Biology and Evolution of the Mollusca

Biology and Evolution of the Mollusca

Volume 1

Winston F. Ponder

David R. Lindberg

Juliet M. Ponder

CRC Press
Taylor & Francis Group
Boca Raton   London   New York

CRC Press is an imprint of the
Taylor & Francis Group, an **informa** business

Front cover photo credits:

1. Ardeadoris rubroannulata © W. B. Rudman; 2. Falcidens crossotus © N. Mikkelsen; 3. Sepioteuthis lessoniana © D. Cowdery; 4. Nassarius glans © L. Charles; 5. Lima sp. © L. Charles; 6. Noctepuna mayana © W. B. Rudman.

CRC Press
Taylor & Francis Group
6000 Broken Sound Parkway NW, Suite 300
Boca Raton, FL 33487-2742

First issued in paperback 2021

© 2020 by Taylor & Francis Group, LLC
CRC Press is an imprint of Taylor & Francis Group, an Informa business

No claim to original U.S. Government works

ISBN-13: 978-0-8153-6169-5 (hbk)
ISBN-13: 978-1-03-217660-4 (pbk)
DOI: 10.1201/9781351115667

Library of Congress Cataloging-in-Publication Data

Names: Ponder, W. F., author. | Lindberg, David R., 1948- author.
Title: Biology and evolution of the mollusca / Winston Frank Ponder, David R. Lindberg and Juliet Mary Ponder.
Description: Boca Raton : Taylor & Francis, 2019. | Includes bibliographical references and index.
Identifiers: LCCN 2019008523 | ISBN 9780815361695 (volume 1)
Subjects: LCSH: Mollusks. | Mollusks--Evolution.
Classification: LCC QL406.7 .P66 2019 | DDC 594--dc23
LC record available at https://lccn.loc.gov/2019008523

Visit the Taylor & Francis Web site at
http://www.taylorandfrancis.com

and the CRC Press Web site at
http://www.crcpress.com

Contents

Volume 2

Introduction to Volume 2

Chapter 12 Molluscan Relationships

Chapter 13 Early History and Extinct Groups

Chapter 14 The Polyplacophora, Monoplacophora and Aplacophora

Chapter 15 The Bivalvia

Chapter 16 The Scaphopoda

Chapter 17 The Cephalopoda

Chapter 18 Gastropoda I - Introduction and the Stem Groups

Chapter 19 Gastropoda II - The Caenogastropoda

Chapter 20 Gastropoda III - The Heterobranchia

Chapter 21 Molluscan Research - Present and Future Directions

Appendix

References

Index

Foreword

There are many books written on molluscs ranging from monographic treatments of a particular group, to physiology texts, aquaculture manuals, shell collector handbooks, and even cook books. It has, however, been some time since an overview of the phylum has been produced. In the two volumes comprising this work, we have made an attempt at doing that. Most of the previous overviews have been edited volumes, with individual sections written by experts in particular systems, processes, or taxa. Of note are (*The Mollusca*, volumes 1–12 (Wilbur, 1983–1988), and *Mollusca*: The Southern Synthesis (Beesley, 1998a). While this approach provides an expert review of a topic, synergies with other topics and taxa in the same volume are often limited. In contrast to edited volumes, co-authorship of the entire work provides opportunities to synthesise across disparate topics and provide profuse cross-referencing throughout the work.

However, any treatment of a group of animals will always reflect the bias of the authors. In our case, we give an overview with, where possible, an evolutionary focus and have attempted, within our limitations, to cover as many aspects relevant to the biology and evolution of molluscs as possible. Given the vast and rapidly expanding literature on this major group of animals, this objective may not always be realised. We also state the obvious in acknowledging that the information presented here is based on the work of thousands of scientists and naturalists, as is obvious from a glance at the bibliography; our final EndNote™ database contained over 30,000 titles. Unlike the practice in taxonomic texts, we have omitted giving the authors of species and other taxa as these are readily available on online databases such as WoRMs and Fossilworks.

To quote the late, great Alan Solem 'Shells have an intrinsic beauty. However, the animals that make the shells are far more beautiful and amazing in their diversified ways of making a living'. (Solem, 1974). Nevertheless, it was shells that started two of us (WFP and DRL) on our long journey with molluscs from an early age. WFP began collecting shells at around 11 years old and decided to become a malacologist soon after. DRL's interest in shell collecting and marine biology also began at an early age, nine years old, although the journey was indirect. JMP has always had an interest in graphic art and painted animal replicas for ten years at the Australian Museum. She became familiar with her main tool, Adobe Illustrator, after retiring from teaching. Many of her illustrations in the book employ colour coding of particular structures to assist with their interpretation.

The genesis of this book came about while WFP and DRL were doing an edited book on molluscan phylogeny (Ponder & Lindberg, 2008, p. viii) and has been a little over a decade in the making. The germ of the idea materialised when one of us was teaching a summer course on molluscs at the University of Wollongong (NSW, Australia) with the identification of the need for a textbook. It has, however, grown considerably beyond the original goal. Another incentive was the realisation and experience that many graduates who end up working with molluscs have had little opportunity to learn basic information about the phylum, much of which is widely scattered in a vast and often difficult to obtain literature or limited to single chapters in general invertebrate zoology texts.

A project such as this would not be possible without significant input from numerous colleagues, photographers, editors, and helpers for which we are most grateful (see Acknowledgements for details).

This volume is concerned with the introductory material for Mollusca, providing an overview of physiology and genomics, and focuses on reviews of the main systems. We also provide a chapter dealing with 'Natural History' which covers a brief review of 'ecology' as well as a review of the relationships that molluscs have with other organisms, including predation, commensalism, and so on, while feeding interactions are mostly covered in Chapter 5. Human interactions with molluscs are summarised in Chapter 10, and the last chapter in this volume looks at past research efforts and potential directions.

Molluscs first appeared in the Cambrian, over 500 million years ago, and waxed and waned over the subsequent eons. Studies of this rich fossil diversity and the changes it encapsulates have provided important insights into evolutionary biology. Many of these studies have been particularly illuminating when combined with morphological, embryological, and molecular studies of living taxa. This integration of the past and present is perhaps more important now than ever before given the rate at which ever-increasing human populations are driving unprecedented global change. It is likely now that every biological system will face major upheavals in the coming decades. For example, the number of known mollusc extinctions is already about equal to that of tetrapod vertebrates, despite there being far fewer mollusc workers to assess diversity changes than those working with vertebrates, and many regional faunas remain poorly known.

We hope that this work might provide some of the tools necessary to train and inspire some of the biologists, palaeontologists, environmentalists, and systematists necessary to continue to investigate and document this amazing group of animals.

Acknowledgements

In particular, we thank our volunteer helpers Rosemary Coucouvinis and Doris Shearman (Figure 1), without whose efforts this project would probably have taken several years longer to complete. Both Rosemary and Doris spent many hundreds of hours cleaning up and maintaining our bibliographic EndNote™ database and proof reading. Doris also carried out general editing of text, spelling, and word checks using various programs, and assisted greatly with producing the index.

We also gratefully acknowledge the contributions of our chapter reviewers. They were:

Chapter 2: Eric Armstrong and Don Colgan. Chapter 3: Janice Voltzow (early draft), Gonzalo Giribet, and Geerat Vermeij. Chapter 5: Sandra Shumway, Alexandre Lobo-da-Cunha, and Peter Beninger. Chapter 6: Elizabeth Andrews (early draft) and Gerhard Haszprunar. Chapter 7: Lauren Sumner-Rooney and Gerhard Haszprunar. Chapter 8: Michael Hadfield, Rachel Collin, Peter Beninger, and Bernhard Ruthensteiner. Chapter 9: Geerat Vermeij, Andy Davis, Jim Carlton, Bruce Marshall, George McKay (birds and mammals), Michael Frese (viruses) and Tom Cribb (platyhelminths). Chapter 10: Geerat Vermeij, Jim Carlton, Sandra Shumway, Wayne O'Connor, and Tom Cribb (flukes). Chapter 11: Eric Armstrong.

Photographs have generously been provided by several people and acknowledgement for these is given in the figure captions. Graham Mayor assisted by providing macros for Microsoft® Word which made our task more efficient.

This work would have been impossible without access to the superb library facilities and online resources of the University of California Berkeley, the University of Sydney, and The Australian Museum, for which we are most grateful.

Lastly, we thank Chuck Crumly and the staff and associates of CRC Press for their assistance and support in producing this work.

FIGURE 1   The two dedicated volunteers that helped make this book a reality—Doris Shearman (left) and Rosemary Coucouvinis (right).

About the Authors

Winston F. Ponder MSc, PhD, DSc is a Senior Fellow at the Australian Museum (Sydney), and an Honorary Associate at the University of Sydney. Before retirement in 2005 he was a Principal Research Scientist at the Australian Museum, where he worked in the malacology section for 37 years and was instrumental in organising and building up the extensive collection. While at the museum he also held several honorary positions at a number of Australian universities. In addition to pursuing research programs, he has also been active in transferring expertise to new generations of young zoologists by way of training research students, hosting overseas collaborators, running workshops and organising a number of conferences. It was when running an intensive mollusc course at the University of Wollongong over several years, that the idea for this book dawned. Winston's research extends broadly across marine and freshwater molluscs and includes observations of living animals, microanatomy, biometrics, cladistics and molecular analysis. He, like his collaborator David Lindberg, has been involved in the revolution in molluscan phylogenetics that has taken place over the last 20 years. His publications include 250 peer-reviewed papers and book chapters, two edited books on molluscs and one on invertebrate conservation. Recently he produced (with two others) an online interactive key and information system of Australian freshwater molluscs, has been involved with the molluscan part of the online Australian faunal directory, and was editor of the journal *Molluscan Research* for 14 years. He organised or co-organised 12 national or international meetings and presented papers at many more. He has undertaken study trips and field work in various parts of the world over more than 40 years but field studies have mainly focused on Australian marine and freshwater habitats. His awards include the Hamilton Prize (Royal Society of N.Z., 1968), Fellow of the Royal Zoological Society, NSW, Silver Jubilee award, Australian Marine Sciences Association (2008), and the Clarke Medal, Royal Society of NSW (2010).

David R. Lindberg is Professor Emeritus of Integrative Biology, Curator Emeritus in the UC Museum of Paleontology, and a former member of the Center for Computational Biology at UC Berkeley. He has authored over 125 peer-reviewed papers and edited or authored three books on the evolutionary history of nearshore marine organisms and their habitats. At Berkeley, he served as major advisor to 21 PhD graduate students and six post-doctoral researchers. He also served as Chair of the Department of Integrative Biology, Director of the UC Museum of Paleontology, and Chair of the UC Berkeley Natural History Museums. In addition to providing graduate seminars in evolution and organismal biology, he regularly taught a marine mammal course, an invertebrate zoology course with laboratories, and two semester principles of phylogenctics course. Prof. Lindberg has conducted research and field work along the rocky shores of the Pacific Rim for over 45 years. In addition to his research and teaching, Prof. Lindberg was actively involved in K–16 outreach projects at the UC Museum of Paleontology, and focused on the use of the internet to increase access to scientific resources, and the training of teachers in principles of evolutionary biology, science, and global change.

Juliet M. Ponder MA Grad Dip Ed has spent 55 years married to the senior author of this book and developing a variety of interests. She excelled at school and began studying science at Auckland University where she met Winston. She was blown away by his passion for molluscs and decided that, if such dedication was required to be a scientist, she didn't have it. However, she was happy to support his work and to this end began developing competence in scientific writing and illustrating and even scuba diving. She and Winston have two children and, while rearing them, Juliet acquired qualifications in fine art, community welfare and education. She worked part-time for 10 years painting lifelike models in the Australian Museum but, as her children grew older, she moved into full time community work and then teaching for 15 years. Retirement from paid work in 2004 gave her the opportunity to learn computer graphics and develop a method of colour coding illustrations to clarify similarities and differences between groups of animals. The drawings in this book combine her love of art, science and education.

1 Introducing Molluscs

This work is devoted to the Mollusca,[1] one of the most diverse and significant groups of animals. It summarises their diversity, utility, physiology, and functional morphology, and their evolutionary history and relationships. Also, we highlight some current areas of research and flag some areas which urgently need work. Due to its size, this work has been produced in two complementary volumes. The first includes an overview of physiology and genomics and covers the anatomy, function, and physiology of the body systems, and their natural history and interactions with humans. In the second volume, we cover the relationships of molluscs with other animal groups and of the class level taxa within the Mollusca, their early fossil history, and details of each of the main taxa, including outlines of their fossil history and evolutionary relationships. We also include an Appendix summarising the classification of each group.

In this introductory chapter, we provide a guide to the terminology and an outline of the rationale used in systematics (classification and phylogeny), and how ecology and genomics are dealt with in the book.

1.1 THE MOLLUSCA: AN INTRODUCTION

Molluscs (= mollusks) are conspicuous and ecologically important, especially in the sea where they often dominate communities. They are also a source of our food and ornaments and are an increasingly utilised resource for pharmaceutical products and experimental animals. Despite comprising a significant component of animal diversity, their study is often neglected, particularly in those parts of the world that have the richest molluscan faunas.

Molluscs include familiar animal groups, such as scallops, cockles, whelks, limpets, slugs, snails, squid, and octopuses, although a casual observer might see little similarity in these rather disparate animals. Their great variety has led to their importance to humans as food and cultural objects, as well as some being pests or hosts for human parasites. While many marine molluscs have wide distributions, some marine and many non-marine species have very limited ranges. These latter taxa are often susceptible to environmental changes and exploitation (see Chapter 10). The impacts of climate change and associated ocean acidification on molluscs are well documented in the fossil record (see Chapter 13), but their effects on extant molluscs have only recently been considered (see Chapter 10).

Molluscs are bilaterally symmetrical eumetazoans and are one of the major groups (phyla). Based on our current knowledge of described animal diversity, the Mollusca is the second largest animal phylum after the Arthropoda (the huge group containing insects, crustaceans, millipedes, spiders, mites, and related animals). Estimates of the number of species of living organisms on earth vary widely from just a few million to tens of millions, largely because of our profound lack of knowledge of many groups that contain small-sized or microscopic organisms such as nematodes, bacteria, viruses, protists, and fungi. Consequently, while our current understanding of the kingdom Animalia includes some hugely diverse groups, notably the insects where much of the described diversity lies, other groups such as nematodes have vast numbers of undescribed taxa and may eventually be shown to even outnumber insects. Molluscs are one of the better-known animal phyla and are estimated to have up to 200,000[2] living species, of which about 52,500 extant and 35,000 fossil species are named[3] (see Section 1.4.2). By way of comparison, the vertebrates (mammals, birds, reptiles, amphibians, fish) comprise the major subgroup of the phylum Chordata (it also includes the tunicates and cephalochordates), and the whole phylum has only about 45,000 living species.

Besides their diversity in body form, molluscs also exhibit a vast range of physiological, behavioural, and ecological adaptations. In size, they range from giant squid (~12 m with some estimates over 14 m in length and one species weighing up to 494 kg) (McClain et al. 2015) and giant clams (up to 115 cm in length and weighing over 300 kg), down to species with adult body sizes of only about half a millimetre. Most molluscs are marine, but large numbers also occupy freshwater and terrestrial habitats. Some live in high mountain habitats and, at the other extreme, over 10 km deep in the sea. They are extremely diverse in their feeding habits, ranging from grazers and browsers on many different biotic substrata to suspension-feeders, predators, and parasites.

Molluscs have been an important source of food for man from early times, and as objects they were, and are, incorporated or depicted in art and religion. Molluscan commerce was engaged in long before recorded history. Shell collecting is a popular hobby, and mollusc fisheries and aquaculture are important industries and are becoming increasingly important as a source of medicines and other useful compounds. As experimental animals, molluscs have played a major role in our understanding of neural function. Besides their economic importance, molluscs have provided valuable scientific insights into evolutionary biology, behaviour, neural biology, biogeography, genomics, and ecology through studies on their molecular and morphological diversity. As covered

---

[1] Derived from the Greek *mollis* – soft. Scientists that study molluscs are malacologists (see Box 1.1).

[2] The large divergence in the estimates is due to markedly different opinions as to the numbers of undescribed species that are either known or yet to be discovered.

[3] The large difference in these numbers depends on estimates of how many of the named species-group taxa are considered synonyms and how many are valid.

BOX 1.1   THE STUDY OF MOLLUSCS – MALACOLOGY

The branch of biology that studies molluscs is called malacology, and its practitioners are malacologists. Most malacologists can also be referred to by their other areas of speciality – e.g., as ecologists, palaeontologists, anatomists, neurobiologists, molecular biologists, physiologists, taxonomists, conservationists, etc.

The term malacology (and its adjective malacological) is apparently derived from an alternative name for Mollusca – Malacozoa – given by Blainville (1825) (Hyman 1967), while the older term, conchology, is generally reserved for the study of shells.

Some historical aspects of the study of molluscs are outlined in Box 1.2.

BOX 1.2   SOME HISTORICAL MUSINGS

Gould (1993) described aspects of the history of malacology from its conchological beginnings in an article entitled 'Poe's greatest hit'. Gould recounted how falling sales of Thomas Wyatt's (1838) *Manual of Conchology* caused Wyatt to produce a less expensive book which he hoped would increase sales at his public lectures. Because of name recognition, he hired the writer/poet Edgar Allen Poe (1809–1849) to produce the new book. The involvement of Poe, and his appropriations of earlier works into his book, have been controversial (Silverman 1991). Gould concluded that Poe made significant contributions that resulted in a much-improved molluscan classification based on both anatomical and shell characters, primarily derived from the works of the French naturalist Georges Cuvier (1769–1832). But, perhaps, more importantly, Poe rebranded the term conchology by arguing that its Greek derivation from conchylion required the study of both the shell *and* animal (our emphasis). Although the taxon descriptions in Poe's work contained descriptions of both shells and anatomy (where known), this arrangement was not new, but rather an acquisition from Brown (1837) who had translated and included Lamarck's (1818) shell- and animal-based classification in *The Conchologist's Text-book*.

While the attribution of including both shell and anatomical characters in the text and classification, and Poe's explicit transformation of conchology into malacology, remains partially shrouded in the past, there is little doubt that the impetus for this change originated among French naturalists. Brown (1837, p. 10) himself mused that '… we were young both in years and science; we then held the opinions and arrangements of Linnaeus as sacred; and consequently, dreaded the new systems, and widely developed views of the French school of natural history, which threatened to overturn the system that had been the delight of our youthful studies. Since that period, our ideas have gradually expanded, and a conviction of the superior classification of the French School has settled in our mind. Since then, malacology and its associated 'systems' have undergone several major transformations, all of which, as with many other branches of science, were met with some resistance from many practitioners.'

in more detail in Chapter 10, human actions have had some devastating impacts on molluscs, particularly those in non-marine environments, resulting in major declines and many hundreds of extinctions. Despite only a small fraction of the non-marine molluscan faunas of the world being adequately assessed, there are more recorded extinctions of non-marine molluscs than all tetrapod vertebrates (birds, mammals, reptiles, amphibians) combined (e.g., Killeen et al. 1998; Ponder & Lunney 1999). Also, introducing alien species has resulted in the homogenisation of many previously unique biotas, especially on islands (Cowie 2001b). Because public concern and conservation efforts continue to focus on vertebrates, the future of rare and endangered molluscs is generally bleak. It is even more serious due to the lack of information from significant parts of the world, signifying that the officially recognised details of molluscs extinctions are just the tip of the iceberg.

Because their shells preserve well, shell-bearing molluscs have an excellent fossil record. Several putative molluscan groups (gastropods, bivalves, helcionellids, and rostroconchs, the latter two now extinct) first appeared in the fossil record of the early Cambrian (520 million years ago [mya]) (see Chapter 13). Their almost simultaneous appearance suggests that they may have existed in the Precambrian as well and older supposed molluscs have been found in the Ediacaran (635–541 mya) of southern Australia and northern Russia. Other groups appear later, with the cephalopods and chitons found from the late Cambrian and scaphopods from the late Silurian (see Chapter 12 for details).

Studies on molluscan evolution can use the rich fossil record as well as anatomical, ultrastructural, embryological, ecological, genetic, and molecular data obtained from living taxa. Such studies have provided valuable insights into evolutionary biology as well as biogeography and ecology.

Consensus as to the identity of the closest living relatives of molluscs has been hard to obtain. Currently, a variety of molecular lines of evidence including genomic, sequence, and mitochondrial gene order data (see Chapter 12) favour the Brachiozoa. The inability to unequivocally identify which group is the nearest relative may reflect the apparent burst of animal evolution before and in the early Cambrian, the so-called 'Cambrian explosion'. It also significantly affects our ability to convincingly envision an ancestral mollusc (see Chapters 12 and 13).

1.1.1 A Brief Early History of Malacology – The Study of Molluscs

The study of molluscs began with the fascination with shells as curiosities. Comprehensive accounts of the history of malacology (see Box 1.1) and shell collecting are provided by Dance (1966, 1986) and Coan et al. (2007) and will not be repeated here other than to briefly mention some of the key events and innovations and some of the main early players. While molluscs have been utilised by humans from well before civilisation began, recorded history of molluscan investigations and classification commenced with Aristotle in Ancient Greece. Numerous works, some large and lavishly illustrated, were used by the great Swedish naturalist Carl Linnaeus (1707–1778) as the basis for the names listed in *Systema Naturae*. The tenth edition of that work (1758) is the starting point for the introduction of valid binomial names and the modern system of biological nomenclature. The basic classification of Linnaeus was improved by the famous French worker Jean-Baptiste Lamarck (1812–1822). Nearly all studies before Lamarck were based primarily on the shell, but Lamarck, and especially his contemporary Georges Cuvier (1769–1832), were some of the earliest comparative anatomists. Subsequent to these studies, many important investigations on comparative molluscan anatomy appeared during the nineteenth and early twentieth centuries, including those of G. P. Deshayes (1795–1875), H. Milne Edwards (1800–1885), T. H. Huxley (1825–1895), W. M. Keferstein (1833–1870), R. Lankester (1847–1929), J. W. Spengel (1852–1921), H. Simroth (1851–1917), J. Thiele (1860–1935), and P. Pelseneer (1863–1945).

Early voyages of discovery and commerce returned natural history specimens to Europe for subsequent study and description. These included those of James Cook on HMS *Endeavour* (1768–1771), HMS *Beagle* (1831–1836, with Charles Darwin on board), and HMS *Sulphur* (1836–1842, with R. B. Hinds), all being British, the French ship *Astrolabe* (1837–1840, with J. J. C. Quoy and J. P. Gaimard, the naturalists), and the United States Exploring Expedition (1838–1842, with most of the material described by A. A. Gould). These were followed by two additional British voyages, HMS *Samarang* (1843–1846, with Arthur Adams), and HMS *Rattlesnake* (1846–1850, with Thomas Huxley). Both Quoy and Gaimard on the *Astrolabe* and Arthur Adams on the *Samarang* made and published numerous illustrations of living animals. Of the nineteenth century world oceanographic expeditions, the voyage of HMS *Challenger* (1872–1876) is probably the most famous, with gastropods described by R. B. Watson, bivalves by E. A. Smith, and cephalopods by W. E. Hoyle. The voyages of the Austrian *Novara* (1857–1859, described mainly by G. R. von Frauenfeld), the German *Valdivia* (1898–1899, with most

molluscs described by J. Thiele). H. F. Nierstrasz on the Dutch *Siboga* (1899–1900) obtained a greatly improved knowledge of the faunas on the continental shelves and deeper waters. Russian expeditions to the Pacific Northwest led by Otto von Kotzebue (*Rurik*, 1815–1818 and *Enterprise*, 1823–1826) were responsible for the discovery of many North American molluscan taxa later described by J. F. Eschscholtz (1793–1831) and M. H. Rathke (1793–1860). In commerce, the shipping routes of the East India Trading Company brought natural history specimens back to England from both the East Indies and the Caribbean, while The Dutch East India Company played a similar role in the Netherlands (Dance 1966) in shipping specimens from many locations in the Indo-West Pacific.

With the further expansion of the British and other European empires during the nineteenth and early twentieth centuries, numerous new faunas were discovered, and large collections brought back to European (including British) collections, ultimately residing in several museums. This was the age of description and consolidation where major contributions were made mainly by European and North American molluscan workers.

The major contribution of many of the early molluscan workers was the description of new taxa (Bouchet 1997). In Europe, Lovell Reeve published the *Conchologia Systematica* (1841–1842) and *Conchologia Iconica* (1843–1878), describing many new species. Later monographic treatments by the Sowerbys (George Brettingham I, II, and III) also described numerous taxa. The brothers Arthur and Henry Adams produced in three volumes *The Genera of Recent Mollusca, Arranged According to Their Organisation* (1853–1858), in which anatomical characters were used to the family level as well as shell morphology, where available, for generic rank.

Some of the most prolific workers in North America were Philip P. Carpenter who described over 500 taxa exclusively from the west coast of North America, and Henry Augustus Pilsbry, who described about 5,800 new invertebrates, the great majority molluscs. Pilsbry's other contributions included authoring many volumes of the *Manual of Conchology*, founding and editing *The Nautilus*, and producing the four-volume *Land Mollusca of North America*. William Healy Dall also described over 5000 molluscan taxa and was the only American malacologist of the second half of the 1800s who explicitly attempted to reflect evolutionary relationships in his classifications. In Australia, English-born ornithologist and malacologist Tom Iredale (1880–1972) described over 2500 new taxa.

Some of these early workers are listed below by country.

Eurasia: Austria (J. P. R. Draparnaud, 1772–1804 and G. R. von Frauenfeld, 1807–1873); Belgium (P. Dautzenberg, 1849–1935 and P. Pelseneer, 1863–1945); Denmark (R. Bergh, 1824–1909 and O. A. L. Mörch, 1828–1878); France (P. D. de Montfort, 1766–1820; A. E. J. de Férussac, 1786–1836; G. Michaud, 1795–1880; G. P. Deshayes, 1796–1875; A. d'Orbigny, 1802–1857; J. C. Chenu,

1808–1879; P. Fischer, 1835–1893; F. P. Jousseaume, 1835–1921; M. Cossmann, 1850–1924; and G. Coutagne, 1854–1928); Germany (H. C. Küster, 1807–1876; W. Dunker, 1809–1885; F. H. Troschel, 1810–1882; C. E. von Martens, 1831–1904; W. Kobelt, 1840–1916; O. Boettger, 1844–1910; J. Thiele, 1860–1935; and L. H. Plate, 1862–1937); Italy (M. Paulucci, 1835–1919; C. M. Tapparone-Canefri, 1838–1891; T. M. A. Monterosato, 1841–1927; and F. Sacco, 1864–1948); Netherlands (M. M. Schepman, 1847–1919 and H. F. Nierstrasz, 1872–1937); Russia (P. G. Demidoff, 1738–1821; J. F. Eschscholtz, 1793–1831; and M. H. Rathke, 1793–1860); Sweden (S. L. Lovén, 1809–1895 and N. H. Odhner, 1884–1973); Switzerland (J. P. B. Delessert, 1773–1847 and A. Naef, 1883–1949); and United Kingdom (G. Montagu, 1753–1815; G. B. Sowerby I–III, 1788–1854, 1812–1884, 1843–1921; J. E. Gray, 1800–1875; J. G. Jeffreys, 1809–1885; E. Forbes, 1814–1854; L. A. Reeve, 1814–1865; S. C. T. Hanley, 1819–1899; A. Adams, 1820–1878; J. C. Melvill, 1845–1929; E. A. Smith, 1847–1916; C. N. E. Elliot, 1864–1931; and J. R. Le B. Tomlin, 1864–1954).

Asia: India (N. Annandale, 1876–1924 and B. Prashad, 1894–1969) and Japan (Yoichirō Hirase, 1859–1925 and T. Kuroda, 1886–1987).

North America: Canada: (P. P. Carpenter, 1819–1877); USA (I. Lea, 1792–1886; A. A. Gould, 1805–1866; J. G. Cooper, 1830–1902; W. G. Binney, 1833–1909; G. W. Tryon, 1838–1888; A. Hyatt, 1838–1902; A. E. Verrill, 1839–1927; W. H. Dall, 1845–1927; H. A. Pilsbry, 1862–1957; F. C. Baker, 1867–1942; and C. M. Cook, 1874–1948).

South America: Chile (R. A. Philippi, 1808–1904).

Australasia: Australia: (R. Tate, 1840–1901; J. W. Brazier, 1832–1930; J. E. Tennison Woods, 1832–1889; J. C. Cox 1834–1912; J. Verco, 1851–1933; C. Hedley, 1862–1926; and W. L. May 1861–1925); and New Zealand: (F. W. Hutton, 1836–1905 and H. Suter, 1814–1918).

Molluscan palaeontology has also had a long and distinguished role in malacology, and a brief introduction to the study of fossil molluscs is given in Chapter 13 (Volume 2).

1.1.2 Some Major Treatments of the Mollusca

There is a vast literature on molluscs. Relatively few seminal works deal with the entire phylum, although there were several major overviews of molluscs published in the late nineteenth and early twentieth centuries. Notable among these were the molluscan chapters in the first edition of H. G. Bronn's *Klassen und Ordnungen des Tierreichs* – the 'Acephala' (bivalves and brachiopods) (Bronn 1862) and the Cephalophora (gastropods, scaphopods, and cephalopods) (Bronn 1862–1866). The second edition, which continued into

the twentieth century, contained more detailed treatments of the 'Amphineura' and Scaphopoda (Simroth 1892–1894a), 'Prosobranchia' (Simroth 1892–1894b, 1896–1907), 'Pulmonata' (Simroth & Hoffmann 1896), Polyplacophora and Scaphopoda (Hoffmann 1932–1939), Opisthobranchia (Hoffmann 1940), and Bivalvia (Haas 1935–1955).

Probably the most important comprehensive systematic review of Mollusca is the *Handbuch der systematischen Weichtierkunde* by Thiele (1929–1935) which has been translated and printed in full (Thiele 1992a, 1992b, 1998). Two sections of the *Handbuch der Paläontologie* were published which dealt with the genera of fossil and living gastropods (Wenz 1938–1944; Zilch 1959–1960), but other molluscan classes were not covered. Significant works, detailing the anatomy and other aspects of molluscs rather than focussing mainly on taxonomy, include three volumes of the *Traité de Zoologie* (Fischer-Piette & Franc 1960a, 1960b; Franc 1960; Fischer-Piette & Franc 1968; Franc 1968; Mangold 1989). Hyman (1967) covered the gastropods, polyplacophorans, and aplacophorans in considerable detail in what was her last volume on the invertebrates.

The *Treatise on Invertebrate Paleontology* covered the taxonomy and fossil history of chitons, monoplacophorans, scaphopods, 'archaeogastropods', and some Paleozoic gastropods (Moore 1964), the bivalves (Cox et al. 1969; Stenzel 1971), and cephalopods (Moore 1957; Teichert et al. 1964; Moore 1996). Some volumes, including the bivalves and Paleozoic gastropods, are currently being revised, but regrettably, the volume on gastropods has not been completed. Several recently revised sections of the bivalves and cephalopods are available in the continuation of the *Treatise* as the *Treatise Online*. An excellent regional treatment of molluscs which has proved invaluable to molluscan workers and students is *The Southern Synthesis* (Beesley et al. 1998a).

Important reviews of molluscan biology and physiology were covered in the two-volume *Physiology of the Mollusca* (Wilbur & Yonge 1964, 1966) and in a special volume devoted to Mollusca in the *Chemical Zoology* series (Florkin & Scheer 1972). These were superseded by the unparalleled 12 volume series *The Mollusca*, published between 1983 and 1988, and involving many authors and several editors. Other useful texts include Sturm et al. (2006) on the study, collection, and preservation of molluscs, Bottjer et al. (1985), notes for a short course on molluscs, the dissection guide by Mizzaro-Wimmer and Salvini-Plawen (2001), and techniques for dealing with small-sized molluscs by Geiger et al. (2007).

General texts on molluscs include those by Fischer (1880–1887), Cooke (1895), Hescheler (1900), Pelseneer (1906), and several chapters in Dogelya and Zenkevicha (1940). At a more general level, there are many books dealing with molluscs, but *Molluscs* by Morton (1958a, 1967), *The Biology of the Mollusca* (Purchon 1968), *Shell Makers* by Alan Solem (1974), *Living Marine Molluscs* by C. M. Yonge and T. E. Thompson (1976), and *Natural History of Shells* by G. J. Vermeij (1993) deserve special mention.

More recently, three edited books have appeared which deal mainly with molluscan phylogeny – Taylor (1996), Lydeard and Lindberg (2003), and Ponder and Lindberg (2008).

Numerous other texts mainly deal with particular groups, faunas or organ systems, or functions such as reproduction, and the reader should refer to the appropriate chapter for that literature.

1.2 WHAT MAKES A MOLLUSC? AN OUTLINE OF THEIR DIAGNOSTIC FEATURES

Because of the enormous morphological and molecular diversity in the group, a simple definition of Mollusca is difficult to frame. Despite these major differences, some shared features distinguish them as a group. These characters are not always obvious, and modification and loss of some have

occurred in most groups. Several morphological features commonly used to diagnose the Mollusca as a clade are outlined in Box 1.3.

1.3 THE MAJOR GROUPS OF MOLLUSCS

Living molluscs comprise eight major groups or 'classes', with Gastropoda (snails, slugs, whelks, and limpets) by far the largest, Bivalvia (scallops, clams, oysters, and mussels) the next most diverse, and then the Cephalopoda (squid, cuttlefish, octopuses, nautilus). The remaining five classes are substantially lower in diversity; the Polyplacophora (chitons), the Scaphopoda (tusk shells), the Monoplacophora

BOX 1.3   SOME MORPHOLOGICAL FEATURES THAT CHARACTERISE MOLLUSCS

A lack of segmentation (or metamerism), although serial repetition of muscles and organs may occur in some groups (see Chapters 3, 4, and 12).

The molluscan body typically has a head, foot, and visceral mass. Homoplastic reduction and loss of a distinct head have occurred in the Bivalvia and possibly in aplacophorans (see Chapter 3).

The body is covered with a mantle which has a bipartite or tripartite margin (see Chapter 3).

The mantle tissue secretes the molluscan shell, while individual cells produce the spicules. In some groups, the shell may be secondarily lost or internalised (e.g., some heterobranch gastropods and cephalopods). Brachiopods also have a shell-secreting mantle, but its early development differs markedly from the formation of the shell gland and early mantle of molluscs (see Chapter 3).

There are typically one or more pairs of ctenidia (gills) which lie in a posterior mantle (= pallial) cavity or a posterolateral mantle groove surrounding the foot. Ctenidia have alternating ciliated filaments on either side of the gill axis (except for one group of protobranch bivalves where they are opposite, and monoplacophorans and some gastropods where they occur on only one side of the gill axis – see Chapter 4). The gills serve both respiratory and ventilation functions, as in other lophotrochozoans.

A ventral foot utilises muscular waves and/or cilia for locomotion, in combination with mucus. In bivalves and scaphopods, the foot is modified for digging. In the aplacophoran groups, the foot is absent or represented by a vestigial groove, and in cephalopods, it is highly modified (see Chapters 3 and 17).

One or more pairs of dorsoventral pedal retractor muscles are present in all molluscan classes (see Chapter 3).

The kidneys, gonads, and anus open into the mantle cavity or mantle groove (see Chapter 4).

The mantle cavity/groove contains a pair of sensory osphradia (and/or posterior sense organs) which are reduced or sometimes lost in some gastropods, and absent in coleoid cephalopods (see Chapters 4 and 7).

The buccal cavity contains a radula – a bilaterally symmetrical ribbon of teeth supported by a muscular odontophore. The radula has been lost in bivalves and a few gastropods. Some potential molluscan outgroups have rather similar buccal structures (see Chapter 5).

A pair of glandular oesophageal pouches is the plesiomorphic condition, but they have been highly modified in some groups and lost in bivalves and some gastropods (see Chapter 5).

Like other lophotrochozoans, molluscs are coelomate, although the coelom is small and represented by the spaces in the kidneys, gonads, pericardium, and the ducts which interconnect them (see Chapters 6, 8, and 12).

The pericardium encloses the heart, and pericardial and/or auricular structures are involved in ultrafiltration (see Chapter 6).

The aorta(s) from the heart open to a haemocoelic circulation system, with the main body cavity a haemocoel, the respiratory pigment (if present) is typically haemocyanin, rarely haemoglobin (see Chapter 6).

As in most lophotrochozoans, molluscs undergo mosaic development and have spiral cleavage. A trochophore and/or veliger larvae are found in many aquatic taxa, but direct development is also common and is the only mode in the Cephalopoda (see Chapter 8).

Note: In the detailed analysis of the morphological characters of molluscs and outgroups by Haszprunar (1996a), some additional ultrastructural characters were listed.

(primarily deep-sea limpets), and the aplacophorans (spicule worms) comprising two classes, the Solenogastres (or Neomeniomorpha) (solenogasters) and the Caudofoveata (or Chaetodermomorpha) (caudofoveates). A few extinct groups are also often treated as classes (see Chapter 13).

The molluscan body plan shows considerable modification between and within groups since they diversified in the Cambrian, although a few groups retain features shared by their ancient relatives, notably monoplacophorans, some protobranch bivalves, a few gastropod groups, nautiloids, and scaphopods (Lindberg et al. 2004) (see Chapters 12 and 13).

Here, in Box 1.4, we provide an outline of classification, basic morphology, and biology of the major living groups (classes) of molluscs. Detailed information on each group is presented in Volume 2 and summaries of their classification are given in the Appendix in that volume.

1.3.1 RELATIONSHIPS OF THE MOLLUSCAN CLASSES

Our understanding of the relationships of the molluscan classes is still in a state of flux. One phylogenetic scheme is shown in Figure 1.1, but the detail of this arrangement is only one of several that are currently in vogue. An overview of past and current hypotheses of molluscan phylogeny is given in Chapter 12. Groupings of classes in a particular phylogenetic hypothesis are sometimes given names, and some of those we refer to in the text are given in Box 1.5.

BOX 1.4   THE CLASSES OF LIVING MOLLUSCA

Eight higher taxa ('classes') of living molluscs are recognised. These are often grouped into higher level groupings (i.e., above Class) – see Box 1.5.

An outline is provided here to serve as a reference point in the following chapters. A few additional extinct classes are recognised, which are covered in Chapter 13. More detailed information on each living class is provided in the relevant taxon chapter, and details of their classification are provided in the Appendix in Volume 2.

Polyplacophora (= Placophora)

The chitons are exclusively marine, with three main groups recognised, two of which are extinct (the Palaeoloricata and Multiplacophora – the latter with up to 17 shell plates), while all living chitons are included in the Neoloricata. They have eight shell plates surrounded by a chitinous girdle (perinotum), a large adhesive foot, and a small, poorly differentiated head. There are multiple ctenidia arranged in a mantle groove around the foot. In feeding, they range from detritivores to herbivores and micro- or grazing carnivores (see Chapter 14).

Ornithochiton

Solenogastres (= Neomeniomorpha)

These marine 'spicule worms' are small, shell-less, and worm-like, and their body is covered with calcareous sclerites embedded in a chitinous cuticle. They are predatory, feeding mainly on cnidarians, and have a narrow, ciliated gliding sole (foot) in a ventral furrow, the pedal groove. There are no ctenidia in the rudimentary posterior mantle cavity (see Chapter 14).

Phyllomenia

Caudofoveata (= Chaetodermomorpha)

Like the solenogasters, these marine 'spicule worms' are small, shell-less, and worm-like, and their body is covered with calcareous sclerites embedded in a chitinous cuticle. They differ from solenogasters in being micro-omnivores, lacking a pedal groove or any trace of a foot and in having a pair of ctenidia in a small posterior mantle cavity. They have an oral shield used for burrowing (see Chapter 14).

Lepoderma

(Continued)

BOX 1.4 (Continued)   THE CLASSES OF LIVING MOLLUSCA

Monoplacophora (= Tryblidia)

The single shell is limpet-like, and they exhibit serial repetition of the ctenidia, shell muscles, kidneys, gonads, and atria. The members of this small group are typically found in the deep-sea, and they appear to be deposit feeders (see Chapter 14).

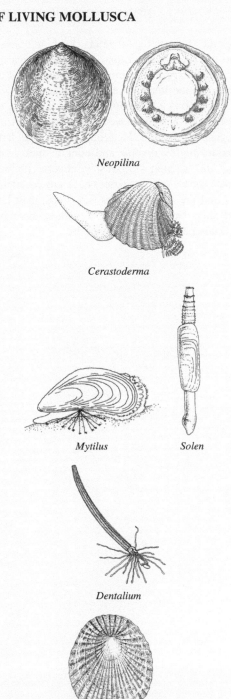

Neopilina

Bivalvia

The bivalves (previously known as pelecypods or lamellibranchs) all have a bivalved shell and include the scallops, oysters, cockles, shipworms, and clams. Bivalves fall into two major groups, entirely marine Protobranchia and the Autobranchia in both marine and fresh-water habitats. The latter contains most of the familiar bivalves and comprises two major groups, the Pteriomorphia (oysters, mussels, scallops, pearl oysters, etc.) and the Heteroconchia (fresh-water mussels, clams, shipworms, etc.) (see Chapter 15).

Cerastoderma

Scaphopoda

The tusk shells are a small entirely marine group that all have a single, tusk-shaped shell open at both ends, a digging foot, and numerous thread-like feeding structures (captacula) that they use for capturing microorganisms. Ctenidia are absent. Two main groups are recognised, Dentaliida and Gadilida, and all are marine (see Chapter 16).

Mytilus   *Solen*

Gastropoda

Comprising the snails, limpets, and slugs, gastropods mostly have a single, often coiled shell or the shell is lost or reduced. The larvae undergo torsion[4] during development so that in many the mantle cavity lies anteriorly. The larvae and some adults have an operculum on the foot. Gastropods are by far the largest class and have been divided into two major lineages, the Eogastropoda with only the true limpets (Patellogastropoda) among living taxa, which are marine, and the Orthogastropoda containing all other gastropods. Within orthogastropods, there are four major groups recognised: the

Dentalium

Cellana

Haliotis

---

[4] See Chapter 8.

(Continued)

BOX 1.4 (Continued)   THE CLASSES OF LIVING MOLLUSCA

Vetigastropoda (top shells, abalone, keyhole limpets, etc.), which are marine, the other three include marine, fresh water, and terrestrial members. These groups are the Neritimorpha (nerites and relatives) (this group, with the Patellogastropoda and Vetigastropoda, comprise the 'lower' gastropods and are covered in Chapter 18) and the Caenogastropoda and Heterobranchia. The latter two groups are highly diverse and contain most living gastropods, the caenogastropods (creepers, periwinkles, whelks, cone shells, cowries, etc.) are covered in Chapter 19 and the heterobranchs (sea slugs, most land snails, and slugs) in Chapter 20. Together, these two groups are known as the Apogastropoda.

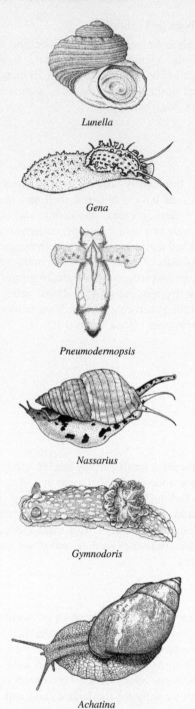

Lunella

Gena

Pneumodermopsis

Nassarius

Gymnodoris

Achatina

(Continued)

BOX 1.4 (Continued)   THE CLASSES OF LIVING MOLLUSCA

Cephalopoda (= Siphonopoda)

Cephalopods have a single, chambered shell, or the shell is reduced or lost. The head is surrounded by arms. The living and extinct cephalopods are covered in Chapter 17. The living cephalopods comprise two major groups, the Nautilida, with only a few species of pearly nautilus, and the Coleoidea, which contains the squid, cuttlefish, and octopuses. A third major group is the ammonites, which are extinct. Nearly all cephalopods are active carnivores, and all are marine.

Nautilus

Nototodaris

Sepia

Octopus

Illustrations from Beesley, P. L., Ross, G. J. B. & Wells, A., Eds. *Mollusca: The Southern Synthesis*. Vol.5 (A & B). Melbourne, CSIRO Publishing, 1998a,. except for *Neopilina, Cerastoderma*, and *Dentalium* which are from Lindberg, D. R., Ponder, W. F. & Haszprunar, G. The Mollusca: relationships and patterns from their first half-billion years. Pp. 252–278 in J. Cracraft & Donoghue, M. J. *Assembling the Tree of Life*. New York, Oxford University Press, 2004.

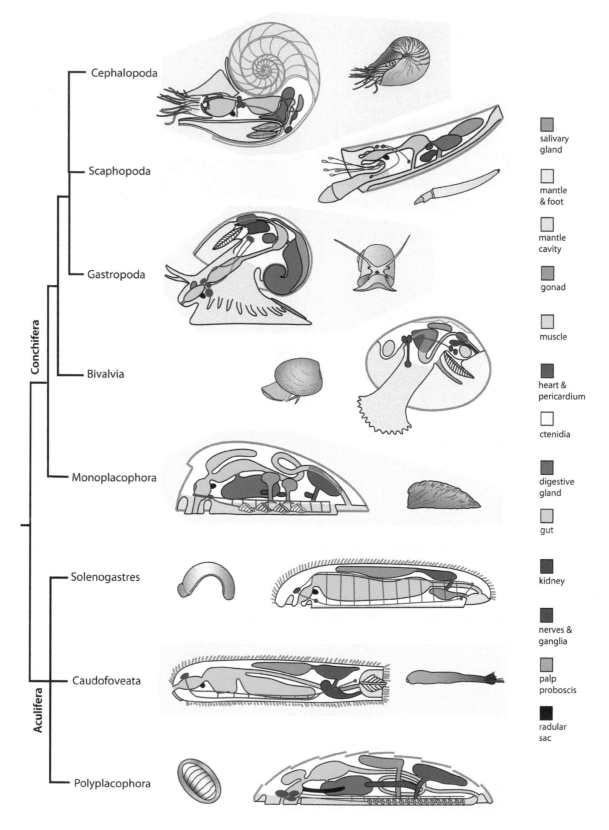

FIGURE 1.1 An example of a hypothesis of relationships (i.e., a phylogenetic tree) of the major groups (classes) of molluscs. In this phylogeny, the aculiferan taxa are shown as unresolved, with a resolved Conchifera. Anatomical cartoons mostly redrawn and modified from Salvini-Plawen, L.V. and Steiner, G., Synapomorphies and plesiomorphies in higher classification of Mollusca, in Taylor, J.D. Ed., *Origin and Evolutionary Radiation of the Mollusca*, Oxford, Oxford University Press, pp. 29–51, 1996.

BOX 1.5 COMMONLY USED GROUPINGS OF MOLLUSCS ABOVE THE CLASS LEVEL

The classes of living molluscs are listed and briefly described in Box 1.4. Below we list the names we use in the text to group those classes. A more detailed list is provided in Chapter 12.

Aculifera – Polyplacophora + Aplacophora (Hatschek 1888–1891).

Aplacophora – Solenogastres + Caudofoveata (Ihering 1876). This group is sometimes treated as a class (see Chapter 14).

Conchifera – Mollusca excluding Aculifera (i.e., Monoplacophora, Cephalopoda, Bivalvia, Scaphopoda and Gastropoda) (Gegenbauer 1878).

Serialia – Polyplacophora + Monoplacophora (Giribet et al. 2006).

Testaria – Conchifera + Polyplacophora (Salvini-Plawen 1972) – equivalent to Eumollusca.

1.4 MOLLUSCAN DIVERSITY

Diversity can be measured in several ways (e.g., taxon, physiological, ecological, and morphological). Here we look at the taxon diversity of living molluscs and, in Table 1.1 and Chapters 3–8 and 13–20, we examine molluscan morphological diversity. Diversity in different habitats

is reviewed in Chapter 9, while fossil molluscan diversity is covered in Chapter 13.

1.4.1 ADAPTATION, ADAPTIVE, AND NON-ADAPTIVE RADIATIONS

A significant source of diversity is accounted for as the result of evolutionary radiations. Terms such as adaptation and radiation have been widely and commonly used and as such have been applied rather loosely. An adaptation, as defined by Gould (2002, p. 662), is 'an emergent character that evolved as a consequence of its value in fitness', while non-emergent characters that contribute to fitness are '*exaptations*'. Given the practical difficulties of teasing out rigorous distinctions, we usually use the term *adaptation* more broadly for any morphological, behavioural, ecological, or physiological trait that appears to have enabled a taxon to gain an evolutionary advantage. This process of evolutionary modification is achieved through natural selection, resulting in improved survival and reproductive success.

Similarly, the term 'adaptive radiation' has been used for a considerable time for many levels of radiation (e.g., Simpson 1944, 1953). Most recent authors follow modifications of the Mayr (1963, p. 663) concept of an adaptive radiation as an 'evolutionary divergence of members of a single phyletic line into a series of rather different niches or adaptive zones'. This restriction of the concept to multiple speciation events driven largely by ecology within a clade has been modified by some to encompass a relatively short time frame, although there is some conjecture over interpretations of 'multiple' and 'relatively short'

TABLE 1.1

Numbers of Named Species of Living Molluscs in Each Molluscan Class According to Five Published Estimates

	Nicol	Boss	Hennig	Van Bruggen	Mizzaro-Wimmer & Salvini-Plawen	Appeltans et al. (Marine Only)
Solenogastres	2000[a]	250[a]	140[a]	100	230	263
Caudofoveata				200	120	133
Polyplacophora		600	600	650	750	930
Monoplacophora		10	6	18	20	30
Scaphopoda		350	300	350	600	572
Cephalopoda		600	600	>1000	650	761
Bivalvia	15,000	7500	31,000	10,000	10,000	9000
Gastropoda	90,000	36,750	95,000	62,000	43,000	32,000–40,000
Total Mollusca	107,000	46,810	128,000	75,000	55,370	43,689–51,689[b]

Sources: Boss, K.J., *Occasional Papers on Mollusks*, 3, 81–135, 1971; van Bruggen, A.C., Biodiversity of the Mollusca: Time for a new approach, in van Bruggen, A.C., Wells, S.M. & Kemperman, T.C.M. (Eds.), *Biodiversity and Conservation of the Mollusca: Proceedings of the Alan Solem Memorial Symposium on the Biodiversity and Conservation of the Mollusca. Siena, Italy 1992. Eleventh International Malacological Congress*, Leiden, the Netherlands, Backhuys Publishers, pp. 1–19, 1995; Hennig, W., *Wirbellose I: Ausgenommen Gliedertiere. Taschenbuch der Zoologie*, Vol. 2, 5th ed., Gustav Fischer, Jena, Germany, 1980; Nicol, D., *Syst. Zool.*, 18, 251–254, 1969; Mizzaro-Wimmer, M. and Salvini-Plawen, L. v., *Praktische Malakologie. Beiträge zur vergleichend-anatomischen Bearbeitung der Mollusken: Caudofoveata bis Gastropoda – Streptoneura*, Springer, Vienna, Austria, 2001.

[a] Total 'aplacophoran' species.

[b] This is the number of species of all molluscs other than gastropods and bivalves.

(Gittenberger 2004). Famous examples of such adaptive radiations are seen in the Galapagos finches and the East African cichlid fish. According to Schluter (2000), in a book on the subject, an adaptive radiation occurs when an ancestral species evolves into multiple species that inhabit several environments and thus, like many earlier authors, he argued for a focus on the 'ecological' aspects of divergence. In an adaptive radiation, if specialisation (niche narrowing) is the norm, the derived taxa should have narrower niches than their ancestors, and while this is often the case, it does not always follow (Schluter 2000; Feranec 2007). 'Adaptive radiation' has also been used much more broadly for radiations of major groups over long periods of time or much more narrowly, such as encompassing the niche divergence of a single species.

There are often so-called non-adaptive radiations where species flocks are mostly allopatric and occupy similar habitats. In those instances, species within the radiation may become sympatric, but may occupy different niches. Thus, in reality, differentiating adaptive and non-adaptive radiations can be difficult, as is the use of terms such as ecological or non-ecological for radiations (e.g., see Gittenberger 1991; Gittenberger & Gittenberger 2005). For these reasons, we usually use the term 'radiation' without a qualifier.

The initial diversifications in biological and physical attributes which enabled the subsequent diversity were driven by many factors. Briefly, these include physical and biological changes in their environment, including food supply, intra- and interspecific competition, temperature, shelter, etc. and genetic factors such as isolation, founder effects, and genetic drift.

The small initial changes at species level eventually led to major innovations which enabled molluscan colonisation and diversification within a huge variety of habitats. The adoption of different feeding habits led to movement into a wider variety of habitats, profoundly influencing molluscan diversification. The change from grazing on hard, shallow-water surfaces is assumed to have occurred in the earliest molluscs and was a huge influence on the radiation of each major taxon, and this happened in different ways in different groups. Structural and physiological changes were essential, including those to the buccal apparatus and alimentary canal (see Chapter 5) and associated and/or parallel modifications to the general body morphology and musculature (see Chapter 3). Other changes included ventilation and respiration (see Chapter 4), reproduction and development (see Chapter 8), circulatory and excretory systems (see Chapter 6), the complexity and centralisation of the nervous system, and the evolution of an array of sense organs (see Chapter 7). Overviews of physiology, including the immune system and genomics, are given in Chapter 2.

1.4.2 Taxon Diversity

Estimates of the number of species of molluscs vary considerably (see below).

Major problems with achieving reliable estimates include very uneven treatments for different faunal groups and regions, very different levels of synonymy from group to group, and the methods of dealing with estimates of undescribed taxa. Various estimates of all living molluscs range from 46,810 to 200,000 (Groombridge & Jenkins 2002; Chapman 2009), while Zhang (2011) estimated there were more than 117,000 described species of living molluscs. Recent estimates for all marine molluscs range from 52,525 (Bouchet 2006) and 43,689 to 51,689 (Appeltans et al. 2012) (see Table 1.1), with, in the latter case, these figures representing 28%–36% of the actual diversity based on advice from experts. Bouchet et al. (2016) estimated there are 46,000 valid extant marine species and around 150,000 undescribed. The number of terrestrial gastropod species approaches 35,000 (Barker 2001a), and there are about 1000 described species of fresh-water bivalves.

1.4.3 Diversity in Habitat

Molluscs have evolved to adopt very diverse lifestyles and occur in almost every habitat except free flying in the atmosphere. They are often one of the more conspicuous groups and are sometimes predominant. They are ubiquitous in marine habitats, inhabiting all zones from the supralittoral to the deepest oceans, where they live on a wide variety of both soft and hard substrata. While gastropods and chitons are normally associated with hard substrata and bivalves with soft substrata, many exceptions occur (see the relevant taxon chapters in Volume 2). Many bivalves (e.g., mussels, oysters) nestle in or bore into hard substrata or are attached to them. Some bivalves burrow into coral, soft rock, or wood, with 'shipworms' (Teredinidae) the most modified, and these make calcareous burrows in wood. Some small bivalves that live commensally with other invertebrates have become slug-like or limpet-like. Various gastropods burrow in soft sediments, while diverse tiny species live interstitially between sand grains. Scaphopods live entirely in soft substrata.

Members of several groups within the chitons, gastropods, and bivalves have successfully adopted the intertidal as their habitat, even though it is a relatively harsh environment (see Chapter 9).

Numerous groups of molluscs are found in the deep-sea, but the most surprising of all are the amazing faunas associated with hydrothermal vents and seeps which consist of many strange molluscs (mostly bivalves and vetigastropods) which often dominate these communities (see Chapter 9).

Most molluscs are marine, and their diversity is highest in nearshore habitats and on the continental shelf and reduces with increasing depth on the continental slope and beyond. As with many other organisms, marine molluscs have their highest diversity in the tropical western Pacific and decrease in diversity toward the poles. Only one comprehensive study on tropical western Pacific molluscan diversity has been carried out (Bouchet et al. 2002), where around 3000 species were found within a single site in coral reef habitat in New Caledonia.

Molluscs are most abundant and diverse in shallow-water marine ecosystems. In the tropics, they numerically comprise 15%–40% of benthic macroinvertebrates and are exceeded only by crustaceans and polychaetes (Longhurst & Pauly 1987).

Only a few families of bivalves and gastropods have successfully invaded fresh water, but several have undergone significant radiations. Some are found in most fresh-water habitats, including streams in caves and other subterranean water such as interstitial groundwater (stygobionts).

Gastropods are the only group of molluscs to have successfully invaded the terrestrial realm, where they are represented by about 35,000 living species (Solem 1984; Bruggen 1995). Terrestrial gastropods can be found in environments ranging from wet tropical rainforests to deserts (see Chapter 9) (Table 1.2).

1.4.4 Diversity in Feeding Types

Many marine chitons and gastropods scrape a living on hard substrata, possibly in much the same way as the earliest molluscs. From these relatively humble beginnings, there has evolved an amazing radiation of marine forms which occupy all marine habitats and access most of the available resources that the marine environment offers. Most marine molluscs are benthic, and many are infaunal, particularly bivalves. Some gastropods and cephalopods live permanently in the water column, while a few gastropods drift on the surface of the ocean, including the violet snails (Epitoniidae) and a few nudibranch slugs (e.g., the 'sea lizard' *Glaucus*). By far the most obvious and successful pelagic species are the squid. These cephalopods have evolved into fast-moving carnivores with most squid being full-time swimmers, as are some octopods, although many of the latter group are benthic. Pelagic gastropods include the planktonic heteropods and pteropods.

A few groups of bivalves and several groups of gastropods live in fresh water and many gastropods are terrestrial. Surprisingly, some non-marine groups are the most basal of the extant taxa in a clade, examples being in the Caenogastropoda (Cyclophoroidea, Viviparoidea, Ampullarioidea) (see Chapter 19) and the heteroconch bivalves (Unionoidea) (see Chapter 15).

Shell morphology can often be correlated with lifestyle and habitat, but this is by no means always the case, as similar shell morphologies do not necessarily indicate similar habits or habitats (see Chapter 3).

A few molluscs are parasites, with ectoparasitic shelled snails evolving in several groups of gastropods and, in one family (Eulimidae), some members have become shell-less internal parasites (see Chapter 19). Although the adults of fresh-water unionoid bivalves are free-living, their glochidia larvae parasitise fish and, more rarely, amphibians (see Chapter 15).

Suspension-feeding is typical of the great majority of bivalves and has also evolved independently in some groups of gastropods (see Chapter 5). The adoption of carnivory has sometimes led to significant radiations (e.g., as in some groups of neogastropods and nudibranchs), while other carnivorous specialists have low diversity, for example, the aplacophoran Solenogastres and a few neogastropod families (e.g., Harpidae).

Some groups of bivalves (e.g., Lucinidae and Solemyidae) have developed symbiotic relationships with bacteria that live in their modified gills, reducing or eliminating the necessity to uptake food by other means. A symbiotic relationship with

---

TABLE 1.2
Habitats Occupied By Living Adult Members of the Major Molluscan Clades

| Taxon | Marine Benthic | | Water Column | Estuarine | Fresh Water | Terrestrial | |
	Shallow	Deep-sea				Damp	Arid
Polyplacophora	●	○					
Solenogastres	○	●					
Caudofoveata	○	●					
Monoplacophora	○	●					
Bivalvia	●	◐	○	◐	◐		
Protobranchia	◐	●		◐			
Autobranchia	●	◐	○	◐	◐		
Scaphopoda	●	●					
Cephalopoda	◐	◐	●	◐			
Gastropoda	●	◐	○	◐	◐	◐	◐
Patellogastropoda	●	○		◐			
Vetigastropoda	●	●		◐			
Neritimorpha	●	○		◐	◐	◐	
Caenogastropoda	●	◐	◐	◐	◐	◐	○
Heterobranchia	●	◐	◐	◐	◐	●	●

● – predominant, ◐ – well represented, ○ – rare, ∘ – occasional or arguably present.

Modified and expanded from Lindberg, D.R. et al., The Mollusca: Relationships and patterns from their first halfbillion years, in Cracraft, J. and Donoghue, M.J. (Eds.), *Assembling the Tree of Life*, pp. 252–278, New York, Oxford University Press, 2004.

zooxanthellae by the 'giant clams' (*Tridacna*), a few other bivalves, and some 'opisthobranch' gastropods similarly provides alternative nutrition. Some members of one group previously included in the 'opisthobranchs', the Sacoglossa, and now part of the Panpulmonata, incorporate plastids from their host algae which function photosynthetically in the cells of these slugs (see Chapters 5 and 20) (Table 1.3).

1.4.5 MORPHOLOGICAL DIVERSITY

As already noted, molluscs are renowned for their incredible morphological diversity ranging from spicule-covered worms to giant snails, clams, and squid. Some major innovations and the main body types are summarised in Table 1.4. Shell morphology can often be correlated with lifestyle and habitat, but this is by no means always the case, as similar

---

TABLE 1.3

Feeding Types in the Major Molluscan Clades

Taxon	Detritivory	Macro Herbivory	Grazing Carnivory	Micro Carnivory	Hunting	Parasitic	Suspension
Polyplacophora	◉	●	◉		◉		
Solenogastres			●				
Caudofoveata	◉			●			
Monoplacophora	●						
Bivalvia	◉			○			●
Protobranchia	●						
Autobranchia	◉			○			●
Scaphopoda	◉			●			
Cephalopoda				○	●		
Gastropoda	●	◉	●	◉	◉	◉	◉
Patellogastropoda	◉	●					
Vetigastropoda	●	◉	●				◉
Neritimorpha	●	◉					
Caenogastropoda	●	◉	●	◉	●	◉	◉
Heterobranchia	●	◉	○	◉	●	◉	◉

● – predominant, ◉ – well represented, ○ – rare.

Modified and expanded from Lindberg, D.R. et al., The Mollusca: Relationships and patterns from their first half-billion years, in Cracraft, J. and Donoghue, M.J. (Eds.), *Assembling the Tree of Life*, pp. 252–278, New York, Oxford University Press, 2004.

TABLE 1.4

Morphological Diversity of Living Adult Members of the Major Molluscan Clades

Taxon	Ano-Pedal Flexure	Worm-Like	Shell(s) Very Reduced or Absent	Number of Shells	Coiled	Slug	Limpet Shaped	Fish-Like
Polyplacophora		○	○	8		∘	●	
Solenogastres		●	●	NA				
Caudofoveata		●	●	NA				
Monoplacophora				1			●	
Bivalvia	◉	○		2	◉	◉	○	
Protobranchia	◉			2				
Autobranchia	◉	○	○	2	◉	◉		
Scaphopoda	●			1				
Cephalopoda	●		○	1	○			●
Gastropoda	●	◉	○	1 (2 in one small group)	●	◉	◉	○
Patellogastropoda	●			1			●	
Vetigastropoda	●		○	1	●	∘	◉	
Neritimorpha	●		○	1	●	◉	○	
Caenogastropoda	●	◉	○	1	●	◉	◉	○
Heterobranchia	●	◉	○	1 (2)	●	●	○	○

● – predominant, ◉ – well represented, ○ – rare, ∘ – occasional or arguably present.

Modified and expanded from Lindberg, D.R. et al., The Mollusca: Relationships and patterns from their first half-billion years, in Cracraft, J. and Donoghue, M.J. (Eds.), *Assembling the Tree of Life*, pp. 252–278, New York, Oxford University Press, 2004.

shell morphologies do not necessarily indicate similar habits or habitats (see Chapter 3).

Some trends have occurred independently in two or more major groups of molluscs, and these trends have fundamental effects on more than one organ system of the body. Briefly, they include a trend to serially repeat organs such as the gills and shell muscles, and to a lesser extent the auricles, and even the kidneys and gonads in monoplacophorans. Serial repetition has been incorrectly equated with segmentation by some workers in the past (see Chapters 3, 4, and 12 for details).

Another trend is the bending of the originally straight gut into a U-shape, a process known as ano-pedal flexure. In extreme cases, such as in scaphopods and cephalopods, it has resulted in a change in body orientation (see Chapters 3, 8, and 12 for details). Reduction and simplification of body organs and structures due to paedomorphosis, where juvenile characters are carried through to the adults, has also been an important evolutionary trend in some groups of molluscs (see Chapters 8 and 12). Another recurring trend in cephalopods

and gastropods is shell reduction and loss, a process that has resulted in profound implications in body morphology, physiology, and habits (see Chapters 3, 17, and 20).

1.5 SOME GENERAL INFORMATION

1.5.1 CLASSIFICATION

Classification and nomenclature are often confused. Nomenclature is the system of naming and is done in the framework of a complex set of rules formulated by the International Commission on Zoological Nomenclature (ICZN) which apply to the names of family-group taxa (i.e., superfamilies, families, subfamilies) and below (genus-group and species-group taxa). Classification is used to organise information about the organisms we study and to facilitate communication about them. Classifications can be represented as hierarchical lists or branching diagrams (trees – e.g., cladograms). The latter communicate hypothesised phylogenetic relationships (Box 1.6).

BOX 1.6   SYSTEMATICS - SOME TERMINOLOGY

Systematics encompasses three distinct activities: taxonomy, classification, and nomenclature (Figure 1.2). Although many systematists practice these three components, they are often amalgamated under the term 'taxonomy' by those outside comparative biology. Not all systematists work across the full breadth of systematics. For example, they can be engaged in the study of molecular phylogenies without applying the results of their studies to the nomenclature of the group. Similarly, resolving nomenclatural issues can be carried out without a phylogenetic study of the species or the generation of a new classification, but usually not without extensive library resources.

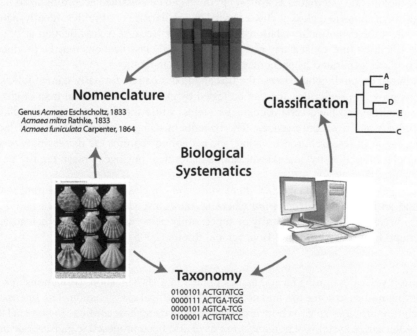

FIGURE 1.2 Biological Systematics – A summary diagram showing the relationship between taxonomy, classification, and nomenclature.

(Continued)

BOX 1.6 (Continued)   SYSTEMATICS-SOME TERMINOLOGY

Sometimes systematics research programs are called 'taxonomy', but this masks the scientific enterprise that modern systematics has become. The three components making up systematics are described below.

Taxonomy

In this process, a primary objective is the study and description of the objects being classified with the ordering (the classification) of an end product. This study of the organisms involves analysis and quantification of the characters they possess, and these may include morphology (hard parts, anatomy, cell ultrastructure), different life history stages (embryonic development etc.), or molecular characters (sequences of genes to genomes). Because of this complexity, character analysis is critical whether it is the homology of morphological structures, alignment of gene fragments, or the physiological state(s) of the individual(s) that produced transcriptome data. Importantly, good taxonomy requires a deep understanding of a taxon and the ability to extract and evaluate the information. Tools used by systematists include microscopy, imaging, histological and molecular facilities, or some subset of them. For extinct taxa, access to isotopic, thin-section, and 3D reconstruction technologies may also be necessary. Unfortunately, and often not from necessity, the taxonomy of many molluscan groups is based on little more than a handful of traditional characters, notably, shell features.

Taxonomy interacts with both nomenclature and classification (Figure 1.2). The taxonomic study describes the characters and states of a taxon. Through interaction with nomenclature, a name can be attached to specimens (grouped as taxa) with unique sets of character states. The interaction of taxonomy with classification requires an additional step – an analysis of the character states, preferably an algorithmic one. There are three major analyses: evolutionary systematics, phenetics, and cladistics. In evolutionary systematics, the analysis depends largely on the systematist and their intimate knowledge of the group to produce an evolutionary scenario. Cladistic techniques can be applied without using computers, but modern phenetic and cladistic analysis uses numerical algorithms which are more computational. Phenetics uses clustering techniques based on overall similarity of the data (e.g., UPGMA [unweighted pair group method with arithmetic mean] and neighbour joining), while cladistic and other phylogenetic reconstruction methods use special similarity (e.g., parsimony) or require an evolutionary model and parameters (e.g., maximum likelihood or Bayesian analysis). Regardless of the method of analysis of the taxonomic data, the process produces a classification.

Classification

Biological classification is where entities (usually species) are grouped into larger groups (taxa). Classifications are also commonly used outside biological systematics as almost anything can be classified according to some criteria or scheme as it is simply the assigning of items to a class or classes based on similarities or affinities. Ideally, a biological classification should attempt to reflect the evolutionary relationships of a group. Because a classification is, of necessity, a hypothesis, more than one classification may exist at any one time. Biological classifications may be produced by evolutionary scenarios or are based on trees generated by phenetic and cladistic computations.

With classifications based on phylogenetic trees, the tips and nodes can be formally named following nomenclatural practices. Classifications provide predictions that can be tested by examining additional taxa or characters. Previously unstudied taxa can be predicted to have certain character states, while the discovery of homoplasy may necessitate re-examination to document putative convergences. The benefits of using classifications that reflect the evolutionary history (phylogeny) of a taxon in research, conservation, and economic ventures are increasingly recognised. Often the replacement of existing classifications by new classifications that reflect phylogeny often require name changes due to nomenclatural requirements (see below).

There are two main types of classification: *ranked classification* – a classification where ranks such as class, order, family, etc. are used and an *unranked classification* where no ranks are used, other than (usually) genus and species. Some classifications are mixed, with ranks to family or superfamily being used, but higher categories are unranked, as for example, by Ponder and Lindberg (1997) and Bouchet and Rocroi (2005).

Nomenclature

Biological nomenclature involves assigning formal names to taxa in a classification. If relationships are represented by a tree, it involves assigning names to some tips and nodes of a hierarchical classification. The International Commission on Zoological Nomenclature provides rules on how taxa will be named and how conflicts in nomenclature (not classification) will be resolved. Alternative systems of nomenclature have also been proposed (e.g., BioCode and PhyloCode), but have yet to gain general acceptance.

(Continued)

BOX 1.6 (Continued)   SYSTEMATICS-SOME TERMINOLOGY

Nomenclature interacts with classification by providing names (and typically ranks[5]) for the different groupings present in the classification and, with taxonomy, by providing unique names to taxa with distinct combinations of character states.

Nomenclature provides relatively stable names governed by a set of rules (unlike the adoption of so-called 'common names') which enables non-specialists (e.g., conservation and economic communities) and specialists to communicate about taxa with minimal confusion. For example, listing species for environmental protection requires a 'scientific name', and the units used to estimate biodiversity are almost always formal scientific names parsed by rank (e.g., species, generic, or familial diversity). For example, names of invasive species must be globally understood to be effective in restricting movement. Likewise, they are needed for the recognition of parasite vectors, patents for natural compounds, and regulating commercial and sport fisheries.

---

[5] Under International Commission on Zoological Nomenclature rules ranks do not need to be assigned for taxa above the family-group. Under PhyloCode, ranks are not assigned except at the species level.

BOX 1.7   SOME COMMON PHYLOGENETIC TERMINOLOGY

Phenetics – taxa are grouped based on overall similarity. Sometimes called numerical taxonomy.

Taxon (plural *taxa*) – any biological entity or group (e.g., species, family, class, phylum). *Crown group taxa* are a clade containing living species and their descendants to the most recent common ancestor. *Stem group taxa* are paraphyletic groups in a clade minus the crown group.

Sister group (or taxon) – is the nearest relative to a particular taxon (or clade) in a phylogenetic tree.

Tree – a branching diagram that represents the relationships between taxa (or any other entities). See https://evolution. berkeley.edu/evolibrary/article/phylogenetics_01 for additional discussion.

Phylogeny – the relationships of organisms, often represented by a tree diagram. The study of phylogeny is phylogenetics.

Phylogenetic methods include:

Cladistics – taxa are grouped based on the possession of shared unique characteristics (synapomorphies) – commonly used with morphological data and less so with molecular data. Computer generated cladistics is usually called *maximum parsimony* as the algorithm employed attempts to find the trees involving the smallest number of changes (and hence the most parsimonious).

Molecular data are most often analysed using model-based approaches such as maximum likelihood and Bayesian inference which use various models of molecular evolution to determine the fit of the data to potential trees.

Maximum likelihood (ML) – assigns probabilities to possible phylogenetic trees using a predetermined model to assess the probability of changes (i.e., mutations). Trees with fewer changes are said to have a higher probability.

Bayesian inference – is similar to ML but is based on the posterior probability distribution of trees. While likelihood methods find the tree that maximises the probability of the data being analysed, a Bayesian analysis finds a tree representing the most likely clades by drawing on the posterior probability distribution.

Total evidence – a cladistic analysis involving both morphological and molecular data.

Some common cladistic terminology:

Clade – a monophyletic group, i.e., a group that shares a common ancestor.

Cladogram (or *phylogram*) – a tree generated using cladistic methods. The branching points are called nodes.

Branch length – the length of a branch on a cladogram where the number of character changes is proportional to the length of the branch.

Apomorphic (also apomorphy) – derived (i.e., modified). A *synapomorphy* is a derived character possessed by two or more taxa in a clade.

(Continued)

BOX 1.7 (Continued)   SOME COMMON PHYLOGENETIC TERMINOLOGY

Plesiomorphic (also plesiomorphy) – primitive (i.e., unmodified). A *symplesiomorphy* is a primitive (i.e., ancestral) character shared by two or more taxa in a clade.

Monophyletic – a group (clade) that shares common descent.

Paraphyletic – a group that does not include all the descendants, but has a common ancestor.

Polyphyletic – a group that does not share a common ancestor.

Grade – a paraphyletic or polyphyletic group united by morphological and/or physiological features. Compare with clade.

There are different philosophical approaches to biological classification. Traditional classifications consist of hierarchical nested groups, labelled with ranked (e.g., class, order, family, genus) names. Monophyly (strict common descent – a clade) is rarely enforced or is untested. Such a classification may contain clades, polyphyletic groups, and grades (paraphyletic groups) which are seldom distinguished so its predictive powers are limited. In contrast, a classification based on phylogeny will (usually) use monophyletic groups and ranks might be used (Box 1.7).

In this book, we mainly use a ranked system for taxa below the family-group level (superfamilies, families, subfamilies, genera, species). We have also used ranks for ordinal-level groups and above and provide a classification for each class of living molluscs in the Appendix in Volume 2. A recent example of a ranked classification and all the complexity it necessitates is that of Carter et al. (2011) for bivalves.

1.5.2 Ecology

Ecology has become a widely used term in different disciplines and can mean different things to different people. In strict biological parlance, it is *the study of the distribution and abundance of organisms* and is often extended to include *the interactions between organisms and their environment.* This book is not a text on molluscan ecology (although one is sorely needed), but we do touch on many aspects that are often considered ecology. For example, the latter part of the definition of ecology just given covers many aspects of biology and physiology, most of which are covered. We do not deal with most aspects of ecology in the strict sense.

To illustrate the broad adoption of the term ecology, some of the recognised subdisciplines of ecology are listed below. The following are covered at least in part in this book in various sections, notably in Chapter 9, but not using these terms.

Ecophysiology – how physiological functions influence interaction with the biotic and abiotic environment. *Behavioural ecology* – how behaviour enables an animal to adapt to its environment. *Evolutionary ecology* – studies ecology considering the evolutionary histories of species and their interactions. *Political ecology* – connects politics and economy to problems of environmental control and ecological change. *Molecular ecology* – examines the relationship between genetics and ecology. *Chemical ecology* – the interaction of organisms with chemicals. *Sensory ecology* – information obtained by an organism from its environment.

We do not cover, or only touch on: *Population ecology* – population dynamics of species. *Community ecology* (or synecology) – interactions between species within an ecological community. *Ecosystem ecology* – examines the flow of energy and matter through the biotic and abiotic components of ecosystems. *Systems ecology* – takes a holistic overview of ecological systems. *Landscape ecology* – examines ecosystems over large geographic areas.

Ecology can also be subdivided according to the area of interest of the researcher – for example, depending on the organisms of interest such as with plant ecology, molluscan ecology, etc. or geographic or habitat focus as with tropical ecology, desert ecology, marine ecology, etc. Other subdivisions relate to the type of science employed, for instance, chemical ecology, genetic ecology, field ecology, theoretical ecology, etc. Thus, the broad use of the term ecology is not to be encouraged as it becomes all-encompassing.

1.5.3 Genetics

The term genetics, like ecology, is broadly used and can cover studies involving heredity, genes, chromosomes, variation, etc. Genetics had its roots in the selective breeding of plants and animals, but modern genetic research is sophisticated and includes gene functions and interactions. We will not provide a synopsis of genetic theories, methods, or terminology here, as these are readily available from many sources (e.g., Silva & Russo 2000; Sunnucks 2000; Deyoung et al. 2005; Karr 2007).

The study of genetic variation is essential in understanding the structure and connections between populations, how speciation occurs, sources of variation, and development. Not only have genetic studies provided much valuable and important scientific data, but there are also many practical applications, especially in fisheries, including commercial shellfish fisheries, managing endangered species, and the control of pest and invasive species. Genetic studies also provide a view of population structure unknowable from other observable traits, providing vital information for management and conservation decision making.

Various techniques have been widely employed in molluscan genetic studies, including allozymes (e.g., Johnson 1976; Johnson et al. 1977; Turner 1977; Colgan & Ponder 2000), DNA-RNA hybridisation (e.g., Karp & Whiteley 1973),

DNA-DNA hybridisation (e.g., Milyutina & Petrov 1989), and radio-immuno-assay (RIA) (e.g., Harte 1992); today most genetic studies use sequences of amino and nucleic acids (e.g., RNA and DNA) and proteomics. Early sequencing of molluscs included the amino acid sequence of histone H2A from *Sepia officinalis* (Wouters-Tyrou et al. 1982), followed later by base pair sequencing from cDNA[6] of *Octopus* rhodopsin (Ovchinnikov et al. 1988) and 18S rRNA sequences representing four molluscan taxa were included in an early molecular estimate of metazoan relationships (Field et al. 1988).

Molluscs have played an important role in our understanding of the principles and properties of genetics, including insights into various evolutionary processes such as speciation, expatriation, vicariance, dispersal, epigenetics, developmental constraints, and species selection. Because of their low mobility and varied dispersal strategies, many molluscs are good subjects for studies on population genetics (e.g., Backeljau et al. 2001; Azuma et al. 2017; LaBella et al. 2017; Proćków et al. 2017; Richling et al. 2017; Rico et al. 2017). This branch of genetics is crucial to our understanding of basic evolutionary processes, as it involves the investigation of changes in the frequencies of alleles, genotypes, and phenotypes within populations, and hence, micro-evolution within species. These changes are driven by selection, mutation, migration, and random genetic drift as well as non-random mating and recombination, some of which are influenced by changes in the abiotic realm.

1.6 HOW TO USE THIS BOOK

In Volume 1, we provide an introductory overview of physiology and genomics and then deal with bodily systems and processes covering structure, function, and physiology in a comparative way. Also included is a general account of various aspects of molluscan biology and natural history

[6] Complementary DNA.

not otherwise dealt with, as well as the significance of molluscs to, and their interactions with, humans, and lastly, aspects of current and future research relevant to the areas covered in this volume.

We have, of necessity, divided the molluscan body into manageable chunks by using the traditional system-based approach. With the focus on systems (digestive, nervous, etc.), the important interrelationships between organ systems can be overlooked. This connectivity through nerves, blood, and chemical messengers is an integral part of the bodily function in molluscs and all other animals.

In Volume 2, we outline the relationships of molluscs to other animals, cover their early fossil history, and then give overviews of the major taxon groups. Where relevant, the taxon chapters include a summary of information covered in the system chapters in this volume and additional detail is provided about some particularly significant systems. Last, we discuss aspects of current and future research relevant to the areas covered in Volume 2. The Appendix in Volume 2 summarises the classifications of all the extant molluscan classes to family level, and the extinct classes are covered in Chapter 13.

We do not provide a glossary, but most terms used are explained in the text where they can be located using the comprehensive index. Most information provided is extensively referenced, and all cited references are listed in the bibliography.

In the book, we give numerous examples using taxon names and, in so doing, we refer to major groups, families, genera, and species. In referring to higher taxa, families, and genera, usually our example is based on one or more species within those taxa. Ideally, such references should be spelled out, for example, as 'a large foot is typical of species of the Turbinidae' or 'a large foot is seen in species of *Turbo*'. For brevity, when higher taxon groups or genera and families are referred to, we generally use shorthand – for example, 'a large foot is typical of turbinids', or 'a large foot is seen in *Turbo*', or '*Turbo* has a large foot'. In these latter cases we, of course, intend that the reference is to species within those groups.

2 Overview of Molluscan Physiology and Genomics

2.1 INTRODUCTION

In this chapter, we give a brief overview of the main aspects of molluscan physiology, immunology, and genomics. We have not attempted to review these topics in detail because from what is known (much of the available information on molluscs is based on relatively few taxa), the general principles are like those in other animals, and detailed information can be obtained from general texts and web sources.

2.2 PHYSIOLOGY

Molluscan physiological studies have a long history because many molluscs are easily obtained and kept in laboratories, and because some are of economic interest. There are, however, only a few comprehensive reviews. The earliest was that of Winterstein (1910–1925), but the two volumes published in the 1960s (Wilbur & Yonge 1964, 1966) were the first dedicated treatment. The 12 volume series, *The Mollusca* (Wilbur 1983–1988), contained many chapters that dealt with aspects of physiology. Most recently an edited two-volume compilation of reviews by Saleuddin and Mukai (2016) covered various aspects of molluscan physiology.

In many respects, molluscs are physiologically analogous to vertebrates, annelids, arthropods, and flatworms, as all those groups have successfully colonised marine, fresh-water, and terrestrial environments. As with all animals, molluscs have a similar basic metabolism, although their metabolic responses to their environments sometimes differ. Besides the information below, the chapters on the organ systems also give, where appropriate, a brief summary of the physiological processes, sources of information, and note where molluscs appear to differ from other animals.

The functions occurring within an animal at the levels of molecules, cells, and organs are interrelated and interdependent and are sometimes difficult to compartmentalise neatly. Thus, of necessity, the separation of Chapters 3 through 7 into organ systems is in part artificial, particularly at the molecular and cellular level. This interrelationship of bodily functions and processes is an important concept in evolution – the body works as a unit, and for it to function well, integration between all the physiological and anatomical systems is required (Figure 2.1). This was demonstrated in a model developed by Menshutkin and Natochin (2007) where they analysed the effects of changes in consumption of food, levels of oxygen, and blood flow to the kidneys and how those changes impacted on the ability to extract and excrete metabolic waste. For example, while an increase in

the efficiency of renal activity has a positive effect on the activity of each cell in an individual, it also increases the energy expenditure of an organism. An awareness of such interrelationships is necessary to not only better interpret physiological and other biological studies, but also to understand their evolution.

Some important aspects of molluscan physiology which do not relate directly or solely to other chapters – namely, metabolism, bodily defences (including the immune system), and the endocrine system, are all briefly outlined below. Explanations of some terms that relate to physiology used elsewhere in this book are provided in the overview below.

2.2.1 EXTERNAL FACTORS

To maintain bodily functions, an animal must uptake various chemical elements and molecules from the external environment, and their concentration in the environment can affect metabolism. The main compounds in cells (including carbohydrates, proteins, lipids, and nucleic acids) require hydrogen (H), oxygen (O), carbon (C), nitrogen (N), sulphur (S), and phosphorus (P). The ionic balance of fluids in cells and the medium external to them is maintained by reactions involving potassium (K), sodium (Na), magnesium (Mg), calcium (Ca), and chlorine (Cl). Several other elements are used in small amounts in a wide range of metabolic reactions. Many of these 'trace elements' are metals (see Section 2.2.5) involved in metalloenzymes, reduction-oxidation (redox), and oxygen transport.

Changes in the external environment may initiate responses in an animal. Such changes may be chemical stimuli such as a change in pH, salinity, oxygen levels, pollutants, or indications of the presence of mates, predators, food, etc.; or physical stimuli such as changes in temperature, light, wave action or currents, change of substrata; or mechanical stimuli, such as being touched, can also have effects. Interestingly, most molluscs do not appear to have any significant response to sound (see Chapter 7). The responses to stimuli include a primary response from the nervous system or may involve changes in hormone levels, followed by secondary responses that are metabolic changes, such as increases in glucose and/or lactate immune function changes, and regulation of protein encoding genes involved in non-immune functions. Tertiary responses include changes in behaviour (such as aggression, mating, feeding) and 'health' (reproduction, digestion, growth, disease resistance). Three of the most important external factors that affect aquatic molluscs are temperature, salinity, and oxygen levels, and these are discussed in more detail below.

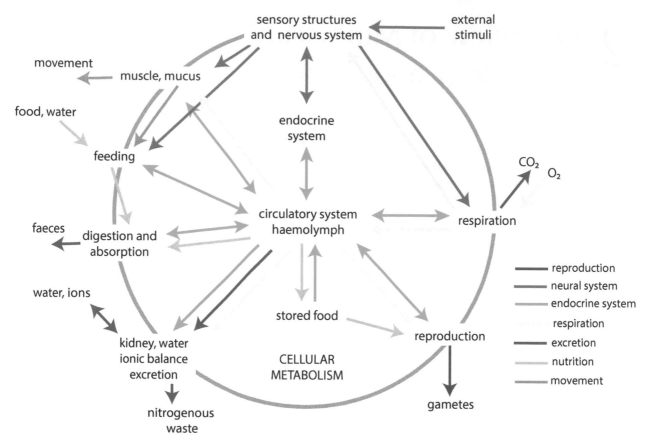

FIGURE 2.1 Diagram showing some of the main physiological interactions in a cell and the body. Original.

Ultraviolet radiation is also an important external factor for intertidal invertebrates, and it has synergistic effects with both salinity and temperature stresses (Przeslawski et al. 2005). Gastropod responses (cellular, physiological, and pharmacological) to these and other stressors have been reviewed by Coutellec and Caquet (2017).

2.2.1.1 Temperature

Molluscs are all *ectothermic*, with their body temperature controlled by the external environment. An overview of how temperature affects the maintenance of metabolic energy balance in marine invertebrates is provided by Newell and Branch (1980) and Brockington and Clarke (2001), some background to physiological adjustments that occur with temperature changes is given in Chapter 9. All animals have a range of temperatures they can tolerate (thermal tolerance) (Kinne 1970; Pörtner 2002), and this is obviously relevant in consideration of the impacts of global warming (see Chapter 10). Animals able to tolerate a wide range of temperatures are termed *eurythermic*, while those that cannot are *stenothermic*.

Physiologically, temperature is linked to many essential metabolic processes including protein function and the rate of many enzymatic reactions. Consequently, temperature has impacts on respiration (see Chapter 4), shell formation (see Chapter 3), digestion (see Chapter 5), excretion and blood circulation (see Chapter 6), reproduction (see Chapter 8),

and neuronal integration, especially neuromuscular control/coordination, such as locomotory capabilities (Clarke et al. 2000). When body temperature approaches the thermal tolerance of an animal, behavioural mechanisms are activated (e.g., decreased locomotion), and increased production of heat shock proteins (HSPs) helps prevent protein damage and disruption of physiological functions (Greenberg et al. 1983; Laursen et al. 1997; Tomanek & Somero 1998, 1999). HSPs are expressed in response to numerous stressors, including cold shock, ultraviolet radiation and injury, as well as high temperature (Sanders 1993; Snyder et al. 2001; Dahms & Lee 2010; Li & Guo 2015) (see also Section 2.4.1). Desiccation, which is highly correlated with environmental temperature and moisture content, is very important for intertidal and terrestrial taxa, and HSPs also have significant roles in both these environments (Dong et al. 2008; Reuner et al. 2008; Evgenèv et al. 2014). Some taxa, mainly non-marine, avoid the worst effects of certain environmental stressors, notably temperature and desiccation, by aestivating (see Section 2.4.6).

Many studies have examined the effects of temperature on molluscs and summaries are provided by Newell and Branch (1980) for marine molluscs, Dillon (2000) for fresh-water molluscs, Gosling (2003) for marine bivalves, and Riddle (1983) for terrestrial gastropods. The thermal ranges of most adult marine bivalve species fall between −3°C and 44°C (Vernberg & Vernberg 1972), but larvae have a narrower range

(Przeslawski et al. 2015). Most bivalves require a higher temperature for spawning than for growth (Gosling 2003). Like other animals, molluscs can adjust to temperature changes (acclimation), although this can take a considerable time (often weeks rather than hours or days) (Segal 1961). Temperature tolerance is typically related to the habitat of the animal, with those undergoing large temperature ranges usually having tolerances reflecting the thermal range. Tomanek and Somero (1999) demonstrated that HSP synthesis in the vetigastropod *Tegula* living in different thermal niches (intertidal versus subtidal) reflected their respective thermal niches and temperate to tropical gradients. They concluded these differences were genetically determined and had an important role in determining thermal tolerances (see also Dong et al. 2008). Thus animals living in sub-Arctic or sub-Antarctic conditions will have a greater tolerance for freezing than those on tropical shores and, conversely, tropical animals will tolerate higher temperatures and have mechanisms for coping with those conditions (e.g., Marshall & McQuaid 2011; Marshall et al. 2011).

Marine sublittoral taxa will typically experience a smaller range of temperature than those in the intertidal or terrestrial environments. These patterns are also present at the molecular level. Stenseng et al. (2005), however, have suggested that it may be the warm-adapted, eurythermic intertidal species under greater threat from warming than subtidal species, which are less heat tolerant. In intertidal and terrestrial environments behaviour is often an important mechanism in mediating temperature fluctuations (e.g., Monaco et al. 2017) (see Chapter 7). While most marine species do not experience wide temperature fluctuations, those found at deep-sea hydrothermal vents can experience changes of more than 13°C over seconds to hours (Johnson et al. 1988; Chevaldonné et al. 1991).

2.2.1.2 Salinity

Salinity is an important factor in aquatic molluscan physiology (Berger & Kharazova 1997; Gainey & Greenberg 1977). As discussed in Chapter 6, marine taxa are typically osmoconformers with the osmolality in their body tissues the same as that of the surrounding water, but those in brackish or fresh water must osmoregulate to maintain an internal osmolality by active regulation, despite that of the external environment. In habitats that experience short-term fluctuations in salinity levels, shelled molluscs can avoid these changes by closing up. Thus, living in fresh-water habitats requires special modifications as does living in high salinities such as occur in evaporated high shore pools, coastal lagoons, and inland saline pools and lakes. Salinity is not only a major environmental factor for adult molluscs, but for the dispersal and settlement of their larvae (e.g., Pechenik 1982; Berger & Kharazova 1997; Montory et al. 2014). Salinity also plays an important role in ocean acidification and has complex interactive effects on molluscan development together with both temperature and pH (Ko et al. 2014).

A general overview of the importance of salinity in molluscan physiology and the terminology involved is given in Chapter 6, and for an overview of high and low salinity habitats see Chapter 9.

The interplay between temperature and salinity is complex and has been investigated in several studies. For example, experiments conducted on the mussel *Geukensia demissa*[1] showed that mortality due to freezing was mainly due to cell dehydration from tissue water being lost as ice and that mortality could be reduced by increasing the salinity up to a point when toxicity occurred from the salt (Murphy & Pierce 1975).

2.2.1.3 Oxygen

As in other animals, the levels of oxygen in the environment surrounding a mollusc are critical to its physiological functioning. As with salinity, there are both conformers and regulators, with the former being respiratory-dependent and the latter being independent. Thus, *oxyconformers* vary their oxygen uptake relative to its availability, while *oxyregulators* largely maintain their oxygen uptake despite any external decrease, although they reach a critical level at which respiration is compromised. In a study on some bivalves, Bayne (1971) was the first to demonstrate that some molluscs were oxygen conformers. More recently, Carey et al. (2012) showed that the subtidal chiton *Leptochiton* oxyconformed and apparently lacked the ability to oxyregulate, while oxyregulation occurred in intertidal chitons which, because of their habitat, experience large fluctuations in oxygen levels. In most terrestrial environments, the oxygen level remains fairly constant, but in water, it fluctuates considerably. Molluscs that can tolerate a low oxygen environment have evolved adaptations such as oxyregulation to cope with these situations. In habitats that experience short-term fluctuations in oxygen levels, shelled molluscs can avoid low levels by closing up. Other solutions are needed in environments with long-term or permanently reduced oxygen levels. Such anoxic habitats are encountered, for example, in the deep-sea, in reducing environments (see Chapter 9), or in stagnant water. Some adaptations to these environments include the use of haemoglobin as a respiratory protein (see Chapter 6), increased respiratory surfaces (see Chapter 4), and adoption of anaerobic metabolism for vital processes (see Section 2.2.4.3).

The oxygen- and capacity-limited thermal tolerance hypothesis (OCLTT) links an animal's oxygen supply strategy to tolerance of other stressors, notably temperature. According to this hypothesis, the relative balance between oxygen supply and demand sets the thermal tolerance range in aquatic organisms, including molluscs. Thus, in this view, oxygen limitation ultimately sets the upper thermal tolerance threshold. For example, the bivalve *Laternula elliptica* exhibits an increased upper thermal tolerance limit under hyperoxic conditions (Pörtner et al. 2006b).

2.2.2 Cell Structure and Cellular Metabolism

Generally, both cell structure (see Box 2.1) and the metabolic pathways in animal cells are highly conserved, being essentially the same in most animals. Some basic details are outlined in Box 2.1 through Box 2.5. Cellular metabolic processes

---

[1] As *Modiolus demissus*.

are essential to body function and involve families of proteins, typically as enzymes (see Box 2.3), that are also conserved, such as adhesion molecules, heat shock molecules, calcium-binding molecules, growth factors, and signalling molecules (e.g., neuropeptides, hormones, and cytokines) responsible for inter-cell communication (see Section 2.3).

2.2.3 ENERGY SOURCES – FOOD SUBSTRATES

Energy sources for molluscs come from all three branches of the tree of life – Archaea, Bacteria, and Eukaryota. While Archaea and Bacteria are important for many grazing, suspension-feeding, and chemosynthetic molluscs, eukaryotes are the preferred food source for most molluscs, especially the 'algae' (Phaeophyceae, Rhodophyta), Plantae, Fungi, and Animalia, although 'algae' and Plantae are richer sources of carbohydrates than fungi and animals, which are generally richer in proteins. All these eukaryote substrates provide the energy sources described below.

2.2.3.1 Carbohydrates

Carbohydrates (saccharides) are large molecules composed of carbon, hydrogen, and oxygen, with the hydrogen:oxygen ratio usually 2:1. They include various sugars (glucose, glycogen, starch, etc.) and range from being simple monosaccharides (e.g., lactate, glucose) to complex polysaccharides (glycogen, amylase or chitin, starch, and cellulose). Sugars such as sucrose and fructose are disaccharides, consisting of two joined monosaccharides. Many carbohydrates found in food items are complex (e.g., starch and cellulose) and need

---

BOX 2.1   CELL STRUCTURE

Cells contain many internal structures, some of which are enclosed in membranes (organelles), and others are not. Here, we summarise the main components in cells and their function, but recommend that more detailed texts also be consulted. Basic cell features are:

Cell membrane (or plasma membrane): this thin, semipermeable membrane surrounds the exterior of the cell and maintains its shape and integrity. It is composed of proteins and lipids and has many functions, including being where the sodium-potassium pumps are located (see Box 2.5).

Cytoplasm (or protoplasm): contains the organelles, the nucleus, and the endomembrane system which are suspended in the cytosol (the liquid component of the cytoplasm), which itself is a complex mixture of ions, proteins, and other molecules, including enzymes.

Cytoskeleton: consists of microfilaments, microtubules, and filaments composed of fibrous proteins. It maintains cell shape, anchors internal structures, and provides tracks for movement of vesicles and organelles.

Cilia and *flagella*: motile filament-like structures composed of microtubules that extend beyond the cell exterior.

Nucleus: contains nucleoplasm, the chromosomes, and one or more nucleoli, and is involved in the synthesis of ribosomes. It is surrounded by a double membrane with pores, and its outer membrane is continuous with the endoplasmic reticulum.

Endomembrane system: within the cytoplasm and includes the endoplasmic reticulum, Golgi apparatus, lysosomes, and peroxisomes.

Endoplasmic reticulum: a network continuous with the outer nuclear membrane. The rough endoplasmic reticulum is involved in protein synthesis and has ribosomes, while ribosomes are absent in the smooth endoplasmic reticulum which is involved in the synthesis of membrane lipids.

Golgi apparatus: consists of stacked flattened vesicles that receive proteins from the endoplasmic reticulum and form secretory vesicles to transport the proteins to other parts of the cell (e.g., vacuoles, lysosomes, or to the exterior as a secretion).

Lysosomes: organelles containing enzymes that break down large molecules.

Peroxisomes: small organelles which degrade amino acids and fatty acids and break down the resulting hydrogen peroxide.

Ribosomes: composed of numerous types of protein and three types of ribosomal RNA (rRNA - 18S, 28S, and 5.8S) and involved in protein synthesis. They lack a membrane, so they are not true organelles.

Mitochondria: found in all cells and are the cell 'powerhouse', being the site of adenosine triphosphate (ATP) (see Box 2.4) production through aerobic respiration (see Section 2.2.4.2). They have their own genomes and are surrounded by a double membrane; the inner membrane forms complex folds (the cristae) which increase surface area for ATP production. The mitochondria are filled with a matrix where carbon metabolism and protein synthesis are carried out, and this is also where the mitochondrial DNA is located. Mitochondria are related to Rickettsiales α-Proteobacteria which, in early eukaryotes, are thought to have entered into an endosymbiotic relationship (Gray et al. 1999).

BOX 2.2   METABOLISM

All animals take up and store nutrients to provide energy for maintenance, growth, and reproduction. In Chapter 5, we review feeding and digestion, but do not describe how food is used to provide energy to carry out metabolic activities in the body. Some of these processes are under hormonal control and are outlined under endocrine systems below (see Section 2.3.1).

Metabolism comprises the chemical reactions needed to maintain life and falls into two main categories:

Anabolism: macromolecules are synthesised in the body – e.g. synthesis of proteins, nucleic acids, or urea and uric acid from ammonia.

Catabolism: metabolic substrates are broken down into simpler compounds and, in that process, energy is released for various functions. The main metabolic substrates are briefly described in Section 2.2.3.

Catabolic metabolism provides the energy used in anabolism and in other metabolic processes (e.g., work by muscles, transport of molecules across membranes, etc.). The energy resulting from catabolic metabolism is stored as high energy phosphate compounds (*phosphagens*) and used to do 'work'. The most important of these is ATP (Box 2.4). The mitochondria provide the majority (about 95%) of ATP by aerobic metabolism of glucose, with glycolysis providing only about 5%. These processes are described briefly below.

BOX 2.3   ENZYMES

Most processes within a cell require enzymes. These proteins are highly specific in their function, such as catalysing the conversion of particular molecules (substrates) into other, often simpler compounds (products). Molecules that act as 'activators' increase enzymatic reactions while others ('inhibitors') decrease them.

Enzymatic reactions in living systems fall into six main classes:

Oxidoreductases: Catalyse the transfer of an electron from one molecule (the electron donor) to another (the electron acceptor). Oxygen may or may not be involved.

Transferases: Move functional groups (i.e., 'group transfer') from one molecule to another. For example, when phosphate groups are transferred between ATP and other compounds, or sugar residues are transferred to form disaccharides, etc.

An example of a transferase is arginine kinase (AK), a kinase that maintains ATP levels. It enables phosphorylation of phosphagens, a high energy source from which ATP can be quickly replenished. Phosphoarginine is the most important phosphagen in many invertebrates and is found, for example, in molluscan muscle. The equivalent compound in vertebrates is phosphocreatine (or creatine phosphate).

Hydrolases: Facilitate hydrolysis, the cleaving of single bonds by adding the elements of water (i.e., OH^- and H^+). For example, phosphatases break the oxygen-phosphorus bond of phosphate esters while other hydrolases function as digestive enzymes, e.g., by breaking the peptide bonds in proteins. Ligases carry out the opposite reaction.

Ligases: Form a single bond by removing the elements of water (c.f. hydrolases) and involve cleaving high energy phosphate bonds in ATP, or some other nucleotide.

Lyases: Form or remove a double bond with group transfer. The functional groups transferred by these enzymes include amino groups, water, and ammonia.

Isomerases: Change the position of a functional group within a molecule, but the number and kind of atoms remain the same in the altered molecule (the isomer).

Some enzyme groups do not fit neatly into the above classification, as for example, the adenylating enzymes (AMPylators) involved in the activation of chemical building blocks (amides, thioesters, and esters) within the cell. Adenylation (also known as AMPylation) occurs when an adenosine monophosphate (AMP) molecule is attached by covalent bonding to a protein side chain. Two important adenylate-forming enzymes are adenylate cyclase (a lyase) and adenylate kinase (a transferase), both of which are found in all animal cells. AMP is involved in the synthesis of ATP.

BOX 2.4   ATP

The ATP molecule is the main source of energy in all cells and is involved in many biochemical pathways. It consists of ribose (a sugar), three phosphate groups, and adenine.[2] There are very high energy bonds between the phosphate groups in ATP and when a phosphate group is removed that energy is released. Cells can make ATP from adenosine diphosphate (ADP) by using the energy (E) released from the breakdown of glucose, and that energy is then stored in ATP until needed. This process occurs in the cytoplasm and mitochondria.

The exothermic reaction (hydrolysis) is reversible and can be summarised as:

$$ATP + H_2O \rightarrow ADP + Pi + E, \text{ where Pi is a phosphate group.}$$

The levels of 'adenylate energy' in molluscs have been used as an indicator of stress, for example, as induced by reduced salinity in estuarine molluscs (e.g., Ivanovici 1960; Rainer et al. 1979).

---

[2] Adenine is one of the two purine nucleobases (the other is guanine) predominantly forming nucleotides of nucleic acids.

BOX 2.5   SODIUM-POTASSIUM PUMPS

The sodium-potassium pumps (Na^+/K^+ pump) (also known as Na^+/K^+-ATPase or sodium-potassium adenosine triphosphatase or just sodium pumps) are found in the plasma membrane of all animal cells and are powered by ATP. They pump sodium out and potassium into the cell, producing an electrical and chemical gradient across the cell membrane. This active transport by the enzyme ATPase results in cells having rather high potassium and low sodium ion concentrations because three sodium ions are moved out for every two potassium ions which move in across the plasma membrane. This facilitates several cellular functions including regulation of the membrane resting potential[3] and cell volume. In addition, the movement of sodium ions out of the cell facilitates the importation of nutrients such as glucose and amino acids into the cell by way of several secondary active transporters in the cell membrane that utilise the sodium gradient. It also functions as a signal inducer in cells through phosphorylation[4] of tyrosine and plays a very important role in nerve cell membranes.

---

[3] The membrane resting potential is the difference in electrical charge between the interior and exterior of a cell, with the interior being more negatively charged than the exterior.

[4] Modification of a protein by the addition of a phosphate group to an amino acid.

---

to be broken down into simple sugars (monosaccharides) to be assimilated. The most important monosaccharide is glucose, which is metabolised by nearly all organisms in the processes described below (see Section 2.2.4.1). Carbohydrate metabolism is described generally for molluscs by Goddard and Martin (1966), for bivalves by Zwaan (1983), gastropods by Livingstone and Zwaan (1983), and cephalopods by Storey and Storey (1983).

Glycogen is a complex polysaccharide utilised as a short-term energy store in animals (and fungi) in the same way that starch is used in plants and is often contained in specialised cells. It is initially broken down by glycogen phosphorylase and, through a series of steps, is converted to glucose 6-phosphate which can be used in some metabolic pathways or reduced to glucose.

Complex carbohydrates are not the only source of glucose, as it is also obtained by the breakdown of lipids and proteins as outlined below.

2.2.3.2   Lipids

Lipids are complex compounds that are oily, greasy, or waxy and are soluble in organic solvents such as chloroform or benzene. The basic components of lipids are fatty acids. Saturated fatty acids have a single bond in the carbon chains while unsaturated fatty acids have double bonds. Fatty acids are rarely found free – they are usually combined with other molecules. Lipids form cell membranes and are also stored as fat. Glycerides are another lipid, and complex triglycerides are often stored. There are several other kinds of lipids found in molluscs, including phospholipids, sphingolipids, steroids, and terpenes.

2.2.3.3   Proteins

Proteins are large macromolecules comprising amino acids and are very important in cellular structures – for example, connective tissue and collagen, in physiological functions such as muscle contraction, and as enzymes.

2.2.3.4 Lipid and Protein Metabolism

Nitrogen metabolism in molluscs is reviewed by Florkin (1966), amino acid metabolism by Bishop et al. (1983), and lipid metabolism by Voogt (1983).

Both lipids and proteins can be broken down to simple carbohydrates that can then be readily metabolised. Complex lipids (triglycerides) are important as sources of energy for ATP synthesis. When they undergo hydrolysis, they produce glycerol (a carbohydrate) and fatty acids. Fatty acid metabolism usually involves a thiokinase and ATP, to form acetyl coenzyme A (acetyl-CoA) which can be acted on by, for example, the citric acid cycle.

Protein metabolism involves the breakdown into constituent amino acids. These metabolic processes release nitrogen (N) and sulphur (S), as well as H, C, and O. The first step in metabolism of amino acids is to remove the amino (NH_2) group so further metabolism can lead to either pyruvate (and then to glucose) or to acetyl-CoA and ketone bodies.

As in vertebrates, molluscs have a complex ketone body metabolism. The enzyme β-hydroxybutyrate dehydrogenase (BHBD) catalyses the interconversion of the ketone bodies, acetoacetate and β-hydroxybutyrate, both of which are important metabolic substrates. The activity of BHBD is relatively high in fresh-water gastropods, bivalves, and terrestrial gastropods, but has not been detected in marine bivalves and gastropods (Stuart & Ballantyne 1996). These authors suggested this may be related to differences in the metabolic organisation of marine and non-marine molluscs, but a detailed explanation has not yet been provided. When they mapped the presence/absence of BHBD onto a phylogeny of bivalves and gastropods, four independent occurrences of the enzyme were indicated, corresponding with the occurrences of non-marine taxa in the phylogenetic tree. Thus, in marine molluscs, ketone bodies are not used (Stuart et al. 1998), and instead, amino acids are the preferred substrates (Ballantyne & Storey 1983).

Most squid use the degradation of muscle and other protein to supply energy when necessary and store little energy as lipid (to about 6% of body weight compared with 15%–25% in fish) or glycogen (less than 0.5% of body weight) (O'Dor & Webber 1986), although some deep-sea species have a higher lipid content (Rosa et al. 2005).

Anaerobic end products accumulate in bivalve muscles during anaerobic metabolism, but these may be converted back to the original substrates when oxygen becomes available. In some, this is done in the muscle tissue while in others at least some of these end products (especially propionate and acetate) are transported by the blood to other parts of the body where they become available as substrates for aerobic metabolism or are converted back to the original substrate (e.g., Zwaan & Wijsman 1976). Gastropods that use the lactate opine (i.e., octopine, strombine, alanopine), and succinate pathways during anaerobic metabolism may distribute the end products to other tissues by way of the blood. The octopine produced anaerobically in muscle by cephalopods is mostly distributed in the blood and used by other tissues as their aerobic metabolic substrate (Storey & Storey 1983), but

at least some is metabolised *in situ*. In the tissues that take up octopine, the arginine and pyruvate portions are metabolised aerobically (e.g., in the kidneys) or resynthesised to glucose (as in the digestive gland) which is recycled to muscle.[5]

Unlike those of most vertebrates, mitochondria in cephalopod heart muscle cannot oxidise lipids, but instead, amino acids such as proline are utilised (Ballantyne et al. 1981), a situation analogous to that in insect flight muscles (Ballantyne 2004). Mitochondria in the hearts of other molluscs, such as bivalves, also preferentially oxidise amino acids (e.g., Ballantyne & Storey 1983).

2.2.4 Aerobic and Anaerobic Cellular Respiration

These metabolic processes convert the biochemical energy in nutrient substrates into ATP. It is the main source of energy in a cell and can occur either in the presence of oxygen (aerobic metabolism) or when oxygen is absent or at very low levels (anaerobic metabolism). Aerobic metabolism is much more efficient than anaerobic metabolism, as during anaerobic respiration only two ATP molecules per molecule of glucose are produced, compared with 36–38 ATP molecules per substrate molecule in aerobic respiration. However, ATP is produced more rapidly in anaerobic respiration, so this source of energy may be utilised as an additional source during bursts of activity by muscle cells. Anaerobic respiration can be used by a cell even before oxygen levels are depleted. Aerobic metabolism is, however, able to support a higher metabolic rate.

Muscular activity, in particular, requires a great deal of energy, but most muscle cells only store enough ATP for a few contractions. Glycogen is another energy source in muscle cells, but, unlike vertebrates, in many molluscs most of the energy needed for contraction is obtained from phosphagens, the most important being creatine phosphate.

2.2.4.1 Glycolysis

This is the first stage in breaking down glucose and involves splitting the glucose molecule into two pyruvate molecules, using ATP and multiple enzymes. This step, which does not require oxygen, occurs in both aerobic and anaerobic metabolism, but from this point on the two processes differ, as outlined below.

2.2.4.2 Aerobic Metabolism

Aerobic metabolism is the release of energy by the oxidation of a carbohydrate substrate, and a common example is the oxidation of glucose to carbon dioxide and water. The first step, glycolysis, occurs in the cytoplasm. The pyruvate then moves into the mitochondria where, after involving the electron transport chain (ETC), two other processes take place; the *citric acid cycle* (= Krebs cycle or tricarboxylic acid cycle – TCA) and oxidative phosphorylation. The complete process involves glucose being broken down into carbon dioxide, water, and energy as ATP.

---

[5] These processes are analogous to the metabolism of lactate in the Cori cycle of vertebrates.

$C_6H_{12}O_6$ (glucose) + $6O_2$ (oxygen) = $6CO_2$ (carbon dioxide) + $6H_2O$ (water) + Energy for 36–38 ATP

The citric acid cycle is a series of enzyme-catalysed reactions in which pyruvate is oxidised and converted to acetyl-CoA and then to CO_2 and water. The first step involves the enzyme citrate synthase in the mitochondria. Energy is acquired by the reduced coenzymes[6] which transfer the energy through the electron transport chain in the inner mitochondrial membrane. The ETC is a series of electron transfers coupled with the transfer of protons (H+ ions) across the mitochondrial membrane. ATP synthesis is driven by the proton gradient resulting from the proton transfer. The energy given to the coenzyme and to succinate[7] by the TCA is transferred via a series of steps in the inner mitochondrial membrane in which ADP is converted (by oxidative phosphorylation) to ATP.

The difference in proton concentration on either side of the mitochondrial membrane produces both an electrical potential and a pH potential across the membrane. The membrane potential is utilised by the enzyme ATP synthase for the phosphorylation of ADP to ATP. An important aspect of the breakdown of carbohydrates is balancing the reduction-oxidation (redox) reactions. In aerobic metabolism 'hydrogen shuttles' in the mitochondria are used to transport reducing agents across the membrane. The two main hydrogen shuttles are the α-glycerophosphate (α-GP) shuttle and the malate-aspartate shuttle (Ballantyne 2004). In both insects and cephalopods, the redox balance between the cytoplasm and the mitochondria is maintained by α-glycerophosphate dehydrogenases (α-GPDH) (Storey & Hochachka 1975).

2.2.4.3 Anaerobic Metabolism

Anaerobic metabolism involves respiration without oxygen, and it occurs when oxygen is absent (*anoxia*) or in short supply (*hypoxia*). An ETC (see above) is used to break down the pyruvate obtained by glycolysis by way of *fermentation*, a process used by a wide range of organisms which involves other substances (with smaller reduction potentials than oxygen) as the electron acceptors. These include sulphate SO_{4-2} nitrate (NO_3^-), sulphur (S), or substrates such as fumarate (in molluscs it is the latter).

In anaerobic metabolism, the ETC occurs in the cytoplasm, not the mitochondria, and produces lactic acid (it is also known as *lactic acid fermentation*), and the ATP generated is produced by substrate phosphorylation. A similar anaerobic fermentation, alcoholic (or ethanol) fermentation, occurs in yeast cells.

Anaerobic metabolism commonly occurs during work in muscle cells, with variations occurring even within different muscle systems in the same species. For example, in *Haliotis* anaerobic energy is supplied from arginine phosphate and from anaerobic glycolysis. In the shell muscle, arginine phosphate levels decreased and arginine levels rose, but neither showed significant changes during work in foot muscle, while both types of muscle had significant increases in the metabolites tauropine and D-lactate (Donovan et al. 1999a). There can also be differences between closely related species with, for example, a small, active, tropical *Haliotis* (*H. asinina*) was more dependent on higher levels of aerobic metabolism than a larger, sedentary, temperate species (Baldwin et al. 2007).

The capacity for anaerobic metabolism differs in different taxa depending on their modes of life. Some animals have periodic oxygen deficits (*facultative anaerobes*), for example, air breathers submerged during high tides. Other species have a highly developed capacity for anaerobic metabolism, as for example, many autobranch bivalves (Zwaan & Wijsman 1976), which might be expected in these and other taxa that lack a respiratory pigment in their blood. Under normal conditions, anaerobic metabolism occurs in patellogastropods, which also lack blood pigments (e.g., Michaelidis & Beis 1990; Marshall & McQuaid 1992) and increases under thermal stress conditions (Bjelde & Todgham 2013; Zhang et al. 2014b). Strict or obligate anaerobes can survive in the absence of oxygen for long periods or permanently. Even usually aerobic molluscs can use anaerobic respiration when needed – for example, while aestivating (see Section 2.4.6). In oysters, heart tissue can live indefinitely without oxygen by using the amino acid aspartate as the major substrate for anaerobic metabolism (e.g., Collicutt & Hochachka 1977).

In a study of abalone, Donovan et al. (1999a) showed that 27% of the energy used during 'fast' movement comes from anaerobic and 73% from aerobic sources, with the anaerobic component only in the foot muscle. These authors compared the energy costs for vertebrates and other animals including other molluscs (see Figure 2.2). They concluded that abalone use more energy than the average similar-sized running vertebrate. Although adhesive crawling is generally thought to be more energetically expensive than running, swimming, or flying, probably due to the production of mucus, it was only when the anaerobic component was included that abalone energy costs exceeded that of running vertebrates. The measurements for the four subtropical marine vetigastropod snails (from Houlihan & Innes 1982) in Figure 2.2 and the temperate marine caenogastropod snail *Littorina* (from Innes & Houlihan 1985) are probably underestimates since no measurement of the anaerobic contribution was made.

Swimming molluscs have higher costs of transport than either medusa or the teleost fish *Oncorhynchus nerka* (Figure 2.3). Within cephalopods, *Nautilus* and cuttlefish have the lowest costs coupled with the lowest speeds, while squid such as *Doryteuthis* and *Illex* have higher costs and are capable of the highest speeds among cephalopods. The highest costs of transport are found in scallops which use rapid

---

[6] The ETC uses the coenzyme nicotinamide adenine dinucleotide (NAD) which accepts and donates electrons. Its low energy form (NAD+) is converted to the high energy (reduced) form NADH by accepting two electrons and a hydrogen atom. Other reduced coenzymes include ubiquinone, the cytochromes, and nicotinamide adenine dinucleotide phosphate (NADP).

[7] Succinate ($C_4H_4O_4$) is an intermediate compound in the citric acid cycle which can donate electrons to the electron transport chain when it is converted to another intermediate compound, fumarate ($C_4H_2O_4$).

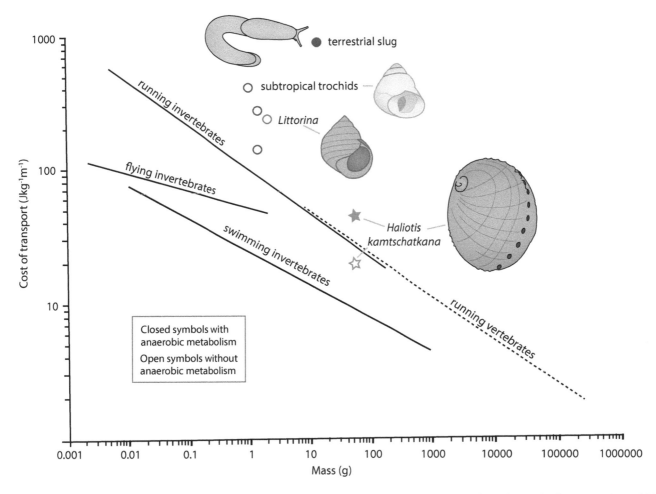

FIGURE 2.2 The minimum cost of transport (COT) of some mainly marine gastropods. Abalone (*Haliotis kamtschatkana*) (stars), trochids (*Phorcus turbinatus, P. articulatus, P. richardi,* and *Steromphala rarilineata*) (red circles), the littorinid *Littorina littorea* (blue circle), and a terrestrial slug (*Ariolimax columbianus*) (brown circle). Regression lines for running, swimming, and flying invertebrates are included for comparison and one for running vertebrates. Redrawn and modified from Donovan, D. A. et al., *J. Exp. Mar. Biol. Ecol.*, 235, 273–284, 1999a.

shell closures to 'swim' (Tremblay et al. 2006; Guderley & Tremblay 2013). Energy costs for scallops are almost three times higher than for squid at similar speeds (Figure 2.3).

In many bivalves, aspartate and glucose metabolism can continue in anaerobic conditions by the amino acid alanine combining with pyruvate to form alanopine. This reaction regenerates NAD$^+$ from NADH/H$^+$ and is catalysed by the enzyme alanopine dehydrogenase. In many sessile littoral bivalves, the anaerobic energy substrate is carbohydrate rather than aspartate. In this way, the aerobic energy pathway can be replaced by alternative anaerobic pathways during short-term or even long-term hypoxia (e.g., Zwaan 1983).

Molluscs possess the enzyme octopine dehydrogenase which assists in maintaining redox balance (see above) under anaerobic conditions. This enzyme (and others of its type), combines an amino acid with pyruvate to produce the anaerobic end product octopine.[8] Molluscs also produce

similar enzymes involved in anaerobic metabolism, including alanopine dehydrogenase (ALPDH), and strombine dehydrogenase (SDH).

Lactate dehydrogenase (LDH) is widely present in animals, but ADH, SDH and octopine dehydrogenase (ODH) are absent in arthropods, echinoderms, and chordates (Livingstone 1983; Livingstone et al. 1990). Except for unionoids, ODH activities are generally absent from freshwater species. These pyruvate oxidoreductase activities also differ between higher taxa of molluscs. Polyplacophorans contain LDH only, while scaphopods and cephalopods contain ODH only. In contrast, most gastropods and bivalves have all three, although some are missing in certain taxa (Livingstone et al. 1990). Other pyruvate oxidoreductases are known from molluscs, including tauropine dehydrogenase (TDH) and β-alanopine dehydrogenase (BDH).

Cephalopods use anaerobic metabolism, notably during strenuous swimming when relative hypoxia occurs. As in other molluscs, the primary phosphagen of cephalopod muscle is arginine phosphate. The arginine is removed by

---

[8] Octopine was first isolated from *Octopus vulgaris* and is found in a wide variety of organisms.

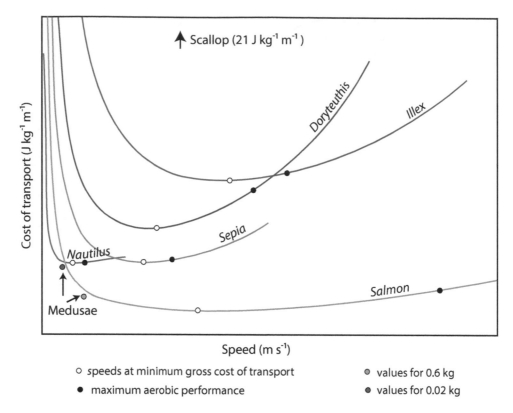

FIGURE 2.3 Cost of transport (COT) in cephalopods in comparison with a scallop, a jellyfish, and a teleost fish. ○ – minimum COT, ● – maximum aerobic performance. Redrawn and modified from O'Dor, R. K. and Webber, D. M., *J. Exp. Biol.*, 160, 93–112, 1991. Taxa – *Nautilus pompilius, Sepia officinalis, Doryteuthis pealeii, Illex illecebrosus, Placopecten magellanicus, Oncorhynchus nerka* (salmon), *Stomolopus meleagris* (jellyfish).

combining with pyruvate to form octopine. As with other molluscs, because anaerobic metabolism is not a steady-state process, the end products (octopine and lactate) accumulate and eventually prevent further glycolysis (see Hochachka et al. 1983a, 1983b; Storey & Storey 1983).

2.2.4.4 Specialisations for High Energy Work

Aerobic glucose metabolism is much more efficient energetically than anaerobic glycolysis, but the latter is used during recovery to refresh carbon and energy reserves. In terrestrial slugs, for example, oxygen uptake increases after movement rather than during it (Denny 1980b), presumably to recharge phosphagen reserves.

In molluscan muscles, arginine phosphate, glycogen, and amino acids are typically the major energy sources. Unlike vertebrates, triglycerides are not a significant energy source for molluscs. Because anaerobic glycolysis is a closed system, how much work it can support depends on the amount of stored glycogen available, while the rate depends on the enzymes involved (Hochachka et al. 1983a). For example, in bivalve adductor muscles, the quick-acting phasic muscle is richer in glycogen than the catch muscle, but the catch muscle can contract for long periods of time, relying more on aerobic metabolism. In scallops, swimming is powered mainly by anaerobic arginine phosphate hydrolysis releasing octopine, with arginine phosphate levels rapidly restored during

recovery (e.g., Chih & Ellington 1986). Arginine phosphate is in higher concentration in the phasic muscle than in the catch muscle. This energy is released by the enzyme phosphoarginine kinase (PAK).

Octopine is also an end product of anaerobic glycolysis in *Nautilus* muscle (Hochachka et al. 1977; Baldwin 1987). While such systems are adequate for low energy work or short bursts of activity, fast-swimming squid (like vertebrates) require additional sources of energy. Hochachka (1994) made interesting comparisons between the ability of vertebrate and cephalopod muscles to maintain a supply of oxygen to muscle tissue. Cephalopods lack at least two of the four main mechanisms used by vertebrates to get oxygen from the blood to the mitochondria (i.e., intracellular myoglobin and mitochondria located only in interfibrillar zones in the muscle cells – with no subsarcolemmal mitochondria). Thus, cephalopods need to obtain maximum ATP yield for the oxygen utilised in sustained activity. They appear to do this by mainly using oxygen-efficient fuels such as carbohydrates and amino acids rather than fatty acids, which are less efficient to oxidise. As noted above, for very quick bursts of activity such as used in prey capture, some squid have been shown to rely mainly on anaerobic mechanisms (e.g., Trueblood & Seibel 2014). The duplicate copies of the genes Cytochrome c oxidase subunits 1-3 (*COX1 - COX3*), ATP synthase membrane subunits 6 and 8 (*ATP6* and *ATP8*) found in the mitochondrial genome of some oegopsid squid may be related to the greater metabolic demands of these cephalopods (Yokobori et al. 2004). These are the only genes duplicated in the mitochondrial genome, where they are involved in the generation of cellular energy.

2.2.5 THE ROLE OF METAL IONS

Many metal ions have a role in biological processes in animals, including playing essential roles in about one-third of enzymes (Da Silva & Williams 1991). Metal ions are usually positively charged, many with a charge greater than one. Atoms or groups of atoms (called ligands) that bond to the metal ion are usually negatively charged or neutral. Specific binding of metal ions occurs with amino acid side chains, particularly the carboxylate groups of aspartic and glutamic acid, and the nitrogen ring of histidine, although other side chains are also involved (Glusker et al. 1999).

Na+ and K+ ions both play essential roles in cell physiological functions such as cellular energy (notably ATP) and water balance, as described in more detail below, and in nerve impulses (see Chapter 7).

The mechanisms involved in the uptake of heavy metals are discussed in Chapter 6. Metal ions such as calcium (Ca^{2+}), magnesium (Mg^{2+}), manganese (Mn^{2+}), zinc (Zn^{2+}), and cooper (Cu^{2+}) have various functions, including helping to stabilise protein configurations, while iron and copper are involved in oxygen transport. Copper binds to the haemocyanin protein molecule to transport oxygen in the blood, as does iron in haemoglobin and myoglobin. Calcium is essential for several processes including building shells. Trace amounts of the metals cobalt (Co^{2+}) and molybdenum (Mo^{2+}) are also important in biological systems. Such ions can modify the electron flow in either an enzyme or a substrate and thus play an important role in controlling a range of biological processes.

Metal ions such as iron (Fe^{2+}), calcium, zinc, copper, molybdenum, and manganese are part of critical enzyme molecules (called metalloenzymes), while certain metals such as zinc and magnesium are essential to some enzyme reactions, but are not part of the enzyme molecule (Simkiss & Mason 1983). Some metal ions also play an important role in protecting the cell against the toxic by-products of the reduction of oxygen to water (Simkiss & Mason 1983).

Metals also play an important role in providing cross-links in some biological materials. Examples include metals such as calcium, copper, and iron joining the proteins in mussel byssus (Reinecke & Harrington 2017), the radular teeth of chitons and patellogastropod limpets (Shaw et al. 2008), and in the glue-like gels that enable the tenacious grip of limpets (Smith 2013).

Most metal ions are toxic in large amounts, for example, by interfering with enzyme functions. Molluscs, like other animals, can accumulate metals, whether essential or not, in amounts exceeding their physiological needs (e.g., Rainbow 2002 and see Chapter 10). Excess or non-essential metal ions are usually stored in a detoxified form. Non-essential toxic metal ions include lead (Pb), cadmium (Cd), mercury (Hg), and arsenic (As). Lead is the least toxic

(e.g., Beeby & Richmond 2010) and does not induce a specific metallothionein (a binding protein) as with cadmium, copper, and zinc (Dallinger et al. 2001). In molluscs, the regulation and excretion of several toxic metals are linked with calcium and phosphorus metabolism as the metals are metabolically isolated, stored, and excreted in calcium or phosphorus-rich intracellular granules (Simkiss et al. 1982; Langston et al. 1998).

2.2.5.1 Calcium

Calcium ions play crucial physiological roles. Calcium is a structural component of some enzymes, for example, amylase, and Ca^{2+} ions are involved in signal transduction, in muscle cell contraction and relaxation, in maintaining the potential difference across excitable cell membranes, in the release of neurotransmitters, in cell division, and as an essential component of the mitotic spindle apparatus. Intracellular Ca^{2+} is stored in mitochondria, the endoplasmic reticulum, and in other parts of cells. In molluscs, calcium is essential for shell and spicule formation (see Chapter 3) and plays a role in heavy metal regulation. Figure 2.4 shows the main movements of calcium and calcium ions in the body of a terrestrial snail. The mechanisms involved in shell formation are outlined in Chapter 3. Conservation of calcium in terrestrial snails is essential, especially on calcium-poor soils (e.g., Fournié & Chetail 1984).

2.2.5.2 Sodium and Potassium

Sodium (Na) and potassium (K) ions both have important roles in cell physiology including in the sodium-potassium pump (see Box 2.5) and in nerve cell membranes, where their electrical charge conducts nerve cell impulses achieved by the ions moving back and forth across the cell membrane to generate a potential.

2.2.5.3 Phosphorus

Phosphorus (P) has an important role in ATP, a critical energy source in the cell (see Box 2.4). It is also a key component of DNA and RNA and the phospholipids that form cell membranes. Like calcium, it also has a role in heavy metal regulation.

2.2.5.4 Iron and Copper

Iron is by far the most widespread and important transition metal and is found in all living organisms. Compounds containing iron are involved in electron transfer (i.e., oxidation-reduction – $Fe^{3+} + e^- \leftrightarrow Fe^{2+}$) reactions and in oxygen transport (haemoglobin and myoglobin). Iron, like other transition metals, can have several oxidation states. In a haemoglobin molecule, for example, there are four haeme[9] groups. Oxygen (O_2) binds with the Fe^{2+} as a ligand in a reversible process.

Succinate dehydrogenase also contains iron. It is the only enzyme involved in both the citric acid cycle and in the electron transport chain. Iron also occurs in other common enzymes like catalase and cytochromes. Catalase converts

---

[9] Haeme is a ferrous ion (Fe^{2+}) surrounded by a porphyrin group, with four nitrogen atoms acting as ligands.

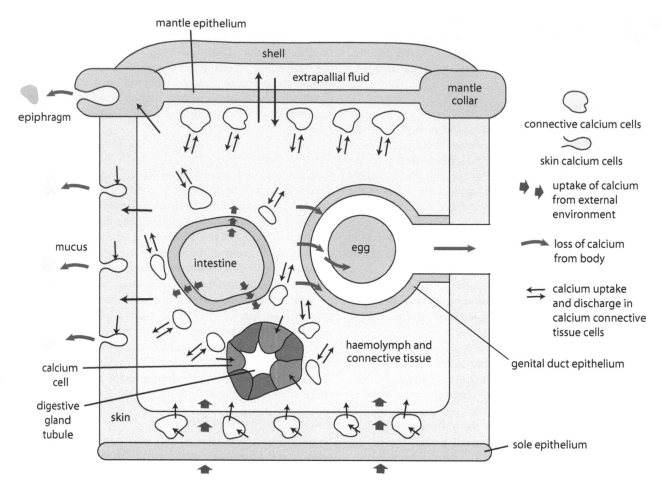

FIGURE 2.4 The movement of calcium in the body of a terrestrial snail. Except for the loss of calcium carbonate granules from the mantle collar and the skin, all other arrows concern only calcium ions. Redrawn and modified from Fournié, J. and Chetail, M., *Am. Zool.*, 24, 857–870, 1984.

hydrogen peroxide (a by-product in many metabolic processes) to water and oxygen while cytochromes are oxidising enzymes found in the inner mitochondrial membrane. They are haemeproteins (containing haeme groups) and are involved in the generation of ATP. Some, such as cytochrome c oxidase, also contain copper.

Copper, like most other transition metals, is involved in electron transfer ($Cu^{2+} + e^- \leftrightarrow Cu^+$) and is found in many proteins. It is also an essential element in molluscs, being necessary for critical bodily processes. Notably, it is a key component of the respiratory protein haemocyanin found in the blood of many molluscs (see Chapter 6 for details). In haemocyanin, one oxygen molecule is bound to two copper atoms (Cu^{2+}) which are oxidised to Cu^{3+} after binding.

Copper is also found in various metalloenzymes such as cytochrome c oxidase, and in luciferase, an enzyme involved in luminescence in the bivalve *Pholas* (Simkiss & Mason 1983).

2.2.5.5 Zinc and Cobalt

Compounds containing zinc and cobalt catalyse the hydrolysis of phosphates and are important in other biological processes. In a shelled mollusc, the shell often contains the largest amount of cobalt. Metallocarboxypeptidases are extracellular enzymes that contain zinc and are involved in many processes, including digestion, blood coagulation, inflammation, and prohormone and neuropeptide processing in many organisms including molluscs (Covaleda et al. 2012). Zinc undergoes rapid ligand exchange, and it is used by several proteins in cell signalling, notably in synaptic vesicles at some glutamatergic nerve terminals. It also regulates some other proteins by changing their concentration and can influence nitric oxide production, and hence, the immune system (Pluth et al. 2011). Examples in molluscs of other metalloenzymes containing zinc include alkaline phosphatase, carbonic anhydrase, lactate dehydrogenase, and malate dehydrogenase (Simkiss & Mason 1983).

2.2.5.6 Molybdenum

Molybdenum is an essential element in many enzymes in animals. Two examples from molluscs are xanthine oxidase and aldehyde oxidase (Simkiss & Mason 1983), the former is involved in the catabolism of purines and the latter converts aldehydes to an acid and hydrogen peroxide.

2.2.5.7 Sulphur

Sulphur (or sulfur), an essential element in all organisms and in most animals, is the third most abundant mineral element in the body. Sulphur is unusual in forming several anions (S^{2-}, S^{22-}, S^{2-}, S^{3-}, S^{32-}) enabling it to form, for example, sulphur dioxide and sulphur trioxide. It generally gains two electrons to form the sulphide ion.

Sulphur is utilised in nucleic acids and proteins and is present in the amino acids cysteine and cystine and in other compounds such as some coenzymes, steroids, and phenols. Glutathione is an important antioxidant that contains sulphur and helps to prevent damage from reactive oxygen products such as free radicals and peroxides. Sulphur is also essential for the electron transport chain and the generation of ATP in oxidative phosphorylation.

There are many types of sulphur-metabolising microbes, including sulphate reducers, and these are utilised in various ways by molluscs in symbiotic relationships with sulphur reducing or sulphur oxidising bacteria (see Chapter 9) in which reduced carbon and nitrogen compounds are provided to the host (see Figure 2.5 for a summary of the processes involved).

2.2.5.8 Magnesium and Manganese

Magnesium is a very important metal ion in some essential bodily processes including the replication and cleavage of DNA and RNA. Magnesium and manganese are both found in the enzyme pyruvate kinase (Simkiss & Mason 1983).

Manganese plays an important role in the body, including the antioxidant system (e.g., as superoxide dismutase – Mn-SOD – in the mitochondria, which prevents the toxic effects of superoxide O^{-2}). It also plays a critical role in development, and is a component of some enzymes (e.g., arginase) and compounds that are cofactors in the enzymatic synthesis of, for example, glycoproteins, fatty acids, and mucopolysaccharides.

2.3 CELL SIGNALLING AND CHEMICAL MESSENGERS

Cell signalling is a critical component of life that allows cells to communicate and coordinate activities throughout the organism. Cells communicate by direct contact (juxtacrine signalling), over short distances (paracrine signalling), and large distances (endocrine signalling). Juxtacrine signalling is especially important during development (e.g., Gerhart 1999), while other forms of signalling become more important as the animal matures and physiological and morphological complexity increase.

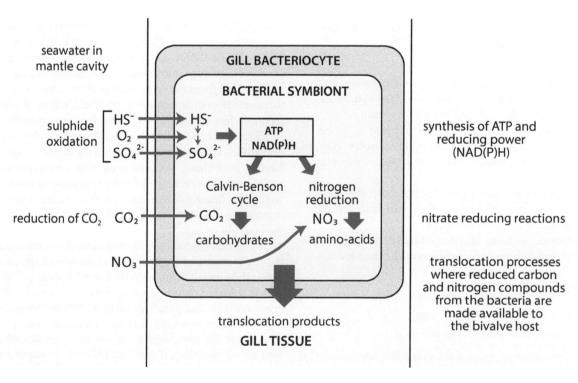

FIGURE 2.5 Diagram depicting the metabolic pathways of sulphur oxidising symbiotic bacteria in the gill of a bivalve. The Calvin-Benson (or Calvin-Benson-Bassham) cycle is utilised to oxidise electron carriers. Redrawn and modified from Felbeck, H. et al., Biochemical interactions between molluscs and their algal and bacterial symbionts, in Hochachka, P.W. (Ed.), *Biochemistry and Physiology. The Mollusca*, Vol. 2, pp. 331–358, Academic Press, New York, 1983.

TABLE 2.1

Chemical Messengers, Other than Nitric Oxide, Employed by Molluscs and Many Other Invertebrates

Compound	Fast Neuro-transmitter	Slow Neuro-transmitter	Neuromodulator	Hormone/paracrine factor
Small polar molecules – e.g.,				
Acetylcholine	•	•		
Glutamate, GABA (gamma aminobutyric acid)	•	•		
Catecholamines		•	•	•
Peptides – e.g.,				
FMRFamide	•			
Enkephalins		•		
RPCH, AKH (red pigment-concentrating hormone and adipokinetic hormones)		•?	•	•
CHH (crustacean hyperglycemic hormone)			•	•
Insulin				•
Lipids – e.g.,				
Juvenile hormones				•
Ecdysteroids			•?	•

Modified from LaFont, R., *Ecotoxicology*, 9, 41–57, 2000.

Regardless of the proximity, chemical messengers transmit the signal to other cells. A chemical messenger is a compound that transmits a 'message' from one cell to another. The main classes of chemical messengers are listed below, although some chemical messengers may have more than one role and could be included in more than one category.

A chemical messenger (Table 2.1) may be:

- Hormones – chemical messengers (including neurohormones) produced by the endocrine system and distributed in the body via the blood system.
- Neurotransmitters – messengers initiated in a neuron and which communicate with adjacent cells. They include nitric oxide and more complex molecules.
- Neuropeptides – small protein-like molecules (peptides) generated in neurons, they can function like hormones or neurotransmitters.
- Pheromones – chemicals released into the environment that initiate responses in members of the same species.

Each category is discussed below.

2.3.1 HORMONES, ENDOCRINE, AND NEUROENDOCRINE SYSTEMS

The *endocrine system*[10] comprises glandular neuroendocrine and/or non-neural cells that release hormones into the blood. The *neuroendocrine system* is made up of modified neurons (neurosecretory neurons or neuroendocrine cells) in the central nervous system (CNS) (see Chapter 7 for more details). As in other animals, the endocrine system in molluscs consists of various ductless glandular tissues, all derived from the ectoderm, which play important physiological roles through their secretion of chemical messengers (hormones) into the blood system. These are very diverse and are usually peptides, proteins, or lipidic molecules. Hormones (including neurohormones) enable general communication with cells and tissues throughout the body without direct contact, in contrast to neuronal communication (see also Chapter 7), although feedback loops expedite hormone release or cessation.

Neuroendocrine cells (NECs) differ from endocrine glands in that they are in intimate contact with standard neurons and interact with them. The endocrine and neuroendocrine systems control a wide range of body functions including bodily well-being (metabolism, osmotic condition, etc.), growth, and reproduction, and more immediate effects such as preparing the body for escape.

Hormones (including neurohormones) are responsible for the maintenance of internal conditions within the animal by controlling metabolism, growth, water balance, etc. This process is achieved by hormonal activity working with the endocrine, nervous, and paracrine systems. While nerves can relay messages very quickly, these are localised. The chemicals released at the synapse are amines and peptides which can also act as hormones if released into the circulatory system, as with neurosecretory cells.

Hormones in the body are generally widespread and in low concentrations. Target cells with receptor molecules specific to a particular hormone can uniformly respond throughout the body to very low concentrations.

---

[10] Glands that secrete into ducts (e.g., salivary glands) are known as the exocrine system.

Most hormones in molluscs and other animals are short polypeptides formed by cleavage (by proteases) from larger precursor proteins (prohormones) produced in the rough endoplasmic reticulum. These are processed via the Golgi apparatus and stored in membrane-bound vesicles. These vesicles appear as granules in the cells that produce peptide hormones (Hartenstein 2006). Bioactive peptides are often short with a low molecular weight, with dipeptides being the shortest.

Instead of the neurotransmitting chemicals produced by 'normal' neurons, neurosecretory neurons produce *neurohormones* that are released into the blood. This neurosecretory system is the major source of hormones in molluscs, whereas the endocrine system is of greater significance in vertebrates. In molluscs, neurosecretory hormones are important in *first-order responses* (where the neurohormone directly affects the target effector organ), also *second-order responses* (where a neurohormone stimulates an endocrine gland that in turn affects the effector organ), and some *third-order responses* (with two intermediate endocrine systems).

While hormones usually affect cells some distance away from the endocrine tissue that secretes them, *paracrine* secretions have their effect on adjacent cells or glands, and an *autocrine* hormone affects receptors in the same cell that produced the hormone. Thus, in contrast to nerve impulses, the messages released by the endocrine system (via hormones) are relatively slow in reaching their destination, but potentially reach all cells in the body. The paracrine system is an intermediate one with its production of local chemical messages sent to adjacent cells.

Hormone production can be controlled simply by negative feedback relating to the hormone levels in the bloodstream, or by physiological parameters which may, in turn, be related to external environmental conditions, including seasonality. Some hormones remain active for long periods, while others lose their activity after a short period.

Numerous hormones have been identified in molluscs, and many have a demonstrated effect on various bodily functions as outlined below, including reproduction, growth of the body and shell, stress responses, and osmoregulation.

2.3.1.1 Molluscan Endocrine Systems

Hormones have been shown to play important roles in many specific activities in molluscs including reproduction (see Chapter 8), osmoregulation (see Chapter 6), and growth of both the body and shell (see Chapter 3).

Molluscan endocrine systems are best known in heterobranch gastropods. An example of a reasonably well-known system is the hormonal control involved in land snail reproduction, which is shown in Figure 2.6.

The morphological aspects of the neuroendocrine system are described in Chapter 7. Unlike most other invertebrates, there are true endocrine glands in molluscs, although endocrine cells are found in annelids.

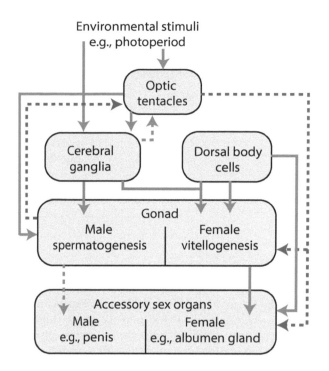

FIGURE 2.6 Diagram showing hormonal interactions in stylommatophoran gastropods. Red dotted lines indicate probable inhibitory links, blue lines show stimulatory links. Redrawn and modified from Flari, V.A. and Edwards, J.P., *Invert. Reprod. Devel.*, 44, 139–161, 2003.

Besides those listed in Table 2.2, some neuropeptides also function as hormones and are secreted by the nervous system (see Chapter 7). In cephalopods, a cardioexcitatory neurohormone is released by neurosecretory cells from neurohaemal areas in the anterior vena cava and pharyngoophthalmic vein.

Most hormones in molluscs, unlike some in arthropods, are not unique to molluscs or even to invertebrates. Similar molecules are found in vertebrates and other invertebrate phyla, suggesting that they were present in ancestral animal taxa. These include several kinds of insulin-like molecules (molluscan insulin-related peptides – MIP), with similar ones found in many invertebrate groups. Others include the oxytocin/vasopressin group of peptides – namely, the molluscan cephalotocin of unknown function, lys-conopressin, involved in osmoregulation and reproduction, and arg-conopressin that is involved in osmoregulation (e.g., Kesteren et al. 1992).

As detailed below, vertebrate-type steroid compounds have been found in most invertebrates, but evidence for their endogenous synthesis is lacking (Scott 2012, 2013), despite most of the vertebrate-type sex hormones (e.g., testosterone, progesterone) being reported in the main molluscan classes (LaFont & Mathieu 2007). An early report of ecdysteroids[11] in molluscs is unconfirmed (LaFont & Mathieu 2007).

---

[11] Insect moulting and sex hormones that include the steroid ecdysone and homologues.

TABLE 2.2

Some Examples of Hormones Other than Neurohormones Identified in Molluscs and Their Source

Group	Hormone	Source
Gastropods	Gonadotropic hormone	Dorsal bodies
	Schistosomin (*Lymnaea*)	Brain/haemocytes?
	'Vertebrate-type' steroids[a]	Gonads
Cephalopods	Gonadotropic hormone	Optic glands (see Chapter 17)

Modified from LaFont, R., *Ecotoxicology*, 9, 41–57, 2000.

[a] These are questionable (see text).

Energy metabolism is also under hormonal control. For example, the storage and breakdown of glycogen is controlled by hormones. Although bivalves and gastropods can tolerate fluctuation in blood glucose levels, high levels of glucose (e.g. after feeding) can stimulate glycogen formation. This is aided by an insulin-like neurohormone that stimulates a glycosyltransferase enzyme (glycogen synthetase) that converts glucose residues into glycogen. Other examples include a neurohormone secreted by the visceral ganglion in mussels (*Mytilus*) which promotes the storage of glycogen, lipid, and protein. A 'hyperglycemic factor' neurohormone secreted by cells in the cerebral ganglia raises blood sugars and has been identified in some bivalves and gastropods (Hemminga et al. 1985; Robbins et al. 1990; Abdraba & Saleuddin 2000).

Peptides similar to vertebrate arginine-vasotocin are found in the CNS of several gastropods and have been shown to enhance the water permeability of the body wall in terrestrial slugs (Deyrup-Olsen et al. 1981).

Some chemical pollutants can disrupt endocrine systems and interfere with various bodily functions, including reproduction (see Chapter 10).

2.3.1.2 Neurosecretion, Neurosecretory Cells, Neurohormones, and Neurotransmitters

Neuroendocrine cells (see also Chapter 7) are derived from nervous tissue and are the basis of most molluscan endocrine systems. The NECs contain extensive rough endoplasmic reticulum and a Golgi apparatus, both typically well developed in peptide-producing cells. The secretory products are usually in the form of granules discharged at the cell margin of one of the neurohaemal endings. They consist of a cell body where synthesis of the neurosecretion (typically a neurohormone) occurs, and an axon, but they differ from typical neurons in several significant ways. Instead of the axon terminating at a dendritic synapse, they typically end in one or more *neurohaemal areas* which terminate in the blood system. In molluscs, the shape and size of the NEC neurons and regular neurons are generally similar.

The NECs are usually at the edges of the ganglia and consistent in number and position in a species. Neurohaemal areas have been described in gastropods and cephalopods (Wendelaar-Bonga 1970; Joosse & Geraerts 1983) and bivalves (e.g., Deaton et al. 2001). These sites are often scattered over the entire nervous system (ganglia, commissures, and nerves), unlike in arthropods, where they are often clustered in special organs.

Neuropeptides are diverse signalling molecules produced in, and released from, neurons, and they may function as transmitters, modulators, or hormones (see Section 2.3.2). As hormones, they are released via neurohaemal organs into the haemolymph, where they may regulate many bodily activities including general metabolism, growth, and reproduction.

Steroid hormones are derived from the lipid cholesterol and include cortisone and oestrogen in vertebrates and ecdysone in arthropods. Like other lipids, they are synthesised in the smooth endoplasmic reticulum and are not stored in vesicles. While a body of literature indicates that sex steroids are important in molluscs, a careful analysis of the data by Scott (2012, 2013) found no evidence for their involvement as reproductive hormones in molluscs. These vertebrate-type steroids carry out similar functions to those in vertebrates and, while they can be detected in molluscan tissues, it is not known for certain if they formed endogenously or are acquired from the environment. Scott (2012, 2013) argued there is no convincing evidence for biosynthesis of vertebrate-type steroids in molluscs, and they do not have the genes for the critical enzymes necessary to progressively transform cholesterol into vertebrate-type steroids or for classical nuclear steroid receptors. It is, however, clear that molluscs can absorb vertebrate steroids from their surroundings and can store them for considerable periods of time. Three steroids, progesterone, testosterone, and 17β-estradiol, which are often thought to be functional in molluscs, are the same as those in humans, and Scott (2012) argued that experimental contamination might account for their presence.

A range of neuronal types exhibit several intermediate modes of neurochemical communication, and thus the distinction between neurosecretory cells and neurons is not clear-cut, with neurosecretory neurons being more similar to the nerve cell precursor than the highly modified neuron. Molluscan NECs are generally similar to neurons (see Chapter 7), although, as noted above, differ because they typically release their secretory products directly into the haemocoel.

The presynaptic neuron terminates at the synapse which is also the site of release of the neurotransmitters. Their structure shows extensive rough endoplasmic reticulum and a well-defined Golgi apparatus typical of cells producing peptides. These secretory products are usually produced as granules stored in small vesicles and are released at the cell margin when the neuron is activated. For the chemical message to be successfully transmitted when liberated into the synapse, the neurotransmitter must bind to a protein receptor on the receiving cell. The receptor cell may be the postsynaptic membrane, or another cell type, or the neuron from which it was released, thus providing feedback. The neurotransmitter may alternatively be modified by an enzyme on release.

The main secretory products from NECs are peptides (neuropeptides) although, in addition to neurohormones, some NECs also make catecholamines (dopamine, octopamine) which are also released into the blood as hormones.

Neurotransmitters are small molecules that act as chemical messengers and are produced by the nervous system. They relay nerve impulses from one cell to another. The receptor cells are not necessarily neurons, but can be other cell types such as gland cells or muscle cells. These small molecules pass through synapses, small gaps (20–50 nanometers) where the cells are in contact.

Two of the important neurotransmitters produced in molluscs are:

Acetylcholine: Acetylcholine (ACh) is a fast neurotransmitter found in many animals and is also expressed in plants and fungi (Horiuchi et al. 2003).

Serotonin: Serotonin (5-hydroxytryptamine or 5-HT) is a monoamine, a neurotransmitter found in serotonergic neurons. These are widespread in molluscan nervous systems – for example, in aplacophorans and chitons (e.g., Faller et al. 2012), bivalves (e.g., York & Twarog 1973; Garnerot et al. 2006), and scaphopods (Wanninger & Haszprunar 2003). Serotonergic cells or clusters of them are known from several groups of gastropods, for example, in abalone (Barlow & Truman 1992; Croll et al. 1995) and in the pteropod *Clione* (Satterlie et al. 1995). In euthyneurans, the serotonergic giant cells in the cerebral ganglia modulate motor output to muscles involved in feeding; others in the pedal ganglia of a nudibranch (*Tritonia*) have been shown to initiate locomotion by activating ciliary cells on the foot sole. Other serotonergic cells in *Aplysia* are cardioexcitors, while some may reset circadian rhythms. While the euthyneuran CNS contains relatively few serotonergic cells – for example, about 120 in *Aplysia* and 200 in *Lymnaea* – the CNS of the caenogastropod *Littorina* contains about 1500

(Croll & Lo 1986). In nudibranchs there are several types of serotonergic cells present that are distributed differently in different taxa (Newcomb et al. 2006).

Neurotransmitters may be an amino acid or a derivative (e.g., the neurotransmitters dopamine and serotonin are made from the amino acids tyrosine and tryptophan, respectively), or small molecules with an amine functional group (e.g., acetylcholine), or peptides (neuropeptides), the latter including the enkephalins, the endorphins, oxytocin, and many others. Neurotransmitters are usually conserved so they can be similar in a wide range of animals, in contrast to neurohormones and hormones which are more diversified.

Some neurotransmitters are fast-acting, forming ion channels in the target cell membrane with almost immediate effects. These effects may cause *depolarisation* resulting in an influx of positively charged ions and the continuation of a nerve impulse. *Hyperpolarisation*, with inward movement of negatively charged ions, inhibits nerve impulses.

Some relatively slow-acting neurotransmitters do not form ion channels on activation, but instead bind to proteins (sometimes called G-protein-coupled receptors – GPCRs). The G-protein is activated when the neurotransmitter binds to a GPCR causing the protein to dissociate into subunits. These either open or close an ion channel and cause an enzyme to be activated or inhibited. The enzyme produces a 'second messenger', activating protein kinase enzymes that phosphorylate certain proteins in the cell, possibly including ion channels.

Neuromodulators are neurohormones that regulate the transmission of multiple synaptic signals, affecting multiple neurons. They include dopamine, serotonin, histamine, and norepinephrine.

Some neurohormones such as serotonin, noradrenaline, and dopamine are found in many animal groups, but do not necessarily have the same effects in different taxa. For example, in molluscs, serotonin and dopamine activated cardiac adenylate cyclase in *Anodonta cygnea* and *Helix pomatia* heart muscle, but noradrenaline increased adenylate cyclase activity in *Helix* and inhibited it in *Anodonta* (Wollemann & Rózsa 1975).

Nitric oxide (NO) is a gas and a free radical[12] that is thought to act as a signalling molecule, including acting as a neurotransmitter in both invertebrates and vertebrates. It has a short half-life and can move rapidly through cell membranes. It has important roles in molluscs (and other animals) and is produced from L-arginine by the enzyme NO synthase in rather complex enzymatic reactions (e.g., Nathan & Xie 1994; Moroz & Gillette 1995; Peruzzi et al. 2004) and plays a role in, for example, the relaxation of blood vessels, behaviour such as feeding, and in the immune system, including microglia communication (see Peruzzi et al. 2004 for literature).

---

[12] An atom, molecule, or ion that has a unpaired valence electron, making them highly reactive.

2.3.1.3 Hormonal Control of Shell Growth

Body and shell growth are controlled by growth hormones. Shell growth is regulated by controlling the transport of calcium from the haemolymph through the mantle edge, and calcium-binding proteins are involved in the movement of calcium through the mantle epithelium. The hormonal control of shell growth has been particularly well studied in the freshwater hygrophilan snail *Lymnaea* where two hormone systems are involved. One is the neurosecretory 'light green cells' (see Chapter 7) in the cerebral ganglia (Wendelaar-Bonga 1970; Geraerts 1976b, 1992) and the other is from the median lip nerve (Saleuddin et al. 1992b). Neurosecretions from the light green cells of *Lymnaea* increase the level of calcium-binding proteins and also stimulate periostracum formation. In some other 'pulmonates', the light green cell function is replaced by other neurosecretory cells such as, for example, the median neurosecretory cells in the planorbid *Helisoma* (e.g., Dillaman et al. 1976).

On the lateral side of each cerebral ganglion is a small lobe, the lateral lobe, which in *Lymnaea* contains three types of neurosecretory cells. The follicle gland is a small epithelial structure on the lateral lobes. The cells in the lateral lobes of the cerebral ganglia have a direct nervous connection to the optic lobes which detect changes in light intensity (Saleuddin et al. 1994) enabling growth (and reproduction) in 'pulmonates' to be closely linked to photoperiod (Figure 2.7). Removal of the lateral lobes (Geraerts 1976a) results in gigantism, suggesting that they inhibit the

hormone secretion by the light green cells (or their equivalents), and thus they appear to balance growth with reproduction. More details regarding these structures are given in Chapter 7. Removal of the gonad also increases growth and results in gigantism, as when some parasitic trematodes cause castration.

2.3.2 Neuropeptides

There are many kinds of neuropeptides in molluscs (e.g., Di Cosmo & Polese 2013) (see also Chapter 7). Their role in acting on the CNS to elicit behavioural responses was first discovered in *Aplysia californica*, in which an egg-laying hormone was shown to result in oviposition (Strumwasser 1984; Smock & Arch 1986). Subsequently, much of the work in this area has been done on *Aplysia* and other heterobranchs such as helicids and *Lymnaea*.

Peptides can either act as neurotransmitters or hormones (Mukai & Morishita 2017). Both the tetrapeptides and small cardioactive peptides (SCP) act as transmitters, while the heptapeptides that circulate in the blood act as hormones (Lesser & Greenberg 1993). Several cardiovascular modulators are found in molluscs and include phenylalanyl-methionyl-arginyl-phenylalanine amide-related peptides (FaRPs), SCP and others (Price et al. 1990; McMahon et al. 1997). Many of the neuropeptides found in molluscs are structurally similar to those in other animals, including vertebrates, such as the peptides similar to FMRFamide   (phenylalanyl-methionyl-arginyl-phenylalanine

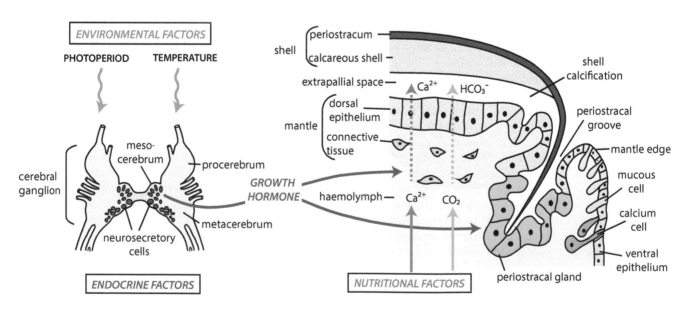

FIGURE 2.7 Environmental and endocrine control of shell growth in a stylommatophoran land snail. Redrawn and modified from Gomot de Vaufleury, Regulation of growth and reproduction, in Barker, G.M. (Ed.), *The Biology of Terrestrial Molluscs*, CABI Publishing, Wallingford, UK, 2001.

amide) (e.g., Higgins et al. 1978; McMahon et al. 1997; López-Vera et al. 2008). Similarly, the pituitary adenylate cyclase-activating polypeptide (PACAP) is found in heterobranch gastropods (Kiss & Pirger 2013), where it has a multifunctional role. Numerous others have been identified by the use of antibodies in the CNS of *Lymnaea*. These include vasopressin, vasotocin, oxytocins, adrenocorticotropin hormone (ACTH), α-melanocyte stimulating hormone (αMSH), dynorphin, met- and leu-enkephalin, glucagon, insulin, vasoactive intestinal peptide, calcitonin, somatostatin, pancreatic polypeptide, secretin, astrin, and several others. Similar but less extensive work has been done on a few other 'pulmonates'.

The tetrapeptide amide FMRFamide was first named from a venerid bivalve (Price & Greenberg 1977) and is only found in molluscs, where it is known from chitons, bivalves, scaphopods, cephalopods, and gastropods (López-Vera et al. 2008) and from aplacophorans (e.g., Faller et al. 2012). Other similar peptides, generally referred to as FaRPs, RFamide, or RF-NH$_2$ peptides are also present in molluscs and in many other invertebrates and vertebrates (López-Vera et al. 2008). FMRFamide is best known as a cardioexcitatory agent and it, and other FaRPs, are employed in many physiological processes involving muscular and nervous tissue including reproduction, locomotion, osmoregulation, feeding, neurogenesis, and modulation of sensory organs (see López-Vera et al. 2008 for review).

The cells that express the FMRFamides are scattered through the CNS and peripheral nervous systems of 'pulmonates', but it is uncertain as to whether they act as neurohormones because no neurohaemal areas have been identified that are associated with this system. FMRFamides are well established as having a role as neurotransmitters and neuromodulators. For example, in *Lymnaea*, the neurons from some of these cells innervate the penial retractor muscle and the heart. In *Cornu aspersum*, a giant neuron (C$_3$) found in the cerebral ganglion is part of this same system and the tentacle withdrawal system (e.g., Prescott et al. 1997).

Neurons that express APGW-amide (Achatina endogenous inhibitory tetrapeptide) are in the right half of the CNS (right cerebral, pedal, pleural, and parietal ganglia) of *Lymnaea* and have axons extending into the penial nerve, mantle nerves, and right cerebropedal connective, the latter innervating the cells of the serotonergic pedal Ib cluster in the right pedal ganglion (Croll & Van Minnen 1992; Koene 2010). These cells do not appear to be neuroendocrine cells, but either act as interneurons or directly innervate peripheral organs. A similar system has been identified in helicid snails.

By 2015, the number of neuropeptides identified in molluscs was 355, and these comprised 12 'families'. In comparison, 2457 neuropeptides in 48 families are known from chordates, but only 24 in five families from annelids (Wang et al. 2015b), no doubt reflecting, at least in part, the very different levels of research activity in these taxa. A summary of the different neuropeptides found in molluscs (and other animals) is provided by Wang et al. (2015b).

Among the numerous other neuropeptides described from molluscs are the cyclic nonapeptides of the vasopressin/oxytocin 'superfamily' found in a wide variety of animals including vertebrates, annelids, arthropods, nematodes, echinoderms, and molluscs. For example, various molluscs and leeches have lys-conopressin, and the neogastropod *Conus* has arg-conopressin (Henry et al. 2013). Cephalopods are unusual because they have two neuropeptides of this 'superfamily', as otherwise only vertebrates are known to possess two. In *Octopus*, these are the neuropeptides octopressin and cephalotocin (Takuwa-Kuroda et al. 2003). Cephalotocin mRNA is mostly expressed in the ventral median vasomotor lobe of the suboesophageal brain, while octopressin mRNA is expressed in both the supraoesophageal and suboesophageal parts of the brain and the buccal and gastric ganglia (Takuwa-Kuroda et al. 2003). These authors assumed that octopressin might be a multifunctional neuropeptide contributing to reproduction, cardiac circulation, and feeding, while cephalotocin may have roles in metabolism, homeostasis, etc. In the cuttlefish *Sepia*, the two neuropeptides are sepiatocin, a neuromodulator very like octopressin, and pro-sepiatocin, a neurohormone similar to cephalotocin (Henry et al. 2013). Sepiatocin is involved in reproduction and blood circulation as it modulates the activity of several muscles such as those of the penis, oviduct, and vena cava, while pro-sepiatocin is liberated into the haemolymph and can potentially target many organs.

2.3.3 PHEROMONES

Pheromones (or ectohormones) are usually proteins or peptides released into the external environment where they can have behavioural or physiological effects on other individuals of the same species. They include alarm, sex, and trail-following pheromones (e.g., Chase 2002; Wyatt 2014). They are best known in terrestrial environments where they can drift in the air, and less well studied in aquatic habitats, although they are relatively well studied in molluscs such as *Aplysia*, *Haliotis*, and certain squid. Some intertidal molluscs, such as *Littorina littorea* (Seuront & Spilmont 2015) produce pheromones that are water-borne and air-borne. If chemicals are released into water, they can diffuse over considerable distances. In relatively calm conditions this can provide directional information for finding mates, including spawning aggregations, as well as locating food, and can also be involved in protective or defensive behaviour (see also Chapter 9). Pheromones can also be released in mucous trails enabling them to diffuse slowly into the surrounding water. In *Haliotis*, for example, such pheromones are released in both pedal mucus (Kuanpradit et al. 2012) and the hypobranchial secretion (Kuanpradit et al. 2010). Water-borne pheromones can also induce aggression, as for example, in male squid (Cummins et al. 2011).

2.3.4 RECEPTORS

Receptors are specialised proteins to which hormones bind. If the hormone can pass through the cell membrane, as with lipid and steroid hormones, the receptors may be in the cytoplasm, and they then may control cell activities, including initiating or inhibiting genes. In contrast, neuropeptides bind to receptor proteins on the surface of the cell membrane. This binding then triggers events in the cell resulting in the transmission of a signal via a second messenger.

2.3.5 GROWTH FACTORS

A growth factor acts as a signalling molecule and, as its name suggests, may stimulate growth, or can be involved in healing wounds, cell differentiation and maturation, and various other processes. Growth factors are usually peptides, but can also be a protein or steroid hormone. They typically act locally (i.e., autocrine or paracrine). Some cells (immunocytes – i.e., haemocytes or lymphocytes acting as macrophages) that secrete these substances can be transported in the blood and the growth factors are then released under specific conditions (e.g., at a wound). They can act as stimulators or inhibitors depending on the target cell and the conditions present (Ottaviani et al. 2001).

Two of the most significant growth factors known are the platelet-derived growth factor (PDGF) and the transforming growth factor-β (TGF-β). PDGF was originally found in vertebrate blood platelets, but it is now known to be synthesised and secreted by several mesenchymal cell types in many animals. It affects various processes including cell proliferation, chemotaxis, and matrix production (Heldin & Westermark 1999). TGF-β has an even wider range of cellular responses, and it is secreted, and responded to, by a variety of cell types (Roberts & Sporn 1996). Like PDGF, TGF-β is released from immunocytes, but it is also found in somatic cells (Ottaviani et al. 1997).

Most of the studies on growth factors (including PDGF and TGF-β), have been on vertebrates with much of the information on invertebrate growth factors based on studies on the fruit fly *Drosophila melanogaster* and the nematode *Caenorhabditis elegans*. Franchini et al. (1996) found PDGF-AB and TGF-β1 in immunocytes in molluscs (Gastropoda – Viviparidae, Planorbidae, and Lymnaeidae, and Bivalvia – Mytilidae), and they have also been found in annelids.

2.4 MOLLUSCAN PHYSIOLOGICAL DEFENCES

The three main kinds of molluscan physiological defences (i.e., other than behavioural and other responses involving the whole animal, including aestivation) are– 1) stress proteins that react mainly to external environmental factors, 2) the immune system that reacts mainly to microbial invasions, and 3) wound repair. We briefly describe each of these below.

As with other animals, molluscs have mechanisms for defensive protection against foreign material, whether non-living (inorganic or organic) or living (viruses, bacteria, fungi, parasites, etc.). External defence by the shell and mucus-covered epidermis is augmented by internal defensive mechanisms with the ability to recognise and kill invaders such as viruses, bacteria, and parasites. These include phagocytosis by haemocytes and various tissues, destruction by digestive enzymes and other gut secretions, encapsulation responses (e.g., as in pearl formation), and proteins in the blood that attach to and destroy foreign material (e.g., antibodies). Immune responses to foreign antigens (involving pattern recognition) also require recognition of the body's own tissue, i.e., to distinguish 'self from non-self'. These responses depend on complex intracellular signalling systems that are still poorly understood in molluscs. There is, however, evidence that suggests there are features in these systems shared with vertebrates and well studied invertebrates such as some arthropods (e.g., Humphries & Yoshino 2003; Ottaviani 2006).

2.4.1 STRESS PROTEINS

Stress proteins (e.g., HSPs) are constantly expressed in cells, but their expression increases in response to stresses such as high and low temperatures, radiation, hypoxia, heavy metal concentrations, desiccation, etc. HSPs help protect tissues from damage by 'chaperoning' proteins changed or damaged by stressors (e.g., Sanders 1993; Liu & Chen 2013). HSPs are named according to their molecular weight. Probably the best known heat shock protein is HSP70 which stabilises proteins during stress by correcting their folding, but there are many other responses. Several studies have demonstrated the roles of HSP responses in molluscs, particularly in intertidal species (e.g., Sanders et al. 1991; Tomanek & Somero 1999; Snyder et al. 2001), and commercial species of bivalves (e.g., Oguma et al. 1998; Li et al. 2007). Stresses due to diverse environmental challenges produce various responses depending on the stage of the life cycle and the tissue type (e.g., Schill et al. 2002; Liu & Chen 2013).

2.4.2 REACTIVE OXYGEN SPECIES (ROS) AND REACTIVE NITROGEN SPECIES (RNS)

These reactive compounds, metabolic by-products of metabolism, contain oxygen and include peroxides, hydroxyl radicals, and two highly reactive forms of oxygen, singlet oxygen and superoxide. These potentially damaging reactive oxygen species (ROSs) are used by immunocytes to kill microbes, and some are involved in signal transduction, cell cycle regulation, and gene expression. Oxidative damage, such as protein misfolding, occurs in macromolecules (e.g.,proteins, DNA, membrane lipids) when the rate of production of ROS exceeds their breakdown. Cellular defences against the harmful ROS involve various enzymes [e.g., catalase conversion of hydrogen peroxide (H_2O_2) to oxygen and water] and the activation of heat shock factors and expression of HSP (see above).

Reactive nitrogen species include NO, nitrogen dioxide (NO_2), and peroxynitrite ($ONOO^-$). Nitric oxide is produced by nitric oxide synthase (NOS) and serves as a signalling molecule in cardiovascular systems. NOS isoforms have been detected in the gastropods *Pleurobranchaea californica* and *Clione limacina* and the bivalves *Azumapecten farreri*, *Crassostrea gigas*, and *Crassostrea virginica* (Donaghy et al. 2015).

2.4.3 THE IMMUNE SYSTEM

There are two types of immune system known in animals – innate and adaptive. Components of the innate immune system react in a non-specific way and are found in all living things.

The molluscan immune system has three components: physical barriers, cellular defences, and humoral mechanisms (Coutellec & Caquet 2017).

- Epithelial mucus forms the initial physical barrier against pathogens and foreign elements, and bioactive, antimicrobial compounds have been found in gastropod mucus (Ehara et al. 2002; Iijima et al. 2003).
- Cellular defences in molluscs are coordinated by immunocytes circulating in the blood. These haemocytes recognise and remove or sequester invading pathogens, using phagocytosis and encapsulation, and through the production of lysosomal enzymes, lipopolysaccharide-induced TNF-a factor, Toll-like receptors, and lectins (Gestal & Castellanos-Martínez 2015).
- Humoral factors are important in the immune responses in molluscs. Besides phagocytosis, haemocytes can secrete soluble host defence peptides (HDPs)[13] and other cytotoxic substances into the haemolymph. Together with other non-specific humoral defence molecules, (e.g., lectins, bactericidins, NO, lysozymes, serine proteases), these form the humoral component of molluscan immunity.

Immune responses common to all invertebrates include the prophenoloxidase pathway (the complement response – see Section 2.4.3.2), phagocytic cells in the blood, cytotoxic effector responses in which foreign cells are destroyed, and antimicrobial compounds (Mydlarz et al. 2006). In molluscs several molecules or mechanisms, both humoral and cellular, have been identified in the innate immune system (e.g., Flajnik & Du Pasquier 2004). Immune responses from the molluscan immune system include those from small molecules circulating in the blood, such as host defence peptides, the complement system, and phagocytosis by cells in tissues or in the blood. Despite being highly advanced in many respects, coleoid cephalopods appear to have a similar complement of immune responses to those of other molluscs (Castillo et al. 2015; Gestal & Castellanos-Martínez 2015).

The adaptive immune response found in vertebrates is not present in molluscs, although there is increasing evidence of some adaptive immunity in bivalves (Ottaviani 2004; Cong et al. 2008; Ottaviani 2011) and maternal transfer of immunity (Yue et al. 2013; Wang et al. 2015a). Thus, invertebrate immune systems can be diverse and complex with both innate and adaptive responses potentially present in molluscs.

Discriminating between self and non-self is essential in any immune system. In many animals, this is facilitated by pattern recognition receptors (PRR) (Medzhitov & Janeway 2002). There are several kinds of PRRs and they apparently function by recognising molecular patterns that characterise particular pathogens. For example, peptidoglycan[14] recognition protein (PGRP) recognises bacterial 'cell walls' in both Gram-positive and Gram-negative bacteria (Steiner 2004). The PGRPs are diverse and may be extra- or intracellular and can activate or inhibit the immune responses, such as the prophenoloxidase (proPO) system well known in arthropods and demonstrated in a few molluscs (e.g., Le Pabic et al. 2014). Although the role of PGRPs in molluscs is not well known, they have been identified in some bivalves (see Wei et al. 2012 for review) and may possibly occur in all other molluscs.

Apoptosis (Type I programmed cell death) also plays a key role in the immune system and appears to be highly conserved in molluscs (Sokolova 2009). It is, for example, involved in protection against parasites by limiting their spread, although some pathogens (especially those that multiply within a cell) can inhibit the apoptosis response.

Signalling pathways in mollusc immune systems are not well understood. They involve mitogen[15] activated protein kinase (MAPK)-like proteins and protein kinase C (PKC) in bivalves (e.g., Canesi et al. 2006) and gastropods (e.g., Lacchini et al. 2006; Coutellec & Caquet 2017).

The process of phagocytosis involves actin polymerisation which utilises energy and thus requires ATP. In *Mytilus*, haemocyte conformational changes and chemotaxis appear to be regulated via protein kinase C which is phosphorylated in response to bacterial lipopolysaccharide. ATP is utilised in protein kinase pathways (Ottaviani et al. 2000) and additional ATPs are supplied when energy demands are high, this process involving phosphoarginine and arginine kinase (Coyne 2011).

2.4.3.1 Cellular Defences (Immunocytes, Amoebocytes, and Haemocytes)

Phagocytosis by immunocytes (amoebocytes in the 'blood' – see Chapter 6 for details) is the main line of defence of the molluscan innate immune response, with these cells engulfing particles such as bacteria to form an internal vacuole (phagosome). The phagosome then fuses with a lysosome and hydrolases break down the particle (phagolysosome biogenesis). Acidification lowers the internal pH of the phagolysosome, enhancing the hydrolytic enzyme activity (Coyne 2011).

---

[13] Also called antimicrobial peptides.

[14] A layer on the outer side of the plasma membrane which serves as a 'cell wall' in most bacteria.

[15] A mitogen is a protein that assists in inducing cell division by triggering mitosis.

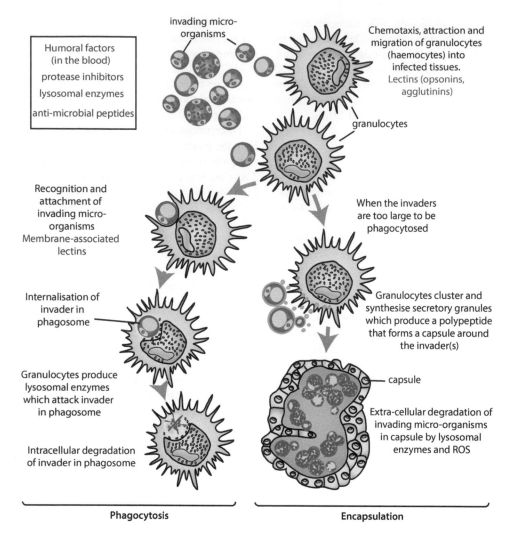

invading micro-organisms

Chemotaxis, attraction and migration of granulocytes (haemocytes) into infected tissues.
Lectins (opsonins, agglutinins)

Humoral factors (in the blood)
protease inhibitors
lysosomal enzymes
anti-microbial peptides

granulocytes

Recognition and attachment of invading micro-organisms
Membrane-associated lectins

When the invaders are too large to be phagocytosed

Internalisation of invader in phagosome

Granulocytes cluster and synthesise secretory granules which produce a polypeptide that forms a capsule around the invader(s)

Granulocytes produce lysosomal enzymes which attack invader in phagosome

capsule

Extra-cellular degradation of invading micro-organisms in capsule by lysosomal enzymes and ROS

Intracellular degradation of invader in phagosome

Phagocytosis   **Encapsulation**

FIGURE 2.8   Humoral and cellular responses involved in bivalve immune defence against infection by a unicellular parasite (the example shown is *Perkinsus*). ROS – Reactive Oxygen Species. Redrawn and modified from Soudant, P. et al., *J. Invertebr. Pathol.*, 114, 196–216, 2013.

The lysosomal enzymes break down the engulfed contents which are either released by exocytosis (i.e., extracellularly) or absorbed intracellularly (Figure 2.8). This process, which is also important in intracellular digestion, is significant in the immune system where amoebocytes (immunocytes) remove pathogens such as bacteria. They also remove dead cells, small inorganic particles, etc.

Once phagocytosed by the haemocyte, the foreign organism is neutralised by a combination of hydrolytic enzymes and ROS (Sokolova 2009; Coyne 2011), as described below. Thus, phagocytosis involves the production of:

- Lysosomal enzymes.
- 'Killing agents' by the immunocytes involving the generation of peroxide and hydroxyl radicals (through reduction of O_2) and their release as an 'oxidative burst' (Mkoji et al. 1988). This can be achieved, for example, by the release of superoxide

by the NADPH[16] oxidase enzyme complex, which occurs in some gastropods and bivalves, and is probably present in other molluscs.

- NOS is another enzyme that reacts with superoxide (peroxynitrite) and damages proteins. NOS has been detected in both bivalves and gastropods (Donaghy et al. 2015).
- Antimicrobial peptides – reported in euthyneuran heterobranchs and bivalves.

Amoebocytes also migrate to the site of injury and agglutinate and transform into fibroblasts synthesising collagen to repair wounds (e.g., Franchini & Ottaviani 2000). Besides wound healing (see Section 2.4.4) and dealing with bacteria and foreign debris that enters at the site of injury,

---

[16] The reduced form of $NADP^+$.

haemocytes also play an important role in body and shell repair. For example, in oysters and abalone haemocytes are involved in shell repair, with those containing calcium carbonate crystals increasing in abundance at the site of shell damage where they release the crystals (Mount et al. 2004; Fleury et al. 2008).

2.4.3.2 The Humoral Component – Antimicrobial Peptides and the Complement System

The humoral immune defences in invertebrates consist mainly of molecules such as host defence peptides and other antimicrobial defence factors that carry out direct attacks on pathogens. In addition, certain small proteins comprise the *complement system* which, as its name suggests, complements the activities of other components of the innate immune system (e.g., Tripp 1975; Mitta et al. 2000a, 2000b; Cellura et al. 2007). Benkendorff (2010, p. 758) suggested that 'a range of different antibacterial, antifungal, antiparasitic, and antiviral secondary metabolites may have evolved in molluscs, for circulation in the haemolymph, as well as for inclusion in mucous secretions on body surfaces'. Humoral factors also include various plasma molecules circulating in the blood (e.g., see Bayne 2003; Ottaviani 2004; Canesi et al. 2006 for reviews). For example, in *Mytilus*, Canesi et al. (2006) reported soluble lectins, hydrolytic enzymes, antimicrobial peptides, and possibly cytokine-like proteins as being relevant in killing bacteria. Agglutinins, proteins which agglutinate particles in the blood, may assist phagocytosis and encapsulation.

Neurosecretory activity plays an essential role in the control of immune responses, with integration between the immune and neuroendocrine systems possible because both utilise cytokines and neuropeptides. Such integration also enables lymphocyte cells to carry out both immune and neuroendocrine responses (Ottaviani 2006).

The antimicrobial peptides, or host defence peptides (or defensins), form part of the innate immune response in all organisms. They are basically broad-spectrum antibiotics effective against many bacteria, viruses, and fungi (e.g., Cellura et al. 2007; Li et al. 2009).

The complement system circulates in the blood, usually as inactive zymogens (enzyme precursors), and is part of the innate immune system. The activation of the zymogens results in a series of chemical reactions which helps in defence against pathogens. The complement system is particularly well-developed in vertebrates, but complement-like activity has also been found in some invertebrates (Dodds & Matsushita 2007), including molluscs (Ma et al. 2005). The process is controlled by the enzyme phenoloxidase which exists in an inactive form as prophenoloxidase. Its activation by microbial compounds results in a series of antimicrobial reactions and initiates other immune responses (Cerenius & Söderhäll 2004).

Neuroendocrine cells associated with the CNS produce immunoreactive substances. These include morphine-like chemicals (e.g., Sonetti et al. 1999), corticotropin-releasing hormone (CRH), and ACTH (Ottaviani 2004), the main mediators of stress response in vertebrates. These substances are released into the blood and stimulate haemocyte activity resulting in their migration and increased phagocytic activity and the release of biogenic amines[17] (Ottaviani et al. 1998; Ottaviani 2006).

Microglia are similar to vertebrate microglial cells and comprise a significant component of nervous tissue in the CNS of molluscs (e.g., Sonetti et al. 1997; Sonetti & Peruzzi 2004). These tiny cells carry out various roles including attending to injuries and controlling pathogens. Communication between microglial cells may involve nitric oxide from neurons, cytokines, growth factors, and peptides (Peruzzi et al. 2004). Activated microglial cells are highly variable in shape, with amoeboid processes. They can proliferate rapidly and migrate quickly toward the site of damage or infection. The microglial cells may then transform themselves into macrophages (Peruzzi & Sonetti 2004; Peruzzi et al. 2004).

2.4.3.2.1 Lectins

Lectins are carbohydrate-binding proteins which can be involved in carbohydrate-recognition as they bind to specific carbohydrate structures, namely, oligosaccharide chains (i.e., glycans). They are important in cell adhesion and apoptosis, and some can recognise non-self-invaders, such as bacteria and parasites, and hence, are important in the innate immune response. Specifically, their roles include non-self-recognition by binding to pathogen-associated molecular patterns (PAMP) on the surface of many pathogens as well as adhering to bacteria and microfungi where they stimulate and enhance phagocytosis and encapsulation. They also appear to have bactericidal effects, although the mechanisms involved are not well known (Wang et al. 2011). Lectins may also recognise senescent cells which are then removed by phagocytes.

Thirteen 'families' of lectins are recognised in animals. Members of one of the most important families are the C-type lectins which are diverse in their structure and function in molluscs (Wang et al. 2011). Lectins occur in many parts of the body, including reproductive structures, eggs, amoebocytes, blood, mantle, body wall, and gut. Several different types of lectins have been isolated from molluscs (e.g., Ozeki 1997; Kim et al. 2006; Fujii et al. 2009; Kawsar et al. 2009; Alpuche et al. 2010; Ito et al. 2010); including C-type lectin (e.g., Espinosa et al. 2010; Ito et al. 2010; Wang et al. 2011), galectin (Tasumi & Vasta 2007; Yamaura et al. 2008), sea urchin egg lectin (SUEL-type lectin) (Naganuma et al. 2006), and D-galactose-binding lectin (Fujii et al. 2011).

Transcripts corresponding to the Toll/NF-kB[18] pathway in some molluscs have also been recorded (e.g., Zhang & Coultas 2011; Collins et al. 2012b; Venier et al. 2016).

---

[17] A biogenic compound with one or more amine groups which can be toxic when in high concentration. They are also precursors in the synthesis of proteins, nucleic acids, etc.

[18] Toll-like receptors recognize protein patterns of invading bacteria leading to the activation of the nuclear factor kappa-light-chain-enhancer of activated B cells.

2.4.3.3 Diverse Responses to Infection

Some remarkably plastic responses to infection occur in molluscs. This has been particularly well studied in the planorbid *Biomphalaria*, a host of the blood fluke *Schistosoma*, in which diverse haemolymph lectins are expressed in response to infection. The responses include:

- Secreted fibrinogen-related proteins (FREP) bind to parasites (e.g., Loker & Kepler 2004; Zhang et al. 2009; Loker 2010; Yang et al. 2014a), and while there appears to be some evidence these may be adaptive, this is not entirely clear (Flajnik & Du Pasquier 2004; Ottaviani 2011). The expression of these proteins is increased following bacterial infection, and they have some features in common with mammalian ficolins. FREP sequences are very diverse among and between individuals. For example, three individuals of *Mytilus galloprovincialis* had no nucleotide or amino acid sequences in common (Romero et al. 2011). This remarkable complexity suggests the potential capacity to recognise and eliminate many pathogens (e.g., Hanington & Zhang 2010; Ottaviani 2011).
- A 'mollusc defence molecule', somewhat similar to the insect haemolin, an immunoglobulin-like molecule (e.g., review by Flajnik & Du Pasquier 2004).

Besides these responses, various experiments, including responses to transplants (Ottaviani 2004) and bacteria (Ottaviani 2006), suggest a short-term memory-type response in at least some molluscan immune responses. Similarly, a range of immune responses to various experimental and natural challenges has been recorded in coleoid cephalopods (see Figure 2.9).

Some parasites can subvert the immune system – examples are *Perkinsus marinus*, the pathogen that causes Dermo disease in oysters (e.g., Soudant et al. 2013), which can inhibit the reactive oxygen intermediates (ROI) that normally kill invading microbes. Infection of *Octopus* by the parasite *Aggregata octopiana* (Apicomplexa) stimulated phagocytic activity of the haemocytes, but it suppressed the oxidative (or respiratory) burst (see Section 2.4.3.1), resulting in a reduction of the expression of antioxidant genes (Castellanos-Martínez 2013).

2.4.3.4 Encapsulation or Nodule Formation

This response is usually initiated against larger foreign bodies such as parasites and involves building a capsule around the invading body using fibroblast-like cells and haemocytes (e.g., Soudant et al. 2013) (Figure 2.8). For example, infection of bivalves with the parasitic perkinsozoan *Perkinsus* produces an inflammatory response from connective tissue with the migration of granulocytes which release a polypeptide to form a capsule. Following encapsulation, lysosomal enzymes and ROS production destroy the parasites (Montes et al. 1995;

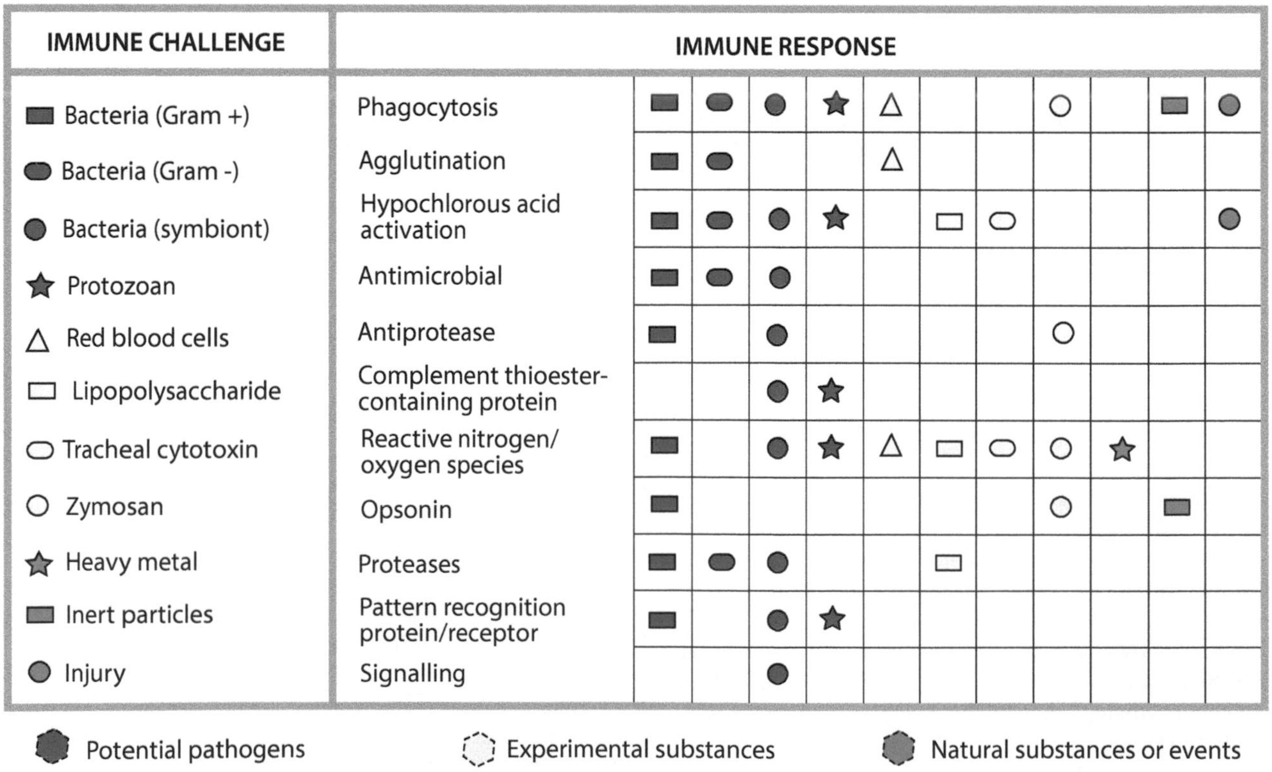

IMMUNE CHALLENGE	IMMUNE RESPONSE											
■ Bacteria (Gram +)	Phagocytosis	■	●	●	★	△			○			■	●
● Bacteria (Gram -)	Agglutination	■	●			△							
● Bacteria (symbiont)	Hypochlorous acid activation	■	●	●	★		▭	○					●
★ Protozoan	Antimicrobial	■	●	●									
△ Red blood cells	Antiprotease	■		●					○				
▭ Lipopolysaccharide	Complement thioester-containing protein			●	★								
○ Tracheal cytotoxin	Reactive nitrogen/oxygen species	■		●	★	△	▭	○	○	★			
○ Zymosan	Opsonin	■							○			■	
★ Heavy metal	Proteases	■	●	●			▭						
■ Inert particles	Pattern recognition protein/receptor	■		●	★								
● Injury	Signalling			●									

⬡ Potential pathogens   ⬡ Experimental substances   ⬡ Natural substances or events

FIGURE 2.9   Immune responses seen in coleoid cephalopods. Redrawn and modified from Castillo, M.G. et al., *Fish Shellfish Immunol.*, 46, 145–160, 2015.

Montes et al. 1996) (Figure 2.8). This reaction is not expressed with other microorganisms such as bacteria (Montes et al. 1997).

2.4.3.5 Environmental Considerations

Adverse environmental conditions (e.g., increased temperature and/or salinity, pollution, pathogen introductions), as well as damage from activities including aquaculture or agriculture, can reduce immune responses and result in disease outbreaks (Mydlarz et al. 2006). For example, various critical processes involving molluscan immunocytes such as chemotaxis and phagocytosis are decreased by abiotic stresses such as pollutants and raised water temperature (Coyne 2011).

2.4.4 Repair Mechanisms in Molluscs

The integrity of the skin is important to maintain bodily health, and this is achieved by blood clotting and wound healing. The mechanisms involved in shell repair are outlined in Chapter 3.

Wound repair includes several responses such as inflammation, matrix formation, and the remodelling of tissue, as well as muscular contraction (e.g., Sminia et al. 1973). Growth factors (Section 2.3.5 above) play a role in this process (e.g., Franchini & Ottaviani 2000; Ottaviani et al. 2001; Ottaviani 2006) by stimulating the migration of cells such as immunocytes into the wound, thus assisting in its closing, cleaning, and healing. Wound healing first involves an inflammation phase, with haemocytes accumulating in the injured area and phagocytising damaged tissue at the margins of the wound. Next, granulation tissue forms and a matrix is formed by haemocytes undergoing fibroblast-like activity. Finally, the epithelial tissues regrow, stimulated by specific growth factors (Ottaviani 2006).

In *Lymnaea*, differentiation of round amoebocytes into flattened cells was observed 18–24 hours after the wound was formed. Collagen started to form between these cells 3–5 days after wound formation, but at 90 days the connective tissue in the wound still differed from the normal tissue (Sminia et al. 1973).

The clumping of haemocytes helps prevent blood loss from wounds. This differs from blood clotting in vertebrates, as there are no extracellular fibrin fibres formed, and it is reversible, with many of the haemocytes re-entering the blood (Chen & Bayne 1995). Sulfhydryl groups play a role in the agglutination of both vertebrate and invertebrate blood cells (Chen & Bayne 1995; Hinzmann et al. 2013).

2.4.5 Secondary Metabolites

Secondary metabolites are used in defence in some gastropods, notably in heterobranchs, and are particularly well utilised by nudibranch sea slugs (see Chapters 9 and 20 for details). They may be sequestered from the diet (animals, fungi, algae, plants) or synthesised *de novo* by the mollusc.

Secondary metabolites are formed by biosynthesis from primary metabolites, with the great majority derived from a few starting compounds created by three main biosynthetic pathways in the food organism or *de novo* in the mollusc as outlined below, based on the account in Cimino and Ghiselin (2009):

- The polyketide pathway commences with acetyl-CoA (acetyl coenzyme A) which provides two carbon units, then links acetate subunits to form linear chains (polyketides, characterised by their oxogroups) which can be cross-coupled to form polyphenols. Polypropionates are polyketides not derived from acetyl-CoA, but from propionyl-CoA. This is very unusual, being found only in 'opisthobranchs' and some bacteria. Fatty acids start with acetyl-CoA and, via a series of reactions, form a linear chain (stearic acid) with two carbon subunits. Stearic acid can be further modified in various ways to form other fatty acids. Shikimic acid derivatives have many hydroxyl groups. One derivative is eicosapentaenoic acid, which forms eicosanoids, while another, arachidonic acid, forms prostaglandins (both are of interest as natural products). Eicosanoids are often used as chemical signals or as defensive chemicals. Acetylenic fatty acids are another important class of biologically active molecules used as defensive metabolites in sponges and algae, and they are sometimes found in 'opisthobranch' taxa that feed on those organisms.
- The isoprenoid pathway also starts with acetyl-CoA, with three units forming mevalonic acid which in turn gives rise to 5-carbon isoprene (2-methyl-1,3-butadiene) units. These isoprene units are linked to form terpenes, steroids, carotenoids, and other similar compounds.
- The amino acid pathway starts from various amino acids produced in the citric acid cycle (see Section 2.2.4.2) and in the glycolytic and shikimic acid pathways. Aromatic amino acids from the shikimic acid pathway are obtained from food (bacteria, fungi, and algae) as they cannot be synthesised by animals. Alkaloids are also derived from these amino acids.

By using the compounds produced by these three pathways and combining, modifying, and rearranging them, a large variety of secondary compounds can be produced (see Chapter 9).

2.4.6 Aestivation (Dormancy, Hibernation)

In unfavourable conditions (usually high or low temperature and/or low water availability in terrestrial or fresh-water taxa) some molluscs, notably terrestrial gastropods, aestivate. During aestivation, the physiology changes to maximise the conservation of water and energy and manage the accumulation of nitrogenous waste. The metabolic rate is much reduced so it is often less than 30% of active animals (e.g., Rees & Hand 1993; Bishop & Brand 2000), and this is accompanied by reduced oxygen uptake (Michaelidis 2002). When favourable conditions reappear, the aestivating individual typically quickly returns to an active state.

Aestivating bivalves close their valves, and operculate snails seal their aperture with the operculum. Stylommatophoran snails (which all lack an operculum) avoid desiccation by either sealing the aperture of the shell to a hard surface using mucus or secreting an epiphragm (see Chapter 20), an operculum-like structure, the latter porous enough for oxygen exchange. In snails, energy during aestivation is mainly obtained from the breakdown of glycogen, although proteins are also important during prolonged aestivation (Livingstone & Zwaan 1983; Rees & Hand 1993). This contrasts with vertebrates, which use lipids as the main source of energy.

Aestivation is accomplished by a complex array of biochemical pathways which together suppress metabolic rate, reprioritise cell function energy use, and increase molecular defence mechanisms such as protein chaperones and antioxidants (Ramnanan et al. 2017). These alterations, in turn, are controlled by intracellular signalling cascades, reversible phosphorylation and enzyme binding, inhibition of transcription and translation, and upregulation of certain genes (Storey & Storey 2010). These changes can occur quickly, as for example, in the stylommatophoran snail *Otala lactea*, where there is a considerable reduction in protein translation 48 hours after aestivation commences and an increase in the expression of HSPs (Ramnanan et al. 2007; Ramnanan et al. 2009). Similarly, arousal from aestivation in land snails, accompanied by a rapid increase in oxygen consumption, increases oxidative stress (e.g., Ferreira-Cravo et al. 2010) from ROS. Arousal is also accompanied by increased levels of antioxidant enzymes. For example, in some helicid snails, levels of enzymes such as catalase, glutathione (GSH), superoxide dismutase (SOD), glutathione S-transferase (GST), and selenium-glutathione peroxidase (GPx) have been shown to increase markedly during arousal (e.g., Ferreira-Cravo et al. 2010), but return to normal levels shortly after. Similar changes have also been observed in the planorbid *Biomphalaria* on arousal from dormancy induced by the drying up of the water body in which they live (Ferreira et al. 2003).

2.5 GENOMICS, INCLUDING CHROMOSOMES AND MOLECULAR STUDIES

Molluscs were included in some of the earliest molecular research (Field et al. 1988) and were chosen because they were easily obtained or raised (especially fresh-water and terrestrial taxa), of economic interest (bivalves), or presented interesting research questions. These investigations addressed questions in biochemistry, development, physiology, genetics, evolution, phylogeny, and systematics. Today, molecular research is both diverse and commonplace and, because of new laboratory techniques and analytical approaches, is much cheaper and more widely accessible, resulting in molecular studies and genomics being among the fastest-growing fields in biological sciences.

2.5.1 CHROMOSOME STUDIES

The earliest genomic data on molluscs came from investigation of gametogenesis (see Chapter 8) at the beginning of the twentieth century. This research documented the reduction in the number of chromosomes in oocytes and sperm in the process later known as meiosis. Many of the study taxa were heterobranch gastropods, including helicid snails (Ancel 1902), the 'sea hare' *Aplysia* (Janssens & Elrington 1904), and several nudibranch sea slugs (Smallwood 1905). The variation reported in these and subsequent studies attracted the attention of biologists, who documented broad patterns first by counting chromosomes visible in cell squashes and later with karyotyping (Jacob 1957, 1958; Burch 1960, 1962; Patterson 1969; Ieyama & Inaba 1974; Patterson & Burch 1978; Thiriot-Quiévreux 2003). The variability and taxonomic distribution of chromosome numbers in molluscs soon caused interest in chromosome duplications and the role of *polyploidy*[19] in molluscan evolution (Burch & Huber 1966; Nakamura 1986; Thiriot-Quiévreux et al. 1989; Ostrovskaya et al. 1996; Hallinan & Lindberg 2011).

Two forms of polyploidy occur in molluscs – one in the germline and one in the somatic line. Thus, chromosome duplications can occur throughout the genome of the organism or can be confined to a specific tissue or cell type (e.g., in the giant neurons of some heterobranch gastropods) (Gillette 1991; Anisimov 2005) (see Chapter 7). While most molluscs have only single pairs of chromosomes (2n or diploid), triploid (3n), tetraploid (4n), and even larger duplication events have been reported in molluscs (Burch & Huber 1966; Patterson 1969; Petkevičiūtė et al. 2007). Tetraploid taxa are commonly artificially produced for fisheries and aquaculture because of their increased vigour (Beaumont & Fairbrother 1991; Okumura et al. 2007).

Polyploidy events are an important mechanism of genetic evolution (Taylor & Raes 2004), and unlike other modes of genome evolution, they can simultaneously affect most or all of the genome. The role of polyploidy in molluscan evolution is contentious. Burch and Huber (1966) thought that polyploidy was important in diversification, but considered it to have little importance in the derivation of higher taxa, while Graham (1985) questioned even its role in diversification. In contrast, Michel (1994) argued that polyploidy might have played an important role in the radiations of endemic lacustrine taxa, and Vinogradov (2000) suggested that the larger genomes of terrestrial eupulmonates were an important factor in their emergence on land.

Polyploidy results from two processes: *autopolyploidy*, which doubles the number of chromosomes in an individual or *allopolyploidy*, where the doubling results from hybridisation between two species. Most molluscan examples appear to have resulted from allopolyploidy (Burch & Huber 1966; Patterson 1969), although instances of autopolyploidy are also suspected (Hallinan & Lindberg 2011). Genes resulting

---

[19] The presence in a cell or organism of more than two sets of homologous chromosomes.

from autopolyploidy are termed *paralogous* because they are derived from common ancestral DNA and *palaeopolyploid*, which are ancient polyploidy events. Many genes produced by polyploidy events eventually become 'silent' and are lost from the genome (Scannell et al. 2007), making the recognition of palaeopolyploidy difficult (Hallinan & Lindberg 2011). Duplicated genes are not always lost and can be diverted to other purposes, thus providing the raw material for evolutionary innovation (Ohno 1967; Kasahara 2007; Scannell et al. 2007).

Hallinan and Lindberg (2011) identified three putative whole genome duplications (WGD) in molluscs – one in caenogastropods, a second early in the history of the stylommatophorans, and a third in cephalopods. Their estimates required that the background rate of chromosome number evolution be relatively low, and this is concordant with observed rates in molluscs, and particularly in gastropods (Chambers 1987). In caenogastropods, Hallinan and Lindberg (2011) identified a WGD between 200 mya and 155 mya in the common ancestor of a clade containing Capulidae, Ranellidae, Cypraeidae, and the Neogastropoda after its divergence from the Strombidae and the other hypsogastropod families. They also found evidence of two possible events in stylommatophorans, either in the early Cenozoic (65 mya) in the common ancestor of Sigmurethra and Orthurethra (after their divergence from Succineidae), or there was a duplication event in the common ancestor of all Stylommatophora after the group diverged from other eupulmonates in the lower Cretaceous (approximately 138 mya), but before they radiated at the beginning of the Cenozoic (65 mya). Steusloff (1942) also suggested a palaeopolypoid event in stylommatophorans, and Burch and Natarajan (1965) suggested the possibility of polyploidy in Succineidae. A third WGD occurred within the Cephalopoda, and although there are several possible positions for the event, Hallinan and Lindberg (2011) found the strongest support for a duplication in the common ancestor of the Coleoidea after it diverged from the Nautiloidea in the lower Ordovician (490 mya), and before the decabrachians diverged from the octobrachians in the Carboniferous (300 mya).

Palaeopolyploidy events are uncommon outside gastropods and cephalopods (Figure 2.10). In chitons, haploid chromosome numbers range between 9 and 12 with a mode of 11 (Hallinan & Lindberg 2011). Odierna et al. (2008) saw no evidence for polyploidy in chitons and instead suggested chromosome loss through a series of Robertsonian fusions[20] (Slijepcevic 1998) and other chromosomal reduction processes. Sampling remains poor in chitons and scaphopods, with only three dentaliid taxa sampled (1n = 10) (Ieyama 1993). Sampling is better in bivalves (Nakamura 1985; Thiriot-Quiévreux 2002), but putative polyploids are uncommon, and variation in chromosome number is substantially less than in gastropods (Nakamura 1985) and cephalopods (Gao & Natsukari 1990; Vitturi et al. 1990) (Figure 2.10).

Three groups of bivalves containing taxa that have undergone polyploidy events are the Cyrenidae, Galeommatoidea, and Sphaerioidea. Polyploidy has been well demonstrated in the galeommatoidean *Lasaea* (Ó Foighil & Thiriot-Quiévreux 1991; Taylor & Ó Foighil 2000) and in the cyrenid *Corbicula* (Park et al. 2000). Based on their observations of variation in chromosome numbers, Burch and Huber (1966) suggested polyploidy was present in some Sphaeriidae, and this has since been confirmed (Lee 1999; Petkevičiūtė et al. 2007; Park 2008).

Burch and Huber (1966) first pointed out that polyploidy in molluscs was often found in parthenogenetic and hermaphroditic taxa capable of self-fertilisation. They argued that the initial polyploid individual would have had difficulty in finding another polyploid mate and/or there would have been complications associated with cellular differentiation during development. Hermaphroditic taxa that lacked sex chromosomes and were capable of self-fertilisation could reproduce, as would parthenogenetic and clonal taxa. This association of polyploidy with reproductive mode appears well supported in bivalves and stylommatophorans, but does not correlate with the polyploidy events seen in caenogastropods or cephalopods, which are predominately gonochoristic.

As noted above, it has been suggested that WGDs can lead to large morphological and physiological innovations, because redundancy can ease constraints on genes throughout the genome (Haldane 1932; Ohno 1967). All three clades identified by Hallinan and Lindberg (2011) fit this model. After having successfully transitioned from aquatic ancestors, stylommatophorans became by far the largest group of land snails and slugs (Vinogradov 2000; Barker 2001a; Mordan & Wade 2008). Coleoid cephalopods are a successful group with disparate size, shape, habits, and ecology and show great changes in anatomical, physiological, and behavioural complexity relative to other molluscs and to their sister taxon, the Nautiloidea (Nishiguchi & Mapes 2008). Last, the WGD event in caenogastropods occurred shortly before the neogastropod radiation and may have had a role in the diversification of this large and diverse clade which has undergone extraordinary radiations as reflected in their anatomical, physiological, behavioural, and ecological disparity (Ponder et al. 2008).

Molecular studies by Yoshida et al. (2011) and Bassaglia et al. (2012) supported a WGD event in coleoid cephalopods. However, a recent study of the genome of *Octopus bimaculoides* by Albertin et al. (2015) found large expansions in only two gene families compared to other invertebrate genomes – the protocadherins and the C_2H_2 zinc-finger transcription factors – known to have also undergone large increases in vertebrates. These genes regulate neuronal development and are involved in neural excitability and in cellular functions associated with the skin, suckers, and nervous system. Albertin et al. (2015) also found evidence of large-scale genomic rearrangements and they suggested that WGDs are not required for key innovations in molluscan evolution, and that repetitive content and gene linkage changes in a handful of gene families may be responsible for the evolution of cephalopods, and especially their large, complex nervous system.

---

[20] A Robertsonian fusion is where a fragment of one chromosome is attached to another, non-homologous chromosome. In Robertsonian translocation, an entire arm is translocated with the break occurring at the centromere.

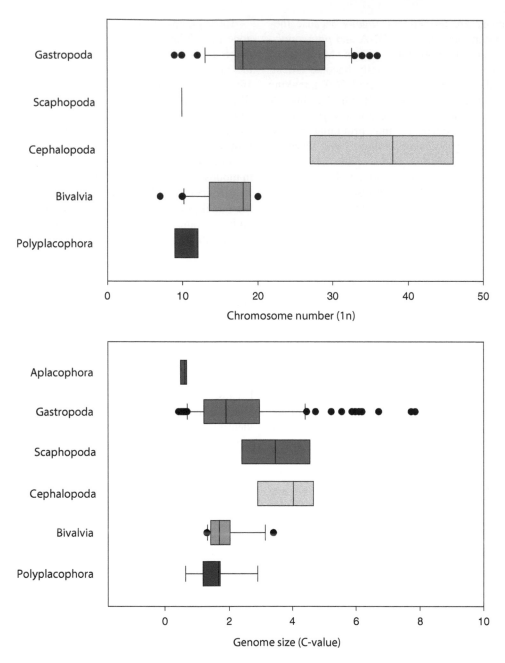

FIGURE 2.10 Chromosome numbers (1n) and genome size in the Mollusca. Chromosome data from Hallinan, N. and Lindberg, D.R., *Genome Biol. Evol.*, 3, 1150–1163, 2011 and genome size data from Gregory (2018). *Animal Genome Size Database*. http://www.genomesize.com.

2.5.1.1 Endopolyploidy

As discussed above, chromosome duplications can also be confined to a specific tissue or cell type. This type of polyploidy, called endopolyploidy (Anisimov 2005), involves the replication of the nuclear genome without cell division or endoreplication. Endoreplication is a disruption of the mitotic cell cycle in which mitosis is aborted before cell division (cytokinesis) or circumvented entirely (Lee et al. 2009). This abbreviated cell cycle still duplicates DNA, but the ultimate cell division is suppressed, producing polyploidy in the cell (Figure 2.11).

There are two types of endoreplication: endocycles and endomitosis. In an endocycle, mitosis is avoided, and the cell only cycles between the G1 and S phases. In endomitosis, the cell begins mitosis after chromosome duplication, but it is not completed.

Endopolyploidy is especially prominent in the Heterobranchia where it occurs in neurons of the CNS ganglia, salivary gland cells, and digestive gland cells (Anisimov 2005). In a study of 25 species of heterobranchs from marine, terrestrial, and fresh-water habitats, Anisimov and Zyumchenko

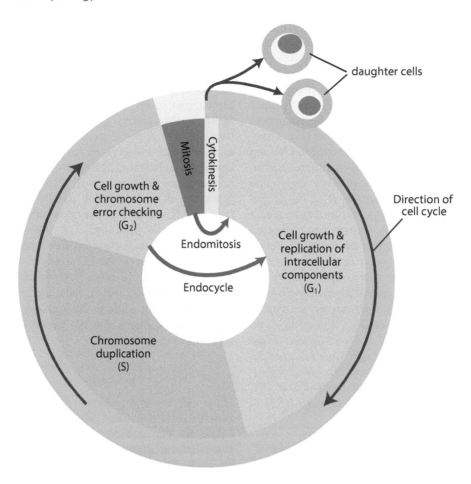

FIGURE 2.11 The mitotic cell cycle showing two mechanisms of endoreplication. Normal cell division consists of five stages, beginning with replication intracellular components in preparation for DNA replication (G_1) and ending with cell division (cytokinesis) and the production of two daughter cells, each with a normal complement of chromosomes. Both endocycles and endomitosis short-circuit the cell cycle and omit the final stages of cell division. Endocycles duplicate the genome without the cell entering mitosis. Successive cell cycles produce multiple copies of chromosomes in these cells. In endomitosis, the chromosomes also replicate, but mitosis is prematurely terminated, producing only a single duplication with each cycle. Redrawn and modified from OpenStax, The Cell Cycle. OpenStax CNX. June 5, 2013, http://cnx.org/contents/d6d279e4-eb7a-4e64-8d53-8710b4c50e81@7.

(2012) found differential polyploidisation in the salivary glands. Cells that secreted proteins (enzymatic) had substantially more chromosomal copies than mucus-secreting cells, 256n versus 32n, respectively. Anisimov and Zyumchenko (2012) attributed the high levels of polyploidy in the enzymatic cells to dietary specialisations that required intense protein syntheses. They found no correlations with habitat, lifespan, or body size. Earlier studies of the CNS of heterobranchs (see Anisimov 2005; Mandrioli et al. 2010) found that while most bivalve and gastropod neurons were 2n, those of some heterobranchs can be as high as 260,000 copies (*Aplysia*). Digestive gland cells in heterobranchs also have elevated DNA content (32–64n), but not as high as either neurons or salivary gland cells. Caenogastropods can also show elevated DNA content in their digestive and salivary gland cells (16n and 32n, respectively), but not in CNS neurons (Anisimov 2005).

2.5.2 Early Molecular Work

In the early 1950s, electrophoresis was used to characterise molluscan proteins (e.g., Jarrige & Henry 1952; Maxfield 1953; Roy 1955). Much of this early work focused on digestive isozymes, but by the middle 1960s electrophoretic data were being used to hypothesise phylogenetic relationships (e.g., Davis & Lindsay 1964; Grant et al. 1988, and references therein). With the advent of DNA sequencing, using electrophoresis to generate characters in determining phylogenetic relationships has waned although it continues to play an important role in molecular biology, including the separation and purification of proteins.

Smith and Quayle (1963) described the base compositions in molluscan DNA and noted that the guanine-cytosine (GC) content of cephalopod DNA (*Octopus vulgaris*) was midway between the values for bivalves and gastropods.

Hinegardner (1974) surveyed the DNA content of 110 molluscan species, including chitons, gastropods, bivalves, and cephalopods, and concluded that DNA content was correlated with chromosome number and that within families, DNA content also correlated with body size. Hinegardner also suggested that the more specialised, and sometimes older, molluscan groups (bivalves and patellogastropods) had relatively lower amounts of DNA than other taxa. Other groups, such as the neogastropods, were identified as having high amounts of DNA, which correlated with increases in body plan complexity and their relatively recent origin and radiation. The pattern of lower amounts of DNA in specialised species was even stronger in bivalves. In a more recent study by Dixon et al. (2001), DNA content was compared in putatively specialised vent species with non-vent species and showed no differences, suggesting that minimal or no changes had occurred since vent colonisation. Bonnivard et al. (2009) have reported the second largest known veneroid genome in the vent species *Calyptogena magnifica*. Our understanding of the patterns of DNA content in invertebrates entered a new phase with the publication of the Animal Genome Size Database (http://www.genomesize.com) (Gregory et al. 2007), although the evolutionary significance and mechanisms of genome size evolution remain controversial (Gregory 2004, 2011; Bainard & Gregory 2013).

2.5.3 Mitochondrial Genomes

The structure of the mitochondrial genome (mtgenome) is broadly conserved across the Metazoa and consists of a circular, double stranded DNA molecule with 13 protein-coding genes, two rRNA, approximately 22 transfer RNAs (tRNA), and a control region (Figure 2.12). As of 1 May 2019, every major group of molluscs, except for the Solenogastres, are represented to varying taxonomic depths in Genbank (http://www.ncbi.nlm.nih.gov/genome/), including two monoplacophorans and four Caudofoveata, six chitons, two scaphopods, 169 gastropods, 156 bivalves, and 44 cephalopods. Within gastropods, there has been a disproportionate amount of study of heterobranch taxa (48%) (Kurabayashi & Ueshima 2000; Grande et al. 2008; Wägele et al. 2008).

Molluscs are one of the few metazoan groups that show substantial coding gene rearrangement and, while this elevated level of change can hamper the reconstruction of relationships, the variation present among molluscan taxa may prove useful in understanding the underlying mechanisms of gene re-ordering (Sun et al. 2017).

Strong conservation of large portions of the gene order seen in the chiton mitochondrial genome (Figure 2.12) is seen in other Lophotrochozoa, suggesting that only one or two rearrangements occurred between brachiopods, chitons, protobranch bivalves, basal gastropods, cephalopods, and caudofoveates (Figure 2.13). There are only two additional rearrangements in most caenogastropods, but most autobranch bivalves and heterobranch gastropods are highly rearranged (Figure 2.13). While it was initially hoped that complete mitochondrial genomes would provide new character sets to help

resolve molluscan relationships, it was soon apparent that characters such as gene order did not provide useful resolution (Grande et al. 2008; Wägele et al. 2008).

In contrast to this pattern of conservation, other groups such as bivalves, scaphopods, patellogastropods, and heterobranch gastropods showed substantial gene rearrangement. MtRNAs also undergo gene order changes, including translocations and duplications, and move at higher rates than changes in the coding genes (Gissi et al. 2008). For example, among heterobranchs, tRNAs are 4.25 times more likely to have changed their position than coding genes. Both coding genes and tRNAs can move rapidly within a genome, and even closely related taxa can show substantial differences in mtgenome order (Rawlings et al. 2001). Based on the time of origin of the major groups (see Chapter 13), gene order changes began during the Paleozoic in several stem lineages (e.g., Cephalopoda, Bivalvia, Heterobranchia). In some clades, such as the Heterobranchia, these changes became fixed for the coding genes, while in other groups (Bivalvia), the rearrangements continued into the Cenozoic, producing unique gene orders in crown taxa.

2.5.4 Nuclear Genomes

Molluscan nuclear genomes are poorly known. Simison and Boore (2008) listed four molluscan taxa selected for whole nuclear genome sequencing; *Aplysia californica*, *Biomphalaria glabrata*, *Mytilus californianus*, and *Lottia gigantea*. Since then four gastropods, *Radix auricularia*, *Colubraria reticulata*, *Conus tribblei*, *Lymnaea stagnalis*, ten additional bivalves, *Crassostrea gigas*, *Crassostrea virginica*, *Bankia setacea*, *Mytilus galloprovincialis*, *Mizuhopecten yessoensis*, *Dreissena polymorpha*, *Corbicula fluminea*, *Bathymodiolus platifrons*, *Modiolus philippinarum*, *Pinctada martensii*, and the octopus *Octopus bimaculoides* have been added. Molluscan genomes tend to be large with many duplication events (see Section 2.5.1). With the continued rapid development of high throughput sequencing and bioinformatic approaches, more molluscan genomes should become available soon.

The patellogastropod *Lottia gigantea* was the first molluscan genome sequenced. The genome consisted of 348 megabase pairs and 23,287 genes, retaining approximately 94% of the inferred ancestral bilaterian gene families (Simakov et al. 2013). Comparison of the *L. gigantea* genome with that of the annelid *Capitella teleta* reveals a high occurrence of conserved linkages between large numbers of orthologous genes (high macrosynteny), and this includes almost half of the protein-coding genes (Simakov et al. 2013). The *L. gigantea* Hox gene complex represents the first intact Hox cluster found in a lophotrochozoan and contains 11 collinear genes – three anterior genes, six central-class genes, plus two posterior-class genes (*Post-1* and *Post-2*). In contrast, the *Crassostrea gigas* Hox gene complex is broken into four clusters, and the central-class gene *Antennapedia* has not been located (Zhang et al. 2012a). The Hox homeodomain contains transcription factors with conserved roles in forming the anterior-posterior

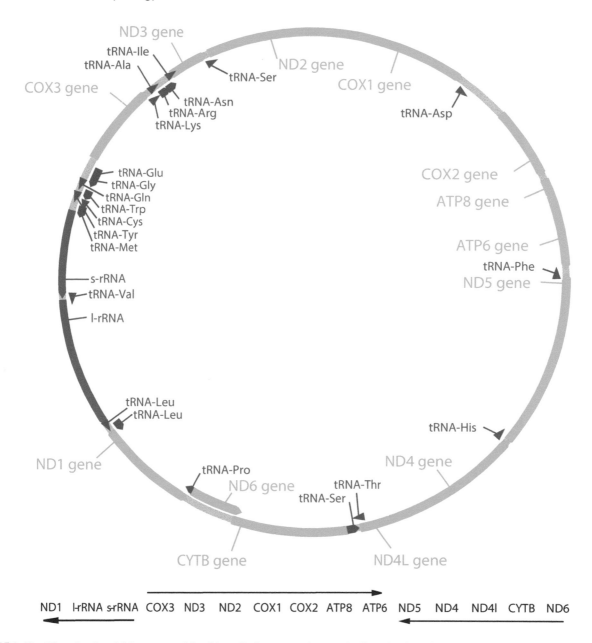

FIGURE 2.12 The mitochondrial genome of the chiton *Katharina tunicata* and a linearisation of the coding genes of the circular genome starting with the position of the NADH dehydrogenase, subunit 1 (ND1). Coding genes are shown in green, tRNAs in purple, and rRNAs in red. Arrows shown in the linearisation indicate the direction of transcription. Reconstructed from GenBank data, https://www.ncbi.nlm.nih.gov/genbank/.

body axis and is thought to have played an important role in molluscan body plan evolution (Callaerts et al. 2002; Lee et al. 2003a; Iijima et al. 2006; Samadi & Steiner 2010). The ParaHox gene cluster, an evolutionary sister of the Hox gene cluster, was not located in the *L. gigantea* genome. One or more genes of this three-gene cluster appear to have been lost in cephalopods (Callaerts et al. 2002), gastropods (Degnan & Morse 1993), and bivalves (Barucca et al. 2003); so far all three genes are present only in chitons (Barucca et al. 2006).

Although limited, the availability of molluscan nuclear genomes has produced insights and new data for the study

of molluscs. Examples include bivalve immunology (Li et al. 2014a), shell formation (Marie et al. 2012, 2013; Zhang et al. 2012a; Suzuki et al. 2013; Suzuki & Nagasawa 2013; Wang et al. 2014), biogeography (Rohfritsch et al. 2013), physiology (Zhang et al. 2012a; Meng et al. 2013), developmental gene expression and signalling (Cho et al. 2010; Yang et al. 2014b), morphogenesis (Sousounis et al. 2013), and neural biology (Veenstra 2010). Many of the studies use the bivalve genomes to address aquaculture-related issues (physiology, development, immunology) given that *Crassostrea gigas* and *Mytilus galloprovincialis* have some of the highest aquaculture production levels of any species (Astorga 2014).

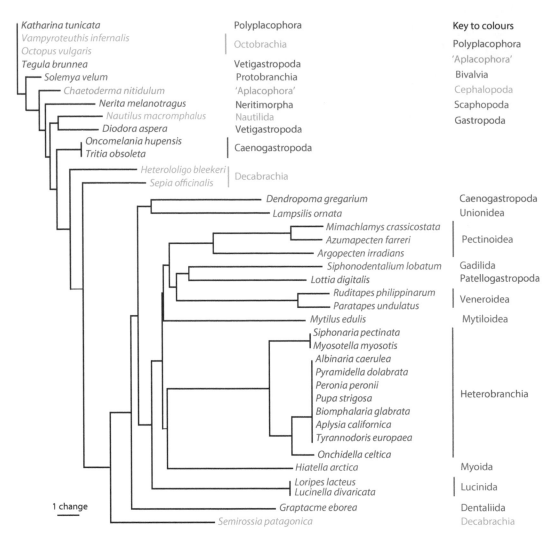

FIGURE 2.13 Breakpoint distance tree based on unique mitochondrial gene orders found in molluscs. Branch lengths represent the change in the number of gene 'adjacencies' (in a linearised genome) present in one genome, but not the other. The colours distinguish the main taxa (see key at top right). The gene order of the chiton *Katharina tunicata* (Figure 2.12) was selected as the initial gene order. Selected taxa from Genbank (GOBASE-Organelle Genome Database) (accessed 2017) and analysis performed with multiple genome rearrangements (MGR) software (http://grimm.ucsd.edu/cgi-bin/mgr.cgi).

Molluscan genomes have also provided important resources for phylogenetic studies which rely on these whole genomes to predict gene identities of transcriptome data (Smith et al. 2011; Zapata et al. 2014).

2.5.5 PROTEOMICS

The complete set of proteins produced or modified by an organism is the proteome, and proteomics is the study of the structure and function of these proteins. Because the metabolic pathways of the cell are regulated by proteins, they vary between organs and systems and reflect local physiology stresses and conditions through time (Tomanek 2014). They are therefore commonly used to assess responses to environmental stresses, including extreme temperatures (Tomanek & Somero 1997, 1998; Tomanek & Sanford 2003),

oxygen and carbon dioxide levels (Sheehan & McDonagh 2008; Tomanek et al. 2011; Wei et al. 2015), salinity; pH and ocean acidification (Kurihara et al. 2007; Kurihara 2008; Amaral et al. 2012b; Tomanek et al. 2012), development and life history (Lopez et al. 2005; Diz et al. 2013), microbes, parasites and diseases (Dolashka & Voelter 2013; Kingtong et al. 2013; Wu et al. 2013; Castellanos-Martínez et al. 2014; Prasopdee et al. 2015), pollutants (Rodriguez-Ortega et al. 2003; Mi et al. 2005; Apraiz et al. 2006; Riva & Binelli 2014), and toxic metals (Cole et al. 2014; Xu et al. 2014). While most of these studies are typically species-specific, investigations comparing sister taxa, and/or populations, provide insights into how cellular processes differ and which are critical to adapting to different settings; insights critical to assessing the response of a taxon to future environmental stresses. Shell formation has also been a topic of proteomic studies because of the diversity of proteins associated with calcium carbonate deposition (e.g., Marie et al. 2009, 2011, 2012, 2013; Mann & Jackson 2014). These studies are also important in fisheries and aquaculture as they seek to maximise environmental conditions for production and reduce disease and parasite loads (Saavedra & Bachere 2006; Huete-Perez & Quezada 2013; Tedesco et al. 2014).

One area of substantial research has been in the proteomics of *Conus* venom, of special interest because of its pharmaceutical potential (Stoecklin et al. 2004; Prashanth et al. 2012; Olivera et al. 2014), especially in pain relief (Lewis et al. 2012). *Conus* venoms or conotoxins are of particular interest because they are a mixture of different toxins (Safavi-Hemami et al. 2011, 2014; Dutertre et al. 2013) that are rapid and specific in their targeting of different cell types (Kaas et al. 2012). Because of this specificity, there are few or no side effects. Ziconotide, the first conotoxin painkiller, was approved for medical use in 2004, and is derived from *Conus magus*; it is about 100 times as powerful as morphine and non-addictive.

2.5.6 RNA Editing

Recent work has demonstrated that cephalopods exhibit a rare post-transcriptional process called RNA editing. During transcription, this process alters specific nucleotide sequences within the RNA molecule and thus changes the amino acid sequence of the encoded protein (Brennicke et al. 1999; Deffit & Hundley 2016). Far more instances of RNA editing occur in cephalopods than is known for any other metazoan. For example, RNA editing in the 'longfin inshore squid' (*Doryteuthis pealeii*) involved more than 57,000 edits (Alon et al. 2015), and a cuttlefish (*Sepia officinalis*), the common octopus (*Octopus vulgaris*), and the 'two-spot octopus' (*Octopus bimaculoides*) had between 80,000 and 130,000 edited sites (Liscovitch-Brauer et al. 2017). In contrast, *Nautilus pompilius* only had about 1000 edited sites suggesting that the extensive recoding evolved in an early stem coleoid lineage. Liscovitch-Brauer et al. also noted that around 1000 of the edited locations were shared between the coleoid species. Most of the recoded genes are associated with the nervous system suggesting that coleoid complexity may be a product of this rare molecular process (Di Cosmo et al. 2018).

2.5.7 Horizontal Gene Transfer

Horizontal gene transfer involves the transfer of functional genes between organisms. This has theoretically occurred with the incorporation of endosymbiotic organelles (such as mitochondria and chloroplasts) and sometimes occurs between bacteria and animals. It has been suggested as a means of animals acquiring genes they would not normally possess, such as those that produce the globin haemeproteins

(Moens et al. 1996) and cellulases (see Davison and Blaxter 2005 for references). The sacoglossan sea slugs *Elysia* spp., like many other sacoglossans, contain functioning chloroplasts obtained from their algal food (see also Chapters 5, 9 and 20), but the nuclear genes required for photosynthesis to be successfully carried out are not normally present in animals. Their presence in *Elysia* spp. was demonstrated and was thought to be an example of horizontal gene transfer (Pierce et al. 2003; Rumpho et al. 2008; Schwartz et al. 2014). Other workers have disputed this, finding no evidence of gene transfer in the sacoglossans they investigated (Wägele et al. 2011; Bhattacharya et al. 2013). No other possible examples of this phenomenon have been demonstrated in molluscs to date.

2.6 NON-GENOMIC MOLECULES

Besides DNA-related data discussed above, other non-genomic molluscan molecules are important sources of both environmental and historical data. Amino acid racemisation studies use proteins from the organic matrix of the shell to date both palaeontological and archaeological sites by comparing the ratio of D:L amino acid isomers (Wehmiller 1990). While most amino acids are in the L form in living animals, after death they are slowly converted to D isomers until an equilibrium of D:L forms is reached. Although this rate is sensitive to temperature, humidity, depositional features, etc., there are corrections for some of these factors such as temperature (Wehmiller & Belknap 1978). Amino acid racemisation dating is used throughout the Quaternary (2.58 mya to present) to date palaeontological deposits and archaeological sites. Molluscs, especially bivalves with thick umbo regions, are particularly useful for analysis. Although newer techniques for dating nearshore deposits are available (e.g., uranium-thorium dating), amino acid racemisation remains an active field for dating deep-sea cores (Huntley et al. 2012), as well as fresh-water (Huntley et al. 2012; Penkman et al. 2013) and terrestrial deposits (Donovan & Paul 2010; Hearty 2010), archaeological sites (shell mounds and middens) (Banerjee et al. 1991; Demarchi et al. 2011), taphonomic processes such as time averaging of fossils (Kosnik et al. 2009), and sea level change (Carr et al. 2010; Switzer et al. 2010), and it remains an important technique for documenting the chronology of previous Holocene climate changes (Wehmiller 2013; Briner et al. 2014).

Molluscan lipids have been used to determine bacterial associations in chemosynthetic vent molluscs. Hot vent community members generally specialise on one of two types of bacteria, either sulphide-oxidising bacteria or sulphate-reducing bacteria. Colaco et al. (2007) have shown that the digestive glands of the mid-Atlantic ridge mussels *Bathymodiolus* spp. and the gastropod *Phymorhynchus* sp. have fatty acid markers indicative of sulphide-oxidising bacteria.

3 Shell, Body, and Muscles

In this chapter, we provide an overview of the external body of molluscs, comprising the mantle, the shell and its formation and growth, the epidermis and associated structures, the foot and operculum (of gastropods), mucoid secretions, locomotion, and general information on cartilage and muscles. Some external structures, such as the suckers found in many coleoid cephalopods, are also important external features, but as they are confined to a single group, we deal with them in the appropriate taxon chapter.

3.1 BODY SYMMETRY AND AXES

Molluscan body axes are complicated, and there can be up to three distinct body orientations which transform during ontogeny. As described in more detail in Chapter 8, in most bilaterally symmetrical animals the embryonic blastula has an animal-vegetal polarity, and during gastrulation, the blastopore forms at the vegetal pole (Biggelaar et al. 2002). In deuterostomes, the blastopore becomes the anus, thus establishing the vegetal pole as the posterior (P) of the embryo and the animal pole as the anterior (A). In protostomes, the initially posterior vegetal blastopore is rotated from its initial position to an anterior-ventral position where it typically forms the mouth opening (Biggelaar et al. 2002). In molluscs, this reorientation of the embryonic A-P axis during the formation of the trochophore larvae is driven by cell proliferation and migration on the embryonic dorsal surface which displaces the blastopore ventrally and then anteriorly (Biggelaar et al. 2002; Wanninger & Wollesen 2015). Thus, by the molluscan trochophore stage, the original embryonic A-P axis is no longer linear, and the originally posterior blastopore has shifted approximately 90° to the ventral surface. In shelled taxa, further cell proliferation and migration resulting from the formation of the dorsal shell gland displaces the secondary anal opening ventrally and anteriorly. This produces the characteristic U-shaped gut indicative of ano-pedal flexure and obscures the original embryonic body axes. A third reorientation of the body axes may occur after metamorphosis when the animal assumes its adult life orientation. Life (or 'ecological') orientation often differs from the anatomical axes in molluscs (Figure 3.1). Some caudofoveates burrow in the sediment with the posterior end dorsal, while scaphopods burrow obliquely to near vertically in the sediment with the head and foot positioned ventrally. In living cephalopods, the ventral axis is typically rotated from 45° to 90° depending on the taxon. In swimming *Nautilus* and crawling octopods, the ventral axis is rotated about 60°, while in cuttlefish and squid, the body axis is rotated a full 90° in its life orientation (Figure 3.1).

Early molluscs were probably[1] bilaterally symmetrical, although the adult orientation of the anterior-posterior axis of the early mollusc is less certain and largely depends on the putative sister taxon and outgroups. If the traditional mollusc-annelid relationship is favored, the adult orientation of the early molluscan body extended along an anterior-posterior axis with a low dorsoventral axis, as in chitons, aplacophorans, and monoplacophorans (Figure 3.1). In most of these animals, the ventral and dorsal orientation of the body is maintained throughout life.

Besides anatomical and ecological changes in body axes, molluscan evolution has been characterised by organ asymmetry and displacement. The rotation of the viscera on the head-foot in gastropod larvae during torsion is a famous example of twisting. Numerous additional changes in body axes occur in gastropods (Figure 3.2), including detorsion, which results in the movement of the mantle cavity posteriorly along the right side of the body. Some 'pulmonate' gastropods have a posterior opening to the modified mantle cavity (as a lung), and this cavity is lost in some other heterobranchs (see Chapters 4 and 20).

Bivalves have also undergone considerable modification. Their shell is divided into two valves, accompanied with considerable lateral compression, and the mantle cavity typically surrounds all or most of the body. Changes in orientation largely result from byssal, as in mussels, or cement attachment, as in oysters (see Figure 3.3 and Chapter 15).

In chitons, aplacophorans, and monoplacophorans the viscera (visceral mass) is distributed along the body, but in bivalves, it is compacted laterally and dorsally, allowing space for an expansive mantle cavity between the enclosing shell valves. In scaphopods and cephalopods, it is elongated dorsally. In *Nautilus* (and many fossil cephalopods such as the ammonites) this elongated tube has been coiled to enable more efficient packing of the viscera. In gastropods, the dorsally extended body is also coiled, but usually asymmetrically into a helical coil.

---

[1] Not necessarily true if Brachiozoa are the sister taxon.

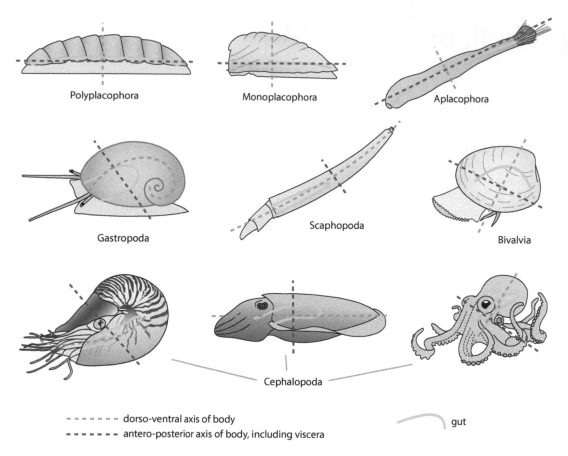

FIGURE 3.1 Anatomical axes in Mollusca. All animals in life orientation. Some aplacophorans are buried in an inclined or near vertical orientation, others lie horizontally. Original.

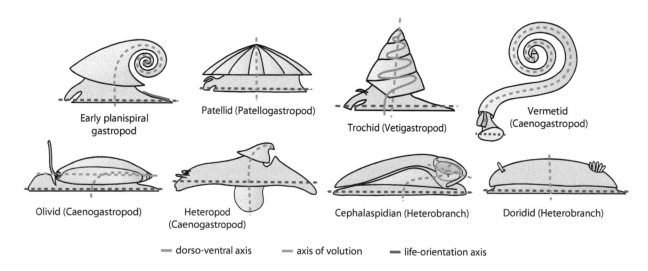

FIGURE 3.2 Axis of symmetry in Gastropoda showing body and shell axes relative to life orientation. Redrawn and modified from Morton, J.E. and Yonge, C.M., Classification and structure of the Mollusca, pp. 1–58, in K.M. Wilbur and Yonge, C.M., *Physiology of Mollusca*, Vol. 1, Academic Press, New York, 1964.

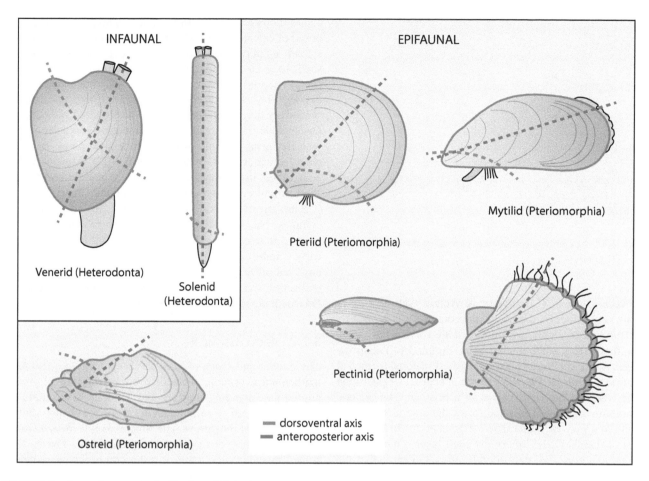

FIGURE 3.3 Axes of symmetry in bivalves. Original.

3.2 SHELLS AND SPICULES

Molluscan shells have attracted the interest of a wide range of scientists from many disparate disciplines as well as naturalists, artists and collectors, and the public.

A calcium carbonate shell or spicules is the most obvious molluscan marker. Whether these shells were formed from fused spicules or not is still debated (see Chapter 13). In living molluscs, only aplacophorans exclusively have spicules, while chitons have eight shell plates surrounded by a girdle covered with spicules, scales, or hairs.

The earliest putative mollusc shells (Rostroconchia) in the fossil record are from the early Cambrian (535–530 mya). Disarticulated plates, cap-shaped, and simple curved or coiled 'shells' are also present during this period and have been interpreted as being shell plates of chiton-like animals, monoplacophorans, or early gastropods. Bivalved shells also appeared in the Cambrian (520 mya) although not all of these were true Bivalvia (see Chapter 13).

Bivalves have two hinged shell valves, while monoplacophorans, scaphopods, gastropods, and cephalopods have a single shell (univalved), and the shell is secondarily lost in some members of the latter two groups.

While most gastropods are univalved, a few secrete an accessory shell valve after the primary shell is formed (e.g., hipponicids and bivalved sacoglossans). Like the operculum, these second shells are not true valves because they are not secreted by the shell gland.

In aplacophorans, the body is covered by spicules embedded in a cuticle covering the epidermis. A cuticle also covers the dorsal surface of chitons and the girdle ornamentation (spines, scales) is embedded in it, and in some taxa it is developed into hairs. Calcareous dermal spicules are also present in some heterobranch slugs (see Chapter 20).

The molluscan shell is a calcareous structure that covers some to all of the upper surface of the animal. It is generally covered by a thin outer organic layer, the periostracum, which is composed of sclerotised proteins and is at least analogous, if not homologous, to the cuticle of chitons and aplacophorans (Checa et al. 2017). The hard part of the shell is comprised of layers of calcium carbonate comprised of the polymorphs calcite and/or aragonite. A small part (0.01%–5% by weight)

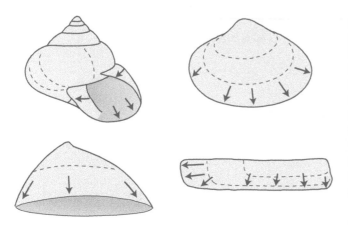

FIGURE 3.4 Accretionary shell trajectories in bivalves and gastropods. Original.

of the shell is organic (Lowenstam & Weiner 1989) and, apart from the periostracum, mainly glycoproteins.

All molluscan shells are produced by incremental additions to their edges and inner surfaces (Figure 3.4). The growing edges of the shell are enlarged by shell material formed at the mantle margin (at the edges of the aperture in a gastropod and the edges of the shell valves in a bivalve). The organic periostracum is secreted first and forms the outermost layer of the shell, and may be mineralised (Checa et al. 2014). Unlike the rest of the shell, the innermost shell layer is not secreted by the mantle edge, but deposited by the outer surface of the thin inner (or dorsal) mantle that underlies the shell. Growth at the edges ceases when the mantle margin is withdrawn from the shell edge, but deposition of the innermost shell layer continues.

The multiple plates of chitons may have been independently derived from a single subdivided shell so cannot necessarily be regarded as primitive (Lindberg & Ponder 1996). Reduction and subsequent loss of the shell, notably in some cephalopods and gastropods, is a derived condition.

While shell morphology can often be correlated with lifestyle and habitat, this is by no means always the case as similar shell morphologies do not necessarily indicate similar habits or habitats (see Section 3.2.3).

3.2.1 Comparison with Outgroups

The ability to secrete an exoskeletal structure composed of calcium carbonate is common in the animal kingdom. Among the lophotrochozoans which secrete shells, the most mollusc-like are the Brachiopoda with their bivalve shell, while a few polychaetes secrete calcareous tubes and some bryozoans (Ectoprocta) have calcareous skeletons. Brachiopod shells superficially resemble bivalves, but differ from most bivalves in having two morphologically different valves which are bilaterally symmetrical through their dorsoventral axis, while most bivalves are symmetrical through their anteroposterior axis. Articulate brachiopods differ markedly from bivalves in having a fragile internal shelly skeleton for the lophophore,

called the brachiophore. In the Inarticulata, the shell is composed of calcium phosphate and chitin, but in the Articulata it is made up of proteins and calcite. As in molluscs, brachiopod shells have an outer organic periostracum which seals off the crystallisation chamber. Many articulate brachiopods have two shell layers, an outer cryptocrystalline layer and an inner secondary fibrous prismatic layer. Other shared shell characters include three calcitic shell microstructures (also shared with Bryozoa), vesicular wall structure, shell pores, and muscle attachment scars (Vinn & Zatoń 2012). Annelid calcareous tubes, like molluscs, have both aragonitic and calcitic shell structures, but typically lack an organic matrix.

The chemical machinery involved in calcification is complex, but, surprisingly, the ability to produce a skeleton evolved in many groups in less than 30 million years, in the early Cambrian (544 mya). This can only be explained with any credibility if the required chemical pathways for biomineralisation were already in place and serving other purposes (Marin et al. 2000) (see Chapter 13).

3.2.2 Shell Geometry

The growth and shape of most shells conform to a few basic mathematical principles. Shells grow by differential growth around an expanding aperture and, while most maintain the same shape from juvenile to adult, others can change slightly or markedly as they grow. The limpet shell form is a simple cone which may or may not be symmetrical. The apex of some limpets is curved and an extension of the curve results in a spiral which may be symmetrical (i.e., isostrophic or planispiral – in the same plane – as in *Nautilus* and many ammonites; i.e., it is bilaterally symmetrical), or asymmetrical (i.e., anisostrophic or conispiral – as in most gastropods). Shell coils approximate a logarithmic spiral (Figure 3.5) in which the distance between adjacent coils increases by a constant factor. A small factor results in a spiral shell, while a

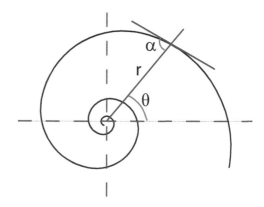

FIGURE 3.5 A logarithmic spiral is generated by a curve determined by r – the distance from the point on the curve to the centre (origin) and θ, the angle between a radius and a horizontal line to the right of the origin. α is a constant, the angle between the radius and a tangent to the curve. Redrawn and modified from Vermeij, G.J., *A Natural History of Shells*, Princeton University Press, Princeton, NJ, 1993.

large factor produces a slight curve, as in a scaphopod shell. If the aperture is not offset, a planispiral coil results, but if it is inclined toward the right of the animal a dextral (right-handed) coiled shell results, while a sinistral (left-handed) coil is produced if the aperture is on the left (Table 3.1; Figure 3.6).

The mathematics of shell coiling has been studied by numerous workers (e.g., Raup 1962, 1966, 1967; Raup & Michelson 1965; Savazzi 1985; Illert 1987, 1989; Ackerly 1989; Stone 1996; Rice 1998; Noshita 2014; Urdy 2015) and much of the early literature is neatly summarised by Vermeij (1993). Most shell shapes result from variations on these themes, and much of that variation is summarised in Table 3.2 and illustrated in Figure 3.7. Attempts to correlate actual shell geometries with theoretical possibilities (e.g., Hickman 1985a; Schindel 1990) have shown that some parts of the 'theoretical morphospace' are not occupied by living taxa, but have been by extinct taxa, whereas other morphospaces have never been occupied (Raup & Michelson 1965; Schindel 1990).

TABLE 3.1

Calculation of Raup's (1962) Four Parameters of Shell Coiling Based on the Measurements in Figure 3.6

Parameter	Measurements from Figure 3.6
Apertural shape (S)	$\dfrac{a1}{a2}$
Whorl expansion rate (W)	$\dfrac{\dfrac{w2}{w1}+\dfrac{w3}{w2}\cdots+\dfrac{wn}{wn-1}}{n-1}$
Position of generating curve (D)	$\dfrac{d1}{d2}$
Whorl translation (T)	$\dfrac{y}{r}$

Source: Lindberg, D.R., *Malacol. Rev.,* 18, 1–8, 1985b.
Note: a = apertural dimension, w = whorl expansion, d = position of generating curve, y = translation, and r = radius.

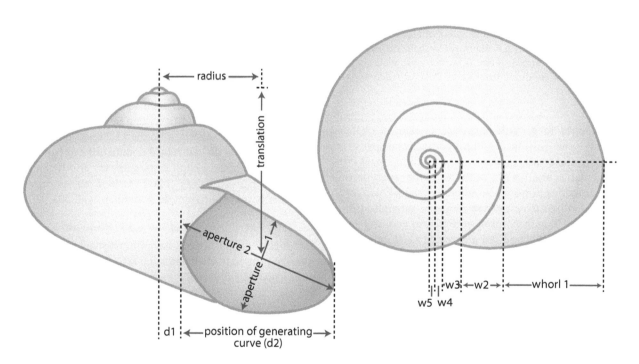

FIGURE 3.6 Shell measurements required for the derivation of the Raup (1962) four parameter model of shell coiling. Also see Table 3.1. Redrawn and modified from Lindberg, D.R., *Malacol. Rev.,* 18, 1–8, 1985b.

TABLE 3.2

Shell Shape and Coiling Parameters

	Simple Cone		Planispiral		Conispiral	
Growth rate	Low	High	Low	High	Low	High
Apertural expansion	Narrow cone	Broad cone	Loose coiling	Tight coiling	Loose coiling	Tight coiling
Curvature	NA	NA	Loose coiling	Tight coiling	Loose coiling	Tight coiling
Aperture moving on Z-axis (spire)	NA	NA	NA	NA	Low spire	High spire

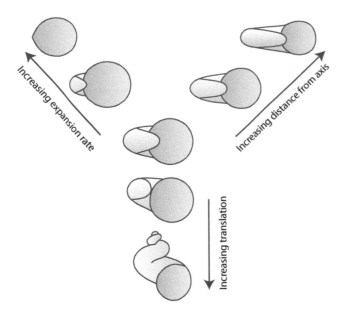

FIGURE 3.7 Shell models generated with different growth parameters – whorl expansion (the shape of the generating curve is equivalent to the shape of the aperture), translation (the height of a generating curve), and distance from the generating curve to the axis. Redrawn and modified from Raup, D.M., *J. Paleontol.*, 40, 1178–1190, 1966.

Small changes in growth parameters result in changes in shell shape and coiling. Tighter coiling results in overlapping sutures and tall, conical shapes, while looser coiling forms a flatter shell with, at the extreme, open coiling (whorls not touching) or even an uncoiled shell. Changes in shell shape also occur as growth changes throughout life. Figure 3.8 shows how the timing of different growth factors can alter the shape of a conical gastropod shell. For example, when growth is relatively rapid in early life, but slows down later, the spire has a dome-shaped profile, while if growth accelerates with increasing age, the spire is concave. A constant growth rate produces a convex cone, while an exponential rate produces a straight-sided cone. Thus, these relationships are far from simple. There is also a relationship between the growth rate and expansion – some parts

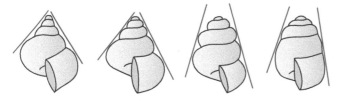

FIGURE 3.8 Variation in spire outline in a gastropod results from different rates of growth. When exponential growth occurs a simple conical spire results, as in the second figure on the left. If the growth rate increases with age, a concave spire results (as in the left-hand figure), while if there is a decrease in growth rate with age, a convex spire will result (as in the two right-hand figures). Redrawn and modified from Rice, S.H., *Paleobiology*, 24, 133–149, 1998.

of the shell edge grow relatively faster, with points further away from the apex usually growing fastest.

In many bivalves, the valves become inflated as growth slows with increasing age so that older, larger individuals are more inflated (Vermeij 1990).

Various shell shapes have evolved multiple times in gastropods. For example, the limpet and conical shell shape have often evolved in many lineages. In most gastropod limpets 'anterior' (i.e., post-torsional anterior) dilation exceeds (or in a few species equals) 'posterior' and lateral aperture expansion, but in patellogastropod limpets the reverse is true with 'posterior' dilation usually exceeding 'anterior' and lateral aperture expansion. Also, uncoiled shells occur in several unrelated groups, such as Truncatelloidea (e.g., Caecidae and some cochliopids), Vermetidae (Vermetoidea), and Siliquariidae (Cerithioidea). Very open coiled shells with the whorls wholly or partially not in contact occur in certain taxa across at least 11 families of living gastropods (Rex & Boss 1976), although no families are exclusively open coiled.

It is generally thought that, once completely lost, coiling cannot be 'reinvented' so an uncoiled limpet, for example, could not revert to a coiled snail. Collin and Cipriani (2003) provided an example in Calyptraeidae showing that coiling was apparently reinvented in uncoiled slipper limpets (*Crepidula*) when a lineage gave rise to the coiled genus *Trochita*. This is at variance with some other results (Simone 2002), and these hypotheses remain untested with new data. In another case involving a vermetid, Gould and Robinson (1994) found that the reinvented coiling differed from normal coiling. Some heteromorph ammonites are also hypothesised to have re-evolved tightly coiled shells, Scaphitidae perhaps being the best known example (Landman et al. 2017).

3.2.2.1 The Direction of the Coil (Handedness)

Although some gastropods have limpet-shaped or planispiral shells, it is more likely that torsion afforded advantages to animals with conispiral shells, enabling tighter coiling and hence a longer spire. When a conispiral shell is viewed with the spire uppermost and the aperture opening facing the observer is on the left, the shell is sinistral, and if it is on the right, it is dextral.

Shell coiling direction (*chirality*) is determined by proteins (e.g., nodal) which determine the direction of cleavage during early development (Figure 3.9) (Levin 2005; Kuroda et al. 2009).

In hyperstrophic snails, the shell coils in the opposite direction along the vertical axis of coiling (the Z-axis) (Figure 3.10). Thus, if a hyperstrophic shell is orientated the same way as a dextral shell, it would appear to be sinistral although anatomically the animal remains dextral, while the shell is apparently sinistral when orientated with the spire uppermost.

Planispiral coiling is rather rare, particularly in living marine gastropods (Vermeij 1975), although it was sometimes common in the Paleozoic. Why the vast majority of modern gastropods exhibit dextral coiling has long puzzled malacologists. The fossil record shows that although both sinistral and dextral coiling were common in early gastropods, dextral taxa

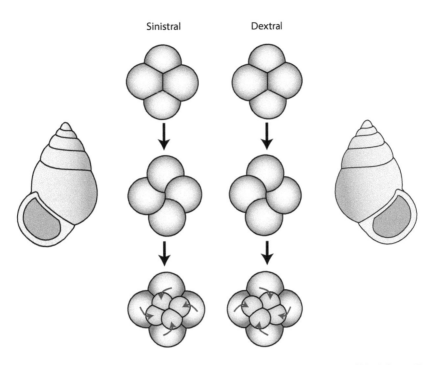

FIGURE 3.9 Chirality in snails is determined by the early cleavage pattern. Redrawn and modified from Grande, C. and Patel, N.H., *Nature*, 457, 1007–1011, 2009.

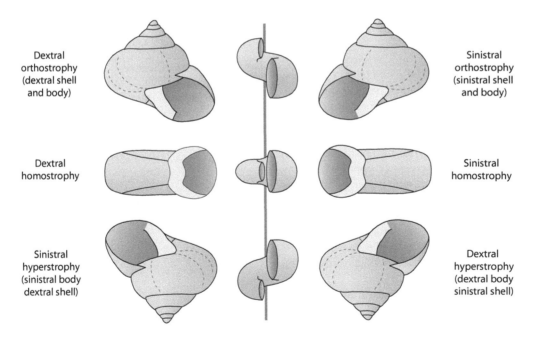

FIGURE 3.10 Shell and body coiling. Diagram illustrating orthostrophy, hyperstrophy, and hypostrophy. Middle vertical row redrawn and modified from Linsley, R.M., *Mem. Nat. Mus. Vic.*, 39, 33–54, 1978c.

later became by far the dominant form with over 90% of living coiled gastropods being dextral (Schilthuizen & Davison 2005). Various authors have related this dextral bias to the effects of torsion on the body, but in sinistral taxa with sinistral animals, torsion occurs in the reverse direction to dextral taxa, and it is difficult to see why a sinistral animal in a sinistral shell would be inherently less efficient than a dextral one in a dextral shell. It is, however, easier to imagine reduction in efficiency where the chirality of the animal and shell differed.

Hyperstrophy occurs in the larval stages of most marine heterobranch gastropods and in the fresh-water caenogastropod genus *Lanistes* (Ampullariidae). Heterostrophy occurs

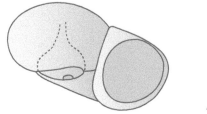
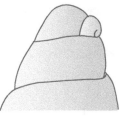

FIGURE 3.11 Examples of heterostrophic protoconchs. Redrawn and modified from Robertson, R., *New York Shell Club Notes*, 329, 12–18, 1993a.

when there is a change in coiling direction during growth, as is obvious in the protoconchs versus the teleoconchs of some heterobranch gastropods (Figure 3.11).

As noted above, most modern gastropods are dextrally coiled but the shells of members of a few gastropod families, namely, Triphoridae,[2] Planorbidae, Physidae, as well as most atlantid heteropods and thecosome pteropods, and some genera and species of stylommatophoran land snails are sinistrally coiled. In addition, some species or genera in other families scattered through the gastropods are also sinistral (Vermeij 1975; Gould 1985; Robertson 1993a, 1993b). In sinistral taxa, the animal within the shell is also sinistral. Freak coiling reversals (to sinistral) occur in many gastropods, but are rare (e.g., Schilthuizen & Davison 2005). Such sinistral sports, or occasional taxa that normally have sinistral shells, usually have a dextral animal.

The distribution of sinistrality in gastropods indicates that change in chirality has occurred repeatedly in the class, particularly in non-marine taxa. Studies on land snails have shown that in some stylommatophorans a single gene may be responsible for chirality and occasionally promotes speciation (e.g., Gittenberger 1988; Asami 1993; Schilthuizen & Davison 2005). In internally fertilising gastropods, mating becomes difficult between partners of opposite chirality because their genitalia are on the wrong side of the body. Thus, selection would normally act against a change in coiling direction (e.g., Schilthuizen & Davison 2005), although exceptions occur, as for example, in the stylommatophoran land snail *Amphidromis* where both left and right coiled snails often occur in approximately equal numbers in interbreeding populations (Schilthuizen et al. 2007, 2014).

Genetic factors are not the only influence affecting coiling patterns in gastropods. Experiments in which near planispiral planorbid shells were weighted on one side, causing the shells to lie at an angle instead of near vertically, resulted in a change in the coiling pattern toward the weighted side (Checa & Jiménez-Jiménez 1997).

Reversal of polarity of bivalve valves can be seen in some asymmetrical bivalves like Chamidae (Odhner 1919) or, uncommonly, in bilaterally symmetrical bivalves where the hinge is transposed (Matsukuma 1996) so the hinge features normally seen on one valve appear on the other.

Asymmetry of bivalve shell valves between left and right occurs rather commonly and is roughly equally divided across the class, but can be constant within a family. Asymmetry may be minimal to extreme, with minimal examples including small differences in inflation or one valve overlapping another. More extreme examples are often associated with cementation. For example, asymmetrical cementing pectinoideans are attached by the right valve despite cementation evolving multiple times within Pectinoidea, while oysters are always cemented by the partially to completely flattened left valve. Within the chamids and rudists, however, either valve is used, depending on the species. So-called torsion in a few bivalves (Arcidae, Mytilidae, the extinct Bakevelliidae, and the hiatellid *Cyrtodaria*) involves twisting of the valves around the anteroposterior axis, the best known example being the arcid *Trisidos* (McGhee 1978; Morton 1983b; Savazzi 1984) (Figure 3.12).

In the Silurian and Devonian, five or six lineages of shelled cephalopods (nautiloids and ammonoids) had conispiral coiled shells with sinistral and dextral taxa about equal in number (Vermeij 1975). Some of these changes in coiling may have been pathological variations resulting from sclerobiont[3] infestation during the life of the ammonites (Stilkerich et al. 2017). Further changes in coiling parameters in ammonoids also occurred in the Mesozoic with the appearance of the uncoiled heteromorph ammonites (see Chapter 17).

3.2.3 SHELL SHAPE AND HABIT

The morphology of a shell is often thought to be correlated with habits and habitat. While many such examples seem to justify this general observation (e.g., gastropods with large apertures clinging to wave-swept platforms, oysters cemented to rocks, burrowing snails with streamlined shells, etc.), often

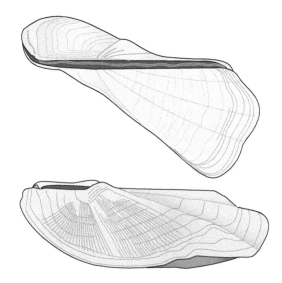

FIGURE 3.12 The 'twisted arc,' *Trisidos tortuosa*, dorsal and lateral views. Orginal

---

[2] There are a few dextral triphorids.

[3] Organisms living on hard substrates such as shells.

shell morphology is not readily correlated. For example, similar shell morphologies frequently occur in different habitats, thus frustrating attempts to correlate them. Limpets are found in a variety of habitats including wave-swept rock platforms and in fast-flowing rivers, but also on various deep-sea biogenic substrata, at hot vents, and in quiet lakes and ponds. Other limpet-like gastropods are parasites on bivalves or starfish. Thus, limpet-shaped shells function well in many very different habitats and suggestions that strong currents or wave action select for limpet-shaped shells are an over-simplification (Ponder & Lindberg 1997). There are many other examples of convergent shell morphology, including cockle-shaped shells of several groups of shallow-burrowing bivalves and the whelk-like shells that have appeared in several lineages of carnivorous caenogastropods. Similarly, elongate tall-spired gastropod shells are often found in shallow sand burrowers such as Terebridae, but other groups with very different shell morphology appear to be equally well adapted to this mode of life (e.g., the globular Naticidae). Also, some tall-spired shells have very different lifestyles, the terrestrial Clausilliidae which live on rocks or tree trunks being one such example.

The orientation of gastropods in relation to the substratum is also related to the nature of the aperture – whether it is radial or tangential (Linsley 1977). An aperture is 'radial' if its plane passes through the axis of coiling, while the plane of a 'tangential' aperture is tangent to the last whorl of the gastropod so the aperture and the ventral-most part of the last whorl lie in the same plane (Figure 3.13). Radial apertures are uncommon in living gastropods, being found in some living vetigastropods (e.g., some trochoideans such as Liotiidae and Angariidae), in architectonicids, and in the tiny omalogyrids, but they are common in older fossil gastropod taxa, including most bellerophontids, and in most coiled cephalopods. A gastropod with a tangential aperture can clamp to the substratum, which is impossible with a radial aperture. These apertural types are further discussed below (see Section 3.2.4).

The lip of a gastropod shell aperture, when viewed from the side, can range from a steep to moderate angle to the axis of coiling (*prosocline*), to parallel to it (*orthocline*), or at a negative angle (*opisthocline*). In conispiral gastropod shells with a steeply inclined prosocline aperture, the snail crawls with the shell apex inclined obliquely. In those where it does not slope (orthocline), the shell lies horizontally on the

substratum surface when the animal is crawling. This latter group often have elongate and/or siphonate apertures, while those with inclined apertures usually have simple rounded to oval openings (e.g., Vermeij 1993).

3.2.4 Gastropod Shell Orientation in Life

Linsley (1977) investigated shell orientation in gastropods in relation to the animal and substratum and proposed that coiled marine gastropods followed several 'laws'. The first two and last two relate to the aperture, the other to the spire. These 'laws' are (from Linsley 1977):

Law 1: The Law of Radial Apertures[4]: Gastropods of more than one volution with radial apertures do not live with the plane of the aperture parallel to the substratum. Most typically it is perpendicular to the substratum.

Law 2: The Law of Tangential Apertures: Gastropods of more than one volution with tangential apertures live with the plane of the aperture parallel to the substratum.

Law 3: The Law of Shell Balance: If the shell of a gastropod is supported above the body, it will be positioned so the centre of mass of the shell and its contents is over the midline of the cephalopedal mass.

Law 4: The Law of Apertural Re-entrants: Angulations or re-entrants on the aperture indicate inhalant or exhalant areas; inhalant areas will be directed anteriorly.

Law 5: The Law of Apertural Elongation: Gastropods having elongated apertures possess only a single gill and develop a water flow through the mantle cavity from anterior to posterior, along the long axis of the aperture; this axis is subparallel to the anterior-posterior axis of the foot.

In gastropods, how the shell is carried on the body is correlated with habits and habitat. If efficiency is achieved when the shell is 'balanced' on the foot as Linsley (1977) theorised, with the centre of mass of the shell balanced over the midline of the head-foot, such 'balance' can be achieved in two main ways (Figure 3.14):

- Lowering the coiling axis
- Swinging the shell so the coiling axis is approximately parallel with the longitudinal axis of the foot – a process called 'regulatory detorsion' by Linsley (1978b).

Snails employ a combination of these two processes to achieve a 'balanced' shell – by adopting a low centre of gravity and maintaining a tangential aperture (Vermeij 1971). Some have also argued that the lower angle of the shell to the foot has

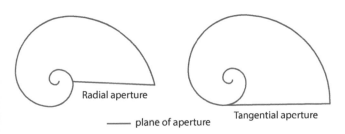

Radial aperture

——— plane of aperture

Tangential aperture

FIGURE 3.13 The two main types of aperture recognised by Linsley (1977). Redrawn and modified from Linsley, R.M., *Paleobiology*, 3, 196–206, 1977.

---

[4] See Section 3.2.3 and Figure 3.13.

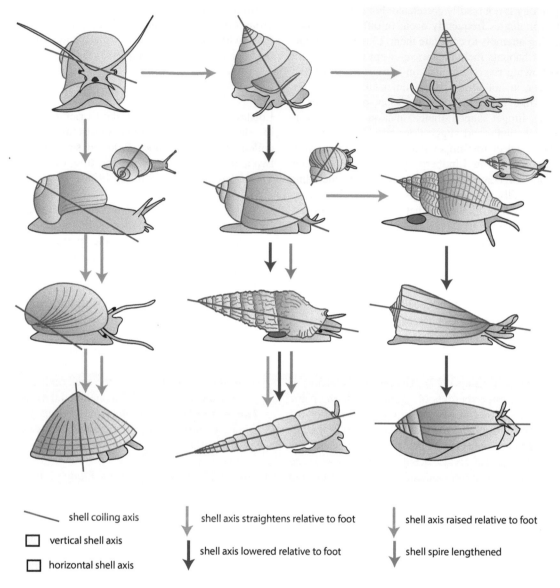

shell coiling axis

□ vertical shell axis

□ horizontal shell axis

↓ shell axis straightens relative to foot

⬇ shell axis lowered relative to foot

↓ shell axis raised relative to foot

⬇ shell spire lengthened

FIGURE 3.14 Some examples of changes in gastropod shell orientation relative to the head-foot. Taxa from left to right are:- top row: Ampullariidae, Pleurotomariidae, Trochidae, second row, Helicidae, Littorinidae, Buccinidae, third row, Neritidae, Cerithiidae, Conidae, bottom row, Patellidae, Terebridae, Olividae. The examples in this figure are not a phylogenetic series, but show trends in morphology. Redrawn and modified from various sources including Hickman, C.S., Comparative morphology and ecology of free-living suspension-feeding gastropods from Hong Kong, pp. 217–234, in B. Morton & Dudgeon, D., *Proceedings of the Second International Workshop on the Malacofauna of Hong Kong and of Southern China, Hong Kong, 6–24 April 1983*, Vol. 1 and 2, Hong Kong University Press, Hong Kong, China, 1985b.

the added benefit of drag reduction and consequently enables more rapid movement (Linsley 1978a, 1978b). Linsley (1977) noted that exceptions to his Law 3 included shells with the aperture placed markedly on one side as in Conidae. High-spired forms such as many cerithioideans and terebrids also do not conform well as the apertural plane is parallel to the axis of coiling. A few groups employ different strategies, such as some Strombidae and Xenophoridae, which use spines to prop up their shells.

The fourth and fifth laws relate to the aperture of caeno-gastropods. Inhalant canals are found in some cerithioideans, triphoroideans, tonnoideans, and neogastropods. Distinct exhalant canals are less common, being especially well-developed in many conoideans, some triphoroideans, and Bursidae. Exhalant shell slits or holes occur in some vetigas-tropods with two ctenidia (Pleurotomariidae, Scissurelloidea, Haliotidae). Apertural elongation is common in neogastro-pods, but uncommon in other caenogastropods, although seen in some such as many Strombidae, and all Ovulidae and Cypraeidae, where it is combined with considerable expan-sion of the last whorl which effectively places the centre of gravity of the shell over the middle of the foot.

3.2.5 Larval Shells and Their Formation

The larval shell is produced initially from the shell gland (see Chapter 8 for details). Larval shells (sometimes called embryonic shells) are called protoconchs in gastropods and prodissoconchs in bivalves. Chitons lack a larval shell and bivalves have paired cap-shaped prodissoconchs aligned in the lateral rather than in the anteroposterior plane. The fossil bivalved group Rostroconchia differ from bivalves in having a larval shell on one valve only and lacking a flexible hinge connecting the valves. In monoplacophorans, scaphopods, and shelled cephalopods the larval shell rests symmetrically on the teleoconch. In most of these taxa it is cap-shaped, although in scaphopods it is tubular; another interesting exception is the very unusual planispiral larval 'ammonitella' of the extinct ammonites. In gastropods, the larval shell is coiled or, in patellogastropods, is caecum-like.

The protoconch is lost in adult scaphopods, but is often retained in adult bivalves and gastropods and can provide an easily accessible marker to determine life history characteristics. For example, in gastropod and bivalve larvae that feed in the plankton (planktotrophic), two separate larval growth stages can be distinguished in the larval shells (protoconch or prodissoconch I and II), the former formed as the initial shell, the latter growing during planktonic feeding. More detailed information on larvae and larval shells is given in Chapter 8 and in the taxon chapters. Aspects of the evolution of gastropod larval shells (protoconchs) through time are given in Chapter 13.

There are significant differences in how chiton shells are formed compared with those of the so-called conchiferan groups prompting Scheltema (1988a) to suggest that their shells were independently derived, although this is rarely accepted (see below).

The chiton larva develops first six and then seven dorsal transverse ridges made up of cells that secrete cuticle, while the depressions between the ridges (the plate fields) secrete the shell plates. The plate fields have cells that develop an enormous flat villus (*stragulum*) which seals each plate field from the external environment and creates a crystallisation chamber in which a rod-shaped shell plate anlage[5] forms (Figure 3.15). After formation of the rod, accretionary growth is then initiated at the edges of the rod, producing the tegmentum layer of the plate, while the ventral hypostracum layer is subsequently added by cells in the central region of the plate field, and the plate is then covered by the organic outer layer, the properiostracum (Kniprath 1980).

Conchiferan larvae usually have the shell-secreting structure, the shell gland, formed early in development from invaginated, thickened ectoderm which subsequently evaginates when the larval shell is secreted. The periostracum is secreted at the distal edge of the shell gland and encloses a crystallisation chamber (the *extrapallial chamber*) closed off from external influences. The shell gland of gastropods is simple,

but in bivalves a groove is formed which develops into the hinge line.

Although they noted these differences, Haas (1981) and co-workers (Haas et al. 1979) concluded that shell formation in both chitons and conchiferans is similar. They suggested that, in chitons, specialised epithelial cells proximal to the shell plates formed a 'properiostracal groove' that produces a thin organic pellicle (properiostracum) which facilitates the deposition of the shell (see below).

3.2.6 Shell Formation/Secretion

The shell is secreted by the mantle (see Section 3.9) and typically covers the viscera and the mantle cavity (see Chapter 4), thus protecting and supporting the soft body. The shell is a complex combination of organic material and different configurations of calcium carbonate crystals that in combination produce a mechanically strong structure. As an example, nacre, a crystalline form of aragonite, is over 1000 times stronger than non-crystalline aragonite (Jackson et al. 1988).

For some time, it was generally assumed that crystals making up the shell crystallised from a supersaturated solution in the extrapallial space between the mantle edge and the periostracum, but recent studies have shown that the process is much more complex than that. It involves an organic matrix (see Section 3.2.6.2) which is elaborated within the extrapallial space using a wide variety of macromolecules, and it is in this matrix that the crystals form.

Scheltema and Schander (2006) reviewed the strategies involved in shell and spicule formation in different groups of molluscs. Their conclusions are summarised below.

Polyplacophora: The mantle secretes a cuticle in which the shell plates and the aragonitic spicules of the girdle and the eight dorsal shell plates are embedded. Spicules are produced in an invagination of either a single cell (the formative cell) or a group of cells (the cell packet) (Figure 3.15), and the proximal region of the spicule is covered with cuticle (Haas 1981; Scheltema 1993). The shell valves are covered by a thin properiostracum secreted by epidermal cells in a groove at the mantle margin, and shell crystals are deposited at the margins and on the ventral surfaces of the shell plates.

Aplacophora: Also have a cuticle covered mantle secreted by the epidermis. As in chitons, the spicules are aragonite, they are secreted extracellularly in the invagination of a single cell (Haas 1981; Okusu 2002) and each has a thin covering of cuticle (Ivanov 1996).

Conchifera: Do not have spicules, and the mantle cuticle is secreted as periostracum in a groove on the mantle edge. Shell growth is by marginal accretion and internal deposition.

Monoplacophora: Calcium carbonate is deposited under the periostracum at the shell edges with a one to one relationship between prisms and mantle cells.

---

[5] A primordium, or part of the embryo that develops into a particular structure.

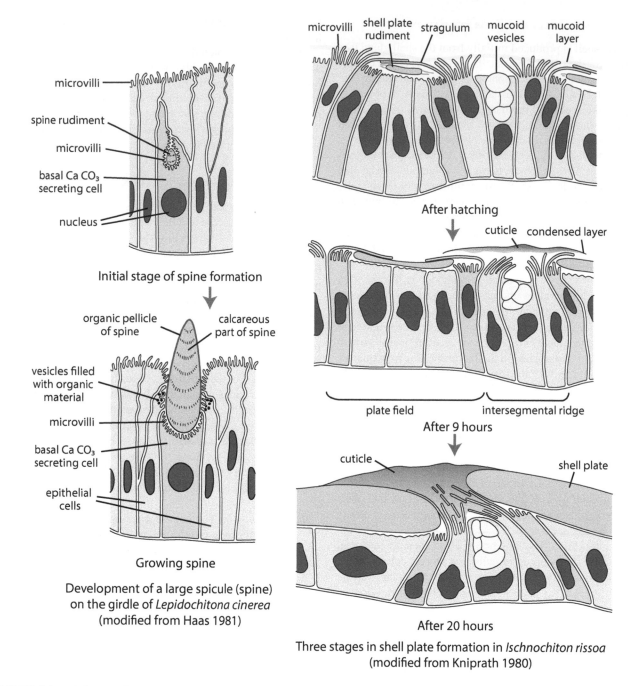

Development of a large spicule (spine)
on the girdle of *Lepidochitona cinerea*
(modified from Haas 1981)

Three stages in shell plate formation in *Ischnochiton rissoa*
(modified from Kniprath 1980)

FIGURE 3.15   Spicule and shell formation in polyplacophorans. The left-hand figures show the development of a spicule. Redrawn and modified from Haas, W., *Malacologia*, 21, 403–418, 1981, and the right-hand figures show the formation of shell plates. Modified from Kniprath, E., *Roux. Arch. Dev. Biol.*, 189, 97–106, 1980.

Further inside, a nacreous layer (foliated aragonite) is secreted by the mantle cells under the prismatic layer (Lemche & Wingstrand 1959; Checa et al. 2009b). No extrapallial cavity has been found.

Nautilus: The shell is composed of aragonite structures that form within individual mantle cells just dorsal to the periostracal groove. Based on an SEM study made on *Nautilus* embryos, the crystals grow out of the mantle edge epithelial cells, fuse together, and then fuse to the periostracum (Arnold 1992). Two forms of crystals occur, proto-prisms and proto-platelets, which once assembled are incorporated into prismatic and nacreous layers, respectively. TEM[6] studies by Westermann et al. (2005) on material from juvenile and early adolescent individuals showed that the prisms were initially formed in intracellular

---

[6] Transmission Electron Microscope.

vacuoles and moved through the cell surface to be assembled extracellularly into their final form. To our knowledge, this method of shell formation has not been confirmed in adults. This finding has been largely ignored in the mainstream literature on shell formation.

Other Conchifera: In most conchiferans, the shell is formed in a fluid-filled extrapallial space between the epithelium of the mantle fold and the periostracum (see Section 3.2.6.1). Crystals form extracellularly in an organic matrix.

In the following sections, we deal firstly with the organic components of the shell and then describe the formation and microstructure of the calcium carbonate component.

3.2.6.1 Periostracum

The periostracum is an organic cuticle-like layer on the outside of the shell (the 'ostracum') that is usually yellow to dark brown. It is composed of *conchiolin* consisting of highly cross-linked quinone-tanned proteins. The periostracum forms a seal at the edges of the shell which isolates the extrapallial space from the external environment, and in that space calcification of the shell occurs, using an organic matrix as a framework (see next Section).

The periostracum of conchiferans is secreted in the periostracal groove in the mantle edge in the form of the soluble protein polymer periostracin which then quickly becomes insoluble via a tanning process involving quinone (Marin & Luquet 2004). It is usually composed entirely of protein, but, in a few taxa, it has calcium carbonate embedded in it.

The periostracum may be smooth or rough and thick and obvious, or it can be barely visible, very thin, and smooth and often eroded away. Some species have regular sculpture or hairs that may be complex and are often species-specific in pattern. Periostracal hairs have often arisen in mollusc evolution and presumably provide adaptive advantages to taxa that possess them (Pfenninger et al. 2005). In species with a heavy, opaque periostracum it obscures the shell beneath so any colour pattern is hidden, as in some species of *Conus*. While the periostracum may be flexible in life, once it has dried out on an empty shell, it can become brittle and flake off. In species where a polished shell surface is secreted by the mantle extending over the outer shell surface, as in cowries, the periostracum is absent.

The periostracum can assist in slowing or preventing shell erosion, notably in fresh-water bivalves and gastropods and, in some bivalves at least, such as some marine mussels (*Mytilus*), it has been shown to assist in reducing biofouling (see Scardino et al. 2003).

The periostracum usually consists of a single layer, but the thick periostracum of fresh-water mussels (Unionidae) consists of two layers. The outer layer is secreted in the periostracal groove as in other bivalves, and the inner layer is secreted by the outer mantle fold epithelium. When shell growth ceases or slows, the outer periostracum continues to be secreted so

it forms folds as it leaves the periostracal groove, and these become conspicuous growth lines (Checa 2000). During long inactive periods, layers of inner periostracum are occasionally added to the interior of the shell. When active shell secretion occurs, both periostracum and shell are secreted (Checa 2000). In some marine species living in cold water, the thick periostracum is also composed of layers. It can be structurally modified, with ridges, hairs, etc. and in some bivalves, and a few gastropods, it is calcified. Chitons lack a true periostracum although a cuticular organic layer (the properiostracum) is present (see Section 3.2.6).

In a departure from the situation in most conchiferans, the unionid bivalve *Amblema* has outer and middle periostracum layers secreted by the periostracal groove and nucleation occurs within the middle layer of the periostracum (Checa 2000). This situation has not been reported in other bivalves, including other unionids, except for a somewhat similar vacuolated middle periostracal layer in a mytilid (Harper 1997b, figure 4a).

3.2.6.2 The Role of Other Organic Compounds in Shell Formation

Despite the organic matrix comprising a minor (less than 5%) part of the shell, it controls many processes involved in shell formation. This organic material acts as a framework on which calcium carbonate[7] is deposited, and some of the acidic proteins are included within the crystal as it grows. This matrix is involved in the synthesis of amorphous minerals from which the crystals form (Marin et al. 2012), and it acts as a point of nucleation of the crystals and determines their growth, position, and form (e.g., Wilt et al. 2003; Marin & Luquet 2005; Jackson et al. 2010; Marin et al. 2012). This organic matrix includes the inner surface of the periostracum from where the outer layer of calcium carbonate crystals is suspended.

The shell matrix proteins increase the resistance of the shell to fractures by several orders of magnitude (Marie et al. 2013). They play a crucial role in determining the type of $CaCO_3$ polymorph, as well as crystal nucleation and their growth and form. The complex organisation of shell matrix proteins is only reasonably well understood in simple shell structures – notably nacre.

Organic compounds secreted into the extrapallial space by the mantle epithelium include proteins, chitin and other polysaccharides, glycoproteins, and proteoglycans that make up the organic shell matrix (e.g., Marin & Luquet 2004, 2005). Of the many proteins expressed during shell formation, some function as enzymes and others in cell signalling (Marin et al. 2012). Some are soluble and others insoluble, the latter apparently involved in the control of crystallisation (e.g., Belcher et al. 1996; Marin et al. 2012). The soluble proteins in the shell matrix inhibit crystallisation until they attach to a suitable

---

[7] All mollusc shells are composed of calcium carbonate except for one questionable record of a phosphatic putative mollusc, the Silurian *Cobcrephora* that was suggested to be a relative of chitons (e.g., Bischoff 1981; Cherns 2004) (see Chapter 13).

insoluble part of the matrix where they act as nucleation points in crystal formation (Marin & Luquet 2004).

Interestingly, for shell crystals to form, the organic matrix needs to be acidic. In contrast, the organic matrix in the non-mineralised gladius of coleoid squid is basic (Dauphin 1996).

Until recently the control of the formation of shell crystals was thought to be through just two mechanisms, crystal nucleation and growth inhibition, but the processes involved are much more complex and under hormonal control. They include interactions between proteins and minerals, protein-protein interactions and interactions with epithelial cells. Some proteins act as enzymes and others may be involved in cell signalling (Marin et al. 2012). In *Lottia*, for example, there are both highly conserved and unique proteins, including enzymes such as peroxidases, carbonic anhydrases and chitinases, acidic calcium-binding proteins, and protease inhibitors (Marie et al. 2013), and this is probably similar in other molluscs. In the pearl oyster *Pinctada fucata*, there are at least 30 kinds of shell matrix proteins, only three of which are shared with gastropods (Miyamoto et al. 2013). In addition, the genes responsible for expressing the proteins are differentially expressed in the larvae, juveniles, and adults, presumably in response to the different environments these stages experience (Jackson et al. 2007). There are also differences in some shell proteins, even between closely related species. For example, in the pearl oysters (*Pinctada* spp.), the shematrin genes, which express proteins with silk/fibroin-like domains, differ between species, suggesting that they may evolve rather rapidly (Jackson et al. 2010; McDougall et al. 2013). Within these biomineralisation genes in abalone, only six were similar to those in the patellogastropod *Lottia*, and only one was shared between the pearl oyster and the abalone (Jackson et al. 2010). The genes in common are probably ancient and may give clues to mineralisation processes in the first molluscan shells. The silk fibroin gel protein has only been identified in nacre, with none known in the superficially similar aragonitic or calcitic prismatic layers (Furuhashi et al. 2009a).

As described above, the shell microstructural elements are formed in an organic matrix. Calcium carbonate precipitates on a scaffold of insoluble chitin. In the model proposed by Suzuki and Nagasawa (2013), the chitin scaffold contains insoluble matrix proteins such as Pif, MSI60, or prismalin-14, which have a hydrophobic region (enabling interaction between protein and chitin) and a hydrophilic-acidic region enabling calcium carbonate-binding. Thus, these proteins provide the connection between the organic (scaffolds) and inorganic (calcium carbonate crystals), form the intercrystalline organic framework, and probably also regulate the nucleation and orientation of the crystals. There are also soluble matrix proteins (e.g., nacrein, MSP-1, or Asp-rich), which, like the insoluble proteins, have a hydrophilic region for calcium carbonate-binding and are also involved in the regulation, formation, and growth of the crystals, and some form the intracrystalline organic matrices. The extracellular organic matrix is secreted by mantle cells.

The most studied shell structure is nacre and, more specifically, the nacre formed in pteriomorphian bivalves. According to Addadi et al. (2006), the first step involves the assembly of the organic matrix in which a chitinous framework is filled with an organic 'hydrophobic silk' gel so it occupies the space in which the nacre crystal will be formed. The orientation and shape of the crystal are determined by the framework, and an amorphous colloidal calcium carbonate replaces the silk gel. This is followed by nucleation on the organic matrix and the commencement of crystal growth utilising some of the acidic proteins included in the crystal.

Three main groups of proteins have been identified in the organic constituents of bivalve nacre (Addadi et al. 2006). These are:

Chitin: Chitin is insoluble and is in the β-form which has parallel chains that extend well beyond individual mantle cells. These chitin fibres lie under individual nacre crystal tablets. The role of chitin in the shell framework appears to vary considerably as some molluscs have several genes involved in the formation of the chitin matrix and others have only one (Jackson et al. 2010).

Acidic proteins: Acidic proteins are highly diverse and are involved in crystal nucleation. They are rich in acidic amino acids like aspartic and glutamic acid (e.g., Tong et al. 2004). Aspartic acid is most abundant in both calcitic and aragonitic shell layers, while glutamic acid is usually associated with amorphous calcium carbonate (Gotliv et al. 2005).

Silk-like proteins: These proteins are similar to arthropod (especially spider) silk and have been found in the nacre of pteriomorphian bivalves such as *Atrina* (Addadi et al. 2006) and in oysters (Takahashi et al. 2012). Addadi et al. (2006) presumed that the silk was a hydrogel before mineral formation, and that water is removed as the mineral forms. The gel probably helps to keep the organic sheets separated before crystallisation. Silk fibroin gel protein has not been identified from aragonitic or calcitic prismatic layers (Furuhashi et al. 2009b) that are superficially similar to nacre and is absent from the nacre of *Haliotis* (Jackson et al. 2010).

Mucin-like proteins: A report of a mucin-like protein in the nacreous shell layer of *Pinna nobilis* was the first evidence that mucins have a probable role in calcification (Marin et al. 2000).

3.2.6.3 Shell Secretion at the Mantle Edge

Shell growth occurs at the edge of the shell when calcium carbonate crystals form in a fluid-filled space between the periostracum and the shell surface, the extrapallial chamber. The extrapallial fluid in the chamber is sealed from the external medium by the periostracum, and ions (mainly calcium) are accumulated by the work of epithelial ion pumps. The calcium is stored in tissues as granules, mobilised as required and pumped (as ions) into the extrapallial space.

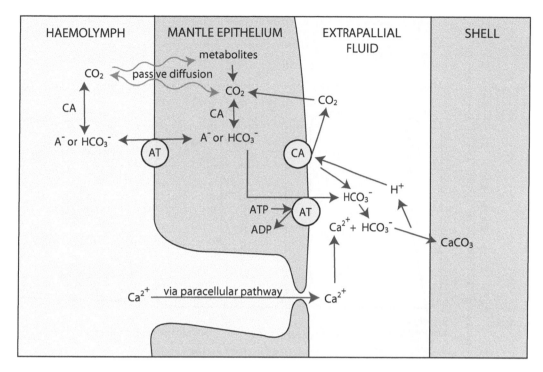

A⁻	Any exchanged anion
ADP	Adenosine diphosphate
AT	Active transport in plasma membrane of mantle cells

| ATP | Adenosine triphosphate |
| CA | Plasma membrane-bonded carbonic anhydrase |

FIGURE 3.16 A postulated metabolism of inorganic shell formation. Redrawn and modified from Wheeler, A.P., Mechanisms of molluscan shell formation, pp. 179–216, in E. Bonucci, *Calcification in Biological Systems*, CRC Press, Boca Raton, FL, 1992.

Crystallisation is possible because the pH of the extrapallial fluid is raised by ammonium ions, providing the conditions for calcium carbonate to be deposited, as in other biomineralisation processes involving carbonates (Loest 1979b). Even so, the process is complex, with one possible scenario illustrated in Figure 3.16. Different possible pathways of the movement of ions via epithelial cells to the extrapallial fluid have been proposed and four are illustrated below (Figure 3.17).

The mineral in mollusc shells is most often aragonite, although layers of aragonite and calcite, and sometimes calcite alone, can be formed, but the factors involved in the production of types of crystalline calcium carbonate in different layers of the shell are still not well known.

Uptake of calcium is essential for shell formation. In seawater, there is no diffusion gradient against calcium uptake because the concentration of Ca^{2+} in seawater and in the haemolymph of a marine mollusc is about $10 \, mM$[8] (Deaton 2008). In contrast, in the haemolymph of fresh-water molluscs the concentrations of calcium range from 3.2 to 8.5 mM, much higher than in the fresh-water medium (0.1–1.0 mM) (Deaton 2008). Thus, calcium must be actively transported from the water against this large concentration gradient in fresh-water

taxa. Regarding sites of calcium ion uptake, the respective roles of the mantle and gills are not known. Calcium ions are transported from the haemolymph to the extrapallial cavity, presumably involving active transport (Weiner & Addadi 2011).

Shell formation requires complex genetic machinery involving a number of genes and transcription factors, and there is also evidence for expansion of gene families being related to mechanisms of shell formation (Takeuchi et al. 2016). On the whole, the transcription factors and signalling genes are conserved, but the expressed proteins are derived and can evolve rapidly (Jackson et al. 2007). The function of some genes is known. Examples are *Hox1* and *Hox4* which control the onset of shell mineralisation in gastropods, *engrailed* which defines the edge of the shell field, and *Dpp* and *nodal* which control shell shape (Hinman et al. 2003; Iijima et al. 2008; Grande & Patel 2009; Samadi & Steiner 2009).

3.2.6.4 Addition of Shell to the Interior of the Shell by the Dorsal Mantle

Most of the studies on shell formation have focused on the processes occurring at the mantle edge, but the addition of material to the inner layers of the shell by the dorsal mantle is less investigated, except in relation to pearl formation (see Section 3.2.6.6) and shell repair (see Section 3.2.6.5). Detailed

---

[8] millimoles.

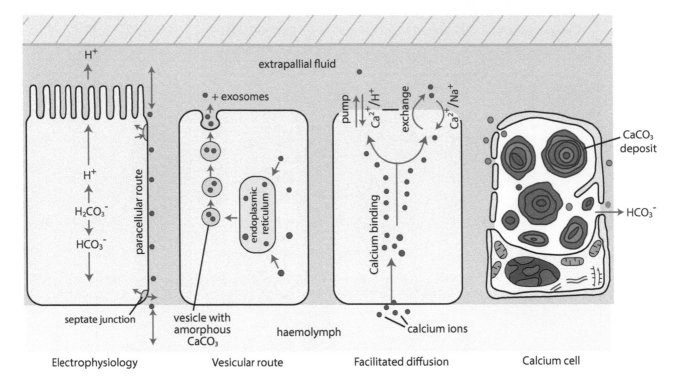

FIGURE 3.17 Four possible systems involving the movement of ions at sites of shell formation. Electrophysiology option involves an increase in acidity that facilitates movement of calcium ions from the haemolymph to the extrapallial fluid. Redrawn and modified from Simkiss, K., Developing perspectives on molluscan shells, Part 2: Cellular aspects, in Saleuddin, S. and Mukai, S., *Physiology of Molluscs*, Vol.1, Apple Academic Press and CRC Press, Palm Bay, FL, pp. 43–76, 2017.

comparative investigations of the mechanisms involved would be of interest because the shell-secreting dorsal mantle differs histologically from the cells on the shell-secreting mantle edge.

3.2.6.5 Shell Repair

Repair of cracked or broken shells is accomplished by deposition of new shell material in the damaged area. The nature of the repair and how quickly it is carried out differs depending on the taxon and the location of the damaged part of the shell, although it is generally carried out more quickly in terrestrial taxa. Much of the information below is from the comprehensive review of Watabe (1983).

Repairs to the shell edge are usually composed of material similar in structure to the rest of the shell, although in some, like the fresh-water planorbid snail *Helisoma*, the outer surface of the repaired area lacks periostracum (Wong & Saleuddin 1972). Such repairs involve little or no change in the mantle, and the repairs have the same structural strength as those of normal shell (e.g., Blundon & Vermeij 1983). When repairs occur away from the edge of the shell (i.e., away from the mantle edge), the shell structure and composition of the organic matrix is often markedly different and also involves changes to the mantle epithelium (Watabe 1983). Shell repairs commence with the formation of organic membranes followed by calcification. The relationship between changes in the organic matrix and resultant shell structure has been experimentally demonstrated in shell repair in the land

snail *Cornu aspersum*. Fernandez et al. (2016) made holes in the last whorl of the shell of *C. aspersum* and these wounds were treated three ways: left uncovered, covered with an inert material, and covered with chicken eggshell membrane with its foreign organic matrix. The initial shell repairs in the presence of the eggshell membrane were calcitic as found in eggshells, but after 24 hours the snail mantle cells were producing their own organic matrix, and the shell repair material was aragonitic. While the repaired shell layers may be the same as in the original shell, in many taxa the shell structure of the repair differs markedly, these differences probably resulting from trauma to the organic matrix (see Watabe 1983 for details). Of special interest are vaterite crystals, along with calcite and/or aragonite, in the repaired shell. This form of calcium carbonate is only very rarely present in normal shells. The mineral composition of the repaired parts of the shell can also be influenced by temperature (Watabe 1983).

Increases in the number of amoebocytes are usually associated with damage, and calcium is removed from storage for the shell repair, mainly from calcium cells in connective tissue and the mantle. Such changes and responses to damage are mediated by hormones, including neurohormones (see Watabe 1983 and Chapter 2).

Shell repairs are usually readily observed as scars on the shell surface. Such scars on recent and fossil shells have often been used as an indicator of predation frequency and failure (e.g., Vermeij et al. 1981; Skovsted et al. 2007), and

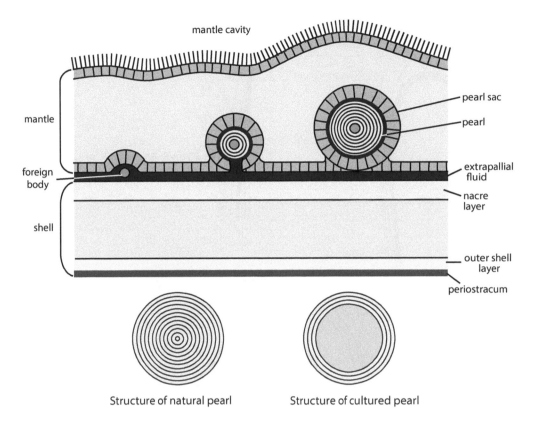

mantle cavity

mantle

foreign body

shell

pearl sac
pearl
extrapallial fluid
nacre layer
outer shell layer
periostracum

Structure of natural pearl   Structure of cultured pearl

FIGURE 3.18 Diagram showing pearl formation in a bivalve and the comparison of the internal structure of a natural and a cultured pearl. Modified from various sources.

Snegin et al. (2016) have used shell repair frequencies of land snails to estimate changes in ungulate (hoofed mammals) populations on Holocene steppe ecosystems.

3.2.6.6 Pearl Formation

A natural pearl is formed as a response to a foreign object, often a parasite, lodged between the mantle and the shell. This may also occasionally be an inorganic irritant (e.g., a sand grain), but usually pearl formation is stimulated by an organic object. The mantle secretes numerous layers of thin shell to cover the invasive object. More broadly, this can be viewed as part of the immune response of the mollusc (see Chapter 2).

Natural pearls can be formed not only in a wide variety of bivalves, but also in gastropods, notably abalone. They are covered with the same shell material that forms the inner side of the shell, whether it is nacre or not. In some pearls, the inner layers may differ from the outer layers, depending on the maturity of the mantle tissue. In pearl oysters, juvenile mantle tissue secretes columnar aragonite while the mature mantle tissue secretes nacre. The pearl is formed in a pearl sac budded off from the mantle (Figure 3.18). Many natural pearls are uneven in shape, with spherical pearls from nacreous bivalves, notably pteriids, being the most prized. Some form 'blisters' on the inner shell surface. Cultured pearls are produced by much the same process as natural ones with various

'seeds' used that form the core of the pearl (see Chapter 10). See Landman et al. (2001) for further discussion of the natural history of pearls.

3.2.7 SHELL MICROSTRUCTURE

Shell material is laid down as crystals of calcium carbonate. It can be either in the form of calcite or aragonite or, very rarely, vaterite[9] (Spann et al. 2010). The other three polymorphs of calcium carbonate (protodolomite, monohydrocalcite, and ikaite) are unknown in molluscs (Marin et al. 2012).

There have been many investigations on the crystallographic features of the shell layers and their physical and chemical properties. Such data on shell microstructure are obtained by a variety of methods, including scanning electron microscopy, and analytical methods including microanalysis, X-ray diffraction, electron diffraction, and infrared spectroscopy (Paula & Silveira 2009).

There may be several different crystalline structures in a shell. In addition, the shell structure at the point of muscle attachments differs from the remainder of the shell and is called

---

[9] Vaterite is also often found in repaired parts of the shell (see Section 3.2.6.5) and in egg capsules of ampullariids and some nudibranch spicules (Watabe 1983).

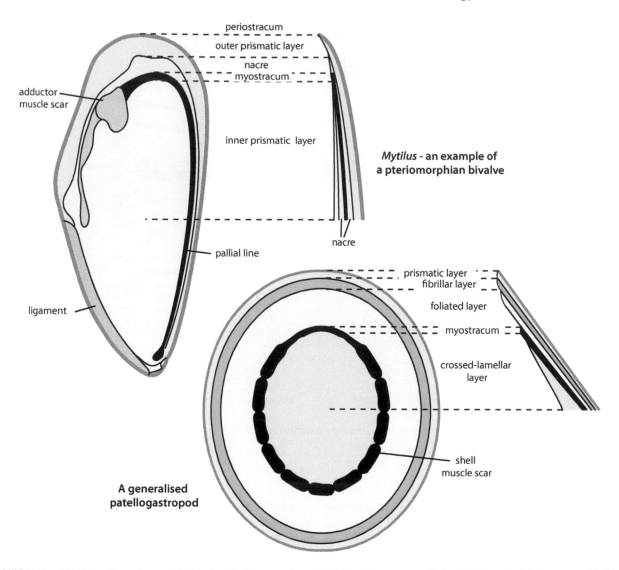

FIGURE 3.19 Shell structure of a mytilid bivalve. Redrawn and modified from Lowenstam, H.A. & Weiner, S., Mollusca, pp. 88–110, in H. A. Lowenstam & Weiner, S., *On Biomineralization*, Oxford University Press, New York, 1989, Weymouth, F.W., *State of California Fish and Game Commission Fish Bulletin*, 1–114, 1921, and a patellogastropod redrawn and modified from Lindberg, D.R., Heterochrony in gastropods, a neontological view, pp. 197–216, in M. L. McKinney, Stehli, F. G. & Jones, D. S., *Heterochrony in Evolution: A Multidisciplinary Approach. Topics in Geobiology*, Plenum Publishing Corporation, New York, 1988a.

the *myostracum*. As the shell grows the muscle attachment areas advance while the retreating portion of the myostracum is covered by the inner shell layer(s), so is retained as a distinct layer within the shell (Figure 3.19). The myostracum is aragonitic and prismatic. $CaCO_3$ spherulites have been reported from the myostracal layer of *Patella vulgata* (Wu et al. 2017).

Diverse shell structures are seen in many bivalves (Taylor et al. 1969; Taylor 1973; Carter & Clark 1985; Watabe 1988; Carter 1990a) and lower gastropods (e.g., MacClintock 1967; Carter & Clark 1985; Carter 1990a; Hedegaard 1990, 1997; Bandel & Geldmacher 1996). The reader is referred to these reviews and to more general reviews by Paula and Silveira (2009) for more detailed information to supplement the general summary below.

The calcium carbonate crystals making up the shell vary in shape, orientation, and size and several types of crystalline structure are recognised. These shell structural types can be subdivided into subtypes resulting in a complex terminology (see Watabe 1988, table 3.1). The formation of these different shell microstructures is under tight biological control.

The main types of molluscan shell microstructure are illustrated in Figure 3.20 as follows[10]:

Prismatic: Uniformly orientated columnar prisms of calcite or aragonite encased in an organic sheath, widely distributed in the Bivalvia, Gastropoda, Scaphopoda, and Polyplacophora. Simple prismatic structures have 'mutually parallel, adjacent structural units (first order

---

[10] The shell structure terminology used here generally follows that of Taylor et al. (1969) and Carter and Clark (1985). Alternative molluscan shell structure taxonomies include Bandel (1990) and Vendrasco and Checa (2015).

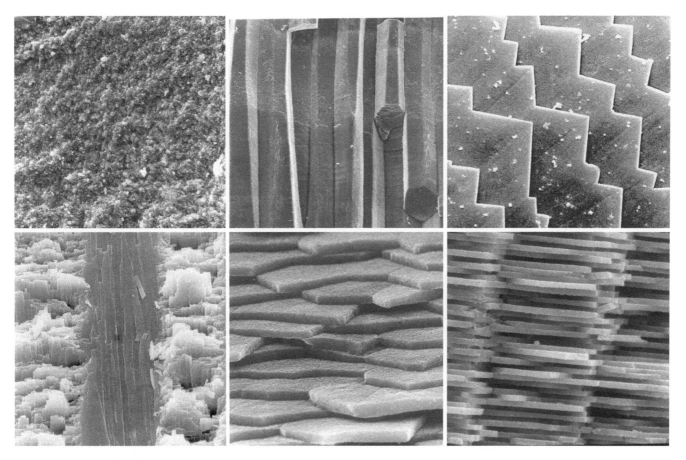

FIGURE 3.20   SEM illustrations of types of shell structure. From left to right: Top row - homogeneous, aragonitic; simple prismatic, calcitic; foliated, calcitic. Bottom row - crossed lamellar, aragonitic; sheet nacre, aragontic; columnar nacre, aragonitic. (Courtesy of J. Carter.)

prisms) that do not strongly interdigitate along their mutual boundaries' (Carter & Clark 1985, p. 52). It resembles a layer of closely packed rods which have their long axes perpendicular to the shell surface. The layer may be thin with almost equidimensional prisms or thicker with elongated prisms, as found in some neritimorph gastropods (Hedegaard 1996).

Several subtypes of prismatic microstructure are recognised (Carter & Clark 1985). Waller (1990) suggested that this shell structure may not be derived from a prismato-nacreous ancestor as often thought, but could have originated as a repair mechanism along the shell margins and may have evolved more than once.

This shell microstructure in Pteriomorphia is calcitic, while in heteroconch bivalves it is aragonitic (Waller 1990). The plesiomorphic state for the molluscan outer shell layer appears to be an aragonitic, prismatic structure (Runnegar & Pojeta 1985). This shell structure overlies inner nacre, crossed lamellar, and foliated structures across the Mollusca (see Taylor 1973).

Nacre: Thin, horizontal aragonitic plates (tablets) or sheets each separated by a thin organic layer and laid parallel to the shell surface are the characteristic features of nacre microstructure. It is present in some Bivalvia, Gastropoda and Cephalopoda. The plates may be overlapping (sheet nacre) or stacked (columnar nacre). Checa et al. (2009a) distinguished sheet nacre from columnar nacre and proposed the term foliated aragonite for sheet nacre. Sheet nacre occurs in both monoplacophorans and some bivalves; columnar nacre in some bivalves, Vetigastropoda, Cephalopoda, and rarely in monoplacophorans.

Nacres in different molluscan classes were thought to be homologous as they shared similarity in composition (aragonite), position (inner shell layers), and appearance (iridescence), as well as their occurrence in putative 'primitive' taxa, but these properties are also shared with some other shell structures. Most shell structures are composed of aragonite; cone complex crossed lamellar structure is also restricted to the interior of shells, and calcitic foliated structures are also iridescent. The distribution

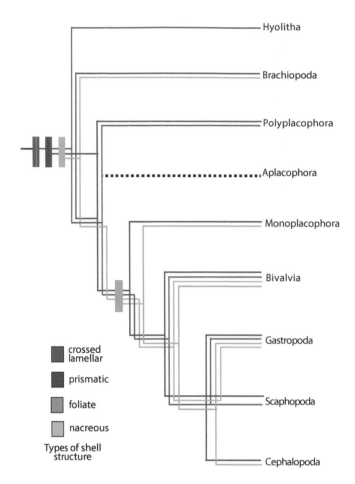

crossed
lamellar

prismatic

foliate

nacreous

Types of shell
structure

FIGURE 3.21 A diagrammatic representation of the distribution of the main types of shell microstructure in molluscan classes. The occurrences of nacre are probably homoplastic. The tree includes the Hyolitha and Brachiopoda as outgroups. Original.

of gastropod, cephalopod, monoplacophoran, and bivalve nacres on the tree in Figure 3.21 suggests that they are probably separately derived, and thus non-homologous.

Nacre properties are presented in Table 3.3, and although they differ between the major molluscan subclades, there appears to be little variation within these groups in the taxa examined. While none of the individual properties differentiates between subclade nacres, unique combinations of these characters clearly distinguish bivalve, cephalopod, and gastropod nacre and the nacre-like structure in monoplacophorans. These findings are supported by Jackson et al. (2010, p. 605) who noted that the nacre-like shell structure of bivalves and gastropods differs in its crystallographic properties, indicating that the 'molecular mechanisms that guide the deposition of the variants of nacre and its derivatives across the Mollusca are fundamentally different'.

Foliate: Foliate structures are sheet-like crystals similar to nacre, but are calcitic. This microstructure is absent in polyplacophorans, scaphopods, and

cephalopods, but present in some pteriomorphian bivalves such as oysters and scallops, and in some patellogastropods and the Paleozoic Platyceratidae. Composed of calcite tile-like crystals arranged parallel to one another, the crystal structure resembles the tiles on a roof. There is less organic matrix than in nacre. Waller (1976) suggested that foliated microstructure was derived from the calcitic prismatic layer, based on the observations that both layers share the same mineralogy and that the foliated layer always developed below a prismatic layer (although in many groups the prismatic layer appears to have been secondarily lost). Carter (1990b) accepted the hypothesis of Waller, but cautioned that, depending on the group, foliated microstructure could have developed either directly or indirectly through intermediate fibrous prismatic, spherulitic prismatic, or homogeneous shell microstructures.

Crossed lamellar: This is the most common shell structure found in polyplacophorans and conchiferans other than cephalopods. The earliest molluscan shell structures are all crossed lamellar derivatives. This shell structure is nearly always aragonitic, although rare calcitic examples have been recognised. Two main types are present: simple and complex. Both types are similar, with elongate crystals running in layers at angles to each other. This shell structure has a little organic material (Dauphin & Denis 2000), including more chitin than in other shell structures (Furuhashi et al. 2009a). Crossed lamellar structures may have preceded the divergence of the extinct Hyolitha (see Chapter 13) and Mollusca (assuming Hyolitha are monophyletic and not molluscs) in the late Precambrian or early Cambrian (570 mya). Crossed lamellar shell structure is resistant to breakage (e.g., Hou et al. 2004) as 'crack energy' is absorbed because fractures are forced to take a convoluted pathway through the criss-cross microstructure (Currey & Kohn 1976).

Aragonitic commarginal crossed lamellar structure occurs in many gastropods and bivalves (Bøggild 1930; Taylor et al. 1969, 1973). It is almost ubiquitous in Patellogastropoda and Fissurellidae (Bøggild 1930; MacClintock 1967; Hedegaard 1990). Aragonitic radial crossed lamellar structure is less common than commarginal. It is common in Lottiidae (Patellogastropoda) (MacClintock 1967; Hedegaard 1990), rare in neritimorphs, but absent in Fissurellidae (Hedegaard 1990).

Wilmot et al. (1992) demonstrated that crystals in aragonitic crossed lamellar structures with different orientation relative to the shell are elongated in different crystallographic directions. Based on only three observations, they concluded that crossed lamellar structures have been derived several times. While elongation in different directions may be an

TABLE 3.3

A Summary of the Properties of the Different Nacres in Four Molluscan Classes

Character	Sources	Monoplacophora	Bivalvia	Gastropoda	Cephalopoda
Crystal axis toward the interior of shell	1	C-axis	C-axis	C-axis	C-axis
Axes in tablet plane	1	Aligned	Aligned	Unaligned	Aligned
Twinning	1	Weak/absent	Absent	–	Prominent
No. of crystals in tablet	1,5,6	1	1	1–20	3
'Pores'	2,3	Absent	Sparse	Abundant	Abundant
'Pore' shape	2,3	–	Small, round	Small, round	Large, oval
Radial membranes	2,3	Present	Absent	Present	Present
Central accumulation	2,3	–	Absent	Present	Present
Chitin	3,7	Absent	Absent or present	Present	Present
Main AA	8	–	Gly, Asx, Ala, Arg, Glx	Gly, Asx, Ala, Ser	Gly, Asx, Ala, Glx
CA activity	8	–	Yes	No	No
Nucleation site	1,5	Edge of lower tablet	Edge of lower tablet	Centre of lower tablet	Centre of lower tablet
Growth zone	1,5	Entire surface	Entire surface	Margin	Margin
Texture	1,2,3	Fibre	Double twin	Fiber	Double twin
Texture strength	1,2,3	–	49–>9999 (10)	88–90 (2)	51–333 (2)
% twinning	1,2,3	–	12–67 (11)	–	50–92 (2)
Angle between a-axis and growth direction	1,2,3	–	90–95 (9)	–	75–100 (2)

Sources of Information: 1 – Hedegaard, C. and Wenk, H.R., *J. Molluscan Stud.*, 64, 133–136, 1998; Chateigner, D. et al., *J. Struct. Geol.*, 22, 1723–1735, 2000; Chateigner, D. et al., *Mater. Sci. Eng.*, 528, 37–51, 2010, 2 – Grégoire, C. et al., *Annales de l'Institut océanographique*, 31, 1–36, 1955; Iwata, K., *J. Fac. Sci.*, 17, 173–229, 1975, 3 – Meenakshi, V.R. et al., Ultrastructure, histochemistry and amino acid composition of the shell of *Neopilina*: Scientific results of the Southeast Pacific Expedition. II, pp. 1–12, in E. Chin, *Scientific Research of the Southeast Pacific Expedition (A. Bruun Report)*, Vol. 2, Texas A & M University, Galveston, TX, 1970, 4 – Mutvei, H., *Zool. Scr.*, 7, 287–296, 1978, 5 – Mutvei, H., The nacreous layer in molluscan shells, pp. 49–56, in M. Ōmori & Watabe, N., *The Mechanisms of Biomineralization in Animals and Plants. Proceedings of the 3rd International Biomineralization Symposium Ago-chō, Japan 1977*, Tokai University Press, Tokyo, Japan, 1980, 6 – Kulicki, C. and Doguzhaeva, L.A., *Acta Palaeontol. Pol.*, 39, 17–44, 1994, 7 – Machado, J. et al., *J. Comp. Physiol. B*, 161, 413–418, 1991, 8 – Marie, B. et al., *ChemBioChem*, 10, 1495–1506, 2009; Data compiled by C. Hedegaard and DRL.

Note: AA = Amino acid; CA = Carbonic Anhydrase activity. Number of taxa observed are given in paraentheses for the last three characters.

inherent property of crossed lamellar structures, it is not necessarily an indication of independent origins. Chateigner et al. (2000, 2010) described different textures (crystallographic orientations) in crossed lamellar structures from numerous taxa, and even from within a single taxon. Belda et al. (1993) reported some deterioration of crossed lamellar shell structures in *Tridacna* related to the availability of increased nutrients. These findings suggest the morphology of crystallites in crossed lamellar shell structures can be heterogeneous and may be subject to some environmental modification.

Cone complex crossed lamellar: Prismatic first order lamellae are composed of second order lamellae which resemble closely spaced, regular stacks of cones, all with their apices pointing toward the exterior of the shell (MacClintock 1967; Carter & Clark 1985; Lindberg & Hedegaard 1996). In the traditional sense of Carter and Clark (1985) and in the expanded sense of Lindberg and Hedegaard (1996), this shell structure occurs in the patellogastropods (MacClintock 1967), cocculinids, most neritimorphs, and in

fissurellids (Hedegaard 1990). It is also found in some autobranch bivalves.

Irregular complex crossed lamellar: This type of crossed lamellar structure has the first order lamellae comprised of 'irregularly shaped, interpenetrating aggregations of parallel second order structural units' (Carter & Clark 1985, p. 62). It is found in some patellogastropods, the lepetelloidean Osteopeltidae, in most neritimorphs (Hedegaard 1990), and in arcid bivalves (Carter & Clark 1985).

Helical: First reported by Bé et al. (1972) in the shell of the pteropod *Cuvierina columnella*, this microstructure was found in other pteropods including the Miocene *Vaginella*, and living *Diacria* and *Cavolinia* (Bandel 1990). Bandel (1977) suggested that helical structures were derived from crossed acicular structures, which are similar to crossed lamellar structure, and reported dendritic to helical structure transitions in secondary shell deposition associated with repair in the shells of species of the pteropod *Cavolinia*. The properties of helical shell microstructure enable the construction of a thin, light shell with resilience to impact (Daraio et al. 2006; Zhang et al. 2011), ideal for organisms that live in the water column. Bandel (1990) found helical structures among crossed lamellar layers in the varices of the neogastropod *Murex*. These varices strengthen the shell, giving it greater resistance to crushing forces (Vermeij 1978, 1979; Paul 1981).

Homogeneous: An aragonitic or calcitic structure that shows a homogeneous structural pattern. Such shell structures are seen in various gastropods and bivalves. The term 'homogeneous' shell structure is unfortunate, as it implies that there is no discernible substructure, although it still retains an arrangement of the crystallographic axes. The term is used for 'aggregations of more or less equidimensional, irregularly shaped crystallites lacking clear first order structural arrangement except for possible accretion banding' (Carter & Clark 1985, p. 63). Thus, when a shell structure has no identifiable elements other than minute granules, it is called homogeneous although homogeneous layers may show some organisation, as with the accretion banding of Carter and Clark (1985). It may also enclose sections of a more distinct structure, where the granules of the homogeneous structure are arranged into the pattern of a well-defined structure. The term 'relics of...' has been used by Lindberg & Hedegaard (1996) to describe the structure of those sections embedded in the homogeneous layer.

Only bivalve and gastropod shell microstructures are particularly diverse and well-studied. The shell structures in some groups of gastropods and bivalves have been extensively used as systematic characters. The high diversity of shell structures in early gastropods and pteriomorphian bivalves strongly contrasts with the more derived taxa, where it is predominantly aragonitic crossed lamellar, sometimes covered by a thin, outer calcitic homogeneous layer (Carter 1990a).

Polyplacophoran shell structure consists of aragonitic prismatic, homogeneous, and crossed lamellar layers (Carter & Hall 1990), while in monoplacophorans, the shell consists of an outer calcitic prismatic layer and an inner aragonitic nacreous layer.

From a historical perspective, inner plesiomorphic crossed lamellar structures have been replaced by several shell structures including fibrillar, foliated, and some others which were all called 'nacre.' All these 'replacement' shell structures are most likely derived from the outer spherulitic prismatic structures and seldom co-occur with one another except in the Bivalvia.

There is a general belief that nacre is plesiomorphic in molluscs and no alternative hypothesis has seriously challenged this traditional view. After examining Cambrian steinkerns for shell structure impressions Runnegar and Pojeta (1985),

repeated by Waller (1998), noted that cone crossed lamellar and nacre structures were both present in these earliest putative molluscs, and thus it was not possible to determine which one was closer to the ancestral state. This view was also supported by Kouchinsky (2000), who considered it likely that nacre and crossed lamellar shell structures evolved independently from the plesiomorphic aragonitic lamellar structure in different molluscan taxa, see also Vendrasco et al. (2011) for a similar conclusion.

The idea that nacre is plesiomorphic is not supported by the distribution of nacre within molluscs (see Figure 3.21), notably with crossed lamellar structures being present in Polyplacophora, and in the extinct, shelled outgroup Hyolitha, and others. Crossed lamellar layers are also present in all conchiferan taxa except for Monoplacophora and Cephalopoda. Nacreous structures are present primarily in non-heterodont bivalves (Protobranchia, Pteriomorphia, Unionida), whereas heterodonts primarily have crossed lamellar shells except for some Anomalodesmata (see Chapter 15, Figure 15.4). In gastropods, nacre is absent in patellogastropods and is only present in some vetigastropod taxa (see Chapter 18, Figure 18.6).

Regrettably the reports of nacre in older fossils are sparse, but there are no demonstrated differences between Paleozoic, Mesozoic, and Recent nacres (Bøggild 1930; Grégoire 1966; Mutvei 1967; Erben et al. 1969; Taylor et al. 1969, 1973; Batten 1972; Mutvei 1983; Dullo & Bandel 1988; Carter 1990a).

In general, the strength of a shell is determined by the mechanical properties of the shell structure and its shape and thickness. Currey (1990) provided an excellent summary of the mechanical properties of shell microstructures so we need not dwell on the details here. Shells built with nacre are stronger than those having other shell structures (Currey 1988, 1990), with foliated and homogeneous structure being the weakest. Crossed lamellar structure is also strong and has the advantage that cracks are restricted due to the criss-crossing lattice of crystals.

Flexibility is increased by incorporating more organic material, as in the shells of the protobranch *Solemya*, the pteriomorphian pinnids, and some stylommatophoran land snails.

Shells are more easily broken by tension (pulling apart) than compression. Nacre is generally the strongest shell structure under tension, compression, or when bent, although because of its high organic composition, nacre is energetically expensive to produce, which is another trade-off.

3.2.8 SHELL COLOUR, PATTERNS, AND PIGMENTS

Colours and patterns are notable features of many molluscan shells. Even fossil shells can reveal colour patterns (not the colour) under ultraviolet light (e.g., Krueger 1974) and traces of patterns in early Paleozoic fossils show that such patterns were developed early in molluscan evolution (Kobluk & Mapes 1989). Shell colours are formed either by pigments or by shell structure, thus falling into two main groups, *schemochromes* (structural colours) and *biochromes* (true pigments) (Fox 1966). Schemochromes include iridescent colours resulting from the interplay of light through very thin layers of calcium carbonate such as in the nacre of shells and pearls (Fox 1953, 1983). Some shells have patches of bright iridescent colouration caused by interference or diffraction from modified surface microstructure. Such structural colouration is relatively rare and differs from that of a normal shell colouration caused by pigments, and it also differs from the iridescent (pearly) appearance of nacre. Examples of such structural colours are seen in some gastropods including the patellogastropod limpets *Patella pellucida* (Figure 3.22) and *Helcion pruinosus* which have bright blue streaks or spots. The structural blue colouration (schemochromes) in these two patellogastropods appears to have developed convergently. In *H. pruinosus*, it is caused by interference or diffraction from highly regular microstructure of crystalline aragonite embedded in the transparent outer shell layer (Brink et al. 2002). In *P. pellucida*, the microstructure of calcite lamellae underlies a thin, transparent, irregular, lamellar layer. These lamellae are underlain by an array of colloidal particles that absorb light and thus provide contrast for the vivid blue colour (Li et al. 2015b). The blue stripes of *P. pellucida* are readily visible through the water at several meters distance, and Li et al. (2015b) speculated that the colouration may indicate Batesian mimicry because two species of noxious nudibranchs with bright blue warning colours also inhabit the *Laminaria* fronds on which *P. pellucida* lives. Interestingly, the spectral range of the blue colour is within the band of minimal light absorption in seawater.

Pigments (biochromes) show as the colour of the reflected wavelengths of light they cannot absorb. They may be laid down in the shell, and/or in the periostracum and the organic matrix. They may exist as a uniform or graded colour, or of regular or irregular patterns made up of contrasting tones or colours.

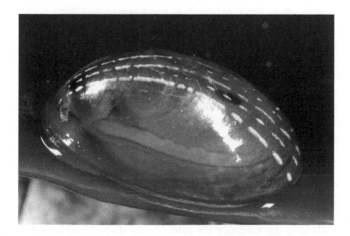

FIGURE 3.22 *Patella pellucida* – East Cornwall, England, on *Laminaria* in the intertidal. (Courtesy of Malcolm Storey.)

In most molluscan shells, the pigmentation is imparted by the mantle edge and is usually found in the outer shell layer, although pigments can also occur in the inner layers. Some shell pigments and their patterns are genetically determined (see below), while others are the by-products of metabolism or derived from food, with the latter changing as diet changes. For example, variable striped colour patterns in the intertidal trochid *Austrocochlea porcata* are mainly composed of the pigment uroporphyrin I, the concentration and abundance of which is related to available algal food and indirectly to wave exposure (Creese & Underwood 1976; Underwood & Creese 1976). Such phenotypic changes have also been demonstrated in several patellogastropods (Lindberg & Pearse 1990; Sorensen & Lindberg 1991), predatory caenogastropods (e.g., Gibbs 1993; Manríquez et al. 2009), and a few other vetigastropods. Whether or not these food-derived patterns are simply a function of metabolic waste disposal (e.g., Comfort 1951a), such phenotypic changes can also impart a selective advantage to some individuals by making them more cryptic (e.g., Sorensen & Lindberg 1991; Manríquez et al. 2009). Crypsis and other mechanisms such as polymorphism, resulting in apostatic selection (see below) by reducing visual predation, may explain the pigmentation colour and/or patterns in some species, although in many other molluscs the utility, if any, of the shell colours and/or patterns, remains unknown. Bauchau (2001) observed that pigment colours and their patterns on shells appear to have no apparent function in many molluscs, but this is difficult to accept given their complexity and the assumed metabolic cost of their production. Thus, while the selective advantage of cryptic and polymorphic shell colouration is readily understood, the reasons (if any) for the great diversity of colours and patterns that exist in many molluscs remain unknown.

The possible functional or selective utility of colours and their patterns is even more enigmatic because many species possessing them live buried in sediment or their shell colouration is hidden by an opaque periostracum or fouling organisms. Nevertheless, selection for cryptic colour may be occurring even in burrowing organisms. For example, Savazzi (1994) noted that the shell colour of some species of *Oliva* was correlated with the sediment colour. For these same reasons, interspecific recognition is an unlikely reason for shell colouration or patterning because of the limited, or lack of, visual capacity of most molluscs.

Although their reason for existence is often mysterious, some theories have been proposed to account for the formation of pigmentation patterns. Waddington and Cowe (1969) proposed a computer model for simulating pigmentation patterns, but did not suggest a definitive physiological mechanism to account for the process. A subsequent computer model by Meinhardt and Klingler (1987) was proposed in the light of a physiological model of Ermentrout et al. (1986), which involved the control of pigment secreting cells by neural activity. This was later developed further by Boettiger et al. (2009). Gong et al. (2012) used this neural network model to explain the remarkable diversity of colour patterns in species of *Conus*. The evidence for such neural networks and neuroendocrine control is not strong, being based mainly on work done on the planorbid *Helisoma* (Saleuddin & Dillaman 1976; Saleuddin & Kunigelis 1984). In cowries (Cypraeidae), colour patterns are found in most species. While these appear to be two-dimensional (as in other molluscs), they are actually three-dimensional in structure due to the shell having been secreted by the overlapping mantle, and visible patterns are often the result of different thicknesses of shell. While the shell patterns are not cryptic (Savazzi 1998), the colour and form of the mantle that often covers the shell may be. When this is withdrawn, it exposes the markedly contrasting, brightly coloured, often patterned shell which Savazzi (1998) suggested may cause predators to confuse them with nudibranchs or other brightly coloured noxious animals.

The models that explain how pigment patterns might be generated do not explain *why* they exist. Bauchau (2001) proposed that the pigmentation patterns created a heterogeneous shell edge which assisted the mantle edge neural network to regulate the growth of the shell. He proposed that uniformly coloured or non-pigmented shells used 'other substances' to mark the shell surface. To our knowledge, this hypothesis remains untested.

The colour of many molluscs is probably largely genetically determined although this has only been demonstrated in relatively few taxa, examples being some land snails (e.g., Cain & Sheppard 1950; Cook & King 1966; Owen 1969; Murray & Clarke 1976; Roth & Bogan 1984; Chiba 1999; Hayashi & Chiba 2004), some marine snails (e.g., Palmer 1984, 1985b; Ekendahl & Johannesson 1997), and some bivalves (e.g., Adamkewicz & Castagna 1988; Winkler et al. 2001; Evans et al. 2009; Zheng et al. 2013). Where known, colour is usually determined by one to only a few genes (e.g., see Zheng et al. 2013). Change, or lack of it, in the frequencies of particular colour morphs over time, has been demonstrated in some long-term studies on helicid snails (Cowie 1992a; Cowie & Jones 1998; Cook et al. 1999; Cameron et al. 2013a). Different immune responses in different colour morphs have also been demonstrated (Scheil et al. 2013).

Some molluscs exhibit distinct colour polymorphism, but the nature of this phenomenon has been investigated in only a few species, the most famous being the European land snail *Cepaea nemoralis* (e.g., Jones et al. 1977), but there are also numerous other land snail examples (e.g., Comfort 1951b; Parkin 1972; Goodhart 1987a), although no comprehensive overviews are available to date. Besides land snails, there are many examples with numerous colour morphs among both marine and fresh-water molluscs (Clarke 1978). Examples include the sandy beach bivalves *Donax variabilis* (Moment 1962) and *Donacilla cornea* (Whiteley et al. 1997), sublittoral bivalves such as many pectinids, sandy beach gastropods such as the trochoideans *Umbonium* (Grüneberg 1980) and *Bankivia* (Clarke 1978; Ermentrout et al. 1986), some of the mangrove-saltmarsh species of *Littoraria* (e.g., Reid

1987), mudflat batillariids *Batillaria* (Miura et al. 2007) and *Velacumantus* (Ewers & Rose 1966), the neritid *Clithon oualaniensis* (Grüneberg & Nugaliyadde 1976), some rocky shore species of *Littorina* (e.g., Reimchen 1979; Atkinson & Warwick 1983) and *Nerita* (e.g., Safriel 1969), and a few chitons (e.g., Rodrigues & Absalão 2005). Some species of freshwater neritids also exhibit pattern and colour polymorphism (e.g., Clarke 1978; Bunje 2007).

Studies investigating colour polymorphism in shelled molluscs have involved a range of approaches including inheritance, genetic drift, frequency-dependent and directional selection, and biochemistry (e.g., Goodhart 1987a; Luttikhuizen & Drent 2008) with several concepts invoked to explain the findings including non-adaptive and adaptive effects. Among the latter are crypsis, visual, disruptive and apostatic selection, climatic effects, assortative mating, genetic drift, area effects, coadaptation, and pseudo-polymorphism (see Goodhart 1987a for review). The latter term was introduced by Grüneberg (1980) to cover examples where there were many, often intergrading, colour forms unlikely to have primarily evolved through visual predation.

Colour polymorphism is thought to have a selective advantage with visual predators because the predators cannot develop a search image for all the morphs. While some studies have demonstrated selection related to visual predation, the colours of some polymorphs are hidden from predators, for example, in some bivalves they are only visible in the interior of the shell (e.g., Cain 1988a; Luttikhuizen & Drent 2008), and in the past, colour patterns evolved before eyes were sufficiently well-developed to see them (e.g., Kobluk & Mapes 1989). Similarly, some complex and colourful shell patterns are hidden beneath a thick periostracum, as noted above. In the tellinid bivalve *Macoma balthica*, there are four, apparently genetically determined, colour morphs that are expressed only in the interior of the opaque shell, although in small floating juveniles this can be visible through the shell (Luttikhuizen & Drent 2008).

Several workers have suggested that colour polymorphism in rocky shore molluscs, such as patellogastropod limpets, is a form of camouflage against predators (Giesel 1970; Hockey et al. 1987; Byers 1989). Crypsis is particularly important. Sorensen and Lindberg (1991) showed that individuals that did not match their background were more likely to be taken, suggesting that selective visual predation can maintain differences in spatial distributions of colour morphs. For some individuals such enhanced crypsis is an obvious selective advantage of colour polymorphisms in heterogeneous environments (e.g., Cain & Sheppard 1950, 1952, 1954; Smith 1975; Jones et al. 1977; Heller 1979; Palmer 1984; Hughes & Mather 1986; Reid 1987; Cain 1988b; Whiteley et al. 1997; Cook 1998; Johannesson & Ekendahl 2002; Rodrigues & Absalão 2005).

Usually the nature of the selection on shell colour morphs is a matter of speculation and may not only be a result of visual predation. Non-biological factors may also be important in maintaining polymorphism, including those determined by colour. For example, the effect of temperature in some snails (e.g., Heath 1975; Jones et al. 1977; Etter 1988b; Stine 1989; Sokolova & Berger 2000), where dark-shelled morphs heat more rapidly and thus may have an advantage in cold areas, while light-coloured shells may be of advantage in warmer locations (Jones 1973; Stine 1989; Miura et al. 2007) or in alpine grasslands (Burla & Gosteli 1993). Similar effects have been observed in dark-shelled morphs of *Mytilus* (Mitton 1977) and darker shells may offer more protection from UV radiation (e.g., Heller 1979). Parasites or parasitoids may also favor particular morphs, invoking a form of selection (e.g., Ewers & Rose 1966; McKillup et al. 2000).

Interactions between two sympatric species may result in colour divergence through frequency-dependent selection from visual predation (e.g., Clarke 1962; Allen 1988; Chiba 1999). Sometimes, different sympatric colour forms regarded as morphs of the same species appear to be different taxa (e.g., Joseph et al. 2014), while allopatric colour forms may be interpreted as representing genetically and/or morphologically distinct forms of the same species (e.g., Nakano et al. 2010; Mariottini et al. 2013).

The chemical composition of shell pigments is rather poorly known with an early study (Comfort 1950, 1951a) showing that porphyrins[11] were present. In lower bivalves, patellogastropods, vetigastropods, and 'opisthobranchs', the shell pigments are acid-soluble, but in many caenogastropods and 'pulmonates' they are not (Comfort 1951a), and the composition of these latter pigments remains elusive. Recent research using Resonance Raman microspectrometry have shown that polyenes, including carotenoids, are common (Barnard & Waal 2006; Hedegaard et al. 2006). The summary below largely follows a recent review by Williams (2017).

The main shell pigments known are carotenoids, melanin, and tetrapyrroles, including porphyrins and bile pigments.

Tetrapyrroles are the main acid-soluble pigments in molluscan shells. They occur in two main forms – cyclic structures (porphyrins) or linear structures (bilins) and produce red, orange, green, blue, violet, yellow, or brown colouration (Fox 1976). *Porphyrins* are the most common shell pigments. In these, the pyrrole rings form a ring, with the most common being uroporphyrin I and conchoporphyrin. They are probably derived from chlorophyll obtained via food and can vary depending on the available chlorophyll (e.g., Underwood & Creese 1976). These compounds produce red, brown, or purple shell colouration (Williams 2017) and have also been identified as responsible for the colour of some pearls and pearl-producing bivalves (see Hedegaard et al. 2006 for references).

Bilins, where the pyrrole rings form a chain, are much less common than porphyrins and are responsible for blue, green, brown, or red colouration (e.g., the green colour in some haliotids, trochids, and turbinids (Jones & Silver 1979).

Quinones may be responsible for the non-acid-soluble pigments in some shells (Comfort 1951a).

Polyenes[12] include *carotenoids*. Hedegaard et al. (2006) used Resonance Raman microspectrometry to identify polyenes as shell pigments in 13 gastropods, one cephalopod, and four bivalve taxa, but found there was no clear relationship between colour, pigment, and taxon with, for example, the same colour being due to a different pigment in different taxa. Although carotenoids are not uncommon in molluscan tissues (e.g., Goodwin 1972; Vershinin 1996; Matsuno 2001), Comfort (1951a) did not find these compounds in shells, although they have since been identified as shell pigments in some taxa. Carotenoids are responsible for the yellow colouration in a species of cowry (Cypraeidae) (Hedegaard et al. 2006) and were identified in *Strombus* shells (Dele-Dubois & Merlin 1981). Carotenoids may also be responsible for the yellow colour of some pearls (Hedegaard et al. 2006; Li et al. 2014b) and have been found in shells of pectinids and some neogastropods by Barnard and Waal (2006) using Raman spectra, suggesting these pigments may be widespread. The striking blue-violet colouration of the pelagic 'violet snail' *Janthina* is like that of its prey, the siphonophores *Physalia* and *Velella*. Analysis of the pigment in *Velella* and *Janthina* showed that the carotenoids responsible for the colour differed in some respects, suggesting that the gastropod may be chemically altering the chemical structure (Barnard & Waal 2006). Blue pigments are also found in a few other gastropod shells, but these have not been investigated.

Indoles have a bicyclic structure and are found as pigments in molluscs in two forms, melanins and indigoids. Melanins come in two forms – eumelanin and pheomelanin (Williams 2017). Melanins have been reported from the shells of *Lymnaea* and *Mytilus*, but are presumably more widespread; eumelanin is an important component of cephalopod ink (Derby 2014) and has also been detected in fossil ink sacs (Glass et al. 2012). Melanins are probably responsible for some yellow, red, brown, and black colours. Indigoids are thought to be present in some shells, such as *Haliotis cracherodii* (Comfort 1949), but Comfort (1951a) later suggested these supposed indigoids may be pyrroles, and there do not appear to have been subsequent reports of indigoids in shells.

The shells (and often the animals) of many molluscs adopt the same or similar colouration as their food and/or their substrata, rendering them cryptic. Most of these cases have not been investigated as to whether the pigments are being sequestered from the food or whether the colours have a genetic basis and are derived *de novo*. Some are clearly derived from pigments in the food (e.g., Comfort 1951a). For example, *Haliotis* fed on different algal foods change colour, sometimes dramatically (e.g., Leighton 1961; Tajima et al. 1980). The shell of *Nucella lapillus* may be white or grey with a diet of barnacles and brown to purple when *Mytilus* is the main food, with an abrupt change in colour if the diet is changed (Moore 1936). Subsequent work, however, has shown that the situation is more complex as the ground colour of the shell in *Nucella* (Spight 1976a; Castle & Emery 1981; Palmer 1984)

---

[11] Heterocyclic or macrocyclic organic compounds comprised of four pyrrole subunits connected via their carbon atom and methine (+CH–). This group of compounds includes pigments, one of which is heme–the red in haemoglobin.

[12] A group of polyunsaturated conjugated organic compounds that are often colored due to their low energy state resulting from multiple carbon to carbon double and single bonds.

and *Haliotis* (Liu et al. 2009) has a genetic basis, and this can be modified by pigments derived from the food.

3.2.9 SHELL PORES

Chiton plates, monoplacophoran shells, and some bivalve and gastropod shells are penetrated by minute pores, and the cellular extensions that enter them are called caeca. Similar structures occur in some brachiopods, bryozoans, ostracods, and barnacles. They have a patchy distribution within both gastropods [some fissurellids, some neomphaloideans (Sasaki et al. 2010b), neritoideans, a few glacidorbids], and some bivalves [some pteriomorphians (all arcoids and limopsoids), cyrenoideans, and Chamidae], but most taxa lack them. In bivalves, the pores are formed by projections from the mantle penetrating the formed shell layers (Waller 1980; Reindl & Haszprunar 1994, 1996a). In most bivalves, they are sparse and do not penetrate right through the shell or are restricted to the juvenile shell (as in Cyrenoidea) (Waller 1998). In the Arcoidea, however, they are dense on the inner side of the pallial line (e.g., Waller 1980).

The function of many of the different caeca is uncertain – the shell limits the exchange of materials and sensory information between the organism and its environment, and the caeca may assist in such exchange. In chitons, the caeca are called aesthetes and have sensory functions including imaging (see Chapter 14), but the cellular extensions that enter the pores in gastropods are less modified. The detailed structure of the caeca inside the pores in arcoid (and some other Pteriomorphia) bivalves, vetigastropods, chitons, and brachiopods are all very different (Reindl & Haszprunar 1996b), and it is unclear as to whether or not they share a common ancestry.

3.2.10 SHELL ORNAMENT AND ITS FUNCTION

Shells can bear a wide range of ornament ranging from enhanced growth lines to long spines or heavy knobs. While sculpture is usually arranged collabrally,[13] which implies some discontinuity in growth, there are also often radial elements. In some taxa, the sculptural elements can be markedly divergent, as for example, in the divaricating sculpture of some bivalves (Seilacher 1972). Sculpture can be continuous as in lines, ridges, frills, or discontinuous as in granules, nodules, or spines.

In extant molluscs with shells, the most elaborate sculpture is seen in bivalves and gastropods, while polyplacophorans either lack sculpture or it is subdued, with ridges and/or granules developed in some taxa. Scaphopods are generally smooth or have radial sculpture. All extant monoplacophoran and cephalopod shells lack sculpture. A different picture presents itself with fossil cephalopods, with some nautiloids and belemnites having developed sculptural elements and with an amazing and, sometimes, bizarre array of sculpture in ammonoids.

In coiled molluscs that develop elaborate projecting sculpture (many gastropods and ammonoids), as the shell grows over the sculptured part of the preceding whorl the protruding sculpture is dissolved by the mantle (e.g., Carriker 1972; Signor 1982b), enabling recycling of the calcium. Vetigastropods tend to cover any sculpture with a thick glaze or have reduced whorl overlap (Vermeij 1973b, 1977b).

It is usually assumed that shell sculpture is functional. Some sculpture roughens the surface, increasing the ability for sediment and fouling biota to lodge and thus facilitating crypsis. Ribs and spines on shells should logically aid in deterring predators, but there have been relatively few attempts to test this. Experiments with epifaunal bivalves tested the effectiveness of spines against predation by muricids and starfish. Spines deterred predation by muricids, but were not effective against predation by starfish (Stone 1998). While evidence suggests that spines on bivalves may discourage borers (Beatty & Rollins 2002), experiments using artificial spines on *Mytilus* found that predation by *Nucella* was not reduced (Willman 2007). Palmer (1977) experimentally demonstrated the role of the shell varices in the muricid *Ceratostoma foliatum* in providing hydrodynamic destabilisation and increasing the probability that it lands aperture down if detached from the substratum. In follow-up work, Palmer (1979) concluded that shell-crushing fish have been important selective agents in the evolution of molluscan shell sculpture, especially in the tropics. In addition to antipredation defenses, it has also been suggested that spines have other adaptive roles such as providing camouflage or for stabilisation on soft substrata. Other adaptive functions for certain sculptural elements include assisting with burrowing, notably the so-called ratchet sculpture (Signor 1982a) in some gastropods and bivalves (e.g., Stanley 1981; Signor 1982a, 1993; Seilacher 1985; Savazzi 1989) and the file-like sculpture seen in the anterior part of the shell in bivalves which bore using mechanical means (e.g., pholadids and teredinids) (e.g., Seilacher 1985; Savazzi 2005).

The 'ratchet sculpture' is common in shallow-water bivalve burrowers that live in sand. It has ridges with a rounded edge toward the direction of burrowing that slip through the sediment while a steep face on the opposite side resists backward movement, thus assisting by reducing slippage when the foot is extruded. This type of sculpture is found in several bivalve families (Vermeij 1993) and some have a modified version – divaricating sculpture, as seen for example, in the lucinid *Divaricella* and the tellinid *Strigilla*. This consists of oblique ridges with steep dorsal edges; the anterior ridges are at right angles to the posterior ones, so they grip the sediment as the shell rotates when burrowing. Such oblique ridges not only enable efficient burrowing, they also strengthen the valves (Checa & Jiménez-Jiménez 2003). Some burrowing gastropods also develop spirally arranged ratchet sculpture which has the steep side toward the apex, while in others, the shell is smooth, but there is a protruding suture (Vermeij 1993).

---

[13] Collabral – shell ornament or growth lines that conform to the growing edge of the shell (the outer lip of the aperture of a gastropod or the shell margin in a bivalve).

Sculpture in gastropods is not confined to the outer part of the shell. The aperture may contain elaborate developments of teeth, ridges, and other structures (Paul 1991) that are often thought to serve as barriers to predators or to strengthen the shell. In the ellobiid *Pedipes*, it has been suggested these teeth prevent the foot of the snail from blocking the pneumostome (Örstan 2010). Such internal sculptural elements are absent in chitons, scaphopods, and extant cephalopods, as well as most bivalves other than some scallops and cockles.

3.2.11 SHELL WINDOWS

Many species of small coiled gastropods have transparent or semitransparent shells, but some of those with an opaque shell have a transparent patch in the last part of the last whorl of the shell behind the aperture. The eyes can be seen through this part of the shell. We are not aware of any experimental studies on coiled gastropods, but a similar situation in a patellogastropod limpet was investigated, and it was demonstrated that it responded to light through translucent patches on the anterior part of the shell which lack further modification (Lindberg et al. 1975; Lindberg & Kellogg 1982). More apomorphic 'shell windows' are found in a few bivalves with symbiotic zooxanthellae where the shell structure is modified to direct light onto the tissues containing the photosynthesising symbionts. The best known example is the cardiid *Corculum cardissa* which has anterior and posterior patches of the shell forming the 'shell windows' that funnel light through the shell in a way analogous to fibre optics (Watson & Signor 1986; Seilacher 1990). These 'windows' are formed mainly from the outer prismatic shell layer, and their shape may help to focus light on the zooxanthellae bearing tissues (Carter & Schneider 1997). Similar, but less developed, windows are also found in some other members of the cardiid subfamily Fraginae.

3.2.12 SECONDARY SHELL STRUCTURES

Adventitious tubes formed by watering pot shells (Clavageloidea), ship worms (Teredinidae), and Gastrochaenidae, are secreted by glands on the siphon (e.g., Savazzi 1982; Morton 1984, 2006c; Morton et al. 2011; see Chapter 15 for more detail). The shell material comprising these tubes is simpler in structure than true shell secreted by the mantle edge.

In gastropods, *Magilus* (Muricidae, Coralliophilinae) and vermetids form long tubes secreted by the mantle edge, and the sole of the foot of hipponicids secretes a basal calcareous plate attached to the substratum (e.g., Knudsen 1991, 1993). In bivalved sacoglossans (the 'bivalved gastropods'), the mantle divides at settlement to produce right and left shell 'valves' with the larval shell remaining attached to the left valve (Kawaguti 1959; Grahame 1969). Subdivision of the mantle presumably also occurs in the formation of the accessory shell plates in pholadid bivalves. A plate, the callum, covers the pedal gape in mature individuals of the wood-boring pholadid *Martesia* and is presumably secreted by the mantle, although the details of this process have not been investigated. Similarly, the calcareous siphonal pallets in Teredinidae are secreted by the siphonal tissue, but the details are, again, not known.

3.2.13 AGGLUTINATION

In many shelled molluscs, epizootic organisms growing on the shell provide camouflage and hence protection against predation. In a few bivalve and gastropod taxa, sand grains and sometimes other objects are agglutinated on the shells. The most famous examples are some members of the Xenophoridae which attach dead shells, stones etc. to the outer surface of their shells. Other examples are the cerithioidean gastropod *Scaliola* (Scaliolidae) and the venerid bivalve *Granicorum* which attach sand grains to the exterior of their shells. Some land snails attach dirt, leaf fragments, lichens etc. to their shells (e.g., Allgaier 2007).

Objects are typically cemented largely through periostracal and shell formation at the mantle edge. In some watering pot shells (Clavagellidae), the adventitious shell produced by siphonal glands has agglutinated sand grains (e.g., Morton 1984). Multicellular arenophilic glands are found on the mantle margins of several other anomalodesmatan bivalves. These secrete mucoid substances which enable the attachment of sand grains and other particles to the periostracum (Sartori et al. 2006).

3.2.14 OTHER SHELL STRUCTURES

Special taxon-specific shell structures (e.g., bivalve hinges and ligament; cephalopod shell chambers) are discussed in the chapters in Volume 2 dealing with groups possessing these features.

3.2.15 SHELL REDUCTION AND LOSS

Adult shell loss in conchiferans occurs only in gastropods and cephalopods, and although extreme shell reduction has occurred in a few bivalves (a few galeommatoideans and Teredinidae), complete shell loss in bivalves is unknown. Examples of adult shell loss are scattered through the gastropods, and in most groups it is rare. Vetigastropod examples include some trochids (e.g., *Gena*), and some fissurellids. Shell loss has occurred in neritimorphs (the slug-like *Titiscania*) and, in caenogastropods, the parasites *Entocolax*, *Entoconcha*, and *Enteroxenos* (Eulimidae) (Lützen 1968). A group of slug-like caenogastropods, the Velutinidae, have thin internal shells. No shell loss is known in patellogastropods.

In heterobranch gastropods, shell reduction and loss is typical of many groups (Morton 1963; Gosliner & Ghiselin 1984; Gosliner 1985, 1991; Tillier 1989) (see Chapter 20). For example, there are no extant 'nudibranch' taxa with a shelled adult, although their sister group, the pleurobranchs, retain the larval shell and often have a reduced internal shell. Developmental studies on the shelled pleurobranch *Berthella* show that the mantle fold extends back over the outside of the larval shell during development, enclosing it (LaForge & Page 2007). In developing nudibranchs, the mantle fold is also reflexed and spreads dorsally after metamorphosis, but it does not enclose the larval shell (see Chapter 20, Figure 20.67), which is discarded (Thompson 1976). Discarding the larval shell also occurs in onchidiids (Fretter 1943) and vaginulids (G. Barker, pers. comm. Nov. 2014).

Shell reduction and shell loss are also predominant in the coleoid cephalopods. While there appears to have been no shell reduction or loss in nautiloids, the external shell has been reduced and internalised in coleoids (belemnoids, cuttlefish, octopods, and squid) or completely lost in some octopods, although a rudimentary shell may be present in some taxa. In both gastropods and coleoids, the adult shell is lost, not the larval shell, which is still produced in early development and then shed at metamorphosis (gastropods) or subsumed by the adult mantle (octopus).

In most cases of shell reduction, the shell is first surrounded by tissue and then reduced. These internal shells may be much reduced and even disappear, but in squid and a few heterobranch gastropods, they are chitinous (Furuhashi et al. 2009a), having minimal or no calcareous material. In some taxa, the shell remains external, but is much reduced in size relative to the rest of the animal – as in teredinid and clavagellid bivalves (see Chapter 15), some 'opisthobranchs' (see Chapter 20), terrestrial 'semi-slugs', and *Testacella* (see Chapter 20). In some gastropods, the shell is covered and eventually enclosed and partly or completely absorbed by either lateral folds of the foot (e.g., naticids or *Akera*) or the mantle encloses the shell as in some fissurellids (e.g., *Scutus*), Cypraeidae and Velutinidae (e.g., *Lamellaria*), and various shelled 'opisthobranchs' (e.g., *Philine, Pleurobranchus*). A similar enclosure of the shell valves by the mantle occurs in some galeommatoidean bivalves such as *Phlyctaenachlamys* and *Devonia* and a few chitons such as *Cryptochiton*. Whether the shell enclosing tissue is pedal or pallial, the shell becomes reduced and may eventually be lost in the adult, but the derivation of the new dorsal integument differs, being either pedal in origin or pallial.

3.2.15.1 Consequences and Opportunities

The advantages of shell reduction or loss include a reduction in weight and musculature and hence energy savings, and enhancing manoeuvrability and the ability to squeeze into small spaces. The shell comprises much of the weight of the mollusc, and moving it contributes significantly to their energy needs (e.g., Hausdorf 2001). A small increase in shell size can markedly increase shell weight. For example, doubling the size of *Nucella lamellosa* required three times the energy to carry the shell (Palmer 1992). Shell formation also expends considerable energy (Palmer 1992). Shelled terrestrial gastropods must contend with greater energy requirements for shell construction and the low calcium levels in many soils. Adopting a slug-like body allows the visceral mass to be lengthened and flattened and the foot to be extended. This process, called 'limacisation',[14] involves the

---

[14] This term, derived from the slug genus *Limax*, is the process of becoming slug-like.

organs of the visceral mass descending into the body cavity. Experiments with semi-slugs and slugs by Hausdorf (2001) showed that slugs were more mobile and flexible, better able to find shelter (by squeezing into tiny crevices) and could grow more quickly because they lacked a shell. These and other advantages have led to about ten independent origins of terrestrial slugs (Pearce & Örstan 2006). In coleoid cephalopods, shell reduction and loss are thought to be related to the development of rapid jet locomotion using a contractile muscular mantle wall (see Section 3.12.3.8.2), an innovation that would be impossible with an external shell (see Chapter 17).

Despite theoretically increasing vulnerability to predation, losing the shell can open opportunities. Besides streamlining, which may facilitate both burrowing and swimming, the much-increased area of skin can provide additional respiratory surface and is often accompanied by reduction or loss of the mantle cavity. Some shell-less slugs (some nudibranchs and acochlidians) have secondarily produced spicules in their mantle (notum) (see Chapter 20). Shell loss has led to, or is correlated with, chemical defence in many nudibranchs and other 'opisthobranchs' where it is often associated with warning colours (Cimino & Ghiselin 1998, 1999) or by the production of defensive mucus as in some terrestrial slugs (see Chapter 20). In a few nudibranch taxa, shell reduction or loss has led to the sequestration of photosynthetic zooxanthellae, or, in many sacoglossans, of plastids (see Chapter 20). Apart from predation-related disadvantages of shell loss, for terrestrial and intertidal taxa it also results in markedly increased susceptibility to desiccation.

3.2.16 Growth

Most studies on molluscan growth have focused on the shell as it is difficult to obtain accurate measurements of body size for soft-bodied organisms such as molluscs. One notable exception is the many growth studies on commercial species of coleoid cephalopods (squid, cuttlefish, octopus) which has also been facilitated by the adoption of standardised measurements for the fishery.

3.2.16.1 Shell Growth and Its Record

Molluscs grow incrementally by adding shell material to the edge of their shell, and the shell also thickens as it grows by way of material added to its inner surface. In some taxa, growth continues through their entire life, this being termed *indeterminate growth*, while others stop growing at a particular stage in their life cycle. This *determinate growth* is obvious in many gastropods where the aperture is thickened or otherwise modified at maturity, in marked contrast to those with indeterminate growth with a thin-edged, undifferentiated aperture throughout their life (Vermeij & Signor 1992).

The elaborations of the last whorl and apertural margin became possible with determinate growth, characteristics that have evolved from indeterminate growth ancestors in many gastropod lineages. Varix formation has appeared more than 40 times during gastropod evolution (Webster & Vermeij 2017), with varices, or at least a terminal varix-like structure in the adult, occurring in many higher caenogastropods (notably tonnoideans, cerithioideans, stromboideans, and neogastropods) as well as some ellobioideans and stylommatophorans. Many of the near basal caenogastropod cerithioideans, and certain 'asiphonate' caenogastropods (e.g., many Rissoinidae and Rissoidae, some Epitoniidae, a few Eulimidae, etc.), also possess apertural varices, indicating this feature has evolved independently in several lineages. The varices are usually thickened externally, but are mainly internal in some families (e.g., Potamididae, Strombidae, Cancellariidae).

As the shell grows, many bivalves and gastropods produce sculptural elements at regular intervals which reflect changes at the mantle margin. Growth lines and some sculpture can be parallel to the growing margin (collabral), producing concentric (bivalves) or axial (gastropods) sculpture, or perpendicular to it, producing spiral elements. These sculptures can be continuous (e.g., ribs) or discontinuous (e.g., granules, spines).

Growth proceeds in bursts (known as episodic growth) in gastropods with intermittent varices, being very rapid between varices. The periostracum and thin shell are quickly formed in the growth spurt and subsequently thickened while the apertural varix is being formed. Varices in most taxa are formed at about the same interval around a whorl, and because the gap between each pair of varices regularly increases, the growth spurts also increase. Episodic growth also occurs in *Siliquaria* (Siliquariidae), which is unique among gastropods because it periodically increases the width of the shell slit and the cross-sectional area of the whorl by breaking the sides of the shell into pieces, moving them outward, and reassembling them (Savazzi 1996).

Even on a daily time scale, shell growth is not continuous, resulting in 'growth lines'. This periodic growth not only occurs in very short time scales (e.g., tidally, diurnally), but also seasonally. Daily growth is recorded in the shell as microscopic growth lines and seasonal growth as macroscopic growth lines (e.g., Lutz 1976; Jones 1980). Such records in the shell have been extensively studied in some bivalves and to a lesser extent in gastropods (e.g., Lutz 1976; Richardson et al. 1980; Jones 1985). For example, Butler et al. (2010) reported a 489 year marine climate chronology in the Irish Sea using data derived from growth increments in the shell of the bivalve *Arctica islandica*, individuals of which can live for over 400 years and show no growth senescence (Ridgway & Richardson 2011).

Microscopic growth lines (microgrowth bands) that form with each tidal cycle have been used mainly in studies of intertidal bivalves (see references above) and some gastropods (e.g., Ekaratne & Crisp 1984). This technique is a more accurate and detailed measure of growth rate than other methods. Some studies show that the variation in growth rates between individuals can be considerable

(e.g., Ekaratne & Crisp 1984). Resorption of the shell may complicate growth rate studies using growth lines, as found in some unionoideans (e.g., Anthony et al. 2001) (see Section 3.2.19).

Shell growth is under hormonal control as discussed in Chapter 2 and is modified by environmental conditions (temperature, food availability, etc.) and the age of the individual. Growth is usually more rapid when young and slower later in life, and ceases in those taxa with determinate growth. Warm conditions also lead to increasing calcification (Graus 1974), and elaborate shell ornamentation tends to increase with increasing temperature (Vermeij 1993). Thus, cold water species typically grow in summer and little or not in winter. Food supply may influence not only growth rate, but also shell shape. For example, in the littorinid *Littorina littorea*, individuals supplied with ample food grew faster and developed low-spired shells with a large aperture, while those with limited food grew more slowly and had higher-spired shells with a smaller aperture (Kemp & Bertness 1984), although this result was reversed (relating to shell shape) in a study on *Littorina saxatilis* (Saura et al. 2012).

Other ways of studying the environmental record in the shell involve the analysis of oxygen and carbon isotopes and other elements as outlined below (see Section 3.2.18).

3.2.17 INTERNAL SEPTA

Internal calcareous partitions that close off the interior of the shell, termed septa, are also seen in some fossil enigmatic gastropods (see Chapters 13 and 19) and a few modern gastropods including the tube-like vermetids, caecids, the coralliophiline *Magilus*, some turritellids (Andrews 1974), and some architectonicids (Bieler 2009). Gastropod and cephalopod septa are analogues, with those in cephalopods differing in being more regular and the chambers between them connected by a siphuncle (see Chapter 17). A similar process to septa formation commonly occurs in many conchiferans where the shell is sealed as the protoconch is lost or the older parts of the shell are removed by wear.

The worm-like shells of most vermetids are cemented to hard substrata and can extend the shell aperture high above the substratum. The production of internal septa in the shell allows it to become much longer than the body. A similar, but unrelated group of gastropods, the siliquariids, are mostly not cemented, and those that live in sponges have a long shell slit and do not produce septa.

Septa-like formations are found in the fresh-water planorbid limpet *Gundlachia* (Basch 1959a) and in a few Melongenidae (Vermeij & Raven 2009). In some land snails, septa are formed because of parasitic fly infection (Knutson et al. 1967). Septa-like formations are also formed in some bivalve shells as foliated layers separated by water-filled spaces, as for example, in some oysters and *Spondylus* (Healy et al. 2001).

3.2.18 Stable Isotopes and Trace Elements in Shells

Many isotopic studies feature molluscan taxa. The accretionary formation of the calcium carbonate shell provides a historical 'time and date stamp' providing data on temperature, habitats, and biological processes (Butler et al. 2010; Stott et al. 2010; Ridgway & Richardson 2011). Traditionally growth increments have often been useful in determining growth rate (see Chapter 3), but the chemical properties of the shell itself can also provide data about the environmental conditions that existed when the shell material was laid down by using carbon or oxygen isotopes (see Chapters 3 and 9) (McConnaughey & Gillikin 2008; Ivany 2012; Gillikin & Dehairs 2013). The ratios of the stable isotopes of carbon ($^{13}C/^{12}C$) and oxygen ($^{18}O/^{16}O$), found in the calcium carbonate of the shell, are useful in estimating palaeosalinity and palaeotemperature. For example, an increase in temperature results in lower $^{18}O/^{16}O$ ratios and salinity tends to increase ^{18}O relative to ^{16}O, while the $^{13}C/^{12}C$ relationship provides estimates of productivity, organic carbon burial, and vegetation type. Isotopes of nitrogen, hydrogen, and sulphur are also commonly used.

Various trace elements are also incorporated in shells by assimilation into the crystal lattice where they replace calcium, absorption on the crystal surface, or between lattice planes, by their incorporation into the organic matrix, or by accidental inclusions (Jones 1985). The relative abundance of calcium and various trace elements can be used to obtain information from fossils about the conditions of deposition and environmental factors such as temperature (see below), although many factors need to be considered, including the affinities of different taxa to different elements.

Each isotopic comparison requires a calibrated standard to which sample results are compared. The standard used to calibrate C_{13} was a mollusc – the Pee Dee Belemnite (PDB), a Cretaceous cephalopod fossil, *Belemnitella americana*, from the Pee Dee Formation, South Carolina, USA. Samples of the standard were depleted during calibration studies which required a new reference sample to be calibrated against the original specimen. Since the laboratory that performed the recalibration was in Vienna, Austria, the standard is called the V-PDB (Vienna PDB). Other isotopes have other standards (or secondary standards now because of depletions). For example, the oxygen and hydrogen standard is Vienna Standard Mean Ocean Water (or V-SMOW), while for nitrogen the standard is atmospheric air and for sulphur, it is the Canyon Diablo meteorite (CD).

Studies of molluscan shell isotopes have led to numerous papers estimating palaeotemperatures and salinities and events such as changes in productivity and hydrography (e.g., Jones 1985). These data, often derived from fossil molluscs, have become important sources of palaeoenvironmental data in this time of unprecedented global climate change. Putative 'molluscan' shell isotopes (^{13}C) have also been used to document the sequence of biomineralisation events during the 'Cambrian explosion' (Kouchinsky et al. 2012) and to reconstruct the carbonate content of Cambrian seas.

Chemosynthetic molluscan communities and taxa have also been studied using isotopes and, besides the shells, tissue of living taxa provides isotopic evidence (^{13}C and ^{15}N) of dietary preferences and chemosynthetic nutrition (e.g., Windoffer & Giere 1997; Naraoka et al. 2008; Soto 2009; Becker et al. 2010; Demopoulos et al. 2010). In terrestrial habitats isotopes are important in determining physical conditions (mainly temperature and rainfall) and food preferences and thereby serve as important palaeovegetation proxies (Yanes et al. 2013). For example, Colonese et al. (2013) have analysed shell carbon isotopes (^{13}C) from Pleistocene and living *Helix figulina* from Greece and were able to identify C_3 vegetation as the main source of carbon for both late glacial and early Holocene snails. Although C_3 plants are typically associated with habitats with moderate sunlight and temperature, plentiful water and high CO_2 concentrations (200 ppm), the ^{13}C values from the shells were similar to snails living in extant Mediterranean woody shrub communities adapted to drought.

Many archaeology studies have combined shell chemistry with the study of growth increments (e.g., Jones 1983a; Hallmann et al. 2009). In the Aleutian Islands, Alaska, growth line studies coupled with stable isotope analyses (^{18}O and ^{13}C) have shown that the cockle *Clinocardium nuttallii* grew faster, larger, and were older in the past (170–400 years BP) than the cockles found there today (Koike et al. 2012). The study was also able to determine that the cockles were collected from the early spring through early autumn and did not correspond to the seasons with the lowest tides. On a more recent time scale, Schöne et al. (2003) used growth increment analysis combined with isotopic studies to document how diversion and management of the waters of the Colorado River in North America have changed growth rates of two species of venerid bivalves in the Colorado River estuary. This study is of particular interest because it demonstrated how the impact of water management policies could be assessed in drainage systems where no pre-impact studies were conducted.

Shells of living and fossil *Tridacna gigas* have the potential to yield reliable records of past changes in seasonality and ENSO (El Niño Southern Oscillation) variability, and mean climate conditions (Welsh et al. 2011). Since the shells are generally more resistant to diagenesis than coral skeletons, they may provide robust estimates of past tropical climate for periods and locations where only altered coral skeletons are available. Molluscan isotopic and sclerochronological time series data have been useful for palaeoclimatic reconstruction (Chauvaud et al. 2005; Schöne & Fiebig 2009; Gilbert et al. 2017), reconstructing life history traits in taxa used in aquaculture (Radermacher et al. 2009), and in reconstructing ENSO events (Carré et al. 2005a, 2005b; Lazareth et al. 2006; Black 2009). They are somewhat limited in their application as they yield relatively short time series (e.g., Carré et al. 2005b). These factors complicate their use for assessing variation in strength and frequency of El Niño events under different conditions, as well as studies of the Western Pacific Warm Pool.

Dense monospecific populations of bivalves in cold seep communities (e.g., the giant white clam *Calyptogena* and *Bathymodiolus* mussels) are of significance because their accretionary mode of growth preserves a detailed record of environmental fluctuations in these unique habitats (Rhoads & Lutz 1980). This record includes chemical and isotopic fingerprints (Rio et al. 1992), while post-mortem accumulations of shells may contain additional taphonomic signatures of fluid venting (Callender & Powell 1992; Callender et al. 1992).

3.2.19 SHELL RESORPTION

Probably all shelled molluscs are capable of resorbing at least some internal shell material after it has been laid down. Most coiled gastropods remove some shell material from the exterior surface adjacent to the inner side of the aperture as the shell grows. This is particularly necessary in highly sculptured taxa such as those with spines or knobs, as resorption is necessary to remove these sculptural elements when the inner lip of the aperture spreads across the parietal wall as it grows (see Section 3.2.10). In some fresh-water mussels resorption of the shell occurs at a similar rate to shell growth with shell reduction recorded in some (Downing & Downing 1993; Anthony et al. 2001).

Resorption of the internal parts of the shell is a characteristic of only a few marine gastropods, notably most neritimorphs, and other examples include a few neogastropods such as *Bullia*, *Conus*, *Conorbis*, and *Olivella*, and in the heterobranchs the cephalaspidean *Mnestia* and ellobiids, with all these possessing elongate apertures (Vermeij 1977b). Species of *Conus* can remove up to around a quarter of the internal shell material. Some terrestrial taxa also resorb internal whorls (Solem 1983); the neritimorph taxa Helicinidae, Ceresidae, and Proserpinidae all have complete resorption of their internal whorls, while ellobiids have partial resorption (e.g., Morton 1955a). Stylommatophorans never completely resorb the internal shell whorls.

3.3 THE BODY

The body of a mollusc consists of the head, foot, and viscera – collectively, and rather quaintly, called the 'soft parts' by some authors. Some primitive molluscs are elongate, with an anterior end and a posterior anus. In others, the dorsoventral axis is extended, and the gut bent into a U-shape (ano-pedal flexure) enabling life in blind tubes and the development of a more compact body shape (see Chapters 1, 5, and 8).

3.3.1 THE EPIDERMIS

The epidermis (or skin or integument) protects the animal and interfaces and interacts with the external world. It covers the entire body and performs many functions including protection, respiration, sensory input, and osmoregulation, while the epidermis of the foot sole facilitates locomotion and tenacity. The epidermis is a single layer of external cells (Figure 3.23) on a basement membrane overlying connective tissue permeated by muscle fibres. It is supplied with nerves and contains sensory cells which, sometimes, have evolved into sense

organs. The epidermis can heal after damage (see Chapter 2) and can also uptake dissolved organic matter (DOM) by way of pinocytosis (see Chapter 5).

The histology of the epidermis has attracted much interest. Epidermal cells are typically columnar in shape and can be ciliated, non-ciliated, or glandular. Their external surface is composed of a microvillous layer which assists in retaining mucoid secretions in place and is sometimes wrongly referred to as a cuticle (see Rieger & Rieger 1976; Rieger 1984). However, a true cuticular layer is found in the mouth region of gastropods and cephalopods.

Microvilli are usually well-developed in the exposed general epidermal regions of the body, but smaller on the outer mantle epithelium.

The epidermis may perform its protective function by the secretion of a shell and, in many gastropods, an operculum. Various chemicals and secretions are also produced by epidermal gland cells, the most important of which is mucus (see Section 3.3.2). Mucus often plays a role in protecting the epidermis in locomotion, and in terrestrial gastropods it assists in protecting against desiccation. With mucus, ciliated epithelial cells help cleanse the body surface and assist with locomotion.

The epidermis may become folded into the connective tissue to form complex glands that may secrete quinone-tanned proteins to produce byssus threads, an operculum, or radula. In contrast, the jaws of many gastropods and the beaks of cephalopods are derived from a simple epithelium.

In addition to epidermal sense organs such as mechanoreceptors, chemoreceptors and light receptors (see Chapter 7), specialised epidermal cells are also involved with antipredator defence, notably repugnatorial glands found in the mantle of some shell-less 'opisthobranchs', and specialised epidermal pigment cells (chromatophores) which enable cephalopods to rapidly change colour for camouflage and displays (see Chapter 17).

In some terrestrial slugs, the epidermal cells produce small buds from their surface during apocrine secretion (Wondrak 1968; Yamaguchi et al. 2000). Non-glandular pigment cells are also found in the epidermis of some slugs (Wondrak 1969, 2012). Ciliated cells are abundant in areas such as the mantle cavity of aquatic snails and the pneumostome of 'pulmonate' gastropods, the mouth, gills, and foot of most molluscs, and the labial palps of bivalves where they assist in maintaining water flow, particle sorting and rejection, and locomotion. Although sparser on the general body surface and outer mantle, they are also important in dispersing mucus over the body surface.

3.3.1.1 Cilia

Epidermal cells often bear cilia that may be immobile and sensory, or motile and involved in moving water, mucus, or in locomotion. Internally each cilium has a ring of microtubules, the axoneme, which passes into the cell as the 'root' which is anchored in the cell by a 'foot' or basal body. The root may be a single or a double structure in molluscs. If double, it consists of a long and short ciliary rootlet. This short rootlet is probably plesiomorphic as it is present in the Polyplacophora, Solenogastres, Caudofoveata, and Scaphopoda, but is absent in Monoplacophora, Gastropoda,

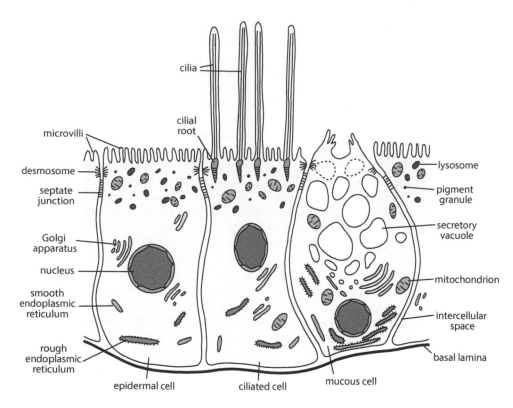

FIGURE 3.23 Three basic types of epithelial cells (epidermal, ciliated, and mucous) found in molluscan epidermis. Sensory cells are not shown. Redrawn and modified from Simkiss, K., Molluscan skin (excluding cephalopods), pp. 11–35, in E. R. Trueman & Clarke, M. R., *Form and Function. The Mollusca*, Vol. 11, Academic Press, New York, 1988 and Voltzow, J., Gastropoda: Prosobranchia, pp. 111–252, in F. W. Harrison & Kohn, A. J., *Microscopic Anatomy of Invertebrates. Mollusca 1*, Vol. 5, Wiley-Liss, New York, 1994.

Bivalvia, and Cephalopoda. There are, however, twin rootlets in autobranch bivalves and cephalopods, but these probably represent a split vertical rootlet (Lundin & Schander 1999, 2001a, 2001b, 2001c; Lundin et al. 2009).

Flattened 'paddle cilia' have been reported from many adult and larval molluscs on structures such as the gills and osphradia. Haszprunar (1985a) reviewed the occurrence of paddle cilia in molluscs and other taxa. They have been variously interpreted as locomotory for either moving water or moving through it, and even for applying byssus threads. They have also been considered as chemoreceptive structures and their presence to be a proof of this function (e.g., Matera & Davis 1982; Haszprunar 1985a, 1987b), but they are now thought to be fixation artefacts (Ehlers & Ehlers 1978; Matera & Davis 1982; Beninger et al. 1995b).

3.3.2 Mucus

The molluscan epithelium typically contains mucus-secreting gland cells which cover the outer surfaces with mucoid secretions. Mucus is a viscous colloidal solution of mostly water plus a protein-polysaccharide complex of high molecular weight and inorganic salts. It may comprise less than 1% to up to over 20% of the wet weight of a mollusc. The inorganic salts comprise about 3% of the wet weight of *Patella* pedal mucus (Grenon & Walker 1980). Two groups of protein-polysaccharide complexes have usually been recognised – the glycosaminoglycans (GAGs; often called mucopolysaccharides or proteoglycans) and glycoproteins (Reid & Clamp 1978; Di et al. 2012).

While the mucoid gels are all called 'mucus', they are generally secretions containing polymers that make large entangled complexes to form slippery gels (Smith 2010). The structure and composition of the polymers differ markedly in different molluscs and in other phyla.

The wide range of secretions is reflected in the diversity of gland cells that produce them in molluscs (see Section 3.3.3). While different kinds of mucus produced, notably by the foot, are presumably derived from different gland cells being activated, secretions from several different types of gland cell may combine to produce a specific type of mucus (Shirbhate & Cook 1987).

The variety and complexity of molluscan mucus, and in particular the legendary slime of terrestrial slugs, is rarely appreciated. The range of different functions (see below) often necessitates different physical and biochemical properties of the mucus which in turn must be produced by a variety of gland cells (Denny 1983; Davies & Hawkins 1998; Smith

2010) (see Section 3.3.3). Despite attention being paid to the biochemical structure of some adhesive gels, largely because of their potential practical applications (e.g., Smith 2006, 2010) (see also Chapter 10), much of the work on molluscan mucus is still in its infancy.

Mucus is dynamic and versatile, functioning as a slippery lubricant, an adhesive, or even a rope. This adhesive substance is a viscoelastic solid which, when stress increases, becomes a liquid, but returns to an effectively solid state when the stress is released. This property enables its use in locomotion (see Section 3.3.2.1). The many functions of mucus in marine molluscs were summarised in a review by Davies and Hawkins (1998), and its many roles in terrestrial molluscs are overviewed by Denny (1983). Mucus is vital in feeding (see Chapter 5), reproduction, and locomotion and adhesion. It also insulates the mollusc from the surrounding environment, acting as a barrier to diffusion (Grimm-Jørgensen et al. 1986) and possibly facilitating ionic regulation (Hillman 1969). Some molluscan mucus has antimicrobial properties (e.g., Otsuka-Fuchino et al. 1992) and can also protect against predators by the addition of distasteful or toxic compounds (see Chapters 9 and 20), the production of copious gelatinous secretions, or by making the surface too slippery to grasp. Mucus has also been shown to protect from pollutants, and even extreme cold (Davies & Hawkins 1998). In those taxa associated with cnidarians, mucus can inhibit the discharge of nematocysts as in some nudibranchs (Chapter 20), the coral living scallop *Pedum spondyloideum* (Yonge 1967), and the coral-boring mytilid Lithophaginae (Morton & Scott 1980). In patellogastropods, foot secretions also trap and stimulate the growth of algae and diatoms on the substratum (Connor & Quinn 1984; Connor 1986).

Mucus plays a vital role in the molluscan gut (see Chapter 5) and is of critical importance in suspension-feeders. In autobranch bivalves, mucus is crucial for the transportation of food from the gills via the labial palps to the mouth and, in some gastropods, sheets or strings of mucus trap food particles. It is also critical in moving (aided by cilia) and binding waste to clean the mantle cavities of most molluscs, where it is secreted by the general epithelial layers and, often in large quantities, by the hypobranchial gland(s).

Once secreted, subsequent dehydration of the mucus can markedly increase its strength and stiffness, as seen in some upper shore and terrestrial taxa. For example, intertidal snails such as littorinids and land snails can use mucus to attach to vertical or overhanging rocks to reduce desiccation. The secretion of a thick layer of mucus that separates the foot from the substratum can also help reduce heat stress, as has been shown in some intertidal snails and limpets (see Davies & Hawkins 1998 for review). Some stylommatophoran snails secrete an epiphragm, a layer of mucus that either seals the aperture to a hard surface or forms an operculum-like sheet across the whole aperture. The epiphragm hardens when dry and prevents moisture loss during aestivation.

It consists mainly of mucoprotein and some calcium (e.g., Campion 1961) and is sufficiently porous to allow some respiration (Barnhart 1983). About half the organic material making up the epiphragm comprises three proteins which, when extracted, have also been shown to cause stiffening in littorinid gel (Pawlicki et al. 2004). Some high intertidal limpets also secrete an epiphragm-like barrier to avoid desiccation (Wolcott 1973).

The mechanisms in marine molluscs are not well understood, but in terrestrial slugs and a few marine molluscs, the mucus is released from secretory cells in packages surrounded by a thin membrane, which immediately absorb water and increase dramatically in volume. Usually a mucous secretion is formed from a mixture of mucin-like secretions from different kinds of gland cells (see Davies & Hawkins 1998; Smith 2010 for reviews). This is particularly the case in the pedal epithelia of gastropods (see Section 3.3.2.1) including terrestrial slugs, the latter being well known for their slime.

Vertebrate mucins are formed from large protein-polysaccharide complexes that tangle together. In molluscs, although structurally and biochemically diverse, mucus is also composed of different protein-polysaccharide complexes (Denny 1983). These polysaccharides are usually called mucopolysaccharides and can be neutral or acidic. Acidic mucopolysaccharides (AMPS) often form sticky adhesive gels, but only after adding specific proteins and/or metal ions sourced from other gland cells. The AMPS are an important component of the mucus and are good lubricants because they are not readily hydrated or separated from the epithelium (Faillard & Schauer 1972). The negative charge facilitates cross-linkage (Smith 2002) and affinity to certain metals (see Section 3.3.2.1). Sulphated AMPS have been found in bivalves and gastropods, and two types were found in a chiton (Höglund & Rahemtulla 1977). This latter report is one of the very few concerned with chiton mucus, despite its production being a large part of their energy budget (Horn 1986). Lectins (see Chapter 2) have also been reported from the mucus of some stylommatophorans and a loliginid squid (see Davies & Hawkins 1998 for references).

Prezant (1990) discussed the evolution of the many and varied roles carried out by mucus in bivalves with most studies concerned with the mucus produced by the gills and its role in feeding (see Chapter 5).

Few details are known regarding the mucus secreted by the foot and mantle in scaphopods. Of particular interest is the adhesive mucus produced by the club-shaped heads of the feeding filaments (captaculae) which are employed in food capture (see Chapter 16).

There are few details available about mucus in cephalopods. Their skin is supplied with gland cells that secrete mucus consisting of a protein-polysaccharide complex (Packard 1988). Coleoid ink (see Chapter 17) is used to deter predators and is bound in mucus to slow dispersion (Denny 1989).

Nautilus tentacles lack suckers or hooks and instead have adhesive papillae that produce a mucoid glue composed mainly of neutral mucopolysaccharides. Gland cells in the non-adhesive epithelium of *Nautilus* produce acidic mucopolysaccharides (Byern et al. 2012).

Some of the biochemical diversity in gastropod mucus presumably reflects phylogeny, but so few taxa have been investigated in detail that generalisations are premature. The three main taxa investigated are described briefly below, based mainly on a review by Smith (2010). They reflect a very limited selection of gastropod diversity with only three of the five major groups of gastropods being investigated in detail and those only from intertidal and terrestrial habitats.

Patellogastropod limpets adhere strongly to the substratum using an adhesive gel, but can alternate between suction and glue, suggesting that the pedal gel properties can change (Smith 1992; Worms et al. 1993). Those investigated include both lottiids and patellids. In *Lottia*, the pedal mucus contains relatively short proteins that cross-link into large aggregates by way of non-covalent bonds to form an adhesive gel that contains relatively little carbohydrate (Smith et al. 1999). This glue-gel contains a protein absent from the mucus used in locomotion and in clamping using suction. These last two gels differ from those of other molluscs in that the non-water component is mainly composed of protein molecules instead of the very large complexes of protein and carbohydrate (Grenon & Walker 1980; Smith et al. 1999; Smith & Morin 2002). This gel stiffening protein may cross-link the gel molecules (Pawlicki et al. 2004).

The caenogastropod 'marsh periwinkle' *Littoraria irrorata* (Littorinidae) (as *Littorina* in much of the literature) produces both locomotory and adhesive gels (Smith & Morin 2002). Like many upper shore littorinids, when not subjected to wetting, they glue the shell aperture to the substratum. This is done by foot movements that transfer gel from the sole of the foot to the edges of the aperture (Bingham 1972b). The adhesive mucus contains protein and carbohydrate in approximately equal amounts, while that used for locomotion has about the same amount of carbohydrate, but much less protein. The carbohydrates are large, complex polysaccharides, and the glue has short proteins not found in the locomotory gel which are implicated in gel stiffening, presumably by cross-linking the carbohydrates (Smith & Morin 2002; Pawlicki et al. 2004).

The mucus of two stylommatophorans, the snail *Cornu aspersum* (as *Helix aspersa* in most literature), and the slug *Arion subfuscus* (Pawlicki et al. 2004) has non-water components mainly consisting of small proteins, and it also contains substantial metal ions involved in cross-linkages (Werneke et al. 2007). Some stylommatophoran slug mucus is renowned for its quantity and stickiness. In *Ariolimax* and *Arion*, the mucus is extruded from both the dorsal surface and the foot, is probably mainly defensive (e.g., Mair & Port 2002), and the total secretion can be over 5% of the body weight of the slug (Martin & Deyrup-Olsen 1986). The dorsal mucus is initially secreted as a viscous slime, but becomes very sticky after several seconds to a couple of minutes. This markedly contrasts with the slippery and viscous mucus these slugs use in locomotion (Deyrup-Olsen et al. 1983) (see Section 3.3.2.1). Both the mucous secretions produced by the foot sole and the dorsal part of the foot contain glycoprotein complexes and smaller proteins, but the glue contains additional short proteins that trigger gel stiffening (Pawlicki et al. 2004). The defensive sticky mucus produced by some terrestrial slugs is rich in metals (calcium, manganese, zinc, iron, magnesium, and copper) (Werneke et al. 2007; Smith 2013).

Mucus production, and particularly that involved in locomotion, involves a considerable cost to the animal, utilising up to 70% of all consumed energy (Davies & Hawkins 1998). While much of the mucus generated in the gut is reutilised, some that binds faecal material is lost. Mucus plays an important role in suspension-feeding molluscs, with much of the mucus utilised being assimilated during feeding, although the mucus binding the pseudofaeces in bivalves is lost to the animal.

The contribution of molluscan mucus to POM[15] and its role in shallow marine ecosystems is virtually unknown, but could be significant (Davies & Hawkins 1998). In a recent study of sedimentary organic matter remineralisation, gastropod mucus in the sediment was found to accelerate the rate of remineralisation of nitrogen by 29% (Hannides & Aller 2016).

Although readily degraded by bacteria, slime trails persist for some time on surfaces where gastropods or chitons have been crawling and may affect microbial film composition and settlement of various marine organisms (e.g., Connor 1986; Holmes et al. 2002).

3.3.2.1 The Role of Pedal Mucus in Locomotion and Adhesion in Gastropods

As noted above, mucus is used for many purposes in molluscs, but here we examine the role of mucus in locomotion. The remarkable properties of pedal mucus enable gastropods to adhere to many surfaces by coating them with a thin (10–20 μm) layer of mucus (Denny 1980a, 1980b, 1981; Shirtcliffe et al. 2012). This layer of mucus between the foot and the substratum has several roles. It provides the liquid in which the propulsive cilia beat, and it can also act as an anchor and/or adhesive. As the animal moves, it leaves a trail of mucus which can often be followed by conspecifics or sometimes by predators (see Chapter 9).

Pedal mucus is extremely important in both cilial and muscular locomotion. In addition, in creeping molluscs, the foot enables the animal to effectively clamp to the substratum or climb or clamp on steep surfaces. The adhesive power of the foot is a combination of the area of attachment and the properties of the pedal mucus. Increased tenacity is associated with low mucous secretion. The adhesive properties of the mucus can be very efficient, with the force required to remove patellid limpets ranging from 1.03 to 5.18 kg/cm² (Branch & Marsh 1978; Grenon & Walker 1981). Little is known about the chemical composition of the pedal mucus of chitons and

---

[15] Particulate organic material.

nothing about that of monoplacophorans, but there is some good information about gastropods (e.g., Trueman 1983).

As a muscular pedal wave (see below) moves across the sole of a gastropod foot, the mucus changes in fluidity. It is much more fluid (and thus not adhesive) beneath the moving wave than beneath the stationary parts of the sole where it is thick and adhesive (see Figure 3.31). The generation of the pedal wave (see Section 3.10.3) results in reduced pressure in that area causing the mucus to become fluid. The mucus then transforms back to its more solid gel phase at the trailing edge of the wave. This transition from fluid to gel states is very rapid and results in mucus acting as a ratchet-like material facilitating moving in the desired direction (Denny 1980a). That part of the sole that is the wave can move with respect to the rest of the foot due to this fluid mucus.

Pedal mucus is a dilute gel made up of a complex mixture of polysaccharides and proteins (Pawlicki et al. 2004), and yet it can produce strong adhesion to wet surfaces, despite being more than 95% water. Trail mucus is mainly composed of large molecules rich in carbohydrates and some relatively small proteins, and the adhesive form of the gel contains 2%–3% more protein than the dilute gel (Smith et al. 1999; Smith & Morin 2002). These proteins are thus an essential component of the glue and probably cross-link polymers in the gel to stiffen it (Pawlicki et al. 2004). Also important are transition metals including zinc, iron, and copper which are necessary for the proteins to act on the gel to form glue. The polymers in the gel are cross-linked with the involvement of metals (Smith 2002, 2006), apparently by catalysing the cross-linking (Werneke et al. 2007).

While the formation and composition of mucous secretions of few species have been examined in detail, there may be significant differences in gastropods from different major groups. For example, in the patellogastropod *Lottia*, there is about six times more protein than carbohydrate in both the trail and adhesive gels, while in the caenogastropod *Littorina* the adhesive form has approximately the same amounts of protein and carbohydrate, with two of the proteins being absent from *Lottia*. In a stylommatophoran slug and snail investigated by Pawlicki et al. (2004) the optimal protein component was between 6% and 27%. These differences, together with structural modifications, have been in part derived from variations in the habits and habitats of these taxa, but there is presumably also a significant evolutionary component, given their long separate histories. It will be of interest to test a wider range of taxa so the taxon-specific components can be separated from those resulting from adjustments to differences in habits.

Davies and Hawkins (1998) noted that the number of types of mucocyte differed in different taxa of marine gastropods. Nine types of mucocyte have been reported from the patellogastropod *Patella vulgata* and six from *Lottia testudinalis* (Grenon & Walker 1978), but only five in the caenogastropod *Littorina littorea* (Shirbhate & Cook 1987). Multiple types of mucins may usually be required to produce mucus with the necessary viscoelastic properties required for locomotion. However, a single kind of mucocyte on the foot sole of the terrestrial slug *Ariolimax columbianus* produced mucus

that can change its viscoelastic properties (Denny & Gosline 1981), although, in stylommatophoran snails and slugs more generally, the proteins necessary to form the glue-like mucus appear to be added from the 'protein glands' on the anterior margin of the foot (Wondrak 2012).

When crawling, gastropods have about a third of the tenacity they have when stationary (Miller 1974a). This is because, in limpets at least, glue-like adhesion is used when immobile for a period of time, and suction is used when they are active (Smith 1991, 1992).

AMPS (and carbonic anhydrase) from the foot and mantle edges of patellogastropod limpets can produce scars on the rock surfaces on which they rest (e.g., Lindberg & Dwyer 1983). These home depressions can be deepened by the limpet scraping the rock with its radula (see Chapter 9).

3.3.3 EPITHELIAL AND SUBEPITHELIAL GLAND CELLS

Gland cells associated with the epithelium can be lodged in the epithelium (*intraepithelial*) or sunk beneath it with a duct that extends through the epithelial cells to reach the surface. These *subepithelial* (or *extraepithelial*) cells may be unicellular or in clusters, and in the latter case, may have a common duct as in, for example, the pedal epithclium of the gastropods *Littorina*, *Veronicella*, and *Siphonaria* (Smith 2010). The epithelial gland cells are usually unicellular, but multicellular epithelial glands do occur. The gland cells that secrete substances onto the exterior of the body are called *exocrine* glands (as opposed to *endocrine* glands whose secretions are supplied to the interior of the body). Gland cells can also be categorised as *holocrine* and *merocrine* depending on their method of secretion. Holocrine secretions are formed in the cell cytoplasm and released by the cell membrane rupturing, destroying the cell. Merocrine secretions are released from the cell from small cytoplasmic vacuoles by way of exocytosis, and the cell continues to function.

The diversity of types of mucoid secretions is related to the diversity of gland cells that produce them. Most molluscan epithelia contain various kinds of secretory glands that produce different materials, but details about the products secreted by specific gland cells are limited to a small number of species. Studies are difficult because several types of glandular cells contribute to mucus production and most gels are the products of more than one type of gland cell mixed together (Smith 2010). Most detailed studies have focused on gastropods, with those on bivalves being mainly concerned with the mucocytes associated with the gills. Consequently, the text below is mainly based on information from gastropod studies.

The diversity of pedal glands in some gastropods was reviewed by Smith (2010). Of the nine kinds of gland cells found in the foot of the European limpet *Patella vulgata*, six are found in the sole, and some of those are subepithelial (Grenon & Walker 1978) (Figure 3.24). Several different gland cells produce either acidic or neutral mucopolysaccharides (see Section 3.3.2.1), proteins, and sulphated and non-sulphated sugars with some magnesium and calcium produced in the secretions (Grenon & Walker 1978, 1980). The subepithelial glands of the sole are the probable source of the adhesive gel (Smith 2010).

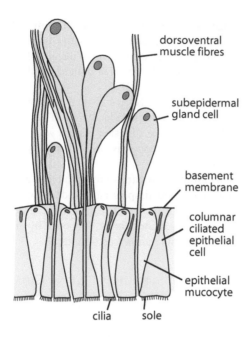

FIGURE 3.24 Epithelial and subepithelial gland cells in a marine gastropod foot sole. Redrawn and modified from Voltzow, J., Gastropoda: Prosobranchia, pp. 111–252, in F. W. Harrison & Kohn, A. J., *Microscopic Anatomy of Invertebrates. Mollusca 1*, Vol. 5, Wiley-Liss, New York, 1994.

Shirbhate and Cook (1987) examined the histochemistry of the pedal and opercular gland cells of three caenogastropod snails from markedly different habitats. The foot of the intertidal *Littorina littorea* has five types of mucus-secreting cells, all subepithelial (Shirbhate & Cook 1987). Two types comprise much of the anterior pedal gland that opens into a groove on the anterior edge of the foot. Each type secretes a different neutral mucoprotein, and one also has neutral mucopolysaccharides. The three types of subepithelial gland cells on the sole of the foot produce sulphated and carboxylated mucopolysaccharide and a mucoprotein. Glands on the dorsal surface of the foot secrete a sulphated mucopolysaccharide and a mucoprotein.

Three gland cell types make up the anterior pedal gland of the terrestrial littorinoidean *Pomatias elegans* and produce mucoprotein, protein, and sulphated mucopolysaccharide. In this terrestrial species, the sole of the foot lacks gland cells except in a median furrow where two cell types produce a neutral and a sulphated mucopolysaccharide. The dorsal surface of the foot has five cell types producing secretions that aid in reducing desiccation.

In the fresh-water truncatelloidean caenogastropod *Bithynia tentaculata*, the single cell type of the anterior pedal gland produces neutral and weakly acidic mucoprotein, while the foot sole has two types of gland cell that secrete AMPS and protein. There is also another cell type that produces carboxylated mucopolysaccharide that is restricted to a transverse band halfway down the sole of the foot. Dorsally, three types of gland cell produce various mucosubstances.

Mucus-secreting cells are also present in the epithelia of the mantle edge and foot of bivalves. Three types of secretory cells were found in the foot of the arcid *Tegillarca granosa*, which are all unicellular and merocrine, being typical of most secretory cells reported in other bivalves (Lee et al. 2012). A study of the venerid *Mercenaria* (Eble 2001) reported pedal gland cells containing acid glycosaminoglycans rich in sulphate and carboxylate groups. In another venerid, *Gomphina*, the pedal mucous cells contained mainly AMPS with carboxylate groups abundant in the acidic material (Park et al. 2012). Specialised mantle glands are found in the pedal apertures of some burrowing and boring bivalves (see Chapter 15).

Specialised defensive glands on the mantle are found in some gastropods including a few caenogastropods, notably the slug-like velutinids (see Chapters 9 and 19) and heterobranchs including several nudibranchs and marine 'pulmonates', notably siphonariids and trimusculids (see Chapter 20).

3.3.4 MODIFICATIONS FOR TERRESTRIAL LIFE

Terrestrial snails and slugs suffer osmotic stress less than fresh-water snails, but their main challenge is preventing water loss. This occurs through evaporation via the skin, locomotion (mucus), and also when discharging waste products (urine and faeces) (see Chapter 6). Terrestrial eupulmonates have developed mechanisms to prevent water loss via the body including various physiological, behavioural, and structural adaptations.

The composition of the slime of terrestrial slugs ranges from dense mucus produced by vesicular mucous cells to clear watery fluid containing little mucus. It also contains high molecular weight proteins too large to diffuse through the skin. Thus the secretions of the skin of stylommatophoran snails and slugs are a combination of mucous secretion from gland cells, large protein molecules, and watery products derived directly by way of ultrafiltration from the blood (Burton 1965; Deyrup-Olsen & Martin 1982). This water and the large molecules are released via large (~0.5 mm long) specialised *channel cells* that penetrate the skin in terrestrial stylommatophoran slugs and snails where they open to the exterior (Luchtel et al. 1984), thus differing from pore cells (see Chapter 6). These cells have a central channel and assist in regulating the body fluid by allowing water, large particles, and macromolecules to be passed directly to the environment (see Figure 3.25). They achieve this by having cell membranes with permeability greater than normal cells – a function facilitated by large pores that allow the big molecules, including haemocyanin, to pass through the cell walls (Simkiss & Wilbur 1977; Deyrup-Olsen & Martin 1982). The channel cells assist in restoring the volume of body fluid to normal levels following hyperhydration and, possibly, providing fluid for mucus secretion and body cleansing. The activity of channel cells depends on neural and/or endocrine[16] control of both

---

[16] The neurohormone arginine vasotocin stimulates the movement of both fluid and particles through the channel cell, and this is inhibited by norepinephrine (Luchtel et al. 1994).

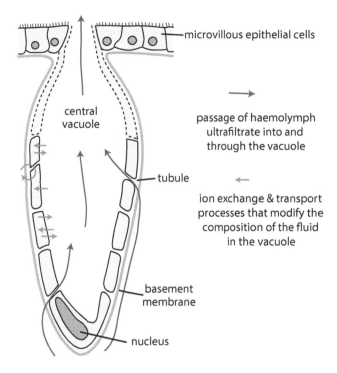

central
vacuole

→

passage of haemolymph
ultrafiltrate into and
through the vacuole

microvillous epithelial cells

tubule

←

ion exchange & transport
processes that modify the
composition of the fluid
in the vacuole

basement
membrane

nucleus

FIGURE 3.25 Channel cell from the dorsal integument of the terrestrial slug *Ariolimax*. Redrawn and modified from Luchtel, D.L. and Deyrup-Olsen, I., Body wall: Form and function, pp. 147–178, in G. M. Barker, *The Biology of Terrestrial Molluscs*, CABI Publishing, Wallingford, UK, 2001.

haemocoelic pressure and the permeability of the channel cells (Martin & Deyrup-Olsen 1982; Luchtel et al. 1984).

Eight types of gland cell occur in different regions of the skin of *Cornu aspersum* (Smith 2010). Of these, three types on the collar (mantle edge) appear to secrete the epiphragm (Campion 1961) (see also Section 3.3.2). They are all subepithelial and large, one of the most common being a mucous gland which secretes an acidic, probably sulphated, mucopolysaccharide and the gland contents appear to be reticulate. Another, less common, type of mucous cell which also secretes AMPS has granular contents and is like cells found on the foot sole. Calcium glands are also common on the ventral part of the mantle collar and contain rod-shaped or spherical granules in a protein matrix (Campion 1961). Also present on the collar are 'protein glands' whose watery contents contain protein, and these may actually be channel cells (Luchtel et al. 1984; Smith 2010).

The skin of slugs (Figure 3.26) has been better studied than other stylommatophorans. In *Ariolimax columbianus* there are three main types of gland cells in the dorsal surface – mucous cells, calcium cells, and channel cells (Luchtel et al. 1984), with the mucous and calcium cells secreting the adhesive mucus (Martin & Deyrup-Olsen 1986).

Less sticky watery mucus is found on the dorsal surface of slugs which lack calcium cells, suggesting that calcium may be the catalyst for rendering the mucus adhesive (Smith 2010) (see Section 3.3.2). Examples of slugs with only mucous and channel cells (i.e., no calcium cells) are the philomycid

Meghimatium (= *Incilaria*) *fruhstorferi* (Yamaguchi et al. 2000) and the limacid *Lehmannia valentiana* (as *L. poirieri*) (Arcadi 1967). In contrast, Cook and Shirbhate (1983) found in the dorsal epithelium of the body surface of another limacid, *Limacus ecarinatus* (as *Limax pseudoflavus*), two kinds of secretory glands (other than channel cells) out of 13, one of which secreted AMPS and an unusual type that secreted protein. All these gland cells are subepithelial, with their narrow necks extending between the epithelial cells (Smith 2010). In stylommatophorans, the peripodial groove surrounds the foot and is richly supplied with glands which, in the slug *Limacus ecarinatus* at least, consist of three types of gland cells (Cook & Shirbhate 1983). In addition, there are three distinct groups of secretory cells on the anterior edge of the foot, one on the dorsal anterior edge where the suprapedal gland duct opens, one a median gland, and the other an inferior gland. The suprapedal gland is a greatly enlarged anterior mucous gland found in all stylommatophorans. It extends posteriorly along the dorsal side of the foot as a tubular suprapedal gland. The whole gland comprises identical mucus-secreting cells (Wondrak 2012).

The systellommatophoran *Veronicella* has 11 types of gland cell in its skin and also, in response to irritation, produces a very sticky mucus on its dorsal surface, where there are two types of gland cells (Cook 1987), but at other times the surface appears to be dry. Unlike the stylommatophoran slugs mentioned above, the subepithelial gland cells secrete into a common duct that opens to the surface. The most abundant type of gland cell secretes neutral and weakly acidic, non-sulphated carboxylated mucopolysaccharides while the second cell type secretes protein. It is presumably the combination of these two secretions that results in the adhesive secretion released on the surface from the inflated common gland ducts when the slug is irritated (Davies & Hawkins 1998; Smith 2010). The foot mucus, secreted by gland cells on transverse ridges on the sole, is, as usual, a mixture of mucopolysaccharides and protein. There is also a very small suprapedal gland which produces a weakly acidic and a neutral mucopolysaccharide and a protein (Davies & Hawkins 1998).

In stylommatophorans, clusters of small gland cells are also often associated with chemoreceptive epithelia on the tentacles (e.g., Wondrak 1981) and Semper's organ, associated with the mouth (see Chapter 7), has three types of gland cells (Lane 1964).

3.3.5 Acid-Secreting Tissues

Some molluscs can produce strong sulphuric acid secretions such as seen in the skin glands of some 'opisthobranchs' (Edmunds 1968b) (see also Chapter 20), and cypraeoideans (Thompson 1969), in the acid glands of *Philine* (Thompson 1986), and in the salivary glands of tonnoideans and pleurobranchoideans (see Chapter 5). The secretory mechanism appears to be the same in all these cases and similar to the secretion of hydrochloric acid by mammalian oxyntic cells (Andrews et al. 1999). The epithelium in contact with free

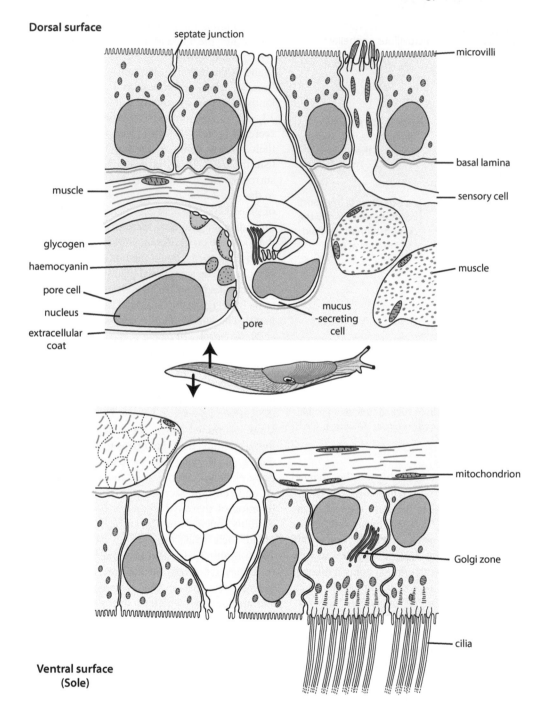

FIGURE 3.26 Diagrammatic sections through the dorsal and ventral (sole) epidermis of terrestrial slugs (based on *Arion and Deroceras*). Redrawn and modified from Newell, P.F., *Malacologia*, 16, 183–195, 1977.

acid secretion is protected by various possibly molecular devices and a coating of mucus, or cirri, or cilia that prevent the acid contacting the cell surface (Andrews et al. 1999).

3.4 MUSCULAR HYDROSTAT

Most movements made by molluscs involve forces generated by antagonistic muscle elongation and contraction against a hydraulic skeleton comprising the blood in the haemocoelic spaces. The dilation and contraction of parts of the body in this way is typically a fairly slow process, although contraction or withdrawal can be rapid if it involves retractor muscles. In many molluscan structures there are retractors, but obvious antagonists are often lacking (Russell-Hunter 1968). Thus, blood flow is reliant on antagonists that are often at some distance from the organ concerned and this feature, together with a constant blood volume in the haemocoel, results in some unusual features and limitations (Russell-Hunter &

Russell-Hunter 1968). Examples include limits to the number of structures that can be extended at any given time. For example, a copulating stylommatophoran land snail displaces a large proportion of its blood to expand the male genital structures, resulting in other parts of the body becoming flaccid.

The total volume of blood in most gastropods and bivalves is limited to that which can be contained in the animal when withdrawn into the shell, but inefficiencies can be overcome, at least in part, by dividing the haemocoel into compartments. If the blood in a particular structure is isolated from the rest of the haemocoel, the antagonistic muscles can be used more efficiently. Examples include the mantle wall of coleoid cephalopods where the radial muscles act against the main circular muscle system in the mantle wall (see Section 3.12.3.8.2). Similar systems occur, for example, in some bivalve siphons (e.g., Chapman 1995) and the proboscis introverts of higher caenogastropods (Golding et al. 2009b).

Mechanisms that sidestep these basic limitations are found in some molluscs. Bivalves with very large siphons and fused mantle edges use seawater contained in the temporarily sealed cavities of the mantle and siphons as an accessory hydrostatic skeleton. In these cases, the siphonal musculature is the antagonist of the shell adductor muscles (Russell-Hunter 1949; Chapman & Newell 1956; Russell-Hunter & Grant 1962). Another very different mechanism is the water-sinus system in naticids (see Section 3.8.2), where the resulting hydraulic

skeleton can be three to four times the capacity of the shell (Russell-Hunter & Russell-Hunter 1968).

3.5 ATTACHMENT OF THE BODY TO THE SHELL

The molluscan body is attached to the shell by muscles. These are mainly the foot or pedal muscles of chitons, bivalves, scaphopods, monoplacophorans, *Nautilus*, and gastropods (columellar muscles), and mantle muscles which attach to the shell (e.g., pallial line in bivalves). Other muscles such as the buccal retractors, head and gill muscles, etc. are also typically attached to the shell which provides a stable anchor point. Unusually, in the palaeoheterodont bivalves (Unionida and Trigoniida), small patches of the mantle also directly attach to the shell (Smith 1983).

It was thought that the muscles attached to the shell by way of an adhesive epithelium (Hubendick 1958), but when closely examined using a transmission electron microscope, it was seen that there was no direct attachment of the muscle to the shell. In gastropods and bivalves, a collagenous intercellular matrix and an epithelium (tendon cells) lie between the muscle and the shell. An extensive network of extracellular organic fibres extends from the microvilli of the epithelial tendon cells to the inner surface of the shell (often called a brush border in earlier literature) (Hubendick 1958; Mutvei 1964; Nakahara & Bevelander 1970; Plesch 1976; Tompa & Watabe 1976; Watabe 1988) (Figure 3.27). More recently Song et al.

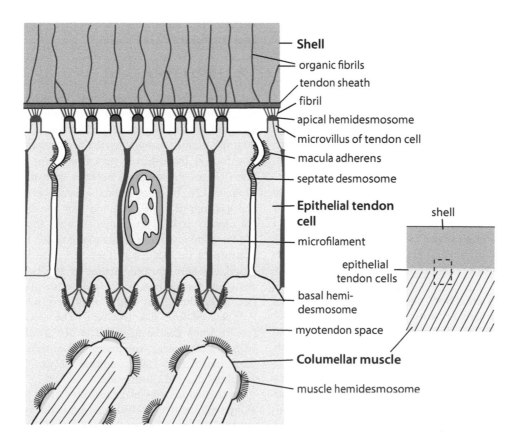

Shell
organic fibrils
tendon sheath
fibril
apical hemidesmosome
microvillus of tendon cell
macula adherens
septate desmosome
Epithelial tendon cell
microfilament
basal hemi-desmosome
myotendon space
Columellar muscle
muscle hemidesmosome

shell
epithelial tendon cells

FIGURE 3.27 Diagrammatic section through the attachment zone of the columellar muscle of a gastropod snail. Redrawn and modified from Tompa, A.S. and Watabe, N., *J. Morphol.*, 149, 339–352, 1976.

(2013) suggested that proteins might act as a glue binding the adductor muscle to the outer calcite layer of the muscle scar in the mussel *Mytilus*.

The mantle also adheres to the shell by tendon cells that contain tonofilaments (Bubel 1984). In bivalves, the attachment scar is indicated as the pallial line on the interior of the shell, although the attachment of muscles and mantle is not always marked on the shell – as for example, on the columella of neogastropods (Price 2003).

3.6 PIGMENTS IN TISSUES

In this section, we briefly outline the main pigments identified in molluscan tissues. For some explanation of the pigment types in shells, see Section 3.2.8 and Williams (2016). Useful treatments of pigments in molluscan tissues include the earlier reviews of Fox (1966), Goodwin (1972), Fox (1983), and the more recent overview of Bandaranayake (2006).

The main pigment classes recognised by Bandaranayake (2006) are carotenoids derived from plants and are responsible for yellow to red colours. They are present in most molluscs (e.g., Kantha 1989; Vershinin 1996). Carotenoproteins produce many colours, including blue, green, purple, and red and have been found in various tissues including the ovary and eggs of some molluscs (e.g., Heras et al. 2007).

Tetrapyrroles include the haemoglobins, haematins, and porphyrins. They play a primary role in oxidation and, secondarily, as pigments (Bandaranayake 2006), with porphyrins also being important as shell pigments (see Section 3.2.8). Bilichromes (or bilins) are metabolic products of porphyrins with colours varying between yellow, green, red, and brown.

Other pigments include the pheomelanins and eumelanins, diverse melanins that range from red to yellow and brown, and brown to black, respectively, and melanoprotein which comprises the ink of some coleoid cephalopods. Ommochromes are found in the chromatophores of coleoid cephalopods (Branden 1976), and indoles include the famous Phoenician dye Tyrian purple, which was produced from the hypobranchial gland of *Stramonita haemastoma* and other muricoidean taxa. Eumelanins are probably involved in the dark brown and black pigment patterns seen on the sides of the foot, mantle edge, and head of many gastropod taxa.

Pigments in the external body obtained from food can assist with crypsis and perhaps aposematism, as with shell colouration (see Section 3.2.8). For example, the aforementioned dark pigmentation of exposed body regions of gastropods probably reduces the threat of visual predators. Pigments in the external body tissues are also considered important in aposematism in some nudibranchs (Edmunds 1991; Haber et al. 2010) and in octopuses (Huffard et al. 2010; Mäthger et al. 2012). Pigment-containing chromatophores under nervous

and muscular control are responsible for rapid colour changes in coleoid cephalopods (see Chapter 17).

3.7 NON-CALCAREOUS STRUCTURAL ELEMENTS IN THE BODY

3.7.1 CHITIN – A STRUCTURAL POLYSACCHARIDE

Chitin, a long-chain polymer of the monosaccharide *N*-acetylglucosamine, which is derived from glucose, is found in the cell walls of fungi, the exoskeleton of arthropods, and in various molluscan structures including the radulae and jaws, and inside epithelial cells lining the body and parts of the gut of aeolidioid nudibranchs (see Chapter 20).

3.7.2 CARTILAGE – STRUCTURE AND HOMOLOGY

Cartilage tissue occurs in several invertebrate groups other than Mollusca, including the Chelicerata, Annelida, and Brachiopoda (Cole & Hall 2004; Schmidt-Rhaesa 2007). In molluscs, cartilage occurs in the buccal or odontophoral cartilages in all classes other than bivalves, and it is also found encapsulating the brain of cephalopods. Although in general the cellular structure of cartilage is rather similar in all invertebrates that possess it, there are some differences. For example, there are differences in the size and number of vesicles in the cells, and squid head cartilage cells have long cytoplasmic extensions into the matrix (Philpott & Person 1970) that are not found in odontophoral cartilages or in the cartilage seen in other invertebrates (e.g., Kryvi 1977). Odontophoral cartilages are described in Chapter 5.

3.8 HEAD-FOOT

Like most bilaterally symmetrical organisms, molluscs have undergone cephalisation with a concentration of nervous tissues and sensory organs at the anterior of the animal, along with the mouth, comprising the head region of the mollusc. The foot is immediately posterior to the head region and often shares dorsoventral muscles with the head (see also Chapter 12). In three groups – scaphopods, cephalopods, and gastropods – the head and foot are distinctly separate, but closely linked, structures – hence the term head-foot. We introduce two new terms to recognise the different relationship of the head with the shell. The *ligocephalic* condition is where the head region is attached dorsally to the shell, and buccal muscle scars are often present on the interior of the shell. This condition is seen in Monoplacophora, Polyplacophora, and Bivalvia. The *apocephalic* condition is when the head region is not attached to the shell. Buccal muscle scars are absent except those integrated with shell muscle(s). This condition occurs in Cephalopoda, Scaphopoda, and Gastropoda.

In shelled taxa (bivalves excepted), it is the head and foot, along with the mantle edge, that are in direct contact with the external environment.

3.8.1 The Head Region

Aplacophorans have a poorly differentiated, eyeless head region with a terminal or subterminal mouth. Solenogasters have a sensory vestibule located anterodorsally to the mouth and, when moving, the head region is held above the substratum and is motile. Caudofoveates have a sensory oral shield that may assist with food detection, but is also involved in burrowing.

In both chitons and monoplacophorans, the head region is constrained by its attachments to the shell. It is small, eyeless, and situated anterior to the foot, from which it is little more than a protrusion to carry the ventral mouth. The head is generally simple in chitons. An oral disc surrounds the mouth and may have extended lips or lappets (e.g., the carnivorous chiton *Placiphorella*). In *Lepidopleurus* there is a dorso-lateral hood surrounding the oral disc similar to the 'velum' structure found in monoplacophorans (Lemche & Wingstrand 1959). Monoplacophorans also have a posterior pair of tentaculate post-oral appendages and a small preoral tentacle between the 'velum' and the anterior mantle fold.

Bivalves are the only molluscan group which lacks a distinct head region. The only remaining vestiges of the head region are the mouth and its associated labial palps, with ciliated tracts which assist in sorting food particles captured either directly (protobranchs) or indirectly via the ctenidia (autobranchs). Because of the lack of a head and associated sensory structures with which to assess the external world, bivalves mainly utilise the mantle edge to gain sensory information. Many have sensory papillae or tentacles along the mantle edge and/or siphons, and in a few taxa (e.g., many scallops) there are eyes, some of which have double retinas and guanine crystal mirrors (Palmer et al. 2017) (see Chapter 7). Eyes are also developed on the ends of siphons in a few taxa. For further details on the sensory equipment of bivalves see Chapters 7 and 15.

The remaining three taxa – Scaphopoda, Gastropoda, and Cephalopoda – all have distinct, detached heads, those of gastropods and cephalopods being especially well-developed. The detachment of the head from a dorsal exoskeleton reorients head musculature, facilitates withdrawal into the shell, and appears correlated with the development of additional cephalic sense organs.

The scaphopod head consists of a large 'snout' (often called a 'proboscis') that is not homologous with the snout seen in gastropods because it is formed anterior to the jaw. A series of long, hair-like tentacles (captaculae) that emerge laterally from the head is unique to scaphopods (see Chapters 5 and 16). Steiner and Dreyer (2003) compared the captaculae to cephalopod arms, and Wanninger and Haszprunar (2002a) concluded that they also resemble the cephalic tentacles of gastropods.

The gastropod head bears a short to elongate snout with the jaws just inside the distal mouth (a synapomorphy of gastropods), a pair of tentacles (cephalic tentacles), and a pair of laterally located small eyes. In a few groups of gastropods one of the cephalic tentacles is modified as a penis, but, in internally fertilising taxa other penial structures are usually present. There are sometimes additional tentacles associated with the foot and mantle edge. The various types of tentacles found in gastropods are discussed in more detail in Chapters 7 and 18–20. In some higher gastropods, the snout is inverted to form a retractile proboscis (introvert) (see Chapter 5).

Cephalopods have a pair of large lateral eyes, and the mouth is surrounded by a series of tentaculate 'arms' (up to 90 in *Nautilus* or 8–10 in coleoids). In coleoids, these may bear suckers or hooks and, if there are ten, a pair are modified as retractile tentacles. *Nautilus* 'tentacles' or cirri lack suckers. The derivation of the arms appears to be from the foot (Shigeno et al. 2008), as discussed in more detail in Chapters 8 and 17. In male coleoid cephalopods, one arm (the hectocotylus) is modified as a copulatory structure.

3.8.2 The Foot

The aplacophoran Caudofoveata lack a foot while the Solenogastres have a very narrow foot (retracted within the ventral pedal groove in preserved specimens) which consists of a non-muscular fold or folds and is ciliated and richly supplied with mucous glands. Movement is by ciliary creeping on mucus produced by a ciliated, eversible pedal pit that lies at the anterior end of the pedal groove. There is evidence that the foot reduction or loss in the aplacophoran groups is secondary (Ivanov 1996; Scheltema 1996).

In polyplacophorans and monoplacophorans the foot has a broad, creeping sole capable of clamping to hard surfaces. In both groups it is surrounded by a mantle groove containing the gills, and the groove is covered externally by a fleshy girdle in chitons and the shell in monoplacophorans.

The foot in gastropods is particularly well-developed, mobile, usually flexible and, in orthogastropods, divided into an anterior propodium and a larger metapodium. Dorsally, there is usually an opercular lobe which secretes the operculum (see Section 3.8.3). A lateral fringe, the epipodial fringe, is present in many vetigastropods and typically bears tentacles and epipodial sense organs (see Chapters 7 and 18). In caenogastropods, other tentacles are associated with the foot, and an opercular lobe is sometimes also present. Lateral extensions of the foot form the parapodia of several groups of euopisthobranchs such as Atyidae, Hydatinidae, Akeridae, Aplysiidae, and Gastropteridae. The wings of the pelagic pteropods are modified parapodia. The other completely pelagic gastropod group, the heteropods, have the foot modified as a keel-like 'fin'. The hydrothermal vent snail *Chrysomallon squamiferum* has developed iron-mineralised scales on the sides and top of the foot (see Chapters 9 and 11).

The foot can, particularly in some burrowing caenogastropods, be markedly expanded. In naticids, this is achieved by taking up external seawater into an extensive pedal watersinus system to expand the foot (Morris 1950; Bernard 1968; Russell-Hunter & Apley 1968; Russell-Hunter &

Russell-Hunter 1968). The external pores of the water-sinus system are in the mesopodium so that during expansion, seawater must be pumped from the mesopodial water-sinuses to those of the propodium. Thus, pedal expansion cannot occur in naticids on dry surfaces, and even in water expansion is slow, taking several minutes, although retraction takes only a few seconds (Russell-Hunter & Russell-Hunter 1968). It is possible that this unique water-sinus system in the foot of naticids could be derived from epidermal invaginations such as seen in the infolded pedal mucous glands of some other caenogastropods (Russell-Hunter & Russell-Hunter 1968). The very large foot of the surf beach nassariid *Bullia* is expanded largely due to haemocoelic pressure, but some minor water uptake may also occur (Brown 1964). Mangum et al. (1978) showed that the foot in *Busycon carica* is expanded by seawater uptake, and that blood and water mix freely in the foot, greatly diluting the blood. The blood is reconcentrated (presumably by the kidney) before returning to the gill and mantle. When the foot is fully retracted into the shell, blood and water are both expelled to enable the body of the whelk to fit within the shell.

In scaphopods and many bivalves, the foot differs from the broad creeping surface typical of chitons, monoplacophorans, and gastropods in being capable of digging in sediment using hydrostatic pressure generated through haemocoelic spaces by enlarged anterior and posterior pedal retractor muscles. The bivalve foot is often large, typically laterally compressed (modified for digging), rarely expanded laterally for crawling, or very reduced or even lost (in adults). In protobranch bivalves, the foot is divided ventrally to form two ventral keels fringed with papillae in most species. The keels can be expanded laterally to act as an anchor in the sediment. In autobranch bivalves, the foot is often hatchet-shaped or rounded and sometimes a small sole-like flattening is present ventrally, especially in the posterior part of the foot, while in most cementing bivalve species the foot is reduced.

In scaphopods, the foot lacks pedal glands and differs in the two main groups, Dentaliida and Gadilida. In the former, the conical foot has well-developed smooth epipodial lobes laterally and ventrally that anchor the foot in the sediment. In the Gadilida, the terminal disc of the foot is often fringed with papillae.

In cephalopods, the foot is so highly modified it cannot be recognised as such, although developmental studies have shown that the arms are derived from the foot (Shigeno et al. 2008) (for further information see Chapters 8 and 17). The muscular funnel directs a stream of water expelled from the muscular mantle cavity, notably during jet propulsion (see Chapter 17 for details).

3.8.2.1 Pedal Glands

Besides epithelial sole glands, glands near the anterior end of the foot in aplacophorans, chitons, monoplacophorans, and gastropods secrete mucus to lubricate the sole during locomotion (see Section 3.3.2.1). Anterior pedal glands are also found in a few bivalves, but these are probably not homologous with those of other molluscs or even within bivalves (Waller 1998) (see Chapter 15).

The anterior pedal gland is particularly well-developed in many gastropods and in some extends deep into the foot. In stylommatophorans, it is developed as the suprapedal gland – a long tubular gland that runs between the viscera and the muscular part of the foot.

While all gastropods have gland cells opening to the sole of the metapodium, in some these are concentrated in a large posterior pedal (or metapodial) gland that opens via an aperture or slit in the sole.

A modified posterior pedal gland, the byssal gland, is found in the foot of many adult autobranch bivalves and in all autobranch bivalve larvae (for more details see Section 3.11.1 and Chapter 15).

The musculature of the foot is described below (see Section 3.12.3.4).

3.8.3 OPERCULUM

Opercula occur in Mollusca, Hyolitha, Bryozoa, and Annelida, and appear to represent multiple derivations of this protective structure, even within phyla (e.g., Brinkmann & Wanninger 2009). In the Mollusca, opercula occur in Cephalopoda – nautiloids and ammonites – and in the Gastropoda, all apparently independently derived. In the cephalopods, ammonites had a pair of horny, and sometimes calcified, lower jaw plates (the aptychi), which closed the aperture of the shell when the animal withdrew. In living *Nautilus*, two modified cirri form a leathery hood structure which closes the shell aperture when the animal withdraws into the body chamber (also see Chapter 17). The gastropod operculum is generated on the dorsal surface of the foot and is composed of conchiolin and sometimes calcified. Some early workers considered the gastropod operculum a second shell valve and equivalent with the bivalve shell (see review in Fretter & Graham 1962, p. 76), but the production of the supposed second valve by the foot epidermis, not the mantle, shows it is not a homologue.

The gastropod operculum is a readily identified synapomorphy of the group. It usually first appears in the larval stage; it is absent in the larvae of pleurobranchs, but present in most nudibranch larvae, although it is lost at metamorphosis (Gibson 2003). An operculum is also absent in direct-developing embryos of some gastropods, including stylommatophorans and vaginulids (G. Barker pers. comm. Nov. 2014). In adult gastropods, an operculum is present in most vetigastropods and caenogastropods, in lower heterobranchs, and some basal euthyneurans. It is absent in adult patellogastropod, cocculinoid and lepetelloid limpets, most euthyneurans, and some taxa in a number of vetigastropod and caenogastropod families. The operculum is also absent in some terrestrial neritimorphs (Proserpinidae and Ceresidae), while limpet-like neritimorphs (e.g., *Phenacolepas* and *Septaria*) have a reduced operculum embedded in a fleshy pocket.

Three basic types of gastropod opercula are usually recognised: multispiral (polygyrous) with numerous spirals and a central nucleus, paucispiral (oligogyrous) with few spirals, and concentric. This latter type is not spiral, and its nucleus can be situated internally, marginally, or terminally. Spiral opercula (whether multispiral or paucispiral) are secreted at the internal edge of the opercular disc (see below), and the operculum rotates as it grows. Based on the outside spiral pattern, the rotation is always clockwise in dextral and counter clockwise in sinistral taxa. Concentric opercula grow differently – new material is added along all or part of the periphery of the operculum, but these seemingly neat distinctions are not as clear-cut as often thought. For example, some opercula change from an initially spiral pattern to become concentric, as in bithyniids. There is also considerable variation in the number of spirals and how tightly they coil so there is not a clear distinction between multispiral and paucispiral opercula.

Calcification of the operculum has occurred multiple times in gastropod evolution. The earliest fossils sometimes interpreted as calcified opercula are the Early Ordovician fossils named *Ceratopea* (Yochelson & Bridge 1957), but these cap-shaped shells are probably not opercula (Yochelson & Wise 1972). Typical multispiral calcified opercula are known as early as the Middle Ordovician, while *Maclurites* from the same period have a paucispiral operculum (Yochelson & Linsley 1972).

The operculum is attached to a disc-like epithelium, on the dorsal surface of the foot, called the opercular disc. It is attached to only part of the operculum, not the whole under surface. Secretory cells in the opercular groove and disc generally secrete mucopolysaccharide and at least two differently staining proteins (Shirbhate & Cook 1987).

The operculum is composed of conchiolin, mainly a homogeneous scleroprotein hardened by quinone tanning. All opercula examined show a similar pattern of amino acids with a high glycine and aspartic acid content, although the varnish has higher levels of glycine than the other layers (Hunt 1976).

The operculum consists of several layers (Figure 3.28), which are typically – from outer to inner side – a thin outer hyaline layer, a thicker main layer, and the adventitious and varnish (shiny) layers (Checa & Jiménez-Jiménez 1998). While the main and varnish layers are always present, a hyaline layer is found in spiral and concentric opercula with the nucleus on the labral edge, and adventitious layers are only found in concentric opercula. In some taxa, one or more calcareous (calcitic or aragonitic) layers are added.

The different layers are produced in different ways. The hyaline layer (when present) is secreted in a slit-like structure of the foot, the opercular groove. On the extended foot, this structure lies at the anterior edge of the opercular disc and runs across the main body axis, and the material it secretes bears similarities to the periostracum. When the foot is retracted, the opercular groove coincides with the inner edge of the aperture. The main layer of the operculum lies below the thin hyaline layer (when present) and is secreted by tall cells at the edge of the opercular disc. Varnish may be secreted by a second groove that lies immediately behind the opercular groove, as in some trochids, but in others (including concentric opercula), it is secreted by a flap on the posterior edge of the opercular disc. In concentric opercula, layers are added posteriorly from the surface of the opercular disc, each added at the bottom of the operculum (see Checa & Jiménez-Jiménez 1998 for more details and references).

In the only detailed investigation of the operculum in recent years, Checa and Jiménez-Jiménez (1998) recognised two main kinds of spiral opercula – those that closely match the aperture in shape (*rigiclaudent*) and those that do not (*flexiclaudent*), as in opercula of vetigastropods like Pleurotomariidae and Trochidae where the operculum fits the aperture by the edges bending outward. These authors distinguished three main types: (1) flexiclaudent spiral (mostly multispiral) operculum, the shape of which does not fit the aperture, but fits by flexing into the aperture, (2) rigiclaudent spiral (fitting the aperture and may be multispiral or, more usually, paucispiral), and the aperture is used as a template when growing, and (3) rigiclaudent concentric, which also typically fits the aperture unless it is reduced in size. The flexiclaudent spiral type is secreted when the foot is extended. The other two types which do not flex upon retraction, except perhaps at their outermost margins, grow when the operculum is held in the aperture, and this growth may or may not be spiral (i.e., the operculum rotating as it grows). As already noted, there are many intermediate situations between the distinctions in form, structure, and formation described above.

Opercular calcification can occur only in rigiclaudent opercula, but is achieved in several ways. For example, in turbinids a thick calcareous layer is added on the outer surface of a corneous paucispiral operculum. This is achieved by a fold of epithelium that lies at the anterior (when the foot is extended) edge of the opercular groove (Figure 3.28). The precipitation of calcium occurs when this fold partially or completely covers the operculum when the animal is crawling, but it is retracted and the operculum exposed when the foot is retracted. Other types of calcified spiral opercula occur in some neritoideans and naticids. In naticids with calcareous opercula, the labial mantle margin, not the foot, deposits the calcium (as aragonite) onto the otherwise corneous operculum (Checa & Jiménez-Jiménez 1998). Some concentric opercula are calcareous (e.g., Bithyniidae, Pomatiidae), where organic and calcareous layers are closely packed together.

A flexiclaudent (mostly multispiral) and rigiclaudent spiral are generally the types present in vetigastropods, with the former predominant. Neritimorphs have a rigiclaudent operculum, and this is also the usual type in lower caenogastropods, although a few cerithioideans have flexiclaudent spiral opercula. Higher caenogastropods mostly have concentric opercula, and this is the only kind found in neogastropods. Rigiclaudent spiral opercula are found in nearly all lower heterobranchs and the few operculate euthyneurans (Checa & Jiménez-Jiménez 1998), with exceptions being concentric opercula in Rissoellidae and *Retusa*.

While the distribution of the different kinds of opercula in gastropods has a phylogenetic basis, this is modified by changes imposed by shell morphologies. For example, the depression of the spire leading to planispiral or near planispiral shell morphology will often result in oval paucispiral opercula becoming circular with more spirals, while an elongated aperture with a canal will result in a concentric operculum (Checa & Jiménez-Jiménez 1998).

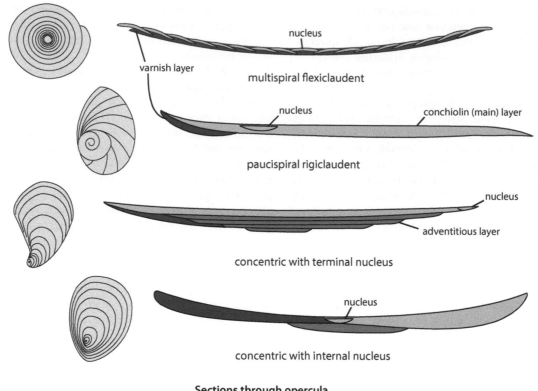

Sections through opercula

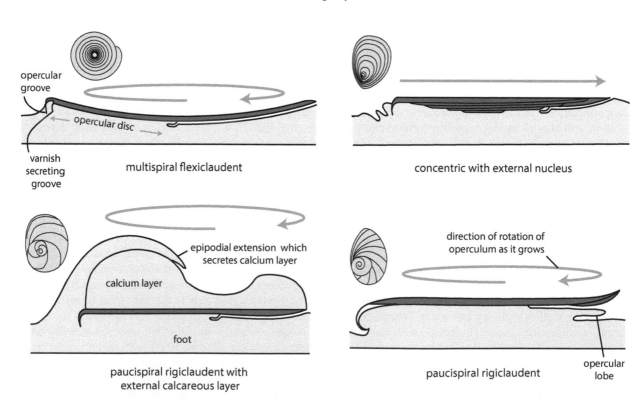

FIGURE 3.28   The structure of various types of gastropod opercula, as seen in cross-section, and its relationship to the foot and opercular disc. Uppermost section redrawn and modified from Checa, A.G. and Jiménez-Jiménez, A.P., *Paleobiology*, 24, 109–132, 1998; next three and lower right redrawn and modified from Fretter, V. and Graham, A.L., *British Prosobranch Molluscs: Their Functional Anatomy and Ecology*, Ray Society, London, UK, 1962; based on Houssay, F., *Archives de zoologie expérimentale et générale*, 2, 171–288, 1884 and Kessel, E., *Zeitschrift fur Morphologie und Okologie der Tiere*, 38, 197–250, 1942; three of lower four redrawn and modified from Checa, A.G. and Jiménez-Jiménez, A.P., *Paleobiology*, 24, 109–132, 1998.

Various theories have been proposed to account for the operculum on the surface of the foot, and how it obtained the ability to secrete the conchiolin which forms the operculum. This substance is chemically like the periostracum and Checa and Jiménez-Jiménez (1998) proposed a 'periostracum shaving model' to account for the origin of the operculum based on the morphology of the flexiclaudent spiral operculum, which appears to be the most primitive type. They argued that in the parietal segment of the mantle the periostracal groove became independent from the rest of the mantle and migrated on to the foot providing the necessary organic material to form an early form of the operculum. With the development of the opercular disc, the flexiclaudent spiral operculum could be formed from the periostracal strip produced by the opercular (i.e., modified periostracal) groove. These opercula do not fit the aperture shape, but this was rectified by the subsequent independent evolution of the rigiclaudent type in several lineages. The evolution of rigiclaudent opercula also preceded their calcification. A greater range of apertural modifications (such as a siphonal canal) became possible and greatly increased with the evolution of the concentric operculum and its wide range of possible shapes (Checa & Jiménez-Jiménez 1998).

3.9 THE MANTLE

Along with the head and foot, the mantle edge has direct contact with the external environment and thus plays a significant role in the interactions of the animal with that environment. The mantle, sometimes called the pallium, is one hallmark of molluscs. It covers the dorsal body, including the mantle cavity, and carries out several functions – the formation of the shell and/or spicules, the reception of sensory information, and often a significant respiratory surface. It is also responsible for regulating water currents entering or exiting the mantle cavity or mantle groove.

The mantle consists of two main parts: the part that typically lies beneath the shell, or in those molluscs where the shell has been reduced or lost, the part that lies over the surface of the body. There is some confusion as to the terminology associated with the mantle, and its extent – for example, in gastropods some (e.g., Hyman 1967) restrict the mantle to the last whorl of the shell, while many others (e.g., Fretter & Graham 1962) also use the term for the dorsal epithelium covering the viscera. Here we treat the mantle as the epithelium that forms the entire dorsal covering of a mollusc body (except for the head and foot) together with, if they are present, the protruding edges. The periostracum and the growing edge of the shell are secreted by the mantle edge. The mantle underlying the shell secretes additional material to thicken or repair the shell from the inside. We use the terminology outlined below for the parts of the mantle.

3.9.1 DORSAL MANTLE

The thin epithelial layer ('skin') covering the visceral part of the body. In shelled molluscs, it lies against the inner surface of the shell. It is in contact with the extrapallial fluid between the mantle and the shell (i.e., that part lying beneath the shell) and can secrete shell material to thicken the overlying shell, or in some taxa, dissolve the shell. For example, in taxa such as ellobiid and neritid gastropods, the inner chambers of the shell are dissolved, making the calcium available to the animal (see Section 3.2.19). In shell-less sea slugs, the dorsal surface (plus the extending edges) is often referred to as the *dorsum* or *notum* (which, depending on its derivation, may be homologous with the dorsal mantle), and the ventral surface beneath the protruding edges is called the *hyponotum*. In systellommatophoran slugs, the hyponotum, along with the foot, forms part of the ventral creeping surface. This is demarcated from the notum by a sharp angulation (or sometimes a groove) called the *perinotum*. In some shell-less taxa the dorsal surface is formed from a reflected mantle edge or parapodial flaps from the foot (see Section 3.2.15).

Other modifications of the dorsal mantle include epidermal papillae in chitons and aplacophorans which are possibly plesiomorphic for Mollusca (see Scheltema 1993), while in some shell-less gastropods (slugs), it can be produced into elaborate projections. The tubules from the dorsal mantle in some gastropod and bivalve taxa extend into the shell forming shell pores (see Section 3.2.9), while in chitons, the aesthetes that penetrate the shell are formed from nervous tissue (see Chapter 14).

3.9.2 MANTLE SKIRT

This term is used for the parts of the mantle that extend beyond the visceral part of the body, particularly that cape-like part of the mantle that forms the external roof of the mantle cavity or, as in nudibranch slugs, the overhanging part of the dorsum. The inner epithelial layer of the mantle skirt, which is often modified as structures such as hypobranchial glands and/or ciliated fields or ridges (see Chapter 4), is not part of the mantle.

3.9.3 MANTLE EDGE

The free edge of the mantle skirt and any protrusions comprise the *mantle edge*. It usually has two or three lobes, one of which may bear tentacles, and, in some taxa, one lobe may be expanded to cover some or all of the shell. The mantle edge consists of two layers of epithelia separated by a space containing some muscle, connective tissue, and blood spaces, and the lower epithelial layer is in contact with the external medium. Parts of the mantle edge are responsible for the secretion of the periostracum and the initial shell material deposited at the growing edge of the shell. The mantle edge is also responsible for the control of water currents entering and leaving the mantle cavity by way of controlling the size of apertures with folds, flaps, or siphons. In some bivalves, the closure of much of the mantle edge confines and directs the water flow (see Chapter 15 for details). Because it is in contact with the external environment, the mantle edge is an important sensory surface and provides a major part of the sensory equipment in many molluscs. Many elaborations, including tentacles and eyes, enhance this function (see Chapter 7), particularly in bivalves because they lack a head. It also has other secretory functions including being a source of mucus secretion in many molluscs and, in some, has developed repugnatorial secretory structures that deter predators. For example, assumed distasteful secretions of the mantle edge in the limpet *Cellana* deter predators and competitors (Branch & Branch 1980) and those from the repugnatorial glands of *Trimusculus* anesthetise potential predators (Rice 1985). There are also examples in other gastropods, and notably nudibranchs, where noxious chemicals are sometimes concentrated on the mantle edge (see Chapter 20).

The mantle edge has one or more folds which are elaborated differently in the major groups of molluscs (Figure 3.29), as outlined by Beedham and Trueman (1967) and Stasek and McWilliams (1973).

In polyplacophorans, the mantle skirt and edge form a girdle around the lateral edges of the eight valves. It is composed of three regions. A small fold on the proximal upper surface secretes the shell and is separated from the main mantle rim by the periostracal groove. The main mantle rim comprises the girdle and secretes the spicules, hairs, or scales. The third component is a low ridge on the underside of the mantle edge that contains mucous glands. It has been suggested that the part of the mantle edge in chitons between the proximal wall of the properiostracal groove and the shell edge is homologous with the conchiferan outer mantle fold (Haas 1972; Stasek 1972).

Aplacophorans are unlike other molluscs in having a spicule-producing mantle surrounding a worm-like body with no distinct mantle edges. Although this condition has been considered primitive for molluscs (e.g., Salvini-Plawen 1980a; Haas 1981), it is only one of several possible evolutionary scenarios (see Chapter 12).

The monoplacophoran mantle edge has a periostracal groove situated ventrally some distance inside the mantle (and hence shell) edge and an inner ventral fold, while the middle fold is less distinct (Haszprunar & Schaefer 1997).

In all bivalves, the mantle edge is comprised of three folds (Stasek & McWilliams 1973) (Figure 3.30). In some galeommatoideans, the middle lobe is expanded over the outer shell surface. In other bivalves these folds fuse in various combinations to close the mantle and form siphons (see Chapter 15 for more details).

An indication of the plesiomorphic state of the mantle edge in cephalopods can be seen in *Nautilus* where it is trilobed, like the situation in bivalves, with the periostracal groove situated between the outer lobe (that secretes the shell) and a middle lobe or fold. There is also an inner fold (Westermann et al. 2005). In coleoids, the mantle has covered the body by an extension of the mantle rim. Thus, the exposed dorsal surface of coleoid cephalopods is not derived from the dorsal mantle, but from the mantle edge.

Despite the size and diversity of the group, gastropods have, with a few exceptions, a relatively uniform arrangement of the mantle edge (Figure 3.30). In most, including patellogastropods, the mantle edge forms a single lobe that secretes the shell, and the periostracal groove is situated adjacent to the shell edge. In many limpet-like orthogastropods, there is

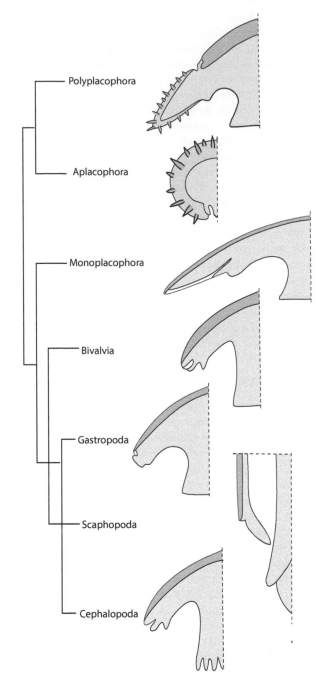

FIGURE 3.29 A diagrammatic comparison of the morphology of the mantle edge of a typical bivalve and a typical gastropod. Redrawn and modified from Stasek, C.R. and McWilliams, W.R., *Veliger*, 16, 1–19, 1973. Monoplacophoran redrawn and modified from Schaefer, K. and Haszprunar, G., *Zool. Anz.*, 236, 13–23, 1997.

a prominent inner fold or swelling, but the largest deviation from the normal gastropod pattern is seen in fissurellids, where the mantle edge is distinctly trilobed. In some species, a well-developed middle fold envelops part of the shell, and a larger ventral fold forms a fleshy curtain below the shell. In other gastropods with the mantle enveloping part or all of the shell, it is the single mantle lobe that is extended (Figure 3.30).

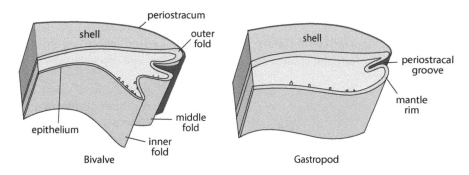

FIGURE 3.30 Mantle edge morphology in molluscs redrawn and modified from Stasek, C.R. and McWilliams, W.R., *Veliger*, 16, 1–19, 1973.

Scaphopods have a large, fleshy lobe-like extension to their anterior mantle edge that assists in closing off the aperture. The periostracal groove lies adjacent to the shell edge.

3.10 LOCOMOTION

Molluscs move by crawling, burrowing, or swimming. In those that crawl or burrow, and in some that swim, the foot is the main organ of locomotion. As deftly expressed by Kappes and Haase (2012), 'The Mollusca provide proverbial examples for "sluggishness" and time-consuming movements "at a snail's pace"'. Chase (2002) argued that the famously slow movements of gastropods are largely due to the lack of a hard, internal skeleton, thus limiting the amplification and antagonism of muscular action. While there is some truth in this, faster movements do occur in some molluscs, such as the leaping movements of strombid gastropods, the aggressive territorial responses of some limpets, the swimming of some shell-less heteropods and pteropods, the rapid jetting of many cephalopods, and the energetic burrowing rates of some bivalves (e.g., *Donax*). Simple rate-based values of distance travelled do not adequately describe the breadth of the roles of locomotion, all of which depend on a complex and flexible system of behaviours. For example, the four basic drives (Four Fs) of *Aplysia* all include locomotion (Dickinson 1996; Moroz 2011).

The broad, creeping foot of chitons, limpets (and presumably monoplacophorans), provides for both adhesion and locomotion. Chitons, limpets, snails, and slugs creep slowly by cilia or waves of muscular activity (pedal waves) that move along the foot. The eight independent shell plates of chitons allow them to adhere and move over highly irregular surfaces that would be difficult for a single-shelled gastropod. Backward locomotion is fastest in polyplacophorans (e.g., *Chiton, Ischnochiton, Cryptochiton*) (Parker 1914; Boyle 1977; Kuroda et al. 2014). Miller (1974b) found backward locomotion in gastropods to be rare, and when present it was slow and of limited duration. Backward locomotion has been reported in both patellogastropods and vetigastropods (e.g., *Patella, Lottia, Fissurella, Monodonta*) (Olmsted 1917; Jones & Trueman 1970; Kuroda et al. 2014).

3.10.1 CILIARY AND MUSCULAR LOCOMOTION USING A CREEPING FOOT SOLE

An early study by Vlés (1907) provided a classification and terminology for the different modes of locomotion in gastropods which were subsequently modified by others (e.g., Olmsted 1917; Miller 1974b) (see Table 3.4). Much of the review below is derived from the detailed studies of gastropod locomotion conducted by Susanne Miller (Miller 1974a, b), who investigated modes of locomotion, speed, tenacity, and foot form of nearly 300 species in 52 families of marine gastropods (other than Heterobranchia). Her work showed that the greatest diversity in locomotory modes occur within the caenogastropods, and that foot size and shape, and small differences in locomotory mode, affect the speed of movement and the strength of adhesion to the substratum.

3.10.2 CILIARY AND ARRHYTHMIC LOCOMOTION, INCLUDING LEAPING

Solenogasters have a long, narrow, reduced foot (see Section 3.8.2) and the animal moves by ciliary action along a thick mucus strand, which has been likened to a monorail (Salvini-Plawen 1968b; Scheltema & Jebb 1994).

Locomotion in some gastropods can be entirely due to the action of pedal cilia, which are highly coordinated and thought to be under nervous and/or neurosecretory control (e.g., Jékely 2011), and they produce a smooth, uniform, and rapid gliding. This type of locomotion is seen in most tiny gastropods, but also occurs in some large taxa, e.g., in some naticids, neogastropods, land and fresh-water 'pulmonates,' and many 'opisthobranchs'. Compared to muscular propulsion (see below), ciliary locomotion is relatively ineffective, especially in larger species, although the weight carried by the foot is reduced in aquatic habitats, enabling larger species to use this means of locomotion. In terrestrial habitats, gastropod locomotion using cilia alone appears to be restricted to small species (Jones 1975).

The speed of locomotion depends on the rate of ciliary beating and on environmental factors (e.g., temperature, substrata). A longer cilium has a faster tip speed, so those with

TABLE 3.4

Classification and Higher Taxon Occurrences of Ciliary and Muscular Locomotory Modes in Polyplacophora and Gastropoda

I. Muscular
  a. Rhythmic
    i. Monotaxic

1. Retrograde	Polyplacophora, Vetigastropoda, Neritimorpha
	Caenogastropoda, Heterobranchia
2. Direct	Caenogastropoda, Heterobranchia

    ii. Ditaxic
      1. Retrograde

a. Transverse	Patellogastropoda (forward), Vetigastropoda, Caenogastropoda
b. Diagonal	Caenogastropoda

      2. Direct

a. Transverse	Patellogastropoda (backward), Vetigastropoda, Caenogastropoda
b. Diagonal	Caenogastropoda

    iii. Composite   Caenogastropoda
  b. Arrhythmic
    i. Distinct Terminating Waves

1. Continuous	Caenogastropoda
2. Discontinuous Type 2	Caenogastropoda

    ii. Indistinct

1. Continuous	Caenogastropoda
2. Discontinuous Type I	Caenogastropoda
3. Discontinuous Type 2	Caenogastropoda

  c. Leaping   Caenogastropoda
II. Ciliary   Solenogastres, Caenogastropoda, Heterobranchia

Sources: Miller, S.L., *J. Molluscan Stud.*, 41, 233–261, 1974b; Salvini-Plawen, L. v. *Sarsia*, 31, 105–126, 1968a; Kuroda, S. et al., *J. Royal Soc. Interface*, 11, 20140205, 2014.

long pedal cilia (as in many aquatic taxa) will move faster than those with shorter cilia (Jones 1975).

While ciliary locomotion is employed by many small-sized molluscs (and other animals), it is also the main locomotory power in some large gastropods such as the nearly 30 cm long and heavy *Cassis tuberosa* (Cassididae) (Miller 1974b). This species and other large gastropods employing ciliary locomotion have a very large foot sole that spreads the weight per unit foot area (Miller 1974b). For example, Copeland (1922) calculated that the pedal cilia of *Polinices duplicata* could transport weights of 4 g/cm^2, although this species, like some neogastropods, uses muscular waves (see below) when greater speed is necessary.

Non-heterobranch gastropods that move by ciliary means differ from those that use muscular waves (see below) in having a broader, longer foot (longer than the shell). For example, the Nassariidae have a much larger foot area than most buccinoideans with non-ciliary locomotion (Miller 1974b).

Some species with ciliary locomotion, such as those in the caenogastropod Nassariidae, can thrash their foot effecting a leaping movement (but in a different way from that seen in Stromboidea and Xenophoridae – see below) to avoid predators.

If the first molluscs were small (see Chapter 12), it is likely that they first moved with cilial action alone. It is thus interesting that chitons and some primitive gastropods move by pedal waves (see below). Many apogastropods[17] move by cilial action, this perhaps being a consequence of their assumed paedomorphic origin (Ponder & Lindberg 1997).

The groups of apogastropods with various types of locomotory methods alternative to cilial movement have presumably developed these independently. Some cilial movers may show irregular, rather unpredictable muscular movements of the foot sole – a phenomenon called *arrhythmic locomotion* (Miller 1974a, b) which was divided into those with distinct terminating waves and those with indistinct waves. The latter type was further subdivided into continuous or discontinuous type 1 and type 2 locomotion (Table 3.4). These indistinct pattern-less movements differ greatly from the regular undulations that form pedal waves described below, and they are grouped with cilial movement in Figure 3.35. These arrhythmic waves can begin and terminate anywhere on the sole, these being areas where the sole epithelium elongates (as in retrograde waves – see below). One type of arrhythmic locomotion (the 'indistinct discontinuous type 2' of Miller) is found only in the Stromboidea and Xenophoroidea and facilitates leaping behaviour. This type is distinguished in Figure 3.35 and is discussed further below.

Arrhythmic waves occur when the movement of the sole is discontinuous and can occur in species with indistinct movement (see above) and in some with a distinctive pedal gait. There are two main types (Miller 1974a, b): in the first, the foot is extended while the shell rests on the substratum and the foot is then attached or anchored as the shell is pulled forward by contraction of the columellar muscles. In the second, the shell is thrust forward, and then the foot is brought forward and attached or anchored at a point in front for the next forward movement of the shell. Leaping is a specialised form of this latter type of discontinuous locomotion and occurs in members of the caenogastropod families Strombidae (Berg 1972, 1974; Field 1977), Struthiolariidae (Crump 1968), and Xenophoridae (Morton 1958c). The operculum is modified as a lever and when the foot twists, the operculum is thrust into the substratum to provide the leverage to vigorously thrust the shell forward or even backward (Miller 1974a). Besides leaping, arrhythmic pedal locomotion is usually less efficient than rhythmic pedal waves (Miller 1974a). A similar, specialised mode of locomotion involves the use of the snout together with the foot, as described below.

Species with either type of arrhythmic locomotion have a wide range of foot sizes and shapes; they mainly have a short, broad foot although some can have a long, narrow foot (mostly Conidae). Foot shape can vary considerably within

---

[17] Caenogastropoda + Heterobranchia.

clades. For example, although all Conoidea examined have indistinct arrhythmic locomotion, species of Terebridae and 'Turridae' have a shorter, broader foot than do members of the Conidae, and these differences are not entirely correlated with substratum preference (see also Section 3.10.6). In caenogastropods (where most variation is seen), the foot area relative to size tends to be most diverse in those taxa with retrograde ditaxic locomotion.

Those species with type 1 discontinuous locomotion (most have indistinct arrhythmic pedal movement) have a comparatively short and broad foot. Miller (1974a) examined the personid *Distorsio anus* which had non-leaping type 2 discontinuous locomotion (arrhythmic terminating waves) and a broad, short foot, but had a broader shell than species with type 1. The leaping strombids have a foot smaller and narrower than the shell than in other species (Miller 1974a).

Some species with type 1 discontinuous locomotion (notably Terebridae) are very effective burrowers (Miller 1974a). Type 2 discontinuous locomotion, including leaping, is characteristic of species with a large, broad shell and a small foot (e.g., Strombidae, Xenophoridae, *Distorsio*). The shell provides a wide, flat, stable base which reduces the possibility of it tipping over. Animals without these shell modifications, such as immature *Strombus gigas*, are unstable on sand, making their movements much less efficient than in mature individuals (Miller 1974b).

3.10.3 RHYTHMIC MUSCULAR LOCOMOTION AND ADHESION

Rhythmic muscular locomotion is broadly distributed among gastropods (Table 3.4). In many taxa, locomotory waves on the creeping sole are used together with adhesion (see Section 3.3.2.1), which also provides anchorage during the movement of other parts of the foot (see Figure 3.31). Chitons and gastropods use the special properties of their pedal mucus in conjunction with pedal waves – a locomotory method known as adhesive locomotion (Denny 1980a,

1980b, 1981; Denny & Gosline 1981; Chan et al. 2005). This method of locomotion is very effective, but is also energetically expensive (Denny 1980b; Tyrakowski et al. 2012). It is achieved by a succession of muscular waves moving along the foot which is connected to the substratum by a thin layer of mucus. This mucus becomes more solid under pressure and more fluid when that pressure is relaxed (see Section 3.3.2.1). The biomechanical properties of mucus combined with the muscular waves produces a very effective means of locomotion that can be mathematically modelled (e.g., Chan et al. 2005; Iwamoto et al. 2014).

Chitons have a broad foot and move using muscular waves, but in gastropods there is a considerable variety of foot shape and the type of movement employed. While some creep on a broad foot, in others the foot is narrow and more agile. In a few cases, the operculum is used to assist locomotory movements, the most famous being the leaping of strombids (see below).

Pedal waves may be *retrograde* (moving in the opposite direction to the animal) or *direct* (moving in the same direction as the animal). The waves may be *monotaxic* (waves extending horizontally across the foot) or *ditaxic* (in two series which alternate on either side of the foot). These waves can be formed either solely by differential contraction of muscles or a combination of muscles acting in concert with blood spaces.

Retrograde waves begin with the extension of the anterior end of the foot, and the wave then moves backward along the foot by the anterior muscles which form the wave contracting and those behind stretching backward (see Figure 3.32). In direct waves, the opposite occurs, with the wave commencing posteriorly and contraction at the anterior edge of the wave. For these reasons, retrograde waves are known as 'waves of elongation' and direct waves are known as 'waves of compression' (Lissmann 1945; Miller 1974a; Trueman 1983). Thus with direct waves, movement is achieved by parts of the sole being attached to the substratum at maximum extension, and forward movement occurs when the remainder of the foot is shortened. With retrograde waves, the foot is anchored by regions at their shortest length, and forward movement occurs with the sole maximally extended.

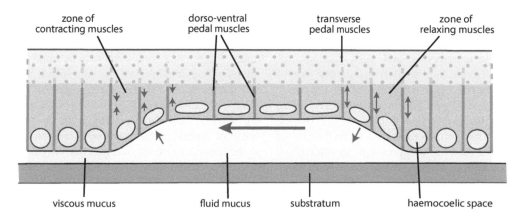

FIGURE 3.31 A diagrammatic representation of the employment of muscles in the formation of a retrograde wave in the patellogastropod *Patella*. Note the compression of the haemocoelic spaces and the mucus changing from viscous to fluid state under the raised part of the pedal wave. Redrawn and modified from Trueman, E.R., *The Locomotion of Soft-Bodied Animals*, Edward Arnold, London, UK 1975.

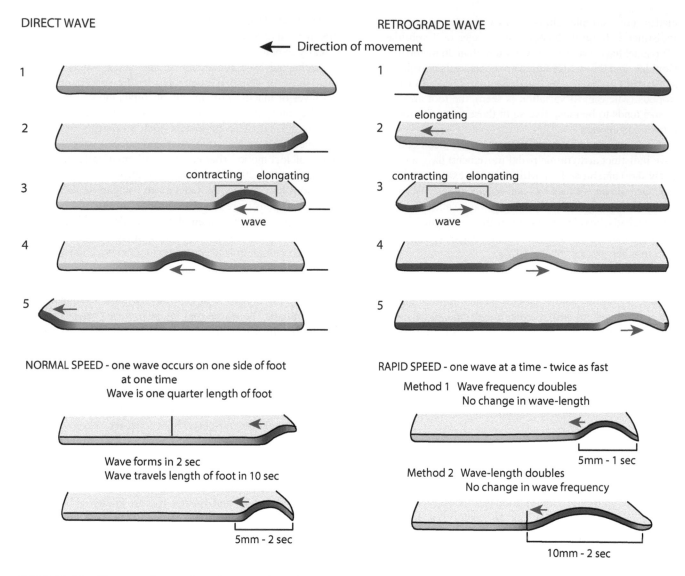

FIGURE 3.32 The upper figures show the two main pedal wave patterns in gastropods, direct waves (waves of compression), and retrograde waves (waves of elongation). The lower figures show two methods of acceleration in vetigastropods. Redrawn and modified from Miller, S.L., *J. Molluscan Stud.*, 41, 233–261, 1974b.

Chitons and some gastropods have retrograde monotaxic waves while in a few other gastropods the waves are direct (Table 3.4). Retrograde ditaxic pedal waves are seen in patellogastropods, some vetigastropods, and many caenogastropods. In the caenogastropod land snail *Pomatias*, one side of the foot remains anchored while the other moves forward – producing an almost bipedal gait. The retrograde ditaxic waveform is associated with a large range of foot sizes and shapes, and these waves are at least one third as long as the foot, while direct waves (and other waves of compression) are often much smaller and the range of foot shape is restricted, usually broad (for high tenacity) or long and narrow. Compression waves can maintain high tenacity even when the animal is moving quickly because the waves can be very small and can travel rapidly. Experiments by Miller (1974a) showed these small waves had greater adherence against shear forces than large pedal waves. Those gastropods with direct waves appear to generate waves smaller than those with retrograde waves and can thus achieve greater gripping power with a smaller foot (Miller 1974a). Small direct waves are, for example, seen in many stylommatophorans and may be important when climbing as only a small part of the foot sole is released from the substratum at any one time. Large waves do have advantages, despite reducing tenacity. Speed can be increased by increasing wave velocity or wavelength. In the first instance, the speed of contraction of the sole musculature must increase, so the energetic cost is greater. If the wavelength is increased, then it is unnecessary to increase the speed of contraction (Miller 1974a).

In species with retrograde ditaxic waves the size and shape of the foot is variable, but none have a foot longer than the shell (Miller 1974a). Taxa with direct ditaxic waves have a large foot area, and some (e.g., some trochids) have a foot longer than the shell.

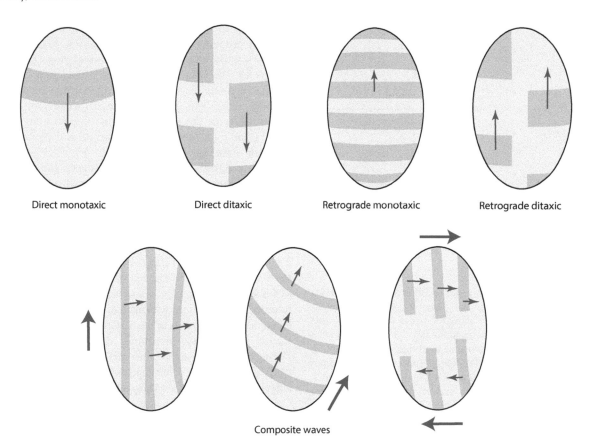

Direct monotaxic   Direct ditaxic   Retrograde monotaxic   Retrograde ditaxic

Composite waves

FIGURE 3.33 The main kinds of pedal wave seen in the gastropod foot. Redrawn and modified from Trueman, E.R., Locomotion in mollusks, pp. 155–198, in Salauddin, A.S.M. and Wilbur, K. M., *Physiology, Part 1. The Mollusca*, Vol. 4., Academic Press, New York, 1983, the bottom row is redrawn and modified from Miller, S.L., *J. Exp. Mar. Biol. Ecol.*, 14, 99–156, 1974a. See Figure 3.35 for their distribution in gastropods. Composite waves are found only in cowries (Cypraeoidea). The small red arrows show the direction that the pedal waves are moving, and the large red arrows show the direction of movement of the animal. In the figures in the top row, the animal is moving toward the top of the page.

So-called *composite* waves are less common and seen in a few gastropods such as cowries where the waves can move in various directions. Modifications of these basic types can be recognised, and Miller in her study of 'prosobranch' locomotion, recognised 14 kinds (Miller 1974a, b). Of all the muscular patterns observed in gastropods by Miller (1974a), she considered the composite waves of cypraeoideans the most complex, enabling them to move quickly and turn easily without reducing the amount of sole attached to the substratum. Taxa with composite wave patterns have similar foot sizes and shapes when compared to species with direct ditaxic waves (Figure 3.33).

If the main shell muscles in the foot cross, the locomotory wave is ditaxic, whereas if they remain separate the wave is monotaxic (Figure 3.34).

The type of locomotion reflects both phylogeny (see Figure 3.35) and habits. In addition, a species may use one pattern for forward movement, but, for an increase in speed, it can modify the normal pattern. Some species use a different pattern of locomotion for escape, reversing, or climbing, but for normal

forward movement, a single pattern of locomotion is usually employed, as is the number and size of muscular waves.

Some burrowing gastropods, notably Naticidae and Olividae, use monotaxic waves to burrow or increase speed (Copeland 1919, 1922; Miller 1974a).

Polyplacophorans, patellogastropods, vetigastropods, and neritimorphs use only *rhythmic pedal waves* which move in regular parallel bands from one end of the foot to the other. The amplitude of the rhythmic waves during normal forward locomotion is low, and all parts of the sole not within a wave at any moment remain stationary and are usually firmly attached to the substratum (Miller 1974a).

Lower caenogastropods have a wide diversity of wave types, but diagonal ditaxic waves are only found in some neogastropods. Other neogastropods use a variety of wave types, but many groups primarily use cilia.

Small-sized 'pulmonates' use either ciliary movement or monotaxic waves and the larger basal 'pulmonates', onchidiids, and stylommatophorans all have direct monotaxic waves.

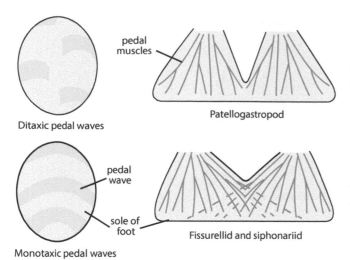

Ditaxic pedal waves

Monotaxic pedal waves

Patellogastropod

Fissurellid and siphonariid

pedal muscles

pedal wave

sole of foot

FIGURE 3.34 A diagrammatic representation of how shell muscle interactions determine whether the pedal waves will be mono-taxic or ditaxic. Figures on left show the pedal waves on the sole, those on the right are transverse sections through the foot showing the musculature (blue). Redrawn and modified from Voltzow, J., Gastropod pedal architecture: Predicting function from structure and structure from function, pp. 83–84, in C. Meier-Brook, *Proceedings of the 10th International Malacological Congress, Tübingen, 27 August – 2 September 1989*, University of Tübingen, Tübingen, Germany, 1992.

In detached caenogastropods, the propodium (with its associated anterior pedal gland) is used to initiate reattachment with the rest of the foot being applied subsequently. In contrast, some vetigastropods (e.g., *Haliotis*, *Trochus*) and patellogastropods can readily attach any part of the sole when righting themselves (Miller 1974a).

Looping using the foot and snout is seen in species of Truncatellidae, and in some Assimineidae and Pomatiopsidae. In this mode of progression, the anterior end of the snout is used along with the foot to practise a looping mode of progression (Figure 3.36) (Fretter & Graham 1962; Davis 1967). Some strombids also practise a similar mode of progression that may sometimes involve the snout (Berg 1974; Solem 1974) (Figure 3.36).

3.10.4 Adhesion and Tenacity

Tenacity is important if chitons, snails, or limpets clamping onto rocks are to withstand wave action and currents and to discourage predators. Two mechanisms are involved in tenacity – suction and adhesion (see also Sections 3.26 and 3.26.1). With suction, a space is formed under the foot to create a negative pressure when pull is exerted. Adhesion is the force holding together two closely applied surfaces with a thin layer of liquid between them. Adhesion is generally possible because the foot sole can closely conform to the shape of most surfaces, and the thin layer of adhesive pedal gel greatly increases its adhesive capabilities.

Tenacity in gastropods is not just passive adhesion, but is an active process involving muscular action. Maximum adhesion is usually attained when an animal stops moving and clamps its shell tightly against the substratum. Stationary caenogastropods usually have the propodium retracted and, although tenacity is greatly enhanced because of a change in the nature of the secreted mucus (see Sections 3.26 and 3.26.1), the total foot area in contact with the substratum is less than when crawling.

Some patellogastropods, once dislodged, will not reattach for several minutes. Once reattached, tenacity is poor at first, but increases with the time that the limpet remains stationary. The strength of pedal adhesion in patellogastropod limpets can exceed the strength of attachment of the muscles to the shell. The highest stationary tenacity recorded by Miller (1974a) was 2.8 kg/cm^2 for *Lottia pelta*, while Aubin (1892) found that a *Patella vulgata* with a foot area of 2.9 cm^2 resisted pulls of up to 4.8 kg/cm^2, but figures of as much as 5.2 kg/cm^2 have been recorded for other limpets (Branch & Marsh 1978). Smith (1992) found that lottiid limpets in the eastern Pacific switched between suction and adhesion attachment depending on tidal height. During high tide about three-quarters of the limpets used suction, while about the same percentage used adhesion during low tides, although mean tenacity was less at high tide than at low tide.

Miller (1974a) experimentally investigated the factors determining the tenacity of a snail or limpet. These include: (1) direction of the dislodging force; (2) whether the animal is moving or stationary; (3) whether the force is sudden or gradual; (4) how long it is applied; and (5) the substratum type and texture. Differences between taxa include the size of the foot and the locomotory patterns. Those with continuous pedal muscular locomotion have high tenacity, especially when stationary, while those with ciliary action or discontinuous locomotion (including leaping) have very low tenacity. Similarly, species with broad feet tend to have tenacity higher than those with narrow feet. The optimal foot for high tenacity is broad and slightly less than the length of the shell so the shell can be clamped tightly against the substratum for maximum adhesion. Other factors such as musculature and the nature and distribution of pedal glands presumably also play an important role.

The requirements for speed and tenacity conflict, and maximum tenacity is usually achieved only when the animal is stationary, but experiments with the trochoidean vetigastropods *Calliostoma* and *Tegula* by Miller (1974a) imply that, at least in those rather narrow-footed taxa, speed can increase without affecting tenacity against shearing forces if wave pattern or the method of movement do not change.

The experiments of Miller showed that the optimal wave pattern is a compromise between speed, energy efficiency, and tenacity. Direct waves and other waves of compression have greater tenacity than retrograde ditaxic waves of elongation because they are typically smaller and closer together. Thus, where both speed and tenacity are needed, rhythmic pedal waves are optimal because the waves move quickly over the sole while much of it remains firmly attached, enabling good adhesion. In contrast, long rhythmic waves (which move a large proportion of foot area at once) probably require less

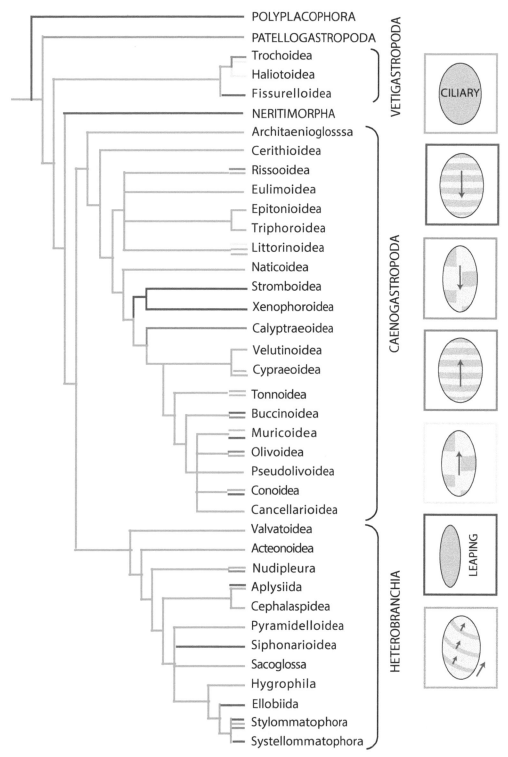

FIGURE 3.35 Pedal wave patterns in polyplacophorans and gastropods based mainly on data from Miller, S.L., *J. Exp. Mar. Biol. Ecol.*, 14, 99–156, 1974a. See Figure 3.33 for key to locomotor types. Note: Rissooidea = Rissooidea + Truncatelloidea.

energy than small, rapid waves achieving the same speed, but tenacity is much reduced.

Because the modes of locomotion differ considerably in the mechanism employed, they might be expected to differ in mechanical power, but no one major type of locomotion is restricted to species of small or large size. The experiments conducted by Miller (1974a) determined that the optimal foot form for high tenacity were ratios of shell length/foot length of about 1.5 and foot length/foot width of between 1.1 and 1.5. This appeared to be due to two factors. Tenacity is approximately proportional to foot area, but if foot size is equal to or exceeds shell size, tenacity is much reduced. Limpets have

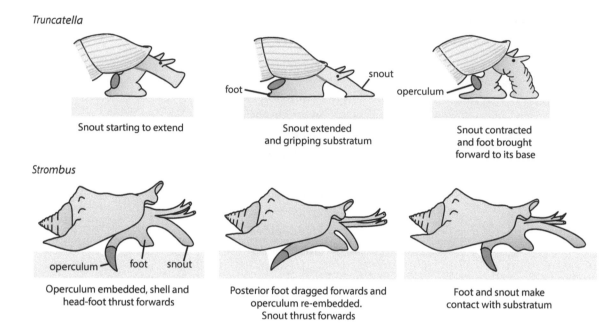

FIGURE 3.36 Looping and utilisation of the snout in locomotion in the caenogastropods *Truncatella* and *Strombus*. *Truncatella* redrawn and modified from Fretter, V. and Graham, A.L., *British Prosobranch Molluscs: Their Functional Anatomy and Ecology*, Ray Society, London, UK, 1962; *Strombus* redrawn and modified from Solem, A.C., *The Shell Makers. Introducing Mollusks*, John Wiley & Sons, New York, 1974.

by far the greatest strength of adhesion when stationary, and Miller (1974a) suggested that tenacity may be increased by the short shell muscles. If, as in snails, shell (i.e., columellar) muscles must curve to reach the outer edges of the foot, much of the pull would be parallel to the substratum, not vertical to it as in limpets. This may in part explain the adoption of a limpet-shape by many intertidal gastropods. The maximum pull by the shell (or columellar) muscles can only be exerted if the foot is smaller than the aperture, allowing the shell to be pulled against the substratum.

While optimal foot size and shape, and the type of locomotion employed for high speed and high tenacity, are not completely compatible, both attributes can be maximised to a high degree because the foot size and shape optimal for both are similar. Thus, most species in exposed habitats studied by Miller (1974a) moved rapidly and clung tightly. Miller showed experimentally that the type of foot ideal for clamping is less flexible and at a marked disadvantage on more complex topography compared with those with a long, flexible foot.

Miller (1974a) demonstrated that many gastropods, unlike most other animals, have no difficulty climbing Teflon coated vertical surfaces whether submerged or in air.

3.10.5 Speed of Locomotion

Cephalopods are the fastest molluscs and are discussed under jet propulsion below (see Section 3.10.9.1).

In gastropods, the speed of movement with normal pedal locomotion (for alternatives see below) differs considerably between taxa. Some have little need to move much and some (vermetids and hipponicids) are permanently cemented to a hard substratum. Others need to move rather quickly for feeding, to reach shelter, to find mates, or to avoid predators.

In gastropods with ciliary locomotion, speed is correlated with the rate of ciliary activity (Copeland 1919), but, in taxa employing muscular locomotion, speed is a function of the size and frequency of the pedal muscular waves (see above), or, in a few, the whole foot can be employed in a 'step' with discontinuous (including leaping) locomotion (Miller 1974a).

Many species with rhythmic pedal waves merely increase the size of waves at greater speeds, while others can increase speed without altering the pattern. For example, the speed of two trochoidean vetigastropods studied by Miller (1974a) doubled when they made contact with a starfish (from about 0.8 to 1.4 mm/sec and 1.2 to 2.9 mm/sec, respectively). They did this by increasing the number and frequency of waves and the size of the steps. While the sizes of steps increased linearly with speed in both species, *Calliostoma* with retrograde ditaxic waves, whose mean foot length was greater, took larger steps than *Tegula* (with direct ditaxic waves) at any speed, while *Tegula* had a greater wave frequency. An increase in wavelength with an increase in speed is common in gastropods, especially in those with retrograde waves. Wave frequency probably increases mainly because of an increase in wave velocity, and not by a change in the spacing of the waves. Similar changes occur in a terrestrial slug (*Limax*), which has 11–19 direct monotaxic waves on the sole at any one time (Crozier & Pilz 1924). Thus, this method of increasing speed may be typical of gastropods with rhythmic pedal waves (Miller 1974a).

TABLE 3.5

Heat Map of Maximum Gastropod Speeds Parsed by Locomotory Mode and Shell Size. Median Values for Size and Mode also Given

Locomotory Mode	<15mm	15–40 mm	41–80 mm	>80 mm	Mode Medians
Monotaxic					
Retrograde					
Direct					
Ditaxic					
Retrograde					
Direct					
Composite					
Arrhythmic					
Terminating waves					
Indistinct					
Discontinuous					
Type 1					
Type 2					
Leaping					
Ciliary					
Size medians					

Slow   Fast

Source of data: Miller, S.L., *J. Molluscan Stud.*, 41, 233–261, 1974b.

Miller (1974a) recorded speeds ranging from 0.1 to 37.2 mm/sec (*Strombus*) within gastropods (Table 3.5). Overall, leaping locomotion was by far the fastest. Leaping is dynamic because the power of the entire head-foot and columellar muscle complex (dorsal-ventral) are involved in each step, and the distance travelled with a single contraction is potentially much greater than that with muscular pedal waves. In general, taxa with ditaxic waves are capable of fairly high speeds, and their tenacity exceeds that of other taxa regardless of the type of locomotion. Larger taxa with ditaxic waves may be slower than those that use leaping locomotion, and small-sized ditaxic movers (<15 mm) are also slower than those of similar size with ciliary locomotion.

Ciliary movement was performed consistently well over all sizes. Rhythmic wave patterns, especially ditaxic, also performed well at larger sizes, as did those with a composite pattern unless size exceeded 80 mm. In contrast, arrhythmic muscular locomotion (other than leaping), was slow, with arrhythmic terminating waves generally performing better than indistinct arrhythmic locomotion and either type of discontinuous locomotion (Table 3.5).

Direct and retrograde ditaxic patterns were close in the maximum speeds attained for animals of similar size. Taxa with retrograde ditaxic waves exhibited a wide range of speeds, and those with direct ditaxic waves more consistently showed higher speeds (Miller 1974a). This is true

even between distantly related taxa – for example, both types occur in the Trochidae and Muricidae in species of similar size, and those with direct waves moved significantly faster than those with retrograde waves. Relatively high speeds were also regularly found in species with direct monotaxic waves which employed waves of compression, so they had a composite pattern.

Several factors were identified by Miller (1974a) as important in determining crawling speed. These were wavelength and foot size and shape and its relationship with shell size, both of which are discussed below.

Wavelength – rapidly moving species with retrograde ditaxic waves usually had a wavelength equal to or greater than half the foot length, while in species with direct ditaxic, the wavelength did not increase more than the normal 1/5 to 1/3 foot length. In theory, if the speed of contraction + contraction per unit sole area + the time between successive waves are all equal, an animal with longer waves should move faster as each wave would cause a larger forward step, but such factors are usually not equal. Thus, among trochids and some muricids, Miller (1974a) found that species with direct waves moved faster even though many had shorter wavelengths.

Loping (sometimes called galloping) occurs intermittently in some stylommatophoran land snails and at least one species of arionid slug (Pearce 1989). In this method of locomotion, the head is lifted from the substratum and thrust forward. It touches the substratum, but a low arch is formed behind and the rest of the body is then moved forward to this new point of contact (Figure 3.37). More than one arch may be present at any one time. This method of progression is not markedly faster than regular gliding. It thus differs from galloping in the sea hare *Aplysia* which is a distinct type of retrograde locomotion faster than crawling (Parker 1917; Pearce 1989).

Foot size and shape and its relationship with shell size – in medium to large species with continuous muscular pedal locomotion, those with a foot length equal to shell length had the highest speeds. As the foot becomes shorter than the shell or when foot length exceeds shell length, maximum speed is reduced. Species with discontinuous locomotion (leaping) are different – in *Strombus* the narrow foot is 1/2 to 1/3 the shell length, yet locomotion is rapid. In those that rely on ciliary action, the foot in the most rapid movers is at least as long as the shell and is often (but not always) broad.

There is relatively little data for the 'minor' classes on locomotory speed. Aplacophorans are very slow-moving and either crawl on their cnidarian food or burrow in soft sediment (see Chapter 14). Chitons are generally slow crawlers, moving rather like a gastropod limpet. Some species of *Ischnochiton* that live on the undersides of rocks can move relatively rapidly away from light when the rock is overturned and move at speeds around 10 cm/min (Boyle 1977).

Scaphopods and many burrowing bivalves can quickly burrow downward into the sediment, but cannot move efficiently laterally, and some shallow-burrowing bivalves can readily move laterally by ploughing through the sediment. Some cockles (Cardiidae) use their long muscular foot to actively leap away from predatory starfish (Morton 1964; Ansell 1967b;

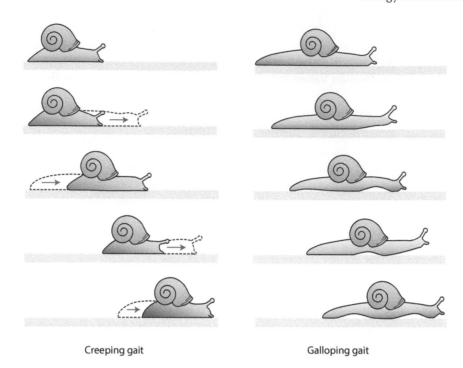

Creeping gait   Galloping gait

FIGURE 3.37 Different modes of locomotion in a terrestrial helicid snail. Redrawn and modified from Lissmann, H. W. *J. Exp. Biol.*, 22, 37–50, 1945 (figures on left), and Gray, J., *Animal Locomotion*, Norton, New York, 1968 (figures on right) as reproduced in Kandel, E.R., *Behavioral Biology of Aplysia: A Contribution to the Comparative Study of Opisthobranch Molluscs*, W. H. Freeman & Co., San Francisco, CA, 1979.

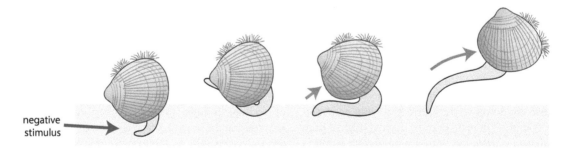

negative stimulus

FIGURE 3.38 Leaping movement of the cardiid bivalve *Laevicardium crassum*. Redrawn and modified from Morton, J.E., Locomotion, pp. 383–423, in Wilbur, K.M. and Yonge, C. M., *Physiology of Mollusca*, Vol. 1, Academic Press, New York, 1964.

Yonge & Thompson 1976) (Figure 3.38). Similar leaping movements have been observed in several other bivalve taxa (Ansell 1969), including *Gari* (Ansell 1967a), and are assumed to be possible in *Neotrigonia* (e.g., Ansell 1969) because of similar foot morphology. The fastest-moving bivalves are the swimming scallops (see Section 3.10.9 and Chapter 2).

The responses of patellogastropod limpets to starfish and other predators have been well studied with some moving away (e.g., Phillips 1975), others twisting (e.g., Bullock 1953), and some even attacking the predator (e.g., Branch 1979).

3.10.6 THE EFFECT OF SUBSTRATUM AND TOPOGRAPHY ON THE LOCOMOTION OF GASTROPODS

Substratum type and topography are often variable and changes experienced by an animal can present problems for

locomotion. Hard and soft substrata provide very different surfaces and topographical complexity can necessitate complex movements such as climbing and turning.

Experimental work carried out on marine gastropods by Miller (1974a) investigated how substratum and topography influenced their form and locomotion.

Differences in the substratum. The experiments that Miller (1974a) conducted showed that taxa living on soft substrata tend to use ciliary locomotion or arrhythmic discontinuous waves and those on hard substrata use rhythmic pedal waves. Those that live on both tend to use arrhythmic waves. Species with ciliary, leaping, or discontinuous locomotion move efficiently on soft substrata and on horizontal hard surfaces, although with their leaping movement strombids are inefficient on smooth hard surfaces, as are species with discontinuous locomotion and poor tenacity (Miller 1974a). To climb vertical, smooth, hard surfaces, species with discontinuous and leaping locomotion usually adopt an indistinct arrhythmic movement. In strombids, this is possible only on the adhesive anterior part of the foot, but large individuals are typically unable to climb. While species employing ciliary movement can usually climb hard surfaces, smooth surfaces can cause slippage.

Most species with muscular pedal waves generally move most effectively on hard surfaces. Miller (1974a) tested the speed on sand of several gastropod species which normally live only on hard substrata and found those with high locomotor tenacities (see below), such as patellogastropods and the neogastropod *Nucella* manage poorly on sand, whereas most of the others from hard substrata had no difficulty, irrespective of locomotory wave pattern. Thus, various muscular wave patterns can be employed for movement on sand, except for those that employ firm adhesion.

Topology. Species with a relatively long, narrow foot generally proved to be the most capable of manipulating themselves over complex surfaces. Species with a broad foot rarely turn more than 90 degrees at once and to achieve even that angle much of the foot area had to be detached. A large shell can be a hindrance on complex topography unless the foot is very flexible, and it is certainly a disadvantage on vertical or overhanging surfaces due to the weight relative to the area of adhesive foot surface.

3.10.7 Correlation with Habitat

Miller (1974a) showed that different locomotor types were strongly correlated with the environment in which they live, as summarised below.

In relation to the substratum, retrograde, direct ditaxic, and composite patterns are mainly found in species on hard substrata, while those with ciliary and discontinuous locomotion are mostly on soft substrata. Species with monotaxic waves present on soft substrata are members of families primarily using ciliary locomotion (Naticidae and Olividae). Almost half the species with distinct arrhythmic terminating waves and indistinct arrhythmic locomotion occurred in sand or mixed sand/rock habitats.

Ciliary and discontinuous locomotion are primarily found in vetigastropods and caenogastropods associated with soft substrata and both have low tenacity. Taxa with arrhythmic terminating waves and continuous indistinct arrhythmic pedal movement tend to inhabit mixed substratum habitats or soft sediments exclusively. Some of those with continuous indistinct arrhythmic pedal movement have a large foot, like that seen in ciliary or discontinuous movers, yet these foot shapes are not essential for effective movement. Similarly, most species with ciliary locomotion have a much larger foot area than necessary for effective propulsion. This suggests that the enlarged foot serves other functions, for instance, burrowing and/or for stability on loose substratum. It has been suggested that the massive foot of some Naticidae is used in prey capture and even transport (Gonor 1965). A similar enlargement of the foot is seen in the mud-living nudibranch *Dendronotus iris*

and contrasts with narrow-footed relatives living on hydroid colonies (Robilliard 1970). The foot of several species of ciliary movers that burrow (e.g., the caenogastropods Olividae, Naticidae, and several shelled 'opisthobranchs') envelops some or all of the shell, and burrowing is assisted by pedal cilia moving sand dorsally over the shell (see also below).

Taxa with high tenacity are found on hard substrata, and those with minimal tenacity usually on soft substrata, while taxa with intermediate tenacity are less restricted. All ciliary movers whose tenacity was measured by Miller (1974a) were sand-dwellers; taxa associated with hard substrata that presumably employ ciliary motion (e.g., Velutinidae, Triviidae) were not tested.

There are also some patterns of locomotory types with a preference for intertidal or subtidal environments and wave exposure. Dislodgement by the forces generated by breaking waves is probably an important selective agent for intertidal snails (Denny 1985; Denny et al. 1985; Wright & Nybakken 2007). The risk of dislodgement can be reduced by a smaller or flatter shell, streamlining, and reduction of the projected surface so that lift is reduced, and increased adhesive strength by way of a larger foot area or increased tenacity (see also Denny 1988).

In the lower littoral or sublittoral, ciliary and discontinuous movers and closely related species with monotaxic patterns dominate, while on wave-exposed sandy beaches the taxa present (often Olividae) are usually burrowers. Most taxa on soft substrata are found in sheltered shores where reduced wave action and inland drainages allow for the accumulation of deep sediments. Those that leap and those with other discontinuous locomotion were also consistently found low on the shore or in the sublittoral.

Species in the higher tidal zones of the more exposed shores are relatively fast-moving and have rhythmic, ditaxic pedal waves. Small pedal waves are found in only a few species in such situations with most taxa in the highest zones (littorinids, planaxids, neritids, and several muricids) typically moving with a large pedal wave ($\geq 1/3$ foot length). The marine panpulmonate limpet *Siphonaria* has initially small waves, but the entire back of the foot then moves forward at once. The patellogastropod lottiids and the vetigastropod *Tegula* have a normal wavelength about 1/3 of the foot length, but can increase this to increase speed. Various muricids living on exposed hard shores have short direct waves, thus maintaining a large area of the foot stationary to increase adhesion. This same pattern is seen in abalone (*Haliotis* spp.) which also live in exposed lower littoral and sublittoral habitats. Others with narrow waves include many species of cypraeids which often hide underneath rocks etc., and trochids which have a long, relatively narrow foot and thus a low tenacity.

Foot size and shape also show some correlation with environmental factors. Many species with foot proportions for high tenacity occur on the upper parts of exposed shores, while species with proportions outside the range for high tenacity usually occur in lower, less exposed zones. While a larger sole size can equate with increased adhesion, improved adhesion can also be increased by the properties of the pedal mucus (see Section 3.3.2.1). Species with arrhythmic locomotion are found only in the low zones and often confined to sheltered habitats and microhabitats. These and other species with unsuitable foot proportions for effective tenacity are mainly found in protected microhabitats.

Despite some generality with the patterns found by Miller (1974a), intraspecific differences always occur. In a more recent study of two intertidal snails in South Australia, Prowse and Pile (2005) found no differences in individuals from sheltered and exposed sites other than a reduction in overall size in both species, possibly indicating that smaller individuals may gain access to refuges more effectively.

3.10.7.1 Burrowing in Soft Sediments

Caudofoveate aplacophorans burrow in soft sediment, but their movements are slow. Burrowing is achieved by the anterior end (with the oral shield) being pushed into the sediment hydrostatically and, with the long posterior spicules acting as an anchor, contraction of the longitudinal muscle bands along the body bring the posterior end forward (Scheltema 1998). Most solenogasters do not burrow, but some species of *Neomenia* can do so using their protrusible pharynx (Scheltema 1998).

The foot is modified for burrowing in many bivalves, some gastropods, and in all scaphopods. In burrowing gastropods, the foot is usually dorsoventrally flattened into a plough-shape, often with extensions from the sides of the foot or the mantle edges to streamline the body so that the animal can plough into and through the substratum, lubricated by mucus.

Although many burrowing gastropods use ciliary action, muscular pedal waves appear to be more effective, especially in larger species such as in many naticids (Copeland 1922) (Figure 3.39), the nassariid *Bullia melanoides* (Ansell & Trevallion 1969), and the olivid *Agaronia* (Cyrus et al. 2012). Pedal waves are developed in some caenogastropod taxa where close relatives have ciliary action alone, and they differ from the waves seen in many other caenogastropods in being long (1/3–1/2 foot length) and monotaxic. Some other burrowing gastropods such as the Terebridae are very effective burrowers (Miller 1974a).

In both scaphopods and bivalves the foot is highly adapted to effectively and efficiently penetrate the sediment using hydrostatic pressure. In burrowing bivalves (Figures 3.39 and 3.40), the movements involve the adductor and foot muscles and, while basically similar (Trueman et al. 1966), there are differences between taxa resulting from structural (including the shell), physiological, and behavioural differences, and differences in the properties of the sediment. Burrowing bivalves usually have a laterally compressed hatchet-shaped or plunger-shaped foot. A hydraulic system is employed, involving haemocoelic fluid being forced into the foot by muscle contraction. This expands the foot distally and anchors the anterior part, enabling the animal to draw down on it (Figures 3.39 and 3.40). The animal may also be prevented from being pushed upward by dilation of the upper part of the body, usually by the shell valves being partially opened.

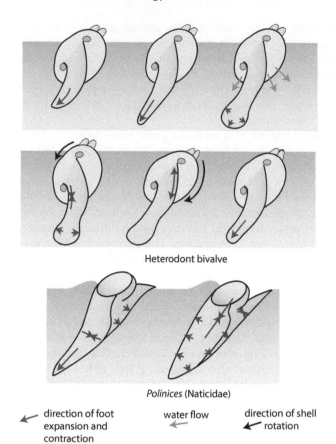

Heterodont bivalve

Polinices (Naticidae)

← direction of foot expansion and contraction   water flow ←   direction of shell ↙ rotation

FIGURE 3.39 Comparison of the burrowing mechanism in a heterodont bivalve and a naticid gastropod. Bivalve redrawn and modified from Trueman, E.R. et al., *J. Exp. Biol.*, 44, 469–492, 1966; naticid from Trueman, E.R., *J. Exp. Biol.*, 48, 663–678, 1968.

Downward movement can also be facilitated by a jet of water from the mantle cavity into the surrounding sediment as the shell is rapidly closed (Figure 3.40). The shell is then dragged through the softened sediment by the foot. Digging can also be assisted by a rocking motion caused by the alternate contraction of the pedal retractor and protractor muscles. Rocking does not occur in bivalves with elongate shells, but can be pronounced in some with rounded shells. Upward movement is accomplished by pushing down with the foot.

Besides the similar muscular hydrostatic burrowing mechanisms seen in living molluscs, convergent features in some bivalves and gastropods include the development of large haemocoelic cavities. In addition to copious blood spaces, naticid gastropods have an 'aquiferous pedal system', which takes in seawater to inflate foot sinuses and is expelled when the animal contracts into its shell (see Section 3.8.2). Many bivalves and gastropods have also developed asymmetrical 'ratchet' sculpture that aids in preventing back slippage and thus makes burial more efficient (see Section 3.2.10). Scaphopods have a unique foot morphology with a fringed epipodium which is expanded to obtain anchorage, followed by its retraction. Octopods and cuttlefish also burrow, and like bivalves, use sediment liquefaction to facilitate their burrowing. Octopods

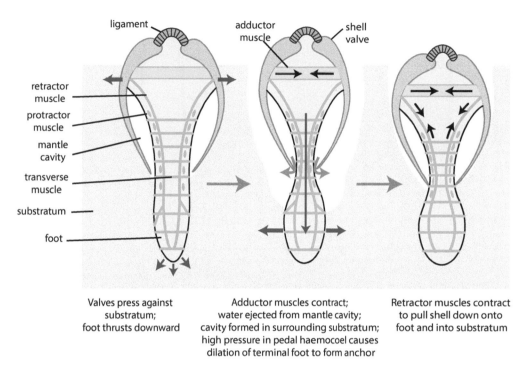

ligament adductor muscle shell valve

retractor muscle
protractor muscle
mantle cavity
transverse muscle
substratum
foot

Valves press against substratum; foot thrusts downward

Adductor muscles contract; water ejected from mantle cavity; cavity formed in surrounding substratum; high pressure in pedal haemocoel causes dilation of terminal foot to form anchor

Retractor muscles contract to pull shell down onto foot and into substratum

FIGURE 3.40 Diagrammatic representation (in transverse section) of the mechanisms involved in bivalve burrowing. Redrawn and modified from Trueman, E.R., Locomotion in mollusks, pp. 155–198, in Salauddin, A.S.M. and Wilbur, K. M., *Physiology, Part 1. The Mollusca*, Vol. 4., Academic Press, New York, 1983.

use water jets from their funnel to liquify the sediment before pulling themselves into the softened sediment with their arms, which function as a muscular hydrostatic burrowing mechanism. Cuttlefish and sepiolids use water jets and their fins to create a depression in which they cover themselves with sediment.

3.10.7.2 Burrowing in Hard Substrata

Some bivalves burrow into hard substrata (wood, clay, soft rock, or in calcareous substrata such as living or dead coral, molluscan shells, or limestone). They achieve this by either mechanical boring or chemical dissolution, or a combination of the two mechanisms, chemical dissolution being only possible on calcareous substrata.

Chemical borers such as date mussels (the mytilid Lithophaginae) do this using acidic secretions from the mantle, while mechanical borers use file-like sculpture on the shell to rasp the substratum. Thus, pholadids and teredinids file the soft rock or wood by rotating the shell back and forth while clamping onto the side of the borehole with their sucker-like foot (Figure 3.41). More details on bivalve boring are given in Chapters 9 and 15.

A few gastropods, such as the coralliophilid *Magilus* and the large vermetid *Ceraesignum maximum* live embedded in massive corals. They do not bore, rather, the coral grows around them, and they produce a tube that maintains a connection with the exterior. A few other limpet-like gastropods (other than patellogastropods) including abalone, the hipponicid *Sabia*, some siphonariids, and *Trimusculus* also construct

home site depressions in calcareous substrata. The excavation process appears to involve both chemical dissolution and radular rasping (Lindberg & Dwyer 1983) (see also Chapter 9).

3.10.8 Bipedal Locomotion

Some benthic octopuses move over sandy bottoms on two of their arms using a rolling gait (Huffard et al. 2005; Huffard 2006). These octopuses bipedally 'walk' backward by alternating the placement of each arm on the substratum and then rolling along the sucker edge of the distal half of the arm. Walking behaviour also occurs in a few cuttlefish taxa, including *Metasepia* (Roper & Hochberg 1988). Bipedal locomotion is the fastest benthic mode of locomotion in the molluscs with observed rates of 60 to 140 mm/sec, equal to or faster than the same individuals crawling (60 mm/sec) (Huffard et al. 2005).

3.10.9 Swimming

Swimming occurs in some adult gastropods and bivalves, and most cephalopods; most bivalves and cephalopods swim by 'jetting' (see Section 3.10.7.1). In bivalves, it is achieved by rapid closures of the valves in a clapping motion with the water being expelled from the mantle cavity. Cephalopods also rapidly expel water from the mantle cavity, but much more efficiently. In coleoids, the rapid contractions of the muscular mantle wall enable these animals to swim using jet propulsion, and they can also swim by flapping their fins. Fins

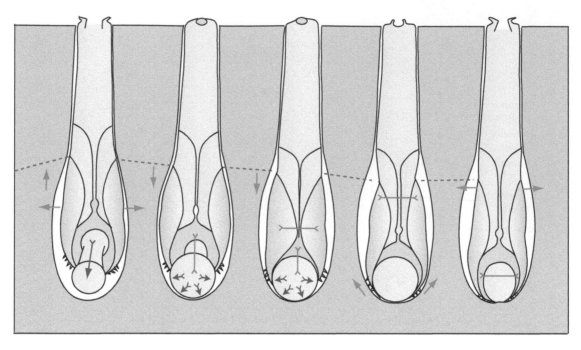

FIGURE 3.41 Burrowing movements of a pholadid bivalve in soft rock. The green arrows show the movements of the shell, while the red arrows indicate the movements of the foot. Redrawn and modified from Nair, N.B. and Ansell, A.D., *Proc. Malacol. Soc. Lond.*, 38, 179–197, 1968.

provide balance and propulsion in squid, cuttlefish, and some finned octopuses. Fin swimming is often used in conjunction with jet propulsion.

Swimming gastropods (Figure 3.42) use parapodia or a modified foot. Heteropods live permanently in the water column and are efficient swimmers, using body movements like a fish. The gymnosome and thecosome pteropods are also permanent members of the water column. The naked gymnosomes are predators on the shelled thecosomes, and *Clione limacina* and probably other gymnosomes can rapidly change their normal swimming wing beat to much higher speeds when involved in prey capture (Satterlie et al. 1985). Pteropods all use a pair of parapodial 'wings' for swimming (see Gilmer & Harbison 1986 for references and details). In thecosome pteropods, differences in the external part of the mantle appear to be related to their flotation abilities (Gilmer & Harbison 1986). Most cavoliniids have reduced external mantles and sink slowly while feeding, but *Diacria* and *Cavolinia* have complex mantles used for flotation and in feeding. *Limacina* has no external mantle, but flotation is assisted by their mucus-feeding webs. Swimming in the pseudothecosome *Peraclis reticulata*, which has an external mantle that covers the shell, is assisted by it being neutrally buoyant.

3.10.9.1 Jet Propulsion, Including 'Flight'

As noted above, in most bivalves jetting is achieved by rapid closures of the valves in a clapping motion with the water being expelled from the mantle cavity. A few bivalves can swim for short periods, but no adult bivalves are long-term swimmers. Most swimming bivalves are pteriomorphians with a single large adductor muscle which rapidly closes the valves. Limids such as *Limaria fragilis* (Donovan et al. 2004) can swim for extended periods (15 min or more), while others, such as *Limaria hians* (Gilmour 1967) are capable of swimming for only a few minutes at most. Physiologically, the energy supplied for swimming is mostly aerobic, with a small anaerobic contribution from arginine phosphate, glycolysis, and ATP (Baldwin & Morris 1983). By way of contrast, scallops swim for short periods, and their swimming is largely powered anaerobically (Gäde et al. 1978; Grieshaber 1978; Livingstone et al. 1981), although their swimming abilities vary considerably. Some such as *Chlamys* can only swim for short distances in a 'zig-zag' fashion, while others such as *Amusium* and *Placopecten* are 'long-range gliders'. Species of *Amusium* and *Ylistrum* are the most efficient bivalve swimmers. Their thin, smooth shells are strengthened by internal ribs, and they are capable of speeds up to 1.6 m/sec and can cover distances of over 23 m in a single swimming event (Joll 1989). The varying swimming abilities of scallops are determined by their shell morphology, including thickness, streamlining, and differential convexity in the upper and lower valves (particularly pronounced in *Pecten*) to generate lift (e.g., Hayami 1991). Allometric changes in shell thickness, shape, and in adductor muscle size also play a role (e.g., Gould 1971).

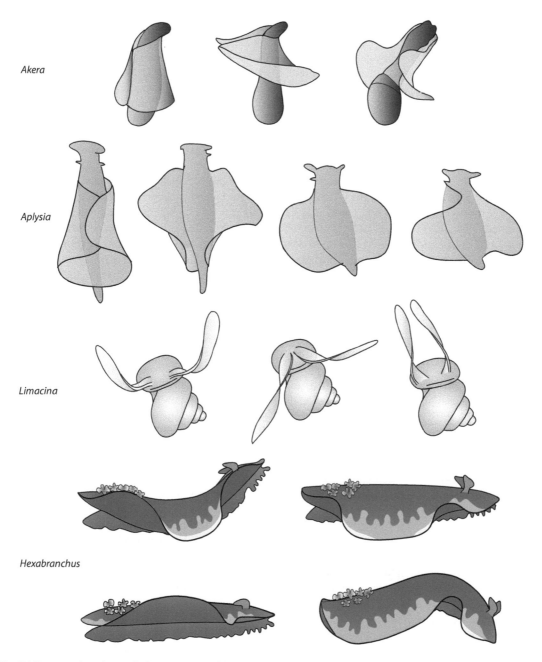

FIGURE 3.42 Different swimming techniques employed by some heterobranch gastropods. Redrawn and modified from the following sources – *Akera* Morton, J.E., Locomotion, pp. 383–423, in Wilbur, K.M. and Yonge, C. M., *Physiology of Mollusca*, Vol. 1, Academic Press, New York, 1964; *Aplysia* Thompson, T.E., *Biology of Opisthobranch Molluscs*, Vol. 1, Ray Society, London, UK, 1976; *Limacina* Morton, J.E., *J. Mar. Biol. Assoc. U. K.*, 33, 297–312, 1954; *Hexabranchus* Edmunds, M., *Proc. Malacol. Soc. Lond.* 38, 121–133, 1968b.

Swimming in limids is less well understood, but the behaviour and mechanics of swimming in *Limaria fragilis*, and paddling by its long mantle tentacles, was described by Donovan et al. (2004).

Cephalopods propel themselves using jet propulsion achieved with a powerful contraction of the mantle muscles. Water flows through the anterior opening into the mantle cavity and over the gills. The mantle builds internal pressure by sealing off the opening except for the funnel. The thick muscles of the mantle wall (see below) contract by as much as 21% in circumference (Ward 1972; Ward & Wainwright 1972; Packard & Trueman 1974) with no change in length. This action squeezes water out of the narrow funnel with enough force to propel the animal at speed. Squid are the fastest marine invertebrates and, using jet propulsion, the squid *Illex* can achieve maximum speeds of around 1.4 m/sec, compared with *Sepia* at 0.65 m/sec, and *Nautilus* at 0.3 m/sec (O'Dor & Webber 1991).

While most of the propulsion and steering in cephalopods originates with the funnel, some propulsion and fine adjustments are made with the fins. Fins provide balance and propulsion in squid, cuttlefish, and some finned octopuses, and are often used in conjunction with jet propulsion from the funnel. The funnel is directed forward to escape a predator, so that the animal propels itself backward. By pointing the funnel backward, forward movement is achieved, such as may be required in prey capture. A few species of squid escape from predators by propelling themselves out of the water and travelling through the air for distances up to 50 m (Heyerdahl 1950; O'Dor et al. 2013). Fins assist in stabilising 'flight' and allow the squid to glide above the surface of the water. A so-called 'flying squid', *Ommastrephes bartrami*, which often swim in schools, have been observed 'flying' in groups of more than one hundred. The 'hooked squid' (*Onychoteuthis*) and the 'Humboldt squid' (*Dosidicus gigas*), also known as the 'jumbo squid', are also capable of becoming airborne, with large adults of the latter able to swim briefly at speeds approaching 30 m/sec (Benoit-Bird & Gilly 2012), and *Sthenoteuthis pteropus* up to 24.5 m/sec (O'Dor et al. 2013).

3.11 PERMANENT ATTACHMENT TO THE SUBSTRATUM

Bromley and Heinberg (2006) reviewed attachment to the substratum by living and fossil molluscs. Benthic animals in the wave-swept intertidal zone require at least temporary grip on the substratum. For most gastropods and chitons, this is achieved by foot clamping (see Section 3.10.3), but for bivalves this is not an option. Many, especially intertidal, bivalves attach using their byssus or by cementing their shell to the substratum (see Chapter 15). The latter method is also used in a few gastropods.

3.11.1 BYSSAL ATTACHMENT

Byssal threads are secreted by a gland in the foot of autobranch bivalve larvae, but this ability is lost in many adults. Byssal attachment is important for many adult pteriomorphians and some heterodont bivalves. Unlike cementation, byssal attachment is usually flexible and can often be transitory. Even in species that will normally remain attached (such as mussels), they are capable of movement along the shore by detaching and reattaching their byssal threads, including reattachment if dislodged, if conditions allow (Paine 1974). This is achieved by the byssus being released at the byssal gland, allowing the animal to creep to a new location using the (usually reduced) foot where it reattaches by secreting a new byssus. There is evidence to suggest that byssal attachment strength may increase relative to wave energy (Witman & Suchanek 1984). More details on the structure and function of the byssus and its role in bivalve evolution are given in Chapter 15.

3.11.2 CEMENTATION

A number of bivalves and gastropods, mostly in the marine environment, can cement their shells to the substratum. A few fresh-water taxa also cement, including a cyrenid from

Sulawesi (Bogan & Bouchet 1998), three genera of unionoideans, *Acostaea* (from Columbia, South America), and *Etheria* (from Africa and Madagascar) in the Mycetopodidae and *Pseudomulleria* (from India) in the Unionidae, indicating that cementation has evolved at least twice in fresh-water mussels (Bogan & Hoeh 2000). Also a very poorly known small fresh-water caenogastropod snail *Helicostoa* (Helicostoidae, ?Truncatelloidea) from the Yangtze River, China, apparently cements its shell to the substratum (Pruvot-Fol 1937).

Oysters are the best known cementing bivalves (e.g., Yonge 1960b; Stenzel 1971), with the settling oyster juvenile using a tanned mucopolysaccharide secreted by the foot for its initial attachment (e.g., Cranfield 1975). Almost immediately, spherulitic calcite cement is then deposited between the periostracum and the substratum (Harper 1992). A similar mechanism of cementation is rather widespread in bivalves, being found in spondylids, dimyids, cleidothaerids, and myochamids (Harper 1992, 1997a, 2012). Some may become secondarily detached (e.g., some oysters and chamids). The extinct rudists (Late Jurassic to the Late Cretaceous), formed extensive carbonate reefs along the shores of the Supertethys (see Chapter 15).

Most vermetid gastropods cement to the substratum, but this is the only gastropod group that does so other than a few siliquariids (e.g., *Stephopoma*) and the supposed truncatelloidean *Helicostoa* mentioned above, although hipponicids lay down a calcareous plate on the substratum to which they are attached by their foot (e.g., Knudsen 1991). Cementation also occurs in a wide variety of other invertebrates, including scleractinian and octocoral corals, craniid brachiopods, and serpulid polychaetes, although the mechanisms differ in detail.

An increase in cementing bivalve taxa in the Early Mesozoic may be related to a marked increase in predation pressure (e.g., Vermeij 1987; Harper 1991). In experiments carried out by Harper (1991), predators preferred byssate bivalves to cemented ones, apparently because cemented prey was more difficult to manipulate.

3.12 MUSCLES AND THE MUSCLE SYSTEMS

Here we deal with the main body muscular system, which is derived from the mesoderm. The muscular system varies considerably from class to class, but is usually related to the head-foot, and its interaction with the shell(s). Another important set of muscles is associated with the buccal mass (except in bivalves which lack a buccal mass), and these are dealt with in Chapter 5.

Before looking at the main muscle systems found in molluscs, we briefly outline below the structure of muscle cells and some of their physiological features. Molluscan muscle systems are both diverse and distinctive and have been a research focus for over 150 years. Besides detailed morphological and physiological studies, molluscan muscles are being investigated for insights into the molecular basis of contraction and its regulation (Hooper & Thuma 2005; Hooper et al. 2008).

Movement nearly always requires muscular action with the energy needs scaling with increases in body size, velocity, and mode of movement. Generally, large fast-moving animals require large amounts of energy, and small, slow-moving animals require small amounts. Muscle tissue occupies a significant component of the body mass – e.g., over 60% in *Octopus* (Hochachka et al. 1983a), and in most other molluscs, it also accounts for a large proportion of the body mass, with muscles in the mantle, head, buccal mass, and many parts of the gut, heart, etc., and particularly the foot. Thus, for molluscs, as for many other animals, muscular work represents one of the main, if not the main, energy demand. Energy and oxygen sources for muscles are crucial for their efficient work, and this aspect of muscle physiology is outlined in Chapter 2.

3.12.1 Muscle Cell Morphology

Chantler (1983) described the structure and biochemistry of molluscan muscles. They are unicellular, composed of elongate cells, and their thick and thin filament organisation is less ordered than that seen in most vertebrate muscles. In all molluscan muscles, the core of the thick filaments is paramyosin which is surrounded by a layer of myosin molecules. It is this layering that gives the wide range of fibre diameters and produces fibres much longer than those of vertebrates. Paramyosin is broadly distributed among invertebrates and is thought to stabilise and extend the myosin filaments (Schmidt-Rhaesa 2007). Paramyosin may also be important in maintaining contraction at a low energy cost, as required, for example, in the opaque white catch muscle of bivalve adductors (see Section 3.12.3.6) and heart muscle (Paniagua et al. 1996).

There are also 'vesicles' that can be seen in sections of the muscle cells (Figure 3.43) that are sections through various intracellular tubules and plasma membrane invaginations (e.g., Heyer et al. 1973; Silva & Hodgson 1987).

Invertebrate muscle cells are generally divided into three types – smooth, transversely (or crossed) striated, and obliquely striated, which are broadly distributed among the Lophotrochozoa (Table 3.6).

These distinctions are an over-simplification of the actual variation present in molluscan muscle. Two classification systems of molluscan muscle are outlined below, both of which rely on finer scale TEM micrographs, although the gross features such as the difference between striated muscles and smooth muscles can be seen using a light microscope. Traditional schemes classifying muscle types were primarily based on the Z-lines or bodies and their alignment. All features do not necessarily covary, and each component of the muscle cell is related to an aspect of muscle function. Another example of a classification of muscle types was developed by Nicaise and Amsellem (1983) and involved scaling six individual components of molluscan muscle. It is summarised as follows.

- 'Z' parameter – ranges from Z1 with continuous Z-bands, as in cross-striated muscle, to Z5, the random arrangement of dense bodies seen in smooth muscle.
- 'L' parameter – the length of the thick filaments (an indicator of contraction speed). This parameter is difficult to measure using TEM because of the length of the filaments (Voltzow 1994).
- 'F' parameter – diameter of the thick filaments, a feature readily measured in TEM micrographs.
- 'M' – the organisation of mitochondria ranging from rare (M1), generally present in the plane of a TEM section (M2), or arranged in packages (M3). Differences in the location of the mitochondria within the cell are not distinguished.
- 'T' parameter – describes three categories of the penetration of tubular infoldings of the plasma membrane into the myoplasm (the contractile portion of the cell) and is only superficially similar to the T system in vertebrates (Nicaise & Amsellem 1983).
- 'R' parameter – describes the condition of the sarcolemma (the cell membrane of the muscle cell), R1 where subsarcolemmal cisternae (flattened membrane discs that are part of the Golgi apparatus) are connected to a sarcoplasmic reticulum tubule system extending into the myoplasm, and R2, where the sarcoplasmic reticulum is restricted to the cell periphery.

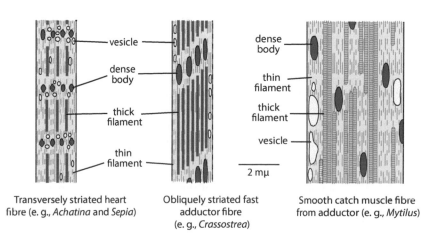

Transversely striated heart fibre (e. g., *Achatina* and *Sepia*)

Obliquely striated fast adductor fibre (e. g., *Crassostrea*)

Smooth catch muscle fibre from adductor (e. g., *Mytilus*)

FIGURE 3.43 Representative molluscan muscle fibres in each of the three main muscle types. Each diagram represents a small portion of a fibre. Redrawn and modified from Twarog, B.M., *J. Gen. Physiol.*, 50, 157–169, 1967.

TABLE 3.6

Distribution of the Three Main Muscle Types in the Lophotrochozoa

Taxon	Smooth	Transverse (or crossed)	Oblique
Brachiopoda	○	○	○
Phoronida	○	○	○
Mollusca	●	○	●
Echiura			○
Annelida	○	○	●
Sipuncula			○
Nemertea	○		○
Platyhelminthes	●	○	
Bryozoa	○	○	
Cycliophora		○	○
Entoprocta	○		○
Acanthocephala	○		
Rotifera	○	○	○
Micrognathozoa		○	○
Gastrotricha	○	○	○
Gnathostomulida		○	

Source of data: Schmidt-Rhaesa, A., *The Evolution of Organ Systems*, Oxford, UK, Oxford University Press, 2007.

Note: ● = Predominant, ○ = Present.

These parameters differ between and within taxa. For example, the Z parameter can vary between 2 (*Rossia*, cephalopod), 2–4 (*Achatina*, gastropod), and 5 (*Venus*, bivalve), whereas in the heterobranch gastropod *Lymnaea,* the head retractor, shell, and diagonal muscles differ in four of the six parameters (Nicaise & Amsellem 1983).

The invaginations of the sarcolemma and elaborations of the sarcoplasmic reticulum described by the T and R parameters are two of the most interesting and variable features of molluscan muscles (Nicaise & Amsellem 1983). Thus the ultrastructure of molluscan muscle cells reveals a much more complex situation than just the three categories discussed below (Nicaise & Amsellem 1983), with differences in the striations due to the ultrastructural organisation of Z-discs and dense bodies. The latter are the functional equivalent of the Z-line or Z-discs in vertebrate striated muscles and are found in all molluscan muscles. The speed of contraction of the muscle cell is linked to the number of contractile filaments in the cell. When moving along a gradient of striated to smooth, the contraction speed decreases, but the ability to maintain sustained tension increases.

Importantly, a single muscle mass may comprise different kinds of muscle cells so the mix of muscles determines the performance (Voltzow 1994). For example, Matsuno and Kuga (1989) reported that the translucent part of the adductor muscle of *Tridacna crocea* consisted of obliquely striated muscles and the opaque part of two types of smooth muscle. Parallel structures are seen in brachiopod adductor muscles where articulated taxa have adductor muscles with both smooth and transverse muscle types, while lingulid brachiopods have adductors with smooth and obliquely striated muscle types (Kuga & Matsuno 1988; James 1997). Only smooth muscles are reported from adductor muscles of most other bivalves (e.g., Matsuno et al. 1993) and in the columellar muscles of a range of gastropods (Frescura & Hodgson 1992). The shell muscles of limpet-like gastropods – patellids, *Haliotis*, and *Siphonaria* – have smooth muscle unusually rich in intercellular collagen (Frescura & Hodgson 1990, 1992).

Most treatments of molluscan muscles categorise them into three main muscle types, as seen in Figure 3.43. The following account is largely based on the review by Paniagua et al. (1996).

3.12.1.1 Smooth Muscle Cells

Smooth muscle is the main type found in molluscs, brachiopods, and many other animals, but is absent from arthropods. The cells are fusiform with a central nucleus. Invertebrate smooth muscle differs from that of vertebrates in having more and thicker myofilaments, but is otherwise superficially similar with the thick and thin filaments showing little order in their arrangement (Paniagua et al. 1996). The force exerted by the muscle cell is related to the number of thin and thick filaments per unit area and the endurance to the number of mitochondria, while the speed of relaxation is determined by calcium availability (Plesch 1977). Smooth catch muscle cells from the anterior adductors of the brachiopod *Lingula unguis* range between 45 and 60 nm in width and about 108 mm in length in *Terebratalia* (Eshleman et al. 1982; Kuga & Matsuno 1988). Widths of myofilaments range from around 25 to 37 nm in the columellar muscle of stylommatophoran snails and 125 to 231 nm in bivalve adductor muscles (Paniagua et al. 1996). The cell diameter is related to the thickness of the thick myofilaments, those with thicker filaments being broader, at least in bivalve adductor muscle (Morrison & Odense 1974). The thin myofilaments are 6 nm wide actin filaments with associated tropomyosin. Regions of cross-linking between myofilaments are indicated by dense (or Z) bodies, rather than the Z-lines seen when these regions are aligned. The thin filaments attach to these bodies.

The morphology of smooth muscles varies not only between species, but also between different muscle systems in the same species (e.g., Plesch 1977). Matsuno (1987) classified smooth muscle into at least four types (A–D), which have been found in molluscs. They are based mainly on the diameter of the thick myofilaments and the density and arrangement of the dense (or Z) bodies that anchor the small filaments (Matsuno 1987).

- In type A, the thick myofilament diameter is less than 14 nm and there are a few irregularly arranged dense Z-bodies. This type of smooth muscle is typical of vertebrates, but has been recorded from the adductor and foot muscle of a few bivalves and gastropods (Matsuno 1987; Lee et al. 2012) and from some other invertebrates.
- The B type is found in various invertebrates including many molluscs and echinoderms and differs from the A type in having thicker myofilaments (diameter about 40 nm) and larger irregularly arranged Z-bodies. In the foot of the caenogastropod *Bullia*, the Type A muscle is found only in the propodium, while Type B is found in the metapodium (Silva & Hodgson 1987).
- The C type cell has even thicker myofilaments (diameter 60–120 nm) and few, but large, Z-bodies. This type of muscle cell has been reported from the adductor muscles of several bivalves (Matsuno 1987).
- D type cells have thick myofilaments of 14–40 nm diameter, and the dense bodies are small, numerous, and more regularly arranged (Matsuno 1987), giving them a similar appearance to obliquely striated muscle, and it appears to be intermediate between these two types of muscle (Paniagua et al. 1996). This type of muscle is found in several groups of invertebrates, notably annelids, and occurs in the siphonal and mantle retractor muscles of cephalopods and in some muscles associated with the buccal mass in vetigastropods and caenogastropods, but is otherwise apparently uncommon in molluscs. In cross sections using a light microscope, this type of muscle appears to be striated.

Smooth muscle cells have been reported to be the only type found in polyplacophorans (Eernisse & Reynolds 1994) and monoplacophorans (Haszprunar & Schaefer 1997). Lemche and Wingstrand (1959) reported striated muscle in the radular retractors of *Neopilina galatheae*.

The function of the smooth muscle cells is regulated by their morphological characteristics, as the speed of contraction is directly related to the cell length, filament length, and width.

3.12.1.2 Transverse Striated Muscle Cells

In transverse or cross-striated muscle, the thick and thin filaments are organised into arrays of distinct A-bands, I-bands, Z-lines, etc. (see above). This most closely resembles vertebrate skeletal muscle, but differs from the striated muscle of vertebrates and arthropods in its poorly defined Z-lines which are discontinuous in appearance, consisting of multiple small electron-dense patches. Because the sarcomeres are poorly defined, this has sometimes resulted in an incorrect interpretation of transverse striated muscles as either obliquely striated or smooth (Paniagua et al. 1996). Like smooth muscles, these cells have a single central nucleus. These muscles are found in the adductor muscles of some bivalves, in the hearts of many bivalves, and in the heart (e.g., Plesch 1977), radular retractors, and foot of some heterobranch gastropods (e.g., Huang & Satterlie 1989). Each thick filament in striated muscle is surrounded by 10–12 thin filaments, with thick filaments ranging from 17 to 20 nm in diameter in bivalve adductor muscles and 24–29 nm in heart muscle (Paniagua et al. 1996). In bivalves, mitochondria are more abundant in heart muscle than in adductor muscle (Paniagua et al. 1996). Interestingly, as noted by Voltzow (1994), typical striated muscle has not been found in 'prosobranchs'. Fretter and Graham (1962: 614) commented that in a scissurellid (Vetigastropoda) all muscles were composed of striped fibres except the columellar muscle, which they noted was 'remarkable'. This observation, however, needs ultrastructural verification.

The Z-band in transverse muscle is composed of multiple electron-dense bodies made up of thin filament attachment plaques intermingled with tubules and vesicles of sarcoplasmic reticulum. Although less developed, there is a transverse and longitudinal component to the 'sarcotubular' system similar to that seen in vertebrate striated muscles. Some tubules are T-tubules that originate as invaginations of the plasma membrane. The sarcoplasmic reticulum tubules and vesicles join in the Z-lines forming dyads comprising the transverse component. The longitudinal component comprises sinuous tubules running parallel to the myofilaments, and they connect the sarcoplasmic reticulum tubules of transverse components (Paniagua et al. 1996).

3.12.1.3 Oblique Striated Muscle Cells

In obliquely striated muscle cells, the myofilaments are helically arranged sarcomeres (Figure 3.44), and they have a few large mitochondria at either end of the cell. Like smooth and transversely striated muscles, they have a single centrally located nucleus, and as in other striated muscles, there are

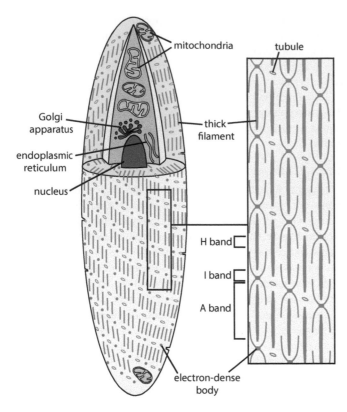

FIGURE 3.44 A diagrammatic representation of an obliquely striated muscle cell of a coleoid cephalopod. Redrawn and modified from González-Santander, R. and Garcia-Blanco, E.S., *J. Submicrosc. Cytol.*, 4, 233–245, 1972; Paniagua, R. et al., *Histol. Histopathol.*, 11, 181–201, 1996.

thick and thin myofilaments forming sarcomeres delimited by Z-lines. The myofilaments are not perpendicular, but instead are oblique to the Z-lines because of the coiling of the fibres, the myofilaments remain aligned with the longitudinal axis of the cell. Because the sarcomeres spiral through the cell, this type of muscle has also been called spiral striated muscle or, in cephalopods, helical smooth muscle by Chantler (1983). Obliquely striated muscle occurs in several other invertebrate groups, notably brachiopod adductor muscles, nematodes, some pelagic tunicates, annelids, and chaetognaths, although each has its own characteristics (Paniagua et al. 1996).

Obliquely striated muscle was first described in cephalopods, where it is a dominant muscle type, and was later shown to be common in the translucent part of the adductor muscles of several bivalves and brachiopods and in various muscle systems in *Lymnaea* (see Paniagua et al. 1996 for references).

The thick filaments are fusiform in shape, and in many molluscs each is surrounded by 12 thin filaments, although nine thin filaments for each thick filament have been observed in *Lymnaea* (Plesch 1977). The maximum width of thick filaments is usually around 20–30 nm, but in some bivalve adductor muscle can be up to 60 nm (Paniagua et al. 1996). The thick filaments are comprised of a paramyosin core covered by myosin (Elliot 1979), while the thin filaments are composed of actin, like those in other muscle types.

As in the transversely striated muscle of molluscs, the Z-lines in molluscan obliquely striated muscle are discontinuous (Matsuno & Kuga 1989), which has sometimes resulted in these muscles being wrongly interpreted as smooth muscles (Paniagua et al. 1996). The sarcotubular system is very similar to that in molluscan transversely striated muscle described above, with the oblique system equivalent to the transverse component of the transversely striated muscle (Paniagua et al. 1996).

Some striated muscles do not fit neatly into the above categories. A type of pseudostriated muscle cell in the thecosome pteropod *Clione* differs from those in other molluscs. The dense bodies form small sarcomere-like units (called pseudosarcomeres) with A-bands and I-bands, which are partially surrounded by sarcoplasmic reticulum (Huang & Satterlie 1989). Another variation, which appears to be intermediate between the helical and oblique forms of striated muscle has been recorded in the propodial foot muscle of *Bullia* (a caenogastropod) (Silva & Hodgson 1987).

Cephalopod muscles are typically obliquely striated, enabling them to contract quickly and powerfully, and parallels have been made between them and the striated muscles of vertebrates (e.g., Mommsen et al. 1981). Unlike most vertebrates, they rely mainly on carbohydrates and amino acids rather than fats for their metabolism (Hochachka 1994) (see Chapter 2). Also, in vertebrates, the mitochondria tend to be grouped centrally within each sarcomere.

In squid and cuttlefish, all mantle muscles are obliquely striated (Bone et al. 1995), but there are significant differences in their attributes. The circular muscle of the mantle wall of cuttlefish and squid contains two main types of fibres. Thin inner and outer layers are comprised of larger cells, with more mitochondria and better blood supply than the smaller cells in the wider middle layer (Bone et al. 1981; Milligan et al. 1997).

The ultrastructure of the arm transverse muscle fibres of decabrachian coleoids are like those known in most other muscular systems in cephalopods, being obliquely (i.e., helically) striated and having relatively long thick filaments. The myofilaments are arranged parallel to the long axis of the fibre, but are staggered, forming a helical alignment. In the tentacles, the transverse muscles have more elements in series and have much shorter myofilaments and sarcomeres compared with those in the arms, although the muscle biochemistry is similar (Kier & Schachat 2008 and references therein). While the change in myofilament length appears to account for the increased contraction speed, the change in striation pattern is less obvious (Kier & Schachat 2008). The evolution of different structural attributes to provide differences in performance contrasts with the situation in vertebrates, where muscle structure is relatively similar and variances in contractile performance relate to biochemical modifications, with different myofilament proteins being expressed (Kier & Schachat 2008). Transcriptome studies of coleoids, cuttlefish, and octopods have failed to identify muscle-specific myosin isoforms involved in the adjustment of contractile velocities, and only alteration of sarcomeric ultrastructure appears responsible for contractile velocity adjustments (Shaffer & Kier 2016).

3.12.2 Metabolism of Contraction

The underlying chemistry of muscle contraction is complex (see Box 3.1). At least two main kinds of muscle contraction are found in molluscs.

In the first type of contraction, typical of smooth muscle in molluscs and other invertebrates, contraction is initiated with calcium ions binding directly to the myosin head and then rapidly cycling cross-bridges to generate force. This involves myosin control of ATPase activity. In these cells, calcium ions (Ca^{2+}) initiate contraction by their interaction with calmodulin[18] to stimulate phosphorylation of the light chain of myosin by way of myosin light chain kinase. The calcium ions act directly on the myosin head forming cross-bridges which bind with the actin filament. These rapidly cycling cross-bridges generate force, allowing contraction to begin. This process is facilitated by energy released from ATP by myosin ATPase activity. Relaxation of the muscle occurs when Ca^{2+} ions are removed and myosin phosphatase is released, which inhibits the dephosphorylation of the myosin. Calcium ions are removed from the muscle cell by the Ca^{2+}ATPase pump in the sarcoplasmic reticulum and/or by similar pumps in the plasma membrane. As with vertebrate smooth muscle, there is also a sustained 'catch' phase with low calcium and where contraction can be maintained with low energy utilisation (see Section 3.12.2.1).

The second type of contraction is found in vertebrate striated muscle and in some molluscan muscles, and differs markedly from that seen in smooth muscle in having an actin-linked regulatory system. On the thin filaments, a protein (tropomyosin) covers the myosin binding sites of the actin molecules. For contraction to occur, calcium ions are pumped from the sarcoplasmic reticulum into the sarcomere. The calcium binds to the protein troponin (Tanaka et al. 2008), and then alters the structure of tropomyosin, so the binding sites on the actin myofilament are uncovered. These exposed binding sites can then bind to the 'heads' of the myosin molecules. This process is powered by ATP binding to each myosin head resulting in the chemical energy in that molecule being converted into mechanical energy by hydrolysis of ATP (involving an ATPase, myosin head ATPase) to ADP and phosphate. This results in each myosin head ratcheting its way along the actin filament. While most of the physiological details have been worked out on vertebrate muscles, myosin head ATPase has been demonstrated, for example, in some molluscan muscles including squid muscle (e.g., Shaffer & Kier 2012; Yamada et al. 2013). In contrast to the myosin in vertebrate striated muscle, molluscan myosin specifically binds two calcium ions, and these sites are essential for the regulatory activity of contraction (Bagshaw 1993).

The sarcoplasmic reticulum (a form of endoplasmic reticulum) controls the concentration of calcium ions. When calcium is released it activates muscle contraction and when removed relaxation occurs. Magnesium ions, although not directly involved in muscle contraction, play a role in the enzymes involved (e.g., Yamada et al. 2013).

3.12.2.1 Catch Muscle and Twitchin

Smooth muscle may contract and relax rapidly (phasic contraction) or may exhibit slow, sustained contraction (tonic contraction), the latter using low calcium and low energy to maintain contraction over long periods. Molluscan catch muscle, such as occurs in the dense white part of bivalve adductor muscles and in bivalve byssus retractor muscles, is unusual in maintaining contraction for very long periods of time while using little energy. This property has been known for over a century (see Bayliss 1927 for an early review).

Both cholinergic and serotonergic nerves[19] control molluscan catch muscle. Stimulation by a cholinergic nerve causes an increase in intracellular Ca^{2+} which results in phasic contraction, and the subsequent decrease in Ca^{2+} concentration initiates the catch state. Catch tension is maintained until serotonin is released, accumulating cAMP in the muscle cell (Funabara et al. 2005) (see Figure 3.46).

Catch is initiated by cholinergic nerve stimulation and occurs with a temporary increase in calcium in the nerve cell which is quickly reduced to near baseline levels. This sustained phase of contraction, or catch phase, is attributed to a protein similar to titin and connectin, called *twitchin* (Butler & Siegman 2010). Twitchin is also known in nematodes, while in insects a different protein, projectin, functions in much the same way. Projectin also occurs in scallops and *Aplysia* (Benian et al. 1996).

Activation of the muscle by stimulating a serotonergic nerve activates a calcium-dependent phosphatase. This dephosphorylates twitchin and allows catch force to be maintained when the level of intracellular calcium decreases. During the catch phase, the rate of ATP utilisation and cross-bridge cycling is extremely slow and calcium levels are low (Funabara et al. 2005; Butler & Siegman 2010).

Twitchin binds to both thick and thin filaments, but binding to the thin filaments depends on phosphorylation, and its uncoupling occurs with dephosphorylation. Regulating catch is thus based on both the attachment of myosin cross-bridges to actin and the state of phosphorylation of twitchin. The process proceeds as follows: on stimulation of the catch muscle, calcium increases in the muscle cell and activates the formation of cross-bridges and twitchin is dephosphorylated. Myosin to actin cross-bridges displace twitchin from the thin filaments and phasic muscle contraction proceeds, but as intracellular calcium concentration is reduced catch ensues, with the myosin cross-bridges detaching from the actin filaments. At this stage, there is a three-way complex between myosin, twitchin, and actin which tethers the thick and thin filaments. Catch is released when serotonergic nerves are stimulated, causing an increase in cAMP and activation of protein kinase A. This enzyme activates phosphorylation of twitchin and the twitchin tethers detach from the thin filaments (Funabara et al. 2005; Butler & Siegman 2010).

The catch muscle has a high proportion of paramyosin to myosin (2–10:1 in mass) and thick filaments that are very long

---

[18] A calcium-binding messenger protein.

[19] See Chapters 2 and 7.

BOX 3.1   THE STRUCTURE AND FUNCTION OF MUSCLE CELLS

Muscle cells have similar functions and properties in animals generally, although invertebrate muscles have single nuclei while skeletal striated muscle in vertebrates comprises multinucleate cells. Muscle cells can be very elongate and contract actively and elongate passively. Their contractile properties are conferred by contractile elements (*myofibrils*) that run the length of the cells and are comprised of bundles of thin and thick filaments composed of elongate complex proteins. The thin filaments are composed of *actin* and the thicker filaments of *myosin*. Each myosin molecule has a globular 'head' and an elongate 'tail'.

The filaments are organised into repeated subunits (*sarcomeres* – the contractile units) along the myofibril. If the sarcomeres are lined up across the myofibrils in the muscle cell, it appears to be striped, and that type of muscle cell is called 'striated'. In smooth muscle cells there is no alignment.

Some details of the morphology of each sarcomere are provided here, and, although largely based on vertebrate muscle, the same terms are used in accounts of molluscan muscle. Each sarcomere is delimited by thin dark bands of dense protein, the Z-material (or *Z-bodies* or *dense bodies*), which is arranged in *Z-lines* (or Z-discs). Both types of filaments are attached to the Z-bodies by the giant protein *titin* (or connectin), with the myosin filaments being linked across the *I-band* and the actin filaments cross-linked with titin in the Z-line via the protein α-actinin. The structure of the Z-material is thus related to the probable tension within the muscle fibre. Abutting the Z-lines in a relaxed state are the pale I-bands, which contain the actin filaments. The remainder of the sarcomere is the slightly darker *A-band* which contains both the thicker myosin filaments and actin filaments. A narrow, paler *H-band* lies in the centre of each A-band when the muscle is relaxed, and it does not contain the actin filaments (see Figure 3.45).

When the muscle contracts, the actin and myosin filaments do not change length, but slide past each other – this is known as the sliding filament concept. The actin filaments are pulled along the myosin filaments toward the middle of the sarcomere where they overlap. This results in the H-band becoming narrower and disappearing when the muscle is fully contracted, and the two Z-lines (discs) are pulled closer together (Figure 3.44).

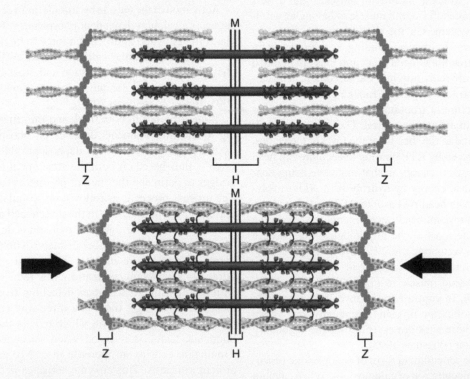

FIGURE 3.45   A diagrammatic representation of the sliding filament model of muscle contraction. Creative Commons Attribution 4.0 International.

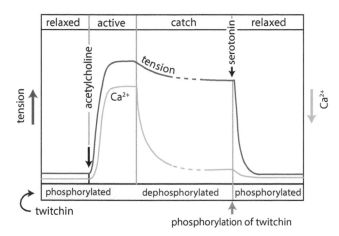

FIGURE 3.46 Schematic representation of the contraction of catch muscle. Redrawn and modified from Funabara, D. et al., *J. Muscle Res. Cell Motil.*, 26, 455–460, 2005.

(25–100 nm) and very thick (diameter 20–100 nm). This type of muscle is unlike others found in molluscs, although paramyosin is found in all invertebrate muscles (Funabara et al. 2005). Its content varies depending on species and tissues, and correlates with the tension which the muscle can develop. Because it is most abundant in catch muscles, it has been suggested that paramyosin plays a role in the thick filaments being able to stand very high tensions (Funabara et al. 2005). As noted above, paramyosin is not found in vertebrate muscles.

3.12.2.2 Muscle Activation and Control

Muscle fibres (cells) connect to nerve, muscle, and glio-interstitial tissue. The significance of the glio-interstitial connections is not certain, but these cells, which have granular contents, may play a role in the ionic regulation of the nerve and/or muscle cells (Nicaise 1973; Muneoka & Twarog 1983).

Muscles can be controlled extrinsically, or by hormones or other humoral factors (see Chapter 2), other muscles or nerves, or intrinsically by autoregulation or local humoral control (either autocrine or paracrine – see Chapter 2).

Three kinds of muscle-muscle junctions (or *myomuscular junctions*) occur in molluscs and are generally similar to those seen in other animals. These are: (1) gap-junctions (nexus)[20], (2) junctions with wider gaps and no basal lamina, and (3) attachment via desmosomes[21] (Nicaise & Amsellem 1983).

Smooth muscle contraction is initiated by the release of membrane-bound vesicles (neurotransmitter vesicles) into the space between the cells. Sympathetic fibres cause both constriction and relaxation of smooth muscle via different receptors; α adrenergic receptors primarily cause contraction, and β adrenergic receptors cause relaxation. The simultaneous stimulation of smooth muscle by both

cholinergic and serotonergic nerves causes a phasic contraction of smooth muscle, while a serotonergic nerve impulse alone initiates the catch phase (Butler & Siegman 2010) (see also Chapter 6).

The point where a nerve and muscle connect is the neuromuscular junction (or myoneural junction) and is a kind of synapse. Striated muscle contractions are initiated from nerves ending in neuromuscular junctions. The typical situation in vertebrate skeletal muscles is that nerve impulses (action potentials) pass from the neuron to the muscle fibre (cell) where they terminate and cause the muscle to contract. The terminal part of the nerve contains many vesicles which typically contain acetylcholine (ACh). A nerve impulse causes vesicles to discharge ACh onto a special membrane on the muscle fibre. This area contains a cluster of transmembrane channels that ACh opens and allows sodium ions (Na^+) to diffuse in to activate the fibre, stimulating the release of calcium ions which interact with troponin (see Section 3.12.2). This process is terminated by the enzyme acetylcholinesterase breaking down the ACh in the neuromuscular junction allowing the sodium channels to close, and the resting potential of the muscle fibre is restored by an outflow of potassium ions. This sequence of reactions also occurs in molluscs (and other animals). Many molluscan neuromuscular junctions employ neurotransmitters other than ACh, such as glutamate, dopamine, serotonin, and FMRFamide (Muneoka & Twarog 1983). This diversity of neurotransmitters means that the nervous control of muscles is often complex.

An example of a well-studied molluscan muscle system is the anterior byssus retractor muscle of *Mytilus*. Both excitatory and relaxing nerves innervate this muscle with the principal neurotransmitters released from these nerves being acetylcholine and serotonin, respectively. Various other monoamines (e.g., dopamine, octopamine) and peptides may also be involved in regulating this muscle as neurotransmitters or neuromodulators (Muneoka et al. 1991). The control by multiple neurotransmitters and neuromodulators of the anterior byssus retractor muscle of *Mytilus* appears to be typical of many molluscan and other invertebrate muscles.

3.12.3 Some Muscular Systems in Molluscs

Eight main muscle systems were recognised in molluscs by Haszprunar and Wanninger (2000). These are the body wall musculature, the buccal musculature, dorsoventral musculature (including the shell muscles), mantle retractor muscles, gut muscles, adductor muscles, enrolling muscles (in chitons and aplacophorans), and extraocular eye muscles. Below we look at some of these and other categories of muscles.

3.12.3.1 Retractor Muscles

The foot retractor muscles carry out various retraction responses. They are robust in burrowing bivalves and in gastropods where pedal manipulation and retraction is well developed. Retractor muscles perform a variety of functions including locomotion, feeding actions, reproductive behaviour, and retraction of a body part away from tactile stimuli.

---

[20] Gap junctions (= nexus or septate junction) are intercellular cytoplasmic connections that allow communication between cells.

[21] Desmosomes are protein structures that serve to adhere cells to each other.

Some of the studies on their structure in gastropods include those on head retractors (Plesch 1977), buccal retractors (Dorsett & Roberts 1980), gill withdrawal muscles (Carew et al. 1974), pteropod wing retractors (Huang & Satterlie 1989), and penis retractors (Wabnitz 1975). The byssal retractors in bivalves are derived from pedal retractor muscles.

The complex muscle system of cephalopods also includes retractor muscles such as the large retractors of the head.

3.12.3.2   Antagonistic Muscles

These are sets of muscles which work against each other by acting in opposite directions. They work in a variety of ways, not the least of which is forcing (stretching) muscles back to their original size and shape after contraction (because muscle cells can only contract). This term is commonly applied to opposing sets of muscles attached to bones in the vertebrate body, but the same functional arrangement is common in molluscan muscular systems ranging from pairs of muscles attached to the shell to sheets of muscles attached to connective tissue.

3.12.3.3   Body Wall Musculature

This basic molluscan body musculature found in most molluscs consists of outer circular, medial diagonal, and inner longitudinal muscles. In addition, dorsoventral fibres stabilise the body (Haszprunar et al. 2008). Variations exist in different groups – for example, in both classes of the shell-less aplacophorans four thick bands of muscle run the length of the body with an extra two ventral bands in Solenogastres. These muscles, along with the body wall muscles, maintain body shape. In the caudofoveates the body musculature is responsible for body movement as these animals lack a foot. The details of body wall musculature in gastropods has not been well studied, but usually consists of an outer layer of circular muscles and inner longitudinal muscles (Voltzow 1994). In the hygrophilan *Lymnaea* there is an outer layer of circular muscle, a middle layer of diagonal muscles, and an inner layer of longitudinal muscles (Plesch et al. 1975) (Figure 3.47).

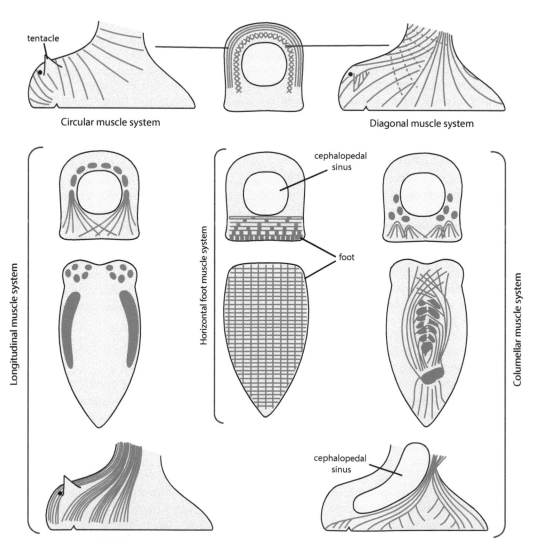

FIGURE 3.47   Major muscle systems in the hygrophilan *Lymnaea stagnalis*. Redrawn and modified from Plesch, B. et al., *Neth. J. Zool.*, 25, 332–352, 1975.

The most developed body wall muscles in molluscs are seen in coleoid cephalopods (see Section 3.12.3.8.2).

3.12.3.4 Foot (Pedal) Musculature

The molluscan foot is highly modified from the assumed plesiomorphic arrangement seen in chitons (see Section 3.8.2), but even the plesiomorphic state is highly complex with the shell muscles (Figure 3.49) comprising much of that complexity (Wingstrand 1985). The foot musculature in monoplacophorans has been described in *Neopilina* (Lemche & Wingstrand 1959; Wingstrand 1985), where there is an outer layer of circular muscles running around the circumference of the foot, this outer part is thickened and the shell muscles are embedded in it. Unlike the situation in chitons, the inner part of the sole is thin and lacking in substantive musculature. In solenogasters the rudimentary foot is not muscularised, however, oblique retractor muscles extend from the lateral sides of the body to the pedal groove (Thiele 1894). In *Proneomenia*, retractor muscle fibres from either side of the body cross and intersperse over the pedal groove (Hoffmann 1949).

The column-like foot in scaphopods is surrounded by an outer layer of circular muscles, inside which are the blocks of longitudinal muscles and radial muscles that make up the bulk of the pedal musculature (Shimek & Steiner 1997) (see Chapter 16 for details). Well-developed circular muscles are also present in the foot of protobranch bivalves (Heath 1937), but are comparatively weakly developed in the autobranch arcid *Tegillarca* and the venerid *Gomphina* (Lee et al. 2007; Park et al. 2012), while the foot of the tellinoidean *Donax* has only oblique and dorsoventral muscle fibres (Trueman & Brown 1985). In cephalopods the foot is markedly modified as the arms (Shigeno et al. 2008) and their complex musculature is briefly described below (see Section 3.12.3.8.1).

In gastropods, the highly muscular foot is very flexible and although well studied externally, surprisingly, the detailed internal muscular structure has only been studied in a few taxa (see Voltzow 1994 for review). In general, the foot of a snail is composed almost entirely of muscle which can be divided into two distinct sections – the dorsal columellar muscle and the ventral 'tarsos' (Voltzow 1985, 1986, 1994). In both, the muscles are packed in bundles bound in connective tissue, but the fibres in the tarsos are smaller in diameter and radiate to connections in connective tissue just internal to the sole epithelium. There is no outer circular muscle layer. The columellar muscle of snails is the shell muscle and attaches to the shell on the columella. It is also attached to the operculum (if present) and/or the dorsal part of the foot. Interspersed with these fibres are bundles of fibres at right angles, and this combined muscle functions as an antagonistic muscular hydrostat (Voltzow 1994). The gastropod foot does not, however, function primarily as a hydrostatic organ, and shows variation in the relative importance of fluid spaces, ranging from being a solid muscular hydrostat to a more fluid-dependent system (Voltzow 1994). Two of the most detailed descriptions of the complex head-foot musculature in gastropods are those of the panpulmonates *Lymnaea* (Plesch et al. 1975) (Figure 3.47) and *Helix* (Trappmann 1916) (Figure 3.51).

Voltzow (1988) noted that in patellogastropod limpets, which have ditaxic pedal waves, there was little crossing of muscle fibres between the left and right lateral pedal muscle blocks, as in monoplacophorans. This contrasts with other kinds of limpets [such as fissurellids (Vetigastropoda), *Crepidula* (Caenogastropoda), and *Siphonaria* (Heterobranchia)] and chitons, which all have monotaxic pedal waves and interspersed fibres from the left and right pedal muscle blocks (Figures 3.34 and 3.48).

3.12.3.5 Shell Muscles

Both polyplacophorans and monoplacophorans have up to eight pairs of composite shell attachment muscles (Figure 3.49). The dorsoventral and oblique muscles which make up the shell muscles are thought to be homologous in conchiferan taxa although there are different patterns of loss in different groups. In bivalves, the plesiomorphic shell muscles are the dorsoventral and oblique pedal muscles, while the adductor muscles are new structures necessitated by the two-part shell (see Section 3.12.3.6 and Chapter 15).

The earliest bivalves such as *Pojetaia* have multiple muscle attachment scars and living protobranchs have up to seven pairs of dorsoventral and oblique pedal muscles. Scaphopods have two pairs of dorsoventral shell muscles as does *Nautilus*, although some fossil nautiloid shells have multiple muscle scars (see Chapter 17).

Thus, in living molluscs, it may seem that multiple, paired shell muscles are plesiomorphic (Figure 3.50) with apomorphic reduction or, possibly, fusion of muscles, ultimately leading to single pair or single muscle, as in cephalopods and gastropods. This is in contrast to the condition in the earliest supposed molluscan ancestors (the small shelly fossils) which have been reported to have only a single pair of muscle scars. However, at least some of these are probably not molluscs (see Chapter 13).

Some gastropods have a pair of shell muscles, including the patellogastropods, haliotids, scissurellids, the phasianellid *Tricolia* (Marcus & Marcus 1960), and most neritimorphs, but in most others (i.e., vetigastropods, caenogastropods, and heterobranchs), there is a single shell dorsoventral attachment muscle, the columellar muscle (Figures 3.51 and 3.52). The few instances of paired shell muscles in heterobranchs (such as Rissoellidae) and the caenogastropod Velutinidae and Triviidae may be explained by heterochrony (Ponder & Lindberg 1997).

In some gastropods, (Patellogastropoda, Neritimorpha, some Cocculinoidea, and Lepetellidae in Lepetelloidea) the shell muscles are superficially divided by blood sinuses passing through them. Given the distribution of this phenomenon within gastropods, it has probably arisen independently in the four main groups (Ponder & Lindberg 1997).

In coiled gastropods, the shell muscle is attached to the columella (Figure 3.51) which, in some, bears folds thought to increase the surface area for the attachment of that muscle, but other ideas have also been advanced. These columellar folds were reviewed in 'turritelliform' gastropods by Signor and Kat (1984), who noted two additional hypotheses to explain the existence of such folds: (1) that they were deposited by

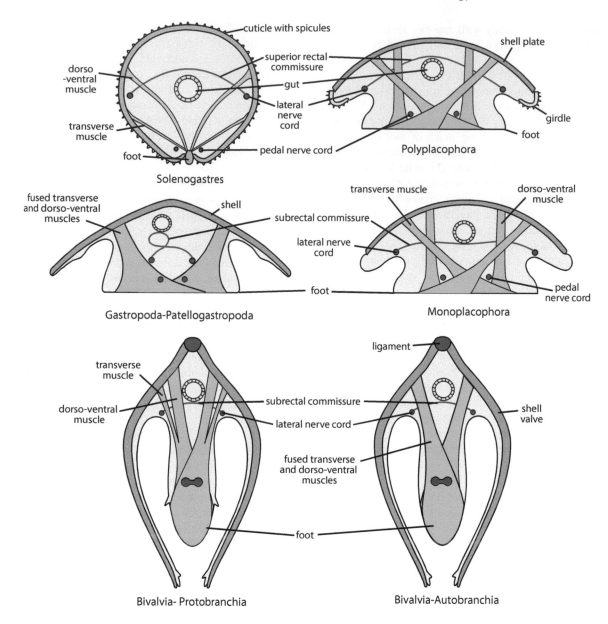

FIGURE 3.48 Diagrammatic transverse sections showing the relationship between shell muscles (i.e., the dorsoventral muscles) (in blue), intestine and rectum, and the main nerve cords in Solenogastres, Polyplacophora, Gastropoda (Patellogastropoda) and Bivalvia. Redrawn and modified from Haszprunar, G., *Z. Zool. Syst. Evol.*, 27, 1–7, 1989a.

an enlarged, folded mantle, and (2) that they were related to a siphonal notch in the aperture. Based on their observations, they rejected all three of these hypotheses. They also noted that they were highly correlated with burrowing tall-spired taxa. Price (2003) investigated four parameters of the columellar muscle in taxa with and without columellar folds. These were the area and length of attachment, the total area of contact between the muscle and the columella, and the depth of attachment inside the shell. She found no significant differences between species with and without folds. Thus, the functional significance of columellar folds is still not understood and, given that they have certainly arisen independently in several groups of gastropods, their utility may differ across taxa.

In extinct rostroconchs, the putative foot retractor muscles appear as a circular or horseshoe-shaped grouping around the upper part of the shell interior, while in most bivalves, they are divided into paired anterior and posterior groups (Waller 1998). Protobranch bivalves have multiple pairs of anterior shell muscles and usually three posterior pairs (Heath 1937; Driscoll 1964). The number of muscles has been reduced in autobranch bivalves which often have only a single pair of anterior and posterior oblique retractors and a single anterior oblique protractor (Figure 3.52).

There are parallel trends in the reduction in the number of shell muscles in bivalves, gastropods, and cephalopods. In some extinct nautiloids with conical shells there were

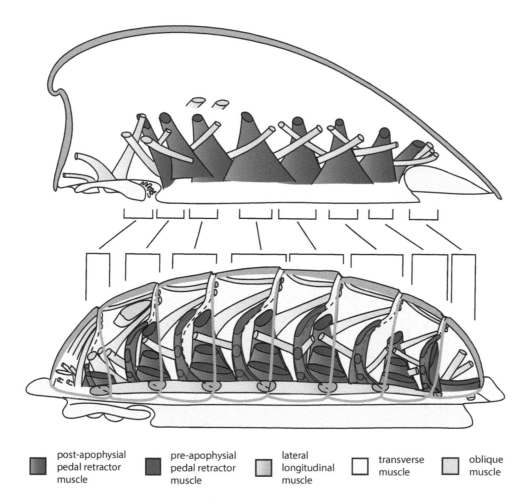

post-apophysial pedal retractor muscle   pre-apophysial pedal retractor muscle   lateral longitudinal muscle   transverse muscle   oblique muscle

FIGURE 3.49 Comparison of the shell and foot musculature of a monoplacophoran and a chiton. Redrawn and modified from Wingstrand, K.G., On the Anatomy and Relationships of Recent Monoplacophora. *Galathea Report*, 16: 1–94., 1985.

multiple shell muscles (see Chapter 17), which may have been reduced to two pairs (as in living *Nautilus*) or a single pair as in coleoids. Fossil nautiloids have three main muscle attachment areas, all in the posterior part of the body chamber (Mutvei 1957). These consist of two continuous bands for attachment of the body and mantle muscles and two pairs of scars for the attachment of the dorsoventral retractor muscles. In some Cretaceous ammonites, there was a pair of muscles dorsally and a single ventral muscle (Jones 1961).

Adult and larval shell muscles are not necessarily homologous (see Chapter 8 for an overview of their ontogeny).

3.12.3.6 Adductor Muscles

Adductor muscles are the muscles that pull the valves together in a bivalved shell. In living molluscs they have been independently evolved in Bivalvia and in the bivalved sacoglossan heterobranch gastropods (see Chapter 20).

The adult adductor muscles of bivalves have undergone many transformations in bivalve evolution (see Chapter 15). Adductor muscles are a 'functional requirement of a bivalved shell' (Waller 1998, p. 10), as they are present in bivalved animals as diverse as various crustaceans, brachiopods, and

bivalved gastropods, but not the extinct bivalved group the rostroconchs. The anterior adductor muscle develops first in ontogeny. All bivalve juveniles are dimyarian, even if monomyarian as adults (Waller 1998), and the rectum always lies above the posterior adductor muscle.

The adductor muscles of bivalves have been the subject of many ultrastructural and physiological studies, in part through interest in the diversity of muscle fibres, especially the smooth catch muscle (e.g., see Section 3.12.2.1).

3.12.3.7 Odontophoral Muscles

The muscles involved in the movement of the mouth, buccal cavity, and the odontophoral apparatus (including the radula) comprise the most complex set of muscles in the body. Bivalves have less complexity than other molluscs because of the lack of a radula, although multiple pairs of mouth muscles, large buccal muscles, and palp musculature are present in protobranchs (Heath 1937) since they are deposit feeders, but are lacking in suspension-feeders (autobranchs). In radula-bearing groups, these muscles are usually red because of the myoglobin they contain. Details of the arrangement and functioning of the buccal and odontophoral muscles are given in Chapter 5.

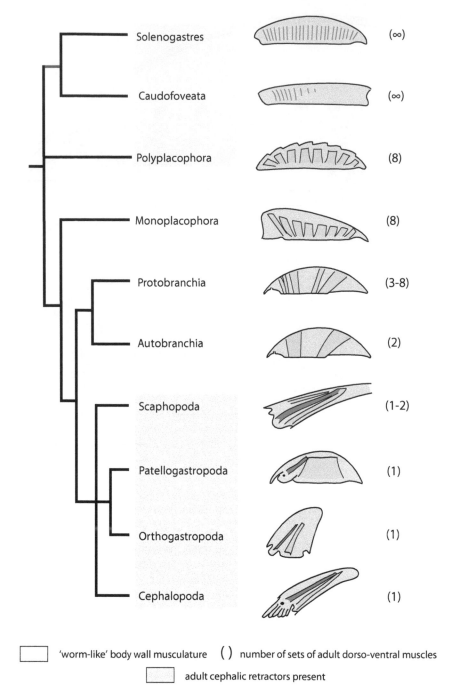

| | 'worm-like' body wall musculature | ( ) | number of sets of adult dorso-ventral muscles |

adult cephalic retractors present

FIGURE 3.50 The main muscles in molluscs. Redrawn and modified from Wanninger, A. and Haszprunar, G., *J. Morphol.*, 254, 53–64, 2002a. The dark blue muscle indicates the cephalic retractor muscles.

3.12.3.8 Specialised Hydrostatic Muscular Systems in Coleoids

There are several molluscan systems that are muscular hydrostats. Bivalve siphons are a good example (see Chapter 15) as is the proboscis of certain caenogastropods (Golding et al. 2009b) (see Chapter 5).

Among the most specialised are the muscular hydrostat systems of coleoid cephalopods (e.g., Kier 1988), which we briefly review below. More details on these structures can be

found in Chapter 17. These include the muscles in the mantle, fins, and arms and tentacles.

3.12.3.8.1 Arms and Tentacles

The arms and tentacles of coleoids are almost entirely composed of muscle arranged in a complex series of layers (Figure 3.53), and they typically bear suckers (see Section 3.12.3.8.4). The arm musculature functions as a muscular hydrostat (Kier 1985, 1988). An outer layer of

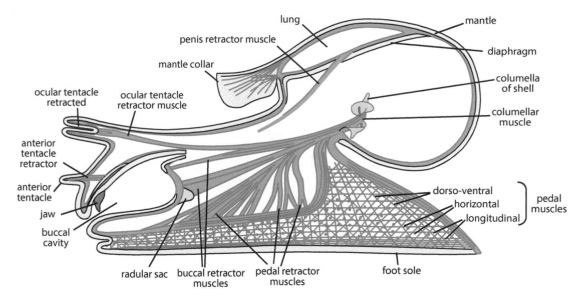

Muscles in *Helix pomatia*, seen in longitudinal section from the left side of the body with the shell removed (other than a small part of the columella). Odontophore not shown.

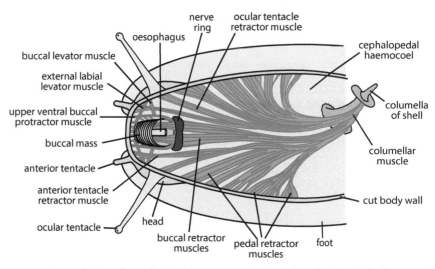

Dorsal view of muscles in *Helix pomatia*, with the dorsal wall and shell removed.

FIGURE 3.51 Musculature of *Helix pomatia*. Redrawn and modified from Trappmann, W., *Z. wiss. Zool.*, 115, 489–585, 1916.

longitudinal muscle is underlaid by connective tissue and oblique muscle layers. This relatively thin sheath of muscle encases a thick ring of bundles of longitudinal muscles that surround a thick core of transverse muscle with the fibres running at right angles to each other and extending between the longitudinal bundles. The axial nerve cord and an artery run through the centre of the transverse muscle. The cirri of *Nautilus* are comprised of similar layers of muscles, but the longitudinal bundles are relatively larger and the transverse muscle mass smaller.

The arms are used in feeding and mating, and sometimes can be involved in locomotion or signalling (Hanlon & Messenger 1996) and even burying (e.g., sepiolids, Boletzky & Boletzky 1970; Huffard et al. 2005). The paired

tentacles of decabrachians are used only for prey capture and are capable of very rapid extension. In the squid *Doryteuthis pealei*, this involves the eight arms flaring open and the two tentacles extending very rapidly (about 20–40 ms) (Kier & Leeuwen 1997). The terminal part of the tentacles has suckers and grasps the prey. The transverse muscles are primarily responsible for this remarkably rapid extension of the tentacles. Very similar, and probably homologous, muscles bend and support the arms, but those in the tentacles are modified and differ in several ways from the arm muscles (Kier & Schachat 2008).

The arms of octopuses are extremely flexible – much more so than 'limbs' in other animals. Recent studies have investigated the system of movement in some detail. It is only

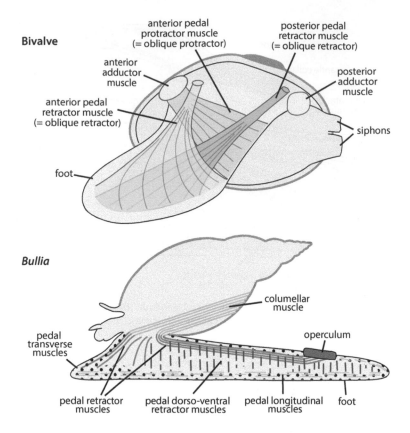

FIGURE 3.52   Comparison of the main shell and foot musculature in a heterodont bivalve and a gastropod (*Bullia*). Bivalve redrawn and modified from Trueman, E.R., Locomotion in mollusks, pp. 155–198, in A. S. M. Salauddin & Wilbur, K. M., *Physiology, Part 1. The Mollusca*, Vol. 4, Academic Press, New York, 1983, and *Bullia* redrawn and modified from Trueman, E.R. and Brown, A.C., *J. Zool.*, 178, 365–384, 1976.

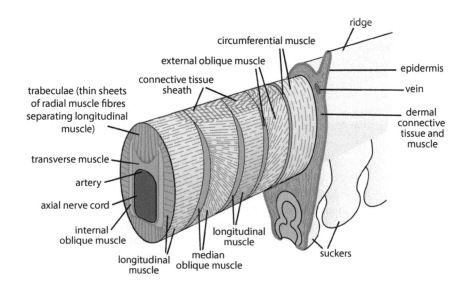

FIGURE 3.53   Details of the musculature of an arm of *Octopus*. Redrawn and modified from Kier, W.M., The arrangement and function of molluscan muscle, pp. 211–252, in Trueman, E.R. and Clarke, M.R., *Form and Function. The Mollusca*, Vol. 11, Academic Press, New York, 1988.

possible because of the highly organised and complex system of motor control. Octopuses also can transform the arm that is carrying out precise movements into a pseudo-jointed structure (Yekutieli et al. 2002, 2005a, b; Sumbre et al. 2005), with many of these actions independent of the brain (Sumbre et al. 2001, 2005).

3.12.3.8.2   Mantle Wall and Funnel

The contraction of mantle muscle fibres in the mantle wall powers the jet propulsion that is a hallmark of cephalopods, and the whole system acts as a muscular hydrostat (Ward & Wainwright 1972) with the mantle wall muscles pressurising the water in the muscle tissue (Gosline & Demont 1985). In all cases there is an antagonistic action of circular and radial muscles. Contraction of the circular fibres expels the water (as the locomotory jet) from the mantle cavity while tension in the longitudinal muscles prevents lengthening of the mantle, and the radial muscles are stretched as the walls thicken (Gosline & Demont 1985). Contraction of the radial fibres makes the mantle wall thinner (and increases the volume of the mantle cavity) and water is drawn in. As in octopods, there are radial and circular muscles in decabrachians, but the longitudinal muscles are replaced by outer and inner layers of stiff tunic composed of layers of collagen fibres that run at right angles to each other. Connective tissue fibres also run diagonally through the muscles (e.g., Budelmann et al. 1997) (Figure 3.54). The tunics prevent the mantle from lengthening when either the radial or circular muscles contract and also provide a firm insertion surface for the radial muscles. The arrangement in octopods enables greater flexibility and movement of the mantle than is possible in the much more rigid arrangement in decabrachians.

In decabrachian coleoids, the radial muscles are entirely fast twitch, glycogenic fibres, contracting during jetting. The thick layer of circular muscles consists of a central mitochondria-poor zone of fast twitch fibres used in jetting and an outer and inner layer of mitochondria-rich slow-twitch, aerobic fibres used in slow swimming. The complex arrangement of collagen fibres consists of some that are stretched on contraction of the circular muscles and others that stretch when the radial muscles contract when the mantle cavity fills, with their elastic rebound helping to expel the water (Gosline & Demont 1985).

The mantle muscle is markedly thicker in coleoid cephalopods that utilise powerful jet propulsion than in those that do not (Trueman & Packard 1968). The pulsating mantle also has an important role in respiration when the animal is at rest. In *Sepia* at least, this is due to the alternating activity of the radial muscles of the mantle and the musculature of the collar flaps, the circular muscle fibres of the mantle are not involved (Bone et al. 1995).

In *Nautilus*, the funnel is a mobile fold of tissue that can be rolled into a tube, while in coleoids it is a tubular structure. The jet stream of water from the mantle cavity is directed through the funnel and is used in steering the animal by it pointing in a particular direction. Other functions include squirting a jet of water at a predator, and (for female octopuses) aerating their egg mass, while some benthic octopuses direct it downward to blow a cavity in soft substratum so they can bury themselves (Hanlon & Messenger 1996).

3.12.3.8.3   Coleoid Fins

Fins are present in all decabrachians, in cirrate octopods, and *Vampyroteuthis*, and they vary in size, shape, and function (Hanlon & Messenger 1996). In sepioids and squid the transverse and dorsoventral muscle bundles work antagonistically and are divided horizontally in the middle section of the fin by a layer of connective tissue (the median fascia) which provides elastic support. There are also strands of connective tissue that run diagonally through the muscles. Again, the whole system acts as a muscular hydrostat as do the arms and tentacles (Kier 1985).

3.12.3.8.4   Suckers

Coleoids often have suckers arranged in rows on their arms (see Chapter 17). Each sucker can be divided into an inner cavity (acetabulum) and the outer part that attaches, the infundibulum, both separated by a narrow section ringed with a sphincter muscle. The infundibulum may be attached broadly to the arm or by a narrow stalk. Its rim forms the outer part of the sucker and may be lined with chitin which can be denticulate or toothed or form one or more hooks.

Three sets of muscles act for both the acetabulum and infundibulum. Radial muscles lie perpendicular to the inner surface of the sucker and there are circular muscles (including the sphincter already mentioned) around the circumference of the sucker. In addition, meridial muscles run at right angles to the two other sets of muscles and with the circular muscles presumably act antagonistically with the radial muscles. The coordination of this complex musculature is assisted by a small ganglion that lies below each sucker.

3.12.3.8.5   Skin Papillae and Chromatophores

Both sepioids and octopods have complex muscles that change the skin texture – from smooth to spike-like papillae that can be 10 mm or more high (Hanlon & Messenger 1996).

Chromatophores have a muscular system that enables them to rapidly expand or contract, thus quickly changing the skin colour (see Chapter 17).

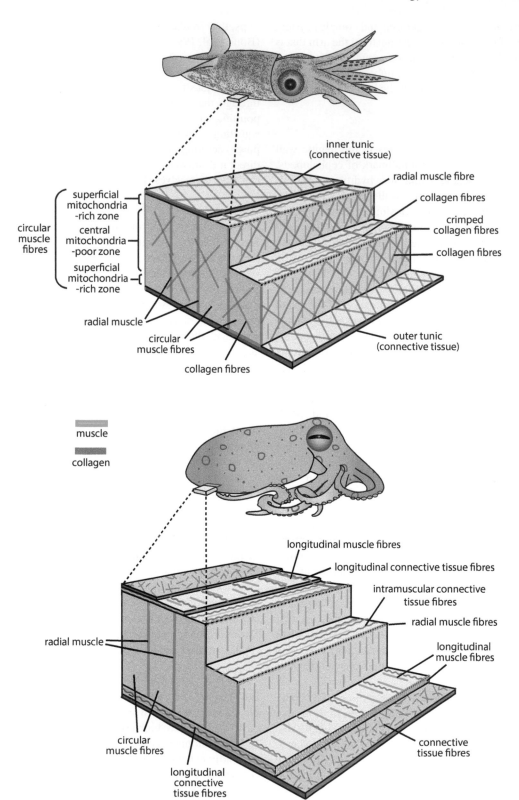

FIGURE 3.54 Diagram of the musculature and connective tissue in a ventral region of the mantle of a squid and an octopus, with outer (skin) layer at the bottom. Redrawn and modified from Kier, W.M. and Thompson, J.T., *Berl. Paläobio. Abh.*, 3, 141–162, 2003.

4 The Mantle Cavity and Respiration

4.1 INTRODUCTION

This chapter deals with the mantle cavity and the structures it contains, and with respiration. The mantle itself, and the head and foot (described in Chapter 3) are in direct contact with the external environment and thus play significant roles in the interactions of the animal with that environment, including at least part of the respiratory role in many taxa. The structures in the mantle cavity are also exposed to the external environment and carry out respiratory and sensory functions, while the cavity itself provides a space into which gametes and waste can be emptied.

Besides the important functional aspects of the mantle cavity, we also outline the very significant role it has played in molluscan evolution.

4.2 THE MANTLE CAVITY

The mantle cavity (also often called the pallial cavity) is a hallmark of the Mollusca. It contains a characteristic array of structures and their associated blood vessels, nerves, and muscles. These structures are involved in respiration (gills or lung) and sometimes feeding (gills), sensory reception (osphradia and other taxon-specific sensory structures), reproduction (gonopores), and excretion (anus and kidney openings). Other structures (rectum, pallial gonoducts, part or all of the kidney and pericardium) may extend into the haemocoel of the roof of the mantle cavity (Fretter & Graham 1962; Lindberg & Ponder 2001).

The mantle cavity of the Solenogastres (Figure 4.1) is often considered a true cloaca, although this putative homology is unlikely. A cloaca is a vestibule that serves as the common opening for the alimentary, reproductive, and excretory tracts (Hejnol & Martindale 2009). Developmentally, the true cloaca

is formed when the reproductive and excretory tracts fail to separate from the alimentary tract. In molluscs, these three tracts are never conjoined during ontogeny (Raven 1966) (see Chapter 8). In Solenogastres, there are no excretory organs, and the reproductive opening is ontogenetically separate from the alimentary system (Thompson 1960; Okusu 2002).

A 'true' cloaca is present in several deuterostome groups including the holothurian echinoderms, elasmobranchs, sarcopterygians, amphibians, birds, reptiles, and monotremes. In the Ecdysozoa, a cloaca occurs in the Nematoda, Tardigrada, and Nematomorpha. Although a true cloaca is found in the Rotifera, there is no true cloaca in any other lophotrochozoans (Hejnol & Martindale 2009). Cloaca-like vestibules are also present in some arachnids.

The molluscan mantle cavity is one of several enclosed spaces or vestibules found in metazoan taxa that combine respiration with the typical structures and openings seen in cloacae. True cloacae with respiratory structures include those in holothurian echinoderms which have a combined respiratory/excretory tree-like structure that opens into a posterior cloacal chamber (Brusca et al. 2016b) and in some diving turtle taxa which have accessory air bladders attached to the cloaca (FitzGibbon & Franklin 2010). In deuterostomes, the Urochordata have an atrium (or atrial cavity) that houses the openings of the alimentary, reproductive, and excretory tracts and encloses a pharynx where respiration and feeding occurs. The Brachiopoda also have a shell-enclosed cavity that houses the openings of the alimentary and excretory tracts and the lophophore – a respiration/feeding structure. Cloaca-like structures in the Lophotrochozoa are present only in aplacophorans, but because ectodermal tissue lines the cavity, it is not formed from the terminal portion of the alimentary tract as in a true cloaca. Like the cavities in

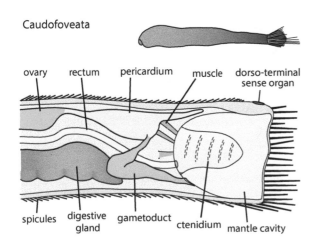
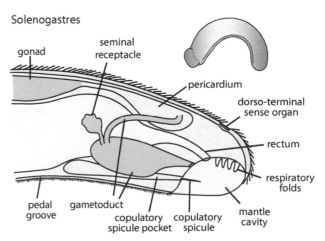

FIGURE 4.1 Aplacophoran mantle cavities. Redrawn and modified from Salvini-Plawen, L.v., *Malacologia*, 9, 191–216, 1969.

Urochordata and Brachiopoda, the molluscan mantle cavity typically houses respiratory structures and sometimes feeding structures.

These cloaca-like cavities have undoubtedly formed independently in Mollusca, Brachiopoda, and Urochordata (Dunn et al. 2008, 2014) and confirm that the molluscan mantle cavity is an apomorphy. Given the convergence among these taxa, and the general similarity of these cavities with the 'cloaca' in Ecdysozoa and Deuterostomia, it is clear that vestibules such as the mantle cavity are common evolutionary solutions

built utilising a variety of developmental pathways (Hejnol & Martindale 2009).

The mantle cavity is typically formed around the central body region (visceral mass and foot) by an expansion of the dorsal mantle epidermis which envelopes the head and foot. In shell-bearing groups, this expansion extends beyond the perimeter of the body and produces a space beneath the shell – the mantle cavity. In limpet-like taxa (polyplacophorans, monoplacophorans, and some gastropods), it is presented as a narrow cleft, the *mantle groove*, surrounding the foot (Figure 4.2).

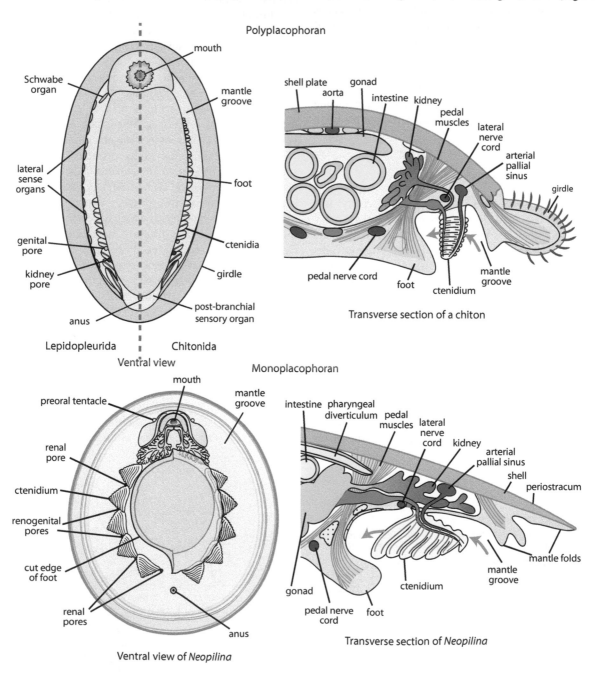

FIGURE 4.2 Details of the mantle cavities (as mantle grooves) of a polyplacophoran (chiton) and the monoplacophoran *Neopilina*. The left and right sides of the chiton show the general configuration in Lepidopleurida (left) and Chitonida (right). *Neopilina* and chiton TS (transverse section) redrawn and modified from Lemche, H. and Wingstrand, K.G., *Galathea Rep.*, 3, 9–71, 1959 and ventral view of the chiton redrawn and modified from Yonge, C.M., *Q. J. Microsc. Sci.*, 81, 367–390, 1939a.

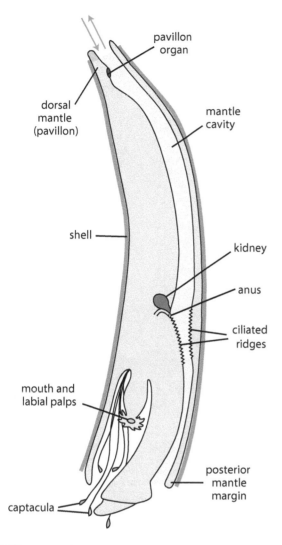

FIGURE 4.3 A diagrammatic longitudinal section of a scaphopod showing the mantle cavity. Redrawn and modified from Reynolds, P.D., *Adv. Mar. Biol.*, 42, 137–236, 2002.

In polyplacophorans, the mantle groove epithelium, unlike the ventral mantle epidermis, lacks a cuticle (Hyman 1967).

The mantle cavity varies in size and position across the Mollusca. In scaphopods (Figure 4.3) and bivalves (Figure 4.4), it surrounds much of the body. In cephalopods, it is a long muscular cavity lying on the functionally ventral (but morphologically posterior) side of the body. In gastropods, it lies anteriorly (above the head) or to the right side, due to torsion (see Chapter 8). While we cannot be certain, the original gastropod mantle cavity may have been shallow and wide, with deepening occurring independently in different gastropod lineages because of changes such as an increase in body size and/or activity levels, exposure to aerial conditions, exposure to lower oxygen tensions, or the adoption of suspension-feeding. While the mantle cavity organ arrangement in caenogastropods shows relatively little variation, there is a variety of arrangements in vetigastropods

and even more in heterobranchs, as outlined below and in Chapters 18 through 20. For example, in some gastropods, mainly shell-less forms, the mantle cavity is reduced and even lost and, in 'opisthobranchs,' the mantle cavity became smaller as it moved along the right side of the body during the detorsion (see Chapters 8 and 20) that occurred in that group (e.g., as in Aplysiidae) or is almost absent (e.g., as in pleurobranchs), and, in the nudibranchs that have become secondarily bilaterally symmetrical, there is no mantle cavity (see Chapter 20).

Just as the mantle cavity varies in size and position across molluscan groups, the associated structures also vary in their occurrence, position, and number both between and within groups. Further complicating the patterns of these shared structures are taxon-specific structures, especially among the sensory surfaces (see Section 4.2.1.8 below). This variability has contributed to extensive discussions of the ancestral state of the molluscan mantle cavity, primarily whether it was metameric with multiple sets of correlated gills, pores, muscles, vessels, and nerves (Lemche & Wingstrand 1959; Wingstrand 1985) or whether the ancestral state was a mantle cavity with only a single pair of gills, pores, etc. (Huxley 1853; Lankester 1883; Yonge 1947; Morton & Yonge 1964). In the former case, the evolutionary trend would be losing mantle cavity structures, while the latter scenario would require the addition of structures. Regardless of the pathway, losses or additions have not been metameric, and the positional correlation between the serial sets of the various mantle cavity structures is difficult to contain within a single developmental hypothesis or process, even in the iconic Monoplacophora (Haszprunar & Schaefer 1996) (see also Chapter 12).

The mantle cavity of *Nautilus* is illustrative of this uncertainty in mantle cavity evolution. Lemche and Wingstrand (1959) and Wingstrand (1985) considered two pairs of gills and kidney openings in the *Nautilus* mantle cavity as primitive, although Willey (1902) had earlier argued that it was not. Although there are two pairs of gills, the two pairs of sensory organs treated as osphradia in *Nautilus* are innervated differently than would be expected from serial replication or deletion (Willey 1902). In molluscan classes where both ctenidium and osphradium are present, a single shared nerve is plesiomorphic. This indicates that the posterior sense organ in *Nautilus* is not an osphradium as it is viscerally innervated, and it should be regarded as a *de novo* structure (see Section 4.2.1.8.5). As discussed in Chapter 12, the serial repetition of organs is probably an autapomorphy of monoplacophorans, and a few examples of independently derived, serially repeated, mantle cavity structures are also seen in other molluscan classes (Wanninger et al. 1999b; Haszprunar & Wanninger 2000), as described below.

4.2.1 Mantle Cavity Structures

4.2.1.1 Ctenidia

Molluscan gills are called ctenidia, a term used to distinguish the 'original' molluscan gills thought to be homologous throughout the phylum. In some taxa, gill-like structures have

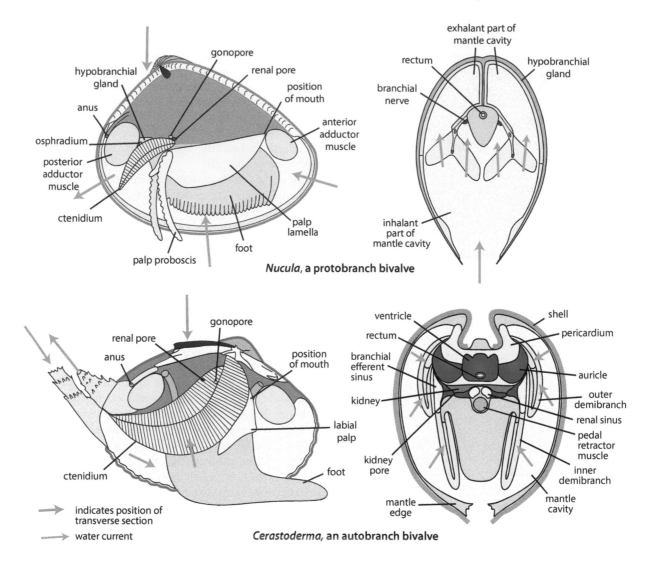

FIGURE 4.4 Protobranch and autobranch bivalve mantle cavities. Cardiid lateral view redrawn and modified from Yonge, C.M., *Proc. Zool. Soc. Lond.*, 123, 551–561, 1953b, Transverse section redrawn and modified from Johnstone, J., *LMBC Mem.*, 2, 1–84, 1899, lateral and TS view of *Nucula* redrawn and modified from Yonge, C.M., *Philos. Trans. Royal Soc. B*, 230, 79–147, 1939b.

evolved that are not homologous with ctenidia, and the term 'secondary gill' is used to denote secondarily evolved respiratory surfaces that function as gills and sometimes resemble ctenidia. In many molluscan taxa, the ctenidial gill has functions other than just respiration, notably the creation of a water current by a band of cilia on the filaments making up the gill, waste disposal or, sometimes, food collection.

A ctenidium is characterised by: (1) a central axis (Figure 4.5) which bears numerous leaflets (or filaments), (2) afferent and efferent vessels or sinuses which, respectively, move haemolymph to and from the leaflets, (3) one or more retractor muscles or fibres, and (4) one or more nerves. The leaflets are typically (5) offset on either side of the axis and their outer surfaces covered by (6) distinct bands of long cilia (the lateral cilia), which generate water currents and (7) shorter cilia along the leaflet edge that facilitate rejection of potentially

clogging particles. In several, sketetal rods (8) along the length of each filament are present. Some of these characters are lost in taxa with reduced or vestigial ctenidia (e.g., *Rhodopetala* and *Cocculina* [Lindberg 1981; Haszprunar 1987a]). It is probable that the common molluscan ancestor had gills with rounded filaments lacking supporting rods and bearing ciliary tracts that created a water current for ventilation. This configuration resembles the ctenidia seen in chitons and Caudofoveata.

The lateral cilia on the surfaces of the filaments create a water current that passes from the outer side to the inner side and, as it does, oxygen is taken up by blood passing from the afferent to the efferent sinuses that run along the median outer and inner surfaces of the ctenidia, respectively. The direction of blood flow and water flow is diametrically opposed, setting up a counter-current exchange system (Figure 4.20) that facilitates the uptake of oxygen by the haemolymph passing

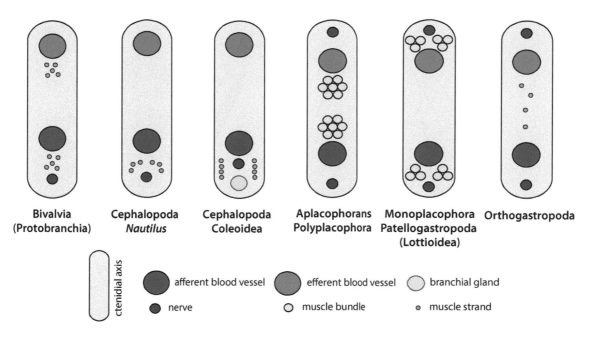

FIGURE 4.5 Schematic drawings of cross sections through ctenidial axis. Original, with data from the following sources: Bivalvia (Yonge, C.M., *Philos. Trans. Royal Soc. B*, 230, 79–147, 1939b), Cephalopoda (Haas, W., *Abh. Geol. B.*, 57, 341–351, 2002), Aplacophorans and Polyplacophora (Salvini-Plawen, L.v., Early evolution and the primitive groups, pp. 59–150, in Trueman, E.R. and Clarke, M.R., *Evolution. The Mollusca.*, Vol. 10, Academic Press, New York, 1985a), Monoplacophora (Lemche, H. and Wingstrand, K.G., *Galathea Rep.*, 3, 9–71, 1959), Lottioidea (Thiem, H., *Jena. Zeitsch. Naturwiss.*, 54, 405–630, 1917b) and Orthogastropoda (Bernard, F., *Bull. biol. Fr. Bel.*, 22, 253–361, 1890a).

through the leaflet. The epithelium of the afferent portion of the leaflet is the thinnest, while the efferent sinus of the leaflet lies in juxtaposition with the thickened epidermis from which the lateral cilia arise.

4.2.1.1.1 Polyplacophora

In chitons, multiple ctenidia are arranged serially in the mantle groove on either side of the foot, and water currents pass from anterior to posterior through the grooves (Yonge 1939a; Fischer et al. 1990) (Figure 4.6). The number of gills differs between chiton taxa (Figure 4.2) and can even vary between individuals and sides. The largest gills are located posteriorly, the smallest anteriorly. The filaments lack supporting rods and are short and semicircular in shape (Figure 4.6).

Each chiton ctenidium is suspended from the roof of the mantle groove and consists of tapering filaments that alternate along an axis (Figure 4.6). The filaments are 'free' – the axis is not attached to the mantle by a membrane as in many other molluscs (see below), and multiple pairs of muscles manipulate each gill (Figure 4.5).

4.2.1.1.2 Aplacophorans

Of the two aplacophoran groups, only the Caudofoveata have a pair of ctenidia in a small posterior mantle cavity (Figures 4.1 and 4.7). These lack supporting rods and resemble chiton gills in general morphology – they are free, with a long axis from which several short, rounded, alternating filaments arise. The ctenidia are attached to the proximal end of the mantle cavity, and water passes over the filaments from their

dorsal (with the efferent sinus) to their ventral (with the afferent sinus) surface (Salvini-Plawen 1985a), the same direction of flow seen in chitons. Members of the other aplacophoran group, Solenogastres, lack ctenidia, but instead the walls of the mantle cavity bear lamella-like folds or papillae thought to serve as respiratory surfaces (Figure 4.1).

Both groups are often illustrated with respiratory structures extending beyond the confines of the mantle cavity (Hyman 1967; Salvini-Plawen 1985a), although this is most likely a fixation or preservation artefact. In life, the mantle cavity is typically wide open with just the tips of the gills, papilla, or folds visible within the cavity, although in low oxygen environments, these structures can be extended beyond the mantle cavity (C. Todt *pers. comm.*, January 2015).

4.2.1.1.3 Monoplacophora

Depending on the size of the species, monoplacophorans have between three and six pairs of ctenidia. Each is a single row of rounded, muscular filaments hanging in the mantle groove surrounding the foot, as in chitons. The ctenidia resemble those in chitons in having free filaments and lacking supporting rods, but they do not have organised bands or patches of cilia. Instead, the leaflet surfaces are densely covered with cilia (Lemche & Wingstrand 1959; Haszprunar & Schaefer 1997). The thickened epithelium that underlies these cilia suggests that the leaflets are not primary respiratory surfaces, and it is probable that respiration is mainly via the surface of the mantle groove (Lindberg & Ponder 1996). In monoplacophoran gills, the internal gill muscles

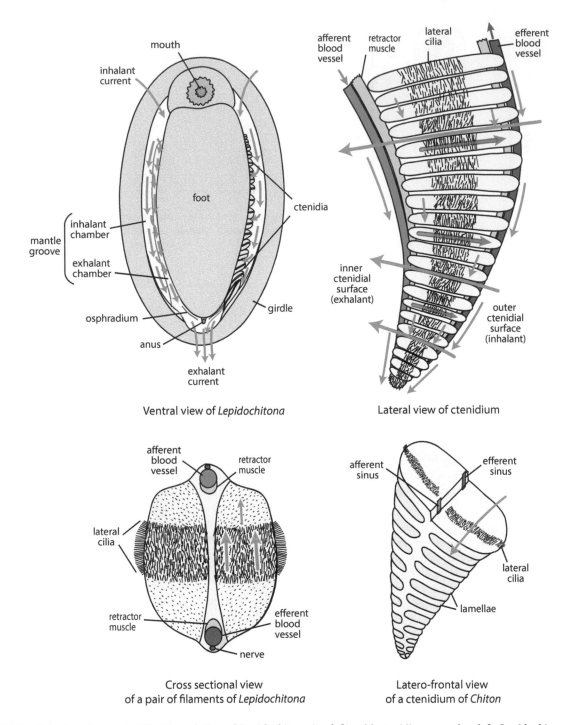

FIGURE 4.6 Polyplacophoran ctenidia. Ventral view of *Lepidochitona* (top left), with ctenidia removed on left. *Lepidochitona* ctenidium (top right) showing water currents and blood flow, and a pair of filaments (bottom left) showing water currents. Bottom right, a 3D diagram of a ctenidium of *Chiton*. Figures of *Lepidochitona* redrawn and modified from Yonge, C.M., *Q. J. Microsc. Sci.*, 81, 367–390, 1939a with the blood flow and water currents on the ctenidium redrawn and modified from Petersen, J.A. and Johansen, K., *J. Exp. Mar. Biol. Ecol.*, 12, 27–43, 1973. *Chiton* ctenidium redrawn and modified from Fischer, F.P. et al., *J. Morphol.*, 204, 75–87, 1990.

are also denser and more defined compared to those of chitons (Lemche & Wingstrand 1959), presumably due to their use as ventilators (fans) rather than primarily as respiratory surfaces. If this is correct, the main purpose of the relatively well-developed blood supply in the filaments may be to facilitate muscular activity. We are not aware of any direct

observations having been made on the water currents in the mantle grooves of monoplacophorans, but they have been reconstructed (Lemche & Wingstrand 1959) as being similar to those in chitons, with water moving from anterior to posterior and across the gills from the outer side to the inner side (see Chapter 14, Figure 14.12).

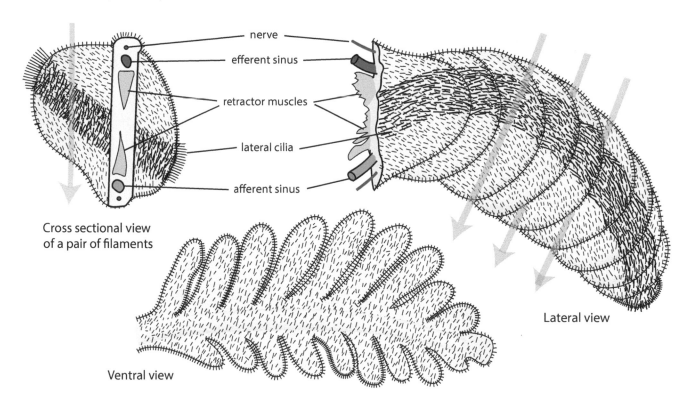

nerve

efferent sinus

retractor muscles

lateral cilia

afferent sinus

Cross sectional view
of a pair of filaments

Lateral view

Ventral view

FIGURE 4.7   Ctenidium and a pair of filaments of the caudofoveate *Falcidens crossotus*. Redrawn and modified from Salvini-Plawen, L.v., Early evolution and the primitive groups, pp. 59–150, in Trueman, E.R. and Clarke, M.R., *Evolution. The Mollusca.*, Vol. 10, Academic Press, New York, 1985a.)

4.2.1.1.4   Bivalvia

Bivalves have a single pair of ctenidia enclosed in a very expansive mantle cavity that generally surrounds the entire body and is convergent with the brachiopod condition. Ctenidial morphology differs considerably between the two major groupings of bivalves – the protobranchs and the autobranchs (Figure 4.8).

Protobranch gills are used primarily for respiration and are typically rather small and short; the filaments are broad and sickle-shaped, with lateral cilia and laterofrontal cilia fields. The gill axes run longitudinally along the body and are attached to the body dorsally in the Nuculida and laterally in the Solemyoida (Yonge 1939b); the leaflets are also substantially larger in the solemyoids than in the nuculoids. Water flows into the mantle cavity ventrally and passes over the surface of the ctenidia to exit dorsally. In the Nuculida, the gill filaments lie opposite each other on either side of the axis, while in the other protobranchs they alternate, as in other molluscan ctenidia (Reid 1998). The respiratory currents are weak compared to the autobranchs, but that is not surprising given the considerable enlargement of the ctenidia in that taxon.

In contrast to protobranchs, the gills of autobranch bivalves are the primary feeding surfaces and are greatly modified for that function, both by their considerable enlargement over most of the body and the modification of the leaflets as filaments (Figure 4.8). In autobranchs, the leaflets are greatly altered by considerable elongation and reflection which results in the formation of ascending and descending

branches – the demibranchs. There is considerable variation in the morphology and complexity of ciliary patterns, and their function within the autobranchs is generally more complex than in other molluscs (see Chapter 15 for more detail), and they also share many features consistent across the phylum. As in protobranchs, there is only one pair of gills, which are attached dorsally to the cavity. The very elongate filaments have supporting rods on the efferent side and are reflected in most taxa. Depending on the group, either entangled cilia or tissue junctions hold the long, thin filaments together. The structure of bivalve gills is described in more detail in Chapter 15, and the feeding mechanisms they employ in Chapter 5.

4.2.1.1.5   Gastropoda

In gastropods, the viscera, together with the mantle cavity, are rotated 180° on the head-foot in a process called torsion (see Chapters 3 and 8). In fully torted gastropods, this results in the ctenidia lying in an anterior mantle cavity above or alongside the head. The ancestral gastropod state is reconstructed as one pair of ctenidia – the *dibranchiate* state, like that illustrated for fissurellids (Figure 4.9), but in most living gastropod taxa (the lottioid patellogastropods, most vetigastropods, neritimorphs, and caenogastropods), there is only a single ctenidium (the left). In some gastropods, this remaining ctenidium may be reduced, lost, or replaced by a secondary gill. The lottioid gill is attached only by its base on the upper left posterior wall of the mantle

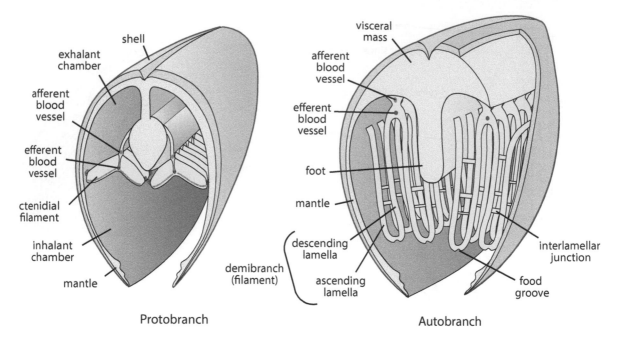

FIGURE 4.8 A generalised diagram of the two main types of bivalve ctenidia, showing the plesiomorphic molluscan gill morphology in protobranchs, and the highly modified gills of the suspension-feeding autobranchs. Redrawn and modified from Westheide, W. and Rieger, R., Eds., *Spezielle Zoologie – Erster Teil: Einzeller und Wirbellose*, Gustav Fischer, Stuttgart, Germany, 1996.

cavity, while in vetigastropods and neritimorphs the ctenidia are also attached to the mantle cavity by a ventral membrane. Some lottioid patellogastropods also have serial secondary gills in the mantle groove around the foot. The ctenidia have apparently been lost in patelloidean patellogastropods as there are only secondary gills in the mantle groove, although putative ctenidial rudiments associated with the osphradia were thought to be present in the mantle cavity by Stützel (1984) (See Chapter 18). In both groups, the roof of the mantle cavity also serves as an important respiratory surface (Davis & Fleure 1903; Fisher 1904; Fretter & Graham 1962; Lindberg 1981).

The reconstructed ancestral gastropod state with two ctenidia is found only in some vetigastropods (Fissurelloidea, Haliotoidea, Pleurotomarioidea, and Scissurelloidea). The leaflets in vetigastropods have discrete bands of cilia, and the two rows alternate on either side of the gill axis (the *bipectinate* condition) as in most molluscan ctenidia (Figure 4.9), but they are not necessarily mirror images, and in some taxa, one leaflet may be substantially larger than the other. Like the Bivalvia (and in contrast to cephalopods), the supporting rods are found on the efferent side of the filaments in most groups of gastropods, although these rods are lacking in patellogastropods, cocculinids, and neritimorphs. The gills in vetigastropods and neritimorphs are attached by both their base and a ventral membrane that extends along the junction of the sides of the mantle cavity with the floor. In the dibranchiate vetigastropods, water enters the mantle cavity ventrally and leaves it dorsally, often via a hole or slit in the mantle roof (and shell). In other vetigastropods, there is an additional dorsal membrane attached to the roof of the mantle cavity that partially divides the cavity into exhalant

and inhalant chambers. In caenogastropods, the single left ctenidium is fused to the mantle roof along the gill axis, and only the right row of leaflets (*monopectinate* condition) (with supporting rods) remains (Figure 4.9). Heterobranchs have lost or substantially modified the original ctenidium (see Section 4.2.1.3.1 and Chapter 20).

While gastropod ctenidia are utilised for respiration in most taxa, in some minute species the ctenidial epithelium is fully ciliated with no respiratory surface, suggesting that in those cases the ctenidial filaments act only as ventilators (Haszprunar et al. 2016).

4.2.1.1.6 Cephalopoda

There are two pairs of ctenidial gills in *Nautilus*, but only one pair in coleoid cephalopods (Figure 4.10). The ctenidia in *Nautilus* are bipectinate and are only attached to the mantle cavity by their bases (Griffin 1900). The leaflets are similar to those previously described, but their surface lacks cilia, and they are folded into secondary lamellae (Eno 1994). In coleoids (Figure 4.11), the base of the ctenidium is attached to the body wall and by a membrane along the afferent axis – as in bivalves, and in contrast to gastropods. There are also supporting rods in the leaflets, but these are on their afferent side, not the efferent sides as they are in bivalves and gastropods, suggesting the possibility of independent derivation (Yonge 1947). The complexity of the leaflets increases among coleoids, both phylogenetically and ontogenetically (Joubin 1885; Yonge 1947), as they become highly folded and richly supplied with a complex set of blood vessels (see below for more details and also Chapter 17). In addition, a haemocyanin synthesising branchial gland is in the gill membrane attached to the body.

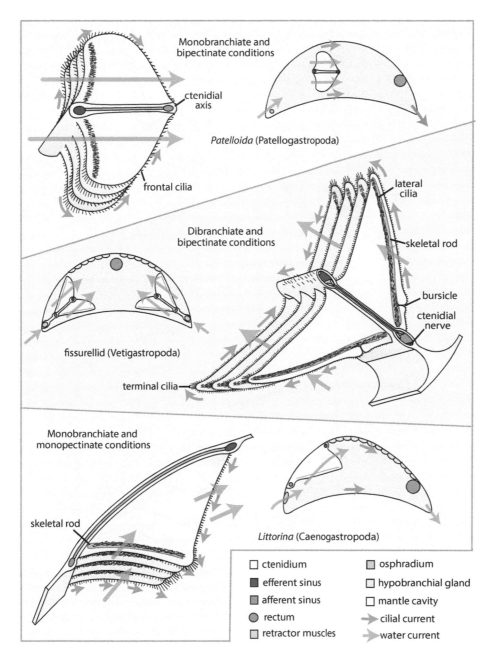

FIGURE 4.9 Some examples of gastropod ctenidia. Centre left figure redrawn and modified from Sasaki, T., *Bull. Univ. Mus.*, 38, 1–223, 1998, arrows added from Yonge, C.M., *Philos. Trans. Royal Soc. B*, 232, 443–518, 1947 and transverse sections and top left figure also redrawn and modified from that publication.

4.2.1.1.7 Modifications for Suspension-Feeding

Except for cephalopods, molluscs have ciliated ctenidial leaflets, with the cilia typically grouped into distinctive bands and patches. These cilia serve two purposes: (1) generating a flow of water that ventilates the mantle cavity and (2) collecting and removing the particles that accumulate and congest the respiratory surfaces. Besides the surface cilia, there are also scattered mucous cells opening onto the leaflet surface. The cleansing cilia and mucous glands on the gills are exapted for capturing suspended food particles enabling the gills to be modified as food collecting surfaces. An increase in ciliation to augment sorting and the movement of mucus involved in food capture is readily achieved with modifications of existing rejection mechanisms. The change in the ctenidia from a purely respiratory/ventilation function evolved once in bivalves (the autobranch bivalves) and several times in gastropods. (See also Chapter 5).

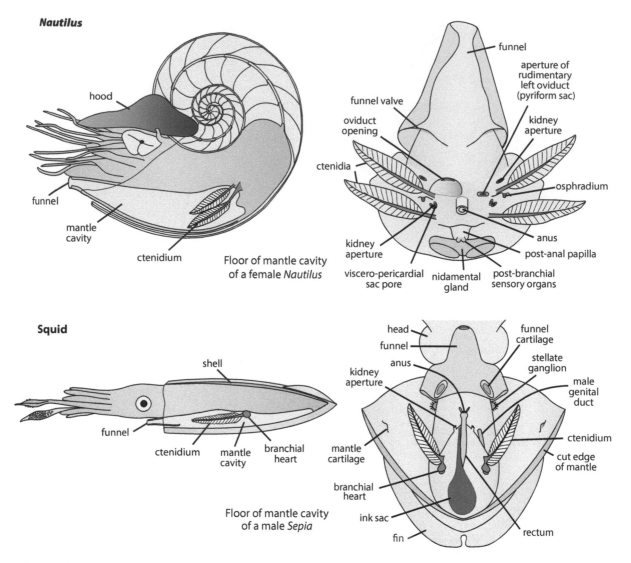

FIGURE 4.10 Comparison of the mantle cavities of *Nautilus* and the cuttlefish *Sepia*. *Nautilus* dorsal view redrawn and modified from a drawing by A.G. Bourne in the *Encyclopedia Britannica* (1900). *Sepia* dorsal view redrawn and modified from Tompsett, D.H., *Sepia*, pp. 1–191, in R. J. Daniel, *L. M. B. C. Memoirs on Typical British Marine Plants and Animals*, University Press of Liverpool, Liverpool, UK, 1939.

The elongation of the ctenidial leaflets into filaments is a general trend in suspension-feeding taxa. This has been carried to the extreme in bivalves where the broad, sickle-shaped leaflets found in protobranch bivalves have been elongated into filaments in all autobranch bivalves. These are so long that they are bent back on themselves enabling them to be wholly contained within the mantle cavity. The vast majority of autobranchs are suspension-feeders (Figure 4.12), and the modifications to the gills have not only greatly altered their morphology, but also produced greater variation and complexity in cilial form and function. Mucus secreted by the gills also plays an important role in particle capture in autobranch bivalves (see Chapter 5 for details). Because the filaments are fragile, they are interconnected by specially modified cilia or tissue interconnections which interlock them and so strengthen the gill and help to maintain its shape.

The nature of these interconnections varies depending on the taxon (see Chapter 15 for details).

Once the food entangled in mucus is collected, cilia move the mucus-entrapped particles to the free edges of the gill demibranchs, where it is rolled into strings and passed along food grooves (and in some, also in the junction of the ascending and descending lamellae) to the labial palps (Figures 4.12 and 4.13). In protobranchs, their large labial palps are used in food gathering and sorting, but the smaller palps in autobranch bivalves are sorting organs only, being involved in the initial screening of material before it enters the mouth.

In suspension-feeding gastropods, the ctenidial filaments are prominent only on the right side of the ctenidium. This morphology results from the loss, reduction, or realignment of the leaflets on the left side of the ctenidium, and the elongated filaments on the right side are never folded over on

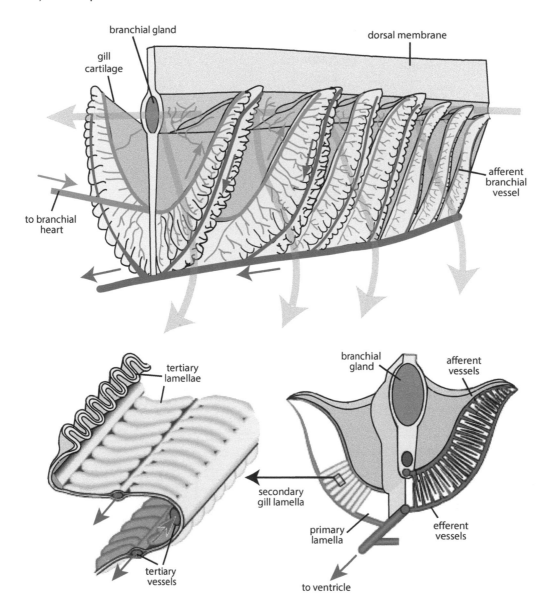

FIGURE 4.11 *Sepia* (Coleoidea) ctenidium. Top figure redrawn and modified from Wells, M.J. and Wells, J., *J. Exp. Biol.*, 99, 315–330, 1982. Lower two figures slightly modified from http://tolweb.org/articles/?article_id=4200 with permission from R. Young. Modified from Schipp, R. et al., *Zoomorphology*, 93, 193–207, 1979.

themselves as in bivalves. Cilia pass the mucus-bound food particles into a food groove on the floor of the mantle cavity which then runs to the mouth. Mucus is secreted by an endostyle, a narrow glandular strip of tissue on the inhalant side of the gill found only in suspension-feeding taxa (see Chapter 5).

Larger and denser cilia and mucus-bound particulate matter on the surface of the gill presumably reduce its respiratory potential, and thus there is probably a need for additional respiratory surfaces in highly modified suspension-feeding taxa (Lindberg & Ponder 2001).

4.2.1.2 Reduction and Loss of Ctenidia

For a complex molluscan structure, the anatomical uniformity of the ctenidium across the major taxa is unusual, but such uniformity can so constrain the definition of a character that

any modification or variation is seen as a non-homologous structure shaped by convergence. Most characters that diagnose the ctenidium are also found in fish gills (Hughes & Morgan 1973) – a gill axis with alternating lamellae on either side, efferent and afferent vessels, a retractor muscle, innervation, and a supporting rod. There is also similar variation in leaflet morphology in fish (Hughes & Morgan 1973), which appears to parallel the variation found in molluscs (Yonge 1947), although the similarity in these structures is undoubtedly the result of convergent evolution.

When mantle cavity gills deviate from or lack the ctenidial character suite, they are usually considered to be *secondary gills* rather than modified ctenidia (Haszprunar 1985f; Ponder & Lindberg 1997). The complete loss of the ctenidial gills has occurred in solenogasters, scaphopods, some

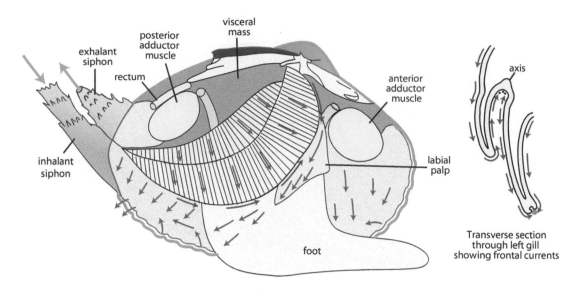

FIGURE 4.12   An example of a heteroconch bivalve (*Cerastoderma*) showing the main feeding currents on the ctenidia (red arrows), and cilial tracts on the foot and mantle (blue arrows). Redrawn and modified from Yonge, C.M., *Proc. Zool. Soc. Lond.*, 123, 551–561, 1953b.

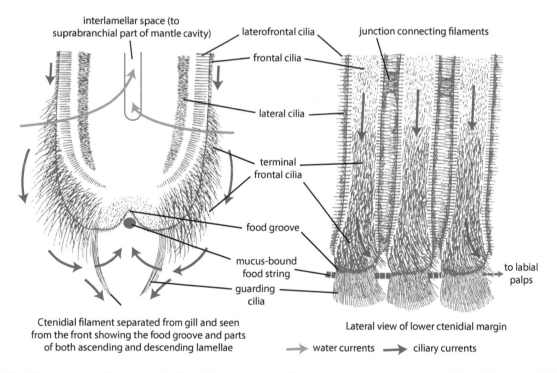

FIGURE 4.13   Water currents (blue arrows) and particle movement and capture (orange arrows) by specialised cilia on ctenidial filaments of an autobranch bivalve (*Corbula*). Redrawn and modified from Yonge, C.M. and Thompson, T.E., *Living Marine Molluscs*, Collins, London, UK, 1976.

septibranch bivalves, and some gastropods. Ctenidial loss has occurred in all major gastropod subclades – Patellogastropoda, Vetigastropoda, Neritimorpha, Caenogastropoda, and Heterobranchia. Loss of the ctenidia is typically associated with one or more of the following: (1) mantle cavity transformations, including loss, (2) mantle cavity brooding, (3) small

body size, and (4) becoming terrestrial (Haszprunar 1988a; Ponder & Lindberg 1997; Lindberg & Ponder 2001).

Examples illustrating the diversity of mantle cavity structure in gastropods are shown in Figure 4.14 and particularly the evolution of different ctenidial configurations in non-heterobranch gastropods. Various modifications occurred

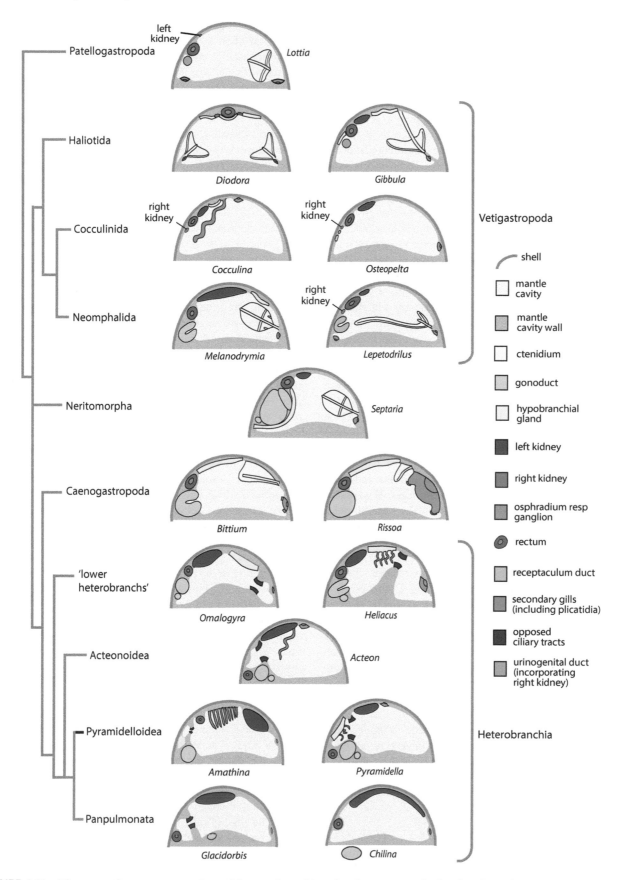

FIGURE 4.14 Diagrammatic transverse sections of the mantle cavities of various gastropods showing the main organ systems and their phylogenetic relationships. Redrawn and modified from Haszprunar, G., *J. Molluscan Stud.*, 54, 367–441, 1988a.

in the Heterobranchia, notably, replacement with a different gill (see Section 4.2.1.3) and/or the development of opposed ciliated ridges (see Section 4.2.1.4). In gastropods, ctenidial loss is correlated in part with mantle cavity transformations, including lung development in terrestrial taxa (Neritimorpha, Caenogastropoda, Heterobranchia), ciliary tract development (Heterobranchia), and shell loss (e.g., Nudipleura) (Ponder & Lindberg 1997). Patellogastropods which use their mantle cavities to brood their young have lost the ctenidium (Lindberg 1987b; Sasaki 1998). Such loss does not occur in bivalves, probably due to the interlinked ctenidial functions of respiration and feeding, but it may have played a role in the loss of the ctenidia in solenogasters, some of which brood their young in the mantle cavity (Hadfield 1979) (see Chapters 8 and 14).

Reductions and modification of the ctenidia are common and have occurred throughout the major gastropod clades. These reductions and modifications include the loss of one row of filaments (bipectinate to monopectinate), loss of, or fusion of, the central axis with the mantle cavity, reductions in size and number of filaments, loss of distinct ciliary groupings, and the loss of skeletal rods. These modifications can occur in isolated species or across an entire group and were first noted by Pelseneer (1896). For example, in the vetigastropod *Lepetodrilus nux*, the right ctenidium has lost the central axis and is represented by a few small, arbitrarily ciliated lappets (Sasaki 1998), while the left ctenidium is identical to that in other *Lepetodrilus* spp., in which only the left is present (Fretter 1988). Also, in the small bodied Scissurellidae and various minute skeneiform trochoideans and sequenzioideans there are transformations from bipectinate to monopectinate gills, sometimes including the reduction of skeletal rods (*Temnocinclis* and *Temnozaga*) (Haszprunar 1989b). Reduced gills are also common in the Cocculinoidea and the Lepetelloidea which show varying degrees of axis, leaflets, and skeletal rod reduction and variation (Haszprunar 1988b). Many caenogastropods are small (<10 mm) (Ponder et al. 2008), and among the smallest taxa reduced ctenidia are common. For example, in some minute Cingulopsoidea, Rissooidea, and Truncatelloidea the ctenidium can be reduced to just a few stubby leaflets or filaments, as in the amphibious to terrestrial Assimineidae where the gill is represented by very few rudimentary filaments. Small, marine, non-heterobranch gastropods that do not brood young in the mantle cavity seldom lose the entire ctenidium, although an exception is seen in at least one member of the minute marine Rastodentidae in the caenogastropods (Ponder 1966).

The reduced ctenidial gills in patellogastropod *Rhodopetala* (Lindberg 1981) and the peltaspirid *Nodopelta* (Fretter 1989) lack skeletal rods and ciliary patches and show a variety of leaflet morphologies, and thus, by way of reduction and simplification, somewhat resemble the secondary gills described in the next section.

4.2.1.3 Secondary Gills and Plicatidia

Secondary gills are recognised as not being ctenidia mainly by their anatomical positions and absences – they lack skeletal rods, distinct ciliary patches, and the characteristic ctenidial leaflets. They also show greater morphological diversity than

ctenidia and range from simple to highly complex. For example, gill variation within euthyneurans is so great that Dayrat and Tillier (2002) concluded that secondary gill characters were too variable to be used in their character analysis.

In the Solenogastres, the posterior mantle cavity in many species is lined with lamella-like folds or papillae. These structures are ciliated, overlay large haemocoelic spaces and sometimes resemble simple ctenidia (Salvini-Plawen 1985a). The mantle cavity and posterior sense organs[1] are innervated by separate nerves from the suprarectal commissure which bridges the lateral nerve cords in both the Solenogastres and the Caudofoveata (Salvini-Plawen 1978b, 1985a; Lindberg & Sigwart 2015); there do not appear to be unique nerves associated with the mantle cavity folds or papillae (Heath 1911b).

In some bivalves (Lucinoidea), secondary gills are found between the anterior adductor muscle and the ventral mantle margin. They are not innervated by the osphradial/ctenidial nerve, but instead by mantle nerves (Taylor & Glover 2000) originating from the cerebropleural ganglia. These gills can be complex with highly folded, ciliated surfaces which may have a central vessel and appear bipectinate (Taylor & Glover 2000). They are positioned in the incoming water flow, follow the course of the substantial mantle vessel, and are interpreted as secondary respiratory surfaces (Pelseneer 1911a; Allen 1958; Taylor & Glover 2000). The haemocoelic spaces are crossed by trabeculae and nerve fibres, but there are no distinct afferent or efferent vessels or associated muscles and nerves. Secondary gills have also been reported in the vesicomyid *Isorropodon mackayi* where there is a triangular patch of heavily folded mantle tissue adjacent to the inhalant aperture which is also thought to serve as a respiratory surface (Oliver & Drewery 2014). Innervation of this area is unknown, but its placement in the inhalant aperture suggests mantle innervation from the visceral ganglia at the end of the lateral nerves.

The mantle cavity and ctenidia are lost in the pelagic shell-less pteropods (Gymnosomata). Instead, there are epidermal surfaces underlain by large haemocoelic spaces. These may occur on the body surface and also as projecting folds and raised ridges at the posterior end of the animal (Hyman 1967). There are no afferent and efferent vessels associated with the blood spaces and only scattered nerve and muscle fibres.

The morphology of the secondary gills of patellogastropods is superficially similar to individual leaflets of ctenidial gills (Davis & Fleure 1903; Nuwayhid et al. 1978; Akşit & Falakali 2011). The leaflets are triangular and hang from the roof of the mantle groove, with large and small ones alternating. The surface of each leaflet has anastomosing ridges unevenly covered by cilia groups, with the greatest concentration of cilia along the edge of the leaflet (Akşit & Falakali 2011). The edge of each lamella is expanded into afferent and efferent regions that communicate directly through the central

---

[1] These structures have also been referred to as the 'gill ganglion' (Heath 1911b) or, in more recent literature, as osphradia.

haemocoelic region. Although both muscle and nerve fibres are present in each lamella, they are not organised at the afferent and efferent margins of the leaflet (Davis & Fleure 1903; Nuwayhid et al. 1978) and are vascularised by the mantle vessel and innervated by mantle nerves as in the aforementioned bivalve secondary gills.

Mantle groove secondary gill lamellae that resemble those of patellogastropods are present along the mantle groove in arminid nudibranchs. Blood flows from these lamellae to the auricle via lateral mantle vessels (García & García-Gómez 1990), and their innervation is from the cerebropleural ganglion (García & García-Gómez 1988).

In molluscs, both the ctenida and osphradia are typically innervated by a common nerve from a lateral nerve or its associated ganglia (e.g., Haszprunar 1987b; Sigwart & Sumner-Rooney 2015), but the final nerve connecting the ctenidium to the lateral nerve or ganglion varies in different groups. In taxa with an osphradium, the ctenidia are innervated by the ctenidial-osphradial nerve (Bivalvia, *Nautilus*, and gastropods) (see Haszprunar 1987b; Lindberg & Sigwart 2015). In contrast, ctenidia in chitons and caudofoveates are innervated by individual nerves from the lateral nerves and/or suprarectal commissure (which bridges the posterior ends of the left and right lateral nerves) depending on the position of the ctenidium. In the Monoplacophora, the ctenidia are also connected directly to the lateral nerve by a pair of nerves, osphradia being absent (Lemche & Wingstrand 1959), and the scaphopods lack both a gill and osphradia (Reynolds 2002). In taxa where either the osphradia or gill is lost, a single connection with the ctenidial-osphradial nerve remains with the surviving structure. This contrasts with *de novo* secondary gills, such as those discussed above in a handful of gastropods and bivalves, which are not innervated by a ctenidial-osphradial nerve.

The association of the ctenidium and osphradia also exists developmentally. In bivalves, cephalopods, and gastropods, the gill rudiment first appears as a raised ciliated epidermal site either within or outside the mantle cavity. If outside, it is subsequently enveloped by the mantle cavity (Arnold 1965; Raven 1966). In chitons, the larval mantle groove is first covered by ciliated epithelium from which the first pair of ctenidia appears, and then these are replicated both posteriorly (few) and anteriorly (many) in the mantle groove (Russell-Hunter & Brown 1965). In osphradium-bearing groups, the osphradia also arise from the ciliated surfaces in juxtaposition with the formation of the ctenidia. The timing of the formation of both is variable, but osphradia often form after the gill rudiments appear, although the osphradial nerve typically appears before this (Raven 1966; Ellis & Kempf 2011). In some bivalve and gastropod taxa, this close association of the gill and osphradia remains into the adult, with the osphradium on the ctenidial axis or immediately adjacent to it (Haszprunar 1987b, 1988a); in other taxa they may separate (e.g., *Nautilus*, Lottioidea) (Griffin 1900; Salvini-Plawen 1969; Haszprunar 1988a). One or both may also be lost, and osphradia and osphradial ganglia can also appear in the ontogeny of taxa that lack osphradia as adults (e.g., Stylommatophora, Nudipleura) (Hammarsten &

Runnström 1926a; Babor & Frankenberger 1930; LaForge & Page 2007).

Taxa with apparently rudimentary ctenidial structures in adults include the Patelloidea (if Stützel [1984] is correct in her interpretation of the 'wart organs') and possibly some reduced gills in heterobranchs. In both of these cases, these states are seen in the ontogeny of other ctenidia, such as the plicate gill state present in developing oysters (Cannuel & Beninger 2006).

4.2.1.3.1 Plicatidia

The distinction of *de novo* secondary gills from highly modified ctenidia has been controversial in molluscs – especially in heterobranch taxa. While a few taxa within the heterobranchs have gills positioned and innervated like ctenidia and superficially resembling a ctenidial gill, others are positionally and morphologically very different (e.g., the leaflet rows in the Mathildidoidea and Architectonicoidea and the mantle groove gills of some nudibranchs). The heterobranch gill is usually considered secondary, i.e., the ctenidium has been lost and secondarily replaced by a *de novo* gill (e.g., Brace 1977b), some of which may somewhat superficially resemble ctenidial gills both positionally and structurally. Often the structure of these putative secondary gills is complex (Figure 4.15) and variable, even within a small clade (e.g., Valvatidae, Rath 1988). Others (e.g., Schmekel 1985; Gosliner 1996) have treated the heterobranch gill as homologous with the ctenidium, even homologising the anal gills of nudibranchs with ctenidia, a controversy with pre-Darwinian roots (Huxley 1853).

The osphradia (when present) and many heterobranch gills are innervated by a nerve (the ctenidial-osphradial nerve) from the supraoesophageal ganglion (Hyman 1967). Gills that are positionally comparable with ctenidia are found in the Valvatoidea (Bernard 1890b; Rath 1988), Acteonoidea (Brace 1977a), Cephalaspidea, Aplysiida, and Pleurobranchida (e.g., Hoffmann 1932–1939). In other taxa, the innervation and position are not as straightforward. For example, the anal gills of nudibranchs (Figure 4.15) are innervated by three nerves; one each from the left and right 'branchial' ganglia and one from the single, left supraoesophageal ganglion (Hancock & Embleton 1852). Because the 'branchial' ganglia innervate both the anterior and posterior mantle regions, their connection with the anal gill plexus is viewed here as secondary. In the planorbid limpet *Laevapex*, Basch (1959b) reported that the osphradium is innervated by a nerve from the left mantle nerve that arises from the left pleural ganglion, while the pseudobranch is innervated by the ventromantle nerve that arises from the left visceral ganglion. In another planorbid, the innervation is similar except that a separate branch from the left mantle nerve also innervates the pseudobranch (Lacaze-Duthiers 1872).

The evidence does not exclude the possibility that heterobranch gills include both reduced and modified ctenidia and *de novo* respiratory structures. If so, it presents a conundrum regarding the homology of some of these gills with the molluscan ctenidium. Many share location, innervation, and similar vascular systems (e.g., Jörger et al. 2010), but

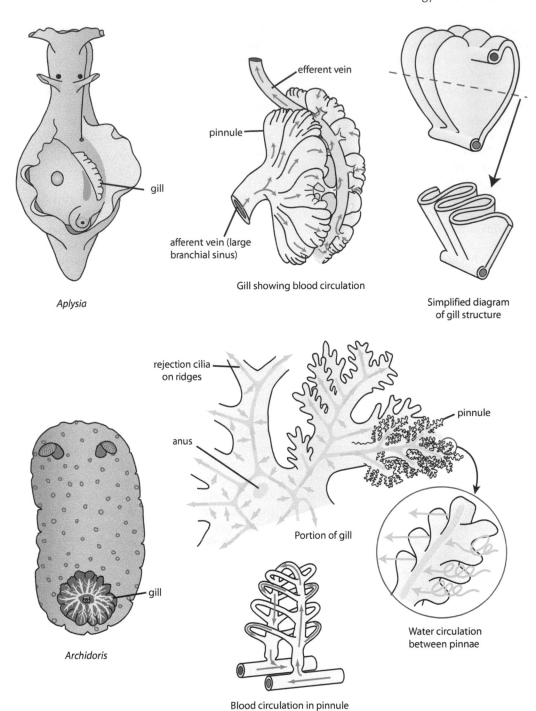

FIGURE 4.15 Gills of two heterobranchs, the plicatidium of the euopisthobranch *Aplysia* and the circum-anal gills of the nudipleuran *Archidoris*. Redrawn and modified from the following sources: top left, Guiart, J., *Contribution à l'étude des Gastéropodes, opistho-branches et en particulier des Céphalaspides*, Bigot Frères, Lille, France, 1901, middle and right top, Carew, T.J. et al., *J. Neurophysiol.*, 37, 1020–1040, 1974, lower right, original, lower middle and right, Potts, G.W., *J. Mar. Biol. Assoc. U. K.*, 61, 959–982, 1981.

lack other ctenidial characters such as a ctenidial axis and characteristic filaments. Rather than being completely *de novo* structures, the development of the ancestral ctenidium may have been curtailed because of heterochrony, possibly associated with paedomorphic events involving initial size reduction in the first heterobranchs (see Chapter 8). Given the long debate of gill homologies in heterobranchs, it is likely

that this scenario will not be universally accepted, but as Hyman (1967, p. 5667) first stated, in relation to planorbid and siphonariid 'pseudobranchs,' 'It seems odd that these families should abandon the ancestral gill and then replace it with sub-stitute structures.' Although future study is needed to resolve this issue, we use the term *plicatidium* (Morton 1972) for het-erobranch gills innervated in the same way as ctenidia, but that are structurally different, notably in lacking a ctenidial axis and filaments. Thus, this is a likely case of atavism – the reappearance of a structure after it has been substantially reduced or lost (Tomić & Meyer-Rochow 2011).

Morton (1972) coined the term plicatidium to distinguish the folded 'opisthobranch' gills from ctenidia, and although not all plicatidia have this folded structure, most do. Morton (1972, p. 345) stated 'the opisthobranch gill would show a reasonable topographic correspondence with a ctenidium. But as soon as its detailed structure and functioning are examined, it becomes clear that it has undergone a very special evolution upon its own lines.' And later, 'Freed from the stereotyped requirements of the ctenidium, the opisthobranch gill has been able to exploit a more plastic form-range than is found among all the rest of the Gastropoda.' Brace (1977b, p. 21) claimed that the ctenidium was lost in the ancestor and the plicatidium 're-developed' in the '…same location as the ctenidium;…' and Haszprunar (1988a) made a similar argument. Hoffmann (1940) earlier noted that the opisthobranch gill (i.e., plicatidium) had the same origin as the ctenidium.

We do not know whether the plicatidium evolved once or multiple times, but it seems to share the ctenidial bauplan, and both may share the same gene expression patterns possibly involving the gene *distal-less* (Lee & Jacobs 1998). The question remains as to whether or not the plicatidia represent *de novo* structures that have replaced the lost ctenidium or whether they are modified ctenidia either: (1) arrested at an earlier developmental stage or (2) with disparate morphologies resulting from a developmental shift that removed adult constraints (Gould 1977; Lindberg 1988a; Haszprunar 1992b; Ponder & Lindberg 1997). In the latter case, the ctenidial-osphradial innervation remains and unites all these diverse structures and, if so, the plicatidium is a convergent morphological state rather than a homologous structure.

Heterobranch gills that are probably not plicatidia include the planorbid pseudobranch, the anal gills of nudibranchs, and the mantle cavity gills of siphonariids. While the planorbid 'pseudobranch' is somewhat similar structurally to plicatidia, its location differs, and the siphonariid gill consists of a row of triangular lamellae on the roof of the mantle cavity which co-exists with ciliated dorsal and ventral ridges, creating the respiratory current (Yonge 1952), although the innervation of this gill has apparently not been clarified in the literature. Indeed, nerves within the gill lamellae, if present, are so inconspicuous they have not been found (Villiers & Hodgson 1987).

4.2.1.4 Opposed Ciliated Ridges

While the mantle cavity in many euthyneuran heterobranchs is markedly modified, reduced, or even entirely lost, many shelled groups have a well-developed mantle cavity which includes a pair of opposed ciliated tracts, or *raphes*, one of which is dorsal, the other ventral, and together they produce an exhalant current. They occur together with plicatidia in many of the larger-sized taxa. They lie on either side of the mantle cavity; on the left side in Omalogyridae and Architectonicidae, and on the right side in Acteonoidea,

Cephalaspidea, Aplysiida, Pyramidelloidea, Sacoglossa, Glacidorboidea, Siphonarioidea, and Amphiboloidea (Haszprunar 1988a) (Figure 4.14). Besides providing ventilation, they can also assist in carrying faecal material from the anus to the edge of the aperture, and in *Amphibola*, they are sites of ionic exchange (Pilkington et al. 1984). In some heterobranchs, these tracts are extended posteriorly into a *pallial caecum*. This substantially increases the length of the ridges posteriorly, thereby probably increasing ventilation within the mantle cavity (Brace 1977b; Mikkelsen 1996; Lindberg & Ponder 2001).

Except for Valvatoidea and Orbitestellidae, opposed ciliated ridges are present in most 'lower heterobranch' taxa (Ponder & Lindberg 1997) and most shelled 'opisthobranchs' including the Aplysiida, most cephalaspideans, and shelled sacoglossans. They are also present in panpulmonates including Siphonarioidea, Glacidorboidea, Amphiboloidea, Pyramidelloidea, and in *Chilina* (Hygrophila). In the remaining heterobranchs, mantle cavities are absent in most acochlidians with a small, simple cavity known in two species (Fahrner & Haszprunar 2002b). In the eupulmonates, the mantle cavity is a lung (see below). Distributing the opposed ciliated ridges and pallial caecum on the phylogenetic tree of Jörger et al. (2010) suggests the ridges may be plesiomorphic for most heterobranch taxa, while the pallial caecum independently originated at least four times (Acteonoidea, Aplysiida, Cephalaspidea, and Hygrophila).

Ponder and Lindberg (1997) suggested that the opposed ciliated ridges in heterobranchs may be homologous with the ciliated ridges often present in the mantle cavity of vetigastropod veligers, juveniles, and in some adult skeneimorphs (A. Warén, pers. comm.). In *Haliotis*, the dorsal ciliated ridge is the ctenidial rudiment with a corresponding ciliated field below (Crofts 1929). While the origin of the ciliated ventral field was not described for *Haliotis* (Crofts 1937, p. 255), Boutan (1895) ascribed the origin of the paired ventral cilia tracts in *Fissurella* to the incorporation of velum cilia into the left and right sides of the deepening mantle cavity. Dorsal and ventral ciliated bands have also been described in the larvae of the nudibranch *Phestilla sibogae* (Bonar & Hadfield 1974). The scenario for the derivation of the heterobranch opposed ciliated ridges from a larval character (the ventral tracts) and the early adult ctenidial rudiment (dorsal tracts) further supports a major paedomorphic reorganisation of ancestral gastropod development patterns at the origin of the Heterobranchia.

Recently, Kollmann (2014) suggested that the extinct Mesozoic 'lower heterobranch' Ptygmatididae modified its internal shell configuration through parietal and palatal plaits that isolated and increased the respiratory surface area in the penultimate 1.5 whorls, which he argued corresponded to the pallial caecum in living groups.

4.2.1.5 Modification as a Lung

Modification of the mantle cavity into a pulmonary sac or 'lung' for aerial respiration is found only in gastropods. As noted above, lungs occur in three of the five major

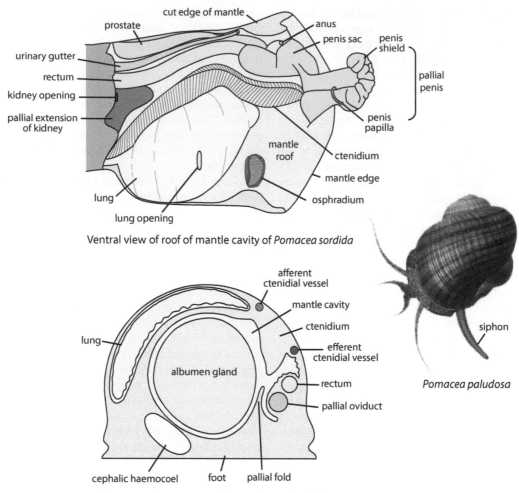

Ventral view of roof of mantle cavity of *Pomacea sordida*

Diagrammatic transverse section through middle region
of mantle cavity of female *Pomacea canaliculata*

Pomacea paludosa

FIGURE 4.16 The roof of the mantle cavity and the lung of the amphibious ampullariid *Pomacea*. Upper figure redrawn and modified from de Simone, L.R.L., *Arquivos do Museu Nacional (Rio de Janeiro)*, 62, 387–504, 2004 and transverse section redrawn and modified from Andrews, E.B., *J. Zool.*, 146, 70–94, 1965b. Living *Pomacea* from Haldeman, S.S., *A Monograph of the Freshwater Univalve Mollusca of the United States, Including Notices of Species in Other Parts of North America*, Conchological Section of the Academy of Natural Sciences, Philadelphia, PA, 1845.

gastropod clades – Neritimorpha, Caenogastropoda, and Heterobranchia. The gill and osphradum are lost in the terrestrial neritimorph taxa and a highly vascularised mantle cavity functions as a lung (Baker 1926). In caenogastropods, the ampullariids have a bilobed mantle cavity with the main part containing a well-developed gill for aquatic respiration and an adjacent section that functions as a lung for aerial respiration (Figure 4.16). In another architaenioglossan group, the terrestrial cyclophorids, the gill is lost, and most lack an osphradium, although it is retained in some species (e.g., Kasinathan 1975).

High intertidal marine littorinoideans have a well-developed ctenidium, but can respire in aerial conditions at higher rates than when submerged (Sandison 1967), while terrestrial littorinoidean taxa such as *Pomatias* have lost the gill and have a vascularised mantle roof. The kidney and heart are also present in the mantle roof, and the former also probably serves as a respiratory surface; both the osphradium and

hypobranchial gland are retained. To keep the respiratory surface moist in aerial conditions, liquid may be provided from the kidney (Fretter & Graham 1962). The amphibious to terrestrial assimineids have a rudimentary ctenidium, but the osphradium is present (Fukuda & Ponder 2003).

4.2.1.5.1 The Panpulmonate Lung

The homology of the lung that occurs in many panpulmonates, including nearly all eupulmonates, has been controversial. Based on its overall similarity in adults, some authors considered the lung to be a modified mantle cavity (Fretter & Peake 1975; Brace 1983; Morton 1988b), but developmental evidence from direct-developing hygrophilan and stylommatophoran taxa showed that the initial formation of the mantle cavity and lung was from separate ectodermal invaginations (Meisenheimer 1896; Heyder 1909; Ghose 1963; Moor 1977), suggesting that the lung was an apomorphy of 'pulmonates.'

There remained the possibility that these developmental patterns were modified by the large amount of yolk in the eggs in these direct-developing taxa. Ruthensteiner (1997) addressed this issue by documenting the formation of the lung in two eupulmonates with indirect development and veliger larvae – *Ovatella myosotis* (Ellobioidea) and *Onchidium branchiferum* (Systellommatophora). Both formed the lung from a single invagination, ontogenetically identical to the formation of the mantle cavity in several outgroups including patellogastropods and vetigastropods. Lung development in the more basal panpulmonate amphiboloideans appears to be a single origin as well (Little et al. 1985), but there does not appear to be information on *Siphonaria*.

The dual invaginations observed in the earlier investigations of development in Hygrophila and Stylommatophora were summarised by Raven (1966) as follows: (1) the lung invagination appears before the mantle cavity invagination, (2) the subsequent invagination of the mantle cavity is in juxtaposition with the first and pushes the lung invagination back into the body and opens into it, (3) the merged lung tissue continues backward and envelops the pericardium-kidney complex (in the terrestrial slug *Arion*), and (4) the pneumostome forms from a slit in the mantle edge and the outer edges of the lung and mantle cavity fuse. This summary from studies of three stylommatophorans and one hygrophilan suggests there is an additional ectodermal contribution in the formation of their lungs. The dual invaginations in the hygrophilan *Lymnaea* (Cumin 1972), which has indirect development, argues against these differences being related to the amount of yolk in the egg. Depending on which recent heterobranch phylogeny is used, dual invaginations in Hygrophila and Stylommatophora are either monophyletic apomorphy (e.g., Dayrat et al. 2011) or homoplastic (e.g., Dinapoli et al. 2011) or reverted to single invaginations on at least two occasions in the Eupulmonata (Ellobioidea, Systellommatophora) (e.g., Jörger et al. 2010). This second (lung) ectodermal contribution may provide additional volume and/or provide additional and alternative connections to other systems (circulatory, nervous) or organs (kidneys, hearts). Developmental data from key taxa are necessary to resolve this further.

In most 'pulmonates,' the mantle cavity is on the right side of the body and opens mid-laterally via the pneumostome (Figure 4.17). The dorsal wall of the cavity is heavily vascularised, and the usual openings to internal systems (gut, gonoduct, kidney) are repositioned on or near the outside of the body. In the stylommatophoran lung, air is pumped in and out by muscular movements of the floor (see below). It passes over the highly folded respiratory epithelium in the inner part of the roof of the mantle cavity. This epithelium is covered by a layer of mucus supplied by subepithelial gland cells, to keep it moist for gas exchange. Capillaries line the thin epithelium which underlies the respiratory epithelium (Figure 4.18). A triangular kidney is located to the left side of the lung with its base abutting the viscera, and the pericardium is located along the left side of the kidney. The roof of the lung between the kidney and pneumostome is extensively vascularised by a pulmonary vein, forming the primary respiratory surface of the lung. The pulmonary vein then extends to the auricle of the heart. A pulmonary vein also occurs in Ampullariidae, but because of the presence of both a lung and ctenidium in that group, it enters the auricle adjacent to the efferent vessel from the ctenidium (Prashad 1925).

Active ventilation was initially thought not to occur in stylommatophorans, but Ysseling (1930) demonstrated these snails carry out breathing movements regulated by the pneumostome and contractions of the floor of the lung. Breathing proceeds as follows: (1) the pneumostome, which is controlled by a sphincter muscle, opens; (2) simultaneously, the floor of the mantle cavity contracts, enlarging the cavity and drawing air in; (3) the pneumostome closes and the muscles of the diaphragm relax, reducing the volume of the cavity and increasing lung pressure; (4) gas exchange in the lung is facilitated by the increase of air pressure in the lung; and (5) the pneumostome opens, the contents of the lung are expelled and the next inhalation commences (Ghiretti 1966; Dale 1974; Barker 2001b).

4.2.1.6 Hypobranchial Glands

The roof of the mantle cavity is often provided with one to a few groups of mucus-secreting cell tracts known as hypobranchial glands. These tracts have a noticeably taller epithelium (often elevated above the adjacent mantle tissue) consisting of large glandular cells (mucocytes) and other secretory cells that alternate with interstitial ciliated cells and some neurosensory cells (Hyman 1967). Hypobranchial glands appear to be plesiomorphic for molluscs, and they are found in the mantle grooves of polyplacophorans and monoplacophorans, in the mantle cavity of protobranch bivalves and many gastropods (Ponder & Lindberg 1997). They have not been reported in the aplacophorans, scaphopods, and cephalopods, although Hoffmann (1949) noted strong similarities between the epithelium of the wall of the shell gland in the solenogaster mantle cavity and the hypobranchial glands of chitons. Salvini-Pawen (1981a) suggested that the nidamental glands of cephalopods may possibly be hypobranchial gland homologues.

In the polyplacophorans, hypobranchial glands are widely distributed in both the Lepidopleurida and Chitonida. Depending on the species, there are between one and four glandular tracts in each mantle groove (Plate 1901; Knorre 1925). Lemche and Wingstrand (1959) described the hypobranchial glands of monoplacophorans as highly developed secretory tissue patches at the base of the gills in the mantle groove.

In protobranch bivalves, a pair of hypobranchial glands run posteriorly and unite at the anus; they are presumably involved in waste consolidation before its removal via the exhalant aperture. In *Nucula delphinodonta*, the hypobranchial gland also secretes the brood pouch which is attached to the exterior of the shell (Drew 1901). Hypobranchial glands have also been recorded from some autobranch bivalves, but are usually absent, presumably due to the reconfiguration of the mantle cavity necessitating waste removal via the inhalant aperture (Morton 1977a) (see Chapter 15). In protobranchs, the gland contains both mucus and ciliated cells, and a somewhat

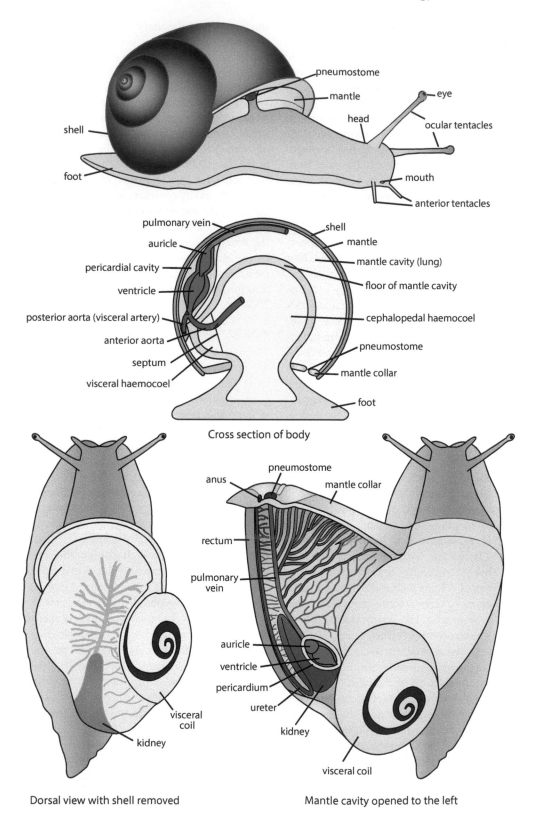

Cross section of body

Dorsal view with shell removed

Mantle cavity opened to the left

FIGURE 4.17 The lung and pneumostome of a helicid snail. Upper figure original; middle figure redrawn and modified from Sommerville, B.A., *J. Exp. Biol.*, 59, 275–282, 1973, lower left figure from Meisenheimer, J., Helix pomatia *Monographien einheimischer Tiere*, 4, 1–140, 1912, lower right figure composite from various sources.

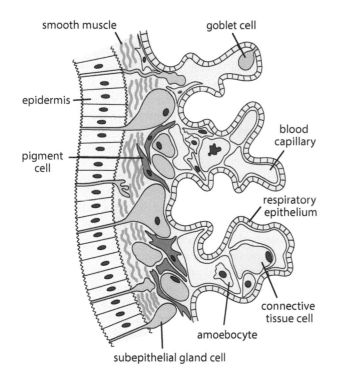

smooth muscle

goblet cell

epidermis

pigment cell

blood capillary

respiratory epithelium

connective tissue cell

amoebocyte

subepithelial gland cell

FIGURE 4.18 Cross-section of lung tissue from the roof of the mantle cavity in *Helix pomatia* (shell removed). Redrawn and modified from Pohunkova, H., *Folia Morphol.*, 15, 250–257, 1967.

similar glandular structure is known in some autobranchs, including the arcoid *Philobrya*, the anomiid *Monia*, a lucinoidean *Fimbria*, the veneroid *Corbicula*, the galeommatoideans *Chlamydoconcha* and *Montacutona*, and the anomalodesmatan *Periploma* (Morton 1977a; Morton & Scott 1980; Morton 1981a). In some of these taxa, the function of the gland has changed. For example, in *Corbicula* and possibly *Fimbria*, the hypobranchial gland appears to function as a nutritional organ for developing larvae (Morton 1977a), as it possibly does in *Chlamydoconcha* (Morton 1981b).

The plesiomorphic condition in gastropods is a pair of hypobranchial glands lying between the gill and anus. They are absent in patellogastropods. In vetigastropods, they are paired or single, with considerable variation in some groups, especially in trochoideans (Fretter & Graham 1994), where the organs on the right side are often reduced in size and the right ctenidium is lost. In contrast, in the genera *Anatoma* (Scissurelloidea) and *Margarites* (Trochoidea), only the right gland is retained. Hypobranchial glands are absent in fissurellids and the vent gastropod *Neomphalus*. Many neritimorphs have a reduced right hypobranchial gland attached to the gonoduct at the posterior end of the mantle cavity; *Theodoxus* and Phenacolepadidae lack a hypobranchial gland. Caenogastropods have a single left hypobranchial gland, while in lower heterobranchs, the single gland is anteriorly positioned and might be homologous with those in other gastropods. Most authors consider a hypobranchial gland to be present in lower heterobranchs, acteonoideans, cephalaspideans, aplysioids, sacoglossans, and shelled pteropods

(Hoffmann 1932–1939, 1940; Wägele et al. 2006), which all have a well-developed shell. In the Panpulmonata, the loss of the hypobranchial gland corresponds with the modification of the mantle cavity as a lung, while in the Nudipleura and Euopisthobranchia, it is correlated with reduction and loss of the shell and mantle cavity.

In some heterobranch taxa, *de novo* glands were formed from the epithelium of the mantle cavity or by modifying the reduced hypobranchial gland, although possible homologies among the various glands in the mantle cavity of 'opisthobranchs' are difficult to establish convincingly (Jensen 2011) (see Chapter 20). For example, two different glands are found in the mantle cavity of aplysioids – the purple gland on the roof and the opaline gland (= gland of Bohadsch) on the floor (Nolen & Johnson 2001), repugnatorial glands in Acteonoidea and Cephalaspidea (Fretter & Graham 1954), and the 'pigmented mantle organ' (see below) in 'lower heterobranchs' (see Section 4.2.1.7).

The plesiomorphic function of the hypobranchial gland(s) appears to be to discharge mucoid secretions in response to irritation, and to play a role in binding waste material entering the mantle cavity, but in some caenogastropods, the hypobranchial gland secretions serve other functions including producing a toxic secretion for immobilising prey and chemical attractants for aggregated spawning. Most of the experimental studies have looked at muricid secretion toxicity (Dubois 1909; Keyl et al. 1957; Hemingway 1978; Srilakshmi 1991; Westley et al. 2006), although the effectiveness of a waterborne toxin has been questioned (e.g., Jullien 1948).

The compounds produced by the hypobranchial glands of muricids include choline esters (urocanylcholines) (Keyl et al. 1957; Roseghini et al. 1996; Shiomi et al. 1998) and halogenated alkaloids, the majority of which contain bromine (Benkendorff 2010; Pauletti et al. 2010). Recent transcriptome studies demonstrate that muricid hypobranchial glands produce a mucous secretion as well as being a site of chemical interaction and biosynthesis (Laffy et al. 2013). The best-known muricid hypobranchial gland secretion is 'royal' or Tyrian purple dye which has been harvested in the Mediterranean region for over 3500 years (Baker 1974; Stieglitz 1994) (see Chapter 10). The major component of the Tyrian purple secretion is 6,6'-Dibromoindigo (Friedländer 1909; Cooksey 2001), a bromine halogenated alkaloid. When initially secreted, the fluid is colourless or pale, but exposed to light it darkens to a deep purple colour (Cooksey 2001). The presence of Tyrian purple precursors in muricid egg capsules suggests that bioactive intermediates are being used to provide a chemical defence for the spawn, although the mode of transfer between the hypobranchial gland and gonoduct remains unknown (Benkendorff et al. 2000; Westley & Benkendorff 2008). Purple hypobranchial secretions are also produced by various other neogastropods and epitoniids (Robertson 1985a).

4.2.1.7 Pigmented Mantle Organ (PMO)

A *pigmented mantle organ*, sometimes called an anal gland (Pelseneer 1914) or larval kidney (Schaefer 1996), occurs in many adult lower heterobranchs (Robertson 1985a; Haszprunar 1985f; Ponder 1987) (see Chapter 20). It consists of a single ectodermal gland positioned near the anus and comprises at least four types of secretory cells (LaForge & Page 2007). The ontogeny and homology of this structure have been controversial. Thorson (1946) described a pigmented structure in the mantle cavity of developing Epitoniidae which Robertson (1983, 1985a) thought would become the hypobranchial gland in these taxa, and he considered that it was a homologue of the pigmented mantle organ (PMO) in heterobranchs. Richter and Thorson (1975) argued that the dark pigment of this structure resulted from fixation.

4.2.1.8 Sensory Structures

Although sensory structures are reviewed in Chapter 7, we outline below those relevant to the mantle cavity as it houses, or is adjacent to, a variety of them, especially in gastropods. They are most often innervated by the lateral nerves arising from the pleural ganglia and are typically associated with the gills or exhalant respiratory flows. Multiple sensor types are often present in these structures, and they are thought to include both chemo- and mechanoreceptors.

The molluscan osphradium is the most ubiquitous of the mantle cavity sense organs and is treated in detail in Chapter 7.

Sensory structures can also be taxon-specific such as Schwabe, lateral, and branchial sense organs in chitons (Thiele 1895; Sigwart et al. 2014), the abdominal and adoral sense organs in some bivalves (Haszprunar 1983, 1985c), or 'rhinophores' in heterobranch gastropods (Thompson 1976). Like the osphradia, these autapomorphic structures range from a simple sensory papilla to gill-like structures and sensory streaks and lines.

4.2.1.8.1 Monoplacophora

Living monoplacophorans are unique among the major groups in lacking mantle groove sense organs (Haszprunar & Schaefer 1997). In the original description of *Neopilina galatheae*, Lemche and Wingstrand (1959), discussed and illustrated a single pair of preoral tentacles on either side of the head above the velum (see Chapter 7, Figure 7.41). Based on their position, innervation, and cell types, they homologised them with the cephalic tentacles of gastropods and suggested that they might be chemoreceptors.

4.2.1.8.2 Scaphopoda

Scaphopods, like monoplacophorans, lack osphradia, but have at least two unique mantle cavity sensory structures. In Dentaliida, the ciliary organ lines the edge of the mantle cavity just inside the ventral opening of the shell, and in both Gadilida and Dentaliida a putative sensory papilla is positioned at the dorsal mantle opening (Reynolds 1992; Steiner 1992; Shimek & Steiner 1997; Reynolds 2002). Both sensory structures consist of three cell types (Reynolds 1992; Shimek & Steiner 1997), and Reynolds (1992) noted similarities of these structures with the chemo- and mechanoreceptors in other molluscs. Unlike other molluscs, water flow through the dorsal mantle opening is not unidirectional, but alternates

between inflow and outflow. The ciliary bands of the pavillion organ were suggested to be sensory (Boissevain 1904), but Reynolds (1992) has argued that they are engaged in ventilation. Based on the morphology of the scaphopod nervous system (Shimek & Steiner 1997), the ventral ciliary organs are innervated by mantle nerves from the pleural ganglia, while the sensory papilla of the dorsal opening is innervated by the visceral ganglia on the lateral nerves.

4.2.1.8.3 Polyplacophora

Three kinds of mantle cavity sensory structures are found in chitons: lateral sense organs and *Schwabe organs* in the Lepidopleurida, and the post-branchial sensory organ in Chitonida (Haszprunar 1987b; Sigwart et al. 2014). In lepidopleurids, the most primitive group of living chitons, the lateral sense organs are on the outer margin of the mantle groove, while the Schwabe organs are on the roof of the mantle cavity on either side of the mouth and, in living animals, are visible as pigmented strips (Sigwart et al. 2014). The Schwabe organs are innervated by the lateral nerve cord, and their epithelia include many ciliated and pigmented cells assumed to be sensory, but their precise function is unknown (Sigwart et al. 2014). In the lepidopleurids, 'branchial sense organs' were recognised at the efferent base of each ctenidium at the point where the gill nerve enters the gill (Haszprunar 1987b), but these are now considered an artefact (J. Sigwart, pers. comm., 2015).

In the Chitonida, the post-branchial sensory organs have been considered to represent osphradia (Yonge 1939a; Haszprunar 1987b), although they are not innervated by the osphradial/ctenidial nerve, but instead by nerves from the supra-anal commissure. These structures have also been called adanal sensory stripes. A second type of putative chitonid sensory structure associated with the mantle grooves, the anterior sense organs, was reported in several taxa (Knorre 1925; Yonge 1939a), but more recent work has failed to confirm these structures (Sigwart et al. 2014).

4.2.1.8.4 Aplacophorans

In both the Caudofoveata and Solenogastres, the medially fused dorsal sensory organs are situated just outside the mantle cavity (Haszprunar 1987c). Their origin as paired structures is based on their bilateral symmetry and separate innervation on each side (Salvini-Plawen 1969). In caudofoveates, the dorsal sense organ forms a longitudinal groove bordered by swellings, and in solenogasters, it is represented by a variable number (up to six) erect or sunken knobs (Haszprunar 1987c). These sensory organs are innervated from the supra-anal commissure. In caudofoveates, ctenidial nerves are separate and are not associated with the nerves to the dorsal sensory organs. This innervation indicates these sensory organs are not osphradia, but similar (probably analogous) to the paired post-branchial sense organs in chitons and *Nautilus* (see below). In the aplacophoran groups, these sensory organs are pre-branchial (inhalant flow) rather than post-branchial (exhalant flow) as in chitons and *Nautilus* (Lindberg & Sigwart 2015).

4.2.1.8.5 Cephalopoda

In nautilids, there are two pairs of mantle cavity sense organs, both of which were thought to be osphradia, and both are ultimately connected to the visceral nerve (Willey 1902), but only the outer osphradia are innervated by the osphradial/ctenidial nerves. The inner (or medial) pair is separately innervated from the visceral nerve loop, as are the post-branchial organs in chitons and the pre-branchial organs in aplacophorans. Because of this innervation pattern, Lindberg and Sigwart (2015) regarded these medial structures as post-branchial sensory organs.

Mantle cavity sense organs are absent in coleoids, this being correlated with the mantle cavity being a muscular sac, and with an extraordinary increase in visual acuity (Muntz 1999). Both nautiloids and coleoids share unique sensory structures (mechanoreceptors and chemoreceptors) on their heads, arms and tentacles, and bodies (Hanlon & Budelmann 1987; Hanlon & Shashar 2003) (see Chapters 7 and 18).

4.2.1.8.6 Bivalvia

Four mantle cavity sensory organs are found in Bivalvia, but of those, only the osphradium is shared outside the group, and this is absent in higher bivalves (Haszprunar 1987b). Two sensory structures are limited to protobranchs (Haszprunar 1985d; Schaefer 2000). *Stempell's organ*, found in the protobranch genus *Nucula*, is situated dorsal to the anterior adductor muscle and innervated from the cerebral ganglia. It is thought to serve as a mechanoreceptor (see Chapter 7). Also, in protobranchs, the *adoral sensory organ* (AdSO), is a pair of raised sensory ridges of unknown function on either side of the labial palps (see Chapter 7).

The *abdominal sense organ* (AbSO), a presumed mechanoreceptor, is limited to the pteriomorphian and palaeoheterodont (unionoidean and trigoniid) bivalves (Haszprunar 1983) (see Chapter 7). The AbSOs are positioned on either side of the anus and innervated by nerves from the visceral loop. In some they are asymmetrically developed, this being related to the asymmetric body and water currents of some of these bivalves (e.g., scallops). Depending on their position relative to the gill axes, the AbSOs may be positioned in the inhalant current (mytilids) or in the exhalant flow (arcids and trigoniids) (Haszprunar 1983). The position and innervation of the AbSOs are reminiscent of the post-branchial sensory organs in chitons and *Nautilus*, and possibly the dorsal sensory organs in aplacophoran groups.

Bivalves have most of their other sensory structures along the mantle edge (eyes, mantle, tentacles).

4.2.1.8.7 Gastropoda

Despite the diversity of mantle cavity morphologies and configurations in gastropods, sensory organs are relatively conservative. Other than sensory streaks in some patellogastropods, and ctenidial sense organs (bursicles) in vetigastropods (see below and Chapter 7), mantle cavity sensory structures are limited to the osphradia (Haszprunar 1985a, 1985g, 1986a). There is, however, substantial variation in size, shape, and

position of the osphradia, especially in caenogastropods (see Chapters 7 and 19). Additional sense organs are present on the gastropod head (cephalic tentacles, eyes, 'rhinophores,' Hancock's organ) and sometimes the foot (epipodial tentacles and sense organs) or mantle (papillae, tentacles or, rarely eyes).

Thiele (1892) first described subpallial *sensory streaks* along the sides of the foot of *Patella*. These anterior pallial sensory streaks, found in patelloideans and nacellids, originate in the mantle cavity near the ventral osphradia and wrap around the anterior ends of the shell attachment muscles (Thiem 1917a). In the patelloideans, these streaks are extended along the side of the foot into the mantle groove as the lateral pallial sensory streaks (Sasaki 1998). In both the patelloideans and lottioids, there is also a streak of sensory epithelium on the left side of the roof of the mantle cavity (Thiele 1895; Haszprunar 1985a). Nerves from the osphradial and pleural ganglia innervate both the roof and mantle groove sensory streaks. Sensory streaks are lacking in the Lepetidae (Angerer & Haszprunar 1996).

Ctenidial sense organs (or *bursicles*) are found in most vetigastropods and consist of small ciliated pockets in the inhalant flow near the base of each leaflet of the ctenidium (see Chapter 7 for details).

4.2.1.9 Openings to the Mantle Cavity

Besides the gills, glands, and sensory structures found in the mantle cavity, three organ systems also typically open into this cavity. They are: (1) the gonopore(s) (reproductive system), (2) the kidney opening(s) (excretory system), and (3) the anus (digestive system).

4.2.1.9.1 Anus

In most molluscs, faecal material is discharged into the mantle cavity by way of the anus which lies between the kidney opening(s) and the opening of the gonoduct. In both polyplacophorans and monoplacophorans, the anus opens into the mantle groove in the mid-posterior position. This presumably plesiomorphic configuration has been secondarily attained in some gastropod slugs, notably the systellommatophorans (e.g., Onchidiidae). In bivalves, the anus opens into the dorsal part of the posterior section of the mantle cavity, near the entrance to the exhalant aperture or siphon. In caudofoveates, the anus opens below the openings of the gonoducts, while in the solenogasters, the anus opens above them. In scaphopods, it opens into the posterior half of the long, enclosed mantle cavity, while in *Nautilus*, it opens to the posterior end of the mantle cavity, but in other cephalopods, it has moved further anteriorly.

The anus is posteriorly located in the mantle cavity of vetigastropods that possess shell slits, grooves, or exhalant apertures, and faecal material is swept away by the outgoing water flow. The loss of the right ctenidium is accompanied by the relocation of the rectum on the right side of the mantle cavity. In taxa retaining the right hypobranchial gland and/or right kidney (most patellogastropods and vetigastropods), the rectum is usually not so markedly displaced to the right as in gastropods that have lost those structures.

In most caenogastropods and neritimorphs, the rectum extends along the right side of the mantle cavity so the anus opens near the mantle edge. In 'lower heterobranchs,' the anus is usually posteriorly located within the mantle cavity, but in nudipleurans and euopisthobranchs, the anal opening is variously positioned – correlated with the reduction or loss of the shell and mantle cavity. In detorted taxa with a reduced mantle cavity on the right side of the body, the anus opens in that position, while in those that lack a mantle cavity, such as doridine nudibranchs (Nudipleura) or have a substantially reduced one as in onchidioidean slugs (Systellommatophora), the anus has reverted to its primitive position in the midline at or near the posterior end. In eupulmonates, the anus has moved anteriorly and lies near the pneumostome.

4.2.1.9.2 Kidney Openings

In chitons, bivalves, cephalopods and scaphopods, and basal gastropods (patellogastropods, vetigastropods), the kidney openings from the right and left kidneys open into the mantle cavity on either side of the anus. Aplacophorans lack differentiated kidneys, but in monoplacophorans, there are multiple openings, each located at the base of a gill and associated with a single kidney. In patellogastropods and most vetigastropods, there are two kidneys, but most living gastropods have a single (left) kidney. In some gastropods, the kidney opening may be elaborated into a papilla or extended as a long ureter (as in many terrestrial taxa) which carries urine to the edge of the lung. In patellogastropods and some vetigastropods, the right kidney opening also serves as the reproductive opening.

4.2.1.9.3 Reproductive Openings

Reproductive openings (apertures, gonopores) are openings from the gonoduct(s). They are typically paired and found on either side of the anus. In chitons, they open above the kidney openings. In monoplacophorans, there are no gonopores as the gonoducts open into a kidney and are discharged into the mantle groove through the kidney opening. In bivalves, the paired gonoducts may open separately to the mantle cavity or are fused with the kidneys near their opening into the mantle cavity. In aplacophorans, the openings from the two gonoducts may be fused into a single opening (many Solenogastres) or separate as in the Caudofoveata. In the remaining three groups, Scaphopoda, Cephalopoda, and Gastropoda, there is a single gonad, and hence a single gonoduct. In scaphopods and some primitive gastropods, the gonoduct enters the right kidney and gametes are discharged through the right kidney opening. All cephalopods have a single reproductive opening in the mantle cavity. In all cephalopods and many gastropods, the gonoduct opens into glandular structures involved in producing secretions that encase the eggs or form spermatophores. In most gastropods, the reproductive openings and pallial parts of the gonoducts (if present) lie within the mantle cavity, but in euthyneuran heterobranchs, the gonoducts largely lie within the haemocoel, and the openings typically lie outside the mantle cavity. Some gastropod reproductive systems have multiple reproductive openings, as seen in the relatively complex systems of the hermaphroditic

euthyneuran heterobranchs and the gonochoric neritimorphs (see Chapters 8, 18, and 20).

4.2.2 Respiration, Ventilation, and Waste Management

4.2.2.1 Respiration

Molluscs require oxygen for their metabolism. When they are small much of this need can be met by diffusion from the water or air through the epidermis into the body. Because moist tissues are permeable to gases, such as oxygen and carbon dioxide, the entire epidermis of many molluscs is typically capable of at least limited cutaneous respiration. The effectiveness of simple diffusion, however, quickly decreases with distance from the body surface, tissue type, and previous consumption by adjacent tissues (Schmidt-Rhaesa 2007). With increasing size, it becomes incumbent on the circulatory system to circulate oxygen throughout the body. This is generally accompanied by the formation of a specialised part of the body for gas exchange (either gills or lungs) and an oxygen carrier in the vascular fluid. Ghiretti (1966) noted three types of respiratory systems in molluscs: (1) cutaneous respiration, (2) branchial respiration, and (3) lung respiration. Some degree of cutaneous respiration is probably present in all molluscs, whether or not specialised respiratory structures exist (Box 4.1).

Most molluscs are aquatic and use gills as their main respiratory surface, but complications arise in those that also use their gills for suspension-feeding. Intertidal taxa, particularly those that live in the highest intertidal zones, are capable of aquatic and limited aerial respiration, as are amphibious taxa living in fresh-water environments. Fully terrestrial taxa are capable of continuous aerial respiration, and molluscs living in hypoxic environments (see Chapter 9) often show additional respiratory adaptations.

Two regions are of particular importance for diffusion because of their intimate association with the ctenidia and inhalant and exhalant water currents – the roof of the mantle cavity or mantle groove, and the epidermis of the head and foot. The kidney can also act as a respiratory surface in those taxa where it is located outside the body cavity, particularly if it has migrated into the roof of the mantle cavity.

Considerable increases in the available respiratory surface on the mantle result from shell reduction or loss (see Section 4.2.2.1.1.6) and also, to a lesser extent, from the adoption of a limpet shape. The plethora of limpet-shaped (patelliform) taxa (regardless of ancestry) at hydrothermal vents, for example (see Chapter 9), may in part result from an ancestral response to low oxygen tension, along with the homoplastic enlargement of ctenidia for respiration, suspension-feeding, or housing symbiotic bacteria. These changes may also be coupled with the provision of secondary respiratory surfaces, with increasing body size (Lindberg & Ponder 2001). Terrestrial gastropods have lost the gill and instead most use a modified mantle cavity as a lung as well as highly vascularised body surfaces for respiration. This latter mechanism is especially common in some slug taxa.

BOX 4.1 RESPIRATION

Respiration involves the exchange of gases (O_2, CO_2) between the animal and the environment. Most molluscs (and other animals) have specialised organ systems for the exchange of gases and a blood system that transports these gases through the body (see Chapter 6).

Oxygen and other dissolved gases diffuse across the thin epithelium covering the respiratory surface (e.g., gills, lung) and into the blood, while the waste product carbon dioxide diffuses out into the water or air.

Oxygen is needed for cellular respiration, which involves a carbohydrate substrate (derived from food) to drive ATP production by the mitochondria (see Chapter 2 for details).

$$C_2H_{12}O_4 \text{ (glucose)} + O_2 \rightarrow CO_2 + H_2O + \text{energy (ATP[2])}.$$

Size matters, as the more cells there are (body mass) (see Figure 4.19), the more oxygen is needed (oxygen demand). Increased activity also leads to more oxygen demand.

The efficiency of diffusion of gases through the respiratory surface depends on the thickness of the membrane, the surface area, the temperature, the difference in concentration (partial pressure) of the gas across the membrane and, in a terrestrial organism, the amount of moisture.

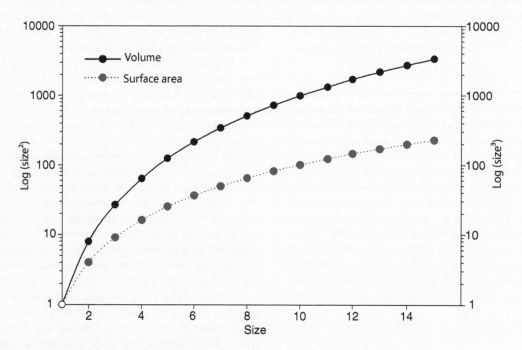

FIGURE 4.19 Graph illustrating the relationship between size and volume. Redrawn and modified from Lindberg, D.R. and Ponder, W.F., *Org. Divers. Evol.*, 1, 273–299, 2001.

---

[2] Adenosine triphosphate (see Chapter 2).

Cephalopods are particularly active and require higher rates of oxygen delivery to their tissues than other molluscs. An increase in the rate of swimming by jet propulsion automatically increases the rate of water flow over their gills, but with haemocyanin as an oxygen carrier as little as 8% of available oxygen is utilised. Thus, cephalopods are less efficient than fish, and consequently a cephalopod needs about four times the rate of blood flow as a typical fish to swim half as fast (Deaton 2008).

As the body mass of the animal becomes larger (see Section 4.2.2.1.1.5), diffusion through the integument becomes less efficient as a respiratory mechanism, and special respiratory structures such as gills or a lung become essential. The surface areas of a selection of various molluscan gills (Pelseneer 1935; Ghiretti 1966; Brown et al. 1989) show that the values of surface area to body weight were similar, except in suspension-feeding bivalves where the gills are mainly used for food collection, and sometimes the respiratory surfaces are enhanced by enlarged mantle folds (see Section 4.2.1.3). Because exchange can occur through most exposed body surfaces, gill size alone is not necessarily an accurate measure of respiratory efficiency. For example, in most gastropods, gas exchange can occur through delicate, moist external surfaces within the mantle cavity (including the gills) or along the inner side of the mantle skirt with its rich blood supply. Blood drains directly to the auricle(s) from these mantle cavity blood spaces and related sinuses and from there it passes to the ventricle, which pumps blood throughout the body. Although the blood system in most molluscs is considered an open one with venous sinuses connecting the arterial and venous systems rather than capillaries, some gastropods have blood pressures between one and three orders of magnitude higher than mammals, and the dimensions of some of the minor sinuses are similar to those found in vertebrate capillary systems (Jones 1983b). Thus, whether a system is open or closed does not appear necessarily to determine the expected physiological outcomes and is more complicated than the simple terms 'open' or 'closed' would suggest (see Chapter 6).

4.2.2.1.1 Gas Exchange

As already outlined, gas exchange takes place by diffusion across a respiratory surface, which serves as the boundary between the environment and the organism. To diffuse, gases must be in solution, with moist, thin, permeable tissues providing the highest rates of diffusion. Diffusion is a passive process, and thus the rate is influenced by the area of the respiratory surface, the thickness of the diffusion surface through which the gas molecules must pass, and the concentration gradient (because diffusion is a passive process, the gas molecules must move from high to low concentration).

The major atmospheric gases are present in seawater and fresh water: nitrogen (N_2), oxygen (O_2), argon (Ar), and carbon dioxide (CO_2) (Table 4.1), although their concentration is determined by their temperature, salinity, and partial pressures in solution.

TABLE 4.1

The Four Most Abundant Gases in the Atmosphere and Seawater

Gas		% Air	% Seawater	Solubility mL/Litre	Volume mL/Litre
Nitrogen	N_2	78.08	62.6	16.9	13.35
Oxygen	O_2	20.95	34.3	34.1	7.16
Argon	Ar	0.934	1.6	0.062	0.06
Carbon dioxide	CO_2	0.033	1.4	1019.0	0.31

Source: Sverdrup, H.U. et al., *The Oceans: Their Physics, Chemistry, and General Biology*, Prentice Hall, New York, 1942.

Note: These gases account for 99.9% of all atmospheric gases at 15°C and 1 atmosphere.

Of these gases, oxygen and carbon dioxide have been most studied because their exchange across a respiratory surface is required to fuel cellular respiration and remove metabolic and other by-products. Because oxygen is consumed in metabolism, body tissues have lower concentrations than the ambient environment, thereby diffusing oxygen across the respiratory surface. Carbon dioxide follows the opposite gradient with higher concentrations in the body tissues than in the ambient environment. In molluscs, the environment gradient is maintained by ciliary and muscular actions that ventilate the respiratory surface, while the organismal gradient is maintained by the circulation of the haemolymph throughout the body. The ambient gradient can be manipulated by varying the ventilation rate or volume, while the internal gradient can be altered by heart rate and stroke volume (Houlihan et al. 1982).

Most molluscs have an oxygen carrier pigment (usually haemocyanin or, rarely, haemoglobin – see Chapter 6) in the haemolymph which can increase oxygen delivery between 20 and 40 times (Bettex et al. 2014) (see also Chapter 6), but their oxygen affinity is also modulated by dissolved carbon dioxide. The high partial pressure of carbon dioxide (pCO_2) results in an increase in blood acidity, which in turn increases the oxygen affinity of haemocyanin. In cephalopods which inhabit the oxygen minimum zone (OMZ), this ability may be an important adaptation to hypoxic conditions (Seibel 2013). Increases in blood acidity also require more buffering with calcium bicarbonate ($CaHCO_3$). Air-breathing gastropods respire at higher oxygen levels than aquatic gastropods because of the higher concentration of oxygen in the atmosphere. This has been accompanied by an increase in internal pCO_2, which increases blood acidity and produces a compensatory increase in blood bicarbonate for buffering. This results in pCO_2 and bicarbonate levels in both terrestrial and aquatic air-breathing gastropods being at least five times higher than in aquatic snails (Barnhart 1992). For terrestrial snails, the lack of ambient water makes pH adjustments using ion exchange more difficult, and respiratory control of pH is favoured (Barnhart 1992).

As noted above, there has been little study of the exchange of other gases in molluscs, and nitrogen and argon in the shell chambers of *Nautilus*, *Spirula*, and *Sepia* contradict the idea that these cephalopods have a unique gas diffusion process. When the new chambers are first formed, they are filled with liquid – a body fluid which provides resistance to the hydrostatic pressure of the ambient environment (Denton & Gilpin-Brown 1966). As the fluid is evacuated from the chamber, a vacuum forms filled by nitrogen, oxygen, argon, carbon dioxide, and other rare gases that diffuse out of solution via the vascularised siphuncle tissue. In these gas-filled chambers, nitrogen and argon occur at atmospheric partial pressures, while the partial pressure of oxygen is lower and pCO_2 higher due to tissue metabolism. This unique gas diffusion between the blood and newly formed chambers establishes the gas-reliant buoyancy mechanism of these cephalopods (Greenwald & Ward 1987) (see also Chapter 17).

4.2.2.1.1.1 Surface Inspiration The inspiration of surface air into a mantle cavity lung occurs in some hygrophilans (Williams 1856) and in the amphibious caenogastropod Ampullariidae (McClary 1964). In both taxa, the snails move to the water/air interface where they use a retractable siphon to fill the lung with air. In some planorbids, the pulmonary siphon consists of a short spout-like projection of the mantle cavity along its ventral edge, while in ampullariids the highly extendable siphon is utilised and when fully deployed can exceed the shell length of the snail. These siphons are both on the left side of the mantle cavity. In some hygrophilans, the tentacles and pulmonary siphon are used to locate the surface film (Hunter 1953). In the ampullariids, the lung is a separate chamber from the main mantle cavity housing the ctenidium. In the hygrophilan family Planorbidae, a secondary gill, the pseudobranch, lies externally near the pulmonary opening (Basch 1963; Hubendick 1978; Burch 1982, 1988). When they do not have access to the water surface, hygrophilans can also fill their lung with water and respire through its vascularised surface (Berg & Ockelmann 1959).

4.2.2.1.1.2 Effect of Temperature The solubility of the respiratory gases oxygen and carbon dioxide decreases as water temperature increases. Thus, colder waters have more dissolved oxygen than warm waters because oxygen is about twice as soluble in seawater at 0°C than at 30°C (Kinne 1970). Temperature also affects the amount of oxygen required for life functions; a temperature rise of 10°C can double the rate of internal chemical reactions. This requires both physiological (e.g., heart rate, absorption efficiency) and behavioural (e.g., grazing or filtering rates) responses to temperature (Newell & Branch 1980). Increasing activities thus require more oxygen and generate more carbon dioxide requiring its removal in animals in both aquatic and terrestrial habitats. One outcome of the relationship between temperature and respiratory gases is the link between thermal tolerance and the availability of oxygen. Because of the physical limitations of ventilating mechanisms and the circulatory system as well as the oxygen carrying capacity of blood pigments, high temperatures can cause a transition from aerobic to anaerobic metabolism (Melzner et al. 2007b). For example, in their study of *Sepia officinalis*, Melzner et al. (2007a) found that as the extremes of temperature tolerance are reached, anaerobic metabolism is initiated in the muscles of essential organs. High temperatures resulted in the failure of the circulatory system and the saturation of haemocyanin, while at low temperatures haemocyanin is desaturated. Responses to heat vary depending on the taxon. The haemocyanin of some eurythermic cephalopods such as *Sepia* can decrease oxygen affinity with temperature changes and enable metabolic rates to be maintained, but more stenothermal species do not have this flexibility (Melzner et al. 2007a). Thermal limits imposed by insufficient oxygen in the tissues have been shown to limit intertidal *Littorina saxatilis* (Sokolova & Pörtner 2003), and these same respiratory features can also be altered, enabling some molluscs to move into other extreme habitats. Seibel

(2011) has shown that organisms in the OMZs must balance their oxygen demand with the potential for suppression of oxygen extraction and transport, anaerobic ATP production, and metabolism. For squid such as *Dosidicus gigas*, anaerobic metabolism and metabolic suppression are used during diurnal (daytime) incursions into the OMZ, while *Vampyroteuthis infernalis*, which lives in the OMZ, has a metabolic rate over 100 times lower than *D. gigas* (Seibel 2013).

4.2.2.1.1.3 Intertidal and Amphibious Molluscs The transition from aquatic to terrestrial habitats is not common, and not simple (Little 1983; Vermeij & Dudley 2000). No fossil molluscs identified as amphibious can be linked to any transitional aquatic to terrestrial habitats, although they undoubtedly exist somewhere in the record (Gordon & Olson 2013). The interface between these two extremes is a gradient of wet and dry conditions surrounding the edges of marine, brackish, and fresh-water habitats. Molluscs that live in this interface are amphibious or semi-amphibious. We define amphibious taxa as those that are active in both aquatic and terrestrial habitats, while the habitat activities of semi-amphibious taxa are typically controlled by the environment. For example, the amphibious Ampullariidae with both a lung and gill are enabled to exploit both aquatic and terrestrial habitats. Most semi-amphibious taxa are found in intertidal habitats where tidal height determines time spent in aquatic and aerial conditions. Semi-amphibious taxa are typically not highly modified, and their ability to remain active in aerial conditions is limited to specific circumstances, such as patellid limpets grazing in moist conditions at night during low tides (Lorenzen 2007). Only gastropods have terrestrial taxa and are therefore potentially amphibious. Intertidal semi-amphibious molluscs are more diverse and include gastropods, bivalves, polyplacophorans, and cephalopods (*Octopus*). Taxa that occur in the intertidal and are not active during either their submergence or emersion are not considered to be semi-amphibious, but are categorised as either terrestrial or aquatic depending on which state they are active in. This definition approximately corresponds to the respiratory states of aerobic and anaerobic respiration, where amphibious and semi-amphibious taxa can sustain activity under aerobic respiration, while terrestrial or aquatic taxa become anaerobic during the alternative phase.

Besides the ampullariids, only a few other gastropods are amphibious as defined here, including some neritids, members of the caenogastropod families Assimineidae (Ponder 1988a), Truncatellidae (Rosenberg 1996), Littorinidae (Reid 1985; Lee & Williams 2002), and members of the heterobranch families Aitengidae (Swennen & Buatip 2009), Ellobiidae (Price 1980), and possibly Onchidiidae; which are all regularly active in both aquatic and terrestrial conditions (see also Chapter 6). Only the ampullariids have separate gill and lung chambers (Andrews 1965b; Jurberg et al. 1997), but other modifications seen in the remaining taxa include gill loss or reduction (Rosenberg 1996), enlarged or elongated mantle cavities (Lindberg & Ponder 2001), and life history changes (Reid et al. 2010).

As noted above, hygrophilans can fill the lung with a bubble of air which serves as a tissue-water interface for transferring oxygen through passive diffusion. Because nitrogen also diffuses to the water, the size of the bubble is diminished over time and must be replenished from the surface. To what extent the bubble functions as a diffusion surface appears to be species dependent (Russell-Hunter 1954). While the ampullariid lung is an autapomorphy for the group, the lung in the Hygrophila is plesiomorphic, being present in basal eupulmonates. Bubble respiration also occurs in aquatic insects and arachnids (spiders) (Seymour & Matthews 2013).

Intertidal panpulmonates such as siphonariids and amphibolids, and to a lesser extent, ellobiids, are capable of respiration in air and water, but siphonariids, which have a gill in their lung, are more efficient at aquatic respiration (Innes et al. 1984).

In semi-amphibious molluscs such as intertidal gastropods, water retained in the mantle cavity during low tides and under certain conditions may allow some respiratory function by the gill to continue for a time (Houlihan et al. 1981). In lottiid limpets, the shallow mantle cavity loses water readily and the ctenidium, lacking skeletal rods, collapses, but the moist, highly vascularised roof of the mantle cavity continues to act as a respiratory surface. Thus the ctenidium is the main respiratory surface in the immersed animal, but the roof of the mantle cavity is used when the animal is exposed at low tide (Kingston 1968). Patellid and nacellid limpets lack a ctenidium, but the highly vascularised roof of the mantle groove has secondary gills which enable them to use aerial respiration, continue feeding activity, and maintain metabolism (thus avoiding anaerobiosis) while exposed at low tide (Brinkhoff et al. 1983; Abele et al. 2010). This behaviour occurs primarily at night when temperatures are low and desiccation minimal. Similar nocturnal feeding activities are exhibited by some chitons, which also have mantle grooves with gills (Cretchley et al. 1997). Some mangrove littorinids have mainly nocturnal activities, while others are active during rain rather than just at night (Ohgaki 1993; Lee & Williams 2002; Bates & Hicks 2005). In these cases, ambient moisture content appears to be an important driver of activity during aerial exposure.

Certain aquatic species can survive out of water in a resting state for long periods with minimal activity (e.g., some high intertidal gastropods and various fresh-water bivalves, notably some unionoideans). Some may be out of the water for longer periods of time than they are submerged and are capable of respiring in air (McMahon 1988), while others have been shown to survive for 1–12 months, or even longer, out of water (Deaton 2008) simply by closing their shell.

4.2.2.1.1.4 Anoxia Molluscs undertake anaerobic respiration when either external or internal environments inhibit or prevent normal aerobic respiration. They experience anoxia[3]

(and hypoxia) in two forms – environmental anoxia and functional anoxia. Environmental anoxia occurs when there is insufficient oxygen in the ambient environment to sustain aerobic respiration, while functional anoxia occurs internally (e.g., in muscular tissues) during high energy expenditure and augments aerobic respiration (Zwaan 1991) (see Chapter 2). Environmental anoxia is most common in aquatic settings because the solubility and diffusion rate of oxygen is substantially lower than in air. Many marine molluscs can tolerate environmental anoxia for days and even weeks, and even though terrestrial molluscs live in high ambient oxygen, they often encounter anoxia during aestivation (Meenakshi 1957; Wieser 1981; Brooks & Storey 1997).

Environmental and functional anoxia use a variety of pathways and compounds in the production of energy (ATP) (See Chapter 2 for more details of the underlying biochemistry).

Besides metabolic pathway adaptations to counteract low oxygen concentrations in the environment and tissues, molluscs also exhibit an array of morphological and behavioural adaptations to low oxygen concentrations, including increasing respiration and ventilation capacity, increasing mantle cavity size, increasing gill size or number, reducing tissue thickness, and increasing the area of respiratory surface by becoming limpet-like (Lindberg & Ponder 2001). Behavioural adaptations include vertical migration (pteropods, cephalopods) (Childress & Seibel 1998; Maas et al. 2012) and posture changes that increase respiratory surface exposure (Levin 2003; Williams et al. 2005).

4.2.2.1.1.5 The Influence of Body Size While the volume of an animal increases as the cube of its linear dimensions (Figure 4.19), the body surface increases only as the square of the body length. Thus, as noted above, small body size reduces the volume to surface area ratios and the need for specialised respiratory structures. The shrinking of the mantle cavity and ctenidium also decreases the Reynolds number[4] of the water flow through the chamber and gill (Vogel 1994). To compensate for this increased viscosity, complex structures like a ctenidium may undergo a reduction in complexity (e.g., the loss of the central axis and reduction in the size and number of filaments) (Eno 1994).

The small size can make specialised respiratory structures unnecessary while larger size necessitates greater respiratory activity and increased efficiency of the ventilation and respiratory functions. If the mantle cavity is enlarged to accommodate larger ctenidia, it not only increases the ctenidial surface area available for respiration, but it also increases the area of the mantle cavity epithelium.

As chitons increase in size, they simultaneously increase both respiratory and ventilation surfaces by adding more gills, from posterior to anterior. A similar increase in ctenidia with size is seen in Monoplacophora, but in that group the ctenidia appear to be mainly ventilating structures, and the mantle

---

[3] Anoxia – absence of oxygen in tissues or environment; hypoxia – deficient oxygen in the body or environment.

[4] The Reynolds number is the ratio of inertial forces to viscous forces for given flow conditions, i.e., the point at which laminar flow becomes turbulent flow.

groove itself provides the respiratory surface (Lindberg & Ponder 1996; Haszprunar & Schaefer 1997).

The two pairs of ctenidia in *Nautilus* are ventilated by currents produced by muscular contractions of the funnel. Thus, although nautiloids had relatively inefficient respiratory equipment compared to coleoids, ctenidial duplication may have enabled them to achieve a large size (as evidenced particularly by some fossil taxa). In coleoids, respiratory problems have been solved in different ways; auxiliary hearts increase the rate of blood flow through the gills and, because they are not encumbered with an external shell, powerful mantle contractions (not cilia) drive the water across the gill and mantle cavity surfaces. These modifications have enabled some coleoids to become the largest and most active molluscs.

The plesiomorphic state of gastropods was paired gills with a small, shallow mantle cavity, although this configuration is highly modified in most extant taxa. With increasing size, some gastropods improved their respiratory efficiency by increasing the length of the mantle cavity, as seen by its extension posterior to the gills in pleurotomarioideans and trochoideans (Lindberg & Ponder 2001). This extension envelops the kidney, and with its high blood flow, provides an additional respiratory surface in the cavity.

Attainment of small size through developmental processes such as paedomorphosis or miniaturisation (Hanken & Wake 1993) appears to have occurred repeatedly. It is important at this point to distinguish between miniatures and paedomorphs. For example, the gastropod taxa Scissurellidae, Skeneidae, and many Rissooidea and Truncatelloidea appear to have attained small body size through miniaturisation. They are characterised by having the full anatomical configuration seen in the adult characters of their larger sister taxon, albeit downsized. In contrast, paedomorphs obtain small body size through juvenilisation and have structures and organ systems reduced or simplified relative to their sister taxon, so that they often exhibit 'new' ground plans such as loss of ctenidia in the mantle cavity. Unique new structures are thought to accompany the release from the developmental constraints of adult morphology *via* heterochrony.

4.2.2.1.1.6 Shell Loss and Its Consequences Shell loss has occurred in the cephalopods, gastropods, and perhaps in the aplacophoran groups. In living cephalopods, shell loss is seen in several groups of coleoids (see Chapter 17). In gastropods, shell loss has rarely occurred in vetigastropods, neritimorphs, and caenogastropods, but is common in heterobranchs (see Chapter 3). While a shell is entirely missing in coleoid 'larvae' a calcified shell may be present or is usually represented as an internal rudiment in adults. Conversely, a shell is typically present in gastropod larvae (Ponder & Lindberg 1997; Dayrat & Tillier 2002), and in those lacking an adult shell, it is lost during metamorphosis and settling.

While there is little doubt the reduction and internalisation of the shell in coleoids facilitated ventilation and locomotion in that group, shell loss *per se* was not a prerequisite for these innovations as they are shared with the externally shelled nautiloids (see Section 4.2.2.1.1.5 and Chapter 17).

The coleoid mantle cavity innovations are more likely to be part of an array of modifications for pelagic life, which have also permitted the secondary return to the benthos by some octobrach taxa.

In gastropods, the patterns of shell loss are more complicated. In vetigastropods, marked shell reduction and loss has occurred only in some fissurellids, notably in the keyhole limpet genera related to *Fissurellidea*. This group is characterised by shell reduction due to overgrowth of the shell by the mantle, with complete loss of the shell in the South American *Buchanania onchidioides* (McLean 1984). The only example of shell loss in the Neritimorpha is *Titiscania*, a genus comprising two species of small (<2 cm) slug-like animals found sporadically throughout the tropical and subtropical Indo-Pacific (Templado & Ortea 2001; Kano et al. 2002). In both of these examples, the mantle cavity is unmodified, except for the thickening of the mantle epidermis over the dorsal surface of the animal. In caenogastropods, complete adult shell loss is limited to a small group of parasitic, worm-like eulimid gastropods found in holothurians where they attach to the walls of the cloaca (Bouchet & Lützen 1980). As with most internal parasites, these gastropods have a highly modified body plan. The mantle cavity and most organs are absent, and the entire animal consists of ovaries and resident dwarf males (Lützen 1968) (see Chapter 19). Another caenogastropod group, the velutinids, are externally slug-like, but retain a reduced internal shell. Shell reduction, decalcification, and enveloping by the mantle are common traits within heterobranchs, and in all members of some taxa the entire adult shell has been lost, including all Rhodopoidea, Gymnosomata, Acochlidia, Systellommatophora, and Nudibranchia, with several additional clades having members that have lost the adult shell or have only an internal rudiment (Cephalaspidea, Aplysiida, Pleurobranchida, Sacoglossa, Trimusculoidea, Stylommatophora). This pattern suggests strong parallelism, occurring more than once in several of these ten clades. This multitude of pathways from reduction to complete shell loss is also reflected in the diversity of morphological outcomes for the exposed mantle and its structures. For example, in Gymnosomata, the mantle cavity is lost, and posterior secondary gills and respiratory patches are present (Hyman 1967). In Acochlidia, a mantle cavity is usually absent, as is the ctenidium and osphradium (Fahrner & Haszprunar 2002b). In the worm-like Rhodopemorpha, the mantle cavity, gill, and osphradium have all been lost, although an osphradium and osphradial ganglion briefly appear during development (Riedl 1960). The greatest variation in gill morphology occurs in the Nudipleura. In the Pleurobranchida, the exposed elongate gill extends along the right side of the body to the posterior portion of the animal, while respiratory provisions in the Nudibranchia include pallial branchia (e.g., *Tritonia*), circum-anal gills (e.g., *Archidoris*), and diverse forms of 'cerata,' which often combine respiratory, digestive, and even defensive functions (Thompson 1976; Thompson & Brown 1984). The cerata are so diverse across nudibranch taxa that Wägele and Willan (2000) identified seven distinct morphotypes.

Although there is tremendous variability in mantle cavity traits associated with shell loss, two general patterns emerge: (1) shell loss is independent and does not affect mantle cavity structure and function (Vetigastropoda, Neritimorpha), and (2) shell loss is linked to whole animal developmental reorganisations. In the shell-less parasitic eulimids, this appears to be a progenetic event in which there is a substantial reduction in body size, and numerous adult organ systems fail to develop; such progenesis is commonly associated with the origin of parasites (Gould 1977). In heterobranchs, there also seems to be whole animal developmental changes, but rather than straight deletions of organs and structures there have been numerous parallel experimentations starting from different combinations of juvenile and unique traits. This is further complicated by detorsion, perhaps the largest developmental change in heterobranchs (see Chapters 8 and 20). Developmental patterns in which new morphologies arise from juvenile states subject to selection in the adult habitat are also considered progenetic, but rather than being a simple arresting of development, they are releases from the constraints of ancestral adult morphologies (Gould 1977; Lindberg 1988a). A similar argument could also be made for cephalopods, which have the most apomorphic development of all molluscs (see Chapters 8 and 17).

4.2.2.2 Ventilation

Ventilation is critical for respiration. As water or air passes through the mantle cavity or lung, diffusion across the respiratory surfaces depletes the fluid of oxygen and enriches it in carbon dioxide. Without ventilation, aerobic respiration would not be possible, and it is referred to as ventilation-limited. If ventilation is too fast or fluid velocity too high, gas exchange will also be inhibited. The exchange of oxygen and carbon dioxide is greatly facilitated by a counter-current system, where blood flows over the respiratory surfaces in the opposite direction to the water flow (Figure 4.20). Although aerial ventilation was initially thought to be absent in 'pulmonate' gastropods, it was later shown that, in stylommatophoran snails at least, they carry out breathing movement associated with the pneumostome and contractions of the floor of the mantle cavity (see Section 4.2.1.5).

Ventilation of the mantle cavity or groove provides a fluid stream which brings the animal oxygen for respiration, chemical and physical information on the local environment and, in some taxa, food, and removes metabolic wastes, sediment, and sometimes gametes from the cavity. Ventilation currents are typically generated by cilia which cover the surface of the mantle cavity and its associated structures (e.g., ctenidia, osphradia), although

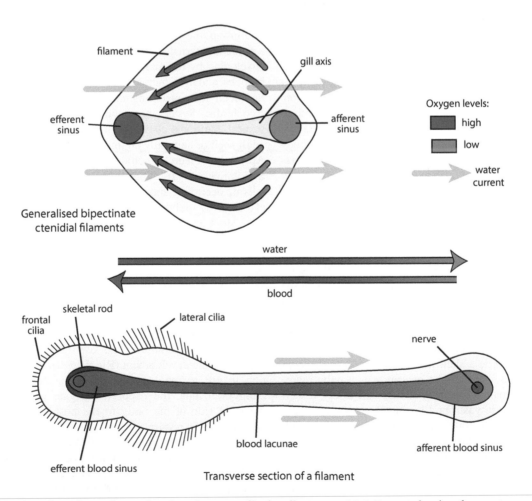

FIGURE 4.20 A surface view and a section through a generalised molluscan ctenidial filament showing the counter-current between internal blood and external respiratory water flow. Original.

in coleoid cephalopods ventilation is controlled by muscular mantle contractions (see Section 4.2.1.1.6). Both ventilation mechanisms (ciliary and muscular) can also be augmented by induced flow[5] (Vogel 1994), which has been demonstrated in gastropods (Murdock & Vogel 1978; Voltzow & Collin 1995) and cephalopods (Vogel 1987). While ciliary ventilation may support resting aerobic respiration in abalone, the induced flow may be important for higher activity levels and recovery from anaerobic events (Taylor & Ragg 2005). Induced flow through the mantle cavity is probably present in other groups such as bivalves and caudofoveates, based on the placement of their exhalant siphons or mantle cavity opening, respectively, just above the sediment-water interface (Stanley 1970; Salvini-Plawen 1971).

While ventilation and respiration are inevitably linked, there is also conflict between these two processes (Lindberg & Ponder 2001). The ciliated cells are larger and denser than the thinner, smaller cells that line the respiratory surface, and therefore the thickness of the epidermis that the gases must diffuse across is increased, which in turn reduces diffusion (Jobling 1994). Thus, except for cephalopods, increasing ventilation capacity on the respiratory surface decreases the respiratory surface area. Lindberg and Ponder (2001) examined the distribution on a gastropod morphological tree of some ctenidial traits associated with this functional contradiction and found this conflict was resolved in different groups by moving all or part of the ventilation and/or respiratory functions to non-ctenidial surfaces. They suggested that the evolution of ctenidial cleansing and sediment rejection systems selected for triangular leaflets with concentrations of cilia at the apex. Triangular respiratory surfaces have only 64% of the surface area of a semicircular surface of the same dimensions. There appear to be multiple responses to this reduction, including smaller body size, increased ventilation, and mantle cavity/ctenidial enlargement. The latter two outcomes are associated with the appearance of skeletal support structures and gill attachment membranes which provide for more efficient respiratory flow between leaflets and further enhance the sediment rejection system. The modification of ventilation traits also occurred, suggesting control by the mantle edge of the passage of fluid through the mantle cavity, the induction of passive flow, and the addition of ventilating surfaces and structures such as the osphradium or ciliated tracts or ridges. There is an elongation of the osphradium in the Neritimorpha, followed by its repositioning next to the ctenidium and a marked increase in size in the Caenogastropoda, and lastly the development of ciliated osphradial leaflets (some Sorbeoconcha). In the 'lower heterobranchs', opposed ciliated ridges in the mantle cavity create flow. These ridges are sometimes extended into an elongate caecum, markedly increasing the length of these ridges (see Section 4.2.1.4).

Besides cilia-powered ventilation, some heterobranchs use muscular contractions to ventilate and discharge debris from the mantle cavity. For example, when *Aplysia* contracts

its gill and siphon and closes the parapodia, water is forced from the mantle cavity through the siphon. As the muscles relax, water is drawn back into the mantle cavity via the parapodia, thus ventilating the mantle cavity. The flow of haemolymph is coordinated with this pulsing which further increases gas diffusion across the gill surfaces (Croll 1985).

Patterns of respiration/ventilation conflict are also present in other molluscs. Four groups (Caudofoveata, Bivalvia, Gastropoda, and Polyplacophora) have gills or ctenidia that function in their plesiomorphic state as both ventilators and respirators. Although Solenogastres lack ctenidia, the mantle cavity with its respiratory folds is ciliated (Salvini-Plawen 1972). In the gills of protobranch bivalves, ventilation and respiration are clearly linked, while in the remaining bivalves the patterns are confused because the gills are highly modified for suspension-feeding and additional respiratory surfaces are furnished by mantle cavity tissue and in a few cases secondary gills (Yonge 1939b, 1947). In living monoplacophorans, the respiratory currents appear to be generated by the beating gills, while the surface of the mantle groove serves as the primary respiratory surface (Lindberg & Ponder 1996; Schaefer & Haszprunar 1997). Scaphopods lack gills, but ciliated ridges circulate water through the elongate cavity, and this circulation is further facilitated by muscular contractions caused by the movements of the foot (Reynolds 2002).

4.2.2.2.1 Direction of Water Flow

Water flows into the mantle grooves of chitons from near the anterior end and exits posteriorly (Yonge 1939a), and similar flow patterns were reconstructed for the monoplacophoran *Neopilina* by Lemche and Wingstrand (1959) (see Chapter 14, Figure 14.12). In chitons, the individual ctenidia hang from the roof of the mantle groove, to which they are attached by the ctenidial axis. The anterior-posterior flow moves from the outer side of the ctenidia to the inner side, which functionally divides the mantle groove into inhalant outer and inner (next to the foot) exhalant channels. The gills are aligned with their axes perpendicular to the cavity ensuring that the water flow crosses each filament from the efferent to afferent sides, creating a counter-current with ctenidial blood flow and thus increasing respiratory efficiency. In both aplacophoran groups, water enters the mantle cavity dorsally and exits ventrally (Salvini-Plawen 1968a, 1985a). In the Caudofoveata the ctenidia are attached to the back wall of the mantle cavity with the efferent side of the gill axis dorsal and the afferent side ventral. Water passes over the laterally extended leaflets, creating dorsal inhalant and ventral exhalant chambers (Salvini-Plawen 1985a).

The paired ctenidia of protobranch bivalves are aligned longitudinally in the spacious mantle cavity that surrounds the body. They are attached to the roof of the mantle cavity by the afferent side of the central axis except in *Solemya*, where the gill axis has rotated clockwise approximately 90°. This rotation creates a substantial vertical space enabling ctenidial enlargement (Yonge 1939b), probably driven by the presence of bacterial symbionts in the gill (Stewart & Cavanaugh 2006)

---

[5] Induced flow occurs when an internal current (e.g., in the mantle cavity) is hydrodynamically coupled with the ambient current (e.g., external water current) to augment flow through the mantle cavity.

(see Section 4.2.3 and Chapters 9 and 15). In protobranchs, the leaflets extend laterally from the axis, dividing the mantle cavity into two chambers. In nuculids, water enters anteriorly into the ventral chamber, but in nuculanoids it enters through the posterior inhalant siphon, with lesser flow from the anterior region of the shell. From the ventral inhalant chamber, it passes vertically across the leaflets into the dorsal exhalant chamber and is then expelled (Atkins 1936; Purchon 1977a). In autobranch bivalves, the adoption of suspension-feeding has brought about a multiplication, elongation, and reflection of the filaments (see Section 4.2.1.1.4 and Chapter 15). As in protobranchs, the filaments hang below their dorsal attachment. Water flow is primarily created by the lateral cilia of the ctenidial filaments, but the rate and direction of water flow through the mantle cavity is controlled by the muscular mantle edges, or, if present, the siphons. Most bivalves have the inhalant stream entering the cavity posteriorly just below the more dorsally located exhalant stream. A few have an anterior inflow into the inhalant chamber (flow-through ventilation), including the protobranch Nuculoidea and the autobranch Galeommatoidea, a highly derived group that includes ectosymbionts, endosymbionts, and burrow associates, as well as free-living forms (Goto et al. 2012). Another autobranch group with anterior inflow is the highly specialised Lucinoidea, recognised for the bacterial symbionts housed in its gills. Most bivalves are infaunal (Stanley 1970), with the inhalant and exhalant openings (often as siphons) at the posterior end for the animal and negating the functionality of the flow-through ventilation assumed to have occurred in early bivalves (see Chapter 15). In autobranch bivalves, the water flow has changed from the simple cross leaflet flow to one that is more efficient. It involves oxygenated water passing across the faces of both the inner and outer filaments into the exhalant chambers between the ascending and descending limbs of the gill filaments from where it is passed out through the exhalant aperture or siphon.

In gastropods, the ctenidial axis is attached to the mantle cavity roof or posterior wall. When present, the gill in patellogastropods extends freely from the posterior wall, while in Cocculinoidea and Heterobranchia the gill extends from the roof of the mantle cavity. In most vetigastropods, the ctenidial axis is attached to the mantle cavity by a ventral (efferent) membrane; dorsal (afferent) membranes are also present in some taxa, although the anterior part of the gill is typically free. In caenogastropods, the right row of leaflets is lost, and the axis of the single left gill is fused to the mantle cavity wall. Despite these various arrangements in gastropods, the flow patterns within the cavity are relatively conservative. Water enters ventrally and flows from left to right (and right to left when the gills are paired) over the leaflet surfaces and is expelled to the top or right. In contrast, there have been significant changes in how water enters or leaves the mantle cavity, partly because of torsion rotating the mantle cavity anteriorly. Some groups have opted to control water flow by regulating the direction and rate of water entering the mantle cavity, while in others, the focus is on the water leaving the cavity – referred to as inhalant and exhalant control,

respectively (Lindberg & Ponder 2001). In vetigastropods the deep mantle cavity, shell slit, or hole are involved in exhalant control and are the hallmarks of several vetigastropod groups, but others have lost these 'hardware' solutions. Some basal caenogastropods (the 'architaenioglossan' Ampullarioidea) also have exhalant control, but in most caenogastropods flow is controlled on the inhalant side, culminating in the extensible, mobile anterior siphon typical of higher caenogastropods. If an anal notch (adapical channel) is present, both inhalant and exhalant control may take place. Heterobranchs also primarily use exhalant control, with many shelled euopisthobranchs and a few sacoglossans extending the posterior portion of the mantle as a siphon.

A major departure in ventilation currents is seen in the highly modified cephalopod mantle cavity which uses muscular pumping to drive water through the gills (see Section 4.2.2.2.2), and their pleated gills lack cilia. Another major difference is that, compared with most other molluscs, the inhalant water current is reversed as it runs over the dorsal surface of the gill and out between the filaments laterally, being discharged from the lateroventral part of the mantle cavity via the funnel (see Section 4.2.2.2.2). Cephalopod ctenidia have the same general configuration of blood vessels as in other molluscs with afferent vessels on the dorsal edges and efferent vessels on the lower, but they have a dorsal to ventral flow rather than the ventral to dorsal seen in all other molluscs except in Caudofoveata. This reversal in flow negates the counter-current system and would seem to lessen respiratory efficiency. Cephalopods appear to have overcome this problem of water flow reversal by second and third order folding of the surface of each filament (see Figure 4.11 and Chapter 17 for more detail on cephalopod gills).

In scaphopods, the dorsal shell aperture is primarily used to circulate water in and out of the mantle cavity. There are no gills, but the respiratory flow is driven by opposed ciliated ridges located midway on the posterior portion of the body. Some water from the dorsal opening passes over the body and mantle before being expelled out of the ventral aperture (Reynolds 2002). This flow is augmented by dilation of the foot and movement of the epipodial lobes (Yonge 1937; Steiner 1991). Periodic contractions of the foot expel waste and gametes from the mantle cavity through the dorsal aperture (Reynolds 2002).

4.2.2.2.2 The Mantle Cavity of Coleoid Cephalopods and Its Modification for Jet Propulsion

The mantle cavity of coleoid cephalopods, as already noted, is highly modified and uses muscular contractions rather than cilia for ventilation. The loss (or internalisation) of the shell removed anatomical constraints imposed by shell secretion, resulting in a muscular hydrostat which provides locomotion and mantle cavity ventilation. The coleoid mantle cavity consists of three layers – outer and inner layers rich in collagen and a middle muscle layer made up of circular fibres (see Chapter 3, Figure 3.54). The central fibres of the middle layer are associated with mantle cavity contractions that produce high-velocity water bursts used in jetting to catch prey or evade predators, while the outer fibres are primarily involved in prolonged aerobic work associated with sustained activity such as swimming (Zwaan 1991) (See also Chapter 17).

This linkage between ventilation and metabolically expensive locomotion produced a unique internal conflict in cephalopod physiology, which Shadwick (1994, p. 69) described as '…a progression towards higher blood and jet pressures…correlated with increasing activity and power density…' Jet propulsion requires a high rate of water movement into and out of the mantle cavity. This passes more water over the gills, but the limiting factor in the cephalopod oxygen delivery system is the low oxygen-carrying capacity of haemocyanin. Thus, to deliver sufficient oxygen to the tissues for sustained swimming and other activities, evolution proceeded along three courses: (1) the modification of the ctenidia to increase surface area, (2) a closed, high pressure circulatory system for more efficient delivery, and (3) (in coleoids) additional hearts to generate greater blood flow (O'Dor et al. 1990; Shadwick 1994). The separately pressurised respiratory and circulatory systems of coleoids are analogous to the condition found in mammals and enable some to pump blood at rates similar to human athletes (O'Dor & Shadwick 1989).

4.2.2.3 Waste Removal

Cilia in the mantle grooves or cavities facilitate the removal of waste (faeces and urine) and potentially clogging particles that enter with the inhalant water stream. These latter particles are caught up in mucus on the gills which are cleansed with the rejection currents on the filaments. The mucus-bound waste is then moved to the exterior by mantle rejection currents. While these currents do not play a direct role in ventilation, they are crucial to its maintenance and performance. In suspension-feeding gastropods, the rejection currents on the floor of the mantle cavity are developed into food grooves, and in bivalves into complex rejection tracts involved in the accumulation of pseudofaeces (see Chapter 5 for more detail).

In scaphopods, periodic contractions of the foot expel faeces through the dorsal aperture (Reynolds 2002).

4.2.3 Symbionts

Epibiotic, endosymbiotic, and intracellular chemosynthetic bacteria are found in the tissues and gills of the mantle cavity (see Chapter 9 for more detail). These thin tissues effectively enable direct access to the molluscs' circulatory system, while the continuous water flow that ventilates the mantle cavity and gills provides a ready supply of reductants and oxidants to the symbionts (Dubilier et al. 2008; Taylor & Glover 2010). The mantle cavity and gills may be secondarily modified to house the bacteria. Modifications include elongated gill filaments (Hawe et al. 2014), disproportionately large gills and gill filament fusion (Taylor & Glover 2010), mantle gills (Allen 1958; Taylor & Glover 2000), and mantle fusion (Kuhara et al. 2014). Anderson (2014) has shown that the mantle gills in lucinid bivalves are detectable using geometric morphometric analysis of shell characters alone, thus suggesting that the acquisition of chemosymbiotic bacteria by this taxon is recognisable in the fossil record.

Other modifications to the mantle cavity are seen at the tissue and cellular level. Bacteria are often housed in epidermal structures (*bacteriocytes*) in the mantle or gill epidermis. Bacteriocytes have been reported in the Monoplacophora (Tryblidia) (Haszprunar et al. 1995), Bivalvia (e.g., Krueger et al. 1996b; Duperron 2010; Brissac et al. 2011), and Gastropoda (Windoffer & Giere 1997; Suzuki et al. 2005; Sasaki et al. 2010b; Hawe et al. 2014; Judge & Haszprunar 2014). In *Tridacna*, mantle tissue with symbiotic algae is further modified by a layer of iridescent cells called iridocytes, which focus primarily red and blue light on the symbiotic algae which are arranged in pillars, while backscattering the potentially damaging yellow light (Holt et al. 2014). This biophotonic system enables photosynthesis to take place deep in the mantle tissue, protects the photosystem from damage, and gives the mantle its distinctive colouration (see Chapter 9 for more information on symbionts).

5 Feeding and Digestion

5.1 GENERAL INTRODUCTION

In this chapter, we examine molluscan feeding and digestive systems, the morphological innovations associated with different feeding strategies, and the physiological aspects of digestion.

Molluscs exhibit one of the most diverse ranges of feeding strategies seen in the Metazoa. From a simple beginning scraping deposits and microbial films, they diversified to exhibit almost all possible feeding strategies. They have become suspension-feeders, grazers on rock surfaces, plants and colonial animals, suctorial parasites and active predators. The evolution of each feeding behaviour required changes in physiology and morphology. Besides modifications to the digestive system itself, other parts of the body related directly or indirectly to a feeding strategy, such as the head, foot, and shell, may also change. In some lineages these changes increased morphological complexity, but in others resulted in simplification. There are also some significant organ asymmetries associated with the digestive system (see Section 5.3.5).

The digestive system of molluscs generally follows a common pattern, although in some it is highly modified. Like most higher animals, most[1] molluscs have a mouth, an oesophagus, a stomach, a digestive gland, an intestine, and a rectum opening to the exterior by way of an anus. In all molluscs except bivalves the mouth opens to a buccal area that may contain (typically paired) jaws. Plesiomorphically, the buccal cavity houses a muscular odontophore, which bears the radula, and a pair of salivary glands usually open to it. Except for the mouth, all of these buccal structures are lost in bivalves.

Odontophoral muscles, along with supporting structures such as odontophoral cartilages, operate the radula. The radular teeth interact directly with the substratum on which the animal is feeding and carry food to the buccal cavity. The oesophagus, sometimes with enzyme-secreting glandular pouches, transports food from the buccal region to the stomach. The oesophagus, much of the stomach, and the intestine are ciliated, allowing transport throughout. A pair of digestive glands associated with the stomach (sometimes called the hepatopancreas) is involved in digestion.

There are two ways food is digested in molluscs:

Extracellular digestion – enzymes break down food in the gut enabling absorption by the gut epithelium and the digestive cells in the digestive gland.
Intracellular digestion – food particles and metabolites are pinocytosed by the digestive cells in the digestive

gland and digested within those cells. Some intracellular digestion may also occur in the hindgut in some bivalves at least (Beninger et al. 2003).

The stomach of many molluscs is complex, typically asymmetrical, with ciliated sorting areas and cuticular areas which protect the stomach wall from abrasion. Food material enters the stomach via the ciliated oesophagus. In most autobranch bivalves and some gastropods, a derivative of the intestine, the style sac, contains a crystalline style, a rotating rod of mucoprotein that releases digestive enzymes (see Section 5.4.2.3.1). The intestine is often long, looped, or coiled, and opens to the rectum from which faecal material is discharged via the anus, which is typically located in the mantle cavity (see Chapter 4).

This arrangement of the gut differs considerably from that seen in the molluscan outgroups. Annelids have a straight, bilaterally symmetrical tube, generally no digestive gland, and digestion and absorption occur within the tubular gut. Members of the Lophophorata (Phoronida, Brachiopoda, Bryozoa, Entoprocta) and Sipuncula have a U-shaped gut, but lack many molluscan features including an odontophore and radula, salivary glands, and digestive glands (although a single dorsal 'digestive gland' occurs in brachiopods and terebellid annelids). Given the anterior-posterior orientation of the gut in chitons, monoplacophorans, and aplacophorans, we assume this is the plesiomorphic adult molluscan condition, and that the U-shaped gut present in several of the outgroup taxa listed above has evolved independently.

It is by no means clear whether a straight or looped hindgut represents the plesiomorphic molluscan condition. Looped hindguts occur in the Polyplacophora and Monoplacophora, while straight hindguts are present in both aplacophoran groups. The presence of both states renders the adult plesiomorphic state at the base of the Aculifera equivocal. Looped hindguts also occur in the inarticulate brachiopods, hyoliths, and sipunculids.

5.2 FEEDING

Many feeding strategies have evolved in molluscs. They probably originally grazed, either indiscriminately or selectively, on biofilms on firm surfaces, and possibly on encrusting animals or detritus (Lindberg et al. 2004). The switch to other kinds of food acquisition was a major feature of molluscan adaptive radiation (Ponder & Lindberg 1997; Vermeij & Lindberg 2000), which had profound influences on the entire body morphology and physiology.

Although we assume that early molluscs were surface film grazers, that mode of feeding is not seen in the outgroups, other than in some polychaetes. In early molluscs, the buccal

---

[1] A few highly modified taxa have lost the gut, including some gastropod internal parasites and a few protobranch bivalves with symbiotic bacteria.

apparatus, including the radula, presumably enabled them to scrape surfaces, and it is possible that some of the earliest radulae were hardened by mineralisation and impregnated with metals increasing abrasion resistance (Dutta et al. 2010), as in some extant taxa (see Section 5.3.2.4.3), although the molluscan radula is probably not a *de novo* structure among the lophotrochozoans. Complex pharyngeal structures are widespread in lophotrochozoan taxa, including the Annelida, Rotifera, Micrognathozoa, and Gnathostomulida. In the Annelida, these structures may be hardened by tanned proteins, calcium carbonate, or other elements, and metals such as iron and copper occur in the structures of some polychaetes (Böggemann 2006). Unlike the molluscan radula, polychaete structures contain little chitin, but instead are composed of amino acids (predominately glycine and histidine).

The combination of the belief that the cnidarian-feeding solenogasters were basal molluscs (see Chapter 12), and the adoption of carnivory in many invertebrate groups, led to the idea that the first molluscs may have been carnivorous (Salvini-Plawen 1988). This is doubtful, although some early molluscs may have incidentally fed on small animals as part of their general grazing activities on films and mats. Changes in feeding modes no doubt occurred slowly, presumably via mixed modes of feeding, with different lineages slowly and independently acquiring the features enabling new feeding strategies.

5.2.1 MAJOR FEEDING INNOVATIONS

We assume that the primitive molluscan diet was probably microphagous, with food particle size small enough for intracellular digestion without the necessity for significant processing. Many living chitons and gastropods scrape detritus and microorganisms from the substratum surface, but most extant molluscs employ quite different modes of feeding. Suspension-feeders, notably most autobranch bivalves and some gastropods, acquire minute, mainly unicellular, organisms and detrital particles from the water column. Deposit feeders also take in particulate food including organic detritus, bacteria, fungi, and protists, along with sediment, and this feeding mode is seen in monoplacophorans, some bivalves (including nearly all protobranchs), caudofoveate aplacophorans, and some gastropods. Scaphopods feed selectively on protists and/or bacteria living among sediments.

The change from the non-selective scraping of surfaces and extraction of minute organisms to grazing on plants, macroalgae, fungi, or encrusting colonial animals for food, occurred in many molluscan lineages. This shift to macrophagy involved the modification of some structures (such as jaws, radula) and, in some lineages, the evolution of new ones (such as a gizzard to break up food).

Grazing carnivory is commonly found in some chitons, solenogasters, and several groups of gastropods. Except for a few anomalous carnivorous ventures such as capturing and feeding on small crustaceans (septibranch bivalves, a few chitons), several gastropod groups and most cephalopods are highly specialised predators.

Although many chitons and marine, fresh-water, and terrestrial gastropods are surface scrapers and detritus feeders, it is perhaps surprising that truly herbivorous grazers are limited to some polyplacophorans and a few groups of gastropods (e.g., Vermeij & Lindberg 2000), with only a few taxa capable of feeding directly on macroalgae or vascular plant tissue. Microfungi and bacteria break down plant detritus, making it more nutritious and more readily digestible than living plant tissue (e.g., Bärlocher et al. 1989b) which may account for the great majority of land snails and slugs utilising decaying vegetation and fungi with only a small percentage preferentially feeding on living plants. This may be because living plant and algal tissues often have low concentrations of essential nutrients. Also, some contain toxins and plant and algal cell walls are hard to digest, although these limitations can be overcome in part by selection of the more nutritious and less toxic plants and/or algae and by physiological adaptations.

An increase in the quality of food intake can be achieved by changing to an animal diet, as this provides more essential nutrients than biofilms or algal or plant tissue. Many molluscs occasionally ingest some animal tissue, even if they normally feed on detritus, algae, fungi, or plants. Minute or small animals can be accidentally or deliberately swallowed with food scraped from the substratum, and some may feed on carrion when it is available. This is particularly true of terrestrial gastropods which are often otherwise constrained in their sources of nutrients and/or calcium (Barker & Efford 2002). Sometimes, the distinction between detrital or herbivorous/fungal/algal-feeding and carnivory is blurred when scavenging, or even predation, is a relatively common source of food. In some taxa, transitional series from herbivory to facultative carnivory and then to obligate carnivory can be recognised (e.g., Barker & Efford 2002), while in other lineages carnivores can revert to herbivory or deposit-feeding, as in some neogastropods (e.g., see Section 5.2.1.13 and Chapter 19).

Each feeding mode is discussed in more detail below.

5.2.1.1 Microphagous Feeding on Biofilms

The plesiomorphic mode of feeding in molluscs was probably unselective microphagous feeding[2] (e.g., Vermeij & Lindberg 2000). Many marine herbivorous vetigastropods, neritoideans, and caenogastropods mainly eat filamentous algae, protists, and other microorganisms, the minute algae and other food items being raked up by the radular teeth. In vetigastropods and neritimorphs, the numerous marginal teeth of their rhipidoglossan radula (see Section 5.3.2.4) act like brooms that sweep the substratum, but exert little force.

Many marine molluscs which live on seagrass or macroalgae are not necessarily feeding on the seagrass or alga itself, but on films of diatoms, bacteria, and cyanobacteria (the *periphyton*) or other attached organisms that cover their surfaces. Similarly, most fresh-water Hygrophila living on macrophytes feed on the film of microalgae and other organisms that cover them rather than the macrophytes themselves.

---

[2] This mode of feeding is sometimes referred to as deposit feeding, but we use that term in a more restricted sense (see Section 5.2.1.2.1).

This has resulted in some interesting mutualistic interactions. The macrophytes benefit from the removal of periphyton by the snails grazing on the leaves and increased nutrient turnover, and the snails not only have a grazing surface, but increased oxygen and shelter from predators (Thomas 1982). For example, the lymnaeid *Radix peregra* is positively attracted to dissolved organic matter excreted by the macrophyte *Ceratophyllum demersum,* although it does not feed on the plant directly, but instead on the periphyton covering its leaves. The removal of the periphyton has a positive effect on the growth rate of the macrophyte (Bronmark 1985).

5.2.1.2 Particle-Feeding

The particulate food obtained by both suspension and deposit-feeding is typically very heterogeneous and is dilute relative to the volume of water or sediment that contains this nutritious food material. Consequently, deposit and suspension-feeders have numerous adaptations with respect to extracting, sorting, and digesting food.

Suspended particulate food consists of either plankton, which is highly variable in size and nutritive value, or re-suspended bottom material which may comprise organic matter, faecal material, and bacteria. In contrast, benthic particulate matter consists largely of various sized mineral particles, organic matter, microalgae, faeces, detritus, protozoans, and bacteria (Ward & Shumway 2004).

All particle-feeding molluscs rely heavily on mucus and cilia for food collection (Beninger & St-Jean 1997a; Davies & Hawkins 1998; Beninger et al. 2007). While most autobranch bivalves are highly specialised suspension-feeders, some bivalves (mainly tellinoideans and protobranchs) are deposit feeders (see Section 5.2.1.2.1). Some suspension-feeding gastropods use mucous nets or threads externally or within the mantle cavity to collect suspended food material (see Section 5.2.1.2.3).

In autobranch bivalves, the distinction between suspension-feeding and deposit feeding is somewhat blurred. Many suspension-feeders collect food at or near the sediment-water interface and so ingest suspended bottom material. Similarly, while vacuuming the sediment surface, some deposit feeders (notably tellinoideans) will uptake some suspended particles via their inhalant siphon.

Particle-feeding in bivalves requires little energy; for example, the energy required to capture and transport suspended particles was less than 2% of the metabolic energy expended by *Mytilus edulis* (Jørgensen et al. 1986). While some infaunal species use extra energy to move about and/or burrow, most of this activity is unrelated to feeding.

Particle selection determines the quality of ingested material and reduces the uptake of harmful substances. In autobranch bivalves, selective removal of non-nutritive particles occurs (reviewed by Ward & Shumway 2004).

While suspension-feeding occurs mainly in autobranch bivalves (see Section 5.2.1.2.2.1), it has also evolved independently in some lineages of vetigastropods and caenogastropods and in at least two groups of heterobranchs (Declerck 1995) (see Table 5.1). However, even highly modified, suspension-feeding gastropods are much less specialised than autobranch bivalves. An important difference is that particle sorting is mostly done within the gut, not partly before ingestion as in bivalves.

Cilial collection of food particles also occurs in planktotrophic veliger larvae (Figure 5.1).

TABLE 5.1
Gastropods Which Use Suspension-Feeding in Adults

Subclass	Superfamily	Family	Suspension-Feeding	Mucous Net Feeding
Vetigastropoda	Neomphaloidea	Neomphalidae		
	Trochoidea	Trochidae (Umboniinae)		
Caenogastropoda	Ampullarioidea	Ampullariidae		
	Viviparoidea	Viviparidae		
	Cerithioidea	Turritellidae		
		Siliquariidae		
	Truncatelloidea	Bithyniidae		
		Hydrobiidae		
	Stromboidea	Struthiolariidae		
	Vermetoidea	Vermetidae		
	Calyptraeoidea	Calyptraeidae		
	Capuloidea	Capulidae		
	Olivoidea	Olividae		
Heterobranchia	Cavolinioidea	Cavoliniidae		
	Cymbulioidea	Cymbulidae		
	Ellobioidea	Trimusculidae		
	Lymnaeoidea	Lymnaeidae		

Source of most data: Declerck, C.H., *Biol. Rev.,* 70, 549–569, 1995.

Note: Each instance involves a separate evolutionary event.

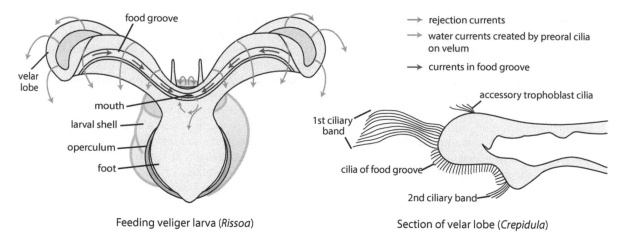

FIGURE 5.1 Suspension-feeding in a gastropod veliger larva using cilia on the velar lobes and transverse section of velar lobe showing detail of cilia. Left figure redrawn and modified from Fretter, V. and Montgomery, M.C., *J. Mar. Biol. Assoc. U.K.*, 48, 499–520, 1968 and right figure redrawn and modified from Henry, J.Q. et al., *Integr. Comp. Biol.*, 47, 865–871, 2007.

A few reports of cephalopods either suspension-feeding or deposit-feeding are apparently erroneous, although some do feed on minute animals (see Section 5.2.1.7.1).

5.2.1.2.1 Deposit-Feeding

The change from feeding on surface films to unselectively taking quantities of sediment into the gut to extract nutritive detritus is seemingly a relatively simple step, although requiring modification to the gut and feeding structures. It has only occurred in some monoplacophorans (Lemche & Wingstrand 1959), to some extent in caudofoveates, and some bivalves and gastropods. In surface scrapers, the gut already copes with the removal of unwanted particles, so the adoption of deposit-feeding with much-increased sediment uptake presumably led, through selection, to modifications to efficiently sort food particles from the sediment and to deal with greater quantities of waste material. These changes included enhancing sorting mechanisms in the stomach and the protection of the delicate gastric epithelia from abrasion, as well as elongation of the intestine for increased opportunity for absorption.

While deposit-feeding in protobranch bivalves is plesiomorphic, deposit-feeding in autobranchs has been secondarily acquired. Protobranchs use ciliated palp proboscides they extend into the sediment, and the great majority of this material is rejected before ingestion. Some suspended particles can be captured by the small ctenidium and carried to the palps, where they are sorted before ingestion (e.g., Davenport 1988), but this is a small component of their total food intake (Stead et al. 2003).

Deposit-feeding autobranchs (mainly tellinoideans) deal with a broader range of particle size than the suspension-feeding taxa from which they evolved. Initial sorting is undertaken on the gills and large labial palps before ingestion, and there are also some gut modifications. Tellinoideans indiscriminately suck up deposits using their long mobile inhalant siphon; the material is transferred to the ctenidia or directly to the labial palps for sorting (e.g., Yonge 1949) and the gill cilia are modified to deal with this (Atkins 1937a,

1937b). The pre-ingestion sorting by the gills and palps generates the production of much pseudofaecal material which can be 30–300 times that of the volume of the faeces (e.g., Hylleberg & Gallucci 1975). Comparative analysis of the material in the pseudofaeces and the stomach shows that both quantitative and some qualitative sorting occurs before ingestion (see Ward & Shumway 2004 for review and references). Many tellinoideans use deposit-feeding as their main source of food, but in addition to feeding on deposits, some also carry out suspension-feeding.

A few adult bivalves use cilia on their foot to collect deposits, as for example, the lucinoidean *Fimbria fimbriata* (Morton 1983a). This habit is much more commonly seen in juveniles of a wide taxonomic range of autobranch bivalves, both marine and fresh water, which feed with cilia on the foot and palps until their gills are functional. This may be the typical means of feeding in all post-larval bivalves, including protobranchs, and was probably the original mode of feeding in the earliest bivalves (Reid et al. 1992; Morton 1996; Waller 1998).

Some studies have suggested that at least some sphaeriids may deposit feed mainly on organic matter rather than suspended particles, but in their size range, particles suspended over the substratum and those deposited on the substratum are probably indistinguishable (Dillon 2000).

Gastropods that ingest surface deposits include members of the Xenophoridae and some cerithioideans, notably Potamididae, Batillariidae, many Cerithiidae, and some fresh-water members of that superfamily.

Scaphopods are sometimes considered to be deposit feeders (e.g., Ward & Shumway 2004), using their ciliated captaculae in food gathering, but they are highly selective in their capturing of food items and do not qualify as deposit feeders in the sense used here.

5.2.1.2.2 Suspension-Feeding Using the Ctenidia

Suspension (or filter) feeding consists of three main processes: (1) direction of water to, and its movement through, the suspension-feeding structure (usually the gill, sometimes mucus); (2) removal of potential food particles from the water; (3) transport of the potential food particles to the mouth. In the sections below, we examine how these processes are achieved, particularly in the autobranch bivalves, the predominant suspension-feeding group among the molluscs.

5.2.1.2.2.1 Bivalves

In bivalves, most of the extraction of food occurs before ingestion with the particles sorted based on their physical and, possibly, their chemical properties (see Section 5.2.1.2.2.2). This process results in the better-quality material being ingested. Much of the following information on bivalve feeding is based on comprehensive reviews of feeding in autobranch bivalves by Ward and Shumway (2004) and Rosa et al. (2018).

Autobranch bivalves are very efficient suspension-feeders because their highly modified gills utilise a combination of cilia, mucus, hydrodynamics, and particle charge to capture and transport particles. The lateral cilia on the ctenidial filaments create a current of water entering the mantle cavity through the inhalant aperture, siphon, or mantle margin. This water flows between the filaments into the suprabranchial chambers and leaves the mantle cavity through the exhalant aperture or siphon. Particles captured by the ctenidium are moved either ventrally (most homorhabdic filibranchs and eulamellibranchs) or either dorsally (initial acceptance, hydrodynamic transport) or ventrally (initial rejection, muco-ciliary transport) to the gill margins. In the ventral marginal tracts (usually a food groove), the particles are carried in a mucous string of variable length, but in the dorsal groove between the gill demibranchs, they are in a watery mucous slurry (Ward et al. 1993; Ward 1996; Beninger et al. 1997a; Beninger & St-Jean 1997b). Although the particles from the dorsal tracts arrive at the palps in a state suitable for sorting, the mucous strings from the food grooves on the gill margins are quickly broken down on the palps enabling sorting to occur (e.g., Beninger et al. 1997b). After sorting on the palps, food is passed to the mouth in a mucus-water slurry (Ward et al. 1994).

Details of the movement of particles on the labial palps are complex and differ with differing palp morphology, which itself can be quite complicated. Generally, material to be ingested is passed anteriorly and perpendicular to the palp crests, while material to be rejected is passed within the palp troughs first to the ventral palp margins, and then posteriorly to the ventro-posterior palp extremity (Beninger et al. 1997a; Beninger & St-Jean 1997b). From here, various means of transporting particles out of the infrabranchial cavity are employed, depending on the particle processing system, thus ensuring that they are not continually resuspended and reprocessed (Beninger & Veniot 1999; Beninger et al. 1999). The size of the palps is correlated with sorting efficiency and, in some taxa, is related to sediment loads in the habitats they occupy (Dutertre et al. 2009).

Some details of autobranch bivalve feeding outlined above have only recently been elucidated, mainly by using the newly developed techniques of endoscopy, mucocyte mapping, and cilia mapping.

5.2.1.2.2.2 Particle Sorting in Bivalves

The efficient selection of suitable food material before ingestion increases the efficiency of sorting and subsequent digestion in the stomach. There is ample evidence of qualitative particle sorting by either or both the palps and gills, depending on the particle processing system (see Ward & Shumway 2004 for review and references). For example, there is good evidence showing that the gills, palps, and stomach of bivalves can select different types of particulate matter, including preferentially selecting living and organic items in favour of inorganic particles and detritus (e.g., Beninger et al. 2008). While the actual separation of accepted and rejected particles is well documented, it is not yet very well understood how bivalves evaluate particle quality, although it may involve interactions between the mucus employed in feeding and the surface properties of the particles, including their charge (e.g., Rosa et al. 2017, 2018).

Ward and Shumway (2004) divided possible bivalve sorting mechanisms into two main categories, passive and active, as follows.

> *Passive (or mechanical) selection* is possible because of the structure of the sorting organ. It includes selection by size, weight, and shape and other mechanical interactions between the surfaces of particles and feeding organs (e.g., 'sticky' particles vs 'non-sticky' particles).
>
> In deposit feeders sorting by size is efficient, as food value is inversely related to particle size in sediments, down to the size of bacteria (Jumars 1993), but in phytoplankton, food value tends to be greater in larger cells, so strict size selection in suspension-feeders is not necessarily the best strategy. Although there are relatively few studies on suspension-feeding bivalves, they suggest that selection is not biased towards the smallest particles, but that particles outside an optimum size range are rejected (see Ward & Shumway 2004 for review and references).
>
> *Active (or behavioural) selection* occurs when changes are initiated by sorting structures (e.g., muscular contractions, ciliary activity) because of stimulation by qualitative features of the particles, notably chemical signatures. Thus, the acceptance or rejection of a particle depends on such cues.
>
> While chemical cues to distinguish particles have been reported in bivalves, in most of those studies differences in size, shape, or surface properties have not been completely removed. In one study, however, Ward and Targett (1989) showed that identical microspheres treated with different phytoplankton-derived chemicals were discriminated. Despite this, the surface properties of the microspheres resulting from the different coatings may have allowed passive rather than active selection as electrostatic charge, sugar, or other organic coatings on particle surfaces have all been shown to affect mechanical particle selection (Ward & Shumway 2004). Thus, while passive (i.e., mechanical) selection is a major

mechanism used by bivalves to distinguish inorganic from organic particles, active selection is also occurring (e.g., Rosa et al. 2018).

Particle retention is typically an asymptotic curve, with retention efficiency decreasing at some point as particle size decreases. Most autobranch bivalves efficiently (>60%) retain particles over 4 μm, while the retention efficiency of smaller particles varies with gill ciliation. For example, some bivalves, including unionids (Dillon 2000) and a tropical marine mytilid (Yahel et al. 2009), efficiently retain particles as small as 1–2 μm; some taxa, including oysters and scallops (with reduced or absent laterofrontal cirri), retain ~100% of 5–6 μm particles, but are much less efficient at capturing smaller particles (Møhlenberg & Riisgård 1978; Yahel et al. 2009). Most autobranch bivalves retain particles over 4 μm, although the retention of smaller particles varies with gill ciliation. For example, some bivalves, including unionids (Dillon 2000), efficiently retain particles as small as 1–2 μm; some taxa, including oysters and scallops (with reduced or absent laterofrontal cirri), retain ~100% of 5–6 μm particles, but are much less efficient at capturing smaller particles.

5.2.1.2.2.3 Gastropoda Several groups of gastropods including the vetigastropod Umboniinae (Trochidae) (Fretter 1975a; Hickman 1985b, 1996, 2003b), the caenogastropod viviparids, turritellids, siliquariids, vermetids, bithyniids, calyptraeids, struthiolariids, and some capulids, have independently evolved ctenidial suspension-feeding (Declerck 1995) (Figures 5.2 and 5.3; Table 5.1). Unlike the situation in bivalves, there is no prior sorting of captured particles, although in umboniine trochids and turritellids long tentacles in the incurrent part of the aperture probably help prevent the largest particles entering (e.g., Hickman 1985b).

The evolutionary step to gastropod ctenidial feeding was a relatively simple one, as it involved only slight modifications to the ciliation that collects waste already present on the filaments, and to the mantle ciliation that rejects this and other waste from the mantle cavity. The ctenidial filaments are elongated (but are never folded as in bivalves), and most have also developed a mucus-secreting strip or ridge, the *endostyle*, which runs along the base of the gill on the left side, over which the incoming water passes.

The most studied gastropod ciliary feeders are species of *Crepidula* with several detailed accounts on feeding in *C. fornicata* (Orton 1912, 1914; Werner 1951, 1953; Beninger et al. 2007; Shumway et al. 2014b) and one on *C. fecunda* (Chaparro et al. 2002). In *Crepidula*, it has usually been suggested that the endostyle produces a mucous sheet that extends across the incurrent side of the gill and entangles the food particles and this sheet, with the entrapped particles, is carried by cilia to a food groove on the neck where it is rolled into a string and transported to the mouth. In addition, there are reports of a porous mucous net suspended in the inhalant water stream.

Thus, under these scenarios, ctenidial feeding in these gastropods is different to the primarily ciliary-based particle capturing in autobranch bivalves. Beninger et al. (2007) found no evidence of a mucous net on the gill or in the inhalant part of the mantle cavity opening, and they also noted that *Crepidula* lacked any ability to qualitatively select food particles before ingestion. Shumway et al. (2014b) found, using video endoscopy, that captured food particles are moved by the frontal cilia of the gill and incorporated into fine mucous strings carried both distally and obliquely across the frontal surface of the filaments to the distal edge of the gill. These are incorporated into the anteriorly moving food string in the food groove which drags them across the gill. The overlapping mucous strings gave the mistaken impression of the 'mucous net' described by earlier workers. Ctenidial mucous sheet feeding has been reported in *Bithynia tentaculata* (Schäfer 1952), but requires confirmation.

5.2.1.2.3 Mucous Net Feeding

Mucous net feeding is rare in gastropods and is known from only four groups (Table 5.1), but particularly in the sessile vermetids (Caenogastropoda) and pelagic thecosome pteropods (Heterobranchia) (Figure 5.4).

Many vermetids secrete a food-entangling mucous thread by way of the foot (e.g., Yonge 1932; Yonge & Iles 1939; Morton 1950, 1955c; Hadfield 1970; Hughes & Lewis 1974; Hughes 1978; Nelson 1980; Kappner et al. 2000). Some vermetids also filter food from the water column using their gill while a few feed with both the gill and mucous threads. The mucus feeders of this family have a smaller gill and very large pedal glands that extend back into the body. A pair of pedal (propodial) tentacles assist in forming the mucous threads. The gill feeders in this group also have propodial tentacles and smaller, but still well developed, pedal glands. In these taxa, the mucus produced appears to be mainly involved in managing waste that collects on the foot and thus is presumably an exaptation to food collection. Ctenidial feeders are better adapted to areas with strong wave action and currents, while mucus feeders are more suited to quiet water, as turbulence would quickly destroy the delicate threads (e.g., Morton 1955c).

The other major group of gastropod mucus feeders are the planktonic thecosome pteropods. Previously, thecosomes were thought to feed in much the same manner as feeding veliger larvae, by collecting particles on their enlarged parapodial 'wings'. All Cavolinioidea examined *in situ* feed using a large, spherical mucous web, while *Peraclis* and other pseudothecosomes use a funnel-shaped mucous sheet (Gilmer & Harbison 1986). Their nets are many times the size of the animals that produce them and these animals have been likened to marine spiders that drift in the plankton. For example, *Gleba cordata* and *Corolla spectabilis*, the adults of which are about 40 mm long, have two-metre diameter nets (Gilmer 1974). The nets are secreted by glands on the edge of the parapodial wings, and the web, with entangled food, is moved to the mouth by cilia. There are few obvious modifications related to this mode of feeding, although in *Diacria* and *Cavolinia*,

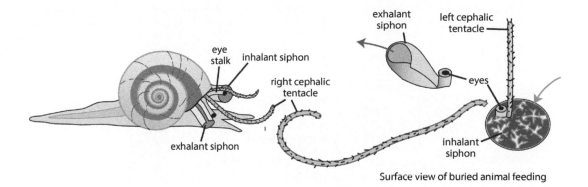

Umbonium (Trochoidea)

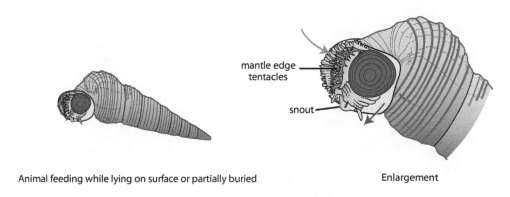

Turritella (Cerithioidea)

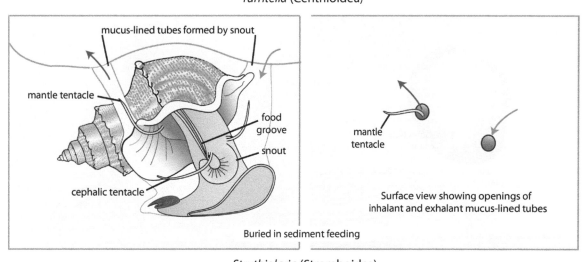

Struthiolaria (Stromboidea)

FIGURE 5.2 Suspension-feeding vetigastropod and caenogastropods that live on soft substrata. Redrawn and modified from the following sources: *Umbonium* Hickman, C.S., Comparative morphology and ecology of free-living suspension-feeding gastropods from Hong Kong, pp. 217–234, in Morton, B. and Dudgeon, D., *Proceedings of the Second International Workshop on the Malacofauna of Hong Kong and of Southern China*, Hong Kong, 6–24 April 1983, Vol. 1 and 2, Hong Kong University Press, Hong Kong, China, 1985b; *Turritella* Fretter, V. and Graham, A.L., *British Prosobranch Molluscs: Their Functional Anatomy and Ecology*, Ray Society, London, UK, 1962; *Struthiolaria* Morton, J.E., *Q. J. Microsc. Sci.*, 92, 1–25, 1951.

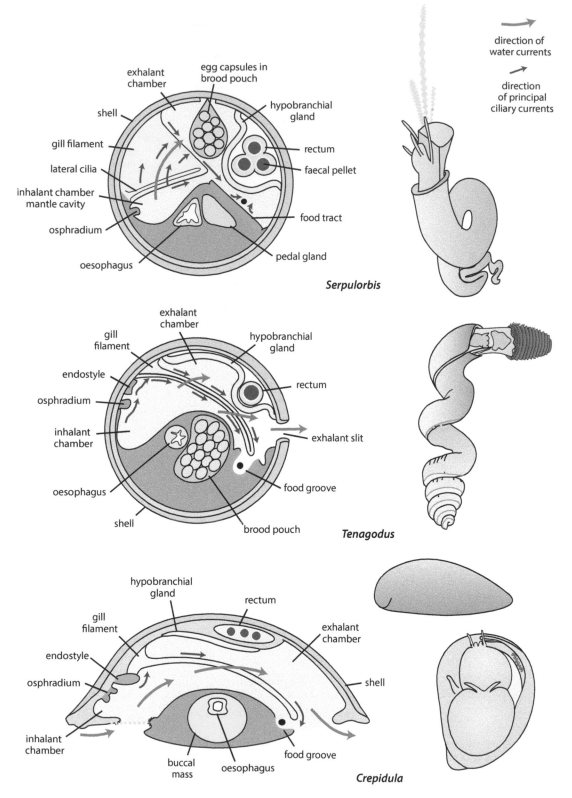

FIGURE 5.3 Some suspension-feeding caenogastropods showing the structure of their mantle cavities in transverse section (TS). Redrawn and modified from the following sources: *Serpulorbis* and *Tenagodus* Morton, J.E., *Pac. Sci.*, 9, 3–15, 1955c, *Tenagodus* whole animal Beesley, P.L. et al., Mollusca: The southern synthesis. Part B., pp. i–viii, 565–1234, in Beesley, P.L. et al., *Fauna of Australia*, CSIRO Publishing, Melbourne, Australia, 1998b; *Crepidula* TS Fretter, V. and Graham, A.L., *British Prosobranch Molluscs: Their Functional Anatomy and Ecology*, Ray Society, London, UK, 1962; *Crepidula* shell original and animal from Yonge, C.M. and Thompson, T.E., *Living Marine Molluscs*. London, Collins, 1976.

Aletes - a pair with portion of
the communal mucous sheet

Serpulorbis with mucous trap
extended during feeding

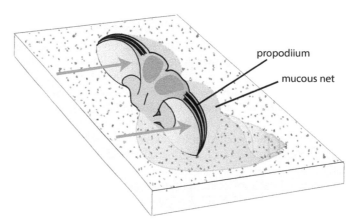

propodiium

mucous net

Olivella crawling over sand

Olivella body oriented towards beach - propodium above
surface holds mucous nets to catch particles in retreating water

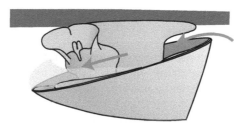

Limacina spreading mucous net

Trimusculus mucous net attached to mantle

FIGURE 5.4   Mucous net feeding in different gastropods. Redrawn and modified from the following sources: *Serpulorbis* and *Aletes* Morton, J.E., *Pac. Sci.*, 9, 3–15, 1955c; *Olivella* Seilacher, A., *Natur und Volk*, 89, 359–366, 1959; *Trimusculus* Walsby, J.R. et al., *J. Zool.*, 171, 257–283, 1973; *Limacina* original, based on data in Gilmer, R.W. and Harbison, G.R., *Mar.* Biol., 91, 47–57, 1986.

the external part of the mantle is complex and may be used for sorting particles before ingestion (Gilmer & Harbison 1986). The food includes small planktonic animals and phytoplankton, so they are effectively omnivorous.

In some species of the normally carnivorous neogastropod genus *Olivella*, the foot secretes mucous sheets that capture particles in the surf of their sandy beach habitats. A sheet emerges from the foot on either side between the pro- and mesopodium, and these large sheets become inflated and cup-like as they fill with water as it recedes down the beach. They are then quickly pulled in and ingested. These animals often live in dense aggregations and can form mucous sheets up to several metres across (Anonymous 2011), but surprisingly, this behaviour has not been well studied. Other species of *Olivella* and related taxa feed carnivorously or omnivorously (e.g., Morris et al. 1980) and at least one species may feed selectively on foraminiferans (Hickman & Lipps 1983). It is probable that mucous trap feeding is used by some species to supplement their diet and may account for these animals achieving high population densities.

A net-like mucous sheet is secreted by the anterior mantle edge in the intertidal marine eupulmonate limpet-like trimusculids (Walsby et al. 1973; Walsby 1975). These animals live on wave-swept hard shores in caves or crevices and orientate themselves to the waves to maximise their feeding (Walsby et al. 1973; Walsby 1975). As the wave approaches, they lower their shell and the water sweeps around the body which is orientated with the head towards the shore. The mucous net fills with water and captures suspended particles. The animal checks the food content of the net with the outer edge of the oral lobes, and it is then hauled in by the radula. Earlier reports (Yonge 1958; Haven 1973) on feeding in these animals are incorrect.

Fresh-water snails, such as some ampullariids and lymnaeids, and the estuarine hydrobiid *Peringia ulvae* collect particles with their pedal mucus and ingest it while floating on surface films (see McClary 1964 for references). This feeding method may be much more common than the literature suggests as floating on surface films is common in small gastropods and the fresh-water hygrophilans. It does not appear to have been investigated recently and is not mentioned, for example, in the otherwise detailed review of feeding in fresh-water snails by Dillon (2000).

5.2.1.3 Scraping Hard Surfaces

While similar to the plesiomorphic method of feeding (see Section 5.2.1), rasping hard surfaces such as rocks requires a large, hard radula and a powerful odontophore to drive it. In addition, the gut must be able to process large particles, both of food and substratum. This feeding method is mainly employed by some chitons, many patellogastropods, and some littorinids. Hardening of the radular teeth typically involves the incorporation of metal ions (see Section 5.3.2.4.3). Gut modifications are needed to sort and dispose of the large, coarse particles generated by this feeding method.

Most rock scrapers are feeding on algae as well as protists and bacteria on the surface, but must also take up some minute or encrusting animals.

5.2.1.4 Omnivorous Feeding

Many generalised scrapers and particle feeders are effectively omnivores, but this term is normally used for animals that typically consume and can digest both macro animal and plant tissue. Relatively few molluscs fall into this category, although, as already noted, animal matter food is sometimes consumed accidentally by herbivores, and carnivores or scavengers may sometimes accidentally consume plant material. Some fresh-water taxa (hygrophilans and ampullariids) are commonly omnivorous, as are some land snails and slugs (Speiser 2001), although in most cases there appear to be very unequal quantities of plant and animal tissue consumed.

5.2.1.5 Herbivorous Feeding

Macroalgae and vascular plants, or the detritus derived from them, are the major source of potential food in many aquatic and most terrestrial ecosystems. The downside of this food resource is its low nutritive value, usually with low nitrogen, and most of the carbon is bound in structural polymers (e.g., cellulose) which cannot be digested by many molluscs. This carbon source is readily available to many fungi and bacteria, and the digestion of cropped plant material can be assisted by the microbial 'flora' in the gut (see Section 5.4.2.3.2).

Although some herbivores produce enzymes that can break down the cell walls of 'algae' and plants (see Section 5.4.2.3), many molluscs eat 'algae' that are minute, including sporelings (microherbivores). These microalgal feeders may be general biofilm or rock scrapers, or suspension-feeders, and while they can be considered at least in part 'herbivores', we use the term 'macro-herbivores' for those that feed directly on large kelp-like or crustose algae or plants, and these taxa have different functional adaptations reflecting these differences in diet.

5.2.1.5.1 Macro-herbivorous Feeding

Relatively few taxa are able to feed on macroalgae or plants, and this ability has evolved independently in different taxa (Vermeij & Lindberg 2000). Erect algae of intermediate size (1–10 cm tall) are eaten less, perhaps being too large to rasp from the substratum and too small to be a substratum to occupy as habitat (Steneck & Watling 1982). Macro-herbivory is found in many chitons (some of which are specialists on various algae or seagrasses) and has independently evolved in several clades of gastropods (some patellogastropods and trochoideans, littorinids, strombids, aplysiids, some stylommatophoran land snail and slug lineages, and the amphibious fresh-water Ampullariidae).

Many chitons, patellogastropods and some trochoideans feed on large (including crustose) 'algae' or plants, requiring penetration of their leathery or hard surfaces. Adopting this mode of feeding requires several adaptations, including the radula being able to cut through the tough cell walls of the algae or plants, and the jaw(s) may also be modified to play a significant role. Differences in the size, structural, and chemical composition of host algae and plants limit their availability. For example, chitons and patellogastropods have the rasping capabilities needed to penetrate plant material such as large, leathery algae (Steneck & Watling 1982). Physiological adaptations may also be required because many algae contain noxious compounds (see Chapter 9). Another limitation, especially for smaller herbivores, is the need to crawl on, and adhere to, the surface of larger algae, seagrasses, or larger land plants, requiring modifications to the foot (e.g., Hickman 2005b). In addition, some algae provide better refuge from predation than others. Thus, it is often a balance of these factors that leads to food choice, rather than simple availability, nutritional value, or ease of digestion (e.g., Cox & Murray 2006).

Feeding preferences may also be more specialised than it first appears. For example, although temperate abalone clamp down on drifting macroalgae to feed on it, they still show preference for particular taxa (e.g., McShane et al. 1994), and while strombids are associated with seagrasses and macroalgae beds, they actually specialise in eating delicate red and green algae (Yonge 1932; Robertson 1961).

Coralline algae, both foliose and encrusting, are an important food source for some intertidal and shallow sublittoral herbivorous gastropods and chitons (e.g., Maneveldt et al. 2006), with some taxa specialising on it. Despite their appearance, these algae have an organic content as high as or higher than fleshy algae (Hawkins 1981).

The caenogastropod Cypraeidae and Columbellidae are primarily carnivorous groups, but also include some secondarily herbivorous taxa. Associated modifications to the alimentary canal have been investigated in columbellids (deMaintenon 1999), but not in cypraeids, and physiological changes have not yet been investigated in either group.

Some herbivorous euopisthobranchs such as the aplysiids and bullids have an oesophageal gizzard to assist in breaking up the plant material on which they are feeding (see Section 5.3.2.7).

While many hygrophilan snails consume macrophytes only if they are dead and decaying or as litter, experiments on *Physa* and the planorbid *Helisoma* have shown that the latter will occasionally eat living plant tissue (Lombardo & Cooke 2004). A physid has been shown to preferentially eat yellow senescent watercress, despite a reduction in nutrients, in preference to the fresh leaves, apparently because of much lower levels of deterrent chemicals (Newman et al. 1992, 1996).

Some stylommatophoran lineages (e.g., many Helicoidea and Limacoidea) can graze directly on living plant tissue, although, in some, this may only comprise a relatively small proportion of the diet (Speiser 2001). Senescent leaves or leaves infested by fungi are often preferred over healthy leaves (e.g., Ramsell & Paul 1990), although sometimes secondary metabolites from fungal infections make the plant distasteful to herbivores (Barker 2008).

While 'plant' food is often detected just by contact, its detection from at best a few centimetres away has been shown to occur in *Aplysia* (e.g., Teyke et al. 1992).

Rather than indiscriminate grazing on plants or algae, there can be preferential selection of food items. Steinberg (1985) demonstrated that the trochid gastropod *Tegula funebralis* selected brown algal food sources based primarily on

phenolic content to avoid algal chemical defences (see also Hay 2009). Experiments on fresh-water gastropods have shown that at least some hygrophilans prefer algae that will provide the most potential energy (Calow 1975; Calow & Calow 1975), and some stylommatophoran slugs and snails show preferences for particular plants (e.g., Mølgaard 1986; Iglesias & Castillejo 1999; Cook et al. 2000).

5.2.1.5.2 Algal Suctorial Feeders

Some heterobranch gastropods, the sacoglossans (e.g., Gascoigne 1977; Jensen 1997; Williams & Walker 1999), and tiny omalogyrids (Fretter 1948), feed suctorially on green algae by piercing individual cells with a specialised radular tooth and then sucking out the cell contents.

5.2.1.6 Fungal-Feeding

Many terrestrial gastropods preferentially feed on the larger fungi (ascomycetes and basidiomycetes), even those that are toxic to mammals (Speiser 2001), although plant material may also be consumed (Davis 2004). Myxomycetes (slime moulds) are eaten by (at least) terrestrial slugs (Keller & Snell 2002).

Many microfungi are ubiquitous in marine, fresh-water, and terrestrial environments and probably provide an important (although poorly documented) component of the diet of a wide range of molluscs.

5.2.1.7 Carnivorous Feeding

Many molluscs feed on animals, either by scraping or biting their tissue, ingesting them whole or by sucking body fluids as outlined below.

5.2.1.7.1 Microcarnivory

Some molluscs feed on animals very much smaller than their body size. Scaphopods are mainly microcarnivores, selectively feeding on small animals such as ostracods, or juvenile molluscs, and 'protists' such as foraminifera using their long, thin captaculae to capture their tiny prey selectively (e.g., Salvini-Plawen 1988; Shimek & Steiner 1997) (see Chapter 16).

Septibranch bivalves (see Chapter 15) capture small crustaceans using muscular septa to suck in prey through the enlarged inhalant siphon (Figure 5.5). The pectinoidean Propeamussiidae also feed on small crustaceans, foraminiferans, and larval bivalves (Morton & Thurston 1989; Dijkstra & Knudsen 1998) which they suck into their mantle cavity by opening the shell valves. Living specimens of the related Cyclochlamydidae (*Cyclochlamys* sp., from New Caledonia) 'crawled about relatively rapidly on a long, narrow foot, and on reaching a fragment of coral or shell, attached themselves with a single byssal thread, which they pulled up against for orientation. The valves and mantle were then held partly open to reveal the withdrawn foot constantly wriggling within, mimicking a worm to attract prey' (B. Marshall in Dijkstra & Knudsen 1998, p. 89). A pteriomorphian bivalve, the limid *Acesta oophaga*, lives in cold seep habitats in the Gulf of Mexico and feeds on the eggs of a vestimentiferan tubeworm (Järnegren et al. 2005).

Cuspidaria   *Poromya*

FIGURE 5.5   The septibranch bivalves *Cuspidaria cuspidata* and *Poromya granulata* feeding on small crustaceans, capturing them by sucking them into the inhalant siphon. *Cuspidaria* redrawn and modified from Reid, R.G.B. and Reid, A.M., *Sarsia*, 56, 47–56, 1974 and *Poromya* from Morton, B., *Sarsia*, 66, 241–256, 1981c.

A few chitons – *Placiphorella*, family Mopaliidae (McLean 1962), *Loricella*, family Schizochitonidae (Ludbrook & Gowlett-Holmes 1989), and *Craspedochiton*, family Acanthochitonidae (Saito & Okutani 1992), capture small crustaceans to supplement food obtained by grazing. They have all apparently evolved this behaviour independently and have an expanded anterior girdle which they raise to capture their prey.

The diet of the vetigastropod genus *Calliostoma* is varied, with some feeding on sponges and others on hydroids, but that of *C. coppingeri* apparently includes small bivalves (Dornellas & Simon 2011). Carnivorous land snails and slugs feed on a variety of prey, although they may preferentially take particular taxa and/or size ranges (e.g., Meyer & Cowie 2010).

Members of the dendronotoidean nudibranch genus *Melibe*, which lack a buccal mass, capture small crustacean prey in a large oral veil. The prey is transferred to the mouth with coordinated movements of the tentacles surrounding the veil and the lips and is then swallowed whole (Watson & Trimarchi 1992).

While many of the mytilid subfamily Bathymodiolinae utilise chemosymbiotic bacteria, a deep-water wood-associated species, *Idas argenteus*, feeds on small *Xyloteredo*, a wood-boring bivalve (Ockelmann & Dinesen 2011).

Some cirrate octopods, and probably some large deep-sea squid such as *Mastigoteuthis*, trap small pelagic or benthic animals that are minute relative to the body size of the predator. *Mastigoteuthis* dangles its long arms downwards, sometimes into sediment on the seafloor, and uses its numerous tiny suckers to capture small animals (e.g., Dilly et al. 1977; Roper & Vecchione 1997; Vecchione et al. 2002). Cirrate octopods use their tent-like webs to enclose the minute to small prey (mainly copepods and other small crustaceans) as they drift and the cirri on the arms probably assist in sweeping them to the mouth (Gibson et al. 2006 and references therein). Some are permanently pelagic and others hover above the sea floor, with benthic opisthoteuthids known to feed on bottom-living crustaceans and polychaetes (Villanueva & Guerra 1991). There are large glands in the lips of *Stauroteuthis syrtensis*, a pelagic cirrate, and some have suggested that, at least in that species, mucus could be involved in capturing small animals (Vecchione & Young 1997).

5.2.1.7.2   Feeding on Colonial Animals

This mode of feeding is seen in some chitons, solenogasters (on cnidarians), and several groups of gastropods, including some vetigastropods, caenogastropods, and heterobranchs. Many colonial animals possess chemicals that deter predation so the predators must be able to assimilate these (see Chapter 9), while others possess tough skeletal material, spicules or, with cnidarians, stinging cells.

Solenogasters are usually found associated with cnidarians, and their diet is mostly inferred from their gut contents and their associations (Scheltema & Jebb 1994). Feeding on sea anemones has been observed in *Neomenia* (Sasaki & Saito 2005) which sucks the anemone tissue into its buccal cavity. *Epimenia australis* has been observed feeding on soft corals where the radula may hook onto the host while pieces of host tissue are sucked off (Scheltema & Jebb 1994). Ingested nematocysts are somehow prevented from discharging during feeding and are often found intact in the gut. Secretions from the foregut glands may assist in neutralising the nematocysts once they are ingested (Salvini-Plawen 1988; Todt & Salvini-Plawen 2004), and the densely packed spicules and a thick subepidermal matrix provide protection while feeding.

Several gastropod taxa are obligate cnidarian feeders, preying on hydroids, stylasterine hydrocorals, discophores, siphonophores, zoanthids, actiniarian sea anemones, black corals (antipatharians), gorgonians, soft corals (alcyonarians), and stony corals (Robertson 1966, 1970a). Although the appearance of coral reef habitat in the early Cenozoic coincided with increased mollusc diversification, few groups of molluscs have adopted corals as a food source (e.g., Robertson 1970a). Feeders on scleractinian corals include a few muricids (see below); some nudibranchs such as a few *Phestilla* spp. feed on corals (see below), and the fissurellid *Diodora galeata* has recently been shown to feed on living coral (*Pocillophora*) (Stella 2012). No other coral feeding vetigastropods are known, although some feed on cnidarians; for example, some species of *Calliostoma* feed on hydroids (Perron & Turner 1978).

The caenogastropod family Ovulidae specialises in feeding on anthozoans (soft corals). Food switches in Ovulidae have been investigated by Schiaparelli et al. (2005). They found that most preyed on octocorals, while predation on the more primitive Hexacorallia was limited. Epitoniids also feed exclusively on cnidarians, as do the basal heterobranch families Architectonicidae and Mathildiddae and many nudibranchs, the most well-known group being the Aeolidioidea. Aeolidioids feed on a variety of cnidarians including hydroids. *Phestilla melanobrachia* feeds on the non-zooxanthellae-bearing 'sun' or 'tube' corals including *Tubastraea*, while some other species of *Phestilla*, as noted above, feed on hard corals, including massive corals such as *Porites* (e.g., Rudman 1981; Ritson-Williams et al. 2003). The nudibranch family Arminidae feed on octocorals including sea pens (Pennatulacea) and soft corals. Some nudibranchs that feed on cnidarians can influence populations of pelagic scyphozoan and hydrozoan medusae by predating their benthic stages. For example, the aeolidiid *Coryphella verrucosa* is the main predator of polyps of the jellyfish *Aurelia* in Swedish fjords (Hernroth & Gröndahl 1985).

The pelagic hydrozoan cnidarians *Velella* and *Porpita* are preyed on by 'violet snails' (*Janthina*), which build bubble rafts and float on the surface (Lalli & Gilmer 1989). The floating glaucid nudibranchs *Glaucus atlanticus* and *G. marginatus* also feed on hydrozoans, including *Velella*, *Porpita*, and *Physalia* (Thompson & Bennett 1970). *Fiona pinnata*, another nudibranch that feeds on *Velella*, accumulates the hydrozoans' blue pigment (Bayer 1963). A few molluscs swimming in the water column (rather than floating on the surface of the sea) also include cnidarians in their diet. The heteropod *Carinaria cristata japonica* includes small siphonophores in its diet (Seapy 1980), and the nudibranch *Cephalopyge trematoides* feeds on the siphonophore *Nanomia* (Arai 2005), while another pelagic nudibranch, *Phylliroe* (Dendronotina), feeds on certain hydrozoan medusae such as *Zanclea*. Part of the diet of the squid *Doryteuthis opalescens* is *Velella* (Brodeur et al. 1987), but there are few other reports of cnidarian predation by coleoids (Arai 2005).

Cnidarian feeding requires modifications to deal with nematocysts. Species of the caenogastropod genus *Drupella* (Muricidae) feed on coral (Moyer & Emerson 1982; Claremont et al. 2011) and have an unusual radula with long, slender lateral teeth with hooked tips, and its proboscis is covered externally with cuticle. During feeding a large pedal gland secretes a mass of mucus that is thrust ahead of the mouth, presumably reducing or preventing injury, while the coral tissue is partially digested externally by salivary secretions before being ingested (Robertson 1970a). Members of the muricid subfamily Coralliophilinae are suctorial coral feeders (e.g., Brawley & Adey 1982; Oren et al. 1998).

Ascidians, which are often protected by a strongly acidic internal fluid, are fed on by some gastropods including members of the Velutinoidea (Triviidae and Velutinidae), as well as a few marginellids, tonnoideans, and nudibranchs.

Some chitons (e.g., Klitgaard 1995; Todt et al. 2009) and several vetigastropods, notably pleurotomariids, calliostomatids, and several fissurellids (e.g., Fretter & Graham 1962), feed on sponges, as do all triphoroideans and some cypraeids. Chromodoridid nudibranchs feed exclusively on sponges (Rudman & Bergquist 2007), as do a number of other groups of nudibranchs (e.g., Thompson 1976). Some nudibranchs feed on bryozoans (e.g., Todd et al. 1991), hydroids (e.g., Lambert 1991), corals (e.g., Faucci et al. 2007), and ascidians (e.g., Paul et al. 1990).

Becerro et al. (2003) showed that the Mediterranean euopisthobranch *Tylodina perversa* strongly preferred sponges with high concentrations of cyanobacterians and avoided those without them. They suggested that *Tylodina* was probably assimilating the cyanobacterians rather than the sponge as only a small fraction of the ingested material is sponge tissue. While this may be an isolated case, it is nevertheless a caution that all sponge feeders may not be strict carnivores. Such an example may also help explain some transitions between carnivory and other food sources, but we do not accept their characterisation of *Tylodina* as a herbivore.

5.2.1.7.3 Scavenging

Some gastropods specialise in scavenging – i.e., eating recently dead animal tissue. Most scavengers feed opportunistically, as carrion is a food resource that is unpredictable and may be infrequent (Britton & Morton 1994) and is, therefore, a feeding strategy that is usually facultative rather than obligate. Because it requires the ability to detect food at a distance, this group of feeders mainly comprises some groups of neogastropods, notably nassariids, which have, as do many other members of that group, a long, often mobile, inhalant siphon and a very well developed chemosensory osphradium. Although little work has been done on specific responses, an example includes a strong feeding response in the nassariid *Tritia obsoleta* that is induced by several compounds including lactic acid (Carr 1967). Most Nassariidae are scavengers (Britton & Morton 1993; Morton 2011), and it is a common feeding mode in the related Buccinidae and Melongenidae (Morton & Jones 2003). There are also records of opportunistic scavenging in a few species in other neogastropod families such as Muricidae, Columbellidae, Marginellidae, Volutidae, and Olividae (e.g., Britton & Morton 1993; Morton 2006b), and in a few heterobranchs. Such opportunistic feeding is probably more common than published records suggest.

Scavengers must be able to locate their food efficiently, feed quickly, and rapidly process and digest large meals on an occasional basis. These features are shared with carnivores, and it is therefore not surprising that all marine gastropod scavengers are derived from predatory lineages, although this is often not the case with scavenging terrestrial snails and slugs, some of which also feed on faeces, and which tend to be omnivorous taxa (e.g., Speiser 2001). The tiny Australian fresh-water gastropod *Glacidorbis hedleyi* appears to feed only on freshly damaged or killed animals, avoiding intact living and moribund individuals (Ponder 1986).

While most cephalopods are predators of live prey, some deep-sea octopods are attracted to fish bait (Isaacs & Schwartzlose 1975).

5.2.1.7.4 Active Predation

Active predation involves selectively seeking prey animals that might be mobile. It is seen in several groups of heterobranchs and caenogastropods and most coleoid cephalopods (see Chapter 9 for an overview). The prey may vary considerably in size relative to the size of the predator; feeding involving capturing relatively minute prey is treated separately above (Section 5.2.1.7.1). Some predators may not eat the entire prey, but feed on different parts of the body selectively. For example, some muricids preferably eat the soft tissues of the gonad and digestive gland, which are rich in glycogen and lipids, and may leave the muscles (e.g., Hughes & Dunkin 1984).

The detection of prey is essential and active predatory molluscs have enhanced sensory capabilities. Most predatory molluscs detect their prey using chemosensory means either through the water column or by following mucous trails in the case of some molluscivores (e.g., some cephalaspideans and carnivorous land snails). Sensory structures (see Chapter 7) involved include the osphradium in combination with the inhalant siphon in caenogastropods, and Hancock's organ or rhinophores in 'opisthobranchs'. Specific chemical cues are no doubt involved in feeding responses. Visual detection of prey is confined to cephalopods and heteropods.

In marine molluscs, various specialised techniques and structural modifications are involved in prey capture and/or penetration and are correlated with the kind of prey selected and how it is eaten. Such modifications are described in more detail below and include a proboscis, modification of the radula, the development of accessory boring organs, the use of toxins to paralyse prey, or the development of grasping structures such as tentacles, suckers, and hooks. Some predators are active swimmers, namely, many cephalopods, and the gymnosome 'pteropods' and 'heteropods', which pursue their prey in the water column.

Toxins to either anaesthetise or kill the prey may be secreted in the salivary gland (e.g., octopods, buccinids, tonnoideans), accessory salivary glands (e.g., muricids), a venom gland (= 'poison gland') developed from the midoesophagus (e.g., conoideans, marginellids), or secreted by the hypobranchial gland (e.g., Andrews et al. 1991), although in muricids it is unlikely that hypobranchial gland secretions play a role in predation (Westley et al. 2006).

The main groups of non-colonial animal prey items available to benthic marine gastropods are polychaete and other 'worms' such as sipunculans, other molluscs, echinoderms, and fish. Some taxa are specialised in their choice of such prey (e.g., many conoideans), while others are generalists. Carnivorous land snails and slugs also consume a variety of organisms including other molluscs, earthworms, and small arthropods which may or may not be swallowed (see Barker & Efford 2002 for details). A summary of feeding preferences for gastropod taxa is given in Table 5.2.

For many terrestrial gastropods, calcium is a limiting resource, and many supplement their supply by scraping shells or calcified eggs of other snails, and this may have led to shell penetration and access to the animal tissue within (Barker & Efford 2002). Changes in the body and gut morphology of terrestrial stylommatophorans that correlate with the adoption of a carnivorous diet are generally not so marked as they are in many marine molluscs (see review by Barker & Efford 2002). Thus, in stylommatophoran land snails there are no significant differences between herbivores and carnivores, other than in the buccal area (see Chapter 20). In stylommatophoran and rathouisiid slugs, there is a reduction in stomach size and shortening of the intestine (e.g., Tillier 1984, 1989), possibly because of the greater opportunities afforded by the lack of visceral coiling or the constraints imposed by the adoption of a slug body.

Sometimes, the prey is eaten in pieces, in other cases swallowed in large chunks or whole. For example, molluscivorous species of *Conus* can take gastropod prey with shells as large as or a little larger than themselves. They remove the animal from the shell and swallow it whole, a feat perhaps made possible by the conotoxins (see Section 5.3.2.7.1.3) relaxing the columellar muscle (Kantor 2007).

In the sections below, we examine some of the main modes of prey capture employed in gastropods. More details on cephalopod feeding are covered in Chapter 17.

5.2.1.7.4.1 Capturing Prey with the Foot

Some gastropods capture prey and hold it with their enlarged foot, including some members of the Volutidae (e.g., Ponder 1970a; Morton 1986; Wolff & Montserrat 2005; Dias 2009; Bigatti et al. 2010) which feed on gastropods and/or bivalves, as do at least some olivids (Cyrus et al. 2012). Naticids (e.g., Carriker 1981; Hughes 1985b; Huelsken 2011), often envelop their prey in their large foot before drilling it, and it has been suggested that at times some may use the foot to suffocate their prey without drilling (e.g., Ansell & Morton 1987; Kabat 1990). Bigatti et al. (2010) have shown that in volutes at least, even though the prey is enveloped in the foot, it is not suffocated, but instead narcotised by salivary secretions. The buccinoidean *Fasciolaria hunteria* envelops small gastropod prey in the foot which helps prevent the operculum of the prey snail closing (Wells 1958). Some cassids capture their echinoid prey by extending their foot over the top of the test (Hughes & Hughes 1981).

5.2.1.7.4.2 Levering and Chipping Shelled Prey

Some predatory marine gastropods have a labral spine (or tooth), a projection extending from the outer lip of the shell aperture. This structure can be used as a lever or wedge to prise open the shells of bivalve prey (e.g., by some buccinids and muricids) or to break into the shell by chipping at the edges. Similarly, barnacles can be entered by using the spine to force the opercular valves apart, or gastropod opercula can be prised open (Marko & Vermeij 1999; Vermeij 2001b). Snails that possess a labral spine can use this to enter their

TABLE 5.2

Feeding Types in the Major Gastropod Clades

Taxon	Detritivory	Macro Herbivory	Grazing Carnivory	Suctorial Herbivore	Hunting or Scavenging	Parasitic	Suspension
Patellogastropoda	○	●					
Vetigastropoda	●	◉	●				○
Pleurotomariida	?		●				
Fissurellida	●	○	◉				
Trochida	●	○	◉				○
Neophalida	◉						●
Neritimorpha	●	◉					
Caenogastropoda	●	◉	●		●	○	◉
Cyclophorida	●	◉					
Viviparida	●						●
Ampullariida	◉	●	○?				
Campanilida	●	○					
Cerithida	●						
Triphorida	●	○	◉			○	○
Strombida	●	◉					○
Calyptraeida							●
Cypraeida		○	●				
Tonnida			◉		◉		
Neogastropoda		○	○		●	○	○
Heterobranchia	●	◉	○	◉	●	○	○
Orbitestellida			●?				
Ectobranchia	●						
Allomorpha	◉?		◉?				
Architectonida			●	◉			
Actonida					●		
Rissoellida	●						
Ringiculida					●?		
Pleurobranchida			●		?		
Nudibranchia			●		●		
Umbrachulida			●				
Pleurocoela	○	○	?		◉		
Amphibolida	●				◉		
Pyramidellida						●	
Siphonariida	●	◉					
Acochlidia	?		◉		◉		
Sacoglossa				●			
Hygrophila	●	◉					
Ellobiida	●	○?					○
Onchidida	●	◉?					
Veronicellida	●	●			○		
Helicida	●	●			◉		

●– predominant, ◉– well represented, ○– rare,

Note: Information from various sources.

shelled prey much faster than by boring. Thus, the evolution of labral spines represents a significant innovation which probably explains why this feature is found in some members of at least ten families and is estimated to have evolved independently at least 58 times since the late Cretaceous (Vermeij 2001b). Most taxa possessing this structure are found in the muricid subfamily Ocenebrinae, but within that group, they have evolved separately in at least four lineages (Marko & Vermeij 1999). This feeding strategy is, however, often not the only one used. For example, the muricid *Acanthina angelica* has labral spines of variable length, and the method it uses to enter the barnacle prey depends on the spine length and the size of the barnacle, with the snail drilling larger prey (Malusa 1985).

Some species of predatory gastropods that lack a labral spine – for example, *Fasciolaria hunteria* – can use the aperture edge to wedge open bivalves (Wells 1958).

5.2.1.7.4.3 Drilling Holes in the Shells of Prey with Calcareous Shells or Exoskeletons

Drilling holes in shelled prey is a predation technique that has evolved independently in several groups of gastropods and also in octopods. While hole drilling occurs throughout the Phanerozoic, attributing the older fossil drill holes to particular predators has often proved challenging (see Deline et al. 2003 for a review of the literature). For this reason, we focus here on living taxa.

Many muricids feed on shelled prey (barnacles, molluscs, serpulid polychaetes) using the radula and secretions from an *accessory boring organ* (ABO) on the anterior part of the foot sole (Carriker 1981), and some also feed on a wide range of additional prey, including polychaetes, sipunculans, etc. (e.g., Taylor 1978). Naticids feed mainly by drilling holes in bivalves, but they also drill gastropods and even crabs (Huelsken 2011) and probably a tubicolous polychaete (Morton & Harper 2009). They also have an ABO (Carriker 1981), but it is on the ventro-anterior part of the proboscis, so these structures in muricids and naticids are not homologous (Figure 5.6).

Boring of bivalves has also been recorded in a few buccinids (Peterson & Black 1995; Morton 2006a), in a newly settled nassariid (Morton & Chan 1997; Chiu et al. 2010), in one genus of marginellids (Ponder & Taylor 1992), and in a few capulids that live on scallop shells (Orr 1962; Matsukuma 1978), but the details of the mechanisms involved are not known. In the case of the capulids, they are not engaging in carnivory, but are kleptoparasites (see Section 5.2.1.8).

Shell drilling in naticids, and especially muricids, has been well studied. It is a slow process, involving alternate use of the ABO and radula (e.g., Carriker 1981). The ABO chemically softens the shell by dissolving the organic matrix, and the radula scrapes away the shell crystals, a process that can take up to 70% of the entire time spent feeding (Connell 1970). Some muricids can take several days to a week or more to drill a hole to access their prey. Thus, this feeding method is not highly efficient, and while drilling, the predator is not feeding for most of the time and is exposed to predators so alternative strategies have evolved in some muricids (see Section 5.2.1.7.4.2).

Marginal (or edge) drilling is employed by some muricids and is reasonably common in naticids (Deline et al. 2003). In a closed bivalve, a hole is often drilled at the point where the valves come together (Ansell & Morton 1987; Vermeij 2001a) because the shell is typically thinner there and drilling is quicker. For example, the muricid *Chicoreus dilectus* can drill the edges of smaller *Chione* shells three times faster than by drilling directly through the valve (Dietl & Herbert 2005). Other positional strategies include drilling through the shell to the most nutritious part of the animal. Thus, the shell of the host may be preferentially drilled where it overlies the gonad

and/or digestive gland, while another strategy focuses on the adductor muscles (e.g., Hughes & Dunkin 1984; Thomas & Day 1995).

The Tonnoidea include the only gastropods that are predators of echinoderms, and many tonnoideans have modified salivary glands (see Section 5.3.2.5.2) that secrete an acid used to assist with boring the shells of their echinoderm prey (see Chapter 19). Some parasitic eulimids drill holes in echinoid tests (Lützen & Nielsen 1975; Crossland et al. 1991; Warén et al. 1994), and the small doridine nudibranch *Okadaia elegans* bores holes in the calcareous tubes of serpulid polychaetes (Young 1969). There are also instances of stylommatophorans drilling holes. For example, some oleacinids, rhytidids, and zonitids have been reported to rasp holes in the shell of their prey (Bromley 1981; Barker & Efford 2002), while others, such as philomycids and some zonitids, scrape a hole in the shell of snail eggs on which they are feeding.

Hole drilling in gastropod, bivalve, and/or crab shells is also carried out by several species of octopuses which use the radula and the rasp-like salivary papilla (see Section 5.3.2.1.6) along with secretions from the buccal glands (e.g., Boucaud-Camou & Boucher-Rodoni 1983; Nixon 1987, 1995). In contrast to hole-drilling gastropods, octopuses inject venom through the hole and wait for the animal to weaken so they can more easily access its tissues.

5.2.1.7.4.4 Reaching Prey at a Distance – The Proboscis

While many carnivorous caenogastropods have a proboscis (see Section 5.3.2.1.5.1; Chapter 19) that facilitates either reaching prey or probing into their bodies, some have a particularly long proboscis that enables them to extend their feeding structures deep into burrows, or otherwise reach prey difficult to approach closely. While the proboscis in many buccinids and nassariids can be extended to more than the length of the shell, in the polychaete feeding turbinellids such as *Vasum* and *Columbarium* and personids (*Distorsio*), the proboscis is extremely long and when retracted lies coiled like string in the proboscis sac. In some cancellariids, colubrariids, and in pyramidellids that feed suctorially (see Section 5.2.1.7.5) on larger prey, the proboscis is very long, enabling feeding to take place without the snail necessarily even being in contact with the host.

5.2.1.7.4.5 Radular Harpooning and Poisoning

Harpooning prey using a hollow, venom-charged harpoon tooth (the 'toxoglossan tooth') held at the tip of the proboscis is a strategy seen only in Conoidea, but within that group, it has evolved at least six times (Taylor 1998; Kantor & Taylor 2000) (see also Chapter 19). Detailed studies of feeding in this group have mainly been undertaken in species of *Conus* which possess the classic elongate, barbed, harpoon-like, venom-charged toxoglossan teeth (e.g., Kohn 1990; Kohn et al. 1999). Toxoglossan teeth are formed from flattened marginal teeth, such as those seen in Drillidae, being progressively rolled up to form a hollow, barbed spear, with some taxa showing intermediate conditions

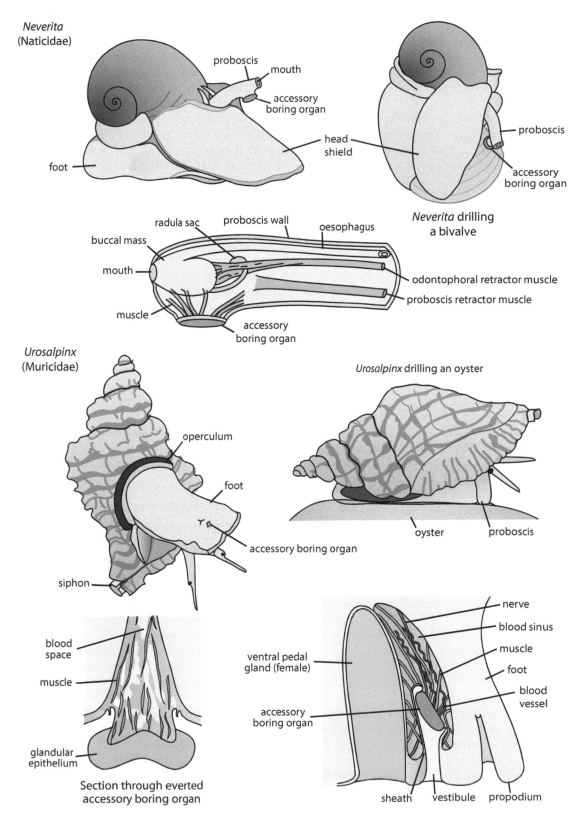

FIGURE 5.6 Comparison of the accessory boring organ of a naticid and a muricid gastropod. Naticids redrawn and modified from Ankel, W.E., *Biologisches Zentrablatt*, 57, 75–82, 1937; ventral view of *Urosalpinx* and lower left figure redrawn and modified from Fretter, V. and Graham, A.L., *British Prosobranch Molluscs: Their Functional Anatomy and Ecology*, Ray Society, London, UK, 1962; second row and lower right figures redrawn and modified from Carriker, M.R., *Malacologia*, 20, 403–422, 1981; *Urosalpinx* lateral view original.

(Taylor 1998; Kantor & Taylor 2000) (see Figure 5.7 and Chapter 19, Figure 19.21). The formation of the harpoon hypodermic tooth is coupled with the reduction and loss of the radular membrane and the possession of a venom gland (see Section 5.3.2.7.1.3) which secretes the toxins and delivers them to the tooth.

In conoideans such as Conidae with hypodermic teeth, one tooth is held at the proboscis tip, where it is gripped by a special sphincter muscle and used to inject venom into their prey. This habit of holding an individual tooth at the tip of the proboscis is also seen in other conoideans, even in those with multiple teeth in each row (Sysoev & Kantor 1987, 1989; Kantor & Taylor 1991; Taylor 1998). In these cases, a marginal tooth is detached from the radular ribbon and then moved to the tip of the proboscis (Taylor 1998). Thus, those conoideans lacking true 'toxoglossate' teeth can still use a radular tooth to stab the prey, allowing venom to enter.

Prey captured by toxoglossans includes polychaete worms, gastropods and, in some species of Conidae, small fish, and the prey is swallowed whole. In some conid species, an ontogenic shift in feeding between juveniles and adults (for example from polychaetes to fish) has been recorded with associated changes in the size, shape, and structure of the harpoon tooth (Nybakken & Perron 1988; Nybakken 1990). Several conoidean taxa have lost the radula (see Chapter 19) and engulf whole prey (Figures 5.8 and 5.9).

5.2.1.7.4.6 The Use of Tentacles, Suckers, or Hooks

Cephalopods employ grasping arms and a sharp beak to capture active prey, and these attributes, together with good eyesight, make them highly effective predators. Coleoid cephalopods have suckers, but these are lacking in *Nautilus*. Arms and, in some taxa, suckers, have also evolved in the shell-less pelagic gymnosome gastropods. This group of 'pteropods' has a pair of arms covered with suckers and/or hooks which aid in capturing their prey. Unlike the visually acute heteropods (which lack arms), gymnosomes lack eyes, but a second more posterior pair of short tentacles may be light sensitive (Lalli & Gilmer 1989).

Scaphopods use numerous thread-like captaculae to feed on minute organisms (see Section 5.2.1.7.1).

5.2.1.7.5 Parasitism and Suctorial Feeding

Parasitism is a form of symbiosis (see Chapter 9). Many so-called parasites are ectoparasites, and there is an arbitrary dividing line between what may be regarded as a parasite and a predator feeding externally on a much larger host. As with many ectoparasites, such external feeders are often host-specific and rarely found except on the preferred host. Such host specificity, when coupled with fixed or poorly mobile adults, requires their larvae to find and settle on a suitable host.

While many gastropods regarded as ectoparasites are suctorial feeders, and grazing predators typically scrape tissue from the host, exceptions occur. For example, some large nudibranchs (Dendrodorididae and Phyllidiidae) lack a radula and are suctorial feeders on sponges. Triphoroidean gastropods feed exclusively on sponges using their radula and acrembolic proboscis. Some are highly specific and others more generalist and often gain access via the oscula to the interior of the sponge to feed on the soft tissues, although some scrape through the outer tissue (Marshall 1983). Because they are typically host-specific and much smaller than their hosts, they could be regarded as ectoparasites. Indeed, many triphorids and cerithiopsids burrow into the sponges they feed on (Fretter 1951; Kosuge 1966; Marshall 1978, 1983). Marshall (1980) suggested that triphorids with long tubular exhalant and inhalant canals may use them as snorkels to assist in feeding on sponges with noxious secretions.

Some cancellariid gastropods feed suctorially using a piercing radula, most spectacularly on resting rays (O'Sullivan et al. 1987), but others feed on body fluids from polychaetes, gastropods, and bivalves, and the contents of cephalopod egg capsules (see Modica et al. 2011 for references). Some colubrariids and a few marginellids feed on resting coral reef fish (see Chapter 19), and these can be regarded as parasites.

Internal parasitism is rare among molluscs. It is only seen in some highly modified eulimid gastropods (most of which are ectoparasites on echinoderms), with some having become internal and thus lost their shell (see Chapter 19). Of these, some lack a gut and absorb food through their body surface. The galeommatoidean bivalve *Entovalva* lives inside holothurians, but is, as far as known, not truly parasitic, but rather an endosymbiont, using the host as a safe residence (Lützen et al. 2011).

5.2.1.8 Kleptoparasitism

Kleptoparasitism (or cleptoparasitism) occurs when an animal takes food from another. This strategy has been well documented in birds in particular (Iyengar 2008), but in molluscs, this form of feeding is less well known, although there are a few interesting examples.

Several capulids, at least some of which are suspension-feeders, are also kleptoparasites. The snail-like capulids *Trichotropis cancellata* and *T. conica* are both kleptoparasites and suspension-feeders. Most individuals steal food from suspension-feeding tube worms (*Serpula*) (Iyengar 2004, 2005, 2007, 2008) using their pseudoproboscis to scoop up the food string of the host. Some species of the limpet-like *Capulus* are also kleptoparasites. *Capulus subcompressus* steals food from tube worms in a similar manner to *Trichotropis* (Schiaparelli et al. 2000). *Capulus ungaricus* is a suspension-feeder (Yonge 1938), but also steals food from its hosts, which include bivalves, gastropods, tube worms, and brachiopods, especially when suspended food is scarce (Iyengar 2008). It obtains food from the gill of a bivalve host by living at the edge of the host shell, in which it sometimes makes a chink to obtain access (Orton 1949; Sharman 1956). Other species of *Capulus* drill a hole through the shell and mantle of their pectinid hosts and steal food strings from the gill and labial palps (Orr 1962; Matsukuma 1978). Another capulid, *Separatista helicoides*, is found on worm tubes and

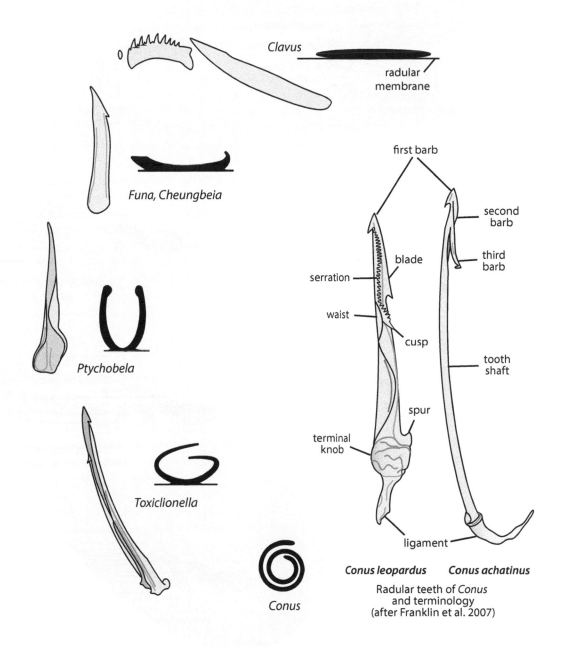

Clavus

radular membrane

first barb

second barb

third barb

blade

serration

waist

cusp

tooth shaft

spur

terminal knob

ligament

Funa, Cheungbeia

Ptychobela

Toxiclionella

Conus

Conus leopardus   Conus achatinus

Radular teeth of *Conus*
and terminology
(after Franklin et al. 2007)

FIGURE 5.7   Diagram showing a transition series of marginal teeth rolling in toxoglossan radulae. Transverse sections (TSs) of the teeth are shown in black. Illustrations redrawn and modified from the following sources: TSs of teeth Kantor, I. and Taylor, J.D., *J. Zool.*, 252, 251–262, 2000; *Conus* teeth Franklin, J.B. et al., *Molluscan Res.*, 27, 111–122, 2007; *Ptychobela* Taylor, J.D. and Wells, F.E., A revision of the crassispirine gastropods from Hong Kong (Gastropoda: Turridae), pp. 101–116, in B. Morton, *The Malacofauna of Hong Kong and Southern China, III: Proceedings of the Third International Workshop on the Malacofauna of Hong Kong and Southern China*, Hong Kong, Vol. 3, Hong Kong University Press, Hong Kong, China, 1994; remainder Powell, A.W.B., *Bull. Auckland Inst. Mus.*, 5, 1–184, 1966.

may also be a kleptoparasite (Iyengar 2008). Not all capulids are kleptoparasites, with *Trichotropis insignis* an obligate suspension-feeder (Iyengar 2008).

While most epitoniids apparently feed on anemone (or other cnidarian) tissue or the mucus which they produce,

Epitonium clathratulum is a kleptoparasite, at least occasionally. Shortly after a feeding event by an anemone (*Bunodosoma biscayensis*), the snail crawls to the mouth and, using its proboscis, steals food from the gastric cavity (Hartog 1987).

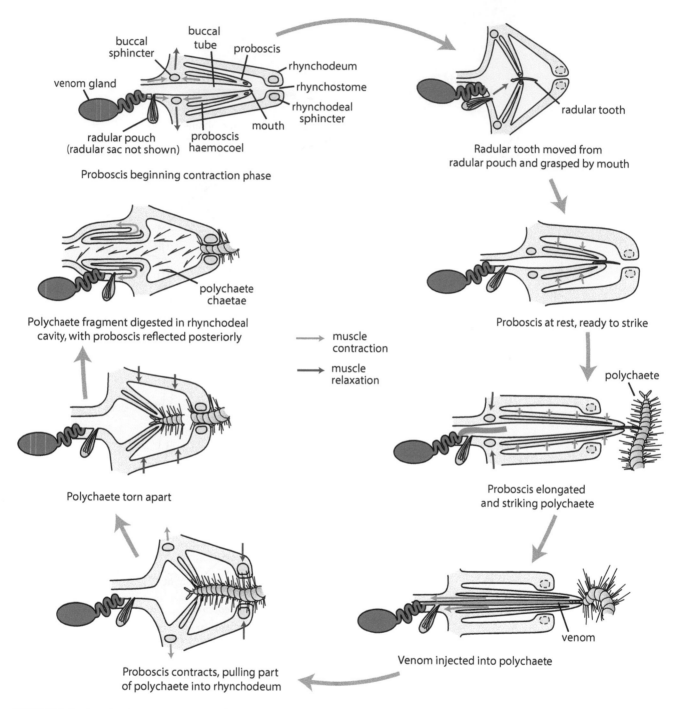

FIGURE 5.8 Feeding in a mangeliid conoidean gastropod (based on *Oenopota* spp. and *Propebela* spp.). Redrawn and modified from Bogdanov, I.P., *La Conchiglia*, 246, 39–47, 1989.

Some predatory buccinids also steal food to supplement their normal diet. *Buccinum undatum* (Rochette et al. 1995) and *Kelletia kelletii* (Rosenthal 1971) both steal food from starfish that are also their potential predators.

Suspension-feeding Calyptraeidae attached to the shells of bivalves may compete with their host for food, even intercepting their feeding currents, a situation described as 'indirect kleptoparasitism' (Iyengar 2008). Some hipponicids were also thought to be 'indirect kleptoparasites' by Iyengar (2008), but this does not appear to be the case as when these limpet-like animals are attached to shells of living gastropods they are often distributed around the exhalant

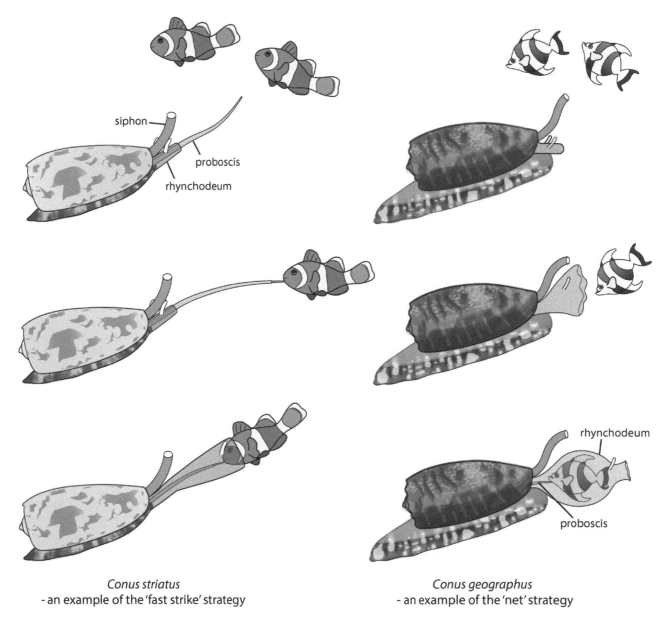

Conus striatus
- an example of the 'fast strike' strategy

Conus geographus
- an example of the 'net' strategy

FIGURE 5.9 Two feeding strategies used by two species of Conidae for catching fish. Redrawn and modified from Olivera, B.M., *Annu. Rev. Ecol. Syst.*, 33, 25–47, 2002.

current feeding on waste material, thus being at least partial coprophages (e.g., Laws 1971). Others do not compete for food or do not appear to intercept food or waste as they are distributed generally over the surface of their host (Morton & Jones 2001) or are attached to rock surfaces (Yonge 1953a, 1960a).

The aeolidioid nudibranch *Cratena peregrina* preys on the hydroid *Eudendrium racemosum* after it has captured planktonic prey, thus taking the planktic organism in addition to the hydroid polyp (Willis et al. 2017), a form of kleptoparasitism these authors term *kleptopredation*.

5.2.1.9 Coprophagy

Coprophagy, the eating of faecal material, is not uncommon in animals. Frankenberg and Smith (1967) showed experimentally that faecal material was ingested by several marine animals, including some molluscs. There have been relatively few studies on this feeding mode in molluscs to indicate the significance of faecal material in molluscan diets, although faecal material must comprise a proportion of the ingested material collected by deposit and suspension-feeders. Coprophagy has been shown to occur in some terrestrial (Speiser 2001), fresh-water (e.g., Brendelberger 1997c), estuarine, and marine (e.g., Frankenberg & Smith 1967) snails as a supplement to normal diets. It has also been suggested as a possible feeding mode in Devonian platyceratids (Sutton et al. 2006).

5.2.1.10 Specialised Feeding on Deep-Sea Biogenic Substrata

Various animal (e.g., whale and fish bones, squid beaks, and shark and skate egg cases), algae, and plant (wood, seagrass rhizomes) substrata that fall to the ocean floor are used by some specialist taxa as both substrata and food sources. Whether the associated molluscs are actually ingesting the decaying material, or just the bacteria on its surface (or incorporated in it), is not well resolved for most of these taxa, although many are apparently highly specialised for particular substrata (see Chapter 9 for details).

5.2.1.11 The Use of Solar Power and Chemoautotrophy

Symbiosis (see Chapter 9) includes many kinds of relationships, such as parasitism (see Section 5.2.1.7.5); those described here involve associations with chemotrophic bacteria and zooxanthellae. They are mutualistic and endosymbiotic with nutrition being provided for the mollusc host. Such associations remain uncommon in molluscs, with several examples only discovered relatively recently. A useful early review is provided by Felbeck et al. (1983).

Some 'opisthobranchs' have functional photosynthetic systems – the so-called 'solar powered slugs' (Rudman 1987; Rumpho et al. 2000). They do this in two very different ways: by incorporating chloroplasts in their tissues (kleptoplasty), this involving a symbiotic relationship with the host plant and the herbivore, or by a symbiotic relationship with zooxanthellae obtained from cnidarian hosts. These relationships are explored in more detail below and in Chapter 9.

In an example rarely cited, the shells of the North American fresh-water mussel *Anodonta grandis* can be eroded so they are translucent or even perforated, exposing the mantle to light, and the mantle tissue in these areas was sometimes green. These green areas contained viable zoochlorellae that could photosynthesise (Pardy 1980; Dillon 2000).

5.2.1.11.1 Kleptoplasty

Many plakobranchian (= elysioidean) sacoglossan sea slugs ingest chloroplasts from their algal hosts and incorporate them in the digestive gland epithelium where, depending on the species, they can use them for months to photosynthesise and fix carbon (Cruz et al. 2013; Christa et al. 2014). This makes it possible for these sea slugs to live for many weeks on the metabolites produced by the chloroplasts (see Wägele & Johnsen 2001 for review). Other sacoglossans may have chloroplasts in their gut, but these show no photosynthetic activity, and the organelles are simply digested for food; some others may have only short-term benefits from the chloroplasts (Jensen 1997). Retention success varies considerably between taxa and may also depend on the algal food.

For example, 50% of the chloroplasts in *Elysia crispata* were still functioning after 58 days of starvation compared with 50% after only 15 days in *Oxynoe antillarum* and five days in *Elysia tuca* (Clark & Busacca 1978). Plastids can function for up to eight months in *Elysia chlorotica* (Mujer et al. 1996). This relationship is thought to have originally evolved by the sacoglossans utilising the chloroplasts to colour their body for camouflage.

The plastids lack the full complement of genes required for photosynthesis and, within a plant cell, the plastids depend on the nuclear genome for over 90% of the proteins needed for their metabolism (Rumpho et al. 2008). Thus, it is very surprising that the chloroplasts function in the absence of an algal nucleus, and it became apparent that the proteins needed for the plastids to function are provided by the slug (Mujer et al. 1996; Pierce et al. 1996; Rumpho et al. 2000). The nuclear gene (*PsbO*) essential for photosynthesis is expressed in *Elysia* and is inherited. This gene has an identical sequence to the same gene in the host alga (*Vaucheria litorea*), and although some have suggested that horizontal transfer of the gene has occurred between the host alga and the slug (Rumpho et al. 2008; Schwartz et al. 2014), this does not appear to be the case (Wägele & Martin 2014; Rauch et al. 2015).

5.2.1.11.2 Zooxanthellae

Symbiotic dinoflagellates (zooxanthellae) occur in several groups of animals including corals and many sea anemones, and in a number of bivalves and gastropods. The best-known examples are the giant clams (*Tridacna*; Cardiidae) and some nudibranch sea slugs.

The giant clams (*Tridacna*, *Hippopus*), have a symbiotic relationship with numerous zooxanthellae (dinoflagellates of the genus *Symbiodinium*) that live in the blood spaces of the siphons and visceral mass (Kirkendale & Paulay 2017). The amoebocytes in the blood remove degenerate zooxanthellae where they are hydrolysed via lysosomes (Goreau et al. 1973; Fankboner & Reid 1990). Residues accumulate in vacuoles in the amoebocytes and are subsequently excreted as nephroliths by the very large kidneys (Reid et al. 1984b). The metabolites from the zooxanthellae are utilised by these bivalves (Fankboner & Reid 1990; Hawkins & Klumpp 1995), and the zooxanthellae may also be broken down in the digestive gland tubules (see below). *Tridacna* can also undertake suspension-feeding and absorb nutrients directly from the seawater by way of its mantle tissues (Fankboner 1971). The products of photosynthesis (mostly glucose) are accumulated and used by the host, being mainly incorporated in mucus and other secretions utilised in feeding and digestion (Fankboner & Reid 1990). Thus, giant clams obtain benefits from the photosynthetic products (e.g., Yonge 1936; Fankboner 1971), but also appear to utilise zooxanthellae cells as a food source (e.g., Maruyama & Heslinga 1997), thus supplementing the normal bivalve suspension-feeding and digestion also carried out by these animals (Fankboner & Reid 1990) (Figure 5.10).

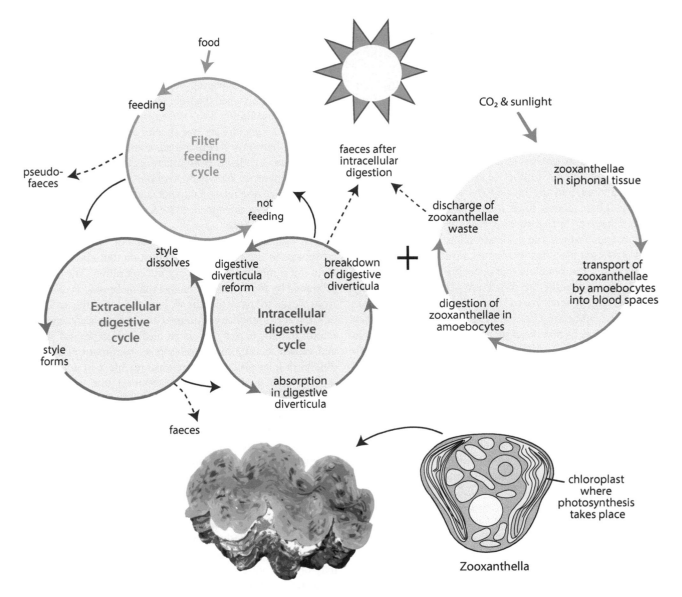

FIGURE 5.10 Diagrammatic feeding cycle of the giant clam *Tridacna* which has symbiotic zooxanthellae in its greatly expanded mantle tissue. Redrawn and modified from Morton, B.S., *J. Zool.*, 185, 371–387, 1978c and zooxanthella from Fankboner, P.V. and Reid, R.G.B., Nutrition in giant clams (Tridacnidae), pp. 195–209, in B. Morton, *The Bivalvia. Proceedings of a Memorial Symposium in Honour of Sir Charles Maurice Yonge (1899–1986)*, Edinburgh, 1986, Hong Kong University Press, Hong Kong, China, 1990.

In *Tridacna*, zooxanthellae are apparently transported from the mantle to the digestive glands by a system of tubes (e.g., Morton 1978c, 1983a; Norton et al. 1992; Hirose et al. 2005), although this has been disputed by some (e.g., Fankboner & Reid 1990). Morton (2000c) provided evidence of a similar tube system between the digestive glands and the kidneys (but not the mantle) in *Fragum erugatum*, another cardiid with symbiotic zooxanthellae. Morton (1978c) showed in *Tridacna* that the movement of the zooxanthellae into the digestive tubules where they are digested resulted in a diurnal cycle of breakdown and re-formation of the digestive tubules (and the crystalline style). That zooxanthellae pass from the haemocoel into the digestive tubules is shown by finding undigested zooxanthellae in the gut and faeces of tridacnines maintained in filtered seawater (Trench et al. 1981). Reade and Reade (1972) injected carbon particles into the adductor muscle of *T. maxima*, and these were taken up by haemocytes, transported to the digestive organs and kidney tubules, and eventually excreted via the faeces and urine, respectively. This experiment supported the existence of two mechanisms of waste removal and a link between the kidneys and digestive glands (Morton 2000c).

The relationships between body size, mantle area, and numbers of symbiotic zooxanthellae differ in different species of giant clams as adults and during their ontogeny (Griffiths & Klumpp 1996). Also, the zooxanthellae can be affected by various environmental stresses, and the clams can undergo bleaching in a similar way to corals (Norton et al. 1995; Buck et al. 2002), with the numbers of zooxanthellae in clams dropping as much as 30-fold (Leggat et al. 2003).

Hawkins and Klumpp (1995) calculated the relative contributions of nitrogen for tissue growth and metabolism from suspension-feeding and from the zooxanthellae in *Tridacna gigas*, with the latter supplying the greater amount. Ammonium was also taken up from seawater, and the zooxanthellae assimilated most of the nitrogen excreted by the host.

The advantages of zooxanthellae symbiosis are maximised through an increase in the host of the surface area exposed to sunlight. Giant clams have undergone a rotation of the body (see Chapter 15, Figure 15.35) so the greatly expanded siphonal tissues (where the zooxanthellae live) lie dorsally. The siphonal area also has iridocytes, special cells containing stacks of regular platelets that probably assist in redirecting light to the zooxanthellae. There are also hundreds of tiny eyes that are highly sensitive to light and shade (Land 2003).

The zooxanthellae are acquired from the environment – they are absent in the eggs and developing larvae. They are taken in via the mouth in juvenile *Tridacna* and move from the stomach into the developing siphonal tissues (Fitt & Trench 1980, 1981).

In the cardiids, there are two lineages containing zooxanthellate taxa: *Fragum*, *Lunulicardia*, and *Corculum* (all members of the Fraginae), and the giant clams (*Tridacna* and *Hippopus*) (Schneider 1995; Maruyama et al. 1998; Kirkendale & Paulay 2017), suggesting independent origins of this symbiosis. All these cardiid bivalves with photosymbionts live in oligotrophic coral lagoons and can be the dominant bivalves found in those habitats. A molecular analysis by Kirkendale (2009) showed there was a single, fairly recent origin of photosymbiosis in the Fraginae and that the genus *Fragum*, with simple triangular shells, was not monophyletic. The genera *Lunulicardia* and *Corculum* have not only modified their shell shapes to increase the surface area available to the zooxanthellae, but also their shell structure, so that translucent 'windows' are built into the shell (Kirkendale 2009). These photosymbiotic cardiids suspension-feed and are generally very like other cardiids (cockles) except that they have reduced labial palps which allow only the smallest food particles to enter the mouth (Morton 2000c).

The reported possession of zooxanthellae by the small Australian estuarine bivalve, *Fluviolanatus subtortus* (Trapeziidae) (Morton 1982) is erroneous (Kirkendale & Paulay 2017).

Some nudibranch sea slugs that feed on cnidarians (e.g., alcyonarians) have a mutualistic symbiotic relationship with the zooxanthellae (*Symbiodinium*) taken from their cnidarian prey (mainly soft corals or, sometimes, hard corals). This symbiotic relationship has been known since the 1930s and has occurred independently in several clades of nudibranchs (Rudman 1991b; Wägele 2004). In a few nudibranchs, such as the dendronotoidean *Melibe* which feeds on small crustaceans, the source of their zooxanthellae is unknown.

In nudibranchs, the zooxanthellae are housed in the digestive gland cells. Like the giant clams, the 'solar powered' sea slugs have undergone bodily changes to increase the surface area available to the zooxanthellae. These include branching of the digestive gland, the provision of dorsal outgrowths (cerata), or expansion of the dorsal body (notum). There are also behavioural adaptations such as orientation towards the light and shading when the light is too intense. Most other sea slug taxa avoid light.

Besides zooxanthellae, one species of nudibranch, *Aeolidia papillosa*, harbours zoochlorellae, another kind of photosynthetic microorganism. Both microorganisms are symbiotic with the anemone on which *Aeolidia* feeds, but, unlike the zooxanthellae, the relationship between the zoochlorellae and the sea slug does not appear to be a stable one (McFarland & Muller-Parker 1993).

The symbiosis with zooxanthellae not only provides an advantage to the nudibranch in nutrition, but also the colour from the symbionts provides cryptic coloration. The nutrients produced by the symbionts enable the nudibranchs to survive from weeks to many months without food. During such periods, the zooxanthellae continue to divide and photosynthesise. Photosynthetic activity has been measured in several species and shown to vary considerably (e.g., Burghardt et al. 2005, 2008b) despite all having long-term retention of zooxanthellae. Different morphological, behavioural, and physiological adaptations all contribute to these differences.

Zooxanthellae symbiosis presumably came about through the host originally ingesting and digesting the zooxanthellae, then their digestion was delayed with the zooxanthellae coloration used for crypsis, and eventually, a stable mutualistic symbiosis evolved, which involved both nutrition and crypsis (e.g., Burghardt et al. 2008b). Because of this symbiosis, the nudibranchs also gain some independence from their primary food source.

In members of the aeolidioidean genus *Phyllodesmium* (Figure 5.11), the less derived species feed on soft corals (that may or may not contain zooxanthellae), while more derived 'solar powered' taxa depend on certain soft corals to harvest their zooxanthellae, and there are also transitional species (Rudman 1991b). The derived taxa not only have enlarged cerata and transparent bodies, some of the more advanced ones have reduced the number of cerata to minimise shading and have enlarged and flattened them to increase their surface area. In the non-specialised taxa, the digestive gland diverticula are unbranched, but they are branched in the symbiotic taxa, providing more space for the zooxanthellae, and the increased surface area enhances exchange of metabolites and respiratory gases. In some species of *Phyllodesmium*, increases in the surface area for occupation by zooxanthellae include narrow tubules lined with cells that store zooxanthellae which permeate the cerata and body wall and, in one species (*P. longicirrum*), zooxanthellae lie in circular chambers in the digestive gland (Burghardt et al. 2008b). In the aeolidioid *Pteraeolidia ianthina*, the digestive gland inside the cerata is little branched, but most of the zooxanthellae are in narrow tubules emanating from the digestive gland to penetrate the body, including the cerata (Burghardt et al. 2008b). Adaptations to zooxanthellae symbiosis similar to

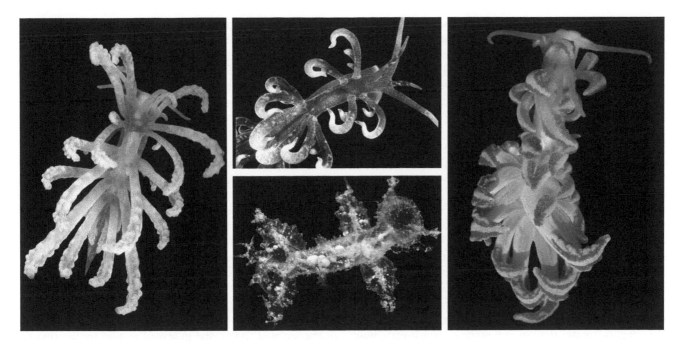

FIGURE 5.11   Some examples of 'solar powered' nudibranchs. Left, *Phyllodesmium lizardense*; top middle, *P. macphersonae*; right, *P. jakobsenae*, lower middle, *Melibe engeli*. (Courtesy of I. Burghardt.)

those in acolidioids are found in *Melibe engeli* (Tritonioidea) (Burghardt et al. 2008b). It also has narrow tubules derived from the digestive gland extending throughout the whole body, and the digestive gland has 'cisterns' packed with zooxanthellae. When individuals of *Melibe* are kept away from any food source, they show no decrease in photosynthesis over 270 days and generate sufficient photosynthetic product to regularly lay egg masses (Burghardt et al. 2008b). *Pteraeolidia ianthina* and *Phyllodesmium longicirrum* apparently derive most of their nutrients from the zooxanthellae as they are rarely found on potential food sources, and the source of their zooxanthellae is unknown. In *Pteraeolidia ianthina*, the zooxanthellae can provide most of the respiratory carbon needs of the host in winter (79%) and excess in summer (173%) (Hoegh-Guldberg & Hinde 1986; Hoegh-Guldberg et al. 1986).

Nudibranch-zooxanthellae symbiosis has not only occurred independently in several taxa, the convergent adaptations such as the fine tubules and cerata are seen in taxa as different as *Melibe* (Dendronotida) (Figure 5.11) and aeolidioideans such as *Phyllodesmium* and *Pteraeolidia*. Another zooxanthellae-bearing nudibranch is the arminid *Dermatobranchus* (Euarminida), which lacks cerata, but has the fine tubules (Wägele & Johnsen 2001; Burghardt et al. 2008b).

5.2.1.12   Bacterial Symbionts

Mutualistic symbiosis with bacteria[3] is common in the gut, with the bacteria providing indirect nutrition as a result of producing enzymes that aid in the breakdown of cellulose and other otherwise indigestible compounds (see Section 5.4.2.3.2).

The mutualistic symbiotic chemoautotrophic bacteria that provide major sources of nutrition for some molluscs, notably several bivalves (see Section 5.2.1.12.1 below), are discussed in this section.

These mutualistic associations are adaptations to life in many kinds of reducing environments. The two main types of chemosynthetic bacterial endosymbionts are methane-oxidising (*methanotrophic*) and sulphur-oxidising (*thioautotrophic*) chemoautotrophic Gamma-proteobacteria. The bacteria obtain energy for fixing organic carbon by oxidising either sulphur compounds or organic compounds, notably methane. The sulphur-oxidising bacteria are anaerobes and fix carbon autotrophically using the Calvin-Benson-Bassham (CBB) cycle,[4] the methanotrophic bacteria use methane as their carbon source, and the aerobes use oxygen to obtain carbon from various substrates.

The host facilitates the supply of the reduced compounds and oxygen from the environment and in return utilises the fixed carbon compounds. The bacteria live intracellularly in specialised epithelial cells (*bacteriocytes*) (see below) usually located in the lateral parts of the gill filaments where they get maximum exposure to the circulating water. Sulphur-oxidising bacteria also assist the host by detoxifying the sulphide from the often oxygen-poor or anoxic environments in which they live, such as deep-sea vents and cold seeps, and also in some bivalves (notably lucinids), in shallow-water reducing environments. Sulphide from the water enters the cells and is detoxified by oxidation in special organelles (sulphur-oxidising bodies) (e.g., Powell & Somero 1985). Other organelles in these cells digest some of the symbiotic bacteria (e.g., Liberge et al. 2001).

---

[3] See Chapter 9 for a more detailed overview of bacterial associations with molluscs.

[4] See Chapter 2 for more details.

For the chemoautotrophy to support growth in the host, there must also be inorganic nitrogen assimilation. Ammonia is primarily assimilated (by way of the enzyme glutamate dehydrogenase) by the bacterial symbionts in the vent taxa *Calyptogena magnifica* and *Bathymodiolus thermophilus*, whereas in the shallow-water *Solemya velum*, ammonia is first assimilated by the host (Lee et al. 1999).

5.2.1.12.1 Bivalves

Bacterial chemosymbionts (see also Chapter 9) are found in the gills in several families of bivalves where these associations have evolved independently (e.g., Williams et al. 2004). In these families, the ctenidial filaments are modified in a similar way to house the bacteria despite there being three different gill types involved – protobranch, homorhabdic filibranch, and eulamellibranch. Increased surface area of the gills is one modification that enhances the exposure of the bacteria to inhalant water containing essential compounds and oxygen. In *Solemya*, the surface area of these hypertrophied gills, in proportion to the rest of the bivalve, is greater than in any other marine invertebrate (Scott 2005).

While the frontal part of the gill filament is unmodified, the abfrontal part is much enlarged by the numerous bacteriocytes that surround blood spaces. As well as water and blood flow, more efficient packing is achieved by adjacent filaments being fused to form cylindrical channels lined with bacteriocytes, as seen in lucinids, vesicomyids, and possibly in one thyasirid (Taylor & Glover 2010).

Associations with sulphur-oxidising bacteria have been reported from the protobranch families Solemyidae and Nucinellidae, the autobranch Mytilidae (Bathymodiolinae only), Lucinidae (Figure 5.12), Thyasiridae, and Vesicomyidae (Duperron 2010; Taylor & Glover 2010) (see also Chapter 15). In Solemyidae, Vesicomyidae, and Lucinidae it has been shown that haemoglobin is involved in transporting the sulphur to the bacteria (see Chapter 6). While in most of these bivalves the sulphide for the bacterial symbionts comes from the water column, the amount that is available is increased in vesicomyids and some lucinids by the foot 'mining' for sulphide in the sediment. Unlike those in lucinids (Dando et al. 1986), the symbionts in mussels do not deposit sulphur granules (Duperron 2010).

Methane-oxidising (methanotrophic) Gamma-proteobacteria have been found in the thyasirid bivalve *Conchocele bisecta* (Kamenev et al. 2001), and in several mytilid species (mostly species of *Bathymodiolus*) from cold seeps, hydrothermal vents, and deep-sea organic substrata (Figure 5.13). Both thyasirids and bathymodioline mussels also contain species with no bacterial symbionts.

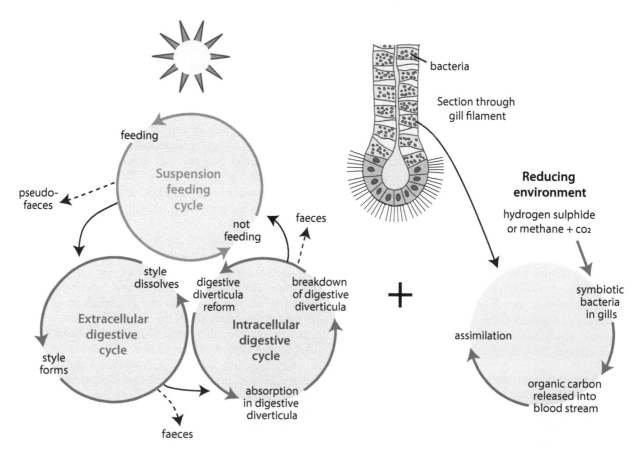

FIGURE 5.12 Feeding cycle of a typical lucinid bivalve. Cycle on left redrawn and modified from Morton, B., *Malacologia*, 14, 63–79, 1973a. Figure in middle is a transverse section through a lucinid gill filament. Redrawn and modified from Giere, O., *Zoomorphology*, 105, 296–301, 1985.

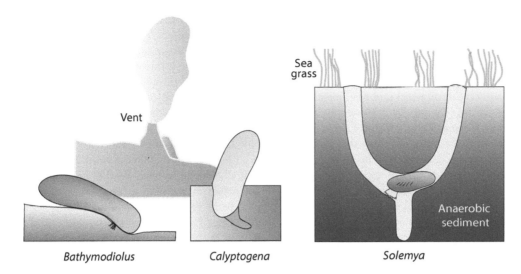

FIGURE 5.13 Some bivalves that live in reducing environments and utilise symbiotic bacteria. Modified from various sources.

Several types of bacterial symbionts are found in some bathymodioline mytilids. These include not only sulphur- and methane-oxidisers, which can co-occur in a single bacteriocyte, but other types of bacterial symbionts may also occur, as recently shown in two cold seep species (Duperron et al. 2007, 2008). Both these mussels have one methanotroph and two distinct thiotrophs as well as methylotrophs which utilise mono-carbon compounds such as methanol. One mussel, *Idas* sp., has two additional bacterial symbionts, one of unknown function and the other related to cellulose-degrading bacteria. This ability to utilise metabolically diverse bacteria may explain the success of these mytilids in a wide range of habitats in the deep-sea. Vent and seep bathymodioline mussels may be derived from ancestors associated with deep-sea organic substrata (e.g., wood, whale carcasses) (Distel et al. 2000), probably multiple times (Jones et al. 2006; Duperron 2010). Three types of bacterial symbionts have also been recognised in a species of *Anodontia* (Lucinidae) (Ball et al. 2009), but their functional differences are unknown.

The morphology of the bacteriocytes is similar in the different bivalve families. They have microvilli covering their dome-shaped outer surface and are separated by narrow intercalary cells. The bacteria lie in numerous vacuoles within each cell. In some taxa, there is only a single bacterium per vacuole, as in Lucinidae and many Bathymodiolinae, while in thyasirids, masses of bacteria are contained in single vacuoles (Taylor & Glover 2010). In solemyids, each bacteriocyte is packed with aligned rod-shaped bacteria lying in single vacuoles (Krueger et al. 1996a; Stewart & Cavanaugh 2006; Taylor 2008).

In some less modified species, the symbiotic bacteria are nestled in the microvilli of the bacteriocytes with this probable intermediate stage found in some Bathymodiolinae living on organic substrata. A further intermediate stage is seen in some thyasirids with taxa having bacterial symbionts in 'extracellular vesicles' beneath a cuticle of modified microvilli (Dufour 2005). Symbiotic methanotrophs are always in vacuoles inside the bacteriocytes.

Bacterial symbiosis appears to be obligate in all Solemyidae, Lucinidae, Vesicomyidae, and probably Mazanellidae. In these bivalves, the gut is simplified or, in some solemyids and mazanellids, lost altogether. Other modifications in solemyids include reduced labial palps and the gills that house the bacteria are unusually (for a protobranch) large and fleshy. Lucinids are modified in several ways in response to their bacterial symbionts. Their ctenidia are unusually large with thick filaments and comprised of only the inner demibranchs. The labial palps are very small, and the foot is elongate and cylindrical, used for making a mucus-lined inhalant tube. Because of the reduced respiratory capability of the ctenidia, the mantle near the anterior adductor muscle is folded and acts as a secondary gill (Taylor & Glover 2000) (see Chapters 4 and 15).

This symbiosis appears to have been established early in solemyids and lucinids, both of which have a fossil record back to the lower Paleozoic (Taylor & Glover 2000, 2006; Taylor 2008). The other three groups are younger – Thyasiridae range from early Cretaceous and Vesicomyidae and Bathymodiolinae early Cenozoic (Taylor & Glover 2010). Some bivalves adopting these symbiotic relationships with bacteria can live in what would normally be regarded as hostile environments, including extremely deep water. For example, some thyasirids and vesicomyids are dominant organisms at 7,326 m (Fujikura et al. 1999).

Labelling has demonstrated the transfer of carbon fixed by both sulphur- and methane-oxidising bacteria to the host tissue (Duperron 2010) with estimates indicating 60% of the organic carbon in one species of *Bathymodiolus* was obtained from methane-oxidisers and 40% from sulphur-oxidisers (Dover et al. 2003). Exactly how the carbon is transferred is uncertain, but may be at least partly due to digestion of the bacteria by phagolysosome-like bodies in the basal part of the bacteriocytes (Fisher et al. 1987; Fiala-Médioni et al. 1994). Detached bacteriocytes have been observed in the food groove of mussels, suggesting that autophagy may also provide a, probably minor, means of carbon transfer.

Vesicomyids (*Calyptogena* spp.) live in deep-sea seeps and vents (Figure 5.13), and bacteria provide all their food (e.g., Goffredi et al. 2004a; Saito & Osako 2007). Some lucinoideans do not rely entirely on the carbon from their bacteria, but are mixotrophic in that their gills can transport captured particles to the mouth which are digested and assimilated in the gut (Duplessis et al. 2004) (Figure 5.12). Generally, mussels with symbiotic bacteria can also suspension-feed, and some bathymodiolines get most of their nutrition this way, while others mainly rely on carbon derived from bacteria.

Although bacterial symbionts do not appear to cospeciate with their hosts (e.g., Won et al. 2008), closely related mussel species harbour closely related bacteria (e.g., Duperron et al. 2006, 2007; Lorion et al. 2012).

Ctenidial bacteriocytes, very similar to those seen in bivalves housing chemosymbiotic bacteria, are found in the wood-boring bivalves Xylophagaidae and Teredinidae. They house cellulolytic, nitrogen-fixing Gamma-proteobacteria thought to produce enzymes which enable these bivalves to break down cellulose (Distel & Roberts 1997; Luyten et al. 2006) (see Section 5.4.2.3.2.1). A report that *Spisula subtruncata* (Mactridae) by Bouvy et al. (1989) had chemoautotrophic bacteria in its gills is incorrect (Le Pennec et al. 1995).

Taylor and Glover (2010, p. 127) suggested that '…symbiosis enables bivalves to live in adaptation to environments where normal food sources for suspension-feeding animals are poor or non-existent'. Thus, some bivalves (notably Solemyidae [Figure 5.13] and Lucinidae) can utilise the sulphur in the shallow anoxic marine sediments common in coral reefs and tropical seagrass beds (Ott et al. 2004; Stewart et al. 2005; Glover & Taylor 2007; Taylor 2008; Taylor & Glover 2010). Suspension-feeding bivalves in these same habitats are often relatively uncommon because the seawater is often oligotrophic, while bivalves with chemosymbiotic bacteria can exploit otherwise unavailable energy sources in the deep-sea, such as in cold seeps, hydrothermal vents, on organic debris, and also in deep-sea sediments (Thyasiridae).

Bacterial symbionts are much less common and much more poorly known in other groups of molluscs. In some monoplacophorans, bacteriocytes have been found in the lateral roof of the mantle groove, the head, the proximal areas of the ctenidia, sides of the foot, and on the post-oral tentacles, but a nutritional role has not been confirmed (Haszprunar et al. 1995; Haszprunar & Schaefer 1997). There are only a few credible reports of gastropods having symbiotic bacteria, and as in bivalves, many of these associations involve the gastropod ctenidia. Chemoautotrophic bacteria have been found in the gills of one of the vetigastropod vent 'limpets', *Lepetodrilus* (Lepetodrilidae), which also collects food by suspension-feeding (Limén et al. 2007). The episymbiotic bacteria are 'farmed' on the gill and ingested via the food groove (see Sasaki et al. 2010b for references). Another unrelated vetigastropod vent limpet, *Hirtopelta* (Peltospiridae), has bacteria in bacteriocytes in the gill (Beck 2002), and bacteriocytes have also been reported in the mantle groove of *Lepetella sierra* (Judge & Haszprunar 2014). Other vetigastropods suspected of having bacterial symbionts include

Gigantopelta chessoia, *Chrysomallon squamiferum*, and *Cyathermia naticoides* (Sasaki et al. 2010b; Chen et al. 2015). *Chrysomallon squamiferum* does not use the gill surfaces and instead houses γ-proteobacteria within the cells of its oesophageal gland (Goffredi et al. 2004b). In the caenogastropods, methane-oxidising (methanotrophic) Gamma-proteobacteria have been found in the provannid *Ifremeria nautilei* (Borowski et al. 2002). While some provannids have both chemoautotrophic and methanotrophic endosymbiont bacteria in their gills (e.g., Suzuki et al. 2006a; Sasaki et al. 2010b), others house only chemoautotrophic (Windoffer & Giere 1997). The only heterobranch reported to be associated with chemoautotrophic bacteria is the orbitestellid *Lurifax vitreus* (Hawe et al. 2014) (see Chapter 9 for a discussion of chemosynthesis in molluscs). Although not bacterial, Zande (1999) reported a filamentous fungal commensal on the gills of the neritimorph *Bathynerita naticoidea* from petroleum seeps in the Gulf of Mexico. The nature of this relationship is not thought to be nutritional; instead, it is suggested that the fungi may detoxify hydrocarbon and sulphide compounds.

5.2.1.12.2   Origin of the Bacterial Symbionts

It is not well understood how the molluscan hosts acquire their bacterial symbionts, although a few species of Lucinidae and the mytilid Bathymodiolinae have been shown to obtain their bacteria from the sediment (Gros et al. 2003; Won et al. 2003, 2008). Recently, settled dual symbiotic species of bathymodioline mytilids have sulphur- and methane-oxidising bacteria (Salerno et al. 2005), but these bacteria were not seen in the eggs, which would suggest that they are quickly acquired from the environment. This is supported by the extracellular location of the bacteria in some mussel taxa, and by observations showing apical bacteriocyte vacuoles engulfing sulphide-oxidising bacteria (Duperron 2010). Similar observations on some lucinids have shown that newly settled individuals that lack symbiotic bacteria take up free-living ones from the sediment, by endocytosis into undifferentiated abfrontal cells in the gill epithelium, followed by rapid proliferation of bacteriocyte cells (Gros et al. 1998a, 1998b, 2003). The mechanisms that involve the recognition by the bivalve host of the bacterium and its entry into the host tissues, and the triggering of the modifications to the gill epithelium to form bacteriocytes, are poorly understood.

An external origin of symbiotic bacteria is also suggested because most symbionts of unrelated shallow-water bivalves (*Solemya*, Lucinidae, Thyasiridae) are members of the same clade of Gamma-proteobacteria, while those found in unrelated deep-water bivalves (*Calyptogena*, Thyasiridae, and *Bathymodiolus*) belong to another clade (Cavanaugh et al. 2005; Dubilier et al. 2008).

In Solemyidae, there is evidence suggesting transmission of the bacteria through the embryo (Krueger et al. 1996b), and this has also been shown for Vesicomyidae (Hurtado et al. 2003), although there is evidence of at least occasional environmental or host to host transmission (Stewart et al. 2008).

5.2.1.13 Dietary Diversification

While some molluscan clades appear rather uniform in their basic diet (e.g., Scaphopoda, Solenogastres, Monoplacophora, Cephalopoda), others are much more diverse. For example, while most chitons are microherbivorous grazers, feeding on diatoms, filamentous algae, algal sporelings etc., some species feed on larger algae or seagrasses, others are restricted to sunken wood, and a few supplement their diet by trapping small animals such as crustaceans under their expanded anterior girdle (see Section 5.2.1.7.1). Some species of *Mopalia* have a mixed diet of algae and animal tissue (e.g., algae, crustaceans, bryozoans), while others feed only on algae (Demopulos 1975; Fulton 1975; Robb 1975). This apparent pattern may also be the result of limited observations of members of the supposedly 'uniform' groups which also have the fewest foraging studies (Fontoura-da-Silva et al. 2013).

Gastropods are the most diverse feeders, both in the breadth of prey types and the morphological diversity required to exploit it (Table 5.2). Patellogastropods are similar to chitons with the majority being microherbivorous grazers, while others graze on carbonate surfaces and calcareous algae, large macroalgae, seagrasses, wood, vestimentiferan tubes, and bacterial mats (Lindberg 2008; Govenar et al. 2015). Both chitons and patellogastropods share the docoglossan type radula (see Sections 5.3.2.4.2.2 and 5.3.2.4.2.6).

It is impossible to state with any certainty what the early vetigastropods fed on, although given their shared rhipidoglossate radular structure, they were probably surface detritus and microorganism feeders. The first members of extant vetigastropod herbivores (which include haliotids and some trochids, turbinids, and fissurellids) are thought to have first arisen in the lineage after the Paleozoic (Vermeij & Lindberg 2000), but it is possible that grazing carnivory on colonial animals may have arisen earlier as the rhipidoglossate radula and gut would only have needed to undergo relatively minor changes. The trochoid Umboniinae are suspension-feeders (see Section 5.2.1.2.2.3) and presumably also evolved from particle feeders, although relatively recently as the group has only a Cenozoic fossil record.

Pleurotomariidae (Harasewych 2002) and, apparently, the Trochaclididae (Marshall 1995), are specialised spongivores. The Calliostomatidae are mainly (but not exclusively) cnidarian feeders (Marshall 1988a; Ferro & Cretella 1993), although some ingest at least some algal material (e.g., Fretter & Graham 1962; Morris et al. 1980), and others feed on sponges and even small bivalves. The slit-bearing pleurotomarioids go back to the latest Cambrian or Lower Ordovician, but whether these early taxa were carnivores is questionable as this may be a feeding mode they relatively recently adopted, given that all living members are found in rather deep water. The later pleurotomariids were common in shallow-water habitats from the Triassic through the Mesozoic (Geiger et al. 2008), but later moved into the deep-water rocky habitats they occupy today (Harasewych 2002). The antiquity of the trochoid grazing carnivore lineages is more obscure as the relationships of the Mesozoic members of this group are not well understood. For example, undoubted calliostomatids are only found

in the Cenozoic, although there are some (e.g., species of *Ueckerconulus*[5] from the Jurassic) that may possibly be ancestors of the family.

Fissurellids have a fossil record back to the Triassic and possibly Carboniferous (Geiger et al. 2008). Their feeding is varied, including microphagous, sponge-feeding, and herbivory (Aktipis et al. 2011), and at least one species of *Puncturella* feeds on forams (Herbert 1991). The latter taxon is basal in the family (Aktipis et al. 2011).

Besides some variation in food in a few taxa (e.g., omnivores), there have been some rapid (in an evolutionary sense) switches in feeding type in some lineages. The caenogastropod cowries (Cypraeidae), although not thoroughly investigated, include herbivores, grazing carnivores, and some with mixed diets (Hayes 1983; Osorio et al. 1993). The diets of the basal cypraeids (Meyer 2003, 2004) are unknown. If mixed grazing is basal in this group, one might expect to see multiple lineages specialising in either herbivory or carnivory.

Many transitions to different feeding types may have occurred slowly and gradually. Nassariids are predominantly scavengers (see Section 5.2.1.7.3), but several have become deposit feeders, even developing a crystalline style (Morton 1960c; Curtis 1980; Curtis & Hurd 1981). In the South African sand beach nassariid *Bullia digitalis*, the shells are colonised by various algae which supplement the predominantly carnivorous diet of this scavenger; cellulolytic symbiotic bacteria in the gut assist with algal digestion (Harris et al. 1986). This is not the only example of herbivory in neogastropods; a possibly rapid switch from carnivory to predominant herbivory has occurred in some clades of columbellids (deMaintenon 1999).

In 'opisthobranchs' there is disagreement as to what the original feeding mode was – herbivorous (Mikkelsen 1996; Cimino & Ghiselin 1999; Göbbeler & Klussmann-Kolb 2011) or carnivorous (Haszprunar 1985f; Rudman & Willan 1998). According to Cimino and Ghiselin (1999), the ancestral herbivorous 'opisthobranch' had some kind of diet-derived chemical defence, and they argue that the ancestral nudibranchs switched to sponge-feeding and by so doing gained protection from the sponge metabolites. Ancestral herbivory is assumed in cephalaspideans, with carnivory occurring independently in two clades (Malaquias & Reid 2008). The phylogeny produced by Medina et al. (2011) is equivocal regarding whether herbivory or carnivory is basal. The lower heterobranchs also exhibit various feeding modes, with some carnivores (Architectonicoidea and probably others such as *Ebala*), microphagous feeders (Rissoellidae, Valvatidae), and algal cell piercers (*Omalogyra*). Acteonoideans and many cephalaspideans are also carnivorous, while basal 'pulmonates' are mostly generalist browsers or, in amphibolids, deposit feeders (Juniper 1986).

Many stylommatophoran land snails and slugs are either herbivorous or omnivorous, but the majority are microphagous, feeding on bacteria and fungi associated with living and decaying vegetation (Graham 1955; Chatfield 1976; Speiser 2001). Carnivory is also present in a few taxa (*Euglandina*,

---

[5] Currently placed in Ataphridae (Gründel 2008).

Haplotrema, Powelliphanta, Testacella), and in many other taxa, it is unclear what is taken in active hunting versus incidental ingestion. Prey taxa include other molluscs, oligochaetes, arachnids, and insects. Coprophagy of mammalian faeces and soil ingestion occurs in many species, and the latter is thought to be a source of humic acid (Speiser 2001). The original feeding mode in stylommatophorans is difficult to estimate as the earliest extant clade ('achatinoid clade') includes predatory snails (*Gonaxis, Rumina*), living plant material feeders, (*Achatina, Archachatina*), and microphagous taxa (Ferussaciidae) (Wade et al. 2001). Large proportions of carnivorous snails are characteristic of Afrotropical snail assemblages (Winter & Gittenberger 1998), and hatchlings of various species of herbivorous land snails cannibalise sibling eggs (Baur 1992b), indicating a possible ontogenetic dietary polarity.

5.2.1.14 Some Behavioural Aspects of Feeding

Molluscan feeding behaviour ranges from simple to complex. Simple behaviours are seen in suspension-feeders, grazers, and deposit feeders, while more complex behaviours are mainly seen in active predators.

Suspension-feeders may respond to variations in the quantity and quality of suspended food in the water column (e.g., Bayne et al. 1993), as well as having their feeding activities adjusted to tidal or other rhythms (see Section 5.4.4.1.1). Adoption of some apparent rhythms may simply be a response to food availability – for example, suspension-feeding by intertidal bivalves is only possible when the tide is in. In other cases, it may be an adaptation to avoid predators, as in *Macoma* (Tellinidae) extending its long siphons only at night to deposit feed on the surface of the sediment, apparently to avoid visual predation by fish (Levinton 1971). Similarly, many intertidal foraging species and other aquatic and many terrestrial taxa feed mainly at night. Other environmental factors may result in changed feeding behaviour – for example, *Macoma* varies its feeding (via the inhalant siphon) with changing water velocity and sediment loads (Levinton 1991), and *Ecrobia ventrosa* (Hydrobiidae) slows its feeding rate in response to increased population density (Levinton 1979).

Some supposedly simple behaviours have hidden complexity. For example, two intertidal limpets on rocky shores in South Africa, one a patellogastropod (*Patella*) and the other a marine 'pulmonate' (*Siphonaria*), both forage during low tide, mostly at night. Both species return to the 'home' areas, with *Siphonaria* returning to a home depression. While *Patella* foraged randomly, *Siphonaria* was non-random with movement in an up shore direction, and it was suggested that there might even be learning involved in relation to their returning to areas of optimal feeding (Gray & Hodgson 1997).

Some patellogastropods 'cultivate' algal gardens near their home depressions (e.g., Branch 1981). Most (but not all) gardening limpets are large species, such as some species of *Scutellastra* (Patellidae), which live on coralline algal substrata and maintain a garden around the periphery of the shell. Usually the main food source is non-coralline filamentous or finely branched red algae growing on the surface of the corallines. Other species maintain algal patches over which they graze (Lindberg 2007b). These limpets increase algal productivity in their gardens by adding nutrients through the release of nitrogenous excretion (Plagányi & Branch 2000) or through nutrient-rich mucus from the foot (Connor 1986). Periphery gardeners mainly use nitrogenous excretions, while patch gardeners are more reliant on mucus enrichment. Patch and periphery gardeners also have different shell morphologies – in most periphery gardeners, the front of the shell is extended and has an angular rather than rounded profile, reflecting an elongated head region that forms a distinctive 'neck', with the extreme example being the South African *Scutellastra cochlear*. This modification enables these limpets to graze their gardens without leaving the vicinity of their home depressions. Patch gardeners typically have a rounded anterior end of the shell (Lindberg 2007b).

Feeding behaviour may differ between juveniles and adults – for example, larger sized temperate species of abalone (Haliotidae) feed on drift algae, while their juveniles scrape bacteria and encrusting algae from the rock surface (e.g., Shepherd 1973; Garland et al. 1985; Tutschulte & Connell 1988).

Extensive study into the feeding behaviour of the herbivorous euopisthobranch *Aplysia* has demonstrated the underlying neural and hormonal drivers of the reflex and fixed pattern behaviours involved in feeding. These are driven by both tactile and chemical stimuli and some basic learning capability (e.g., Preston & Lee 1973; Kandel 1979; Rosen et al. 1982) (see also Chapter 7). Similar, but generally less detailed, studies have been carried out on a few other 'opisthobranchs' such as the carnivorous nudibranch *Tritonia* (e.g., Willows 1980) and *Pleurobranchaea* (e.g., Davis et al. 1980; Gillette & Gillette 1983), and several hygrophilans and stylommatophorans (see Elliott & Susswein 2002 for review).

Many studies have demonstrated that predators preferentially select their prey. Grazing carnivores may be permanently associated with their host, as with ovulids on their anthozoan host. Others may move about seeking the sponge, hydroid, or ascidians on which they feed, thereby requiring sensory equipment to detect their prey at a distance. This ability to detect prey is more acutely developed in active predators that often have complex behaviours and associated structural modifications.

Benthic marine gastropod predators move slowly over the substratum relying on chemical cues obtained via the surrounding water. To be effective, they must be able to determine the direction of the food. Predatory caenogastropods do this by using their often mobile inhalant siphon to 'sniff' the incoming water, and nudibranchs detect chemical cues in water currents passing over their bodies and follow them upstream (e.g., Tyndale et al. 1994; Megina & Cervera 2003; Wyeth et al. 2006). Some are specific in their choice of prey species, as for example, some conids (e.g., Kohn 1966; Duda et al. 2008; Remigio & Duda 2008), while others are more opportunistic. Predatory land snails find their prey mainly by olfaction, or by tracking the mucous trails of prey snails or slugs (Barker & Efford 2002).

All heteropods are highly modified swimming carnivores, feeding on a variety of planktonic animals and hunting mostly by sight. This markedly contrasts with the pelagic gymnosome pteropods which are mostly active predators, but some 'sit-and-wait'. Gymnosomes feed exclusively on shelled pteropods (thecosomes) and are highly species-specific in their prey choice, relying on tactile recognition. Once detected, the prey is captured using a unique array of feeding structures including, in some, 'buccal cones', eversible grasping structures or arms with or without suckers and, in most, grasping hooks that lie in a pair of lateral sacs on either side of the mouth (see Chapter 20, Figure 20.28). The buccal cones or tentacles are used to grasp the thecosome, and the hooks and radula remove the prey from its shell before it is swallowed and quickly digested (Conover & Lalli 1972, 1974; Lalli & Gilmer 1989). The gymnosome pteropod *Clione limacina* captures its prey, the thecosome *Limacina helicina*, in a rapid movement which involves opening the mouth and extruding three pairs of buccal cones in 60–90 ms. If the buccal cones do not contact prey they quickly retract, but if they do a viscous, sticky material is released from papillae on the cones to hold the prey (Hermans & Satterlie 1992). There is a strong correlation between the metabolic rate and hence locomotory performance of the gymnosome and its thecosome prey (Seibel et al. 2007).

Prey specialisation can lead to some spectacular radiations in both grazing carnivores and active predators. Colonial animals such as anthozoans, sponges, and bryozoans are highly diverse, as are some of the molluscs that feed on them; many are specialised to feed on a single species or a few similar species. Notable examples include the sponge-feeding Triphoroidea (Triphoridae and Cerithiopsidae) and various groups of sponge-feeding dorid nudibranchs (e.g., Chromodorididae) and the anthozoan-feeding Epitoniidae and Ovulidae. Such specialisation required the development of sensory equipment capable of detecting species-specific chemical clues for prey detection.

One of the better-studied groups of predatory gastropods are members of the Conidae (the cone shells). Most species are specialised predators on polychaetes (vermivorous), hemichordates, or molluscs (molluscivorous), while the most specialised are some larger species that, as adults, kill and swallow small fish (piscivorous) (e.g., Olivera 2002). Conids only radiated extensively as recently as the Miocene. A conservative view of the classification of the group was accepted until recently, so what was then considered a single genus, *Conus*, was the most species-rich living genus of marine animals (Duda et al. 2001). This single genus has been divided into several genera (e.g., Puillandre et al. 2011), although this nomenclature is not universally accepted. Like some other conoideans, conids have hollow harpoon-like radular teeth (see Section 5.2.1.7.4.5) that are charged with venom from a large venom gland. A phylogenetic study indicated that marked shifts in feeding strategies occurred only a few times early in the radiation of conids (Duda et al. 2001) and were associated with the evolution of a large array of peptide neurotoxins (e.g., Remigio & Duda 2008).

Activities such as drilling can take a long time, with increased energy expenditure and risk of predation being issues and, in an intertidal animal, also the possibility of desiccation. Thus, there are behavioural mechanisms that increase predation efficiency. Predators typically select a preferred prey size that maximises both efficiency and reward – for example, the muricid *Thais mutabilis* preys on a preferred size of the prey snail (*Cerithidea cingulata*) relative to the size of the predator, but when the preferred prey size was not present, they selected larger prey and, only when larger individuals were rare or absent, did they prey on small-sized *Cerithidea* (Ekaratne & Goonewardena 1994).

Predators also attack their prey in ways likely to result in success. For example, *Thais* preying on *Cerithidea* usually prefer to drill the thin, horny operculum rather than the shell. Conditioning can alter this behaviour; *Thais* experimentally conditioned to drill bivalve shells drilled *Cerithidea* shells for several days before drilling the operculum (Ekaratne & Goonewardena 1994).

Another muricid, *Hexaplex trunculus*, drills the thicker bivalve shells and uses its labral spine on thinner ones (see Section 5.2.1.7.4.2) to chip at the shell margin. The method of attack is also related to predator and prey size; for example, young *Mytilus* are chipped and older ones drilled by larger individuals of the muricid *Hexaplex trunculus*, but small individuals only used drilling (Morton et al. 2007).

Experiments with the naticid *Euspira lewisii* have shown that they preferentially select thinner shelled individuals of the same prey species of bivalve (Grey et al. 2005), but this was not the case with another naticid *Neverita* (as *Polinices*) *duplicata* (Boggs et al. 1984). The actual site on the prey shell where drilling occured was often not randomly chosen, but equated with the area where the shell was thinnest and/or where the most nutritious part of the prey animal (typically the gonad or digestive gland) lay. For example, whelks feeding on limpets drilled over the gonad and digestive gland rather than over the shell muscle (Black 1978; Palmer 1988; Thomas & Day 1995). In an interesting contrast, the muricid *Lepsiella* (as *Haustrum*) *baileyanum* drilled holes in the thickest part of *Haliotis* shells to access the nutritious underlying shell muscle, but this may be in part related to the subtidal habitat of the prey where desiccation and wave action are not relevant (Thomas & Day 1995). Whelks may avoid the foot/shell muscle tissue when feeding on intertidal limpets, presumably to avoid them detaching (Black 1978; Palmer 1988). When the prey was smaller than the predator the foot was eaten, as the predator is able to grip the substratum independently (Palmer 1988).

Other factors probably influencing the selection of a drill site by predatory snails include shell ornamentation, the orientation of the prey and its relationship with the substratum, and defensive behaviour.

Cephalopods exhibit sophisticated feeding behaviours. These are outlined in Chapter 17 and are reviewed by Hanlon and Messenger (1996).

5.3 DIGESTIVE SYSTEM

Previous overviews of the digestive system of molluscs have been provided by Owen (1966) and Salvini-Plawen (1981b, 1988), and the detailed treatments for the major taxon groups in the *Microscopic Anatomy of Invertebrates* series.

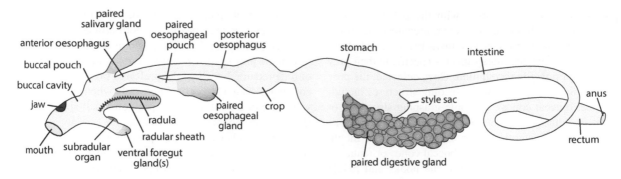

FIGURE 5.14 A generalised molluscan gut. Redrawn and modified from Mizzaro-Wimmer, M. and Salvini-Plawen, L.v., *Praktische Malakologie. Beiträge zur vergleichend-anatomischen Bearbeitung der Mollusken: Caudofoveata bis Gastropoda – Streptoneura*, Springer, Vienna, Austria, 2001.

5.3.1 Development and Evolution of the Molluscan Digestive System

The molluscan digestive system (Figure 5.14) typically consists of four developmentally independently derived components – the foregut, midgut, hindgut, and rectum. Structures of the foregut include the buccal cavity, jaws, salivary glands, and oesophagus and are typically formed from the ectodermal stomodaeum. Midgut structures include the stomach, style sac, digestive gland, and (if present) the gastric caecum, are endodermal and typically derived from the archenteron. The hindgut or intestine is also derived from endoderm or, in gastropods, from a third tissue group – the endomesoderm (see Chapter 8). The rectum and anus are typically formed from the ectodermal proctodaeum (Raven 1966).

The development of the molluscan digestive system follows a similar pattern to that seen in the putative outgroups, although with several modifications. First, there are additions to the foregut, with outpocketings forming the radular sac and various glands. Midgut modifications include the formation of the prominent digestive gland and, in some groups, a style sac and gastric shield. Developmentally, the molluscan hindgut is most like the lophophorate hindgut in being derived from the endo- and mesoderm rather than ectoderm as in annelids, although ectodermal tissue may contribute to the terminal portion of the hindgut in some polyplacophoran and aplacophoran groups (see Chapter 8). In chitons, bivalves, and gastropods, the hindgut differs from molluscan outgroups and aplacophorans in being plesiomorphically highly coiled but has become secondarily simplified in the more derived members of some groups. As discussed in Chapter 8, a U-shaped gut is seen in some tube-dwelling annelids, in most lophophorate groups, and in three molluscan taxa (Cephalopoda, Scaphopoda, and Gastropoda).

For the sake of simplicity, in the review below, we treat the intestine and rectum as a unit (the hindgut), as they are morphologically and functionally similar. The division into fore-, mid and hind gut is largely concordant with the developmental origins of the digestive system, as outlined above. It differs from that of some authors who recognise the midgut as the stomach through to and including the intestine, with the rectum alone being the hindgut.

5.3.2 The Foregut

The foregut (= anterior gut) commences at the mouth and associated structures, which open to the buccal cavity. The buccal mass and the operational part of the radula lie within the buccal cavity, and the salivary glands open to it. The oesophagus commences at the posterior edge of the buccal cavity and runs to the stomach (part of the midgut).

The foregut ingests food and commences its mechanical breakdown (by the radula and jaws), as well as mixing the food with saliva for lubrication and, often, preliminary enzymatic action. The oesophagus may have an oesophageal gland which provides additional enzymes, or it may be expanded into a crop for holding food or, in some 'opisthobranchs', a muscular gizzard has developed to further physically break down food (see Section 5.3.2.7 and Chapter 20).

5.3.2.1 The Mouth, Jaws, Buccal Cavity, and the Acquisition and Ingestion of Food

The mouth is the interface between the external environment and the alimentary canal. Some molluscs have sensory organs for detecting food at a distance, and there are usually sensory epithelia or structures around the mouth. The mouth opens to an oral area (sometimes constricted as a tube) that often bears a pair of jaws, a single dorsal jaw or, in cephalopods, a dorsal and ventral 'beak'. The oral tube is continuous with the buccal cavity in which the odontophore and the anterior working part of the radula lie. Behind the opening of the radular sac (where the radula is formed), the buccal cavity constricts and the oesophagus, and often the paired salivary glands, open to it. This space is sometimes called the pharynx, but we treat it as the posterior part of the buccal cavity as it is continuous with it. Below, we briefly survey these structures in the major groups of molluscs.

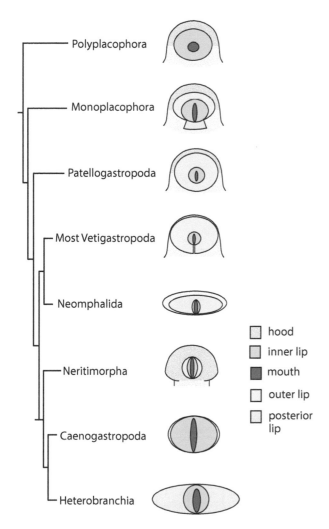

Polyplacophora

Monoplacophora

Patellogastropoda

Most Vetigastropoda

Neomphalida

Neritimorpha

Caenogastropoda

Heterobranchia

☐ hood
☐ inner lip
■ mouth
☐ outer lip
☐ posterior lip

FIGURE 5.15 Diagram showing the mouth structure in different gastropod groups, chitons, and monoplacophorans. Original.

The mouth is located ventrally at the anterior end of the animal. It may be covered with a hood and have various lip and other structures associated with it which are generally consistent in each major group (Figure 5.15).

Jaws are lacking in aplacophorans except in Prochaetodermatidae (Caudofoveata) (see below). Jaws are absent in chitons, and there is a dorsal jaw with lateral flanges in monoplacophorans and patellogastropods. Cephalopods have dorsal and ventral beak-like elements, and it is probable that the dorsal 'beak' is derived from the dorsal jaw from a shared ancestor with monoplacophorans. Scaphopods also have a single dorsal jaw in some taxa (Shimek & Steiner 1997), but it is often absent. In all the above conchiferan taxa, the jaw is composed of thick, uniform cuticle, which is tanned in cephalopods, some of which have calcium carbonate added. In contrast, orthogastropods have a pair of jaws that are plesiomorphically dorso-laterally located and composed of chitinous rods secreted by individual epithelial cells (Figures 5.16 and 5.21).

5.3.2.1.1 Aplacophorans

In Caudofoveata, the mouth opening is anterior and partially or completely surrounded by the oral shield and the mouth opens into a folded, expandable buccal cavity, which typically contains the odontophore and radula (see Section 5.3.2.4.2.1) and which, in the Chaetodermatidae, is protrusible. In solenogasters, the mouth is anteroventral to ventral and lies inside the preoral sensory vestibule. The buccal cavity in solenogasters may be a simple tube, a muscular sucking pump, or, in a few, the buccal epithelium forms an invaginated sheath around the foregut to allow it to be protruded as a short 'proboscis' (Salvini-Plawen 1988). There is usually an 'oral cavity' and a 'buccal cavity' (= pharynx). These may be lined with similar epithelium, or they may be quite different. Jaws are lacking in Caudofoveata except in Prochaetodermatidae where they are cuticular and paired (Figure 5.16), lying laterally in the oral cavity where they are inserted lateroventrally and connected by muscle (Scheltema & Ivanov 2000). Presumably these jaws are not homologous to conchiferan jaws. They are assumed to hold food while the radula macerates it, and when the associated muscles relax, the jaws help to hold the mouth open (Salvini-Plawen 1988).

In caudofoveates, feeding has only rarely been observed. In *Prochaetoderma*, it consists of sweeping microorganisms or other organic material into the widely open mouth using the radula (Scheltema 1981; Salvini-Plawen 1988). While radular protrusion also occurs in some other caudofoveates, this is not possible in the Chaetodermatidae, where the probable main function of the two forceps-like teeth of the radula is to move food posteriorly in the foregut and, possibly, to play a role in food selection (Scheltema 1981; Salvini-Plawen 1988). The pedal shield is thought to be involved in the actual uptake of food.

Solenogasters feed on cnidarians. *Neomenia*, which lacks a radula, uses well-developed foregut sphincter muscles to suck anemone tissue into the posterior buccal cavity where enzymatic secretions from unicellular glands begin digestion (Sasaki & Saito 2005). Other solenogasters have multiple radular teeth which are thought to be protruded to feed on their prey.

5.3.2.1.2 Polyplacophora

The simple mouth lies ventral to the poorly differentiated 'head' and is surrounded by an unpaired preoral veil of mantle tissue. The anterior oral cavity has a cuticular coating but no jaw and, behind it, the short buccal cavity contains the openings to the subradular pouch and radular sac. Behind the opening of the radular sac, the salivary glands open into the posterior part of the buccal cavity (Figure 5.17).

The ventral subradular organ (see Section 5.3.2.2.1) is extended and pressed against the substratum before feeding, presumably to 'taste' the potential of the food. Food, together with inorganic particles, is scraped from the substratum when the powerful radula (see Section 5.3.2.4.2.2) is extended through the mouth. The material is carried into the mouth by the radula and mixed with secretions from

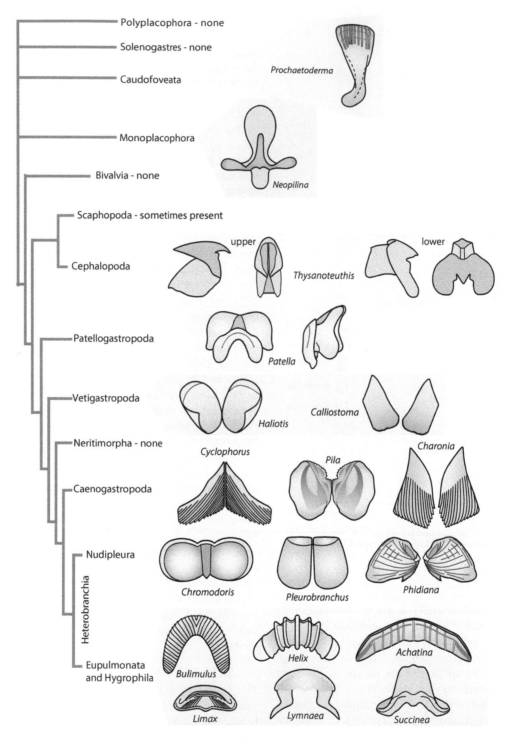

FIGURE 5.16   An overview of the diversity of jaw structure in molluscs. Jaw drawings redrawn and modified from many sources.

the buccal and salivary glands. Food is then transferred by ciliary action to the oesophagus.

5.3.2.1.3   *Monoplacophora*

The ventral mouth is bordered dorsally by the 'velum' which continues to form a flattened extension on each side. Lemche and Wingstrand (1959) suggested that the velum was homologous to the larval velum, but there is no evidence to support this idea (Haszprunar & Schaefer 1997), although it may be homologous to the preoral veil in chitons. A pair of clusters of short post-oral tentacles, of unknown function, lies on either side behind the mouth and immediately in front of the foot. Lemche and Wingstrand (1959) assumed that they are involved in feeding and suggested that they are homologous to the inner labial palps of bivalves (disputed by Waller 1998, p. 8), the arms of cephalopods, and the captaculae of scaphopods.

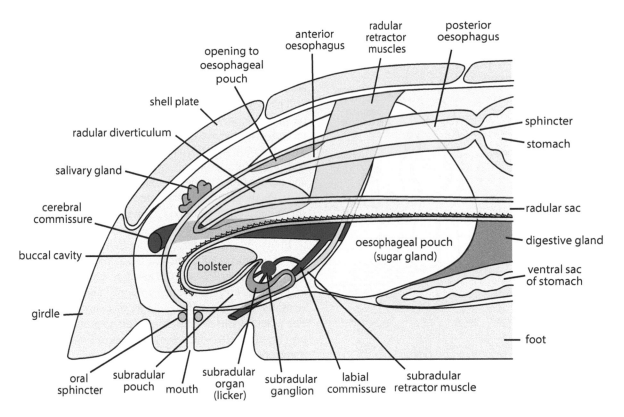

FIGURE 5.17 Longitudinal section of the anterior end of a generalised chiton showing the main features of the anterior gut with most of the musculature omitted. Redrawn and modified from Fretter, V., *Trans. Royal Soc. Edinburgh*, 59, 119–164, 1937, and Wingstrand, K.G., *Galathea Report*, 16, 1–94, 1985.

The cuticle-lined buccal cavity contains the odontophore and radula (see Section 5.3.2.4.2.3), a subradular pouch (see Section 5.3.2.2.1) ventro-posteriorly, and a trilobed dorsal jaw (Figure 5.16) latero-dorsally. The salivary glands open into the lateroventrally compressed posterior buccal cavity.

Based on radular musculature, Lemche and Wingstrand (1959) proposed that food was gathered by 'food grooves' rather than the radula, and (Wingstrand 1985) once in the buccal cavity, the food was transported back into the posterior buccal cavity by the radula and transported through the oesophagus by cilia. Graham (1964) argued that even with the reduced musculature relative to chitons, the monoplacophoran radula could interact with the substratum, and feeding marks were subsequently reported on the xenophyophore *Stannophyllum* (collected from a *Neopilina* locality) by Tendal (1985). Subsequently feeding traces have been observed involving other monoplacophorans (see Chapter 14).

5.3.2.1.4 Scaphopoda

The mouth is at the end of a mobile snout-like 'proboscis' or buccal tube and is generally incapable of extending beyond the opening of the shell (Shimek & Steiner 1997). Although this structure resembles a gastropod snout, it is not homologous because it is formed anterior to the jaw. The mouth opening is surrounded by lips (oral lappets) that

are sometimes extended into elaborate frills, especially in dentaliids (Figure 5.18), and sort material before it enters the mouth (Dinamani 1964). Peristaltic movements in the buccal tube move food to the storage pouches and eventually to the radula (see Section 5.3.2.4.2.4).

At the base of the buccal tube, attached to a pair of lateral bulges, are few to many thin, contractile tentacles, the captaculae (see Chapter 16). Each has a ciliated terminal expansion (bulb), and a ciliated tract runs along each tentacle. The captaculae are used in feeding and are extended by ciliary action and retracted by muscles (Shimek & Steiner 1997). The bulbs on the ends of the captaculae collect the food from the surrounding sediment and hold it by adhesion. Larger items are brought to the mouth directly by the captaculae, often with two or more captaculae working together, while smaller particles move to the mouth along the ciliated tracts on the captaculae (Dinamani 1963; Byrum & Ruppert 1994; Shimek & Steiner 1997; Reynolds 2002).

Prey is stored in a pair of lateral pouches in the buccal tube for up to 30 hours before it is moved to be macerated by the radula (Shimek 1990). A single, dorsal, rather large, horseshoe-shaped jaw in the buccal cavity may protect the underlying tissues and nerves from damage by the massive radula, especially in gadilids (Shimek & Steiner 1997). There is a ventral subradular organ, but no salivary glands.

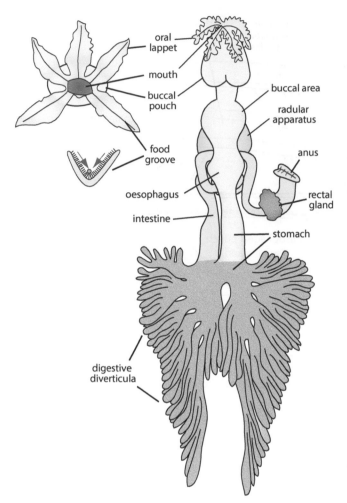

FIGURE 5.18 The alimentary canal of the scaphopod *Dentalium* showing the oral lappets. Anus and rectal gland have been extended to the right for clarity. Oral lappet details redrawn and modified from Gainey, L.F., *Veliger*, 15, 29–34, 1972; main figure redrawn and modified from de Lacaze-Duthiers, H., *Zoologie*, 6, 225–228, 319–385, 1856–1857.

5.3.2.1.5 Gastropoda

The overall trend in the gastropod alimentary canal is one of increasing simplification from their ancestral state, although some increased complexity is seen in the foregut, notably in those taxa with a proboscis and accessory foregut glands.

The mouth (Figure 5.19, 5.20, 5.25) mouth is often not a simple aperture, and several distinct morphological types can be recognised, especially in the 'lower' gastropods. These include a hood over the mouth (in Patellogastropoda and Neritimorpha), with scale-like structures present in some patellogastropods and neritimorphs, variously shaped openings including round (as in Patellogastropoda and some Vetigastropoda) or a vertical slit (in most other gastropods). In many heterobranchs, there is a lateral expansion of the sides of the mouth area.

In many gastropods, the mouth is at the anteroventral end of a protrusion from the head, the 'snout', a structure lacking in other molluscs. This may be short and broad as in patellogastropods, vetigastropods, and neritimorphs, or ventrally elongate and mobile as in many caenogastropods. The mouth opens to allow the radula to protrude during feeding (Figure 5.20), which works in concert with the jaw(s) and, if present, the licker.

The snout has become introverted in some caenogastropods and heterobranchs to form a proboscis (see Section 5.3.2.1.5.1).

The mouth opens either directly to the buccal cavity or to a narrow oral tube, which is elongated in some derived taxa. The buccal cavity may be lined with cuticle or an epithelium with gland cells, and there is usually a pair of jaws (see below). The salivary glands and radular sac open into this cavity and, in Fissurellidae, a median subradular diverticulum (Fretter & Graham 1962, fig. 96). A subradular organ is present in some lower gastropods and basal caenogastropods (see Section 5.3.2.2.1). The radula is very diverse (see Section 5.3.2.4.2.6) and sometimes lacking. Posterodorsally, the buccal cavity opens to the oesophagus with a sheet of tissue (the oesophageal 'valve') lying between the radula and the dorsal buccal cavity/oesophagus.

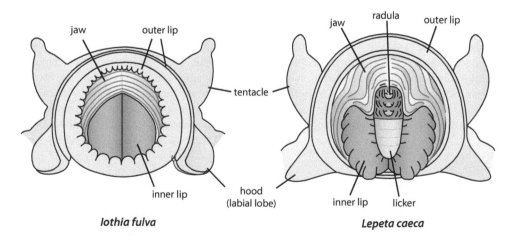

Iothia fulva   *Lepeta caeca*

FIGURE 5.19 The open mouths of the patellogastropods *Iothia fulva* and *Lepeta caeca*. Redrawn and modified from Fretter, V. and Graham, A.L., *British Prosobranch Molluscs: Their Functional Anatomy and Ecology*, Ray Society, London, UK, 1994.

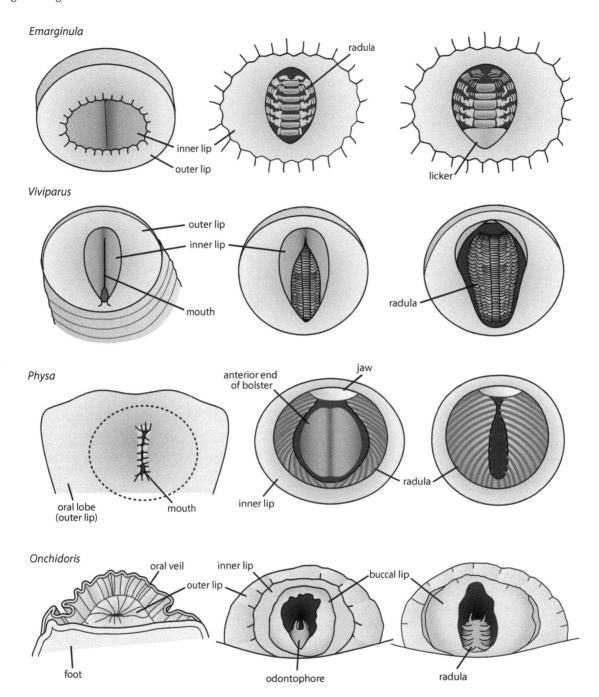

FIGURE 5.20 Mouth action in a vetigastropod (*Emarginula*), a lower caenogastropod (*Viviparus*), and two heterobranch gastropods (*Physa* and *Onchidoris*) to show the protrusion of the odontophore and radula. The illustrations are redrawn and modified from Eigenbrodt, H., *Zeitschrift für Morphologie und Ökologie der Tiere*, 37, 735–791, 1941, except for *Onchidoris* which is redrawn and modified from Crampton, D.M., *Trans. Zool. Soc. Lond.*, 34, 45–86, 1977.

In patellogastropods, the jaw consists of anterior and posterior paired lobes fused dorsally (Sasaki 1998) (Figure 5.16) and is composed of a homogeneous material as in all other jaw-bearing molluscs, except most other gastropods. In cocculinids, there is only a single vestigial dorsal jaw (Haszprunar 1987a), and neritimorphs have only very thin lateral plates. As noted above, the jaws in orthogastropods have, as their plesiomorphic condition, a single pair of dorsolateral jaw elements that are sometimes fused dorsally and composed of small rod-like elements (Ponder & Lindberg 1997) (Figures 5.16 and 5.21). Jaws are sometimes lost, as in most littorinoideans, and some assimineids. With few exceptions, neogastropods also lack jaws, although muricids (Carriker 1943; Wu 1965) have dorsal and ventral elements of doubtful homology with the jaws in other gastropods. Cancellariids have a tubular jaw to stabilise and direct the elongate, piercing radular teeth (Petit & Harasewych 1986) (see Chapter 19, Figure 19.18).

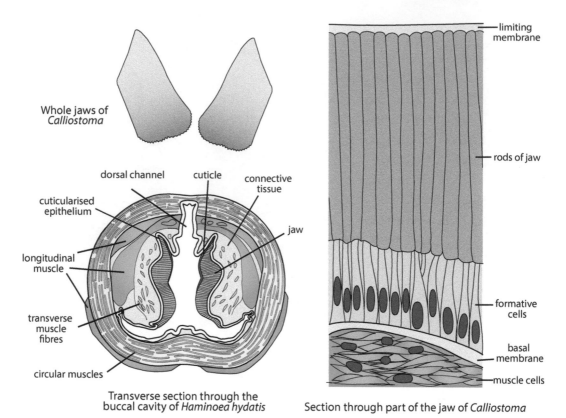

Whole jaws of *Calliostoma*

dorsal channel   cuticle   connective tissue

cuticularised epithelium

longitudinal muscle

transverse muscle fibres

circular muscles

jaw

Transverse section through the buccal cavity of *Haminoea hydatis*

limiting membrane

rods of jaw

formative cells

basal membrane

muscle cells

Section through part of the jaw of *Calliostoma*

FIGURE 5.21   Jaws of *Calliostoma* with a section through part of the jaw showing rods and a transverse section through the jaws of *Haminoea*. Redrawn and modified from Randles, W.B., *Q. J. Microsc. Sci.*, 48, 33–78, 1905 (*Calliostoma*) and Fretter, V., *Trans. Royal Soc. Edinb.*, 59, 599–646, 1939 (*Haminoea*).

In heterobranchs, the jaws are diverse and in some 'lower heterobranchs' are highly modified, as for example, in orbitestellids, where the multiple elements making up each lateral pair of jaws are modified into a few large plates (Ponder 1990b). The eupulmonates have a single dorsal element probably derived from the fusion of the two lateral jaws. In the suctorial pyramidellids, the jaw is modified as a piercing stylet (Fretter & Graham 1949). Some 'opisthobranchs' (Gosliner 1994), valvatoideans (Ponder 1990a), and rissoelloideans often have the dorsal ends of the rods elaborated, or the rods may fuse to form tooth-like structures (some 'opisthobranchs', Gosliner 1994) or, in the case of hygrophyilans, eupulmonates and siphonariids, a single dorsal element (Figure 5.16). Jaws are lost in sacoglossans and a few other panpulmonates such as Amphiboloidea, Glacidorboidea, Latiidae, Smeagolidae, some onchidiids, and many carnivorous stylommatophorans (Barker 2001a). In eupulmonates where the jaw is dorsal (presumably the two lateral elements have fused), it is still composed of fused rods. A few other orthogastropods in various groups have lost the jaws. While they are mostly dorso-lateral to lateral on the wall of the buccal cavity/oral tube, in the tonnoideans, they are often connected dorsally (e.g., several examples illustrated by Warén & Bouchet 1990). In Succinoidea, the 'elasmognathous' jaw has an accessory plate (Figure 5.16) to which buccal muscles attach (Barker

2001a). Such differences in jaw morphology inspired a classification of stylommatophorans based on jaws (Fischer 1880–1887).

In stylommatophorans, the food is squeezed between the jaw and the radula to cut off pieces, and it is not further broken up by the radula (Märkel 1957; Mackenstedt & Märkel 2001).

During feeding, the jaws remain open until the radula is retracted, and the neurological and muscular mechanisms involved have been extensively studied in *Aplysia* (e.g., Nagahama & Takata 1988).

5.3.2.1.5.1 The Proboscis   There has been an independent evolution and elongation of a proboscis in several major groups of predatory marine caenogastropods (e.g., Kantor 1991; Golding et al. 2009b) (see also Chapter 19), and several types of proboscis can be recognised. Most fall into two main functional types: the *acrembolic* type in which the retractor muscles are at the tip of the proboscis, and the *pleurembolic* type in which the retractors are inserted in the middle of the proboscis (Figure 5.22). When retracted, the acrembolic type has the tip of the proboscis fully inverted because of the distal attachment of the retractor muscles. Because the pleurembolic type of proboscis has the retractor muscles inserted on the sides of the proboscis, the distal part is not inverted

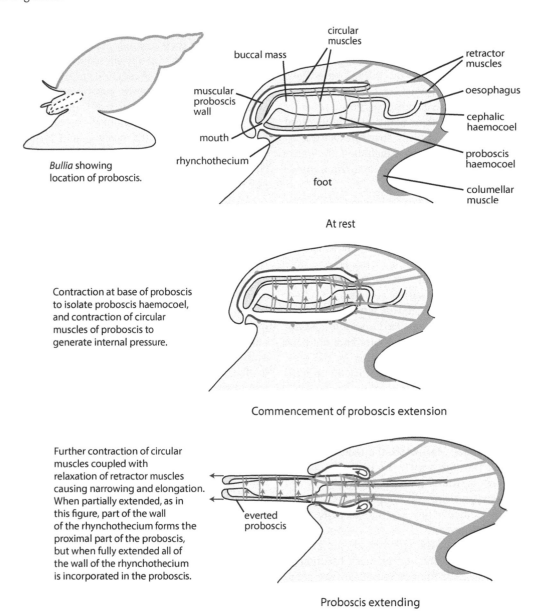

FIGURE 5.22 Extrusion of a pleurembolic proboscis in a buccinoidean neogastropod (*Bullia* – Nassariidae). The head and foot are shown in longitudinal medial section. Redrawn and modified from Trueman, E.R. and Brown, A.C., *J. Zool.*, 211, 505–513, 1987.

when the proboscis is retracted. While the acrembolic type of proboscis has independently evolved several times in both caenogastropods and heterobranchs, the pleurembolic type may have evolved only once, in caenogastropods (Golding et al. 2009b; Simone 2011). Another proboscis type, found in many tonnoideans and recognised by Day (1969) as the 'argobucciniform' type, was renamed *teinoembolic* by Golding et al. (2009b). This is characterised by being non-introvertable, and the proboscis is retracted solely by longitudinal muscle contraction and, when retracted, is enclosed in an external rhynchodeum.

In an *intraembolic* proboscis, the buccal mass is at the base of the proboscis, and the elongate oral tube extends to the tip. This is the basic type in the Conoidea and a few other neogastropods. The intraembolic type was considered

by Kantor (2002) as the most primitive in Neogastropoda, with the primitive neogastropods having a short proboscis with a basal buccal mass. Such a proboscis is seen in several conoidean families such as Drillidae and Turridae (in the strict sense) (see Chapter 19) and also in the non-conoidean Pseudolividae and Ptychatractidae. The structure of this type would not allow the radula to interact with the food directly, suggesting that they feed suctorially, perhaps involving external digestion. There are a variety of other kinds of proboscis recognised in conoideans (see Chapter 19, Figures 19.20 to 19.22) that are developmental modifications and simplifications of the same basic type.

The mechanisms involved in the extrusion of the proboscis differ depending on the type of proboscis, but most involve hydrostats (Trueman & Brown 1987; Golding et al. 2009b).

5.3.2.1.5.2 Accessory Structures Associated with the Proboscis in Some Neogastropods Branch-like or club-like accessory proboscis structures (see Chapter 19, Figure 19.22), or rhynchodeum outgrowths (Fedosov 2008), have been described in some conoideans. In some terebrids at least, they can be elaborate and may be chemosensory (Taylor & Miller 1990; Taylor et al. 1993) rather than being directly involved in feeding. Similar accessory proboscis structures have been described in some other conoideans including *Zemacies* (Borsoniidae), *Horaiclavus* (Horaiclavidae), and *Tritonoturris* (Raphitomidae) (Fedosov 2008).

An unusual structure, the *epiproboscis* (Ponder 1972; West 1990, 1991; Harasewych 2009) (Figure 5.23), is found only in the neogastropod family Mitridae. It consists of a long, retractile, muscular papilla housed in the proboscis, but which can be extended considerably beyond it. The extremely narrow salivary ducts are contained within the epiproboscis, and it is assumed this structure delivers salivary secretion (Ponder 1972) to the sipunculan prey on which this family specialises (Kohn 1970; Fukuyama & Nybakken 1983; Loch 1987; Taylor 1989, 1993) before the prey is engulfed. There have not been studies on the effect of the salivary secretion on the prey, although observation of feeding indicates that the epiproboscis is extended and inserted into the prey (Loch 1987). West (1990, 1991) provided the most detailed observations of feeding behaviour. After locating the prey, it was grasped by the peristomal rim at the distal end of the proboscis and, if it could not be immediately extracted, the radula rasped a hole in the sipunculan through which the epiproboscis was extended. It then wrapped around the viscera of the prey and dragged it to the mouth, or, using a pumping action, brought fluids and other body contents into the mouth, with these actions repeated several times.

The origin and homology of the epiproboscis are controversial. Ponder (1972) argued that the primary function of the epiproboscis is the targeted application of salivary gland secretions and resulted from the ventral migration of the salivary ducts and their opening at a papilla at the inner ventral edge of the mouth with the subsequent elongation and modification of the papilla. West (1991) noted similarities with the subradular organs of lower gastropods and other molluscs although, given the distribution of this structure in gastropods (see Section 5.3.2.2.1), we doubt any relationship. Harasewych (2009) observed that the position and arrangement of the ducts were like those of accessory salivary glands, and Ponder (1972) noted that mitrid salivary glands have different histology from most neogastropod acinous salivary glands.

An analogous ventral, but much shorter, salivary papilla is found in octopods (see next Section).

5.3.2.1.6 Cephalopoda

In cephalopods, the mouth is surrounded by the characteristic arms seen in all members of the group (see Chapter 17 for details) and contains a pair of parrot-beak-like jaws (see below) surrounded by two circular lips, the innermost bearing papillae and the outer smooth. In the decabrachian coleoids and *Nautilus*, there is also a circle of miniature webbed arms, the buccal (or peribuccal) membrane (or crown), which in decabrachian coleoids sometimes bear minute suckers, and in some taxa is differentiated in females into a copulatory pouch.

The cephalopod jaw is beak-like (Figure 5.16), with sharply pointed large dorsal and ventral elements used for biting prey. The lower jaw closes around the slightly smaller upper jaw, the latter being the homologue of the jaw in other conchiferans (Salvini-Plawen 1988). In *Nautilus*, the beak tips are reinforced with calcareous material. Broad inner and outer lateral extensions of the jaws form attachment surfaces for the powerful muscles that lie between them.

The buccal cavity of coleoids contains the massive odontophore and, in octopods, a 'salivary papilla' (also called the 'tongue' or 'subradular organ'); it is not homologous

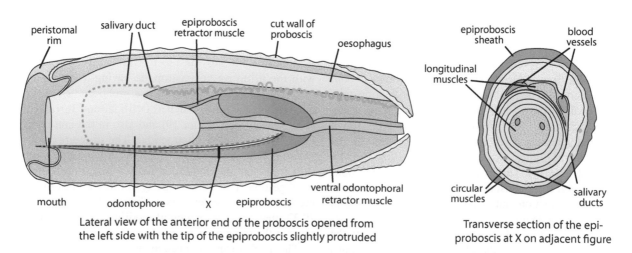

Lateral view of the anterior end of the proboscis opened from the left side with the tip of the epiproboscis slightly protruded

Transverse section of the epiproboscis at X on adjacent figure

FIGURE 5.23 The proboscis of the mitrid *Strigatella paupercula* opened laterally to show the epiproboscis, and a transverse section through that structure. Redrawn and modified from Ponder, W.F., *Malacologia*, 11, 295–342, 1972.

with the subradular organ of other molluscs, although a sub-radular pouch (radular diverticulum) is present in coleoids (see Section 5.3.2.2.1) (Salvini-Plawen 1981b).

Cephalopod salivary glands are described below (see Section 5.3.2.5.4). In coleoids, the salivary papilla lies at the anteroventral edge of the odontophore and is a muscular, highly innervated, eversible structure mainly enclosed by the submandibular gland, a mass of tissue of unknown function. A muscular sheath surrounds the anterior part of the papilla (Young 1965b; Nixon 1979a, 1979b, 1980). The salivary duct runs through the papilla and opens at the anterior end where the venomous salivary secretion is released (Ballering et al. 1972). In some octopods, the salivary papilla has a rasp-like surface and acts as an accessory boring organ (e.g., Nixon 1980) (see also Section 5.2.1.7.4.3).

Dorso-laterally there is a pair of lateral lobes (also called 'palatine lobes' or 'buccal palps') which lie on either side of the food passage above the radula. In *Nautilus*, these lobes are glandular, but in coleoids, they are covered with a chitinous layer with irregular teeth-like projections orientated towards the oesophagus.

In cephalopods, prey capture may involve several steps – grasping the prey with the arms, envenomation, and sometimes predigesting the food (see below) and, before swallowing, biting and rasping with the radula. In some octopuses, it may also involve drilling a hole through the shell of the prey.

5.3.2.1.7 Bivalvia

As described in other sections, in most autobranch bivalves food is acquired by suspension-feeding via the gills and is pre-sorted by the labial palps or, in protobranch bivalves, by deposit-feeding using the elongate appendages on the labial palps. In all bivalves, the food enters the mouth largely by ciliary action.

The mouth opens directly to the simple oesophagus, there being no buccal cavity, odontophore, or jaw. There are two pairs of labial palps on either side of the mouth, one inner and one outer. Both have complex ciliated sorting ridges on their opposed surfaces. On each side, the inner demibranch of the gill lies between the two palps and delivers food to their ridged surfaces.

The labial palps seen in bivalves are unique and found in all bivalves except the protobranch Solemyoidea which have a highly specialised relationship with bacterial symbionts (see Section 5.2.1.12.1) and do not ingest food directly. The labial palps are probably derived from inner and outer lips of the mouth (e.g., Beninger et al. 1990), and the groove between the palps leads to the side of the mouth.

In protobranch bivalves, the palps are very large and receive food mainly from the palp appendages, although small amounts can be derived from the gills (Reid et al. 1992). Although the palps differ considerably in size in different autobranch bivalve taxa, they are never as relatively large as in protobranchs.

In all bivalves, the labial palps are more or less triangular in shape and have broad bases attached to the visceral mass. Cilia in a groove running between the palp bases also carry some food directly to the mouth. The ciliated ridges and grooves on the palps lie at either right angles or obliquely to this groove and run to the free edge of each palp. Complex ciliation on these ridges and grooves assists in sorting the food particles before they enter the mouth. The ridges may have secondary folds (Figure 5.24) which separate opposed waste currents in the grooves from the lighter food particles moving across the summits of the ridges towards the mouth. The waste moves along the grooves to the outer edges of the palps where it is passed onto the mantle surface and is eventually rejected as part of the pseudofaeces. The structure of the ridges and folds and their ciliary patterns ranges from highly complex as in some Solenidae, to very simple and reduced as in Teredinidae (Purchon 1968).

Material leaves the palps and enters the mouth aided by cilia. Direct observations have shown that particles in the gut of some bivalves are suspended in a slurry (Ward et al. 1994), not bound in mucus as suggested in earlier accounts (e.g., Morton 1960a; Purchon 1977b), and this may be the normal situation in bivalves.

The mouth of most bivalves is surrounded by two lips which vary in morphology (see Chapter 15). In a study of the mouth structures of bivalves, Bernard (1972) showed that most bivalves have simple lips, but in some pteriomorphians (Pectinidae [but not the related Propeamussiidae], Spondylidae, and Limidae [see also Morton 1979]), the lips are hypertrophied into elaborate structures. While these are separate structures from the labial palps, these elaborated lips, like the palps, appear to be involved in particle sorting (Bernard 1972). In taxa with complex lips, the mouth entrance is reduced to numerous fine passages. Fusion has occurred in some, with the openings of the mouth reduced to a series of ostia.

5.3.2.2 Sublingual (Subradular) Pouches and Associated Structures

A space below the odontophore, the *sublingual* (or *subradular*) *pouch* (or sac), opens to the anterior buccal cavity. This space may contain a median sensory organ called a *licker* or *subradular organ* and/or lateral paired 'glands' or may be unmodified.

5.3.2.2.1 The Subradular Organ (Licker)

A subradular (or sublingual) pouch with a dorsal subradular organ is found in polyplacophorans, monoplacophorans, *Nautilus*, scaphopods, and some basal gastropods. It is absent in aplacophorans (Salvini-Plawen 1988).

In chitons, the subradular organ is bilobed and lined with microvillous ciliated cells and mucous cells (Boyle 1975). In chitons and patellogastropods (see below), the subradular organ can be protruded from the mouth and is apparently used to sense potential food, presumably using chemical stimuli. This structure has not been well studied – for example, the fine structure has only been described in detail in one chiton (Boyle 1975), where the epithelium is a mix of ciliated and microvillous cells and nerve connections to the subradular ganglia.

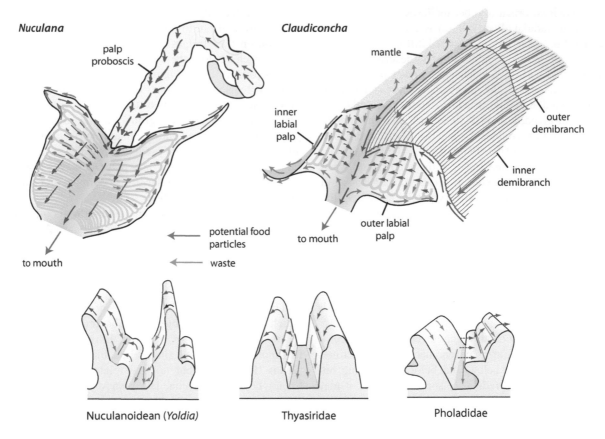

FIGURE 5.24 Labial palps of a protobranch (*Nuculana*) and an autobranch (*Claudiconcha*) bivalve, with three examples of the ciliation and ridge structure in three bivalve taxa. Illustrations redrawn and modified from the following sources: *Claudiconcha* Morton, B.S., *J. Zool.*, 184, 35–52, 1978b, *Nuculana* Atkins, D.G., *Q. J. Microsc. Sci.*, 79, 181–308, 1936, *Yoldia* Stasek, C.R., *Malacologia*, 2, 349–366, 1965, thyasirid Allen, J.A., *Philos. Trans. Royal Soc. B*, 241, 421–484, 1958, pholadid Purchon, R.D., *Proc. Zool. Soc. Lond.*, 124, 859–911, 1955.

There is a large subradular organ in monoplacophorans, but its function, in what is apparently an unselective deposit feeder, is unclear. In scaphopods, unlike the other groups that possess one, the subradular organ (like the radula) cannot be everted and thus is never in contact with the substratum. It is probably involved in the initial processing of food, being brought into contact with prey while it is being masticated by the radula and perhaps 'tasting' it before swallowing (Shimek & Steiner 1997).

The subradular organ is plesiomorphic in gastropods, and its simplification or loss probably occurred independently in several groups (Ponder & Lindberg 1997). This 'licker' is best-developed in patellogastropods, but is also found in a less developed state in some vetigastropods (including peltospirids) and ampullariids (Architaenioglossa, Caenogastropoda) (Haszprunar 1988a; Ponder & Lindberg 1997; Strong 2003). In patellogastropods, its surface is smooth or variously folded or ridged in different taxa. Its surface in these and other gastropods bears cuticular scales. It is thought to be sensory and is usually associated with the labial commissure. Strong (2003) described a non-sensory structure with many goblet cells that lies at the anterior end of the odontophore in many lower caenogastropods which she equated with the subradular organ, although there remains some doubt as to its homology.

The licker of patellogastropods (Figure 5.19) is applied to the substratum just before the radular teeth and moves along it immediately behind the radula collecting anything left on the substratum by pulling it into the mouth on retraction (Fretter & Graham 1994, p. 608). Thus, the licker, at least in patellogastropods, appears to have a mechanical feeding function, perhaps in addition to a chemosensory one. The labial commissure, sometimes with ganglia, innervates the subradular organ in most gastropods that possess one, but is absent in the vent-living peltospirids, ampullariids, and other caenogastropods (Strong 2003).

In cephalopods, there are paired lateral lobes below the buccal mass that are glandular in *Nautilus*, and this structure has been called a subradular organ (e.g., Salvini-Plawen 1988; Sasaki et al. 2010a), although its homology has not been tested. A possibly equivalent structure in coleoid cephalopods is covered with small chitinous teeth (Budelmann et al. 1997) and is separate from the median salivary papilla which can also have small teeth (see Section 5.3.2.1.6).

5.3.2.2.2 Paired Glands Associated with the Sublingual Pouch

Paired glands associated with the sublingual pouch are present in Caudofoveata and Solenogastres, Polyplacophora, Monoplacophora, and Scaphopoda (where they are rudimentary) which were thought to be homologous by Salvini-Plawen (1972, 1988). Similar paired structures are also seen in two groups of gastropods, the cocculinids (Haszprunar 1987a; Salvini-Plawen 1988), and neritimorphs (Bourne 1909; Whitaker 1951; Fretter 1965; Strong 2003). The homology of these structures in each of these two groups is doubtful (Strong 2003), and because patellogastropods lack these structures, they may be separate innovations within both these gastropod groups.

5.3.2.3 The Buccal Mass

The molluscan buccal mass[6] or apparatus is composed of multiple discrete elements that operate together during feeding. They include numerous muscles, often one or more pairs of underlying odontophoral cartilages, and the radula, which is drawn over the cartilages during feeding. There is no buccal apparatus in bivalves, and it has been lost in a few gastropods.

The plesiomorphic condition of the buccal mass in gastropods is an ovoid muscular mass, but the shape can change substantially depending on function. For example, in herbivorous 'opisthobranchs' and 'pulmonates', the buccal mass is more or less spherical. These taxa ingest relatively small pieces of food by way of the radula and jaw. In contrast, an enlarged, highly muscular, elongate buccal mass has been thought to be characteristic of carnivores in stylommatophorans so they can ingest large food items and hold struggling prey (e.g., Tillier 1989). While this is sometimes true, Barker and Efford (2002) pointed out that a small, spheroidal buccal mass is found in some carnivorous taxa. They note that the large, elongate buccal mass is only found in obligate carnivores which have long, dagger-like radular teeth, and several stylommatophoran families show transitional conditions.

In some taxa, the buccal area is modified as a buccal pump. This has occurred in some solenogasters and in several groups of gastropods where they have independently evolved, and sometimes a separate pumping chamber has budded off. Examples of gastropod taxa with highly modified buccal pumps include many Eulimidae, some nudibranchs (e.g., Forrest 1953) (see Figure 5.25), and Pyramidellidae (see Figure 5.26).

5.3.2.3.1 Muscles of the Buccal Mass

In the Caudofoveata, the buccal apparatus bearing the radula consists of a pair of supporting bolsters with both protractor and retractor muscle bundles. Most solenogasters have a weakly differentiated bolster with muscles attached, but some lack a radula, and in those, the muscular buccal cavity acts as a suctorial pump.

---

[6] The terms buccal mass, buccal bulb, pharyngeal bulb, and pharynx are used interchangeably in the literature.

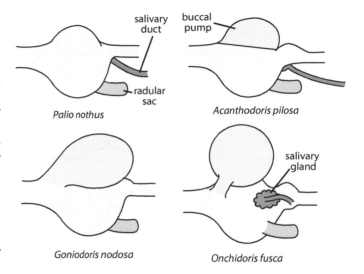

FIGURE 5.25 Buccal pumps in some dorid nudibranchs. Redrawn and modified from Forrest, J.E., *Proc. Linn. Soc. Lond.*, 164, 225–235, 1953.

Plate (1897) found that the complex buccal musculature of the chiton *Acanthopleura* consisted of 38 paired and six unpaired muscles, and the musculature of other chiton species has been subsequently described (Graham 1973; Wingstrand 1985). Graham (1973) described how these muscles work together, and noted that the odontophore is under greater control in chitons than in gastropods. The buccal muscles of chitons can be reduced to a few functional sets (Graham 1973; Wingstrand 1985) (see modified summary in Table 5.3).

Wingstrand (1985) carefully compared the buccal muscles of monoplacophorans with those of chitons and found a great deal of similarity, although with some simplification. He also noted, as did Graham (1973), the general similarity and homology of the buccal musculature of chitons and monoplacophorans with those of gastropods, although often individual muscles are modified. One such difference is that many radular muscles are attached to the shell in chitons and monoplacophorans, but not in gastropods, except for the median protractor of the subradular membrane in some gastropods where it connects to the shell or columellar muscle.

The complex musculature of the gastropod buccal mass has been studied in some detail. In the vetigastropod *Monodonta*, Nisbet (1973) recognised 33 muscles, but some simplification in buccal musculature is evident in higher gastropods – in the neogastropods *Nassarius* (Burton 1971) and *Urosalpinx* (Carriker 1943) have only 24 and 15, respectively.

Although details of these complex muscles differ between major groups of gastropods, they have many elements in common. The retractors of the subradular membrane provide the feeding stroke of the radula and are thus the most powerful muscles in the buccal system. The odontophore moves backwards and forwards within the buccal cavity and oral tube in coordination with the protraction and

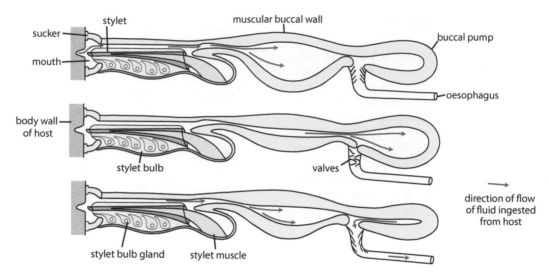

FIGURE 5.26 Diagrammatic longitudinal section of the buccal apparatus of *Odostomia plicata* (Pyramidellidae) showing the action of the buccal pump. Redrawn and modified from Maas, D., *Z. Morphol. Ökol. Tiere*, 54, 566–642, 1965.

retraction of the subradular membrane and hence the radula. During protraction of the odontophore, the radular teeth are pulled outwards over the bending plane and inwards over it on retraction. The odontophoral protractors and the membrane retractors are the most powerful muscles in the buccal mass complex.

Based on the published accounts noted above, some of the main muscles found in the buccal masses of chitons, monoplacophorans, and gastropods are listed below and diagrammatically illustrated in Figure 5.27.

The protractor muscles of the odontophore: The odontophoral protractors comprise a group of muscles that originate in the anterior body wall and oral lips and attach to the posterior end of the cartilages. The median protractor muscles run back through the nerve ring to join the columellar muscle in gastropods.

The retractor muscles of the odontophore: The odontophoral retractors are a pair of rather narrow muscles that extend from the posterior ends of the cartilages (the posterior odontophoral muscles) or radular sac (posterior radular retractor muscles) to the body wall behind the buccal mass.

Muscles associated with the radula and radular membrane: The radular tensor muscles play a role in the coordination of movements of the radular membrane and the radular sac. The ventral approximator muscle is responsible for the erection of the radular teeth as they move over the bending plane by separating the dorsal edges of the front ends of the cartilages when it contracts. The lateral protractor muscles protract the subradular membrane and

extend from the subradular membrane to the posterior ends of the cartilages.

The radular membrane is retracted by two sets of muscles: the subradular membrane retractor muscle(s) – these massive muscles run from the cartilage to the subradular membrane and the radular sac. They retract the membrane and bring the cartilages together and, in so doing, fold away the radular teeth.

Muscles associated with the buccal wall: The act of swallowing food requires movements of the buccal wall and the transverse fold which are achieved by the interaction of several muscles – the buccal dilators, buccal circular muscles, and a retractor of the transverse fold. Some additional muscles are present in polyplacophorans which perform minor functions, but many of these are lost in higher gastropods (Graham 1973).

5.3.2.3.1.1 Evolutionary Trends All the primary muscles are found in chitons and gastropods and simplification involves loss of only ancillary muscles (Graham 1973). According to Graham, the reasons for this include an increasing organisation of the head, functional changes in the radula, and simpler use of the buccal mass, notably with the evolution of a proboscis in higher Caenogastropoda.

In gastropods, the musculature associated with radular functioning differs depending on the group. In patellogastropods, the radula undertakes powerful scrapes in a direct line. Thus, the muscles responsible for this movement of the odontophore are strong, and ancillary muscles involved in adjusting the odontophore are absent. In addition, some of

TABLE 5.3
The Main Muscles in the Buccal Musculature of Chitons and Their Function

Function	Muscle(s)
Dilation of buccal cavity	*Buccal dilator muscles*: Muscle bands extending from the mouth and buccal regions to adjacent parts of the body wall.
Protraction of odontophore	*Lateral protractors of odontophore*: The outer and inner protractor muscles which originate at the posterior ends of the cartilages and insert in the anterior wall of the head.
	Ventral protractors of odontophore: A pair of strong muscles extends from the posterior end of the cartilages to the ventral side of the subradular pouch and the wall of the head and mouth. They retract the pouch and also protract the cartilages.
	A group of muscles that extend from the ventral surface of the cartilages to the buccal wall.
Retraction of the odontophore	*Odontophoral retractor muscles*: Attached to the posterior end of the bolsters to insert in the foot behind.
	Retractors of subradular pouch: A few slender muscles attached to the subradular pouch are fixed in the body wall.
Retracting, anchoring, or stabilising odontophore	An unpaired muscle extending from the midventral sublingual pouch (radular diverticulum) to the midventral body wall.
	A pair of *odontophoral retractor muscles* which run from the posterior ends of the cartilages to the body wall.
	A pair of slender muscles that run from the posterior side of the subradular pouch to the ventral anterior ends of the bolsters, binding them together.
	Stabilising muscles, including a pair running from the radular diverticula to the upper side of the radular sheath and a rather weak pair of muscles that run from the lower side of the oesophagus to the upper side of the radular sheath.
Retraction of the subradular membrane (and the radula)	A group of muscles which extend from the radular diverticulum to the ventral surface of the bolster on each side and move the radula back and forth, assisted by another which inserts in the body wall.
	The main pair of radular retractors (the *dorsal radular retractors*) is attached to the ventral radular sheath and runs to the second shell plate.
	The *posterior radular retractors* connect to the third shell plate.
	A pair of muscles run from the posterior end of the cartilages to the radular sheath.
Protraction of the subradular membrane (and the radula)	*Median protractors*: These run backwards from the anterior end of the subradular membrane at the anterior end of cartilage and insert on the third shell plate.
	Lateral protractors: Originate at the anterior end of the cartilage and run to join the lateral pedal muscles.
	Outer protractors: Inserted laterally on the sublingual pouch and run to the ventral muscles.
	A muscle that moves the radula to the mouth is attached to the anterior end of each bolster and inserted in the sides of the mouth.
Spreading of radular teeth	Assisted by the *ventral approximator muscle* which connects the two main cartilages ventrally. Its contraction spreads apart the dorsal edges of the cartilages.
	A short, strong muscle runs from the posterior end of each cartilage to the adjacent body wall and on contraction pulls the posterior ends of the cartilages apart, which helps to bring their anterior ends together.

Modified form Graham, A.L., *J. Zool.*, 169, 317–348, 1973, and Wingstrand, K.G., *Galathea Report*, 16, 1–94, 1985.

the muscle hypertrophy results from multiple cartilages being present, requiring that they are locked together during radular use to achieve the necessary rigidity and stability needed for radular rasping.

In contrast, in vetigastropods, the rhipidoglossan radula brushes rather than scrapes the feeding surface. While there is no need for powerful strokes, many minor adjustments are required. Thus, ancillary muscles for adjusting the odontophore are present.

In caenogastropods, which have taenioglossan or stenoglossan radulae, fine control is unnecessary, but greater scraping power is typically required than in vetigastropods. In neogastropods, the levator and depressor odontophoral muscles are lost (Graham 1973). Instead, the muscles involved in odontophoral movements are strengthened, and the ancillary muscles for adjusting the odontophore are absent. Those typically present in lower caenogastropods are the buccal dilators and constrictors, protractors and retractors of the odontophore, protractors and retractors of the subradular membrane, and a ventral approximator of the cartilages (Graham 1973). A reduction in the number of cartilages in caenogastropods (see next Section) also strengthens the odontophore.

The radula of chitons behaves more like that of a gastropod taenioglossan radula (Graham 1973; Guralnick & Smith 1999), although it is more similar in morphology to the patellogastropod docoglossan radula. The chiton buccal mass has more muscles than found in any gastropod.

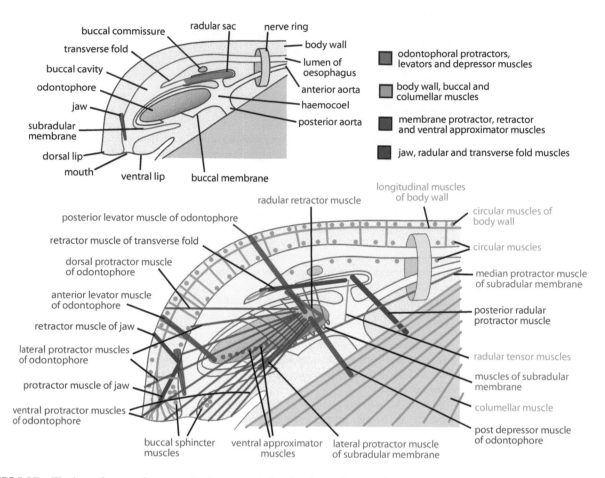

FIGURE 5.27 The buccal mass of a generalised gastropod showing the main musculature. Redrawn and modified from Graham, A.L., *J. Zool.*, 169, 317–348, 1973.

Graham (1973) argued this is not due to a need for accurate control of radular and odontophoral movements, but that the larger number of muscles may be a primitive feature, with gastropods showing greater simplicity reflecting improved efficiency, or that the larger number of ancillary muscles in chitons may be related to the poorly developed head region which is much less mobile and flexible than in gastropods (the ligocephalic condition, see Chapter 3). In contrast, in gastropods, the head is flexible and muscular (the apocephalic condition) and is involved in part in the manipulation of the odontophore. Similarly, monoplacophorans also have a poorly developed head and, like chitons, have several extra ancillary muscles. The buccal musculature of monoplacophorans closely resembles that of chitons (Lemche & Wingstrand 1959; Graham 1973; Wingstrand 1985), including the possession of 'ancillary muscles' that assist in steadying parts of the 'feeding apparatus' (Graham 1973).

Of the remaining 'conchiferans', scaphopods, despite having a relatively very large radula, have much simplified odontophoral musculature with only two lateral pairs of muscles attached to the bolster (see next Section) with a strong horizontal muscle anterodorsally between the bolsters.

The buccal mass of cephalopods is highly modified, being a large ovoid bulb surrounded by a sheath of connective tissue, with much of the musculature involved in operating the beaks. This musculature forms the walls of the buccal mass. The buccal bulb is surrounded by a blood sinus that allows it to rotate freely while biting, and decabrachian coleoids have muscles surrounding the sinus allowing the buccal mass to be protruded (Bidder 1966). Detailed descriptions of the buccal muscles of cephalopods are provided by J. Z. Young (1994), Tanabe and Fukuda (1999), and Uyeno and Kier (2005).

5.3.2.3.2 *Odontophoral Cartilages, Bolsters and Vesicles*

Both groups of aplacophorans lack true odontophoral cartilages, but have paired bolsters composed of either interwoven muscle fibres and possibly collagen or, in a few solenogasters, large clear cells (probably chondroid tissue) with attached muscle fibres (Scheltema 1981; Scheltema et al. 1994).

There are two pairs of odontophoral cartilages in polyplacophorans and in larger monoplacophorans and only one pair in the minute *Micropilina*. In chitons and monoplacophorans, the cartilages are associated with a pair of large, fluid-filled radular vesicles (Wingstrand 1985; Haszprunar & Schaefer 1997) which, although closely associated with the cartilages, are not homologous. In monoplacophorans, the vesicles are bordered anteriorly (and posteriorly in *Laevipilina antarctica*) by cartilage tissue (Haszprunar & Schaefer 1997).

Scaphopods have a single pair of odontophoral cartilages connected by a dense rod of connective tissue (Shimek & Steiner 1997). Cephalopods lack cartilages, but have at least analogous 'bolster rods' that contain a 'semi-fluid, gelatinous substance' (Messenger & Young 1999).

Gastropod cartilages have been investigated by Katsuno and Sasaki (2008) and Golding et al. (2009a). All patellogastropods have two or more pairs of cartilages, with up to five pairs in Patellidae, and this condition is possibly the result of secondary subdivision (Salvini-Plawen 1988). Some authors (Graham 1964; Wingstrand 1985) have suggested that the anterior cartilage of patellids includes the homologue of the radular vesicle seen in chitons and monoplacophorans; this structure is otherwise unknown in gastropods.

Thus, two pairs of odontophoral cartilages and a pair of radular vesicles can be regarded as plesiomorphic in molluscs, with multiplication, reduction, or loss having occurred independently in several groups. Within non-heterobranch gastropods, there are five to two pairs of cartilages, or the two are fused into a single bilobed piece (e.g., as in *Janthina*). The 'lower' vetigastropods have three pairs of cartilages, the third pair probably being formed by the subdivision of one of the original pairs. 'Higher' vetigastropods have two pairs. Neritimorphs have two pairs and a small pair of median cartilages that are sometimes fused. Cocculinids, the neomphaline vetigastropods, and nearly all caenogastropods (the buccal apparatus is lost in a few taxa) have a single pair of cartilages (Figure 5.28).

The plesiomorphic condition for gastropods is debatable, with multiple pairs in patellogastropods and lower vetigastropods. Sasaki (1998) argued that a single pair was basal with multiplication having occurred in some groups, but we

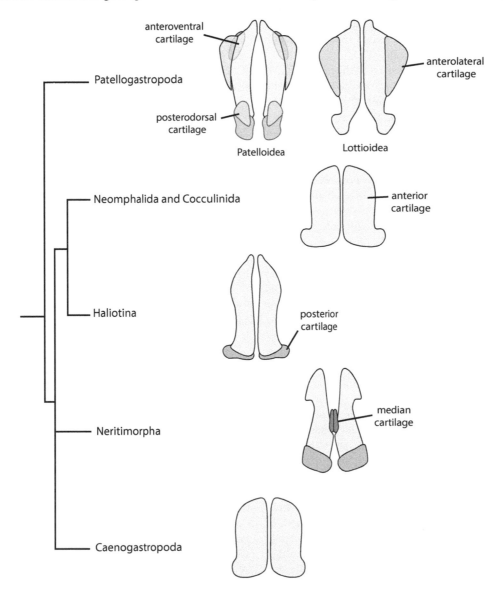

FIGURE 5.28 Gastropod odontophoral cartilages. Redrawn and modified from Sasaki, T., *Bull. Univ. Mus.*, 38, 1–223, 1998.

consider this unlikely given two pairs in polyplacophorans and monoplacophorans, suggesting this may be the plesiomorphic state in gastropods.

The odontophoral cartilages consist of strongly vacuolated turgor cells, and each is ensheathed in a layer of collagenous tissue (Katsuno & Sasaki 2008). Processes on the cartilages provide anchorage points for the buccal muscles. In contrast, the odontophore of 'pulmonates' consists of two pieces of modified cartilage connected by muscles at the posterior end. In many 'pulmonates', turgor cells are intermingled with muscles, but in Stylommatophora muscles replace the turgor cells. Due to its muscular nature, the 'pulmonate' odontophore can change shape during feeding. While a pair of odontophoral cartilages are commonly reported in 'pulmonates', including a single one reported in *Chilina* (Brace 1983), it is sometimes assumed that these 'cartilages' are not homologous to those in other gastropods (e.g., Haszprunar 1985f, 1988a; Ponder & Lindberg 1997; Katsuno & Sasaki 2008). This latter view is supported by the absence of buccal cartilages of any kind in basal heterobranchs and 'opisthobranchs', as well as in their structural differences (Katsuno & Sasaki 2008), although these suppositions need further testing.

In development, the cartilages arise from the mesoderm, in contrast to the rest of the foregut which is ectodermal. In the limpet *Patella*, there is only a single pair of cartilages in the metamorphosing larva (Smith 1935). Similarly, in *Haliotis*, the two dorsal cartilages are formed during the late veliger stage as derivatives of the mesoderm below the buccal cavity, and even in the two month old post-larva, the extra cartilages have not developed (Crofts 1937). It is thus possible that cartilage reduction (and their assumed loss in heterobranchs) may have resulted from heterochrony.

The cartilages that were plesiomorphically integrated into the vesicles in monoplacophorans and chitons have become separated in patellogastropods where the dorso-lateral and medial cartilages remain, but the connective tissue linking the two is lost. In patellogastropods, only the anterior cartilages (i.e., the medial cartilages of Guralnick & Smith 1999) support the radula at the bending plane, as in orthogastropods. Unlike the situation in orthogastropods, the subradular membrane does not become associated with the anterior cartilages posterior to the bending plane; instead, it rests on the dorsal cartilages or musculature. The dorsal cartilages and dorsal muscles (and hence radula) appear to move anteriorly and even further dorsally relative to the anterior cartilages during the feeding stroke, causing even greater dissociation from the anterior cartilages and increasing the distance the radula moves over the bending plane (Guralnick & Smith 1999).

A single pair of cartilages is present in most caenogastropods, except for the cases mentioned below and for a few in which the odontophore is lost. These cartilages are considered homologous with the anterior cartilages (medial cartilages of Guralnick & Smith 1999) of lower gastropods (Wingstrand 1985; Sasaki 1998; Golding et al. 2009a). They often overlap and can vary considerably in length, with long, narrow cartilages being found in some neogastropods – their shape perhaps related to being contained within a narrow proboscis

(Graham 1973), although many proboscidate taxa have short cartilages (Golding et al. 2009a). Basal caenogastropods ('Architaenioglossa') have short cartilages relative to those seen in vetigastropods and neritimorphs (Golding et al. 2009a). Dorsal processes are present on the lateral margins of the cartilages in at least some basal caenogastropods, and these serve as additional sites for muscle attachments (Golding et al. 2009a).

While it has been generally accepted that caenogastropods have only one pair of cartilages, small 'subradular cartilages' (Figure 5.30) have been found in many taxa (Golding et al. 2009a), although these were absent in all the carnivorous taxa examined in that study. These subradular cartilages are composed of tissue like that of the main cartilages and are located below the subradular membrane. They may be homologous with the dorsal cartilages in patellogastropods which are located anteriorly in the buccal mass (Guralnick & Smith 1999), not above and posterior to the anterior cartilage to which they are often connected. Golding et al. (2009a) suggest that the main function of both the dorsal and subradular cartilages is to support the lateral surfaces of the radula during feeding.

An additional pair of thin, flat 'accessory' cartilages was recognised in the buccal mass of *Carinaria cristata* (Carinariidae) by Golding et al. (2009a), where they lie dorsal to both the radula and the food groove.

5.3.2.3.3 Radular Diverticulum (Caecum)

A radular caecum (or radular diverticulum) is found in polyplacophorans (Graham 1973), monoplacophorans (Lemche & Wingstrand 1959; Haszprunar & Schaefer 1997), scaphopods, *Nautilus* (Salvini-Plawen 1988), and some 'lower' gastropods. In lower gastropods, it is a blind, often anteriorly paired, pouch of the anterodorsal radular sheath that separates the radular sheath from the oesophagus (Salvini-Plawen 1988).

Unique to the vetigastropod family Fissurellidae, a ventral radular sac or diverticulum extends ventrally between the two halves of the buccal mass (Fretter & Graham 1962, fig. 96).

5.3.2.4 Radula and Radular Sac

The radula is one of the most diagnostic molluscan characters. It lies on the subradular membrane, a flat strip of chitinous material that lies dorsally on the odontophore and is secreted in the radular sac. The radular apparatus is made up of the numerous rows of radular teeth, the supporting subradular and supraradular epithelium, often called the radular sheath, which move with the radular teeth from the radular sac to the functional part of the radula in the buccal mass. The radular sac (not to be confused with the sheath), is a blind pouch composed of specialised cells. The radula is secreted in the posterior end of the sheath from membranoblast (which continually secrete the radular membrane) and odontoblast cells, the latter responsible for intermittently secreting the radular teeth.

The radula consists of (usually many) horizontal rows of backwards pointing radular teeth, and in their basic form consist of a *basal plate* and one or more sharp projections (*cusps*). A complex set of muscles (see Section 5.3.2.3.1) works the radula and the associated supporting cartilages (if present) (see Section 5.3.2.3.2). The radula is protruded, and the radular membrane is bent over the anterior edge of the underlying supporting structure resulting in the teeth being elevated so they can function by scraping, rasping, sweeping, or tearing the food that is then passed back into the oesophagus. The working teeth get worn or broken and are continuously replaced from behind, new ones being secreted and hardened in the radular sac, as described below. Worn teeth are dislodged and may pass through the gut with the food.

5.3.2.4.1 The Radular Sac

The radula is formed in the tubular radular sac (Figure 5.29), lying initially dorsal to the odontophore. The radular sac may extend behind the odontophore into the cephalic haemocoel and may be short and straight or curved, or looped or coiled, with different taxa showing different configurations. Thus, the radular sac may curve dorsally in some taxa, or ventrally in others. In some taxa, the radular sac is located anterodorsal, over the radular cartilages, while in others it is located mid-posterior in the visceral mass. It is U-shaped in section and lined by a ventral and dorsal epithelium. Chondroid tissue (the *collostyle*) fills the space between the U-shaped layer of dorsal epithelium (e.g., Mischor & Märkel 1984), making the sac look rounded or oval in section. The collostyle is particularly obvious in stylommatophorans, and a cuticular 'collostylar hood' closes the anterior end of the radular sac (see Figure 5.30) (Mackenstedt & Märkel 1987, 2001).

The blind posterior end of the radular sac ranges from simple to distinctly bifid, the latter being the usual condition in vetigastropods.

In all gastropods except euthyneurans, the odontoblasts are grouped in cushions comprising several hundred cells – each giving rise to a longitudinal row of teeth. The mechanism of radular formation is most readily understood in 'pulmonates', where the odontoblasts form a subterminal girdle of large cells arranged in small groups at the end of the radular sac, and an additional cell type secretes the basal plates of the teeth (see Luchtel et al. 1997 for review).

Hardening of the initially fibrous teeth (proteins and mucopolysaccharides, mainly chitin) occurs when the newly-formed teeth are in contact with the supraepithelium (upper epithelium) of the radular sac, which enables tanning (incorporation of organic compounds) and incorporation of inorganic salts, including mineralisation in some (see Section 5.3.2.4.3).

5.3.2.4.1.1 Radular Teeth and Tooth Fields

In traditional descriptions of radulae, a tooth at the centre of each row of radular teeth is called the central (or rachidian) tooth. Teeth on either side of this are called the lateral teeth, and marginal teeth lie outside the laterals, being demarcated by a distinct change in tooth morphology and, often, by a change in position and/or orientation on the ribbon.

The *docoglossan* type of radula is seen in polyplacophorans, scaphopods, and patellogastropods. The few plate-like marginal teeth are reduced from six in polyplacophorans to three in monoplacophorans, to even fewer (0–3) in patellogastropods, and they are the least robust teeth in these radulae. In vetigastropods and neritimorphs, the marginal teeth are elongate, delicate, numerous, and serrated, functioning as brush-like structures. In contrast, many caenogastropods have only seven teeth in each row (the taenioglossate condition), while the caenogastropod neogastropods have between five and one teeth in each row (the rachiglossan condition), or the radula is lost (see Section 5.3.2.4.2.6 and Chapter 19). Heterobranch gastropods exhibit wide variation in their radulae (see Section 5.3.2.4.2.6 and Chapter 20).

The fusion of two odontoblast precursors during development (Raven 1966) gives rise to the bilateral symmetry of the molluscan radula. The ontogeny of the lateral teeth in gastropods indicates that they belong to two distinct ontogenetic tooth fields, the inner lateral and outer lateral. The central (= rachidian) teeth results from the fusion of the innermost teeth within the inner lateral tooth field (Ivanov 1990a; Ponder & Lindberg 1997) (Figure 5.31). In some gastropods, such as in some species of the vetigastropod *Tricolia* (Robertson 1985b) and the panpulmonates *Physa* (Eigenbrodt 1941) and *Amphibola* (Farnie 1919; Golding et al. 2007), the relatively large central teeth appear to be the result of further fusion involving the central and two innermost lateral teeth. Several additional inner lateral teeth may occur on either side of the central tooth, or, as in most caenogastropods, the lateral

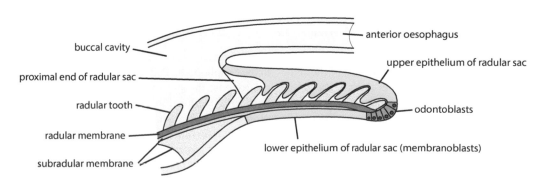

FIGURE 5.29 Diagram of a longitudinal section of a generalised gastropod radular sac showing the main structures and associated parts of the anterior gut. Redrawn and modified from Fretter, V. and Graham, A.L., *British Prosobranch Molluscs: Their Functional Anatomy and Ecology*, Ray Society, London, UK, 1962.

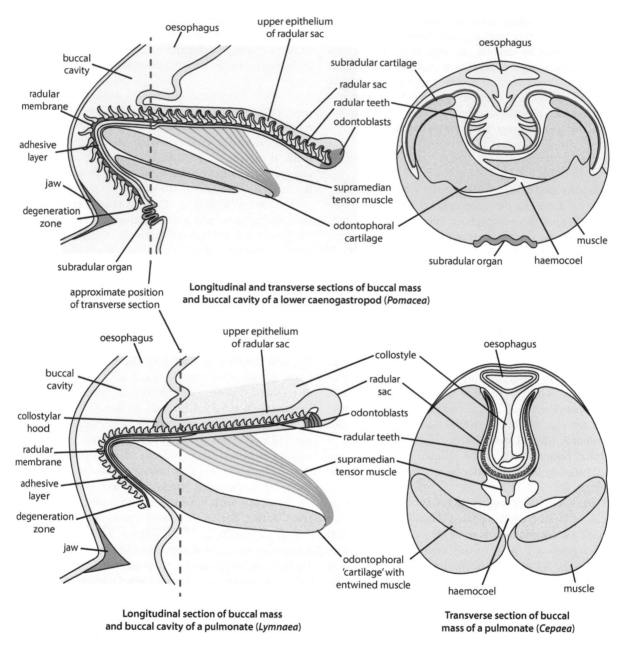

FIGURE 5.30 Comparisons of the buccal region of a caenogastropod (*Pomacea*) and the 'pulmonates' *Lymnaea* and *Cepaea*. Redrawn and modified from the following sources: longitudinal sections Mackenstedt, U. and Märkel, K., *Zoomorph.*, 107, 209–239, 1987, TS of *Pomacea* Mischor, B. and Märkel, K., *Zoomorph.*, 104, 42–66, 1984 and *Cepaea* Mackenstedt, U. and Märkel, K., Radular structure and function, pp. 213–236, in Barker, G.M., *The Biology of Terrestrial Molluscs*, CAB Publishing, Wallingford, UK, 2001.

tooth field is reduced to a single tooth on either side of the central tooth. In gastropods with multiple lateral teeth, the outer lateral teeth are immediately adjacent (and typically posterior to) the inner lateral teeth. In patellogastropods, neritimorphs and some vetigastropods, the basal attachment plates and teeth are often fused in the outer lateral tooth field forming a large, multicuspid tooth sometimes called the pluricuspid. The marginal teeth lie peripheral to the outer lateral teeth and are typically demarcated by a change in tooth morphology, a gap in the radular membrane, or different basal plate structure. In contrast to the highly conserved radular configuration

in the Scaphopoda, Polyplacophora, and taenioglossate caenogastropods, the presence or absence of tooth fields and the number of teeth per field is highly variable in some gastropod groups and recognising tooth homologies is more complicated than it may first appear.

In higher gastropods, the rudiment of the radular sac arises around the first phase of torsion, but appears earlier in patellogastropod and vetigastropod larvae. In proboscis-bearing caenogastropods, the radula, although forming beneath the oesophagus, is not connected with the buccal area until metamorphosis (e.g., Page 2000). The radular teeth are typically

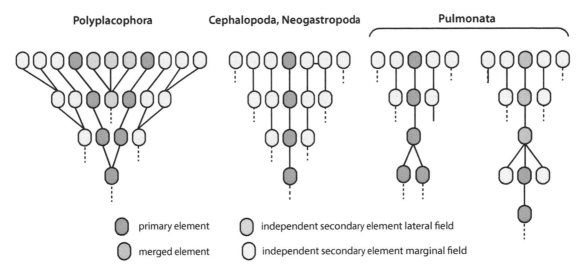

FIGURE 5.31 Some of the different types of radular morphogenesis, with development from bottom to top of page. Redrawn and modified from Ivanov, D.L., *Evolutionary Morphology of Mollusca. Patterns of Morpho-functional Changes to the Radular Apparatus*, Moscow, Russia, 1990a.

not formed until shortly before hatching. They are secreted by transverse bands of odontocytes proliferated from odontoblasts in the distal end of the radular sac. The bands of odontocytes are separated by transverse bands of cells which produce the radular membrane. Many gastropods and polyplacophorans initially form the outer lateral teeth, but in some gastropods, the central tooth (i.e., the inner lateral field) appears first. Each part of the odontoblast undergoes progressive differentiation, producing increasing numbers of teeth laterally until the adult arrangement is achieved.

The cells (or odontoblasts) making up these tooth fields and generating the radular teeth are small and numerous in Polyplacophora, Cephalopoda, and 'prosobranchs', but are large and few in number in euthyneurans (i.e., 'opisthobranchs' and 'pulmonates'), although too little information is available about lower heterobranchs to draw conclusions. In 'pulmonates', the shape of the radular teeth is determined by the profile of the odontoblasts rather than the odontoblastic cushions (Mischor & Märkel 1984).

Radulae are sometimes called *monostichous* (a single tooth per row), *distichous* (two teeth per row), or *polystichous* (several teeth per row), but which arrangement is plesiomorphic is unknown. An ancestral distichous condition has been suggested by Boettger (1956), Minichev and Sirenko (1974), Ivanov and Tsetlin (1981), Scheltema (1981), and Eernisse and Kerth (1988), but Salvini-Plawen (1972, 1985a) argued for an original monostichous radula. The original 'conchiferan' radula was polystichous according to Minichev and Sirenko (1974) or distichous (Ivanov & Tsetlin 1981).

The radulae of patellogastropods, polyplacophorans, and monoplacophorans are similar in having relatively few, generally plate-like marginal teeth and large, typically heavily mineralised outer lateral teeth.[7] This general similarity has resulted in these radulae being called docoglossate. In marked contrast, the rhipidoglossate radulae of neritimorphs and vetigastropods

have less well mineralised and relatively smaller outer laterals, and usually numerous, often long, finely serrate marginals.

Polyplacophorans and monoplacophorans, in each tooth row, have a single central tooth flanked by a single cusped inner lateral teeth on either side (see Chapter 14). The outer lateral teeth are multicuspid in polyplacophorans, each with either two or three cusps, while the outer lateral teeth of monoplacophorans each have two outer lateral cusps. Polyplacophorans have three pairs of marginal teeth, each unicuspid, and monoplacophorans have two pairs of marginal teeth. Polyplacophorans and monoplacophorans do not have tooth bases that overlap within a row, but both do have tooth bases in one row adjacent to cusps in the row behind; both groups lack outer lateral tooth bases independent of the membrane. Ventral tooth plates and lateral tooth plates are found only in patellogastropods (Moskalev 1964).

In polyplacophorans and monoplacophorans, the inner and outer lateral teeth are mineralised, and central teeth are present. McLean (1990) noted that, in polyplacophoran and monoplacophoran radulae, the teeth articulate within a row, but this articulation is not homologous in the two groups. Thus, although McLean (1990) argued that the medial-lateral process found on the first inner lateral of neolepetopsids is similar to the articulation seen in chitons, the articulation in that group is formed from a medial extension of the base along its whole length, while in neolepetopsids, it is from a strongly medially directed process off a mostly laterally directed base. Although most neritimorphs and vetigastropods have three or more inner lateral tooth cusps per row, neritids have only one outer lateral cusp, while species of *Fissurella* have four cusps on the outer lateral tooth. Both taxa have central teeth and numerous marginal teeth. The bases of the outer lateral teeth in both taxa

---

[7] Exceptions are lepetids, neolepetopsids, and monoplacophorans which lack mineralisation of the outer lateral teeth.

are attached to the membrane and do not overlap the bases of the inner lateral teeth. In an unusual condition in *Fissurella,* the outer lateral teeth bases in one radular row extend into the row behind where they overlap and fuse with the bases of the outer laterals in that row. Medial-lateral processes on the inner laterals are absent in *Nerita* and *Fissurella,* although processes involved in tooth articulation occur on the outer laterals. Ventral plates are absent in both taxa (Guralnick & Smith 1999).

5.3.2.4.2 The Radula – Function and Value as a Taxonomic Tool

Radular characters are important in traditional gastropod systematics where they have been used at all levels. Early studies focussing on the radula included those by Troschel (1856–1879) and Thiele (1891–1893), and these were refined by Thiele (1929–1935). Radular-based names such as Docoglossa, Stenoglossa, Rachiglossa, Taenioglossa, Toxoglossa, and Ptenoglossa became entrenched in the literature. A more recent review of molluscan radulae was provided by Ivanov (1990a). The prominence of radular characters in gastropod systematics was premised on the lack of variability in the radula. With molecular techniques new characters became available to test this assumption and have revealed evidence for plasticity in radular dentition (Padilla 1998, 2001; Reid & Mak 1999; Simison & Lindberg 1999).

Early workers could only use light microscopy and work with largely two-dimensional images but, with the availability of scanning electron microscopy from the late 1960s, radular images became three dimensional and provided much more information (Figure 5.32). Important taxonomic characters

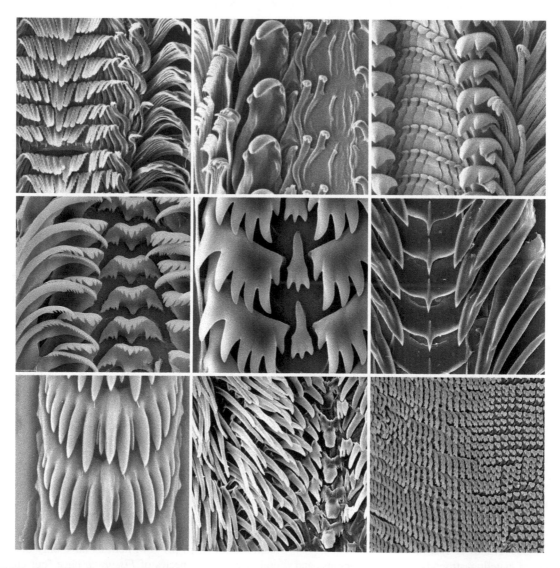

FIGURE 5.32 Scanning electron micrographs of examples of orthogastropod radular teeth. Top left *Clyposectus* (Vetigastropoda, Clyposectidae) (Courtesy of C. Hickman). Top centre *Cocculina* (Vetigastropoda, Cocculinidae) (Courtesy of C. Hickman). Top right *Hemitoma* (Vetigastropoda, Fissurellidae) (Courtesy of C. Hickman). Middle left *Gabbia* (Caenogastropoda, Bithyniidae) (Courtesy of J. Studdert). Middle centre *Granulifusus* (Caenogastropoda, Fasciolariidae) (Courtesy of Y. Kantor). Middle right *Sibogasyrinx* (Caenogastropoda, Cochlespiridae). (Courtesy of Y. Kantor). Bottom left *Cuthona* (Heterobranchia, Cuthonidae) (Courtesy of J. Avern). Bottom middle *Amphibola* (Heterobranchia, Amphibolidae) (Courtesy of R. Golding). Bottom right *Glyptophysa* (Heterobranchia, Planorbidae) (Courtesy of J. Studdert).

include the number of lateral and marginal teeth, the shape and number of cusps on each tooth, and the morphology of the tooth, including the base. The main characteristics of the radula in the major groups of molluscs are outlined in more detail below, and illustrations of a range of examples of the radulae in each major group are provided in the relevant taxon chapters in Volume 2.

The functioning of the molluscan radula has been discussed by numerous authors from various points of view. In radulae with markedly different teeth across a row (e.g., in the rhipidoglossan radulae), different teeth may function in very different ways. The lateral and central teeth, for example, cut and scrape while the marginal teeth sweep food into the middle part where it can be passed into the gut (Figure 5.33). The number, size and shape of radular teeth vary considerably, especially in gastropods, and sometimes these different morphologies can be readily linked to diet (e.g., see Chapter 20), while in others they are not so obviously related (e.g., see Chapter 19). The radular teeth may also have other structures such as interlocking sockets or abutting shoulders that assist in supporting the teeth during the feeding stroke with those behind taking some of the strain (e.g., Solem 1974; Hickman 1980). Additional aspects of radular function are described below (see Section 5.3.2.4.2.7).

In life, the radula is orientated so the cusps of the teeth point posteriorly, not anteriorly – i.e., they are upside down, contrary to the way they are usually depicted in illustrations.

In radulae with multiple teeth in each row, the configuration of the rows is significant, with the basic patterns depicted in Figure 5.34. In the plesiomorphic condition, the lateral and marginal teeth form an inverted-V with the marginal teeth lying posterior to the lateral and central teeth. We term this condition *alveate* – in this state the inner teeth (central and lateral) contact the substratum first. Examples include chitons, monoplacophorans, some patellogastropods, most vetigastropods and caenogastropods, acteonids, Aplysiidae, Tylodinidae, Onchidiidae, Amphibolidae, and Latiidae. In some taxa, the reverse configuration occurs, with the central and lateral teeth lying anterior to the central teeth. This was called *sagittate* by Marshall (1984) for the condition in the newtoniellid Adelacerithiinae (Triphoroidea) in which the outer teeth make contact first. Examples of this condition are seen in various gastropods including gastropterids, trimusculids, physids, achatinellids, some bothriembryontids, and in some carnivorous taxa such as pleurobranchs, testacellids, and rhytidids. Sometimes, particularly in some stylommatophorans, the radular row is straight, or approximately so. We term this condition *lineate*.

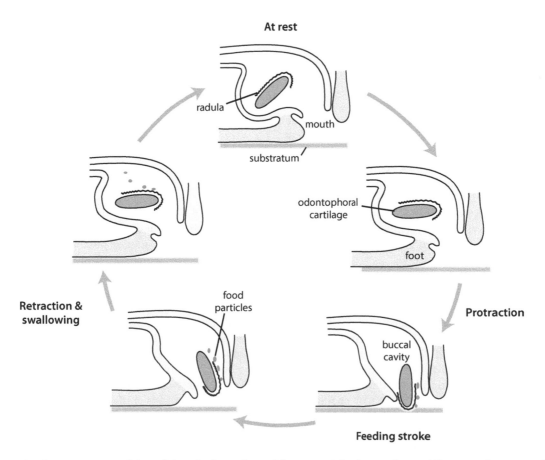

FIGURE 5.33 Feeding movements of the radula and odontophore of *Lymnaea*, a fresh-water hygrophilan panpulmonate – the food particles are shown in green. Redrawn and modified from Kater, S.B., *Integr. Comp. Biol.*, 14, 1017–1036, 1974.

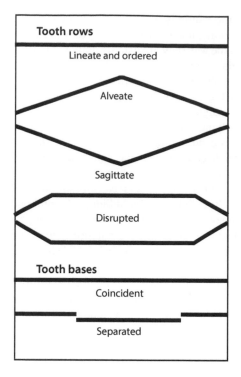

FIGURE 5.34 Basic patterns of radular tooth rows (primarily based on gastropods). See text for explanation. Original.

In most radulae, the row of teeth in each half row forms a line that runs in the same direction – we term this *ordered*. In some radulae, the radular row may undergo a change in direction so the outer teeth are on a different trajectory from the inner ones in each half row, a condition we term *disrupted*, as in some patelloideans, helicids, and planorbids.

Some vetigastropods have the inner teeth in a sagittate configuration, and the outer teeth are alveate – as in most Phasianellidae and Addisoniidae.

In a few rhipidoglossate taxa, the bases of the marginal teeth do not lie in the same row as the lateral teeth, a state described by Hickman (1984) in Fissurellidae and in Cocculinidae. The normal condition is here named *coincident*, while the state in fissurellids and cocculinids is called *separated*.

Guralnick and Lindberg (1999) examined the ontogeny of row orientation in patellogastropods and concluded that similar configurations in different taxa were built in different ways. In *Patella* and *Scutellastra*, V-shaped (alveate) and stepped (disrupted) configurations resulted from changes in the initial secretion parameters (the shape, size, position, and secretion rate of odontoblasts), while in *Acmaea* and *Lottia* differences between alveate and disrupted rows were driven by post-secretional shape changes, but before maturation.

Ontogenetic changes occur during the development of the radula. In groups with simple radulae, these may be relatively minor, but in others, such as those of many euthyneurans, they may be profound and usually involve the size of the teeth, an increase in the number of tooth rows, and, in many cases, increases in the number of teeth per row. How these changes

relate to the odontoblasts is not entirely clear as few studies have been carried out, but they do involve increases in the number of odontoblast cells. Such modifications, based on relatively few studies, mainly in stylommatophorans, were reviewed by Ivanov (1990a) who recognised five main patterns of radular morphogenesis. These were:

- *A gradual change in the shape of the tooth*. For example, in many stylommatophoran embryos, an initial row of non-specific protuberances transforms into tooth plates with several cusps which gradually grow into teeth with fewer cusps (e.g., Sterki 1893), perhaps reflecting a dietary shift.
- *Fusion of teeth*. The central tooth of some stylommatophorans is initially a pair of tubercles, which fuse to form a single tooth (e.g., Kerth 1979).
- *Tooth reduction or loss*. The juvenile teeth become rudimentary or are lost, as for example, the central tooth in juvenile patellids, which is lost or rudimentary in adults (e.g., Smith 1935).
- *Fragmentation*. In this type of development, a single tooth plate is divided into two or more, as in polyplacophorans (e.g., Minichev & Sirenko 1974) (Figure 5.31).
- *Change in the location of the teeth*. In the earliest radula of *Ancylus fluviatilis*, there are two longitudinal rows of hook-like teeth, but these later change into a typical 'pulmonate' radula (Kerth 1983).

Warén (1990) described ontogenetic changes in the radulae of several trochoidean taxa that included an increase in the number of marginal and lateral teeth and changes in the shape and ornamentation of the teeth. An interesting finding was that many of the juvenile radulae he studied looked very similar, with the changes that characterise each group occurring later in ontogeny, although some small-sized taxa retain the juvenile configuration.

5.3.2.4.2.1 Aplacophorans In the aplacophoran groups, the radula is simple, but diverse in structure. Most have a radular membrane (as in other molluscs), which may be divided longitudinally. This membrane is absent in the caudofoveate chaetodermatids where the number of radular teeth is reduced to two forceps-like teeth (Scheltema 1981). A subradular membrane is usually absent.

In most caudofoveates, the distal teeth emerge from the radular sac into the buccal cavity, while in solenogasters, the radula is bent into a ventral pocket, so the teeth exposed in the buccal cavity are in the middle section. In most caudofoveates, the radula consists of two rows of teeth (distichous condition), but in the Chaetodermatidae, there are just two teeth that function as forceps, moving food through the foregut and, possibly, playing a role in food selection (Scheltema 1981; Salvini-Plawen 1988). In solenogasters, the radula may be monostichous, distichous, or polystichous (Ivanov 1990b). In all cases, only the lateral field is present.

The radula of aplacophorans does not function as a rasp except possibly in the Prochaetodermatidae (Caudofoveata), as the teeth show some wear (Scheltema 1981). The radula of at least one solenogaster (*Epimenia*) has been observed hooking onto the soft coral prey while the muscular buccal cavity sucks in polyps (Scheltema & Jebb 1994). The radula has been lost in about 20% of solenogasters, which all have the buccal mass modified as a buccal pump (Scheltema et al.1994).

5.3.2.4.2.2 Polyplacophora The radula in chitons is of the docoglossate type (see below) and is regarded as primitive in molluscs (with the possible exception of aplacophoran radulae), as this type of radula is also found in monoplacophorans and patellogastropods. In all these taxa, the rows are markedly alveate (some patellogastropods excepted). Subsequent multiplication and loss of teeth fields, and variation in the number of teeth and tooth fusion within these fields, has led to the evolution of the other types of radulae.

A pair of the major outer lateral teeth in each row of the chiton radula is highly mineralised (see Section 5.3.2.4.3), and their radulae are fairly uniform throughout the group. They usually have 17 (rarely 13 or 15) teeth in each row which consists of both inner and outer lateral and marginal fields. The radular sac is long and straight with the pair of large mineralised lateral teeth in each row alternating so they are closed up like a zip. The radula usually functions as a combination of backwards and forwards scrapes which pull the opposed teeth together (Eernisse & Reynolds 1994). These actions scrape and tear off potential food items that are moved to the oesophagus by the backwards movement of the radula.

The ontogenetic sequence of radular tooth multiplication in some chitons starts from an initially single bilobed tooth in the larva (Minichev & Sirenko 1974) (Figure 5.35), which becomes the large, outer lateral teeth, followed by the plate-like outermost marginal tooth, and subsequent appearance of the inner marginal and inner lateral and central teeth (Minichev & Sirenko 1974; Sirenko & Minichev 1975; Eernisse & Kerth 1988). Based on these findings, Eernisse and Kerth (1988) argued that this is not an initially monostichous radula, but, because it is bilobed and quickly divides in two, it is distichous. In some other chitons, the initial teeth are the main functional teeth – the second lateral, and the fifth and eighth pairs of teeth (marginals) (Eernisse & Kerth 1988), and is thus a polystichous condition.

The initial appearance of the large outer lateral teeth was likened to the usual single or double teeth seen in the aplacophoran groups by Salvini-Plawen (1988), who argued this was evidence for the aplacophoran radula being primitive, although it could also suggest that the aplacophoran radular morphologies are heterochronic and apomorphic.

5.3.2.4.2.3 Monoplacophora The monoplacophoran radula is docoglossate, being generally similar to that seen

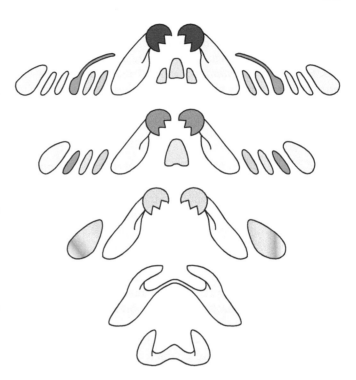

FIGURE 5.35 Development of a chiton radula. Redrawn and modified from Minichev, S. and Sirenko, B.J., *Zool. Zhurnal*, 53, 1133–1139, 1974.

in chitons and patellogastropods. There is little variation (Warén & Gofas 1996b), with 11 teeth in each alveate row, even in the tiny *Micropilina*. Each row consists of a small central, an inner small lateral, and an outer large, broad lateral that appears to correspond to the large, mineralised outer lateral teeth in chitons. The two inner marginal teeth have broad, finely cuspate cutting edges, while the outermost pair of marginal teeth in each row lack cusps.

5.3.2.4.2.4 Scaphopoda The scaphopod radula is the most massive relative to body size of any molluscs (Morton 1959). It consists of five, or rarely seven, mineralised teeth in each row. The lateral field typically consists of a large central tooth and (usually) a single pair of large lateral teeth in a sagittate configuration. The marginal field consists of a single, large plate-like marginal tooth.

The radula of scaphopods is not eversible, and in dentaliids functions like a ratchet, being used to move food from the buccal tube to the oesophagus, while in gadilids, the radula, together with the muscular buccal wall, breaks up the food (Shimek & Steiner 1997).

5.3.2.4.2.5 Cephalopoda In cephalopods, the arrangement of radular teeth consists of a central and (usually two) lateral teeth and one or more marginal teeth. Nixon (1995) reviewed cephalopod radulae, both living and fossil, and provided a scheme for naming the teeth. As in chitons and scaphopods, the outermost lateral tooth is plate-like, although this is lost in some taxa.

The tooth rows range from alveate in most coleoids to weakly sagittate in *Nautilus*, with 13 teeth in each row in *Nautilus* and typically nine or seven in Coleoidea, with five in Gonatidae. There is a general reduction in radular size from *Nautilus* through octopods to squid and sepioids. Cirrate octopods and *Spirula* lack a radula.

5.3.2.4.2.6 Gastropoda

The radula of gastropods is highly diverse and, in this group in particular, has been used extensively in classification and taxonomy (see Section 5.3.2.4.2), particularly within the Orthogastropoda. Here we provide a new, simpler way of recognising this variation in orthogastropods (see below).

The main types of gastropod radulae largely conform to the taxonomic groupings of gastropods. While major radular types are generally recognised among the 'prosobranch' groups, those of heterobranchs are also very diverse.

Patellogastropod radulae are of the docoglossan (= stereoglossan) type. In many characteristics, these resemble the radulae of the outgroup polyplacophorans and monoplacophorans, and, as noted above, are considered close to the ancestral radular type in gastropods. Each row of teeth consists of 2–4 inner lateral teeth with a central tooth sometimes present. On each side are 2–4 outer lateral teeth that are typically fused into a large multicuspid tooth in patelloideans, and from 0 to 3 marginal teeth. The teeth are hardened by the incorporation of metal ions (see Section 5.3.2.4.3) and are firmly fixed to the radular ribbon so that the radula acts like a rasp.

Orthogastropod radulae are treated in some detail below. We introduce some new ways of dealing with the variation observed in this system, resulting from the recognition of the two radular fields (see Section 5.3.2.4.1.1). We suggest the following terminology to classify the radular configurations in the Orthogastropoda:

Description of fields
- *Lateromarginal* – both lateral and marginal fields occupied.
- *Haplolateral* – only the lateral field occupied.
- *Haplomarginal* – only marginal field occupied.
- *Field unknown* – teeth of uncertain or unknown origin.

Number of teeth per row
- *Multidentate* – many (ten or more) teeth in each row (central tooth present or absent).
- *Enneadentate* – nine teeth in each row (central tooth present).
- *Octodentate* – eight teeth in each row (central tooth absent).
- *Heptadentate* – seven teeth in each row (central tooth present).
- *Hexadentate* – six teeth in each row (central tooth absent).

- *Pentadentate* – five teeth in each row (central tooth present).
- *Tetradentate* – four teeth in each row (central tooth absent).
- *Tridentate* – three teeth in each row (central tooth present).
- *Bidentate* – two teeth in each row (central tooth absent).
- *Monodentate* – one tooth in each row (the central).

Thus, a radula could be described as a lateromarginal multidentate radula (as in the rhipidoglossan condition), haplolateral multidentate (as in many 'pulmonate' or 'opisthobranch' radulae), haplolateral monodentate (as in the monodentate rachiglossans or sacoglossans), or haplomarginal bidentate (typical toxoglossan teeth).

Radular formulae are often used to describe radular configurations in gastropods. For example, Fretter and Graham (1962) encoded the patellid radula (docoglossan) as $3 + D + 2 + R + 2 + D + 3$; with R = rachidian (i.e., central [C]) tooth, followed by the number of lateral teeth with a dominant (D) tooth being separately annotated, and then the marginal teeth. This approach is often simplified to count only the number of teeth – e.g., $3 + 3 + C + 3 + 3$ to reflect the number of teeth and plates present in the inner and outer lateral tooth fields.

The patellid radula could be encoded as:

$$\underbrace{3}_{\substack{\text{Marginals}}} + \underbrace{2 + 2}_{\substack{\text{Outer} \\ \text{laterals}}} + C^7 + \underbrace{2 + 2}_{\substack{\text{Inner} \\ \text{laterals}}} + 3$$

Vetigastropods and neritimorphs have lateromarginal multidentate radula with alveate rows, a condition known as *rhipidoglossan*. Each row consists of a central tooth and, on each side, several (usually five) lateral teeth, the outermost sometimes largest, and numerous slender marginal teeth (sometimes called uncini). A typical formula would be $\infty + D + 4 + C + 4 + D + \infty$ or $\infty + 1 + D + 3 + C + 3 + D + 1 + \infty$, with '$\infty$' indicating a large number of marginal teeth, and D the 'dominant' pluricuspid tooth. The long, flexible marginal teeth may act like a broom, sweeping up minute food particles. This type of radula is often asymmetrical, notably in Trochoidea, Fissurelloidea, Cocculinoidea, and Pleurotomarioidea with one side slightly to significantly more anterior on the ribbon than the other (Hickman 1981). The central tooth in some vetigastropods may also be asymmetrical. Some euthyneuran radulae are also asymmetrical – for example, in many pleurobranchs and in the nudibranch Aegiridae where the central teeth are missing, the rows alternate in the middle of the ribbon much like a zipper.

In Pleurotomariidae, the *hystrichoglossan* radula is a modification of the rhipidoglossan type with numerous elongate lateral teeth and many long marginal teeth bearing a broom-like cluster of filaments distally (Woodward 1901; Hickman 1984) (Chapter 18, Figure 18.26). An example is the radula of the pleurotomariid *Entemnotrochus rumphii* with a formula $\infty + 14 + 27 + C + 27 + 14 + \infty$, thus possessing 41 lateral teeth in inner and outer groups. There are other highly modified rhipidoglossan radulae. For example, a considerable reduction in tooth numbers per row is seen in seguenzioideans (e.g., Quinn 1991) which may reflect heterochronic changes. Thus, some seguenzioideans have only seven teeth in each row, the outermost pair being the marginals, and so converge on the taenioglossate condition seen in many caenogastropods. In trochaclidids, the central and lateral teeth are reduced (or apparently lost in *Acremodonta*), and the marginals are finely and repeatedly branched distally to form broad, fan-like ends (Marshall 1995). Another strange modification of the rhipidoglossan type of radula is seen in *Thysanodonta* (Thysanodontinae, Calliostomatidae) which has only 10–11 exceedingly narrow, elongate, barbed marginal teeth in each row (Marshall 1988a). Some of the deep-sea 'cocculiniform' (cocculinoidean and lepetelloidean) limpets also have reduced and highly modified rhipidoglossate radulae, notably with the loss of the marginal teeth, but typical rhipidoglossate radulae are also found in some families.

Most caenogastropods, other than neogastropods, have lateromarginal heptaserrate radula (the *taenioglossan* type) with seven teeth in each of the alveate rows, consisting of a symmetrical central tooth, and, on each side, one lateral and two marginal teeth (formula $2 + 1 + 1 + 1 + 2$). The central and lateral teeth act as rakes, and the marginals sweep the material into the mouth (e.g., Steneck & Watling 1982). The taenioglossan radula is remarkably uniform across a wide range of feeding types within caenogastropods (see Chapter 19, Figures 19.14, 19.16), but considerable modification from what is presumably a taenioglossan ancestral type is seen in the cnidarian-feeding Epitoniidae with the *ptenoglossan* radula. Here, the multidentate radular row consists of numerous similar hook-shaped pointed teeth of uncertain homology and alveate to near linear configuration. Ontogenetic change involving the loss of subsidiary cusps has been described in at least one epitoniid (Page & Willan 1988). The sponge-feeding triphoroideans include assumed basal taxa with taenioglossate radulae and others that show tooth multiplication (see Section 15.1.1.5.1 and Chapter 19, Figure 19.15). Some cerithiopsids have brush-like ends to the marginal teeth convergently resembling those in the Pleurotomariidae (see above) (e.g., Nützel 1998), both of which are sponge feeders.

Radular-bearing neogastropods have five to one teeth in each row and are collectively known as *stenoglossan*. The term 'stenoglossan' is often treated as a synonymous with *rachiglossan*, but the former was originally introduced (as a higher taxon Stenoglossa) to include neogastropods having rachiglossan and toxoglossan radulae. Ontogenetic changes in tooth morphology have been shown in some neogastropods and may reflect dietary changes from juvenile to adult. Examples include muricids (Herbert et al. 2007) and conids (Nybakken & Perron 1988; Nybakken 1990).

The rachiglossan type is tridentate to monodentate and is found in neogastropods such as Buccinoidea and Muricoidea. Each (typically) sagittate row has a central tooth and often one lateral tooth on each side (i.e., three teeth per row). Some families (e.g., Marginellidae and most Volutidae) have just the central teeth. Only a few taxa in the rachiglossate groups have lost the radula.

Many conoideans have the toxoglossan type, with the typical condition being bidentate, but with pentadentate and tridentate conditions also found, and radular loss has occurred in several lineages. A pentadentate radula is found in most Drillidae, while a tridentate radula is seen in a few lineages, with these radular rows being alveate, unlike the sagittate condition in most other neogastropods. The typical toxoglossan condition has the marginal teeth rolled into hollow spears with other teeth lacking (see Figure 5.7) and has independently evolved in several lineages. In the typical toxoglossan condition, each tooth is not firmly fixed to the radular ribbon. Each row has two teeth of which only one is used at a time. The teeth ready for use are typically stored in a radular pouch, individually filled with venom and transferred to the proboscis to be used for spearing and the envenomation of prey (see Section 5.2.1.7.4.5). The radula has been independently lost in several conoidean taxa.

Another modification of the stenoglossan radula is the *nematoglossan* type (Olsson 1970) of the Cancellariidae, which consists of a single row of very long central teeth, usually with three distal cusps. These teeth are protruded through a tubular jaw and used for piercing the prey before suctorial feeding (Harasewych & Petit 1986) (see Chapter 19, Figure 19.18). Some cancellariids (e.g., some species of Admetinae, Oliver 1982) lack a radula, but retain the tubular jaw.

Heterobranch radulae have not been categorised in the literature like those of 'prosobranchs', but show much diversity in form between groups. Within the lower heterobranch groups, there is a considerable range of radular diversity (see Chapter 20), although typically the radular rows are alveate. The family Hyalogyrinidae (*Hyalogyrina* and *Xenoskenea*) (Marshall 1988b) (see Haszprunar et al. 2011 for anatomical treatment) has a multidentate radula that closely resembles the rhipidoglossate condition with three laterals and ~ten marginals in *X. pellucida* (Warén et al. 1993). A few taxa have enneadentate radulae, such as the cornirostrids *Cornirostra* and *Noerrevangia* (Warén et al. 1993), while another cornirostrid *Tomura* is heptadentate (i.e., taenioglossate-like) (Warén et al. 1993), as are species of *Valvata* (Valvatidae), and *Jeffreysiella* (Rissoellidae) (Ponder & Yoo 1977), and at least one mathildidid (*Gegania*). The radula is pentadentate in other Mathilididae and in Architectonicoidea (Bieler 1988), tridentate in Orbitestellidae (Ponder 1990b), and monodentate in at least some species of *Omalogyra*, while other species of *Omalogyra* and *Ammonicera* (Sleurs 1985) have a tetradentate radula. While a marginal field is probably present in Hyalogyrinidae and arguably present in some (Rissoellidae, Valvatoidea), the lateral and marginal teeth are not clearly morphologically distinct in Architectonicoidea and Mathildidoidea. Thus, the lower heterobranchs may contain taxa with lateromarginal and haplolateral radulae. Pyramidelloideans have lost the radula, but members of the Ebalidae have a multi-element jaw apparatus which Warén (1994) suggested might be a combination of the radula and the jaw, although we consider this very unlikely.

The radulae of 'euthyneurans' (see Chapter 20) are even more diverse than those of the 'lower heterobranchs', ranging from multidentate with numerous similar teeth (polyglossate condition) to monodentate; radular rows can be alveate, lineate, or sagittate with the latter condition evolving a number of different times in carnivorous lineages.

There can be considerable variation in the number of tooth rows present in some taxa. Philinoideans, for example, have a pair of large teeth (usually interpreted as laterals) adjacent to the central tooth (if present) and a series of smaller simple teeth (usually interpreted as marginals). The 'marginals' are lost in some lineages, as is the central tooth (Rudman 1978), resulting in a tridentate or bidentate radula. In many other 'opisthobranch' radulae, the transition between laterals and marginals is often rather gradual so the division becomes an arbitrary one. In some chromodoridid radulae (Rudman 1984, fig. 4), the tooth next to the central (if present) is sometimes large and different, while in others, it is not distinguishable from the other teeth near it. For these reasons, the homology of the teeth in euthyneuran radulae is uncertain and, although they are generally interpreted as comprising both lateral and marginal fields, there is often no clear division between them. The difficulty of interpreting the homology of the teeth is compounded by a paucity of ontogenetic studies, and none of these has addressed this question. We favour the idea that the euthyneuran radula is haplolateral as suggested by Ponder and Lindberg (1997), but comparative ontogenetic studies are needed to resolve this question. A haplolateral monodentate radula is found in sacoglossans and most aeolidioid nudibranchs.

The number of teeth in polyglossate radulae often exceeds 100 per row and can be as many as 1100 in *Umbraculum* (Peile 1939). The numerous uniform teeth typical of many 'pulmonate' polyglossate radulae is a secondary phenomenon. During development in stylommatophorans, usually only two longitudinal rows of teeth (the innermost lateral teeth) appear initially (the so-called 'distichous phase'), with additional rows of lateral teeth added later as well as the central teeth (Schnabel 1903; Kerth 1979) (Figure 5.31). Sterki (1893) showed three teeth appearing initially in some stylommatophorans. Ontogenetic changes in cusp morphology can be marked in stylommatophorans (Figure 5.36) and have also been noted in the ellobiid subfamily Melampinae (Martins 1996). In these cases, the youngest teeth have multiple cusps, but adults have only a single cusp.

There have been few detailed studies on the ontogeny of the radula in 'opisthobranchs'. The number of radular teeth increases with age in those with many radular teeth in each row in adults, but in some nudibranchs such as *Polycera* with few teeth in adults, the number of teeth decreases as the animal grows (see Martínez-Pita et al. 2006 for review). In species of *Tritonia*, which have many teeth per row in adults, the radula increases from 0 + C + 0 teeth per row to 4 + C + 4 in one day after settlement, and the shape and denticulation of the teeth change with growth (Thompson 1962).

5.3.2.4.2.7 Radular Function Radular teeth can laterally flex outwards and then sweep inwards (the *flexoglossate* condition), or the teeth can remain fixed (the *stereoglossate* condition) during the feeding stroke. An important distinction between stereoglossate and flexoglossate feeding is the force produced during the feeding stroke and thus the impact on the substratum. For example, patellogastropods can penetrate and even excavate hard substrata, while other taxa can only penetrate softer substrata or brush the surface of a hard substratum. This is probably because flexing teeth exert force laterally and downwards, and thus do not have as strong a downwards force as the stereoglossate patellogastropods (Guralnick & Smith 1999). Thus, patellogastropods differ from other molluscs in the details of their feeding mechanism (Guralnick & Smith 1999) in having an entirely stereoglossate feeding stroke, and all other radular-bearing molluscs are, at least to some degree, flexoglossate. These differences can be best seen in their feeding traces, with the stereoglossate radula producing parallel strokes while the flexoglossate radula scrapes the teeth inwards (Figure 5.37).

The lack of flexure may partially account for the strength of the stroke of the patellogastropod radula, but other factors accounting for this include the hypertrophied muscles, the use of the radula during odontophore protraction, and possibly the shape of the medial odontophoral cartilages. Guralnick and Smith (1999) pointed out that it is only in patellogastropods that the 'medial' (i.e., the anterior) cartilages change markedly in shape along their lengths as it is small at the bending plane and taller towards the posterior end before levelling. The evolutionary change to stereoglossy in the patellogastropod radula allowed them to utilise new rocky shore intertidal resources, although flexoglossy supports more flexible feeding habits and diets than stereoglossy. Flexoglossy allows the animal to grasp and hold food items while pulling them towards the oesophagus and keeping ingested food from moving back towards the mouth (Guralnick & Smith 1999).

Garland et al. (1985) illustrated the grazing tracts of juvenile abalone (*Haliotis rubra*) on coralline algae. These tracts show the individual teeth (of the central-lateral field) parallel to one another as they are dragged across the surface (like the stereoglossate condition) closely resembling the parallel tooth tracts produced by adult patellogastropods. The role of the marginal teeth in juvenile abalone is uncertain, but adults are typically flexoglossate.

The workings of the buccal apparatus and its role in radular function have been described in detail by Guralnick and Smith (1999). They involve the muscles, radula, and odontophoral cartilages (over which the radula is drawn), working in concert. The feeding action (or 'feeding stroke') requires the coordinated movements of both the odontophore and the radula (Figure 5.38). Odontophore protraction and retraction moves the buccal mass downwards out of the mouth towards

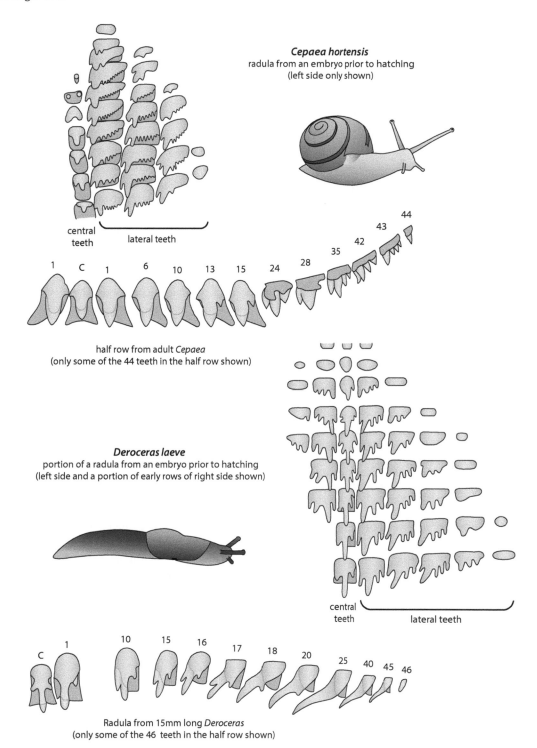

FIGURE 5.36 Radular development in stylommatophorans; the snail *Cepaea* and the slug *Deroceras*. *Cepaea* redrawn and modified from Schnabel, H., *Z. wiss. Zool.*, 74, 616–655, 1903 (juvenile), adult from Taylor, J. W., *Monograph of the Land and Freshwater Mollusca of the British isles*. Leeds, Taylor Brothers, 1894–1914. *Deroceras* from Sterki, V., *Proc. Acad. Nat. Sci. Phila.*, 45, 388–400, 1893.

the feeding surface, and at the same time, radular protraction and retraction move the radula over the cartilages. During the feeding stroke, the radula is pulled over the anterior end of the cartilages by the radular protractor muscles. At the bending plane at the most anterior part of the odontophore, the radular

teeth point downwards instead of upward, and it is this part of the radula that contacts the feeding surface.

The feeding stroke of patellogastropods differs from those of other molluscs not only in being stereoglossate, but also because their radular teeth make contact with the substratum

FIGURE 5.37 Radula and feeding traces in a stereoglossate patellogastropod (*Lottia*) and a flexoglossate caenogastropod (*Littorina*). Redrawn and modified from Guralnick, R.P. and Smith, K., *J. Morphol.*, 241, 175–195, 1999.); radulae: *Lottia paradigitalis* courtesy of B. Simison. *Littorina littorea* courtesy of D.G. Reid.

be infolded, if it is not, the teeth cannot flex out as they pass over the bending plane. Folding of the radula occurs when it lies in a groove between the supporting cartilages that is narrower than the radular width (Figure 5.39). Third, the subradular membrane must be closely associated with the underlying odontophoral supporting structures anteriorly (i.e., near and at the bending plane) allowing the teeth to unfold. Last, the radula must be partially or fully flattened as it passes over the bending plane and the radular teeth then flex laterally. Part of the flattening is passive and is partly achieved by lateral muscles attached to the subradular membrane.

Polyplacophorans were considered stereoglossate, but Guralnick and Smith (1999) showed them to be flexoglossate as they conform to the above conditions. The radula sits more shallowly in the groove between the supporting vesicles in chitons and monoplacophorans than it does in orthogastropods, suggesting a less modified flexoglossate condition in those taxa. In patellogastropods, there is a groove between the supporting cartilages, but the radula is not folded into it (Figure 5.39).

Guralnick and Smith (1999) argued that the apomorphic patellogastropod condition is largely achieved by the lack of close association of the radular apparatus and the underlying cartilages near the bending plane, resulting in the teeth not being able to curve around the outer lateral edges of the cartilage before the bending plane.

The configuration of the monoplacophoran radula most closely resembles that of lepetid patellogastropods, but its supporting apparatus is like that of chitons. Both have fluid-filled vesicles and a radular apparatus resting in the groove formed by the vesicles, resulting in folding of the radula before the bending plane. It is thus assumed that monoplacophorans use their radula in a flexoglossate manner.

In the caudofoveate aplacophoran *Limifossor* (Heath 1905), the radula lies deep within the groove formed by the supporting structure anterodorsally; the two teeth separate along the midline and, as they move anteriorly, their tips become more inclined outwards which, in the opinion of Guralnick and Smith (1999), appears to be flexoglossate.

In Scaphopoda, the radula is folded between medial cartilages, and the anterior portion of the radular apparatus can flex outwards.

during odontophoral protraction, not, as in other groups, during retraction and because several tooth rows, not one, makes contact with the feeding surface. Their contact can be forceful, enabling penetration even in hard surfaces.

Unlike earlier interpretations, Guralnick and Smith (1999) argued that the flexoglossate condition is plesiomorphic. They suggested four conditions were needed for the radula to function in a flexoglossate manner. Firstly, a sturdy anterior structure is needed to support the membrane as it is pulled over the end (the bending plane). Secondly, the radula must

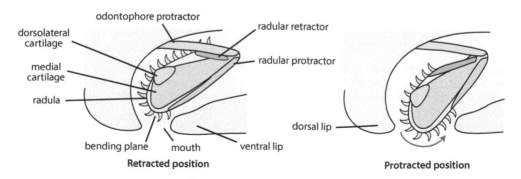

FIGURE 5.38 Diagram showing the main components of the odontophore involved in the feeding stroke of a generalised mollusc. Redrawn and modified from Guralnick, R.P. and Smith, K., *J. Morphol.*, 241, 175–195, 1999.

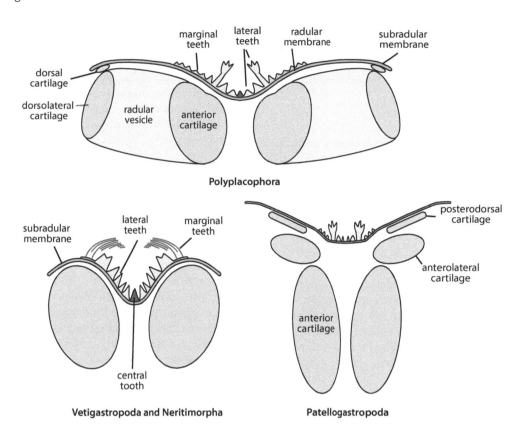

FIGURE 5.39 Diagram showing the position and shape of the radula and underlying cartilages in chitons and lower gastropods. Muscles are not shown. Redrawn and modified from Guralnick, R.P. and Smith, K., *J. Morphol.*, 241, 175–195, 1999.

In cephalopods, at least some coleoid squid such as species of Loliginidae have medial cartilage-like structures, but there are only muscle and connective tissue in *Nautilus* and *Octopus* (Young 1994; Uyeno & Kier 2005). In *Sepia*, the radula is folded anteriorly into a groove formed by muscle and cartilage, although muscle appears to predominate at the bending plane. In at least some squid and octopuses, the teeth have been observed to point backwards; as they are pulled over the bending plane, they point first upwards and then forwards, indicating rotation during feeding. *Nautilus* lacks cartilages, but has a pair of muscular structures that contain 'membrane-enclosed rods' (Sasaki et al. 2010a, p. 41), and the long lateral teeth are erected so that they do not cover the foremost central teeth during feeding (Guralnick & Smith 1999).

Grasping and holding onto food items would be essential for carnivores feeding on active prey and give extra flexibility in food choice for herbivores – for example, some polyplacophorans can not only scrape the microbial film, but also snip off and consume algal blades, an action only possible in flexoglossate feeders. The stereoglossate condition does not enable grabbing, holding, cutting, or biting, but only rasping and collecting the dislodged scrapings.

The cartilages and radula do not necessarily change in concert during evolution. Thus, even though the patellogastropod radula is relatively plesiomorphic, the cartilages are not.

Barker (2001a) suggested the evolutionary changes in the broad, multidentate (polyglossate) radula found in 'pulmonates' first occur in the teeth at the margins (i.e., the so-called

marginal teeth), especially near the transition between those with the 'lateral' teeth, because, as the radular ribbon moves over the anterior margin of the odontophore during feeding, these teeth undergo most elevation and rotation and are the first to make contact with the food. Somewhat controversially, he regarded this type of radula as essentially rhipidoglossate as there are some functional similarities. The central and lateral teeth directly rasp the substratum, while the food is gathered by the teeth at the margins (the 'marginal' teeth or, more correctly, the outer laterals) when they fold back over the central and 'lateral' (i.e., inner lateral) teeth (Mackenstedt & Märkel 2001).

The radular teeth of obligate carnivorous stylommatophorans have become elongated and narrow and the tooth row markedly sagittate. Solem (1974) recognised two basic types – 'stabbers' and 'slicers', although Barker and Efford (2002) found that this distinction was not fundamental, but rather gradual evolutionary stages that occur independently in several lineages.

It is also important to consider the support structures of the radular teeth themselves. Each tooth has three main parts, a cusp-bearing head, a shaft, and a base, the latter attached to the radular membrane or a plate. All three components can be modified in radular evolution, but it is modifications to the base that are particularly important when considering the mechanical functions of the radula (e.g., Hickman 1980). Hickman noted that an enlarged base might have several adaptive functions – mechanical support and distribution of stress forces, as well as providing overlap and interlock within and between rows. Such interlocking systems help to keep teeth aligned during radular movements over the odontophore and as it passes over uneven surfaces.

5.3.2.4.3 Radular Mineralisation

Radular tooth mineralisation occurs in several groups of molluscs, but different elements are utilised. In general, differences in mineralisation correlate with differences in the hardness of the teeth (e.g., Runham et al. 1969). While most studies have focused on mineralisation in one or a few species, Okoshi and Ishii (1996) studied the concentration of 17 elements in the radulae of 24 molluscs. Aktipis et al. (2008) used these and other available data as a set of characters in their analysis of gastropod phylogeny.

In the caudofoveate aplacophorans, the teeth are chitinised and strengthened with calcium phosphate and iron (Salvini-Plawen 1988) or, more specifically, with amorphous iron oxide and a crystalline form of calcium phosphate, hydroxyapatite ($Ca_{10}(PO_4)_6(OH)_2$) (Cruz et al. 1998). The latter is a common component of vertebrate teeth and bones, but uncommon in invertebrates, although it is also found in chiton radular teeth.

The most heavily mineralised radular teeth in molluscs are seen in chitons and patellogastropods, and in these groups mineralisation commences from the start of radular formation. Polyplacophorans use iron and calcium-based minerals to harden some of their radular teeth, enabling them to graze on algae on hard substrata, although tooth wear still necessitates a rapid turnover of teeth. The only pair of teeth mineralised using iron and calcium are the major (i.e., outer) lateral teeth, and in these, the posterior surface of each cusp is hardened with magnetite.[8] When these teeth are formed in the posterior end of the radular sac, the odontoblasts secrete an extracellular organic matrix consisting mainly of the polysaccharide α-chitin (Evans et al. 1990), which is critical in the mineralisation of the tooth (Wal et al. 2000). This fibrous organic material is one of the three major components in the tooth cusp, together with acicular crystals of goethite (hydrated iron oxide – α-FeOOH), and silica (SiO_2). The goethite crystals lie parallel to the fibres making up the organic framework (Wal et al. 1989). The hardness of the front edge of the cusp is about twice the hardness of the back edge, probably due to the more compact mineral material in the front part (Wal et al. 1989). This is one of the remarkable characteristics of chiton teeth mineralisation – they have the various minerals separated in different parts of the tooth cusp (Figure 5.40). This greatly enhances the capacity of the teeth to withstand wear, shocks, and cracking and also provides some self-sharpening ability (Wal et al. 2000).

The incorporation of magnetite, a magnetic ferrous mineral (FeO Fe_2O_2), in the radula of chitons is very unusual and is known in very few instances in nature – namely, in very

small amounts in birds and insects where it is utilised in magnetoreception (Eernisse & Reynolds 1994). In chiton teeth, it is first deposited as ferrihydrite and then transformed into magnetite (Eernisse & Reynolds 1994).

Studies on the mechanisms involved in this biomineralisation show two main stages: transfer of minerals via the dorsal epithelium of the radular sac that surrounds the tooth cusps and also by way of the *stylus canal*, a structure only found in the base (*stylus*) of the mineralised teeth of chitons. This latter mechanism, while suspected for some time, has only recently been verified (Shaw et al. 2009). It involves moving elements stored at the *junction zone*, which lies between the tooth cusp and the tooth base (stylus), and where there is an initial accumulation of 'precursor elements' (Macey & Brooker 1996), that reach there via the stylus canal, and from there the minerals diffuse to the tip of the cusp.

The tube-like stylus canal is filled with columnar epithelial tissue that contains ferritin particles and is similar to the dorsal epithelium that surrounds the mineralised cusps within the radular sac. The stylus canal transports iron (and perhaps other elements) to the junction zone for its later delivery into the tooth cusps. Above the stylus canal, a plume of mineral elements between the junction zone and the posterior edge of the cusp coincides with iron appearing in the dorsal epithelium of the radular sac and the beginning of mineralisation. The plume may also later move phosphorus and calcium into the tooth core, the last part of the cusp to be mineralised (Shaw et al. 2009).

Patellogastropods (Figure 5.40) have similar patterns of mineralisation to chitons, but instead of magnetite, the iron oxide goethite is used with hydrated silica ($SiO_2 \cdot nH_2O$) in the mineralisation of their teeth (e.g., Liddiard et al. 2004); this is one of the strongest biological materials known (Barber et al. 2015). Also, as in chitons, the initial influx of iron occurs at the junction zone in limpets (Liddiard et al. 2004; Sone et al. 2007), but they lack a structure like a stylus canal.

The hydrothermal vent patellogastropod *Paralepetopsis ferrugivora* differs from its shallow-water relatives in lacking any crystalline phase in tooth mineralisation, with amorphous silica in the teeth cusps and amorphous iron oxide in the junction zone (Cruz & Farina 2005).

Radular teeth incorporating metal ions is the plesiomorphic condition in gastropods (Ponder & Lindberg 1997), but the minerals incorporated differ from group to group. For example, while iron and silica are utilised in patellogastropod (and chiton) teeth, calcium is utilised in at least some 'pulmonates' and species of the ampullariid genus *Pomacea* (Mackenstedt & Märkel 2001). Loss of mineralisation has occurred within vetigastropods, possibly neritimorphs (although neritid teeth contain silica), and also with some patellogastropods (Neolepetopsidae), and is commonly lost in other gastropod groups. While there is some mineralisation recorded in other gastropods, for example, even a species of the scavenging neogastropod *Cominella* has Si, Ca, Fe, and Pb present (Meyer-Rochow & West 1999), there is much less than in chitons and limpets.

Monoplacophoran teeth are only weakly mineralised (composition unknown). The radular teeth of a species of *Nautilus* are strengthened with calcium and silica, while those

---

[8] An oxide of iron (Fe_3O_4) that is magnetic.

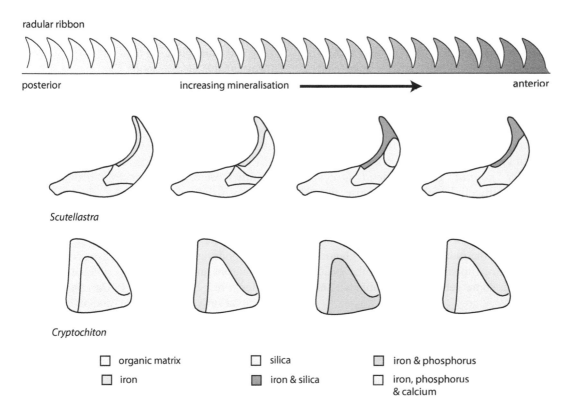

radular ribbon

posterior

increasing mineralisation

anterior

Scutellastra

Cryptochiton

□ organic matrix   □ silica   ■ iron & phosphorus

□ iron   ■ iron & silica   □ iron, phosphorus & calcium

FIGURE 5.40 Tooth mineralisation in a patellogastropod (*Scutellastra*) and a polyplacophoran (*Cryptochiton*). Middle row redrawn and modified from Liddiard, K.J. et al., *Molluscan Res.*, 24, 21–31, 2004 and lower row from Lowenstam, H.A. and Weiner, S., Mollusca, pp. 88–110, in Lowenstam, H.A. and Weiner, S., *On Biomineralization*, Oxford University Press, New York, 1989.

of scaphopods contain iron and calcium, with silica, barium, zinc, and copper in trace amounts (Shimek & Steiner 1997).

5.3.2.5 Foregut Glands

This section deals with the glands associated with the foregut, including the salivary glands, although some other glandular foregut structures are dealt with above (see Section 5.3.2.2). Although there are subepithelial mucocytes in the oesophagus of species of *Mytilus* (Beninger et al. 1991) that secrete mucopolysaccharides (Beninger & Le Pennec 1993), we do not regard these as salivary glands.

We deal with the aplacophorans in some detail below as these are well studied and the glandular elements in their foregut are not readily homologised with those of other molluscs.

5.3.2.5.1 *Aplacophorans*

Lateroventral glandular organs which open ventrally to the radula are present in both the aplacophoran groups. These are considered the homologues of the paired glands associated with the sublingual pouch in other molluscs (Salvini-Plawen 1988) (see Section 5.3.2.2.2). There are two additional main foregut glands in both caudofoveates and solenogasters (Salvini-Plawen 1988). Most of these glands are called 'salivary glands' by Scheltema et al. (1994, p. 38), but we do not use this terminology because of their doubtful homology with the salivary glands of other molluscs. We treat them separately in the two groups because of this difficulty. They have

been described in detail and are important in the systematics of the two groups (particularly Solenogastres).

In the particle-feeding caudofoveates, the foregut glands provide enzymes that assist in the initial breakdown of food, while in the mainly cnidarian-feeding solenogasters, the secretions from the foregut glands probably aid in preventing the discharge of the nematocysts of the cnidarian prey as well as providing enzymes that aid in the initial breakdown of prey tissue (Salvini-Plawen 1988).

In caudofoveates, epithelial or subepithelial glands are scattered through the buccal epithelium, but are concentrated near the radula where they form distinct lateral and/or dorsal glands, or lateral lobular organs in a few taxa (e.g., Chaetodermatidae). Ventrally, immediately in front of the radula, some unicellular glands open by way of one or two papillae and some have stalked dorsal gland cells which may form paired groups above or behind the radula (Salvini-Plawen 1988; Todt & Salvini-Plawen 2004).

The several types of 'foregut glands' recognised in solenogasters (Salvini-Plawen 1988; Todt & Salvini-Plawen 2004; Handl & Todt 2005; Todt 2006; Garcia-Alvarez & Salvini-Plawen 2007, L.v. Salvini-Plawen pers. comm., March, 2008) are listed below.

Buccal (or *Pharyngeal*) *glands*: These unicellular, subepithelial (or rarely epithelial) glands discharge into the buccal cavity. They are very common in most solenogasters and are the only foregut glands present in Hemimeniidae and Neomeniidae. The buccal glands are especially well developed in some taxa with a muscular buccal wall. This type of gland is also represented by subepithelial glands in the anterior part of the oesophagus or, in proboscidate taxa, at the posterior end of the 'proboscis' sheath.

Preoral glands: Only in Gymnomeniidae; single gland cells (some with a subepithelial cell body) are arranged preorally in the roof between the atrium and the mouth (Todt & Salvini-Plawen 2004).

Dorso-pharyngeal papilla gland (*dorsal foregut gland*): These multicellular glands comprise elongate gland cells with subepithelial cell bodies which open together as a papilla-like projection into a mid-dorsal narrow pouch in the buccal cavity, just anterior to the radula. This type of gland is found in some Pararrhopaliidae, and similar structures are seen in some genera in Pruvotinidae and Dondersiidae.

Lateroventral foregut glands: These paired, multicellular organs are the most obvious glands associated with the foregut and are taxonomically important. They open just in front of, or at both sides of, the radula. Several types of lateroventral foregut glands can be recognised (Handl & Todt 2005; Todt & Salvini-Plawen 2005; Todt 2006; Garcia-Alvarez & Salvini-Plawen 2007) as outlined below:

- Clustered gland cells with subepithelial cell bodies open close together into the buccal cavity (present in Meiomeniidae and three genera of Simrothiellidae). In Gymnomeniidae, a clustered type (type 3 of Todt & Salvini-Plawen 2004) is restricted to the posterior buccal cavity on both sides of the radula.
- Subepithelial gland cells open into a pair of ducts (which may be surrounded by a muscle layer).
- Epithelial gland cells that open to a paired duct and are surrounded by musculature. In *Simrothiella* these are bulbous organs.
- A paired ramified duct with each branch having a cluster of terminal gland cells surrounded by muscles is found only in the Amphimeniidae.
- Circumpharyngeal glandular organs with subepithelial cell bodies arranged in several (globular) bundles around the buccal cavity are found only in the Unciherpiinae (Pruvotinidae).

5.3.2.5.2 Salivary Glands of Polyplacophora, Monoplacophora, and Gastropoda

Distinct glandular masses (salivary glands) are found in chitons, monoplacophorans and gastropods, but are absent in bivalves and scaphopods. Besides these structures, other foregut glands in those groups include epithelial goblet cells that secrete lubricatory mucus into the buccal cavity, but these are not dealt with further. Some gastropods have 'accessory salivary glands', which are discussed below (see Section 5.3.2.5.3).

Paired salivary glands lacking ducts are present in polyplacophorans (Fretter 1937) and monoplacophorans (Wingstrand 1985), and while these are thought to be homologous with the salivary glands of gastropods, they are questionably homologous with so-called salivary glands of cephalopods (Salvini-Plawen 1988) (see Section 5.3.2.5.4).

In chitons, the dorsal salivary glands (= buccal glands) are compound in larger species and lie on the anterior part of the buccal region. They provide lubrication during feeding (Salvini-Plawen 1988). Monoplacophorans have lateroventrally located, ductless paired salivary (or buccal) glands. Each consists of several glandular lobes.

In gastropods, the salivary glands open dorsally or laterodorsally to the buccal cavity. In vetigastropods, the glands are ductless, but ducts are present in most patellogastropods and most other gastropods except for most neritimorphs and cocculinids which lack salivary glands (Fretter 1965; Haszprunar 1987a; Salvini-Plawen 1988). Gastropod salivary glands are often called acinous salivary glands to distinguish these 'true' salivary glands from other 'accessory salivary' glands that occur in some gastropods (see Section 5.3.2.5.3), typically in addition to the acinous glands.

In gastropods, the position of the nerve ring varies relative to the buccal mass – in some, it is in front of the buccal mass, in others behind. In the latter situation, found in caenogastropods, the acinous salivary glands, or their ducts, either pass through the nerve ring or over it. In rissooideans, truncatelloideans, and neogastropods, the salivary ducts do not pass through the nerve ring (Ponder 1974, 1988a), but in many other caenogastropods, they do (see Ponder & Lindberg 1997 for exceptions).

The acinous salivary glands of gastropods appear rather late during ontogeny as dorso-lateral diverticula on the buccal mass and then usually extend back along the oesophagus (see Ponder & Lindberg 1997 for references).

In most gastropods, the acinous salivary glands secrete 'mucus', which mainly consists of acid mucopolysaccharide (glycosaminoglycan) that lubricates the passage of food (Andrews 1991), and in some taxa, serous cells secrete proteoglycan (a glycoprotein) which probably contains some enzymes (see Section 5.4.2.3).

In patellogastropods, there is considerable variation in salivary gland morphology, ranging from ductless tubular glands to large glands with long ducts, or two pairs of ducts in Patellidae (Sasaki 1998). In *Patella*, the salivary glands contain two types of cells (Pugh 1963; Benmeradi & Benmeradi 1992) and produce several enzymes (see Section 5.4.2.3). Many other gastropod salivary glands contain three types of cells.

The salivary glands of vetigastropods and lower caenogastropods mostly do not secrete digestive enzymes – only mucus and occasionally protein (Fretter & Graham 1962; Voltzow 1994), although enzymes are produced in at least some caenogastropods (see Section 5.4.2.3). The salivary glands of some tonnoideans and neogastropods produce toxins utilised to paralyse their prey, as described below.

Tonnoideans have a pair of large salivary glands; each consists of a small anterior acinous lobe and a large posterior lobe, the latter sometimes called the 'proboscis gland' or 'acid gland' (Barkalova et al. 2016). Separate ducts from these two lobes join behind the nerve ring and run as a single duct to enter the buccal cavity dorso-laterally, as in other caenogastropods. These glands are covered with a thin layer of muscle which contracts to force their secretions through the ducts into the buccal cavity where they open. The glands are comprised of tubules lined with large cells that produce sulphuric acid (Day 1969; Houbrick & Fretter 1969; Nüske 1973; Fänge & Lidman 1976; Hughes & Hughes 1981). This is utilised in boring holes in the tests of echinoderm prey and may also assist with digestion (Andrews et al. 1999).

While the salivary gland complex of tonnoideans has sometimes been thought to include accessory salivary glands, both parts are modifications of the acinous salivary glands (Andrews et al. 1999; Barkalova et al. 2016). The acinous lobe secretes mucins, while the 'proboscis gland' produces the acidic secretion. The salivary glands of some tonnoideans have also been shown to produce toxins and enzymes (Andrews et al. 1999). A chelating agent is present in the 'acid gland' secretion that makes its penetration of calcareous shell more effective (Day 1969; Andrews et al. 1999). A peptide toxin is probably secreted by cells in the salivary ducts, while the small anterior acinar lobes may produce proteolytic enzymes (Andrews et al. 1999). The first reported tonnoidean toxin was from *Cassis tuberosa* and was unidentified (Cornman 1963). A neurotoxin that targets nicotinic-like acetylcholine receptors was reported in *Cymatium intermedius* (West et al. 1998), and a salivary toxin from *Charonia saulae* induced haemoptysis in mice (Shiomi et al. 1994). Proteinaceous toxins ('echotoxins') from the salivary glands of *Monoplex parthenopeus* have been rather well studied and at least five different components have been identified that react with the erythrocyte membrane when injected into mice, causing haemolysis (Shiomi et al. 2002; Kawashima 2003; Gunji et al. 2010). One of the echotoxin components is very similar to actinoporins from sea anemones (Kawashima 2003; Gunji et al. 2010).

The salivary glands of at least some buccinoidean neogastropods secrete toxic substances such as tetramine and smaller amounts of histamine, choline, and choline ester (Endean 1972) and, besides tetramine, three additional unidentified toxins that appear to inhibit neuronal Ca^{2+} channels were identified in the salivary glands of the buccinid *Neptunea antiqua* (Power et al. 2002).

Tetramine (tetramethylammonium) ($(CH_3)_4N^+$) is a paralytic neurotoxin that affects the autonomic nervous system by blocking nicotinic acetylcholine receptors. There are high levels present in the acinous salivary glands of species of *Neptunea* (e.g., Fänge 1960; Power et al. 2002) and some species of *Buccinum*, but other species of this latter genus have insignificant levels (Shiomi et al. 1994). Another buccinoidean with high levels of this toxin is *Hemifusus tuber* (Melongenidae) (Kawashima et al. 2002).

No pharmacological activity was found in the acinous salivary glands of *Nucella* (Andrews et al. 1991), but in another muricid, *Stramonita haemastoma*, these glands contain a substance toxic to mice that is a powerful vasodilator and a CNS depressant (Huang & Mir 1972).

These differences in the products of the acinous salivary glands in neogastropods may be reflected in their histology as they have, in at least one nassariid, three types of cells, while in the muricid *Nucella* there are only two (Andrews 1991), but it is not known how general these differences are. Like the pancreatic acinar cells of mammals, the enzyme-secreting cells of the (acinous) salivary glands of *Nucella* are specialised for regulated secretion (Andrews 1991). One type of nassariid salivary gland cell produces a glycoprotein secretion rich in disulphide groups similar to that expressed in the accessory salivary glands of *Nucella* (Martoja 1964; Minniti 1986; Fretter & Graham 1994) (see next Section). Extracts of salivary glands of *Colubraria reticulata*, a suctorial feeder on the blood of fish, have anticoagulant properties (Modica & Holford 2010; Modica et al. 2015). While most toxins in species of *Conus* are produced in the venom (or 'poison') gland (see Section 5.3.2.7.1.3), the salivary glands also secrete substances that closely resemble conopeptides and are different from those in the venom gland (Biggs et al. 2008).

As with other major groups of gastropods, the salivary glands of heterobranchs have only been investigated in ultrastructural detail in a few taxa. In 'opisthobranchs', early histological studies on the salivary glands have been supplemented by recent studies of the herbivorous euopisthobranchs *Aplysia* (Lobo-da-Cunha 2001, 2002) and *Bulla* (Lobo-da-Cunha & Calado 2008), and the carnivorous *Philinopsis* (Lobo-da-Cunha et al. 2009), all of which have ribbon-shaped glands. Whereas the salivary glands of *Bulla* contain two types of secretory cells, those of *Aplysia* and *Philinopsis* contain three types. While all produce acidic mucopolysaccharides, the salivary glands of *Philinopsis* also secrete substantial amounts of protein, probably including digestive enzymes, although little is known of the enzymes involved in 'opisthobranch' salivary secretion (see Section 5.4.2.3). The salivary glands of other cephalaspideans were investigated by Lobo-da-Cunha et al. (2015).

Some nudibranchs lack salivary glands, but instead have glands known as *ptyaline glands* (or oral glands) associated with the anterior part of the oesophagus. These glands have not yet been studied in detail (Gosliner 1994; Lobo-da-Cunha et al. 2009).

The salivary glands of the marine eupulmonate *Otina otis* have a single type of secretory cell, and small ciliated cells (Morton 1955b), but four or more cell types are found in other 'pulmonates' (e.g., Walker 1970b; Serrano et al. 1996; Luchtel et al. 1997). The glands may be simple and tubular (as in planorbids), or with ramified ducts (as in Lymnaeidae and Stylommatophora) (Luchtel et al. 1997; Moura et al. 2004). As in other gastropods, the saliva of 'pulmonates' mainly has a lubricatory function, but in some species, enzymes have been reported (see Section 5.4.2.3).

5.3.2.5.3 Accessory Salivary Glands of Gastropoda

Several kinds of 'accessory salivary glands' have been described in gastropods (e.g., Fretter & Graham 1962, 1994). Some lottioids have a second type of gland that opens to the buccal pouches or the oesophagus (Fretter & Graham 1962), but these are unlike those seen in other gastropods (Andrews 1991), and little is known about them. Other groups possessing accessory salivary glands are epitonioideans, and several groups of neogastropods (Ponder 1974; Andrews 1991). Reports of accessory glands in neritoideans refer to glands in the sublingual pouch (see Section 5.3.2.2.2).

The accessory glands in the Epitonioidea (Fretter & Graham 1962; Graham 1965; Collin 2000; Strong 2003) are tubular and open labially at the base of the proboscis sheath as cuticular stylets (Strong 2003). There have been no studies on their secretion, although it is likely that they are toxic and involved in narcotisation of their prey. Andrews (1991) suggested a possible homology of the epitonioidean accessory salivary glands with those of neogastropods because they both open labially and are tubular in structure, but they are histologically distinct, so homology is unlikely (Ball et al. 1997).

The accessory salivary gland(s) found in several groups of neogastropods have a single, anterior opening at the lower lip and are structurally similar throughout the group (Ponder 1974; Andrews 1991). They are absent in buccinoideans and are present in most (but not all) other neogastropod superfamilies (Ponder 1974; Taylor & Morris 1988; Kantor 1996, 2002; Modica & Holford 2010), including some conoideans (e.g., Schultz 1983; Taylor et al. 1993). The accessory salivary glands develop from invaginations of the lower lip (Ball et al. 1997), and their narrow ducts join (when a pair is present) just before discharging at the lower lip of the mouth. In the few taxa actually tested, their secretions can contain pharmacologically active components that are toxic and used to paralyse prey, but enzymes are lacking (Andrews et al. 1991). For example, the paralytic neurotoxin serotonin is present in these glands in the muricid *Nucella* (West et al. 1994). Serotonin, a powerful muscle relaxant in the venom of many arthropods, is found in some buccinoidean acinous salivary glands (see previous Section), and in *Conus* venom (McIntosh et al. 1993), but there is about eight times the concentration of this toxin in the smaller accessory salivary glands of *Nucella lapillus* (West et al. 1994). West et al. pointed out that the accessory salivary glands have structural similarities to the non-homologous *Conus* venom gland, mainly in that both possess an external muscle layer, and they also share some histochemical characteristics. The muscular coating of the accessory salivary glands may enable the secretion to be forcefully injected into prey. West et al. (1994) also noted other effects from the accessory salivary gland extract that could not be attributed to serotonin, suggesting that other bioactive compounds may be present. A toxin has also been demonstrated in the costellariid *Mitromica foveata*, and this may originate in the accessory salivary glands (Maes & Raeihle 1975). Clearly, much remains to be done in this area of research.

The toxic secretions from the accessory salivary gland(s) presumably largely assist in relaxing the muscles of the prey to enable easy access to the tissue. Although accessory glands are absent in some neogastropods, some of those appear to have compensated for this lack by the acinous salivary glands producing toxins (see previous Section).

While the accessory salivary glands in muricoideans open labially, as they do in some conoideans (Marsh 1971; Schultz 1983), in other conoideans that possess these glands the proboscis is formed by the elongation of the oral tube (Taylor & Morris 1988), so that the opening of the accessory salivary gland(s) is some distance from the buccal cavity. How the secretions from the glands are utilised in these taxa, and how they interact with those produced by the venom gland, is not known.

5.3.2.5.4 Buccal Glands of Cephalopods

There are three types of glands associated with the buccal area of cephalopods (Bidder 1966; Boucaud-Camou & Boucher-Rodoni 1983; Budelmann et al. 1997) that have been loosely described as 'salivary glands' (e.g., Budelmann et al. 1997) and include the submandibular gland and the posterior venom glands.

The single submandibular gland (or 'sublingual' or 'subradular' gland) lies below the salivary papilla through which the posterior glands open. This lobulate gland is represented by two small folds in *Nautilus* and is especially well developed in octopods and *Vampyroteuthis*. The function of this gland is unclear, but may assist with lubrication.

Paired anterior buccal glands comprise ramifying tubules at the back of the dorsal buccal cavity. In sepioids, they lie within the lateral glandular lobes of the buccal mass, while in octopods they lie freely on either side of the buccal mass. The glands open via a narrow opening from the lateral lobes into the buccal cavity and produce large quantities of mucus which may assist in liquefying the secretion from the posterior buccal glands. They are missing in the cirrate octopods. According to Budelmann et al. (1997), these glands are present in the lateral pharyngeal lobes in *Nautilus*, but according to Bidder (1966) and Salvini-Plawen (1988), they are not, although the lateral lobes in *Nautilus* have a glandular epithelium. These glands are possibly the homologues of the salivary glands in polyplacophorans, monoplacophorans, and gastropods, but this is by no means certain.

Large, paired, or single posterior buccal glands are often called salivary or venom glands, although their homology with salivary glands of other molluscs is unlikely (e.g., Salvini-Plawen 1988). These large glands are only found in coleoids, where they lie well behind the buccal mass, although they may be represented by 'lateral glands' on the buccal bulb in *Nautilus*. They are paired in octopods and sepioids, but are single in teuthoids and *Cirroteuthis*. These posterior buccal glands are well developed in octopods and are larger in cuttlefish than in squid, but absent in cirrate octopods and *Spirula* (Bidder 1966). In at least some octopods, the salivary duct (from the posterior buccal gland[s]) opens to a ventral salivary papilla. In coleoids, the salivary glands are composed of numerous tubules clustered into many lobules. They produce a viscous secretion that is passed through a long, single duct which runs alongside the oesophagus through the nerve ring ('brain') to the opening (which may be on the muscular salivary papilla) in the floor of the anterior buccal cavity just behind the jaws. In octopuses, this papilla is covered with a thin cuticle formed into small tooth-like projections. Larger teeth present at the opening of the salivary duct can be everted to drill shells (Nixon 1980). The secretion from the posterior buccal gland(s) is a mixture of mucus and various biologically active substances, the function of which is to rapidly immobilise the prey. Certain enzymes it contains may have a predigestive function with some taxa showing evidence of external predigestion (e.g., Nixon 1984; Kasugai 2001).

The posterior buccal gland(s) and their ducts are under nervous control, being well supplied with nerves from the subradular ganglion and the superior buccal lobe of the brain.

Budelmann et al. (1997) listed substances detected in the posterior buccal glands of coleoids as:

- Neutral and acid proteoglycans.
- Toxins including eledoisin (a polypeptide), cephalotoxin (a neurotoxic glycoprotein), and maculotoxin (a tetrodotoxin found in the blue-ringed octopus).
- Proteolytic enzymes and hyaluronidase.
- Cardioexcitatory and vasodilatory transmitters such as acetylcholine (ACh), 5-hydroxytryptamine, (5-HT) and some catecholamines. These substances facilitate the rapid distribution of the toxins within the vascular system of the prey.

5.3.2.6 Buccal and Oesophageal Pouches

Lateral pockets of the buccal cavity and/or the most anterior part of the oesophagus are present in some basal gastropods and caenogastropods. In lower gastropods (patellogastropods, vetigastropods, cocculinids, and neritimorphs) and in some caenogastropods, the *buccal pouches* form simple pockets on either side of the dorsal food channel in the buccal cavity (Salvini-Plawen & Haszprunar 1987). In some caenogastropods, such as Ampullariidae and Littorinidae, a pair of glandular *oesophageal pouches* with narrow openings arise from the most anterior part of the oesophagus. The function of these pouches is obscure, although Andrews (1964) noted that in an ampullariid they produced an acidic mucopolysaccharide.

The terminology relating to buccal and oesophageal pouches across gastropods and their homology is confused in the literature. While both buccal and oesophageal pouches have been treated as putatively homologous (Sasaki 1998; Strong 2003), the latter, at least in some taxa, appear to be related to the oesophageal gland (Ponder 1983) (see Section 5.3.2.7.1.1).

5.3.2.7 The Oesophagus

Food particles are transported to the stomach via the oesophagus, which is variously modified in different molluscs and, in some, glandular structures are associated with it. In all molluscs where the nerve ring is behind the buccal mass, the oesophagus narrows as it passes through this ring.

The oesophagus of aplacophorans may be short to moderately long. In solenogasters, there is a distinct oesophagus immediately behind the radula and, while this is mostly provided with gland cells with subepithelial cell bodies, there is no distinct oesophageal gland.

In chitons, the short oesophagus is divided into anterior and posterior sections. The anterior part is laterally extended into pouch-like glandular structures which, posteriorly, receive the openings of the oesophageal glands (= pharyngeal diverticula or sugar glands) (see below). A pair of dorsal ciliated tracts lie on either side of the food groove, and there is also a ventral ciliated tract. The short, ciliated, simple posterior oesophagus is separated from the stomach by a sphincter (Fretter 1937).

As in chitons, the anterior part of the monoplacophoran oesophagus contains a dorsal food channel. The middle part receives the ducts of the two pairs of oesophageal pouches (see below), and the ciliated posterior oesophagus runs to the stomach.

In cephalopods, the anterior ectodermal portion of the oesophagus lies anterior to the nerve ring ('brain') and passes through it. Posterior to the nerve ring, the endodermal cuticularised part of the oesophagus has, in most octopods and *Nautilus*, an enlarged anterior portion – the crop or 'prestomach'. Muscle fibres in the wall of the oesophagus create peristaltic movements.

The oesophagus of autobranch bivalves is a simple ciliated tube, but in many protobranchs, the anterior portion has a dorsal food channel bordered by a pair of ridges, as in many other molluscs (Salvini-Plawen 1988).

In heterobranchs, the oesophagus is usually a simple ciliated tube throughout, but in some may be modified as a crop or as a gizzard (see below) and show some diversification, as for example, in *Philinopsis depicta* (Cephalaspidea) the dilated oesophagus forms a crop with two large internal folds creating a food channel (Lobo-da-Cunha et al. 2011) (Figure 5.42), while in *Aglaja tricolorata* oesophageal pouches contain secretory cells (Lobo-da-Cunha et al. 2014).

In gastropods other than heterobranchs, the oesophagus can be divided into anterior, mid, and posterior sections. The anterior oesophagus is usually simple apart from bearing dorsal and, sometimes, ventral ridges. The midoesophagus may or may not have a simple to highly modified oesophageal gland and, in neogastropods, often a valve of Leiblein. These structures are described in more detail below (see next Section). The primitive dorsal and ventral folds in the oesophagus are lost in nearly all heterobranchs and most caenogastropods, although they are present in architaenioglossans, with rudiments in some other lower caenogastropods (Strong 2003).

A pair of dorsal and ventral glandular folds bordering ciliated food-conducting channels is the plesiomorphic condition in the gastropod anterior oesophagus, and their subsequent development, modification, and loss are significant in gastropod evolution (Fretter & Graham 1962; Salvini-Plawen & Haszprunar 1987; Salvini-Plawen 1988; Fretter & Graham 1994; Ponder & Lindberg 1997; Strong 2003). Ventral folds are present in patellogastropods, vetigastropods, and neritimorphs and have generally been regarded as mostly absent from caenogastropods, although Strong (2003) recognised putative rudimentary ventral folds in many non-neogastropod caenogastropods (see also Chapter 19). In vetigastropods and patellogastropods, the ventral folds are well developed, and a ciliary tract runs between them in the anterior oesophagus. This may become a prominent T-shaped structure in the anterior to midoesophagus due to the expansion and meeting of the oesophageal glands ventrally (Strong 2003).

While the posterior oesophagus is typically a simple narrow tube, in several groups of gastropods it is expanded into a crop. An oesophageal crop is seen in some 'opisthobranchs'. Most Acteonidae and Hydatinidae have a large crop used to briefly store ingested polychaete prey. A smaller crop and a lateral expansion, the oesophageal caecum, are found in *Toledonia* and sacoglossans. In some euopisthobranch, notably cephalaspideans, thecosome pteropods, and aplysioids, the crop is highly muscular and has internal chitinous plates forming an oesophageal gizzard (Figure 5.41 and Chapter 20). In some cephalaspideans, there are three equal-sized plates (most Bulloidea and Philinoidea) (Figure 5.41); most thecosomes have four equal-sized plates, and sea hares (Aplysioidea) have many plates of several sizes arranged in rows. A gizzard formed from the anterior stomach is also found in some 'pulmonates' (see Section 5.3.3.1.6 and Chapter 20).

5.3.2.7.1 Structures Associated with the Midoesophagus

In polyplacophorans, two large, lateral glandular pouches open narrowly to the anterior part of the short, simple oesophagus. These structures have been variously called sugar glands, pharyngeal[9] diverticula, pouches, or glands in monoplacophorans, polyplacophorans, aplacophorans, and gastropods. Internally, the pouches contain numerous villi lined with a glandular epithelium which produces carbohydrate digesting enzymes (carbohydrases) (Meeuse & Fluegel 1958, 1959) and may also be involved in nutrient absorption (Andrews & Thorogood 2005). Monoplacophorans also have large oesophageal pouches, but they open to the middle part of the oesophagus. Unlike other 'conchiferans' and chitons, there are two pairs of these pouches in all species examined except the minute *Micropilina arntzi* (Haszprunar & Schaefer 1997). They were thought, incorrectly, to be a dorsal coelom by Lemche and Wingstrand (1959). These large pouches run posteriorly into the viscera and are lined with squamous to cuboidal ciliated cells with a dense microvillar border. The pouches in chitons and monoplacophorans are probably homologous with the oesophageal glands of gastropods and scaphopods (Wingstrand 1985; Salvini-Plawen 1988). Somewhat similar pouches, although much reduced, also occur in protobranch bivalves (Salvini-Plawen 1988).

5.3.2.7.1.1 Oesophageal Glands of Gastropoda

Strong (2003) noted the confusion surrounding the use of the terms buccal and oesophageal pouches (see Section 5.3.2.6) and oesophageal glands in gastropods, and restricted 'oesophageal gland' to the lateral glandular pouches of the midoesophagus (see Figure 5.42). In some lower gastropods, there is ventral glandular tissue associated with the ventral fold(s), and this may be paired, as in the neritimorph *Theodoxus fluviatilis* or a 'solid disc' as in the cocculinid *Macleaniella*. This ventral glandular tissue has been interpreted as being continuous with the glands of the oesophageal pouches (although they are histologically distinct), and all these glandular structures have been treated as the oesophageal gland (Salvini-Plawen & Haszprunar 1987; Sasaki 1998). This 'ventral glandular mass' is sometimes confined to the anterior oesophagus (e.g., *Macleaniella*, Strong 2003) and is absent in caenogastropods and heterobranchs, while the oesophageal gland proper (= midoesophageal gland of Strong 2003) is present in lower gastropods and caenogastropods, and absent in heterobranchs.

5.3.2.7.1.2 The Oesophageal Gland Proper (= Midoesophageal Gland)

In 'prosobranch' gastropods, the midoesophageal gland(s) have undergone a complex evolution (Amaudrut 1898; Fretter & Graham 1962; Ponder 1970b, 1974; Salvini-Plawen & Haszprunar 1987; Salvini-Plawen 1988; Ponder & Lindberg 1997; Sasaki 1998; Strong 2003; Andrews & Thorogood 2005). Some gastropods such as patellogastropods, vetigastropods, and many caenogastropods have large oesophageal pouches with internal septa, and in vetigastropods these septae are modified, with the lining forming papillae (Salvini-Plawen & Haszprunar 1987). In the vent vetigastropod *Chrysomallon squamiferum*, the midoesophageal glands are hypertrophied and house chemosynthetic bacteria (Warén et al. 2003; Goffredi et al. 2004b; Chen et al. 2015). In neritids, the oesophageal glands have two lobes with different histology and are probably a combination of buccal pouches and oesophageal glands (Whitaker 1951; Fretter 1984b; Strong 2003) (see also Section 5.3.2.6). In those oesophageal glands investigated, cilia beat from the glands into the oesophagus.

At least five main states can be recognised for the midoesophageal gland. The most plesiomorphic condition is an internally simple gland confined to lateral pouches (as in cocculinids and the neritimorphs), or septate (as in patellogastropods). An expanded, internally papillate oesophageal gland is found in most vetigastropods, and an internally septate gland is typical of many 'asiphonate'[10] caenogastropods (Strong 2003). The oesophageal gland is stripped from the oesophagus to form a gland of Leiblein or a venom gland in many neogastropods and is confined to a central glandular strip in the midoesophagus in cancellariids (Graham 1966). The oesophageal gland has been lost in many

---

[9] This term (and buccopharyngeal) suggests the presence of a pharynx in the molluscan digestive system; an anatomical region of the digestive tract well delineated in vertebrates, but poorly defined in invertebrate organisms and therefore not used here.

[10] We use the informal name 'asiphonate clade' to encompass several superfamilies of caenogastropods including Littorinoidea, Naticoidea, Triphoroidea, Rissooidea, and Truncatelloidea following Colgan et al. (2007).

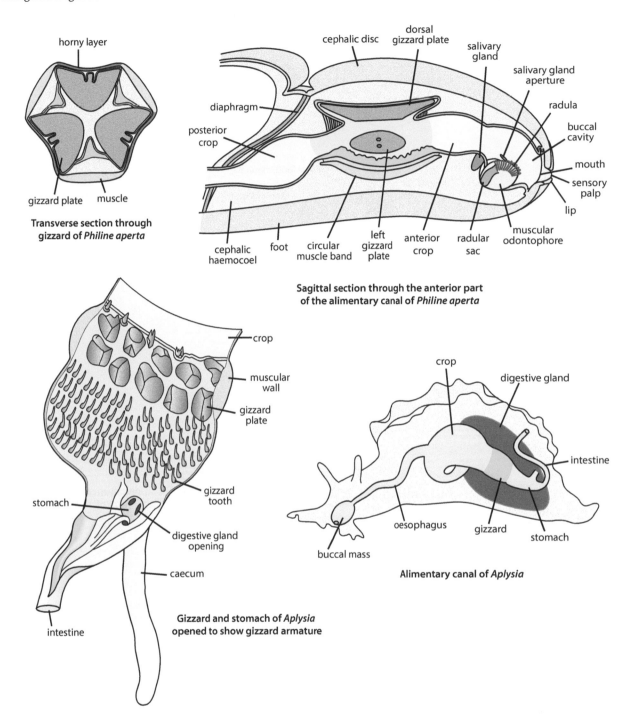

FIGURE 5.41 Highly modified oesophageal gizzards of the euopisthobranchs *Philine* and *Aplysia*. Illustrations redrawn and modified from the following sources: *Philine* transverse section Brown, H.H., *Trans. Royal Soc. Edin. Earth Environ. Sci.*, 58, 179–210, 1934, sagittal section Fretter, V., *Trans. Royal Soc. Edin.*, 59, 599–646, 1939; *Aplysia* crop Cuvier, G., *Annales du Muséum d'histoire naturelle Paris*, 2, 287–314, 1803, diagrammatic view of gut modified from http://www.asnailsodyssey.com/LEARNABOUT/MOLLUSCA/nudiFood.php).

caenogastropods, as in most (but not all) of those with a crystalline style (e.g., Graham 1939) and in some groups of neogastropods (Ponder 1974).

In *Patella*, the oesophageal glands take up radioactive glucose and amino acids and may also absorb dissolved organic material, perhaps taking over some functions of the reduced stomach (Bush 1989). *Littorina* has an epithelium in the oesophageal gland like that seen in *Patella*, suggesting a similar function, and this may be a plesiomorphic condition of the gastropod midoesophagus (Andrews & Thorogood 2005). The ability to absorb nutrients in the oesophagus has interesting implications, including the ability to uptake food via incoming water as shown in the muricid *Reishia clavigera* (Lau & Leung 2004) (see Section 5.5).

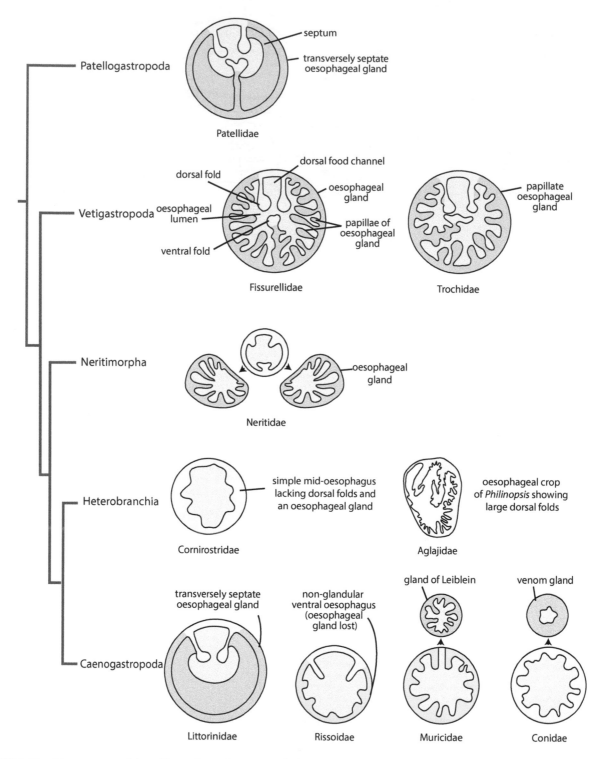

FIGURE 5.42 The structure of the midoesophagus in gastropods shown by diagrammatic transverse sections (TSs). Redrawn and modified in part from Fretter, V. and Graham, A.L., *British Prosobranch Molluscs: Their Functional Anatomy and Ecology*, Ray Society, London, UK, 1962. Left-hand heterobranch TS redrawn and modified from Ponder, W.F., *J. Molluscan Stud.*, 56, 533–555, 1990a and right hand-figure redrawn and modified from Lobo-da-Cunha, A. et al., *J. Molluscan Stud.*, 77, 322–331, 2011.

In carnivorous caenogastropods, there is a tendency for the dorsal folds to separate the dorsal and ventral parts of the midoesophagus. This has occurred independently, along with developing a proboscis, in neogastropods and some other carnivorous caenogastropods, by the dorsal folds meeting or overlapping (see Andrews & Thorogood 2005, fig. 19). The evolution of an acrembolic proboscis in the 'ptenoglossan' families and the Naticoidea has not resulted in a major displacement of the dorsal folds, nor has separation of the dorsal and ventral parts of the oesophagus occurred in tonnoideans to the same extent it has in neo-gastropods. In naticoideans, epitoniids, and tonnoideans, the oesophageal gland remains connected to the oesophagus along its whole length, but is separated from the food channel by overlapping dorsal folds, the left overlapping the right.

While most tonnoideans have a well-developed oesophageal gland, the large holothurian-eating tonnoidean *Tonna* lacks this gland, but has a so-called 'Delle Chiaje' organ (Weber 1927). This resembles the oesophageal glands, but arises from the anterior oesophagus, opens to the posterior buccal cavity, and has a posterior caecum ending in a bulb. Some have suggested that it is a modified buccal gland (Day 1969; Hughes & Hughes 1981), but Andrews and Thorogood (2005) argued that it is probably an oesophageal gland that has migrated anteriorly. Indeed, it has been suggested that all foregut glands (other than the salivary glands and accessory salivary glands) are derived from the same glandular tracts in the oesophagus (Fretter & Graham 1994), including the buccal pouches, oesophageal glands, and their derivatives.

The oesophageal gland in neogastropods has been stripped from the oesophagus and, in many taxa, so have the glandular dorsal folds which join ventrally to form a duct; together they form the gland of Leiblein. This structure is large in some neogastropods, particularly in muricids, and may be important not only in digestion and absorption, but also in storing food reserves and in the uptake of heavy metals (see Chapter 6) (Andrews & Thorogood 2005) (see also Section). The gland of Leiblein is associated with the valve of Leiblein in neogastropods (see next Section). Like many muricids, some naticids and tonnoideans bore holes in their prey, but lack a valve of Leiblein and a separated oesophageal gland. This indicates these modifications to the oesophagus are not functionally required if a proboscis is present, or for feeding activities such as boring, although it may be related to the type of proboscis.

5.3.2.7.1.3   The Valve of Leiblein and the Gland of Leiblein of Neogastropoda   The separation of the dorsal food channel and the dorsal folds from the oesophagus, along with the oesophageal gland, to form a gland of Leiblein, occurs to a varying degree in many neogastropods. Similarly, in many neogastropods, there is a valve of Leiblein, a glandular pyriform swelling that lies at the junction of the anterior oesophagus and midoesophagus. Its structure varies a little between different neogastropod taxa, and because the valve and its function in the muricid *Nucella* (Graham 1941; Andrews & Thorogood 2005) has been described in some detail, it can serve as a model for many other neogastropods. Internally, it has a ciliated valve-like cone, and there is much glandular epithelium giving the valve a whitish appearance in life. The siphon-like structure within the valve effectively prevents back-flow from the midoesophagus (Andrews & Thorogood 2005), thus preventing regurgitation of the oesophageal contents, especially during the extension of the proboscis.

The homology of the valve of Leiblein with foregut structures in other caenogastropods has long been debated (see Chapter 19), but it has been recently suggested (Golding & Ponder 2010) that this structure may be derived from glandular dorsal folds of the anterior oesophagus, similar to those seen in tonnoideans.

In many neogastropods, the valve of Leiblein is reduced or lost, in others it is hypertrophied. The valve is most prominent in species with a large gland of Leiblein, and it has been suggested that the function of the valve is to retain the enzymatic secretion of the gland in the midoesophagus (Ponder 1974). Andrews and Thorogood (2005) argued that its larger size in muricids is probably linked with the large amounts of partially digested material in the midoesophagus. A valve of Leiblein is not necessarily correlated with the disposition of the gland of Leiblein. In most neogastropods, the duct of the gland of Leiblein opens to the oesophagus behind the valve, but in marginellids and conoideans, it opens to the buccal area and such taxa typically lack a valve of Leiblein. However, a valve of Leiblein, or a rudiment of it (an oesophageal caecum), is present in at least some marginellids (Ponder 1970b; Coovert & Coovert 1995; Strong 2003) and a valve is also found in some raphitomid Conoidea (Kantor & Taylor 2002). As particles enter the valve, they are bound up in a sticky secretion and squeezed into a string by thick lateral pads. As the string passes into the midoesophagus, it is covered with secretions from the gland of Leiblein (Graham 1941). Buccinoideans have reduced valves and glands of Leiblein, or have lost them, possibly because the food they swallow is in larger pieces and these pass rapidly through the muscular oesophagus (e.g., Payne & Crisp 1989; Andrews & Thorogood 2005).

The gland of Leiblein is modified in several families where the equivalent of the duct forms a long tubular structure with a small, often muscular, gland of Leiblein attached distally. This modification has occurred independently in Marginellidae (Ponder 1970b) and possibly the ancestral conoidean (see also Chapter 20). The most extreme example of this tubular type is the toxoglossan venom gland which opens directly to the buccal cavity. The long tube (the venom gland) is considered the homologue of the glandular dorsal folds of the midoesophagus, while some have suggested that the posterior bulb may be the homologue of the gland of Leiblein (Amaudrut 1898; Ponder 1970b, 1974). While there is support from developmental studies for this hypothesis for the origin of the venom gland, the homology of the bulb is equivocal (Page 2011).

While Volutidae, some Muricidae (Ponder 1974), and some Costellariidae (Fedosov & Kantor 2010) have achieved a somewhat similar stripping of the dorsal folds to form a tubular duct (Figure 5.43), in these groups, the duct opens just behind the valve of Leiblein so it cannot be used to envenomate prey.

The venom gland of conoideans, and in particular Conidae, produces a complex array of neurotoxic peptides (conotoxins) that consists of 10–40 amino acid residues held together by disulphide bonds. Different conotoxins have different inhibitory activities; alpha-conotoxins (α-conotoxins) inhibit the

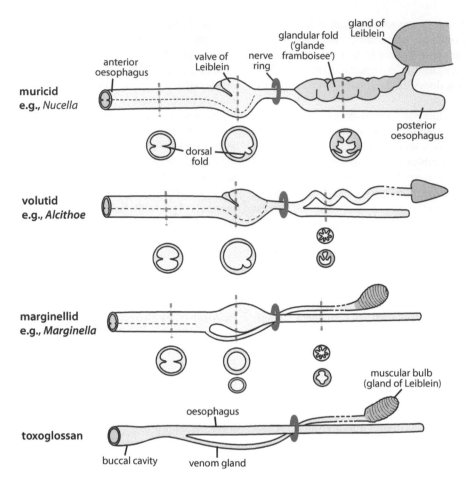

FIGURE 5.43 A series illustrating the possible derivation of the toxoglossan venom gland from the 'glande framboisee' of a rachiglossan in Neogastropoda. Redrawn and modified from Ponder, W.F., *Malacologia*, 12, 295–338, 1974, and Fretter, V. and Graham, A.L., *British Prosobranch Molluscs: Their Functional Anatomy and Ecology*, Ray Society, London, UK, 1994.

nicotinic acetylcholine receptors of nerves and muscles, delta-conotoxins (δ-conotoxins) inhibit the inactivation of voltage-dependent sodium channels, kappa-conotoxins (κ-conotoxins) inhibit potassium channels, mu-conotoxins (μ-conotoxins) inhibit voltage-dependent sodium channels in muscles, and omega-conotoxins (ω-conotoxins) inhibit N-type voltage-dependent calcium channels. This latter type of conotoxin has medical implications because this kind of channel is related to pain sensitivity (algesia) (see Chapter 10). Conantonkins are a sixth type of conotoxin that blocks nerve impulses which use glutamic acid (instead of acetycholine) as the neurotransmitter. Injection of *Conus* venom can cause death in humans.

5.3.3   The Midgut

5.3.3.1   The Stomach and Style Sac

The stomach and associated structures, including the style sac and digestive glands (see below), comprise the midgut. In general, food particles enter the stomach from the oesophagus. The beating cilia on the ridges of sorting areas on the floor of the stomach keep light food particles suspended where they can be acted upon by digestive enzymes before being moved into the digestive glands via narrow openings or ducts. Cilia in the grooves of the sorting area carry the heavier particles to the intestinal groove (which runs alongside the major typhlosole), where they are moved into the intestine to form one component of the faecal material. The stomach may sometimes have an outpocketing generally called a 'caecum' or 'gastric caecum', but these structures are not all homologous. The stomach wall also consists of cuticularised areas, with one area, the gastric shield, being especially thickened and standing apart.

The gastric cuticular lining, and especially the gastric shield, is thought to protect the epithelium of the stomach wall from abrasion, particularly from the rotating head of the protostyle or crystalline style, although the gastric shield is often found where there is no style and also in some carnivorous taxa where coarse food particles are not present. While it is probably retained in these taxa as a plesiomorphy, it may also serve some other function (see Section 5.4.2.7).

The style sac is associated with, or wholly or partially separated from, the intestinal opening. Two prominent ridges, the typhlosoles, run into this then extend across the stomach in various ways, depending on the group. The style sac has on its walls long brush-like cilia which beat to maintain a circular motion, revolving either mucus-bound waste material (the 'protostyle') or a crystalline style.

A crystalline style is found in autobranch bivalves, some caenogastropods, and at least one basal heterobranch (Salvini-Plawen 1981b; Ponder & Lindberg 1997). In autobranch bivalves, it is secreted by special cells of the typhlosoles, not the style sac epithelium (see Section 5.3.3.1.5.2). It is a hyaline mucoprotein rod rotated by long cilia on the walls of the style sac, and enzymes (see below) and carbonic acid are released from its dissolving head as it rotates within the gastric chamber, which has a relatively high pH. The carbonic acid buffers the stomach contents so pH levels are optimal for enzymatic activity (e.g., Yonge 1930). The crystalline style is often not a permanent structure in the stomach in many bivalves, and its size can vary with tidal or other environmental cues in a cyclical way (see Section 5.4.4.1.1). Rotation of the style may also assist in circulating the particles in the stomach and help in bringing them into contact with the sorting area (Graham 1949; Morton 1952a, 1953; Reid 1965; Owen 1966).

The style sac is otherwise involved in faeces formation, and in those with a so-called protostyle, this is continuous with the faecal string. Crystalline styles have evolved independently from a protostyle in autobranch bivalves and in several groups of gastropods.

Figure 5.44 shows an overview of the general structure of the post-buccal gut in molluscs and, in particular, the stomach.

5.3.3.1.1 Aplacophora

The stomach of caudofoveates has the ventral digestive gland opening to it posteriorly. This contrasts with solenogasters, where the stomach has the digestive gland cells opening directly into it. In this latter group, the stomach is often extended dorso-anteriorly into a single or paired caecum. In both aplacophoran groups, a dorsal ciliary tract (as a band, fold, or groove) runs down the stomach and leads to the intestine. This structure appears to be at least analogous to the typhlosole(s) seen in the stomach of other molluscs (Scheltema 1981).

A protostyle is present in the more advanced caudofoveate stomachs (as in some of the carnivorous Chaetodermatidae) (Scheltema 1981). In the most derived members of the family, the protostyle rotates in a style sac formed from the anterior end of the intestine, and a gastric shield is present in the stomach. However, a protostyle and gastric shield are lacking in most aplacophorans (Salvini-Plawen 1988; Scheltema et al. 1994).

5.3.3.1.2 Polyplacophora

Primitive chitons have a small, simple stomach while in others the stomach is large, muscular, and asymmetrical (Fretter 1937; Salvini-Plawen 1988). This latter stomach type has a cuticle-lined or ciliated ventral sac with two ciliated bands running longitudinally along the dorsal wall and extending into the intestine, which leaves the stomach posteriorly to the left. The oesophagus opens anteriorly to the narrow stomach, from which it is separated by a sphincter. The two digestive gland ducts open

posteriorly, as does the intestine. The ventral sac may partly represent the gastric shield area (Graham 1949; Owen 1966) of the stomach in bivalves and gastropods. In chitons, it forms an expanded chamber for food storage; its muscular contractions also help to break up food particles facilitating the solely extracellular digestion mainly by way of proteolytic enzymes from the digestive glands (Fretter 1937; Salvini-Plawen 1988). From there food is passed into the anterior part of the intestine by muscular contractions and ciliary action, there being no rotation within the stomach (Salvini-Plawen 1988). This long part of the intestine is at least analogous to the style sac of other molluscs (Graham 1949; Owen 1966; Salvini-Plawen 1988), in having ciliated, typhlosole-like ridges running longitudinally and the contents rotating within it. A muscular valve, which pinches off pellets from the amorphous faecal material in the anterior part (Fretter 1937), separates it from the long posterior intestine.

5.3.3.1.3 Monoplacophora

Monoplacophorans have a small, almost bilaterally symmetrical, triangular stomach. In the larger species, the digestive glands open laterally, and in the smaller species, they open anteriorly at a common opening. The oesophagus opens anterodorsally, and the intestine leaves posteriorly to the left. Large monoplacophorans have a short protostyle (Haszprunar & Schaefer 1997), incorrectly called a crystalline style by Lemche and Wingstrand (1959) and Wingstrand (1985), which lies in a short blind caecum that runs posteriorly with its opening near the oesophageal opening. The smaller species apparently do not have a protostyle (Haszprunar & Schaefer 1997), but they do have the caecum. There is no sorting area or gastric shield and no typhlosoles (Lemche & Wingstrand 1959; Wingstrand 1985; Salvini-Plawen 1988; Haszprunar & Schaefer 1997). Digestion is both intracellular and extracellular (see Section 5.3.3.2.3).

5.3.3.1.4 Scaphopoda

Scaphopods have a small simplified stomach containing a cuticular area which may represent a relictual gastric shield, but there is no style sac (Morton 1959; Salvini-Plawen 1988), which Waller (1998) suggested may be a secondary loss. The intestine leaves the stomach anteriorly, and two large digestive glands open to it laterally. Digestion is both intracellular and extracellular (see Section 5.3.3.2.4).

5.3.3.1.5 Bivalvia

Bivalve stomachs are complex (Figure 5.45) and have been rather well studied. They receive the food already partly presorted by the labial palps (see Section 5.2.1.2.2.2 and Chapter 15). Because protobranch bivalves are deposit feeders, they have a less modified stomach than autobranchs (see below), and it is similar to that of some primitive gastropods. There are a simple sorting area and a style sac that contains a loose, gelatinous protostyle (Morton 1983a) and a major and minor typhlosole. Most other bivalves have a crystalline style (see Section 5.3.3.1.5.2), and its absence in protobranchs probably primary, although it has also been suggested to be secondarily absent (Waller 1998).

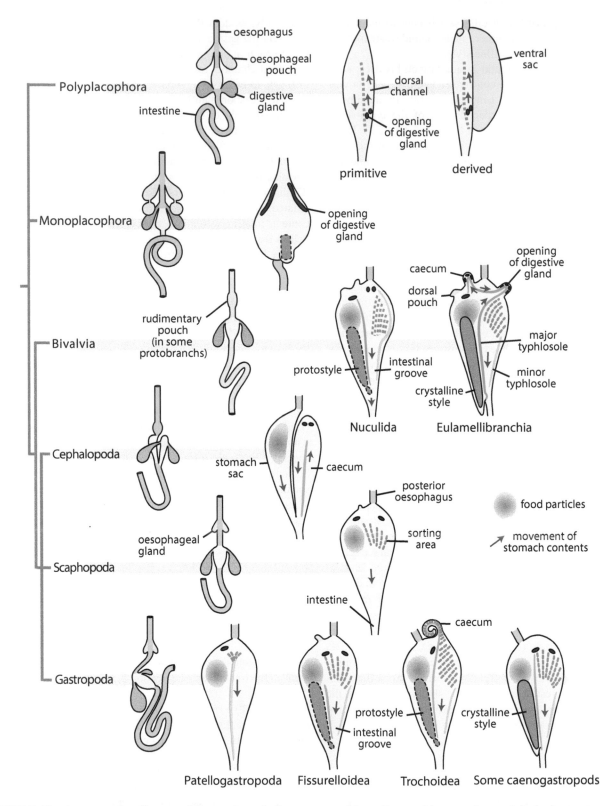

FIGURE 5.44   A comparative diagram of the post-buccal alimentary canal in molluscs, with comparisons of the basic structure of the stomach. Redrawn and modified from Salvini-Plawen, L.v., The structure and function of molluscan digestive systems, pp. 301–379, in Trueman, E.R. and Clarke, M.R., *Form and Function. The Mollusca*, Vol. 11, Academic Press, New York, 1988.

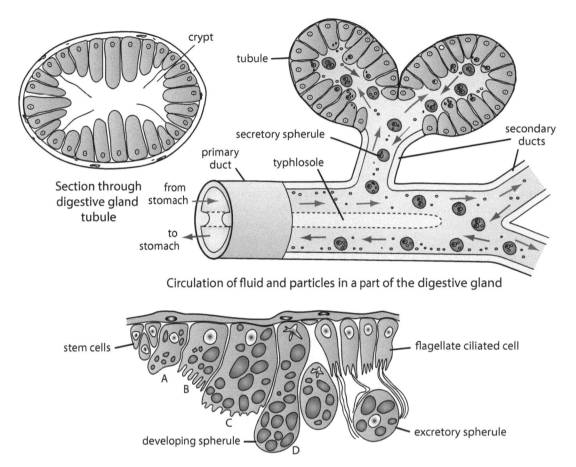

Section through digestive gland tubule

crypt

tubule

secretory spherule

primary duct

from stomach

to stomach

typhlosole

secondary ducts

Circulation of fluid and particles in a part of the digestive gland

stem cells

flagellate ciliated cell

A
B
C
D

developing spherule

excretory spherule

Cells lining walls of tubule in *Tridacna*, A-D stages in development of holocrine secretory cells

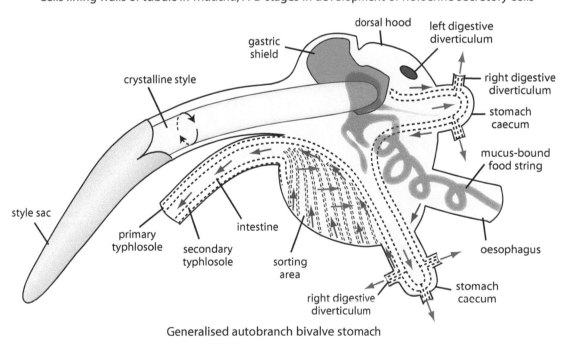

dorsal hood

left digestive diverticulum

gastric shield

crystalline style

right digestive diverticulum

stomach caecum

mucus-bound food string

style sac

primary typhlosole

secondary typhlosole

intestine

sorting area

right digestive diverticulum

oesophagus

stomach caecum

Generalised autobranch bivalve stomach

FIGURE 5.45 Stomach and digestive gland morphology of autobranch bivalves. Redrawn and modified from the following sources: top right Owen, G., *Q. J. Microsc. Sci.*, 96, 517–537, 1955, top left and middle Fankboner, P.V. and Reid, R.G.B., Nutrition in giant clams (Tridacnidae), pp. 195–209, in B. Morton, *The Bivalvia. Proceedings of a Memorial Symposium in Honour of Sir Charles Maurice Yonge (1899–1986)*, Edinburgh, 1986, Hong Kong University Press, Hong Kong, China, 1990, lower Morton, J.E., *Molluscs*, 1st ed., Hutchinson, London, UK, 1958a.

Stomach morphology has been used in higher classification (Purchon 1959, 1987b; Bieler et al. 2014). Purchon (1959) classified the bivalve stomach into five main types (see Chapter 15 for details), based on the pattern of ciliated waste grooves and how they were covered by the extension of the intestinal major typhlosole, modifications which presumably evolved to prevent clogging of the ducts to the digestive glands.

5.3.3.1.5.1 Aspects of Stomach Function in Bivalves

The rotating crystalline style and cilia on the stomach wall circulate particles in the stomach lumen, enhancing extracellular digestion and contact with sorting areas on the walls and floor of the stomach. The oesophagus and stomach contents are mildly acid (pH 5.0–6.0), rendering mucus less viscous and maintaining a slurry-like mixture with suspended particles – the conditions needed for efficient sorting (Figure 5.46).

Particle sorting in the stomach is the last opportunity for particle selection. Detailed early studies of bivalve stomach morphology and function (e.g., Yonge 1923, 1926a, 1939b, 1949; Graham 1949; Owen 1955, 1956; Reid 1965; Purchon 1987a) required opening the stomach which has resulted in some observational artefacts (Ward & Shumway 2004). The rotating style and gastric shield triturate material which enters the stomach from the oesophagus, breaking apart larger and aggregated particles. As the head of the style dissolves, enzymes are released and mixed with the stomach contents. Sorting of the particles occurs on the ridged sorting areas and in the stomach pouches. Small, light organic particles and fragments are passed into the ducts of the digestive glands, while the larger, denser material is moved to the intestinal

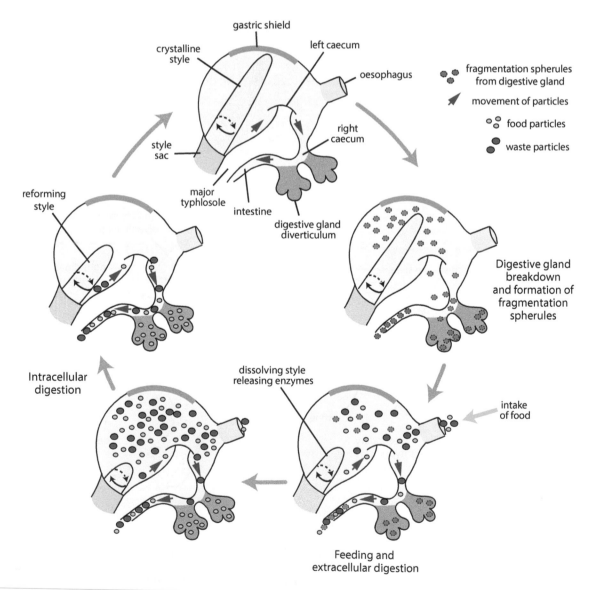

FIGURE 5.46 The digestive cycle in the midgut of an intertidal bivalve showing the crystalline style dissolving and reforming. Redrawn and modified from Morton, B., *Malacologia*, 14, 63–79, 1973a.

groove and then to the intestine. Extracellular digestion in the stomach reduces larger food particles to a size where they can be passed to the digestive glands for intracellular digestion.

Processing on the gastric sorting areas is probably affected mostly by particle density, with the heavier ones falling into rejection grooves. Thus, not only larger particles are removed, but also small, dense particles (e.g., silt) ingested with the food. Size and weight are particularly significant in sorting, but other factors such as extracellular digestibility or other qualitative factors may also be important. For example, gastric sorting in at least some bivalves is able to distinguish between otherwise identical particles with organic and inorganic coatings and live and dead phytoplankton cells (see Ward & Shumway 2004 for review and references).

Purchon (e.g., Purchon 1958) devised a classification of bivalve stomachs for which details are given in Chapter 15. Within the stomach, three main types of sorting mechanism were recognised by Reid (1965). The basic type (A) is found in all groups of autobranch bivalves except septibranchs, but the derived type (B) is found in only Purchon's type III stomachs (e.g., many pteriomorphians), and type C in type V stomachs (many heterodonts). Type (A) is widespread in bivalves (and other molluscan stomachs) and consists of a ridged surface with heavier particles carried by cilia in the grooves towards the intestine, and the cilia on the ridges keeping lighter particles in suspension. This basic type is present in most stomachs, even if more complex sorting mechanisms are also present. The more complex types (B and C) have evolved from this basic type. In the type B stomach, a narrow tongue derived from the major typhlosole extends into the intestinal groove (involved in rejection) and parts of the digestive gland duct tracts adjacent to that groove. This is typically associated with a sorting caecum and effectively moves the sorting area away from the openings to the digestive gland ducts. Type A and B sorting systems have the disadvantage of the ducts from the digestive glands sharing the same space where unsorted material from the oesophagus is received. In Purchon's type III stomachs, this conflict is resolved by having the digestive gland ducts opening into entirely separate spaces (right and left caeca), this being the C type sorting system. These stomach types are illustrated and described in more detail in Chapter 15.

The stomachs of the deposit-feeding tellinoideans have modifications that enable them to deal with much particulate material including larger particle sizes, of which only a small proportion is suitable for food (e.g., Lopez & Levinton 1987). The stomach of these bivalves has a large caecum for the temporary storage of particles, a larger crystalline style, and a thicker gastric shield than most other bivalves (Morton 1960a), as well as some having reduced sorting areas.

5.3.3.1.5.2 Style Sac and Crystalline Style The connection between the intestine and style sac differs in the different stomach types in bivalves. In Purchon's type I and type II stomachs the style region is open to the intestinal groove. In most bivalves with stomach type III, some with type IV, and a few with type V, the style is in a tube connected to the intestine via a narrow slit formed by a double (stomach type IV) or single (stomach type V) typhlosole. The closure of the slit between the style sac and the intestine results in the style sac being a separate caecum, although a groove remains where the slit has closed over. This has occurred in a few taxa with a type III stomach and many with the type V stomach (Morton 1983a).

The crystalline style of bivalves has been intensively studied. As it rotates within the style sac, it stirs the stomach contents, mediating extracellular digestion. The style is secreted from behind and dissolves at the free end which protrudes into the stomach lumen. Several enzymes are secreted within the style sac, incorporated in the style, and then released as it dissolves in the stomach lumen. The style sac itself does not have an absorptive function, as once thought.

The histology of the autobranch style sac is distinctive, with non-secretory columnar cells bearing long, rather stiff cilia which rotate the style. Taller cells on the typhlosole(s) are ciliated and secretory and may produce the style enzymes (Morton 1983a). They facilitate style rotation and may also assist in keeping it elevated from the opening to the intestine. Kato and Kubomura (1954) introduced a terminology for the cells associated with the style sac (Figure 5.47). The ciliated cells in the style sac that rotate the style are non-secretory and are termed the 'A cells'. Those on the typhlosoles are secretory, producing the style enzymes, and their cilia also assist in style rotation (Morton 1983a). There are a further two to three groups of cells in the intestinal groove below the typhlosole(s). The 'D cells' secrete the material that forms the style and are especially active in the basal end of the style sac (Judd 1979), while the 'C cells' opposite with short, rather stiff, cilia, help to transfer this secretion to the lumen of the style sac. A similar, but additional, type of secretory cell was described by Judd (1979) in the base of the intestinal groove in some bivalves with a separated style sac (see Morton 1983a, p. 92 for additional discussion and references).

The bivalve crystalline style is composed of protein, including high molecular weight glycoproteins (Judd 1987). Some of this protein is bound to sugars such as glucose, mannose, and galactose, but it is not known how these substances form the solid hyaline structure of the style. Digestive enzymes are liberated from the dissolving crystalline style (see Section 5.4.2.3.1).

5.3.3.1.6 Gastropoda

Gastropods (other than patellogastropods) are like scaphopods and cephalopods in having the intestine departing the stomach anteriorly. Although the patellogastropod stomach is highly simplified, from outgroup comparison the plesiomorphic gastropod stomach probably had a style sac, gastric shield, and sorting areas.

An outpocket, the *gastric caecum*, has been developed in different gastropod taxa, while others lack this. It is a derived feature, not a primitive one as has been suggested (e.g., Graham 1949, 1985). In vetigastropods, a caecum lies close to the gastric shield and may be small and crescent-shaped to large and spiral when developed as a sorting

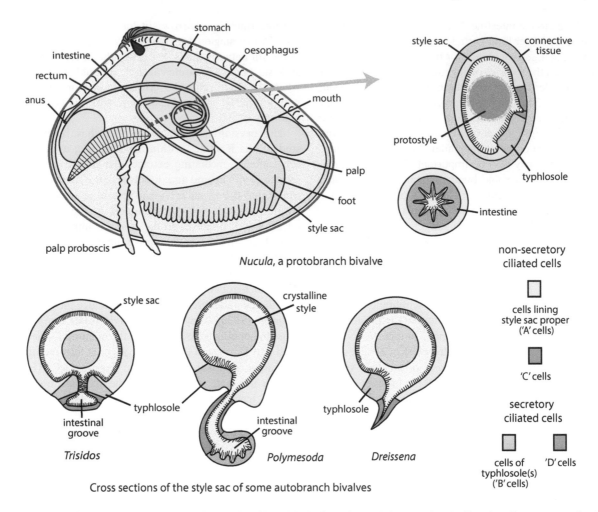

FIGURE 5.47 The bivalve style sac showing its relationship with the intestine and the associated ciliated and secretory cells. *Nucula* redrawn and modified from Yonge, C.M. and Thompson, T.E., *Living Marine Molluscs*, Collins, London, UK, 1976. Transverse sections as follows: top right figures based on unpublished figures by B. Morton with permission; lower figures redrawn and modified from Morton, B., Feeding and digestion in Bivalvia, pp. 65–147, in Saleuddin, A.S.M. and Wilbur, K.M., *Physiology, Part 2. The Mollusca*, Vol. 5, Academic Press, New York, 1983a.

area, as in Haliotidae, Pleurotomariidae, and Trochoidea (Figure 5.48), or it is absent, as in cocculinids. Extensions of the major and minor typhlosoles and intestinal groove run into the large vetigastropod gastric caecum which opens ventrally behind the gastric shield near the posterior end of the gastric chamber. In those taxa where the caecum is large and spiral, the sorting area that lies to the left of the minor typhlosole also extends into the caecum (Graham 1949; Morton 1955d; Strong 2003). Gastric caeca found in caenogastropods are usually small and not homologous with those in vetigastropods and, even within that group, are not all homologues (Strong 2003). There are also superficially similar gastric caeca in some lower 'pulmonates' as well as other molluscs (cephalopods, some chitons, and the rudimentary structures seen in some scaphopods and protobranch bivalves).

Comparison with outgroups suggests that the first gastropods probably had a stomach with a sorting area, a gastric shield, and a style sac with a protostyle, indicating that the crystalline style evolved independently in bivalves and in some gastropod taxa.

In most gastropods, the gastric sorting areas are largely unmodified, but in a few, they form a series of ridged leaflets that can almost fill the stomach lumen. Such structures are found in the caenogastropod *Alcithoe* (Volutidae) (Ponder 1970a) and in the heterobranch *Onchidium* (Awati & Karandikar 1940). The lower caenogastropod *Campanile* (Campanilidae) has leaflets around the opening to the digestive gland and these spiral down into the digestive gland ducts (Houbrick 1989).

When present in gastropods, the gastric shield typically forms a tooth-like projection of the cuticular lining. A gastric shield is absent in patellogastropods, and some neogastropods

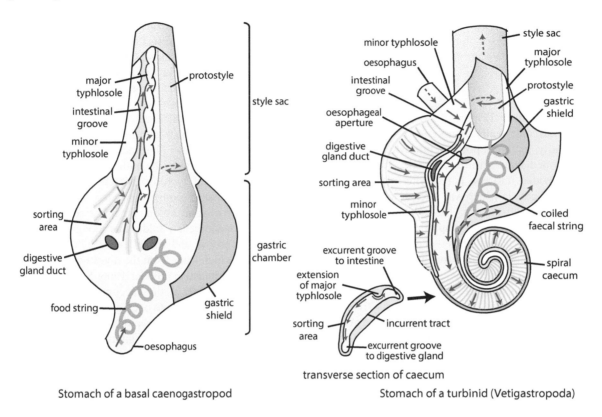

FIGURE 5.48 The stomach and style sac of a basal caenogastropod and a vetigastropod (a turbinid). They are shown opened dorsally with the cut dorsal wall reflected. Left figure redrawn and modified from Strong, E.E., *Zool. J. Linn. Soc.*, 137, 447–554, 2003 and right from Morton, J.E., *Proc. Malacol. Soc. Lond.*, 31, 123–137, 1955d and Morton, J.E., *Biol. Rev.*, 35, 92–140, 1960a.

and heterobranchs, other than some valvatoideans. Strong (2003) noted that the position of the gastric shield varied in the gastropods she examined. In the cocculinid, it was dorsal and left of the midline, while in the neritoidean, it was dorsal and to the right. In caenogastropods, the shield is ventrolateral, and in vetigastropods, it wass often large and occupies much of the right side of the stomach (Sasaki 1998). While it is possible that these different positions may indicate convergent origins of the shield, it is more likely that they are responses to different functional and/or developmental factors (Lindberg & Ponder 1996).

Distinctive tall cells with very uniform cilia line the style sac in the taxa possessing a crystalline style, and these cilia are instrumental in its rotation. These same ciliated cells are also found in the style sac in many gastropods that lack a crystalline style (but may have a protostyle), although in some they are lacking, and the style sac is instead lined with shorter cilia similar to those in some other parts of the stomach (Strong 2003). This latter arrangement is seen in cocculinids and in some caenogastropods such as naticids, cypraeids, epitoniids, and some neogastropods among the taxa examined by Strong (2003). It is also characteristic of patellogastropods (e.g., Graham 1932), whereas neritoideans and some neogastropods (Strong 2003), and vetigastropods, have the tall uniform cilia in the style sac. The style sac is not well

differentiated in most heterobranchs; of those investigated, only the primitive *Orbitestella* (Orbitestellidae) has the tall uniform cilia and a short crystalline style (Ponder 1990b).

In patellogastropods, the part of the gut behind the openings to the digestive glands is typically wider than the part anterior to it and has been interpreted as part of the stomach (e.g., Walker 1968; Lindberg 1981) or as the intestine or style sac (Graham 1949; Fretter 1990; Ponder & Lindberg 1997). We regard it as the homologue of the style sac.

In stylommatophorans, the stomach is a small, simple sac, but in many 'lower pulmonates' (e.g., Siphonarioidea, Hygrophila, Amphiboloidea, Ellobioidea), the anterior stomach forms a muscular gizzard, but its position is variable, with some having the gizzard just behind the oesophageal opening and, in a few of these, the muscular area is strongly bilobed (see Chapter 20, Figure 20.40). In others, where the oesophagus and rectum open close together, the stomach forms a separate bag which may be developed into a gizzard. In *Onchidella* (Onchidiidae) (Fretter 1943), the gizzard is formed in the posterior stomach. These gizzards form an interesting contrast to the oesophageal gizzards found in several euopisthobranchs (see Section 5.3.2.7 and Chapter 20), although a few 'opisthobranchs' (Ringiculidae and many Dendronotoidea) have thin chitinous plates in the stomach, but these are not a proper gizzard as there is no significant muscular development.

The stomach has been effectively lost in a few gastropods, including some eulimids, stylommatophorans, and nudibranchs where it is just the point where the digestive gland(s) open.

5.3.3.1.7 Cephalopoda

The cephalopod stomach (Figure 5.49) is highly modified with its thick muscular walls composed of mainly circular muscles. Internally, its initial portion, into which the oesophagus opens, is lined with chitinised cuticle. The stomach is longitudinally divided into two parts, with a dorsal sac-like muscular part with cuticle which was interpreted as a vestigial gastric shield by Salvini-Plawen (1988). This functions as a gizzard and a ventral part forms a spiral caecum, into which the digestive glands open, and it has two grooves separated by a longitudinal typhlosole (the 'columellar ridge'). The intestine departs the stomach anteriorly.

The spiral caecum is a novel structure and nothing to do with the gastric caeca found in some gastropods. Internally, the caecum surface area is greatly increased by many densely ciliated leaflets analogous with those found in a few gastropod stomachs (see previous Section). The caecum and its leaflets are complex both structurally and functionally. The coiled

caecum has a half to one turn in *Nautilus*, while in coleoids such as *Sepia* and *Octopus,* there are several coils. In many squid, the caecum has a large, sac-like extension in which the leaflets are spread out like a fan (Bidder 1966) (Figure 5.49).

The anterior-most part of the stomach in *Nautilus* forms a thin-walled vestibule lined with a glandular epithelium, while the main stomach is a thick walled, cuticle-lined muscular sac that forms a blind pouch and receives the intestine near its opening to the vestibule. The small caecum opens to the most proximal part of the intestine, rather than the stomach proper.

In Coleoidea, a typhlosole separates the digestive (or 'hepatopancreatic') groove and the mucous (intestinal) groove, which continues along the intestine. A fold, the digestive ('hepatopancreatic') fold, runs along the other side of the mucous groove and can effectively cover the groove to form a tube which leads from the digestive gland through the caecum to the intestine (Bidder 1966; Salvini-Plawen 1988). In *Nautilus*, not only is the caecum much smaller, but the digestive groove is a simple open groove.

On the side opposite the typhlosole, the mucous groove is covered by numerous ciliated leaflets which sort and transport the indigestible heavier particles to the mucous groove and from there they are transported to the intestine. In coleoids,

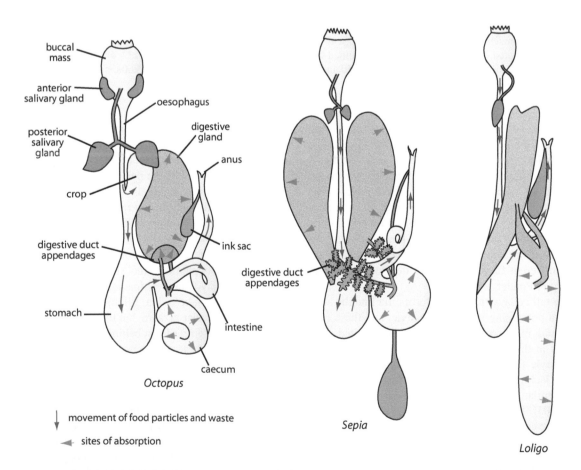

FIGURE 5.49 Comparative diagrams of the gut in *Octopus*, *Sepia*, and *Loligo*. Redrawn and modified from Bidder, A.M., Feeding and digestion in cephalopods, pp. 97–124, in Wilbur, K.M. and Yonge, C.M., *Physiology of Mollusca*, Vol. 2, Academic Press, New York, 1966, and Boucaud-Camou, E. and Boucher-Rodoni, R., Feeding and digestion in cephalopods, pp. 149–187, in Saleuddin, A.S.M. and Wilbur, K.M., *Physiology. Part 2. The Mollusca*, Vol. 5, Academic Press, New York, 1983.

the leaflets have primary and secondary folds and in *Sepia* are covered with glandular and ciliated cells, but there are only ciliated cells in squid. In *Nautilus*, the leaflets have only primary folds (Budelmann et al. 1997).

5.3.3.2 Digestive Glands

The digestive gland (= midgut gland, liver, or hepatopancreas), plesiomorphically consists of two large lobes made up of tubules that coalesce to communicate with the stomach by way of one, two, or several ciliated ducts. The cytology of molluscan digestive glands is generally similar in chitons, bivalves, scaphopods, cephalopods, and gastropods, consisting of digestive and secretory (basophilic) cells. There are, however, some important differences in the morphology of these glands which are briefly outlined below, in the accounts of the major groups.

These organs are a major site of absorption, intracellular digestion, phagocytosis, secretion of enzymes, nutrient storage, excretion and, often, detoxification. The tubules of the digestive gland(s) are typically surrounded externally with only a thin connective tissue, and lie within the visceral haemocoel, enabling nutrient exchange.

There are two basic gland cell types in the digestive gland – the absorptive columnar *digestive cells* and the broad-based, somewhat triangular *secretory cells* or basophilic cells (Figure 5.50). The former ingests food particles at the bases of the microvilli by pinocytosis, and these are contained within large vesicles (phagosomes). Intracellular digestion takes place in lysosomes, an organelle that contains (or receives) acid hydrolytic enzymes from the Golgi apparatus, although it is produced in the rough endoplasmic reticulum.

In a recent study, Lobo-da-Cunha et al. (2018) described the digestive cells in several euopisthobranchs which they concluded resembled those of most other gastropods. The distal microvillous ends of the cells formed endocytic vesicles in which calcium was detected, apparently forming endosomes showing arylsulphatase activity. This enzyme was also found in lysosomes, some of which fused to form vacuoles. The secretory cells contained large secretory vesicles containing glycoprotein, while small vacuoles contained calcium concretions in some shelled taxa.

The products of digestion are transferred from the basal part of the digestive cells into the haemolymph in adjacent blood spaces or to amoebocytes in the surrounding connective tissue, (e.g., Fankboner 1971). Waste products are contained in insoluble residual bodies (usually spherules) within the digestive cells, and these cells also play a role in detoxification, for example in removing potentially harmful metal ions. The waste spherules are eventually discharged into the lumen of the digestive tubule and are passed along the ciliated digestive gland ducts to the stomach, and then to the intestine where they are added to the faecal material. The connective tissue around the tubules also contains cells that store glycogen. A few muscle fibres may also surround the tubules and enable them to carry out weak contractions that, with cilia and muscles in the stomach and digestive gland ducts, enable the movement of material in and out of the tubules.

The digestive gland is a very important source of digestive enzymes (discussed below – see Section 5.4.2.3), and these are largely produced by the secretory (also called basophilic, crypt, or calcium) cells. These cells are pyramidal in shape, and their distal surface may have microvilli and can also be flagellate. Their cytoplasm contains rough endoplasmic reticulum and prominent Golgi apparatus, indicating protein synthesis. While the exact role of secretory cells in the digestive gland is not well understood, they are generally assumed to secrete enzymes for extracellular digestion. They are apparently absent in cephalopods (see Section 5.3.3.2.7).

As demonstrated, notably in bivalves, digestive patterns in the digestive gland tubules are often related to extrinsic factors such as tides (see Section 5.4.4.1) or the availability of food. These patterns can be determined by examining the morphological changes in the digestive cells that line the tubules and are responsible for the intracellular digestion of food (e.g., Yonge 1926b; Morton 1956b; McQuiston 1969; Owen 1970; Mathers 1972). These studies have resulted in four phases in the intracellular digestive processes of the digestive cells being recognised – absorptive, disintegrative, reconstitutive, and holding (Morton 1973a, 1983a; Langton 1975; Robinson & Langton 1980). Morphological changes of the digestive gland are also sometimes correlated with food availability (Langton & Gabbott 1974; Wilson & La Touche 1978).

Peroxisomes are small organelles involved in the oxidation of fatty acids and other processes related to lipid metabolism and other important functions. They are present in the digestive gland cells of bivalves and gastropods (e.g., Lobo-da-Cunha et al. 1994; Cancio et al. 2000) and are probably present in digestive gland cells of all molluscs. They have been shown to proliferate in response to environmental pollution (e.g., Cajaraville et al. 2003).

5.3.3.2.1 *Aplacophora*

The single digestive gland in the Caudofoveata contains two types of digestive cells. Cells on the dorsal side of the diverticulum are packed with granules, while the laterally and ventrally located cells have a large vacuole and granules. The vacuole contents are glandular and are released by the rupturing of the cell (Scheltema et al. 1994). Food particles are not found in the gland, indicating that it is a source of enzymes for extracellular digestion in the stomach and intestine (Salvini-Plawen 1988). In contrast, the digestive cells lining the 'stomach' in solenogasters are much more variable and both intra- and extracellular digestion occurs. These cells are expanded into a ventrolateral pouch on either side of the 'stomach' (i.e., midgut). Phagocytosis of nematocysts occurs, although they are only rarely digested (Salvini-Plawen 1988).

5.3.3.2.2 *Polyplacophora*

The digestive gland in chitons consists of two very large, initially symmetrical lobes which become asymmetrical during development (Salvini-Plawen 1988). They open separately and asymmetrically to the stomach and extend posteriorly to lie among the coils of the intestine. The right lobe extends anteriorly over and around the anterior part of the stomach,

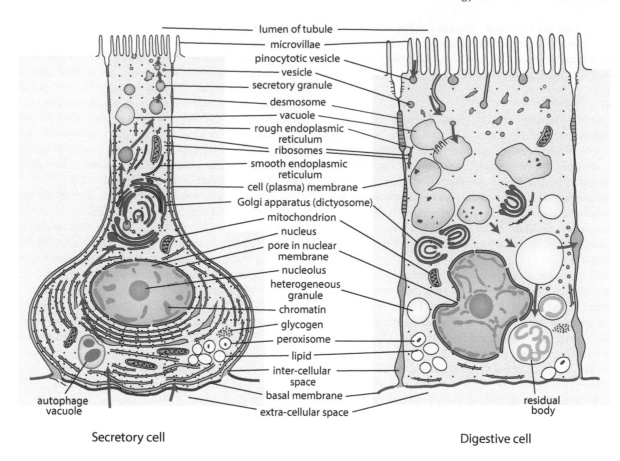

Secretory cell   Digestive cell

FIGURE 5.50 Digestive gland cells of a venerid bivalve (*Ruditapes*) – see text for an explanation of some of the main cellular processes and Chapter 2, Box 2.1, for a brief description of the main organelles. Redrawn and modified from Henry, M., *Vie marine*, 6, 7–15, 1984a; Henry, M., *Vie marine*, 6, 17–23, 1984b.

while the left lobe is on the posterior part of the stomach and among the highly coiled intestine. There are two main cell types, digestive and secretory cells, the latter with calcareous granules (Eernisse & Reynolds 1994). The digestive cells secrete proteolytic enzymes (Salvini-Plawen 1988).

5.3.3.2.3 Monoplacophora

The two pairs of lateral lobes of the digestive gland open to the stomach by long slits on either side or, in the smaller species, by a single opening. The digestive glands are more compact than in chitons, they lie on either side of the stomach and do not extend posteriorly among the coils of the intestine. The two main cell types (secretory and digestive) are present, the latter having a vacuolate outer edge with the vacuoles containing food particles (Haszprunar & Schaefer 1997) indicating intracellular digestion. Muscles in the walls of the digestive glands show they are capable of at least minor contractile movements, perhaps aiding the movement of enzymes into the stomach.

5.3.3.2.4 Scaphopoda

The single (left) (Gadilida) or paired (Dentaliida) digestive glands open laterally to the stomach by way of slits. The glands consist of several lobes composed of tubules lined with two cell types, triangular secretory cells and

digestive cells, and the latter undergo a cycle of secretion of enzymes and absorption by phagocytosis and pinocytosis (Shimek & Steiner 1997).

5.3.3.2.5 Bivalvia

The digestive glands and the process of digestion in bivalves have been described in more detail than in other molluscs. There is a more complex system of digestive gland ducts than in other groups, with primary ducts leading to the digestive gland branching into smaller ducts that may again subdivide before opening to the blind-ended digestive tubules. Different branching patterns are characteristic of different bivalve taxa. Muscles in the duct walls and cilia assist the passage of particles through the ducts, and this may also be aided by contractions of the adductor muscles (Morton 1983a).

Good summaries of digestive gland structure and histology can be found in Morse and Zardus (1997) and Gosling (2003). As in other groups, the tubules contain two main cell types (Figure 5.50), the numerous digestive cells and less abundant secretory cells, the latter sometimes flagellate. In most bivalves, the digestive cells produce excretory spherules, but in *Tridacna*, the entire digestive cell is packaged and released into the lumen of the tubule at the end of its cycle (Fankboner & Reid 1990). Other types of cell are sometimes recognised – for example, a columnar cell with a single long flagellum lacking conspicuous organelles is common in the digestive gland of *Crassostrea* (Weinstein 1995). In protobranchs, the two cell types are randomly organised, while in autobranchs they are grouped.

The primary digestive gland ducts of protobranch bivalves are unciliated, but the secondary ducts are ciliated. This contrasts with the primary ducts of the digestive tubules of autobranch bivalves which are lined with ciliated columnar cells that, at least in some (e.g., *Tridacna*), secrete proteolytic enzymes used in extracellular digestion in the stomach (Fankboner & Reid 1990). Cilia on these cells move the enzymes towards the stomach and move food into the tubules. The secondary ducts in autobranch bivalves are ciliated or unciliated, and the epithelial cells apparently do not have a role in absorption or secretion. The secondary ducts open to the digestive tubules where intracellular digestion occurs.

The digestive tubules of bivalves can undergo cyclical changes induced by external stimuli, usually daylight or tides (see Section 5.4.4.1.1 and Figure 5.54), but in permanently submerged bivalves, these changes may occur simultaneously within the tubules.

5.3.3.2.6 Gastropoda

In gastropods, the digestive gland occupies much of the visceral mass. In most gastropods, the gland consists of a large, seemingly single mass usually composed of two unequal lobes, the left lobe typically much larger in coiled taxa. In a few taxa, the right lobe is very tiny or lost. This asymmetry is presumably largely a result of shell coiling, although it may also be, at least in part, a carryover from the asymmetry necessitated by visceral packing as in chitons (see Section 5.3.3.2.2). In secondarily bilaterally symmetrical species (e.g., many nudibranch slugs), the digestive glands remain unequal in size, although in some nudibranch slugs they have become subequal and may also become branched. In Aeolidioidea and some Sacoglossa, the branches extend into dorsal processes (cerata) or other body extremities (Wägele & Willan 2000) (Figure 5.51). The compact, normal digestive gland is termed *holohepatic*, while the branched type is called *cladohepatic*.

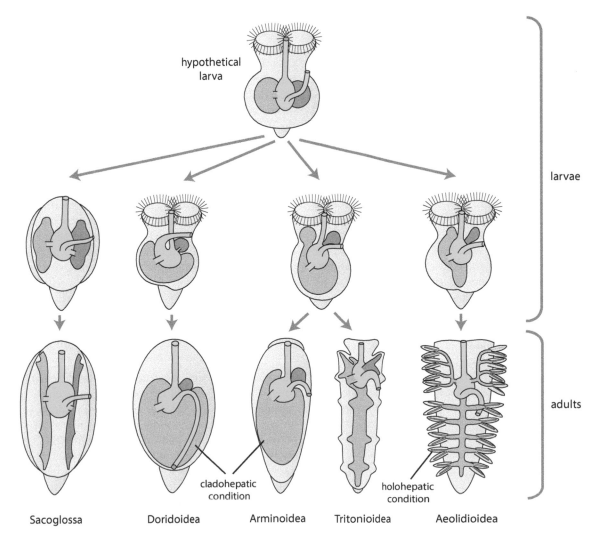

FIGURE 5.51 Differences in the differentiation of the digestive gland in some 'opisthobranchs'. Redrawn and modified from Schmekel, L. and Portmann, A., *Opisthobranchia des Mittelmeeres, Nudibranchia und Saccoglossa*, Springer-Verlag, Berlin, Germany, 1982.

Such branching is often associated with other specialisations such as the storage of zooxanthellae (see Section 5.2.1.11.2) or chloroplasts (see Section 5.2.1.11.1) or, in Aeolidioidea, the storage of cnidocysts in special cnidosacs at the distal ends of the digestive gland diverticulae which are housed in the finger-like cerata (see Chapters 9 and 20). In Doridoidea, the right lobe of the digestive gland is lost or very small and, if present, differs from the left histologically, being lined only with ciliated cells (Wägele & Willan 2000).

Each lobe of the digestive gland is typically connected by a single duct to the stomach, but in some taxa (e.g., Patellogastropoda, Trochoidea, some Hygrophila), the ducts fuse before reaching the stomach, while in others the ducts branch so that several ducts enter the stomach. In Onchidiidae, the gland is composed of three lobes (two derived from the left lobe), and there are three ducts entering the stomach.

The cytology of gastropod digestive glands has been the subject of many studies (Fretter & Graham 1962, 1994; Andrews 2010; Lobo-da-Cunha et al. 2018). As in other molluscs, there are two basic types of cells lining the tubules, the digestive cells and secretory cells. These have activity cycles during which they change morphologically, resulting in the recognition in earlier literature of more types of cell. The digestive cells are columnar, with microvilli on their outer surfaces, and contain numerous vesicles, along with other inclusions and a basal nucleus. These cells absorb nutrients and/or ingest particles using pinocytosis. The secretory cells are microvillous distally, much shorter, less common, pyramid shaped, and occur in small clusters in the corners of tubules. They often contain one or two large spherules along with other inclusions, and a large nucleus. In many caenogastropods, the spherules contain yellowish, probably excretory, concretions. Gros et al. (2009) found three types of cells in the tubule epithelium of the digestive gland of *Strombus*, with vacuolated lipid-storing cells besides the usual two, this being a unique type of cell (probably excretory) not seen in other gastropods.

Taïb and Vicente (1999) and Taïeb (2001) described the two cell types found in *Aplysia*. These both change morphologically during the digestive cycle, the changes in the secretory (calcium) cells being related to ion metabolism and those in the digestive cells to intracellular digestion.

A detailed account of the function and structure of the digestive gland in 'pulmonates' is given by Luchtel et al. (1997). Four to five types of cell are recognised in stylommatophorans (e.g., Sumner 1965, 1966; Walker 1970b), the three additional cell types being excretory cells with large excretory vacuoles probably derived from either the digestive or secretory cells, narrow columnar 'thin cells' that probably give rise to digestive or secretory cells, and mucous cells; the latter are also known from *Lymnaea*.

5.3.3.2.7 Cephalopoda

In coleoid cephalopods, the digestive glands are the largest gland in the body. In early development, they are initially paired, equal-sized lobes, but these fuse together, in some completely, and are enveloped by a thin sheath of muscles and connective tissue. The digestive gland of *Nautilus* is composed of three to (usually) five main lobes. In coleoids, the digestive gland comprises smaller lobules made up of numerous tubules. These tubules collect into a lumen that runs the length of the gland in each of the two original lobes, each draining into a duct joined with the common duct which opens to the apex of the stomach caecum. In *Nautilus* each of the lobes comprises several lobules which consist of a mass of tubules that open to a central muscular chamber. The lobules are each connected by narrow ducts which unite just before their entry to the apex of the caecum.

The cephalopod digestive gland synthesises and secretes digestive enzymes, resorbs and metabolises nutrients, synthesises and stores lipids, lipoproteins, vitamins and protein-bound metal ions, and excretes waste products (Budelmann et al. 1997). In *Nautilus*, it also synthesises haemocyanin, a function taken over by the branchial heart complex in coleoids (e.g., Dilly & Messenger 1972; Beuerlein et al. 1998) (see also Chapter 6).

The histology of the digestive gland of coleoids has been described by Budelmann et al. (1997). While several types of cell have been described in the literature, these are functional stages of digestive cells, and separate secretory cells are absent. The progenitors of the digestive cells (basal or replacement cells) are pyramid shaped and grow into the columnar digestive cells. The digestive cells have a microvillous border and contain waste in vacuoles, and also secrete enzymes. These enzymes are bound up in 'boules' – large proteinaceous inclusions (primary lysosomes) which release the enzymes (chymotrypsin, cathepsin, and amylases) when discharged from the cell. In the early stages of feeding, the number of boules increases in many cells, and these are released several hours after feeding, reaching the stomach via the digestive ducts. Other digestive cells actively take up particles by way of endocytosis.

In *Nautilus*, each tubule is divided into a terminal alveolar part, middle excretory, and inner absorptive sections. The terminal alveolar part is lined with pyramid shaped cells like the replacement cells in coleoids, but many of these apparently differentiate into haemocyanin synthesising cells. Other replacement cells differentiate into digestive cells, which have resorptive, storage, and secretory functions. The digestive cells line the other parts of the tubules.

5.3.3.3 Digestive Duct Appendages ('Pancreas') of Coleoid Cephalopods

These structures are associated with the digestive gland in coleoid cephalopods. In octopods, they are enclosed in the capsule surrounding the digestive gland, while in squid they are made up of creamy-white tubules or alveoli on the walls of the paired digestive gland ducts, into which they open separately. In all squid and cuttlefish, the appendages lie within the kidney, lined with a single layer of columnar epithelial cells which play a role in the absorption of nutrients (such as amino acids and carbohydrates) from the fluid in the digestive gland ducts. They also secrete several enzymes and have an osmoregulatory function (see Budelmann et al. 1997 for references). There is no similar structure in *Nautilus*, or in other molluscs.

5.3.4 The Hindgut – The Intestine and Rectum

The intestine is sometimes included as part of the midgut, but, as indicated above, because it and the rectum are a functional unit and often hardly distinguishable, we treat them together as components of the hindgut. Some workers (e.g., Morton 1983a) refer to the typhlosole-bearing part of the intestine as the midgut.

The intestine passes through the ventricle of the heart in some molluscs (see Chapter 6) including monoplacophorans, many bivalves, and some vetigastropod clades, but does not do so in most gastropods, including patellogastropods, or chitons, or aplacophorans.

Plesiomorphically, the rectum opens posteriorly at the anus into a mantle cavity or mantle groove. The anus is often extended as a papilla.

5.3.4.1 Aplacophorans

In the caudofoveate aplacophorans, the long, ciliated intestine bends once or twice as it leaves the stomach, but is otherwise not looped or coiled, and opens to the mantle cavity between the ctenidia. Solenogasters lack an intestine and have a short ciliated rectum which empties into the reduced posterior mantle cavity above or below the gametoduct openings (Scheltema et al. 1994).

5.3.4.2 Polyplacophora

In chitons, the intestine is long and coiled with the anterior and posterior parts separated by a valve. Two ciliated ridges

from the stomach continue in the anterior intestine which closely resembles the stomach in its histology (Fretter 1937) and, as noted above, may be homologous with the style area of other molluscs (see Section 5.3.3.1.2). The valve at the commencement of the posterior intestine nips the faecal material into pellets. In most chitons the posterior intestine is long with multiple loops (fewer in Lepidopleurida), including highly symmetrical anterior-posterior and dorsal-ventral coiling (Figure 5.52). Internally, the ciliated epithelium contains mucous cells. The rectum is short, passing through the muscles of the posterior-most part of the body and opening at the anus in the midline in the mantle groove. The anterior intestine, along with the stomach, is the main area where (extracellular) digestion occurs (Salvini-Plawen 1988) facilitated by amylolytic enzymes from the oesophageal glands and proteolytic enzymes from the digestive glands. Just why the intestine is so long and coiled in most chitons does not appear to have been satisfactorily addressed.

5.3.4.3 Monoplacophora

The intestine of monoplacophorans is like that of chitons in having multiple coils (Lemche & Wingstrand 1959; Wingstrand 1985; Haszprunar & Schaefer 1997) and is lined by simple ciliated epithelium. Based on their ultrastructural investigation of the intestinal epithelium, Haszprunar and Schaefer (1997) concluded that it is specialised for transport rather than absorption. The rectum is short and straight and has a similar epithelium to the intestine. It opens at the anus in the posterior midline in the mantle groove.

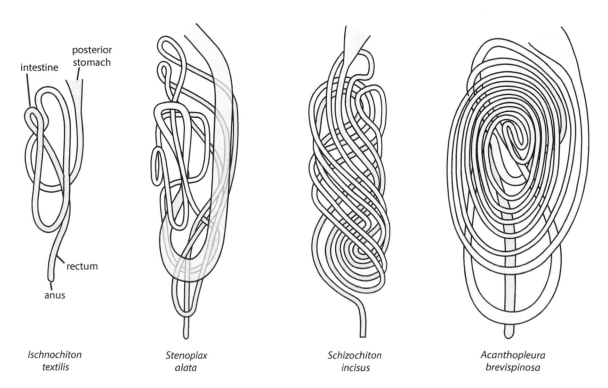

FIGURE 5.52 Variation in gut coiling in chitons. Redrawn and modified from Hoffmann, H., *H. G. Bronn's Klassen und Ordnungen des Tier-Reichs wissenschaftlich dargestellt in Wort und Bild*, 3, 1–511, 1929–1930.

5.3.4.4 Scaphopoda

In scaphopods, the intestine lies anterior to the stomach where it is looped, usually three times. Internally, the intestine is indistinctly ridged and is surrounded by muscles. Ciliated and glandular cells line the distal intestine, and there are only ciliated cells proximally (Shimek & Steiner 1997). The rectum bends ventrally just anterior to the stomach, is lined with ciliated cells, and coated with a muscle layer. A rectal gland is present (see Section 5.3.4.9) and the anus opens to the mantle cavity on a papilla just behind the foot.

5.3.4.5 Bivalvia

In bivalves, the intestine is usually long and coiled and typically extends ventrally into the viscera and haemocoel in the dorsal part of the foot. It is rather uniform throughout the bivalves, having a columnar epithelium with microvilli and/or cilia on the outer surface (Morse & Zardus 1997) and is involved in absorption. The rectum runs just above the posterior adductor muscle, and the anus opens near the exhalant aperture.

5.3.4.6 Gastropoda

The plesiomorphic condition in gastropods is a looped intestine (Ponder & Lindberg 1997), but this condition is simplified within most gastropod clades. In some vetigastropods and patellogastropods, the intestine can be several times as long as the body (e.g., Sasaki 1998). Some workers have divided the intestine into several regions based mainly on differences in the epithelial cells (Campbell 1965; Ward 1966; Lutfy & Demian 1967). The trend in gastropods towards a shorter, straighter intestine is not necessarily associated with feeding, although it has typically been attributed to moving from herbivory to carnivory, and a reduced need to consolidate faecal matter (Fretter & Graham 1962; Owen 1966). Similar intestinal morphology is seen in groups of caenogastropods and heterobranchs which exhibit a wide range of feeding strategies. In more derived patellogastropod taxa the intestine is shortened (Lindberg 1988a, 1988b), but this is not correlated with the nature of the food ingested, instead appearing to be due to paedomorphosis (Lindberg 1988a). In stylommatophorans, the intestine is secondarily coiled, probably in response to their terrestrial habits and the need to absorb water from the faecal material. In caenogastropods, some intestinal coiling is seen in 'architaenioglossans', while in most 'asiphonate' caenogastropods, the intestine is uniquely looped back around the style sac. In most carnivorous caenogastropods (e.g., epitoniids, many neogastropods and cypraeids), the intestine runs straight from the stomach to the mantle cavity (Ponder & Lindberg 1997; Strong 2003).

There have only been a few detailed descriptions of the intestinal epithelium of gastropods which include ultrastructure and histochemistry, but one example is outlined below to illustrate the potential complexity. In their study on *Haliotis*, Cui et al. (2001b) confirmed that the intestine and the rectum have an important role in digestion and absorption as well as being tracts for passing food and faeces. The long intestine in *Haliotis* can be divided into roughly equal anterior, middle, and posterior sections.

The epithelia of both the intestine and rectum are composed of five cell types with different functions. Epithelial cells with microvilli absorb nutrients and water and are the main type in the mid-intestine, but are also common in the anterior intestine. The ciliated epithelial cells assist with moving material through the gut, but are otherwise structurally similar to the microvillar cells. They are the main cell type in the anterior intestine, the intestinal groove typhlosoles, and the rectum. There are two types of glandular cells: the first type is narrow distally with short microvilli, and is more abundant towards the posterior end of the intestine and much less common in the rectum. These cells release enzymes into the gut lumen (see Section 5.4.2.3). The second type of gland cell is found mainly in the rectum and typhlosoles. Its distal and middle parts are full of tiny secretory granules, much smaller and denser than in the first type. These do not show enzymatic activity and are probably involved in faecal consolidation. Goblet cells are found mainly in the rectum and contain mucoprotein and secrete mucopolysaccharide to lubricate the rectum and cement the faeces.

The main function of the rectum is the transfer of faecal pellets to the anus, a process assisted by lubrication from goblet cells and movement by muscles in the rectal wall. A study of *Littorina saxatilis* by Brough and White (1990b) showed that the folding and histology of the rectal epithelium suggest an absorptive and/or excretory function, and Brough and White (1990a) showed heavy metal excretion by some of these cells.

Structurally, the intestine in many 'pulmonates' can be differentiated into four sections – a pro-, mid-, and post-intestine, and the rectum, the pro-intestine having a ventral typhlosole. Limacid slugs differ from other stylommatophorans in having a small caecum between the intestine and rectum.

The location of the anus varies in gastropods, being plesiomorphically located posteriorly in the mantle cavity and apomorphically near or at the edge of the mantle cavity, except in some heterobranchs where the mantle cavity is lost (see Chapter 4 for more details).

5.3.4.7 Cephalopoda

In cephalopods, the intestine is short and straight, departing from the caecum with the typhlosole ('columellar ridge') and the intestinal groove running along it. It runs anteriorly and is looped in some taxa. In coleoids, the rectum is short and has the ink gland (see Section 5.3.4.11) opening to it. In *Nautilus*, the proximal intestine is divided longitudinally into two sections – one section with transverse folds that receives the main mucous groove (see Section 5.3.3.1.7) from the caecum and a longitudinally folded section leading from the stomach. Both parts open into a long intestinal loop through which the main mucous groove from the caecum continues. In all cephalopods, the intestinal walls are muscular, and the epithelium consists of ciliated and glandular cells, and the short rectum is even more muscular. The anus lies just posterior to the funnel in coleoids, but in *Nautilus*, it is at the posterior end of the mantle cavity.

5.3.4.8 Faecal Material

The faecal material in molluscs varies considerably from group to group, but within groups, it is rather consistent, as shown by studies by the Japanese malacologist Kohman Y. Arakawa in the 1960s and 1970s (Arakawa 1963, 1965, 1968, 1970, 1972) and by Bandel (1974). Faecal pellets are produced in Polyplacophora (Fretter 1937; Arakawa 1968), but the plesiomorphic condition is probably a simple faecal string like that seen in monoplacophorans (Ponder & Lindberg 1997). The solenogasters also produce a loose faecal string, but the caudofoveate aplacophorans produce a string of faecal pellets (Salvini-Plawen 1988). Protobranch bivalves produce firm, mostly grooved faecal rods (Yonge 1939b; Arakawa 1970), while autobranch bivalves produce rods with or without one or more grooves; in some, the rod is constricted into pellet-like lumps (Arakawa 1965). Scaphopods produce unconsolidated mucus-bound faecal strings (Shimek & Steiner 1997).

The lower gastropods (patellogastropods and pleurotomariids) produce a faecal string (Bandel 1974; Arakawa et al. 1978) and a similar, probably convergent condition is seen in many 'pulmonates'. In some vetigastropods, and also in some bivalves, one or more typhlosoles run along inside the rectum and result in the faecal rod becoming longitudinally grooved (Arakawa 1965; Bandel 1974). This type of faecal sculpture appears to be unknown in other gastropods except for *Valvata* (Cleland 1954, fig. 7). In most 'higher' gastropods, the faecal material is simple in section, and the string may be separated into pellets. Thus, many caenogastropods produce pellets, but some, including some neogastropods, produce faecal strings (e.g., Arakawa 1965, 1968; Bandel 1974). Similarly, pellets are produced by some 'pulmonates', but most, like 'opisthobranchs', produce strings (e.g., Arakawa 1968).

The pellet formation in polyplacophorans is convergent with that in gastropods as the pellets are produced by a special intestinal valve (Fretter 1937) (see Section 5.3.4.2) not found in gastropods.

Cephalopods, including *Nautilus*, bind tangled strings of waste with mucus which they eliminate at intervals (Bidder 1966).

5.3.4.9 Rectal Gland of Scaphopods

In scaphopods, a rectal gland opens into the right side of the middle part of the rectum. The gland consists of several tubules lined with a cuboidal epithelium with long cilia (Shimek & Steiner 1997) which unite to form a short duct. While many functions of this gland have been suggested, excretion of lipoids is the most likely (Sahlmann in Shimek & Steiner 1997).

5.3.4.10 Anal (= Rectal) Gland of Gastropods

In some neogastropods, the rectal (= anal) gland is a caecum-like structure near the distal end of the rectum and typically opens to it near the anus by way of a narrow duct. In neogastropods, this organ is involved in excretion and protein metabolism and is discussed in more detail in Chapter 6.

Structures referred to as anal glands are also known in some vetigastropods, strombids, and naticids (see Schaefer 1996 for references). Certainly the anal gland of trochids is not homologous with that of neogastropods (Andrews 1992), and it is likely that the others are also convergent (Ponder & Lindberg 1997). In the trochid *Gibbula*, the anal gland is a simple glandular structure that provides mucins to assist in binding faecal material (Andrews 1992).

5.3.4.11 Ink Gland of Cephalopods

During development, the ink gland found in nearly all coleoids but not in *Nautilus*. It is derived from the hindgut and consists of a reservoir (the ink sac) and a duct, the latter opening to the rectum (near the anus) via a short ampulla and controlled by a sphincter muscle. It produces a dark brown to dark blue – almost black – ink used in defence and escape. Muscular contractions under nervous control drive the ink from the gland. The melanin pigments making up the ink are secreted by cells lining the gland (see Chapter 17 for more details).

5.3.5 OVERVIEW OF THE EVOLUTION OF THE MOLLUSCAN DIGESTIVE SYSTEM

The digestive system provides important insights into molluscan evolution. The system has been used to show evolutionary trends in molluscs (e.g., Salvini-Plawen 1988), but others have argued that feeding and digestive structures are strongly convergent because of similar diets. While this emphasis on a correlation between structure and function in the digestive system has obscured the phylogenetic usefulness of several gut characters, many are actually useful in cladistic analyses, as long as their structural complexity is adequately interpreted (Strong 2003).

A distinct digestive gland is absent in the solenogasters, and most of the digestive system consists of the large stomach lined with digestive cells. They also have an anterior extension of the stomach (caecum) which is sometimes paired and extends dorsally from the point at which the short oesophagus enters the stomach. The hindgut is short in solenogasters and opens to the rectum which then opens to the mantle cavity. The polyplacophoran gut is somewhat similar to that of the Caudofoveata in its general configuration, but differs in the hindgut being looped, sometimes complexly, in the possession of paired digestive glands which open dorsally into the posterior region of the stomach and are interspersed among the stomach and hindgut, and the rectum extends below the digestive gland.

The Polyplacophora gut shares numerous plesiomorphies with the remaining molluscan groups, including several foregut structures (distinct buccal and salivary glands, subradular organ), digestive gland position, configuration and ducts, and hindgut development and looping.

It is unclear as to which of these three aculiferan configurations best represents the plesiomorphic molluscan condition because they show a somewhat confused pattern of transformations. Based on comparisons with annelids, solenogasters might represent the plesiomorphic condition in lacking a distinct digestive gland (although the sometimes paired dorsal outpocketings [caeca] may represent reduced digestive gland[s]) and an ectodermal (rectum) connection between the midgut and anus. The configuration seen in the Caudofoveata with its unique posterior and ventral digestive gland is probably not plesiomorphic. The organogenesis of the hindgut in the Caudofoveata is not known. The aplacophoran digestive systems thus have some unique states (ventral digestive gland, lack of hindgut), suggesting the strong possibility of developmental heterochrony operating in the evolution of these worm-like molluscs, rather than these conditions indicating ancestral states (see Chapters 12 through 14). In polyplacophorans and solenogasters, the midgut and hindgut are derived from endoderm.

Unique character states (apomorphies) of the Conchifera include an anteroposterior (A-P) compression and dorsoventral (D-V) elevation of the visceral mass, as seen in the placement of the hindgut and digestive gland over the stomach and often overlapping the posterior portion of the foregut. Also, the position of the rectum has been reversed relative to its position in polyplacophorans, with the rectum now extending over the top of the hindgut loops rather than below them (Figure 5.52). While this compression and elevation may not be unexpected in bivalves and gastropods, it is unexpected in Monoplacophora which are thought to have a primitive body plan little modified from their early Cambrian origin. Instead, the state of their viscera suggests a previously unsuspected configuration in the conchiferan ancestor (see Chapters 12 and 13). Also, in bivalves and monoplacophorans, the openings of the digestive glands are no longer at the posterior end of the stomach, but have moved anteriorly and are now in the midsection of the stomach. The basic configuration of the bivalve digestive system and that of living monoplacophorans is virtually identical, and this may represent the ancestral conchiferan condition.

Although monoplacophorans and bivalves are respectively dorsoventrally and laterally compressed, respectively, they, along with the Polyplacophora, Solenogastres, and Caudofoveata, are the only molluscan taxa in which the digestive system lies in a definitive anterior-posterior body axis. In the remaining taxa, Cephalopoda, Scaphopoda, and Gastropoda, the anterior-posterior axis of the body underwent a U-shaped bending that placed the anus and mouth roughly in the same plane. This bending is called *ano-pedal flexure* (or simply 'flexure') and results in the juxtaposition of the anterior and posterior body components and the elongation of the dorso-ventral (D-V) axis (also see Chapter 8).

Although ano-pedal flexure acts along the A-P axis of the body, it can be overlooked in cephalopods and scaphopods because their orientations in life are along their D-V axis (Chapter 3, Figure 3.1), not their A-P axis, and the anus is located under the viscera rather than behind in both these groups. Whether the ancestral conchiferans shared a more D-V life axis is not known, but based on outgroup comparisons with the other molluscan taxa, a more A-P lifestyle would be supported. Also, the openings of the digestive glands moved further anteriorly and are typically at or near the anterior end of the stomach in conchiferans (Figure 5.44).

In gastropods, ano-pedal flexure is sometimes thought to result from torsion, but it actually occurs before torsion begins, with the larval gut bent into a U-shape because of the differential growth of the shell gland (see Chapter 13). Compared to cephalopods and scaphopods, torsion in gastropods places the rectum on top of the viscera rather than underneath. Other changes in the digestive system in gastropod evolution included the reduction in length and coiling of the hindgut, reduction of the digestive glands, and adding specialised foregut glands, and pouches, especially in caenogastropods (see Section 5.3.2.5). The two 'higher' gastropod clades, caenogastropods and heterobranchs, are thought to be heterochronic (Ponder & Lindberg 1997), which may, at least partly, account for their shortened intestines.

In both cephalopods and scaphopods, there appears to be a major reversal of the A-P compression seen in the conchiferan lineage, with a posterior expansion of midgut structures associated with the stomach. The configuration of the *Nautilus* digestive system suggests these patterns are convergent in these two groups. Except for the modified foregut, a small caecum, and the concentration of intestinal and digestive gland openings at the anterior region of the stomach, the basic configuration of the *Nautilus* digestive system is like that of the Monoplacophora and Bivalvia, except for reduced hindgut looping. In the coleoid cephalopods, and in scaphopods, there was a strong posterior displacement of structures associated with the stomach as the body underwent posterior elongation. Although the midgut structures moved posteriorly in both groups, hindgut looping, although reduced, remained beneath the stomach. In scaphopods, hindgut compression resulted in a substantial overlap of the hindgut loops under the foregut.

The connections between the foregut and stomach and the stomach and digestive glands are broad and not distinct in scaphopods, especially in Dentaliida. Anterior movement of the digestive glands in Gadilida placed these organs under and around the stomach and hindgut – an arrangement also found convergently in coleoid cephalopods and may be related to both scaphopods and the ancestral coleoids having a tubular shell.

In the coleoid cephalopods, there is a posterior elongation of the small gastric caecum seen in *Nautilus*, with it often being coiled (thereby maintaining its length, but not its volume) in the rather compressed bodies of octopods and sepiids (cuttlefish) (Figure 5.49). In contrast, in squid (Oegopsida and Myopsida) with their more elongate bodies, the caecum is not coiled. All three coleoid groups have an anterior and dorsal digestive gland and show a reduction of hindgut length. In squid, the hindgut is reduced to the point of being straight with no looping – a condition convergent with higher gastropods.

The plesiomorphic condition of the digestive system of *Nautilus* further supports the view those cephalopods are little changed since their origin in the early Paleozoic. The palaeontologist John Peel recognised the secondary anterior-posterior elongation of various molluscs and coined the term 'scaphopodisation' for this condition (Peel 2006), which he pointed out has happened multiple times in different molluscan groups and was probably driven by environmental and ecological changes (see Chapter 13 for further discussion). Thus, for example, we consider convergence rather than shared descent the most likely explanation for the similarity of digestive system configuration seen in coleoid cephalopods and scaphopods.

5.4 PHYSIOLOGICAL PROCESSES

5.4.1 FOOD DETECTION

Food detection is usually achieved either by chemosensory receptors or organs, or visually in coleoid cephalopods and heteropod gastropods (see Chapter 7). In suspension-feeding taxa, food is separated from waste largely based on its size and weight, besides other characteristics (see Section 5.2.1.2).

In solenogasters, food is detected by chemoreception with the protrusible *atrial sense organ* and mechanoreception with the *periatrial stereocirri* (Haszprunar 1986b). Chitons, patellogastropods, and vetigastropods use their subradular organ to 'taste' food and, although the latter two groups have an osphradium, it is unlikely this is involved in food detection, contrary to the situation in caenogastropods. Carnivorous chitons also have an array of mantle tentacles sensitive to the touch of prey items beneath the expanded and elevated anterior girdle (McLean 1962).

In some euopisthobranchs and hygrophilans, an osphradium (when present) may play a minor role in food detection, but other sensory structures are involved and are more significant. 'Opisthobranchs' have a number of different sensory structures associated with the anterior part of the body which are used for distance (rhinophores, Hancock's organ) or touch (oral tentacles, lip organ) chemoreception (Göbbeler & Klussmann-Kolb 2007). Hygrophilans such as *Lymnaea* do not have the ability to detect food at a distance and must rely on 'taste' (e.g., Bovbjerg 1968; Tuersley & McCrohan 1987). In stylommatophorans, both the eye-bearing (ocular) tentacles and the anterior (oral or inferior) tentacles are involved in distance olfaction in these terrestrial snails and slugs (e.g., Chase 2001) (see also Chapter 7) and their lips in 'taste' (Salánki & Bay 1975).

Heteropods visually hunt prey using their very efficient eyes, but the carnivorous gymnosome pteropods rely on touch, having tactile tentacles that make direct contact with the prey (Lalli & Gilmer 1989).

In scaphopods, the terminal bulbs of the tentacle-like captaculae locate prey, but the sensory mechanisms involved are unknown (Shimek & Steiner 1997).

Suspension-feeding rates in bivalves are stimulated by factors such as temperature, but are slow in response to increasing phytoplankton density (e.g., Winter 1978) (see also Section 5.4.4). Sensory tentacles on the siphon of microcarnivorous septibranch bivalves are stimulated by the touch of a member of the zooplankton, resulting in the rapid extension of the voluminous siphon which captures the prey (Reid & Crosby 1980).

The whole feeding process in cephalopods is under complex neural control. Many coleoids visually locate their prey, with motion a key stimulus. In squid and cuttlefish, there are three steps involved (Messenger 1968) – *attention* (involves changes such as body and arm position and colour patterns), *positioning* (movement in relation to prey), and *seizure* (tentacles in squid or cuttlefish, or arms in octopods, grasp prey). In octopuses, *approach* and *jump* stages are generally more applicable (Hanlon & Messenger 1996). Many octopods rely heavily on tactile, rather than visual stimuli, for example, extending an arm into holes or crevices, or in many deep-water taxa due to reduced or absent vision (Hanlon & Messenger 1996). Hanlon and Messenger (1996) have provided a detailed review, and more information is given in Chapters 7 and 17.

5.4.2 DIGESTION AND ASSIMILATION

In molluscs, the passage of food through the gut involves several processes. Generally, in many molluscs, ingested food is bound in mucus from the buccal area and mixed with saliva (which may also contain enzymes), and in some at least, is subjected to enzymes from glands in the oesophagus. The food is quickly moved to the stomach. Enzymes produced in the style sac are released from the crystalline style if one is present (see Section 5.4.2.3.1) and additional enzymes may be produced in the digestive gland. Small particles are carried into the digestive gland by cilia, and larger ones moved into the intestine. While most of the final digestion and absorption occur in the digestive gland tubules, these same processes also occur to a varying extent throughout the gut, including the oesophagus, the crop (if one is present), digestive gland, and intestine, all of which have been shown to absorb both soluble and particulate material. Such absorption of nutrients can be experimentally determined using tracers such as iron saccharate, graphite, titanium dioxide, tritiated D-glucose, or ferritin.

The energy content of a particular food does not necessarily indicate whether a particular animal can use it because it may not be readily digestible; this depends on gut morphology and the available complement of enzymes. For example, phytoplankton cells and algal and plant material can be digested more readily if the thick cell walls are dissolved using enzymes such as cellulases or if they are physically ruptured to allow access to the cell contents. While some rupturing may be achieved by the radula, additional mechanisms have evolved in some lineages to assist with this, such as the gizzards of some 'opisthobranchs' and 'pulmonates' (see Sections 5.3.2.7 and 5.3.3.1.6, and Chapter 20).

The energetic cost of digestion can be high – for example, it amounts to about 17% of the total energy expenditure in *Mytilus* (Widdows & Hawkins 1989).

5.4.2.1 The Gut Environment

Little is known of the internal conditions in the gut of most molluscs, other than some basic data on pH in some and, in a few taxa, enzymes and oxygen levels. The pH levels can have a significant effect on enzyme activity, and pH and oxygen levels are also important for providing a suitable environment for symbiotic gut bacteria to thrive (e.g., Plante & Jumars 1992) (see also Section 5.4.2.3.2). The gut microenvironment influences microbial composition, but it can also be modified by the metabolic processes of that community.

The gut environment is important for optimal enzymatic reactions – in particular pH and temperature, with the requirements differing depending on the enzyme involved. Over a limited range (the 'biokinetic range') enzymatic reactions increase with temperature, which extends from about 0°C to 45°C, above which activity typically falls. For example, Banerjee and Sur (1991) examined the optimal temperatures for some hydrolysing enzymes in three very different gastropods, a terrestrial systellommatophoran (a veronicellid), and two caenogastropods (a fresh-water ampullariid and a marine melongenid). The activity of acid protease from the digestive gland of the veronicellid was highest at an extraordinary 70°C. The optimal activity of the other enzymes tested (cellulase, invertase, alkaline protease) were in more 'normal' ranges (37°C–45°C).

Yonge (1925b) examined the pH of the gut of a range of gastropods (*Patella*, *Crepidula*, *Buccinum*, and *Doris*) and found it ranged as follows: salivary glands 5.7–6.4, crop – *Patella* 5.4–6.0, *Aplysia* 4.4–4.6, oesophagus 5.6–7.0, gizzard (*Aplysia*), 4.4–5.4, stomach 5.4–6.2, style (*Crepidula*) 5.8, digestive gland 5.6–6.3, intestine 6.7–8.0, and rectum 5.9–8.6. The pH of the gut contents of *Strombus gigas* varied between pH 5.25 and 6.65 (Horiuchi & Lane 1966).

Studies on the gut environment in bivalves rarely include the whole gut, and oxygen levels are apparently unstudied. The stomach contents have a lower pH (down to pH 5.5) because of the dissolving style and acid secretions from the digestive gland (Owen 1974). In early studies (e.g., Yonge 1925a, 1926a), the activity of enzymes such as amylase and glycogenase released from the crystalline style was shown to require acid conditions, and this is true for the whole gut. In oysters, this ranges between 5.4 and 5.5, with the style having a pH of 5.2 (Yonge 1926a; Galtsoff 1964) and is the most acid part of the gut. In contrast, Dean (1958) found that extracts from the style of an oyster ranged in pH from 5.8 to 6.0. In the absence of the style, the stomach pH increases. Yonge (1925b) showed in a range of bivalves that pH in the gut ranged from 5.6 to 6.8 in the oesophagus, 5.4–6.0 in the stomach, the style 4.4–5.4, digestive gland 5.5–5.8, intestine 5.8–6.2, and the rectum 6.4–7.0.

The gut of an abalone (*Haliotis laevigata*) is effectively anaerobic, while the pH profile of the gut was 6.2, falling to 5.3 in the crop and slightly increasing through the stomach (5.5) and intestine to reach 6.6 in the rectum (Harris et al. 1998b). In contrast, the pH in the gut of helicid snails ranged from very acidic (pH ~ 1.4) in the crop to neutral or alkaline in the intestine (Charrier & Brune 2003). This same study also looked at oxygen levels and showed that the whole gut was anoxic, but their data indicated high uptake of oxygen via the gut epithelium. In addition, it was shown there was an accumulation of hydrogen in the intestine and digestive gland.

Walker et al. (1996b) investigated the pH of the gut of the slug *Deroceras reticulatum* using ion-selective microelectrodes, and Charrier and Brune (2003) used microsensors to study pH, hydrogen, and oxygen levels in four helicid snails. This latter study showed an increase in gut pH from the crop (average pH 5.3–6.1) to the distal intestine (average pH 5.6–7.4) with a slight decrease in the rectum of the phyllophagous (leaf feeding) species. Studies on the saprophagous slugs *Arion ater* (James et al. 1997) and *D. reticulatum* (Walker et al. 1996a, 1999) showed that the major part of the gut involved in the digestion of proteins was the digestive gland, with a pH of ~6.5.

5.4.2.2 Extracellular and Intracellular Digestion

Food particles must be broken down through digestive processes before they can be utilised. The digestive cells take up food in two ways – in part by absorption of nutrients (these are largely the products of extracellular digestion) through their outer surfaces which are covered in microvilli or, alternatively, food particles are ingested by endocytosis and digested in lysosomes within the cell (intracellular digestion) (Loboda-Cunha et al. 2018), this being essentially the same process as in unicellular organisms. Intracellular digestion occurs in the digestive cells of many molluscs, but by its very nature is limited to dealing with very small particles. The undigested remains from intracellular digestion are accumulated in the vesicles and then discarded and carried via the digestive gland ducts and stomach to the intestine.

The evolution of an alimentary canal in multicellular organisms enabled extracellular digestion to occur. This method of digestion enables larger food items to be broken down and assimilated via either absorption or intracellular digestion. The gut also allows for food to be processed in a conveyer belt fashion with different parts of the digestive processes being separated. This allows not only for different physical processes to occur, but also for the creation of different chemical environments – for example, protein digestion requires an acid environment, while carbohydrate and lipid digestion need an alkaline or neutral one.

As both intracellular and extracellular digestion occur in many molluscs, the question arises as to which was the likely mode of digestion in the first molluscs? Arguably, the most primitive living molluscs are the chitons, and the nearest putative outgroups might be expected to provide some clues. Brachiopods have midgut diverticula that are at least analogues of the molluscan digestive gland and, like that gland in molluscs, these diverticula are involved in intra- as well as extracellular digestion. The posterior part of the gut has cilia somewhat reminiscent of the style sac in many molluscs that rotate the contents, although in articulate brachiopods that part of the gut is a blind tube, with the faecal material being periodically expelled through the mouth. In inarticulate brachiopods, the gut is a U-shaped tube with the anus opening into the lophophoral cavity. The closely related phoronids have a U-shaped gut and no digestive gland – digestion is partly intra- and partly extracellular within the gut. In polychaetes, digestion is extracellular with enzymes secreted by the gut epithelium, and the food is absorbed through the intestinal epithelium. Sipunculans have a U-shaped, coiled gut, and probably also mainly extracellular digestion. Thus, within the outgroups, there is considerable variation with a mixture of both intracellular and extracellular digestion. Extracellular digestion might be the primitive method in molluscs, as there is mainly extracellular digestion in caudofoveates and chitons, although both methods occur as they do also in solenogasters.

5.4.2.3 Digestive Enzymes

Besides a few detailed investigations in a handful of taxa, digestive enzymes in molluscs have been rather superficially studied. Most of the older research focused on the presence or absence of the major classes of enzymes, but the relatively few modern studies have had access to much more sophisticated detection techniques and a greater understanding of the diversity and complexity of the enzymes. In our brief overview below, we indicate some findings in the older literature, while focussing in more detail on some recent studies. Digestive enzymes have also been identified in the haemolymph where they play a protective role against invading microorganisms and possibly metazoan parasites (Cheng 1978; Cheng & Downs 1988) (See Chapter 2 for further details).

Enzymes are liberated from cells mainly in the digestive glands, but also the salivary glands in some taxa (see below and Section 5.3.2.5.2), oesophageal glands (when present) and, at least sometimes, from cells in the lining of other parts of the gut.

Enzymes are formed in cells through transcription of DNA to messenger RNA (mRNA). The mRNA is transported via the cytoplasm to ribosomes in the endoplasmic reticulum where it is translated to the amino acid sequence of the protein (the enzyme). Minute vesicles containing the enzyme break off and are accumulated and coalesced in the Golgi apparatus. From there, they accumulate in the apical end of the cell as secretory granules awaiting release.

In general terms, three modes of release of enzymes from the secretory cells occur in molluscs and other animals – (1) a pinched off piece of the apical part of the cell breaks down and liberates the secretory granules containing the enzymes (*apocrine secretion*), (2) the whole secretory cell disintegrates to release its contents (*holocrine secretion*), or (3) the enzymes are released in tiny vesicles by exocytosis through the cell membrane (*merocrine secretion*). This latter mode seems to be the secretory process in the basophilic digestive gland cells and has been detected in some other cells, for example, in the anterior oesophagus of *Bulla striata*, where the apical portions of the cells being released into the oesophageal lumen did not contain secretory vesicles or any other organelle (Lobo-da-Cunha et al. 2010).

The digestive enzymes fall in two very different categories: those employed in extracellular digestion and secreted by basophilic cells and the lysosomal acid hydrolases utilised in intracellular digestion (such as cathepsins) in digestive cells.

The six classes of enzymes are described in Chapter 2. One of the most important classes of digestive enzymes is the hydrolases, which are found throughout the gut in molluscs (e.g., Livingstone & Zwaan 1983).

Enzymes are very specific to particular substrate types, but within those, some can act on a range of compounds. There are three main categories of hydrolytic digestive enzymes – proteases, carbohydrases, and esterases. Several types of enzyme within each main category act on a more specific range of substrates. We do not explain the biochemistry involved in specific enzymatic reactions, but this information is readily available elsewhere.

- *Proteolytic enzymes* (proteases). All animals, including strict herbivores, need to have these enzymes to hydrolyse proteins to release amino acids.
- *Carbohydrases.* These enzymes hydrolyse carbohydrates – from disaccharides (e.g., sucrose) or polysaccharides (e.g., starch) into smaller and simpler carbohydrates (e.g., glucose). *Cellulases* are an important group of carbohydrases which hydrolyse forms of cellulose (a long-chain polymeric polysaccharide carbohydrate, of beta-glucose) and other complex polysaccharides to beta-glucose molecules which can be digested by animals. For the complete degradation of cellulose, three classes of enzymes are required: endocellulases, exocellulases, and cellobiase (Vonk & Western 1984). An enzyme acting as an endocellulase breaks internal bonds, creating exposed chain ends which the exocellulase breaks into small units (tetrasaccharides and disaccharides). Cellobiase (or beta-glucosidase) hydrolyses these small units into monosaccharides (e.g., Payne et al. 1972; Wood & Bhat 1988).
- *Esterases* – including *lipases*. These enzymes hydrolyse lipids (including fats and related substances) into fatty acids.

The proportions and mix of digestive enzymes in a particular mollusc are largely related to its feeding mode. Carnivorous taxa have high protease and esterase activities, whereas detritivores and herbivores show high cellulase activity (e.g., Boetius & Felbeck 1995).

Cellulases are needed to digest plant and algal material. Their distribution in molluscs was reviewed by Stone and Morton (1958) and in bivalves by Crosby and Reid (1971). Cellulases are mainly, but not exclusively, produced in the molluscan gut by symbiotic bacteria (see Section 5.4.2.3.2). In a few cases, endogenous production of cellulases has been reported in the vetigastropod *Haliotis* (Suzuki et al. 2003), the caenogastropod *Ampullaria* (Wang 2003; Wang et al. 2003a), and the bivalve *Corbicula* (Sakamoto et al. 2007; Sakamoto & Toyohara 2009) and has been found in a few other animals. It is possible that this may be more widespread in molluscs. For example, independent cellulolytic activity has been found in the caenogastropod *Strombus* (Horiuchi & Lane 1965), the panpulmonates *Helix pomatia* (Strasdine & Whitaker 1963) and *Radix peregra* (Brendelberger 1997a) and the bivalve *Scrobicularia plana* (Payne et al. 1972).

The enzymes found in bivalves were reviewed by Reid (1982) and summarised by Gosling (2003). They are produced by the digestive gland, stomach wall, and the crystalline style (see next Section).

Enzymes in (mainly autobranch) bivalve stomachs (from the summary in Gosling 2003, p. 114–115), although not all of which are necessarily endogenous), include:

- Esterases – lipid digestion.
- Acid and alkaline phosphatases – probably involved in absorption and phagocytosis of material from the stomach.
- Endopeptidases, e.g., trypsin – enable the breakdown of proteins.
- Carbohydrases – carbohydrate-splitting enzymes including amylase, glucosidase, galactosidase, maltase, chitinase (needed to break down the chitin, e.g., in diatom skeletons) and cellulases (needed to break down cellulose), the last being particularly active in bivalve stomachs. Endogenous cellulase has been shown to be produced in a few molluscs (see above), but more often is produced by symbiotic bacteria in the gut (see Section 5.4.2.3.2).

Acid enzymes such as cathepsins, carboxypeptidases, and acid phosphatases have been reported from bivalve stomachs and digestive glands (Reid 1982).

The enzymes alkaline phosphatase and maltase are present in many regions of the gut, but are most likely to be involved with membrane transport as they are associated with epithelial brush borders (Payne & Crisp 1989; Boetius & Felbeck 1995).

As might be expected, the microcarnivorous septibranch *Cardiomya* has much higher proteolytic activity in the stomach than that in suspension-feeding or deposit-feeding bivalves (Reid 1977).

Changes in digestive enzyme activities in bivalves have been shown to be induced by temperature changes (e.g., Seiderer & Newell 1979; Newell & Branch 1980), food availability (Reid & Rauchert 1976), or reproductive state (Kreeger 1993).

Digestive enzymes are produced in the salivary glands in some gastropods. In the patellogastropod *Patella*, the salivary glands have been shown to produce peptidases, glucosidases, phosphatases, and, unusually, amylase (Bush 1989; Benmeradi & Benmeradi 1992), which are secreted into the oesophagus. This markedly contrasts with what is thought to be the case with the salivary glands of vetigastropods and lower caenogastropods as apparently most do not produce digestive enzymes (e.g., Fretter & Graham 1962; Ward 1966) (see also Section 5.3.2.5.2). Protease activity has been reported in the salivary glands of abalone (Edwards & Condon 2001), and the salivary glands in the herbivorous fissurellid *Megathura* produce cystine arylamidase and trypsin and small amounts of naphthol-AS-GI-phosphohydrolase, alpha-glucosidase (α-glucosidase), cellulase, and amylase (Martin et al. 2011) hinting that, for herbivorous species at least, salivary enzymes may be more widely available in vetigastropods.

The salivary glands of the lower caenogastropods *Pomacea* (Andrews 1965a) and *Littorina* (Jenkins 1955 cited by Fretter & Graham 1962) produce amylases, and proteases are produced in these glands in at least some muricids (Andrews 1991), but it is not known how commonly the acinous salivary glands of muricids and other neogastropods secrete these enzymes.

Little is known of the enzymes involved in 'opisthobranch' salivary secretion, but strong amylase and weak protease activity were detected in *Aplysia punctata* (Howells 1942). In 'pulmonates', digestive enzymes are found in the saliva, and they are apparently important in the preliminary digestion of food, unlike the situation in many other gastropods. For example, in the planorbid *Biomphalaria straminea*, amylase has the highest activity, then cellulase, maltase, and aminopeptidase (trypsin was not detected) (Moura et al. 2004). Enzymes, with amylase dominating, have also been recorded in *Lymnaea stagnalis* (Carriker & Bilstad 1946; Boer et al. 1966), *Deroceras reticulatum* (Walker 1970a), and helicids (Charrier & Rouland 1992; Flari & Lazaridou-Dimitriadou 1996) with cellulase also present in helicids (Stone & Morton 1958).

The salivary gland enzymes begin the extracellular digestion of food in the oesophagus, with their optimum pH probably higher than required by enzymes in the stomach (Andrews & Thorogood 2005). In the neogastropod *Nucella*, the non-salivary enzymes are mainly produced in the gland of Leiblein and the digestive gland (Andrews & Thorogood 2005). The oesophageal gland of some vetigastropods and caenogastropods has also been shown to produce digestive enzymes (e.g., Fretter & Graham 1962), and in a patellogastropod (*Patella*) its cells produce an amylase by apocrine secretion in blebs (Bush 1989), but surprisingly few studies have been undertaken. The midoesophageal region (including the oesophageal gland) of the herbivorous fissurellid *Megathura* contained cellulase, amylase, lysozyme, lipase, and protease (Martin et al. 2011). Two of these enzymes, lysozyme and lipase, occur in molluscan haemocytes (e.g., Foley & Cheng 1977; Cheng 1978), so they may not be produced by the oesophageal gland cells themselves.

A suite of enzymes has been found in abalone and includes lipases, proteases, and glucosidase hydrolases (e.g., cellulase, laminarinase, alginase, carrageenase, and agarase) (see Serviere-Zaragoza et al. 1997 for references). While some enzymes are produced in the gut, enteric bacteria in abalone also produce enzymes that can hydrolyse various seaweed polysaccharides (Erasmus et al. 1997) (see also Section 5.4.2.3.2). The abalone *Haliotis midae* can synthesise the endogenous polysaccharases agarase, carrageenase, alginate lyase, carboxymethylcellulase (CMCase), and laminarinase with their expression regulated by diet (Erasmus et al. 1997). In addition to this suite of endogenous enzymes, others produced by bacteria isolated from the gut of *H. midae* can hydrolyse carrageenan, laminarin, alginate, carboxymethylcellulose, and agarose, all of which occur in the algal food of the abalone, suggesting these play a role in digestion (Erasmus et al. 1997). Changes in enzyme activity in different parts of the gut were investigated in juvenile abalone by

Garcia-Esquivel and Felbeck (2006), who assayed 19 digestive enzymes in different gut locations. Their results showed high activities of cellulase and another carbohydrase, lysozyme,[11] lipase, and the proteases chymotrypsin and/or aminopeptidase. Such a complement of enzymes is to be expected in a herbivore. The digestive gland-stomach region was characterised by high activities of cellulase and lysozyme, and proteases such as chymotrypsin, with most enzymes in the digestive gland. Both the buccal cavity and the intestine typically had high activity of lipase and aminopeptidase. Epithelial cells in the intestine of *Haliotis* contain vacuoles with many secretory granules that release enzymes into the intestinal lumen. The granules contained protease, non-specific esterase and lipase, and 3–4 kinds of polysaccharide-degrading enzymes which facilitate extracellular digestion within the intestine (Cui et al. 2001b).

Proteases are very important in animals as most cannot synthesise half of the 20 L-amino acids required to build proteins, and hence determine growth (Serviere-Zaragoza et al. 1997). Generally, the proteases that hydrolyse proteins in the food are pepsin in the stomach, and trypsin, chymotrypsin, and carboxy- and aminopeptidases in the intestine (Whitaker 1994). These enzymes enable herbivorous gastropods to utilise the protein content in macroalgae. Protease in abalone (*Haliotis rubra*) has dual pH optima (pH 3 and 10) and, surprisingly, an optimal temperature of 45°C, much higher than the temperature to which the animal is actually exposed (Edwards & Condon 2001).

In naticids, a protease is secreted by the oesophageal glands and amylolytic, lipolytic, and proteolytic enzymes by the digestive glands. Neogastropods have a more diverse array of enzymes involved in extracellular digestion. In many, the acinous salivary glands (Andrews 1991) and the gland of Leiblein secrete proteolytic enzymes, but not amylolytic and lipolytic enzymes.

The digestive gland is the main site of digestion and absorption in most gastropods and is the major source of enzymes in molluscs (e.g., Voltzow 1994) with its acidity providing optimal conditions. Picos-García et al. (2000) provided a list of enzymes known from the digestive gland of *Haliotis*; the secretory gland cells contain protease, non-specific lipase, esterase, and four types of polysaccharide-digesting enzymes (Cui et al. 2001a). Molecular studies have shown that abalone can produce an endocellulase and a glucanase in the digestive gland (Suzuki et al. 2003), and earlier work showed several kinds of endogenous carbohydrases (alginate lyase, CMCase, laminarinase, agarase, and carrageenase) in that organ.

Both *Haliotis* (Garcia-Esquivel & Felbeck 2006) and *Megathura* (Martin et al. 2011) have high levels of phosphohydrolases and medium to low levels of esterase and lipase. The peptide hydrolases included high levels of leucine arylamidase, medium levels of valine arylamidase, medium to low levels of cystine arylamidase, and low levels of trypsin, while alpha-chymotrypsin (α-chymotrypsin) was high in *Megadenus* and low in *Haliotis*. Levels of the glycosidases also varied, with beta-galactosidase (β-galactosidase), beta-glucuronidase (β-glucuronidase), and alpha-fucosidase (α-fucosidase) being high in both, and alpha-galactosidase (α-galactosidase), alpha-glucosidase, (α-glucosidase) and beta-glucosidase (β-glucosidase) medium. Levels of *N*-acetyl-beta-glucosaminidase (β-glucosaminidase) and alpha-mannosidase (α-mannosidase) were higher in *Haliotis* than in *Megadenus* (Martin et al. 2011). The presence of these enzymes show these vetigastropods can digest complex polysaccharides.

In a detailed study of the hydrolytic enzymes in the digestive gland in *Haliotis discus hannai*, Cui et al. (2001a) found that 'vesicles' within the digestive cells (i.e., lysosomes) that contained acid phosphatase and non-specific esterase are involved in intracellular digestion. The digestive cells also secrete protease and contain many lipid globules (stored food reserves) and basal vesicles (residual bodies) that accumulate undigested residues. The basophilic cells have a well-developed rough endoplasmic reticulum and numerous refractile spherules that contain metals and may play a role in detoxification. These cells also secrete protease. *In vitro* tests for enzymes in the digestive gland of *Haliotis* by Cui et al. (2001a) also showed diastase, cellulose, alginase, and laminarinase.

The intestine of the fissurellid *Megathura* has a wide range of enzymes present including beta-glucuronidase, amylase, cellulase, lysozyme, lipase cysteine, arylamidase, trypsin, naphthol-AS-GI-phosphohydrolase, and alpha-galactosidase, with more activity in the anterior region than the posterior indicating that digestion continues in that part of the gut (Martin et al. 2011).

The gut of the 'saltmarsh periwinkle' *Littoraria irrorata* was found to contain 19 enzymes active on lipids, peptides, and a wide range of carbohydrates, but not chitin, and protease activity was weak (Bärlocher et al. 1989a).

The digestive gland of *Aplysia* produces a wide range of enzymes including alkaline and acid phosphatase, esterases, arylamidases, trypsin, galactosidases, mannosidase, and fucosidase (Taïeb 2001).

In the terrestrial stylommatophoran helicids, enzymes in the oesophagus are produced by the stomach and digestive gland, but not the salivary glands (e.g., Dimitriadis & Hondros 1992). Of approximately 30 enzymes obtained from the gut of helicid snails, over 20 are carbohydrases (Dimitriadis 2001). *Helix pomatia* has enzymes that hydrolyse numerous carbohydrate substrates, including sucrose, maltose, the plant polysaccharides starch, cellulose, xylan and mannan, the animal and fungal polysaccharide chitin, and glycogen. This ability to hydrolyse both plant and animal polysaccharides is surprising in what is thought to be a strictly herbivorous species (e.g., Pollard 1975), but many other helicids include fungi in their diets. Helicids have a wide spectrum of carbohydrases including mannan-degrading and cellulolytic enzymes that act in their crop (Charrier & Rouland 1992; Charrier et al. 2001). Their enzymes hydrolyse numerous carbohydrate substrates, including sucrose, maltose, xylan, mannan, starch, glycogen,

---

[11] An enzyme that attacks peptidoglycans, a polymer consisting of sugars and amino acids that forms part of the cell wall of many bacteria.

chitin, and cellulose. Xylans are polymers of xylose and are constituents of plant cell walls, while mannans are plant polysaccharides that are polymers of mannose. The ability to hydrolyse plant polysaccharides (xylan, mannan, starch, and cellulose) as well as chitin, an animal and fungal polysaccharide, may indicate the potential for an omnivorous diet (Myers & Northcote 1958). Proteolytic activity in 'pulmonates' has been found in the crop, stomach, digestive gland, and intestine.

While there has been controversy as to whether stylommatophoran snails, particularly Helicidae, have endogenous cellulase and chitinase in their gut, studies show that some of these enzymes are produced endogenously and are active in the crop and digestive gland (Charrier & Rouland 1992; Flari & Charrier 1992; Dimitriadis 2001; Charrier et al. 2006) (see also Section 5.4.2.3.2), although bacteria obtained from their food also produce similar enzymes. Cellulolytic bacteria have not been found inhabiting the crop or the rest of the gut (Charrier et al. 2001). The acidic crop fluid (pH 5.3–6.1, Charrier & Brune 2003) is optimal for the most powerful glucanases from the crop (e.g., Charrier & Rouland 2001).

A broad range of carbohydrases has been found in the gut of autobranch bivalves (e.g., Kristensen 1972; Wojtowicz 1972; Mathers 1973; Hameed & Paulpandian 1987; Palais et al. 2010), the crystalline styles of bivalves and gastropods (see next Section), and the digestive glands of many molluscs (e.g., Onishi et al. 1984). Endogenous amylases in bivalves are usually alpha-amylases (Owen 1966; Morton 1983a; Palais et al. 2010). Laminarinase, a type of carbohydrase, was the most active in most species, with the highest activity in bivalves, especially in the crystalline style, rather than gastropods (Onishi et al. 1984). Degradation of structural polysaccharides also occurs in herbivorous gastropods lacking a crystalline style – for example, in the patellogastropod Lottiidae (on alginic acid, in brown seaweed) and Patellidae (on xylan, which replaces cellulose in some green algae), and others, such as *Patelloida*, *Littorina*, and *Turbo* on microcrystalline cellulose (Onishi et al. 1984). Fucoidan (a sulphated polysaccharide found in many brown algae) was also commonly degraded in these taxa, but not as efficiently (Onishi et al. 1984).

Enzymatic activity can differ considerably between taxa, even when food is not fundamentally different. For example, in a study of two fresh-water gastropods, the hygrophilan *Radix peregra* and the truncatelloidean *Bithynia tentaculata*, Brendelberger (1997b) identified the extracellular enzymes cellobiase, chitobiase (possibly of bacterial origin), and protease in the gut, and their activities were shown to change markedly with food conditions. While cellobiase activity in *R. peregra* was ten times higher than in *B. tentaculata*, the other enzymes were of the same order of magnitude in both species. This was related to feeding, as *R. peregra* is a herbivore (hence the high cellobiase activity), but also feeds on carrion, for which protease is necessary. *Bithynia* is much more selective, feeding on minute green algae and diatoms (Brendelberger 1995), requiring high chitobiase activity for digesting microorganisms on substrate surfaces and high

protease activity when feeding on diatoms (Brendelberger 1997b). Esterase and protease were detected in the whole digestive tract of both species. Lipase was not detected in this study, and phosphatase found in the stomach of *Bithynia* was thought to have originated from bacteria associated with the food. Intracellular phosphatases are present in the digestive cells of some gastropods and bivalves (e.g., Summer 1969; Dimitriadis & Liosi 1992). This class of enzymes is rarely recorded in the gut of gastropods, but it does occur in at least some gastropod neurosecretory cells (e.g., Lane 1966), haemocytes, and haemolymph (e.g., Franchini & Ottaviani 1990; Suresh et al. 1993).

Given the findings in the most recent studies such as that of Martin et al. (2011), digestive enzymes in gastropods are obviously more diverse than previously shown. Apparently, the ability to secrete many digestive enzymes may reside in secretory cells lining many parts of the gut, but the enzymes expressed vary with the region of the gut and the quantity and type of food (e.g., Garcia-Esquivel & Felbeck 2006), although external factors, such as diurnal and/or tidal cycles, may also have a role.

The digestive enzymes in cephalopods were reviewed by Boucaud-Camou and Boucher-Rodoni (1983). The enzymes (proteases, lipases, and amylases) are mainly derived from the digestive gland and digestive duct appendages and transported to the caecum by the digestive gland ducts. The foregut glands also produce some enzymes.

In a study of octopod and oegopsid squid paralarvae, Boucaud-Camou and Roper (1995) recognised 14 enzymes involved in digestion (esterases, glycosidases, and peptidases). Overall protease activity was high, while amylase activity was very low or undetectable, confirming that paralarvae were carnivorous and had similar enzymatic activity to adults. Also, as in adults, the digestive gland had the most enzyme activity, notably proteolytic enzymes. Acid phosphatase, dipeptidyl aminopeptidase II, (DAP II) and acetylglycosaminidase showed high activity in the digestive gland, reflecting intracellular digestive processes. Besides the midgut enzyme activity, the posterior buccal glands had high proteolytic activity, suggesting that these glands may be more involved in digestion in the paralarvae than in adults (where they are mostly venom glands).

5.4.2.3.1 The Role of the Crystalline Style

As noted above, cellulases are produced in the crystalline style of most (if not all) style-bearing bivalves (e.g., Brock & Kennedy 1992; Sakamoto & Toyohara 2009). Cellulases and carbohydrases have also been recorded from the crystalline style of *Strombus* (Horiuchi & Lane 1965), a herbivorous caenogastropod. The truncatelloidean *Oncomelania*, which has a crystalline style like all members of that group, can digest and survive on a diet of filter paper, but the source of the cellulases is unknown (Winkler & Wagner 1959). Successful digestion of an algal diet requires a wider range of enzymes than just cellulase and the enzymes found in algal-feeding gastropods (e.g., *Haliotis*, *Littorina*, and *Aplysia*) include agarase, cellulase, and alginate lyase (Gomez-Pinchetti & Garcia-Reina 1993).

Several studies have provided quantitative data on crystalline style enzyme activity (e.g., Newell et al. 1980; Lucas & Newell 1984; Seiderer et al. 1984; Palais et al. 2010). Palais et al. (2010) reviewed amylase and CMCase activity in the crystalline style (and digestive gland) of a range of bivalves.

The style sac in the protobranch bivalve *Nucula*, which does not have a crystalline style, secretes amylase and lipase (Owen 1956). The crystalline style of some autobranch bivalves has been shown to contain carbohydrases (including cellulase), amylase, lipase, alginase, and chitobiase. The style sac of the vetigastropod *Scutus* secretes amylase (Owen 1958).

A study of the herbivorous fissurellid *Megathura* showed that both the stomach and style sac (with a crystalline style, unlike many fissurellids) have high levels of alpha-amylase, cellulase, and lysozyme. The first two enzymes break down plant material, while the lysozyme catalyses the hydrolysis of bacterial cell walls (Martin et al. 2011). Similar enzymes were found in the crystalline style of the caenogastropod *Telescopium* where two enzymes (laminarinase and glucosidase) were found only at the proximal end of the style (Alexander et al. 1979), although the crystalline styles in caenogastropods are surprisingly under studied. *Strombus* has carbohydrase activity in the style (cellulase and maltase), but most enzyme activity is in the digestive gland (Horiuchi & Lane 1965, 1966). Another stromboidean, the suspension-feeding *Struthiolaria*, showed strong amylolytic enzyme activity in the style, but it did not produce enzymes that could hydrolyse disaccharides or cellulose. Besides amylase activity, digestion was intracellular (Morton 1951).

Carbohydrase activity has been determined in the crystalline styles and digestive glands of many molluscs. Laminarinase, a type of carbohydrase, was the most active in most species. Activity was generally higher in bivalves than in gastropods, and especially in the crystalline style of some bivalves (Onishi et al. 1984).

Protease is also present in the style sac in bivalves (Reid & Sweeney 1980) and was at a high level in that region in *Megathura* (Martin et al. 2011); this may assist with the periodic degradation of the style.

At least some enzymes produced in the crystalline style have often been attributed to symbiotic bacteria as discussed in the next section below.

5.4.2.3.2 *The Role of Gut Bacteria in Digestion*

In many herbivores, especially terrestrial taxa, parts of the digestive tract are modified to accommodate a microbial population involved in food digestion by producing enzymes that do not originate from the host and also provide other benefits (e.g., McBee 1971; Harris 1993). In turn, the host provides a microenvironment that is ideal for the bacteria to flourish, with suitable pH (see Section 5.4.2.1). Structural modifications of the host are important. For example, in bivalves, the (usually long) intestine has the highest populations of bacteria (Harris 1993) with food taking up to three days to traverse it, giving ample time for bacterial populations to adjust and interact (Prieur et al. 1990).

Seillière (1906) first isolated bacterial cellulases in *Helix pomatia*, and based on this and other studies, the digestion of plant material was generally attributed to the activities of the gut microbiota. Indeed, it is well known that bacteria (and possibly also microfungi) in the gut can play an important role in digestion in many animals, notably in breaking down complex polysaccharides such as cellulose. Earlier studies (e.g., Galli & Giese 1959; Crosby & Reid 1971; Harris 1993; Erasmus et al. 1997; Harris et al. 1998b) often included the use of carbohydrate substrates to determine the role of bacteria in digestion. Both aerobic and anaerobic bacteria have been shown to be involved. Although many cases of the proven or suspected involvement of bacteria in digestion in molluscs have been reported, there are relatively few detailed studies on the composition of gut bacteria and their role in producing digestive enzymes.

Without suitable bacteria to produce enzymes such as cellulases, it is generally thought that molluscs (like most animals) could not use most of the energy in plant material. The source of cellulase in some molluscs was thought to originate mainly from spirochaete bacteria found in the crystalline style of some bivalves (e.g., Stone & Morton 1958; Judd 1979; Harris 1993) and a gastropod (Morton 1952b) (see below). While this does appear to be the case for some bivalves, some recent studies have shown the cellulases can be expressed endogenously in some taxa (Wang et al. 2003a; Sakamoto et al. 2007; Sakamoto & Toyohara 2009). This has also been found to be the case in a few other animals (Watanabe & Tokuda 2001), and it is possible that endogenous cellulase expression may be more widespread in molluscs than currently recognised. For example, cellulases are produced in the crystalline style of the fissurellid *Megathura* (Martin et al. 2011) where bacteria have not been located. In addition, high levels of lysozyme in that species (Martin et al. 2011) and in the gut of juvenile *Haliotis* (Garcia-Esquivel & Felbeck 2006) suggest that bacteria are probably being digested. Despite this, the role of bacterial enzymes is important.

Seiderer et al. (1984) noted a bacteriolytic enzyme in the crystalline style of the mussel *Choromytilus meridionalis* that is capable of lysing the majority of free-living bacterial taxa in the local water column. They suggested that bacteria may serve as a complementary food source, especially when phytoplankton is scarce. Similar bacteriolytic activity has also been observed associated with the style of some other mytilids (e.g., Seiderer et al. 1987).

The polysaccharolytic enzymes of the abalone *Haliotis midae* and its resident gut bacteria were investigated by Erasmus et al. (1997). Bacteria isolated from the abalone gut degraded the polysaccharides laminarin, carboxymethylcellulose (CMC), alginate, agarose, and carrageenan. Alginate lyase, CMCase, laminarinase, agarase, and carrageenase were found in the digestive gland where bacteria were absent, indicating that a range of polysaccharases is produced endogenously by *H. midae*, and these varied in response to diet. Nonetheless, gut bacteria apparently helped to digest alginate, laminarin, agarose, carrageenan, and cellulose (Erasmus et al. 1997).

A study using molecular markers on the bacterial communities in the gut of *Haliotis* showed considerable diversity, with the *Vibrio* group dominant (Tanaka et al. 2004). A shift in gut bacteria as a result of a change in diet was demonstrated by Tanaka et al. (2003) in cultured *Haliotis*. They found that in juvenile abalone, the bacteria matched those cultured from seawater, but they changed when the food of the cultured abalone was switched from microalgae to algal pellets. Thus, at about four months, the ambient gut microbes were replaced by algal polysaccharide-degrading bacteria which were almost entirely facultative anaerobes, including species of *Vibrio*.

In general, the importance of bacteria in contributing enzymes for digestion remains unclear. Some studies have used antibacterial agents to remove gut bacteria to test their necessity. For example, in abalone, while bacteria from the gut can break down a range of substances, after antibacterial treatment they can still degrade polysaccharides (Erasmus et al. 1997), suggesting that while the bacteria assist in digestion, they are not essential.

Where microbial enzymes play a role in digestion, the successful breakdown of food particles may be limited by the density of microbes and the size of the food particle, as the smaller food particles have a greater relative surface area for microbial interactions.

Since the initial discovery of spirochaetes in the crystalline style in the 1880s, they have rarely been investigated in molluscs despite their supposed importance as a source of cellulases. A recent study on spirochaetes in the crystalline styles of some bivalve taxa used molecular markers and showed that the diversity in each species was low, but that the taxa differed in different bivalves and encompassed three genera, *Cristispira*, *Spirochaeta*, and *Brachyspira* (Husmann et al. 2010). In two cases, these workers found similar spirochaetes in the gill, and they assumed that the bacteria were picked up from the environment although they also found that only about half the styles of the species investigated had spirochaetes, so they concluded that they are not essential symbionts.

Studies by M. Charrier and colleagues have focused on the gut microbial biota of helicids, in particular those in the common garden snail *Cornu aspersum*. For example, the bacteria live free in relatively stable populations within the gut (Lesel et al. 1990; Charrier et al. 1998) and can digest living plant tissue, with plant fibres being digested with 60%–80% efficiency (Davidson 1976; Charrier & Daguzan 1980), and the bacteria increase in numbers in the presence of food (Charrier et al. 2001). Successful plant digestion requires a range of polysaccharide depolymerases and glycoside hydrolases. Studies on helicid snails have shown that some complex polysaccharides (cellulose, lamarin, and mannan) are efficiently degraded while others (starch, pullulan, alpha-glucosides, xylan, and beta-xylosides) are only weakly hydrolysed (Charrier et al. 2006). It is, however, unclear as to the exact role of the bacteria, as no cellulolytic or methanogenic bacteria occur in the gut (Charrier et al. 2001), although some appear to assist in the fermentative processes (Charrier et al. 1998). For example, some bacteria may break down soluble

cellulose and chitin (Lesel et al. 1990; Brendelberger 1997a), and some are apparently transient in the gut, interacting with trace metals (Simkiss 1985; Simkiss & Watkins 1990).

In helicids, the breakdown of polysaccharides and oligosaccharides mainly occurs in the oesophagus, crop, and stomach where the bacteria are mainly aerobic (Charrier et al. 1998), with the dominant bacteria that could be cultivated being Enterobacteria (Lesel et al. 1990; Watkins & Simkiss 1990) and Enterococci (Charrier et al. 1998). Bacteria in the intestine are at even higher levels and are found in both fed and starved animals and, based on culturing and sequence analysis, comprise several genera of both Gamma-proteobacteria and Firmicutes (Charrier et al. 2006). These intestinal bacteria can also help to digest plant material.

Charrier et al. (2001) reported 14 bacterial strains in ten genera (determined by sequence data) in the anaerobic intestine of *Cornu aspersum*, with a similar diversity in *Helix pomatia*. These comprised a single strict anaerobe (*Desulfotomaculum guttoideum*, a sulphate-reducer) and 13 facultative anaerobes. They concluded that helicid snails are 'hindgut fermenters'.

In the helicids studied by Charrier and Brune (2003), oxygen is absent throughout the intestine lumen, a surprising finding given that aerobic bacteria are involved in digestive processes, although the anoxic conditions may be due in part to oxygen consumption by the bacteria. These authors suggest that the gut bacteria may be avoiding the anoxic conditions by resting against the oxic gut epithelium. Anaerobic bacteria (family Enterobacteriaceae) are most abundant in the intestine of these snails (Lesel et al. 1990; Charrier et al. 1998), with *Enterococcus casseliflavus* the most abundant. The same bacteria are also commonly found in the gut of wood-feeding arthropods, such as termites. Although there are high levels of cellulase activity in the crop of helicids (Charrier & Rouland 1992; Flari & Charrier 1992), cellulolytic bacteria were not found in that region or in the rest of the gut by Charrier et al. (1998); Charrier et al. (2001) suggesting an endogenous origin of the suite of cellulolytic enzymes. The acidic crop fluid (pH 5.3–6.1, Charrier & Brune 2003) is optimal for the most powerful glucanases from the crop (e.g., Charrier & Rouland 2001). Nevertheless, it is probable that a mix of endogenous and exogenous enzymes break down complex polysaccharides in the crop, digestive gland, and stomach rather than enzymes produced solely by the host or by the bacteria. Studies of the digestive gland of *Cornu* found few bacteria (Charrier 1990), and they were absent in the cytoplasm of the cells of the digestive gland (Charrier et al. 1998); when snails were fed sterilised food their growth increased, although cellulase activity was significantly reduced in the crop, suggesting a possible deleterious effect of the enzymes from exogenous bacteria (Charrier et al. 1998). This finding also suggested that microorganisms ingested with the food may provide cellulases in the foregut. Starvation considerably reduced bacterial populations in the crop, but not in the intestine, suggesting that crop microbial populations originate from the environment and are probably lysed after ingestion. In contrast, the stable intestinal bacteria remained in high densities (Charrier et al. 1998).

No matter what the nature of the food was, the activity by cellulolytic bacteria degrading microcrystalline cellulose in the intestine was insignificant. The presence of *Enterococcus* as the dominant lactic bacterium in the intestine of *C. aspersum* suggests that lactic acid may be important in the digestive processes (Charrier et al. 1998).

While hydrogen production is typical of the termite hindgut and also the rumen of herbivorous mammals such as cattle, it was only recently reported from terrestrial gastropods (Charrier & Brune 2003). Hydrogen accumulates in the proximal intestine, presumably produced by the many bacteria present, and also in the digestive gland where there are few microorganisms, although it may diffuse there from the intestine.

5.4.2.3.2.1 Symbiotic Bacteria in Teredinid Bivalves In the wood-boring teredinids, particles of wood are phagocytosed by cells in specialised digestive tubules after being partially hydrolysed by cellulase derived from symbiotic bacteria (Waterbury et al. 1983). The more derived teredinids, which have short gills and consequent reduced ability to suspension-feed, ingest wood fragments, the larger ones being stored in a large caecum-like appendix of the stomach and eventually released as faecal material, while smaller fragments are digested. In the less specialised genera, the gills are much larger and the caecum is small (Turner & Johnson 1971); it is absent in *Kuphus*, the only non-wood-boring teredinid. The large gills in *Kuphus* house chemoautotrophic bacteria (Distel et al. 2017).

The digestion of wood is probably commenced in the stomach by way of extracellular digestion through enzymes released into the oesophagus by the bacteria in the so-called 'glands of Deshayes', structures found only in Teredinidae (Morton 1978a). These 'glands' are actually bacteriocytes which contain a densely packed Gram-negative bacterium (Popham & Dickson 1973), the only marine bacterium currently known which is able to both produce cellulase and fix dinitrogen (Waterbury et al. 1983), and is also responsible for proteolytic activity (Greene et al. 1989). The bacteriocytes occur in the duct inside each afferent branchial vessel of the gills and near the 'calcareous gland' in the mantle (Moraes & Lopes 2003). It is probably the same bacterium in all teredinids (Distel et al. 1991), and it was identified as a proteobacterium (Distel et al. 1991), and named *Teredinibacter turnerae* (Distel et al. 2002b). More recently, three additional bacteria were identified in the bacteriocytes of *Lyrodus pedicellatus* (Distel et al. 2002a). It is thought that enzymes produced by the bacterial symbiont(s) facilitate wood digestion and provide a source of nitrogen to the shipworms (Waterbury et al. 1983; Greene et al. 1988). A diagrammatic representation of the feeding cycle in teredinids is given in Figure 5.55.

5.4.2.4 Absorption

Food digested extracellularly by enzymes is broken down into amino acids (from protein), monosaccharides (from carbohydrates), and monoglycerides, glycerol, and fatty acids (from lipids). These substances, and ions, vitamins and water, are absorbed through epithelial cells lining the gut and, in particular, the digestive cells in the digestive gland, and then transferred from those cells to the haemolymph ('blood'). These processes are complex and essentially similar to those that occur in most animals, so are not dealt with here in more detail.

Important physical aspects that increase the efficiency of absorption are a large absorptive surface area and mechanisms facilitating the movement of the fluid containing the material to be absorbed across that surface. Various strategies are employed to increase the absorptive surface such as intestinal coiling and epithelial folding within it, multiplication of digestive gland tubules at the tissue or organ level, and microvilli at the cellular level. Passage across the cell membrane can be achieved by passive diffusion, facilitated diffusion, active transport, or pinocytosis.

5.4.2.5 Digestion in Aplacophorans

Little is known about the detail of the digestive processes in aplacophorans. Digestion is extracellular in caudofoveates, although picnosis may occur (Salvini-Plawen 1988), and both intracellular and extracellular digestion occurs in at least some solenogasters (Baba 1940; Salvini-Plawen 1988). Secretions from the foregut glands are responsible for some enzymes involved in extracellular digestion, but most are from the digestive glands.

5.4.2.6 Digestion in Polyplacophorans

In chitons, some particles are ingested by amoebocytes, but the bulk of the food is digested extracellularly (Salvini-Plawen 1988). Carbohydrases are secreted by the 'sugar glands' (= oesophageal glands), which are sometimes wrongly referred to as salivary glands (e.g., Greenfield 1972). These empty into the stomach, thus making the stomach, and the digestive glands, the main parts of the gut involved in carbohydrate digestion. The stomach also contains lipases and proteases which are probably secreted by cells of the digestive gland (Greenfield 1972). Movements of the stomach and the ventral sac mix the enzymes and food together, and absorption occurs in the digestive cells of the digestive gland and the long, coiled intestine (Greenfield 1972), especially the very elongate, looped posterior portion.

5.4.2.7 Digestion in Bivalves

Protobranchs mainly use extracellular digestion while intracellular digestion is most important in many autobranchs. There is also a tendency for increased reliance on extracellular digestion in bivalve evolution with the intestine becoming more important for digestion and absorption (Morton 1983a; Gosling 2003). Intracellular digestion is suited for processing the nearly continuous stream of fine particles resulting from suspension-feeding. While the phases of the cycle of digestive gland cell maturation may occur simultaneously within the tubules in many permanently submerged bivalves, changes in these cells can also be externally induced, usually in a cyclical way, by extrinsic factors such as light or tides (see Section 5.4.4.1.1 and Figure 5.54).

In protobranchs, intracellular digestion occurs in the digestive glands, but they rely more on extracellular digestion in the stomach and intestine than autobranch bivalves (Owen 1974). The processes of bivalve digestion have been best studied in autobranch bivalves, showing they sort particles in the stomach before the digestible fraction is transferred to the digestive glands where intracellular digestion, (including absorption and phagocytosis), and extracellular digestion occur.

Various digestive enzymes are liberated from both the crystalline style and digestive glands, those from the former and some from the latter ending up in the stomach lumen to facilitate extracellular digestion (see Section 5.4.2.3). The stomach wall and possibly even the gastric shield also secrete digestive enzymes. The gastric shield is a layer of thick cuticle-like chitinous material and was thought to function solely in protecting the stomach wall from the grinding action of the crystalline style. It overlies an epithelium with the outer cell surfaces covered with long microvilli which extend through the cuticle of the gastric shield, and these cells may produce some enzymes thought to originate from the crystalline style (Halton & Owen 1968; McQuiston 1970; Mathers 1973; Morton 1983a).

The complex stomachs of suspension-feeding autobranchs show considerable structural differences in different groups (see Chapter 15). They all have a crystalline style, well developed sorting regions, several typhlosoles, large digestive glands with an often complex series of ducts, and, in many, one or two food-sorting caeca. These modifications are mainly to enhance particle processing to select tiny particles suitable for ingestion, which is an important element of suspension or deposit-feeding (see Section 5.2.1.2). At least in some bivalves, relatively large particles can be ingested; for example, *Corbicula fluminea* can ingest particles from 5 to 20,000 μm^2 (Wallace et al. 1977). Assimilation of ingested particles in fresh-water bivalves was reviewed by Dillon (2000). Evidence for assimilation could be inferred by comparison of carbon taken in and lost from bivalves, by radio-labelling and by using controlled diets to monitor growth or reproduction of individuals.

The epithelium of the primary ducts of the digestive glands in *Tridacna* (and other autobranch bivalves) comprises ciliated columnar cells that secrete proteolytic enzymes used in extracellular digestion in the stomach (Fankboner & Reid 1990). Cilia on the cells move enzymes towards the stomach and food into the tubules.

5.4.2.8 Digestion in Monoplacophorans and Scaphopods

Little is known about digestion in monoplacophorans. The stomach lumen contains relatively large food particles such as diatoms, dinoflagellates, etc. (Tendal 1985; Haszprunar & Schaefer 1997) which presumably undergo extracellular digestion.

Scaphopods undertake extracellular digestion in the stomach and absorption mostly occurs in the digestive gland (e.g., Salvini-Plawen 1981b).

5.4.2.9 Digestion in Gastropods

Although a comprehensive review of digestion in gastropods is long overdue, there is an early one by Owen (1966) and Andrews and Thorogood (2005) provided a very useful update of many aspects.

As with most other molluscs, gastropod digestion is typically partly extracellular and partly intracellular. Extracellular digestion occurs mainly in the stomach, but also in other parts of the gut, while intracellular digestion takes place in the digestive glands and sometimes by amoebocytes in the stomach. In several gastropods the walls of the stomach and intestine have been shown to be absorptive (e.g., Forester 1977; Nelson & Morton 1979; Bush 1988). In gastropods, as in other molluscs, the main source of extracellular gastric enzymes is the digestive gland (see Section 5.4.2.3), and the acid conditions in the stomach provide an optimal environment for most enzymatic activity, although these conditions are not needed for all digestive enzymes. In the account below, we focus on a few aspects of gastropod digestion that have come to light in recent years. Other aspects are covered in other sections.

In most taxa, the food is mixed with saliva as it enters the dorsal food groove in the buccal cavity and some extracellular digestion may occur as it moves through the oesophagus.

Diet mainly determines the digestive processes. Herbivorous vetigastropods and patellogastropods generally have extracellular digestion. The enzymatic activity in some herbivorous gastropods is very effective in breaking down the cell walls of plants and algae (see Section 5.4.2.3).

A study of *Haliotis* (McLean 1970) showed that the crop and intestine, and the digestive gland, were important areas of absorption and that digestion is largely extracellular (e.g., Edwards & Condon 2001). The large spiral gastric caecum of trochoideans sorts the material it receives from the oesophagus and waste and enzymes from the digestive glands, and extracellular digestion occurs there. Waste material is moved to the style sac and then the rectum (e.g., Morton 1955d).

Phagocytosis by the cells of the digestive glands is important in many microherbivorous caenogastropods that feed fairly continuously. In these taxa, oesophageal glands are nearly always reduced or lost (e.g., Graham 1939), and they have a crystalline style that releases amylolytic and lipolytic enzymes, which are usually the only extracellular enzymes found in the stomach (Owen 1966). In contrast, extracellular digestion predominates in herbivorous and carnivorous caenogastropods that feed discontinuously.

In the stomach, the food is mixed with enzymes secreted by the digestive glands and, in most euthyneurans, also by the salivary glands. There is also phagocytosis by the digestive cells (Owen 1966), but in the nudibranch *Jorunna* (Millott 1937), digestion is supposedly entirely intracellular.

In gastropods with a large crop, such as abalone, the oesophagus can be an important region of absorption, while digestion and absorption continue to occur as food passes along the intestine. In many gastropods, the products of this digestion are absorbed, at least in part, in the oesophageal gland or gland of Leiblein, if one of these is present, and possibly also by the epithelium of the posterior oesophagus (Andrews & Thorogood 2005).

While phagocytosis has been shown to occur in stylommatophorans (e.g., Weel 1950), they also possess a large variety of extracellular enzymes. The 'pulmonate' gut appears to be capable of secretion, nutrient storage, and absorption throughout most of its length (Luchtel et al. 1997). Some enzymes (see Section 5.4.2.3) may be secreted by the salivary glands, but most originate in the digestive gland and some in the epithelium lining the gut. Other enzymes are produced by bacteria (see Section 5.4.2.3.2). In stylommatophorans, absorption occurs in the crop and intestine as well as the digestive gland, in part by amoebocytes (Luchtel et al. 1997).

The oesophageal gland of patellogastropods, vetigastropods, neritimorphs, and many caenogastropods (see Section 5.3.2.7.1.1) secretes enzymes for extracellular digestion. In neogastropods, the oesophageal gland is separated from the oesophagus as the gland of Leiblein and, anterior to that, a valve of Leiblein (see Section 5.3.2.7.1.3).

The oesophageal pouches (glands) of *Patella* are lined with cells that secrete enzymes and absorptive cells involved in intracellular digestion (Bush 1989), where liquid material is squeezed into the pouches by muscular contraction. The apparent ability of the oesophageal pouch cells to absorb dissolved organic matter may be correlated with the simple stomach in *Patella* (Andrews & Thorogood 2005). A similar absorptive epithelium is found on the ventrolateral walls of the oesophageal gland of caenogastropods such as littorinids, naticids, and tonnoideans (Andrews & Thorogood 2005), suggesting that the absorptive capacity of the oesophageal gland may be typical of non-neogastropod caenogastropods.

The main site of extracellular digestion and absorption in neogastropods may be the foregut or midgut, but details vary depending on the group (Andrews & Thorogood 2005). For example, studies show that there are significant differences in the way that muricoideans and buccinoideans process food. In muricids, extracellular digestion begins in the oesophagus with enzymes from the acinous salivary glands, facilitated by the relatively slow passage of food through both the anterior and midoesophagus. Absorption and further digestion occur in the gland of Leiblein. In their detailed study of the gland of Leiblein in a muricid (*Nucella lapillus*) and a nassariid (*Tritia reticulata*), Andrews and Thorogood (2005) showed that the epithelium of the gland is absorptive and secretory. Apart from occasional mucous cells, they showed that there were two main types of cells and that both absorbed solutes through pinocytosis and carried out intracellular digestion in lysosomes. They also showed that ciliated cells were involved in protein metabolism and unciliated cells took up and stored lipids and carbohydrates. The enzymes in the epithelial cells are lysosomal and mainly concerned with intracellular digestion through absorptive activity, not through phagocytosis. A mucous string that emerges from the gland of Leiblein in *Nucella* does not enter the midoesophagus, and the residual

bodies released from the epithelial cells in the gland remain intact at least until the stomach, where they may release some enzymes (Andrews & Thorogood 2005). There is also evidence that the gland of Leiblein absorbs cadmium, and possibly some other toxic substances, from the blood and the lumen of the gland.

The digestive cycle in the gland of Leiblein cells was described in some detail by Andrews and Thorogood (2005). Digestion triggers apocrine secretion by the cells of the gland of Leiblein in preparation for the next round of absorption and intracellular digestion. The indigestible material is retained in blebs and not shed until the food appears in the gland. This is in contrast to what happens in the digestive gland (see Section 5.3.3.2) and may enable the release of any remaining active enzymes into the food. A few hours after shedding their apical blebs, most of the cells are receptive to absorption. Despite the narrow tubular connection with the oesophagus, the gland of Leiblein (of *Nucella*) uptakes carbohydrates and proteins by endocytosis and absorbs lipids.

Although extracellular digestion is predominant in neogastropods, some intracellular digestion also occurs following absorption. As in other gastropods, absorption takes place via epithelial cells in the fore-, mid-, and hindgut. Dimitriadis and Andrews (2000) demonstrated endocytosis in the digestive gland cells of *Nucella*. In contrast, it has been argued that the buccinid digestive gland is not a site of absorption because the gland is isolated by ciliary currents running out of it and by a valve-like fold on the ducts (Kantor 2003). There is, however, absorption by epithelial cells throughout the gut (as in other gastropods). In muricids, the foregut and the midgut are the two main sites of extracellular digestion and absorption (and hence intracellular digestion). In some neogastropods, there is no gland of Leiblein, or it is very reduced, and the digestive gland is the predominant site of digestion and absorption.

In nassariids (Buccinoidea) food passes more rapidly through the oesophagus where little extracellular digestion occurs. This may be associated with a general trend for reduction or even loss of both the valve and gland of Leiblein in some buccinoideans. Buccinids have a small gland of Leiblein which is probably involved in enzyme production and possibly absorption (Andrews & Thorogood 2005), and their tendency to swallow large pieces of food may be correlated with this reduction in size while their stomach is typically well developed and is the main site of digestion. In contrast, in neogastropods where the food is largely liquid, either through external predigestion, as in the Harpidae (Hughes & Emerson 1987), or suctorial feeding as in Colubrariidae and (probably) Cancellariidae (see Chapter 19), both the midoesophagus and stomach are greatly simplified. Some partial external digestion may also occur in some naticids (Reid & Friesen 1980).

In conoideans extracellular digestion occurs either in the buccal cavity or via external digestion in the proboscis sac (Andrews & Thorogood 2005). They do not undertake extracellular digestion and absorption in the midoesophagus because the gland of Leiblein is either lost or incorporated in the venom gland (see Section 5.3.2.7.1.3).

5.4.2.10 Digestion in Cephalopods

Digestion in cephalopods has been reviewed by Bidder (1966) and Boucaud-Camou and Boucher-Rodoni (1983). In most cephalopods (including *Nautilus*), the prey is reduced to small pieces using the beak; the pieces are coated with mucus in the buccal cavity then transported to the stomach by way of rapid peristaltic movements of the oesophagus (Bidder 1950; Boucher-Rodoni & Mangold 1977; Westermann et al. 2002). Although *Nautilus* and octopods can store food in the oesophageal crop while the first part of a meal is digested (Westermann et al. 2002); a crop is lacking in other cephalopods.

There is partial external predigestion in some coleoids, notably octopods, prior to ingestion. Proteolytic enzymes, probably from the posterior buccal glands (which contain a large suite of these enzymes) are injected into the prey, and the partially predigested tissues are then ingested (Boucaud-Camou & Boucher-Rodoni 1983) with digestion continuing in the oesophagus. A major function of this predigestion is probably to loosen muscle attachments, with the result that mainly intact crab exoskeletons, for example, are left empty. Cuttlefish do not have this ability as their posterior buccal glands lack suitable enzymes, although they do contain toxins.

In the stomach, the food is digested when mixed with enzymes (see Section 5.4.2.3) in the acid (pH 5.0–5.8) conditions, and it is then passed through the caecum to the digestive gland. There are counter currents of particulate and dissolved food entering the digestive gland and enzymes being moved into the stomach. These are maintained by a coordinated opening and closing of sphincters at the stomach opening and the openings of the digestive gland ducts into the stomach, along with movements by the muscular stomach walls. Nerves from the gastric ganglion regulate these processes. Within the digestive gland, the largely liquid food undergoes intracellular digestion and absorption.

5.4.3 Storage of Food Reserves

The products of digestion can be stored in various parts of the body and in various forms (see Chapter 2). While structural and other proteins can also be catabolised to form amino acids during times of stress (e.g., Bishop et al. 1983), the main stored reserves are in the form of carbohydrates or lipids.

Carbohydrate-based stored food reserves are typically glycogen which can be readily converted to glucose. Glycogen can be stored in a variety of tissues, notably the digestive gland, but also in connective tissue and in various parts of the body such as the mantle, foot, gonad, and muscle (e.g., Livingstone & Zwaan 1983).

Lipids, particularly triacylglycerols, are mainly stored in the digestive gland and gonad, and these are particularly important reserves in chitons, some gastropods, and cephalopods, while in other gastropods and bivalves carbohydrate reserves are more significant (Voogt 1983). The sterols in chitons differ from those in other molluscs, while in gastropods and cephalopods the dominant sterol is cholesterol. Protobranch bivalves are also rich in cholesterol, but autobranch bivalves are not (Voogt 1983).

5.4.4 Intrinsic and Environmentally Induced Feeding Rhythms

Feeding is often regulated by an environmental and/or intrinsic (endogenous) rhythm. Some of the most obvious environmental rhythms include the tidal cycle for intertidal animals, diurnal (day-night) changes, and phases of the moon (lunar).

5.4.4.1 Cyclical Feeding and Digestion

Phasic changes in the digestive cells of the digestive gland have been studied in several taxa (e.g., J.E. Morton 1956b; B. Morton 1971, 1977b; Merdsoy & Farley 1973; Boghen & Farley 1974; Nelson & Morton 1979), while Zaldibar et al. (2004) found this rhythm not only in the digestive gland, but also in the epithelium of the stomach as it underwent synchronous renewal. The phagocytic digestive cells within the tubules of the digestive glands show phased uptake of food and release of excretory products (Morton 1983a), and the accompanying cytological changes are reflected in their cytology. Indeed, the lining of the digestive tubules undergoes significant and quite rapid change in the number, shape, and contents of both digestive and secretory cells as they respond to the availability of food and the environment (e.g., Morton 1983a) and may even undergo complete regeneration in each feeding cycle (Nelson & Morton 1979). The absorptive, digestive, and fragmentation phases of the digestive cells are synchronised with the feeding cycle that is in turn often correlated (at least in part) with environmental changes (e.g., the tidal cycle).

The digestive gland epithelium also responds quickly to stress of various kinds, possibly as a result of an increase in the basophilic cells, but more often due to an apparent increase in those cells because of the loss of many of the digestive cells (see Zaldibar et al. 2004 for summary of literature). Such changes are not seen in molluscs that rely on more continuous, extracellular digestion (e.g., some neogastropods, Owen 1966).

Cyclic feeding activity with associated changes in digestive activities has been observed in some gastropods such as *Littorina* (Merdsoy & Farley 1973; Boghen & Farley 1974), the cerithioid *Telescopium* (Alexander et al. 1979), the calyptraeid *Maoricrypta* (Nelson & Morton 1979), the neogastropod *Tritia* (Curtis 1980), and in the stylommatophoran land snail *Helix pomatia* (Krijgsman 1925, 1928).

Cyclic feeding rhythms have been studied in detail in many bivalves as outlined below.

5.4.4.1.1 Bivalves

Figure 5.53 schematically shows the rhythmic and cyclical nature of feeding and digestion in many shallow-water autobranch bivalves as described by Morton (1973a, 1983a). In many intertidal and some shallow-water bivalves ctenidial feeding is cyclical, being switched on and off depending on the state of the tide or the day. Thus, food enters the stomach only during feeding (i.e., when the tide is in, covering the animal), and the crystalline style largely or completely dissolves (e.g., Bernard 1973; Morton 1983a). For example, in the oyster *Crassostrea*, when the animal is out of the water during

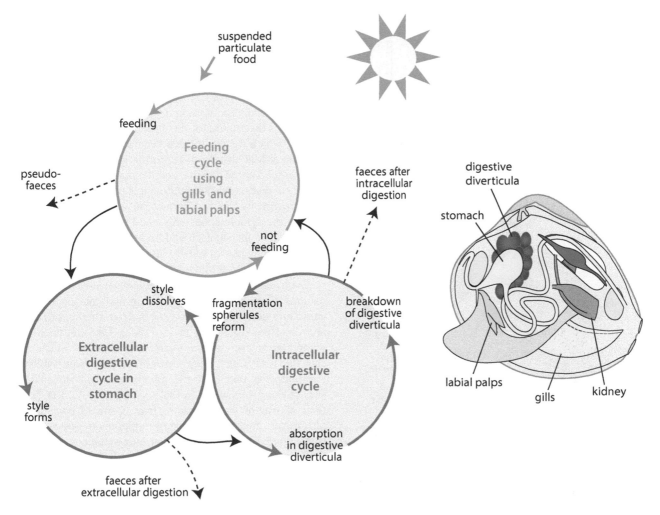

FIGURE 5.53 A generalised feeding cycle for a shallow-water autobranch bivalve. Redrawn and modified from Morton, B., *Malacologia*, 14, 63–79, 1973a.

low tide, the style disappears in about two hours, but once resubmerged can reform completely within about half an hour (Bernard 1973). In sublittoral and fresh-water bivalves, daily cycles of activity of valve opening followed by inactivity are often seen (see Morton 1983a for examples and references). During periods of non-feeding the valves will shut, and the remaining waste in the stomach is passed to the intestine. Any food remaining in the stomach is moved to the diverticula. The crystalline style reforms and the cells lining the digestive gland tubules break down, producing fragmentation spherules. These are moved to the stomach, aided by contractions of muscle fibres around tubules. Assimilated food is transported around the rest of the body via amoebocytes. The cells of the digestive gland then reform ready for the next cycle (Morton 1983a).

J. E. Morton (1956b) first demonstrated feeding and digestion cycles linked to tidal activity in the small intertidal bivalve, *Lasaea* (Figure 5.54). Since then many other examples of rhythmic feeding and digestion have been reported (see Morton 1983a for references). In intertidal bivalves, feeding occurs over the high-tide period and

ceases during the lower part of the tidal cycle when extracellular digestion occurs in the stomach. Food is passed to the digestive gland during the next high-tide period. Owen (1974) argued there is little or no evidence for rhythmic digestion as material once ingested was rapidly passed to the digestive gland for intracellular digestion (e.g., Mathers 1972) – an assertion repeated by Gosling (2003). Morton (1983a) countered this contention by demonstrating there was indeed substantial evidence for digestive cycles in animals in environments where either tidal or diurnal rhythms were important, and that the phases in the digestive tubules within the same animal can vary depending on, for example, their proximity to the stomach. In some sublittoral species, the style remains a constant length, and different phases occur simultaneously in different digestive tubules (Morton 1983a).

In most cases where rhythms occur they are not endogenous, because experimental animals kept in constant conditions often rather quickly lose their cyclic behaviour and adopt a steady state. As pointed out by Morton (1983a), constant conditions are rare in shallow aquatic ecosystems.

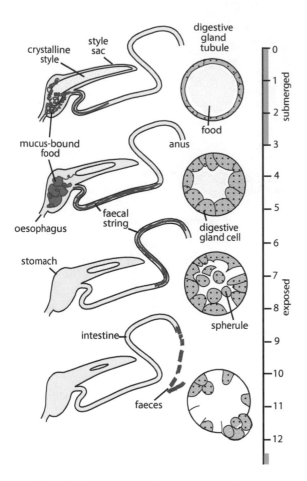

FIGURE 5.54 The relationship between periodicity in the digestive system and the tidal regime in the small intertidal autobranch bivalve *Lasaea rubra*. The scale on the right indicates hours. Redrawn and modified from Morton, J.E., *J. Mar. Biol. Assoc. U. K.*, 35, 563–586, 1956b.

The duration of the intracellular digestive cycle in the digestive cells of the digestive gland varies among different bivalves, ranging from two to 20 hours (e.g., Owen 1966, 1974; Morton 1977b).

Where cyclical feeding is possible due to constant inundation, in bivalves it may have advantages over continuous feeding, because largely intracellular digestion occurs. In bivalves fed increasing concentrations of seston, the efficiency of absorption is decreased, suggesting that the digestive tubules can approach saturation with the subsequent loss of potential food (Ward & Shumway 2004).

In the fresh-water unionids, rhythmic activity has been reported, but the correlations are not clear, although at least in some species (see Dillon 2000 for review), they may be related to light and temperature, while another fresh-water bivalve, *Dreissena*, feeds at night, at least in laboratory conditions (Morton 1969).

5.4.4.1.2 Cyclic Feeding in Shipworms and Giant Clams
Some bivalves utilise food sources in addition to those gained via the gills and palps. The integration of two

feeding mechanisms has been well studied in shipworms and in *Tridacna*, and both are cyclical, as described below.

Shipworms (*Teredo* and related genera) feed like normal bivalves, but, unlike other less specialised wood borers (*Martesia* and *Xylophaga*), they also utilise the wood fragments. These are obtained when they burrow into the wood in which they live (See Section 5.4.2.3.2.1) by using bacteria in specialised diverticulae of the digestive gland. In the few species for which the feeding has been described in some detail (Morton & McQuiston 1974; Morton 1978a), shipworms suspension-feed in the daytime and the normal part of the digestive gland digests this food during this phase. Digestion of the wood fragments obtained from boring occurs at night with the help of the symbiotic bacteria (see Figure 5.55). Thus shipworms utilise the stomach and intestine at different times for the two feeding phases and also use different parts of the digestive gland (Morton & McQuiston 1974; Morton 1978a, 1983a).

Giant clams (*Tridacna* and *Hippopus*) inhabit coral reef habitats where the waters are low in nutrients such as inorganic nitrogen, and they have a circadian rhythm of feeding and digestion (e.g., Reid et al. 1984a). Like the teredinids, giant clams use daytime suspension-feeding which they shut down at night. They also obtain food from symbiotic zooxanthellae that pack the greatly enlarged siphonal tissues of these very large animals (see Section 5.2.1.11.2). Via their zooxanthellae, the clams take up dissolved inorganic nitrogen in daylight, but release ammonium waste at night (see Figure 5.10). In comparison, non-symbiotic larvae and newly settled juveniles with few zooxanthellae release ammonium both during the day and at night (Fitt et al. 1993). Studies on *Tridacna gigas* suggest that they may obtain nutrition from their zooxanthellae during periods of emersion (during which most other bivalves cannot feed), although at a lower level of activity (Mingo-Licuanan & Lucas 1995).

5.5 ABSORPTION OF EXOGENOUS DISSOLVED AND PARTICULATE CARBON

Huge quantities of *dissolved organic material* (DOM) are contained in seawater and in many fresh-water systems, and may be available as a source of nutrition for aquatic organisms including adult molluscs (e.g., Gorham 1990) and their larvae (e.g., Manahan 1990).

Marine mussels (Mytilidae) have been shown to uptake amino acids and sugars (see review by Hawkins & Bayne 1992), and the sugar glands in chitons might absorb dissolved organic matter from seawater (Andrews & Thorogood 2005); these latter authors suggest this ability may be correlated with the simple stomach.

In *Tridacna*, Goreau et al. (1973) demonstrated that tritiated leucine and carbon dioxide with a radioactive carbon isotope ($^{14}CO_2$) were quickly taken up through the siphonal epithelium, while Fankboner (1971) showed that dissolved and particulate material from the seawater entered by way of micropinocytotic channels that form at the bases of epithelial microvilli covering the siphonal epithelium.

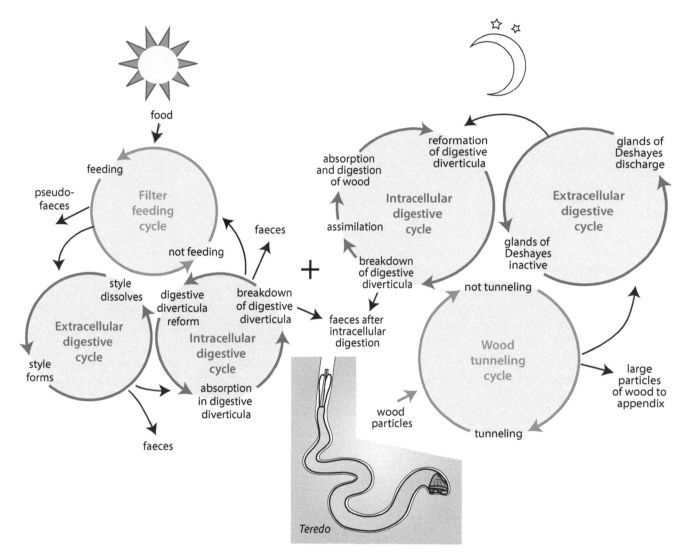

FIGURE 5.55 A diagrammatic representation of the day and night feeding cycles of the wood-boring bivalve *Teredo*. Redrawn and modified from Morton, B. and McQuiston, R.W., *Forma et Functio*, 7, 59–80, 1974.

Dillon (2000) reviewed non-particulate feeding in freshwater bivalves. An early experiment with unionids showed uptake through the gut and outer epithelium when they were fed solutions of fat, starch, and albumen (Churchill 1916). Other experiments included those of Efford and Tsumura (1973) where they showed that sphaeriids could assimilate labelled glucose and glycine from low concentration solutions and they calculated that, given typical levels of dissolved organic material in mud bottoms where the sphaeriids live, at most 4% of energy requirements could be obtained in this way. They did, however, suggest that such uptake of dissolved nutrients may be important for marsupial young in the brood chambers in the mantle cavity.

Absorption of DOM through the external body epithelium may be more common than reported in the literature. For example, Boucaud-Camou and Roper (1995) observed a high level of alkaline phosphatase activity in the skin of squid and octopod paralarvae, suggesting active uptake of nutrients from seawater to provide part of their energy requirements.

While absorption and intracellular digestion occur in the oesophagus of at least some gastropods (see Section 5.3.2.7.1.2), Lau and Leung (2004) showed that the muricid *Thais clavigera* could provide 10% of its energy requirements by 'drinking' seawater and absorbing suspended and soluble organic matter from it. This ability may assist these animals, and possibly many other similar taxa, to obtain essential nutrients to sustain them between feeding events. These authors did not identify the site of absorption, but Andrews and Thorogood (2005) found that absorption and intracellular digestion occurred in the gland of Leiblein (or oesophageal gland), which appears to mesh well with the findings in muricids.

6 Circulatory and Excretory Systems

6.1 INTRODUCTION

Animals more than a few cells thick require organ systems that can bring oxygen and nutrients to their internal cells and carry away products and waste. Because of their high surface to volume ratio, in small animals this may be accomplished solely by bathing the cells in fluids, but in larger and generally more active animals, a circulatory system provides the internal transport for distributing nutrients and oxygen and carrying away waste products. As in other animals, the molluscan circulatory system is typically made up of a network of blood spaces, channels, and vessels, with blood (haemolymph) to carry the nutrients and dissolved gases, and a pump (heart) to move this fluid through the network. There are also specialised cells present in the blood to assist in carrying waste and nutrients and fighting pathogens.

In molluscs, the heart, pericardium, and kidneys (the renopericardial complex) are closely interrelated. In most, these structures and the gonad arise from a common mesodermal cell group (Raven 1966), and in many taxa, this shared ontogeny is preserved by various connections (e.g., renopericardial ducts) and shared functions (e.g., excretion, respiration) of these structures. While the development of the renopericardial complex in cephalopods is highly modified (e.g., formation of both branchial hearts and a systemic heart), the complex is still primarily mesodermal in origin, and the development of these components and their connections (often secondary rather than primary) remain highly integrated (Ranzi 1931a, 1931b; Raven 1966). Most molluscs have an open circulatory system. Unlike the familiar 'closed' vertebrate circulation system with its arteries, capillaries, and veins, the molluscan circulatory network is typically not fully contained in vessels, but rather is an 'open' system with the blood moving through and between open, interconnected sinuses (spaces) among connective tissue. Conversely, most cephalopods have a closed circulatory system. In both closed and open systems, an anterior vessel ('aorta') is present in most taxa except the 'aplacophorans' where a dorsal sinus extends anteriorly from the ventricle, and capillary-sized vessels are found in a few structures in some gastropods (see Section 6.3.5). In most molluscs, a respiratory protein, usually haemocyanin, and occasionally haemoglobin, circulates in the haemolymph ('blood').

The molluscan heart is enclosed within a sac-like pericardium and consists of a single ventricle with usually two or more auricles (one in several gastropod lineages). The chambered heart is unique among lophotrochozoans which typically have only contractible vessels circulating the blood throughout the body (Schmidt-Rhaesa 2007; Brusca et al. 2016b).

In adult molluscs, excretion takes place by means of ultrafiltration by podocytes through the atrial wall into the pericardium and the filtrate is further modified by the kidneys (also called nephridia or renal organs). The urine is released into the mantle cavity, although diffusion of some wastes such as ammonia (NH_3) occurs through the epidermis and/or the gill(s) in aquatic taxa. The kidneys are paired and connected to the pericardium, as are the gonads in some taxa. Each cavity (lumen) of these mesodermal structures (pericardium, kidneys, and gonads) is plesiomorphically connected and together are thought to represent the coelom of the Mollusca (see Chapters 8 and 12). In larvae, excretion takes place via a pair of protonephridia that are located anteriorly.

The circulatory system moves respiratory gases and nutrients around the body and also collects wastes and moves them to locations where they can be excreted or sequestered. Mechanisms to prevent the loss of beneficial nutrients and water from the body, and for excreting harmful metabolites, are essential roles of the excretory system. The circulatory and excretory systems are thus linked, and they are also integrated with the respiratory surfaces where gases are exchanged (see Chapter 4), and with the digestive system, which supplies the nutrients. These processes are influenced by environmental parameters, both externally (e.g., temperature, oxygen, salinity, and pH in all habitats, and humidity in terrestrial habitats), and internally (osmotic pressure, oxygen levels, nutrients, hormones, etc.).

6.2 MAINTAINING SALT AND WATER BALANCE

6.2.1 OSMOSIS AND THE MOVEMENT OF MOLECULES ACROSS CELL MEMBRANES

Osmotic balance (see Box 6.1 for teminology) is critical to maintain homeostasis of the body fluids, preventing them from becoming too diluted or too concentrated. A change in osmotic concentration of the extracellular fluid ('blood') causes changes in the diffusion of water and ions across cell membranes which could cause cells to rupture if the osmotic pressure is too low, or to shrink if it is too high. This maintenance of the osmotic balance (*osmoregulation*) prevents diffusion of water by *osmosis* from the side containing the lower concentration to the hypertonic (higher concentration) side by utilising a selectively permeable cell membrane. Thus, if the concentrations are not equal, osmosis will cause the cell to shrink or expand as water or solutes move in or out of the cell until the concentrations are balanced on either side of the cell membrane. As in all animals, the fluids (and the salts and ions they contain) in molluscan cells are also controlled by a combination of osmosis and ion pumps, the latter a process requiring energy expenditure in which ions are 'pumped' across the cell membrane. Ion

BOX 6.1   SOME TERMINOLOGY RELATED TO OSMOSIS

Osmosis – the movement of dissolved molecules and ions through a selectively permeable membrane (typically a cell wall) that equalises the solute concentrations on both sides of the membrane.

Osmotic pressure – the pressure needed to maintain an equilibrium with no net movement of solvent.

Osmolytes – dissolved compounds that affect osmosis.

Osmoconformers – animals with the same osmotic pressure as the external medium.

Osmoregulators – animals that actively regulate the osmotic pressure of their internal fluids.

Homoosmotic – those osmoregulators able to maintain a more or less constant osmotic pressure different to that of the external medium.

Heteroosmotic – variable osmotic pressures in body fluids that in some may include osmoregulation.

The concentration of salts, ions etc. in the external medium compared with body tissues

Hypertonic – the external medium is a higher concentration than the body fluids – results in a *hyperosmotic* state.

Isotonic – the external medium is the same concentration as the body fluids – results in an *isoosmotic* state.

Hypotonic – the external medium is a lower concentration than the body fluids – results in a *hypoosmotic* state.

Heterotonic – the external medium is of varying concentrations (e.g., in a typical estuary) relative to body fluids – *heteroosmotic* regulators can adapt to different osmotic conditions.

pumps make osmoregulation possible, enabling the cell contents to be maintained at a different osmotic pressure from the fluid surrounding the cells.

Osmotic regulation involves the maintenance of water content and ionic composition of both the cytoplasm and the extracellular fluid. Even when these are in osmotic equilibrium, their ionic composition differs. Cells have relatively high concentrations of potassium ions (K^+) and low sodium (Na^+), the reverse of the extracellular fluid, a difference maintained by the sodium-potassium pump (see below and Chapter 2) which maintains the concentrations of these ions at optimal levels.

The passage of molecules across a cell membrane can be achieved by:

- Passive diffusion by way of osmosis. Water molecules pass through the lipid bilayer of the plasma membrane, facilitated by hydrophilic channels formed by transmembrane proteins. These enable the diffusion of water from a higher to lower concentration (as distinct from the concentration of solutes in the water). Water is never actively transported against its concentration gradient, and its concentration can only be altered by the active transport of solutes. For example, reabsorption of water from the kidney back into the blood depends on the active transport of sodium ions (Na^+) (see below).
- Facilitated diffusion (or 'indirect active transport' or 'down-hill' transport) uses an ion to pump some other molecule or ion (usually Na^+) against its gradient. The gradient is established by the 'sodium-potassium pump' (Na^+/K^+ pump or simply 'sodium pump') driven by the enzyme sodium-potassium adenosine triphosphatase (Na^+/K^+-ATPase) in the plasma membrane.

- Active transport or 'pumping' ('uphill' transport involving the expenditure of energy) usually involves the transport of Na^+ or K^+ using Na^+/K^+-ATPase and typically uses energy from the hydrolysis of ATP.[1] Much of the energy generated by the mitochondria in animal cells is used to run the Na^+/K^+-ATPase pump.
- Pinocytosis, a process in which fluid is captured at the cell surface by cell membrane invaginations to form small fluid-filled vesicles.

6.2.2   Salinity and Its Importance in Aquatic Habitats

Molluscan aquatic habitats can differ greatly in water and solute chemistry, with salinity (see Box 6.2 for terminology) being one of the main factors determining the suitability of a particular habitat for a particular species.

The majority of molluscs are marine species (salinity ~ 35‰). Many species live exclusively in fresh-water habitats, and many gastropods are terrestrial. There is a relatively low diversity in brackish water where salinity, in particular, fluctuates considerably from near fresh water (~4‰) to ~80‰ through evaporation. Very few species can tolerate hypersaline (>35–300‰) conditions (see also Chapter 9). The great variation in water chemistry, particularly salinity, necessitated the evolution of mechanisms to maintain salt and water balance in aquatic molluscs. The mechanisms used by molluscs to cope with maintaining water balance in different environments have been reviewed by Deaton (2008).

From an economic viewpoint, salinity is critical for the establishment and management of shellfisheries and the success or failure of invasive species (Tettelbach & Rhodes 1981; O'Connor & Lawler 2004; Sarà et al. 2008) and molluscan pests such as shipworms (Shipway et al. 2014).

---

[1] Adenosine triphosphate (see Chapter 2).

Fresh water – very low salinity (<0.5‰).
Oligohaline – low salinity (0.5–5.0‰).
Brackish water – salinity ranging from 0.5 to 29‰.
Euhaline – normal seawater (salinity 30 to 35‰).
Metahaline – salinity from 36 to 40‰.
Brine – very saline (>50‰) water from which salts crystallise.
Thalassic – salinity derived from the ocean.
Homoiohaline – salinity remains rather constant.
Poikilohaline – salinity variable.
Euryhaline – an organism that can tolerate a wide range of salinities.
Stenohaline – an organism that can tolerate only a narrow range of salinities.
Haline – salts derived from the sea.
Saline – salts derived from the land.

6.2.2.1 Cell Volume Regulation

In marine osmoconformers, the volume of each cell in the body changes with fluctuations in the salinity of the external water, and the amount of water in each cell must be controlled by intracellular mechanisms to avoid cell death. These mechanisms change the amount of osmotic solute in the cell, and this is often achieved by adjusting the levels of free amino acids (which act as osmotic solutes) (Pierce & Amende 1981; Pierce 1982). When salinity is low, water enters the cells which then swell, and free amino acids leave the cell with water, enabling the cell volume to return to normal. In high salinity, cells shrink, and free amino acid concentrations are increased, thus preventing further water loss because of the increased osmotic pressure in each cell. The details of the mechanisms involved can be complex. In a study on the red blood cells (erythrocytes) of an arcoidean bivalve (*Noetia ponderosa*) Smith and Pierce (1987) found that, following hypoosmotic stress, cell volume regulation began immediately with an efflux of intracellular K^+ and Cl^- (but not Na^+) and that, after many minutes, this was followed by an efflux of the organic sulphonic acid taurine. In addition, the volume regulation of the erythrocytes depended on extracellular Ca^{2+} levels as taurine efflux is increased when external levels are high, but does not affect the efflux of K^+ and Cl^- ions. Given the complexity of the processes involved and the energy required, not surprisingly only a few species can control osmotic pressure (of the blood and hence cell volume) over the whole non-lethal range of salinity.

6.2.2.2 Hormonal Control of Osmoregulation

Body volume and osmoregulation in molluscs is controlled in part by hormones. The rate of filtration depends on the local pressure of the haemolymph, which is controlled by the heart rate, and that is controlled by cardioactive neuropeptides (FMRFamide and related peptides [FaRP family]).

There are several neuroendocrine centres in the ganglia which control ionic balance and osmoregulation, and details differ between taxa. For example, a marine gastropod such as the sea hare *Aplysia* loses water to the environment (the opposite problem to fresh-water taxa). This is controlled by large neurosecretory cells (R15 cells) in the abdominal ganglion which continuously produce an antidiuretic small peptide neurohormone that causes water uptake. Production only ceases in a hypoosmotic medium (e.g., in a tide pool diluted by rain). In the fresh-water snail *Lymnaea*, the neurosecretory dark green cells (DGC) of the pleural and parietal ganglia secrete a diuretic neurohormone[2] that increases urine flow in fresh water, but is inactive in saline conditions (e.g., With et al. 1994). Yellow cells (YC) in the parietal and visceral ganglia of *Lymnaea* also react to osmotic changes and produce a sodium influx stimulatory peptide (SISp). This peptide acts on the sodium pump in both the epidermis and the kidney which compensates for losing sodium ions in a hypoosmotic environment (With et al. 1994).

Other neurohormones may also be involved in the control of ion reabsorption, as in the fresh-water planorbid *Helisoma* (e.g., Saleuddin et al. 1992a; Madrid et al. 1994). In that genus, there are two types of neurosecretory cells involved, most of which are in the visceral ganglion and a few in the parietal ganglia. The nerve fibres emerging from these cells run to the smooth muscle surrounding the epithelium of the kidney. One of these neurosecretory cell types becomes active in distilled water, producing a diuretic hormone that increases the spaces between the kidney cells, enabling movement of fluid from the haemolymph into the kidney. The other type of cell becomes active in salty water, producing an antidiuretic factor, an FMRFamide-related peptide. Such peptides play a role in osmoregulation of other molluscs, including the marine venerid bivalve *Mercenaria* (Deaton 1990). In stylommatophoran slugs and *Aplysia*, an arginine-vasotocin-like peptide secreted by the pleuropedal ganglia causes the release of fluid through the body wall (Sawyer et al. 1984).

6.2.2.3 Marine Environment

The haemolymph in most marine molluscs has the same osmotic concentration as the seawater in which they live (i.e., they are osmoconformers). Many osmoconforming marine molluscs are stenohaline, tolerating only a small range of salinity, while euryhaline taxa are more tolerant of salinity variation. All aplacophorans, cephalopods, scaphopods, monoplacophorans, and polyplacophorans are stenohaline, as are many bivalves and gastropods, although these latter classes also contain many euryhaline taxa. Some euryhaline osmoconformers can live within a considerable salinity range and produce urine isotonic to the haemolymph.

In marine osmoconforming molluscs, the maximum K+ concentration in cell cytoplasm is about 200–300 mM, but with other cations and anions, the total osmotic pressure is

[2] This hormone is similar to the thyrotropin releasing hormone (TRH) of vertebrates.

about 500–600 mOsm[3] (Deaton 2008). This osmotic concentration in cells is thus considerably less than that of both seawater and the extracellular fluid, which is usually around 1100 mOsm. The difference is made up of organic osmolytes – amino acids and amine compounds. A change in the osmotic concentration of the surrounding medium, and consequently the extracellular fluids, causes rapid changes to cell size (see above), although these effects can be mitigated by changing the organic osmolyte concentration. Indeed, the range of salinities tolerated is correlated with the availability of free amino acids necessary for intracellular volume regulation and the ability of a particular species to osmoregulate. The amino acids involved in regulating the cytoplasmic osmotic changes are mainly taurine, alanine, glutamic acid, and glycine (Deaton 2008), but the concentrations of different amino acids change over time in response to the osmotic changes (e.g., Baginski & Pierce 1978). The biochemistry behind these and other related changes in the cell has only been studied in bivalves and is summarised in Deaton (2008). There are also data suggesting that the CNS may play a role, as some neurotransmitters (e.g., Deaton 1990) and neuropeptides (e.g., Weiss et al. 1989) influence water content, at least in some bivalve taxa. These generalities are, however, based on observations of only a few taxa, nearly all bivalves, and virtually nothing is known regarding details of these mechanisms in most other molluscs.

Molluscs use various behavioural and structural strategies to avoid short-term osmotic stress caused by, for example, the flood plume at a river mouth, or a change in salinity due to evaporation or rain in an intertidal pool. Many gastropods close their shell with the operculum, limpets clamp their shell onto the rock surface, and bivalves close their shell valves or burrow into the sediment. Such mechanisms effectively isolate the animal from the external medium for relatively short periods, such as those experienced between tides in the intertidal or in an estuary. Thus, such behaviour in waters of rising salinity maintains the osmotic pressure of the body fluids at a lower level than that of the external medium, as exchange is prevented or much reduced (Figure 6.1). Consequently, when transferred to fresh water or to high salinity water, shelled marine molluscs can typically survive for much longer than those lacking shells (e.g., Berger & Kharazova 1997). Simply closing the shell to survive for long periods brings its own physiological challenges, and shelled estuarine molluscs also typically have adaptations to survive without respiration for rather long periods and to deal with the accumulation of the acidic byproducts of anaerobic metabolism (e.g., Aliakrinskaia 1972) (see Chapter 2).

The reactions to salinity changes are in response to detectors on the cephalic tentacles, mantle edges, or siphons (Freeman & Rigler 1957; Davenport 1979, 1981). Two kinds

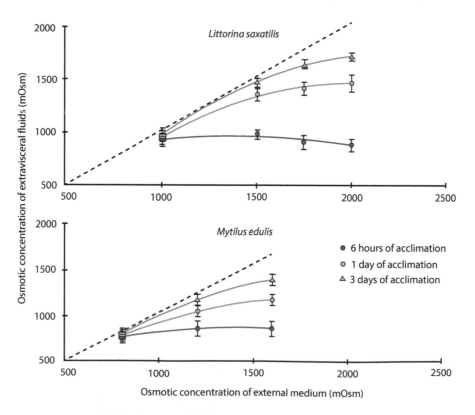

FIGURE 6.1   Osmotic concentration of extravisceral fluids of *Littorina saxatilis* and *Mytilus edulis* in water of high salinity. The 95% confidence intervals are indicated by vertical lines. Redrawn and modified from Berger, V., *Biologiya Morya*, 6, 30–35, 1989.

---

[3] Milliosmole – 1/1000 of an osmole (Osm) – the number of moles of solute that make up the osmotic pressure of a solution.

of detector have been identified; those that respond to sodium ion concentration and others that respond to osmotic pressure (Natochin 1966; Berger & Kharazova 1997).

In osmotic and ionic conformers, there is little need for the kidney to play a role in salt and water balance. Perhaps surprisingly, urine produced in marine molluscs such as *Octopus*, *Strombus*, and *Haliotis* is similar in quantity to fresh-water bivalves. The reason for this is unclear (Deaton 2008) and markedly contrasts with the much higher urine output by many fresh-water gastropods (e.g., see Section 6.7.7.4.5.1).

6.2.2.4 Upper Littoral and Supralittoral Environment

The adaptations that allow some gastropods, notably littorinids, to live successfully in the upper and supralittoral fringe include the ability to aestivate, tolerate temperature extremes, minimise water loss, and produce uric acid as a waste product (e.g., McMahon 1990; Britton 1992; Lang et al. 1998).

The amphibious ampullariid *Pomacea* can aestivate for up to 400 days losing around 20% of body water, with the osmotic pressure of the haemolymph doubling in some individuals (Little 1968). Many marine patellogastropod and siphonariid limpets can tolerate increased osmolality of the haemolymph caused by either desiccation or hypersaline water (see Williams & Morritt 1995 for review). Experiments on the supralittoral littorinid *Cenchritis* (previously *Tectarius*) *muricatus* showed that its blood concentration could increase by up to 250%, although some regulation of haemolymph concentration occurs (Emson et al. 2002). This markedly contrasts with another littorinid, the littoral *Littorina littorea*, in which its haemolymph is isoosmotic with seawater over a limited salinity range; it can tolerate considerable cell volume changes (Taylor & Andrews 1988).

6.2.2.5 Estuarine Environment

Estuaries and similar habitats with brackish water have smaller faunas than nearby marine habitats because relatively few molluscs can succeed in environments with rapidly fluctuating salinities. Consequently, many estuarine molluscs cannot cope with continuous fully marine or fully fresh-water conditions. These oligohaline molluscs can be both conformers and hyper-regulators, responding to moderate increases or decreases in salinity. In brackish water with higher salinities, they are osmoconformers, and in low salinities, they are hyperosmotic regulators. On the limited data available, the switch from conformity to hyper-regulation is 60–70 mOsm in bivalves and 125–150 mOsm for gastropods (Deaton 2008). When acclimated to fresh water, the blood of oligohaline species has an osmotic concentration the same as fresh-water species (lower limit about 40 mOsm).

When the osmotic concentration of the haemolymph is maintained above that of the aquatic environment, there is a continuous influx of water and efflux of ions. Hyper-regulating aquatic animals must produce large quantities of urine to maintain their body volume and actively take up ions to compensate for those lost in the urine. We have been unable to find studies on details of urine production and sodium and chloride fluxes in oligohaline molluscs. While it is possible

for oligohaline molluscs to live in fresh water, they cannot reproduce in it and, unlike fresh-water species, they cannot survive for long periods in deionised water, probably because of their inefficient ion uptake mechanisms (Deaton 2008).

The greatest range of salinities (full seawater to fresh water) is tolerated by only a handful of taxa, including the bivalves *Rangia* and *Polymesoda*, and the gastropods *Melanopsis* and *Potamopyrgus* (Burton 1983). This range is also found in neritimorphs which move from marine to fresh water during their lives (Estabrooks et al. 1999; Kano 2009).

Experiments on some saltmarsh snails have shown that they can tolerate rather large salinity changes. The assimineid *Assiminea grayana* can survive up to twice the concentration of normal seawater, and the osmotic pressure of the haemolymph is maintained at slightly above that of the external medium over the 50%–200% range (Little & Andrews 1977). The saltmarsh eupulmonate *Ovatella myosotis* (Ellobiidae) can tolerate more extreme saltwater concentrations, up to 257% normal seawater (Seelemann 1968a).

Larvae may be less tolerant of low salinity than adults, as shown, for example, in the caenogastropod *Tritia* (as *Ilyanassa*) *obsoleta* (Kasschau 1975; Vernberg & Vernberg 1975) and the marine panpulmonate *Amphibola crenata* (Little et al. 1984).

Of the few taxa investigated, most brackish-water caenogastropods remain isoosmotic with the water over most of their salinity range (Little 1981), although some are capable of osmotic regulation. These latter include the truncatelloideans *Peringia* (as *Hydrobia*) *ulvae*, an estuarine hydrobiid (Todd 1964; Avens 1965; Negus 1968), the tateid *Potamopyrgus antipodarum*, a primarily fresh-water species that can also survive in estuarine conditions (Todd 1964), and the saltmarsh *Assiminea grayana* (Seelemann 1968b; Little & Andrews 1977). If this is true of other truncatelloideans in marginal or non-marine environments, it could be an important factor in the success of this group in estuarine, fresh-water and, to a lesser extent, terrestrial environments. Some marine 'pulmonates' occupy estuarine habitats (amphiboloideans, many ellobiids, most onchidiids), while others are found only in fully marine habitats (trimusculids, siphonariids, otinids, smeagolids, some ellobiids, few onchidiids). The estuarine amphiboloidean *Amphibola crenata* can tolerate a wide range of salinities and can even survive in fresh water for several days (Shumway & Marsden 1982; Little et al. 1984), but, like many other estuarine gastropods, remains isoosmotic with the medium over most of its range (Little 1981; Orange & Freeman 1988). This is in marked contrast to ellobiids, such as *Ovatella myosotis*, which also tolerate a wide range of salinities, but maintain the osmotic pressure of their blood (Seelemann 1968a).

6.2.2.6 Hypersaline and Hydrothermal Environments

Virtually nothing is known about the physiology of molluscs that inhabit hypersaline (salinity >46‰) habitats, although some species of bivalves live in salinities up to 80‰ in saline habitats in Texas and the Yucatan Peninsula (Britton & Morton 1989). The pomatiopsid gastropod *Coxiella* lives in salt lakes in Australia and is active in salinities between about 6 and 124‰, although living (but inactive) *Coxiella* is found in sites with salinities above and below these levels (Williams & Mellor 1991). *Coxiella* is an osmoconformer over a wide range of salinities, but the haemolymph takes 20–28 hours to reach equilibrium with the external environment (Williams & Mellor 1991).

Bivalves and gastropods are found in hypersaline pools near hydrocarbon seeps on the floor of the Gulf of Mexico; little is known of their physiology although the neritoidean *Bathynerita naticoidea* can tolerate salinities of up to 85‰ and has been shown to osmoconform over a range of 35–75‰ (Gaest et al. 2007). At hydrothermal vents, environmental conditions are characterised by high sulphide concentrations, fluctuating osmolality, and irregular temperature regimes (Nakamura-Kusakabe et al. 2016). Like the hypersaline environments, physiological studies are uncommon in this extreme environment.

6.2.2.7 Fresh-Water Environments

Both bivalves and gastropods have independently invaded fresh water multiple times. Fresh-water species are typically limited to salinities below 2‰, and most can also osmoregulate to about 3‰ salinity. Like other fresh-water animals, they usually produce urine with a lower osmotic concentration than the haemolymph, although the osmotic concentration of the haemolymph of fresh-water bivalves is generally lower than that of fresh-water gastropods and, consequently, these gastropods produce more urine than the bivalves. In many bivalves bicarbonate ions are equal or higher in concentration in the haemolymph than chloride ions, while in gastropods bicarbonate is lower than chloride (Deaton 2008).

Fresh-water molluscs must extract ions from the surrounding water against an electrochemical gradient. This extraction occurs in the kidney and also, at least in bivalves, via the gills. In unionids, the epithelium lining the water channels of the gills has clusters of non-ciliated cells with microvilli. These are morphologically typical of ion-transporting cells and contain numerous mitochondria, but lack other organelles such as Golgi apparatus and endoplasmic reticulum (Kays et al. 1990).

Ions may be taken up independently, as for example, with sodium and chloride in unionids, *Corbicula*, and the fresh-water hygrophilans (Deaton 2008), but in the 'zebra mussel' *Dreissena*, chloride uptake is dependent on sodium (Horohov et al. 1992). Unionid bivalves have many serotonergic synapses in their gills, and ion uptake is stimulated by the hormones serotonin and cAMP (e.g., Scheide & Dietz 1986), while prostaglandins inhibit the uptake of sodium (Graves & Dietz 1979).

Most fresh-water bivalves survive for months in deionised water, but for long-term survival, *Dreissena polymorpha* requires water with low concentrations of Na^+, K^+, Mg^{2+}, and Cl^- because its epithelium is unusually 'leaky' (Dietz et al. 1996).

With very few exceptions, fresh-water molluscs have a limited tolerance of brackish water and die with continued exposure to greater concentrations than about 200 mOsm (Deaton 2008).

Based on the limited data available (Jordan & Deaton 1999), the capacity of cells to accumulate amino acids in hyperosmotic water is apparently lower in fresh-water gastropods than in fresh-water bivalves, and oligohaline bivalves show greater rates of accumulation than unionids (Deaton 2008).

6.2.3 THE TERRESTRIAL ENVIRONMENT

Some groups of gastropods are the only molluscs that can live permanently in terrestrial habitats, which they do very successfully. A major problem is overcoming desiccation, as water evaporates continuously from their exposed body and is the major cause of water loss (Machin 1975). Water is also lost from the respiratory surface, through mucus secretion (which is about 98% water) and waste excretion (urine and faeces). Urine flow can cease during dehydration and inactivity, when there is increased reabsorption of water from the filtrate, and increase again with hydration.

In terrestrial gastropods, the state of hydration is related to activity, with dehydrated animals being inactive, and the most active being well hydrated. Some stylommatophorans can tolerate considerable water loss. A number of helicids survive around 50% loss of mass and a limacid slug 80% (Burton 1983), but some terrestrial gastropods can survive a reduction in normal body weight of up to 90% through water loss (Prior 1984). Although terrestrial caenogastropod or neritimorph snails can close their shell aperture with an operculum, they are also subject to considerable variation in water content. Their urine is either hypoosmotic (e.g., the helicinid *Alcadia* [Neritimorpha] and the cyclophoroidean *Poteria*) or isoosmotic (e.g., the littorinoideans [Pomatiasidae] *Pomatias* and the *Tropidophora*) (e.g., Rumsey 1972). In all terrestrial snails, dehydration is accompanied by corresponding changes in osmotic and solute concentrations.

Slugs cannot live for more than a few days unless they have access to water and can lose 30%–40% of their body weight within a two hour period of activity (Prior 1985). Snails retracted into their shells and in an inactive state can survive for long periods of time, some even for years. In aestivating snails, the haemolymph significantly increases in osmotic concentration due to water loss through evaporation.

For these reasons, terrestrial gastropods have evolved not only physiological responses to water loss but also behaviours to avoid it. These include choosing suitable habitats, nocturnal activity, avoidance of activity in dry, hot conditions, modification of respiratory activity and behaviours such as homing and 'huddling' (reviews in Machin 1975 and Prior 1985).

Aestivation is typically triggered by environmentally stressful conditions such as dehydration, which results in the gastropods choosing sheltered places or burying themselves in the soil. Activity recommences following the resumption of suitable (typically moist) conditions or mechanical stimulation.

Active slugs usually remain in the region of highest humidity, but some slugs (e.g., *Limax maximus*) increase activity in dry conditions (Hess & Prior 1985), suggesting that after an initial short-term reduction in activity, dehydration in snails and slugs may initiate a response to seek out moisture (Prior 1985). Body size is an important factor, with small slugs and snails dehydrating more rapidly than larger species or individuals because of their high surface area to volume ratio.

Typical water loss through evaporation for terrestrial eupulmonate snails and slugs generally ranges from 2–20 mg per gram of body weight per hour (mg g^{-1} h^{-1}) when active, but was greatly reduced (<0.05–5 mg g^{-1} h^{-1}) when aestivating (Machin 1968; Withers et al. 1997). Largely, the marked decrease in water loss in aestivating snails is because the animal has withdrawn into the shell. Other factors include reduced water permeability of the mantle during aestivation, sealing the aperture with an epiphragm, and sealing to hard surfaces (e.g., Machin 1968, 1972, 1975; Appleton et al. 1979). Body weight fluctuates by 10%–50% even if conditions are constant (Deaton 2008). Depending on humidity and the taxon, active eupulmonate snails and slugs lose water at 2–45 mg g^{-1} h^{-1} (Machin 1975).

Over-hydration of slugs results in large amounts of fluid (including haemolymph) being expelled through the body wall and pneumostome, a process apparently controlled by neurotransmitters (Sawyer et al. 1984). The adverse effects of over-hydration may explain why snails and slugs tend to avoid direct exposure to rainfall (Prior 1985).

Slugs use the pneumostome to assist in water balance control – when overhydrated, it is kept wide open, but when dehydrated, it is reduced in diameter and is rhythmically opened and closed. Slugs which have the pneumostome kept fully open experimentally lose 7% more water than slugs with normal pneumostome function (Prior 1985).

Water can be obtained by uptake directly from the environment (from food or by absorption through the body wall by osmosis) or via metabolism. An active terrestrial slug must access more water than can be obtained from its food (Deaton 2008). Stylommatophoran snails and slugs do not appear to actually drink water directly (although this is not yet a certainty), but they can absorb water through the body wall. Thus rehydration in slugs and snails can occur rapidly by the absorption of water directly through the foot when it is in contact with a moist surface. This is done by way of osmotic uptake through the bottom of the foot and in sufficient quantity to make drinking unnecessary. This uptake occurs via a *paracellular pathway* – intercellular spaces between epithelial cells in the foot (e.g., Ryder & Bowen 1977) that enable ionic diffusion by allowing water and even relatively large molecules to pass through (Prior & Uglem 1984) (see also Section 6.6.4.1). Slugs dehydrated to 60%–70% of their normal body weight move to water, and when they reach it, their body assumes a flattened posture which is maintained for around 10–20 minutes while sufficient water is absorbed through the surface of the foot. This behaviour is controlled by changes in osmolality of the haemolymph (Prior 1985).

The body water content and haemolymph solute composition of rehydrated active stylommatophoran snails and those of resting or aestivating individuals generally show similar patterns (e.g., Withers et al. 1997). Haemolymph concentration of terrestrial gastropods ranges from about that seen in fresh-water gastropods to about twice that level (Deaton 2008), fluctuating depending on the state of hydration.

6.2.3.1 Amphibious Molluscs

A few gastropods are the only true amphibious molluscs (i.e., being regularly active in both aquatic and terrestrial conditions) and include some neritids, ampullariids, assimineids, truncatellids, littorinids, and possibly onchidiids (see also Chapter 4). Some aquatic species can survive out of water for long periods in a resting state, as for example, some bivalves and gastropods at higher intertidal levels, but these are not strictly amphibious. For example, some bivalves have been shown to survive for up to 12 months out of water (Deaton 2008).

Responses of amphibious species to prolonged exposure in the air are varied. The ampullariid *Pomacea lineata* lost 50% of its body weight after aestivating for 200 days, and the osmotic pressure of the haemolymph more than doubled (Little 1968). Similarly, non-amphibious bivalves exposed to air show marked increases in the osmotic concentration of their haemolymph. Water loss through evaporation results in an increase in inorganic ions in some molluscs, but in others, there is also an increase in amino acids (Deaton 2008).

6.3 THE VASCULAR SYSTEM

The main function of the vascular system is the distribution, by way of the haemolymph ('blood'), of oxygen and nutrients to tissues around the body and the removal of carbon dioxide and other wastes. There are also other important functions, such as providing hydrostatic pressure and distributing chemical messengers such as hormones.

Like most other invertebrates and fish (but not tetrapod vertebrates), molluscs have a single circulatory system[4] (Figure 6.2). The heart has a ventricle which pumps oxygenated blood through the body via a network of vessels and sinuses – the arterial system. In molluscs, deoxygenated blood is gathered in the vessels (veins) and sinuses of the venous system and eventually passes through a respiratory surface, usually the gill(s), before being collected in the auricle(s) of the heart. From there it is pumped into afferent vessels (aortae, arteries). This is the reverse of the pattern in fish (Figure 6.2), in which the deoxygenated blood is pumped by the heart to the respiratory surface.

Vessels may connect sinuses, and several arterial and venous vessels are typically associated with the heart and gills. Closed systems have also evolved in the Mollusca, most notably in stylommatophoran gastropods and coleoid cephalopods where capillary systems link the arteries and veins (see Section 6.3.5). Beds of minute capillary-like vessels sometimes occur in tissues and organ systems in some other molluscs (see Section 6.3.5). Although the closed blood vessels of cephalopods are lined with endothelial cells (Budelmann et al. 1997), the blood spaces and 'vessels' in most other

---

[4] The vascular system of coleoid cephalopods is somewhat analogous to a double system as described in 6.3.1.

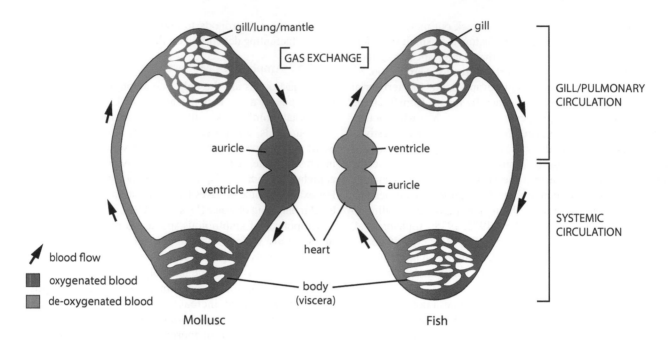

FIGURE 6.2 Diagrammatic comparison of the basic circulatory pattern in a typical mollusc and a fish. From various sources.

molluscan groups are surrounded by a layer of connective tissue and are not lined with endothelium, although some endothelial cells may be present in some caenogastropods (Voltzow 1994) and 'pulmonates' (Pentreath & Cottrell 1970; Luchtel et al. 1997).

Except in cephalopods, respiration (see Chapter 4) in shelled taxa is typically mainly via the gills and the mantle tissue lining the mantle cavity. This is notably true of autobranch bivalves where, in *Mytilus*, for example, about 15% of respiration is via the gills, compared with 85% via the inner mantle. In naked marine and terrestrial taxa, it is mostly via the outer mantle surface, including in air-breathing species. Land snails have the roof of the mantle cavity highly vascularised, and it functions as a lung (see Chapter 4). Sometimes respiration can also occur (at least in part) through the walls of the kidney. Changes in the respiratory surface utilised has led to corresponding modifications in the vascular supply.

Because many molluscs have complex vascular systems, a heart is necessary to pump blood around the body. Thus, despite their large haemocoel spaces, molluscs do not have particularly low-pressure vascular systems.

The volume of blood relative to that of the body decreases with increasing sophistication of the vascular system. Thus, coleoid cephalopods have a much lower blood to body volume (~6%) than bivalves (~55%), gastropods (34%–66%), and *Nautilus* (20%) (McMahon et al. 1997). In tetrapod vertebrates, a double circulatory system has evolved where an additional auricle and ventricle have developed in the heart, enabling a separate system for oxygenating the blood before it is pumped into the rest of the body. Although there is no true double circulatory system in molluscs, coleoid cephalopods have an analogous one (see next Section), as they have evolved not only a closed

circulatory system, but also additional pumping structures (the branchial hearts) that push blood through the gills (Schipp 1987b).

Most molluscan hearts have a pair of auricles, but nautiloids and larger monoplacophorans have two pairs, some polyplacophorans three pairs, and many gastropods have a single auricle. In gastropods, the number of auricles typically reflects the number of gills present as each auricle receives oxygenated blood from its adjacent gill. Some gastropods have two auricles and a single gill (e.g., in trochids) or a single auricle and no gill; the four auricles in monoplacophorans each serve more than one 'gill,'[5] and in polyplacophorans, there are also multiple gills per auricle. Each auricle opens to the ventricle by way of a single opening, except in chitons where there are two to four pairs of openings (e.g., Hyman 1967). Molluscan hearts all have a well-developed inner layer of the pericardium, the *epicardial* layer, and lack an endothelium (see Section 6.3.2).

In most molluscs, low-pressure blood passes through the gills and/or mantle and then to the heart. Some or all of this blood may also pass through the kidneys on the way to the gills. There is considerable variation in gastropods, from taxa where most of the blood goes to the gills, to those where the kidney is the only respiratory surface (Hyman 1967; Haszprunar 1988a). In contrast, in cephalopods the blood is under pressure as it passes through the gills.

Vascular systems in molluscs are best known in gastropods, bivalves, and cephalopods, and the main features of these systems are outlined below. The relatively simple and less studied vascular systems of the other groups are briefly discussed.

---

[5] These structures are involved in ventilation, not respiration (see Chapter 4).

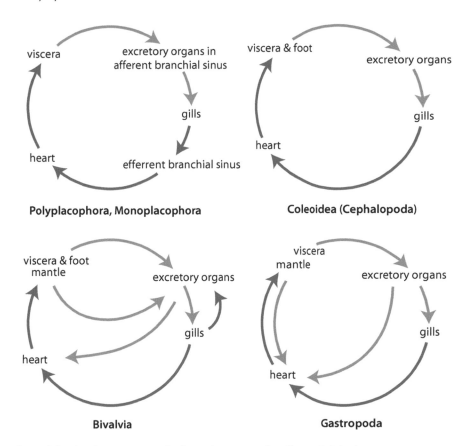

FIGURE 6.3 Comparison of the circulatory patterns in the main groups of molluscs. Original.

6.3.1 OVERVIEW OF THE CARDIOVASCULAR SYSTEM AND CIRCULATION PATTERNS

There are substantial differences in the details of circulatory patterns (Figure 6.3) and heart morphologies among molluscan taxa. Most of these differences are associated with the addition of ancillary respiratory surfaces or the number of gills present in the taxon. Deviations from the general patterns are also most common in animals with small body size or in larger, active taxa.

In most molluscs, the heart is the main organ responsible for blood propulsion. Composed of a muscular ventricle and auricles, the heart lies in a coelomic sac, the pericardium, which is typically connected to the kidneys by way of renopericardial ducts. Like other organ systems, the molluscan heart and pericardium show considerable morphological diversity (see Figure 6.4).

The pericardial cavity and heart are formed by a pair of embryonic mesodermal extensions that later fuse. Thickenings on the walls of the pericardium invaginate to give rise to the ventricle and auricle(s). During development, the ventricle anlagen engulf the intestine (so it finally passes through the ventricle) and, in some taxa (Monoplacophora, Scaphopoda, many bivalves, and some gastropods), this configuration is maintained in the adult. This incorporation of the intestine in the pericardium may be heterochronic as that is the way it commences in the development of some conchiferan molluscs (e.g., in *Nucula* [Drew 1901] where the adult intestine

does not pass through the ventricle, unlike the situation in other protobranch groups [Yonge 1939b]). The penetration of the pericardium and ventricle by the intestine appears to be plesiomorphic for conchiferans, but may not be for molluscs because in chitons the heart develops above the intestine, which never penetrates the pericardium (Moor 1983). Also, in solenogasters, the mesodermal cell mass proliferates above the early gut causing the intestine to pass below the pericardium in adults (Thompson 1960) (Figure 6.5).

Oxygenated blood is received by the auricle(s), where ultrafiltration (see Section 6.6.1) typically takes place before it is passed to the ventricle. The ventricle then pumps the oxygenated blood around the main mass of the body via large aorta(e) – one in chitons, monoplacophorans, and some bivalves, two in gastropods, *Nautilus*, and most bivalves, and several in coleoid cephalopods. Blood flow is controlled by a system of valves or similar devices, which prevents the reflux of blood from the ventricle back into the auricle(s) during ventricular systole (the contraction and ejection phase), and from the aorta(e) to the ventricle during ventricular diastole (the expanding and filling phase) (see Figure 6.14).

The polyplacophoran circulatory system is mostly an open system and, besides the pumping of the heart, body movement also plays a role in assisting blood flow (McMahon et al. 1991). The heart has a single pair of auricles, opening to the ventricle by way of posterior openings on each side. In the family Lepidopleuridae there is a single opening between each

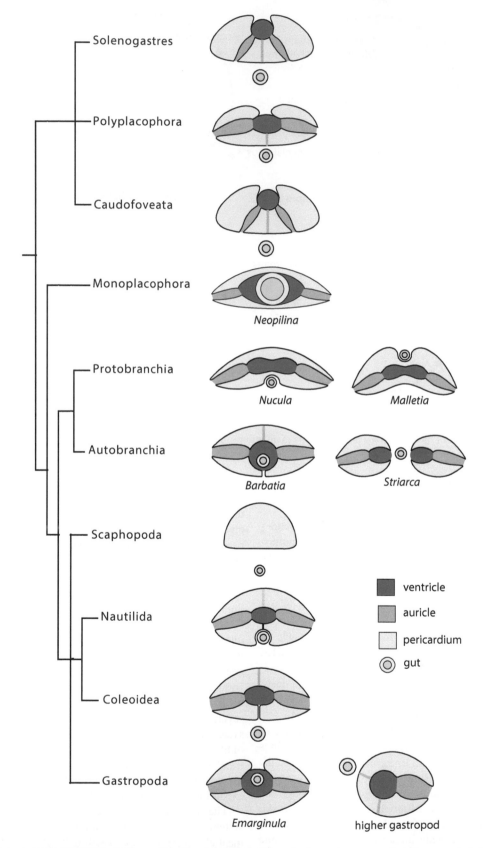

FIGURE 6.4 Diagrammatic transverse sections of the heart and pericardium in the main groups of molluscs, and their relationship with the gut. Redrawn and modified from Naef, A., *Ergebnisse und Fortschritte der Zoologie*, 6, 27–124, 1926.

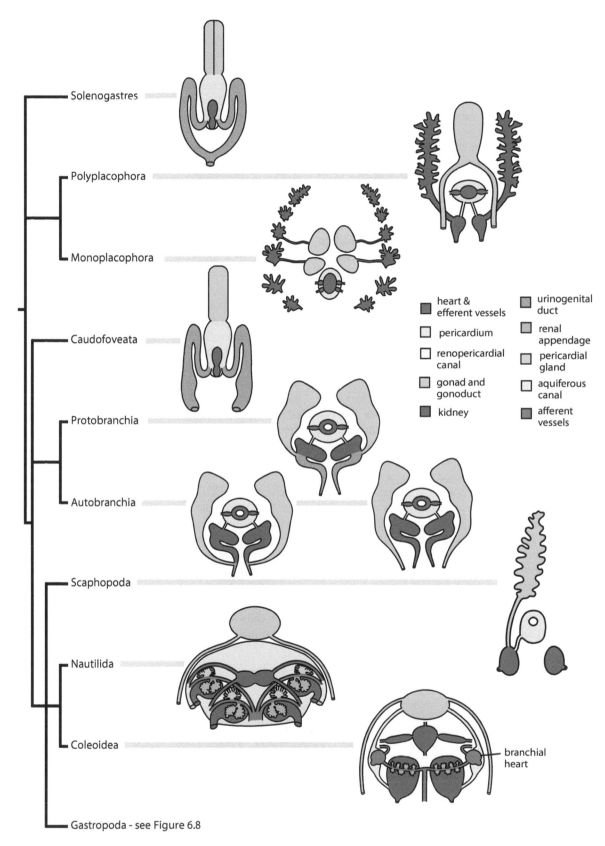

FIGURE 6.5 Comparative schematic diagrams of the renopericardial system of molluscs (excluding gastropods), and the relationship of that system to the genital system. Redrawn and modified (in part) from Martin, A.W. and Harrison, F.M., Excretion, in Wilbur, K.M. and Yonge, C.M. (eds.), *Physiology of Mollusca*, Vol. 2, pp. 353–386, Academic Press, New York, 1966 and various other sources.

auricle and the ventricle, but in other polyplacophorans there are two or sometimes four openings (ostia), leading to the suggestion that each auricle represents a fused pair (Hyman 1967; Eernisse & Reynolds 1994). The auricles receive blood from the efferent branchial sinuses that connect with the multiple gills in the mantle groove. The ventricle pumps blood through the only aorta, the anterior. This vessel carries blood to sinuses in the head and viscera from where it passes through several sinuses to the gills, then to the auricles via the efferent branchial sinuses. Blood moves through the body to collect in ventral sinuses, and ultimately those of the foot. Transverse sinuses direct the blood to the afferent sinuses from which each of the numerous gills is supplied by a branch. After passing through the gill and the vascularised mantle roof, the oxygenated blood is collected and returned to the heart via the efferent branchial sinuses, which lie in the outer portion of the roof of the mantle groove. The efferent branchial sinus runs along either side of the body and connects through several vessels with the right and left auricles.

The typical monoplacophoran circulatory system is similar to that of chitons except there are two pairs of auricles. As in chitons, these lie in the same position relative to the foot retractor muscles. Lemche and Wingstrand (1959) incorrectly interpreted the heart of *Neopilina* as having two ventricles with the rectum lying between, but the rectum is actually surrounded by a single ventricle (Haszprunar & Schaefer 1997). Oxygenated and deoxygenated blood is separated in afferent and efferent systems. The two pairs of auricles receive oxygenated blood through short vessels from ventral mantle sinuses, which then collect the oxygenated blood from the gills and kidneys. The ventricle discharges into a single aorta, where most of the blood passes anteriorly to the foot and opens directly into anterior body sinuses. The blood is collected in the pericardial sinus before passing into the dorsal venous mantle sinuses and the gills (Lemche & Wingstrand 1959). The minute *Micropilina* lacks a heart (Haszprunar & Schaefer 1997; Ruthensteiner et al. 2010).

In the two aplacophoran groups, the large pericardium contains a ventricle and, usually, a pair of auricles, although in some taxa they are fused into a single auricle. There are no blood vessels. The auricle receives oxygenated blood from the gills (when present) and sinuses which surround the posterior

mantle cavity or chamber. The ventricle opens directly into the dorsal sinus, and blood moves from there to the anterior sinuses. It then drains through the open system and moves to the posterior end of the animal where the deoxygenated blood flows into the respiratory surfaces. Body movements help circulate the blood through the body sinuses.

Scaphopods lack a heart, but Reynolds (1990b) suggested that, in a dentaliid (*Dentalium rectius*), a muscular sinus surrounding the rectum, the perianal sinus, is homologous with the pericardium of other molluscs because ultrastructurally it is similar to that of bivalves in having podocytes. Simone (2009) identified a pericardium in the dentaliids *Coccodentalium* and *Paradentalium*, but not in the gadilidan *Gadila* and *Polyschides*. There are no aortic or other true blood vessels (Shimek & Steiner 1997). Blood moves from the perianal sinus both anteriorly and posteriorly through a series of relatively continuous sinuses (Figure 6.6), first to the head and viscera, then the blood is reoxygenated along the ventral surface of the mantle before being returned to the perianal sinus.

Bivalve hearts have a ventricle and a single pair of auricles. The ventricle is often thin-walled with a thin outer muscle layer and trabeculae which run across the lumen. The intestine penetrates the ventricle to at least some extent in many bivalves, including some protobranchs, and that is the assumed plesiomorphic condition. This is modified independently in several taxa, e.g., the protobranch *Nucula* and the pteriomorphians *Lima*, some modiolines, and *Vulsella*, where the intestine lies ventral to the heart. In the protobranch *Malletia* and several pteriomorphians including oysters (other than *Pycnodonte*), *Pinctada*, *Isognomon*, *Malleus*, and some modiolines, the unionoidean *Mulleria*, and the myoidean *Teredo*, the intestine lies above the heart (White 1942; Narain 1976). An aortic bulb may occur on the posterior or anterior aorta (see Section 6.3.4). Because of the important role the gills play in feeding in autobranch bivalves, the circulation patterns have been modified to exploit additional organs and surfaces for respiration, including the mantle and kidneys (Figure 6.7).

Because of torsion, the ctenidia in gastropods are plesiomorphically located anteriorly, and the heart lies anterior to the visceral mass except in the secondarily bilaterally symmetrical

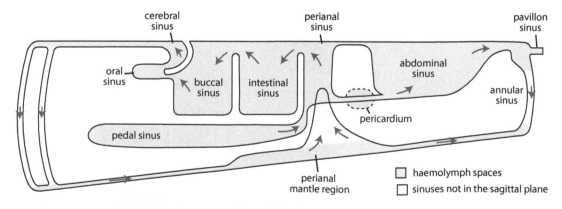

FIGURE 6.6 Diagram of the haemolymph spaces in a scaphopod with their assumed connections and haemolymph circulation. Redrawn and modified from Shimek, R.L. and Steiner, G., Scaphopoda, in Harrison, F.W. and Kohn, A.J. (eds.), *Microscopic Anatomy of Invertebrates. Mollusca 2*, Vol. 6B, pp. 719–781, Wiley-Liss, New York, 1997.

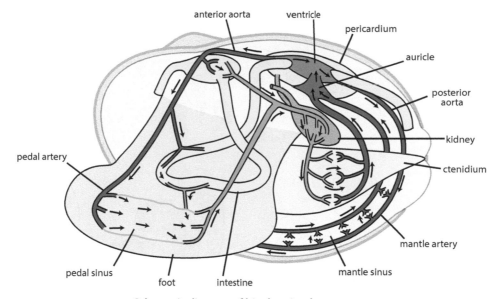

Schematic diagram of bivalve circulatory system

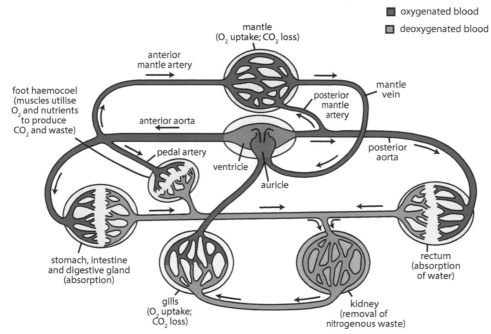

Circulatory system of a generalised bivalve

FIGURE 6.7 A schematic diagram of a bivalve (a unionid) circulatory system and a diagrammatic representation of the circulatory system of a generalised autobranch bivalve. Top figure redrawn and modified from Buchsbaum, R., *Animals Without Backbones: An Introduction to the Invertebrates*, Penguin Books, Harmondsworth, UK, 1951; lower figure redrawn and modified from Guthrie, M.J. and Anderson, J.M., *General Zoology*, Wiley, New York, 1957.

fissurellids, sacoglossans, and nudibranchs. In most gastropods, the heart differs from other molluscs in being on the post-torsional left side of the body rather than in the midline. By outgroup comparison, the plesiomorphic condition of the heart in gastropods was a ventricle and two auricles which receive blood from a pair of gills. This condition survives in only some vetigastropods, the so-called 'Diotocardia' (Pleurotomarioidea, Haliotioidea, Fissurelloidea, and Scissurelloidea). The left and right auricles receive oxygenated blood from the efferent veins from the left

and right mantle and gills, respectively, which indirectly receive blood from the left and right kidneys, respectively (e.g., Russell & Evans 1989; Fretter & Graham 1994) (Figure 6.9). If the right gill is reduced in size relative to the left, the right auricle is also normally smaller. As already noted, the number of auricles is usually correlated with the number of gills, but this is not the case in some taxa. Vetigastropods that have lost the post-torsional right gill (e.g., Lepetodriloidea, Trochoidea) have two functional auricles and retain both kidneys (Figure 6.8). The blood entering the right

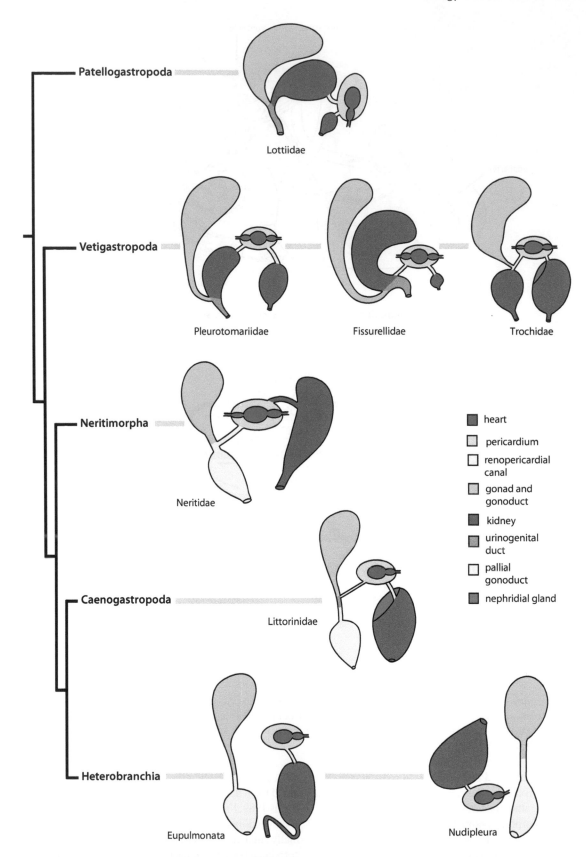

FIGURE 6.8   Comparative schematic diagrams of the renopericardial system of gastropods and the relationship of that system to the genital system. Redrawn and modified from various sources.

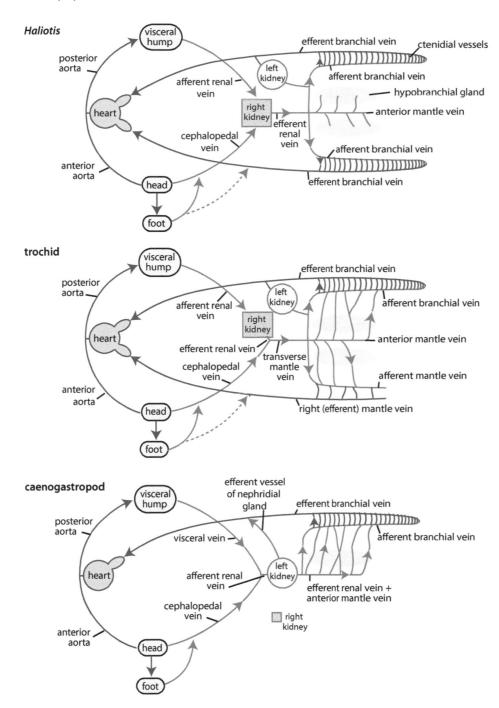

FIGURE 6.9 Schematic comparisons of the vascular systems of a haliotid, a trochid, and a typical caenogastropod. Redrawn and modified from Fretter, V. and Graham, A.L., *British Prosobranch Molluscs: Their Functional Anatomy and Ecology*, Ray Society, London, UK, 1962.

auricle is collected from the efferent mantle vein, while that on the left receives blood from the efferent ctenidial and renal veins (e.g., Fretter & Graham 1994; Sasaki 1998) (Figure 6.9).

The loss of the right gill and kidney in neritimorphs is accompanied by a marked reduction in the size of the right auricle which becomes non-functional in some taxa, while in others it is connected to the remaining (left) kidney (Sasaki 1998). In cocculinoideans, neomphaloideans, and caenogastropods, the right auricle is lost along with the right gill (Figures 6.8 and 6.9). Heterobranch gastropods also have a single auricle (Figure 6.14) and may or may not have a plicatidium (see Chapter 4), secondary gill or lung (Figure 6.12).

The ventricle surrounds the intestine in many vetigastropods and neritimorphs, but in other gastropods, including nearly all patellogastropods, the intestine lies outside the pericardium. One assumed progenetic patellogastropod, *Erginus apicina*, has the intestine surrounded by the pericardium, not the ventricle (Lindberg 1988b).

In gastropods, the ventricle is more muscular than in bivalves and contains thick trabeculae that traverse the lumen.

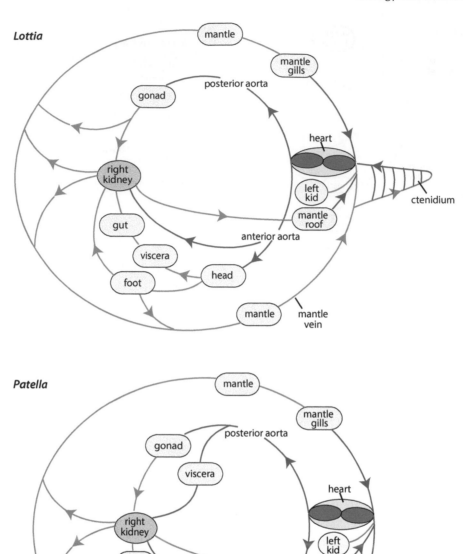

FIGURE 6.10 Schematic comparison of the circulatory patterns in a representative of the two main groups of patellogastropods. Note that only some species of *Lottia* have mantle gills. *Lottia gigantea* based on Fisher, W.K., Zool. Jahrb. Abt. Anat. Ontog. Tiere, 20, 1–66, 1904; *Patella* redrawn and modified from Davis, J.R.A. and Fleure, H.J., *Liverpool Mar. Biol. Comm. Mem.* 10, 1903.

There is no separate blood supply for the heart, oxygen being supplied via diffusion from the blood in the heart.

In vetigastropods, the anterior and posterior aortae arise from the left side of the pericardium, but in patellogastropods (Figure 6.10), the heart is rotated so the aortae emerge from the posterior side of the heart (Ponder & Lindberg 1997). This rotation is also shown by the post-torsional left and right kidneys both being located on the right of the pericardium, and the left renopericardial duct lies above the right (Figure 6.8).

A similar, but less marked, rotation of the heart is seen in *Neomphalus* (Fretter et al. 1981, figure 15), but other limpet-like gastropods do not show such rotation. In these and coiled gastropods, the pericardium is on the left side of the body, and the (post-torsional left) kidney is to its left. Exceptions are helicinids (Neritimorpha) and cocculinids which show evidence of a rotation in the opposite direction, as the left auricle lies on the right side of the ventricle (Ponder & Lindberg 1997). In the fissurellid *Megathura crenulata*, the pericardium

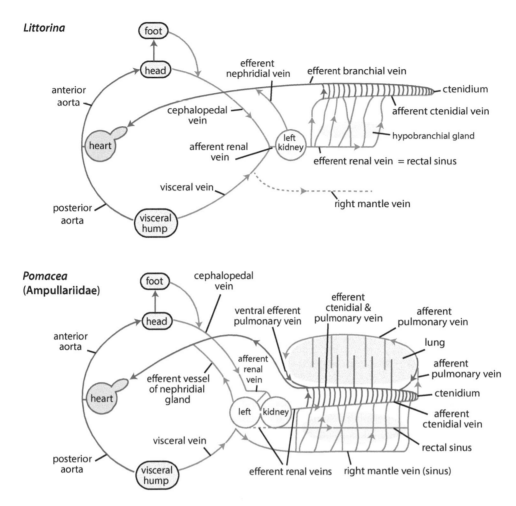

FIGURE 6.11 Schematic comparison of the vascular system of a typical intertidal caenogastropod such as *Littorina* and the amphibious ampullariid *Pomacea canaliculata* with an additional respiratory surface (lung). Redrawn and modified from Andrews, E.B., *J. Zool.*, 146, 70–94, 1965b.

lies at the midline of the body with the left kidney above and slightly to the left (Illingworth 1902).

The addition or subtraction of respiratory surfaces necessitates changes to the circulatory system - as in patellogastropods with or without a ctenidium (Figure 6.10) or the addition of a lung (Figures 6.11 and 6.12).

Cephalopods are active aerobic predators with a large brain, and the basic molluscan circulatory system is 'pushed to its limits in order to satisfy oxygen demand' (McMahon et al. 1997, p. 947). The circulatory system of *Nautilus*, with its extended peripheral and dorsal sinuses and lack of true capillaries (Bourne 1987c), is intermediate between the open system as seen in other molluscs, and the closed system in coleoid cephalopods. Coleoids have a cardiovascular system that suits their active, predatory lifestyle, in marked contrast to the more open circulatory organisation in nautiloids which have a lower cardiac performance. The nautiloid heart has a single bilobed ventricle, as in coleoids, but there are two pairs of auricles, one for each of the four gills. Blood is pumped to the body through five aortae and from there via small vessels to a series of sinuses (Figure 6.13) (Bourne 1987c; Schipp 1987a, 1987b). The coleoids have not only reduced the blood sinuses, but they also have a more developed venous system,

extensive arteries, and an extended capillary area, all consisting of true endothelial vessels. There is a largely separate respiratory circulatory system, where the blood is pumped through the gills by two accessory branchial hearts (see Section 6.3.3), and from there it flows to the true, or systemic, heart. Because this system is largely separated from the circulation around the rest of the body, it is, as noted above, somewhat analogous to the double circulatory system of tetrapod vertebrates (see Section 6.3.2). This unique molluscan circulatory system may be a product of early modifications to the larval circulatory system necessitated by increased yolk and provided circulation to the yolk sac (see Section 6.3.3 and Chapter 8).

The cephalopod systemic heart, as in most other molluscs, consists of a ventricle and one (in coleoids) or two (in *Nautilus*) pair(s) of 'auricles,'[6] corresponding to the number of gills present, although it differs in a number of ways from those of other molluscs. The pericardium is enlarged in *Nautilus*, squid, and sepioids and contains the branchial hearts, part of the intestine, and part of the ink sac (Young & Vecchione 1996). The intestine

---

[6] These structures are not homologous with the auricles in other molluscs - see 6.3.2.

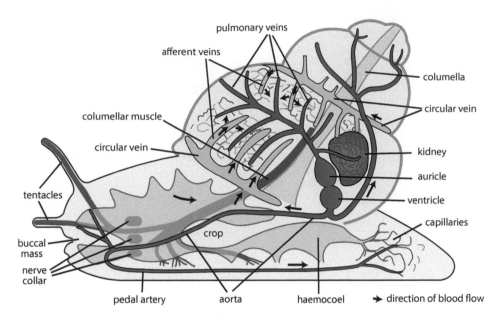

FIGURE 6.12 The blood system in the eupulmonate *Helix pomatia* showing the circulation pattern and the haemocoelic spaces. Only a few arteries and a small part of the capillary network in the posterior foot are shown. Redrawn and modified from Borradaile, L.A. et al., *The Invertebrata. A Manual for the Use of Students*, Cambridge University Press, Cambridge, UK, 1958.

does not pass through the ventricle, but lies within the pericardium just below the heart. *Nautilus* has peripheral sinuses and lacunae (Figure 6.13) where gas exchange takes place, notably in the gills. Blood leaves the heart through three or four vessels, the two main ones being a cephalic artery (= anterior aorta) and a posterior aorta, in addition to smaller genital and renal arteries.

The coleoid systemic heart consists of a single powerful, but rather compact, ventricle fed by the two 'auricles' with muscular valves that receive blood from a gill (Figure 6.13). Three aortae, with valves, leave the ventricle, the largest (dorsal or cephalic) running anteriorly to the head region mainly to supply the brain and the majority of the body. There are no internal trabeculae crossing the lumen of the ventricle, but there is a partial septum in the ventricle (Figure 6.15) which may direct blood toward the two smaller aortae lying to one side (McMahon et al. 1997).

The pericardium is much reduced in octobrachians (i.e., octopods and *Vampyroteuthis*). In octopods, the pericardium is represented only by a pair of lateral, thin-walled, semi-transparent, flask-shaped pouches. These lie behind the bases of the ureters, and each opens anteriorly into the corresponding kidney by way of a long, narrow renopericardial duct. Each pouch contains the branchial heart appendage (pericardial gland) thought to be at least analogous to the pericardial gland of gastropods and bivalves (e.g., Isgrove 1909). Thus, unusually, in octopods the heart itself is not contained in the pericardium, but lies within the body cavity. Cephalopod hearts obtain their blood supply from branches of the posterior aorta (Wells 1983), and it has been demonstrated in *Octopus* that oxygen is sourced directly from the lumen of the ventricle via the capillary network in the ventricle wall with the flow occurring during diastole, oxygenating the myocardium, and enabling increased performance (e.g., Agnisola 1990; Agnisola et al. 1990). The coronary blood collects in veins

and drains into the lateral venae cavae, and the coronary vessels are highly innervated (Smith & Boyle 1983).

There are valves to regulate inflow and outflow into the main arteries in cephalopods. The cephalic artery is on the left side in *Nautilus* and on the right in coleoids, but is probably derived from the same central vessel in ancestral cephalopods (Naef 1913). Other major vessels are the abdominal artery in coleoids (which is equivalent to the mantle artery and siphonal artery in *Nautilus*), the genital artery, and renal artery, the latter being lost in octopods (Budelmann et al. 1997) (Figure 6.13).

Most coleoid veins are contractile, especially the anterior vena cava, and their contractions can aid in the movement of blood, while muscle fibres in the lacunae and peripheral sinus can also facilitate blood flow. The muscles involved with the circulatory system are under nervous control, both autonomous and non-autonomous, and are also under the control of neurohormones (see Chapter 2).

Squid such as *Doryteuthis pealei* can swim at rates as high as 7 m/sec (Cole & Gilbert 1970) placing great demands on their cardiovascular system. It has been shown that the systemic heart output in coleoids can rival or exceed that of mammalian hearts in work and power (Shadwick et al. 1990; O'Dor & Webber 1991), but only for a very short time. Studies on *Sepia* have indicated that alternating contractions of the radial and circular mantle muscles in the mantle wall may assist blood flow through the capillaries in the mantle (King & Adamo 2008). Thus, besides heart action, mantle contractions during respiration and/or jetting presumably assist blood circulation.

6.3.2 HEART STRUCTURE AND FUNCTION

The myocardium (and epicardium) of polyplacophoran hearts are primitive compared to other molluscs (Økland

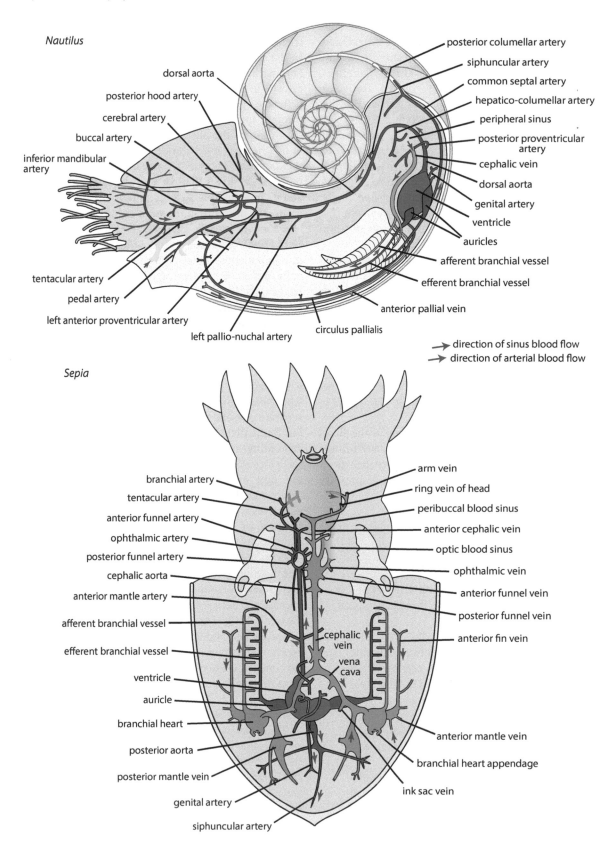

FIGURE 6.13 Schematic diagrams of the circulatory systems of *Nautilus* and *Sepia*. Redrawn and modified from Schipp, R., *Experientia*, 43, 474–477, 1987b.

1980). The myocardium is loose, consisting of muscle cells and nerve processes embedded in a collagen matrix and, unlike vertebrates, there is no endocardium (Box 6.3). Muscular trabeculae are absent in the basal *Lepidopleurus*, but present in the more 'advanced' chiton *Tonicella*. The epicardium of the ventricle is flat and simple, but that of the auricle has podocytes (see Section 6.6.2).

Molluscs, other than coleoid cephalopods, also lack a compact myocardium and vascularisation of the heart. Even in coleoids, the muscle layer making up the myocardium is not complete, with some blood able to contact the basal lamina.

A major function of the heart is to pump blood, but it also has an additional important role in many molluscs (but not cephalopods) as the epicardium is often the site of filtration and primary urine production (see Section 6.6.1).

A few molluscs lack a distinct heart, including all scaphopods and some others with very small body size such as a few heterobranch gastropods, some very small protobranch bivalves (Sanders & Allen 1973), and the monoplacophoran *Micropilina*. In these cases, and in other molluscs, circulation of body fluids is assisted by muscular contractions associated with body movement. In some molluscs, circulation is further assisted by accessory pumping structures (see Sections 6.3.3 and 6.3.4).

The heart rate, and hence blood flow, are under hormonal control with cardioactive neuropeptides being widespread in molluscs (see Chapter 2).

As mentioned above (Section 6.3), the role of the heart in molluscan circulation is substantially different from the tetrapod heart. In the three or four-chambered hearts of tetrapods, the auricle *pushes* the blood to the respiratory

surface where it is oxygenated, and then it flows to the ventricle, from which the blood is pumped through the body. In the molluscan heart (cephalopods excepted), the auricle (and perhaps the ventricle) *pulls* the blood from the respiratory surface(s) where it has been oxygenated, and then it flows to the ventricle, from where the blood is pumped through the body. In molluscs, the mechanics involved in the filling of the auricle(s) and the role of the pericardium in this process are still not well known, although one idea, the 'constant volume hypothesis,' has gained some acceptance. It supposes that the total volume of the pericardium and its contents is constant so that when the ventricle contracts, the pressure in the pericardial cavity is reduced, and blood moves into the auricle(s) (see Jones 1983b for review). Berg (1992) proposed a model that lends general support to this hypothesis. This 'constant volume' model was originally based on studies of the heart of stylommatophoran land snails. It assumed that both the auricular and ventricular volumes are the same and that the pericardium itself does not change in volume. This hypothesis, however, is found wanting for hearts involved in filtration and in those where the auricle and ventricle volumes are not equal (e.g., Fretter & Graham 1994). This latter point is demonstrated in a detailed study of *Littorina* by Andrews and Taylor (1988), who showed that the auricle is functionally divided into two separate chambers so that, on contraction, blood is pumped into both the ventricle and the nephridial vein, the latter generating a 'tidal flow' with the blood moving back and forth on each contraction/relaxation of the auricle. Although this flow model has only been demonstrated in *Littorina littorea* to date, in many caenogastropods, the nephridial gland vessel is connected to the efferent ctenidial vessel at the base of the auricle suggesting that this mode of heart operation is typical of many members of that group (Fretter & Graham 1994), except for neogastropods (Andrews 2010) (see Section 6.3.8). Thus, the constant volume hypothesis may be an over-simplification for caenogastropods, at least because the actual atrial volume must be in excess of that of the ventricle (Andrews & Taylor 1988). Similarly, the functional role of the pericardium in the control of cardiac output is rather poorly understood, despite being an important component of the constant volume hypothesis (McMahon et al. 1997). While most pericardia are assumed to maintain the same volume as required by the constant volume hypothesis, in *Busycon*, for example, the pericardial wall is contractile (Jones 1988).

There are four muscular contractile auricles in *Nautilus*, and two in coleoids. Developmental studies on coleoids have shown that the 'auricles' are actually expanded efferent branchial veins which connect with the ventricle late in development. Earlier in development, the so-called branchial hearts are connected to the ventricle and are presumably the homologues of the auricles (Portmann 1926; Raven 1966; Andrews 1988).

In octopods, the ventricle lumen is partially divided internally by a septum which directs most of the incoming blood from the right side of the 'auricle(s)' into the dorsal aorta, while that from the left side flows mainly into the abdominal and gonadal arteries (Smith 1981b) (Figure 6.15).

BOX 6.3   THE STRUCTURE OF THE HEART WALL

The molluscan heart consists of a ventricle and one or more auricles. Its wall can have up to three basic layers, these being:

- The **epicardium** on the outer (pericardial) side is a thin coelomatic epithelium resting on the basal lamina (see Section 6.3.1). This layer may be differentiated into podocytes (see Section 6.6.2).
- The **basal lamina** – an extracellular matrix (usually collagen) separating the epicardium and myocardium.
- The **myocardium** is a layer of muscle cells on the inside of the basal lamina. This layer is also the source of irregular trabeculae that cross the lumen if these are present. It is also incomplete, enabling blood to make contact with these structures and the basal lamina.

An endocardium, the innermost lining of the heart in vertebrates, is absent in molluscs.

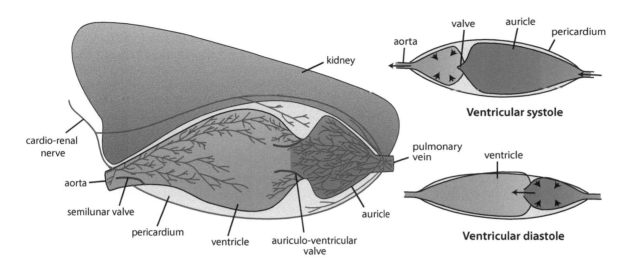

FIGURE 6.14 Heart structure and function in the stylommatophoran *Helix*. The figure on the left shows the structure and innervation of the heart and those on the right show the function. Left figure redrawn and modified from Aseyev, N. et al., *Peptides*, 31, 1301–1308, 2010; the right figures redrawn and modified from Jones, H., *Comp. Biochem. Physiol. A*, 39, 289–295, 1971.

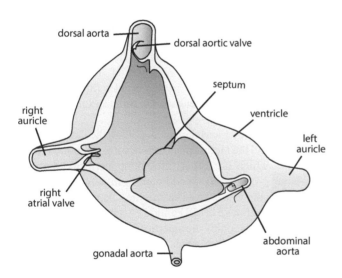

FIGURE 6.15 The anatomy of the systemic heart of an octopus with part of the ventral surface removed to show the partial septum and the valves. Redrawn and modified from Smith, P.J.S., *J. Exp. Biol.*, 93, 243–255, 1981b.

6.3.2.1 Heart Musculature

Heart muscle ultrastructure (see Chapter 3 for details) has been studied in relatively few molluscan taxa. One important difference in the structure of molluscan cardiac muscle fibres is the arrangement of the Z-material. It varies from relatively simple smooth muscle as in bivalves to complex oblique or helical striated muscles like those in coleoid cephalopods (González-Santander & García-Blanco 1972). The number of mitochondria in heart muscle also varies greatly between different molluscan classes, averaging from 80% of the cross-sectional volume in squid heart tissue to only 30% in a bivalve (Dykens & Mangum 1979), reflecting differences in work demand.

In chitons, the muscle fibres (cells) are simple, unbranched, with a single nucleus, and mitochondria are located peripherally. The Z-material consists of dense bodies and attachment plaques; sarcomeres, myofibrils, transverse tubules, and intercalated discs are all absent. Muscle fibres in the ventricle of *Lepidopleurus* have only septate junctions, but those in the more advanced chiton *Tonicella* show both septate junctions and desmosomes (Økland 1980). The myocardium contractile material in chitons has a very primitive arrangement, more so than in other molluscs and, according to Økland (1980), is generally more similar to the myoepithelial cells in the hearts of the polychaete *Arenicola marina* (Jensen 1974). Unusually, heart muscle fibres have been observed in the pericardium in chitons (Økland 1981) and in the presumed homologue of the pericardium of scaphopods (Reynolds 1990b).

There are insufficient data to reliably compare details of cardiac muscles in the different molluscan classes, but some generalities can be noted. In bivalves, the Z-line (see Chapter 3) is irregular, and in the normal state of contraction in chitons, no such structures are found (Økland 1980). The Z-material in gastropods and cephalopods is fairly regular, although disrupted. In *Patella*, the muscle fibres are short and narrow. Compared to the myocardial ultrastructure in Polyplacophora, the heart muscles of *Patella* are more complex in structure. In chitons, the myocardial fibres lack T-tubules, and the Z-material is unfused, features seen in *Patella* (Økland 1980, 1982). The fine structure of monoplacophoran heart muscle has not been described. The heart muscle of some bivalves, gastropods, and cephalopods is striated and helical (see Chapter 3), although there is a gradation between the fully striated helical muscle seen in coleoids and the spiral arrangement of 'smooth' cardiac muscle such as that seen in bivalves. The structure of the helical heart muscles in gastropods and coleoids has Z-arrangements 'like steps in a spiral staircase' (McMahon et al. 1997). The development of striations is correlated with an increase in performance and power output (McMahon et al. 1997). The myocardium in the heart of the neogastropod *Busycotypus* (as *Busycon*) *canaliculatus* is well-organised, the contractile units are aligned, and the Z-material is not fused. While intercalated discs are present, T-tubules are apparently lacking (Sanger 1979).

The ventricle of cephalopods is complex (Wells 1983), with the muscle cells comprising the myocardium obliquely striated (Kling & Schipp 1987). The myocardium also contains many nerve fibres which, with various chemical transmitters, control the heart rhythm.

Modification of the heart muscles may have been a major innovation that facilitated the move from a sedentary to a highly active lifestyle (Dykens & Mangum 1979). Thus, heart muscle ultrastructure of a squid differs notably in the much greater density of mitochondria and tubules when compared with, for example, that of a bivalve. It also differs from gastropods and bivalves in the Z-material and the arrangement of myofibrils. While some gastropods also have a helical organisation in the heart muscle, this is less pronounced than in coleoids because their sarcomere arrangement is looser, and it is looser still in bivalves, becoming parallel only during contraction (e.g., Dykens & Mangum 1979; McMahon et al. 1997). Muscle fibres similar to the striated muscle found in the hearts of gastropods and coleoid cephalopods occur in the walls of the veins and arteries of coleoids. Interestingly, *Nautilus* heart muscle ultrastructure is more like that of bivalves than coleoids (Dykens et al. 1982).

6.3.2.2 Heart Regulation and Control

Some factors influencing heart (i.e., ventricle) output include:

Blood pressure in the auricle(s). It was discovered in early experiments on *Aplysia* (Straub 1901, 1904) that the output of the ventricle is directly related to the atrial filling pressure (i.e., the preload pressure). This was subsequently also demonstrated in other taxa, including a neogastropod (*Busycon*), a bivalve (*Mercenaria*) (Smith 1985), and an octopod (*Eledone*) (Smith 1981b).

Heart rate (the number of heartbeats per unit time) is typically regulated by a *pacemaker*. There is good evidence for a neuronal pacemaker region that governs the contraction of the ventricle (Jones 1983b), but this may be confined to a small area (e.g., the atrioventricular valve) or is more diffuse across the myocardium (see McMahon et al. 1997 for discussion and references) (see Chapter 7 for more details).

Stroke volume, the volume of blood pumped from the ventricle with each beat.

Work of the heart muscle fibres determines both stroke volume and heart rate which are regulated by the permeability properties of the membrane and excitation-contraction coupling (see Section 6.3.2.4).

Nervous control and neurohormones (see below, and for general information on neurohormones, see Chapter 2).

Information on heart performance is given below (Section 6.3.2.4).

Interactions with heartbeat and neurons and hormones have been studied in several taxa and are reviewed by McMahon et al. (1997), for heterobranch gastropods (Kodirov 2011) and, for cephalopods, by Kling and Schipp (1987) and Kling and Jakobs (1987). Nerves may directly influence the heartbeat, or carry inhibitory or excitatory neurotransmitters to the heart. Four major neurotransmitters are known to influence heart activity (see McMahon et al. 1997; Kodirov 2011 for reviews): (1) cholinergic neurons with the neurotransmitter acetylcholine (ACh), which usually acts as a cardioinhibitor and can also sometimes be excitatory, (2) serotonergic neurons release serotonin, (5-hydroxytryptamine [5-HT]) which usually acts as a cardioexcitor, but can also sometimes be inhibitory, (3) peptidergic neurons release the FMRFamide-related neuropeptides (FaRPs) and various small cardioactive peptides (SCPs), and (4) catecholaminergic (CAergic) neurons which produce the neurohormones epinephrine (adrenaline), norepinephrine (noradrenaline), and dopamine which are released in the haemolymph when the animal is stressed (Kodirov 2011). For example, in *Helix*, a glutaminergic motor neuron (V43) is thought to relax the ventricle, while dopamine and octopamine may also regulate the performance of the heart (Rózsa 1987; Buckett et al. 1990). In cephalopods, noradrenaline is an important cardio-excitatory compound affecting the systemic heart (Kling & Jakobs 1987; Versen et al. 1999), although in *Nautilus* it is present only in the auricles (Springer et al. 2005), while in coleoids other catecholamines excite the branchial heart (Fiedler & Schipp 1990).

As pointed out by McMahon et al. (1997), the data on heart performance and the factors that influence it are obtained from experiments which are often difficult to perform, making comparisons problematic. In addition, in immunohistochemistry studies, there are cross-reactivities between peptides. For example, although FMRFamide is apparently confined to molluscs (Price 1986), immunoreactivity to the antibody occurs in all major animal phyla (McMahon et al. 1997).

6.3.2.3 Heart Metabolism

In bivalves and gastropods (Ellington 1985) and *Nautilus* (Kling & Schipp 1987), oxygen reaches the ventricular muscles by diffusion from the blood in the ventricular lumen and the metabolism utilises glycogen. Coleoid cephalopod hearts differ from other molluscs in the following ways: (1) they have a very high rate of oxygen consumption (Houlihan et al. 1987), necessitating an oxygenated blood supply; (2) metabolism is aerobic and carbohydrate-based; (3) they oxidise amino acids as a source of fuel; and (4) increased energy demands are met by the activities of the enzymes phosphofructokinase and citrate synthase (see McMahon et al. 1997 for further details).

For reviews of the metabolic pathways employed in molluscan hearts see Ellington (1985), McMahon et al. (1997), and in cephalopods, Driedzic et al. (1990).

The action potential of molluscan heart muscle is very similar (especially in gastropods) to that of vertebrates, consisting of a prepotential, spike, and plateau (e.g., Jones 1983b; McMahon et al. 1997), but although similar in pattern, the ions involved differ. From the *resting potential*, the slowly rising *prepotential* is a result of a sodium leakage current, then the rapidly rising *spike* is due to calcium, which is quickly inactivated. An increase in sodium conductance maintains a relatively long-lasting (sometimes up to several seconds) *plateau* before a return to the resting potential. The resting potential is due mainly to potassium ions and, to a minor extent, chloride ions (for more details see Jones 1983b; Brezden & Gardner 1992; McMahon et al. 1997). So-called 'ion channels' are enzyme-like protein molecules that, when 'open,' catalyse the transfer of specific ions through cell membranes (e.g., Brezden & Gardner 1992) and are common to animal and bacterial, plant, and fungal cells (e.g., Morris & Sigurdson 1989). In some cells, the open state is affected by cell membrane deformation, or 'stretch.' Thus, in muscle cells, elongation of the muscle cell will cause the activation of some ion channels, but will shut down others (e.g., Morris & Sigurdson 1989).

6.3.2.4 Heart Performance

The heart performance (rate or stroke volume) can only be improved by improving the response of all the individual fibres because the heart muscles contract as a unit. Thus, muscle performance depends on the permeability of the muscle cell membrane to particular ions and *excitation-contraction coupling* – the conversion of an electrical stimulus to a mechanical response (see Chapter 3). Other intrinsic or extrinsic factors also act on the muscle fibres, either in the myocardium generally or via a pacemaker. McMahon et al. (1997) noted some similarities in performance and configuration of the molluscan cardiovascular system with that of vertebrates, including the action potential, despite the differences in muscle cell structure and some physiological aspects.

Overviews of the performance of molluscan hearts include general reviews (e.g., Smith 1987b; McMahon et al. 1997) and details of gastropods and bivalves (Jones 1983b) and cephalopods (Wells 1983; Wells & Smith 1987). As mentioned above, heart output is a combination of heart rate and stroke volume. Although the latter is more difficult to measure experimentally, it plays an important, and sometimes dominant, role in regulating cardiac output.

Considerable, apparently natural, *in vivo* variation in heart rate can occur, for example in *Busycon* it is markedly erratic, ranging from 5–15 beats min^{-1} (Jones 1988), although heart rate can usually be influenced by external and internal factors. For example, it has been induced in bivalves by opening and closing the shell valves (Colman 1974). In *Aplysia*, the heart rate increases in response to external stimuli such as noxious substances, food and feeding, temperature increase, moderate hypoxia, or overall activity, with some of these effects under direct neuronal control (reviewed by Koester & Koch 1987). Such control by the nervous system has been well studied in a few species of euthyncuran gastropods. Two are stylommatophorans – *Helix pomatia* (Rózsa 1979, 1983, 1987; Zhuravlev & Safonova 1984) and *Achatina fulica* (Furukawa & Kobayashi 1987), and the aquatic hygrophilan *Lymnaea* (Buckett & Dockray 1990; Buckett et al. 1990).

The neurons involved in the control of heartbeat are in the visceral (= abdominal) ganglia. The most extensive studies of neuronal heart control have been on *Aplysia* where at least ten types of motor neurons innervate the cardiovascular system, all but one pair in the visceral ganglia (one pair of neurons are in the pedal ganglia), and three innervate the heart (Koester & Koch 1987; Skelton & Koester 1992; Skelton et al. 1992).

Temporary cessation of heartbeat (acardia) has been reported in various gastropods (Bourne et al. 1990). In certain conditions, octopods can stop their heart beating. In *Octopus vulgaris,* it can stop for a few seconds in anoxic conditions (Wells 1979), but in the much larger and colder water inhabitant *O. dofleini* (Johansen & Martin 1962), it can stop for up to two hours under seemingly normal conditions. In the pteropod *Clione*, Arshavsky et al. (1990) showed that heartbeat is controlled by four neurons in the left visceral ganglion. In swimming activity, the heart stops beating during natural pauses lasting a few minutes. Cessation of heartbeat (and locomotion) also occurred when the head was touched, but if the tail was touched, both locomotion and heartbeat accelerated. Heartbeat also increased when hunting.

In an abalone, *Haliotis cracherodii*, cardiac output is correlated with body size and ranges from 100 to 150 mL kg^{-1} min^{-1}, a similar output to that seen in mammals, but with a stroke volume of 5 mL kg^{-1} body weight, much larger than that of a mammal (Jorgensen et al. 1984). Jones (1988) noted systolic ventricular pressures recorded for gastropods, the highest to that date being from terrestrial slugs which were typically around 74.5 cmH$_2$O, but up to 90 cmH$_2$O, while diastolic pressures were about 52 cmH$_2$O (Duval 1983). Thus, the ventricle is increasing pressure by about 23 cmH$_2$O. In comparison, the ventricular systolic pressures of helicid snails are at about 25 cmH$_2$O with diastolic levels about 6 cmH$_2$O, with the ventricle increasing pressure by about 19 cmH$_2$O (Jones 1971). In the neogastropod *Busycon*, systolic pressures were usually 15–30 cmH$_2$O with very brief periods of over 50 cmH$_2$O (maximum 80 cmH$_2$O). As the diastolic level was only 1–3 cmH$_2$O, the ventricle was responsible for most of the systolic pressure (Jones 1988). Cephalopods show sustained levels of systolic pressure, varying from about 30 to about 70 cm H$_2$O (Bourne 1987b; Wells & Smith 1987), with rather high diastolic levels in some. Thus *Busycon* is capable of producing pressures in the ventricle for one to a few contractions as great as the sustained pressures in some cephalopods (Jones 1988).

Significant increases in stroke volume during activity have been recorded in some gastropods and bivalves (McMahon et al. 1997). For example, the pectinid bivalve *Placopecten magellanicus*, can increase stroke volume five times and the heart rate two to three times when escaping (Thompson et al. 1980). Cephalopods also regulate cardiac stroke volume during exercise, as in *Nautilus* which has a two to threefold increase in stroke volume, with heart rate increases up to approximately 30% (Wells et al. 1992). *Octopus* shows no marked increase in heartbeat frequency during exercise, or in recovery from low oxygen conditions, although an increase of up to three times in stroke volume substantially increases the supply of oxygen (Wells 1979), with the power output increasing by about 700% during activity (Armstrong & Smith 1992). The cardiovascular system of squid is even more impressive as they have 'pushed molluscan cardiovascular capability to an extreme' (McMahon et al. 1997, p. 949). During exercise, they show a similar stroke volume increase to that of *Octopus*, but at double the heart rate (McMahon et al. 1997). The open-water squid have increased the relative heart size to increase output because both stroke volume and heart rate have probably reached their limit (Wells 1992a, 1992b). Thus, *Illex* has a heart nearly twice as large as that of *Doryteuthis* with a similar body mass. *Illex* has a similar cardiac performance to teleost fish, although it uses much more energy than a fish such as a salmon, with their hearts pumping at approximately 400 mL per kg^{-1} of blood with a stroke volume of 2.6 mL per kg^{-1}, and yet the heart is only 3 g per kg^{-1} (Wells & Smith 1987).

The hearts of active decabrachian coleoids pump blood at rates that may approach those of a human athlete (O'Dor & Shadwick 1989). Thus the coleoid cephalopods '… are a dramatic example of what can be developed from the basic molluscan cardiovascular design. Saddled with a poor respiratory pigment carried in solution and a vascular system which is inevitably compromised by the constriction of arteries and veins during activity' (McMahon et al. 1997, p. 948), coleoids represent a 'triumph of engineering over design' (Wells & Smith 1987, p. 487).[7]

Other differences seen in coleoid cephalopod heart muscle are correlated with their high metabolic rate and limited anaerobiotic capacity (unlike bivalves, see Chapter 2). According to Dykens and Mangum (1979), the performance of squid heart muscles at 20°C is equalled only in hummingbird muscle at 36°C.

6.3.3 THE BRANCHIAL HEARTS OF CEPHALOPODS

A unique feature of coleoid cephalopods is a pair of globular accessory branchial hearts lying near the base of each gill, between the anterior vena cava and the afferent branchial vessel. Their derivation, which is discussed in Chapter 8, is from the original auricles, while the adult auricles are derived from the paired vena cava (Naef 1909). Together, the systemic heart and branchial hearts provide a 'double' circulation pattern analogous to that of tetrapod vertebrates (see Section 6.3.2).

Compared with the systemic heart, the branchial hearts are weakly muscular, although sufficiently so to raise the returning venous blood pressure, forcing the blood quickly through vessels in the gill on its return to the 'auricles' (Wells 1983; McMahon et al. 1997). In octopods at least, the contractile lateral venae cavae may also contribute to blood pressure before blood reaches the gill (Smith 1982b). The branchial hearts in octopods do not have a valve directing flow into the 'auricles,' but one is present in decabrachian coleoids (Smith 1982b; Fiedler & Schipp 1990). This unique vascular

system may result from modifications to the larval circulatory system necessitated by increased yolk, and to provide its distribution to the developing embryo (see *Nautilus* below and Chapter 8).

A layer of muscle surrounds each branchial heart, and trabeculae extend internally among the peculiar vacuolate cells lining the interior. Besides assisting with blood circulation, the appendages of the branchial hearts (the pericardial appendages) which lie within the pericardial cavity and contain podocytes (see Section 6.6.2), are responsible for the production of primary urine. Other functions involve excretion by way of the vacuolate cells and the absorption of carbohydrates (Martin 1983; Budelmann et al. 1997) (see Section 6.7.6.2).

In octopods and decabrachian coleoids, a 'cardiac ganglion' lies on the ventral surface of the branchial hearts and innervates them. Internally, these ganglia have an 'intraganglionar body,' a sac of connective tissue and muscle fibres of unknown function, which contracts in rhythm with the branchial hearts, lateral venae cavae, and the ventricle (Budelmann et al. 1997; McMahon et al. 1997).

Contractile 'branchial hearts' in embryonic *Nautilus* (Arnold 1987) are the homologues of the coleoid branchial hearts. but these are reduced in adults to contractile vessels (the venae cavae) which, like the coleoid branchial hearts, increase the blood pressure to cope with the resistance in the gill (Bourne 1987a, 1987c). *Nautilus* also has pericardial appendages with podocytes that are homologous with the branchial heart appendages.

The muscles in the coleoid branchial hearts have a helical structure (Fiedler & Schipp 1987) as in the ventricle, further indicating they are the original auricles. There are differences in the way the muscles are arranged and sometimes in their colour, which, in octopods is deep red to yellow-brown, while the ventricle is pale. This colour difference is due to iron-rich pigments in the spongy tissue of the branchial hearts, which may play a role in oxygen metabolism (Martin 1983). In contrast, the branchial hearts in *Sepia* are white (Beuerlein et al. 2002).

6.3.4 THE AORTIC SYSTEM AND AORTIC BULB

Monoplacophorans, polyplacophorans, and aplacophorans have only a single (anterior) aorta, while cephalopods have an anterior and a posterior aorta that emerge from either end of the ventricle. Most bivalves, including many protobranchs, also have an anterior and a posterior aorta, but the posterior aorta is reduced or lost in some protobranchs and pteriomorphians (Narain 1976). Scaphopods lack both the ventricle and aorta (see below). The gastropods (some patellogastropods excepted) have a single vessel that emerges from the posttorsional posterior end of the ventricle and immediately bifurcates into a posterior and anterior aorta, both these vessels being homologues of the anterior aorta in other molluscs.

Blood circulation is assisted in some bivalves by parts of the posterior blood vessels being modified as accessory pumps quite separate from the heart. These include contractile pumping by the posterior aorta of siphonate bivalves (Narain 1976). In some siphonate bivalves, the posterior aorta

---

[7] See O'Dor and Webber (1986) for a contrary view.

can develop a large bulb contained within the pericardium which may possibly act as a receptacle for blood during rapid contraction of the siphon(s) (Narain 1976), but in some heterodont bivalves at least, this muscular aortic bulb (or *bulbus arteriosus*) is thought to act as a neurohaemal structure that also plays a role in the regulation of blood circulation (Deaton et al. 2001). Some oysters have pulsating vessels in the mantle folds, and a pair of 'accessory hearts' in the cloacal region has also been described in other oysters (Galtsoff 1960; Yonge 1960b; Galtsoff 1964). Contractile bulbs have also been recorded from the anterior aorta of a few bivalves across several autobranch groups (Narain 1976). A possible function of these anterior aortic bulbs is to help prevent the backflow of blood into the ventricle during diastole by accommodating it.

In gastropods, the aorta is often swollen into an aortic bulb where it first emerges. These structures differ anatomically and histologically in different groups of gastropods, so they are probably analogues and may even differ in function (Økland 1982). In *Patella*, there are cardiac muscle fibres in the contractile aortic bulb, but not in the aortae themselves, and the epicardium resembles that of the rest of the heart. This suggests that the patellid aortic bulb[8] serves as an accessory pump (Jones 1968), and is part of the heart complex (Økland 1982), although confirmation is needed from developmental studies. Heart muscle fibres have also been observed in the proximal aorta in *Lymnaea stagnalis* (Plesch 1977).

An ampulla, which may function as a compensation chamber, is found in the initial part of the anterior aorta of ampullariid gastropods and lies within the pericardial cavity. Its lumen contains numerous amoebocytes which might be produced from specialised cells on its inner wall (Andrews 1965b).

An aortic bulb is present in some aplacophorans (Scheltema et al. 1994), but not in chitons.

6.3.4.1 Blood Flow through the Arteries

In vertebrates, the elasticity of the artery walls smooths the pulsating flow of blood from the heart to peripheral vessels, a well-known property known as the *Windkessel effect*. This effect appears to be present in some gastropods such as *Haliotis* (Bourne et al. 1990) and *Aplysia* (Brownell & Ligman 1992), but is best known in cephalopods where arterial elasticity has a significant haemodynamic influence on the vascular system. While both the aortae and venae cavae of *Octopus* are highly distensible, elastic structures, they have different properties and functions. The aorta functions as an elastic storage reservoir, in contrast to the vena cava which is a low-pressure capacitance[9] vein (Shadwick & Nilsson 1990). The blood pressure in the body increases markedly during locomotion (Wells 1979; Wells et al. 1987) and is compensated for by the elastic aorta (Shadwick & Nilsson 1990). Thus, differences in the 'stiffness' and thickness of aortae

in different cephalopods are correlated with activity levels. *Nautilus* has thin-walled high volume arteries, whereas in an active squid like *Nototodarus*, they are thick-walled, low-volume vessels (Gosline & Shadwick 1982). In other active squid, and to a lesser extent in *Sepia* and *Octopus*, the arteries are reinforced with a dense layer of elastic longitudinal fibres (e.g., Shadwick 1994).

A unique innervated, muscular, elastic accessory organ surrounds the mantle vessels of *Loligo* and *Sepia* and is thought to prevent blood being ejected from the mantle during high-pressure contraction of the mantle wall (Alexandrowicz 1962). This modification is absent in octopods, which have a less active lifestyle.

Many neurotransmitters and about half the cells in the suboesophageal lobes of the brain of *Octopus* control the vascular system, indicating that peripheral circulation and its control is of importance in cephalopods (Young 1971; Schipp 1987a; Schipp et al. 1991).

Detailed studies of the heterobranch *Aplysia* have shown how body movements can interfere with circulation, and how this is mitigated (Brownell & Ligman 1992). Muscular arterial valves control flow, and the stiffness of various vessels is intrinsically and extrinsically regulated (by nerves and peptides). In addition, the characteristics of the blood passing the abdominal ganglion can affect the performance of the cardiovascular system (Furgal & Brownell 1987).

6.3.4.2 Arterial Haemocoels

In all groups with a radula, the muscular and highly active odontophore lies within the buccal haemocoel, a space surrounded by a membrane continuous with the wall of the anterior aorta. A cephalic artery supplies blood to the 'brain' in cephalopods. Gaps in this membrane, especially anteriorly and ventrally, communicate with the venous haemocoel of the head and also allow muscles to pass through (Graham 1973).

6.3.5 Exchange Vessels – Capillaries

Capillaries are best developed in coleoid cephalopods (Smith 1963; Bourne et al. 1990), which have an effectively closed vascular system. *Nautilus* does not have fine capillaries comparable to those in coleoids, but instead has fine lacunae (Bourne 2010). The capillaries of coleoids are endothelial vessels of about 8.0 μm diameter (Browning & Casey-Smith 1981; Browning 1982). Vessels of similar diameter occur in at least some gastropods (Russell & Evans 1989; Bourne et al. 1990). For example, in *Haliotis*, there is a mesh of capillary-like sinuses over the posterior visceral mass (Bourne et al. 1990). In gastropods, the exchange system is best developed in some 'pulmonates.' Vessels from the anterior and posterior aortae channel blood into beds of capillary-like, but non-endothelial, sinuses associated with the various organs of the body (Fretter 1975b), such as sinuses of 10 μm diameter or less on the gut of *Lymnaea*. These end in ostioles surrounded by muscle fibres which can regulate the flow of blood into the haemocoel by closing the openings (Carriker & Bilstad 1946). It is not only the gut that has such networks. A mesh of small arterial vessels

---

[8] If the 'aortic' bulb is actually part of the heart and not the aorta, then both an anterior and posterior aorta are present in patellids. Lottiids do not have an 'aortic' bulb, and there are anterior and posterior aortae present.

[9] Capacitance is the opposite of elasticity in terms of the ability of a vessel to distend.

and capillaries has been described over the CNS in *Helix* (Pentreath & Cottrell 1970) and in other areas (Figure 6.10). Even in bivalves, the peripheral sinuses can be complex, with blood from the arteries circulating through progressively smaller branching vessels in the mantle where it passes into small haemal spaces which collect into the venous vessels running to the kidneys and the gills (McMahon et al. 1997).

The thickness of diffusion surfaces (including the basement membrane) is usually around 0.1 μm, similar to that in mammals (McMahon et al. 1997). The density of vessels in cephalopod muscle tissue is generally much less than that of mammals (e.g., Browning & Casey-Smith 1981), although a similar density is achieved in brain tissue and continually active muscle in cephalopods. Vessel densities are low in muscles that are not in continuous use, such as those involved in burst activity (Abbott & Bundgaard 1987), and hence there is an oxygen debt accumulated during exercise (Wells et al. 1983a; Houlihan et al. 1986).

6.3.6 The Venous System and Haemocoelic Spaces

Large blood sinuses form the venous blood system of polyplacophorans, monoplacophorans, aplacophorans and scaphopods and, to a lesser extent, bivalves and gastropods. They are the basis of the hydrostatic skeleton and 'hydraulic anchors' used in locomotion (e.g., Trueman 1983) (see Chapter 3). In some gastropod body parts and in the siphons of bivalves, muscular hydrostats may assist the hydraulic function of the blood or are even predominant (e.g., Voltzow 1985; Bourne et al. 1990; Golding et al. 2009b).

In gastropods, the transverse septum separating the visceral from the cephalopedal haemocoel enables the build-up of differential pressures (e.g., Fretter & Graham 1994). Stylommatophorans have more complete septa in the anterior haemocoel than other gastropods. For example, in *Helix pomatia*, there are, in addition to the transverse septum, two horizontal septa and muscular ostia controlled by muscle fibres. The more dorsal one encloses a blood space, while the second isolates the buccal mass and anterior oesophagus from the reproductive tract lying above (Kisker 1923).

Moderately large blood spaces are also found in cephalopods where they function as reservoirs, but the hydrostatic functions are performed by arrays of antagonistic muscles in the arms, funnel, and mantle wall. Like the arteries, the veins in coleoid cephalopods are also contractile, facilitating the return of blood to the hearts. This is particularly so in the highly contractile lateral venae cavae which supply blood to the branchial hearts in coleoids or, in nautiloids, directly into each gill, but even in these veins the blood pressure is low (Wells & Smith 1987).

6.3.7 Blood Flow to the Respiratory Surface(s) and Kidney(s)

Typically in molluscs, after supplying the body tissues, deoxygenated blood enters the respiratory plexus (in the gill[s], mantle, or lung), where it is reoxygenated before being collected in

the auricles and passed to the ventricle, from where the blood is pumped through the arterial system. As previously pointed out, this sequence is the opposite to that in the vertebrate circulatory system (Figure 6.2).

Some gastropods have convergently modified their circulatory patterns to use other body surfaces for respiration such as the outer mantle or kidneys, in addition to, or instead of, the gills or lung.

In gastropods, as in other molluscs, the heart receives oxygenated blood and, in vetigastropods and caenogastropods, most passes through the kidney(s); much less comes from the kidney in heterobranchs (Fretter 1975b). In caenogastropods, venous blood from the visceral haemocoel supplies the kidney and, on leaving the kidney, goes to the mantle and then the gill. In heterobranchs, the heart and kidney lie in the roof of the mantle cavity, and blood to the kidney arrives from the mantle. In all gastropods, the mantle cavity roof is highly vascularised and becomes a 'lung' in terrestrial taxa and some larger hygrophilans, although in most hygrophilans the lung is water-filled. Blood to the 'lung' is derived from the 'venous circle,' a series of interconnecting sinuses along the edge of the pulmonary sac, which itself receives blood from the visceral haemocoel. Most of the blood from the lung drains via the efferent pulmonary vein to the auricle, but some goes via the kidney (Figure 6.12).

6.3.8 Compensation Mechanisms

Compensation mechanisms are necessary for the blood displaced in sinuses when sudden contraction occurs. In snails with a large foot, a much larger volume of blood is displaced when the animal retracts into its shell. In carnivorous caenogastropods, displacement also occurs with proboscis retraction or, in bivalves, with the sudden withdrawal of large siphons or the foot. Spaces are required to accommodate the displaced fluid, and this occurs in different ways in different groups.

In caenogastropods, the blood sinuses in the nephridial gland appear to compensate for blood displaced from the mantle skirt (see Sections 6.6.3 and 6.7.7.4.2) when a snail retracts into its shell. Andrews and Taylor (1988) showed that the superficial sinuses that connect the nephridial gland vein and the dorsal kidney wall in *Littorina* open only under stress, notably when the snail retracts into its shell. Blood from the auricle, via the nephridial gland vein, enters the efferent renal vein and rectal sinus. On complete retraction of the animal into the shell, blood may escape to the mantle cavity through a 'blood pore' (see also Section 6.3.9 below) on the anal papilla (Fretter 1982; Andrews 2010). Stress on the ventricle is prevented by the expansion of the 'bulbous aortae' and the slowing or cessation of heartbeat.

The 'superficial sinuses' seen in a caenogastropod such as *Littorina* are absent in muricids and buccinids (Andrews 2010). In these neogastropods, the blood expelled from the cephalic haemocoel during proboscis retraction or retraction of the snail into its shell is accommodated mainly in secondary folds on the kidney dorsal wall. Blood from the heart and nephridial gland is rerouted, avoiding possible back pressure on the heart. In a muricid such as *Nucella*, a 'pressure valve' on the transverse branch of the dorsal afferent renal vein opens and shunts excess blood flows into the efferent renal vein and rectal sinus. Thus, the rectal sinus provides alternative storage, at least in *Nucella* and probably other muricids. At extreme pressures, some blood may be released from the rectal sinus into the mantle cavity through a 'blood pore' near the anus, as in some other gastropods (see next Section). Buccinids appear to differ from muricids in having a narrow rectal sinus with thick walls (Andrews 1992) and, on retraction, blood accumulates in a large vessel that connects with extensive blood spaces beneath the hypobranchial gland. The burrowing nassariid *Bullia* uses this same mechanism and also provides space in the upper coils of the shell (Andrews 2010). There is no blood pore known in buccinoideans, and this is thought to be the reason that some (such as *Buccinum*) cannot withdraw completely into their shell (Andrews 2010); some intertidal buccinids can do so, but these have not yet been studied.

As noted above, at least some neogastropods compensate when blood fills the secondary folds of the kidney. Andrews (2010) argued that the development of two different types of folds in the kidney of neogastropods (and other carnivorous caenogastropods) (see Section 6.7.7.4.3) was related to the development of a retractile proboscis housed in the cephalic haemocoel. When the proboscis or head-foot retracts, the blood displaced from the cephalic haemocoel moves into the dorsal afferent renal vein and is accommodated in the secondary kidney folds. In neogastropods, a septum separates the two branches of the afferent renal vein, so blood from the cephalopedal haemocoel flows into the dorsal vein and from there into the secondary folds which become distended when the head-foot is completely retracted. This blood drains back into the cephalopedal haemocoel when the proboscis (or head-foot) is extended.

While complex kidneys are also found in carnivorous groups of non-neogastropod caenogastropods such as Cypraeoidea, Velutinoidea, and Tonnoidea (Strong 2003; Andrews 2010), there are no detailed studies to date on their role, if any, in compensation. *Strombus*, which is sometimes considered to have a proboscis, actually has an extensible snout (Golding et al. 2009b) and is usually recognised as having undifferentiated kidney folds (e.g., Strong 2003). Nevertheless, there are two sets of folds in *Struthiolaria* (Morton 1956a) and *Strombus* (Little 1965a; Strong 2003). One set resembles the secondary folds and is supplied by a branch of the afferent renal vein. It is likely that these are not involved in compensating for blood displaced during the shortening of the snout as it does not retract into the cephalic haemocoel, but the head-foot is well-developed and active in these animals, and additional compensatory blood spaces may be required when it is retracted.

In neogastropods and naticids, haemocyanin aggregates form in the secondary folds of the kidney. In neogastropods, these aggregates only occur under high pressure after retraction of the proboscis, and then disperse when the blood pressure falls during protraction (Andrews 2010).

Architectonicoideans and pyramidelloideans are exceptional among heterobranch groups in developing a long proboscis (Fretter & Graham 1949; Maas 1965; Haszprunar 1985d; Ponder 1987; Andrews 2010). The large visceral haemocoel acts as the 'compensation sac' receiving the blood when the proboscis is retracted. A blood vessel connecting the visceral and cephalopedal haemocoels protects the heart from excess pressure during rapid retraction. When the proboscis is extended, blood is forced through this vessel and blood from the visceral haemocoel is carried in a second vessel to the sinuses in the roof of the mantle cavity. These latter sinuses are interconnected with vascularised folds in the kidney which in turn open to the efferent mantle vein and heart. This oxygenated blood is then supplied to the viscera. Muscular and proboscis activity is important in pushing blood around the body and facilitating the role of the heart (Andrews 2010).

6.3.9 Blood Release Mechanisms and the Addition of Seawater

Some gastropods possess a 'blood pore' that allows the escape of blood when a sudden contraction of the body may cause pressure which would otherwise damage the heart (Morris 1950; Mangum 1979; Fretter & Graham 1994). First shown in the fresh-water hygrophilan *Lymnaea stagnalis* (Lever & Bekius 1965), 'blood pores' have since been demonstrated on the rectal sinus in the caenogastropods *Littorina* and *Nucella*, and there is little doubt that these or similar mechanisms will be found in many other gastropods. Helicid snails and terrestrial slugs also lose blood through the pneumostome (e.g., Martin & Deyrup-Olsen 1982), and 'blood venting' via channels in the skin has been demonstrated in some terrestrial slugs (see Chapter 3); it can be induced by certain stimuli including several neural transmitters (Deyrup-Olsen & Martin 1982) (see also channel cells, Section 6.6.4.1).

At least two groups of caenogastropods (naticids and melongenids) have independently evolved the ability to enhance the expansion of the foot by sucking seawater through pores into the haemocoelic spaces in the foot tissue and then expelling it on retraction. In melongenids, water assists in the enlargement of the propodium during burrowing (Voltzow 1985). Mangum (1979) found blood in this expelled fluid in *Busycotypus canaliculatus* suggesting blood 'overflow' occurs. Water and blood mix in the foot in both melongenids and *Polinices* and haemocyanin concentrations are much lower in this pedal fluid than in other parts of the vascular system, suggesting that there is a haemocyanin filter. This filter appears to be in the secondary folds of the kidney of naticids (Andrews 1981).

6.3.10 Comparison of Vertebrate and Coleoid Cephalopod Vascular Systems

There are many parallels between the circulatory system of coleoid cephalopods and that of tetrapod vertebrates in anatomy, vascular structure, and the elasticity of the arterial walls, as well as increased heart size and arterial pressure with increased activity (Shadwick 1994). Some mechanisms that increase performance include the anatomical separation of the respiratory and systemic systems; this converges on the tetrapod vertebrate double circulatory system, and their pumps (see Section 6.3 above). Other specialisations include the combining of locomotory and circulatory pumping functions in the mantle wall of actively swimming squid, interruption of blood flow from the heart during intense jetting, a shift to anaerobic metabolism, and other anatomical specialisations that assist in reducing circulatory problems in the mantle wall that would normally be caused by the high pressures generated during locomotion (Shadwick 1994).

6.4   MOLLUSCAN BLOOD

Molluscan blood (more accurately referred to as haemolymph) typically contains nutrients, respiratory pigments, and blood cells (haemocytes). It circulates through the body, distributes nutrients and oxygen to cells, and removes their waste products. It also carries hormones and the haemocytes aid in bodily defence. Each of these functions is discussed in more detail below.

6.4.1   RESPIRATORY PIGMENTS

Three types of respiratory proteins that can reversibly bind oxygen occur in animals. These are globins (myoglobin, haemoglobin, and chlorocruorin), haemocyanin, and haemerythrin. While haemocyanin is common in molluscs and arthropods, haemerythrin does not occur in molluscs, although it is utilised by some other lophotrochozoans (see below). Globins are widely distributed, even occurring in some bacteria, but their distribution in animals, including molluscs, is patchy.

The two major kinds of respiratory proteins are haemoglobin and haemocyanin. When haemoglobin is oxygenated, it changes colour from dark red to bright red, while haemocyanin changes from colourless to blue. These differences are due to the metal incorporated in the protein which binds a dioxygen molecule; in haemoglobin, the metal is an iron atom located in a porphyrin ring (the haeme group), and in haemocyanin, oxygen is bound between two copper atoms.

In vertebrates and some invertebrates, including a few molluscs, haemoglobin is transported in the blood within cells (*erythrocytes*) (see Section 6.4.2.1). In most invertebrates with haemoglobin (including some molluscs), it is extracellular, being freely dissolved in the haemolymph, and each protein molecule is large, with up to 144 oxygen-binding sites. Extracellular haemocyanin molecules can be even larger, with up to 160 oxygen-binding sites (Decker et al. 2007).

The different types of respiratory pigments differ markedly in their physical properties, structure, molecular weight, their capacity for oxygen binding and the site where binding occurs. They all share the ability to bind oxygen reversibly and transport, store, or transfer it. The colour of respiratory compounds is often pronounced (hence the term respiratory pigments) and is due to specific absorption of wavelengths of light that change with the degree of oxygenation.

While both molluscs and arthropods utilise haemocyanin, haemoglobins are the primary oxygen carriers in annelids and in phoronids. Molluscan and arthropod haemocyanins are different structurally so may have evolved independently. Surprisingly, only a few molluscs use haemoglobin for oxygen transport, but several taxa (Polyplacophora, Monoplacophora, Gastropoda and Protobranchia) use myoglobin in specific musculature.

The presence of more than one respiratory protein in any one animal is generally rare, but occurs in many molluscs, as they have haemocyanin in the haemolymph and myoglobin in some tissues, notably the muscles of the buccal mass. Two types of respiratory protein in the same individual is not unique to molluscs; some annelids, for example, have haemoglobin in the blood and haemerythrin in the body wall muscles.

Phenoloxidases (enzymes that oxidise phenolic compounds to quinones) are found in all organisms. They are involved in many critical functions, including wound healing, and immunological processes such as antioxidative protection, using reactive quinones to attack and encapsulate invaders, as well as providing protection against ultraviolet light (Decker et al. 2006). Phenoloxidases and haemocyanins are similar, being in the same copper type-3 family and both have a very similar active site – one oxygen molecule bound between two copper atoms (Holde et al. 2001). Under certain conditions haemocyanin can function as a phenoloxidase (tyrosinase/catecholoxidase) (Terwilliger & Ryan 2006; Decker et al. 2007).

6.4.1.1   Haemocyanin

Haemocyanins have between 6 and 160 oxygen-binding sites, and some 'have the highest molecular cooperativity observed in nature' (Decker et al. 2007, p. 631). Haemocyanin is present in many molluscs, although in bivalves it is found only in protobranchs. Most autobranch bivalves lack an oxygen transport agent in their blood and may be oxyconformers. Haemocyanin is also absent in scaphopods and solenogasters (Lieb & Todt 2008). While present in most gastropods, it is absent in patellogastropods (e.g., Lieb & Todt 2008).

Molluscan haemocyanin is complex and varied in structure, more so than in arthropods. Recent studies on molluscan haemocyanins have revealed a great diversity of structure that has provided insights into their evolution within the phylum (Markl 2013). These studies have suggested that mollusc haemocyanin probably evolved during the late Precambrian (ca. 740 Ma, Tonian) (Decker et al. 2007). Recent studies have shown that the haemocyanin is more than an oxygen carrier, being linked to certain innate immune responses (Coates & Nairn 2014).

Each haemocyanin unit consists of multiple subunits grouped into cylindrical structures (Bonaventura & Bonaventura 1985), with most molluscan haemocyanins consisting of eight functional subunits. Two different haemocyanin isoforms are expressed in many molluscs, while others have only a single form (Markl 2013). Recently it was shown that a third isoform was present in the eggs and early embryos of *Sepia* (Thonig et al. 2014). Cephalopod haemocyanin typically consists of seven subunits as they lack the C-terminal functional unit found in other taxa, although in *Sepia* eight subunits have been reported, due to a duplication of FU-d (Markl 2013). Environmental conditions, notably pH and temperature, influence oxygen binding of haemocyanins (e.g., Decker et al. 2007). Thus, the upper and lower limits of the thermal tolerance of haemocyanin are limits for the aerobic metabolism and thus can fail during thermal extremes. Survival in such extremes requires shut down or a switch to anaerobic metabolism. It has, for example, been shown that *Sepia officinalis* switches to anaerobic energy production in critically high or low temperatures (Melzner et al. 2007b).

The haemocyanin of the protobranch bivalve *Nucula* differs from other molluscan haemocyanins, notably in being much less glycosylated (see Bergmann et al. 2007). Interestingly, analysis of the haemocyanin molecular structure suggests a closer relationship between gastropods and protobranchs than to cephalopods, with the common ancestor of protobranchs and gastropods estimated to have lived 494 million ± 50 million years ago (late Cambrian)[10] (Bergmann et al. 2007) adding support to recent genomic phylogenies (Kocot 2013). Bergmann et al. (2006) also analysed the complete gene structure of *Nautilus* haemocyanin and showed that it is generally similar to that of *Octopus*. Based on these data, it was calculated (using a 'molecular clock') that their common ancestor lived 415 ± 24 million years ago (early Devonian), which is in good agreement with the fossil record.

A special type of connective tissue cell (pore cells or rhogocytes) (see Section 6.5 below) synthesises, stores, and releases haemocyanin.

6.4.1.2 The Globins: Myoglobin and Haemoglobin

6.4.1.2.1 Myoglobin

Myoglobin is a form of, and functionally equivalent to, haemoglobin and is widely distributed in animals. In molluscs, it is present in muscle cells, particularly in the buccal mass, stomach, heart, nerve cells, and in adductor muscles of bivalves (Tentori 1970). Myoglobins have a greater oxygen affinity than haemocyanin and act as an oxygen storehouse.

Myoglobin contains a haeme group at the centre of a single-chain globular protein that typically consists of 140–160 amino acids. It differs from haemoglobin in that it does not have 'cooperative binding' of oxygen, where binding at one site in the molecule causes changes in the binding affinities at other sites. In addition, oxygen binding is not affected by the oxygen concentrations in surrounding tissue.

Myoglobins have been reported from chitons, scaphopods, gastropods, and bivalves and may be present in all molluscan classes. Molluscan myoglobins are diverse, with the chiton *Liolophura* having only 19%–20% sequence identity with caenogastropod myoglobins and a little higher (26%–29% identity) with 'opisthobranchs' (Suzuki et al. 1993).

The buccal muscles and muscular stomachs of gastropods have 'one of the most remarkable occurrences of red

muscle in invertebrate animals' (Suzuki & Imai 1998, p. 986). Myoglobin can be very abundant in gastropod buccal mass muscles, two or three times more so than in human heart or skeletal muscle (Read 1966). They are rather well known, with reviews including those by Terwilliger and Terwilliger (1985) and Suzuki and Imai (1998).

A peculiarity of molluscan myoglobin is the presence of a dimeric form which usually coexists with the monomeric form (Tentori 1970). In most chitons, both monomer and dimer forms are present, with the latter being disulphide-bonded and unique to polyplacophorans. These two kinds of myoglobin differ in their oxygen dissociation curves, with monomeric myoglobin (or haemoglobin) showing a hyperbolic curve, while the dimeric myoglobin shows a sigmoid curve similar to that seen in vertebrate and polychaete haemoglobins (Suzuki & Imai 1998). It has been suggested that these two types of myoglobins are related to the often intertidal habitat of chitons, with both aquatic and aerobic conditions (Smith et al. 1988). The monomeric myoglobin has a 20 times slower auto-oxidation rate than other molluscan myoglobins (Suzuki et al. 1993). Both monomeric and dimeric myoglobins have been recorded from one tonnoidean and some neogastropods, but the sampling is too poor to discuss their distribution in these groups. The sampling to date indicates that all other gastropods have only monomeric myoglobins (e.g., Suzuki & Imai 1998), although the myoglobin of the amphibious cerithioidean *Cerithidea rhyzophorarum* is unusual in that it converts from monomer to dimer on oxygenation (Matsuoka et al. 1996).

Several vetigastropods (haliotids, turbinids, trochids) have myoglobin that expresses an enzyme similar to indoleamine dioxygenase,[11] (IDO) found in vertebrates. Although unstable, IDO can take an oxygenated form with an absorption spectrum similar to that of haemoglobin and myoglobin (Suzuki & Imai 1998; Suzuki et al. 1998b; Suzuki 2003; Yuasa et al. 2007). It is possible that IDO is present in all or most vetigastropods, but sampling to date is insufficient to be definitive. This type of myoglobin is not present in other gastropods, including patellogastropods and neritimorphs (Suzuki et al. 1998a).

In bivalves, myoglobin has been reported from the adductor muscle, heart, and foot of a number of taxa (Alyakrinskaya 2003), with two kinds present in the adductor muscle of some (Suzuki & Imai 1998), and apparently absent from others (Alyakrinskaya 2003). In scaphopods, what is presumably myoglobin was identified in the buccal mass as haemoglobin (Manwell 1963).

6.4.1.2.2 Haemoglobin

Haemoglobin is found sporadically in molluscs. It may occur within tissues, or within cells, and extracellularly. It can also occur in erythrocytes (as in vertebrates; see Section 6.4.2.1) or as extracellular molecules of haemoglobin free in the blood. Molluscan haemoglobins can be relatively simple such as

---

[10] This is not in accord with the fossil record (see Chapter 13).

[11] A tryptophan-degrading enzyme containing haem.

those found in nerve tissue in *Aplysia*, to very complex like that in planorbid snails. The largest known intracellular haemoglobin molecule in any animal is from the bivalve *Barbatia* (Arcidae), where it occurs in erythrocytes (Grinich & Terwilliger 1980).

In bivalves, haemoglobin has evolved *de novo* in a few unrelated groups of autobranchs. Why these bivalves have haemoglobin instead of haemocyanin is not clear. The functional role of haemoglobin is typically to carry oxygen to tissues, but in some bivalves, and a gastropod with bacterial symbionts in the gills, a form of haemoglobin is involved in sulphur transfer. For example, in the protobranch bivalve *Solemya velum*, the gills (which contain sulphur-oxidising bacteria) have approximately equal quantities of two kinds of haemoglobins. When sulphide is lacking, the haemoglobin is oxygenated and deoxygenated in the normal way. When sulphide is present, approximately half of the haemoglobin is reversibly converted from the aquoferric form to the ferric form which binds sulphide, while the remainder functions normally. This apparently enables sulphur to be carried to the sulphur-oxidising symbiotic bacteria in the gill (Doeller et al. 1988). Haemoglobin has also been found in the gills of another protobranch, *Yoldia* (Stead & Thompson 2003). Another bivalve with haemoglobin and gill bacterial symbionts is the large vent and seep bivalve *Calyptogena*, which typically has two forms of haemoglobin, one in erythrocytes binds oxygen, and another which is dissolved in the blood is involved in binding sulphur (Zal et al. 2000). More recently, Decker et al. (2017) have reported four different intracellular, monomeric haemoglobin molecules in three other vesicomyid species (*Calyptogena valdiviae*, *Elenaconcha guiness*, and *Abyssogena southwardae*) all of which differ between species in their molecular weight. This variation, coupled with patterns of abundance in sulphide-rich sediments, led Decker et al. to suggest a role for haemoglobin in determining clam dominance at sites. Lucinids also contain sulphur-oxidising bacteria and have three types of haemoglobin in their gills (Read 1965), one of which binds sulphur (Kraus & Wittenberg 1990; Rivera et al. 2008), although the haemoglobin is confined to special bodies in the bacteriocytes in the gills (Frenkiel et al. 1996).

The provannid caenogastropod *Alviniconcha hessleri* is the only gastropod in which haemoglobin and bacterial symbionts have been recorded. This large-sized snail from hydrothermal vents in the western Pacific has chemoautotrophic, sulphide-oxidising bacteria and haemoglobin in bacteriocytes in the gills (Wittenberg & Stein 1995). The concentrations of gill haemoglobin for most bivalves with bacterial gill symbionts, and for *Alviniconcha*, is between 20–250 μmol/kg wet weight of gill tissue, but in some lucinids, it is more than 500 μmol/kg (Wittenberg & Stein 1995). It thus appears that bacteriocyte haemoglobin is a feature of both gastropod and bivalve bacterial symbioses (see Chapters 5, 9 and 15 for more details on bacterial symbiosis in bivalves).

Haemoglobins have been reported as the blood oxygen transport agent mainly in the pteriomorphian Arcidae (e.g., *Tegillarca granosa*) (Bao et al. 2013), and in some species of the arcoidean *Glycymeris* (Glycymeriidae) (Thomas 1975). It is also found in the archiheterodonts *Cardita* (e.g., Yonge 1969), *Astarte* (e.g., Alyakrinskaya 2003), and at least some crassatellids (Taylor et al. 2005). In gastropods, blood haemoglobin is found only in the heterobranch Planorbidae and the neritimorph Phenacolepadidae.

Haemoglobin can be intracellular and confined to the gills as in the non-symbiont containing protobranchs *Nucula* (Doeller et al. 1988) and *Yoldia* (Angelini et al. 1998; Alyakrinskaya 2003). In the Antarctic *Yoldia eightsi*, the haemoglobin is adapted to cold conditions and differs from other molluscan globins (Dewilde et al. 2003).

The haemoglobin molecule is made up of subunits of a globular protein. Each of these subunits consists of a protein chain tightly associated with a non-protein haeme group of an iron (Fe) ion held in a porphyrin ring. Oxygen binds to the charged iron ion. In invertebrates, the number of subunits can vary considerably, while in vertebrates it is always four. Structurally, the haemoglobin in different molluscs is quite diverse. That found in the blood of arcoidean bivalves differs from that found in planorbids in the latter being structurally similar to the muscle myoglobin of planorbids and other 'pulmonates.' This suggests that the blood haemoglobin in planorbids is derived from myoglobin, although the molecule has markedly increased in size (Lieb et al. 2006). This switch from haemocyanin to haemoglobin in planorbids is unique in higher gastropods, but has also occurred in the neritimorph family Phenacolepadidae, although the haemoglobin, present (in erythrocytes) in the latter group (Kano et al. 2002), has not been examined in detail to date. In both these cases, haemoglobin may have assisted their ancestors in oxygen-poor environments. Planorbids often live in stagnant fresh water and have haemoglobin with high oxygen affinity (Lieb et al. 2006), while phenacolepadids typically live in anoxic sediments beneath wood, coral, and rock. In the latter case, these animals have also invaded deep-sea chemosynthetic habitats which are poor in oxygen (Warén & Bouchet 1993, 2001; Kano et al. 2002). Other gastropods live in oxygen-poor environments such as in hydrothermal vents or deep under stones, but these have yet to be investigated. As noted above, another deep-sea vent gastropod, *Alviniconcha hessleri*, has the haemoglobin confined to gill bacteriocytes, although other provannids do not appear to have obvious haemoglobin.

The haemoglobins of autobranch bivalves fall into three main groups, two of them analysed by Suzuki et al. (1993). One group, the arcid erythrocyte haemoglobins, contains homodimeric, heterodimeric, tetrameric, and didomain chains and, in the second group, the haemoglobins are made up of three haemoglobin chains. This latter group is found in *Calyptogena* and lucinids, both of which have bacterial symbionts in their gills (Suzuki et al. 1993). The third group, not included in the Suzuki et al. (1993) analysis, are the very large haemoglobin molecules of archiheterodonts such as *Cardita* (Terwilliger et al. 1978) and *Astarte* (Yager et al. 1982) with molecular weights of $8–12 \times 10^6$ g·mol^{-1} and which have an unusual quaternary structure. In comparison, haemoglobins of other bivalves are much smaller; arcids, for example, having molecular weights of $34{,}000–64{,}000 \times 10^3$ g·mol^{-1} (Nagel 1985). These smaller haemoglobins are closely related to the myoglobin in the same taxa (Lieb et al. 2006).

Some other bivalves, particularly those living in low oxygen environments, may also be found to have haemoglobin. For example, the colour of the galeommatid *Barrimysia* suggests that haemoglobin may be present in its tissues (Kano & Haga 2006).

Haemoglobin occurs in Solenogastres, but not Caudofoveata, and occurs in erythrocytes in at least some taxa (e.g., Salvini-Plawen 1978a) and has undergone only preliminary investigation (Lieb & Todt 2008).

6.4.2 Haemocytes

In addition to blood pigments, haemolymph also contains several kinds of blood cells (haemocytes) which have immunogenic or phagocytic functions (Xu et al. 2017) (see also Chapter 2). They are sometimes distinguished on the basis of cytoplasm structure and morphology (e.g., Cavalcanti et al. 2012), but their uniqueness is suspect, as different haemocyte morphology can reflect different physiological states of cells (Martin et al. 2007). Two main kinds of haemocytes are usually recognised – *granulocytes* and smaller *hyalinocytes*, the former mainly engaged in phagocytosis. Many definitions rely on the amount of cytoplasm relative to the nucleus, their general morphology which can be amoeboid (amoebocytes) or spherical, their colour, or the number of granules present (granulocytes have many granules, while hyalinocytes have few).

Some seasonal and sexual differences in the abundance of haemocytes have been noted in some taxa.

Haemocytes are produced (by haematopoiesis) in the blood (by cell division) and in the connective tissue and epithelia associated with the blood vessels and sinuses, or within the pericardium and some other parts of the body (Fornůsek & Větvička 1992; Pila et al. 2016). Serous cells produced in the pericardial gland (also known as Keber's gland) are usually recognised as a different kind of haemocyte in bivalves (e.g., Kuchel et al. 2010; Li et al. 2015c), or are sometimes thought to be rhogocytes (Haszprunar 1996b; Hine 1999) (see next Section).

Cephalopods only have a single type of haemocyte, and these are produced in 'white bodies' or Hensen's glands (Claes 1996) located behind the eyes in the orbital pit area. Nevertheless, two subtypes of haemocytes were recognised by Castellanos-Martínez (2013) in *Octopus*, large and small granulocytes. Although both had defensive abilities, the large form had greater phagocytic ability and a higher oxidative burst (see Chapter 2) than the smaller ones.

6.4.2.1 Erythrocytes

Some taxa contain haemoglobin carried in the blood in erythrocytes (see also Section 6.4.1.2.2), which may or may not have nuclei. Erythrocytes are typical of vertebrates (but not other chordates), and while uncommon in invertebrates, are present in some molluscs, as well as in some annelids, brachiopods, echiurans, priapulids, sipunculans, and echinoderms (Cohen & Tamburri 1998). These erythrocytes vary in morphology from group to group. In molluscs, they have been reported from only one group of gastropods, the phenacolepadids (e.g., Fretter 1984b), several bivalve taxa, including arcids, vesicomyids (see Section 6.4.1.2.2), *Xylophaga* (Ansell & Nair 1968), and most species of *Astarte* (Alyakrinskaya 2003), and solenogasters (e.g., Salvini-Plawen 1978a).

6.4.2.2 Immune Response and Other Functions

The circulatory system also serves as part of the immune system, and haemocytes play an important role, as outlined in Chapter 2. Amoebocytes in the haemocoel attack invasive microbes, and they encapsulate larger invading microorganisms, while invaders such as parasites too large to be phagocytised by single cells are encapsulated by groups of cells (Ottaviani 2006). Amoebocytes also phagocytose damaged cells, particles, and stray bacteria etc.

6.5 PORE CELLS (RHOGOCYTES)

Pore cells (or *rhogocytes*) (Figure 6.16) are unique to molluscs and occur in all groups. They are found throughout the haemocoel, connective tissue, and, especially, in the mantle skirt (Haszprunar 1996b). They are sometimes called 'brown cells,' and have filtration sites on their surface. These often solitary cells are responsible for the synthesis and recycling of haemocyanin (e.g., Sminia & Boer 1973; Albrecht et al. 2001) as well as haemoglobin in planorbids (Sminia et al. 1972; Lieb et al. 2006). In addition, pore cells also have an important role in the physiology of heavy metal accumulation and its detoxification (see Haszprunar 1996b for review) (Figure 6.16).

Other functions attributed to pore cells include glycogen storage, phagocytosis, and collagen production (e.g., Luchtel et al. 1984). Filtration pressure in the pore cells is probably caused by endocytosis (compared to muscular activity in the somewhat similar podocytes) (Haszprunar 1996b). Pore cells resemble podocytes (see Section 6.6.2), but differ in being a mesenchymate cell type surrounded by an extracellular matrix, lacking cell processes, and in their location. The pore cells and podocytes also resemble terminal cells (cyrtocytes) seen in larval protonephridia, and the arthropod nephrocytes, and these four types of cells may have a common origin (Haszprunar 1996b). So-called calcium-cells may be a specific type of pore cell (G. Haszprunar, pers. comm. 2018).

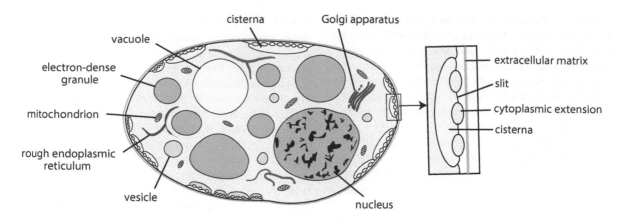

FIGURE 6.16 A generalised molluscan rhogocyte. Redrawn and modified from Haszprunar, G., *J. Molluscan Stud.*, 62, 185–211, 1996b, and Luchtel, D.L. and Deyrup-Olsen, I., Body wall: Form and function, in Barker, G.M. (ed.), *The Biology of Terrestrial Molluscs*, pp. 147–178, CABI Publishing, Wallingford, UK, 2001.

6.6 THE EXCRETORY SYSTEM

Excretion has three main roles: (1) ridding body fluids of metabolic wastes that would be toxic if allowed to accumulate, (2) retaining vital macromolecules and water in the body, and (3) maintaining osmotic pressure. To undertake those roles, the cells making up the molluscan excretory systems carry out three basic functions: (1) filtration of the primary urine from the haemocoel, (2) transport of waste products from the haemolymph into the urine, and (3) reabsorption of useful metabolites (and water if necessary) from the urine.

There are two general kinds of excretory organs, epithelial and mesodermal, the latter being tubular structures generally referred to as kidneys or nephridia.

Most epithelia restrict the flow of solutes and water, but some are specialised to transport solutes actively (e.g., ions, glucose, amino acids, urea), and even wastes such as heavy metals, via organs such as the gills or skin, as well as the kidneys. Thus, although we concentrate on pericardial and kidney structures here, it is important to bear in mind that epithelia on the gills and other surfaces may also be involved in excretion to a greater or lesser extent in different taxa. Excretory activities in molluscs can also sometimes be undertaken by organs in the body other than the epithelia or kidneys, as for example, the digestive gland and the anal gland in some neogastropods (see Section 6.7.7.4.4).

The heart-kidney complex (the heart, kidneys, and associated blood vessels) carry out the related tasks of filtration, reabsorption, and excretion, each accomplished by specialised structures and cells that filter the haemolymph and then transport, accumulate, and excrete the waste products.

Within the pericardium, the molluscan heart is covered by an epithelium (the epicardium). Beneath this lies an extracellular matrix on which rest the podocytes (see Section 6.6.2) involved in filtration (see Section 6.6.1), although important exceptions occur, as described below.

The filtrate, the primary urine, accumulates in the pericardial cavity and drains through the renopericardial duct(s) into the kidney(s), where reabsorption of filterable materials (e.g., water, glucose) occurs, and secretions from the kidney cells are added to form the final urine. Resorption in lower gastropods is confined to the right kidney. In apogastropods, which only have a left kidney, absorption in most caenogastropods occurs in the nephridial gland (see Sections 6.6.3 and 6.7.7.4.2) and excretion in the remainder of the kidney, while in heterobranchs absorption is not confined to a particular part of the kidney and most nitrogenous excretion occurs in the digestive gland (see Section 6.7.7.5).

Primitively, filtration probably took place in the pericardial wall of the auricles, and the filtrate was transferred via a pair of coelomic ducts into the mantle cavity (Haszprunar et al. 2008). This is essentially the system found in aplacophorans (although few details of the processes of excretion in those groups are known) and chitons (see below). These coelomic ducts are metanephridia and are homologous with the kidneys in other molluscs, which are plesiomorphically tubular and structurally similar to those found in annelids and sipunculans. In molluscs they connect to the pericardium through the renopericardial ducts. The kidneys (nephridia, renal organs) open into the mantle cavity or mantle groove through small apertures, the kidney openings (or *nephridiopores*) through which the excretory products are released. These mesodermal tubular excretory structures probably originally evolved for the excretion of solutes rather than nitrogenous waste because ammonia is readily excreted across epithelia (e.g., on the gills).

Molluscan kidneys are associated with the reproductive system in Monoplacophora, Scaphopoda, and patello- and vetigastropods and some bivalves, and members of these taxa generally release their gametes through the kidney opening into the environment (see also Chapter 8). Rudiments of the right kidney are incorporated in the genital ducts of neritimorphs and caenogastropods (see Chapter 8).

In some heterobranch gastropods and bivalves the protonephridia of the larvae (see Box 6.4 and Chapter 8) are said to be derived from both ectoderm and mesoderm, while in the remaining molluscs, they are ectodermal in origin (Raven 1966). The protonephridia are ultimately reabsorbed and replaced in adults by a mesodermal coelomic kidney, a metanephridium (Box 6.4) (Andrews 1988), which is typically derived from the mesodermal pericardial mass. An exception to this was thought to be the 'kidney' in the tiny heterobranch slug *Rhodope* which has features similar to a protonephridium (Riedl 1960). Haszprunar (1997) considered it to probably be a modified metanephridium because of the lack of auricular filtration sites. Protonephridia undertake filtration via terminal flame cells, while the metanephridia require filtration (by podocytes) from the blood to a coelomic system. Metanephridia use muscular pressure, and processing of the filtrate occurs in the coelomic duct which forms the kidney (Ruppert & Smith 1988; Haszprunar 1992b).

There is considerable diversity in the morphology of kidneys in molluscs. Although their anatomy and function are reasonably well understood for some taxa (reviews by Andrews 1981, 1988), many gaps still exist regarding significant taxa, particularly in marine gastropods.

Monoplacophorans alone have three to seven pairs of kidneys, and chitons, bivalves, and scaphopods have a pair of kidneys which are the same in appearance and function. In gastropods, there are either two kidneys that are morphologically and functionally very different, or only one, the left (see Section 6.7.7). The kidneys in cephalopods are highly modified (see Section 6.7.6).

In many molluscs, the kidneys are differentiated into two (sometimes more) distinct areas. Usually one has epithelia typical of resorptive cells (i.e., with microvilli and basal infolding), and the other is lined with excretory cells with numerous vacuoles containing granules (or concretions) of varying size, shape, and number (see Figure 6.17). Excretory products are removed from the blood by the excretory cells which have a large base with long basal processes to increase the surface area in contact with the blood. Their basal cytoplasm is also rich in mitochondria, and uric acid crystals are precipitated in excretory vacuoles (Andrews 1988).

Simple tubular kidneys are seen in the early ontogeny of bivalves, but in adult bivalves, as their length increases, they become U-shaped (Raven 1966). Typically the two limbs are structurally and functionally different – the proximal limb being resorptive and the distal one excretory. While U-shaped tubular kidneys occur in all bivalves, sac-shaped kidneys are seen in scaphopods, gastropods, and cephalopods (Figure 6.18).

In monoplacophorans, the location of the excretory organs in the mantle skirt rather than within the visceral mass as in polyplacophorans, is a significant change. Similarly, in many gastropods, they are placed in the roof of the mantle cavity, and this enables the kidneys to serve as respiratory surfaces (see Chapter 4).

The course of blood circulation and urine formation in *Littorina* is shown in Figure 6.24. The structure and function of the nephridial gland vein and the auricle in *Littorina* has been described in detail by Andrews and Taylor (1988). This vein dilates when the auricle contracts, as does the ventricle, and contracts when the auricle expands. The auricle is subdivided by muscles into two compartments, one of which has the 'filtration chambers' containing podocytes, where the primary urine is produced. This compartment connects with both the efferent ctenidial and nephridial gland veins, but not the ventricle. The other compartment opens to the nephridial gland vein and the ventricle, but not the efferent ctenidial vein (Andrews 1988; Andrews & Taylor 1988). The tubules of the nephridial gland (see Sections 6.6.3 and 6.7.7.4.2) are lined with mucous cells and ciliated resorptive

---

BOX 6.4 LARVAL AND ADULT MESODERMAL EXCRETORY STRUCTURES

Two main types of mesodermal excretory structures are recognised (mainly from Ruppert & Smith 1988):

Protonephridia – blind tubules that do not have a direct connection to the lumen of a coelomic cavity. Fluid is drawn into the tube by terminal cilia or flagella (flame cells, cyrtocytes, or solenocytes), and then it passes along the tube and exits via a pore. There are several different kinds recognised, depending on the complexity of the terminal ciliated or flagellate cells. Protonephridia are found in the larval stages of phoronids, annelids, and molluscs (except cephalopods), and in adult Platyhelminthes, Gnathifera, scalidophoran Nemathelminthes, Entoprocta, and a few polychaetes. The fluid inside the tubule is at a negative pressure relative to the fluid of the primary body cavity.

Metanephridia – found in animals such as molluscs, annelids, arthropods, and deuterostomes, which have both a coelomic and vascular space. They have a direct connection with the coelom via a ciliated opening (nephridiostome) which draws in fluid which is then expelled via the kidney opening (= kidney or renal pore, nephridiopore). As described in Box 6.5, the primary urine (or pro-urine) is formed by filtration, under hydrostatic pressure, of vascular fluid into the coelomic space, and secondary urine is formed by the movement of coelomic fluid into the metanephridium. This is the type of excretory structure (kidney) seen in adult molluscs (and in other groups such as brachiozoans, annelids, arthropods, and deuterostomes).

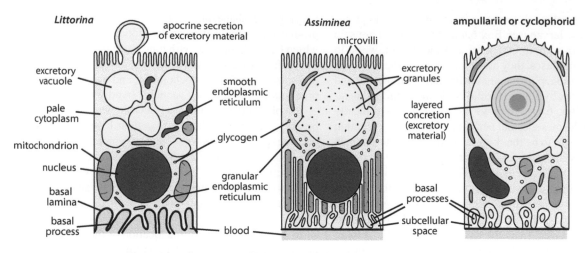

Some types of excretory cells found in the dorsal wall of the kidney of caenogastropods

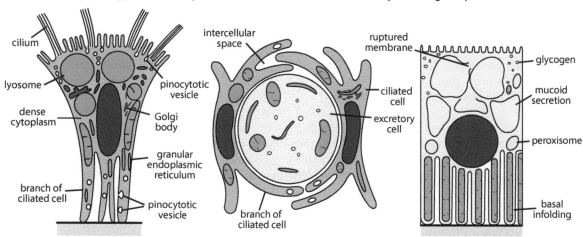

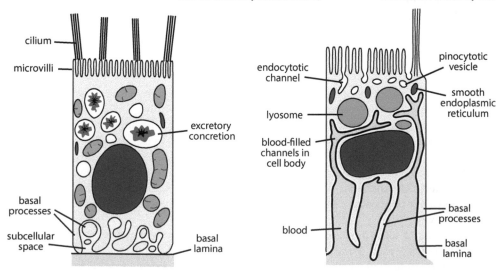

FIGURE 6.17 Kidney cells of some caenogastropods and vetigastropods. Redrawn and modified from Andrews, E.B., *J. Molluscan Stud.*, 47, 248–289, 1981.

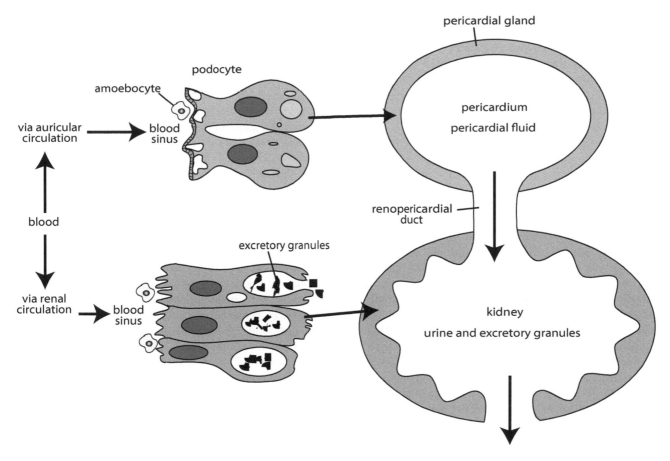

FIGURE 6.18 A generalised diagram of bivalve excretory pathways. Redrawn and modified from Morse, M.P., *Am. Zool.*, 27, 737–746, 1987b.

BOX 6.5   URINE FORMATION

Urine – the primary urine forms in molluscs by filtration of the blood (haemolymph) through a fine sieve and is modified to a greater or lesser extent to form the secondary urine before it is expelled.

> *Filtration* – the formation of primary urine (or pro-urine, or filtrate) is a process where fluid passes through pores of a membrane filter, usually formed by podocytes (see Section 6.6.2), in response to hydrostatic or osmotic pressure; notably the arterial blood pressure is higher (due to the pumping of the ventricle) than the blood osmotic pressure. In molluscs, filtration typically occurs through the walls of the heart (usually the auricle[s]). with the primary urine collecting in the pericardium. It is carried from the pericardium to the kidney, via the renopericardial duct, for modification, and eventual elimination.
>
> Filtration occurs when the pore size is small enough to prevent colloids and large solutes (e.g., proteins) passing through. Apart from size (molecular weight), this process is mostly non-selective.

The processes of urine modification include:

> *Reabsorption* – removal of water and nutrients (glucose, amino acids) and other solutes (urea, chlorine, and sodium ions etc.) from the blood or urine. Water can be passively reabsorbed by osmosis in some taxa, but in others, the structures carrying the urine are relatively impermeable to water with the urine being hypoosmotic to the blood. Concentration to produce hyperosmotic urine may occur by osmosis or by pumping ions or other solutes into it, or both.
>
> *Secretion* – the liberation of specific waste products into the urine including nitrogenous waste (ammonia, urea, uric acid, some organic acids), ions, organic acids, or bases, resulting in hyperosmotic urine.
>
> *Excretion* – the elimination of waste products, and other materials not needed, from the body.

cells (Andrews 1981, 1988). There is experimental evidence for resorption in *Littorina* in both the nephridial gland and the dorsal wall of the kidney (Andrews 1988).

The move from the marine to non-marine environments necessitated major structural and physiological changes in the excretory system, some of which will be revisited in the taxon chapters in Volume 2, but the general aspects are outlined below (see Section 6.6.4.1).

6.6.1 FILTRATION

The primary filtrate has a similar osmotic pressure to the blood, including the concentrations of Na^+ and Cl^-, and it contains organic molecules, including glucose and amino acids. The secondary urine is altered through reabsorption and secretion and thus differs from the blood. Resorption by the kidney(s) removes useful solutes from the urine, and excretory products are released into the urine from vacuoles in the renal *excretory cells*. In terrestrial or amphibious gastropods, there is control over the amount of water being excreted through reabsorption, and in fresh-water bivalves and gastropods, the urine is rendered hypoosmotic to the blood by the removal of ions.

There may be three different filters involved in producing the filtrate from the blood. In the auricle, some fluid is forced through the basal lamina (an extracellular matrix), which thus acts as a fine filter, due to the difference in pressure in the auricle and the pericardium. A coarser filter is provided by the podocytes (see next Section) which allows fluid from the blood to pass into the subcellular space between the pedicels of the podocytes and their main bodies. The third type of filter is provided by narrow channels that allow fluid to move through the heart wall. Some channels penetrate the podocytes (or other epicardial cells) enabling the primary urine to pass into the pericardial cavity.

In some taxa, urine formation occurs without filtration. This is apparently the case with the right kidney of the vetigastropod lepetelloid limpets (Haszprunar & McLean 1996), the left kidney of fissurellids (Andrews 1985), and presumably the minute monoplacophoran *Micropilina* that lacks a heart (Haszprunar & Schaefer 1997). Similarly, in the heart-less sacoglossan *Alderia modesta*, the urine is apparently formed directly in the kidney without filtration (Fahrner 2002).

6.6.2 PODOCYTES

Podocytes are special sieve cells involved in filtration associated with the heart and/or pericardium where they lie on the basal lamina. Pedicels are processes that extend from the cell and intertwine with those from adjacent cells. The spaces in the pedicel mesh can be subdivided by diaphragms that extend from the sides of the pedicels. The mesh formed from both kinds of processes lies beneath the podocytes and extends over the basal lamina in contact with the blood (typically in the auricle).

Podocytes are widespread in invertebrate animals, being present in annelids, phoronids and brachiopods

(Kuzmina & Malakhov 2015), phylactolaematan Bryozoa, and sipunculans, as well as in crustaceans and deuterostomes (Haszprunar 1996b). In most molluscs, podocytes are found only on the epicardium of the auricle(s), and this condition is probably plesiomorphic (Andrews 1988; Fretter & Graham 1994), being found in chitons (Økland 1980), solenogasters[12] (Reynolds et al. 1993), monoplacophorans (Haszprunar & Schaefer 1997), lower bivalves, and many gastropods (Morse & Reynolds 1996). They are part of the pericardial epithelium in scaphopods (Reynolds 1990b); in patellogastropods (Økland 1982), they are on both the ventricle and auricles, and in Cyclophoridae, the wall of the ventricle is the main site of filtration (Andrews & Little 1972). Podocytes may also be clustered in specialised structures, including the *'pericardial glands'* (Keber's organs). In protobranch and pteriomorphian bivalves, the pericardial glands are situated on the auricles (Figure 6.20), while in autobranch bivalves they are located in an anterodorsal position to the pericardial cavity (Figure 6.19). In cephalopods, convergent 'pericardial appendages' also occur in nautiloids and in the branchial heart of coleoids (see below).

In gastropods, podocytes have been well studied in a few taxa such as the vetigastropod *Gibbula* (Trochidae) (Andrews 1976b) and the fresh-water caenogastropod *Viviparus* (Andrews 1979a). In *Gibbula*, the numerous pedicels are around 500 nm long and 100 nm in diameter with narrow gaps between them (about 33 nm) with diaphragms projecting from the pedicels to make the gaps even smaller.

The sieves of podocytes may have very small openings (approx. 20 nm in bivalves) (Morse 1987b) that may be edged by cell surface proteins that further block large macromolecules, ensuring that they remain in the blood. Charges on the surface of the filter are also thought to either attract or repulse solutes (Martin 1983; Meyhöfer et al. 1985). Podocytes in bivalves have also been shown to be possible sites of absorption and secretion in addition to filtration (e.g., Jennings 1984; Morse 1987b; Andrews & Jennings 1993). The absorbed material is either degraded or sequestered in lysosomes, which may be stored before being shed into the urine (Andrews & Jennings 1993).

In most molluscs, the podocytes lie in oxygenated blood, but in cephalopods they do not, and are associated with veins (see below). The cephalopod 'auricles' are actually expanded efferent branchial veins, while the branchial hearts are the presumed homologues of the auricles (see Section 6.3.2).

In a few tiny taxa where the heart and pericardium have been lost, as in the monoplacophoran *Micropilina* (Haszprunar & Schaefer 1996, 1997) and in some 'opisthobranchs' (see below), podocytes are, not surprisingly, absent.

Podocytes in the outer pericardial wall are known only in scaphopods, heterodont bivalves, cephalopods, and in one nudibranch gastropod. In *Nautilus*, they are found in the 'pericardial glands' (see Section 6.7.6.1), and in coleoid cephalopods, in the homologous branchial heart appendages (see Section 6.7.6.2). In bivalves, groups of podocytes on the auricular wall are usually called 'auricular glands'

---

[12] Podocytes have not yet been found in Caudofoveata.

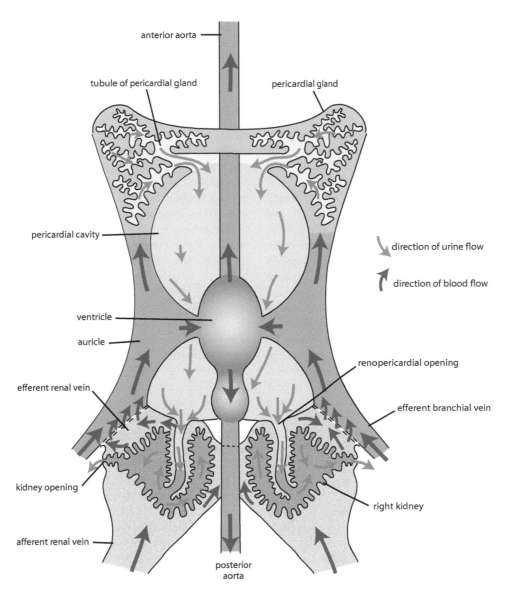

anterior aorta

tubule of pericardial gland

pericardial gland

pericardial cavity

direction of urine flow

direction of blood flow

ventricle

auricle

renopericardial opening

efferent renal vein

efferent branchial vein

kidney opening

right kidney

afferent renal vein

posterior aorta

Heart-kidney complex of *Scrobicularia plana* - with rectum removed

FIGURE 6.19 The renopericardial complex of the estuarine bivalve *Scrobicularia plana*, with the rectum removed. Redrawn and modified from Andrews, E.B. and Jennings, K.H., *J. Molluscan Stud.*, 59, 223–257, 1993.

and those in the pericardial wall 'pericardial glands.' The pericardial glands are formed from much-folded auricular epicardium that creates complex blind-ending ducts (Meyhöfer et al. 1985; Andrews & Jennings 1993; Meyhöfer & Morse 1996) (Figure 6.19). Podocytes in the outer pericardial epithelium may coexist with those on the auricle(s), as in many bivalves (Andrews & Jennings 1993; Morse & Zardus 1997), and in some 'opisthobranchs' (see Section 6.7.7.5.1). While the podocyte-containing 'pericardial glands' of bivalves are thought to be an adaptation to estuarine or fresh-water habitats (Andrews & Jennings 1993), the function of the pericardial wall podocytes in the marine nudibranch *Hypselodoris* is not known (Fahrner & Haszprunar 2002a).

While the plesiomorphic condition of auricular podocytes is seen in patellogastropods, vetigastropods, neritimorphs,[13] and many caenogastropods, there is a convergent evolutionary tendency to separate the functions of filtering and pumping blood (Andrews 1981, 1985). Thus, there are special filtration chambers in some vetigastropods such as *Emarginula* and *Haliotis* and more complex filtration pouches in some caenogastropods (Fretter & Graham 1994). In these latter structures, there is an enlarged subcellular space under groups of podocytes, the floor of which is made up of the pedicels of the podocyte, and the roof of squamous epithelial cells is raised above the basal lamina. Although

---

[13] Not confirmed in terrestrial neritimorphs

details are only known in a few taxa of marine caenogastropods, these pouches are more common and less specialised in lower caenogastropods than in higher groups (Fretter & Graham 1994). The filtration pouches also involve modifications to the muscles of the auricle so that the peripheral part involved in filtration is isolated from the central part involved in moving blood to the ventricle. Part of this change also involved incorporating blood flow from the nephridial gland in the auricular circulation (see next Section), as described in detail in *Littorina* (Andrews & Taylor 1988) (see Section 6.7.7.4.2).

Adaptations to life in fresh water include podocytes being lost in some non-marine gastropods (see Section 6.6.4.1, and relevant subsections under 6.7.7.4.5 and 6.7.7.5.2), although the patterns of loss appear to be haphazard. For example, in the fresh-water 'architaenioglossans' the podocytes are in chamber-like appendages on the inner wall of the auricle in Viviparidae (Andrews 1979a), while in Ampullariidae, the podocytes are lost, and filtration occurs via extracellular channels (Andrews 1976a). In the marine interstitial cephalaspidean *Philinoglossa*, podocytes are also lacking, and filtration is facilitated by 'slashed cells' (podocyte-like cells), located on a part of the outer pericardial wall adjacent to the kidney (Bartolomaeus 1997). In contrast, estuarine and fresh-water bivalves appear to have concentrated podocytes into pericardial glands (Andrews & Jennings 1993) (see Section 6.7.4.1). These and other excretory adaptations of fresh-water and terrestrial gastropods are discussed in the taxonomic survey below.

6.6.3   The Nephridial Gland

An important structure associated with the kidney in marine caenogastropods (see Section 6.7.7.4.2) and some vetigastropods (see Section 6.7.7.2) is the nephridial gland, which lies between the kidney and pericardium. This structure is comprised of connective tissue, muscle fibres, and blood spaces and is lined with ciliated cells. The invaginations of the gland contain blood spaces which come into close contact with urine in the kidney lumen.

There is a 'tidal flow' of blood from the auricle to the blood spaces of the nephridial gland, but, contrary to many earlier accounts, this does not pass into the kidney proper (Andrews 2010) (see Section 6.7.7.4.2). In caenogastropods which possess a nephridial gland, the primary urine passes down the renopericardial duct to the nephridial gland where organic solutes are resorbed from the urine by the epithelium lining the nephridial gland tubules (e.g., Andrews & Taylor 1988, 1990; Taylor & Andrews 1988). Additional solutes are extracted by the dorsal wall of the kidney. In addition, some metabolites and toxins are removed by phagocytes in the nephridial gland sinuses (the so-called 'blood gland'). The phagocytes may also play a role in regulating the ionic composition of the blood and may also be involved in the catabolism of toxins and macromolecules such as haemocyanin (Andrews 1988, 2010). These processes ensure that most solutes are returned to the blood before it enters the aortic system (Andrews 1988). In some caenogastropods, the large

nephridial gland vein and 'blood gland' accommodate displaced blood when the animal retracts quickly, a role also performed by secondary folds in the roof of the kidney in some taxa (see Section 6.3.8).

6.6.4   Excretory Challenges

As molluscs evolved, their excretory systems adapted to new challenges. For example, the evolution of carnivory (i.e., resulting in a protein-rich diet) in cephalopods and many groups of gastropods, resulted in high levels of toxic nitrogenous waste in the body, which in turn required a more efficient excretory system (see Sections 6.6.4.2 and 6.6.4.2.1). The move from the marine environment, where blood and urine are close to isotonic, to fresh-water and terrestrial systems, posed different challenges (see next Section).

6.6.4.1   Estuarine, Fresh-Water, Supralittoral, and Terrestrial Molluscs

Here, we briefly explore some general issues, and details of particular groups are covered in the comparative section below. Estuarine, fresh-water, supralittoral, and terrestrial environments all provide their own particular suite of challenges which must be addressed by the animals living in them. Because many different evolutionary lineages of mostly bivalves and gastropods moved into such environments, their responses to osmotic and excretory issues encountered have been independently met in a variety of ways.

Marine molluscs, like most marine invertebrates, are isoosmotic to seawater and thus, in general, have low filtration rates. In contrast, fresh-water molluscs have high filtration rates, and the rates in terrestrial taxa are somewhat intermediate. Estuarine habitats are particularly challenging because they often undergo rapid and considerable changes in salinity.

Gastropods and bivalves that live in fresh-water environments undergo osmotic stress because the osmolarity of their haemolymph is greater than their environment due to the greater internal concentration of ions (mainly K^+, Ca^{++}, Mg^{++}, Cl^-, CO_3^-). The influx of water from the environment by osmosis is compensated for by producing copious amounts of urine. The loss of ions, because of this increased urine production, is compensated for by active uptake (resorption) by the kidney. Similarly, the blood in fresh-water taxa is at lower osmotic pressures than marine molluscs, but the mechanisms by which such physiological adaptations have been achieved vary in different groups. Not all of the water taken into the body of fresh-water gastropods is due to osmosis, with as much as 20%–30% of the urine production in *Lymnaea* due to water taken in orally while feeding (With 1996).

As with other terrestrial animals, the excretion of ammonia is not a viable option for terrestrial gastropods, but instead, nitrogenous waste is usually excreted as uric acid. The most successful by far of the terrestrial gastropod groups is the Stylommatophora. These snails and slugs possess behavioural and physiological adaptations (see Section 6.7.7.5.2.3) to avoid or impede water loss, as well as structural ones such as the contractile pneumostome, which reduces evaporation from the mantle cavity. The pneumostome serves a double purpose as stylommatophorans can take up water through the pneumostome which enters the gut via the anus,[14] by way of a rectal pump. This serves as a reserve for dry conditions to help stabilise the blood, or water can be extruded from the anus and used to help keep the external body surfaces moist, a habit found in many stylommatophoran taxa (Neuckel 1985). Urine stored in the mantle cavity can provide an additional source of fluid that can act as a water reservoir for subsequent resorption (e.g., Blinn 1964; Smith 1981a).

Terrestrial gastropods can be vulnerable to excess hydration, for example, during flooding or heavy rain. Excess water can be removed from the body by increasing urine output (Luchtel & Deyrup-Olsen 2001), and in some terrestrial slugs, it is removed through secretion via epithelial channel cells (Luchtel et al. 1984; Martin & Deyrup-Olsen 1986; Martin et al. 1990) (see Chapter 3). Just how widespread these cells are in stylommatophoran slugs is unclear, with only a few reports of their occurrence (e.g., Yamaguchi et al. 1999; Smith 2010). Fluid from the channel cells probably also helps to remove dirt and other debris from the skin of the slug. Channel cells may be homologous with a type of protein gland cell found in the skin of some land snails (Luchtel et al. 1984; Smith 2010).

In stylommatophorans and some terrestrial caenogastropod taxa (e.g., *Pomatias*), and in the amphibious freshwater ampullariids, the production of primary urine is under greater control because of the loss of the auricular podocytes. In those taxa, the podocytes have been replaced by extracellular tubules in the auricle and/or ventricle. Some upper littoral and supralittoral taxa such as the truncatelloidean *Assiminea* (Little & Andrews 1977) and the littorinids *Melarhaphe neritoides* (Andrews 1988) and the supralittoral

Cenchritis muricatus (Emson et al. 2002) have also lost the podocytes, thereby reducing the filtration rate (see also Section 6.7.7.4.5.2). Podocytes are also absent in fresh-water hygrophilans (see Section 6.7.7.5.2.2).

Little (1983) argued that terrestrial eupulmonates were probably derived from saltmarsh taxa and retain a relatively high osmotic pressure as do the littorinoidean pomatiasids and terrestrial assimineids, which are also thought to have a marine origin. In contrast, the helicinids (Neritimorpha) were derived from fresh-water taxa and have a lower osmotic pressure and produce hypoosmotic urine (Little 1983). Cyclophorids are also thought to have fresh-water ancestors, possibly related to ampullariids and/or viviparids (Little 1983).

6.6.4.2 Nitrogen Metabolism

Metabolism of carbohydrates and lipids produces only carbon dioxide waste, but amino acid metabolism produces waste nitrogen and sulphur. Whereas sulphur is readily excreted as sulphur dioxide (SO_2), nitrogen (obtained from the synthesis of free amino acids) is usually converted to ammonia and then to urea or uric acid. Most animals secrete their nitrogenous waste as one of three main products – ammonia, uric acid, or urea, as outlined in Box 6.6.

Many terrestrial species of caenogastropods are ureotelic, but some can accumulate significant levels of uric acid comprising as much as one-third of the total nitrogen from amino acid metabolism (Bishop et al. 1983). Nitrogen excretion by terrestrial eupulmonates is a complex mixture of ammonotely, ureotely, and uricotely, and the relative importance of a mode of excretion can vary with activity, season, or other environmental conditions (see Section 6.7.7.5.2.3).

BOX 6.6   EXCRETION OF NITROGENOUS WASTE

Ammonotely. Ammonia excretion is the simplest and probably the plesiomorphic form of nitrogenous excretion and is typical of aquatic species, although its relative importance may vary. Ammonia is the final breakdown (involving several specific enzymes) product of proteins and purines. It is highly soluble and easily diffuses because of its low molecular weight and is thus readily excreted across epithelia such as those of the gills or the body surface. It is toxic so levels must remain low in the body. Ammonia excretion requires large amounts of water so is not suitable for terrestrial animals. It can, however, be expelled from the body as gas as, for example, demonstrated in some 'pulmonate' snails (e.g., Martin 1983). Ammonia is the principal excretory product in many marine molluscs including chitons, bivalves, and cephalopods, although urea, uric acid, and amino acids are also excreted in small amounts (e.g., Bishop et al. 1983).

Uricotely. Involves the excretion of mainly uric acid (a purine) which is insoluble and thus has low toxicity. It is an ideal substance for nitrogen excretion for terrestrial taxa as it enables water conservation.

Ureotely. Involves the excretion of mainly urea, and is common in terrestrial animals, including 'pulmonates' and caenogastropods. Ammonia is converted to urea via a complex urea cycle involving input from enzymes and energy. Urea is less toxic than ammonia, but, unlike uric acid, is soluble.

---

[14] Neuckel (1985) noted that anal uptake of water has also been observed in 'larval' squid (Bidder 1950) and apparently occurs in *Strombus gigas* (Little 1967).

6.6.4.2.1 Carnivorous Molluscs and Nitrogenous Waste

Carnivores eat animal tissue of one form or another which has a high protein and water content. The digestion of protein results in nitrogenous waste, typically urea, which requires a considerable volume of water for its excretion. This is not an issue for aquatic taxa, including cephalopods, but is in terrestrial carnivorous snails and slugs, although putative adaptations in those latter groups are not well studied. Nitrogenous excretion is handled by the right kidney in lower gastropods, all of which are aquatic, but in higher gastropods, it is the modified, folded right dorsal wall of the left kidney that extracts nitrogenous waste. This part of the kidney receives nearly all the blood from the head-foot and visceral hump en route to the gill or lung and is more complex in the almost exclusively carnivorous neogastropods (e.g., Andrews 1988; Fretter & Graham 1994) than in other caenogastropods. In 'opisthobranchs,' the excretory changes involving carnivory are unclear, given the large range of morphological changes to the kidney which have occurred in that group (Andrews 1988).

6.6.4.2.2 Utilisation of Ammonia

Some cephalopods sequester ammonia to aid in buoyancy (see Section 6.7.6.5 and Chapter 17), and certain molluscs such as *Tridacna* that contain abundant symbiotic algae provide nitrogen for them by taking it up from the seawater during the day, but excrete ammonia at night while no photosynthesis is occurring (Fitt et al. 1993).

Ammonia gas (NH_3) is also thought to benefit shelled taxa by its conversion to ammonium (NH_4^+) due to the removal of protons in the extrapallial fluid which occurs during calcium carbonate deposition. This thereby increases the rate of shell formation as indicated by the equation $2NH_3 + H_2CO_3 + Ca^{2+} = 2NH_4^+ + CaCO_3$ (Campbell & Boyan 1974; Loest 1979b).

While gaseous ammonia is excreted by at least some terrestrial snails, a few terrestrial slugs absorb significant quantities of this gas from the air through their epidermis, perhaps allowing them to conserve nitrogen used in the synthesis of amino acids that would otherwise have to be replaced in their diet (Loest 1979a).

6.6.5 HEAVY METAL UPTAKE AND EXCRETION/DETOXIFICATION

Like other animals, molluscs take up metal ions from their food or from the environment via the gills or epidermis, with their concentration in the body increasing with exposure (e.g., Simkiss & Mason 1983). In some cases, these trace metals are essential, or at least beneficial, and in others they can be toxic. Most metals, including the toxic ones, form stable complexes with metallothioneins, low molecular weight proteins. These occur in the Golgi apparatus of cells involved in detoxification, including the digestive cells of the digestive gland of gastropods and bivalves.

Because the metals in the body are available metabolically, if they are toxic they must either be excreted or detoxified to avoid

deleterious effects. Some metals, such as zinc, are beneficial in small quantities but can become toxic at higher levels (Rainbow & Luoma 2011). Other toxic 'heavy metals' include arsenic, lead, mercury, cadmium, iron, and aluminium. It is not necessarily the total amount of a particular metal accumulated in the body, but its total bioavailability that is critical, as toxic metals are typically stored in the body in a detoxified form (see reviews in Marigómez et al. 2002; Vijver et al. 2004; Amiard et al. 2006; Rainbow & Luoma 2011). Toxicity can occur when one or more metals are taken up at a higher rate than they can be excreted or detoxified, although these levels differ between species and within species under different circumstances (e.g., Rainbow 2002, 2007; Vijver et al. 2004; Luoma & Rainbow 2005, 2008; Rainbow & Luoma 2011).

The mechanisms of metal uptake and accumulation, their localisation and transport within the body, and their elimination have been the subject of many studies. Marigómez et al. (2002) extensively reviewed the cellular processes involved in metal metabolism in molluscs and their distribution in the body. They concluded that the processes of metal bioaccumulation might be similar in different groups of molluscs at cellular and subcellular levels, but noted that the properties of the surrounding environment might affect the uptake and excretion of the metals.

Cells involved in metal uptake, storage, and detoxification include digestive cells, haemocytes, and pore cells (see Section 6.5). While metal uptake can occur by diffusion, active transport, or endocytosis, it can be enhanced by the synthesis of metallothioneins or by the formation of granules. Such granules (e.g., phosphate or calcium granules) can contain metals such as zinc, lead, etc. Mineralised excretory concretions in the kidney(s) are also common, and these may be intracellular or extracellular. Granules and other forms of metal accumulation are also found in other tissues such as the digestive gland, gills and connective tissue of many molluscs (e.g., Marigómez et al. 2002), and in the gland of Leiblein of neogastropods (Andrews & Thorogood 2005).

The gills are a significant point of uptake of metal ions in aquatic molluscs. The ions are bound to metallothioneins prior to being incorporated into lysosomes and are then released to the blood where haemocytes may ingest them. In contrast, particulate metal is mostly taken up in the digestive system by endocytosis. The particles are moved to lysosomes and then to residual bodies which are mainly housed in digestive cells in the digestive gland (Marigómez et al. 2002).

Metals can also be accumulated selectively in specific cell types because the pools of potential bonding agents (ligands) these cells contain may differ in their binding characteristics. Oxygen donor ligands are in basophilic cells in the digestive gland and calcium cells in connective tissue, and these bond metals such as Ca, K, and Mg with carbonate, oxalate, phosphate, and/or sulphate to form insoluble granules. For example, pyrophosphate forms insoluble salts with Ca, K, Mg, Mn, and Zn as well as some other metal ions (Marigómez et al. 2002). In *Cornu aspersum*, for example, phosphate granules are comprised of Ca, Mg, P, and trace elements such as Ag, Al, Cd, and Zn (Almendros &

Porcel 1992). On the other hand, metals such as Cd, Cu, and Fe are bound in cells rich in sulphur and nitrogen ligands by sulphur donor metal-binding proteins, such as metallothioneins, found particularly in digestive cells in the digestive gland (e.g., Simkiss & Mason 1983; Gibbs et al. 1998; Andrews & Thorogood 2005), the kidney cells, and also in podocytes and pore cells (Marigómez et al. 2002). Metals are transported between tissues by haemocytes, while pore cells mobilise metals as well as accumulate them (Marigómez et al. 2002).

The insoluble granules that bind the potentially harmful metals may be retained through life or excreted via the faeces. Useful elements can be resorbed as required. The composition of the metals in these granules can also vary considerably across molluscan taxa (e.g., Gibbs et al. 1998; Marigómez et al. 2002). Despite a widespread assumption that the main function of the phosphate granules is toxic metal sequestration, this may not be a primary function (Marigómez et al. 2002) because they are also important stores of calcium and/or phosphate (e.g., Gibbs et al. 1998).

While the digestive gland, in particular, is responsible for much of the regulation, detoxification, and sequestering of metals in many molluscs (e.g., Simkiss & Mason 1983; Marigómez et al. 2002; Andrews & Thorogood 2005), in grazing gastropods such as *Littorina*, the kidney is the major site of excretion and production of metallothioneins involved in detoxification (Mason & Nott 1981; Bebianno & Langston 1998).

In neogastropods, the proboscis associated with their carnivorous diet necessitated a major change in the role of the kidney, with the secondary folds becoming blood pressure compensation sacs (see also Section 6.3.8) and a resulting reduction in its role in nitrogenous excretion (Andrews 2010) (see Section 6.7.7.4.2). While some of the excretory function in neogastropods may be undertaken in the digestive gland, which has a unique type of cell (probably excretory) not seen in other gastropods (Fretter & Graham 1962, 1994; Andrews 2010), the gland of Leiblein can also be an important site of sequestration and storage, as in *Nucella*, of some metals such as cadmium (Leung & Furness 1999). Cadmium uptake in the gland of Leiblein is by unciliated cells via the blood and also the lumen of the gland (Andrews & Thorogood 2005). The rectal gland in *Nucella* has also been shown to be involved in detoxification (Andrews 1992) (see Section 6.7.7.4.4).

Vacuolar cells in the branchial hearts of coleoid cephalopods contain large amounts of iron associated with adrenochrome molecules and may have an excretory (e.g., Witmer 1974) or detoxifying role (Martin 1974; Fiedler & Schipp 1987).

6.6.6 EXCRETION VIA THE DIGESTIVE GLAND

In most molluscs, excretory material that is a by-product of digestion is voided via the digestive gland, and this important process is described in Chapter 5. In 'pulmonates,' the digestive gland replaces the kidney as the main excretory organ (see Section 6.7.7.5.2).

6.7 COMPARATIVE KIDNEY MORPHOLOGY, THE SITE OF FILTRATION AND EXCRETION

6.7.1 APLACOPHORANS

While kidneys have not been unequivocally shown to be present in aplacophorans, a pair of large coelomic ducts (also serving as gametoducts) are present that may well be their homologues. These arise posteriorly from the pericardium and bend anteriorly and then posteriorly (thus folded in a C-shape) to open into the mantle cavity or its remnant. Developmentally, positionally, and functionally (i.e., they receive the filtrate from the pericardium), these ducts are usually regarded as homologous with kidneys in other molluscan groups (e.g., Martin 1983; Andrews 1985; Morse & Reynolds 1996), although their role as kidneys is yet to be demonstrated. They are derived from mesoderm and have been reported as containing cells which are structurally similar to kidney tissue in other molluscs (Scheltema 1978a; Morse & Reynolds 1996). The name gonoduct or gametoduct is often applied to these ducts because gametes pass through them, rather than because of any assumed homology with the gonoducts of other molluscs. This possible absence of functional kidneys in aplacophorans is either primary, as argued by Salvini-Plawen (1985a) and Haszprunar (1988a), or secondary (Scheltema 1993, 1996). The presence of kidneys (metanephridia) in many of the putative outgroups, including annelids and brachiopods, argues against their primitive absence in molluscs.

Excretion in aplacophorans has been reported as occurring via the epidermis, although this has not yet been unequivocally demonstrated.

6.7.2 POLYPLACOPHORA

The pericardium contains podocytes on the auricles, but, unlike those present in other classes, these are said not to have slit diaphragms (Økland 1981; Eernisse & Reynolds 1994), although this seems improbable (G. Haszprunar, pers. comm. 2018). The pericardial wall is contractile and probably assists in forcing the filtrate through the pair of renopericardial ducts.

The large, paired kidneys of chitons lie ventrally. In lepidopleurids, each kidney consists of only a single nephridial canal which extends along the ventral surface of the body cavity close to the pedal sinuses. The outer surface is digitate, and the renopericardial duct enters the kidney near the opening of the nephridiopore. In other chitons, each kidney is a U-shaped tube, consisting of an outer arm near the mantle groove and an inner one near the midline. Thus, although it is U-shaped, like those in many bivalves and the putative kidneys of aplacophorans, the chiton kidneys differ in that the two arms lie in a horizontal plane instead of a vertical plane as in aplacophorans and bivalves.

The outer nephridial canal bears the kidney opening and is probably homologous with the single nephridial canal of the basal lepidopleurids. The inner nephridial canal originates posteriorly from the renopericardial duct and extends anteriorly before turning back posteriorly and joining the outer nephridial canal. In some taxa, posterior extensions of the outer nephridial canal occur which track the transverse pedal sinuses.

The nephrocytes lining the kidney are cuboidal to columnar and vacuolate, with some vacuoles containing granules also found in the kidney lumen. The cell bases are infolded, and the apical edges have microvilli. Thus these cells appear to be involved in both active transport and excretion (Andrews 1988; Eernisse & Reynolds 1994).

6.7.3 Monoplacophora

The pericardium contains auricular podocytes (Haszprunar & Schaefer 1997). There is no direct communication (i.e., no renopericardial ducts) between the kidneys and the pericardium (Haszprunar & Schaefer 1997), although renopericardial ducts were incorrectly reported by Lemche & Wingstrand (1959) and doubtfully reported by Wingstrand (1985). There are several (3–7) pairs of non-interconnected kidneys with pores opening to the mantle groove (Wingstrand 1985; Haszprunar & Schaefer 1997). The implications of the lack of renopericardial ducts to carry the urine from the pericardium to the kidneys for absorption from the filtrate, or even the role of the kidneys, have not been investigated.

As in bivalves, the kidneys lie outside the shell muscles, thus differing from chitons where they occur inside and ventrally. It has been suggested that the monoplacophoran condition may have been derived from a polyplacophoran-like kidney where the lobules of the elongate kidney extend between the shell muscles (Haszprunar 1988a).

The nephrocytes are uniform throughout the kidneys, relatively large, vacuolate and their apices are lined with microvilli. There are basal infoldings where the few mitochondria are located (Haszprunar & Schaefer 1997).

6.7.4 Bivalvia

Podocytes in many lower bivalves (i.e., many protobranchs and pteriomorphians) are housed in so-called auricular 'glands' on the surface of the auricles (Figure 6.20), and these are the sites of filtration in those taxa. In contrast, in heteroconch bivalves and some protobranchs and pteriomorphians, 'pericardial glands' (Figure 6.19) are the principal site containing podocytes (Meyhöfer et al. 1985; Morse 1987b; Andrews 1988; Andrews & Jennings 1993). The pericardial glands are pouches on the pericardial wall formed from modified auricular epicardium with many complex blind-ending ducts lined with podocytes, greatly increasing the surface area available for filtration (Andrews & Jennings 1993). Blood from the auricles is moved into the glands where filtration into the pericardial cavity occurs, and the primary urine passes from there via the renopericardial canals to the kidneys.

Apart from the podocytes, pore cells are involved in excretion in some bivalves. These cells can be found on the auricles of *Pecten* and *Crassostrea*, and completely replace the podocytes in the former. They appear to phagocytose particles of low molecular weight (Andrews 1988).

Because ammonia excretion occurs via the extensive epithelial surfaces of the gills and mantle, the role of the kidneys in excretion is reduced. The pair of tubular, usually U-shaped (straight in some taxa such as *Mytilus* and *Pecten*) kidneys lie under or behind the pericardium, and the details of their morphology are highly variable. There is little epithelial differentiation within the kidneys in protobranchs, but there is between two limbs of the U-shaped kidneys of most autobranch bivalves, although the difference between these two sections may not be obvious, and their relative length can vary (Andrews 1988). Typically, in autobranch bivalves, the proximal limb is resorptive and the distal limb excretory with concretions produced in the epithelial cells. When the kidneys are ventral to the pericardium, the renopericardial ducts open anteriorly, but if they are posterior to the pericardium, the renopericardial ducts open posteriorly. Each kidney opens to the suprabranchial chambers by way of a simple aperture. In bivalves, there is no pallial elaboration or extension of the kidney.

6.7.4.1 Estuarine and Fresh-Water Bivalves

To meet the need for an increased filtration rate, estuarine and fresh-water bivalves tend to increase the surface area available

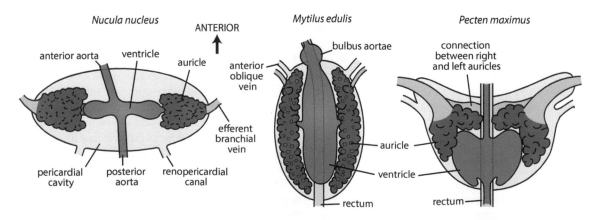

FIGURE 6.20 Dorsal views of the pericardium and heart showing examples of auricular glands in marine bivalves. Redrawn and modified from Andrews, E.B. and Jennings, K.H., *J. Molluscan Stud.*, 59, 223–257, 1993.

for filtration by the so-called pericardial glands (see above). In estuarine taxa, such as *Scrobicularia* (Figure 6.19) and *Polymesoda*, both auricular and pericardial glands are found. In the more specialised fresh-water taxa, only pericardial glands are found, and the auricular walls are thin and simple. These 'glands' can lie either inside or outside the pericardium. In the entirely fresh-water Unionoidea, the pericardial glands can be particularly well developed, as in *Anodonta*, where they form a separate part of the pericardial cavity anterior to the main chamber. The renopericardial ducts open to this chamber so the primary urine can flow directly to the kidney, bypassing the main pericardial lumen (Andrews 1988; Andrews & Jennings 1993). Because bivalve podocytes can also carry out excretory activity and absorb organic and inorganic (e.g., heavy metals) particles from the blood or primary urine (see Section 6.6.2), their role in fresh-water bivalves is particularly important.

Because excretion is also carried out by the digestive gland, the kidneys in estuarine and fresh-water bivalves can become elaborated for the effective resorption of ions. In the estuarine tellinid *Scrobicularia* (Andrews 1988), the proximal limb is narrow and simple internally, with a resorptive epithelium containing glycogen. The distal limb has a vacuolated excretory epithelium; part of it has folded walls and is expanded as a bladder.

In the fresh-water unionoid *Anodonta*, the major function of the kidney is resorption involving active transport. The urine is hypoosmotic to the blood and produced in large quantities. The osmotic work required is reduced by the low osmotic pressure of the blood (and the tissues) (Potts 1954). With exposure to air, blood osmolality in fresh-water bivalves can double (Dietz 1974; Byrne et al. 1989).

6.7.5 SCAPHOPODA

The pair of kidneys are sac-like and lie just above the anus in the middle section of the body; they discharge into the long mantle cavity. There is a questionable (Shimek & Steiner 1997) renopericardial connection to the right kidney (which also connects to the gonad), and both kidneys are lined with two types of highly vacuolated cells (Reynolds 1990a). In the few well-studied species, the pericardium partially encloses the putative rudimentary ventricle ('perianal sinus') with podocytes, but no auricles (see Morse & Reynolds 1996 for references; Reynolds 2002). Given the questionable presence of renopericardial ducts, it is uncertain how the filtrate enters the kidney lumen (Shimek & Steiner 1997).

6.7.6 CEPHALOPODA

Cephalopod excretory systems diverge considerably from those of other molluscs and involve several distinct organs and tissues with different functions. As outlined above, the coelomic spaces (pericardium, kidney, and gonad) are greatly expanded in most cephalopods (see Section 6.3.2), the haemocoel is replaced by a closed blood system in coleoids (mainly closed in nautiloids), and the renal portal system has been

lost. These changes are thought to be correlated with a high metabolic rate and much more efficient circulation of blood through the gills than seen in other molluscs.

Coleoid cephalopods have two 'kidneys' while *Nautilus* has four. Although cephalopod kidneys are homologous with those of other molluscs, they are highly modified. They lie in the mantle roof and extend behind the posterior wall of the mantle cavity in *Nautilus*, but are entirely within the mantle cavity of coleoids. Unlike other molluscs, there is no filtration associated with the heart. Instead, the pericardial coelom envelopes venous vessels which have been co-opted for filtration (see Chapter 17, Figure 17.33 and 17.34).

The pericardial coelom of *Nautilus* is much like that of many decabrachian coleoids. In *Nautilus*, the palliovisceral ligament partially separates the dorsal gonadal coelom from the ventral pericardial coelom. The latter contains the heart and pericardial glands and opens to the mantle cavity via two pores at the bases of the posterior gills.

Although the excretory systems of the nautiloid and coleoid lineages have some basic similarities, they also show substantial differences. In nautiloids, the pericardial and renal structures are found on the four contractile afferent branchial veins (the venae cavae) and are outpocketings of those blood vessels. These large veins collect blood from the sinuses, and their muscular contractions pass the blood into the gills under some pressure. The pericardial appendages (also referred to as 'glands' or 'sacs') (see next Section) lie dorsally within the pericardium while the renal appendages (see Section 6.7.6.3.1 and Figure 6.21) are ventrally located on the afferent branchial veins. The contractile pericardial appendages fill with blood with each contraction of the venae cavae. These pericardial appendages are the homologues of the branchial heart appendages in coleoids and, like them, contain podocytes and are involved in filtration. They lie slightly closer to the gills than the renal appendages.

Each renal appendage (Figure 6.21) is a sacculate organ that lies within each of the four renal sacs ('kidneys') associated with each of the venae cavae. Each renal sac is a coelomic space and opens separately to the mantle cavity by way of a kidney opening. Blood from the vena cava passes in and out of the sacculations comprising the renal appendages, this flow being assisted by muscle fibres within the appendages.

In coleoids, each branchial heart (see Section 6.3.3) has a small branchial heart appendage (see Section 6.7.6.2), the latter being the site of filtration and containing podocytes (Schipp et al. 1971). These appendages are the functional equivalent and homologue of the pericardial glands in *Nautilus*.

The marked differences between nautiloid and coleoid 'kidneys' are outlined below. In coleoids, the filtrate produced by the branchial heart appendage bathes the renal appendages in the urinary sac of the renal coelom, and in *Nautilus*, the urine is released into the mantle cavity through a pair of openings of the pericardial coelom, not via the 'kidneys.' The renal sacs of *Nautilus* do not contain filtrate, but instead minute granules of calcium phosphate (see Section 6.7.6.3.1).

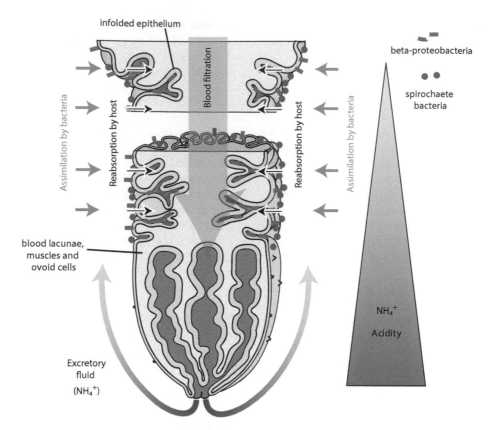

Schematic longitudinal section of a pericardial villus

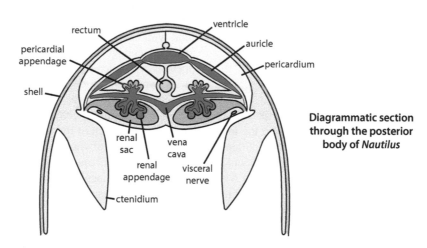

Diagrammatic section through the posterior body of *Nautilus*

FIGURE 6.21   The pericardial appendages of *Nautilus*. Upper figure redrawn and modified from Schipp, R. et al., *Zoomorphology*, 105, 16–29, 1985, and Pernice, M. and Boucher-Rodoni, R., *Environ. Microbiol. Rep.*, 4, 504–511, 2012; lower figure redrawn and modified from Naef, A., *Ergeb. Fortschri. Zool.*, 3, 329–462, 1913.

Unlike other molluscs, in many decabrachian coleoids (squid, cuttlefish) the ventricle (systemic heart) and much of the gut lie within the pericardial coelom which communicates with the dorsal gonadal coelom. The branchial heart appendages also lie in the pericardial coelom and produce the filtrate, which has a similar ionic composition to the blood and contains glucose and amino acids. In decabrachians, the digestive gland ducts are externally lobulated, forming the digestive gland duct appendages (the so-called 'pancreas'), which lie within the pericardial cavity. These structures, along with the renal appendages on the walls of the venae cavae, are responsible for reabsorption (e.g., Martin 1983). Externally, the surface of both these structures is lobulated and covered with an absorptive epithelium richly supplied with blood. The modified filtrate leaves the pericardial coelom via the paired, funnel-like renopericardial ducts into the renal sac, another coelomic space. In cranchiids, the coelom is enlarged and contains ammonia to provide neutral buoyancy (see Chapter 17).

In octopods, the filtrate reabsorbing area has been modified, along with a considerable reduction in the extent of the coelom, the heart, and branchial hearts (but not their appendages) lying outside it. The filtrate from the pericardial glands is contained within a pair of narrow channels and is only accessible to the external surface of the branchial heart appendages. The filtrate passes through the renopericardial ducts into the kidney lumen. In contrast to the rather large paired structures in decabrachians, in octopods the digestive gland duct appendages are reduced to a small mass of tissue in the posterior part of the digestive gland (Martin 1983; Young & Vecchione 1996). The renopericardial duct in octopods differs from that in decabrachians in being important in resorbing glucose and amino acids (Martin 1983) from the filtrate on its passage to the renal sacs, where further resorption occurs in the renal appendages. There may also be absorption in narrow 'water canals' which link the gonadal coelom and the much reduced pericardial coelom (Budelmann et al. 1997).

A major difference between cephalopods and most other molluscs is the configuration of the 'kidneys.' In other molluscs, the urine is inside the kidney, and the blood is outside, while in the cephalopod 'kidney' (the renal appendages), the blood is inside and the urine is outside. Thus, the cephalopod 'kidneys' are analogous to the pericardial glands of some bivalves and gastropods, which also have the blood inside and the urine outside, and are surrounded by a coelomic membrane (Martin 1983). Martin (1983) suggested that the excretory structures became specialised in coleoid cephalopods, with one part retaining filtration and its relationship with the pericardial coelom, while the other part moved along the vena cava, where it became incorporated in the renal coelom, taking over the secretory activity of the normal molluscan kidney. In Nautilus, the renal appendages became specialised for calcium retention, so the pericardial glands carry out some of the secretory functions performed by the renal appendages in coleoids. It is not clear as to which of these configurations is most like the original plesiomorphic condition in cephalopods.

Symbiotic dicyemid mesozoans and chromidinid ciliates are found between the infoldings of the renal epithelia of many coleoids, more rarely in the digestive gland duct appendages, and in one case in the branchial heart appendage (of a species of Rossia) (Furuya et al. 2004) (see Chapter 9). They may be involved in acidification, and their ciliated bodies may assist the local transport of urine (Lapan 1975).

The pericardial and renal appendages are described in more detail below.

6.7.6.1 Nautiloid Pericardial Appendages and Their Symbiotic Bacteria

The pericardial appendages of Nautilus (Figure 6.21) consist of a series of contractile, finger-shaped villi, each with a microvillous 'apical' region and a 'basal-medial' region with crypt-like invaginations. In the 'apical' part of each, several channels run to the tip to open at a pore through which the filtrate flows into the pericardium. This filtrate is produced by the podocytes which lie in the region of the villus above the channels. The filtrate is an acidic, ammonia-rich fluid and is modified by

the resorptive microvillous epithelium of the channels, which also excretes waste material (Schipp & Martin 1981; Martin 1983). These latter functions are facilitated by the invaginations being surrounded by blood spaces. Thus, the pericardial appendages of Nautilus have adopted not only the functions of the equivalent structures in coleoids (filtration), but also those carried out by the coleoid renal appendages (secretion and absorption), leaving the renal appendages in Nautilus solely involved in calcium metabolism (Martin 1983).

The pericardial appendages in Nautilus are thus highly specialised and differ considerably in structure and function from those of the same name in coleoids. They have also been shown to harbour symbiotic bacteria in high densities (Schipp et al. 1990; Pernice et al. 2007). These bacteria do not occur in the channelled distal region of each appendage, but occur in large numbers in the crypt-like invaginations and outer epithelium in the basal-medial section (Schipp et al. 1985) (Figure 6.21). Two kinds of bacteria have been identified; a betaproteobacterium and a coccoid spirochaete, neither of which are closely related to any other known bacterium (Pernice et al. 2007). The spirochaetes are associated with the epithelium in peripheral areas; the betaproteobacterial symbionts are less closely associated with the epithelium and are in peripheral and invaginated areas.

The betaproteobacteria belong to an ammonia-oxidising lineage not otherwise known in a symbiotic association with animals (Pernice et al. 2007). Although these bacteria live in very dense concentrations in an ammonia-rich environment, the exact role they play is still not clear (Pernice & Boucher-Rodoni 2012). They could play a crucial role by transforming ammonia, thus detoxifying the body tissue, and/or by providing nitrogen gas for filling the chambered shell responsible for its neutral buoyancy (the gas within these chambers is >90% nitrogen) (Denton 1974; Boucher-Rodoni & Mangold 1994). Also, nitrogen may also enter the chambers by simple diffusion from the blood, as suggested by Denton and Gilpin-Brown (1966), or perhaps both mechanisms are involved. Another possibility is that the symbiotic bacteria may degrade proteins present in the coelomic fluid into amino acids facilitating reabsorption in the basomedian region of the pericardial appendages, enabling Nautilus to conserve nitrogen (Pernice et al. 2007). It is not yet known how these bacteria are transferred through the generations.

Although symbiosis for recycling nitrogenous waste products is not known to have evolved in any other molluscs, it is known in a few terrestrial arthropods and also in a marine and a terrestrial annelid (Pernice et al. 2007).

6.7.6.2 The Branchial Heart Appendages in Coleoids

The appendages of the branchial hearts in coleoids are homologues of the pericardial appendages in Nautilus. They are attached to the branchial hearts (see Section 6.3.3) and connect to the other coelomic spaces. Unlike Nautilus, symbiotic bacteria are not present.

The branchial hearts and their branchial or pericardial appendages (Figure 6.22) are closely connected and originate as a single structure in ontogeny. Both are contractile and the

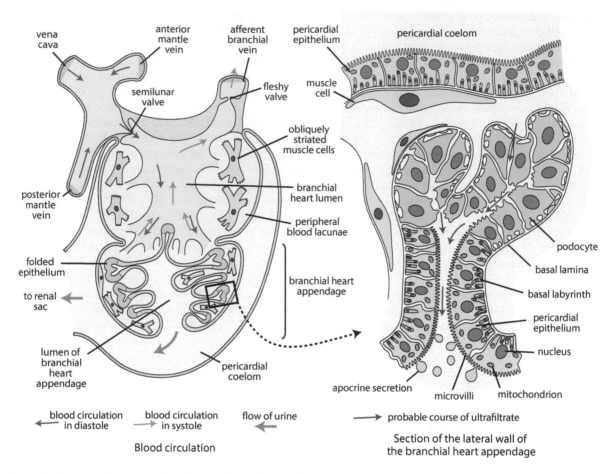

FIGURE 6.22 Branchial heart complex of the cuttlefish *Sepia officinalis*. Redrawn and modified from Schipp, R. and Hevert, F., *J. Exp. Biol.*, 92, 23–35, 1981.

blood in them is under relatively high pressure, passing in and out of both in a shuttle movement. In *Octopus* at least, most of the venous blood passes through the main part of each branchial heart back into the branchial afferent vessel. Some of it remains in the spongy tissue inside the branchial heart, and it is part of this blood that enters the rather small channel into the branchial heart appendage (Johansen & Martin 1962).

As noted above, the branchial heart appendages contain numerous podocytes (Figure 6.22) and are the main site producing the filtrate. This fluid is similar to the blood in ionic composition and is passed into the pericardial coelom (e.g., Schipp & Hevert 1981). In addition, the epithelium of the branchial heart appendages is responsible for secretion and reabsorption, including carbohydrates (Budelmann et al. 1997). Ovoid-polygonal cells in the wall of the pericardial appendages are thought to be involved in detoxification and breakdown of haemocyanin, in storing some substances such as proteins containing Cu and Fe, and in immunological defence mechanisms (Budelmann et al. 1997).

6.7.6.3 The Renal Appendages in the 'Kidneys'

6.7.6.3.1 Nautiloids

The renal appendages inside the renal sacs in *Nautilus* are primarily secretory. *Nautilus* can concentrate up to several

grams of extracellular spherical concrements (uroliths) in the kidneys. These uroliths contain mainly Ca, but also Mg, P, and strontium (Sr), and are thought to be calcium storage vesicles for use in shell growth or repair, and, in particular, the formation of new septa (e.g., Martin et al. 1978), although some may be excreted. Septal formation occurs approximately once a month over nearly three years of growth (Martin et al. 1978). The process requires rapid calcium mobilisation, and the cells involved are microvillous and rich in phosphatase (Schipp & Martin 1981).

6.7.6.3.2 Coleoids

As in *Nautilus,* the renal appendages in coleoids are located inside the renal sacs, but, unlike *Nautilus,* the renal sac receives urine via the renopericardial ducts, and the renal appendages are bathed in this fluid. Also, unlike *Nautilus,* the renal appendages are the main sites of the synthesis and secretion of ammonia and may also be involved in reabsorption and mineral metabolism (Martin 1983). The renal appendages, in both nautilids and coleoids with calcified shells, store uroliths in their tissues. In *Sepia* these contain Ca, Mg, Na, P, Na, K, Cl, and S (Schipp & Boletzky 1975) and are thought to be storage vesicles for Ca used for shell growth or repair; although some may be excreted. These uroliths are 40–60 μm in diameter and are formed in crypt-like cavities in the epithelium of the appendages (Budelmann et al. 1997).

As in *Nautilus*, the renal appendages of coleoids are bud-like mesodermal pouches that arise from the vena cava and a few veins such as the abdominal veins that branch from the vena cava (see Chapter 17). They have a single-layered excretory epithelium like that of the branchial heart appendages, which is also involved in active transport. Below the epithelium are blood lacunae, and muscle fibres producing rhythmic contractions (Budelmann et al. 1997).

6.7.6.4 The Digestive Gland Duct Appendages ('Pancreas') of Coleoids

In decabrachian and octopod coleoids, the digestive gland duct appendages differ in morphology and, in part, in function. No similar structure is present in *Nautilus*. In decabrachians, these structures consist of lobulations on the two digestive gland ducts which lie within a dorsal lobe of the renal coelom (Young & Vecchione 1996). In *Sepia*, it has been shown that the outer (mesodermal) epithelium of the appendage (which lies exposed in the renal coelom) is involved in ammonia excretion and the inner epithelium, exposed inside the ducts, plays an important role in water reabsorption (Schipp & Boletzky 1976).

In contrast, in octopods the homologues of the decabrachian digestive gland appendages lie inside a capsule of connective tissue surrounding the digestive gland and are associated with the bases of the digestive gland ducts. Their single layer of epithelial cells is involved solely in water resorption and plays an important role in maintaining blood volume (Wells & Wells 1989). They are referred to as a single structure by Young and Vecchione (1996) or two closely applied lobes (Wells & Wells 1989). The bases of the epithelial cells are in contact with blood lacunae, and their microvillous outer surfaces are exposed in the digestive gland duct which contains the 'chyme' from the caecum (Wells & Wells 1989).

6.7.6.5 Utilisation of Ammonia for Buoyancy in Cephalopods

In general, most tissue has a greater density than seawater, so a pelagic animal must expend a considerable amount of energy to maintain its position in the water column. Coleoid cephalopods are ammonotelic, with more than 70% of their nitrogenous waste being excreted as ammonia (e.g., Boucher-Rodoni 1989), which is lighter than seawater. At least ten squid families have overcome the buoyancy problem by adding ammonia to special coelomic compartments or vesicles scattered through their tissue, thus decreasing their density (Voight et al. 1994). By maintaining a lower pH in the areas where ammonia is stored than in the haemolymph, the ammonia is maintained as ammonium ions, and concentrations can exceed 500 mM (Voight et al. 1994).

In cranchid squid, the coelom has enlarged considerably with about 40% of the nitrogen which is processed during the lifetime of the animal retained as ammonium chloride (Denton et al. 1969). While ammonium chloride is isotonic with seawater, it is less dense than sodium chloride, and therefore provides buoyancy.

As in coleoids, *Nautilus* is ammonotelic, with ammonia representing more than 70% of the organic nitrogen excreted (Boucher-Rodoni 1989). Compared with the shell-less cephalopods, ammonia excretion is low in *Sepia* and *Nautilus* (Boucher-Rodoni & Mangold 1994), as both use nitrogen gas in their shell for buoyancy, suggesting that they may convert much of their nitrogenous waste from ammonia to nitrogen gas. In *Nautilus*, symbiotic bacteria in the pericardial gland may be involved in this process (see Section 6.7.6.1), although there is no direct evidence of this. Over 90% of the gas contained in the chambers of the shell of *Nautilus* and in the internal shells of *Sepia* and *Spirula* is nitrogen (Denton & Taylor 1964; Denton & Gilpin-Brown 1966), but small quantities of oxygen and other atmospheric gases are also present. The gas diffuses as liquid from the tissues into the chambers of the shell via the siphuncle, and ions are pumped out (see also Chapter 17).

6.7.7 Gastropoda

The site of filtration in gastropods is usually in the wall of the auricle(s), although in the patellogastropod *Patella*, the whole heart is covered by podocytes (Økland 1982). While there is some porosity in the ventricle of some gastropods, it contributes very little to the total filtrate (Fretter & Graham 1994). Higher gastropods show modifications from the plesiomorphic condition seen in vetigastropods, where the epicardium of the auricles is composed entirely of podocytes; this is also likely to be the plesiomorphic condition for molluscs (see Section 6.6.2). Nevertheless, there is some separation of filtration and pumping in the auricles of a number of vetigastropods, with filtration chambers being developed, for example, in *Haliotis* and *Emarginula* (Andrews 1981, 1985). In caenogastropods, filtration chambers are more localised and elaborate, and they can form filtration pouches (Fretter & Graham 1994) (see Section 6.6.2). These pouches become fewer in higher caenogastropods, and the musculature of the auricle is modified to enable more effective separation of the pumping and filtration functions (Fretter & Graham 1994).

In some taxa where the epicardium lacks filtration pouches, it is penetrated by extracellular channels that may also enable fluid to pass through the auricular wall (e.g., Andrews 1976a). In some gastropods, only a small part of each auricle is involved with filtration, but this is supplemented by podocytes in the epithelium overlying the vein connecting the auricle to the kidney (Andrews 1988).

Unlike the situation in other classes, in gastropods that possess paired kidneys (patellogastropods and most vetigastropods), they are asymmetric. This may have been a consequence of the originally longitudinally orientated heart at the inner end of the mantle cavity becoming transverse due to torsion, resulting in the kidneys, which retained their original relationship with the pericardium, becoming asymmetrical. Early in gastropod evolution, the ancestral gastropod kidneys took on different functions and hence different morphologies, as reflected in patellogastropods and vetigastropods. In patellogastropods the small left kidney lies at the base of the visceral mass behind the mantle cavity, while the larger right kidney extends posteriorly under the alimentary system and gonad. In some taxa the left kidney may extend into the mantle roof (Lindberg 1981). In vetigastropods, the left kidney (called the 'papillary sac'), which is very small in fissurelloideans, lies in the mantle skirt anterior to the pericardium and the right kidney lies in the viscera behind the mantle cavity. In these taxa, the right kidney performs excretory functions and also provides a route for gametes through its lumen (usually in a separated channel) to the kidney opening. In contrast to this situation, some vetigastropod lineages and other orthogastropods have retained only the left kidney, with this modification occurring independently several times in the course of gastropod evolution (Ponder & Lindberg 1997). Thus, a single (left) kidney is found in the vetigastropod Cocculinoidea and Neomphalida, as well as the Neritimorpha, Caenogastropoda, and Heterobranchia.

Gastropod excretory systems are highly variable, especially in the form and function of the kidney(s). This is in part a reflection of their habitat diversity (marine, brackish, fresh water, terrestrial) in neritimorphs, caenogastropods, and heterobranchs which necessitated independently acquired solutions to excretory issues. As is so often the case, much of what is outlined below is based on studies of only a handful of species, with little being done since the reviews by Andrews (1988) and Fretter and Graham (1994).

In other classes, each of the paired kidneys typically has one section for resorption and another for excretion. The ancestral gastropods uniquely separated these different functions between the two kidneys, with the right kidney undertaking the excretory functions and the left being the absorptive area. Later in gastropod evolution, when only a single kidney remained, these two functions were combined (independently in several lineages) in the single kidney (Andrews 1988). Because the single kidney is involved in resorption, there is a reduction in excretory activity in the kidney usually associated with increased excretory activity by the digestive gland (Fretter & Graham 1962, 1994) or an increase in storage of calcareous waste granules in connective tissue, mainly around blood vessels (Andrews 1979a).

The different functions of the right and left kidneys in patellogastropods and vetigastropods are reflected in the source of their respective blood supplies. The right kidney (or the excretory portion of the left kidney when only one kidney is present) receives deoxygenated blood primarily from the visceral sinuses, while the resorptive left kidney (the 'papillary sac' in vetigastropods or the 'nephridial gland' analogue in taxa with just the left kidney), receives partially oxygenated blood primarily from the head, mantle, and foot. A distinct 'nephridial gland' is absent in Cocculinoidea, Neritimorpha, and Heterobranchia, whereas a 'nephridial gland' is found in at least some 'higher vetigastropods,' namely, some trochoideans including the phasianellid *Tricolia* (Marcus & Marcus 1960), where it lies in a specialised part of the papillary sac (the left kidney) (see Section 6.7.7.2). In the Neomphalida it is

incorporated in the single (left) kidney (Fretter et al. 1981), as it is in most caenogastropods (Andrews 1985, 1988; Ponder & Lindberg 1997) (see Section 6.7.7.4.2). The distribution of 'nephridial glands' in gastropods indicates that this structure has evolved independently in these three groups.

The left kidney of patellogastropods is involved in resorption and thus functions in a similar way to that of the vetigastropod 'papillary sac' and the nephridial gland seen in many other gastropods (Andrews 1985). In caenogastropods and in those vetigastropods in which a nephridial gland occurs, the nephridial gland vein has an oscillating 'tidal flow' which moves oxygenated blood back and forth between the auricle and the nephridial gland (see Sections 6.6.3 and 6.7.7.4.2).

Developmental studies of neritimorphs and caenogastropods have shown that a rudiment of the right kidney is incorporated into the reproductive system, but this lacks any excretory function. Such an incorporation could not be verified in heterobranchs (Ponder & Lindberg 1997) (see also Chapter 8). In all these taxa, the remaining (left) kidney is usually responsible for both excretion and reabsorption, but details differ in the major groups, as outlined in the systematic overview below.

Invasions of land and fresh water by gastropods presented major challenges for the excretory and circulatory systems (see above), and they have been surmounted in different ways in different groups. All successful terrestrial and fresh-water groups of gastropods are monocardiate with a single kidney. Whether or not the combination of secretory and absorptive functions in a single kidney was one of the key innovations enabling non-marine living is unclear.

6.7.7.1 Patellogastropoda
Podocytes are uniquely present on the auricle and ventricle in *Patella* (Økland 1982), but the site(s) of filtration in other patellogastropods has not yet been investigated.

There is an important difference between the location of the right kidney in patellogastropods and vetigastropods. The large right kidney of patellogastropods extends posteriorly under the digestive gland and gut as in Polyplacophora, but in vetigastropods, this kidney lies on top of the digestive gland.

The small left and large right kidneys are distinct histologically. The left kidney epithelium is columnar and has, as in vetigastropods, no excretory vacuoles and many lysosomes and is involved in resorption (Andrews 1985). While this small kidney has some similarities to the 'nephridial gland,' there is no nephridial gland vein. In contrast, the right kidney is lined with ciliated vacuolated cells with excretory spherules, which contain a high proportion of melanin, and tall microvilli, their structure indicating that they are involved in active transport (Andrews 1985). The renal epithelia in both kidneys in patellogastropods differ in their fine structural details from those of other gastropods leading Andrews (1988, p. 404) to note that they 'must have diverged from other archaeogastropods at a very early stage in phylogeny.'

Only a couple of patellogastropod species are found in brackish waters; there are no fresh-water taxa. Kidney function and morphology have not been investigated in the brackish-water taxa.

6.7.7.2 Vetigastropoda

In studied vetigastropods, podocytes are found only on the auricles (see Section 6.6.2).

In the right kidney, the renal epithelium occurs only on branches of the efferent renal vein, while other parts of the walls of the kidney are lined with squamous epithelium (Andrews 1985, 1988). In contrast, the left kidney is compact, with the lumen filled with many blood-filled papillae (hence its alternative name, papillary sac), which provide a very large surface area and is a unique feature of vetigastropods.

In vetigastropods (e.g., Trochoidea and some hot vent taxa) that have lost the right gill, the left kidney has a nephridial gland, a specialised part of the papillary sac. It overlies the pericardium posteriorly and the main blood vessel that links the 'gland' and the efferent ctenidial vein. Andrews (1985) showed that this linking vessel (the posterior renal vein) might facilitate flow through the nephridial gland in a way similar to the 'tidal flow' in caenogastropods (see Sections 6.6.3 and 6.7.7.4.2). In this area, invaginations from the bases of the papillae are in contact with the blood and the renal epithelium. The tubule-like invaginations are lined with cells whose structure suggests that they are involved in resorption of organic solutes (see Sections 6.6.3). In the externally bilaterally symmetrical fissurellids, the left kidney is vestigial, but still functional. The papillae are not developed, but the epithelial cells are similar to those in the papillary sac of other vetigastropods (Andrews 1988).

Andrews (1985) summarised the differences in the left and right kidneys in vetigastropods as shown below (Table 6.1).

In the cocculinid *Macleaniella moskalevi*, the single (left) kidney lies in the mantle roof, is a simple sac without papillae and possesses a uniform excretory epithelium (Strong 2003).

No living vetigastropods are found in terrestrial or freshwater habitats.

6.7.7.3 Neritimorpha

Neritimorphs have podocytes only on the auricles (Estabrooks et al. 1999). Uniquely, the pericardium is considerably expanded posteriorly and to the right in a long pericardial caecum. As in other gastropods with only a single (left) kidney, this carries out the functions of both the right and left kidneys in vetigastropods and patellogastropods. Unlike the situation in those groups, the neritimorph kidney abuts the posterior wall of the mantle cavity. It is ovoid and divided into two distinct sections; a 'glandular' section and a large, thin-walled bladder which opens to the mantle cavity (Little 1972; Delhaye 1974a; Andrews 1981; Estabrooks et al. 1999). Together these form a tight U-shape (Figure 6.23), giving rise to comparison with the U-shaped kidneys of bivalves (Andrews 1981, 1988). The 'glandular' proximal part of the kidney has folded walls and is lined with excretory cells like those of the right kidney in vetigastropods. These cells have numerous excretory vacuoles containing fine granules with a very low uric acid content (Delhaye 1974a). The renopericardial duct opens to the proximal 'glandular' part of the kidney. In marine neritids, this smooth distal 'bladder' is partly excretory and, unusually, the only likely resorptive cells it contains are near the kidney opening (Andrews 1988).

There is no nephridial gland and, unlike the situation in trochoidean vetigastropods, a vein does not connect the kidney with the efferent ctenidial vein, except, convergently, in the limpet-shaped *Phenacolepas* (Fretter 1984b). Instead, they show the plesiomorphic configuration where the efferent

---

TABLE 6.1

A Summary of the Functions of the Right and Left Kidneys of Vetigastropods and the Suggested Differences in the Composition of Their Blood and Urine

Left kidney		Right kidney	
Secretes basic dyes, extracts ferritin from blood and resorbs solutes from urine (Na⁺-glucose transport); small storage capacity.		Secretes acid dyes, PAH[a] and PSP,[b] and extracts cations like Fe³⁺ and ferritin from blood; large storage capacity.	
Blood	*Urine*	*Blood*	*Urine*
Post-branchial oxygenated pH ~ 8.04; haemocyanin high concentration, dissociates and polymerises, perhaps adjusting colloid osmotic pressure and lowering viscosity. Low concentration of purines and ammonia.	No nitrogenous excretion; pH as in post-branchial blood and primary filtrate (pH ~8.04).	Prebranchial deoxygenated with pH ~ 7.85; haemocyanin low concentration, no polymerisation. High concentration of amino acids, purines and ammonia.	Excretion of purines and ammonia, possible acidification of urine and trapping of NH₃ as NH₄⁺. High concentration of NH₃ likely to inhibit some types of active transport.
Conditions favour active transport.	Conditions favour active transport.	Conditions favour diffusion of NH₃.	

Modified from Andrews, E.B., *Philos. Trans. Royal Soc. B*, 310, 383–406, 1985.

[a] Polycyclic Aromatic Hydrocarbon.

[b] Phenolsulfonphthalein.

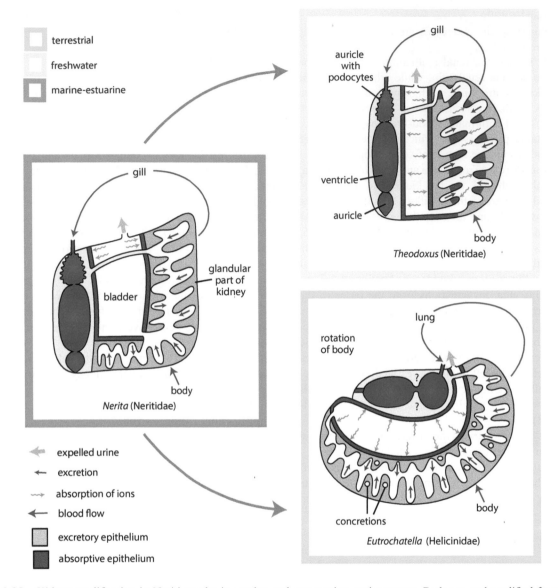

FIGURE 6.23 Kidney modification in Neritimorpha in marine and non-marine environments. Redrawn and modified from Little, C., *J. Exp. Biol.*, 56, 249–261, 1972, and Delhaye W., *Archives de Biologie*, 85, 235–262, 1974a.

renal vein flows to the afferent ctenidial vein. This configuration led Andrews (1988, p. 407) to postulate that the neritimorphs are an ancient group that evolved 'parallel to the archaeogastropods.'[15]

In contrast to non-marine neritimorphs (see next section), the few marine neritids studied have blood isoosmotic with the urine, suggesting that they osmoconform (Little 1972; Andrews 1988). Some marine neritoideans can also tolerate a wide range of conditions. The limpet-like phenacolepadids live in largely anoxic conditions, while the brine-pool neritoidean *Bathynerita naticoidea*, which lives in salinities up to 85‰ in deep-water cold seep mussel beds in the Gulf of Mexico, has been shown to osmoconform over a range of 35–75‰ (Gaest et al. 2007).

6.7.7.3.1 *Non-Marine Neritoideans*

Unlike the marine taxa, fresh-water neritoideans resorb ions from the urine (Little 1972), this being mainly carried out in the 'glandular' excretory limb of the kidney and also, to a lesser extent, in the smooth-walled distal limb (Delhaye 1974a; Andrews 1988).

As shown in Figure 6.23, the heart lies horizontally in the terrestrial helicinids (it is longitudinal in other neritimorphs) across the posterior wall of the mantle cavity. Other than the kidney opening having moved to the right, the kidney is similar in construction to other neritimorphs (Little 1972). The excretory cells in the brackish and fresh-water neritid *Theodoxus* have large granules containing proteins and phospholipids, and there is no trace of the lipofuscins, uric acid, urates, or calcium salts found in the excretory cells of the fresh-water hygrophilans. Terrestrial helicinids have large

---

[15] Presumably vetigastropods, not patellogastropods.

excretory granules containing proteins, lipofuscins, urates, and calcium salts (Andrews 1981). There is some ultrastructural evidence that the excretory cells in helicinids may also be involved in absorbing solutes (Andrews 1988).

Resorption occurs in the helicinid kidney in the distal 'bladder,' while in *Theodoxus*, it takes place in the proximal 'glandular' section, these differences reflecting their independent origins from marine taxa (Delhaye 1974a; Andrews 1981, 1988).

6.7.7.4   Caenogastropoda

6.7.7.4.1   The Heart and Pericardium

In marine caenogastropods such as *Littorina*, filtrate accumulates in the pericardium and is passed to the kidney via the renopericardial duct where, in the nephridial gland (see below), the urine is modified through manipulation of ions, resorption of solutes, and the addition of nitrogenous waste from the dorsal kidney epithelium before being released into the mantle cavity. Various strategies have been employed by non-marine caenogastropods to enable them to live in either fresh-water or terrestrial environments (see Section 6.7.7.4.5).

In caenogastropods, the main chamber of the auricle (see Section 6.3.2) receives oxygenated blood from the efferent branchial vein during diastole. During systole, some of this blood is moved to the nephridial gland vein and its sinuses, where some resorption occurs before being moved back to the dorsal channel of the auricle during the next diastole, and from there it passes into the ventricle (Andrews 1988, 2010; Andrews & Taylor 1988). The rest of the blood in the auricle is filtered through podocytes in the filtration chambers on the sides of the auricle (Andrews 1976b). The filtrate then passes down the renopericardial duct into the nephridial gland tubules, and from there into the lumen of the kidney proper.

6.7.7.4.2   The Nephridial Gland

Most caenogastropods have a nephridial gland on the left side of the kidney which resembles the convergent nephridial gland in the left kidney (i.e., 'papillary sac') of higher vetigastropods (see Section 6.7.7.2). Like the vetigastropod structure, it is connected to the left auricle by a vein (nephridial gland vein) which moves post-branchial blood back and forth between the left auricle and the kidney (Figure 6.24). While cerithioideans have been reported to lack a nephridial gland, the batillariid *Lampanella* has a small one, and a similar structure may be present in other cerithioideans (Strong 2003). A nephridial gland is absent in architaenioglossans, although the blood circulation patterns are similar to those seen in other caenogastropods, with some blood flowing directly to the auricle from the kidney (Strong 2003).

Neogastropods have a much larger and more complex nephridial gland than that in lower caenogastropods such as *Littorina,* and it is more distinctly separated from the rest of the kidney (Andrews 1988, 2010) (Figure 6.25). Resorption occurs in the nephridial gland of neogastropods, but the renal epithelium in the main part of the kidney is involved only in excretion (Andrews 1988).

The blood flow from the heart to the nephridial gland, as described above, has only been worked out in detail in *Littorina* (Figure 6.24). Previously it was thought that the main 'efferent' vein of the nephridial gland transported prebranchial (i.e., unoxygenated venous) blood to the auricle because of superficial sinuses that link the nephridial gland with the dorsal wall of the kidney (see Fretter & Graham 1994, for review), an arrangement shown to be incorrect in *Littorina* (Andrews & Taylor 1988), and it cannot occur in neogastropods such as *Nucella* and *Buccinum* which lack such connections (Andrews 2010) (Figure 6.25).

Organic solute resorption occurs in both the nephridial gland and the dorsal wall of the kidney (see below). The final composition of the blood is modified by amoebocytes and other cells in the connective tissue in the blood spaces of the nephridial gland ('blood gland') before it circulates around the body (Andrews 2010).

The nephridial gland epithelium is predominantly composed of resorptive epithelia; in lower caenogastropods, resorptive cells are also found on the ridges of the folds on the dorsal wall of the kidney, and on the primary folds in carnivorous taxa, as described below. Andrews (2010) argued that the nephridial gland is probably more important than the dorsal wall of the kidney with regard to solute resorption, but this has yet to be demonstrated experimentally.

The fresh-water caenogastropod *Bithynia* has a vestigial nephridial gland and a hypertrophied blood gland (E. B. Andrews, pers. comm. 2012). Strong (2003) stated that the nephridial gland receives blood from the afferent renal vein in *Bithynia* (not confirmed by Andrews 2010), and *Strombus*. While these observations may be an exception to the 'tidal flow' of blood between the auricle and nephridial gland vein, they require confirmation.

6.7.7.4.3   The Kidney Proper

In caenogastropods, the single (left) kidney has developed specialised areas for resorption as well as excretion, processes facilitated by the afferent renal vein which ramifies in the dorsal wall of the kidney, delivering blood from the visceral venous sinuses.

The epithelium in the dorsal kidney wall of a marine lower caenogastropod such as *Littorina* is composed mainly of excretory cells, among which are some pigmented cells. It has been suggested that the pigmented cells might at least be analogues of the resorptive cells of the 'papillary sac' of vetigastropods, the homologue of the caenogastropod kidney (Andrews 1981, 1988). Andrews also showed that the excretory cells had similarities with the 'papillary sac' epithelium in having long basal processes, indicating that these cells may have been secondarily modified to take on a new role. Whether these assumptions are correct or not needs to be further tested.

Nitrogenous waste is extracted in the highly folded right dorsal wall of the kidney, where most of the blood from the head-foot and viscera are received on route to

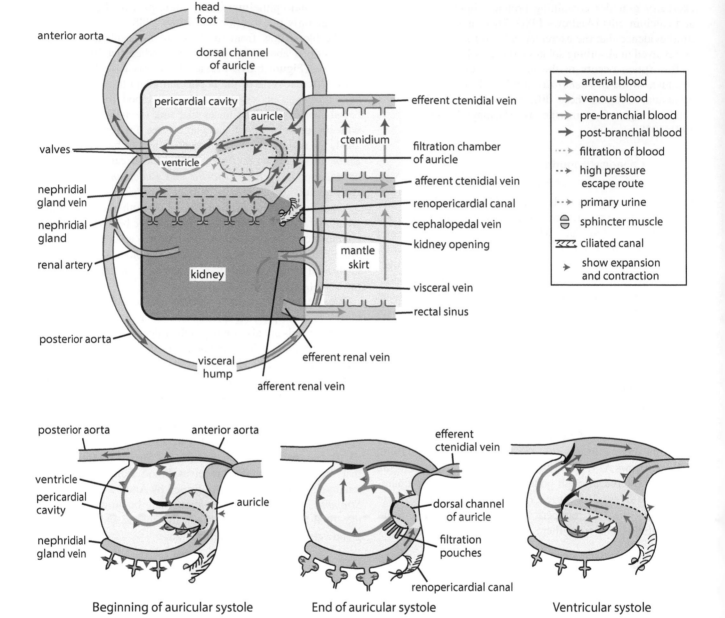

FIGURE 6.24 The function of the heart and nephridial gland in *Littorina*. Lower figures show different phases of heartbeat (dorsal view). Redrawn and modified from Andrews, E.B. and Taylor, P.M., *J. Comp. Physiol. B*, 158, 247–262, 1988.

the ctenidium. Filtration occurs in the wall of the auricle (Andrews 1985, 1988), but unlike vetigastropod and archi-taenioglossan taxa, the podocytes in *Littorina* and *Nucella* are found in appendages on the inner wall of the auricle (Andrews 1976b, 2010) (Figures 6.24 and 6.25).

In most non-carnivorous caenogastropods, the excretory tissue of the kidney is uniform, but in most carnivorous taxa, primary and secondary lamellae are formed from two distinct kinds of excretory tissue. Thus, within caenogastropods, the kidneys of naticoideans and cypraeoideans are more complex than those in grazing or browsing taxa such as *Littorina* (Fretter & Graham 1962; Andrews 1981; Strong 2003).

In non-neogastropod proboscis-bearing caenogastropods (cypraeids, naticids), and in the neogastropod cancellariids, there is no clear separation of the blood supply to the primary and secondary folds. According to Strong (2003), the dorsal afferent renal vein in a cypraeid, a naticid, and a cancellariid also supplies some excretory (i.e., primary) folds, but Andrews (2010) found that the secondary folds in a naticid had a separate vein. It appears that in all neogastropods, the large primary and smaller secondary folds have a different blood supply and differ in histology and function.

The primary folds of neogastropods are covered with excretory and ciliated resorptive cells and receive venous

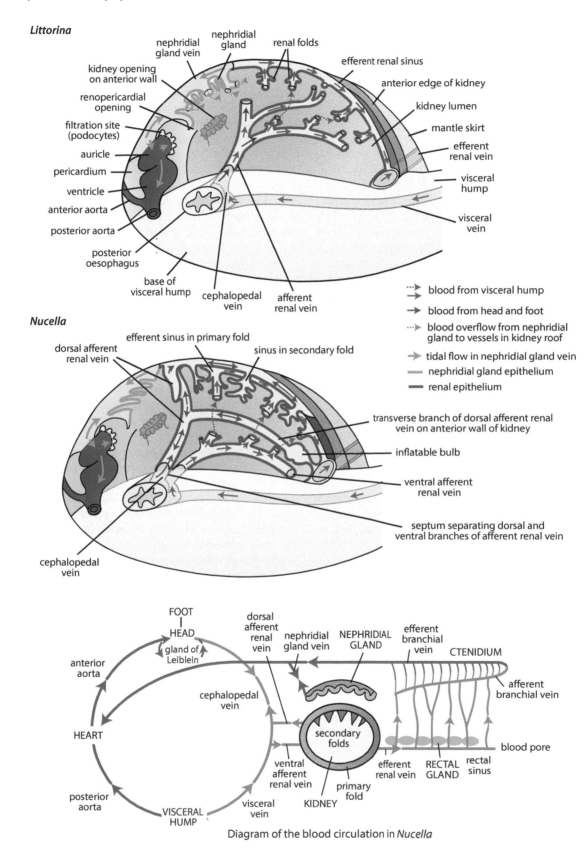

FIGURE 6.25 A comparison of the kidney and heart structure and function in the caenogastropods *Littorina* and *Nucella* as shown by diagrammatic transverse section through the base of the visceral hump and looking anteriorly through the kidney. Redrawn and modified from Andrews, E.B., *J. Molluscan Stud.*, 76, 211–233, 2010 (*Nucella*); *Littorina* based on a drawing provided by E. B. Andrews, 2011.

blood from the digestive gland and other parts of the viscera. The resorptive cells store glycogen and absorb organic solutes and are located mostly on the crests of the folds, while the sides are mostly covered with vacuolate excretory cells which lie over the blood spaces (Dimitriadis & Andrews 1999; Andrews 2010). In contrast, the secondary folds receive blood from the head-foot (Andrews 1988), which has low nitrogenous waste and may be more oxygenated (e.g., Brown 1984). The secondary folds on the left (adjacent to the nephridial gland) are lined with a resorptive epithelium of microvillous cuboidal cells that lack excretory vacuoles and are histologically similar to those in the papillary sac of trochids (Andrews 1988). While it could be argued that the secondary folds are the equivalent of the papillary sac in vetigastropods and the primary folds on the right are equivalent to the right kidney in vetigastropods (and the renal folds in lower caenogastropods such as *Littorina*), the histology of the cells of the neogastropod secondary folds indicate that they have only a minor role in absorption. Indeed, their main function is to compensate for blood displaced by the retraction of the proboscis (Andrews 2010) (see Section 6.3.8).

These two kinds of folds often interdigitate in neogastropods. If they do, the condition is called *pycnonephridial*, and this is the condition seen in most neogastropods. If they do not interdigitate, it is referred to as *meronephridial*; although these conditions can overlap a little, they are rather consistent in different families (Ponder 1974; Strong 2003). The carnivorous non-neogastropod caenogastropods and a few neogastropod families (e.g., Volutidae, Marginellidae, Mitridae, Olividae and most Conoidea) have a meronephridial condition (Strong 2003). Andrews (2010) speculated that these differences might be related to shell shape (e.g., long narrow aperture) and other factors involved in proboscis structure and its retraction, as well as in head-foot retraction.

6.7.7.4.4 The Rectal Gland of Neogastropods

In some neogastropods the rectal (= anal) gland is a diverticulum near the distal end of the rectum and typically opens by a narrow duct near the anus. Strong (2003) found that the rectal gland of a cancellariid opens to the rectum along its entire length, a considerable departure from the normal neogastropod configuration, and she also found that the rectal gland of the marginellid *Prunum* opens separately to the mantle cavity.

The structure and function of the rectal gland in two muricids has been investigated by Andrews (1992). It is involved in the extraction (from the blood of the rectal sinus) and metabolism of cations and macromolecules absorbed by the gland of Leiblein. While it is a major site of protein metabolism and contains glycogen and lipids, it does not accumulate purines (Andrews 1992).

At least in muricids, the rectal gland houses extracellular symbiotic bacteria in the lumen and in a pit in each epithelial cell (Andrews 1992, 2010). These (sulphide-oxidising?) bacteria break down molecules, and the metabolites derived from the bacteria are absorbed by the rectal gland epithelium (Andrews 1992).

The rectal gland appears early, prior to torsion, at least in *Stramonita* (D'Asaro 1966), and is a distinct larval structure suggesting it might have originally evolved as a larval organ (Ponder & Lindberg 1997). The absence of this gland in some neogastropods (Fretter & Graham 1962; Ponder 1974; Strong 2003) is correlated with a reduction in the size of the rectal sinus and an increase in renal complexity, suggesting that, in those taxa where it is lacking, its functions may be carried out by the kidney (Andrews 1992, 2010). The lack of a rectal gland in buccinoideans is correlated with a relatively small gland of Leiblein. A rectal gland may be present or absent in those Olivellinae (Olividae) which lack a gland of Leiblein (Kantor 1991) and absent in Harpidae, which have also lost the gland of Leiblein (Andrews 2010). A rectal gland occurs in some conoideans, all of which have the gland of Leiblein modified as a venom gland.

Rectal (or anal) glands are reported from a few other gastropods, but these are not homologous with those in neogastropods, and their function is either not understood or is not excretory (see Chapter 5).

6.7.7.4.5 Non-Marine Caenogastropods

While most caenogastropods are marine or estuarine, some groups live successfully in fresh-water and a few in terrestrial environments. We briefly explore below some of the changes to the renal system that enable these snails to live in non-marine environments.

6.7.7.4.5.1 Fresh-Water Caenogastropods
Fresh-water caenogastropods include the architaenioglossan Viviparoidea and Ampullarioidea, several groups in both the Cerithioidea and Truncatelloidea, and minor incursions in a few other groups (see Chapter 19). An essential prerequisite is the ability to produce copious urine hypoosmotic to the blood, and this has been achieved through different structural adaptations in each group, reflecting the multiple independent incursions into fresh water. An increase in the rate of urine production and a corresponding increased rate of resorption of salts from it are both required. The former requires an increase in the area where filtration occurs, and the latter an increase in absorptive surfaces. Additional efficiency can be obtained by muscles being involved in helping to move the urine, rather than just cilia (Little 1985).

There are major differences between the two fresh-water architaenioglossan groups, the viviparids and ampullariids. Viviparids have well-developed filtration chambers and associated podocytes (Boer et al. 1973; Andrews 1979a), and this remains the only group of fresh-water caenogastropods where the filtration site has been studied in detail. Andrews (1988) noted that the truncatelloidean *Bithynia* had similar filtration pouches to viviparids and that most other examined fresh-water caenogastropods employ auricular podocytes, although podocytes and filtration chambers are lost in the amphibious ampullariid *Marisa* (Andrews 1981), and presumably other ampullariids. In this taxon, filtration occurs through extracellular channels in the epicardium. These channels probably open when blood pressure is high, allowing filtrate to escape (Andrews 1976a). The urine passes quickly from the

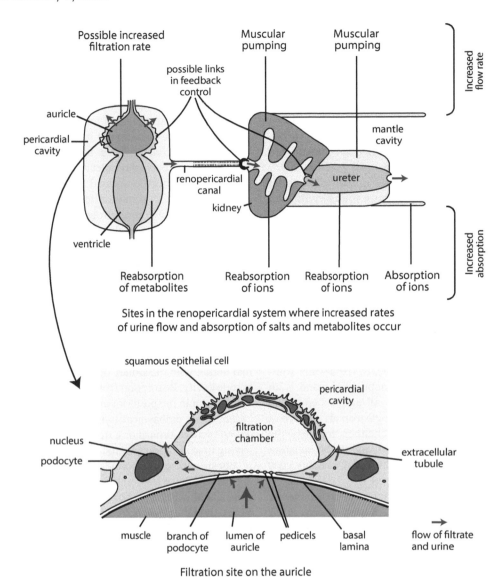

FIGURE 6.26 Renopericardial system of the fresh-water caenogastropod *Viviparus* showing some of the adaptations for living in fresh water. Upper figure redrawn and modified from Little, C., *Am. Malacol. Bull.*, 3, 223–231, 1985; lower figure redrawn and modified from Andrews, E.B., *J. Molluscan Stud.*, 47, 248–289, 1981.

small pericardial cavity into the large kidney, which lacks a nephridial gland, but the epithelium on the left wall of the kidney retains some resorptive function (Andrews & Little 1982; Andrews 1988), although most resorption occurs in the pallial ureter (Andrews 1965b, 1976a).

In viviparids (see Figure 6.26), the kidney lies in the mantle roof and opens via a long, pallial, ectodermal ureter near the anus, anteriorly on the right side of the mantle roof. In marked contrast to ampullariids, the kidney consists of two parts, a much-enlarged portion that consists of a spongy mass full of tubules where excretion occurs,[16] and a small section that lacks dorsal folds. Urine, which has low concentrations of uric acid,

is circulated through the kidney by both ciliary and muscular action. Uric acid is stored in connective tissue associated with the kidney (Andrews 1981), and ammonia is probably excreted via epithelial tissues (e.g., the gill and mantle) or the digestive gland. The ureter is large, contains more urine than the kidney and holds it for longer, enabling ions to be resorbed by the epithelium. It is lined with cells specialised for active transport and which contain glycogen reserves (Andrews 1979a).

In *Viviparus*, there is a very efficient uptake of water and ions, with around 94% of the ions removed from the urine by the kidney and ureter (Little 1965c, b; Burton 1983), resulting in the osmotic pressure of the final urine being about 23% of that of the blood. Initially, glucose (Little 1979) and amino acids (Taylor & Andrews 1987) are resorbed from the primary urine by cells in the walls of the ventricle. The output of urine

---

[16] This part of the kidney in *Viviparus* was incorrectly called a nephridial gland by Andrews (1979a).

in *Viviparus* is very high, being 10–20 times more than that in marine caenogastropods such as *Strombus* (Little 1985).

The ampullariid kidney is subdivided into two chambers, with the posterior chamber largely embedded in the viscera, the anterior being an ectodermal ureter in the mantle roof. The posterior chamber is involved with excretion and water storage, while resorption occurs in the anterior part (Andrews 1979a). The renal excretory cells contain vacuoles with layered concretions containing uric acid which accumulate during aestivation and are released when activity resumes (Little 1968). Most of the ion resorption occurs in the pallial ureter (i.e., anterior kidney chamber) which has a folded epithelium rich in glycogen and is supplied by blood from a branch of the afferent renal vein (Andrews 1965b, 1976a). This separation of the resorptive and excretory functions allows nitrogenous waste to accumulate in the kidney during aestivation and thus is in contrast to the arrangement in viviparids (Andrews 1988).

It is clear that the absorptive parts of the kidney have evolved differently in different groups of fresh-water caenogastropods. Both the ampullariid and viviparid kidneys have an ectodermal pallial ureter, while in some fresh-water cerithioideans the visceral kidney is divided into resorptive and excretory compartments (Delhaye 1977). In fresh-water truncatelloideans generally, the modified nephridial gland is an ovoid structure on the dorsal kidney wall and has epithelial cells like those seen in marine caenogastropods (Andrews 1988). Only remnants of the tubular channels found in the nephridial glands of marine taxa remain, and the blood space under the gland is continuous and contains many amoebocytes. Underlying the gland, the kidney lumen is a large bladder-like space, and an anterior lobe has extended into the mantle roof in some taxa, although the kidney opening remains in the posterior wall of the mantle cavity. This modification is particularly marked in bithyniids where the excretory part of the kidney lies in the mantle roof enabling the resorptive parts, together with the large storage chamber ('bladder'), to expand in the visceral portion[17] (Andrews 1988).

6.7.7.4.5.2 *Terrestrial Caenogastropods* Relatively few caenogastropods have become terrestrial, these including all members of Cyclophoroidea, some Littorinoidea and Truncatelloidea. Each of these groups has invaded terrestrial habitats independently, and consequently, their modifications for terrestrial life differ.

Cyclophoroideans are mostly tropical snails that live in humid habitats and usually aestivate in dry conditions, thus 'avoiding, rather than overcoming, the problems of terrestrial life' (Andrews 1988, p. 420). Their blood has a low osmotic pressure suggesting a fresh-water ancestry (Andrews 1988), and they are uricotelic, with their kidney, which lies mostly in the mantle roof, being a 'kidney of accumulation.' Other adaptations of the cyclophoroidean kidney are shared by other non-marine caenogastropods, namely, its partial migration into the mantle cavity roof and the lack of a nephridial gland.

All the cyclophoroideans examined had two kinds of renal epithelium; the resorptive yellow-green pigmented ciliated cells and vacuolated excretory cells (Andrews & Little 1972, 1982). In nearly all species, these latter cells have a single, very large vacuole containing a concretion made up of layers of excretory material, including uric acid, and it appears as though these are retained for life. Evidence of a kidney of accumulation was also found in the New Zealand cyclophoroidean *Murdochia pallidum* by Morton (1952b) and in the tiny European cyclophoroidean, *Acicula* (Creek 1953), but these have not been examined ultrastructurally. The kidney of *Acicula* does not contain uric acid, but instead concretions composed mainly of mucoprotein (Delhaye 1974b).

Cyclophoroideans reduce water loss by lessening the filtration rate. This is accomplished by a reduction in the number of auricular podocytes in all but the most primitive taxa. The kidney is divided into distinct anterior and posterior sections. The anterior (pallial) part is internally simple and holds urine for resorption of water, ions, and organic molecules. The walls of the posterior (visceral) part of the kidney have internal lamellae, and the epithelium excretes lipids and purines (Andrews & Little 1972, 1982). There is no ectodermal ureter, with the kidney opening to the posterior end of the mantle cavity. Resorption from the released urine occurs in an embayment in the posterior end of the mantle cavity below the kidney. This embayment is lined with an ectodermal cuboidal epithelium typical of a transporting epithelium, including having basal infoldings (Andrews & Little 1972).

Because the cyclophoroidean kidney is modified for osmoregulation, the waste is uniquely secreted by the hypobranchial gland which also performs its usual secretory function. Subepithelial tissue beneath the gland secretes the waste which is passed to the edge of the mantle by a groove on the right side of the mantle cavity (Andrews & Little 1972). The urine from the kidney helps to keep the epithelia of the mantle cavity moist, thus assisting with respiration.

Terrestrial littorinoideans such as the European *Pomatias* (Pomatiasidae) can generally tolerate drier conditions than cyclophoroideans, and differ from other terrestrial snails in having a high blood osmotic pressure which indicates a marine origin (Andrews 1988). This is not surprising given the number of supralittoral taxa found in the related Littorinidae. *Pomatias elegans* exhibits some unusual terrestrial adaptations (Andrews 1988). The auricular filtration chambers are small, thus reducing the filtration rate. The pericardium and kidney lie in the roof of the mantle cavity and, as in other terrestrial gastropods, there is no gill, and the mantle cavity roof is highly vascularised and acts as a lung. In order to minimise water loss, this highly modified kidney has its opening to the mantle cavity at the end of a tube invaginated into the lumen of the kidney (Andrews 1981). In addition, the dorsal wall is expanded into an anterior chamber lined with excretory epithelium and some ciliated cells. Near the renopericardial duct and the kidney aperture, this chamber opens to a smooth-walled posterior bladder partly lined with ciliated resorptive cells. The bladder is homologous with the left kidney wall of other caenogastropods except that, as in other terrestrial

---

[17] A similar modification is seen in the salt-marsh assimineids (Little & Andrews 1977).

caenogastropods, there is no nephridial gland. Ultrastructural examination of the bladder epithelium by Andrews (1981) showed that it was composed of cells specialised for organic and possibly water resorption and that ion-transporting cells were absent. Because the kidney cannot effectively resorb ions, this essential task is uniquely performed by the modified anterior pedal mucous gland. A tube within the foot arises from this gland and is lined with an epithelium capable of taking up ions from water in the soil (Delhaye 1974c, 1974d). This enables these snails to maintain a higher osmotic pressure of their blood than other terrestrial gastropods, even though the osmotic pressure of the urine is only slightly lower than that of the blood (Rumsey 1972).

Like other terrestrial caenogastropods, *Pomatias* lacks a ureter or an exhalant flow from the mantle cavity so, as in most cyclophorids, it stores nitrogenous waste for long periods, even for life (Kilian 1951; Morton 1952b; Martoja 1975). Unlike cyclophorids, this waste is not stored in the kidney (Martoja 1975), but instead substantial amounts of uric acid are deposited for storage in connective tissue in the blood spaces of the digestive gland just behind the kidney (the so-called 'concretion gland') (Creek 1951; Kilian 1951; Little 1981). There is a lack of uric acid in the kidney because the nitrogenous waste has been removed from the blood to the concretion gland prior to it reaching the excretory part of the kidney. Instead, the kidney in *Pomatias* accumulates melanins and lipofuscins in the excretory cells (Delhaye 1974b; Martoja 1975), although small amounts of uric acid (produced by an epithelium on the anterior pericardial wall and around the renopericardial duct[18]) remain in the urine (Andrews 1988). The blood supply to the anterior pericardial wall is via the homologue of the efferent vein of the nephridial gland. Andrews (1988) hypothesised that ammonia would normally diffuse from the pulmonary vessels into the mantle cavity, but that, during aestivation, it would accumulate in the blood and be removed by the pericardial cells. Unusually, the deposits in the concretion gland may be used as a source of nitrogen when needed, a process that may be aided by enzymes from both intracellular and extracellular symbiotic bacteria in the connective tissue (Kilian 1951; Andrews 1988).

The supralittoral semi-terrestrial littorinid *Cenchritis muricatus* is modified even more than *Pomatias* by the replacement of the filtration chambers with extracellular tubules that penetrate both the auricular and ventricular walls, thus reducing the amount of primary urine formed. Also, the nephridial gland is reduced, and the excretory cells contain single large concretions of phospholipids and calcium salts similar to those found in some other caenogastropods (Emson et al. 2002). In the terrestrial assimineid *Pseudocyclotus* (Truncatelloidea), podocytes are also absent and, as in *Cenchritis* and ampullariids, filtration is achieved by way of extracellular channels through the lining of the auricular wall (Little & Andrews 1977) (see Section 6.7.7.4.5.1). The kidney of *Pseudocyclotus* has a reduced nephridial gland and the left side, which is expanded into the mantle roof, is involved with

resorption of organic solutes. The excretory area of the kidney is large and layered excretory concretions containing purines are accumulated (Little & Andrews 1977).

6.7.7.5 Heterobranchia

Ancestral heterobranchs were small, shelled, and had probably lost the ctenidial gill (Ponder & Lindberg 1997). Although Andrews (1988) suggested that podocytes were lost in the euthyneuran ancestor, they have been shown to be present in some nudibranchs (see next Section), as well as Sacoglossa and Acochlidia, but are apparently absent in Hygrophila and Stylommatophora. There are no studies to date on the fine structure of the heart complex in 'lower heterobranchs'.

All living 'lower heterobranchs' have the kidney located in the mantle roof (Ponder & Lindberg 1997) or dorsal body if the mantle cavity is very reduced or lost, where it acts in part as a respiratory surface. As in most other gastropods, the kidney has an anterior opening. Heterobranchs do not have a nephridial gland, and the pigmented ciliated cells that resorb organic solutes in other gastropods are also absent. Instead, the heterobranch kidney has a resorptive function, and the digestive gland is the main site of excretion of uric acid and purines (Andrews 1988) (Figure 6.27).

6.7.7.5.1 'Opisthobranchs'

In 'opisthobranchs,' the single kidney is typically present in the roof of the mantle (as in lower heterobranchs), but, due to detorsion, is rotated posteriorly and many have a plicatidium (see Chapter 4). The trend toward shell loss has enabled the use of the general body surface for gas exchange and excretion of ammonia. Detorsion, shell, and mantle cavity reduction and loss, and the growth of the mantle skirt have resulted in major changes in the blood supply to the kidney compared with the patterns in other gastropods (Brace 1977a; Andrews 1988). Notably, there has been a change of blood flow, from right to left instead of posterior to anterior; most of the blood flowing to the kidney is partially to well oxygenated, and little of it has passed through the viscera, including the digestive gland. This latter change resulted in the marked reduction or loss of the kidney as a site of nitrogenous excretion (Andrews 1988).

Despite earlier reports that podocytes were absent in 'opisthobranchs,' more recent investigations have shown that they can be present. They have been reported on the auricular wall of the pelagic gymnosome and thecosome 'pteropods' (Fahrner & Haszprunar 2000), the sacoglossans *Bosellia* (Fahrner & Haszprunar 2001) and *Elysia* (Neusser et al. 2018), the interstitial Acochlidia (Fahrner & Haszprunar 2002b), and the nudibranch *Hypselodoris tricolor* (Fahrner & Haszprunar 2002a). *Hypselodoris* differs from other 'opisthobranchs' (and all other molluscs) in that, in addition to auricular podocytes, the outer pericardial endothelium is also made up of flat podocytes. Although very few taxa have been examined, it is apparent that not all nudibranchs share this arrangement as it is not present in aeolidiids, but it is probable that it may be present in at least closely related dorid-like taxa (Fahrner & Haszprunar 2002a). In most examined 'opisthobranchs' that have podocytes, the auricular epicardium is

---

[18] Andrews (1988) noted that a similar arrangement occurs in *Haliotis* and *Truncatella*.

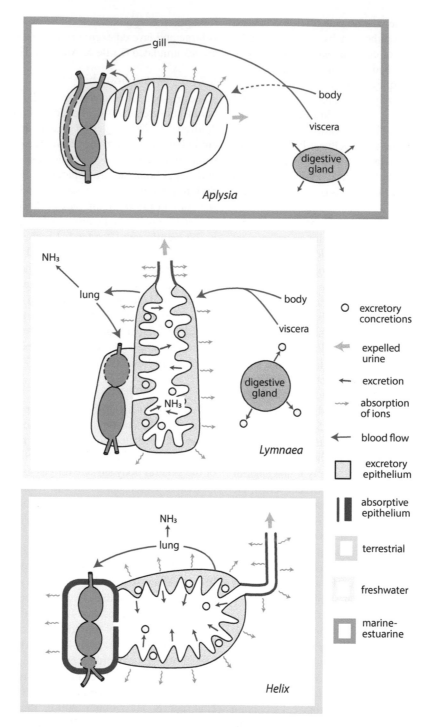

FIGURE 6.27   Kidney modification in higher heterobranchs comparing a marine euopisthobranch (*Aplysia*), a fresh-water hygrophylian (*Lymnaea*), and a terrestrial stylommatophoran (*Helix*). Redrawn and modified from various sources.

the site of filtration. Some 'opisthobranchs' differ, especially small ones that have lost the heart. These include the probably paedomorphic, interstitial cephalaspidean *Philinoglossa helgolandica* which has podocyte-like cells in the pericardial wall that are the probable site of filtration, and it is likely that these are homologous to the podocytes in other molluscs (Bartolomaeus 1997).

All examined 'opisthobranchs' have solitary pore cells (rhogocytes) (see Section 6.5) in the haemocoel and in connective tissue. Because they are structurally similar to podocytes, Fahrner and Haszprunar (2002a) suggested that these may provide additional filtration sites. This is, however, unlikely as it is unclear as to how they might actually function in this role (E. B. Andrews pers. comm. 2011). The minute sacoglossan *Alderia modesta* lacks not only a heart, but also the pericardium, and there is no evidence of any cells capable of filtration in this species (Fahrner & Haszprunar 2001; Fahrner 2002).

As in other gastropods, the renopericardial duct in 'opisthobranchs' connects the pericardial cavity with the kidney. The renal epithelia and the histology of the renopericardial duct appear to be rather uniform (Fahrner & Haszprunar 2002a). Notably, the renal epithelium has only a single cell type, with apical microvilli and basal infoldings. The cells contain large vacuoles as well as endosomes, lysosomes, residual bodies, and a large number of mitochondria and dense granules (glycosomes) (Andrews 1988; Fahrner & Haszprunar 2002a). This epithelium is apparently mainly responsible for reabsorption of organic solutes and, although being lined with vacuolated cells containing simple granules, excretory activity is low (Andrews 1988). Another indication that the kidney is not a major site of nitrogenous excretion is, as noted above, due to its isolation from the digestive gland (Andrews 1988), and most such excretion occurs in the digestive gland (Fretter 1939). There is also no absorptive surface in the pericardial cavity (Andrews 1988). On the basis of these observations, Andrews (1988) suggested that the renal epithelium, as well as having some excretory function, resorbs organic solutes, but lacks histological specialisation because of the low rate of urine production. This is similar to the situation in 'pulmonates' (see below).

The so-called 'tectibranch opisthobranchs' have the primitive vascular connection between the kidney and the auricle (the main efferent renal vein), although it is not known whether there is any tidal flow in it, as demonstrated in some other gastropods (see Sections 6.6.3 and 6.7.7.4.2). Where a plicatidium is present, the gill muscles may supplement the pumping action of the heart (Andrews 1988) with neural inhibition of the heartbeat when the gill contracts, as demonstrated in *Aplysia* (Kandel 1976).

'Opisthobranchs' show a diverse array of vascular patterns and kidney development made possible by the reduction and eventual loss of the mantle cavity in several lineages (see Chapter 20). This has resulted in the original pallial kidney being located in the dorsal body wall or in the main haemocoel, and its originally compact form becoming diffuse or branched.

The mechanisms facilitating osmoregulation are in place in at least some 'opisthobranchs.' For example, some species of *Aplysia* will stop producing an antidiuretic neurohormone from the visceral ganglion if it is exposed to fresh water (e.g., in an intertidal pool in the rain) (e.g., Weel 1957; Skinner & Peretz 1989) (see also Chapter 2), via a response from the osphradium (Stinnacre & Tauc 1969), while others apparently do not osmoregulate (Scemes et al. 1991). The only freshwater 'opisthobranchs' are some acochlidians.[19] While marine acochlidians have a simple sac-like kidney with a short duct (Neusser et al. 2006; Jörger et al. 2008; Neusser et al. 2009),

the fresh-water taxa have a well-developed kidney divided into two chambers, and there is a long pallial duct where resorption of ions takes place (Neusser & Schrödl 2009).

6.7.7.5.2 'Pulmonates'

There have been several reviews of the 'pulmonate' excretory system (Delhaye & Bouillon 1972a–c; Andrews 1988; Luchtel et al. 1997). The heart and kidney both lie in the mantle roof, but in contrast to 'opisthobranchs,' the kidney in 'pulmonates' is associated with the digestive gland and the excretory part receives venous blood from it (Andrews 1988). The epithelium of the excretory part of the kidney is lined with cuboidal to columnar vacuolated cells similar to those in other gastropod taxa, but they carry out both resorptive and excretory functions, as in 'opisthobranchs.' The vacuoles in the renal cells contain layered concretions that accumulate deposits of uric acid and purines, with uric acid predominant in slugs and purines in snails (Riegel 1972). Blood vessels in the ends of the renal lamellae drain the blood from the kidney, and these connect either directly with the pulmonary vein or by way of a pair of efferent renal veins. Thus, the nitrogenous waste has been removed from the blood reaching the auricle from the kidney and is likely to be oxygenated because the kidney is in close contact with the lung and is thus probably also oxygenated (Andrews 1988).

In the few well-studied 'pulmonate' taxa (all non-marine), the site of filtration varies, with four different sites identified, all in the renopericardial system (Andrews 1988; Luchtel et al. 1997; Fahrner 2002). These are: (1) in the epicardium of the auricle or (2) ventricle via paracellular filtration, (3) by way of paracellular or transcellular filtration in parts of the kidney, or (4) restricted to a small specialised area of the kidney with arterial haemolymph supply, the latter with podocyte-like cells found on a small part of the kidney supplied with arterial blood (Matricon-Gondran 1990).

6.7.7.5.2.1 Marine and Estuarine 'Pulmonates' Some families of marine 'pulmonates' are restricted to fully marine habitats including the Siphonariidae, Trimusculidae, Otinidae, Smeagolidae, and some Ellobiidae. Almost all are intertidal to supralittoral, an exception being the sublittoral *Williamia* (Marshall 1981). Many members of the Ellobiidae and all Amphiboloidea occupy estuaries and can be extremely abundant in the high intertidal zone of mangroves and saltmarsh habitats.

The kidney epithelium is composed entirely of vacuolated excretory cells. These cells may be able to resorb electrolytes and water, and possibly organic solutes; they show alkaline phosphatase activity in the cell bases and around the excretory vacuole (Delhaye & Bouillon 1972a; Barker 2001a). Although the kidneys of marine 'pulmonates,' including Amphiboloidea and some Ellobioidea, lack a ureter, the kidney of some ellobiids has a simple distal pouch-like extension at least analogous with the orthureter (see Section 6.7.7.5.2.3) in some non-marine eupulmonates (Delhaye & Bouillon 1972a, 1972b; Nordsieck 1985; Tillier 1989).

Surviving in intertidal habitats necessitates the ability to cope with fresh-water influx and salinity increases due to

---

[19] Although traditionally thought to be 'opisthobranchs,' in recent molecular phylogenies this group nests within the taxa treated traditionally as 'pulmonates' and are now included in the clade Panpulmonata (see Chapter 20).

evaporation. *Siphonaria* has some ability to osmoregulate by increasing the pool of soluble amino acids (Bedford 1969), but these limpets mainly tend to escape salinity changes by clamping to the rock surface (McAlister & Fisher 1968). Ellobiids that live in the supralittoral zone on fully marine shores must cope with desiccation, but the ways they manage this have not yet been studied. 'Pulmonates' living in brackish water must be able to osmoregulate. The New Zealand amphibolid, *Amphibola crenata*, is an effective osmoregulator (Little et al. 1984), but surprisingly, very few ellobiids have been studied in this regard. The saltmarsh western Atlantic *Melampus bidentatus* is the most investigated in which osmoregulation (Price 1980) appears to be under the control of neurohormone peptides (Khan et al. 1999). A traditional scenario is that estuarine members of the Ellobiidae, or its ancestors, are related to the ancestors of non-marine 'pulmonates' which first colonised the land and then fresh-water habitats (e.g., Morton 1955a). While it is possible that the ancestors of stylommatophorans and the hygrophilans occupied estuarine habitats, recent molecular phylogenies do not support the idea of ellobiids being basal (Holznagel et al. 2010; Dayrat et al. 2011).

6.7.7.5.2.2 Fresh-Water 'Pulmonates' (Hygrophila)

The excretory system of hygrophilans is similar to that of terrestrial stylommatophorans (see next Section) with additional adaptations to their fresh-water life. As in other 'pulmonates,' both the heart and kidney lie in the mantle skirt. As with terrestrial 'pulmonates,' podocytes are lacking in the heart, leading to the incorrect suggestion that the kidney is the site of urine formation (Khan & Saleuddin 1981a, 1981b). As discussed by Andrews (1988), lymnaeids and planorbids appear to have independently evolved the means of filtration via the auricle. In lymnaeids, it is by way of modified epicardial cells with a sieve formed from basal processes somewhat convergent with those of podocytes, while in planorbids, it is via channels through the epicardial epithelium. Because the auricle is the site of filtration, and urine production is high, the renopericardial duct is much larger than in stylommatophorans (Andrews 1988).

The greatly increased osmoregulatory demands of hygrophilans result in the production of copious hypoosmotic urine, this probably being the major osmoregulatory mechanism (Khan & Saleuddin 1979). They have a simple mesodermal primary ureter (referred to as a 'nephridial pouch' by Delhaye & Bouillon 1972a), or orthureter, but, unlike many stylommatophorans, there is no ectodermal secondary ureter. The epithelia of both the kidney and primary ureter are specialised for the resorption of ions (Andrews 1988).

Unlike fresh-water caenogastropods, uric acid is accumulated in both the kidney and digestive gland in hygrophilans, which they excrete in layered concretions (as in stylommatophorans – see next Section), perhaps suggesting a terrestrial or semi-terrestrial ancestry (Andrews 1988). Ammonia is the main nitrogenous excretory product in hygrophilans, although the details vary. For example, when the planorbid *Biomphalaria* is living in water, it excretes

about four times the amount of nitrogen as ammonia than urea, but when it is aestivating, starved, or parasitised, urea production is markedly increased, probably to detoxify the ammonia (Becker & Schmale 1978). Similarly, lymnaeids are ammonotelic when aquatic, but, when aestivating, become uricotelic (e.g., Newman & Thomas 1975). Thus urea can also be a major part of the nitrogen excreted in the freshwater hygrophilans under certain circumstances (Freidl 1974; Schmale & Becker 1977; Freidl 1979).

6.7.7.5.2.3 Terrestrial Eupulmonates

The stylommatophorans, and to a lesser extent, systellommatophorans, successfully invaded a wide range of terrestrial habitats. To achieve this, they evolved a suite of behavioural and physiological adaptations, with the latter particularly involving the excretory system.

Many stylommatophorans can withstand great variation in body water content. Such fluctuations may occur during dry spells or aestivation, but also during daily activities. To enable rehydration, water can be taken in rapidly via food, the pneumostome, or the integument.

Water conservation can also be enhanced by behaviour (see Section 6.2.3) and, when aestivating, the aperture can be sealed with dried mucus to form an epiphragm that functions rather like an operculum, or the aperture is sealed with mucus to a hard surface such as wood or rock (Riddle 1983; Solem 1985) (see Chapter 20).

As with hygrophilans, it is often incorrectly thought that there is no pericardial filtration in stylommatophorans because the heart lacks podocytes. It has been stated that filtration occurs directly from the renal capillaries into the kidney (Vorwohl 1961; Martin et al. 1965) suggesting that there is high blood pressure in these capillaries. While some experimental evidence has been presented for such filtration by way of intercellular spaces (e.g., Newell & Skelding 1973), this is, at best, equivocal (Andrews 1988). Andrews (1988) discussed the evidence for filtration via the heart. The ventricle of *Cornu aspersum* has a thin-walled 'bulbous appendage' that can be isolated from the ventricle by muscular contraction. The epicardium of this structure was 'permeated by channels,' providing a possible means of filtration. In an unpublished experiment, E. B. Andrews injected a vital dye (methylene blue in Ringer solution) into the efferent pulmonary vein of *Cornu aspersum* and a few minutes later part of the auricular epicardium and the pericardium near the renopericardial opening was stained, indicating that it had crossed the auricular epicardium (E. B. Andrews, pers. comm. 2012). These data suggest that filtration may be occurring in the heart in addition to the kidney, as shown in Figure 6.28.

As might be expected in such a large group as the stylommatophorans, there is some diversity in their excretory systems (Delhaye & Bouillon 1972a–c). The kidney consists of three morphologically and functionally distinct sections – the broad kidney proper that lies proximally in the mantle cavity roof is lamellate internally, and connects with the pericardium by way of a renopericardial opening. A simple pouch (*orthureter*) is present in only the primitive orthurethran taxa, and in most

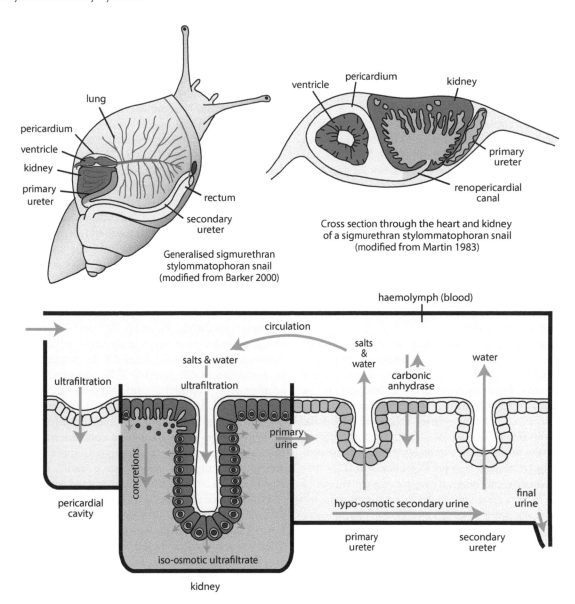

FIGURE 6.28 The kidney structure and the physiological processes involved in the renopericardial system of a helicoid snail. Figure on upper left redrawn and modified from Barker, G.M., Gastropods on land: Phylogeny, diversity and adaptive morphology, in Barker, G.M. (ed.), *The Biology of Terrestrial Molluscs*, pp. 1–146, CABI Publishing, Wallingford, UK, 2001a; upper right redrawn and modified from Martin, A.W., Excretion, in Saleuddin, A.S.M. and Wilbur, K.M., *Physiology, Part 2. The Mollusca*, Vol.5, pp. 353–405, Academic Press, New York, 1983; lower figure redrawn and modified from Vorwohl, G., *Z. Vgl. Physiol.*, 45, 12–49, 1961.

an ectodermal secondary ureter may be represented by either an open ciliated groove or a closed tube of various lengths (see below). The closed ureters have internal transverse lamellae (Delhaye & Bouillon 1972c), and an absorptive function.

In stylommatophoran snails, the kidney is supplied with oxygenated blood by a renal artery, but in slugs, there is no direct arterial supply to the kidney (Duval & Runham 1981). Instead, both ureters are bathed in oxygenated blood from the pulmonary veins, and then this blood (in slugs) or part of it (in snails) passes through the excretory folds of the kidney, where nitrogenous waste is removed before it drains into the auricle (Andrews 1988).

The excretory folds in the kidney are lined with vacuolated excretory cells, which also have a resorptive function as shown by their cell structure, which includes well-developed microvillae. In addition, glycogen deposits suggest that glucose absorption occurs (Andrews 1988). On the basis of the characteristics of the basal infolding, Andrews (1988) considered resorption of ions to be unlikely. A large apical vacuole in each cell contains a layered concretion made up of calcium salts, uric acid, and purines. These concretions are eventually expelled from the cell into the kidney lumen (Bouillon 1960; Runham & Hunter 1970; Andrews 1988). The layers of the concretions are progressively built up by way of contributions from the smooth endoplasmic reticulum (Skelding 1973), in a very similar way to those in the Cyclophoridae and Ampullariidae (see Sections 6.7.7.4.5.1 and 6.7.7.4.5.2).

The ureter is involved in the reabsorption of water. The primary ureter is derived from the nephridial sac mesoderm and the secondary from invaginated ectoderm. The primary ureter is lined with an epithelium comprised of ciliated cells, cells with a brush border, and extensive basal interdigitations with mitochondria (Runham & Hunter 1970). The cells lining the secondary ureter also have a brush border and have dense cytoplasm and mitochondria. Most resorption of solutes and water from the primary urine occurs in the primary ureter (e.g., Martin et al. 1965), the rest in the secondary ureter. The resorbed fluid is hyperosmotic to the blood plasma, and the final urine is more alkaline than the blood due to insoluble nitrogenous materials being excreted into it (Andrews 1988).

In addition to resorption in the ureters, some resorption of organic solutes probably also occurs in the pericardial cavity, as indicated by microvilli and glycogen deposits in the epicardium of *Helix* (Andrews 1988).

The differing condition of the ureter in stylommatophorans has been used as an important character in their classification (see Chapter 20 for details). The most primitive stylommatophorans have usually been considered to be the orthurethran taxa (Pupilloidea, Chondrinoidea, and Partuloidea), but in recent phylogenies this is no longer the case (see Chapter 20). In the orthurethran group, the kidney opens to a primary ureter (the orthureter), similar to the condition in some hygrophilans (see previous Section). This ureter lacks ciliated cells and typically opens into an excretory 'uretal groove' which runs anteriorly to terminate near the pneumostome. In the more modified stylommatophorans, this uretal groove is replaced by an ectodermal secondary ureter. Unlike the simple orthureter, the primary ureter in these taxa is tube-like with folded walls and lies along the margin of the kidney. The secondary ureter is long and straight and runs parallel to the rectum, opening near the pneumostome. Mesurethra (including the Clausiliidae and Strophocheilidae) lack the primary ureter (orthureter), and the secondary ureter is represented by an open groove. In the Sigmurethra, both the primary and secondary ureters are closed and have folded walls, thus increasing the absorptive surface area. This latter group includes most of the typical stylommatophoran snails and slugs. The Succineidae (the so-called 'Heterurethra') (Delhaye & Bouillon 1972b) have ureters like those of the Sigmurethra and are often included with them.

While it is generally argued that the simple, pouch-like orthureter is found only in the orthurethran taxa, a pouch-like differentiation in the kidney occurs in a number of non-orthurethran families, and the folding of the walls of the primary ureter may be related to body size (Tillier 1989). Some issues relating to the use of kidney structure in the classification of stylommatophorans are discussed further in Chapter 20.

Stylommatophorans are not the only terrestrial panpulmonates, with some in the Systellommatophora and Ellobioidea. A secondary ureter which runs posteriorly has evolved independently in the terrestrial Veronicellidae and Rathousiidae

(Systellommatophora), but there is no secondary ureter in the mainly estuarine onchidiids (Delhaye & Bouillon 1972b), a group which also contains a small number of terrestrial taxa (Dayrat 2009).

Stylommatophoran snails and slugs are primarily purinotelic, enabling nitrogen excretion with reduced water loss (Riddle 1983). Experiments have shown that the levels of ammonia and uric acid in tissues are controlled at least in part by neurohormones (see Chapter 7). Slugs use both the purine and arginine pathways for removal of ammonia and often produce more urea than purines for excretion. To minimise water loss, slugs excrete semi-solid nitrogenous waste, with uric acid comprising about half of the nitrogen excreted in this way. About 15% is xanthine, and there are smaller quantities of guanine, hypoxanthine, and adenine (e.g., Horne & Beck 1979). Slugs are able to synthesise urea as they have the necessary urea cycle enzymes. Consequently, they excrete about 50% of their total nitrogen as urea and ammonia release is very small. In *Limacus flavus*, for example, urea and purines, respectively, make up 59% and 41% of the excreted nitrogen, with urea excreted in the urine while the purines are deposited in a semi-solid form. Of the purines excreted, uric acid makes up 64%, guanine 22%, and xanthine 14%, while adenine or hypoxanthine were also sometimes detected (e.g., Horne & Beck 1979). Helicid snails excrete most nitrogenous waste matter as semi-solids or solids in the form of uric acid, as well as guanine and xanthine concretions, and generally excrete little ammonia (e.g., Speeg & Campbell 1968) or urea, although this changes in aestivating snails, with substantial quantities of ammonia gas being excreted (see below).

Excreta from the ureter is voided through the pneumostome, and in some stylommatophoran land snails, this is in the form of firm pellets deposited at sometimes lengthy intervals. Thus solid crystals can be moved through the system, but can also be flushed when water is available. In the slug *Deroceras*, 92% of the total nitrogenous waste is made up of uric acid and xanthine, and this largely solid waste is expelled from the kidney daily in the early evening (Runham & Hunter 1970), while in contrast, *Helix pomatia* sheds its excreta in anything from two to 20 days (Vorwohl 1961). *Mesomphix vulgatus*, a small, short-lived stylommatophoran, does not void excretory material and presumably stores all its nitrogenous waste (Badman 1971); this may be more common than currently recognised to date as most studied species are relatively large.

The urea synthesised in stylommatophorans may accumulate and reach high levels during aestivation or fasting (e.g., Horne 1971; Hiong et al. 2005). Thus, for example, in *Helix pomatia*, the relative proportion of uric acid in the kidneys of hibernating individuals is nearly twice that of active individuals (Florkin 1966), while urea levels in the body increase considerably in *Bulimulus*, perhaps helping to reduce water loss by evaporation (Horne 1971).

The purines are synthesised *de novo* rather than by degradation of nucleotides (Bishop et al. 1983), and urea is also formed *de novo* from ammonia (NH_3) and by arginine metabolism (e.g., Campbell & Speeg 1968; Horne & Boonkoom 1970). The general low to very low levels of urea in stylommatophoran snail tissues and its virtual absence in the excreta (Campbell & Speeg 1968; Tramell & Campbell 1970; Weiser & Schuster 1975) is apparently due to high activity of the enzyme urease that breaks down urea into ammonia and carbon dioxide as soon as it is formed (Campbell & Speeg 1968). Thus, as in many other animals, the arginine-ornithine-urea cycle is important, especially during aestivation, playing an important role in ammonia detoxification (e.g., Hiong et al. 2005) and reducing evaporative water loss (Horne 1971). In stylommatophorans where urease activity is absent or low, especially in slugs (Horne 1977a, 1977b), and in snails during aestivation (Horne 1973a, 1973b), urea can contain a major part of the nitrogen excreted, as it can in the fresh-water hygrophilans (see Section 6.7.7.5.2.2).

Experiments on the land snail *Bulimulus dealbatus* have shown that urea accumulates more rapidly during aestivation (Horne 1971), suggesting that urea, in that case at least, was involved in ammonia detoxification because protein degradation had increased as a result of fasting (Horne 1973a, 1973b). This view is also supported by recent experiments on *Achatina fulica* that have shown the rate of urea synthesis increases, as do the activities of the enzymes involved in the ornithine-urea cycle, in response to an increase in ammonia as a result of fasting or aestivation (Hiong et al. 2005). Thus at least some of the physiological functions of urea may be unrelated to nitrogenous excretion or osmotic water retention in animals capable of aestivation.

Part of the nitrogenous waste in stylommatophorans is expelled as gaseous ammonia via the body surface and the lung, and some diffuses through the shell. Ammonia gas may be excreted in small quantities, as in the slug *Deroceras reticulatum*, where less than 1.5% of the total nitrogenous waste is lost this way, but in active snails it can be around 5%, and in aestivating snails as much as 30% (Andrews 1988). Andrews suggested that these differences may be due to the different arrangement of the kidney blood vessels in snails and slugs as outlined above.

6.7.7.5.3 Why Are There No Terrestrial 'Opisthobranchs?'

Andrews (1988) asked an interesting question – given that there are many terrestrial 'pulmonates,' why have the 'opisthobranchs'[20] not produced any terrestrial lineages, especially given the number of terrestrial eupulmonates with reduced or absent shells? The mantle cavity is commonly reduced or lost in 'opisthobranchs,' but is retained as a lung in most 'pulmonates.' In addition to other necessary changes, such as a reduction in permeability of the external body surface (e.g., Machin 1975) coupled with an increase in the thickness of the body wall muscles that serves to reduce the amount of blood near the surface (Andrews 1988), 'pulmonates' have a smaller blood volume compared with 'opisthobranchs' (Burton 1983; Jones 1983b). In addition, Andrews (1988) argued that two major factors relating to the renal system in 'pulmonates' are also responsible for their success in non-marine environments; (1) the production of waste from nitrogen metabolism that does not interfere with resorption in the kidney, and (2) the development of a ureter that undertakes additional resorption. Like the lung, the pulmonate ureter lies within the mantle cavity. Facilitating these changes is a vascular system that supplies oxygenated blood to the sites of resorption. In contrast, the absence of both renal and pallial sites of ion uptake in 'opisthobranchs' may at least partly explain their restriction to the marine environment.

---

[20] The semi-terrestrial genus *Aiteng* in the Acochlidia (Neusser et al. 2011), is an exception. The Acochlidia are mostly marine but also have a few fresh-water members. Acochlidians have traditionally been treated as 'opisthobranchs,' but along with Sacoglossa and Pyramidelloidea, now form part of the Panpulmonata.

7 Nervous System, Sense Organs, Learning and Behaviour

7.1 GENERAL INTRODUCTION

In this chapter, we cover the nervous system and sensory structures and outline aspects of molluscan behaviour and learning. Sensory receptors (which may be clustered and specialised as sense organs) collect external cues and the neurons, comprising the nervous system, transmit this information to relevant parts of the body with responses such as reproductive behaviour, escape, or feeding being initiated. The nervous system is also involved in initiating innate responses and maintaining and coordinating essential bodily functions such as digestion, respiration, excretion, and reproduction. The nervous system also underpins the behaviours, memory, and learning demonstrated in molluscs.

The molluscan nervous system is more variable and diversified than that of other protostomes, and, unlike annelids and arthropods, the brain is often not clearly divided into compartments (Faller et al. 2012). The variety of configurations of the nervous system is paralleled by a diverse range of sense organs, many of which have independently evolved in each taxonomic group.

As in vertebrates, the nervous system has two main components, the central nervous system (CNS) – the 'brain' with both sensory and motor cells – and a peripheral nervous system (PNS). The latter includes a somatic nervous system involving voluntary muscle movements and an autonomic nervous system responsible for involuntary movements of the heart, gut, blood vessels, and some glands.

The basic nervous system of molluscs consists of a concentration of nerves around the head region, the cell bodies of which are typically clustered in *ganglia*. The molluscan CNS usually consists of four main paired ganglia, the cerebral, pleural, pedal, and buccal; a fifth pair, the visceral ganglia, is present in scaphopods, gastropods, and bivalves. Pairs of ganglia are connected by *commissures*, while different ganglia are joined by *connectives*. The PNS plesiomorphically includes paired lateral and pedal nerve cords that extend through the mantle and foot, respectively, but this system has been modified in cephalopods, bivalves, scaphopods and many gastropods. Ladder-like cross connections[1] between the pedal nerve cords are present in aculiferans and many vetigastropods, but are absent in monoplacophorans and patellogastropods.

The CNS and the main components of the sensory equipment are in the head region in gastropods and cephalopods, where eyes and tactile organs such as tentacles are present. In conchiferans, statocysts used for orientation are associated with the pedal ganglia, and a pair of chemosensory osphradia is typically found in the mantle cavity. Epidermal unicellular photoreceptors (*phaosomes*) are nearly always present, except in cave or deep-water inhabitants. Complex, eye-like structures occur on the dorsal surface of chitons and onchidiid slugs and, in some bivalves, on the mantle edge. Most gastropods have small cephalic eyes, while most cephalopods have large, complex cephalic eyes that parallel those of vertebrates.

The nervous system is composed of two main cell types – neurons (including both neurosecretory and excitatory neurons) and glial cells. These are described below.

7.2 THE MOLLUSCAN NERVOUS SYSTEM

Molluscan nerve cords are made up of clusters of axons with their cell bodies more or less interspersed along their lengths. They differ from those of vertebrates in lacking a hollow embryonic nerve cord, and the axons are not surrounded by a myelin sheath. While there is no such sheath in molluscs, independently developed myelinated nerves are found in some arthropods and annelids (Roots 2008). Some major molluscan nerves are single axons.

The basic arrangement of the nervous system and CNS ganglia in various molluscs is shown in Figure 7.1. Nerve cords run from the anterior nerve clusters or ganglia into the viscera and through the foot where additional ganglia are usually located. Plesiomorphically, there is an anterior *circumbuccal* (or, if behind the buccal mass, a *circumoesophageal*) nerve ring with nerves to the head region, a pair of pedal nerve cords that innervate the foot, and a pair of visceral nerve cords that innervate the viscera and mantle. The anterior nerve ring (typically comprised of the cerebral, pleural, and pedal ganglia, and including the buccal ganglia) lies at the anterior end of the buccal mass in most molluscs. In all gastropods, the cerebral ganglia are initially anterior to the buccal mass during development, but move behind in cephalopods, a few vetigastropods, most caenogastropods, and in a few 'opisthobranchs' (e.g., *Haminoea*, *Adalaria*, and aplysiids) (Ponder & Lindberg 1997).

In more advanced ganglionated nervous systems, such as in many gastropods or cephalopods, the cerebral ganglia innervate the head region, the buccal or supraoesophageal ganglia innervate the buccal area, salivary glands, oesophagus, and stomach, the pedal ganglia the foot, the pleural ganglia the lateral body wall and mantle, and the visceral ganglia the viscera, posterior gut, and posterior body.

Earlier studies on nervous systems relied on standard dissection and histology, but recent studies, for example on the nervous systems of aplacophorans and polyplacophorans, have

---

[1] We refer to these as 'connectives', but they are sometimes called commissures.

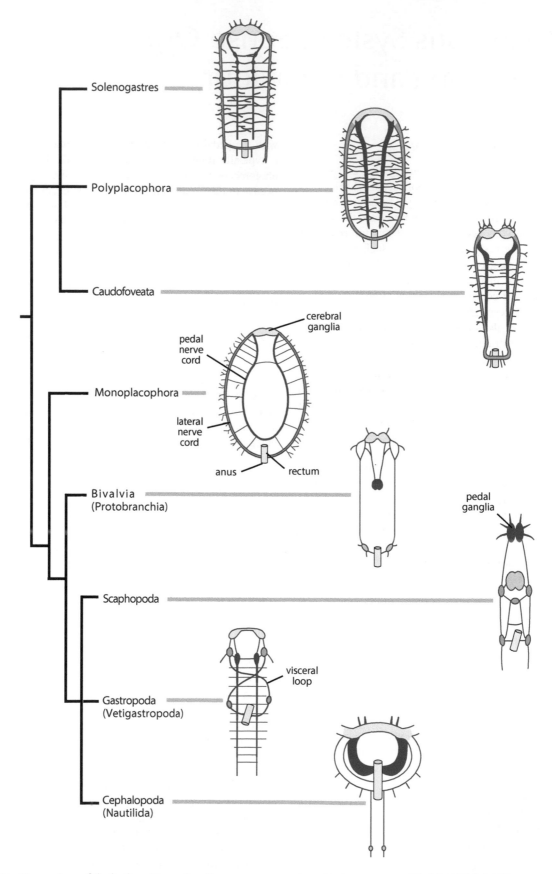

FIGURE 7.1 Comparison of the basic patterns of molluscan nervous systems. Redrawn and modified from Salvini-Plawen, L. v., *Z. wiss. Zool.*, 184, 205–394, 1972.

used new techniques such as 3D reconstruction and fluorescent markers for the neurotransmitters serotonin and FMRFamide (phenylalanine-methionine-arginine-phenylalanine amide), as well as acetylated α-tubulin, a structural protein in neurons (e.g., Shigeno et al. 2007; Todt et al. 2008; Faller et al. 2012). Sometimes acetylated α-tubulin is not present in molluscan neurons (e.g., Jackson et al. 1995).

Initially, the origin of the gastropod nervous system was described as the delamination of cells from the ectoderm (see Chapter 8), with the cerebral ganglia arising from the initially formed cephalic plates and initiated by the neurons of the apical sense organ (ASO) of the larva (Raven 1966). However, Croll and Voronezhskaya (1996) reported that the first features of the nervous system of *Lymnaea stagnalis* were three posteriorly located cells with anteriorly directed fibres which provided the scaffolding for the formation of ganglia and their interconnections. These cells are called EFAP cells (early Fa-LIR,[2] anteriorly projecting). The number and location of EFAP cells differs between taxa, and their origin in the caudal region (telotroch) rather than from the ASO has now been reported in the caenogastropod *Crepidula fornicata* and the heterobranchs *Aplysia californica*, *L. stagnalis*, and *Biomphalaria glabrata* (Nezlin & Voronezhskaya 2017). In polyplacophorans, the earliest components of the nervous system were seen in peripheral neurons which projected into the apical organ. Among these cells, two pairs of exceptionally large lateral cells determine the pathways of the developing adult nervous system (Voronezhskaya et al. 2002). The pedal and lateral nerve trunks grow backwards from the cerebral ganglia, and the pedal trunks are connected by commissures. The more peripheral ganglia develop later, including the buccal ganglia and, if present, the subradular ganglia (Friedrich et al. 2002). In gastropods, each of the circumoesophageal (or circumbuccal) ganglia is usually formed from separate primordia.

There is a question as to the homology of the visceral (or mantle [= palllial] or lateral) nerve cord in different classes because it lies between the dorso-ventral muscles in Solenogastres, outside the shell muscles in monoplacophorans, polyplacophorans, bivalves, and scaphopods and inside those muscles in cephalopods and gastropods. In ancestral gastropods and cephalopods, this transition from outside to inside the shell muscles might have been achieved by later ontogenetic development of the shell muscles relative to the nerve cords, enabling a shift in the position of the nerve cord before the shell muscles were formed (Ponder & Lindberg 1997).

Potential outgroups have both similar and markedly different nervous system patterns. The brachiozoan nervous system differs substantially from the molluscan nervous system. A dorsal and ventral ganglion are connected by a nerve ring around the oesophagus, and their associated nerves innervate shell muscles, mantle, and lophophore where it may form a nerve ring around the base of the lophophore; an intraepidermal nerve plexus is also present (Schmidt-Rhaesa 2007; Lüter 2016; Santagata 2016). This simple nervous system is thought to result from the sessile habits of these taxa, and as pointed out by Schmidt-Rhaesa (2007), whether it is plesiomorphic or derived from other patterns is not known.

Annelids have dorsal cerebral ganglia, paired connectives which form a nerve ring around the foregut, and merge ventrally to form a pair of 'subpharyngeal' ganglia. These ganglia give rise to paired, ganglionated nerve trunks of the ventral nerve (Rouse 2016). Besides the ventral nerve, there are 0–17 longitudinal nerve cords (Schmidt-Rhaesa 2007). The body wall is innervated by the segmental ganglia along the ventral nerve, and an intraepidermal PNS is also present in some basal taxa (Rouse 2016).

Like molluscs, Entoprocta have a tetraneurous nervous system and a complex serotonin-expressing apical organ (Wanninger et al. 2007; Haszprunar & Wanninger 2008; Merkel et al. 2012).

We briefly describe below the main features of the nervous system in each major group of molluscs and these are summarised in Figure 7.1.

7.2.1 POLYPLACOPHORANS AND MONOPLACOPHORANS

The polyplacophoran nervous system is the most conservative in molluscs and its condition has been called 'amphineurous' or 'tetraneurous', being comprised of paired lateral and ventral nerve cords[3] that arise from a paired cerebral centre (Salvini-Plawen 1981a; Sigwart & Sumner-Rooney 2016) (Figure 7.1). Such lateral (mantle/pallial) and pedal (ventral) nerve cords are found in both chitons and monoplacophorans, and the former are the homologues of the visceral (or pleural) cords in gastropods, scaphopods, and bivalves. In polyplacophorans, the pedal cords are cross-connected while the lateral cords, which lie outside the pedals, are connected to the pedal cords in some taxa, but not in others (e.g., Hyman 1967 figs 43A, B), but never directly to each other. In monoplacophorans, unlike chitons, the two lateral cords join posteriorly above the rectum, and the pedal cords join below the rectum. There are also numerous cross connections between the pedal and lateral nerve cords, but none between the two pedal nerve cords, except for an interpedal commissure and the posterior linkage (Lemche & Wingstrand 1959; Wingstrand 1985). This monoplacophoran arrangement of the pedal nerve cords is somewhat similar to that seen in patellogastropods. In both monoplacophorans and chitons, the lateral nerve cords innervate the mantle, the mantle cavity, and all the sensory organs other than the subradular organ (licker). However, the statocysts in monoplacophorans are cerebrally innervated. In addition, these cords innervate most of the internal organs, while the pedal cords innervate the foot. In chitons, the shell muscles are innervated by nerves from the lateral nerve cord, while in monoplacophorans, it is by the latero-pedal connectives.

---

[3] These cords are sometimes referred to as 'medullary', likening them to the vertebrate medullary system.

In both groups, the pedal and lateral cords are connected to the lateroventral part of the supraoesophageal ganglionic cord (the cerebral commissure) in the head region which is usually thought to be equivalent to the cerebral + pleural ganglia (Figure 7.3), but this has recently been questioned (Sumner-Rooney & Sigwart 2018). Innervation to the head region arises from this part of the nervous system, including a pair of buccal and subradular ganglia which are both connected to the lateroventral parts of the supraoesophageal cord. A narrow subcerebral commissure completes the ring around the anterior gut.

7.2.2 Aplacophorans

Both aplacophoran groups have a typical tetraneurous system (Salvini-Plawen 1985a) with some differences as described below and illustrated in Figures 7.1, 7.2, and 7.3.

In both groups, the most anterior cross connective between the pedal cords is often treated as the equivalent of the subcerebral commissure, but it is unclear if it is homologous.

7.2.2.1 Caudofoveata

The cerebral ganglia, which may be separated or partially or completely fused, give rise to a pair of lateral (mantle/pallial) and a pair of ventral (pedal) nerve cords and, usually, a pair of buccal and pleural ganglia (Sigwart & Sumner-Rooney 2016). Both pairs of cords form a ganglionic swelling in their anterior part and join posteriorly, forming a ganglionic suprarectal commissure that innervates the ctenidia and posterior sense organ(s) (see Figure 7.3 and Section 7.6.1). Ventral and lateroventral cross connections may also be present. In some 'advanced' taxa, the cerebral ganglia have several separate or fused precerebral ganglia (see Section 7.2.2) which innervate the sensory head shield and oral region. The lateral cords innervate the body wall and viscera while the ventral cords have few nerves. Shigeno et al. (2007) and Sigwart and Sumner-Rooney (2016) described caudofoveate nervous systems in detail.

Caudofoveates have precerebral ganglia (see Figure 7.3) which have been considered to resemble the mushroom bodies

in the brains of annelids and arthropods (Faller et al. 2012). Although the mushroom bodies differ structurally from anything in molluscs, they have small, densely packed cells called *globuli cells*. These are somewhat similar to the small, densely packed cells in the precerebral ganglia of caudofoveates and have also been observed in the posterior part of the brain of a scaphopod (*Antalis entalis*) and in the procerebrum of eupulmonate gastropods. The arthropod globuli cells may be involved in olfaction, and are convergent with the similar cells in molluscs, given their absence in most molluscan taxa (Faller et al. 2012). In caudofoveates, nerves from the precerebral ganglia connect with sensory cells in the oral shield which are assumed to be chemosensory (Shigeno et al. 2007).

7.2.2.2 Solenogastres

The cerebral ganglia are fused and, as in caudofoveates, give rise to a pair of lateral and a pair of ventral nerve cords (Figure 7.3) and, usually, a pair of buccal and pleural ganglia. Both pairs of cords have weak ganglionic swellings along their length and are cross-connected to one another. The lateral cords join posteriorly where they form a ganglionic suprarectal commissure that innervates the mantle cavity and posterior sense organ. The cerebral ganglia, and sometimes there may be precerebral ganglia, innervate the atrial sense organ and the oral region, while the lateral cords innervate the body wall and viscera, and the ventral (pedal) cords innervate the pedal groove. Todt et al. (2008) and Sigwart and Sumner-Rooney (2016) gave detailed accounts of the nervous systems of solenogasters.

7.2.3 Bivalves

A diagrammatic illustration of the nervous system of an autobranch bivalve is shown in Figure 7.4. The CNS consists of three pairs of ganglia; the cerebral ganglia that lie on either side of the oesophagus are actually cerebropleurals as they consist of fused cerebral and pleural ganglia, a pair (often fused) of pedal ganglia at the base of the foot within the visceral mass, and a pair of closely adjacent visceral (or visceroparietal or posterior)

Caudofoveata

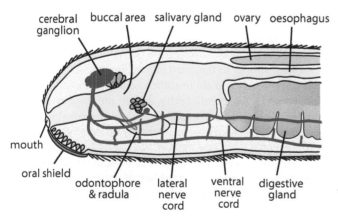

Solenogastres

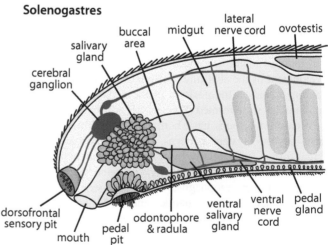

FIGURE 7.2   A diagrammatic comparison of the anterior nervous system of the two aplacophoran groups, shown in lateral view. Redrawn and modified from Salvini-Plawen, L. v., *Malacologia*, 9, 191–216, 1969.

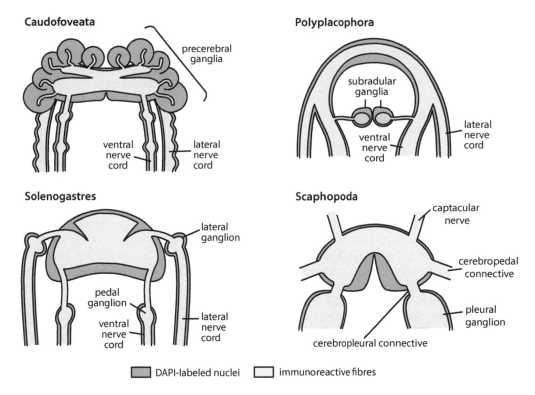

FIGURE 7.3 Diagrams of examples of the central nervous systems of four molluscan classes. DAPI is a fluorescent stain that binds to adenine-thymine rich regions of DNA. Redrawn and modified from Faller, S. et al., *Zoomorphology*, 131, 149–170, 2012.

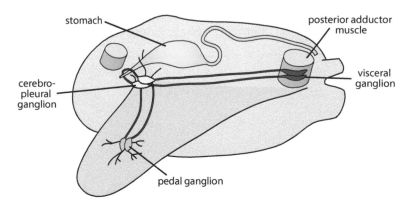

FIGURE 7.4 A diagram of a typical autobranch bivalve nervous system. Redrawn and modified from Buchsbaum, R., *Animals Without Backbones: An Introduction to the Invertebrates*, Penguin Books, Harmondsworth, UK, 1951.

ganglia ventral to the posterior adductor muscle. There are various modifications in the development of these ganglia, but generally the bivalve nervous system is fairly uniform. The fusion of the cerebral and pleural ganglia is usually complete, but is partial in some protobranch bivalves such as *Nucula* and *Leda*, and in a few autobranch bivalves. Buccal ganglia have been lost. The (fused) pleural ganglia in bivalves (and scaphopods) are at least analogous to those in gastropods because they innervate some of the same areas (shell muscles, visceral nerve loop).

The visceral ganglia are usually the largest of the three pairs of ganglia and extend laterally through the body from the vicinity of the kidneys to the region of the posterior adductor muscle. In burrowing bivalves especially, there appears to be greater centralisation of nervous coordination in the visceral

ganglia. Fusion also occurs in pteriomorphian bivalves. In some, the cerebropleural ganglia are fused with the visceral ganglia, and a long cerebral commissure is present, as in *Spondylus* and *Lima* (Pelseneer 1931). *Pecten* and *Spondylus* have the most complex visceral ganglionic mass known in a bivalve, comprising several discrete lobes and two accessory ganglia which connect with the visceral ganglion and the branchial nerves in *Pecten*.

Nerves from the cerebropleural ganglia innervate the anterior adductor muscle and the anterior mantle. The anterior mantle nerves anastomose with the posterior mantle nerves from the visceral ganglion, and the combined nerve cord forms the circummantle nerve which runs around the edge of the mantle. Nerve(s) from the cerebropleural ganglion also innervate the labial palps, oral area, and oesophagus. A cerebral nerve runs to the so-called branchial (or 'cephalic') eye located on the base of the first gill filament in many pteriomorphians (see Section 7.9.8.2). The statocyst nerves also run to this ganglion along the cerebropleural/pedal connectives.

The pedal ganglion innervates the foot via only a few (usually 3–6) nerves, as well as the anterior and part of the posterior musculature and the byssal retractors.

The large visceral ganglion gives off many nerves. These include the branchial nerves to the gills, dorsal, and ventral mantle (pallial) nerves and the mantle nerves to the posterior parts of the mantle (including those to the siphon if one is present). Other nerves supply the posterior adductor muscle, the anal region, and the posterior foot retractor muscles. Nerves from this ganglion, or from the cerebrovisceral connective, supply the stomach and the rest of the gut and digestive gland, the gonads, kidney, and venous sinus. The pericardium, pericardial glands, and heart are innervated from the posterior or dorsal mantle nerves.

In most bivalves, the osphradia are innervated by the ctenidial nerve from the visceral ganglion or directly by a nerve from that ganglion, but in several heterodont clades they are also innervated in part by the cerebral ganglion (Salvini-Plawen & Haszprunar 1982; Haszprunar 1987b).

7.2.4 SCAPHOPODS

A diagrammatic illustration of the nervous system of a scaphopod is shown in Figure 7.5. There is a pair of closely associated cerebral ganglia and a separate, but close, pair of pleural

ganglia. The rounded pedal ganglia are some distance from the pleural ganglia to which they are connected by apparently fused cerebropedal + pleuropedal connectives. One or two pairs of widely separated visceral ganglia connect to the pleural ganglia by a pair of visceral nerves, and an anterior pair of small subradular ganglia are present, with connectives to the cerebral ganglia. The connectives to the larger, more posteriorly placed buccal ganglia arise from the subradular ganglia. A mantle nerve arises from the pleural and cerebral ganglia, and there are small peripheral ganglia associated with the captaculae which are innervated from the cerebral ganglion. The cerebral ganglia also innervate the snout. The pedal ganglia give off a few nerves to the foot.

The posterior parts of the connectives to the visceral ganglia are cord-like, with cell bodies scattered along them. The visceral complex innervates the posterior mantle, stomach, digestive gland, rectum, and kidney. A detailed account of the scaphopod nervous system is provided by Sigwart and Sumner-Roonery (2016).

7.2.5 GASTROPODS

The nervous system of gastropods varies considerably from group to group. A diagrammatic illustration of the nervous system of a generalised basal gastropod is shown in Figure 7.6. This basic gastropod nervous system consists of a pair of dorsal cerebral ganglia connected via the cerebral commissure. These are connected to two ventral pedal ganglia and smaller pleural ganglia located close to the pedal ganglia. As in other molluscs,

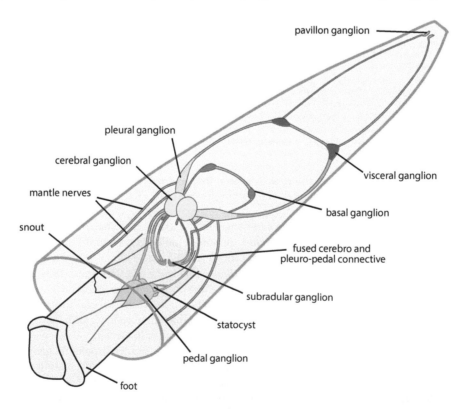

FIGURE 7.5 Diagram of a scaphopod nervous system. Redrawn and modified from Shimek, R.L. and Steiner, G., Scaphopoda, in Harrison, F.W. and Kohn, A.J., *Microscopic Anatomy of Invertebrates. Mollusca 2*, Vol. 6B, pp. 719–781, Wiley-Liss, New York, 1997.

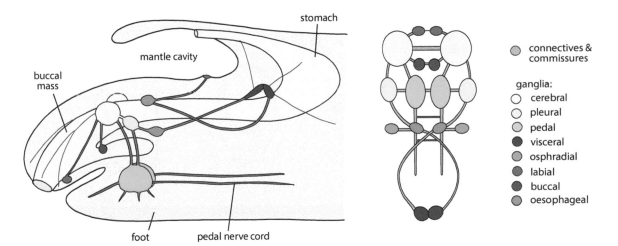

FIGURE 7.6 Lateral diagrammatic view of the arrangement of ganglia in a generalised epiathroid caenogastropod. Right figure a schematic dorsal view of the ganglia in a generalised lower gastropod. Left figure redrawn and modified from Fretter, V. and Graham, A.L., *British Prosobranch Molluscs: Their Functional Anatomy and Ecology*, Ray Society, London, UK, 1962; right figure original.

the pleural ganglia differ from the other circumoesophageal ganglia in not being connected to each other by a commissure. Pleuropedal connectives and cerebral-pleural connectives typically link these ganglia unless they are fused. In basal gastropods, the paired pedal ganglia have elongate pedal cords and there is one, or sometimes two or more, commissures.

The innervation of organs from the main ganglia is typically:

- Cerebral ganglia – the eyes, statocysts, head tentacles, skin and some muscles of the lips, head and neck and, in some, the penis. The tentacular nerve can be single or bifurcated. This nerve is homologous with the *nervus clypei capitis* of cephalaspidians, aplysiids, and sacoglossans which innervates the head shield or head, but is lost, along with the cephalic tentacles, in nudipleurans, umbrachulids, and *Rhodope* (Huber 1993).
- Buccal ganglia – buccal mass, salivary glands, and anterior oesophagus.
- Pleural ganglia – the mantle in non-heterobranch gastropods (see above) and the shell muscles. Where both post-torsional left and right shell retractors comprise the shell muscle, innervation is from both left and right pleural ganglia, with the nerves crossed due to torsion. Where there is a single muscle (the situation in most gastropods), the nerves arise from the left pleural ganglion (Haszprunar 1985e).
- Pedal ganglia – the foot muscles and skin. In most gastropods, there are few anterior pedal nerves from the pedal ganglia, while in caenogastropods there are many (Ponder & Lindberg 1997). Many caenogastropods also have accessory pedal ganglia.
- Supraoesophageal and suboesophageal ganglia – innervate the gills (ctenidium or plicatidium), osphradium, and mantle.
- Parietal ganglia (present only in some euthyneurans) – the lateral body wall and mantle.

- Visceral ganglion – the stomach, digestive gland, intestine, rectum, the anal region and adjacent skin and body wall, reproductive organs, kidney, and heart.

The posteriorly located, usually unpaired, visceral ganglion is connected to the pleural ganglia by typically long connectives which form the visceral loop, the homologue of the lateral nerve cord in other molluscs. Unlike the pedal nerve cords, the visceral nerve cord is affected by torsion. In *Patella*, the twisted visceral loop is formed *after* torsion (Smith 1935), but *during* torsion in *Haliotis* (Crofts 1937) and *Marisa* (Demian & Yousif 1975). The arrangement of the visceral loop has been important in gastropod classification – the higher taxon names Streptoneura and Euthyneura were based on its configuration (see next Section).

Typically, an additional pair of ganglia lies within the visceral loop, and these become asymmetrically positioned due to the twisting of the loop. The ganglion connected to the left pleural ganglion is located ventrally (the suboesophageal or 'subintestinal' ganglion), while that attached to the right pleural ganglion is located dorsally (the supraoesophageal or 'supraintestinal' ganglion). In many euthyneurans, there is an additional pair of ganglia on the visceral loop, the parietal ganglia (the pentaganglionate condition). The primitive hygrophilan *Chilina* has an extra ganglion in addition to the usual five (Brace 1983). Haszprunar (1985f) argued that the parietal ganglia were budded off the pleural ganglia because the nerve supplying the columellar muscle arises in the left parietal ganglion in pentaganglionate taxa and in the left pleural ganglion of other gastropods.

The ganglia in the visceral loop vary considerably in position and are often fused in various combinations, generally associated with shortening of the visceral loop. These changes have occurred independently many times (see Figures 7.7 through 7.9). Extreme concentrations are seen in many 'opisthobranchs' and stylommatophorans, as well as in some higher caenogastropods.

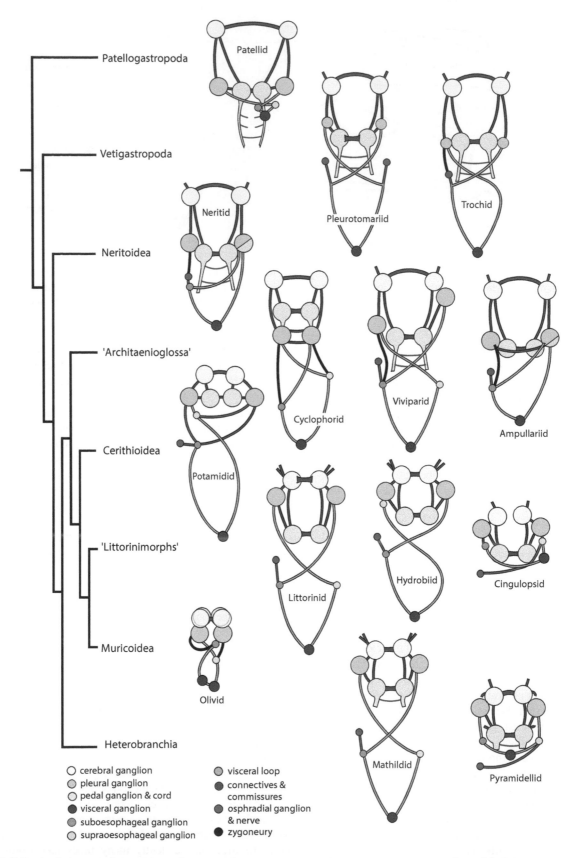

FIGURE 7.7 A diagrammatic overview of gastropod nervous systems. Most figures redrawn and modified from Haszprunar, G., *J. Molluscan Stud.*, 54, 367–441, 1988a, patellid, cyclophorid and hydrobiid original. For additional heterobranch nervous systems see Figures 7.8 and 7.9.

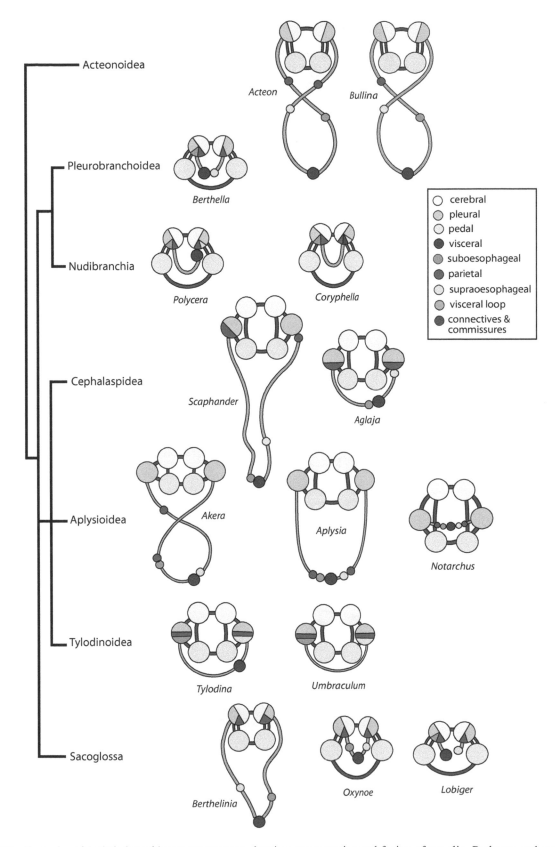

FIGURE 7.8 Examples of 'opisthobranch' nervous systems, showing concentration and fusion of ganglia. Redrawn and modified from Schmekel, L., Aspects of evolution within the opisthobranchs, in Trueman, E.R. and Clarke, M.R., *Evolution. The Mollusca*, Vol. 10, pp. 221–267, Academic Press, New York, 1985, with the addition of *Bullina* redrawn and modified from Rudman, W.B., *Zool. J. Linn. Soc.*, 51, 105–119, 1972.

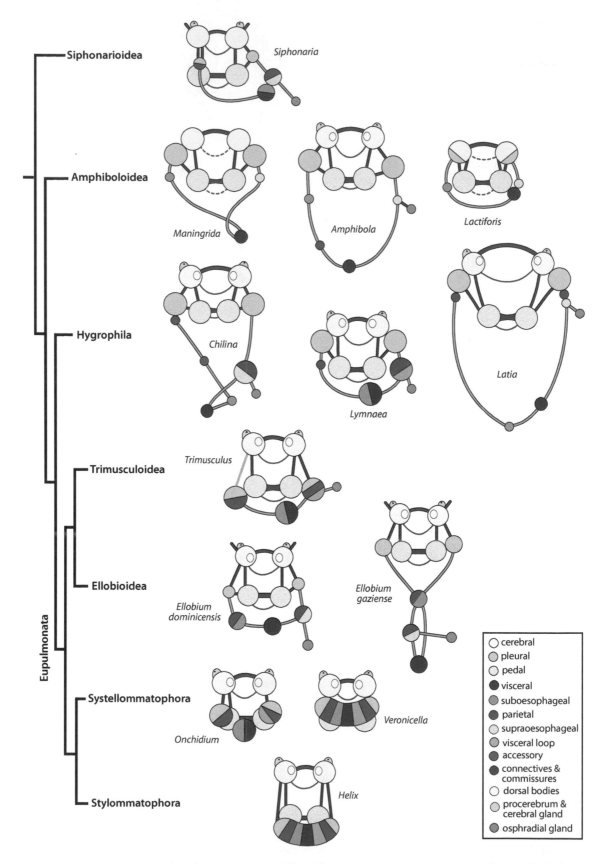

FIGURE 7.9 'Pulmonate' nervous systems, showing concentration and fusion of ganglia. Redrawn and modified from Haszprunar, G. and Huber, G., *J. Zool.*, 220, 185–199, 1990, with the addition of two amphiboloideans redrawn and modified from Golding, R.E. et al., *Zootaxa*, 50, 1–50, 2007.

The plesiomorphic connective to the supraoesophageal ganglion is from the right pleural, and to the suboesophageal, it is from the left pleural ganglion, although secondary connections are often developed. A connection between one of the pleural ganglia and a nerve from the branchial or oesophageal ganglion on the same side (post torsion) of the visceral commissure, is a condition known as *dialyneury*, and is typical of many vetigastropods. A similar, but direct, connection between the pleural and oesophageal ganglia is called *zygoneury*. A zygoneurous connection may occur between the supraoesophageal and the left pleural ganglia, or between the suboesophageal and the right pleural ganglia, or both. Such zygoneurous connections are common in caenogastropods and are thought to provide a more direct route for nerve impulses.

Additional ganglia include the buccal ganglia that connect to the cerebral ganglia and have a commissure below the anterior gut, and paired or a single osphradial ganglion. In some basal gastropods, there is also a pair of labial ganglia and in others anterior accessory pedal ganglia.

Patellogastropods, and those vetigastropods with two ctenidia (haliotids, pleurotomariids, fissurellids), have the nerve cells distributed along the nerve cords and lack true separate ganglia. They have branchial ganglia, and a supraoesophageal ganglion is found only in fissurellids. Other gastropods (and most other molluscs other than chitons and monoplacophorans) have the nerve cells concentrated in ganglia. Page (1994) noted that the development of ganglia in 'opisthobranch' larvae is retarded relative to caenogastropods, vetigastropods, and patellogastropods.

Topological relationships between the pleural, cerebral, and pedal ganglia are important markers in the evolution of the molluscan nervous system. These include concentration of the ganglia resulting eventually in their fusion. Also, the position of the pleural ganglia relative to the cerebral and pedal ganglia is notable. If the pleural ganglia are near the pedal ganglia and distant from the cerebral ganglia, the condition is called *hypoathroid*, while if the pleurals are located close to the cerebral ganglia and well separated from the pedal ganglia, it is called *epiathroid*. Because the hypoathroid condition is mostly found in basal gastropods, it is probably the original (plesiomorphic) condition for gastropods, although there are convergent epiathroid conditions also found in scaphopods and bivalves.

A hypoathroid condition is found in a few basal caenogastropods (cyclophorids and ampullariids), while viviparids have an intermediate condition (*dystenoid*). The remaining caenogastropods are epiathroid. A hypoathroid condition also occurs in some euthyneuran heterobranch gastropods, as, for example, in aplysiids and many eupulmonates, including some ellobiids. At least some go through an epiathroid stage in early ontogeny, and the more basal 'opisthobranchs' and the lower heterobranchs have a near epiathroid nervous system. These observations suggest that one or more reversals to a hypoathroid condition have occurred in heterobranch evolution (Ponder & Lindberg 1997).

The buccal ganglia are associated with the buccal mass and found in all gastropods; they are connected to each other by a commissure that runs beneath the buccal mass (it is dorsal in chitons) and to the cerebral ganglion by the cerebrobuccal connective. In heteropods and some neogastropods, they are fused with the cerebral ganglia.

Labial ganglia are found only in patellogastropods and some vetigastropods, although a labial commissure is present in some neritimorphs, architaenioglossan caenogastropods, and in some heterobranchs. There is a 'labial lobe' associated with the cerebral ganglion from which the buccal connectives arise, which may represent a fused labial ganglion in some cocculinid and lepetelloidean limpets. Small subradular ganglia occur in some patellogastropods.

The innervation of the mantle and associated structures in gastropods is quite variable. In patellogastropods and *Haliotis*, a mantle (= pallial) nerve originates from each of the pleural ganglia and runs around the mantle edge; this is not the homologue of the mantle nerve cord in chitons and is separate from the visceral loop. In *Haliotis* and *Trochus*, the left mantle nerve forms a dialyneurous connection with the visceral loop. The ctenidium and osphradia are innervated from the visceral loop in patellogastropods, but various changes in the innervation of those structures have occurred in other gastropod groups; in most gastropods, the supraoesophageal ganglion innervates the mantle skirt, gill, and osphradium, although there are exceptions as outlined below.

In patellogastropods, branches of the mantle nerves innervate the secondary gills (if present) and the mantle skirt. The visceral loop arises from the pleural ganglia and contains sub- and supraoesophageal ganglia which innervate the right and left osphradia, respectively. A nerve from the left osphradial ganglion innervates the ctenidium (when present), while in *Patella* a nerve from each osphradial ganglion innervates the presumed ctenidial rudiments, the wart organs (Stütsel 1984).

There are no ganglia formed in the visceral loop in some large-sized 'lower' vetigastropods such as *Mikadotrochus* (Woodward 1901) and *Haliotis* (Crofts 1937). This suggests that the formation of these ganglia in vetigastropods may be convergent with that in both patellogastropods and apogastropods. As in patellogastropods, the mantle (pallial) nerves arise from the pleural ganglia, but vetigastropods have branchial ganglia arising from the visceral loop from which the ctenidia and osphradia are innervated. These differ from the supra- and suboesophageal ganglia in caenogastropods because they lie on short branches off the visceral loop, and may be homologues of the osphradial ganglia.

The neritimorph nervous system is hypoathroid, and the two pleural ganglia are, unusually, connected by a thick commissure and are fused with the pedals (e.g., Bouvier 1887; Sasaki 1998), an arrangement unique in gastropods, as is the origin of the visceral loop nerves which both originate from the right pleural ganglion. There is a zygoneury with the left pleural ganglion that connects with the 'suboesophageal ganglion'.

In the lower caenogastropod Ampullariidae, the supraesophageal ganglion innervates the ctenidium and the osphradium in part, with the latter mainly innervated by the left pleural ganglion (Prashad 1925; Berthold 1988), and this arrangement appears to be a synapomorphy of this group and viviparids (Annandale & Sewell 1921; Ponder & Lindberg 1997). This configuration appears to result from zygoneury with the mantle nerve from the pleural ganglion. Many other caenogastropods have a zygoneury formed with the mantle nerve arising from the left pleural ganglion and the osphradial nerve. In most, the supraoesophageal ganglion has two separate nerves innervating the ctenidium and osphradium.

A true ctenidium is absent in heterobranchs, but in some a modified structure, the plicatidium, is present (see Chapter 4), and many have lost the osphradium. Taxa with a plicatidium and an osphradium feature plesiomorphic innervation, with both innervated by a single nerve from the supraoesophageal ganglion. In some, a mantle nerve is given off from the osphradial ganglion, a situation never seen in caenogastropods. The sinistral fresh-water limpet *Laevapex* has a very concentrated nerve ring with the left mantle nerve from the left pleural ganglion also supplying the osphradium, and two right mantle nerves emerge from the right 'visceral ganglion' (presumably fused with the supraoesophageal ganglion). Another mantle nerve from the 'left visceral ganglion' supplies the posterior part of the mantle edge and gill (Basch 1959b). Thus, in this small limpet, there are several nerves from three ganglia which supply the mantle edges. The mantle nerves are less conspicuous in coiled gastropods than they are in the limpet-like taxa. In *Littorina*, the mantle nerves arise from the pleural ganglia and the left forms a dialyneury with a nerve from the supraoesophageal ganglion. In the lepetelloidean pseudococculinid limpets, a mantle nerve arises from both pleural ganglia, but in another lepetelloidean limpet, *Cocculinella*, the left mantle nerve arises from the pleural ganglion, but the right emerges from the suboesophageal ganglion. There is a different pattern in the superficially similar cocculinid limpets where the left mantle nerve arises from the pleural ganglion or the visceral loop near that ganglion and the right from the pedal ganglion. The different patterns of the innervation of the mantle by the mantle nerve reflect the different origins of these secondarily limpet-shaped taxa (Ponder & Lindberg 1997). Other variations exist; for example, in the neogastropod *Buccinum*, the left mantle nerve arises from the supraoesophageal ganglion and the right from the suboesophageal ganglion, while the siphonal nerves arise from the left pleural ganglion and form a dialyneury with the left mantle nerve (Dakin 1912). In *Aplysia*, the mantle nerves arise from the pleural ganglia, but the situation in 'opisthobranchs' is confused, partly owing to the combining of pedal and mantle tissues. In neritimorphs, the mantle nerves arise from the pleural ganglia, the left also innervating the ctenidial gill.

The condition seen in patellogastropods (long pedal cords, no complete cross connections other than the posterior connection, a condition possibly correlated with a disc-like foot) is like that seen in monoplacophorans. In some of the other lower gastropod groups (Vetigastropoda and Neritimorpha), the pedal nerve cords have numerous cross connections (commissures), a condition that may be related to the narrower, more mobile foot. In higher gastropod groups, reduction in the number of commissures and the concentration of

the cords into pedal ganglia occurred independently, with cypraeids, for example, having pedal cords with many commissures (Riese 1930). The pedal ganglia of some heterobranchs have two commissures, one (the posterior parapedal commissure) possibly being a relict. Caenogastropods generally have only the pedal commissure, although littorinoideans can have two (Fretter & Graham 1962), ampullariids have two or three (Berthold 1991), and cypraeids have several (Bouvier 1887).

7.2.5.1 Streptoneury and Euthyneury

The visceral loop is twisted into a figure eight, a condition known as streptoneury (= chiastoneury), a consequence of torsion. This is the plesiomorphic condition in gastropods, but in some, the visceral loop is secondarily untwisted – the euthyneurous condition (= orthoneury). There has been debate on how the euthyneurous condition evolved, and it is not homologous in all gastropod taxa in which it occurs.

In most streptoneurous gastropods, the visceral loop is twisted into a figure eight, with the twist occurring anterior to the oesophageal ganglia so that the supraoesophageal and suboesophageal ganglia lie on the left and right sides, respectively. In some caenogastropods where the connectives between the pleural ganglia and the oesophageal ganglia are short, the oesophageal ganglia lie anterior to the twist so that the supraoesophageal is on the right and the suboesophageal on the left. As far as the positions of the oesophageal ganglia are concerned, this latter configuration emulates that seen in euthyneurous gastropods, such as many 'opisthobranchs' (Figure 7.8) and all 'pulmonates' (Figure 7.9), where the visceral loop is not twisted. Streptoneury is also present in most 'lower heterobranchs', acteonoids and some cephalaspideans (Figure 7.8).

The process known as 'detorsion' (see Chapters 8 and 20) is often invoked to explain euthyneury, but there are other suggestions. For example, Krull (1934) explained the configuration in many 'pulmonates' by invoking the loss of the supraoesophageal ganglion and its connectives so that the ganglion usually identified as the supraoesophageal is actually the right parietal ganglion (which is otherwise missing). According to the hypothesis of Krull, the connection from the suboesophageal ganglion and the right parietal ganglion is zygoneurous, and the osphradium is innervated by the right parietal ganglion. These ideas have not gained general acceptance. Instead, most workers accepted the pentaganglionate theory (Schmekel 1985; Haszprunar 1988a) which regards the right parietal ganglion of 'pulmonates' to be the fused right parietal plus the supraoesophageal ganglion, while the 'visceral ganglion' is the fused visceral and suboesophageal ganglion.

7.2.6 CEPHALOPODS

Cephalopods have the largest, most complex invertebrate nervous system. They have an unusually large CNS ('brain'), in large part because they possess very elaborate sense organs and therefore process much more information than other molluscs. A cephalopod brain works in a very different way from human brains because much of the processing happens in ganglia distributed in various parts of the body. This distributed processing facilitates peripheral sensory input and output and, in coleoids, allows rapid communication with the arms, suckers, and skin (including the chromatophores).

The cephalopod CNS is more concentrated, with fusion of ganglionic elements, than in all other molluscs. In addition, there are brachial, optic, olfactory, and peduncle lobes in the brain, and additional peripheral ganglia, notably the stellate ganglia. They are one of the very few invertebrate groups where much of the nervous system is concentrated in the brain.[4] The squid giant axons connect the stellate ganglia with the brain. These nerves are among the best studied nerve cells in neuroscience, and there is a great deal of information on the morphology of some cephalopod brains. Little is known about the operation of the neural networks that underlie the behaviours seen in cephalopods. The better-studied ones were reviewed by Williamson and Chrachri (2004), notably those networks underlying the operation of the giant fibres, chromatophores, statocysts, eyes, learning, and memory.

Because the cephalopod nervous system has been the subject of considerable interest and investigation, we describe it in some detail below.

7.2.6.1 The Cephalopod Brain

As in other molluscs, the CNS in cephalopods, the so-called brain, is derived from a circumoesophageal ring of nervous tissue where parts have become enlarged. It is difficult to align the parts of the cephalopod brain with the ganglia of other molluscs because they may have evolved independently from an essentially non-ganglionated condition (Shigeno et al. 2015). The CNS of *Nautilus* (Figure 7.10) shows no sign of a ganglionated condition and Young (1988a) suggested that it might have been derived from a similar nervous system to that seen in chitons by shortening and concentration around the oesophagus. This idea is not in conflict with the more usually accepted idea that cephalopods were derived from a 'monoplacophoran'-like ancestor. In *Nautilus*, many nerve cells are located around the periphery below the skin and in the brachial nerve cords in the arms. Thus, the nervous systems of *Nautilus* and coleoids have a sophisticated system of peripheral reflexes upon which is superimposed processing and 'decision-making' by the CNS. It contains about twice the number of nerve cells as the CNS, and all the ganglia associated with it are lower motor centres (Budelmann 1995).

Whereas the brain of *Nautilus* has 13 identifiable lobes (Young 1965a), there are 23 in the cirrate octopod *Cirrothauma* and 38 in *Loligo* (Young 1988a). The simpler

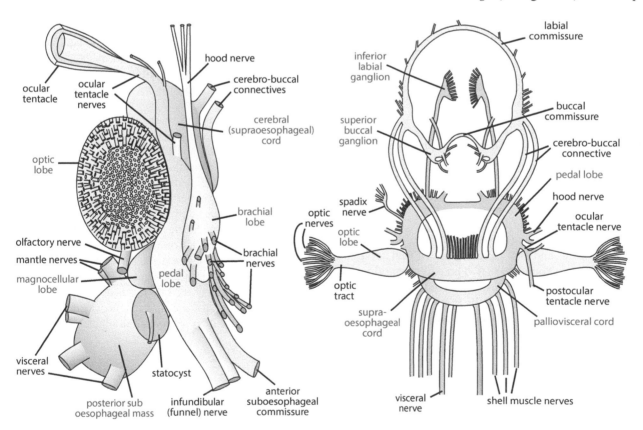

FIGURE 7.10 Dorsal and lateral views of the brain of *Nautilus*. Red labels identify parts of the brain. Dorsal view redrawn and modified from Griffin, L.E., *Mem. Natl. Acad. Sci.*, 8, 101–230, 1900, lateral view redrawn and modified from Young, J.Z., *Philos. Trans. Royal Soc. B*, 249, 1–25, 1965a.

---

[4] Also in brachyuran crabs.

brain in *Nautilus* is correlated with simpler eyes not capable of detailed discrimination and a statocyst that lacks cristae to monitor angular acceleration (see Section 7.5.4.1). This latter feature in coleoids allows eye movements that stabilise the image on the retina, but *Nautilus* is incapable of this (Young 1988a), and the *Nautilus* brain also differs from that of coleoids in having a more developed olfactory lobe.

7.2.6.1.1   Nautilus CNS

In *Nautilus*, three major ganglionic cords make up the brain (Figure 7.10). Two ventral cords (the palliovisceral and the more anterior suboesophageal cord) join laterally before joining the dorsal cord, the supraoesophageal (or cerebral) cord. The main ganglionic masses are at the point where these three cords join laterally. They consist of the cerebral, brachial, and pedal lobes (together with the brachiopedal lobes) and the more lateral optic lobes. These paired lateral masses represent the main part of the brain and receive the nerves from the eyes, lips, arms, statocysts, olfactory organs, and cephalic tentacles. Here, two pairs of lobes (called the magnocellular lobes because their position is similar to that of the magnocellular lobes in coleoids) receive the nerves from the two pairs of ocular tentacles (absent in coleoids) and the statocyst nerves. These lobes are in the dorsal part of the lateral brain, and a more anterior brachial lobe lies at the lateral part of the pedal cord. These two pairs of lobes are comparable to the anterior and middle suboesophageal masses of coleoids (see below), respectively. The magnocellular lobes are relatively smaller than those in coleoids and do not receive giant nerve cells as in decabrachians, but they do contain the largest nerve cells in the CNS (Young 2010).

The suboesophageal (or brachiopedal) cord consists of more than just the equivalent of the pedal ganglia as it not only gives off the nerves to the funnel (derived from the foot), but also the tentacle (brachial) nerves which connect directly to the cerebral part of the brain (the magnocellular region). This latter area also innervates the ocular tentacles, arms, and the hood (Young 1965a, 1988a).

The ventral posterior (palliovisceral) cord gives off mantle and visceral nerves and may represent the visceral nerve cord of other molluscs (Young 1988a). The anteroventral part of the supraoesophageal cord coordinates feeding and gives rise to the buccal ganglia, while laterally it connects with the eyes and tentacles. This part of the brain is equivalent to the frontal and vertical lobes of coleoids.

The large optic lobes join the supraoesophageal cord laterally. A pair of olfactory lobes lie at the back of the optic tract as in coleoids, but are larger. Both the optic and olfactory lobes are continuous with the supraoesophageal cord.

The superior buccal ganglia lie anterior to the nerve ring at the sides of the buccal mass and are joined to the supraoesophageal cord by two pairs of connectives. *Nautilus* lacks the basal lobes and peduncular lobe seen in coleoids.

The general organisation of the brain in *Nautilus* is sufficiently like that of coleoids to offer clues as to what the brain of their common ancestor may have looked like.

7.2.6.1.2   The Coleoid Brain

There are several general accounts of coleoid brain structure and function including those of Wells (1962), Bullock and Horridge (1965), and Young (1988a), as well as some overviews and insights in the more recent literature by Hanlon and Messenger (1996), Nixon and Young (2003), Hochner et al. (2006), and Zullo and Hochner (2011). The account below is based on these sources.

As in *Nautilus*, the brain in coleoid cephalopods (Figures 7.11 and 7.12) can be divided into two main regions: the supraoesophageal mass which, as the name suggests, lies above the oesophagus, and the ventral suboesophageal mass. These are connected laterally by the dorsal magnocellular lobes and the basal lobes that represent the fusion of two lateral cords, as seen in *Vampyroteuthis*, the likely living sister group to the modern coleoids.

In all coleoids, the suboesophageal mass is subdivided into three masses as well as the magnocellular lobe. The pair of magnocellular lobes runs around the sides of the brain joining the dorsal cerebral and optic regions of the brain with the ventral pedal and palliovisceral lobes. These lobes (unlike the equivalent in *Nautilus*) have two ventral commissures (anterior and posterior) in decabrachians and one (posterior) in octopods. These lobes are the main higher motor system for the control of the arms and mantle, giving rise to the first-order giant nerve cells in decabrachian cephalopods.

The posterior masses of the brain are joined laterally by the basal lobe complex and the dorsal magnocellular lobes. These lateral structures are fused in coleoids, but were originally separate cords, as shown by their partial separation in *Vampyroteuthis* (Young et al. 1999).

In *Octopus*, the central part of the brain contains about 50 million neurons, and this part includes the vertical lobe, the centre of learning and memory. This central brain processes information from the much larger optic lobes which contain about 120 million neurons, as well as from the largely autonomous peripheral nervous system in the arms, containing about 300 million neurons (Zullo & Hochner 2011).

The basal and peduncle lobes have distinctive parallel fibres and are found in all coleoids, and they all have a superior frontal/vertical lobe system. They also all possess the endocrine optic glands (see Chapter 17) that lie on the optic tract, but are not part of the brain.

The main lobes making up the coleoid brain are briefly described below and are illustrated in Figures 7.11 (*Octopus*) and 7.12 (*Sepia*).

7.2.6.1.2.1   Sub- and Supraoesophageal Lobes

Suboesophageal

The suboesophageal lobes represent the joined anterior and posterior suboesophageal cords of *Nautilus*. This mass consists of anterior, middle, and posterior sections (often called brachial, pedal, and palliovisceral 'ganglia', respectively) which are subdivided as detailed below. There are also vasomotor

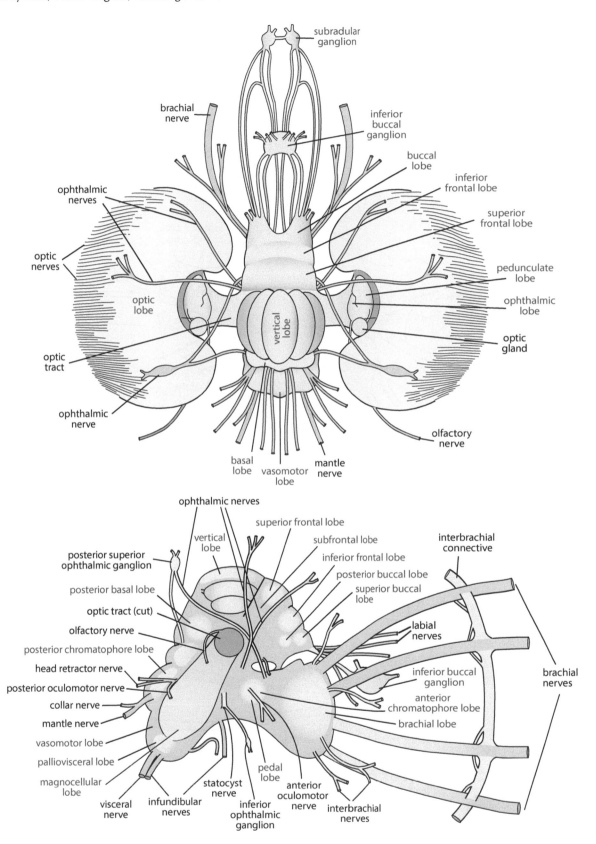

FIGURE 7.11 Dorsal and lateral views (the latter with the optic lobes removed) of the brain of *Octopus vulgaris*. Red labels identify parts of the brain. Redrawn and modified from Young, J.Z., *The Anatomy of the Central Nervous System of Octopus vulgaris*, Clarendon Press, Oxford, UK, 1971.

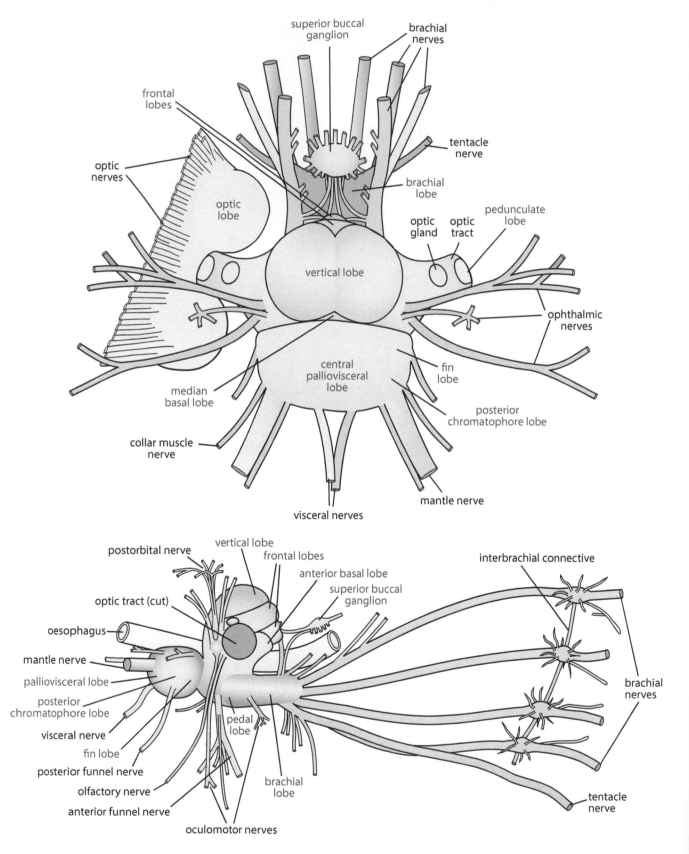

FIGURE 7.12 Dorsal and lateral (the latter with optic lobes removed) views of the brain of *Sepia officinalis*. Red labels identify parts of the brain. Redrawn and modified from Tompsett, D.H., *Sepia*, in Daniel, R.J., *L. M. B. C. Memoirs on Typical British Marine Plants and Animals*, pp. 1–191, University Press of Liverpool, Liverpool, UK, 1939.

and chromatophore lobes. All these lobes receive input from various receptors and coordinate motor functions.

Anterior and middle suboesophageal mass: There are prebrachial and brachial lobe with connections to the arms, and an anterior lobe of the middle suboesophageal (pedal) mass also has additional nerves that connect to the arms and suckers and are involved in their actions. The middle oesophageal mass comprises the intermediate and lower motor centres involved in most actions of the animal. The anterior lateral pedal lobes of the middle suboesophageal mass control eye movements and the posterior lateral pedal lobes probably initiate an attack. Anterior chromatophore lobes lie near the front of the middle suboesophageal mass and the posterior part, the posterior pedal lobes, control the funnel.

Posterior suboesophageal mass: This area includes the palliovisceral lobe centrally and laterally, part of the magnocellular lobes. These centres are important in controlling the mantle and its respiratory and locomotory functions and various activities of the viscera. Posterior chromatophore lobes and fin lobes are found in decabrachians and cirrate octopods. All these sections are closely coordinated by cross-linking fibres.

Supraoesophageal lobes

These consist of two functionally distinct systems – the basal lobes and the frontal/vertical lobe system.

Basal lobes: These contain the higher motor centres, including the anterior and medial dorsal lobes and the peduncle (= pedunculate) lobes. They are generally similar in all coleoids and represent about 20% of the brain excluding the optic lobes (Maddock & Young 1987). Inputs from the eyes and statocysts are received by these lobes, and they regulate movements, including eye movements. The fine parallel fibres in these lobes are similar to those in the cerebellum of the vertebrate brain and, like that structure, appear to be involved with the timing of movements as experiments show that motor activity is maintained, regulated, or suppressed by the lobes.

In *Octopus*, the anterior basal lobe controls the movement of the head and eyes, while the median basal lobe is involved in the control of the movements of the mantle and funnel. The dorsal basal lobe probably controls defensive and avoidance actions, and the lateral basal lobes control the chromatophores and skin muscles. The peduncle lobes coordinate motor activity, including colour changes and inking (Nixon & Young 2003).

Frontal and vertical lobes: These lobes show much more variation in development than the basal lobes – for example, in *Loligo*, they represent over 20% of the brain volume, but only 5% in cirrate octopods (Young 1988a). The superior frontal and vertical lobes are involved with vision and, in octopods, with touch. Together with the lateral superior frontal lobes and subvertical lobe, they regulate the exploratory behaviour, learning, and memory.

In *Octopus*, the median inferior frontal and lateral inferior frontal lobes, the subfrontal lobes, and posterior buccal lobe are functionally integrated and process chemotactile information from the arms and suckers.

The frontal and vertical lobes connect with the superior buccal lobes, and this system regulates prey capture and is involved in food selection and learning. The learned responses involve the tiny *amacrine*[5] nerve cells that make up the subfrontal and vertical lobes (Young 1991). Thus, an animal with the vertical lobe removed behaves normally apart from its memory functions.

When the brains of an octopus and a decabrachian such as *Sepia* are compared, the most obvious difference is the sizes of the vertical lobe system, inferior frontal lobes, and optic lobes. The superior frontal and vertical lobes are largest in decabrachians living in shallow water with, for example, 21% of the total brain volume in loliginids and 22% in *Sepia* (Young 1988a), but they are also large (22%) in the deep-sea pelagic *Vampyroteuthis*. They are smaller in oceanic decabrachians (means of 14% for six mesopelagic species and 12%–13% for buoyant species), although each category has some variation (figures from Young 1988a). The smallest vertical lobes (7%–8%) in decabrachians are in some *Mastigoteuthis* and *Joubiniteuthis* which have, respectively, very long tentacles and arms, but these taxa have large inferior frontal lobes and magnocellular lobes. *Chiroteuthis* has very long tentacles, and has larger vertical and smaller magnocellular lobes.

While the vertical lobe system is smaller in volume in epibenthic octopods than in many decabrachians (average 14.8%), it is folded in octopods in a way analogous to the folding in the cerebral cortex of mammal brains, so the actual area is greater. The lobes are smaller in bathybenthic and pelagic octopods (11%–12%), while the cirrate octopods have the smallest vertical lobes of any coleoid (3.6%).

7.2.6.1.2.2 Magnocellular Lobes These lobes lie lateral to the oesophagus and link the higher supraoesophageal and lower suboesophageal motor centres. They are concerned with rapid escape and defence. In decabrachians, giant nerve fibres originate from these lobes.

---

[5] Nerve cells with branches, but without an axon.

7.2.6.1.2.3 The Buccal Lobes and Inferior Frontal System This system shows more variation than most others. The *superior buccal lobe* in decabrachians (and *Nautilus*) lies anterior to the brain, but is joined with it in octopods. In *Octopus*, this lobe is concerned with the motor control of feeding.

The *inferior frontal lobe* in decabrachians is comparable to the *posterior buccal lobe* that lies behind the superior buccal lobe of octopods. Octopods also have an extra system of lobes not present in decabrachians – the large *lateral* and *median inferior frontal* and *subfrontal lobes*. The subfrontal and median inferior frontal system of octopods is concerned with chemotactile memory (Young 1991). It presumably evolved when octopods became benthic and increasingly used their arms for assessing their environment. This idea is supported by the fact that secondarily pelagic octopods (e.g., *Argonauta* and *Tremoctopus*) have a poorly developed subfrontal system (Figure 7.13).

In decabrachians, the arms handle prey, while in octopods, they also discriminate it. Experiments show that *Octopus* establishes memories of touched objects in the inferior frontal system. These lobes are more developed in a few decabrachians with numerous suckers on their arms as they may also be involved in chemotactile discrimination (Young 1988a).

Bathypelagic and epipelagic octopods have small to vestigial inferior frontal systems (Figure 7.13). Thus in *Argonauta*, it comprises only 0.7% of the brain, although cirrate octopods have large (2.3%) inferior frontals suggesting that touch is important, perhaps needed for groping for food on the sea floor.

7.2.6.1.2.4 The Optic Lobes These large lobes act as higher motor centres and analyse visual information. In coleoids, a chiasm between the eye and the optic lobe conducts information from the eyes and presumably rectifies the inverted retinal image prior to it being processed by the optic lobe (Young 1962a). The chiasm is lacking in *Nautilus* and *Cirrothauma*, both of which lack a lens (Hanlon & Messenger 1996). The optic lobes are where visual memories are stored and thus function as a 'high brain centre'.

The optic lobes are generally similar in structure in all cephalopods, but some differences occur. For example, the cell structure of the outer layers (deep retina) is simple in *Nautilus* and cirrate octopods.

The relative size of the optic lobes also varies considerably. For example, in *Nautilus*, which has a pinhole eye, the optic lobes are equivalent to 27% of the volume of the rest of the brain. In the 'blind' finned cirrate octopod *Cirrothauma*

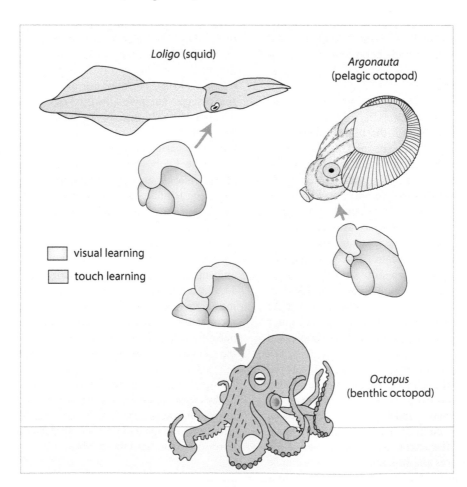

FIGURE 7.13 Comparison of the touch learning and visual learning brain centres in a benthic and a secondarily pelagic octopod (*Argonauta*) with a squid. Redrawn and modified from Hanlon, R.T. and Messenger, J.B., *Cephalopod Behaviour*, Cambridge University Press, Cambridge, UK, 1996.

(with an open-cup eye), it is only 14% of the volume of the rest of the brain. In other cirrates with eyes containing lenses, it is 40%–50%. In most octopods, the optic lobes are relatively smaller (150% of the volume of the rest of the brain in *Octopus*) than in decabrachians (340% in *Loligo*; 260% in *Sepia*), but pelagic octopods living in relatively shallow water all have large optic lobes. Those of *Argonauta* are 220% of the volume of the rest of the brain, while the optic lobes in *Tremoctopus* are 250% and *Alloposus* 370%. Deep-water octopods mostly have small optic lobes, e.g., *Bathypolypus* 42% of the volume of the rest of the brain and *Benthoctopus* 49%. Some deep-water decabrachians have the largest optic lobes of any cephalopods – for example, in *Cranchia*, the optic lobes are 610% of the brain volume, *Taonius* 460%, and *Grimalditeuthis* 720%. *Histioteuthis* has unequal-sized eyes, and this is reflected in the optic lobes with the left optic lobe 640% and the right 300% of the volume of the rest of the brain (Maddock & Young 1987; Young 1988a).

7.2.6.1.2.5 The Ganglia of the Peripheral Nervous System

Many bodily functions in cephalopods are controlled by peripheral ganglia (Figure 7.14) and nerve cells. Dispersed nerve cells are found in the mantle, gut, and cardiac organs, and while most are sensory, some, notably a gut plexus system, are probably involved in motor control (Boyle 1986). Other nerve cells are distributed along nerves where they can be clumped as a *ganglionic chain*, such as in the arms and gills, although most of the peripheral nerve cells are massed in the ganglia listed below.

Brachial ganglionic chain: Formed in the nerves which run from the brachial brain lobes into each arm. They are cross-linked by the interbrachial commissure before they enter the arms. In *Octopus*, these are the largest nerves in the body, and this part of the nervous system represents the 'greater part of the nervous system of an octopus' (Young 1963a, p. 249). The cells in the nerves running along the centre of the arms cluster to form a *brachial* (or *sucker*) *ganglion* above each sucker. Lateral nerves are given off from the main brachial nerve and supply the arm muscles and also a small *subacetabular ganglion* that lies just above each sucker.

The branchial ganglionic chain runs along the base of each gill and originates from the visceral nerve.

Stellate ganglia: The pair of mantle nerves arises from the palliovisceral lobe of the brain and terminates at the large stellate ganglia, one on each side of the mantle. Smaller nerves radiate from the stellate ganglion to supply the mantle muscles. In decabrachians, this ganglion is where the second and third order neurons synapse (Llinás 1999), and the ganglion also acts as a lower motor centre for mantle movements.

Ganglia in the buccal region are innervated by the superior buccal lobe. The *inferior buccal ganglia* lie just behind the buccal mass below the oesophagus, and they control buccal movements. The pair of *subradular ganglia* lies at the base of the salivary papilla and innervates its musculature.

The large, unpaired *gastric ganglion* lies at the proximal end of the intestine and is connected to the brain via the visceral nerve which supplies the crop, stomach, caecum, and intestine. In octopods, paired *fusiform ganglia* arise from the visceral nerves near the base of the renal papillae. These nerves supply parts of the cardiac system. The *cardiac ganglia* lie on, and innervate, each of the two branchial hearts.

7.2.6.1.2.6 Function of the Cephalopod Brain and Nervous System

There have been many studies on the structure, function, and organisation of the cephalopod brain, particularly on *Octopus*, *Loligo*, and *Sepia*, employing anatomy,

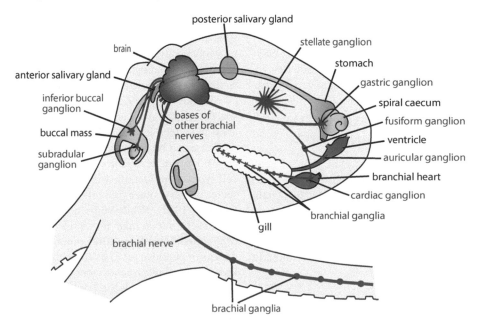

FIGURE 7.14 A diagrammatic plan of the peripheral nervous system of *Octopus*. Purple labels indicate parts of the nervous system. Redrawn and modified from Boyle, P.R., Neural control of cephalopod behaviour, in Willows, A.O.D., *Neurobiology and Behavior, Part 2. The Mollusca*, vol. 9, pp. 1–99, Academic Press, New York, 1986.

BOX 7.1   SOME SIMILARITIES AND DIFFERENCES IN *OCTOPUS* BRAINS AND THE BRAINS OF VERTEBRATES

While some aspects of the brain of a coleoid cephalopod like *Octopus* are similar to those of vertebrate brains, others are not.

- The nervous system is a distributed, rather than a centralised, system in coleoids. In *Octopus* there are more neurons in the arms than in the CNS (Young 1963a).
- The vertical, superior frontal, and inferior frontal lobes of the brain are involved in the consolidation of memory and are thus functionally, but not structurally, similar to the mammalian cortex (Young 1995). The vertical lobe, in particular, has similarities with the mammalian hippocampus in function and structure (Hochner 2008).
- Many neurotransmitters (see Section 7.3.2) are similar – notably dopamine (DA), noradrenaline (NA), and serotonin (5-HT) receptor subtypes resemble those found in vertebrates (Messenger 1996). *Octopus* and vertebrates are the only known taxa to have two members of the vasopressin/oxytocin 'superfamily' of peptides – these are expressed in the brain in *Octopus* (Takuwa-Kuroda et al. 2003).
- The optic lobe in coleoids is organised in a similar way to the deeper layers of the retina in vertebrates.
- The peduncle lobe is similar cytologically to the vertebrate cerebellum.
- Brain size relative to body weight is smaller in octopuses than in birds and mammals, but larger than in fish and reptiles. The number of neurons in an octopus brain (~500 million) is more than in amphibians and small mammals and only a little less than mammals such as cats and dogs (Hochner 2008).

---

histology, and experiments involving electrical stimulation and lesions, as well as learning and behaviour.

Bodily reactions and functions are mainly controlled by motor neurons that originate in the lower and intermediate motor centres mostly in the suboesophageal lobes of the brain. This part of the brain has been likened to the vertebrate spinal column. These lower and intermediate centres are controlled by the higher motor centres – primarily, the peduncle and basal lobes, which are in turn controlled by the optic lobes (Hanlon & Messenger 1996). Thus sensory input (mainly from the eyes, statocysts, and the arms and their suckers) is processed in the motor centres (controlled by the optic lobes) and in the higher centres (the vertical and superior frontal lobes) of the brain where memories are formed (Young 1977; Messenger 1983; Hanlon & Messenger 1996). See also Section 7.16. See Box 7.1 for a comparison of the octopus brain with that of vertebrates.

7.3   MOLLUSCAN NEUROSECRETORY CELLS AND THE NEUROENDOCRINE SYSTEM

While neurotransmitters enable the communication of neurons with other cells via synapses, some cells communicate via the endocrine system. This involves chemical signals (hormones) affecting the target cells less directly than by way of cell to cell contact (see Chapter 2 for a general introduction). In molluscs, most of this research has been conducted with cephalopods and heterobranch gastropods. As already noted, both groups have very large neurons.

7.3.1   NEUROSECRETORY CELLS IN THE CNS

Neurosecretory cells are discussed in Chapter 2. The secretory product is synthesised in the body of the cell. The process commences with the rough endoplasmic reticulum secreting the

neurohormone (peptides) from where it is transported in tiny vesicles to the Golgi zones. Here, the vesicles are packaged into granules and transported through the axon to the neurohaemal endings (Figure 7.15) from where the hormone is either released indirectly to the blood by diffusing through the perineurium and connective tissue sheath or directly into the bloodstream. In the latter cases, such as the cerebral arteries of stylommatophoran land snails or the vena cava of cephalopods, the axons line the walls of blood vessels. Sometimes, where the secretory cell axons lie within peripheral nerves, they may deliver the hormone directly to a target cell or tissue. This *peripheral neurosecretion* is commonly seen in 'pulmonates'; for example, the 'dark green cells' of *Lymnaea* have axons that reach into the kidney and are probably involved in regulating ion transport. A large number of neurohaemal endings may be present with estimations of 80,000 endings in the ovulation hormone-producing cells in *Lymnaea* (Wendelaar Bonga 1971), with about 800 per cell. These widely spaced release areas are in contrast with the more restricted areas in vertebrates and arthropods.

There can be many types of neurosecretory cells in a species. For example, by the early 1980s, as many as 25 different cell types were identified in *Lymnaea* based on their peptide secretions and other properties (Schot et al. 1981; Joosse & Geraerts 1983). While some are molluscan specific, others are structurally similar to peptides found in vertebrates. As Mukai and Morishita (2017) point out '...there are probably dozens of peptides yet to be fully characterised at a functional level *in any single animal*' (our emphasis).

Among molluscs, heterobranch gastropods have been most studied (Mukai & Morishita 2017). Experiments done in the late 1970s and 1980s showed that neurohormones influenced a wide range of body processes including those involved with reproduction (such as egg-laying), heart rate, respiration, growth and metabolism, excretion, and immune responses.

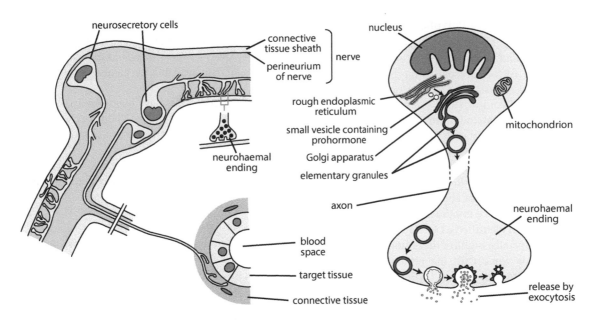

FIGURE 7.15 A diagrammatic representation of neurosecretory cells in neurohaemal areas in a stylommatophoran gastropod with a diagram of the detail of a neurosecretory cell on the right. Figure on left redrawn and modified from Joosse, J., Endocrinology of molluscs, in Durchon, M. and Joly, P., *Actualités sur les hormones d'invertébrés*, pp. 107–123, Colloques internationaux du Centre national de la recherche scientifique, Villeneuve-d'Ascq, France, 1976 and figure on right redrawn and modified from Joosse, J. and Geraerts, W.P.M., Endocrinology, in Saleuddin, A.S.M. and Wilbur, K.M., *Physiology, Part 1. The Mollusca*, Vol. 4, pp. 317–406, Academic Press, London, UK, 1983.

For example, a neurohormone produced by cells associated with the parietal ganglia influences the production of urea and uric acid in *Lymnaea* (Hanumante et al. 1978, 1980), and one in the pleuvrovisceral complex of *Onchidium* controls ammonia levels (Hanumante & Deshpande 1980). Similarly, various neurohormones produced in the so-called 'bag cells' of the visceral ganglia of *Aplysia* have been shown to be involved in the control of the renal aperture, respiratory movements of the mantle cavity, constriction of arteries and heartbeat (Alevizos et al. 1991), as well as ovulation and egg-laying (the egg-laying hormone – ELH) (see Chapter 8). The control and initiation of neurohormone secretions can be complex. The egg-laying hormone in *Aplysia* is initially released from small cells in the cerebral and pleural ganglia; this activates its release from the bag cells in the visceral ganglia and is thus an example of a distributed neurosecretory network (Hatcher & Sweedler 2008).

In contrast to heterobranchs, other gastropod taxa are poorly studied. Early studies on a few caenogastropods (e.g., Gabe 1966), including *Viviparus* (Gorf 1961, 1963) and *Bithynia* (Andrews 1968) showed neurosecretory activity in most ganglia except, in some species, the pedal and osphradial ganglia, and low activity in the suboesophageal and cerebral ganglia. *Littorina* has been shown to have large numbers of small serotonergic cells in all the major ganglia, with most in the pedal ganglia (Croll & Lo 1986).

7.3.2 Neurotransmitters

The main neurotransmitters are briefly described in Chapter 2. Several have been identified in the nervous systems of both euthyneuran gastropods and coleoid cephalopods, including acetylcholine, dopamine, octopamine, histamine, adrenaline,

noradrenaline, peptides, and some amino acids. In cephalopods, additional neurotransmitters include serotonin (5-HT), glutamate, γ-aminobutyric acid (GABA), alanine, aspartate, glycine, taurine, and tyramine (Budelmann et al. 1997, p. 329). In addition, FMRFamide-like, substance P-like, and somatostatin immunoreactives (ISRIF) have been demonstrated in several parts of the brain. Budelmann et al. (1997) suggested that the diversity of neurotransmitters is correlated with variation in form and content of the synaptic vesicles seen in neuronal processes and synaptic clefts in the brain and optic lobes.

Neurons are often divided into different types based on their neurotransmitters. Thus, *cholinergic neurons* use the neurotransmitter acetylcholine (ACh), *serotonergic neurons* release 5-HT (serotonin), and *peptidergic neurons* release the FMRFamide-related neuropeptides (FaRPs) and other peptides.

Nitric oxide (NO) is also important in molluscan nervous systems, as outlined in Chapter 2. There are many reports in molluscs, including gastropods (e.g., Liu et al. 1996; Inoue et al. 2001; Peruzzi et al. 2004), that the microglia produce NO which acts as a messenger molecule in neuron-microglia communication. It is also critical in memory formation (see Section 7.6.2).

7.3.3 Neuroendocrine System

The role and some of the general functions of the molluscan neuroendocrine system are outlined in Chapter 2 and are reviewed elsewhere (e.g., Lever & Boer 1983; Joosse 1988; Hartenstein 2006). Peptides are the most diverse hormones, but others are lipids and amino acid derivatives.

Here, we briefly overview the physical structure of the neuroendocrine system and its relationship to the rest of the nervous system in the major molluscan taxa.

7.3.3.1 Aplacophorans and Polyplacophorans

The nervous system of solenogasters has the neurotransmitters FMRFamide and serotonin (Todt et al. 2008; Faller et al. 2012), with the former present throughout the nervous system, but serotonin is restricted to the longitudinal nerve cords, the cerebropedal commissure, and part of the cerebral ganglion. It is particularly abundant in an unpaired bundle of neurites adjacent to the ciliary cells of the pedal fold (Faller et al. 2012). Neurotransmitters were not detected in the buccal system (Todt et al. 2008). In the caudofoveate *Chaetoderma*, serotonin-like and FMRFamide-like immunoreactivity is mainly localised in the posterior section of the brain (Shigeno et al. 2007; Faller et al. 2012), not the anterior lobed section.

Amano et al. (2010b) identified a GnRH[6]-like peptide in the cerebral, pleural, and pedal ganglia of the chiton *Acanthopleura japonica*. Subsequent work with the chiton *Lepidochitona* demonstrated immunoreactivity for FMRFamide and serotonin in the cerebrobuccal and subradular ganglia, the paired lateral nerve cords and ventral and lateroventral commissures (Faller et al. 2012).

7.3.3.2 Scaphopods

The cerebral, pleural, and pedal ganglia have FMRFamide-like immunoreactive somata on the outer side of the central mass of fibres, and serotonin is also present in the cerebral and pedal ganglia and in the visceral nerves (Wanninger & Haszprunar 2003; Faller et al. 2012).

7.3.3.3 Bivalves

Some commercially important bivalves such as oysters, mussels, and clams are relatively well studied (e.g., Mathieu et al. 1991). Stewart et al. (2014b) identified 74 putative neuropeptide genes from genome and transcriptome data of two pteriomorphian bivalves, the pearl oyster *Pinctada fucata* and the Pacific oyster *Crassostrea gigas*. They also identified over 300 encoding precursors for potential bioactive peptide products, a gene for the GnRH, and two egg-laying hormones.

Most bivalves studied have serotonin and dopamine in their ganglia, and these hormones regulate various functions such as heart rate, movements of the foot, reproduction, and ciliary activity. GABA, an inhibitory neurotransmitter, is found in low amounts in the ganglia of bivalves (Cochran et al. 2012a, 2012b).

Neuropeptides involved in regulating growth, gametogenesis, and glycogen metabolism have been identified, for example in mytilids, and are released from several types of neurosecretory cells, mostly in the cerebral ganglia (Gosling 2003).

7.3.3.4 Gastropods

7.3.3.4.1 Patellogastropods, Vetigastropods, and Neritimorphs

Based on the investigation of the genome of *Lottia gigantea*, about 59 genes that encode for putative neuropeptides have been identified (Veenstra 2010). Most of these have been previously identified in other molluscs or invertebrates, and the genes that produce at least some neuropeptides are thought to have evolved in early metazoans. Other than this study, there is little data available for patellogastropod hormones or neuroendocrines.

Neuroendocrine cells have been identified in *Haliotis* and *Trochus* (e.g., Hahn 1994a, b; Thongkukiatkul et al. 2001), some at least of which appear to be associated with reproductive activity.

There are apparently no relevant studies on neritimorphs to date although Pfeiffer (1992) described neuroendocrine cells in the gut epithelium of a species of *Nerita*.

7.3.3.4.1.1 Juxtaganglionar Organ
Endocrine structures are located within the connective tissue sheath of the cerebral commissures of some patellogastropods (*Patella*), vetigastropods (*Haliotis*), and trochids (*Gibbula*, *Phorcus*, and *Monodonta*) (e.g., Martoja 1965a; Herbert 1982; Clare 1987). In vetigastropods, this juxtaganglionar organ is a light yellow structure that runs along the cerebral commissure and is thus similar to the dorsal bodies of some hygrophilans. The juxtaganglionar cells in vetigastropods are structurally similar to the same organ in the gymnosome *Hydromyles* and in *Aplysia*, with nerve axons containing neurosecretory granules involved with female reproduction (e.g., Martoja 1965b, 1965c; Clare 1987). Interestingly, no equivalent structures have been identified in neritimorphs, caenogastropods, or basal heterobranchs, raising doubts about their homology.

The control mechanisms of the juxtaganglionar organs of vetigastropods and 'opisthobranchs' are not understood, but the equivalent structures in 'pulmonates', the dorsal bodies, have been shown to be under photoperiodic, neuroendocrine, and nervous control (reviewed by Saleuddin et al. 1994).

7.3.3.4.2 Caenogastropods

While a considerable body of work has been carried out on the endocrinology of *Crepidula* (Joosse & Geraerts 1983), particularly in relation to reproduction, there are only a few studies on other caenogastropods, most of which simply demonstrate the presence of neurosecretory cells. For example, Croll and Lo (1986) identified serotonin-like immunoreactivity in the CNS of *Littorina*, noting that the neurosecretory cells were smaller and much more numerous (>1500) than those in euthyneuran gastropods. They are found in all the major ganglia, but most were in the pedal ganglia. Cardioinhibitory peptides including the catch-relaxing peptide (CARP) have been shown to inhibit the heartbeat of *Rapana thomasiana* and are closely related to myomodulin, another neuropeptide found in at least some neogastropods (Fujiwara-Sakata & Kobayashi 1992).

7.3.3.4.3 Heterobranchia

Mukai and Morishita (2017) have recently reviewed peptides and neurotransmitters in the heterobranchs *Aplysia*, *Lymnaea*, *Helisoma*, *Biomphalaria*, *Helix*, and *Achatina*.

---

[6] Gonadotropin-releasing hormone. See Chapter 8.

7.3.3.4.3.1 'Opisthobranchs' The GnRH system of *Aplysia* is particularly well studied (e.g., Zhang et al. 2000; Sun et al. 2012; Mukai & Morishita 2017) and is produced in the CNS, mainly in the pedal ganglia, but also in the cerebral and abdominal ganglia (Jung et al. 2014).

The bag cell neurons of *Aplysia* (see also Section 7.3.1) secrete several neurohormones (peptides) some of which mediate egg-laying behaviour, and these are all derived from the egg-laying prohormone (proELH). The ELH induces ovulation and egg-laying behaviour (Conn & Kaczmarek 1989) and is detailed by Mukai and Morishita (2017) (see also Chapter 8).

In 'opisthobranchs' such as *Aplysia*, the endocrine juxtaganglionar organs are located in the connective tissue sheath of the cerebral commissure with the cells generally dispersed (Herbert 1982; Switzer-Dunlap 1987). The cells making up the organ are elongated or irregular in shape and often have thin processes like those seen in the dorsal bodies (Switzer-Dunlap 1987). They are otherwise structurally similar except for larger and more numerous granules in the cell body. Nerve axons containing neurosecretory granules enter the juxtaganglionar organs. The function is less certain than that of the

dorsal bodies (see below), although Saleuddin et al. (1994) posit gonadotrophic hormone production in *Aplysia* similar to the dorsal body cells of 'pulmonates' and the optic glands of cephalopods.

7.3.3.4.3.2 'Pulmonates' Neuroendocrine cells have been extensively studied in the hygrophilan *Lymnaea stagnalis* (Figure 7.16). There are several groups of neurosecretory cells recognised in the 'pulmonate' CNS (except for the pedal ganglia, where none are present). They are distinguished by their staining properties with various vital stains. The summary below is focused on *Lymnaea* (Figure 7.16), but is relevant for 'pulmonates' in general and is largely based on the detailed account in Luchtel et al. (1997).

7.3.3.4.3.3 Light Green Cells (LGC = MDC, LDC), Cerebral Green Cells, and Canopy Cells There is an interconnected mediodorsal and latero-dorsal cluster of LGCs on each cerebral ganglion (Figure 7.16); they are often called mediodorsal (MDC) and latero-dorsal (LDC) cells and are involved in growth regulation and related processes. Besides contact with the median lip nerve, the axons from these cells do not appear

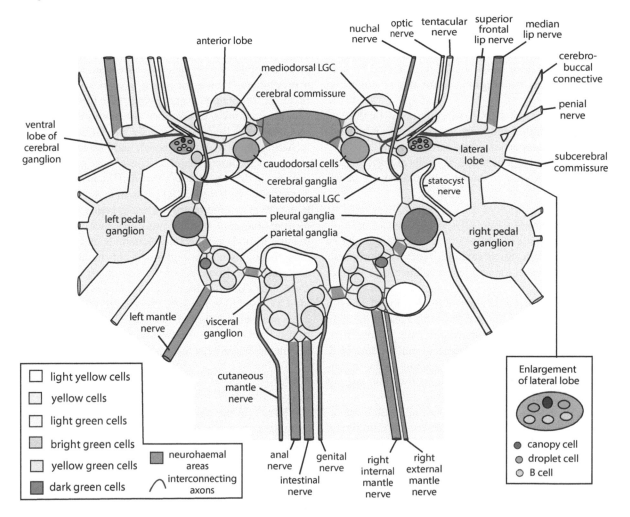

FIGURE 7.16 Neurosecretory cell groups and their neurohaemal areas in the CNS of the hygrophilan panpulmonate *Lymnaea stagnalis*. Redrawn and modified from Wendelaar-Bonga, S.E., *Z. Zellforsch. Mikrosk. Anat.*, 108, 190–224, 1970.

to innervate other neurons or organs. They receive sensory input from neurons at the base of the cephalic tentacles. These are equivalent to the *cerebral green cells* (CeGC) (= dorsal green cells) in stylommatophorans such as helicids (e.g., Wijdenes et al. 1987). The axon branches of these cells innervate the non-nervous endocrine dorsal bodies which produce gonadotropic hormones (see Section 7.3.3.4.3.8).

There is a single large cell, the *canopy cell* (CC) in each lateral lobe on the cerebral ganglia of *Lymnaea* (e.g., Minnen et al. 1979) (Figure 7.16). These two cells are coupled between the two ganglia and are histochemically and structurally like the LGC cells; they are also connected to an axon in contact with the median lip nerve. Both the LGC and CC produce four insulin-related neuropeptides (e.g., Smit et al. 1993; Benjamin & Kemenes 2013).

7.3.3.4.3.4 Caudodorsal Cells (CDC)
These cells are in the caudodorsal part of the cerebral ganglia in *Lymnaea*, with more (about 70 cells) on the right ganglion. A small group of these cells is also sometimes found on the pleural ganglia. There is a dorsal cluster of CDCs with only one axon, and a ventral cluster with two axons, these axons running through the cerebral commissure and discharging at the neurohaemal area. Two neuropeptide genes (*CDCH-1* and *CDCH-II*) are expressed in the CDCs to produce two related hormones, both concerned with ovulation (e.g., De Lange et al. 1994). These genes are also expressed by neurons in the skin in the head region, on the mantle edge, and in the genital tract.

In stylommatophorans, CDC-like cells occur in the visceral and parietal ganglia, and there are small cells with similar immunoreactive properties in the cerebral ganglia.

The CDCs are homologous to the bag cells of *Aplysia*. Like the bag cells, the CDCs are involved in regulating egg-laying and associated behaviour.

7.3.3.4.3.5 Dark Green Cells (DGCs), Yellow-Green Cells (YGCs), and Yellow Cells (YCs)
The visceral ganglia of *Lymnaea stagnalis* have three separate types of neuropeptidergic cells involved with ion and water balance (e.g., With & Berg 1992).

In *Lymnaea*, the DGCs (Figure 7.16) are found in both pleural ganglia, but mostly in the right, with a few in the right and left parietal ganglia and visceral ganglion. Despite their separation, these cells are interconnected and synchronised. Their axons discharge in various neurohaemal areas, including at the peripheries of the ganglia connectives, the layer of connective tissue surrounding the ganglia, the nuchal nerves, and (mainly via pedal nerves) to the foot and mantle integument. They can be activated by low ionic concentrations, and some have suggested that the DGCs regulate mucus secretion. The neuropeptide released by these cells appears to be related to the thyrotrophic releasing hormone of vertebrates.

The YGCs (Figure 7.16) are intermingled with the YCs in several small groups of cells in the ganglia of the visceral ring and are near the points of origin of nerves and connectives. The axon endings of these cells are in the peripheries of

several connectives and nerves, some reaching the kidney and ureter, and some branches ramify in the connective tissue surrounding the CNS. The neuropeptide secreted by these cells has not been characterised, and details of the function of these cells are unknown. They are, however, thought to be involved in osmoregulatory processes. The YGCs and YCs are innervated by a single interneuron – the 'visceral white cell' which belongs to the group of FMRFamide neurons.

The YCs are very like the YGCs with similar axon patterns and neurohaemal areas. They produce the sodium influx stimulating (SIS) peptide which regulates sodium uptake, both through the skin and kidney. The YC system innervates several organs, including the pericardium, ureter, aorta, and the ventral buccal artery, suggesting involvement in the regulation of water and electrolyte balance processes and blood pressure (Hatakeyama et al. 2004). Innervation of the ovotestis suggests a possible role in maintaining the blood-testis barrier, and nerves to the ventral buccal artery suggest a role in regulating blood pressure to the buccal mass.

YC type cells have been found in planorbids and in the semi-amphibious *Succinea*, but not in other (terrestrial) stylommatophorans which appear to lack this system.

7.3.3.4.3.6 Light Yellow Cells (LYC)
Clusters of LYCs (Figure 7.16) are found in the right parietal ganglion and visceral ganglion of *Lymnaea* and have been identified in other hygrophilans (Planorbidae) and several stylommatophorans (Boer & Montagne-Wajer 1994). The LYCs form a large neurohaemal area, as do other systems in the visceral complex. Axons of this system ramify into all the central ganglia including the cerebral and pedal, and into the muscles of the aorta and the ureter papilla. These cells produce a protein containing three peptides (LYCP-I-III) and are assumed to be homologous with the R3–R14 system in *Aplysia*. They appear to be involved indirectly with egg-laying, copulation, and feeding, and may regulate processes such as blood pressure (related to body posture). In planorbids and stylommatophorans, the axons innervate the veins of the lung and the sinus containing the blind sac of the secondary ureter in the slug *Deroceras* (Boer & Montagne-Wajer 1994), suggesting that the LYC system is involved in pressure regulation related to urine production. The homologous R3–R14 system in *Aplysia* is also involved in cardiovascular processes (e.g., see Boer & Montagne-Wajer 1994 for references).

7.3.3.4.3.7 VD1 and RPD2
Two giant neurons (not shown in Figure 7.16) named VD1 (in the visceral ganglion) and RPD2 (in the right parietal ganglion) are white in living animals (rather than the orange-yellow of most neurons) and are probably involved in regulating respiratory and cardiovascular processes. Dopamine acts as an inhibitor and histamine as an excitor. Branches to the neuropil extend into the visceral and right parietal ganglia. Second order neurons belonging to this system are found in the same ganglia in *Lymnaea* and in the cerebral ganglia, and third order neurons are found in the pedal ganglia (Kerkhoven et al. 1992).

VD1 runs to the skin near the pneumostome and osphradium via the right parietal nerve and the skin of the lips via the lip nerve. VD1 and RPD2 axons are both in the pericardial branch of the intestinal nerve, and they innervate the auricle. Several neuropeptides are produced by these nerves, and other cells in the cerebral, visceral, and right parietal ganglia produce similar ones.

This VD1/RPD2 system in *Lymnaea* and *Achatina* is probably homologous to the R15 of *Aplysia*, with similar neurons and neuropeptides in stylommatophorans with 1–4 giant neurons in the visceral complex. In *Lymnaea*, these nerves innervate the auricle, but not the ventricle (Kerkhoven et al. 1993), and they produce a peptide that excites the auricle (Bogerd et al. 1994). Luchtel et al. (1997, p. 693) suggested this system is common to all gastropods, but this seems unlikely as the homologous cells have only been observed in euthyneurans (Heterobranchia) – i.e., in stylommatophorans (Kerkhoven et al. 1993) and in *Aplysia*. They are lacking in the pelagic heterobranch *Clione limacina*, although the 'HE' cell in the left pedal ganglion of *Clione* is a cardio excitor and three 'HI' cells in the left visceral ganglion are cardiac inhibitors (Arshavsky et al. 1990). Antibodies of the VD1/RPD2 α-neuropeptide are, however, expressed in bivalves, vetigastropods, caenogastropods, stylommatophorans, *Aplysia*, and some coleoids suggesting that related neuropeptides are expressed in some neurons in a wide variety of molluscs (Kerkhoven et al. 1993). Genetics of the VD1/RPD2 system in *Lymnaea* have been reviewed by Benjamin and Kemenes (2013).

7.3.3.4.3.8 Dorsal Bodies in Panpulmonates

With the juxtaganglionar organs, the panpulmonate dorsal bodies are the only non-nervous endocrine structures identified in gastropods. In *Siphonaria* and hygrophilans, these bodies are on the mediodorsal part of the cerebral ganglia and extend on to the dorsal surface of the cerebral commissure (Saleuddin et al. 1997). In stylommatophorans, these cells are scattered through the connective tissue sheath of the ganglia making up the CNS, although concentrated on the dorsal surface of the cerebral commissure (Joosse 1988; Saleuddin et al. 1991).

The cells comprising the dorsal bodies appear to be derived from mesoderm (Boer et al. 1968; Boer & Joosse 1975), not the ectoderm from which the nervous system is derived. The cells making up the disc-like dorsal bodies are separated from the cells of the nervous system by a basement membrane, except in the panpulmonate *Siphonaria*. In that taxon, the dorsal bodies are located entirely on the cerebral ganglia and well separated by the long cerebral commissure. Unlike other 'pulmonates', in which all the dorsal body cells lie outside the basement membrane, in *Siphonaria* some of the dorsal body cells lie immediately underneath the basement membrane of each cerebral ganglion. The other cells making up the dorsal body form a simple mass (Saleuddin et al. 1997). The dorsal body cells show differences in structure depending on the reproductive state of the individual. In hygrophilans, there is no direct contact between the dorsal body cells and neuroendocrine cells, but in at least one planorbid, the dorsal body

cells penetrate the perineurium (Boer et al. 1968; Khan et al. 1990). The close proximity of neurosecretory axons suggests neuronal control of dorsal body activity is likely. In contrast, in stylommatophorans, there appears to be direct nervous control, with the dorsal body cells innervated by axons from the cerebral ganglia (reviewed by Luchtel et al. 1997) with the control being inhibitory.

In hygrophilans, the dorsal bodies (or mediodorsal bodies) are two yellowish clumps of cells located on the dorsoposterior surface of the cerebral commissure. They are well separated in lymnaeids and in the planorbid *Ancylus*, but in other planorbids (*Australorbis*, *Planorbarius*, and *Helisoma*), they lie close together or are partly fused (e.g., Boer et al. 1968; Saleuddin et al. 1994). Besides these dorsal bodies, smaller paired latero-dorsal bodies with similar cells are found in some species. The ellobiid *Melampus* has only the mediodorsal bodies, which are on the cerebral commissure (Price 1977).

The dorsal body cells in hygrophilans are arranged in lobules, usually in groups of 6–12 cells separated by connective tissue etc. Each dorsal body cell is elliptical, with fine processes that interdigitate with those from other dorsal body cells in the medulla, which contains blood spaces, neurosecretory axons, and connective tissue (Saleuddin et al. 1994). The dorsal body cell processes are closely associated with nerve axons from the cerebral ganglia that penetrate the basement membrane.

The dorsal body cells of stylommatophorans lie in the connective tissue surrounding the cerebral ganglia and the cerebral commissure. These cells are usually irregular in shape and occur singly or in small groups. *Succinea* is an exception, with the dorsal bodies forming compact structures attached to the cerebral ganglia (Boer & Joosse 1975). In some other stylommatophorans such as the helicid *Cornu aspersum*, the dorsal bodies are innervated by FMRFamide-containing axons in the cerebral ganglia (e.g., Marchand et al. 1991). These neurosecretory cells regulate growth by inhibiting the activity of the dorsal body cells. Mating causes significant changes in the dorsal body cells, and their activity is controlled by the gonad. The dorsal body cells control oocyte maturation and growth, and the differentiation of the female accessory sex organs. The hormone produced by the dorsal bodies of the planorbid *Helisoma* is not well understood, but may be an ecdysteroid[7]-like substance (Mukai et al. 2001).

7.3.3.4.3.9 Lateral Lobes of 'Pulmonates'

The lateral lobes (Figure 7.16) are neuroendocrine centres in 'pulmonates'. They are a small lateral projection on each of the cerebral ganglia, which are connected to the cerebral ganglion by a nerve, and are beneath the origin of the tentacular and optic nerves (Luchtel et al. 1997). Each lateral lobe receives axons from the optic nerve. Experiments have shown that they are involved in the maturation of gametes, oviposition, and growth control.

---

[7] Ecdysteriods include insect sex and moulting hormones, but also occur in some other invertebrates.

The lateral lobes are functionally closely associated with the dorsal body cells, and they also contain the *follicle gland* (see below). Also, in each lateral lobe, a large (~100 μm) canopy cell is connected to the cerebral ganglion via a nerve and is part of the 'light green cell' neurosecretory system (see Section 7.3.3.4.3.3).

7.3.3.4.3.10   The Procerebrum of Stylommatophorans

The *procerebrum* is the equivalent structure to the lateral lobe in Hygrophila. It lies on the frontolateral part of each cerebral ganglion and contains a *follicle* or *cerebral gland*. It is thought to be homologous to the rhinophoral ganglion in 'opisthobranchs'. In Ellobioidea and Stylommatophora (Eupulmonata), globineurons are present and there are typically two connectives to the cerebral ganglion (Mol 1974).

The procerebrum arises from a separate ectodermal invagination from the remainder of each cerebral ganglion, and they fuse late in development (reviewed by Mol 1967). The extensive study by Mol showed that the procerebrum is present in all stylommatophorans, and although details differ from group to group, it is thought to be an adaptation to terrestrial life.

Each procerebrum consists of small neurons, and it has been calculated that it has as many or more neurons than the rest of the CNS. This structure is involved in the chemoreception of odours (i.e., olfaction) via the tentacles. The olfactory nerve from the optic tentacle arises from the apex of the procerebrum. The tentacle ganglion, at the distal end of the olfactory nerve, is similar histologically to the procerebrum in possessing many small neurons and a finely textured neuropil (Ratté & Chase 2000).

Olfactory information is carried to the procerebrum by the olfactory nerve and the medial lip nerve (from the optic and anterior tentacles, respectively). Some nerve cells from the pedal and buccal ganglia also send axons into the procerebrum, at least in some species.

7.3.3.4.3.11   The Follicle (or Cerebral) Gland of 'Pulmonates'

This epithelial structure, of uncertain function, consists of a central lumen having epithelial cells interspersed with gland cells and neuronal processes bearing cilia (Brink & Boer 1967). It arises from the same tube-like ectodermal invagination as the procerebrum. The lumen of this embryonic tube forms the follicle(s) after it closes off. In at least one ellobiid the follicle does not close off, but remains connected by a tube to the external environment (Lever et al. 1959; Mol 1967). In hygrophilans, there is a single follicle gland associated with each lateral lobe (the equivalent of the procerebrum), but in stylommatophorans, there may be several follicles with parts inside and outside the procerebrum.

The function of the follicle gland is unknown as there is apparently no evidence that neuroendocrines are produced (Luchtel et al. 1997), although some have suggested that it is possibly endocrine in nature or even a receptor organ (e.g., Boer & Joosse 1975).

7.4   SENSORY STRUCTURES

Sensory cells (i.e., sensory receptors) detect information from both the external and internal environments and send that information by way of the nervous system, typically to the CNS where it is processed to generate a response (e.g., muscle contraction). The sensory information is detected by a specialised cell membrane and is converted (transduced) into an electrical signal (the receptor potential).

If the (primary) sensory receptor is a neuron, the receptor potential is converted to an action potential for transmission in the neuron, but if it is a (secondary) sensory cell, the receptor potential passes through a synapse to a connecting neuron where it is converted into an action potential.

The different kinds of sensory cells and systems that respond to different classes of stimuli are outlined in Box 7.2.

BOX 7.2   THE MAIN TYPES OF RECEPTOR CELLS

- Chemoreceptors – detect specific chemicals (often proteins). They also include chemicals involved in interspecific and intraspecific interactions – e.g., mate recognition, predators, and food recognition (taste). Distance chemoreception is called *olfaction*, while contact chemoreception is *gustation*.
- Thermoreceptors – detect changes in temperature. These are usually of two kinds: cold or heat receptors, which initiate an action potential with a fall or rise in temperature, respectively.
- Mechanoreceptors – detect mechanical pressure (touch) and/or displacement (distortion). The simplest forms are naked neurons connected either to a bristle or to ciliated epidermal cells with underlying nerves, as in statocysts. In coleoid cephalopods, the statocysts are elaborated to an angular acceleration system.
- Photoreceptors – respond to light. These may be concentrated into light-sensing or even visual organs, or dispersed across the body. Dermal photoreceptors may be photosensitive nerve endings in the epithelium with photoreceptive pigments in them, but these are still poorly understood (Ramirez et al. 2011). Also, some ganglia have photosensitive neurons (see Section 7.9.2). Accumulation of photoreceptors into eyes has often occurred in molluscs.
- Magnetoreceptors – while found in some animals, notably some birds and fish, to date they are unknown in molluscs, although a few chitons and gastropods have been shown to respond to magnetic stimuli (see Section 7.8).
- Electroreceptors – are known in some animals (mainly vertebrates), but as yet are unknown in molluscs.

7.4.1 EPIDERMAL RECEPTORS

Various sensory cells in the epidermis include touch and chemosensory receptors (Bubel 1984) and sometimes photoreceptors. Epidermal temperature receptors, like those in vertebrates (e.g., Spray 1986), are unknown in molluscs, although responses to temperature occur in some terrestrial gastropods (see Section 7.7).

The epidermal receptors (Figures 7.17 and 7.18) are bipolar nerve cells with their cell bodies within or just below the epidermis. Each receptor has a single dendrite extending to the surface of the epidermis and a single axon connecting to other nerves. The function (i.e., chemo-, mechano-, or photoreceptor) of the receptor cells cannot be readily determined by TEM examination as they are all similar in structure (Bubel 1984), although chemosensory cells are typically not ciliated, while mechanoreceptor cells are generally ciliated, and photoreceptors may have long cilia or microvilli. These simple dispersed receptor cells in the epidermis have been little studied

in molluscs, although where aggregated, they form the basis of the better-known sensory organs such as the osphradia and eyes. All such sensory organs are derived from the ectoderm during development.

In bivalves, receptors are mainly in the siphons, the mantle edge, or on tentacles on the mantle edge, these being the parts of the animal interfacing with the external environment. In gastropods, they are found in the general epidermis of the head-foot, but are mainly in the cephalic tentacles (Figure 7.18), epipodial tentacles (mainly vetigastropods), around the mouth, on the foot, and on the mantle edge. In cephalopods, they are also found in the epidermis, but are mainly concentrated on the arms, especially on the rims of the suckers, and the mouth. A detailed account is provided by Bubel (1984).

Pre-oral and post-oral tentacles in monoplacophorans are usually considered sensory, but Haszprunar and Schaefer (1997) could not locate sensory cells in a species they examined (but see Section 7.6.7).

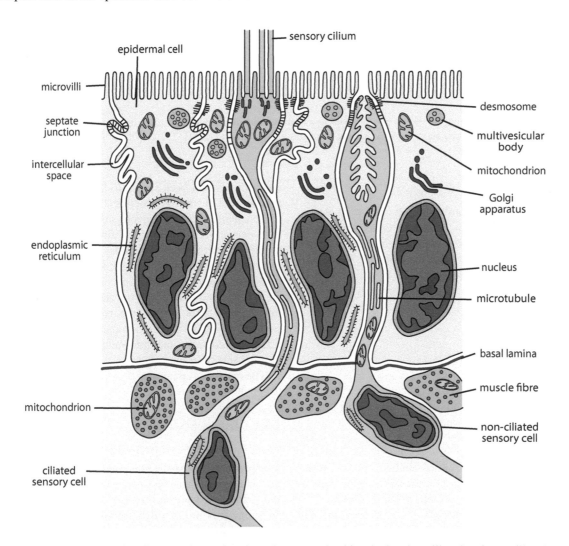

FIGURE 7.17 A diagrammatic representation of a section through gastropod epidermis showing ciliated and non-ciliated sensory cells. Redrawn and modified from Bubel, A., Mollusca: Epidermal cells, in Bereiter-Hahn, J. et al., *Biology of the Integument: Invertebrates*, Vol. 1, pp. 400–447, Springer Verlag, Berlin, Germany, 1984, which was based on figures in Crisp, M., *J. Mar. Biol. Assoc. U.K.*, 51, 865–890, 1971 (a nassariid) and Newell, P.F., *Malacologia*, 16, 183–195, 1977 (stylommatophoran slug).

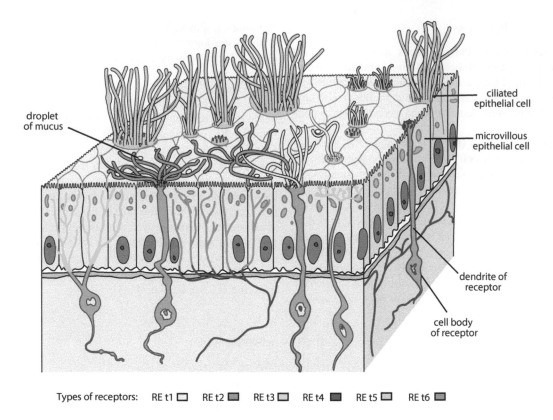

Types of receptors: RE t1 ☐   RE t2 ▨   RE t3 ☐   RE t4 ▪   RE t5 ☐   RE t6 ▨

FIGURE 7.18   A diagrammatic representation of a small section of the epithelium in the sensory region on a cephalic tentacle of *Lymnaea stagnalis* showing six types of epidermal receptors. Redrawn and modified from Zaĭtseva, O.V. and Bocharova, L.S., *Cell Tissue Res.*, 220, 797–807, 1981.

7.4.2   Ampullae or Ciliary Bottles

These distinctive cells are neurons that have invaginated to form a flask or bottle-shape with the interior containing numerous cilia. Although their homology is unlikely, such structures are found in the osphradia of vetigastropods, the apical sensory organ of many gastropod larvae (see Section 7.5.1.2), olfactory organs of cephalopods, the ampullary organs of chiton larvae, and the nuchal organs of some polychaetes (Haszprunar et al. 2002).

7.5   SENSE ORGANS

Sense organs can respond to simple mechanical or tactile stimuli, to chemicals, light, or gravity and, in some cases, temperature. They comprise sensory receptor cells which respond to a specific kind (or kinds) of stimulus. Some sensory receptors may themselves be neurons (and thus primary receptors), others are not, and they activate neurons via neurotransmitter signals (secondary receptors). While the function of most sense organs is established, there remain a few where it is equivocal or unknown.

Chemoreception is vital, not only for food detection, but for the detection of mates, eggs or sperm (with free-spawning taxa), finding suitable substrata during larval settlement, the recognition of predators or prey, and, in some taxa, trail following and homing behaviour. The cephalic and mantle tentacles, the osphradium and various areas of the mouth, mantle,

and foot have all been shown to have a chemosensory function, at least in some gastropods (Croll 1983).

Touch and chemical receptors or simple shadow responses from dermal light receptors all directly monitor critical environmental information that does not need a sophisticated nervous system to process. In most animals, including molluscs, the main light receiving organs are clusters of variously organised photoreceptors called eyes (see Section 7.9). Simple light-dark or shadow receptors do not require a sophisticated CNS and do not qualify as 'eyes', but information gathered from image-forming eyes requires processing in some sort of brain.

The main molluscan sense organs, and their distribution in the phylum, are listed in Table 7.1.

7.5.1   Larval Sense Organs

7.5.1.1   Water Pressure Reception

A response to hydrostatic pressure was observed in *Mytilus* larvae, where it is considered a form of mechanoreception (Bayne 1963). Increased pressure results in the veliger larvae swimming upwards in the water column, perhaps to facilitate dispersal. As the pediveliger approaches settlement, the pressure response fades, and the larva moves downwards to the substratum. Pressure-related changes have also been observed in other bivalve larvae (Cragg & Gruffydd 1975; Mann & Wolf 1983), but do not seem to have been noted in other molluscan larvae.

TABLE 7.1

Summary of Distribution of Some of the Main Sense Organs in the Molluscan Classes

Sense Organ	Caudo-foveata	Solenogastres	Poly-placophora	Monoplaco-phora	Bivalvia	Scaphopoda	Gastropoda	Cephalopoda
Cephalic Eyes[a]	○	○	○	○	● Some pteriomorphians	○	●	●
Statocysts	○	○	○	●	●	●	●	●
'Lateral line'	○	○	○	○	○	○	○	Coleoids only
Osphradium	○	○	○	○	●	○	●	*Nautilus* only
Posterior SenseOrgan	●	● (Chitonida)	●	○	○	○	○	*Nautilus* only
Cephalic Tentacles	○	○	○	○?	○	○	●	○
Mantle (edge) Eyes[a]	○	○	●?	○	● (evolved separately in several groups of autobranchs)	○	◐ (very rare)	?
Dorsal Mantle Eyes[a] **and Shell Eyes**	○	○	● (shell eyes-aesthetes)	○	○	○	● (onchidioidean slugs)	○
Subradular Organ	○	○	●	●	○	●	● (patello-gastropods, vetigastropods and lower caenogastro-pods)	● (*Nautilus* only)
Specialised Olfactory Organ[a]	○	○	○	○	○	○	● (some euthyneuran gastropods - Hancock's organ and optic tentacles)	●

● = present, ◐ = present in a few members, ○ = absent.

[a] Several independent derivations.

Note: Organs restricted to one small group are not included in the Table.

7.5.1.2 Apical Sensory Organs in Larval Gastropods and Some Bivalves

The apical sensory organ (or apical organ, frontal organ, etc.) (ASO) (see Kempf et al. 1997) is found in both gastropod and bivalve larvae, but is best known in gastropods, where it differs in different groups (Schaefer & Ruthensteiner 2001; Page 2002a; LaForge & Page 2007; Kristof & Klussmann-Kolb 2010). It typically consists of neuronal cell bodies clustered into a neuropil (the apical ganglion). Some cells have two or three clumps of long cilia, others (presumed sensory neurons) have a dendrite-like extension that lies parallel to the surface of the epidermis. Another distinctive type is the 'ampullae' or 'ciliary bottles' (see Section 7.4.2), these being flask-shaped, with cilia within the invaginated lumen. Some of these cells show serotonin-like immunoreactivity (e.g., Page & Parries

2000). Multiple ampullae are found in the apical pits of caenogastropods and heterobranchs (Page 2002a), but only one (Page 2002d) in a lottiid patellogastropod, and apparently none in *Patella* (Haszprunar et al. 2002). The ASO of *Nerita* is, in many respects, like that of other gastropods, but lacks the ampullae (Page & Kempf 2009). Instead, there is a pocket-like cavity derived from a supporting cell and filled with cilia provided by several sensory cells (Figure 7.19). The ASO of *Nerita* also differs from that of other gastropods in lacking sensory neurons with an apical dendritic process that expresses serotonin-like immunoreactivity. Instead, the ciliary tuft in hatching larvae forms an elongate strip of cilia that extends along the dorsal margin of the apical ganglion (Page & Kempf 2009).

There are, to date, no published reports of ampullae associated with the ASO of vetigastropod larvae. Differences

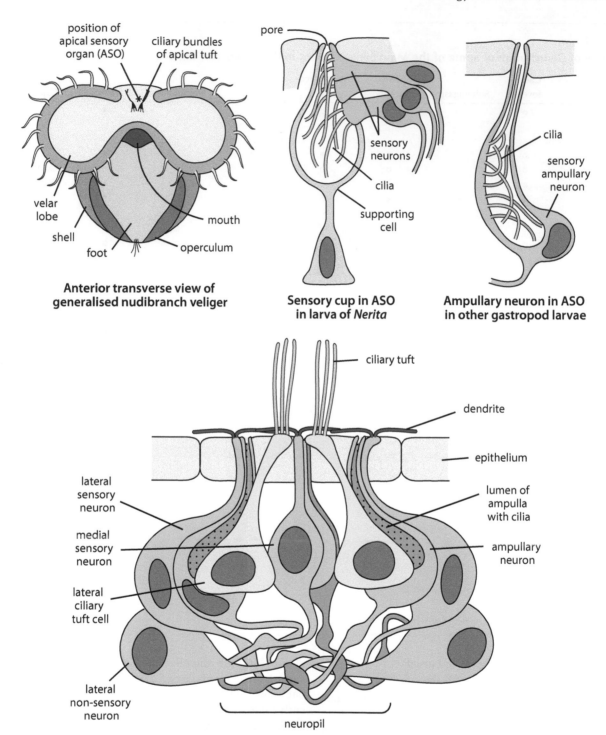

Anterior transverse view of generalised nudibranch veliger

Sensory cup in ASO in larva of *Nerita*

Ampullary neuron in ASO in other gastropod larvae

Major neuronal types in a nudibranch apical sensory organ

FIGURE 7.19   Apical sensory organ (ASO) of gastropod veliger larvae. Larvae redrawn and modified from Kempf, S.C. et al., *J. Comp. Neurol.*, 386, 507–528, 1997, details of cells redrawn and modified from Page, L.R. and Kempf, S.C., *Zoomorphology*, 128, 327–338, 2009, and nudibranch ASO redrawn and modified from Page, L.R., *Invertebr. Reprod. Dev.*, 41, 193–200, 2002a.

may be associated with the larvae of neritimorphs, caeno-gastropods, and heterobranchs often being planktotrophs, while those of patellogastropods and vetigastropods are non-feeding.

The ASO is lost at metamorphosis. Studies on nudibranch larvae have suggested that a role of the ASO appears to be in perceiving settlement cues at metamorphosis (Hadfield et al. 2000), and it may also detect stimuli relevant to locomotory activity by stimulating cilia of the velum (Kempf et al. 1997; Page 2002a).

While many bivalve larvae have an apical tuft of cilia, associated invaginated ampulla-like cells containing cilia have only been described in a few species, including a scallop (*Pecten*) (Cragg & Crisp 1991), a venerid (*Ruditapes*) (Tardy & Dongard 1993), and *Dreissena* (Pavlicek et al. 2018). It is possible that the development of these cells in bivalves is convergent with those in gastropods, although both share a pretrochal position in the larvae. They are not present in examined protobranch bivalves or in scaphopods (Haszprunar et al. 2002).

7.5.1.3 Ampullary System in Polyplacophoran Larvae

Chitons have a larval 'ampullary system' that is lost at metamorphosis. It consists of ampullary cells arranged in four pairs lying dorsolaterally and ventrolaterally in the pretrochal part of the larva (Haszprunar et al. 2002). Their axons enter the cerebral commissure.

7.5.2 Nociception

Nociceptors respond to noxious stimuli which may be chemical, mechanical, or thermal. They are primary receptors that are free nerve endings in the integument and may be present in some molluscs (Bullock & Horridge 1965). They are highly conserved across phyla, and it has been suggested that molluscs would provide good models for studies on pain in mammals including humans, although the molluscan system is much simpler (Crook & Walters 2011). These authors reviewed the physiology of nociceptors in molluscs and the behavioural responses caused by noxious stimulation. They suggested that pain-like states may be experienced by molluscs with more complex nervous systems, although how pain is interpreted in this context is highly subjective.

Nociceptors have been studied in some gastropods and cephalopods, and there is little or no evidence for their occurrence in the other classes, or in non-heterobranch gastropods. There have been relevant studies on some stylommatophoran land snails and on *Lymnaea*, but experiments on *Aplysia*, in particular, have provided evidence of nociception backed by supporting anatomical data. It has also been shown that the nociceptors in *Aplysia* can be desensitised, and sensitisation can be quite complex with, for example, site-specific responses and specific responses to particular stimuli (e.g., Walters 1991; Mason et al. 2014). Despite the numerous studies on cephalopod nervous systems and behaviour, knowledge of their nociceptor abilities (i.e.,

pain response) is poor (e.g., Hanlon & Messenger 1996) and is largely limited to a few recent studies.

It has long been known that coleoid cephalopods can learn to modify their behaviour in experiments using electric shocks. In the squid *Doryteuthis* (as *Loligo*), it has only recently been demonstrated that behavioural responses can also be modified for at least a couple of days duration following experimentally inflicted minor injury (Crook et al. 2011). Squid nociceptors selectively respond to noxious mechanical stimuli (e.g., injury), but not to heat, with anecdotal accounts of octopuses not responding to intense heat (Crook & Walters 2011). Mechanical stimuli cause spontaneous activity and responses occur on both sides of the body, not mainly the affected side as in mammals (Crook et al. 2013). Also, injured squid respond more quickly to threats than uninjured ones (Crook et al. 2014). Thus, while it is almost certain that cephalopods respond to what might be called pain, the physiological basis for these responses is still being worked out.

The evidence that some molluscs can perceive pain-like stimuli raises animal welfare issues, which are discussed in Chapter 10.

7.5.3 Mechanoreceptor Organs

Cilia-bearing epithelial mechanoreceptors are found dispersed across the external epithelium, but are often morphologically indistinguishable from chemoreceptive cells (e.g., Crisp 1971). Such cells may be modified and grouped into sensory organs, such as in the sensory hair system in a pit at the ends of the siphonal tentacle of some cardiids (Zugmayer 1904).

Experiments have demonstrated that the rhinophores and tentacles of some 'opisthobranchs' have mechanosensory receptors important in detecting the direction of water currents (e.g., Bicker et al. 1982a, 1982b).

A probable mechanoreceptor is found dorsal to the anterior adductor muscle in *Nucula*. This tube-like sense organ is called Stempell's organ (see Section 7.12.2.2).

Some epithelial mechanoreceptors are highly modified, such as the hair cells in the lateral line system of some coleoids (see Section 7.5.8) and in statocysts (see below). *Nautilus* responds to vibration, but how it achieves this is unknown (Soucier & Basil 2008).

7.5.4 The Statocysts – Gravity, Balance, and Direction

Statocysts (= otoliths) function as organs of balance and orientation and are found in all conchiferans. They are paired sacs filled with fluid; their internal walls have ciliated sensory receptor cells, and they contain solid, usually calcareous, structures which are a single *statolith* or multiple *statoconia*. In many species, the statoconia act individually, but in some, they act as a single mass. Gravity causes the statolith or statoconia to lie against part of the statocyst wall, activating the hair cells in that region. The hair cells signal the CNS, conveying information about the direction and magnitude of gravity. The pair of statocysts lie near the pedal ganglia and are innervated by the cerebral ganglia. They are cyst-like structures that appear early in development, usually from ectodermal invaginations, although in Hygrophila, they arise from solid ingrowths from the ectoderm (Raven 1952). They are thus present in larvae and are retained in adults.

Monoplacophoran statocysts open to the anterior mantle groove by way of a narrow duct (Haszprunar & Schaefer 1997). The statocyst nerve is penetrated by a narrow duct in some bivalves (Morse & Zardus 1997), and an open duct is present in protobranch bivalves, the pteriomorphian bivalves *Mytilus* and *Pecten*, and also in *Nautilus*. There is a closed duct in coleoids (see next Section). Such ducts have not been reported associated with gastropod or scaphopod statocysts. In the bivalves with open ducts, externally derived particles are utilised as statoconia, but in all other taxa, the inclusions are calcareous and secreted by the statocyst epithelium (Markl 1974).

The inclusions (statolith or statoconia) have a higher specific gravity than the fluid in which they are contained. As noted above, these inclusions interact with the sensory hairs and, as shown in experiments, they detect the orientation and directional movements of the animal. Coleoid cephalopods can also detect angular acceleration, an ability shared in invertebrates only by some crustaceans (Budelmann 1988).

Statoconia are found in the statocysts of monoplacophorans, scaphopods, cephalopods, in lower gastropods, and in protobranch and most other bivalves. Considerable diversity in the inclusions has been reported in anomalodesmatan bivalves (Morton 1985). In many higher caenogastropods, the statocysts each contain a single statolith although during development, there is often initially a single statolith, and statoconia are added progressively, suggesting that in adults, single statoliths may arise as the result of heterochrony (Ponder & Lindberg 1997). There are, however, exceptions, for example, the developing statocysts of the planorbid *Biomphalaria* have numerous statoconia (Gao et al. 1997). In *Aplysia*, the statoconia have been shown to develop in the supporting cells lining each statocyst and increase in size and number as the animal grows (Wiederhold et al. 1990).

Statocysts are broadly found throughout metazoan phyla, including Cnidaria and Ctenophora, most lophotrochozoan taxa (e.g., Acoelomorpha, Platyhelminthes, Annelida, Brachiopoda, Mollusca), ecdysozoan taxa (e.g., Euarthropoda), and deuterostomes (e.g., Holothuria, Tunicata) and are thought to have evolved several times in parallel (Schmidt-Rhaesa 2007). In molluscs, statocysts are absent in polyplacophorans and aplacophorans, but are present in the other classes (Conchifera), where they lie laterally near the pedal ganglia, except in scaphopods, where they lie posterior to the pedal ganglia (Figure 7.20). They are present in the larvae of scaphopods, the veliger larvae of bivalves and gastropods, and in the paralarvae of cephalopods. Because they are absent in the adults and larvae of chitons and aplacophorans, it is likely that statocysts evolved in molluscs independently

from those seen in other phyla, despite their general similarity. Nevertheless, the same or similar genes (e.g., *Pax258*) appear to be expressed in a range of metazoans during their development (O'Brien & Degnan 2003).

The statocysts are lined with supporting cells and numerous mechanosensory cells with multiple sensory hairs ('hair cells') (Bubel 1984). The cilia arising from the hair cells have been likened to kinocilia on the hair cells in the vertebrate inner ear (e.g., Barber & Dilly 1969). Hair cells are involved in the gravity receptor systems of most invertebrates and all vertebrates. They are usually primary receptor cells (which have an axon extending from the base), but secondary sensory cells (without an axon) are found in coleoid cephalopods and ctenophores (Budelmann 1988).

The ciliated sensory cells (hair cells) in the statocysts of most conchiferan molluscs are distributed evenly over the statocyst wall, although the number of sensory cilia (kinocilia) and their size differ in different groups. While in most examined conchiferans the bases of the cilia are evenly spread, but in some, they are oriented in a particular direction, and this latter condition has been described as polarised. Polarised hair cells have been reported in the basal caenogastropod *Pomacea* (Stahlschmidt & Wolff 1972) and in coleoid cephalopods. The individual cilia are polarised in a single direction on polarised hair cells in cephalopods, while in gastropods and bivalves, the cilia are polarised in a radial or spiral pattern, and the cell is described as unpolarised (Kuzirian et al. 1981; Budelmann 1988). In most bivalves and scaphopods, these cilia are polarised radially towards the centre of the distal surface of the cell, while in gastropods, they are usually polarised towards the periphery (Budelmann 1988). Uniformly polarised hair cells are not present in the relatively complex statocyst of the pelagic, carnivorous heteropod *Pterotrachea* (Barber & Dilly 1969; Markl 1974) (Figure 7.21), although the individual cilia are morphologically and physiologically polarised.

The number of cilia per hair cell differs considerably in different gastropod taxa, with *Pomacea paludosa* having only 30–40 polarised cilia, while there are many more in the unpolarised hair cells of *Aplysia californica* (720) and *Limacus flavus* (500–700) (Barber 1968; McKee & Wiederholt 1974; Bubel 1984). In cephalopods, the hair cells of octopods and decabrachians have a relatively small number (50–150) of polarised cilia (Burighel et al. 2011), while in *Nautilus* there are only 10–15 unpolarised cilia (Neumeister & Budelmann 1997).

In cephalopods, gastropods, scaphopods, and bivalves, the statocyst nerve originates from the cerebral ganglion, but usually runs through the cerebropedal connective, giving the impression that the statocysts are innervated from the pedal ganglion. They are innervated directly from the cerebral ganglion in a few caenogastropods and at least one neritimorph (Fretter & Graham 1962). In the monoplacophorans *Neopilina* and *Vema*, the statocyst nerve arises not from the pedal ganglion, but from the second latero-pedal connective, which runs between the lateral nerve cord and the cerebropedal connective, both of which are connected to the cerebral ganglion (Wingstrand 1985).

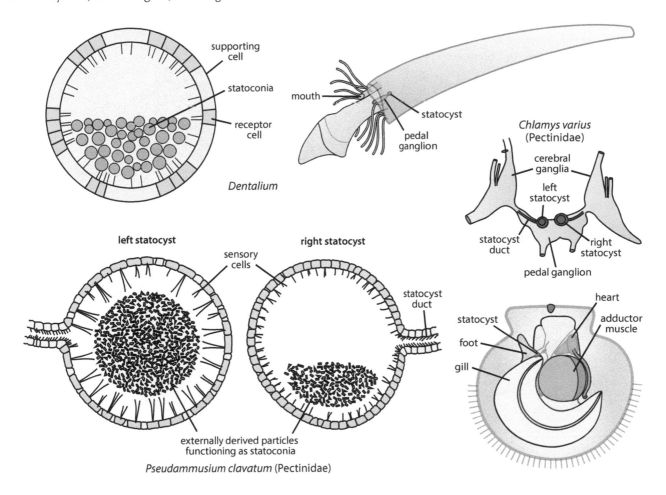

FIGURE 7.20 Diagrammatic sections of the statocysts of a scaphopod and a pectinid bivalve. Scaphopod statocyst redrawn and modified from Budelmann, B.U., Morphological diversity of equilibrium receptor systems in aquatic invertebrates, in Atema, J. et al., *Sensory Biology of Aquatic Animals*, pp. 757–782, Springer-Verlag, New York, 1988 and the pectinid redrawn and modified from Buddenbrock, W., *Zool. Jahrb. Abt. Allg. Zool. Physiol. Tiere*, 35, 301–356, 1915. Whole animal figures original.

While statocysts are typically present in adults and larvae of bivalves, they become reduced or even disappear in some adult burrowing taxa such as *Panopea* (Bower & Blackbourn 2003). The statocysts are normally symmetrical, but can be unequal in strongly inequivalve bivalves such as pectinids, with one side (the right [lower]) larger and more complex and containing statoconia and the other statocyst (the left [upper]) is smaller (Figure 7.20), and, in some pectinids, has a single statolith (Barber & Dilly 1969). Little is known about the neurotransmitters employed in statocysts, although histamine is indicated as a neurotransmitter in the hair cells in some studies on heterobranch gastropods, and may be more widespread in molluscs (Braubach & Croll 2004).

There have been experiments on the effects of altered gravity on statocyst development. Thus, embryos of *Aplysia californica* reared in simulated increased gravity resulted in the size and number of statoconia decreasing (Pedrozo et al. 1996), while in aquatic and terrestrial snails reared in microgravity in space, the statoconia increased in size (e.g., Balaban et al. 2011).

7.5.4.1 Cephalopod Statocysts

In cephalopods, the statocysts (or, as they are often known in coleoids, the angular acceleration receptors) are in the cartilage below the brain. In decabrachians at least, they provide information for the maintenance of orientation by way of gravity and angular acceleration, thus enabling bodily movements to compensate for any deviations. This has been demonstrated in coleoids by ablation of the statocysts which results in the animal spiralling and zigzagging as it swims. As well as disrupting eye orientation, octopuses can no longer discriminate or learn differences between horizontal and vertical shapes or the plane of polarised light (see Hanlon & Messenger 1996 for references). The statocysts also control countershading reflexes and are sensitive (in some coleoids at least) to vibrations, this latter sensitivity giving rise to the suggestion they may 'hear' (see next Section).

The statocysts in *Nautilus* are much simpler than in coleoids and are generally similar to those in other molluscs. They are simple, ovoid, small (2.0–2.5 mm in diameter) ectodermal sacs lined with numerous receptor cells and contain fluid and

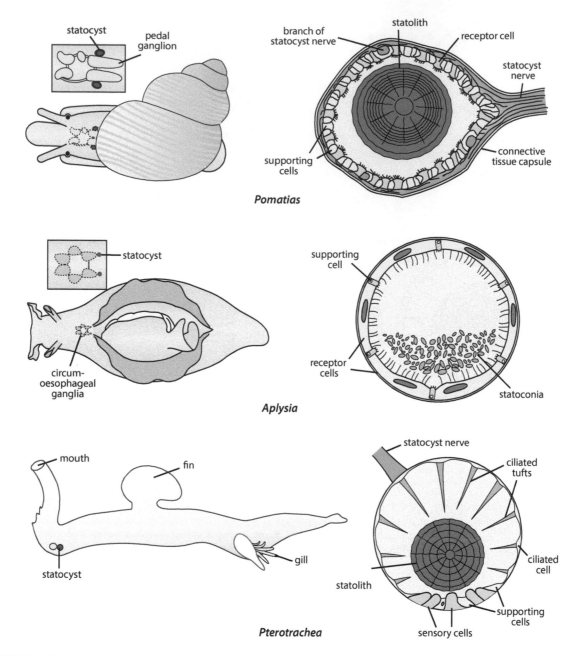

FIGURE 7.21   Diagrammatic sections of examples of gastropod statocysts. Statocysts redrawn and modified from the following sources; *Pomatias* Zaĭtseva, O.V., *Neurosci. Behav. Physiol.*, 27, 83–89, 1997; *Aplysia* Wiederhold, M.L. et al., *Hearing Res.*, 49, 63–78, 1990; *Pterotrachea* Markl, H., The perception of gravity and of angular acceleration in invertebrates, in Kornhuber, H.H., *Handbook of Sensory Physiology. Vestibular System Part 1: Basic Mechanisms*, Springer-Verlag, Berlin, Germany, pp. 17–74, 1974. Whole animal figures original.

numerous small, angular calcareous statoconia. They connect to the exterior by a narrow ciliated duct (Kölliker's canal) that opens below each eye (Young 1965a; Neumeister & Budelmann 1997). This canal is the embryological remnant of the invagination that formed the statocyst. It is also present in coleoids, but is closed in adults (Budelmann et al. 1997). The hair cells in *Nautilus* statocysts are primary receptors (in coleoids they are secondary receptors), and there are no nerve cells in the statocyst (Young 1988b). There are two types of hair cells in *Nautilus*. In the ventral half of each statocyst, the hair cells have a single row of 10–15 kinocilia-like cilia, while in the upper

half, the hair cells have 8–10 scattered kinocilia-like cilia with the cilia in both halves polarised (Neumeister & Budelmann 1997). Experiments have shown that the statocysts in *Nautilus* not only function as gravity receptors, but can detect rotatory movements (angular accelerations) despite lacking the sophisticated receptor systems described below for coleoids (Neumeister & Budelmann 1997). Compensatory eye movements occur that are statocyst controlled (Hartline et al. 1979), and inputs from both the statocysts and eyes control compensatory funnel movements (Neumeister & Budelmann 1997). Thus, *Nautilus* statocysts differ from those of most gastropods, scaphopods, and bivalves in having many more hair cells and in having the receptor cells divided into different regions, suggesting that they detect not only gravity, but also direction and rotation (Neumeister & Budelmann 1997).

Coleoid statocysts are much more complex than those of *Nautilus*, and their structure shows marked modifications in different taxa related to their lifestyles (Young 1989). Large irregularly shaped statoliths are associated with the morphologically anterior wall of the statocyst chamber, not ventrally as in other molluscs (Budelmann 1975; Budelmann 1988). In cephalopods, the anterior wall of the statocyst in a horizontally oriented coleoid is the equivalent of the ventral wall in a vertically oriented mollusc (see Chapter 3 and Figure 3.1).

The statocysts of octopods and *Vampyroteuthis* differ from those of decabrachians and other molluscs in that each has an outer and inner wall. They are in cavities in the ventral cranial cartilage (Figure 7.22). The outer wall is separated from the inner sac by a fluid-filled (perilymph) space crossed by many strands containing blood vessels (Young 1988b). The inner sac contains fluid (endolymph), and its inner surface has the sensory systems (see below) and contains the statolith. The statocysts of decabrachians are also in the ventral cranial cartilage (Figure 7.23).

As in other molluscan statocysts, in coleoids the hair cells are the main receptors. These cells interact with accessory statoliths or layers of statoconia. They are secondary sensory cells in synaptic contact with afferent and efferent nerve endings. The numerous hair cells in coleoids are polarised and arranged in two receptor systems within the statocyst, the *macula* and the *crista*.

The macula system: Comprises the gravity receptor system with interactions between the hair cells and the statolith providing information about the direction of gravity and linear acceleration. This input is used to orientate the eyes, head, and body (Budelmann 1970). It also regulates countershading (e.g., Ferguson & Messenger 1991; Ferguson et al. 1994) to adjust for changes in orientation. In octopods, there is only one macula (which carries the single statolith) in each statocyst, but in decabrachians, there are three, with the largest (the *macula princeps = macula statica*) having a large statolith partially attached to it. The two smaller maculae (*maculae neglectae*) are covered with a layer of small statoconia (Budelmann et al. 1997). These maculae are arranged at right angles to each other (Figure 7.23).

The statolith is on the vertical anterior wall of the statocyst – the anatomical ventral wall in ancestral cephalopods. It is irregular, with the shape being species-specific. Although a flange is attached to the macula princeps, the rest can make small movements that are registered by the hair cells. Each part of the irregularly shaped statolith is responsible for the detection of different directions of movement of the animal (Arkhipkin & Bizikov 2000). The statoliths are calcareous (aragonite), with an organic matrix, and often show growth rings which, in some squid at least, can be used to determine age (Jackson 1994).

The movements of the statolith not only stimulate the hair cells, but they can also alter the pattern of flow of fluid in the statocyst, which is also registered (Arkhipkin & Bizikov 2000). The hair cells communicate with large neurons around the statocyst. These neurons end in various parts of the brain (magnocellular lobe, lateral pedal lobe, peduncle lobe, and anterior and probably median basal lobes) (Budelmann & Young 1984). A first-order giant nerve cell lies above the macula princeps and closely connects with the macular nerve fibres.

The crista-cupula system: This is the angular acceleration receptor system that consists of a crista epithelium to which *cupulae* are attached. The crista is a ridge that runs in three dimensions on the inner wall of the statocyst and is subdivided into four segments set at right angles to each other in decabrachians (Figure 7.23), while there are nine segments in octopods. A flap (the cupula) is attached to each segment that protrudes into the lumen of the statocyst, and this swings back and forth in response to movements of the fluid (endolymph). The cristae are covered with hair cells, which are polarised at right angles to the long axis and, depending on the species, they range from a few hundred to thousands on each segment (Burighel et al. 2011). The hair cells on the cristae are activated by the movement of fluid in the statocyst, and they send information to the brain about rotational acceleration (Figure 7.25), enabling the animal to control the position of the eyes, head, and funnel (Budelmann et al. 1987). The hair cells (which are secondary sensory cells) are accompanied by primary sensory cells and are separated by supporting cells. The hair cells are not uniformly arranged on the cristae. In the middle of the ridge, large hair cells are arranged in two to four rows, and on either side, there are two to four rows of smaller hair cells. These large hair cells are in afferent contact with large, first-order neurons, and the small hair cells make contact with second order neurons (Burighel et al. 2011).

The statocyst wall in fast-swimming squid (loliginids and ommastrephids) has several projections (*anticristae* or trabeculae) that divide part of the statocyst interior into canal-like spaces, lined with hair cells, that have been likened to the semicircular canals of vertebrates (Stephens & Young 1978; Budelmann & Tu 1997) (Figures 7.24 and 7.25). These canal-like structures slow the movement of fluid, enabling the effects of rapid turns to be registered. These 'canals' are absent in slow-moving squid such as the cranchiids and in octopods. An octopus statocyst and a human inner ear are compared in Figure 7.24.

In faster-moving squid, the statolith is relatively larger in proportion to the statocyst than in the slower moving taxa, notably cranchiids, making the former more sensitive to low angular accelerations. The horizontal channel is particularly

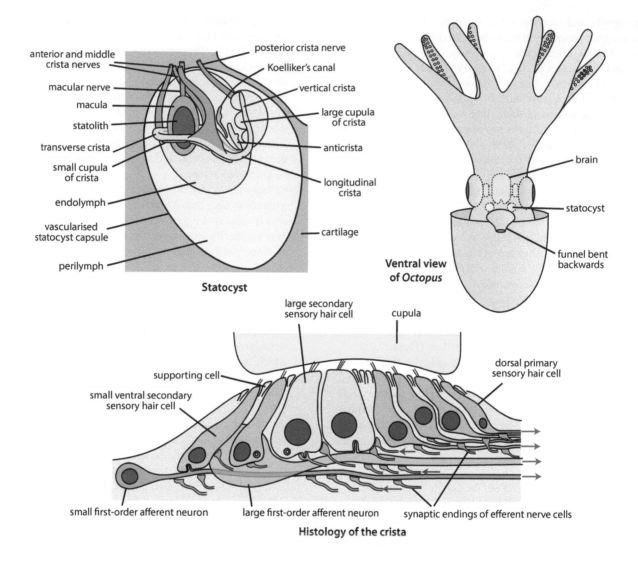

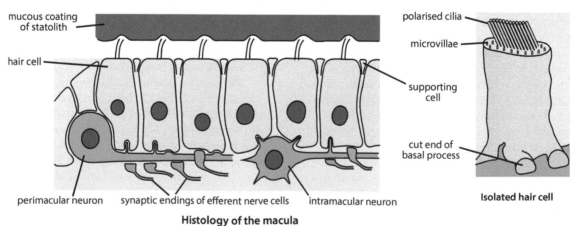

FIGURE 7.22 Statocyst of *Octopus* showing its location and general anatomy. Upper left figure redrawn and modified from Young, J.Z., *Proc. R. Soc. B*, 152, 3–29, 1960, upper right original. Cytology of the crista and macula redrawn and modified from Budelmann, B.U., Morphological diversity of equilibrium receptor systems in aquatic invertebrates, in Atema, J. et al., *Sensory Biology of Aquatic Animals*, pp. 757–782, Springer-Verlag, New York, 1988, and hair cell redrawn and modified from Colmers, W.F., *J. Comp. Neurol.*, 197, 385–394, 1981.

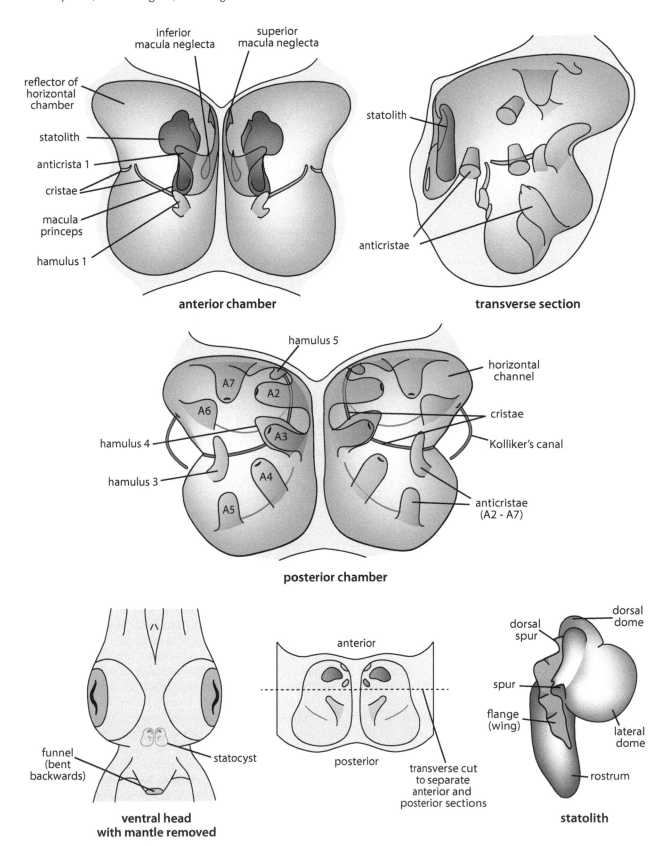

FIGURE 7.23 Statocysts and statolith of the squid *Berryteuthis*. Redrawn and modified from Arkhipkin, A.I. and Bizikov, V.A., *J. Zool.*, 250, 31–55, 2000 except for the two lower left figures of a 'squid' which are redrawn and modified from Budelmann, B.U., The statocysts of squid, in Gilbert, D.L. et al., *Squid as Experimental Animals*, pp. 421–439, Plenum Press, New York, 1990.

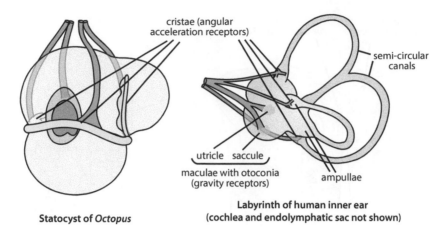

cristae (angular
acceleration receptors)

semi-circular
canals

utricle   saccule

maculae with otoconia
(gravity receptors)

ampullae

Statocyst of *Octopus*

Labyrinth of human inner ear
(cochlea and endolymphatic sac not shown)

FIGURE 7.24   Comparison of an octopus statocyst and the inner ear of a human. Redrawn and modified from Wells, M.J., *Am. Sci.*, 49, 215–227, 1961.

well-developed in decabrachians, with the endolymph displaced during linear movement by the statolith flowing via this channel to the *crista verticalis* in the posterior chamber (Figure 7.25) (Arkhipkin & Bizikov 2000). Of the other cristae, the *crista transversalis* anterior and the crista transversalis posterior, in the anterior and posterior chambers, respectively, are sensitive to accelerations in the pitch plane, while the *crista longitudinalis* and crista verticalis are more sensitive to changes in the rolling and yawing planes (Arkhipkin & Bizikov 2000) (Figure 7.25).

Coleoid statocysts can also detect low-frequency sounds (see next Section).

7.5.4.2 Sound Reception

Apparently no mollusc can detect vibrations caused by sound in a similar way to that of terrestrial vertebrates, but some have been shown to detect low-frequency sound, presumably as vibrations, via the statocysts or the substratum the animal is in contact with. Thus, molluscs are not known to use sound to communicate (Vermeij 2010), or even to detect sound from predators. For example, squid cannot detect the acoustic signalling of whales hunting them, although they have compensated by developing very large eyes (see Section 7.9.10.1). There has been debate as to whether cephalopods can 'hear', but the answer, as implied above, depends on the definition of 'hearing' used, as some definitions require an air-filled cavity and the ability to detect pressure changes (Budelmann 1996).

Experiments from the 1980s showed that some coleoids, presumably by way of their statocysts, could detect low-frequency sounds (i.e., external low-frequency vibrations >10 Hz) (e.g., Hanlon & Budelmann 1987; Packard et al. 1990) inaudible to humans (sometimes called 'infrasound'). It is also known that the nerves from the statocyst cristae and maculae respond to low-frequency sound (Hanlon & Messenger 1996). The first direct evidence that statocysts were responsible for sound detection was obtained from experiments with *Octopus ocellatus*, which was able to detect sounds ranging from 50–150 Hz (Kaifu et al. 2008). Also, the macula/statolith complex was specifically implicated in detecting sound waves in a

squid and an octopus. *Sepioteuthis lessoniana* was capable of detecting sounds ranging from 400 to 1500 Hz and *Octopus vulgaris* from 400 to 1000 Hz (Hu et al. 2009).

7.5.5 Cephalopod Arms and Suckers

Coleoids have a well-developed sense of touch, particularly via their arms and suckers, with *Octopus* having at least four kinds of mechanoreceptors in the sucker rims. *Nautilus* also has a well-developed sense of touch with its many arms and ocular tentacles, but these lack suckers so do not have the discriminatory abilities of coleoids, particularly octopods.

The highly sophisticated neural coordination of arms and suckers in *Octopus* has been reviewed by Grasso (2008). Young (1971) observed that the nervous system of the arms contains more neurons than the central brain and many functions are largely independent of the brain. While separated arms can function, they are coordinated as a unit in the intact animal. Similarly, suckers in octopods have some neural autonomy, but are also coordinated with relevant arm movements. They have neural connections to and from the arm and the brain, and each sucker has its own ganglion. Apart from their obvious gripping functions, octopod suckers are important sensory receptors. The sucker rims have high densities of mechano- and chemoreceptors – around 10,000 per sucker (Graziadei & Gagne 1976). The sensory axons from the sucker rim and proprioceptors (muscle sensors) from the sucker muscles are received by the sucker ganglion which communicates with the nervous system of the arm via the brachial ganglion that lies over each sucker. These form a chain of interconnected ganglia along each arm (Figure 7.14). Nerves from these ganglia also directly connect to the sucker, bypassing the sucker ganglion. This all results in a mix of brain, arm, and independent control of each sucker (Grasso 2008).

While they explore surfaces, the suckers of *Octopus* can make complex movements which can be independent of arm movements and visual cues. Tactile discriminatory abilities in octopuses are considerable and well known following many experiments (summarised by Wells 1978). The sensory

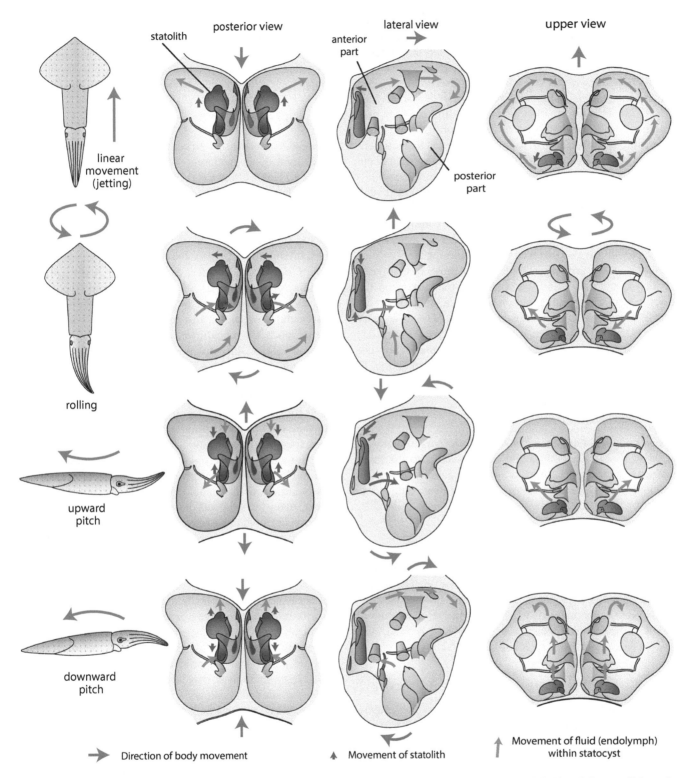

posterior view   lateral view   upper view

statolith

anterior part

posterior part

linear movement (jetting)

rolling

upward pitch

downward pitch

Direction of body movement   Movement of statolith   Movement of fluid (endolymph) within statocyst

FIGURE 7.25 Statocyst function in the squid *Berryteuthis* – the figures indicate the movement of internal fluid and the statolith as the squid moves and changes orientation. Redrawn and modified from Arkhipkin, A.I. and Bizikov, V.A., *J. Zool.*, 250, 31–55, 2000.

information from the suckers in octopuses is processed in the frontal, subfrontal, and posterior buccal lobes. Decabrachians lack these higher centres in the brain, although their inferior frontal lobe is comparable (Young 1988a). Experiments with *Octopus* have shown that objects can be discriminated based on shape and texture, and this information can be learned (e.g., Wells 1978). This information, however, is limited. They cannot distinguish by touch the orientation of grooves on a cylinder or discriminate differences in weight. The reason for these limitations appears to be related to the lack of information about the orientation of the suckers in space because of the mobility and flexibility of the arms (Wells & Wells 1957). Although there is evidence of arm-eye coordination (Byrne et al. 2006) this is rather poor (Hanlon & Messenger 1996). Interestingly, it is likely that the parts of the brain responsible for touch learning in *Octopus* are also involved in taste learning.

Squid suckers are less complex than those of octopuses (see also Chapter 17), and the control of the suckers on the retractile tentacles is much simpler. From a study of *Loligo* (Kier & Leeuwen 1997; Leeuwen & Kier 1997), it appears that the extension of the retractile paired tentacles (and its terminal suckers) is largely a reflex action. The tentacles are shot out at great speed, too fast for tentacle-sucker coordination, so sucker action is triggered on contact with the prey.

7.5.6 Pressure Receptors

Little is known about general pressure sensitivity in molluscs and what, if any, receptors might be involved. There is some evidence to suggest that *Nautilus* and some squid modify their behaviour in response to increased water pressure (Hanlon & Messenger 1996), as do some bivalve larvae (see Section 7.5.1).

7.5.7 Scaphopod Captaculae

These long thread-like appendages are unique to scaphopods and are described in some detail in Chapter 16. The terminal bulb contains a ganglion, and the captaculae respond to various prey items differently. Although the terminal bulb is ciliated, no specialised receptor structures have been located (Sigwart & Sumner-Rooney 2016).

7.5.8 Coleoid Cephalopod Lateral Line Analogue

Squid and cuttlefish are sensitive to local water movements (e.g., Budelmann & Bleckmann 1988; Komak et al. 2005) due to a 'lateral line system' comprised of lines of sensory hair cells on the head and arms, a system analogous to the lateral line of amphibians and fish. Cuttlefish have four lines on each side of the head, three of which (not the most lateral) extend onto the arms. In squid, there is also a fifth line on the ventral side of the head. Lines are also present in octopus hatchlings, but not in adults (Budelmann et al. 1997).

The lines contain thousands of hair cells that respond to water movements within the 0.5–400 Hz range and especially between 75 and 100 Hz (Bleckmann et al. 1991). These findings suggest that a cuttlefish should be capable of detecting a potential

predator (e.g., a 1 m long fish) about 30 m away (Budelmann et al. 1991). Presumably, the lines may also assist with prey detection. The hair cells in the lines are primary sensory cells, each with a few to many polarised cilia, and each is surrounded by supporting cells with microvilli (Burighel et al. 2011).

7.6 CHEMOSENSORY STRUCTURES

A range of chemosensory structures to detect chemical stimuli has evolved in molluscs – the equivalents of taste and smell (e.g., Kohn 1961a; Croll 1983). The chemosensory cells, whether they are simple receptors in the epidermis or clustered in organs, usually possess an invaginated distal surface bearing numerous sunken sensory cilia.

Some of the chemosensory organs in molluscs include the osphradium (see below), the rhinophores and Hancock's organ of 'opisthobranchs' (see Section 7.6.8.1.1), the tentacles of stylommatophorans (see Section 7.6.8.1.2), and the cephalopod olfactory organs (see Section 7.6.8.2). Others include the subradular organ, various tentacles and the bursicles in vetigastropods, the cephalic tentacles and lips of most gastropods, and various other structures in heterobranch gastropods (see Section 7.6.8.1).

Octopuses and probably all cephalopods are sensitive to chemicals over their entire bodies and especially the suckers (see Section 7.5.5). Cephalopods, particularly *Nautilus* and octopuses, use chemoreception to help locate food, and it may also be used in communication and mating. Experiments with *Sepia* show it also detects and responds to chemical signals (Boal & Golden 1999), but given the complement of chemosensory structures in cephalopods (skin, suckers, olfactory organs), many experiments involving chemosensory behaviour have not unambiguously distinguished their individual roles (Hanlon & Messenger 1996).

7.6.1 Osphradia and Posterior Sense Organs

The distribution of the osphradium in mollusc taxa and its location within the mantle cavity or mantle groove is covered in some detail in Chapter 4. Osphradia, which are innervated by the branchial nerve and often associated with their own ganglion, are present only in bivalves, gastropods, and *Nautilus* (Figure 7.26). In bivalves, they lie in the exhalant water stream, while in *Nautilus* and gastropods, they lie in the inhalant stream. They are absent in scaphopods, some heterobranch gastropods, coleoid cephalopods, chitons, monoplacophorans, and aplacophorans. Sensory structures (posterior sense organs) found in the posterior-most parts of the mantle grooves/cavities of chitons and aplacophorans were previously thought to be osphradia (see Chapter 4). In both aplacophoran groups, this sense organ lies at the dorsal edge of the mantle cavity in the incoming water current (Salvini-Plawen 1968a). Their function has been thought to be chemosensory, perhaps related to reproduction. Their dorsal position, rather than ventral, and innervation from the suprarectal commissure suggests that they are not osphradia and, following Lindberg and Sigwart (2015), they are called *posterior sense organs*. They are possibly homologous with

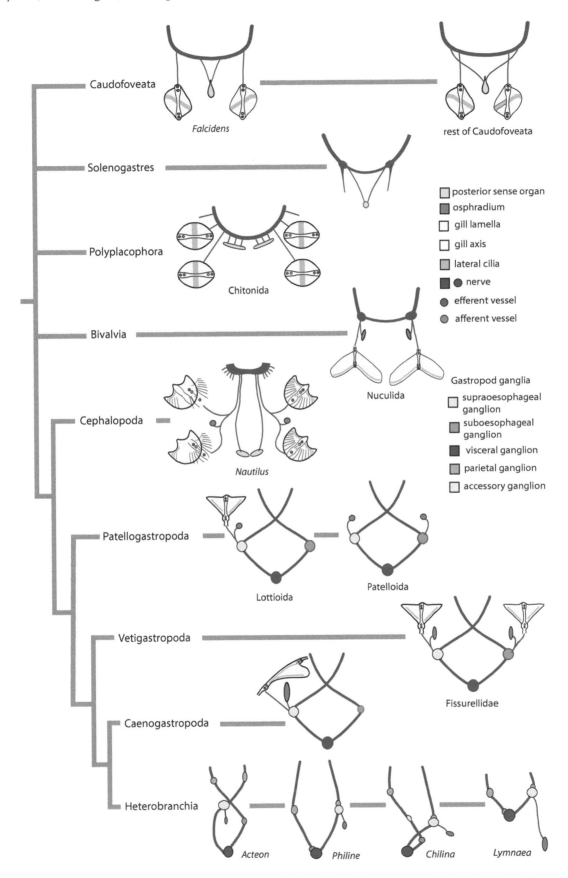

FIGURE 7.26 Innervation of osphradia and posterior sense organs in the major molluscan taxa and showing their position relative to the ctenidia. Redrawn and modified from Haszprunar, G., *Philos. Trans. Royal Soc. B*, 315, 37–61, 1987b, Patellogastropoda original.

the post-branchial sensory organs in chitons which share the same innervation (see also Chapter 4). Similarly, the posterior pair of sensory structures in *Nautilus*, previously thought to be osphradia, are also considered to be posterior sense organs and are innervated from the visceral nerve (see Chapter 4), but it is not clear if they share the same cytology. There are no posterior sense organs in scaphopods, coleoid cephalopods, monoplacophorans, bivalves, or gastropods. Based on the structure of the receptor cells, Haszprunar (1987b, 1987c) suggested a chemosensory role in reproduction in the posterior sense organs of the aplacophoran groups and chitons.

7.6.1.1 Osphradial Structure and Function

There have been several important studies on the structure of the osphradium (Crisp 1973; Haszprunar 1985a, g, 1986a, 1987b; Taylor & Miller 1989) and osphradial sensory cells are described by Haszprunar (1985a, g, 1986a, 1987b, c, 1992a). Ciliated lateral fields are present on the osphradia of most caenogastropods (the Sorbeoconcha – i.e., other than architaenioglossans) and neritimorphs, but, whereas they are formed by cylindrical ciliated cells and some sensory cells in neritimorphs, in sorbeoconch caenogastropods, the lateral ciliated areas also contain specialised cells (the microvillous Si1 and ciliated Si2 cells), with both types having 'column microvilli' (Haszprunar 1985a). Another unique sorbeoconch cell type, the Si4 cells, are found at the edges of the central zone of the osphradium and have a distinctive layered endoplasmic reticulum. Another type of specialised cell, 'cilia bottles' or ampullae (see Section 7.4.2), are only found in the osphradia of higher vetigastropods (Haszprunar 1985a).

The osphradium is an important chemoreceptor organ in aquatic gastropods (e.g., Brown & Noble 1960; Downey & Jahan-Parvar 1972; Townsend 1973; Sokolov et al. 1980; Wedemeyr & Schild 1995; Lindberg & Ponder 2001) and unionoidean bivalves (Sokolov & Zaĭtseva 1982). Earlier, it was (probably incorrectly) thought to be a mechanoreceptor in some lower caenogastropods (Hulbert & Yonge 1937). The evidence shows that the chemoreceptive role is the predominant one (Lindberg & Ponder 2001; Lindberg & Sigwart 2015), although detailed physiological studies on most molluscan osphradia have not yet been undertaken. Haszprunar (1987b, c) suggested a chemosensory function associated with reproduction in chitons, bivalves, and primitive gastropods, while in higher caenogastropods, the osphradium sensory functions are enhanced to enable food detection.

A role in reproduction has also been suggested for the osphradium in some lower caenogastropods such as *Viviparus* (Wölper 1950), in heterobranch gastropods such as the sea hare *Aplysia* (e.g., Tsai et al. 2003) and *Lymnaea* (Nezlin 1997), and in various bivalves (e.g., Haszprunar 1987b; Beninger et al. 1995a) (see also Chapter 8).

It has recently been shown that the osphradium of lymnaeids has an important role in the enhancement of long-term memory associated with chemicals (kairomones) emitted by a predator (in this case crayfish) (Karnik et al. 2012a), while in the marine panpulmonate *Siphonaria*, it plays a role in homing (Kamardin 1988). Another suggested osphradial chemoreceptive function includes, in *Lymnaea*, detection of levels

of oxygen and carbon dioxide (Kamardin 1976; Wedemeyr & Schild 1995), a role disputed by Karnik et al. (2012b). The osphradium of *Aplysia* can detect differences in osmolality (Downey & Jahan-Parvar 1972). Such diversity in known functions is not surprising, given the histological differences in gastropod osphradia in particular (see above).

7.6.2 Lip Receptors

The lips are important chemosensory structures that facilitate responses to chemicals in potential food items, either at a distance (as odours) or on contact. Based on their innervation, lips are homologous in all gastropods (Klussmann-Kolb et al. 2013), and probably in all molluscs.

The distinction between taste (gustation or contact chemoreception) and smell (olfaction or distance chemoreception) is less apparent in aquatic animals than in those on land, although it is still useful if they are distinguished by distance vs contact, and low vs high stimulus (Kohn 1961a). Taste receptors (i.e., close-range chemoreceptors) are likely to be present in the oral structures of many molluscs and have been demonstrated in some – for example, they have been reported in *Aplysia* (Emery 1976a), land snails (e.g., Salánki & Bay 1975), and other gastropods (e.g., Kohn 1961a), octopuses and cuttlefish (Graziadei 1965), and squid (Emery 1975). Putative taste receptors have been suggested on the smooth palp surface of *Crassostrea gigas* (Dwivedy 1973), but have been questioned by Beninger et al. (1990).

7.6.3 Adoral Sense Organ

Found only in protobranch bivalves, this small structure consists of a pair of thickened epithelial ridges containing three types of bipolar receptor cells innervated by the cerebral ganglion. Each ridge lies on either side of the mouth on the outer side of the base of the labial palp. Based on a detailed study of this structure, Schaefer (2000) suggested that the adoral sense organ was chemosensory, but noted that it is apparently not involved in feeding as it is separated from the mouth by the axis of the labial palps and thus from the furrow that transports food to the mouth. The function of these organs remains unknown.

7.6.4 Preoral Sense Organ

The preoral (or atrial) sense organ is found in the aplacophoran solenogasters and is associated with the detection of their food (mainly Cnidaria, but also polychaetes and other invertebrates). The Caudofoveata mainly feed on microorganisms and detritus and lack this structure, their only anterior sense organ being the highly sensitive pedal shield (Salvini-Plawen 1981b).

7.6.5 Ctenidial Sense Organs (Bursicles) of Vetigastropods

Ctenidial sense organs or 'bursicles' (Szal 1971; Haszprunar 1987d) (Figure 7.27) are present in most vetigastropods, including at least some neomphalidans (Geiger et al. 2008). They are small elongate sacs that lie within the gill filament just above the base of the efferent side of each ctenidial

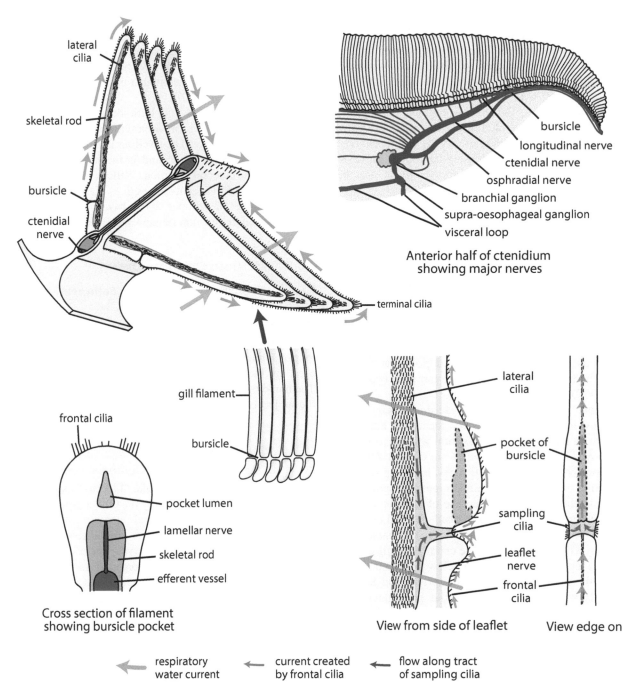

lateral
cilia

skeletal rod

bursicle

ctenidial
nerve

terminal cilia

bursicle
longitudinal nerve
ctenidial nerve
osphradial nerve
branchial ganglion
supra-oesophageal ganglion
visceral loop

Anterior half of ctenidium
showing major nerves

gill filament

bursicle

frontal cilia

pocket lumen

lamellar nerve

skeletal rod

efferent vessel

Cross section of filament
showing bursicle pocket

lateral
cilia

pocket of
bursicle

sampling
cilia

leaflet
nerve

frontal
cilia

View from side of leaflet   View edge on

← respiratory   ← current created   ← flow along tract
  water current   by frontal cilia   of sampling cilia

FIGURE 7.27 Location and detail of vetigastropod bursicles. Redrawn and modified from the following sources: upper left Sasaki, T., *Bull. Univ. Mus.*, 38, 1–223, 1998, upper and lower right Szal, R.A., *Nature*, 229, 490–492, 1971, lower left Haszprunar, G., *J. Molluscan. Stud.*, 53, 46–51, 1987d, middle original.

lamella and thus lie in the inhalant water current. They are open to the exterior, form a swelling, and extend inside the filament to the skeletal rods. The bursicles are innervated by the efferent nerve in the filament – i.e., the ctenidial nerve.

Each bursicle has inwardly beating cilia around the opening, and the inside of the pocket-like cavity is lined with ciliated cells, with a sensory epithelium in the innermost part which bears paddle-like processes and microvilli (Haszprunar 1987d). In the trochoidean *Tegula*, the bursicles have been shown to detect predatory starfish (Szal 1971).

7.6.6 Cruciform Muscle Sense Organ in Tellinoidean Bivalves

The X-shaped cruciform muscle (Graham 1934) is found on the ventral margin of tellinoidean bivalves at the base of the siphon (see Chapter 15, Figure 15.17). A narrow tubular sense organ lies in the posterior arms of the cruciform muscle and, in most species, opens to the exterior, but is closed in *Tagelus* and *Donax* (Graham 1934). Its structure has been described in some detail in *Scrobicularia plana* by

Odiete (1978). The tubular sense organ consists of a narrow distal section, and the other end opens to a larger cavity, which in turn communicates via a narrow opening with the siphonal lumen. The epithelium of the tube is penetrated by long ciliated dendrites (Frenkiel & Mouëza 1980) from neurons that connect with a small ganglion associated with the pair of sense organs. These sense organs probably open when the siphons are extended and close when the valves are closed. Chemoreceptor (Odiete 1978) and mechanoreceptor (Pichon et al. 1980) functions have been suggested for this organ, the latter function supported by the presence of collar receptors.

7.6.7 CEPHALIC TENTACLES

A pair of cephalic tentacles is present in gastropods and perhaps monoplacophorans. They are innervated by the cerebral ganglion, provide chemosensory and tactile information and accommodate the cephalic eyes at their bases in most gastropods. Cerebrally innervated head appendages occur in other classes, including the 'velum' of monoplacophorans, the labial palps of bivalves, the captaculae of scaphopods, and the arms of cephalopods, but they are unlikely to be homologues of cephalic tentacles (Ponder & Lindberg 1997).

Lemche and Wingstrand (1959) described a pair of 'preoral tentacles' in *Neopilina galatheae* located on either side of the mouth, at the junction of the velum with the roof of the mantle groove. These small tentacles are ciliated with their distal portions bearing several cell types, including secretory cells as well as several undetermined types distinguished by differences in their nuclei. At the base of each tentacle is a small concentration of 'dark, granulate cells' (Lemche & Wingstrand 1959, figure 70) (Figure 7.41). The tentacles are innervated by a robust nerve from the cerebral ganglion which terminates in a plate-like plexus just below presumed sensory epithelia at the tip of the tentacle; Haszprunar and Schaefer (1997) could not locate sensory cells in the 'pre-oral tentacles' of two species they examined. Whether these small tentacles are homologous with the cephalic tentacles of gastropods is not known. Innervation and position suggest similarities, but their vestigial condition masks their actual affinity, and they remain worthy of further study.

7.6.7.1 Cephalic Tentacles in Gastropods

Cephalic tentacles are found in most gastropods, although they are secondarily lost in some heterobranchs. Typical cephalic tentacles are found in most lower heterobranchs, but their homologues in euthyneurans are sometimes modified. Thus, the rhinophores of nudibranchs and the head shield (in part) of cephalaspidians are homologous to the cephalic tentacles of other gastropods (Huber 1993; Staubach 2008). Stylommatophorans have two sets of tentacles, the posterior pair is the homologue of the cephalic tentacles of other gastropods, and the anterior tentacles are a new feature derived from the oral lobes, as are the anterior tentacles in some 'opisthobranchs' (e.g., aplysiids) (Staubach 2008).

The tentacle epithelium has sensory cells embedded in it. While there are no detailed comparative studies published on caenogastropods or heterobranchs, patellogastropods, vetigastropods, and neritimorphs have been contrasted, and each group has a characteristic combination of epithelial cell types (Künz & Haszprunar 2001).

In both patelloidean and lottioidean patellogastropods, there is an additional sensory area at the base of each cephalic tentacle next to the eye. The swelling is rich in sensory cells (Choquet & Lemaire 1969) and is innervated by a unique nerve from the cerebral ganglion (Willcox 1898; Fisher 1904; Thiem 1917a). More recent studies (e.g., Marshall & Hodgson 1990; Künz & Haszprunar 2001) have not addressed this structure, and its function remains undetermined.

7.6.8 SPECIALISED OLFACTORY ORGANS

7.6.8.1 Cephalic Sense Organs of Euthyneuran Heterobranch Gastropods

The euthyneuran heterobranchs possess several sensory organs in the cephalic region that have mixed functions, being in part olfactory and mechanosensory. Many also possess eyes, but these are considered separately (see Section 7.9.4). The homology of these non-optic sensory structures has been the subject of recent studies, as outlined below.

7.6.8.1.1 *Hancock's Organ, Lip Organs, Oral Veils, Oral Tentacles, Rhinophores, and Cephalic Shields of 'Opisthobranchs'*

These structures comprise the suite of 'cephalic sense organs' of some 'opisthobranchs', notably the Cephalaspidea.

There are two anterolateral sensory structures in Cephalaspidea divided into anterior and posterior parts, each innervated by different cerebral nerves. These were called the anterior and posterior Hancock's organ (Edlinger 1980), but more recently that name has been restricted to the posterior folded structure, and the simple anterior part has been treated as a posterior extension of the lip organ (Staubach & Klussmann-Kolb 2007; Staubach 2008). Both the lip organ and Hancock's organ are found in most shelled 'opisthobranchs' (Cephalaspidea and Acteonoidea), although the latter is apparently secondarily lost in *Acteon tornatilis* (Staubach & Klussmann-Kolb 2007).

The Hancock's organ is characterised by a distinctively folded epithelium. Although Huber (1993) considered it homologous with the labial tentacles of Aplysiida and Pleurobranchida, detailed studies of its innervation have shown that it is actually homologous with the rhinophores of nudipleurans and aplysioids (Staubach 2008; Klussmann-Kolb et al. 2013), as suggested by Schmekel (1985), and probably the cephalic tentacles of other gastropods (Staubach 2008).

The ultrastructure of the cephalic sense organs has been described by Göbbeler and Klussmann-Kolb (2007) who concluded that Hancock's organ is mainly used for distance chemoreception and detection of water currents (rheoreception), whereas the oral tentacles and lip organ are possibly involved in contact chemoreception and mechanoreception.

The sensory oral veil of Pleurobranchida, the lip organ (= anterior Hancock's organ) of the Cephalaspidea and Acteonoidea, and the labial tentacles of Aplysiida and Nudibranchia are probable homologues as they are innervated by homologous nerves (Klussmann-Kolb et al. 2013). Collectively, these structures can be called the anterior sense organs (Staubach 2008).

Experiments and other studies have demonstrated the chemosensory nature of the rhinophores (e.g., Bicker et al. 1982b; Chia & Koss 1982; Levy et al. 1997; Wertz et al. 2007; Cummins et al. 2008), although in some taxa such as *Tritonia* (Wyeth & Willows 2006), there may also be mechanoreceptors functioning as rheoreceptors in detecting water currents. Response to light has also been documented in *Aplysia* rhinophores (Chase 1979).

Sacoglossans have only a single pair of tentacles, which are similar to rhinophores, although not homologous as they appear to represent fused oral and cephalic tentacles, being innervated by three cerebral nerves (Hoffmann 1932–1939; Huber 1993).

7.6.8.1.2 *The Tentacles of Stylommatophorans*

The ability to detect and distinguish odours (olfaction) is the principal means that stylommatophoran land snails and slugs have of sensing environmental cues at a distance because they cannot detect sound and have poor vision (Chase 2001). Each tentacle (of the two pairs) on the head has an olfactory organ at the tip, and most of the knowledge of the functioning of these organs is based on those on the posterior tentacles (also called the *ommatophores*, or optic tentacles). These tentacles have long been known to be olfactory organs. The bulging distal end of the tentacle contains the olfactory organ which lies just below the eye (Figure 7.28). It is a rather complex structure, with an

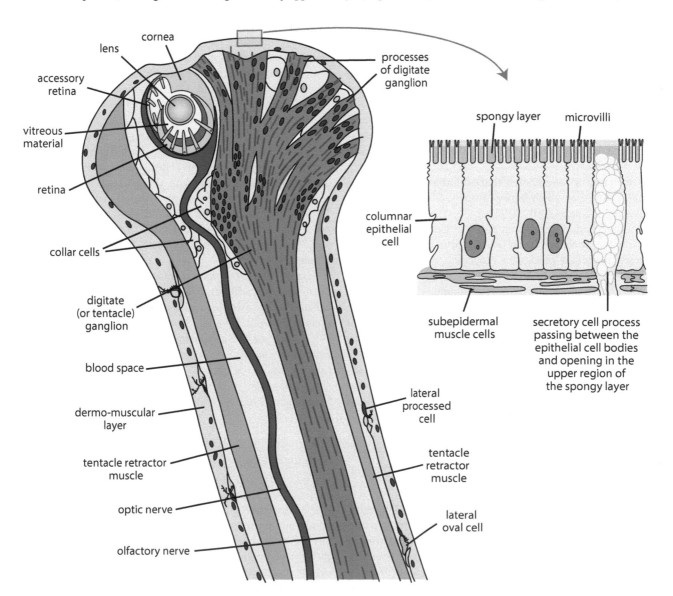

FIGURE 7.28 Longitudinal section of a posterior (optic) tentacle of the stylommatophoran slug *Deroceras reticulatum*. Details of the olfactory epithelium based on *Cornu aspersum* redrawn and modified from Rogers, D.C., *Z. Zellforsch. Mikrosk. Anat.*, 114, 106–116, 1971; tentacle section redrawn and modified from Runham, N.W. and Hunter, P.J., *Terrestrial Slugs*, Hutchinson University Library, London, UK, 1970.

outer epithelial layer below which lie muscles and then sensory neurons. The dendrites of these sensory cells extend through the brush border of the epithelium. The receptors are similar in morphology to olfactory receptors in vertebrates (Chase & Rieling 1986). The nerves of the olfactory organ originate in the tentacle (=digitate) ganglion that lies just below it. This ganglion also supplies the main nerve in each optic tentacle, the olfactory (or tentacle) nerve, which runs down the tentacle into the procerebrum. The procerebrum and tentacle ganglion are similar in cytological structure, having small neurons and a finely textured neuropil. The procerebrum is thought to be the area where olfactory information is processed (see Section 7.3.3.4.3.10) and where olfactory memories are stored in stylommatophorans (Ratté & Chase 2000). The smaller anterior tentacles also have an olfactory organ at their tips, and the medial lip nerve communicates signals from them to the procerebrum.

Various experiments have shown that slugs and snails use olfactory cues to follow slime trails of other snails. This is used, for example, by snail predators to track prey and by conspecifics to find mates. Other experiments have shown that some terrestrial slugs, for example, can return to a home site upwind over a distance of 90 cm by a direct route where cues from mucous trails or vision are not involved. Instead cues are apparently obtained from chemical odours deposited at the home site (Gelperin 1974; Cook 1979a) (see also Section 7.15.3.4.3).

Several studies have shown that stylommatophoran slugs and snails can distinguish between new and familiar odours (see reviews in Chase 2001, 2002).

In some terrestrial stylommatophoran gastropods, the epidermis on the posterior tentacles has primary sensory cells with long branched processes emerging from them (e.g., Rogers 1971; Wright 1974; Kataoka 1976). The processes are embedded in a mass of long, fine filaments that may have a protective function and form the so-called spongy layer (Bubel 1984). This spongy layer (Figure 7.28) of terrestrial stylommatophoran snail and slug tentacles is thought to be involved in olfaction by serving as a fluid trap for airborne chemicals. These modifications are lacking in hygrophilans (Zylstra 1972) and other aquatic gastropods, but a somewhat similar modification is seen in the cephalic tentacle epithelium of the terrestrial caenogastropod *Pomatias* (Wondrak 1981).

The posterior tentacles, along with their eyes and olfactory organs, can regenerate following removal, as for example, demonstrated in *Achatina* (Chase & Kamil 1983).

7.6.8.2 Olfactory Organs in Cephalopods

Because many cephalopods, including *Nautilus*, are nocturnal or live in deep water, chemoreception may be more important than vision in those taxa. *Nautilus* can detect and follow waterborne odours (Basil et al. 2005), and coleoids are capable of both contact and distance chemoreception (Budelmann et al. 1997).

Olfactory organs (rhinophores) in *Nautilus* have only one kind of sensory cell, flask-shaped cells which open to the exterior by a narrow pore. The lumen of the flask contains numerous cilia and some microvillae (Barber & Wright 1969a). Only two other cell types are present, ciliated epithelial cells and some mucous cells.

In coleoids, the olfactory organs are small pit-like structures below and behind the eyes where they lie in the inhalant respiratory current. They are capable of distance chemoreception, being lined with chemoreceptors able to detect low concentrations of chemicals (see Budelmann et al. 1997 for review). They are extremely developed in some mid-water cephalopods where they project from the head on stalks.

Coleoid olfactory organs are more specialised than those in *Nautilus*, having at least five kinds of assumed receptor cells present, including several cilia-filled flask-shaped cell types (e.g., Emery 1975, 1976b). These receptor cells are similar to other chemoreceptor cells in structure and experiments have shown that they are sensitive to coleoid ink and its precursors. These substances cause a jet response in squid, and some have suggested that the olfactory organs may function in the detection of alarmed individuals in these schooling animals. They may also function in other ways including the detection of sex pheromones (Hanlon & Messenger 1996).

The chemoreceptive abilities of *Octopus* suckers are well known, with early experiments showing that *Octopus* can discriminate tastes (Wells 1963). Octopuses suckers have many receptors, with far fewer found in cuttlefish and squid suckers (Budelmann et al. 1997). In *Octopus*, the receptors are mostly in the sucker rims with about 10,000 on each sucker, compared with about 100 on each sucker in *Sepia* (Graziadei 1964). This highlights the tactile detection of food by octopuses out of visual range using their arms, while cuttlefish and squid use their arms to grasp visually located prey (Hanlon & Messenger 1996).

Experiments by several researchers have shown that *Octopus* can readily discriminate between a range of chemical substances using its suckers and is sensitive to concentrations at up to 1,000 times lower than detectable by humans. While some of these experiments have shown that octopuses can use distance chemoreception in food detection, the relative importance of suckers, lips, and olfactory organs in this ability depends on a variety of circumstances (Hanlon & Messenger 1996; Di Cosmo & Polese 2018).

7.6.9 Subradular Organ (Licker)

In chitons, monoplacophorans, scaphopods, *Nautilus* (but not coleoids), and some gastropods, the ventrally located sublingual (or subradular) pouch, which opens to the anterior buccal cavity, contains a sensory organ called a licker or subradular organ. The structure (and therefore the homology) of the subradular organ is poorly understood, but it is thought to be involved in 'tasting' food. In chitons, large nerves extend from the epithelium to the paired subradular ganglia (Boyle 1975). The location and taxonomic distribution of the subradular organ are described in more detail in Chapter 5.

7.7 TEMPERATURE – THERMORECEPTION

Little is known about temperature receptors in molluscs, although many are sensitive to temperature, as shown by both behavioural and physiological responses. This lack of information about the neural mechanisms involved and how animals can distinguish temperature gradations and choose optimal temperatures is also the case for most animals (e.g., Hamada et al. 2008).

Examples of nerve activity being related to temperature include experiments involving the giant neurons of *Aplysia* (Chalazonitis & Chalazonitis-Arvanitaki 1961) and the lip, tentacle and pedal nerves, and the cerebral ganglia of some stylommatophoran land snails (Voss et al. 2000; Voss & Schmidt 2001). In the former case, the giant neurons were very sensitive to rapid changes of temperature, but the details of the mechanisms underlying these responses are unclear. There may be a direct influence of temperature on the central neurons involved in various behaviours and the physiology of the nerve impulses themselves (e.g., Murray 1966; Voss et al. 2000). For example, the gymnosome pteropod *Clione* is typically oriented vertically as it swims in the water column, but if the water is warmed this orientation changes. This occurs because motor neurons involved in muscular movements of the 'tail' and 'wings' responsible for orientation are mediated by interneurons that transmit signals from the statocysts. These interneurons are strongly depolarised when the animal is warmed (Panchin 1997; Deliagina et al. 2000). These responses are inactivated during defensive behaviour (Deliagina et al. 2000).

Temperature is well known to be the cue for gonad maturation and spawning in various molluscs (see Chapter 8), and also strongly influences other physiological processes (e.g., Innes & Houlihan 1981; Gosling 2003). Temperature is implicated in the migration and reproduction of some cephalopods (Hanlon & Messenger 1996), although day length may be more important in some (Mangold & Froesch 1977). While day length is the main stimulant for reproduction in some gastropods, a combination of day length and temperature is the trigger in others (e.g., Wayne & Block 1992; Udaka & Numata 2008).

In stylommatophoran land snails, it has been demonstrated that responses to temperature are delayed by exposing them to odours of peppermint extract (containing menthol) or vegetable juice (Kavaliers & Tepperman 1988).

7.8 PERCEPTION OF MAGNETIC AND ELECTROSTATIC FIELDS

Unlike the proven ability of some birds, sharks, marine turtles, isopods, and lobsters to detect magnetic fields (Cain et al. 2005), there is little evidence to show some molluscs may have this ability. While a few gastropods apparently respond to magnetic fields in experiments, possibly via magnetite crystals in certain cells (Chase 2002), to date these responses to magnetism are not completely convincing (see Chase 2002 for

discussion). Experiments conducted in the early 1960s hinted at this ability in *Tritia* (=*Nassarius*) *obsoleta*, but the results were not clear-cut, and the experiments have not been replicated (Chase 2002). Kavaliers and Ossenkopp (1991) reported an apparent influence by magnetism on opioid signalling in the nervous system of the land snail *Cepaea nemoralis*. Later experiments showed that light modulated these apparent magnetic field effects with the effect being greater in darkness (Prato et al. 1998).

Because of the presence of magnetite in the radular teeth of most polyplacophorans, there has been interest in their responses to magnetic fields (Ratner & Jennings 1968; Ratner 1976; Tomlinson et al. 1980; Sumner-Rooney et al. 2014). These studies have included both field observations and experimental manipulations, the latter of which have demonstrated statistically significant movement or orientation responses to magnetic fields by several chiton species, all belonging to the families Ischnochitonidae, Lepidochitonidae, and Mopaliidae. In contrast, *Leptochiton rugatus* (Lepidopleurida) also showed some propensity to orient, but the results were not statistically significant (Sumner-Rooney et al. 2014).

The nudibranch *Tritonia* is one of the most carefully investigated cases, with experiments in the laboratory and field showing responses to magnetism (see Cain et al. 2005 for summary and references). In that animal, it seems that stimuli from the magnetic field of the earth are collected from as yet unidentified peripheral receptors (Pavlova et al. 2011) in a single neuron in each pedal ganglion and that in turn stimulates pedal ciliary beat influencing the direction of movement of the animal (e.g., Lohmann & Willows 1987; Popescu & Willows 1999; Willows 1999; Wang et al. 2003b; Beron & Murray 2014). Neurobiological work done in *Tritonia* is one of the few studies on any animal that have attempted to investigate the neural mechanisms involved in magnetic orientation. It is not known if the findings summarised below are likely to be found in other 'opisthobranchs'.

In *Tritonia*, three pairs of neurons in the pedal ganglia respond to changes in the magnetic field. Two (Pd5 and Pd6) are large neurons which are excited by magnetic field changes, presumably triggered by as yet unidentified peripheral sensor(s), and this induces pedal activity. The third pair (Pd7) is inhibited by the same magnetic changes. Although some details still need to be worked out, Pd5 and Pd6 neurons, which innervate the foot, control the pedal cilia, and hence the direction of locomotion. The Pd7 neurons have branches to the cerebral ganglia and from there to the mouth, oral veil, and rhinophores. They may function in suppressing behaviour (e.g., feeding) that is incompatible with reorientation or locomotion (Cain et al. 2005).

While the above results from experiments on two euthyneuran gastropods and one caenogastropod indicate that the magnetic field of the earth may have some influence on behaviour in those animals, we are a long way from fully understanding the mechanisms involved and whether the magnetic field of the earth is commonly employed by a range of molluscs in their behavioural responses.

7.9 EYES AND OTHER LIGHT RECEPTORS

Molluscs have various responses to light ranging from orientation (e.g., movement away from strong light) to defence (e.g., shadow inciting a withdrawal reflex), and light also can trigger hormonal release that initiates reproductive development. These responses may be mediated by single-celled dermal photoreceptors, simple photoreceptors made up of clusters of dermal sensory cells (often called eyespots or ocelli), or invaginated (subdermal) groups of photoreceptors which form eyes (see Section 7.9.4 below).

Dermal sensory cells are probably found in all molluscs and enable photosensitivity, i.e., light responses not mediated by an eye or eye-like structure, although rather little is known about these structures in molluscs. These responses can also be due to light-sensitive neurons within the CNS, as described below.

7.9.1 Dermal Receptors

Light sensitive epidermal organelles known as *phaosomes* are present in molluscs as well as Annelida, Cnidaria and many other taxa. Typically, there is a sac-like vacuole filled with densely packed microvilli, which is either closed or connected via a narrow pore to the external environment (Crisp 1971; Haszprunar & Kunz 1996). In chitons, phaosomes are often developed as part of the macraesthetes.

Shadow responses are widespread in molluscs with such responses possibly mediated by dispersed photoreceptor cells in the epidermis. Dermal receptors have been reported in scaphopods (Messenger 1991), which lack eyes. Some species of chitons not only have numerous 'shell eyes', but also have light-sensitive cells in the girdle and ventral surfaces. Dermal receptors have been reported in the mantle and siphons of some bivalves (e.g., Kennedy 1960), and are probably more common than the few reports indicate. Similarly, the mantle and head-foot of some gastropods are sensitive to shadows. Such shadow responses in molluscs have long been known (Willem 1892) and occur in all major groups of gastropods. Hodgson et al. (1987) reported electron-dense structures in the mantle edge tentacles of several south African patellogastropods which they suggested facilitated light scattering as part of the dermal receptor system. Dermal light receptors that respond to shadows have long been known from 'pulmonate' gastropods such as *Lymnaea*, *Helix*, and *Onchidium* (e.g., Gotow & Nishi 2008) and 'opisthobranchs' such as *Aplysia* (e.g., Stoll 1972; Duivenboden 1982). Shadow responses have been observed in the siphon of nassariids (Crisp 1972), and another nassariid, the blind whelk *Bullia digitalis*, is also sensitive to light (Brown & Webb 1985). Similarly, in the sacoglossan *Elysia*, the zooxanthellae-containing parapodia are spread in response to light, in part due to three sets of photoreceptors, two of which are not the eyes (Rahat & Monselise 1979). Numerous dermal photoreceptors on the papillae that bear the eye-like ocelli in *Onchidium* (see Figure 7.40) can

cause contraction of the papillae independent of the CNS (Shimatani et al. 1986). Such responses by dermal receptors in gastropods are probably much more common than indicated in the literature.

The phototransduction pathway genes involved in dermal receptors have only been identified in *Lymnaea* and *Sepia*. In *Lymnaea stagnalis*, experiments indicate that a phototransduction cascade using CNG[8] ion channels is utilised by the dermal receptors (Pankey et al. 2010), while in *Sepia*, it is the TRP[9] channels (Mäthger et al. 2010). This latter channel is used in the retinal receptor cells in both *Lymnaea* (Sakakibara et al. 2005) and *Sepia* (e.g., Chrachri et al. 2005), suggesting a common origin for the dispersed and retinal photoreceptor cells in *Sepia*, but separate origins in *Lymnaea* (Ramirez et al. 2011).

7.9.2 Nerve Cells as Photoreceptors

Intrinsic sensitivity of parts of the nervous system to light has been recorded in some bivalves, some heterobranch gastropods, and *Sepia*, but is probably more widespread than the literature suggests. Shadows cause a reflex action in some bivalves (e.g., Kennedy 1960; Wiederhold et al. 1973). In the mactrid bivalve *Spisula*, the light-sensitive area is a part of the mantle nerve close to the visceral ganglion. Similar neural light responses have been observed in the heterobranch gastropods *Aplysia*, *Helix*, and *Onchidium* and in the coleoid *Sepia*, and in some other groups of invertebrates (Kennedy 1960; Charles 1966). This sensitivity is mediated through specific photosensitive pigments within the cell body of the neuron. In *Aplysia*, the photosensitive neurons are on the cerebral and visceral ganglia (e.g., Block & Smith 1973; Brown & Brown 1973). Experimental work on the visceral (abdominal) ganglion of *Aplysia* has shown that some nerve cells respond to light and light-sensitive pigments are present (see Gotow & Nishi 2008 for review). Large light-sensitive neurons are present in the dorsal part of the abdominal ganglion of *Onchidium verruculatum* (Hisano et al. 1972; Gotow 1986; Nishi & Gotow 1992). These are, however, much slower to respond than eye photoreceptors in *Onchidium* 'stalk eyes' (Gotow & Nishi 2008). The ganglionic photoreceptor neurons are not morphologically distinguishable from non-photoreceptive ones, even at the ultrastructural level, and have none of the structural modifications seen in eye photoreceptors (Gotow & Nishi 2008). In *Helix pomatia*, the left parietal ganglion is light-sensitive (Pasic & Kartelija 1995). A photosensitive nerve in *Spisula* (Kennedy 1960) is located peripherally and responds directly to illumination, and apparently mediates the response of siphon retraction.

The function of these photosensitive neurons is not well known, although some have suggested that they respond to ambient light in their natural habitat or help to maintain locomotory rhythms (e.g., Messenger 1991).

---

[8] Cyclic nucleotide-gated ion channel.
[9] Transient receptor potential channel.

The mechanoreceptive hair cells in the statocysts of *Lymnaea* have also been shown to respond to light (Tsubata et al. 2003).

7.9.3 LARVAL EYES

Although present in putative outgroups, most mollusc classes lack larval photoreceptors. In annelids and some brachiopods, larval pretrochal photoreceptors originate as post gastrulation invaginations in the region of the apical plate. In the brachiopod *Terebratalia transversa*, the larval photoreceptors consist of a lens cell, a pigment cell with a large, modified cilium, and connection with the larval brain. Gene expression data indicate these are cerebral eyes with ciliary photoreceptors for directional light detection (Passamaneck et al. 2011; Passamaneck & Martindale 2013). Larval photoreceptors also appear in most annelid trochophores and may persist in juveniles and adults, but are typically replaced by eyes during later development (Rhode 1992; Purschke et al. 2006).

Polyplacophoran trochophore larvae have larval (trochophore) photoreceptors, and although these may persist for a short period after metamorphosis in Chitonida, they are supplanted by the photoreceptors of the shell aesthetes. In Lepidopleurida, the larval photoreceptors appear to be retained in the adult Schwabe organ (see Chapter 4; Sumner-Rooney & Sigwart 2016). Polyplacophoran larval eyes are unique among trochozoans because they are post-trochal and innervated by the lateral nerve cord, rather than pretrochal as in other taxa (Henry et al. 2004). Larval eyes (or ocelli) of the chiton *Katharina* have several pigment cells and one sensory cell with microvilli and a cilium (Rosen et al. 1979), and the larval eyes of *Lepidochiton* have four sensory cells and six pigment cells, both kinds with 2–3 cilia and many microvillae (Bartolomaeus 1992). The trochophore larvae of caudofoveates, patellogastropods, and vetigastropods lack photoreceptors. Taxa with a test-cell larva (pericalymma) (Solenogastres, protobranch bivalves, and scaphopods) also lack larval eyes, but this highly modified type of trochophore makes potential comparisons difficult.

In the remaining groups (cephalopods, gastropods, and bivalves), the earliest photoreceptors are not larval eyes, but anlagen of the adult eyes. They differ from typical larval eyes in appearing at more advanced developmental stages (e.g., veliger stage of gastropods and bivalves or stage 16 of *Octopus* development) and are homologous with the eyes of the adult. They are often thought to be an example of an adult organ accelerated into a larval phase. Indeed, in a review of gastropod and bivalve larval eyes, Bartolomaeus (1992) noted that gastropod larval eyes have the same ultrastructural features as the adult eyes although the number of photoreceptor, pigment, and corneal cells is increased in adults. For example, the early eyes of the littorinid *Lacuna* each consist of one pigment cell, one rhabdomeric[10] cell, one ciliary sensory cell, and one corneal cell that acts as a lens (Bartolomaeus 1992). Some

other caenogastropod larvae have small lens-like structures in their eyespots, and these have a role in directional photoreception during their planktonic phase (Blumer 1994).

7.9.4 ADULT EYES

The first light receptor organs were probably formed by the clustering of a few photosensitive cells, and once they reach a certain level of complexity, they are termed 'eyes'. To be termed an eye, it must have multiple photoreceptor cells, often arranged in a cup shape, and associated with surrounding or interspersed pigment cells, which ensure that light from only one direction is registered, and the information from multiple photoreceptors can be compared simultaneously to build a spatial picture (Land & Nilsson 2012). Eye formation occurred independently many times in animals and has occurred multiple times within molluscs, where an astounding variety of eye types is found, more so than in other phyla. Some are simple, cup-shaped eyes with or without a lens, pinhole eyes or enclosed eyes with a lens, and scanning eyes and compound eyes have also evolved. Some are cephalic eyes, and there are also different kinds of complex non-cephalic eyes, particularly in bivalves, but also in chitons and in a few gastropods. Primitively, these eyes initiate a shadow response, this being facilitated by ciliary rather than microvillar photoreceptors (Messenger 1991). More complex non-cephalic eyes can provide information about movement and direction, with examples being the arcid compound eye and the pectinid mirror eye (Messenger 1991).

As with other receptors, the eyes develop from ectodermal invaginations that form optic vesicles differentiated as eyes. While eyes are placed on the head in gastropods and cephalopods, in bivalves they are typically on the mantle edge or siphons, and in chitons, they are located within the shell valves. Molluscan eyes show an amazing range of diversity from simple structures that lack a cornea and lens to the highly organised mantle eyes seen in some bivalves (notably some pectinids), the cephalic eyes of many gastropods, and, particularly, the large complex eyes of coleoid cephalopods. While the photoreceptor cells themselves may be homologous within molluscs and probably across animal phyla (see below), it is clear the photoreceptor organs (the eyes) have evolved their complex structures independently in different groups.

Living monoplacophorans, aplacophorans, and scaphopods lack eyes – possibly in part due to their deep-water habitats (most monoplacophorans and solenogasters) or burrowing habits (scaphopods and caudofoveates).

7.9.5 THE CORNEA AND LENS

The lens focuses light on the retina. While not all molluscan eyes have a lens, most do. Sometimes, these cannot form an image, while those that do form images have non-homogeneous lenses which allow refraction to occur (Messenger 1991).

Several gastropods and cephalopods lack a cornea, and the lens is in direct contact with the surrounding water. If a cornea is present in aquatic animals, it does not play a significant role in

---

[10] See Section 7.11.2 for a description of the types of photoreceptor cells.

image formation because of the similar refractive indices of the liquid inside and outside the eye. This is not the case with eyes of terrestrial animals, notably tetrapod vertebrates, where the cornea can have a role in focussing (e.g., Land 2012) because of the difference in the refractive index of air and water. In terrestrial gastropod eyes, the lens has the main role (Gál et al. 2004).

Lenses are typically spherical in shape, because the radii of curvature of its surfaces have to be small so they maximise refraction. In a lens of uniform transparent material, the image is formed too far from the lens to focus on the retina (spherical aberration). Many lens eyes cannot focus a sharp image on the retina, but they capture light more efficiently than eyes which lack a lens (Land & Nilsson 2012; Nilsson 2013).

Eye lenses that can form a focused image have a gradient of refraction. Such lenses have evolved at least eight times in coleoid cephalopods, at least four times in gastropods, and several times in bivalves, as well as in fish, marine mammals, the carnivorous alciopid polychaetes, in one genus of copepod crustaceans, and even in cubozoan jellyfish (Land 2012). These lenses, called Matthiessen lenses, have focal lengths of about 2.5 lens radii (Matthiessen's ratio), an observation originally made by Ludwig Matthiessen in the 1880s on fish eyes. These lenses have a refractive index gradient that varies from about 1.52 in the centre to 1.33 at its surface and usually lack spherical aberration so they form good quality images (Warrant & Locket 2004). The lenses of some gastropods (*Littorina*, *Strombus*, heteropods, and *Lymnaea*) and coleoid cephalopods have been shown to achieve Matthiessen's ratio (i.e., focal length about 2.5 times lens radius), enabling a good image to be produced on the retina in water (Land 1984; Seyer 1992, 1994).

Unusually, the lens of *Strombus* has a Matthiessen's ratio of about 2.0, markedly lower than the usual 2.5 found in the lens of many marine animals that correct for spherical aberration (i.e., are aplantic) (Seyer 1994).

Most fish and mollusc eyes are wide-angle (180°) because the spherical lens forms an image of similar quality across the hemispherical retina (Land 2012). In deep water (>300 m), downwelling light is restricted to a cone about 60° across, and this limitation has resulted in structural changes to the eyes of many cephalopods (and fish) which have become directed upwards and are somewhat tubular (Land 2012). At less than 1000 m there is sufficient light to distinguish the surfaces and shapes of animals, but below 1000 m no surface light penetrates, with bioluminescence being the only light source. These differences in available light require modifications to the eyes. In a recent study, Gagnon et al. (2013) examined the optics of lenses from animals living at different depths, including a heteropod gastropod and cephalopods and found they compensated for depth.

The scanning eyes of heteropod gastropods have a spherical, highly specialised lens with unidirectional aberrations that correspond with a ribbon-shaped retina (Land 1982; Gagnon et al. 2013) (see Section 7.9.9.1). Convergent scanning eyes are also seen in jumping spiders (salticids), some crustaceans, and some insect larvae (Land 2012).

In animals, lenses are constructed from various materials, most of which occur in other parts of the body where they perform different functions (Land 2012). Thus, lenses are comprised of transparent substances otherwise present in lower concentrations in cells, but can be expressed *en masse* and densely packed in the cells making up the lens. These substances vary according to the group of animals (Land 2012), although they are usually proteins called crystallins. The lens is covered with layers of living epithelial cells with very high concentrations of crystallins, and there is a core of dead cells. The transparent cornea may also have a high proportion of crystallins. Several different crystallins may be employed in building the lens to produce a refractive index gradient. Vertebrate lenses are composed of α-crystallins related to heat shock proteins. *Octopus* lenses comprise proteins including aldehyde dehydrogenase and glutathione S-transferase (Land 2012); the latter is also found in mammal lenses (Tomarev & Piatigorsky 1996). Thus, the crystallins in the lenses of coleoid cephalopods have evolved independently of those found in vertebrates (Brahma 1978) and possibly involved gene duplication (Piatigorsky 2003). Similarly, amino acid sequences of the crystallin in the lens of scallops is about 55% similar to that of cephalopods, but is surprisingly 64% to 67% identical to the two forms of crystallin found in the lens of human eyes (Piatigorsky et al. 2000).

The derivation of Hygrophila from terrestrial taxa (see Chapter 20) is supported by their eye structure. Many share a fixed lens with other gastropod lens eyes, and some semi-amphibious lymnaeids have modified the retinal surface so it lies in two pits, with the deeper one suitable for receiving images under water and the shallower one suitable for vision in the air (Gál et al. 2004). The amphibious caenogastropod *Pomacea* (Ampullariidae) has relatively poor vision (Seyer et al. 1998), but a better image is formed in the aquatic *Viviparus* (Viviparidae) (Zhukov et al. 2006).

7.9.6   THE MAIN TYPES OF EYE

Until recently, the independent evolutionary origin of the various types of eye was often advocated (see Section 7.11.2). More recently, work on developmental genes has shown that a single gene is responsible for initiating eye development in several animal groups (see Section 7.10), suggesting that the basic dermal photoreceptor structures, at least in several groups, may be monophyletic (Gehring 2001, 2005), sharing a Precambrian ancestry with subsequent adaptive radiations generating the great diversity of eye types found.

There are four main types of eyes:

'*Cup eyes*' with a simple cup-like structure lined with photoreceptors and pigment. The amount of light entering these eyes is rarely controlled (e.g., in patellogastropods), but in *Nautilus*, an iris can achieve this, creating a '*pinhole camera eye*'.

'*Lens eyes*' with a single lens projecting onto a retina (found in vertebrates, cephalopods, some gastropods, and some chitons). There may or may not be a cornea, a thin transparent layer separating the lens from the exterior. A specialised type found only in coleoid cephalopods and vertebrates is called a '*camera eye*', where the lens can move relative to the retina. Thus, the camera eye can accommodate, maintaining a focused image on the retina.

'*Mirror eyes*' which, in the case of the scallops (e.g., *Pecten*), use both a lens focussing the light onto a distal retina and a reflecting parabolic mirror projecting the light onto a proximal retina.

'*Compound eyes*' which have multiple ommatidia, each with a group of photoreceptor cells. These are typically found in arthropods, where each eye possesses a lens, although a somewhat similar type that lacks lenses is found in certain polychaetes and in some arcoid bivalves.

The evolution of different eye types has ramifications for other aspects of nervous system evolution. A lens is required for focussing an image on the retina, and high numbers of relatively small receptors are necessary to convey that image to the brain. The need for complex visual processing in cephalopods has led to the evolution of the largest invertebrate optic lobes. In coleoid cephalopods, further control is obtained by muscles that move the lens for focussing and the eye itself relative to the environment, thus paralleling vertebrate eyes.

We describe below the basic structure of the various types of eyes found in molluscs. This is followed by a comparison of the structure of the retinas found in various types of eyes, and an account of the cellular structure of the light-sensitive (photoreceptor) cells they contain.

7.9.7 CHITON SHELL EYES

Aside from the phaosomes in the macraesthetes, many chitons also possess so-called shell eyes (ocelli or eyespots) that have differentiated from the aesthetes, dorsal nerve strands in pores in the shell plates (see Chapter 14, Figure 14.8). Only the dorsal surface of the chiton is exposed to light, as the head is hidden beneath that surface. Chiton ocelli comprise a lens, vitreous body, pigmented sheath, and retina with structural differences in different groups of chitons (see Eernisse & Reynolds 1994 for further information). The eyespots (also termed intrapigmented aesthetes) are clusters of pigmented photoreceptors found in various taxa and have, in *Chiton*, been demonstrated to form images (Kingston et al. 2018). The innervation for these eyes and eyespots is from the lateral nerve cord.

Unusually, the birefringent lenses in the shell eyes of the chiton *Acanthopleura granulata* are composed of aragonite (Speiser & Johnsen 2011) and could potentially facilitate image formation in both air and water (Li et al. 2015a).

7.9.8 BIVALVE EYES

Cerebrally innervated eyes appear on the outer side of the first gill filament in the veligers of many Pteriomorphia (Pelseneer 1900). In most species, they are lost in mature stages. For example, in *Ostrea edulis*, a pair of eyes develop in the body wall, dorsal to the attachment point of the paired gill rudiments, just before settlement. Each eye consists of a pigmented cup filled with a gelatinous matrix and closed by a lens-like structure (Cole 1938). After settlement, these eyes quickly degenerate. In some pteriomorphian taxa (e.g., *Philobrya, Pteria, Anomia, Mytilus, Modiolaria*, and *Avicula*), these eyes remain present in the adult where they still lie on the first gill filament (see Section 7.9.8.2), and in many species, bivalves appear to have compensated for the lack of an eye-bearing head by evolving eyes on the mantle edges (mantle eyes), as described below.

7.9.8.1 Bivalve Mantle Eyes

The diversity of eye types in bivalves ranges from simple cup-like structures to complex lens-bearing eyes with an everse or inverse, single or double, retina, to simple compound eyes. Bivalve eyes are an excellent example of the independent evolution of complex structures. Each of the main types is briefly described below. Eyes are absent or degenerate in fresh-water and deep-sea bivalves (e.g., Malkowsky & Götze 2014).

Mantle (=pallial, including siphonal) eyes differ considerably in structure and location in different groups of bivalves (Morton 2001, 2008a), indicating that they are (as organs) probably not homologous structures. They are on the outer mantle fold of representatives of the Arcoidea, Limnopsoidea, Pterioidea, and Anomioidea, the middle fold in the Pectinoidea and Limoidea, and the inner fold in the Cardioidea (including *Tridacna*).

In the Arcoidea (e.g., *Arca, Anadara, Barbatia, Glycymeris*), mantle eyes are found in shallow-water taxa and are located on the outer mantle fold. They range from simple photosensory cells to pits containing both photosensory and pigmented cells, to protruding eyes ('caps') composed of groups of photosensory and pigmented cells (Nilsson 1994). The 'caps' may each have 10–80 receptors surrounded by pigment cells that form simple, ommatidium-like structures (Figure 7.29) which have been likened to arthropod compound eyes (e.g., Charles 1966; Morton 2001), but are not homologous. There is no lens, but dense distal cilia were mistaken for a lens in early works. These eyes can apparently detect movement (Braun 1954), which triggers valve closure.

Pearl oysters (Pterioidea) have single photoreceptor cells on the outer mantle fold. Members of the Limidae have simple cup-shaped eyes on the middle fold, with or without a lens, and a simple retina (Figure 7.29). These eyes are not homologous with the more complex eyes in pectinids (Waller 1998).

Pectinoideans (e.g., *Pecten, Amusium, Spondylus*) have eyes on the middle mantle fold which differ in size and number between species. Their rather complex structure consists of a vesicle surrounded by a layer of pigment cells, except for a thin, transparent epithelial cornea, below which lies a large, multi-cellular lens. A thin septum composed of connective tissue separates the lens from the most unusual feature of these eyes, the double retina (see below for a description of its structure), a feature otherwise known only in the anomalodesmatan *Laternula* (see below), although it differs in structure. The lateral lobes of the parieto-visceral ganglion function as optic lobes (Spagnolia & Wilkens 1983). Many pectinid eyes

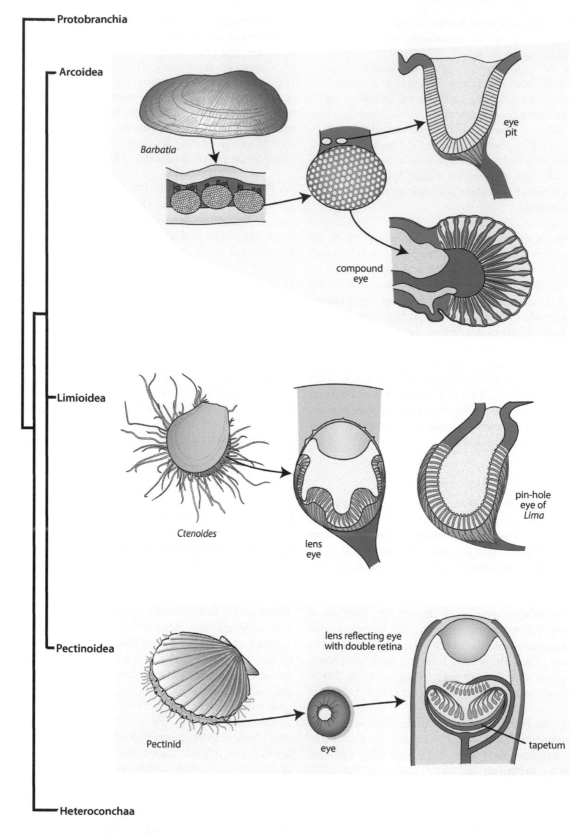

FIGURE 7.29 Eyes on the mantle edges of some pteriomorphian bivalves. Redrawn and modified from Salvini-Plawen, L.v., and Mayr, E., *Evol. Biol.*, 10, 207–263, 1977 and the following sources: arcids Nowikoff, M., *Zool. Anz.*, 67, 277–289, 1926; Nilsson, D.E., *Philos. Trans. Royal Soc. B*, 346, 195–212, 1994 and limids Morton, B., *J. Molluscan Stud.*, 66, 449–455, 2000a.

are so-called *mirror eyes*, in which the image is formed by a reflector (see Section 7.11.2.2.2).

Cardiidae (e.g., *Cerastoderma*) have small eyes (Figure 7.30) on the sensory tentacles (derived from the middle mantle fold) of both siphons. These are covered by a cornea, have a multi-cellular lens, and a single layer of about 12–20 receptor cells in the retina and a reflecting layer (*tapetum*) behind it (Barber & Wright 1969b). In the cardiid Tridacninae, large

Tridacna have several thousand 'hyaline organs' on the mantle over the expanded siphonal area. These multi-cellular, lens-like structures (Figure 7.30) do not have a retina, and they let light through the mantle tissue to enhance the growth of the symbiotic zooxanthellae it contains (see Chapters 9 and 15). They may represent modified eyes (Yonge 1953b), and it has been suggested they could facilitate predator avoidance (Fankboner 1977, 1981) or even collectively perceive a

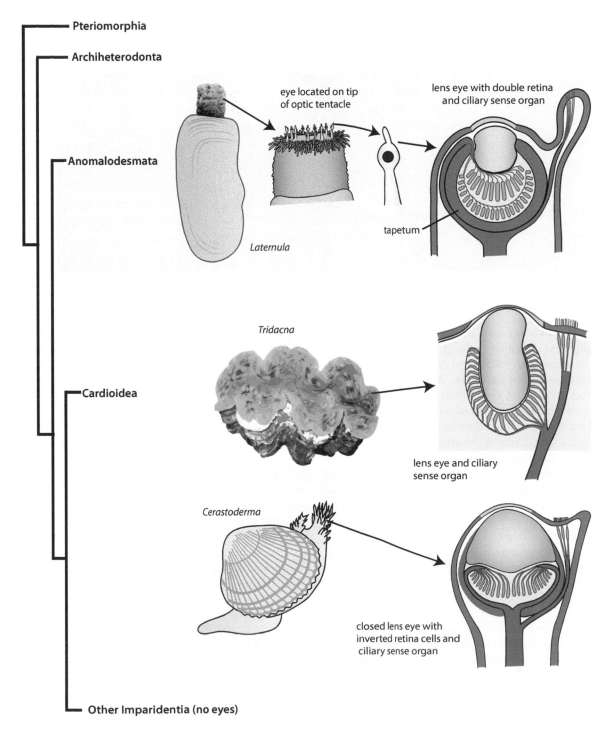

FIGURE 7.30 Mantle and siphonal eyes of some heteroconch bivalves. Redrawn and modified from Salvini-Plawen, L.v. and Mayr, E., *Evol. Biol.*, 10, 207–263, 1977 and *Laternula* siphon from Adal, M.N. and Morton, B., *J. Zool.*, 170, 533–556, 1973.

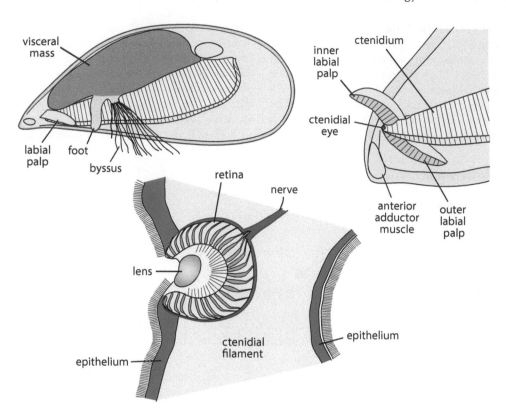

FIGURE 7.31 Location of branchial eye in *Mytilus* (above) and below a section through the branchial eye in *Musculus marmoratus* (Mytilidae). *Mytilus* redrawn and modified from Rosen, M.D. et al., *The Veliger*, 21, 10–18, 1978 and *Musculus* redrawn and modified from List, T., *Die Mytiliden des Golfes von Neapel und der Angrenzenden Meeres-Abschnitte*, Friedländer & Sohn, Berlin, Germany, 1902.

mosaic image (Fankboner 1977). Both *Tridacna*, which has hyaline organs, and the closely related *Hippopus* which lacks them, have an immediate withdrawal response to shadow, and they also close their valves at night. *Tridacna gigas* has the largest number of 'eyes' of any animal and is also the second largest invertebrate in body size and weight.

In the Anomalodesmata, *Laternula* has one of the most complex bivalve eyes (Adal & Morton 1973). Each eye lies in an optic tentacle on the siphon. The epithelial cornea lies above the large globular multi-cellular lens composed of two cell types, and there is a double, ciliary-based, retina (Figure 7.30). The eye is surrounded by a pigmented, non-reflective tapetum, and an appendage with sensory cilia may help to protect the eye from contact damage (Adal & Morton 1973). The double retina mantle eyes seen in Laternulidae and Pectinidae are convergent (Morton 2001, 2008a).

7.9.8.2 Bivalve Branchial or 'Cephalic' Eyes

Paired branchial (often called 'cephalic' because of their innervation) eyes occur on the anterior end of the base of the first ctenidial filament of the inner demibranch and are seen only in some pteriomorphian bivalves, namely, representatives of the Arcoidea, Limnopsoidea, Mytiloidea, Pterioidea, and Anomioidea (Morton 2001, 2008a). In *Anomia* and some species of *Pteria*, there is only one branchial eye, usually on the left side.

These branchial eyes appear late in the development of these groups, and in Ostreoidea (although absent in adults),

where they are thought to play a role in settlement. As shown in Figure 7.31, they are open, pigment-lined cups composed of photosensory cells innervated from the cerebral ganglia. The interior of the open-cup is filled with a crystalline material which functions as a simple lens. Opposite each 'eye' in some taxa there is a translucent spot in the shell valve through which diffused light can pass. Rosen et al. (1978) described in some detail the structure of these eyes in *Mytilus*.

The homology of these cerebrally innervated ocelli-like organs with the cephalic eyes of gastropods and cephalopods is unlikely. Although protobranch bivalves have no eye-like structures, they also lack a comparable developmental morphology to help determine the plesiomorphic state in bivalves. Ultrastructurally, cephalic bivalve photoreceptor cells are a fairly unmodified mixed type (microvilli plus an intact cilium).

7.9.9 Gastropod Light Receptor Organs

7.9.9.1 Cephalic Eyes in Gastropods

Gastropod eyes vary in complexity (Figure 7.32), with the most primitive condition seen in patellogastropods. In that group, the eyes are open pits lacking a lens, a condition also seen in some vetigastropods (e.g., *Trochus*). In many vetigastropods, they remain narrowly open, but have a lens (e.g., *Haliotis*) or are closed with a lens as in Fissurellidae, *Tricolia* (Phasianellidae), and some turbinids (Marcus & Marcus

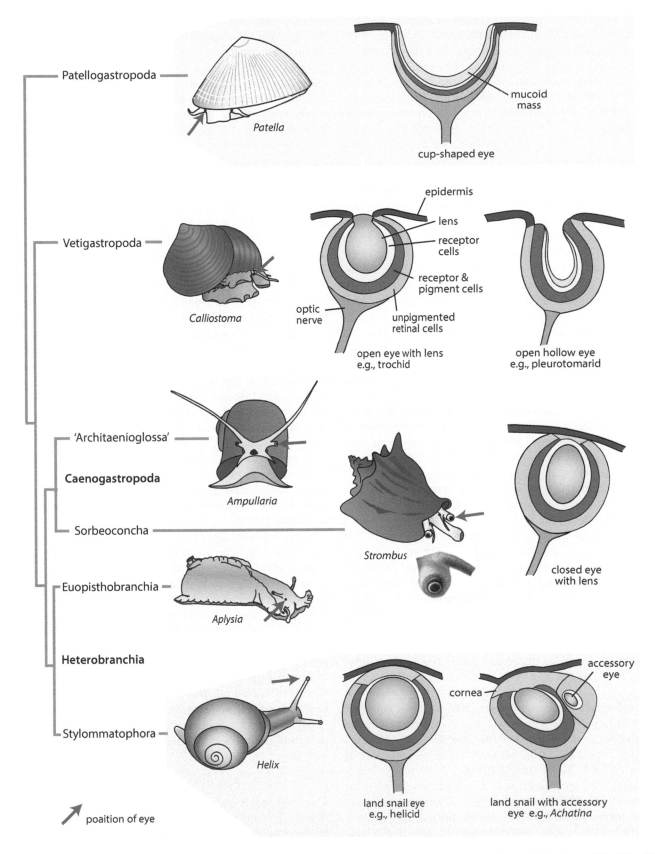

FIGURE 7.32 Gastropod cephalic eyes. Redrawn and modified from Salvini-Plawen, L.v. and Mayr, E., *Evol. Biol.*, 10, 207–263, 1977 and Sasaki, T., *Bull. Univ. Mus.*, 38, 1–223, 1998, with the eupulmonate eyes of *Helix* redrawn and modified from Hesse, R., *Das Sehen der niederen Tiere*, Fischer, Jena, 1908 and *Achatina* from Tamamaki, N. and Kawai, K., *Zoomorphology*, 102, 205–214, 1983.

1960; Sasaki 1998). Neritimorphs, caenogastropods, and heterobranchs all have closed eyes with a lens, except for some deep-sea and subterranean taxa where the eyes are reduced or lost, including some Neritiliidae (Neritimorpha) that live in dark submarine caves and possess 'open pit' eyes (Kano & Kase 2002). Given the long independent histories of the groups involved, these differences in eye morphology should not be thought of as an evolutionary sequence. The pelagic gymnosomes and *Janthina* also lacks visible eyes, as do some naticids and some other burrowing gastropods.

In *Patella* and *Haliotis*, the eyes develop during torsion as small pigment patches which become invaginated following torsion, while in most 'higher' gastropods (neritimorphs, caenogastropods, and heterobranchs) the ectodermal invaginations become detached to form two closed vesicles in which the retina and lens differentiate.

Although they are all on the head, the location of the cephalic eyes differs in detail between groups of gastropods. In patellogastropods, vetigastropods, neritimorphs, and most caenogastropods, the eyes are at or near the outer bases of the cephalic tentacles. In vetigastropods, neritimorphs, and the caenogastropod architaenioglossans, the eyes are typically on peduncles on the outer sides of the tentacle bases, and a convergent condition is found in some mathildidids (Heterobranchia: Architectonicoidea).

The eyes of caenogastropods such as littorinids have a spherical, non-homogeneous lens, and some evidence (including behavioural studies), indicates that they can form rudimentary images on their retina (e.g., Newell 1965; Hamilton et al. 1983). During development, the lens of gastropods is formed by secretions from corneal and supportive cells, this material being applied to the lens surface by microvilli or cilia (e.g., Blumer 1999).

In neogastropods and some other caenogastropods, the tentacle bases are extended to place the eye along the tentacle well above the base and the rest of the head. In strombids, which, besides heteropods, have the most efficient eyes in caenogastropods, the eyes are on long peduncles, and the tentacle bases are also extended, with the distal tentacle either small or lost. Strombid eyes (Figure 7.32) are up to about 2 mm in diameter, with a spherical lens with good resolving power that produces an image apparently free of aberrations (Seyer 1994). The efficient, mainly microvillar, retina has four cell types (Gillary 1974; Gillary & Gillary 1979; Land 1984) and has more photoreceptors (nearly 100,000) than any other gastropod. Their good visual ability is complemented by the eye stalks being able to move in unison or separately although how these apparent visual abilities relate to behaviour is not well understood (Messenger 1991).

One type of specialised lens eye is the so-called scanning eye (Land 1982; Messenger 1991) of heteropods, which employ their mobile eyes in hunting prey. These actively swimming carnivorous caenogastropods have (for a gastropod) large eyes with a large spherical lens and a unique ribbon-shaped retina (see Section 7.11.2.2.3). Heteropod eyes are almost constantly moving in a regular fashion – they move rapidly downwards and more slowly upwards through about 90°. This is assumed to be an effective means of extracting information from a stationary scene, and these animals are presumably seeking objects that glisten in the light from above (Land 1982). Such scanning eye movements are very rare in other animals and unknown in other molluscs. Despite the unusual and interesting features of these eyes, there have been no detailed studies of their structure since that of Hesse (1900). Heteropod eyes are also unusual because they are the only gastropod eyes with a well-developed 'optic ganglion'.

In some lower heterobranchs the eyes lie in the middle of the tentacle bases, but in others, they lie on the inner sides of the tentacles. In many 'opisthobranchs', the small eyes are embedded below the body surface and lie on, or near, the cerebral ganglia. Their small lens is probably unable to form an image, and the degenerate retina can have as few as five photoreceptors (Chase 2002).

In marine and fresh-water panpulmonates the eyes are at the inner bases of the cephalic tentacles, while in stylommatophorans they lie at the tips of invertible tentacles.

Zieger and Meyer-Rochow (2008) reviewed the data on the image-forming capabilities of 'pulmonate' eyes. Their eyes function using fixed focal length optics, and while fairly focused images can be formed in at least some aquatic taxa (due to retinal modification), terrestrial snails and slugs cannot form a sharp image on their retina, but rather gauge the light intensity and at best form a blurred image.

The cephalic eyes of terrestrial 'pulmonate' snails and slugs probably register light intensity and direction (Eakin & Brandenburger 1975) for use in orientation as it has been shown, for example, in the helicid *Otala*, that following eye removal, the gastropod no longer moves away from light (Hermann 1968).

Some stylommatophorans have an accessory eye within the primary cephalic eye (Figure 7.32), but slightly separate from it both morphologically and functionally (Newell & Newell 1968; Tamamaki & Kawai 1983; Tamamaki 1989a, 1989b). Both eyes have a similar arrangement of microvilli and spectral sensitivity. Although the lens of the accessory eye cannot form an image, this eye probably monitors light intensity, when the tentacle is partially retracted and the primary eye cannot access light.

In many gastropods, the eyes are normally located under the lip of the shell, and that part of the shell covering the eyes may be modified as a translucent window. Some species of lottiid limpets, for example, have been shown to respond to light through the translucent patches in the anterior part of the shell (e.g., Lindberg et al. 1975).

7.9.9.2   Other Eyes in Gastropods

There are only a few examples of non-cephalic eyes in gastropods. In caenogastropods, a closed mantle eye is present in the amphibious cerithioidean *Cerithidea* (Potamididae) (Houbrick 1984) (Figure 7.33). It is at the inner edge of the inhalant siphon, has a spherical lens, and a well-developed retina and pigment and is innervated from a ganglion lying in the siphon.

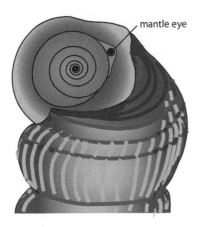

FIGURE 7.33 Mantle eye of *Cerithidea scalariformis*. Redrawn and modified from Houbrick, R.S., *Am. Malacol. Bull.*, 2, 1–20, 1984.

In the Heterobranchia, the pteropod *Corolla* has lost the cephalic eyes and has numerous secondary eyes, each with lens and retina, around the edges of the wings (Lalli & Gilmer 1989).

The systellommatophoran onchidiid slugs not only have dermal photoreceptors and cephalic eyes borne on optic tentacles, but dorsal eyes as well (Figure 7.40). The dorsal lens eyes are on papillae that also contain photoreceptive cells lacking microvilli; instead, the photoreceptive cells have cilia-derived membranous whorls and the eyes detect shadows (Yanase & Sakamoto 1965).

7.9.10 Cephalopod Light Receptor Organs

Besides dermal receptors, there are two main photosensitive organs in cephalopods; the eyes and the photosensitive vesicles.

7.9.10.1 Cephalic Eyes

Complex, large, and conspicuous cephalic eyes in cephalopods are situated laterally on the head. While the eyes in all cephalopods are large, in some they are extremely large as, for example, the huge eye of giant squid. There are two main kinds, the pinhole eye found in *Nautilus*, and the lens eyes in coleoids.

Nautilus has a simple, large, cup-shaped eye with a small pupil-like opening which lacks a cornea and lens (Figure 7.34). The eye works on a 'pinhole camera' principle, with the pupil responding in size to changes in light intensity. While it is open to the surrounding seawater, protective mucoid material is secreted by the retina. Despite

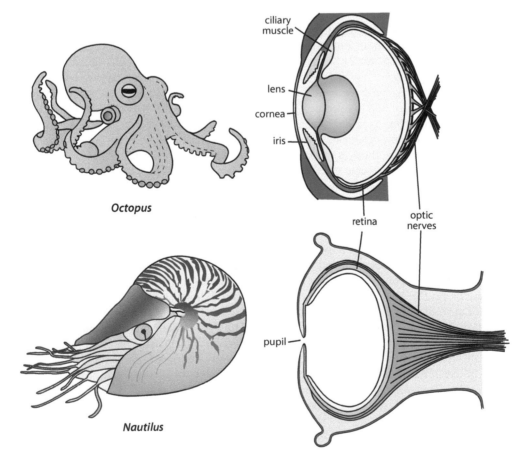

FIGURE 7.34 A diagrammatic comparison of the eyes of *Octopus* and *Nautilus*. *Octopus* eye redrawn and modified from Wells, M.J., Cephalopod sense organs, in Wilbur, K.M. and Yonge, C.M., *Physiology of Mollusca*, Vol. 2, pp. 523–545, Academic Press, New York, 1966 and *Nautilus* redrawn and modified from Mugglin, F., *Vierteljahresschr. Naturforsch. Ges. Zürich*, 84, 25–118, 1939.

having a rather complex retina, it cannot form good quality images (Muntz & Raj 1984; Muntz 1986), because there is no lens; it cannot stabilise an image on the retina as the simple statocyst is unable to effectively monitor angular acceleration (unlike coleoids) (Young 1988a) (see Section 7.5.4.1).

Thus, the resolution and sensitivity in the *Nautilus* eye are poor (Land 1984; Muntz 1986), much less than in coleoids, with the image being about 13 times dimmer than in *Octopus* (Land 1984). Relatively advanced features shared with coleoid cephalopods include compensatory eye movements mediated by the statocysts, and a pupil which can expand and contract, although much more slowly than in coleoids (and vertebrates). These features are not found in the eyes of non-cephalopod molluscs.

Nautilus spends much of its life in low light intensity, but it is positively phototactic, an unexpected ability in an animal with such habits. The eyes may be involved in hunting bioluminescent prey and/or be involved in vertical migration (Ward et al. 1984; Muntz 1987, 1994). Although *Nautilus* appears to use its eyes primarily as phototactic organs, some features suggesting that they may be degenerate, as they have four million receptors and the ability to make compensatory eye movements (Messenger 1991). Although there are many fewer receptors in the *Nautilus* retina than in *Octopus* (Young 1962c), it still has two orders of magnitude more receptors than in *Strombus* which has the largest gastropod lens eyes (Messenger 1991). Thus, there may have been some secondary simplification of the eyes in *Nautilus* (Land 1984; Ogura et al. 2013) in much the same way that the eyes in a few deep-water coleoids have become secondarily simplified (e.g., *Cirrothauma*). It is, therefore, possible that a closed lens eye was present in ancient nautiloids and that such eyes did not evolve first in coleoids.

Coleoid cephalopods have complex eyes that are often cited as an example of convergent evolution with those of vertebrates as they are superficially very like the eyes of fish (see also Section 7.9.11). As in vertebrates, coleoid eyes have a large posterior chamber, a retina, lens, iris, and cornea, as well as a choroid (a vascular layer surrounding the eyeball) with an outer coat (sclera) and elaborate intrinsic and extrinsic eye muscles, the latter moving the eyes in the orbit (Budelmann & Young 1984; Hanlon & Messenger 1996). Such eye movements occur in vertebrates, but in molluscs are found only in coleoid cephalopods, although the contraction of their muscles is slower than in vertebrates (McVean 1984). They enable the eye to compensate for movement of the body or to respond to visual cues (Budelmann & Young 1984). The eye muscles respond to information from the macula of the statocysts, enabling the eye to remain steady despite variation in the position of the head or body (for reviews see Budelmann 1977; Budelmann et al. 1987). In some coleoids, eye movements may be used for depth perception (Messenger 1977a), with special muscles involved (Budelmann & Young 1993).

There are some significant differences between coleoid and vertebrate eyes in the retinal structure (Figure 7.35). In cephalopods, the photoreceptors are microvillar (rhabdomeric – see Section 7.11.2) as in many other molluscs, are usually of a single type, and point towards the light (everse type). In contrast,

vertebrate eyes have ciliary photoreceptors of two types (rods and cones) and point away from the light (inverse type). The cephalopod retina is also much simpler than that in vertebrates, and visual processing is undertaken in the optic lobes, while in vertebrates the processing is in the brain itself. Cephalopods do not have the blind spot (see Section 7.9.11.1) found in vertebrates because of the everse construction of the retina.

The lens consists of two parts, front and back, separated by a septum, the posterior part being larger. It has a graded refractive index (high in the centre and progressively lower towards the periphery) that results in a focal length of about 2.5 times the lens radius (see Section 7.9.5), enabling the formation of a good quality image, without accommodation, of objects located a few centimetres away and into the far distance (Hanlon & Messenger 1996). The lens colour can differ with water depth, with deep-water species having a transparent lens, but in some surface-living species, the lens is yellow and absorbs blue light (Denton & Warren 1968).

Many coleoid cephalopods undertake diel vertical migration (also known as diurnal vertical migration), spending daylight hours in the dark, deep sea and migrating into shallow water at night (Chapter 17 and Figure 17.18). In response to darkness, light entering the eye is maximised by dilation of the pupil and retraction of the pigment in the supporting cells of the retina. Mesopelagic coleoids increase their ability to capture light by having very large eyes with wide pupils. The giant deep-sea squid *Architeuthis* and *Mesocychoteuthis* have the largest eyes known for any animal, with a diameter up to 40 cm (Nilsson et al. 2012), although given their body length, these are not exceptional for coleoids, with many others having very large eyes relative to their body length (Schmitz et al. 2013a, 2013b). Nilsson et al. (2012, 2013) argued these large (and therefore highly sensitive) eyes might be important in spotting the bioluminescent glow created by the movements of predatory sperm whales.

Some mesopelagic coleoids, such as *Sandalops* and *Amphitretus*, have dorsally-directed tubular eyes that are adaptations related to capturing down-welling daylight (Herring 2002). The deep-sea squid *Histioteuthis meleagroteuthis* has one eye much larger than the other and which projects out from the head. It was thought to point upwards to capture light while the squid swam horizontally, but observations from submersibles show these squid suspended vertically in the water column, although this may be due to disorientation because of the lights of the submersible (Warrant & Locket 2004).

The eyes of bathypelagic cephalopods are often reduced in size, but functional. The bathypelagic squid *Bathyteuthis* has a well-developed fovea suggesting its use in detecting point source bioluminescent prey (Hanlon & Messenger 1996). *Cirrothauma* lives below 3000 m and has secondarily simplified eyes which lack an iris, ciliary muscles, and lens. They consist of only a simple cup, a cornea, and a large pupil.

7.9.10.2 Photosensitive Vesicles

Besides their eyes, many coleoid cephalopods have photosensitive vesicles (they are absent in *Nautilus*), with considerable differences in their structure between the three major groups

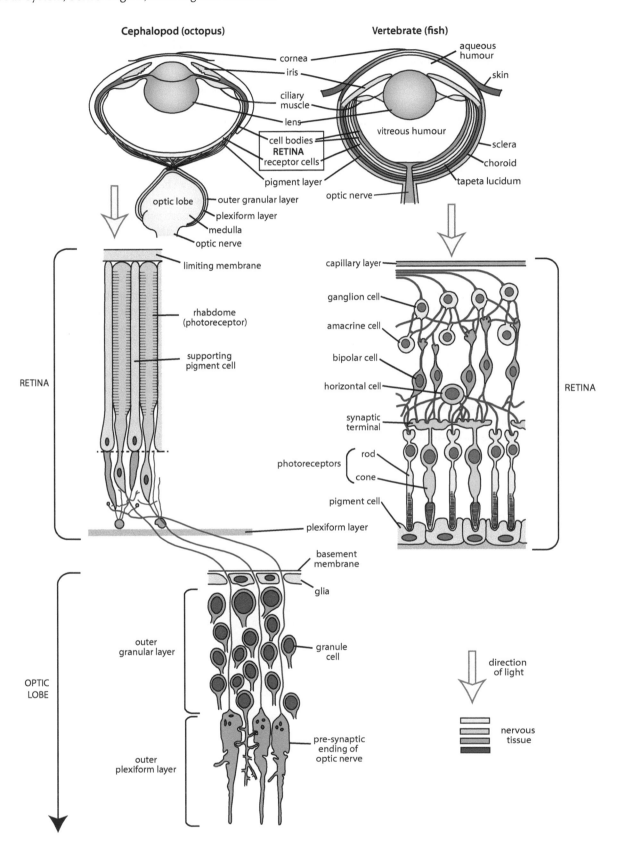

FIGURE 7.35 Comparison of the eyes and retinal structure of an octopus and a fish. *Octopus* eye redrawn and modified from Wells, M.J., Cephalopod sense organs, in Wilbur, K.M. and Yonge, C.M., *Physiology of Mollusca*, Vol. 2, pp. 523–545, Academic Press, New York, 1966 and fish eye from several sources. *Octopus* retinal structure redrawn and modified from Young, J.Z., *Philos. Trans. Royal Soc. B*, 245, 1–19, 1962c, and the optic lobe redrawn and modified from Dilly, P.N. et al., *Proc. Royal Soc. B*, 158, 447–455, 1963.

(Budelmann et al. 1997), suggesting that they are probably independently acquired. Like the eyes, these light-sensitive structures consist of rhabdomeric receptor cells. They also contain the visual pigment rhodopsin and the accessory photopigment retinochrome (Messenger 1991). They are probably mainly involved in regulating countershading by photophores through the detection of daylight (Young et al. 1979). They may also be utilised in diurnal migration as they are well-developed in many mid- and deep-water cephalopods. In some pelagic decabrachians, the dorsal and ventral photosensitive vesicles are differentiated, with light from above detected by the former and the ventral photophores being monitored by the latter (Young 1973).

In octopods, the photosensitive vesicles are attached to the posterior side of the stellate ganglia. In contrast, in decabrachians, one or more pairs of vesicles lie on either side of the head close to the olfactory lobe of the brain and the optic tract. *Vampyroteuthis* has two pairs of vesicles, one close to the eye (thought to be a luminescent organ by early workers), the other on the front edge of the ventral mantle (Budelmann et al. 1997).

7.9.11 CEPHALIC EYES – COMPARISON OF GASTROPODS, CEPHALOPODS AND VERTEBRATES

In molluscs, true cephalic eyes are present only in gastropods and cephalopods. They range in size from less than 1 mm to 10 cm or more and are the most similar in their position to vertebrate eyes – that is, they are paired and located on the head. It is not certain whether cephalic eyes evolved in the common ancestor of gastropods and cephalopods because it developed a distinct head or whether they evolved independently, although the latter scenario is most likely (e.g., Salvini-Plawen 1982).

The cephalic eyes of gastropods typically have relatively few (<10,000) retinal cells in each eye that function mainly as light detectors. Cephalopods (and vertebrates) have much larger numbers of retinal cells (>1 million) in each eye, and the eye can form a sharp image on the retina which the CNS interprets. The retina of vertebrate eyes, especially in the fovea, has many more cells than *Octopus*, although fish, for example, have about the same visual acuity as *Octopus* (Hanlon & Messenger 1996).

Similarities and differences between coleoid and fish eyes are outlined in Table 7.2. The major difference in retinal construction occurs due to different embryological origins (see next Section). In the vertebrate retina, light must pass through layers of cells before reaching the photoreceptors at the back. The cephalopod retina is not only arranged differently, with the light striking the photoreceptors first, but the retinal construction is also simpler, largely because some of the equivalent neuronal components found in the vertebrate retina are located in the outer layers of the optic lobe (Wells 1978) (see Figure 7.35). The optic lobe consists of four to five parts – an outer cell body layer, a neuropil layer, another cell layer, and

TABLE 7.2
Comparison of Coleoid Cephalopod and Fish Eyes

	Coleoid	Fish
Developmental origin	Ectodermal	Outgrowth from brain (which is of ectodermal origin)
Lens	Spherical, bipartite, with gradient construction	Spherical, with gradient construction
Refractive index of lens	About 1.58 in centre – changes towards periphery and thus avoids spherical aberration	About 1.58 in centre – changes towards periphery and thus avoids spherical aberration
Focal length	About 2.5 radii	About 2.5 radii
Aperture	About f/0.8	About f/0.8
Focussing	Lens moved relative to retina by muscles	Change in lens shape by attached muscles
Retinal shape	Inverted half sphere	Inverted half sphere
Number of types of photoreceptor cells	Usually one	Several – comprising rods (dim light receptors) and cones (colour receptors)
Retina	Simple, non-inverted	Inverted, multi-layered
Orientation of photoreceptor cells	Direct	Inverted
Retinal processing	Very little	Considerable
Nerve connection	From each retinal cell to the optic lobe	Optic nerve with markedly fewer axons than there are retinal cells
Colour vision	None – with one exception	Present
Detection of polarised light	Yes	Some
Adaptation to changes in intensity of light	An iris present; pigment migration slow	No iris; pigment migration fast

Compiled from various sources.

then another neuropil layer followed by a cell body layer, or a central medulla (Williamson et al. 1994). It is very large and consists of more than half of the nervous tissue of the brain (Wells 1962).

As noted above, the cytology of the deeper layers of vertebrate retinas is somewhat similar to that of the cortex of the optic lobes of coleoids (Figure 7.35) (see Section 7.11.2.2.4), notably in both having inner and outer granule cell layers and a layered plexiform zone between, and both containing amacrine and horizontal cells (e.g., Packard 1972; Budelmann 1995) (Figure 7.35). Packard (1972) also noted that second order visual cells of the coleoid retina lead to a system of columns similar to the situation in the fish tectum (=optic lobe), and that the peduncle lobe (located where the optic nerve exits the optic lobe) differs in structure from other parts of the coleoid CNS, while it is convergently somewhat similar to that of the folia of the vertebrate cerebellum.

7.9.11.1 Development of the Cephalopod Eye and a Comparison with Vertebrates

Despite being superficially similar in adult morphology, cephalopod and vertebrate eyes develop very differently (Figure 7.36), emphasizing their independent origins. In cephalopods, the anlage of the retina forms from a layer of ectodermal cells; the nuclei in the cells move to either the end or the middle of the cell, and the anlage develops into two layers. The edges of the retinal anlage curve inwards to form the optic vesicle from which the retina, lens, and ciliary body are formed. The developing eye vesicle is sealed off by three cell layers – the outer and inner ectodermal layers and a thin layer of mesoderm sandwiched between them from which the sclera and lens ligaments are derived. The outer ectodermal layer becomes the iris, the outer lens segment, and the eye covering, while the inner ectodermal layer gives rise to the inner lens segment. The cornea is formed by an extra epidermal fold (Harris 1997; Tomarev et al. 1997).

In vertebrates (Figure 7.36), the optic vesicle is formed from an evagination of the forebrain (diencephalon), and the lens develops from the overlying ectoderm. As the optic vesicle grows, it remains connected to the forebrain (by the optic stalk). The vesicle then invaginates, forming the optic cup, resulting in the inner layer of cells becoming the neural retina and the outer layer the pigmented epithelium. After the ectodermal lens forms, it separates, and the epithelium above differentiates as the cornea.

As noted above, these strikingly different origins of the retina account for the differently located cephalopod photoreceptors being at the front, facing the incoming light, while in vertebrate eyes they are on the inside of the retina, pointing away from the light entering the eye. Thus, in vertebrates, light must pass through several cell layers before it can activate the photoreceptors. The part of the vertebrate retina where the nerve fibres come together to form the optic nerve lacks photoreceptors (the 'blind spot'). In cephalopods, the optic nerve is on the outer side of the eye, and consequently, they do not have a blind spot.

The differences between cephalopods and vertebrates are also reflected in the genes expressed during eye development (see next Section).

7.10 GENE EXPRESSION AND EYE DEVELOPMENT

It is clear from embryology and phylogeny that the similarity of the eyes in coleoids such as octopods and vertebrates such as fish or humans is due to convergence, but how is this played out in the genes that express these eyes? Ogura et al. (2004) compared the gene expression in the eyes of *Octopus* and humans and found that 729 (69.3%) genes were commonly expressed between the human and *Octopus* eyes, but, of those, a large proportion was found in the common bilaterian ancestor. They suggested that given the number of conserved genes and their similar gene expression, this may be responsible for the convergent similarity of the eyes.

There is some evidence that the visual system evolved from chemoreceptors (Gaines & Carlson 1995). The gene *eyeless* (or *Pax6* – paired box gene 6) is a 'master control' gene (see Chapter 8). It switches on a cascade of many hundreds of genes involved in eye morphogenesis and has been identified in most groups of animals, including molluscs. This discovery has led some to argue for a monophyletic origin of eyes. Even the very simple eyes of flatworms are formed by a cascade initiated by the control genes *Pax6* and *sine oculis* and finishing with the photoreceptor genes (rhodopsin family). More advanced eyes presumably evolved by adding genes to the developmental pathway (Gehring & Ikeo 1999). As well as the eyes, *Pax6* is also involved in the development of chemosensory organs, and in the morphogenesis of the brain (Gehring 2001).

The development of vertebrate eyes is controlled by a suite of four *Pax6* genes which 'control' eye development and the development of some other sensory organs derived from the ectoderm. In squid, five forms (splicing variants) of *Pax6* have been identified as involved in eye formation and, unlike insects, there is no duplication of the *Pax6* gene (Yoshida et al. 2014). The *SIX3* and *SIX6* genes are expressed in eye development in some molluscs such as squid (*Idiosepius*), but not in *Nautilus*, which lacks a lens (Ogura et al. 2013). Pax family genes have been identified in aplacophorans, chitons, bivalves, gastropods, and cephalopods, but in molluscs, *Pax6* is only known at present from cephalopods, polyplacophorans, and solenogasters (Scherholz et al. 2017).

Apart from *Pax6*, other genes involved in cephalopod eye formation include *eya* (*eyes absent*), *so* (*sine oculis*), and *dac* (*dachshund*) (Peyer et al. 2014). Studies on the squid *Euprymna scolopes* have shown that all four genes (*Pax6*, *eya*, *so*, and *dac*) are expressed in developing not only the eyes, but also the photophore which houses bioluminescent bacteria (*Vibrio fischeri*), even though these structures arise from different tissues in the embryo (Peyer et al. 2014). In the few gastropods studied, the situation

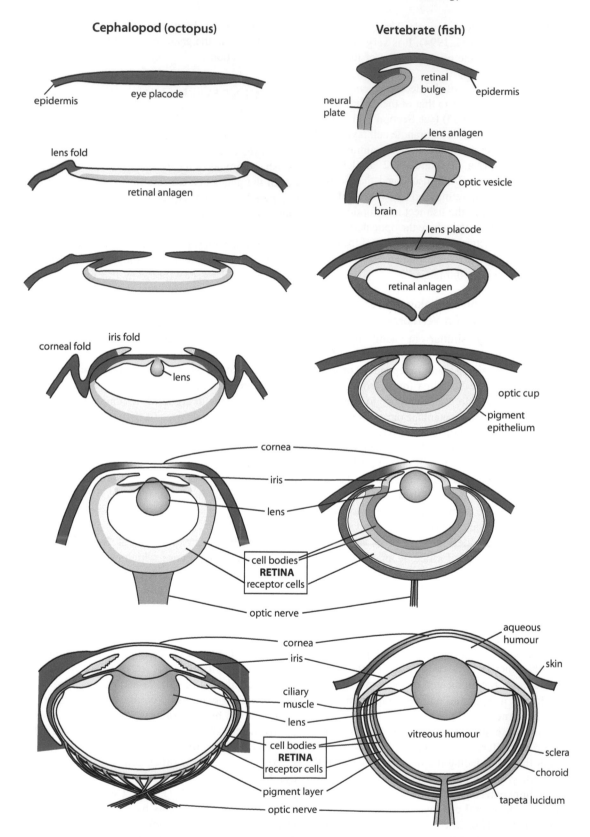

FIGURE 7.36 Comparison of the developmental stages in the convergently similar lens eyes in vertebrates and a coleoid cephalopod (*Octopus*). Developmental stages redrawn and modified from Harris, W.A., *Proc. Natl. Acad. Sci. USA*, 94, 2098–2100, 1997. For source of adult eyes see Figure 7.35.

appears to be more complex. Two Pax genes (*HasPax*-258 and *HasPax*-6) are expressed in eye formation in *Haliotis* (O'Brien & Degnan 2000), and differ from those expressed in cephalopods. Also, Hox genes have been identified in *Aplysia* and *Haliotis* (Ruddle et al. 1994), and the *POU-III* and *POU–IV* genes are expressed in eye formation in *Haliotis* (O'Brien & Degnan 2000, 2002). Otherwise, the Pax and Hox genes are involved in the formation of neural tissue (O'Brien & Degnan 2000; Hinman et al. 2003) and the statocysts (O'Brien & Degnan 2003). These results are in accord with the structural differences, possibly indicating that cephalopod and gastropod eyes were independently evolved from an eyeless ancestor.

The *Pax*6 gene is, no doubt, only part of the story of gene expression in eye formation, but, at the time of writing, that story is largely untold for molluscs. Some clues as to other classes of genes involved include, for example, the POU domain genes which are widely represented in animals. They play critical roles in the development and functioning of the nervous and neuroendocrine systems with the *POU–IV* gene regulating neuronal development, including various sensory neurons. These genes have been reported in only a few molluscs (Lozano et al. 2014), notably abalone (*Haliotis*) (O'Brien & Degnan 2000, 2002), where they are found in a variety of sensory tissues including the eyes, in the developing stellate ganglion of the squid *Doryteuthis* (Burbach et al. 2001), and are expressed in several tissues in the eyeless oyster *Crassostrea* (Zhang et al. 2012a), and the scallop *Pecten*, where it is also expressed in the mantle eyes (Lozano et al. 2014).

7.11 VISUAL ABILITY

In order to form an image, the lens and retina must be separated (Land 1981). Separation of these two structures is usually aided by the presence of a clear fluid-like 'vitreous body', or, in lens-less vetigastropod taxa, the vitreous body may act like a lens (L. Sumner-Rooney, pers. com. 2018). The properties of the lens are also important (see Section 7.9.5).

As noted above, the simplest eyes only detect light, enabling orientation relative to light, and/or responding to shadows as a defence against predation. This level of capability is the case with most bivalve eyes, although some, such as the mirror eyes of scallops, are capable of forming reasonable images (Land & Nilsson 2012).

Gastropod eyes range from simple pits to the rather sophisticated lens eyes of *Strombus* and the scanning eyes of heteropods (see Section 7.9.9.1). There are, however, relatively few studies on the visual abilities of gastropods, although in many the main purpose may be orientation relative to light sources. We do not know of any studies relating to the eye function in patellogastropods or the simple cup eyes of certain vetigastropods. The relatively simple lens eyes of the hygrophilan *Lymnaea* invoke positive phototactic behaviour to reach the water surface to breathe (Stoll 1973). *Lymnaea* also uses its eyes for orientation (Stoll 1973), and they can probably form simple images (Land 1984).

The lens eyes of the caenogastropods *Littorina littorea* and *Strombus raninus* have rather high visual acuity, and they can produce a good quality image because of the separation between the lens and the retina (Seyer 1992, 1994). *Strombus* also appears to be able to detect small objects (Seyer 1994). As already noted, it is thought to be the best-developed gastropod eye, and its abilities are comparable to the functionality of the eyes of small fish (Seyer 1998). As in most gastropods, because the rhabdomes lack shielding between them which allows light to spread, the visual acuity of the eye is limited (Seyer 1994; Seyer et al. 1998). Although the lens and retina in the lower caenogastropod *Ampullaria* are hardly separated, the eye does appear to form a poorly focused image on the retina (Seyer et al. 1998).

The relatively simple eyes of 'opisthobranchs' are capable of registering changes in light intensity and, in some at least, their responses to light are subject to a circadian rhythm induced by a pacemaker in the eye (see Section 7.14).

7.11.1 VISION IN COLEOID CEPHALOPODS

Sweeney et al. (2007) compared the visual acuity of a range of coleoid eyes by determining the resolving capabilities of their lenses. They found those of some squid compared favourably with those of vertebrates, but their most surprising result was that the optical acuity of the lens of the deep-water vampire squid *Vampyroteuthis infernalis* was comparable to that of the human eye.

Most of the work on determining what coleoids can see has been carried out on *Octopus* and to a lesser extent the cuttlefish *Sepia*, using their ability to learn. They are trained to make different responses to different visual stimuli. These tests have shown, for example, that *Octopus* can discriminate between vertical, horizontal, and oblique markings, black and white, some shapes, and discern length, area, outline, and form (e.g., Boycott & Young 1955b, 1957; Messenger & Sanders 1972) and the direction of polarised light (see next Section). Different orientations of objects can be discerned because the eye is normally horizontal in the orbit. This is controlled by the statocysts, the removal of which disables eye orientation and the ability to discriminate the orientation of objects.

Experiments have thus demonstrated that octopuses have similar visual abilities to vertebrates (Sutherland 1957a, 1957b, 1959, 1960), and most of these capabilities in octopods appear to extend, at least in part, to other cephalopods.

7.11.1.1 Discrimination and the Ability to Perceive Polarised Light

Experimental work shows that octopuses discriminate on brightness, size, orientation, form, and plane of polarisation. Although they cannot discriminate shapes on differences in their hue (colour), brightness discrimination allows them to distinguish not only between black and white, but

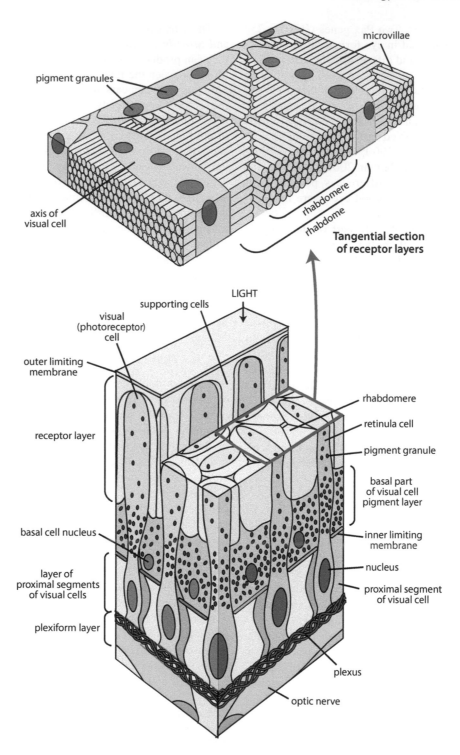

FIGURE 7.37 Diagrammatic section of the retina of *Octopus* (lower figure) and detail of a rhabdome (upper figure). Figures redrawn and modified from Moody, M.F. and Parriss, J.R., *Nature*, 186, 839–840, 1960 (upper figure), and Moody, M.F. and Parriss, J.R., *Z. Vergl. Physiol.*, 44, 268–291, 1961 (lower figure).

also different intensities of colours (Messenger et al. 1973) (Figure 7.37). The absence of colour vision in octopuses has been demonstrated in several experiments (e.g., Messenger et al. 1973; Roffe 1975; Messenger 1977a) and is in accord with the structure of the retina, which contains a single type of photoreceptor cell and one type of visual pigment (see Section 7.11.2.2.4).

It has been known that octopuses were sensitive to polarised light since the 1960s (Moody & Parriss 1960, 1961), but can only readily distinguish between vertical and horizontal e-vectors[11] (see below for other cephalopods and for the mechanisms behind this ability). Experiments have shown that polarisation vision can be used by *Octopus* for object discrimination, and it may provide information similar to that from colour vision in other animals (Shashar et al. 1996).

There has been much less experimental work done to test the abilities of decabrachian coleoids. Although they have a somewhat similar retinal structure (Figure 7.38), it differs in detail from that of octopods. Experiments on the squid *Lolliguncula* show it can discriminate horizontal from vertical rectangles, black spheres from white, and a white horizontal rectangle from a white sphere (Allen et al. 1985). Messenger (1977b) also found that *Sepia officinalis* could discriminate between small and large squares regardless of distance.

While polarisation sensitivity is well known in shallow-water coleoids (Varjú 2004), the structure of the retinas of some deep-water taxa suggest that they also possess it. In relation to orientation to polarised light, *Euprymna* and *Sepioteuthis* can orient themselves in four directions (0°, 45°, 90°, 135°) relative to the e-vector (Jander et al. 1963). These results differ from the results from experiments on *Octopus* (Tasaki & Karita 1966; Sugawara et al. 1971) and squid (Saidel et al. 1983; Saidel et al. 2005) and are due to different orientations of the microvilli within the retina (see also Section 7.11.2.2.4). While it is not known for certain how cephalopods apply this ability to detect the plane of polarised light, it may assist in retaining orientation during navigation (e.g., Jander et al. 1963). For example, experimental

work on *Sepia* suggests that they may also use this ability to see through the camouflage of the mirror-like sides of silvery fish as the light reflecting from their sides is partially polarised (Shashar et al. 2000). Thus, sensitivity to polarised light enhances detection of opaque, silvery-reflecting, and transparent prey (Shashar et al. 1998).

Some coleoids show polarisation patterning in their skin to facilitate intra- and perhaps interspecific communication. It has been suggested that this technique would not be detected by predators such as many marine mammals, because they are insensitive to polarised light (Shashar et al. 2002), although at least some fish can detect it.

The extent to which coleoids are colour blind is unknown because only a few species have been investigated. Colour vision is absent in octopuses, cuttlefish, and some squid, such as the Pacific squid *Todarodes pacificus* (Flores et al. 1978; Flores 1983), although at least one squid (*Watasenia scintillans*) may detect colours (see below). If most coleoids are colour blind, it is surprising that they demonstrate such striking colour changes and patterns and can accomplish precise colour matches to backgrounds.

7.11.2 Photoreceptor Cells and Comparative Retinal Structure

Photoreceptor cells are specialised sensory cells on or near the exterior of the animal so they can receive light. How the photoreceptor cells evolved from undifferentiated cells is unclear.

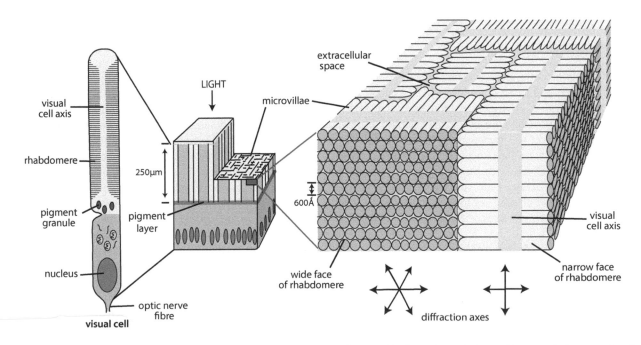

FIGURE 7.38 Schematic diagram of a slice of the retina (centre figure) of the squid *Alloteuthis subulata* with expanded views of a visual cell (left figure) and the arrangement of photoreceptor membranes in the rhabdomes (right figure). The diffraction axes are obtained by X-ray diffraction and calculated by Fourier transformation. Redrawn and modified from Saibil, H.R. and Hewat, E., *J. Cell Biol.*, 105, 19–28, 1987.

[11] The light vector.

Eyespots have even evolved in some protists, and these have rhodopsin-like photosensitive pigment, although these are organelles and different from the receptor cells. It has been suggested that the properties of photosensitive cells that make up eyes may have originated from intracellular symbionts, as did mitochondria and chloroplasts (Gehring 2001).

In animals, photoreceptor cells have generally increased their outer membranous surface area to maximise their ability to intercept light. The photoreceptor molecules (opsins) are housed in that membrane, so an increase in its area also increases the amount of opsins. This increased surface area has been achieved in three ways in animals (Salvini-Plawen 1982) and involves different signalling pathways.

1. The ciliary type – by the modification of the plasma membrane of cilia or flagella to increase the surface area by forming lamellae or discs (Figure 7.39). The highly specialised cilium may only be recognisable from remnants such as the basal body. Found in cnidarians, ctenophores, bryozoans, some molluscs, chaetognaths, urochordates, larval brachiopods, and chordates.
2. The rhabdomeric (sometimes called rhabdomal) type – increase the surface area by microvillar elaborations of the distal cell membrane (Figure 7.39). This kind is found in the eyes of most molluscs, and rotifers, platyhelminths, sipunculans, annelids, and arthropods.
3. Ganglionic or diverticular type – these photoreceptors are found only in a few animals, e.g., rotifers and some protochordates, and certain neurons in some molluscs are photosensitive (see Section 7.9.2).

Some photoreceptor cells are a mixture of both ciliary and rhabdomeric types, and these combined types occur in some pteriomorphian bivalve eyes and in chiton ocelli (Salvini-Plawen 1982), and possibly the bilaterian ancestor (Arendt 2003). As noted above, the two main types of photoreceptor cells are found in molluscan eyes. The photoreceptors in the mantle eyes of bivalves such as *Pecten*, *Cardium*, and *Tridacna* are the ciliary type, and these usually possess 75–200 modified cilia flattened out to form a membrane array for the attachment of a photosensitive pigment. They also occur in *Laternula*, where they are more complex than in other bivalves (Adal & Morton 1973).

The shell eyes of chitons have rhabdomeric photoreceptors, and these are also the main type found in the cephalic eyes of most gastropods and cephalopods.

There are varying theories as to the origin of the different kinds of photoreceptor cells. They may all be derived from ciliary cells, thus having a monophyletic origin (Vanfleteren 1982). Conversely, Salvini-Plawen and Mayr (1977) and Salvini-Plawen (1982) argued that some photoreceptors evolved from simple sensory cells independently in different animal groups, based on fundamental differences in their position, innervation, and structure. Salvini-Plawen (1982, p. 149) summed up by stating that 'there is no factually

supported argument in favour of hereditary transmission of molluscan photoreceptors, either adult eyes or larval ocelli, from their non-molluscan ancestors. In contrast, all available comparative-anatomical, ontogenetic, and evolutionary-functional data distinctly point to an independent acquisition of cephalic eyes in adult forerunners of Gastropoda and Siphonopoda (cephalopods), and to a secondary differentiation of lateral ocelli in placophoran[12] larvae successive to the original Pericalymma'. He further argued that the similar microvillar receptor structures in the cerebral eyes of molluscs and annelids result from similar functional requirements.

Despite the obvious differences, some argue that the more recent evidence that a single gene initiates eye development in many animals supports a single origin of photoreceptors (e.g., Gehring 2001, 2005) (see also below and Chapter 8), but if this is actually the case, it remains unclear as to what a common ancestral photoreceptor cell type might have been like.

The highly folded ciliary membranes of each photoreceptor cell contain photosensitive protein pigments (chromoproteins), which undergo a photochemical reaction when stimulated by light. In a similar way, in several lineages, the plasmalemma of sensory cells became photosensitive to evolve into the rhabdomeric type of photoreceptors. Because there are many ultrastructurally distinct photoreceptor cells, these structures probably evolved several times, most likely from sensory cells with existing limited photosensitive ability in both cilia and simple microvilli (Salvini-Plawen & Mayr 1977; Salvini-Plawen 1982). Subsequent evolution resulted in modification of these structures to produce cells with mixed receptors, or the hypertrophic development of one or the other to produce either the ciliary or rhabdomeric type.

All animal eye photoreceptors investigated have a light-sensitive photopigment based on vitamin A combined with opsin (a protein) and a chromophore (the part of the molecule responsible for its colour). As indicated above, the rhabdomeric and ciliary kinds of photoreceptors differ in how they increase the membranous surface area, but they also differ in the signal transduction pathways they utilise. These are based on pathways found in all animals – the phosphodiesterase (PDE) or the inositol (tri) phosphate (IP3) (e.g., Arendt 2003). The signalling pathway in ciliary photoreceptors involves the activation of PDE which alters the amount of cyclic guanosylmonophosphate (cGMP) in the cell, while in rhabdomeric photoreceptors, the signalling pathway involves activation of phospholipase C (PLC) and the IP3 pathway (Arendt 2003).

The ciliary and rhabdomeric photoreceptors also have different physiological responses. Rhabdomeric cells discharge impulses when stimulated by light ('ON' response), while the reverse is true for ciliary receptors which discharge when illumination is reduced in intensity or ceases ('OFF' response). It appears that the basal body found in the ciliated cell types is linked with the 'OFF' response. The possession of the two different types of photoreceptor is believed to account for defensive and orientation behaviour in molluscs. Those that

---

[12] Polyplacophoran.

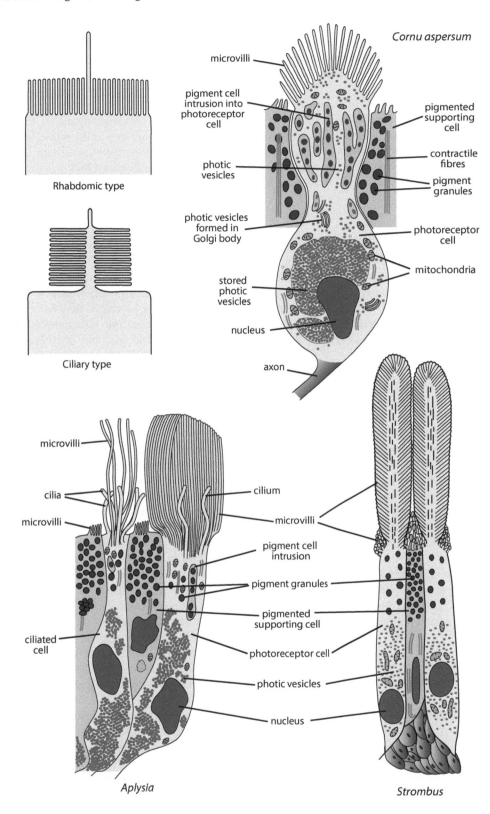

FIGURE 7.39 Diagram (top left) showing the basic differences between the rhabdomic and ciliary types of photoreceptor cells, and showing three rhabdomic examples from gastropods. Redrawn and modified from the following sources: upper left Arendt, D., *Int. J. Dev. Biol.*, 47, 563–571, 2003, *Cornu* Brandenburger, J.L. and Eakin, R.M., *Vision Res.*, 10, 639–653, 1970, *Aplysia* Hughes, H.P.I., *Z. Zellforsch. mikrosk. Anat.*, 106, 79–98, 1970, *Strombus* Hughes, H.P.I., *Cell Tissue Res.*, 171, 259–271, 1976.

show shadow responses possess ciliary photoreceptors that respond to a decrease in illumination, while those using light for orientation have rhabdomal photoreceptors that respond to illumination (Land 1968). However, chiton shell eyes are rhabdomeric and detect shadows.

The retraction response to changes in light intensity seen in many gastropods is mediated by ciliary epidermal photoreceptors on the mantle edge or, as in *Lymnaea*, for example, in the cephalic tentacles (Bubel 1984). These ciliary receptors are also in the siphons of caenogastropods such as nassariids, and if that structure is removed, the shadow response is markedly reduced (Crisp 1972).

7.11.2.1 Visual Pigments

At least three kinds of photoreceptors containing different pigments are required to distinguish colour. Most cephalopods investigated have a single visual pigment (Seidou et al. 1990; Messenger 1991), which is more sensitive in mesopelagic species. Some octopuses and squid have two, or rarely three visual pigments (e.g., the bathypelagic squid *Bathyteuthis* and *Watasenia*) (Seidou et al. 1990; Kito et al. 1992; Michinomae et al. 1994).

The eyes of some cephalopods (including octopuses, some squid, and *Nautilus*), and at least some gastropods have two visual pigments, rhodopsin and retinochrome (e.g., Hara & Hara 1976; Ozaki et al. 1986; Hara et al. 1995). Both of these photopigments are also present in coleoid extraocular photoreceptors and the dorsal eyes of *Onchidium* (Katagiri et al. 2001).

In *Octopus*, adaptation to light comes with a decrease in rhabdome size and a redistribution of these pigments, while in the dark, the rhabdome increases in size (Torres et al. 1997). These changes in size may be due to the degradation or formation of rhabdomere microvilli, apparently mediated by Rho GTPase (Miller et al. 2005) and heat shock proteins (HSPs) (Ochoa et al. 2002).

The crystal structure of squid rhodopsin (*Loligo forbesi*) has been determined (Davies et al. 1996; Davies et al. 2001; Murakami & Kouyama 2008) and *Octopus* rhodopsin was compared with vertebrate visual pigments by Nakagawa et al. (1999).

7.11.2.2 Comparison of the Receptor Cells and Retinal Structure in Molluscan Classes

An overview of the retinal structure and receptor cells in each group of molluscs is given below.

7.11.2.2.1 Polyplacophora

In some chitons, the modified aesthetes that can form ocelli (the macraesthetes) have a cup-shaped retina composed of up to around 100 microvillous receptors with the microvilli lying parallel to the lens surface. Associated with the macraesthetes are smaller micraesthetes which comprise a cluster of pigmented rhabdomes and are capped with a simple lens-like structure. They are present in the tegmental layer of the shell, and the shell may be transparent above them, thus acting like a lens (Hyman 1967). The details of the processes involved in light reception in chitons are not well understood although the

shell eyes and eyespots (both elaborated macraesthetes) can form images (Li et al. 2015a; Kingston et al. 2018).

The photosensory cells in polyplacophoran eyes vary, even in the same family. In the chitonids, *Chiton* ocelli have numerous long villi and a few cilia, while in *Onithochiton*, both rhabdomeric and ciliary photosensory cells are present. In *Acanthochitona* (Acanthochitonidae), the median aesthetes have rhabdomeric sensory cells, while lateral aesthetes are much smaller and have lamellate ciliary cells assumed to be photoreceptive (Vanfleteren 1982).

7.11.2.2.2 Bivalvia

Some bivalve siphons have epithelial photoreceptors that mediate withdrawal responses (see Kennedy 1960). The photoreceptor cells on the siphons of *Mya* and *Pholas* were used in classic studies that produced, for the first time, photochemical rate constants in a light receptor (Hecht 1920; Hecht 1927).

Significant structural differences in the photoreceptors occur between bivalve taxa – for example, in arcoids (e.g., *Arca*, *Barbatia*, and *Glycymeris*), the receptor cells have closely packed disc-like membranes derived from cilia, while those of *Cardium* have numerous lamellate cilia (Vanfleteren 1982). In the branchial (or cephalic) eyes at the base of the anterior filaments of the inner demibranchs in mytilids and in some other pteriomorphians (Morton 2001) (see Section 7.9.8.2), the photoreceptors are of the mixed type.

The double retina of the mantle eyes of pectinids is of interest in that the retina nearest the lens (proximal) contains receptor cells with modified lamellate cilia which point towards the lens, while the distal retina contains photoreceptors with microvilli and only one or two cilia which face away from the lens (Barber et al. 1967). The image is formed by reflection from the mirror-like tapetum, comprised of a crystalline array in the cells which lie behind the double retina and imparts the distinctive metallic blue-green sheen of these eyes. The image is reflected to the distal retina which has been shown to respond to movements of stripes. The lens forms the image behind the back of the eye, and its shape suggests it is not focussing light, but rather correcting for the hemispherical mirror. The mirror consists of alternating layers of cytoplasm and guanine crystal and is one of the few examples of an invertebrate tapetum other than in arthropods; they are the only true mirror eyes in molluscs (Land 1984; Messenger 1991), although a reflecting layer is also present in cardiid eyes (see Section 7.9.8.1).

The function of the proximal retina in pectinids appears to be to detect increases in light levels. Experiments have shown that proximal receptors are depolarised in light (the 'ON' response), while the distal retinal photoreceptors are hyperpolarised (the 'OFF' response) (Wilkens 2008). This latter response is seemingly similar to that seen in vertebrate ciliary receptors, but there is a major difference, at least in some bivalve eyes (e.g., *Pecten* and *Lima*), where it is the result of an increase in membrane permeability to K+, while in vertebrates it results from decreasing Na+ permeability (Messenger 1991).

Ciliated receptors in the eyes of *Cardium* and *Tridacna* are much simpler than in *Laternula* (Adal & Morton 1973).

7.11.2.2.3 Gastropoda

The 'larval' eyes of gastropods develop into the adult eyes (Salvini-Plawen 1980b, 1982; Salvini-Plawen 2008). In a neritoidean (Blumer 1995), and in most caenogastropods, the larval eyes differ from the adult eyes in possessing only ciliary receptors (Blumer 1994, 1995, 1996, 1998, 1999), while in the metamorphosed adult eye the receptors are rhabdomeric or the mixed type. In contrast, the larval eyes in 'opisthobranchs' have only rhabdomeric photoreceptors (e.g., Buchanan 1986), as in most of the adult eyes. Only the larval eyes of the caenogastropod *Lacuna divaricata* are known to possess both types of photoreceptors (Bartolomaeus 1992).

The rhabdomeric retina in gastropod eyes has the photoreceptor elements directed towards the source of light, thus differing from some bivalve eyes (see above). In most gastropods, the retina is composed of microvillous photoreceptors, with some having cilia and microvilli. Some marine gastropods have four or five types of photoreceptors located in different parts of the retina, but stylommatophorans have only two (Chase 2002).

The rhabdomeric photosensory cells of most adult gastropod cephalic eyes vary, with different taxa possessing well-developed, reduced, or no cilia (Salvini-Plawen 1982). The photoreceptors of the abalone, *Haliotis* (Vetigastropoda), have extremely regular microvilli (Kataoka & Yamamoto 1981), while those of the heterobranchs *Aplysia*, *Helix*, *Onchidium*, and the caenogastropod littorinids and the nassariid *Tritia* have both types (rhabdomeric and ciliated) while nudibranchs have only rhabdomeric receptors (Bartolomaeus 1992).

The retina of the patellogastropod *Scutellastra cochlear* is apparently composed of a single cell type (Marshall & Hodgson 1990), but related patellogastropods (*Patella* spp.) have two types (pigment and sensory, the latter with a microvillous border) (Davis & Fleure 1903; Stützel 1984). There is no data available on the retinal cytology of other patellogastropods. The retinas of many other gastropods and cephalopods also contain pigment and sensory cells.

The scanning eyes (see Section 7.9.9.1) of the swimming carnivorous caenogastropod heteropods *Pterotrachea* and *Oxygyrus* have a ribbon-shaped retina only 3–6 receptors wide, but several hundred long (Land 1982). Adult heteropods have the retina divided into two parts – an anterior part with rhabdomeric photoreceptors and a posterior part with ciliary photoreceptors (Blumer 1998).

The eyes of *Aplysia* are small, although larger than in many 'opisthobranchs', and have a relatively large lens which abuts the retina, so it does not form a sharp image. The retina is unusually complex, even more so than in cephalopods, with five kinds of receptors (Figure 7.39) and two types of neurons (Herman & Strumwasser 1984). The largest photoreceptor has long microvilli, while the other four kinds are small, and while they only occupy a small part of the retina, they comprise about half the receptors. These cells have cilia and microvilli. There is also some specialisation within the retina, with the microvillar receptors and one of the small 'mixed' receptors distributed through the retina, while the three types of small receptors are only located ventrally. Despite being unable to form a clear image, this is not a simple eye as it also contains the *neuronal circadian oscillator system* or pacemaker. Pacemakers are also found in some other 'opisthobranch' eyes (see Section 7.14).

In contrast to the eyes of *Aplysia*, nudibranchs have very small eyes (e.g., about 40 μm diameter in *Glaucus*) with only a few photoreceptor cells (e.g., seven in *Rostanga* and five in *Tritonia* and *Hermissenda*). The extremely simple eyes of *Hermissenda* have been used as a model for studies of memory (e.g., Alkon 1987).

The eyes of some 'pulmonates' have also been investigated in some detail. The cephalic eyes of *Onchidium* contain microvillar (rhabdomeric) photoreceptors and show depolarising 'ON' responses to illumination (Fujimoto et al. 1966). Thus they do not mediate shadow responses, unlike the dorsal eyes which have ciliary lamellate sensory cells and a reversed retinal structure similar to that of vertebrates (Katagiri et al. 1985) (Figure 7.40).

The cephalic eyes of other 'pulmonates' (stylommatophorans and hygrophilans) also have typical microvillar (rhabdomic) photoreceptors (Figure 7.39), and the retina also contains pigmented cells. There are two types of microvillar photoreceptors, one adapted for dim light, the other for bright light. An exception is the aquatic planorbid *Planorbarius corneus* which lives in somewhat deeper water and has only the dim light receptors (Bobkova et al. 2004). Stylommatophorans have simple cup-shaped retinas, but hygrophilans (lymnaeids, planorbids, physids) have their retinas divided into dorsal and ventral depressions separated by a ridge. These separate parts presumably enable the eye to function both in air and water in these often partially amphibious taxa (Gál et al. 2004).

There are a few reports of polarisation sensitivity in marine gastropods, notably in littorinids, but these have yet to be convincingly proven (Varjú 2004) (see also Section 7.15.3.4).

A dual photopigment system like that in cephalopods (see below) may be common in gastropod eyes, although the two pigments are present in different proportions (Messenger 1991). In *Onchidium*, for example, two visual pigments (rhodopsin and retinochrome) are found (Katagiri et al. 2001). These pigments are located separately in the retina of the cephalic eyes, but in the dorsal eyes they are not distinctly localised (Katagiri et al. 2002).

7.11.2.2.4 Cephalopoda

Cephalopod eyes contain only microvillous (rhabdomeric) photoreceptors which are long and narrow and lack cilia. The retina has been studied in several coleoids and is generally similar and rather complex in structure. As described above (Section 7.11.1.1), it differs markedly from that of vertebrate eyes in having the receptor cells oriented towards the light instead of away from it, and there is only one kind of photoreceptor. The summary below is largely based on the

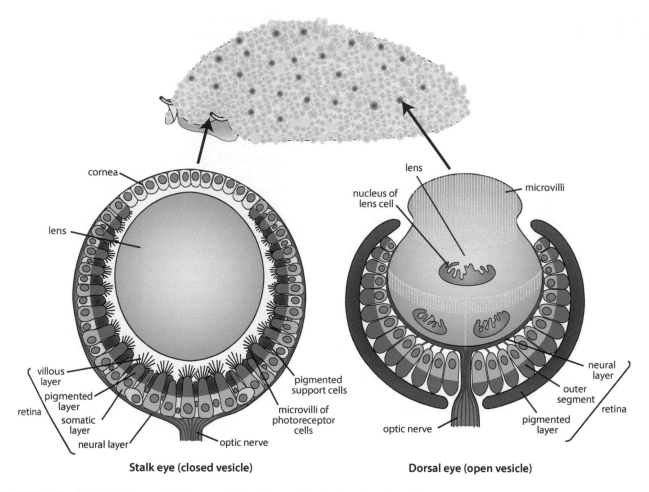

FIGURE 7.40 'Stalked' (i.e., cephalic) eye and dorsal eye of *Onchidium* showing their profound structural differences. Redrawn and modified from Katagiri, Y. et al., *Neurosci. Res.*, 2, S1–S15, 1985.

overviews of Messenger (1991) and Budelmann et al. (1997) to which the reader is referred for more detailed information. The inner surface of the retina is covered with a limiting membrane, below which lies an 'outer segment' layer comprised of the outer part of the elongate receptor cells. Below this lie the receptor cell bodies which form a narrower 'inner-segment' layer and below that lies a plexiform layer. Between each receptor cell are narrow supporting cells that contain pigment and so help to collect light. The receptor cells vary from around 20,000 to 250,000 cells per mm^2 depending on the species, and in *Octopus*, the retina has about 20×10^7 photoreceptors (Budelmann et al. 1997). Each photoreceptor cell contains pigment granules which, when stimulated by light, move towards the lens, and when it is dark, retract to the base of the outer segment. These cells also become shorter in the light. Studies have been made on the biochemistry involved in vision in coleoids, and these have aided our understanding of the fundamentals of photoreception. Cooper et al. (1986) noted that the biochemistry of photoreception in squid and vertebrates was closely convergent. Although there are two photopigments involved in cephalopod vision, only rhodopsin is a visual pigment and is found only in the outer segment of the retina. While similar, it is not identical to rhodopsins

in other animals. The second pigment is retinochrome, an accessory pigment involved in the regeneration of rhodopsin, found in the inner retinal segment composed of the bodies of the receptor cells. In these cell bodies lie the nucleus, mitochondria, and sheets of thin, ordered membranes, the somal or myeloid bodies, which contain retinochrome.

While rhodopsin is the visual pigment in all cephalopods, the absorption properties of this molecule vary, mainly relating to habitat with, for example, deep-water species having different absorption properties to those living in shallow water (e.g., Muntz & Johnson 1978).

The distal end of each photoreceptor cell has a central axis from which there are numerous microvillar projections which make up a vertical stack, the *rhabdomere* (Figure 7.38). Four rhabdomeres from four sensory cells are grouped to form a unit, the *rhabdom* (or rhabdome) (Young 1971). The microvilli make up an orthogonal arrangement, being oriented horizontally or vertically relative to the plane of the eye. They contain not only the visual pigment rhodopsin, but other signal transduction proteins such as Gq (a G protein) that activates the enzyme phospholipase C (e.g., Saibil & Hewat 1987).

The organisation of the two plane rhabdomere system with orthogonal (intersecting at right angles) arrays of microvilli (see Figures 7.37 and 7.38) provides the alignment of the rhodopsin molecules needed for the discrimination of polarised light. As with some vertebrates, octopuses can distinguish differences in polarisation as small as 4.4–9.7 minutes of arc (Muntz & Gwyther 1988).

Proximally, an axon from the end of each photoreceptor cell extends below the inner retinal segment through the plexiform layer, which also contains glial cells (Figure 7.35). These axons unite in the choroid, then enter the sclera (Wells 1978; Ali 1984), and group together to form the optic nerve which makes a complex dorso-ventral chiasma before entering the optic lobe (Young 1962a). The chiasma presumably 'corrects' the inverted retinal image produced by the lens, and it has been suggested that it may also project a visual map relative to a gravitational map (Young 1962b).

The density and morphology of the receptor cells in the retina is not always uniform. For example, the receptor density may be higher in some areas. In the benthic *Octopus* and the near benthic *Sepia*, there is a horizontal strip of longer, thinner receptor cells across the middle of the retina that is assumed to be a particularly sensitive area. In *Sepia*, the posterior part of this strip is particularly sensitive and is where the prey image is formed during its capture (Young 1963b). Epipelagic squid such as loliginids lack this strip.

There is good evidence that most coleoids cannot discriminate wavelength because their retina has only one type of rhodopsin (visual pigment) and thus cannot see colours (hue), only brightness. It may still be possible to discriminate wavelength with a single visual pigment if the retina contains multiple banks of receptors. Despite the retinae of many cephalopods being studied, banked receptors have only been found in the bioluminescent deep-sea firefly squid *Watasenia scintillans*, which, like some other bathypelagic squid, also has three visual pigments in the retina (Matsui et al. 1988; Seidou et al. 1990; Kito et al. 1992), one type found throughout the retina, and the other two in the ventral region, which suggests this species may have at least rudimentary colour vision. This raises the interesting question as to why these abilities have apparently not evolved in (most?) other coleoids. In addition, *Watasenia* possesses a specialised ventral zone, another apparently unique feature of this unusual squid, although a ventral sensitive area in the deep-water *Bathyteuthis* is used for resolution of light from above.

7.11.3 Visual Processing

How the information from the photoreceptors is passed to the CNS and processed is poorly understood in most molluscs, although there are some data for a few gastropod taxa, but in some cephalopods it is reasonably well understood.

Detailed studies on the very simple eye of the nudibranch *Hermissenda* have produced a fairly complete picture of visual processing in this positively phototaxic species (Alkon 1987). Each eye has only five photoreceptors associated with 14 optic ganglion cells and other second order cells in the central ganglia. Different optic ganglion cells give different levels of response – for example, some mediate weak arousal, while stimulation of others can result in reorientation of the animal. Even this simple system can initiate a complex range of behaviours (Alkon 1987).

At the other end of the visual complexity scale, cephalopod visual processing has been rather well studied. Much of the information comes from detailed anatomical and behavioural studies and, to a lesser extent, physiological studies (Messenger 1991). The large optic lobes lie on either side of the brain and can be several times larger in volume than the brain itself (see Section 7.2.6.1.2.4). They process visual input and, based on that input, discriminate, learn, and regulate behaviour and chromatophore displays (Budelmann et al. 1997). As with the brain itself, the optic lobes have an outer cortex and a central medulla. The outer cortical region of the optic lobe receives the axons of the retinal receptor cells and consists of three (in *Nautilus*) or four (in coleoids) layers: an outer granule cell layer, a plexiform zone (where the optic nerves mostly terminate), and an inner granule cell layer. In the squid *Loligo*, there is a palisade layer between the inner granule cell layer and the center of the optic lobe, the medulla. The cell bodies of the second order visual cells lie in the granule layers. This arrangement is like that in the deep layers of vertebrate retinas. The inverted image from the eye is re-inverted in the dorso-ventral chiasma of the optic nerves before being received by the plexiform layer. The plexiform zone of coleoid optic lobes also contains the processes of amacrine cells, whose cell bodies lie in the granule layers, as well as fibres that proceed well into the medulla. The medulla contains many millions of cells and is one of the most complex areas of the brain, being part higher motor centre, visual memory storage, and 'association' area.

The peduncle lobe of the brain has an array of long parallel fibres and, as noted above, receives input from both the optic lobes and statocysts, and controls locomotion, posture, and muscle tone (Messenger 1967). Visual input is recorded as memories in the vertical lobe of *Octopus*. This lobe consists of five lobules, the cortex of which contains very small amacrine nerve cells with short axons and numerous presynaptic vesicles which synapse with large nerve cells (Gray 1970). Both long and short-term memories are acquired and retained in this part of the brain (Sanders 1970; Young 1991; Shomrat et al. 2008).

7.12 OTHER SENSORY ORGANS

We briefly describe below some sense organs of bivalves, gastropods, and cephalopods that have not been dealt with so far. Some are poorly known and of unknown function, as for example, a small, protrusible, anterior, ciliated structure found in some solenogaster aplacophorans. Others associated with the mantle cavity are described in Chapter 4, including the Schwabe organ in lepidopleuran chitons, possible sensory areas in scaphopods, and the sensory streaks in the mantle grooves of patellogastropods.

7.12.1 Monoplacophora

A patch of 'dark granulate cells' lies on the outer base of each of the pair of preoral tentacles in *Neopilina* (see Section 7.6.7 and Figure 7.41), perhaps representing a rudimentary cephalic eye (Lemche & Wingstrand 1959).

7.12.2 Bivalvia

7.12.2.1 Mantle Edge Tentacles

In bivalves, the sensory structures are concentrated around the mantle edge, as is obvious in swimming pteriomorphians such as *Pecten* and *Lima*. In these taxa, stimulation of mantle

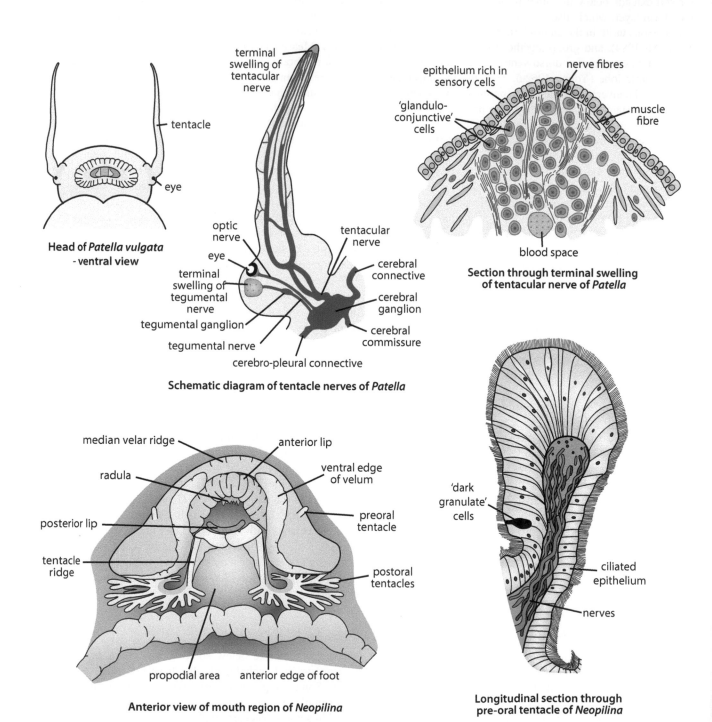

FIGURE 7.41 Possible tentacular sense organs of unknown function in a patellogastropod and a monoplacophoran. *Patella vulgata* figures redrawn and modified from Choquet, M. and Lemaire, J., *B. Soc. zool. Fr.*, 94, 39–53, 1969; *Neopilina galatheae* redrawn and modified from Lemche, H. and Wingstrand, K.G., *Galathea Report*, 3, 9–71, 1959.

edge sensory receptors (eyes and/or tentacles) triggers contraction of the large single adductor muscle and muscular movements of the 'velum', the wide, curtain-like inner mantle fold that controls water outflow, which together enable swimming. Experiments using pieces of mantle edge (and its tentacles) of *Lima* showed that extracts of a predatory starfish elicit strong impulses in the mantle nerves and associated movements of the velum and tentacles, showing that these responses are controlled by peripheral neurons, not the CNS (Stephens 1978).

7.12.2.2 Stempell's Organ and Stenta's Organ

Stempell's organ is a tube-like, paired structure situated immediately above the anterior adductor muscle in the protobranch bivalve *Nucula* (Haszprunar 1985c). It is innervated from the cerebral ganglion (which it lies just below) and contains collar receptors, suggesting a mechanoreceptor function. It is assumed that it detects the level of contraction of the anterior adductor, and that this organ may be related to the anterior inhalant water current in the Nuculidae (Haszprunar 1985c), a feature not shared by other protobranchs.

A somewhat similar sensory organ, Stenta's organ, is found in nuculanids and related protobranchs. It is a tubular structure which lies in a ciliated depression in the inner mantle fold just above the anterior adductor muscle (Zardus 2002).

7.12.2.3 Abdominal Sense Organs

The abdominal sense organs (AbSOs) are a probable mechanoreceptor found on the ventral surface of the posterior adductor muscle in pteriomorphian and palaeoheterodont (both Trigonida and Unionida) bivalves where they are located on either side of the anus and innervated by nerves from the visceral loop. It was thought that the AbSOs were a synapomorphy of these two groups (Haszprunar 1983, 1985b), but Waller (1998) suggested instead that the AbSO was lost in heterodonts with the evolution of siphons and more effective incurrent control. The AbSOs are probably mechanoreceptors that indirectly regulate water currents (Haszprunar 1985b). Several detailed studies on the AbSOs of scallops have been undertaken by P. M. Zhadan and colleagues (e.g., Zhadan & Semenkov 1984; Zhadan & Sizov 2000) (see also Chapter 4).

7.12.3 Patellogastropoda

A terminal swelling of the tentacular nerve has been described in *Patella*, and it may well have a sensory function. The tegmental nerve just below the eye, also has a terminal swelling (Figure 7.41).

7.12.4 Vetigastropoda

The sense organs described below are unique to vetigastropods (see also Chapter 18).

7.12.4.1 Sensory Papillae

Sensory papillae (Figure 7.42) are found only in vetigastropods. These typically numerous small projections occur on the cephalic tentacles, epipodial tentacles (see below), and sometimes the surface of the foot. They have a pit at the end of the papilla with a protrusible ciliary bundle extending from it.

Somewhat similar cilial bundles are also found on the cephalic tentacles of patellogastropods (Marshall & Hodgson 1990) and, while they are possibly homologous with the ciliary bundle of vetigastropods, they are not associated with a papilla.

7.12.4.2 Cephalic Lappets and Tentaculate Neck Lobes

Cephalic lappets, folds on the inner sides of the bases of the cephalic tentacles, occur in many vetigastropods and some neritimorphs. They are lost in some vetigastropods, including some (but not all) minute trochoideans and are also absent in scissurellids and fissurellids. Tentaculate neck lobes (Figure 7.42), at least partly sensory, are found in 'higher' vetigastropods and may be present on one side or both. Nontentaculate neck lobes are found in some of the hot vent gastropod groups and may be homologous. Analogous structures in some caenogastropods, namely some architaenioglossans (the exhalant siphon of viviparids and ampullariids), and turritellids and bithyniids, are not primarily sensory.

7.12.4.3 Epipodial Skirt and Tentacles

The pedally innervated epipodium reaches its greatest development in 'higher' vetigastropods, notably trochoideans. The possession of long epipodial tentacles (Figure 7.42) with sensory papillae and epipodial sense organs (see below) is a characteristic feature of many vetigastropods (Haszprunar et al. 2017). An epipodial skirt is found in many vetigastropods, and there are possible homologues in a few patellogastropods and some neritimorphs. Cerithioideans such as *Bittium* and allies sometimes have a narrow ridge and/or epipodial tentacles, the former perhaps homologous with the epipodial skirt of vetigastropods. Particularly long epipodial tentacles are developed in some members of the cerithiimorph family Litiopidae. A skirt-like structure identified as an epipodium is also present in *Janthina*, but is absent in the less derived Epitoniidae, suggesting it is a *de novo* structure related to the pelagic habits of that genus.

No adult patellogastropod has epipodial tentacles, however, Haszprunar et al. (2017) argue that *epipodial sense organs* (ESOs) are present in juveniles. Epipodial tentacles are also absent in adult neritimorphs and in Heterobranchia and most caenogastropods. The parapodia derived from the sides of the foot in some euopisthobranchs (mainly cephalaspidians and aplysioids) are presumably *de novo* structures as the more plesiomorphic heterobranchs lack them.

Haszprunar et al. (2017) considered the 'epipodial tentacles' of Cocculinidae, Neomphalina, and some vetigastropod (e.g., Fissurellidae, Haliotidae, Lepetodrilidae) to be ESOs (see next Section).

7.12.4.4 Epipodial Sense Organs

ESOs (Figure 7.42) lie at the bases of the 'epipodial tentacles' in vetigastropods and are innervated by pedal nerves (MacDonald & Manio 1964; Crisp 1981). They are

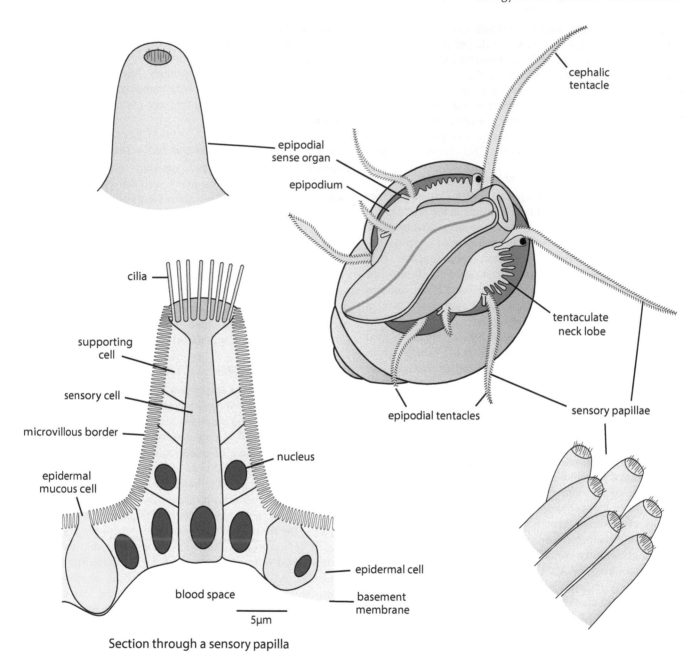

cilia

supporting
cell

sensory cell

microvillous border

epidermal
mucous cell

nucleus

epidermal cell

basement
membrane

blood space

5μm

Section through a sensory papilla

epipodial
sense organ

epipodium

cephalic
tentacle

tentaculate
neck lobe

epipodial tentacles

sensory papillae

FIGURE 7.42   External sensory structures in vetigastropods. All redrawn and modified from Künz, E. and Haszprunar, G., *Zool. Anz.*, 240, 137–165, 2001, except *Tricolia* animal redrawn and modified from Fretter, V. and Graham, A.L., *British Prosobranch Molluscs: Their Functional Anatomy and Ecology*, Ray Society, London, UK, 1962.

characterised by the possession of collar receptor cells with a long cilium and microvilli and are assumed to function as mechanoreceptors. They are described and contrasted with other innervated molluscan sense organs with collar receptors by Haszprunar (1985c). An evolutionary scenario was presented for epipodial structures by Haszprunar et al. (2017). They considered a posterior pair of ESOs to be an apomorphy of the gastropod stem lineage. This character is retained in early juveniles of several groups which lack them as adults, including patellogastropods and many cocculinids and juvenile vetigastropods.

The varying morphologies and combinations of ESOs, epipodial tentacles, and sensory papillae in lower gastropods and especially vetigastropods is thought to result from the modularity of the ESOs, cephalic/epipodial tentacles, and sensory papillae, with each module having the potential to become serially repeated, thus accounting for the various combinations of these structures among the basal gastropod clades (Haszprunar et al. 2017).

Lindberg and Ponder (2001) suggested this complement of epipodial-related sense organs arranged as lateral arrays on either side of the body in lower gastropods may be associated with the lack of inhalant control of water into the mantle cavity in these groups, which must rely to a much greater extent on external water-borne signals and direct contact with the body to assess the direction of stimuli (Burke 1964; Yarnall 1964).

7.12.5 HETEROBRANCH GASTROPODS

7.12.5.1 Semper's Organ in Stylommatophorans
The lips of stylommatophorans contain a highly developed sensory epithelium with many sensory neurons opening to the surface as well as a complex arrangement of glands opening to the lip surface. This sensory epithelium is connected to a small ganglion which is part of Semper's organ, the other part being a small gland which discharges onto the surface of the mouth lobe. While Semper's organ was originally considered an endocrine organ (e.g., Lane 1964), it has been shown to be an exocrine gland (Mol 1967; Renzoni 1968).

7.12.6 CEPHALOPODS

The large range of well-developed sense organs in cephalopods (eyes, arms, suckers, statocysts, rhinophores, lateral lines) are described above. These are complemented in deca-brachians (squid and cuttlefish) by the *nuchal proprioceptive organ* on their dorsal neck region underneath the nuchal cartilage. This organ is comprised of hair cells innervated by a branch of the postorbital nerve. On each side of the nuchal ridge that forms the midline, the hair cells are divided into an anterior and a posterior group, and each cell has up to about 300 kinocilia arranged in up to seven rows. Both the left and right anterior hair cells have the kinocilia polarised in the medial direction (i.e., parallel to the transverse axis), while those of the two posterior groups are polarised in the anterior direction (parallel to the longitudinal axis). They detect and enable the control of pitch and roll body movements (Budelmann 1994; Preuss & Budelmann 1995).

Embryos, paralarvae, and newly settled incirrate octopods have bristle-like *Kölliker organs* of uncertain function over much of the outer surface of the animal, but these are lost in subadults and adults. They consist of a cluster of rodlets secreted by special ectodermal cells that are spread into cilia-like rodlets by associated muscles (Boletzky 1973; Brocco et al. 1974; Villanueva & Norman 2008).

7.13 RHYTHMIC ACTIVITY, PACEMAKERS, AND BIOLOGICAL CLOCKS

Rhythmically recurring abiotic factors affect the activities of living organisms, with day and night (light and temperature) and the tidal cycle being obvious ones. In response to these cycles, most animals are more active at certain times – day or night, high tide or low tide, etc. This rhythmic behaviour is partly controlled by endogenous 'clocks', the existence of which can be observed experimentally by placing the animals

in controlled environmental conditions. Such rhythms or 'clocks' regulate many aspects of physiology and behaviour in most animals, including metabolism, hormone secretion, locomotor activity, feeding, and various other processes (e.g., Bell-Pedersen et al. 2005; Gillette & Sejnowski 2005).

Several studies have demonstrated rhythmic activity in both marine and fresh-water bivalves, with the latter largely related to tides. These include valve opening, metabolic and cytological changes in the crystalline style and digestive gland (e.g., see Chapter 5, Figure 5.54), byssal formation, shell secretion, and heart function. These changes are, however, often directly related to external cues rather than being endogenous (e.g., Gnyubkin 2010). Numerous other studies have demonstrated tidal rhythms (e.g., Palmer 1995; Naylor 2001). In gastropods, some involve *circadian* (daily) rhythms – a repeated pattern every 24 hours. This may be manifested as locomotory activity that is diurnal (e.g., in the sea hares *Aplysia* [Jacklet 1972] and *Bursatella* [Block & Roberts 1981]), nocturnal (e.g., the garden snail *Cornu* [Bailey 1975] and the terrestrial slug *Limax maximus* [Sokolove et al. 1977]), or crepuscular (evenings or dawn) (e.g., in the fresh-water caenogastropod *Melanoides tuberculata* [Beeston & Morgan 1979]), or both nocturnal and crepuscular (e.g., in the fresh-water planorbid *Helisoma trivolis* [Kavaliers 1981]). *Cornu* also has an approximately annual cycle (linked to the time to hibernate) when kept in constant conditions (Bailey 1981), while others show a rhythmic pattern of temperature selection as in *Helisoma* (Kavaliers 1980). Interactions between different endogenous rhythms occur – for example, in a species of the intertidal gastropod *Nerita* there is both a tidal clock and a circadian rhythm modified by environmental interactions. Thus locomotory and feeding activity was suppressed during daytime low tide because it fed at night (Zann 1973). Similar rather complex interactions have been investigated in other intertidal molluscs (e.g., Vannini et al. 2008).

The detail of how these rhythms are generated has progressed in recent years through studies on invertebrates, including a few gastropods, notably *Aplysia*. Various pacemaker neurons associated with a few different organ systems have been identified, but among the best studied are those in the heart, where signals originate from a few specialised neurons located in a ganglion and are transmitted by their neurotransmitters (e.g., serotonin) to the heart cells (see Chapter 6). Both excitatory and inhibitory neurons have been mapped in the ganglia of some species, the best known being the heterobranch gastropods *Achatina*, *Aplysia*, *Helix*, and *Lymnaea* (e.g., Kodirov 2011).

Central pattern generators (see Section 7.15.2) also produce endogenous rhythmic output without exogenous input.

7.14 PACEMAKERS

Pacemakers can be neurons which spontaneously discharge in the absence of inputs from chemical or mechanical stimuli. The two common patterns of such endogenous discharge are: (1) the 'beating pattern' which is exhibited as regular discharge spikes with constant intervals between, or (2) clusters of spikes separated by intervals – the so-called 'bursting pattern'. Neurons that produce the latter pattern are known as 'bursting neurons'.

Certain pacemakers are not neurons, the best known of these being the pacemaker myocytes associated with the heart or blood vessels. These undergo regular contractions initiated by the muscle cell (myocyte) itself rather than by an outside stimulus or nerve. Hearts are usually categorised as myogenic, with contractions initiated intrinsically, as in molluscs, chordates, and at least some brachiopods, or neurogenic with nerves initiating the muscular contractions as in many annelids and arthropods. Exceptions occur with, for example, some arthropods and annelids being myogenic (e.g., McMahon et al. 1997).

Myogenic pacemaker activity involved in maintaining heartbeat has been studied in various molluscs, including some chitons, bivalves, several gastropods, and coleoid cephalopods. The pacemaker can be intrinsic in the musculature of the heart, diffusely distributed within it, or more localised, as for example, in gastropods and cephalopods, it is thought to be localised near the atrioventricular valve (Jones 1983b; McMahon et al. 1997). Numerous studies on the effects of various neurohormones and other organic and non-organic compounds and ions on heartbeat have demonstrated that nervous, humoral (including neurohormones and neuropeptides) (see Deaton 2009 for a recent overview), metabolic, and peripheral influences on blood flow can all affect heartbeat frequency and amplitude.

The most extensive studies of neuronal mechanisms controlling the heart have been carried out on a few species of heterobranch gastropods including *Aplysia californica* (e.g., Kandel 1976; Koester et al. 1979; Koester & Koch 1987), the pteropod *Clione limacina* (Arshavsky et al. 1990), and the land snails *Helix pomatia* (Rózsa 1979; Zhuravlev & Safonova 1984; Rózsa 1987) and *Achatina fulica* (Furukawa & Kobayashi 1987). In stylommatophorans (*Helix* and *Achatina*), the cerebral, abdominal, and parietal ganglia contain a few neurons which either accelerate or decelerate heartbeat. In the euopisthobranch *Aplysia*, four neurons are found in the visceroparietal (i.e., the fused parietal and abdominal) ganglion, two of which are excitory and two inhibitory. In the nudibranch *Archidoris montereyensis* heart regulation is by way of a few neurons in the right pleural and visceral ganglia (Wiens & Brownell 1990). In *Clione*, three heart inhibitors are found in the visceroabdominal ganglion, and an excitor neuron in the left pedal ganglion (Arshavsky et al. 1990).

Cellular pacemakers and how they express circadian and other rhythms have been studied using gastropods, especially euthyneurans, as models. Experiments with excised eyes kept in darkness show a circadian rhythm, and these pacemakers have been found in several euopisthobranchs, notably the aplysiids *Aplysia* and *Bursatella* (Block & Roberts 1981; Roberts et al. 1987), and the cephalaspideans *Bulla*, *Navanax* (Eskin & Harcombe 1977), and *Haminoea* (McMahon & Block 1982), with most studies conducted on *Aplysia* and *Bulla* (see Blumenthal et al. 2001 for details). Pacemakers are presumably present in at least some other euopisthobranch taxa, but it is not known if they occur in other 'opisthobranchs' such as

nudibranchs, although a circadian pacemaker is thought to be located in or near the eyes in the dendronotoidean nudibranch *Melibe* (Newcomb et al. 2014). The 'brain' also plays a role in pacemaker activity because the neurons from the pacemakers in the eye enter the CNS (Messenger 1991).

There is a circadian rhythm in locomotory behaviour in *Aplysia*, which is diurnally active, while *Bulla*, which also exhibits a circadian rhythm in locomotion, is nocturnally active. Additional non-ocular pacemakers are apparently present (e.g., Blumenthal et al. 2001), presumably related to the presence of photoreceptors other than the eyes, for example, dermal and neural photoreceptors (see Section 7.9.1 and 7.9.2). In *Aplysia*, there are neural photoreceptors in the cerebral ganglia (Roberts & Block 1982). Thus, eyeless *Aplysia* and *Bulla* show some circadian activity. Non-ocular photoreception (after eye removal) has been observed in experiments on the planorbid fresh-water snail *Helisoma* (Kavaliers 1981), terrestrial limacid slugs (Beiswanger et al. 1981), and a nudibranch *Melibe* (Newcomb et al. 2014).

Within the eye of *Bulla* and *Aplysia*, different cells carry out different aspects of pacemaker function. At the base of the retina in each eye in *Bulla*, there are about 130 'basal retinal neurons' surrounding the neuropil which make up the circadian pacemaker (Block et al. 1984; McMahon et al. 1984; Blumenthal et al. 2001; Block & Colwell 2014) (Figure 7.43). They continue to produce a circadian rhythm even when the

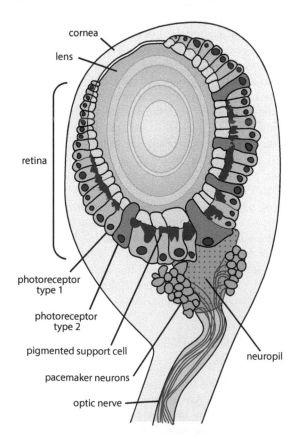

FIGURE 7.43 A longitudinal section through the eye of the cephalaspidean *Bulla gouldiana*. Redrawn and modified from Jacklet, J.W. and Colquhoun, W., *J. Neurocytol.*, 12, 673–696, 1983.

photoreceptor layer of the eye is removed experimentally. When that has been done, the rhythm can be altered by light pulses, suggesting these neurons are photoresponsive (e.g., Block & McMahon 1984). These neuronal cells produce a protein similar to opsin (Geusz et al. 1997), and even an isolated basal retinal neuron can produce a circadian rhythm (Michel et al. 1993). A similar pacemaker is in the basal retinal cells in the eye of *Aplysia*, where it is comprised of about 1000 secondary neurons (e.g., Herman & Strumwasser 1984) and these are also photoresponsive (Block & McMahon 1983; Jacklet & Barnes 1993).

As reviewed by Blumenthal et al. (2001) and Block and Colwell (2014), the timing of the ocular circadian pacemaker can be modified, both by light and by efferent input from the CNS, with calcium regulating the pacemaker phase. Although the details of the role that calcium plays are uncertain, light causes the pacemaker cell membranes to depolarise, allowing calcium ions to enter. The effect of light depends on the time in the circadian cycle when exposure occurs (e.g., Ralph & Block 1990). In *Aplysia*, at least, both cyclic guanosinemonophosphate (cGMP) (e.g., Eskin et al. 1984) and/or cAMP (cyclic adenosine monophosphate) are involved in transmitting the light response (e.g., Eskin & Takahashi 1983). The efferent signals from the CNS are controlled by a neurotransmitter (serotonin) with the fibres reaching the eye via the optic nerve (Takahashi et al. 1989). In *Bulla*, the efferent neurotransmitter is the tetrapeptide, FMRFamide (see Blumenthal et al. 2001 for review).

Modulation of long-term memory by circadian clocks can occur and has been demonstrated in *Aplysia* (Fernandez et al. 2003; Lyons et al. 2005; Lyons et al. 2006; Lyons 2011). *Aplysia* is diurnally active and long-term memories are formed in experimental animals during the daytime, but not at night. These results can also be obtained in eyeless sea hares, suggesting that extraocular circadian oscillators associated with extraocular photoreceptors can also modulate long-term memories (Lyons et al. 2006).

Although several stylommatophoran land snails and slugs have been shown to have circadian rhythms (see Zieger & Meyer-Rochow 2008 for references), an internal ocular pacemaker has not yet been demonstrated.

Pairett and Serb (2013) produced evidence there may be a circadian clock in scallop eyes as they have all six core circadian clock genes present in the house mouse *Mus musculus* and the fruit fly *Drosophila melanogaster*, although it is not yet known which cells in the scallop function as the circadian clock.

7.15 BEHAVIOUR

In this section, we summarise aspects of the behavioural repertoire in the main groups of molluscs and some of the neural and sensory aspects involved. Other aspects of behaviour are covered elsewhere, mainly in the taxon chapters and in Chapter 9, while reproductive behaviour is discussed in more detail in Chapter 8.

Bullock (1990) argued that the nervous system and behaviour were together most responsible for large evolutionary leaps in complexity, although, as pointed out almost three decades

ago by Hanlon (1991), there is little interaction between neurobiologists and evolutionary biologists in relation to the study of molluscan eyes. A meeting organised by the American Malacological Union in 1990 started to address this gap, and papers were published from it (Hanlon 1991). As always, there was a considerable bias towards euthyneuran gastropods with virtually all papers dealing with those taxa. Thus, while molluscan behaviour has been studied in detail in certain euthyneuran gastropods and cephalopods, it is relatively poorly understood in most other molluscs except for some intertidal taxa such as various patellogastropod limpets, littorinids, and a few muricid whelks and chitons.

Molluscan behaviour ranges from rather limited, as in chitons, bivalves, and patellogastropods, to the very complex, seemingly intelligent behaviour exhibited by coleoid cephalopods. Behavioural complexity and the complexity of the nervous system and sensory organs appear intimately linked. Throughout this broad range of behavioural responses, sense organs provide the necessary information to maintain life, most notably facilitating avoidance of predators and unsuitable habitat conditions and finding food and mates. Different sense organs are important in different groups, depending on their development. For example, with gastropod eyes, some are capable of object recognition, and together with dermal light receptors, they initiate simple shadow reflexes and movement away from (*negative phototaxis*) or towards (*positive phototaxis*) light. Most gastropods rely primarily on chemoreception (olfaction) for distance perception, while nearby objects are perceived by a combination of chemoreception and mechanoreception (Chase 2002).

In molluscs, as in other animals, behaviour can be *stereotyped* – i.e., it has a genetic basis and is performed in the same way despite the situation, such as many responses to stimuli. This contrasts with *plastic* behaviours, which are variable responses typically involving learning.

Some stereotyped behaviours have been studied extensively at the cellular level. These include the touch responses in *Aplysia* and stimulation of the chemosensory cells on the rhinophores, oral tentacles, and lips in *Aplysia* and in other euthyneuran gastropods, including the carnivorous marine *Pleurobranchaea*, *Tritonia*, and *Navanax* and the fresh-water *Lymnaea* (e.g., Audesirk & Audesirk 1985).

7.15.1 NEURAL NETWORKS AND BEHAVIOUR

A neural network[13] (or pathway) consists of interconnected neurons activated in a linear fashion. Both simple and complex neural networks are known in molluscs, and some examples are given below. The neurons involved in neural networks are often complex and have specific functions which may only be understood in relation to the complete circuit. Some may even code specific pieces of information and are known as 'grandmother cells' or 'gnostic' units (e.g., see Gross 2002).

---

[13] This term is also used for artificial neural networks in computer programming, so it is sometimes specified as a biological neural network.

Neural networks involved in feeding in gastropods were reviewed by Elliott and Susswein (2002). They found that many herbivorous gastropods had generally similar behaviour and underlying neural anatomy, and similarities in pharmacology and physiology. While carnivorous gastropods differ in their methods of capturing prey, some features of neural control in feeding are conserved. They observed that feeding was flexible in all the taxa in the study, with both behaviour and physiology able to adapt to changes in the external environment or the internal state of the animal because of past experience. Such flexibility may involve many neural sites.

7.15.2 Central Pattern Generators

Central pattern generators (e.g., Marder et al. 2005; Dickinson 2006) are neural networks that produce rhythmic patterned outputs without the sensory feedback from muscles or other organs that control rhythmic behaviours such as feeding (e.g., Perrins & Weiss 1996; Straub et al. 2002), respiration (e.g., Taylor & Lukowiak 2000), as demonstrated in *Lymnaea stagnalis* and in *Aplysia*, and in locomotion including escape responses (e.g., Katz 1998; Jing & Gillette 1999).

An example of a central pattern generator that has been well studied in molluscs is one in the nudibranch *Melibe* that underlies swimming and is controlled by an endogenous circadian clock. It consists of eight neurons in the circumoesophageal ganglia, including the pedals (Thompson & Watson 2005; Lillvis et al. 2012; Sakurai et al. 2014). A similar example controlling swimming in the gymnosome pteropod *Clione* has been identified (Panchin et al. 1995). Newcomb et al. (2012) explored the relationship between neural circuits and behaviour by examining swimming in various nudipleuran sea slugs which swim mainly by way of either dorso-ventral or left-right flexing. Central pattern generators have been demonstrated in the dorso-ventral swimming seen in the nudibranch *Tritonia diomedea* (Katz et al. 2001) and the pleurobranch *Pleurobranchaea californica* (Jing & Gillette 1999), while left-right central pattern generators have been investigated in the unrelated nudibranchs *Melibe leonina* (see above) and *Dendronotus iris* (Newcomb & Katz 2007). They found that the left-right and dorso-ventral swimming behaviours involved different sets of neurons and there were differences in the details of the circuits within each swimming category, signifying the separate origin of these behaviours in different clades.

7.15.3 Summary of Behaviour in Each Class

7.15.3.1 Chitons

Photoreceptors in the girdle and valves (aesthetes) are utilised in shadow responses (to avoid potential predators) (e.g., Arey & Crozier 1919; Boyle 1972, 1977) and in phototaxic movements into lighter or darker areas. In *Chiton tuberculatus*, which lacks shell eyes, but has eyespots, young individuals are negatively phototactic, while older ones are positively phototactic (Arey & Crozier 1919). Negative phototaxis is commonly seen in many adult chitons, and chitons also respond to gravity and physiological stress, often have diurnal

patterns, and some intertidal species show homing behaviour (Boyle 1977) (see Section 7.15.3.4.3).

7.15.3.2 Bivalves

The mantle eyes and/or dermal light receptors of bivalves initiate simple reflexes from shadows, which are important in avoiding predators, with responses such as valve closure, rapid burrowing, or siphon retraction (Ramirez et al. 2011). They may also enable the animal to move away from, or towards, light. Some show phototaxic behaviour, as for example, in the negative phototaxis of the fresh-water invasive 'zebra mussel' *Dreissena polymorpha* (Kobak 2001) and in the small crevice-living intertidal bivalve *Lasaea rubra* (Morton 1960b) which is modified by an accompanying strong negative geotactic response. Although juvenile scallops make phototaxic responses to light (Howell 1989), the function of the sophisticated eyes in adult scallops remains obscure as, for example, predator detection appears to be related more to mechanical and chemical stimuli (Morton 2000b), although scallops have also been shown to respond to visual stimulus (Speiser & Johnsen 2008). In those scallops and limids that swim, this behaviour can be elicited by various stimuli such as movement, shadows, bright light, and vibration.

Giant clams (*Tridacna*) have thousands of small mantle eyes associated with their symbiotic zooxanthellae (see Section 7.9.8.1 and Chapters 9 and 15) and can produce surprisingly complex responses (Wilkens 1986). A strong shadow or movement results in rapid contraction of the siphon and closure of the valves so that a water jet is sent towards the disturbance, but a shadow falling on the mantle without marked changes in intensity or movement results in a slow retraction. The mantle and siphon also respond to increased light intensity by expanding, thus providing a larger surface area for the symbiotic zooxanthellae.

7.15.3.3 Scaphopods

Scaphopods show only a relatively limited range of feeding and burrowing behaviours which are reviewed by Reynolds (2002).

7.15.3.4 Gastropods

In gastropods, basic behaviours are feeding, mating, spawning, orientation, locomotion, aggression, defence, and resting. Common defensive behaviours usually involve shutting down (e.g., a limpet clamping) or a snail retracting into its shell. Sometimes noxious chemicals are released, or flight reactions occur, or autotomy is performed (see Chapter 3). Reproductive behaviours usually involve mating, spawning, or brood protection (see Chapter 8).

Responses to light can initiate simple orientation responses or behaviours which can vary depending on the conditions; for example, circadian locomotory activity (see Section 7.13), or they may differ in different stages in the life cycle. Some responses are rather sophisticated; for example, a sea hare, *Aplysia brasiliana*, can maintain a straight direction in swimming, apparently by using, at least in part, celestial cues (Hamilton & Russell 1982a, 1982b). Many larvae respond to light, although this may be complicated by other sensory inputs. For example, there is a positive phototaxis in early *Strombus* larvae and a strong negative geotaxis, but the former decreases with the age of the larvae (Barile et al. 1994). Similar changes have been observed in some other gastropod larvae, as for example, in the nudibranch *Phestilla* (Miller & Hadfield 1986). In the salt marsh littorinid *Littoraria irrorata*, factors such as salinity, elevation, food, light, gravity, temperature, water, etc. all influence behavioural responses (Bingham 1972a), while the behaviour of the rocky shore *Littorina littorea* is influenced by waves and secondarily by light (Gendron 1977). A trochid, *Phorcus turbinatus*, like several other littoral gastropods, migrates down-shore in the daytime (positive geotaxis and negative phototaxis) and upwards at night (negative geotaxis) (Chelazzi & Focardi 1982).

Although in some gastropods their limited visual capability may be important, most mainly rely on chemoreception to locate food or mates (Messenger 1991). Vision can play a role, as for example, in the marsh periwinkle, *Littoraria irrorata* (Littorinidae), which can locate the plant stems which they climb to avoid predation (Hamilton 1977) because it responds preferentially to vertical rather than horizontal lines (Hamilton & Winter 1982). These same authors later tested a subtidal vetigastropod, *Turbo castanea*, a supralittoral littorinid *Cenchritis muricatus*, and the nocturnal terrestrial snail *Cornu aspersum*. They found that both *Turbo* and *Cenchritis* had better vision than *Cornu*, and that *Turbo* was less able to discriminate shape differences, possibly because its eyes do not have a cornea (Hamilton & Winter 1984). Two of the best-developed gastropod eyes are those in the benthic caenogastropod Strombidae and in the pelagic heteropods, the latter with a scanning retina (see Section 7.9.9.1). The heteropod eyes contrast sharply with the lack of eyes in another group of pelagic predatory gastropods, the gymnosome pteropods such as *Clione limacina*, which use a set pattern of movements to detect their thecosome pteropod prey (Varona et al. 2002).

Clione is rather well studied experimentally. It tends to stay at a fairly constant depth by using its 'wings', but every few minutes, as the wings stop moving, the animal sinks 20–50 cm. The wings then recommence their beating and the animal returns to its initial depth. When it touches its prey (a thecosome), it is grasped through the eversion of three pairs of buccal cones, and the wing beats increase by two to three times, but return to normal when the prey has been swallowed. Other stimuli can change the wing beat – touching the tail results in an increase, but touching the head stops them (e.g., Arshavsky et al. 1990). Newcomb et al. (2012) compared the neural circuits involved in swimming in a range of nudibranchs and pleurobranchs (i.e., nudipleurans) (see Section 7.15.2).

Some early experiments with a nassariid (Baylor 1959) and littorinids (Charles 1961a, 1961b, 1961c) suggested that they could, like coleoid cephalopods, detect the plane of polarisation in light, but little has been done to examine this phenomenon further and some argue that this ability is not yet proven (Varjú 2004). Littoral gastropods may possibly use this apparent ability in orientation along with a 'sun compass' and/or 'moon compass' (Newell 1958a, 1958b; Evans 1961; Charles 1966;

Messenger 1991). It has also been suggested that sometimes visual cues (i.e., involving a memory for topographic features) may also be used (e.g., Edelstam & Palmer 1950; Evans 1961; Hamilton 1977, 1978; Chelazzi & Vannini 1980), although this is doubted by others (e.g., Cook 1979b).

Chase (2002) reviewed seasonal and daily cycles in gastropods and their endogenous circadian clocks (see Section 7.13) and the underlying controlling neuronal processes. These can range from being daily to yearly. Gastropod responses will often vary at different times; for example, response to a particular stimulus may sometimes produce a certain behaviour, while at other times it might produce a weak response or none.

7.15.3.4.1 Trail Following

Many gastropods and chitons engage in trail following using chemosensory means. Following intraspecific or interspecific pedal mucous trails may have originally evolved from utilising the mucus as a supplementary food source (Davies & Hawkins 1998), especially as these trails can entangle microbes and other food (see Ng et al. 2013 for review). Interestingly, the mucus of the homing limpets *Lottia gigantea* and *Lottia scabra* grows microalgae more effectively than the mucus of the non-homing limpet *Lottia digitalis*. This enhancement is possible because *L. gigantea* mucous trails can last from four to 15 days on the intertidal rock surface (Connor & Quinn 1984; Connor 1986). Other mucous trails can persist even longer, with the trail of *Littorina littorea* having a half-life of about 12 days, and pedal mucus produced by resting *Patella vulgata*, a temperate intertidal limpet, had an incredible half-life of approximately 40 days (Davies et al. 1992). This markedly contrasts with the mucus produced by a resting tropical limpet, *Cellana grata*, which persisted for only about six days (Davies & Williams 1995). In an unusual adaptation, possibly derived from trail following, the nassariid *Tritia obsoleta* derives a proportion of its amphipod prey from individuals trapped in its mucous trails (Coffin et al. 2012).

Carnivorous gastropods track their gastropod prey by following their mucous trails. This behaviour is seen in naticids, some cephalaspideans such as *Navanax*, and stylommatophoran carnivores, the most famous of which is probably *Euglandina rosea* (Cook 1985; Ng et al. 2013). As far as known, no predatory gastropods can detect the direction in which the prey snail has moved (Chase 2002), although some species at least can follow conspecifics directionally (e.g., Clifford et al. 2003).

7.15.3.4.2 Territoriality

Some intertidal gastropods are territorial, with the best-known examples being among the patellogastropod limpets. For example, limpets such as *Lottia gigantea* (Wright 1982) and some species of *Scutellastra* (Branch 1975a) defend their territory against both conspecifics and other species, and this behaviour can be modified by experience (Wright & Shanks 1993). Territorial patellogastropod limpets are also often 'gardeners', an algal 'garden' being maintained around the periphery of their home scar by some species of *Scutellastra* (see next Section) with algal growth encouraged by nutrient enrichment by way of either nitrogenous waste or mucus (Lindberg 2007b).

7.15.3.4.3 Homing Behaviour in Gastropods and Chitons

Homing behaviour, where an individual returns to a specific location after feeding, is often associated with territoriality and is well known in intertidal chitons and in a range of both intertidal and terrestrial gastropods (reviews by Underwood 1979; Branch 1981; Chelazzi et al. 1988; Chelazzi 1990; Cook 2001). Homing and non-homing sometimes occur among members of a co-existing population, for example, in some intertidal patellogastropod (e.g., Underwood 1977) and siphonariid (Creese & Underwood 1982) limpets. This phenomenon raises interesting questions about how the animals that exhibit homing behaviour can relocate their 'homes' after feeding excursions. The 'homes' themselves provide refuges that can help reduce physical and biological constraints. In some intertidal rocky shore molluscs, the homes may be deep scars in the rock surface called 'home scars' (Figure 7.44). The shell fits neatly into these scars making the owner more difficult to dislodge and reducing desiccation (Cook 1969, 1971; Branch 1975a; Hahn & Denny 1989; Lindberg 2007b).

Homing terrestrial gastropods usually shelter in nooks or crannies that reduce desiccation and the likelihood of predation. Homing can, sometimes, involve groups of individuals of a species. Such collective homing (e.g., Cook 1979a, 1979b; McFarlane 1980) has been observed in a few chitons and marine and terrestrial gastropods where clustering in crevices or holes can reduce desiccation, heating, and predation more effectively than with individuals (Chelazzi 1990). This clustering to reduce desiccation is particularly important in terrestrial slugs where interspecific clusters can occur, as they can also do with land snails (Chelazzi 1992).

After foraging for food, gastropods and chitons relocate their 'homes' mainly by olfactory cues. These are usually by way of mucous trail following, or for some terrestrial gastropods, distance chemoreception appears to play an important role (Cook 1979b; Chase & Croll 1981; Chelazzi et al. 1985; Chelazzi et al. 1988; Cook 1992). Thus, when snails or slugs are removed tens of metres from their 'home', they can return without trail following (e.g., Cook et al. 1969; Thomas 1973; Dunstan & Hodgson 2014), although following old trails can occur if they fail to locate the home (e.g., Cook 1979a, 1979b) or to help locate the home (Cook 1969, 1971; Chelazzi et al. 1985). Thus, for example with the European slugs *Limacus maculatus* and *L. flavus*, if they cannot locate 'home' using an 'olfactory beacon' detected by their posterior (optic) tentacles, they will, as a backup, follow trails (using their anterior tentacles) made by the individual involved or by conspecifics, and will sometimes briefly follow the trails of other limacids, but not those of other slugs (Cook 1977). Similar trail following has been observed in intertidal gastropods where an individual may respond to its own trail if it is a solitary 'homer', or to conspecific trails in the case of 'collective homers'. The intertidal snail *Nerita textilis* is a collective homer which uses long-lasting trails to mark the home area (Chelazzi et al. 1985). In an early experiment involving the European limpet *Patella vulgata*, crossing the trails of conspecifics did not confuse the individual limpets (Funke 1968). Success in locating 'home' is markedly improved in some species at least by the trails being polarised, and the gastropod can detect the direction in which it was laid down and whether or not it is an outwards or inwards trail (e.g., Cook & Cook 1975). Chitons

FIGURE 7.44   Two examples of patellogastropod home scars. Upper middle figure shows the home scar of the Californian *Lottia scabra* on *Lottia gigantea*, and the left figure shows the exterior of *L. scabra in situ*. (Courtesy of D. Lindberg.) An example of *L. gigantea* with *L. scabra* on its back is shown in the lower middle figure. (Courtesy of National Park Service/Channel Islands.) On the lower right an eroded *Patella caerulea* from Greece with the home scar of a larger individual above. (Courtesy of W. Ponder.)

(*Acanthopleura*) removed from the rock surface and placed on the trail of a conspecific do not return to their home, and only a few (~8%) go to the home of the conspecific individual. Those removed and replaced in the reverse direction on their own trail had ~76% success in returning home (Chelazzi et al. 1987). If a section of the trail is experimentally removed, the animal will search for the trail and follow it again when found (e.g., McFarlane 1980; Chelazzi et al. 1987). An individual can, at least in some species, recognise its own trail (Funke 1968), and many homing species can determine the direction of movement when the trails were formed (Ng et al. 2013).

It is not known what chemicals are involved in the 'olfactory beacon' emanating from home sites, although there are suggestions that it may be particular chemicals released by the resting animal(s). *Achatina fulica* releases a family-specific chemical (Chase et al. 1980), while in others, it may be species-specific. There is no reliable evidence to suggest that any gastropod utilises memorised visual cues when homing.

Many intertidal and terrestrial molluscs do not exhibit homing behaviour. For example, those patellogastropod limpets which return to a specific site on the substratum after feeding excursions are typically territorial species (Branch 1975a; Lindberg 2007b) or are species that live and feed on large brown algae (Branch 1975a; Carlton 1976). Sometimes, there may be multiple 'home sites' within the territory or home range of the limpet (Fisher 1904). Even closely related species exhibit differing homing abilities, as shown with comparisons of three species of the chiton genus *Acanthopleura*, where 'homing performance' differed considerably (Focardi & Chelazzi 1990).

7.15.3.5 Cephalopods

Cephalopod behaviour has been reviewed by Hanlon and Messenger (1996). Their large brains, their ability to learn, and their superior sense organs enable cephalopods to have a remarkable repertoire of behaviours.

With only pinhole eyes, *Nautilus* can make only simple phototactic responses, in marked contrast to coleoids, which use their lens eyes to find and capture prey (Messenger 1968). However, despite their sensory sophistication and large, well-developed eyes, most coleoids are colour blind, although they can discriminate brightness and the plane of polarisation of light (see Section 7.11.1.1.1), and their remarkable body colour and skin pattern changes enable intraspecific signalling/communication and camouflage (e.g., Barbato et al. 2007). Some of these aspects are further discussed below.

Some benthic octopuses exhibit homing behaviour, and experiments have shown that they use the memory of visual landmarks to relocate their 'dens' (e.g., Mather 1991), although no true territorial behaviour has been observed in coleoids (e.g., Hanlon & Messenger 1996).

7.15.3.5.1 Colour Changes and Camouflage

Despite lacking colour vision, by using their chromatophores coleoids can generate rapid and striking colour changes for communication and for camouflage. They rapidly change their appearance using colour, skin, and body morphology responding to visual stimuli and can do this multiple times in a short space of time. They can also use these colour patterns and morphology changes to communicate with conspecifics (Hanlon & Messenger 1996).

Experiments on cuttlefish and octopuses have shown that various background features are assessed before generating a particular pattern. The chromatophores change brightness and pattern to match the background (as determined visually), but the match of the background colour is assisted by the iridophores and leucophores, special skin structures that reflect light from the environment (see below). Tests with spectrometer measurements of the cuttlefish skin and the substratum showed that they were very similar (Mäthger et al. 2008), and these authors argued that cuttlefish chromatophore colours (and presumably those of other coleoids) have evolved to match the colours of their habitats. Cuttlefish display some of the most striking colour patterns of any coleoids and these can be divided into three main categories (Hanlon & Messenger 1988; Barbosa et al. 2008), uniform pattern (or uniformly stippled) with minimal contrast, mottled pattern, with fine to medium contrast distributed fairly uniformly, and disruptive patterns with contrasting dark and light colours. Besides colour patterns, skin projections of various shapes can be displayed. Thus, within each of categories, many variations are possible.

As already noted, skin colour is mainly the product of the chromatophores, but reflectors (leucophores and iridophores) in the skin (see Chapter 17 for details) also play an important role. Leucophores reflect ambient light and matching wavelength and intensity (Froesch & Messenger 1978) and affect how the overlying chromatophores appear, but they may be important in deeper water where the shorter wavelength colours (blue and green) dominate. Iridophores reflect specific wavelengths, and can also alter the appearance of chromatophores (e.g., Mäthger & Hanlon 2007).

Important cues used by cuttlefish while producing their camouflage have experimentally been shown to be contrast and object size. Thus, the area of a light object on a dark background is visually assessed, not its shape or aspect. Substrate edges and contrast, and the interactions of various substrate features, are also important (see Mäthger et al. 2008 for summaries of the literature).

While coleoid camouflage studies are mostly carried out in daylight, Hanlon et al. (2007) reported that the 'Australian giant cuttlefish' *Sepia apama* showed different day and night camouflage body patterns, suggesting that they have good night vision.

Some squid have been shown to split their displays longitudinally so they can display completely differently on their right and left sides (e.g., if a same-sex individual is on one side, and one of the opposite sex on the other) (Byrne et al. 2003).

Antipredator responses of the squid *Sepioteuthis sepioidea* may include specific displays or a fleeing response. These vary according to the level of threat indicating that they can modify an otherwise simple decision based on the perceived threat that a particular fish species presents (Mather 2010).

Other displays that involve intraspecific interactions are described in the next section.

7.15.3.5.2 *Intraspecific Communication in Coleoids*

Among molluscs, only the coleoid cephalopods use vision for the location and selection of mates, and some have elaborate sexual and other intraspecific displays. Body patterns are used to communicate sex and maturity, as well as species identity, the latter being especially important in schooling squid (Hanlon 1982; Sugimoto & Ikeda 2012).

Messenger (1991) and Hanlon and Messenger (1996) emphasised the role played by light in the life and behaviour of cephalopods. Sexual maturation can be related to light intensity via the optic gland (e.g., Nixon 1969; Goff & Daguzan 1991). Many mid-water cephalopods have diel rhythms, presumably influenced by light. Light and temperature may also influence the long migrations undertaken by some species before mating or spawning. Visual cues, perhaps involving sensitivity to polarised light, may aid the navigation of schooling squid, and although this has not yet been experimentally tested, short distance navigation aided by polarised light has been demonstrated in cuttlefish (Cartron et al. 2012). There is also evidence that certain camouflage patterns used by some squid can only be seen by eyes that detect polarised light. These patterns are achieved by the polarised aspect of the iridescent colours from the iridophores beneath the chromatophores (Mäthger & Hanlon 2006). Intraspecific visual communication of cuttlefish using body pattern signals involves a minimal change in achromatic or polarised patterns when observers are females, but the females may use polarised patterns when interacting with male conspecifics. This polarisation patterning may be used to complement the zebra colour patterning sexual display used by both males and females (Boal et al. 2004).

A general similarity in some display patterns (Moynihan 1975, 1985) suggests that they may have evolved in early coleoid evolution, although they are overlain by species-specific displays. The more general patterns are those that are very visual and used in relation to other species, including predators, while intraspecific signalling, including sexual displays, are more species-specific. Moynihan and Rodaniche (1982) suggested that the Caribbean reef squid (*Sepioteuthis sepioidea*) uses patterns as a 'visual language' that evolved because of their schooling behaviour and that their repertoire of signals was greater than other non-human animals. Similarly, squid movements and postures can also provide a means of intraspecific communication (Mather et al. 2010).

7.15.3.5.3 *Correlation of Behavioural Ability and Brain Morphology*

The vertical lobe of the coleoid brain is the centre of learning and memory function, with this converging on the mammalian hippocampus and cerebellum. In size, brains of coleoids lie between those of fish and birds or mammals of similar body size, although they have evolved entirely independently as their common ancestor lacked a brain.

Brain size may be influenced by metabolic rate as in the actively swimming squid *Loligo* which has a relatively larger brain than the benthic *Octopus* (e.g., Packard 1972). Brain size might also be influenced by the amount of information being processed. For example, in gastropod eyes, the number of receptor cells vary from around five in some 'opisthobranchs' to about 4,000 in *Helix*, while in the most complex bivalve eye (*Pecten*) there are as many as 10,000. In comparison, there are about four million receptors in each eye in *Nautilus* and about 20 million in *Octopus*. Thus, the visual areas in the cephalopod brain are much larger than those in gastropods. Each optic lobe in *Octopus* contains about 65 million cells (Young 1963a). Visual learning is correlated with the vertical lobe system, while the great complexity of the coleoid optic lobe is equated with form vision. Other sensory inputs are also considerable – for example, those from the complex statocysts and the arms and suckers. Touch learning in *Octopus* is associated with the subfrontal and inferior frontal lobes and, to a lesser extent, the vertical lobe (Wells 1959; Wells & Young 1975). The brain is also responsible for the highly sophisticated motor systems involved in locomotion, feeding, and blood supply and the control of the many thousands of cells making up the chromatophore system, with the chromatophore lobe size in *Octopus* related to the complexity of the body patterns.

Maddock and Young (1987) studied the brain size of many cephalopods and found that the relative size of the higher centres in the brain showed unexpected patterns which they could not explain. For example, the supraoesophageal lobes, involved in 'higher level processing' including learning and motor control, vary from a quarter to half of the volume of the brain, but the relative size of these centres is not always obviously related to behavioural abilities. While they found values approaching 50% in genera such as *Sepia*, *Loligo* and *Sepioteuthis*, which have sophisticated behaviours, higher values were found in the pelagic, slow-moving, *Spirula* and *Cranchia*, while *Octopus vulgaris*, the subject of so many behavioural and learning experiments, has only 33%. Another centre, the vertical lobe system, varies from 3%–27% of the total brain volume. *Sepia* and loliginids were among the highest (20%–27%), but, once again, the largest vertical lobe was in a deep-water species, *Vampyroteuthis*. In comparison, *Octopus vulgaris* was only 12.6%.

7.15.3.5.4 *Do Cephalopods Have Consciousness?*

Mainly based on studies of octopuses, the idea has been raised (Mather 2008b) that cephalopods have personality, can play, think, and are even conscious. Individual differences in reactions that are strong and enduring enough to be called 'personalities' have been observed in octopuses (Mather & Anderson 1993). Mather (2008b) also suggested that they can form basic concepts – as evidenced in their assessment of complex sensory information and their choice of motor response. Early experiments with shape discrimination showed that octopuses did not apply simple rules (Mather 2008a). Similarly, octopuses have been shown to choose one or several methods to enter their bivalve prey, each using a different effector and prey orientation underneath the arm web, thus not involving vision (Anderson & Mather 2007), and it has been shown that at least one species of octopod can recognise individual humans (Anderson et al. 2010). Using observations such as these, it has been argued that cephalopods have a 'primary consciousness', with the capacity to 'integrate perception and learned information with motivation to make decisions about complex actions' (Mather 2008b, p. 51). The reasoning is that cephalopods, especially octopuses, learn from both developed touch sensitivity (suckers) and vision, and may have a form of 'domain generality' (domain-general learning – a cohesive impression forged from experience) and can form simple concepts. They also know their position (orientation and location in space) coupled with the memory of recent foraging excursions. Thus, Mather (2008a, p. 37) concludes that if 'using a global workspace' which evaluates memory input and focuses attention is the criterion, cephalopods appear to have primary consciousness' (see Box 7.3 for definition).

7.16　LEARNING

The types of learning demonstrated in some molluscs (mainly cephalopods) include:

Classical conditioning (also called *Pavlovian conditioning*) is a form of associative learning where the subject is presented with objects (often food) along with a stimulus. This learning has been used in most experiments with cephalopods and results in acquiring associative memories. It was discovered in the 1950s and 1960s that, using this experimental method, octopuses could learn to visually distinguish shape, orientation, size, and brightness, including various combinations, with the acquired memories retained for, sometimes, months. Blinded octopuses could also distinguish and remember different objects using their suckers. Other senses such as taste can also involve learning through conditioning – for example, cuttlefish have been shown to learn quickly in taste aversion experiments (Darmaillacq et al. 2004), and *Aplysia* can learn to modify its feeding behaviour in response to food edibility (e.g., Susswein et al. 1986).

Imitation learning (or *observational learning*) occurs because of observing and replicating novel behaviour

---

BOX 7.3　LEARNING, MEMORY, INTELLIGENCE, AND CONSCIOUSNESS

Learning is the acquisition of new information and using it to modify subsequent behaviour. This involves *memory*, the ability to recall past experiences. *Explicit memory* is the conscious acquisition of knowledge about individuals, places, and things. It is related to intelligence (see below) and is only found in higher vertebrates. *Implicit memory* is the non-conscious learning of various skills and tasks undertaken by some molluscs. This form of learning can be involved in various activities such as navigational abilities or prey capture.

Memory is essential to learning, it enables the storage and retrieval of the learned information. The experiences retained in memory are a framework in which to associate new information.

Learning and memory are closely interrelated, but the processes underlying them are different. Because of this, learning and memory are dealt with separately below.

Some molluscs have played a fundamental role in our understanding of how these processes operate in animals in general (see Section 7.17).

There are several types of memory:

Associative memory – recalls facts, events, and situations from the past. *Spatial memory* is a type of associative memory and records information about the environment and its spatial orientation.

Non-associative memory – includes two simple forms of learning called habituation and sensitisation. *Habituation* is a decrease or cessation of response to a repeated stimulus. *Sensitisation* is the opposite – increased responsiveness to a repeated stimulus.

Cognition is the assimilation and integration of learned information, while *intelligence* is the assimilation of relevant parts of that knowledge and applying that information.

Primary consciousness is an experience made up of basic perceptions and motor stimuli and is sometimes called perceptual or phenomenal consciousness. This contrasts with *higher order consciousness* in mammals, where primary consciousness is interpreted as including a sense of self.

in another individual, but reports of this in octopuses are somewhat controversial. Fiorito and Scotto (1992) trained octopuses to choose a red or white ball, and then untrained octopuses watched them and were later given a choice. They selected the same ball as the observed group more often and learned more rapidly than the original individuals had under Pavlovian conditioning, but attempts to replicate those experiments have been mixed or negative (e.g., Boal et al. 2000).

Spatial learning is remembering landmarks and/ or distances travelled as a means of navigating in the environment. Coleoids have good navigational abilities using their visual powers, as shown in field observations on *Octopus* (Mather 1991) and in laboratory studies using cuttlefish (Karson et al. 2003) (see also spatial memory below).

Conditional discrimination[14] is a form of complex learning usually associated only with vertebrates. Maze experiments using coleoid cephalopods (octopods and cuttlefish) have shown that they are capable of conditional discrimination. Their ability to navigate through the maze involved discriminating between useful landmarks and their context (context or condition sensitivity) (Hvorecny et al. 2007).

7.16.1 Differences within Coleoids

Octopus and *Sepia* possess the same three neuronal elements within the vertical lobes although there are some anatomical differences in the synaptic organisation. Both also use the same neurotransmitters for fast synaptic conduction. Thus, glutamate facilitates the connection between the amacrine interneurons (AI) and the superior frontal lobe, while the AI connection to the large efferent neurons is cholinergic; memory is localised at different synaptic sites with different neurotransmitters involved. Accordingly in the octopus short- and long-term synaptic plasticity is at the glutamatergic synaptic connection, but in cuttlefish, it occurs at the cholinergic synaptic connection (Shomrat et al. 2011, 2015). Given the lack of detailed comparative studies on other coleoid taxa, it is unclear as to the evolutionary significance of these differences which may be related either to different lifestyles or to the independent evolution of advanced learning and memory systems (Shomrat et al. 2015).

7.16.2 Memory

Both sensitisation and habituation are important behaviours for survival. There have been many experiments using *Aplysia* to demonstrate both of these non-associative memory mechanisms. As an example, a reflex involving local contraction of the body is caused by a nip from a crab and learning a rapid response gives *Aplysia* a better chance of survival. *Aplysia*

has ten major ganglia which, in *Aplysia californica*, contain about 20,000 nerve cells (Scheller 1985). Many of the neurons are large and can be seen with the naked eye. This enables the identification and monitoring of changes in the individual nerve cells involved in a behaviour and in memory acquisition. Similarly, the hygrophilan snail *Lymnaea*, which also has large neurons, can be shown to acquire short-, medium-, and long-term memories in response to conditioning or sensitisation experiments (e.g., Benjamin & Kemenes 2010).

In sensitisation experiments, as in the famous Pavlovian conditioning, the animal is taught to respond strongly to an otherwise neutral stimulus, although for this sensitisation to succeed, a previous negative stimulus (such as an electric shock) must be remembered. How long that memory lasts is related to the number of repetitions of the negative stimulus. While a single shock results in a memory that lasts minutes, multiple shocks result in longer-lasting memory. Thus, in gastropods like *Aplysia*, learning, as in vertebrates, consists of very short-term transient memory and a longer-term memory.

The gill and siphon withdrawal reflex in *Aplysia* is an involuntary, defensive reflex involving direct synaptic connections between sensory and motor neurons. Eric Kandel and his co-workers (e.g., Kandel 1979) found this reflex could be modified by three types of learning: habituation, sensitisation, and classical conditioning. Similar experiments with *Lymnaea* involve feeding and respiration responses (e.g., Benjamin & Kemenes 2010) (see below). For example, in *Aplysia*, a light tactile stimulus that would not normally produce a defensive reflex is repeatedly applied along with a strong noxious stimulus (usually an electric shock) and, after a few treatments, the animal shows a defensive reflex to the light touch. Frost and Kandel (1995) found that a single stimulus to the tail resulted in short-term sensitisation, while repeated stimulation produced long-term sensitisation. Thus, following sensitisation, the animals react more strongly to a noxious stimulus, whereas following habituation they learn to ignore a neutral stimulus. Both forms of learning give rise to a short-term memory lasting minutes to hours and a long-term form lasting days to weeks (Kandel 2001). The memory for sensitisation results in an increased strength of the connection between the sensory and motor neurons, while habituation memory reflects a decrease. Bailey and Chen (1983) first showed that long-term sensitisation resulted in an increased number of synapses in *Aplysia* in each sensory neuron and these persisted as long as the gill-withdrawal reflex lasted. As the memory decays, the new synaptic terminals are lost. Conversely, after long-term habituation, the number of synapses between sensory and motor neurons decreases. Such changes are facilitated by several neurotransmitters such as serotonin (5-HT). Thus, memories are stored by structural changes in synaptic connections in individual neurons.

Our early knowledge of the cellular processes involved in memory was in large part a result of studies on *Aplysia*. They clarified the role of individual synapses in learning and memory and provided a model for vertebrate memory systems. Details of the chemical and other processes involved were shown in later studies on *Aplysia* (Manseau et al. 1998; Manseau et al. 2001; Barbas et al. 2005; Marinesco et al. 2006; and many

---

[14] A discrimination in which two stimuli are presented, and the correct stimulus is determined based on which of the two stimuli is present or was presented recently (Bouton 2007).

others). Serotonergic neurons are activated during this sensitisation training, and serotonin is released, inducing either short- or long-term memories (for detailed reviews of the biochemistry involved see Barbas et al. 2003; Hawkins et al. 2006).

In long-term memory memory (days or weeks), the new new synaptic connections form in a process involving altered gene expression and new protein and mRNA synthesis. Short-term memory involves modification of proteins which then modify synaptic connections. This is a process where phosphates are added to proteins (thus modifying them) to make existing nerve connections stronger. This is achieved by forming pores for ions involved in nerve signals to pass through the nerve cell membrane allowing future signals to pass more easily. Long-term memory involves the activation of gene expression and synthesis of new protein for building new connections or extending old ones. Stimuli causing short-term changes prepare the synapse for later long-term change, but those changes will not occur without the protein synthesis required to make the structural changes for the memory to persist. Some of the receptors involved have been characterised (e.g., Hai 2006), and the reader is referred to this and other recent studies for details on the biochemistry underlying memory acquisition.

Relatively few memory experiments have been carried out on gastropods other than *Aplysia*. The fresh-water hygrophilan *Lymnaea* has been the subject of some studies. For example, Lukowiak et al. (1996) showed that aerial respiratory activity could be conditioned with the animals being taught not to use their pneumostome, but instead to respire entirely via the skin. These memories lasted for about a month with the length of retention depending on the length and intensity of the training (Lukowiak et al. 2000). More recently, Lukowiak (2017) has argued the importance of *Lymnaea stagnalis* as a model system for studying the causal neuronal mechanisms which form long-term memory. Other studies have shown that feeding responses in *Lymnaea* can be conditioned to respond to chemical, touch, and visual stimuli (e.g., Whelan & McCrohan 1996; Staras et al. 1999; Andrew & Savage 2000; Benjamin & Kemenes 2010; Ito et al. 2013; Spencer et al. 2017). Such feeding-related experiments have also been carried out on various gastropods including *Aplysia* and other 'opisthobranchs' (*Hermissenda* and *Pleurobranchaea*) and some terrestrial slugs and snails (see Chase 2002 for review). For example, *Achatina fulica* was shown to develop long-term memory for particular food odours (Croll & Chase 1977).

An extensive body of work has shown that the large brains of coleoid cephalopods contain dedicated learning and memory centres (for reviews, see Mather 1995; Hanlon & Messenger 1996; Hochner et al. 2006; Alves et al. 2007; Mather & Kuba 2013) with the vertical and subfrontal lobe complexes the centres of visual and tactile memory, respectively (e.g., Boycott & Young 1955a; Maddock & Young 1987; Young 1991; Hochner et al. 2006). The relatively simple brain of *Nautilus* lacks specialised regions dedicated to learning and memory (Young 1965a), but Crook and Basil (2008) demonstrated that *Nautilus* is capable of both short- and long-term memory, the latter for up to 24 hours, despite lacking these centres. The memory profile is similar to that of *Sepia*

(Messenger 1973), although the duration of long-term memory is much shorter in *Nautilus*. Thus the absence of these dedicated memory centres does not seem to limit the ability of *Nautilus* to perform simple cognitive tasks (Crook & Basil 2008; Crook et al. 2009; Basil & Crook 2017).

As already noted, learning in coleoids is mainly concentrated in the vertical and superior frontal lobes. The vertical lobe is the main centre of learning and memory as these abilities can be impaired by lesions to this part of the brain, as shown in experiments on *Octopus* (e.g., Fiorito & Chichery 1995; Shomrat et al. 2008). These experiments show that the learning and memory systems of *Octopus* are separated into short- and long-term memory sites, as in mammals (Shomrat et al. 2008). These two memory sites are not independent because acquiring long-term memory is mediated by the vertical lobe, which is also involved with modulating short-term learning and behaviour.

Learning in many animals is often associated with social habits, but octopuses are generally solitary animals that do not rear their young, so parental influence does not occur. In contrast, schooling squid exhibit social behaviour, but there is surprisingly little detail available on their learning abilities (Hanlon & Messenger 1996; Boal 2006; Ikeda 2009).

Nitric oxide is important for memory formation and hence learning, where its role is to initiate various essential pathways (second-messenger cascades) required for both short- and long-term memory (reviewed by Susswein et al. 2004). For example, blocking the synthesis of NO will stop *Aplysia* learning that a particular food is inedible (Katzoff et al. 2006).

7.16.3 Cognition and Intelligence in Cephalopods

Much of the complex behaviour of cephalopods involves memory and some form of decision-making (Hanlon & Messenger 1996), including using learned information to assess threats, navigation, locating and capturing prey, finding and signalling mates, and interactions with conspecific individuals of the same sex.

Depending on the definition used, some coleoid cephalopods might be considered intelligent. Despite numerous references on the web and in non-scientific literature to the intelligence of cephalopods, there are very few in the scientific literature. Statements such as 'the octopus, an extremely successful and highly intelligent cephalopod' (Scheibel & Schopf 1997, p. xvi) are not backed up with convincing arguments. There is, however, no doubt that cephalopods are capable of cognition (see Box 7.3), and interested readers are referred to two recent publications on the subject (Darmaillacq et al. 2014; Gutnick et al. 2017).

7.17 THE MOLLUSCAN NERVOUS SYSTEM IN EXPERIMENTAL PHYSIOLOGY

Behaviour can be simple to complex, although complex behaviours can be compounded with simple behaviours, many of which are encoded genetically but can still be modified through learning.

The sea hare *Aplysia*, the hygrophilan *Lymnaea*, the stylommatophoran slug *Lehmannia* (often referred as *Limax*), octopuses, squid, and cuttlefish have been used as model organisms to study neural biology and physiology including such phenomena as circadian rhythms, behaviour, learning, and memory. *Aplysia* is used by neurobiologists because of its large neurons – a soma of a motor neuron can be up to 1 mm in diameter, making it relatively easy to study their physiological responses, and thus how they accomplish learning. *Aplysia* is also used in operant conditioning, where a spontaneous behaviour is modified by reinforcement. Thus, for example, aspects of feeding behaviour can be experimentally modified by taste (gustatory) reinforcement via sensory receptors in the buccal cavity that transmit taste stimuli to the buccal ganglion via the buccal nerve. Like *Aplysia*, *Lymnaea* has a well understood and relatively simple neuronal network that mediates tractable behaviours and possesses long-term memory.

Contributions of fundamental general importance have been made by scientists using molluscan nervous systems as models. These mainly involve work on some coleoid cephalopods (squid, octopuses, and cuttlefish) and some euthyneuran gastropods, notably the sea hare *Aplysia*, the pond snail *Lymnaea stagnalis*, some land snails such as *Helix pomatia* and *Achatina fulica*, and some 'opisthobranchs' such as *Bulla*, *Navanax*, *Pleurobranchaea*, and *Hermissenda*. Their findings include such famous examples as:

1. Electrophysiological experiments using the giant nerve fibres in squid that run from the stellate ganglia were carried out from the late 1930s by Alan Hodgkin, Andrew Huxley, and their colleagues. Their work was the basis of our understanding of nerve transmission, and they received the Nobel Prize in Physiology or Medicine in 1963 (along with Sir John Eccles) 'for their discoveries concerning the ionic mechanisms involved in excitation and inhibition in the peripheral and central portions of the nerve cell membrane'.

2. In the 1960s, Sir Bernard Katz and Ricardo Miledi and their colleagues first discovered the mechanisms of how nerve synapses function by using the squid giant synapse. This work showed the role that calcium plays in the release of neurotransmitters. Katz shared the Nobel Prize for Physiology or Medicine in 1970 with Ulf von Euler and Julius Axelrod 'for their discoveries concerning the humoral transmitters in the nerve terminals and the mechanism for their storage, release and inactivation'.

3. Ruth Hubbard and Robert St. George published pioneering work on the rhodopsin system of the squid in 1958, and this work was carried on by others (see Messenger 1991 for review), which was relevant to our current understanding of phototransduction.

4. The cellular processes resulting in memory were largely worked out on studies on *Aplysia* by Eric Kandel and his colleagues in the 1990s. Kandel was awarded the Nobel Prize in Physiology or Medicine in 2000 along with two others (A. Carlsson and P. Greengard). More recent studies that related electrical and molecular properties of neurons to behaviour and to memory formation in *Lymnaea* have also contributed.

5. The cellular mechanisms in pacemakers through studying those associated with the eye in *Bulla* and *Aplysia*.

These and many other findings have not only contributed greatly to our understanding of molluscan biology and physiology, but have contributed in important ways to neurobiology and neurophysiology in general.

8 Reproduction and Development

8.1 GENERAL INTRODUCTION

Molluscan reproductive strategies are some of the most diverse in the Metazoa, and their varied reproductive modes are a rich field of inquiry for ecological and evolutionary studies. Most molluscan clades are *gonochoristic*, with gonochorism being both plesiomorphic and dominant; gender is genetically determined and does not change during the lifetime of the individual. *External fertilisation* is plesiomorphic, although *internal fertilisation* has evolved independently in the solenogasters and cephalopods, and in several groups of gastropods. *Hermaphroditism* is also broadly distributed among molluscs and includes both sequential (*protandrous* or *progynous*) and *simultaneous* conditions. In some groups such as the Heterobranchia, hermaphroditism appears to have evolved once in the common ancestor, while in other groups, it is found haphazardly distributed among species (e.g., venerid bivalves). Molluscs contain both *iteroparous* (most molluscs) and *semelparous* taxa (e.g., most coleoid cephalopods and some euopisthobranchs). *Asexual* reproduction occurs in some gastropods (*parthenogenetic*) and a few bivalves (both parthenogenetic and *androgenic*) (Box 8.1).

In externally fertilising aquatic species, the gametes are typically released into the water column where fertilisation occurs. This type of spawning requires high fecundity to ensure fertilisation and reproductive success. Numerous behaviours have evolved which increase fertilisation probability, including aggregation, synchronised spawning, pseudo-copulation, and both external and internal egg retention. The spawn may also be placed on the substratum in a compact mass and fertilised either by sperm present in the water column or by males directly interacting with the egg mass. Pseudo-copulation occurs when the male and female are in very close proximity, enabling direct sperm transmission. Spermatophores are present in several groups, and they serve to increase fertilisation probability by delivering a concentrated packet of sperm that can be released near the eggs. Egg capsules are found in some gastropods and cephalopods, but this reproductive mode requires copulation, followed by internal fertilisation, to encapsulate one or more fertilised eggs before the deposition of the capsule. Encapsulation also enables the provision of additional nutrients for developing embryos, as well as providing protection and potentially shortening a pelagic development phase or the retention of the larvae until they can emerge as 'crawl-away' juveniles.

Development in the Mollusca is mosaic; that is, the blastomeres have a specific spatial orientation to one another, limited developmental potency, and early specification of cell fate in the developing embryo. Cell cleavage pattern is spiral and typically dextral. Molluscs are protostomes, and the developing blastula undergoes gastrulation, plesiomorphically producing a trochophore larval stage with its distinctive bands of locomotory cilia – the *prototroch*. Three types of blastulas are found in molluscs: the *coeloblastula* in polyplacophorans, bivalves, scaphopods, patellogastropods, and vetigastropods, the *placula-like larvae* in caenogastropods and heterobranchs, and the *stereoblastula* in cephalopods, gastropods, and bivalves with yolk-rich eggs (Verdonk & Cather 1983). These three types differ primarily in the shape of the blastocoel. In the placula type, the blastocoel forms a narrow slit, in the coeloblastula, the blastocoel forms a wide, broad cavity, and in the stereoblastula (sometimes called the *discoblastula* in cephalopods), the blastocoel is almost non-existent, being represented by the space between the small micromeres and the large yolk-filled macromeres. The pattern of early development in molluscs is characteristic across the entire group, although it can be highly modified in taxa such as cephalopods that have exceedingly yolky eggs (see Chapter 17). In gastropods and bivalves, a later larval stage, the veliger, is also present. The veliger is formed by further growth and development of the prototroch into the wing-like velum, providing enhanced locomotion for the larvae. In planktotrophic taxa, the velum has been co-opted to capture food. Such feeding larvae are found only in the more derived gastropods and in autobranch bivalves, other molluscan larvae being lecithotrophic and sustained by yolk reserves during their pelagic phase. Nutritive dissolved organic material (DOM) may also be taken up from the surrounding water by the larvae.

Some molluscs abandon pelagic development and undergo what is usually termed direct development, in which the dispersal phase is omitted because the embryos develop on or within the body of the female (brooding) or within an egg mass or capsule (see also Section 8.10.1). In gastropods and bivalve taxa which undergo direct development, the typical veliger phase may be present, truncated, or absent.

Compared to pelagic taxa, the dispersal capability of direct-developing taxa is often limited. While this correlation appears to be true in ecological time (Hadfield & Strathmann 1996; Hoskin 1997; Kyle & Boulding 1998), at deeper, evolutionary time scales there appears to be little difference in dispersal potential between pelagic and non-pelagic taxa (Ó Foighil 1988; Vermeij et al. 1990; Marshall et al. 2012).

Regardless of the mode of development, most molluscs undergo a transition from the larval stage to the juvenile or adult stage. In pelagic developing taxa, this transition is marked by changes in shell morphology and reflects the change in habitat as the larval animal typically moves from the water column to the benthos. In direct-developing taxa that crawl away from the refuge of their mother or spawn, this change is also often reflected in a marked change in shell morphology (e.g., shape, sculpture, or colour). Besides the shell and ecological changes that accompany settlement,

---

BOX 8.1   SOME BASIC REPRODUCTIVE TERMINOLOGY

Gonochoristic – one individual has one sex, either male or female.

Hermaphroditic – the same individual has both sexes. *Simultaneous hermaphrodites* produce gametes of both sexes at the same time. *Protandrous hermaphrodites* are male first and female second, while *protogynous hermaphrodites* are female first and male second. In most taxa this change occurs once, in others the sequence may repeat multiple times. Sex changing hermaphrodites are also sometimes referred to as *consecutive hermaphrodites*.

Asexual reproduction – occurs when the offspring inherit a copy of the genome from only one parent and because the genome is unmodified by recombination, the offspring are clones.

Parthenogenetic – an asexual mode of reproduction in which unfertilised eggs develop into offspring.

Androgenic (= *paternal apomixis*) – the nuclear chromosomes in the embryo are entirely paternal and derived from one or more sperm nuclei.

Semelparous – the offspring are produced in a single reproductive event followed by the death of parents. Typically the lifespan is annual.

Iteroparous – offspring are produced in several reproductive events and lifespans are usually several years.

External fertilisation – fertilisation occurs outside the body of the animal with reproductive interactions being between sperm and eggs. Fertilisation success rate is typically lower than with internal fertilisation.

Internal fertilisation – fertilisation occurs within the body of the animal, and the reproductive interactions are between individual animals. Fertilisation success rate is typically higher than with external fertilisation.

---

internal anatomical changes also accompany this transition, and the juvenile quickly enters a period of often rapid growth. Such growth is commonly indeterminate (continuous growth until death) or may be determinate (growth terminated at a particular time or size), the latter often marked by distinctive shell features, for example, in gastropods, reflected, flared, or thickened shell apertures (Vermeij & Signor 1992) (see Chapter 3).

All characteristics and generalities discussed above have their exceptions and amendments which are developed and discussed in more detail below. The diverse array of reproductive strategies and modes in the Mollusca has attracted many researchers beginning with Aristotle in 350 BCE and continue to be explored in the context of evolutionary developmental biology (e.g., Perry et al. 2015). Because some molluscs (e.g., bivalves, cephalopods, and gastropods) preserve their egg size and subsequent ontogeny in their shells, some reproductive parameters of those molluscs are discernible in the fossil record (e.g., Jablonski & Lutz 1983; Nützel 2014).

8.2   REPRODUCTIVE ANATOMY

The molluscan reproductive system consists of a gonad for gamete production and ducting to transfer the gametes to the environment or another individual. The contact between these two components is typically well defined histologically – the former being derived from mesodermal tissue and the latter from ectodermal tissues. The gonad is part of the coelomic system, which also includes the pericardium and kidney. The reproductive system may be integrated with these other systems, most commonly the excretory system (see also Chapter 6). Thus, in some taxa, the kidney is incorporated into the reproductive system or, much more rarely, the pericardium.

In most groups there is also an associated ectodermal component that makes up the glands, ducts, and pouches of the distal part of the reproductive system. This ectodermal section shows the greatest modification and diversity, especially in gastropods and to a lesser extent in cephalopods.

The general characteristics and configuration of molluscan reproductive systems are summarised below and in Figure 8.1. More details on each group are provided in the taxon chapters in Volume 2.

8.2.1   POLYPLACOPHORA

Although thought to be one of the most plesiomorphic groups of molluscs, the polyplacophoran (chiton) reproductive systems do not incorporate the excretory organs. Polyplacophorans typically have a single fused median gonad with bilateral gonoducts that open into the mantle grooves. In a few species, the gonads may be only partially fused or, in some Chitonida, still paired (Andrews 1979b). Chitons are mostly gonochoristic broadcast spawners, but a few are simultaneous hermaphrodites (Eernisse 1986), and several mantle groove brooders are known (Pearse 1979). The gonoduct in females differs from that of males in being lined with secretory cells referred to as shell glands, and these secrete the often elaborate egg hulls that form a protective coat on the unfertilised eggs (see Section 8.3.1.3 and Chapter 14, Figure 14.6).

8.2.2   CAUDOFOVEATA AND SOLENOGASTRES

Aplacophoran reproductive systems are the least differentiated in molluscs. In a departure from the configuration seen in all other molluscs, in both caudofoveates and solenogasters the gonad(s) open directly into the pericardium. Typically, the gonads are paired in solenogasters and juvenile caudofoveates,

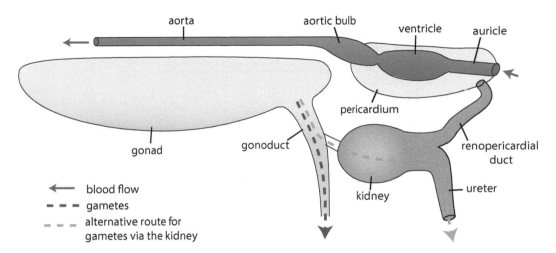

FIGURE 8.1 Generalised molluscan reproductive system. Reproductive system in purple, circulatory system in red, and excretory system in brown. Redrawn and modified from Mizzaro-Wimmer, M. and Salvini-Plawen, L.v., *Praktische Malakologie. Beiträge zur vergleichend-anatomischen Bearbeitung der Mollusken: Caudofoveata bis Gastropoda—Streptoneura*, Springer, Vienna, Austria, 2001.

although they are fused in adult caudofoveates (Hadfield 1979). After traversing the pericardium, the gametes travel through the coelomoducts to the mantle cavity. In solenogasters, the paired coelomoducts fuse before entering the mantle cavity, and they have seminal receptacles at the junction of the proximal and distal portions of the coelomoducts and paired copulatory spicule glands. Caudofoveates are broadcast spawners, with external fertilisation (Nielsen et al. 2007). In contrast, solenogasters are simultaneous hermaphrodites with internal fertilisation and show a diversity of spawning strategies, including full pelagic development, benthic egg masses, and complete and partial brooding (Hadfield 1979; Okusu 2002).

8.2.3 MONOPLACOPHORA

Monoplacophorans have between one (in tiny species) and three (in large species) pairs of gonads. These are connected to the kidneys through a gonoduct and the gametes pass through the kidneys, which in turn open via kidney openings into the mantle groove. The gonads may be discrete or diffuse, sometimes interpolating into the digestive gland or extending into the lateral mantle regions. Sexes are separate in larger species and they are thought to be broadcast spawners, while one of the smallest taxa, *Micropilina arntzi*, is a hermaphroditic brooder (Haszprunar & Schaefer 1996).

8.2.4 BIVALVIA

Bivalve reproductive system structures are conservative. The gonads are typically paired, but may be fused in some taxa, and open into the mantle cavity through either shared kidney ducts or separate gonoducts (Morse & Zardus 1997) (Chapter 15, Figure 15.27). Bivalves may be gonochoristic or hermaphroditic, and many smaller taxa brood their embryos (Sastry 1979). Fertilisation is typically external; however, spermatophore-like sperm balls and spermatozeugmata

(see Section 8.3.1.2) which facilitate internal fertilisation have been described in a few taxa.

8.2.5 SCAPHOPODA

Scaphopods, like bivalves, are conservative in their reproductive anatomy. A single gonad connects to the right kidney, and the gametes are released into the water for external fertilisation through the kidney aperture (Steiner 1992). Sexes are usually separate, although rare hermaphrodites have been reported (McFadien-Carter 1979).

8.2.6 CEPHALOPODA

In cephalopods, there is a single gonad and the sexes are separate (Budelmann et al. 1997). In males (Figure 8.2), the gonad opens in close juxtaposition with the 'accessory organ' in which the spermatophores are formed. Sperm is transferred from the testis to the accessory organ via a vas deferens. The accessory organ is divided into distinct regions and includes the *spermatophoric organ*, where sperm are packaged into spermatophores, and the *spermatophoric* or *Needham's sac* where they are stored. In *Nautilus*, the spermatophoric sac leads directly to a penis (Arnold 1984), and there are other secondary sexual structures derived from modified arms – the spadix and anti-spadix – that function in spermatophore transfer during copulation. In coleoid cephalopods, a modified arm (the hectocotylus) is used to transfer spermatophores from males to females. The female system (Figure 8.2) consists of a single ovary that connects through one or two oviducts with oviducal and paired nidamental glands involved in capsule and spawn formation. Accessory nidamental glands are present in some taxa and host a consortium of bacteria (Collins et al. 2012a). In some squid, the buccal pouches serve as seminal receptacles for spermatophores, while in other taxa, the spermatophores are stored on the single oviduct (Arnold &

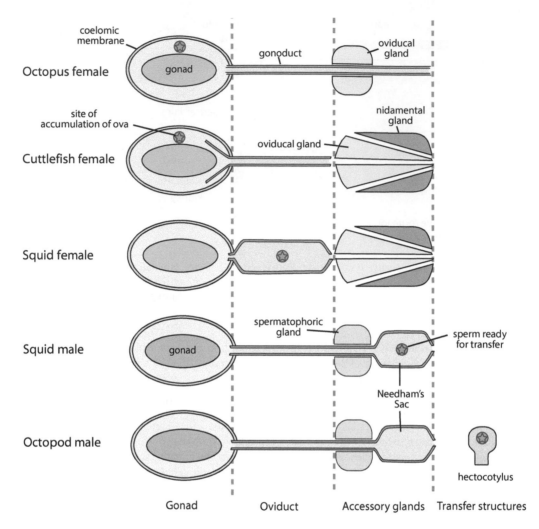

FIGURE 8.2 Schematic representation of the structure of male and female reproductive systems in coleoid cephalopods. Redrawn and modified from Arkhipkin, A.I., *J. Northwest Atl. Fish. Sci.*, 12, 65–74, 1992.

Williams-Arnold 1977). Fertilisation may be external or internal and development is direct, taking place within the egg capsules. Egg capsules may be unattended or brooded, and young are released as either 'crawl-away' juveniles or, in many octopods and squid, into the water column as swimming '*paralarvae*'. The paralarvae are not a larval form, but instead a specialised planktonic juvenile stage (see Chapter 17).

8.2.7 GASTROPODA

Gastropod reproductive systems are highly diverse, with the greatest diversity in a single clade – the Heterobranchia (Gosliner 1994; Voltzow 1994; Luchtel et al. 1997). All gastropods possess a single post-torsional right gonad. In patellogastropods and most vetigastropods, the gonad opens either into the right kidney or the renopericardial duct between the pericardium and kidney, and gametes pass through the kidney opening into the mantle cavity. Exceptions to this basic configuration are found in some neomphaloideans which are associated with chemosynthetic habitats and show modifications for both internal fertilisation

and provisioning of the eggs with nutritive and protective membranes (Fretter 1989).

Reproductive system morphology in patellogastropods and vetigastropods resembles the plesiomorphic configuration in other molluscan classes (Figure 8.1), except that a separate gonoduct is absent. In the Neritimorpha, Caenogastropoda, and Heterobranchia, the reproductive systems are much more elaborate. Caenogastropods and neritimorphs have a rudiment of the right kidney incorporated in the oviduct, but there is no indication of this in heterobranchs (Fretter & Graham 1964; Webber 1977; Ponder & Lindberg 1997). In some neritimorphs, the female system may have up to three separate openings into the mantle cavity, and along with the other higher gastropod groups, there are large glands associated with the ectodermal portion of the female reproductive system, including albumen and capsule glands, as well as a bursa copulatrix and a seminal receptacle (see Chapter 18). In caenogastropods (Chapter 19), there is typically a single opening in the female from which the eggs are released, and it also receives sperm during copulation. In a few truncatelloideans these two functions are separated, with copulation via the kidney opening (Ponder 1988a).

In both neritimorphs and caenogastropods, the albumen gland provides nutrition and the capsule gland secretes the coating that encapsulates the eggs. The sperm sacs include the seminal receptacle where sperm is stored and the bursa copulatrix which receives sperm, the latter usually involved in resorbing excess sperm or its secretions and breaking down spermatophores (e.g., Eales 1921; Kunigelis & Saleuddin 1986; Fretter & Graham 1994). The homology of these different structures (i.e., seminal receptacle, bursa copulatrix, albumen, and capsule glands) between higher gastropod taxa is unlikely, but they are probably homologous within subclades.

Male systems in the Neritimorpha, Caenogastropoda, and Heterobranchia are independent of the right kidney, and a prostate gland and penis are typically present.

Broadcast spawning is common in both patellogastropods and vetigastropods, whereas copulation and internal fertilisation is the rule in neritimorphs, caenogastropods, and heterobranchs. Brooding occurs in all groups, as does both direct and indirect development. In taxa with encapsulated development, the young may be released either as 'crawl-away' juveniles or as veliger larvae into the water column. While the majority of patellogastropods, vetigastropods, neritimorphs, and caenogastropods are gonochoristic, hermaphroditism occurs in some members of these groups as well, but is an apomorphy for heterobranchs.

8.2.7.1 Heterobranchs

Unlike the great majority of other gastropods, heterobranchs are nearly all simultaneous hermaphrodites with a single ovotestis. Their great diversity in reproductive anatomy is primarily expressed in the secondary sexual characteristics of additional glands, ducting, chambers, and copulatory structures (Gosliner 1994; Luchtel et al. 1997). For example, in nudibranchs such as *Doris* (Figure 8.3), there are separate openings and ducts for eggs,

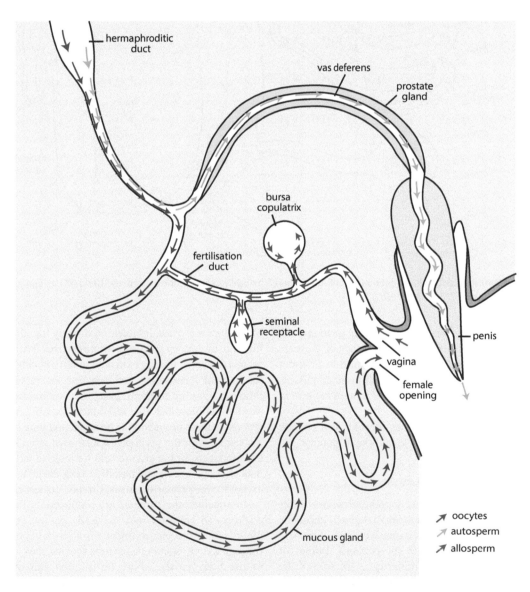

FIGURE 8.3 Diagrammatic reproductive anatomy of the nudibranch *Doris pseudoargus*. Redrawn and modified from Thompson, T.E., *Biology of Opisthobranch Molluscs*, Ray Society, London, UK, 1976.

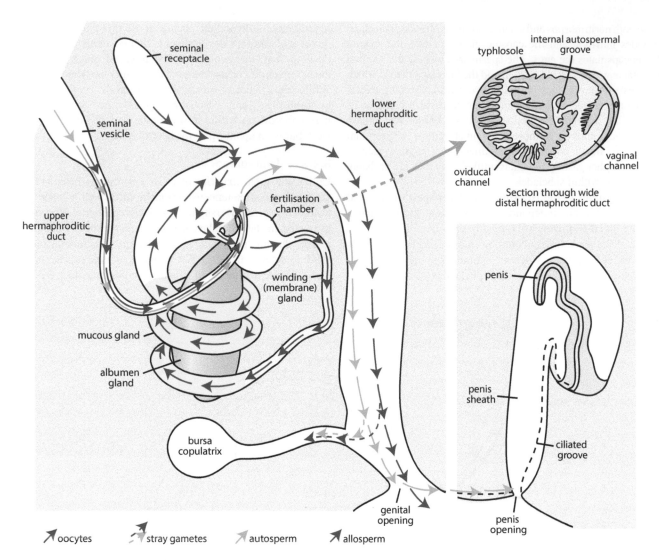

FIGURE 8.4 Diagrammatic reproductive anatomy of an aplysiid heterobranch. Redrawn and modified from Thompson, T.E. and Bebbington, A., *Malacologia*, 7, 347–380, 1969.

exosperm (sperm from a partner), and *autosperm* (own sperm), while in the aplysiids (Figure 8.4), all three kinds of gametes utilise much of the same ducting, but are confined to different channels within the ducts. Both the examples illustrated possess seminal receptacles for exosperm storage and a bursa copulatrix for sperm storage and digestion of excess sperm. The eggs receive a coating of mucus after fertilisation, but aplysiids also provide albumen for development from a distinct additional gland.

8.2.8 COMPARISONS WITH OUTGROUPS

The reproductive systems of all other lophotrochozoans differ substantially from those of molluscs. Most lophotrochozoans lack distinct gonads and instead have transient sites of gamete production, typically associated with the coelomic linings of the body cavity. The only major similarity is the role of the kidney openings in the release of the gametes from the body. In brachiopods and phoronids, the gametes are produced by transient gonads that form as thickened areas of the peritoneum.

The gametes are usually released directly into the environment (through the kidney opening), where external fertilisation occurs, or, in a few taxa, the eggs are held internally in brood chambers for internal fertilisation and brooded for varying lengths of time before being released. Ectoprocts reproduce primarily by asexual budding, but also reproduce sexually. As in the other groups, gamete production is transient and derived from tissues associated with the peritoneum. Gametes are released through coelomoducts, or the ova are held for internal fertilisation. Both gonochoristic and hermaphroditic taxa occur in these groups, except for the primarily hermaphroditic ectoprocts.

In annelids, the gametes are produced on the walls of the coelom and released into the body cavity, and from there directly into the environment through the coelomoducts, or through kidney openings. In some species, they are maintained in the body cavity where fertilisation can occur, and the zygotes are either released or brooded. In sipunculans, they are selectively removed from the coelomic fluid by the kidneys and stored until being free spawned into the water column.

8.3 THE MOLLUSCAN GONAD

The discrete gonad is a significant molluscan character compared to the transient nature of gamete-producing tissue in the outgroups. The molluscan gonad is coelomic (mesodermal) in origin and initially was a paired structure. It still shows varying degrees of midline fusion in aplacophorans, polyplacophorans, and bivalves; but only a single gonad remains in scaphopods, cephalopods, and gastropods. The position of the gonad within the visceral mass varies. It is generally dorsal to the stomach and therefore occupies the highest position within the visceral mass, as aplacophorans, polyplacophorans, protobranch bivalves, cephalopods, scaphopods, and most gastropods all display this topological relationship. The reversed position occurs in monoplacophorans, autobranch bivalves, patellogastropods, and *Cocculina*, with the gonad ventral to the digestive system. Except for bivalves, these are limpet-like and secondarily flattened. In some vetigastropod limpets, the gonad is arguably intermediate in position, being located more posteriorly, but it is still mostly dorsal to the alimentary system.

8.3.1 GAMETES AND GAMETOGENESIS

Molluscs are plesiomorphically gonochoristic, and the great majority reproduce sexually from the fertilisation of eggs by sperm, with only a few gastropods and bivalves able to reproduce asexually (see Section 8.4.3). Molluscan sperm and eggs are produced by meiosis in the gonads (Figure 8.11) and the gametes have a haploid number of chromosomes (1n) – half the number of chromosomes found in the original parent cell (2n), and with *crossing over*, each gamete is genetically different. For more detail on molluscan chromosomes, see Chapter 2.

8.3.1.1 Spermatozoa

Schmidt-Rhaesa (2007) recognised three general sperm forms – round-headed (*aquasperm*), filiform (*introsperm*), and ancillary sperm (*spermatozeugmata*) (see Section 8.3.1.2), this latter type lacks a cilium (flagellum). These different types appear to be correlated with the mode of fertilisation (Franzén 1955; Rouse & Jamieson 1987), and several synonymous terms for sperm types have resulted from different taxonomies emphasising either functional, ecological, or morphological criteria. Externally fertilising aquasperm are plesiomorphic in Mollusca (Figure 8.5) and are present in all putative outgroups (Schmidt-Rhaesa 2007). Franzén and Rice (1988) considered spermatogenesis to be determined by ancestry and fertilisation biology, thus having both a phylogenetic and a functional component. Where it has been investigated, convergence in sperm morphology appears to be due to functional selection. This includes the overall similarity of sperm morphology in taxa with internal fertilisation, such as for example, filiform gastropod and solenogaster sperm (Buckland-Nicks & Scheltema 1995). Although it has long been known that testes and their primary germ cells are derived from stem cells of the coelomic pericardial epithelium (Raven 1966), the genetic

components of testes formation and spermatogenesis have remained obscure. Klinbunga et al. (2009) and Amparyup et al. (2010) have reported suites of genes linked to sex determination and gonad development in *Haliotis*, and further studies, driven in part by aquaculture interests, are likely to extend our knowledge of the underlying genetics associated with spermatogenesis in molluscs. Spermatogenesis and sperm function in molluscs is further reviewed by Maxwell (1983).

8.3.1.1.1 Eupyrene (Normal Chromatin)

Molluscan sperm have three distinct regions – the head, midpiece, and tail (or axoneme). In aquasperm the short rounded head consists of a nucleus and acrosome, the midpiece has four mitochondria arranged in a circle below the nucleus, and the tail has a cilium (or flagellum) about 50 μm long (Figure 8.6). Aquasperm can be further divided into *ectaquasperm* for external fertilising species and *entaquasperm* for species where the sperm fertilise eggs held within a space such as the mantle cavity. Aquatic sperm are thought to be adapted for efficient swimming in an aquatic environment.

In molluscs with internal fertilisation, sperm morphology has converged to produce introsperm (Figure 8.7). Introsperm have elongated heads and midpieces along which the mitochondrial spheres are condensed as sheaths around the axial filament. It has long been recognised that, with a change from external to internal fertilisation, sperm changes from aquasperm to introsperm both within and between taxa. The rearrangement and the increase in the number of mitochondria along the axoneme has been suggested to be an adaptation to provide more energy for movement through the higher viscosity found within reproductive tracts. Both aquasperm and introsperm also occur in annelids.

The sperm acrosome complex typically forms in early spermatids and is produced by the Golgi apparatus. It forms above the sperm nucleus and facilitates fertilisation by penetrating the outer layers of the egg (e.g., the vitelline membrane) (see 8.5 below). The acrosome complex consists of two parts: the membrane-bound acrosomal vesicle and the underlying subacrosomal material. In octopods, the acrosome is screw-shaped. In the anomalodesmatan bivalve *Laternula*, a Golgi apparatus forms a temporary acrosome and then migrates posteriorly as the sperm develops, leaving no acrosome complex in the mature sperm (Kubo 1977). Acrosomes are also absent in some polyplacophoran taxa (Chitonina and Acanthochitonina), but members of the basal group Lepidopleurina appear to have more typical aquatic sperm with acrosomes (Hodgson et al. 1988).

The nucleus containing the paternal chromatin underlies the acrosomal complex. During spermatogenesis, it condenses and may change its shape from a spherical mass to more elongate shapes to conform with the morphology of the mature sperm. Microtubules (the manchette) are associated with this transformation, a process well demonstrated in cephalopods (Maxwell 1974; Arnold & Williams-Arnold 1978) and gastropods (Buckland-Nicks & Chia

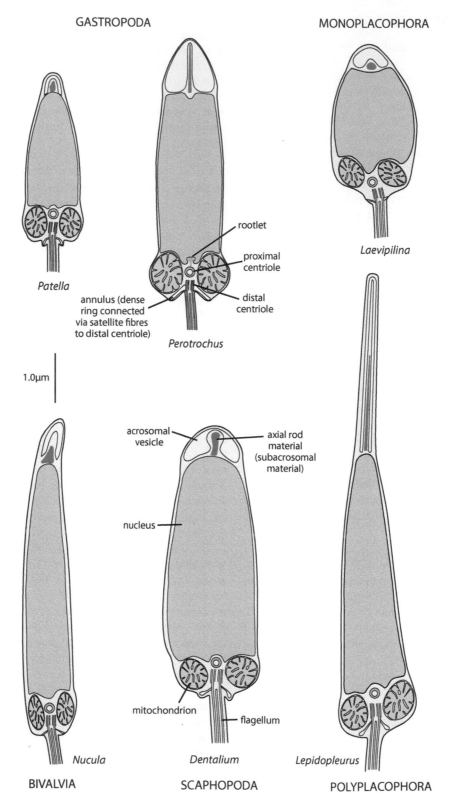

FIGURE 8.5 Generalised aquasperm morphology among molluscan classes. Redrawn and modified from Healy, J.M., Sperm morphology and its systematic importance in the Gastropoda, in Ponder, W.F., Eernisse, D.J. and Waterhouse, J.H. (Eds.), *Prosobranch Phylogeny. Malacological Review Supplement*, Malacological Review, Ann Arbor, MI, pp. 251–266, 1988c (*Bayerotrochus*); Healy, J.M. et al., *Mar. Biol.*, 122, 53–65, 1995 (*Laevipilina* and *Lepidopleurus*); Hodgson, A.N. et al., *Philos. Trans. Royal Soc. B*, 351, 339–347, 1996 (*Patella*); Lamprell, K.L. and Healy, J.M., *Rec. Aus. Mus. Suppl.*, 24, 1–189, 1998 (*Dentalium*); and Popham, J.D. and Marshall, B.A., *Veliger*, 19, 431–433, 1977 (*Nucula*).

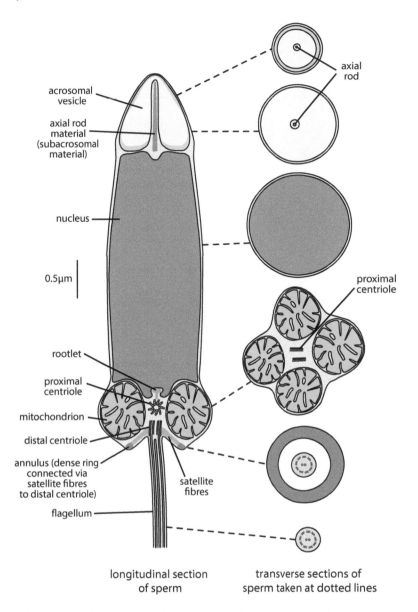

acrosomal
vesicle

axial rod
material
(subacrosomal
material)

nucleus

0.5μm

rootlet

proximal
centriole

mitochondrion

distal centriole

annulus (dense ring
connected via
satellite fibres
to distal centriole)

flagellum

axial
rod

proximal
centriole

satellite
fibres

longitudinal section
of sperm

transverse sections of
sperm taken at dotted lines

FIGURE 8.6 Detailed semi-diagrammatic illustration of the structure of the aquasperm of *Bayerotrochus westralis* – a pleurotomariid vetigastropod. Redrawn and modified from Healy, J.M., *J. Molluscan Stud.*, 54, 309–316, 1988b.

1976b). Two centrioles (proximal and distal) lie at the base of the nucleus with the proximal one often in a small invagination of the nucleus. The distal centriole forms the basal body of the sperm flagellum or axoneme. Particularly elongated sperm heads are seen in bivalves with large eggs (Franzén 1983).

Mitochondrial provisioning in sperm is conservative in molluscs. During spermatogenesis, the scattered mitochondria are concentrated at the base of the nucleus into 4–5 spheres around the centrioles and do not undergo further condensation. Thus, the midpiece of aquatic sperm typically contains four large, oval mitochondria which represent the fusion of multiple, smaller mitochondria. In bivalves, scaphopods, polyplacophorans, caudofoveates, and basal gastropods, these mitochondria are large and unmodified. In neritimorphs, caenogastropods, heterobranchs,

solenogasters, and cephalopods, further modifications reflect the independent evolution of internal fertilisation in these groups. In neritimorphs, caenogastropods, and heterobranchs, the mitochondrial spheres condense to two and then extend as ribbons down the sperm tail to form the mitochondrial sheath (Figure 8.7). A single pair of spherical mitochondrial condensations is found in solenogasters and several taxon-specific patterns are seen in cephalopods (Figure 8.8). In *Nautilus*, the paired mitochondria lie on either side of the nucleus (Arnold & Williams-Arnold 1977). In octopods, condensation produces a mitochondrial sheath similar to that found in higher gastropods. In decabrachian cephalopods, the mitochondria migrate posteriorly and form a spur-like extension of the head of the mature sperm that may facilitate penetration of the jelly coat of the egg (Fields & Thompson 1976). In *Vampyroteuthis*, there are four spherical

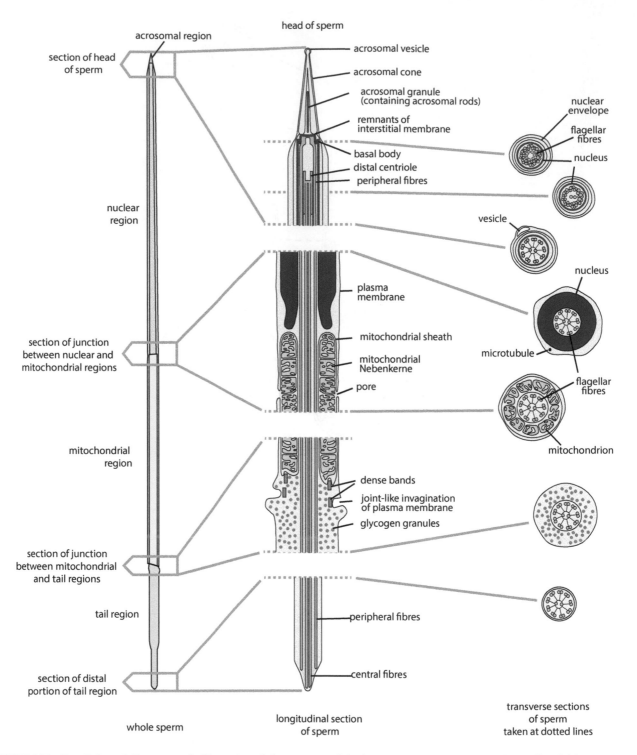

FIGURE 8.7   Detailed semi-diagrammatic illustration of the structure of the introsperm of *Littorina scutulata*, a littorinid caenogastropod. Redrawn and modified from Buckland-Nicks, J.A. and Chia, F.-S., *Cell Tissue Res.*, 172, 503–516, 1976a.

condensations around the centrioles and the sperm resembles the aquasperm of other molluscs (Figure 8.8).

The sperm tail, (flagellum, or axoneme) is a cilium, typically with a 9 + 2 microtubule structure (Figure 8.9). It is attached to the sperm by the posterior centriole, often called the basal body. In cephalopods, accessory fibres are also present on the outside of the axoneme.

In the final stages of spermatogenesis, the bulk of the cytoplasm is lost, including the ribosomes, Golgi apparatus, and endoplasmic reticulum.

8.3.1.1.2   Paraspermatozoa

Paraspermatozoa (or more simply, parasperm) are sterile gametes found in neritimorphs and caenogastropods

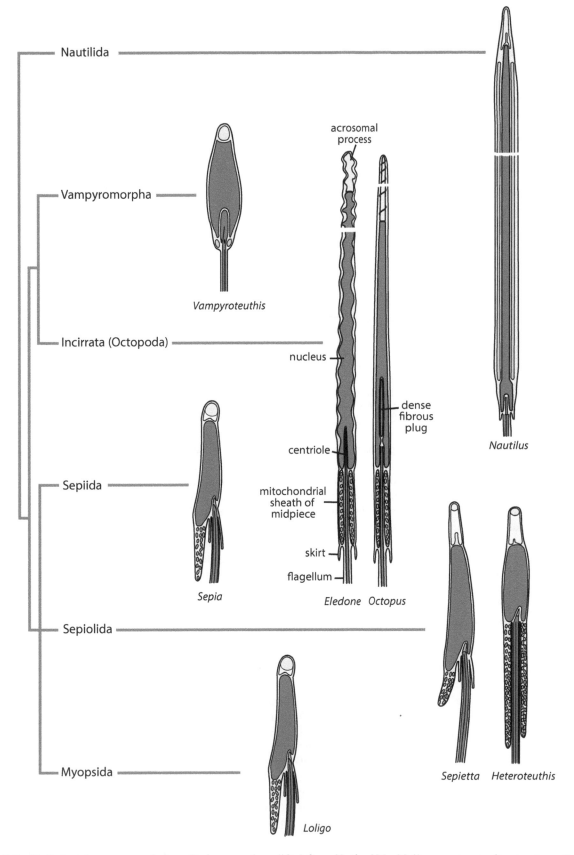

FIGURE 8.8 Cephalopod sperm morphology. Redrawn and modified from Healy, J.M., Molluscan sperm ultrastructure: Correlation with taxonomic units within the Gastropoda, Cephalopoda and Bivalvia, in Taylor, J.D. (Ed.), *Origin and Evolutionary Radiation of the Mollusca*, Oxford University Press, Oxford, UK, pp. 99–113, 1996.

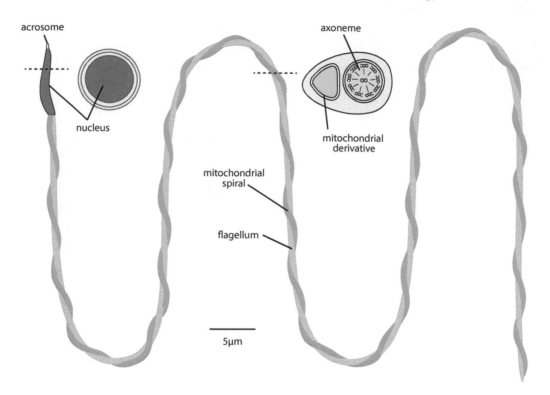

FIGURE 8.9   An introspermatozoan of the nudibranch *Doris pseudoargus*, showing the highly modified flagellum. Redrawn and modified from Thompson, T.E., *Malacologia*, 14, 167–206, 1973.

(Figure 8.10), and a few bivalves have spermatophore-like sperm bundles (see next Section). Most paraspermatozoa have reduced amounts of chromatin relative to the euspermatozoa and generally lack an acrosome, although there are exceptions, including those in the caenogastropod taxa Cerithioidea and Campaniloidea. Parasperm are termed *oligopyrene* (reduced chromatin) or *apyrene* (no chromatin), while those with more chromatin than found in euspermatozoa are referred to as *hyperpyrene*. Parasperm range from simple flagellated cells typically lacking a nucleus, as in neritimorphs, to large structures carrying hundreds of eusperm as in Cerithiopsoidea, these being termed spermatozeugmata (e.g., Jamieson 1987). A variety of secretion products are released by exocytosis from gastropod parasperm. *Cerithium* parasperm release dense granules into the testicular fluid and those in *Littorina* release a fine granular secretion. The role of these secretions is not known.

Formation of parasperm is similar in the taxa studied so far. Both euspermatozoa and paraspermatozoa are formed from the spermatogonium. Different authors have reported the ability to recognise parasperm precursors at different stages of spermatogenesis, giving rise to terms such as paraspermatogonia or spermatoblasts. Developing paraspermatozoa are readily distinguishable from euspermatozoa after differentiation of the secondary spermatogonia, giving rise to distinctive paraspermatocysts. The variability in chromatin content in parasperm of different taxa results from abnormal meiotic cell divisions, but the greatest morphological changes occur during the maturation of the paraspermatids. During this stage,

the nuclear material often degenerates and is lost or forms glycoproteins. The centriole gives rise to individual flagella, and mitochondria migrate and form the parasperm midpiece, while glycogen accumulates in the body of the parasperm. The control mechanisms causing the differentiation of the spermatogonium into either eusperm or parasperm are not well understood, with arguments for both genetic and environmental control suggested. Gamarra-Luques et al. (2009) suggested that the unique parasperm found in *Pomacea canaliculata* were produced by the truncation of normal euspermiogenesis in that species.

8.3.1.2   Spermatophores and Spermatozeugmata

Although some parasperm (spermatozeugmata) transfer sperm between individuals, they are not to be confused with spermatophores, which are sperm storage containers with a solid wall and serve as a mechanism to increase the number of sperm transferred between mating individuals. Spermatophores occur in gastropods and cephalopods, although analogous structures have been described in a few bivalves (see below). In gastropods, spermatophores occur in neritimorphs and some caenogastropods and heterobranchs (Robertson 2007) (Table 8.1).

Spermatozeugmata differ from spermatophores in being intracellular in origin and are parasperm with numerous attached eusperm (Hodgson et al. 1997; Buckland-Nicks et al. 2000). In contrast, spermatophores are formed within the reproductive tract external to the gonad and encase eusperm into capsule-like structures, often with elongate tails.

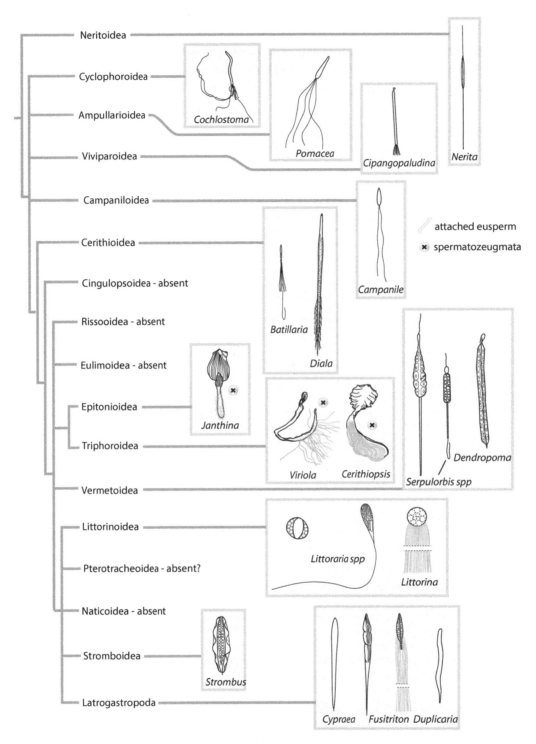

FIGURE 8.10 Examples of neritimorph and caenogastropod parasperm. Redrawn and modified from the following sources: Nishiwaki, S., *Sci. Rep. Tokyo Kyoiku Daigaku Sec. B*, 11, 237–275, 1964 (*Nerita, Batillaria, Janthina,* and *Strombus*); Selmi, M.G. and Giusti, F., *Atti Accad. Fisiocr. Siena*, 115–147 (*Cochlostoma*); Gamarra-Luques, C. et al., *Biocell*, 30, 345–357, 2006 (Pomacea); Tochimoto, T., *Sci. Rep. Tokyo Kyoiku Daigaku Sec. B*, 13, 75–109, 1967 (*Cipangopaludina* and *Duplicaria*); Healy, J.M., *Helgoländer Meeresun.*, 40, 201–218, 1986c (*Campanile*); Healy, J.M., *Helgoländer Meeresun.*, 40, 177–199, 1986b (*Diala*); Healy, J.M., Sperm morphology and its systematic importance in the Gastropoda, pp. 251–266, in Ponder, W.F., Eernisse, D.J. and Waterhouse, J.H. (Eds.), *Prosobranch Phylogeny. Malacological Review Supplement*, Malacological Review, Ann Arbor, MI, 1988c (*Virola*); Fretter, V. and Graham, A.L., *British Prosobranch Molluscs: Their Functional Anatomy and Ecology*, Ray Society, London, UK, 1962 (*Cerithiopsis*); Healy, J.M., *J. Molluscan Stud.*, 54, 295–308, 1988a (*Serpulorbis* and *Dendropoma*); Reid, D.G., *The Littorinid Molluscs of Mangrove Forests in the Indo-Pacific Region: The Genus Littoraria (Mollusca, Gastropoda, Littorinidae)* (*Littoraria*), Butler & Tanner, London, UK, 1986a; Buckland-Nicks, J.A., *Micron*, 29, 267–280, 1998 (*Littorina* and *Fusitriton*); Healy, J.M., *J. Molluscan Stud.*, 52, 125–137, 1986a (*Cypraea*).

TABLE 8.1

Presence of Paraspermatozoa, Spermatozeugmata, and Spermatophores in Select Gastropod Taxa and the Presence or Absence of a Copulatory Organ (Penis) in Those Taxa

Taxon	Paraspermatozoa	Spermatophore	Penis
Vetigastropoda	●		
Neritimorpha			
Neritoidea	●	●	●
Caenogastropoda			
Cyclophoroidea			●
Viviparoidea	●		●
Ampullarioidea	●		●
Cerithioidea	●	●	○
Campaniloidea	●	●	○
Abyssochrysoidea	■		○?
Littorinoidea	■		●
Pterotracheoidea		●	●
Vermetoidea		●	○
Cypraeoidea	■?		●
Stromboidea	●		●
Triphoroidea	■	●	○
Epitonioidea	■	○	○
Tonnoidea	■	○	●
Neogastropoda	●	○	●
Heterobranchia			
Architectonicoidea		●	○
Rhodopoidea		●	○
Cephalaspidea		●	●
Runcinida		●	●
Cavolinioidea		●	●
Clionoidea		●	●
Pyramidelloidea		●	●
Parahedyloidea		●	○
Aeolidioidea		●	●
Siphonarioidea		●	●
Stylommatophora		●	●

● = present ■ = function as spermatozeugmata ○ = absent ? = uncertainty

Updated and modified from Robertson, R., *Malacologia*, 30, 341–364 1989; Healy, J.M., Sperm morphology and its systematic importance in the Gastropoda, in W. F. Ponder, Eernisse, D. J. & Waterhouse, J. H. (Eds.), *Prosobranch Phylogeny. Malacological Review Supplement*, Malacological Review, Ann Arbor, MI, pp. 251–266, 1988c; Healy, J.M., *Mar. Biol.*, 105, 497–507, 1990; Robertson R., *Am. Malacol. Bull.*, 23, 11–16, 2007.

Note: Taxa lacking parasperm or spermatophores are not listed. Presence indicates a published record from one or more species in that group and does not necessarily suggest that they are present in all members of that taxon.

In gastropods, spermatozeugmata occur within the Abyssochrysidae, Vermetoidea, Triphoroidea, Littorinidae, Campaniloidea, Tonnoidea, Epitonioidea, and possibly in the Cypraeidae (Robertson 2007) (Table 8.1). Spermatophores occur sporadically in many gastropod groups, including the Neritimorpha, Cerithioidea, Vermetoidea, Pterotracheoidea, Triphoroidea, Architectonicoidea, Pyramidelloidea, Cephalaspidea, Runcinoidea, Thecosomata, Acochlidiomorpha, Aeolidioidea, Siphonarioidea, and Stylommatophora (Robertson 1989, 2007). The types of spermatophore transfer identified in heterobranchs are: (1) transfer via copulation – the common method; (2) transfer via placing the spermatophore on the body, followed by (2a) migrating to the female opening, as in *Aeolidiella glauca* (Haase & Karlsson 2000), or (2b) entering by way of dermal insemination, as in some Acochlidia (Jörger et al. 2009) and *Rhodope* (Cuervo-González 2017).

Although most groups that rely on spermatozeugmata and spermatophores to transfer sperm have copulatory structures, some do not (Table 8.1). In some caenogastropods, a modified sperm called a lancet parasperm forms a plug in the reproductive system of the female which may prohibit later mating (Buckland-Nicks 1998). Their sporadic distribution in gastropods (Figure 8.10 and Table 8.1) suggests that, as suspected by Robertson (1989), both spermatophores and spermatozeugmata are convergent rather than plesiomorphic.

Spermatophores are present throughout cephalopods. Copulatory structures are also present and may transfer the spermatophore either directly to the female or, in many coleoids, via a hectocotylised tentacle. In neither case is the sperm deposited internally; rather the spermatophore is placed in juxtaposition with the female opening (Arkhipkin 1992).

In bivalves, large paraspermatozoa occur in some members of the Galeommatoidea (see below), while in others, sperm is transferred to the gills of conspecifics within gelatinous non-encapsulated sperm balls, sometimes called spermatophores, although they do not strictly qualify as such (Ó Foighil 1985; Lützen et al. 2001a, 2005; Jespersen & Lützen 2006). Spermatozeugmata have been reported in the Unionidae (*Anodontoides*, *Lampsilis*, *Lasmigona*, and *Utterbackia* [Lynn 1994], *Tuncilla* [Waller & Lasee 1997]), *Ostrea* (Ostreidae) (Coe 1931; Ó Foighil 1989), and *Nutricola* (Veneridae) (Falese et al. 2011). In some galeommatoid taxa, sperm transfer may involve an interaction between normal euspermatozoa and sterile and often rather large paraspermatozoa (Jespersen et al. 2001, 2002; Jespersen & Lützen 2006, 2009) or the transfer of sperm packets consisting of minute portions of the testis (Lützen et al. 2004; Jespersen & Lützen 2006).

8.3.1.3 Eggs

As with the testis, the origin of ovaries and primary germ cells (oocytes) from stem cells of the coelomic pericardial epithelium has long been known (Raven 1966). Two forms of localised oogenesis are found in molluscs – a well defined ovary typically with a genital duct, or a simple invagination of germinal epithelium of the coelom. Oogenesis may be *alimentary* (egg develops in association with accessory cells) or *solitary* (egg develops without accessory cells, as in some bivalves). Alimentary development can be either *follicular* or *nutrimentary*. In follicular development, the developing eggs are associated with follicle cells which provide nutrients to the developing egg(s), while in nutrimentary development, the eggs receive nutrients from non-follicle accessory cells.

Oocytes are typically large cells requiring a substantial maternal investment. The eggs of free spawning molluscs are relatively small, but are larger in those with internal fertilisation and/or engaging in brood protection. In bivalves, oocytes develop within acini or tubules that make up the female gonad (e.g., Jong-Brink et al. 1983; Pipe 1987; Morse & Zardus 1997). Although it is a continuous process, six phases of oogenesis are generally distinguished – germinal vesicle formation, RNA synthesis, vitellogenesis, cortical granule formation (not known in gastropods) (Jong-Brink et al. 1983), attachment of auxiliary cells to the oocyte, and in some taxa, in the last phase after ovulation the oocyte produces a polar body. The increased volume of the ovary is typically associated with a milky appearance due to vitellogenin (a precursor of yolk), glyco-lipo-phosphoproteins, and a lipovitellin (a component of yolk), in the spaces surrounding the oocytes (Puinean et al. 2006). Developing oocytes usually remain attached to the acini walls by a stalk as they develop. Mature oocytes have a germinal vesicle, an outer vitelline layer, and the cytoplasm mainly contains numerous yolk granules in addition to Golgi bodies, mitochondria, and

free ribosomes. The yolk can be either lipid or proteinaceous and can have an endogenous or exogenous origin (see Morse and Zardus [1997] for more details). When auxiliary cells are present, they appear to have a mainly nutritive function and disappear when the vitelline layer forms late in development, which is followed by the addition of an outer jelly coat. Different roles for oocyte-follicle cell associations have been recorded in molluscs, including vitellogenesis, the formation of egg coats, nutritive loading, and the determination of egg polarity (Jong-Brink et al. 1983; Collier 1997). Although only one type is known in bivalves (Morse & Zardus 1997), the distribution and significance of these within molluscs is too poorly understood to know if there is any phylogenetic significance.

Follicle cells are somatic in origin and surround the developing egg. Non-follicle accessory cells are typically derived from the germ line and include sibling cells arising from mitotic divisions, abortive oocytes, and some non-germ line cells distinct from follicle cells. In some taxa, oogenesis begins in association with accessory cells, but these have disappeared by the time the eggs have completed development. Follicle cells also occur in association with developing oocytes in polyplacophorans and cephalopods (Selman & Arnold 1977) and, in gastropods, the oocytes are also associated with numerous cells typically called follicle cells. These cells enclose the oocyte and isolate it from the lumen of the ovary (Pal & Hodgson 2005). As the oocyte matures and enlarges, the follicle cells are pushed aside, but may persist up to the late vitellogenic stage (Pal & Hodgson 2002). In the veneroid bivalve *Gaimardia trapesina,* the follicle cell may persist into spawning and assists in the attachment of the ova and developing embryos to the brood chambers of the inner and outer demibranchs of the gill (Ituarte 2009).

In *Octopus*, follicle cells are involved in the synthesis and transfer of proteins and other molecules to the developing ovum. In the squid *Loligo*, the follicle cell wall penetrates the developing ovum and facilitates yolk transport, a process that probably contributes to the rapid growth of the yolk-rich egg. Whereas oocyte secretions produce the plesiomorphic jelly egg hulls of the lepidopleurid chitons, follicle cells produce the egg hulls (chorions), which are well-developed in the Chitonidae and other higher chiton taxa (Buckland-Nicks & Reunov 2010).

Accessory cells associated with nutrient provision include nurse cells connected to the developing oocyte by cytoplasmic bridges, which result from incomplete cytokinesis during mitosis. Unlike nurse cells, nutritive eggs are not mitotic siblings and therefore do not remain attached to the developing egg through cytoplasmic bridges. Instead, nutritive eggs are abortive oocytes that either transfer nutrients or are phagocytised by the developing ovum. Nutritive eggs and other extracellular components are also present in the spawn of some gastropods, particularly in caenogastropods (see Chapter 19).

Primary oogonia multiply through mitosis (Figure 8.11). In different taxa, the number of mitotic divisions may be fixed or variable. Secondary oogonia undergo three stages: (1) premeiotic, (2) previtellogenic, and (3) vitellogenic (Wourms 1987). Meiotic prophase begins during the premeiotic period.

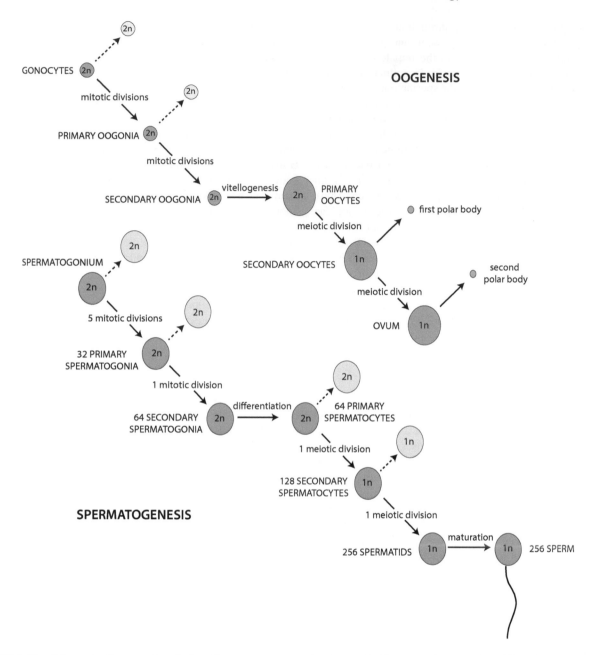

FIGURE 8.11 Schematic diagram of typical molluscan gametogenesis. Oogenesis diagram redrawn and modified from Wourms, J.P., Oogenesis, in Giese, A.C., Pearse, J.S. and Pearse, V.B. (Eds.), *Reproduction of Marine Invertebrates. General Aspects: Seeking Unity in Diversity*, Blackwell Scientific Publications & Boxwood Press, Palo Alto, CA, Vol. 9, pp. 49–178, 1987 and spermatogenesis diagram redrawn and modified from Dohmen, M.R., Gametogenesis, in Verdonk, N.H., Biggelaar, J.A.M., van den and Tompa, A.S. (Eds.), *Development. The Mollusca*, Vol. 3, Academic Press, New York, pp. 1–48, 1983.

Nuclear events predominate during the previtellogenic stage, including formation of the germinal vesicle, the appearance of lampbrush chromosomes, and RNA synthesis. Mitochondria are also duplicated during this period. Finally, the developing egg undergoes vitellogenesis, during which yolk is added to the egg through either auto- or heterosynthesis. In autosynthesis, the yolk is synthesised within the egg, while in heterosynthesis, the yolk is synthesised outside the egg and transported into it (Schechtman & Knight 1955). Autosynthesis has been reported in several molluscan groups

(bivalves, gastropods, and polyplacophorans) (Selwood 1986; Pipe 1987; Hodgson & Eckelbarger 2000; Chung 2007). As discussed above, heterosynthesis of yolk by follicle cells and its subsequent transfer to the developing oocyte has been demonstrated in several cephalopods (O'Dor & Wells 1975; Arnold & Williams-Arnold 1977).

During vitellogenesis, the egg substantially increases in volume as yolk and cell organelles increase in volume and number, respectively. During this period, there is also differentiation of the outer egg cortex – the cell membrane with its outer surface or jelly coat and the underlying cortical cytoplasm – and the inner ooplasm. The extracellular surface coat or egg envelope provides egg protection and controls and facilitates fertilisation (Wourms 1987). The cell membrane provides the interface between the developing ovum and its environment, including accessory cells, while the cortical cytoplasm provides developmental information for the differentiation of specific cell and tissues during development (Sardet et al. 2002).

Molluscan eggs may be enclosed in up to three distinct envelopes. Egg envelopes are primarily defined by where and how they are formed in the reproductive system (Table 8.2). Primary and secondary egg envelopes occur in external fertilising species, while internal fertilisation is required for some secondary envelopes (cephalopods) and all tertiary ones. The possession of only individual egg membranes is most likely the plesiomorphic state in the Mollusca.

The inner ooplasm contains the endomembrane system, which includes the nuclear envelope, endoplasmic membrane, Golgi complex, lysosomes, cortical granules, etc. During oogenesis, the ooplasm also accumulates ribosomal precursors which will be used by the developing zygote. These include 5S, 18S, 28S rRNAs, and ribosomal proteins (Long & Dawid 1980; Simison et al. 2006). Yolk, the nutritive component of the egg that provides energy for the developing embryo, also accumulates in the ooplasm. There are two main types, lipid and proteinaceous, the former stored as membrane-less droplets or globules and the latter as granules or platelets with membranes; both may be present in the same egg. Thus, yolk comprises lipids, proteins, and carbohydrates, all of which vary both between taxa and between oogenesis stages. Yolk protein concentrations range from 50% (by volume) in *Lymnaea*, 66%–75% in the Nudibranchia, and 75% in *Crepidula* (Raven 1972). Lipid components are considerably lower. Volumetric estimates range from 2% to 5% in nudibranchs, 5% in *Lymnaea*, and 14% in the mussel *Mytilus* (Raven 1972). In contrast, Kessel (1982) reported that in the patellogastropod *Lottia digitalis,* lipid droplets were the most numerous component in the cytoplasm. Carbohydrates in molluscan eggs are predominately glycogen and galactogen. Previous studies (Giese 1969) have shown carbohydrate content to be variable, ranging from 1.3% (gonad dry weight) in a coleoid (*Loligo*) to 4.3%–9.8% in vetigastropods (*Haliotis* and *Megathura*),

13.3% in a polyplacophoran (*Katharina*), and 45.3% in a venerid bivalve (*Tivela*). Carbohydrates can also occur in the intracapsular fluid of gastropod egg capsules (Raven 1972). Mollusc eggs are often pigmented, with the most common pigments being carotenoids which range in colour from yellow and orange to purple, blue, and green, depending on protein binding. These pigments (biochromes) originate from the maternal diet and are thought to have multiple possible functions, including providing protein reserves and protection from UV radiation and desiccation (Wourms 1987).

The developmental stage at which the eggs are fertilised in molluscs is at metaphase of the first maturation (meiotic) division except in bivalves, where it occurs at the formation of the germinal vesicle (Figure 8.11) (Giese & Kanatani 1987). Thus, the egg is arrested in an intermediate meiotic phase (e.g., prophase 1 in *Spisula* or metaphase 1 in *Mytilus*), and the fertilisation event itself serves as the stimulus for completion of egg maturation and polar body formation. In *Dentalium*, fertilisation occurs during the formation of secondary oocytes, but before the formation of the first polar body.

The energy needed for egg production, particularly yolk formation (vitellogenesis), is derived from food and is often utilised via stored reserves, mainly glycogen (see Chapter 2). This is facilitated in some chitons and in cephalopods by the formation of a special vascular system around each oocyte or juxtaposition with the digestive gland (Jong-Brink et al. 1983). In coleoids, vitellogenesis occurs at the end of the life cycle and involves catabolism of muscle tissue. Egg size generally reflects the amount of yolk added to the egg, with all living cephalopods, and some gastropods, producing large yolky eggs. Lecithotrophic development typically requires more yolk than planktotrophic development because energy resources for development must be included in the egg. This correlation is reflected in the size of the initial part of the protoconch (gastropods, cephalopods, monoplacophorans) or prodissoconch (bivalves), because these initial shells enclose the developing larvae and its yolk reserves. Protoconch morphology (see 8.10.2.1 and Figure 8.45) is characteristic of major clades so this correlation has enabled palaeobiologists to recognise feeding types in the fossil record (Jablonski & Lutz 1983; Nützel & Frýda 2003; Nützel 2014) (see also Chapter 13).

Yolk protein (vitellin) is mostly synthesised in the ovary (Jong-Brink et al. 1983). The formation of yolk granules in

---

TABLE 8.2

The Nature of Egg Cases and Capsules

Type of Egg Case	Source	Taxa	References
Primary (e.g., egg membrane; jelly coats; fertilisation membranes)	Oocyte within ovaries	Most molluscs	Raven (1961)
Secondary (e.g., hulls; chorion)	Follicle cells within ovaries	Polyplacophora, Cephalopoda	Arnold (1971); Pearse (1979)
Tertiary (e.g., egg capsule; ribbons; eggshell)	Outside of ovaries (oviducal, nidamental); post-fertilisation	Neritimorpha, Caenogastropoda, Heterobranchia	Ponder and Lindberg (1997)

Modified and updated from Wourms, J.P., Oogenesis, in Giese, A.C., Pearse, J.S. and Pearse, V.B. (Eds.), *Reproduction of Marine Invertebrates. General Aspects: Seeking Unity in Diversity*, Vol. 9, Blackwell Scientific Publications & Boxwood Press, Palo Alto, CA, pp. 49–178, 1987.

molluscs is mainly endogenous and is a complex intracellular process involving mitochondria, Golgi apparatus, and rough endoplasmic reticulum, and in some taxa there are also exogenous controlling processes involved (Jong-Brink et al. 1983). Because of the high metabolic cost of producing them, any oocytes remaining in the ovary after spawning degenerate and are reabsorbed by the adult.

8.4 SEXES

Gonochorism is plesiomorphic in Mollusca, although a wide diversity of sexual systems is present (Collin 2013). Many gastropod and bivalve families are primarily gonochoristic, simultaneous hermaphrodites, or sequential hermaphrodites, while in a few families almost every sexual strategy occurs. The multiple evolutionary transitions between sexual systems in molluscs should be ideal for comparative studies of associated ecological factors. Unfortunately data for all but a few taxa do not allow rigorous testing, although some preliminary generalisations are possible. For example, the phylogenetic distribution of sexual systems in molluscs shows that gastropods and bivalves demonstrate different patterns, possibly associated with the presence/absence of copulation. The analysis by Collin (2013) of sex change in gastropods suggests that sequential hermaphrodites do not evolve from simultaneous hermaphrodites, and that sex reversal mostly occurs in free spawning, not copulating taxa. Moreover, three of the best studied protandrous taxa (calyptraeids, coralliophilines, and patellogastropods) show similar, convergent responses to environmental conditions. These conditions include sedentary life styles, occurrence in groups or aggregates, and variation in the size at which sex change occurs among different populations. Experimental studies conducted with members of these three groups suggest that the presence of females or larger individuals represses growth and sex change of males, but the behaviour of small males can mediate this influence (Collin 2013).

8.4.1 SEX DETERMINATION

The debate of endocrine versus metabolic control of sex determination was first investigated by Orton (1927) using the protandrous hermaphrodite *Crepidula fornicata* and the sequential hermaphrodite *Ostrea edulis*. Orton concluded that rhythmical changes in the general metabolism of the animal were responsible for sex change in these species. Yusa (2007) reviewed sex determination in molluscs and concluded that two primary categories of mechanisms were involved – environmental sex determination and genetic sex determination. These could act separately or combined to determine the sex of an individual. Environmental factors which lead to sex determination include temperature, food abundance, (quality or scarcity), day length, or interaction with conspecifics (Bull 1983). Bivalves, such as oysters, are thought to have trophic-related sex determination, while some scallops respond to both temperature and food abundance (Sastry 1968, 1997). In gastropods, the slipper snail *Crepidula* provides the classic example of how interactions with conspecifics promotes sex determination (Hoagland 1978).

Sex determination in bivalves has been recently reviewed by Breton et al. (2019).

Genetic factors in sex determination are divisible into three forms, *heterogamety* (two factor systems), *oligogenic*, and *polygenic*. Heterogamety relies on two genetic factors, such as the XY sex determination system of many mammals. It has been reported for some patellogastropods, vetigastropods, and caenogastropods (Coe 1944; Patterson 1969; Nakamura 1986; Thiriot-Quiévreux 2003). Within some clades there is a diversity of two factor systems of sex chromosomes (e.g., *Viviparus*) (Baršiene et al. 2000; Thiriot-Quiévreux 2003), as well as multiple independent originations of sex chromosome determination in distantly related gastropod clades (Collin 2019). Among gastropods, the Neritidae were thought to be unique with an XX (females) or XO (males) sex determination system. This system, in which the males end up with an odd number of chromosomes, has also been found in the vetigastropod *Monodonta vermiculata* and the caenogastropods *Neotricula aperta*, *Melarhaphe neritoides*, and *Pterosoma planum* (Thiriot-Quiévreux 2003). Sex chromosomes in heterobranchs have not been reported (Thiriot-Quiévreux 2003). Heterogamety has also been reported in some bivalves (Allen et al. 1986; Guo & Allen 1994; Breton et al. 2010; Chávez-Villalba et al. 2011). In a recent study, Capt et al. (2018), using comparative transcriptomics, reported three deeply conserved sex determining genes (*Sry*, *Dmrt1*, *Foxl2*) in the unionioid bivalves *Venustaconcha ellipsiformis* and *Utterbackia peninsularis*, these genes being shared with vertebrates. Capt et al. (2018) also concluded, in support of Breton et al. (2010), that sex determination in bivalves with *doubly uniparental inheritance* (DUI) (see Section 8.5.1.1) may involve both nuclear and mitochondrial genomes (see also Breton et al. 2019).

In contrast to two factor systems, oligogenic and polygenic sex determination rely on several or many genes, respectively. Examples include oligogenic sex determination in the apple snail *Pomacea canaliculata* (Yusa & Kumagai 2018) and possibly the bivalves *Crassostrea* and *Mytilus* (Yusa 2007). Polygenic systems in molluscs remain undocumented, although the recent report of a multiple gene system in the unionids by Capt et al. (2018) may represent one such system, and Webber (1977) suggested that polygenic systems might underlie the presence of balanced (same age of sex change) and unbalanced (variable age of sex change) hermaphroditism in closely related non-heterobranch gastropods.

Sex determination in other groups of molluscs remains poorly known, if at all. From the limited data available, molluscs, and especially gastropods and bivalves, appear to be evolutionarily flexible and provide a diverse array of sex determination mechanisms in much need of further study.

8.4.2 HERMAPHRODITISM

In hermaphroditic taxa, both sexes are present in the same individual, and this may occur *contemporaneously* (simultaneous hermaphrodites) or *sequentially* (consecutive hermaphrodites); consecutive hermaphrodites may change sex once or multiple times (*alternating*). Sequential sex change in molluscs is also almost exclusively protandric (male to female) and has only been reported among gastropods and bivalves (see below). Sometimes, protandric sex change occurs because of the presence of other individuals of the same species, such as occurs with *Crepidula*. The distinction between sequential and simultaneous hermaphroditism is not always clear-cut, with the transitional phase between sexes sometimes having a functional gonad producing both eggs and sperm.

In a study of molluscan hermaphroditism, Heller (1993) concluded that about 40% of molluscan genera are hermaphroditic. The vast majority of these are found in the Heterobranchia, where simultaneous hermaphroditism is an apomorphy for this diverse and disparate clade. Hermaphroditism also appears to be an apomorphy in Solenogastres, but has been independently derived multiple times in most other molluscs. Protandric hermaphroditism has evolved independently in several groups within both bivalves and gastropods, while protogyny has only been reported in a few bivalves (see Section 8.4.2.2). The patchy distribution of protandric hermaphroditism among patellogastropod, vetigastropod, and caenogastropod gastropods, and pteriomorphian and heterodont bivalves, further suggests that it is unlikely that a single causal mechanism can explain its evolution among such disparate taxa.

Heterobranchs are simultaneous hermaphrodites except for the gonochoric acochlidian families Microhedylidae and Ganitidae (Schrödl & Neusser 2010). Given the phylogenetic position of these taxa within heterobranchs (Jörger et al. 2010), gonochorism in these taxa presumably represents a secondarily derived state, possibly associated with reduction in body size and the specialised habits of these taxa.

Building on the critique of hermaphroditism nomenclature by Hoagland (1984), Ponder and Lindberg (1997) categorised hermaphroditism using a time/space relationship which they argued better described the timing and anatomy of hermaphroditic reproduction and introduced the terms *synhermaphroditism* and *apohermaphroditism* (Table 8.3). With simultaneous synhermaphroditism, sperm and eggs are present in the same gonad simultaneously (+ +) (Figure 8.12), while in simultaneous apohermaphroditism, eggs and sperm are present in different gonads simultaneously (+ –). Although the reproductive outcome is the same, the morphology differs.

In comparison, a sequential protandric hermaphrodite has eggs and sperm in the same gonad at different times (– +) (sequential synhermaphroditism) or in separate gonads at different times (sequential apohermaphroditism) (– –). The pattern for the plesiomorphic state in stem groups analysed by Ponder and Lindberg (1997) also shows multiple originations of these different forms of hermaphroditism across molluscs, as also concluded by Collin (2013).

8.4.2.1 Simultaneous Hermaphroditism

Simultaneous hermaphrodites are broadly distributed throughout the Mollusca (Fretter & Graham 1964; Fretter 1984a; Heller 1993, 2001), with all solenogasters and almost all heterobranchs being simultaneous hermaphrodites. In non-heterobranch gastropods, simultaneous hermaphrodites are found in the patellogastropods (*Erginus*) (Lindberg 1983) and the vetigastropod groups Cocculinoidea and Lepetelloidea (Haszprunar 1988b). Another vetigastropod, the fissurellid *Puncturella noachina*, has also been reported to have an hermaphrodite gonad (Rammelmeyer 1925) and may also be a simultaneous hermaphrodite. Simultaneous hermaphrodites are also found in the caenogastropod families Velutinidae and Eulimidae (Warén 1983; Fretter 1984a), with additional scattered occurrences among the neogastropods (e.g., muricids, cancellariids, turrids) (Heller 1993; Collin 2019). In bivalves, simultaneous hermaphrodite taxa include several Pectinidae and Cardiidae, most members of the Anomalodesmata, most Galeommatoidea (e.g., Sphaeriidae, Erycinidae, Montacutidae), several unionoideans (e.g., *Margaritifera, Anodonta, Carunculina*), some venerids (e.g., *Gemma, Tivela*), and some teredinids (*Lyrodus, Teredo*) (Sastry 1979; Heller 1993). In Polyplacophora, simultaneous hermaphroditism has been reported in two brooding chitons, *Lepidochitona caverna* and *L. fernaldi* (Eernisse 1988a, 1988b). Simultaneous hermaphroditism has not been reported in monoplacophorans, scaphopods, caudofoveates, or cephalopods.

Simultaneous hermaphroditism potentially permits three reproductive options: self-fertilisation, female-acting outcrossing, and male-acting outcrossing. A fourth option, sperm-sharing, was reported by Monteiro et al.

TABLE 8.3

Time and Space Relationships in Different Types of Hermaphroditism

Form of Hermaphroditism	Timing	Location	Morphology
Simultaneous apohermaphrodite	+	–	Both ovary & testis present
Simultaneous synhermaphrodite	+	+	Ovotestis
Protandrous or progynous synhermaphrodite	–	+	Single gonad
Protandrous or progynous apohermaphrodite	–	–	Both ovary & testis present

Source: Ponder, W.F. and Lindberg, D.R., *Zool. J. Linn. Soc.*, 119, 83–265, 1997.

Timing: + at the same time, – at different times; Location: + occupying the same gonad, – in separate gonads.

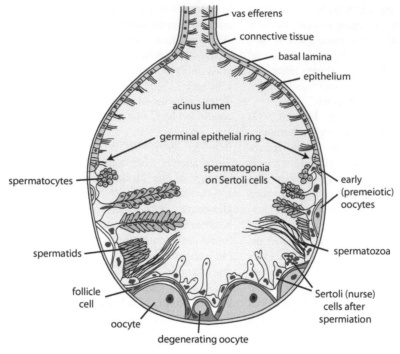

Lymnaea stagnalis (Hygrophila)

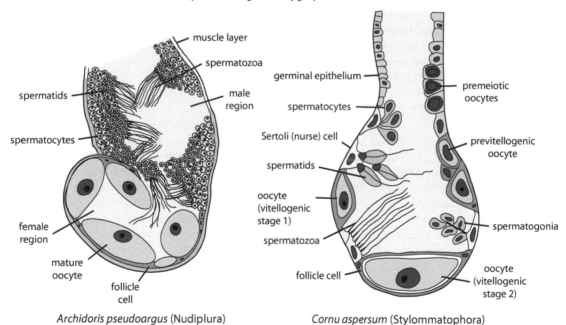

Archidoris pseudoargus (Nudiplura)　　　*Cornu aspersum* (Stylommatophora)

FIGURE 8.12 Simultaneous production of sperm and eggs in the ovotestes of some representative euthyneuran heterobranchs. Redrawn and modified from the following sources: Joosse, J., Structural and endocrinological aspects of hermaphroditism in pulmonate snails, with particular reference to *Lymnaea stagnalis* (L.), in Reinboth, R. (Ed.), *Intersexuality in the Animal Kingdom*, Springer-Verlag, Berlin, Germany, pp. 158–169, 1975 (*Lymnaea*); Thompson, T.E., *Philos. Trans. Royal Soc. B*, 250, 343–374, 1966 (*Archidoris*); and Griffond, B. and Bolzoni-Sungur, D., *Reprod. Nutr. Dév.*, 26, 461–474, 1986 (*Cornu*).

(1984) in laboratory-reared *Biomphalaria*, where some cross-fertilising individuals impregnated partners with exosperm received from an earlier mating. Thus, while the sperm-sharing individual functions physically as a male, it is not functioning genetically as one because the sperm transferred are not autosperm.

8.4.2.2 Sequential Hermaphroditism

Besides protandry (male → female) and protogyny (female → male), some molluscs may also display alternating sexuality (male → female → male →…) in which an individual functions first as one sex and then as the other. In protandry and protogyny, sex change typically occurs once and the timing of sex change is often environmentally determined by the presence of the opposite sex (Heller 1993). Although protandry is by far the dominant form of sequential hermaphroditism in molluscs, possible cases of protogyny have been reported in a few bivalves including a Malaysian unionid *Contradens ascia* (Berry 1974), the mussel *Mytella charruana* (Stenyakina et al. 2010), and *Panopea generosa* (Calderon-Aguilera et al. 2014). Sequential hermaphroditism has not been reported in scaphopods, caudofoveates, or cephalopods.

In chitons, only *Cyanoplax dentiens* is a protandric sequential hermaphrodite (Pearse 1979), and in the Monoplacophora, the minute *Micropilina arntzi* is a protandric hermaphrodite (Haszprunar & Schaefer 1996). Several bivalve taxa are sequential hermaphrodites, including oysters (Ostreidae), pearl oysters (*Pinctada*), fresh-water mussels (*Elliptio*), the giant clams *Tridacna* and *Hippopus*, the cyrenid *Corbicula*, the venerid *Mercenaria*, and the teredinid *Bankia*. In unionids and pteriomorphians, gonochoristic taxa often have a few hermaphrodites in the population as well (Heller 1993; Dillon 2000). Morton (1972a) reported that when a veliger of the commensal bivalve *Borniopsis* (as *Pseudopythina*) *subsinuata* settles near a female on its host (typically a stomatopod), it becomes male and fertilises the female. When the female dies, the male changes to female. Settling veligers that are not in the presence of a female immediately mature as females without a male phase. Sequential alternating hermaphroditism is common in some bivalve taxa (e.g., Ostreidae and Pholadidae) (Andrews 1979b; Sastry 1979; Hoagland 1984).

Sequential hermaphroditism is predominant in a few non-heterobranch gastropods. In patellogastropods, protandry has been demonstrated in *Patella* spp. (Orton 1920, 1927, 1928; Le Quesne & Hawkins 2006), *Atalacmea fragilis* (Willcox 1898), *Erginus sybariticus* (Golikov & Kussakin 1972), *Lottia gigantea* (Wright & Lindberg 1982; Lindberg & Wright 1985), and *Scutellastra* spp. (Branch 1974; Lindberg 2007b). Examples of sequential hermaphroditism in vetigastropods are rare (*Diodora* [Fretter 1984a]), and documenting its presence in broadcast spawners is difficult because of the lack of secondary sexual characteristics. In caenogastropods, sequential hermaphroditism occurs primarily in the Calyptraeidae, Eulimidae, Capulidae, Epitoniidae, Hipponicidae, Coralliophilinae, and Triviidae (Warén 1980b; Gosliner & Liltved 1982; Warén et al. 1984; Chen & Soong 2002; Collin 2013, 2019). Many of these taxa are limpet-like and sex change is socially and size mediated; others are ectoparasites.

There are scattered unique occurrences in other families including littorinids (Reid 1986b), assimineids (Fukuda & Ponder 2004), tornids (Bieler & Mikkelsen 1988), and vermetids (Calvo & Templado 2005). Sequential, alternating hermaphroditism is uncommon in caenogastropods and has been reported in epitoniids (Melone 1986), some eulimids (Warén 1980a, 1983), *Janthina* (Laursen 1953), and the vermetid *Thylacodes arenarius* (Calvo & Templado 2005).

Some panpulmonate heterobranchs are protandrous, including the fresh-water *Glacidorbis* where the males do not change into females until they have received sperm from another male, and virgin males can grow to be as big as females

(Ponder 1986). Thecosome pteropods are also protandrous (Hadfield & Switzer-Dunlap 1984; Lalli & Gilmer 1989).

In the well-studied and distantly related gastropod superfamilies Calyptraeoidea and Patelloidea, individuals of some species respond to local ecological changes by altering the age at which they change sex, although the critical ecological changes appear to be different. The adaptive value of protandry in these two groups most probably relates to the limited availability of females, and the consequent size-independent nature of male reproductive success versus the size-dependent nature of reproductive success in females (Charnov 1982).

Several theories relate to the development of sequential hermaphroditism. It is thought that sex change may occur if a species can produce more offspring as one sex when small and as the opposite sex when large (the size-advantage hypothesis) (Ghiselin 1969; Warner 1975; Iwasa 1991). Thus several models have been developed in an attempt to determine the time (size) of sex change in sequential hermaphrodites (Charnov 1982; Iwasa 1991; Collin 1995; Muñoz & Warner 2003). Most imply that the life history transition is an *evolutionary stable strategy* (ESS) and predict that the optimal time (size) of sex change yields maximum lifetime reproductive success. The original ESS model of Charnov (1982) suggests that the optimal time of sex change depends on the existing sex ratio in the population and the relative fitness of each sex at that time. It predicts a critical age (size) above which all individuals should be the second sex and below which all individuals should be the first sex, although it assumes a constant environment and similar growth and mortality rates for both sexes.

8.4.3 ASEXUAL REPRODUCTION

Asexual reproduction is rare in molluscs (Runham 1993). The only molluscs known to reproduce asexually include two members of the Bivalvia – the androgenic cyrenid *Corbicula* (Hedtke et al. 2008; Pigneur et al. 2012) and the parthenogenetic galeommatoidean *Lasaea* (Crisp et al. 1983). In caenogastropods, some Viviparidae (Hubricht 1943; Johnson & Bragg 1999), Tateidae (*Potamopyrgus antipodarum*) (Wallace 1992), and Thiaridae (Jacob 1957; Ben-Ami & Heller 2005) have been described as parthenogenetic. Parthenogenesis in these taxa appears to be facultative as sexually reproducing individuals and rare males have been reported in some populations (Patil 1958; Schalie 1965; Heller & Farstey 1990; Wallace 1992) (see also Chapter 19).

8.4.3.1 Schizogamy

In a few of the pelagic cavoliniid pteropods, a form of asexual reproduction is said to co-occur with two forms of hermaphroditism (protandry and simultaneous) and is most similar to *schizogamy*, which occurs in some annelid worms. This form of asexual reproduction was referred to as strobilation. These asexual divisions of the body have not yet been demonstrated to produce viable offspring (Lalli & Gilmer 1989) and may prove to be artefacts.

Cavoliniids have long been known to have aberrant forms often referred to as 'skinny' or 'minute' (Spoel 1967). While some of these morphological oddities were shown to be preservation artefacts (Gilmer 1986; Lalli & Gilmer 1989), Spoel (1973, 1979) reported that some of these aberrant forms represented individuals that had reproduced asexually by splitting into two, and he referred to this form of reproduction as strobilation. Later studies of both preserved and living individuals (Pafort-van Iersel & Spoel 1986; Pokora 1989) appeared to reveal a complex form of reproduction in which asexual reproduction is part and also includes both protandry and self-fertilisation (Figure 8.13). It is argued that when oceanic conditions are favourable, cavoliniids are outcrossing (protandrous) hermaphrodites. When oceanic conditions become unfavourable, some mature female specimens apparently split transversely across their bodies, producing a shell-less anterior region with parapodia and digestive system, but initially lacking a gonad (the primary individual), while the posterior shelled region develops two 'wings' *de novo* and maintains the original gonad with both ovary and testis present (the metamorphosed individual) (Pafort-van Iersel & Spoel 1986). The lack of a penis in the metamorphosed individual led these workers to conclude these individuals are self-fertilising. In the primary individual, a band of interstitial cells on the left side of the body gives rise to a new testis and a seminal groove and penis redevelops in these individuals (Spoel 1967).

Thus, it appears that when environmental conditions are favourable, the reproductive strategy of members of the Cavoliniidae consists of the outcrossing protandry. When oceanographic conditions are unfavourable, the females of some taxa undergo fission, producing two individuals – the primary individual, which once again returns to the male stage of the protandric cycle, and the hermaphroditic metamorphosed individual which self-fertilises (Figure 8.13). This form of asexual reproduction is otherwise unknown in molluscs. The origins of the fission component of this reproductive strategy might be derived from autotomy (see Chapter 3). Janssen (1985) reported similar 'skinny' and 'minute' forms in the Miocene of the North Sea and Aquitaine Basins, which, it was suggested, argues for the existence of this reproductive strategy throughout much of the Neogene.

8.5 FERTILISATION

In outcrossing taxa, contact between the egg and sperm occurs either externally or internally (Brahmachary 1989). In external fertilising gastropods (mostly patellogastropods and vetigastropods), the eggs are typically misshapen from compaction in the gonad, but on contact with seawater, the egg quickly becomes spherical as the outer glycoprotein jelly coat swells; in patellogastropods, the jelly layer is sloughed off after fertilisation (Smaldon & Duffus 1985). The jelly coat agglutinates sperm on the egg surface and, in conspecifics, increases sperm motility, while endocrine-like substances released by sperm can produce self-inactivation or can inactivate other sperm (Tyler 1949). During the acrosomal reaction, a lysine released from the sperm head dissolves the vitelline layer of the egg prior to fertilisation (see below). These compounds were mostly called 'fertilizin' in the early literature. In Polyplacophora (Miller 1977) and *Haliotis* (Himes et al. 2011), other compounds released with the eggs can stimulate chemotactic responses by conspecific sperm.

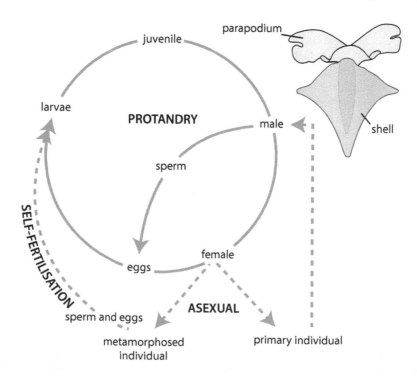

FIGURE 8.13 Possible reproductive modes of *Clio pyramidata* during favourable (solid) and unfavourable (dashed) oceanic conditions. Redrawn and modified from Pokora, Z., *Prz. Zool.*, 33, 397–410, 1989. Note: the asexual component of this hypothetical life cycle is disputed by some workers.

Micropyles or openings for sperm entry have been reported in Polyplacophora, Cephalopoda, and Vetigastropoda. In the Polyplacophora and Cephalopoda, the micropyle has been identified in the egg hull and chorion (Hall 1970; Sakker 1984; Villanueva 1992), although Buckland-Nicks et al. (1990) showed that fertilisation in the chiton *Tonicella lineata* does not require a micropyle. In the Vetigastropoda, micropyles have been illustrated extending through the egg membranes in Trochidae (Lebour 1937; Holmes 1997) or through the albumen (jelly) layer in *Fissurella* (Huaquin et al. 2004). In neither case does sperm penetration of the egg membrane appear to be constrained to these sites, and thus the role of the micropyle in fertilisation appears undetermined in vetigastropods. A micropyle was reported in fresh-water unionid bivalves, but was later shown to represent the attachment area of the developing egg on the ovarian wall (Focarelli et al. 1988).

Molecular processes mediating fertilisation are complex and have been investigated and elucidated only recently in a few taxa, although they have been especially well studied in oysters (Brandriff et al. 1978; Moy & Vacquier 2008; Wu et al. 2011). Several molecules are involved in gamete recognition and are thought to have critical roles in diversification, especially in broadcast species (Moy et al. 2008; Krug 2011). Two of the better understood molecular processes are the role of the protein bindin, which bonds the sperm to the egg during fertilisation, and lysin, which dissolves the vitelline egg membrane. Bindin is found in the sperm acrosomes and binds the sperm to glycoprotein receptors on the surface of eggs in oysters at least (Moy et al. 2008).

Studies by Victor Vacquier and his collaborators have contributed greatly to our understanding of the role of bindin

in bonding the sperm to the egg, and lysin in the acrosomal reaction. Both molecules have rapidly evolved and play a major role in species-specific fertilisation and diversification in broadcast spawning bivalve and gastropods (Talbot et al. 1980; Vacquier 1998; Hellberg & Vacquier 1999; Kresge et al. 2001; Swanson & Vacquier 2002). Our knowledge of these and other molecules associated with fertilisation in molluscan groups outside gastropods and bivalves is virtually non-existent and is a potentially rich research area. In addition, the role of these molecules in internally fertilising molluscs is not well studied but, given their role in other internally fertilising groups such as arthropods and mammals, they will probably be important (Krug 2011).

Penetration of the egg by the sperm is facilitated by the acrosomal reaction, which consists of two stages: (1) the opening of the acrosomal vessel and release of species-specific lysin, followed by (2) the dissolution of the vitelline egg membrane and penetration of the egg by the acrosomal process (Longo 1983) (Figure 8.14). Penetration of the sperm into the egg is also assisted by movement of the flagellum (Wada et al. 1956; Hylander & Summers 1977; Smaldon & Duffus 1985). In coleoid cephalopods, the 'hooking action' of the mitochondrial spur on the sperm head may facilitate passage through the jelly layer and movement down the micropyle (Fields & Thompson 1976).

After the acrosomal reaction has opened and then fused the egg and sperm membranes, the head of the sperm migrates into the egg, initiating activation (Figure 8.14). Besides the sperm nucleus, Longo (1983) reported that in *Spisula* and *Mytilus*, the mitochondria, centrioles, and a portion of the axonemal complex enters the egg. Arnold and Williams-Arnold (1977)

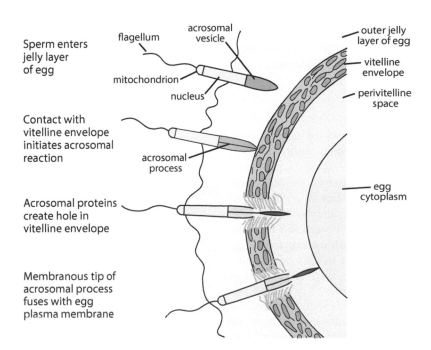

FIGURE 8.14 Fertilisation sequence in a vetigastropod (*Haliotis*). Redrawn and modified from Kresge, N. et al., *BioEssays*, 23, 95–103, 2001.

state that no portion of the flagellum enters the egg in coleoid cephalopods. Egg activation includes the disappearance of the cortical granules followed by the appearance of a space between the vitelline membrane and the egg (sometimes called the perivitelline space), and the breakdown of the germinal vesicle of the primary oocyte.

Depending on whether the egg was arrested as a primary oocyte (Bivalvia), or at metaphase of the first maturation (meiotic) division (Polyplacophora, Scaphopoda, Cephalopoda, Gastropoda) (Figure 8.11), egg activation will differ in details of the breakdown of the germinal vesicle and the formation and organisation of the meiotic spindle (Raven 1966; Longo 1983). The breakdown of the germinal vesicle marks the initiation or resumption of meiotic maturation of the egg, and promotes the development of the male *pronucleus* (Longo & Kunkle 1978). The first polar body is generated at the completion of the first meiotic division, and a second polar body is formed by the second meiotic division (Figure 8.11). These meiotic divisions are followed by the formation of the female pronucleus.[1] Development, migration, and fusion of the male (sperm) and female (oocyte) pronuclei have been well documented in both gastropods and bivalves and are summarised by Raven (1966) and Longo (1983). The conclusion of fertilisation is demarcated by the alignment of the male and female chromosomes in preparation for the first cleavage of the zygote (Longo 1973).

The above sequence of events associated with fertilisation in molluscs requires that a single successful sperm enters the egg and fuses with the female pronucleus. In internally fertilising species, there are multiple behavioural and morphological instances to control sperm abundance and ensure optimal fertilisation opportunities. In externally fertilising species, the interaction is between gametes in the water column, necessitating the release of substantial numbers of gametes to ensure fertilisation success. Large numbers of sperm, however, potentially increase the probability of *polyspermy*[2] (see next Section), an event almost always fatal for the zygote.

8.5.1 BROADCAST SPAWNING

Fertilisation and the subsequent fusion of an egg and sperm DNA to produce a zygote are often overlooked as a critical component of reproductive strategies. As discussed above, fertilisation may be *external* or *internal*, but, as is often the case, the distinction between these lies on a gradient of structures and strategies rather than at a sharp boundary.

In externally fertilising species, the interaction is between gametes and not individuals as it is with internal fertilising taxa. Usually this interaction takes place in the water column and is often called *broadcast spawning*. Broadcast spawning typically involves large numbers of gametes increasing the probability of fertilisation success, which is also determined by such things as gamete dilution and dispersion, organism traits (e.g., abundance, sex ratio, spawning synchrony), and

gamete characteristics (Denny & Shibata 1989; Levitan 1995; Levitan & Petersen 1995; Levitan 2006). Some behaviours also increase the likelihood of fertilisation, including the timing of gamete release and the aggregation of either individuals or their eggs (see Section 8.7).

Several reproductive strategies have been reported for broadcast spawners: a single synchronous spawning event, a protracted spawning period (with or without partial spawning), and continuous spawning, also known as trickle-spawning (Levitan 2006). The advantage of a single synchronous spawning event is that the chance of sperm limitation due to dilution and dispersal is decreased, promoting higher fertilisation success (Oliver & Babcock 1992). The disadvantage of this strategy is that all eggs are literally in the same basket, and subsequent unfavourable environmental conditions will compromise all the propagules.

Generally, in marine invertebrates, successful fertilisation of oocytes is achieved when spawning coincides with optimal environmental conditions suitable for larval development (Sastry 1979; Strathmann 1987a; Olive 1992). Fertilisation success depends on the proximity of potential mates and whether dispersing clouds of eggs and sperm intersect during a critical interval, possibly lasting only a matter of seconds (Picken & Allan 1983; Denny & Shibata 1989). Two strategies appear to have emerged which maximise fertilisation probabilities. The first is *massive synchronised spawning* (MSS) events in which individuals in a population synchronously release their entire gonadal contents. The other is the *partially synchronised spawning* (PSS) strategy, which corresponds to the synchronous release of only a fraction of the gonadal contents of an individual. PSS events are repeated throughout the reproductive cycle until the gonads are depleted, redeveloped, or resorbed (e.g., Barber et al. 2005; Cardoso et al. 2009). The MSS strategy is like a boom or bust one, while the PSS strategy is more of a bet-hedging strategy.

The stochastic nature of fertilisation success, larval survival in the water column and subsequent settlement success, has given rise to several hypotheses regarding the genetic makeup of populations of broadcast species. For example, the *sweepstake reproductive success* hypothesis states that animals producing many eggs (or embryos) (as in MSS) have extremely large variances in individual reproductive success (often by chance alone), leading to a complex pattern of genetic patchiness among annual cohorts of recruits (Hedgecock & Pudovkin 2011). Moreover, individuals surviving recruitment (see Chapter 9) probably represent only a small fraction of the adult breeding population, with these representing those that effectively matched their reproductive effort with optimal conditions favouring gonad development, successful fertilisation, larval survival, and development and ultimately metamorphosis.

The PSS strategy found in some other broadcasting taxa can lead to distinct pools of larvae within a single reproductive cycle. In the marine bivalve *Mya arenaria*, microsatellite markers suggest the absence of a pattern of sweepstake reproductive success, and instead suggest that the first larvae produced in the season are the most successful to survive recruitment. Results also show direct evidence for

---

[1] The gamete nuclei prior to their fusion at fertilisation.
[2] The fertilisation of the egg by multiple sperm.

larval retention and demonstrate larval and post-larval kin aggregation in this species (St-Onge et al. 2015).

Continuous or 'trickle' spawning is associated with the PSS strategy and may increase the chances of sperm and egg collisions over time. This is thought to be a bet-hedging strategy that ensures that at least some larvae will find favourable conditions for survival (Levitan 1995; Levitan & Petersen 1995). Continuous spawning also has been argued to be a strategy of infaunal taxa such as bivalves which counters limited sperm longevity and gamete dilution (Pennington & Chia 1985; Denny & Shibata 1989; Havenhand 1991; Levitan 1995; Levitan & Petersen 1995).

Among marine invertebrates that freely release their gametes into the water, fertilisation success is highly variable, ranging between 0% and 100% (Levitan 1995; Levitan & Petersen 1995). Individual traits of gametes can also greatly influence fertilisation success. These traits include the velocity and behaviour of sperm (Gray 1955; Vogel et al. 1982; Levitan 1993, 2000; Farley 2002; Kupriyanova & Havenhand 2002), and the size and receptivity of the oocyte (Rothschild & Swann 1951; Vogel et al. 1982; Cox & Sethian 1985; Levitan 1993; Duda & Palumbi 1999; Evans & Marshall 2005). Experiments show that differences arise among many marine invertebrate species; for example, those with smaller eggs and faster, but shorter-lived sperm achieve relatively fewer fertilisations compared to species with larger eggs and slower, but longer-lived sperm (Levitan 1998). Although many of the above studies have focused on echinoderms, they are generally applicable to molluscs and provide a rich field of inquiry.

Much of the data covering sperm concentrations which ensure fertilisation while avoiding polyspermy (see previous Section) come from aquaculture research, primarily bivalves (Stephano & Gould 1988; Clotteau & Dubé 1993; Rampersad et al. 1994; Gould & Stephano 2003; Dong et al. 2012) and abalone (Stephano 1992; Grubert et al. 2005; Suphamungmee et al. 2010). Ginzburg (1975) reported that optimum fertilisation occurred at sperm concentrations between 10^6 or 10^7 sperm/cm^3 in *Crassostrea gigas*, *Spisula sachalinensis*, and *Mactra chinensis* (as *M. sulcataria*), and concluded that the drop-off in fertilisation success at higher concentrations was related to reduced sperm activity levels. Clotteau and Dubé (1993) reported optimal fertilisation events when sperm:egg ratios were between 50:1 and 100:1 for the surf clam *Spisula solidissima*, while sperm:egg ratios of 100:1 were also found to be optimum in the abalone *Haliotis asinina* (Suphamungmee et al. 2010). These and other studies suggest that higher sperm concentrations increased polyspermy, while at low concentrations there may be insufficient sperm to fertilise available eggs. A variety of mechanisms to avoid polyspermy are known in free spawning molluscs (Gould & Stephano 2003). These mechanisms fall into two types—global and local (Moy et al. 2008). Global mechanisms are associated with the entire egg and are not species-specific. They include modification of the extracellular layer of the egg to prevent the sperm binding, micropyle access, changes to the egg membrane potential, and formation of the perivitelline space (Gould & Stephano 2003). In contrast, local mechanisms act on individual sperm based

on compatible molecular recognition systems and are thus species-specific. Local mechanisms can act on components of sperm activation, chemoattraction, and molecular mechanisms associated with sperm-egg binding and the acrosomal reaction (Moy et al. 2008). Contact with seawater often makes eggs more resistant to polyspermy (Togo et al. 1995). Abalone eggs also appear to be resistant to polyspermy, even when sperm:egg ratios are 100-fold higher than those required for 100% fertilisation (Stephano 1992).

Generally, for broadcast spawners, theoretical and experimental evidence indicates that fertilisation probabilities are extremely low. Denny (1988), building on the earlier calculations of Vogel et al. (1982), added water movement to the calculation of fertilisation efficiency and suggested that, even at low velocities, only a few percent of the eggs are likely to be fertilised, and the overall number of fertilisation events was not substantially improved by close proximity of males and females. In experiments using sea urchin eggs, Pennington (1985) determined that fertilisation efficiency within 20 cm of spawning males was between 60% and 95%, while at distances greater than 20 cm less than 15% of the eggs were fertilised. These values were higher than those later calculated by Denny and Shibata (1989), who attributed the difference in the two results to uncertainties regarding the mean velocity of water in the experiments of Pennington. Denny and Shibata (1989) concluded that for organisms along wave-swept shores, between 99% and 99.9% of the eggs released are never fertilised. If true, this finding substantially changes the generally accepted view of larval mortality in broadcast spawning taxa, which typically assumes that the lack of reproductive success is due to high predation rates on larvae, larval starvation, or their loss through offshore transport: thus these mortality rates could all be substantially lower than traditionally thought. Last, from an evolutionary perspective, Chaffee and Lindberg (1986) suggested that the marked decrease in fecundity of gastropod taxa with internal fertilisation compared with external fertilising taxa, when both had pelagic development, represented the cost of external fertilisation. Based on their survey, fertilisation efficiency would be between 3% and 15% in external fertilising gastropod taxa and consistent with both theoretical and experimental results. Multiple trends in the evolution of molluscan reproductive systems have apparently addressed this problem, the most common being the independent transition from external to internal fertilisation, most notably in the cephalopods, solenogasters, and gastropods.

8.5.1.1 Doubly Uniparental Inheritance in Bivalves

While most animals transmit their maternal mitochondrial genome, paternal mitochondrial DNA (mtDNA) transmission is known in some invertebrates and vertebrates (e.g., Korpelainen 2004). The transmission of mtDNA from both parents is known only within bivalves, a phenomenon known as doubly uniparental inheritance (DUI). This has been reported in eight families of bivalves (Mytilidae, three families of Unionidea, Trigoniidae, Veneridae, Donacidae, and Solenidae) (e.g., Hoeh et al. 1996, 2002; Zouros 2000; Theologidis et al. 2008;

Glavinic et al. 2010; Alves et al. 2012). In DUI taxa, there are two lineages of mtDNA, one transmitted via the eggs (maternal or type 'F genome') to both male and female offspring, the other (paternal or type 'M genome') transmitted in the sperm only to male offspring. The F genome predominates in the somatic tissue and the M type in the gonads in males (Stewart et al. 1995; Sutherland et al. 1998). In *Mytilus* species, there is also evidence of paternal mtDNA in somatic female and male tissues (Stewart et al. 1995; Garrido-Ramos et al. 1998; Dalziel & Stewart 2002), as well as in eggs (Obata et al. 2006), however, the sperm has been reported to be free of the maternal genome of the male (Venetis et al. 2006). The result is there are two distinct mtDNA lineages in female and male tissues, and while sometimes these are very distinct (up to around 50%), this is not always the case, and in some mytilids, genomes may become similar through recombination (Burzynski 2003). This situation is further complicated in mytilids by mitochondrial genome *masculinisation* in which a maternal or type F genome gives rise to a new paternal or type M genome (Hoeh et al. 1996). Subsequent extinction of the ancestral type M genome(s) and its replacement by the new one (Hoeh et al. 1997; Quesada et al. 1999) resets the time of divergence between conspecific gender-associated mtDNA lineages and probably accounts for low divergence levels in some species of *Mytilus* (Hoeh et al. 1996). In contrast, there is no evidence of this occurring in unionoideans which consistently show a high level of divergence (Hoeh et al. 1996).

Comparative phylogenetic and genomic analyses, conducted by Doucet-Beaupré et al. (2010), hypothesised a single origin of DUI in an early autobranch lineage with its subsequent loss in descendant lineages, including the Pectinoidea, Ostreoida, and numerous veneroid groups. Alternatively, it could have independent origins in each higher group (Mytiloidea, Palaeoheterodonta, Veneroidei, and possibly Tellinoidea) (Passamonti & Ghiselli 2009).

8.5.2 Internal Fertilisation

In some species, sperm is transferred between individuals by copulatory or transfer structures (penis, spermatophore, hectocotylus, etc.) and the eggs are fertilised internally; sperm are often stored following sperm transfer and before fertilisation.

Internal fertilisation is absent in scaphopods and rare in bivalves, polyplacophorans, and patellogastropods where it is associated with the retention of the eggs or brooding, generally in the mantle cavity (see below). Copulation is also rare in Vetigastropoda, but dominant in Neritimorpha, Caenogastropoda, and Heterobranchia, and in Cephalopoda and Solenogastres. In contrast to broadcast spawners, the interaction is between individuals rather than between gametes. In internal fertilising species, the sperm received from the partner is called *exogenous* sperm or *exosperm* and that produced by the individual is termed *endogenous* sperm and is referred to as *autosperm*. While this distinction is not very significant in gonochoristic species, it is important in hermaphroditic taxa where sperm with different origins is often present in the same individual.

Although copulation is the most common form of sperm transfer in internal fertilising species, special structures, such as spermatophores and spermatozeugmata, can facilitate the transfer of sperm masses and are found widely distributed among unrelated molluscan groups (see Section 8.3.1.2 and Table 8.1).

In internally fertilising gastropods, the ova are fertilised at a point immediately before the junction of the oviduct with the oviducal glands, and thus before they receive the nutritive and protective coatings that together make up the egg (see Section 8.5.4). They receive sperm that has been stored, usually in one or more seminal receptacles. The area where fertilisation takes place is called the fertilisation chamber or, in eupulmonates, the *carrefour*, although the latter structure often has sperm-storing structures associated with it and has a differentiated fertilisation sac or pouch (see also Chapter 20).

Internal fertilisation substantially increases fertilisation probability over externally fertilising taxa. The ability for an individual to bulk transfer sperm (copulation, spermatozeugmata, spermatophores, sperm balls, etc.) should be highly advantageous to an individual (Pemberton et al. 2003a, 2003b). Copulatory structures have evolved only in cephalopods and gastropods, having evolved separately in every major gastropod clade, including patellogastropods and vetigastropods. In these two latter groups, copulatory organs are relatively rare and mostly appear in taxa living in difficult habitats, such as hydrothermal vents or boreal regions, or in some with small body size. Similarly, the appearance of gelatinous egg masses in bivalves is also associated predominantly with taxa facing similar latitudinal challenges, and appear sporadically on the bivalve phylogenetic tree (Collin & Giribet 2010). The appearance of these innovations has not led to any substantial radiations in patellogastropods, vetigastropods, or bivalves.

8.5.2.1 Sperm-Casting

Sperm-casting has not been as well studied as other forms of fertilisation methods. It is defined as the entrainment of sperm from the water column by the female and subsequent internal or *semi-internal*[3] fertilisation in the absence of copulatory or transfer structures (penes, spermatophores, hectocotyli, etc.). It occurs when males broadcast sperm, while females retain eggs and brood the developing embryonic stages (Pemberton et al. 2003a, 2003b; Falese et al. 2011). A few bivalves produce gelatinous egg masses (Collin & Giribet 2010) (see Chapter 15) that are deposited externally on the substratum where embryos develop outside the body of the animal.[4] The eggs are probably fertilised before their encasement in the gelatinous mass, and there is some observational evidence to support this (e.g., *Abra tenuis*) (Gibbs 1984). We also regard this strategy as a form of sperm-casting, a phenomenon which may have played an important role in the evolution of internal fertilisation (Fretter 1988) by selecting for the ability to retain entrained sperm in

---

[3] Fretter (1988) distinguished internal fertilisation (within the reproductive tract) from semi-internal fertilisation (within the mantle cavity).

[4] Gelatinous egg masses are also present in a few vetigastropods, including *Diodora* (Strathmann 1987b).

the mantle cavity. A scenario where tissue folds in the mantle cavity may have provided early sperm storage/concentration is plausible; further invagination of these folds into the genital tract may have given rise to sperm storage and fertilisation chambers. Pseudo-copulation occurs when the male and females are in very close proximity, enabling direct sperm transmission. In the bivalve *Oivariscintilla yoyo* the foot may play a role in assisting sperm transfer (Mikkelsen & Bieler 1992).

8.5.2.2 Copulation

Copulatory structures have evolved independently in gastropods and cephalopods and have led to accompanying changes in sperm morphology as well as modifications to both female and male systems in these groups. There are probably two sources of selection acting here: (1) factors associated with storing sperm including *sperm competition* (see next Section), the potential for delayed or repeated spawning, and the energetic advantage of digesting excess sperm and (2) internal fertilisation, making possible the encapsulation of the spawn and the addition of secondary nutrition such as albumen for the developing embryo(s). Copulatory structures in gastropods are typically derived from foot, mantle, or head tissue and are commonly called a penis or verge. In the female system, these two potential selection factors are manifest as elaborations of the reproductive tract with multiple sperm sacs and glands (Runham 1988). Copulatory structures in cephalopods include a penis, and in some coleoid taxa (e.g., octopods, decabrachian squid), a specialised arm(s) for spermatophore transfer known as the *hectocotylus* (see below and Chapter 17). When present, the hectocotylus provides more accurate placement of the spermatophore on the female than the penis alone (Arkhipkin & Laptikhovsky 2010).

Internal fertilisation and the associated increased complexity of reproductive systems are derived characters within gastropods and cephalopods (e.g., Arkhipkin 1992; Ponder & Lindberg 1997; Giribet et al. 2006). While the penis (and the associated Needham's sac) appears to be homologous across coleoids, the penis in gastropods has been formed independently in many groups and thus is not homologous. Although copulatory structures and more complex reproductive systems are generally associated with caenogastropods and heterobranchs, copulatory structures are present in all major gastropod groups, including a few patellogastropods (Golikov & Kussakin 1972; Lindberg 1987b), vetigastropods (e.g., Kano 2008), and most neritimorphs (Holthuis 1995; Barker 2001a; Kano et al. 2002), but are absent among the more basal members of those latter groups. Like most gastropods, the copulatory structures of patellogastropods and vetigastropods are associated with the head. Examples include: *Gorgoleptis* – left wall of snout (gonochoristic), *Lepetodrilus* and *Clypeosectus* – ventral base of right cephalic tentacle and right side of inner pallial wall (gonochoristic), Lepetelloidea – right cephalic tentacle, left side of inner pallial wall (both gonochoristic and hermaphroditic), Seguenziidae – right neck lobe or left side of inner pallial wall (gonochoristic), *Bathymargarites* – right

subocular peduncle (gonochoristic), and Skeneidae s.s. – anterior right corner of foot (simultaneous hermaphrodite). In patellogastropods, the copulatory structure is of the pipette type with no ciliated groove between the urogenital papilla and the copulatory structure (Golikov & Kussakin 1972; Lindberg 1987b). It transfers sperm into the female mantle cavity where the young are brooded. In vetigastropods and neritimorphs, the presence of a seminal receptacle is correlated with the presence of a penis, and the sperm are often modified for semi-internal fertilisation. In vetigastropods, the penis is also pipette-like, with a ciliated groove, rolled flap, or an enclosed tube (Kano 2008) connecting the urogenital papilla with the penis. Of interest is that much of the morphological diversity of cephalic copulatory structures in non-heterobranch gastropods has been reported as cephalic teratologies in patellogastropods (Pelseneer 1928).

In lower heterobranchs, the penis lies on the right side of the head. In the more plesiomorphic shelled taxa, it is contained within the mantle cavity, as in neritimorphs and caenogastropods, but in more derived heterobranchs, the copulatory organ is invaginated on the right side of the animal (see Chapter 20). Sperm transfer is from the genital opening by way of an open ciliated sperm (or seminal) groove, which extends to the tip of the penis. This open sperm groove (e.g., in aplysiids, Figure 8.5) is similar to the convergent arrangement found in some caenogastropods such as *Littorina* (e.g., Fretter & Graham 1962). See Chapter 19 for further discussion of caenogastropod copulatory structures.

Speculation as to the drivers of the invagination of the penial apparatus into a penial sheath (or preputium) which occurred in euthyneuran heterobranchs includes the reduction of the shell and mantle cavity that potentially resulted in the exposure of the penis on the head in 'opisthobranchs' (Ghiselin 1965). This would have been particularly disadvantageous in burrowing taxa, but would not explain the inversion of the penial apparatus in 'pulmonates' where the basal taxa are shelled. The invagination of the penis into the body wall may, of course, have occurred independently in more than one euthyneuran lineage, but this possibility does not seem to have been rigorously investigated.

Copulation in cephalopods is accomplished by transferring spermatophores from the male to the female and is typically accompanied by complex courtship behaviours which facilitate copulation (Figure 8.15). In this clade, besides the presence of a penis, there are other morphological features, typically derived from tentacles, that assist with the transfer. In *Nautilus*, the penis lies within the mantle cavity and the actual transfer of spermatophores relies on the spadix, a complex erectile organ derived from modified small tentacles (cirri) (Arnold 1984). Many extant coleoids (octopods, cuttlefish, and nearshore squid) typically have short penes contained within the mantle cavity. Offshore squid typically have longer penes, which are erectile in some taxa (Arkhipkin & Laptikhovsky 2010).

Transfer of the spermatophores to the females during mating may be facilitated by the hectocotylus (Arnold 1984) (Figure 8.15). In the pelagic octopod *Argonauta,* the hectocotylus may detach from the male and remain

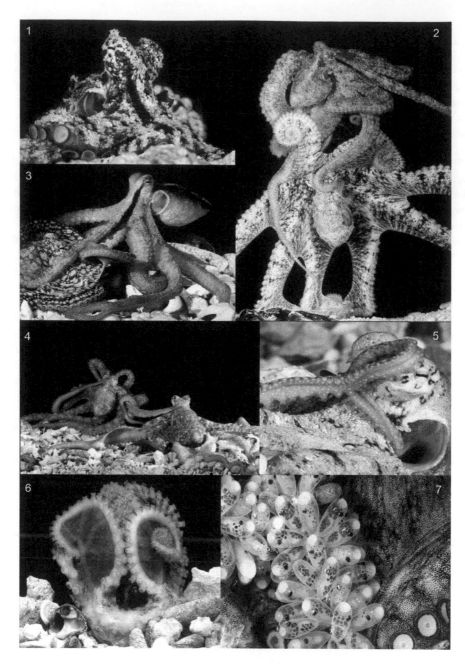

FIGURE 8.15   Mating and egg-laying sequence of *Abdopus aculeatus*. 1. Male display, 2. First contact, 3. Continued courtship behaviour, 4. Insertion of hectocotylus by male, 5. Close-up of hectocotylus insertion, 6. Female brooding eggs, 7. Close-up of eggs being brooded. Courtesy of R. Caldwell.

in the mantle cavity of the female. Spermatophores may be found on the external mantle, head, and neck and in some large, deep-water squid (e.g., *Taningia danae*) they are found implanted in cuts thought to have been made by the males (Hoving et al. 2010). *Vampyroteuthis* and the cirrate octopod cirroteuthids lack the hectocotylus, and the mechanism of spermatophore transfer is unknown (Pickford 1957). In female *Vampyroteuthis*, the transferred gametes are stored in a preocular pouch which serves as a seminal receptacle (Arnold 1984). See Chapter 17 for additional discussion of cephalopod reproductive morphology and behaviour.

8.5.2.3   Multiple Paternity and Sperm Competition

Many internally fertilising gastropods mate multiple times and sperm can be stored in sperm storage structures (seminal receptacle, bursa copulatrix, etc.), often for long periods. For example, in *Littorina*, sperm is placed in the bursa copulatrix during copulation and can be stored there for up to three months (Buckland-Nicks et al. 1999). It migrates to the seminal receptacle, where the sperm of more than one partner may accumulate (Paterson et al. 2001). Viable sperm may be stored in the seminal receptacle for a considerable time; in some stylommatophoran snails for up to a year (Baur 1994a). Recent studies using genotyping have demonstrated that the offspring of a single littorinid brood can be sired by as many as 20 males, although fewer are generally the rule in this and other taxa (Mäkinen et al. 2007; Panova et al. 2010; Xue et al. 2014). In contrast, in *Crepidula fornicata*, Proestou et al. (2008) found that 83% of the offspring from a single female were fathered by the largest and closest of the males stacked on the back of the female's shell, although the success of the largest male decreased as the number of mature males increased.

When sperm of more than one partner can be stored, sperm competition can occur, i.e., competition for fertilisation between the sperm of different individuals. For example, in *Crepidula fornicata*, spermatozoa which first anchor onto the walls of spermathecal lobes have a greater chance of fertilising the oocytes than later arrivals, because when the walls are covered, spermatozoa from subsequent fertilisations cannot attach and decompose in the lumen of the lobules (Beninger et al. 2016).

While fertilisation success is determined by many factors, some of these may be influenced by the sperm donor. Such factors may include the number and quality of the sperm or prevention or reduction in the success of matings of the female with other males. Structural features of the reproductive tract may also make a difference. For example, Garefalaki et al. (2010) showed that mating order and epiphallus length were traits affecting the outcome of sperm competition in *Cornu aspersum*. In some buccinids and muricids, studies have shown that the last male to mate with a female sires the largest proportion of the brood (Lombardo et al. 2012; Xue et al. 2016).

In simultaneous hermaphrodites, gender conflicts can arise when one individual tries to maximise the import of its preferred gender role – typically male (Bateman's principle) (Bateman 1948). As an example, an individual provides sperm to fertilise the eggs of a mate, while at the same time does not allow its own eggs to be fertilised by the partner. The exchange of sperm (sperm trading) through reciprocal copulation in simultaneous hermaphrodites is thought to be one way to solve such gender conflicts (Leonard 1991). To work optimally, partners should provide equal access to their sperm and eggs (Anthes et al. 2005, 2006b), although this ideal situation is further complicated by some studies showing that equal access does not necessarily occur even during reciprocal copulation (Anthes & Michiels 2005; Anthes et al. 2006b).

8.5.2.4 Self-Fertilisation

Self-fertilisation ('selfing') should theoretically be possible in any simultaneous or protandric hermaphrodite provided that, in the latter case, sperm is retained until the female phase. There are advantages to self-fertilisation – notably there is a sperm transmission advantage since the sperm is maintained in a single individual. There is also assurance of reproductive success, even in an environment with little or no chance of mating. There are also costs, most notably inbreeding depression. Even in taxa capable of self-fertilisation, it is almost always not the preferred option. This may well have been a trend driven by selection, as selfing leads to a loss of genetic diversity and fitness due to inbreeding depression (Jarne & Charlesworth 1993). Selfing, however, has a strong advantage for founders in new habitats following a dispersal event or for survivors of a disaster. Self-fertilisation is

common in stylommatophoran land snails and slugs and in freshwater hygrophilans (see Chapter 20) and has been recorded in the protandrous caenogastropod *Calyptraea* (Wyatt 1960).

8.5.3 Mating Aggregations

As in broadcast species, mating aggregations usually increase reproductive success by bringing potential mates into close contact, although such aggregations of internal fertilising taxa produce new selective pressures including male-male competition, multiple paternity, sperm competition, and cryptic female choice (Ghiselin 1987; Baur 1998; Hall & Hanlon 2002). Perhaps not surprisingly, mating aggregations of internal fertilising species appear to be less common than in external fertilising species (see Section 8.7).

8.5.3.1 Eggs, Capsules, and Egg Masses

Egg masses produced by externally fertilising gastropods are restricted to a handful of vetigastropods which produce gelatinous masses or ribbons affixed to the benthos (see Chapter 18). These include the fissurellid *Diodora apertura* and several trochid species among the *Calliostoma*, *Cantharidus*, *Gibbula*, *Margarites*, and *Margarella* (Webber 1977; Strathmann 1987b; Hickman 1992; Przeslawski 2004b). Two sources for the adhesive mucus that binds the spawn have been identified. The first is the jelly coat of the eggs themselves, which in *Calliostoma* and *Cantharidus* may be supplemented by secretions from tissues surrounding the urogenital papilla (Webber 1977). Vetigastropod spawning masses lack the organisation of masses and capsules produced by internal fertilising gastropods. Instead, they typically have the eggs surrounded by three to four distinct layers which serve either to provide nutrients to or protect the developing embryo. In heterobranchs, the egg capsules consist of three layers: (1) an inner nutritive 'albumen' layer (which consists of the carbohydrate galactogen and sometimes protein [e.g., Ghiselin 1965]); (2) a thin membrane layer (or egg covering) defining the individual capsules which contain one or more eggs, and in many heterobranchs the membrane layer may form a connecting strand (*chalazae*) between capsules; and (3) an outer gelatinous layer of mucus that lies outside the membrane. This latter layer may be single or a series of layers, composed of material produced by the mucous or nidamental glands. In neritimorphs, most caenogastropods, and many eupulmonates one or more additional firm layers form the egg capsule (Figures 8.16 and 8.17).

In aquatic environments, gelatinous egg masses are thought to confer several advantages, besides providing an extra protective layer (e.g., against abrasion, desiccation in the intertidal) (Pechenik 1979). Egg masses also contain embedded chemical defences to discourage predators (Ebel et al. 1999), sunscreens against UV (Przeslawski 2004a; Przeslawski et al. 2004), and antibacterial compounds which prevent infection (Benkendorff et al. 2000, 2001b; Ramasamy & Murugan 2007). The egg masses of the fresh-water hygrophilan *Biomphalaria glabrata* contain at least 20 polypeptides, 16 of which are identified as defensive (Hathaway et al. 2010). Several of these peptides have been isolated from the albumen gland, and the immunoprotection these proteins provide is thought to protect against sexually transmitted

FIGURE 8.16 Some examples of gastropod spawn. Top left, *Nerita atramentosa* (Neritidae), New South Wales, Australia. Courtesy of R. Przeslawski. Top middle, *Dicathais orbita* (Muricidae), New South Wales, Australia. Courtesy of K. Benkendorff. Top right, *Nucella lamellosa* (Muricidae), California, USA, egg capsule mass (approx. 30 cm diameter) from communal capsule deposition. Courtesy of D. Lindberg. Middle left, *Tonna galea* (Tonnidae), Greece, spawning egg ribbon. Courtesy of G. Sangiouloglou. Middle right, *Dolabrifera dolabrifera* (Aplysiidae), New South Wales, Australia. Courtesy of K. Benkendorff. Lower row, left, *Rostanga rubicunda* (Discodorididae), New Zealand; middle, *Siphonaria funiculata* (Siphonariidae), New South Wales, Australia, pelagic egg mass; right, *Fastosarion freycineti* (Helicarionidae), New South Wales, Australia, eggs of this terrestrial semislug; all courtesy of W. B. Rudman.

pathogens and to prevent their vertical transmission to the embryo (Hathaway et al. 2010). Gelatinous coatings also increase spacing between embryos, which in turn increases oxygen diffusion (Chaffee & Strathmann 1984; Moran & Woods 2007).

Several hypotheses have been put forward as to the reasons gelatinous egg masses have not evolved in certain groups of marine invertebrates. These include physiological constraints on packing embryos too tightly together, limits on safe deposition sites, and difficulties in transferring sperm to eggs. By forming artificial egg masses with normally free living embryos, Strathmann and Strathmann (1989) demonstrated that special embryological adaptations are not necessary for development within a gelatinous egg mass. In addition, physiological factors

appear to constrain the shape and size of egg masses, and the distribution inside the mass, but any constraints seem fairly permissive as many forms of egg masses occur in nature. With bivalves, it is not immediately clear why gelatinous egg masses are so uncommon as limitations on bringing eggs and sperm together inside the female are overcome in a fairly diverse array of brooding bivalves.

Bivalves rarely produce externally attached spawn, but this occurs in a few phylogenetically diverse taxa (Collin & Giribet 2010) (Chapter 15). The largest known bivalve eggs are produced by the protobranch *Acharax alinae* (600–660 µm) (Beninger & Le Pennec 1997). The protobranchs *Solemya velum* and *S. reidi* deposit individual eggs in adhesive capsules

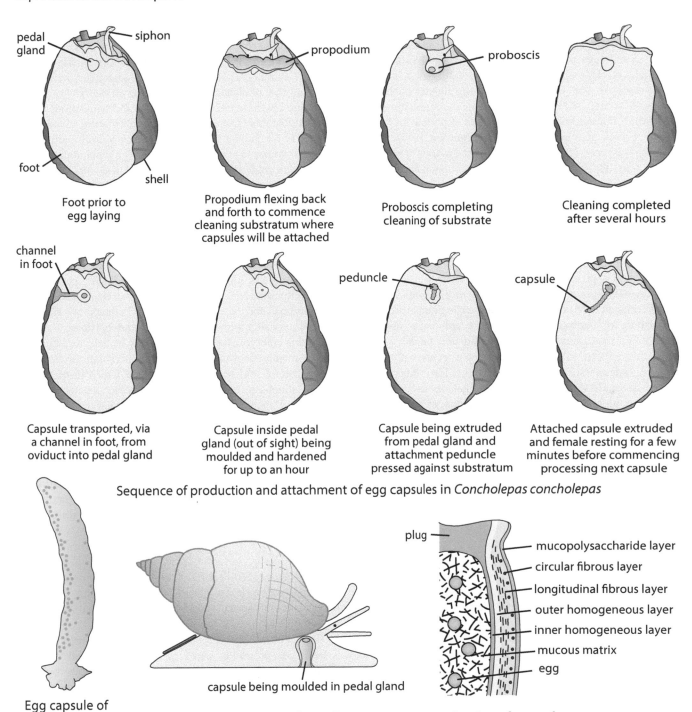

FIGURE 8.17 Egg capsule morphology, and production and attachment of capsules by the muricid caenogastropods *Concholepas concholepas* (as observed through aquarium glass), and *Nucella lapillus*. *Concholepas* sequence and capsule redrawn and modified from Castilla, J.C. and Cancino, J., *Mar. Biol.*, 37, 255–263, 1976.) *Nucella* animal original; egg capsule section redrawn and modified from Bayne, C.J., *Proc. Malacol. Soc. Lond.*, 38, 199–212, 1968a.

on the substratum (Gustafson & Reid 1986; Gustafson & Lutz 1992), and, in *Nucula delphinodonta*, the eggs develop in a gelatinous sac attached to the posterior valves (Drew 1901).

The venerid *Turtonia minuta* attaches a single egg capsule with 8–19 embryos to its byssus (Ockelmann 1964). Strings

of oocytes enclosed in a mucous sheath and attached to the byssus of the bivalve or substratum are produced by a few mytilids, including *Musculus discors* (Ockelmann 1958) and *Musculus niger* (Thorson 1946). *Musculus niger* has large eggs, measuring over 500 × 350 μm, which are embedded in a mucous string and attached to the byssus (Ockelmann 1958). In the Carditida, *Astarte elliptica* and *A. sulcata* attach their mucus coated eggs to the substratum by way of an adhesive membrane (Ockelmann 1958). In the tellinoideans, Gibbs (1984) reported that *Abra tenuis* laid egg masses containing between 400 and 1800 eggs directly on the substratum, and the cardiid *Parvicardium exiguum* was reported to lay its eggs in a thick gelatinous covering (Jørgensen 1946). In the lucinids, Pelseneer (1926a) reported that *Loripes orbiculatus* (as *L. lacteus*) deposited stalked egg capsules containing multiple eggs on the substratum, although other workers have suggested that the capsules are actually released into the water column (see Collin & Giribet 2010). The tropical lucinid *Phacoides pectinatus* deposits spherical egg masses that resemble the egg cases of arenicolid polychaetes in sea grass beds (Collin & Giribet 2010). Fertilisation of the eggs before encasement or encapsulation is probably facilitated by sperm-casting or other mechanisms of sperm transfer (see Sections 8.5.2 and 8.5.2.1).

Ockelmann (1958) observed that increasing latitude resulted in a pattern of increased maternal care and larger eggs for bivalves. Further examination by Collin and Giribet (2010) found this pattern to be equivocal for species with either gelatinous egg masses or encapsulated eggs. They did point out that two of the six species known to have gelatinous egg masses are hosts to sulphur-oxidising bacteria, and that members of two of the species with sticky, encapsulated eggs, *Solemya velum* and *Solemya reidi*, also host such bacteria. The symbiotic bacteria in individuals of *S. velum* are vertically transmitted, while those in the lucinid *Codakia orbicularis* are environmentally transmitted (Gros et al. 1997). It would be interesting to further examine the possible link between chemosymbiotic bacteria and the evolution of egg masses in bivalves.

All cephalopods produce complex spawn masses as described in more detail in Chapter 17.

8.5.4 Oviducal Glands

Oviducal glands associated with the oviduct that package eggs are found in many molluscan taxa. These glands are often seen as thick folds composed of gland cells interspersed with ciliated cells. As the cilia move the eggs through the oviducts, glandular secretions are deposited over them, producing the corresponding layered components of the spawn. In more complex systems such as those found in cephalopods and gastropods, the glands are separated from the lumen of the oviduct, and the secretions are deposited on the fertilised eggs as they pass by the opening to the gland. Various combinations of these glands may be present in different taxa, and there is a long history of confused terminology and homology of these structures, between and within groups (e.g., Ghiselin 1965).

Oviducal glands are diverse across the molluscs. In some polyplacophorans, the proximal portions of the gonoducts have large diverticula that apparently coat the eggs with mucus before spawning (Pearse 1979). 'Shell glands' have been reported in Solenogastres, which are thought to secrete an egg membrane (Heath 1911a; Hadfield 1979); a similar glandular

epithelium is also found surrounding the openings of the gametoducts of the Caudofoveata (Scheltema 1973; Hadfield 1979). Oviducal glands are rare in bivalves (Runham 1988), although Purchon (1941a) described a glandular accessory organ in the pholadoidean *Xylophaga dorsalis*. Oviducal glands are absent in monoplacophorans and scaphopods.

Most coleoids have a well-developed oviducal gland (see Chapter 17 for details), but this is absent in *Nautilus*, although the proximal tissue of the gonoducts is highly glandular and a single nidamental gland is also present (Haven 1977). Paired nidamental and accessory nidamental glands are present in most decabrachians (Arnold & Williams-Arnold 1977), but in octopods there are paired oviducts, each with an oviducal gland and nidamental glands. As in other molluscan taxa, the oviducal glands coat the eggs with jelly layers, the outermost forming the outer egg capsule (Runham 1988).

In gastropods, distinct oviducal glands are present in some vetigastropods (e.g., *Calliostoma*), caenogastropods, neritimorphs, and heterobranchs (see Chapters 18, 19, and 20) and typically provide protection and nutrients for the developing embryo. These structures may include: an albumen gland, a membrane gland that secretes a membrane around the albumen, a mucous or muciparous gland which provides an outer gelatinous covering of the egg mass, and an outer capsule or oothecal gland which produces the outer egg capsule. Sometimes there is a second smaller membrane gland (e.g., the winding gland in aplysiidans) which lies between the albumen and membrane glands in most euopisthobranchs (Ghiselin 1965) (Figure 8.4). In some groups, the eggs pass through the albumen gland, while in some heterobranchs (e.g., Sacoglossa and Aplysiida), a secretion from the blind gland is accumulated in a proximal area where it then coats the eggs (Ghiselin 1965).

8.5.5 Encapsulation

As noted above, fertilisation of the eggs before encasement or encapsulation is facilitated by sperm-casting or other mechanisms of sperm transfer, including copulation (see Section 8.5.2).

Those species that produce benthic eggs must provide a means of protecting them as they lie exposed on the substratum, such as encapsulation and/or protection by the parent. In some neritid and tateid gastropods, the capsule itself is reinforced with calcareous spherulites, sand grains, diatoms, and other small particles incorporated into the outer covering of the capsule (Tan & Lee 2009). In addition to the production of gametes and any additional resources provided as a food source, energy is also expended in capsule formation (Perron 1981; Perron & Corpuz 1982). The cost of constructing capsule walls can be very high, e.g., 39%–45% of the total biomass of the brood (Stickle 1973) or 20%–50% of the energy invested in reproduction (Perron 1981; Perron & Corpuz 1982). In those gastropods which abandon the egg mass, the degree of investment in the capsule wall may vary with the number of eggs in the capsule or the length of the intracapsular developmental period (Gibson et al. 1970; Spight 1975; Perron 1981). In many caenogastropods, the capsule is transferred to the pedal gland of the foot where it is shaped and hardens before being affixed to the substratum (Figure 8.17).

Egg capsule size is related to body size of the female, at least in some neogastropods (Spight et al. 1974). Thus with increasing age (and hence size), there may be changes in the allocation of resources to capsule production, which require correlated changes in the production of ova and intracapsular fluid (Perron & Corpuz 1982). As capsule size increases, surface/volume ratios are reduced and capsule wall thickness increases. The number of eggs per capsule may not change much as respiratory constraints may limit the number of embryos due to capsule surface area and/or wall thickness. Where this is the case, per ovum costs of encapsulation increase with increasing female age and size.

In many gastropods with early intracapsular development, there are both embryonic and nurse eggs in the capsules. The nurse eggs are normally consumed by the embryos as an energy source to supplement yolk during intracapsular development (Fretter 1984a; Pechenik 1986). Fecundity is determined by the number and/or size of the capsules and the number of eggs each contains (Castilla & Cancino 1976; Spight 1976b; Gallardo 1977, 1979, 1981; Perron 1981; Perron & Corpuz 1982; Deslous-Paoli & Heral 1986). Many internal fertilising marine gastropods provide their embryos with extra-embryonic food sources other than nurse eggs, such as nutritive fluids including albumen and yolk. This supplementary food is packaged in the capsules with the fertilised eggs and can influence the hatching size of the embryos (Gallardo 1979, 1981; Hoagland 1986; Pechenik 1986). In the absence of such extra-embryonic material, the size of the embryo at hatching may show little intraspecific variation (Fioroni 1966). This variation may result from differences in food storage among capsules or spawn masses, due to different numbers of embryos per capsule (Spight 1976b). According to Rivest (1983), differences in hatching size resulting from variation in the nurse egg/embryo ratio may be an adaptive mechanism when the environment is unpredictable and when there is a relatively long period between egg deposition and the emergence of the juvenile. This outcome is thought to be adaptive because there are always some larger juveniles in the brood that have a greater chance of survival should conditions be poor at the time of hatching.

The role of variability in the allocation of eggs and nutritional resources to capsules extends to other life history traits, including breeding seasons and cycles, egg and larval sizes, larval duration, etc. – which all have substantial adaptive significance – and are thought to increase fecundity and survivorship of individuals as well as increasing recruitment across populations and their persistence through geological time (Hadfield & Strathmann 1996).

Lesoway et al. (2016) provided an interesting approach to understanding the origin of nurse eggs in caenogastropods. Because nutritive embryos, or nurse eggs, serve as nutrition for their viable capsule mates, they are an example of an alternative developmental phenotype. *Crepidula navicella* produces broods composed primarily of nurse eggs and a small number of viable embryos. Lesoway et al. (2016) compared the transcriptomes of viable and nutritive embryos and found that while viable embryos expressed high levels of transcripts associated with known developmental events, nutritive embryos expressed high levels of *apoptosis* (programmed cell death) transcripts. This result suggests that apoptosis has an important role in the formation of nutritive embryos in *Crepidula*.

8.5.6 Parental Care

Parental care can be defined as parental behaviour likely to increase the survival and/or fitness of the offspring. In its broadest sense, it could include the production of large, yolky eggs, the provision of unfertilised nurse eggs, brooding, or the preparation of sites for egg-laying, but in another sense, it can refer to the care of eggs or young detached from the body of the parent (e.g., Baur 1994b). Parental care strategies for marine gastropods (Fretter & Graham 1962, 1964; Purchon 1968; Sastry 1979; Brahmachary 1989) include eggs being retained within the female reproductive tract, or on the body, or within special brood chambers. Some terrestrial oviparous gastropods protect their eggs by depositing them in excavations which they dig in the soil. Species in which the eggs are retained within the female reproductive tract for the entire developmental period are called *ovoviviparous*. In *viviparous* species, nutritional material is provided by the parent to the developing embryo. Variation across these three modes is nearly continuous in gastropods and bivalves. In cephalopods, most female octopuses guard their eggs until hatching (Wells & Wells 1977a; Boal 2006) (see Chapter 17).

Protection of offspring is thought to be energetically expensive; in gonochoristic species, these costs are generally incurred by the female, with egg capsules and other protective structures representing up to 50% of the investment in a brood (Perron & Corpuz 1982). The process of egg-laying and moulding the capsules is also expensive, sometimes taking several days, during which the females usually do not feed and may be vulnerable to predation (Brokordt et al. 2003). In soft-bottom habitats, caenogastropods often attach their eggs to their own shells, as, for example, in the nassarids *Bullia melanoides* and *Buccinanops* spp. (Averbuj et al. 2014; Averbuj & Penchaszadeh 2016), the collumbellids *Bifurcium bicanaliferum* and *Strombina* spp. (Fortunato et al. 1998; Fortunato 2002), and the olivid *Olivella puelcha* (as *O. plata*) (Pastorino 2007). Some, such as the muricid *Hexaplex nigritus*, the hydrobiid *Peringia ulvae*, and the tateid *Ascorhis* also carry conspecific egg capsules, but it is not known if they are their own (Thorson 1946; Ponder & Clark 1988; Cudney-Bueno et al. 2008). Female *Solenosteira macrospira* (Buccinidae) deposit capsules on their male partners (Kamel & Grosberg 2012). Parental care often comes with significant fitness costs and Kamel and Grosberg (2012) have shown reduced growth in both male and female snails with attached capsules compared to controls without egg capsules. Miglavs et al. (1993) reported that at least one species of *Oenopota* (Mangeliidae) places its eggs amongst the brooded eggs of the boreal shrimp *Sclerocrangon boreas*. How this is accomplished remains unknown, but the shrimp's ventral brood space provides protection and is well ventilated by the crustacean.

Some other caenogastropod females guard their benthic egg masses until they hatch. For example, female *Buccinum isaotakii* guard their eggs until another female deposits on the mass (Ilano et al. 2004), and in *Fusitriton* spp. and some other cymatiids, females guard their eggs by sitting on or adjacent to the spawn (Ramon 1991; Beu 1998; Canete et al. 2012). Female cowries (Cypraeidae) also guard their spawn by pushing potential predators away with their shells, lifting the shell up and bringing it down on the intruder, or biting with the radula (Ostergaard 1950; Wilson 1985; Osorio et al. 1999). While most species leave the mass when the eggs have hatched, one species of cowrie has been observed to apparently assist hatching by chewing off the tops of the capsules (Katoh 1989). Parental care is less common in heterobranchs (see Baur 1994b).

8.5.6.1 Brooding

Brooding is defined here as the retention of fertilised eggs, on or in a parent, and through their subsequent development, until their release as veligers or crawl-away juveniles. In broadcast spawning taxa, brooding is thought to increase fertilisation frequency by concentrating eggs, while in others, fertilisation may be achieved through sperm-casting, or more rarely, copulation. The majority of brooded young complete development near the parent, and sibling competition is thought to be initially high, although this may be offset to some extent by the likely favourable habitat encountered by the brood upon release. Thus, brooding is also associated with reduced dispersal potential and (perhaps) diversification (Cohen & Johnston 1987; Pearse et al. 2009).

Brooding can occur on the external shell or within the body or, as with the pelagic octopod *Argonauta*, in a shell-like external egg case secreted by specialised arms of the female (see Chapter 17). Internal brooding may occur in the mantle cavity, in specialised chambers, or in the ectodermal parts of the reproductive system. Specialised brood chambers may be derived from the mantle cavity or, in bivalves, in the gills, but can also include a modified kidney in a patellogastropod (see below) or the cephalic chamber of some cerithioideans and the pedal gland in a hot vent provaniid gastropod (Reynolds et al. 2010). Sometimes the shell is deformed to accommodate the brood (see Section 8.6 below). While most theories argue that brooding shields larvae from pelagic mortality (Strathmann & Strathmann 1982; Pechenik 1999), in estuarine species, specialised brood chambers may isolate developing embryos from physiological dangers such as salinity stress (Chaparro et al. 2009) or ammonia accumulation (Chaparro et al. 2011). But as with all parental investments, there is also a cost. For example, females of the gastropod *Crepipatella dilatata* brood their egg capsules in the mantle cavity for several weeks until the offspring hatch as juveniles. When the female is tightly clamped to the substratum during periods of low salinity, it is isolated from the environment, limiting the available oxygen for developing embryos (Segura et al. 2016).

In bivalves, chitons, and patellogastropods, brooded eggs are typically retained in the mantle cavity (Golikov & Kussakin 1972; Turner 1978; Andrews 1979b; Lindberg 1979;

Sastry 1979; Eernisse 1988b). All have external fertilisation and depend on sperm-casting, although the patellogastropod *Erginus rubella* has a distinct brood chamber formed from a modified left kidney (Thorson 1935; Lindberg 1982, 1983, 1988a). Most brooders have relatively small body sizes. In neritimorphs, caenogastropods, and heterobranchs, brooding is less common, presumably because they produce egg capsules or ribbons which are analogous to brooding, but without the cost of the parent maintaining the brood internally. Exceptions include some caenogastropod limpet-like taxa (e.g., Calyptraeidea), all vermetids, a few littorinids, and fresh-water groups such as all viviparids, thiarids, and some truncatelloideans (Vail 1977; Fretter 1984a; Lively & Johnson 1994; Haase 2005; Glaubrecht 2006; Golding et al. 2014). In the Eupulmonata, occurrences of brooding have been documented by Tompa (1979), while in other heterobranchs brooding is rare and limited to pteropods, the acteonoidean *Pupa affinis* (as *P. kirki*) (Hadfield & Switzer-Dunlap 1984), and some glacidorbids (Ponder & Avern 2000).

Bivalve brooding is widely distributed across the class (Sellmer 1967; Sastry 1979), with their broods contained in demibranchial or epibranchial chambers, or, more rarely, the mantle cavity, on the outside of the shell, or specialised brood chambers (Figure 8.18) (Pelseneer 1935; Hansen 1953; Oldfield 1964; Chanley & Chanley 1970; Mackie 1984; Russell & Huelsenbeck 1989; Graf & Ó Foighil 2000; Ó Foighil & Taylor 2000; Guralnick 2004). In some fresh-water bivalves, the embryos develop in a 'marsupium', such as in the sphaeriid *Musculium heterodon* (Figure 8.19), which forms around the developing embryos after fertilisation and is thought to provide nourishment (Okada 1935). In some astartiids, broods can also be attached to the exterior of shells with adhesive mucus (Ockelmann 1958). Brooded larvae may undergo synchronous development (all at the same stage) or asynchronous development (different stages present).

Nautilus does not brood, and, in decabrachian cephalopods, there is no evidence for brooding (Arnold & Williams-Arnold 1977). Instead egg capsules are attached individually to the substratum or in common spawning masses. In contrast, external brooding is common in octopods where, in benthic species, the eggs are attached to the substrata and are guarded and cared for by the female throughout their development. In pelagic species, the brood may be held in a specialised egg case (*Argonauta*), attached to the webbing between the arms (*Tremoctopus*), within the mantle cavity (*Vitreledonella*), or internally in the oviducts (*Ocythoe*) (Wells & Wells 1977a). The brooding period of the deep-sea octopus *Graneledone boreopacifica* has been measured *in situ* at over four years (53 months), the longest egg-brooding period reported for any animal (Robison et al. 2014).

Because many brooded embryos are released as crawl-away juveniles, they have the potential to become fossilised along with the adult. Examples of fossil brooding have been documented in caenogastropods, namely, viviparids (Ashkenazi et al. 2010) and calyptraeids (Herbert & Portell 2004), the venerid bivalve *Transennella* (Lindberg 1984b; Russell et al. 1992), and some ammonites (Mironenko & Rogov 2015).

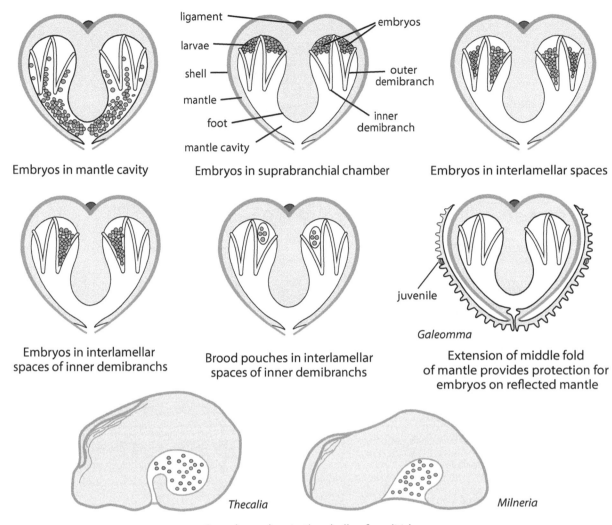

Embryos in mantle cavity

Embryos in suprabranchial chamber

ligament · embryos

larvae

shell · outer demibranch

mantle

foot · inner demibranch

mantle cavity

Embryos in interlamellar spaces

Embryos in interlamellar spaces of inner demibranchs

Brood pouches in interlamellar spaces of inner demibranchs

juvenile

Galeomma

Extension of middle fold of mantle provides protection for embryos on reflected mantle

Thecalia

Milneria

Brood pouches in the shells of carditids

FIGURE 8.18 Modes of brooding in bivalves. Redrawn and modified from Mackie, G.L., Bivalves, in Tompa, A.S., Verdonk, N.H. and Biggelaar van den, J.A.M. (Eds.), *Reproduction. The Mollusca*, Academic Press, New York, Vol. 7, pp. 351–418, 1984.

8.6 SEXUAL DIMORPHISM

Sexual dimorphism is used to describe a wide range of features that differentiate the sexes in molluscs, including shell morphology, shell size, anatomy, body colour, karyotypes (Andriychuk & Garbar 2015), and steroid levels (Janer et al. 2006). One of the most common examples of sexual dimorphism is the presence of copulatory organs (e.g., penes, hectocotyli) to aid in sperm transfer from males to females (and during the male phase in protandric species). Specialised copulatory structures are present only in gastropods and cephalopods, where they are especially diverse in both taxa. Similar structures are unreported in polyplacophorans, monoplacophorans, caudofoveates, bivalves, and scaphopods. However, apparently at least two bivalve taxa have co-opted existing anatomy to facilitate fertilisation. Mikkelsen & Bieler (1992) reported that two species of the galeommatid genus *Divariscintilla* may have co-opted the foot to transfer sperm into the mantle cavity

of conspecifics, while in the shipworm *Bankia setacea* the male's exhalant siphon is inserted into the female's inhalant siphon and sperm passed between individuals (Townsley et al. 1966). The exhalant siphons are sexually dimorphic with the male siphon bearing rows of papilla lacking in females. In a study of sperm morphometrics, Popham (1974) showed that shipworms which internally fertilised had different acrosomal morphologies, but did not provide information as to the manner of internal fertilisation. Solenogasters, all of which are simultaneous hermaphrodites, sometimes have putative copulatory spicules (or stylets) in the mantle cavity; however, these may be deciduous and their absence temporary in an individual (Scheltema 1992), and their function during mating has not been documented. Besides external reproductive anatomy, radulae sometimes show sexual dimorphism (e.g., Arakawa 1958; Mutlu 2004; Matthews-Cascon et al. 2005). While genomic and physiological differences between sexes can also be thought of as examples of sexual dimorphism,

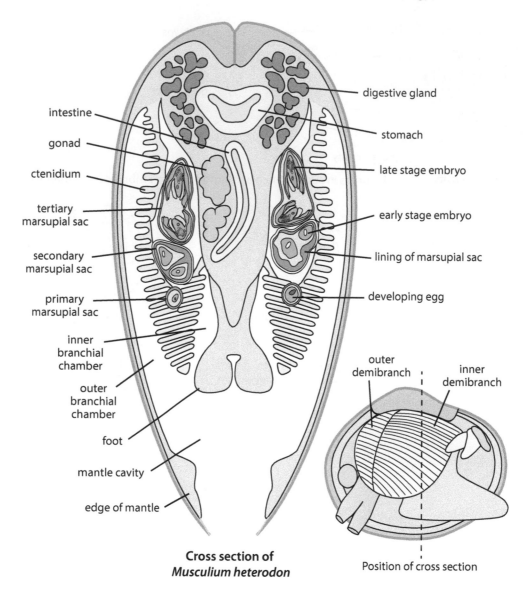

intestine

gonad

ctenidium

tertiary
marsupial sac

secondary
marsupial sac

primary
marsupial sac

inner
branchial
chamber

outer
branchial
chamber

foot

mantle cavity

edge of mantle

digestive gland

stomach

late stage embryo

early stage embryo

lining of marsupial sac

developing egg

outer
demibranch

inner
demibranch

**Cross section of
*Musculium heterodon***

Position of cross section

FIGURE 8.19 Marsupial brooding in the hermaphroditic fresh-water sphaeriid bivalve *Musculium heterodon*. Redrawn and modified from Okada, K., *Sci. Rep. Tohoku Imp. Univ.*, 9, 315–328, 373–391, 1935.

we focus here primarily on anatomical and morphological modifications and especially those features thought to increase fecundity or survival of the spawn. Alternative ecological drivers of sexual dimorphism have been reviewed by Shine (1989).

Protandrous molluscs with internal fertilisation are almost always sexually dimorphic as copulatory structures are reduced or lost during the female phase. Increased female body size is usually necessary for egg production or the brooding of young. However, it is not always the females that are larger in gonochoristic species (Table 8.4). While females can be significantly larger than males, as for example, in the muricids *Urosalpinx cinerea* and *Eupleura caudata* (Griffith & Castagna 1962), and the littorinid *Littoraria variegata* (Riascos & Guzman 2010), males can also be substantially larger than females in both gastropods and bivalves. Examples

include the cypraeids *Umbilia hesitata* (Griffiths 1961), the olivid *Olivella biplicata* (Edwards 1968), and the unionid *Lampsilis siliquoidea* (Jass & Glenn 2004). The most extreme examples of sexual size-dimorphism are found in taxa that have dwarf males (see Section 8.6.1).

It is also often difficult to ascribe sexual dimorphism to differences in body size (or age) because differences between males and females in populations can also result from differential mortality (Paine 1969a), sex change, or growth with no sexual selection (e.g., Abbott 1960; Mills and Cote 2003). The complexity in the interactions of these parameters can be seen in a study of growth rates, maturation sizes, and shell sexual dimorphism in the closely related and co-occurring mangrove snails *Littoraria zebra* and *L. variegata*. In *Littoraria zebra*, Riascos and Guzman (2010) found no differences in growth rate, maximum size,

TABLE 8.4

Some Examples and Distinguishing Feature(s) of Shell Sexual Dimorphism in Gonochoristic Gastropods

Major Group	Example Genus	Reference	Shell Feature		
			Size	Volume	Morphology
Patellogastropoda	*Scutellastra*	Branch (1985)	•		
	Lottia	Wright & Lindberg (1982)	•		
	Patella	Le Quesne & Hawkins (2006)	•		
Vetigastropoda	*Margarites*	Lindberg & Dobberteen (1981)			•
	Turbo	Amio (1955)			•
	Larochea, Larocheopsis	Kase & Kano (2002); Marshall (1993)			•
	Liotipoma	McLean (2012)			•
Neritimorpha	*Helicina*	Richling (2004)		•	
	Neritina, Septaria, Helicina	Govindan & Natarajan (1974); Haynes (1992)	•		
Caenogastropoda	*Pomacea, Cochlostoma*	Cazzaniga (1990); Baur et al. (2012)	•		
	Viviparus	Minton and Wang (2011)			•
	Strombus	Cotton (1905); Abbott (1960); Minton & Wang (2011)	•		•
	Olivella	Edwards (1968)	•		
	Hydrobia	Quick (1920)			•
	Nucella	Pelseneer (1926b); Son and Hughes (2000)			•
	Buccinanops	Soledad Avaca et al. (2013)			•

or shell morphology between males and females; minimum size at maturity was similar, but the mean size of mature males was smaller than that of females. In *L. variegata*, females grew almost twice as fast as males, reached larger maximum size, attained sexual maturity at a larger size, and had more globose shells than males. They concluded that the marked sexual dimorphism present in *L. variegata* can be explained by increased fecundity due to selection by way of different growth rates in males and females (Riascos & Guzman 2010).

Differences in shell morphology are also known in *Nautilus*, with females having larger shells (Vayssière 1895; Saunders & Spinosa 1978; Tajika et al. 2015). As in all extinct molluscs, the recognition of sexual dimorphism in the shells of fossil cephalopods is difficult (Westermann 1969). The co-occurrence of large and small ammonites which share identical early whorl morphology, but differ in ornamentation of the later whorls, or peristomal modifications (Makowski 1962; Callomon 1963, 1969), has been thought of as evidence of sexual dimorphism (see Chapter 17). These criteria are also applied to extinct groups of gastropods and bivalves (Linsley 1978c) and have been found to also be associated in extant taxa (e.g., Dobberteen & Ellmore 1986). While the diagnosis of sexual dimorphism may be correct, assigning the sex associated with each morphology remains problematic. Because of the space occupied by eggs, volumetric arguments favour small males and large females. This reasoning led Makowski (1962) to consider the smaller ammonoids with the more complex peristome to be males and larger individuals to be females. Callomon (1963) was more conservative and argued that the determination of which sex was larger would always be speculative.

The trochid *Margarites vorticiferus* exhibits a marked sexual dimorphism in the size of the shell umbilicus used by females to brood developing embryos (Lindberg & Dobberteen 1981). The morphological difference between males and females is so apparent that they were described as different species (Lindberg 1985b), with the difference in coiling parameters seen in the earliest teleoconchs (Dobberteen & Ellmore 1986). Umbilical brooding has also been reported in some other vetigastropods, including *Arene*, *Munditia*, and *Liotipoma*, in the lower heterobranch *Philippia* (Architectonicidae) (Shasky 1968; Robertson 1970b; Burn 1976; McLean 2012), and in a few stylommatophoran land snails such as *Cavellia marsupialis* (Powell 1941). This use by gastropods of the umbilicus as a brood chamber is considered one of the classic examples of the exaptation of non-adaptive morphology (Gould 1997).

Internal brooding may also result in differences in shell morphology between males and females. In the scissurellid genus *Larochea*, the young are brooded in the right subpallial cavity of females, which is supported by a large internal inner lip septum; this septum is substantially smaller in males (Kase & Kano 2002). In another genus, *Larocheopsis*, an internal inner lip septum is absent, and the smaller male is found attached to the last whorl of the female outside the parietal area (Marshall 1993). Bivalves also exhibit shell sexual dimorphism. For example, some species of the bivalve *Astarte* are both protandric and sexually dimorphic in shell shape, with the shell of the female more elongated than the male shell (Saleuddin 1964). In some other bivalves, morphological characters, including greater shell thickness, crenulated lower margins, greater shell inflation, and coarser ornamentation have been used to distinguish female from male shells (Gardner & Campbell 2002). Kauffman and Buddenhagen (1969) noted these characters coincided with sex change and suggested that shell inflation provided more room for the gonad and the subsequent development of large eggs. Gardner and Campbell (2002) confirmed this morphological transformation (and therefore sexual dimorphism) was also common in Middle Jurassic Astartinae. In the chemosynthetic bivalve *Calyptogena pacifica*, females have larger and more expanded posterodorsal shells (Coan et al. 2000), and this was later argued for the entire genus (Krylova & Sahling 2006), while Parra et al. (2009) also confirmed sexual dimorphism in *Calyptogena gallardoi*. Shell shape sexual dimorphism in the Unionoidea was first reported by Kirtland (1834). He described an inflated ventral-posterior region in females, which he suggested provided more space in the ctenidial brood chambers (marsupia) containing the eggs and developing embryos. Male shells were less inflated and had more acute posterior margins. This morphological difference has been reported in two groups, the Lampsilini, and in one species of the Amblemini (Burch 1975).

8.6.1 DWARF MALES

Dwarf males are often found among small, sedentary marine invertebrates that occur at low densities in habitats where potential mates or resources are rare and limited (Ghiselin 1974; Collin 2013; Yamaguchi et al. 2013). Besides their small body size, dwarf males often appear to have resulted from progenetic heterochronic changes so they have juvenilised somatic structures combined with adult reproductive structures. A few bivalves, cephalopods, and gastropods have dwarf males.

In cephalopods, the pelagic argonaut octopods exhibit extreme sexual dimorphism in size and lifespan. Females may be up to 10 cm in length and form egg cases (often called 'paper nautilus' shells) up to 30 cm in length. In comparison, males are rarely greater than two cm in length. In addition, males are semelparous (mate only once and die), whereas the females (very unusually for coleoids) are iteroparous, having multiple spawning events during their longer lives. In another argonautoidean octopod, *Tremoctopus violaceus*, females may be up to 2 m long, and males less than 2.5 cm in length; a difference of almost two orders of magnitude. Weight ratios between the sexes are about 10,000:1 and can reach 40,000:1 (Norman et al. 2002).

Among bivalves, dwarf males are known in several commensal galeommatoideans where, depending on the species, they live attached either internally or externally to their host (Figure 8.20) (see Heller 1993 for review). Dwarf males are also found in the wood-ingesting pholadidoidean *Xylophaga* (Ockelmann & Dinesen 2011; Haga & Kase 2013) and in the teredinid shipworm *Zachsia* (Figure 8.20) (Turner & Yakovlev 1983; Voight 2015). In gastropods, dwarf males also occur predominantly in parasitic taxa. In the limpet-shaped eulimid *Thyca crystallia*, which lives affixed to its starfish host, the dwarf males are attached under the mantle of the female (Elder 1979). Not all members of

the genus have dwarf males; in *Thyca stellasteris*, males are about half the size of females and are in close association, while in *T. ectoconcha*, the males and females are of similar size (Warén 1980a). The most extreme known case of dwarf males in eulimids is in *Enteroxenos*, a shell-less endoparasite of holothurians in which the minute male is fused inside the female body cavity and consists of little more than a testis (Lützen 1979).

8.7 REPRODUCTIVE BEHAVIOUR

Reproductive behaviour in molluscs ranges from the recognition of simple environmental or conspecific chemical cues to induce spawning or mating, to more complex behaviours involving body contact or displays.

Whether a copulating or broadcasting species, gonochoristic, and hermaphroditic, in terrestrial or aquatic habitats, reproductive synchronicity begins with the gametogenic cycle, which must be coordinated between individuals to ensure the availability of mature gametes for spawning. In taxa with continuous gamete production, some portion of the population is always gravid and often undergoing partial spawning as, for example, in the bivalves *Mya arenaria* and deep-sea protobranchs (Sanders & Hessler 1969; Rokop 1974). For most free spawning molluscs, the environmental conditions and cues under which they spawn in the field remain largely unknown, but for synchronised spawning to occur at all, gametogenesis must be coordinated between individuals. Once this criterion is met, two other factors become important, coordinating simultaneous gamete release and proximity for successful fertilisation of eggs (see Section 8.5.1), both of which may be mediated by chemical cues. Spawning aggregations increase interactions between individuals and gametes and are present in most molluscan groups. In internally fertilising coleoids, aggregations occur in cuttlefish (Hall & Hanlon 2002) and loliginid squids (Moltschaniwskyj et al. 2002; Forsythe et al. 2004). Spawning aggregations have not been reported in *Nautilus*. In 2018 E/V Nautilus (https://www.youtube.com/channel/UC1KOOWHthbQVXH2kZue3_xA) posted images of large aggregations (>1000) of brooding octopus (*Muusoctopus robustus*) at seeps off Monterey, California. This appears to be the first report of spawning aggregations in octopuses. Olfactory clues are often hypothesised for producing aggregations, but details are poorly known. Boal et al. (2010) have demonstrated that cuttlefish are attracted to spawned eggs of conspecifics, the data suggesting that *Sepia* eggs could be the source of the hypothesised reproductive pheromones. Nhan et al. (2010) experimentally demonstrated in the laboratory the release of attraction pheromones by male abalone (*Haliotis asinina*).

Gamete release in broadcasting species may be synchronised over short or long periods or be continuous (Sastry 1979; Levitan 2006). Thorson (1950) argued that synchronicity of spawning was important, to coordinate it with larval food resources and maximise survival and growth of larvae (see also review by Olive 1992). Because most molluscan clades have plesiomorphic, non-feeding larvae, the availability of larval

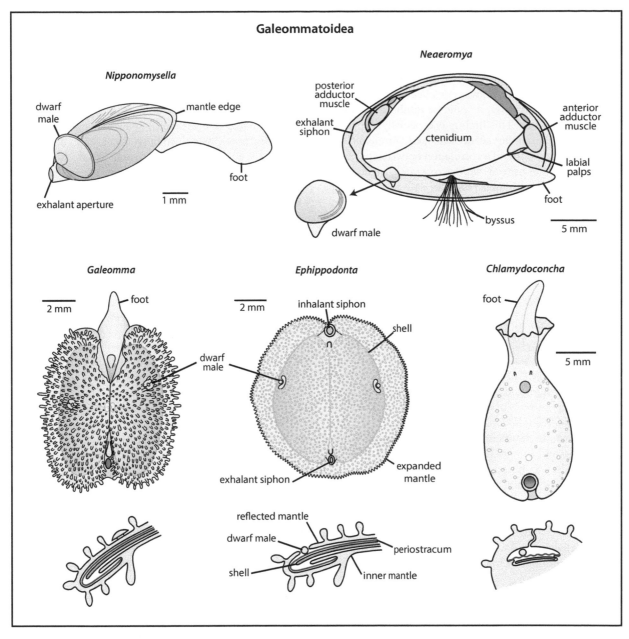

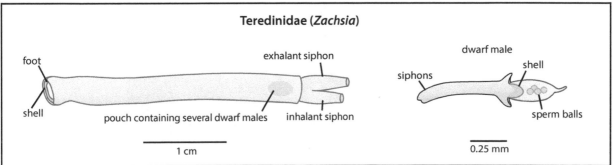

FIGURE 8.20   Examples of dwarf males in bivalves. Redrawn and modified from the following sources: Lützen, J. et al., *J. Zool.*, 254, 429–440, 2001b (*Nippomysella*); Narchi, W., *The Veliger*, 12, 43–52, 1969 (*Neaeromya* – general figure), dwarf male from photograph courtesy of Jingchun Li; Morton, B.S., *J. Zool.*, 169, 135–150, 1973c (*Galeomma*);   Morton, B.M., *J. Conchol.*, 29, 31–39, 1976 (*Ephippodonta*); Morton, B., *J. Zool.*, 195, 81–121, 1981b (*Chlamydoconcha*); Turner, R.D. and Yakovlev, M., *Science*, 219, 1077–1078, 1983 (*Zachsia*).

food resources is probably not an important issue in the overall history of the taxon (Hodgson 2010). Correlations between spawning events and productivity in the water column (as well as temperature) are often identified by researchers. For example, in chitons, Himmelman (1975) reported that spawning in two of the polyplacophoran *Tonicella* was correlated with the spring phytoplankton bloom, although it could be inhibited by colder than normal temperatures. This finding is especially surprising because polyplacophoran larvae do not feed. Arsenault and Himmelman (1998) also found a correlation between spawning and phytoplankton blooms for *Chlamys islandica*, but not with temperature, photo period, lunar phase, or salinity. Starr et al. (1990) reported that a heat-stable metabolite released by phytoplankton taxa triggered spawning in mussels (Mytilidae) and noted that phytoplankton blooms were better than single factor triggers because they integrated numerous local abiotic and biotic factors.

Gamete-associated chemical cues have also been found to initiate spawning in bivalves, including *Mytilus* (Young 1942), *Tridacna* (Wada 1954), *Turtonia* (Oldfield 1964), *Pecten* and *Placopecten* (Beninger et al. 1995a), and oysters (Andrews 1979b), and the pearl oyster *Pinctada fucata* (Yu et al. 2008). In the scallops *Pecten maximus* and *Placopecten magellanicus*, Beninger et al. (1995a) demonstrated these chemical signals are sensed through the osphradia. Fujikura et al. (2007) presented two hypotheses for spawning cues in the deep-sea vent and seep-dwelling species of *Calyptogena*. They proposed two factors, a chemical cue related to sperm in the water in conjunction with reduced current speed, while the second hypothesis was based solely on the concentration of sperm present in the water column. Current speed has also been suggested as a spawning cue in *Placopecten magellanicus* (Desrosiers & Dubé 1993).

Studies of chitons and other molluscs have shown that sperm cells can sense eggs in the water and will move up an egg-water gradient, thus creating higher sperm concentrations at the egg source (Miller 1977; Kaupp et al. 2006). Sperm-attracted agents have also been identified in *Dreissena* species (Miller et al. 1994) and abalone (Krug et al. 2009; Himes et al. 2011). In patellogastropods and other rocky shore vetigastropods, coordinated spawning occurs following physical disturbances, such as storms (Orton et al. 1956; Branch 1974; Creese 1980; Onitsuka et al. 2010). Temperature, both high and low, has also been implicated in coordinated spawning in bivalves (Sastry 1979; Minchin 1992), including *Tresus* (Breed-Willeke & Hancock 1980), *Spondylus* (Parnell 2002; Enriquez-Diaz et al. 2009; Kim et al. 2010), and in the vetigastropods *Tegula* (Velez-Arellano et al. 2009) and *Haliotis* (Counihan et al. 2001). Chitons have also been shown to coordinate spawning based on diurnal light-dark cycles and a semi-diurnal tidal immersion cycle (Yoshioka 1989). This latter behaviour results in gametes being released at high water and concentrated over the entire intertidal population. Water currents have also been suggested to trigger spawning in estuarine oysters as there are important ecological consequences associated with gamete encounters and the dispersal of fertilised eggs (Bernard et al. 2016). Collin et al. (2017) documented and reviewed reproductive cycles in tropical intertidal gastropods and found that tidal amplitude cycles rather than lunar cycles were synchronised with reproductive periods. Photoperiod has also been identified in coordinating spawning in *Mytilus* (Dominguez et al. 2010), however Babcock et al. (1992), found no correlation between spawning and lunar or diel cues in *Arca* species on the Great Barrier Reef, Australia.

Aggregation behaviour or pseudo-copulation reduces the distance between individuals and increases the probability of fertilisation events. In broadcasting species, there is limited evidence of spawning aggregations in polyplacophorans (Okuda 1947; Pearse 1979), while in gastropods, numerous examples have been reported. One of the most spectacular aggregations is the behaviour seen in the patellogastropod *Nacella concinna* in the Antarctic (Figure 8.21) (Picken & Allan 1983;

FIGURE 8.21   Spawning aggregation of *Nacella concinna* at Signy Island, South Orkney Island, Antarctica showing release of sperm (solid) and egg (dotted) streams. Reprinted from *J. Exp. Mar. Biol. Ecol.*, 71, Picken, G.B. and Allan, D., Unique spawning behaviour by the Antarctic limpet *Nacella* (*Patinigera*) *concinna* (Strebel, 1908), Figure 1, Copyright 1983, with permission from Elsevier.

Stanwell-Smith & Clarke 1998). Aggregation behaviour has been reported in some large vetigastropods such as *Haliotis* (e.g., Seamone & Boulding 2011) and *Trochus* (Maboloc & Mingoa-Licuanan 2013), and Hickman and Porter (2007) reported a water column spawning aggregation of over 20,000 *Scissurella spinosa* and *Sinezona plicata* that had been attracted to a light trap suspended approximately 2 m above the bottom in Moorea, French Polynesia. Within copulating caenogastropods, aggregations have been reported in muricids (Webber 1977), including *Nucella* (Stickle 1973), *Urosalpinx* (Hancock 1956), and *Hexaplex* (Wolfson 1968), buccinids (e.g., Martel et al. 1986), and in the littorinid *Lacuna* (Gallien & Larambergue 1939). In heterobranchs, spawning aggregations have been reported in *Onchidoris bilamellata* (Claverie & Kamenos 2008) and *Aplysia* (e.g., Pennings 1991). Spawning aggregations have also been suggested for fresh-water unionid bivalves (Amyot & Downing 1998). Studies of scaphopods are rare and no spawning cues have been identified (Rokop 1977), and data for Caudofoveata and Monoplacophora are lacking.

Multispecies synchronous broadcast spawning involving several phyla, including some molluscs, occurs on coral reefs (Babcock et al. 1992; Babcock 1995).

Examples of the more complex behaviours are found in coleoids and include complex courtship and male-male aggression (Hanlon & Messenger 1996). Because of their exceptional visual acuity, visual cues and communication are often highly important and reproductive body patterning in squid can involve as many as 19 chromatic, five locomotor, and 12 postural components (Postuma & Gasalla 2015). There is also evidence for the use of achromatic (neutral greys, white, or black) and polarised light by cuttlefish (Boal et al. 2004) and bioluminescent signals for sexual recognition. For example, Herring (2000) noted a correlation between the late development of secondary light organs and the onset of sexual maturity in adult females of the squid *Liocranchia* and the octopod *Japetella*. Additional complex mating behaviours are also seen in some other octopods (Huffard et al. 2008), but complex behaviours can also be absent in both squid and octopods (Cheng & Caldwell 2000; Rodrigues et al. 2012). In contrast to many coleoids, *Nautilus* appears to be limited to chemical cues like the majority of other molluscs (Hanlon & Shashar 2003).

Mating aggregations also occur in internally fertilising taxa. In cephalopods, they have been reported for squid (Sauer et al. 1992, 1997) and cuttlefish (Hall & Hanlon 2002), although purported mating aggregations in octopods were dismissed by Wells & Wells (1977a). Examples of mating aggregations in gastropods include those in some muricids (e.g., Feare 1971; Fotheringham 1971; Cudney-Bueno et al. 2008), Strombidae (e.g., Catterall & Poiner 1983), *Conus* (e.g., Kohn 1961b), Buccinidae (e.g., Martel et al. 1986), and Nassariidae (e.g., Straw & Rittschof 2004) and range from hundreds to thousands of individuals. In the heterobranchs, mating aggregations have been reported in the Aglajidae, Gastropteridae (Anthes & Michiels 2007), in Aplysiidae (Carefoot 1987), *Siphonaria* (Borland 1950), *Melampus*

(Russell-Hunter et al. 1972; Price 1979), and *Onchidoris* (Claverie & Kamenos 2008). In an innovative study of the evolution of cephalaspidean reproductive strategies in the Aglajidae and Gastropteridae, Anthes et al. (2008) reported several correlations between the formation of mating aggregations and both behavioural and morphological traits. For example, species that formed mating aggregations engaged in shorter copulation periods and therefore less sperm investment per mating. This trait was negatively correlated with the size of the seminal receptacle, but positively correlated with the size of the bursa copulatrix, leading the authors to conclude there was selection to decrease space for sperm storage (seminal receptacle), while increasing sperm digestion capabilities (bursa copulatrix).

Other than the gonochoristic cephalopods, some of the most complex reproductive behaviours are found in the simultaneous hermaphroditic heterobranchs, especially in terrestrial eupulmonates. These complex courtship behaviours include stereotypical movements (dancing), body contact, shell mounting, tentacle touching, penis eversion, nuptial secretions, and exchanging 'love darts' (Lipton & Murray 1979; Leonard 1992; Pomiankowski & Reguera 2001; Reise 2007; Benke et al. 2010) (see Chapter 20). These complex behaviours are thought to result from sexual selection driven by different combinations of reciprocal mating behaviour. In taxa where each individual is functioning as both male and female during mating bouts (reciprocal mating), male-female (sexual) conflict is likely as each individual attempts to fertilise as many as possible of the eggs of their partner, while only giving sperm of the fittest partners access to their eggs. For taxa that perform one sexual role at a time during a mating bout, conflict can arise over their respective sexual roles. If both animals prefer to mate in some role and simultaneous reciprocal insemination is impossible, the conflict can be resolved by sex role swapping after the first insemination, potentially resulting in a form of copulation called *conditional reciprocity* (Koene & Maat 2005). These conflicts are probably responsible for the evolution of a large suite of the reproductive behaviours and morphologies present in euthyneuran heterobranchs (Leonard & Lukowiak 1991; Leonard 2006).

One of the strangest mating behaviours is exhibited by terrestrial slugs of the genus *Ariolimax*, which will sometimes bite off the penis of their partner after copulation (*apophally*) (Heath 1916; Leonard et al. 2002, 2007). The interpretation and hypotheses for this behaviour are varied. For example, Jones (2001) suggested that individuals bite off their own penes after copulation which reduces subsequent sperm competition in the partner. However, a sphincter muscle in the female reproductive tract, which may serve to grip the penis of the partner (Mead 1943), requires alternative hypotheses that would benefit the recipient of the penis rather than the amputee (Reise & Hutchinson 2002), including a potential nutrient source and reducing the subsequent matings of the partner as a male. Penis biting has also been reported in some other agriolimacid slugs with external sperm exchange (Reise 2007). Rymzhanov (1994) reported that in one population of *Deroceras laeve*, some individuals bite off their own penis after copulation, and it is then eaten by the partner. In other populations, aphallic individuals never produced a penis (Reise & Hutchinson 2002). Much work remains to be done with these systems.

The complexity of molluscan reproductive behaviours has also resulted in the use of euthyneuran heterobranchs as models for the study of the evolution of behaviour and life history strategies (Kandel 1979; Chase 2002). Molluscan reproductive behaviour has also contributed to sex allocation theory (Carrillo-Baltodano & Collin 2015) and game theory. The *Hermaphrodite's Dilemma* serves as a game theory model in psychology with a premise similar to the *Prisoner's Dilemma* (i.e., why two individuals might not cooperate, even if it is in their best interests to do so) (Smith 1982a; Leonard 1990). Because of the evolution of complex copulatory behaviours with different fitness outcomes, euthyneuran heterobranchs (especially Stylommatophora) are ideal organisms in which to explore these and other hypotheses of the adaptive influences of sexual conflict and selection.

Besides complex courtships, copulatory behaviour also appears to be influenced by the presence of both male and female genitalia in the same individual. Determining and maintaining alignment during copulation is likely to be substantially more complicated for heterobranchs than for gonochoristic species. Asami et al. (1998) explored the relationship between stylommatophoran shell morphology and mating and concluded that low-spired snails tended to mate side to side and had reciprocal sperm exchange, while high-spired snails mated by shell mounting and had unilateral sperm exchange. In their investigation into the evolutionary role of the aforementioned love darts in stylommatophoran mating, Davison et al. (2005) also investigated mating positions and shell morphology. They concluded that dart-bearing species with simultaneous reciprocal, face to face mating were restricted to two superfamilies – the Helicoidea and Limacoidea, which either have predominantly low-spired shells or were slugs, whereas shell mounting taxa were generally high-spired. A subsequent analysis by Jordaens et al. (2009) concluded, however, that information about gastropod mating behaviour is too sparse to detect general evolutionary patterns. They observed that: (1) some species displayed considerable intraspecific variation in mating behaviour; (2) mating position did not predict reciprocity of sperm exchange; and (3) reciprocal intromission of penes did not always lead to reciprocal transfer of sperm. In the 'common garden snail' *Cornu aspersum*, Chase et al. (2010) reported a small protruding structure associated with the female opening that played an important role in positioning of snails before copulation and showed that functionally equivalent structures often occurred in other reciprocally copulating taxa.

8.7.1 MATE CHOICE

In copulating taxa, assortative mating (or assortative pairing) occurs between individuals with similar (positive assortative mating) or dissimilar (negative assortative mating) characteristics or traits and often indicates either male and/or female sexual

selection. Such behaviours can result in reproductive isolation. Assortative mating has been shown to occur in some gastropod and coleoid taxa and may be more common than so far demonstrated. The most detailed studies are on *Littorina* species (e.g., Hull 1998; Johnson 1999; Zahradnik et al. 2008) (see below). Size-assortative mating has been demonstrated in the fresh-water caenogastropods *Viviparus* (Staub & Ribi, 1995) and *Pomacea* (Estebenet & Martín 2002), and in the heterobranch intertidal limpet *Siphonaria* (Pal et al. 2006). Although much of this literature has focused on male/female size, other factors have also been identified. For example, in the caenogastropod *Neptunea arthritica*, males mate preferentially with females that have previously copulated (Miranda et al. 2008; Lombardo et al. 2012). This was a surprising result as reproductive theory suggests that mating with females that had already copulated would risk both increased sperm competition and dilution for subsequent partners, especially in taxa with the ability to store sperm. By monitoring paternity with microsatellite markers, Lombardo et al. (2012) showed that males could achieve considerable reproductive success with a mated female by removing sperm (method unknown) from earlier copulation events.

8.7.1.1 The Role of Body Size in Reproductive Behaviour

There is a considerable body of literature and theory dealing with the relationship of body size and behaviours associated with courtship, mating frequency, sperm transfer, etc. (Calder 1996; Blanckenhorn 2000). Sexual selection is often invoked to explain body size-dimorphism in gonochoristic taxa (see Section 8.6). Although female choice of mating partners has been thought to be the normal situation, males may selectively allocate their sperm to, for example, large females with the highest fecundity. Much of the work addressing size-assortative mating in molluscs has focused on gonochoristic caenogastropods (especially the intertidal littorinids) and hermaphroditic euthyneuran heterobranchs. In the former groups, where sexual dimorphism is common, female fecundity is positively correlated with body size and eggs can be fertilised by stored sperm from several males.

Littorinid examples include the study by Zahradnik et al. (2008), who experimentally showed that males of *Littorina subrotundata* preferentially copulated with larger females and sometimes aggressively competed with smaller males for access to females (see also Ng et al. 2016). Male *Strombus pugilis* and *Conomurex* (as *Strombus*) *luhuanus* have also been observed fighting near groups of females (Bradshaw-Hawkins & Sander 1981; Kuwamura et al. 1983). In both these cases, the larger more aggressive males are thought to engage in more copulations than smaller individuals. Zahradnik et al. (2008) found that males preferentially preferred virgin females and appeared to discriminate against, or show modified behaviour towards, non-virgin females. Males of *Littorina littorea* and *L. saxatilis* also prefer larger females and will copulate for significantly less time with parasitised females compared to non-parasitised individuals (Saur 1990). Erlandsson and Johannesson (1994) found that small and large males of *Littorina littorea* had equal success in accessing females, however, copulations with large females lasted longer, suggesting the possibility of greater sperm transfer. A small form of *Littorina saxatilis* (previously called *L. neglecta*) also displays size-assortative mating, with males preferring larger and more fecund females, but not necessarily the largest females; this behaviour has been suggested as a possible reproductive barrier between the *L. neglecta* form of *L. saxatilis* and the larger typical form of *Littorina saxatilis* (Johnson 1999). In the nassariid *Buccinanops deformis* (as *B. globulosus*), males are smaller than females and medium and large size males copulate with larger females than small-sized males (Avaca et al. 2012). Rolán-Alvarez et al. (2015), Zahradnik et al. (2008), and Riascos and Guzman (2010) have provided recent overviews of these and other caenogastropod studies.

There has been interest in the role that body size has in sex role preferences in simultaneous hermaphrodites, and it is predicted that there is a preference for the sex role with a lower mating cost (Charnov 1979; Michiels 1998). Conflict arises when numerous individuals have a preference for the same sex role, resulting in potential mates having incompatible interests (i.e., gender or sexual conflicts). It is the origin and determination of these conflicts, and their resolution, that are thought to have determined the evolution of mating behaviour (see review by Anthes et al. 2006a). In heterobranchs, the situation is further complicated by the possibility of selection acting independently on male and female traits in the same individual. In most models, simultaneous hermaphrodites are assumed to divide resources between male and female function, which can cause both size-dependent mating behaviours and conflict with potential mates over their respective roles. Evolutionary outcomes may include size-assortative mating, conditional exchange of gametes, and mating patterns where relative body size affects the investment in each sexual role. Yusa (2008) demonstrated some mating advantage of larger individuals over smaller in *Aplysia kurodai*, perhaps due to preference by those assuming the male role for large 'female' partners due to their greater reproductive capacity. The preference for large 'female' partners was also consistent with their finding that mating periods were longer with larger 'females' than with smaller. In the Cephalaspidea, where most copulations involve conditional reciprocity, Anthes et al. (2006a) demonstrated that in *Mariagiaja tsurugensis* (as *Chelidonura sandrana*) male copulation duration increased with partner body size, but decreased when the partner had previously copulated. In the pond snail *Lymnaea stagnalis,* there is no evidence of mate choice based on body size (Koene et al. 2007). In laboratory experiments, the cephalaspidean *Bulla gouldiana* revealed complicated effects of body size on mate choice and duration (Chaine & Angeloni 2005). Pairs were more likely to mate if they included at least one large animal, with the larger typically inseminating the smaller. When both individuals were large, they were more likely to each mate in both sexual roles by switching roles once during copulation. This suggests some level of reciprocity, which is unlikely to be conditional given the rate of unilateral mating. When the larger member of the mating pair inseminated the smaller, the duration of insemination increased with the size of the smaller sperm recipient. Copulations lasted longer in pairs that switched sexual roles than in those that did not switch roles (Chaine & Angeloni 2005).

The relative body size in paired individuals has been suggested as a potential factor that influences the preferred sex role. Consequently, in theory, for species with unilateral penis intromission, smaller individuals (i.e., those with less reproductive resources) should invest more in sperm than in eggs and therefore should adopt the male sex role more often, whereas larger individuals should preferentially adopt the female sex role. While some experimental studies (see Chaine & Angeloni 2005 for review) have supported this idea, others have not. Several studies have been carried out on euthyneurans, some of which support it (e.g., Angeloni 2003; Angeloni et al. 2003; Anthes et al. 2006a; Yusa 2008; Dillen et al. 2010), but several others show no effect (e.g., Switzer-Dunlap et al. 1984; Baur 1992a; Koene & Maat 2007). Some, such as the study on the sacoglossan *Oxynoe* (Gianguzza et al. 2004) and the cephalaspidean *Bulla* (Chaine & Angeloni 2005), showed that small individuals often assumed the female role when copulating. Because size is related to age, possibly age could be a confounding effect in some of the above cited experiments and few have considered age. In experiments with *Lymnaea*, Hermann et al. (2009) demonstrated that age, not size, was the main determinant of mating behaviour, and Webster et al. (2003) showed that mate choice in *Biomphalaria glabrata* is not controlled by one individual, but depends upon both mating partners, and females can actively reject a potential partner.

Size-dependent mating behaviours in coleoids differ from those in other molluscs and can feature different copulation strategies rather than a simple preference for larger individuals. This may be due to a combination of spawning in large aggregations and the semelparous life history strategies of most. In some decabrachians, females are accompanied to spawning sites by 'consort' males (Hanlon 1998; Hanlon et al. 2002). Larger lone males and smaller sneaker males occupy the periphery around spawning grounds. Lone males fight consort males to obtain access to females. Outcomes of most fights often depend on body size, based on observations in both captivity (Wada et al. 2005) and in the field (Hanlon et al. 2002). Sneaker males are not aggressive towards consort males, but instead rely on stealth to gain access to females. Sneaker males use a different copulatory position to the larger consort and lone males and also have larger sperm (Iwata et al. 2011; Hirohashi & Iwata 2013). These alternative mating behaviours and sperm size appear to depend on relative body size (Wada et al. 2005). Other factors in mate choice are probably present also. For example, Hooper et al. (2016) examined sperm investment in the sepiidan *Sepiadarium austrinum* and found that when sequentially presented with small versus large females, sperm investment was determined by mating order, rather than partner size, and in subsequent matings males consistently decreased sperm investment. The octopus *Abdopus aculeatus* also incorporates sneaker matings, mate guarding, and male-male aggression, making it more similar to aggregating decabrachians than any documented octopus reproductive social system (Huffard et al. 2008).

8.8 ENDOCRINE CONTROL OF REPRODUCTION

Synchronisation is important in reproduction. In gonochoristic species, male and female gametes must mature simultaneously to prepare for fertilisation and spawning, and in hermaphroditic species, reproductive systems must be controlled sequentially (protandry and protogyny) or simultaneously in the same individual. This is achieved by the endocrine system which releases hormones produced by neurosecretory cells in response to external (exogenous) and/or internal (endogenous) stimuli (Giese & Pearse 1974). External cues include: temperature, photoperiod, food, potential mates, chemicals, mechanical stimuli, etc. which have been shown to activate receptors in the nervous system which stimulate neurohormonal secretions (Dogterom et al. 1983). Internal cues come from body fluid pressure, oxygen level, tissue hydration, etc. and are often associated more with maintaining homeostasis of the animal rather than responses to external stimuli. Natural and human-produced chemicals in the environment can disrupt molluscan endocrine systems (Porte et al. 2006; Ketata et al. 2008), including pharmaceuticals, dioxin, polychlorinated biphenyls, pesticides, plasticisers, and tributyltin (TBT). Tributyltin compounds, which were used to control fouling communities on boat hulls, have been shown to severely disrupt the endocrine system in several caenogastropods, notably muricids, to produce *imposex* individuals, for example, development of male genitalia (penis, vas deferens) by females (e.g., Matthiessen & Gibbs 1998) (see Chapter 10).

Although studies of molluscan nervous systems have identified neurosensory cells in all major taxa (see Chapters 2 and 7), the rather few studies of endocrine control of reproduction include a small number of taxa, of mainly bivalves, cephalopods, and gastropods (Landau et al. 1997; Morishita et al. 2010). Perhaps surprisingly, hormonal regulation of reproduction in coleoid cephalopods is relatively simple, while in heterobranchs, it is almost as complex as that found in vertebrates (Geraerts 1987). Here, we review the role of the endocrine system in molluscan reproduction.

8.8.1 MORPHOLOGY, MESSENGERS, AND RESPONSES

As outlined in Chapter 2, the molluscan endocrine system consists primarily of neurosecretory cells, endocrine glands, and the neurohormones and steroids which are thought to act directly on the target tissues (Ketata et al. 2008). Despite this relative simplicity, the molluscan endocrine system is the most diverse hormonal system of any invertebrate phylum (LaFont & Mathieu 2007). There is also substantial variation among the classes, and especially within gastropods.

The neurohormonal secretions are typically released into the haemolymph from cells associated with the central nervous system (CNS) (Tsai et al. 2003); most often the cerebral, pleural, and in some heterobranchs, the visceral (= abdominal)

ganglia (see Chapter 7 for more details). Most of the work on molluscan reproduction has focused on the role of the sources of these secretions. In most gastropods, these are the juxtaganglionar organs (Martoja 1965a) – a group of large membrane-bound cells surrounding the cerebral ganglia, and the dorsal bodies, found in Stylommatophora and Hygrophila, and probably homologous with organs of the same name in 'opisthobranchs' (Switzer-Dunlap 1987), vetigastropods, and patellogastropods (see Chapter 7). The juxtaganglionar organs are symmetrical and found on the dorsal surface of the cerebral ganglia except in patellogastropods, where they are more lateral and at the base of the optic nerve and cerebral ganglion (Martoja 1965a). In cephalopods, studies have focused on the optic glands on the optic nerves, and unlike the juxtaganglionar organs and dorsal bodies, these have a neural origin (Wells & Wells 1977b). Additional cellular sources of neurohormonal secretions have been identified in heterobranch gastropods including the caudodorsal cells, bag cells, and atrial glands (see Chapter 7). Bag cells are cell clusters on the left and right pleuro-abdominal connectives near the abdominal ganglion (Kupfermann & Kandel 1970), while caudodorsal cells, like the juxtaganglionar organs, are associated with the cerebral ganglia, and the atrial gland is marked by a yellowish swelling on the hermaphroditic duct near the gonopore of *Aplysia* (Beard et al. 1982). Knowledge of molluscan endocrine systems has predominately come from studies of *Lymnaea* and *Aplysia*, two heterobranchs that have provided model systems for behavioural and endocrine research (Chase & Goodman 1977; Kriegstein 1977b; Kandel 1979; Wayne 2001; Chase 2002).

Neuropeptides and peptide hormones play critical roles in neural and neurohormonal regulation of reproduction (Morishita et al. 2010). These messengers affect gametogenesis, vitellogenesis, oocyte maturation, albumen synthesis, and the growth of the accessory sex organs (for reviews see Boer & Joosse 1975; Joosse & Geraerts 1983; Joosse 1988; Saleuddin et al. 1990). Molluscan neurohormones involved in reproduction include APGWamide (amidated tetrapeptide Ala-Pro-Gly-Trp-NH2), caudodorsal cell hormone, dorsal body hormones, egg-laying hormone (ELH), and insulin-like peptides (Geraerts et al. 1988; Nagle et al. 1989; Smit et al. 1996; Di Cosmo & Di Cristo 2006). Other neurohormones (e.g., gonadotropin-releasing hormone, GnRH) are produced by the CNS ganglia (cerebral, visceral, or pedal). Besides neuropeptides, other messengers such as sex steroids and eicosanoids are thought to be involved in regulating reproductive processes in molluscs. Eicosanoids, including prostaglandin E2 (PGE2), mediate several aspects of reproduction in molluscs such as the induction of gamete release (Martínez et al. 2000) and egg production (Kunigelis & Saleuddin 1986). The role of eicosanoids in molluscan reproduction is reviewed by Stanley (2014). The nervous system also produces aminergic neurosecretions (dopamine, noradrenaline, serotonin).

Molluscs were thought to be unique among the lophotrochozoans in synthesising *de novo* vertebrate-type steroids (LaFont & Mathieu 2007). Sternberg et al. (2010) have argued that the absence of androgen and progesterone receptors and the presence of an oestrogen-unresponsive oestrogen receptor in gastropods suggests that the sex steroids detected in gastropods, if functional, would act via some other signalling pathway. However, Scott (2012) argued that there is no convincing evidence for biosynthesis of vertebrate-type steroids in molluscs, and that experimental contamination may account for their reported presence (see Chapter 2 for more discussion on hormones).

GnRH directly regulates gonadal functions and is central to the initiation and maintenance of reproduction in the Metazoa. It was originally investigated in vertebrates, but GnRH-like molecules are also present in multiple invertebrate phyla, including Cnidaria, Platyhelminthes, Annelida, Arthropoda, and Mollusca, suggesting that GnRH is an ancient peptide, and its role in reproduction arose before the divergence of lophotrochozoans, ecdysozoans, and deuterostomes (Tsai 2006; Roch et al. 2011; Treen et al. 2012). In molluscs, GnRH peptides have been identified in a range of taxa including polyplacophoran *Acanthopleura japonica* (Amano et al. 2010b), the gastropods *Patella* and *Lottia* (Patellogastropoda) (Morishita et al. 2010; De Lisa et al. 2013), *Haliotis* (Vetigastropoda) (Amano et al. 2010a; Nuurai et al. 2010), *Helisoma* (Hygrophila) (Goldberg et al. 1993), *Aplysia* (Aplysiida), in the bivalves *Mizuhopecten* and *Crassostrea* (Pteriomorphia) (Nakamura et al. 2007; Rodet et al. 2008), and in the cephalopods *Octopus vulgaris* (Iwakoshi et al. 2002), *Sepia officinalis* (Morishita et al. 2010), and *Uroteuthis edulis* (Onitsuka et al. 2009). Amino acid sequences for active GnRH peptides are highly conserved in molluscs (Table 8.5). GnRH receptors have been demonstrated in both bivalves and polyplacophorans by Pazos and Mathieu (1999) and Gorbman et al. (2003), respectively. In chitons, Vicente and Gasquet (1970) showed that the release of unidentified neurosecretory product(s) into the haemolymph of the chitons *Lepidochitona cinereus* and *Chiton olivaceus* was correlated with gonadal development. Similar correlations have been noted in a bivalve (*Perna canaliculus*) (Mahmud et al. 2014).

8.8.2 Gastropods

More than ten hormones and steroids have been reported to be involved in gastropod reproduction, however, the diversity of sampled taxa remains small (Table 8.6). While most patellogastropod limpets are gonochoristic, the protandric species have received the most attention, primarily from the perspective of endocrine control of sex change (e.g., *Patella*, *Scutellastra*, *Lottia*) (Lindberg 2007b). The work of Choquet (1964, 1965, 1967, 1970, 1971) first demonstrated a cerebral ganglion neurohormone associated with the initial formation of the testis, followed by its transformation into an ovary in the protandric *Patella vulgata*. Highly conserved regions of molluscan gonadotropin sequences were later reported in *Patella caerulea* and *Lottia gigantea* by De Lisa et al. (2013), who also showed that GnRH mRNA expression was widespread in both male and female germ lines during gametogenesis, as inferred by the experiments of Choquet. Choquet (1970) later identified a cephalic tentacle factor that exerted an inhibitory influence on spermatogenesis and was responsible for quiescence of spermatogenesis after spawning. Production of the tentacular messenger was progressively reduced during the subsequent female phase and appeared to have no influence on oogenesis, whereas the cerebral hormone was essential to egg production. Thus, sex change was conditioned by the persistence of the tentacular inhibition on the male line and dependent on the cerebral ganglia hormone to initiate oogenesis in the gonad. Based on the work of Choquet, Wright (1988) proposed that control of sex change in *L. gigantea* actually resided in the spermatogonia and oogonia themselves, with the oogonia changing their sensitivity to the cerebral factor as the animal aged. Thus, the mitogenic factor caused the spermatogonia to differentiate and multiply initially, but did not affect oogonia until they reached a threshold later in life. After this threshold was reached, proliferating oogonia exerted an inhibitory effect on spermatogonia, such that sperm production was prevented by egg production. Choquet assumed that the sensitivity of the oogonia was ontogenetically set, but hormonal or metabolic modulation of the switch *in vivo* is also a possibility (Wright 1988). Field experiments suggested food supply as a possible proximal cue of sex change in *Lottia gigantea* (Lindberg & Wright 1985; Wright 1989), while density may be a proximal cue in *Patella vulgata* (Borges et al. 2016).

TABLE 8.5

Dodecapeptide Sequences for Molluscan Gonadotropin-Releasing Hormone (GnRH)

Taxon	Amino Acid Sequence
Mizuhopecten yessoensis	-Glu-Asn-Phe-His-Tyr-Ser-Asn-Gly-Trp-Gln-Pro-Gly-Lys-G-Arg-
Crassostrea gigas	-Glu-Asn-Tyr-His-Phe-Ser-Asn-Gly-Trp-Gln-Pro-Gly-Lys-G-Arg-
Lottia gigantea	-Glu-His-Tyr-His-Phe-Ser-Asn-Gly-Trp-Lys-Ser-Gly-G-G-Arg-
Patella caerulea	-Glu-His-Tyr-His-Phe-Ser-Asn-Gly-Trp-Lys-Ser-Gly-G-G-Arg-
Aplysia californica	-Glu-Asn-Tyr-His-Phe-SSer-Asn-Gly-Trp-Tyr-Ala-Gly-Lys-Lys-Arg-
Octopus vulgaris	-Glu-Asn-Tyr-His-Phe-Ser-Asn-Gly-Trp-His-Pro-Gly-Gly-Lys-Arg-
Sepia officinalis	-Glu-Asn-Tyr-His-Phe-Ser-Asn-Gly-Trp-His-Pro-Gly-Gly-Lys-Arg-
Uroteuthis edulis	-Glu-Asn-Tyr-His-Phe-Ser-Asn-Gly-Trp-His-Pro-Gly-Gly-Lys-Arg-

Source: De Lisa, E. et al., *Zool. Sci.*, 30, 135–140, 2013.

Note: Conserved regions shown in grey, substitutions coloured, and gaps = - G -

TABLE 8.6

Reproductive Hormones Reported in Gastropods

Hormone	Source	Function	Exemplar Taxa	References
Melatonin, 5-methoxytryptophol	Cerebral ganglia	Transduce photoperiodic signals	*Cornu aspersum*	Blanc et al. (2003)
APGWamide	Cerebral and pedal ganglia, male reproductive tract	Control of male reproductive behaviour and ejaculation	*Aplysia, Tritia obsoleta, Cornu aspersum, Lymnaea*	Fan et al. (1997); Koene et al. (2000); Morishita et al. (2010)
Conopressin	Vas deferens	Control of ejaculation	*Lymnaea*	Golen et al. (1995)
Caudodorsal cell hormone	CNS	Control of egg-laying	*Aplysia, Lymnaea*	Hatcher and Sweedler (2008)
FMRFamide	CNS	Control of egg-laying	*Lymnaea*	Saunders et al. (1992)
GnRH (gonadotropin-releasing hormone)	CNS; cerebral, pedal, and abdominal ganglia; gonads	Control of egg-laying	Patellogastropoda, Heterobranchia	Zhang et al. (2008); De Lisa et al. (2013)
ELH (egg-laying hormone)	Bag cells	Control of egg-laying	*Aplysia*	Arch et al. (1976); Chiu et al. (1979)
20-hydroxyecdysone	Dorsal bodies	Stimulation of oogenesis and oocyte maturation	*Helisoma, Cornu aspersum*	Mukai et al. (2001); Bride et al. (1991)
Progesterone	Unknown	Stimulation of oogenesis	*Helix pomatia, Littorina littorea*	Csaba and Bierbauer (1979); Lehoux and Williams (1971)
Testosterone	Unknown	Maturation of the male reproductive tract and spermatogenesis; development of male secondary sex characteristics; stimulation of oocyte maturation	*Tritia obsoleta*	Gooding and LeBlanc (2001)
9-cis retinoic acid	Unknown	Male reproductive tract development	*Reishia clavigera, Biomphalaria glabrata, Nucella lapillus, Tritia obsoleta*	Nishikawa et al. (2004); Bouton et al. (2005); Castro et al. (2007); Sternberg et al. (2008)

Adapted from: Sternberg, R.M. et al., *Ecotoxicology*, 19, 4–23, 2010.

In the gonochoristic vetigastropod *Haliotis rufescens* (Miller et al. 1973) and the trochids *Gibbula* (Herbert 1982; Clare 1987) and *Monodonta* (Clare 1987), cellular activity in the juxtaganglionar organ has an annual cycle related to female reproduction, but this correlation has not been observed in males. Griffiths (1961) reported similar annual control in the Cypraeidae. Saitongdee et al. (2005) identified an egg-laying hormone secreted by the cerebral, pleuropedal, and visceral ganglia in *Haliotis asinina* and York et al. (2012) demonstrated a complex succession of neuropeptide expression which contributed to regulating the diurnal spawning cycle of *H. asinina*. A putative neuropeptide from the parietal ganglion, also called an egg-laying substance, has been shown to induce egg-laying in the neogastropod *Busycon* (Ram 1977).

As with patellogastropods, the best studied caenogastropod is also a protandric species – the slipper limpet *Crepidula*. Species of *Crepidula* often live in close association with one another, sometimes forming stacks of individuals with the smallest at the top (males) and the largest at the base (females) (e.g., *Crepidula fornicata*) (Coe 1936; Fretter & Graham 1962; Collin 1995). In this species, both spermatogenesis and oogenesis are under the influence of a factor secreted by the cerebral ganglion and, as in patellogastropods, a hormone produced in the tentacles suppresses spermatogenesis. Based on external cues regarding their position in the stack, small individuals sequentially reduce and terminate spermatogenesis and commence oogenesis in the gonad (Figure 8.22) (Coe 1936).

Thus, the CNS control of sex change in *Crepidula* differs markedly from the patellogastropod *Patella* where control of sex change appears to reside with the spermatogonia and oogonia in the gonad (see above). In addition, sex reversal in *Crepidula* and other copulating taxa is not a simple replacement of spermatogenesis by oogenesis in the gonad as it is in broadcast spawners such as patellogastropods. In copulating taxa, it also involves the reduction of the male reproductive system (vas deferens, seminal vesicles, prostate, sperm groove, penis) and its partial or complete replacement by female secondary sex organs (oviduct, seminal receptacle, pallial oviduct, etc.) (Figure 8.22). In some species, components of the male system may be directly co-opted into the female reproductive system. For example, in *Crepidula fornicata*, the seminal vesicle of the male becomes the seminal receptacle of the female (Beninger et al. 2016). Some changes are controlled by neurohormones not associated with the gonad (Le Gall & Streiff 1975; Joosse & Geraerts 1983). For example, a factor from the pedal ganglion differentiates the penis (Moffett 1991); this putative hormone is apparently not specific to *Crepidula* as it is also present

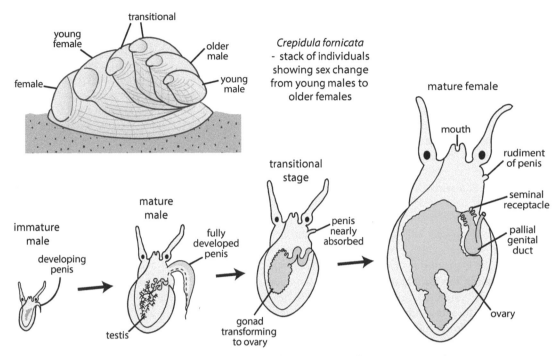

Crepidula onyx - animals removed from shell, showing changes in reproductive system with increasing size

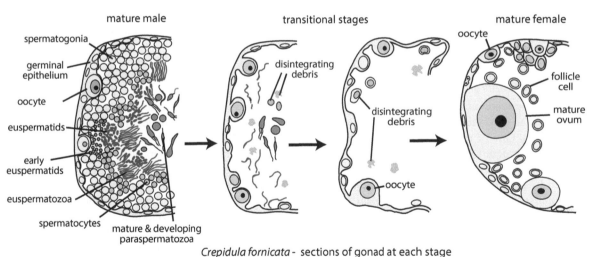

Crepidula fornicata - sections of gonad at each stage

FIGURE 8.22 Sex change in the caenogastropod *Crepidula*. Large females are at the base of a stack, followed by transitional individuals, and males. The morphological and gonadal transition from male and to female is under hormonal control (see text). Top left original; second and bottom row figures redrawn and modified from Coe, W.R., *Q. Rev. Biol.*, 19, 85–97, 1944.

in the pedal ganglia of gonochoristic male caenogastropods. The pedal factor appears to be activated by a factor produced by neurosecretory cells associated with the cerebropleural ganglion in *Crepidula*. In contrast, the degeneration of the penis during the onset of the female phase is controlled by a second hormone produced by neurosecretory cells in the pleural ganglia (Moffett 1991).

Because these factors are not water soluble and thus require the direct contact of individuals (Carrillo-Baltodano & Collin 2015), the habit of *Crepidula* to stack and live in close association with one another is not unexpected. Because of

their close proximity, tissue/tissue contact can be maintained by both the cephalic tentacles and the mantle margins. Le Gall and colleagues (Le Gall & Streiff 1975; Le Gall 1981; Le Gall et al. 1983) documented a masculinising factor in female mantle tissue sensed by males through tentacular contact. The presence of this factor produces neural and hormonal changes in the cerebral ganglion which then initiates morphogenetic activity in male pedal ganglia and the release of neurohormone that accumulates near the right tentacle (and penis) and prevents sex change; a second neurohormone inactivates somatic growth of male individuals. Males release a feminising factor from their cephalic tentacles which is sensed by the mantle edges of individuals in the lower part of the stack. This factor causes the pleural ganglion in the lower individuals to release a neurohormone that stimulates the reduction of the male reproductive system and promotes formation of the female reproductive system. A neural signal also goes to the cerebral ganglion, producing a neurohormone that inactivates locomotion and triggers somatic growth of the lower individuals. Thus, in *Crepidula* at least, large individuals suppress growth and sex change in smaller animals, while small animals prompt sex change and increase the growth rate of larger animals (Carrillo-Baltodano & Collin 2015).

8.8.2.1 Heterobranchs

We know more about aspects of the hormonal control of reproduction in euthyneuran gastropods, notably the euopisthobranch *Aplysia*, the hygrophilans *Lymnaea* and *Helisoma*, a few stylommatophoran land snails (e.g., *Helix*, *Theba*) than in all other gastropods combined. All heterobranchs are hermaphroditic and therefore individuals can provide insights into both male and female reproductive endocrinology; however, these systems may be more complicated than in gonochoristic taxa because of potential interactions between the two systems and the production and maintenance of two sets of reproductive anatomies. As mentioned above, in addition to the juxtaganglionar organ (i.e., dorsal bodies), some heterobranchs have additional neurosecretory glands, including caudodorsal cells, bag cells, and atrial glands.

8.8.2.1.1 *Euopisthobranchs*

Members of the euopisthobranch genus *Aplysia* (sea hares) have long provided an important model system for neurobiology (Kandel 1979). One of the first invertebrate neurohormones to be studied in detail was the ELH of *Aplysia* (Toevs 1970; Arch et al. 1976; Chiu et al. 1979; Rothman et al. 1983). ELH is secreted by neurosecretory cells in 'bag cells' which straddle the junction of the left and right connectives from the pleural to the visceral ganglia. The cells are covered in a fibrous sheath, and in addition to neurosecretory cells, they contain neurons and are bathed in blood from the adjacent aorta (Kupfermann et al. 1974). The main nerves extending from this plexus of neurosecretory cells and ganglia are the branchial (and osphradial) nerve, the siphon nerve, and the genital-pericardial nerve.

Neurohormones are stored in the bag cells as granules and are released in response to bursts of electrical activity, thereby producing a sudden, coordinated release of both bag cell peptides and ELH. Bag cell peptides provide autoexcitation of the ELH release, contraction of the hermaphroditic duct, and the release of additional peptides by the atrial gland (Brown & Mayeri 1989). The release of ELH induces the complete behavioural and physiological repertoire for ovulation of oocytes, their transport through the reproductive tract, and their fertilisation and packaging in the egg string, followed by the extrusion of the egg mass and its fixation to the substratum. The stimulus for this hormonal release appears to be driven, in part, by temperature and perhaps a

pheromone sensed by the rhinophores (Morishita et al. 2010). Early studies showed that bag cell neurons respond to vertebrate GnRH, suggesting the presence of an endogenous GnRH peptide that can bind and activate receptors associated with the bag cells. More recent work has shown there are at least two GnRH systems in *Aplysia* (Tsai et al. 2003), and that each system has a different form of GnRH with a distinct pattern of distribution. One system is widely distributed throughout the CNS and is detectable by an anti-serum against tunicate I-GnRH. The second system appears to be exclusively in the osphradium (detectable by an anti-serum against mammalian gonadotropin-releasing hormone, mGnRH), suggesting a direct relationship between the chemosensory osphradium and GnRH function in *Aplysia*.

8.8.2.1.2 *Panpulmonates*

In panpulmonates, most research has centred on a few hygrophilans and stylommatophorans (Geraerts et al. 1978, 1988; Joosse & Geraerts 1983; Vincent et al. 1984; Geraerts 1987; Wijdenes et al. 1987; Adamson et al. 2015). Male and female gamete maturation is more simultaneous in Hygrophila than in Stylommatophora and, not surprisingly, endocrine control of reproduction differs substantially between the two groups, although there are some shared pathways that remain intact (Figure 8.23).

The dorsal bodies (juxtaganglionar organs) in hygrophilans consist of cells associated with the cerebral ganglia near the cerebral commissure; there may also be medial dorsal bodies and lateral dorsal bodies or lobes in some groups (Figure 8.23) (Joosse & Geraerts 1983) (see also Chapter 7). The lateral lobes of the cerebral ganglia are thought to initiate simultaneous development of both male and female gametes through pathways that include the caudodorsal and dorsal body cell groups. In stylommatophorans, gametogenesis is successive, with sperm production preceding oocyte development. While totipotent gametes can differentiate directly into eggs in the presence of the dorsal body hormone, sperm production requires a second neurohormone from the optic tentacle; this latter factor also suppresses differentiation of oocytes. Thus, the initial onset of spermatogenesis is controlled by both the dorsal body hormone and an optic tentacle factor. The decrease in optic tentacle factor over time allows for the onset of oogenesis. As in hygrophilans, vitellogenesis is initiated by the dorsal body hormone, while development of the male and female accessory sex organs results from release of male and female factors from the gonads (Figure 8.23) (Joosse & Geraerts 1983). Saleuddin et al. (1990) reported that the interaction of dorsal bodies with the neuroendocrine pathway that controls somatic growth may indicate an inverse relationship between growth and reproduction in both stylommatophorans and hygrophilans. APGWamide has also been shown to regulate male reproductive activity in *Lymnaea* (Morishita et al. 2010).

The caudodorsal cells of hygrophilans are similar to the bag cells of *Aplysia* and produce a hormone similar in function to the egg-laying hormone, stimulating development of the albumen gland (Joosse & Geraerts 1983). In *Lymnaea*,

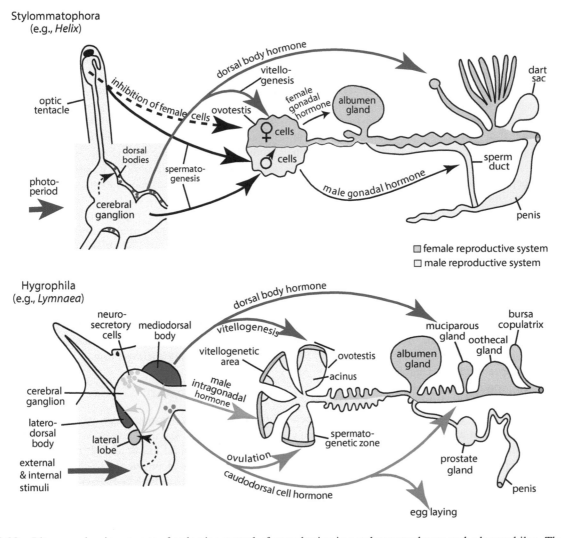

FIGURE 8.23 Diagrams showing aspects of endocrine control of reproduction in a stylommatophoran and a hygrophilan. The red, blue, orange and black arrows represent hormonal effects. Redrawn from Joosse, J. and Geraerts, W.P.M., Endocrinology, in Saleuddin, A.S.M. and Wilbur, K.M., *Physiology, Part 1. The Mollusca*, Vol. 4, pp. 317–406, Academic Press, London, UK, 1983.

the two caudodorsal cell groups are asymmetric with almost double the number of cells on the right as on the left. Axons project from these cell groups into the cerebral commissure, and stimuli are thought to include photoperiod, temperature, oxygen level, and food availability. Like bag cells, the caudodorsal cells discharge after electrical stimulation, with egg-laying beginning 5–10 minutes after discharge. Also, like the bag cells, different peptides are released that contribute to different aspects of egg-laying; caudodorsal cells control ovulation and oviposition, while the dorsal bodies induce oocyte maturation, vitellogenesis, albumen synthesis, and development of accessory sex organs and functioning, as it does in stylommatophorans. In contrast to the caudodorsal cells, the lateral lobes of the cerebral ganglia are thought to process photoperiod information from the optic nerve and coordinate growth and reproduction (see reviews in Boer & Joosse 1975; Joosse & Geraerts 1983; Joosse 1988; Saleuddin et al. 1990).

In stylommatophorans, there may be as many as six putative endocrine cell groups in the connective tissue surrounding the cerebral commissure. These cells contain neurosecretory granules that are released by exocytosis. Dorsal bodies in stylommatophorans control oocyte growth, vitellogenesis, differentiation, and growth of accessory sex organs. Cell morphology suggests both peptide and steroid secreting functions. In contrast to stylommatophorans, there does not appear to be gonadal control of male or female accessory sex organs in hygrophilans (Joosse & Geraerts 1983).

8.8.3 BIVALVES

A GnRH is present in bivalves (e.g., *Mizuhopecten, Crassostrea, Mytilus, Ruditapes*), where it appears to act directly on the gonad by stimulating cell division (Pazos & Mathieu 1999; Song et al. 2015). In a study of *Crassostrea gigas* and *Mytilus edulis,* Pazos and Mathieu (1999) demonstrated increased DNA synthesis and mitogenic activity in gonadal tissue following treatment with five vertebrate GnRHs. Extracts of the cerebropedal ganglia also increased cell division and mitogenic activity in the gonad, similar to that produced by the vertebrate GnRH treatments. Nakamura et al. (2007) used a GnRH antibody to identify neurons in the cerebropedal ganglia containing GnRH-like peptides; no nerve endings were found in the gonad, suggesting that GnRH is probably produced in the cerebral ganglia and transported by the haemolymph to the gonad.

FMRFamide peptides were first identified in the venerid clam *Macrocallista nimbosa*, and these help to regulate cardiac activity (Price & Greenberg 1977). Henry et al. (1995), Too and Croll (1995), and Pani et al. (2005) later demonstrated FMRFamide in the CNS, haemolymph, and in tissues surrounding the gonad in pectinids, but their function in bivalves remains unclear. APGWamide, another amide peptide, thought to regulate reproduction in *Lymnaea*, has been identified in the CNS and gonad of *Mytilus* by Smith et al. (1997). There is, however, little evidence for APGWamide involvement in the neural regulation of reproduction in bivalves (Morishita et al. 2010).

As discussed above and in Chapter 2, evidence for sex steroids in the control of reproduction has also been controversial (Scott 2012, 2013). Steroids, including estradiol-17, testosterone, and progesterone have been reported in bivalves (Liu et al. 2014 and references therein); their association with reproduction is typically based on their concentrations, which vary with the reproductive cycle (Alon et al. 2007; Ketata et al. 2008; Liu et al. 2014). Experimental exposure to steroids stimulates oogenesis and spermatogenesis, induces spawning, and can also promote sexual differentiation and shift sex ratios. It may be mediated by intracellular sex steroid receptors (Varaksina & Varaksin 1991; Wang & Croll 2003; Croll & Wang 2007). Vitellogenesis has been suggested to be under control of another steroid, estradiol-17 (Li et al. 1998; Saavedra et al. 2012). The work of Croll (2007) and Croll and Wang (2007), suggests that steroids may play an important role in the reproductive control of scallops, although the absence of known steroid receptors (Sternberg et al. 2010) complicates this conclusion.

8.8.4 Cephalopoda

Control of reproduction in coleoid cephalopods is relatively simple and appears to involve only one GnRH-like hormone from one pair of endocrine glands, the neural optic glands. In all coleoids, these are on the optic nerve and often bright orange in colour in mature individuals. *Nautilus* has no distinct optic gland, but there are groups of cells, positioned at the junction of the optic and olfactory lobes, which have abundant cytoplasm and argentophilic inclusions and may have a neurosecretory function (Young 1965a, 2010).

The optic glands secrete GnRH under nervous control (Wells 1960; Nishioka et al. 1970; Wells & Wells 1977b;

Kanda et al. 2006). Light has been shown to inhibit optic gland function in octopods, and this function is mediated by a complex neural pathway from the retina to the subpedunculate lobe of the brain and then to the optic gland (Wells & Wells 1977b; Di Cristo 2013); severing these connections in juveniles results in premature development of gonads. Cephalopod GnRH is not sex specific. It regulates yolk production by follicle cells and stimulates growth and synthetic activity of accessory sex organs in females and spermatophore production and development of the male reproductive ducts (O'Dor & Wells 1973, 1975). It also appears to promote protein breakdown and mobilisation of amino acids for use by reproductive structures.

While the gonads of cephalopods have been shown to synthesise steroids, their role in reproduction has not been confirmed (LaFont & Mathieu 2007). Carreau and Drosdowsky (1977) observed *in vitro* steroid biosynthesis in male and female gonads of *Sepia officinalis*. Seven enzyme systems were demonstrated in the cuttlefish including pregnenolone, progesterone, testosterone, androstenedione, dehydroepiandrosterone, and estradiol-17β. Avila-Poveda et al. (2015) further showed that gonadal progesterone levels were elevated during vitellogenesis.

8.9 DEVELOPMENT

Like most lophotrochozoan taxa, molluscan development is characterised by spiral cleavage, a D-quadrant-dependent transition from spiral to bilateral cleavage and gastrulation, followed by the development of a trochophore larva. Molluscs are coelomate animals with body cavities typically formed by the splitting of embryonic mesodermal masses (*schizocoely*). They also exhibit protostomous development – the mouth typically develops before the anus (Figure 8.24). These characteristics are shared with several other phyla (e.g., Nemertea, Platyhelminthes, Annelida), and these are grouped as the Trochozoa within the Spiralia.

The animal-vegetal axis of the developing embryo (and earlier oocyte) is defined by a line passing through the nucleus of the egg. The hemisphere which includes the nucleus is defined as the animal pole, while the opposite hemisphere is defined as the vegetal pole; in molluscs the majority of the yolk is concentrated in the vegetal pole. Spiral cleavage is a stereotypic pattern of cell division that segregates cytoplasmic determinants in specific regions of the embryo and creates a highly organised developmental architecture, which is radially symmetrical to the animal-vegetal axis (Lambert & Nagy 2002). In spiral-cleaving embryos, each cell after the four cell state can be uniquely identified using a nomenclature of quadrants (A, B, C, D) and quartets (1, 2, 3, 4) (Wilson 1892) (Figure 8.27 and Section 8.9.2 below). Each cell lineage makes a distinct contribution to adult morphology and some cell lineage contributions are conserved across phyla (Wilson 1898; Guralnick & Lindberg 2001). Because of the large yolky eggs of some molluscs, especially cephalopods, the spiral cleavage pattern can be highly modified (Naef 1928) (see Section 8.9.2).

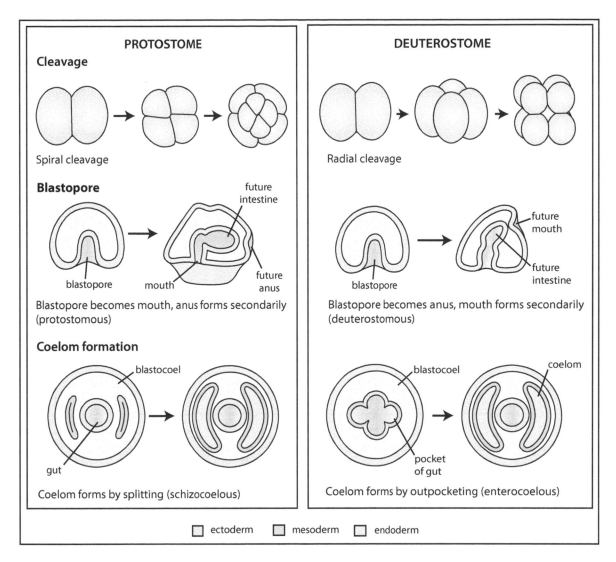

FIGURE 8.24 Comparison of generalised bilaterian cleavage patterns and blastopore and coelom formation. Redrawn and modified from various sources.

8.9.1 Spiral Cleavage

In most lophotrochozoans, the early cleavage pattern is spiral, meaning that the animal pole blastomeres are rotated with respect to those of the vegetal pole. In some molluscs, the handedness of the spiral offset is maternally inherited. Also, present during spiral cleavage, but independent of it, is a cellular pattern derived from the first and second quartet of cells that form a transitory cross-shaped pattern on the animal surface of the blastomeres at the 64-cell stage. The molluscan and 'sipunculan' crosses are formed by cells $1a^{12}$–$1d^{12}$, while in non-sipunculan annelids, the cross pattern is formed by the cells $1a^{112}$–$1d^{112}$ (Scheltema 1996; Lindberg & Guralnick 2003a). In molluscs, two cleavage modes are present: (1) the discoidal segmentation of cephalopods and (2) typical spiral cleavage (all other molluscs); these are two distinct modes with no intermediate forms (Fioroni 1966).

8.9.2 Cleavage and Early Development

Upon fertilisation, the first two divisions of the egg are meridional and produce four macromeres (A, B, C, D) (Figure 8.25). Beginning with the third cleavage cycle, the first quartet of micromeres (1a, 1b, 1c, 1d) are produced from each macromere, however, the divisions are not along the meridian of each cell. Instead, the cleavage spindles are oblique, and the new micromeres are displaced to the right (clockwise) and lie in the furrows between the macromeres. In sinistral gastropods, the spindle direction is reversed and the macromeres are displaced to the left and lie in a counter-clockwise configuration relative to their respective macromeres when viewed from the animal pole. In subsequent cell divisions, the spindles are oriented at right angles to the previous divisions, and therefore the following group of micromeres alternate between clockwise and counter-clockwise placements. The micromeres

EQUAL CLEAVAGE

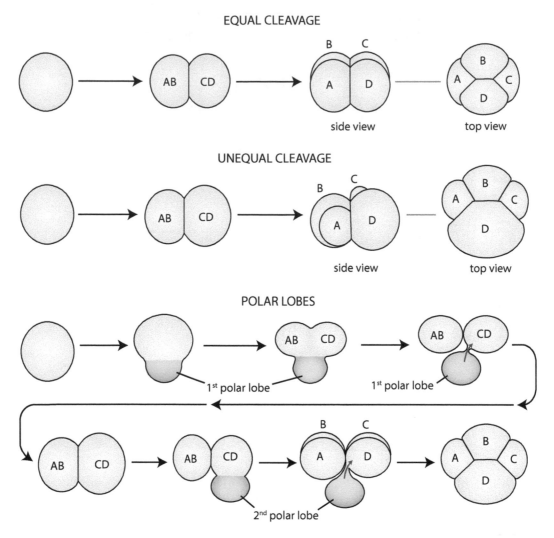

UNEQUAL CLEAVAGE

POLAR LOBES

FIGURE 8.25 Cleavage patterns in non-cephalopod molluscs (see text). Because of the large yolk content in cephalopod eggs, their cleavage patterns (see Figure 8.26) are not comparable with other. The red arrows show transfer of cytoplasm from the polar lobes to the D macromere. Redrawn and modified from Verdonk, N.H. and Van den Biggelaar, J.A.M., Early development and the formation of the germ layers, in Verdonk, N.H., Biggelaar, J.A.M., Van den and Tompa, A.S., *Development. The Mollusca*, Vol. 3, pp. 91–122, Academic Press, New York, 1983.

begin dividing at the fourth cleavage cycle and also show an alternation between clockwise and counter-clockwise placement of their descendants.

Early cleavage patterns in cephalopods are markedly different because of the large amount of yolk present in the cephalopod egg (Figure 8.26). Instead of the entire egg undergoing cleavage, superficial cleavage occurs at the animal pole, and a *discoblastula* is formed on the yolk surface during early cleavages (Raven 1966; Boletzky 1989). There is no trace of spiral cleavage as in other molluscs, and instead the early divisions are bilaterally symmetrical. The first cleavage furrow divides the discoblastula into left and right sections, while the second cleavage furrow is perpendicular to the first and demarcates anterior and posterior regions of the embryo. In the anterior portion of the disc, the third cleavage furrow is radial and divides the anterior region into four sections, while in the posterior region, the third cleavage furrow is parallel

with the first (Boletzky 1989). The fourth cleavage furrow is the first to actually form blastomeres, and subsequent divisions increase the number of cells in the central portion of the disc (Raven 1966). This process produces a single layer of cells resting on top of the uncleaved yolk sac.

Variation in cleavage patterns is found throughout molluscs. These include variation in the symmetry of cell divisions, the presence of polar lobes, and variation in the overall rate of cell cleavage, the relative rates of division among different cell lineages, and the total number of cell divisions for specific cell lineages (Freeman 1979; Wray 1994; Lindberg & Guralnick 2003b) (Figure 8.27). Much of this variation appears associated with accelerating the establishment and organisation of the D-quadrant blastomeres (Lindberg & Guralnick 2003b and references therein).

In polyplacophorans, solenogasters, scaphopods, and most gastropods, the first two cell divisions are typically equal, but in

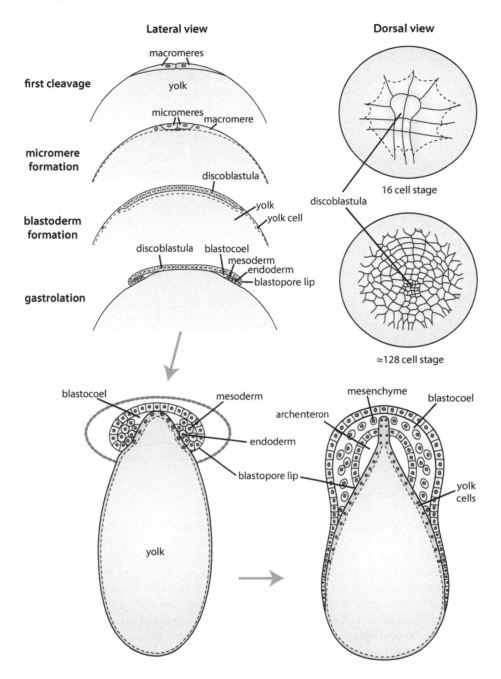

FIGURE 8.26 Coleoid cephalopod early development. Redrawn and modified from Naef, A., *Die Cephalopoden. Embryologie*, R. Friedländer & Sohn, Berlin, Germany, 1928 and Singley, C.T., *An analysis of gastrulation in Loligo pealei*, Ph. D., University of Hawaii, 1977.

many bivalves the cell divisions are unequal and the D-quadrant is often the largest cell at the four cell stage (Raven 1966; Freeman & Lundelius 1992) (Figure 8.25). The formation of *polar lobes*[5] (Figure 8.25) is another mechanism by which morphogenetic factors can be partitioned into a specific cell lineage, and they are also thought to facilitate the early specification of cell fates (Henry et al. 2006). Polar lobes occur in the solenogasters, caenogastropods, scaphopods, and pteriomorphian bivalves (Freeman & Lundelius 1992). Basal caenogastropods

have small polar lobes, whereas in more derived taxa, they are relatively large (Verdonk & Cather 1983).

Variation in the relative rates of division among different cell lineages which produces an apparent acceleration of the 4d cell lineage relative to cell number shows a strong phyletic pattern. For example, in the Polyplacophora, 4d mesentoblast formation occurs when the embryo is composed of about 73 cells (Heath 1899), in the Scaphopoda it appears at 64 cells (Dongen & Geilenkirchen 1974), and in basal gastropod taxa (e.g., Patellogastropoda and Vetigastropoda), mesentoblast formation occurs at about the 63 cell stage (Biggelaar & Haszprunar 1996).

---

[5] Outpocketings of cytoplasm on the vegetal pole of the oocyte.

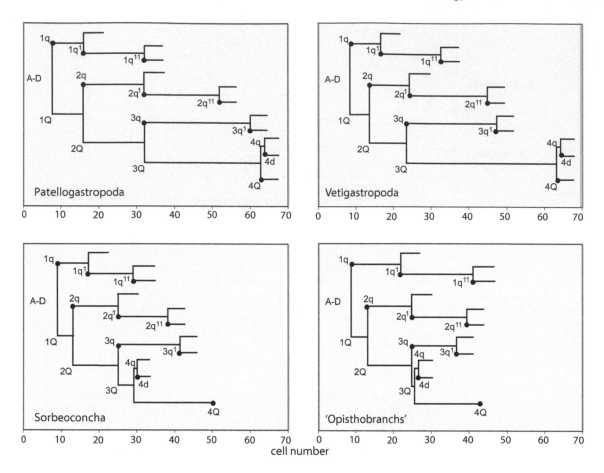

FIGURE 8.27 Varying rates of cell division and differentiation of cell lineages in four groups of gastropods. Data from Lindberg, D.R. and Guralnick, R.P., *Evol. Dev.*, 5, 494–507, 2003b.

In higher gastropod taxa (Caenogastropoda and Heterobranchia), mesentoblast formation occurs at the 24 cell stage (Verdonk & Biggelaar 1983; Biggelaar & Haszprunar 1996). This acceleration appears to be due to an earlier onset of 3D macromere division relative to the divisions of the first, second, and third quartets (Verdonk & Biggelaar 1983). The 3D vegetal macromere is the developmental signalling centre or organiser in molluscan development. Its role is to transition the developing embryo from a spiral to a bilateral cleavage pattern and establish the secondary axis, called either the anterior-posterior axis (D-quadrant posterior), or the dorsal-ventral axis (D-quadrant dorsal) (Biggelaar 1977; Edsinger-Gonzales et al. 2007).

In some cases, the acceleration of 4d mesentoblast formation appears to be related to life history evolution, especially in gastropods (Freeman & Lundelius 1992; Biggelaar & Haszprunar 1996). In other taxa, changes in cell lineage differentiation appear to be associated with diversification of body plans irrespective of life history (Ponder & Lindberg 1997; Guralnick & Lindberg 2001). Regardless of the ultimate causality, changes in the timing of formation of cell lineages can also change the spatial and temporal patterns of cells within the developing embryo, and these new alignments undoubtedly affect both subsequent specific inductive interactions between cells while shifting other specifications from local to global patterning signals or vice versa (Biggelaar 1978; Freeman 1979; Davidson 1990; Wray

1994) (Figure 8.28). Moreover, temporal changes in cell lineage formation might disrupt regulatory gene activity that previously constrained variation in certain quartets by delimiting specific periods during which gene expression or signalling occurred.

Continued cell divisions produce the blastula. Like most metazoan embryos, the molluscan blastula has a clear animal-vegetal polarity (Biggelaar et al. 2002). Cells associated with the vegetal pole contain most of the yolk of the embryo and have little cytoplasm, while the cells that make up the animal pole have less yolk and exhibit faster cell divisions. In taxa with little yolk, a wide cleavage cavity is present within the blastula, forming a hollow sphere of blastomeres, or a *coeloblastula*. A coeloblastula is formed in the Polyplacophora, Solenogastres, Scaphopoda, and Patellogastropoda. It also occurs in the bivalve *Dreissena*, the hygrophilans *Physa* and *Lymnaea*, and in the caenogastropod *Bithynia*, all of which live in fresh water, and in the terrestrial stylommatophorans *Succinea* and *Limax* (Raven 1966). In taxa with larger amounts of yolk, such as the many vetigastropod, neritimorph, and caenogastropod species, a *stereoblastula*, an almost solid mass of blastomeres, is formed (Raven 1966). In cephalopods, there is no blastula, but rather a discoblastula or *blastoderm*—a single layer of ectodermal cells that spread over the surface of the yolk and denote the animal pole (Figure 8.26).

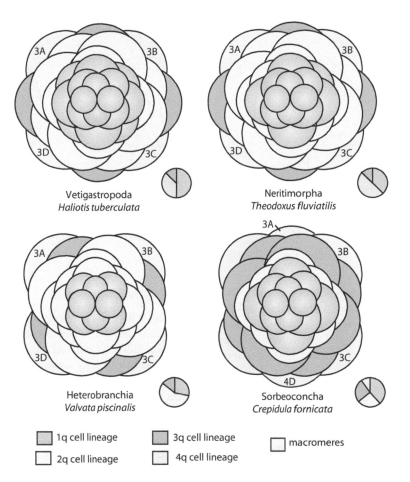

FIGURE 8.28 Stylised embryo composition at the 32–36 cell stage in four gastropod subclades – Vetigastropoda (36 cells), Neritimorpha (36 cells), Heterobranchia (32 cells), and Sorbeoconcha (34 cells). Cells are coloured to show cell lineage affinities, and pie charts summarise overall embryo composition for each taxon at this stage. Note changing cell lineage compositions in different gastropod clades at similar cell numbers. Cell lineage data from Van den Biggelaar, J.A.M. and Haszprunar, G., *Evolution*, 50, 1520–1540, 1996.

Gastrulation in molluscs with a coeloblastula involves the invagination of the vegetal pole cells into the interior of the embryo. The opening created by the invagination of these cells is called the *blastopore* and becomes the mouth in most taxa (Biggelaar et al. 2002). The invagination of the vegetal pole cells into the blastula forms the archenteron, or primordial gut (endoderm), and also gives rise to the mesoderm (Lyons et al. 2015). Proliferation and movement of the animal pole cells (epiboly) during gastrulation produces the outer epithelial layer (ectoderm) of the embryo. In taxa that have stereoblastulae, gastrulation occurs primarily by epiboly (Raven 1966). Gastrulation in cephalopods is markedly different (Figure 8.26). Instead of a spherical blastula undergoing gastrulation, the process is initiated at the periphery of the discoblastula. Ectodermal cells of the blastoderm layer delaminate from the yolk sac and fold under the periphery of the discoblastula (Naef 1928). This folding, followed by cell proliferation, produces a thickened, multicellular layer around the margin, occurring around the entire margin of the discoblastula in octopods (Sacarrao 1952), while in *Loligo*, the folding is limited to a horseshoe shaped area (Teichmann 1903).

8.9.3 DEVELOPMENT OF GERM LAYERS

Molluscs are triploblastic metazoans, and their tissues and organs are derived from one of three germ layers (ectoderm, endoderm, and mesoderm) or combinations thereof (Figure 8.29). Differentiation of these three germ lines occurs early in development and generally accompanies gastrulation (Figure 8.30). Ectoderm in the head region (pretrochal) is derived from the first quartet of micromeres (1a–1d), while trunk ectoderm (post-trochal) is derived from the second quartet (2d), and with a lesser contribution from the third quartet (Raven 1966; Verdonk & Biggelaar 1983). The third quartet also gives rise to the ectomesoderm in gastropods (Dictus & Damen 1997). Endoderm differentiates with the formation of the blastopore and forms the epithelial lining of the archenteron; participating cells include 4A–4D and 4a–4c. Lastly, mesoderm has two origins within the lophotrochozoans: endomesoderm which forms from daughter cells of 4d (teleoblasts) which proliferate into mesodermal bands on either side of the archenteron, and ectomesoderm which is derived from the second and third quartets of micromeres (Boyer

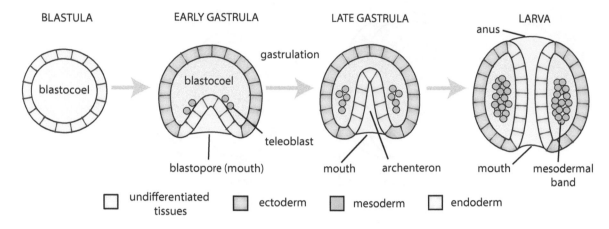

FIGURE 8.29 Schematic diagram of molluscan tissue formation. Original.

et al. 1996). Traditionally, these two sources of mesoderm have been considered to produce adult and larval mesoderm. Boyer et al. (1996) have argued that both contribute to adult mesoderm in Polycladida and Annelida, and Verdonk and Biggelaar (1983) have also pointed out that both may form larval and adult structures in molluscs. Lyons et al. (2015) have shown that only five genes were exclusively expressed in gastropod ectomesoderm, and these may contribute to the epithelial-to-mesenchymal transition. Moreover, shared expression of genes in both mesodermal sources suggests that components of the conserved endomesoderm program were either co-opted for ectomesoderm formation or that ecto- and endomesoderm are derived from a common mesodermal precursor that became subdivided into distinct domains during evolution (Lyons et al. 2015; Perry et al. 2015).

8.9.3.1 Germ Cells

Formation of the germ line or *primordial germ cells* (PGCs) in molluscs results from both epigenetic (internal inductive signals in the developing embryo) and preformation mechanisms (maternally inherited determinants). This is uncommon in lophotrochozoans, where most germ lines form by epigenesis (Extavour & Akam 2003; Extavour 2007). Other groups which share a dual mechanism are the Platyhelminthes and Annelida. Based on the limited data available (reviewed by Extavour & Akam 2003), we provide the following summary. Solenogasters germ lines are mesodermal in origin and are formed by epigenesis after metamorphosis. In polyplacophorans, they are post-embryonic and also epigenetic in origin. In cephalopods and bivalves, the germ line is reported to be specified by preformation mechanism(s). In cephalopods, this occurs at the blastoderm stage and originates from the superficial layer of the blastoderm, while in bivalves, germ line formation occurs during early cleavage and is derived from the 4d cell. Lastly, in gastropods, both epigenesis and preformation mechanisms have been identified (Raven 1966). Germ cells are formed during late embryogenesis and early cleavage stages and are derived from mesoderm or the early blastomeres.

8.9.4 Gene Expression

In his classic paper on gene expression, Haldane (1932) suggested that torsion and body asymmetry of gastropods was not a larval adaptation as suggested by Garstang (1928), but rather an evolutionary novelty caused by changes in the temporal expression of genes. He also proposed these could be combined effects of multiple genes. This early insight into gene expression in molluscs is prophetic not just for gastropods, but also the other classes as gene expression studies have provided important insights into the genetic mechanisms of molluscan diversification. However, until very recently, identification of key developmental genes has been limited (de Oliveira et al. 2016).

8.9.4.1 Signalling Pathways

Signalling pathways activate specific genes through signal regulated transcription factors, while preventing their expression in other cells. This control allows these pathways to produce a diversity of gene expression patterns which control most cell fate decisions in developing bilaterian animals (Gerhart 1999; Barolo & Posakony 2002). In the Bilateria, there are seven major cell-cell signalling pathways: Wnt/β-Catenin, TGF-β (transforming growth factor beta 1), Hedgehog (*Hh*), receptor tyrosine kinase (RTK), nuclear receptor, Jak/STAT (signal transducer and activator of transcription proteins), and Notch. Each pathway is used repeatedly during development and activates different subsets of target genes in different developmental contexts. In molluscs, Hedgehog, Wnt, and Notch signalling pathways have been identified (de Oliveira et al. 2016), and have been shown to have diverse roles in molluscan development and morphological diversity. Wnt signalling genes in the pearl oyster (*Pinctada fucata*) probably participate in shell formation and early embryonic and larval development (Gao et al. 2016). Wnt signalling has also been implicated in regulating early oocyte development and is of possible importance in understanding oocyte infertility in bivalve aquaculture (Pauletto et al. 2014). In the patellogastropod *Patella vulgata*, Hedgehog is expressed in the ventral midline of the larva between the mouth and telotroch (see Section 8.9.5) and is hypothesised to be involved in development of the

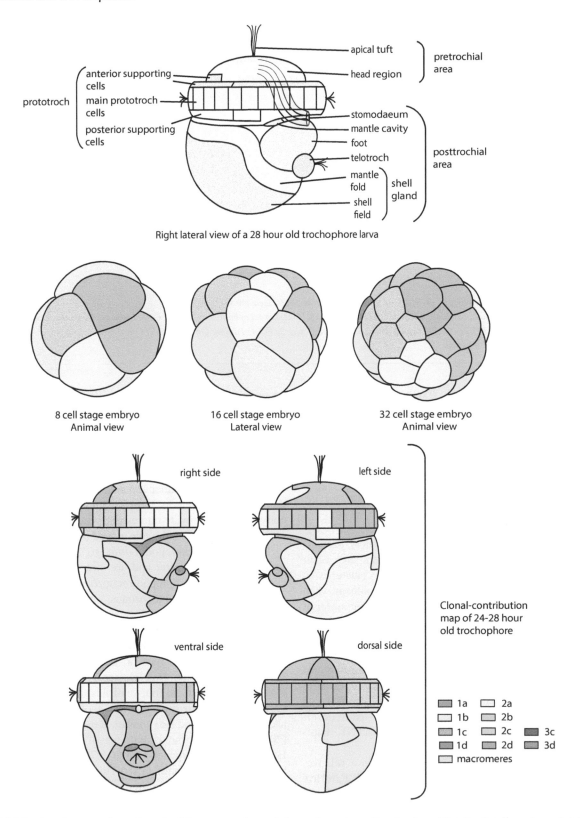

FIGURE 8.30 Development, formation of cell lineages, and morphology of a gastropod trochophore (*Patella*, Patellogastropoda). Redrawn and modified from Dictus, W.J.A.G. and Damen, P., *Mech. Dev.*, 62, 213–226, 1997.

TABLE 8.7

Complete and Incomplete Hox Clusters Reported in the Mollusca and Putative Outgroups

Higher Taxon	Species	PG-1	PG-2	PG-3	PG-4	PG-5	Lox5	Antp	Lox2	Lox4	Post-1	Post-2
Brachiopoda	*Lingula anatina*	•	•	•	•	•	•	•	•	•	•	•
Annelida	*Capitella teleta, Nereis virens*	•	•	•	•	•	•	•	•	•	•	•
Mollusca												
Polyplacophora	*Acanthochitona crinita*	•	•	•	•	•	•	•	•	•	?	•
Solenogastres	*Wirenia argentea*	?	?	•	•	•	•	•	•	•	?	?
	Gymnomenia pellucida	•	•	•	•	•	•	•	•	•	•	•
Caudofoveata	*Scutopus ventrolineatus*	?	?	?	?	•	?	?	•	•	?	?
Bivalvia	*Nucula tumidula*	•	•	•	?	•	?	•	?	•	•	•
	Pecten maximus, Pinctada fucata	•	•	•	•	•	•	•	•	•	•	•
Scaphopoda	*Antalis entalis*	•	•	•	•	•	•	?	?	•	•	•
Gastropoda	*Lottia gigantea, Gibbula varia*	•	•	•	•	•	•	•	•	•	•	•
Cephalopoda	*Nautilus*	•	•	•	•	•	•	•	•	•	•	•
	Idiosepius notoides	?	?	•	?	•	•	•	•	?	?	•

Source: Barucca, M. et al., *J. Dev. Biol.*, 4, 12, 2016; de Oliveira, A.L. et al., *BMC Genomics*, 17, 905, 2016.

Gene Abbreviations: Antp: Antennapedia; lab: labial; Lox2: Lophotrochozoa Hox2; Lox4: Lophotrochozoa Hox4; Lox5: Lophotrochozoa Hox5; pb: Proboscipedia; PG-1: Paralog Group-1; PG-2: Paralog Group-2; PG-3: Paralog Group-3; PG-4: Paralog Group-4; and PG-5: Paralog Group-5; Post-1:Posterior-1; Post-2: Posterior-2. ? = not reported.

nervous system (Nederbragt et al. 2002), while in cuttlefish, the Hedgehog signalling pathway is probably involved in the differentiation of striated muscle fibres in mantle development (Grimaldi et al. 2008). Lastly, Notch has been implicated in shell colouration patterns in the bivalves *Meretrix meretrix* and *Crassostrea gigas* (Feng et al. 2015; Yue et al. 2015).

8.9.4.2 Hox, ParaHox, and Other Developmental Homeobox Clusters

Hox genes are a family of transcriptional regulatory proteins governing embryonic development in molluscs and other Bilateria. One of the special features of the Hox cluster is its spatial collinearity. Hox clusters are said to be collinear when the order of the genes on the chromosome reflects the order of gene expression and function along the anterior to posterior axis of the embryo. Spatial and temporal collinearity is a noted feature of Hox gene expression and is probably plesiomorphic in the last common ancestor of deuterostomes and protostomes. In Lophotrochozoa, collinearity has been reported in only a few annelids and molluscs (Fritsch et al. 2015; Barucca et al. 2016). Other homeobox gene families involved in developmental processes include ParaHox, EHGbox, and NK (natural killer) genes.

The ParaHox cluster is an evolutionarily close relative, or paralog, of the Hox cluster. Both Hox and ParaHox clusters appear to have arisen by duplication of an ancestral ProtoHox cluster early in metazoan evolution. More distantly related homeobox genes are in the NK cluster, these being evolutionary relatives of both Hox and ParaHox genes and are found in clusters in some bilaterian taxa. Comparative genomics studies strongly suggest that Hox, ParaHox, and NK clusters were originally close together in the genome. The expression patterns of these three related sets of homeobox

genes appears to correspond broadly with one of the three embryonic germ layers: Hox genes are expressed in all germ layers, but predominantly in the neuroectoderm[6] (Holland 2005), ParaHox genes are expressed primarily in endodermal derivatives (Brooke et al. 1998), and NK genes are expressed mostly in mesodermal derivatives (Jagla et al. 2001).

Changes in the number and sequence of Hox genes, and in their expression patterns, have been related to the evolution of body plans and provide interesting insights into the evolution of the Hox cluster and the role played by Hox genes.

In the patellogastropod *Lottia gigantea*, 11 Hox genes occur as a single cluster and are structurally collinear with clusters found in non-lophotrochozoan genomes (including sponges, cnidarians, placozoans, and humans) (Simakov et al. 2013). In putative molluscan outgroups, the most intact cluster is found in the annelid *Capitella teleta*, which shares synteny[7] with *L. gigantea*, however, the genes are distributed among three chromosomes or scaffolds. In the brachiopod *Lingula anatina*, the Hox cluster is non-syntenic with *L. gigantea* and occurs on four scaffolds, leading Simakov et al. (2013) to conclude that the last mollusc-annelid ancestor had a single 11-gene Hox cluster. Recently, de Oliveira et al. (2016) confirmed the orthology[8] of numerous molluscan Hox and ParaHox genes (Table 8.7), and also reported a specific molluscan motif in the Hox paralog group 5 and a lophotrochozoan ParaHox motif in the *Gsx* gene.

---

[6] Ectodermal tissue which gives rise to the nervous system.

[7] Synteny occurs when groups of genes are found in the same order in the genome of two species being compared. Macrosynteny refers to the equivalence of a large portion of genes, while microsynteny refers to preservation of synteny of only a few genes.

[8] Orthologs are genes in different species that evolved from a common ancestral gene by speciation.

Data on Hox expression in molluscs are known from gastropods (Degnan & Morse 1993; Giusti et al. 2000; Hinman et al. 2003; Samadi & Steiner 2009), cephalopods (Lee et al. 2003b; Pernice et al. 2006), bivalves (Zhang et al. 2012a; Paps et al. 2015), and polyplacophorans (Fritsch et al. 2015, 2016). These data suggest a role of Hox genes in CNS development, shell formation, and, in cephalopods, tentacle and funnel development. Hox expression has also been identified in sensory organs such as the apical organ, statocysts, and in the light organ in cephalopods. Lee et al. (2003a) reported novel recruitment and spatial expression patterns of Hox genes in the coleoid *Euprymna scolopes*. In this study, Hox genes were expressed in the developing intrabrachial ganglia of the brachial nervous system which contrasted with the lack of Hox expression in the development of gastropod pedal nerve cords (Hinman et al. 2003). This difference suggests that at least two, and possibly seven, Hox genes were co-opted in the reorganisation of the plesiomorphic molluscan pedal nervous system into the discrete ganglia that make up the cephalopod brachial nervous system. Moreover, each arm pair expresses a unique combination of these seven genes and is not collinear (Lee et al. 2003a). Analysis of early development in the oyster *Crassostrea gigas* has shown that, as in gastropods and cephalopods, Hox genes are not activated in temporal collinearity. Instead, *Lox4* expression occurs before gastrulation, while *Lox5* and *Post-2* are expressed during the trochophore stage, and the remaining genes in late development (Paps et al. 2015). This lack of temporal and spatial collinearity suggests that Hox genes were co-opted for the formation of some of the body plan novelties found in gastropods, bivalves, and cephalopods (Fritsch et al. 2015) rather than their ancestral role of patterning structures along the anterior-posterior body axis as in most other bilaterians. Temporal and spatial collinearity in Hox gene expression in brachiopods is also lacking (Schiemann et al. 2017).

Patterning of the body along its anteroposterior (A-P) axis involves striking differences in the deployment of Hox genes between studied Aculifera and Conchifera reviewed by Wanninger and Wollesen (2018). The putative ancestral role of patterning structures along an anterior-posterior body axis is present in the staggered expression patterns of early and mid-trochophore larva of the polyplacophoran *Acanthochitona crinita* (Fritsch et al. 2015), where genes *Acr-Hox1-5*, *Hox7*, and *Post-2* are expressed in a co-linear pattern along the anteroposterior axis. In contrast, the gastropod *Gibbula varia* exhibits structure-specific expression of the Hox genes, with expression in, for example, the apical organ, the foot, shell field, and the prototroch (Samadi & Steiner 2009). Likewise in Cephalopoda, Hox genes are not obviously employed in A-P patterning, but are expressed in particular organ systems, including the gills, arms, and funnel (Lee et al. 2003a).

Molluscan ParaHox genes have been investigated to a lesser extent than Hox genes. Some have suggested that the ParaHox gene *Gsx* patterned the foregut of the last common bilaterian ancestor, based on its foregut expression pattern in some lophotrochozoan, ecdysozoan, and deuterostome taxa. *Gsx* is also expressed in the bilaterian CNS, and in the apical organs of larval gastropods and annelids (Wollesen et al. 2015; Fritsch et al. 2016), but is not expressed in the developing digestive tract in both the scaphopod *Antalis entalis* and the coleoid cephalopod *Idiosepius notoides*. Instead, in the scaphopod, it is expressed in cells of the apical organ and, in older trochophores, in the developing cerebral and pedal ganglia (Wollesen et al. 2015; Fritsch et al. 2016). In cephalopods, *Gsx* is expressed in the cerebral, palliovisceral, and optic ganglia, while in later embryos, it is also expressed close to the eyes and in the supraoesophageal and posterior suboesophageal masses and optic lobes. Just before hatching, *Gsx* is only expressed near the eyes (Wollesen et al. 2015; Fritsch et al. 2016).

The NK cluster was first identified in bivalve molluscs by Mesías-Gansbiller et al. (2013) and probably consists of six or seven genes. In other Bilateria, the NK homeobox gene cluster appears to play a primary role in mesoderm specification and formation and nervous system development. Six NK cluster genes have been reported in the oyster *Crassostrea gigas* (Zhang et al. 2012a) and three highly conserved NK homeobox sequences in the venerid bivalve *Venerupis corrugata* (as *V. pullastra*) and two from the oyster *Ostrea edulis* (Perez-Parallé et al. 2016). Single NK genes have been identified from the gastropod *Lymnaea stagnalis* (Iijima et al. 2006) and the cephalopod *Sepia officinalis* (Navet et al. 2008).

The EHGbox cluster is a group of three genes – engrailed (*en*), motor neuron restricted (*Mnx*), and gastrulation brain homeobox (*Gbx*). Engrailed has been reported in bivalves, gastropods, cephalopods, scaphopods, chitons, and bivalves and appears to be involved in shell formation (Wray et al. 1995; Jacobs et al. 2000; Kocot et al. 2017). The gastrulation brain homeobox has been reported from gastropods, cephalopods, solenogasters, and bivalves (Degnan & Morse 1993; Iijima et al. 2006; Perez-Parallé et al. 2016). Phylogenetic analysis of bivalve *Gbx* gene sequences enabled Perez-Parallé et al. (2016) to conclude that the gene was highly conserved among the Solenidae, Pectinidae, Veneridae, Ostreidae, and Mytilidae. There remain no reports of the presence of the third gene (*Mnx*) in molluscs.

The Pax gene family encodes transcription factors involved in myogenesis, neurogenesis, biomineralisation, and the development of sensory systems and excretory organs in the Bilateria (Scherholz et al. 2017). In molluscs, *Pax6* appears involved in central nervous system development, especially the development of the cerebral ganglia and the ventral nerve cords, but not in developing the lateral nerve cords. The Pax2/5/8 gene family have also been implicated in spicule development in both polyplacophorans and solenogasters (Scherholz et al. 2017). See Chapter 7 for the role of *Pax6* in sensory structure development in molluscs.

8.9.5 LARVAE

Larvae are typically recognised as distinct morphological forms of a developing organism. Besides a discrete external morphology, larvae may also have organs or systems not found in adults, or that are replaced by adult structures with their own separate ontogenies (Strathmann 1993).

Larval transformations are often associated with transitions between habitats, or during metamorphosis, and therefore have unique gene expression patterns associated with the different larval stages (Lambert et al. 2010; Heyland et al. 2011; Huan et al. 2012; Huang et al. 2012; Bassim et al. 2014; Xu et al. 2016). Different transcription factors are also expressed at the same larval stage, but in different phenotypes. For example, Lesoway et al. (2016) have shown that the transcriptomes of nutritive embryos or 'nurse eggs' of *Crepidula navicella* have high levels of cell death related gene expression compared to viable embryos which show normal developmental transcript patterns.

Molluscan larvae are morphologically diverse (Figure 8.31–8.33, 8.35, and Table 8.8). After gastrulation, many aquatic molluscan taxa pass through a trochophore larval stage, a larval type found in a variety of lophotrochozoan taxa including the Annelida, Entoprocta, and Nemertea. The trochophore typically consists of an oval cell mass with a dorsal apical tuft, one or more circumferential bands of cilia (*prototroch* and *metatroch*), and a ciliated ventral patch (*telotroch*); the prototroch provides propulsion, while the telotroch marks the putative site of the anal opening. The prototroch divides the body into an upper pretrochal region and a lower post-trochal region, and the mouth is located on the ventral surface of the post-trochal region. Three components are often present in the developing pretrochal region: (1) a head vesicle, (2) apical plate and tuft, and (3) cephalic plates (Raven 1966). The head vesicle forms as a space between the ectoderm and endoderm on the dorsal surface of the pretrochal region and is penetrated from below by mesodermal cells (Raven 1966). Head vesicles are best developed in stylommatophorans and are common in

caenogastropods (Raven 1966; Collin 2004). In the fresh-water Unionidae, the head vesicle transforms into the thread gland, which may assist the glochidia larvae in attaching to its host (Howard & Anson 1922). A sensory apical tuft arises from the apical plate in most molluscan taxa; an apical sense organ (see Chapter 7) is also present in most taxa. Cephalic plates are on either side of the apical plate and give rise to the cerebral ganglia, tentacles, and eyes (Raven 1966). A trochophore larval stage is absent in cephalopod, neritimorph, caenogastropod, and heterobranch development, and instead the gastrula develops directly into a paralarva in coleoid cephalopods and into a shelled veliger larva in most gastropods and bivalves. In the veliger larva, the prototroch gives rise to the velum which provides propulsion and is sometimes also used as a feeding organ (Figure 8.32). In Solenogastres and Protobranchia, the prototroch is modified into test cells that enclose the developing embryo forming a pericalymma larva (see Chapters 14 and 15).

Downstream particle capture by the prototroch and metatroch is used by trochozoans with feeding larvae (Nielsen 1995). Derived gastropods and bivalves are the only molluscs with feeding larvae (planktotrophic) and feeding occurs only in the veliger stage, not the trochophore. In some groups (protobranch bivalves and solenogasters), the hyposphere (the region below the prototroch) of the trochophore is covered from above by a ciliated layer of cells – the pericalymma (or test cell larvae) (Figure 8.31). Pericalymma larvae are also found in some annelids, leading Salvini-Plawen (1980b) to suggest that this type of larva was plesiomorphic for the Mollusca + Annelida and that the 'true' trochophore was only found in annelids. Nielsen (2004) also supported a Mollusca + Annelida clade based on shared trochophore features, including cell lineages, ciliary bands, and other developmental and anatomical features.

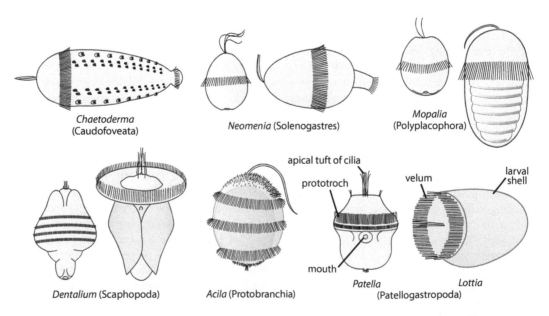

FIGURE 8.31 Representative molluscan larvae. Redrawn and modified from the following sources: *Chaetoderma, Mopalia*, and *Lottia*, Nielsen, C., *Animal Evolution: Interrelationships of the Living Phyla*, 2nd ed., Oxford University Press, Oxford, UK, 2001; *Dentalium*, Lacaze-Duthiers, F.J.H., *Zoologie*, 4, 319–385, 1858; *Neomenia*, Thompson, T.E., *Proc. Royal Soc. B*, 153, 263–278, 1960; *Patella*, Fretter, V. and Graham, A.L., *British Prosobranch Molluscs: Their Functional Anatomy and Ecology*, Ray Society, London, UK, 1962 (left); and *Acila*, Zardus, J.D. and Morse, M.P., *Invertebr. Biol.*, 117, 221–244, 1998.

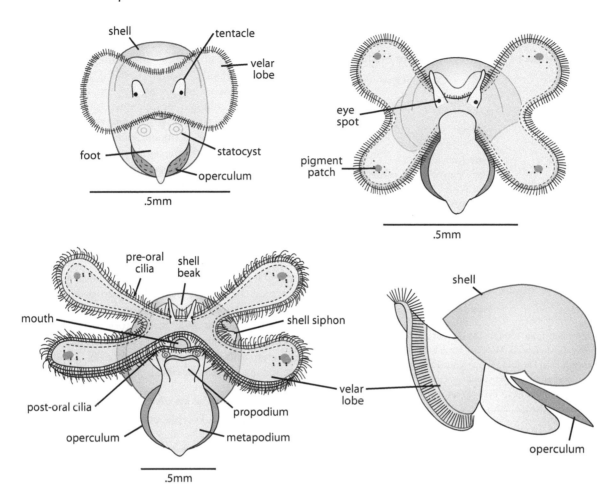

FIGURE 8.32 Ontogeny and development of the velum of the veliger larva of the caenogastropod *Tritia incrassata* (Nassariidae). The early stage (upper left) is reminiscent of velum morphology found in patellogastropods and vetigastropods. Redrawn and modified from Fretter, V. and Graham, A.L., *British Prosobranch Molluscs: Their Functional Anatomy and Ecology*, Ray Society, London, UK, 1962; figure on lower right a stylised lateral view of the veliger larva of *Crepidula*, redrawn and modified from Nielsen, C., *Animal Evolution: Interrelationships of the Living Phyla*, 2nd ed., Oxford University Press, Oxford, UK, 2001.

Whole genome and other molecular analyses support a Mollusca + Brachiopoda clade, with the Annelida as their sister taxon (Paps et al. 2009; Luo et al. 2015) (see Chapter 12).

In polyplacophorans and caudofoveates, there is no secondary larval form beyond the trochophore; instead, the trochophore undergoes a ventral elongation and develops directly into the adult morphology (Buckland-Nicks et al. 2002b; Nielsen et al. 2007). In solenogasters, the trochophore larva is succeeded by a pericalymma larva and in the scaphopods by a stenocalymma larva before transitioning into the adult morphology. The stenocalymma larva (Figure 8.31) is characterised by its covering of 'test cells' derived from the trochoblast lineage and several ciliated rings. The homology of the pericalymma larva and the stenocalymma larva remains uncertain.

In lower gastropods, a veliger stage typically follows the trochophore (Buckland-Nicks et al. 2002a) (Table 8.8), while the veliger stage sometimes develops without going through a well-defined trochophore stage in higher gastropods. Direct development, sometimes with modified larvae, is found in caenogastropods and heterobranchs; modified larvae include

echinospira larvae (e.g., Velutinidae, Eratoidae, Capulidae), polytrochous larvae of pelagic pteropods, and the uniformly ciliated Warén's larvae from the hydrothermal vent provannid *Ifremeria nautilei* (Reynolds et al. 2010). In bivalves, a veliger stage generally follows the trochophore. In the fresh-water mussels (Unionida), the veliger has been substantially modified for larval parasitism (Zardus & Martel 2002; Barnhart et al. 2008), with the glochidium, lasidium, and haustorial larvae enabling both transmission and parasitism (see Chapter 15). The late bivalve veliger stage often becomes demersal before metamorphosis and is known as a *pediveliger*. In both gastropods and bivalves, the veliger and other stages may be absent and the trochophore develops directly into the adult body plan, this often being referred to as *direct development*.

As previously mentioned, protobranch bivalves have a pericalymma larva (also called an endolarva). While Salvini-Plawen (1980b) argued for this type of larva being ancestral in bivalves because it also occurs in polychaetes, sipunculids, aplacophorans, and scaphopods, others (Ivanova-Kazas 1985a, 1985b; Wingstrand 1985) have argued these larvae evolved independently and somewhat differently in each group.

TABLE 8.8
General Distribution of Larval Types in the Mollusca

Taxon	Larval Type
Caudofoveata	Trochophore
Solenogastres	Pericalymma
Polyplacophora	Trochophore
Bivalvia	
Protobranchia	Pericalymma
Autobranchia	Trochophore, veliger, pediveliger
Unionidea (fresh water)	Glochidium, lasidium, or haustorium
Scaphopoda	Trochophore, (veliger), stenocalymma
Gastropoda	
Patellogastropoda, Vetigastropoda	Trochophore, veliger
Neritimorpha, Veliger Caenogastropoda	Veliger, or Warén's larva, or echinospira
Heterobranchia	Veliger, or poytrochous larva
Cephalopoda	Paralarva

Adapted and modified from Young, C.M., A brief history and some fundamentals, in Young, C.M., Sewell, M.A. and Rice, M.E. (Eds.), *Atlas of Marine Invertebrate Larvae*, San Diego, CA, Academic Press, pp. 1–19, 2002.
Note: Stages in brackets are abbreviated, and direct development after formation of the embryo may occur in some taxa.

8.9.5.1 Larval Organs

Larval organs and structures appear in the larval phase and are lost in metamorphosis. They are typically associated with the locomotory, sensory, and nutritional needs of the developing embryos (e.g., velum, apical organs, larval hearts, protonephridia, etc.). Larval organs should not be confused with precocious adult structures that may appear in the larval phase, but may then be lost in the adult of some taxa, examples including the opercula in patellogastropods and many heterobranchs (Ponder & Lindberg 1997), byssal gland in bivalves (Yonge 1962), the mantle cavity of some heterobranchs (Thompson 1976; Nielsen 2015), and some nervous system components (Nielsen 2015). Larval organ loss may be rapid, such as the casting off of the velum in gastropods, bivalves, and scaphopods or test cells in protobranch bivalves (Drew 1899), or they may disappear slowly by means of gradual reduction such as the protonephridia in polyplacophorans (Baeumler et al. 2011), or by absorption; e.g., test cells in solenogasters (Thompson 1960).

Protonephridia occur in larval polyplacophorans, caudofoveates, scaphopods, and bivalves, while in the gastropods, larval protonephridia have been identified in patellogastropods, caenogastropods, and heterobranchs (Haszprunar & Ruthensteiner 2000; Ruthensteiner et al. 2001; Stewart et al. 2014a). Protonephridia are typically formed by an invagination of ectoderm in the anterior region of the larva, although it has been suggested that they originate from mesoderm in the bivalve *Sphaerium* (Okada 1939) and several hygrophilans, including *Planorbis*, *Physa*, and *Lymnaea* (Raven 1964).

Larval digestive systems can also show substantial modification depending on the larval food source (Fioroni 1966). In gastropods, one of the most prominent larval structures is the enlarged velum for feeding (Figure 8.32). While nutrient absorption in feeding larvae is centred in digestive glands, Fioroni (1982b) listed several other digestive system modifications in larval gastropods, including an enlarged oesophagus for added absorption in *Fusinus* and a 'specialised gut region' in *Buccinum*. Some molluscan larval organs such as the apical organ and protonephridia are symplesiomorphic traits and shared with putative outgroups, including annelids, brachiopods, and phoronids (Kempf et al. 1997; Baeumler et al. 2011, 2012), while others such as the anal organs of protobranch bivalves and the configuration of the larval musculature of gastropods and bivalves appear to be autapomorphic in specific taxa (Zardus & Morse 1998; Haszprunar & Wanninger 2000). Larval structures that probably evolved in association with particular functions may also be reallocated when life history strategies change. For example, gastropods, which develop in large, dense egg masses without free swimming larvae, use their plesiomorphic locomotory cilia to induce water movement through the egg masses which then facilitates gas exchange within the capsule (Chaffee & Strathmann 1984; Strathmann & Chaffee 1984).

8.9.5.2 Ano-Pedal Flexure

Monoplacophora, Polyplacophora, and both aplacophoran clades have linear alimentary systems as adults (i.e., the mouth and anus lie in a saggital plane), but the remaining molluscs (and extinct hyoliths) have U-shaped guts (i.e., the mouth and anus lie in a transverse plane) and are said to display ano-pedal flexure as adults. Ano-pedal flexure in molluscan embryos begins when the developing gut folds into a U-shape as the posterior telotroch (anus) is displaced from its ancestral posterior position to a ventral position under the proto-mouth and foot (Figure 8.33 [21 hours] and Figure 8.34). This alignment is obligatory for life in a capsule-like shell with a single opening. The displacement of the telotroch appears to be driven by cell proliferation and migration associated with the appearance of the shell gland on the dorsal surface of the larvae (Figure 8.33, nine hours). This developmental process occurs in all living molluscan classes except for Caudofoveata (Nielsen et al. 2007); in *Neomenia* (Solenogastres) this folding results from the formation of the pericalymma (Figure 8.34). In the non-placophoran groups, it is maintained in adults, although it is not as evident in bivalves.

Molluscan evolution features displacements and asymmetries of other organ systems as well. Torsion, the rotation of the viscera on the head-foot axis in gastropod larvae, is probably the best known molluscan example of organ and system rotation (see Section 8.9.5.6), but similar organ rotations are found in most bivalves, gastropods, and cephalopods (Lindberg & Ponder 1996; Ponder & Lindberg 1997). For example, Seydel (1909) first noted an unusual orientation of the byssus in the limid bivalves *Limaria hians* and *Lima inflata* and suggested that it may have been produced by a 180° horizontal rotation of the foot. Pelseneer (1907) observed twisted pedal nerves and fusion of the cerebral and

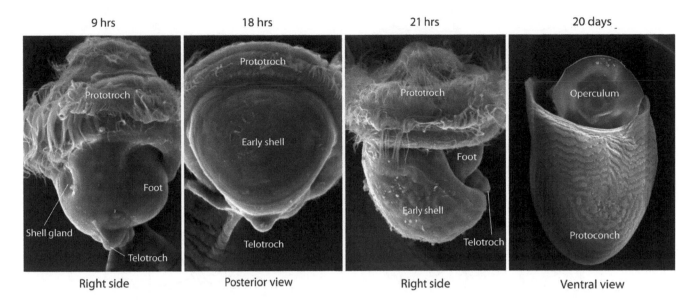

FIGURE 8.33 Early development of *Lottia gigantea* (Gastropoda). From left: trochophore showing early invaginated shell gland and foot rudiment; trochophore with evaginated shell gland with early shell covering the dorsal and posterior surface of the trochophore; 90° displacement (posterior --> ventral) of the telotroch as the early shell envelopes the posterior region of the trochophore; and lastly, the retracted veliger showing the operculum. Scanning electron micrographs courtesy of E. Edsinger.

visceral ganglia in *L. inflata* and *Lima vulgaris*. Odhner (1914) observed the rotation of the foot during the development of *L. inflata,* but provided no insight into the mechanics of the rotation. More recently, Gilmour (1990) noted similar foot rotation during settlement and early post-larval development of *Limaria parafragile*.

In monoplacophorans, scaphopods, and nautiloids, the alimentary system is above the gonad(s), while in aplacophorans, bivalves, and polyplacophorans, the alimentary system lies below the gonad(s) (Lindberg 1985a). In gastropods, the only group with the gonad directly under the alimentary system is the patellogastropods (Lindberg 1985a).

One obvious result of ano-pedal flexure is the compaction of the body mass, and although it is often associated with living in closed tubes and coiling, its distribution in molluscs (see above) casts doubt on a simple functional relationship. For example, scaphopods exhibit ano-pedal flexure, but live in a tube open at both ends. Did their ancestors live in closed tubes? Supposedly, they shared a common ancestor with rostroconchs and these groups were included with the bivalves in the Diasoma (= through body; i.e., lacking ano-pedal flexure) (Pojeta & Runnegar 1976). Although the proposed ancestral linear gut of rostroconchs would have been an ideal exaptation for scaphopod shell morphology, we find marked ano-pedal flexure in scaphopods a strong argument against this derivation. Ano-pedal flexure has also been confused with coiling (e.g., Knight 1952); it allows the visceral mass to elongate and it follows that an efficient way to further compact this mass is to coil it. Thus, although ano-pedal flexure and coiling are functionally linked, they are different processes. *Nautilus* and many extinct shelled cephalopods, such as many other nautiloids and most ammonites, had coiled shells, but their body only occupied a small, relatively straight portion of the shell. Coiling of their tubular, septate ancestor, such as proposed by Pojeta and Runnegar (1976), appears to have evolved independently of visceral mass compaction.

In gastropods, ano-pedal flexure represents a plesiomorphic condition in their development and, as noted above, is unrelated to the synapomorphy of torsion (Cox 1960; Fretter & Graham 1962). Pelseneer's (1911b) description of teratological gastropod embryos that underwent ano-pedal flexure but failed to undergo torsion, is consistent with these developmental events being separate.

Ano-pedal flexure and the formation of the detached head may be a synapomorphy that unites Gastropoda, Cephalopoda, and Scaphopoda and may be an argument in favour of a common origin of these taxa – the Cephalomalacia of Keferstein (1862) (see Chapters 3 and 12 for further discussion). Termier and Termier (1968) recognised ano-pedal flexure as a synapomorphy uniting gastropods and cephalopods, although scaphopods were inexplicably omitted from their phylogeny, and have even been included with non-flexed taxa by Pojeta and Runnegar (1976). Alternatively, the relatively simple process of folding the body during development may be homoplastic. For example, similar differential growth phenomena occur in several tube dwelling annelid groups as well as sipunculans, and in brachiopods (Nielsen 1991), and may not be developmentally difficult to achieve. Comparative studies of the development of ano-pedal flexure in gastropods, cephalopods, and scaphopods might provide insights into whether or not this process is homologous.

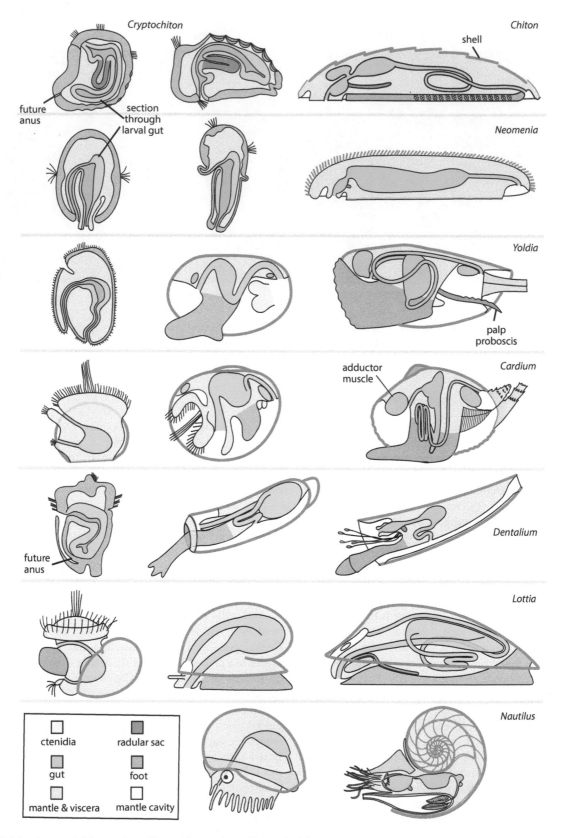

FIGURE 8.34 Ano-pedal flexure in molluscan larvae, juveniles, and adults. Original.

8.9.5.3 Left-Right Asymmetry

Besides shuffling the ancestral anterior-posterior Hox cluster, molluscs, especially gastropods, have also broken ancestral bilateral symmetry. Gastropods are unique in undergoing ano-pedal flexure, torsion, and then strong left-right asymmetry in the development of their mesoderm-derived tissue and organ systems (Raven 1966). There are few groups with such marked left-right asymmetry in both the larval and adult morphologies. It is also a deep character suite dating from the Cambrian. Left-right asymmetries in gastropods extended from cleavage patterns to the adult morphology (Palmer 1996). One of the most marked left-right anatomical asymmetries in gastropods is the loss of mesoderm-derived organs from the post-torsional right side of the body (i.e., kidney, heart, gill, shell muscle) in the most derived members of the group. Other common asymmetries such as gut coiling, asymmetrical stomach, and organ placement are present in gastropods; gut coiling and asymmetrical stomachs are also present in chitons, cephalopods, and bivalves (Lindberg & Ponder 1996).

In deuterostomes, left-right asymmetries (such as the displacement of the heart to the left side of the body in humans, gut coiling in vertebrates) use the signalling molecule Nodal (Grande 2010). Nodal control of body chirality was first demonstrated in molluscs by Grande and Patel (2009) who found that *nodal* (and its target gene *Pitx*) were expressed on the right side of the embryo in the dextral species *Lottia gigantea* and on the left side in the sinistral species *Biomphalaria glabrata*. The role of nodal expression in molluscs now includes gastrulation regulation, germ layer specification, and left-right asymmetries, while in the putative outgroup Brachiopoda, *nodal* is expressed after gastrulation and germ layer specification (Namigai et al. 2014; Grande et al. 2015). The phylogenetic distribution of the *nodal-Pitx* cascade suggests that it may represent an ancestral feature of the Bilateria (Namigai et al. 2014; Grande et al. 2015).

8.9.5.4 Initial Shell Formation

In all molluscs, shell formation takes place on the dorsal surface of the post-trochal region of the trochophore. In polyplacophorans, shell formation differs from that in conchiferans, beginning with the formation of seven transverse ridges of goblet cells across the dorsal surface of the larva. This division produces eight grooves which become the plate fields. The cells in the depression formed between the ridges are the presumptive mantle tissue. Around the edges of each field are specialised cells with cellular extensions – the stragula – which extend inward over the groove, sealing it off from the environment and forming the crystallisation or mineralisation chamber in which the nucleus of the shell plate forms (Watanabe & Cox 1975; Kniprath 1980) (see also Chapters 3 and 14). The initial shell is rod-like, conforming to the transverse grooves and becoming the tegmentum layer of the plate; the ventral hypostracum layer is subsequently added by cells in the central region of the plate field (Kniprath 1980). Spicules also appear over the head region and extend around the

margin of the dorsal region as part of the developing girdle. Jacobs et al. (2000) demonstrated abundant engrailed expression around the edge of each plate field and spicule formation in the developing girdle. The shell plates continue to enlarge and thicken as the perimeter of each plate field expands and deepens with overall body growth. After the formation of the initial rod-like structure in the grooves of the early plate fields, the plates grow by accretionary growth at the margins and ventral surfaces.

In most conchiferans, there is a single undivided shell field induced by cell-cell contact between the ectoderm and the endomesoderm of the developing digestive tract. The ectoderm cells in the contact area thicken and divide. In *Lymnaea*, the earliest shell field cells are paired and bilaterally symmetrical and test positive for alkaline phosphatase (AP) activity; they also secrete AP positive extracellular material (Hohagen & Jackson 2013). As the two groups of AP positive cells divide, another band of AP positive cells forms around the original central pair. The central area, surrounded by the second band of AP positive cells, invaginates, and the central cells further divide forming a deepening chamber which then evaginates and forms the shell field. An initial periostracum forms over the invaginated central cells, providing the isolated crystallisation chamber in which calcification occurs. Subsequent accretionary growth begins after the central region evaginates and flattens, forming a fold (periostracal groove) around the periphery of the shell field.

In *Nautilus* and monoplacophorans, the initial shell (apical cap) is shield-like (Arnold et al. 1987; Warén & Gofas 1996b) and shows an accretionary growth pattern after the initial calcification of the shell field (*cicatrix*). In gastropods and scaphopods, the apical cap extends into a more bulbous or tubular protoconch. In molluscs with internal shells (coleoids and some heterobranchs), the shell field is subsequently overgrown by the body and forms an internal shell sac (Kniprath 1979, 1981).

In bivalves, the original shell field is single, as in other conchiferans, but rather than remaining approximately circular throughout its development, it first expands transversely across the dorsal region of the trochophore (Kin et al. 2009; Hashimoto et al. 2015). The developing shell field invaginates as engrailed expression on the periphery. The shell field then evaginates and emerges dumbbell-shaped. The expression of the signalling molecule Dpp (Decapentaplegic) on either side of the central region of the shell field is thought to produce the dumbbell shape by inhibiting cell proliferation in the middle of the shell field (Kin et al. 2009). After evagination, engrailed expression is absent within the presumptive shell hinge which lies across the narrow, central region of the dumbbell-shaped shell field. Instead, Dpp expression activity is present along the hinge line (Kin et al. 2009), culminating in the two-valved shell of bivalve veligers (Figure 8.35). In gastropods, Dpp may also have a role in limiting the shell field as it circumscribes the engrailed-delimited shell field (Kin et al. 2009). Based on experimental isolation of blastomeres in *Mytilisepta* (as *Septifer*) *virgata*, Hashimoto et al. (2015) argued that the bilateral division of the shell field in bivalves was autonomously

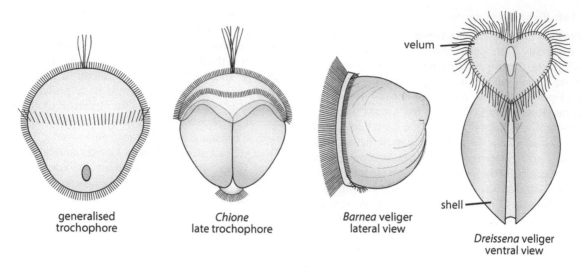

FIGURE 8.35 Examples of autobranch bivalve larvae showing the main developmental stages. Redrawn and modified from the following sources: trochophore original; *Chione* Mouëza, M. et al., *Invertebr. Biol.*, 125, 21–33, 2006), *Barnea* veliger Nielsen, C., Phylum Mollusca, pp. i, 110–123, *Animal Evolution: Interrelationships of the Living Phyla*, Oxford University Press, Oxford, UK, 1995, *Dreissena* Korschelt, E., *Sber. Ges. naturf. Freunde Berl.*, 21, 131–146, 1891.

controlled by intrinsic mechanisms within the D lineage and not dependent on cell-cell interactions with descendants of the A, B, or C blastomeres.

The formation of the shell field and the initial shell shows these cell lineages, cell movement, and proliferation patterns are deeply conserved (Hohagen & Jackson 2013). There also appears to be deep conservation of the gene expression patterns associated with shell field ontogeny. Seemingly homologous engrailed expression patterns, with modest modification, have been demonstrated in polyplacophorans, scaphopods, gastropods, bivalves, and cephalopods (Wray et al. 1995; Moshel et al. 1998; Jacobs et al. 2000; Wanninger & Haszprunar 2001; Baratte et al. 2007; Kin et al. 2009). In the putative outgroup Brachiopoda, engrailed expression is found on the dorsal surface of the bilobed larva and is thought to be associated with early shell deposition (Vellutini & Hejnol 2016). Hox genes have also been demonstrated to be involved with early shell formation in molluscs (Hinman et al. 2003; Samadi & Steiner 2009; Kocot et al. 2017), and likewise may also indicate deep conservation.

Spicule production is intracellular in aplacophorans and polyplacophorans (Hoffmann 1949) (see Chapter 3). In polyplacophorans, spicule formation occurs before the initiation of plate calcification and first appears as a row of spicules just anterior of the prototroch before extending around the body on the outside of the developing mantle (Watanabe & Cox 1975). In solenogasters, spicule formation occurs haphazardly over the surface of the body, with the exception of larval *Nematomenia banyulensis* (Pruvot 1890), in which the spicules are grouped in seven transverse rows. Scheltema and Ivanov (2002) described six spicule groups on an unidentified neomenioid larva as possibly similar to the report by Pruvot and reflective of the primitive state, but as pointed out by Todt and Wanninger (2010), their pattern results from a very different

growth form and occurs at a much later stage than that of Pruvot's larva. Early spicule production in the Caudofoveata has not yet been observed, however, Nielsen et al. (2007) suspected that the epidermal papillae were the source of the spicules. In *Chaetoderma*, the epidermal papillae occur in seven rows as in chitons, however in chitons the papillae are not involved with either plate or spicule formation and are thought to secrete cuticle (Hyman 1967), as they also do in solenogasters (Okusu 2002). In the solenogaster *Epimenia babai*, when the papillae were subjected to programmed cell death staining they were found to be positive, suggesting a secretory role with high cell turnover (Okusu 2002).

8.9.5.5 Control of Shell Coiling

Cleavage direction has long been noted to be closely associated with the direction of shell coiling in gastropods and is thought to be controlled by a single maternally inherited gene expressed as two alleles, one determining right handed patterns and the other left (Kuroda 2015). The right handed allele is generally dominant in right handed taxa. In the heterobranchs *Partula* and *Laciniaria*, the left-handed allele is dominant over the right (Freeman & Lundelius 1982). Kuroda et al. (2009) demonstrated that the direction of spiral cleavage at the third cell division determined the direction of shell coiling, as well as defining which side of the embryo expressed nodal (Grande and Patel 2009). Besides nodal and Pitx, Dpp has been shown to be expressed on the right side of the shell gland and later in the mantle tissue of the heterobranch *Lymnaea stagnalis* by Shimizu et al. (2011). When Dpp expression is blocked before the formation of the trochophore stage, the protoconch is misshapen and unmineralised. When Dpp expression is inhibited at the trochophore or veliger stage, the shells remain uncoiled and are cone-shaped. In subsequent work with patellogastropods and both dextral and sinistral lineages of *L. stagnalis*, Shimizu et al. (2013) demonstrated that patellogastropods had symmetrical expression patterns of Dpp corresponding to their symmetrical cap-shaped shells, while in the dextral and sinistral forms of *L. stagnalis*, mirror image, asymmetric Dpp expression patterns were present.

While uncommon, sinistral coiling became fixed in some taxa with normally dextral snails and survived for long periods of geological time (e.g., Vermeij 1975), with examples including the conoidean *Antiplanes* and the buccinid *Pyrulofusus* which date to the Miocene (16 Ma). Vermeij further noted that 12 of 13 reversals in Cenozoic marine taxa occurred in those taxa with crawl-away juveniles, and that reversals were more common in land and fresh-water snails which also have direct development.

8.9.5.5.1 Enantiomorphy

This term is normally used for a pair of crystals, molecules, or compounds that are mirror images of each other, but are not identical. It has also been used to describe sinistral-dextral dimorphism in stylommatophoran land snails (e.g., Sutcharit et al. 2007) and a few other groups of animals based on a primary whole-body polymorphism in development. Such *chiral dimorphism* (enantiomorphy) is relatively rare and can only be seen in groups with asymmetrical bodies (Asami et al. 1998).

8.9.5.6 Torsion in Gastropods

Torsion is the 180° rotation of the gastropod head-foot (cephalopodium) relative to the visceropallium (shell and viscera) during development and is a significant gastropod synapomorphy (Figures 8.36 and 8.37). The torsion event in basal gastropods has been traditionally described as a relatively rapid 90° rotation of the visceropallium by muscle contraction, followed by the second 90° of the rotation through differential growth (Wanninger et al. 2000). This rotation moves the mantle cavity from its primitive posterior position to the derived anterior position, and the shell also rotates from the exogastric to an endogastric state (whether the shell coil is away from or over the visceral mass, respectively; Figure 8.37). The rotation also twists any organs or structures that cross the rotational plane between the visceropallium and cephalopodium at the time of torsion – e.g., larval musculature, digestive, and nervous systems. Torsion also reverses the orientation of structures relative to their pre- and post-torsional positions (e.g., 'left' and 'right' kidneys). Detorsion occurs after metamorphosis in many heterobranchs and can be a subsequent 90° or more clockwise rotation of the visceropallium which returns the mantle cavity to the right side or a more posterior position on the animal (Figure 8.38).

In many discussions of torsion, it has been assumed that the various anatomical structures and organs affected by rotation are a tightly associated unit with the mantle cavity being the most prominent feature (e.g., Naef 1911; Garstang 1928; Crofts 1955; Morton 1958b) (Figure 8.37). More recent work by others (e.g., Voltzow 1987; Page 1997b, 2002b, 2006a; Ruthensteiner 1997; Wanninger et al. 1999a, 2000; Kurita & Wada 2011) has challenged this view (see Jenner 2006). Instead, torsion

is now recognised as a compilation of developmental events that may be largely or wholly ontogenetically disassociated from one another in different taxa. Variation in such a critical and unique process in gastropod development should not be surprising given there have been over 500 million years of subsequent evolution since this event first became a fixture of gastropod ontogeny. What we observe today are modifications of a shared ancestral developmental pathway that produced an endogastric shell, anterior mantle cavity, rotated oesophagus, and streptoneurous nervous system, either synchronously or via the concordance of different ontogenetic trajectories.

Here, we update previous descriptions of the torsion process and review previous hypotheses regarding its adaptive significance in gastropods. The following description is based primarily on studies by Ruthensteiner (1997), Wanninger et al. (1999a, 2000), Page (2003, 2006b), Bondar and Page (2003), and Kurita and Wada (2011).

The torsion narrative has traditionally been based on two distinctive events, although variations within and between events and phases are present in different taxa. The role of muscular contractions of the larval muscle system has been disputed in more recent work by Page (2002b), who demonstrated that torsion can occur without larval retractor muscle activity (see Section 8.9.5.6.2). Instead of contractions, Kurita and Wada (2011) identified cell proliferation on the right side of mantle epithelium as the 'driving force' of the rotation of the cephalopodium in a patellogastropod. This asymmetric cell proliferation is initiated by TGF-β signalling (which includes the nodal pathway); cell proliferation and torsion did not occur when the TGF-β signal was blocked. This result highlights yet another important role of asymmetric cell proliferation and movement in molluscan development, and suggests that the two major reorganisations of the primitive, bilaterally symmetrical molluscan bauplan – ano-pedal flexure (see Section 8.9.5.2) and gastropod torsion – both appear to be driven by this mechanism.

8.9.5.6.1 Mantle Cavity Complex

As described in Chapter 4, the adult mantle cavity complex typically contains the anal and kidney openings, hypobranchial glands, reproductive tract openings, osphradia, and ctenidia. While thought to be highly integrated in the adult gastropod, this complex is assembled during development and not all of these structures originate as part of the mantle cavity or are associated with the mantle cavity during torsion (although torsion cartoons often show many, if not all, as present). The assembly of the pre-torsional mantle cavity begins with ano-pedal flexure.

The presumptive opening of the anus is initially at the posterior end of the embryo where it is often surrounded by a distinctive ring of cilia, the telotroch. During the course of trochophore development, when the shell gland has developed, the putative anus migrates from its posterior position to a more ventral position adjacent to the foot (Figure 8.36). It remains there until the mantle cavity forms. The migration of the anus is associated with the bending of the anterior-posterior axis. Raven (1966) referred to this bending as ventral flexion (= ano-pedal flexure), and suggested that the invagination/

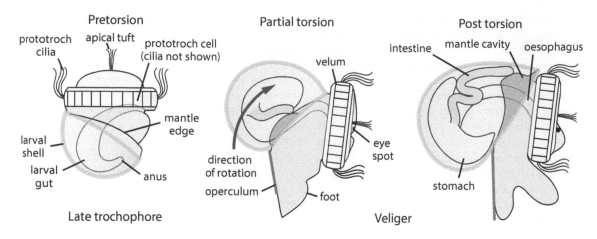

FIGURE 8.36 Stages of torsion in a gastropod larva. Redrawn and modified from Smith, F.G.W., *Philos. Trans. Royal Soc. Edin.*, 225, 95–125, 1935 and Crofts, D.R., *Proc. Zool. Soc. Lond.*, 125, 711–750, 1955.

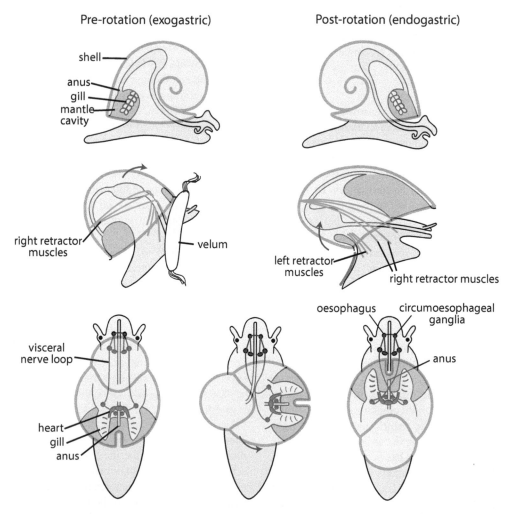

FIGURE 8.37 Some examples of gastropod torsion scenarios from the literature. Redrawn and modified from the following sources: upper line Page, L.R., *Integr. Comp. Biol.*, 46, 134–143, 2006a, middle line Morton, J.E. and Yonge, C.M., Classification and structure of the Mollusca, pp. 1–58, in Wilbur, K.M. and Yonge, C.M., *Physiology of Mollusca*, Vol. 1, Academic Press, New York, 1964, and bottom line Naef, A., *Ergeb. Fortschr. Zool.*, 3, 329–462, 1913.

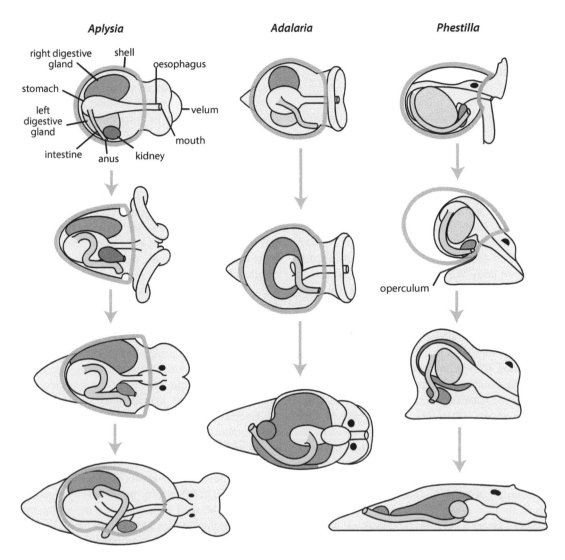

FIGURE 8.38 Stages of detorsion in the euopisthobranch *Aplysia* and two nudipleurans. Note that cilia on the velar lobes is not shown. Redrawn and modified from the following sources: *Aplysia* Saunders, A.M.C. and Poole, M., *Q. J. Microsc. Sci.*, 55, 497–539, 2010, and Kriegstein, A.R., *J. Exp. Zool.*, 199, 275–288, 1977a, *Adalaria* Thompson, T.E., *Philos. Trans. Royal Soc. B*, 242, 1–58, 1958, and *Phestilla* Bonar, D.B. and Hadfield, M.G., *J. Exp. Mar. Biol. Ecol.*, 16, 227–255, 1974.

evagination of the shell gland as well as subsequent apical cap and protoconch growth is responsible for this displacement. Biggelaar et al. (2002) pointed out this developmental event is not limited to some molluscs, but that embryonic growth of the anterior-posterior axis in protostomes in general is irregular and primarily limited to cells on the dorsal side of the embryo, as noted by Raven. In addition, they point out that cells on the ventral side of the embryo migrate into the interior of the embryo and into the lateral and dorsal regions. Because of this uneven cell propagation and migration, the ventral region of the embryo does not extend and the original anterior-posterior axis is bent. This asymmetry is subsequently corrected in most protostome groups, thereby producing bilaterally symmetrical worms like nemerteans and annelids, as well as molluscs that have an anterior mouth and posterior anus (see Thompson 1960 for solenogaster example). In the ancestor of cephalopods, scaphopods, and gastropods, this deep protostome developmental character appears to have been co-opted to produce ano-pedal flexure and set the stage for these three diversifications, one of which includes gastropod torsion.

The mantle cavity typically appears as an epithelial thickening on the right side of the trochophore (Ruthensteiner 1997; Page 2006a), not posteriorly as illustrated in most torsion cartoons. Even this generalisation varies, even within the same group. For example, in the patellogastropod *Patella vulgata*, the mantle cavity first appears as a symmetrical invagination on either side of the stomodaeum (Lespinet et al. 2002), while in *Patella caerulea*, it appears under the foot on the ventral side of the embryo (Wanninger et al. 2000), and only in *Cellana sandwicensis* does it first appear on the right side of the body (Ruthensteiner 1997). As the mantle cavity broadens and deepens by increased cell divisions, it subsumes the anal opening. Further expansion brings the mantle cavity into contact with the developing paired kidneys. Thus, just before torsion, the putative mantle cavity complex typically consists of a partially invaginated mantle cavity, anal opening, and paired kidney openings.

8.9.5.6.2   Rotation

The rotation of the cephalopodium relative to the visceropallium is variable and relatively rapid. Different mechanisms have been thought responsible for this initial rotation. Crofts (1955) proposed that the straightening of asymmetrical larval retractor muscles was responsible for the rapid rotation. Other workers have noted that torsion still occurs even when the larval retractor muscles are unattached to the protoconch (Hickman & Hadfield 2001; Page 2002b) or molecularly blocked (Kurita & Wada 2011). Wanninger et al. (2000) noted muscular contractions followed by foot movements that peristaltically pumped body fluid from the visceropallium to the cephalopodium. This sequence of contractions, followed by peristaltic pumping, rotated the cephalopodium 180° in about two hours in *Patella caerulea*. In other basal taxa with an initial 90° rotation, the second phase of the rotation typically takes longer and is accomplished through differential growth.

The axis of rotation is typically said, or illustrated to be, dorsal-ventral with the mantle cavity rotating counter-clockwise from the posterior to the anterior of the animal. As Wanninger et al. (2000) pointed out, ano-pedal flexure has placed the anus on the right ventral side of the animal. Thus, the rotation of the mantle cavity is from ventral to dorsal rather than posterior to anterior. The size and position of the mantle cavity during rotation varies in different taxa and, as noted above, the final enlargement and assembly of the mantle cavity complex typically occurs after it is in its final position, allowing the withdrawal of the cephalopodium (head-foot) into the mantle cavity (and, therefore, an endogastric shell orientation). In several taxa, the larval shell (i.e., the protoconch) rotation and visceropallium rotation are not synchronous, but disassociated. Page (1997a) observed that in *Haliotis kamtschatkana*, the shell obtained a full 180° rotation, while the mantle cavity rudiment had rotated only 90°. Subsequent differential growth expanded the mantle cavity into a more central and dorsal position. Similar disassociation of shell and mantle cavity rotation has been observed in some caenogastropods (Page 2006a), and in the heterobranch *Ovatella* (Ellobiidae), the mantle cavity does not even invaginate until well after torsion is completed (Ruthensteiner 1991).

8.9.5.6.3   Final Development of the Mantle Cavity Complex

The gill rudiments typically form in the mantle cavity after rotation and mantle expansion is complete, although there is substantial variability. In some caenogastropods, the gill rudiments are present on the mantle epithelia before invagination of the mantle cavity rudiment and, like the anus, these rudiments are subsumed by the mantle cavity and then rotate during torsion (Raven 1966). The hypobranchial glands appear during the second stage of, or after, torsion. In *Haliotis*, left and right hypobranchial glands develop asynchronously, the left during the differential growth phase, and the right after

torsion is completed (Crofts 1937), while in the heterobranch *Amphibola*, the hypobranchial gland develops 3–4 weeks after metamorphosis (Little et al. 1985). Lastly, the osphradia develop within the mantle cavity, and in all reported studies appear after torsion. With the appearance of the osphradia, the mantle cavity complex is complete in both position and complement, except in some heterobranchs, as described below.

8.9.5.6.4   Detorsion

Detorsion occurs in 'opisthobranchs' and in a limited number of eupulmonates, and is often characterised as a 90° or more clockwise rotation of the visceropallium which returns the mantle cavity to the right side or a more posterior position on the animal. Like torsion, detorsion is highly variable within and between taxa. What is broadly referred to as detorsion is the disassociation of the mantle cavity components – anus, mantle cavity, gill, and renal openings. These structures, which become closely associated during torsion, now become disassociated and produce a new suite of character states and combinations which influence other structures and organs. In the systellommatophoran *Onchidella*, the post-torsional mantle cavity (with anus and kidney pore) moves posteriorly along the right side of the body to a posterior position (Fretter 1943). Displacements of the mantle cavity, anus, and shell are even more disassociated and therefore variable in the Euopisthobranchia (Figure 8.38). In some taxa, the full 180° rotation occurs and is maintained into adulthood (e.g, in the nudibranch *Aeolidiella alderi*) (Tardy 1991), while in other taxa, the mantle cavity complex only goes through a partial rotation or none at all (Thompson 1976; Brace 1977b). For example, in *Philine*, the mantle cavity remains on the right side, while the shell is displaced to the left and the anus and kidney openings to the right (Raven 1966). In the nudibranch, all of the organs originate in their 'post-torsional' positions (Thompson 1958). The extent and striking morphological outcomes of different detorsional pathways have even been proposed as diagnostic of ordinal rank (Baranetz & Minichev 1994) in the Nudibranchia with Doridoxida = right handed position of the mantle complex, Phyllida = terminal and ventral position of the anus and gills, Corambida = ventral 'ctenidium', and Doridida = dorsal mantle complex.

Like torsion, detorsion is not a single developmental event, but rather an adult character suite achieved through a variety of different developmental pathways, and often, further dissociation of the elements of the mantle cavity complex. Whereas there is an early dissociation of the mantle cavity and anus, these structures, along with the kidney openings, are the earliest assembled components of the mantle cavity complex and have remained developmentally associated since gastropods originated. Many descriptions of detorsion (see above) are examples of dissociation of these formerly concomitant characters, and thus the viscera (or mantle cavity if formed), shell, anus, and renal openings exhibit independent ontogenetic trajectories. Thus, detorsion *per se* is not a synapomorphy of 'opisthobranchs', but rather the developmental release that produced the subsequent dissociation of these formerly coupled characters. Regardless of its ultimate cause, the resultant character states seen in detorted taxa illustrate some of the variation that can be produced when formerly cohesive developmental pathways are released.

8.9.5.6.5 *Historical Explanations for Torsion*

There have been many attempts to explain the evolution of torsion in gastropods. These theories have focused either on the supposed (1) advantages to the adult, (2) advantages to the larva, or (3) a combination of both.

The first theories were concerned with how torsion was an adult adaptive response to adult morphology and were also explanations for the development of asymmetry in gastropods. In one of the earliest treatments, Lankester (1883) proposed that gravity caused the shell and viscera to lean to the right side of the metamorphosed crawling snail. Pressure and strain then caused differential muscle development which brought about torsion, spiral growth of the viscera (and ultimately shell), and the atrophy of organs on the right side of the body. Lang (1900) supposed that the ancestors of gastropods developed narrowly conical uncoiled shells while increasing the size of the visceral mass. Like Lankester, he argued that to facilitate movement, the shell most likely flopped to the side of the animal, leading both to torsion and to the asymmetry seen in gastropods. In contrast, Naef (1911) observed that the oldest untorted snails in the fossil record were bilaterally symmetrically coiled and exogastric with an assumed posterior mantle cavity. He argued that the earliest members of this lineage were pelagic (an idea that has not gained acceptance), but when they became benthic the heavy shell lay on the head of the animal. This difficulty was overcome by rotating the body, resulting in the mantle cavity facing forwards and the shell resting on the foot. A modification of the Lang theory was proposed by Solem (1974).

Garstang (1928), or perhaps the more famous poem distributed as part of his presidential address to the Zoology section of the British Association for the Advancement of Science (1929), was the first to suggest that torsion provided an adaptive advantage to the larva. He proposed that torsion resulted from a larval mutation that enabled the withdrawal of the head and velum into the mantle cavity to be protected by the operculum, rather than leaving them exposed to (unspecified) dangers as in the pre-torsional state. This theory was widely accepted (e.g., Crofts 1937, 1955; Yonge 1947, 1960c; Eales 1950; Morton 1958a, 1958b; Fretter & Graham 1962; Morton & Yonge 1964; Franc 1968; Goodhart 1987b). Experimental work, however, has questioned the supposed anti-predator defence role of torsion. For example, Pennington and Chia (1985) concluded that withdrawing into the larval shell and closing the aperture with the operculum provides little or no protection from a variety of planktonic predators. For an alternative conclusion, see Hickman (2001).

Despite potentially conferring advantages to the larva, difficulties arise for the adult by the new location of the mantle cavity – largely the famous sanitation issues – the problem of the releasing of waste over the head. Garstang suggested this stream of waste inhibited the growth of the mantle, resulting in an embayment in the shell which developed into the slit seen in some lower vetigastropods. Criticisms of this seemingly elegant theory have been advanced, and Voltzow and Collin (1995) and Lindberg and Ghiselin (2003) showed that the supposed sanitation issue is fallacious and that the anus is located below the gills in the outgroups, not above as shown in the hypothetical ancestor by Pelseneer (1894) and later envisioned by Garstang.

Garstang's idea of torsion originating from a single larval mutation was seemingly supported by the results of Crofts (1937), which suggested that torsion was the result of an asymmetrical larval muscle. These findings were modified by subsequent observations in the 1980s (e.g., Bandel 1982; Voltzow 1987) (see discussion above). The fact that an operculum is found in all gastropods with swimming larvae, even if absent in the adults (e.g., most limpets, slugs), suggests that it may have functional significance for the larva, particularly as the larval animal can retract into the shell (Bandel 1982), despite some observations to the contrary (e.g., Underwood 1972). Morton (1958b) suggested that torsion, and the resulting anterior mantle cavity, had significant advantages to adult gastropods because they could utilise water currents to detect chemical stimuli in front of the animal. While this is true in more advanced gastropods, it was unlikely to have been a significant factor in the earliest gastropods which were non-selective grazers with poorly developed osphradia.

Ghiselin (1966) suggested that torsion occurred as a response to the difficulties that an untorted snail would have balancing the shell during locomotion at larval settlement. Underwood (1972) and Salvini-Plawen (1984) suggested that torsion might have resulted from equilibrium problems in the early larva. The partial rotation in the larva may have improved the balance of the velum relative to the shell, while the second half of the rotation favoured the adult (Underwood 1972). This idea, that the first phase of torsion results in advantages for the larva (protection by the operculum and possibly balance) with the second phase being adaptations of benefit for the adult, was also favoured by Haszprunar (1988a). Lever (1979) reviewed the various theories of torsion and concluded that most did not adequately explain the fact that most gastropods coil dextrally – thus, like several others before him, he confused coiling with torsion rather than treating them as two separate processes. He argued that an asymmetrical pre-gastropod was the ancestor and that torsion was simply an extension of this asymmetry. Lastly, Edlinger (1988a, 1988b) proposed that the ancestors to gastropods had visceral coiling, and enlargement of the posterior mantle cavity resulted in the formation of a waist between the viscera and the foot. This, combined with the reduction in the number of shell muscles, enabled torsion to occur. This biomechanical model was rejected by Haszprunar (1989a) and Ruthensteiner et al. (2010).

Perhaps even more puzzling has been the practice of showing torsion with cartoons of adult snails. In these, the foot axis remains constant and the visceropallium rotates relative to the cephalopodium (e.g., the top and bottom rows in Figure 8.37). In actuality, the larval retractor muscles are attached to the shell, which is part of the visceropallium, and in the developing larvae, the muscular-hydrostatic contractions rotate the cephalopodium relative to the visceropallium. In addition, using terms such as post- and pre-torsional for organs and mantle cavity structures (e.g., Fretter & Graham 1962; Haszprunar 1988a; Ponder & Lindberg 1997) presupposes the presence of these structures or their precursors in the developing larvae. In many instances, this is not the case. For example, the gill rudiments only form in the mantle cavity after it achieves its dorsal position and therefore have no pre-torsional orientation.

Since Spengel's (1881) original concept of torsion and his division of gastropods into major groupings based on its effects (or lack thereof) such as Streptoneura (twisted nerves) and Euthyneura (straight nerves, due to detorsion), our understanding of the process of torsion, especially over the last 20 years, has undergone major revision because of dramatically improved developmental and imaging techniques and methods. These improvements have probably been responsible for the fact that studies of the process of torsion have now finally exceeded the number of papers arguing selective advantages for torsion.

8.9.5.7 Asymmetry in Gastropods

The diverse array of asymmetries in the gastropod body has long been appreciated by developmental biologists, many of whom included torsion as just one of several asymmetries (Raven 1958, 1966; Thompson 1958; Page 2006a). Workers studying heterobranchs, including Thompson (1958), Brace (1977a), Tardy (1991), and Page (1994) have also been critical of traditional descriptions of torsion, probably due to the highly variable patterns of torsion and detorsion observed in their study taxa. These and other studies of both larval and adult morphologies have decoupled torsion, shell coiling, and some left-right asymmetries. The separation of coiling and torsion into two separate processes was an important step in understanding the limits of torsion, although the traditional view of torsion has persisted.

Page (2006a) pointed out that the traditional view of torsion rests on a circular argument; namely, the presence of an anterior mantle cavity complex in living taxa is assumed to be evidence of a 180° rotation from a posterior position in the adult pre-torted common ancestor of gastropods[9]. Page (2006a) concluded that without independent evidence for a posterior mantle cavity in the common ancestor, gastropod torsion is a grand tautology. Page referred to the traditional view as the *rotation hypothesis* of torsion and argued that while it is consistent with aspects of gastropod development and adult anatomy, more recent developmental work combined with robust hypotheses of gastropod relationships required a new hypothesis. She proposed the *asymmetry hypothesis*, which views the asymmetries of the gastropod body plan as an accumulation of gradual change in the dimensions and positioning of mantle epithelium that delineates the mantle cavity and does not require shell rotation. Also, the asymmetry hypothesis does not require a maladaptive intermediate stage to justify subsequent stages, nor does it require a major macromutation as proposed by Garstang

(1929). The rotational hypothesis of torsion is obviously too simple and mixes larval processes and adult morphologies indiscriminately. While less generalised, the asymmetry hypothesis appears to be a more accurate characterisation of what is referred to as torsion and potentially brings other asymmetries into the mix.

Non-torsion rotation or displacement of organ systems in most gastropods has not been well documented. One exception is the gastropod *Neomphalus fretterae*, where displacements beyond torsion were proposed to account for the anatomical peculiarities seen in this hydrothermal vent taxon. These include a leftward rotation of the visceral mass around an anteroposterior axis, best seen in the heart and stomach, and also present in parts of the nervous system (Fretter et al. 1981). Using Fretter et al. (1981) criteria (i.e., the position of the gastric shield and digestive gland ducts of the stomach and the relative positions of the aorta and kidneys to the pericardium), Ponder and Lindberg (1997) reviewed heart and digestive systems in several basal gastropod clades. In the patellogastropods, they identified rotation of the pericardium and kidneys (Lindberg 1983, 1988a), while in the cocculiniform limpets, the positions of the gastric shield and digestive gland ducts indicate a 180° rotation of the stomachs in the Cocculinoidea compared to the Lepetelloidea (Haszprunar 1988b). A possible rotational asymmetry was also seen in the position of the left osphradial nerve in the Cocculinoidea (Haszprunar 1987a; Strong et al. 2003). While most of these examples are from groups that have become secondarily flattened, it is unlikely these rotations are restricted to taxa that become limpet-like (Fretter et al. 1981). Nor are they mandatory for becoming a limpet, as seen in the two alternative stomach arrangements in the cocculiniform limpets. Whether the asymmetry hypothesis can be extended to include other asymmetries in gastropod development remains to be determined.

8.9.6 Metamorphosis

The term metamorphosis refers to the phase at which an often environmentally mediated, irreversible commitment is made by an individual to transform from the larval to the juvenile stage (Hadfield et al. 2001; Bishop et al. 2006), and typically represents a change in habitat, morphology, and function. This requires that the larva be at a stage in its development where it is capable or competent to undertake this transformation. Just because a larva has become competent does not mean it will immediately undergo metamorphosis, as metamorphosis may be delayed until specific cues or other factors are sensed by the larva (Thorson 1950; Scheltema 1974; Hadfield 1978; Miller & Hadfield 1990; Pechenik 1990; Pires et al. 2000; Bishop et al. 2006). These cues may be physical, chemical, or biological (Pawlik 1992); some of these factors are broadly shared across taxa, while others appear to be unique to specific taxa (Bishop et al. 2006).

In a study of the timing of metamorphosis in marine invertebrates, Bishop et al. (2006) noted substantial variation in larval competence in the absence of settling cues and/

---

[9] Wagner (2001) discussed the implications of the interpretation of torted versus untorted muscle scars.

or food reserves (both maternal and external), and how the responses to settling cues changed over time. They parsed this variation into three hypotheses: (1) desperate larvae, (2) variable retention larvae, and (3) 'death before dishonour' larvae. The desperate larvae hypothesis predicts that as food reserves decline, a competent larva becomes less subject to settling cues and will settle without any cue. The variable retention hypothesis is similar to the former hypothesis, except the response is less likely to be associated with larval food limitations. Lastly, in the 'death before dishonour' hypothesis competent larvae do not undergo metamorphosis unless a specific settling cue is provided, regardless of food and time (see also Krug 2009). Combinations of these categories may occur in a single species. For example, Krug et al. (2007) demonstrated that in the sacoglossan *Alderia willowi*, the lecithotrophic larvae either metamorphose between 0 and 48 hours of hatching, or delay metamorphosis for up to 14 days while waiting to receive carbohydrate cues from their host algae *Vaucheria* spp. Those that do not receive that cue during the two week component period die, thus combining aspects of variable retention and 'death before dishonour' hypotheses in a single cohort.

Molluscs, especially heterobranchs and commercially important taxa such as abalone (*Haliotis* spp.), *Concholepas* and bivalves (*Crassostrea, Mercenaria, Mytilus*) have figured prominently in studies of settling cues, but data for other taxa are patchy. Lord (2011) reported that the lecithotrophic larvae of the polyplacophoran *Cryptochiton stelleri* began settling on coralline algae, or in the presence of coralline extracts, three days after hatching, and remained competent for up to two months. Settlement cues have been investigated in patellogastropods by Kay and Emlet (2002), in the Vetigastropoda, with most involving *Haliotis* spp. (Morse et al. 1984; Bryan & Qian 1998; Daume et al. 1999a, 1999b, 2000; Roberts & Lapworth 2001; Takami et al. 2002; Gapasin & Polohan 2005; Viçose et al. 2010) and *Trochus* (Heslinga 1981). Studies of caenogastropod settlement cues include those on *Crepidula* (Pechenik et al. 2002), *Littorina* (Hohenlohe 2002), *Strombus* (Davis & Stoner 1994; Boettcher & Targett 1996, 1998), *Polinices* (Kingsley-Smith et al. 2005), *Chorus* (Gallardo & Sanchez 2001), and *Concholepas* (Manríquez et al. 2004). In most of these studies, biofilms and adult habitat cues (conspecifics, algae, barnacles, or other prey items) were found to elicit settling responses, but abiotic factors (sediment type, hydrodynamics) can also be involved (Reidenbach et al. 2009; Koehl & Hadfield 2010). In heterobranchs, much of the settling responses work has focused on nudipleuran taxa, including studies by Thompson (1958), Gibson (1995), Pechenik et al. (1995), Krug and Manzi (1999), Pires et al. (2000), Croll et al. (2003), Hadfield & Koehl (2004), Ritson-Williams et al. (2009), and Marshall (2012). There is also an extensive literature on morphological changes at metamorphosis (e.g., Thompson 1976; Bonar 1978; Gibson 2003). Although studies of neritimorphs are lacking, Dattagupta et al. (2007) suggested that habitat clues (deep-sea mussel beds) may be important in larval settlement given the behaviour of adult *Bathynerita naticoidea* in response

to seawater conditioned with *Bathymodiolus childressi*. Lastly, Yin et al. (2015) identified seven core genes in the metamorphosis of *Lottia gigantea* larvae which were involved in nerve development, tissue differentiation, secondary shell formation, and energy metabolism, and revealed a possible molecular mechanism of metamorphosis.

8.9.7 HETEROCHRONY

Heterochronic events are changes in the timing of the development of organs and structures and can produce two forms of morphological expression: (1) paedomorphosis, or the retention of ancestral juvenile characters by later ontogenetic stages of descendants, and (2) peramorphosis, where new descendant characters are produced by additions to the ancestral ontogeny (Love 2015). The role of heterochrony in biotic evolution received renewed interest and study with the publication of the seminal treatment of ontogeny and phylogeny by Steven J. Gould (1977). Many of Gould's examples were molluscan and included the study by Ockelmann (1964) of small paedomorphic bivalves, progenetic transitions in bivalve habits by Stanley (1972), dissertation work on life history evolution in *Crepidula* by Hoagland (1975), as well as Gould's own work on *Poecilozonites* and *Cerion* (Gould 1968, 1969, 1970, 1984). Heterochronic processes were invoked in early molluscan evolutionary studies, including the study of fossil ammonoids by A. Hyatt (1894), investigations of bivalve phylogeny by R. T. Jackson (1890), and the patellogastropod classification of W. H. Dall (Lindberg 1998). More recent studies invoking heterochronic processes in molluscs are numerous. Some focus on insights on molluscan phylogeny (Haszprunar 1992b; Scheltema 1993; Lindberg & Ponder 1996; Jacobs et al. 2005), gastropod phylogeny (Ponder & Lindberg 1997), or anatomy (Lindberg 1988a; Zaitseva et al. 2015). Others have applied these findings to fossils (Geary 1988; Landman 1988; Allmon 1994; Jones & Gould 1999; Steuber 2003; Goodwin et al. 2008; Teichert & Nützel 2015), bivalves (Morton 2004a, 2007; Collins et al. 2016; Echevarria 2016), cephalopods (Boletzky 1989, 1997; Shigeno et al. 2001b), or radular development (Warén 1990; Guralnick & Lindberg 1999; Herbert et al. 2007), and developmental patterns (Freeman & Lundelius 1992; Page 1998; Tills et al. 2013; Page & Hookham 2017). Molluscan examples are also prominent in the general treatments of heterochrony by McKinney (1988) and McKinney and McNamara (1991).

8.9.8 ECOLOGICAL DEVELOPMENT (ECO-DEVO)

Traditionally, gene-based differences were thought to be wholly responsible for determining ontogeny, but there has long been evidence that the ontogeny of an organism can be adaptively modified by its environment (Nilsson-Ehle 1914). And while the genetic ability to respond to environmental factors is inherited, it remains the environment that produces the specific phenotype (Gilbert & Epel 2008). Research in this area is often called ecological developmental biology or eco-devo (Gilbert 2001; Sultan 2007; Gilbert & Epel

2008). The ability to develop into more than one phenotype is known as phenotypic plasticity or developmental plasticity when observed in embryos and larvae. There are two types of phenotypic plasticity: (1) reaction norms and (2) polyphenisms. In 'reaction norms', the genome encodes a continuous range of potential phenotypes, whereas in 'polyphenisms' the phenotypes are discontinuous (Woltereck 1909; West-Eberhard 2003). The prevalence of multiple phenotypes in many molluscan species, often called ecophenotypes or ecotypes, might suggest a wealth of eco-devo studies on members of the phylum, but this area of research in molluscs is in its infancy (Nakano & Spencer 2007; Krug et al. 2012; Lesoway et al. 2015, 2016). These studies will probably expand as the environmental factors resulting from global climate change increasingly impact molluscs.

In both plants and animals, exposure to a host of environmental factors (e.g., temperature, nutrition, light, salinity, pH, heavy metals, predators, conspecifics) can influence ontogenetic pathways and produce distinct phenotypes (Gilbert & Epel 2008). Environmental regulation of developmental gene expression occurs in a variety of ways, including the regulation of gene transcription by *cis*-regulatory elements and methylation patterns, by the transmission of environmental signals through the neuroendocrine system and transgenerational predator-induced polyphenisms (Gilbert & Epel 2008). In a study of the effect of temperature (heat shock proteins—HSP, see Chapter 2) on population structure and phenotypic diversity within and among ten populations of the Mediterranean helicoidean land snail *Xeropicta derbentina*, Di Lellis et al. (2014) demonstrated that Hsp70 levels varied between populations, and were correlated with population genetic structure, and shell colour. Specifically, populations that exhibited little variation in shell colour were found to have higher Hsp70 levels, both constitutively and under heat stress. Population structure (based on an analysis using COI) was not correlated with phenotypic diversity, but with Hsp70 expression levels. These and many other results support the idea that Hsp70 functions as a 'molecular chaperone' to buffer potentially destabilising mutations as well as being a protein folding catalyst (e.g., Fink 1999). In response to waterborne cues from the presence of shell-crushing predators (e.g., crabs), some gastropods respond by altering shell thickness and sculpture which make the shell more resistant to crushing (Appleton & Palmer 1988; Palmer 1990; Selden et al. 2009) (see Chapter 9).

Besides phenetic plasticity, physiological plasticity in early developmental stages has also been documented in gastropods. Rudin-Bitterli et al. (2016) have shown that encapsulated embryos of *Littorina obtusata* respond to moderate hypoxia by slowing their developmental rate and modifying the ontogeny of their developing cardio-respiratory system. The velum, which may play a role in gas exchange, was larger in hypoxic conditions, and developed more slowly. They concluded that the surviving embryos had an earlier onset of the expression of adult structures and also came from eggs with higher albumen content, while embryos that died retained their larval cardio-respiratory features longer.

8.9.9   ORGANOGENESIS

The differentiation of the three metazoan tissue types (mesoderm, ectoderm, endoderm) and the subsequent development of structures and organs from these tissues occurs very early in molluscan development. This type of development is called mosaic development, with each tissue type being derived from specific cell lineages and maintaining its integrity through development and growth. Regulative development, in which cell-cell interactions determine cell fates, also occurs in molluscs, but does not override the basic organisation or fates of the early cells. Thus, the larva and adult are mosaics of these three tissue lineages. This type of development is typical of most lophotrochozoan groups, except for lophophorates (Valentine 1997).

8.9.9.1   Ectoderm-Derived Organs

Ectoderm forms the molluscan integument, nervous system, and some components of both the digestive and reproductive systems. Tissues and structures derived from the ectoderm secrete the shell (see below) and spicules and form external structures such as tentacles and ctenidia, as well as the nerves and sensory structures. Ectoderm may also invaginate into the body and, in conjunction with either mesoderm (reproductive tracts), or the endodermal and endomesodermal gut, give rise to the external openings of these systems (see also Chapters 3 and 4). Many larval organs are ectodermal in origin, including the larval shell, larval tentacles, the apical plate, protonephridia, and prototroch (Lyons & Henry 2014; Lyons et al. 2015).

Two distinctive molluscan features – the foot and the shell-secreting tissues – have ectodermal origins. The foot first appears as an epidermal thickening below the mouth (the former blastopore) on the ventral side of the post-trochal region. In gastropods, the early foot may be bilobed, while it is trilobed in scaphopods (Raven 1966). In polyplacophorans and solenogasters, the foot rudiment extends ventrally over most of the post-trochal region. The shell gland first appears as an ectodermal thickening on the dorsal surface of the post-trochal region, opposite the developing mouth and foot. This epidermal cell thickening then invaginates below the surface before evaginating back to the surface where it secretes the larval shell over the central region. Following evagination, the periphery of the shell gland is bordered by the mantle fold which enlarges the initial larval shell through accretionary shell growth. Polyplacophorans do not undergo an invagination followed by evagination; instead, seven transverse ectodermal ridges divide the dorsal surface of the post-trochal region into eight shell fields between the ridges (see Section 8.9.5.4 and Chapter 14 for further discussion).

8.9.9.1.1   Nervous System

Ectodermal development of the nervous system is referred to as *neurulation* and is initially specified in the apical region of the trochophore (Raven 1966; Verdonk & Biggelaar 1983). The first structure to form is the apical organ, which consists of an apical tuft of long cilia and sensory and neurosecretory cells (see Chapter 7). Apical organs are not unique to molluscs, also being found in cnidarians, lophophorates, and some deuterostomes (Marlow et al. 2014), although their homology has not been convincingly established. They are thought to integrate sensory information (e.g., light, touch, and chemical cues), which in turn controls ciliary locomotion, and these cues may also initiate metamorphosis (Page 2002a; Jékely 2011).

In most gastropods, bivalves, and scaphopods, the ganglia originate from independent ectodermal primordia. For example, the cerebral ganglia may form as invaginations of cells (the cephalic plates) on either side of the apical organ, while the pedal ganglia originate from foot ectoderm, the pleural and parietal ganglia from the pleural groove behind the velum, and the buccal ganglia from the stomodaeum. The visceral ganglion originates from an unpaired ectodermal thickening at the back of the developing mantle cavity (Raven 1966) and, lastly, commissure and connectives between ganglia are formed by ganglionic outgrowths or cell migration. In gastropods that lack a trochophore stage (e.g., heterobranchs), development of the CNS is modified. For example, in the heterobranch *Ovatella myosotis,* the first component of the CNS to form is the cerebral commissure followed by the cerebral and pleural ganglia, and the pedal ganglia continue to be derived from foot ectoderm (Ruthensteiner 1999). As pointed out by Croll and Dickinson (2004), the adult central nervous system forms independently of the larval nervous system which precedes it in development.

In polyplacophorans and aplacophorans, the cerebral ganglia also arise as invaginations of the cephalic plates. Cells from the cerebral ganglionic cell masses migrate posteriorly, forming the buccal and subradular ganglia and pedal and lateral nerve cords and commissures (Hammarsten & Runnström 1926b; Baba 1938).

In cephalopods, the cephalic ganglia are formed from ectodermal cell proliferations of the cephalic lobes which delaminate into several layers (Raven 1966). The delaminating cells form four pairs of clusters. The anterior-most cell masses give rise to the cerebral ganglia, which subsequently fuse, while the other ganglionic pairs become the optic ganglia, the pedal ganglia, and the visceral ganglia (Marthy 1987; Shigeno et al. 2001a, 2001c). While previous workers have postulated that the cephalopod nervous system develops from the fusion of multiple ganglia, as found in scaphopods, gastropods, and bivalves, Shigeno et al. (2015), using neurogenetic gene expression, have shown that the octopus nervous system follows a medullary cord pattern of development as in polyplacophorans and aplacophorans, rather than one based on multiple spherical ganglia as in scaphopods, gastropods, and bivalves – a surprising result given the complex brains of cephalopods.

8.9.9.1.2 *Foregut*

While the stomach, digestive gland, and style sac (when present) are products of the endoderm, foregut structures (radular sac, salivary glands, buccal cavity, oesophagus, buccal glands, etc.) are derived from the ectodermal stomodaeum which forms the mouth on the ventral side of the embryo just behind the prototroch (Raven 1966; Moor 1983). Hammarsten and Runnström (1926b) stated that the oesophagus and 'sugar glands' in polyplacophorans were

derived from endoderm and not ectoderm, but this has been disputed by Raven (1966). In solenogasters, the foregut gives rise to the buccal ganglia which ultimately fuse with the cerebral ganglia (Thompson 1960). Developmental studies of the foregut in bivalves are rare (e.g., Gustafson & Reid 1988; da Costa et al. 2008). In contrast, foregut development in gastropods has been well studied, especially in caenogastropods where the foregut is diverse and important in systematics (see Chapter 19).

Foregut development in gastropods begins with a ventral outpocketing of the stomodaeum. In patellogastropods, vetigastropods, and heterobranchs, it typically forms the radular sac, and the buccal cavity is formed from the stomodaeum just in front of the radular sac outpocketing (Smith 1935; Crofts 1937; Thompson 1958). In some caenogastropods at least, the outpocketing almost simultaneously produces both the radula and buccal cavity (Page 2002c) (Figure 8.39).

Mesodermal cells later aggregate around the radular outpocketing and ultimately form the buccal mass musculature. In addition, bilaterally symmetrical outpocketings of the buccal cavity form the salivary glands. Additional foregut derived glands are also produced in some taxa, for example, the paired pharyngeal diverticula in monoplacophorans (Wingstrand 1985), the paired buccal pouches in patellogastropods, vetigastropods, neritimorphs, and some caenogastropods (Hyman 1967; Ponder & Lindberg 1997) and the accessory salivary glands present in some neogastropods (Ponder 1974). Foregut ontogeny is more complicated in neogastropods with proboscis development.

Foregut development in vetigastropods may be specified by *Gsx* expression (Samadi & Steiner 2010), but this is not implicated in foregut differentiation in other classes (see Section 8.9.4.2). Page (2005) noted that a ventral outpocketing from the larval oesophagus also gives rise to the anterior oesophagus and valve of Leiblein in a nassariid neogastropod, therefore suggesting that they are also derivatives of the buccal cavity tissues and not the larval gut system in the feeding larvae.

This suggests these two distinct gut systems (larval and adult) are independent as exemplified by the loss of the larval oesophagus and mouth during metamorphosis. Page (2005) concluded her analysis with the hypothesis that the gastropod buccal cavity and buccal mass are a 'developmental module' and that the canalised development of this module may have been an important feature in the evolution of the neogastropod foregut, because it enabled the elaborate foregut structure of adult neogastropods to arise late in development without compromising larval feeding structures. Page and Hookham (2017) further developed the idea of modularity in neogastropods by considering foregut development to be divisible into a ventral module consisting of the outpocketed buccal cavity and radular sac and a dorsal module consisting of the stomodaeum derived larval gut. They linked their modularity hypothesis with the evolution of feeding larvae in gastropods and suggested that the uncoupling and heterochronic offset of the two modules (dorsal and ventral foregut modules) enables the larval foregut to be co-opted as a functional oesophagus in feeding larvae.

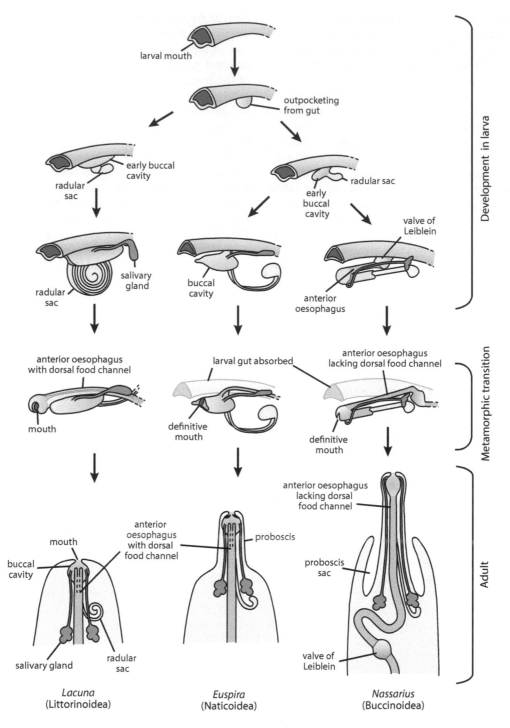

FIGURE 8.39 Foregut development in three caenogastropod taxa (odontophore cartilages and muscles not shown). Redrawn and modified from Page, L.R., *Evol. Dev.*, 2, 25–34, 2000.

8.9.9.2 Endoderm-Derived Organs

The endoderm gives rise to the stomach (and associated structures) and the digestive gland (Raven 1966). The early stomach (archenteron) is composed of the progeny of the 4A–4D macromeres which make up the roof of the archenteron, while the fifth quartet of micromeres make up the sides (Verdonk & Biggelaar 1983). Small cell endoderm forms at the tip of the archenteron and connects with the ectodermal oesophagus or buccal cavity, while large cell endoderm gives rise to paired lobes – the nascent digestive glands. The endomesodermal hindgut originates as an outgrowth of 4d cell progeny at the posterior part of the archenteron (Verdonk & Biggelaar 1983; Lyons et al. 2015) and elongates to the ectoderm where it forms the anal opening (Raven 1966). Conklin (1897) first noted the contribution of the endomesoderm to the formation of the hindgut. Using a high-resolution fate map, Hejnol et al. (2007) identified the involvement of the 4d micromere in hindgut formation in *Crepidula fornicata*, while Lyons et al. (2012) further resolved the contributions of the early 4d cell lineages, identifying four sublineages of 4d that contributed to the hindgut in *Crepidula*. Larval gut modifications are primarily ectodermal in origin at the anterior and posterior regions of the endoderm derived portion (see Section 8.9.5.1).

8.9.9.3 Mesoderm-Derived Organs

The mesoderm gives rise to connective tissue, cartilage, muscles, the heart walls, blood cells and lymph vessels, kidneys, gonads, and primary genital ducts. Mesodermal larval structures are predominately the larval retractor muscles.

Mesoderm arises in two regions of the embryo – ectomesoderm is derived from the second and third quartet micromeres, while endomesoderm is derived from the 4d micromere or mesentoblast (Boyer et al. 1996; Lambert 2008; Lyons et al. 2012). While the specific micromeres which produce ectomesoderm vary among taxa, the derivation of endomesoderm from the 4d cell appears to be highly conserved across spirally dividing lophotrochozoans. Division of the 4d micromere typically produces paired endodermal precursors and mesodermal bands from which the visceral mesoderm originates (Lyons et al. 2012), while the anterior mesoderm is associated with the blastopore and may contribute to the mesodermal mouth structures (Lartillot et al. 2002). Based on expression patterns of *fork head* and *goosecoid* genes in *Patella vulgata* and other metazoan taxa, Lartillot et al. (2002) suggested that specification of the A-P axis of the embryo utilised both an anterior and posterior organiser linked with the dipolar mesoderm, and that this form of specification was ancestral in the Bilateria. Specification of the 4d cell may occur when the embryo consists of as few as 25 cells in some caenogastropods and heterobranchs (Biggelaar 1996; Biggelaar & Haszprunar 1996), although the ancestral state in molluscs appears to be later (>60 cells) (Lindberg & Guralnick 2003b).

The visceral mesoderm produces the pericardium, heart, kidneys, most of the musculature, mesenchymal cells, often germ cells and gonads, and contributes to the hindgut (Raven 1964; Baeumler et al. 2012; Lyons et al. 2012). The mesodermal bands are typically paired and lie on either side of the endoderm. The bands give rise to paired cell masses or a single cell mass to the right of the developing hindgut (some 'littorinimorphs' and Stylommatophora) (Raven 1966). In taxa where there is only a single band, the cell mass divides in two with the posterior portion forming the coelomoduct rudiments and the anterior portion giving rise to the pericardium and heart. In molluscs with two separate rudiments, they fuse and the pericardium forms as a lumen in the coelomic anlage, which often expands to envelop the hindgut (Raven 1966). The coelomoducts appear as buds on either side of the cell mass and, with continued growth and differentiation, separate into kidneys and pericardium + heart, although connections (the renopericardial ducts) between these two organs often remain. In aplacophorans, this developmental history is well displayed as the gonads, pericardium and coelomic ducts form

a pair of long, continuous tubes (Chapter 14, Figure 14.18). Although their coelomic ducts are poorly differentiated, they still perform the functions (ultrafiltration, reabsorption) of the molluscan kidney, including providing a passage for the gametes from the gonad to the mantle cavity (Reynolds et al. 1993; Scheltema et al. 1994; Baeumler et al. 2012). In the squid *Loligo,* the kidneys and pericardium differentiate at the same time, forming separate lumina in the formerly solid cell masses (Naef 1909).

Mesodermal structures in solenogasters evidently form very late in development, but this has not been directly observed. Thompson (1960) reported that the heart, coelomoducts, excretory tissue, and germ cells were not present ten weeks after metamorphosis, perhaps explaining why, of all molluscan groups, the solenogaster mesodermal structures show the least morphological differentiation.

Heart formation in molluscs typically begins with a thickening of the dorso-posterior surface of the pericardium mesodermal cell mass over the hindgut. This ventrally expanding tissue mass surrounds the hindgut and forms a lumen between the hindgut and pericardial wall (Raven 1952). In polyplacophoran and patellogastropod adults, the dorsal surface of the ventricle remains attached by the mesocardium to the dorsal wall of the pericardium along its longitudinal axis (Fleure 1904). Subsequent development of two smaller cell masses from the mesoderm anlagen, or invaginations of the pericardial walls, form the right and left atrial cavities which fuse laterally with the ventricle and the haemocoelic spaces at the base of the gill rudiments. At this point, the heart is a transverse tube across the pericardium and opens to the body cavity on either side. It is fused with the atrial cell masses to form auricles, arteries, and veins which arise from the heart rudiments (Raven 1966). In the few 'zygobranch' vetigastropods studied (e.g., *Haliotis*), both left and right auricles form from the heart rudiment (Crofts 1937), while in gastropods with a single auricle (e.g., *Patella, Viviparus, Bithynia, Planorbis, Arion*), the dorsal thickening invaginates, forming the tube across the pericardium that constricts medially to form the auricle and ventricle (Raven 1966). In both bivalves (e.g., *Nucula*) and gastropods (e.g., *Lottia*), subsequent growth of the kidneys and other organs can displace the pericardium and push the rectum outside the ventricle and pericardium (Heath 1937; Lindberg 1988a) (also see Section 8.9.5.7).

Little is known of heart formation in *Nautilus*. The branchial hearts in embryonic *Nautilus* are large and develop relatively early (14 days) in order to provide nutrients from the external yolk sac to the developing embryo (Arnold 1987). In adult *Nautilus*, they are reduced to contractile vessels which, like the coleoid branchial hearts, increase pressure prior to blood entering the gills (Bourne 1987a, 1987c). By six months of age, the renopericardial system of *Nautilus* is fully developed (Shigeno et al. 2010). Heart formation in *Sepia* shows some similarities to heart development in gastropods and originates from the pericardial-intestinal sinus, expanding into the developing pericardial cavity (Distaso 1908). The outpocketing differentiates into heart musculature, but remains connected with the hindgut through a membrane (the mesocardium) as in patellogastropods and polyplacophorans. Naef (1909) described both the branchial hearts and paired ventricle heart rudiments arising from folds in the wall of the pericardium. The branchial hearts connect with the developing venous system returning from the external egg sac and the developing gill. The ventricle rudiments first formed as a pair of independent parallel tubes, but became united posteriorly to form a single chamber, each with a separate anterior aorta. The posterior aorta then forms in the median region of the fused ventricle and the left aorta subsequently atrophies. Lastly, *de novo* 'auricles' are formed from branchial veins that secondarily connect the gills and ventricle and lie outside the pericardium (Naef 1909). Development appears similar in octopuses, although the gills and pericardial glands appear much earlier than the ventricle rudiments, which again fuse at the midline to form the single chambered ventricle and connect with the anterior and posterior aortas (Marthy 1968). Overall, the development of the heart appears to be more variable in cephalopods than in other molluscs, and is probably the product of early modifications to the larval circulatory system necessitated by increased yolk content and the need to provide circulation for retrieval of nutrients from the external yolk sac (Portmann 1926; Boletzky 1987).

The reproductive system of molluscs forms late in development (Raven 1966; Moor 1983); in most taxa, the reproductive system does not develop until after metamorphosis, including Solenogastres (Thompson 1960), Polyplacophora (Pearse 1979), and Gastropoda (Moritz 1939). In the 'queen conch' *Strombus gigas*, the ontogeny of the reproductive system is not completed until the third year (D'Asaro 1965), while in many heterobranch gastropods, the formation of the reproductive tract is entirely a juvenile developmental event (Tardy 1970; Moor 1983; Visser 1988). The belated organogenesis of the reproductive system often excludes this system from larval developmental studies (e.g., Smith 1935; D'Asaro 1965), and thus its ontogeny is more readily observable in juveniles (Schulhof & Lindberg 2013). Heterobranch gastropods have the added study value of a hermaphroditic reproductive system providing both male and female anatomies (Rouzaud 1885; Hoffmann 1922; Tardy 1970; Griffond & Bride 1985).

Where known, gonads are derived from the primary mesoblasts which also form the kidneys and pericardium (Dohmen 1983; Moor 1983). The gonad rudiments are initially paired, but can become fused or partially fused into a single structure to varying degrees in different taxa. Gonads are generally paired in aplacophoran taxa although some have a single gonad (e.g., *Chaetoderma*) (Nierstrasz & Hoffmann 1929); in polyplacophorans, the gonad is originally paired, but becomes fused into a single gonad in most species (Hyman 1967). Bivalve gonad development was reviewed by Sastry (1979) and in that group the gonads typically originate late in development from mesodermal cell masses near the pericardium and kidneys. In *Sphaerium,* the early paired gonads fuse under the pericardium to form a single gonad.

In oysters, gonads do not persist during the resting stages between spawnings and are diffuse and transient (Andrews 1979b). Scaphopods, cephalopods, and gastropods have a single gonad, which does not result from fusion, but forms from a single mesodermal anlage. Moor (1983) identified three origins for the gonad: (1) cells from the pericardium; (2) an isolated mesodermal cell mass in the body cavity; or (3) from differentiation of the PGCs. Differentiation of the PGC triggers gonad development in some gastropods (heterobranchs) and also in a few bivalves. Depending on the timing of PGC differentiation, the gonad is formed from cells either associated with the mesoderm bands, or if differentiation occurs later, from the early pericardium (Raven 1966; Moor 1983).

In cephalopods, gonad development is poorly known despite numerous fishery studies of reproductive cycles of commercially important species (e.g., Avila-Poveda et al. 2009; Arizmendi-Rodriguez et al. 2012; Cuccu et al. 2013). Naef (1909) described and illustrated the ontogeny of the gonad in *Loligo* where it is derived from a mesodermal cell mass that also gives rise to the pericardium and kidneys. It first appears as a medial cell mass between the developing heart and shell gland, but development of the pericardium moves the gonad cell mass dorsally before enveloping the developing heart with its posterior aorta. The gonad remains undifferentiated until late in development when it connects with the pericardium. Gonoducts are mesodermal in origin in cephalopods, forming as an outgrowth of the pericardial epithelium, and they open into the ectodermal oviduct gland; the nidamental glands are also ectodermal in origin (Figure 8.40).

Single gonoducts occur in scaphopods, most cephalopods and gastropods; they are paired in octopuses. In bivalves, the mesodermal gonoducts open into the kidneys in basal taxa (e.g., *Solemya, Nucula, Pecten*), while in most autobranchs, the paired gonoducts open individually into the mantle cavity with little, if any, ectodermal contribution (see also Chapter 15). Tardy (1970) has suggested that the 'vésicules séminales' in the bivalve genera *Montacuta* and *Jousseaumiella* were derived from ectoderm. Gonoducts of chitons are also paired mesodermal outgrowths that open into the mantle groove, with little ectodermal invagination in the elaboration of the reproductive system. In basal gastropods, in which the gametes pass through the left kidney, there is also little ectodermal contribution to the reproductive system. In internally fertilising taxa (Caenogastropoda and Heterobranchia),[10] the role of ectodermally derived structures in the elaboration of the reproductive duct is more pronounced (Moor 1983), and in euthyneuran heterobranchs, the forming gonad only becomes recognisable after it contacts the invaginated ectodermal components of the reproductive system (Ghiselin 1965; Tardy 1970). In neritimorphs and caenogastropods, the right kidney is incorporated into the gonoduct, but the majority of the duct is ectodermal (also referred to as 'pallial') (Ghiselin 1965; Tardy 1970; Thiriot-Quiévreux & Martoja 1976; Moor 1983; deMaintenon 2001).

---

[10] The Neritimorpha, with their diaulic (two openings) and triaulic (three openings) reproductive systems, have not been adequately studied to determine the origin of these complex reproductive systems.

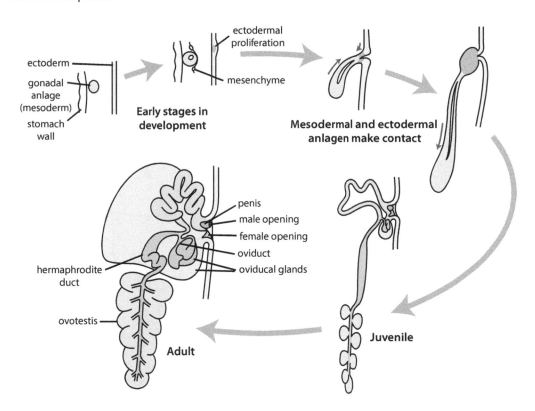

FIGURE 8.40 Ectodermal and mesodermal contributions to the development of the reproductive system in the aeolidioid nudibranch *Aeolidiella alderi*. Redrawn and modified from Tardy, J., *Bull. Soc. Zool. Fr.*, 95, 407–428, 1970.

8.9.9.3.1 Shell Muscles

In polyplacophorans, the larval prototroch musculature and adult longitudinal muscles appear slightly before the dorsoventral shell and ventrolateral muscles (Wanninger & Haszprunar 2002b); in some taxa, the dorsoventral shell muscles appear before the shell plates are formed (Hammarsten & Runnström 1926b). Lastly, the transverse muscles appear. The dorsoventral shell muscles do not appear to become functional until after the completion of metamorphosis (Wanninger & Haszprunar 2002b). In caudofoveates, the lateral longitudinal muscles appear first, followed by the prototroch muscle ring, and lastly the circular muscles (Nielsen et al. 2007). In solenogasters, muscle fibres form between the developing gut and ectoderm and ultimately give rise to the lateral longitudinal and circular muscles; dorsoventral muscles from the foot rudiment to the body walls also form (Baba 1938; Thompson 1960). This early work was greatly enhanced by Scherholz et al. (2015), who meticulously followed myogenesis in *Wirenia argentea* and showed that ontogenetic components were shared with polyplacophorans. In scaphopods, the cephalic and dorsoventral retractor muscles appear almost simultaneously in the early trochophore larva (Wanninger & Haszprunar 2002a).

In bivalves, the larval musculature is lost at metamorphosis, except for the larval adductors which often give rise to the adult adductor musculature. In unionids, however, the larval adductors degenerate and the adult adductors are *de novo* adult structures

formed from scattered mesenchyme. Anterior and posterior dorsoventral retractor muscles form from paired groups of muscle cells that extend ventrally into the foot, followed by the transverse musculature of the foot (Raven 1966).

In basal gastropods with two shell muscles (patellogastropods, some vetigastropods, and neritimorphs), the separate adult muscles are derived from the left and right pre-torsional mesodermal bands (Smith 1935; Crofts 1955) (Figure 8.41). In most gastropods (e.g., higher vetigastropods, most caenogastropods, and heterobranchs), there is a single shell attachment muscle, the columellar muscle, derived from the right pre-torsional mesodermal band. The presence of paired muscles in the caenogastropod velutinoideans is possibly due to retention through heterochrony; paired shell muscles have also been reported in the heterobranch *Rissoella* (Fretter & Graham 1962; Haszprunar 1988a).

In the heterobranch *Aplysia*, the adult musculature consists of pedal, anterior, and metapodial retractor muscles; only the anterior retractor muscles appear after metamorphosis, the other two first appear in the veliger phase and co-occur with the larval musculature (see below) (Wollesen et al. 2007).

Larval muscles occur in polyplacophorans, caudofoveates, gastropods, and bivalves (Haszprunar & Wanninger 2000; Kurita et al. 2016). Most are associated with the foot and prototroch (and velum when present). Chitons also have a unique larval pretrochal muscle grid, which is lost upon completion of metamorphosis. Although there is no unique

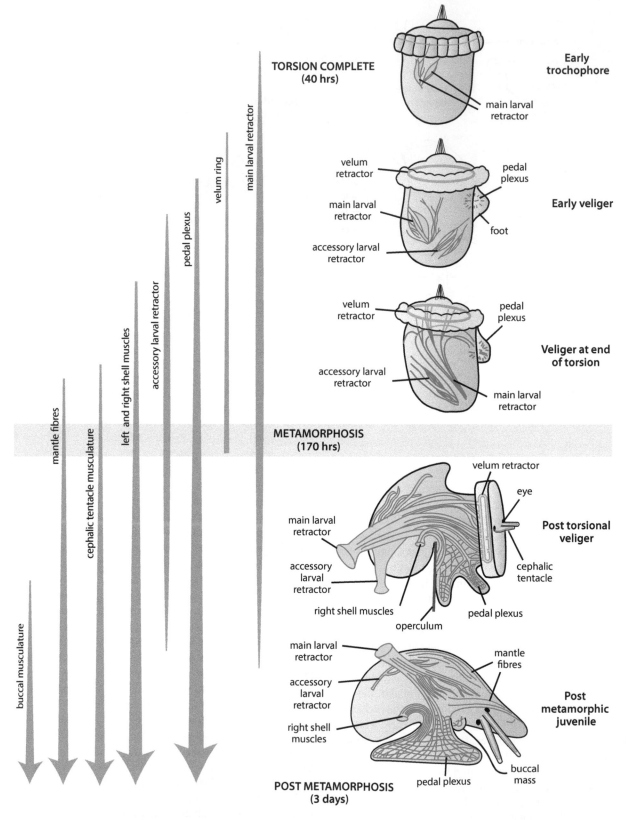

FIGURE 8.41 Myogenesis in the patellogastropod *Patella*. Transverse muscle not shown. Redrawn and modified from Wanninger, A. et al., *Dev. Genes Evol.*, 209, 226–238, 1999b.

larval musculature in scaphopods, the developing adult retractor muscles give rise to fibres that allow for the retraction of the prototroch, and are lost during metamorphosis (Wanninger & Haszprunar 2002a). In caudofoveates, the lateral longitudinal muscles appear to give rise to the prototroch muscle ring (Nielsen 2015). Such a ring was not reported in solenogasters by Thompson (1960), possibly due to the presence of the test larva in that group.

In bivalves and gastropods, the larval musculature is usually independent of the adult musculature and is derived from ectomesoderm (Haszprunar & Wanninger 2000). In bivalves, the apomorphic adductor muscles are typically derived from myoblasts that also produce the larval musculature. The anterior adductor muscle forms from the fusion of left and right dorsal groups of myoblasts and typically before the posterior adductor muscles, which are derived from the ventral larval retractor. In patellogastropods, these myoblasts give rise to the main larval retractors (Kurita et al. 2016).

Bivalves typically have three pairs of larval retractors: (1) the ventral larval retractors, (2) the dorsal, and (3) ventral velum retractors; four pairs have been reported in some pteriomorphians (Kakoi et al. 2008). In gastropods, patellogastropods have three larval shell muscles – the unpaired main larval retractor (Figure 8.41) and accessory larval retractor and the paired transverse muscle, although only the first two were reported in earlier accounts (Damen & Dictus 2002). Page (1995) also described three pairs of larval retractor muscles in several nudibranch species. There are two larval retractor muscles in vetigastropods and *Nerita* (Page 1997a; Page & Ferguson 2013), while most other groups have one or two (Page 1998). In all taxa studied, the larval retractor muscle arises first from the pre-torsional right mesodermal block. If the pre-torsional left muscle forms at all, it is after the formation of the right muscle (as in vetigastropods, patellogastropods, and some 'opisthobranchs'). The pre-torsional left muscle may develop to equal size with the right, or perhaps even exceed it. The adult columellar muscle of the caenogastropod *Neverites lewisii* (Naticidae) is derived from a portion of the larval pedal muscle (Page 1998), but in euthyneurans the contribution of the larval retractor muscles to the adult muscular system is unclear. For example, in *Aplysia,* the larval musculature consists of the larval, accessory larval, and metapodial retractor muscles, all of which appear before and during the veliger phase, but degenerate during metamorphosis (Wollesen et al. 2008). *De novo* originations of the adult musculature have also been documented in the cephalaspidean *Retusa* (Smith 1967) and the nudibranch *Aeolidiella* (Kristof & Klussmann-Kolb 2010). In contrast, Thompson (1962, p. 195) stated that in *Tritonia,* 'The adult musculature…is derived from the embryonic and larval cephalopedal muscle complex'.

The advent of computer-assisted three dimensional reconstructions of muscles from transmission electron microscopy and staining with fluorescently labelled phalloidin for actin labelling, combined with confocal laser scanning microscopy, has provided amazing detail of myogenesis and related phenomena in molluscs (Page 1998; Wanninger et al. 1999b; Scherholz et al. 2015). These advances have also been accompanied by molecular investigations, such as those of Degnan et al. (1997), who provided some of the first insights into the developmental genetics of molluscan gastropod musculature, although questions remain on the persistence of the larval musculature into the adult stage in many taxa.

8.10 LIFE HISTORY EVOLUTION

The study of molluscan life history patterns focuses on patterns of reproduction, development, growth, maturation, and survival. These traits are important because they largely determine the survival and reproductive output of an individual and are therefore major components of fitness (Stearns 1976, 1977, 1980; Ghiselin 1987). Molluscan life history traits are diverse (Figure 8.42) and can be compared across different biological categories, including morphology (e.g., test cell larval, trochophore), ecology (e.g., pelagic, non-pelagic development), physiology (e.g., feeding, non-feeding larvae), behaviour (e.g., external fertilisation, copulation), or life cycle stage (e.g., larval, juvenile, adult), etc. The biological consequences of different modes of larval development in molluscs have evolutionary significance at both the ecological (Thorson 1950; Salvini-Plawen 1980b; Fioroni 1982a; Krug 2009) and geological time scales (Shuto 1974; Scheltema 1977, 1978b, 1979; Hansen 1983; Jablonski 1986; Nützel 2014). The diversity of traits provides opportunities to explore life history evolution at micro- and macroevolutionary levels although interactions may not necessarily be simple (Russell et al. 1988). Many life history traits are also phylogenetically informative at various taxonomic levels.

As discussed earlier in this chapter, molluscan spawning patterns generally fall into three main categories, successive, intermittent, and cyclic spawning. Successive ('dribble') spawning is where oogenesis is not synchronised and oocytes in various stages of maturation are found in a single ovary with relatively few eggs being released at any one time over a long breeding period, often the entire year. Intermittent spawning also has asynchronous oogenesis and more than one spawning event with large batches of eggs released. Lastly, in cyclic spawning egg production occurs periodically, usually following a seasonal cycle, with most eggs being released at one time or during a short period. Different reproductive strategies are imposed by particular environments, notably extreme environments, and it is interesting to note that all three patterns exist in molluscs from abyssal depths (Scheltema 1994).

8.10.1 DEVELOPMENTAL STRATEGIES

Two critical criteria for larval characterisation in molluscs have been defined as: (1) the source of the nutrition that powers development and (2) the dispersal capability of the larvae (Figure 8.43). Different classifications have emphasised different aspects of this division although they are often thought to be positively correlated (i.e., more nutrition = more

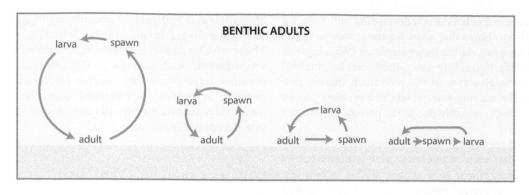

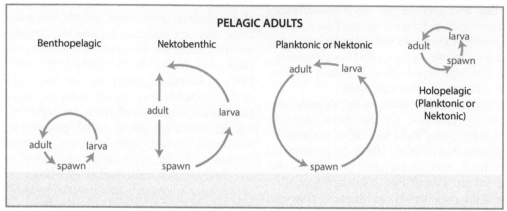

FIGURE 8.42 Molluscan life cycle stages sorted by adult habitat and pelagic duration (when present). Benthic adult life cycles are present in most molluscs, while pelagic adult life cycles are characteristic of pelagic cephalopods and two groups of gastropods – Gymnosomata and Thecosomata (holopelagic). 'Spawn' may consist of either gametes or fertilised eggs. Original.

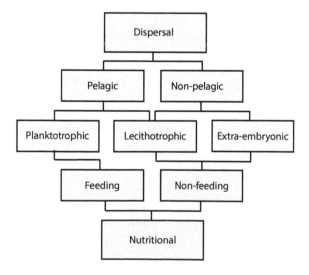

FIGURE 8.43 Primary dichotomy and relationships of larval developmental classifications. Original.

dispersal). Only the primary relationships are shown as these categories (especially nutritional source) are not mutually exclusive. For example, some caenogastropods with *mixed development* begin development using maternal yolk (lecithotrophic), then switch to nurse eggs (extra-embryonic) before hatching from the egg capsule, and completing development as a planktotrophic feeding larva. These taxa also have both pelagic and non-pelagic (capsular) phases (mixed) with the dispersal potential determined by the duration of the former relative to the latter. Lucas et al. (1986), in a study of larval development in *Mytilus edulis*, referred to these three nutritional sources as endotrophy, followed by mixotrophy, and exotrophy. Additional subdivisions can also be added to these basic categories (e.g., extra-embryonic could be further broken down to distinguish nurse eggs from cannibalism of developing siblings, or extra albumen, or viviparity).

We prefer to use the term pelagic or nektonic rather than planktonic. Planktonic organisms are assumed to be unable to swim against a current and can only drift in the plankton, which is typically limited to the top 150–250 m of an aquatic environment This view of passive planktonic larvae is being challenged by studies on larval behaviour. Larvae often undertake diurnal migrations, which exploit the use of nearshore wind, upwelling cells, and other oceanographic features to actively move in the water column (Ebert & Russell 1988; Grantham et al. 2003; Shanks & Brink 2005). In addition, documentation by Bouchet and Warén (1994) of ontogenetic vertical migration and dispersal of deep-sea gastropod larvae from hydrothermal vents, as well as the palaeontological and extant occurrences of sustained cold water faunas in subtropical climes (Valentine 1955; Emerson 1956), suggests that larval dispersion is not passive. Organisms which live in the pelagic realm and are motile are called nektonic (Figure 8.42).

8.10.2 CLASSIFICATION OF LARVAL STRATEGIES

The diversity of molluscan life history traits has produced a descriptive vocabulary with its share of synonyms and homonyms (Fioroni 1982a; Jablonski & Lutz 1983; Turner et al. 1987). The variety and distribution of these traits has also produced a rich mosaic of characters and states, as well as terminology (Table 8.9), all of which have contributed to multiple hypotheses of life history evolution in the Mollusca (Salvini-Plawen 1980b; Chaffee & Lindberg 1986; Haszprunar 1992b; Hickman 1999). Most analyses have supported an ancestral lecithotrophic (non-feeding) pelagic larvae for the molluscan crown group (see Rouse 2000), and based on their small body sizes, it has been suggested that Cambrian stem groups may have been brooders (Chaffee & Lindberg 1986; see also Haszprunar 1992b). Many of the categories are often paired and used as dichotomous comparisons or trade-offs in life history evolution (e.g., benthic versus pelagic development, feeding [planktotrophic] versus non-feeding [lecithotrphic] larvae, direct versus indirect development).

The hypothesised plesiomorphic traits (gonochoristic, external fertilisation, non-feeding larvae) in the molluscan life cycle remain present in at least some members of all but two classes; the Cephalopoda and Solenogastres. These plesiomorphic traits are also present in all putative outgroups, including the Annelida, Brachiopoda, and Entoprocta (Giribet 2008). Internal gamete fertilisation followed by direct development (brooding) has convergently evolved in all classes except the Caudofoveata, although details for that taxon are scant (Nielsen et al. 2007). Given this high occurrence of internal fertilisation across the Mollusca, it is surprising that only gastropods and cephalopods have evolved copulatory structures. Instead, the remaining taxa

facilitate internal fertilisation with different forms of sperm-casting (Pemberton et al. 2003b; Falese et al. 2011).

As outlined above, earlier studies (e.g., Thorson 1950; Scheltema 1978b) contrasted pelagic and non-pelagic development. Thorson further recognised five subcategories within this basic dichotomy, based primarily on the source of nutrition. Scheltema emphasised broader division between pelagic and non-pelagic development. Subsequent schemes have generally followed emphasis by Thorson on nutrition source and dispersal capability, although some have added categories based on the secondary morphology of the larva (trochophore, veliger, pericalymma, rotifer, etc.).

Turner et al. (1987) also noted that different larval development classifications emphasised ecological settings of development (e.g., egg membrane, capsule, pelagic, benthic), while others emphasised morphological definitions (e.g., trochophore, veliger, pediveliger), and they presented a dichotomous key to developmental types which incorporated both. Their first couplet distinguished feeding from non-feeding larvae. Within the feeding category they distinguished the morphological stage at which the larvae enter the pelagic realm. Lastly, they distinguished whether the developing embryos were brooded or encapsulated before their release. Two ecological categories were recognised under non-feeding – free living versus encapsulated or brooded. Under the encapsulated or brooded categories, taxa were further divided by dispersal capability as crawl-away juveniles or a partial pelagic mode. Lastly, the non-feeding embryos were categorised by whether they were brooded or encapsulated, losing the ability to estimate dispersal capability in this nutrient-based scheme (Table 8.9).

Fioroni (1982b) distinguished 19 main larval developmental modes in molluscs in a scheme that differentiated the various stages of development using their morphological diversity, where they occurred in the environment, and the nutritional source(s) for development. The classification of Fioroni has been modified and expanded to include at least 20 different molluscan life histories (Figure 8.44).

8.10.2.1 Larval Dispersal, Morphology, and Ecology

As evidenced above, one of the most discussed dichotomies in molluscan life history modes is the presence or absence of a larval dispersal phase. This dichotomy has been contrasted utilising several terminologies, including brood or broadcast, direct or indirect development, benthic or pelagic development, etc. These classifications are neither synonymous nor mutually exclusive, and the source of trophic resources for development, which can often be highly correlated with dispersal capability (see Jablonski & Lutz 1983), is sometimes confounded in these classifications (see Chapter 9 for more on dispersal mechanisms).

Molluscan larval stages occur in both aquatic and terrestrial habitats. In terrestrial habitats, the larvae typically remain immersed in fluid within desiccation resistant eggs, capsules, or parental brood chambers (Tompa 1979). The aquatic realm is home to both pelagic larvae and non-pelagic larvae, the latter housed in capsules, egg ribbons, parental brood chambers, and 'nests' placed on the benthos. Not surprisingly, correlations

TABLE 8.9

Information Content of Molluscan Larval Terminology

Term	Characteristic		
	Dispersal	Nutrition	Larval Morphology
Direct development – lacking larval stages	No	No	Embryo > juvenile
Juveniparous – eggs hatch within parent, offspring emerge as juveniles	No	No	Embryo > juvenile
Larviparous – eggs hatch within parent, offspring emerge as larvae	No	No	Embryo > larvae
Oviparous – egg-laying	No	No	N/A
Ovoviviparous – eggs hatch within parent, supplied with yolk and albumen	Yes	Yes	N/A
Planktotroph – veliger feeds on plankton	Yes	Yes	Embryo > ? > veliger
Lecithotroph – non-feeding veliger	No	Yes	Embryo > ? > veliger
Non-pelagic or benthic – on the benthos	Yes	No	N/A
Pelagic or nektonic – active swimming	Yes	No	Embryo > ? > veliger

Terms modified from the Biotic Database of Indo-Pacific Marine Mollusks (http://clade.ansp.org/obis/mollusc_keywords.php).

with poor dispersal potential are strongest in terrestrial habitats, but dispersal potential cannot be assumed to be good in aquatic habitats. The plesiomorphic condition is typically considered to be a benthic adult with pelagic spawn and larvae (Figure 8.42), but Chaffee and Lindberg (1986) provided an alternative argument that the small body size of early putative molluscs is more characteristic of taxa with non-pelagic larvae.

Morphological features which correlate with life history traits include the presence or absence of certain larval types and copulatory structures, embryo encapsulation, brooding structures, and protoconch size and number of whorls; only this last morphological feature can be reliably determined in the fossil record (Shuto 1974; Jablonski & Lutz 1983; Nützel & Frýda 2003). The criteria of Shuto (1974) for the recognition of planktotrophic versus non-planktotrophic gastropod species was based on the relationship between the diameter and the number of whorls of the protoconch. In summary, large diameter protoconchs are produced by species with large, yolk-rich eggs, while small diameter protoconchs indicate eggs with little yolk (Figure 8.45). The feeding larvae construct additional whorls on the larval shell (also referred to as the formation of protoconch II) (Figure 8.45) the longer it spent in the pelagic realm during feeding. In his analysis of protoconch morphology and developmental type, Shuto (1974, figure 4) plotted two larval types, planktotrophy (pelagic) and non-feeding (non-pelagic) species, and established measurement ranges which corresponded to these categories.

Jablonski and Lutz (1983) expanded the relationship between planktotrophy/non-planktotrophy and egg size in gastropods to include other molluscan groups. Their conclusions were similar to those of Shuto (1974). Like the classification of Shuto, that of Jablonski and Lutz also confounded pelagic/non-pelagic dichotomy by including within their non-planktotrophic category both pelagic and non-pelagic lecithotrophic species. Moreover, unless egg sizes of planktotrophic species do not overlap with non-planktotrophic species, there exists the possibility for the confounding of pelagic and non-pelagic development in the planktotrophic category as well.

Because dispersal capability affects the evolutionary potential of a species (e.g., radiation, persistence, extinction), the pelagic/non-pelagic dichotomy is usually considered of primary importance, however the planktotrophic/non-planktotrophic dichotomy is used to estimate dispersal capability. Moreover, do these dichotomies correlate with egg size, the primary determinate of protoconch size correlation? In two well documented studies of Indo-Pacific *Conus* species Kohn and Peron (1994) and Kohn (2012) examined species with egg sizes ranging from 125 μm to nearly 1 mm, and showed that species with the smaller egg sizes had shorter intracapsular development, grew more slowly after hatching, had longer pelagic stages as feeding larvae, and were more likely to be widely dispersed as theory suggested.

The terms *direct development* and *indirect development* can be used to denote different larval morphologies or even pelagic or non-pelagic development (i.e., 'crawl-away' versus pelagic larvae). In most scenarios, the ancestral molluscan larval morphologies are typically thought to be indirect and consist of an embryo (gastrula) > trochophore > juvenile. In gastropods and bivalves, an additional larval type is inserted between trophophore and juvenile (e.g., veliger, Warén's larva). In numerous taxa, the larva are often modified or bypassed (e.g., caenogastropod gastropods, cephalopods, solenogasters). Here, we have restricted our use of the terms direct and indirect development to denote differences in larval morphology, primarily the absence or modification of the plesiomorphic trochophore larvae (direct development). In species with direct development the larvae transform after gastrulation to the veliger phase with a velum in gastropods and bivalves – a typical trochophore stage is not present. In some brooded gastropods and bivalves, even the veliger larval phase is truncated, but this is rare even in encapsulated caenogastropod development (Figure 8.46).

Brooding structures typically involve some anatomical modifications, and even shell modifications are sometimes associated with brooding behaviour in some bivalve and gastropod taxa (Lindberg & Dobberteen 1981; Kotrla & James

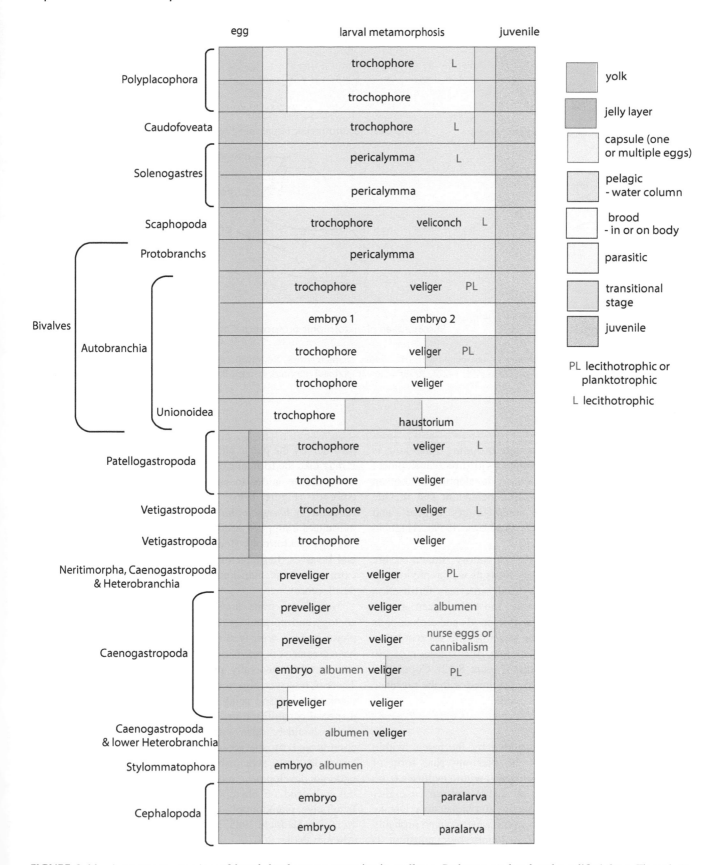

FIGURE 8.44 A summary overview of larval development strategies in molluscs. Redrawn, updated and modified from Fioroni, P., *Malacologia*, 22, 601–609, 1982a.

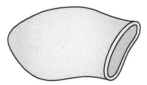

Uncoiled larval shell (Patellogastropoda)

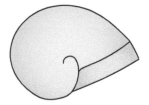

Trochoidean lecithotrophic larval shell (Vetigastropoda)

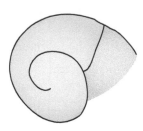

Littorinoidean lecithotrophic larval shell
(Caenogastropoda)

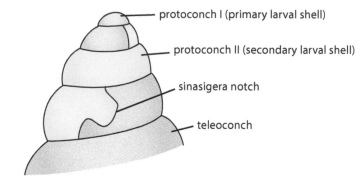

protoconch I (primary larval shell)

protoconch II (secondary larval shell)

sinasigera notch

teleoconch

Buccinoidean planktotrophic protoconch (Caenogastropoda)

FIGURE 8.45 Protoconch morphologies associated with life history characteristics. Redrawn and modified from the following sources: patellogastropod, Sasaki, T., *Bull. Univ. Mus.*, 38, 1–223, 1998; vetigastropod and littorinid, Bandel, K., *Fazies*, 7, 1–198, 1982; buccinoidean, original.

1987) (see Section 8.6). Although brooding behaviour is typically thought of as being correlated with direct development and 'crawl-away' larvae (non-pelagic development) (Thorson 1946, 1950), examples of brooding behaviour that include a pelagic phase are found in solenogasters, bivalves, and gastropods (e.g., Webber 1977; Andrews 1979b; Hadfield 1979; Sastry 1979).

In an evolutionary context, copulatory structures are correlated with several life history traits as well as phylogeny. While copulatory structures and internal fertilisation have evolved numerous times in most molluscan groups (see Section 8.5.2), it was in the gastropods and cephalopods that internal fertilisation made possible the encapsulation of embryos followed by a tremendous diversification of life history strategies in these groups (Figure 8.42). Copulation and internal fertilisation are also correlated with the presence of planktotrophic larvae in gastropods.

8.10.2.2 Larval Nutrition

Yolk is intercellular and is provided by the parent within individual eggs. It represents the plesiomorphic form of developmental nutrition (lecithotrophy) within molluscs (e.g., Nützel et al. 2006; Wilson et al. 2010) and can be highly variable between individuals, species, and higher taxa. Uptake of dissolved organic material from the water column probably supplements yolk nutrition. Extra-embryonic nutrition is provided by the parent and exists outside the egg membrane. It may take the form of extracellular albumen, or 'nurse' eggs and embryos added to egg capsules containing developing embryos (Figure 8.46). Extracellular albumen may also function in *haemotrophic viviparity* or *matrotrophy* where nutrients are supplied directly by parent tissue (Ostrovsky et al. 2016). Planktotrophy (feeding) is convergent in gastropods and bivalves. It supplements both yolk and, where present, extra-embryonic nutrition, and requires a pelagic phase of development.

Nutritional traits are correlated with larval dispersal, ecology, and morphology; for example, there is always a pelagic phase in taxa with feeding larvae, and lecithotrophic eggs have greater yolk provisioning than the eggs of feeding larvae and therefore generally produce larger initial protoconchs. There is, however, also substantial overlap in egg sizes among planktotrophic and pelagic lecithotrophic taxa, so inferring dispersal capability from the initial whorl of the protoconch alone should be done with caution. This dichotomy is also complicated by the presence of facultative planktotrophic larvae in some taxa (especially heterobranchs) which can undergo development without feeding (i.e., lecithotrophy), but will feed on phytoplankton when it is available (i.e., facultative planktotrophy) (Thompson 1958; Kempf 1981; Kempf & Hadfield 1985; Miller 1993; Botello & Krug 2006; Allen & Pernet 2007).

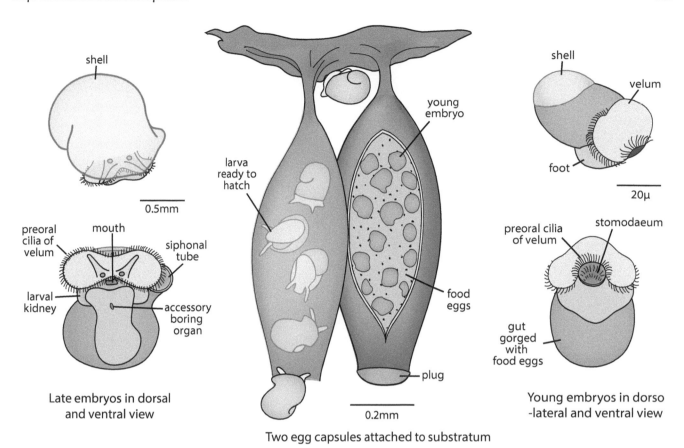

FIGURE 8.46 Larval development of the caenogastropod *Nucella lapillus*. Redrawn and modified from Fretter, V. and Graham, A.L., *British Prosobranch Molluscs: Their Functional Anatomy and Ecology*, Ray Society, London, UK, 1962.

It is generally thought that transition from feeding to non-feeding larvae occurs when taxa become direct-developing brooders and truncate the pelagic phase of their life history. Collin (2004) presented data that argued that species which ancestrally had planktotrophic development and had acquired a lecithotrophic mode of development, including loss of a larval stage, had reverted back to planktotrophic development. In a study of slipper limpets (Calyptraeidae), Collin mapped developmental type on a phylogenetic tree and found three cases where reversal may have occurred from 'direct' development to feeding larvae. In all cases, the direct development involved nurse eggs as nutrient so the eggs were not excessively yolky, and the developing larvae in the egg capsule retained the velum. In cases where direct development occurred from large yolky eggs, no reversals were found. In an even larger study of sacoglossan heterobranchs, Krug et al. (2015) found 27 origins of lecithotrophy in the group, with most in lineages which had invested in extra-embryonic yolk as planktotrophs.

9 Natural History and Ecology

9.1 INTRODUCTION

This chapter covers aspects of molluscan biology not covered, or only touched on, in the preceding chapters. It also deals with aspects of what is often called 'ecology' (see Chapter 1 for discussion). Although we deal with many topics below, it is important to remember that many aspects of the biology and 'ecology' of an animal are interrelated and that their expression is, to a greater or lesser extent, the result of evolutionary 'trade-offs'.

9.2 LONGEVITY AND GROWTH RATES

The length of time individuals of a particular species live is important biological information and can be determined by a variety of methods. These include examining population structure, tagging experiments, and observation of captive animals or, if it has a shell, by counting growth increments, and isotope methods such as carbon dating, although reliable age data for molluscs are generally scant in the literature.

Traditionally, growth lines on shells have been counted, but this is not reliable unless the time intervals can be validated using independent means. Such validation can be obtained using mark and recapture, or by using isotope 'sclerochronologies'.[1]

For example, variation in oxygen ($\delta^{18}O$) and carbon ($\delta^{13}C$) isotope ratios have been used to determine growth rates and age. These methods are especially applicable when there are marked seasonal temperature differences because they vary with temperature (e.g., Richardson 2001), although they can also vary with salinity (e.g., Epstein & Lowenstam 1953). This technique has been used successfully in a number of molluscs, including abalone (Gurney et al. 2005; Roussel et al. 2011), which had proved difficult to age (e.g., Day & Fleming 1992), in part because their growth lines are not necessarily annual (e.g., McShane & Smith 1992). This methodology can also be used with fossil shells (e.g., Goewert & Surge 2008).

Powell and Cummins (1985) and Heller (1990) provided overviews of molluscan longevity. The former authors compiled molluscan longevity data to look for correlations between age and long-term abiotic cycles, and how composition and structure in benthic communities might be affected. Philipp and Abele (2010) briefly reviewed bivalve longevity. Heller (1990) used age and reproductive frequency to distinguish between short-lived (one to two years, or longer with only one reproductive event; i.e., *semelparous* or the so-called R-strategists) and long-lived species, those that breed more than once and live for more than two years (*iteroparous*,

or K-strategists). Heller found that short-lived taxa in terrestrial and marine habitats (1) lacked an external shell, (2) had an external shell that was semitransparent, (3) occurred in a habitat exposed to high solar radiation and temperatures (although short-lived species in cold temperatures may counteract this), (4) had highly predictable environments, and (5) had relatively small body size (<10 mm).

A summary of the data for longevity for each major group analysed by Heller (1990) is shown graphically in Figure 9.1. Only two polyplacophoran species (see also Boyle 1977) had reliable data – one (*Cryptochiton stelleri*) living for at least 25 years and the other for four years, and these are not shown in Figure 9.1. Longevity estimates for scaphopods and aplacophorans appear to be non-existent. Except for *Nautilus*, which lives for over 20 years (Saunders 1984), most shallow-water coleoid cephalopods are short-lived (<3 years), dying after one reproductive season (e.g., Arnold & Williams-Arnold 1977), but some deep-water taxa may live longer and have multiple reproductive seasons (see Chapter 17). For smaller taxa (e.g., *Pterygioteuthis gemmata*), longevity is often measured in months and even the 'colossal squid' *Mesonychoteuthis hamiltoni*, which can reach mantle lengths of nearly 3 m and weigh as much as 494 kg, is estimated to only live between one and three years (Arkhipkin 2004).

Sclerochronology, which involves a painstaking study of microgrowth lines in shells, usually bivalves, has often provided detailed and reliable results (Jones 1983a; Richardson 1989). Using this and other methods showed that the oldest mollusc, a marine bivalve, *Arctica islandica* from Iceland, lives 450–507 years (Butler et al. 2013). This more than doubles the previously reported age (220 years – shown in Figure 9.1), making it the oldest known non-colonial animal. Another bivalve, the large, deep-burrowing 'geoduck' *Panopea generosa*, lives for up to 168 years (Orensanz et al. 2004). The 'North Pacific cockle' *Clinocardium nuttallii* has been estimated to grow between 0.66 and 0.775 mm per year and live up to 16 years (Koike et al. 2012). American freshwater mussels (Unionidae) grow rapidly for their first and second years, but after about six years only very small increases in size can be measured (Chamberlain 1931). Some unionoideans have been shown to live for several decades, and one specimen of the European fresh-water mussel *Margaritifera margaritifera* survived for 116 years (Heller 1990); a specimen of the same species from Finland was estimated to be over 160 years old and some may reach 200 years (Helama & Valovirta 2008).

In bivalves, large size does not necessarily equate with a long life. For example, the deep-sea protobranch *Tindaria callistiformis* is about 100 years old at around 8 mm, as determined by radium-228 concentrations (Turekian et al. 1975). Growth rates in bivalves vary enormously from extremely

---

[1] Sclerochronology is the study of physical and chemical variation in the shell (or other accretionary hard tissue).

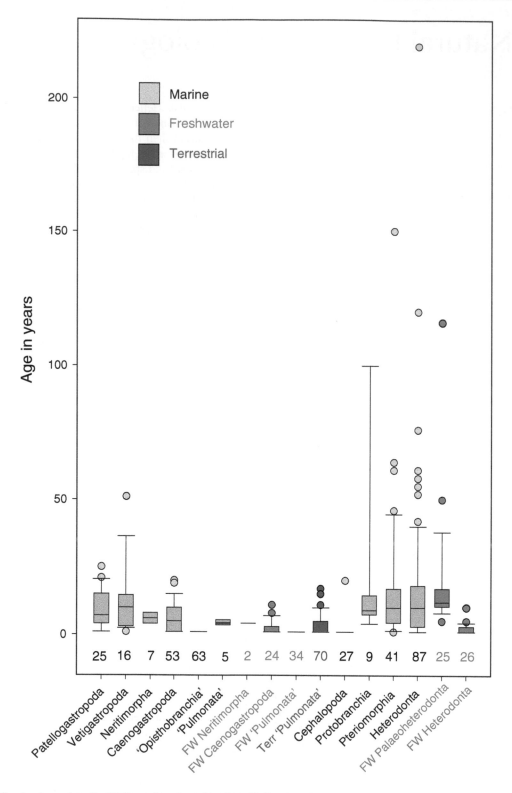

FIGURE 9.1 Graph of age data for Mollusca based on data from Heller, J., *Malacologia*, 31, 259–295, 1990. FW – fresh water, Terr – terrestrial. The numbers at the bottom of the graph are the number of species included. The box contains the middle 50% data. The upper edge of the box is the 3rd quartile, and the lower edge is the 1st quartile with the line in the box marking the median. The ends of the 'whiskers' indicate the minimum and maximum data values other than outliers, which are indicated by small circles.

slow (as in *Arctica islandica*) to very rapid. *Tridacna gigas* is the largest and fastest-growing bivalve (Bonham 1965), growing 8–12 cm per year (Beckvar 1981), which is about 16 times the shell accretion rate of oysters (Goreau et al. 1979). *T. gigas* can reach up to around 115 cm in length and attains sexual maturity at around 6–8 years, but may live for over 100 years (Rosewater 1965), although analyses of these massive shells rarely record any beyond 60 years old (Jones et al. 1986; Watanabe et al. 2004). Zooxanthellae in its tissues have been suggested to facilitate the high growth rates, although some other species of *Tridacna* which also have zooxanthellae are not as large, with *T. maxima*, for example, only reaching around 20 cm in length.

Among marine gastropods, with the probable exception of many small-sized species, most of the non-heterobranch gastropods are in the 'long-lived' category, as are the few marine 'pulmonates' for which there are data. In contrast, 'opisthobranchs' are nearly all short-lived and die once they have reproduced, although they may undertake several spawnings during their one reproductive season. They die as a result of the atrophy and breakdown of the digestive gland (known as the senescent syndrome). Their life spans vary from a few weeks to a year or more, with larger species usually living longer than small ones.

Territorial patellogastropod limpets, such as *Lottia gigantea* and the South African *Scutellastra*, are long-lived species and have been estimated to live for 25–30 years (Branch 1974; Morris et al. 1980). Espinosa et al. (2007) estimated the age of a European limpet, *Patella ferruginea*, as between nine and 35 years, depending on the environment, while another, *Patella vulgata*, lives for about 3–15 years (Fretter & Graham 1962). In comparison, non-territorial *Scutellastra*, including *S. granatina*, *S. oculus*, and *S. granularis*, live between two and eight years (Branch 1974). In the Nacellidae, *Cellana radiata* lives up to 4–5 years (Rao 1976; Ismail & Elkarmi 1999), and Creese (1981) examined growth and longevity of Australian lottiid limpets and calculated that *Notoacmea petterdi* can live for over ten years, *Patelloida latistrigata* for three years, and *Patelloida alticostata* for between five and six years.

Regarding vetigastropods, the fissurellid *Montfortula rugosa* was estimated to live for more than three years (Creese 1981), and the large Chilean *Fissurella crassa* may live for up to ten years (Bretos 1980). Mark and recapture studies suggest a maximum age of about 50 years for at least one species of *Haliotis*, *H. kamtschatkana* (Breen 1980), although other data for that species give an age of 19 years (Shepherd et al. 2000). Maximum ages recorded for other species of *Haliotis* suggest that they mostly live for less than 20 years – e.g., 14 years for *H. fulgens* (Mexico) (Shepherd & Turrubiates 1997), 18 years for *H. corrugata* (California) (Shepherd & Avalos-Borja 1997), and 13 years for *H. gigantea* (Japan) (Inoue & Oba 1980). The abalone in temperate Australia show marked differences in longevity with the maximum age of the large *H. laevigata* being about 30 years and the smaller *H. rubra* 20 years, while *H. roei*, *H. scalaris*, and *H. cyclobates* (which are smaller) again live for only ten, eight, and six years, respectively (Shepherd 2008). In Oman, the fast-growing tropical *H. mariae* lives up to only

seven years (Shepherd et al. 1995). In some species of abalone, a growth line indicates a single year, while in others there are two growth lines per year (Shepherd et al. 1995).

Some trochoideans are also rather long-lived. *Tegula funebralis* lives to around 30 years in the colder northern parts of its range (Frank 1975), while the large, tropical *Tectus niloticus* lives about 6–15 years (e.g., Smith 1987a; Lemouellic et al. 2008).

Fresh-water neritimorphs and some larger fresh-water caenogastropods are relatively long-lived. For example, a specimen of the fresh-water neritid *Neritodryas cornea* was recaptured eight years after being released (Kosuge 2007). By comparison, fresh-water hygrophilans are all short-lived, as are some (mainly small-sized) fresh-water caenogastropods. Hygrophilans may have one to three generations per year depending on the locality and the species. Although regarded as short-lived, a few hygrophilans can reproduce in a second year, with some species showing considerable variation in their reproductive patterns that may have environmental, phenotypic, or genetic causes (Russell-Hunter 1978; Brown 1985).

The large South American volutid gastropods *Odontocymbiola magellanica* and *Adelomelon brasiliana* live to about 20 years (Bigatti et al. 2007) and another volute, *Zidona dufresnei*, for 17 years (Giménez et al. 2004), but, as far as known, other large-sized marine gastropods are shorter lived. For example, the whelk *Buccinum undatum* lives for up to 12 years (Gendron 1992), the strombs *Strombus gigas* and *S. costatus* live for seven and five years respectively (Wefer & Killingley 1980), although the relatively small tropical strombid *Gibberulus gibberulus* is probably an annual (Vermeij & Zipser 1986). The temperate, small-sized turritellid *Gazameda gunnii* lives for about seven years (Carrick 1980).

9.3 DISPERSAL, RECRUITMENT, AND MIGRATION

The movement of individuals through migration or dispersal between populations facilitates gene flow, and thus inhibits genetic divergence, if the newly arrived individuals establish, interbreed, and reproduce. While free-swimming molluscs such as cephalopods, heteropods, and pteropods can readily disperse, most molluscs are benthic and rely on larvae or accidental transport for long-distance dispersal, this being especially so for sessile species. Larval dispersal is not necessarily advantageous in all circumstances (e.g., Hedgecock 1986) and in many taxa it has been superseded by 'direct' means of development (see also Chapter 8).

In non-marine habitats, the chance of dispersal is reduced by greater distances between suitable habitats. Such dispersal may be by accidental attachment to mobile animals or by wind or other weather conditions such as flooding. The chances of successful dispersal are related to other factors such as the reproductive condition of the individual, the nature of the habitat, and the habits of the species concerned. For example, the access to individuals by potential dispersive agents will be much greater in species that live on the surface of the substratum, or on algae or plants, than buried species, or those that hide under rocks, logs, and so on. Hermaphrodites that can self-fertilise or parthenogenetic species have a greater chance of successful colonisation following dispersal than dioecious species.

In marine taxa, distances between suitable settlement substrata will dictate dispersal success – for example, islands, seamounts, coastal rocky headlands, estuaries, and deep-sea hydrothermal vents. In a study of a chain of seamounts off the coast of Brazil, Leal and Bouchet (1991) showed that species diversity of 'prosobranchs' fell with distance from the coast, and that the species that were present had approximately equal numbers of those with lecithotrophic and planktotrophic larvae. This suggested that both can be effectively dispersed, probably by way of island hopping across the relatively small distances (100–250 km) between the individual seamounts. In contrast, the number of direct-developing taxa (those lacking a swimming larval stage) fell off rapidly away from the coast.

The lack of dispersal opportunities can lead to genetic differentiation and eventually speciation. Isolated relictual non-marine habitats or isolated marine habitats containing species that have little chance of dispersal often contain endemic taxa. Such is the case in remote limestone outcrops, rainforest remnants, desert springs, and oceanic islands and seamounts. In the deep sea, seeps and hot vents are often separated by considerable distances along and between deep oceanic ridges, although the animals present in these habitats appear to have a mix of reproductive strategies similar to those in shallow water. Genetic analyses of populations found at vents along oceanic ridges sometimes show patterns suggestive of isolation by distance (IBD), although about half of the studies failed to detect IBD and the lack of a statistically robust pattern in some of these studies may be due to inadequate sampling (Audzijonyte & Vrijenhoek 2010). Recent modelling of larval transport of deep-sea chiton larvae also suggests the possibility of inadequate sampling (Yearsley & Sigwart 2011). Accurate modelling provides quantitative frameworks not only in which to evaluate hypotheses, but can also predict both the existence and probable location of intermediate populations (Yearsley & Sigwart 2011). Alternatively, the lack of population structure could result from widely dispersing larvae (e.g., Lutz et al. 1986; Gage & Tyler 1991; Craddock et al. 1997; Vrijenhoek 1997; Tyler & Young 1999).

Thorson (1946) argued that pelagic lecithotrophic larvae (which are produced with intermediate-sized eggs) were a compromise between planktotrophy (with small eggs, high fecundity, high dispersal and colonisation potential, and low survival rate) and non-pelagic lecithotrophy (large eggs, low fecundity, minimal dispersal and colonisation potential, and high survivorship). While many molluscs have lecithotrophic larvae, planktotrophic larvae are found in many autobranch bivalves and many neritimorph, caenogastropod, and heterobranch gastropods (see Chapter 8).

The evolutionary importance of dispersal is thought to range from the fitness of an individual (e.g., Christiansen & Fenchel 1979) to properties associated with species selection (Jablonski 1986; Vermeij 1996a), thereby cutting across several spatial and temporal scales and potentially providing

insights into micro- to macroevolutionary processes. At the population level, gene flow and genetic differentiation between populations have been argued to be directly tied to these different reproductive strategies (e.g., Palumbi 1994), and molecular investigations have confirmed the importance of either brooded or broadcast larvae in producing either genetically homogeneous or heterogeneous populations (e.g., Johannesson 1988; Hellberg 1996; Kyle & Boulding 2000). At the macroevolutionary level, speciation and extinction rates of marine invertebrates have been argued to be correlated with dispersal mode and its relationship with geographic range (Jackson 1974; Hansen 1978; Scheltema 1978b; Jablonski & Lutz 1983; Collin 2001). Even the evolution of suspension-feeding communities in the Ordovician has been linked to macroevolutionary changes in dispersal capabilities of Paleozoic taxa (Chaffee & Lindberg 1986; Signor & Vermeij 1994), although, obviously, long-distance dispersal must have individual advantages and macroevolutionary consequences are incidental. It is also interesting to note that long distance dispersal seems to have been a general disadvantage during mass extinctions, when the plankton on which planktotrophic larvae depend was devastated.

While the highly correlated and often mosaic nature of dispersal capability combined with trophic provisioning in marine invertebrate development (see below) provides a rich array of evolutionary outcomes and patterns for interpreting life history strategies, it also makes comparisons with adequate trophic, geographic, and phylogenetic control more difficult. For example, the early contribution of Menge (1975) to understanding the evolutionary consequences of the brood or broadcast dichotomy lacked phylogenetic control by comparing two co-occurring, yet distantly related, species with contrasting plesiomorphic states. In another example, the analyses by Jablonski (1982) of evolutionary rates and modes in late Cretaceous gastropods confounded dispersal vs no dispersal with the presence or absence of feeding larvae (Russell & Lindberg 1988) (see Chapter 8).

Pelagic larvae cannot metamorphose until they become competent, even if they encounter suitable habitat. For example, dispersal is a consequence of the pelagic phase of both lecithotrophic and planktotrophic larvae, as the swimming larva must feed (e.g., yolk reserves or by suspension-feeding) and develop before it reaches a competent state, a process that may take weeks to months. Some planktotrophic gastropod larvae can survive for long periods in the plankton and are called teleplanic larvae (see Section 9.3.3).

Fresh-water neritoideans and a few thiarids are the only fresh-water gastropods with a marine larval stage that enables them to disperse between rivers and streams along the coast (e.g., Hodges & Allendorf 1998), and presumably even between streams and rivers on different island groups, given the wide distributions of some of these taxa.

Fresh-water mussels (Unionoidea) have larvae that are ectoparasites on fish (see Chapter 15) to facilitate dispersal within the same water body. Mussel species richness and fish species richness are positively correlated within at least some North American rivers (Vaughn & Taylor 2000).

9.3.1 Recruitment

Recruitment, the addition of new individuals (typically juveniles) to a population, may be by settlement of larvae or by juveniles that have emerged from an egg capsule or a brood pouch after direct development, or, much more rarely, by adding mature individual(s). For terrestrial and most fresh-water molluscs, recruitment occurs by way of juveniles following direct development, but, because many marine molluscs undergo a free-swimming larval stage, recruitment is much more haphazard and open to receive juveniles from other populations via the plankton, including those from far away. While recruitment is important in determining population densities, it is only one of a range of factors, including the survival of recruits (Caley et al. 1996). According to some studies (e.g., Powell et al. 1984), survival is usually markedly overestimated (i.e., mortality is very high).

There is considerable variation in larval settlement and recruitment in the marine environment (e.g., Underwood & Keough 2001) or the 'supply' of recruits (Lewin 1986; Underwood & Fairweather 1989), due to a wide range of factors, not least of which is the supply of larvae (e.g., Miron et al. 1995). Planktonic larvae have higher mortality than direct developers, and long-lived planktotrophic larvae more than the shorter-term lecithotrophic larvae (e.g., Thorson 1950); their availability is also influenced by the proximity of adult populations and by ocean current patterns. Factors determining settlement of larvae include the availability of suitable substrata, temperature, salinity, food availability, and chemical cues that initiate metamorphosis. Such factors may be complex – for example, grazing gastropods can markedly affect recruitment by altering the substratum (Anderson & Underwood 1997). Large-scale differences in recruitment rates and times also occur, as shown, for example, in an eight-year study along the west coast of North America by Broitman et al. (2008) of *Mytilus* spp. and by Harris et al. (1998a) in South Africa. At a local scale, the consequences of biological interactions such as predation are important (Underwood & Fairweather 1989). For example, Fairweather (1988) showed in experiments that successful recruitment and survival of barnacles depended very much on the absence of a predatory muricid whelk (*Tenguella*, as *Morula*).

Relatively few studies have followed larval development and subsequent settlement. Because of their minute size, accurately identifying and following molluscan larvae is often very difficult if not impossible. Thus, studies are based on work using genetic markers (see Section 9.3.4), assumptions made from field collections, or by observing larvae in the laboratory. In a rare example of a detailed field study, the larvae of a venerid bivalve, *Ruditapes decussatus*, were studied in a coastal lagoon in Portugal. Their development was equated with various environmental factors and showed a significant relationship between planktonic and recruitment abundances (Chícharo & Chícharo 2001).

9.3.2 Non-Larval Dispersal

Many organisms, including many molluscs, disperse by means other than by larval dispersal. This may be by way of an adaptation or event that enhances the dispersal of the adult, or the larval stage, or it may be entirely 'accidental' by way of currents, wind, or flooding. Some examples are given below.

The post-larvae of many bivalves produce a byssal thread (Sigurdsson et al. 1976) or mucous cord (Nakamura 2013), which they can use to drift in the sea, analogous to the way spiders use gossamer threads to float in the air. Byssal drifting has been observed in all major groups of bivalves, including many in which the adults lack a functional byssus.

Some small adult gastropods have large pedal mucous glands and some of these have been reported as drifting by using mucous threads (e.g., Martel & Chia 1991; Martel & Diefenbach 1993).

Swimming or drifting is a means of dispersal in some adult molluscs such as most cephalopods (see Chapter 17), heteropod and pteropod gastropods (see Chapters 19 and 20, respectively), and, for short distances, some pectinid and limid bivalves (see Chapter 15), and 'opisthobranchs' (see Chapter 20). The epitoniid snails *Janthina* and *Recluzia*, and the aeolidioid nudibranch *Glaucus* (Glaucidae), spend their life drifting upside down on the surface of the ocean. This ability to swim or drift enables dispersal, with some of the permanent members of the pelagic zone (some coleoids, all heteropods and pteropods, and a few nudibranchs) having trans-oceanic distributions.

Some scissurellids, a small-sized benthic group, can swim (Haszprunar 1988c) and sometimes form large spawning aggregations in the water column (Hickman & Porter 2007) where they can presumably be swept along by currents.

In some species, occasional 'rafting' of adults or spawn on drifting algae, either on their fronds or holdfasts, provides a potential mechanism for dispersal (e.g., Highsmith 1985). For example, some cold water bivalves such as *Gaimardia* have large distributions because they are byssally attached to drifting algae (Helmuth et al. 1994). Others, such as *Mytilus*, may also disperse attached to floating algae, wood, or similar surfaces, but have most often been distributed on the hulls of ships. Such human-assisted dispersal is discussed in Chapter 10. Naturally occurring floating objects such as pumice or, particularly, wood are a potential means of dispersal for small gastropods, oysters, and wood-boring bivalves, notably teredinids (Edmondson 1962; Turner 1966), and today, plastic is, regrettably, also a very significant means of dispersal. The March 2011 tsunami, generated by the Tōhoku earthquake in Japan, provided a spectacular example of rafting with masses of debris including man-made objects, much of it plastics, providing a dispersal opportunity for boring bivalves, clinging limpets and chitons, and byssally attached or cemented bivalves. In all, 289 species of living Japanese marine organisms (protists, invertebrates, and fish) travelled ~7000 km from Japan across the Pacific to Hawaii and the Pacific coast of North America over five years (Carlton et al. 2017). Given that floating man-made debris is so abundant, the authors convincingly argued that such dispersal could only increase in future.

Passive dispersal is important in fresh-water molluscs (Kappes & Haase 2012), and besides drifting, dispersal is often by chance and successful introductions can be rare. Dispersal of some fresh-water molluscs by larger fresh-water insects such as water beetles has been observed. Kew (1893) provided numerous examples of molluscs (sphaeriids, fresh-water limpets, and bithyniids) attached to aquatic bugs (Hemiptera), dragonflies (Odonata), and beetles (Coleoptera). In a spectacular example, Driscoll (2011) provided photographic evidence of many fresh-water limpets (20 *Ancylus* and one *Acroloxus*) attached to the elytra of a single water beetle (*Discus*) and, in another, Cotton (1934) provided a photograph of a large fresh-water mussel (Hyriidae) attached to the foot of a duck.

Dispersal of small or juvenile terrestrial molluscs by wind may be rather common (Kirchner et al. 1997), and dispersal by cyclonic winds is probably how many oceanic islands were colonised (Vagvolgyi 1975; Peake 1981; Cowie & Holland 2006). While hurricanes or cyclones are a possible means by which some fresh-water gastropods may have arrived on oceanic islands, it is more likely that they came attached to birds (e.g., Ponder 1982), although successful colonisation of remote habitats by either means of dispersal must be extremely rare.

At a more local scale, passive dispersal within a habitat can occur by wave action, currents, and sediment movements, as described, for example, for the small infaunal venerid bivalve *Gemma gemma* (Commito et al. 2013), an often abundant species on low-energy sand-mud flats and in estuarine lagoon soft bottoms of eastern North America.

Upstream movement of fresh-water snails (and other aquatic invertebrates) is a rather common phenomenon and has been recorded in many groups of gastropods (Huryn & Denny 1997). The reasons for this are partly to counter the down-stream flow and accidental dislodgement, but other reasons may include predator avoidance and searching for space or food. Because this behaviour has been reported from some members of all major groups of fresh-water gastropods, Huryn and Denny (1997) argued that it has been independently acquired in different lineages and may result from constraints imposed by shell morphology rather than a specific adaptation to life in lotic fresh-water systems. They tested this idea invoking a model in which torque is generated by the hydrodynamic drag on the shell surface of species of *Elimia* (Pleuroceridae), found in North American streams. They concluded that body size, food limitation, and hydrodynamics may all interact to influence movement patterns of snails.

While most fresh-water snail migrations in streams are relatively small scale, this is not the case with many tropical fresh-water neritids that spend part of their larval life in the marine environment and migrate upstream to breed (e.g., Schneider & Frost 1986; Schneider & Lyons 1993; Blanco & Scatena 2007). These migrations can comprise huge numbers of individuals in clusters or long lines (some >5000 m²) and have been recorded particularly from Costa Rica, Puerto Rico, Hawaii, and French Polynesia, but are probably more widespread, particularly in the Indo-Pacific. Fresh-water neritids are generally rheophilic (prefer fast-flowing water) and the migrations typically occur within five days of a flood and end after about a week (see Blanco & Scatena 2007 for references and details).

Some migrations comprise mixed species of neritids and others include more than one group of gastropods. For example, Schneider and Lyons (1993) showed that a species of *Neritina* (Neritidae), together with a small-sized species of *Cochliopina* (Cochliopidae), migrate upstream in a Costa Rican river in huge, mixed aggregations (>500,000 individuals) during the dry season. While *Neritina*, like other fresh-water neritids, has pelagic larvae swept down-stream to the sea and the young individuals migrate upstream, *Cochliopina* has direct development. The size distribution of the snails suggested that in this case the migration takes more than one year for *Neritina* and perhaps less than a year for *Cochliopina*. In this instance, a mollusc-eating fish was common down-stream, and the authors argued that the 'distance migrated may be a balance between size-dependent energetic constraints on up-stream movement and increased predation pressure down-stream' (Schneider & Lyons 1993, p. 3). It is unclear as to what role predation may play in similar migrations undertaken by *Neritina* in many tropical streams and rivers (see above), and it seems unlikely that predation is the predominant driver in this behaviour.

In an interesting divergence from the normal migration of juveniles upstream in some Pacific islands, a few species 'hitchhike' a ride on the back of subadults of another species. During their ride upstream they hardly grow, but do so after they abandon their ride (Kano 2009).

The term *diadromy* was introduced for fish that migrate back and forth between seawater and fresh water at various stages of their life (Myers 1949). Three categories of diadromy were recognised by Myers and reviewed by McDowall (1992, 2007), but only one applies to molluscs. The term *amphidromy* is used where spawn hatch in fresh water, their larvae move to the sea where they live for a relatively short period, and then return to fresh water as small juveniles that then migrate up the rivers or streams, grow and mature, and eventually spawn. This is the situation seen in many fresh-water neritids (see above) and some thiarids (e.g., Glaubrecht et al. 2009), but not in other groups of fresh-water gastropods that undergo their entire life history in fresh water.

9.3.3 Teleplanic Larvae

By delaying metamorphosis, some planktotrophic veliger larvae can spend up to a year or more in the plankton, and these larvae can travel long distances in the open sea using transoceanic surface currents. These teleplanic-type larvae are known from several marine invertebrate groups (some echinoderms, sipunculids, crustaceans, polychaetes, and cnidarians) as well as gastropods (Scheltema 1968, 1986). Their ability to travel long distances enables some to even traverse from the western to the eastern Pacific Ocean, a feat thought to be impossible for ordinary larvae (Scheltema 1988b). Teleplanic larvae are an extreme case of an R-strategy (see Section 9.2), as large quantities of larvae never reach an appropriate substratum on an island or continental shelf to metamorphose.

Within gastropods, teleplanic larvae are found in tonnoideans (e.g., Beu 2008) and some other caenogastropod families (Naticidae, Triphoridae, Ovulidae, Cypraeidae, and Muricidae), and in the 'lower heterobranch' Architectonicidae.

While most field and laboratory observations suggest teleplanic larvae typically live for about a year in the plankton, and some up to about 14 months or a little more, the larvae of a tonnoidean, *Fusitriton oregonensis*, reared in a laboratory took 4.5 years from hatching to metamorphosis, and the metamorphosed juveniles reached sexual maturity in 3.5 years (Strathmann & Strathmann 2007). This species has a range from California to northern Japan. Whether tonnoidean teleplanic larvae actually live this long in the ocean is not known, and in this amount of time, oceanic currents could conceivably transport larvae around 14,000 km (Strathmann & Strathmann 2007). Larvae of a species of *Aplysia* have been experimentally shown to live for nearly a year (Kempf 1981), but no records of naturally occurring long-lived larvae in 'opisthobranchs' are known to us.

The ability of the larvae of some gastropods to travel extraordinary distances across oceans has resulted in a few species having circum-tropical distributions (e.g., some tonnoideans such as *Monoplex parthenopeus*, *M. exaratus*, *Charonia lampas*, and *Turritriton labiosum*). Other circum-tropical tonnoideans include some species of *Bursa*, *Eudolium* and *Oocorys*, and *Malea pomum* (A. G. Beu, pers. comm., April, 2019). This said, genetic studies would be of value in these and other groups to confirm that cryptic sibling species are not involved, especially for inter-oceanic distributions with inevitable reduced gene flow.

Most bivalves do not have teleplanic larvae, but those of coastal species can be carried out to sea by currents and even dispersed many hundreds of kilometres across oceans (Scheltema 1971), although larvae of some Pinnidae and Teredinidae are considered teleplanic (Scheltema & Williams 1983). Experiments have shown that, in the presence of suitable food, bivalve planktotrophic larvae will grow rapidly, but in the open ocean, where plankton is often much less abundant, larvae may survive over much longer periods and show little growth (e.g., Millar & Scott 1967).

A larval bivalve, *Planktomya*, was originally described as a North Atlantic pelagic species (e.g., Allen & Scheltema 1972), but was later recognised as the larva of a benthic Caribbean species,[2] with other species known from southern Angola (Gofas 2000) and the eastern Atlantic (Aartsen & Engl 2001).

9.3.4 Gene Flow as a Measure of Dispersal

While it is very difficult to obtain direct measures of levels of dispersal and interbreeding success, this can be measured indirectly by estimating gene flow using analyses of allelic population data. This is done by using a range of molecular

methods including microsatellite data (e.g. Neigel 1997; Broquet & Petit 2009; Yamamichi & Innan 2012).

Restriction of gene flow is commonly present in non-marine taxa, resulting in marked genetic differentiation with numerous examples in land snails, slugs, and fresh-water molluscs. In the marine environment, such restrictions are often not as apparent, although they do occur, even at local scales (e.g., Ponder 2003).

The ability of the larval swimming stage to disperse, coupled with many larvae being able to delay metamorphosis to extend the pelagic phase and hence their dispersal potential, is often cited as the main selective advantage of this type of life cycle (see reviews by Strathmann 1980, 1985, 1993; Jablonski & Lutz 1983; Grahame & Branch 1985; Scheltema 1986; Pechenik 1990). The success of this means of dispersal is reflected in the genetic makeup of populations over varying geographic scales (Hellberg et al. 2002; Hellberg 2009).

In marked contrast to species with larval dispersal, the direct-developing (i.e., non-pelagic) lecithotrophic species such as muricid whelks, in particular *Nucella*, show marked population differentiation (Grant & Utter 1988; Goudet et al. 1994; Marko 1998; Carro et al. 2012; Colson & Hughes 2007; Pascoal et al. 2012) as do other gastropods (e.g., Parsons 1996; Hoskin 1997; Kyle & Boulding 2000; Collin 2001). These findings suggest that such larvae may be 'behaviourally adapted to avoid dispersal in the water column and thereby recruit locally' (Todd 1998, p. 2) and raise the question as to whether the larval stage is always primarily for dispersal and/or gene flow (e.g., Strathmann 1982). The relationship between larval type, gene flow, range size, and speciation is not simple and is subject to many factors (e.g., Hellberg 2009). For example, in Cretaceous gastropods, the use of larval type to explain differences in range size and diversity (Jablonski & Lutz 1980; Jablonski 1986) has been criticised (Hedgecock 1986).

Although small-scale genetic differentiation is often present, in some species that show phenotypic variation, there is little or no corresponding genetic differentiation as in the seagrass-living patellogastropod limpet, *Tectura paleacea* (Begovic & Lindberg 2011) and differing rates of evolution in different phenotypes of some oysters have also been investigated (Hedgecock & Okazaki 1984).

As might be expected, in a study contrasting two coexisting species of *Littorina*, Janson (1987) found that the direct-developing species *L. saxatilis* had greater genetic and morphological differences between populations than *L. littorea* with planktonic larvae. Planktonic larvae generally facilitate dispersal and gene flow between marine populations (Selkoe & Toonen 2011), and there are many studies of molluscs that show this. Some examples include mussels (Mytilidae) (Levinton & Suchanek 1978; Colgan 1981), *Siphonaria* (Johnson & Black 1982, 1984a, 1984b), a muricid (Liu et al. 1991), and a venerid bivalve (Borsa et al. 1994). The possession of a planktonic larva does not necessarily reduce genetic structuring. Todd (1998) compared two nudibranchs with swimming larvae, *Goniodoris nodosa* with planktotrophic larvae, and *Adalaria proxima* with shorter-term lecithotrophic larvae. While *Goniodoris* populations showed little genetic differentiation, there was considerable

---

[2] The taxonomic position of this genus was changed from Mesodesmatidae to 'Montacutidae' (i.e., Lasaeidae) (Gofas 2000).

genetic structuring in *Adalaria*, despite it having a larval swimming stage. Similarly, other studies have shown high levels of genetic structuring in some species with lecithotrophic larvae, for example, the intertidal trochid, *Austrocochlea* (Parsons 1996; Colgan & Schreiter 2011). Not surprisingly, there are also many studies on species with non-pelagic development with marked heterogeneity between populations, such as in direct-developing littorinids (e.g., Ward & Warwick 1980; Janson & Ward 1984; Janson 1987; Ward 1990) and the muricid *Nucella* as mentioned above.

The patterns discussed above are by no means universal. Some largely immobile species that lack swimming larvae can readily colonise new habitats as in the brooding *Littorina saxatilis* (e.g., Johannesson 1988) and others with long-term swimming larvae have well-differentiated local populations, as in *Siphonaria* (Johnson & Black 1984a) and *Littorina littorea* (Johannesson 1992). In cases like that of *Siphonaria* (Johnson & Black 1982, 1984b), populations are genetically differentiated, but over time the cohorts of recruits in any one location are also genetically different. Such a pattern is called *chaotic* (or *fluctuating*) *genetic patchiness* and may have multiple causes, including larvae originating from different source populations, differential selection of the larvae or newly settled juveniles, or selective reproductive success for particular genotypes (Hellberg 2009).

In summary, most genetic studies on marine species with pelagic larvae have shown that genetic connectivity exists over rather wide geographic areas, such as between islands and archipelagos in the Pacific Ocean (e.g., Palumbi 1997; Meyer 2003; Williams & Reid 2004), but species with lecithotrophic swimming larvae can show marked geographic genetic differentiation (e.g., Meyer et al. 2005). Historical factors may also be important in interpreting present-day patterns of population differentiation. In a broad-scale study of *Tridacna maxima* in the western and central Pacific, Benzie and Williams (1997) found a high level of genetic exchange within island groups, but little between them. Interestingly, they found that the patterns of differentiation did not conform to the present-day patterns of ocean currents and suggested that historical events, such as previous patterns of dispersal during times of lower sea levels, might account for their findings. Similarly, marked differentiation was found by Benzie and Ballment (1994) in western Pacific populations of the 'black-lipped pearl oyster' *Pinctada margaritifera*.

9.4 BIOLOGICAL INTERACTIONS

9.4.1 PREDATION

Molluscs are important prey and predators in both terrestrial and aquatic food chains. They interact with trophic levels ranging from those of the primary producers (e.g., chemosynthetic bacteria, angiosperms, and algae) to higher-level consumers such as vertebrates. Some reviews of predators of molluscs are those of Vermeij (1987), Boyle & Rodhouse (2005)

for cephalopods, Barker (2004b) for terrestrial molluscs, and Dillon (2000) for fresh-water molluscs.

9.4.1.1 Molluscs as Prey

Predators of molluscs include representatives of almost every co-occurring major group of organisms. In marine systems, elasmobranch and bony fish, marine mammals, birds, crustaceans (e.g., crabs, lobsters, crayfish), echinoderms (starfish), and even other molluscs are especially voracious. In fresh-water systems, they include fish and other vertebrate predators (birds, mammals, turtles), crayfish, and crabs. Terrestrial gastropods are eaten mainly by vertebrates (mammals, birds, reptiles, the latter including specialised snail-eating snakes), but invertebrate predators such as carnivorous snails and slugs, beetles, ants, planarians, and flatworms can also be significant (Barker 2004b). Some flies, including members of the Sciomyzidae, lay their eggs on intertidal, terrestrial, and some amphibious fresh-water snails, and the larvae parasitise the living snail. Because some of these flies are restricted to certain snail taxa, they have been used as biological control agents. Other arthropod predators include spiders, centipedes, and mites. These groups of predators are discussed in more detail below.

Many predators eat molluscs opportunistically and they typically consist of only a small part of their prey, while some mainly target molluscs. Such species are described as *molluscivores* and many have special adaptations for dealing with their molluscan prey. The main strategies adopted by predators of shelled molluscs when extracting the fleshy body from the shell are shell crushing, shell peeling, and body extraction, the latter method leaving the shell more or less intact, with the animal pulled from it by sucking or grasping. Shell drilling is another method employed by some predatory gastropods and some octopods. Dropping the shelled mollusc onto a hard surface is a strategy adopted by some birds (e.g., gulls, some corvids) and other birds (e.g., thrushes, pittas) crack open shells by hitting them on a hard surface. Several mammals such as sea otters and mongooses also use hard surfaces to break the shells of mollusc prey.

The predators of molluscs have, of course, changed dramatically over time as different groups evolve and others decline and become extinct (Figure 9.2).

Molluscs have adopted several strategies to reduce predation. For shelled molluscs, these have been reviewed by Vermeij (1983) and include:

- Attaining a large size – although with shells being dropped by birds to crack them open, larger sizes are often chosen.
- Increase in shell thickness including apertural strengthening modifications.
- Development of spines, ribs, knobs, varices.
- Adoption of a more globular shape is a good defence against some crushing predators found typically in warmer hard-bottom environments. For defence against peelers a tall-spired shell into

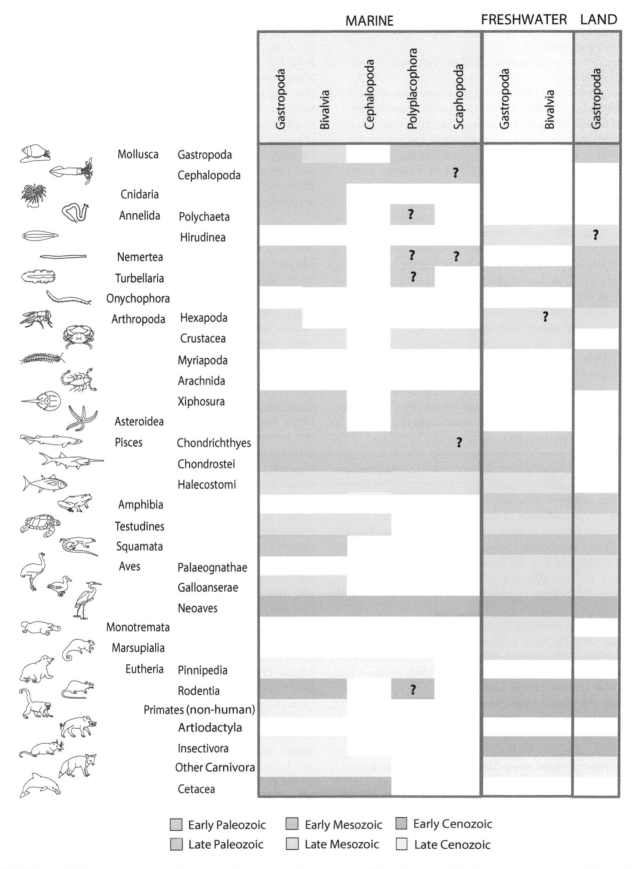

FIGURE 9.2   Molluscan predators and how they change over time (original). The colours used indicate the origination of the predator group. All groups listed are still extant.

which the animal can retract deeply is a useful adaptation and is most commonly seen in soft-bottom gastropods.

- Narrowing of the aperture and development of apertural barriers.
- Adoption of an efficient shell microstructure. Nacre is most able to withstand compression and has the greatest tensile strength. The crossed lamellar structure is next, and more resistant to abrasion and penetration (Currey 1988, 1990), while crystal size is also important, with tropical taxa tending to have larger, stronger crystal structure (Vermeij & Currey 1980).

Predatory shell-crushing through time has been reviewed by Alexander (2003). Planispiral and 'trochiform' gastropod shells were much more common in the Paleozoic than later in the fossil record or among living gastropods (Vermeij 1975; Cain 1977), as were umbilicate and loosely coiled shells (Vermeij & Currey 1980). Major changes were seen by the mid-Mesozoic with more modern shell morphologies dominating, including those with thickened outer lips, narrow, elongate apertures, and increasingly defensive shell sculptural modifications. A similar change also occurred in ammonoid shells, with Mesozoic species showing stronger sculpture than those in the Paleozoic, and Cretaceous taxa had the most strongly developed spines and tubercules (Ward 1981). Ammonites were fed on by ichthyosaurs and mosasaurs, and those reptiles, like their prey, were extinct by the end of the Cretaceous.

Crustaceans (Eumalacostraca) diversified after the Devonian, and some may have been capable of crushing shells. The Jurassic saw the beginning of some possible shell-breaking crustaceans including the stomatopods in the early Jurassic, when brachyuran crabs and spiny lobsters (palinurids) appeared. The first crabs with claws dedicated to shell-crushing (by having 'molariform teeth') appeared in the Paleocene and diversified through the early Cenozoic (Vermeij 1977a). In the Middle Triassic, cephalopods developed calcified jaws capable of crushing shelled prey.

The Devonian and early Carboniferous saw the appearance of diverse shell-crushing predatory fish such as placoderms and lungfish (Dipnoi) and later certain chondrichthyans, including rays (Batoidea) in the Jurassic. Shell-crushing teleost fish first appeared in the Eocene.

Among the reptiles, the Triassic placodonts had large flat teeth used for crushing molluscs and brachiopods, while in the Cretaceous, ichthyosaurs and mosasaurs had modified dentition for crushing molluscs. Birds arose in the Jurassic, but the diets of these early birds are not well understood. Shell-crushing mammals first appear in the Miocene.

9.4.1.1.1 Vertebrate Predators

While predatory vertebrates have substantial impacts on marine, terrestrial, and fresh-water molluscs, the number of taxa of mammals and birds known to feed on molluscs is relatively small, and there are very few that specialise in eating molluscs. For example, Allen (2004) noted that only eight of the 102 species of mammals in Britain are known to include gastropods in their diet (Corbet & Southern 1977) as do only 19 of the 648 terrestrial bird species in Australia (Blakers et al. 1984). As discussed by Allen (2004), these figures may in part reflect a lack of data, but they demonstrate the general observation that very few species of tetrapod vertebrates are molluscivores. Nevertheless, significant numbers of predatory vertebrates may at least occasionally feed on molluscs; for example, Molloy et al. (1997) recorded 176 predators of zebra mussels, 36 of which were bird species and 38 fish, and Wild and Lawson (1937) documented 20 vertebrate predators of *Planorbis*. Aspects of predation by each of the major groups are outlined below.

9.4.1.1.2 Mammals

Molluscs, notably benthic species and squid, form a substantial part of the diet of some marine mammals (Pauly et al. 1998). Some whales, notably the 'gray whale' (*Eschrichtius robustus*), feed partly on benthic invertebrates, also, to a lesser extent, the 'bowhead whale' (*Balaena mysticetus*), 'sperm whale' (*Physeter macrocephalus*), 'pigmy sperm whale' and 'dwarf sperm whales' (*Kogia* spp.), some 'beaked whales' (Ziphiidae), the 'white whale' (*Delphinapterus leucas*), and the 'narwhal' (*Monodon monoceras*). Some dolphins (Delphinidae) and porpoises (Phocoenidae), sea lions (Otariidae), most seals (Otariidae and Phocidae) (Klages 1996), and the 'La Plata dolphin' (*Pontoporia blainvillei*) also include molluscs in their diet. The 'ribbon seal' (*Phoca fasciata*) and 'bearded seal' (*Erignathus barbatus*) (both Phocidae) and the 'South American marine otter' (*Lontra felina*, Mustelidae) also consume large amounts of benthic organisms, and benthic invertebrates make up the majority of the diet of the walrus (*Odobenus rosmarus*) and the 'sea otter' (*Enhydra lutris*). While the sea otter eats a wide variety of molluscs (and other invertebrates) and either crushes the shell with its jaw or breaks it by hammering them on a hard surface, the 'walrus' mainly eats bivalves, and its mouth and palate are adapted to efficiently suck the animal out of its shell.

Squid make up a major component of the diet of various marine mammals (Lindberg & Pyenson 2006), including the beaked whales, sperm whales, pigmy and dwarf sperm whales, white whales, narwhals, most porpoises and dolphins (Clarke 1996), all sea lions, and most seals (Boyle & Rodhouse 2005). 'Elephant seals' (*Mirounga leonina*) on South Georgia Island alone have been estimated to eat at least 2.3 million tonnes of cephalopods (mainly squid) yearly (Boyd et al. 1994). Various squid make up about 95% of the food of the 'sperm whale' (Boyle & Rodhouse 2005), and that species also includes 'giant squid' (*Architeuthis dux*) and the 'colossal squid' (*Mesonychoteuthis hamiltoni*) as prey. Other squid-eating whales include 'pilot whales' (*Globicephala* spp.) and 'southern bottlenose whales' (*Hyperoodon planifrons*), and small squid are consumed in low numbers by the 'fin whale' (*Balaenoptera physalus*) and the 'sei whale' (*B. borealis*).

Over evolutionary time, the interactions between cetaceans (whales, porpoises, and dolphins) and cephalopods may qualify as a form of escalation (in the sense of Vermeij 1987), with the most spectacular example observed in the adult body sizes of the deepest diving and largest echo-locating cetacean, the 'sperm whale' (*Physeter*) and its largest prey item, the 'giant squid' (*Architeuthis*) (Kubodera & Mori 2005) (see also Chapter 17).

Although the intertidal zone is frequented by terrestrial mammals that feed on molluscs (and other intertidal animals), there are few studies documenting this. Such mammals have been termed 'maritime mammals' by Carlton and Hodder (2003), who reviewed the literature to that time. One such study in Chile (Navarrete & Castilla 1993) showed that an introduced rat (*Rattus norvegicus*) was the primary mammalian predator in the rocky shore intertidal. The rats mainly targeted 'keyhole limpets' (*Fissurella crassa*), as well as some other molluscs and crabs. In that study area, only four other species of mammals were foraging in the intertidal. In the review by Carlton and Hodder (2003), 45 species of mammals in eight orders were identified as feeding in the intertidal. Although they consumed many prey items, about two thirds of these were molluscs, crabs, and fish. The largest number of maritime mammals (21 species) occurs on the eastern North Pacific coast compared with only eight species on the eastern South Pacific coasts. Besides various rats, examples of other mammals known to forage in the intertidal and that include molluscs in their diet are mice, raccoons, otters, coyotes, foxes, jackal, mink, black bears, pigs, baboons, and long-tailed macaques.

Fresh-water bivalves and snails are sometimes consumed by various mammals including rats, mice, and other rodents. In the Americas, Asia, and Europe, predation is by otters and mink (Mustelidae) (Dillon 2000) and the insectivore desmans (Talpidae), while in Australia, the 'water rat' (*Hydromys chrysogaster*)[3] and the 'platypus' (*Ornithorhynchus anatinus*) include molluscs in their diet. Despite being primarily herbivorous, other rodents such as the 'muskrat' (*Ondatra zibethicus*) and 'water vole' (*Arvicola amphibius*) (both Cricetidae) sometimes feed on unionoid mussels (Neves & Odom 1989). A few seals and cetaceans also live in large fresh-water bodies and presumably include molluscs in their diets.

Terrestrial slugs and snails are frequently consumed by small mammals such as 'hedgehogs' (Erinaceidae), 'shrews' (Soricidae), rats and mice (Muridae), 'voles', 'dormice' (Cricetidae), and several others (see review by Allen 2004), some of which peel shells, while others crush them (Vermeij 1987). 'Mongooses' (Herpestidae) break land snail shells by flinging them against a hard surface (Vermeij 1987). In one study, it was shown that over 70% of the deaths of the introduced *Cornu aspersum* in marginal environments in California were due to small mammal predation (Potts 1975), while introduced rats (*Rattus* spp.) have been implicated in the extinction or near extinction of several native land snails in New Zealand (Daniel 1973) and on Lord Howe (Murphy 2002) and Norfolk (Neuweger et al. 2001) islands. In New Zealand, rat eradication on offshore islands has been followed by an increase in numbers of fauna previously consumed by rats (e.g., Towns 2009).

We have only located one reference to bats feeding on molluscs: 'Bats have been found feeding on *Paludina*, *Planorbis*, and *Anodonta*, but I think that they must have been pressed by hunger to do so.' (Masefield 1899, p. 162).

9.4.1.1.3 Birds

Oceanic seabirds are important predators of squid, which can be abundant at the surface at night, making them significant items in the diet of non-diving, nocturnal feeding seabirds such as albatross (e.g., Croxall & Prince 1994), shearwaters, and petrels (e.g., Warham 1990). The diurnal vertical migration patterns of many squid enable even some 'deep-water' species to be included in their diet. 'Penguins' (Spheniscidae) also include squid as a significant prey item, but 'puffins' (*Fratercula* – Alcidae) only occasionally eat molluscs. Other seabirds such as 'gannets' (*Morus*) feed primarily on fish, but also eat squid, while the related (both Sulidae) 'booby' (*Sula*) are fish eaters.

Some aquatic birds such as terns and cormorants are mainly fish eaters, but others commonly feed on molluscs. In the intertidal, molluscs are mainly preyed upon by 'seagulls' (Laridae) (e.g., Bahamondes & Castilla 1986) and 'oystercatchers' (*Haematopus* – Haematopodidae). Oystercatchers are voracious mollusc predators, feeding broadly on cockles, mussels, limpets and other gastropods, and chitons (Lindberg et al. 1987). They often access their shelled prey by hammering it against hard surfaces or compact sand (e.g., Vermeij 1987). These predation events can have substantial impacts on the communities in which they occur. Wintering flocks of oystercatchers have been estimated to sometimes contain between six and 12 thousand birds (Drinnan 1957), and each bird has been estimated to consume over 350 cockles per day from cockle beds in Great Britain (Horwood & Goss-Custard 1977). Along the two-kilometre shoreline of Traeth Melynog in Wales, Griffiths (1990) estimated that oystercatchers consumed over 18 million cockles per winter. In one case of oystercatcher predation on rocky shores, Lindberg et al. (1987) documented the consumption of over 380 limpets by six black oystercatchers (*Haematopus bachmani*) in less than three hours. Similar observations have also been made in southern Africa (Hockey & Branch 1984). A falconid, the 'chimango caracara' (*Milvago chimango*), the most abundant bird of prey in coastal areas of South America, feeds extensively on a variety of invertebrates in the intertidal zone including robbing prey from oystercatchers (García & Biondi 2011). On the Antarctic Peninsula and sub-Antarctic islands, 'sheathbills' (Chionidae) eat intertidal chitons, limpets, and other gastropods (e.g., Burger 1982).

Wading birds such as storks (Ciconiidae), herons (Ardeidae), curlews, snipes, sandpipers and woodcocks (Scolopacidae), avocets and stilts (Recurvirostridae), 'jacana' or 'Jesus birds' (Jacanidae), rails, coots and their relatives (Rallidae), spoonbills (Threskiornithidae), at least one curlew (Burhinidae), and a few plovers and lapwings (Charadriidae) feed opportunistically on a variety of animals including molluscs in both marine and fresh-water habitats. The 'limpkin' (*Aramus guarauna*, Aramidae) is similar to a rail in

---

[3] *Hydromys* is also an intertidal predator.

appearance and, like the 'snail kite' (*Rostrhamus sociabilis*, Accipitridae), specialises in feeding on ampullariid snails in tropical parts of the Americas (e.g., Michelson 1957; Dillon 2000). It breaks the shell by using its large beak as a hammer (Snyder & Snyder 1969). One recent experimental study shows the potentially complex interactions involved in wading bird predation. When the 'red knot' (*Calidris canutus*) was prevented from feeding on its preferred prey, the venerid bivalve *Dosinia*, that species flourished at the expense of a lucinid bivalve *Loripes*, and the reduced uptake of sulphide by the lucinid led to increased toxification of the sediment (van Gils et al. 2012).

Some other waterfowl such as ducks, and to a lesser extent geese and swans (Anatidae) and loons (Gaviidae), include both fresh-water and terrestrial molluscs in their diet (e.g., Michelson 1957; Dillon 2000), with the 'eider duck' (*Somateria mollissima*) feeding preferentially on marine mussels (Mytilidae) (e.g., Bustnes & Erikstad 1990). The dippers (Cinclidae), small passerines that live in fresh-water habitats, include molluscs in their diet.

Terrestrial molluscs are consumed by a variety of bird families (Allen 2004). Of these, the main ones include Accipitridae (hawks and kites, including the 'snail kite'), some falcons (Falconidae), owls (Strigidae), pigeons and doves (Columbidae), and ravens and crows (Corvidae). While crows and their relatives open snail shells by hammering with the beak, many of the passerine thrushes and their relatives (Turdidae), the starlings (Sturnidae), and pittas (Pittidae) open shells by breaking them on a hard surface (sometimes called an anvil). Other passerines that target snails include the 'great tit' (Paridae), the 'pied flycatcher' (Muscicapidae), 'pipits' (Montacillidae), and the 'po'o-uli' of Hawaii (Drepanididae). The galliforms (chickens, grouse, turkeys, and pheasants – Phasianidae) feed on a wide variety of invertebrates including molluscs. In New Zealand, two native parrots (Psittacidae) eat large native snails, and New Zealand 'kiwis' (Apterygidae) sometimes also eat snails.

9.4.1.1.4 Reptiles and Amphibians

Various reptiles are mollusc predators, and for some species, they are a conspicuous part of their diet. This is not new, for as noted above, the Mesozoic marine plesiosaurs and ichthyosaurs fed on belemnites and ammonites besides other prey.

Among modern reptiles, the Squamata is the most diverse group. The main subgroup, the Lacertilia (lizards, monitors, skinks, geckos) are mainly terrestrial and some eat land snails (reviewed by Laporta-Ferreira & Da Graca Salomão 2004). These authors note that, despite the great diversity of lizards, gastropod predation is known to occur in only a few genera in six families. These include Agamidae, genus *Agama* in the Middle East and *Gehyra* in the Gekkonidae which is widespread in the Indo-West Pacific region. Species of *Varanus* (Varanidae) sometimes include gastropods in their extensive range of food items, while the mainly northern hemisphere lizards of the Anguidae include some snail and slug-eating genera, such as *Anguis* from Europe, *Elgaria* from North

America, *Diploglossus* from Brazil, and *Ophisaurus* from the USA and northwest Africa. There are several genera of skinks (Scincidae) that eat gastropods, including the Australian 'blue tongue' *Cyclodomorphus*, the Indo-Pacific *Emoia*, the central and eastern North American, northwest African and western Asian *Eumeces*, *Oligosoma* from New Zealand, and *Tiliqua* from Indonesia, New Guinea, and Australia. Two genera of the 'whiptail lizards' (Teiidae) feed on gastropods – *Dracena* from South America are semi-aquatic and have a modified jaw enabling them to feed on ampullariid snails, while *Tupinambis* from the same region feeds on slugs. The only living rhynchocephalian, *Sphenodon*, the New Zealand 'tuatara', also commonly feeds on snails and slugs. Vermeij (1987) listed Iguanidae as feeding on terrestrial molluscs, and in northern Chile, the iguanid *Microlophus atacamensis* feeds on intertidal limpets, littorines, and mussels (Fariña et al. 2013). Other records of modern reptiles feeding on intertidal molluscs are rare. Of the remaining Squamata, the 'worm lizards' (Amphisbaenia) contain some species of Amphisbaenidae, which are known to feed on land snails, using their jaws to crush the shells.

Some marine turtles (Cheloniidae), notably the 'loggerhead' (*Caretta caretta*), 'Kemps ridley' (*Lepidochelys kempii*), and the 'flatback' (*Natator depressus*) eat molluscs besides other food items, while an east Pacific 'green turtle' (*Chelonia mydas*) was recorded as including 'sea hares' (*Aplysia*) in its diet (Seminoff et al. 2002), although these may have been accidentally consumed in this otherwise herbivorous species. Many fresh-water turtles also include molluscs in their diet and, although their jaws are well adapted to crushing shells, only a few choose molluscs as their preferred food items, the most notable being the Southeast Asian Malayan or Mekong 'snail-eating turtle' (*Malayemys subtrijuga* – Geoemydidae). In North America, species of mud and musk turtles (Kinosternidae) eat molluscs besides other benthic invertebrates (Dillon 2000), as do the 'pig-nosed turtle' (*Carettochelys insculpta* – Carettochelyidae) and 'snapping turtles' (Chelydridae). A few species of Australian 'long-necked turtles', *Chelodina* and *Emydura* (Chelidae), have been recorded as including molluscs in their diet, as do species of the South American chelid genera *Hydromedusa* and *Phrynops* and the Central and South American *Platemys* (Laporta-Ferreira & Da Graca Salomão 2004). A 'hidden-necked turtle' *Pelomedusa subrufa* (Pelomedusidae), from Africa and Madagascar, also eats gastropods along with other prey. In the mainly aquatic Emydidae, several genera include gastropods in their diet, but species of *Graptemys* preferentially feed on molluscs. In the terrestrial tortoises (Testudinidae), members of the Central and North American genus *Gopherus*, the African *Kinixys*, and the Eurasian *Testudo* feed on gastropods (Laporta-Ferreira & Da Graca Salomão 2004).

Crocodilians (alligators, caimans, and crocodiles) generally prey on a wide range of animals, including molluscs occasionally. Two fresh-water species, the 'Chinese alligator' (*Alligator sinensis*) from China and the 'broad-snouted caiman' (*Caiman latirostris*) of South America, have blunt teeth and broad snouts and specialise in eating shelled molluscs. Other species of *Caiman* also take some molluscs, as do the related *Melanosuchus niger* and *Paleosuchus* spp. in tropical South America. Two species of *Crocodylus* are also known to sometimes feed on molluscs, the South American *C. intermedius* and the Philippine *C. mindorensis* (Laporta-Ferreira & Da Graca Salomão 2004), both of which are endangered species. Thus, in general, fresh-water molluscs form only a minor part of the diet of most aquatic or semi-aquatic reptiles.

The prey of a few marine, fresh-water, and terrestrial snakes may sometimes include molluscs. For example, a marine file snake (*Acrochordus granulatus*) eats mainly small fish, but sometimes also crustaceans and snails (Heatwole 1999). In various parts of the tropics, several species of terrestrial snakes have specialised in eating snails and slugs, with the highest number (17 species) of these predators known from Brazil (Agudo-Padrón 2012, 2013). Because these snakes cannot crush the shell, snail-eating species of the colubrid subfamily Pareatinae, including the Japanese *Pareas iwasakii*, are morphologically specialised in having asymmetrical jaws to more readily extract the animal from dextral shells (Hoso et al. 2007; Lillywhite 2014) (Figure 9.3). The Javanese *Pareas carinatus* has been recorded feeding on young *Achatina fulica* (Mead 1961). Some species of snakes from four families in two major groups, Scolecophidia (blind snakes and thread snakes – Leptotyphlopidae and Typhlopidae) and Alethinophidia (Uropeltidae, Atractaspididae, and Colubridae), are known to feed on gastropods (Laporta-Ferreira & Da Graca Salomão 2004), with blind snakes eating snails and slugs besides other prey (mainly insects). Species of *Cylindrophis* are the only uropeltids known to sometimes feed on gastropods as do two genera (*Aparallactus* and *Chilorhinophis*) of the Atractaspididae, a family found mainly in Africa and the Middle East. Colubrid snakes are a diverse group found in tropical regions around the world. Several genera in different subfamilies include gastropods as part of their diet, but in the Pareatinae, *Pareas* feeds mainly on gastropods (snails and slugs) and *Aplopeltura* feeds exclusively on snails. Some genera (*Calamodontophis*, *Tomodon*, and *Sibynomorphus* commonly feed on vaginulid slugs, while *Siphlophis* and several other colubrid genera include gastropods among their food. One species of *Sibon*, *S. sanniola* from Mexico, feeds mainly on operculate land snails, while species of the Central and South American semi-arboreal genus *Dipsas* feed on snails and/or slugs and are the most specialised gastropod-feeding snakes. Other exclusively gastropod-eating colubrids include species of *Duberria* from Africa and *Contia* from the western USA, while *Storeria* from eastern North America feeds mainly on slugs. The presence of snail-eating snakes in the tropics has led to the suggestion that not only changes in snail handedness,[4] but the evolution of bizarre apertural shapes were developed to avoid snake predation (Hoso & Hori 2008).

In areas in Japan where right-hand asymmetry is present in the jaw of the snail-eating snake *Pareas iwasakii* (Figure 9.3), sinistral snails are more prevalent than in other areas. It was demonstrated that evolution of sinistrality had occurred at least six times in the genus *Satsuma* (Camaenidae), a preferred prey item, with the sinistral snails surviving predation attempts (Hoso et al. 2010). This form of selective predation may help to account for the unusual number of sinistral land snails in Southeast Asia (Vermeij 1975; Hoso et al. 2007).

Many frogs and toads (Anura) include fresh-water snails and terrestrial snails and slugs in their diet (e.g., Speight 1974), as do salamanders and various species of newts (Michelson 1957; Hanlin 1978). They generally comprise a relatively minor part of their diet (Hamilton 1948; Michelson 1957; Dillon 2000), although two Ethiopian species of *Paracassina* (Hyperoliidae) have a specialised skull enabling them to swallow whole snails without crushing them (Drewes & Roth 1981). The 'cane toad' (*Bufo marinus*) has been introduced to several areas including Florida, northern Australia, the Philippines, and Fiji where it is a major pest and eats a wide variety of animals including molluscs (e.g., Hinckley 1963; Bailey 1976; Grant 1996; Pearson et al. 2009; Shine 2010). The toxins in this toad are deadly to a range of animals and even eating their eggs can kill lymnaeid snails (Crossland & Alford 1998). Some other toads, such as *Xenopus laevis victorianus*, have also been reported feeding on molluscs (Vercammen-Grandjean 1951).

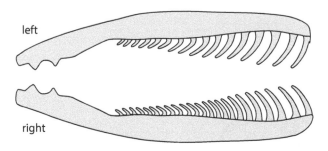

left

right

assymetrical lower mandibles of *Pareas*

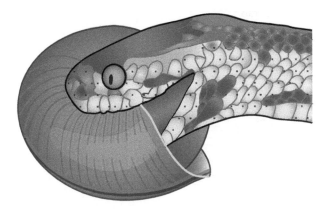

FIGURE 9.3   *Pareas iwasakii*, a snail-eating colubrid snake eating a land snail (*Satsuma*, Camaenidae) and detail of the asymmetrical lower mandibles. Upper figure drawn and modified from Hoso, M. et al., *Biol. Lett.*, 3, 169–172, 2007; lower figure drawn and modified from https://en.wikipedia.org/wiki/Iwasaki%27s_snail-eater.

---

[4] The direction in which the shell coils – i.e., sinistral or dextral (see Chapter 3).

9.4.1.1.5 Fish

Fish are very significant predators on marine and fresh-water molluscs, but it is in only a relatively small proportion of species that they are the main component of their diet. For example, in a study of the diets of 88 fish species in Western Port, Victoria, Australia, 29 included benthic molluscs in their diet, but only seven fed predominantly on molluscs (Edgar & Shaw 1995). Of those, only two primarily ate benthic molluscs. Similarly, in the West Indies, of over 212 species investigated, only 22 primarily consumed shelled invertebrates such as molluscs (Randall 1967). In the United States, Baker (1916) estimated that molluscs comprise about 6% of the food consumed by fresh-water fish.

Most fish that eat molluscs consume them along with a variety of other prey, but there are a few that prefer molluscs,

including cephalopods (Smale 1996). These molluscivorous fish include both teleosts (bony fish) and chondrichthyans (sharks and rays) and a third class, the Sarcopterygii (the lobe-finned fish), also include molluscs in their diet. Those mollusc eaters that target shelled taxa such as bivalves and snails have convergently evolved crushing mouthparts (e.g., Grubich 2003). A brief review of the fish groups that target molluscs is given below.

9.4.1.1.5.1 Class Chondrichthyes (Cartilaginous Fish)

Mollusc-eating (and crab-eating) sharks, skates, and rays have hard, crushing molar-like teeth or plates to grind up shells. Stingrays (Dasyatidae), 'cow-nosed rays' (Rhinopetridae) (Figure 9.4), and 'eagle rays' (Myliobatidae) (Kolmann et al. 2015) have crushing teeth fused into plates; some 'skates'

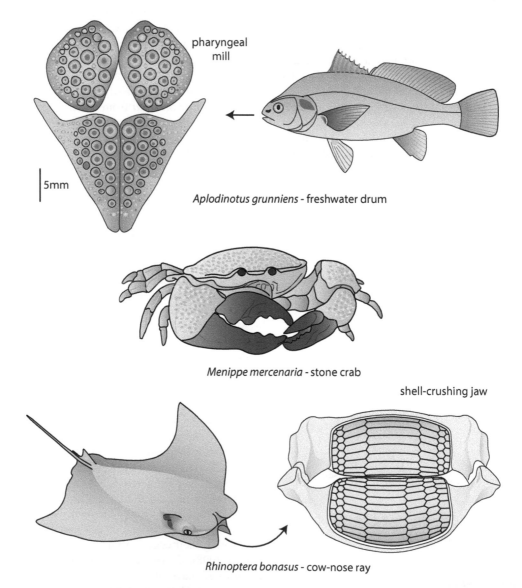

FIGURE 9.4   Three examples of shell-crushing predators; upper row the pharyngeal mill of a mollusc-eating fresh-water teleost fish (the drum, *Aplodinotus*); middle row the intertidal crab *Menippe* which has strongly asymmetrical claws that enable shell crushing and peeling; bottom row the ray *Rhinoptera* with details of its shell-crushing jaw. Top left figure redrawn and modified from French, J.R.P., *J. Freshwater Ecol.*, 12, 495–498, 1997; bottom left from Summers, A.P., *J. Morphol.*, 243, 113–126, 2000. Others original.

(Rajidae) have molariform crushing teeth, while a few swallow whole bivalves (e.g., Du Buit 1978). These bottom feeders flap their wings or jet water from the buccal cavity, moving the sediment to reveal buried prey.

While most sharks swallow their food whole, a few like the 'horn shark', 'bullhead shark', and 'Port Jackson shark' (all *Heterodontus* spp., Heterodontidae, order Heterodontiformes) are mollusc feeders and have pointed front teeth for grasping and grinding back teeth for crushing. Some 'hound sharks' (Triakidae) also include molluscs in their diet. The order Orectolobiformes contains the 'collared carpet sharks' (Parascylliidae), 'bamboo sharks' (Hemiscylliidae), and 'wobbegongs' (Orectolobidae) which also include molluscs in their diet. While all the typical sharks and rays belong to the subclass Elasmobranchii, the other small group of living cartilaginous fish, the subclass Holocephali, contains the chimaeras ('ghost sharks' and relatives, Chimaeriformes, including the Callorhyinchidae), which feed on molluscs and have three pairs of grinding plates. 'Cat sharks' (Scyliorhinidae – order Carcharhiniformes) include large gastropods (such as 'whelks') and small cephalopods amongst their prey (Eales 1949).

Many sharks consume cephalopods as a small part of their diet (e.g., Follesa et al. 2010), including some other carcharhinids (order Carcharhiniformes) such as the 'bull shark' (*Carcharhinus leucas*), the 'school shark' (*Galeorhinus galeus*, Triakidae), 'thresher shark' (*Alopias* spp., Alopiidae), 'hammerhead shark' (*Sphyrna* spp., Sphyrnidae), 'weasel shark' (Hemigaleidae), 'dogfish' (Squalidae, Centrophoridae, and Etmopteridae – order Squaliformes), and some 'carpet sharks' (order Orectolobiformes). 'Sleeper sharks' (Somniosidae – order Squaliformes) consume large octopuses and squid in addition to other prey items (e.g., Orlov 1999; Sigler et al. 2006), while squid and other coleoids make up a large part of the diet of the 'blue shark' (*Prionace glauca*, Carcharhinidae) (Nakano & Stevens 2008). 'Tiger sharks' (*Galeocerdo cuvier*) preferentially eat gastropods when juvenile, but rarely when adult (Aines et al. 2018).

9.4.1.1.5.2 Class Actinopterygii (Ray-Finned Fish) The ray-finned fish are by far the largest group of fish and are divided into several higher groups including many orders. There are two subclasses, the Chondrostei containing the most primitive members and the Neopterygii comprising the teleosts and Holostei, the former group containing the great majority of living fish.

The Chondrostei comprise ancient fish that inhabit brackish and fresh water. The group contains the sturgeons (order Acipenseriformes, Acipenseridae) which include molluscs in their diet, as does another member of this group, the paddlefish (Polyodontidae). The few other taxa are apparently not mollusc eaters.

Another ancient group, the holosts, has only a few living species, including the 'blowfin' (*Amia calva*; Amiidae), a member of a family that dates back to the Jurassic. They include molluscs in their varied diet, but the other living group, the gars (order Lepisosteiformes, Lepisosteidae) apparently do not.

The bonefish (Albulidae, order Albuliformes) are brackish-water and marine fish that swallow their prey of various benthic invertebrates (including molluscs), which they grind up using a pharyngeal mill (e.g., Colton & Alevizon 1983).

The remaining members of the class belong to the Teleostei, the true bony fish. One order, Tetraodontiformes, contains significant mollusc eaters with crushing mouthparts including the Diodontidae ('porcupine fish', 'blowfish') and Ostraciidae ('boxfish', 'trunkfish'). 'Pufferfish' and their relatives (Tetraodontidae) are mostly marine and estuarine and typically include molluscs in their diet, with some species that can be regarded as primarily molluscivores. One of the few freshwater species is *Tetradon schoutedeni* from the Congo. Their teeth are fused into upper and lower plates used for crushing the shells of molluscs and crustaceans. In 'porcupine fish' (Diodontidae), the teeth are fused into a beak-like structure.

The order Batrachoidiformes contains a single living family, Batrachoididae, which comprises the 'toadfish', some of which feed on molluscs and can crush shells in their jaws (Vermeij 1987).

The order Perciformes contains about 40% of all living bony fish and includes marine and fresh-water groups such as the 'drums' (e.g., *Aplodinotus*) (Figure 9.4) and 'croakers' (e.g., *Genyonemus* and *Plagioscion*) (Sciaenidae), the 'grunts' (Haemulidae), the 'jacks' and relatives (Carangidae), and the fresh-water 'perches' (Percidae), which all include mollusc feeders with modified crushing teeth (Vermeij 1987; Grubich 2003). The Sciaenidae includes *Aplodinotus grunniens* (Figure 9.4), a specialised fresh-water mollusc feeder from North and Central America with a crushing pharyngeal mill (Figure 9.4). The Cichlidae is an important fresh-water perciform group that contains mollusc eaters and is one of the largest families of fish, with particularly rich faunas in South America and Africa. Although cichlid diets are very varied across the family, some, including various species of *Serranochromis*, *Haplochromis*, and *Trematocranus* from Africa, specialise in mollusc-eating (e.g., McKaye et al. 1985; Slootweg 1987; Brodersen et al. 2003; Makoni et al. 2005). Cichlids possess a pharyngeal mill that can be employed in crushing prey (e.g., Smits et al. 1996). The marine Embiotocidae ('surf perches') and Labridae ('wrasses', 'gropers'), also have pharyngeal mills and some members of a related family, the Odacidae ('cales' and 'weed whitings'), feed on small molluscs. Another large marine to fresh-water group are the 'gobies' (Gobiidae), some of which include molluscs in their diet, with one molluscivorous Eurasian species (*Neogobius melanostomus*) introduced to the North America Great Lakes. Several other perciform families eat molluscs by swallowing them whole, including Mullidae ('goatfish'), Nototheniidae, and the 'sand perches' and relatives including the New Zealand 'blue cod' (Pinguipedidae) and some 'blennies' (Blennioidei; e.g., families such as Tripterygiidae and Clinidae). Some other groups of Perciformes commonly include molluscs in their diet, such as the Sparidae ('porgies', 'bream'), Lethrinidae ('emperors'), the fresh-water 'sunfish' (Centrarchidae), the latter including species of *Lepomis* (Sadzikowski & Wallace 1976; Dillon 2000), the marine

Malacanthidae ('tilefish'), Leiognathidae ('ponyfish'), and Sillaginidae (certain 'smelts' and 'whitings') (Vermeij 1987).

Scorpaeniform fish also include many mollusc eaters, including Scorpaenidae ('scorpionfish'), Cottidae ('sculpins'), Psychrolutidae ('flatheads'), Hexagrammidae ('greenlings'), and Triglidae ('gurnards') (e.g., Castriota et al. 2012).

The order Cypriniformes, a diverse group of mostly fresh-water fish, includes 'loaches' (Cobitidae), 'carp' and 'minnows', and their relatives (Cyprinidae), with varied diets, but many species often consume molluscs, as do many members of the allied 'catfish' (order Siluriformes). Both groups include several primarily molluscivorous species, with some cyprinids using a pharyngeal mill to crush mollusc shells. The 'black carp' (*Mylopharyngodon piceus*, Cyprinidae) is a molluscivore from Asia that has been introduced into the United States. This species, along with two native species of molluscivorous catfish translocated outside their normal ranges, have impacted native fresh-water bivalves in the USA (Bogan 2006). Several cyprinids are important molluscivores in fresh-water systems in Europe and North America (e.g., the 'roaches' [*Rutilus*]) and Africa, including the famous lakes Malawi and Tanganyika, where some studies have suggested evolution has resulted in changes to shell morphology related to fish predation (e.g., Vermeij & Covich 1978; Palmer 1979; Evers et al. 2011). In contrast to the shell-crushing cyprinids, loaches are 'slurpers' using their modified snout to grasp the animal in the shell. 'Carp' (*Cyprinus carpo*) usually have a generalist diet, but have been shown to selectively consume benthic molluscs in a Yugoslavian lake (Stein et al. 1975). Many 'catfish' (order Siluriformes) include molluscs in their diet, and some consume their prey whole. For example, a New Guinea estuarine species, *Cinetodus froggatti* (Ariidae), was recorded with 14 species of intact molluscs in its gut that comprised its only food, while some other 'catfish' (family Plotosidae) crush shelled molluscs in their jaws (Turner & Roberts 1978).

The order Cyprinodontiformes includes the Profundulidae, the 'killifish', which feed on fresh-water gastropods which they crush with their jaws (Vermeij 1987), while the 'sticklebacks' (order Gasterosteiformes), another fresh-water group, include some molluscs in their diet (Maitland 1965).

The order Gadiformes are mostly marine and include the true 'cod' (Gadidae), which swallow molluscs along with other animals, and the 'eelpouts' or 'wolf eels' (Zoarcidae), the latter family being mollusc crushers (Vermeij 1987). Another order, Salmoniformes, comprises only the Salmonidae, including salmon, trout, char, and their relatives, all of which spend at least part of their life in fresh water. Some salmonids, in particular various species of trout, consume a diverse range of prey items that may include molluscs (e.g., Dillon 2000), which they often swallow whole without crushing. The order Esociformes includes species of *Umbra* (Esocidae), the 'mudminnows', which are mollusc predators, and the 'pikes' (Esocidae) which are not. The fresh-water 'eels' (order Anguilliformes, Anguillidae) sometimes include molluscs in their diet (e.g., Sinha & Jones 1967; De Nie 1982). The order Lophiiformes includes the 'anglerfish', and members of one family, Ogcocephalidae ('batfish'), feed on

bottom-dwelling animals including molluscs, the shells of which they crush in a pharyngeal mill (Randall 1967). The 'squirrelfish' (order Beryciformes, Holocentridae) also include molluscs in their diet and the 'clingfish' (order Gobiesociformes, family Gobiesocidae) are significant mollusc predators in the intertidal (see below). 'Flatfish' (Pleuronectiformes) bite off bivalve siphons (a habit also practised by various other fish), as well as consuming small molluscs (e.g., Tyler 1972).

Although there are apparently no specialised cephalopod-feeding bony fish, many include some cephalopods in their diet (Boyle & Rodhouse 2005; Hunsicker et al. 2010), including 'cod' (*Gadus* spp., Gadidae), 'groupers' (Serranidae), Sparidae and some Carangidae (Perciformes), 'flatfish' (flounders, soles, halibut, which comprise several families in the order Pleuronectiformes), 'salmon' (Salmonidae), 'tuna' and 'mackerels' (Scombridae), 'marlins' (Istiophoridae), 'swordfish' (*Xiphias gladius*, Xiphiidae), 'scorpionfish' (Scorpaenidae), 'javelin fish' (Macrouridae), 'lancetfish' (Alepisauridae), 'cuskeels' (Ophidiidae), and some true eels (order Anguilliformes). Cephalopods are of particular importance for many large pelagic fish (Boyle & Rodhouse 2005; Logan et al. 2012), and they can have a substantial impact on cephalopod populations. In the Azores, swordfish and blue sharks are estimated to consume between 4500 and 10,811 tonnes of cephalopods yearly (Santos et al. 2001). It has been suggested that decreasing fish stocks may result in an increase in coleoid cephalopod populations (Lipiński et al. 1998; Chotiyaputta et al. 2002).

Fish predation on coleoids is not new as there is an early Jurassic actinopterygian fossil, *Pachycormus* (Order Pachycormiformes), with a preserved gut containing hooks from *Phragmoteuthis*, an early coleoid.

Pteropods are commonly eaten by a diverse range of zooplankton-feeding fish. For example, in a study of West Indian fish, the perciform 'mackerel scad' (*Decapterus macarellus*, Carangidae) had over 90% of the volume of its gut contents composed of pteropods, while the beloniform 'halfbeak' (*Hemiramphus balao*, Hemiramphidae) had 31.4%, the tetraodontiform 'triggerfish' (*Canthidermis sufflamen*, Balistidae) 21.2%, and the labrid 'Creole wrasse' *Clepticus parrae* 19.2% (Randall 1967). Some 'lantern fish' (Myctophidae) feed largely on pteropods (Paxton 1972; Van Noord 2012).

Fish predators of fresh-water snails have occasionally been used as biological control agents in the control of schistosome vector snails, although with questionable success (Michelson 1957; Chiotha et al. 1991; Slootweg et al. 1994) and the invasive 'quagga mussel' *Dreissena bugensis* is consumed by native mollusc-eating 'sunfish' in the Lower Colorado River (Karp & Thomas 2014). Generally, however, the impact of fish predators on fresh-water mollusc populations is mixed (see Dillon 2000); although deliberately or accidentally introduced molluscivorous fish can cause significant declines and even extinction (see Chapter 10).

In Chile and southern Africa, intertidal 'clingfish' (Gobiesociformes, Gobiesocidae) are important predators on molluscs (Paine & Palmer 1978; Lechanteur & Prochazka 2001). In Chile, a 'clingfish', *Sicyases sanguineus*, feeds on chitons, mussels, siphonariids, fissurellids, littorinids, tro-chids, patellogastropods, and the neogastropod *Concholepas*. The fish is amphibious and attaches to wave-swept vertical walls where it feeds. Molluscs, as well as barnacles and algae, are removed by the clingfish using its chisel-like teeth. It is unusual for both the breadth of this diet and its ability to feed in the high intertidal (Munoz & Ojeda 1997). Predation by *Sicyases* has effects throughout the intertidal on both the abundance and size of prey species and even parasite trans-mission, with human harvesting subtly affecting the system (Loot et al. 2005) (Figure 9.5). In southern Africa, another large 'clingfish' (*Chorisochismus dentex*) is also a vora-cious molluscan predator feeding on chitons, fissurellids, siphonariids, and urchins, but its primary prey is the diverse fauna of patellid limpets that occur on the shore, especially *Helcion* spp. (Stobbs 1980; Lechanteur & Prochazka 2001).

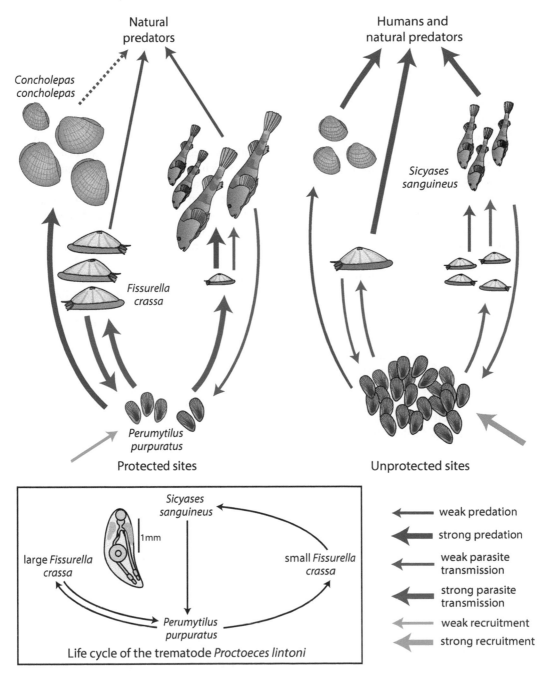

FIGURE 9.5 Effects of exclusion of human predation on parasitism in an intertidal food web in Chile. Redrawn and modified from Loot, G. et al., *Conserv. Biol.*, 19, 203–212, 2005, and a diagram of the life cycle of the trematode parasite *Proctoeces lintoni*. Image of the parasite redrawn and modified from Cortés, Y and Muñoz, G., *Rev. Biol. Mar. Oceanog.*, 43, 255–263, 2008.

Gray and Hodgson (1998) have suggested that activity patterns in *Helcion pectunculus* are shaped, in part, by clingfish predation.

9.4.1.1.5.3 Class Sarcopterygii (Lobe-Finned Fish, Including 'Lungfish')

'Lungfish' are fresh-water fish that occur in Africa, Australia, and South America. They eat a variety of aquatic animals including molluscs, the shells of which are crushed by their jaws.

9.4.1.1.6 Echinoderm Predators

Those starfish (Asteroidea) that are carnivorous are mostly opportunistic, consuming a wide range of prey including molluscs (Sloan 1980; Shivji et al. 1983; Harper 1994a). Many starfish are predators of marine molluscs and other invertebrates, including intertidal taxa. The more primitive asteroids (order Paxillosida), such as *Astropecten* (Asteropectinidae) and *Luidia* (Luidiidae) swallow their prey whole; preliminary digestion occurs in their cardiac stomach and is completed in their pyloric stomach. Members of this group regularly feed on molluscs (e.g., Sloan 1980; Chiu et al. 1983) and indigestible material such as the shells are expelled via the mouth. Other asteroids have suckered tube feet; the cardiac stomach is everted and surrounds the food item that is then partially digested externally by enzymes before being engulfed, thus enabling the consumption of larger prey. Members of the order Forcipulatida forcibly open bivalve shells using their tube feet.

Most 'sea urchins' (Echinoidea) are algal feeders, but the Cidaridae (subclass Perischoechinoidea), which includes the 'pencil urchins' (and several extinct groups), are the most primitive living echinoids and at least some are carnivorous (Mortensen 1928). They feed mainly on a variety of encrusting animals such as bryozoans, sponges, and barnacles, as well as some molluscs (e.g., McClintock 1994), which they crush with powerful teeth in their Aristotle's lantern.

Ophiuroids are usually detritus feeders, but may also prey on small invertebrates such as crustaceans and worms, and a few, such as some species of Ophiuridae (e.g., *Ophiura* and *Ophionotus*), on small molluscs (e.g., Vermeij 1987). Holothurians are also detritus feeders and are not known to feed on molluscs except perhaps by accidentally ingesting small species.

9.4.1.1.7 Xenoturbellidan and Acoelan Predators

The subphylum Xenoturbellida, an enigmatic and rare group related to the acoelan flatworms, that probably include eggs and larvae of bivalves in their diet (Bourlat et al. 2003). Members of another subphylum, the Acoelomorpha, include the Acoela which feed on minute organisms, including larval molluscs (Brusca et al. 2016a).

Xenoturbella was originally thought to be related to turbellarian flatworms, but molecular results published in the 1990s indicated a bivalve relationship. This was subsequently shown to be due to contamination from bivalves in the diet. More recent work using both morphology and molecular data places *Xenoturbella* in the deuterostom Xenacoelomorpha, which also includes the Acoelomorpha and Nemertodermatida (Bourlat et al. 2003; Bourlat et al. 2006; Telford 2008; Philippe et al. 2011).

9.4.1.1.8 Arthropod Predators

Three major groups of arthropods include species that prey upon molluscs – the Chelicerata (arachnids and pycnogonids), Crustacea, and Insecta.

9.4.1.1.9 Chelicerates

This subphylum contains the Xiphosura ('horseshoe crabs') and the Arachnida (scorpions, 'whip spiders', and other spiders).

The marine 'horseshoe crabs' (Xiphosura, Limulida, Limulidae) are found along the Atlantic and Gulf coasts of North America and in Southeast Asia. They use the bases (the gnathobases) of their five pairs of legs to grind up food (mainly shelled molluscs and worms) (Carmichael et al. 2004), which is then passed to the mouth. The anterior four pairs of legs each have a small claw.

'Harvestmen spiders' differ from most arachnids in being able to swallow chunks of food, and in addition to other food items, most species sometimes eat both terrestrial slugs and snails. Some members of Trogulidae and at least one species of Ischyropsalididae are specialised gastropod feeders (Nyffeler & Symondson 2001), and a few trogulids use their chelicerae to extract the animal from the shell of their land snail prey.

The only record of a scorpion feeding on terrestrial snails appears to be that of Lamoral (1971), where *Opisthophthalmus carinatus histrio* preyed upon the subulinid *Xerocerastus* in the Kalahari Desert.

Some mygalomorph spiders also prey on gastropods, besides their mainly arthropod prey, and there are also very sparse records of araneomorph spiders including an occasional mollusc in their diet (Nyffeler & Symondson 2001; Pollard & Jackson 2004). No spiders are known to feed exclusively on gastropods.

The subclass Acari (mites) contains some predaceous species parasitic on molluscs (e.g., Michelson 1957; Fain 2004) (see Section 9.4.4.9.6).

Pycnogonids are entirely marine and carnivorous, with few records of them feeding on molluscs. In one case, they have been observed feeding on the autotomised cerata of an aeolidioid nudibranch (Piel 1991).

9.4.1.1.10 Crustaceans

Two of the three subclasses of the class Malacostraca contain mollusc predators. The subclass Hoplocarida contains only the Stomatopoda, while the Eumalacostraca comprises the majority of crustaceans. Most of the molluscan predators are within the superorder Eucarida, order Decapoda. The most significant predators are members of the Brachyura (the true crabs), with others found among the Anomura ('hermit crabs', 'squat lobsters', and 'porcelain crabs'),

Astacidea (lobsters, crayfish), Achelata (spiny lobsters, etc.), and Caridea (snapping shrimps, etc.).

Stomatopods ('mantis shrimps') are aggressive visual predators. Members of Gonodactylidae and related families open mollusc shells by hammering them with the exceptionally robust dactyl bases of the claw-like second maxillipeds (Burrows 1969; Full et al. 1989; Patek & Caldwell 2005; Weaver et al. 2012).

The Astacidea include the marine lobsters (Nephropidae and Enoplometopidae) and the fresh-water crayfish (the Northern Hemisphere Astacidae and Cambaridae, and the Southern Hemisphere Parastacidae). These crustaceans are omnivorous, feeding on a variety of animals including molluscs, and use their powerful chelae to crush or peel the shell (Ennis 1973; Vermeij & Covich 1978; Elner & Campbell 1981; Carter & Steele 1982). North American fresh-water crayfish have been shown to consume numbers of molluscs (Dillon 2000) and one species, *Procambarus clarkii* (Cambaridae), was introduced to Kenya, where it is thought to be largely responsible for the disappearance of *Biomphalaria* and *Bulinus* from some regions (e.g., Hofkin et al. 1991).

The claw-less spiny lobsters (Achelata, Palinuridae) are also omnivorous feeders and include molluscs in their diet (e.g., Jernakoff et al. 1993). In South Africa the local 'rock lobster' *Jasus lalandii* is a dominant predator in the kelp beds on the west coast where it feeds mainly on smaller individuals of two species of mussels (Griffiths & Seiderer 1980).

Alpheidae (Decapoda, Alpheoidea) ('snapping shrimps', 'pistol shrimps') feed on a range of animals including molluscs (e.g., Beal 1983) and use their large chela to crush prey. One species has been observed lifting the operculum of a retracted whelk with the large chela and nipping off pieces of the foot with the smaller cheliped (Goldberg 1971).

A fresh-water prawn, *Macrobrachium* (Decapoda, Caridea, Palaemonidae), includes fresh-water molluscs in its diet. It has been suggested that they assist in controlling schistosome intermediate hosts (Alkalay et al. 2014).

Among the Anomura, molluscs are eaten by some hermit crabs (Paguroidea) including those that crush shells using their chelae (various species of Paguridae, Diogenidae, and Lithodidae). Terrestrial hermit crabs such as *Coenobita cavipes* and the 'coconut crab' (*Birgus latro*) (both Caenobitidae) will kill and eat land snails, including *Achatina fulica* (Mead 1961, 1979). Of the other Anomura, king crabs (Lithodoidea) include molluscs in their diet (e.g., Jewett & Feder 1982; Anisimova et al. 2005) as does the South American *Aegla longirostri* (Aegloidea, Aeglidae), one of only two fresh-water anomurans (Santos et al. 2008). While molluscs may occasionally be included in the food of some other anomuran taxa, they are not considered to be significant mollusc predators.

Crabs (Brachyura) are among the most important invertebrate predators of benthic molluscs, and their activity is thought to have driven phenotypic changes in shell morphology and behaviour in at least some gastropods (e.g., Vermeij 1977a, 1978; Bertness & Cunningham 1981; Vermeij 1987; Fisher et al. 2009) (see Section 9.4.3.1.1).

Many crabs (families Cancridae, Matutidae, Carpiliidae, Grapsidae, Ocypodidae, and Parthenopidae) crush their prey, using their strong toothed chelae that are capable of exerting considerable force (Warner & Jones 1976; Elner 1978; Brown et al. 1979; Taylor 2000). In contrast, the Calappidae (box crabs) have chelae modified to peel the shell (Ng & Tan 1984; Bellwood 1998). This shell-opening technique involves grasping the aperture edge and breaking it back along the line of the shell coil, often making both modern and fossil shells that have been attacked (successfully or, by evidence of repaired breaks, unsuccessfully) easily recognisable (see Section 9.4.3.1). These crabs usually have the right chela much larger than the left and the suggestion has been made that this asymmetry may be functionally related to opening dextral snail shells (Ng & Tan 1985). At least some Portunidae, Xanthidae, and Menippidae (Figure 9.4) also have asymmetrical chelae modified for peeling dextral snails (Bertness & Cunningham 1981; Ng & Tan 1985; Shigemiya 2003), although these families are usually considered shell crushers (e.g., Vermeij 1987). Experiments with the xanthid crab *Eriphia smithii* showed those with large right chelae were more effective in opening the spired shells of *Planaxis*, but not the more globular shells of *Nerita* (Shigemiya 2003). Some fresh-water crabs are also molluscan predators, for example, *Syntripsa* (Parathelphusidae) in lakes in Sulawesi (Schubart & Ng 2008). Some Cancridae, Portunidae, Grapsidae, Xanthidae, Menippidae, and the fresh-water Thelfusidae extract the animal from the shell (Vermeij 1978). Spider crabs (Majoidea) have more delicate chelae and are not usually shell crushers, although the Inachidae, which include the giant spider crabs, may include molluscs in their diet.

Several studies have shown how crab predation on molluscs can influence intertidal zonation. For example, studies on littorinids on the west coast of the USA suggest that members of that family could inhabit the lowermost intertidal and subtidal if it were not for predation by crabs (and presumably fish) (Yamada & Boulding 1996).

9.4.1.1.11 Insects

Despite being the most speciose group of animals, there are relatively few groups of insects that prey on molluscs, as shown below.

9.4.1.1.11.1 Hemiptera Large fresh-water bugs will eat aquatic snails (Dillon 2000; Jackson & Barrion 2004; Turner & Chislock 2007). Examples include some Belostomatidae, including *Belostoma* in the Americas, with other snail-eating belostomatids occurring elsewhere. In tropical Australia and East Asia, the largest taxa (*Lethoceras* spp.) reach about 12 cm in length.

Strangely, there are no records of assassin bugs (Reduviidae) eating snails or slugs and predation of terrestrial gastropods by bugs is rare (Jackson & Barrion 2004). An exception is a few species of 'harlequin bugs' (*Dindymus*, Pyrrhocoridae) from Australia, New Zealand, the Philippines, and Asia that eat gastropods and their eggs (Jackson & Barrion 2004).

9.4.1.1.11.2 Diptera While several fly families (mainly members of the calyptrate Muscoidea) feed on dead molluscs, a few opportunistically feed on living molluscs (notably some horseflies – Tabanidae). Some flesh flies, including two species of *Sarcophaga* (Sarcophagidae), live on *Littoraria* in mangrove habitats in Queensland, Australia (McKillup et al. 2000), while others attack terrestrial snails and slugs (e.g., Mead 1979). A few species of *Melinda* (Calliphoridae) have been reared from larvae feeding on snails (Kurahashi 1970).

Many members of the acalyptrate marsh flies, Sciomyzidae, have a close predaceous relationship with living molluscs. The biology of this group was reviewed by Knutson and Vala (2011). Some feed on dead or moribund gastropods, but the larvae of most feed on a wide range of fresh-water bivalves and gastropods, many terrestrial slugs and snails (Barker et al. 2004) (Figure 9.6), and some intertidal gastropods such as littorinids (Knutson & Vala 2011). The relationship of these flies to molluscs is variously considered predatory or parasitic. They occur in most parts of the world, although less diverse in Australasia

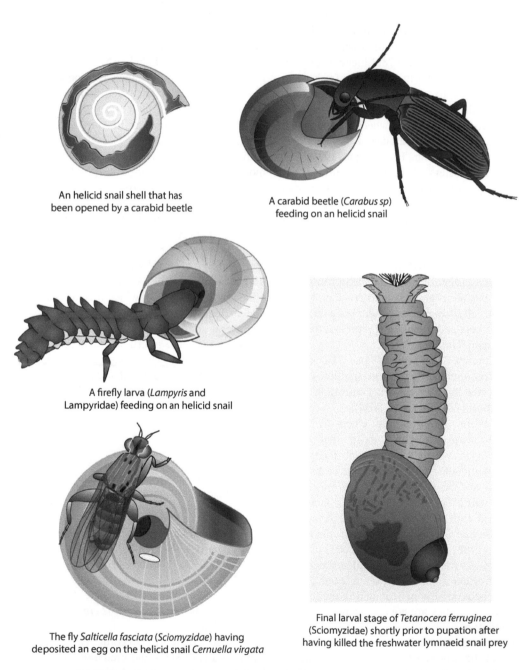

An helicid snail shell that has been opened by a carabid beetle

A carabid beetle (*Carabus sp*) feeding on an helicid snail

A firefly larva (*Lampyris* and Lampyridae) feeding on an helicid snail

The fly *Salticella fasciata* (Sciomyzidae) having deposited an egg on the helicid snail *Cernuella virgata*

Final larval stage of *Tetanocera ferruginea* (Sciomyzidae) shortly prior to pupation after having killed the freshwater lymnaeid snail prey

FIGURE 9.6 Some examples of insect predators of molluscs. Drawn and modified from the following sources: upper left drawn from http://www.molluscs.at/gastropoda/terrestrial.html?/gastropoda/terrestrial/enemies.html, middle left from https://zandvleitrust.org.za/archive/art-ZIMP%20biotic%20-%20insects%20-%20beetles.html, lower right redrawn and modified from Knutson, L.V. and Vala, J.C. *Biology of Snail-killing Sciomyzidae Flies*, Cambridge University Press, Cambridge, UK, 2011; others original.

and Oceania, in damp areas around swamps or marshes, and less so in drier areas. Some species lay their eggs on living molluscs and others lay eggs on vegetation, and the young larvae then find a mollusc host. Some that lay eggs in riparian habitats produce larvae that consume fresh-water molluscs.

Sciomyzids can be divided into five biological groups (Berg & Knutson 1978): (1) predators of hygrophilan fresh-water snails, (2) predators of fresh-water sphaeriid bivalves, (3) terrestrial parasitoids, (4) slug predators, and (5) eaters of gastropod eggs and embryos.

Because of the host specificity of some sciomyzids, some have been suggested for use in the biological control of exotic pest snails (e.g., Coupland & Baker 1995).

The larvae of species that feed on aquatic snails have modifications to enable them to breathe, such as having hydrofuge hairs posteriorly that keep their posterior spiracles at the water surface. These larvae kill and consume their prey quickly and move from snail to snail, often consuming several to many individuals before pupation. Some sciomyzids (species of *Ilione* and *Renocera*) are specialised feeders of sphaeriid bivalves, having elongated processes on their bodies enabling them to respire in water. These larvae kill their host and then move to several other individuals before pupation. Those that predate land snails lay their eggs on vegetation or directly on the shell (Figure 9.6). The larvae live in a single land snail that is eventually killed as the larva consumes the body before pupating, usually within the shell. Those that feed on slugs usually kill their host and move to other individuals before pupation. Some slug specialists, notably *Tetranocera elata*, produce a neurotoxin in the salivary glands that immobilises the prey (Rozkošný 1984). Species of *Antichaeta* feed on egg masses of lymnaeids and succineids (Knutson 1966).

Some sciomyzids attack marine molluscs, including the 'flesh fly' *Liosarcophaga choudhuri*, a parasitoid on the mangrove littorinid *Littoraria* (*Palustorina*) *melanostoma* in mangrove forests in India (Sinha & Nandi 2010). There are few other reports of insect parasitoids of intertidal gastropods. These include two species of *Sarcophaga* that 'parasitise' *Littoraria filosa* in Australian mangrove forests (McKillup et al. 2000), a report of an unidentified dipteran larva that 'parasitises' the intertidal patellogastropod limpet *Scurria* in Chile (Uribe & Otaïza 1996; Camus & Barahona 2002), and various intertidal gastropods that are parasitised by some species of *Dictya* and one of *Hoplodictya* (Knutson & Vala 2011).

Some Chironomidae live in the mantle cavities of fresh-water snails or fresh-water mussels (e.g., Michelson 1957), some burrowing into the viscera; a few others have been recorded feeding on egg masses of aquatic snails (Knutson & Vala 2011).

Members of the genus *Wandolleckia* (Phoridae) live ectoparasitically inside the mantle cavity of *Achatina fulica* in Africa (Mead 1961). The larvae of a few other phorids feed on snails and the eggs of both slugs and snails (Knutson & Vala 2011).

9.4.1.1.11.3 Coleoptera Some beetles are significant predators of terrestrial snails and slugs (Symondson 2004; Harper et al. 2005), notably Carabidae which feed by inserting their head into the aperture of a snail shell, others nip at the shell with their mandibles breaking the shell away until they reach the retracted animal (e.g., Brandmayr & Brandmayr 1986; Symondson 2004) (Figure 9.6). Some carabids have been used, with little or no success, as biological control agents against *Achatina fulica* and certain other snails (Mead 1979). Some members of a few other beetle families also feed on gastropods, such as some Staphylinidae, Silphidae, Lampyridae, Elateridae, and, in fresh water, the diving beetles (Dytiscidae), although all these beetles, including most carabids, are polyphagous and thus do not feed exclusively on gastropods (Symondson 2004).

The bioluminescent so-called 'lightning bugs' or fireflies (Lampyridae) are beetles whose glow-worm larvae feed on terrestrial and fresh-water gastropods (both snails and slugs) (Michelson 1957; Dillon 2000; Symondson 2004) (Figure 9.6). One large Indian species (*Lamprophorus tenebrosus*) has been unsuccessfully used in attempts at biological control of *Achatina fulica* in Mauritius, Indonesia, Guam, and Hawaii (Mead 1979).

The larvae of the 'click beetle' genus *Drilus* (Elateridae – previously Drilidae) bore holes in the shells of various land snails (Örstan 1999; Symondson 2004), including members of the clausiliid genus *Albinaria* (Baalbergen et al. 2014). The larvae of the 'silphid beetle', *Ablattaria arenaria*, also feed on snails (Sekeroglu & Colkesen 1989).

Some diving beetles (Dytiscidae), a very diverse family, sometimes consume fresh-water molluscs in both their larval and adult stages (Michelson 1957; Dillon 2000; Klecka & Boukal 2012).

9.4.1.1.11.4 Other Insects Certain ants will take terrestrial gastropods occasionally (e.g., Wild & Lawson 1937; Mead 1979), as do various 'fire ants' including the 'little fire ant' (*Wasmannia auropunctata*) and the 'red fire ant' (*Solenopsis invicta*) (e.g., Wojcik et al. 2001). Both species have become established in several countries including the USA and Australia (Ascunce et al. 2011). *Solenopsis invicta* has also been observed attacking the eggs of *Pomacea* exposed during dry conditions (Stevens et al. 1999) and is implicated in the extinction of a subspecies of the 'tree snail' *Orthalicus reses* in Florida (Forys et al. 2001). Other tropical invasive ant species, such as the 'crazy ant' (*Anoplolepis gracilipes*) and *Wasmannia auropunctata* have become widely established in the tropics, including on some oceanic islands, but their impact on native snail faunas has not yet been investigated.

Dragonfly nymphs (Odonata) are important predators of aquatic snails, especially when fish and crayfish are absent (Turner & Chislock 2007). One species of *Gomphus* (Gomphidae) has been observed inside the mantle cavities of gravid unionid mussels where it was eating the gills (Levine et al. 2009).

In an unusual departure from other lepidopterans, the caterpillar of the moth *Hyposmocoma molluscivora* and some congeneric species, all from Hawaii, trap small land snails with their silk and then consume them (Rubinoff & Haines 2005).

9.4.1.1.12 Chilopoda (Centipedes) and Myriapoda (Millipedes)

Centipedes are carnivorous, and there are a few reports of them attacking terrestrial and even intertidal molluscan prey (e.g., Barker 2004a), but none are known to specialise in eating molluscs.

A few species of millipedes stray from their normally detritivorous habits, with some carrion-feeding species sometimes eating snails (Barker 2004a).

9.4.1.1.13 Annelid Predators

These segmented worms contain two groups, polychaetes and leeches, that sometimes prey on molluscs.

9.4.1.1.14 Polychaetes

Various polychaetes eat small molluscs, including some members of the orders Eunicida (families Eunicidae, Lumbrineridae) and Phyllodocida (families Nephtyidae, Polynoidae) (Fauchald & Jumars 1979; Brönmark & Malmqvist 1986; Vermeij 1987; Aditya & Raut 2005).

9.4.1.1.15 Leeches

Several species of the family Glossiphoniidae (mainly species of *Glossiphonia*) feed on the haemolymph of freshwater snails (Michelson 1957; Brönmark 1992; Dillon 2000; Aditya & Raut 2005), and they have been suggested for use in biological control programs (see Aditya & Raut 2005).

It has been considered that terrestrial leeches do not prey on snails and slugs (Barker 2004b), but some recent observations indicate that some do so occasionally (Shikov 2011) and https://en.wikipedia.org/wiki/Leech.

9.4.1.1.16 Nemertean Predators

Nemerteans are mostly predatory and inhabit marine intertidal and subtidal zones. They feed on molluscs and other invertebrates (Vermeij 1978; McDermott & Roe 1985).

A species of terrestrial nemertean from New Guinea, *Platydemus manokwari*, is a land snail predator (e.g., Sugiura 2010) and has been introduced into various Pacific islands, primarily in an attempt to control *Achatina fulica*, and also Queensland, Australia. It has failed in controlling *Achatina*, but has contributed instead to the decline and extinction of native land snails.

9.4.1.1.17 Platyhelminth Predators

Certain marine polyclad 'worms' of the family Leptoplanidae eat small snails, limpets, and nudibranchs (e.g., Phillips & Chiarappa 1980), and some members of the family Panoceridae, one of which has been shown to utilise tetrodotoxin to kill its prey (Ritson-Williams et al. 2006), also eat gastropod molluscs. Several are predators of bivalves such as oysters, for example, the so-called 'oyster leech', *Stylochus ellipticus* (family Stylochidae) and other species of *Stylochus* have been documented feeding on various bivalves (e.g., Newman et al. 1993; Kraeuter 2001).

A few terrestrial planarians (Tricladida) such as *Geoplana*, *Freemania*, and *Dugesia* are voracious predators of terrestrial gastropods (Vermeij 1978; Mead 1979; Winsor et al. 2004).

9.4.1.1.18 Cnidarian Predators

Sea anemones such as some Actinidae and Stichodactylidae will sometimes feed on gastropods, including various snails, 'sea hares' and nudibranchs (Vermeij 1987; Meij & Reijnen 2012).

Coral polyps capture and consume molluscan larvae as part of the zooplankton component of their diet.

The 'Portuguese man o' war' *Physalia* feeds mainly on fish and their larvae, but can also capture and consume small cephalopods (Purcell 1984).

9.4.1.1.19 Ctenophore Predators

Ctenophores feed on zooplankton and are significant predators of molluscan larvae.

9.4.1.1.20 Molluscs as Predators of Other Molluscs

An overview of molluscan feeding is given in Chapter 5. Here, we briefly look at the only two classes, gastropods and cephalopods, which are predators of other molluscs.

9.4.1.1.20.1 Gastropoda

Some omnivorous species feed unselectively on newly settled young and juveniles as they browse; for example, *Pomacea canaliculata* has been recorded as eating both juveniles and adults of some fresh-water snails (Kwong et al. 2009). Here, we focus on those species that selectively feed on molluscs. Neogastropod 'whelks', including many murcids and volutids, and some buccinids, melongenids, fasciolariids, olivids, and costellariids feed on other molluscs (Vermeij 1978). Some soft-bottom taxa (volutes, olivids, naticids) often wrap their prey in their large foot while feeding. Hole drilling through shelled prey, characteristic of Naticidae and Murcidae, has also been observed in one genus of Marginellidae (*Austroginella*) (Ponder & Taylor 1992) and reported in *Cominella* (Buccinidae) (Morton 2006a) (see also Chapter 5).

Some Conidae feed on shelled gastropods (e.g., Kohn 1959; Kohn & Nybakken 1975) by swallowing them whole, and there is a report of them feeding on nudibranchs (Valdés et al. 2013).

Several species of ranellids and related taxa (Tonnoidea; Caenogastropoda) eat molluscs, while among the heterobranch gastropods, cephalaspideans including Philinidae, Retusidae, and Aglajidae feed on small molluscs, the shells of which are crushed in their gizzards. Some nudibranchs also prey on molluscs, including the cladobranch nudibranch *Dirona albolineata* (Dironidae), which consumes small shelled gastropods as part of its rather varied diet, crushing the shells with its jaws (Robilliard 1970). Some nudibranchs feed on other sea slugs, including some species of *Gymnodoris* that feed on certain other nudibranchs (Kay & Young 1969; Hughes 1985a). Several species of the aeolidioid genus *Phidiana* eat other aeolidioids and dendronotids (Lance 1962; Thompson 1976; Goddard et al. 2011), and species of the polycerid genus *Roboastra* feed on other nudibranchs and sacoglossans (Robilliard 1970; Megina & Cervera 2003). The planktonic gymnosomes are specialist feeders on shelled pteropods (thecosomes), a prey item sometimes also included in heteropod diets (Lalli & Gilmer 1989).

The terrestrial stylommatophorans include voracious predators in the families Streptaxidae, Haplotrematidae, Rhytidae, and Oleacinidae, all of which feed by inserting their head into the aperture of the prey snail (Mead 1961, 1979; Barker 2004b). The Oleacinidae includes the infamous *Euglandina rosea*, a species largely responsible for the extinction of many snails on islands in the Pacific and Indian Oceans where they were imported as a 'biological control' agent (see Chapter 10). The Central and South American rhytidoid Scolodontidae (=Systrophiidae) are assumed to be carnivorous, but this has not been confirmed (Barker 2004b). In contrast, most freshwater snails are herbivores or detritivores, although some can act as scavengers and predation on other molluscs is only rarely observed (Michelson 1957).

9.4.1.1.20.2 Cephalopoda

Most juvenile cephalopods feed on small crustaceans and, when larger, take fish and other cephalopods, with cannibalism relatively common (Boyle & Rodhouse 2005). Planktonic coleoids (mainly squid) undertake diel migration following prey and/or avoiding predation. The mainly benthic octopods feed on gastropods and bivalves by forcing open their shells or, sometimes, drilling through them using the salivary papilla (see Chapters 5 and 17). An octopus (*Graneledone*) on hydrothermal vents crushes the shells of gastropods and bivalves with its heavy beak before ingesting them (Voight 2000). Cuttlefish may also include shelled molluscs in their diet (e.g., Boyle & Rodhouse 2005).

9.4.1.2 Molluscs in Food Webs

Molluscs can occupy any trophic level within a food web except for that of primary producer. This diversity in food web location reflects their wide range of feeding strategies (see Chapter 5). The molluscs closest to primary producers are those that form symbiotic relationships with bacteria, protists, and chloroplasts, such as deep-sea chemosynthetic mussels (*Bathymodiolus*), tropical giant clams (*Tridacna*), and sacoglossan gastropods (e.g., *Elysia*), respectively. Although these molluscs are not autotrophs, they do provide their associated primary producers with some of the components to produce food through either photosynthesis or chemosynthesis, which in turn is consumed by the mollusc. These food chains are therefore short (Figure 9.7).

In a simple food chain, molluscs often function as primary consumers by filtering phytoplankton from the water column (bivalves), grazing on algal or bacterial mats on various substrata (chitons, aquatic and terrestrial gastropods), or feeding directly on plants (aquatic and terrestrial gastropods). These primary consumers are then consumed by secondary consumers such as starfish, other molluscs, fish, birds, and mammals in intertidal, shallow fresh water, and terrestrial habitats (Figure 9.7). In terrestrial communities, gastropods become secondary consumers when they feed directly on fungi and bacteria, which serve as primary consumers as they decompose

plant material and recycle it (Speiser 2001). In some food chains, carnivorous molluscs can reside at or near the top, although they rarely function as apex predators because they may be prey items for larger predators. Examples include venomous tropical *Conus* species and octopuses (Figure 9.7). Cephalopods, such as squid, often occupy high trophic levels, but they too are still not apex predators because they are consumed by apex predators including sea birds (e.g., albatross), odontocete whales (e.g., the 'sperm whale', *Physeter*), and fish (e.g., 'tuna' – *Thunnus*) (Figure 9.7).

One of the more bizarre trophic relationships involves a gastropod mollusc and a spiny lobster in Africa. In one subtidal community on an island off the west coast of South Africa, lobsters dominate the community and prey on mussels. A 'whelk', *Burnupena papyracea*, survives in low numbers among the lobsters because its shell is covered by a protective bryozoan (Barkai & McQuaid 1988). On another island, lobsters are rare and three species of *Burnupena* occur in high densities. An attempt to introduce spiny lobsters to the second island failed when the introduced lobsters were attacked and eaten by the whelks. Within a week of the introduction, none of the 1000 lobsters released at the island were found alive (Barkai & McQuaid 1988). In this extraordinary example, the typical trophic relationship between spiny lobsters and gastropods appears to have been reversed, and both food chains appear to be stable and driven by the numbers of whelks present on each island.

From the preceding examples and earlier sections in this chapter, it is evident that molluscs are important both as prey items and predators in both aquatic and terrestrial food chains. While they are often eaten by other invertebrates and vertebrates, they also feed on a wide variety of animal prey. Also, greater species diversity generally results in greater numbers of predators (Paine 1966, 1980).

While food chains and trophic levels are important concepts in ecology, evolutionary theory predicts that predators and prey will be subjected to an evolutionary 'arms race' in which, through natural selection, gradually increasing defences in the prey are met with an increased ability in the predators to meet those defences (Vermeij 1987). For example, an increase in shell thickness by a prey species may be met by an increased ability by the predator to crush shells. Similarly, an increase in the size of the prey may lead to an increase in the size of the predator. In such cases, these modifications are typically limited by other constraints.

9.4.1.2.1 Cephalopods

Coleoid cephalopods are an important component of marine food webs. They are active predators that feed on shrimps, crabs, fish, other cephalopods, and with octopods, on other molluscs. This is, in part, because the short life span of coleoids means they grow rapidly with high food conversion rates, and they provide a critical link between lower and higher trophic levels in ocean food webs (Boucher-Rodoni et al. 1987; O'Dor & Wells 1987). For example, 90% of all odontocete (toothed) whales feed on squid (Clarke 1996). It has been

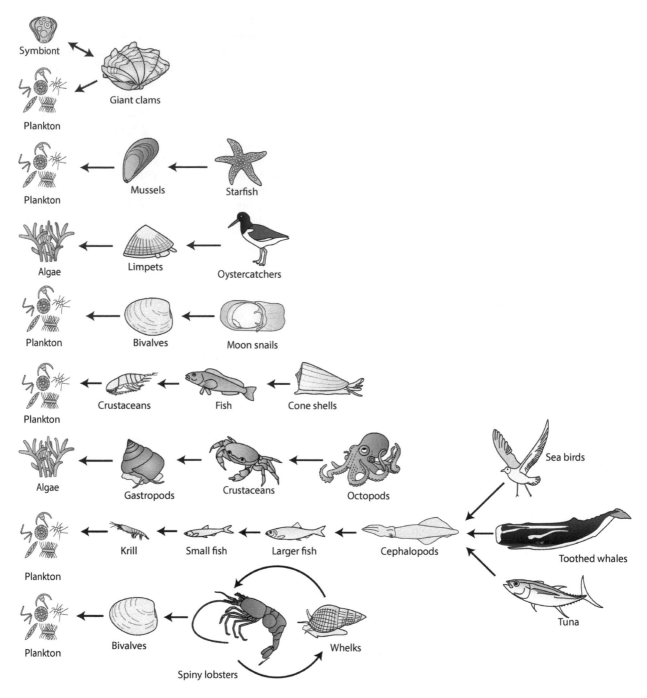

FIGURE 9.7 Some examples of simple trophic interactions involving marine molluscs. Original.

estimated that between 12.5 and 24 million tonnes of squid are consumed annually in the Antarctic by whales, pinnipeds, and seabirds, and 'sperm whales' (*Physeter microcephalus*) alone may consume as much as 320 million tonnes globally (Santos et al. 2001). Squid predation by 'sperm whales', 'swordfish' (*Xiphias gladius*), and the 'blue shark' (*Prionace glauca*) is summarised in Figure 9.8.

As cetacean populations recover from past over-hunting, their impact on cephalopod taxa can only be predicted to increase. Add to this the increasing human predation on cephalopod stocks, rising from approximately 3.7 million tonnes

in 2009 to over 4 million tonnes in 2012, although this represented only about 3% by value of the world fish trade (FAO 2014) (see also Chapter 10). The potential for unforeseen trophic repercussions or 'ecological surprises' (sensu Doak et al. 2008) may well be in the future for this critical taxon in marine food webs.

9.4.2 Trophic Status

Over the last 20 years, there has been increasing interest and sophisticated use of stable isotope analysis to trace pathways

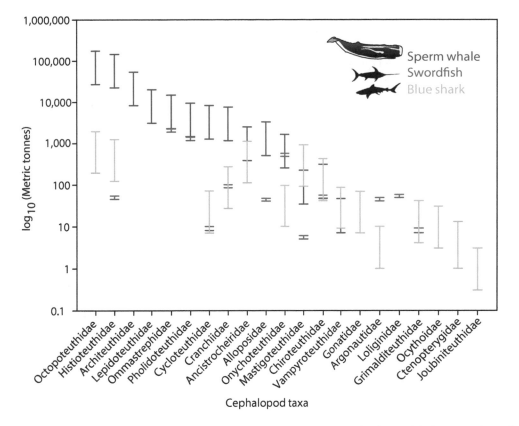

FIGURE 9.8 Maximum and minimum estimates of cephalopod taxa annually consumed by three of their top predators in the vicinity of the Azores. Data from Santos, M.B. et al., *Fish Res.*, 52, 121–139, 2001.

of organic matter (primarily carbon and nitrogen) in food webs, and to determine the contributions of various food items to the diet of an organism (Fry 2006). Questions like this have typically been addressed with approaches such as stomach content analysis, which provides information only on recently ingested food items, whereas stable isotope analysis provides information on the long-term diets of organisms enabling estimation of their positions in food webs. Taxa investigated include bivalves (Riera & Richard 1996; Sauriau & Kang 2000; Christian et al. 2004; Kasai et al. 2004; Dubois et al. 2007), gastropods (Sitnikova et al. 2012a; Fedosov et al. 2014; Raw et al. 2016), and cephalopods (Cherel & Hobson 2005; Cherel et al. 2009; Navarro et al. 2013; Guerreiro et al. 2015). Sampled tissues include muscle, mantle, digestive gland, foot, cephalopod beaks, as well as whole-body samples; the amino acids of an individual can also be used (Ng et al. 2007). When used in conjunction with environmental sampling, stable isotope analysis can be used to evaluate potential food sources for molluscs in various habitats, as, for example, in woodland streams in midwestern North America (Raikow & Hamilton 2001). Results of such studies often show that terrestrial input of particulate matter is more important in fresh-water and estuarine bivalve diets than aquatic-based nutrition (Kasai & Nakata 2005; Antonio et al. 2010). Stable isotope studies have also been used to interpret trophic relationships of molluscs in habitats such as the Kairei hydrothermal vent field of the central Indian Ridge (Dover 2002) or within kelp forest communities (Fredriksen 2003). Stable isotope analysis can also identify the trophic diversity of an individual, as has been done in fresh water and estuarine systems for lymnaeids and hydrobiids (Doi et al. 2010; Riera 2010).

9.4.3 Defence Mechanisms

Molluscs, like all organisms, are constantly surrounded by threats from competitors, predators, parasites, diseases, and the physical environment. Here, we examine some of the common strategies that molluscs adopt to reduce the impact of threats. Given the huge variety of form and habits within molluscs, these are highly varied. Although some of the main mechanisms are outlined here, more specific details are provided, where necessary, for particular taxon groups in the chapters dealing with those groups in Volume 2.

Adaptation by way of natural selection is undoubtedly the usual process involved in the evolution of defensive traits. Adaptations must be a balance between coping with the physical environment and biological interactions, particularly predation. Our knowledge of such matters has largely been gleaned from experiments and observations carried out on intertidal rocky shores. There, predation by whelks, starfish, crabs, oystercatchers, and fish is often more important than wave action in intertidal mollusc mortality (e.g., Dayton 1971; Paine & Palmer 1978; Robles 1987; Hahn & Denny 1989).

Adaptations can occur relatively quickly when significant changes in the environment greatly increase selection pressure, as with interactions with new parasites or predators, although some responses can be induced quickly in individuals because of their existing genetic makeup. Such phenotypic plasticity can invoke rapid responses in sometimes unexpected ways, as for example, sometimes where introduced predators interact with native species. An example of this latter type of interaction was the introduction of the 'green crab' (*Carcinus maenas*) to the Atlantic coast of North America (see next Section).

9.4.3.1 The Shell in Defence

For many molluscs, the shell is the main defence against predators. Modifications to resist predation typically include overall thickening and strengthening by way of ridges, spines, or varices, and in gastropods, by apertural barriers and thickened apertural lips. In chitons, the scaly, leathery, or spiky girdle acts with the shell to provide protection, and when dislodged they curl up, protecting the foot and mantle. In aplacophorans, the body spicules presumably provide some protection from small predators.

Spines in shell ornament (e.g., in some gastropods, as in many muricids and bivalves such as spondylids) are relatively uncommon, possibly because developing spines is physiologically expensive and, for crawling gastropods, can result in a cumbersome shell that is also susceptible to being dislodged by water currents or waves. Spines may be useful in deterring predators as in some spondylids and oysters, but they also provide a complex scaffold to cover with sponges and other encrusting organisms, which provide excellent camouflage and may increase survivorship (Pitcher & Butler 1987).

Fresh-water bivalves and gastropods typically lack many of the shell modifications to deter predators seen in marine counterparts (Vermeij & Covich 1978; Vermeij & Dudley 1985, 2000). While their generally thinner shells may result from the lower concentrations of calcium carbonate in fresh water, the absence of many predators found in marine habitats may also be a factor (Vermeij & Dudley 2000).

Inner organic layers (conchiolin sheets) are sometimes formed in the shell of corbulids, unionoids, and oysters (Harper 1994b) and in some small to minute caenogastropod snails (Eatoniellidae, Cingulopsidae, Anabathridae, and Barleeidae) (Ponder 1988a). These sheets may provide a barrier against predatory drillers that use shell chelating agents (see Chapter 5) or, in the case of unionoids, protection against the corrosive action of acidic fresh water.

The frequency of shell repair has been considered a measure of the success of shell protection (Vermeij et al. 1980). Crabs peel back the outer lip of the aperture and these attacks are often unsuccessful, as measured by the often numerous repairs to the shell aperture. Vermeij et al. (1980) examined terebrid gastropod shells to gain a measure of the rates of predation by crabs. They found that the frequency of repair was highest in this largely tropical group in the Indo-West Pacific, with tropical America next, and warm temperate areas the least. These patterns coincided with the presence

and diversity of *Calappa*, a highly specialised lip-peeling crab genus. Similarly, Dudley and Vermeij (1978) found that the frequency of drilling of turritellid shells by gastropod predators was about three times higher in the tropics than in temperate areas.

In a bizarre departure from other molluscs, a deep-sea hot vent gastropod *Crysomallon squamiferum* uniquely has protective magnetised iron-sulphide scales over the sides of its foot (Goffredi et al. 2004b; Suzuki et al. 2006b).

Defences other than the shell include chemical and behavioural defences that are also briefly discussed below.

9.4.3.1.1 Shell Defences and the Arms Race

The interactions between predators such as crabs and their molluscan prey have resulted in an evolutionary 'arms race'. One manifestation of this is that quite rapid phenotypic changes can occur in response to predation, as outlined below.

Experiments have shown that shelled molluscs can help to deter attack by shell-crushing fish by developing stout spines or nodules (Palmer 1979). There are more such fish in tropical than in temperate seas, and this is correlated with many shells in the tropics bearing spines or nodules (Vermeij 1974, 1978). As noted above, changes in shell morphology in relation to predation pressure by fish have also been suggested in African ancient lake faunas (Vermeij & Covich 1978).

Some melongenid whelks such as *Busycon carica* and *Sinistrofulgur sinistrum* chip pieces from the edges of bivalves until they can gain purchase to wedge open the valves with their shell aperture. Dietl (2003) used the fossil record to show that the bivalve prey *Mercenaria mercenaria* of the whelk *Sinistrofulgur sinistrum* evolved a larger shell and there was a corresponding increase in the size of the predator. In the presence of *Nucella lapillus*, the 'blue mussel' *Mytilus edulis* increased in shell thickness when compared with mussels not in contact with that whelk (Smith & Jennings 2000), as has also been shown to occur in the presence of crab predators (see below).

Different crab taxa have claws specialised to open shells in different ways and, because smaller-sized prey are preferred by most crabs (Seed & Hughes 1995), the molluscan prey can potentially escape predation by increasing in size. Shell adaptations other than size can also reduce predation, including shell thickness (e.g., Palmer 1985a), shape (e.g., globose vs conical), strong external sculpture including ribs, lamellae, spines, etc. (e.g., Donovan et al. 1999b), narrow or armoured apertures in gastropods and reducing gape in bivalves (e.g., Boulding 1984; Seed & Hughes 1995), although such differences are not evenly distributed (e.g., Vermeij 1976).

Many studies have examined the interrelationships of crabs and their molluscan prey, with different crab predators inducing different responses as shown, for example, in *Mytilus* (Smith & Jennings 2000). Sedentary molluscs, particularly bivalves, have been shown to respond to waterborne predator cues by enhancing defences against predation – notably by increasing shell thickness; in mussels (*Mytilus*) (Leonard et al. 1999; Smith & Jennings 2000; Freeman & Byers 2006; Fässler & Kaiser 2008), the muricid whelk *Nucella* (e.g., Vermeij 1978, 1982; Palmer 1985a; Donovan et al. 1999b), and *Littorina* (Appleton & Palmer 1988; Palmer 1990; Trussell 1996; Rochette et al. 2007; Edgell et al. 2008). This effect is caused by crabs (native or introduced) that have been in an area for many years, but is generally not induced by recently introduced crabs (e.g., Edgell et al. 2008). Other shell features can be affected by predatory crab as, for example, in *Littorina obtusata* (Rochette et al. 2007; Edgell et al. 2008) and include reduced spire height and increased thickness (Seeley 1986). In *Nucella* spp., these affects include smaller apertures and larger apertural teeth (Appleton & Palmer 1988; Palmer 1990), or increased length (Fisher et al. 2009), or both (Vermeij 1982), or overall size (Fisher et al. 2009). The crabs can respond to the changes in the shells of their prey by strengthening their claws (Smith & Palmer 1994), with these changes in concert likened to an 'arms race' (e.g., Vermeij 1977a; Vermeij 1983; Seed & Hughes 1995; Yamada & Boulding 1998).

A study of 14 species of gastropods showed that taller spired species could retract further into their shells, offering them some protection against peeling crabs, although this relationship was not apparent when closely related species were compared (Edgell & Miyashita 2009). Variation in shell morphology is common and some of this may be the result of interactions with shell-breaking (or other) predators, but it may also result from other responses to the environment. While there are relatively few examples of inducible defences rigorously demonstrated in molluscs, they may be common. Thus, because considerable phenotypic plasticity in the shell has been demonstrated in some taxa, some earlier studies where differences in shell features were interpreted as evolutionary change or genetic variation among populations may need to be re-examined (Leonard et al. 1999). In a parallel fashion, crab claw strength and armature have increased over time.

A reduction in the feeding rates of the prey organisms in the presence of crabs (i.e., chemical cues) can change community structure. Thus, for example, where crabs were present in northeastern North America, barnacles and fucoid algae were more abundant, because *Nucella lapillus* consumed 29% fewer barnacles in the presence of crabs than when crabs were absent, and *L. littorea* consumed more than 4.5 times fewer fucoids than controls where crabs were absent (Trussell et al. 2003).

Another interesting example of the 'arms race' between predator and prey is being fought in the water column between shell-less gymnosome pteropods and their shelled thecosome prey. The gymnosomes are highly species-specific in their choice of thecosome prey with, for example, *Clione limacina* feeding exclusively on *Limacina helicina* (Conover & Lalli 1972). The coevolution of gymnosomes and thecosomes has resulted in highly specialised prey capture mechanisms in gymnosomes, which rely on tactile recognition (see Chapters 5 and 20), and on predator avoidance in the thecosomes. Seibel et al. (2007) showed that, in these highly coevolved taxa, not only did the complex feeding structures vary with the prey, but also their metabolism and locomotion.

9.4.3.2 Responses to the Chemical Defences of Food Items

Chemical defence is thought to be a major force in evolution, with modification of protective chemicals (biotransformation) and responses to those by the herbivore or predator are part of the 'arms race' between predators and their animal prey and herbivores and their algal or plant food. Chemical defences are commonly used by organisms to limit predation and herbivory, but some taxa's modifications enable them to feed on poisonous or distasteful organisms and detoxify and/or excrete the noxious compounds. Responses against these defences may be behavioural, physiological, or biochemical. A biochemical response typically includes detoxification by the transformation of the poisonous chemicals by way of enzymes. Such chemicals are called *allelochemicals* – a term that encompasses biologically synthesised chemicals with many functions and includes, for example, pheromones, venom, and antimicrobial compounds.

These defensive chemicals are found in many terrestrial plants, marine 'algae' (e.g., Steinberg 1985; Estes & Steinberg 1988; Hay & Fenical 1988; Hay & Steinberg 1992; DeBusk et al. 2000) and in seagrasses (Vergés et al. 2007; Vergés et al. 2011). Herbivorous gastropods and chitons utilise enzymes such as glutathione transferase to enable them to modify the defensive chemicals in their algal food (e.g., Bornancin et al. 2017). Some recent studies with herbivores have shown that the food plant or alga may respond to being fed on by increasing the levels of noxious compounds in their cells (Borell et al. 2004). For example, a carbohydrase enzyme (α-amylase) in the saliva of *Littorina obtusata* has been shown to cause the alga *Ascophyllum nodosum* to increase the levels of phlorotannins, compounds associated with defence against herbivory, thus increasing its resistance to the herbivore (Coleman et al. 2007). It has also been demonstrated that the phlorotannins in the brown alga (*Eisenia*) can deter the feeding of the vetigastropod herbivore *Turbo cornutus* by inhibiting its glycosidase enzymes (Shibata et al. 2002).

Chemical deterrents are well known in many terrestrial plants, but their relationship with terrestrial gastropods is much more poorly understood than that for insects.

Poisonous allelochemicals are also found in many colonial animals such as sponges and gorgonian corals. Gorgonians, for example, are rich in defensive allelochemicals and are fed on, and the allelochemicals detoxified by, ovulid snails (see Section 9.4.3.4.13). Similarly, numerous toxic chemicals are obtained by nudibranchs and some other euthyneurans from their animal or algal food and coopted in a defensive role (see Section 9.4.3.4.1).

9.4.3.3 Chemical Signalling

Most intra- and interspecific communication between molluscs is chemical (see Chapter 7). Some other aspects of intraspecific signalling (or 'communication'), notably the visual signalling seen between shallow-water coleoid cephalopods, have been briefly covered in Chapter 7.

So-called alarm pheromones have been reported in some gastropods, for example, in the vetigastropod *Tegula* (Jacobsen & Sabell 2004) and the caenogastropods *Littorina littorea* (Jacobsen & Stabell 1999) and *Tritia obsoleta* (Nassariidae) (Atema & Stenzler 1977). In the neogastropod *Buccinum undatum*, 'alarm chemicals' may assist in inducing predator-recognition learning (Rochette et al. 1998). In 'opisthobranchs', the best known 'alarm chemicals' are the 3-alkylpyridine alkaloid haminols in Mediterranean species of *Haminoea* and there are similar compounds in other cephalaspideans (Cimino & Ghiselin 2009). These animals follow the mucous trails of conspecifics and an escape response is elicited by species-specific haminols in the mucus. The haminol is a pheromone secreted when *Haminoea* is attacked, although it does not deter the predator. Similar behaviour is seen in conspecifics of the aglajid *Navanax* when it secretes its alarm pheromone (navenone, a polypropionate) on its trail. In *Philinosis*, another aglajid, a similar pheromone is used to locate mates. The carnivorous aglajids probably acquire these chemicals from herbivorous cephalaspidean prey (Cutignano et al. 2011).

Aplysia species can detect proteins (such as the appropriately named attractin, enticin, temptin, and seductin) from other individuals of the same species (Painter et al. 1998, 2004), sex pheromones involved in mate attraction and influencing the duration of egg-laying.

The detection by chemoreception of chemical cues from food or conspecifics also plays an essential role in foraging and larval settlement (e.g., Chia & Koss 1978, 1982; McGee & Targett 1989; Pawlik 1989; Avila 1998; Ritson-Williams et al. 2009).

9.4.3.4 Molluscan Chemical Defences

Some gastropods can sequester and concentrate chemicals from their food and use these secondary metabolites for their own defence. Notably, shell-less sea slugs have often utilised such distasteful *antifeedant* molecules, which they obtain from their food (and/or the symbionts within the food) and move them to sacs or glands in their skin. These chemicals are mainly alkaloids, many of which are terpenes and their derivatives. There are many examples in dorid nudibranchs and other 'opisthobranchs' that feed on colonial animals, notably sponges, and utilise their secondary metabolites (see next Section). Although much of the attention on chemical deterrents has been concentrated on nudibranch sea slugs, some other gastropods, including some with shells, possess noxious chemicals as outlined below.

Noxious defensive chemicals can also be obtained from algae – as in some sacoglossans (Cimino & Ghiselin 1998). The 'sea hares' (*Aplysia* and other aplysiidans) are well known for using chemicals obtained from their algal food to deter predators by way of secretions in the skin, opaline glands, and ink gland (see Section 9.4.3.4.4.1), the latter analogous to the ink sac in many coleoid cephalopods (see Chapter 17). Other chemicals are obtained from the food of grazing carnivores that feed on sponges, soft corals, or tunicates, although the origins of the chemical compounds sequestered by the mollusc may not necessarily be a product of the organism being consumed, but may be derived from symbiotic bacteria or Cyanobacteria associated with the prey species.

The effectiveness of these noxious chemicals in deterring predators has been mainly shown in laboratory experiments (e.g., see review by Pawlik 1993), but has been demonstrated in relatively few field experiments. In one such experiment, the ability of reef fish to discriminate between nudibranch models was tested by Giménez-Casalduero et al. (1999). Various coloured models were used, but only one colour was associated with a feeding deterrent. The fish learned to avoid the colour associated with the feeding deterrent, but when the colour association was changed the fish failed to form an association during the experiment. Similarly, feeding experiments showed that both crabs and fish rejected the nudibranch *Doriopsilla* (Long & Hay 2006).

9.4.3.4.1 Secondary Metabolites in Euthyneurans

In euthyneurans, the secondary metabolites are usually found in the epithelial and subepithelial layers, or they can be concentrated in special glands associated with the dorsal or lateral epithelium, and some can be stored in the digestive gland. For example, in *Aplysia*, the cholinergic agent aplysin and other metabolites are stored in both the digestive gland and epithelium (e.g., Stallard & Faulkner 1974).

The treatment here provides details pertaining mainly to the secondary metabolites of 'opisthobranchs', as they are a vital part of the defence mechanism of those animals (Faulkner & Ghiselin 1983; Karuso 1987; Avila 1995; Cimino & Ghiselin 1998, 1999; Marin et al. 1999). It also touches on the relatively small amount of information on 'pulmonates'. In general, 'opisthobranchs' use aposematic signals as toxicity is generally related to the conspicuousness of the nudibranch (Cortesi & Cheney 2010), with aposematic species comprising about 50% of all opisthobranch species according to Gosliner and Behrens (1990).

There is a large body of literature on the secondary metabolites of 'opisthobranchs' (see reviews by (Proksch 1994; Cimino & Ghiselin 1998, 1999, 2009; Cimino & Gavagnin 2006; Benkendorff 2010), and here, we only outline some of the main findings that are directly relevant to the biology of these animals. The account given is largely a summary from the detailed overviews given by Cimino and Ghiselin (2009) and Bornancin et al. (2017). The reader should refer to these works for more information and for most of the reference sources for this body of work. A more general introductory account of secondary metabolite synthesis is given in Chapter 2.

Secondary metabolites are mainly found in sponge, bryozoan, or algal eating taxa, but not in most aeolidioids (all cnidarian feeders), which have an alternative means of defence (see Chapter 20 and Section 9.4.3.12.1). The main relevance for organisms of some metabolites is their toxicity and their potential in chemical defence for either the prey or, as with many nudibranchs, their secondary use for defence. The original adoption and subsequent shifts in the animals and plants or algae utilised as food require the ability to tolerate, detoxify (by biotransformation), and, sometimes, exploit by sequestering and utilising metabolites from their prey for defence. The defensive deployment of accumulated metabolites may require their modification (reduced or increased toxicity). Sometimes, the metabolites are synthesised *de novo* by the mollusc, thus releasing it from its dependence on a specific food as the source of the defensive compound. Cimino and Ghiselin (1999) postulated that enzymes involved in detoxifying metabolites can be modified to carry out other transformations. Thus, the ability to modify a food-sourced metabolite (e.g., by oxidation, reduction, or degrading) is, they argued, relatively easy to evolve as these processes are already undertaken in the body, and the pathways for *de novo* synthesis are present and used in the synthesis of other compounds.

Cimino and Ghiselin (1999) suggested simple models to explain how particular biosynthetic capacities might have evolved. They argued that the synthesis of defensive compounds *de novo* may have been achieved by a preadaptive phase followed 'by evolution in a retrosynthetic mode, with selection favouring enzymes that enhance the yield of end products already present in the food' (Cimino & Ghiselin 1999, p. 187). This process presumably began with the ability to make a metabolite obtained from food either more, or less, toxic. Once the enzyme that facilitated this change had evolved 'evolution by gene duplication could occur, working backward to establish an entire pathway' (Cimino & Ghiselin 2009, p. 190).

De novo synthesis of metabolites originally obtained from sponge prey is well documented in doridoids (Cimino et al. 1983; Faulkner et al. 1990). This ability enables them to become liberated from their sponge prey while remaining chemically defended, but the details of the mechanisms involved in such synthesis are currently unknown. The *de novo* synthesis of particular compounds can give clues to the previous dietary habits of a particular taxon. For example, the dendronotoidean *Tethys fimbria*, which eats crustaceans, makes use of prostaglandins as defensive chemicals, suggesting that its ancestor ate octocorals, as do many of its relatives (Cimino & Ghiselin 1999).

When a nudibranch that derives its defensive metabolites from its prey shifts to a different food source, whether it be a different group of sponges, or bryozoan or ascidian prey, the sequestered metabolites must also change, resulting in considerable physiological challenges. There are exceptions – for example, some phanerobranch Doridina exploit bryozoans defended with alkaloids, and this facilitates some members of that group switching to ascidians, which have similar defensive chemicals (Cimino & Ghiselin 1999). In theory, those that synthesise defensive chemicals *de novo* should be able to exploit a broader range of food, including food that lacks defensive compounds, and because they are not tied to a particular food item, they should also be able to expand their range geographically (Cimino & Ghiselin 1999).

Many display what appears to be warning colouration (i.e., aposematic colouration), while some are cryptic. While there appears to be a correlation between conspicuous colouration and the presence of toxic compounds (Cortesi & Cheney 2010), this does not translate into a correlation with increased body size (Cheney et al. 2014).

The nudibranchs have undergone more chemical studies than any other group of molluscs, and within that taxon, the most widely investigated group are the doridine nudibranchs. Many of their secondary metabolites have antifeedant properties and are obtained directly from their food or synthesised by the slug.

Lists of the main secondary metabolites found in euthyneurans are given in Table 9.1, and are briefly discussed below.

9.4.3.4.2 Acteonoidea

Acteonoideans are polychaete feeders. The aplustrid *Micromelo undata* contains polypropionates (Napolitano et al. 2008).

9.4.3.4.3 Cephalaspidea

This group includes both herbivores and carnivores. The herbivorous cephalaspidean *Bulla* has toxic polypropionates (as do the acteonoidean *Micromelo*, some sacoglossans, and marine 'pulmonates'), where they are found in glands along the edge of the mantle. These compounds have also been detected in aglajids that feed on *Bulla*. Another herbivorous group, the haminoeids, also produce an array of secondary chemicals. One of these, an antifeedant halogenated polyacetate (kumepaloxane), is released in mucus when the animal is disturbed (Polner et al. 1989). Similar compounds have been found in some red algae and, rarely, in some sponges. Several polyproprionates have also been found (e.g., Nuzzo et al. 2016). Cephalaspideans also produce phenolic and alkaloid 'alarm pheromones'. For example, *Haminoea* species biosynthesise species-specific haminols and similar compounds which are secreted as part of the mucous trail. The carnivorous cephalaspidean *Scaphander* produces lignarenones that also probably function as 'alarm pheromones', as does a similar compound (navenone-B) in the aglajid *Navanax*, the latter secreted from a 'yellow gland' in the mantle cavity. *Philinopsis* has been shown to use pheromones for locating mates (Cimino & Ghiselin 2009).

In the carnivorous philinoideans, some, like *Philine*, secrete sulphuric acid on their body surface as a predator deterrent (Thompson 1986). Aglajids such as *Philinopsis* and *Navanax* feed on other 'opisthobranchs' and derive some of their noxious compounds, mainly various polypropionates, from their herbivorous cephalaspidean (e.g., *Bulla* and *Haminoea*) or aplysiidan (e.g., *Stylocheilus*) prey. Similarly, several cytotoxic peptides in *Philinopsis* are probably derived from Cyanobacteria, most likely via their 'opisthobranch' prey (Nakao et al. 1998). Another philinoidean cephalaspidean, *Sagaminopteron* (Gastropteridae), is found on sponges, and the metabolites it stores are probably acquired from its food. One of these (a phenol, oxy-polybrominated diphenyl ether) is concentrated in its parapodial wings. Some gastropterids have vivid colours, consistent with warning colouration, while others are cryptic, but both have similar secondary metabolites, which are at more than twice the concentration in their parapodia than in the sponge (Becerro et al. 2006).

TABLE 9.1

The Major Secondary Compounds Described from Euthyneuran Heterobranchs

Compound	Taxa	Notes
Polyketides (including fatty acids)	This large group of compounds is comprised of acetate (ethanoate) units arranged in a line. They are produced by nearly all organisms and include compounds used as antifungals and antibiotics. Fatty acids are significant sources of energy as they produce ATP	
Fatty acids	Aplysiida Haminoeoidea Sacoglossa	Fatty acids have the carbonyl group of the acetate units reduced. They are medium to long chain saturated and unsaturated monocarboxylic acids, with an even number of carbons
Polyacetylenes	Doridoidea	Polymers derived from polyketides. They are a long chain of carbon atoms with alternating single and double bonds, and each with one hydrogen atom
Prostaglandin lactones	Tritonioidea	The prostaglandins are lipids derived from fatty acids. They contain 20 carbon atoms, including a 5-carbon ring. A lactone is a cyclic ester based on shorter chain fatty acids
Cyclic acetogenins	Aplysiida Nudibranchia (Doridoidea)	Acetogenins are typically found in plants and red algae. They have 32 or 34 linear carbon chains that contain various oxygenated functional groups
Macrocyclic fatty acid lactones	Aplysiida Nudibranchia (Doridoidea)	A lactone is a cyclic ester based on shorter chain fatty acids. A macrocyclic molecule is a cyclic part of a molecule that contains multiple (nine or more) atoms
Macrolides	Nudibranchia (Polyceroidea) Cephalaspidea (Philinoidea) Aplysiida Eupulmonata (Onchidioidea)	Large ring lactones
Aromatic Polyketides	Cephalaspidea (Philinoidea, Bulloidea, Haminoeoidea)	Aromatic polyketides are polyketides with oxygen atoms attached to alternate carbon atoms
Glyceride Esters	Glycerides (or acylglycerols), are esters formed from glycerol and fatty acids. They include animal fats and vegetable oils. The following four categories are grouped as glyceride esters	
Fatty acid esters	Nudibranchia (Tritonioidea) Umbrachuloidea	An ester formed from the combination of a fatty acid with an alcohol (which may be a glycerol)
Sesquiterpenoid esters	Nudibranchia (Doridoidea)	Terpenoids are lipids derived from 5-carbon isoprene units. Sesquiterpenes consist of three isoprene units and have 15 carbon and 24 hydrogen atoms. Sesquiterpenoids are modified sesquiterpenes and esters are derived from them
Diterpenoid esters	Nudibranchia (Doridoidea)	Diterpenoids consist of four isoprene units (20 carbon atoms) and the esters are derived from them
Polypropionates	Acteonoidea Gymnosomata (Clionoidea), Pleurobranchida, Sacoglossa (Limapontioidea, Plakobranchoidea), Cephalaspidea (Philinoidea, Bulloidea, Haminoeoidea) Aplysiida Panpulmonata (Siphonarioidea,[a] Onchidioidea)	A polymer of propionates (a salt or ester of propionic acid), a short chain fatty acid
Ethers	Ethers contain an ether group, an oxygen atom connected to two alkyl or aryl groups. Ethers commonly form links in carbohydrates	
Polyethers	Aplysiida	Compounds with more than one ether group. Includes some dinoflagellate toxins
Terpenoids (and steroids)	The majority of compounds isolated from heterobranchs to date are terpenoids (sometimes called isoprenoids) (e.g., Andersen et al. 2006). They are lipids formed by 5-carbon isoprene units and classified by the number of C_5 units that they contain. The largest group of natural products and precursors to steroids and sterols; some added to proteins. A number are highly noxious	
Monoterpenoids	Nudibranchia (Tritonioidea) Aplysiida	Modified from monoterpenes – terpenes consisting of two isoprene units

(Continued)

TABLE 9.1 (*Continued*)

The Major Secondary Compounds Described from Euthyneuran Heterobranchs

Compound	Taxa	Notes
Sesquiterpenoids	Nudibranchia (Tritonioidea, 'Metaminoidea', Onchidoridoidea, Doridoidea, Phyllidioidea, Arminoidea, Bathydoridoidea) Cephalaspidea (Haminoeoidea) Aplysiida Sacoglossa (Oxynooidea, Plakobranchoidea) Panpulmonata (Onchidioidea)	Modified from sesquiterpenes – terpenes consisting of three isoprene units
Halogenated sesquiterpenoids	Aplysiida	Sesquiterpenoids are a class of 15-carbon isoprenoid compounds found in many organisms. Halogenation occurs when a halogen is added
Isocyanosesquiterpenoids	Nudibranchia (Phyllidioidea, Doridoidea, Polyceroidea)	This sesquiterpene (3-isoprenes) contains an isomer of cyanide (nitrogen triple bonded to a carbon atom) (i.e., an isocyanide)
Furanosesquiterpenoids	Nudibranchia (Phyllidioidea, Tritonioidea, Doridoidea)	Furanosesquiterpenoids are sesquiterpenes (3-isoprenes) that also contain a five membered aromatic ring with four carbons and one oxygen
Diterpenoids	Pleurobranchida Nudibranchia (Tritonioidea, Arminoidea, Aeolidioidea, Phyllidioidea, Doridoidea, Polyceroidea) Aplysiida Sacoglossa (Plakobranchoidea, Limapontioidea), Eupulmonata (Ellobioidea)	Terpenoids with 4-isoprene units (20 carbons)
Sesterterpenoids	Nudibranchia (Doridoidea, Phyllidioidea)	Sesquiterpenes have 25 carbon atoms (5-isoprene units). Derived from them, sesterterpenoids are found widely in plants
Cyclic sesterterpenoids	Nudibranchia (Doridoidea)	Sesterterpenoids consist of 5-isoprene units (25 carbon atoms). The cyclic form contains one or more rings
Triterpenoids	Pleurobranchida Nudibranchia (Polyceroidea, Onchidoridoidea) (Also in *Lottia*, a patellogastropod)	Triterpenes are terpenes consisting of six isoprene units. Triterpenoids are derived from triterpenes
Steroids	Nudibranchia (Onchidoridoidea, Doridoidea, Aeolidioidea) Aplysiida Eupulmonata (Ellobioidea)	Steroids are derived from tetracyclic triterpenes. They have four cycloalkane rings (each consisting of a ring of carbon atoms) joined to each other
Phenols and Quinones	Umbrachuloidea Nudibranchia (Doridoidea, 'Metaminoidea', Phyllidioidea, Aeolidioidea) Cephalaspidea (Philinoidea) Aplysiida Sacoglossa (Limapontioidea)	Phenols have a hydroxyl group (–OH) bonded to an aromatic hydrocarbon group. The simplest is phenol or carbolic acid Phenols undergo oxidation to quinones by conversion of an even number of –CH = groups into –C(=O)- groups with any necessary rearrangement of double bonds
Nitrogenous compounds	Cimino and Ghiselin (2009) include in this category compounds derived from the amino acid pathway, including alkaloids, specialised amino acids, and peptides – the latter dealt with separately below Alkaloids are common in terrestrial plants and include substances such as strychnine, caffeine, and nicotine	
Nitrogenous compounds other than peptides	Pleurobranchida Nudibranchia (Polyceroidea, Doridoidea, 'Metaminoidea', Arminoidea, Aeolidioidea) Umbrachuloidea Aplysiida Sacoglossa (Oxynooidea)	See above
Peptides	Cephalaspidea (Philinoidea) Pleurobranchida Aplysiida Sacoglossa (Plakobranchoidea)	Peptides are short chains of amino acid monomers linked by peptide (amide) bonds

Source of data: Cimino, G. and Ghiselin, M.T., *Proc. Calif. Acad. Sci.*, 60, 175–422, 2009.

[a] Some polyprionates from siphonariids are not natural products, but probably isolation artifacts (e.g., Beye & Ward 2010).

9.4.3.4.4 *Aplysiida*

Most aplysiidans are herbivores that feed on macroscopic algae. While nothing is known of chemicals in Akeridae, a wide variety of compounds have been isolated from aplysioideans, some of them toxic. Indeed, poisoning of humans can occur if 'sea hares' or their eggs are eaten without special treatment (see Cimino & Ghiselin 2009 for examples, and Chapter 10). Many of these toxic and/or noxious compounds are terpenes, but there are also sterols, fatty acid derivatives, alkaloids, cytotoxic peptides such as dolastatins, and glycoproteins, and several proteinaceous sex pheromones (see Cimino & Ghiselin 2009; Benkendorff 2010; Bornancin et al. 2017 for references), although some could be bacterial tetrodotoxins associated with their food (W. B. Rudman, pers. comm.). Some noxious compounds derived from their algal food, including several halogenated compounds, are stored in the digestive gland (e.g., Stallard & Faulkner 1974) rather than the skin, ink, or eggs (Pennings & Paul 1993) and, presumably because of this, the digestive gland is rejected by most predators. While the digestive gland can detoxify at least a few of the noxious chemicals, to be effective as deterrents, the chemicals must reside on the outer surface of the body. The body wall in some aplysioids contains polyhalogenated monoterpenes, degraded sterols, steroids, and lactonised dihydroxy fatty acids (aplyolides), some of which have antifeedant properties (e.g., Hay et al. 1987; Bornancin et al. 2017). The result is a noxious mix that would presumably deter most potential predators. As well as noxious chemicals, the eggs contain unsaturated fatty acids that prevent bacterial growth (Benkendorff et al. 2005). Different species of *Aplysia* feed preferentially on red, green, or brown algae; some change from one to the other at different growth stages, and consequently their metabolites can differ markedly (Cimino & Ghiselin 2009).

Dolastatins (see Pettit 1997 for review) are various cytotoxic peptides produced in *Dolabella* (some of which have anticancer properties), and are probably derived from cyanobacterians consumed along with their algal food. *Stylocheilus* and *Bursatella* are primarily cyanobacterial feeders, and consequently have a different set of metabolites, including polyacetates and polyethers (e.g., Pennings et al. 1996; Suntornchashwej et al. 2005).

Dolabrifera contains the unusual polypropionate dolabriferol, and similar compounds have been found outside the Aplysioidea in *Pleurobranchus membranaceus*, *Micromelo undata*, and some marine 'pulmonates'. In *Dolabrifera*, it is only found in the outer integument so is probably biosynthesised *de novo*, possibly because of diet change (Cimino & Ghiselin 2009). The related *Dolabella*, which has exogenous metabolites such as caulerpenyne (Bornancin et al. 2017) to defend itself, also has polypropionates, although these are found in low concentration in the internal organs. It also contains dolabellin, a heterocyclic compound (a bis-thiazole) (Sone et al. 1995) similar to compounds with antifungal properties. In addition, diterpenoids and, as in *Aplysia*, cyclic acetogenins, macrocyclic fatty acid lactones, polyethers, and sesquiterpenoids have been identified.

The aplysioidean *Syphonota* feeds on the seagrass *Halophila* rather than algae and thus has a different suite of metabolites derived from their food, the main one being syphonoside, a macrocyclic glycoterpenoid. This is sequestered and some is biotransformed to the steroid syphonosideol and then to fatty acid derivatives (Bornancin et al. 2017).

9.4.3.4.4.1 *The Purple and Opaline Glands of Aplysioids*

Aplysioids have two unique glands in the mantle cavity; the purple gland on the roof and, beneath the floor, the opaline gland (=gland of Bohadsch), which opens to the cavity by a single duct. These two glands operate to provide an effective predator deterrent. The release of the secretions of each gland is independently controlled by different ganglia – the ink gland by the abdominal ganglion and the opaline gland by the right pleural ganglion. They can be released together or separately following provocation (Carew & Kandel 1977; Tritt & Byrne 1980).

The opaline secretion is white and viscous and is released less readily than the fluid ink, and in smaller amounts. The secretion tastes bitter and smells unpleasant to humans. It contains a protein apparently toxic to crabs and bacteria (Kamiya et al. 1989) and which suppresses feeding. Opaline secretion also contains mycosporine-like amino acids, some of which act as alarm signalling molecules and others that do not (Derby 2007; Derby et al. 2007; Kicklighter et al. 2007; Kamio et al. 2011). The mycosporine-like amino acids are derived from the algal food and can perform various functions in different parts of the body including acting as sunscreens as they absorb UV (e.g., Carefoot et al. 2000), and as alarm cues (Kamio et al. 2011). While some chemicals in the opaline secretion are obtained from the algal food, others are produced *de novo* (Derby 2007; Kamio et al. 2011).

The ink gland is mainly composed of muscular 'red-purple vesicles' that contain the purple ink, but smaller numbers of 'amber vesicles' and 'clear vesicles' are also present (Prince et al. 1998). The ink release is controlled by a valve that, when opened, releases the ink from each red-purple vesicle through a duct that opens at a pore on the ventral surface of the purple gland. When the vesicles are discharged, the ink is carried out either anteriorly through the siphon or posteriorly through the parapodial margins.

Although all aplysioids have ink and opaline glands, not all release purple ink from the ink gland as a defence strategy when disturbed (e.g., Kandel 1979; Johnson & Willows 1999). *Aplysia juliana*, which normally feeds on green algae, can only produce purple ink when fed on red algae (Prince & Johnson 2006). *Aplysia parvula* can produce both white and purple ink, but *Dolabrifera dolabrifera* cannot produce purple ink, no matter what algae it eats (Prince & Johnson 2013). When *Aplysia californica* is fed only on brown algae, it loses its ability to produce ink, but this ability is restored when fed red algae (Chapman & Fox 1969). In *Aplysia juliana*, the secretion from the purple gland is not purple, is toxic to crabs, and has antibacterial properties (Kamiya et al. 1989). The purple secretion of *Aplysia californica* had mixed effects on predatory lobsters (Kicklighter et al. 2005) and negative effects on large sea anemones (Nolen et al. 1995; Kicklighter & Derby 2006).

The secretions from both the opaline glands and the ink sac typically mix together on release and comprise several amino acids and ammonium (Kicklighter et al. 2005; Johnson et al. 2006) and hydrogen peroxide (Aggio & Derby 2008). This cocktail adversely affects predators such as lobsters and fish by massively stimulating their chemosensory systems, thus preventing normal function and resulting in confusion and cessation of attack (Johnson & Willows 1999; Kicklighter et al. 2005; Sheybani et al. 2009; Nusnbaum & Derby 2010). This latter effect, sensory disruption, is complemented by another, *phagomimicry* (see below).

In the most studied species, *Aplysia californica*, the ink and opaline gland secretions react together when the secretions mix on their release. This results in a range of products (notably hydrogen peroxide, ammonium ions, α-keto acids, and carboxylic acids), which in turn undergo other changes. The ink is mainly composed of the purple (or violet) pigment aplysioviolin (about two thirds of the dry weight) and a protein called escapin. The colour of aplysioviolin is due to an ester of a pigment obtained from red algae, and thus species that do not feed on red algae do not produce purple ink (Faulkner 1992). Escapin is an L-amino acid oxidase[5] (Yang et al. 2005), while the L-amino acid substrate on which the enzyme acts (L-lysine) is in the opaline gland secretion. Aplysioviolin is derived from the photosynthetic pigment r-phycoerythrin in red algal and cyanobacterial food. This conversion occurs in the specialised vacuoles in digestive gland cells that have been called 'rhodoplast digestive cells', a cell type found only in aplysioids, and is even found in those that do not produce coloured ink (Prince & Johnson 2013). The resulting compound phycoerythrobilin is then carried via the haemolymph to the ink gland where it is methylated to form aplysioviolin (Kamio et al. 2010 and references therein). The pigment phycoerythrobilin is also found in the skin of *Aplysia* (Kamio et al. 2010). The ink protein is synthesised in the ink gland in cells rich in rough endoplasmic reticulum (Prince & Johnson 2006). These compounds in the ink have been shown to have effects on an anemone and a crab (see Bornancin et al. 2017).

The ink chemicals deter feeding by crustaceans, but apparently not other likely predators such as anemones and fish (Derby 2007). In particular, it has been shown that escapin, and similar proteins in other aplysiidans, induce antipredator responses and also have antimicrobial (and antitumour) activity (Kicklighter et al. 2005; Prince & Johnson 2006). Other compounds in the secretions from the ink gland generate the feeding response (phagomimicry) by predators (Kicklighter et al. 2005; Derby & Aggio 2011), which are attracted to the

ink cloud rather than the body of the 'sea hare'. The active compounds involved in eliciting the feeding response include free amino acids and ammonium, which are concentrated in the ink and opaline fluids of *Aplysia* (Kicklighter et al. 2005; Derby 2007; Derby et al. 2007).

The ink of 'sea hares' is functionally like that of coleoid cephalopods (cuttlefish, squid, and octopuses), which release ink when attacked by predators (see Chapter 17). Coleoid ink was thought to function mainly to visually deter predators as 'smoke screens' or decoys (Caldwell 2005), but Derby et al. (2007) showed that it resembles the ink of *Aplysia* in containing free amino acids (mainly taurine, aspartic acid, glutamic acid, alanine, and lysine) and ammonium in these secretions. Both large decapod crustaceans and fish have specific receptor systems for these amino acids. Thus, Derby et al. (2007) argued that both aplysioid and coleoid ink appeared to have the potential to cause sensory disruption and/or phagomimicry as a means of defence. However, experiments with the squid *Sepioteuthis* by Wood et al. (2010) suggested that coleoid ink is not involved in phagomimicry, but is instead unpalatable to predators, a finding agreeing with numerous anecdotal observations.

9.4.3.4.5 Pteropods

Nothing is known about possible defensive compounds in pteropods other than a polypropionate (pteroenone) in the gymnosome *Clione* (Yoshida et al. 1995). This chemical is not present in *Limacina*, the thecosome on which *Clione* preys. *Clione* is sometimes carried by the Antarctic hyperiid amphipod *Hyperiella dilatata*, apparently to deter predators (McClintock & Janssen 1990).

9.4.3.4.6 Umbrachulida

The Umbrachulida comprises the genera *Umbraculum* and *Tylodina*, both sponge feeders. They contain several antifeedant metabolites derived from the sponges on which they feed. In *Tylodina*, these include brominated alkaloids and a phenolic pigment (uranidine). Some parts of their food sponges contain Cyanobacteria, and the slugs preferentially feed on those (Becerro et al. 2003).

Umbraculum contains fatty acid esters (diacylglycerols and a bis-hydroxybutyric acid ester) in its skin that are toxic to fish (Cimino et al. 1988, 1989; Gavagnin et al. 1990).

9.4.3.4.7 Nudipleura – Pleurobranchida

The pleurobranchs have glands in the skin that produce sulphuric acid (Thompson & Slinn 1959; Gillette et al. 1991), and these glands may be plesiomorphic for Nudipleura. A modified salivary gland, the acid gland, opens into the buccal area near the mouth, and secretions from this gland (pH 1.2) are used to immobilise prey (Morse 1984). *Pleurobranchaea*, a generalist predator and scavenger with a preference for sea anemones, has labdane diterpene-type aldehydes in the skin. In contrast, the mainly ascidian-feeding *Pleurobranchus* contains polypropionates (membrenones) (Ciavatta et al. 1993), which may be synthesised *de novo* (Cimino & Ghiselin 1999),

---

[5] L-amino acid oxidases are enzymes that oxidatively deaminate L-amino acids. This type of enzyme is found widely in nature and a number of varieties are common in 'opisthobranchs' (Derby 2007).

and they also contain chlorinated diterpenes and a cyclic peptide that occur in ascidians (e.g., Fu et al. 2004). One species has been found to contain two triterpenoids in its skin and mucus like those from sponges (Spinella et al. 1997).

9.4.3.4.8 Nudipleura – Nudibranchia – Doridina

Several groups of Doridina have independently evolved the ability to biosynthesise a variety of metabolites. The most primitive living nudibranch is the deep-sea and Antarctic *Bathydoris* (Bathydoridoidea) that feeds on a wide variety of prey. One species has been shown to contain in its skin a sesquiterpene (hodgsonal) which repels a sympatric starfish. This compound is not present in the gut or digestive gland (Avila et al. 2000) and, although a closely similar compound is known from sponges, whether it is synthesised *de novo* is unclear.

In the Polyceroidea, the tropical, shallow-water, conspicuously coloured *Hexabranchus* (Hexabranchidae) is, like *Bathydoris*, using a wide range of animals (sponges, tunicates, worms, gastropods, and echinoderms) as food (Gosliner 1987). Probably because of this, it is very unusual among Doridina in containing a wide range of metabolites that seem to protect this common, conspicuous animal. These include antifeedant compounds derived from sponges, including isocyanosesquiterpenoids, di- and triterpenoids, and macrolides including kabiramide, dihydrohalichondramide, and tetrahydrohalichondramide, the latter biotransformed from the sponge metabolite halicondramide (Bornancin et al. 2017). These compounds are found in various parts of the body, including the skin and some on the egg mass (Pawlik et al. 1988). Other metabolites found in *Hexabranchus* include an esterified carotenoid pigment (hurghadin, a conjugated dialdehyde), found in parts of the mantle involved in visual signalling and chemical defence (Cimino & Ghiselin 1999).

The polycerids *Triopha* and *Polycera* synthesise an alkaloid otherwise found in bryozoans, while the onchidoridid *Acanthodoris* biosynthesises sesquiterpenoids, and the dendrodoridids *Dendrodoris* and *Doriopsilla* also synthesise sesquiterpenoids like those in sponges (Cimino & Ghiselin 2009). The sesquiterpenoid dialdehyde polygodial from the mantle tissue of *Dendrodoris* deters fish feeding (Pawlik 1993).

Most other nudibranchs are specialised feeders, containing only a few noxious compounds. While some Doridina derive metabolites from their food, others biosynthesise similar metabolites *de novo*. Many nudibranchs feed on sponges, which contain spicules, or noxious compounds, or both, while others have neither. An antifeedant compound is often restricted to a particular group of sponges or may be produced by symbiotic bacteria (including Cyanobacteria) living in the sponge – in some of those latter cases, the compound may be relatively widespread.

The other polyceroids, like *Hexabranchus*, are often conspicuously coloured and include the Aegiridae, which obtain alkaloids such as naamidine-A[6] from the calcareous sponges on which they feed, while polycerids contain defensive alkaloids derived from their ascidian or bryozoan food.

In at least one species, these alkaloids may be synthesised *de novo* (Cimino & Ghiselin 1999). Bryozoan-feeding polycerids such as *Tambja* and *Nembrotha* sequester the antifeedant tambjamines obtained from their prey for defence. *Tambja* is fed on by another polycerid, *Roboastra tigris*, which also uses these compounds (Pawlik 1993). The gymnodoridid *Nembrotha* contains tambjamines, obtained in part from its ascidian prey, which deter fish feeding (Pawlik 1993).

Some members of the family Onchidorididae (Onchidoridoidea) contain terpenoids, probably derived from their bryozoan food. These include a triterpenoid in *Adalaria* and several sesquiterpenoids in *Acanthodoris* synthesised *de novo*. The compounds from these somewhat cryptic sea slugs impart a rather pleasant (to humans) odour.

The remaining Doridina are mainly sponge feeders, comprising the Phyllidioidea, Doridoidea, Onchidoridoidea, and Polyceroidea. The suctorially feeding (i.e., they lack a radula) phyllidioideans comprise the dendrodoridids and phyllidiids. Dendrodoridids are often brightly coloured and can feed on various sponges, perhaps because they biosynthesise protective terpenoids *de novo*, mainly noxious drimane sesquiterpenoids (Cimino & Ghiselin 1999). They are combined with fatty acid esters in the gonads and transferred to the eggs where they provide a source of energy for the developing larva and also protection. A number (up to 16 in one species of *Dendrodoris*) of other sesquiterpenoids in the body are also biosynthesised *de novo*, one of which (polygodial) deters predators. While several compounds are identical to some that occur in sponges, they are sometimes not found in the sponges used for food, although ancestrally that may have been the case (Cimino & Ghiselin 1999).

Similar dialdehydes to those found in *Hexabranchus* are found in dendrodoridids along with other sesquiterpenes, one diterpene, isocyanides (like those in phyllidiids and some sponges – see below), and 15-20 carbon terpenoids like those in Chromodorididae (e.g., Cimino & Ghiselin 1999).

In contrast to dendrodoridids, all phyllidiids appear to be restricted to sponges that contain isocyanide terpenoids (Axinellida and Halichondrida), which they sequester and use defensively (Fusetani et al. 1992; Yasman et al. 2003). Some also use other metabolites derived from sponges, including several sesquiterpenoids, with two forms of isocyanopupukeanane from the sponge *Hymeniacidon* (Hagadone et al. 1979). Some of the secondary metabolites in *Phyllidia* have antifouling properties (Hirota et al. 1998) and one, axisonitrile-1, while not deterring feeding, is toxic to fish in low concentrations in seawater (Pawlik 1993).

Species of *Cadlina* (Cadlinidae) are drab-coloured, protected by dermal spicules (Penney 2006) and chemical defences. They feed on various spiculate sponges; they contain and sequester many metabolites (mostly terpenoids), which they derive from these sponges. They can also generate some *de novo* (e.g., Dumdei et al. 1997; Kubanek et al. 1997, 2000). The eastern Pacific *Cadlina luteomarginata*, which has been shown experimentally to be noxious to predators (Penney 2004), feeds on at least ten species of sponge and produces several secondary metabolites stored in the dorsum,

---

[6] This compound has anticancer properties.

at least five of which have antifeedant activity (Thompson et al. 1982). One chemical (luteone) is aromatic (Hellou & Andersen 1981). The compounds found in *Cadlina* include isocyanides, and terpenoids synthesised *de novo* include at least three sesquiterpenoids. While the metabolites derived from food vary geographically, the *de novo* sesquiterpenoids do not. The most noxious compound found in *Cadlina* is albicanyl acetate, a sesquiterpenoid, two forms of which are found only in the eggs, while another two forms are found mainly in the skin. Others include luteone and cadlinaldehyde (Bornancin et al. 2017).

The sponge-feeding Chromodorididae contains over 300 species (Johnson & Gosliner 2012) and have modifications of the dorsum called *mantle dermal formations* (MDFs) in which concentrated secondary metabolites from the sponge prey are stored (Avila & Durfort 1996). Each genus has a unique arrangement of these MDFs (Rudman 1984).

Besides their use in defence, the ability to store toxic metabolites obtained from food in these special organs might assist in isolating those toxins from the rest of the body. This may have enabled the chromodoridids to feed on even more toxic prey, opening up new adaptive advantages (Wägele 2004). Although sometimes considered a synapomorphy of chromodoridids, Wägele (2004) pointed out that MDFs were not restricted to that family, but were present in a polycerid, *Limacia clavigera*, a non-specific bryozoan feeder, and in the sacoglossan *Plakobranchus ocellatus* (Plakobranchidae), an algal feeder. Nevertheless, MDFs presumably played a significant role in the evolution and diversification of chromodoridids (e.g., Rudman 1984; Gosliner & Johnson 1999), which can be highly specific regarding their sponge prey (Rudman & Bergquist 2007). Their bright aposematic colours and patterns contrast with the more cryptic colouration of other families of cryptobranchiate nudibranchs (Gosliner 2001; Wägele 2004).

Most chromodoridids feed upon non-spiculate dictyoceratid and dendroceratid sponges that contain an array of defensive secondary metabolites, notably terpenoids (Rudman & Bergquist 2007). In contrast, the related Actinocyclidae are drab-coloured and feed on halisarcid sponges that lack defensive metabolites and spicules.

Members of each chromodoridid genus specialise in utilising particular terpenoids. Amongst the array of metabolites reported from sponges, some are unique to particular sponge taxa and often found in chromodoridid mantles. Of the dictyoceratid sponges, three families (Spongiidae, Thorectidae, and Irciniidae) contain sesterterpenes, but sesquiterpenes are found in Dysideidae. While the Spongiidae and Irciniidae have similar compounds, the thorectids have a unique sesterterpene molecule (scalaradial) that is often found in the mantle glands of the chromodoridids that feed on them. In contrast, the dendroceratid sponges (Darwinellidae and Dictyodendrillidae) have spongian diterpenes and are fed on by many chromodoridids.

As far as known, all species of the conspicuously coloured chromodoridid genus *Glossodoris* feed on thorectid (Demospongiae) sponges (Rudman & Bergquist 2007) with

G. hikuerensis and *G. cincta* sequestering the sesterterpene heteronemin from the sponge *Hyrtios* (Bornancin et al. 2017). Some chromodoridid species contain metabolites (all terpenoids) characteristic of Spongiidae (diterpenoids), such as pongiatriol (Han et al. 2018), rather than those typical of Thorectidae (scalarane sesterterpenoids), although some of the latter have been modified by the slugs to detoxify them. These metabolites are often in the mantle glands, and sometimes other parts of the body. Compounds found in the mantles of two species of *Glossodoris* deter fish feeding (Rogers & Paul 1991) as does the furanosesquiterpenoid metabolite longifolin extracted from *G. valenciennesi*; this compound has also been found in mantle tissues of species of *Hypselodoris* (Pawlik 1993). Other sponge-derived metabolites found in species of *Hypselodoris* include the furanosesquiterpenes tavacpallescensin, spiniferin, microcionin, furodysinin, and nakafuran (Bornancin et al. 2017). Some of these metabolites are also found in *Ceratosoma gracillimum* (furodysinin, nakafuran) and *Hypselodoris maridalidus* (nakafuran), while the sponge-derived furanoterpene tetradehydrofurospongin has also been found in a species of *Hypselodoris* (Bornancin et al. 2017).

Species of what was previously considered a single genus, *Chromodoris*, are brightly coloured and lack spicules in their integument. They are now thought to comprise two main groups (the genera *Chromodoris* and *Goniobranchus*) (Johnson 2010; Johnson & Gosliner 2012). In typical *Chromodoris*, nitrogenous macrolides (latrunculins) are characteristic. These fish-repelling compounds are probably produced by sponge symbionts. The sesquiterpenoid dendrolasin is also obtained from the sponge prey, as are macrocyclic fatty acid lactones. Four chlorinated homoditerpenes, a sesterterpene, and spongian diterpene lactones have also been found in species in this group.

In the large group of species now placed in *Goniobranchus* (Johnson & Gosliner 2012), most contain rearranged diterpenoids (for a review of diterpenoids in 'opisthobranchs' see Gavagnin & Fontana 2000). Aplysillins, aplyroseol norditerpene, and norrisolide similar to those in the host sponges (Darwinellidae), have also been located in some species, as have toxic sesterterpenoids and polybrominated biphenyl esters. A species of another genus of chromodoridids, *Tyrinna*, has been shown to have a sesquiterpenoid and other terpenoids (Fontana et al. 1998). The conspicuous chromodoridid slug *Ceratosoma* has projections near the gill armed with repugnatorial glands. The colour and prominence of the projections aim to divert attention away from the gill (e.g., Rudman 1991a). Chemicals noxious or toxic to fish are concentrated in these glands and include sesquiterpene furans and thiosesquiterpenes in one species and furanosesquiterpenoids in other species. These compounds are derived from dysideid sponges. Other genera of Chromodorididae, *Thorunna*, *Mexichromis*, *Risbecia*, *Felimare*, and *Hypselodoris*, also mostly feed on dysideid sponges and all are colourful. The first two genera have sesquiterpenoids sequestered from their dysideid sponge food, while *Risbecia* and *Felimare* have furanosesquiterpenoids. Some species of *Hypselodoris* contain terpenoids derived from their demosponge (Thorectidae) food including sesterterpenes and sesquiterpenes, with the former being modified to an apparently inactive form and stored in the mantle glands in *H. orsini*.

The chromodorid *Felimida norrisi* feeds on the sponge *Aplysilla* (Darwinellidae) and contains several spongian diterpenes, two of which (shahamin C and polyrhaphin C) inhibit fish feeding (Bobzin & Faulkner 1989). Polyrhaphin has also been found in another species of *Felimida*, *F. luteorosea* (Gavagnin et al. 1992). Two species of another chromodoridid genus, *Miamira*, sequester oxy-polybrominated diphenyl ethers in their mantle rather than terpenes (Dewi et al. 2016).

The Dorididae – the true dorids – contains several genera that can be brightly coloured or cryptic on their similarly coloured host sponge. The primitive doridid *Aldisa* contains steroids in the integument, one of which repels predators. These steroids are not found in the sponge on which one species feeds, although another species appears to modify a non-repellent steroid obtained from the sponge (Cimino & Ghiselin 2009).

The remaining members of Dorididae utilise terpenoid glycerides for defence with the cadlinid, *Cadlina*, and the doridid genera *Archidoris*, *Anisodoris*, and *Doris* synthesising diterpenoic acid diglycerides *de novo*, and one species of *Doris* (*D. tanya*) producing diterpene and sesquiterpene glycerides. While most of the terpenoids used are typical sponge metabolites, these slugs are the only nudibranchs to synthesise both the terpenoid and the glycerols. Thus, species of *Doris* contain noxious compounds that may, depending on the species, be terpenoid acid glyceride esters, sesquiterpenoid, or diterpenoid glycerides, and all are synthesised. In at least some species, a nucleoside used to defend the eggs is also synthesised.

While two compounds from *Archidoris* (a dimenoic acid glyceride and a glyceryl ether) deterred fish feeding, three of the other tested compounds did not (Gustafson & Andersen 1985). *Anisodoris nobilis* sequesters N-methylnucleoside doridosine originating from the sponge *Tedania* on which it feeds (Kim et al. 1981).

The Discodorididae are broad, flattened, mostly pale-coloured slugs with conspicuous large dark spots in several species. They contain a variety of metabolites derived from sponges. There are two groups, one with sensory organs (caryophyllidia) of unknown function on the dorsal surface (Valdés & Gosliner 2001). The eastern Pacific *Diaulula sandiegensis* feeds on a variety of sponges and sequesters polyacetates with nine chlorinated polyacetylenes identified at San Diego, California, some or all of which may be involved in chemical defence (Walker & Faulkner 1981). In another location (British Columbia, Canada), supposedly the same species contained two steroids, presumably because a different sponge prey was being consumed (Williams et al. 1986). At least one metabolite (diaulusterol A) has the polyacetate portion of the molecule synthesised *de novo*, but there was no evidence of acetate incorporation into the steroid part (Kubanek & Andersen 1999). *Jorunna funebris* contains isoquinoline quinones obtained from a petrosiid sponge. Isoquinoline alkaloids, possibly also derived from a sponge, have been recovered from the integument and mucus of this species (Cimino & Ghiselin 2009) and a derivative of one of these was trialled as an anticancer drug.

Discodoridids that do not have caryophyllidia include *Peltodoris*, *Paradoris*, *Asteronotus*, and *Halgerda*. *Peltodoris* contains various sterols and high-molecular-weight polyacetylenes derived from petrosiid and chalinid sponges, but these metabolites are only found in the digestive gland and are thus not used in defence. Instead, the integuments of these slugs are very spiculose. Some species of *Paradoris* are cryptic. *P. indecora* lives on irciniid sponges from which it obtains and sequesters furanosesterterpenoids in dorsal tubercules and which are released when it is disturbed (Marin et al. 1997).

The rather drab *Asteronotus caespitosus* feeds on dysideid sponges and contains metabolites including phenols, quinones, and sesquiterpenoids. The digestive gland (but not the integument) contains cytotoxic polybrominated diphenyl ethers and a hexachlorinated alkaloid. Sesquiterpenes are found in the integument of specimens from some areas, presumably also obtained from their food (Cimino & Ghiselin 2009).

Most species of *Halgerda* are of cryptic appearance. Of those surveyed for secondary metabolites, these cryptic species lack defensive compounds, unlike two conspicuously coloured species, *H. gunnessi* and *H. aurantiomaculata*, which contain toxic compounds obtained from their food. *H. aurantiomaculata* contains several alkaloids, while *H. gunnessi* contains mixtures of acylated tetrasaccharides.

9.4.3.4.9 Nudipleura – Nudibranchia – Cladobranchia

Of the Tritonioidea, only members of Tritoniidae and Tethydidae have been studied. *Tritonia hamnerorum* feeds on sea fans (gorgonians) and sequesters from them a furano germacrene sesquiterpenoid. In experiments, the intact slugs are not attacked by fish. Another tritoniid, *Tochuina tetraquetra*, feeds on a soft coral (*Alcyonium*) from which it obtains cuparane sesquiterpenoids. In addition, diterpenoids were found in this slug and one of these was also found in the soft coral. The Antarctic tritoniid *Tritoniella belli* feeds on an octocoral (*Clavularia*) and contains the glyceride ester chimyl alcohol, which protects them from starfish (McClintock et al. 1994).

The highly modified Tethydidae have paddle-shaped cerata and catch small crustaceans. *Tethys fimbria* contains prostaglandin lactones (prostaglandins are defensive metabolites in many octocorals), and these stimulate the regeneration of their cerata, which are readily autotomised. *Melibe viridis* has similar prostaglandin lactones to those in *Tethys*, but other species of *Melibe* lack these defensive compounds and instead have terpenoids that are probably produced *de novo* (Barsby et al. 2002). Like prostaglandins, terpenoids are defensive metabolites in many octocorals, the assumed ancestral food source of these two genera.

The arminid (Arminoidea) *Armina maculata* contains diterpenoids and similar compounds probably derived from the pennatulaceans (Octocorallia) on which they feed (Guerriero et al. 1988, 1990). Another species of *Armina* (*A. babai*) contains a waxy lipid (ceramide) that has also been isolated from a gorgonian (Ishibashi et al. 2006). *Dermatobranchus ornatus* contains different diterpenoids that are also found in gorgonians (Zhang et al. 2006; Cimino & Ghiselin 2009), while *D. otome* contained three sesquiterpenoids of unknown origin (Ishibashi et al. 2006). *Leminda millecra* also contained sesquiterpenes and quinones typical of gorgonians on which they feed (Pika & Faulkner 1994). *Antiopella* (as *Janolus*) *cristatus* (Janolidae, Proctonotoidea) is a bryozoan feeder and consequently has different metabolites, notably a lipotripeptide (janolusimide) (Cimino & Ghiselin 2009).

Some aeolidioideans use aposematic colours and it has been suggested that this is associated with the stored nematocysts (Aguado & Marin 2007) in their cerata (see Section 9.4.3.12 and Chapter 20).

Although most aeolidioids do not use defensive chemicals, there are exceptions. Some hydrozoans contain secondary metabolites such as modified steroids and these are found in the aeolidioids that prey on them, notably species of *Hervia* and *Flabellina*. A species of *Cratena* contains prenylphenols in its skin and two anthozoan feeders, *Phyllodesmium* and *Phestilla*, sequester secondary metabolites (Cimino & Ghiselin 2009; Bogdanov et al. 2014). A species of *Phestilla* contains alkaloids derived from a compound found in the hard coral *Tubastraea* on which it feeds. Species of *Phyllodesmium* feed on soft corals (Octocorallia) and utilise their symbiotic zooxanthellae to photosynthesise (see Chapters 5 and 20). Their cnidae are apparently not used in defence, and these animals are very cryptic when on their hosts. One species uses terpenoids for defence, and another contains a diterpenoid derived from their food that deters fish.

Relatively few experimental studies have been carried out on the effectiveness of the cnidocysts in the cnidosacs in deterring predators (see Chapter 20). One study suggested that the aeolidioidean nematocysts are not functional, but that chemical defences are more likely (Penney 2009). The relative use of chemicals and nematocysts in aeolidioidean defence requires much more study before definitive conclusions can be drawn (Edmunds 2009).

9.4.3.4.10   *Panpulmonata – Sacoglossa*

Sacoglossan chemical defences have been reviewed by Cimino and Ghiselin (1998, 2009) and Marín and Ros (2004). The shelled sacoglossans *Ascobulla* and *Oxynoe* feed on *Caulerpa*. They modify a sesquiterpenoid (caulerpenyne) derived from the alga for defence by turning it into more toxic forms (oxytoxins) that are stored in the body and released in mucus when the animal is disturbed. In addition, *Oxynoe* sequesters palmitic acid (a fatty acid), beta-sitosterol (a phytosterol), and caulerpin (Bornancin et al. 2017). *Lobiger*, another shelled *Caulerpa* feeder, does not contain caulerpenyne, but does have the toxic derivative oxytoxin (see Cimino & Ghiselin 2009 for references) as well as caulerpin.

The shell-less lineages of sacoglossans exploit a variety of green algae as food and defensive metabolites, often polypropionates. The switch to a different algal food was facilitated by *de novo* synthesis of polypropionates for defence and their modification as photoactive sunscreens for the 'solar powered' taxa (see Cimino & Ghiselin 2009 for discussion and

details), or involvement in the regeneration of cerata. Cimino and Ghiselin (2009) speculated that these *de novo* compounds in sacoglossans either originated as defensive metabolites that were then coopted to stimulate regeneration or whether the opposite occurred, with the compounds originally involved in wound healing. Polypropionates are not present in shelled sacoglossans, but are found in some species of *Elysia* and *Plakobranchus* (Plakobranchidae). Taxa that practise autotomy (see Section 9.4.3.8) of the cerata for defence, such as *Ercolania* (Limapontiidae) and *Cyerce* (Caliphyllidae), often synthesise propionates *de novo*, and sometimes these compounds have been shown to stimulate ceratal regeneration (Di Marzo et al. 1993). *Placida* (Limapontiidae) relies on polypropionates for defence, but does not autotomise.

Species of *Elysia* produce several cytotoxic peptides (kahalalides and diterpenes) derived from feeding on the chlorophyte *Bryopsis* (see Benkendorff 2010; Bornancin et al. 2017). Those that feed on the chlorophyte *Caulerpa* utilise the toxic sesquiterpene caulerpenyne, produced by the alga for their defence, which, in some species of *Elysia*, is converted to oxytoxin and at least one species also utilises caulerpin (Bornancin et al. 2017). Some which feed on other green algae utilise different secondary metabolites. For example, some species of *Elysia* and *Bosellia* (Boselliidae) live on the calcareous green alga *Halimeda*, which produces the toxic trialdehyde secondary metabolites halimedatrial and halimeda tetraacetate, which certain species utilise for their defence (Paul & Alstyne 1988; Marín & Ros 2004; Bornancin et al. 2017).

Unlike the cryptic colouration of many sacoglossans, most species of the plakobranchid genus *Thuridilla* are brightly coloured, suggesting that they employ chemical defence. Two species have been shown to contain diterpenoids derived from their green algal food which are modified as several detoxified forms of the diterpenoid thuridillin (Bornancin et al. 2017).

A species of *Costasiella* (Stiligeridae) preferentially feeds on the green alga *Avrainvillea*, which contains a brominated diphenylmethane derivative, avrainvilleol, a fish antifeedant sequestered by the slug (Hay et al. 1990). This sea slug contains γ-pyrone polypropionates (which may be synthesised *de novo*), α-pyrones, and an unusual hydroperoxide. The limapontiid *Ercolania* eats *Chaetomorpha* and also contains γ-pyrone polypropionates. Endogenous polypropionates are also found in the caliphyllid *Cyerce nigricans* which feeds on another green alga, *Chlorodesmis*, and contains the diterpenoid chlorodesmin, as does a species identified as *Elysia* sp. (Hay et al. 1989). Chlorodesmin has been shown to deter fish feeding (Pawlik 1993). Similarly, *Mourgona*, another caliphyllid, obtains prenylated bromohydroquinones (such as cyclocymopol) from its food source *Cymopolia*, a calcareous green alga. This metabolite is secreted by the slug in a viscid toxic mucus (Jensen 1984). This compound is structurally similar to a brominated diphenylmethane derivative found in the limapontiid *Costasiella* (Cimino & Ghiselin 2009).

Cyerce autotomises its cerata when attacked, but they regenerate rapidly and, like other caliphyllids, placidenes and cyercenes are used in defence and regeneration (Cimino &

Ghiselin 2009). Polypropionates have not been detected in the cryptic species *Caliphylla mediterranea*, and its cerata do not readily autotomise. Pyrones are present in the hermaeid *Aplysiopsis formosa*.

9.4.3.4.11 Panpulmonata — Marine 'Pulmonates'

Polypropionates are produced by some marine 'pulmonates', namely, the siphonariids[7] and *Onchidium* (e.g., Manker et al. 1988, 1989; Manker & Faulkner 1989; Rodríguez et al. 1992; Garson et al. 1994; Norte et al. 1994; Paul et al. 1997; Beukes & Davies-Coleman 1999). In addition, some novel fatty acid derivatives have been isolated from *Siphonaria* (Carballeira et al. 2001), pyrone esters and a novel peptide from *Onchidium* (Ireland et al. 1984; Fernandez et al. 1996), sesquiterpenoids from *Onchidella* (Ireland & Faulkner 1978), and *Trimusculus* produces sterols and terpenes (Manker & Faulkner 1996; San-Martin et al. 1996). Of the large variety of compounds in siphonariids, only one (the polypropionate vallartanone B) has been shown to deter fish (e.g., Pawlik et al. 1986; McQuaid et al. 1999).

The metabolites of some siphonariid and trimusculid limpets show surprising convergences with pleurobranchs (Cimino & Ghiselin 2009). For example, *Pleurobranchaea meckelii* has two diterpenes that resemble one from *Trimusculus reticulatus*, and *Pleurobranchus membranaceus* has a polypropionate (membrenone-C) that resembles vallartanone-D from *Siphonaria maura*.

Regarding non-marine 'pulmonates', the lack of regular reviews of the natural products of terrestrial molluscs (in contrast to the situation with marine taxa) may account for the small number of studies on stylommatophorans (Benkendorff 2010), despite their potential. For example, in helicids, there has been work on mucoproteins and bioactive lectins (the latter reviewed by Bonnemain 2005), while new terpenes have been found in *Achatina fulica* that appear to inhibit HIV-1 reverse transcriptase, an enzyme essential for HIV infection (Patil et al. 1993). At least one species of the stylommatophoran slug *Arion* can sequester and detoxify alkaloids from their food plants (Aguiar & Wink 2005) and Hesbacher et al. (1995) showed that some snails (*Balea* [Clausiliidae], *Chondrina* [Chondrinidae], and *Helicigona* [Helicidae]) sequester bioactive terpenes from their lichen food. Also, cardioexcitatory peptides have been identified in the fresh-water hygrophilian *Lymnaea stagnalis* (Tensen et al. 1998).

9.4.3.4.12 Caenogastropoda–Cypraeoidea

Some ovulids have been investigated, including *Ovula ovum* that feeds on the soft coral *Sarcophyton* and transforms the terpene found in the soft coral tissue into a less toxic (to fish) compound (Coll et al. 1983). *Cyphoma gibbosum* feeds on gorgonian octocorals and has enzymes that modify a suite of toxic allelochemicals obtained in its food (Vrolijk & Targett 1992; Whalen et al. 2010). Another ovulid, *Simnialena uniplicata*, lives on the gorgonian *Leptogorgia virgulam*, which has secondary metabolites that deter fish predation on the gorgonian, but this is not the case with *S. uniplicata*, which has apparently not acquired the antipredator defence from its prey item, as this snail is eaten by fish. In this species, extracts from both the ovulid and the gorgonian contained diterpenoid hydrocarbons that inhibited the settlement of barnacle larvae (Gerhart et al. 1988).

Antimicrobial properties have been observed in extracts from species of cowries (Cypraeidae) (Anand & Edward 2002), but there do not appear to be credible reports of the mantle or other parts of the head-foot containing noxious chemicals.

9.4.3.4.13 Caenogastropoda–Velutinoidea

Lamellarins, pyrrole-derived alkaloids, were first isolated from the slug-like velutinid *Lamellaria* (Andersen et al. 1985), but subsequently other related lamellarins have been obtained from ascidians (on which velutinids feed) and the sponge *Dendrilla* (Fan et al. 2008). A staurosporine analogue was isolated from another velutinid, *Coriocella nigra*, while *Coriocella hibyae* had two lamellarins and that have shown antitumour activity (Cantrell et al. 1999). Similar staurosporines have been obtained from an ascidian (Reyes et al. 2008).

9.4.3.4.14 Vetigastropoda and Other Molluscs

There are few examples of chemical deterrents from other molluscs. The only patellogastropod known to have chemical defence is *Lottia limatula*, with the triterpene limatulone, which deters feeding by 'kelpfish' (*Gibbonisia elegans*) (Albizati et al. 1985; Pawlik et al. 1986). Despite this, Lindberg et al. (1987) have shown that *L. limatula* is a preferred prey of oystercatchers.

Some abalone (*Haliotis* spp.) respond to starfish attack by releasing mucus from the exhalant shell pores. The mucus release is thought to have repellent properties and to obscure the direction in which the abalone escaped (Bancalà 2009). There is also some evidence that the same secretions also act as an intraspecific alarm signal (Bancalà 2009). A chemical deterrent in the epipodium of *Haliotis rubra* has also been reported (Day et al. 1995), but similar strategies do not appear to have been reported in other vetigastropods.

In some bivalves, an unusual adaptation against vertebrate predators, such as the venerids *Saxidomus* spp. and the mactrid *Spisula solidissima*, is that they appear to sequester paralytic shellfish poisoning toxins (PSPTs) in their siphonal tissues. These toxins, notably saxitoxin, are obtained from planktonic dinoflagellates such as *Protogonyaulax* spp. The nervous system of *Saxidomus* is unusual in being tolerant of high concentrations of saxitoxin (Kvitek & Beitler 1991). Consumption rates of the toxin-containing bivalves were much lower than normal levels when fed to gulls (Kvitek 1991a), sea otters (Kvitek et al. 1991), and siphon-nipping fish (Kvitek 1991b) (see also Section 9.4.3.10.3).

Enzymatic detoxification of certain toxic chemicals in algae has been demonstrated in an abalone and a chiton (Kuhajek & Schlenk 2003), and certain bivalves can detoxify paralytic shellfish toxins (Bricelj & Shumway 1998).

---

[7] See earlier note on siphonariid polyproprionates.

The properties of the ink of coleoid cephalopods are discussed in Chapter 17.

9.4.3.5 Antimicrobial Secondary Metabolites and Sunscreens Associated with Gastropod Spawn

Gastropod egg capsules provide protection against predation and environmental stresses such as desiccation and osmotic stress, but the physical barrier provided by the capsule wall may not offer sufficient protection from microbial infection or UV radiation (e.g., Przeslawski 2004b).

Antimicrobial and/or antifungal protection of spawn by way of secondary metabolites has been demonstrated in several instances. For example, antifungal activity has been shown in nudibranch egg masses (e.g., Matsunaga et al. 1986) and antibacterial activity in those of *Aplysia* (e.g., Kamiya et al. 1984; Kisugi et al. 1989) and nudibranchs (e.g., Matsunaga 2006).

Benkendorff et al. (2001b) investigated antibacterial activity in the egg masses of 38 species of gastropods including members of Neritimorpha (1), Caenogastropoda (12) and Heterobranchia (25), and the cephalopod *Sepioteuthis*, with the gastropods mostly marine, but including two fresh-water and one terrestrial species. Three human pathogenic bacteria (*Escherichia coli*, *Staphylococcus aureus*, and *Pseudomonas aeruginosa*) were used in the tests. The cephalopod and all the gastropods showed some antimicrobial activity. Benkendorff et al. (2005) later tested the antimicrobial properties of free fatty acids and sterols in the spawn of some aquatic gastropods and *Sepioteuthis*. They found significant antimicrobial activity in the spawn of aplysiids, *Philine*, amphiboloideans, siphonarioideans, and a planorbid, and in both the gelatinous spawn and leathery capsules of caenogastropods other than neogastropods, and *Sepioteuthis*. The egg capsules of at least some muricids have antimicrobial protection provided by the precursors of Tyrian purple from the hypobranchial gland (Benkendorff et al. 2000, 2001a; Westley & Benkendorff 2008). Antifouling activity has also been observed in extracts from three muricid egg masses (Ramasamy & Murugan 2007).

Increased ultraviolet radiation because of ozone depletion has led to research in sunscreen activity in egg masses (e.g., Przeslawski 2004a; Przeslawski et al. 2004). Some taxa avoid this problem by depositing egg masses in shaded locations, but others, such as several species of siphonariids, lay their spawn in full sunlight and suffer high levels of mortality despite natural sunscreens, including the mycosporine-like amino acids (Shick & Dunlap 2002), being present (Russell 2008).

9.4.3.6 Luminescence

A few taxa use flashes of light (bioluminescence) to startle predators (see Section 9.4.8).

9.4.3.7 Aggressive Movements, Speed and Agility

Most mobile molluscs are famed for slow movement, so 'speed and agility' hardly seem applicable, although not all are 'sluggish'. Some squid, for example, are among the fastest-moving

animals, and they and their relatives use their speed to avoid predators and to capture prey.

Most molluscs rapidly retreat into their shells when threatened – limpets and chitons clamp tightly to the substratum and bivalves retract their siphons and close their shells. Some gastropods and bivalves move quickly in response to touch by, or chemicals from, predators, for example, the leaping response of cardiids and strombs or the twisting behaviour of limpets in response to contact with predatory starfish (e.g., Phillips 1975, 1976) (see Chapter 3), the thrashing of the foot by some snails (e.g., Fishlyn & Phillips 1980), or, in the intertidal, their movement out of the water (e.g., Phillips 1978). Other responses include the very rapid burrowing by some bivalves or olivid snails (Phillips 1977) or small gastropods living on seagrass dropping off if they detect a predatory starfish (e.g., Fishlyn & Phillips 1980).

Gymnosome pteropods are efficient swimmers that can rapidly change their wing beat rate to enable higher speeds when chasing prey (Seibel et al. 2007).

9.4.3.8 Autotomy

Some animals can discard a part of their body that has been grasped or when the animal is very disturbed. This is known as autotomy or self-amputation and usually involves breakage at a zone of weakness. The discarded body part often continues to move independently so the potential predator is distracted by it. Skinks and geckos that drop their tails are a well-known example of this phenomenon, but there are also numerous invertebrate examples, including arthropods, echinoderms, and molluscs. In molluscs practising autotomy, regrowth of the discarded structure occurs in a matter of weeks or months (see Stasek 1967; and Fleming et al. 2007 for reviews).

When the foot of the caenogastropod fig snail *Ficus ficus* is mechanically stimulated, a portion of the mantle on the side of the inner lip swells then autotomises. This part of the mantle has an extensive network of muscle fibres and connective tissue, which, during autotomy, expands three-fold on the dorsal side and 15-fold on the ventral side (Liu & Wang 2002).

Stomatelline trochids such as *Gena* are well known for the autotomy of their foot. This, and the subsequent regeneration of the foot, have been studied in some detail in *Gena varia*. An autotomy line separates the metapodium from the propodium and enables autotomy. The animals can regenerate the discarded part of the foot and re-autotomise and re-regenerate several times (Fishelson & Qidron-Lazar 1966; Fishelson & Kidron 1968). Some caenogastropods can also autotomise the posterior end of the foot, while others such as some cypraeids (Burgess 1970), *Harpa* (Liu & Wang 2002), and the olivid *Agaronia* (Rupert & Peters 2011) autotomise the posterior part of the foot, as do some stylommatophoran slugs (arionids, agriolimacids, and limacids), when threatened by a predator (e.g., Pakarinen 1994). In an arionid, autotomy is achieved by the fast contraction of a ring of muscles (Luchtel & Deyrup-Olsen 2001). It is also known in a species of camaenid land snail, a member of the genus *Satsuma*, that is the prey of a snail-eating snake (Hoso 2012). The large Cuban *Polydontes* snails can discard their 'tail', which is reported to then twitch for up to 54 hours (Solem 1974) providing an alternative target for a predator.

Some sea slugs, such as the nudibranch *Discodoris fragilis* and the pleurobranch *Berthella martensi*, can cast off the mantle skirt. Autotomy of the cerata has been described in *Melibe leonina* and involves complex interactions of muscles and nerves (Bickell-Page 1988, 1989). The cnidosac-bearing cerata of some aeolidioid nudibranchs also undergo autotomy (e.g., Piel 1991; Miller & Byrne 2000). *Phyllodesmium*, an aeolidioid that does not store nematocysts, has developed, in place of the cnidosac, a large gland that produces a sticky secretion. When disturbed by a potential predator, the animal drops several cerata that move about and become sticky from the secretion (e.g., Burghardt et al. 2008a); cerata are also discarded when the animal is starved. *Oxynoe* (a sacoglossan) can autotomise the tail of the foot (Lewin 1970). Limid bivalves can readily autotomise their long, sticky mantle edge tentacles (Gilmour 1963, 1967), and some other bivalves can autotomise their siphons (e.g., Hodgson 1984).

Some octopuses and squid undergo autotomy of their arms when attacked (e.g., Norman 1992; Norman & Finn 2001; Norman & Hochberg 2005; Bush 2012). Arm autotomy typically occurs within an 'autotomy zone' or 'fracture plane' – an anatomically weak area. In the octopus *Abdopus aculeatus*, this occurs between proximal suckers 4 and 8 (Alupay 2013). In the squid *Octopoteuthis deletron*, 'fracture planes' were associated with arm autotomy in only two of the 15 individuals examined (Bush 2012), suggesting that autotomy in this species occurred at potential, rather than at pre-formed, fracture planes. The autotomised arm of *O. deletron* typically writhes about after separation and the terminal photophore bioluminesces, possibly to distract the attacker. In some octopuses, the hectocotylus arm is detached during mating and remains within the mantle cavity of the female.

9.4.3.9 Crypsis

Avoidance of detection by a predator can be achieved by various means, such as transparency, mimicry, and camouflage (including mimesis), or by evolving behaviours such as being nocturnal, burrowing, hiding in crevices, under rocks or logs, and so on.

Many molluscs are adept at camouflage, the employment of a cryptic colour, and/or shape that blends with the background. This may be done by the selection of an appropriate background that suits the body form and/or colour, or by colour/shape changes, provided the taxon concerned has the visual competence to discern backgrounds. Other methods include disguising the body with sediment, epizooids, or epiphytes. The thorny oysters (Spondylidae) are good examples where the spines on the shell, in addition to the possible antipredator role discussed above, provide excellent surfaces for the growth of sponges and other animals and attract algae and trap sediment.

Many patellogastropods (notably lottiids) often come to resemble the background on which they occur (e.g., barnacles, algae, other molluscs, etc.) (Giesel 1970; Lindberg &

Pearse 1990; Sorensen & Lindberg 1991; Espoz et al. 2004). Colour correspondence appears to be driven by the ingestion of the algae on which they graze, although this mechanism does not explain accompanying changes in shell sculpture seen in some taxa (Lindberg & Pearse 1990). In some subtidal species (*Acmaea*, *Erginus*), the shell is overgrown by the substratum (bryozoans, coralline alga) on which they are living (e.g., McLean 1966). Very good examples of visual crypsis are also seen in some specialised caenogastropods, notably ovulids living on soft corals and velutinids living on ascidians, where their animal has the colour and texture of the host. Xenophorid caenogastropods attach empty gastropod shells aperture up and the inner sides of empty valves of bivalves, and sometimes small pebbles, to their shells (e.g., Morton 1958c). In an unusual example, chemical camouflaging is used by the seagrass lottiid limpet *Tectura paleacea*, which incorporates flavonoid compounds from *Phyllospadix* on which it lives into its shell, making it largely undetectable by the small predatory starfish *Leptasterias hexactis* which occupies the same habitat (Fishlyn & Phillips 1980).

Many nudibranchs are difficult to see against their food, such as sponges, and sacoglossans blend perfectly with their green algal food. These animals not only match their host in colour, but often also in texture. In *Aplysia*, and presumably other aplysiidans, chemicals obtained from their algal food provide cryptic pigmentation, and these are stored in cellular structures in the skin (Prince & Young 2010).

Non-marine molluscs are often drab-coloured and match their backgrounds well, but a few land snails are brightly coloured and may be polymorphic (see Section 9.4.3.11). One land snail, *Napaeus barquini*, from the Canary Islands, lives on rock faces covered with lichens, and it covers its shell with lichens fixed in place by the snail (Allgaier 2007).

9.4.3.9.1 Changing Colour

Benthic coleoid cephalopods (cuttlefish and octopus) are famous for their extraordinary ability to rapidly change colour and shape (see Chapter 17).

A few 'opisthobranchs' can also change colour to match their surroundings, although more slowly than coleoids. The 'bubble shell' *Haminoea navicula* has pigmented epithelial and subepithelial cells; when these cells are extended or contracted, the external colour can change from black or dark brown to white within a few hours (Edlinger 1982).

9.4.3.9.2 Countershading

Countershading is the deployment of different colours or shades on the upper and lower surfaces of animals to make them more cryptic and reflects different adaptive pressures from the dorsal versus the ventral views (Ruxton 2004). This is most common in pelagic or surface floaters (some squid, *Glaucus*, *Janthina*), which have dark colour above and pale below, although similar traits are seen in some arboreal snails.

As in pelagic fish, pelagic squid are counter-shaded (ventral surface lighter than dorsal surface) to avoid predation. Pelagic squid also use photophores (light organs) to achieve countershading (see Chapter 17). Countershading is also seen in some swimming bivalves such as the pectinids *Amusium* and *Placopecten*, where the ventral valve is markedly paler than the dorsal (Thayer 1971).

9.4.3.10 Warning Colours and Behaviour – Aposematism

Distinctive colouration or behaviour is used by some species to 'warn' potential predators that they are dangerous or distasteful. The concepts and theory of warning colouration are beyond the scope of this work, but the interested reader can refer to discussions in the literature (e.g., Guilford & Cuthill 1991; Guilford & Dawkins 1993; Pawlik 2012).

As described above (see Section 9.4.3.4.1), many nudibranch slugs often have distasteful or poisonous chemicals in their bodies, and these species typically have very distinctive colours and shapes that serve to 'educate' and 'warn' potential predators (Figure 9.9). Such signals can be reinforced by *Müllerian mimicry* where two or more species are very similar in their aposematic signalling (e.g., colour) and are also poisonous or distasteful. For example, there are several rather restricted geographic groups of chromodoridid species with very similar colour patterns and Rudman (1991a) suggested these colour groups are evidence of Müllerian mimicry.

Examples are the red-spotted species from southeastern Australia and the 'blue chromodoridids' of the Mediterranean (Ros 1977). In these cases, the similarly coloured sea slug species are common enough for potential predators to learn that they are distasteful. Other common colour patterns are found over large geographic areas and these seem unlikely to be explained by aposematism.

The limid bivalve *Ctenoides ales* produces a bright, flashing light display using light scattering silica nanospheres embedded in their tissue, as well as rapid mantle movement (Dougherty et al. 2014). Dougherty et al. (2017) demonstrated a significant increase in the flash rate of *C. ales* related to both increases and decreases of light (shadow response) and suggested that flashing in this species may function as an aposematic signal (see also Section 9.4.3.10.2).

9.4.3.10.1 Batesian Mimicry

In Batesian mimicry, the mimic does not share the attribute that makes it unpalatable to predators. This type of mimicry is rather common, with some well-known examples where a harmless species has evolved to closely resemble a dangerous one, such as flies that look like wasps. In the marine realm examples include polyclad flatworms that

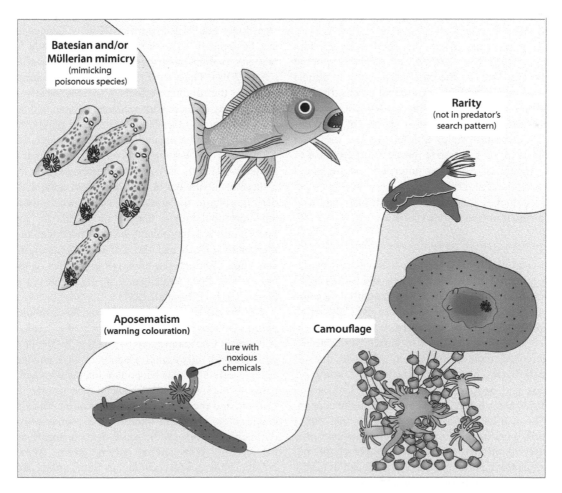

FIGURE 9.9 Examples showing how colour provides sea slugs with protection from predators. Redrawn and modified from Rudman, W.B., *J. Molluscan. Stud.*, 57, 5–21, 1991a.

mimic, toxic phyllidiid dorid nudibranchs; a good example being *Pseudoceros imitatus* that imitates the noxious nudibranch *Phyllidiella pustulosa* (Newman et al. 1994). Besides flatworms, some nudibranchs and even a holothurian also mimic phyllidiid nudibranchs (Rudman 2004). An extraordinary cephalopod example is the 'mimic octopus' *Thaumoctopus mimicus* (Norman & Hochberg 2005) that alters its body shape and colour to resemble venomous taxa such as sea snakes or lionfish. Other examples of mimicry are not so readily explained, such as amphipods mimicking the columbellid *Alia carinata*, and shell patterns of juveniles of *Littorina mariae* mimicking a tube-building serpulid polychaete that co-occurs on algal fronds (Reimchen 1989).

9.4.3.10.2 *Deimatic Display*

Sudden flashes of colour or patterns can startle predators. These deimatic displays are best seen in some cephalopods, notably blue ring octopuses and cuttlefish (e.g., Adamo et al. 2006).

This behaviour is also seen in a few nudibranchs where bright colours that are usually hidden can be rapidly displayed to startle potential predators. For example, the 'Spanish dancer', *Hexabranchus sanguineus*, suddenly displays a brilliant red and white colour pattern on the dorsal side of its mantle when it is disturbed and begins to swim, and many glaucid aeolidioids erect their often brightly coloured cerata when disturbed.

When on its green algal food, the shelled sacoglossan *Lobiger* is very cryptic. If disturbed, the brightly coloured tentacles around the edges of the parapodia unroll to almost double the size of the animal. In extreme situations, the tentacles can break off and writhe (see Section 9.4.3.8).

The long dorsal 'horn' seen in most species of the nudibranch genus *Ceratosoma* acts as a defensive lure to attract potential predators to the location (Figure 9.9) where most of the distasteful chemicals (furanosesquiterpenoids) are stored (Rudman 1998; Mollo et al. 2005).

The brightly coloured long tentacles of limids may also have a warning function (see also 9.4.3.10).

9.4.3.10.3 *Prey Avoidance of Temporarily Toxic Molluscs*

Shellfish toxins can modify the feeding behaviour of predators (including humans) of bivalves. Kvitek and Bretz (2005) showed that harmful algal bloom toxins could change the dietary preferences and foraging behaviour of five Californian shorebirds feeding on molluscs and crabs. For example, oystercatchers decreased their intake of molluscs with ingested toxins and increased their consumption of those that do not accumulate the toxins, notably limpets (mostly *Lottia* spp.). Similarly, Kvitek and Bretz (2004) showed that Californian sea otters modified their behaviour dependent on the concentration of PSPTs in their preferred bivalve prey (the 'butter clam', *Saxidomus giganteus*). In areas of intermediate prey toxicity, the otters continued to forage on butter clams, but discarded the most toxic parts of the body, although in highly toxic locations they avoided butter clams and other large toxic bivalve prey, and consumed smaller non-toxic species. These authors also noted that 'butter clams' were larger and more

abundant in highly toxic feeding areas, suggesting that areas with high concentrations of paralytic shellfish poisoning provide a refuge from sea otter predation for these bivalves.

9.4.3.11 Spreading the Threat: Polymorphisms

A polymorphism exists where more than one morph is found in a population. Different morphs can be effective in reducing predation pressure and are typically a phenotypic expression of a genetic polymorphism. For example, where there is considerable small-scale variation in habitat, a different colour/pattern variation will have different selective advantages. In such cases, polymorphisms would have a selective advantage. A famous example is the different colour and banding patterns in the European land snail *Cepaea nemoralis* that was used to test early ideas on the adaptive advantages of polymorphisms. The main predator of this snail is the 'song thrush' (*Turdus philomelos*) and, because they more easily detect snails that stand out from the background, it was shown to be important in maintaining the polymorphism (Cain & Sheppard 1954; Clarke 1962). This is complicated by predation also being frequency dependent – the predators favour the most common morphs[8] – another factor that might assist in maintaining the polymorphism. There are other nonpredator-related factors, such as the effect of shell colour on thermoregulation (Jones et al. 1977) and, of course, genetics (e.g., Jones et al. 1980; Davison 2002). Changes in frequency can be rapid in *Cepaea nemoralis*, in as little as two generations, perhaps contributing to the success of that species (Oz'go 2011). Thus, most polymorphisms are probably maintained by the combined effects of various selective factors and the interplay of these with genetics, habitat, and behaviour.

A number of other land snails show polymorphisms, but few have been investigated. *Euhadra peliomphala* has different patterns of polymorphisms on different islands (Hayashi & Chiba 2004) and, in *Arianta arbustorum*, pale morphs have a thermal advantage in absorbing less solar radiation than dark morphs in the alpine grasslands where they occur and are more abundant in that habitat (Burla & Gosteli 1993). Spectacular polymorphisms such as those seen in the Florida tree snails *Liguus* and the Cuban *Polymita* remain largely unstudied although the underlying genetics of the polymorphism have been investigated in *Liguus fasciatus* (Roth & Bogan 1984; Hillis et al. 1987).

Colour differences in some taxa, for example, in *Nerita polita* (Grüneberg 1981), are largely the result of pseudopolymorphism, mainly due to changes in colour and pattern with age with only a small fraction of these changes visible in a single individual. Such a population includes a mix of colour patterns that intergrade with each other. Another neritoidean, *Clithon*, and the trochoidean *Umbonium* both exhibit considerable colour variations interpreted as pseudo-polymorphisms (Grüneberg 1980, 1982); some discrete forms are probably the result of conventional polymorphism. Apparent colour polymorphisms are common in some other shallow-water

---

[8] Apostatic selection is where a particular morph is targeted by a predator and is usually dependent on the higher frequency of the particular morph.

umboniines such as the Australian *Bankivia*, but have not been investigated. The neritoideans are hard-shore taxa, but the shallow-water umboniines are suspension-feeders that live in large numbers on sandy beaches at or below low tide. Both groups are presumably highly susceptible to visual predation by birds and fish, respectively.

Other investigated gastropod polymorphisms include the intertidal littorinid *Littorina mariae* where the different colour morphs match different coloured backgrounds (Reimchen 1979). Another group of littorinids, species of *Littoraria*, often show colour polymorphism (Reid 1986a, 1987) that may be accounted for in part by background matching, but might also relate to other factors such as differential heating properties (Cook & Freeman 1986; Reid 1987). McKillup and McKillup (2002) investigated the predation of *Littoraria filosa* by two parasitoid flies, both species of *Sarcophaga*, one of which attacked and killed more snails that did not match their background than those that did. The authors suggested that greater attention should be given to the importance of parasitoid insects in selection of morphs of terrestrial snails. Crab predation on the colour polymorphic tropical intertidal bivalve *Donax faba* has been shown to be non-random (Smith 1975).

9.4.3.12 Associations with Cnidarians to Ward Off Threats

A number of molluscs coopt the nematocysts of cnidarians for protection. These associations range from cnidarians living on the shells of living molluscs, molluscs living within corals, or molluscs preying on cnidarians sequestering the stinging cells and reusing them.

The relationship between sea anemones and shells occupied by hermit crabs is well known (e.g., Gusmão & Daly 2010). Some species of gastropod have one or more anemones attached to their shells and, for some, the anemones are host specific (see Ates 1997 for review), while in others they are more generalist (e.g., Mercier & Hamel 2008; Goodwill et al. 2009), although the anemones involved tend to be specialised in this habit. Some evidence suggests this relationship provides mutual protection against predators and more food resources (e.g., Mercier et al. 2011). The columbellid *Nitidella nitida* is associated with an anemone, but it is not known whether it simply uses it for protection or is also feeding on it (Robertson 1966). Relationships between scaphopods and anemones are also known (see Section 9.4.4.8.3.1).

Some gastropods and bivalves associated with living corals and other colonial cnidarians are presumably utilising these animals because they provide shelter and defence by way of their nematocysts. Such examples include the coralliophiline muricids and bivalves such as *Pedum*, while several other bivalves find shelter in the coral skeleton by boring into it (see review by Morton 1990) (see also Section 9.4.4.8.3.1 and Chapter 15).

Gastropods and bivalves living in association with hard corals have received particular attention. The gastropods are variously described as parasites or predators (the distinction between these relationships is often blurred) and several

caenogastropods and heterobranchs are obligate cnidarian feeders (see Chapter 5). Some members of three caenogastropod families have independently adopted symbiotic relationships with stony (scleractinian) corals – some epitoniids, ovulids, and muricids (*Drupella* and Coralliophilinae). Ovulids are mainly associated with gorgonians and alcyonarians, but *Jenneria* and *Pedicularia* feed on hard corals (e.g., D'Asaro 1969; Robertson 1970a).

Members of the muricid subfamily Coralliophilinae lack jaws and a radula and are all associated with cnidarians – a few with gorgonians, zoanthids, antipatharians, and alcyonarians – but most are associated with stony corals (Robertson 1970a). Some are embedded within the body of the host, others live attached to the outside where they form a scar on the coral surface. Three genera (*Leptoconchus*, *Magilus*, and *Magilopsis*) bore mechanically into living stony corals as juveniles and remain connected to the exterior by way of a small aperture. The coral tissue, partly broken down by salivary secretion, and the surface mucus, is sucked up by the proboscis and pumping action of the muscular oesophagus (Ward 1965; Oren et al. 1998; Baums et al. 2003).

Among the lower heterobranchs, members of the Architectonicidae feed on cnidarians, mostly anemones, with *Heliacus* associated with, and feeding on, *Zoanthus* and *Palythoa*, both colonial zoanthid sea anemones (Robertson 1967), but species of *Philippia* feed on stony corals (Robertson et al. 1970).

At least some coralliophilid larvae appear to be immune to the stinging cells of the coral polyps (Gohar & Soliman 1963) in contrast to species of *Philippia*, which have no immunity (Robertson et al. 1970).

Many nudibranchs feed on various cnidarians (see Chapters 5 and 20) and may have intimate relationships with them. Of note is the ability of some to sequester the stinging cells from their cnidarian prey and reuse them for their own defence purposes (see Chapter 20 for details). While this ability is not unique to nudibranchs, as it is also found in some ctenophores and platyhelminths, it is best studied in the nudibranch groups Aeolidioidea (Greenwood 1988, 2009; Putz et al. 2010) and Cladobranchia (*Hancockia* and *Embletonia*) (Martin et al. 2009, 2010).

The examples above relate to gastropods, but two interesting examples of cephalopods apparently utilising cnidarians for protection have been reported. Small individuals of the pelagic 'blanket octopus' *Tremoctopus violaceus* have been reported to carry pieces of *Physalia* tentacle on their dorsal arms for defence, these being held by the suckers (Jones 1963; Norman et al. 2002). A 'paper argonaut', *Argonauta hians*, has been observed to 'ride' jellyfish, but it is not entirely clear as to whether or not they are feeding on it, as suggested by Heeger et al. (1992), or simply clasping it. Strickland (in Norman 2000) observed an *A. hians* orienting the much bigger jellyfish for defence when danger approached, and it used the jellyfish as a platform from which to hunt smaller comb jellies and other prey items, thus conserving energy as *Argonauta* is negatively buoyant (Finn & Norman 2010).

9.4.4   Relationships with Other Organisms

Besides food sources and predators, there are many relationships with other organisms, including microorganisms, that are well known to be important to the well-being of particular animals and that have contributed significantly to their evolution (e.g., Ruby et al. 2004; Hickman 2005c; McFall-Ngai et al. 2005). Such relationships are generally called symbiotic (see Box 9.1). We cover some important examples of symbiosis (including parasitism) in Chapter 5, but here we look at the broader concept and terminology, examples of how the symbionts are transferred to the host, and the various kinds of parasitism (Box 9.1).

9.4.4.1   Symbionts of Molluscs (Including Parasites)

We provide below some examples of relationships in molluscs with other organisms, including commensals and parasites, the latter sometimes causing diseases. Often the line between parasites and commensals is blurred. Diseases caused by parasites significant in molluscan aquaculture are reviewed in Chapter 10.

Molluscs, like other animals, have evolved immune responses (reviewed in Chapter 2) to parasites and microorganisms causing disease.

We have not attempted to be comprehensive in the following overview, but instead offer examples to illustrate the range of relationships that occur. In some groups, notably 'bacteria' and 'protists', classifications have changed markedly since the introduction of molecular techniques. Often, for example, particularly in the older literature, 'bacteria' are not identified adequately enough to place them within a particular phylum. Thus, the examples given are mainly from the recent literature where there is greater certainty regarding the identification of the symbionts.

9.4.4.2   Associations with Viruses

Viruses are the most abundant life forms, but must infect a living cell to reproduce. The great majority infect 'bacteria', but all life forms can serve as hosts. Little is known about viruses in molluscs, except for some that cause diseases.

Viruses cause several diseases in both adult and larval molluscs, but details are not known for some commercially significant taxa, notably abalone and some bivalves. While many reports are not specific as to the virus encountered, the following have been reported from molluscs, some of which can cause diseases in humans. Of those in the latter category, it must be stressed these human pathogens do not harm the mollusc harbouring them, nor do they replicate in them. They are obtained from the environment, and their presence is often a result of sewage contamination. The classification below

---

BOX 9.1   SYMBIOSIS

Symbiosis is a close interaction between two species. The terminology surrounding the various symbiotic relationships is sometimes confusing. Broadly, symbiotic relationships may be mutualistic, commensal, or parasitic. We define some of the terms used as follows.

A *mutualistic* relationship is one in which both organisms derive a benefit.

A *commensal* relationship is one where two organisms live together and one benefits, but the other is not particularly harmed or benefited. Commensal relationships may involve transport (*phoresy*), shelter or housing (*inquilinism*), or using skeletal or other remains (*metabiosis*), such as gastropod shells being used by hermit crabs.

A *parasitic* relationship is where the relationship benefits one symbiont, but harms the other. Although some parasites can be quite large, they are always smaller than their host. Distinctions between 'normal' predation and parasitism are often difficult to make in real-life situations. A predator typically kills and consumes all or part of the prey, while a parasite rarely kills the host, at least initially.

An *obligate* symbiotic relationship is critical for the survival of one or both of the symbionts; if not it is *facultative*. It is possible for the relationship to be obligate for one of the symbionts and facultative for the other.

An *ectosymbiont* lives on the exterior of the other symbiont, while an *endosymbiont* lives inside. If the relationship is parasitic, these are described as *ectoparasites* and *endoparasites*, respectively.

Symbiotic relationships may also involve the acquisition of organelles such as plastids during feeding (*kleptoplasty*) as seen in sacoglossans, or the transfer of microorganisms (e.g., zooxanthellae) from food as seen in some aeolidioids.

Obligate parasites require a host for their survival during at least some stage of their life cycle, while a few are *facultative* parasites, being able to live freely or to parasitise. Some parasites are temporarily attached to the host (e.g., ticks, fleas, leeches), while others are associated with a host for most of their life cycle. Some are parasitic only as larvae.

Parasites of parasites are called *hyperparasites*.

Parasites with *direct life cycles* infect a single host, but those with *indirect life cycles* have two or more hosts.

The *primary* or *definitive* host is the final host in an indirect life cycle, while a *secondary* or *intermediate* host is a transitional host in an indirect life cycle.

When a parasite afflicts its host so it is impaired, the condition is called a *disease*, although diseases can also be caused by invasion of bacteria or viruses by way of ingestion or a wound.

follows the International Committee on Taxonomy of Viruses (ICTV) (https://talk.ictvonline.org/taxonomy/ – accessed 2 August 2018).

9.4.4.2.1   Order Herpesvirales

These DNA viruses, called 'herpes-like viruses' (Herpesviridae) in much of the literature, have been implicated in the high mortality of larval oysters (e.g., Elston 1997), and *Pecten* and *Ruditapes* in hatcheries (e.g., Renault & Novoa 2004). Herpes-like viruses are also known from adult oysters and abalone (McGladdery 1999; Savin et al. 2010; McGladdery 2011) and are associated with disease. A herpes-like virus has also been reported from the Chinese fresh-water mussel *Hyriopsis cumingii* (Liu et al. 1993).

One herpes-like virus (OsHSV-1, or OsHV) found in oysters is not a notable pathogen in *Ostrea*, but has been responsible for massive mortality in *Crassostrea gigas* (e.g., Segarra et al. 2010; Renault et al. 2012). An abalone herpes virus (AbHV) is related to the oyster herpes virus (Savin et al. 2010) and causes abalone viral ganglioneuritis (AVG), which, as its name suggests, affects the nervous system of abalone in several parts of the world (e.g., Corbeil et al. 2012). AVG has severely infected farmed abalone in China and Taiwan and in Victoria, Australia, and has also been found in wild populations.

9.4.4.2.2   Order Picornavirales

Picornavirus-like viruses (Picornaviridae) and other small virus-like particles have been reported from various bivalves including mytilids, *Pecten*, *Pinctada*, *Paphies*, *Ruditapes*, and *Cerastoderma* (McGladdery 1999; Renault & Novoa 2004; McGladdery 2011; Bateman et al. 2012) and also from a unionid (Ip & Desser 1984).

Raw oysters are infamous for being a source of the hepatitis A viruses that infect humans and that probably mainly originate from sewage, but other viruses in raw oysters and other bivalves can cause human health problems. Enteroviruses and aichiviruses (both Picornaviridae) have been identified from *Crassostrea gigas* by Le Guyader et al. (2008).

The viruses in this group comprise those with one or a few single stranded RNAs.

9.4.4.2.3   Order Ortevirales

Retro-like viruses (?Retroviridae) have been reported from some bivalves including *Mytilus*, *Cerastoderma*, and *Mya* where they may cause haemic neoplasia.

This group contains both single stranded RNA and DNA viruses.

9.4.4.2.4   Families Unassigned to an Order

Family Iridoviridae: These viruses are responsible for gill necrosis, haemocyte infection, and the mortality of pediveligers in oysters (e.g., Elston 1997; McGladdery 1999, 2011; Renault & Novoa 2004). They have also been implicated in causing tumours in the arm muscles of *Octopus* (Hanlon & Forsythe 1990). Members of this family have a double stranded DNA genome.

Family Papillomaviridae: Papovavirus-like viruses can cause neoplasia[9] and have been reported from some bivalves (*Crassostrea*, *Pinctada*, and *Mya*); similar viruses are known from several other bivalves, including infections of the labial palps (Elston 1997; McGladdery 1999, 2011). A 'papilloma-like' virus is responsible for gametocyte hypertrophy in oysters and possibly some other bivalves (Garcia et al. 2006).

Family Birnaviridae: Birnavirus-like viruses have been reported from several marine bivalves and the fresh-water *Corbicula*, and in *Littorina* and *Patella* (McGladdery 1999, 2011; Grizzle & Brunner 2009).

Families Togaviridae, Reoviridae, Adenoviridae, and Astroviridae: Members of these families have been recorded from bivalves (Renault & Novoa 2004; Le Guyader & Atmar 2007). A *Reovirus* and *Rotavirus* (both Reoviridae) have been identified in *Crassostrea gigas* by Le Guyader et al. (2008), and a reovirus-like virus has also been reported in cuttlefish (Hanlon & Forsythe 1990; McGladdery 1999).

Family Arenaviridae: An Arenaviridae-like virus has been reported from cultured *Hyriopsis cumingii*, a Chinese fresh-water mussel (Unionidae) (Shao et al. 1995).

Family Nodaviridae: *Betanodavirus* was recorded from skin lesions on cultured and wild *Octopus vulgaris* (Fichi et al. 2015).

Family Caliciviridae: *Norovirus*, primarily responsible for shellfish borne gastroenteritis in humans, is known from the gut of oysters (Le Guyader & Atmar 2007; Le Guyader et al. 2008) and some other bivalves (Terio et al. 2010).

9.4.4.2.5   Other Viruses

Viral diseases also occur, but are mainly not identified. For example, the virus causing acute virus necrobiotic disease (AVND) infected the scallop *Azumapecten farreri* in China (Tang et al. 2010). Other viruses have been found in oysters, *Pinctada*, and *Tellina* (see Renault & Novoa 2004 for details). A leukaemia-like disease of many bivalve species (termed disseminated neoplasia) may be related to retroviral infections, but this is unconfirmed (Elston 1997; Renault & Novoa 2004), and a few diseases of internal organs in coleoid cephalopods are suspected to be caused by unidentified viruses (e.g., Hanlon & Forsythe 1990). Another virus causes oyster velar virus disease (OVVD), and is found in *Crassostrea gigas* (Elston & Wilkinson 1985).

9.4.4.3   Associations with 'Bacteria'

We briefly review the many associations that molluscs have with 'bacteria', using a modern classification as a framework. Older works distinguished Gram[10]-negative and Gram-positive bacteria based on staining properties, and this method is still used today.

---

9 An abnormal growth on or in the body.
10 Named after H. C. Gram (1853–1938), a Danish bacteriologist.

What were regarded previously as 'bacteria' fall into two major 'domains' – the Archaea and the Bacteria (see Box 9.2). The remainder of life (other than viruses) falls into the domain Eukaryotes. The modern classification of bacteria was only developed in recent years, notably with the classifications proposed by Cavalier-Smith (e.g., Cavalier-Smith 2006) and Ruggiero et al. (2015a, 2015b) (Box 9.2).

Bacteria are generally *aerobic* (require oxygen), *anaerobic* (cannot tolerate gaseous oxygen), or they are *facultative anaerobes* (prefer oxygen, but can live without it).

They also have a wide range of nutritional types including *heterotrophs* that derive energy from breaking down complex organic compounds they must take in from the environment – this includes saprobic bacteria found in decaying material and those that rely on *fermentation* or *respiration*. Thus, many bacteria are saprophytic, decomposing dead matter or organic waste and often producing methane as a by-product. Some are *autotrophic*, fixing carbon dioxide to make their own food source. Energy may be from light (*photoautotrophic*) using photosynthesis to produce complex organic compounds

BOX 9.2   CLASSIFICATION OF BACTERIA

An outline of the classification of 'bacteria' used here is based on Ruggiero et al. (2015a, 2015b). Groups referred to in the text are in bold and indicated by an asterisk (*)

Domain Archaea (=Archaebacteria) (prokaryotes with some genes and metabolic pathways resembling those in eukaryotes; live in a wide range of habitats; reproduce by dividing asexually, and do not produce spores)
Phylum *Crenarchaeota (found in most environments, some thermophilic; includes some sulphur reducers)
   ***Euryarchaeota** (includes methanogenic bacteria, halophiles (salt), and some thermophilic; also includes some sulphur reducers).
Domain Bacteria (=Eubacteria) (prokaryotes, living in all environments; cell wall differs from that of Archaea; reproduce by dividing asexually, and some produce spores).
Subkingdom Negibacteria
   Phyla *Acidobacteria (common in soils, some acidophilic, diverse, but not well understood)
      Aquificae (in thermal environments)
      Armatimonadetes (includes some aerobic chemoheterotrophic bacteria)
      ***Bacteroidetes** (Gram-negative rod-shaped bacteria)
      Caldiserica (contains a distinct thermophilic bacterium)
      ***Chlamydiae** (contains obligate intracellular pathogens such as *Chlamydia*)
      Chlorobi (green sulphur bacteria)
      Chrysiogenetes (contains a single species that respires using arsenate)
      ***Cyanobacteria** (blue-green 'algae')
      ***Deferribacteres** (contains bacteria that metabolise iron)
      ***Deinococcus-Thermus** (=Hadobacteria) (include highly resistant bacteria that can modify toxic materials including nuclear waste)
      Dictyoglomi (contains a single genus that can live at high temperature and metabolises organic molecules)
      Elusimicrobia (live in various contaminated habitats and in termite guts)
      Fibrobacteres (includes the rumen bacteria that break down cellulose)
      ***Fusobacteria** (anaerobic, Gram-negative bacteria)
      ***Gemmatimonadetes** (important soil-living bacteria)
      ***Lentisphaerae** (includes some mammalian gut and marine bacteria)
      ***Nitrospira** (contains nitrogen-fixing bacteria)
      ***Planctomycetes** (aquatic bacteria with a holdfast)
      ***Proteobacteria** (Gram-negative purple bacteria and relatives)
      Classes *Alphaproteobacteria (purple non-sulphur bacteria)
         ***Beta-proteobacteria** (aerobic, some nitrogen fixers, a few pathogenic)
         ***Delta-proteobacteria** (largest group of sulphate reducing bacteria)
         ***Epsilon-proteobacteria** (includes endosymbionts and pathogens, some from hydrothermal vents; oxidise sulphur, formate,[11] or hydrogen)
         ***Gamma-proteobacteria** (purple sulphur bacteria)
         ***Zeta-proteobacteria** (iron oxidisers in marine and estuarine habitats and hydrothermal vents)

---

[11] Derived from formic acid.

(Continued)

BOX 9.2 (Continued)   CLASSIFICATION OF BACTERIA

Phyla *Spirochaetae (spiral or corkscrew-shaped bacteria – many are free-living and anaerobic)
  Synergistetes (anaerobic Gram-negative rod-shaped bacteria; some responsible for diseases)
  Thermodesulfobacteria (thermophilic, sulphur-reducing bacteria)
  Thermotogae (Gram-negative, anaerobic, thermophilic bacteria)
    ***Verrucomicrobia** (a small group in fresh water and soil etc.)
Subkingdom Posibacteria
  Phyla *Actinobacteria (one of the Gram-positive group)
    ***Chloroflexi** (=Chlorobacteria) (green non-sulphur bacteria)
  ***Firmicutes** (Gram-positive bacteria that produce highly resistant spores – some (heliobacteria) can photosynthesise and some reduce sulphur)
    ***Tenericutes** (includes Mollicutes; lack a cell wall; parasites of animals and plants)

from simple substances from the environment or *chemotrophic* using inorganic chemical reactions (chemosynthesis). Chemotrophic bacteria obtain energy from chemical reactions involving oxidation of electron donors in their environments and synthesise all necessary organic compounds from carbon dioxide. Some chemotrophs obtain their energy through fermentation, where electrons from a reduced substrate are transferred to oxidised intermediate substances to form reduced fermentation products such as lactate, ethanol, or hydrogen. Because the energy in the substrates is higher than that of the products, fermentation provides energy to synthesise ATP for metabolic processes.

Chemoautotrophic bacteria use various *inorganic* energy sources as listed in Table 9.2.

Many of the Bacteria or Archaea using inorganic substrates live in extreme environments such as deep-sea vents where they are the primary producers. Some produce oxygen as a by-product and various Archaea produce methane (methanogens). Some bacteria are *halorespirers* and these anaerobes use halogenated compounds (e.g., chlorinated phenols) as terminal electron acceptors (various Proteobacteria, Chloroflexi, and Clostridia). *Chemoheterotrophic* bacteria use organic carbon for growth as they cannot fix carbon. *Chemolithoheterotrophs*

obtain energy from inorganic sources (e.g., sulphur), while *chemoorganoheterotrophs* use organic energy sources (e.g., protein, carbohydrate, or fats). *Syntrophic* bacteria live off the products of another bacterium.

Given this amazing array of lifestyles, it is not surprising that many groups symbiotic with molluscs that live in a wide variety of habitats (see Figure 9.10), as detailed below.

Many free-living Bacteria (Actinobacteria, Aquificae, Bacilli, Chloroflexi, Chlorobi, Spirochaetae, and Proteobacteria) and the Sulfolobales (Archaea) fix carbon using reduced sulphur compounds (*thiotrophy*). Nitrous oxide (N_2O) is produced as a by-product of ammonia oxidation by nitrifiers (ammonia-oxidising bacteria and Archaea), especially when oxygen is limited. In contrast, denitrifiers have N_2O as an intermediate product in anaerobic respiration.

Photoautotrophs are common and diverse, but chemoautotrophs are less so. The Cyanobacteria (blue-green algae), Chlorobi (green sulphur bacteria), Gamma-proteobacteria (purple sulphur bacteria), and Alpha-proteobacteria (purple non-sulphur bacteria) are all chemoautotrophs. While some, like Cyanobacteria, use water as a hydrogen donor and oxygen is a by-product, others like the sulphur bacteria use hydrogen sulphide (H_2S) as the hydrogen donor and sulphur is a by-product.

As with other animals, bacteria form a wide range of very significant relationships with molluscs, including symbionts that provide nutrition, others in the gut that provide enzymes to aid in digestion, and some that cause diseases. Most diseases caused by bacteria in molluscs are recorded from bivalves and a few gastropods, notably abalone. Many older accounts (e.g., Michelson 1957) often do not accurately identify the bacteria involved. There are few instances where bacteria cause disease in cephalopods, except in culture. This does not, however, suggest that potentially pathogenic bacteria are absent. For example, sampling the skin of a species of healthy squid, Ford et al. (1986) showed a rich bacterial flora with 1700–6770 viable cells per cm^2 on the dorsal mantle compared with 340–1100 cells mL^{-1} in the seawater. Those on the mantle comprised at least 13 different bacteria, of which more than half were potential pathogens.

TABLE 9.2

Examples of Inorganic Substrates Utilised by Bacteria and Archaea

Substrate	Type of Bacterium and Action
Hydrogen sulphide	Sulphur oxidisers – reduce sulphur compounds to sulphuric acid
Sulphur	Sulphur reducers – e.g., reduce sulphate to hydrogen sulphide
Ferrous iron	Iron oxidisers – oxidise soluble iron to ferric oxide
Molecular hydrogen	Hydrogen oxidisers – convert hydrogen to water
Gaseous nitrogen	Nitrogen fixers – convert gaseous nitrogen to ammonium or nitrogen dioxide
Ammonia	Ammonia oxidisers – convert ammonia to nitrate
Methane	Methanotrophs – oxidise methane

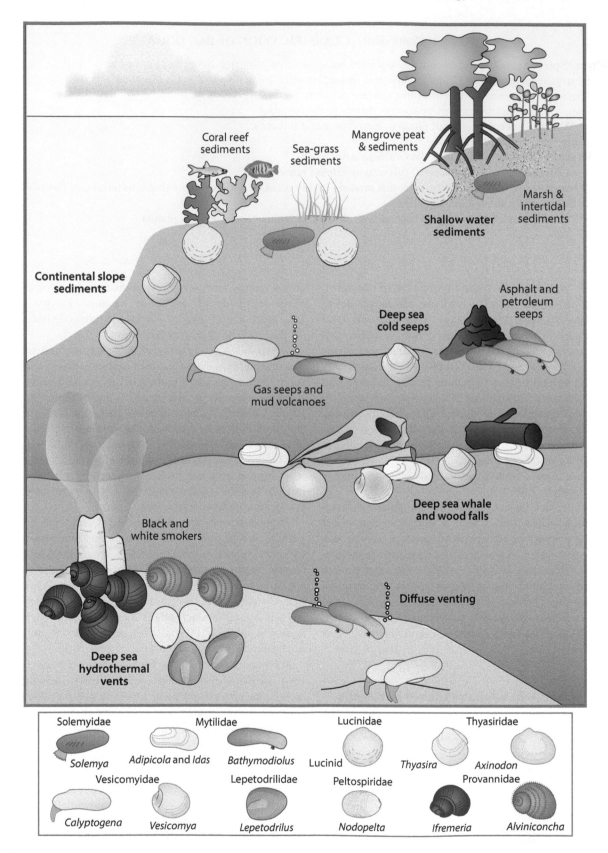

FIGURE 9.10 Examples of molluscs with chemosynthetic symbioses with various bacteria. Redrawn and modified from Dubilier, N. et al., *Nat. Rev. Microbiol.*, 6, 725–740, 2008.

Due to their suspension-feeding habits, bivalves accumulate large quantities of bacteria, notably Gram-negative species (Lauckner 1983b). Also, symbiotic bacteria are known from many molluscs as summarised below.

The summary of mollusc-associated bacteria that is given below is based on data from a relatively small number of recent studies as older studies did not use the modern groupings.

9.4.4.3.1   Domain Archaea

Although members of this group were originally discovered in extreme environments, such as hot springs and hydrothermal vents, they are also found in many 'normal' habitats such as seawater and soil as well as in anaerobic habitats, or in highly saline or acidic ones. These microbes have no membrane-bound organelles (including a nucleus) in their cells.

Wrede et al. (2012) reviewed symbioses in Archaea, but gave no specific molluscan examples. The group includes two main phyla: Euryarchaeota and Crenarchaeota, which are briefly described below.

9.4.4.3.2   Phylum Euryarchaeota

The 'class' Halobacteria (=Halomebacteria) are found in highly saline water, and their blooms can cause salt lakes to turn pink or reddish, as they are purplish in colour. They use the pigment to absorb light and the resulting energy creates ATP, a process quite different from photosynthesis. There are apparently only rare relationships with molluscs, although they have been found in the gut of the oyster *Crassostrea* (King et al. 2012).

9.4.4.3.3   Phylum Crenarchaeota
             (=Crenarchaea, Eocytes)

These Archaea are abundant in soil and comprise about a fifth of the microorganisms found in the oceanic picoplankton. Some oxidise ammonia and play an important role in nitrogen cycling. Others are anaerobic methanogens found in the gut of animals, as well as in swamps and hot springs, that produce methane gas as a by-product. The group also includes sulphur-dependent thermophiles (some living in >100°C). They come in a wide variety of shapes and stain Gram-negative.

Relationships between molluscs and Crenarchaeota are rarely recorded, but they commonly form associations with sponges and are known to be symbionts of ascidians. They can also be significant components of biofilms on marine mollusc shells (e.g., Pfister et al. 2010) and have been found in the gut of the oyster *Crassostrea* (King et al. 2012).

9.4.4.3.4   Domain Bacteria (=Eubacteria)

Most known bacteria are Gram-negative 'diderms',[12] but the 'phyla' Firmicutes (low GC[13] Gram-positive, monoderm[14]), Actinobacteria (high-GC Gram-positive, monoderm), Deinococcus-Thermus (Gram-positive, diderm), and Chloroflexi (mostly Gram-negative, monoderm) are exceptions.

Some higher groupings ('Subkingdoms') have been proposed as shown in Box 9.2. Other classifications have been suggested that include, for example, a higher clade, the 'Eobacteria', composed of two 'phyla', Deinococcus-Thermus and Chloroflexi, both of which lack the lipopolysaccharide in the outer membrane of Gram-negative diaderm bacteria. Other major groupings no longer used include the 'FCB group' (or Sphingobacteria) comprising the phyla Bacteroidetes, Chlorobi, and Fibrobacteres. The 'PVC group' (or Planctobacteria) contains the 'phyla' Chlamydiae, Lentisphaerae, Planctomycetes, and Verrucomicrobia.

We have not found recorded associations with molluscs for the following bacterial phyla – Caldiserica, Synergistetes, Chlorobi (green sulphur bacteria), Thermotogae, Chrysiogenetes, Dictyoglomi, Elusimicrobia, Fibrobacteres, Thermodesulfobacteria, and Aquificae.

9.4.4.3.5   Phylum Deinococcus-Thermus (or
             Xenobacteria or 'Hadobacteria')

The Deinococcus-Thermus are a small group of spherical bacteria (cocci) comprising two groups ('classes'), one highly resistant to radiation (Deinococcales) and the other to heat (class Deinococci). The former is famous for its ability to 'eat' nuclear waste and other toxins, and they can also survive in space and endure temperature extremes. The cells have thick walls and they stain with Gram-positive stains, although, as in Gram-negative bacteria, the walls possess a second membrane.

Recorded relationships with molluscs are few. They have been found associated with the gut of planorbid snails (Horn et al. 2012) and the oyster *Crassostrea* (King et al. 2012).

9.4.4.3.6   Phylum Chloroflexi (=Dehalococcoides,
             Chlorobacteria, Green Non-Sulphur Bacteria)

The members of this group are diverse, including aerobic thermophiles that use oxygen and thrive in high temperatures, anoxygenic phototrophs that carry out photosynthesis using light, and anaerobic halorespirers that use halogenated organics as energy sources. One 'class' is recognised, the Thermomicrobia.

One species is recorded from the digestive gland of *Saccostrea* (Green & Barnes 2010), and members of the group have been found associated with the gut of the oyster *Crassostrea* (King et al. 2012) and planorbid snails (Horn et al. 2012).

9.4.4.3.7   Phylum Firmicutes

This is one of the two phyla of monoderm, mostly Gram-positive bacteria, with this group characterised by a low guanine-cytosine (GC) content. Some are pathogens and most form spores. Three 'classes', Bacilli, Clostridia, and Negativicutes are recognised. Within Clostridia, Heliobacteriaceae (including *Heliobacterium*) are photosynthetic and Negativicutes have Gram-negative walls.

Reported molluscan associations with Firmicutes include the arcid *Anadara* (Romanenko et al. 2008), the gills of bivalves from mud volcanoes (Rodrigues et al. 2010), and (bacilli and cocci) with a galeommatoidean (Oliver

---

[12] Diderm – the cell wall having a double membrane.
[13] guanine-cytosine content.
[14] Monoderm – the cell wall having a single membrane.

et al. 2013). Firmicutes have also been recorded from the gill and gut of deep-sea chitons associated with sunken wood (Duperron et al. 2012).

Two taxa (a Bacilli and a Clostridia) were found in *Crassostrea* (Fernandez-Piquer et al. 2012) and four taxa in the digestive gland of *Saccostrea* (Green & Barnes 2010). Firmicutes have been found associated with the gut of abalone (Tanaka et al. 2004), helicid (Charrier et al. 2006) and achatinid (Cardoso et al. 2012; Oyeleke et al. 2012) land snails, fresh-water planorbids (Horn et al. 2012), and the oyster *Crassostrea* (King et al. 2012), where they may sometimes provide useful enzymes.

Toxins from some species of *Bacillus* have been shown to kill zebra mussels and the planorbid *Biomphalaria* (Singer et al. 1997). A *Bacillus*-like bacterium and *Streptococcus* were found on healthy skin in octopuses, while both *Bacillus* and *Staphylococcus* were associated with skin ulcers on octopuses (Hanlon & Forsythe 1990). *Lactococcus* (Streptococcaceae) was recorded from skin lesions on cultured and wild *Octopus vulgaris* (Fichi et al. 2015).

Some species of *Bacillus* (with the gamma-proteobacterium *Shewanella*) have been shown to produce tetrodotoxin in blue ring octopuses (Hwang et al. 1989), as they do in puffer fish and some other marine animals.

9.4.4.3.8   Phylum Actinobacteria (High-GC Gram-Positive Bacteria)

This is one of the two phyla of monoderm Gram-positive bacteria, the other being Firmicutes, which has a lower GC content. Many Actinobacteria are significant because they produce secondary metabolites. One species has been recorded from the digestive gland of *Saccostrea* (Green & Barnes 2010), and other Actinobacteria have been found associated with the gut of *Achatina* (Cardoso et al. 2012), planorbids (Horn et al. 2012), the *Crassostrea* (King et al. 2012), and a deep-sea wood-associated chiton (Duperron et al. 2012). Members of this group have also been reported from the arcid *Anadara* (Romanenko et al. 2008) and the gills of bivalves from mud volcanoes (Rodrigues et al. 2010). Mycobacterial infections ('subclass' Actinobacteridae, order Actinomycetales) have been reported from fresh-water snails (Michelson 1961; Bean-Knudsen et al. 1988), and *Mycobacterium* has been recorded from some bivalves such as zebra mussels (Winters et al. 2011) and oysters (Beecham et al. 1991).

9.4.4.3.9   Phylum Acidobacteria

Acidobacteria are diderm Gram-negative bacteria that are very abundant in many soils and very diverse, including acidophiles and non-acidophiles, but are difficult to culture and consequently not well studied. Usually their metabolism slows when nutrients are low, and they are tolerant of fluctuations in soil moisture. Several 'classes' are recognised including Solibacteres, Holophagae, and Acidobacteria.

Acidobacteria have been found associated with the gut of planorbid snails (Horn et al. 2012) and the oyster *Crassostrea* (King et al. 2012).

9.4.4.3.10   Phylum Planctomycetes

These bacteria are found in aquatic habitats and reproduce by budding. They are ovoid, with a stalk-like holdfast. They have been found in *Crassostrea* (Fernandez-Piquer et al. 2012; King et al. 2012), where they can be abundant in the gut, and in deep-sea mussels (Rodrigues et al. 2013) and the gut of planorbid snails (Horn et al. 2012). They have also been found in the gills and gut of deep-sea chitons associated with sunken wood (Duperron et al. 2012).

9.4.4.3.11   Phylum Spirochaetes (Spirochetes)

These chemoheterotrophic bacteria are diderm and most have long, spirally coiled cells. They have a unique arrangement of the flagella and reproduce asexually by transverse fission. Many, but not all, are free-living and anaerobic.

Spirochaetes have been found in oysters (Green & Barnes 2010; Fernandez-Piquer et al. 2012; King et al. 2012) and also associated with the gills of bivalves from mud volcanoes (Rodrigues et al. 2010). Most well-known is the association of the spirochaete *Cristispira* with the crystalline style of bivalves and some gastropods, although the nature of this association is not clear (e.g., Sitnikova et al. 2012b).

9.4.4.3.12   Phylum Bacteroidetes

These very diverse, important diderm Gram-negative bacteria include opportunistic pathogens and gut and skin commensals as well as free-living taxa in the soil, sediment, and sea water. They are anaerobic, rod-shaped, and do not form spores. There are four classes recognised. By far the best studied are in the class Bacteroidia, as members of this class include some human pathogens. Other classes are Flavobacteria, Cytophagia, and Sphingobacteria.

Bacteroidetes have been recorded as associates of several molluscs including deep-sea mussels (Rodrigues et al. 2013), in the digestive gland of *Pectinodonta*, a wood-associated patellogastropod (Zbinden et al. 2010), the accessory nidamental glands of coleoid cephalopods (Pichon et al. 2005), and the gill and gut of deep-sea chitons associated with sunken wood (Duperron et al. 2012). Bacteroidia and many kinds of Flavobacteria are found in *Crassostrea* (Fernandez-Piquer et al. 2012). A *Flavobacterium*-like bacterium has been found on the skin of squid (Hanlon & Forsythe 1990). Several Bacteroidetes have also been found associated with the gut of abalone (Huang et al. 2010; Zhao et al. 2012), the 'giant African snail' *Achatina* (Cardoso et al. 2012), planorbid snails (Horn et al. 2012), and the oyster *Crassostrea* (King et al. 2012). Sphingobacteria and Flavobacteria have been found associated with the arcid *Anadara* (Romanenko et al. 2008).

9.4.4.3.13   Phylum Proteobacteria (Purple Bacteria and Relatives)

Purple bacteria were the first bacteria discovered to photosynthesise without having oxygen as a by-product. Some are related to the mitochondria in the cells of eukaryotes. Many are free-living and include the nitrogen-fixing bacteria, but there are also many pathogens (e.g., *Escherichia*, *Salmonella*,

Vibrio, *Helicobacter*, and *Pseudomonas*). They are all Gram-negative and engage in a range of strategies – many produce sulphur as the by-product of photosynthesis, but the members of the 'order' Rhodospirillales, which include the family Acetobacteraceae, are the acetic acid bacteria that produce acetic acid as a by-product of photosynthesis. Together these two 'orders' comprise the purple non-sulphur bacteria. Members of the 'order' Burkholderiales use oxalic acid as their source of carbon.

Proteobacteria are divided into several classes, as outlined below.

9.4.4.3.13.1 'Class' Alpha-proteobacteria (α-Proteobacteria)

This diverse group includes the 'purple non-sulphur bacteria', which are phototrophic and use light energy in a photosynthetic process that has ATP as an end product. This process does not involve water and no oxygen is produced. They include many that live in the open ocean plankton. Other Alpha-proteobacteria include some that metabolise chlorine compounds, many are symbionts of plants (the nitrogen fixers) and animals, while others are pathogens, including some Rickettsiaceae. Ancestors of this latter group likely gave rise to mitochondria.

The 'order' Rhizobiales comprises several families that are symbiotic with plant roots and fix nitrogen. One family in this group, the Methylocystaceae, comprises methanotrophs, utilising methane or methyl alcohol as their source of energy and carbon.

Alpha-proteobacteria can be significant members of the biofilm on mollusc shells (e.g., Pfister et al. 2010) and vent Solenogastres (species of *Helicoradomenia*) have been shown to harbour both epibiotic and endocuticular members of this group (Katz et al. 2006). Many have been recorded as associates of *Crassostrea* (Fernandez-Piquer et al. 2012) and other oysters (Pujalte et al. 1999), including in their digestive gland (Green & Barnes 2010). They have also been found in deep-sea mussels and *Thyasira* (Rodrigues et al. 2013), in *Pectinodonta*, a wood-associated patellogastropod (Zbinden et al. 2010), and in the gills and gut of deep-sea wood-associated chitons (Duperron et al. 2012). They are also associated with oysters (e.g., Pujalte et al. 1999) and the arcid *Anadara* (Romanenko et al. 2008). This group has also been found in the gills of bivalves from mud volcanoes (Rodrigues et al. 2010) and the gut of abalone (Tanaka et al. 2004; Huang et al. 2010; Zhao et al. 2012), planorbids (Horn et al. 2012) and *Crassostrea* (King et al. 2012).

Intracellular *Rickettsia* have been found in several bivalves (e.g., Azevedo & Villalba 1991; Wu & Pan 2000; Costa et al. 2012) and abalone (e.g., Azevedo et al. 2006). An extracellular giant *Rickettsia* closely associated with bacteria with a Gram-negative type wall has been reported from the gill epithelium of *Crassostrea gigas* (Azevedo & Villalba 1991). Rickettsiales-like bacteria also occur in gill epithelia and the ova of the bivalve *Mercenaria* (Fries & Grant 1991). The rickettsiales-like prokaryote, *Xenohaliotis californiensis*, is a pathogen of *Haliotis* spp. in the northeastern Pacific where it causes 'abalone withering syndrome' – an infection of the digestive tract, which causes the abalone to starve to death (Crosson et al. 2014). Besides abalone die-offs in California, rickettsiales-like bacteria have been associated with mass mortalities of scallops in the USA and France (Mortensen 2000).

Some species of Alpha-proteobacteria have been recorded as endosymbionts in the cerata of nudibranchs and may have originated from their cnidarian prey (Doepke et al. 2012). One species is known from the ctenidium of a galeommatoidean (Oliver et al. 2013), and others were found in the accessory nidamental glands of coleoid cephalopods (Pichon et al. 2005; Collins et al. 2012a).

9.4.4.3.13.2 'Class' Beta-proteobacteria (β-Proteobacteria)

This group contains aerobic or facultative[15] bacteria that utilise a wide range of substrates and includes chemolithotrophic, chemoautotrophic, and some phototrophic taxa. Some are also involved in nitrogen fixation and the group also includes pathogens (Neisseriaceae). They are found in aquatic and terrestrial environments and are often endosymbionts of 'protists'.

Members of this group of bacteria have been found associated with oysters (Fernandez-Piquer et al. 2012) and (genus *Alcaligenes*) with reared squid (Hanlon & Forsythe 1990). They are also found in the gills of bivalves from mud volcanoes (Rodrigues et al. 2010), deep-sea mussels (Rodrigues et al. 2013), and the gut and gills of deep-sea chitons associated with sunken wood (Duperron et al. 2012). They have also been found in the gut of abalone (Huang et al. 2010), *Achatina* (Cardoso et al. 2012), planorbids (Horn et al. 2012), and *Crassostrea* (King et al. 2012).

9.4.4.3.13.3 'Class' Delta-proteobacteria (δ-Proteobacteria)

These bacteria are mostly aerobic, but the group contains some anaerobic taxa, including most of the sulphate and sulphur-reducing bacteria and others such as the ferric iron-reducing bacteria, and a number are syntrophic.

Delta-proteobacteria have been recorded as associates of several molluscs including oysters (Fernandez-Piquer et al. 2012), in the gills of *Pectinodonta* (Zbinden et al. 2010) and deep-sea chitons on wood (Duperron et al. 2012). Members of this group have also been found in the gut of planorbids (Horn et al. 2012) and *Crassostrea* (King et al. 2012).

The mineralised scales of the 'scale-footed' gastropod *Crysomallon squamiferum* are coated with iron sulphides (pyrite and greigite) and covered with Delta-proteobacteria (and Epsilon-proteobacteria), which are probably involved in the mineralisation of the very robust scales (Goffredi et al. 2004b; Yao et al. 2010).

---

[15] Facultative bacteria can use either dissolved oxygen or oxygen from sulphate or nitrate ions. Thus, they can live under aerobic, anoxic, or anaerobic conditions.

9.4.4.3.13.4 'Class' Gamma-proteobacteria (γ-Proteobacteria)

These bacteria are of ecological and medical importance with many pathogenic taxa (e.g., Enterobacteriaceae, Vibrionaceae, and Pseudomonadaceae). It also includes the purple sulphur bacteria that live in stagnant water or hot springs and oxidise hydrogen sulphide to sulphur using photosynthesis, but do not produce oxygen. Marine biofilms, which provide important substrates for molluscan recruitment, can have high concentrations of the human pathogens *Vibrio cholerae* and *Escherichia coli* (Shikuma & Hadfield 2010).

Gamma-proteobacteria can be significant members of the biofilm on mollusc shells (e.g., Pfister et al. 2010) and have been recorded as epibiotic associates of several molluscs including oysters (e.g., Green & Barnes 2010; Fernandez-Piquer et al. 2012). Vent Solenogastres (species of *Helicoradomenia*) have been shown to harbour both epibiotic and endocuticular gamma-proteobacterians (Katz et al. 2006).

Symbiotic gamma-proteobacterians are associated with a cold seep lucinid and *Thyasira* (Brissac et al. 2011), deep-sea mussels (Zielinski et al. 2009; Rodrigues et al. 2013), the wood-boring teredinids (e.g., Distel et al. 2002b; Distel 2003), some of which produce cellulases (see also Chapter 5), the wood-feeding *Pectinodonta* (Zbinden et al. 2010), and the gills and gut of deep-sea chitons from wood (Duperron et al. 2012). They are also one of the main groups of bacteria associated with the gut of abalone (Tanaka et al. 2004; Huang et al. 2010; Zhao et al. 2012), helicids (Charrier et al. 2006), *Achatina* (Cardoso et al. 2012), planorbids (Horn et al. 2012), *Crassostrea* (King et al. 2012), and *Anadara* (Romanenko et al. 2008).

This group also includes the symbiotic chemoautotrophic (sulphur-oxidising) bacteria associated with both shallow-water and deep-sea bivalves (*Calyptogena*, *Vesicomya*, lucinids, *Solemya*, the mytilids *Bathymodiolus* and *Idas*, and thyasirids) (Rodrigues et al. 2010, 2013), the vent provaniid gastropods *Alviniconcha*, *Ifremeria*, and *Crysomallon squamiferum* (Goffredi et al. 2004b; Dubilier et al. 2008; Nakagawa & Takai 2008), and the vent vetigastropod *Lepetodrilus* (see also Vrijenhoek 2010; Bates et al. 2011).

The vent-associated molluscs house the Gamma-proteobacteria in their gills, except for the 'scale-footed' gastropod *Crysomallon squamiferum*, which houses them in its much enlarged oesophageal gland and the rest of the gut is reduced. This morphology suggests *Crysomallon* is reliant on these chemoautotrophic bacteria for its nutrition (Goffredi et al. 2004b).

Symbiotic bacteria of the family Vibrionaceae (species of *Photobacterium* and *Vibrio*) provide the light in the light organs of some coleoid cephalopods (e.g., Nishiguchi et al. 2004), with nitric oxide (NO) playing a role in their establishment (Davidson et al. 2004). These bacteria colonise the light organ in juveniles (see also Section 9.4.8 and Chapter 17). Other species of *Vibrio* are associated with bivalves such as oysters (e.g., Pujalte et al. 1999) and coleoids, with six species of *Vibrio* found in wild-caught octopuses along with other bacteria, including other Gamma-proteobacteria (*Proteus* and *Aeromonas*) (Farto et al. 2003). Some pathogenic species of

Vibrio are responsible for human health issues, including diseases associated with eating molluscs, notably bivalves such as oysters (e.g., Fernandez-Piquer et al. 2012). *Photobacterium* (Vibrionaceae) was recorded from skin lesions on cultured and wild *Octopus vulgaris* (Fichi et al. 2015).

Vibrio, *Pseudomonas*, *Acinetobacter*, *Klebsiella*, *Plesiomonas*, and *Areomonas* were cultured from reared coleoids (Hanlon & Forsythe 1990). A disease in laboratory-reared cuttlefish that affects the blood sinuses is associated with two species of *Vibrio* and one of *Pseudomonas* (Hanlon & Forsythe 1990). Generally wild squid have fewer bacterial populations than tank-reared individuals (Castellanos-Martínez & Gestal 2013). Gamma-proteobacteria have also been found in the accessory nidamental glands of coleoid cephalopods (Pichon et al. 2005).

The mobile gamma-proteobacterians *Aeromonas* and *Pseudomonas* were the predominant bacteria isolated from healthy unionids (Starliper & Morrison 2000). Some species of the former genus are pathogenic in zebra mussels, and in the latter genus some have been shown to produce lethal toxins that can kill fresh-water bivalves (Grizzle & Brunner 2009). Several Gamma-proteobacteria have been recorded as endosymbionts from the cerata of nudibranchs and may have originated from their cnidarian prey (Doepke et al. 2012). The gamma-proteobacterium *Shewanella* (previously *Alteromonas*) on the substratum can induce oyster larvae to settle (Weiner et al. 1989), and some gamma-proteobacterians are used in aquaculture because they produce antibiotic substances (e.g., Riquelme et al. 1996, 2001; Prado et al. 2009). Some species of *Shewanella*, together with *Bacillus* (Firmicutes), have been shown to produce tetrodotoxin in blue ring octopuses (Hwang et al. 1989), as they do in puffer fish and some other marine animals.

Intracellular members of the order Oceanospirillales are found in bathymodioline mussels and the limid *Acesta* (Jensen et al. 2010).

9.4.4.3.13.5 'Class' Epsilon-proteobacteria (ε-Proteobacteria)

This small group contains some significant intestinal symbionts and pathogens (e.g., *Helicobacter*). Others are common in hydrothermal vents and cold seeps and are chemolithotrophic. A member of this class is an endosymbiont in the gill of the vent gastropods *Alviniconcha* (Suzuki et al. 2005) and *Lepetodrilus* (Bates et al. 2011).

Epsilon-proteobacteria have been recorded as associates of oysters (Romero et al. 2002; Fernandez-Piquer et al. 2012), *Thyasira* (Brissac et al. 2011) and several deep-sea gastropods including in the gills of *Pectinodonta* (Zbinden et al. 2010) and a deep-sea chiton, both associated with sunken wood (Duperron et al. 2012). Sulphur-oxidising bacteria associated with the vent provaniid gastropod *Alviniconcha* and the 'scale-footed' gastropod *Crysomallon squamiferum* (see below) also belong to this group. In addition, they have been found associated with the gills of bivalves from mud volcanoes (Rodrigues et al. 2010), the gut of abalone (Tanaka et al. 2004), *Achatina* (Cardoso et al. 2012), and planorbids.

The mineralised scales of the 'scale-footed' gastropod *Crysomallon squamiferum* are covered with Epsilon-proteobacteria (and Delta-proteobacteria), which, as noted above, are probably involved in the mineralisation of the scales.

Novel strains of *Campylobacter* (order Campylobacterales) were found in oysters and mussels (Endtz et al. 1997; Debruyne et al. 2009); some species are pathogens of mammals, including humans.

9.4.4.3.13.6 Other Proteobacteria

The remaining Proteobacteria comprise two small 'classes', one of which is the Acidithiobacillia, containing a small number of bacteria that metabolise iron and/or sulphur to produce sulphuric acid. Another 'class', Zeta-proteobacteria, was named for an aerobic, chemolithoautotrophic iron-oxidising bacterium. These are not known from molluscs.

9.4.4.3.14 Phylum Fusobacteria

Fusobacteria are anaerobic, Gram-negative bacteria implicated in several human diseases. They have been found associated with oysters (Green & Barnes 2010; Fernandez-Piquer et al. 2012; King et al. 2012) and the gut of abalone (Tanaka et al. 2004; Huang et al. 2010).

9.4.4.3.15 Phylum Cyanobacteria (=Cyanoprokaryota, Blue-Green Algae)

This major photosynthetic clade is thought to be responsible for the oxygen in the atmosphere of the early Earth. Cyanobacteria, initially through an endosymbiotic relationship, gave rise to plastids (chloroplasts).

Cyanobacterians can be significant members of the biofilm on mollusc shells (e.g., Pfister et al. 2010). Several taxa can corrode gastropod and bivalve shells (Nielsen 1972; Lauckner 1983b), and some bore into scaphopod shells (Reynolds 2002, 2006). This group of bacteria has also been found associated with the gills of a deep-sea wood-associated chiton (Duperron et al. 2012), the gills of bivalves from mud volcanoes (Rodrigues et al. 2010), as well as in *Crassostrea* (Fernandez-Piquer et al. 2012), and in the digestive gland of another oyster, *Saccostrea* (Green & Barnes 2010). An obligate endocytobiotic cyanobacterium has also been found in the digestive gland of the ampullariid *Pomacea* (Vega et al. 2006). Even though this cyanobacterium has chlorophyll-like pigment(s), photosynthesis would not be possible as light would not penetrate through the shell and body of the snail.

Biologically active products such as dolastatin, originally isolated from the aplysiid *Dolabella*, are produced by some cyanobacterians such as *Symploca* and sequestered by the aplysiid (Piel 2004).

9.4.4.3.16 Phylum Chlamydiae (Chlamydiales)

These diderm, weakly Gram-negative bacteria are obligate intracellular pathogens that replicate inside eukaryotic host cells. Many are as small, or even smaller, than viruses.

Chlamydiae, includes *Chlamydia*, a virus that causes chlamydia infections, are found in some bivalves (e.g., Fryer & Lannan 1994) and have been reported from the gill of a deep-sea chiton from wood (Duperron et al. 2012).

9.4.4.3.17 Phylum Gemmatimonadetes

This small group are Gram-negative and aerobic. They have been found in the gut of planorbids (Horn et al. 2012).

9.4.4.3.18 Phylum Lentisphaerae

These bacteria can be aerobic or anaerobic and are marine, being found in sediments, fish, and coral, as well as in the gut of terrestrial mammals and birds. They have been found in the gut of *Crassostrea* (King et al. 2012).

9.4.4.3.19 Phylum Nitrospira

These Gram-negative nitrite-oxidising bacteria have been found in the gut of planorbids (Horn et al. 2012).

9.4.4.3.20 Phylum Tenericutes

These bacteria lack a cell wall and include the Mollicutes (previously included in Firmicutes), members of that class being abundant in the gut and gills of one species of chiton associated with sunken wood (Duperron et al. 2012). Mollicutes have also been reported from the gill and gut of abalone (Tanaka et al. 2004; Huang et al. 2010; Zhao et al. 2012). Three taxa of these Gram-negative bacteria have been found in *Crassostrea* (Fernandez-Piquer et al. 2012).

9.4.4.3.21 Phylum Verrucomicrobia

These bacteria live in soil and water as well as being symbionts with eukaryotes. They have been found in *Crassostrea* (Fernandez-Piquer et al. 2012; King et al. 2012), in the gut of planorbids (Horn et al. 2012), in the gut and gills of deep-sea chitons from wood (Duperron et al. 2012), in deep-sea mussels (Rodrigues et al. 2013), and in the accessory nidamental glands of 'bobtailed squid' (*Euprymna*) (Collins et al. 2012a).

9.4.4.3.22 Phylum Deferribacteres

This small group of bacteria contains a single family. They have been found in some deep-sea mussels (Rodrigues et al. 2013).

9.4.4.4 Associations with 'Protists'

Protists are no longer considered a monophyletic assemblage of organisms. The classification followed below is largely based on that of Walker et al. (2011) and Adl et al. (2012). Organisms generally referred to as the Protista are common, diverse, and occur in most environments. Many are associates with animals and some cause disease. We briefly outline below the main groups known to have relationships with molluscs. For more detailed reviews of the marine 'parasitic' groups see Rohde (2005).

9.4.4.5 Alveolata

This major grouping of 'protists' includes the dinoflagellates and some parasitic taxa.

9.4.4.5.1 Dinoflagellata

Some dinoflagellates, a mostly free-living group, are important food items for many molluscs, particularly suspension-feeders. Some are parasites on a variety of marine animals, including molluscs. The parasitic dinoflagellates typically have a feeding stage (trophont) which lives on the host, and an asexually dividing tomont stage away from the host, followed by a free-swimming infectious stage (dinospore), which is typically encased in cellulose and contains chloroplasts or pigments. These pigments are sometimes neurotoxic to mammals when concentrated in suspension-feeding bivalves or fish (e.g., Wang 2008).

Blooms of free-living dinoflagellates, mainly of the genera *Karenia*, *Alexandrium*, and *Gymnodinium*, are sometimes responsible for mass mortalities of marine molluscs because of the toxic metabolites they release. Other dinoflagellates ingested by bivalves or gastropods can cause shellfish poisoning (see also Chapter 10). One of these, *Prorocentrum lima*, is an epiphytic species living as an epibiont on *Mytilus* shells and is one of the dinoflagellates responsible for diarrhetic shellfish poisoning (Levasseur et al. 2003).

Dinoflagellate parasites have been recorded in some molluscs (e.g., an octopod, Forsythe et al. 1991) and an unnamed symbiotic dinoflagellate related to the order Syndiniales occurs in the gills of the hydrothermal vent bivalves *Calyptogena* and *Bathymodiolus* (Noguchi et al. 2013), but whether it is an endocommensal or an endoparasite is unknown. A dinoflagellate similar to *Protodinium* has been found embedded in the skin of *Octopus* where it causes lesions (Hochberg 1990).

Zooxanthellae (*Symbiodinium*) are dinoflagellates of the class Dinophyceae. They form important associations with a few molluscs and several other marine invertebrates, mainly corals. In molluscs, this association occurs in a few bivalves, notably *Tridacna*, and in some nudibranchs (see Chapters 5, 15, and 20).

9.4.4.5.2 Perkinsida

This non-photosynthetic alveolate group contains *Perkinsus*, an intracellular parasite that causes disease (perkinsosis or 'dermo') in a wide variety of molluscs (abalones and many bivalves including oysters, scallops, pearl oysters, mussels, and venerids) (Villalba et al. 2004; Carnigie in Rohde 2005; Bower 2006; Joseph et al. 2010; Costa et al. 2012). For more information on the mollusc-related human diseases caused by this group see Chapter 10.

9.4.4.5.3 Paramyxida

The paramyxids, a group related to dinoflagellates, include significant parasites of bivalves, annelids, and crustaceans. Of these, *Marteilia* and *Marteilioides* are noteworthy parasites of oysters (including QX disease) and a wide range of other bivalves (Berthe et al. 2004; Bower 2006).

9.4.4.5.4 Ciliophora (Ciliates)

Ciliates differ from other alveolates in having two kinds of nuclei. Although most are free-living, many are commensals, and some are parasitic. The ectoparasitic forms can be irritants that cause fouling, or even lesions, and the endoparasitic taxa can also cause damage.

Some ciliates, such as the stalked *Vorticella* and *Epistylis*, can be part of the epibiotic covering of the shell in the fresh-water gastropod *Pomacea*, and some mobile ciliate taxa are associated with this covering (e.g., Vega et al. 2006).

Symbiont ciliophorans have been reported from a wide variety of marine, fresh water, and terrestrial molluscs, with most being reported from gastropods and bivalves, a smaller number from cephalopods, and a few from chitons. They are most commonly found in the mantle cavity and/or on the gills, but some are known from inside the gut and other organs. These ciliates are diverse and include representatives of four of the eight classes of ciliophorans (Polyhymenophorea, Colpodea, Phyllopharyngea, and Oligohymenophorea) (As & Basson 2004).

Ciliates are reported from the shell and operculum of living *Pomacea* (Dias et al. 2006, 2008), and the gills of chitons (Kozloff 1961; Lauckner 1983c), gastropods (Lauckner 1980; As & As 2000), the planktonic thecosomate gastropod *Corolla*, and cephalopods (Hochberg 1990; Forsythe et al. 1991). They are common in the gills and gut of autobranch bivalves (e.g., Fenchel 1965; Sindermann & Rosenfield 1967; Nair & Saraswathy 1971; Beninger et al. 1988; Fokin et al. 2003), including fresh-water taxa (e.g., Antipa & Small 1971; Karatayev et al. 2000; Grizzle & Brunner 2009), fresh-water gastropods (e.g., Jarocki 1935; Michelson 1957; Lauckner 1983c; Vega et al. 2006), and terrestrial snails and slugs (e.g., Mead 1979), including some pathogens (e.g., Segade et al. 2016). In terrestrial gastropods, several types of ciliates have been reported in the genital system, kidney, gut, and mantle cavity, but usually the nature of the association is unknown (Runham & Hunter 1970; Mead 1979; South 1992; As & Basson 2004).

Trichodine ciliates occur in the mantle cavity of some scaphopods, presumably as commensals, and others have been found in the blood (Reynolds 2002; Reynolds 2006).

In bivalves, ciliates can be abundant in the gills, but the exact nature of the relationship is unclear. Some of these gill ciliates are incorporated in the food string and ingested (Beninger et al. 1988). Of the Scandinavian bivalves investigated up to the 1960s, 22 species lacked ciliates, 17 species had one species of ciliate, and seven had two species. Eight species had three to five, six species were known from *Macoma balthica*, and seven from *Mytilus edulis* (Fenchel 1965). Three species of ciliate have been reported from the fresh-water bivalve *Dreissena*, with one, the endosymbiont *Conchophthirus acuminatus*, being in higher numbers than all other associated 'animals' (Karatayev et al. 2000). Other members of that genus have an obligate association with other fresh-water bivalves such as unionids (Michelson 1957), where they can occur in large numbers in the mantle cavity (Grizzle & Brunner 2009). Another ciliate (*Ophryoglena* sp.) in *Dreissena* only occurs in the digestive gland (Karatayev et al. 2000).

At least five families of ciliates are parasites in the kidney, gut, and digestive glands of a range of coleoid cephalopods, but those in the kidney are reduced and even eliminated if large numbers of dicyemids (see Section 9.4.4.8.5) are present (Hochberg 1990). Some ciliates have more than one host in their life cycle; for example, *Chromidina* inhabits a crustacean host and a cephalopod in parts of its life cycle (Hochberg 1990). Members of the genus *Opalinopsis* (Opalinopsidae) inhabit the midgut and digestive glands of coleoids and heteropods (Hochberg 1990).

Fokin et al. (2003) reported a multilevel system with bacteria in the cytoplasm of ciliates inhabiting the 'zebra mussel' *Dreissena*, and viruses were present in some of the bacteria.

9.4.4.5.5 Apicomplexa (Sporozoans)

This large group of spore-forming parasites are obligate intracellular parasites for most of their life. Most species are found in the gut or body cavity of a single host, but a few cycle between molluscan and crustacean hosts.

Apicomplexans are recorded from gastropods (e.g., Lauckner 1983c; Azevedo & Padovan 2004; Volland et al. 2010; Aldana-Aranda et al. 2011; Azmi et al. 2015), cephalopods (e.g., Hochberg 1990; Pascual et al. 1996; Gestal et al. 2002; Ibáñez et al. 2005; Castellanos-Martínez & Gestal 2013), scaphopods (Lauckner 1983a), chitons (Lauckner 1983c), and bivalves (Lauckner 1983b; Azevedo & Matos 1999; Bower 2006). The life cycle of some involve a second host, as for example, in the species of the genus *Aggregata* recorded from coleoid cephalopods which have, or are likely to have, a crustacean host (Hochberg 1983, 1990). To date there are nine named and several unnamed species of *Aggregata* known from coleoids that are responsible for intestinal coccidiosis in both wild and cultivated populations. One such species infects a deep-sea vent octopod (Gestal et al. 2010).

Some eugregarine 'sporozoans' infect both bivalves and gastropods where they infest gills and blood sinuses (Lauckner 1980, 1983b). Members of another group of 'sporozoans', the Coccidia, are known to infest scaphopods (Lauckner 1983a), chitons (Lauckner 1983c), and gastropods. One (*Merocystis*) infests the kidney of *Buccinum*, and another, *Pseudoklossia*, the digestive gland of the limpet *Patella* (Lauckner 1980).

The porosporid *Nematopsis legeri* occurs in various gastropods, bivalves, and at least one chiton (Lauckner 1980, 1983b, 1983c) with decapod crustaceans the definite hosts. Species of *Isospora* and *Pfeifferinella* infect non-marine gastropods, including some slugs (South 1992), but appear to have only a minor impact on the host (Kudo 1954).

9.4.4.5.6 Excavata

This major group of single-celled eukaryotes includes free-living and symbiotic forms. They lack typical mitochondria and are flagellate.

9.4.4.5.6.1 Euglenozoa This diverse phylum of unicellular flagellate 'protists' includes many free-living species, some of which contain chloroplasts, and others are significant parasites, including a few that infect humans.

Euglenoidea: Euglenoids are commonly found in fresh water, and there are also some marine and endosymbiotic members. Many have chloroplasts, but others feed by diffusion or phagocytosis. They comprise an important part of the food of some suspension-feeding and browsing molluscs (e.g., Dazo & Moreno 1962; Molina et al. 2010). Vega et al. (2006) reported motile euglenoids associated with the 'algal' epibionts on the shell of *Pomacea*, along with other mobile 'protists'.

Kinetoplastea: Kinetoplastids are single-celled flagellates that include both free-living and parasitic taxa. Hochberg (1990) reported flagellate (Bodonidae) infestations of mainly the gills in laboratory populations of octopods, which, when severe, caused mortality. They have also been reported from the external body and the gills in wild octopuses (Hochberg 1990; Forsythe et al. 1991).

A kinetoplastid pathogen *Cryptobia* (Bononidae) infects the reproductive system of land snails (e.g., Jordaens et al. 2007) and is apparently transmitted to the next generation via the eggs (Current 1980). It is also known in fresh-water snails (e.g., Kitajima & Paraense 1983) and in abalone (Chen et al. 2004).

9.4.4.5.6.2 Rhizaria This group includes many pseudopod-forming amoeboids and flagellates as well as Foraminifera and Radiozoa (including radiolarians). They include both aquatic and terrestrial taxa and most are free-living, but with a few gymnamoebae parasitic in animals.

Mytilus has been shown to act as a reservoir for *Neoparamoeba*, which causes amoebic fin disease in salmon (Tan et al. 2002), and another fish gill amoebic parasite, *Paramoeba*, has been reported to cause mortality in molluscs (Sprague 1978).

9.4.4.5.7 Heterokonta (=Stramenopiles, Heterokontophytes)

This grouping includes the multicellular brown and golden algae and the unicellular diatoms, in addition to Labyrinthulomycetes.

Many of the multicellular heterokont have a motile larval stage with two kinds of flagella. Their chloroplasts, when present, are surrounded by four membranes. While most are photosynthetic, some are heterotrophic.

9.4.4.5.7.1 Labyrinthulomycetes (Slime Nets) Labyrinthulomycetes (=Labyrinthulea, Labyrinthulomycota, Labyrinthomorpha) do not form pseudopodia, but instead produce networks of 'slime channels'. Previously classified with the slime moulds, they are now considered a separate group. A number are saprobic or parasitic on marine molluscs and other organisms. There are two main groups, the Labyrinthulida (slime nets) and Thraustochytrida.

A marine bivalve parasite that causes QPX disease in *Mercenaria* is a thraustochytrid labyrinthulomycete (Grizzle & Brunner 2009) which was originally identified as a fungus. A similar parasite was found in *Ruditapes* (McGladdery 1999). These parasites infect the connective tissue throughout the body. There are a few reports of similar organisms causing disease in coleoid cephalopods (Hanlon & Forsythe 1990). *Labyrinthuloides haliotidis* causes mortality in juvenile abalone in western North America (Bower 1987, 2006).

9.4.4.5.7.2 *Bacillariophyceae* (*Diatoms*)

This important group of mostly unicellular (some are colonial) 'algae' are abundant in the phytoplankton and, uniquely, their cell wall is composed of silica.

Diatoms can be significant food items for many browsing and suspension-feeding molluscs. They can also be significant epibionts in both marine and fresh-water ecosystems (e.g., Vega et al. 2006).

9.4.4.5.7.3 *Peronosporomycetes* (*Oomycota, Oomyceta*)

This group contains some pathogens that were considered in the past to be fungal diseases. Larval mycosis of *Crassostrea* and *Mercenaria* is caused by *Sirolpidium*, a peronosporomycete. Similar 'fungus-like' diseases have been reported infecting muricid egg capsules (Lauckner 1980).

9.4.4.5.7.4 *Phaeophyceae* (*Brown Algae*)

This very diverse group of mostly marine algae includes many of the familiar 'seaweeds' that are important as habitats and food for many marine animals, including molluscs. Their brown colouration comes from the pigment fucoxanthin. Unlike other heterokonts, brown algae are multicellular with differentiated tissues.

Brown algae are important food items for many herbivorous gastropods; one group, the kelps (Laminariales), are especially important because of their weak chemical defences. Brown algae also occur as part of the epibiont community, particularly on molluscs in the mid to lower intertidal and shallow subtidal. Associations between particular taxa of brown algae (often Laminariales) and particular mollusc species (notably patellogastropod limpets and some trochoidean vetigastropods), are not necessarily old, despite the antiquity of the two groups. For example, North American cool-temperate associations between algae and gastropods are mostly no older than the Pliocene (Vermeij 1992). These large kelps often form beds or 'forests' that support diverse molluscan communities throughout the temperate world (Steneck et al. 2002).

9.4.4.5.7.5 *Endomyxa: Haplosporidia* (*haplosporidians*)

The Haplosporidia is a small group of parasitic protists often treated as a separate phylum (see Burreson & Ford 2004 for review). The life history of most consists of two stages – a multinucleate plasmodium and a resistant spore. Most are marine, with only one known fresh-water species, and they infect a range of 'invertebrates'. *Haplosporidium* species are significant parasites of oysters (e.g., *Haplosporidium nelsoni* causes MSX disease in oysters and *H. costale* causes SSO disease, also in oysters) (e.g., Bower 2006) and are also found in some venerids (e.g., Costa et al. 2012) (see also Chapter 10). Haplosporidians have been recorded from fresh-water snails (Physidae) (Burreson 2001), scaphopods (Lauckner 1983a; Reynolds 2002, 2006), and chitons (Lauckner 1983c). The group includes the genus *Bonamia*, apparently lacking spores, which infects the haemocytes of oysters and is often lethal. *Mikrocytos*, another parasite of oyster haemocytes, is closely related to *Bonamia*. Some haplosporidians are hyperparasites.

9.4.4.5.8 *Incertae Sedis*

Laboratory colonies of species of *Lymnaea*, *Planorbis*, *Bithynia*, and *Bulinus* have been killed due to the flagellate *Dimoeriopsis destructor* (Hollande & Chabelard 1953) described from *Lymnaea* (Hollande & Pesson 1945), and it was suggested that this parasite might be effective as a biological control agent. The taxonomic position of this taxon is uncertain.

9.4.4.6 Associations with Fungi

Fungi have been variously regarded as 'plants', 'protists', or even 'animals', but they are now treated as a separate kingdom that is sister to Animalia. Like animals, fungi are heterotrophic, absorbing nutrients from surrounding organic material by way of their hyphae, and are thus saprotrophs. They have a unique cellular structure and the cell walls are chitinous. They have spores and sometimes produce spectacular fruiting bodies (e.g., mushrooms). Both the hyphae and fruiting bodies are utilised as food by many gastropods, but fungi can also cause diseases in some molluscs, although details are almost non-existent (e.g., Mead 1979).

Fungal diseases have been reported from oysters where they cause larval mortality (McGladdery 1999) and are also reported from coleoid cephalopods where they can affect gills and damage skin (Hanlon & Forsythe 1990; McGladdery 1999).

A fungus was reported to be invading and destroying spawn of a planorbid and embryos of *Physopsis*, although in the latter case at least it may have been due to contamination (Michelson 1957). Unidentified 'fungal infections' were noted in a number of fresh-water snails by McCaffrey and Johnson (2017). Reports of boring fungi in mollusc shells (Lauckner 1980) are usually mis-identifications of lichens.

9.4.4.6.1 *Lichens*

Lichens are complex organisms that are actually symbiotic relationships between fungi and algae or cyanobacterians (Henssen & Jahns 1974). They occur in both terrestrial and aquatic environments. In terrestrial environments, they can be an important food source for snails (see Chapter 5). In marine intertidal environments, lichens commonly infect molluscs, barnacles, and other calcium carbonate secreting organisms by boring through the shell with the fungal mycelium. Because of this erosion, the shell is weakened and colour patterns and morphologies of the infected taxa are often highly modified (Bonar 1936; Test 1945; Lindberg 1978). In Chile, the lichen *Thelidium litorale* actually mediates cryptic mimicry in several intertidal patellogastropods (*Scurria* spp.) by exfoliating exterior shell layers, which exposes inner pigmented layers and causes the limpet to closely resemble an acorn barnacle (Espoz et al. 1995, 2004). Which in turn reduces bird predation on the infected limpets (Hockey et al. 1987).

9.4.4.6.2 Zygomycota: Mucoromycotina

The fungus *Rhizopus*, in the Mucorales, was identified in the gut of *Archachatina marginata* (Oyeleke et al. 2012). Most *Rhizopus* species are saprobic on plants or specialised parasites of animals.

9.4.4.6.3 Ascomycota: Eurotiomycetes

The fungal pathogens *Aspergillus flavus* and *A. niger* were identified in the gut of *Archachatina marginata* where it was suggested that *A. niger* may produce useful enzymes (Oyeleke et al. 2012). These cause diseases in plants and some strains produce mycotoxins.

9.4.4.6.4 Ascomycota: Pezizomycotina

The fungus *Fusarium* was identified in the gut of *Archachatina marginata* (Oyeleke et al. 2012). Most of these fungi are found in soil or associated with plants. A few produce mycotoxins.

9.4.4.6.5 Microsporidia

Microsporidians are a very large group of tiny eukaryotic intracellular parasites within the kingdom Fungi. They form small unicellular spores, but unlike typical fungi they lack mitochondria. Previously treated as a group of protists or a separate phylum, they are recognised as a group of fungi following Walker et al. (2011).

Microsporans are found associated with a wide range of animals, including marine and estuarine bivalves (Lauckner 1983b; Bower 2006) and fresh-water (McClymont et al. 2005) and terrestrial (Selman & Jones 2004) gastropods. One common microsporidian, *Steinhausia mytilovum*, parasitises the ova of *Mytilus* while other species parasitise eggs of certain other bivalves including some oysters (Lauckner 1983b). Some species infect the intestine and digestive gland of bivalves. There are a few records from coleoid cephalopods (Hochberg 1990) and a few reports from other molluscs (e.g., Richards & Sheffield 1970; Maurand & Loubes 1979; Sagrista et al. 1998; Matos et al. 2005). Microsporidia infect various fresh-water snails (Michelson 1963; Cunningham & Daszak 1998), while others are hyperparasites of trematodes of both land and fresh-water snails (Canning & Basch 1968; Canning et al. 1974). In at least one case, a microsporidian (*Steinhausia* sp.) has been implicated in the extinction of a species of the land snail *Partula* (Cunningham & Daszak 1998). Molecular studies have shown that the microsporidians infecting fresh-water snails are more diverse than previously thought (Cunningham & Daszak 1998).

9.4.4.7 Associations with 'Plants', Including Red and Green 'Algae'

The common concept of plants (Plantae) that includes 'algae' is not a natural grouping, although they all utilise photosynthesis. 'Algae' are a polyphyletic assemblage with three main higher groups: brown (Phaeophyceae), red (Rhodophyta), and green (Chlorophyta), and all contain both unicellular and multicellular taxa (see below). The green algae are included in the clade ('phylum') Viridiplantae (or Chloroplastida), which also includes the land plants, but brown and golden algae are included in Heterokonta above.

9.4.4.7.1 Rhodophyta (Red Algae)

The red algae are one of the largest groups of 'algae' and the most ancient. The majority are multicellular and many of these are 'seaweeds'. Their chloroplasts were probably derived from endosymbiotic cyanobacterians, but different pigments are used in photosynthesis, resulting in their different colour. They include the coralline algae, which secrete calcium carbonate and are important in reef building.

Although the red algae are some of the most chemically defended, they are important food items for many herbivorous gastropods. They are also frequently an important component of the epibiont community, living on the shells of molluscs, particularly in the lower intertidal and subtidal. For example, the encrusting red coralline alga *Lithothamnion* often encrusts the shells of a variety of intertidal and shallow sublittoral living molluscs.

9.4.4.7.2 Viridiplantae (Green 'Plants')

The green plants (green algae and terrestrial plants) belong to the 'phylum' Viridiplantae. The terrestrial plants (Embryophyta or Metaphyta), which also include some aquatic taxa (macrophytes, seagrasses) are a monophyletic group (Puttick et al. 2018).

Green plants have cell walls that contain cellulose, they store food as starch and their chloroplasts, which contain chlorophylls *a* and *b*, probably originated from endosymbiotic Cyanobacteria.

Green algae and plants are important food items for many herbivorous gastropods. Green algae can also be an important component of the epibiont community on the shells of molluscs living in fresh water and marine mid to lower intertidal and shallow subtidal habitats, while plants can form dense seagrass beds with diverse molluscan communities and, of course, are critically important in terrestrial habitats.

9.4.4.7.2.1 Chlorophyta
This division includes some of the 'green algae', but the name has also been used to encompass all green algae. The group includes such characteristic green 'seaweeds' as *Ulva*, green filamentous algae, and many less well known green algae.

Members of the class Chlorophyceae are distinguished from other chlorophytes by several differences in their ultrastructure. Some chlorophyceans form epibiotic communities on fresh-water gastropods (e.g., Vega et al. 2006).

Several taxa of chlorophyceans have been implicated in corroding shells (e.g., Nielsen 1972). Some common small gastropods (Hydrobiidae) on mudflats in parts of the Northern Hemisphere provide hard substrata on which *Ulva* can settle and grow where they would otherwise have difficulty doing so (Schories et al. 2000). Members of another class, Zygnematophyceae, including *Spirogyra*, are also significant epibionts on some fresh-water gastropods (e.g., Vega et al. 2006). The single-celled green alga *Chlorella* can infest unionoids so they appear to be green in colour (Michelson 1957).

The class Bryopsidophyceae contains well-known macro-green algae such as *Codium* and *Caulerpa*. These and various other green algae are fed on by sacoglossan sea slugs that acquire plastids from them, which are coopted to produce photosynthetic products within the slugs (see Chapters 5 and 20).

9.4.4.7.3 Associations with Angiosperms

Green plants are of great significance to non-marine and some marine molluscs in terms of providing a substratum, food, and shelter, but we will not consider them further in this section as they are not strictly in symbiotic relationships although some arboreal snails and various marine seagrass and mangrove specialists are obligate plant associates. Patellogastropods have had a long association with marine angiosperms with notable radiations in the Paleogene of Europe and the Neogene of the Caribbean (Lindberg 1987a, 1990c). Non-limpet seagrass obligates have evolved adaptations to cope with the challenges of living on flexible leaves and the water currents that swirl around them (e.g., Hickman 2005a), and similar adaptations are, of course, needed for taxa that live on algal fronds.

The bark and/or leaves of bushes, palms, and trees are used as essential habitat by some arboreal gastropods. Similarly, some arid-zone snails burrow into or beneath grass or similar plants to survive.

9.4.4.8 Associations with Animals

9.4.4.8.1 Associations with Poriferans (Sponges)

Sponges are a food source for various molluscs. Burrowing sponges such as *Cliona* of the family Clionidae (Demospongiae) penetrate the shells of dead and living molluscs as well as other calcareous structures such as coral, and they are a significant bioerosion agent. These boring sponges are also preyed on by various invertebrates including gastropods (Guida 1976; Stefaniak et al. 2005).

Non-boring sponges are often epibionts on molluscan shells (e.g., Wulff 2006; Schejter et al. 2011; Mackensen et al. 2012) and can afford protection mainly by way of crypsis, although associations with non-boring sponges on the shells of living bivalves and gastropods can be more complex. The shells of some species of scallop are often encrusted with sponges which may provide some protection from predation and colonisation by boring sponges, while the sponges benefit from the inhalant water currents of the host (e.g., Forester 1979;

Pitcher & Butler 1987). A similar relationship has been shown with *Arca noae* (Corriero et al. 1991; Marin & Belluga 2005). Sponge-covered gastropods are rare, such as some individuals of certain species of *Euchelus* (Hickman & McLean 1990) and *Herpetopoma* (Chilodontidae), although the significance of these associations has not been investigated.

The siliquariid gastropods *Siliquaria* and *Tenagodus* are obligate commensals of certain sponges, living embedded within the host sponge, but not feeding on it, as they are suspension-feeders (Pansini et al. 1999). Other small gastropod species are often found within sponge canals, but also occur broadly in the environment and are therefore not obligates (Koukouras et al. 1996). Similarly, the pteriid bivalves *Vulsella* (Tsubaki & Kato 2012) and *Crenatula modiolaris* (Reid & Porteous 1980) are obligate sponge associates, being permanently encased in their host sponges.

9.4.4.8.2 Associations with Cnidarians

We discuss the defence-related aspects of molluscan associations with cnidarians above (see Section 9.4.3.12).

Useful reviews of associations between cnidarians and molluscs have been provided by (Robertson 1966, 1970a, 1980; Rees 1967). They are utilised for food (see Chapter 5), shelter and protection, and in some their stinging cells are coopted for defence (see Section 9.4.3.12 and Chapter 20).

Below, we provide examples of relationships of molluscs with members of the three main cnidarian classes. There are no known molluscan relationships with Cubozoa (box jellyfish).

9.4.4.8.2.1 Hydrozoa (Hydroids, 'Hydromedusae', Siphonophores)
Some hydroids use mollusc shells (bivalves, gastropods, and scaphopods) as substrata (i.e., are part of the fouling community). Others that live on the shells are symbiotic with the host mollusc as, for example, with some nassariids (e.g., Hiroshi 1991) and nuculid bivalves (Edwards 1965). Similar associations occur in other gastropods, including several thecosome pteropods, and some bivalves, with several families involved (Puce et al. 2008).

Among the most modified hydroids are the naked family Eirenidae that reside, in an apparently mutualistic relationship, in the mantle cavity of some bivalves, including species of mussels, oysters and venerids (Rees 1967; Piraino et al. 1994; Kubota 2000, 2004; Govindarajan et al. 2005), and teredinids (Nair & Saraswathy 1971). The hydroid receives shelter, protection, and food while the bivalve may benefit from the hydroid ingesting sporocysts, as trematode infection is low when the hydroids are present (Piraino et al. 1994). The planula larvae settle in a host bivalve and develop into hydroids, which eventually bud medusoids that exit the bivalve and undergo the sexual part of their life cycle in the water column. The hydroids attach to the mantle, labial palp, foot, and sometimes the gill of the host bivalve by way of a sucker-like disc at the base of their stalk-like body (Govindarajan et al. 2005).

The athecate hydroids, *Klinetocodium* (Hydractiniidae) and *Pandea* (Pandeidae) are probably obligatory associates with a few thecosome pteropods, but the nature of their association is uncertain (Lalli & Gilmer 1989).

Phylliroe bucephala is symbiotic with the medusa stage of the 'hydromedusa' (i.e., Anthomedusa) *Zanclea costata*. The medusa is attached to the nudibranch ventrally just behind the mouth. The post veliger initially settles on the medusa and feeds on the prey captured by it, then the remains of the medusa are carried around by the adult nudibranch, which feeds on siphonophores (Ankel 1952; Lalli & Gilmer 1989).

9.4.4.8.3 Scyphozoa (Jellyfish)

There is a report of possible imitation of jellyfish by squid paralarvae of the family Gonatidae (Arkhipkin & Bizikov 1996). Certain jellyfish are reported to be prey items for *Argonauta* (Heeger et al. 1992).

9.4.4.8.3.1 Anthozoa (Sea Pens, Sea Anemones, Hard and Soft Corals)

Epitonids and architectonicids feed on anemones and some other cnidarians, with some of these gastropods being much smaller than their host and thus could be regarded as ectoparasites. A few galeommatoid bivalves are commensal with anemones, including burrowing species (e.g., Ponder 1971; Morton 1980a, 1988a; Goto et al. 2012). Some relationships with anemones provide protection (see Section 9.4.3.12), and they also are fed on by many epitoniids and some aeolidioid nudibranchs.

A symbiotic (mutualistic) relationship between a deep-sea scaphopod and a sea anemone, which lives on the dorsal (concave) side of the shell has been described (Shimek 1997; White et al. 1999). Similar relationships are known in a few other scaphopods and a relationship with a solitary coral and a scaphopod is also known (Reynolds 2002).

Sea pens (Pennatulacea) are food items and a source of secondary metabolites for some nudibranchs (see Section 9.4.3.4.8). Similarly, many gastropods and some bivalves are associated with octocorals, such as gorgonians and soft corals (Alcyonacea), and may utilise them for food (e.g., ovulids, the nudibranch *Tritonia*, and solenogaster aplacophorans), shelter, or substrata (e.g., some species of *Pteria*). Among the Hexacorallia, the Antipatharia (black corals) have some species of *Pteria* attached to them, as has the muricid (Coralliophilinae) *Rhizochilus antipathum*.

Relationships between coral and boring bivalves have been documented by Morton (1990) and Kleemann (1996). The mytilid *Fungiacava eilatensis* lives within *Fungia* in cavities chemically excavated by secretions from the mantle that envelops the delicate shell (Goreau et al. 1969, 1972) (see Chapter 15, Figure 15.36). This bivalve lives in a commensal relationship with the coral, its enlarged inhalant siphon opening into the coelenteron of the coral and its mobile foot assisting with food collection (Goreau et al. 1970). It is unrelated to the other coral-boring mytilids, the lithophagines, which bore into hard coral by chemical means. Other coral-boring bivalves such as *Gastrochaena* use mechanical means. The giant clams *Tridacna crocea* and *T. fossor* are also mechanical coral borers and can assist in breaking down coral boulders. A galeommatoidean bivalve is also known to be associated with living coral (Morton 1980a).

One of the best known hard coral associates is the pectinid bivalve *Pedum* (Yonge 1967; Kleemann 1990; Scaps 2011), which lives embedded in living scleractinian corals such as *Porites*. It is attached byssally and is surrounded by the living coral. Its presence is thought to reduce coral predation by the 'crown of thorns starfish' (*Acanthaster planci*) (DeVantier & Endean 1988).

While all members of the Coralliophilinae (Muricidae) are coral associates, young *Leptoconchus* and *Magilopsis* are mechanical borers and the adult flask-shaped burrows in living coral have very small apertures (Gohar & Soliman 1963).

9.4.4.8.4 Associations with Acoelomorpha (Acoela)

Members of this phylum were, until recently, included as part of the platyhelminths. Apart from the acoelan *Convoluta japonica* being implicated in the death of a venerid in Japan (Kato 1951), we have not found other reports of acoel and molluscan relationships.

9.4.4.8.5 Associations with Dicyemida

Dicyemid mesozoans are a phylum of highly specialised microscopic, worm-like marine hermaphroditic organisms that live exclusively in the kidneys of coleoid cephalopods. They spend most of their life cycle attached to the renal appendages of the coleoid hosts. The dicyemid is attached by its ciliated head, the rest of the body hanging free in the renal coelom absorbing its nutrients from the urine of the host. The body consists of about 14–40 cells surrounding an elongate axial cell and the organism grows from the early stage simply by enlarging its cells. There are no organs and their simple body form is probably a consequence of the 'parasitic' life these animals lead, although there is no agreement on the exact nature of the relationship. Most authors consider them to be commensal with little positive or negative consequence to the host (e.g., Hochberg 1990). Alternatively, it has been suggested that dicyemids facilitate the excretion of ammonia by the host (Lapan 1975), and there is also evidence they can cause some damage to the epithelium of the renal appendages (Finn et al. 2005).

While most are host specific, some coleoids contain two or three species of dicyemids. There is a correlation between host size and the size of the adult dicyemids, with the size of the adult constrained by the size of the organs they inhabit (renal sac volume, the diameter of renal tubules, and the extent of internal folding in the renal appendages) (Furuya et al. 2003a, 2003b).

Although often present in high densities, infection rates are sometimes low, and they are absent from some coleoids. Infection appears to be higher in temperate latitudes and is restricted to benthic and epibenthic coleoid taxa with pelagic taxa apparently not infected (Hochberg 1990; Finn et al. 2005). The mode of infection of the host is unknown, but is probably due to a minute motile stage entering the renal papillae (Finn et al. 2005).

The phylogenetic position of this taxon is controversial. It has been considered the outgroup to metazoans (as one of the two members of the 'Mesozoa'), or, alternatively, as a basal lophotrochozoan related to platyhelminths. They have spiral cleavage, and recent phylogenetic studies suggest that dicyemids may be more closely related to the higher lophotrochozoans (Petrov et al. 2010; Suzuki et al. 2010) (see also under Orthonectida below).

9.4.4.8.6 Associations with Orthonectida

This group, like the Dicyemida, was previously included in the Mesozoa. Recent phylogenetic analyses have found them to be either a separate phylum within the 'Spiralia', and unrelated to the Dicyemida (Sliusarev 2008), or that they form a monophyletic group with the Dicyemida (i.e., Mesozoa) within the Lophotrochozoa (Petrov et al. 2010).

Although very reduced, orthonectids are more complex animals than dicyemids. They have circular and longitudinal muscles; their body is ciliated and has a cuticle. They also have a very simple nervous system and a multicellular receptor in adults. Their life cycle includes sexual, asexual, and parthenogenetic stages (Sliusarev 2008).

An orthonectid has been recorded from the mantle cavity, blood sinuses, kidney, and ovotestis of the cephalaspidean *Philine* (Lang 1954) and also from a rissoid, a columbellid, *Solariella* and *Lepeta* (Shtein 1954), and some bivalves (Atkins 1933; Kozloff 1992). They are mainly found in the gonads and sometimes also the haemocoel and blood vessels of the mantle and gill.

9.4.4.8.7 Associations with Platyhelminthes

Traditionally, the platyhelminths were divided into three classes, the free-living 'Turbellaria' and two parasitic classes, the Cestoda and the Trematoda. As with many other groups, the classification of platyhelminths has recently undergone major changes and is now quite complex. The major vertebrate parasitic taxa (Cestoda, Monogenea, and Trematoda) are grouped together in the Neodermata (Riutort et al. 2012), and all include at least some involvement with molluscs in their life-cycles. Other mainly free-living taxa, the triclads, rhabdocoels, and polyclads, are grouped in the Archoophora and also include some taxa associated with molluscs.

9.4.4.8.8 Archoophora: Polycladida

This major group includes the polyclad flatworms that can be associated with the epibionts on encrusted mollusc shells, but some have closer relationships with molluscs. Commensal relationships between polyclads and intertidal gastropods are not uncommon where they live in the mantle cavity, including a planktonic species that lives with *Janthina* (e.g., Lauckner 1980; Faubel et al. 2007). Most polyclads associated with bivalves are members of the acotylean family Stylochidae. These polyclads, sometimes called 'oyster leeches', are predators of various bivalves, including several commercial taxa (e.g., oysters, mussels, pearl oysters, giant clams) and, particularly, juveniles (Galleni et al. 1980; Littlewood & Marsbe

1990; Newman et al. 1993). A polyclad, *Stylochoplana parasitica*, lives in the mantle groove of some chitons in Japan (Kato 1935), and some polyclads mimic toxic nudibranchs, such as *Pseudoceros imitates*, that imitates *Phyllidiella pustulosa* (Newman et al. 1994) (see also Section 9.4.3.10.1).

9.4.4.8.9 Neoophora

This grouping includes the triclads, rhabdocoels, and the parasitic clades. The Neodermata encompasses the parasitic groups (trematodes, cestodes, and Monogenea).

9.4.4.8.9.1 Rhabdocoela
The subclass Rhabdocoela contains a few taxa parasitic in molluscs. For example, two genera of Graffillidae are parasites of bivalves, including *Graffilla* found in the digestive gland of a teredinid (Nair & Saraswathy 1971). Species of the genus *Paravortex* usually live in the gut, but also the mantle cavity, kidney, and heart, of some marine bivalves with sometimes high levels of infestation (e.g., Lauckner 1983b; Brusa et al. 2006). *Urastoma cyprinae* (Urastomidae), a supposedly cosmopolitan species, infests bivalves, but is also found free-living (Robledo et al. 1994; Cáceres-Martínez et al. 1998).

Members of the 'order' Temnocephalida are all parasitic or commensal. They can be associated with the gill and mantle cavity of their hosts, and some are parasites in the gut. They differ from other platyhelminthes in having a ventral adhesive disc with which they attach to the host and there are finger-like projections around the head. *Temnocephala* (Temnocephalidae) are, for example, common in the mantle cavity of ampullariid snails (e.g., Damborenea et al. 2006; Vega et al. 2006), although they are mainly associated with crustaceans.

9.4.4.8.9.2 Neodermata
The Neodermata encompasses the parasitic groups (trematodes, cestodes, and Monogenea).

Cestoda (Tapeworms): Members of this large group of internal parasites have no mouth or gut, nutrients being absorbed through the skin. Almost all tapeworms have at least two hosts.

There are two groups, the eucestodes, which include most tapeworm taxa; a chain of segments (proglottids) are budded off from the 'neck' and the head is a holdfast with variable attachment structures. In the other group, the Cestodaria, the body is unsegmented, the 'head' is not modified as a holdfast, and larvae are at the posterior end. This latter 'subclass' contains only two orders, and they are not known to occur in molluscs.

Besides one report of an adult cestode in an octopus (see Hochberg 1990), all other records of cestodes in molluscs are secondary larval stages, and include reports from bivalves (Lauckner 1983b; Bower 2006) and gastropods (Lauckner 1980), including terrestrial slugs (Runham & Hunter 1970; South 1992). None appear to have been reported from chitons or scaphopods.

Eucestoda: This is the largest 'subclass' of cestodes and contains a dozen orders. Members of the orders Diphyllidea and Cyclophyllidea and several unidentified cestodes have been recorded as larvae from various marine gastropods (Lauckner 1980). In bivalves, larvae of members of the orders Trypanorhyncha, Lecanicephalidea, Tetraphyllidea, and Diphyllidea (all of which have elasmobranchs as definitive hosts), as well as unidentified taxa, have been recorded. Some of the larval cestodes are implicated in initiating pearl formation (Lauckner 1983b).

Several cyclophyllidean cestodes have been found in land snails with their definitive hosts being mammals or birds, although they are generally rarely encountered (see Mead 1979 for a summary). The best known is *Davainea proglottina*, which infects poultry and certain other birds and has as its intermediate host a number of species of terrestrial slugs and snails (e.g., Abdou 1956).

Several eucestode taxa are recorded from coleoid cephalopods, although only as larvae (e.g., Hochberg 1990; Pascual et al. 1996; Shukhgalter & Nigmatullin 2001) with their incidence suggesting that they are important intermediate hosts. Members of the orders Tetraphyllidea and Trypanorhyncha are the most commonly encountered (Hochberg 1990), both of which are usually parasites in the gut of cartilaginous fish. The first host is normally a small crustacean and the larvae can be acquired either directly by these being ingested or indirectly by feeding on small fish, or even other cephalopods.

Several genera of the family Phyllobothriidae (Tetraphyllidea) occur in the gut of a range of coleoids, often in large numbers, with the definitive host being elasmobranchs or occasionally cetaceans (Hochberg 1990). Similarly, larval cestodes that may be members of the Tetrabothriidae (Tetrabothriidea) have been found in cephalopods, and these have their final stage in sea birds, cetaceans, and pinnipeds (Hochberg 1990).

Members of the family Tentaculariidae (order Trypanorhyncha), including the genera *Nybelinia* and *Tentacularia*, have four tentacles with hooks and are found in a wide variety of coleoids where they can occur in large numbers. They are acquired from feeding on the crustacean primary hosts, and the final (definitive) hosts are usually sharks.

Monogenea: Monogeneans are small ectoparasites usually found on the skin or gills of fish, but some are internal parasites. They have only one larval generation in their life cycle (hence their name). Both ends, and especially the posterior end, bear hooks or other attachment devices (but not suckers) with which they attach to their hosts. Some taxa have been found associated with squid where they infect large blood vessels or are attached to the mantle cavity, gills, or arms (Hochberg 1990).

9.4.4.8.9.3 Trematoda The most important molluscan parasites are digenean trematodes, with most trematode flukes utilising snails as obligatory first intermediate (primary) hosts (Figure 9.11). Thus, gastropods can be considered keystone species for trematodes and their communities (Esch et al. 2001), supporting the development of over 18,000 species of these parasites. While most gastropods that harbour trematodes are aquatic, they are also widely found in terrestrial snails

and slugs (reviewed by Mead 1979). Most shallow-water and intertidal marine snails harbour trematode parasites, most of which have birds or fish as their definitive host. Bivalves are also important as primary hosts for trematodes and a very few use scaphopods, and there is only one unconfirmed record of a trematode infection of a chiton (Lauckner 1983c). The other molluscan classes do not act as primary hosts, although trematodes parasitise cephalopods, which are commonly used as the secondary or definitive host (Hochberg 1990).

Trematode flukes typically have holdfasts (sucker-like structures) and are active feeders with an intestine, and a complex lifecycle (see Box 9.3 and Figure 9.11) with one to three intermediate hosts. Most of these parasites have, as definitive hosts (i.e., the host with the adult stage), vertebrates such as fish (but rarely chondrichthyans), amphibians, reptiles, birds, and mammals. Several important species infect humans (see Chapter 10).

With very few exceptions, molluscs (see above) are the primary (i.e., first intermediate) hosts of trematodes. The gastropod groups encompass a wide range of taxa, but are clustered mainly in the 'asiphonate' caenogastropods and, for some trematode groups, in the fresh-water Hygrophila and intertidal vetigastropods and patellogastropods.

Trematode sporocysts and rediae infections typically result in castration and may change the host growth rate by either reducing it or increasing it, sometimes resulting in gigantism. These changes may correlate with the level of castration (Gorbushin 1997), but are also the result of a complex interplay of factors, including the species of trematode, the longevity of the snail host, the nature of the habitat, and other environmental parameters (Gorbushin & Levakin 1999). It seems that short-lived species may often be more inclined to gigantism than longer lived ones (Sousa 1983; Gorbushin & Levakin 1999). Similarly, trematodes negatively affect survival and usually fecundity, although sometimes there is a positive short term effect (e.g., Sorensen & Minchella 2001; Jordaens et al. 2007).

Although trematode parasites have an intimate relationship with their gastropod intermediate hosts, the nature of this relationship at the molecular level is not well understood. Host specificity is typically high and sometimes this relationship is so specialised that cospeciation is indicated. Similarly, trematodes will be rejected by an unsuitable snail host. The types of signalling that may occur between the parasite sporocyst and its snail host include (Lockyer et al. 2004) snail receptor molecules interacting with the sporocyst secretions, direct contact between the parasite and host tissue, or receptors and ligands on the sporocyst surface interacting with soluble host molecules. The immunological responses of the mollusc to parasitic infections are discussed in Chapter 2.

The class Trematoda is divided into two subclasses, Aspidogastrea and Digenea, with two orders and several suborders recognised within the latter subclass, as detailed below, where we follow the phylogenetic classification of Olson et al. (2003). Much of the information summarised below is obtained from the valuable reviews by T. H. Cribb

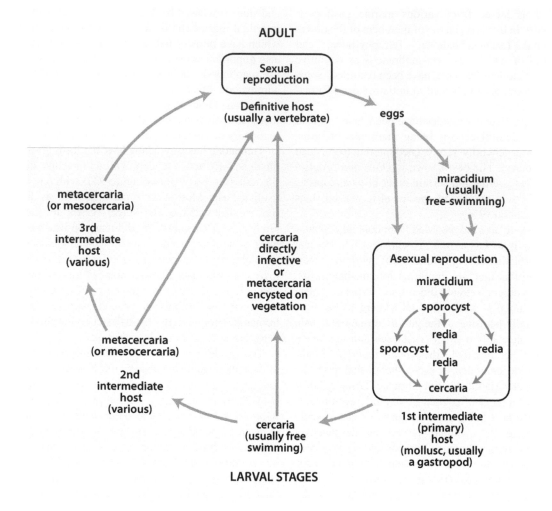

FIGURE 9.11   Summary of trematode life cycles. From various sources.

BOX 9.3   THE STAGES IN THE LIFE CYCLE OF A TREMATODE

Eggs – produced by the adult in the definitive host. Develop and hatch as a miracidium.

Miracidium – usually a free-swimming ciliated larva that infects the mollusc primary host. In a few taxa, the eggs are directly ingested by the mollusc primary host.

Asexual stages within the first mollusc host

    Sporocyst – a sac in which multiple offspring develop that, depending on the species, may develop into a redia or another sporocyst. If a redia is not developed, the sporocyst produces the cercaria.

    Redia – an asexual stage, with mouth, pharynx, and gut, which produces cercaria.

Actively swimming stage that emerges from the secondary intermediate host

    Cercaria – has a tail that may be simple or forked.

Stages that infect the definitive host

    Metacercaria – a form present in most digenean life cycles, either in the tissues of a second or third intermediate host or on vegetation where it may be encysted.

Sexual adult stage – In definitive host. Most are hermaphrodite, with sexual reproduction, although sometimes self-fertilisation occurs. Blood flukes have separate sexes, but the male and female are held together. This stage is formally called the sexual adult to draw attention to the fact that sporocysts and rediae are also adults, although they resemble larvae.

See Figure 9.11.

BOX 9.4   SOME TREMATODE TERMINOLOGY

There are several kinds of adult body forms recognised in trematodes:

Amphistome – large, with large sucker at posterior end
Distome – mouth surrounded by a sucker and also a ventral sucker
Echinostome – oral sucker surrounded by spinose collar – ventral sucker also present
Monostome – usually only the oral sucker is present. Some lack suckers
Holostome – body divided into two regions
Gasterostome – mouth near centre of body and gut very simple
Schistosome – sexes separate, male usually larger, with female in groove.

and coworkers (Cribb et al. 2001a, 2001b, 2003). Some terminology is outlined in Box 9.4.

Subclass Aspidogastrea (Aspidobothrea, Aspidocotylea): This small group (~80 species) comprises four families divided into two orders, Aspidogastrida and Stichocotylida. They have either a single divided ventral sucker or a row of suckers. The Stichocotylida contains a single species which is a parasite of elasmobranchs.

For the Aspidogastrida, the first (and usually the only) host is a marine or fresh-water gastropod or bivalve, although in some, there is a second host (both cartilaginous and bony fish, and turtles). Unlike digeneans, there is no asexual reproduction at any stage. Host specificity is often low, but variable, with some species infesting both bivalves and gastropods as well as tortoises and fish.

The largest family is the Aspidogastridae, and members of that family are unusual in parasitising fresh-water bivalves, including unionids, *Corbicula* and *Dreissena* (Grizzle & Brunner 2009), as well as some gastropods. Some members of the family may carry out their life cycle entirely within the mollusc (e.g., species of *Aspidogaster*), although if fish eat the mollusc, they can become infected (Molloy et al. 1997), while others mature when eaten by either a fish or turtle.

Subclass Digenea: Members of this large group (~18,000 species comprising over 150 families) (Cribb et al. 2001b) usually have two or more hosts in their lifecycles (Esch 2002). The first intermediate (primary) host is nearly always a mollusc. The sporocyst and redia stages in the mollusc reproduce asexually. The adult stages are found in vertebrates (including fish, amphibians, reptiles, birds, and mammals), the definitive host, where they can cause serious disease. Most are found in the gut of the definitive host, but they can occur in most other internal organs (liver, lungs, bladder, and blood system).

The traditional divisions of the Echinostomida, Plagiorchiida, and Strigeida are not monophyletic (Olson et al. 2003), but instead the digeneans can be split into two main groups, the Diplostomida and Plagiorchiida. The former contains three superfamilies, the latter 19. The Plagiorchiida are further subdivided into 13 higher groupings (see Olson et al. 2003 for further details).

All groups of vertebrates are important definitive hosts except for chondrichthyans which are surprisingly poor hosts (T. Cribb, Personal communication). Gastropods commonly serve as intermediate hosts, with bivalves being the next

most important. At least 66 digenean families utilise gastropods compared with only nine (including Gymnophallidae, Fellodistomatidae, Bucephalidae) utilising bivalves (e.g., Vazquez et al. 2006) and some Lecithasteridae and Ptychogonimidae have been found in scaphopods (Køie et al. 2002). The Aporocotylidae is exceptional in using gastropods, bivalves, and polychaetes as first intermediate hosts.

Digeneans usually have two suckers, one ventral and one oral ('distomes'), although some ('monostomes') have a single sucker. Most are hermaphroditic.

Superorder Diplostomida: The Diplostomida contains a single order, Diplostomata, the sister group to the rest of the digeneans.

Order Diplostomata: This order includes trematodes having cercariae with mostly forked tails. It includes three superfamilies. The Diplostomatoidea contains six families, and all have their adult phase in tetrapods. Their primary hosts are gastropods and the metacercariae mostly encyst in vertebrates (fish or amphibians) as secondary intermediate hosts, although some use molluscs or annelids as their second host. The second host is then eaten by the vertebrate definitive host. The Brachylaimoidea all have molluscs as their second intermediate hosts and three of the four families have entirely terrestrial life cycles, utilising terrestrial 'pulmonates' with these cercariae having very reduced tails. The primary hosts are birds or mammals. The fourth family, Leucochloriodiomorphidae, has an aquatic life cycle in fresh water, utilising snails as primary and secondary intermediate hosts.

The Schistosomatoidea, which includes the human blood flukes (*Schistosoma* spp.), comprise three families and the Clinostomidae, which is sometimes included in a separate superfamily. Schistosomes are unusual among digeneans in having separate sexes. Schistosomatoids use gastropods as primary hosts, except for one of the blood fluke families, Aporocotylidae, which also utilise polychaetes and bivalves as primary hosts. The fork-tailed cercaria produced by the rediae or sporocysts directly infect the definitive host, except clinostomids which differ in having fish or amphibians as intermediate hosts and hygrophilan 'pulmonates' as primary hosts.

Many snails are hosts of schistosomes (Schistosomatidae) that infect birds and mammals, with the most significant being the snail hosts of human schistosomes. These are either pomatiopsids (*Oncomelania* spp.) or planorbids

(*Bulinus* spp. and *Biomphalaria*) (see also Chapter 10). When *Biomphalaria glabrata* is infected with *Schistosoma mansoni* (Cooper et al. 1996) or *Bulinus truncatus* is infected by *S. haematobium* (Schrag & Rollinson 1994), egg production by the snail host is much reduced. A similar reduction in fecundity has been recorded in *Lymnaea stagnalis* infected with the avian schistosome *Trichobilharzia* (Joosse & Elk 1986).

Other families in this group include the Spirorchiidae, which infect marine turtles, the Aporocotylidae, which infect both chondrichthyan and teleost fish, and the Clinostomidae, which infect fish-eating reptiles (crocodiles) and birds (Cribb et al. 2001b).

Several marine snail species are known as carriers of the larval stages of avian schistosomes, whose emerging cercariae can produce dermatitis (e.g., 'Pelican itch') in humans in various parts of the world.

Superorder Plagiorchiida: This group encompasses the great majority of digeneans, containing 13 orders consisting of 19 superfamilies. We treat each order separately below.

Order Bivesiculata: This small group, the sister to the remaining Plagiorchiida, contains only the Bivesiculidae, a family that uses cerithioid gastropods as the primary host and a fish as the definitive host where they infect the small intestine (Cribb et al. 2003). The cercariae have forked tails.

Order Transversotremata: The single small family in this order, Transversotrematidae, have caenogastropod primary hosts and no secondary intermediate host; their definitive hosts being teleost fish where they live beneath their scales as ectoparasites. The cercariae have forked tails.

Order Hemiurata: This group includes twelve families grouped in two superfamilies (sometimes treated as separate orders), and the cercariae of both groups have forked tails. One superfamily (Azygioidea), contains a single family, and these parasites have a vertebrate as the definitive host and various aquatic gastropods as the primary hosts. The other superfamily, Hemiuroidea, contains several families and primary hosts include a range of aquatic gastropods, and (rarely) bivalves and scaphopods. Secondary hosts are mostly crustaceans or other arthropods, but coleoid cephalopods, gastropods, bivalves, chaetognaths, vertebrates, or even ctenophorans are also sometimes utilised. In one family, the Didymoxoidae, the metacercaria parasitise the stomach of coleoid cephalopods, which are infected via their crustacean prey. Didymozoids have three to four hosts in their life cycle, culminating in fish, with pelagic gastropods the probable primary intermediate hosts (Hochberg 1990). Larvae of some members of the Hemiuridae infest *Argonauta*, Hirudinellidae infect squid, and Derogenidae are parasites of cuttlefish (Hochberg 1990).

The definitive hosts of azygioideans are teleosts and chondrichthyans, and Hemiuroidea also parasitise these two groups of fish and a few tetrapods. For example, some species of *Halipegus* (Hemiuridae) have frogs as the definitive host and a snail as the secondary host, with arthropods (ostracods, insects) as the tertiary hosts (e.g., Macy et al. 1960). Infection of *Helisoma anceps* by *Halipegus occidualis* results in reduced fecundity (Keas & Esch 1997).

Order Heronimata: This order contains a single family, Heronimidae, containing only a single species for which the primary host is a hygrophilan, and the definitive host fresh-water turtles, in which the parasite lives in the lungs. Unusually for Plagiorchiida, there is no secondary intermediate host. The cercariae have a simple tail.

Order Bucephalata: This order contains two superfamilies, Bucephaloidea and Gymnophalloidea, the former containing a single family, the latter five families. These trematodes have cercariae with forked tails and are unusual in having bivalves as the primary host (e.g., Bower 2006), including fresh-water taxa, notably *Dreissena* and unionids (Grizzle & Brunner 2009). The Bucephalidae have fish as their secondary hosts with only teleost fish as their definitive host, where they infect the intestine. The Gymnophallidae have birds as their definitive host, and the other five families infest teleost fish. Two families do not have secondary hosts, the remainder use a wide range of animals as their second hosts, such as molluscs (including cephalopods), echinoderms and, more rarely, vertebrates, crustaceans, annelids, or cnidarians.

Order Pronocephalata: These trematodes comprise several families in two superfamilies, the cercariae of which have simple tails. The Paramphistomatoidea and Pronocephaloidea have various gastropods as their primary hosts, and there is no secondary host. The definitive hosts are teleosts and tetrapods. The Paramphistomatoidea are characterised by the absence of an oral sucker and the possession of a posterior ventral sucker. The better known of the four included families, the Paramphistomatidae, are usually large flukes that parasitise the gut of herbivorous mammals causing paramphistomatiasis and have various fresh-water hygrophilans as their intermediate hosts. The Pronocephaloidea contains six families, members of which are found in marine mammals such as dugongs, while one family, Pronocephalidae, are parasites of marine and fresh-water turtles with various caenogastropod and hygrophilan snails as primary hosts.

Order Haplosplanchnata: This small group contains only the family Haplosplanchnidae, which utilises cerithioid snails as the primary host, and there is no secondary intermediate host. The cercariae have simple tails, and the definitive hosts are mainly herbivorous teleost fish.

Order Echinostomata: This group comprises a few families grouped in a single superfamily. The primary hosts are various gastropods that may be marine, fresh water, or terrestrial. The secondary intermediate hosts may be gastropods, crustaceans, chordates, or lacking entirely (i.e., two-host cycles). The definitive hosts are usually tetrapods, and the cercariae have simple tails.

The group includes some significant parasites, including the families Fasciolidae (adults large, thin-bodied, and leaf-shaped) that parasitise mammals (e.g., the liver flukes, *Fasciola*, and the intestinal fluke *Fasciolopsis*), and Echinostomatidae (elongate, with spinose collars) found in the gut of birds, reptiles, and mammals.

Order Opisthorchiata: This small group consists of a single superfamily for which the primary hosts are caenogastropod snails, and the secondary intermediate hosts are fish or rarely amphibians. Definitive hosts are teleost fish and tetrapods.

Included in this group are the families Heterophyidae (small flukes that infect birds and mammals including humans) and Opisthorchiidae (in the gall bladder and bile duct of mammals, birds and reptiles – e.g., *Opisthorchis sinensis*, the 'southeast Asian liver fluke', a human parasite). The cercariae have simple tails.

Order Apocreadiata: This order contains a single family of teleost parasites. The primary hosts are caenogastropods (often truncatelloideans), and the secondary intermediate hosts are gastropods or annelids. The cercariae have simple tails.

Order Lepocreadiata: This group contains several families of teleost parasites grouped in a single superfamily. The primary hosts are caenogastropods, the secondary can be cnidarians, platyhelminths, molluscs, crustaceans, annelids, echinoderms, or vertebrates. The cercariae have simple tails.

Order Monorchiata: This group contains two families of teleost parasites in a single superfamily. The primary hosts may be gastropods or bivalves, and the secondary intermediate hosts include molluscs, crustaceans, annelids, platyhelminths, and echinoderms. The cercariae have simple tails.

Order Xiphidiata: This large, diverse group may not be monophyletic. Cercariae typically have simple tails. The many families are grouped in four superfamilies. The Gorgoderoidea, Allocreadioidea, and Microphalloidea all of which have both gastropods and bivalves as the primary hosts, but Plagiorchioidea has only gastropods (marine, fresh water, and terrestrial).

The Allocreadioidea utilise various aquatic gastropods and bivalves and the Microphalloidea only caenogastropods and stylommatophorans. The secondary intermediate hosts are mostly arthropods or molluscs (including octopods), but also include annelids (not in Gorgoderoidea) or vertebrates, and sometimes echinoderms in Microphalloidea. The definitive hosts include teleost fish and tetrapods, besides chondrichthyan fish in Gorgoderoidea alone.

The Gorgoderoidea includes the families Dicrocoeliidae (small parasites of gut and associated organs) and Troglotrematidae (infest various internal organs of birds and mammals). This latter family includes the human lung fluke *Paragonimus westermani*. The Dicrocoeliidae contains *Dicrocoelium dendriticum*, known as the lanceolate liver fluke, which infects ruminants, as well as pigs and, rarely, humans. It has as its first intermediate host a land snail or slug and the second is an ant that is driven to suicidal behaviour by the parasite (Manga-González et al. 2001). Cercariae from the ant then infect the definitive host.

The Plagiorchioidea includes eight families, including the Brachycoeliidae, which contains taxa that infect slugs and have amphibians as their definitive host.

Trematode Life Cycles

The parasite life cycles (Figure 9.11) range from simple to complex. Some have just the intermediate molluscan host, such as the schistosomes (blood flukes), where the swimming cercariae are shed from the snail and directly infect the definitive host. In others, such as the liver fluke *Fasciola hepatica*, the cercaria emerge from the snail primary host, encyst on plants, and only enter the definitive host (such as cattle or humans) when the vegetation is ingested.

Some other trematodes require two intermediate hosts, and in almost all cases, the first (primary) host is a mollusc. For example, in species of *Paragonimus*, the cercariae from the snail enter the second intermediate host, often a crustacean, in which they encyst in the tissues of the host as a metacercaria. The definitive host is infected when it eats the second intermediate host.

In general, the miracidia larvae swim and are host specific, but the adult flukes are often less host specific. Once inside the mollusc (usually a snail) first intermediate host, the parasite undergoes a series of asexual divisions. This asexually dividing stage may be a simple sac (sporocysts) or have a mouth, pharynx, and a rudimentary gut (redia). In some, there are two generations of sporocysts, with the second producing the cercariae. In others, such as *F. hepatica*, there is a single generation of sporocysts, followed by rediae that produce the cercariae.

Some fresh-water bivalves are significant hosts of trematodes. Seven genera are reported from *Dreissena* spp., and they may be the first or second intermediate host or the only host (*Aspidogaster* sp.) depending on the species of trematode (Molloy et al. 1997).

Impact on Molluscan Hosts: The common consequence of trematode infection is castration of the host gastropod or bivalve, and the infection may also alter the host behaviour (Moore 2002). For example, the normal foraging behaviour of the tateid *Potamopyrgus antipodarum* is altered to greater activity in daylight hours when predation is more likely (Levri & Lively 1996; Levri 1999), and similar behavioural changes have also been shown to occur in other gastropods such as *Littorina littorea* (Lambert & Farley 1968), *Physa integra* (Bernot 2003), *Peringia ulvae* (Huxham et al. 1995), and *Cerithidea scalariformis* (Belgrad & Smith 2014). Bivalves also have behaviours altered by trematodes. For example, the common 'European cockle' *Cerastoderma edule* (Cardiidae) infested with species of *Himasthla* (Plagiorchiida, Echinostomata, Himasthlidae) and/or *Gymnophallus* (Plagiorchiida, Bucephalata, Gymnophallidae) have reduced burrowing ability so they are more susceptible to predation and, because they are on the surface, can suffer mass mortality events (e.g., Thieltges 2006). Similarly, individuals of the venerid bivalve *Austrovenus stutchburyi* infected with another himasthlid, *Curtuteria australis*, have a smaller foot and consequently do not burrow as effectively, making them more susceptible to oystercatchers (Thomas & Poulin 1998). One of the most bizarre examples is caused by species of the trematode *Leucochloridium* (Strigeidida, Leucochloridiidae) that parasitise succineid snails. The sporocysts move into the tentacles, the behaviour of the snail is modified causing it to move into the open and the swollen tentacles, which cannot be retracted, pulsate, their movement attracting birds (Wesołowska & Wesołowski 2014).

Gymnophallid trematodes are common in bivalves and sometimes leave shell blisters or other traces seen in fossils and that can be followed through time (Ruiz & Lindberg 1989; Huntley & De Baets 2015). For example, scars on the interior of the valves of Neogene specimens from North America suggested that these trematode infections originated in the Atlantic and were introduced to the Pacific at the Pliocene-Pleistocene boundary (2.6 Ma) (Ruiz & Lindberg 1989). Holocene fossils of the Chinese corbulid *Potamocorbula amurensis* enabled infestations to be tracked (Huntley et al. 2014). Other changes can be induced – for example, warts are formed on the hinge plate of individuals of the tellinid *Macoma balthica* infected with gymnophallid trematodes (Gantsevich et al. 2016).

It is probable that the molluscan hosts have adapted in various ways because of their interactions with trematodes. For example, in the bivalve *Nutricola*, infections by the gymnophallid trematode *Telolecithus* and the monorchiid (Plagiorchiida) *Parvatrema* have been suggested to result in size reduction and contribute to the maintenance of sexual dimorphism for size (Ruiz 1991).

Trematode infection can cause reduced survival under normal (e.g., Fredensborg et al. 2005) or extreme environmental conditions (e.g., Sousa & Gleason 1989). The effect of trematodes on their snail host can lead to gigantism, typically because of castration, although this is not always the case. For example, in hydrobiids infections from microphallid and heterophyid trematodes cause gigantism, whereas infections from notocotylids and bunocotylids do not (Gorbushin 1997). While trematode infection often causes gigantism in such relatively short-lived snails, long-lived taxa rarely show gigantism and instead may exhibit stunted growth (Sousa 1983; Huxham et al. 1993). Shell shape changes can also occur (Hay et al. 2005; e.g., Buckland-Nicks et al. 2013; Bordalo et al. 2014).

9.4.4.8.10 Associations with Rotifera

Rotifers are microscopic multicellular aquatic animals found in both aquatic and moist terrestrial non-marine environments, only rarely in brackish and marine habitats. They feed using a terminal crown of cilia and are attached by way of a stalk-like 'foot'.

Besides being included as food items, there are few recorded associations of rotifers with molluscs, other than them attaching to the exterior of fresh-water mollusc shells (e.g., Vega et al. 2006; Bołtruszko 2010). Vega et al. (2006) reported a rotiferan in the gut and faeces of cultured *Pomacea* (Ampullariidae) that remained alive after being removed from the snail. One species, *Proales gigantean*, attacks the eggs of fresh-water snails (Hygrophila), destroying the embryo, and laying its own eggs in the capsule (Michelson 1957; Rao & Ramakrishna 1993).

A group of modified rotifers, previously considered a separate phylum, are the Acanthocephala, which are a sister taxon to the true rotifers. These small parasitic 'worms' are commonly known as spiny or thorny headed worms, and there are over a thousand species. Their life cycle involves at least two hosts, the definitive (final) host almost always a vertebrate, and the intermediate host almost always an arthropod. A few species are known to parasitise squid (Hochberg 1990), and one has been recorded in a teredinid (Nair & Saraswathy 1971).

9.4.4.8.11 Associations with Nematoda (Roundworms, Nemathelminthes)

This group has around 40,000 described species, of which nearly half are parasitic, but the majority of taxa have yet to be named. Recent studies (Holterman et al. 2006) recognise up to 12 clades within nematodes, but older classifications recognised fewer.

Nematodes are associated with most, if not all, groups of molluscs, although relatively few such associations are reported in the literature, probably mainly owing to the dearth of taxonomic expertise in the group. For the same reason, many accounts regarding nematodes are not very specific about the taxa involved.

Nematodes are present in the epibiont layer on the exterior of at least some fresh-water snail shells (e.g., Vega et al. 2006) and can also be common in epibiont coverings of marine taxa, including chitons (e.g., Connelly & Turner 2009). They are common in the mantle cavities of fresh-water bivalves such as *Dreissena*, and, although they are probably free-living species without any obligate association (Molloy et al. 1997), they have been reported as symbionts (Karatayev et al. 2000). Others have been recorded as parasites as, for example, *Daubaylia potomaca* in fresh-water hygrophilan snails (Zimmermann et al. 2011).

Many of the reported nematode associations with molluscs are infections from nematode larvae with adults infecting vertebrates. There are a few records of parasitic larval nematodes in marine gastropods, including infections from larvae of spiny headed (Order Spirurida) nematodes of the genus *Echinocephalus* (Gnathostomatidae) in abalone and *Conuber* (Naticidae) (Lauckner 1980). In contrast, Grewal et al. (2003) reviewed the parasitic associations of nematodes with non-marine snails and slugs, of which 108 described nematode species were recorded. From that analysis, it was concluded that in several instances nematodes utilised mollusc hosts independently. Comparatively poor data are available for marine gastropods and other molluscan classes, although bivalves are considered a rather uncommon host (Lauckner 1983b). Some species of nematodes infest a wide range of bivalves and gastropods. For example, to 1990, the third and fourth stage larvae (the first and second undergo their development in the egg) of the turtle parasite *Sulcascaris sulcata* (Ascaridoidea, Anisakidae) (Figure 9.12) had been reported from nine species of bivalves and five gastropods, where they infest various parts of the body and organs, including muscles (Lauckner 1980). A species of *Urosporidium* is a hyperparasite on the nematode larvae in the mactrid *Spisula* (Perkins et al. 1975). Neither parasite is considered harmful to humans. In contrast, *Echinocephalus sinensis* is a parasite of 'eagle rays', and the larval stages infect the gonoduct of the oyster *Crassostrea gigas* in Asia. If consumed alive, it can penetrate the gut wall in mammals and possibly humans (Ko et al. 1975; Ko 1976). Other species of *Echinocephalus* infect various other bivalves and gastropods (see Lauckner 1980, 1983b; and Bower 2006 for reviews). Larval trematodes (identified only to order Spirurida) encyst mainly in muscles of the

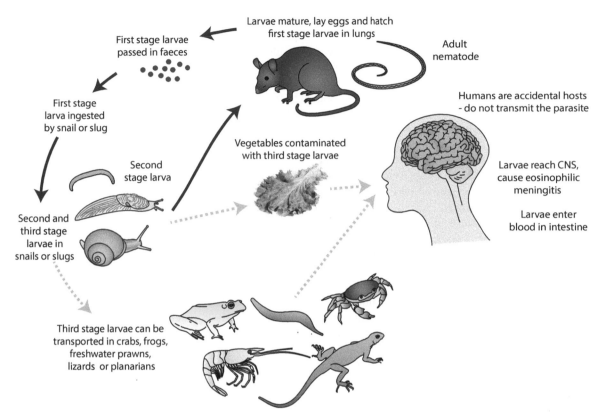

Life history of a non-marine (mainly terrestrial) parasitic nematode, *Angiostrongylus cantonensis*

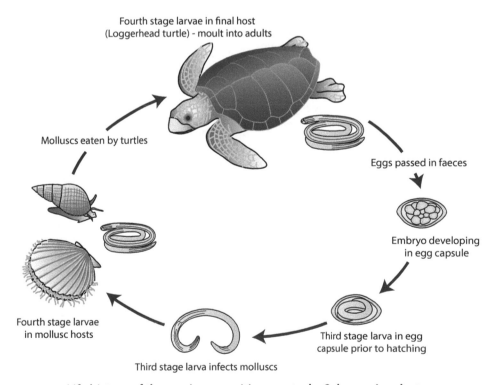

Life history of the marine parasitic nematode, *Sulcascaris sulcata*

FIGURE 9.12 Life history of a terrestrial and a marine nematode involving molluscan intermediate hosts. Life cycle of *Angiostrongylus cantonensis* from various sources and *Sulcascaris sulcata* life cycle redrawn and modified from Berry, G.N. and Cannon, L.R.G., *Int. J. Parasitol.*, 11, 43–54, 1981.

marine bivalve *Tagelus* (Vazquez et al. 2006). Endoparasitic nematodes are known from scaphopods (Reynolds 2002).

Larval nematodes are reported in coleoid cephalopods, with the majority being larval Ascaridae, but also Acanthocheilidae and Anisakidae (e.g., Hochberg 1990; Pascual et al. 1996). The adults reside in vertebrates such as fish and cetaceans that feed on the coleoids. Some, such as *Anisakis*, where third stage larvae encyst in the gut and gonads, can cause human health issues if the squid (or fish) is consumed raw (Hochberg 1990).

Nematode parasites are associated with terrestrial snails (Mead 1979; Morand et al. 2004) and slugs (Runham & Hunter 1970; South 1992). Some can be transmitted sexually or by way of infected eggs (e.g., Anderson 1960; Morand & Faliex 1994; Jordaens et al. 2007). Terrestrial gastropods are significant hosts for two major groups of nematodes; metastrongyloids, where they are intermediate hosts, and rhabditids, where they are definitive hosts (Grewal et al. 2003). Some nematodes have been used as biological control agents for terrestrial slug and snail pests (Grewal et al. 2005). For example, infection by *Nemhelix bakeri* (Cosmoceroidea, Cosmocercidae) decreases egg production in *Cornu aspersum* (Morand 1989), while *Phasmarhabditis hermaphrodita* (Rhabditoidea, Rhabditidae) is used in slug control (Speiser et al. 2001; Morand et al. 2004).

Various terrestrial slugs and snails are the intermediate hosts of mammalian parasites such as *Parelaphostrongylus tenuis* (Strongylida, Protostrongylidae), which infect the brain of deer, moose, etc. (Cervidae), and other ungulates in North America. One of the mollusc-associated nematodes most dangerous to humans is the 'rat lungworm' *Angiostrongylus cantonensis* (Strongylida, Metastrongylidae) (Figure 9.12), which in parts of Asia and the Pacific, causes human eosinophilic meningoencephalitis, a sometimes fatal disease (Figure 9.12). Normally, the intermediate hosts are terrestrial or amphibious gastropods, but non-marine mollusc-eating crustaceans can act as paratenic[16] hosts. Humans are infected by ingesting raw or under-cooked intermediate or paratenic hosts. Even the marine bivalves *Crassostrea virginica* and *Mercenaria mercenaria* are capable of acting as intermediate hosts for *A. cantonensis* (Cheng & Burton 1965).

Some nematodes such as the ascaridoid *Sulcascaris sulcata* (Ascaridoidea, Anisakidae) infect marine bivalves (Lauckner 1983b; Sindermann 1990), and bivalves can also act as a paratenic host for another anisakid, the 'codworm' *Phocanema decipiens* (Lauckner 1983b).

9.4.4.8.12 Associations with Annelida

9.4.4.8.12.1 Polychaeta
Polychaetes are prey items for the Ficidae and various neogastropods, notably conoideans, vasids and turbinellids, including the largest living gastropod, *Syrinx aruanus* (Taylor & Glover 2003). Some pyramidellids are ectoparasites of various polychaetes.

The polychaetes *Polydora* and *Boccardia* (Spionidae) bore into the shells of both dead and living molluscs, sometimes riddling the shell so it readily disintegrates. After penetrating

the shell of a living bivalve such as an oyster or mussel, it perforates the shell and enters the extrapallial cavity. As a result, blisters are formed on the inner surface of the shell in which these 'mud worms' reside (Lauckner 1983b; Bower 2006). After the worm dies other inhabitants can occupy the blister, including other polychaetes (e.g., Lauckner 1983b; Machkevsky 1997). For example, certain cirratulids and sabellids may use the burrow then enlarge it, and some nereidids may occupy the burrows made by boring sponges, while others form internal blisters (Lauckner 1983b). One sabellid is a pest on abalone (see Chapter 10).

Lang (1932) noted that the openings of suspension-feeding serpulid polychaete tubes attached to ammonite shells were always directed towards the current opening of the body chamber and suggested a symbiotic relationship in the past between worm and ammonite.

Some polynoid polychaetes have commensal associations with gastropods, living in the mantle cavities of mostly limpet-like taxa (Davenport 1950; Hickok & Davenport 1957; Ansell et al. 1998), and this relationship is also reported in bivalves (Lauckner 1983b). Members of the polynoid genus *Branchipolynoe* are commonly found in the mantle cavity of *Bathymodiolus* species living at hydrothermal vents and cold seeps (Zhang et al. 2017). Some galeommatoidean bivalves have a commensal relationship with a polychaete (Ponder 1965; Wear 1966), while some species of capitellids apparently live only in the egg masses of loliginid squid (Hochberg 1990).

9.4.4.8.12.2 Oligochaetea and Aphanoneura
These two, mainly non-marine, classes can both be associated with the epibiotic coating on the shells of fresh-water molluscs (e.g., Vega et al. 2006) and are sometimes reported from the mantle cavities of fresh-water bivalves such as Dreissena, but these are normally free-living species presumed to have inadvertently entered the bivalve, although they are sometimes considered commensals (Karatayev et al. 2000). *Tubifex* has been recorded as a 'symbiont' of *Helisoma* and *Physa* (McCaffrey & Johnson 2017).

The tubificid oligochaete *Chaetogaster limnaei* is a common commensal of fresh-water snails and bivalves in the Northern Hemisphere. It is of interest because it consumes trematode larvae (Conn et al. 1996; Rodgers et al. 2005; Ibrahim 2007), and it has also been recorded from Australia living on a lymnaeid host of *Fasciola hepatica* (Rajasekariah 1978). High levels of infestation of this tubificid negatively affects reproductive output in the snail hosts (McCaffrey & Johnson 2017).

9.4.4.8.12.3 Hirudina (Leeches)
Fresh-water leeches (Glossiphoniidae) can be associated with the epibiotic coating on the shells of fresh-water molluscs (e.g., Vega et al. 2006). Leeches are also sometimes found in the mantle cavities of fresh-water bivalves such as *Dreissena*, but these are normally free-living species that have presumably inadvertently entered the bivalve (Karatayev et al. 2000). In South America, several species of glossiphoniid leeches are found associated with ampullariid gastropods where they live mainly in the mantle cavity (e.g., De-Carli et al. 2014). One species, *Helobdella ampullariae*, is symbiotic in ampullariids and

---

[16] A paratenic host is required for completion of the life cycle, often as a means of dispersal, but the parasite does not develop in it.

Chilina and is not found in the external environment (Vega et al. 2006). *Helobdella* has also been found in the lung of planorbids (Negrete et al. 2007), while other species of *Helobdella* are considered predators of fresh-water molluscs (see Section 9.4.1.1.15).

Three species of leeches of the family Piscicolidae have been found in *Octopus dofleini*. Piscicolids are normally fish parasites, with some also found on crustaceans, and it is likely that transfer occurs when the octopus feeds on the crustacean (Hochberg 1990).

9.4.4.8.13 Echiurans (Spoon Worms)

Several species of galeommatoidean bivalves and a tornid gastropod are associated with echiurans, living in their burrows (Morton 1988a; Anker et al. 2005). This and the following taxon were previously considered as separate phyla, but, based on molecular data, they are now thought to be unsegmented annelids (e.g., Struck 2017).

9.4.4.8.14 Sipunculids (Peanut Worms)

Sipunculids are the exclusive food of members of the Mitridae, while burrowing sipunculids have several galeommatoidean commensal associates (Morton 1988a; Goto et al. 2012).

9.4.4.9 Associations with Arthropoda

9.4.4.9.1 Associations with Crustacea

The subphylum Crustacea contains 6–7 recognised classes, some of which have members with associations with molluscs. These are listed below.

9.4.4.9.2 Maxillopoda

The class Maxillopoda contains, besides other groups, the 'subclasses' Copepoda, Thecostraca, and Branchiura, all of which contain taxa that have relationships with molluscs.

9.4.4.9.2.1 Copepoda
A detailed, but dated, list of copepods associated with molluscs is provided by Monod and Dollfus (1932). Associations between copepods and bivalves are common and several copepod families also parasitise gastropods, especially 'opisthobranchs' (Lauckner 1980). Most of the mollusc-associated copepods are members of the order Poecilostomatoida, while only a few members of the orders Caligoida, Harpacticoida, Lernaeoida, Cyclopoida, and Monstrilloida are recorded as mollusc parasites. Parasitic copepods have not been reported from fresh-water molluscs. Various 'commensal' copepods live in the mantle cavity of bivalves, scaphopods, and gastropods and are little modified (Bocquet & Stock 1963), in contrast to some of the parasitic taxa. An ectoparasitic copepod of unknown relationship is known from scaphopods (Reynolds 2002), and one was described from a deep-sea cocculinid *Coccopigya hispida* (Jones & Marshall 1986), while an unidentified internal crustacean parasite was noted by Haszprunar (1987a) in the same species.

In some gastropod species, copepod egg sacs have probably been misinterpreted as brooded gastropod young (Huys et al. 2002).

Order Poecilostomatoida

Members of the Mytilicolidae are common in bivalves, with *Mytilicola* spp. having highly modified bodies. They live in the intestine of bivalves and may sometimes cause mortality (Bower 2006). Other highly modified copepods include *Axinophilus*, which attaches to the anterior adductor muscle of *Thyasira*, and *Teredoika* which lives in the stomach of *Teredo* (Lauckner 1983b).

Pectinophilus ornatus is a highly modified mytilicolid that was originally thought to be a rhizocephalan (Cirripedia). It is considered a pest of scallops in Japan (Nagasawa et al. 1993). *Teredicola typica* is associated with Teredinidae, while the clausidiid *Leptinogaster major* has been found in the mantle cavity of several heterodont bivalves (Humes & Cressey 1960). Many other taxa are also associated with bivalves including *Anthessius* (Anthessiidae), *Paranthessius* and *Doridicola* (Rhynchomolgidae), *Lichomolgus*, *Gelastomolgus*, *Herrmannella* and *Modiolicola* (Lichomolgidae), *Myicola*, *Ostrincola* and *Pseudomyicola* (Myicolidae), and *Conchyliurus* (Clausidiidae) (Lauckner 1983b). The nature of the association of many of these is poorly understood.

Trochicola enterica, a mytilicolid copepod, is a parasite of trochoideans (Kleeton 1961) with the females living in the intestine and males and juveniles living in the mantle cavity. Other species of *Trochicola* are found in various gastropods. A related genus *Panaietis* is found in abalone and various other vetigastropods, while the anthessiid *Neanthessius renicolis* is found in the kidney of fasciolariids (Lauckner 1980).

Doridicola longicauda (Rhynchomolgidae) lives in the mantle cavity and on the gills of cuttlefish where it moves about feeding on mucus. Members of the genus *Octopicola* (Octopicolidae) are found in the gills of octopuses and also move about freely (Hochberg 1990).

Order Monstrilloida

A monstrillid (*Monstrilla helgolandica*) parasitises the small ectoparasitic gastropod *Odostomia*, while the philoblennid *Philoblenna arabici* lives attached to the mantle in cowries (Cypraeidae), and the lichomolgid *Epimolgus trochi* lives in the mantle cavity of trochids. Various genera of lichomolgids, including some that live endoparasitically within the body cavity, are associated with 'opisthobranchs'.

Order Cyclopoidea

Two species of *Ozmana* (Cyclopoidea; Ozmanidae) are endosymbionts of *Pomacea*. Although these can occur in large numbers in the haemocoel, ctenidium, and mantle cavity, they do not seem to harm the host snail. They appear to be transferred during copulation (Vega et al. 2006).

Cyclopoid copepods (Chitonophilidae) parasitise chitons, cocculinoidean and lepetelloidean limpets, and muricids (Lamb et al. 1996; Huys et al. 2002; Avdeev & Sirenko 2005; Schwabe et al. 2014), and a species of *Lepetodrilus*, a hydrothermal vent limpet (Tunnicliffe et al. 2008). These parasites can cause extensive tissue damage, castration, and death. In cocculiniform limpets, the copepod eggs have been wrongly interpreted several times as evidence of brooding (see Huys et al. 2002 for review).

Order Siphonostomatoidea

Larval stages of pennellid copepods (*Pennella* spp.) are found in the gills of several coleoid cephalopods, mostly from the Mediterranean (Hochberg 1990; Pascual et al. 1996), and species of lichomolgids have also been reported from squid and octopuses (Pascual et al. 1996).

Some thecosome pteropods may act as intermediate hosts of parasitic pennellid copepods that have fish as their final host (Lalli & Gilmer 1989).

Order Caligoidea

Among caligoid copepods, which are normally fish parasites, four genera are associated with cephalopods, including *Anchicaligus nautili* (Caligidae) associated with *Nautilus*, where it lives in the mantle cavity. The other three genera live in the mantle cavities of coleoids (Hochberg 1990).

Order Harpacticoidea

Some harpacticoid copepods are associated with bivalves where they live in the mantle cavity (Lauckner 1983b), and one is recorded from chitons where they cling to the girdle (Glynn 1968). They are also associated with deep-sea octopods (e.g., Humes & Voight 1997), usually attached to the gills, but some species have been found on the arm webs or ventral head. Unlike some parasitic species in this group, they are all relatively unmodified (Hochberg 1990).

9.4.4.9.2.2 *Thecostraca: Cirripedia (Barnacles)* Thoracican barnacles attach themselves to hard substrata, and their body is enclosed in shell plates. They are all marine and suspension-feeders with two swimming larval stages. One shell-less cirripede group are crab parasites. Barnacles provide shelter for juvenile and small molluscs and a food source for some predatory gastropods, notably some muricids, and they compete for space with molluscs such as limpets, oysters, and mussels in the intertidal zone. They are often epibionts on the shells of intertidal and subtidal or deeper water molluscs. The balanid barnacle *Amphibalanus improvisus* on living *Mytilus edulis* shells grow significantly faster than those on empty shells, which may indicate a commensal relationship (Laihonen & Furman 1986). Other experiments have shown that barnacle epibionts can reduce mussel growth (Buschbaum & Saier 2001), reduce body size of co-occurring taxa such as limpets on rocky shores (Dunmore & Schiel 2003), reduce burrowing success, or even prevent closure of the valves in bivalves (Lauckner 1983b), thus increasing their mortality. Barnacle epibionts decrease reproductive success in *Littorina* (Buschbaum & Reise 1999).

Some barnacles (Acrothoracica) bore into mollusc shells (e.g., Batham & Tomlinson 1965), including those of the extinct swimming belemnite cephalopods (Seilacher 1968). Some of these boring barnacles are naked, and a naked pedunculate barnacle, *Malacolepas conchicola* (Malacolepadidae), lives on the inside of the arcoidean *Cucullaea* (Cucullaeidae) and a venerid from Japan. The peduncle is attached to the interior of the shell and, in *Cucullaea*, is enclosed in a shelly tube secreted by the bivalve (Lauckner 1983b).

9.4.4.9.2.3 *Branchiura* These small crustaceans are external parasites of teleost fish and there is also one record from the skin of a cuttlefish (*Sepia*) (Hochberg 1990).

9.4.4.9.3 *Malacostraca*

This class contains most of the familiar crustaceans in the subclass Eumalacostraca. There are three 'superorders' recognised in Eumalacostraca, one of which is the 'superorder' Peracarida containing the amphipods and isopods, the Eucarida, containing the larger crustaceans such as crabs, lobsters, and shrimps (order Decapoda), and lastly the shrimp-like Syncarida.

9.4.4.9.3.1 *Peracarida* Sphaeromatid isopods have been found associated with intertidal molluscs around the world. In Puerto Rico, some are found living in the mantle groove of large chitons (Glynn 1968; Lauckner 1983c), while in Japan, they occur in the mantle groove of the limpet *Cellana* and on the dorsal foot and mantle cavity of neritid gastropods (Nishimura 1976). Two sphaeromatid species are found associated with limpets in southern Africa (Branch 1975b).

A few cymothoid isopods have been recorded from coleoid cephalopods, mainly in the mantle cavity, but are rare (Hochberg 1990; Pascual et al. 1996). This group of isopods are typically sexually dimorphic protandrous hermaphrodites and are commonly found on fish. Isopods are not known to be symbiotic associates of bivalves.

Amphipods are often associated with the outer surface of larger shelled molluscs as part of the epibiont fauna. There are several amphipod taxa commensal with molluscs (Vader 1972), such as the hyalid amphipod *Parhyale hawaiensis*, which occupies the mantle groove of a Puerto Rican chiton, *Chiton tuberculatus* (Glynn 1968), and *Calliopiella michaelseni*, which occurs in pairs and completes its life cycle under limpets in southern Africa (Branch 1975b). A few amphipods inhabit the mantle cavity of marine bivalves where they get protection and can use some of the food of the bivalve. One of the most modified is the gammarid *Cardiophilus baeri* (Behningiellidae), which lives in the mantle cavity of the cardiid *Didacna baeri* in the Caspian Sea (Vader 1972). A looser positive association can also be found, as between the intertidal limpet *Cellana tramoserica* and the hyalid amphipod *Apohyale media* (as *Hyale*), where grazing by the limpet improved the habitat of the amphipod, enabling it to increase in numbers (Underwood & Verstegen 1988).

Hyperiid amphipods attach to, and feed on, various zooplankton (including heteropods and pteropods). There is one report of a hyperiid on a laboratory-reared loliginid squid where it was attached to the skin by its mouthparts (Hochberg 1990), but until then at least, no reports were known from wild coleoids.

Some terrestrial isopods and an amphipod feed on the surface mucus of *Achatina fulica* in Hawaii (Mead 1979).

9.4.4.9.3.2 *Eucarida* Crabs are important predators of molluscs (see Section 9.4.1.1.10), especially gastropods, and they in turn are significant food items for coleoids, notably octopuses.

Undoubtedly the most well-known commensal association between crustaceans and molluscs is the relationship between pinnotherid crabs (pea crabs) and their bivalve hosts. Members of the Pinnotheridae, comprising about 50 genera including *Pinnotheres*, live in the mantle cavity of many bivalves, including mytilids, oysters, scallops, pinnids, and various heterodonts including venerids, *Mya*, and others. There is generally a low level of host specificity (see Lauckner 1983b for review). These small, soft-bodied crabs have also been recorded from a few larger gastropods such as *Haliotis* and *Strombus*, and the mantle grooves of the very large chiton *Cryptochiton stelleri*, as well as echinoids, ascidians, the rectum of certain holothurians, and the tubes or burrows of various invertebrates. It is only the females in residence, the much smaller males being free-living. The relationship is variously described as commensal and parasitic. Protection is afforded by the shelled host and the crab steals food from the host, also feeding on the mucus of the mollusc. Damage to the host can occur. The mature females produce larvae that go through two larval stages before leaving the host. The larvae grow into small adults of both sexes, which, having left the host, mate, and look like ordinary crabs with hard shells. After mating, the males die, the females take up residence in a host and go through several moults before reaching maturity. The female then produces larvae from eggs fertilised by sperm from the single mating.

The 'watchman shrimps' *Conchodytes* and *Anchistus* (Palaemonidae) are commensals in the mantle cavity of *Pinna* (Morton 1988a).

Commensal associations of galeommatoidean bivalves with Eucarida such as crabs (Morton 1988a, 2008b; Boyko & Mikkelsen 2002), mantis shrimps (e.g., Morton 1988a; Mikkelsen & Bieler 1992), ghost shrimps (Thalassinidea) (Kato & Itani 1995), and hermit crabs (Goto et al. 2007) are not uncommon.

There is a report of a megalopa crab larva 'commensal' in a loliginid squid (Seréne 1957), but this needs confirmation.

Some species of slipper limpet (Calyptraeidae) live inside shells occupied by hermit crabs (e.g., Karlson & Cariolou 1982; Morton 1988a).

9.4.4.9.4   Associations with Insects
Chironomid larvae are sometimes reported from among epibiont 'algae' on fresh-water molluscs (e.g., Vega et al. 2006). Others found in the mantle cavities of fresh-water bivalves such as *Dreissena* are presumably free-living species that have inadvertently entered the bivalve, but they have sometimes been reported as commensals (Ricciardi 1994; Karatayev et al. 2000). Chironomid larva have been recorded living on *Chilina* (Prat et al. 2004) and with several other hygrophilan snails (McCaffrey & Johnson 2017).

A small land snail, the subulinid *Allopeas myrmekophilos*, is found in colonies of an army ant *Leptogenys distinguenda*[17] (Formicidae), where they reside only inside the host bivouacs and move about without interference by the ants. When the ants migrate, the snails are carried by the workers to the new location along with the brood and prey items. This association is possible because of a foamy substance secreted by the snail when in contact with the ants, and the snail shares food with the carnivorous ant (Witte et al. 2002). The snail secretion was usually produced only during contact with *L. distinguenda* and not other ponerine ants. *Allopeas myrmekophilos* is the only known molluscan ant associate.

Many species of insects, especially dipterans, are parasites or parasitoids of terrestrial, fresh-water and, more rarely, marine molluscs. The most significant group is the snail-parasitising 'marsh flies' (Diptera: Sciomyzidae) (see Section 9.4.1.1.11), the larvae of which range from being predators, parasitoids, or saprophages of a wide variety of gastropods in mainly terrestrial and fresh-water habitats, as well as feeding on molluscs cast up on the shoreline, fresh-water sphaeriid bivalves, and a few marine intertidal species (Knutson & Vala 2011; Chapman et al. 2012). Many sciomyzids are host specific and have the potential to be used as biological control agents for pest snails, including fluke-carrying lymnaeids. In addition to sciomyzids, over a hundred genera in several additional dipteran families attack terrestrial molluscs (Barker et al. 2004; Coupland & Barker 2004) (see also Section 9.4.1.1.11).

9.4.4.9.5   Associations with Arachnida- Acari (Acarina, Mites)
The order Prostigmata contains four families with members parasitic on either fresh-water or terrestrial molluscs. They contain several genera that are parasites of fresh-water bivalves, including *Unionicola* (Unionicolidae) with several species parasitic on Holarctic Unionidae (e.g., Edwards & Vidrine 2013). They are found in the mantle cavity where they feed on mucus and haemocytes (see Fain 2004 for review), often damaging the epithelium and causing inflammation. A species of *Unionicola* lives in an unknown relationship in the mantle cavities of some ampullariids (Vega et al. 2006).

Other mites have been reported from the mantle cavities of fresh-water bivalves such as *Dreissena*, and while these may be free-living species that have inadvertently entered the bivalve, they have been reported as commensals (Molloy et al. 1997; Karatayev et al. 2000).

Larvae and nymphs of the halacaroid mite *Halixodes chitonis* live attached to the gills of the large New Zealand chiton *Cryptoconchus porosus* and similar larval mites have been found in the mantle cavity of a New Zealand calyptraeid (Bartsch 1986).

Terrestrial snails and slugs, particularly the latter, have parasitic associations with several taxa of mites (Fain 2004) that live externally and in the lung, including species in Ereynetidae and two other families of Prostigmata. Species of the ereynetid genus *Riccardoella* are common parasites of several stylommatophoran land snails and slugs (e.g., Barker & Ramsay 1978; Mienis 1990; Baur & Baur 2005; Zabludovskaya & Badanin 2010) and feed on their blood. They live in the lung or externally, are transmitted during copulation or other body contact (Schüpbach & Baur 2008), and can markedly decrease fitness (e.g., Graham et al. 1996).

---

[17] Sometimes treated as a subspecies of *L. processionalis*.

9.4.4.9.6   Associations with Pycnogonida

Adult and juvenile pycnogonids are occasionally found in the mantle cavities of 'opisthobranchs' where they are thought to feed on the faeces of the gastropod (Arnaud 1978; Lauckner 1980). Pycnogonids have also been recorded from the mantle cavity of some bivalves (Lauckner 1983b).

9.4.4.9.7   Associations with Bryozoa (Ectoprocta, Polyzoa) and Entoprocta

Bryozoans, especially the ctenostome *Alcyonidium* (Alcyonidiidae), are common epibionts on the shells of gastropods and bivalves (e.g., Ryland & Porter 2006) and at least one chiton (Dell'Angelo & Laghi 1980). Cazzaniga (1988) observed a plumatellid bryozoan (*Hyalinella vahiriae*) living in the suture of the fresh-water, amphibious gastropod *Pomacea canaliculata*.

Some members of the orders Cheilostomata and Ctenostomata burrow into various substrata including gastropod and bivalve shells (Lauckner 1980, 1983b; Smyth 1988). Bryozoans are also food items for gastropods such as various nudibranchs, some Cystiscidae, and juvenile *Urosalpinx* (Franz 1971). Fresh-water bryozoans are eaten by introduced *Pomacea canaliculata* in Thailand, but not by the native *Pila* (Wood et al. 2006). The cheilostome bryozoan *Membranipora membranacea* (Membraniporidae) produces spines in response to predation by several species of nudibranchs (Iyengar & Harvell 2002).

Pagurid hermit crabs that occupy empty gastropod shells appear to have an association with some bryozoans (e.g., Taylor et al. 1989), but, despite an apparent symbiosis between gastropods and a trepostome bryozoan recorded from the Upper Ordovician (McNamara 1978), there are few recorded symbiotic associations of bryozoans and living marine gastropods. The subtidal western North American patellogastropod limpet *Acmaea funiculata* is always encrusted by living bryozoans, possibly a case of crypsis (Lindberg 2007a). In another example, the buccinid whelk *Burnupena papyracea* is protected by the alcyonidiid bryozoan *Alcyonidium nodosum* against predation by the rock lobster *Jasus lalandii*, in part by the encrustation strengthening the shell, but also by way of chemical defences produced by the bryozoans (Gray et al. 2005).

An epibiotic (ectoparasitic?) bryozoan *Pseudobathyalozoon profundum* is found attached ventrally to the girdle of the deep-sea Pacific lepidopleurid chiton *Nierstraszella andamanica*. The zooids feed within the mantle groove among the gills (Sigwart 2009).

The entoproct family Loxosomatidae are solitary and often associated with various animals, including the external shell of bivalves (Iseto 2005). Fresh-water entoprocts such as *Urnatella* (Barentsiidae) live attached to the shells of fresh-water bivalves such as *Dreissena*.

9.4.4.9.8   Associations with Brachiopoda

The galeommatoidean bivalve *Koreamya* has a commensal relationship with lingulid brachiopods (Savazzi 2001a; Sato et al. 2011). Otherwise, we are not aware of other associations between molluscs and brachiopods, besides them being a source of food for some carnivorous taxa, especially in ancient seas.

9.4.4.9.9   Associations with Phoronida

This small group of marine, suspension-feeding, worm-like animals sometimes bore into mainly dead molluscan shells (Lauckner 1983b). Phoronids are usually treated as a separate phylum closely related to brachiopods (Santagata 2015).

9.4.4.9.10   Associations with Echinoderms

Echinoderms are preyed on by relatively few gastropods. Some tonnoideans exclusively eat echinoderms, and eulimids parasitise them, with some being highly specialised shell-less internal parasites (see Chapter 19). Some echinoids also make depressions in rock in which molluscs can shelter.

Some galeommatoidean bivalves live externally or internally on holothurians (e.g., Kato 1998; Middelfart & Craig 2004) and heart urchins (Spatangoidea) (e.g., Goto et al. 2012).

The Paleozoic limpet-like Platyceratoidea lived affixed to the calyx of crinoids, but the detailed nature of the association is unknown, although a coprophagous habit has been suggested (Bowsher 1955).

9.4.4.9.11   Associations with Other Molluscs

Molluscan epibionts that live on the shells of larger molluscs include calyptraeids, vermetids, lottiids, and bivalves such as mytilids, zebra mussels, and oysters. Exotic bivalves (zebra and quagga mussels, and the mytilid *Limnoperna*) sometimes form such dense aggregations on the shells of fresh-water mussels (Unionoidea) that they smother them. Some lottiid limpets preferentially live on the shells of other molluscs (e.g., Morton 1980b; Mapstone et al. 1984; Lindberg 1990a; Minchinton & Ross 1999; Espoz et al. 2004).

Boring bivalves attack the shells of some larger bivalves and gastropods. For example, *Lithophaga* spp. (Mytilidae) and *Penitella conradi* (Pholadidae) bore into the shells of abalone (Smith 1969; Alvarez-Tinajero et al. 2013).

Various molluscan predators feed on other molluscs, including many muricids that prey on bivalves, other gastropods, and sometimes chitons. Naticids feed mainly on bivalves, as do some Volutidae, while others such as *Melo* feed on predatory gastropods (Morton 1986; Shau-Hwai et al. 2010).

Pyramidellids are suctorial ectoparasites on various bivalves and gastropods, but rarely chitons. The related family Amathinidae are ectoparasites on bivalves (Ponder 1987) as are some cancellariids (e.g., *Trigonostoma*) (Garrard 1975; Hemmen 2007). Species of Capulidae steal food from the gastropods or bivalves they are attached to (i.e., are kleptoparasites) (Iyengar 2008) (see Chapter 5). A few species bore holes in living scallops where they presumably steal food from the gills (Orr 1962; Morton 1988a).

Galeommatoidean bivalves only rarely have associations with other bivalves – one example being a species of *Arthritica* living with the rock boring pholad *Anchomasa similis* in New Zealand (Ponder 1965; Morton 1973b).

Some coleoid cephalopods are predators of other molluscs, such as squid feeding on other coleoids and octopods feeding on gastropods and bivalves, often by drilling them (see Chapters 5 and 17).

9.4.4.9.12 Associations with Ascidians

Many ascidians are fouling organisms and are often epibionts on larger benthic marine molluscs. Some families of gastropods are specialist predators of ascidians, including nudibranchs (Thompson 1976; Thompson & Brown 1984), some marginellids (Ponder 1970b), ranellids (Laxton 1971), triviids, and velutinids (Dalby 1989). Others include ascidians in their diet, including some grazing vetigastropods, some cypraeids, columbellids, muricids, fasciolariids, and *Melongena* (Dalby 1989).

9.4.4.9.13 Associations with Vertebrates

Molluscs provide an important source of food for many vertebrates (including humans) and some fish provide food for molluscs – for example, the blood sucking colubrariids and marginellids that feed on resting teleost fish and the cancellariids that feed on resting rays (see Section 9.4.1.1.1 and Chapters 5 and 19), and some species of *Conus* harpoon small fish.

Several small species of the family Carapidae (pearlfish) have commensal relationships with bivalves, living inside the mantle cavity, and some can produce sound pulses within the bivalve host (Kever et al. 2014).

9.4.5 CANCER-LIKE DISEASES

Cancer-like neoplastic diseases producing tumours have been found in some bivalves (Sindermann & Rosenfield 1967; Rosenfield 1976; Lauckner 1983b; Twomey & Mulcahy 1989; Elston et al. 1992; Cheng 1993; Barber 2004), including blood cancers (Walker & Böttger 2008). There have been fewer reports of tumours in gastropods (e.g., Michelson & Richards 1975; Tascedda & Ottaviani 2014), and none in cephalopods, perhaps due to peculiarities of their physiology (e.g., Lee et al. 1994).

The neoplastic tumours in bivalves (Barber 2004) are associated with haemocytes that are enlarged two to four times their normal diameter and circulate anaplastic cells with characteristic cytology. These cancers can be transmitted between individuals and can occur in the majority of individuals in some populations of bivalves. Such occurrences have usually been reported in the visceral mass, muscle and blood sinuses, and vessels in the clam *Mya arenaria*, the mussel *Mytilus trossulus*, and the cockle *Cerastoderma edule*. The occurrence is often associated with environmental degradation. Neoplasia is also known in the gonads in some bivalves. The causes of tumours in bivalves have not been determined, but have been suggested to include stressors such as viruses, biotoxins, and environmental contaminants.

9.4.6 TRANSFER OF SYMBIONTS TO THE HOST

Acquiring symbiotic gut bacteria is generally via ingested food as discussed in Chapter 5. Other symbionts are acquired in various ways, some of which are outlined below, but for many, details are obscure.

The chemoautotrophic sulphur-oxidising bacteria housed intracellularly in the gills of lucinids are not transmitted by the gametes and are absent in the veliger larvae. Instead, they appear to be taken up by the juveniles after metamorphosis with acquisition continuing throughout their lifetime (Gros et al. 2012). Zooxanthellae in giant clams (*Tridacna*) are not associated with the gametes and when veliger larvae feed on them they are digested, so they are not acquired by larval ingestion. Instead, several days after metamorphosis, a tubular structure develops from the stomach to the mantle. This becomes packed with intact zooxanthellae obtained by juvenile feeding and develops in the mantle as the clam grows and the zooxanthellae actively reproduce to populate the mantle and siphonal tissues (Hirose et al. 2006). Thus, individual clams may have more than one kind of zooxanthellae (Carlos et al. 2000; Ishikura et al. 2004). Whether or not subadult or adult clams continue to take up zooxanthellae from the environment and add them to the symbiotic population, or whether they just use them as food, is unclear.

The acquisition of zooxanthellae by some nudibranchs is discussed in Chapters 5 and 20. In at least one nudibranch, the aeolidioid *Pteraeolidia ianthina*, there may be several genetically distinct zooxanthellae symbionts in any one individual (Ishikura et al. 2004; Loh et al. 2006), with this diversity perhaps conferring advantages to the host species if it spans a wide geographic range or a variety of habitats.

Various symbiotic bacteria are associated with the accessory nidamental glands in coleoid cephalopods (e.g., Pichon et al. 2005; Collins et al. 2012a), some of which are horizontally transmitted by their deposition on the jelly coat of the eggs, while others colonise from seawater, at least in some squid (Kaufman et al. 1998).

Symbiotic bacteria (*Vibrio fischeri*) associated with the light organs of some coleoid cephalopods (see Section 9.4.8 and Chapter 17) are acquired post embryonically (Nyholm & Nishiguchi 2008). For example, in the sepiolid *Euprymna scolopes*, the nascent light organ has crypts colonised by the bacterial symbionts from the seawater. This involves a complex process of elimination of the non-target bacteria, in part by the epithelial cells of the crypts secreting mucus to which only living Gram-negative bacteria can attach. There are also other factors that prevent invasion by non-target bacteria, including characteristics of the host tissues and of the bacterium (e.g., Davidson et al. 2004).

9.4.7 PARASITIC MOLLUSCS

As noted earlier, the distinction between parasitism and predation is not clear-cut, with some taxa variously described as parasites or predators in the literature. The brief review below covers those groups that conform to our definition of parasites. There are no known parasitic bivalves, although some galeommatoideans have been suggested to be parasites. For example, species of *Entovalva* and similar galeommatoideans live externally or internally on or in the cloaca of holothurians, but are apparently not harming the host (see also Chapter 15).

Some gastropods are the only 'true' molluscan parasites. Eulimids (Caenogastropoda) are ecto- or endoparasites on echinoderms, and these range from shelled, free moving ectoparasites to others permanently attached externally or embedded in the epidermis of the host. Others are internal with their shells reduced or lost, and some are worm-like (see Chapter 19). A number display considerable sexual dimorphism and some are protandric, while others have tiny 'parasitic' males, comprised mainly of the testis, embedded in their bodies so early workers thought these animals were simultaneous hermaphrodites. Other gastropods include two parasitic families in the Pyramidelloidea (Heterobranchia), the Pyramidellidae, which are ectoparasites mainly on annelids and other molluscs, and the Amathinidae, which are bivalve parasites. These closely-related families have a highly modified anterior gut with a very long proboscis and a jaw modified as a piercing stylet.

A few gastropods (some Colubrariidae, Cancellariidae, and Marginellidae) feed suctorially on sleeping fish via their long proboscis (see Chapter 5), and Coralliophilinae (Muricidae) are associates of either hard or soft corals where they feed on the surface mucus and polyps, with some being predatory, others parasitic (e.g., Martin et al. 2014).

Small size grazers on colonial animals are sometimes regarded as ectoparasites. For example, the Triphoroidea (Triphoridae and Cerithiopsidae) are sponge feeders, the Ovulidae graze soft corals, the Triviidae ascidians, and the Architectonicidae and Epitoniidae cnidarians. Similar examples of feeding on large prey are seen in many other gastropods (e.g., some nudibranchs, calliostomatids), which are usually treated as carnivorous grazers.

9.4.8 BIOLUMINESCENCE

Bioluminescence is the emission of visible light by animals or their symbionts and has evolved independently many times in different groups of organisms (Haddock et al. 2010). It is produced by a complex organic molecule ('luciferin'), which, when catalysed by a specific enzyme ('luciferase'), releases light. The terms luciferin and luciferase are general ones and refer to a wide range of different molecules and enzymes that reflects the wide distribution of bioluminescence among animals. In metazoans, there are two types of luminescence – via symbiotic bacteria by way of a luminous secretion, or by luminous cells. Luciferases probably evolved from 'mixed-function oxygenases' (Bolstad 2003) and two or more kinds are sometimes present (Haddock et al. 2010). Bioluminescence in molluscs was reviewed by Tsuji (1983), Herring (1987), and briefly by Haddock et al. (2010).

Nearly all bioluminescence produced by marine animals is blue-green light (470–490 nm), the optimum wavelengths for visibility in seawater. The eyes of deep-sea squid are most sensitive to the blue-green light, and in deep-sea loliginids the parolfactory vesicles (on the optic tracts of decabrachians) are photoreceptive (Bolstad 2003).

Bioluminescence has been suggested to serve several purposes including countershading, startling predators, lures to attract food, communication, and attracting mates.

Bioluminescence is important in finding mates in sparsely distributed deep-sea animals (Herring 2000), or simply detecting other animals (e.g., Johnsen 2005).

Many oceanic squid have luminescent organs, some complex (see Chapter 17 for details). Other luminescent molluscs include a few members of the caenogastropod family Planaxidae (Haneda 1958; Ponder 1988c; Deheyn & Wilson 2011), one species of Eulimidae (Marshall 1997), and the bivalves *Pholas* (Dunstan et al. 2000) (Pholadidae), and *Rocellaria* (Haneda 1939) (Gastrochaenidae). The limid *Ctenoides ales* (Okutani 1994) produces bright flashes of light often confused with bioluminescence, but this is actually the result of scattered light being mirrored by a broadband reflector built from silica spheres in the mantle (Dougherty et al. 2014).

All major groups of coleoid cephalopods have some taxa that produce luminescence. It is produced by bacterial symbionts (family Vibrionaceae – see Section 9.4.4.3.13.4) in sepiolids and loligids (e.g., Nishiguchi et al. 2004) and by the secretion of intrinsic chemicals (a luciferin with a luciferase) in other coleoids. There are about 70 genera of light-producing squid alone (Herring 1977). Coleoid light organs are often species-specific in their arrangement, typically produce ventrally directed light, and are involved in countershading (e.g., Johnsen et al. 2004) (see Section 9.4.3.9.2). While squid use their luminescence for counter-illumination, some squid can even modify the colour of their luminescence to compensate for daylight or moonlight (Young & Mencher 1980; Herring et al. 1992).

Of those bacteria in coleoid light organs, the most studied is *Vibrio fischeri*, found in the luminescent organs of the sepiolid squid *Euprymna scolopes* (see Chapter 17). In the Sepiolidae that contain light organs, these structures are found in the mantle cavity and are associated with the ink sac and intestine. These complex structures have evolved to enhance and direct light from the symbiotic bacteria they contain. Nishiguchi et al. (2004) suggested that the sepiolid light organ was most likely derived from an accessory gland open to the environment so bacterial colonisation could occur. This structure became more and more specialised to eventually become their now complex light organs. Light organs with symbiotic bacteria are also found in some Loliginidae, where they are associated with the ink sac and intestine, and possess a lens and reflector. Light control is managed by chromatophores and a screen involving the ink sac (Tsuji 1983, 2002).

Oceanic teuthoid (mainly oegopsid) epi- and mesopelagic squid have light organs that do not house symbiotic bacteria (Herring 1977; Tsuji 1983), although the chemical reactions producing the bioluminescence in these pelagic squid can differ. For example, that in the squid *Symplectoteuthis oualaniensis* differs from other bioluminescent animals and reacts in a different way (Tsuji 1983; Fujii et al. 2002). The small 'firefly squid' (*Watasenia scintillans*) found in Japanese waters has ventral photogenic organs on the head, arms, funnel, and eyes. These produce a bluish luminescence, the brightest from the tips of the fourth pair of arms, and the chemical nature of this luminescence has been well studied (Tsuji 1983, 2002).

The deep-sea *Vampyroteuthis infernalis* produces bioluminescence from organs on the arm tips and the fluid released forms a luminescent cloud surrounding the animal (Robison et al. 2003). Bioluminescence is produced by modified suckers along the length of the arms of the rather slow-moving cirrate octopod *Stauroteuthis syrtensis* and possibly related taxa, and may function as a lure to attract prey (Johnsen et al. 1999).

Some nudibranch genera of Polyceridae, Tethydidae, and Phylliroidae are luminous, including two genera of polycerid nudibranchs (subfamily Triophinae), *Kaloplocamus* and *Plocamopherus*, have knobs that can flash light when disturbed (Kato 1949; Vallès & Gosliner 2006). In contrast, the pelagic nudibranch *Phylliroe* (Phylliroidae) has luminescent epithelial cells scattered over its transparent body (Lalli & Gilmer 1989).

The New Zealand hygrophilan fresh-water limpet *Latia* is the only known fresh-water luminescent gastropod. The luminescence is produced chemically, bound in mucus secreted from cells in various parts of the head-foot and mantle, and discharged through the pneumostome when the animal is disturbed, presumably to distract predators (Tsuji 1983; Moore & Meyer-Rochow 1988; Nakamura et al. 2004). The only caenogastropods definitely known to be luminescent are some planaxids, which produce bioluminescence from a gland on the mantle edge when disturbed (Haneda 1955, 1958; Ponder 1988c; Deheyn & Wilson 2011). Haneda (1958, p. 152) reported that when *Tonna galea* 'moves through water with its foot well extended, it emits a greenish-white light', although this observation does not appear to have been confirmed.

The only known terrestrial mollusc that produces bioluminescence is the helicarionoidean *Quantula striata* (often known as *Dyakia striata*) (Dyakiidae) from Cambodia, Malaysia, and Singapore (Haneda 1981; Tsuji 1983). Its luminous organ, in the anterior head-foot, produces flashes of yellow-green light by way of an intracellular reaction, especially in juveniles (Tsuji 1983; Counsilman & Ong 1988; Isobe et al. 1991; Daston & Copeland 1993), and is thought to be from a modified glandular tissue. This luminescence may have a social role (Counsilman et al. 1987), such as assisting in the aggregation of snails around food sources.

The pholad bivalve *Pholas dactylus* releases a luminescent secretion from its siphon when disturbed (Tsuji 1983). This has been known for many years, and it was the first organism in which the luciferin-luciferase reaction was described in the late 1880s. The substances that produce the luminescence are produced in light organs in the mantle and siphon. Luminescence has also been reported in a coral-boring bivalve, the gastrochaenid *Rocellaria*. When the animal is irritated, a strip of cells along the mantle produces a luminous secretion squirted out through the siphon (Haneda 1955).

9.5 ECOLOGY

Ecology (see Chapter 2 for a general discussion of the scope) is the study of the interaction of organisms with their environment to understand their patterns of distribution and abundance. Molluscan ecology has a rich history and has provided significant contributions in both basic and applied research. As well as biotic interactions (e.g., competition, predation), they also experience a wide variety of abiotic interactions as well with temperature, light, salinity, precipitation, turbulence, fire, pollutants, etc.

There have been three main approaches used in ecological studies – descriptive, experimental, and evolutionary, and, more recently, a fourth has been added, molecular ecology. Molluscan studies have been important in all four. As its name implies, descriptive ecology is concerned with describing patterns. Hypotheses as to what drives those patterns can be tested with experiments (experimental ecology), and the historical reasons observed patterns exist are hypothesised by evolutionary ecologists. Molecular ecologists use molecular (including genetic) methods to address ecological questions, and there is an obvious overlap with evolutionary ecology. Other branches of ecology are often recognised (see Introduction to this volume), such as behavioural ecology, the relationship between ecology and observed animal behaviours.

When investigating the interactions of an organism with its biotic and abiotic environment, the evolutionary history of the organisms should not be ignored (although it often is!), as many attributes may not have evolved in the habitat that an organism currently occupies. Information about species, such as their life history, longevity, physiology, and feeding habits are essential for a proper understanding of the role that a species has in an ecosystem, as are data on their abundance, interactions with other species (including intraspecific competition), predators, parasites, and dispersal ability.

Molluscs have been important organisms in understanding and testing some of the fundamental principles of ecology, especially along rocky shores, for nearly 70 years (e.g., Lodge 1948; Connell 1961a, 1980, 1983; Paine 1966, 1969a; Underwood 1979, 1985; Underwood et al. 1983), but overviews of molluscan ecology are surprisingly rare. The contribution by Russell-Hunter (1983) in the edited *Ecology* volume of *The Mollusca* series is an exception; other chapters in the volume address ecological aspects of deep-sea molluscs, fresh-water, mangrove and coral-associated bivalves, nudibranchs, fresh-water and terrestrial gastropods, and cephalopods. Molluscan ecology is also the subject of several books. Although restricted to fresh-water molluscs, *The Ecology of Freshwater Molluscs* (Dillon 2000) provides an important overview of the processes that have shaped and accompanied this invasion of fresh water by some gastropod and bivalve taxa. Earlier, a book by Mozley (1954) on fresh-water molluscan ecology was primarily a treatment of gastropods as vectors of schistosomiasis. *Ecology of Marine Bivalves: An Ecosystem Approach* by Dame (1996) provides excellent coverage of processes and interactions that shape marine bivalve populations as well as their contributions to ecosystem services. While these contributions are both significant and substantial, their limited focus under-samples molluscan diversity and interactions, and ultimately limits comparative study and discussion across the phylum.

9.5.1 ECOLOGICAL FUNCTIONS

Molluscs, with other 'invertebrates', perform important functions or 'services' (e.g., Kellert 1993; Ponder & Lunney 1999; Ponder et al. 2002), as outlined below (modified from Ponder et al. 2002; de Bello et al. 2010).

- *Breakdown of plant and macroalgal matter and detritus.* For example, in coastal marine habitats where large amounts of plant material are produced, molluscs and other marine 'invertebrates' play a significant role in breaking this material down, facilitating recycling. Although much of the decomposition of wood and other plant material results from fungal or microbial activity, and much wood in oceanic, coastal marine, and estuarine systems is broken down by teredinids (shipworms). Within sediment, for example, infaunal macrofauna play an important role through bioturbation that generally increases the degree of microbial action.
- *Recycling of nutrients.* Suspension and deposit-feeding molluscs recycle large quantities of particulate matter and dissolved nutrients in the environment (e.g., Ward & Shumway 2004). The large populations of bivalve and gastropod molluscs found in coastal waters, including estuaries and inland rivers and lakes, cycle large quantities of organic material and convert some of it into tissue that is released as gametes, or is broken down after death. Waste material from the animals is utilised by coprophages or is readily broken down by bacteria or fungi.
- *Energy and nutrient transfer between trophic groups.* Molluscs play a significant role in transferring carbon and nutrients assimilated by plants, algae, fungi, protists, and bacteria. Many molluscs feed directly or indirectly on such organisms and serve (as summarised earlier in this chapter) as food for a wide range of predators. In terrestrial, fresh water, and shallow marine ecosystems, 'plants', macroalgae, fungi, and many microorganisms are the major sources of primary food material, but in the open ocean, phytoplankton is the primary source of carbon. A large proportion of phytoplankton falls to the sea bed either directly, or indirectly as faecal material from zooplankton herbivores, where it is available to benthic deposit feeders.
- *Ecological interactions.* Interactions such as herbivory, predation, parasitism, mutualism, and competition perform essential roles in ecosystems. For example, molluscan (and other) herbivores can prevent algae from dominating shallow-water ecosystems, while molluscan (and other) carnivores play an important role in regulating populations.
- *Maintaining environmental health.* Suspension-feeding bivalves remove bacteria and other microorganisms, suspended sediment, and phytoplankton

from the water column, thus improving water quality. The resulting increase in water clarity allows more light penetration that may stimulate algal, macrophyte, or seagrass growth (e.g., Ward & Shumway 2004). This applies not only to marine ecosystems, but to fresh-water systems where bivalves are abundant (e.g., Vaughn et al. 2004). Infaunal molluscs (along with other infaunal invertebrates) also play important roles in reworking sediments (bioturbation) when burrowing, thus reducing effects such as anoxia. Scavenging molluscs play an important role in consuming the remains of dead animals.

- *Ecosystem stabilisation.* Each species plays a role in the ecosystem and in general it is believed that the more diverse the ecosystem is the more stable it becomes. While losing key species will have a clear impact and may cause cascade extinctions, it also raises controversial issues such as 'functional redundancy' (e.g., Levin & Lubchenco 2008; Cadotte et al. 2011) that we will not deal with further.
- *Provision of habitat for other biota.* Dense beds of attached molluscs such as mussels and oysters provide habitat for a wide range of other animals and algae, and larger benthic molluscs with an exposed shell provide potential habitat (see Section 9.5.9.1).

9.5.2 BIOLOGICAL INTERACTIONS

Populations and individuals are affected by abiotic (physical) factors as discussed above and by the interactions between the organisms in the ecosystem. These latter interactions are central to ecology and the main kinds are listed in Box 9.5. Such interactions may be between individuals or groups of the same species (intraspecific interactions) or different species (interspecific interactions).

The concept of *competition* is important in understanding these interactions. This is where different individuals, populations, or species attempt to use the same limited resources, thus lowering the fitness of one or both competitors. In theory, natural selection will cause those species less able to compete successfully either to adapt or becoming extinct.

While competition for mates is always intraspecific, competition for resources (mainly food and space) can be inter- or intraspecific, as can predation (if intraspecific, it is usually called cannibalism).

Predators of molluscs are dealt with above (see Section 9.4.1), but to better understand those relationships, we outline in this section some of the key aspects of predator-prey relationships and the role that molluscs have played. Predation, in its broadest definition, is any organism consuming any other organism and thus includes carnivory, herbivory, and parasitism. Here, we use it equated with carnivory – animals consuming animals.

Other types of interactions comprise symbiotic relationships including mutualism, commensalism, and parasitism, all of which are discussed above (see Section 9.4.4 and

BOX 9.5   KINDS OF BIOLOGICAL INTERACTIONS

Biological interactions can be grouped as follows:

Amensalism – These interactions occur when the actions of one individual or group of one species causes detriment to another, but receives no benefit. An example would be grazing of algae by an herbivorous fish, destroying habitat for a small gastropod relying on the algae for shelter or food.

Antagonism – In these interactions, one species benefits at the detriment of another. Such interactions include predation, herbivory, parasitism, and various types of competition. Studies of these interactions have largely dominated ecology.

Neutralism – Interactions between two species that do not affect either of them. Such interactions are difficult to demonstrate and are probably rare.

Facilitation – Those interactions are positive to at least one species and include:

Commensalism – where one species benefits from an interaction and the other is neither positively nor negatively affected.

Mutualism – where both species benefit from an interaction. This includes, for example, most cases of bacterial symbiosis.

Symbiosis – close relationships between different species (see Box 9.1).

Box 9.1), and the broader relationships between animals are listed in Box 9.5.

Understanding the effects of interactions has been an important goal of ecology, and ecologists using intertidal molluscs (and other animals) have played an important role in experimentally demonstrating the outcomes of both predation and competition, as for example, in the classic studies of Robert Paine (1969a, 1977), Joseph Connell (1961a, 1980, 1983), Tony Underwood (1979), and Bruce Menge and John Sutherland (1976). Similarly, studies on molluscs have also been important in developing ecological theory in relation to various other interactions.

9.5.2.1   Niche Specialisation

In a review of niche theory, Vandermeer (1972) defined the niche of an organism as 'ecological position in the world'. More precisely, a niche is a description of how an organism 'fits' into a community or ecosystem. It is the product of the evolutionary history and adaptations of the organism to its surroundings, involving morphology, physiology, and behaviour, and the resulting suite of habitat requirements for shelter, feeding, reproduction, etc. Theoretically, a niche is a 'multiple

variant space', and describing the niche of a particular organism is not a simple process. Niche concepts differ in detail, but ideas such as *niche breadth* are about the habitats and resources used by a species, while *niche partitioning* clarifies the niche differences between coexisting species. Literature on niche theory has also promoted the idea of *niche construction*, where some argue that the relationship between the organism and its environment results in influences on both, resulting in changes to the environment and to the organism, the latter reflected in genetic changes (e.g., Vandermeer 1972).

Niches can also be looked at from the viewpoint of being simple habitats that may either be occupied by an organism or available to occupy. A body of literature has argued that many niches in many ecosystems remain vacant (e.g., Cornell 1999). For example, Walker and Valentine (1984) estimated that between 12% and 54% of niches suitable for marine 'invertebrates' are empty, and the same is true for parasites (Rohde 2006).

Examples of supposed unoccupied niches are seen in tropical rainforests in northeastern Australia. These habitats have a diverse land snail fauna, but very few arboreal species compared to other similar rainforests in nearby Papua New Guinea or in other parts of Asia. Similarly, rainforest streams in tropical northeastern Australia have depauperate neritid and thiarid faunas compared with other nearby tropical areas. Do these habitats represent unoccupied niches not populated because of evolutionary or environmental history, or are there some other reason(s) such as unique physical factors that make them unsuitable?

Coral reefs provide a wide range of niches for the high diversity of plants and animals for which these habitats are renowned, including perhaps one third of fish species, and a great molluscan diversity (Goreau et al. 1979). Tropical coral reef diversity differs greatly in different parts of the world (e.g., Karlson & Cornell 1999; Knowlton et al. 2010), although the basic habitat is much the same. Similarly, intertidal communities in most tropical to temperate parts of the world are diverse, but this is not always the case. For example, a small oceanic island off the northeast coast of Brazil, Fernando de Noronha, has a single littorinid and a nerite. Unlike those on most tropical shores where there are several species, these are highly variable and occur throughout the intertidal zone, combining the features of upper shore and lower shore specialist species seen elsewhere (Vermeij 1972a). Thus, these species have expanded their niches compared to their relatives on other shores, although higher diversity does not necessarily result in tighter niche specialisation. In a study of the sacoglossan *Placida* that feeds on *Codium*, Trowbridge et al. (2009) found low niche differentiation despite the study area having the highest known number of *Codium* feeding sacoglossans globally.

There are numerous other instances of molluscs being used to address questions regarding the niche concept and testing theory, some including examples involving general marine molluscs (Lavm & Nevo 1981), bivalves (Green 1971; McClain et al. 2011), intertidal gastropods (Haven 1971; Kitching 1977; Noy et al. 1987), coral reef gastropods

(Reichelt 1982), fresh-water gastropods (Calow & Calow 1975), land snails (Barker & Mayhill 1999; Gauslaa 2008), and cephalopods (Voight 1995; Gowland et al. 2002).

9.5.2.2 Aggressive Conspecific and Intraspecific Behaviour

While it is well known that coleoid cephalopods engage in sometimes complex, agonistic behaviour, particularly during mating (Hanlon & Messenger 1996), some gastropods also show aggression towards both conspecifics and other species. For example, several patellogastropod limpets will attack potential predators (e.g., Branch 1979), and other limpets and chitons are aggressive towards competing conspecifics (e.g., Branch 1975a; Chelazzi et al. 1983). Intraspecific agonistic behaviour has also been observed in some sea slugs (e.g., Zack 1975), and both intra- and interspecific aggression occurs in limacid, agriolimacid, and arionid terrestrial slugs (Rollo & Wellington 1979). In this latter example, fatal wounds are sometimes inflicted by the aggressor using the radula and jaw, and the aggressor may even eat the victim. Such behaviour was only seen in mature individuals of a few particularly aggressive species and was marked in summer when shelter and food are scarce. Rollo and Wellington (1979) suggested that the agonistic behaviour they observed may be significant in regulating population size. Interestingly, aggressive behaviour in these slugs does not appear to occur in winter, when there are often aggregations of conspecifics, but the same species show aggressive behaviour in summer.

9.5.2.3 Parasites

As outlined earlier in this chapter, molluscan bodies may have both internal and external parasites from a wide range of phyla (see Section 9.4.4.1) that interact and are themselves the subject of ecological studies (e.g., Poulin 2007). These diverse parasite communities are sometimes used to assess the origins of stocks of fish and other vertebrates, and this technique has been suggested as a tool in assessing stocks of coleoids (Pascual & Hochberg 1996) and scallops (Oliva & Sánchez 2005).

9.5.3 Populations and Communities

Molluscs occupy every trophic level in marine communities, from primary producers to top carnivores. As already noted above, a few can be equated with primary producers as they have symbiotic relationships with plant plastids (obtained from algal food) or contain unicellular organisms embedded in their tissues that can photosynthesise, or otherwise convert simple chemicals in the environment (see Section 9.4.4.1 and Chapter 5), but the great majority of molluscs depend directly or indirectly on primary producers.

Like other biotic communities, molluscan communities (see Box 9.6) are structured through the interplay between physical, extrinsic biotic, and intrinsic biotic factors. In aquatic systems, these primarily include nutrient levels in the water, which phytoplankton, algae, and aquatic plants depend upon (e.g., Birkeland 1987; Fabricius 2005), as well as factors such as temperature, salinity, and, especially in fresh water, pH.

BOX 9.6   POPULATIONS AND COMMUNITIES

Population – A group of interbreeding individuals in a circumscribed area.

Metapopulation – A group of spatially separated populations of the same species. It is assumed there is sufficient gene flow between populations to maintain genetic continuity.

Community – The collection of different species simultaneously occupying the same geographical area. Sometimes called a *biocoenosis*.

In terrestrial systems, soil pH, moisture, calcium availability, ground cover, and other shelter etc. are important factors that determine gastropod communities.

Phytoplankton is directly utilised by suspension-feeders either in the water column (such as many larvae and Thecosomata) or, more importantly, by benthic bivalves. Molluscan herbivores are relatively uncommon (see Chapter 5), and include marine gastropods such as some trochids, turbinids, strombids, and aplysiids, with a number of chitons, limpets, and neritids also being important algal grazers on rocky shores, while in freshwater, ampullariids and some hygrophilans graze on macrophytes. In the terrestrial realm, species of stylommatophoran snails and slugs are also herbivores (see Chapter 5).

Many molluscs also utilise the detrital material in sediments, which is largely composed of material derived from primary producers (including bacteria, fungi, and other microorganisms), while many others specialise in scraping the microbiota (protists, bacteria etc.) and detritus from the surface of substrata (see Chapter 5). Most cephalopods and many gastropods are carnivores, but their distribution and abundance depends heavily on available food, which largely depends on the primary producers and, ultimately, nutrients.

Where nutrient levels are generally low (oligotrophic) as in many tropical seas, phytoplankton productivity is also low. These conditions are favoured by corals, with the photosynthetic symbionts in the coral tissues greatly exceeding the primary productivity of the surrounding waters. Other similar primary producers with photosynthetic hosts include ascidians and sponges, as well as *Tridacna* and a few other bivalves and gastropods. Such animals provide a wealth of food for carnivores, which are also abundant and diverse in coral reef habitats.

In marine systems with intermediate levels of nutrients (mesotrophic), algae and marine angiosperms (mainly seagrasses) dominate and, in the tropics, may out-compete corals. Molluscs are often abundant, but lower in diversity, and bivalves are a larger component of the communities than in oligotrophic habitats. Herbivores are prominent and carnivores typically less abundant.

In high nutrient level (eutrophic) marine systems, suspension-feeding bivalves such as mussels and oysters often predominate on hard substrata, while burrowing bivalves are common in soft sediments.

Scavengers and/or carnivores such as nassariids, naticids, and buccinids can be common in eutrophic habitats. Taylor (1998) investigated these relationships from the perspective of the diet of tropical predatory gastropods. In oligotrophic habitats, such as coral reefs, the main prey items are polychaetes and herbivorous gastropods. In mesotrophic habitats where algae and seagrasses typically dominate, bivalves become more important prey, while in eutrophic habitats, bivalves and barnacles are the predominant prey.

9.5.3.1 Some Factors Involved in the Maintenance of Molluscan Populations

The main factors involved in maintaining molluscan populations are the same as those maintaining populations of other animal taxa. They include recruitment, competition for food and other resources, including space, mates, and shelter, with predation (see Section 9.4.1), parasitism, and disease also important.

Numerous studies have been made on inter- and intra-specific competitive interactions in molluscs and also those between molluscs and other 'invertebrates'. These topics are well covered elsewhere with overviews including those of Russell-Hunter (1983) on molluscs in general, Allen (1983) on deep-sea molluscs, Underwood (1977, 1979, 1998) on intertidal rocky shore molluscs, Dillon (2000) on fresh-water molluscs, and (Barker 2001b, 2002) on terrestrial molluscs. The reader is referred to these and similar works for further general information.

The factors maintaining particular populations are not necessarily well understood, but are largely a balance between recruitment (see Section 9.3.1) and predation (9.4.1), as well as other factors such as the availability of food and shelter. For example, Runham and Hunter (1970) thought that the main density-dependent factors controlling terrestrial slug populations in Europe were predation, parasites, and disease, but competition for food and shelter and associated agonistic behaviour were later stressed (Rollo & Wellington 1979; Barker & Efford 2002) (see Section 9.5.2.2), although it was previously thought that aggressive interaction did not occur in terrestrial gastropods (Lomnicki 1969).

The feeding behaviour of grazing gastropods, in particular larger patellogastropod limpets, can markedly influence settlement and community structure on intertidal rocky shores (Underwood 1979; Branch 1981; Lubchenco & Gaines 1981; Lindberg et al. 1998). Thus, the Californian intertidal limpet *Lottia gigantea* limits incursions by other grazing species and recruitment by encrusting species by repeatedly grazing the surrounding area (e.g., Stimson 1973; Shanks 2002). When large grazers, such as limpets, chitons, and large trochids and turbinids are absent, other grazers and sessile species (algae, mussels, barnacles, etc.) increase in abundance (e.g., Underwood 1980; Underwood & Jernakoff 1981; Creese 1982; Lindberg et al. 1998; Kido & Murray 2003).

9.5.4 Chance Environmental Impacts

Many impacts such as fires, earthquakes, tsunamis, and meteor impacts are rare and may cause high mortality,

even extinction, although such effects are usually localised. Species better adapted to surviving extreme conditions may have a higher chance of survival and these are presumably more common in environments that often experience such disturbances (e.g., from ice, floods, or storms).

Fires can be devastating to terrestrial molluscs which are severely reduced or even eliminated from fire-prone habitats (e.g., Stanisic & Ponder 2004).

While markedly disturbed habitats may have reduced faunal diversity compared with stable ones, evidence suggests those experiencing intermediate levels of disturbance result in higher diversities than those experiencing low levels (e.g., Connell 1978).

9.5.5 Sea Shore Ecological Studies

Molluscs figure predominantly in the descriptive phase of ecology. Some very early studies related to seashore zonation patterns (e.g., Audouin & Milne Edwards 1832–1834), but more detailed early studies on the intertidal were not made until the early part of the 20th century. One of the earliest studies that described patterns of zonation in the intertidal zone was that of Oliver (1923) on the intertidal communities of New Zealand shores. Others soon followed, including the seminal study of Stephenson and Stephenson (1949). There have also been many studies on soft shores and, although the zonation patterns of the infauna are hidden, they nevertheless occur and, as with rocky shores, can be experimentally examined (e.g., Reise 1985). Descriptive accounts of seashore ecology have been provided by numerous authors and in several books (Morton & Miller 1968; Stephenson & Stephenson 1972; Morton & Morton 1983; Ricketts et al. 1985; Britton & Morton 1989; Morton 2004b).

While zonation exists on hard shores, the causal factors involved in these generally harsh environments were not clear. Initially, they were thought to be related to critical tidal levels (i.e., the time immersed) (e.g., Colman 1933; Southward 1958; Newell 1970). This idea was refuted by Connell (1961b), Paine (1966), and Underwood (1978), who showed that intertidal distributions of organisms such as gastropods and barnacles were not strictly related to the critical levels, but that several other factors were involved including wave exposure, food availability, predation, settlement patterns, substratum, etc. Testing the importance of these factors has been done experimentally, often involving molluscs, in several parts of the world (e.g., reviews by Paine 1977; Underwood et al. 1983; Underwood 2000; Morton 2004b). Reviews of molluscs in intertidal ecology include Underwood (1979) on intertidal gastropods, Branch (1981) on limpets, Boyle (1977) on chitons, and Dame (2012) on bivalves.

Physical factors such as wave exposure can have compounded effects. Molis et al. (2015) showed that high wave exposure increased the toughness of the algal food (*Fucus*) of *Littorina obtusata* inducing phenotypic changes in the radula of the snail.

In the 1940–1970s experimental manipulations of intertidal molluscs addressed some of the fundamental questions in ecology, including prey-predator relationships, ecological succession, and competition (Jones 1948; Lodge 1948; Burrows & Lodge 1950; Connell 1961a; Paine 1966, 1969a). Although some argued that many early studies lacked robust experimental design (Hurlbert 1984; Creese 1988; Johnson et al. 1997; Underwood 1997), subsequent studies have generally supported the directionality and outcomes of these interactions (e.g., Menge et al. 1994; Scheibling 1994; Boaventura et al. 2002; Robles et al. 2009).

Movement and feeding of intertidal animals affect the environment in which they live and, although often negligible, these impacts can sometimes be marked, as with predators impacting on the populations of their prey or the actions of grazers altering the composition of algae. For example, the selective grazing of the Caribbean chiton *Choneplax lata* provides conditions that encourage a reef-building crustose coralline alga to grow (Littler et al. 1995), and some large limpets produce algal 'gardens' (see Chapter 18).

A useful source of information on many aspects of rocky shore animals and their ecology is Denny and Gaines (2007), but, surprisingly, there is no general, up-to-date review of hard-shore molluscan ecology, However, a number of textbooks, including Little and Kitching (1996), Raffaelli and Hawkins (1996), and Knox (2000), and some other general texts on marine ecology, deal with the subject in some detail.

Soft-shore studies involving molluscs have also resulted in a substantial literature, but have been less significant than hard-shore studies. While many of these have been in marine conditions, there is also a substantial body of work involving estuarine (including mangrove and saltmarsh) molluscan ecology. General reviews of soft shore-ecology can be found in Little (2000) and McLachlan and Brown (2006), and in various more general textbooks on marine ecology. There are also some detailed regional accounts that provide valuable descriptive information (e.g., Morton & Miller 1968; Morton & Morton 1983; Britton & Morton 1989; Morton 2004b).

9.5.6 Deep-Sea Ecological Studies

The discovery in 1976 of the hydrothermal vent bivalve populations at the Galapagos Rift (Lonsdale 1977) fundamentally changed our understanding of deep-sea ecology and the diversity of molluscs associated with chemosynthetic communities (see Section 9.7.1.6). Ecological studies on other deep-sea mollusc communities are rather scant. They include an early review (Allen 1983) and a more recent one (Rex & Etter 2010).

9.5.7 Fresh-Water Ecological Studies

There are many texts dealing with fresh-water ecology in general (e.g., Wetzel 2001; Dodds 2002; Lampert & Sommer 2007; Moss 2010). The ecology of fresh-water molluscs has been reviewed by Dillon (2000) and that of fresh-water mussels by Strayer (2008). Also, detailed accounts relate to some invasive bivalves and the invasive ampullariid gastropods (*Pomacea*) (Cowie & Hayes 2012; Hayes et al. 2015; Joshi

et al. 2017), and there are numerous papers on the ecology of intermediate hosts of human schistosomes, notably the planorbids *Biomphalaria* and *Bulinus* spp. and the pomatiopsids *Oncomelania* spp. Other relatively well-known taxa include some lymnaeids, as they are the intermediate hosts of other economically important diseases such as liver fluke (*Fasciola*). Studies on these intermediate hosts of parasites of importance in human or livestock health produced some of the earliest ecological accounts of fresh-water molluscs. These investigations went under the general categories of 'medical malacology' or 'veterinary malacology', which flourished during the mid to late 20th century (e.g., Malek & Cheng 1974). Such studies remain important and relevant in some parts of the world, notably Africa (e.g., Stothard & Kristensen 2000; Brown 2002), Asia, and central and South America.

Invasive bivalves have also spawned many studies, notably on the 'Asian clam' *Corbicula* (Sousa et al. 2008), the 'zebra mussel' and 'quagga mussel' *Dreissena* spp. (Nalepa & Schloesser 2014), and the 'golden mussel' *Limnoperna*.

Many fresh-water molluscs must be able to withstand a highly variable habitat – with, for example, flooding in the wet season (resulting in turbulence, reduction in salinity, etc.) and drought (drying, increase in salinity) in the dry season. To withstand these conditions, strategies such as aestivation and rapid breeding (R-selection) are needed. In more stable environments, species are less resistant to drying, have low dispersal ability, and tend to have reproductive strategies towards the K-selected end of the spectrum.

9.5.8 Terrestrial Ecological Studies

Terrestrial gastropods occur on all continents except Antarctica, and live in most habitats. Many species favour damp conditions, such as in leaf litter or under logs or rocks. Disturbed habitats (gardens, fields, etc.) are usually favoured by exotic species, and these are generally more resistent to disturbance than many native species.

Reviews of the ecology of terrestrial gastropods include a general account by Sallam and El-Wakeil (2012). Cook (2001) reviewed behavioural ecology, and their physiological ecology has been reviewed by Riddle (1983). A major focus has been on the ecology and control of pest species such as the helicid snails, various slugs, and the 'giant African snail' (*Achatina fulica*) (see Chapter 10).

Besides historical factors, the distribution and diversity of terrestrial molluscs is determined largely by climate (temperature and rainfall), which in turn influences vegetation. Vegetation and other shelter (logs, litter, rock piles, crevices, etc.) positively affect the conditions (temperature, light, moisture) for terrestrial molluscs. Changes in available moisture, vegetation cover (e.g., by land clearing), and fire have major influences on snail populations, often causing local extinction. Factors such as drought and fire are critically important, with some terrestrial species not being drought resistant, although many others can survive drought or other unfavourable conditions by aestivating (see Chapter 2 and Section 9.7.5).

The effects of fire can be devastating to land molluscs with only those species or individuals whose behaviour includes habits such as burrowing in the soil, hiding under rocks, or in crevices surviving. Thus, fire-prone habitats generally have low gastropod diversity.

Soil and rock type are important, particularly for snails, as low calcium content can be detrimental. Land snails scrape dead shells, limestone, and cement to obtain calcium for their shells and, because this is passed up the food chain, they can be important in calcium cycling. The availability of calcium is limiting, as shown when available calcium in the soil is artificially increased and there is an increase in individuals and species (Johannessen & Solhøy 2001). Snails also uptake other metals and consequently can be used in environmental monitoring for toxins such as cadmium. Their faeces can also be significant sources of nitrogen in poor soils (e.g., Zaady et al. 1996), and their feeding activities can assist with soil formation (Shachak et al. 1987, 1995; Jones & Shachak 1990).

Because land snails move little and have relatively narrow environmental tolerances, they can be very good indicators of the environmental changes that have occurred at a particular location. Thus subfossil or fossil shells can be used for palaeoenvironmental reconstruction (Goodfriend 1992).

While attention is often focused on the impact of pest herbivorous species on crops, the consumption of seedlings by native herbivorous snails and slugs can presumably also influence native plant populations. Most terrestrial gastropods are not herbivorous, but instead consume detrital material, bacteria, fungi, etc. Many terrestrial snails and slugs play a role as decomposers, though their contribution is small compared with other decomposing organisms (e.g., Wolters & Ekschmitt 1997).

Predators of land snails and slugs include various insects and some reptiles, amphibians, certain birds, and small mammals (see Section 9.4.1.1). Snails and slugs are also intermediate hosts of a number of parasites that infect tetrapods.

9.5.9 KEYSTONE SPECIES

The keystone species concept was introduced by Paine (1969b) as a result of his studies on the predation of intertidal animals by the starfish *Pisaster ochraceus* (Paine 1966) and is now a well-established primary concept in ecology and conservation biology (e.g., Mills et al. 1993; Bond 1994; Paine 1995; Power et al. 1996; Simberloff 1998; Smith et al. 2015). The concept recognises that the removal of certain species (i.e., keystone species) from an ecosystem results in dramatic changes to the communities. Keystone species are thus, relative to the rest of the community, thought to be exceptional in their ability to maintain the diversity and organisation of the community. These species have more profound impacts on the communities in which they live than might be expected from species of similar biomass and/or abundance. Molluscs considered keystone species include the limpet-like predatory muricid *Concholepas* (e.g., Navarrete & Manzur 2008; Manríquez et al. 2009), the bivalve *Limaria hians* (Trigg et al. 2011) and large limpets including *Cellana tramoserica*

(Fraser et al. 2015), and patellids (Henriques et al. 2017). Epifaunal bivalves that live in large aggregations (e.g., oysters, mussels) have sometimes been considered keystone species (e.g., Dame & Prins 1997), but given the concept of keystones outlined above, we do not consider that these examples are confirmed. Similarly, shallow-water or intertidal infaunal bivalves living in large aggregations ('beds') including various 'cockles' are also an essential ecological component of their communities, and it could be argued that they are keystone species (e.g., Smith et al. 2015). Not all keystone species need to be large. It has been suggested that *Limacina helicina* could be considered a keystone species in Arctic pelagic ecosystems as it comprises about 63% of the total zooplankton in the Ross Sea and is an important component of the food chain in that area (Hunt et al. 2008). This thin-shelled species, along with other thecosomes, may be vulnerable to ocean acidification.

Esch et al. (2001) argued that snails are keystone species for trematodes, and it could be equally well suggested that coleoid cephalopods would be for mesozoans, but this use of the term 'keystone species' differs somewhat from what is normally accepted.

Other examples of molluscan keystone species that have been suggested (see Power et al. 1996) include the role of *Nucella lapillus* (Menge 1976) in mussel predation on northwest Atlantic rocky shores and the role of the land snail *Eucondrus* in lichen consumption in relation to weathering of rock.

9.5.9.1 Molluscs as Habitat Generators and Ecosystem Engineers

Molluscs are significant habitat for a wide range of biota from microbes to other molluscs. As detailed in the earlier sections of this chapter, organisms calling an individual mollusc home include numerous bacteria and other microorganisms, parasites (mainly trematodes, nematodes, and various protozoans), commensals that live either in the mantle cavity or in tubes or burrows created by the host mollusc, and a wide range of epibionts such as sponges, hydroids, bryozoans, and tunicates.

Ecosystem engineers can be defined as organisms that create, modify, and maintain habitats in a significant way (Jones et al. 1994, 1997) and include services as varied as trees creating shade to minute organisms altering soil. Some molluscs live in dense beds that form significant habitats, particularly in intertidal zones – these include bivalves such as oysters and 'mussels' (e.g., Seed 1996). Masses of oyster shells, either as reefs or where they form zones on the shore, are very significant habitat for a wide range of epibionts and nestling animals including many other molluscs. Oyster reefs may be intertidal or shallow subtidal. They provide important ecological services as invertebrate and fish habitat (Harding & Mann 2001; Beck et al. 2009, 2011), processors of nutrients and suspended material (Dame et al. 1984, 1989), and stabilising shorelines. These reefs have deteriorated or even disappeared in many parts of the world due to human disturbance (see Chapter 10).

Reef-like aggregations of other bivalves, notably mussels (Mytilidae), are also common on shores and provide similar services to oyster reefs. Similarly, aggregations of large bivalves such as pinnids, *Tridacna* and non-reef-forming oysters can provide substratum and structuring, adding complexity to habitats. The extinct rudist bivalves were significant reef-forming organisms (see Chapter 15), and their substantial fossil deposits are now important as oil reservoirs (see Chapter 10). Aggregations of some fresh-water bivalves, mainly various unionoideans (Gutierrez et al. 2003), create 'shoals'.

Intertidal vermetid gastropods sometimes crowd together to form distinct zones on the shore and sometimes minor reefs in parts of the Caribbean, eastern Mediterranean, the Atlantic islands such as Bermuda and Cape Verde, and, in the Pacific, Hawaii, and some islands off northern New Zealand and western Mexico (e.g., Safriel 1975).

Similar habitats are also provided by barnacles, tube worms, etc., and these are also well utilised by molluscs. The byssal nests of the limid bivalve *Limaria hians* form benthic reefs in the waters around the United Kingdom that are important habitats for many species (Hall-Spencer & Moore 2000).

Accumulated drifts of dead mollusc shells (notably bivalves, including fresh-water mussels) also provide habitat and shelter for a variety of animals and an attachment surface for 'algae' and 'plants', as do even individual shells. Broken down shell can form a significant, and sometimes the major constituent of, sediment. These calcareous sands provide habitat for numerous animals, and can also comprise a major component of beach rock in tropical environments.

Empty snail shells not only provide a home for hermit crabs, but also a variety of other animals including small octopuses, sipunculid, and various polychaete worms. Empty microsnail shells are inhabited by hermit crab-like tanaid crustaceans, while 'straight' or 'symmetrical' hermit crabs (Pylochelidae) occupy empty scaphopod shells.

9.6 ABIOTIC INTERACTIONS – THE PHYSIO-CHEMICAL ENVIRONMENT

The physical environment includes potentially damaging forces such as wind, wave, or current action, as well as temperature and chemical variables such as salinity and alkalinity/acidity. Combinations of these variables and the important interactions between them are critical to understanding ecological observations (e.g., McLusky et al. 1986; Przeslawski et al. 2005; Sarà et al. 2008). Here, we briefly review the main variables.

9.6.1 TEMPERATURE

Physical factors such as wind, temperature, humidity, substratum nature and colour, and available potential cover determine the rate of desiccation and heating in intertidal and terrestrial molluscs, although this can be greatly mediated by the behaviour of the animal (e.g., seeking shade or clustering) if they are mobile. Temperature (especially temperature extremes) is especially important in the intertidal zone, terrestrial habitats, and in shallow coastal or inland waters, largely for physiological reasons (see Chapter 2). Investigation of the temperatures in living animals *in situ* has been greatly enhanced by thermal imaging techniques (e.g., Caddy-Retalic et al. 2011) and biomimetic temperature sensors (Helmuth et al. 2010). Sea temperature is also a limiting factor for most species and serves as a trigger for reproduction in many taxa, as do lunar cycles (see Chapter 8).

Responses to temperature include high levels of evaporative cooling in active terrestrial snails and slugs, but inactive snails show no such effect, while some non-siphonate bivalves can maintain a 1°C–2°C temperature gradient by having their valves slightly gaping (Withers 1992).

9.6.2 SALINITY

Salinity also has a significant impact on physiological processes (see Chapter 2) and is a major factor in estuaries, coastal lakes and lagoons, slow-flowing inland rivers and the intertidal, etc. where evaporation or rainfall can cause major and rapid changes in salinity. Relatively few species can survive in hypersaline environments (see Section 9.7.3), and diversity also declines in hyposaline systems such as most estuaries.

9.6.3 PHYSICAL FORCE–WAVE ACTION ETC.

In the intertidal and shallow subtidal wave action is important and can also be significant in large lakes. Intertidal wave-swept habitats are among the most violent physical environments, with the animals that live in them having to either contend with these extreme conditions or be swept away. Such forces are important in structuring molluscan diversity and abundance in intertidal and reef crest communities (Denny 1995), with drag being the hydrodynamic force most likely to dislodge intertidal gastropods (Denny et al. 1985; Trussell et al. 1993). In many oceanic systems and rivers, currents, and the scouring they can cause, can also have very important effects on habitats, such as sediments, and the animals in them. Mechanisms to avoid being swept away by waves or currents include methods to firmly attach to the substratum and to reduce the surface area affected by drag. In molluscs, these adaptations include a broad foot (chitons, limpets, abalone) and increased foot tenacity by the pedal mucus (see Chapter 3), byssus threads (e.g., mussels and some other bivalves – see Chapter 15), or cementation (e.g., oysters, vermetid gastropods). Behavioural adaptations – hiding in crevices, creating depressions (scars) in the rock, etc. are also important. These include activity patterns, as the adhesive properties of pedal mucus are reduced when the animal is crawling.

Biomechanics, the study of how the form, function, and biology of organisms is influenced by physical forces in the environment, started with D'Arcy Thompson (1917) *On Growth and Form*. Biomechanics examines how organisms withstand the physical forces imposed by the environment, in particular, in relation to their size and shape.

Common trends in the shell morphology of molluscs on wave-swept shores (e.g., Denny 1985, 1988; Gibbs 1993; Denny & Gaylord 2010) include:

- Reduction in body size – the forces generated during the acceleration of a wave increase with the volume of the object being struck, and thus have a relatively greater impact on large-sized animals.
- Flattening the body will reduce impact of drag from waves, but will increase the lift forces. Flattening is seen in surf zone gastropods (e.g., limpets, *Concholepas*, abalone) and rock oysters. Limpets in the surf zone with an intermediate ratio of shell height to diameter will undergo lower lift forces and impact from wave acceleration than taller or flatter limpets – another trade-off situation (see Section 9.9).

Limpets attached to seagrasses or kelps have the most streamlined body shapes because their substratum moves with the current. Other limpets living on rock must contend with water forces from many directions, requiring a more generalised body form (Denny 1988).

Intraspecific changes in shell morphology with exposure can occur in different degrees of wave exposure, with individuals on exposed shores typically smaller and squatter than those in more sheltered areas (Etter 1988a; Sundberg 1988; Trussell et al. 1993; Trussell 1997). This is not always the case, as, for example, Prowse and Pile (2005) found no morphological correlation with wave exposure in an intertidal trochid and a neritid.

9.6.4 HABITATS

Molluscs have evolved very diverse lifestyles enabling them to live in almost every habitat except free flying in the atmosphere. They are often one of the more conspicuous groups and are sometimes predominant. They are ubiquitous in marine environments, inhabiting all zones from the supralittoral to the deepest oceans, where they live on a wide variety of both soft and hard substrata. While gastropods and chitons are normally associated with hard substrata and bivalves with soft substrata, many exceptions occur. Many bivalves (e.g., mussels, oysters) nestle in or bore into hard substrata, or are attached to them. Some bivalves burrow into coral, soft rock, or wood, with 'shipworms' (Teredinidae) the most modified (see Section 9.6.5.1.1 and Chapter 15). Various gastropods burrow in soft sediments (see Chapter 3), while a diverse assortment of tiny species live interstitially between sand grains (see Section 9.6.5.3).

Members of several groups within the chitons, gastropods, and bivalves have successfully adopted the intertidal as their habitat (see Section 9.7.1.1), even though it is a relatively harsh environment.

Numerous groups of molluscs are found in the deep-sea, but the most surprising of all are the amazing faunas associated with hydrothermal vents and seeps; these consist of many strange molluscs (mostly bivalves and vetigastropods) that often dominate these communities (see Section 9.7.1.6).

Molluscs originated as epibenthic animals in shallow-water marine habitats, and similar habitats today accumulate high diversities because of their rich niche diversity, long-term stability, and minimal physiological adaptation. Others are outside the basic molluscan comfort zone and required significant adaptation, either structural, morphological, or physiological. Thus molluscan habitat diversity was extended by colonisation of a wide range of substrata, and this was paralleled by movement into the intertidal and ultimately non-marine habitats on the one hand, and into deeper water on the other. Some acquired the ability to leave the benthos and move into the water column as adults. All of these 'ecological' adjustments resulted in physiological, behavioural, and structural modifications that have led to the amazing molluscan diversity we see today.

9.6.5 SUBSTRATA

9.6.5.1 Hard Substrata

Hard substrata are both abiotic and biotic in their origin and can extend over hundreds of kilometres and hectares or be less than 1 cm^2 in area. They occur throughout marine, fresh water, and terrestrial habitats. Besides tropical coral reefs (see Section 9.7.1.2), some of the best known marine hard substrata are the nearshore and rocky intertidal habitats of temperate regions (Section 9.7.1.1) (Branch et al. 1981; Ricketts et al. 1985; Underwood & Chapman 1995). Approximately one third of the coastlines of the world are rocky shores (Johnson 1988), and they are renowned for both their organismal diversity and productivity (Darwin 1839; Grosberg et al. 2012). Hard substrata also enhance diversity and productivity by providing attachment for macroalgae, especially the large, fast-growing subtidal kelps and intertidal species.

Rocky substrata are more common on active continental margins than on passive ones. This distribution was recognised and contrasted as 'Atlantic' versus 'Pacific' coastal types in the 19th century (Suess 1888)[18] and in the 1970s was reframed in the context of plate tectonics. Simply stated, active margins and the dynamic properties associated with them (subduction, faulting, accretion) have resulted in both a higher abundance of rocky shores and a greater diversity of rock types along these shores than are found along the quieter and less dynamic passive margins of continents. Not unexpectedly, living biodiversity patterns of rocky shore species generally reflect this 'active' and 'passive' dichotomy.

Beyond the intertidal and nearshore, and below fairweather and storm wave-base[19] water depths (15–40 m), sediments tend to rapidly accumulate, thus making soft-bottom communities predominant on continental shelves, bathyal depths, and

---

[18] Eduard Suess (1831–1914) also recognised and coined the terms 'biosphere', 'Gondwanaland' and the 'Tethys Ocean'.

[19] The depth at which the sea bed is disturbed by wave action during fair and inclement weather.

abyssal plains. Even in these extensive tracts of soft substrata, hard substrata can be found as large-scale geological features and include rocky seamounts (mostly volcanoes), plutons, batholiths, spreading, and fracture zones. Besides these geological hard substrata, biogenic hard substrata also find their way onto the sea bottom (see Section 9.7.1.7). These include marine mammal skeletons, wood, sea turtle carapaces, and even cephalopod beaks.

The molluscan fauna that inhabit hard substrata usually have anatomies and behaviours that enable them to remain firmly attached to the substratum. This includes the large expansive foot of polyplacophorans, arm suckers of octopuses, cementation in bivalves and gastropods, byssus attachment by bivalves, nestling behaviour in both gastropods and bivalves, and the evolution of the limpet morphology in every major group of gastropods (Vermeij 2016). Thus, not surprisingly, most rocky shores of the world have impressive chiton, limpet, abalone, mussel, and littorinid faunas. These communities are seldom similar at lower taxonomic ranks and instead reflect historical and current ecological interactions. For example, Lindberg (1988b) showed that although southern Africa, Chile, Europe, Australia, and western North America all have noteworthy patellogastropod faunas, each is different in its combination of members from up to three different groups.

Some hard substrata are made up of smaller elements (e.g., gravel, cobbles, and boulders) (see Section 9.6.5.3), and although these habitats have many properties of hard substrata, they, like soft substrata, are unstable and capable of shifting subject to the size and frequency of disturbance. Depending on the size of the elements that make up the substratum, organisms (including molluscs) may nestle between them, much like smaller interstitial organisms.

9.6.5.1.1 Boring Into Hard Substrata (Including Wood)

Burrowing into soft sediments and boring into hard substrata is a common means of bivalves reducing predation. Few molluscs are true rock borers, although many are nestlers, living in crevices or holes made by other animals. Several bivalves are borers in coral and calcareous rock (e.g., limestone), notably some mytilids such as *Lithophaga*, and pholadids, most of which bore into soft rock or clay (see Chapter 15 for details). Burrowing bivalves are significant in the bioerosion of corals and a number of taxa can be involved (e.g., Kleemann 1996). For example, at Low Isles, Queensland, the important coral burrowers were the mytilids *Lithophaga* (five species) and *Modiolus* (one species), two species of gastrochaenid (*Gastrochaena*), an arcid, and two species of *Tridacna* (Otter 1937).

Some vermetids excavate a trench in the surface of calcareous rocks, coral, or shell using chemical etching. Others have their bodies encased partly within growing coral, such as *Ceraesignum maximum* (Golding et al. 2014). *Dendropoma* sp. (Savazzi 2001b) etches its way deeper into the substratum as the rock erodes. This species, and the clavagellid bivalve *Bryopa* (Morton 2005), are the only species known to move deeper into the substratum while part of the shell is cemented.

Three related families of bivalves have adopted wood boring; the Pholadidae ('piddocks') (subfamilies Pholadinae, Martesiinae), Xylophagaidae, and Teredinidae[20] ('shipworms'). In the Pholadinae, some genera are occasionally found boring in wood, but mostly they bore in peat, firm mud or clay, or soft rock. The widely distributed *Martesia* and the central Indo-West Pacific *Lignopholas* are both members of the Martesiinae and bore into wood, including the woody husks of large nuts, but do not feed on it. The Xylophagaidae are all wood-borers and replace teredinids in the deep-sea. They may also utilise cellulose (Purchon 1941b), but do not have the modifications seen in teredinids (see Chapter 15 for more details).

Teredinids are important in marine systems as they are one of the main groups of marine organisms that break down wood in the sea. The large mud-living teredinid *Kuphus* commences life as a wood-borer (Shipway et al. 2018), but then harnesses nutrients from symbiotic bacteria in its gills (Distel et al. 2017). The wood-boring teredinids line their burrows with calcium carbonate and can digest and utilise wood for food aided by symbiotic bacteria in their gut (see Chapter 5). Their wood-boring activities in man-made structures such as wharf piles and boats have made them pests (see Chapter 10).

9.6.5.2 Living on and in Soft Substrata

Much of the marine benthic and intertidal habitat is dominated by soft sediments, as are many fresh-water habitats. While some molluscs living on soft sediments are epifaunal, the majority are infaunal, burrowing into the sediment. So-called 'soft' sediments range from coarse sands to fine muds and silts. Here, we briefly look at the epibenthic forms living in these environments and the burrowers that live within the sediments.

9.6.5.2.1 Epibenthic Molluscs in Muddy Environments

Fine-grained sediments such as mud can be very soft and any epifauna living on them may require special modifications for sanitation and to prevent sinking. Sinking into the substratum can be avoided, for example, by decreasing shell weight, the development of a large foot, or, in some, developing flanges or other processes such as spines as, for example, seen in species of *Murex* or in some strombids and xenophorids. In others, the shell has become wide and flat, as in the window pane shell (*Placuna*) and various scallops (Pectinidae).

Suspended particles must be largely excluded from the mantle cavity to avoid clogging the gills. To this end, various mechanisms are employed, including mesh-like tentacles, ciliated rejection tracts, and by restricting or controlling the inhalant area. Similar sanitation adaptations are also needed for any infaunal species in these habitats.

---

[20] A molecular study (Distel et al. 2011) suggested that all the wood boring groups may form a monophyletic group with only a single family recognised.

9.6.5.2.2 Burrowing in Soft Sediments

Some aspects of burrowing in sediments are discussed in Chapter 3. This habit involves several modifications to the shell and animal, as discussed in more detail in Chapter 15 for bivalves, Chapter 16 for scaphopods, and in Chapters 18 through 20 for gastropods.

Sediment scour is a problem for shallow-water burrowing animals and several strategies may be employed for them to remain buried. In some bivalves, these include shell ornamentation, including ridges and spines that help anchor the animal in the sediment. The ability to rapidly burrow again if dislodged is important, especially in high energy environments. Various shallow-burrowing gastropods such as naticids, terebrids, olivids, and *Bullia* can rapidly burrow again, as can bivalves such as donacids and mesodesmatids. This ability is usually associated with a streamlined shell and a powerful foot. Another way to avoid predation or exposure from scouring is to burrow deep in the sediment. Deep-burrowing gastropods have not evolved, but several lineages of bivalves have adopted this habit. These usually have smooth shells with long siphons and often have a permanent gap in the posterior end of the shell, and sometimes the anterior end. Bivalve families that have adopted deep burrowing include the Solenidae, Cultellidae and Myidae, and some Mactridae and Hiatellidae (*Panopea*) (see also Chapter 15).

9.6.5.2.3 High Energy Beaches and the Surf Zone

The problems of living on the surface of unstable soft sediment (sand) in higher energy areas are such that few molluscs have achieved it, other than occupying this habitat for short periods after emerging from beneath the surface (naticids, terebrids, *Oliva*, *Olivella*, some *Conus*, *Bullia*, and a few cephalaspideans and acteonoideans. Most molluscs living in such habitats avoid the problems of dealing with living on the surface of high energy soft substrata by burrowing below the surface.

Some taxa, notably *Bullia digitalis* and some species of *Donax* and *Olivella*, have been reported to 'surf', with the gastropods expanding their large foot and parapodia, respectively, to provide a larger surface area. *Olivella* flaps the parapodia and effectively swims, while *Bullia* uses its large foot as a sail. Some of the surf zone sandy beach gastropods are carnivores or scavengers, but others, living in the lower intertidal or subtidally, are suspension-feeders – the umboniine trochids such as the flattened *Umbonium* and *Zethalia*, and the tall-spired *Bankivia* and *Leiopyrga* feed using their gills, but some species of *Olivella* produce mucous nets with which they capture suspended particles (Seilacher 1959) (see Chapter 5). The ability of these gastropods to suspension-feed probably accounts for the high population densities they can sometimes achieve.

9.6.5.3 Gravel, Cobble, and Boulder Substrata and Interstitial Habitats

Gravels, cobbles, and boulders are substrata that are typical of some intertidal areas and many rivers. They differ from soft substrata in the larger particle size of the substratum (sand particles are less than 2 mm in diameter, gravel 2–44 mm, pebbles 4–64 mm, cobbles 64–256 mm, and boulders over 256 mm). Most such substrata consist of a mixture of most or all of these categories. These substrata differ from hard surfaces in being unstable, and depending on the level of this disturbance, their faunas generally less diverse than the more stable rocky reefs (e.g., Wilson 1987). Molluscs associated with these habitats include, for example, in New Zealand, the ellobiid *Marinula*, the assimineid *Suterilla*, and a small slug, *Smeagol*, in the upper part of the shore. Boulder fields in the lower parts of the shore have chitons, especially the mobile *Ischnochiton*, and various gastropods, notably trochids, which survive the frequent disturbance due to movement, with these species also found on rocky shores. In the western Mediterranean, the small truncatelloideans *Botryphallus* and *Caecum* live among gravel in the lower intertidal along with the tiny trochoidean *Skenea* (Ponder 1990c). These shelled gastropods occupy interstitial spaces among the gravel and coarse sand as do some tiny heterobranch slugs found on beaches in many parts of the world (e.g., Swedmark 1967; Challis 1969; Morse & Norenburg 1983; Poizat 1985; Arnaud et al. 1986; Morse 1987a; Poizat 1991; Seaward 2001). Similarly, small interstitial snails (mainly truncatelloideans) are found in coarse sand and gravel in fresh-water systems in many parts of the world.

9.6.6 INTO THE WATER COLUMN–SWIMMERS AND FLOATERS

The move from a benthic to a pelagic swimmer or a surface floater requires significant changes in body form and lifestyle to overcome instability and sinking, and achieve locomotion.

Passive floating has rarely evolved in molluscs – gastropod examples are the 'violet snails' *Janthina* and the 'sea lizard' *Glaucus*, although many gastropods can take advantage of the surface tension to float for short periods, typically using a mucous thread, while tiny or juvenile bivalves use a byssal thread. Some gastropod shells, notably some Eulimidae, have a hydrophobic surface, as do some larval shells (Collin 1997), enabling them to float in the surface film.

With swimming animals, the tendency to sink depends on the density of the body and body size; small animals, including larvae, can remain in the water column, especially if they have a relatively large surface area. Larger animals must either reduce the density of the body or use force to create an upward thrust (i.e., swim or jet). Swimming animals must also contend with drag and flow forces that can be counteracted by streamlining the body.

Among larger adult molluscs, swimming has evolved in members of three molluscan classes, the most adept being the cephalopods. Primitively, cephalopods used gas (mostly nitrogen) in their shell chambers to achieve buoyancy, but most living coleoids have reduced or absent shells and stay afloat by swimming or by using ammonia in their tissues to achieve a reduced density approaching that of seawater (see Chapter 17).

A few groups of gastropods have become efficient swimmers, with the heteropods and pteropods being exclusively so. Shell reduction and loss has occurred in both of these unrelated groups, with some members of both having long, almost fish-like bodies. Modified paddles aid in swimming and, when present, shells are thin and light. Some other gastropods can swim using flapping parapodia as in some aplysioids, or undulating body movements as in the large nudibranch *Hexabranchus* (see Chapter 3), but these are both benthic for much of their existence.

A few bivalves can swim for short periods, but no adult bivalves are full-time swimmers. The monomyarian pteriomorphian pectinids and limids have a few species that can swim for short periods of time by rapidly clamping their valves to expel water (Donovan et al. 2004). This is achieved by the fast contraction of the large, single adductor muscle in the middle of the valves. The most efficient bivalve swimmers are scallops of the genera *Amusium* and *Ylistrum*. These have thin, smooth shells strengthened by internal ribs. They are capable of speeds up to 1.6 m/sec and can cover distances of over 23 m in a single swimming event (Hayami 1991).

9.7   THE MAJOR DOMAINS

As summarised below in Tables 9.3 and 9.4, molluscs occupy all three major domains. Each of these is briefly discussed below.

Gastropods occupy almost every possible habitat with bivalves as a close second, co-occurring with gastropods in most habitats except for the terrestrial biomes. The remaining classes are all limited to marine biomes (Table 9.3). The distribution of major subgroups of gastropods and bivalves in habitats is given in Table 9.4.

9.7.1   MARINE HABITATS–LIVING IN THE COMFORT ZONE

As discussed above, most marine molluscs are benthic, and many are infaunal, particularly bivalves, but some gastropods and cephalopods live permanently in the water column, while a few gastropods drift on the surface of the ocean. By far the most obvious and successful pelagic species are the squid. These cephalopods have evolved into fast moving carnivores with most squid being full-time swimmers as are some octopods, although many of the latter group are benthic. Pelagic gastropods include the planktonic heteropods and pteropods.

Molluscs first evolved in a shallow-water marine environment where they probably scraped a living on hard substrata in much the same way as done today by chitons and some primitive gastropods. From these relatively humble beginnings, there evolved an amazing radiation of marine forms that occupy all marine habitats and access most of the available resources that the marine environment offers. As they radiated into different environments during their evolution, they adapted to a wide range of conditions imposed by these environments.

The sea is a relatively benign environment, but is far from uniform. For example, while light and temperature decrease with increasing depth, salinity often increases. Dissolved oxygen is at near saturation at the surface, although its solubility is decreased at higher temperatures. Thus colder surface waters hold more dissolved oxygen than warmer parts of the sea. The level of dissolved oxygen falls rapidly with depth and is often lowest at between 400–2000 m (the oxygen minimum zone), increasing a little in deeper water. Similarly, pH may fall between 2000–3000 m and then rise again. Oxygen depletion can also occur in many estuaries and bays because of the large quantities of nutrients entering these waters from the land, stimulating phytoplankton and algal growth. The decomposition of this organic material can deplete the oxygen and result in hydrogen sulphide being produced.

The intertidal zone is the interface between the marine and terrestrial environments, providing unique opportunities and posing special challenges (see next Section). The terminology used for marine zones is by no means settled. The term littoral zone is often used for just the intertidal and shallow sublittoral (as we use it here), but it has also been defined as that part of the marine environment including the shore and nearshore habitats. In the latter, sunlight (to about 1% of surface levels) can penetrate the water to the bottom enabling prolific growth of seagrasses and algae. There are two main components; the 'eulittoral zone', more commonly called the intertidal zone, and the shallow water below low tide mark that is often called the 'sublittoral'. While this latter term is commonly used just for the shallow inshore water, in some textbook definitions, it is used to include the continental shelf (i.e., to around 200 m), while in others, it includes the shore and sublittoral (shallow nearshore) habitats. We have adopted a scheme that recognises intertidal and subtidal zones and the continental shelf (Figure 9.13).

9.7.1.1   Intertidal Zone

The intertidal zone, often called the littoral (or eulittoral) zone is that part of the shore regularly covered and uncovered by tides (Figure 9.13). To live in the intertidal zone animals must be able to withstand physiological extremes (e.g., Tomanek & Helmuth 2002), including desiccation, high and low salinities, high and low temperatures, turbulence and scouring, and predation from both terrestrial and aquatic predators. Despite the rigours of this environment, a rich assemblage of molluscs have successfully adapted to it (Table 9.5).

A common strategy by bivalves and operculate gastropods (e.g., littorinids and most other caenogastropods) is to close up during low tide to prevent water loss, while limpets clamp on to the substratum. Other behavioural adaptations include hiding in crevices or under rocks, while burrowing bivalves and gastropods bury during low tide in sediment between rocks, in pools, or on soft shores. Some species on rocky shores are restricted to tide pools that remain after the tide retreats. In the middle to upper shore such pools can become

TABLE 9.3

Major Biome, Region, and Zonal Distribution of Adult Members of the Extant Molluscan Classes

	Monoplacophora	Polyplacophora	'Aplacophora'	Bivalvia	Cephalopoda	Gastropoda	Scaphopoda
Aquatic Biomes	•	•	•	•	•	•	•
Marine	•	•	•	•	•	•	•
Oceans	•	•	•	•	•	•	•
Pelagic					•	•	
Benthic	•	•	•	•	•	•	•
Intertidal		•		•	•	•	•
Vents & Seeps		•	•	•	•	•	
Coral Reefs	•	•		•	•	•	
Estuaries				•	•	•	
Fresh water				•		•	
Lakes & Ponds				•		•	
Springs & Seepages				•		•	
Aquifers						•	
Swamps & Wetlands				•		•	
Rivers & Streams				•		•	
Terrestrial Biomes						•	
Tundra						•	
Rainforest						•	
Savanna						•	
Taiga						•	
Temperate Forest						•	
Temperate Grasslands						•	
Alpine						•	
Chaparral						•	
Desert						•	

hypersaline in hot weather or hyposaline in rain, restricting the species present in them to those that can tolerate these conditions for a few hours.

Avoiding extremes of heat and cold is also necessary for the animals living in this zone. Thus the avoidance of heat loss is required in very cold conditions and adaptations that assist in heat avoidance are necessary, not only in tropical areas, but on hot summer days in temperate climates. Heat avoidance can be achieved in several ways including:

- Reduction of direct contact with the substratum (e.g., attach by sticky mucus on the edge of the shell or by mucous strands – e.g., some littorinids). This markedly contrasts with some gastropods or chitons that must remain attached by their foot, which allows heat transfer from the substratum to the animal.
- Evaporation from the foot and/or mantle for cooling – not an option for snails that seal their aperture with an operculum. This strategy is commonly adopted by middle shore taxa, but is generally not available to supralittoral fringe species (McMahon 1990).

- Modifications of shell shape and posture – reduction of surface area perpendicular to the rays of the sun. High-spired snails that lie at right angles to the rays of the sun present a smaller surface area than broad species. A tall-spired species of littorinid with its spire pointed towards the sun on a tropical shore will have a temperature significantly less than one lying horizontally (Garrity 1984; Lim 2008), although high-spired shells are probably more prone to being dislodged by wave action, suggesting that trade-offs occur. Similarly, limpets (patellogastropods and siphonariids) on the upper shore often have relatively taller shells than the lower shore limpets, while upper shore neritids tend to be more globose than lower shore species (Vermeij 1973a, 1993).
- Shell size – with species that characteristically live in the upper part of the intertidal, their shell size increases up shore, while in species more typical of the lower intertidal, it often decreases up shore (Vermeij 1972b).

TABLE 9.4

Habitats Occupied by Living Adult Members of Subgroups of the Two Major Molluscan Clades, Gastropods and Bivalves

| Taxa | Marine | | | | Fresh Water | Terrestrial | |
| | Marine Benthic | | Water Column | Estuarine | | | |
	Shallow	Deep-Sea				Damp	Arid
Bivalvia	●	●	○	•	•		
Protobranchia	•	•		◉			
Pteriomorphia	•	•		•	◉		
Palaeoheterodonta	•			•			
Heterodonta	•	•		•			
'Anomalodesmata'	•	•		•			
Gastropoda	•	•	○	•	•	•	•
Patellogastropoda	•	◉		◉			
Vetigastropoda	•	•		•			
Neritimorpha	•	•		•	•	•	
Caenogastropoda	•	•	•	•	•	•	◉
Heterobranchia	•	•	•	•	•	•	•

● – predominant, • – well represented, ◉ – rare, ○ – occasional or arguably present.

Data modified and expanded from Lindberg, D.R. et al., The Mollusca: Relationships and patterns from their first half-billion years, in Cracraft, J. and Donoghue, M.J., *Assembling the Tree of Life*, Oxford University Press, New York, pp. 252–278, 2004.

- Pale shell colour – tropical taxa tend to have pale colours. Conversely, to aid in absorbing heat, temperate and colder water intertidal taxa often have dark coloured shells.
- Shell sculpture – upper shore littorines and nerites are generally more highly sculptured than lower shore taxa. The adaptive advantages of this sculpture may include increased surface area for heat loss (Vermeij 1973a).

Throughout the world, members of the family Littorinidae make up many of the caenogastropod species encountered in the upper intertidal (eulittoral fringe) (Stephenson & Stephenson 1972). High intertidal species have several adaptations (McMahon 1990), including small size, reduction of metabolic rate, shells that reflect heat, and lethal limits greater than most marine gastropods. They can also cement the edge of the aperture to the rock surface allowing them to maintain their position while being retracted. This latter habit is also known in a few tropical Pacific species of the cerithioidean Potamididae that often live on mangrove trees (McMahon 1985; McMahon & Britton 1985; McMahon & Cleland 1990).

In littorinid snails, the contribution to reduction in temperature by way of shell colour, shape, and sculpture is small (less than 2°C), while behavioural modifications (orientation, foot removed from substratum – which enables the aperture to be sealed with the operculum) can achieve 2°C–4°C lowering of temperature (Miller & Denny 2011).

In regions with diurnal tides (a single high and low tide daily), animals living in the mid and lower littoral are hydrated with each high tide, while those in the supralittoral undergo emergence, and hence desiccation, for long periods. It is thus to be expected that the animals living in these two zones should exhibit very different adaptive strategies (McMahon 1990). Those in the middle littoral maintain activity in the air, utilise evaporative cooling, and have low thermal tolerances and few or no adaptations to reduce the absorption of radiant heat. While it might seem logical that aerial respiration rates increase with increasing height on the shore, this is not the case, as those caenogastropods exhibiting maximum gas exchange live in the mid-shore, and not the high shore (McMahon 1988). In regions with mixed tides (two unequal high and low tides a day), the patterns and periods of emergence are more complex especially in the middle littoral which may or may not be submerged by the second high tide of the day. Mixed tides create the inverse complication in the lower littoral where the possibility of two exposures per day are possible (Ricketts et al. 1985).

Tolerance of water loss by gastropods frequenting the upper intertidal reaches extremes in some supralittoral taxa. While operculate littorinids and neritids are well adapted to avoid heavy water loss, the non-operculate, salt marsh ellobiid *Melampus bidentatus* can survive losing up to almost 80% of its body water (Price 1980).

Besides macroalgae or seagrasses, the lower intertidal provides abundant food either in suspended material or deposits on

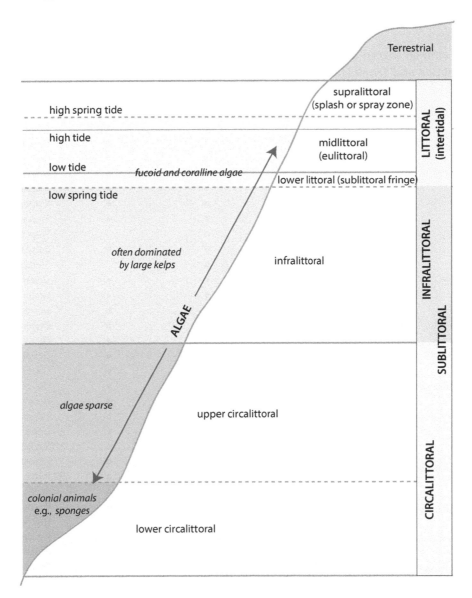

FIGURE 9.13 A scheme of intertidal and subtidal zonation. Adapted from Stephenson, T.A. and Stephenson, A., *Life Between Tidemarks on Rocky Shores*, Freeman, San Francisco, CA, 1972 with terminology from Connor, D.W. et al., *The Marine Habitat Classification for Britain and Ireland Version 04.05*, Joint Nature Conservation Committee, Peterborough, UK, 2004.

soft shores, while on hard shores, rich biofilms, an excellent food resource for grazing molluscs, are found throughout the intertidal. Most species that live in this zone, and some of those in the mid-intertidal, have substantial populations in the sublittoral.

Many (but not all) intertidal molluscs feed only when covered with water. Suspension-feeders can only filter when covered by the tide, and some bivalves have been shown to have digestive cycles tied to tidal rhythms (see Chapter 5).

It is not only the larger, more obvious species that are abundant in the lower parts of the intertidal. Many micromolluscs (mainly gastropods) occur, often in high densities and/or diversity in algal turf. For example, sampling of coralline turf in several parts of the world shows molluscan species diversity ranging from 85 in southeastern Australia to 22 in

Japan (Kelaher et al. 2007). A few tiny bivalves (e.g., *Lasaea, Neolepton*) are also commonly found in these habitats.

The animals living in the lowermost intertidal are usually only uncovered at low spring tides so are little different from those in the shallow subtidal zone. In contrast, the uppermost part of the intertidal, the supralittoral, is usually wet only by spray at high tide and by seawater only during spring tides and storms, The molluscs in this zone are semi-terrestrial, and some extend into true terrestrial habitats behind the littoral zone (some ellobiids, assimineids, truncatellids, littorinids, etc.) and face special challenges due to the near terrestrial conditions. Various true terrestrial animals (mostly arthropods) can also be present in this zone, but terrestrial molluscs are rarely present because of salt intolerance.

TABLE 9.5

Some Common Macro-Molluscan Epifaunal Families Associated with Hard Shores in Temperate and Tropical Marine Habitats

Class	Family	Temperate	Tropical	Trophic Role
Polyplacophora	Ischnochitonidae	●	○	Grazer
	Chitonidae	●	●	Grazer
	Mopaliidae	●	●	Grazer
	Acanthochitonidae	○	○	Grazer
	Cryptoplacidae	○	○	Grazer
Bivalvia	Mytilidae	●	●	Suspension
	Ostreidae	●	●	Suspension
	Anomiidae	●	●	Suspension
	Chamidae	○	●	Suspension
	Carditidae	○	○	Suspension
	Cleidotheridae	○		Suspension
Gastropoda	Patellidae	●	●	Grazer
	Lottiidae	●	●	Grazer
	Nacellidae	●	○	Grazer
	Fissurellidae	●	○	Grazer
	Trochidae	●	●	Grazer
	Turbinidae	●	●	Grazer
	Neritidae	●	●	Grazer
	Cerithiidae	○	●	Grazer
	Planaxidae	○	●	Grazer
	Batillariidae	○	○	Grazer
	Littorinidae	●	●	Grazer
	Vermetidae	○	○	Suspension
	Hipponicidae	○	○	Deposits
	Cymatiidae	○	○	Predator
	Nassariidae	○	○	Scavenger
	Buccinidae	○	○	Predator
	Columbellidae	○	●	Herbivore
	Muricidae	●	●	Predator
	Fasciolariidae	○	●	Predator
	Turbinellidae	○	●	Predator
	Costellariidae	○	●	Predator
	Mitridae	○	●	Predator
	Conidae	○	●	Predator
	Siphonariidae	●	●	Grazer
	Trimusculidae	○	○	Suspension
	Onchidiidae	○	○	Grazer
	Aplysiidae	○	○	Herbivore

○ – rare to uncommon on most shores, ● – moderately common to common on some shores, ● – abundant on many shores.

Some intertidal bivalves, notably some mytilids and ostraeids, form reef-like habitats (see Section 9.5.9.1) that provide shelter for a wide range of organisms.

9.7.1.2 Coral Reefs

Coral reefs have been likened to the rainforests of the sea. They are primarily built by scleractinian (stony) corals that serve as a base for myriads of other organisms. They grow in shallow, nutrient-poor tropical and subtropical waters and only occupy about 0.1% of the area of the oceans yet support perhaps up to a third of known marine diversity. The zooxanthellae in the coral require light for photosynthesis so the reefs are usually less than about 60 m deep, well within the subtidal zone.

Tropical cays and islands derived from coral reefs lack hard substrata other than the coral itself and 'beach rock' formed from agglutinated sand and rubble. On many tropical shores, this latter environment is important as an intertidal hard-shore habitat for molluscs.

While coral reefs are renowned for their biotic diversity, including molluscs, only a few molluscan taxa actually live on living coral, the great majority live instead in the habitats provided by dead coral. These include nestling among dead coral, under slabs of dead massive coral, or in coral borings, buried in coral rubble, and in or on coral sand.

Some of those that live associated with living coral include coralliophilines that feed on the mucus secreted by both hard and soft corals and/or on their polyps. The muricid *Drupella* feeds directly on the polyps of hard corals, and some nudibranchs, including the aeolidioideans *Phestilla* and *Cuthona* (both Tergipedidae) and the arminoidean *Pinufius*, feed on stony corals, while some tritonids and a genus of arminids eat soft corals.

Other molluscs such as *Pedum* (Pectinidae) live embedded in living coral, but suspension-feed in the normal way as does the similarly embedded large vermetid *Ceraesignum maximum* (previously *Dendropoma*). This mode of feeding is also employed by a few species of the coral-boring bivalve genus *Lithophaga* that bore into living coral (most bore dead coral). Members of the Ovulidae all live and feed on soft corals, including sea whips and sea fans (Alcyonacea).

Coral reefs are threatened by global warming, acidification, and direct human activities (see Chapter 10).

9.7.1.3 Continental Shelf

The continental shelf is the zone that borders continental land masses and a similar shelf is found around many islands. The shelf break is usually at about 140 m where the much steeper continental slope begins (Figure 9.14).

These habitats are home to rich molluscan faunas in many parts of the world, but comparative figures are rare. Some areas are particularly diverse. For example, in New Zealand, a single dredge haul on the West Norfolk Ridge, west of Cape Reinga, northern North Island, contained over 600 species (B. Marshall, Personal Communication, March 2009), mostly undescribed or new records, this representing about 13% of the known (described and undescribed) New Zealand marine fauna (4702 species and subspecies – B. Marshall, Personal communication, May 2016).

9.7.1.4 Deep-Sea

The deep sea is permanently dark and comprises the bathyal zone (1000–4000 m) (Figure 9.14) characterised by cold water (around 4°C) with the pelagic component called bathypelagic. Although the extreme conditions of abyssal depths (4000–6000 m) result in reduced faunal diversity, there are many molluscs found only in these habitats characterised by very cold water (around 2°C–3°C), great pressure, and reduced oxygen. Many parts of the abyssal ocean floor are swept by currents and the sediments are mostly ooze. Abyssal plains are flat parts of the sea floor at 3000–6000 m and can be very extensive.

These areas have low productivity and the molluscan diversity is low, decreasing with increasing distance away from continental margins into the abyssal plain (Rex 1973). Even lower diversity is encountered in the deeper (>6000 m) hadal zone found only in deep trenches, the deepest of which (the Mariana Trench in the western Pacific) extends to nearly 11,000 m.

The movement of early molluscs into deeper water probably occurred early in their evolution and has been repeated in various groups numerous times over the course of molluscan evolution. In the fossil record, and with living faunas, there is a faunal turnover from shallow water, across the continental shelf and slope, and into the deep sea and the abyssal plain. Protobranch bivalves are one group that is more diverse in the deep sea than in shallower water (Allen 1978) (Figure 9.22), and most living monoplacophorans are found in this environment.

9.7.1.4.1 Extreme Marine Environments

Some marine habitats occupied by molluscs can be considered 'extreme' in the sense that few species are adapted to withstand the challenges these habitats pose. Examples include supralittoral habitats (see Section 9.7.1.1), abyssal and hadal depths (see previous Section), and those listed below.

9.7.1.5 Hypoxic and Anoxic Habitats

Water with low dissolved oxygen levels is described as hypoxic, and when entirely depleted of oxygen is anoxic. There are several types of hypoxic and anoxic marine habitats mostly found in pockets of water that do not mix with surrounding water masses, although the reasons for this lack of mixing are varied. Hypoxic or anoxic sediments are common where oxygen is prevented from moving into the deeper sediments by relatively impervious layers of overlying fine sediment or by oxygen being depleted by microorganisms, notably bacteria, at a faster rate than it can be replenished.

Some marine basins are anoxic due to lack of mixing of deeper water with the oxygenated surface waters. Examples include the Bannock and L'Atalante Basins in the Mediterranean Sea, and the Caspian Sea and Black Sea are anoxic below 100 m and 50 m, respectively. The Permian-Triassic mass extinction events probably resulted from widespread anoxia in the oceans.

While molluscs cannot live in anoxic habitats, some are adapted to cope with hypoxic conditions. For example, the 'vampire squid' *Vampyroteuthis infernalis* is one of the few large animals that can live in the 'oxygen minimum zone'. It can cope with these conditions in part due to unusually efficient haemocyanin in the blood.

The bivalve families Solemyidae and Lucinidae are typically found in hypoxic sediments and have several adaptations, including possessing symbiotic bacteria that utilise hydrogen sulphide from the sediments (see Chapter 5). Lucinids have secondary gills formed from the mantle and long siphons that extend to the surface. Some other marine molluscs often

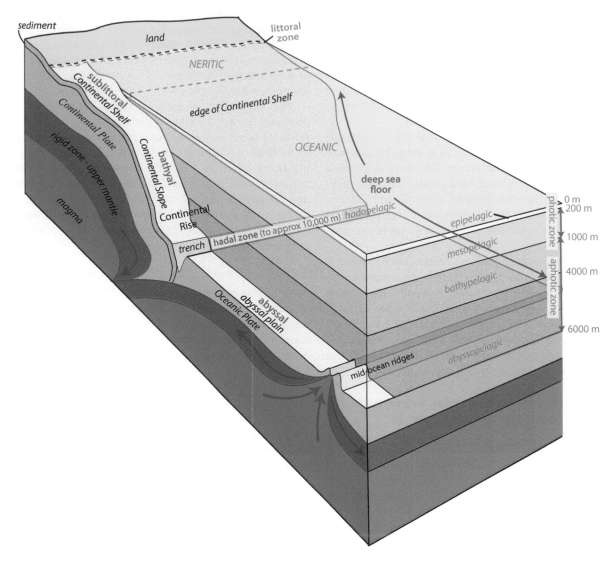

FIGURE 9.14   Benthic and oceanic zonation and sea-floor morphology. From various sources.

found in hypoxic conditions have haemoglobin in their blood, including the Phenacolepadidae (Neritimorpha) and some Arcidae (Pteriomorphia).

9.7.1.6   Hydrothermal Vents and Seeps

These relatively recently discovered structures in the deep sea are fissures through which superheated mineral rich water escapes. While this activity was known in the Red Sea in the 1960s, the discovery of vents on the East Pacific Rise by using a submersible in 1977 resulted in a flurry of research on these structures (Grassle 1987; Lutz & Kennish 1993; Desbruyères et al. 2006; Sasaki et al. 2010b; Taylor & Glover 2010). Since then, hydrothermal vents have been discovered in most areas associated with tectonically active oceanic ridges. The vent structures are diverse and, because minerals precipitate from the hot water after it is released,

they may form chimney-like vents that can reach tens of metres in height.

Although they make up only a tiny fraction of the total area occupied by the deep-sea floor, they are biologically much more productive and have vastly higher densities of organisms. Chemosynthetic bacteria form the basis of the system – either free-living or, commonly, in symbiotic relationships. The bacteria obtain their primary energy from the dissolved chemicals, notably methane or hydrogen sulphide. These abundant bacteria provide a resource base for animal grazers and suspension-feeders and have chemosynthetic relationships with others. The animal communities are surprisingly diverse, with numerous molluscs, giant tube worms, crustaceans, and fish. They are unlike the vast majority of biological communities in that they do not obtain their primary energy from the sun. Other deep-sea communities rely on the 'rain' of debris from the upper parts of the ocean so are ultimately dependent on solar-based energy.

While some species are tolerant of the hot, chemical-rich water issuing from the vents, others are not, resulting in a zonation of taxa around each vent. The species living in these hostile conditions have also evolved physiological adaptations to deal with the toxic compounds and, in taxa with symbiotic bacteria, those needed by the symbionts are channelled to them (e.g., Vetter 1985, 1991; Somero et al. 1989).

Hundreds of species of animals are associated only with hydrothermal vents, including over 200 species of gastropods and more than 20 species of bivalves (Table 9.6). Each group of hydrothermal vents (see Figure 9.15) is geographically separated and has its own characteristic fauna (e.g., Dover et al. 2002) that can be divided up into six major provinces (Bachraty et al. 2009). Many of the specialised taxa in vent systems around the world are closely related and their evolution has been driven by the dynamics of continental drift, as well as more contemporary events, and their reproductive and dispersal abilities. Data from observations on reproductive biology and genetics suggest that considerable mixing of fauna occurs between groups of vents on the same ridge (e.g., Tyler & Young 1999). There are some surprises, with, for example, the Indian Ocean vent fauna being related to the fauna of western Pacific vents (see Dover et al. 2002). The biogeographic relationships of these systems remain incomplete with new and significant discoveries undoubtedly awaiting further exploration.

Molluscan groups found in the vents and seeps include solenogasters, polyplacophorans, and several bivalve and gastropod families (see Table 9.6 and Figure 9.16). Some molluscs associated with the vents are so distinct that they have been allocated their own families and even superfamilies.

Besides the families restricted to vents and/or seeps, Sasaki et al. (2010b) listed several additional family-group taxa with a few species in seeps and vents (see Table 9.6).

Hot vents are often rich in minerals and are of interest to some mining companies. Such activities are likely to pose a major threat to these fascinating and unique communities.

In some areas, cold water escapes from deep-sea vents and these are called cold seeps. The organisms they support are similar, but different to those in hydrothermal vent communities and the composition of these depends on the chemicals (e.g., hydrogen sulphide, methane, iron, silica, manganese) that dominate the water issuing from the vents. Thus, some vents are known as sulphide or methane vents and are surrounded by dense communities of molluscs. Others form 'pools' of brine on the sea floor where the water percolates through sediments containing salt deposits, which also contain significant molluscan communities (e.g., Cary et al. 1989; MacDonald et al. 1990; Gaest et al. 2007).

Mud volcanoes found along the Mediterranean Ridge also have dense communities, mainly composed of molluscs such as lucinids and vesicomyids (Olu-Le Roy et al. 2004). These habitats all have their own unique communities and, as with the hydrothermal vents, the environmental conditions form a gradient from the source of the discharge to the perimeter of the system that is reflected in the associated communities (e.g., Barry et al. 1997). Bivalves, some of considerable size, are often the dominant animals in these communities with five families (Solemyidae, Mytilidae, Vesicomyidae, Lucinidae, and Thyasiridae) having independently evolved adaptations for life in these habitats. In particular, dense stands of the large bivalves, notably *Calyptogena* or *Bathymodiolus*, can surround these habitats, which in turn provide habitat for a range of smaller invertebrates including other molluscs (e.g., Turnipseed et al. 2003). The molluscs in these communities, as in others, also harbour symbionts and parasites (e.g., Terlizzi et al. 2004; Ward et al. 2004).

Several hydrothermal vent communities are known from the fossil record from sites as far apart as the Antarctic and Greenland and extend back to the Silurian (Kelly et al. 1995; Little et al. 2002). Kiel and Little (2006) have shown that some living seep genera have significantly longer geological ranges than generally seen in living marine molluscs, but with similar ranges to deep-sea taxa. The antiquity of the vent systems has raised the question as to whether the animals associated with them are 'living fossils' (Newman 1985), but on evidence from the fossil record and molecular divergence estimates, it appears that most modern vent taxa are relatively recent and that the kinds of animals making up vent communities have changed considerably over time (Little & Vrijenhoek 2003). The oldest communities contain brachiopods and monoplacophorans as well as bivalves and gastropods. Bivalves and gastropods dominated by the end of the Mesozoic, but the chemosymbiotic solemyids and mytilids found in modern vent communities first appeared in these systems around that time. Similarly, provannid snails appear in seep communities in the late Jurassic and vesicomyid bivalves in the early Cretaceous (Little et al. 2002). Some families found in modern vent and seep communities are yet to be identified in the fossil record.

Many of the modern vent animal taxa have apparently moved to vent habitats via biogenic substrata such as dead animal carcasses or sunken wood (e.g., Samadi et al. 2007 and references therein).

9.7.1.7 Deep-Sea Biogenic Substrata

Biogenic habitats serve as both complex substrata and as food sources for a wide variety of molluscs, including polyplacophorans, bivalves, and gastropods. Biogenic habitats are typically found in low oxygen environments, provide chemosynthetic nutritional resources, are patchy in their distribution, and are relatively short-lived in comparison with other marine habitats (Warén 2011). Many of these same characteristics are also found in hydrothermal vent and seep habitats, but vents and seeps tend to have more rapid fluctuations in temperature and oxygen, as well as

TABLE 9.6

Molluscs Recorded from Deep Sea Hydrothermal Vent and Seep Communities

Major Group	Family	Number of Genera	Number of Species
Solenogastres	Simrothiellidae	1?	~9
Polyplacophora	Ischnochitonidae	1	1
	Leptochitonidae	1	1
Gastropoda			
Patellogastropoda	Neolepetopsidae*	3	10
	Pectinodontidae	3	8
Vetigastropoda	Lepetodrilidae*	4	30
	Pyropeltidae*	1	7
	Family?	6	9
	Sutilizonidae*	3	5
	Fissurellidae	3	8
	Trochidae	1	1
	Calliostomatidae	3	3
	Neomphalidae*	5	8
	Melanodrymidae	2	5
	Turbinidae	8	13
	Peltospiridae*	9	19
	Pseudococculinidae	2	2
	Seguenziidae	1	1
	Cataegidae	1	1
	Chilodontidae	2	2
	Collonidae	1	4
	Solariellidae	1	1
Neritimorpha	Phenacolepadidae	2	5
	Neritidae	1	1
Caenogastropoda	Provannidae[a]	5	25
	Cerithiopsidae	1	1
	Elachisinidae	1	1
	Rissoidae	3	3
	'Tornidae'	1	1
	Capulidae	1	1
	Ranellidae	1	1
	Buccinidae	6	15
	Muricidae	1	3
	Volutidae	2	2
	Cancellariidae	3	3
	Conidae	7	18
Heterobranchia	Hyalogyrinidae	2	5
	Orbitestellidae	1	3
	Xylodisculidae	1	1
	Pyramidellidae	1	1
	Dendronotidae	1	1
Total gastropods	37+	101	218
Bivalvia			
Protobranchia	Solemyidae	1	1
Pteriomorphia	Mytilidae	2	12
	Pectinidae	2	2

(*Continued*)

TABLE 9.6 (*Continued*)

Molluscs Recorded from Deep Sea in Hydrothermal Vent and Seep Communities

Major Group	Family	Number of Genera	Number of Species
Heterodonta	Vesicomyidae	1	7
Total bivalves	4	6	22
Cephalopoda			
Octopoda	Cirroteuthidae	2	2
	Grimpoteuthidae	1	1
	Octopodidae	3	3
Total cephalopods	3	6	6
Grand totals	47+	116	257

Sources: Sasaki, T. et al., Gastropods from Recent hot vents and cold seeps: Systematics, diversity and life strategies, in Kiel, S., *The Vent and Seep Biota: Aspects from Microbes to Ecosystems. Topics in Geobiology*, Springer, Dordrecht, the Netherlands, 2010b; Desbruyères, D. et al., Eds., *Handbook of Deep-sea Hydrothermal Vent Fauna*, Linz, Austria, Land Oberösterreich, Biologiezentrum der Oberösterreichische Landesmuseen, pp. 169–254, 2006.

* = families restricted to vents and seeps.

a Most members of the Provannidae are vent or seep species, but a few are known from sunken wood.

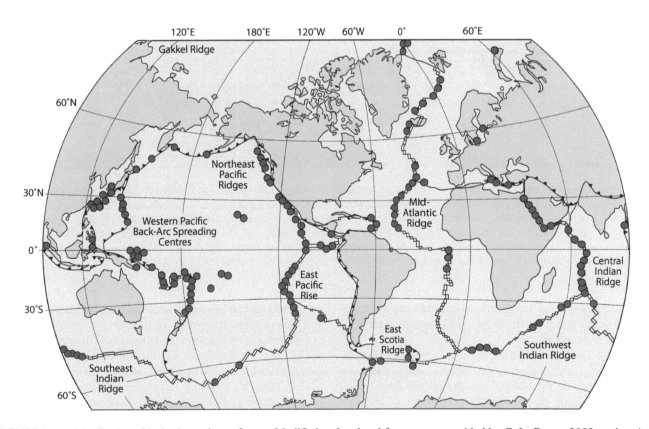

FIGURE 9.15 Distribution of hydrothermal vent faunas. Modified and updated from a map provided by C. L. Dover. 2002, and various other sources.

FIGURE 9.16 Representative hydrothermal vent taxa. (a). *Alviniconcha hessleri* (Provannidae), Western Pacific and Indian Ocean vents. (b). *Cyathermia naticoides* (Neomphalidae), East Pacific Rise: 9-21°N. (c). *Lepetodrilus atlanticus* (Lepetodrilidae), Mid-Atlantic Ridge. (d). *Bathymodiolus azoricus* (Mytilidae), Menez-Gwen vent field. (e). *Lepetodrilus fucensis* (Lepetodrilidae) among *Ridgeia picesae*, Juan de Fuca Ridge. (f). *Ifremeria nautilei* (Provannidae), Kilo Moana, Lau Back-Arc Basin. (g). *Calyptogena magnifica* (Vesicomyidae), Southern East Pacific Rise:17°S. (h). *Vulcanoctopus hydrothermalis* (Enteroctopodidae) on *Riftia* tubes, East Pacific Rise:13°N. (i). *Lepetodrilus cristatus* (Lepetodrilidae) on an active chimney, East Pacific Rise: 13°N. Images courtesy of P. Briand © Ifremer.

high concentrations of toxic substances such as hydrogen sulphide, heavy metals, petroleum, and supersaturated brine (McMullin et al. 2000).

There are two sources of biogenic substrata – those that originate within the marine realm (animal, algal, and plant elements) and those introduced into the oceans from terrestrial habitats (mostly plant components). Marine invertebrate substrata utilised by molluscs include cephalopod beaks, polychaete worm tubes, crustacean carapaces, sponges, and the shells of other molluscs (e.g., Haszprunar 1988b; Warén & Gofas 1996a). Marine vertebrate substrata include whale and fish bone, elasmobranch egg cases, and turtle shells (Haszprunar 1988b; Smith & Baco 2003; Warén 2011).

Terrestrial input is primarily wood and to a lesser extent fruits such as coconuts. Regardless of their origin, these substrata, except perhaps for the plants and algae, have little direct nutritional value for their associated molluscs. Instead, bacterial decomposition directly or indirectly provides the nutritional source. These bacteria may be on or embedded in the substratum or in the digestive system of the associated mollusc. With large, animal substrata (i.e., many marine mammals) seepage of organics from the carcass enriches the sediment around the fall. So the skeleton not only provides a substratum, but also enhances the underlying and adjacent sediments (Smith & Baco 2003; Treude et al. 2009).

Specialised molluscs occupying biogenic substrata in the ocean have been known since at least 350 BCE when 'shipworms' were first referred to by the Greeks, but it was not until Sellius (1733) that the molluscan affinities of shipworms were revealed. Shipworms were regarded as relatively shallow-water taxa, attacking floating ships, pier pilings, and dikes, and their presence in the deep sea was often interpreted as the sinking of waterlogged wood that had become infested at the surface (e.g., Smith 1885). The association of molluscan taxa with biogenic substrata was rarely highlighted in early descriptions of deep-sea taxa. Exceptions were Jeffreys (1882), who noted the association of the bivalve *Teredo* with *Cocculina* and *Idas* on waterlogged wood recovered from 940 m off Scotland, and Verrill (1884), who reported a similar group of taxa associated with wood in 3707 m off the east coast of North America and remarked on the similarity of the wood-associated fauna with that reported two years earlier by Jeffreys. In contrast, Dall (1882) first described the wood-associated taxa *Cocculina*, *Lepetella*, and *Pectinodonta*, and the elasmobranch egg case taxon *Addisonia*, but with no reference to their biogenic substrata. Thus, by the late 1880s recurrent associations of bivalve and gastropod taxa with wood substrata had been noticed, but it would take the discovery of the hydrothermal vent communities almost 100 years later to understand these habitats and place them in a chemosynthetic context.

Many of the biogenic substrata occupied today occur deep in the fossil record. Fossil cephalopod beaks are first found in the Carboniferous (Tanabe & Fukuda 1999), and polychaete worm tubes and crustacean carapaces are older, first appearing in the early Paleozoic. Sharks and bony fish have been around since the Devonian, providing, respectively, egg cases and bone. Although whale falls did not occur until the Eocene (Lindberg & Pyenson 2007), the diversification of marine reptiles (plesiosaurs, monosaurs, ichthyosaurs) provided analogous fall habitats during the Mesozoic (Kaim et al. 2008), and sea turtles began to diversify during the Cretaceous, providing not only bone, but their carapace. Wood as a biogenic substratum dates from the Middle Devonian with the appearance of progymnosperms, followed by the Carboniferous appearance of lycopod and conifer forests (DiMichele et al. 1992; Taylor et al. 2009). Angiosperms began their input during the Cretaceous, adding both hardwoods and marine plants (notably sea grasses) to the biogenic community. Lastly, although 'algae' date from the Paleozoic, the Laminariales, or kelp, with their robust stipes and holdfasts are thought to date from the Upper Cretaceous (Silberfeld et al. 2010).

Polyplacophorans associated with biogenic substrata include at least two lineages of Lepidopleurida, these being associated with deep-sea wood in the western Pacific (Sigwart et al. 2010). This group of chitons has not been reported from hydrothermal vents. In the Bivalvia, lineages clearly associated with biogenic substrata are limited to the Bathymodiolinae and Teredinidae (Taylor & Glover 2010; Warén 2011). The bathymodioline *Idas* (Mytilidae) is found in association with wood and whale falls, while the Teredinidae ('shipworms') and Xylophagaidae are wood boring and ingesting bivalves, with

the latter family mainly in deep water. Gastropods represent by far the greatest diversity associated with biogenic substrata, and the vast majority of that diversity is found in the basal lineages, especially the Vetigastropoda. In the Patellogastropoda, members of the genus *Pectinodonta* (Pectinodontidae) are wood specialists that have an associated bacterial community including polymer-degrading and denitrifying bacteria, which appear similar to those in termites and some plant-ingesting aquatic insects (Zbinden et al. 2010). The closely related *Serradonta* occurs on worm tubes that it ingests. Although the Neolepetopsidae are typically associated with cold seeps and hydrothermal vents, *Neolepetopsis nicolasensis* and *Paralepetopsis clementensis* have also been found associated with whale bones (McLean 2008). The Cocculinoidea are also wood specialists although a few taxa have been reported from whale bone, and the Bathysciadiidae occur predominantly on squid beaks (Lindberg 2008; Warén 2011). Numerous vetigastropod taxa associated with biogenic substrata are listed in Table 9.7.

Besides many vetigastropods, other orthogastropod taxa that live on wood include neritimorph phenacolepadid limpets and a few provaniids. Truncatelloidean taxa found on wood substrata include members of the Iravadiidae and Tornidae (Warén 2011) and also the lower heterobranch families Xylodisculidae (wood and marine angiosperm debris) and Hyalogyrinidae (wood and coconuts) (Warén 2011). Provannids are found associated with whale bone and wood. They have been associated with biogenic substrata since at least the Cretaceous when they occurred on plesiosaur remains (Kaim et al. 2008).

While various other caenogastropod taxa have been reported near or on wood, they are primarily carnivores and therefore would not be directly obtaining nourishment through the chemosynthetic pathways associated with biogenic substrata. Instead, they are probably attracted to the increased biomass that the biogenic substrata support.

9.7.2 ESTUARIES AND OTHER BRACKISH-WATER ENVIRONMENTS

Estuaries are characterised by both low and fluctuating salinities depending on the volume of fresh water entering them, while other coastal habitats, such as lagoons, can have drastically variable salinity changes, ranging from near fresh water after flooding, to highly saline following periods of little or no fresh water input. The frequency of these fluctuations may be diurnal, lunar, seasonal, annual, or even decadal in areas that experience long-term droughts.

Estuarine habitats can have a range of substrata available to molluscs ranging from fine, soft mud to hard shores, and include seagrass, mangrove, and saltmarsh communities.

These brackish-water areas contain a variety of molluscs with differing abilities to adapt to fluctuating salinities – i.e., the ability to osmoregulate and volume regulate (e.g., Gainey & Greenberg 1977) (see Chapter 6). Thus, most estuaries have gradients in species diversity with more taxa at the saline coastal end than at the fresh water end. Also, taxon diversity is lower

TABLE 9.7

Vetigastropod Taxa Associated with Biogenic Substrata

Cocculinoidea	Cocculinidae	Wood, Algal Holdfasts, and Whale Bone
Lepetelloidea	Bathysciadiidae	Cephalopod beaks
	Lepetellidae	Polychaete worm tubes
	Addisoniidae	Elasmobranch egg cases
	Bathyphytophilidae	Marine angiosperms
	Caymanabyssiidae	Wood
	Cocculinellidae	Fish bone
	Osteopeltidae	Whale bone
	Pseudococculinidae	Wood
	Tentacoculus (Pseudococculinidae)	Algal holdfasts, crab carapaces
	Pyropeltidae	Whale bones
Seguenzioidea	Choristellidae	Elasmobranch egg cases
	Cataegidae	Wood
	Chilodontidae	Wood
Fissurelloidea	Fissurellidae	Wood
Scissurelloidea	*Anatoma, Scissurella*	Wood
Lepetodriloidea	Lepetodrilidae	Wood
Neomphaloidea	*Leptogyra, Xyleptogyra, &* *Leptogyropsis*	Wood
Trochoidea	Trochidae	Wood
	'Skeneidae'	Wood
	Turbinidae	Wood, whale bone

Sources: Haszprunar, G., Superorder Cocculiniformia, in Beesley, P.L. et al., *Mollusca: The Southern Synthesis. Part B. Fauna of Australia*, CSIRO Publishing, Melbourne, Australia, pp. 653–664, 1998; Marshall, B.A., *New Zeal. J. Zool.*, 12, 505–546, 1985; Sasaki, T. et al., Gastropods from Recent hot vents and cold seeps: Systematics, diversity and life strategies, in Kiel, S., *The Vent and Seep Biota: Aspects from Microbes to Ecosystems. Topics in Geobiology*, Springer, Dordrecht, the Netherlands, pp. 169–254, 2010b; Warén, A., Molluscs on biogenic substrates, in Bouchet, P. et al., *The Natural History of Santo*, Vol. 70, Muséum national d'Histoire naturelle, Paris, France, pp. 438–448, 2011.

because many estuarine areas are unstable, often small in area and volume, and are short-lived geologically (Croghan 1983). Richer estuarine faunas are found in areas with relatively stable conditions over long periods of time (e.g., tropics, temperate Australia) than in areas affected by Pleistocene glaciation, or on narrow, tectonically active continental margins (e.g. west coast of the Americas).

9.7.2.1 Inland Seas

Some inland seas such as the Baltic, Caspian, Aral, and White Seas in Eurasia are hyposaline (<34‰) – i.e., with low levels of salinity, and while similar to the lower levels experienced in estuaries, they are much more constant. Their faunas are generally depauperate (e.g., Filippov & Riedel 2009), with only those marine taxa that can tolerate low salinities or are specialised for low salinity, as are many true estuarine taxa.

9.7.3 Hypersaline Environments

In some other enclosed water bodies, fresh water input does not keep pace with evaporation and the result is hypersaline (>34‰) conditions. While some enclosed or semi-enclosed coastal habitats become hypersaline, most inland hypersaline habitats are classed as salt lakes (see below). These habitats usually have abundant microorganisms and cyanobacterial mats so food is not a limiting factor.

Relatively few molluscs (some bivalves and a few gastropods) are halophilic and thus able to cope with hypersaline conditions (40‰ to 80‰), and some can even thrive in hypersaline conditions, sometimes in high densities, but none live in brine (i.e., at salinities >80‰) (Hickman 2003a).

9.7.3.1 Hypersaline Coastal Lagoons

Hypersaline coastal lagoons are found in parts of the world that have relatively low annual rainfall or where fresh water input from rivers or streams is low and evaporation is high. Animals in these environments must be able to cope with osmotic stress and mollusc diversity is usually low (e.g., Bayly 1972). The Laguna Madre in coastal Texas and the Sivash between the Black and Azov Seas in Russia are well-known examples of hypersaline lagoons, although salinities in the former range from less than that of seawater to 86‰ and in the latter 36‰ to 132‰ (Hickman 2003a). Both are dominated by bivalves. The Laguna Madre contains several bivalve species with the mactrid *Mulinia lateralis* and the venerid *Anomalocardia auberiana* dominating, with the latter, and the cyrenid *Polymesoda floridana*, capable of surviving salinities of up to 80‰ (Britton & Morton 1989).

If there are occasional or seasonal fresh water inputs, these lagoons can be dynamic systems. For example, the Coorong is a large, mostly hypersaline embayment at the mouth of the Murray River in South Australia and becomes increasingly saline (as much as 80‰ to 100‰) in times of low river flow (sometimes over several years), but when river flow is high, these same areas are 35‰ to 70‰, resulting in lower faunal diversity (Geddes & Butler 1984). Diversity is low with the bivalves *Notospisula trigonella* (Mactridae) and *Arthritica semen* (Lasaeidae) and the gastropods *Ascorhis tasmanicus* (Tateidae) and *Salinator fragilis* (Amphibolidae) dominating.

The hypersaline waters of Lharidon Bight and Hamelin Pool, Shark Bay, northwest Australia, have huge numbers of a small cockle (*Fragum erugatum*) (Cardiidae). Piles of shells wash ashore and comprise the majority of the beach sediment (Berry & Playford 1997; Morton 2000c).

In the hypersaline Keta Lagoon, a large coastal lagoon in Ghana, Africa, again the community is dominated by bivalves – a species of *Tellina* and two species of the venerid *Tivela* (Lamptey & Armah 2008). The hypersaline Bardawil Lagoon in Egypt, near the Suez Canal, connects to the Mediterranean Sea and contains only six molluscan species (Barash & Danin 1982).

Some species that can tolerate hypersaline environments are also found in estuaries – for example, the cockle *Cerastoderma glaucum* is widely distributed in Europe and North Africa in estuaries, coastal lagoons, and inland seas, but it can only tolerate salinities up to 40‰ (Barnes 1994). The European hydrobiid gastropods *Peringia ulvae* and *Ecrobia ventrosa* can live in both estuarine and hypersaline conditions.

Bivalves have been shown to cope with the osmotic stress of hypersalinity by releasing intracellular free amino acids into the blood (Pierce & Greenberg 1972). *Fragum erugatum* maintains symbiotic zooxanthellae, perhaps necessary for its increased energy demands from osmoregulation (Morton 2000c).

The adoption of mutualisms by animals as a way of coping in these harsh environments prompted one malacologist to suggest that they could provide clues about life elsewhere in the universe (Hickman 2003a).

9.7.3.2 Inland Hypersaline Salt Lakes

The Dead Sea (between Israel and Jordan), Great Salt Lake in Utah, Mono Lake in California, and Lake Eyre in Australia are famous examples of large hypersaline lakes, but they occur in most continents and can be small. An overview of world salt lakes is provided by Williams (1986).

Hypersaline lakes rarely contain molluscs, but a few have become adapted to these harsh conditions. For example, some species of pomatiopsids can tolerate hypersaline conditions with the Australian *Coxiella* living exclusively in those habitats (Williams & Mellor 1991), while species of the southern

African *Tomichia* live in a variety of habitats including salt lakes (Davis 1981). The potamidid *Batillariella estuarina* is found in a saline cut off arm of Lake Eliza in South Australia (Bayly & Williams 1966). While no other potamidids inhabit salt lakes today, a few did in the past (Plaziat 1993).

9.7.3.3 Deep-Sea Brine Pools and Lakes

Most deep-sea seeps are rich in sulphides or methane, but some are brine as a result of water percolating through salt deposits in the sediments. These undersea brine 'pools', 'lakes', or 'pockmarks' are a feature of many cold seeps in the Atlantic, Mediterranean, and Caribbean, and they have characteristic faunas, notably of bivalves (see Section 9.7.1.6).

The L'Atalante, Urania, and Discovery Basins are anoxic brine lakes at the bottom of the Mediterranean Sea over 3000 m below the surface and only microbes are known to inhabit them.

9.7.4 Fresh Water

Most fresh-water habitats are connected to the oceans, but the majority of marine phyla never moved into fresh water. In molluscs, this was achieved by only a few groups of bivalves and some gastropods that made the necessary physiological changes. The most important of these are adaptations for osmotic changes because the body fluids of marine and fresh-water animals have a similar ionic composition to that of seawater (e.g., Deaton 1981; Croghan 1983; Miller & Labandeira 2002) (see Chapter 6). The resulting osmotic gradient between the low concentration in the external medium and the high internal body concentration results in water diffusing into the body across all the permeable surfaces. Not all fresh-water snails have evolved from marine taxa as the fresh-water Hygrophila may have evolved from terrestrial ancestors. Their modified mantle cavity can function as a lung, although some have a 'pseudobranch' for extracting oxygen from water (see Chapter 4).

Strategies adopted by fresh-water molluscs are generally energetically expensive and include the excretion of large quantities of dilute urine and limiting the permeability of surfaces, although the latter solution is not used by most molluscs. Associated modifications include being able to reabsorb the needed ions and salts from the urine before it is excreted (see Chapter 6). Other adaptations include modifications to reproductive strategies with direct development becoming the norm (although not universal). Thus, relatively few large eggs are produced and laid attache d to aquatic plants or hard substrata, or are brooded (see Chapter 8). Direct development brings its own limitations, especially in reducing dispersal opportunities. Fresh-water mussels (Unionoidea) have overcome dispersal issues by developing a larva that is temporarily parasitic on fish, while some fresh-water neritids retain a swimming veliger stage and some carry hitchhiking juveniles upstream (Schneider & Frost 1986).

As in the sea, water is able to dissolve many substances including chemicals, minerals, and nutrients. These dissolved substances change the properties of fresh water with changes being reflected in measures such as pH and 'hardness'. The pH indicates the acidity of water – it is a measure of the abundance of hydrogen ions (H^+) on a logarithmic scale, with acidic waters less than pH 7 and alkaline greater than seven. Most natural fresh water is typically slightly acidic to slightly basic, with pH around 6.5–8.5.

9.7.4.1 Ancient Lakes

Ancient lakes are large, old lakes found on several continents and were mostly formed through tectonic activity. They have had a continuous history for hundreds of thousands of years or, depending on the definition used, for a million years or more. Most other lakes are relatively short-lived. Ancient lakes have accumulated and formed their biota over many thousands or even millions of years and their faunas can be diverse. The evolution of 'species flocks' in these lakes has long given them recognition as 'hot spots' of molluscan diversity (e.g., Boss 1978; Michel 1994; Strong et al. 2008). Many studies on the mechanisms that drive this speciation have been carried out in molluscs and other taxa. Background information, reviews, and overviews of studies can be found in Rossiter and Kawanabe (2000) and Wilke et al. (2008). There is evidence that both allopatric (or peripatric) and parapatric speciation occurs in these lakes, with most allopatric speciation probably being explained by so-called microallopatric separation due to ecological differences (e.g., Glaubrecht 2011).

Ancient lakes containing significant endemic molluscan faunas are Lakes Baikal, Ohrid, Tanganyika, and some lakes in Sulawesi. Certain families of cerithioideans and truncatelloideans are the most speciose in ancient lakes.

Lake Baikal in Siberia is the largest (31,494 km^2), deepest (1637 m), and oldest (>25 million years) of the ancient lakes, being the largest fresh-water lake in the world by volume. Most of its endemic gastropods are truncatelloideans (40+ species, most are Amnicolidae belonging to the endemic subfamily Baicaliinae), but there are also 12 hygrophilans (Planorbidae and Acroloxidae) (Boss 1978). There are 35 species of bivalve recorded from this lake, 20 endemic (Prozorova & Bogatov 2006; Slugina 2006), including five species of unionids, but most are sphaeriids.

East African Great Lakes (see below) are famous for their fish and mollusc faunas and many evolutionary studies have been conducted there. The faunas of some of the other lakes are also receiving some attention. For example, the mollusc faunas of the European sister lakes Ohrid and Prespa of the Balkans and the two central lake systems of Sulawesi (Indonesia) have recently been studied (Albrecht et al. 2012; von Rintelen et al. 2012).

The Great Rift Valley of Africa extends for about 6000 km and contains several ancient lakes including the second largest fresh-water lake in the world by volume, Lake Tanganyika, which is divided between four countries. It is about 32,900 km^2 in area, reaches 1470 m in depth, and came into existence about 9–12 million years ago. This lake is famous for its fish fauna with 176 endemic species. There are also 60 gastropod species, with perhaps 50 endemics (Brown 1994; Glaubrecht 2008), and these are dominated by a remarkable radiation of cerithioideans. Many of the cerithioideans superficially resemble marine taxa, and the term 'thalassoid' was coined for these. They comprise the family Paludomidae and a few species of the small-sized Syrnolopsidae once grouped with truncatelloideans. The bivalves of this lake comprise 15 species (Coulter 1991), including three unionoidean families.

There are several other ancient lakes in the Great Rift Valley of Africa, including Lake Victoria, the second largest lake in the world by surface area (25,600 km^2), but it is only up to 83 m deep. It was formed about 400,000 years ago, and contains endemic fish and molluscs, but is now badly damaged environmentally. Lake Malawi (29,600 km^2, 560 m deep and about 4.5 million years old) has a large number of endemic fish and thiarids – mostly *Melanoides* (Genner et al. 2007) and ampullariid (*Lanistes*) snails. Other huge lakes in the Rift Valley are Lake Albert, Lake Kivu, and Lake Turkana, the latter being the largest alkaline lake in the world.

The highest and one of the largest (8562 km^2) is Lake Titicaca in the Andes, straddling Peru and Bolivia. It is deep (284 m), over a million years old, and has 30 gastropods, many of which are endemic, and all are small truncatelloidean snails (Cochliopidae) (Kroll et al. 2012). Bivalves are represented by three species of sphaeriid, one of them endemic.

Lake Ohrid, in the Balkans of Europe is about 288 m deep, 2–5 million years old, and this 358 km^2 lake is the most biodiverse lake in the world by size. It is rich in gastropods with 68 species, of which 73.5% are endemic (Hauffe et al. 2011), these comprising mostly hydrobiids (s.l.), planorbids, and valvatids. There are 13 bivalve species known from the lake, but only one sphaeriid is endemic (Albrecht & Wilke 2008). The nearby sister lake, Lake Prespa (273 km^2, 54 m deep), is at a higher altitude and the source of some of the water of Lake Ohrid; it also contains a possible endemic bivalve and several gastropods (e.g., Albrecht et al. 2012).

Lake Biwa, Honshū, Japan, is about 670 km^2 in extent and about 103 m deep, and is the largest lake in Japan. It contains about eight endemic gastropods, including a viviparid, four pleurocerids, a lymnaeid, and two planorbids (Boss 1978). There are 16 species of bivalves (unionids, cyrenids, and sphaeriids), nine of which are endemic (Slugina 2006).

Ancient lakes in Sulawesi, Lake Poso (450 m deep, 323.2 km^2) and the Malili lakes (Danau Matano, Mahalona and Towuti) have rich mollusc faunas, particularly radiations of cerithioidean snails (Pachychilidae) (Glaubrecht & Rintelen 2008), tateids (Ponder & Haase 2005; Haase & Bouchet 2006; Zielske et al. 2011), and planorbids. Lake Poso also contains an extraordinary cemented cyrenid bivalve (Bogan & Bouchet 1998) and another endemic cyrenid, and two other endemic cyrenids occur in the Malili lakes (Glaubrecht et al. 2003).

Lake Lanao in Mindanao, the Philippines, is another ancient lake with a surface area of 340 km^2, and it reaches 112 m in depth. The molluscan fauna of this lake is not well known. Other putative ancient lakes include those in Myanmar (Lake Inlé) with ten endemic gastropods (Boss 1978), Greece (Lake Pamvotis – with two endemic molluscs) (Frogley & Preece 2004), and Asia Minor (Lake Egirdir) with only one endemic gastropod (Wilke et al. 2007). Some other old lakes, such as the Californian Mono Lake, are saline.

Relatively long-lived lakes such as Great Lake and the sister lakes Crescent and Sorrell on the central plateau in Tasmania also contain a few endemic molluscs (Fulton 1983).

Many of these ancient lakes are under pressure from human activities (see Chapter 10), and the impacts of deliberately introduced fish or invasive species including molluscs.

Not only are the living faunas of lakes important in evolutionary studies, a very comprehensive record of fossil gastropods from the Miocene Lake Steinheim in southwest Germany (Janz 1999; Rasser & Covich 2014), the Late Miocene Lake Pannon in eastern Europe (Muller et al. 1999), and the Plio-Pleistocene Lake Turkana in northern Kenya (Williamson 1981) have also inspired several evolutionary studies.

9.7.4.2 Springs and Spring-Fed Streams

Spring-fed systems comprising permanent springs, streams, and seepages in some parts of the world contain impressive radiations of mostly allopatric species where there have been continuous environments for long periods of time. Some of the better-known examples include those below.

The Balkans, including much of the Balkan Peninsula, is a large area in southeastern Europe that lies to the east of Italy. It is comprised of several countries including Albania, Bosnia and Herzegovina, Croatia, North Macedonia, Montenegro, and Serbia, but does not include Greece. The largely mountainous terrain contains numerous spring habitats that contain a rich fauna of 'hydrobiid' gastropods, many of which are only known from springs and typically have very narrow distributions, often a single spring (e.g., Radoman 1983; Falniowski et al. 2012). Greece also has a rich spring fauna, and smaller spring faunas are found in other unglaciated parts of southern Europe including Italy and Spain.

There are important areas of springs, seepages, and streams in North America, including a large percentage of the USA fauna found in the southeast of the country (Neves et al. 1997), including Florida (Thompson 1968). There are important truncatelloidean snail faunas in arid-zone springs, notably those of the Great Basin Region (Hershler 1998, 1999), including Ash Meadows and Death Valley springs (Hershler & Sada 1987; Hershler et al. 1999; Hershler & Liu 2008). Another important arid-zone spring radiation of mainly truncatelloidean snails is found in Coahuila, Mexico (Taylor 1966; Hershler 1985).

Significant radiations of truncatelloidean snails occur in spring-fed streams in southeastern Australia, including Tasmania (Ponder et al. 1993, 1994, 2005; Miller et al. 1999; Clark et al. 2003) and in arid-zone springs in northern South Australia (Ponder et al. 1989) and western Queensland (Ponder & Clark 1990). Other significant stream and spring radiations have been described from New Zealand (Haase 2008) and New Caledonia (Haase & Bouchet 1998; Zielske & Haase 2015).

All these small stream and spring habitats are fragile and very susceptible to changes in ground water levels and the endemic fauna they contain are highly vulnerable (e.g., Ponder & Colgan 2002).

9.7.5 TERRESTRIAL

Relatively few groups of organisms have transitioned from the aquatic environment to land (Vermeij & Dudley 2000); this is also true within the Mollusca where only a handful of gastropod taxa have managed it. Terrestrial lineages evolved independently in different groups and via different routes (Little 1983), with many frequenting wet tropical forests, while, at the other extreme, some can withstand arid conditions. By far the most successful group are the stylommatophorans, which have developed numerous adaptations to terrestrial life. The success of this group is reflected in its species-level diversity with over 25,000 species. They have been reported to have a fossil record back to the Carboniferous (Solem & Yochelson 1979), but those fossils were probably cyclophoroideans and the first undoubted stylommatophorans did not appear until the Jurassic-Cretaceous boundary (Bandel 1997, 2000). Both the terrestrial caenogastropod cyclophoroideans and terrestrial neritimorphs are among basal living members of their respective clades (e.g., Ponder & Lindberg 1997).

Several substantial modifications are needed for a gastropod to survive in a terrestrial environment and these typically include:

- Protection against desiccation and/or behavioural changes to avoid desiccation (see below and Chapter 6).
- Modifications to the respiratory and circulatory system to enable respiration in air (see Chapter 4).
- Modification to sense organs so they function in air (see Chapter 7).
- Locomotory modifications (see Chapter 3).
- Adoption of altered or new diets and modification of digestive processes (see Chapter 5).
- Reproductive modifications to protect developing embryos from desiccation (see Chapter 8).

Many of the changes presumably evolved before the adoption of a truly terrestrial life. The transition was a gradual one as living in marginal aquatic environments, where an amphibious lifestyle is essential, necessitates several exaptations to terrestrial life (Little 1983). Thus, routes to the land from the aquatic environment were varied; some probably from the sea via river estuaries into fresh-water bodies and then to land, while others transitioned directly to land from the sea via mangrove or saltmarsh in estuaries or from the supralittoral zone on marine shores. Living in the supralittoral requires adaptations to these harsh environments in both shell morphology (Vermeij 1973a) and physiology. Several modern groups that frequent the upper intertidal and supralittoral have semi-terrestrial or terrestrial members including neritoideans, the caenogastropod Littorinoidea, the truncatelloidean families Truncatellidae, Assimineidae and Pomatiopsidae, and the eupulmonate Ellobiidae. A group showing multiple, relatively recent, terrestrial invasions is the Caribbean Truncatellidae (Rosenberg 1996), a group typically found in the supralittoral.

While the adaptations listed above are all necessary for successful terrestrial life, some are more important than others. It would seem logical that mechanisms to avoid desiccation or, alternatively, develop tolerance of water loss, were essential, although certain terrestrial snails, including some cyclophorids, assimineids and many stylommatophorans, are confined to permanently damp habitats and cannot survive in dry conditions. Others are highly tolerant of drying as are some supralittoral species. Although the eggs of most terrestrial snails and slugs are poorly resistant to desiccation (e.g., Bayne 1968b; Riddle 1983), at least some adult stylommatophoran slugs such as *Deroceras reticulatum*, can tolerate up to about 80% weight loss (i.e., water loss) (Bayne 1969). Some land snails are so well adapted to arid conditions, they can survive in the driest deserts where they may only be active for a few hours a year when the rare rain falls.

While such mechanisms to avoid water loss are important, reducing the loss of water and ions through excretion (as urine) is also vital. Thus, resorption from the kidney and reduction in urine output are essential modifications achieved in various ways including, convergently in several groups, the reduction or loss of podocytes on the auricular wall with replacement by other means of filtration such as developing extracellular perforations in the heart (see Chapter 6).

Another necessary adaptation is internal fertilisation – no vetigastropod is terrestrial, but all those gastropod groups with a well-developed reproductive system involving internal fertilisation (Neritimorpha, Caenogastropoda, Heterobranchia) (see Chapter 8) have given rise to terrestrial taxa. This reproductive mode includes associated sensory and behavioural (locating mates, mating behaviour) modifications, and while eggs laid in aquatic settings have water, calcium, and trace elements directly available to the developing embryos, most land snail eggs must contain all the necessary nutrients and trace elements needed for development (that is they are *cleidoic*) (e.g., Tompa 1980; Baur 1994b).

In terrestrial gastropods, the dorsal wall of the mantle cavity has lost the ctenidium and is transformed into a lung supplied with a network of capillary blood vessels. In terrestrial eupulmonates, water loss is minimised by the mantle cavity opening being reduced to a small breathing aperture (pneumostome) controlled by a muscular sphincter.

Losing water through evaporation is a critical problem for terrestrial gastropods and must be minimised. In caenogastropods and neritimorphs, the shell aperture can be sealed with the operculum, but some terrestrial cyclophoroideans and neritimorphs have developed special breathing tubes that enable air to pass in and out of the aperture via very narrow openings (Figure 9.17). Some of these tubes are closed externally, as in *Alycaeus conformis* (Alycaeidae), except for many microtunnels that open to it from the shell surface (Páll-Gergely et al. 2016). Terrestrial stylommatophorans do not have an operculum, but instead seal the shell aperture using various adaptations, including with hardened mucus

that in some taxa is calcified to form a shelly operculum-like structure. Others seal the shell aperture with dried mucus glued against a hard surface (see Chapter 20).

A major source of water loss is in the production of the mucus secreted during locomotion and for the maintenance of body moisture, especially with slugs, although this is minimised because the slime of terrestrial molluscs is generally hygroscopic (attracts water rather than loses it).

The activity of terrestrial snails and slugs may increase/decrease with a decrease in temperature, humidity/moisture, time of day, food availability, and other factors and some of this activity is endogenous (see Cook 2001 for review). Behavioural adaptations include using suitable hiding places during dry periods. Many are confined to permanently damp habitats such as beneath litter or rotting logs because they are relatively poorly adapted to withstand drying. Small snails, especially those with flattened shells, and slugs can hide in narrow spaces of rock or bark where moisture is maintained and there is good protection from predators. Some burrow into the soil to aestivate, while others climb plant stems, tree trunks, or fence posts and aestivate. The climbing strategy is related to the fact that temperature decreases above the ground. This is counteracted by a general increase in air movement that will increase evaporation, but because evaporation depends mainly on the temperature of the evaporating surface, both evaporation and daytime temperature tend to decrease with increasing height from the ground in conditions where there is a relatively sparse canopy (Jaremovic & Rollo 1979). Similar climbing behaviour is seen in the salt marsh *Littorina irrorata* for the same reasons (McBride et al. 1989). The shell aperture in these climbing taxa is closed by a membrane formed from dried mucus that can also firmly attach the shell to the substratum.

Body size is critical as smaller species and juveniles are more prone to desiccation because of their relatively larger surface area. Thus juveniles sometimes have different behaviours from adults and/or live in different microhabitats.

The ability to crawl into suitable narrow hiding places and also maintain a reasonable body size may well have been a factor in what otherwise might appear to be an enigmatic situation – the shell reduction and even loss of an external shell seen in several eupulmonate lineages. Another significant advantage includes escaping the dependence on calcium for shell building, enabling slugs to live in calcium-poor areas that would be a challenge for snails.

The osphradium is lost from the mantle cavity as it is not adapted to respond to airborne chemical signals, these functions being taken over by the tentacles in stylommatophorans, which have several sensory adaptations that make them well suited to life on land. These include olfactory organs in their eye-bearing tentacles, while the anterior (oral) tentacles enhance the senses of touch and smell. All four tentacles are used for the orientation of the snail and there are taste receptors around the mouth used for slime trail following, etc. (see Chapter 7 for more information).

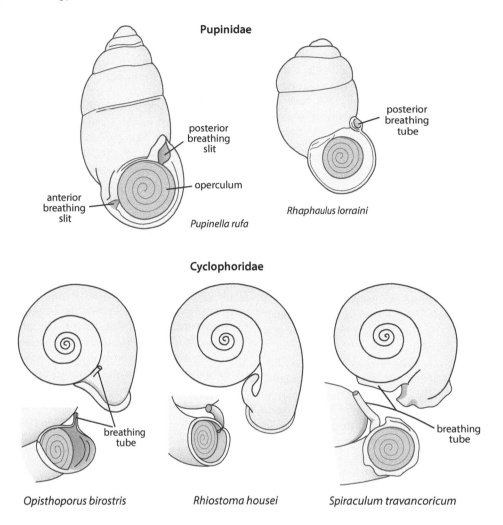

Pupinidae

posterior breathing slit

operculum

anterior breathing slit

Pupinella rufa

posterior breathing tube

Rhaphaulus lorraini

Cyclophoridae

breathing tube

breathing tube

Opisthoporus birostris   *Rhiostoma housei*   *Spiraculum travancoricum*

FIGURE 9.17   Examples of apertural breathing slits and tubes in the terrestrial operculate Pupinidae and Cyclophoridae (Cyclophoroidea). Top row drawn from photographs, bottom row redrawn and modified from Cooke, A.H. et al., *Molluscs, Brachiopods (Recent) and Brachiopods (Fossil)*, MacMillan, London, UK, 1927.

Terrestrial caenogastropods and neritimorphs lack modified eyes and tentacles.

Like their marine relatives, terrestrial caenogastropods and neritimorphs have separate sexes and internal fertilisation. As noted above, it is the more primitive clades in both these groups that have successfully invaded the land and fresh water (see Chapters 19 and 20). 'Pulmonates' differ mainly in being hermaphrodites. Both lay eggs in capsules that sustain and protect the embryos during direct development. Some are brooders, as for example, many clausiliids, subulinids, and microcystids and some oreohelicids.

Being an hermaphrodite doubles the chance of encountering a mating partner, a useful circumstance in unfavourable environmental conditions (Heath 1977). An additional advantage is that if 'pulmonate' snails do not find a mate they can self-fertilise (see Chapter 8). This latter ability enhances the success of colonisation following dispersal.

Rich terrestrial gastropod faunas associated with limestone outcrops are known from many parts of the world, with these often being entirely or largely endemic. The richest

limestone associated molluscan faunas known are in Malaysia (e.g., Clements et al. 2008b).

Radiations of terrestrial molluscs on oceanic islands, often involving species in a few genera or family-group taxa, can lead to impressive species numbers considering the relatively small size of these habitats (e.g., Solem 1984; Cowie 1996; Cameron et al. 2013b; Triantis et al. 2015, 2016). Some examples include Hawaii (752 species) (Solem 1990; Cowie 1995), the Canaries (227 species), the Maderian archipelago (Alonso et al. 2006; Cook 2008) (161 species), Samoa (67 species), Society Islands (161 species), Mascarene (162 species), and the Galapagos (97 species) (Parent & Crespi 2006, 2009). Because many of these taxa evolved in the absence of the range of predators and competitors that would normally be present in continental habitats, they are prone to extinction when human induced changes occur (see Chapter 10), with some famous examples being the achatinellids in Hawaii (Hadfield 1994) and the Partulidae in a number of islands in the Pacific (Cowie 1992b).

9.8 MOLLUSCAN DIVERSITY

While there are many studies on molluscan species diversity, these are usually regional in focus, or concerned with change-over in faunas through time. There are no published comprehensive studies of extant molluscan diversity that cover all taxa and all regions. The following section highlights some of the general findings and trends.

9.8.1 Marine Molluscan Diversity

Marine molluscan species diversity is generally highest in nearshore habitats and on the continental shelf and reduces with increasing depth on the continental slope and beyond. As with many other organisms, marine molluscs have their highest diversity in the tropical western Pacific and decrease in diversity towards the poles (Figures 9.18 and 9.19). Results from only one comprehensive sampling effort of tropical western Pacific molluscan diversity have been published (Bouchet et al. 2002), where almost 3000 species were found while intensely sampling 42 stations in a 295 km² area of a coral reef habitat in New Caledonia.

Diversity in the tropical Indo-West Pacific generally decreases away from continental margins, as for example, with cypraeid taxa (Figure 9.20).

Molluscs are most abundant and diverse in shallow-water marine ecosystems. In the tropics, they numerically comprise 15%–40% of benthic macroinvertebrates and are exceeded only by crustaceans and polychaetes (Longhurst & Pauly 1987). There are only a few reasonably reliable estimates of species numbers in marine molluscan faunas around the world – some are listed in Table 9.8.

Overall diversity also decreases with depth, as for example, shown in eastern North Atlantic neogastropods (Figure 9.21) and some groups of bivalves, although protobranch bivalves and anomalodesmatans are more diverse in deeper water (Figure 9.22).

While biogeographic provinces are well known in coastal marine waters (e.g., Spalding et al. 2007), the deep-sea benthos is often regarded as rather uniform. There are, however, biogeographic provinces recognisable in the lower bathyal and abyssal benthos (>800 m) that have characteristic faunal and physical (temperature, salinity, etc.) characteristics (Watling et al. 2013).

9.8.2 Non-Marine Molluscan Diversity

Non-marine molluscs, although being relatively restricted in terms of the major groups involved, are specifically diverse largely due to their fragmented habitats. Human impacts on these habitats, combined with their often restricted distributions has resulted in many being threatened, endangered, or extinct (see Chapter 10).

Some non-marine groups are the most basal of the extant taxa in a clade, these being in the Caenogastropoda (Cyclophoroidea, Viviparoidea, Ampullarioidea) (see Chapter 19) and the hetero-conch bivalves (Unionoidea) (see Chapter 15).

9.8.2.1 Fresh-Water Molluscs

A few groups of bivalves and several groups of gastropods live in fresh water. Many are found in surface waters, but others are subterranean, living in streams in caves and other habitats such as interstitial ground water or aquifers (stygobionts) (e.g., Hershler & Holsinger 1990).

Although only a few families of bivalves and gastropods have successfully invaded fresh water, some have undergone significant radiations. These include all members of the bivalve superfamilies Unionoidea (the fresh-water mussels or clams), Sphaerioidea (the pea clams) and some members of the families Cyrenidae (including the genus *Corbicula* – the 'basket clam'), and Dreissenidae (the 'zebra mussel' and relatives). The significant gastropod families include the heterobranch Valvatidae, the fresh-water hygrophilans such as Chilinidae, Lymnaeidae, Physidae, and Planorbidae, and four important groups of caeno-gastropods, the truncatelloidean families related to Hydrobiidae (including Amnicolidae and Tateidae) and Pomatiopsidae, the cerithioidean Thiaridae, Pachychilidae, Semisulcospiridae, Pleuroceridae, Melanopsidae, and Paludomidae, and the archi-taenioglossan Viviparidae and Ampullariidae. Several other families of gastropods, and a few of bivalves, also have fresh-water representatives in some, mainly tropical, parts of the world (Strong et al. 2008).

Some ancient lakes contain rich endemic molluscan faunas (see Section 9.7.4.1), and high levels of endemicity also occur in some fresh-water stream systems and in some arid-zone springs (see Section 9.7.4.2).

As with marine faunas, reliable comparative data on fresh water molluscan faunas are rather sparse (Table 9.9). However, it is clear that the richest recorded fresh-water molluscan faunas inhabited some rivers of southeast North America, with 97 species of unionids recorded from the Tennessee River (Hughes & Parmalee 1999) and 118 species of gastropods found in the Mobile River Basin alone (Neves et al. 1997). Unionid bivalve densities can reach nearly 700 per square metre in southeast USA and 1800 in the Ukraine (Protasov et al. 2015). These communities and other fresh-water mollusc habitats have been severely affected by human activities such as river modification and pollution (Neves et al. 1997; Strong et al. 2008). Exotic invasive species also pose threats to fresh-water molluscs, particularly those with restricted distributions. One of the most aggressive is the 'zebra mussel' (*Dreissena polymorpha*), which in some areas in the USA have been recorded as exceeding 30,000 per square metre (Dermott & Munawar 1993) (see Chapter 10).

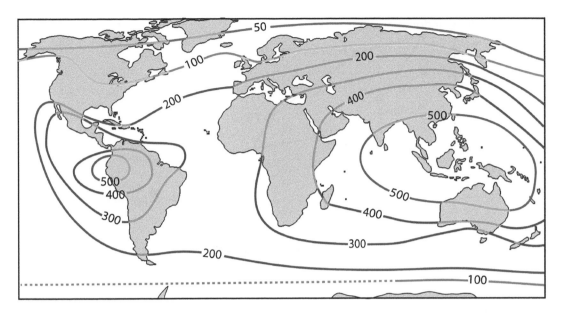

FIGURE 9.18 Marine bivalve diversity contours based on the number of species. Data from Crame, J.A., *Paleobiology*, 26, 188–214, 2000, including the addition of contours (Courtesy of P. Middelfart).

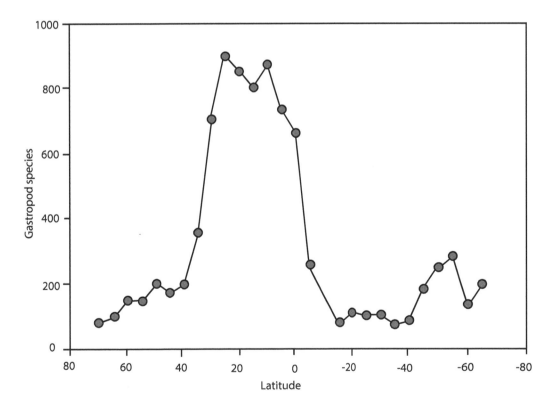

FIGURE 9.19 Latitudinal variation in gastropod species richness on the continental shelf of the eastern Pacific Ocean and Antarctica. Redrawn and modified from Rex, M.A. et al., *Ecology*, 86, 2288–2297, 2005.

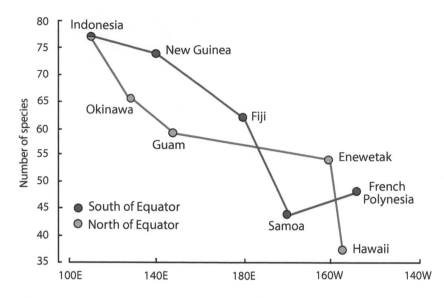

FIGURE 9.20 Numbers of cowry (Cypraeidae) species from east to west across the Pacific Ocean. Redrawn and modified from Kay, A.E., *B. Mar. Sci.*, 47, 23–34, 1990.

TABLE 9.8

Some Numbers of Marine Molluscs Based on Regional Checklists

Region	Number of Species	Source
Northeast Atlantic	3641	CLEMAM Check List of European Marine Mollusca http://www.somali.asso.fr/clemam/index.clemam.html
West Africa	2500	R. von Cosel, unpublished
Indo-Pacific	32,000	24,269 in Biotic database of Indo-Pacific marine mollusks http://data.acnatsci.org/obis/, estimated to be 2/3 complete
Panama region (Gulf of California to Columbia)	2535	Keen (1971)
South Africa	2788	Kilburn and Herbert (1999)
North Pacific	1744	Kantor and Sysoev (2005)
New Zealand	2091	Spencer et al. (2009)
Antarctic and Magellanic	800	Bouchet (2006) estimate
Western Atlantic	6170 (gastropods only)	Rosenberg (2005) Malacolog 4.0 http://data.acnatsci.org/wasp

Data from *Bouchet, P., The magnitude of marine biodiversity, in Duarte, C.M., The Exploration of Marine Biodiversity: Scientific and Technological Challenges*, Fundación BBVA, Bilbao, Spain, pp. 31–62, 2006.

9.8.2.2 Terrestrial Molluscs

Gastropods are the only group of molluscs to have successfully invaded the terrestrial realm, where they are represented by about 35,000 living species (Solem 1984; Bruggen 1995). Terrestrial gastropods can be found in environments ranging from wet tropical rainforests to Again, reliable diversity data are scant but some comparative figures are given in Table 9.10.

Relatively few gastropod lineages have become terrestrial, these being the neritimorph superfamilies Hydrocenoidea and most Helicinoidea, the caenogastropod Cyclophoroidea, some Truncatelloidea and Littorinoidea, the heterobranch Rathouisioidea, some Onchidioidea and Ellobioidea, and all the highly diverse Stylommatophora. In all, this represents about 112 families with over ten separate invasions of land being involved (Barker 2001a), although this may have occurred several times in a few families including the Truncatellidae (Rosenberg 1996) and Ellobiidae (Martins 2001). While in some tropical regions the terrestrial gastropod faunas comprise high diversities of non-stylommatophoran groups, in most parts of the world stylommatophorans dominate with more than 30,000 species estimated (Barker 2001a). Some land snails adopt bizarre shell morphologies, such as several from Cuba (e.g., Aiken 2009) and Malaysia (Clements et al. 2008a). Areas such as limestone outcrops, isolated rainforest relics, and

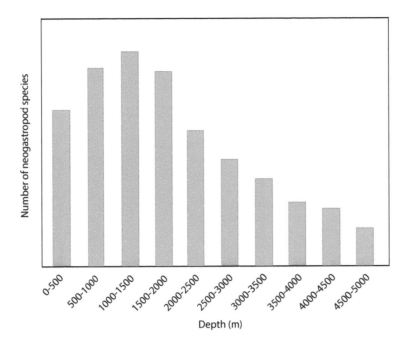

FIGURE 9.21 Bathymetric pattern of species richness of deep-sea neogastropods in the eastern North Atlantic. Redrawn and modified from Rex, M.A. et al., *Ecology*, 86, 2288–2297, 2005.

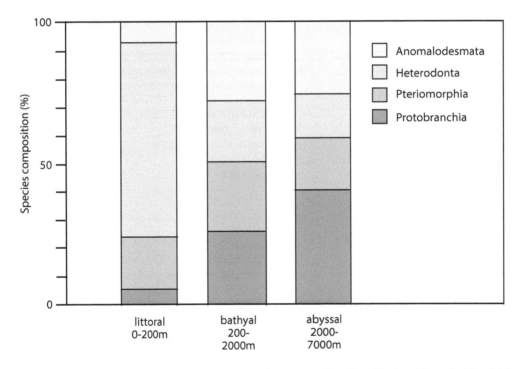

FIGURE 9.22 Bathymetric patterns of species richness of the major bivalve taxa. Data from Zardus, J.D., *Adv. Mar. Biol.*, 42, 1–65, 2002; after Hickman, C.S., *Annu. Rep. West. Soc. Malacol.*, 7, 41–50, 1974.

oceanic islands often have high endemism, and those faunas are under threat from invasive predators, land clearing, and other human activities (see Chapter 10). One of the most diverse land snail faunas is that found on the limestone hills of Perak, Malaysia with 122 species, 34 of which were restricted to one hill (Foon et al. 2017).

While numbers of species are often low, this is not always the case, with up to 95 species living in a square kilometre of rainforest in Cameroon (Winter & Gittenberger 1998), 60 species have been found in the same size area in a Borneo rainforest (Schilthuizen & Rutjes 2001), and in a four ha. temperate rain forest in New Zealand (Solem et al. 1981).

TABLE 9.9

Comparisons of Some Fresh-Water Molluscan Faunas

Area	Number of Species		Source
	Bivalves	Gastropods	
Europe	53	~397	Estimated from Welter-Schultes (2012) and CLECOM (http://www.weichtiere.at/clecom/index.html?/clecom/clecom.html)
North America	300	650	http://molluskconservation.org/index.html
Australia	41	306	http://keys.lucidcentral.org/keys/v3/freshwater_molluscs/ (native, described species)
New Zealand	6	83 (8)	B. A. Marshall, pers. comm., January (2014)
Africa	–	341[a]	Brown (1994)
Brazil	117	256	Simone (2006)
Mascarene Islands	2	21	Griffiths and Florens (2006)

Source: Strong, E.E. et al., *Hydrobiologia*, 595, 149–166, 2008.
Note: Figures in brackets are the number of introduced species included in the total.
[a] Excluding hydrobiids from North Africa.

TABLE 9.10

Comparisons of Some Terrestrial Faunas (Described Species)

Area	Number of Described Species	Source
Europe	~1970	Estimated from Welter-Schultes (2012)
North America	1200	Nekola (2014)
Cuba	1400 (97% endemic)	Aiken (2009)
Australia	1500 (50)	F. Köhler, pers. comm., November (2017)
Lord Howe Island	59	Hyman and Köhler (2016) (https://australianmuseum.net.au/blogpost/amri-news/a-quest-for-endangered-land-snails-on-lord-howe-island)
New Zealand	502 (32)	B. Marshall, pers. comm., January (2015)
Brazil	700	Simone (2006)
Mascarene Ids	209	Griffiths and Florens (2006)
American Samoa	50	Cowie (2001b)
Madeira	117 (36)	Cook (2008)

Note: Figures in brackets are the number of introduced species.

9.9 PUTTING IT ALL TOGETHER – EVOLUTIONARY TRADE-OFFS

Species are products of evolution over time and comprise an integrated set of systems, shaped through selective forces, which are adequate to meet the particular circumstances of an organism in the time and place where it occurs. They are sometimes compromises of conflicting demands from external factors such as energy sources, chemical constituents, substratum, and intrinsic traits such as size, mobility, and reproductive mode. A modification of a trait in a particular system may have costs and impacts on one or more other systems that require associated changes to continue integrated functions. Successful evolutionary modifications can only occur within the constraints imposed by the construction of the body and its physiology and the availability of any additional energy and material inputs. An analogy with commercial endeavours is drawn by G. Vermeij (1993) in that the cost of undertaking a 'project' must be feasible. Carrying his analogy further, he argued that the 'market place' in which organisms compete for resources (i.e., for food, mates, and shelter) dictates the costs and benefits. Risks to survival from predators, competitors, and the physical environment must be balanced against the potential benefits and, ultimately, reproductive success (see also Ghiselin 1974).

Vermeij (1993) examined the costs and benefits of building shells, and we will use this as an example to illustrate the above points. On the cost side, he pointed out that building a shell requires time and energy. Calcium may be limited. Also, time and energy must be devoted to prevent the deterioration of the shell through damage or erosion and transporting the shell requires additional energy with increase in mass. While calcium carbonate is rarely limited in the oceans, except in very cold water and the deep sea, it may be in short supply in many terrestrial and fresh-water habitats. Thus, animals living in those habitats have to expend more energy obtaining the calcium and incorporating it in their shells. Also, shells in calcium-poor aquatic environments are subjected to greater etching through dissolution and often must compensate for this by building a thick periostracum.

Shell construction involves the laying down of a protein matrix upon which calcium carbonate crystals form, in addition to a protective outer organic periostracum. The metabolic 'cost' of building the organic matrix is seven to 30 times higher than that of calcification, perhaps accounting for the independent, evolutionary loss in several groups (e.g., within the vetigastropods) of microstructures (such as nacre) with a high organic component (Palmer 1983, 1992). For a shell with only 1.5% organic matrix, the cost of building that component would account for 22% of the total cost of shell production compared with a shell with 5% organic matrix where it would account for nearly 50% of the cost of building the shell in marine gastropods (Palmer 1992). Thus the cost of calcification in a trochid such as *Tegula* would be much higher than a caenogastropod, equivalent to about 75% of the total energy invested in somatic growth and 410% of the energy expended on reproduction (Paine 1971). Thus, shells with more protein matrix are more energetically expensive and probably slower to repair than other shell structures with a smaller organic component.

Other economies in shell building include:

- Approximation of a spherical shape: The most efficient shape is a sphere (maximum volume enclosed), and some gastropods and bivalve shells approximate this.
- Reduction in shell thickness to produce thin shells and/or shell reduction or loss.

Given the difficulties of forming shells in cold water, one might expect to see evidence of globose shell shapes, thinner shells, and shell reduction in polar regions, but these trends are hardly apparent (Vermeij 1993), suggesting that the costs of shell formation have not imposed absolute limits on molluscs. Living in calcium-poor aquatic environments imposes other costs – notably shell dissolution. This is particularly so in cold and in deep marine waters, and in acidic fresh waters. Strategies to combat this include:

- Building the shell with calcite rather than aragonite – calcite is 36% less soluble. Some cold water groups have adopted this strategy. Many littorinids living in cold waters have calcitic shells, while their tropical relatives have aragonitic shells (Taylor & Reid 1990).

Many cold water taxa have aragonitic shells and many tropical taxa have calcitic shells. In the tropics, calcitic shells are only seen in some fresh-water nerites.
- Building a thick periostracum. Many cold water taxa have engaged in this strategy, but, again, it is not unique to these environments. Similarly, some cold water taxa have a thin periostracum. Some fresh-water taxa also have a thick periostracum.
- Build the shell in layers separated by organic sheets. Seen in some bivalves, e.g., some oysters, corbulids, and Unionoidea.
- Encourage epiphytic or epizoic growth. Encrusting sponges, hydroids, ascidians, algae, etc. can also provide protection by shielding the shell surface from direct contact with seawater-encrusting organisms and can also prevent settlement of bioeroding organisms (Smyth 1989). For mobile taxa, however, such adornment might well hamper movement.

Shells can also be attacked by borers – either predatory (whelks, octopus) or non-predatory (some sponges, fungi, forams, algae, worms, barnacles, bivalves, etc.) organisms that can excavate the shell surface. Shell boring or excavation of both kinds is commonest in warm marine waters. The defence mechanisms that have evolved include some of those listed immediately above. For example, the organic layers in oysters and corbulid bivalves provide resistance against predatory drillers (e.g., Lewy & Samtleben 1979).

Calcium carbonate is about three times denser than seawater. Thus the larger and heavier the shell, the more energy that must be expended to move it. Many modifications have occurred to increase the efficiency of dragging a shell around. These include streamlining and reduction in weight by thinning the shell, and/or reducing the overall shell size. Some weight reduction is also obtained by increasing the organic component. Protein is about the same density as seawater, thus increasing the relative proportion of protein in the shell will reduce its density, but this too has its costs (see above). Calcite is about 7% less dense than aragonite and swimming scallops usually have calcite shells, although other swimmers (limids, heteropods, pteropods, and cephalopods) have aragonitic shells.

The above examples using the shell provides some insights into the complexity of the constraints and effects of evolutionary modifications not only relating to the shell, but to all organ systems and parts of the body. Most of these constraints and interactions remain poorly understood and unravelling them is one of the many challenges for future biologists.

10 Molluscs and Humans

10.1 INTRODUCTION

Humans have been interacting with molluscs from their earliest history. They were used for food, ornaments, building materials, tools, dyes, medicines, and money. Today they are significant as food items, objects of fascination for collectors and artists, religious symbols, decoration, and a source of drugs and natural medicines. They are of considerable economic importance, particularly in aquaculture, and are also intermediate hosts of some of the most serious human parasitic diseases. Humans have also affected molluscan habitats and populations in many direct and indirect ways causing major conservation issues.

In this chapter, some ways humans interact with molluscs, both directly and indirectly, are briefly explored.

10.2 MOLLUSCS AS FOOD

More than half of all seafood eaten worldwide today is farm raised, compared to only about 9% in 1980, primarily from the expansion of aquaculture in China (Food and Agriculture Organization of the United Nations [FAO 2006]). In the United States in 1975, aquaculture produced approximately 151,365 tonnes of molluscs, while in 2016, it produced 173,725 tonnes. This relatively modest increase markedly contrasts with what has occurred in China where molluscan aquaculture rose from approximately 277,500 tonnes in 1975 to 14,473,630 tonnes in 2016.

Ancient peoples primarily collected molluscs for food. It has been suggested that early humans in the East African Rift Valley fed on fish and molluscs, a diet rich in long-chain polyunsaturated fatty acids, and this may have facilitated the growth of the cerebral cortex in their brains (Broadhurst et al. 1998). The most visible evidence of shellfish diets are shell middens found throughout the world on coastlines, estuaries, and river banks. These contain many species of medium-sized to larger intertidal or shallow subtidal bivalves, gastropods, and sometimes chitons. They are typically composed of marine species, but some contain mainly or entirely terrestrial or fresh-water molluscs as, for example, in the late Palaeolithic, middens from in and around Franchthi Cave in Greece (Whitney-Desautels & Beer 1999) and fresh-water molluscs from sites in the Guanzhong Basin of northwest China of mid-late Neolithic age (Li et al. 2013). The oldest known shell middens in the world are about 150,000 years old, from the middle Stone Age of South Africa (Avery et al. 2008; Steele & Klein 2008). Many younger shell middens are found around the world, and some can be quite recent, with some in Australia, for example, being accumulated as recently as the early to mid-1800s (e.g., Colley 1997) (Figure 10.1).

A great variety of molluscs were, and are, collected for food from wild populations, including scallops, oysters, cockles, abalone, trochus, whelks, winkles, limpets, chitons, squid, cuttlefish, and octopus. This is in part because they are often relatively easy to gather, but they also have considerable nutritional value. Mollusc flesh is low in saturated fats and provides proteins, essential elements (notably iron, zinc, and copper), vitamins such as vitamin B-12, and unsaturated fats (omega-3 fatty acids). This applies not only to marine molluscs (e.g., Miletic et al. 1991; Dong 2009; Wright et al. 2018a), but also to those living in fresh water (e.g., Baby et al. 2010) and on land (e.g., Fagbuaro et al. 2006). In parts of Africa where malnutrition is prevalent, it has been suggested that the meat of terrestrial snails such as giant African land snails *Achatina* and *Archachatina* (Achatinidae) could be a cheap way of supplementing diets because it is not only rich in protein and polyunsaturated fatty acids, but contains high levels of the amino acids arginine and lysine, minerals such as iron, calcium, magnesium, phosphorus, copper, zinc, vitamins such as A, B-6, B-12, K, and folate, and essential fatty acids such as linoleic and linolenic acids (Fagbuaro et al. 2006; Malik et al. 2011).

Harvesting of edible molluscs can also deplete populations and affect community structure, leading to severe reductions in numbers and even local extinction (see Section 10.9.2.1). Depletion of wild stocks has resulted in an upsurge in aquaculture, especially bivalves such as oysters and mussels, but also other molluscs such as abalone, octopuses, and scallops (see Section 10.2.2). In some parts of the world many molluscs harvested or grown in commercial aquaculture enterprises are mostly high-end, 'luxury' species (e.g., oysters, scallops, abalone), although this is generally not the case for many aquaculture enterprises, notably in Asia and South America.

Given the great variety of molluscs eaten as food around the world (Figure 10.2), with many specialist cookbooks for 'shellfish' such as oysters and mussels, it is perhaps surprising there appears to be only a few published so far that provide recipes for a wide range of molluscs, one being Shumway et al. (2014a),

FIGURE 10.1 Australian aboriginal man and child collecting oysters, Port Macquarie area, New South Wales. State Library of New South Wales, call number 04719, file number FL1701046, with permission, photograph by T. Dick, ca. 1905.

and another Ivanov and Sysoev (2009), the latter featuring 1600 recipes and covering over 1100 species.

10.2.1 Some Commercially Important Species Caught in Wild Fisheries

Harvesting from wild populations has been the traditional method of collecting molluscs for food, ornaments, or other purposes such as specimen shells (for the collector market). This is usually done by hand, or by using nets, dredges, trawls, rakes, or digging. Some of these methods can be damaging to the habitat and local populations can be quickly depleted. Management of wild stocks is strictly enforced in some areas to prevent overfishing (e.g., Hauck & Gallardo-Fernández 2013; Semmens et al. 2018) but, in some parts of the world, such management is minimal. One well regulated scallop fishery, the world's largest commercial molluscan fishery, is the sea scallop fishery in the Gulf of Maine (NEFMC, 2018).

Other than abalone, bivalves generally have an economic value much greater than gastropods and many pteriomorphians (mainly Mytilidae, Ostreidae, Pectinidae) are of particular importance as high-end human food items and comprise the great bulk of bivalves consumed. Many other species of bivalves are notable as items of harvested human food including arcoideans (e.g., *Anadara, Glycymeris*), Pinnidae (*Atrina, Pinna*), Cryenidae (e.g., *Batissa, Geloina, Polymesoda,*

Corbicula), Cardiidae (e.g., *Cerastoderma*), Myidae (e.g., *Mya*), Veneridae (e.g., *Mercenaria, Chione, Austrovenus, Ruditapes, Tivela, Katylesia*), Arcticidae (*Arctica islandica*), Solenidae (e.g., *Solen, Ensis*), Mesodesmatidae (*Paphies, Mesodesma*), Mactridae (e.g., *Spisula, Mactra*), Donacidae (*Donax*), and Hiatellidae (*Panopea*).

Popular names can be confusing, with, for example, many kinds of bivalves being called 'clams' in North America and various shallow-burrowing taxa often referred to as 'cockles' – including cardiids in Europe and arcids (e.g., *Anadara*) in Asia and Australia. 'Cockles' are caught by hand, rake, or shallow dredge, and numerous venerids are also fished in this way, as are some donacids and mesodesmatids.

Oysters are among the few animals that many people eat alive. They are still collected around the world from wild populations on the shore, or by dredging, but almost all of the commercially harvested oysters are now cultivated (see Section 10.2.2.2). Significant oyster species are *Crassostrea gigas*,[1] originally from Asia, but now introduced to many other parts of the world, the closely related (or synonymous) *C. angulata* which is the dominant species

---

[1] The generic name *Magallana* has been recently introduced for this and other Pacific species of *Crassostrea* (Salvi & Mariottini 2016). This name is not adopted here pending further studies as suggested by Bayne et al. (2017).

FIGURE 10.2 Some examples of molluscs being used as food. Middle top, two left figures in middle row and left and middle in lower row courtesy of J. M. Ponder. Top right, middle left and lower right public domain.

used in aquaculture in Europe and also in Taiwan and southern China (Hsiao et al. 2016), *Crassostrea virginica* from the east coast of USA (introduced to several areas and has become established in O'ahu, Hawaii, and British Columbia), *Ostrea edulis* in Europe (introduced populations established on the east coast of the USA and a record from southwestern Australia), *Ostrea angasi* in southern Australia, *Ostrea chilensis* in New Zealand and Chile, *Saccostrea cucullata* in the tropical Indo-Pacific including northern Australia, and *Saccostrea glomerata* in eastern Australia and New Zealand.

Several commercially significant species of scallops, including *Placopecten magellanicus*, are fished on the Atlantic coast of North America, with *Chlamys islandica*, *Argopecten gibbus*, and *A. irradians* on the east coast and *Patinopecten caurinus* on the west coast making up the majority of scallop fisheries in North America. *Pecten*

maximus, *Aequipecten opercularis*, *Chlamys islandica*, and *Mimachlamys varia* are the main species fished in Europe. *Amusium pleuronectes* is fished in much of Asia, but in China, *Chlamys farreri* is by far the most important commercial species. In Japan, *Amusium japonicum*, along with *Patinopecten yessoensis* and *Pecten albicans*, are the main species fished, with *P. yessoensis* also important in Russian waters. In Australia, *Pecten fumatus*, *Mimachlamys asperrima*, and *Ylistrum* (previously *Amusium*) *balotti* are fished and, in New Zealand, *Pecten novaezelandiae*. In Brazil, *Nodipecten nodosus* and *Euvolva ziczac* are the most important species, while in Argentina, it is *Zygochlamys patagonica* and *Aequipecten tehuelchus* (Shumway & Parsons 2006, 2016).

Fished scallops are trawled or dredged and together comprise a valuable industry, although overfishing has severely depleted stocks in some areas. For example, in the South

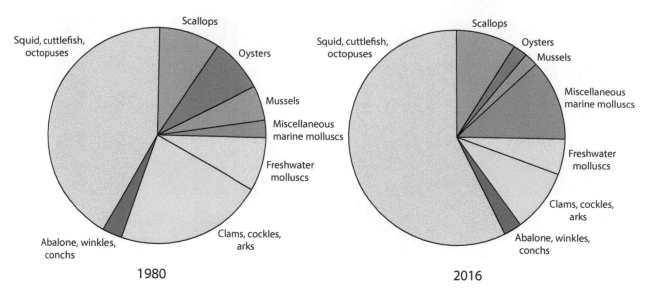

FIGURE 10.3 Marine mollusc wild fisheries catch based on FAO Fisheries statistics for 1980 and 2016. The total catch for 1980 was 3,671,997 tonnes and for 2016 was 6,325,971 tonnes. This figure does not include aquaculture (see Figure 10.6).

Island of New Zealand, scallop catches (*Pecten novaezelandiae*) fell from 1246 tonnes in 1975 to only 41 tonnes in 1980. Following the closure of the fishery, it was reopened, but the catches fell from 850 tonnes in 1994 to 65 tonnes in 2006 (Mincher 2008). This small fishery is dwarfed by the *Placopecten magellanicus* fishery along the Atlantic coast of North America which peaked at 325,125 tonnes in 2004 and was 206,177 tonnes in 2016 (FAO Fisheries 2018). A fishery for *Argopecten ventricosus* in the Gulf of Panama collapsed in 1991 (Medina et al. 2007).

Deeper burrowing bivalves such as Solenidae ('razor clams', 'razorfish') are fished in Europe, the Pacific Northwest region of America, and some other parts of the world and 'soft-shelled clams' or 'sand gaper' (*Mya arenaria*) in eastern North America and in parts of the eastern Atlantic coast of Europe. The largest commercially collected bivalve is the deep-burrowing 'geoduck' (*Panopea generosa*) in northwest North America and Mexico. These can weigh up to 4.5 kg with a shell length of over 200 mm and massive siphons up to about a metre in length. The value of this fishery for these long-lived animals, up to 168 years (Orensanz et al. 2004), is about $US80 million annually, and there is some farming of this species. A smaller species of *Panopea* is harvested from wild populations in New Zealand.

Giant clams (*Tridacna* and *Hippopus*) are still fished in parts of the Indo-Pacific despite their CITES (Convention on International Trade in Endangered Species of Wild Fauna and Flora) listing. They are also 'farmed' for food and ornaments in some parts of the Pacific (e.g., Hart et al. 1998).

Although the fishing catch of cephalopods is small compared with 'fin fish', it is slowly increasing, rising from ~3.9 million tonnes in 2005, to over 4 million tonnes in 2014 (compared with the total fisheries catch of 68 million tonnes), it nevertheless comprised 20.6% of the total mollusc

catch of 23.16 million tonnes in the same year (Figure 10.3). The cephalopod catch steadily increased each year until the mid 2000s (Figure 10.4), totalling (in millions of tonnes) 2.4 in 1990, 3.4 in 1999, 3.9 in 2005, and nearly 4.8 million in 2014, but has dropped a little in more recent years. It has increased as a percentage of the total fisheries catch from less than 3% in the early 1990s to about 7% in 2014 (FAO 2016).

Figures given by Boyle and Rodhouse (2005) for 1999 indicate that octopus comprised about 18.4% of the total catch, cuttlefish about 24.1%, with the remainder being squid. In 2012, FAO results indicated the following proportions: squid 72%, octopuses 9%, cuttlefish 1%, and undetermined 18%. For details regarding cephalopod fisheries and their management, see Boyle and Rodhouse (2005).

Abalone (Haliotidae) have a world-wide distribution in the coastal waters of every continent except southern South America[2] and northeastern North America. Most commercially important species are found on the coasts of New Zealand, South Africa, Australia, western North America, and Japan with one species being found along the southern Atlantic coast of Europe and in the Mediterranean. Wild abalone stocks are depleted in many areas with, for example, the sportfishing season being cancelled in California in 1997 and again in 2018 due to the threat of population collapses of some species. About a quarter of the world wild catch is fished in Tasmania, and Australia as a whole has about half of the world wild catch. The value of these animals is such that there is significant poaching and illegal trafficking of abalone meat, with the amount poached in some areas thought to be greater

---

[2] Although wild abalone do not occur along the southern coasts of South America, there is a robust abalone aquaculture enterprise in Chile (1200 tonnes in 2014) (FAO 2017).

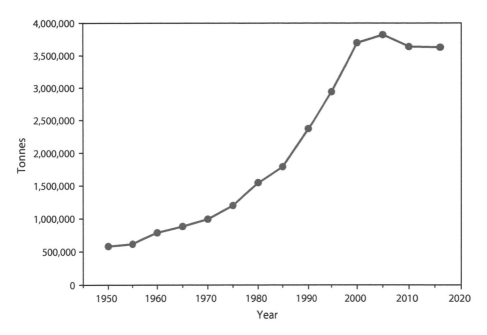

FIGURE 10.4 Coleoid cephalopod catches from 1950 to 2014. Data from FAO Fisheries, Food and Agriculture Organisation of the United Nations, Fisheries and Aquaculture Department, Statistics, http://www.fao.org/fishery/statistics/en, November 2018.

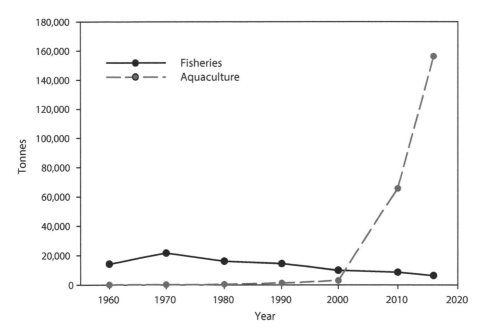

FIGURE 10.5 Abalone fisheries catch and aquaculture production 1960–2014. Data from FAO Fisheries, Food and Agriculture Organisation of the United Nations, Fisheries and Aquaculture Department, Statistics, http://www.fao.org/fishery/statistics/en, 2018.

than the legal take. In recent years, most abalone meat sold comes from aquaculture (Figure 10.5). Abalone shell is also used for jewellery and ornaments, notably the New Zealand *Haliotis iris* ('paua').

While abalone is undoubtedly the most valuable gastropod fishery, various other gastropods also comprise significant fisheries. These include whelks (notably Europe and Japan) and patellogastropod limpets (notably Hawaii, Azores) and the

'loco', an abalone-like muricid (*Concholepas concholepas*) in Chile. Some of these fisheries are substantial. For example, a 'whelk', *Buccinum undatum,* is fished in several countries around the North Atlantic with 2016 global production of around 41,000 tonnes (FAO Fisheries 2018).

Vetigastropods such as some trochids are fished widely in the tropical Indo-West Pacific for the aquarium trade, for their 'meat' and for their nacreous shells that are popular as ornaments; some are still used in the button trade. The most commonly fished trochid, *Tectus niloticus*,[3] is a large, widely distributed tropical Indo-West Pacific species that has been overharvested in many parts of its range.

Because of their large size and intertidal habitat, limpets have been important components of human diets for over 150,000 years, but there is currently little economic interest. Larger marine and fresh-water neritids have been used by subsistence gatherers as food items throughout prehistoric time. Caenogastropods are also a source of food, including species such as the conch (*Strombus* spp.), winkles (*Littorina* spp.), potamidid 'mud whelks', and various buccinoidean 'whelks' such as *Buccinum* (see above), *Neptunea*, *Busycon*, and *Busycotypus*. Other neogastropods fished commercially in various parts of the world include *Babylonia*, various muricids, and some Volutidae, with members of the latter family being harvested particularly in South America. Various cerithioidean snails are also eaten; notably the large mangrove-living potamidids such as *Telescopium*, *Terebralia*, and *Cerithidea*, which are used as food in some tropical areas, notably parts of Asia and Africa.

Some fresh-water mussels (Unionida) are significant as food items in some cultures, but more so in earlier times. China accounts for the great majority of reported world catches of fresh-water molluscs.

10.2.2   Aquaculture and Mariculture

Farming aquatic molluscs (aquaculture – the cultivation of aquatic organisms, mainly for food) is responsible for the production of a very significant proportion of molluscs utilised for food and also for pearls and pearl shell. Some taxa forming valuable fisheries, notably squid (calamari), do not lend themselves to aquaculture so are obtained entirely from wild catches.

Aquaculture underwent rapid growth, particularly through the last few decades of the 1900s, and molluscs represent a large proportion of aquaculture production. Marine aquaculture (Figure 10.6), especially mariculture, accounts for most of the production of molluscs which includes mostly bivalves, notably several pteriomorphians such as oysters (see Section 10.2.2.2), scallops (see Section 10.2.2.5), mussels (see Section 10.2.2.3), pearl oysters (see Section 10.2.2.4), and some species of venerid 'clams', including the 'Manila clam' *Ruditapes* (previously *Venerupis* or *Tapes*) *philippinarum*.

Much has been written on bivalve aquaculture (e.g., Quayle & Newkirk 1989; Spencer 2002; Shumway et al. 2003; Shumway 2011; FAO 2014), and only a brief overview is provided here.

The major aquaculture activity involving gastropods concerns abalone (see Section 10.2.2.7), which are farmed in California, Asia (Japan, China, Taiwan, Korea, Thailand), Australia, New Zealand, Chile, South Africa, and a small amount in Europe.

Cultivated species are often prone to disease (see Section 10.2.2.10), particularly because the animals are in such high densities. There are also environmental problems with the aquaculture activities themselves which sometimes generate pollution, introduce accompanying invasive taxa (see Section 10.7.1.4), and the location of the facilities may often cause considerable environmental damage. On the positive side, it can be argued that shellfish aquaculture not only increases food production, but can serve other purposes including the provision of habitat and amelioration of effluent, such as that resulting from fish farming, and reducing phytoplankton densities (e.g., Caddy & Defeo 2003; Shumway 2011).

10.2.2.1   Stock Enhancement by 'Seeding'

Aquaculture methods include rearing juveniles of certain species and then either growing them to marketable size or transplanting them back into the 'wild' once they have reached a suitable size, effectively avoiding the high levels of mortality in larvae and juveniles. Thus, for example, in Japan, most abalone production is by way of juveniles 'planted' on the sea floor for later harvesting. In some tropical areas in the Pacific, juveniles of vulnerable species such as *Tridacna* spp., *Tectus niloticus*, and *Turbo marmoratus* are reared, and then the reefs are restocked with them when they reach a size where predation levels are acceptably low (Crowe et al. 2002; Purcell 2004). Cultured 'giant clams' (*Tridacna* spp.) are 'farmed' in several parts of the tropical Pacific, including Australia, mainly for restocking reefs and for the aquarium trade.

Commercially significant bivalve fisheries are sometimes 'seeded' with reared spat. This is common practice with oysters (see next Section), but is also done with, for example, cockles (*Cerastoderma edule*), which are sometimes seeded in Holland and France to enhance wild stocks. In the Pacific Northwest USA and western Canada young 'geoduck' *Panopea generosa* (Hiatellidae) are 'planted' on suitable tidal flats, as these large, deep-burrowing bivalves fetch high prices. Seeding the 'soft shell clam' *Mya arenaria* is also carried out on the east coast of Canada and northeastern USA. 'Razor clams' (Solenidae and Pharidae) are mostly fished from wild stocks, but the cultivation of the 'Chinese razor clam' or 'Agemaki clam', *Sinonovacula constricta*, a pharid, occurs in China where spat is seeded on estuarine mud flats.

10.2.2.2   Oyster Mariculture

Oysters are by far the major mollusc resource in many areas. This is also reflected in the value of the resource.

The largest mariculture production (Figure 10.7) for an individual species (including non-molluscs) is that of the 'Pacific oyster' (*Crassostrea gigas*) with production being about 587,289 tonnes in 1980 and 573,696 tonnes in 2016. This production, which was only 149,163 tonnes in 1950, increased dramatically in the mid 1970s to around present levels (FAO Fisheries 2019). Production of this species mainly occurs in Asia, particularly in China and Korea, but it has also been introduced into the Pacific coast of the

---

[3] Previously known as *Trochus niloticus*.

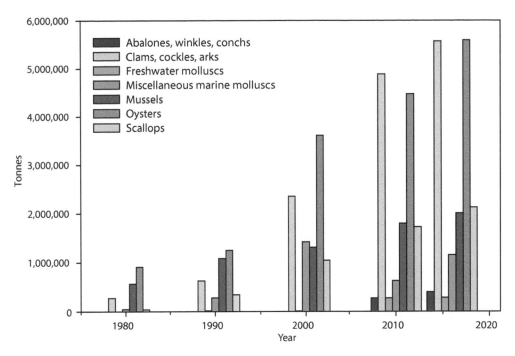

FIGURE 10.6 Marine mollusc aquaculture production (in tonnes) based on FAO statistics for 1980, 2000, and 2016. The total 2016 aquaculture production was 17,139,140 tonnes.

FIGURE 10.7 Left, oysters (*Crassostrea gigas*) being harvested from suspended cages in Prince William Sound, Alaska. (Courtesy of D. R. Lindberg.) Upper centre, oyster stick culture, Hiroshima, Japan. (Courtesy of S. E. Shumway.) Lower centre, commercial mussel (*Mytilus*) rafts, Ria de Arosa, Spain. Courtesy of S. E. Shumway.) Right, a raft and a line used in *Crassostrea angulata* culture in Vietnam. (Courtesy of W. A. O'Connor.

USA, Europe, temperate Australia, and New Zealand where it is also cultivated. In New Zealand, eastern and southern Australia, and the American Pacific Northwest, it has established in the wild. Introduced oysters make up the bulk of the industry in some areas, including Europe, where over three-quarters of oysters produced are Pacific oysters. The native European oyster, *Ostrea edulis*, is also farmed in Europe and in the USA and South Africa.

Oysters start maturity as males or females and then change sex, with more than one sex reversal possible. An oyster can spawn up to about 20 million eggs, and the gonads are the major portion of the body tissue when mature. After spawning, the oyster is 'out of condition', and not suitable for harvesting.

Some oysters spawn freely into the water column, while others brood the fertilised eggs in their gills and later release them as shelled veliger larvae. Either way, the veliger settles and attaches to the substratum using the byssus gland in their foot. They then permanently attach and cement to the substratum and the foot atrophies and is lost. For many cultured species, hatcheries are used to supply oyster 'spat', but in some areas farmers collect the spat in good 'spatting' locations by setting out suitable substratum(e.g., wooden stakes, polyvinyl chloride [PVC] slats) for the larvae to settle on. They then place these substrata in growing areas, where conditions are ideal for growth and fattening, until they reach a marketable stage. Some, such as the rock oysters, may remain attached to hard surfaces such as stakes through their growing period, while others are transferred to trays and cages suspended in the water column to grow (Figure 10.7). In Asia, settlement on shells and grow-out under rafts is very common.

Oyster culture, like many agricultural and aquacultural activities, is subject to fluctuations. For example, the oyster fisheries were markedly reduced along eastern North America (*Crassostrea virginica*), western North America (*Ostrea lurida* on the northern Pacific coast and *O. conchaphila* in Mexico) and eastern Australia (*Saccostrea glomerata*). In the USA, the value of oysters produced in 2009 was $136.5 million, although the value of the 2010 'crop' was less, at $117.6 million, reflecting a decline in recent years. This decline was because of such factors as overfishing, deteriorating water quality and increased sedimentation, impacts from hurricanes, oil spills, and diseases such as Dermo and MSX (see Section 10.2.2.10), and the state of the global economy. Thus, while diseases have been partially responsible for the decline of the oyster industry in some areas, reductions in some locations is often due to ecosystems having changed dramatically from overfishing, habitat disturbance, sedimentation, and nutrient loading.

10.2.2.3 Mussel Mariculture

As with oysters, in some parts of the world, mussel culture has been important for hundreds of years and today occurs in most maritime countries. Various methods are employed using growing surfaces such as ropes, baskets, rocks, and stakes (Figure 10.7).

In cool-temperate waters, species of *Mytilus* are mostly used, while in warmer waters, generally species of *Perna* are cultured. Mussels are relatively low-value seafood, but are hardy and have the advantage of rapid growth, ease of cultivation and harvesting, and good survival during distribution to markets. 'Blue mussels' (*Mytilus edulis*) and 'black mussels' (*M. galloprovincialis*) are grown mainly in Europe and North America, and the global harvest is increasing – the total (FAO Fisheries statistics) in 2002 was 571,377 tonnes and in 2014 was 655,886 tonnes. The 'green mussel' *Perna viridis* is cultured through parts of the central Indo-West Pacific; its production is declining (312,634 tonnes in 2002 down to 163,521 tonnes in 2014), but the largest member of the genus, *P. canaliculus*, which comprises the bulk of the mussel crop in New Zealand, increased from 78,000 tonnes in 2002 to 97,438 tonnes in 2014.

10.2.2.4 Pearl Oyster Mariculture

The most important species of *Pinctada* are the 'Japanese pearl oyster' or 'Akoya', *Pinctada fucata*[4] (Red Sea and Persian Gulf, tropical Indo-West Pacific including subtropical Australia, China, and Japan); the 'Atlantic' or 'Gulf pearl oyster', *Pinctada radiata* (Persian Gulf, Red Sea, Mediterranean Sea); the 'black-lipped pearl oyster', *Pinctada margaritifera* (Persian Gulf, southwestern part of Indian Ocean, Burma, Australia, and Pacific including Fiji and Tahiti); the 'white-lip', 'silver-lip', or 'gold-lip' pearl oyster, *Pinctada maxima* (Australia, Fiji, Tahiti, Myanmar, Philippines, China); and the 'La Paz pearl oyster', *Pinctada mazatlanica* (Pacific coast of Mexico). Some members of the genus *Pteria* are also used for pearl culture, notably *P. penguin* in the Indo-West Pacific and *P. sterna* in the eastern Pacific.

In the past, the adult pearl oysters were collected from the ocean floor (Figure 10.8), but aquaculture methods now usually involve obtaining stock from hatcheries and rearing the juveniles. When they are large enough, the bivalve is implanted with a mantle irritant around which a pearl will form while they are suspended in baskets in the sea (Figure 10.8).

10.2.2.5 Scallop Mariculture

Scallops are farmed commercially, mainly in China and Japan, using a variety of methods such as nets, ropes, or trays, or even in fenced off areas of the seabed. Spat is collected or cultured in hatcheries. Some of the main species cultured are *Argopecten irradians* in Japan and China and eastern USA, *A. purpuratus* in Chile, and *Patinopecten yessoensis* are cultured in Japan and China. Other cultured species are *Chlamys islandica* in eastern USA, *Mimachlamys crassicostata* in Japan and China, *Aequipecten opercularis*, *Pecten maximus*, and *Mimachlamys varia* in Europe, *Mizuhopecten yessoensis* in Japan, China, eastern Russia, and Western Canada, *Chlamys farreri* in China, while small numbers of both *Placopecten magellanicus* and *Argopecten irradians* are cultivated in eastern USA. Other cultured species include *Argopecten purpuratus* in Chile and Peru and *Nodipecten nodosus* in Brazil (Shumway & Parsons 2016).

10.2.2.6 'Clam' Mariculture

The global mariculture of clams has been steadily increasing and involves several species, notably venerids. The most important is the so-called 'Japanese carpet shell' or 'Manila clam' (*Ruditapes philippinarum*[5]), which is native to Japan, but widely cultivated in Europe, western North America (where it was previously accidentally introduced), and Asia. Other venerids such as the 'hard clam' or 'northern quahog', *Mercenaria mercenaria*, are farmed in Florida and other eastern seaboard parts of the USA, part of its native range, but it is also farmed in parts of the United Kingdom and Europe and on the west coast of the USA. In parts of the United Kingdom and Europe, the so-called 'grooved carpet shell' or 'native

---

[4] Also called *P. imbricata* in the past.

[5] Sometimes included in other genera such as *Venerupis* or *Tapes*.

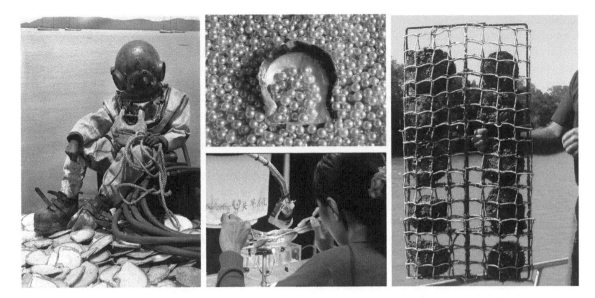

FIGURE 10.8 Left, pearl diver with catch, Thursday Island, Torres Strait, Australia, early 1920s. Australian Museum Archives, with permission, photograph by F. Hurley. Upper middle, cultured Akoya (*Pinctada fucata*) pearls from New South Wales, Australia. Courtesy of W. O'Connor. Lower middle, pearl technician seeding pearls. Courtesy of C. McDougall. Right, cage containing pearl oysters, Broome, Australia. Courtesy of J. M. Ponder.

clam', *Venerupis decussata*,[6] is cultivated, while the 'blood cockle', *Anadara granosa*, is an important aquaculture species in Southeast Asia.

Some other clams have stocks enhanced by 'seeding', where spat is grown in hatcheries and then 'planted' on suitable habitat for growing and later harvesting (see Section 10.2.2.1).

10.2.2.7 Abalone Mariculture

Abalone has been farmed successfully in California since the 1980s thanks to the discovery that hydrogen peroxide-induced spawning (Morse et al. 1978). Farming is done in tanks in land-based operations or in enclosures in the sea mostly using 'red abalone' (*Haliotis rufescens*) and to a lesser extent the 'green abalone' (*H. fulgens*). Abalone is also farmed in China and Taiwan (mainly *H. discus hannai* and *H. diversicolor*), and those countries are now the biggest producers in the world, although Korea and Thailand are also significant producers. Other areas where abalone are farmed include the Philippines (*H. asinina*), South Africa (*H. midae*), Chile (*H. rufescens*), and Australia (*H. rubra*, *H. laevigata*, and a hybrid of these two species). There are also smaller operations in Europe, Iceland, and some Pacific Rim countries.

Farming of abalone in Taiwan and in Victoria, Australia has been severely affected by viral diseases in recent years (see Section 10.2.2.10).

10.2.2.8 Cephalopod Mariculture

The main cultured species are *Sepia officinalis*, *Sepioteuthis lessoniana*, *Octopus maya*, and *O. vulgaris*, both for food

and as a source of experimental animals. Ongoing challenges include the development of a sustainable artificial diet and the control of reproduction (Villanueva et al. 2014).

10.2.2.9 Heliciculture

Many land snails are edible, and some are significant food items in various parts of the world, with a few species farmed, although in relatively small-scale operations. Land snails are farmed commercially in Europe, especially France, Italy, and Greece, and there is also notable production in North America, Taiwan, China, Indonesia and some other Asian countries, and Brazil and Argentina.

Most farmed snails are produced for the restaurant trade because *escargot* is famous in French cuisine. In France, two native species are normally used in heliciculture. One of these, the '*petit-gris*' *Cornu aspersum*, is common in temperate countries where it is often a pest. It is the 'common garden snail', long known as *Helix aspersa*. The other is the 'Roman snail' or 'Burgundy snail' (*Helix pomatia*), which is now uncommon or even rare in many parts of Europe. Some other species of Helicidae are collected from wild populations for eating, and a few are farmed, mainly in Europe.

The large pest species, the 'giant African snail' (*Achatina fulica*) and the related *A. achatina* and *Archachatina marginata* are also used for *escargot*. Other species are also eaten, such as the large New Caledonian native *Placostylus fibratus* (Brescia et al. 2003).

10.2.2.10 Some Diseases Affecting Farmed Molluscs

Pests and diseases can severely affect not only natural mollusc populations, but cultured ones. Examples of some of

---

[6] Sometimes included in other genera such as *Ruditapes* and *Tapes*.

the major pathogens affecting farmed molluscs are briefly described below. See Chapter 9 for additional information on parasitic and disease-causing pathogens of molluscs.

Some spionid polychaetes, notably some species of Polydora, damage and often kill oysters and various other bivalves and abalone. These 'mud worms' or 'blister worms' burrow into the inner surface of the shell, and their tubes are covered by the mollusc with a thin layer of shell, forming a blister. In the 1990s, an introduced sabellid (*Terebrasabella heterouncinata*) infected the California abalone aquaculture industry. This polychaete, which settles inside the shell aperture, arrested shell growth when present in heavy infestations (Kuris & Culver 1999). The sabellid, apparently of South African origin, was not found in wild populations, in California at least, being confined to abalone culture facilities, until a discharge carried larvae onto the shore near Morro Bay. It has been found in California and Baja California, Mexico, Chile, South Africa, and Iceland. It was eradicated from culture facilities (and as far as known, in the wild) in California in 1999.

The bucephalid trematodes *Prosorhynchus* spp. are common in *Mytilus* (Coustau et al. 1993; Francisco et al. 2010) and scallops (Hutson et al. 2004), where they infect the gonad and can cause castration.

The significant diseases of commercial oysters include protist parasites such as *Bonamia ostreae*, an intracellular haplosporidian parasite affecting the haemocytes of *Ostrea edulis* and *O. chilensis* (Lane et al. 2016). Heavily infected oysters are in poorer condition than uninfected oysters. Bonamia disease from a related parasite, *Bonamia exitiosa*, decimated the oyster beds in southern New Zealand (*Ostrea chilensis*) and also infects *Ostrea edulis* and *O. angasi* (e.g., Carrasco et al. 2012b). Winter mortality disease in oysters in New South Wales, Australia, was thought to be caused by a similar parasite '*Mikrocytos*' *roughleyi*. This was assumed to be a species of *Bonamia* (Cochennec et al. 2003), but this has been shown not to be the case, and its relationships remain uncertain (Carnegie et al. 2014). Spiers et al. (2014) have shown this parasite probably does not even cause winter mortality. Another related species, *Mikrocytos mackini*, causes Denman Island disease in *Crassostrea gigas* (Hervio et al. 1993).

MSX disease can devastate *Crassostrea* stocks, although it is not harmful to humans. It has been a serious problem in eastern North America, where the disease was introduced and has caused massive mortalities (e.g., Burreson et al. 2000; Burreson & Ford 2004). Species of *Haplosporidium* can also infect 'pearl oysters' (*Pinctada* spp.), with, for example, *H. hinei* in *Pinctada maxima* (Bearham et al. 2008). QX disease is similar to MSX disease and is caused by another haplosporidian, *Marteilia sydneyi*. It can kill many oysters (up to 80% in some areas) in infected estuaries and was responsible for the decline in the industry in Queensland and northern New South Wales, Australia (e.g., Roubal et al. 1989). Another species, *Marteilia refingens*, also infects various oysters, while *Marteilioides chungmuensis* infects *Crassostrea gigas* in Japan and Korea (e.g., Park & Chun 1989) and *Saccostrea echinata* in northern Australia (Hine & Thorne 2000). These parasites can also infect other bivalves with, for example, *Marteilia refingens*

and *M. maurini* infecting various mussels (Mytilidae), both *Marteilia* and *Marteilioides* infecting *Venerupis philippinarum* (e.g., Itoh et al. 2005), and a species of *Marteilia* is found in *Cerastoderma edule* (Carrasco et al. 2012a). The microsporidian *Steinhausia mytilovum* infects *Mytilus*.

Dermo disease in oysters is caused by a protist (Apicomplexa) parasite, *Perkinsus marinus*. It infects the haemocytes of *Crassostrea virginica*, but is not known to be harmful to humans. A species of *Perkinsus* was thought to be responsible for wiping out large numbers of *Tridacna* in Queensland in the 1980s (Goggin & Lester 1987) and also infecting *Venerupis philippinarum* in Asia (e.g., Park & Choi 2001) and *V. decussata* in Europe (Azevedo 1989). Abalone (*Haliotis* spp.) can also be infected with species of *Perkinsus* (e.g., Goggin & Lester 1995).

Both wild and farmed Californian abalone, such as *Haliotis cracherodii*, can be affected by a fatal wasting disease called 'withering syndrome' caused by a bacterial pathogen (*Xenohaliotis californiensis*) (e.g., Friedman & Finley 2003).

A fungus, *Ostracoblabe implexa*, infects oysters and may sometimes cause their death.

Bacterial diseases of mussels include vibriosis, caused by species of *Vibrios*, and rickettsiosis by *Rickettsia*-like and *Chlamydia*-like bacteria.

Viral diseases also occur, but are mainly not identified. For example, acute virus necrobiotic disease (AVND) infected the scallop *Azumapecten farreri* in China (Tang et al. 2010). This virus is a herpes-like virus, and other herpesviruses include significant pathogens of molluscs (various bivalves and abalone). One of the best studied, oyster herpesvirus (OsHV), infects oysters. It is not a serious pathogen in *Ostrea*, but has been responsible for massive mortality in *Crassostrea gigas* (e.g., Segarra et al. 2010; Jenkins et al. 2013). An abalone herpesvirus (AbHV) is related to the oyster herpesvirus (Savin et al. 2010) and causes abalone viral ganglioneuritis (AVG) which, as its name suggests, affects the nervous system of abalone in several parts of the world (e.g., Corbeil et al. 2012). AVG has severely infected farmed abalone in China and Taiwan and in Victoria, Australia, and in Victoria it has also been found in wild populations. Another virus, 'oyster velar virus disease' (OVVD) is also found in *Crassostrea gigas* (Elston & Wilkinson 1985). Viral diseases in mussels and *Ruditapes* include a picornaviridae-like virus (Bateman et al. 2012).

Other viruses are serious pathogens, including the Akoya viral disease (AVD) with a very serious impact on the Japanese cultivated pearl oyster industry (Kuchel et al. 2011) and the papova-like virus from *Pinctada maxima* (Norton et al. 1993). These and other viruses in bivalves are reviewed by Elston (1997).

10.3 MOLLUSCS IN CULTURE AND SOCIETY

10.3.1 Pearls and Pearl Shell

Pearls have been significant in many human societies and remain so today. Besides often being considered to symbolise human emotions such as purity and love, and sometimes being used for various medicinal purposes, they are objects with considerable intrinsic value as jewellery.

Most commercially valuable pearls are from members of the 'pearl oyster' family Pteriidae, mainly from species of *Pinctada* (e.g., Landman et al. 2001) (Figure 10.8). The pearl oyster shell itself (the 'mother of pearl'), is widely used in ornaments, for art, and for other decorative purposes.

The process of pearl formation is described in Chapter 3. Perfect natural pearls are very rare, and most pearls sold in recent years are cultured. It is widely held that the method used for culturing pearls was developed in Japan in the late 1800s. Nuclei, usually a bead made from fresh-water mussel shell, are implanted surgically near the gonad with a small piece of mantle tissue from another oyster. The cultured pearls can be harvested 1–2 years later, and the same oyster can often have a second nucleus implanted.

Several species are important in the cultured pearl industry (see Section 10.2.2.4), and each produces pearls with different characteristics. Most cultured pearls are from pteriids (pearl oysters), notably species of *Pinctada*, but some species of *Pteria* are also used, especially for mabé (blister pearls) that form on the internal shell rather than in the mantle. They are formed by a mould (usually plastic) being inserted between the shell and the mantle. Eventually, the nacre-covered mould is cut from the shell. Pearls vary in colour although they are usually white, cream, or pinkish, with darker 'black' pearls coming from *P. margaritifera*. Their lustre depends upon the reflection and refraction of light from the translucent layers of nacre making up the pearl – pearls with numerous thin layers have the best lustre, while iridescence is caused by very thin, overlapping layers breaking up light.

Cultured pearls are also obtained from some fresh-water mussels (Unionoidea), notably from China, but also Japan and the USA, and natural pearls are mainly from central Europe, the USA, and China.

Some fresh-water mussels were important as sources of pearl shell for buttons (Figure 10.9) and pearls, notably in the USA (e.g., Neves 1999) and Asia. While these industries still exist in Asia and the USA, today they are mainly harvested for their shells as a source of the nuclei for the marine cultured pearl industry (see above). The impact of the fresh-water mussel industry is visible from the heaps of fresh-water mussel shells with holes punched through them that lie alongside some of the larger rivers of the American Midwest (Figure 10.9).

Natural and cultured nacreous pearls are obtained from abalone. Saleable pearls are sometimes produced by other nacreous gastropods, notably *Tectus*, but are rarely of excellent quality although the nacre of the shell is commonly used for buttons and ornaments. A 'conch', *Strombus gigas*, produces marketable non-nacreous pearls, although mostly of poor quality. Various other molluscs produce pearls (see Landman et al. 2001 for details), but of generally low commercial value.

10.3.2 CULTURAL OBJECTS AND ORNAMENTS

The shells of many molluscs are used for ornaments, in decoration, and as jewellery. These include many gastropods (cone

FIGURE 10.9 The steamboat *City of Idaho* picking up mussel shells at Vincennes, Indiana, headed for a button factory in Muscatine, Iowa. Reproduced from *Vincennes* by Richard Day and William Hopper (1998) with permission of Arcadia Publishing. Insert, a shell of the unionid *Lampsilis teres* from which buttons have been cut. Some examples of unfinished and finished buttons are also shown. Courtesy of R. Warren, with permission from Illinois State Museum.

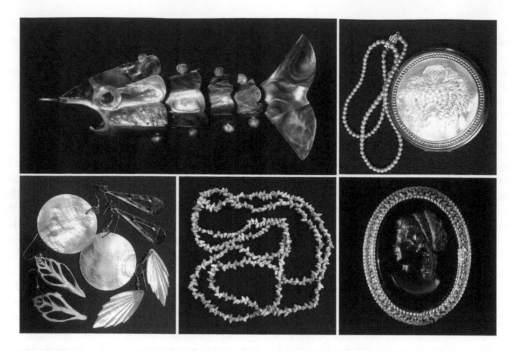

FIGURE 10.10   A few examples of modern jewellery made from shells. Top left, fish pendant with inlaid abalone shell over sterling silver armature, Mexico. Courtesy of D.R. Lindberg. Top right, a pearl necklace and a locket made from pearl shell; bottom left, four examples of shell ear rings; bottom middle, a necklace made from assimineid shells. Courtesy of J. M. Ponder. Bottom right, cameo brooch, Florence, Italy. Courtesy of D.R. Lindberg.

shells, cowries, volutes, trochids, haliotids, and many others) and bivalves, especially some pteriomorphians (pearl oysters, some scallops, *Spondylus* and *Placuna*, the 'window pane shell'). Some heterodonts, notably venerids and some tellinids, are highly coloured and/or sculptured and are also popular as ornaments and in shell art. The chambered, nacreous shell of *Nautilus* is also valued as an ornament (Figure 10.10).

The earliest evidence of hominoid engravings on a unionid shell is from Trinil, Java, which is assumed to be by *Homo erectus* and which has been dated at between 0.54 and 0.43 million years (Joordens et al. 2015). The earliest known pieces of 'jewellery' made by modern humans are two shell beads dated at 100,000–135,000 years old from Skhul Cave on the slopes of Mount Carmel in Israel and one 90,000 years old from Oued Djebbana in Algeria. These finds (Vanhaeren et al. 2006) pre-dated some other ancient examples of 'jewellery' by 25,000 years. They are three shells of the nassariid *Tritia gibbosulus*, a species still living in the Mediterranean Sea, each with a hole made in them to allow them to be strung together as a necklace or bracelet. Other younger prehistoric examples of shell beads using small gastropods like nassariids or littorinids are also known (see D'Errico et al. 2008 for discussion).

Shells provide an excellent line of evidence for trade in prehistoric archaeological settings. For example, marine species in inland sites and the biogeographic patterns of the species provide evidence of contact between different regions. It is especially useful in areas such as the Near East, North Africa, or Central America, where different faunas are within reasonable trading distance.

Pieces of shell, usually cut from marine 'clams' (*Mercenaria*) or 'whelks' (*Busycotypus*), were polished to make beads (called wampum) and were used by many Native North American tribes as money. The Narragansett people from the eastern USA are thought to have first produced these beads, which they created by cutting and drilling the shells of whelks or bivalves. Wampum had a high value because it was time-consuming to make and the beads were widely traded and used as currency over much of early North America. Even in colonial times in New England, it become legal tender between 1637 and 1661. Probably the phrase 'shelling out' money came from this period.

Similar uses of cut-out shell for currency are seen in other parts of the world, as for example, kemetas in New Ireland comprising reddish beads made from a trochid. A similar type of shell currency is found on Malaita Island, Solomon Islands, where strings of shell beads made from a few species of bivalve[7] are still used in exchange for goods (Figure 10.11).

Scaphopods, notably *Antalis pretiosum*, were valued objects of Native Americans, Inuit, and First Nations peoples. The shells were used as decorative material on clothing and accessories, especially breastplates, and were also strung together as a form of currency (Figure 10.13). In addition to scaphopod shells, hair pipes fashioned from the lip of *Strombus gigas* shells were commercially manufactured in New Jersey, USA and traded with

---

[7] Mostly *Chama pacifica*, *Beguina semiorbiculata*, *Anadara granosa*, and *Atrina vexillum*.

FIGURE 10.11 Some examples of shells used as ornaments and money in Oceania. Top left, man wearing a kapkap, Western Province Solomon Islands, 1920–1930. M. Ward Collection, Australian Museum Archives, with permission. Top centre, an ornament (kapkap), used for currency in Melanesia and made from *Tridacna* shell with an overlay of turtle shell. This example is from Malaita, Solomon Islands, 1921. Australian Museum, with permission; photograph by J. M. Ponder. Top right, man, and lower centre, woman, adorned with shell ornaments, Goaribari Village, Gulf Province, Papua New Guinea, 1920–1923. Australian Museum Archives, with permission, photograph by F. Hurley. Middle left, Langalanga shell money, Solomon Islands; beads made from various mollusc shells are used. Courtesy of K. Szabo. Middle centre. Bidayuh priestess belt made from beads cut from shell, Borneo. Courtesy of K. Szabo. Middle right, shell money beads and laoniasi, a disc of *Tridacna* shell, from the Solomon Islands. Courtesy of K. Szabo. Bottom left, portions of two belts ornamented with shell beads and cowries, Trobriand Islands, Papua New Guinea. Australian Museum, with permission; photograph by J. M. Ponder. Bottom centre, see above. Bottom right, shell arm ornament made from *Conus litteratus*, Central Province, Papua New Guinea, 1883, Australian Museum, with permission; photograph by J. M. Ponder.

the Native Americans of the Great Plains, USA between the late 1700s and 1880. During the 1880s, the use of shell hair pipes was superseded by the production of bone 'hair pipes' which were both larger and more uniform.

The 'window pane oyster' (*kapis* shell) industry in the Philippines uses the translucent, flat valves of *Placuna ephippium* (Placunidae) to make lampshades (Figure 10.12), windows, and ornaments for both local use and the tourist trade. Stocks of this species have declined, and efforts have been made to restock some wild populations in the Philippines (Gallardo et al. 1995).

Many shells, mainly various bivalves and gastropods, are widely used for decoration, as ornaments for the body (Figures 10.11 and 10.13), including jewellery, or as artwork, including their use as ornamental objects. 'Mother of pearl' was, and still is, widely used, as was the nacreous shell of abalone and trochids, in decoration and in making buttons and cameos. Such nacreous shell inlay has been widely used to ornament objects

FIGURE 10.12   Some examples of practical uses for shells. Left, *Charonia tritonis* shell used as a trumpet, from Papua New Guinea. Courtesy of K. Szabo. Upper centre, lampshade made from kapis (*Placuna ephippium*). Lower centre, trolling lure made from *Pinctada* shell, Solomon Islands. Courtesy of K. Szabo. Right, carved wooden head, Makira Province, Solomon Islands, ornamented with cowrie shells, Australian Museum, with permission; photograph by J. M. Ponder.

FIGURE 10.13   Scaphopods used as ornaments. Left, Cheyenne chief and medicine man White Thunder wearing scaphopod breast plate. Photographed in 1872 by an unknown photographer. Centre, scaphopod necklace from Pacific Northwest coast of North America. Courtesy of K. Szabo. Right, Daughter of Ogalam chief American Horse wearing scaphopod cape. Library of Congress, Prints & Photographs Division, Edward S. Curtis Collection (LC-USZ62-96974).

both in modern (Figure 10.10) and ancient cultures. Cameo carving of shells dates from Greek and Roman times, but has been prevalent since the fifteenth century with the popular use of cameos beginning around the mid-eighteenth century (Figure 10.10). A wide variety of molluscan shells have been used in making cameos, with the most popular being *Cypraecassis rufa*, but others include other species of *Cassis*, notably *C. madagascariensis*, various cowries, and some strombids.

The nacre on the interior of pteriid (pearl oysters) and unionoidean (fresh-water mussels) bivalves and haliotids (abalone) and some trochoideans (notably *Tectus*) are used commercially for buttons and mother of pearl (see also the previous Section). Similarly, the pearly shell of *Nautilus* is

widely used as ornament. Other uses of objects constructed from shells include fish hooks (Figure 10.12), beads, scrapers, cutters, and bowls, the latter from baler shells and large abalone species.

Some species of Neritidae are, and were, used as decoration, especially the smaller species as 'beads', and some of the flatter patellogastropod limpets are often used as shell jewellery because of their medallion-like shape and the lustre of the shell surface.

Their diversity of shell form and colour has ensured that caenogastropods have played a prominent role in human art, culture, and as currency (see below) and tools. Shell trumpets using large whelks such as *Charonia* (Figure 10.12), *Bursa*,

Strombus, *Fasciolaria*, *Turbinella*, and *Cassis* have been used since Neolithic times and are found in many parts of the world including Polynesia, Japan and other parts of Asia, Australasia, South America, and the Mediterranean. They were used for signalling, in ceremonies, on the battlefield, and as musical instruments.

Many cultures have used shells for money, and they are the most widely and longest used currency in human history. Cowry shells (Cypraeidae) were a common form of ancient money, used in Africa, Asia, including India, and other parts of the world. In China, the Chinese character for 'money' is a representation of a cowrie shell, which they used from about 1200 BCE. In parts of Africa, cowries were used for this purpose as recently as the middle of the twentieth century. The Indo-west Pacific 'money cowry', *Monetaria*[8] *moneta*, was the shell most widely used as currency. 'Chank shells' (*Turbinella*[9] *pyrum*) were used as currency in India until the mid-nineteenth century.

It is not only shells that were important culturally. The Phoenicians produced the royal purple dye from the hypobranchial gland secretion of certain Mediterranean muricids, with references to it dating from about 600 BCE. Variously known as Royal purple, Tyrian purple, and purple of the Phoenicians and ancient Romans, the practice originated in Tyre (hence the name). The dye was used not only for colouring clothing for nobles, but also as ink for writing. The extraction of this dye was arguably the first large-scale chemical industry. The use of the dye declined with the Roman Empire, and large-scale production ceased when Constantinople fell in 1453.

10.3.3 Molluscs and Religion

Cowrie shells have been used as religious objects from prehistoric times in Africa, Egypt, India, and China, and in some early European religions. In particular, they have been associated with human female sexual organs, and seen as fertility and fecundity symbols, but were also important in other religious aspects, including the afterlife.

In India and Sri Lanka, the 'sacred chank' (*Turbinella pyrum*) (Figure 10.14) has been important in the Hindu and Buddhist religions. The chank shell is used in ceremonies as a vessel and as a trumpet and was regarded as sacred in the Hindu religion from before 2000 BCE. It plays a part in both Hindu and Buddhist ceremonies today. In Japan, trumpets made from the shell of the 'giant triton' (*Charonia tritonis*) are still used by Shinto priests for a daily call to prayers.

Other molluscs have also been important as religious symbols, as decorations, and as objects for use in ceremonies; for example, a giant clam shell is sometimes used as a baptismal font, and *Pecten* is used in some churches as a religious symbol (it is associated with Saint James).

FIGURE 10.14 Carved sacred chanks (*Turbinella pyrum*) or Shankha – Left: with carved with image of Lakshmi and Vishnu with silver added; from Bangladesh or West Bengal state, India (Pala period – eleventh to twelfth century). Middle: Pala period, India, ca. eleventh century. Right: India, Pala period, eleventh century or earlier. Courtesy of Clair H. Originally posted to Flickr, uploaded to Commons 25 December 2009.

10.3.4 Myths, Fables, and Legends

While many Western fables feature vertebrates and arthropods (e.g., Aesop), molluscs are poorly represented in this genre. One exception to this trend is the more recent *Limpet Fable*, which was first used by the philosopher and Christian apologist C. S. Lewis to explain why people do not believe in the Christian God. In his book *Miracles* (1947), Lewis posited '...a mystical limpet, a sage among limpets, who (rapt in vision) catches a glimpse of what Man is like. In reporting it to his disciples, who have some vision themselves (though less than he) he will have to use many negatives. He will tell them man has no shell, is not attached to a rock, is not surrounded by water. And his disciples, having a little vision of their own to help them, do get some idea of Man... they build up a picture of Man as a sort of amorphous jelly (he has no shell), existing nowhere in particular (he is not attached to a rock)

---

[8] Previously *Cypraea*.
[9] Previously *Xancus*.

and never taking nourishment (there is no water to drift it towards him'. This story was illustrative of the premise of Lewis that what is learned from experience depends on the kind of philosophy brought to the experience and was the centrepiece of his argument against naturalism and for supernaturalism. The works of Lewis are still widely read and cited.

Cephalopods are part of human legends and folklore, and more recently, featuring as monsters in many movies. Sightings of giant squids and large octopuses presumably gave rise to legends of monsters like the Kraken (Figure 10.15).

In parts of medieval Europe, ammonoid fossils were thought to be petrified snakes and were often fitted with snake-like heads. For example, the legend of Saint Hilda of Whitby who, according to English folklore, removed the heads of snakes with a whip, inspired locals to carve snake heads in ammonite fossils. Such 'snakestones' (Figure 10.16) are known from Viking times. In Nepal, ammonoids are thought by Hindus also to have religious significance.

FIGURE 10.15   A cephalopod monster illustrated in 1871 in Jules Verne's *Twenty Thousand Leagues Under the Sea*.

FIGURE 10.16 A snakestone carved from a specimen of *Hildoceras bifrons* from the early Jurassic of Yorkshire, England. Courtesy of The Trustees of the Natural History Museum, London, UK.

Another common myth involving molluscs includes stories of giant clams (*Tridacna* spp.) capturing divers or reef walkers.

10.3.5 MOLLUSCS IN PAINTINGS

Molluscs feature widely in paintings from early times. For example, they were commonly depicted in the ancient civilisations of the Mediterranean region including sites such as Pompeii. One of the most famous paintings is *Birth of Venus* by Botticelli (ca. 1486) which depicts Venus in a *Pecten* shell. Some other early paintings of note depicting shells include *Peace and the Arts* by Cornelis (1607) and *Cabinet d'amateur* by Francken (ca. 1636), both of which are reproduced in Dance (1986) along with more information on this subject.

Because of their brilliant colours and forms shells often decorate still life paintings, but one of the most famous of the shell depictions is *The Shell* (*Conus marmoreus*) by Rembrandt, etched in black and white in 1650. Drab limpets have also been featured in art, including *Blasting Limpets on the Barbary Coast* by William Heath Robinson (1906), and *War. The Exile and the Rock Limpet* (1842) by J. M. W. Turner; the former a tribute to the tenacity of Mediterranean limpets, while the second contrasted the exile of Napoleon to St. Helena with the solitary existence of the patellogastropod.

10.3.6 RECREATIONAL ACTIVITIES

Molluscs, and particularly their shells, are widely used as the focus of hobbies such as shell and fossil collecting (see Section 10.3.7), shell art, and photography of shells and living animals (see Section 10.3.8). Other activities include,

for example, keeping living molluscs in aquaria or collecting images of molluscs on stamps. Snails and slugs are also often kept as pets. The invasive giant African land snail *Achatina fulica* is popular as a pet in some areas and, although illegal in some countries (e.g., USA, Australia), this must increase the risk of illegal imports.

10.3.7 SHELL COLLECTING

Shell collecting has been a popular hobby for hundreds of years. Early books on shells date from the 1600s (see Dance 1986 for a detailed history of shell collecting). The reason for making collections can range from a fascination for the beauty and form of the objects to scientific objectives. Collecting the shells of both living and extinct molluscs can be a useful window for participants to learn about the natural and evolutionary history of an important group of organisms. At the other extreme, in which commercial interests reign, it can encourage exploitation of rare or threatened species, damage habitat, and, sometimes, deplete populations (see Section 10.9.2.1).

10.3.8 MOLLUSC PHOTOGRAPHY

The often beautiful 'opisthobranch' sea slugs have long attracted the attention of naturalists, but increasingly other living molluscs including marine, fresh-water, and terrestrial snails, octopuses and other cephalopods, and, to a lesser extent, bivalves, are being photographed, often displaying colourful features only seen in living animals. These images are produced by both amateur and professional photographers and are becoming increasingly prevalent on the world wide web. The interest in photography of living marine molluscs has increased greatly since the advent of SCUBA (self-contained underwater breathing apparatus) diving, while digital photography and comparatively inexpensive cameras have made nature photography relatively easy for both amateur and professional.

10.4 SOME MEDICAL AND COMMERCIAL APPLICATIONS

10.4.1 MEDICINES AND MEDICAL APPLICATIONS

The medicinal properties of some molluscs have been recognised for centuries and have recently undergone a resurgence through alternative medicines (Pereira et al. 2016) and in assessing the bioactivity of extracts and secondary metabolites from molluscs (e.g., see Avila 2006; Benkendorff 2010), particularly in their use as anticancer (e.g., Simmons et al. 2005) and antimicrobial agents. These latter substances are mainly peptides from various molluscs which can be safely used in aquaculture and may have wider applications (e.g., Li et al. 2009; Li et al. 2011), including in human medicine as replacements for antibiotics.

Of particular note are the many hundreds of conotoxins (peptides) present in cone shell venom, one of which is now marketed as an effective drug in the relief of chronic pain, and others are in trials. Each species of *Conus* has between 40 and 200 peptides with little overlap between species (Myers et al.

1993; Livett et al. 2004; Terlau & Olivera 2004; Buczek et al. 2005; Han et al. 2008; Twede et al. 2009; Robinson et al. 2014).

Molluscs have also long been used in traditional medicines (e.g., Herbert et al. 2003; Lewbart 2006; Benkendorff 2010). Examples include the medicinal and aphrodisiac properties of oysters since at least Roman times, the fresh-water bivalve *Corbicula* is thought to improve liver function, and extract from the New Zealand green mussel (*Perna canaliculus*) is said to help with joint mobility.

In more conventional medicine, there is recognition of the considerable potential use of chemical compounds produced by molluscs, although this is largely an untapped resource (e.g., Avila 2006; Benkendorff 2010). To cite just a few examples, a compound (acharan sulphate) produced by *Achatina fulica* has antitumour activity (Lee et al. 2003c), while extracts from the potamidid *Telescopium* are reported to have antimicrobial, antiprotozoal, anti-fertility, and immunomodulation properties (Roy et al. 2010), and extracts of galactans from the eggs of *Pomacea* assisted healing of wounds in the small intestine of rats (Cruz et al. 2004).

The chemical substances produced by sea slugs (see Chapter 9), notably nudibranchs, have resulted in them being significant targets for natural products research (e.g., Fontana 2006; Benkendorff 2010), but the potential of bioactive secondary metabolites of molluscs has hardly been realised (Benkendorff 2010). For example, the ink of aplysiids has been shown to have many bioactive molecules (Yamada et al. 2010; Pereira et al. 2016) and dolastatins from the aplysiid *Dolabella auricularia* have been trialled as antitumour agents. Other examples include the antitumour properties of coleoid ink (Derby 2014), *Megathura crenulata* haemocyanin (Harris & Markl 1999) and brominated indole derivatives from the hypobranchial gland of the muricid whelk *Dicathais orbita* (e.g., Esmaeelian et al. 2013). Other cancer-related applications include the assessment of gene expression by cancer-related genes in a mussel (*Mytilus edulis*) and the transmission of a leukaemia-like cancer in the 'soft-shelled clam' *Mya arenaria* in natural situations to assist in modelling contagious cancer (Metzger et al. 2015).

10.4.2 OTHER PRACTICAL USES, INCLUDING BIOMATERIALS

There are many practical uses of molluscs other than those outlined above (see Section 10.3.2). A few examples are given below.

Besides decorative uses, nacre from bivalve shells was used by Mayans in Central America to construct the first dental implants over 2000 years ago (Bobbio 1972). Nacre is also of interest today. The nacreous abalone is very strong due to its construction of tile-like calcium carbonate crystals held together with protein. Impacts result in the 'tiles' sliding in the flexible protein, thus absorbing the energy of the impact. This configuration has inspired the use of this concept to help build stronger body armour (e.g., Huang et al. 2011).

The internal shell of a cuttlefish (cuttlebone) is sold in the pet trade as a calcium source for caged birds and some other pets. In the past, the cuttlebone was ground into a powder for use as a polishing agent, added to toothpaste, or used as an antacid or an absorbent (Norman & Reid 2000).

In another practical application, molluscs have long provided reliable and long-lasting writing ink with one source being Tyrian purple (see Section 10.3.2), another being ink from octopuses and some other coleoid cephalopods.

The byssus of the very large, and now rare, *Pinna nobilis* of the Mediterranean was made into cloth (sea silk) and was highly prized by ancient Greeks, Romans, Arabs, and Chinese, and continued to be traded until the early nineteenth century. More recently, mussel (*Mytilus*) byssus threads have been investigated because they are extremely strong and are firmly anchored to permanently wet substrates. The very adhesive properties of the byssus are unlike synthetic adhesives, in not being degraded by seawater. This realisation has resulted in research including the investigation of the considerable practical potential of these adhesive proteins (e.g., Hagenau & Scheibel 2010). The proteins in byssal threads have also been shown to have self-healing properties with many potential applications (e.g., Vaccaro & Waite 2001; Holten-Andersen et al. 2011; Lee et al. 2011).

Land snail mucus has been long used for various mainly medicinal purposes, while the mucus of various marine gastropods has inspired visco-elastic adhesives (Dodou et al. 2011). Some of these adhesive gels form attachments that can approach the strength of barnacle and mussel byssus glues, yet they are temporary. With the ability to form strong, temporary attachments, these adhesives would have many practical applications.

The radular teeth of chitons and some gastropods, notably patellogastropods, are well-known for their hardness, being impregnated with iron (see Chapter 5). These have been investigated as inspiration for producing very hard materials (e.g., Weaver et al. 2010; Barber et al. 2015; Joester & Brooker 2016).

Many molluscan fossils, particularly ammonites, have value as collectable items or ornaments, but one group of mollusc fossils indirectly have a major economic value. Cretaceous rudist shells formed massive accumulations during the Cretaceous. Today, these carbonate platforms hold substantial oil reservoirs in the Gulf of Mexico and the Middle East (Simo et al. 1993).

Some cephalopod fossils have broad applications. The geochemical signature obtained from fossil belemnite guards from the Cretaceous Peedee Formation in South Carolina, USA is a global standard (VPDB [Vienna Pee Dee Belemnite] – previously PDB [Pee Dee Belemnite]), against which all other such samples are measured for both carbon and oxygen isotopes. Belemnites have also been used to establish palaeotemperatures using oxygen isotope analysis. In addition, many ammonoids and some belemnites (e.g., Doyle & Bennett 1995) are important index fossils enabling geologists to date Mesozoic strata. This is particularly true of ammonoids, due to their complex suture morphologies which enable them to be identified relatively easily. This, coupled with their rapid speciation and widespread distributions, makes them excellent index fossils. Similarly, other molluscan fossils are also important in stratigraphy where they are used for dating rocks through the Phanerozoic and Cenozoic.

10.4.3 MOLLUSCS AS EXPERIMENTAL ANIMALS

Some molluscs have proved exceptional models, providing significant insights into several fundamental processes such as the production and function of neuroendocrines, synaptic transmission in nerves, the generation of the action potential, the neural basis of learning and memory, and the reception of noxious and favourable stimuli and responses to them. Probably the most famous were the studies of Eric Kandel and his colleagues, using *Aplysia*, in which they demonstrated how memory is located in modified synapses, and how it is lost. While *Aplysia* was particularly important, some other heterobranch sea slugs, several land snails (e.g., *Helix* and *Achatina*) and *Lymnaea* have also been important in experimental studies on neuron function, endocrine systems, sensory systems, and studies on the neurophysiological basis of learning and behaviour (see also Chapter 7). Many other molluscs have proved valuable in laboratory experiments addressing a wide range of questions. These include cephalopods, notably species of *Octopus*, *Sepia*, and *Loligo*.

Molluscs are valuable as experimental animals for various reasons. They are often easy to maintain in the laboratory and have sufficiently well-developed nervous systems. Individual taxa provide special advantages. For example, the giant axon of the squid can be 100–1000 times larger than a mammalian axon, and the neurons in the CNS of euthyneuran heterobranchs are of unusually large size and relatively few, enabling them to be targeted individually.

Some molluscs may also be useful in screening for drug effects (e.g., Adriaens & Remon 1999).

10.4.3.1 Animal Welfare Issues When Working with Molluscs

Although molluscs have been widely used as model organisms in experiments, little attention has been paid to animal welfare issues. Crook and Walters (2011) reviewed the evidence for molluscs experiencing pain (nociception) and concluded that pain-like states may be experienced by at least those taxa with complex nervous systems and pointed out the difficulties with actually demonstrating this. They concluded with a recommendation that investigators should attempt to minimise the potential for pain-like sensations in their experimental subjects. Particular concern has been expressed regarding cephalopods (e.g., Moltschaniwskyj et al. 2007).

10.5 MOLLUSCS AS DISEASE VECTORS

An important negative human impact of a few molluscs is through their roles as disease carriers. Gastropods can serve as intermediate hosts for many parasites, particularly trematodes (Platyhelminthes) and, to a much lesser extent, nematodes. Although the vast majority of these parasites are harmless to humans and domestic animals, a few are involved in serious diseases. Work on molluscan intermediate hosts has spawned major research activity called 'medical malacology' (e.g., Malek & Cheng 1974).

The most important group of molluscan parasites are digenean trematodes (see Chapter 9) which are endoparasites of vertebrates, including humans (Table 10.1).

The most serious human disease caused by snail-borne trematodes is schistosomiasis (or bilharzia, named after Theodor Bilharz, a German pathologist who worked in Egypt in the 1850s) caused mainly by three 'human blood flukes' – *Schistosoma mansoni* (Africa, Middle East, and South America), *S. haematobium* (Africa and Middle East), and *S. japonicum* (Far East, particularly China). These 'blood flukes' live in the veins in the mesenteries attached to organs in the body including the intestine (*S. japonicum*, *S. mansoni*) or bladder (*S. haematobium*). Around 200 million people are typically infected, mostly in Africa and Asia. The disease is contracted through contact with water containing the free-swimming larvae which penetrate the skin (Figure 10.17). It can be severely debilitating, can stunt growth in children, and can be fatal. The intermediate snail hosts are the planorbids *Biomphalaria* spp. (*S. mansoni*) and *Bulinus* spp. (*S. haematobium*) and the pomatiopsids *Oncomelania* spp. (*S. japonicum*). Two species of *Schistosoma* are much more localised than the previous three, these being *Schistosoma mekongi*, with the pomatiopsid *Neotricula aperta* as the intermediate host, which is found in parts of Cambodia and Laos, and *Schistosoma intercalatum*, with the intermediate host being two species of the planorbid *Bulinus*, found in parts of western and central Africa (see Table 10.1).

Most fluke infections in humans can be treated with antihelmintics (or anthelminthics) such as praziquantel and various benzimidazoles which, in Africa, have been greatly assisted by funds from the Bill & Melinda Gates Foundation.

There are several genera of 'blood flukes' that infect birds, and while many do not cause problems for humans, some can cause cercarial dermatitis, an affliction known by a variety of names such as 'swimmers itch' or 'duck itch'. It occurs when the cercariae partially penetrate the skin causing an immune reaction and rash-like symptoms. It is known in most parts of the world and occurs in both fresh-water lagoons and lakes (as in Europe and North America) and also in estuaries (e.g., in parts of Australia).

Other significant trematodes include the human 'liver flukes' *Clonorchis sinensis* (Chinese liver fluke) and *Opisthorchis viverrini* (Southeast Asian liver fluke), with several taxa of fresh-water snails, all caenogastropods (bithyniids, thiarids, assimineids, pachychilids), acting as the first intermediate host of the former and bithyniids for the latter. The snail infection is followed by a fish as the second intermediate host (SIH). The parasite is transferred to the human host by eating raw or poorly cooked fish and infects the bile ducts of the liver. The 'cat liver fluke', *Opisthorchis felineus*, also infects humans and a bithyniid is the first intermediate host.

TABLE 10.1

A List of Most Trematodes Which Infect Humans, with Their Definitive Hosts, Intermediate Hosts and Distribution

Fluke	Common Name and Definitive Host	Disease	Intermediate Hosts	Region and Habitat
Echinostomata				
Fasciola hepatica (Fasciolidae)	Sheep liver fluke; sheep, cattle, and other ruminants, and some other mammals including humans	Fascioliasis	Several Lymnaeidae; metacercariae encyst on aquatic plants	Originally from Europe, but now in most of the world. Fresh water
Fasciola gigantica (Fasciolidae)	Ruminants, also occasionally humans	Fascioliasis	Several Lymnaeidae; metacercariae encyst on aquatic plants	Asia, Africa. Fresh water
Fasciolopsis buski (Fasciolidae)	Giant intestinal fluke (the largest human fluke); pigs, humans	Fasciolopsiasis	Planorbidae (species of *Segmentina*. *Gyraulus* and *Hippeutis*); SIH: metacercariae encyst on aquatic plants	South and Southeast Asia. Fresh water
Echinostoma ilocanum (Echinostomatidae)	Intestinal fluke; humans and various mammals and birds	Echinostomiasis	*Gyraulus* and *Hippeutis* (Planorbidae); SIH *Lymnaea*, *Pila*, Viviparidae; *Clea* as well as fish and tadpoles	Asia, including Indonesia. Fresh water
Echinostoma spp (Echinostomatidae)	Intestinal fluke; humans and various mammals and birds	Echinostomiasis	Lymnaeidae, Physidae, Planorbidae, Viviparidae; SIH various snails, fish, tadpoles, and bivalves (*Corbicula*, 'mussels')	World-wide (human infections most commonly in Asia). Fresh water
Acanthoparyphium tyosenense (Echinostomatidae)	Intestinal fluke; birds and humans	Echinostomiasis	Naticidae; SIH Naticidae, and marine bivalves (Mactridae, Solenidae)	Korea. Marine
Artyfechinostomum malayanum (Echinostomatidae)	Intestinal fluke; various mammals and occasionally humans	Echinostomiasis	Planorbidae (*Gyraulus*, *Indoplanorbis*) (another species, *A oraoni*, has been reported with infections in India, has a lymnaeid as its primary host); SIH *Pila*, Lymnaeidae, Bithyniidae	Southeast Asia, India. Fresh water
Echinochasmus spp. (Echinostomatidae)	Intestinal fluke; various birds and mammals, including humans	Echinostomiasis	Viviparidae, Bithyniidae, Lymnaeidae; SIH fish	Japan, China, Korea, and Middle East. Fresh water
Echinoparyphium recurvatum (Echinostomatidae)	Intestinal fluke; birds and mammal including rats and sometimes humans	Echinostomiasis	Physidae, Planorbidae, Lymnaeidae, Valvatidae; SIH tadpoles, frogs, Planorbidae, and Lymnaeidae	Taiwan, Indonesia, Egypt. Fresh water
Episthmium caninum (Echinostomatidae)	Intestinal fluke; various mammals such as dogs, birds, some human infections	Echinostomiasis	First intermediate host unknown; SIH fresh-water fish	India, Thailand. Fresh water
Himasthla muehlensi (Echinostomatidae)	Intestinal fluke; birds, one reported case in humans	Echinostomiasis	*Littorina* (Littorinidae); SIH bivalves (*Mytilus*, *Mya*, *Mactra*)	Brazil, Central America, and USA. Marine
Hypoderaeum conoideum (Echinostomatidae)	Intestinal fluke; birds, sometimes humans	Echinostomiasis	Planorbidae, Lymnaeidae; SIH tadpoles	Europe, Siberia, Japan, and Thailand. Fresh water
Isthmiophora melis (Echinostomatidae)	Intestinal fluke; rodents and carnivores, occasionally humans	Echinostomiasis	Lymnaeidae; SIH tadpoles, fresh-water fish	Romania, Russia, China, Taiwan, and USA. Fresh water
Philophthalmus spp. (Philopthalmidae)	Eye flukes; birds, rare in humans	Philophthalmiasis	Thiaridae, Batillariidae; encyst on aquatic vegetation or various hard surfaces	Asia, southern North America, South America, Africa, Europe, Russia, and Australia. Marine and fresh water
Opisthorchiata				
Clonorchis sinensis (Opisthorchiidae)	Chinese liver fluke; fish-eating mammals including humans	Clonorchiasis	Bithyniidae, one Assimineidae, some Thiaridae, and *Semisulcospira*; SIH fish	Asia, Russia. Fresh water
Opisthorchis viverrini (Opisthorchiidae)	Southeast Asian liver fluke; fish-eating mammals including humans	Opisthorchiasis	Various Bithyniidae; SIH fish	Asia. Fresh water

(Continued)

TABLE 10.1 (*Continued*)

A List of Most Trematodes Which Infect Humans, with Their Definitive Hosts, Intermediate Hosts and Distribution

Fluke	Common Name and Definitive Host	Disease	Intermediate Hosts	Region and Habitat
Opisthorchis felineus (Opisthorchiidae)	Cat liver fluke; fish-eating mammals including humans	Opisthorchiasis	Various Bithyniidae; SIH fish	Europe, Russia, and Turkey. Fresh water
Metagonimus spp. (Heterophyidae)	Intestinal fluke; fish-eating mammals (especially dogs, cats, rats), including humans (especially *M. yokogawai*)	Metagonimiasis or heterphyidiasis	Semisulcospiridae. SIH fish	Parts of Asia, southern Europe, and Russia. Fresh water
Haplorchis taichui (Heterophyidae)	Intestinal fluke; humans and other fish-eating mammals, birds	Heterphyidiasis	Thiaridae; SIH fish	Asia, Middle East. Fresh water
Haplorchis yokogawai and *H. pumilo* (Heterophyidae)	Intestinal fluke; humans, various other fish-eating mammals, and birds	Heterphyidiasis	Thiaridae; SIH fish	Asia, Hawaii, Egypt, and Australia. Fresh water
Haplorchis other spp. (Heterophyidae)	Intestinal fluke; various mammals, including humans	Heterphyidiasis	*Cleopatra* (Paludomidae), Thiaridae; SIH fish	Asia, Egypt, and Australia. Fresh water
Centrocestus armatus (Heterophyidae)	Intestinal fluke; various mammals including cats and dogs; rarely humans	Heterphyidiasis	*Semisulcospira;* SIH frogs, fish	Malaysia, Thailand, Japan and China, and Egypt. Fresh water
Centrocestus spp. (Heterophyidae)	Intestinal fluke; fish-eating birds and mammals, sometimes humans	Heterphyidiasis	Thiaridae; SIH fish	Asia, Tunisia, Hawaii, and Mexico. Fresh water
Apophallus donicus (Heterophyidae)	Intestinal fluke; various fish-eating mammals, and sometimes humans	Heterphyidiasis	*Fluminicola* (Lithoglyphidae); SIH fish	USA
Ascocotyle longa (Heterophyidae)	Intestinal fluke; fish-eating birds and mammals, including humans	Heterphyidiasis	*Heleobia* (Cochliopidae); SIH fish	Europe, Asia, Africa, and the Americas. Estuarine
Cryptocotyle lingua (Heterophyidae)	Intestinal fluke; fish-eating birds and mammals, rarely humans	Heterphyidiasis	Littorinidae; SIH fish	Europe, Russia, North America, Greenland, and Asia. Marine
Heterophyes spp. (Heterophyidae)	Intestinal fluke; carnivorous mammals, sometimes humans	Heterphyidiasis	Potamididae; SIH fish	North Africa, Middle East, eastern Mediterranean, and Asia. Estuarine
Heterophyopsis continua (Heterophyidae)	Intestinal fluke; in cats and birds, sometimes humans	Heterphyidiasis	Snail host unknown; SIH fish	Japan, Korea. Estuarine
Procerovum spp. (Heterophyidae)	Intestinal fluke; dogs, rarely humans	Heterphyidiasis	Thiaridae; SIH fish	Asia, Africa, and Australia. Fresh water
Pygidiopsis summa (Heterophyidae)	Intestinal fluke; cats, dogs, and humans	Heterphyidiasis	Potamididae; SIH fish	Japan, China, and Korea. Estuarine
Stellantchasmus spp. (Heterophyidae)	Intestinal fluke; various mammals (e.g., rats, dogs, cats), sometimes humans	Heterphyidiasis	Thiaridae; SIH fish	Asia, Hawaii, and Palestine. Fresh water
Stictodora spp. (Heterophyidae)	Intestinal fluke; cats, dogs, seagulls, rarely humans	Heterphyidiasis	Batillariidae; SIH fish	Korea, Australia. Estuarine

Other genera of Heterophyidae (intestinal flukes) which occasionally infect humans are *Diorchitrema* and *Phagicola*.

Xiphidiata

Fluke	Common Name and Definitive Host	Disease	Intermediate Hosts	Region and Habitat
Plagiorchis spp. (Plagiochiidae)	Intestinal fluke; birds, bats, rats, and cats	Plagiorchiidiasis	Lymnaeidae; SIH larval insects, snails, and fish	Asia. Fresh water
Paragonimus westermani (Troglotrematidae)	Oriental lung fluke; various carnivores, rodents, pigs, and humans	Paragonimiasis	*Semisulcospira* spp. (Semisulcospiridae), *Oncomelania* (Pomatiopsidae); SIH crabs, crayfish	Mainly SE Asia, China, and Japan. Fresh water
Paragonimus spp. (Troglotrematidae)	Lung flukes; mammals including humans (about ten species infect humans)	Paragonimiasis	Various cerithioidean and truncatelloidean snails; SIH crabs, crayfish	Southeast Asia, West Africa, and Central and South America. Fresh water

(Continued)

TABLE 10.1 (*Continued*)

A List of Most Trematodes Which Infect Humans, with Their Definitive Hosts, Intermediate Hosts and Distribution

Fluke	Common Name and Definitive Host	Disease	Intermediate Hosts	Region and Habitat
Phaneropsolus bonnei (Lecithodendriidae)	Intestinal fluke; humans and presumably other hosts	Lecithodendriidiasis	Bithyniidae; SIH insects (Odonata)	Southeast Asia, India. Fresh water
Prosthodendrium molenkampi (Lecithodendriidae)	Intestinal fluke; humans and insectivores	Lecithodendriidiasis	Bithyniidae; SIH insects (Odonata)	Indonesia, Thailand, Laos, and Vietnam. Fresh water
Paralecithodendrum spp. (Lecithodendriidae)	Intestinal fluke; humans and insectivores	Lecithodendriidiasis	*Thais* (Muricidae); SIH insects (Odonata)	Indonesia.
Spelotrema brevicaeca (Microphallidae)	Intestinal fluke; fish, skinks, birds (terns), and presumably other hosts; rarely humans	Microphallidiasis	Snail host unknown (a congeneric species uses *Bittium*); SIH decapod crustaceans	Philippines
Nanophyetus salmincola (Nanophyetidae)	Intestinal fluke; humans and various other fish-eating mammals and birds	Nanophyetiasis	*Juga plicifera* (Semisulcospiridae); SIH fish (salmonids)	Western North America and eastern Siberia. Fresh water
Dicrocoelium dendriticum and *D. hospes* (Dicrocoeliidae)	Liver fluke; ruminants, rarely humans	Dicrocoeliasis	Various land snails; SIH ants	Europe, North Asia, North America, North Africa, and Australia. Terrestrial
Gymnophalloides seoi (Gymnophallidae)	Intestinal fluke; humans, and various mammals and seashore birds	–	First intermediate host not known; SIH Ostreidae (*Crassostrea*)	Korea.
Metorchis conjunctus (Opisthorchiidae)	Liver fluke; fish-eating mammals and sometimes humans		*Amnicola limosus* (Amnicolidae); SIH fish.	North America. Fresh water
Eurytrema pancreaticum, E. coelomaticum, and *E. ovis* (Heterophyidae)	Pancreas fluke; infects pancreas, liver and intestine; various animals including sheep, goats, pigs, cattle, and also humans	Eurytrematiasis	Land snails such as *Bradybaena, Cathaica*; SIH locusts	Asia, South America. Terrestrial

Pronocephalata

Fischoederius elongatus (Gastrothylacidae)	Intestinal fluke; ruminants, rarely humans	Paramphistomatidiasis	Unknown; Redia encyst on plants etc.	China and India
Watsonius watsoni (Gastrodiscidae)	Intestinal fluke; various primates, rarely humans	Paramphistomatidiasis	Presumably a snail, but identity unknown; Redia encyst on plants etc.	Africa
Gastrodiscoides hominis (Gastrodiscidae)	Intestinal fluke; pigs, various other mammals such as rats, monkeys, and occasionally humans	Gastrodiscoidiasis	*Helicorbis* (Planorbidae); SIH tadpoles, frogs, crayfish, squid(?), aquatic plants	Asia, Kazakhstan. Fresh water

Diplostomata

Alaria americana (Diplostomidae)	Intestinal fluke (can also infect other parts of the body including eyes); usually in dogs and related mammals, rarely humans	Diplostomiasis	*Heliosoma* and related Planorbidae; SIH tadpoles and frogs	Northern North America. Fresh water
Fibricola cratera (Diplostomidae)	Intestinal fluke; various mammals, rarely humans	Diplostomiasis	Physidae; SIH frogs	North America. Fresh water
Neodiplostomum seoulense (Diplostomidae)	Intestinal fluke; rats, mice, etc., and sometimes humans	Diplostomiasis	Planorbidae (*Hippeutis, Segmentina*); SIH tadpoles and frogs. Snakes are paratenic hosts and can transmit the parasite to humans	Korea and northeast China. Fresh water
Brachylaima cribbi (Brachylaimidae)	Intestinal fluke; reptiles, birds, mammals including humans	Brachylaimidiasis	Helicidae	Australia. Terrestrial

(*Continued*)

TABLE 10.1 (*Continued*)

A List of Most Trematodes Which Infect Humans, with Their Definitive Hosts, Intermediate Hosts and Distribution

Fluke	Common Name and Definitive Host	Disease	Intermediate Hosts	Region and Habitat
Cotylurus japonicus (Strigeidae)	Intestinal fluke; birds, mammals, rarely humans	Strigeidiasis	Lymnaeidae, Physidae, Planorbidae	Japan, China. Fresh water
Clinostomum complanatum (Climostomatidae)	Infects throat; usually in wading birds; rarely humans	Clinostomiasis	*Biomphalaria* (Planorbidae); SIH fish.	Asia. Fresh water
Schistosoma mekongi (Schistosomatidae)	Blood fluke (infects mesentery veins of intestine); humans, dogs, pigs	Schistosomiasis	*Neotricula aperta* (Pomatiopsidae)	Laos, Cambodia. Fresh water
Schistosoma japonicum (Schistosomatidae)	Blood fluke (infects mesentery veins of intestine); hosts include a wide range of mammals including humans	Schistosomiasis	*Oncomelania* spp. (Pomatiopsidae)	China, Taiwan, Southeast Asia, Philippines, Sulawesi, and (extinct in Japan). Fresh water
Schistosoma haematobium (Schistosomatidae)	Blood fluke (infects veins around bladder); humans	Schistosomiasis	*Bulinus* spp. (Planorbidae)	Africa, Middle East, and Corsica. Fresh water
Schistosoma mansoni (Schistosomatidae)	Blood fluke (infects mesentery veins of intestine); humans, but can also infect rodents and primates	Schistosomiasis	*Biomphalaria* spp. (Planorbidae)	South America, Caribbean, Africa, and Madagascar. Fresh water
Schistosoma intercalatum (Schistosomatidae)	Blood fluke (infects mesentery veins of rectum); humans	Schistosomiasis	*Bulinus* spp. (Planorbidae)	Western and Central Africa. Fresh water

Data from various sources including Wikipedia and published papers including Grove, D.I., *A History of Human Helminthology*, CAB International, Wallingford, UK, 1990; Fried, B. et al., *Parasitol. Res.*, 93, 159–170, 2004; Chai, J.-Y., Intestinal flukes, in Murrell, K.D. and Fried, B. (Eds.), *Food-Borne Parasitic Zoonoses. Fish and Plant-borne Parasites*, Springer, New York, pp. 53–115, 2007; Chai, J.Y. et al., *Korean J. Parasitol.*, 47, S69–S102, 2009; Madsen, H. and Hung, N.M., *Acta Tropica*, 140, 105–117, 2014.

Abbreviation: SIH, second intermediate host.

Two important 'liver flukes' of stock (such as cattle and sheep), and occasionally humans, are *Fasciola hepatica* in temperate areas and *F. gigantica* in tropical regions – both with species of lymnaeid snails as intermediate hosts (Figure 10.18).

Other 'liver flukes' that occasionally infect humans such as *Dicrocoelium dendriticum*, which is commonly found in sheep in Europe, can infect the bile ducts of humans. The first intermediate host is the small land snail *Conchlicopa lubrica* and the second host is an ant. A related species, *Dicrocoelium hospes*, also sometimes infects humans and is common in cattle in Africa.

Pancreatic flukes *Eurytrema pancreaticum*, *E. coelomaticum*, and *E. ovis* are found in stock and also occasionally infect humans. Their intermediate hosts are first land snails and the second an insect.

The 'Oriental lung fluke' *Paragonimus westermani* infects species of the fresh-water snail *Semisulcospira* (Pachychilidae) as its first intermediate host and decapod crustaceans as the second intermediate host. After being ingested, it becomes encapsulated in the parenchyma of the lung of the definitive host, which includes several mammals, including humans. Two other 'lung flukes', *Paragonimus mexicana* and *P. skrjabini*, also sometimes infect humans.

The 'intestinal fluke', *Fasciolopsis buski*, which is the largest human fluke (up to about 7.5 cm long), is a common parasite of pigs in Asia, and mainly has species of planorbids as its intermediate host. There are several other 'intestinal flukes' that can infect humans (see Table 10.1), and these include *Heterophyes heterophyes*, *Metagonimus yokogawai*, *Echinostoma ilocanum*, *Watsonius watsoni*, and *Gastrodiscoides hominis*.

Some intestinal flukes such as the lecithodendriids *Phaneropsolus bonnei* and *Prosthodendrium molenkampi* have aquatic insect larvae (Odonata) as their second intermediate hosts and infect humans because the nymphs are eaten by some communities.

Some nematodes are also significant parasites, but only a few of those carried by molluscs are of concern in human health. Of these, probably the most important is the 'rat lung worm' *Angiostrongylus cantonensis*, found in the lungs

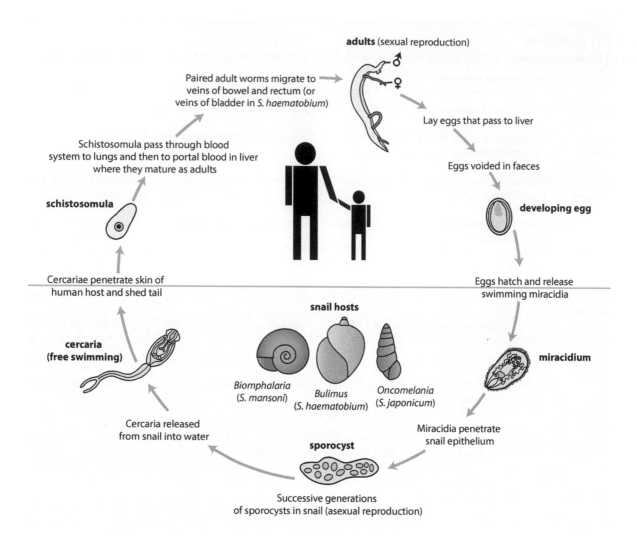

adults (sexual reproduction)

Paired adult worms migrate to
veins of bowel and rectum (or
veins of bladder in *S. haematobium*)

Lay eggs that pass to liver

Schistosomula pass through blood
system to lungs and then to portal blood in liver
where they mature as adults

Eggs voided in faeces

schistosomula

developing egg

Cercariae penetrate skin of
human host and shed tail

Eggs hatch and release
swimming miracidia

snail hosts

**cercaria
(free swimming)**

miracidium

Biomphalaria
(*S. mansoni*)

Bulimus
(*S. haematobium*)

Oncomelania
(*S. japonicum*)

Cercaria released
from snail into water

Miracidia penetrate
snail epithelium

sporocyst

Successive generations
of sporocysts in snail (asexual reproduction)

FIGURE 10.17   The life cycle of *Schistosoma* spp. Redrawn and modified from various sources.

of rodents that pass the eggs via their faeces (Chapter 9, Figure 9.12). The infection is passed on via a molluscan (typically a terrestrial snail or slug) or crustacean paratenic host and the worms migrate from the gut to the brain causing debilitating eosinophilic meningitis or meningoencephalitis which can be fatal. Humans contract the disease mainly through ingesting infected gastropod intermediate hosts. It is most common in the islands of the Pacific, Southeast Asia, and Taiwan. *Achatina* is a vector (Iwanowicz et al. 2015), as are many other introduced snails and slugs (e.g., veronicellids) in the tropics, and other terrestrial snails and slugs in warm temperate areas (Cowie 2013a, 2013b; Kim et al. 2014). Another species, *Angiostrongylus costaricensis* in South America, causes a gastrointestinal syndrome in humans and is also carried by rats with mainly terrestrial slugs as the intermediate hosts.

10.6 POLLUTION AND HEAVY METAL INDICATORS/MONITORING

Shellfish populations suffer negative population impacts from human activities other than fishing (notably coastal pollution) and can serve as indicators of nearshore ecosystem health. Molluscs are widely used for monitoring chemical and bacterial pollutants in marine and fresh-water systems. Such studies are undertaken routinely to monitor, for example, organic pollution, heavy metals (see Chapter 6), and chemicals such as tributyltin (TBT) (see Section 10.9.5.1).

For species to be useful as indicator organisms, they should meet several prerequisites (e.g., Phillips 1980; Rainbow 1995). These include relevant practical considerations such as appropriate distribution and ease of sampling, identification, and handling. Factors influencing the uptake and accumulation of

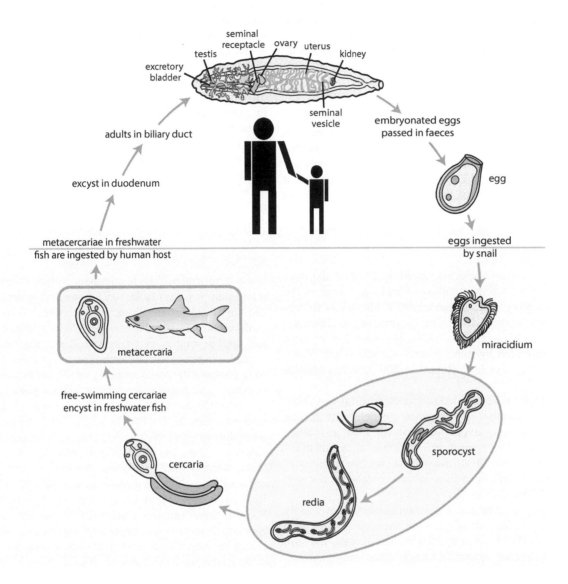

seminal
receptacle ovary uterus
testis kidney
excretory
bladder

adults in biliary duct

seminal
vesicle

excyst in duodenum

embryonated eggs
passed in faeces

egg

metacercariae in freshwater
fish are ingested by human host

eggs ingested
by snail

metacercaria

miracidium

free-swimming cercariae
encyst in freshwater fish

sporocyst

cercaria

redia

FIGURE 10.18 The life cycle of *Opisthorchis viverrini*, the Southeast Asian human liver fluke. Redrawn and modified from various sources.

metals should also be well understood – e.g., the effects of temperature, season, age, body size and diet, and the relationship between background concentrations and uptake. Bivalves have been extensively used as pollution indicators, not only for heavy metals, but for TBT and other pollutants in fresh-water, estuarine, and marine ecosystems (e.g., Byrne & O'Halloran 2001).

Molluscs are also useful for monitoring changes in the composition of animal communities, as they are easily collected, relatively easily identified, and tend not to be markedly seasonal. Examples of molluscs commonly used in biological monitoring studies of rocky shore communities include mytilids, patellids and other large limpets, littorinids, trochids, and muricid whelks.

Some species of molluscs can accumulate metals with the concentrations reflecting the levels of those that are bioavailable in their environment making them useful indicators of metal pollution, including dangerous metals such as lead, mercury, and uranium. This is particularly true of bivalves which are more widely used than any other group of organisms in biomonitoring metals in both marine and fresh-water environments (Phillips 1990; Byrne & O'Halloran 2001), and even historical data can be retrieved. For example, shells of fresh-water mussels (Hyriidae) were used to archivally retrieve levels of heavy metal contaminants (including copper and uranium) associated with a mine in northern Australia (Markich et al. 2002). Some gastropods, including terrestrial species (e.g., Berger & Dallinger 1993), and marine and fresh-water taxa, are also used for such monitoring.

Uptake of toxic heavy metals in some molluscs can cause potential health risks for humans and other animals. For example, lead and arsenic can be accumulated in garden snails in rather high quantities (e.g., Coeurdassier et al. 2010).

10.7 MOLLUSCS AS PESTS

While a few native species of molluscs have become pests in some areas, most molluscan pests are invasive species that have become established in areas outside their native habitats. Relatively few molluscs are highly successful colonisers, but outlined below are some of those which have become pests where they have been introduced.

10.7.1 MOLLUSCAN INVADERS

The spread of invasive species results in some of the most harmful impacts to ecosystems worldwide and some of the most difficult to reverse. It can result in alterations to ecosystems, often to their detriment, biodiversity decline, and homogenisation (Cowie et al. 2009). The spread of invasive species is facilitated by many human activities, both direct and indirect (e.g., Cowie & Robinson 2003), and by climate change.

Some molluscs have invaded marine, terrestrial, and freshwater habitats because of accidental or deliberate human intervention, including importing them for food, as biological control agents (see Section 10.7.2), and as aquarium occupants or even as pets. Most invasive species (the majority of which are small) have been introduced accidentally, not only as adults, but also as juveniles or eggs.

While many species become established, they are rarely regarded as pests, and although their interactions with local molluscs and other biota may still be detrimental, they are subtle and difficult to quantify. Others cause major agricultural and ecosystem impacts.

Important accidental pathways of mollusc introductions include, in the case of marine molluscs, transport on hulls and in ballast tanks of ships, of fresh-water molluscs by the aquarium trade, and terrestrial gastropods with agricultural produce such as cut flowers and potted plants. Some are also transported on equipment, various goods, cars, and boats.

Non-native molluscs may have even been seen on the deep-sea hydrothermal vents (Van Dover 2014) with, for example, some having been accidentally transferred from sampling gear (Voight et al. 2012), but there is, as yet, no record of establishment.

10.7.1.1 Some Characteristics That Make a Successful Invader

A potential invading species must first be transported to its new environment, where it must survive to reproduce and spread to new locations. Invaders appear to establish most successfully in disturbed habitats and ultimately often displace native fauna (e.g., Vermeij 1996b). Some characteristics shared by molluscan invaders are that they are opportunistic and tolerant of a wide range of environmental conditions, have high fecundity, are often hermaphrodites with the ability to self-fertilise (as in many stylommatophorans), or parthenogenetic, and some may brood their young. These latter characteristics enable even a single individual to colonise successfully. The combination of all these features is, perhaps fortunately, only possessed by a relatively small proportion of species. Parthenogenetic brooding caenogastropods include some cerithioidean thiarids and the truncatelloidean *Potamopyrgus* (see Section 10.7.1.3).

10.7.1.2 Terrestrial Invasive Species

It is in terrestrial habitats that invasive molluscs are most often recognised as pests (Table 10.2). Some European land snails and slugs have become virtually cosmopolitan – the 'common garden snail', *Cornu* (previously *Helix* or *Cantareus*) *aspersum*, and several slugs (Arionidae, Limacidae) being good examples in temperate and warm temperate areas. Some helicoideans from the Mediterranean, which favour warmer, drier climates, have become serious agricultural pests in southern Australia (Baker 1989, 2002), these being *Theba pisana*, *Cernuella virgata*, *Cochlicella acuta*, and *C. barbara*. Others are mainly tropical, such as the elongate 'miniature awl snail' *Subulina octona*, the 'giant African snail' *Achatina fulica*, the 'Asian banded snail' *Babybena similaris*, and some veronicellid slugs such as *Sarasinula plebeia* (Rueda et al. 2002), *Veronicella cubensis*, and *Laevicaulis alte* (Kim et al. 2016). Many of these introduced terrestrial molluscs are garden and agricultural pests (e.g., Barker 2002), and these same species can be problems in domesticated habitats (such as gardens) in their original areas. Some local species can become pests, notably in Europe (e.g., Hommay 2002; Port & Ester 2002), the source of many of the pest species in other parts of the world. Introduced snails and slugs may also compete with native species.

One of the most serious invasive pests in the tropics is the 'giant African snail' (*Achatina fulica*), originally from East Africa, because of the damage it does to crops and gardens. This species was originally introduced to Réunion and later Hawaii, apparently because of its supposed medicinal properties (Cowie & Robinson 2003). It has since been introduced to Asia, including Indonesia and the Philippines, several Pacific Islands, New Guinea, the West Indies, and Brazil (e.g., Mead 1979; Raut & Barker 2002; Thiengo et al. 2007). It is subjected to intense quarantine scrutiny in many countries, and early infestations in Florida and Australia were detected and successfully eradicated, although it has since re-established in Florida (Iwanowicz et al. 2015). Ill-informed attempts to control

TABLE 10.2
The Main Species of Invasive Land Snails and Slugs

Species	Original Range	Invasive Range
Neritimorpha		
Alcadia striata (Helicinidae)	Latin America	North America
Caenogastropoda		
'Cyclotropis' sp.	Southeast Asia, Indonesia, Papua New Guinea, and Australia	Hawaii
Heterobranchia		
Carychium minimum (Carychiidae)	Europe	North America
Belocaulus angustipes (Veronicellidae)	Latin America	North America
Sarasinula plebeia (Veronicellidae)	Neotropics	Southern USA, Central America, Caribbean, northern South America, Africa, Australia, New Caledonia, and Oceania
Leidyula floridana (Veronicellidae)	North America	Latin America
Laevicaulis alte (Veronicellidae)	Central Africa	North America, Asia, Australia, and Oceania
Veronicella cubensis (Veronicellidae)	Cuba	California, Florida, Antilles, Mariana Ids, Hawaii, and Samoa
Veronicella sloanii (Veronicellidae)	Jamaica	Florida and widespread in Caribbean islands.
Semperula wallacei (Veronicellidae)	?	Oceania, China
Diplosolenodes occidentalis (Veronicellidae)	Lesser Antilles	Greater Antilles, Central America, and northern South America
Sarasinula linguaeformis (Veronicellidae)	Brazil	Guadalupe, Dominica, and northern South America
Succinea tenella (Succineidae)	Asia	Hawaii
Succinea costaricana (Succineidae)	Central America	Hawaii
Calcisuccinea dominicensis (Succineidae)	Latin America	North America
Succinea putris (Succineidae)	Europe	North America
Cochlicopa lubrica (Cochlicopidae)	North America and Europe	Latin America, Africa, and Australasia
Lamellidea oblonga (Achatinellidae)	Cook Islands (native range south and central Pacific islands)	Oceania
Lamellidea pusilla (Achatinellidae)	Marshall Islands	Oceania
Pacificella variabilis (Achatinellidae)	Easter Island	Oceania
Tornatellides oblongus (Achatinellidae)	Cook, Marquesas, and Society Islands	Polynesia
Elasmias apertum (Achatinellidae)	Polynesia	Oceania, Indian Ocean Islands, and Kermadec Islands
Cochlicopa lubricella (Cochlicopidae)	Europe	Africa
Lauria cylindracea (Pupillidae)	Europe	North America, Africa
Pupilla muscorum (Pupillidae)	North and Latin America	Australasia
Pupoides coenopictus (Pupillidae)	Europe, Asia	Latin America, Africa
Gastrocopta pediculus (Vertiginidae)	Australia?	Oceania
Gastrocopta servilis (Vertiginidae)	Neotropics	Australasia, Oceania
Pupisoma dioscoricola (Vertiginidae)	?	North and Latin America, New Caledonia
Pupisoma orcula (Vertiginidae)	Africa, Asia	Australasia, Oceania
Vertigo ovata (Vertiginidae)	North America	Latin America, Australasia
Vallonia costata (Valloniidae)	North Africa, Europe to Central Asia	Africa, North America, Mexico, and Australasia
Vallonia excentrica (Valloniidae)	North America, Europe	Mexico, Asia, and Australasia
Vallonia pulchella (Valloniidae)	Holarctic	North and Latin America, Africa, Asia, Hawaii, and Australasia
Chondrina amicta (Chondrinidae)	Europe	Latin America
Pleurodiscus balmei (Pleurodiscidae)	Europe	Australasia
Ena montana (Enidae)	Europe	Australasia
Rachistia histrio (Enidae)	Europe	Africa, Asia, Fiji, New Caledonia, and Australasia
Bulimulus guadalupensis (Orthalicidae)	Caribbean	North America (Florida), Hawaii

(Continued)

TABLE 10.2 (*Continued*)

The Main Species of Invasive Land Snails and Slugs

Species	Original Range	Invasive Range
Bulimulus tenuissimus (Orthalicidae)	Latin America	North America
Achatina[a] *fulica* (Achatinidae)	East Africa	North and Latin America, Asia, Oceania, and Indian Ocean Islands
Limicolaria aurorea (Achatinidae)	Africa	Latin America
Allopeas clavulinum (Subulinidae)	East Africa	North and Latin America, Europe, Asia, Australasia, and Oceania
Allopeas gracile (Subulinidae)	Latin America	North and Latin America, Africa, Asia, Australasia, and Oceania
Allopeas micra (Subulinidae)	Latin America	North America, Oceania
Beckianum beckianum (Subulinidae)	Latin America	Oceania
Eremopeas[b] *tuckeri* (Subulinidae)	Australasia	Oceania
Opeas hannense[c] (Subulinidae)	Neotropics	North America, Europe, and Oceania
Opeas opella (Subulinidae)	Asia	Latin America, Oceania
Opeas spp. (Subulinidae)		Oceania (several spp. listed by Cowie (2001a); also one species each of *Prosopeas, Paropeas* and *Beckianum*)
Paropeas achatinaceum (Subulinidae)	Southeast Asia, Australasia	Oceania
Rumina decollata (Subulinidae)	Mediterranean	Europe, North and Latin America
Subulina octona (Subulinidae)	Neotropics	North America, Africa, Europe, Australasia, and Oceania
Subulina striatella (Subulinidae)	Africa	Europe
Cecilioides acicula (Ferussaciidae)	Europe	North and Latin America, Australasia
Cecilioides aperta (Ferussaciidae)	Latin America	North America, Oceania
Ferussacia follicula (Ferussaciidae)	Europe	North America, Australasia
Gonaxis kibweziensis (Streptaxidae)[d]	East Africa	Oceania
Gonaxis quadrilateralis (Streptaxidae)	East Africa	Oceania
Gulella io (Streptaxidae)	Africa	Latin America
Huttonella bicolor (Streptaxidae)	India?, Southern Africa or Mascarene Islands?	North and Latin America, Australasia, and Oceania
Luntia insignis (Streptaxidae)	Africa?	Latin America
Streptostele musaecola (Streptaxidae)	Africa	Latin America, American Samoa
Euglandina rosea[e] (Spiraxidae)	Southeast USA	Oceania
Testacella haliotidea (Testacellidae)	Europe	North America, Australasia
Testacella maugei (Testacellidae)	Europe	Africa
Paralaoma servilis (= *P. caputspinulae*) (Punctidae)	New Zealand[f]	North and Latin America, Europe, and Hawaii
Discocharopa aperta (Charopidae)	Australasia, Oceania	Europe
Helicodiscus paralellus (Endodontidae)	North America	Europe
Helicodiscus singleyanus inermis (Endodontidae)	North America	Europe
Discus rotundatus (Discidae)	Europe	North America
Anguispira alternata (Discidae)	North America	Latin America
Arion ater (Arionidae)	Europe	North America
Arion cicumscriptus (Arionidae)	Europe	Mexico
Arion fasciatus (Arionidae)	Europe	North America
Arion hortensis (Arionidae)	Europe	North America
Arion intermedius (Arionidae)	Europe	Latin America, Africa, Hawaii, and Australasia
Arion lusitanicus (Arionidae)	Europe	Africa, Australasia
Arion rufus (Arionidae)	Europe	North America
Arion subfuscus (Arionidae)	Europe	North America

(Continued)

TABLE 10.2 (*Continued*)
The Main Species of Invasive Land Snails and Slugs

Species	Original Range	Invasive Range
Arion sylvaticus (Arionidae)	Europe	North America
Meghimatium striatum (Philomycidae)	Asia	Oceania
Meghimatium pictum (Philomycidae)	China	Brazil
Macrochlamys indica (Ariophantidae)	Asia	Africa
Parmarion martensi (Ariophantidae)	SE Asia	Hawaii
Ovachlamys fulgens (Chronidae)	Japan	Latin America, Oceania
Euconulus fulvus (Euconulidae)	Europe	Australasia
Guppya gundlachi (Euconulidae)	Latin America	North America, Asia
Liardetia doliolum (Euconulidae)	Philippines, Micronesia	Asia, Australasia
Llardetia samoensis (Euconulidae)	Oceania	Latin America, Asia
Zonitoides arboreus (Gastrodontidae)	North America and Caribbean	Latin America, Europe, Africa, Asia, Australasia, and Oceania
Zonitoides nitidus (Gastrodontidae)	North America?	Latin America, Europe, and Africa
Oxychilus alliarius (Oxychilidae)	Europe	North and Latin America, Africa, Australasia, and Oceania
Oxychilus cellarius (Oxychilidae)	Europe	North and Latin America, Africa, and Australasia
Oxychilus draparnaudi (Oxychilidae)	Europe	North America, Africa, and Australasia
Oxychilus helveticus (Oxychilidae)	Europe	North America
Hawaiia minuscula (Pristilomatidae)	North America?, East Asia?	Latin America, Europe, Australasia, and Oceania
Vitrea contracta (Pristilomatidae)	Europe	Latin America, Australasia
Vitrea crystallina (Pristilomatidae)	Europe	Africa, Australasia
Quantula striata (Dyakiidae)	Asia	Oceania
Coneuplecta calculosa (Euconulidae)	?	Oceania, eastern Australia
Diastole conula (Euconulidae)	Society Islands	Cook Islands
Liardetia discordiae (Euconulidae)	?	Oceania
Liardetia doliolum (Euconulidae)	?	Oceania, East Timor, and Northeast Australia
Liardetia samoensis (Euconulidae)	Fiji, Samoa, and Marquesas	Oceania
Liardetia spp. (Euconulidae)		Oceania (see Cowie 2001a for details).
Kaliella microconus (Chronidae)	?	Malaysia, Oceania, and tropical Australia
Milax gagates (Milacidae)	Europe	North and Latin America, Africa, Hawaii, and Australasia
Tandonia budapestensis (Milacidae)	Europe	Australasia
Tandonia sowerbyi (Milacidae)	Europe	Latin America, Australasia
Lehmannia marginata (Limacidae)	Europe	North America, Australasia
Lehmannia nyctelia (Limacidae)	Europe	Latin America, Africa, and Oceania
Lehmannia valentiana (Limacidae)	Iberian Peninsula	Europe, North and Latin America, Africa, and Hawaii
Limax maximus (Limacidae)	Europe	North and Latin America, Africa, Hawaii, and Australasia
Limacus flavus (Limacidae)	Europe	North and Latin America, Africa, Asia, Hawaii, and Australasia
Limacus maculatus (Limacidae)	Europe	North America
Malacolimax tenellus (Limacidae)	Europe	Hawaii
Boettgerilla pallens (Boettgerillidae)	W Mediterranean	Asia, North and Latin America
Deroceras agreste (Agriolimacidae)	Europe	Latin America
Deroceras invadens (Agriolimacidae)	Italy?, United Kingdom	Europe, Americas, South Africa, Kenya, Middle East, and Australasia
Deroceras laeve (Agriolimacidae)	Holarctic	North America, Mexico, Europe, Africa, Asia, and Hawaii

(*Continued*)

TABLE 10.2 (*Continued*)
The Main Species of Invasive Land Snails and Slugs

Species	Original Range	Invasive Range
Deroceras panormitanum (Agriolimacidae)	Europe	North and Latin America, Africa, Australasia
Deroceras reticulatum (Agriolimacidae)	Europe	North and Latin America, Africa, Asia, and Hawaii, Australasia
Lacteoluna selenina (Sagdidae)	Latin America	North America
Polygyra cereolus (Polygyridae)	Florida	Latin America, Hawaii
Praticolella griseola (Polygyridae)	North America	Latin America
Caracolus marginellus (Pleurodontidae)	Latin America	North America
Zachrysia provisoria (Pleurodontidae)	Latin America	North America
Zachrysia trinitaria (Pleurodontidae)	Latin America	North America
Zachrysia auricomya havanensis (Pleurodontidae)	Cuba	Mexico
Thysanophora caeca (Thysanophoridae)	North America	Latin America
Bradybaena similaris (Bradybaenidae)	East and Southeast Asia	North and Latin America, Africa, Oceania, and Australasia
Arianta arbustorum (Helicidae)	Europe	North America
Cepaea nemoralis (Helicidae)	Europe	North America
Cantareus apertus (Helicidae)	Africa and Europe	North America, Australasia
Cornu[g] aspersum (Helicidae)	Western Europe and Mediterranean	North and Latin America, Africa, Asia, Australasia, and Oceania
Eobania vermiculata (Helicidae)	Mediterranean	North America, Australasia, and Oceania
Helix pomatia (Helicidae)	Europe	North and Latin America
Otala lactea (Helicidae)	Europe	North and Latin America, Australasia
Otala punctata (Helicidae)	Europe	North and Latin America
Theba pisana (Helicidae)	Mediterranean	Latin America, Africa, and Australasia
Candidula intersecta (Hygromiidae)	Europe	Latin America, Australasia
Cernuella cisalpine (Hygromiidae)	Europe	North America
Cernuella virgata[h] (Hygromiidae)	Europe	North and Latin America, Australasia
Cernuella neglecta (Hygromiidae)	Europe	North America, Australasia
Microxeromagna armillata (Hygromiidae)	Europe	Australasia
Monacha cantiana (Hygromiidae)	Europe	North America
Trichia hispida (Hygromiidae)	Europe	North America
Trichia striolata (Hygromiidae)	Europe	North America
Trochoidea elegans (Hygromiidae)	Europe	North America
Xerotricha conspurcata (Hygromiidae)	Europe	North America
Cochlicella acuta (Cochlicellidae)	Europe	North America, Australasia
Cochlicella barbara (Cochlicellidae)	Mediterranean	North America, Africa, and Australasia

From various sources including http://www.invasive.org/species/mollusks.cfm and several published sources, including Robinson, D.G., *Malacologia*, 41, 413–438, 1999; Cowie, R.H., Non-indigenous land and freshwater molluscs in the islands of the Pacific: Conservation impacts and threats, in G. Sherley (Ed.), *Invasive Species in the Pacific: A Technical Review and Draft Regional Strategy*, South Pacific Regional Environment Programme, Apia, Samoa, pp. 143–172, 2000; Cowie, R.H., *Biol. Invasions*, 3, 119–136, 2001a; Hausdorf, B., *J. Molluscan Stud.*, 68, 127–131, 2002; Cowie, R.H. et al., *Int. J. Pest Manag.*, 54, 267–276, 2008; Rumi, A. et al., *Biol. Invasions*, 12, 2985–2990, 2010; Hayes, K.A. et al., *Bishop Museum Occas. Pap.*, 112, 21–28, 2012; Kim, J.R. et al., *Pac. Sci.*, 70, 477–493, 2016; Naranjo-Garcia, E. and Castillo-Rodriguez, Z.G., *Nautilus*, 131, 107–126, 2017).

Species that have invaded nearby areas only are generally not included in the table.

[a] *Lissachatina* is sometimes used as a subgenus or genus for this species.

[b] Sometimes treated as a subgenus of *Pseudopeas*.

[c] *Opeas pumilum* and *O. goodallii* are synonyms.

[d] A few additional streptaxids are listed by Cowie (2001a), mainly from Hawaii.

[e] Recent work suggests that more than one species may be involved, in Hawaii at least (Meyer et al. 2017).

[f] The complex status of this species is unravelled by Christensen et al. (2012).

[g] Previously placed in *Helix, Cantareus* and a few other genera (see Cowie 2011).

[h] *Helicella virgata* is a synonym.

this species using biological control methods have been abject failures and have resulted in the extinction of perhaps hundreds of native species (see Section 10.7.2).

Numerous small-scale introductions have also occurred, for example, three rather large Cuban pleurodontids – *Caracolus marginellus*, *Zachrysia provisoria*, and *Z. trinitaria* – were introduced to Florida during the early 1900s and are now established there, with *Z. provisoria* also being introduced to Puerto Rico, the Virgin Islands, and the Bahamas.

10.7.1.3 Fresh-Water Invasive Species

Some invasive molluscs have had significant impacts on fresh-water ecosystems around the world (Tables 10.3 and 10.4), but particularly so in North America. In particular, the 'zebra mussel', *Dreissena polymorpha* (Dreissenidae), originally from southern Russia and Ukraine, has been a major pest in lakes and rivers in North America and Europe and is a significant fouling organism, clogging power plant intake pipes. It smothers unionids (many of which are threatened species) by dense aggregations attaching to the posterior ends of their shells. Another species of *Dreissena*, the 'quagga mussel', *D. bugensis*, originally from Ukraine, is a pest in the Great Lakes system in North America as well as in the western United States, along with *D. polymorpha*.

The 'Asiatic clam', *Corbicula fluminea* (Cyrenidae), has impacted many North and South American and European aquatic ecosystems, including the Great Lakes, with considerable economic impact due to clogging of power plant intake pipes. It often occurs with a similar Asian species, *Corbicula fluminalis*. In Asia and South America, the 'golden mussel', *Limnoperna fortunei*, originally from China, has spread to some other parts of Asia, including Japan, and to river systems in South America (Boltovskoy et al. 2006).

The large South American fresh-water 'apple snail', *Pomacea canaliculata* (Ampullariidae) and *P. maculata* are

major pests in many of the rice growing areas of Southeast Asia where they were originally introduced for food (e.g., Halwart 1994; Joshi et al. 2017). Some other species of Ampullariidae have also become established outside their normal ranges, mostly because of the aquarium trade. In Florida, for example, there are five species of *Pomacea* established in the wild, only one of which is native, with the introduced taxa potentially causing damage to fauna, agriculture, and human health (Rawlings et al. 2007). In addition, the systematics of these snails has been confused with some species being difficult to distinguish (Rawlings et al. 2007; Hayes et al. 2008).

Some other significant fresh-water invasive species include parthenogenetic, ovoviviparous taxa such as the Asian 'red-rimmed melania' *Melanoides tuberculata* (Thiaridae) which has invaded many tropical and subtropical areas, as has another thiarid, *Tarebia granifera*. The 'New Zealand mud snail' *Potamopyrgus antipodarum* (Tateidae) has invaded fresh-water ecosystems in Australia, Europe, and in the last 30 years, North America (Alonso & Castro-Díez 2012).

Invading species have displaced native taxa with many of the recorded displacements being anecdotal, but nevertheless signal serious impacts on native fauna. For example, in New Zealand and Australia, the species of the native high-spired planorbid *Glyptophysa* have been severely affected to the point of near extinction by the introduction of the 'bladder snail' *Physa*[10] *acuta* (Winterbourn 1980; Zukowski 2001; Zukowski & Walker 2009, pers. observ. WFP). The introduced North American *Pseudosuccinea columella* (Lymnaeidae) has largely replaced the native lymnaeid *Austropeplea tomentosa* in New Zealand as it has other species of *Austropeplea* in southeastern Australia (pers. observ. WFP).

Field observations in Australia suggest that the introduced *Potamopyrgus antipodarum* (Tateidae) appears to displace related native tateid species (Ponder 1988b), although this may be in part due to the apparent favouring by *Potamopyrgus* of disturbed environments (Ponder 1988b; Schreiber et al. 2003). Also, experiments in Australia showed statistically positive interactions between native non-molluscan aquatic invertebrates that are probably eating the faecal pellets of *Potamopyrgus* (Schreiber et al. 2002), indicating that interactions are far from simple. In New Zealand, where *P. antipodarum* is native and very widespread, it is frequently found with the introduced and widespread *Physa acuta*. Experiments carried out by Cope and Winterbourn (2004) showed that reproduction and growth of *Physa* were stimulated by the presence of *Potamopyrgus*, perhaps facilitating the spread of the former. In another experiment, Riley et al. (2008) looked at the impact of exotic *Potamopyrgus antipodarum* on a native snail, *Pyrgulopsis robusta* in the western USA and found that *Potamopyrgus* limited the growth of the native snails which in turn facilitated the growth of *P. antipodarum*.

TABLE 10.3
The Most Significant Invasive Fresh-Water Bivalves

Species	Original Range	Invasive Range
Limnoperna fortunei ('golden mussel') (Mytilidae)	Asia	South America
Dreissena polymorpha ('zebra mussel') (Dreissenidae)	Southern Russia	North America and Europe
Dreissena bugensis ('quagga mussel') (Dreissenidae)	Black Sea	North America
Corbicula fluminea (Cyrenidae)	Asia, Africa?	North and Latin America, Europe, and Oceania (Hawaii)
Corbicula fluminalis (Cyrenidae)	Asia	North and Latin America, Europe

Source: From various sources.

---

[10] Previously known as *Physella* or *Haitia acuta*.

TABLE 10.4
Significant Invasive Fresh-Water Gastropods

Species	Original Range	Invasive Range
Caenogastropoda		
Pomacea canaliculata (Ampullariidae)	South America	North America, Mexico, Asia, and Hawaii
Pomacea diffusa[a] (Ampullariidae)	Latin America	North America, Australia, Hawaii, Guam, and Papua New Guinea
Pomacea maculata (Ampullariidae)	Latin America	North America
Pomacea paludosa (Ampullariidae)	North and Latin America	Hawaii
Pila conica (Ampullariidae)	Asia	Hawaii, Guam
Marisa cornuarietus (Ampullariidae)	Latin America	North America, Africa
Cipangopaludina chinensis (Viviparidae)	Asia	North America, Hawaii
Cipangopaludina japonica (Viviparidae)	Asia	North America, Fiji
Melanoides tuberculata (Thiaridae)[b]	Asia or Middle East and Africa	North and Latin America, Europe, Africa, Australasia, and Oceania
Tarebia granifera (Thiaridae)	Madagascar?, Asia?	North and Latin America, Africa, ad Oceania
Thiara scabra (Thiaridae)	Asia	North America, Africa, and Oceania
Bithynia tentaculata (Bithyniidae)	Europe	North America
Gabbia robusta (Bithyniidae)	Asia	Hawaii
Potamopyrgus antipodarum (Tateidae)	New Zealand	North and Latin America, Europe, and Australia
Pyrgophorus coronatus (Cochliopidae)	Latin America	Hawaii
Pyrgophorus platyrachis (Cochliopidae)	Florida	Singapore
Heterobranchia		
Physa acuta (Physidae)	North America	Latin America, Europe, Africa, Asia, Australasia, and French Polynesia
Physa virgata (Physidae)[c]	North America	Oceania
Amerianna carinata (Planorbidae)[d]	Tropical Australia	Latin America, West Africa, and Asia
Biomphalaria straminea (Planorbidae)	Latin America	Asia
Biomphalaria guadeloupensis (Planorbidae)	Latin America	Africa
Indoplanorbis exustus (Planorbidae)	Asia	Africa, Oceania, and Australasia
Planorbella duryi (Planorbidae)	North America	Mexico, Europe, Africa, Hawaii, and Australasia
Planorbarius corneus (Planorbidae)	Europe	North America, Fiji, and Australasia
Ferrisia noumeensis (Planorbidae)	New Caledonia	Oceania
Galba truncatula (Lymnaeidae)	Europe	North and Latin America, Africa
Lymnaea stagnalis (Lymnaeidae)	Europe, North America	Africa, Australasia
Pseudosuccinea columella (Lymnaeidae)	North America	Latin America, Europe, Africa, Australasia, and Oceania
Radix auriculata (Lymnaeidae)	Europe	North America, Mexico, and Asia
Radix peregra (Lymnaeidae)	Europe	New Zealand
Radix rubiginosa (Lymnaeidae)	Asia	Europe
Radix viridis (Lymnaeidae)	Asia	Australasia, Oceania

From various sources, including Cowie, R.H., *Biol. Invasions*, 3, 119–136, 2001, and Joshi, R.C. et al., *Biology and Management of Invasive Apple Snails*, Philippine Rice Research Institute, Muñoz, Philippines, 2017.

[a] Previously known as *P. bridgesii*.

[b] Cowie (2001a) listed numerous taxa of Thiaridae as alien and cryptogenic, but it is highly likely that many of these are naturally occurring in islands like Fiji, Samoa, New Caledonia, and Papua New Guinea so they are not listed here.

[c] Cowie (2001a) recorded two additional species of *Physa* from Hawaii.

[d] Cowie (2001a) listed three supposed exotic species of *Gyraulus*, mainly from New Caledonia and Fiji and *Physastra nasuta* from New Caledonia. These require further study and some, particularly the latter, may be native.

10.7.1.4 Marine Invasive Species

Some marine introduced species (Tables 10.5 and 10.6) were distributed around the world on the hulls or among ballast of early sailing ships, and sometimes their original distribution is even unclear.[11] This includes nudibranchs transported on the hydroids or other colonial fouling organisms on which they feed, and both the host and the sea slug have become widespread. Most of these species, rightly or wrongly, are not considered pests. One species, *Mytilus galloprovincialis*, is more obvious, but rarely considered a pest and instead is often harvested and/or farmed in areas where it was introduced. This species appears to have largely replaced native *Mytilus* in some areas (e.g., Westfall & Gardner 2010), but the situation is complex, largely because of the difficulty in distinguishing the native and introduced species and through introgression occurring (Colgan & Middelfart 2011). Ballast water and the hulls of ships remain important vectors of potentially invasive species today (e.g., Bax et al. 2003a).

TABLE 10.5
Some of the Significant Invasive Marine Bivalves

Species	Original Range	Invasive Range
Pteriomorphia		
Arcuatula senhousia ('Asian nest mussel') (Mytilidae)	Asia (Japan and China)	Australasia, Mediterranean, west coast of North America, Mexico, Tanzania, and Madagascar
Perna viridis ('green mussel') (Mytilidae)	India, Southeast Asia	Australia, Japan, Caribbean, Venezuela, and Georgia and Florida, North America
Perna perna ('brown mussel') (Mytilidae)	Southern India, Sri Lanka, Africa, and western South America	Gulf of Mexico, Texas
Xenostrobus securis ('black pygmy mussel') (Mytilidae)	Australia and New Zealand	Japan, Mediterranean
Mytilus galloprovincialis ('black mussel') (Mytilidae)	Mediterranean, Adriatic, and Black Seas	West coast North America and Mexico, Japan, Korea, South Africa, Chile, Australia, and New Zealand
Brachidontes pharaonis (Mytilidae)	Red Sea and western Indian Ocean	Mediterranean
Geukensia demissa ('ribbed marsh mussel') (Mytilidae)	Eastern North America	West Coast USA, Baja California
Anadara demiri (Arcidae)	Indian Ocean	Mediterranean, Aegean, and Adriatic Seas
Anadara transversa (Arcidae)	Northwest Atlantic	Mexico
Crassostrea gigas ('Pacific oyster') (Ostreidae)	Asia Pacific coast	North America, Europe, Australia, and New Zealand
Mytilopsis leucophaeata ('false dark mussel') (Dreissenidae)	Gulf of Mexico	Eastern North America and Europe (brackish and fresh water)
Mytilopsis sallei ('black striped mussel') (Dreissenidae)	Caribbean	Numerous Indo-Pacific locations
Heteroconchia		
Ruditapes philippinarum ('Manila clam') (Veneridae)	Japan	West coast of North America, Europe, and Mediterranean
Nuttallia obscurata ('purple varnish clam') (Psammobiidae)	Japan, Korea, and China	Northeast Pacific, North America
Theora lubrica ('Asian semele') (Semelidae)	Asia	Temperate Australia, New Zealand, California, Mexico, and Mediterranean
Rangia cuneata ('common rangia') (Mactridae)	Gulf of Mexico	Atlantic coast of North America, Belgium
Raeta pulchella (Mactridae)	Southeast Asia	Victoria, Australia
Corbula gibba ('European corbula') (Corbulidae)	Europe	Temperate Australia
Corbula amurensis ('Amur river corbula') (Corbulidae)	Asia	North America (California)
Teredo navalis ('shipworm') (Teredinidae)	Uncertain, possibly Mediterranean, East Atlantic	Widely distributed
Teredo bartschi ('shipworm') (Teredinidae)	Southeast coast of USA, Bermuda	Widely distributed
Bankia destructa ('shipworm') (Teredinidae)	Caribbean	Widely distributed
Bankia zeteki ('shipworm') (Teredinidae)	Caribbean	Widely distributed

Sousa, R. et al., Biol. Invasions, 11, 2367–2385, 2009 and various other sources.

---

[11] Species that are not clearly introduced or native are referred to as *cryptogenic*.

TABLE 10.6

Some of the Most Significant Invasive Marine Gastropods

Species	Original Range	Invasive Range
Caenogastropoda		
Littorina littorea ('common periwinkle') (Littorinidae)	Europe	Eastern North America
Cenchritis muricatus ('beaded periwinkle') (Littorinidae)	Gulf of Mexico, Caribbean	Gulf of California
Crepidula fornicata ('slipper shell') (Calyptraeidae)	Western Atlantic coast of North America	Northwest coast North America, Europe, and Japan
Batillaria attramentaria ('Japanese horn snail') (Batillariidae)	Japan	West Coast USA
Tritia obsoleta ('eastern mudsnail') (Nassariidae)	Atlantic coast of North America	West Coast USA
Busycotypus canaliculatus ('channeled whelk') (Buccinidae)	Atlantic coast of North America	West Coast USA
Urosalpinx cinerea ('Atlantic oyster drill') (Muricidae)	Atlantic coast of North America	West Coast USA, northern Europe
Rapana venosa ('rapa whelk') (Muricidae)	Pacific coast of Asia	Black Sea, Atlantic coast of Europe, eastern North America, and southeast coast of South America
Heterobranchia		
Boonea bisuturalis (Pyramidellidae)	Northwest Atlantic coast of North America	West Coast USA and Mexico (via oyster aquaculture)
Cuthona perca (sea slug) (Tergipedidae)	Unknown	Widely distributed
Polycera hedgpethi (Polyceratidae)	Unknown	Widely distributed in Indo-Pacific region; northeast Pacific, northeast Atlantic
Godiva quadricolor (Facelinidae)	South Africa	Widely distributed in Indo-Pacific region, Mediterranean, northeast Atlantic

From various sources.

Some bivalve species have been introduced for food and are thus rarely considered as problems, although sometimes these can be significant local invaders and could be considered pests. They include *Crassostrea gigas* (the 'Pacific oyster'), originally from Asia, but deliberately introduced into many parts of the world and which have subsequently caused problems in some areas. The western Atlantic species *C. virginica* has been deliberately introduced into British Columbia and Hawaii. *Ostrea edulis* from Europe has also been introduced to many parts of the world. Other commercially important mussel species include the mussels *Mytilus* spp. which, as noted above, have been widely distributed by both shipping and deliberate introductions. *Ensis leei* (previously *E. americanus*) (Pharidae) and *Mya arenaria* (Myidae) from the Atlantic coast of the USA were both introduced to Europe, and the latter also to western North America. These species can have considerable impacts on local communities and can often drastically alter the habitats which they invade (Sousa et al. 2009). The 'Manila clam' *Ruditapes philippinarum* is another economically important species widely introduced to other parts of the world for aquaculture and has also caused some local problems where it has become established in the wild (Gosling 2003). Commercially significant bivalves in the tropical Pacific such as giant clams (*Tridacna* spp.), 'pearl oysters', true oysters, and the 'green mussel' *Perna viridis* have been deliberately introduced to islands outside

their normal range (Eldredge 1994). Gastropod examples of deliberate introductions include the large *Turbo marmoratus* which has a native range west of Fiji on continental islands. It was introduced to Tahiti from Vanuatu in 1967 and later established in the Cook Islands (Eldredge 1994). Similarly, *Trochus niloticus* was introduced to many Pacific islands outside its normal range.

Attempts, successful or otherwise, to transport oysters have also unwittingly introduced other marine species, some of which have proved to be pests. These include the American 'slipper limpet' *Crepidula fornicata* (Calyptraeidae) and the 'oyster drill' *Urosalpinx cinerea* (Muricidae) translocated with oysters from eastern USA to Europe and now pests, particularly in the European oyster industry. Other examples include the venerid bivalve *Gemma gemma* and a mytilid, *Geukensia demissa*, both western Atlantic species introduced to the west coast of North America along with oysters from the Atlantic coast in 1890s. The venerid *Venerupis largillierti*, the malletiid *Neilo australis*, and *Chiton* (or *Amaurochiton*) *glaucus* were introduced to Tasmania along with oysters from New Zealand in the early 1900s and are not considered pests, but the turritellid *Maoricolpus roseus*, which was introduced in the same way, has become the dominant organism on the shelf in eastern Tasmania and parts of southeastern Australia where it is displacing native species (Bax et al. 2003b).

Another marine invasive gastropod that may have been moved out of its range by aquaculture activity or (as larvae) in ballast water is the Asian *Rapana venosa*, a large predatory muricid that has invaded eastern USA and, in Europe, the Black, Aegean, and Adriatic seas, the northern Atlantic coast of France, and the South Atlantic coasts of Uruguay and Argentina. Many other species of marine molluscs have been inadvertently introduced to various parts of the world along with ballast, attached to hulls of ships, or among debris associated with transported molluscs, notably oysters, or with fish, but many of these are not recognised as pests.

Some areas such as large ports with favourable conditions are more susceptible to invasive species than others. For example, of 30 species of marine molluscs introduced to North America listed by Carlton (1992), 19 (63%) occur in San Francisco Bay and the reports of invasive species have continued to accelerate (Cohen & Carlton 1998).

Some bivalves are significant invasive species (see Table 10.5). These include the 'green mussel' (*Perna viridis*), originally from Southeast Asia, which has become established in the Caribbean and on the coasts of Venezuela and Florida. This species, as also suspected in many other introductions, was probably transported on the hulls of ships or in their ballast tanks. Another mytilid, the 'nest mussel' *Arcuatula senhousia*, can form dense mats overlying benthic communities, while the 'black striped mussel' *Mytilopsis sallei* is a dreissenid similar to the fresh-water *Dreissena* and is a fouling pest in harbours and estuaries. An outbreak of this latter species was successfully eradicated in Darwin, Northern Australia, in 1999 (Ferguson 1999).

Some nudibranchs have also proved to be successful invaders as they live on colonial epifauna on the hulls of ships.

Many marine molluscs have invaded the Mediterranean Sea via the Suez Canal as well as being carried by ships from the Atlantic by way of the Strait of Gibraltar or through escapees from aquaculture. There are at least 215 alien marine molluscs known from the Mediterranean Sea, over 30 of which are 'opisthobranchs' (Pancucci-Papadopoulou et al. 2005; Zenetos et al. 2005, 2010, 2012; Cervera et al. 2010). While the majority of these are not considered pests, 19 were listed among the 'worst invasive species' in the Mediterranean, although only five were considered to have socio-economic impacts (Streftaris & Zenetos 2006).

Some species were introduced in early American colonial times, as with the European 'common periwinkle' *Littorina littorea* on the Atlantic coast of North America, probably introduced with ships' ballast in the early 1800s to eastern Canada (Brawley et al. 2009), and then spread south along the North American Eastern seaboard, where it profoundly altered rocky shore, mudflat, and salt marsh communities (Steneck & Carlton 2001; Chapman et al. 2007).

In recent times, the marine aquarium trade has been responsible for moving several marine bivalves out of their native ranges (e.g., Williams et al. 2012), and some, such as small species of 'giant clams' (*Tridacna* spp.) and 'pearl oysters' (*Pinctada* spp.) have been flagged as potentially of concern in Florida (Ray 2005; Ferriter et al. 2009) (Table 10.6).

10.7.1.5 Homogenisation of Faunas

The replacement of native faunas with non-native species is now commonplace due to the human-assisted dispersal of exotic species, particularly in urban and rural areas, and is having, and will have, major consequences for ecosystems around the globe (e.g., Olden et al. 2004). This is occurring in all biota, with molluscs no exception, especially with terrestrial faunas (e.g., Cowie 2001a, 2001b).

10.7.2 BIOLOGICAL CONTROL

There have been attempts to address problems caused by invasive species by using biological control methods. Various organisms have been used in this way, and there have been some spectacular failures, such as the 'cane toad' *Bufo marinus* in Australia. These have mainly proved to be ineffective largely because they are not sufficiently specific to the pest that was the intended target, so native species are sometimes severely affected.

The worst example of biological control gone wrong that primarily involves molluscs is the use of *Euglandina rosea*[12] (the 'rosy wolf snail') that is a native of Florida. This predatory species was introduced in an attempt to control the giant African snail *Achatina fulica*. Introducing *Euglandina* directly resulted in the extinction of perhaps hundreds of native snail species (e.g., Clarke et al. 1984; Griffiths et al. 1993; Coote & Loève 2003; Meyer et al. 2017), but failed as a control agent. African streptaxids, *Gonaxis* spp., were also introduced to some Pacific Islands as a predator of juvenile *Achatina* and may also have impacted native species, although little is known (Cowie 2000). In virtually all cases, attempts using snails as biological control agents have had little or no success and often with serious impacts on non-target organisms (Cowie 2001c). In another seriously flawed attempt at controlling *Achatina fulica*, a predatory flatworm from New Guinea, *Platydemus manokwari*, was intentionally introduced to islands in the Indian and Pacific Oceans. As with *Euglandina rosea*, this has proved ineffective against *Achatina fulica*, but has had serious impacts on native snail species (Sugiura & Okochi 2006; Sugiura 2010). This flatworm predator is now established in Guam, Hawaii, the Philippines, Northern Mariana Islands, Australia, Palau, Japan, Maldives, Tonga, and Vanuatu. *Rumina decollata*, a carnivorous North African land snail, is commercially available in Europe and North America where it is used to control pest snails and slugs. Unfortunately, it also preys on native species, but there has been little attempt to gauge

---

[12] It is now known that two species of *Euglandina*, one unnamed, were introduced to Hawaii (Meyer et al. 2017).

its impact. Some thiarids have also been used to control fresh-water snail pests such as *Pomacea canaliculata* and *Biomphalaria* spp., the latter because they are intermediate hosts for schistosomes which causes schistosomiasis. These introduced thiarids are probably also responsible for displacing non-target native species.

Another control method that has been employed is the use of helicoid specific flies (see Chapter 9), with the snails eaten by the fly larvae. Because of the host specificity, this control measure may be more effective than some others in areas where there are no native helicids such as in South Australia, where *Sarcophaga penicillata* was introduced to combat the European pest helicid *Cochlicella acuta* in 2000.

A parasitic nematode and a carnivorous beetle are available commercially to control slugs and snails in Europe, and these presumably also affect non-target species.

10.7.3 Marine Fouling and Boring

Although not as serious as some other marine invertebrates, some molluscs are considered pests because they foul the bottoms of boats. Mussels (notably *Mytilus* spp.) grow rapidly and adhere firmly to the substratum with their byssal threads. Other fouling non-mytilid molluscs include some ostreids (oysters), anomiids, and dreissenids (such as *Dreissena* and *Mytilopsis*).

As a consequence of fouling the hulls of boats, several bivalves have been successfully transported by shipping and have established outside their normal range where some have become pest species (see Section 10.7.1.4 and Tables 10.3 and 10.5).

The wood-boring ship worms (Teredinidae) (e.g., Clapp & Kenk 1963; Turner 1966; Turner & Johnson 1971) (see also Chapters 5, 9, and 15), are pests in estuaries where they destroy submerged wooden structures such as wharf piles. They can be a significant problem for boats with wooden hulls unless protective measures are used. These worm-like bivalves were a challenge for early mariners and remain a problem today. At the same time, teredinids play a very important role in recycling dead wood in estuarine and open ocean habitats. Some pholads, notably *Martesia*, also bore into wood, but generally present much less of a problem.

Some chemical treatments to prevent fouling are themselves problematic. For example, carcinogenic biocide creosote was used to treat pier pilings, but polluted the waterways, and the notorious TBT and triphenyltin (TPT), which were used as fouling deterrents, also caused various environmental problems including imposex in many gastropods (see Section 10.9.5.1).

10.8 POISONOUS AND VENOMOUS MOLLUSCS

A few molluscs are poisonous or venomous to humans. Venom can be administered directly by way of a bite or 'sting', or poisons are ingested when molluscs are eaten. The most serious form of the latter poisoning is so-called 'shellfish poisoning' (see Section 10.8.1.1).

A few fatal bites from blue ring octopuses (*Hapalochlaena* spp.) have been recorded. Most, if not all, octopuses have venom glands, and other species may be harmful to humans, particularly to individuals that exhibit acute sensitivity, so all should be handled with caution.

Approximately 20 species of cone shells are known to be dangerous to humans, with the most significant being those that are fish eaters. *Conus geographus* and *C. textile* have caused most human fatalities and *C. striatus*, *C. aulicus*, *C. gloriamaris*, *C. marmoreus*, *C. omaria*, and *C. tulipa* have also been implicated (e.g., McIntosh & Jones 2001). On a positive note, the peptides in cone shell venom are being used to produce powerful drugs for pain control (see Section 10.4.1).

The toxicity of aplysioids (sea hares and relatives) and their eggs has long been known, although there are records of Pacific Islanders eating *Dolabella* and *Aplysia* after removing the viscera (Cimino & Ghiselin 2009). Some details regarding the chemicals involved in these and other toxic molluscs are discussed in Chapter 9.

10.8.1 Molluscs and Pathogens

There are vast numbers of bacteria, including many pathogens, in the marine environment (e.g., Govorin 2007). Many of these are allochthonous, that is they were transported there from other polluting sources – and from the human perspective, the notable ones are from sewage, stormwater, and other such sources. Many edible molluscs can uptake bacteria, viruses, and heavy metals while feeding. This is particularly the case with suspension-feeding bivalves that remove potentially harmful bacteria and viruses from the water and either eliminate them along with the pseudofaeces or ingest and digest them. The internal bacterial flora of marine bivalves is mainly obtained from the external environment and may only be temporarily held in the gut (see Chapter 5). While primarily composed of Vibrionaceae and Pseudomonadaceae (see Chapter 9), there are also many other species. Consequently, bivalves are well-known accumulators and vectors of bacteria pathogenic to humans, a problem exacerbated by them often being consumed raw or only partially cooked. Consumption of bivalves with these pathogens generally results in minor to more serious gastrointestinal problems and chronic infections like giardiasis or illnesses such as salmonellosis or hepatitis A (see Section 10.8.2).

Estimates of the accumulation of bacteria and viruses are often based on levels of the non-pathogenic gamma-proteobacterium *Escherichia coli*, data typically used as an indicator of sewage contamination, although additional indicators should ideally be used for effective monitoring. Much of the production of edible bivalves occurs in locations where there is at least some sewage contamination. If bivalves are grown in polluted areas, prior to marketing, the growers often transfer the 'crop' to clean water areas for a period. Thus, it is essential that the dynamics of bacterial uptake and elimination are understood. Bacteria are generally too small to be retained in suspension-feeding unless they are aggregated or attached to other particles – which they often are – and can sometimes be an important component of the food of a bivalve. Some bacteria may be retained longer than others (e.g., Janssen 1974), and studies also show there is considerable variation in the rates that bacteria are cleared from bivalves as it depends on the species and the environmental factors involved (e.g., Bernard 1989; Prieur et al. 1990).

10.8.1.1 Shellfish Poisoning

Edible bivalves are also occasionally responsible for various kinds of 'shellfish poisoning' (Shumway et al. 2018). This kind of poisoning is actually not caused by the bivalve itself, but by toxins produced by planktonic algae (usually dinoflagellates) which are ingested, accumulated, and sometimes metabolised by the bivalve. These poisonings are expressed in several ways depending on the toxins involved and the amount consumed. *Paralytic shellfish poisoning* (PSP) is probably the most significant shellfish poisoning, and several toxins responsible for it have been identified, all derivatives of saxitoxin. It results in neurological effects including tingling, burning, numbness, drowsiness, incoherent speech, and respiratory paralysis. It occurs mainly in bivalves, but cases have also been reported in carnivorous and grazing gastropods, the latter including abalone (Shumway 1995; Deeds et al. 2008).

Other shellfish poisonings include *diarrhetic shellfish poisoning* (DSP) which causes gastrointestinal disorders and results from okadaic acid accumulated from certain dinoflagellates. This chemical causes cells to become more permeable to water, causing diarrhoea (Dawson & Holmes 1999). *Neurotoxic shellfish poisoning* (NSP) (Watkins et al. 2008) results from brevetoxin accumulated from a dinoflagellate (*Karenia*) and results in gastrointestinal and neurological symptoms. *Amnesic shellfish poisoning* (ASP) is caused by the toxin domoic acid produced by certain diatoms and results in short-term memory loss, brain damage, or even death (Waldichuk 1989; Jeffery et al. 2004).

10.8.2 Infections Caused by Eating Molluscs

Issues related to bacterial and viral infections in shellfish often result from the effects of coastal pollution. Several bacterial and viral diseases can be acquired by eating molluscs. Bacterial (e.g., *Escherichia coli*) diseases can result from eating raw shellfish. Also, at least 11 species of *Vibrio* bacteria cause diseases in humans, but *Vibrio vulnificus* accounts for most deaths associated with raw oyster consumption in recent years (e.g., Liu et al. 2006). *Vibrio* bacteria grow in muddy areas as the water warms.

Raw oysters are infamous for being a source of the dangerous hepatitis A virus which probably mainly originates from sewage, but other viruses in raw oysters and presumably other shellfish can cause human health problems. Le Guyader et al. (2008) identified five enteric viruses in individuals of *Crassostrea gigas* which caused gastroenteritis in France. These included noroviruses, an enterovirus, an aichi virus, and a rotavirus. Other viruses found in shellfish include rotaviruses, astroviruses, and adenoviruses, although disease-related issues involving these are only rarely reported (Le Guyader & Atmar 2007).

10.8.3 Poisoning due to Toxins

Tetrodotoxin is a potent neurotoxin found in various invertebrates, some amphibians, and fish (notably puffer fish) (e.g., Miyazawa & Noguchi 2001; Chau et al. 2011), and is also sometimes found in the body tissues of some gastropods such as some members of Naticidae (Hwang et al. 1991), Nassariidae (Hwang et al. 1992; Liu et al. 2004), Olividae (Hwang et al. 2003), and a few tonnoideans (e.g., Narita et al. 1981; Noguchi et al. 1984). Most records of poisoning are from various parts of Asia where these gastropods are commonly eaten. The poison is mainly in tissues such as the digestive gland or muscle and is presumably ingested by the gastropods during feeding. Gastropods containing tetrodotoxin are not only insensitive to its effects, but are often attracted to it (Hwang et al. 2004). Tetrodotoxin is also produced as one component of the venom of Conidae and blue-ringed octopuses (*Hapalochleaena*). In these cases, the tetrodotoxin appears to be produced by symbiotic bacteria (e.g., Hwang et al. 1989; Miyazawa & Noguchi 2001; Chau et al. 2011).

Using some buccinoideans (e.g., *Buccinum*, *Neptunea*, *Hemifusus*) and tonnoideans (e.g., *Fusitriton*) as food has resulted in some poisoning incidents due to the poisons (such as tetramine) produced in their salivary glands (see Chapter 5). Such incidents are usually short and rarely fatal (e.g., Kawashima et al. 2004).

10.9 HUMAN IMPACTS ON MOLLUSCS AND THEIR CONSERVATION

Human impacts on the environment are well known and are primarily a consequence of uncontrolled population growth coupled with the seemingly insatiable need for ever-increasing economic growth. They can range from subtle, as in the example illustrated in Chapter 9, Figure 9.5, to devastating. Increasing population results in escalating habitat destruction by way of land clearing for agriculture and urban development, and the generation of an endless array of pollutants, including greenhouse gases and the increasing production of indestructible plastics. Largely because of degradation of terrestrial and aquatic environments and unsustainable loss of habitat, the biodiversity of the world is fast diminishing (e.g., Jenkins 2003; Maxwell et al. 2016) to the extent that many scientists believe that a sixth mass extinction may be underway (e.g., Barnosky et al. 2011; Ceballos et al. 2015), including in the oceans (e.g., Payne et al. 2016).

Two of the most biodiverse habitats of earth are among the most affected – well over half of the rainforests of the world have been destroyed, and coral reefs have a bleak future from human pressures, pollution, and global warming. What is not well appreciated is that molluscs, particularly non-marine species, are being severely affected. There were about the same number of known extinctions[13] of terrestrial and fresh-water molluscs (300) in the last 200 years than all the tetrapod vertebrates combined (297 amphibians, reptiles, birds, and mammals) (IUCN 2018). Mollusc extinctions, almost all of which are non-marine species, comprise 40% of recorded animal extinctions (750). Worse, this shocking figure probably represents only the tip of the iceberg, as unlike vertebrates, only a relatively small proportion of molluscan faunas are well documented (e.g., Régnier et al. 2009), with their geographic distribution (Figure 10.22) highly correlated with areas that have well entrenched malacological traditions. The number of marine extinctions (commonly set aside in the literature as 'missing species') is probably under-represented as well.

Many issues relate to the impacts of various kinds of pollution which may cause death or have more subtle effects, for example on growth, development, or longevity. Atmospheric pollution resulting in increased greenhouse gases has a threefold impact – increasing temperatures, rising sea levels, and increased ocean acidification, all of which, particularly the latter, have potentially profound impacts on many molluscs (see Sections 10.9.5.2 and 10.9.5.3).

Conservation concerns relating to 'invertebrates' and, to a lesser extent, their habitats, are plagued by apathy and ignorance not only from the general public, but also from most of the government and non-government bureaucracies responsible for conservation and environmental protection.

10.9.1   Impacts on Non-Marine Molluscs

As noted above, nearly all recorded mollusc extinctions are of non-marine molluscs, with high numbers of both fresh-water and terrestrial species having become extinct due to human activity in the last 200 years (Lydeard et al. 2004) (Figures 10.19 and 10.22). The current numbers of recently extinct taxa however, probably greatly under-represent actual molluscan extinctions due to lack of studies (e.g., Régnier et al. 2009). All recorded bivalve extinctions are fresh-water mussels (Unionoidea) (Figure 10.21), and most gastropod extinctions are stylommatophorans, although there are also significant numbers of fresh-water taxa (Figures 10.19 and 10.20). The bivalves of conservation concern are all fresh water, other than four species of *Tridacna* which are the only marine bivalves listed as threatened.

Some of the major causes of the declines and extinctions in non-marine molluscs include:

- Habitat modification and destruction – including clearing of native vegetation, reclamation of swamps and marshes, river modifications such as channelling and damming.
- Pollution, acidification, and salination.
- Predation, competition, or displacement by introduced taxa (e.g., rats, pigs, several helicids, various slugs, *Dreissena* the 'zebra mussel' and *Corbicula*, notably in the USA; fish such as carp, trout, and others) (see Section 10.7.1).
- Ill-conceived biological control (e.g., particularly the misguided efforts to control *Achatina fulica* with

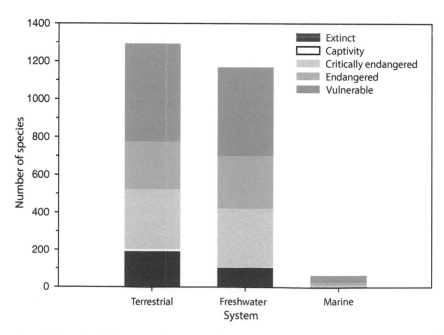

FIGURE 10.19   Comparison of the recognised vulnerability of non-marine and marine molluscs. Data from IUCN, The IUCN Red List of Threatened Species, Version 2018-2, https://www.iucnredlist.org/, November 2018.

---

[13] Excluding the 'extinct in the wild' category.

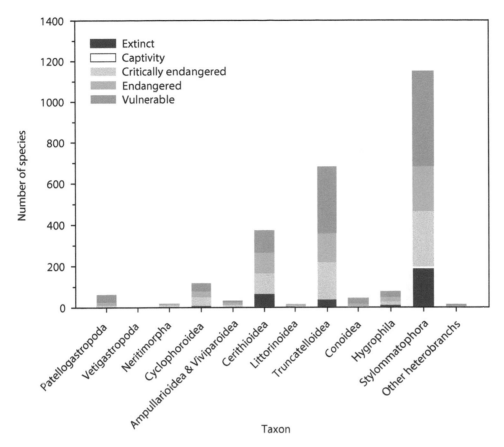

FIGURE 10.20 Comparison of the recognised vulnerability of different gastropod taxa. Data from IUCN, The IUCN Red List of Threatened Species, Version 2018-2, https://www.iucnredlist.org/, November 2018.

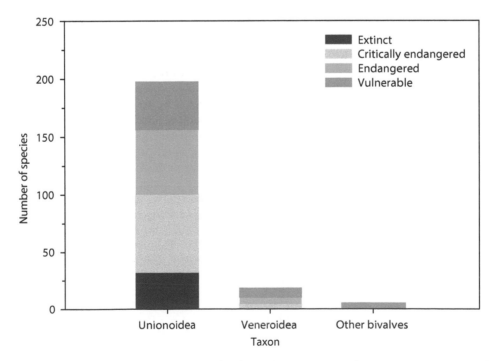

FIGURE 10.21 Comparison of the recognised vulnerability of different bivalve taxa. Data from IUCN, The IUCN Red List of Threatened Species, Version 2018-2, https://www.iucnredlist.org/, November 2018.

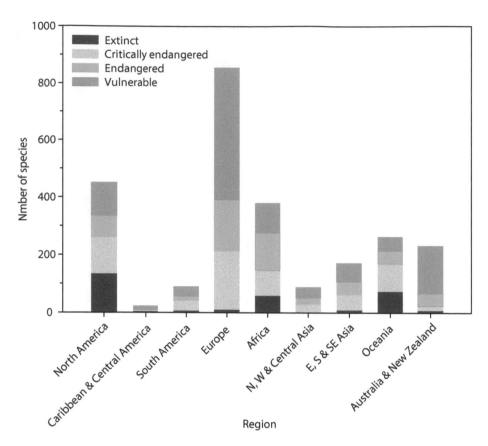

FIGURE 10.22 Comparison of the number of endangered molluscs from different geographic regions. Data from IUCN, The IUCN Red List of Threatened Species, Version 2018-2, https://www.iucnredlist.org/, November 2018.

various snail-eating predators, see Section 10.7.2, but also mosquito fish, and cane toads).

- Many non-marine taxa have very small distributions and hence are highly susceptible to small-scale threats (see Section 10.9.4).
- Over-harvesting is an issue for some edible or commercially significant species, such as some fresh-water mussels.

The removal of native vegetation directly affects terrestrial species, and the destruction of riparian vegetation and an increase in sedimentation affects fresh-water habitats and the species living in them. The ever-increasing demands placed on fresh-water systems have resulted in many modifications, including water diversion, dam construction for reservoirs, weirs, drains, and canals. Removal of water from aquifers and rivers for town supply and irrigation can cause flow reduction and salination, while adding pollutants including sewage and fertilisers from agriculture also have significant impacts.

Some fire control strategies involve regular 'controlled' burning, and this has been shown to have a significant negative impact on many terrestrial invertebrates, particularly gastropods (e.g., Nekola 2002; Stanisic & Ponder 2004; Santos et al. 2009; Bros et al. 2011).

10.9.1.1 Fresh-Water Mussels (Unionoidea)

Many fresh-water unionoideans are threatened, endangered, or recently extinct (Figure 10.21) in North America and, to a lesser extent in many other parts of the world, largely because of habitat destruction, overharvesting, pollution, and, sometimes, invasive species. The unionid fresh-water mussel fauna of the USA is by far the richest in the world. Of the approximately 1000 species of fresh-water mussels worldwide, over 297 occur in the USA – this compares with, for example, only 12 species in all of Europe and 18 in Australia, an area of similar size to the USA. In the USA, 213 of the 297 species (71.7%) are endangered, threatened, or of 'special concern'. Twenty-one (7.1%) are endangered, but possibly extinct, 77 (20.6%) are endangered, but extant, 43 (14.5%) are threatened, and 72 (24.2%) are listed as of 'special concern' (Williams et al. 1993). This marked decline is due to various factors including pollution, river regulation (e.g., dam construction), and overfishing (see Section 10.9.1).

10.9.1.2 Overharvesting of Non-Marine Molluscs

Edible gastropods and bivalves in fresh-water systems have often been overharvested, with the most obvious examples being unionoideans in Europe, Asia, and, especially, the

USA (see previous section). While much of the decline in fresh-water mussels in the USA is due to environmental degradation, overharvesting of the mussels as a source of shell buttons, and, more recently, the nuclei used to seed artificial pearls in pearl oysters, probably also played a role.

Medium-sized to large terrestrial snails are used as food in many parts of the world, and harvesting impacts are generally poorly studied. Examples include the widespread (in Europe) edible terrestrial 'Roman snail' (*Helix pomatia*) where its collection for food has resulted in it becoming rare in some parts of Europe (e.g., Andreev 2006) and native species of *Placostylus* traditionally harvested for food in parts of New Caledonia are now threatened (Brescia et al. 2003).

10.9.2 IMPACTS ON MARINE MOLLUSCS

Marine conservation issues have been gaining more attention in recent years, but for a long time it was a matter of 'out of sight, out of mind'. Many coastal marine habitats are heavily affected by people, and it has become apparent that, despite earlier thinking, the ocean is not an unlimited resource, with pollution from chemicals and plastics apparent in the sea in even the most remote parts of the globe. Most commercial marine species are threatened by overfishing, and most of the benthos of the continental shelf is severely altered by trawl fishing. Many inshore areas are affected by development or pollution, and few coral reefs are pristine.

Impacts of trawling and commercial dredging on benthic habitats and communities are well documented and have had a great effect on most of the benthos of the inshore systems, continental shelves and slopes of the world. In particular, the three-dimensional structure of benthic communities is destroyed, with large emergent invertebrates such as sponges and soft and hard corals being the first to go. These provide shelter and refuge for not only many marine molluscs, but also many adult and juvenile fish. The benthic fauna also provides food for many demersal fish. Improvements in fishing gear are helping, but there is still a long way to go. Inshore systems are being severely affected by development and pollution. For example, highly productive mangrove and saltmarsh communities are increasingly polluted and much reduced in many parts of the world.

Human activities, either direct or indirect, have been implicated in the decline of oyster reefs in many parts of the world. These reefs provide important ecological services such as habitat and improving water quality (see Chapter 9) as well as food for humans and other animals. Their importance has led to efforts to rehabilitate these reefs in some areas (Beck et al. 2009; Beck et al. 2011).

Coral reefs are an important habitat for molluscs and a great deal of serious damage has been done in many parts of the world by overfishing (including the common use of explosives and cyanide) and pollution from human activities. These impacts are exasperated by outbreaks of 'crown of thorns starfish' and, more recently, coral bleaching due to increasing temperatures.

While overharvesting of marine molluscs has caused serious declines in some species and even local extinction (see Section 10.9.2.1), some of these impacts can be complex. For example, on the West Coast of the USA, 'owl limpets' (*Lottia gigantea*) are commonly collected for food or bait, but are also a favourite prey item of the local 'black oystercatcher', *Haematopus bachmani*. These birds are, however, sensitive to the presence of humans and are thus virtually absent from populated coasts. In those areas small individuals dominate the limpet populations because human predation targets the larger limpets (e.g., Lindberg et al. 1998) which, in this protandrous species, are females. Thus oystercatcher predation and human exploitation can affect *Lotia gigantea* populations in different ways (Kido & Murray 2003; Fenberg et al. 2010). Such influences due to human 'harvesting' must be having profound effects on rocky shore intertidal communities generally.

Not all recent extinctions are directly due to human activities, as for example, the disappearance (and likely extinction) of the patellogastropod *Lottia alveus* in New England was probably due to the severe decline of their eelgrass habitat in the 1930s as the result of 'fungal' disease (Lindberg 1984a; Carlton et al. 1991).

The causes of some apparent extinctions are not clear, as in the case of the unique brackish-water species of patellogastropod *Potamacmaea fluviatilis* that is known only from estuaries in India and Burma and has not been seen alive in over 100 years (Lindberg 1990b). The impacts on estuaries and molluscs living in them can be severe due to human pressures exerted on these habitats. For example, in Japan, many species, including numerous molluscs, are listed as threatened or endangered in estuarine habitats (Japanese Association of Benthology 2012).

10.9.2.1 Overharvesting of Marine Molluscs

Many mollusc resources are overexploited. Overharvesting can result not only in declines in numbers, but also local extinction. Harvesting of edible molluscs can also lead to a drop in overall body size and, if excessive, it will not only deplete populations so they become uneconomic to harvest, but also affect community structure (e.g., Keough et al. 1993; Lindberg et al. 1998; Roy et al. 2003). Even prehistoric foraging on shores resulted in local depletion of edible molluscs (Mannino & Thomas 2002), and there is evidence it has reduced and continues to reduce both maximum and mean size of target species such as limpets at some localities (Eekhout et al. 1992; Erlandson et al. 2011). In what is probably the earliest example where this impact has been demonstrated, Steele and Klein (2008) have shown that the size of intertidal patellid limpets and a 'turban' snail, *Turbo sarmaticus*, declined significantly in late Stone Age sites in South Africa, starting as much as 13,000 years ago. Historical declines in size have also been shown in other areas including Southern California (Roy et al. 2003). Overharvesting of intertidal molluscs remains a significant problem in many areas (e.g., Thompson et al. 2002) and remains uncontrolled in many parts of the world, although in some countries it is managed by the creation of reserves and bag limits.

There is a world-wide trend of decreasing global catches from fishing, and, in the absence of proper management, stock collapse occurs (e.g., Jamieson 1993) (see also Section 10.2.1). Sometimes, management options involve licensing, size restrictions, gear restrictions, imposition of bag limits, seasons being declared, or even a complete ban on collecting. These often belated efforts to reverse population declines have had mixed success. In a review of 'shellfish' resources for FAO, Caddy and Defeo (2003) argued that some of the overexploitation might be countered by stock enhancements, while in some parts of the world marine reserves are seen as an answer to provide recruitment.

Such declines are particularly obvious in large, hard-shore, high-value species such as most of the larger species of abalone (*Haliotis*) with wild abalone fisheries having crashed or declined around much of the world (e.g., Tenger 1989; Karpov et al. 2000). But where they are carefully managed, these fisheries continue successfully. One critically endangered Californian species, the 'white abalone' (*H. sorensoni*), was down to about 1% of its original population (Hobday et al. 2000) due to unregulated overfishing. It was listed as an endangered species, and a rescue operation was carried out involving the few, older, widely separated animals left. Some were collected and used to produce young in culture conditions from 2001, although it remains 'critically imperilled' (NatureServe conservation status: http://www.natureserve. org/explorer/ranking.htm. NatureServe Inc. Retrieved 29 April 2018).

In Chile, numbers of the limpet-like muricid, 'loco', *Concholepas concholepas*, and the fissurellid 'lapas', *Fissurella* spp., have significantly declined, but limited harvesting continues under strict restrictions. Similarly, the fishing of conchs, *Strombus* spp., has declined due to overfishing, leading to the listing of the 'Caribbean queen conch' (*S. gigas*) on CITES. This latter species has been an important food item for hundreds of years, and it is only in recent decades that this very valuable fishery has become depleted due to overfishing (both legal and illegal) (e.g., Berg & Olsen 1989; Valle-Esquivel 2003).

Differences in reproductive biology were highlighted by Jamieson (1993) as a reason for the more rapid declines in exploited gastropods (often low fecundity and often short-lived or no larval development, hence more limited dispersal) compared with bivalves (often high fecundity, long larval development). Clearly, exceptions occur with, for example, *Strombus* having larvae that can live in the plankton for weeks (Brito-Manzano & Aranda 2004) and *Concholepas* for months (Disalvo & Carriker 1994).

Even very widely distributed species in the tropical Indo-West Pacific are affected – for example, the commercial harvest of *Tectus* species for the button industry, which began in the early twentieth century, has significantly reduced populations in some areas.

Gastropod harvesting can quickly affect local populations, as for example, with whelks such as *Busycon* in Georgia, USA,

where harvesting three times over just three weeks caused a marked decline in mean shell length (Shalack et al. 2011). Whelk populations (*Buccinum undatum*) in the Wadden Sea have become extinct due to overfishing, TBT pollution, and damage from fishing gear (Cadee et al. 1995), and significant declines have been recorded in other areas.

Some burrowing bivalves have also been seriously overexploited, for example, the once abundant New Zealand 'toheroa' (*Paphies ventricosum*), a large ocean beach burrower, now has a very restricted harvesting season. This was an important food source of Maori in the pre-European times, but was heavily exploited from the early 1900s with commercial production peaking in 1940 and ceasing because numbers had declined to where it was uneconomic only three decades later (Redfearn 1974). More recently, harvesting has been banned or severely restricted (Akroyd et al. 2002). Interestingly, two closely related, but smaller species of *Paphies* in New Zealand remain abundant, even though they are harvested in large quantities.

Many wild scallop fisheries have collapsed and then may sometimes re-establish following closure (e.g., the 'Australian scallop', *Pecten fumatus*) (Semmens et al. 2012) (see Section 10.2.1). The commonly used method of harvesting is dredging, which can cause considerable damage to the seafloor. Much of the scallop meat now marketed comes from aquaculture. A well-known example of a decline in bivalve production is the 'American eastern oyster', *Crassostrea virginica*, where production along the United States East Coast (largely Rhode Island to South Carolina) declined nearly 60% from the 1890s by 1940 and almost 99% by 2004. Although overfishing was a significant cause of this decline, other factors including market demand, habitat deterioration, and disease were involved (MacKenzie 2007). A lesser known example is the 'limestone rock burrowing mussel *Lithophaga lithophaga* which was once popular as a food item in Italy and is now listed as an endangered species. The 'fan mussel' *Pinna nobilis* was once abundant in the Mediterranean, but is now rare. It was harvested mainly for its byssus ('sea silk'), and habitat degradation has contributed to its more recent decline. Giant clams (*Tridacna* spp. and *Hippopus*) were fished to the point of near extinction in some parts of the Indo-Pacific, but are now all listed by CITES, although aquaculture of some species is now occurring successfully (see next Section).

10.9.3 The Shell Trade and Shell Collecting

While marine molluscs are often common and widespread, some are potentially vulnerable through harvesting as human food items (see above) or because of their value (e.g., by way of being rare and having an attractive shell) (e.g., Wells 1989). Vulnerability is related to size and ease of access to the habitat of the species, as well as reproductive attributes such as low fecundity and direct development (Ponder & Grayson 1998). Many species are prized by shell collectors. Some volutes (Volutidae), for example, are eagerly sought by commercial and amateur collectors, and the situation is made worse by the species having very restricted ranges (see Section 10.9.4), low fecundity, and direct development. Local threats (e.g., pollution, development) are more likely to impact on such species and, if sought after, overharvesting is a likely outcome.

Shell collecting is a popular and laudable pastime, and the information some shell collectors acquire can be very useful. Unfortunately, when commercial interests are involved, shell harvesting can be non-sustainable. Local communities collecting for commercial dealers and tourists (as well as for food) have severely depleted intertidal and reef systems in many parts of the Indo-Pacific and Caribbean. The shell trade puts great value on some species of, for example, Pleurotomariidae, Volutidae, Muricidae, Harpidae, Conidae, and Cypraeidae, and collectors can pay hundreds of dollars for exquisite specimens. Despite high prices being paid for some specimens, the vast majority of the millions of shells collected are sold cheaply to the tourist trade as souvenirs or ornaments.

Large species are particularly vulnerable. Giant clams (*Tridacna* and *Hippopus*) naturally occur through much of the Indo-Pacific, but have declined or even disappeared in some areas. For example, the largest species, *Tridacna gigas*, the largest living shelled mollusc, has now disappeared from much of western Indonesia, the eastern Caroline Islands, and Guam (Wells et al. 1983b). Similarly, overcollecting of some large, widespread, Indo-Pacific gastropods such as the large, conspicuous whelks, *Charonia tritonis*, the 'giant triton', and *Cassis cornuta*, the 'horned helmet shell', have been severely depleted throughout much of their range with unknown consequences (e.g., Wells et al. 1983b). Collecting for the shell trade may have been responsible for much of this decline, but with *Cassis* and strombids, for example, presumably also by overcollecting for food by local communities.

Some have suggested that reduction in numbers of *Charonia tritonis*, which has been identified as a predator of the coral-eating 'crown of thorns starfish' (*Acanthaster planci*), may have been in part responsible for outbreaks of that starfish on the Great Barrier Reef, although this idea is generally disputed.

Nautilus shells are harvested for the shell trade (Angelis 2012), but little is known about the size of *Nautilus* populations or their structure, nor their growth rates in their natural environment. This is of concern as fisheries for *Nautilus* in the Philippines and Vanuatu are reportedly being fished unsustainably (e.g., Dunstan et al. 2010).

Some land snails are also affected by the shell trade with, for example, the 'Manus green tree snail' *Papustyla pulcherrima* being listed by CITES (Whitmore 2015). There is also a considerable trade in fossil molluscs, which are collected and sold as ornaments or to specialist collectors. Such fossils include attractive bivalves and gastropods and, particularly, cephalopod fossils, notably ammonites. Such collecting is not sustainable and has led to some fossil outcrops being severely damaged, with fossil collecting banned in some areas.

10.9.4 NARROW-RANGE TAXA

Relictual, isolated habitats, including oceanic islands, often contain many endemics and are consequently of great conservation significance. Non-marine species with small ranges can be confined to small patches of suitable habitat such as limestone outcrops, small rainforest remnants, springs, or small islands. Limestone outcrops are notable for their endemic land snail faunas, especially those in semi-arid or arid areas. These are often restricted in size and potentially highly vulnerable, particularly given the need to mine limestone.

Such narrow-range taxa are extinction-prone because they can be adversely impacted by small-scale threats. Unlike the situation in many other invertebrate groups, most molluscs have a shell that can often leave 'fossil' evidence long after an extinction event and, also, early collections held in museums can provide evidence of subsequent loss. Nevertheless, such evidence gathering is often frustrated by poor sampling in the past in many parts of the world.

Many of the recorded extinctions of non-marine molluscs are species from islands, as there is often historical evidence of their past existence. Those taxa with small distributions in streams or rivers, or in relictual habitats such as isolated springs, have probably experienced far higher extinction rates than the recorded extinctions suggest, usually because of the lack of previous collecting and poor 'fossil' evidence. Thus, most of the probable extinctions in such habitats are unlikely ever to be adequately documented.

Although many benthic marine molluscs have effective larval dispersal and often have wide ranges, many marine taxa are poor dispersers because they may have direct development or very short-term larvae. Some of these have narrow ranges (Ponder 2003), being confined to a small segment of the coast, sometimes through specific habitat or host requirements, an isolated bay or estuary, or a small oceanic island or seamount. Such species are vulnerable to small-scale pollution events or shore modification, or if of interest to fishermen or shell collectors, can have their populations rapidly depleted. In the Pacific, for example, there are many marine endemics associated with isolated islands such as Hawaii (20% of the fauna) and the very isolated Easter Island has 37% endemism. Isolated seamounts also show endemism, but relatively few have been well sampled for molluscs.

A few examples of narrow-range taxa and their declines are briefly outlined below.

Terrestrial examples:

- The tree-living land snail genus *Partula* lives on South Pacific Islands such as Tahiti, Moorea, and Guam. Over 100 species have been named, and many are threatened, endangered, or extinct. The seven species on Moorea were extinct about five years after the introduction of *Euglandina* by the French authorities in a misguided attempt to control *Achatina* (see Section 10.7.2), against the advice of the malacological community (Clarke et al. 1984; Murray et al. 1988). These species were the subject of several detailed genetic studies (see Goodacre 2002 for review). In all, the 61 species in the Society Islands have been reduced to only five by this ill-conceived biological control program (Coote & Loève 2003).

- *Achatinella* is a Hawaiian genus of land snails with 41 species, all endemic to O'ahu. Every species is listed as endangered and 16 are extinct due to a combination of land clearing and exotic predators. Most of the extant species occur only in small numbers and have very restricted ranges.
- The large carnivorous iconic land snails of the genera *Paryphanta* and *Powelliphanta* in New Zealand comprise about 22 species and 51 subspecies (Walker 2003). Many are threatened and several endangered, mainly by the combined impacts of heavy predation from introduced pigs, possums, rats, and land clearing, as well as over-collecting in the past.

Fresh-water examples:

- Many fresh-water and terrestrial caenogastropods have very short ranges (Ponder & Colgan 2002) making them particularly vulnerable to habitat modification or destruction (e.g., Lydeard et al. 2004; Strong et al. 2008). In parts of Eastern Europe (particularly the Balkans), the USA, and Australia, there are significant radiations comprising dozens of species of small fresh-water truncatelloidean snails. Many are highly restricted, sometimes to a single stream, spring, or cave. While there is direct evidence for extinction in a few cases, this is mostly assumed where springs or streams have ceased to flow or where significant habitat modification has occurred in areas where similar habitats support endemic taxa.
- Rich endemic molluscan faunas are found in several ancient lakes (Boss 1978; Wilke et al. 2006) (see also Chapter 9) such as Lakes Ohrid and Prespa in the Balkans, Lakes Malawi and Tanganyika in East Africa, Lakes Poso and Malili in Sulawesi, Lake Biwa in Japan, Lake Titicaca in South America, and Lake Trichonis in Greece. Significant human impacts on some of these lakes include pollution and the introduction of exotics.
- Faunas in arid-zone artesian springs in the USA and Australia contain many endemic species, mainly of small truncatelloideans, nearly all described in the last few decades (e.g., Strong et al. 2008). Most are restricted to a few springs and some to only one. Most of the springs are on private properties and are often badly damaged by stock. Water extraction from the aquifers supplying the springs has caused the extinction of many of the springs and their fauna.

Marine examples:

As noted above, short-range marine endemics are present on isolated islands such as Hawaii, Easter Island, and the Galapagos. Also, oceanic seamounts have endemic taxa, although these are not as well-known as those from islands. Similarly, the upwelling zone in Oman has several short-range endemic species, as do parts of southern Africa and Australia. Some short-range endemics are found at the northern or southern ends of the ranges of their genera (or families) that often coincide with a transition from tropical to temperate zones.

Some specific examples include:

- Some direct-developing, large species may have very short ranges. For example, *Cymbiola thatcheri* is only known from the small Chesterfield Reef, Coral Sea. Other volutes can also have short ranges – for example, the rare *Lyria laseroni* is only known from a small part of the coast of northern New South Wales, Australia.
- Some marine habitats are restricted. For example, the marine eupulmonate slug *Smeagol* (Otinoidea) is found on the upper shore among pebbles in temperate Australasia and Japan. This type of substratum is rather uncommon in southeastern Australia, with pebble shores often widely separated and, probably because of this, species are very restricted (Tillier & Ponder 1992).

10.9.5 Pollution and Its Impacts

Pollution occurs when potentially damaging contaminants are released into the environment. It can be caused by chemical substances, radiation from nuclear waste, or excess heat or light. Many pollutants can cause problems for molluscs, including oil spills, thermal pollution caused by power plants, chemical wastes, plastic particles (e.g., Rochman et al. 2015), heavy metals (e.g., Sobrino-Figueroa et al. 2007; Nuñez et al. 2012; Emilia et al. 2016), and high nutrient loads. Such impacts on animals in general are detailed elsewhere, for example, heavy metals (Boyd 2010) and organic pollutants (Walker 2008). Here, one specific form of organic pollution is examined in more detail – endocrine disruption.

10.9.5.1 Endocrine Disruption

It was discovered in the 1980s that the antifouling biocide TBT used in marine paints mimics a hormone and induces 'intersex' (imposex) in molluscs (e.g., Alzieu 2000a). This is one of the few proven cases of endocrine disruption in invertebrates other than the insect growth regulators used in pest control (e.g., Sharara et al. 1998; Depledge et al. 1999; Pinder et al. 1999; Oehlmann & Schulte-Oehlmann 2003; Matthiessen 2008).

The effects of TBT have been well studied in gastropods (e.g., Lagadic et al. 2007; Oehlmann et al. 2007), for example, whelks (e.g., Ellis & Aganpattisina 1990), hydrobiids and *Littorina* (e.g., Schulte-Oehlmann et al. 1998), but other molluscs and a variety of aquatic invertebrates and vertebrates can also be affected to different degrees (e.g., Alzieu 2000a; Weltje & Schulte-Oehlmann 2007).

The first recorded adverse effects of TBT on molluscs were on *Crassostrea gigas* at the Bay of Arcachon, a major oyster aquaculture centre in France. It resulted in ball-shaped shell deformations in adults and a decline in spat fall resulting in serious consequences for local oyster production (Alzieu 2000b). Since then, the related compound TPT has also been shown to have similar effects (Figure 10.23).

The use of both TBT and TPT has now been banned in many countries and by the International Maritime Organisation. They are thought to function by activating the retinoid X receptor (RXR), one of the nuclear receptors, which appears to play a major role in the induction, differentiation, and growth of male genitalia in female gastropods (Nishikawa 2006).

Other endocrine disrupting compounds (EDCs) are found in runoff that includes sewage (including that from treatment plants) and agricultural effluent, including oestrogenic compounds. These include 17α-ethynylestradiol (a synthetic estrogen used in contraceptive pills) and 4-nonylphenol. Nonylphenol polyethoxylates are released when alkylphenol polyethoxylates (a common chemical constituent of emulsifiers, detergents, etc.) break down, and 4-nonylphenol is the most common of these. These chemicals have been shown to have negative impacts on molluscs including causing feminisation in male bivalves (e.g.,

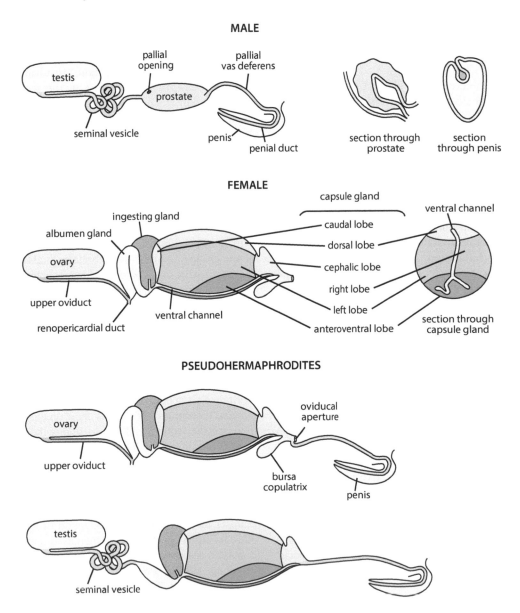

FIGURE 10.23 Morphological expression of imposex in the muricid neogastropod *Nucella lapillus*. Typical male and female morphologies at the top of the figure, followed by two examples of pseudohermaphroditic morphologies induced by exposure to TBT compounds. Redrawn and modified from Oehlmann, J. et al., *J. Molluscan Stud.*, 57, 375–390, 1991.

Langston et al. 2007; Andrew et al. 2008; Andrew-Priestley et al. 2012) and other changes such as effects on burrowing behaviour and antioxidant activity (e.g., Matozzo et al. 2004).

10.9.5.2 Acidification

In general, fresh-water molluscs are not tolerant of acid waters (i.e., pH less than 7) and decreasing pH in fresh-water ecosystems because of acid rain from industrial pollution affected mollusc communities in Canada and Europe (e.g., Økland & Økland 1986). In waters exposed to alkaline soils (e.g., those originating from limestone and dolomite rock), the acid is usually neutralised, but in, for example, soils originating from granite, the surface waters are much more prone to acidification. Consequently, water bodies in northern USA and Canada, for example, are more prone to acidification than those further south. Acidification of estuarine water by leaching from disturbed acid sulphate soils, which are rich in iron sulphide, can affect oysters (Wilson & Hyne 1997; Dove & Sammut 2007) and other molluscs (e.g., Amaral et al. 2012a).

More recently, the increased levels of CO_2 in the atmosphere has raised considerable concern. This will cause increasing ocean acidification as the oceans absorb CO_2 in the atmosphere and it is converted into carbonic acid. As a result, the ocean pH could drop from the current 8.1 to 7.7 by about 2100 with potentially drastic impacts on calcifying organisms (Orr et al. 2005; Feely et al. 2008; McNeil & Matear 2008; Wootton et al. 2008; Gattuso & Hansson 2011).

The acid water will dissolve the calcium carbonate shells of molluscs, in particular, the thin and fragile shells of their larvae and the planktonic shelled heteropods and thecosome pteropods, as well as having impacts on corals, crustaceans, and any other animals with calcium carbonate skeletal material. This is particularly concerning as changes in ocean acidity have triggered mass extinction events in the past with warmer and more acidic oceans probably contributing to the greatest mass extinction event 250 million years ago in the Permian where about 95% of marine life became extinct (see Chapter 13).

Even for those taxa not directly affected, flow-on effects may occur. For example, the highly susceptible thecosome pteropods (e.g., Comeau et al. 2009) are preyed on by shell-less gymnosomes that target single species. Thus, ocean acidification that affects thecosomes will also probably affect the naked gymnosome pteropods.

Experiments have demonstrated negative impacts on larvae (e.g., Gibson et al. 2011) and adults, with even the relatively thick shells of larger taxa probably becoming thinner, rendering them more susceptible to predation (Gazeau et al. 2007; Welladsen et al. 2010). Although it has been demonstrated in an oyster that adaptations to larvae can occur that minimise the effects of acidification (Parker et al. 2011; Parker et al. 2015; Goncalves et al. 2016), it seems unlikely that such changes will occur rapidly enough to make an appreciable difference in most aquatic molluscs.

Physiological impacts may also occur with increasing acidification – for example, the immune response appears to be diminished in mussels (e.g., Bibby et al. 2008) and oysters undergo changes in metabolic pathways (Lannig et al. 2010).

10.9.5.3 Climate Change

The biggest foreseeable challenge for scientists and technologists, now and in the future, is global climate change. All aspects, including basic research, aquaculture, fisheries, invasive species, medical applications, conservation, and biogeography, will be confronted by this global event. Some aspects are only outlined here as considerable readily accessible literature covers this topic.

The numerous predicted consequences of climate change on ecosystems and biodiversity are very serious. It is brought on by anthropogenic pollution causing changes to the relative composition of gases in the atmosphere with increasing levels of greenhouse gases, notably carbon dioxide and methane, resulting in increasing temperatures, greater variance in the weather including more frequent storms and droughts, and rising sea levels These affects will have, and are having, negative impacts on biodiversity (e.g., Foden et al. 2008; Heller & Zavaleta 2009; Willis & Bhagwat 2009; Bellard et al. 2012) and human society.

Inevitably, climate change will cause changes to mollusc populations and the distributions of species. Assessing these will be problematic in many groups because of the lack of reliable historical and contemporary data. Museum mollusc collections contain some of the best distributional information for any group of marine invertebrate animals (see Chapter 11). These and other data could provide a benchmark for monitoring distributional changes brought on by climate change, but to date are largely unutilised. Increased research efforts are needed to assess these challenges.

Some of the ecological and other changes caused by climate change will be complex. For example, some saltmarshes in southeastern USA were turned into barren mudflats, and the few remaining patches of grass were covered in *Littoraria irrorata*. This species feeds on fungi that grow on damaged *Spartina* (cordgrass). The snails chew on the grass, the fungus grows on the wound, and the snail eats the fungus. Stresses on *Spartina* such as pollution and climate change make it less able to deal with the pressures of snail grazing, putting the system out of balance (Silliman et al. 2005).

In another example, the balance of the often overlooked, but important role of parasitism population dynamics could be upset by warming. The production of cercaria in trematodes generally increases with increased temperature leading to marked reductions or even expatriation of hosts (Poulin & Mouritsen 2006). Predator-prey interactions can also be affected as shown, for example, with oysters and their predatory whelks – the oyster shells are thinner making them more susceptible to drilling, and elevated carbon dioxide levels increase the predator's energy levels increasing its consumption rate (Wright et al. 2018b).

Changes to habitat caused by climate change may reach a point at which a species can no longer survive. This is an issue for some marine molluscs, especially for those that are direct developers, or that are restricted to specialised habitats. It is particularly critical for most non-marine molluscs, as many live in discontinuous habitats such as fragmented forests, mountains, lakes, and rivers, and they have no means of dispersing across barriers formed by unsuitable habitat other than by accidental means or by human intervention.

10.9.6 Oxygen Depletion

Because colder waters hold more oxygen than warmer waters, and the oceans have warmed over the past 50 years, it should be expected that the oxygen minimum zones will expand with a reduction in dissolved oxygen levels. The consequences of this for marine ecosystems are likely to be profound with expected impacts on subtropical, temperate, and subpolar regions larger than in the tropics (Stramma et al. 2008).

Human pollution can cause local anoxic conditions, but can also contribute to larger coastal 'dead zones', some of which are extensive (Diaz & Rosenberg 2008). Some of the oceanic 'oxygen minimum zones' have diverse microbial communities and are nitrite-rich and some are essentially anoxic (Ulloa et al. 2012), and there is evidence that these zones are expanding (e.g., Stramma et al. 2008; Stramma et al. 2012) due to warming and other factors with likely serious consequences for some marine ecosystems.

10.9.7 Lack of Knowledge and Unequal Effort

Compared with vertebrates, the effort expended on assessing the status of threatened molluscan taxa is miniscule, despite the larger number of extinctions referred to above.

Information from basic research is essential to make informed conservation decisions. Even the listing, let alone eventual management of threatened taxa, depends on knowledge not readily available for most molluscan species. Ideally, basic biological and distributional information is required, but our present knowledge base falls far short of this ideal because many molluscs (and other invertebrates) are not even named, and little or nothing is known about the distribution, feeding, reproduction, life history, and physiology of even many common species. This paucity of information hinders the ability to list a species as threatened under most endangered species legislation or in the International Union for Conservation of Nature (IUCN) Red List, as certain criteria need to be met. These include distribution, evidence for the decline of numbers or a reduction in range, and the identification of threats. It is also important to identify values, such as phylogenetic uniqueness and/or the biological, ecological, economic, or social importance of a species when addressing the best strategy for its conservation.

11 Research on Molluscs – Some Historical, Present, and Future Directions

11.1 MALACOLOGY

A brief introduction to early malacological workers is given in Chapter 1. The twentieth century saw a massive increase in molluscan studies – as shown below. This increase in malacological studies has not only taken place in the traditional malacological centres, but has now expanded into many parts of Asia, including China, Taiwan, South Korea, Thailand, and South and Central America, and eastern Europe. While many of these earlier studies focused on describing molluscan diversity, molluscs have also been widely used as study organisms in many areas of scientific enquiry, notably in ecological and physiological studies.

11.1.1 TECHNOLOGY

The introduction of new technology, as with other branches of science, greatly assisted researchers. An important early invention included the stereoscopic microscope, which was first marketed by Zeiss in 1897, whereas compound microscopes had been around since the early 1600s. The transmission electron microscope (TEM) was developed in 1932, but only became commonly available in the 1960s, and it gave whole new levels of resolution to the examination of tissues and cellular structure. The scanning electron microscope (SEM) was developed in the late 1930s and came into common use in the 1970s. Modern ecological studies have been mainly carried out during the latter half of the 1900s and molecular studies from the 1970s to 1980s. Other imaging techniques such as confocal microscopy have been useful in examining the development of musculature and other organ systems and several long-standing controversies have been resolved. Other imaging technologies that have been available in recent years such as micro computed tomography or MicroCT scanners are briefly discussed in Chapter 21.

There was a period before, and mainly during, the 1980s in which computer algorithms were used to generate phenetic classifications; a process called numerical taxonomy in which early desktop programs such as NT-SYS were employed. Then the broad application of what Hennig (1950) termed phylogenetic systematics, but which later became cladistic methodology, or cladistics, began in the 1970s. Applications of this methodology were markedly increased by the development of maximum parsimony algorithms and became commonplace in the 1990s. Its wide use became possible with the availability of desktop computers. What started with the mainframe installations of Clad/OS and PHYSYS by S. Farris and M. Mickevich, respectively (Albert 2005), was migrated to the desktop in the 1980s in the form of PHYLIP, PAUP, and HENNIG86 cladistics programs. In the 1990s many other programs and methods were developed (e.g., NONA, MacClade, MESQUITE, PAUP*), followed by the development of non-parsimony-based methods using algorithms such as maximum likelihood (e.g., RAxML) and Bayesian inference (e.g., MrBayes) which are currently preferred for phylogenetic inference using molecular data.

11.1.2 MOLLUSCAN RESEARCH PRODUCTIVITY

Solem (1974) provided estimates of the number of papers published per year on molluscs. We updated this survey using the Clarivate Analytics (Thomson Reuters) Web of Science™ database. The earlier estimate of publication productivity follows the same general trend as the data from the Web of Science. The major difference between the two estimates is that the Solem's data are on average 35% lower than the Web of Science data, except for the 1911–1920 (12%) period. The Web of Science data were produced by an 'all databases' search using the string 'mollus*'. The 'all databases' option includes: the Web of Science Core Collection; the BIOSIS Citation Index; BIOSIS Previews; CABI: CAB Abstracts® and Global Health® databases; Chinese Science Citation Database SM; Current Contents Connect; the Data Citation Index; Derwent Innovations Index; Inspec®; KCI-Korean Journal Database; MEDLINE®; The U.S. National Library of Medicine® (NLM®); Premier Life Sciences database; the Russian Science Citation Index; SciELO Citation Index; and the Zoological Record. The general search string was also combined with four keywords – physiology, ecology, molecular, and evolution – each highlighting a major area of molluscan research. The string 'mollus*' retrieved almost 560,000 items.

The Web of Science and associated databases do not include all publications on molluscs. For example, a Google Scholar search using the same search string recovers an additional 90,000 items. Also, most citation databases underestimate the number of taxonomic and systematic works because many taxonomic journals are not indexed (Wägele et al. 2014). This also substantially underestimates citation indices and hence affects the impact factor of a journal (Bouchet et al. 2016). However, citation indices provide an important and readily available insight into the publication record of our research community, its breadth, and its contributions to our understanding of biodiversity and processes of our world. Lastly, the reduction in the numbers of publications as you approach the 2016 cut-off (general, physiology, ecology) is probably an artefact due to sampling error similar to the Signor-Lipps effect in extinction data, but with missing publications rather than taxa. Unlike missing fossil taxa, many of these 'missed' publications are subsequently entered into the databases, and after a while, the number of publications typically rises above preceding years. The lack of a drop in the molecular and evolution publications suggests that journals addressing these topics are initially better sampled, and fewer papers are subsequently discovered and entered.

Five arbitrary changes in publication rates were recognised: two corresponding to world wars and three with global recession/depressions (Figure 11.1). These reductions in publication rates were also seen in physiology, ecology, and molecular molluscan publications, but were absent in publications including evolution. Recoveries from these events appear to have a lag of about one year (approximately the time to publish a paper) and are typically followed by substantial sustained increases in publication rates. For example, a sustained increase in publications occurred following the Long Depression (1873–1886), World War I, the combined Great Depression (1929–1941) and World War II, and the global depression of the 1990s. The slope of publication rates also shifted radically following World War II and the global depression of the 1990s. Between 1864 and 1945 publication rates were highly variable ($r^2 = 0.4593$), but with a relatively slow increase in productivity (slope = 10.64). After World War II, a sustained increase in publication rates produced a 12.9-fold increase in the slope of publications per year (137.40) between 1946 and 1997, and with much less variability ($r^2 = 0.9515$). The most recent shift occurred in 1998 and represented a five-fold increase in the slope (681.11); also, with little variation ($r^2 = 0.9542$).

The overall pattern of the general search is also present in combined searches featuring the keywords physiology, ecology, and molecular. The publication of *The Origin of Species* by Charles Darwin in 1859 was probably the most revolutionary event in the history of biology. It coincided with the great age of discovery and gave impetus and direction to much of the biological work. However, molluscan studies do not appear to have been part of this revolution (see also Lindberg 1998). The first Web of Science molluscan evolution papers appeared in 1889 and featured studies of variation in both fossil and living molluscs (James 1889; Keyes 1889; Pearce 1889; Williams 1889). The rate of increase in evolution papers remained relatively constant until the early 1950s, which then saw a slight increase through the late 1960s. In the early 1970s publications again increased, followed by exponential growth between 1998 and 2016.

11.2 SOME AREAS OF RESEARCH FOCUS, NOW AND IN THE FUTURE

Here, we present areas of research relevant to the topics covered in this volume. Aspects of research relevant to systematics and phylogeny and palaeontology are discussed in Chapter 21.

11.2.1 PROVISION OF BASIC KNOWLEDGE

As noted in Chapter 10, basic biological knowledge (feeding, reproduction, life history, habitat requirements, etc.) and distributional information are essential to make informed decisions for resource management, ecological studies, assessment of

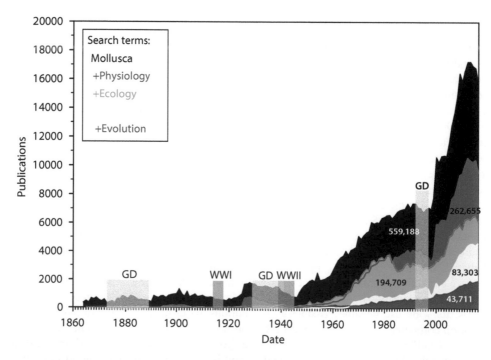

FIGURE 11.1 Area plots of molluscan publications between 1864 and 2016. The search string 'mollus*' was combined with one of four keywords: physiology, ecology, molecular, and evolution. Five reductions in publication rates are recognised; two with world wars (WW) and three with global recession/depressions (GD). The numbers of citations are also given.

communities and their interactions, commercial shellfisheries, biomedical studies, modelling, and assessment of conservation status etc. It is also vital if the 'values' of a particular species are to be assessed. Such assessment would usually include aspects such as economic or social importance, biological or ecological roles, and phylogenetic uniqueness. However, this information is often lacking, even for many common species. Unfortunately, gathering basic biological (natural history) information is not given a high priority in most universities and other research institutions today, and even publishing such data is typically not encouraged by many reviewers and editorial boards of non-speciality journals. Consequently, much of the cited information relies on data gathered from mainly European species in the 1800s to the mid-1900s. The reliability of using such information from other species as a surrogate depends on how well known the relationships are between the taxa in question – the more dependable the phylogenetic relationships, the more reliable the predictions are likely to be. Unfortunately, only generalised data are typically available for the great majority of species. How can this situation be rectified? Ways to achieve rapid increases in knowledge that have been successfully used in recent times include the use of museum collections and research workshops. In the future, better vetting of potential data may be facilitated by deep text mining as discussed below. Obviously, there is much to be done.

The need for a more systematic approach for gathering data became readily apparent to us as we prepared these volumes. While on the one hand we were often amazed by the considerable data that are available, we were also often reminded of just how few taxa some data are based on. Numerous examples of 'data holes' are alluded to in each chapter in these two volumes. They include, for example, our knowledge of digestive enzymes and gut bacteria being restricted to a small number of taxa, studies of the endocrine system and neuroendocrinology are non-existent for most groups of molluscs, as are many detailed histological studies of molluscan organ systems and structures throughout the phylum. Undersampling limits our understanding of molluscan systems biology at all levels of the biological hierarchy – from our overall estimates of molluscan anatomical, physiological, and ecological diversity to our understanding of variation in biochemical pathways and genomics.

11.2.2 Genomics

Molluscan genomics is positioned to make substantial progress because of the numerous technological, theoretical, and computational advances made during the last ten years. In 2007, the rate of cost reduction for DNA sequencing surpassed Moore's Law which states that the cost of computing power drops 50% biennially because of technological and computational advances (Hayden 2014). At this time, it is highly likely that producing and analysing genomic data will continue to become cheaper and faster in the foreseeable future. Cloud-based machine learning applications will also improve the accuracy and speed of the annotation of genomic data, reducing one of the critical bottlenecks for taxa only distantly related to a model organism. However, sampling issues are likely to remain with certain groups being well sampled, while other groups remain unknown. As in other datasets, taxa of economic, medical, and cultural interest are often well represented. While these criteria will continue to be essential, our knowledge of molluscan genomics could be substantially improved by coordinating taxon selection to maximise research initiatives across and between scientific fields and disciplines. An important criterion should be consideration of the phylogenetic distinctiveness of potential taxa.

11.2.3 Evolutionary and Ecological Development

The emerging CRISPR (clustered regularly interspaced short palindromic repeats) genome editing technology will provide a critical new tool for molluscan evolutionary developmental studies. Molluscs, like many lophotrochozoans, have few genomic tools with which to target and silence specific genes as is commonly done in the better-known model organisms with well-known and annotated genomes. With CRISPR technology, it is possible to cut the genome at specific locations which permits the removal and/or addition of new genes, the ability to turn genes on and off in a reversible manner and to change their levels of activity in a developing embryo. Early work on molluscs using CRISPR genome technology includes Perry and Henry (2015), Henry and Lyons (2016), Jackson and Degnan (2016), and Goulding and Lambert (2016).

11.2.4 Ecology

Molluscan ecological research shows two distinct expansions (Figure 11.2). The first sustained productivity increase began at the end of WW II. Publication rates weakened in the early 1980s and would remain below the 1982 rate for the next 20 years. There was a slight upturn in the late 1980s, but it appears to have been terminated with the early 1990s global recession. The substantially lower publication rates of the 1980s are not seen in Web of Science searches of either keyword alone (i.e., mollusc* or ecology), and the driver(s) of the hiatus in molluscan ecology publications remain unknown. In contrast, the 1990s reduction was broadly across molluscan publications (Figure 11.1). The second and the current increase in publication rates began in 1999 with the recovery from the earlier global recession.

11.2.5 Physiology Studies

Molluscan physiology (*sensu lato*) accounts for most of the molluscan research publications recovered from the Web of Science (Figure 11.3). It is not surprising then that physiological research reflects the overall publication productivity seen in the general search for mollusc-related publications shown in Figure 11.1. Bivalves appear to be the taxon of choice except for several years in the late 1970s and early 1980s when they were briefly eclipsed by gastropods; cephalopods remain a distant third in physiological research.

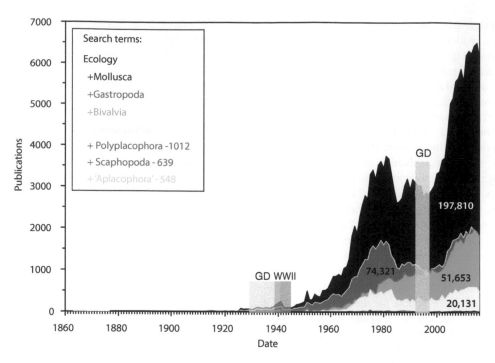

FIGURE 11.2 Area plots of molluscan ecological publications between 1864 and 2016. The search string 'ecology' was combined with one of seven taxon names: Mollusca, Gastropoda, Bivalvia, Cephalopoda, Polyplacophora, Scaphopoda, and 'Aplacophora'. The number of publications is included for the last three taxa as they are too low to be reflected at this scale. Five reductions in publication rates are recognised; one with the second world war (WW II) and others with the 1930s and 1990s global recession/depressions (GD). The numbers of citations are also given.

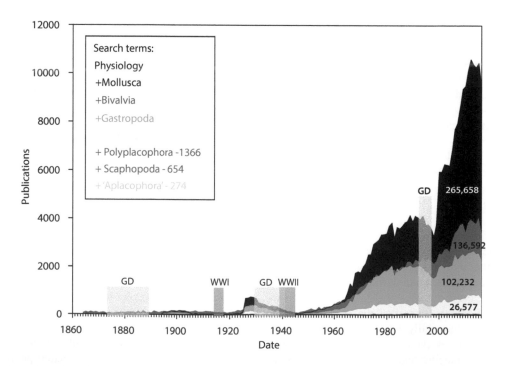

FIGURE 11.3 Area plots of molluscan physiological publications between 1864 and 2016. The search string 'physiology' was combined with one of seven taxon names: Mollusca, Bivalvia, Gastropoda, Cephalopoda, Polyplacophora, Scaphopoda, and 'Aplacophora'. Numbers of publications are included for the last three taxa as they are too low to be reflected in the area plot. Five reductions in publication rates are recognised; two with world wars (WW) and three with global recession/depressions (GD). The numbers of citations are also given.

Molluscan physiology has also been prominent in several major treatments beginning with the (1964–1966) two-volume treatise entitled *Physiology of Mollusca* (Wilbur & Yonge 1964). Three subsequent volumes: S*tudies in the Structure, Physiology, and Ecology of Molluscs* (Fretter 1968), *The Biology of Mollusca* (Purchon 1968), and *Pathways in Malacology* (Spoel et al. 1979) also dealt primarily with molluscan physiology. In the 12-volume series, *The Mollusca* (Wilbur 1983–1988) almost every volume contains contributions on molluscan physiology, and most recently, *Physiology of Molluscs: A Collection of Selected Reviews* (Saleuddin & Mukai 2016) has appeared. Besides these physiological treatments across the entire phylum, several volumes focused at the class level or below have also appeared, including, amongst others, *Behavioral Biology of Aplysia: A Contribution to the Comparative Study of Opisthobranch Molluscs* (Kandel 1979), *Physiology of Cephalopod Molluscs: Lifestyle and Performance Adaptations* (Pörtner et al. 1994), *The Biology of Terrestrial Molluscs* (Barker 2001b), *Behavior and Its Neural Control in Gastropod Molluscs* (Chase 2002), and *Marine Mussels: Their Ecology and Physiology* (Bayne 2009).

Molecular tools have to a large extent revolutionised physiological studies. For example, physiologists can now identify the ecological importance of, for example, habitat preferences, on physiological, biochemical, and molecular processes (e.g., heat shock proteins), enabling the integration of physiological measurements with mechanistic ecology (Tomanek 2002). One new tool is transcriptome data[1] which provides a snapshot of which genes are being expressed in cells and tissues at a given time. Because biotic and abiotic stress produces physiological responses seen as gene expression, transcriptome data provide a high-resolution view of the genetic underpinnings of ecological and behavioural responses. By selectively sampling cells and tissues under different conditions and at different times, it is possible to determine when and where each gene is turned on or off, and determine the number of transcripts and estimate the amount of gene expression.

What is being transcribed at a given time varies between cells and among different tissues and also depends on the physiological condition of the organism. For example, Johnson et al. (2015) have shown that, under high CO_2 and low O_2 conditions, the hepatopancreas transcriptome of the 'palaemon shrimp' (*Litopenaeus vannamei*) contains both known and novel haemocyanin isoforms which have different amino acid sequences and are either encoded by different genes or they are from the same gene, but have had different exons removed.

The neuronal transcriptome of *Aplysia californica* was one of the first molluscan transcriptomes identified (Moroz et al. 2006). Additional transcriptomes now include the neuronal transcriptomes of *Octopus vulgaris* (Zhang et al. 2012b),

transcriptome studies of shell secretions and biomineralisation (Zhang et al. 2012b; Werner et al. 2013; Kocot et al. 2016; Yarra et al. 2016), immunogenomics (Halanych & Kocot 2014; Zhang et al. 2014a; McDowell et al. 2016; Schultz & Adema 2017), and gene expression associated with reproduction and development (Bassim et al. 2014; Paps et al. 2015; de Oliveira et al. 2016).

Transcriptome studies will be crucial to studies related to climate change, both for determining physiological limits to stressors (high CO_2, low O_2, low pH, high temperatures, etc.) and estimating acclimation and adaptation potential for a taxon (Stillman 2003; Pörtner et al. 2006a; Sunday et al. 2014). Thermal stress has been of special interest, especially since the discovery and characterisation of heat shock proteins in molluscs (Greenberg et al. 1983; Laursen et al. 1997; Tomanek & Somero 1998, 1999). Heat shock proteins are expressed in response to numerous stressors, including cold shock, ultraviolet radiation, and injury, besides high temperature (Sanders 1993; Dahms & Lee 2010; Li & Guo 2015). With the threat of global warming, the responses of heat shock proteins are of special interest to researchers working in intertidal habitats. Intertidal organisms have been particularly important in physiological studies because they live in both aquatic and semi-terrestrial environments. Transcriptomes of experimentally stressed intertidal organisms show that widespread changes in gene expression occur in response to environmental differences, such as those related to temperature variation and access to molecular oxygen (Somero 2002). Physiological modifications and limits as a result of increasing temperatures are also of major concern to shellfisheries (e.g., oysters, scallops, clams, abalone) (Park et al. 2015; Jiang et al. 2016; Zhu et al. 2016; Nagata et al. 2017; Nie et al. 2017).

Since the late 1970s, there has been increasing study of the effects of acidification on molluscs and other calcium carbonate-secreting organisms (e.g., Hofmann et al. 2008). Here too, transcriptome data have provided an important tool to experimentally test some hypotheses of the effects of ocean acidification on molluscs, including compromising the ability of bivalves and gastropods to secrete robust shells (Werner et al. 2013; Li et al. 2016) and the regulation of pH associated with high CO_2 and low oxygen in high performance cephalopods (Hu et al. 2013, 2014). Like cephalopods, pteropods subjected to projected ocean acidification levels show differential transcription activity in response to extreme levels of CO_2 and responses to acidic stress, and differential expression of biomineralisation genes (Koh et al. 2015; Maas et al. 2015).

11.2.6 ENDOCRINE SYSTEM

Studies on the endocrine system and neuroendocrinology are lacking for most groups of molluscs. As with most other aspects of physiological research, there are many studies, mainly on a relatively small number of species of euthyneuran gastropods, and to a lesser extent on cephalopods. Little to virtually nothing is known regarding other taxa.

---

[1] DNA must be 'read' and transcribed (copied) into RNA (ribonucleic acid) for 'instructions' to be carried out. The individual gene readouts are transcripts, and the collection of all the transcripts in a cell is the transcriptome. It is possible to count the number of transcripts to determine the amount of gene activity (gene expression) in a cell or tissue.

As detailed in Chapter 7, the unusually large cell size of the neurons in euthyneurans has made them ideal candidates for neuroendocrine studies, but modern molecular approaches should enable sophisticated studies to be made on other molluscs. A very useful line of investigation would be to examine the comparative functioning of endocrine systems in different groups – why, for example, do dorsal bodies or similar structures occur in vetigastropods and euthyneurans, but apparently not in caenogastropods or neritimorphs? In those taxa where they are absent, what is performing the role? Similarly, the situation with other groups such as patellogastropods, polyplacophorans, bivalves, and scaphopods is not clear. A few studies have provided insights into the presence of non-neural endocrine systems, but there is still little known, and the information is very scattered.

11.3   MOLLUSCS AS MODEL ORGANISMS

Given the taxon, habitat, anatomical and functional diversity of molluscs, and their remarkable fossil record of over half a billion years, it is amazing that, with few exceptions, they have only served as model organisms for a handful of biological endeavours (e.g., *Aplysia* in neurobiology, *Biomphalaria* in parasitology, and *Crepidula* for development). There is no question of the potential of molluscs to fundamentally contribute to evolutionary theory, especially the origins of biodiversity, their historical biogeography, and the processes of speciation and radiation. And while numerous papers treating aspects of the evolution of specific taxa exist, it has been shown in other disciplines (e.g., evo-devo, see Chapter 8) that these are '… not enough to encourage a synthesis of this information into an overarching evolutionary framework' (Love 2009, p. 39). This lack of synthesis is thought to date from *The Modern Synthesis* (Huxley 1942) and extends to many other 'invertebrate' groups.

Based on a survey of the taxa featured in the major works of the modern synthesis (e.g., Dobzhansky, Simpson, Waddington, de Beer), Love (2009) noted the overall lack of marine invertebrates as exemplar or model organisms. Love found this exclusion surprising because it impacts the completeness of the synthesis regarding questions involving topics such as evolvability, novelty, and variation. The analysis by Love found only two participants in the modern synthesis, G. G. Simpson and G. R. de Beer, evoked multiple marine invertebrate exemplars in their work. Regarding molluscs, Simpson (1944) focused on the ammonite *Kosmoceras* and the bivalve *Gryphaea* in his discussion of the tempo and mode of evolution. Other bivalves mentioned included *Leda, Lima, Modiola, Nucula, Ostrea,* and *Pteria*. The developmental biologist G. R. de Beer also featured molluscs in his numerous contributions to *The Modern Synthesis*. The dearth of marine invertebrates in *The Modern Synthesis* was probably reflected in taxon choice for the research topics which followed. Between 1947 and 1960 only just over 3% of the papers published in the journal *Evolution* treated marine invertebrates (Love 2009).

The study by Love provided an important insight into the relative absence of marine invertebrates from *The Modern Synthesis* and suggested this absence continues to relate to research today. As Love (2009) argues, it is important that conceptual content and experimental practice in a research topic be evaluated for its generality and completeness. Only with comprehensive sampling across the breadth of biodiversity can we test and expand our understanding of evolutionary and other biological theories. The lack of marine invertebrates in the synthesis is probably still in play today.

Glaubrecht (2009) pointed out that efforts to build model systems are rare in malacology. What is needed are active and sustained efforts to bring molluscan exemplars to the forefront of their respective branches of biology. Adding molluscan taxa will provide additional insights and integration in these fields and will advance and extend our understanding of processes and mechanisms, as argued by Love (2009), producing hypotheses and theories that are better explained, both theoretically and generally.

11.4   MOLLUSCAN ONTOLOGY

Biological ontologies define the basic terms and relations in biological domains and are used as community reference, as the basis for interoperability between systems, and for search, integration, and exchange of biological data (Lambrix et al. 2007). To make progress, there must be shared vocabulary. Examples of biological ontologies include the Gene Ontology (GO) – a bioinformatics initiative to unify the representation of gene and gene product attributes across all species, and the Mammalian Phenotype Ontology (MP), which documents observable morphological, physiological, behavioural, and other characteristics of mammalian organisms that are displayed throughout their development and lifespan.

While not complete, this work provides a beginning for a malacological ontology.

11.5   THE FUTURE

There is no counter-argument to the fact that molluscs are important. Of the 46 commonly recognised branches of biology (http://www.biology-online.org/dictionary/Branches_of_biology), 29 (63%) feature molluscan study organisms. Outlined here, in these two volumes, is a partial record of the evolutionary history of an amazing clade. It includes parallels with our own optic evolution, morphological, and anatomical diversity which exceeds most other phyla, an ecological diversity which includes every terrestrial and aquatic biome and associated regions of the globe, and has been leaving a fossil record for half a billion years. The research potential and options molluscan diversity provides have barely been sampled. Molluscs have important contributions to make to the new biology, but it will take new tools and new approaches to realise this potential.

Malacology, like all branches of biology, is poised for rapid progress in a host of areas due to methodological and computational improvements; however, our ability to organise and share information remains rudimentary or dependent on keyword searches of bibliographic databases.

Social media clients such as ResearchGate, LinkedIn, Frontiers, Academic.edu are available to facilitate interactions, etc., but no mollusc-specific working group or organisation exists to communicate molluscan research efforts and needs in a coordinated and cooperative manner. We tend to focus on specific groups of interest and ignore most of the incredible diversity in the phylum. As a result, certain groups are relatively well-studied, while others remain almost unknown. Moreover, improved communication across the full breadth of biological research would undoubtedly benefit numerous researchers who use molluscs as study organisms, as many of us are perhaps too focused on our own speciality. We must make more and better use of social media and 'the cloud'. A molluscan-based platform could also provide a clearinghouse for potential collaborators, including citizen scientists and volunteers, to assist in research projects. Such an enterprise would probably lead to larger, more active, and inclusive research groups and a more systematic approach to unravelling the biology and evolution of the Mollusca.

In our final chapter of the second volume (Chapter 21) we explore some other potentially important areas of development of knowledge and methods, better access to the information, and the important role of 'citizen science'.

References

van Aartsen, J. J. & Engl, W. (2001). Il genere *Planktomya* nell'Atlantico orientale. *La Conchiglia* 300: 14–22.

Abbott, N. J. & Bundgaard, M. (1987). Microvessel surface area, density and dimensions in brain and muscle in the cephalopod *Sepia officinalis*. *Proceedings of the Royal Society B* 230: 459–482.

Abbott, R. T. (1960). The genus *Strombus* in the Indo-Pacific. *Indo-Pacific Mollusca* 1: 33–146.

Abdou, A. H. (1956). Observations on the life cycle of *Davainea proglottina* in Britain. *Journal of Helminthology* 30: 189–202.

Abdraba, A. M. & Saleuddin, A. S. M. (2000). A putative hyperglycemic factor from the cerebral ganglia of *Otala lactea* (Mollusca: Pulmonata). *Journal of Comparative Physiology B* 170: 219–224.

Abele, D., Kruppe, M., Philipp, E. E. R. & Brey, T. (2010). Mantle cavity water oxygen partial pressure (Po2) in marine molluscs aligns with lifestyle. *Canadian Journal of Fisheries and Aquatic Sciences* 67: 977–986.

Ackerly, S. C. (1989). Kinematics of accretionary shell growth, with examples from brachiopods and molluscs. *Paleobiology* 15: 147–164.

Adal, M. N. & Morton, B. (1973). The fine structure of the pallial eyes of *Laternula truncata* (Bivalvia, Anomalodesmata, Pandoracea). *Journal of Zoology* 170: 533–556.

Adamkewicz, S. L. & Castagna, M. (1988). Genetics of shell color and pattern in the bay scallop *Argopecten irradians*. *Journal of Heredity* 79: 14–17.

Adamo, S. A., Ehgoetz, K., Sangster, C. & Whitehorne, I. (2006). Signaling to the enemy? Body pattern expression and its response to external cues during hunting in the cuttlefish *Sepia officinalis* (Cephalopoda). *Biological Bulletin* 210: 192–200.

Adamson, K. J., Wang, T., Zhao, M., Bell, F., Kuballa, A. V., Storey, K. B. & Cummins, S. F. (2015). Molecular insights into land snail neuropeptides through transcriptome and comparative gene analysis. *BMC Genomics* 16: 308.

Addadi, L., Joester, D., Nudelman, F. & Weiner, S. (2006). Mollusk shell formation: A source of new concepts for understanding biomineralization processes. *Chemistry* 12: 980–987.

Aditya, G. & Raut, S. K. (2005). Feeding of the leech *Glossiphonia weberi* on the introduced snail *Pomacea bridgesii* in India. *Aquatic Ecology* 39: 465–471.

Adl, S. M., Simpson, A. G., Lane, C. E., Lukeš, J., Bass, D., Bowser, S. S., Brown, M. W., Burki, F., Dunthorn, M. & Hampl, V. (2012). The revised classification of eukaryotes. *Journal of Eukaryotic Microbiology* 59: 429–514.

Adriaens, E. & Remon, J. P. (1999). Gastropods as an evaluation tool for screening the irritating potency of absorption enhancers and drugs. *Pharmacological Research* 16: 1240–1244.

Aggio, J. F. & Derby, C. D. (2008). Hydrogen peroxide and other components in the ink of sea hares are chemical defenses against predatory spiny lobsters acting through non-antennular chemoreceptors. *Journal of Experimental Marine Biology and Ecology* 363: 28–34.

Agnisola, C. (1990). Functional morphology of the coronary supply of the systemic heart of *Octopus vulgaris*. *Physiological Zoology* 63: 3–11.

Agnisola, C., Zummo, G. & Tota, B. (1990). Coronary drainage in the *Octopus vulgaris* systemic heart. *Journal of Experimental Zoology Part A* 253: 1–6.

Aguado, F. & Marin, A. (2007). Warning coloration associated with nematocyst-based defences in aeolidioidean nudibranchs. *Journal of Molluscan Studies* 73: 23–28.

Agudo-Padrón, A. I. (2012). Brazilian snail-eating snakes (Reptilia, Serpentes, Dipsadidae) and their alimentary preferences by terrestrial molluscs (Gastropoda, Gymnophila & Pulmonata): A preliminary overview. *Biological Evidence* 2: 2–3.

Agudo-Padrón, A. I. (2013). Snail-eating snakes ecology, diversity, distribution and alimentary preferences in Brazil. *Journal of Environmental Science and Water Resources* 2: 238–244.

Aguiar, R. & Wink, M. (2005). How do slugs cope with toxic alkaloids? *Chemoecology* 15: 167–177.

Aiken, S. (2009). Cuban land snails: The weird and the wonderful. *American Conchologist* 37: 17–19.

Aines, A. C., Carlson, J. K., Boustany, A., Mathers, A. & Kohler, N. E. (2018). Feeding habits of the tiger shark, *Galeocerdo cuvier*, in the northwest Atlantic Ocean and Gulf of Mexico. *Environmental Biology of Fishes* 101: 403–415.

Akroyd, J. A. M., Walshe, K. A. R. & Millar, R. B. (2002). Abundance, distribution, and size structure of toheroa (*Paphies ventricosa*) at Ripiro Beach, Dargaville, Northland, New Zealand. *New Zealand Journal of Marine and Freshwater Research* 36: 547–553.

Akşit, D. M. & Falakali, B. (2011). The external morphology of the gill of *Patella caerulea* L. (Mollusca: Gastropoda). *Turkish Journal of Zoology* 35: 603–606.

Aktipis, S. W., Giribet, G., Lindberg, D. R. & Ponder, W. F. (2008). Gastropoda: An overview and analysis. pp. 201–237 in W. F. Ponder & Lindberg, D. R. (Eds.) *Phylogeny and Evolution of the Mollusca*. Berkeley, CA, University of California Press.

Aktipis, S. W., Boehm, E. & Giribet, G. (2011). Another step towards understanding the slit-limpets (Fissurellidae, Fissurelloidea, Vetigastropoda, Gastropoda): A combined five-gene molecular phylogeny. *Zoologica Scripta* 40: 238–259.

Albert, V. A. (2005). *Parsimony, Phylogeny and Genomics*. Oxford, UK, Oxford University Press.

Albertin, C. B., Simakov, O., Mitros, T., Wang, Z. Y., Pungor, J. R., Edsinger-Gonzales, E., Brenner, S., Ragsdale, C. W. & Rokhsar, D. S. (2015). The *Octopus* genome and the evolution of cephalopod neural and morphological novelties. *Nature* 524: 220–224.

Albizati, K. F., Pawlik, J. R. & Faulkner, D. J. (1985). Limatulone, a potent defensive metabolite of the intertidal limpet *Collisella limatula*. *The Journal of Organic Chemistry* 50: 3428–3430.

Albrecht, C. & Wilke, T. (2008). Ancient Lake Ohrid: Biodiversity and evolution. *Hydrobiologia* 615: 103–140.

Albrecht, C., Hauffe, T., Schreiber, K. & Wilke, T. (2012). Mollusc biodiversity in a European ancient lake system: Lakes Prespa and Mikri Prespa in the Balkans. *Hydrobiologia* 682: 47–59.

Albrecht, U., Keller, H., Gebauer, W. & Markl, J. (2001). Rhogocytes (pore cells) as the site of hemocyanin biosynthesis in the marine gastropod *Haliotis tuberculata*. *Cell and Tissue Research* 304: 455–462.

Aldana-Aranda, D., Frenkiel, L., Brulé, T., Montero, J. & Cárdenas, E. B. (2011). Occurrence of Apicomplexa-like structures in the digestive gland of *Strombus gigas* throughout the Caribbean. *Journal of Invertebrate Pathology* 106: 174–178.

Alevizos, A., Skelton, M. E., Weiss, K. R. & Koester, J. (1991). A comparison of bursting neurons in *Aplysia*. *Biological Bulletin* 180: 269–275.

Alexander, C. G., Cutler, R. L. & Yellowless, D. (1979). Studies on the composition and enzyme content of the crystalline style of *Telescopium telescopium* (L.) (Gastropoda). *Comparative Biochemistry and Physiology Part B* 64: 83–89.

Alexander, R. M. (2003). *Principles of Animal Locomotion*. Princeton, NJ, Princeton University Press.

Alexandrowicz, J. S. (1962). An accessory organ of the circulatory system in *Sepia* and *Loligo*. *Journal of the Marine Biological Association of the United Kingdom* 42: 405–418.

Ali, M. A. (1984). *Photoreception and Vision in Invertebrates [NATO Advanced Study Institute Workshop held in July 1982 at Bishop's University, Quebec, Canada]*. New York, Plenum Press.

Aliakrinskaia, I. O. (1972). Biochemical adaptations of aquatic molluscs to inhabiting air [in Russian]. *Zoologicheskii Zhurnal* 51: 1630–1636.

Alkalay, A. S., Rosen, O., Sokolow, S. H., Faye, Y. P. W., Faye, D. S., Aflalo, E. D., Jouanard, N., Zilberg, D., Huttinger, E. & Sagi, A. (2014). The prawn *Macrobrachium vollenhovenii* in the Senegal River Basin: Towards sustainable restocking of all-male populations for biological control of Schistosomiasis. *PLoS Neglected Tropical Diseases* 8: e3060.

Alkon, D. L. (1987). *Memory Traces in the Brain*. New York, Cambridge University Press.

Allen, A., Michels, J. & Young, J. Z. (1985). Memory and visual-discrimination by squids. *Marine Behaviour and Physiology* 11: 271–282.

Allen, J. A. (1958). On the basic form and adaptations to habitat in the Lucinacea (Eulamellibranchia). *Philosophical Transactions of the Royal Society B* 241: 421–484.

Allen, J. A. & Scheltema, R. S. (1972). The functional morphology and geographic distribution of *Planktomya henseni*, a supposed neotenic pelagic bivalve. *Journal of the Marine Biological Association of the United Kingdom* 52: 19–31.

Allen, J. A. (1978). Evolution of the deep sea protobranch bivalves. *Philosophical Transactions of the Royal Society B* 284: 387–401.

Allen, J. A. (1983). The ecology of deep-sea molluscs. pp. 29–75 *in* W. D. Russell-Hunter (Ed.) *Ecology. The Mollusca*. Vol. 6. New York, Academic Press.

Allen, J. A. (1988). Reflexive selection is apostatic selection. *Oikos* 51: 251–253.

Allen, J. A. (2004). Avian and mammalian predators of terrestrial gastropods. pp. 1–36 *in* G. M. Barker (Ed.) *Natural Enemies of Terrestrial Molluscs*. Oxford, UK/Cambridge, MA, CABI Publishing.

Allen, J. D. & Pernet, B. (2007). Intermediate modes of larval development: Bridging the gap between planktotrophy and lecithotrophy. *Evolution & Development* 9: 643–653.

Allen, S. K., Hidu, H. & Stanley, J. G. (1986). Abnormal gametogenesis and sex ratio in triploid soft-shell clams (*Mya arenaria*). *Biological Bulletin* 170: 198–210.

Allgaier, C. (2007). Active camouflage with lichens in a terrestrial snail, *Napaeus (N.) barquini* Alonso and Ibáñez, 2006 (Gastropoda, Pulmonata, Enidae). *Zoological Science* 24: 869–876.

Allmon, W. D. (1994). Patterns and processes of heterochrony in lower Tertiary turritelline gastropods, U. S. Gulf and Atlantic coastal plains. *Journal of Paleontology* 68: 80–95.

Almendros, A. & Porcel, D. (1992). Phosphatase activity in the hepatopancreas of *Helix aspersa*. *Comparative Biochemistry and Physiology Part A* 103: 455–460.

Alon, G., Shore, L. S. & Steinberger, Y. (2007). Correlation between levels of sex hormones (progesterone, testosterone, and estrogen) and ecophysiological-behavior stages in two species of desert snails (*Sphincterochila zonata* and *Sphincterochila prophetarum*) in the Northern Negev Desert. *General and Comparative Endocrinology* 151: 122–127.

Alon, S., Garrett, S. C., Levanon, E. Y., Olson, S., Graveley, B. R., Rosenthal, J. J. & Eisenberg, E. (2015). The majority of transcripts in the squid nervous system are extensively recoded by A-to-I RNA editing. *Elife* 4: e05198.

Alonso, Á. & Castro-Díez, P. (2012). The exotic aquatic mud snail *Potamopyrgus antipodarum* (Hydrobiidae, Mollusca): State of the art of a worldwide invasion. *Aquatic Sciences* 74: 375–383.

Alonso, M. R., Goodacre, S. L., Emerson, B. C., Ibanez, M., Hutterer, R. & Groh, K. (2006). Canarian land snail diversity: Conflict between anatomical and molecular data on the phylogenetic placement of five new species of *Napaeus* (Gastropoda, Pulmonata, Enidae). *Biological Journal of the Linnean Society* 89: 169–187.

Alpuche, J., Pereyra, A., Mendoza-Hernández, G., Agundis, C., Rosas, C. & Zenteno, E. (2010). Purification and partial characterization of an agglutinin from *Octopus maya* serum. *Comparative Biochemistry and Physiology Part B* 156: 1–5.

Alupay, J. S. (2013). Characterization of arm autotomy in the octopus, *Abdopus aculeatus* (d'Orbigny, 1834). PhD, University of California, Berkeley.

Alvarez-Tinajero, M. D. C., Càceres-Martìnez, J. & Gonzàlez-Avilés, J. G. (2013). Shell boring clams in the blue abalone *Haliotis fulgens* and the yellow abalone *Haliotis corrugata* from Baja California, Mexico. *Journal of Shellfish Research* 20: 889–893.

Alves, C., Chichery, R., Boal, J. G. & Dickel, L. (2007). Orientation in the cuttlefish *Sepia officinalis*: Response versus place learning. *Animal Cognition* 10: 29–36.

Alves, F., Beasley, C. R., Hoeh, W. R., Martins, R., Ricardo, L., de Simone, L. R. L. & Tagliaro, C. H. (2012). Detection of mitochondrial DNA heteroplasmy suggests a doubly uniparental inheritance pattern in the mussel *Mytella charruana*. *Revista Brasileira de Biociências* 10: 176–185.

Alyakrinskaya, I. O. (2003). Tissue hemoglobins in Bivalvia (Mollusca). *Biology Bulletin (translated from Izvestiia Akademii nauk SSSR Seriya biologicheskaya)* 30: 735–746.

Alzieu, C. (2000a). Impact of tributyltin on marine invertebrates. *Ecotoxicology* 9: 71–76.

Alzieu, C. (2000b). Environmental impact of TBT: The French experience. *Science of the Total Environment* 258: 99–102.

Amano, M., Moriyama, S., Okubo, K., Amiya, N., Takahashi, A. & Oka, Y. (2010a). Biochemical and immunohistochemical analyses of a GnRH-like peptide in the neural ganglia of the Pacific abalone *Haliotis discus hannai* (Gastropoda). *Zoological Science* 27: 656–661.

Amano, M., Yokoyama, T., Amiya, N., Hotta, M., Takakusaki, Y., Kado, R. & Oka, Y. (2010b). Biochemical and immunohistochemical analyses of GnRH-like peptides in the nerve ganglion of the chiton, *Acanthopleura japonica*. *Zoological Science* 27: 924–930.

Amaral, V., Cabral, H. N. & Bishop, M. J. (2012a). Effects of estuarine acidification on predator–prey interactions. *Marine Ecology Progress Series* 445: 117–127.

Amaral, V., Thompson, E. L., Bishop, M. J. & Raftos, D. A. (2012b). The proteomes of Sydney rock oysters vary spatially according to exposure to acid sulfate runoff. *Marine and Freshwater Research* 63: 361–369.

Amaudrut, A. (1898). La partie antérieure du tube digestif et la torsion chez les mollusques gastéropodes. *Annales des sciences naturelles. Zoologie* 4: 1–291.

Amiard, J. C., Amiard-Triquet, C., Barka, S., Pellerin, J. & Rainbow, P. S. (2006). Metallothioneins in aquatic invertebrates: Their role in metal detoxification and their use as biomarkers. *Aquatic Toxicology* 76: 160–202.

Amio, M. (1955). On growth and development of spines of the top shell, *Turbo cornutus* Solander. *Journal of the Shimonoseki College of Fisheries* 4: 57–68.

Amparyup, P., Klinbunga, S. & Jarayabhand, P. (2010). Identification and expression analysis of sex-specific expression markers of Thai abalone *Haliotis asinina*, Linnaeus, 1758. *Journal of Shellfish Research* 29: 765–773.

Amyot, J. P. & Downing, J. A. (1998). Locomotion in *Elliptio complanata* (Mollusca: Unionidae): A reproductive function? *Freshwater Biology* 39: 351–358.

Anand, T. P. & Edward, J. K. P. (2002). Antimicrobial activity in the tissue extracts of five species of cowries *Cypraea* spp. (Mollusca: Gastropoda) and an ascidian *Didemnum psammathodes* (Tunicata: Didemnidae). *Indian Journal of Marine Sciences* 31: 239–242.

Ancel, P. (1902). La réduction numerique des chromosomes dans la spermatogenèse d'*Helix pomatia*. *Bibliographie Anatomique* 11: 17–20.

Andersen, R. J., Faulkner, D. J., He, C. H., Van Duyne, G. D. & Clardy, J. (1985). Metabolites of the marine prosobranch mollusk *Lamellaria* sp. *Journal of the American Chemical Society* 107: 5492–5495.

Andersen, R. J., Desjardine, K. & Woods, K. (2006). Skin chemistry of nudibranchs from the west coast of North America. pp. 277–302 *in* G. Cimino & Gavagnin, M. (Eds.) *Molluscs: From Chemo-Ecological Study to Biotechnological Application*. Berlin, Germany, Springer-Verlag.

Anderson, L. C. (2014). Relationships of internal shell features to chemosymbiosis, life position, and geometric constraints within the Lucinidae (Bivalvia). pp. 49–72 *in* D. I. Hembree, Platt, B. F. & Smith, J. J. (Eds.) *Experimental Approaches to Understanding Fossil Organisms. Topics in Geobiology*. Dordrecht, the Netherlands, Springer Science + Business Media.

Anderson, M. J. & Underwood, A. J. (1997). Effects of gastropod grazers on recruitment and succession of an estuarine assemblage: A multivariate and univariate approach. *Oecologia* 109: 442–453.

Anderson, R. C. (1960). On the development and transmission of *Cosmocercoides dukae* of terrestrial molluscs in Ontario. *Canadian Journal of Zoology* 38: 801–825.

Anderson, R. C. & Mather, J. A. (2007). The packaging problem: Bivalve prey selection and prey entry techniques of the octopus *Enteroctopus dofleini*. *Journal of Comparative Psychology* 121: 300–305.

Anderson, R. C., Mather, J. A., Monette, M. Q. & Zimsen, S. R. M. (2010). Octopuses (*Enteroctopus dofleini*) recognize individual humans. *Journal of Applied Animal Welfare Science* 13: 261–272.

Andreev, N. (2006). Assessment of the status of wild populations of land snail (escargot) *Helix pomatia* L. in Moldova: The effect of exploitation. *Biodiversity and Conservation* 15: 2957–2970.

Andrew-Priestley, M. N., O'Connor, W. A., Dunstan, R. H., Van Zwieten, L., Tyler, T., Kumar, A. & MacFarlane, G. R. (2012). Estrogen mediated effects in the Sydney rock oyster, *Saccostrea glomerata*, following field exposures to sewage effluent containing estrogenic compounds and activity. *Aquatic Toxicology* 120: 99–108.

Andrew, M. N., Dunstan, R. H., O'Connor, W. A., Van Zwieten, L., Nixon, B. & MacFarlane, G. R. (2008). Effects of 4-nonylphenol and 17α-ethynylestradiol exposure in the Sydney rock oyster, *Saccostrea glomerata*: Vitellogenin induction and gonadal development. *Aquatic Toxicology* 88: 39–47.

Andrew, R. J. & Savage, H. (2000). Appetitive learning using visual conditioned stimuli in the pond snail, *Lymnaea*. *Neurobiology of Learning and Memory* 73: 258–273.

Andrews, E. B. (1964). The functional anatomy and histology of the reproductive system of some pilid gastropod molluscs. *Proceedings of the Malacological Society of London* 36: 121–140.

Andrews, E. B. (1965a). The functional anatomy of the gut of the prosobranch gastropod *Pomacea canaliculata* and of some other pilids. *Proceedings of the Zoological Society of London* 145: 19–36.

Andrews, E. B. (1965b). The functional anatomy of the mantle cavity, kidney and blood system of some pilid gastropods (Prosobranchia). *Journal of Zoology* 146: 70–94.

Andrews, E. B. (1968). An anatomical and histological study of the nervous system of *Bithynia tentaculata* (Prosobranchia) with special reference to possible neurosecretory activity. *Proceedings of the Malacological Society of London* 38: 213–232.

Andrews, E. B. & Little, C. (1972). Structure and function in the excretory systems of some terrestrial prosobranch snails (Cyclophoridae). *Journal of Zoology* 168: 395–422.

Andrews, E. B. (1976a). The ultrastructure of the heart and kidney of the pilid gastropod mollusc *Marisa cornuarietis*, with special reference to filtration throughout the Architaenioglossa. *Journal of Zoology* 179: 85–106.

Andrews, E. B. (1976b). The fine structure of the heart of some prosobranch and pulmonate gastropods in relation to filtration. *Journal of Molluscan Studies* 42: 199–216.

Andrews, E. B. (1979a). Fine structure in relation to function in the excretory system of two species of *Viviparus*. *Journal of Molluscan Studies* 45: 186–206.

Andrews, E. B. (1981). Osmoregulation and excretion in prosobranch gastropods. Part 2: Structure in relation to function. *Journal of Molluscan Studies* 47: 248–289.

Andrews, E. B. & Little, C. (1982). Renal structure and function in relation to habitat in some cyclophorid land snails from Papua New Guinea. *Journal of Molluscan Studies* 48: 124–143.

Andrews, E. B. (1985). Structure and function in the excretory system of archaeogastropods and their significance in the evolution of gastropods. *Philosophical Transactions of the Royal Society B* 310: 383–406.

Andrews, E. B. (1988). Excretory systems of molluscs. pp. 381–448 *in* E. R. Trueman & Clarke, M. R. (Eds.) *Form and Function. The Mollusca*. Vol. 11. San Diego, CA, Academic Press.

Andrews, E. B. & Taylor, P. M. (1988). Fine structure, mechanism of heart function and haemodynamics in the prosobranch gastropod mollusc *Littorina littorea* (L.). *Journal of Comparative Physiology B* 158: 247–262.

Andrews, E. B. & Taylor, P. M. (1990). Reabsorption of organic solutes in some marine and freshwater prosobranch gastropods. *Journal of Molluscan Studies* 56: 147–162.

Andrews, E. B. (1991). The fine structure and function of the salivary glands of *Nucella lapillus* (Gastropoda: Muricidae). *Journal of Molluscan Studies* 57: 111–126.

Andrews, E. B., Elphick, M. R. & Thorndyke, M. C. (1991). Pharmacologically active constituents of the accessory salivary glands and hypobranchial glands of *Nucella lapillus*. *Journal of Molluscan Studies* 57: 136–138.

Andrews, E. B. (1992). The fine structure and function of the anal gland of the muricid *Nucella lapillus* (Neogastropoda) (and a comparison with that of the trochid *Gibbula cineraria*). *Journal of Molluscan Studies* 58: 297–313.

Andrews, E. B. & Jennings, K. H. (1993). The anatomical and ultrastructural basis of primary urine formation in bivalve molluscs. *Journal of Molluscan Studies* 59: 223–257.

Andrews, E. B., Page, A. M. & Taylor, J. D. (1999). The fine structure and function of the anterior foregut glands of *Cymatium intermedius* (Cassoidea: Ranellidae). *Journal of Molluscan Studies* 65: 1–19.

Andrews, E. B. & Thorogood, K. E. (2005). An ultrastructural study of the gland of Leiblein of muricid and nassariid neogastropods in relation to function, with a discussion on its homologies in other caenogastropods. *Journal of Molluscan Studies* 71: 269–300.

Andrews, E. B. (2010). The fine structure of the excretory and venous systems of the neogastropods *Nucella lapillus* and *Buccinum undatum* and its functional and phylogenetic implications. *Journal of Molluscan Studies* 76: 211–233.

Andrews, H. E. (1974). Morphometrics and functional morphology of *Turritella mortoni. Journal of Paleontology* 48: 1126–1140.

Andrews, J. D. (1979b). Pelecypoda: Ostreidae. pp. 293–341 *in* A. C. Giese & Pearse, J. S. (Eds.) *Reproduction of Marine Invertebrates. Molluscs: Pelecypods and Lesser Classes.* Vol. 5. New York, Academic Press.

Andriychuk, T. V. & Garbar, A. V. (2015). On sexual dimorphism of karyotypes of *Viviparus viviparus* and *V. contectus* (Gastropoda, Viviparidae). *Vestnik Zoologii* 49: 105–112.

Angelini, E., Salvato, B., Di Muro, P. & Beltramini, M. (1998). Respiratory pigments of *Yoldia eightsi*, an Antarctic bivalve. *Marine Biology* 131: 15–23.

de Angelis, P. (2012). Assessing the impact of international trade on chambered *Nautilus. Geobios* 45: 5–11.

Angeloni, L. (2003). Sexual selection in a simultaneous hermaphrodite with hypodermic insemination: Body size, allocation to sexual roles and paternity. *Animal Behaviour* 66: 417–426.

Angeloni, L., Bradbury, J. W. & Burton, R. S. (2003). Multiple mating, paternity, and body size in a simultaneous hermaphrodite, *Aplysia californica. Behavioral Ecology* 14: 554–560.

Angerer, G. & Haszprunar, G. (1996). Anatomy and affinities of lepetid limpets (Patellogastropoda = Docoglossa). pp. 171–175 *in* J. D. Taylor (Ed.) *Origin and Evolutionary Radiation of the Mollusca.* Oxford, UK, Oxford University Press.

Anisimov, A. P. (2005). Endopolyploidy as a morphogenetic factor of development. *Cell Biology International* 29: 993–1004.

Anisimov, A. P. & Zyumchenko, N. E. (2012). Evolutionary regularities of development of somatic polyploidy in salivary glands of gastropod mollusks: Part V. Subclasses Opisthobranchia and Pulmonata. *Cell and Tissue Biology* 6: 268–279.

Anisimova, N., Berenboim, B., Gerasimova, O. V., Manushin, I. & Pinchukov, M. (2005). *On the Effect of Red King Crab on some Components of the Barents Sea Ecosystem. Report PINRO.* Murmansk, Russia, Polar Research Institute of Marine Fisheries and Oceanography: 1–9.

Ankel, W. E. (1937). Wie bohrt *Natica. Biologisches Zentrablatt* 57: 75–82.

Ankel, W. E. (1952). *Phylliroe bucephala* Per. und Les. und die Meduse *Mnestra parasites* Krohn. *Pubblicazioni della Stazione Zoologica di Napoli* 23: 91–140.

Anker, A., Murina, G. V., Lira, C., Caripe, J. A. V., Palmer, A. R. & Jeng, M. S. (2005). Macrofauna associated with echiuran burrows: A review with new observations of the innkeeper worm, *Ochetostoma erythrogrammon* Leuckart and Ruppel, in Venezuela. *Zoological Studies* 44: 157–190.

Annandale, N. & Sewell, R. B. S. (1921). The banded pond snail of India *Viviparus bengalensis. Records of the Indian Museum* 22: 215–292.

Anonymous. (2011). Monterey Bay National Marine Sanctuary site characterization. Shallow soft bottom habitats. II Community zonation. from http://montereybay.noaa.gov/sitechar/shallow2.html Accessed 21 July 2011.

Ansell, A. D. (1967a). Leaping movements in two species of Asaphidae (Bivalvia). *Proceedings of the Malacological Society of London* 37: 395–398.

Ansell, A. D. (1967b). Leaping and other movements in some cardiid bivalves. *Animal Behaviour* 15: 421–426.

Ansell, A. D. & Nair, N. B. (1968). Occurrence of haemocoelic erythrocytes containing haemoglobin in a wood boring mollusc. *Nature* 217: 357.

Ansell, A. D. (1969). Leaping movements in the Bivalvia. *Proceedings of the Malacological Society of London* 38: 387–399.

Ansell, A. D. & Trevallion, A. (1969). Behavioural adaptations of intertidal molluscs from a tropical sandy beach. *Journal of Experimental Marine Biology and Ecology* 4: 9–35.

Ansell, A. D. & Morton, B. (1987). Alternative predation tactics of a tropical naticid gastropod. *Journal of Experimental Marine Biology and Ecology* 111: 109–119.

Ansell, A. D., Gibson, R. N. & Barnes, M. (1998). Symbiotic polychaetes: Review of known species. *Oceanography and Marine Biology Annual Review* 36: 217–340.

Anthes, N. & Michiels, N. K. (2005). Do 'sperm trading' simultaneous hermaphrodites always trade sperm? *Behavioral Ecology* 16: 188–195.

Anthes, N., Putz, A. & Michiels, N. K. (2005). Gender trading in a hermaphrodite. *Current Biology* 15: R792–R793.

Anthes, N., Putz, A. & Michiels, N. K. (2006a). Hermaphrodite sex role preferences: The role of partner body size, mating history and female fitness in the sea slug *Chelidonura sandrana. Behavioral Ecology and Sociobiology* 60: 359–367.

Anthes, N., Putz, A. & Michiels, N. K. (2006b). Sex role preferences, gender conflict and sperm trading in simultaneous hermaphrodites: A new framework. *Animal Behaviour* 72: 1–12.

Anthes, N. & Michiels, N. K. (2007). Precopulatory stabbing, hypodermic injections and unilateral copulations in a hermaphroditic sea slug. *Biology Letters* 3: 121–124.

Anthes, N., Schulenburg, H. & Michiels, N. K. (2008). Evolutionary links between reproductive morphology, ecology and mating behavior in opisthobranch gastropods. *Evolution: International Journal of Organic Evolution* 62: 900–916.

Anthony, J. L., Kesler, D. H., Downing, W. L. & Downing, J. A. (2001). Length-specific growth rates in freshwater mussels (Bivalvia: Unionidae): Extreme longevity or generalized growth cessation? *Freshwater Biology* 46: 1349–1359.

Antipa, G. A. & Small, E. B. (1971). The occurrence of thigmotrichous ciliated Protozoa inhabiting the mantle cavity of unionid molluscs of Illinois. *Transactions of the American Microscopical Society* 90: 463–472.

Antonio, E. S., Kasai, A., Ueno, M., Kurikawa, Y., Tsuchiya, K., Toyohara, H., Ishihi, Y., Yokoyama, H. & Yamashita, Y. (2010). Consumption of terrestrial organic matter by estuarine molluscs determined by analysis of their stable isotopes and cellulase activity. *Estuarine, Coastal and Shelf Science* 86: 401–407.

Appeltans, W., Ahyong, S. T., Anderson, G., Angel, M. V., Artois, T., Bailly, N., Bamber, R. et al. (2012). The magnitude of global marine species diversity. *Current Biology* 22: 2189–2202.

Appleton, R. D. & Palmer, A. R. (1988). Water-borne stimuli released by predatory crabs and damaged prey induce more predator-resistant shells in a marine gastropod. *Proceedings of the National Academy of Sciences of the United States of America* 85: 4387–4391.

Appleton, T. C., Newell, P. F. & Machin, J. (1979). Ionic gradients within mantle-collar epithelial cells of the land snail *Otala lactea*. *Cell and Tissue Research* 199: 83–97.

Apraiz, I., Mi, J. & Cristobal, S. (2006). Identification of proteomic signatures of exposure to marine pollutants in mussels (*Mytilus edulis*). *Molecular and Cellular Proteomics* 5: 1274–1285.

Arai, M. N. (2005). Predation on pelagic coelenterates: A review. *Journal of the Marine Biological Association of the United Kingdom* 85: 523–536.

Arakawa, K. Y. (1958). On the remarkable sexual dimorphism of the radula of *Drupella*. *Venus* 19: 206–214.

Arakawa, K. Y. (1963). Studies on the molluscan faeces (I). *Publications of the Seto Marine Biological Laboratory* 11: 185–208.

Arakawa, K. Y. (1965). Studies on the molluscan faeces (II). *Publications of the Seto Marine Biological Laboratory* 13: 1–21.

Arakawa, K. Y. (1968). Studies on the molluscan faeces (III). *Publications of the Seto Marine Biological Laboratory* 16: 127–139.

Arakawa, K. Y. (1970). Scatological studies of the Bivalvia (Mollusca). *Advances in Marine Biology* 8: 307–436.

Arakawa, K. Y. (1972). Studies on the molluscan faeces (IV). *Publications of the Seto Marine Biological Laboratory* 19: 347–357.

Arakawa, Y., Nakano, D., Tsukuda, O. & Hoshino, T. (1978). On faecal pellets and food habits of emperor's slit shell, *Mikadotrochus hirasei* (Pilsbry). *Venus* 37: 116–120.

Arcadi, J. A. (1967). The two types of mucous gland cells in the integument of the slug, *Lehmania poirieri* (Mabille): A study in metachromasy. *Transactions of The American Microscopical Society* 86: 506–509.

Arch, S., Earley, P. & Smock, T. (1976). Biochemical isolation and physiological identification of the egg-laying hormone in *Aplysia californica*. *Journal of General Physiology* 68: 197–210.

Arendt, D. (2003). Evolution of eyes and photoreceptor cell types. *International Journal of Developmental Biology* 47: 563–571.

Arey, L. B. & Crozier, W. J. (1919). The sensory responses of *Chiton*. *Journal of Experimental Zoology* 29: 157–260.

Arizmendi-Rodriguez, D. I., Rodriguez-Jaramillo, C., Quinonez-Velazquez, C. & Salinas-Zavala, C. A. (2012). Reproductive indicators and gonad development of the Panama brief squid *Lolliguncula panamensis* (Berry 1911) in the Gulf of California, Mexico. *Journal of Shellfish Research* 31: 817–826.

Arkhipkin, A. I. (1992). Reproductive system structure, development and function in cephalopods with a general scale for maturity stages. *Journal of Northwest Atlantic Fishery Science* 12: 65–74.

Arkhipkin, A. I. & Bizikov, V. (1996). Possible imitation of jellyfish by the squid paralarvae of the family Gonatidae (Cephalopoda, Oegopsida). *Polar Biology* 16: 531–534.

Arkhipkin, A. I. & Bizikov, V. A. (2000). Role of the statolith in functioning of the acceleration receptor system in squids and sepioids. *Journal of Zoology* 250: 31–55.

Arkhipkin, A. I. (2004). Diversity in growth and longevity in short-lived animals: Squid of the suborder Oegopsina. *Marine and Freshwater Research* 55: 341.

Arkhipkin, A. I. & Laptikhovsky, V. V. (2010). Observation of penis elongation in *Onykia ingens*: Implications for spermatophore transfer in deep-water squid. *Journal of Molluscan Studies* 76: 299–300.

Armstrong, J. & Smith, P. J. S. (1992). Dynamic responses of cardiac output. pp. 15–21 in R. B. Hill, Kuwasawa, K., McMahon, B. R. & Kuramoto, T. (Eds.) *Phylogenetic Models in Functional Coupling of the CNS and the Cardiovascular System. Comparative Physiology*. Vol. 11. Basel, Karger.

Arnaud, F. (1978). A new species of *Ascorhynchus* (Pycnogonida) found parasitic on an opisthobranchiate mollusc. *Zoological Journal of the Linnean Society* 63: 99–104.

Arnaud, P. M., Poizat, C. H. & von Salvini-Plawen, L. (1986). Marine interstitial Gastropoda (including one freshwater interstitial species). pp. 153–173 in L. Botosaneanu (Ed.) *Stygofauna Mundi. A Faunistic, Distributional and Ecological Synthesis of the World Fauna Inhabiting Subterranean Waters*. Leiden, E.J. Brill.

Arnold, J. M. (1965). Normal embryonic stages of the squid, *Loligo pealii* (Lesueur). *Biological Bulletin* 28: 24–32.

Arnold, J. M. (1971). Cephalopods. pp. 265–310 in G. Reverberi (Ed.) *Experimental Embryology of Marine and Fresh-water Invertebrates*. New York, Elsevier Publishing Co.

Arnold, J. M. & Williams-Arnold, L. D. (1977). Cephalopoda: Decapoda. pp. 243–290 in A. C. Giese & Pearse, J. S. (Eds.) *Reproduction of Marine Invertebrates. Molluscs: Gastropods and Cephalopods*. Vol. 4. New York, Academic Press.

Arnold, J. M. & Williams-Arnold, L. D. (1978). Spermiogenesis of *Nautilus pompilius*. I. General survey. *Journal of Experimental Zoology Part A* 205: 13–25.

Arnold, J. M. (1984). Cephalopods. pp. 419–454 in A. S. Tompa, Verdonk, N. H. & van den Biggelaar, J. A. M. (Eds.) *Reproduction. The Mollusca*. Vol. 7. New York, Academic Press.

Arnold, J. M. (1987). Reproduction and embryology of *Nautilus*. pp. 353–372 in W. B. Saunders & Landman, N. H. (Eds.) *Nautilus: The Biology and Paleobiology of a Living Fossil (Topics in Geobiology)*. New York, Springer.

Arnold, J. M., Landman, N. H. & Mutvei, H. (1987). Development of the embryonic shell of *Nautilus*. pp. 373–400 in W. B. Saunders & Landman, N. H. (Eds.) *Nautilus: The Biology and Paleobiology of a Living Fossil (Topics in Geobiology)*. New York, Springer.

Arnold, J. M. (1992). *Nautilus* embryology: A new theory of molluscan shell formation. *Biological Bulletin* 183: 373–374.

Arsenault, D. J. & Himmelman, J. H. (1998). Spawning of the Iceland scallop (*Chlamys islandica* Müller, 1776) in the northern Gulf of St. Lawrence and its relationship to temperature and phytoplankton abundance. *The Veliger* 41: 180–185.

Arshavsky, I., Deliagina, T. G., Gelfand, I. M., Orlovsky, G. N., Panchin, V., Pavlova, G. A. & Popova, L. B. (1990). Neural control of heart beat in the pteropod mollusc *Clione limacina*: Coordination of circulatory and locomotor systems. *Journal of Experimental Biology* 148: 461–475.

van As, J. G. & Basson, L. (2004). Ciliophoran (Ciliophora) parasites of terrestrial gastropods. pp. 559–578 in G. M. Barker (Ed.) *Natural Enemies of Terrestrial Molluscs*. Oxford, UK/Cambridge, MA, CABI Publishing.

van As, L. L. & van As, J. G. (2000). *Licnophora bassoni* sp. n. (Ciliophora: Heterotrichea) from South African turban shells (Gastropoda: Prosobranchia). *Acta Protozoologica* 39: 331–335.

Asami, T. (1993). Genetic variation and evolution of coiling chirality in snails. *Forma* 8: 263–276.

Asami, T., Cowie, R. H. & Ohbayashi, K. (1998). Evolution of mirror images by sexually asymmetric mating behavior in hermaphroditic snails. *American Naturalist* 152: 225–236.

Ascunce, M. S., Yang, C. C., Oakey, J., Calcaterra, L., Wu, W. J., Shih, C. J., Goudet, J., Ross, K. G. & Shoemaker, D. (2011). Global invasion history of the fire ant *Solenopsis invicta*. *Science* 331: 1066–1068.

Aseyev, N., Zakharov, I. S. & Balaban, P. M. (2010). Morphology of neuropeptide CNP2 modulation of heart activity in terrestrial snail. *Peptides* 31: 1301–1308.

Ashkenazi, S., Klass, K., Mienis, H. K., Spiro, B. & Abel, R. (2010). Fossil embryos and adult Viviparidae from the Early-Middle Pleistocene of Gesher Benot Ya'aqov, Israel: Ecology, longevity and fecundity. *Lethaia* 43: 116–127.

Astorga, M. P. (2014). Genetic considerations for mollusk production in aquaculture: Current state of knowledge. *Frontiers in Genetics* 5: 435 (431–436).

Atema, J. & Stenzler, D. (1977). Alarm substance of the marine mud snail, *Nassarius obsoletus*: Biological characterization and possible evolution. *Journal of Chemical Ecology* 3: 173–187.

Ates, R. M. L. (1997). Gastropods carrying actinians. pp. 11–20 *in* J. C. den Hartog, van Ofwegen, L. P., & van der Spoel, S. (Eds.) *Proceedings of the 6th International Conference on Coelenterate Biology, the Leeuwenhorst, Noordwijkerhout, The Netherlands, 16–21 July 1995*. Leiden, The Netherlands, Nationaal Natuurhistorisch Museum.

Atkins, D. G. (1933). *Rhopalura granosa*, sp. nov., an orthonectid parasite of a lamellibranch, *Heteranomia squamula* L., with a note on its swimming behaviour. *Journal of the Marine Biological Association of the United Kingdom* 19: 233–252.

Atkins, D. G. (1936). On the ciliary mechanisms and interrelationships of lamellibranchs. Part I: New observations on sorting mechanisms. *Quarterly Journal of Microscopical Science* 79: 181–308.

Atkins, D. G. (1937a). On the ciliary mechanisms and interrelationships of lamellibranchs. Part III: Types of lamellibranch gills and their food currents. *Quarterly Journal of Microscopical Science* 79: 375–421.

Atkins, D. G. (1937b). On the ciliary mechanisms and interrelationships of lamellibranchs. Part II: Sorting devices on the gills. *Quarterly Journal of Microscopical Science* 79: 339–373.

Atkinson, W. D. & Warwick, T. (1983). The role of selection in the colour polymorphism of *Littorina rudis* Maton and *Littorina arcana* Hannaford-Ellis (Prosobranchia: Littorinidae). *Biological Journal of the Linnean Society* 20: 137–151.

Aubin, P. A. (1892). The limpet's power of adhesion. *Nature* 45: 464–465.

Audesirk, T. E. & Audesirk, G. (1985). Behaviour of gastropod molluscs. pp. 1–94 *in* A. O. D. Willows (Ed.) *Neurobiology and Behavior, Part 1. The Mollusca*. Vol. 8. New York, Academic Press.

Audouin, J. V. & Milne Edwards, H. (1832–1834). *Recherches Pour Servir a l'Histoire Naturelle du Littoral de la France, ou Recueil de Mémoires sur l'Anatomie, la Physiologie, la Classification et les Moeurs des Animaux des Nos Côtes. Tome 1*. Vol. 1–2. Paris, Crochard.

Audzijonyte, A. & Vrijenhoek, R. C. (2010). When gaps really are gaps: Statistical phylogeography of hydrothermal vent invertebrates. *Evolution* 64: 2369–2384.

Avaca, M. S., Narvarte, M. & Martín, P. (2012). Size-assortative mating and effect of maternal body size on the reproductive output of the nassariid *Buccinanops globulosus*. *Journal of Sea Research* 69: 16–22.

Avdeev, G. V. & Sirenko, B. I. (2005). New and recognized species of copepods (Chitonophilidae): Parasites of chitons of Northern Pacific. *Parazitologiya* 39: 516–543.

Avens, A. G. (1965). Osmotic balance in gastropod molluscs. II: The brackish water gastropod, *Hydrobia ulvae* Pennant. *Comparative Biochemistry and Physiology* 16: 143–153.

Averbuj, A., Rocha, M. N. & Zabala, S. (2014). Embryonic development and reproductive seasonality of *Buccinanops globulosus* (Nassariidae) (Kiener, 1834) in Patagonia, Argentina. *Invertebrate Reproduction & Development* 58: 138–147.

Averbuj, A. & Penchaszadeh, P. E. (2016). Reproductive biology in the South Western Atlantic genus *Buccinanops* (Nassariidae): The case of *Buccinanops paytensis*. *Molluscan Research* 35: 1–9.

Avery, G., Halkett, D., Orton, J. H., Steele, T., Tusenius, M. & Klein, R. (2008). The Ysterfontein 1 Middle Stone Age rock shelter and the evolution of coastal foraging. *South African Archaeological Society Goodwin Series* 10: 66–89.

Avila-Poveda, O. H., Colin-Flores, R. F. & Rosas, C. (2009). Gonad development during the early life of *Octopus maya* (Mollusca: Cephalopoda). *Biological Bulletin* 216: 94–102.

Avila-Poveda, O. H., Montes-Perez, R. C., Koueta, N., Benitez-Villalobos, F., Ramirez-Perez, J. S., Jimenez-Gutiérrez, L. R. & Rosas, C. (2015). Seasonal changes of progesterone and testosterone concentrations throughout gonad maturation stages of the Mexican octopus, *Octopus maya* (Octopodidae: Octopus). *Molluscan Research* 35: 161–172.

Avila, C. (1995). Natural products from opisthobranch mollusks: A biological review. *Oceanography and Marine Biology Annual Review* 33: 487–559.

Avila, C. & Durfort, M. (1996). Histology of epithelia and mantle glands of selected species of doridacean mollusks with chemical defensive strategies. *The Veliger* 39: 148–163.

Avila, C. (1998). Competence and metamorphosis in the long-term planktotrophic larvae of the nudibranch mollusc *Hermissenda crassicornis* (Eschscholtz, 1831). *Journal of Experimental Marine Biology and Ecology* 231: 81–117.

Avila, C., Iken, K., Fontana, A. & Cimino, G. (2000). Chemical ecology of the Antarctic nudibranch *Bathydoris hodgsoni* Eliot, 1907: Defensive role and origin of its natural products. *Journal of Experimental Marine Biology and Ecology* 252: 27–44.

Avila, C. (2006). Molluscan natural products as biological models: Chemical ecology, histology and laboratory culture. pp. 1–23 *in* G. Cimino & Gavagnin, M. (Eds.) *Molluscs. Progress in Molecular and Subcellular Biology*. Berlin, Heidelburg, Springer-Verlag.

Awati, P. R. & Karandikar, K. R. (1940). Structure and bionomics of *Oncidium verraculatum* Cuv. *Journal of the University of Bombay* 8: 3–57.

Azevedo, C. (1989). Fine structure of *Perkinsus atlanticus* n. sp. (Apicomplexa, Perkinsea) parasite of the clam *Ruditapes decussatus* from Portugal. *Journal of Parasitology* 75: 627–635.

Azevedo, C. & Villalba, A. (1991). Extracellular giant rickettsiae associated with bacteria in the gill of *Crassostrea gigas* (Mollusca, Bivalvia). *Journal of Invertebrate Pathology* 58: 75–81.

Azevedo, C. & Matos, E. (1999). Description of *Nematopsis mytella* n. sp. (Apicomplexa), parasite of the mussel *Mytella guyanensis* (Mytelidae) from the Amazon estuary and description of its oocysts. *European Journal of Protistology* 35: 427–433.

Azevedo, C. & Padovan, I. P. (2004). *Nematopsis gigas* n. sp. (Apicomplexa), a parasite of *Nerita ascencionis* (Gastropoda, Neritidae) from Brazil. *Journal of Eukaryotic Microbiology* 51: 214–219.

Azevedo, C., Conchas, R. F., Tajdari, J. & Montes, J. F. (2006). Ultrastructural description of new Rickettsia-like organisms in the commercial abalone *Haliotis tuberculata* (Gastropoda: Haliotidae) from the NW of Spain. *Diseases of Aquatic Organisms* 71: 233–237.

Azmi, N., Ghaffar, M. & Cob, Z. C. (2015). Apicomplexa-like parasites of economically important bivalves from Merambong shoals, Johor. *The Malayan Nature Journal* 66: 13.

Azuma, N., Zaslavskaya, N. I., Yamazaki, T., Nobetsu, T. & Chiba, S. (2017). Phylogeography of *Littorina sitkana* in the northwestern Pacific Ocean: Evidence of eastward trans-Pacific colonization after the Last Glacial Maximum. *Genetica* 145: 139–149.

Baalbergen, E., Helwerda, R., Schelfhorst, R., Cajas, R. F., Castillo, Van Moorsel, C. H. M., Kundrata, R., Welter-Schultes, F. W., Giokas, S. & Schilthuizen, M. (2014). Predator-prey interactions between shell-boring beetle larvae and rock-dwelling land snails. *PLoS ONE* 9: e100366.

Baba, K. (1938). The later development of a solenogastre, *Epimenia verrucosa* (Nierstrasz). *Journal of the Department of Agriculture of the Kyushu Imperial University* 6: 21–40.

Baba, K. (1940). The mechanisms of absorption and secretion in a solenogastre, *Epimenia verrucosa* (Nierstrasz), studied by means of injection methods. *Journal of the Department of Agriculture of the Kyushu Imperial University* 6: 119–168.

Babcock, R., Mundy, C., Keesing, J. & Oliver, J. (1992). Predictable and unpredictable spawning events: *in situ* behavioral data from free-spawning coral-reef invertebrates. *Invertebrate Reproduction & Development* 22: 213–228.

Babcock, R. (1995). Synchronous multispecific spawning on coral reefs: Potential for hybridization and roles of gamete recognition. *Reproduction, Fertility and Development* 7: 943–950.

Babor, J. & Frankenberger, A. Z. (1930). The osphradial ganglion and its metamorphosis in the Stylommatophora. *Spisy lékařské fakulty Masarykovy University v Brně* 9: 147–154.

Baby, R. L., Hasan, I., Kabir, K. A. & Naser, M. N. (2010). Nutrient analysis of some commercially important molluscs of Bangladesh. *Journal of Scientific Research* 2: 390–396.

Bachraty, C., Legendre, P. & Desbruyères, D. (2009). Biogeographic relationships among deep-sea hydrothermal vent faunas at global scale. *Deep Sea Research Part I: Oceanographic Research Papers* 56: 1371–1378.

Backeljau, T., Baur, A. & Baur, B. (2001). Population and conservation genetics. pp. 383–412 *in* G. M. Barker (Ed.) *The Biology of Terrestrial Molluscs*. Wallingford, UK, CABI Publishing.

Badman, D. G. (1971). Nitrogen excretion in two species of pulmonate land snails. *Comparative Biochemistry and Physiology Part A* 38: 663–673.

Baeumler, N., Haszprunar, G. & Ruthensteiner, B. (2011). Development of the excretory system in the polyplacophoran mollusc, *Lepidochitona corrugata*: The protonephridium. *Journal of Morphology* 272: 972–986.

Baeumler, N., Haszprunar, G. & Ruthensteiner, B. (2012). Development of the excretory system in a polyplacophoran mollusc: Stages in metanephridial system development. *Frontiers in Zoology* 9: 23.

Baginski, R. M. & Pierce, S. K. (1978). A comparison of amino acid accumulation during high salinity adaptation with anaerobic metabolism in the ribbed mussel, *Modiolus demissus demissus*. *Journal of Experimental Zoology* 203: 419–428.

Bagshaw, C. R. (1993). *Muscle Contraction*. New York, Chapman & Hall.

Bahamondes, I. & Castilla, J. C. (1986). Predation of marine invertebrates by the kelp gull *Larus dominicanus* in an undisturbed intertidal rocky shore of central Chile. *Revista Chilena de Historia Natural* 59: 65–72.

Bailey, C. H. & Chen, M. (1983). Morphological basis of long-term habituation and sensitization in *Aplysia*. *Science* 220: 91–93.

Bailey, P. (1976). Food of the marine toad, *Bufo marinus*, and six species of skink in a cacao plantation in New Britain, Papua New Guinea. *Australian Wildlife Research* 3: 185–188.

Bailey, S. E. R. (1975). The seasonal and daily patterns of locomotor activity in the snail *Helix aspersa* Müller, and their relation to environmental variables. *Journal of Molluscan Studies* 41: 415–428.

Bailey, S. E. R. (1981). Circannual and circadian rhythms in the snail *Helix aspersa* Müller and the photoperiodic control of annual activity and reproduction. *Journal of Comparative Physiology A: Neuroethology, Sensory, Neural, and Behavioral Physiology* 142: 89–94.

Bainard, J. D. & Gregory, T. R. (2013). Genome size evolution: Patterns, mechanisms, and methodological advances. *Genome* 56: 70–80.

Baker, F. C. (1916). The relation of mollusks to fish in Oneida Lake. *Technical Publication No. 4, New York State College of Forestry, Syracuse University* 16: 1–366.

Baker, G. H. (1989). Damage, population dynamics, movement and control of pest helicid snails in southern Australia. *Monograph-British Crop Protection Council* 41: 175–185.

Baker, G. H. (2002). Helicidae and Hygromiidae as pests in cereal crops and pastures in South Australia. pp. 193–215 *in* G. M. Barker (Ed.) *Molluscs as Crop Pests*. Wallingford, UK, CABI Publishing.

Baker, H. B. (1926). Anatomical notes on American Helicinidae. *Proceedings of the Academy of Natural Sciences of Philadelphia* 78: 35–56.

Baker, J. T. (1974). Tyrian purple: An ancient dye, a modern problem. *Endeavour* 33: 11–17.

Balaban, P. M., Malyshev, A. Y., Ierusalimsky, V. N., Aseyev, N., Korshunova, T. A., Bravarenko, N. I., Lemak, M. S., Roshchin, M., Zakharov, I. S., Popova, Y. & Boyle, R. (2011). Functional changes in the snail statocyst system elicited by microgravity. *PLoS ONE* 6: e17710 (17711–17713).

Baldwin, J. & Morris, G. M. (1983). Re-examination of the contributions of aerobic and anaerobic energy production during swimming in the bivalve mollusc *Limaria fragilis* (Family Limidae) *Australian Journal of Marine and Freshwater Research* 34: 909–914.

Baldwin, J. (1987). Energy metabolism of *Nautilus* swimming muscles. pp. 325–330 *in* W. B. Saunders & Landman, N. H. (Eds.) *Nautilus: The Biology and Paleobiology of a Living Fossil. Topics in Geobiology*. New York, Springer.

Baldwin, J., Elias, J. P., Wells, R. M. G. & Donovan, D. A. (2007). Energy metabolism in the tropical abalone, *Haliotis asinina* Linné: Comparisons with temperate abalone species. *Journal of Experimental Marine Biology and Ecology* 342: 213–225.

Ball, A. D., Taylor, J. D. & Andrews, E. B. (1997). Development of the acinous and accessory salivary glands in *Nucella lapillus* (Neogastropoda: Muricoidea). *Journal of Molluscan Studies* 63: 245–260.

Ball, A. D., Purdy, K. J., Glover, E. A. & Taylor, J. D. (2009). Ctenidial structure and three bacterial symbiont morphotypes in *Anodontia* (*Euanodontia*) *ovum* (Reeve, 1850) from the Great Barrier Reef, Australia (Bivalvia: Lucinidae). *Journal of Molluscan Studies* 75: 175–185.

Ballantyne, J. S., Mommsen, T. P. & Hochachka, P. W. (1981). Studies of the metabolism of the migratory squid, *Loligo opalescens*: Enzymes of tissues and heart mitochondria. *Marine Biology Letters* 2: 75–85.

Ballantyne, J. S. & Storey, K. B. (1983). Mitochondria from the ventricle of the marine clam, *Mercenaria mercenaria*: Substrate preferences and effects of pH and salt concentration on proline oxidation. *Comparative Biochemistry and Physiology Part B* 76: 133–138.

Ballantyne, J. S. (2004). Mitochondria: Aerobic and anaerobic design—Lessons from molluscs and fishes. *Comparative Biochemistry and Physiology Part B* 139: 461–467.

Ballering, R. B., Jalving, M. A., Yen, D. A., VenTresca, L. E., Hallacher, J. T., Tomlinson, J. T. & Wobber, D. R. (1972). Octopus evenomation through a plastic bag via a salivary proboscis. *Toxicon* 10: 245–248.

Bancalà, F. (2009). Function of mucus secretion by lamellose ormer, *Haliotis tuberculata lamellosa*, in response to starfish predation. *Animal Behaviour* 78: 1189–1194.

Bandaranayake, W. M. (2006). The nature and role of pigments of marine invertebrates. *Natural Product Reports* 23: 223–255.

Bandel, K. (1974). Faecal pellets of Amphineura and Prosobranchia (Mollusca) from the Caribbean coast of Colombia, South America. *Senckenbergiana maritima* 6: 1–31.

Bandel, K. (1977). Die Herausbildung der Schraubenschicht bei Pteropoden. *Biomineralisation* 9: 73–85.

Bandel, K. (1982). Morphologie und Bildung der frühontogenetischen Gehäuse bei conchiferen Mollusken [Morphology and formation of the early ontogenetic shells of conchiferan molluscs]. *Facies* 7: 1–198.

Bandel, K. (1990). Shell structure of the Gastropoda excluding Archaeogastropoda. pp. 117–134 in J. G. Carter (Ed.) *Skeletal Biomineralization: Patterns, Processes and Evolutionary Trends.* Vol. 1. New York, Van Nostrand Reinhold.

Bandel, K. & Geldmacher, W. (1996). The structure of the shell of *Patella crenata* connected with suggestions to the classification and evolution of the Archaeogastropoda. *Freiberger Forschungshefte* 3: 1–71.

Bandel, K. (1997). Higher classification and pattern of evolution of the Gastropoda: A synthesis of biological and paleontological data. *Courier Forschungsinstitut Senckenberg* 201: 57–81.

Bandel, K. (2000). The new family Cortinellidae (Gastropoda, Mollusca) connected to a review of the evolutionary history of the subclass Neritimorpha. *Neues Jahrbuch für Geologie und Paläontologie Abhandlungen* 217: 111–129.

Banerjee, S. & Sur, R. K. (1991). Comparative studies on thermal behaviour of some hydrolyzing enzymes in gastropod molluscs. *Comparative Physiology and Ecology* 16: 156–165.

Banerjee, S., Sur, R. K. & Misra, K. K. (1991). Comparative studies on the microanatomy and scanning electron microscopic observations on the gut of three gastropods. *Proceedings of the Zoological Society (Calcutta)* 44: 161–175.

Bao, Y. B., Wang, Q., Guo, X. M. & Lin, Z. H. (2013). Structure and immune expression analysis of hemoglobin genes from the blood clam *Tegillarca granosa*. *Genetics and Molecular Research* 12: 3110–3123.

Baranetz, O. N. & Minichev, S. (1994). The evolution of the mantle complex in nudibranchiate molluscs (Gastropoda, Nudibranchia). *Zoologicheskii Zhurnal* 73: 29–35.

Barash, A. & Danin, Z. (1982). Contribution to the knowledge of mollusca in the Bardawil Lagoon. *Bollettino Malacologico* 18: 107–128.

Baratte, S., Andouche, A. & Bonnaud, L. (2007). Engrailed in cephalopods: A key gene related to the emergence of morphological novelties. *Development Genes and Evolution* 217: 353–362.

Barbas, D., DesGroseillers, L., Castellucci, V. F., Carew, T. J. & Marinesco, S. (2003). Multiple serotonergic mechanisms contributing to sensitization in *Aplysia*: Evidence of diverse serotonin receptor subtypes. *Learning and Memory* 10: 373–386.

Barbas, D., Campbell, A. K., Castellucci, V. F. & DesGroseillers, L. (2005). Comparative localization of two serotonin receptors and sensorin in the central nervous system of *Aplysia californica*. *Journal of Comparative Neurology* 490: 295–304.

Barbato, M., Bernard, M., Borrelli, L. & Fiorito, G. (2007). Body patterns in cephalopods: 'Polyphenism' as a way of information exchange. *Pattern Recognition Letters* 28: 1854–1864.

Barber, A. H., Lu, D. & Pugno, N. M. (2015). Extreme strength observed in limpet teeth. *Journal of the Royal Society Interface* 12: 20141326 (20141321–20141326).

Barber, B. J. (2004). Neoplastic diseases of commercially important marine bivalves. *Aquatic Living Resources* 17: 449–466.

Barber, B. J., Fajans, J. S., Baker, S. M. & Baker, P. (2005). Gametogenesis in the non-native green mussel, *Perna viridis*, and the native scorched mussel, *Brachidontes exustus*, in Tampa Bay, Florida. *Journal of Shellfish Research* 24: 1087–1095.

Barber, V. C., Evans, E. M. & Land, M. F. (1967). The fine structure of the eye of the mollusc *Pecten maximus*. *Zeitschrift für Zellforschung und Mikroskopische Anatomie* 76: 295–312.

Barber, V. C. (1968). The structure of mollusc statocyst, with particular reference to cephalopods. pp. 37–62 in J. D. Carthy & Newell, G. E. (Eds.) *Invertebrate Receptors. Proceedings of a Symposium held at the Zoological Society of London on 30 and 31 May 1967.* New York, Academic Press.

Barber, V. C. & Dilly, P. N. (1969). Some aspects of the fine structure of the statocysts of the molluscs *Pecten* and *Pterotrachea*. *Zeitschrift für Zellforschung und Mikroskopische Anatomie* 94: 462–478.

Barber, V. C. & Wright, D. E. (1969a). The fine structure of the sense organs of the cephalopod mollusc *Nautilus*. *Zeitschrift für Zellforschung und Mikroskopische Anatomie* 102: 293–312.

Barber, V. C. & Wright, D. E. (1969b). The fine structure of the eye and optic tentacle of the mollusc *Cardium edule*. *Journal of Ultrastructure Research* 26: 515–528.

Barbosa, A., Mäthger, L. M., Buresch, K. C., Kelly, J., Chubb, C., Chiao, C.-C. & Hanlon, R. T. (2008). Cuttlefish camouflage: The effects of substrate contrast and size in evoking uniform, mottle or disruptive body patterns. *Vision Research* 48: 1242–1253.

Barile, P. J., Stoner, A. W. & Young, C. M. (1994). Phototaxis and vertical migration of the queen conch (*Strombus gigas* Linne) veliger larvae. *Journal of Experimental Marine Biology and Ecology* 183: 147–162.

Barkai, A. & McQuaid, C. D. (1988). Predator-prey role reversal in a marine benthic ecosystem. *Science* 242: 62–64.

Barkalova, V. O., Fedosov, A. E. & Kantor, Y. I. (2016). Morphology of the anterior digestive system of tonnoideans (Gastropoda: Caenogastropoda) with an emphasis on the foregut glands. *Molluscan Research* 36: 54–73.

Barker, G. M. & Ramsay, G. W. (1978). The slug mite *Riccardoella limacum* (Acari: Ereynetidae) in New Zealand. *New Zealand Entomologist* 6: 441–443.

Barker, G. M. & Mayhill, P. C. (1999). Patterns of diversity and habitat relationships in terrestrial mollusc communities of the Pukeamaru Ecological District, northeastern New Zealand. *Journal of Biogeography* 26: 215–238.

Barker, G. M. (2001a). Gastropods on land: Phylogeny, diversity and adaptive morphology. pp. 1–146 in G. M. Barker (Ed.) *The Biology of Terrestrial Molluscs.* Wallingford, UK, CABI Publishing.

Barker, G. M. (2001b). *The Biology of Terrestrial Molluscs.* Wallingford, UK, CABI Publishing.

Barker, G. M. (2002). *Molluscs as Crop Pests.* Wallingford, UK, CABI Publishing.

Barker, G. M. & Efford, M. G. (2002). Predatory gastropods as natural enemies of terrestrial gastropods and other invertebrates. pp. 279–404 in G. M. Barker (Ed.) *Natural Enemies of Terrestrial Molluscs.* Oxford, UK/Cambridge, MA, CABI Publishing.

Barker, G. M. (2004a). Millipedes (Diplopoda) and centipedes (Chilopoda) (Myriapoda) as predators of terrestrial gastropods. pp. 405–426 in G. M. Barker (Ed.) *Natural Enemies of Terrestrial Molluscs.* Oxford, UK/Cambridge, MA, CABI Publishing.

Barker, G. M. (2004b). *Natural Enemies of Terrestrial Molluscs.* Oxford, UK/Cambridge, MA, CABI Publishing.

Barker, G. M., Knutson, L., Vala, J.-C., Coupland, J. B. & Barnes, J. K. (2004). Overview of the biology of marsh flies (Diptera: Sciomyzidae), with special reference to predators and parasitoids of terrestrial gastropods. pp. 159–226 in G. M. Barker (Ed.) *Natural Enemies of Terrestrial Molluscs.* Oxford, UK/Cambridge, MA, CABI Publishing.

Barker, G. M. (2008). Mollusc herbivory influenced by endophytic clavicipitaceous fungal infections in grasses. *Annals of Applied Biology* 153: 381–393.

Bärlocher, F., Arsuffi, T. L. & Newell, S. Y. (1989a). Digestive enzymes of the saltmarsh periwinkle *Littorina irrorata* (Mollusca: Gastropoda). *Oecologia* 80: 39–43.

Bärlocher, F., Newell, S. Y. & Arsuffi, T. L. (1989b). Digestion of *Spartina alterniflora* (Loisel) material with and without fungal constituents by the periwinkle *Littorina irrorata* (Say) (Mollusca: Gastropoda). *Journal of Experimental Marine Biology and Ecology* 130: 45–53.

Barlow, L. A. & Truman, J. W. (1992). Patterns of serotonin and SCP immunoreactivity during metamorphosis of the nervous system of the red abalone, *Haliotis rufescens. Journal of Neurobiology* 23: 829–844.

Barnard, W. & de Waal, D. (2006). Raman investigation of pigmentary molecules in the molluscan biogenic matrix. *Journal of Raman Spectroscopy* 37: 342–352.

Barnes, R. S. K. (1994). Macrofaunal community structure and life histories in coastal lagoons. pp. 311–362 in B. Kjerfve (Ed.) *Coastal Lagoon Processes.* Vol. 60. Amsterdam, Elsevier Science B. V.

Barnhart, M. C. (1983). Gas permeability of the epiphragm of a terrestrial snail, *Otala lactea. Physiological Zoology* 56: 436–444.

Barnhart, M. C. (1992). Acid-base regulation in pulmonate molluscs. *Journal of Experimental Zoology Part A* 263: 120–126.

Barnhart, M. C., Haag, W. R. & Roston, W. N. (2008). Adaptations to host infection and larval parasitism in Unionoida. *Journal of the North American Benthological Society* 27: 370–394.

Barnosky, A. D., Matzke, N., Tomiya, S., Wogan, G. O. U., Swartz, B., Quental, T. B., Marshall, C., McGuire, J. L., Lindsey, E. L. & Maguire, K. C. (2011). Has the Earth's sixth mass extinction already arrived? *Nature* 471: 51–57.

Barolo, S. & Posakony, J. W. (2002). Three habits of highly effective signaling pathways: Principles of transcriptional control by developmental cell signaling. *Genes & Development* 16: 1167–1181.

Barry, J. P., Kochevar, R. E. & Baxter, C. H. (1997). The influence of pore-water chemistry and physiology on the distribution of vesicomyid clams at cold seeps in Monterey Bay: Implications for patterns of chemosynthetic community organization. *Limnology and Oceanography* 42: 318–328.

Barsby, T., Linington, R. G. & Andersen, R. J. (2002). De novo terpenoid biosynthesis by the dendronotid nudibranch *Melibe leonina. Chemoecology* 12: 199–202.

Baršiene, J., Ribi, G. & Barsyte, D. (2000). Comparative karyological analysis of five species of *Viviparus* (Gastropoda: Prosobranchia). *Journal of Molluscan Studies* 66: 259–271.

Bartolomaeus, T. (1992). Ultrastructure of the photoreceptor in the larvae of *Lepidochiton cinereus* (Mollusca, Polyplacophora) and *Lacuna divaricata* (Mollusca, Gastropoda). pp. 215–236 in P. Ax (Ed.) *Microfauna Marina.* Vol. 7. Stuttgart, Gustav Fischer Verlag.

Bartolomaeus, T. (1997). Ultrastructure of the renopericardial complex of the interstitial gastropod *Philinoglossa helgolandica* (Hertling), 1932 (Mollusca: Opisthobranchia). *Zoologischer Anzeiger* 235: 165–176.

Bartsch, I. (1986). A new species of *Halixodes* (Halacaridae, Acari) and a review of the New Zealand species. *Journal of the Royal Society of New Zealand* 16: 51–56.

Barucca, M., Olmo, E. & Canapa, A. (2003). Hox and paraHox genes in bivalve molluscs. *Gene* 317: 97–102.

Barucca, M., Biscotti, M. A., Olmo, E. & Canapa, A. (2006). All the three ParaHox genes are present in *Nuttallochiton mirandus* (Mollusca: Polyplacophora): Evolutionary considerations. *Journal of Experimental Zoology Part B* 306B: 164–167.

Barucca, M., Canapa, A. & Biscotti, M. A. (2016). An overview of Hox genes in Lophotrochozoa: Evolution and functionality. *Journal of Developmental Biology* 4: 12.

Basch, P. F. (1959a). Status of the genus *Gundlachia* (Pulmonata, Ancylidae). *Occasional Papers. Museum of Zoology, University of Michigan* 602: 1–9.

Basch, P. F. (1959b). The anatomy of *Laevapex fuscus*, a freshwater limpet (Gastropoda: Pulmonata). *Miscellaneous Publications of the Museum of Zoology, University of Michigan* 108: 1–56.

Basch, P. F. (1963). A review of the Recent freshwater limpet snails of North America (Mollusca: Pulmonata). *Bulletin of the Museum of Comparative Zoology* 129: 399–461.

Basil, J. & Crook, R. J. (2017). Learning and memory in the living fossil, chambered *Nautilus.* pp. 103–136 in S. Saleuddin & Mukai, S. (Eds.) *Physiology of Molluscs.* Vol. 2. Oakville, ON, Apple Academic Press Inc.

Basil, J. A., Bahctinova, I., Kuroiwa, K., Lee, N., Mims, D., Preis, M. & Soucier, C. (2005). The function of the rhinophore and the tentacles of *Nautilus pompilius* L. (Cephalopoda, Nautiloidea) in orientation to odor. *Marine and Freshwater Behaviour and Physiology* 38: 209–221.

Bassaglia, Y., Bekel, T., Da Silva, C., Poulain, J., Andouche, A., Navet, S. & Bonnaud, L. (2012). ESTs library from embryonic stages reveals tubulin and reflectin diversity in *Sepia officinalis* (Mollusca: Cephalopoda). *Gene* 498: 203–211.

Bassim, S., Tanguy, A., Genard, B., Moraga, D. & Tremblay, R. (2014). Identification of *Mytilus edulis* genetic regulators during early development. *Gene* 551: 65–78.

Bateman, A. J. (1948). Intra-sexual selection in *Drosophila. Heredity* 2: 349–368.

Bateman, K. S., White, P. & Longshaw, M. (2012). Virus-like particles associated with mortalities of the Manila clam *Ruditapes philippinarum* in England. *Diseases of Aquatic Organisms* 99: 163–167.

Bates, A. E., Harmer, T. L., Roeselers, G. & Cavanaugh, C. M. (2011). Phylogenetic characterization of episymbiotic bacteria hosted by a hydrothermal vent limpet (Lepetodrilidae, Vetigastropoda). *Biological Bulletin* 220: 118–127.

Bates, T. W. & Hicks, D. W. (2005). Locomotory behavior and habitat selection in littoral gastropods on Caribbean limestone shores. *Journal of Shellfish Research* 24: 75–84.

Batham, E. J. & Tomlinson, J. T. (1965). On *Cryptophialus melampygos* Berndt, a small boring barnacle of the order Acrothoracica abundant in some New Zealand molluscs. *Transactions of the Royal Society of New Zealand: Zoology* 7: 141–154.

Batten, R. L. (1972). The ultrastructure of five common Pennsylvanian pleurotomarian gastropod species of Eastern United States. *American Museum Novitates* 2501: 1–34.

Bauchau, V. (2001). Developmental stability as the primary function of the pigmentation patterns in bivalve shells? *Belgian Journal of Zoology* 131 (Suppl): 23–28.

Baums, I. B., Miller, M. W. & Szmant, A. M. (2003). Ecology of a corallivorous gastropod, *Coralliophila abbreviata,* on two scleractinian hosts. II. Feeding, respiration and growth. *Marine Biology* 142: 1093–1101.

Baur, A. & Baur, B. (2005). Interpopulation variation in the prevalence and intensity of parasitic mite infection in the land snail *Arianta arbustorum*. *Invertebrate Biology* 124: 194- 201.

Baur, B. (1992a). Random mating by size in the simultaneously hermaphroditic land snail *Arianta arbustorum*: Experiments and an explanation. *Animal Behaviour* 43: 511–518.

Baur, B. (1992b). Cannibalism in gastropods. pp. 102–127 *in* M. A. Elgar & Crespi, B. J. (Eds.) *Ecology and Evolution among Diverse Taxa*. Oxford, UK, Oxford University Press.

Baur, B. (1994a). Multiple paternity and individual variation in sperm precedence in the simultaneously hermaphroditic land snail *Arianta arbustorum*. *Behavioral Ecology and Sociobiology* 35: 413–421.

Baur, B. (1994b). Parental care in terrestrial gastropods. *Experientia* 50: 5–14.

Baur, B. (1998). Sperm competition in molluscs. pp. 255–306 *in* T. R. Birkhead & Moller, A. P. (Eds.) *Sperm Competition and Sexual Selection*. San Diego, CA, Academic Press.

Baur, H., Reichenbach, F. & Neubert, E. (2012). Sexual dimorphism in shells of *Cochlostoma septemspirale* (Caenogastropoda, Cyclophoroidea, Diplommatinidae, Cochlostomatinae). *ZooKeys* 208: 1–16.

Bax, N., Williamson, A., Aguero, M., Gonzalez, E. & Geeves, W. (2003a). Marine invasive alien species: A threat to global biodiversity. *Marine Policy* 27: 313–323.

Bax, N. J., Gowlett-Holmes, K. L. & McEnnulty, F. R. (2003b). *Distribution and Biology of the Introduced Gastropod Maoricolpus roseus (Quoy and Gamard, 1834) (Caenogastropoda Turritellidae) in Australia*. Hobart, Tasmania, Centre for Introduced Marine Pests.: 1–32.

Bayer, F. M. (1963). Observations on pelagic mollusks associated with the siphonophores *Velella* and *Physalia*. *Bulletin of Marine Science* 13: 454–466.

Bayliss, W. M. (1927). *Principles of General Physiology*. New York, Longmans, Green and Co.

Baylor, E. R. (1959). The responses of snails to polarized light. *Journal of Experimental Biology* 36: 369–376.

Bayly, I. A. E. & Williams, W. D. (1966). Chemical and biological studies on some saline lakes of south-east Australia. *Marine and Freshwater Research* 17: 177–228.

Bayly, I. A. E. (1972). Salinity tolerance and osmotic behavior of animals in athalassic saline and marine hypersaline waters. *Annual Review of Ecology and Systematics* 3: 233–268.

Bayne, B., Ahrens, M., Allen, S., D'auriac, M. A., Backeljau, T., Beninger, P., Bohn, R., Boudry, P., Davis, J. & Green, T. (2017). The proposed dropping of the genus *Crassostrea* for all Pacific cupped oysters and its replacement by a new genus *Magallana*: A dissenting view. *Journal of Shellfish Research* 36: 545–547.

Bayne, B. L. (1963). Responses of *Mytilus edulis* larvae to increases in hydrostatic pressure *Nature* 198: 406–407.

Bayne, B. L. (1971). Oxygen consumption by three species of lamellibranch mollusc in declining ambient oxygen tension. *Comparative Biochemistry and Physiology Part A* 40: 955–970.

Bayne, B. L., Iglesias, J. I. P., Hawkins, A. J. S., Navarro, E., Heral, M. & Deslous-Paoli, J. M. (1993). Feeding behaviour of the mussel, *Mytilus edulis*: Responses to variations in quantity and organic content of the seston. *Journal of the Marine Biological Association of the United Kingdom* 73: 813–829.

Bayne, B. L. (2009). *Marine Mussels: Their Ecology and Physiology*. Cambridge, Cambridge University Press.

Bayne, C. J. (1968a). Histochemical studies on the egg capsule of eight gastropod molluscs. *Proceedings of the Malacological Society of London* 38: 199–212.

Bayne, C. J. (1968b). A study of the desiccation of egg capsules of eight gastropod species. *Journal of Zoology* 155: 401–411.

Bayne, C. J. (1969). Survival of the embryos of the grey field slug *Agriolimax reticulatus*, following desiccation of the egg. *Malacologia* 9: 391–401.

Bayne, C. J. (2003). Origins and evolutionary relationships between the innate and adaptive arms of immune systems. *Integrative and Comparative Biology* 43: 293–299.

Bé, A. W. H., MacClintock, C. & Currie, D. C. (1972). Helical shell structure and growth of the pteropod *Cuvierina columnella* (Rang) (Mollusca, Gastropoda). *Biomineralization Research Reports* 4: 47–79.

Beal, B. F. (1983). Predation of juveniles of the hard clam *Mercenaria mercenaria* (Linne) by the snapping shrimp *Alpheus heterochaelis* Say and *Alpheus normanni* Kingsley. *Journal of Shellfish Research* 3: 1–9.

Bean-Knudsen, D. E., Uhazy, L. S., Wagner, J. E. & Young, B. M. (1988). Systemic infection of laboratory-reared *Biomphalaria glabrata* (Mollusca: Gastropoda) with an acid-fast bacillus. *Journal of Invertebrate Pathology* 51: 291–293.

Beard, M., Millecchia, L., Masuoka, C. & Arch, S. (1982). Ultrastructure of secretion in the atrial gland of a mollusc (*Aplysia*). *Tissue and Cell* 14: 297–308.

Bearham, D., Spiers, Z., Raidal, S., Jones, J. B., Burreson, E. M. & Nicholls, P. K. (2008). Spore ornamentation of *Haplosporidium hinei* n. sp. (Haplosporidia) in pearl oysters *Pinctada maxima* (Jameson, 1901). *Parasitology* 135: 521–527.

Beatty, W. L. & Rollins, H. B. (2002). The role of spinose ornament in predator deterrence: The bivalve *Arcinella*, Pinecrest (Pliocene) of Florida. pp. 356–357 *in Geological Society of America: 2002 Denver Annual Meeting [October 27–30, 2002]*. Denver CO, The Geological Society of America.

Beaumont, A. R. & Fairbrother, J. E. (1991). Ploidy manipulation in molluscan shellfish: A review. *Journal of Shellfish Research* 10: 1–18.

Bebianno, M. J. & Langston, W. S. (1998). Cadmium and metallothionein turnover in different tissues of the gastropod *Littorina littorea*. *Talanta* 46: 301–313.

Becerro, M. A., Turon, X., Uriz, M. J. & Templado, J. (2003). Can a sponge feeder be a herbivore? *Tylodina perversa* (Gastropoda) feeding on *Aplysina aerophoba* (Demospongiae). *Biological Journal of the Linnean Society* 78: 429–438.

Becerro, M. A., Starmer, J. A. & Paul, V. J. (2006). Chemical defenses of cryptic and aposematic gastropterid molluscs feeding on their host sponge *Dysidea granulosa*. *Journal of Chemical Ecology* 32: 1491–1500.

Beck, L. A. (2002). *Hirtopelta tufari* sp. n., a new gastropod species from hot vents at the East Pacific Rise (21 S) harbouring endocytosymbiotic bacteria in its gill (Gastropoda: Rhipidoglossa: Peltospiridae). *Archiv für Molluskenkunde* 130: 249–257.

Beck, M. W., Brumbaugh, R. D., Airoldi, L., Carranza, A., Coen, L. D., Crawford, C., Defeo, O. et al. (2009). *Shellfish Reefs at Risk: A Global Analysis of Problems and Solutions.* Arlington, VA, The Nature Conservancy.

Beck, M. W., Brumbaugh, R. D., Airoldi, L., Carranza, A., Coen, L. D., Crawford, C., Defeo, O. et al. (2011). Oyster reefs at risk and recommendations for conservation, restoration, and management. *BioScience* 61: 107–116.

Becker, E. L., Lee, R. W., Macko, S. A., Faure, B. M. & Fisher, C. R. (2010). Stable carbon and nitrogen isotope compositions of hydrocarbon-seep bivalves on the Gulf of Mexico lower continental slope. *Deep Sea Research Part II: Topical Studies in Oceanography* 57: 1957–1964.

Becker, W. & Schmale, H. (1978). The ammonia and urea excretion of *Biomphalaria glabrata* under different physiological conditions: Starvation, infection with *Schistosoma mansoni*, dry keeping. *Comparative Biochemistry and Physiology Part B* 59: 75–79.

Beckvar, N. (1981). Cultivation, spawning, and growth of the giant clams *Tridacna gigas, Tridacna derasa* and *Tridacna squamosa* in Palau, Caroline Islands. *Aquaculture* 24: 21–30.

Bedford, J. J. (1969). The soluble amino acid pool in *Siphonaria zelandica*: Its composition and the influence of salinity changes. *Comparative Biochemistry and Physiology* 29: 1005–1014.

Beeby, A. & Richmond, L. (2010). Magnesium and the regulation of lead in three populations of the garden snail *Cantareus aspersus*. *Environmental Pollution* 158: 2288–2293.

Beecham, H. J., Oldfield, E. C., Lewis, D. E. & Buker, J. L. (1991). *Mycobacterium marinum* infection from shucking oysters. *The Lancet* 337: 1487.

Beedham, G. E. & Trueman, E. R. (1967). The relationships of the mantle and shell of the Polyplacophora in comparison with that of other Mollusca. *Journal of Zoology* 151: 215–231.

Beesley, P. L., Ross, G. J. B. & Wells, A., Eds. (1998a). *Mollusca: The Southern Synthesis.* Melbourne, Australia, CSIRO Publishing.

Beesley, P. L., Ross, G. J. B. & Wells, A. (1998b). *Mollusca: The Southern Synthesis.* Part B. pp. i–viii, 565–1234. Melbourne, Australia, CSIRO Publishing.

Beeston, D. C. & Morgan, E. (1979). A crepuscular rhythm of locomotor activity in the freshwater prosobranch, *Melanoides tuberculata* (Müller). *Animal Behaviour* 27: 284–291.

Begovic, E. & Lindberg, D. R. (2011). Genetic population structure of *Tectura paleacea*: Implications for the mechanisms regulating population structure in patchy coastal habitats. *PLoS ONE* 6: e18408.

Beiswanger, C. M., Sokolove, P. G. & Prior, D. J. (1981). Extraocular photoentrainment of the circadian locomotor rhythm of the garden slug *Limax*. *Journal of Experimental Zoology Part A* 216: 13–23.

Belcher, A. M., Wu, X. H., Christensen, R. J., Hansma, P. K., Stucky, G. D. & Morse, D. E. (1996). Control of crystal phase switching and orientation by soluble mollusc-shell proteins. *Nature* 381: 56–58.

Belda, C. A., Cuff, C. & Yellowlees, D. (1993). Modification of shell formation in the giant clam *Tridacna gigas* at elevated nutrient levels in sea water. *Marine Biology* 117: 251–257.

Belgrad, B. A. & Smith, N. F. (2014). Effects of predation and parasitism on climbing behavior of the marine snail, *Cerithidea scalariformis*. *Journal of Experimental Marine Biology and Ecology* 458: 20–26.

Bell-Pedersen, D., Cassone, V. M., Earnest, D. J., Golden, S. S., Hardin, P. E., Thomas, T. L. & Zoran, M. J. (2005). Circadian rhythms from multiple oscillators: Lessons from diverse organisms. *Nature Reviews: Genetics* 6: 544–556.

Bellard, C., Bertelsmeier, C., Leadley, P., Thuiller, W. & Courchamp, F. (2012). Impacts of climate change on the future of biodiversity. *Ecology Letters* 15: 365–377.

Bellwood, O. (1998). The phylogeny of box crab genera (Crustacea: Brachyura: Calappidae) with notes on their fossil record, biogeography and depth distribution. *Journal of Zoology* 244: 459–471.

Ben-Ami, F. & Heller, J. (2005). Spatial and temporal patterns of parthenogenesis and parasitism in the freshwater snail *Melanoides tuberculata*. *Journal of Evolutionary Biology* 18: 138–146.

Benian, G. M., Tang, X. & Tinley, T. L. (1996). Twitchin and related giant Ig superfamily members of *C. elegans* and other invertebrates. *Advances in Biophysics* 33: 183–198.

Beninger, P. G., Le Pennec, M. & Salaün, M. (1988). New observations of the gills of *Placopecten magellanicus* (Mollusca: Bivalvia), and implications for nutrition. Part I. General anatomy and surface microanatomy. *Marine Biology* 98: 61–70.

Beninger, P. G., Auffret, M. & Le Pennec, M. (1990). Peribuccal organs of *Placopecten magellanicus* and *Chlamys varia* (Mollusca: Bivalvia): Structure, ultrastructure and implications for feeding: I. The labial palps. *Marine Biology* 107: 215–223.

Beninger, P. G., Le Pennec, M. & Donval, A. (1991). Mode of particle ingestion in five species of suspension-feeding bivalve molluscs. *Marine Biology* 108: 255–262.

Beninger, P. G. & Le Pennec, M. (1993). Histochemistry of the bucco-oesophageal glands of *Mytilus edulis*: The importance of mucus in ingestion. *Journal of the Marine Biological Association of the United Kingdom* 73: 237–240.

Beninger, P. G., Donval, A. & Le Pennec, M. (1995a). The osphradium in *Placopecten magellanicus* and *Pecten maximus* (Bivalvia, Pectinidae): Histology, ultrastructure, and implications for spawning synchronization. *Marine Biology* 123: 121–129.

Beninger, P. G., Potter, T. M. & St-Jean, S. D. (1995b). Paddle cilia fixation artefacts in pallial organs of adult *Mytilus edulis* and *Placopecten magellanicus* (Mollusca, Bivalvia). *Canadian Journal of Zoology* 73: 610–614.

Beninger, P. G., Dufour, S. C. & Bourque, J. (1997a). Particle processing mechanisms of the eulamellibranch bivalves *Spisula solidissima* and *Mya arenaria*. *Marine Ecology Progress Series* 150: 157–169.

Beninger, P. G. & Le Pennec, M. (1997). Reproductive characteristics of a primitive bivalve from a deep-sea reducing environment: Giant gametes and their significance in *Acharax alinae* (Cryptodonta: Solemyidae). *Marine Ecology Progress Series* 157: 195–206.

Beninger, P. G., Silverman, H. G., Lynn, J. W. & Dietz, T. H. (1997b). The role of mucus in particle processing by suspension-feeding marine bivalves: Unifying principles from diverse systems. *Journal of Shellfish Research* 16: 337–338.

Beninger, P. G. & St-Jean, S. D. (1997a). The role of mucus in particle processing by suspension-feeding marine bivalves: Unifying principles. *Marine Biology* 129: 389–397.

Beninger, P. G. & St-Jean, S. D. (1997b). Particle processing on the labial palps of *Mytilus edulis* and *Placopecten magellanicus* (Mollusca: Bivalvia). *Marine Ecology Progress Series* 147: 117–127.

Beninger, P. G. & Veniot, A. (1999). The oyster proves the rule: Mechanisms of pseudofeces transport and rejection on the mantle of *Crassostrea virginica* and *C. gigas*. *Marine Ecology Progress Series* 190: 179–188.

Beninger, P. G., Veniot, A. & Poussart, Y. (1999). Principles of pseudofeces rejection on the bivalve mantle: Integration in particle processing. *Marine Ecology Progress Series* 178: 259–269.

Beninger, P. G., Le Pennec, G. & Le Pennec, M. (2003). Demonstration of nutrient pathway from the digestive system to oocytes in the gonad intestinal loop of the scallop *Pecten maximus* L. *The Biological Bulletin* 205: 83–92.

Beninger, P. G., Decottignies, P., Guiheneuf, F., Barillé, L. & Rincé, Y. (2007). Comparison of particle processing by two introduced suspension feeders: Selection in *Crepidula fornicata* and *Crassostrea gigas*. *Marine Ecology Progress Series* 334: 165–177.

Beninger, P. G., Valdizan, A., Cognie, B., Guiheneuf, F. & Decottignies, P. (2008). Wanted: Alive and not dead—Functioning diatom status is a quality cue for the suspension-feeder *Crassostrea gigas*. *Journal of Plankton Research* 30: 689–697.

Beninger, P. G., Valdizan, A. & Le Pennec, G. (2016). The seminal receptacle and implications for reproductive processes in the invasive gastropod *Crepidula fornicata*. *Zoology* 119: 4–10.

Benjamin, P. R. & Kemenes, G. Y. (2010). *Lymnaea* learning and memory. *Scholarpedia* 5: 4247.

Benjamin, P. R. & Kemenes, I. (2013). *Lymnaea* neuropeptide genes. *Scholarpedia* 8: 11520.

Benke, M., Reise, H., Montagne-Wajer, K. & Koene, J. M. (2010). Cutaneous application of an accessory-gland secretion after sperm exchange in a terrestrial slug (Mollusca: Pulmonata). *Zoology* 113: 118–124.

Benkendorff, K., Bremner, J. B. & Davis, A. R. (2000). Tyrian purple precursors in the egg masses of the Australian muricid, *Dicathais orbita*: A possible defensive role. *Journal of Chemical Ecology* 26: 1037–1050.

Benkendorff, K., Bremner, J. B. & Davis, A. R. (2001a). Indole derivatives from the egg masses of muricid molluscs. *Molecules* 6: 70–78.

Benkendorff, K., Davis, A. R. & Bremner, J. B. (2001b). Chemical defense in the egg masses of benthic invertebrates: An assessment of antibacterial activity in 39 mollusks and 4 polychaetes. *Journal of Invertebrate Pathology* 78: 109–118.

Benkendorff, K., Davis, A. R., Rogers, C. N. & Bremner, J. B. (2005). Free fatty acids and sterols in the benthic spawn of aquatic molluscs, and their associated antimicrobial properties. *Journal of Experimental Marine Biology and Ecology* 316: 29–44.

Benkendorff, K. (2010). Molluscan biological and chemical diversity: Secondary metabolites and medicinal resources produced by marine molluscs. *Biological Reviews* 85: 757–775.

Benmeradi, S. S. & Benmeradi, N. (1992). Ultrastructure et activité de la gland salivaire de *Patella safiana* Lmk (Mollusque, Gastéropode). *Actes de Colloque* 13: 107–112.

Benoit-Bird, K. J. & Gilly, W. F. (2012). Coordinated nocturnal behavior of foraging jumbo squid *Dosidicus gigas*. *Marine Ecology Progress Series* 455: 211–228.

Benzie, J. A. H. & Ballment, E. (1994). Genetic differences among black-lipped pearl oyster (*Pinctada margaritifera*) populations in the western Pacific. *Aquaculture* 127: 145–156.

Benzie, J. A. H. & Williams, S. T. (1997). Genetic structure of giant clam (*Tridacna maxima*) populations in the West Pacific is not consistent with dispersal by present-day ocean currents. *Evolution* 51: 768–783.

Berg, C. J. (1972). Ontogeny of the behavior of *Strombus maculatus* (Gastropoda, Strombidae). *American Zoologist* 12: 427–443.

Berg, C. J. (1974). A comparative ethological study of strombid gastropods *Behaviour* 51: 274–327.

Berg, C. J. & Olsen, D. A. (1989). Conservation and management of queen conch (*Strombus gigas*) fisheries in the Caribbean. pp. 421–442 *in* J. F. Caddy (Ed.) *Marine Invertebrate Fisheries: Their Assessment and Management*. New York, John Wiley & Sons.

Berg, C. O. & Knutson, L. (1978). Biology and systematics of the Sciomyzidae. *Annual Review of Entomology* 23: 239–258.

van den Berg, H. (1992). Model for mechanics of mollusc systemic heart. *Comparative Biochemistry and Physiology Part A* 101: 835–844.

Berg, K. & Ockelmann, K. (1959). The respiration of freshwater snails. *Journal of Experimental Biology* 36: 690–708.

Berger, B. & Dallinger, R. (1993). Terrestrial snails as quantitative indicators of environmental metal pollution. *Environmental Monitoring and Assessment* 25: 65–84.

Berger, V. & Kharazova, A. D. (1997). Mechanisms of salinity adaptations in marine molluscs. *Hydrobiologia* 355: 115–126.

Bergmann, S., Lieb, B., Ruth, P. & Markl, J. (2006). The hemocyanin from a living fossil, the cephalopod *Nautilus pompilius*: Protein structure, gene organization, and evolution. *Journal of Molecular Evolution* 62: 362–374.

Bergmann, S., Markl, J. & Lieb, B. (2007). The first complete cDNA sequence of the hemocyanin from a bivalve, the protobranch *Nucula nucleus*. *Journal of Molecular Evolution* 64: 500–510.

Bernard, F. (1890a). Recherches sur les organes palléaux des Gastéropodes prosobranches. *Annales des sciences naturelles. Zoologie* 9: 89–404.

Bernard, F. (1890b). Recherches sur la *Valvata piscinalis*. *Bulletin scientifique de la France et de la Belgique* 22: 253–361.

Bernard, F. R. (1968). The aquiferous system of *Polinices lewisi* (Gastropoda, Prosobranchiata) *Journal of the Fisheries Research Board of Canada* 25: 541–546.

Bernard, F. R. (1972). Occurrence and function of lip-hypertrophy in the Anisomyaria (Mollusca, Bivalvia). *Canadian Journal of Zoology* 50: 53–57.

Bernard, F. R. (1973). Crystalline style formation and function in the oyster *Crassostrea gigas* (Thunberg, 1795). *Ophelia* 12: 159–170.

Bernard, F. R. (1989). Uptake and elimination of coliform bacteria by four marine bivalve mollusks. *Canadian Journal of Fisheries and Aquatic Sciences* 46: 1592–1599.

Bernard, I., Massabuau, J. C., Ciret, P., Sow, M., Sottolichio, A., Pouvreau, S. & Tran, D. (2016). In situ spawning in a marine broadcast spawner, the Pacific oyster *Crassostrea gigas*: Timing and environmental triggers. *Limnology and Oceanography* 61: 635–647.

Bernot, R. J. (2003). Trematode infection alters the antipredator behavior of a pulmonate snail. *Journal of the North American Benthological Society* 22: 241–248.

Beron, C. & Murray, J. (2014). *Behavioral and Neural Activity During Magnetic Stimulation of Tritonia tetraquetra Imply Conditional Magnetotactic Response*. Student Research Papers. Friday Harbour, Washington, DC, Friday Harbour Laboratories: 1–29.

Berry, A. J. (1974). Freshwater bivalves of Peninsular Malaysia with special reference to sex and breeding. *Malayan Nature Journal* 27: 99–110.

Berry, G. N. & Cannon, L. R. G. (1981). The life history of *Sulcascaris sulcata* (Nematoda: Ascaridoidea), a parasite of marine molluscs and turtles. *International Journal of Parasitology* 11: 43–54.

Berry, P. F. & Playford, P. E. (1997). Biology of modern *Fragum erugatum* (Mollusca, Bivalvia, Cardiidae) in relation to deposition of the Hamelin coquina, Shark Bay, Western Australia. *Marine and Freshwater Research* 48: 415–420.

Berthe, F. C. J., Le Roux, F., Adlard, R. D. & Figueras, A. (2004). Marteiliosis in molluscs: A review. *Aquatic Living Resources* 17: 433–448.

Berthold, T. (1988). Anatomy of *Afropomus balanoideus* (Mollusca, Gastropoda, Ampullariidae) and its implications for phylogeny and ecology. *Zoomorphologie* 108: 149–159.

Berthold, T. (1991). *Vergleichende Anatomie, Phylogenie und Historische Biogeographie der Ampullariidae (Mollusca: Gastropoda) [Comparative Anatomy, Phylogeny and Historical Biogeography of the Ampullariidae (Mollusca: Gastropoda)].* Vol. 29. Hamburg, Berlin, P. Parey.

Bertness, M. D. & Cunningham, C. (1981). Crab shell-crushing predation and gastropod architectural defense. *Journal of Experimental Marine Biology and Ecology* 50: 213–230.

Bettex, D. A., Prêtre, R. & Chassot, P. G. (2014). Is our heart a well-designed pump? The heart along animal evolution. *European Heart Journal* 35: 2322–2332.

Beu, A. G. (1998). Superfamily Tonnoidea. pp. 792–803 in P. L. Beesley, Ross, G. J. B. & Wells, A. (Eds.) *Mollusca: The Southern Synthesis. Part B. Fauna of Australia.* Melbourne, Australia, CSIRO Publishing.

Beu, A. G. (2008). Recent deep-water Cassidae of the world: A revision of *Galeodea, Oocorys, Sconsia, Echinophoria* and related taxa, with new genera and species (Mollusca, Gastropoda). *Mémoires du Muséum national d'histoire naturelle* 196: 269–387.

Beuerlein, K., Schimmelpfennig, R., Westermann, B., Ruth, P. & Schipp, R. (1998). Cytobiological studies on hemocyanin metabolism in the branchial heart complex of the common cuttlefish *Sepia officinalis* (Cephalopoda, Dibranchiata). *Cell and Tissue Research* 292: 587–595.

Beuerlein, K., Löhr, S., Westermann, B., Ruth, P. & Schipp, R. (2002). Components of the cellular defense and detoxification system of the common cuttlefish *Sepia officinalis* (Mollusca, Cephalopoda). *Tissue and Cell* 34: 390–396.

Beukes, D. R. & Davies-Coleman, M. T. (1999). Novel polypropionates from the South African marine mollusc *Siphonaria capensis. Tetrahedron* 55: 4051–4056.

Beye, G. E. & Ward, D. E. (2010). On the origin of siphonariid polypropionates: Total synthesis of baconipyrone A, baconipyrone C, and siphonarin B via their putative common precursor. *Journal of the American Chemical Society* 132: 7210–7215.

Bhattacharya, D., Pelletreau, K. N., Price, D. C., Sarver, K. E. & Rumpho, M. E. (2013). Genome analysis of *Elysia chlorotica* egg DNA provides no evidence for horizontal gene transfer into the germ line of this kleptoplastic mollusc. *Molecular Biology and Evolution* 30: 1843–1852.

Bibby, R., Widdicombe, S., Parry, H., Spicer, J. I. & Pipe, R. K. (2008). Effects of ocean acidification on the immune response of the blue mussel *Mytilus edulis. Aquatic Biology* 2: 67–74.

Bickell-Page, L. R. (1988). Autotomy of cerata by the nudibranch mollusc *Melibe leonina*: A morphological and neurophysiological inquiry. *American Zoologist* 28: 1A-214A.

Bickell-Page, L. R. (1989). Autotomy of cerata by the nudibranch *Melibe leonina* (Mollusca): Ultrastructure of the autotomy plane and neural correlate of the behaviour. *Philosophical Transactions of the Royal Society B* 324: 149–172.

Bicker, G., Davis, W. J. & Matera, E. M. (1982a). Chemoreception and mechanoreception in the gastropod mollusc *Pleurobranchaea californica*. 2. Neuroanatomical and intracellular analysis of afferent pathways. *Journal of Comparative Physiology A* 149: 235–250.

Bicker, G., Davis, W. J., Matera, E. M., Kovac, M. P. & Stormo-Gipson, D. J. (1982b). Chemoreception and mechanoreception in the gastropod mollusc *Pleurobranchaea californica*: Part 1: Extracellular analysis of afferent pathways. *Journal of Comparative Physiology A* 149: 221–234.

Bidder, A. M. (1950). The digestive mechanism of the European squids *Loligo vulgaris, Loligo forbesii, Alloteuthis media,* and *Alloteuthis subulata. Quarterly Journal of Microscopical Science* 91: 1–43.

Bidder, A. M. (1966). Feeding and digestion in cephalopods. pp. 97–124 in K. M. Wilbur & Yonge, C. M. (Eds.) *Physiology of Mollusca.* Vol. 2. New York, Academic Press.

Bieler, R. (1988). Phylogenetic relationships in the gastropod family Architectonicidae, with notes on the family Mathildidae (Allogastropoda). pp. 205–240 in W. F. Ponder, Eernisse, D. J. & Waterhouse, J. H. (Eds.) *Prosobranch Phylogeny. Malacological Review Supplement.* Ann Arbor, MI, Malacological Review.

Bieler, R. & Mikkelsen, P. M. (1988). Anatomy and reproductive biology of two western Atlantic species of Vitrinellidae, with a case of protandrous hermaphroditism in the Rissoacea. *The Nautilus* 102: 1–29.

Bieler, R. (2009). Architectonicidae of the Indo-Pacific. *Abhandlungen des Naturwissenschaftlichen Vereins in Hamburg* 30: 1–377.

Bigatti, G., Penchaszadeh, P. E. & Cledón, M. (2007). Age and growth in *Odontocymbiola magellanica* (Gastropoda: Volutidae) from Golfo Nuevo, Patagonia, Argentina. *Marine Biology* 150: 1199–1204.

Bigatti, G., Sacristán, H., Rodríguez, M. C., Stortz, C. A. & Penchaszadeh, P. E. (2010). Diet, prey narcotization and biochemical composition of salivary glands secretions of the volutid snail *Odontocymbiola magellanica. Journal of the Marine Biological Association of the United Kingdom* 90: 959–967.

Van den Biggelaar, J. A. M. & Haszprunar, G. (1996). Cleavage patterns and mesentoblast formation in the Gastropoda: An evolutionary perspective. *Evolution* 50: 1520–1540.

Van den Biggelaar, J. A. M. (1977). Development of dorsoventral polarity and mesentoblast determination in *Patella vulgata. Journal of Morphology* 154: 157–186.

Van den Biggelaar, J. A. M. (1978). The determinative significance of the geometry of the cell contacts in early molluscan development. *Biologie Cellulaire* 32: 155–161.

Van den Biggelaar, J. A. M. (1996). The significance of the early cleavage pattern for the reconstruction of gastropod phylogeny. pp. 155–160 in J. D. Taylor (Ed.) *Origin and Evolutionary Radiation of the Mollusca.* Oxford, UK, Oxford University Press.

Van den Biggelaar, J. A. M., Edsinger-Gonzales, E. & Schram, F. R. (2002). The improbability of dorso-ventral axis inversion during animal evolution, as presumed by Geoffroy Saint Hilaire. *Contributions to Zoology* 71: 1–8.

Biggs, J. S., Olivera, B. M. & Kantor, I. (2008). α-Conopeptides specifically expressed in the salivary gland of *Conus pulicarius. Toxicon* 52: 101–105.

Bingham, F. O. (1972a). The influence of environmental stimuli on the direction of movement of the supralittoral gastropod *Littorina irrorata. Bulletin of Marine Science* 22: 309–335.

Bingham, F. O. (1972b). The mucus holdfast of *Littorina irrorata* and its relationship to relative humidity and salinity. *The Veliger* 15: 48–50.

Birkeland, C. (1987). Nutrient availability as a major determinant of differences among coastal hard-substratum communities in different regions of the tropics. pp. 45–97 in C. Birkeland. *Comparison between Atlantic and Pacific Tropical Marine Coastal Ecosystems: Community Structure, Ecological Processes and Productivity.* Paris, UNESCO Reports in Marine Science.

Bischoff, G. C. O. (1981). *Cobcrephora* n. g., representative of a new polyplacophoran order Phosphatoloricata with calciumphosphatic shells. *Senckenbergiana lethaea* 61: 173–215.

Bishop, C. D., Huggett, M. J., Heyland, A., Hodin, J. & Brandhorst, B. P. (2006). Interspecific variation in metamorphic competence in marine invertebrates: The significance for comparative investigations into the timing of metamorphosis. *Integrative and Comparative Biology* 46: 662–682.

Bishop, S. H., Ellis, L. L. & Burcham, J. M. (1983). Amino acid metabolism in molluscs. pp. 243–327 *in* P. W. Hochachka (Ed.) *Metabolic Biochemistry and Molecular Biomechanics. The Mollusca.* Vol. 1. New York, Academic Press.

Bishop, T. & Brand, M. D. (2000). Processes contributing to metabolic depression in hepatopancreas cells from the snail *Helix aspersa. Journal of Experimental Biology* 203: 3603–3612.

Bjelde, B. E. & Todgham, A. E. (2013). Thermal physiology of the fingered limpet *Lottia digitalis* under emersion and immersion. *Journal of Experimental Biology* 216: 2858–2869.

Black, B. A. (2009). Climate-driven synchrony across tree, bivalve, and rockfish growth-increment chronologies of the northeast Pacific. *Marine Ecology Progress Series* 378: 37–46.

Black, R. (1978). Tactics of whelks preying on limpets. *Marine Biology* 46: 157–162.

de Blainville, H. M. D. (1825). *Manuel de Malacologie et de Conchyliologie.* Paris, F.-G. Levrault.

Blakers, M., Davies, S. J. J. F. & Reilly, P. N. (1984). *The Atlas of Australian Birds.* Carlton, Victoria, Royal Australasian Ornithologists Union, Melbourne University Press.

Blanc, A., Vivien-Roels, B., Pevet, P., Attia, J. & Buisson, B. (2003). Melatonin and 5-methoxytryptophol (5-ML) in nervous and/or neurosensory structures of a gastropod mollusc (*Helix aspersa maxima*): Synthesis and diurnal rhythms. *General and Comparative Endocrinology* 131: 168–175.

Blanckenhorn, W. U. (2000). The evolution of body size: What keeps organisms small? *Quarterly Review of Biology* 75: 385–407.

Blanco, J. F. & Scatena, F. N. (2007). The spatial arrangement of *Neritina virginea* (Gastropoda: Neritidae) during upstream migration in a split-channel reach. *River Research and Applications* 23: 235–245.

Bleckmann, H., Budelmann, B.-U. & Bullock, T. H. (1991). Peripheral and central nervous responses evoked by small water movements in a cephalopod. *Journal of Comparative Physiology A* 168: 247–257.

Blinn, W. C. (1964). Water in the mantle cavity of land snails. *Physiological Zoology* 37: 329–337.

Block, G. & Smith, J. T. (1973). Cerebral photoreceptors in *Aplysia. Comparative Biochemistry and Physiology Part A* 46: 115–121.

Block, G. D. & Roberts, M. H. (1981). Circadian pacemaker in the *Bursatella* eye: Properties of the rhythm and its effect on locomotor behavior. *Journal of Comparative Physiology A* 142: 403–410.

Block, G. D. & McMahon, D. G. (1983). Localized illumination of the *Aplysia* and *Bulla* eye reveals new relationships between retinal layers. *Brain Research* 265: 134–137.

Block, G. D. & McMahon, D. G. (1984). Cellular analysis of the *Bulla* circadian pacemaker system. 3. Localization of the circadian pacemaker. *Journal of Comparative Physiology A* 155: 387–395.

Block, G. D., McMahon, D. G., Wallace, S. F. & Friesen, W. O. (1984). Cellular analysis of the *Bulla* circadian pacemaker system. 1. A model for retinal organization. *Journal of Comparative Physiology A* 155: 365–378.

Block, G. D. & Colwell, C. S. (2014). Molluskan ocular pacemakers: Lessons learned. pp. 213–232 *in* G. Tosini, Iuvone, P. M., McMahon, D. G. & Collin, S. P. (Eds.) *The Retina and Circadian Rhythms.* New York, Springer.

Blumenthal, E. M., Block, G. D. & Eskin, A. (2001). Cellular and molecular analysis of molluscan circadian pacemakers. pp. 371–400 *in* J. S. Takahashi, Turek, F. W. & Moore, R. Y. (Eds.) *Handbook of Behavioral Neurobiology: Circadian Clocks.* Vol. 12. New York, Kluwer Academic/Plenum Publishers/Springer.

Blumer, M. J. F. (1994). The ultrastructure of the eyes in the veligerlarvae of *Aporrhais* sp. and *Bittium reticulatum* (Mollusca, Caenogastropoda). *Zoomorphology (Berlin)* 114: 149–159.

Blumer, M. J. F. (1995). The ciliary photoreceptor in the teleplanic veliger larvae of *Smaragdia* sp. and *Strombus* sp. (Mollusca: Gastropoda). *Zoomorphology* 115: 73–81.

Blumer, M. J. F. (1996). Alterations of the eyes during ontogenesis in *Aporrhais pespelecani* (Mollusca, Caenogastropoda). *Zoomorphology (Berlin)* 116: 123–131.

Blumer, M. J. F. (1998). Alterations of the eyes of *Carinaria lamarcki* (Gastropoda, Heteropoda) during the long pelagic cycle. *Zoomorphology* 118: 183–194.

Blumer, M. J. F. (1999). Development of a unique eye: Photoreceptors of the pelagic predator *Atlanta peroni* (Gastropoda, Heteropoda). *Zoomorphology (Berlin)* 119: 81–91.

Blundon, J. A. & Vermeij, G. J. (1983). Effect of shell repair on shell strength in the gastropod *Littorina irrorata. Marine Biology* 76: 41–45.

Boal, J. G. & Golden, D. K. (1999). Distance chemoreception in the common cuttlefish, *Sepia officinalis* (Mollusca, Cephalopoda). *Journal of Experimental Marine Biology and Ecology* 235: 307–317.

Boal, J. G., Wittenberg, K. M. & Hanlon, R. T. (2000). Observational learning does not explain improvement in predation tactics by cuttlefish (Mollusca: Cephalopoda). *Behavioral Processes* 52: 141–153.

Boal, J. G., Shashar, N., Grable, M. M., Vaughan, K. H., Loew, E. R. & Hanlon, R. T. (2004). Behavioral evidence for intraspecific signaling with achromatic and polarized light by cuttlefish (Mollusca: Cephalopoda). *Behaviour* 141: 837–861.

Boal, J. G. (2006). Social recognition: A top down view of cephalopod behaviour. *Vie et Milieu* 56: 69–80.

Boal, J. G., Prosser, K. N., Holm, J. B., Simmons, T. L., Haas, R. E. & Nagle, G. T. (2010). Sexually mature cuttlefish are attracted to the eggs of conspecifics. *Journal of Chemical Ecology* 36: 834–836.

Boaventura, D., da Fonseca, L. C. & Hawkins, S. J. (2002). Analysis of competitive interactions between the limpets *Patella depressa* Pennant and *Patella vulgata* L. on the northern coast of Portugal. *Journal of Experimental Marine Biology and Ecology* 271: 171–188.

Bobbio, A. (1972). The first endosseous alloplastic implant in the history of man. *Bulletin of the History of Dentistry* 20: 1–6.

Bobkova, M. V., Gál, J., Zhukov, V. V., Shepeleva, I. P. & Meyer-Rochow, V. B. (2004). Variations in the retinal designs of pulmonate snails (Mollusca, Gastropoda): Squaring phylogenetic background and ecophysiological needs (I). *Invertebrate Biology* 123: 101–115.

Bobzin, S. C. & Faulkner, D. J. (1989). Diterpenes from the marine sponge *Aplysilla polyrhaphis* and the dorid nudibranch *Chromodoris norrisi. The Journal of Organic Chemistry* 54: 3902–3907.

Bocquet, C. & Stock, J. H. (1963). Some recent trends in work on parasitic copepods. *Oceanography and Marine Biology Annual Review* 1: 289–300.

Boer, H. H., Wendelaar Bonga, S. E. & van Rooyen, H. (1966). Light and electron microscopical investigations on the salivary gland of *Lymnaea stagnalis* L. *Zeitschrift für Zellforschung und Mikroskopische Anatomie* 76: 228–247.

Boer, H. H., Slot, J. W. & van Andel, J. (1968). Electron microscopical and histochemical observations on the relation between medio-dorsal bodies and neurosecretory cells in the basommatophoran snails *Lymnaea stagnalis*, *Ancylus fluviatilis*, *Australorbis glabratus* and *Planorbarius corneus*. *Zeitschrift für Zellforschung und Mikroskopische Anatomie* 87: 435–450.

Boer, H. H., Algera, N. H. & Lommerse, A. W. (1973). Ultrastructure of possible sites of ultrafiltration in some gastropods, with special reference to the auricle of the freshwater prosobranch *Viviparus viviparus* L. *Zeitschrift für Zellforschung und Mikroskopische Anatomie* 143: 329–341.

Boer, H. H. & Joosse, J. (1975). Endocrinology. pp. 245–307 *in* V. Fretter & Peake, J. (Eds.) *Pulmonates: Functional Anatomy and Physiology.* Vol. 1. London, Academic Press.

Boer, H. H. & Montagne-Wajer, C. (1994). Functional morphology of the neuropeptidergic light-yellow-cell system in pulmonate snails. *Cell and Tissue Research* 277: 531–538.

Boetius, A. & Felbeck, H. (1995). Digestive enzymes in marine invertebrates from hydrothermal vents and other reducing environments. *Marine Biology* 122: 105–113.

Boettcher, A. A. & Target, N. M. (1996). Induction of metamorphosis in queen conch, *Strombus gigas* Linnaeus larvae by cues associated with red algae from their nursery grounds. *Journal of Experimental Marine Biology and Ecology* 196: 29–52.

Boettcher, A. A. & Target, N. M. (1998). Role of chemical inducers in larval metamorphosis of queen conch, *Strombus gigas* Linnaeus: Relationship to other marine invertebrate systems. *Biological Bulletin* 194: 132–142.

Boettger, C. R. (1956). Beiträge zur Systematik der Urmollusken (Amphineura). *Verhandlungen der Deutschen Zoologischen Gesellschaft Zoologischer Anzeiger Supplement* 19: 223–256.

Boettiger, A., Ermentrout, B. & Oster, G. (2009). The neural origins of shell structure and pattern in aquatic mollusks. *Proceedings of the National Academy of Sciences of the United States of America* 106: 6837–6842.

Bogan, A. & Bouchet, P. (1998). Cementation in the freshwater bivalve family Corbiculidae (Mollusca: Bivalvia): A new genus and species from Lake Poso, Indonesia. *Hydrobiologia* 389: 131–139.

Bogan, A. E. & Hoeh, W. R. (2000). On becoming cemented: Evolutionary relationships among the genera in the freshwater bivalve family Etheriidae (Bivalvia: Unionoida). pp. 159–168 *in* E. M. Harper, Taylor, J. D. & Crame, J. A. (Eds.) *Evolutionary Biology of the Bivalvia.* London, The Geological Society (Special Publication No. 177).

Bogan, A. E. (2006). Conservation and extinction of the freshwater molluscan fauna of North America. pp. 373–383 *in* C. F. Sturm, Pearce, T. A. & Valdes, A. (Eds.) *The Mollusks: A Guide to Their Study, Collection, and Preservation.* Boca Raton, FL, Universal Publishers.

Bogdanov, A., Kehraus, S., Bleidissel, S., Preisfeld, G., Schillo, D., Piel, J., Brachmann, A. O., Wägele, H. & König, G. M. (2014). Defense in the aeolidoidean genus *Phyllodesmium* (Gastropoda). *Journal of Chemical Ecology* 40: 1013–1024.

Bogdanov, I. P. (1989). Il peculiare sistema di alimentazione dei molluschi di Oenopotinae. (Peculiar feeding mechanism of the Oenopotinae molluscs). *La Conchiglia* 246: 39–47.

Bogerd, J., Li, K. W., Jiménez, C. R., Van der Schors, R. C., Ebberink, R. H. M. & Geraerts, W. P. M. (1994). Processing, axonal transport and cardioregulatory functions of peptides

derived from two related prohormones generated by alternative splicing of a single gene in identified neurons VD1 and RPD2 of *Lymnaea*. *Molecular Brain Research* 23: 66–72.

Böggemann, M. (2006). Worms that might be 300 million years old. *Marine Biology Research* 2: 130–135.

Bøggild, O. B. (1930). The shell structure of the mollusks. *Det Kongelige Danske Videnskabernes Selskabs Skrifter. Naturvidenskabelig og Mathematisk Afdeling, ser.* 9(2): 231–326.

Boggs, C. H., Rice, J. A., Kitchell, J. A. & Kitchell, J. F. (1984). Predation at a snail's pace: What's time to a gastropod? *Oecologia* 62: 13–17.

Boghen, A. & Farley, J. (1974). Phasic activity in the digestive gland cells of the intertidal prosobranch *Littorina saxatilis* (Olivi) and its relation to the tidal cycle. *Proceedings of the Malacological Society of London* 41: 41–56.

Boissevain, M. (1904). Beiträge zur Anatomie und Histologie von *Dentalium. Jenaische Zeitschrift für Naturwissenschaft* 38: 553–572.

von Boletzky, S. & von Boletzky, M. V. (1970). Das eingraben im sand bei *Sepiola* und *Sepietta* (Mollusca, Cephalopoda). *Revue suisse de Zoologie* 77: 536–548.

von Boletzky, S. (1973). Structure et fonctionnement des organes de Kölliker chez les jeunes octopodes (Mollusca, Cephalopoda). *Zeitschrift für Morphologie der Tiere* 75: 315–327.

von Boletzky, S. (1987). Ontogenetic and phylogenetic aspects of the cephalopod circulatory system. *Experientia* 43: 478–483.

von Boletzky, S. (1989). Early ontogeny and evolution: The cephalopod model viewed from the point of developmental morphology. *Geobios* 12: 67–78.

von Boletzky, S. (1997). Developmental constraints and heterochrony: A new look at offspring size in cephalopod molluscs. *Geobios* 21: 267–275.

Bolstad, K. (2003). *Deep-Sea Cephalopods: An Introduction and Overview. TONMO. com: The Octopus News Magazine Online.*

Boltovskoy, D., Correa, N., Cataldo, D. & Sylvester, F. (2006). Dispersion and ecological impact of the invasive freshwater bivalve *Limnoperna fortunei* in the Río de la Plata watershed and beyond. *Biological Invasions* 8: 947–963.

Bołtruszko, J. S. (2010). Epizoic communities of Rotifera on freshwater bivalves. . *Oceanological and Hydrobiological Studies* 39: 75–82.

Bonar, D. B. & Hadfield, M. G. (1974). Metamorphosis of the marine gastropod *Phestilla sibogae* Bergh (Nudibranchia, Aeolidiacea). 1. Light and electron microscopic analysis of larval and metamorphic stages. *Journal of Experimental Marine Biology and Ecology* 16: 227–255.

Bonar, D. B. (1978). Morphogenesis and metamorphosis in opisthobranchiate molluscs. pp. 177–196 *in* F. S. Chia & Rice, M. E. (Eds.) *Settlement and Metamorphosis of Marine Invertebrate Larvae.* New York, Elsevier.

Bonar, L. (1936). An unusual ascomycete in the shells of marine animals. *University of California Publications: Botany* 19: 187–192.

Bonaventura, J. & Bonaventura, C. (1985). Physiological adaptations and subunit diversity in hemocyanins. pp. 21–34 *in* J. Lamy, Truchot, J.-P. & Gilles, R. (Eds.) *Respiratory Pigments in Animals: Relation Structure-Function.* Berlin, Springer-Verlag.

Bond, W. J. (1994). Keystone species. pp. 237–253 *in* E.-D. Schulze & Mooney, H. A. (Eds.) *Biodiversity and Ecosystem Function.* Berlin, Heidelberg, Springer-Verlag.

Bondar, C. A. & Page, L. R. (2003). Development of asymmetry in the caenogastropods *Amphissa columbiana* and *Euspira lewisii*. *Invertebrate Biology* 122: 28–41.

Bone, Q., Pulsford, A. & Chubb, A. D. (1981). Squid mantle muscle. *Journal of the Marine Biological Association of the United Kingdom* 61: 327–342.

Bone, Q., Brown, E. R. & Usher, M. (1995). The structure and physiology of cephalopod muscle fibres. pp. 301–329 *in* N. J. Abbot, Williamson, R. & Maddock, L. (Eds.) *Cephalopod Neurobiology, Neuroscience Studies in Squid, Octopus and Cuttlefish*. Oxford, UK, Oxford University Press.

Bonham, K. (1965). Growth rate of giant clam *Tridacna gigas* at Bikini Atoll as revealed by radioautography. *Science* 149: 300–302.

Bonnemain, B. (2005). *Helix* and drugs: Snails for western health care from antiquity to the present. *Evidence-based Complementary and Alternative Medicine* 2: 25–28.

Bonnivard, E., Catrice, O., Ravaux, J., Brown, S. C. & Higuet, D. (2009). Survey of genome size in 28 hydrothermal vent species covering 10 families. *Genome* 52: 524–536.

Bordalo, M. D., Ferreira, S. M. F., Jensen, K. T. & Pardal, M. A. (2014). Impact of trematodes on the population structure and shell shape of the estuarine mud snail *Hydrobia ulvae* from a Southern European estuary. *Marine Ecology* 35: 1–10.

Borell, E. M., Foggo, A. & Coleman, R. A. (2004). Induced resistance in intertidal macroalgae modifies feeding behaviour of herbivorous snails. *Oecologia* 140: 328–334.

Borges, C. D. G., Hawkins, S. J., Crowe, T. P. & Doncaster, C. P. (2016). The influence of simulated exploitation on *Patella vulgata* populations: Protandric sex change is size-dependent. *Ecology and Evolution* 6: 514–531.

Borland, C. (1950). An ecological study of *Benhamina obliquata* (Sowerby) a basommatophorous pulmonate in Otago Harbour. *Transactions of the Royal Society of New Zealand* 78: 385–393.

Bornancin, L., Bonnard, I., Mills, S. C. & Banaigs, B. (2017). Chemical mediation as a structuring element in marine gastropod predator-prey interactions. *Natural Product Reports* 34: 644–676.

Borowski, C., Giere, O., Krieger, J., Amann, R. & Dubilier, N. (2002). New aspects of the symbiosis in the provannid snail *Ifremeria nautilei* from the North Fiji Back Arc Basin. *Cahiers de Biologie Marine* 43: 321–324.

Borradaile, L. A., Potts, F. A., Eastham, L. E. S., Saunders, J. T. & Kerkut, G. A. (1958). *The Invertebrata. A Manual for the Use of Students*. Cambridge, UK, Cambridge University Press.

Borsa, P., Jarne, P., Belkhir, K. & Bonhomme, F. (1994). Genetic structure of the palourde *Ruditapes decussatus* L. in the Mediterranean. pp. 103–113 *in* A. R. Beaumont (Ed.) *Genetics and Evolution of Aquatic Organisms*. London, Chapman & Hall.

Boss, K. J. (1971). Critical estimate of the number of recent Mollusca. *Occasional Papers on Mollusks* 3: 81–135.

Boss, K. J. (1978). On the evolution of gastropods in ancient lakes. pp. 385–428 *in* V. Fretter & Peake, J. (Eds.) *Pulmonates: Systematics, Evolution and Ecology*. Vol. 2A. London, Academic Press.

Botello, G. & Krug, P. J. (2006). 'Desperate larvae' revisited: Age, energy and experience affect sensitivity to settlement cues in larvae of the gastropod Alderia sp. *Marine Ecology Progress Series* 312: 149–159.

Bottjer, D. J., Hickman, C. S. & Ward, P. D., Eds. (1985). *Mollusks: Notes for a Short Course*. Studies in Geology. Knoxville, TN, University of Tennessee Department of Geological Sciences.

Boucaud-Camou, E. & Boucher-Rodoni, R. (1983). Feeding and digestion in cephalopods. pp. 149–187 *in* A. S. M. Saleuddin & Wilbur, K. M. (Eds.) *Physiology. Part 2. The Mollusca*. Vol. 5. New York, Academic Press.

Boucaud-Camou, E. & Roper, C. F. E. (1995). Digestive enzymes in paralarval cephalopods. *Bulletin of Marine Science* 57: 313–327.

Boucher-Rodoni, R., Boucaud-Camou, E. & Mangold, K. (1987). Feeding and digestion. pp. 85–108 *in* P. R. Boyle (Ed.) *Cephalopod Life Cycles. Comparative Reviews*. Vol. 2. London, Orlando FL, Academic Press.

Boucher-Rodoni, R. (1989). Consommation d'oxygène et excrétion ammoniacale de *Nautilus macromphalus. Comptes rendus de l'Académie des sciences Serie III Sciences de la Vie* 309: 173–179.

Boucher-Rodoni, R. & Mangold, K. M. (1994). Ammonia production in cephalopods, physiological and evolutionary aspects. pp. 53–60 *in* H. O. Pörtner, O'Dor, R. K. & Macmillan, D. L. (Eds.) *Physiology of Cephalopod Molluscs: Lifestyle and Performance Adaptations. Marine and Freshwater Behaviour and Physiology, Special Issue*. Basel, Gordon and Breach Science Publishers SA.

Boucher-Rodoni, R. & Mangold, K. M. (1977). Experimental study of digestion in *Octopus vulgaris* (Cephalopoda: Octopoda). *Journal of Zoology* 183: 505–515.

Bouchet, P. & Lützen, J. (1980). Deux gastéropodes parasites d'une Holothurie Elasipode. *Bulletin Muséum National Histoire Naturelle, Paris: Zoologie Biologie Ecologie Animales* 2: 59–75.

Bouchet, P. & Warén, A. (1994). Ontogenetic migration and dispersal of deep-sea gastropod larvae. pp. 98–117 *in* C. M. Young & Eckelbarger, K. J. (Eds.) *Reproduction, Larval Biology, and Recruitment of the Deep-sea Benthos*. New York, Columbia University Press.

Bouchet, P. (1997). Inventorying the molluscan diversity of the world: What is our rate of progress? *The Veliger* 40: 1–11.

Bouchet, P., Lozouet, P., Maestrati, P. & Héros, V. (2002). Assessing the magnitude of species richness in tropical marine environments: Exceptionally high numbers of molluscs at a New Caledonia site. *Biological Journal of the Linnean Society* 75: 421–436.

Bouchet, P. & Rocroi, J.-P. (2005). Classification and nomenclator of gastropod families. *Malacologia* 47: 1–397.

Bouchet, P. (2006). The magnitude of marine biodiversity. pp. 31–62 *in* C. M. Duarte (Ed.) *The Exploration of Marine Biodiversity: Scientific and Technological Challenges*. Bilbao, Spain, Fundación BBVA.

Bouchet, P., Bary, S., Héros, V. & Marani, G. (2016). How many species of molluscs are there in the world's oceans, and who is going to describe them? *Tropical Deep-Sea Benthos* 29: 9–24.

Bouillon, J. (1960). Ultrastructure des cellules rénales des mollusques. I. Gastéropodes pulmonés terrestres (*Helix pomatia*). *Annales des sciences naturelles. Zoologie* 2: 719–749.

Boulding, E. G. (1984). Crab-resistant features of shells of burrowing bivalves: Decreasing vulnerability by increasing handling time. *Journal of Experimental Marine Biology and Ecology* 76: 201–223.

Bourlat, S. J., Nielsen, C., Lockyer, A. E., Littlewood, D. T. J. & Telford, M. J. (2003). *Xenoturbella* is a deuterostome that eats molluscs. *Nature* 424: 925–928.

Bourlat, S. J., Juliusdottir, T., Lowe, C. J., Freeman, R., Aronowicz, J., Kirschner, M., Lander, E. S., Thorndyke, M., Nakano, H. & Kohn, A. B. (2006). Deuterostome phylogeny reveals monophyletic chordates and the new phylum Xenoturbellida. *Nature* 444: 85–88.

Bourne, G. B. (1987a). Circulatory physiology of *Nautilus. Experientia* 43: 484–486.

Bourne, G. B. (1987b). Hemodynamics in squid. *Experientia* 43: 500–502.

Bourne, G. B. (1987c). The circulatory system. pp. 271–280 *in* W. B. Saunders & Landman, N. H. (Eds.) *Nautilus: The Biology and Paleobiology of a Living Fossil. Topics in Geobiology.* New York, Springer.

Bourne, G. B., Redmond, J. R. & Jorgensen, D. D. (1990). Dynamics of the molluscan circulatory system: Open versus closed. *Physiological Zoology* 63: 140–166.

Bourne, G. B. (2010). The circulatory system. pp. 271–279 *in* W. B. Saunders & Landman, N. H. (Eds.) *Nautilus. The Biology and Paleobiology of a Living Fossil. Topics in Geobiology.* Vol. 6. New York, Springer.

Bourne, G. C. (1909). Contributions to the morphology of the group Neritacea of the aspidobranch gastropods: Part 1. The Neritidae. *Proceedings of the Zoological Society of London* 8: 810–887.

Boutan, L. (1895). Recherches sur le byssus des Lamellibranches. *Archives de zoologie expérimentale et générale* 3: 297–338.

Bouton, D., Escriva, H., De Mendonça, R. L., Glineur, C., Bertin, B., Noël, C., Robinson-Rechavi, M., de Groot, A., Cornette, J. & Laudet, V. (2005). A conserved retinoid X receptor (RXR) from the mollusk *Biomphalaria glabrata* transactivates transcription in the presence of retinoids. *Journal of Molecular Endocrinology* 34: 567–582.

Bouton, M. E. (2007). *Learning and Behavior: A Contemporary Synthesis.* Sunderland, MA, Sinauer Associates.

Bouvier, E. L. (1887). Système nerveux, morphologie générale et classification des gastéropodes prosobranches. *Annales des sciences naturelles. Zoologie* 3: 1–510.

Bouvy, M., Soyer, J., Cahet, G., Descolas-Gros, C., Thiriot-Quiévreux, C. & Soyer-Gobillard, M.-O. (1989). Chemoautotrophic metabolism of intracellular gill bacteria in the marine bivalve *Spisula subtruncata* (Da Costa). *Netherlands Journal of Sea Research* 23: 29–34.

Bovbjerg, R. V. (1968). Responses to food in lymnaeid snails. *Physiological Zoology* 41: 412–423.

Bower, S. M. (1987). *Labyrinthuloides haliotidis* n. sp. (Protozoa: Labyrinthomorpha), a pathogenic parasite of small juvenile abalone in a British Columbia mariculture facility. *Canadian Journal of Zoology* 65: 1996–2007.

Bower, S. M. & Blackbourn, J. (2003). Geoduck clam (*Panopea abrupta*): Anatomy, histology, development, pathology, parasites and symbionts: Normal histology—Organs associated with the visceral mass. Retrieved February, 2015, from http://www.dfo-mpo.gc.ca/science/aah-saa/species-especes/shellfish-coquillages/geopath/viscera-eng.html.

Bower, S. M. (2006). Parasitic diseases of shellfish. pp. 629–677 *in* P. T. K. Woo (Ed.) *Fish Diseases and Disorders: Protozoan and Metazoan Infections.* Vol. 1. Wallingford, UK, CABI Publishing.

Bowsher, A. L. (1955). Origin and adaptation of platyceratid gastropods. *University of Kansas: Paleontological Contributions* 5: 1–11.

Boycott, B. B. & Young, J. Z. (1955a). A memory system in *Octopus vulgaris* Lamarck. *Proceedings of the Royal Society B* 143: 449–479.

Boycott, B. B. & Young, J. Z. (1955b). Memories controlling attacks on food objects by *Octopus vulgaris* Lamarck. *Pubblicazioni della Stazione Zoologica di Napoli* 27: 232–249.

Boycott, B. B. & Young, J. Z. (1957). Effects of interference with the vertical lobe on visual discriminations in *Octopus vulgaris* Lamarck. *Proceedings of the Royal Society B* 146: 439–450.

Boyd, I. L., Arnbom, T. A. & Fedak, M. A. (1994). Biomass and energy consumption of the South Georgia population of southern elephant seals. pp. 98–120 *in* B. J. Le Beouf & Laws, R. M. (Eds.) *Elephant Seals: Population Ecology, Behaviour, and Physiology.* Berkeley, CA, University of California Press.

Boyd, R. S. (2010). Heavy metal pollutants and chemical ecology: Exploring new frontiers. *Journal of Chemical Ecology* 36: 46–58.

Boyer, B. C., Henry, J. Q. & Martindale, M. Q. (1996). Dual origins of mesoderm in a basal spiralian: Cell lineage analyses in the polyclad turbellarian *Hoploplana inquilina. Developmental Biology* 179: 329–338.

Boyko, C. B. & Mikkelsen, P. M. (2002). Anatomy and biology of *Mysella pedroana* (Mollusca: Bivalvia: Galeommatoidea), and its commensal relationship with *Blepharipoda occidentalis* (Crustacea: Anomura: Albuneidae). *Zoologischer Anzeiger* 241: 149–160.

Boyle, P. & Rodhouse, P. G. K. (2005). *Cephalopods: Ecology and Fisheries.* Oxford, UK, Wiley-Blackwell.

Boyle, P. R. (1972). The aesthetes of chitons. Part 1. Role in the light response of whole animals. *Marine Behaviour and Physiology* 1: 171–184.

Boyle, P. R. (1975). Fine structure of the subradular organ of *Lepidochitona cinerea* L. (Mollusca, Polyplacophora). *Cell and Tissue Research* 162: 411–417.

Boyle, P. R. (1977). The physiology and behaviour of chitons (Mollusca: Polyplacophora). *Oceanography and Marine Biology Annual Review* 15: 461–509.

Boyle, P. R. (1986). Neural control of cephalopod behavior. pp. 1–99 *in* A. O. D. Willows (Ed.) *Neurobiology and Behavior, Part 2. The Mollusca.* Vol. 9. New York, Academic Press.

Brace, R. C. (1977a). Anatomical changes in nervous and vascular systems during the transition from prosobranch to opisthobranch organization. *Transactions of the Zoological Society of London* 34: 1–25.

Brace, R. C. (1977b). The functional anatomy of the mantle complex and columellar muscle of tectibranch molluscs (Gastropoda, Opisthobranchia). *Philosophical Transactions of the Royal Society B* 277: 1–56.

Brace, R. C. (1983). Observations on the morphology and behaviour of *Chilina fluctuosa* Gray (Chilinidae), with a discussion on the early evolution of pulmonate gastropods. *Philosophical Transactions of the Royal Society B* 300: 463–491.

Bradshaw-Hawkins, V. I. & Sander, F. (1981). Notes on the reproductive biology and behavior of the West Indian fighting conch, *Strombus pugilis* Linnaeus in Barbados, with evidence of male guarding. *The Veliger* 24: 159–164.

Brahma, S. K. (1978). Ontogeny of lens crystallins in marine cephalopods. *Journal of Embryology and Experimental Morphology* 46: 111–118.

Brahmachary, R. L. (1989). Mollusca. pp. 281–348 *in* K. G. Adiyodi & Adiyodi, R. G. (Eds.) *Reproductive Biology of Invertebrates: Fertilization, Development and Parental Care.* Vol. 4 A. Chichester, UK, John Wiley & Sons.

Branch, G. M. (1974). The ecology of *Patella* Linnaeus from the Cape Peninsula, South Africa. 2. Growth rates. *Transactions of the Royal Society of South Africa* 41: 161–193.

Branch, G. M. (1975a). Mechanisms reducing intraspecific competition in *Patella* spp.: Migration, differentiation and territorial behaviour. *Journal of Animal Ecology* 44: 575–600.

Branch, G. M. (1975b). The ecology of *Patella* from the Cape Peninsula South Africa. *Zoologica Africana* 10: 133–162.

Branch, G. M. & Marsh, A. C. (1978). Tenacity and shell shape in six *Patella* species: Adaptive features. *Journal of Experimental Marine Biology and Ecology* 34: 111–130.

Branch, G. M. (1979). Aggression by limpets against invertebrate predators. *Animal Behaviour* 27: 408–410.

Branch, G. M. & Branch, M. L. (1980). Competition between *Cellana tramoserica* (Sowerby) (Gastropoda) and *Patiriella exigua* (Lamarck) (Asteroidea), and their influence on algal standing stocks. *Journal of Experimental Marine Biology and Ecology* 48: 35–49.

Branch, G. M. (1981). The biology of limpets: Physical factors, energy flow, and ecological interactions. *Oceanography and Marine Biology Annual Review* 19: 235–380.

Branch, G. M., Branch, M. L. & Bannister, A. (1981). *The Living Shores of Southern Africa*. Cape Town, Struik Publishers.

Branch, G. M. (1985). Limpets: Evolution and adaptation. pp. 187–220 in E. R. Trueman & Clarke, M. R. (Eds.) *Evolution. The Mollusca*. Vol. 10. New York, Academic Press.

van den Branden, C. (1976). A study of the chromatophore pigments in the skin of the cephalopod *Sepia officinalis* L. *Biologisch Jaarboek (Dodonaea)* 44: 345–352.

Brandenburger, J. L. & Eakin, R. M. (1970). Pathway of incorporation of vitamin a³H₂ into photoreceptors of a snail, *Helix aspersa*. *Vision Research* 10: 639–653.

Brandmayr, P. & Brandmayr, T. Z. (1986). Food and feeding behaviour of some *Licinus* species (Coleoptera: Carabidae: Licinini). *Monitore Zoologico Italiano (Italian Journal of Zoology)* 20: 171–181.

Brandriff, B., Moy, G. W. & Vacquier, V. D. (1978). Isolation of sperm bindin from the oyster (*Crassostrea gigas*). *Gamete Research* 1: 89–99.

Braubach, O. R. & Croll, R. P. (2004). Evidence that histamine acts as a neurotransmitter in statocyst hair cells in the snail, *Lymnaea stagnalis*. *Journal of Gravitational Physiology* 11: 57–66.

Braun, R. (1954). Zum Lichtsinn augenlöser Muscheln. *Zoologische Jahrbücher. Abteilung für Allgemeine Zoologie und Physiologie der Tiere* 65: 194–208.

Brawley, S. H. & Adey, W. H. (1982). *Coralliophila abbreviata*: A significant corallivore! *Bulletin of Marine Science* 32: 595–599.

Brawley, S. H., Coyer, J. A., Blakeslee, A. M. H., Hoarau, G., Johnson, L. E., Byers, J. E., Stam, W. T. & Olsen, J. L. (2009). Historical invasions of the intertidal zone of Atlantic North America associated with distinctive patterns of trade and emigration. *Proceedings of the National Academy of Sciences* 106: 8239–8244.

Breed-Willeke, G. M. & Hancock, D. R. (1980). Growth and reproduction of subtidal and intertidal populations of the gaper clam *Tresus capax* from Yaquina Bay, Oregon USA. *Proceedings of the National Shellfisheries Association* 70: 1–13.

Breen, P. A. (1980). Measuring fishery intensity and annual production in the abalone fishery of British Columbia. *Canadian Technical Report of Fisheries and Aquatic Sciences*: 1–49.

Brendelberger, H. (1995). Growth of juvenile *Bithynia tentaculata* (Prosobranchia, Bithyniidae) under different food regimes: A long-term laboratory study. *Journal of Molluscan Studies* 61: 89–95.

Brendelberger, H. (1997a). Bacteria and digestive enzymes in the alimentary tract of *Radix peregra* (Gastropoda, Lymnaeidae). *Limnology and Oceanography* 42: 1635–1638.

Brendelberger, H. (1997b). Contrasting feeding strategies of two freshwater gastropods, *Radix peregra* (Lymnaeidae) and *Bithynia tentaculata* (Bithyniidae). *Archiv für Hydrobiologie* 140: 1–21.

Brendelberger, H. (1997c). Coprophagy: A supplementary food source for two freshwater gastropods? *Freshwater Biology* 38: 145–157.

Brennicke, A., Marchfelder, A. & Binder, S. (1999). RNA editing. *FEMS Microbiology Reviews* 23: 297–316.

Brescia, F. M., Poöllabauer, C. M., Potter, M. A. & Robertson, A. W. (2003). A review of the ecology and conservation of *Placostylus* (Mollusca: Gastropoda: Bulimulidae) in New Caledonia. *Molluscan Research* 28: 111–122.

Breton, S., Stewart, D. T., Shepardson, S., Trdan, R. J., Bogan, A. E., Chapman, E. G., Ruminas, A. J., Piontkivska, H. & Hoeh, W. R. (2010). Novel protein genes in animal mtDNA: A new sex determination system in freshwater mussels (Bivalvia: Unionoida)? *Molecular Biology and Evolution* 28: 1645–1659.

Breton, S., Capt, C., Guerra, D. & Stewart, D. (2019). Sex-determining mechanisms in bivalves. pp. 165–192 in J. L. Leonard (Ed.), *Evolution of Sexual Systems*. New York, Springer Verlag.

Bretos, M. (1980). Age determination in the keyhole limpet *Fissurella crassa* Lamarck (Archaeogastropoda: Fissurellidae), based on shell growth rings. *Biological Bulletin* 159: 606–612.

Brezden, B. L. & Gardner, D. R. (1992). A review of the electrophysiological, pharmacological and single channel properties of heart ventricle muscle cells in the snail *Lymnaea stagnalis*. *Experientia* 48: 841–851.

Bricelj, V. M. & Shumway, S. E. (1998). Paralytic shellfish toxins in bivalve molluscs: Occurrence, transfer kinetics, and biotransformation. *Reviews in Fisheries Science* 6: 315–383.

Bride, J., Gomot, L. & Saleuddin, A. S. M. (1991). Mating and 20-hydroxyecdysone cause increased galactogen synthesis in the albumen gland explants of *Helix aspersa* (Mollusca). *Comparative Biochemistry and Physiology Part B* 98B: 369–373.

Briner, J. P., Kaufman, D. S., Bennike, O. & Kosnik, M. A. (2014). Amino acid ratios in reworked marine bivalve shells constrain Greenland Ice Sheet history during the Holocene. *Geology* 42: 75–78.

Brink, D. J., van der Berg, N. G. & Botha, A. J. (2002). Iridescent colors on seashells: An optical and structural investigation of *Helcion pruinosus*. *Applied Optics* 41: 717–722.

Brink, M. & Boer, H. H. (1967). An electron microscopical investigation of the follicle gland (cerebral gland) and of some neurosecretory cells in the lateral lobe of the cerebral ganglion of the pulmonate gastropod *Lymnaea stagnalis* L. *Zeitschrift für Zellforschung und Mikroskopische Anatomie* 79: 230–243.

Brinkhoff, W., Stöckmann, K. & Grieshaber, M. K. (1983). Natural occurrence of anaerobiosis in molluscs from intertidal habitats. *Oecologia* 57: 151–155.

Brinkmann, N. & Wanninger, A. (2009). Neurogenesis suggests independent evolution of opercula in serpulid polychaetes. *BMC Evolutionary Biology* 9: 270.

Brissac, T., Rodrigues, C. F., Gros, O. & Duperron, S. (2011). Characterization of bacterial symbioses in *Myrtea* sp. (Bivalvia: Lucinidae) and *Thyasira* sp. (Bivalvia: Thyasiridae) from a cold seep in the Eastern Mediterranean. *Marine Ecology* 32: 198–210.

Brito-Manzano, N. & Aranda, D. A. (2004). Development, growth and survival of the larvae of queen conch *Strombus gigas* under laboratory conditions. *Aquaculture* 242: 479–487.

Britton, J. C. (1992). Evaporative water loss, behaviour during emersion, and upper thermal tolerance limits in seven species of eulittoral-fringe Littorinidae (Mollusca; Gastropoda) from Jamaica. pp. 69–82 in J. Grahame, Mill, P. J. & Reid, D. G. (Eds.) *Proceedings of the Third International Symposium* on Littorinid Biology, Dale Fort Field Centre, Wales, 5–12 September 1990. Vol. 3. London, Malacological Society of London.

Britton, J. C. & Morton, B. (1993). Are there obligate marine scavengers? pp. 357–391 in B. Morton (Ed.) The Marine Biology of the South China Sea. Proceedings of the First International Conference on the Marine Biology of Hong Kong and the South China Sea. Hong Kong, Hong Kong University Press.

Britton, J. C. & Morton, B. (1994). Food choice, detection, time spent feeding, and consumption by two species of subtidal Nassariidae from Monterey Bay, California. The Veliger 37: 81–92.

Britton, J. C. & Morton, B. (1989). Shore Ecology of the Gulf of Mexico. Austin, TX, University of Texas Press.

Broadhurst, C. L., Cunnane, S. C. & Crawford, M. A. (1998). Rift Valley lake fish and shellfish provided brain-specific nutrition for early Homo. British Journal of Nutrition 79: 3–21.

Brocco, S. L., O'Clair, R. M. & Cloney, R. A. (1974). Cephalopod integument: The ultrastructure of Kölliker's organ and their relationship to setae. Cell and Tissue Research 151: 293–308.

Brock, V. & Kennedy, V. S. (1992). Quantitative analysis of crystalline style carbohydrases in five suspension- and deposit-feeding bivalves. Journal of Experimental Marine Biology and Ecology 159: 51–58.

Brockington, S. & Clarke, A. (2001). The relative influence of temperature and food on the metabolism of a marine invertebrate. Journal of Experimental Marine Biology and Ecology 258: 87–99.

Brodersen, J., Chimbari, M. J. & Madsen, H. (2003). Prosobranch mollusc species- and size-preferences of Sargochromis codringtonii (Cichlidae) in Lake Kariba, Zimbabwe. African Journal of Aquatic Science 28: 179–182.

Brodeur, R. D., Lorz, H. V. & Pearcy, W. G. (1987). Food Habits and Dietary Variability of Pelagic Nekton off Oregon and Washington, 1979–1984. Seattle, WA, Northwest Fisheries Science Centre.

Broitman, B. R., Blanchette, C. A., Menge, B. A., Lubchenko, J., Krenz, C., Foley, M., Raimondi, P. T., Lohse, D. & Gaines, S. D. (2008). Spatial and temporal patterns of invertebrate recruitment along the west coast of the United States. Ecological Monographs 78: 403–421.

Brokordt, K. B., Guderley, H. E., Guay, M., Gaymer, C. F. & Himmelman, J. H. (2003). Sex differences in reproductive investment: Maternal care reduces escape response capacity in the whelk Buccinum undatum. Journal of Experimental Marine Biology and Ecology 291: 161–180.

Bromley, R. G. (1981). Concepts in ichnotaxonomy illustrated by small round holes in shells. Acta geológica hispánica 16: 55–64.

Bromley, R. G. & Heinberg, C. (2006). Attachment strategies of organisms on hard substrates: A palaeontological view. Palaeogeography, Palaeoclimatology, Palaeoecology 232: 429–453.

Brönmark, C. (1985). Interactions between macrophytes, epiphytes and herbivores: An experimental approach. Oikos 45: 26–30.

Brönmark, C. & Malmqvist, B. (1986). Interactions between the leech Glossiphonia complanata and its gastropod prey. Oecologia 69: 268–276.

Brönmark, C. (1992). Leech predation on juvenile freshwater snails: Effects of size, species and substrate. Oecologia 91: 526–529.

Bronn, H. G. (1862). Malacozoa, Abt. 1: Malacozoa Acephala [bivalves and brachiopods]. pp. 1–706 in H. G. Bronn (Ed.) Die Klassen und Ordnungen des Thier-Reichs. Vol. 3. Leipzig und Heidelberg, C. F. Winter.

Brooke, N. M., Garcia-Fernandez, J. & Holland, P. W. H. (1998). The ParaHox gene cluster is an evolutionary sister of the Hox gene cluster. Nature 392: 920–922.

Brooks, S. P. J. & Storey, K. B. (1997). Glycolytic controls in estivation and anoxia: A comparison of metabolic arrest in land and marine molluscs. Comparative Biochemistry and Physiology Part A 118: 1103–1114.

Broquet, T. & Petit, E. J. (2009). Molecular estimation of dispersal for ecology and population genetics. Annual Review of Ecology, Evolution, and Systematics 40: 193–216.

Bros, V., Moreno-Rueda, G. & Santos, X. (2011). Does postfire management affect the recovery of Mediterranean communities? The case study of terrestrial gastropods. Forest Ecology and Management 261: 611–619.

Brough, C. N. & White, K. N. (1990a). Localization of metals in the gastropod Littorina saxatilis (Prosobranchia: Littorinoidea) from a polluted site. Acta Zoologica 71: 77–88.

Brough, C. N. & White, K. N. (1990b). Functional morphology of the rectum in the marine gastropod Littorina saxatilis (Olivi) (Prosobranchia: Littorinoidea). Journal of Molluscan Studies 56: 97–108.

Brown, A. C. & Noble, R. G. (1960). Function of the osphradium in Bullia (Gastropoda). Nature 188: 1045.

Brown, A. C. (1964). Blood volumes blood distribution and seawater spaces in relation to expansion and retraction of foot in Bullia (Gastropoda). Journal of Experimental Biology 41: 837–854.

Brown, A. C. (1984). Oxygen-tensions in the pedal sinus of the whelk Bullia digitalis (Dillwyn). Journal of Molluscan Studies 50: 122.

Brown, A. C. & Webb, S. C. (1985). The dark response of the whelk Bullia digitalis (Dillwyn). Journal of Molluscan Studies 51: 351–352.

Brown, A. C., Trueman, E. R. & Stenton-Dozey, J. M. E. (1989). Gill size and respiratory requirement in the Mollusca with special reference to the prosobranch Gastropoda. South African Journal of Science 85: 126–127.

Brown, A. M. & Brown, H. M. (1973). Light response of a giant Aplysia neuron. Journal of General Physiology 62: 239–254.

Brown, D. S. (1994). Freshwater Snails of Africa and Their Medical Importance. London, Taylor and Francis.

Brown, D. S. (2002). Freshwater Snails of Africa and Their Medical Importance. London, Taylor & Francis.

Brown, H. H. (1934). A Study of a tectibranch gasteropod mollusc, Philine aperta (L.). Transactions of the Royal Society of Edinburgh, Earth and Environmental Science 58: 179–210.

Brown, K. M. (1985). Intraspecific life history variation in a pond snail: The roles of population divergence and phenotypic plasticity. Evolution 39: 387–395.

Brown, R. O. & Mayeri, E. (1989). Positive feedback by autoexcitatory neuropeptides in neuroendocrine bag cells of Aplysia. The Journal of Neuroscience 9: 1443–1451.

Brown, S. C., Cassuto, S. R. & Loos, R. W. (1979). Biomechanics of chelipeds in some decapod crustaceans. Journal of Zoology 188: 143–159.

Brown, T. (1837). The Conchologist's Text-book, Embracing the Arrangements [of] Lamarck and Linneaus, with a Glossary of Technical Terms. Glasgow, Archibald Fullarton & Co.

Brownell, P. H. & Ligman, S. H. (1992). Mechanisms of circulatory homeostasis and response in Aplysia. Experientia 48: 818–827.

Browning, J. & Casey-Smith, J. R. (1981). Tissue channel morphology in Octopus. Cell and Tissue Research 215: 153–170.

Browning, J. (1982). The density and dimensions of exchange vessels in *Octopus pallidus*. *Journal of Zoology* 196: 569–579.

Van Bruggen, A. C. (1995). Biodiversity of the Mollusca: Time for a new approach. pp. 1–19 *in* A. C. Van Bruggen, Wells, S. M. & Kemperman, T. C. M. (Eds.) *Biodiversity and Conservation of the Mollusca: Proceedings of the Alan Solem Memorial Symposium on the Biodiversity and Conservation of the Mollusca. Siena, Italy 1992. Eleventh International Malacological Congress.* Leiden, Backhuys Publishers.

Brusa, F., Ponce de León, R. & Damborenea, C. (2006). A new *Paravortex* (Platyhelminthes, Dalyellioida) endoparasite of *Mesodesma mactroides* (Bivalvia, Mesodesmatidae) from Uruguay. *Parasitology Research* 99: 566–571.

Brusca, R. C., Moore, W. & Shuster, S. M. (2016a). Introduction to the Bilateria and the Phylum Xenacoelomorpha. Triploblasty and bilateral symmetry provide new avenues for animal radiation. pp. 345–372 *in* R. C. Brusca, Moore, W. & Shuster, S. (Eds.) *Invertebrates.* Sunderland, MA, Sinauer Associates.

Brusca, R. C., Moore, W. & Shuster, S. M. (2016b). *Invertebrates.* Sunderland, MA, Sinauer Associates.

Bryan, P. J. & Qian, P.-Y. (1998). Induction of larval attachment and metamorphosis in the abalone *Haliotis diversicolor* (Reeve). *Journal of Experimental Marine Biology and Ecology* 223: 39–51.

Bubel, A. (1984). Mollusca: Epidermal cells. pp. 400–447 *in* J. Bereiter-Hahn, Matoltsy, A. B. & Richards, K. S. (Eds.) *Biology of the Integument: Invertebrates.* Vol. 1. Berlin, Springer-Verlag.

Buchanan, J. A. (1986). Ultrastructure of the larval eyes of *Hermissenda crassicornis* (Mollusca: Nudibranchia). *Journal of Ultrastructure and Molecular Structure Research* 94: 52–62.

Buchsbaum, R. (1951). *Animals without Backbones: An Introduction to the Invertebrates.* Harmondsworth, UK, Penguin Books.

Buck, B. H., Rosenthal, H. & Saint-Paul, U. (2002). Effect of increased irradiance and thermal stress on the symbiosis of *Symbiodinium microadriaticum* and *Tridacna gigas*. *Aquatic Living Resources* 15: 107–117.

Buckett, K. J. & Dockray, G. J. (1990). Pharmacology of the myogenic heart of the pond snail *Lymnaea stagnalis*. *Journal of Neurophysiology* 63: 1413–1425.

Buckett, K. J., Peters, M., Dockray, G. J., Van Minnen, J. & Benjamin, P. R. (1990). Regulation of heartbeat in *Lymnaea* by motoneurons containing FMRFamide-like peptides. *Journal of Neurophysiology* 63: 1426–1435.

Buckland-Nicks, J. A. & Chia, F.-S. (1976a). Fine structural observations of sperm resorption in the seminal vesicle of a marine snail *Littorina scutulata*. *Cell and Tissue Research* 172: 503–516.

Buckland-Nicks, J. A. & Chia, F.-S. (1976b). Spermatogenesis of a marine snail *Littorina sitkana*. *Cell and Tissue Research* 170: 455–476.

Buckland-Nicks, J. A., Chia, F.-S. & Koss, R. (1990). Spermiogenesis in Polyplacophora, with special reference to acrosome formation (Mollusca). *Zoomorphology (Berlin)* 109: 179–188.

Buckland-Nicks, J. A. & Scheltema, A. H. (1995). Was internal fertilization an innovation of early Bilateria? Evidence from sperm structure of a mollusc. *Proceedings of the Royal Society B* 261: 11–18.

Buckland-Nicks, J. A. (1998). Prosobranch parasperm: Sterile germ cells that promote paternity? *Micron* 29: 267–280.

Buckland-Nicks, J. A., Bryson, I., Hart, L. & Partridge, V. (1999). Sex and a snail's sperm: On the transport, storage and fate of dimorphic sperm in Littorinidae. *Invertebrate Reproduction & Development* 36: 145–152.

Buckland-Nicks, J. A., Healy, J. M., Jamieson, B. G. M. & O'Leary, S. (2000). Paraspermatogenesis in *Littoraria (Palustorina) articulata*, with reference to other Littorinidae (Littorinoidea, Caenogastropoda). *Invertebrate Biology* 119: 254–264.

Buckland-Nicks, J. A., Gibson, G. & Koss, R. (2002a). Phylum Mollusca: Gastropoda. pp. 261–287 *in* C. M. Young (Ed.) *Atlas of Marine Invertebrate Larvae.* San Diego, CA, Academic Press.

Buckland-Nicks, J. A., Gibson, G. D. & Koss, R. (2002b). Phylum Mollusca: Polyplacophora, Aplacophora, Scaphopoda. pp. 245–259 *in* C. M. Young (Ed.) *Atlas of Marine Invertebrate Larvae.* San Diego, CA, Academic Press.

Buckland-Nicks, J. A. & Reunov, A. A. (2010). Egg hull formation in *Callochiton dentatus* (Mollusca, Polyplacophora): The contribution of microapocrine secretion. *Invertebrate Biology* 129: 319–327.

Buckland-Nicks, J. A., Chisholm, S. A. & Gibson, G. (2013). The living community inside the common periwinkle, *Littorina littorea*. *Canadian Journal of Zoology* 91: 293–301.

Buczek, O., Bulaj, G. & Olivera, B. M. (2005). Conotoxins and the posttranslational modification of secreted gene products. *Cellular and Molecular Life Sciences* 62: 3067–3079.

Von Buddenbrock, W. (1915). Die Statocyste von Pecten, ihre Histologie und Physiologie. *Zoologische Jahrbücher. Abteilung für Allgemeine Zoologie und Physiologie der Tiere* 35: 301–356.

Budelmann, B.-U. (1970). Investigations on the function of the statolith organs of *Octopus vulgaris*. *Verhandlungen der Deutschen Zoologischen Gesellschaft* 64: 256–261.

Budelmann, B.-U. & Young, J. Z. (1984). The statocyst oculomotor system of *Octopus vulgaris*: Extraocular eye muscles, eye muscle nerves, statocyst nerves and the oculomotor center in the central nervous system. *Philosophical Transactions of the Royal Society B* 306: 159–190.

Budelmann, B.-U. (1988). Morphological diversity of equilibrium receptor systems in aquatic invertebrates. pp. 757–782 *in* J. Atema, Fay, R. R., Popper, A. N. & Tavolga, W. N. (Eds.) *Sensory Biology of Aquatic Animals.* New York, Springer-Verlag.

Budelmann, B.-U. & Bleckmann, H. (1988). A lateral line analogue in cephalopods: Water waves generate microphonic potentials in the epidermal head lines of *Sepia* and *Lolliguncula*. *Journal of Comparative Physiology A* 164: 1–5.

Budelmann, B.-U. (1990). The statocysts of squid. pp. 421–439 *in* D. L. Gilbert, Adelman, W. J. & Arnold, J. M. (Eds.) *Squid as Experimental Animals.* New York, Plenum Press.

Budelmann, B.-U., Riese, U. & Bleckmann, H. (1991). Structure, function, biological significance of the cuttlefish 'lateral lines'. pp. 201–209 *in* E. Boucaud-Camou (Ed.) *La Seiche: Actes du Premier Symposium International sur la Seiche, Caen, 1–3 Juin 1989 (The Cuttlefish: 1st International Symposium on the Cuttlefish Sepia, Caen, June 1–3 1989).* Caen, Centre de Publications de l'Université de Caen.

Budelmann, B.-U. & Young, J. Z. (1993). The oculomotor system of decapod cephalopods: Eye muscles, eye muscle nerves, and the oculomotor neurons in the central nervous system. *Philosophical Transactions of the Royal Society B* 340: 93–125.

Budelmann, B.-U. (1994). Cephalopod sense organs, nerves and the brain: Adaptations for high performance and lifestyle. pp. 13–33 *in* H. O. Pörtner, O'Dor, R. K. & Macmillan, D. L. (Eds.) *Physiology of Cephalopod Molluscs: Lifestyle and Performance Adaptations.* Basel, Gordon and Breach Science Publishers SA.

Budelmann, B.-U. (1975). Gravity receptor function in cephalopods with particular reference to *Sepia officinalis*. *Fortschritte der Zoologie* 23: 84–96.

Budelmann, B.-U. (1977). Structure and function of the angular acceleration receptor systems in the statocysts of cephalopods. pp. 309–324 *in* M. Nixon & Messenger, J. B. (Eds.) *The Biology of Cephalopods. Symposia of The Zoological Society of London*. London, Academic Press.

Budelmann, B.-U., Sachse, M. & Staudigl, M. (1987). The angular acceleration receptor system of the statocyst of *Octopus vulgaris*: Morphometry, ultrastructure, and neuronal and synaptic organization. *Philosophical Transactions of the Royal Society B* 315: 305–344.

Budelmann, B.-U. (1995). The cephalopod nervous system: What evolution has made of the molluscan design. pp. 115–138 *in* O. Breïdbach & Kutsch, W. (Eds.) *The Nervous Systems of Invertebrates: An Evolutionary and Comparative Approach*. Basel, Birkhäuser Verlag.

Budelmann, B.-U. (1996). Active marine predators: The sensory world of cephalopods. *Marine and Freshwater Behaviour and Physiology* 27: 59–75.

Budelmann, B.-U., Schipp, R. & von Boletzky, S. (1997). Cephalopoda. pp. 119–414 *in* F. W. Harrison & Kohn, A. J. (Eds.) *Microscopic Anatomy of Invertebrates: Mollusca 2. Mollusca*. Vol. 6A. New York, Wiley-Liss.

Budelmann, B.-U. & Tu, Y. (1997). The statocyst-oculomotor reflex of cephalopods and the vestibulo-oculomotor reflex of vertebrates: A tabular comparison. *Vie et Milieu* 47: 95–99.

Bull, J. J. (1983). *Evolution of Sex Determining Mechanisms*. Vol. 1. Menlo Park, CA, The Benjamin/Cummings Publishing Company.

Bullock, T. H. (1953). Predator recognition and escape responses of some intertidal gastropods in presence of starfish. *Behaviour* 5: 1–11.

Bullock, T. H. & Horridge, G. A. (1965). The Mollusca. pp. 1273–1515 *in* T. H. Bullock, Horridge, G. A. & Freeman, W. H. (Eds.) *Structure and Function in the Nervous Systems of Invertebrates*. Vol. 2. San Francisco, London, W. H. Freeman & Co.

Bullock, T. H. (1990). Goals of neuroethology *Bioscience* 40: 244–248.

Bunje, P. M. E. (2007). Fluvial range expansion, allopatry, and parallel evolution in a Danubian snail lineage (Neritidae: *Theodoxus*). *Biological Journal of the Linnean Society* 90: 603–617.

Burbach, J. P. H., Hellemons, A. J. C. G. M., Hoekman, M., Grant, P. & Pant, H. C. (2001). The stellate ganglion of the squid *Loligo pealeii* as a model for neuronal development: Expression of a POU class VI homeodomain gene, Rpf-1. *Biological Bulletin* 201: 252–254.

Burch, J. B. (1960). Chromosome morphology of aquatic pulmonate snails (Mollusca, Pulmonata). *Transactions of the American Microscopical Society* 79: 451–461.

Burch, J. B. (1962). Chromosome numbers and systematics in euthyneuran snails. *Malacologia* 1: 215–241.

Burch, J. B. & Natarajan, R. (1965). Cytological studies of Taiwan freshwater pulmonate snails. *Bulletin. Institute of Zoology. Acadamia Sinica (Taipei)* 4: 11–17.

Burch, J. B. & Huber, J. M. (1966). Polyploidy in mollusks. *Malacologia* 5: 41–43.

Burch, J. B. (1975). *Freshwater Unionacean Clams (Mollusca: Pelecypoda) of North America*. Washington, DC, Hamburg, MI, U.S. Environmental Protection Agency (Malacological Publications).

Burch, J. B. (1982). North American freshwater snails V-X. V. Keys to the freshwater gastropods of North America. VI. Generic Synonomy. VII. Supplemental Notes. VIII. Glossary. IX. References. X. Index. *Walkerana* 1: 217–365.

Burch, J. B. (1988). North American freshwater snails. I–III. I. Introduction. II. Systematics, nomenclature, identification and morphology. III. Habitats and distribution. *Walkerana* 2: 1–80.

Burger, A. E. (1982). Foraging behaviour of lesser sheathbills *Chionis minor* exploiting invertebrates on a sub-Antarctic island. *Oecologia* 52: 236–245.

Burgess, C. M. (1970). *The Living Cowries*. New York, A.S. Barnes.

Burghardt, I., Evertsen, J., Johnsen, G. & Wägele, H. (2005). Solar powered seaslugs: Mutualistic symbiosis of aeolid Nudibranchia (Mollusca, Gastropoda, Opisthobranchia) with *Symbiodinium*. *Symbiosis* 38: 227–250.

Burghardt, I., Schrödl, M. & Wägele, H. (2008a). Three new solar-powered species of the genus *Phyllodesmium* Ehrenberg, 1831 (Mollusca: Nudibranchia: Aeolidioidea) from the tropical Indo-Pacific, with analysis of their photosynthetic activity and notes on biology. *Journal of Molluscan Studies* 74: 277–292.

Burghardt, I., Stemmer, K. & Wägele, H. (2008b). Symbiosis between *Symbiodinium* (Dinophyceae) and various taxa of Nudibranchia (Mollusca: Gastropoda), with analyses of long-term retention. *Organisms Diversity & Evolution* 8: 66–76.

Burighel, P., Caicci, F. & Manni, L. (2011). Hair cells in non-vertebrate models: Lower chordates and molluscs. *Hearing Research* 273: 14–24.

Burke, W. R. (1964). Chemoreception by *Tegula funebralis* (Mollusca). *The Veliger* 6 Suppl: 17–20.

Burla, H. & Gosteli, M. (1993). Thermal advantage of pale coloured morphs of the snail *Arianta arbustorum* (Helicidae, Pulmonata) in alpine habitats. *Ecography* 16: 345–350.

Burn, R. (1976). Shell with a built-in nest. *Australian Shell News* 16: 3.

Burreson, E. M., Stokes, N. & Friedman, C. (2000). Increased virulence in an introduced pathogen: *Haplosporidium nelsoni* (MSX) in the eastern oyster *Crassostrea virginica*. *Journal of Aquatic Animal Health* 12: 1–8.

Burreson, E. M. (2001). Spore ornamentation of *Haplosporidium pickfordi* (Haplosporidia), a parasite of freshwater snails in Michigan, USA. *Journal of Eukaryotic Microbiology* 48: 622–626.

Burreson, E. M. & Ford, S. E. (2004). A review of recent information on the Haplosporidia, with special reference to *Haplosporidium nelsoni* (MSX disease). *Aquatic Living Resources* 17: 499–517.

Burrows, E. M. & Lodge, S. M. (1950). A note on the inter-relationships of *Patella*, *Balanus* and *Fucus* on a semi-exposed coast. *Journal of the Marine Biological Association of the United Kingdom* 62: 30–34.

Burrows, M. (1969). The mechanics and neural control of the prey capture strike in the mantid shrimps *Squilla* and *Hemisquilla*. *Journal of Comparative Physiology A* 62: 361–381.

Burton, D. J. F. (1971). The functional anatomy of the feeding apparatus of some selected members of Stenoglossa (prosobranch gastropods). PhD, University of Reading.

Burton, R. F. (1965). Relationship between the cation contents of slime and blood in the snail *Helix pomatia* L. *Comparative Biochemistry and Physiology* 15: 339–345.

Burton, R. F. (1983). Ionic regulation and water balance. pp. 291–352 *in* A. S. M. Saleuddin & Wilbur, K. M. (Eds.) *Physiology, Part 2. The Mollusca*. Vol. 5. New York, Academic Press.

Burzynski, A. (2003). Evidence for recombination of mtDNA in the marine mussel *Mytilus trossulus* from the Baltic. *Molecular Biology and Evolution* 20: 388–392.

Buschbaum, C. & Reise, K. (1999). Effects of barnacle epibionts on the periwinkle *Littorina littorea* (L.). *Helgoland Marine Research* 53: 56–61.

Buschbaum, C. & Saier, B. (2001). Growth of the mussel *Mytilus edulis* L. in the Wadden Sea affected by tidal emergence and barnacle epibionts. *Journal of Sea Research* 45: 27–36.

Bush, M. S. (1988). The ultrastructure and function of the intestine of *Patella vulgata*. *Journal of Zoology* 215: 685–702.

Bush, M. S. (1989). The ultrastructure and function of the oesophagus of *Patella vulgata* Linnaeus. *Journal of Molluscan Studies* 55: 111–124.

Bush, S. L. (2012). Economy of arm autotomy in the mesopelagic squid *Octopoteuthis deletron*. *Marine Ecology Progress Series* 458: 133–140.

Bustnes, J. O. & Erikstad, K. E. (1990). Size selection of common mussels, *Mytilus edulis,* by common eiders, *Somateria mollissima:* Energy maximization or shell weight minimization? *Canadian Journal of Zoology* 68: 2280–2283.

Butler, P. G., Richardson, C. A., Scourse, J. D., Wanamaker, A. D., Shammon, T. M. & Bennell, J. D. (2010). Marine climate in the Irish Sea: Analysis of a 489-year marine master chronology derived from growth increments in the shell of the clam *Arctica islandica*. *Quaternary Science Reviews* 29: 1614–1632.

Butler, P. G., Wanamaker, A. D., Scourse, J. D., Richardson, C. A. & Reynolds, D. J. (2013). Variability of marine climate on the North Icelandic Shelf in a 1357-year proxy archive based on growth increments in the bivalve *Arctica islandica*. *Palaeogeography, Palaeoclimatology, Palaeoecology* 373: 141–151.

Butler, T. M. & Siegman, M. J. (2010). Mechanism of catch force: Tethering of thick and thin filaments by twitchin. *Journal of Biomedicine and Biotechnology* 2010: 725207 (725201–725220).

Byern, J., von, Wani, R., Schwaha, T., Grunwald, I. & Cyran, N. (2012). Old and sticky-adhesive mechanisms in the living fossil *Nautilus pompilius* (Mollusca, Cephalopoda). *Zoology* 115: 1–11.

Byers, B. A. (1989). Habitat-choice polymorphism associated with cryptic shell-color polymorphism in the limpet *Lottia digitalis*. *The Veliger* 32: 394–402.

Byrne, P. A. & O'Halloran, J. (2001). The role of bivalve molluscs as tools in estuarine sediment toxicity testing: A review. *Hydrobiologia* 465: 209–217.

Byrne, R. A., McMahon, R. F. & Dietz, T. H. (1989). The effects of aerial exposure and subsequent reimmersion on hemolymph osmolality, ion compositon, and ion flux in the freshwater bivalve *Corbicula fluminea*. *Physiological Zoology* 62: 1187–1202.

Byrne, R. A., Griebel, U., Wood, J. B. & Mather, J. A. (2003). Squid say it with skin: A graphic model for skin displays in Caribbean Reef Squid (*Sepioteuthis sepioidea*). pp. 29–35 *in* K. Warnke, Keupp, H. & von Boletzky, S. (Eds.) *Coleoid Cephalopods through Time*. Vol. 3. Berlin, Berliner Paläobiologische Abhandlungen.

Byrne, R. A., Kuba, M. J., Meisel, D. V., Griebel, U. & Mather, J. A. (2006). *Octopus* arm choice is strongly influenced by eye use. *Behavioural Brain Research* 172: 195–201.

Byrum, C. A. & Ruppert, E. E. (1994). The ultrastructure and functional morphology of a captaculum in *Graptacme calamus* (Mollusca, Scaphopoda). *Acta Zoologica* 75: 37–46.

Cáceres-Martínez, J., Vásquez-Yeomans, R. & Sluys, R. (1998). The turbellarian *Urastoma cyprinae* (Platyhelminthes: Urastomidae) associated with natural and commercial populations of *Mytilus galloprovincialis* and *Mytilus californianus* from Baja California, NW México. *Journal of Invertebrate Pathology* 72: 214–219.

Caddy-Retalic, S., Benkendorff, K. & Fairweather, P. G. (2011). Visualizing hotspots: Applying thermal imaging to monitor internal temperatures in intertidal gastropods. *Molluscan Research* 31: 106–113.

Caddy, J. F. & Defeo, O. (2003). *Enhancing or Restoring the Productivity of Natural Populations of Shellfish and other Marine Invertebrate Resources*. Rome, Food & Agriculture Organization, Fisheries Technical Paper 488.

Cadee, G. C., Boon, J. P., Fischer, C. V., Mensink, B. P. & Ten Hallers-Tjabbes, C. C. (1995). Why the whelk (*Buccinum undatum*) has become extinct in the Dutch Wadden Sea. *Netherlands Journal of Sea Research* 34: 337–339.

Cadotte, M. W., Carscadden, K. & Mirotchnick, N. (2011). Beyond species: Functional diversity and the maintenance of ecological processes and services. *Journal of Applied Ecology* 48: 1079–1087.

Cain, A. J. & Sheppard, P. M. (1950). Selection in the polymorphic land snail *Cepaea nemoralis*. *Heredity* 4: 275–294.

Cain, A. J. & Sheppard, P. M. (1952). The effects of natural selection on body colour in the land snail *Cepaea nemoralis*. *Heredity* 6: 217–231.

Cain, A. J. & Sheppard, P. M. (1954). Natural selection in *Cepaea*. *Genetics* 39: 89–116.

Cain, A. J. (1977). Variation in the spire index of some coiled gastropod shells, and its evolutionary significance. *Philosophical Transactions of the Royal Society B* 277: 377–428.

Cain, A. J. (1988a). The scoring of polymorphic colour and pattern variation and its genetic basis in molluscan shells. *Malacologia* 28: 1–15.

Cain, A. J. (1988b). The colours of marine bivalve shells with special reference to *Macoma balthica*. *Malacologia* 28: 289–318.

Cain, S. D., Boles, L. C., Wang, J. H. & Lohmann, K. J. (2005). Magnetic orientation and navigation in marine turtles, lobsters, and molluscs: Concepts and conundrums. *Integrative and Comparative Biology* 45: 539–546.

Cajaraville, M. P., Cancio, I., Ibabe, A. & Orbea, A. (2003). Peroxisome proliferation as a biomarker in environmental pollution assessment. *Microscopy Research and Technique* 61: 191–202.

Calder, W. A. (1996). *Size, Function, and Life History*. Mineola, NY, Dover Publications.

Calderon-Aguilera, L. E., Aragón-Noriega, E. A., Morales-Bojórquez, E., Alcántara-Razo, E. & Chávez-Villalba, J. (2014). Reproductive cycle of the geoduck clam *Panopea generosa* at its southernmost distribution limit. *Marine Biology Research* 10: 61–72.

Caldwell, R. L. (2005). An observation of inking behavior protecting adult *Octopus bocki* from predation by green turtle (*Chelonia mydas*) hatchlings. *Pacific Science* 59: 69–72.

Caley, M. J., Carr, M. H., Hixon, M. A., Hughes, T. P., Jones, G. P. & Menge, B. A. (1996). Recruitment and the local dynamics of open marine populations. *Annual Review of Ecology and Systematics* 27: 477–500.

Callaerts, P., Lee, P. N., Hartmann, B., Farfan, C., Choy, D. W. Y., Ikeo, K., Fischbach, K.-F., Gehring, W. J. & de Couet, H. G. (2002). HOX genes in the sepiolid squid *Euprymna scolopes*: Implications for the evolution of complex body plans. *Proceedings of the National Academy of Sciences of the United States of America* 99: 2088–2093.

Callender, W. R. & Powell, E. N. (1992). Taphonomic signature of petroleum seep assemblages on the Louisiana upper continental slope: Recognition of autochthonous shell beds in the fossil record. *Palaios* 7: 388–408.

Callender, W. R., Powell, E. N., Staff, G. M. & Davies, D. J. (1992). Distinguishing autochthony, parautochthony and allochthony using taphofacies analysis: Can cold seep assemblages be discriminated from assemblages of the nearshore and continental shelf? *Palaios* 7: 409–421.

Callomon, J. (1969). Dimorphism in Jurassic ammonites. pp. 111–121 *in* G. E. G. Westermann (Ed.) *Sexual Dimorphism in Fossil Metazoa and Taxonomic Implications.* Vol. 1. Stuttgart, E. Schweizerbart'sche Verlagsbuchhandlung.

Callomon, J. H. (1963). Sexual dimorphism in Jurassic ammonites. *Transactions of the Leicester Literary and Philosophical Society* 57: 21–56.

Calow, P. (1975). The feeding strategies of two freshwater gastropods, *Ancylus fluviatilis* Müll. and *Planorbis contortus* Linn. (Pulmonata), in terms of ingestion rates and absorption efficiencies. *Oecologia* 20: 33–49.

Calow, P. & Calow, L. J. (1975). Cellulase activity and niche separation in freshwater gastropods. *Nature* 255: 478–480.

Calvo, M. & Templado, J. (2005). Reproduction and sex reversal of the solitary vermetid gastropod *Serpulorbis arenarius*. *Marine Biology* 146: 963–973.

Cameron, R. A. D., Cook, L. M. & Greenwood, J. J. D. (2013a). Change and stability in a steep morph-frequency cline in the snail *Cepaea nemoralis* (L.) over 43 years. *Biological Journal of the Linnean Society* 108: 473–483.

Cameron, R. A. D., Triantis, K. A., Parent, C. E., Guilhaumon, F., Alonso, M. R., Ibáñez, M., Martins, A. M., de Frias, Ladle, R. J. & Whittaker, R. J. (2013b). Snails on oceanic islands: Testing the general dynamic model of oceanic island biogeography using linear mixed effect models. *Journal of Biogeography* 40: 117–130.

Campbell, J. L. (1965). The structure and function of the alimentary tract of the black abalone, *Haliotis cracherodii*. *Transactions of the American Microscopical Society* 84: 376–395.

Campbell, J. W. & Speeg, K. V. (1968). Arginine biosynthesis and metabolism in terrestrial snails. *Comparative Biochemistry and Physiology* 25: 3–32.

Campbell, J. W. & Boyan, B. D. (1974). On the acid-base balance of gastropod molluscs. pp. 109–133 *in* N. Watabe & Wilbur, K. M. (Eds.) *The Mechanisms of Mineralization in the Invertebrates and Plants.* Columbia, SC, University of South Carolina Press.

Campion, M. (1961). The structure and function of the cutaneous glands in *Helix aspersa*. *Quarterly Journal of Microscopical Science* 102: 195–216.

Camus, P. A. & Barahona, R. M. (2002). Insectos del intermareal de Concepción, Chile: Perspectivas para la investigación ecológica [Intertidal insects from Concepción, Chile: Perspectives for ecological research] *Revista Chilena de Historia Natural* 75: 793–803.

Cancio, I., Völkl, A., Beier, K., Fahimi, H. D. & Cajaraville, M. P. (2000). Peroxisomes in molluscs, characterization by subcellular fractionation combined with western blotting, immunohistochemistry, and immunocytochemistry. *Histochemistry and Cell Biology* 113: 51–60.

Canesi, L., Betti, M., Ciacci, C., Lorusso, L. C., Pruzzo, C. & Gallo, G. (2006). Cell signalling in the immune response of mussel hemocytes. *Invertebrate Survival Journal* 3: 40–49.

Canete, J. I., Gallardo, C. S., Cespedes, T., Cardenas, C. A. & Santana, M. (2012). Encapsulated development, spawning and early veliger of the ranellid snail *Fusitriton magellanicus* (Röding, 1798) in the cold waters of the Magellan Strait, Chile. *Latin American Journal of Aquatic Research* 40: 914–928.

Canning, E. U. & Basch, P. F. (1968). *Perezia helminthorum* sp. nov., a microsporidian hyperparasite of trematode larvae from Malaysian snails. *Parasitology* 58: 341–347.

Canning, E. U., Foon, L. P. & Joe, L. K. (1974). Microsporidian parasites of trematode larvae from aquatic snails in West Malaysia. *The Journal of Protozoology* 21: 19–25.

Cannuel, R. & Beninger, P. G. (2006). Gill development, functional and evolutionary implications in the Pacific oyster *Crassostrea gigas* (Bivalvia: Ostreidae). *Marine Biology* 149: 547–563.

Cantrell, C. L., Groweiss, A., Gustafson, K. R. & Boyd, M. R. (1999). A new staurosporine analog from the prosobranch mollusk *Coriocella nigra*. *Natural Product Letters* 14: 39–46.

Capt, C., Renaut, S., Ghiselli, F., Milani, L., Johnson, N. A., Sietman, B. E., Stewart, D. T. & Breton, S. (2018). Deciphering the link between doubly uniparental inheritance of mtDNA and sex determination in bivalves: Clues from comparative transcriptomics. *Genome Biology and Evolution* 10: 577–590.

Carballeira, N. M., Cruz, H., Hill, C. A., De Voss, J. J. & Garson, M. J. (2001). Identification and total synthesis of novel fatty acids from the siphonarid limpet *Siphonaria denticulata*. *Journal of Natural Products* 64: 1426–1429.

Cardoso, A. M., Cavalcante, J. J. V., Vieira, R. P., Lima, J. L., Grieco, M. A. B., Clementino, M. M., Vasconcelos, A. T. R., Garcia, E. S., de Souza, W., Albano, R. M. & Martins, O. B. (2012). Gut bacterial communities in the giant land snail *Achatina fulica* and their modification by sugarcane-based diet. *PLoS ONE* 7: e33440 (33441–33446).

Cardoso, J. F. M. F., Witte, J. I. & Van der Veer, H. W. (2009). Differential reproductive strategies of two bivalves in the Dutch Wadden Sea. *Estuarine, Coastal and Shelf Science* 84: 37–44.

Carefoot, T. H. (1987). *Aplysia*: Its biology and ecology. *Oceanography and Marine Biology Annual Review* 25: 167–284.

Carefoot, T. H., Karentz, D., Pennings, S. C. & Young, C. L. (2000). Distribution of mycosporine-like amino acids in the sea hare *Aplysia dactylomela*: Effect of diet on amounts and types sequestered over time in tissues and spawn. *Comparative Biochemistry and Physiology Part C* 126: 91–104.

Carew, T. J., Pinsker, H., Rubinson, K. & Kandel, E. R. (1974). Physiological and biochemical properties of neuromuscular transmission between identified mononeurons and gill muscle in *Aplysia*. *Journal of Neurophysiology* 37: 1020–1040.

Carew, T. J. & Kandel, E. R. (1977). Inking in *Aplysia californica*. *Journal of Neurophysiology* 40: 692–707.

Carey, N., Galkin, A., Henriksson, P., Richards, J. G. & Sigwart, J. D. (2012). Variation in oxygen consumption among 'living fossils' (Mollusca: Polyplacophora). *Journal of the Marine Biological Association of the United Kingdom* 93: 197–207.

Carlos, A. A., Baillie, B. K. & Maruyama, T. (2000). Diversity of dinoflagellate symbionts (zooxanthellae) in a host individual. *Marine Ecology Progress Series* 195: 93–100.

Carlton, J. T. (1976). Marine plant limpets of the northeastern Pacific; patterns of host utilization and comparative plant-limpet distributions. *Western Society of Malacologists Annual Report* 9: 22–25.

Carlton, J. T., Vermeij, G. J., Lindberg, D. R., Carlton, D. A. & Dudley, E. C. (1991). The first historical extinction of a marine invertebrate in an ocean basin: The demise of the eelgrass limpet *Lottia alveus*. *Biological Bulletin* 180: 72–80.

Carlton, J. T. (1992). Introduced marine and estuarine mollusks of North America: An end-of-the-20th-century perspective. *Journal of Shellfish Research* 11: 489–505.

Carlton, J. T. & Hodder, J. (2003). Maritime mammals: Terrestrial mammals as consumers in marine intertidal communities. *Marine Ecology Progress Series* 256: 271–286.

Carlton, J. T., Chapman, J. W., Geller, J. B., Miller, J. A., Carlton, D. A., McCuller, M. I., Treneman, N. C., Steves, B. P. & Ruiz, G. M. (2017). Tsunami-driven rafting: Transoceanic species dispersal and implications for marine biogeography. *Science* 357: 1402–1406.

Carmichael, R. H., Rutecki, D., Annett, B., Gaines, E. & Valiela, I. (2004). Position of horseshoe crabs in estuarine food webs: N and C stable isotopic study of foraging ranges and diet composition. *Journal of Experimental Marine Biology and Ecology* 299: 231–253.

Carnegie, R. B., Hill, K. M., Stokes, N. A. & Burreson, E. M. (2014). The haplosporidian *Bonamia exitiosa* is present in Australia, but the identity of the parasite described as *Bonamia* (formerly *Mikrocytos*) *roughleyi* is uncertain. *Journal of Invertebrate Pathology* 115: 33–40.

Carr, A. S., Bateman, M. D., Roberts, D. L., Murray-Wallace, C. V., Jacobs, Z. & Holmes, P. J. (2010). The last interglacial sea-level high stand on the southern Cape coastline of South Africa. *Quaternary Research* 73: 351–363.

Carr, W. E. S. (1967). Chemoreception in the mud snail, *Nassarius obsoletus*. 1. Properties of stimulatory substances extracted from shrimp. *Biological Bulletin* 133: 90–105.

Carrasco, N., Andree, K. B., Lacuesta, B., Roque, A., Rodgers, C. & Furones, M. D. (2012a). Molecular characterization of the *Marteilia* parasite infecting the common edible cockle *Cerastoderma edule* in the Spanish Mediterranean coast: A new *Marteilia* species affecting bivalves in Europe? *Aquaculture* 324: 20–26.

Carrasco, N., Villalba, A., Andree, K. B., Engelsma, M. Y., Lacuesta, B., Ramilo, A., Gairín, I. & Furones, M. D. (2012b). *Bonamia exitiosa* (Haplosporidia) observed infecting the European flat oyster *Ostrea edulis* cultured on the Spanish Mediterranean coast. *Journal of Invertebrate Pathology* 110: 307–313.

Carré, M., Bentaleb, I., Blamart, D., Ogle, N., Cardenas, F., Zevallos, S., Kalin, R. M., Ortlieb, L. & Fontugne, M. (2005a). Stable isotopes and sclerochronology of the bivalve *Mesodesma donacium*: Potential application to Peruvian paleoceanographic reconstructions. *Palaeogeography, Palaeoclimatology, Palaeoecology* 228: 4–25.

Carré, M., Bentaleb, I., Fontugne, M. & Lavallee, D. (2005b). Strong El Niño events during the early Holocene: Stable isotope evidence from Peruvian sea shells. *The Holocene* 15: 42–47.

Carreau, S. & Drosdowsky, M. (1977). The in vitro biosynthesis of steroids by the gonad of the cuttlefish (*Sepia officinalis* L.). *General and Comparative Endocrinology* 33: 554–565.

Carrick, N. (1980). Aspects of the biology of *Gazameda gunni* (Reeve, 1849), a viviparous mesogastropod and potential 'indicator' of perturbation induced by sewage pollution. *Journal of the Malacological Society of Australia* 4: 254–255.

Carriker, M. R. (1943). On the structure and function of the proboscis in the common oyster drill *Urosalpinx cinerea* Say. *Journal of Morphology* 73: 441–506.

Carriker, M. R. & Bilstad, N. M. (1946). Histology of the alimentary system of the snail *Lymnaea stagnalis apressa* Say. *Transactions of the American Microscopical Society* 65: 250–275.

Carriker, M. R. (1972). Observations on the removal of spines by muricid gastropods during shell growth. *The Veliger* 15: 69–74.

Carriker, M. R. (1981). Shell penetration and feeding by naticacean and muricacean predatory gastropods: A synthesis. *Malacologia* 20: 403–422.

Carrillo-Baltodano, A. & Collin, R. (2015). *Crepidula* slipper limpets alter sex change in response to physical contact with conspecifics. *Biological Bulletin* 229: 232–242.

Carro, B., Quintela, M., Ruiz, J. M. & Barreiro, R. (2012). AFLPs reveal different population genetic structure under contrasting environments in the marine snail *Nucella lapillus* L. *PLoS ONE* 7: e49776.

Carter, J. A. & Steele, D. H. (1982). Stomach contents of immature lobsters (*Homarus americanus*) from Placentia Bay, Newfoundland. *Canadian Journal of Zoology* 60: 337–347.

Carter, J. G. & Clark, G. R. (1985). Classification and phylogenetic significance of molluscan shell microstructure. pp. 50–71 in T. W. Broadhead (Ed.) *Mollusks: Notes for a Short Course organized by D.J. Bottjer, C.S. Hickman, and P.D. Ward*. University of Tennessee Studies in Geology. Knoxville, TN, University of Tennessee Department of Geological Science.

Carter, J. G., Ed. (1990a). *Skeletal Biomineralization: Patterns, Processes and Evolutionary Trends [Vol. I]: Atlas and Index [Vol. II]*. New York, Van Nostrand Reinhold.

Carter, J. G. (1990b). Shell microstructural data for the Bivalvia: Part V. Order Pectinoida. pp. 363–389 in J. G. Carter (Ed.) *Skeletal Biomineralization: Patterns, Processes and Evolutionary Trends*. Vol. 1. New York, Van Nostrand Reinhold.

Carter, J. G. & Hall, R. M. (1990). Polyplacophora, Scaphopoda, Archaeogastropoda and Paragastropoda (Mollusca). pp. 29–51 in J. G. Carter (Ed.) *Skeletal Biomineralization: Patterns, Processes and Evolutionary Trends*. Vol. 2. New York, Van Nostrand Reinhold.

Carter, J. G. & Schneider, J. A. (1997). Condensing lenses and shell microstructure in *Corculum* (Mollusca: Bivalvia). *Journal of Paleontology* 71: 56–61.

Carter, J. G., Altaba, C. R., Anderson, L. C., Araujo, R., Biakov, A. S., Bogan, A. E., Campbell, C. et al.. (2011). A synoptical classification of the Bivalvia (Mollusca). *Paleontological Contributions* 4: 1–47.

Cartron, L., Darmaillacq, A. S., Jozet-Alves, C., Shashar, N. & Dickel, L. (2012). Cuttlefish rely on both polarized light and landmarks for orientation. *Animal Cognition* 15: 591–596.

Cary, C., Fry, B., Felbeck, H. & Vetter, R. D. (1989). Multiple trophic resources for a chemoautotrophic community at a cold water brine seep at the base of the Florida Escarpment. *Marine Biology* 100: 411–418.

Castellanos-Martínez, S. (2013). Coccidiosis and molecular basis of the immune response of common octopus (*Octopus vulgaris* Cuvier, 1797). PhD, Universida de Vigo, Spain.

Castellanos-Martínez, S. & Gestal, C. (2013). Pathogens and immune response of cephalopods. *Journal of Experimental Marine Biology and Ecology* 447: 14–22.

Castellanos-Martínez, S., Sheila, Diz, A. P., Alvarez-Chaver, P. & Gestal, C. (2014). Proteomic characterization of the hemolymph of *Octopus vulgaris* infected by the protozoan parasite *Aggregata octopiana*. *Journal of Proteomics* 105: 151–163.

Castilla, J. C. & Cancino, J. (1976). Spawning behaviour and egg capsules of *Concholepas concholepas* (Mollusca: Gastropoda: Muricidae). *Marine Biology* 37: 255–263.

Castillo, M. G., Salazar, K. A. & Joffe, N. R. (2015). The immune response of cephalopods from head to foot. *Fish and Shellfish Immunology* 46: 145–160.

Castle, S. L. & Emery, A. E. H. (1981). *Nucella lapillus*: A possible model for the study of genetic variation in natural populations. *Genetica* 56: 11–15.

Castriota, L., Falautano, M., Finoia, M. G., Consoli, P., Pedà, C., Esposito, V., Battaglia, P. & Andaloro, F. (2012). Trophic relationships among scorpaeniform fishes associated with gas platforms. *Helgoland Marine Research* 66: 401–411.

Castro, L. F. C., Lima, D., Machado, A. L. D., Melo, C., Hiromori, Y., Nishikawa, J., Nakanishi, T., Reis-Henriques, M. A. & Santos, M. M. (2007). Imposex induction is mediated through the Retinoid X Receptor signalling pathway in the neogastropod *Nucella lapillus*. *Aquatic Toxicology* 85: 57–66.

Catterall, C. P. & Poiner, I. R. (1983). Age- and sex-dependent patterns of aggregation in the tropical gastropod *Strombus luhuanus*. *Marine Biology* 77: 171–182.

Cavalcanti, M. G. S., Filho, F. C., Mendonça, A. M. B., Duarte, G. R., Barbosa, C. C. G. S., De Castro, C. M. M. B., Alves, L. C. & Brayner, F. A. (2012). Morphological characterization of hemocytes from *Biomphalaria glabrata* and *Biomphalaria straminea*. *Micron* 43: 285–291.

Cavalier-Smith, T. (2006). Rooting the tree of life by transition analyses. *Biology Direct* 1: 1–135.

Cavanaugh, C. M., McKiness, Z. P., Newton, I. L. G. & Stewart, F. J. (2005). Marine chemosynthetic symbioses. pp. 475–507 *in* M. Dworkin, Falkow, S. I., Rosenberg, E., Schleifer, K. H. & Stackebrandt, E. (Eds.) *The Prokaryotes: An Evolving Electronic Resource for the Microbiological Community*. New York, Springer.

Cazzaniga, N. J. (1988). *Hyalinella vahiriae* (Ectoprocta) en la Provincia de San Juan. *Revista de la Asociación de Ciencias Naturales del Litoral* 19: 205–208.

Cazzaniga, N. J. (1990). Sexual dimorphism in *Pomacea canaliculata* (Gastropoda: Ampullariidae). *The Veliger* 33: 384–388.

Ceballos, G., Ehrlich, P. R., Barnosky, A. D., García, A., Pringle, R. M. & Palmer, T. M. (2015). Accelerated modern human-induced species losses: Entering the sixth mass extinction. *Science Advances* 1: e1400253.

Cellura, C., Toubiana, M., Parrinello, N. & Roch, P. (2007). Specific expression of antimicrobial peptide and HSP70 genes in response to heat-shock and several bacterial challenges in mussels. *Fish and Shellfish Immunology* 22: 340–350.

Cerenius, L. & Söderhäll, K. (2004). The prophenoloxidase-activating system in invertebrates. *Immunological Reviews* 198: 116–126.

Cervera, J. L., Tamsouri, N., Moukrim, A. & Villani, G. (2010). New records of two alien opisthobranch molluscs from the northeastern Atlantic: *Polycera hedgpethi* and *Godiva quadricolor*. *Marine Biodiversity Records* 3: e51.

Chaffee, C. & Strathmann, R. R. (1984). Constraints on egg masses. I. Retarded development within thick egg masses. *Journal of Experimental Marine Biology and Ecology* 84: 73–83.

Chaffee, C. & Lindberg, D. R. (1986). Larval biology of early Cambrian molluscs: The implications of small body size. *Bulletin of Marine Science* 39: 536–549.

Chai, J.-Y. (2007). Intestinal flukes. pp. 53–115 *in* K. D. Murrell & Fried, B. (Eds.) *Food-borne Parasitic Zoonoses. Fish and Plant-borne Parasites*. New York, Springer.

Chai, J.-Y., Shin, E. H., Lee, S. H. & Rim, H. J. (2009). Foodborne intestinal flukes in Southeast Asia. *Korean Journal of Parasitology* 47: S69–S102.

Chaine, A. & Angeloni, L. (2005). Size-dependent mating and gender choice in a simultaneous hermaphrodite, *Bulla gouldiana*. *Behavioral Ecology and Sociobiology* 59: 58–68.

Chalazonitis, N. & Chalazonitis-Arvanitaki, A. (1961). *Thermoreceptor properties of single nerve cells*. Proceedings of the 1st International Biophysics Congress, July 31–August 4, 1961, Stockholm, Sweden, Elsevier.

Challis, D. A. (1969). An ecological account of the marine interstitial opisthobranchs of the British Solomon Islands Protectorate. *Philosophical Transactions of the Royal Society of London* 255: 527–539.

Chamberlain, T. K. (1931). Annual growth of freshwater mussels. *Bulletin of the Bureau of Fisheries* 46: 713–739.

Chambers, S. M. (1987). Rates of evolutionary change in chromosome numbers in snails and vertebrates. *Evolution* 41: 166–175.

Chan, B., Balmforth, N. J. & Hosoi, A. E. (2005). Building a better snail: Lubrication and adhesive locomotion. *Physics of Fluids* 17: 1–10.

Chanley, P. E. & Chanley, P. (1970). Larval development of the commensal clam *Montacuta percompressa* Dall. *Proceedings of the Malacological Society of London* 39: 59–67.

Chantler, P. D. (1983). Biochemical and structural aspects of molluscan muscle. pp. 77–154 *in* A. S. M. Saleuddin & Wilbur, K. M. (Eds.) *Physiology, Part 1. The Mollusca*. New York, Academic Press.

Chaparro, O. R., Thompson, R. J. & Pereda, S. V. (2002). Feeding mechanisms in the gastropod *Crepidula fecunda*. *Marine Ecology Progress Series* 234: 171–181.

Chaparro, O. R., Segura, C. J., Montory, J. A., Navarro, J. M. & Pechenik, J. A. (2009). Brood chamber isolation during salinity stress in two estuarine mollusk species: From a protective nursery to a dangerous prison. *Marine Ecology Progress Series* 374: 145–155.

Chaparro, O. R., Montory, J., Pechenik, J., Cubillos, V., Navarro, J. & Osores, S. (2011). Ammonia accumulation in the brood chamber of the estuarine gastropod *Crepipatella dilatata*: How big a problem for mothers and brooded embryos? *Journal of Experimental Marine Biology and Ecology* 410: 29–38.

Chapman, A. D. (2009). *Numbers of Living Species in Australia and the World*, 2nd ed. Canberra, A Report for the Australian Biological Resources Study: 1–78.

Chapman, D. J. & Fox, D. L. (1969). Bile pigment metabolism in the sea-hare *Aplysia*. *Journal of Experimental Marine Biology and Ecology* 4: 71–78.

Chapman, E. G., Przhiboro, A. A., Harwood, J. D., Foote, B. A. & Hoeh, W. R. (2012). Widespread and persistent invasions of terrestrial habitats coincident with larval feeding behavior transitions during snail-killing fly evolution (Diptera: Sciomyzidae). *BMC Evolutionary Biology* 12: 175 (171–122).

Chapman, G. & Newell, G. E. (1956). The role of the body fluid in the movement of soft-bodied invertebrates. II. The extension of the siphons of *Mya arenaria* L. and *Scrobicularia plana* (Da Costa). *Proceedings of the Royal Society B* 145: 564–580.

Chapman, G. (1995). Versatility of hydraulic systems. *Journal of Experimental Zoology Part A* 194: 249–269.

Chapman, J. W., Carlton, J. T., Bellinger, M. R. & Blakeslee, A. M. H. (2007). Premature refutation of a human-mediated marine species introduction: The case history of the marine snail *Littorina littorea* in the Northwestern Atlantic. *Biological Invasions* 9: 995–1008.

Charles, G. H. (1961a). The mechanism of orientation of freely moving *Littorina littoralis* (L.) to polarized light. *Journal of Experimental Biology* 38: 203–212.

Charles, G. H. (1961b). The orientation of *Littorina* species to polarized light. *Journal of Experimental Biology* 38: 189–202.

Charles, G. H. (1961c). Orientational movements of the foot of *Littorina* species in relation to the plane of vibration of polarized light. *Journal of Experimental Biology* 38: 213–224.

Charles, G. H. (1966). Sense organs (less Cephalopods). pp. 455–521 *in* K. M. Wilbur & Yonge, C. M. (Eds.) *Physiology of Mollusca*. Vol. 2. New York, Academic Press.

Charnov, E. L. (1979). Simultaneous hermaphroditism and sexual selection. *Proceedings of the National Academy of Sciences of the United States of America* 76: 2480–2484.

Charnov, E. L. (1982). Alternative life-histories in protogynous fishes: A general evolutionary theory. *Marine Ecology Progress Series* 9: 305–307.

Charrier, M. & Daguzan, J. (1980). Consommation alimentaire: Production et bilan énergétique chez *Helix aspersa* Müller (Gastéropode Pulmoné terrestre). *Annales de la Nutrition et de l'Alimentation* 34: 147–166.

Charrier, M. (1990). Evolution during digestion of the bacterial flora in the alimentary system of *Helix aspersa* (Gastropoda: Pulmonata): A scanning electron microscopic study. *Journal of Molluscan Studies* 56: 425–433.

Charrier, M. & Rouland, C. (1992). Les osidases digestives de l'escargot *Helix aspersa*: Localisations et variations en fonction de l'état nutritionnel. *Canadian Journal of Zoology* 70: 2234–2241.

Charrier, M., Combet-Blanc, Y. & Ollivier, B. (1998). Bacterial flora in the gut of *Helix aspersa* (Gastropoda Pulmonata): Evidence for a permanent population with a dominant homolactic intestinal bacterium, *Enterococcus casseliflavus*. *Canadian Journal of Microbiology* 44: 20–27.

Charrier, M., Fonty, G., Gillard-Martinie, B., Martin, M.-C. & Andant, G. (2001). Characterization of some intestinal symbionts of edible snails: A first approach of the chemical substances consumed and released by the microflora. pp. 54 *in* L. Salvini-Plawen, Voltzow, J., Sattmann, H. & Steiner, G. (Eds.) *Abstracts: World Congress of Malacology 2001, Vienna, Austria (19–25 August)*. Vienna, Unitas Malacologica.

Charrier, M. & Rouland, C. (2001). Mannan-degrading enzymes purified from the crop of the brown garden snail *Helix aspersa* Müller (Gastropoda Pulmonata). *Journal of Experimental Zoology* 290: 125–135.

Charrier, M. & Brune, A. (2003). The gut microenvironment of helicid snails (Gastropoda: Pulmonata): In-situ profiles of pH, oxygen, and hydrogen determined by microsensors. *Canadian Journal of Zoology* 81: 928–935.

Charrier, M., Fonty, G., Gaillard-Martinie, B., Ainouche, K. & Andant, G. (2006). Isolation and characterization of cultivable fermentative bacteria from the intestine of two edible snails, *Helix pomatia* and *Cornu aspersum* (Gastropoda: Pulmonata). *Biological Research* 39: 669–681.

Chase, R. & Goodman, H. (1977). Homologous neurosecretory cell groups in the land snail *Achatina fulica* and the sea hare *Aplysia californica*. *Cell and Tissue Research* 176: 109–120.

Chase, R. (1979). Photic sensitivity of the rhinophore in *Aplysia*. *Canadian Journal of Zoology* 57: 698–701.

Chase, R., Croll, R. P. & Zeichner, L. L. (1980). Aggregation in snails, *Achatina fulica*. *Behavioral and Neural Biology* 30: 218–230.

Chase, R. & Croll, R. P. (1981). Tentacular function in snail olfactory orientation. *Journal of Comparative Physiology A* 143: 357–362.

Chase, R. & Kamil, R. A. (1983). Morphology and odor sensitivity of regenerated snail tentacles. *Journal of Neurobiology* 14: 43–50.

Chase, R. & Rieling, J. (1986). Autoradiographic evidence for receptor cell renewal in the olfactory epithelium of a snail. *Brain Research* 384: 232–239.

Chase, R. (2001). Sensory organs and the nervous system. pp. 179–211 *in* G. M. Barker (Ed.) *The Biology of Terrestrial Molluscs*. Wallingford, UK, CABI Publishing.

Chase, R. (2002). *Behavior and its Neural Control in Gastropod Molluscs*. Oxford, UK, Oxford University Press.

Chase, R., Darbyson, E., Horn, K. E. & Samarova, E. (2010). A mechanism aiding simultaneously reciprocal mating in snails. *Canadian Journal of Zoology* 88: 99–107.

Chateigner, D., Hedegaard, C. & Wenk, H.-R. (2000). Mollusc shell microstructures and crystallographic textures. *Journal of Structural Geology* 22: 1723–1735.

Chateigner, D., Ouhenia, S., Krauss, C., Hedegaard, C., Gil, O., Morales, M., Lutterotti, L., Rousseau, M. & Lopez, E. (2010). Voyaging around nacre with the X-ray shuttle: From bio-mineralisation to prosthetics via mollusc phylogeny. *Materials Science and Engineering: A* 528: 37–51.

Chatfield, J. E. (1976). Studies on food and feeding in some European land molluscs. *Journal of Conchology* 29: 5–20.

Chau, R., Kalaitzis, J. A. & Neilan, B. A. (2011). On the origins and biosynthesis of tetrodotoxin. *Aquatic Toxicology* 104: 61–72.

Chauvaud, L., Lorrain, A., Dunbar, R. B., Paulet, Y. M., Thouzeau, G., Jean, F., Guarini, J. M. & Mucciarone, D. (2005). Shell of the Great Scallop *Pecten maximus* as a high-frequency archive of paleoenvironmental changes. *Geochemistry, Geophysics, Geosystems* 6: 1–15.

Chávez-Villalba, J., Soyez, C., Huvet, A., Gueguen, Y., Lo, C. & Moullac, G. L. (2011). Determination of gender in the pearl oyster *Pinctada margaritifera*. *Journal of Shellfish Research* 30: 231–240.

Checa, A. G. & Jiménez-Jiménez, A. P. (1997). Regulation of spiral growth in planorbid gastropods. *Lethaia* 30: 257–269.

Checa, A. G. & Jiménez-Jiménez, A. P. (1998). Constructional morphology, origin, and evolution of the gastropod operculum. *Paleobiology* 24: 109–132.

Checa, A. G. (2000). A new model for periostracum and shell formation in Unionidae (Bivalvia, Mollusca). *Tissue and Cell* 32: 405–416.

Checa, A. G. & Jiménez-Jiménez, A. P. (2003). Evolutionary morphology of oblique ribs of bivalves. *Palaeontology* 46: 709–724.

Checa, A. G., Ramirez-Rico, J., González-Segura, A. & Sánchez-Navas, A. (2009a). Nacre and false nacre (foliated aragonite) in extant monoplacophorans (=Tryblidiida: Mollusca). *Die Naturwissenschaften* 96: 111–122.

Checa, A. G., Sánchez-Navas, A. & Rodríguez-Navarro, A. (2009b). Crystal growth in the foliated aragonite of monoplacophorans (Mollusca). *Crystal Growth and Design* 9: 4574–4580.

Checa, A. G., Salas, C., Harper, E. M. & de Dios Bueno-Perez, J. (2014). Early stage biomineralization in the periostracum of the 'living fossil' bivalve *Neotrigonia*. *PLoS ONE* 9: e90033.

Checa, A. G., Vendrasco, M. J. & Salas, C. (2017). Cuticle of Polyplacophora: Structure, secretion, and homology with the periostracum of conchiferans. *Marine Biology* 164: 64.

Chelazzi, G. & Vannini, M. (1980). Zonal orientation based on local visual cues in *Nerita plicata* L. (Mollusca: Gastropoda) at Aldabra Atoll. *Journal of Experimental Marine Biology and Ecology* 46: 147–156.

Chelazzi, G. & Focardi, S. (1982). A laboratory study on the short-term zonal oscillations of the trochid *Monodonta turbinata* (Born) (Mollusca: Gastropoda). *Journal of Experimental Marine Biology and Ecology* 65: 263–273.

Chelazzi, G., Focardi, S., Deneubourg, J. L. & Innocenti, R. (1983). Competition for the home and aggressive behaviour in the chiton *Acanthopleura gemmata* (Blainville) (Mollusca: Polyplacophora). *Behavioral Ecology and Sociobiology* 14: 15–20.

Chelazzi, G., Della Santina, P. & Vannini, M. (1985). Long-lasting substrate marking in the collective homing of the gastropod *Nerita textilis*. *Biological Bulletin* 168: 214–221.

Chelazzi, G., Della Santina, P. & Parpagnoli, D. (1987). Trail following in the chiton *Acanthopleura gemmata*: Operational and ecological problems. *Marine Biology* 95: 539–545.

Chelazzi, G., Focardi, S. & Deneubourg, J. L. (1988). Analysis of movement patterns and orientation mechanisms in intertidal chitons and gastropods. pp. 173–184 in G. Chelazzi & Vannini, M. (Eds.) *Behavioral Adaptation to Intertidal Life*. New York, Plenum Press.

Chelazzi, G. (1990). Eco-ethological aspects of homing behaviour in molluscs. *Ethology Ecology and Evolution* 2: 11–26.

Chelazzi, G. (1992). Invertebrates (excluding arthropods). pp. 19–43 in F. Papi (Ed.) *Animal Homing*. London, Chapman and Hall.

Chen, B., Xu, L., Guo, Z. & Yang, H. (2004). A new species of *Cryptobia* sp. n. (Kineloplastida, Bodinina, Bodonidae) found in the blood of the farmed abalone, *Haliotis diversicolor* Reeve. *Journal of Shellfish Research* 23: 1169–1172.

Chen, C., Copley, J. T., Linse, K., Rogers, A. D. & Sigwart, J. D. (2015). The heart of a dragon: 3D anatomical reconstruction of the 'scaly-foot gastropod' (Mollusca: Gastropoda: Neomphalina) reveals its extraordinary circulatory system. *Frontiers in Zoology* 12: 1–16.

Chen, J.-H. & Bayne, C. J. (1995). Bivalve mollusc hemocyte behaviors: Characterization of hemocyte aggregation and adhesion and their inhibition in the California mussel (*Mytilus californianus*). *Biological Bulletin* 188: 255–266.

Chen, M.-H. & Soong, K. (2002). Estimation of age in the sex-changing, coral-inhabiting snail *Coralliophila violacea* from the growth striae on opercula and a mark–recapture experiment. *Marine Biology* 140: 337–342.

Cheney, K. L., Cortesi, F., How, M. J., Wilson, N. G., Blomberg, S. P., Winters, A. E., Umanzör, S. & Marshall, N. J. (2014). Conspicuous visual signals do not coevolve with increased body size in marine sea slugs. *Journal of Evolutionary Biology* 27: 676–687.

Cheng, M. W. & Caldwell, R. L. (2000). Sex identification and mating in the blue-ringed octopus, *Hapalochlaena lunulata*. *Animal Behaviour* 60: 27–33.

Cheng, T. C. & Burton, R. W. (1965). The American oyster and clam as experimental intermediate hosts of *Angiostrongylus cantonensis*. *Journal of Parasitology* 51: 296.

Cheng, T. C. (1978). The role of lysosomal hydrolases in molluscan cellular response to immunologic challenge. pp. 59–71 in L. A. Bulla & Cheng, T. C. (Eds.) *Invertebrate Models for Biomedical Research*. New York, Springer.

Cheng, T. C. & Downs, J. C. U. (1988). Intracellular acid phosphatase and lysozyme levels in subpopulations of oyster, *Crassostrea virginica*, hemocytes. *Journal of Invertebrate Pathology* 52: 163–167.

Cheng, T. C. (1993). Non-infectious diseases of marine molluscs. pp. 289–318 in J. A. Couch (Ed.) *Pathobiology of Marine and Estuarine Organisms. Advances in Fisheries Science*. Boca Raton, FL, CRC Press.

Cherel, Y. & Hobson, K. A. (2005). Stable isotopes, beaks and predators: A new tool to study the trophic ecology of cephalopods, including giant and colossal squids. *Proceedings of the Royal Society of London B: Biological Sciences* 272: 1601–1607.

Cherel, Y., Ridoux, V., Spitz, J. & Richard, P. (2009). Stable isotopes document the trophic structure of a deep-sea cephalopod assemblage including giant octopod and giant squid. *Biology Letters* 1: 364–367.

Cherns, L. (2004). Early Palaeozoic diversification of chitons (Polyplacophora, Mollusca) based on new data from the Silurian of Götland, Sweden. *Lethaia* 37: 445–456.

Chevaldonné, P., Desbruyères, D. & Le Haître, M. (1991). Time-series of temperature from three deep-sea hydrothermal vent sites. *Deep Sea Research Part A: Oceanographic Research Papers* 38: 1417–1430.

Chia, F.-S. & Koss, R. (1978). Development and metamorphosis of the planktotrophic larvae of *Rostanga pulchra* (Mollusca: Nudibranchia). *Marine Biology* 46: 109–119.

Chia, F.-S. & Koss, R. (1982). Fine structure of the larval rhinophores of the nudibranch *Rostanga pulchra*, with emphasis on the sensory receptor cells. *Cell and Tissue Research* 225: 235–248.

Chiba, S. (1999). Character displacement, frequency-dependent selection, and divergence of shell colour in land snails *Mandarina* (Pulmonata). *Biological Journal of the Linnean Society* 66: 465–479.

Chícharo, L. & Chícharo, M. A. (2001). Effects of environmental conditions on planktonic abundances, benthic recruitment and growth rates of the bivalve mollusc *Ruditapes decussatus* in a Portuguese coastal lagoon. *Fisheries Research* 53: 235–250.

Chih, C. P. & Ellington, W. R. (1986). The role of glycogen phosphorylase in the control of glycolysis during contractile activity in the phasic adductor muscle of the bay scallop, *Argopecten irradians concentricus*. *Physiological Zoology* 59: 563–573.

Childress, J. J. & Seibel, B. A. (1998). Life at stable low oxygen levels: Adaptations of animals to oceanic oxygen minimum layers. *Journal of Experimental Biology* 201: 1223–1232.

Chiotha, S. S., McKaye, K. R. & Stauffer, J. R. (1991). Use of indigenous fishes to control schistosome snail vectors in Malaŵi, Africa. *Biological Control* 1: 316–319.

Chiu, A. Y., Hunkapiller, M. W., Heller, E., Stuart, D. K., Hood, L. E. & Strumwasser, F. (1979). Purification and primary structure of the neuropeptide egg-laying hormone of *Aplysia californica*. *Proceedings of the National Academy of Sciences of the United States of America* 76: 6656–6660.

Chiu, M. Y., Paul, K. S., Shin, L., Wong, K. P. & Cheung, S. G. (2010). Sibling cannibalism in juveniles of the marine gastropod *Nassarius festivus* (Powys, 1835). *Malacologia* 52: 157–161.

Chiu, S. T., Lam, V. W. W. & Shin, P. K. S. (1983). Mollusc predation by *Luidia* spp. (Echinodermata: Asteroidea) in Tolo Harbour and Channel, Hong Kong. pp. 907–933 in B. Morton (Ed.) *The Marine Flora and Fauna of Hong Kong and Southern China II: Taxonomy and Ecology*. Hong Kong, Hong Kong University Press.

Cho, S.-J., Vallès, Y., Giani, V. C., Seaver, E. C. & Weisblat, D. A. (2010). Evolutionary dynamics of the WNT gene family: A lophotrochozoan perspective. *Molecular Biology and Evolution* 27: 1645–1658.

Choquet, M. (1964). Biologie expérimentale: Culture organotypique de gonades de *Patella vulgata* L. (Mollusque Gastéropode Prosobranche). *Comptes rendus hebdomadaires des séances de l'Académie des sciences* 258: 1089–1091.

Choquet, M. (1965). Recherches en culture organotypique sur la spermatogenèse chez *Patella vulgata* L (Mollusque Gastéropode). Rôle des ganglions cérébroïdes et des tentacules. *Comptes rendus hebdomadaires des séances de l'Académie des sciences* 261: 4521–4524.

Choquet, M. (1967). Gamétogenèse in vitro au cours du cycle annuel chez *Patella vulgata* L. en phase mâle. *Comptes rendus hebdomadaires des séances de l'Académie des sciences* 265D: 333–335.

Choquet, M. (1970). Analyse des cycles sexuels naturels chez les mollusques hermaphrodites et gonochoriques. *Bulletin de la Société zoologique de France* 95: 393–406.

Choquet, M. (1971). Étude du cycle biologique et de l'inversion du sexe chez *Patella vulgata* L. (Mollusque Gastéropode Prosobranche). *General and Comparative Endocrinology* 16: 59–73.

Choquet, M. & Lemaire, J. (1969). Étude histologique du complexe tentacule-ganglion cérébroide de *Patella vulgata* L. (Mollusque Gastéropode Prosobranche). *Bulletin de la Société zoologique de France* 94: 39–53.

Chotiyaputta, C., Nootmorn, P. & Jirapunpipat, K. (2002). Review of cephalopod fishery production and long term changes in fish communities in the Gulf of Thailand. *Bulletin of Marine Science* 71: 223–238.

Chrachri, A., Nelson, L. & Williamson, R. (2005). Whole-cell recording of light-evoked photoreceptor responses in a slice preparation of the cuttlefish retina. *Visual Neuroscience* 22: 359–370.

Christa, G., Gould, S. B., Franken, J., Vleugels, M., Karmeinski, D., Händeler, K., Martin, W. F. & Wägele, H. (2014). Functional kleptoplasty in a limapontioidean genus: Phylogeny, food preferences and photosynthesis in *Costasiella*, with a focus on *C. ocellifera* (Gastropoda: Sacoglossa). *Journal of Molluscan Studies* 80: 499–507.

Christensen, C. C., Yeung, N. W. & Hayes, K. A. (2012). First records of *Paralaoma servilis* (Shuttleworth, 1852) (Gastropoda: Pulmonata: Punctidae) in the Hawaiian Islands. *Bishop Museum Occasional Paper* 112: 3–7.

Christian, A. D., Smith, B. N., Berg, D. J., Smoot, J. C. & Findlay, R. H. (2004). Trophic position and potential food sources of 2 species of unionid bivalves (Mollusca: Unionidae) in 2 small Ohio streams. *Journal of the North American Benthological Society* 23: 101–113.

Christiansen, F. B. & Fenchel, T. M. (1979). Evolution of marine invertebrate reproduction patterns. *Theoretical Population Biology* 16: 267–282.

Chung, E.-Y. (2007). Oogenesis and sexual maturation in *Meretrix lusoria* (Röding 1798) (Bivalvia: Veneridae) in western Korea. *Journal of Shellfish Research* 26: 71–80.

Churchill, E. P. (1916). The absorption of nutriment from solution by freshwater mussels. *Journal of Experimental Zoology Part A* 21: 403–429.

Ciavatta, M. L., Trivellone, E., Villani, G. & Cimino, G. (1993). Membrenones: New polypropionates from the skin of the Mediterranean mollusc *Pleurobranchus membranaceus*. *Tetrahedron Letters* 34: 6791–6794.

Cimino, G., De, R. S., De, S. S., Sodano, G. & Vil, G. (1983). Dorid nudibranch elaborates its own chemical defense. *Science* 219: 1237–1238.

Cimino, G., Crispino, A., Spinella, A. & Sodano, G. (1988). Two ichthyotoxic diacylglycerols from the opisthobranch mollusc *Umbraculum mediterraneum*. *Tetrahedron Letters* 29: 3613–3616.

Cimino, G., Spinella, A., Scopa, A. & Sodano, G. (1989). Umbraculumin-B, an unusual 3-hydroxybutyric acid ester from the opisthobranch mollusc *Umbraculum mediterraneum*. *Tetrahedron Letters* 30: 1147–1148.

Cimino, G. & Ghiselin, M. T. (1998). Chemical defense and evolution in the Sacoglossa (Mollusca: Gastropoda: Opisthobranchia). *Chemoecology* 8: 51–60.

Cimino, G. & Ghiselin, M. T. (1999). Chemical defense and evolutionary trends in biosynthetic capacity among dorid nudibranchs (Mollusca: Gastropoda: Opisthobranchia). *Chemoecology* 9: 187–207.

Cimino, G. & Gavagnin, M., Eds. (2006). *Molluscs: From Chemo-Ecological Study to Biotechnological Application*. Berlin, Heidelberg, Springer-Verlag.

Cimino, G. & Ghiselin, M. T. (2009). Chemical defense and the evolution of opisthobranch gastropods. *Proceedings of the California Academy of Sciences* 60: 175–422.

Claes, M. F. (1996). Functional morphology of the white bodies of the cephalopod mollusc *Sepia officinalis*. *Acta Zoologica* 77: 173–190.

Clapp, W. F. & Kenk, R. (1963). *Marine Borers. An Annotated Bibliography*. Washington, DC, US Govt. Printing Office.

Clare, A. S. (1987). Studies on the juxtaganglionar organ of trochids. pp. 342–349 in H. H. Boer, Geraerts, W. P. M. & Joosse, J. (Eds.) *Neurobiology. Molluscan models. Proceedings of the Second Symposium on Molluscan Neurobiology, held at the Department of Zoology of the Free University, Amsterdam, the Netherlands, August 18–22, 1986*. Amsterdam, North-Holland Publishing Company.

Claremont, M., Reid, D. G. & Williams, S. T. (2011). Evolution of corallivory in the gastropod genus *Drupella*. *Coral Reefs* 30: 977–990.

Clark, K. B. & Busacca, M. (1978). Feeding specificity and chloroplast retention in four tropical Ascoglossa, with a discussion of the extent of chloroplast symbiosis and the evolution of the order. *Journal of Molluscan Studies* 44: 272–282.

Clark, S. A., Miller, A. C. & Ponder, W. F. (2003). Revision of the snail genus *Austropyrgus* (Gastropoda: Hydrobiidae): A morphostatic radiation of freshwater gastropods in southeastern Australia. *Records of the Australian Museum* 28: 1–109.

Clarke, A. H. (1978). Polymorphism in marine mollusks and biome development. *Smithsonian Contributions to Zoology* 274: 1–13.

Clarke, A. P., Mill, P. J. & Grahame, J. (2000). Biodiversity in *Littorina* species (Mollusca: Gastropoda): A physiological approach using heat-coma. *Marine Biology* 137: 559–565.

Clarke, B. (1962). Natural selection in mixed populations of two polymorphic snails. *Heredity* 17: 319–345.

Clarke, B., Murray, J. & Johnson, M. S. (1984). The extinction of endemic species by a program of biological control. *Pacific Science* 38: 97–104.

Clarke, M. R. (1996). Cephalopods as prey. III. Cetaceans. *Philosophical Transactions of the Royal Society B* 351: 1053–1065.

Claverie, T. & Kamenos, N. A. (2008). Spawning aggregations and mass movements in subtidal *Onchidoris bilamellata* (Mollusca: Opisthobranchia). *Journal of the Marine Biological Association of the United Kingdom* 88: 157–159.

Cleland, D. M. (1954). A study of the habits of *Valvata piscinalis* (Müller) and the structure and function of the alimentary canal and reproductive system. *Proceedings of the Malacological Society of London* 30: 167–203.

Clements, R., Liew, T. S., Vermeulen, J. J. & Schilthuizen, M. (2008a). Further twists in gastropod shell evolution. *Biology Letters* 4: 179–182.

Clements, R., Ng, P. K. L., Lu, X. X., Ambu, S., Schilthuizen, M. & Bradshaw, C. J. A. (2008b). Using biogeographical patterns of endemic land snails to improve conservation planning for limestone karsts. *Biological Conservation* 141: 2751–2764.

Clifford, K. T., Gross, L., Johnson, K., Martin, K. J., Shaheen, N. & Harrington, M. A. (2003). Slime-trail tracking in the predatory snail, *Euglandina rosea*. *Behavioral Neuroscience* 117: 1086–1095.

Clotteau, G. & Dubé, F. (1993). Optimization of fertilization parameters for rearing surf clams (*Spisula solidissima*). *Aquaculture* 114: 339–353.

Coan, E. V., Valentich-Scott, P. & Bernard, F. R. (2000). *Bivalve Seashells of Western North America [including] Marine Bivalve Mollusks from Arctic Alaska to Baja California*. Santa Barbara, CA, Santa Barbara Museum of Natural History.

Coan, E. V., Kabat, A. R. & Petit, R. E. (2007). 2,400 years of Malacology. 4th edition, revised., from https://www.malacological.org/downloads/epubs/2400-years/2400yrs_of_Malacology_complete.pdf

Coates, C. J. & Nairn, J. (2014). Diverse immune functions of hemocyanins. *Developmental and Comparative Immunology* 45: 43–55.

Cochennec, N., Reece, K. S., Berthe, F. & Hine, M. (2003). *Mikrocytos roughleyi* taxonomic affiliation leads to the genus *Bonamia* (Haplosporidia). *Diseases of Aquatic Organisms* 54: 209–217.

Cochran, T., Brown, C., Mathew, K., Carroll, M. A. & Catapane, E. J. (2012a). A study of GABA in bivalve molluscs. *FASEB Journal* 26: 762–765.

Cochran, T., Matthew, K., Catapane, E. J. & Carroll, M. A. (2012b). The presence of GABA in ganglia of bivalve molluscs. *Journal of Shellfish Research* 31: 270.

Coe, W. R. (1931). Spermatogenesis in the California oyster (*Ostrea lurida*). *Biological Bulletin* 61: 309–315.

Coe, W. R. (1936). Sexual phases in *Crepidula*. *Journal of Experimental Zoology Part A* 72: 455–477.

Coe, W. R. (1944). Sexual differentiation in mollusks. II. Gastropods, Amphineurans, Scaphopods, and Cephalopods. *Quarterly Review of Biology* 19: 85–97.

Coeurdassier, M., Scheifler, R., Mench, M., Crini, N., Vangronsveld, J. & de Vaufleury, A. G. (2010). Arsenic transfer and impacts on snails exposed to stabilized and untreated As-contaminated soils. *Environmental Pollution* 158: 2078–2083.

Coffin, M. R., Barbeau, M. A., Hamilton, D. J. & Drolet, D. (2012). Effect of the mud snail *Ilyanassa obsoleta* on vital rates of the intertidal amphipod *Corophium volutator*. *Journal of Experimental Marine Biology and Ecology* 418: 12–23.

Cohen, A. N. & Carlton, J. T. (1998). Accelerating invasion rate in a highly invaded estuary. *Science* 279: 555–558.

Cohen, A. S. & Johnston, M. R. (1987). Speciation in brooding and poorly dispersing lacustrine organisms. *Palaios* 2: 426–435.

Cohen, W. D. & Tamburri, M. N. (1998). Distinctive cytoskeletal organization in erythrocytes of the cold-seep vesicomyid clam, *Calyptogena kilmeri*. *Biological Bulletin* 194: 7–13.

Colaco, A., Daniel, D. & Guezennec, J. (2007). Polar lipid fatty acids as indicators of trophic associations in a deep-sea vent system community. *Marine Ecology* 28: 15–24.

Cole, A. G. & Hall, B. K. (2004). The nature and significance of invertebrate cartilages revisited: Distribution and histology of cartilage and cartilage-like tissues within the Metazoa. *Zoology* 107: 261–273.

Cole, C., Coelho, A. V., James, R. H., Connelly, D. & Sheehan, D. (2014). Proteomic responses to metal-induced oxidative stress in hydrothermal vent-living mussels, *Bathymodiolus* sp., on the Southwest Indian Ridge. *Marine Environmental Research* 96: 29–37.

Cole, H. A. (1938). The fate of larval organs in the metamorphosis of *Ostrea edulis*. *Journal of the Marine Biological Association of the United Kingdom* 22: 469–484.

Cole, K. S. & Gilbert, D. L. (1970). Jet propulsion of squid. *Biological Bulletin* 138: 245–246.

Coleman, R. A., Ramchunder, S. J., Moody, A. J. & Foggo, A. (2007). An enzyme in snail saliva induces herbivore-resistance in a marine alga. *Functional Ecology* 21: 101–106.

Colgan, D. J. (1981). Spatial and temporal variation in the genotypic frequencies of the mussel *Brachidontes rostratus*. *Heredity* 46: 197–208.

Colgan, D. J. & Ponder, W. F. (2000). Incipient speciation in aquatic snails in an arid-zone spring complex. *Biological Journal of the Linnean Society* 71: 625–641.

Colgan, D. J., Ponder, W. F., Beacham, E. & Macaranas, J. M. (2007). Molecular phylogenetics of Caenogastropoda (Gastropoda: Mollusca). *Molecular Phylogenetics and Evolution* 42: 717–737.

Colgan, D. J. & Middelfart, P. A. (2011). *Mytilus* mitochondrial DNA haplotypes in southeastern Australia. *Aquatic Biology* 12: 47–55.

Colgan, D. J. & Schreiter, S. (2011). Extrinsic and intrinsic influences on the phylogeography of the *Austrocochlea constricta* species group. *Journal of Experimental Marine Biology and Ecology* 397: 44–51.

Coll, J., Tapiolas, D., Bowden, B., Webb, L. & Marsh, H. (1983). Transformation of soft coral (Coelenterata: Octocorallia) terpenes by *Ovula ovum* (Mollusca: Prosobranchia). *Marine Biology* 74: 35–40.

Colley, S. M. (1997). A pre-and post-contact Aboriginal shell midden at Disaster Bay, New South Wales south coast. *Australian Archaeology* 45: 1–19.

Collicutt, J. M. & Hochachka, P. W. (1977). The anaerobic oyster heart: Coupling of glucose and aspartate fermentation. *Journal of Comparative Physiology A* 115: 147–157.

Collier, J. R. (1997). Gastropods: The snails. pp. 189–217 *in* S. F. Gilbert & Raunio, A. M. (Eds.) *Embryology: Constructing the Organism*. Sunderland, MA, Sinauer Associates.

Collin, R. (1995). Sex, size, and position: A test of models predicting size at sex change in the protandrous gastropod *Crepidula fornicata*. *American Naturalist* 146: 815–831.

Collin, R. (1997). Hydrophobic larval shells: Another character for higher level systematics of gastropods. *Journal of Molluscan Studies* 63: 425–430.

Collin, R. (2000). Development and anatomy of *Nitidiscala tincta* (Carpenter, 1865) (Gastropoda: Epitoniidae). *The Veliger* 43: 302–312.

Collin, R. (2001). The effects of mode of development on phylogeography and population structure of North Atlantic *Crepidula* (Gastropoda: Calyptraeidae). *Molecular Ecology* 10: 2249–2262.

Collin, R. & Cipriani, R. (2003). Dollo's law and the re-evolution of shell coiling. *Proceedings of the Royal Society B* 270: 2551–2555.

Collin, R. (2004). Phylogenetic effects, the loss of complex characters, and the evolution of development in calyptraeid gastropods. *Evolution* 58: 1488–1502.

Collin, R. & Giribet, G. (2010). Report of a cohesive gelatinous egg mass produced by a tropical marine bivalve. *Invertebrate Biology* 129: 165–171.

Collin, R. (2013). Phylogenetic patterns and phenotypic plasticity of molluscan sexual systems. *Integrative and Comparative Biology* 53: 723–735.

Collin, R., Kerr, K., Contolini, G. & Ochoa, I. (2017). Reproductive cycles in tropical intertidal gastropods are timed around tidal amplitude cycles. *Ecology and Evolution* 7: 5977–5991.

Collin, R. (2019). Transitions in sexual and reproductive strategies among the Caenogastropoda. pp. 193–220 *in* J. L. Leonard (Ed.) *Transactions Between Sexual Systems*. New York, Springer-Verlag.

Collins, A. J., LaBarre, B. A., Wong, B. S., Shah, M. V., Heng, S., Choudhury, M. H., Haydar, S. A., Santiago, J. & Nyholm, S. V. (2012a). Diversity and partitioning of bacterial populations within the accessory nidamental gland of the squid *Euprymna scolopes*. *Applied and Environmental Microbiology* 78: 4200–4208.

Collins, A. J., Schleicher, T. R., Rader, B. A. & Nyholm, S. V. (2012b). Understanding the role of host hemocytes in a squid/*Vibrio* symbiosis using transcriptomics and proteomics. *Frontiers in Immunology* 3: 91.

Collins, K. S., Crampton, J. S., Neil, H. L., Smith, E. G. C., Gazley, M. F. & Hannah, M. (2016). Anchors and snorkels: Heterochrony, development and form in functionally constrained fossil crassatellid bivalves. *Paleobiology* 42: 305–316.

Colman, J. (1933). The nature of the intertidal zonation of plants and animals. *Journal of the Marine Biological Association of the United Kingdom* 18: 435–476.

Colman, N. (1974). The heart rate and activity of bivalve molluscs in their natural habitats. *Oceanography and Marine Biology Annual Review* 12: 301–313.

Colmers, W. F. (1981). Afferent synaptic connections between hair cells and the somata of intramacular neurons in the gravity receptor system of the statocyst of *Octopus vulgaris*. *Journal of Comparative Neurology* 197: 385–394.

Colonese, A. C., Zanchetta, G., Perlès, C., Drysdale, R. N., Manganelli, G., Baneschi, I., Dotsika, E. & Valladas, H. (2013). Deciphering late Quaternary land snail shell $\delta^{18}O$ and $\delta^{13}C$ from Franchthi Cave (Argolid, Greece). *Quaternary Research* 80: 66–75.

Colson, I. & Hughes, R. N. (2007). Contrasted patterns of genetic variation in the dogwhelk *Nucella lapillus* along two putative post-glacial expansion routes. *Marine Ecology Progress Series* 343: 183–191.

Colton, D. E. & Alevizon, W. S. (1983). Feeding ecology of bonefish in Bahamian waters. *Transactions of the American Fisheries Society* 112: 178–184.

Comeau, S., Gorsky, G., Jeffree, R. A., Teyssie, J. L. & Gattuso, J. P. (2009). Impact of ocean acidification on a key Arctic pelagic mollusc (*Limacina helicina*). *Biogeosciences* 6: 1877–1882.

Comfort, A. (1949). Identity of the shell pigment of *Haliotis macherodii* Leach. *Nature* 163: 647.

Comfort, A. (1950). Biochemistry of molluscan shell pigments. *Proceedings of the Malacological Society of London* 28: 79–85.

Comfort, A. (1951a). The pigmentation of molluscan shells. *Biological Reviews* 26: 285–301.

Comfort, A. (1951b). Observations on the shell pigments of land pulmonates. *Proceedings of the Malacological Society of London* 29: 35–43.

Commito, J. A., Currier, C. A., Kane, L. R., Reinsel, K. A. & Ulm, M. (2013). Dispersal dynamics of the bivalve *Gemma gemma* in a patchy environment. *Ecological Monographs* 65: 1–20.

Cong, M., Song, L., Wang, L., Zhao, J., Qiu, L., Li, L. & Zhang, H. (2008). The enhanced immune protection of Zhikong scallop *Chlamys farreri* on the secondary encounter with *Listonella anguillarum*. *Comparative Biochemistry and Physiology Part B* 151: 191–196.

Conklin, E. G. (1897). The embryology of *Crepidula*, a contribution to the cell lineage and early development of some marine gastropods. *Journal of Morphology* 13: 1–226.

Conn, D. B., Ricciardi, A., Babapulle, M. N., Klein, K. A. & Rosen, D. A. (1996). *Chaetogaster limnaei* (Annelida: Oligochaeta) as a parasite of the zebra mussel *Dreissena polymorpha*, and quagga mussel *Dreissena bugensis* (Mollusca: Bivalvia). *Parasitology Research* 82: 1–7.

Conn, P. J. & Kaczmarek, L. K. (1989). The bag cell neurons of *Aplysia*. A model for the study of the molecular mechanisms involved in the control of prolonged animal behaviors. *Molecular Neurobiology* 3: 237–273.

Connell, J. H. (1961a). Effects of competition, predation by *Thais lapillus*, and other factors on natural populations of the barnacle *Balanus balanoides*. *Ecological Monographs* 31: 61–104.

Connell, J. H. (1961b). The influence of interspecific competition and other factors on the distribution of the barnacle *Chthamalus stellatus*. *Ecology* 42: 710–723.

Connell, J. H. (1970). A predator-prey system in the marine intertidal region. I. *Balanus glandula* and several predatory species of *Thais*. *Ecological Monographs* 40: 49–78.

Connell, J. H. (1978). Diversity in tropical rain forests and coral reefs. *Science* 199: 1302–1310.

Connell, J. H. (1980). Diversity and the coevolution of competitors, or the ghost of competition past. *Oikos* 35: 131–138.

Connell, J. H. (1983). On the prevalence and relative importance of interspecific competition: Evidence from field experiments. *American Naturalist* 122: 661–696.

Connelly, P. W. & Turner, R. L. (2009). Epibionts of the eastern surf chiton, *Ceratozona squalida* (Polyplacophora: Mopaliidae), from the Atlantic coast of Florida. *Bulletin of Marine Science* 85: 187–202.

Connor, D. W., Allen, J. H., Golding, N., Howell, K. L., Lieberknecht, L. M., Northen, K. O. & Reker, J. B. (2004). *The Marine Habitat Classification for Britain and Ireland Version 04. 05.* Peterborough, UK, Joint Nature Conservation Committee.

Connor, V. M. & Quinn, J. F. (1984). Stimulation of food species growth by limpet mucus. *Science* 225: 843–844.

Connor, V. M. (1986). The use of mucous trails by intertidal limpets to enhance food resources. *Biological Bulletin* 171: 548–564.

Conover, R. J. & Lalli, C. M. (1972). Feeding and growth in *Clione limacina* (Phipps), a pteropod mollusc. *Journal of Experimental Marine Biology and Ecology* 9: 279–302.

Conover, R. J. & Lalli, C. M. (1974). Feeding and growth in *Clione limacina* (Phipps), a pteropod mollusc. II. Assimilation, metabolism, and growth efficiency. *Journal of Experimental Marine Biology and Ecology* 16: 131–154.

Cook, A., Bamford, O. S., Freeman, J. D. B. & Teideman, D. J. (1969). A study of the homing habit of the limpet. *Animal Behaviour* 17: 330–339.

Cook, A. (1977). Mucus trail following by the slug *Limax grossui* Lupu. *Animal Behaviour* 25: 774–781.

Cook, A. (1979a). Homing by the slug *Limax pseudoflavus* (Evans). *Animal Behaviour* 27: 545–552.

Cook, A. (1979b). Homing in the Gastropoda. *Malacologia* 18: 315–318.

Cook, A. & Shirbhate, R. (1983). The mucus-producing glands and the distribution of the cilia of the pulmonate slug *Limax pseudoflavus*. *Journal of Zoology* 201: 97–116.

Cook, A. (1985). The organisation of feeding in the carnivorous snail *Euglandina rosea* Férussac. *Malacologia* 26: 183–189.

Cook, A. (1987). Functional aspects of the mucus-producing glands of the systellommatophoran slug, *Veronicella floridana*. *Journal of Zoology* 211: 291–305.

Cook, A. (1992). The function of trail following in the pulmonate slug, *Limax pseudoflavus*. *Animal Behaviour* 43: 813–821.

Cook, A. (2001). Behavioural ecology: On doing the right thing, in the right place, at the right time. pp. 447–487 *in* G. M. Barker (Ed.) *The Biology of Terrestrial Molluscs*. Wallingford, UK, CABI Publishing.

Cook, L. M. & King, J. M. B. (1966). Some data on the genetics of shell-character polymorphism in the snail *Arianta arbustorum*. *Genetics* 53: 415–425.

Cook, L. M. & Freeman, P. M. (1986). Heating properties of morphs of the mangrove snail *Littoraria pallescens*. *Biological Journal of the Linnean Society* 29: 295–300.

Cook, L. M. (1998). A two-stage model for *Cepaea* polymorphism. *Philosophical Transactions of the Royal Society B* 353: 1577–1593.

Cook, L. M., Cowie, R. H. & Jones, J. S. (1999). Change in morph frequency in the snail *Cepaea nemoralis* on the Marlborough Downs. *Heredity* 82: 336–342.

Cook, L. M. (2008). Species richness in Madeiran land snails, and its causes. *Journal of Biogeography* 35: 647–653.

Cook, R. T., Bailey, S. E. R., McCrohan, C. R., Nash, B. & Woodhouse, R. M. (2000). The influence of nutritional status on the feeding behaviour of the field slug, *Deroceras reticulatum* (Müller). *Animal Behaviour* 59: 167–176.

Cook, S. B. (1969). Experiments on homing in the limpet *Siphonaria alternata*. *American Zoologist* 9: 1139.

Cook, S. B. (1971). A study in homing behavior in the limpet *Siphonaria alternata*. *Biological Bulletin* 141: 449–457.

Cook, S. B. & Cook, C. B. (1975). Directionality in the trail-following response of the pulmonate limpet *Siphonaria alternata*. *Marine Behaviour and Physiology* 3: 147–155.

Cooke, A. H. (1895). Molluscs. pp. 1–459 *in* S. F. Harmer & Shipley, A. E. (Eds.) *The Cambridge Natural History*. Vol. 3. London, UK, Macmillan Publishers.

Cooke, A. H., Shipley, A. E. & Reed, F. R. C. (1927). *Molluscs, Brachiopods (Recent) and Brachiopods (Fossil)*. London, UK, MacMillan Publishers.

Cooksey, C. J. (2001). Tyrian purple: 6, 6′-dibromoindigo and related compounds. *Molecules* 6: 736–769.

Cooper, A., Dixon, S. F. & Tsuda, M. (1986). Photoenergetics of *Octopus* rhodopsin. *European Biophysics Journal* 13: 195–201.

Cooper, L. A., Larson, S. E. & Lewis, F. A. (1996). Male reproductive success of *Schistosoma mansoni* infected *Biomphalaria glabrata* snails. *Journal of Parasitology* 82: 428–431.

Coote, T. & Loève, É. (2003). From 61 species to five: Endemic tree snails of the Society Islands fall prey to an ill-judged biological control programme. *Oryx* 37: 91–96.

Coovert, A. G. & Coovert, H. K. (1995). Revision of the supraspecific classification of arginelliform gastropods. *The Nautilus* 109: 43–110.

Cope, N. J. & Winterbourn, M. J. (2004). Competitive interactions between two successful molluscan invaders of freshwaters: An experimental study. *Aquatic Ecology* 38: 83–91.

Copeland, M. (1919). Locomotion in two species of the gastropod genus *Alectrion* with observations on the behavior of pedal cilia. *Biological Bulletin* 37: 126–138.

Copeland, M. (1922). Ciliary and muscular locomotion in the gastropod genus *Polinices*. *Biological Bulletin* 42: 132–142.

Corbeil, S., Williams, L. M., Bergfeld, J. & Crane, M. S. J. (2012). Abalone herpes virus stability in sea water and susceptibility to chemical disinfectants. *Aquaculture* 326: 20–26.

Corbet, G. B. & Southern, H. N. (1977). *The Handbook of British Mammals*. Oxford, Blackwell Scientific Publications for the Mammal Society.

Cornell, H. V. (1999). Unsaturation and regional influences on species richness in ecological communities: A review of the evidence. *Ecoscience* 6: 303–315.

Cornman, I. (1963). Toxic properties of the saliva of *Cassis*. *Nature* 200: 88–89.

Corriero, G., Pronzato, R. & Sarà, M. (1991). The sponge fauna associated with *Arca noae* L. (Mollusca, Bivalvia). pp. 395–403 *in* J. Reitner & Keupp, H. (Eds.) *Fossil and Recent Sponges*. Berlin, Heidelberg, Springer-Verlag.

Cortesi, F. & Cheney, K. L. (2010). Conspicuousness is correlated with toxicity in marine opisthobranchs. *Journal of Evolutionary Biology* 23: 1509–1518.

Costa, P. M., Carreira, S., Lobo, J. & Costa, M. H. (2012). Molecular detection of prokaryote and protozoan parasites in the commercial bivalve *Ruditapes decussatus* from southern Portugal. *Aquaculture* 370–371: 61–67.

Cotton, B. C. (1934). A freshwater mussel attached to a duck's foot. *South Australian Naturalist* 15: 113.

Cotton, H. (1905). Sexual dimorphism in *Strombus pugilis*. *The Nautilus* 3: 139–140.

Coulter, G. W. (1991). Composition of the flora and fauna. pp. 200–274 *in* G. W. Coulter (Ed.) *Lake Tanganyika and Its Life*. Oxford, UK, Oxford University Press.

Counihan, R. T., McNamara, D. C., Souter, D. C., Jebreen, E. J., Preston, N. P., Johnson, C. R. & Degnan, B. M. (2001). Pattern, synchrony and predictability of spawning of the tropical abalone *Haliotis asinina* from Heron Reef, Australia. *Marine Ecology Progress Series* 213: 193–202.

Counsilman, J. J., Loh, D., Chan, S. Y., Tan, W. H., Copeland, J. & Maneri, M. (1987). Factors affecting the rate of flashing and loss of luminescence in a Asian land snail, *Dyakia striata*. *The Veliger* 29: 394–399.

Counsilman, J. J. & Ong, P. P. (1988). Responses of the luminescent land snail *Dyakia* (*Quantula*) *striata* to natural and artificial lights. *Journal of Ethology* 6: 1–8.

Coupland, J. & Baker, G. (1995). The potential of several species of terrestrial Sciomyzidae as biological control agents of pest helicid snails in Australia. *Crop Protection* 14: 573–576.

Coupland, J. B. & Barker, G. M. (2004). Diptera as predators and parasitoids of terrestrial gastropods, with emphasis on Phoridae, Calliphoridae, Sarcophagidae, Muscidae and Fanniidae (Diptera, Brachycera, Cyclorrhapha). pp. 85–158 *in* G. M. Barker (Ed.) *Natural Enemies of Terrestrial Molluscs*. Oxford, UK/Cambridge, MA, CABI Publishing.

Coustau, C., Robbins, I., Delay, B., Renaud, F. & Mathieu, M. (1993). The parasitic castration of the mussel *Mytilus edulis* by the trematode parasite *Prosorhynchus squamatus*: Specificity and partial characterization of endogenous and parasite-induced anti-mitotic activities. *Comparative Biochemistry and Physiology Part A* 104: 229–233.

Coutellec, M.-A. & Caquet, T. (2017). Gastropod ecophysiological response to stress. pp. 303–396 *in* S. Saleuddin & Mukai, S. (Eds.) *Physiology of Molluscs*. Vol. 1. Oakville, Ontario, Canada, Apple Academic Press.

Covaleda, G., Del Rivero, M. A., Chávez, M. A., Avilés, F. X. & Reverter, D. (2012). Crystal structure of novel metallocarboxypeptidase inhibitor from marine mollusk *Nerita versicolor* in complex with human carboxypeptidase A4. *Journal of Biological Chemistry* 287: 9250–9258.

Cowie, R. H. (1992a). Shell pattern polymorphism in a 13-year study of the land snail *Theba pisana* (Müller) (Pulmonata: Helicidae). *Malacologia* 34: 87–97.

Cowie, R. H. (1992b). Evolution and extinction of Partulidae, endemic Pacific island land snails. *Philosophical Transactions of the Royal Society B* 335: 167–191.

Cowie, R. H. (1995). Variation in species diversity and shell shape in Hawaiian land snails: *In situ* speciation and ecological relationships. *Evolution* 49: 1191–1202.

Cowie, R. H. (1996). Pacific island land snails: Relationships, origins and determinants of diversity. pp. 347–372 *in* A. Keast & Miller, S. E. (Eds.) *The Origin and Evolution of Pacific Island Biotas, New Guinea to Eastern Polynesia: Patterns and Processes*. Amsterdam, SPB Academic Publishing.

Cowie, R. H. & Jones, J. S. (1998). Gene frequency changes in *Cepaea* snails on the Marlborough Downs over 25 years. *Biological Journal of the Linnean Society* 65: 233–255.

Cowie, R. H. (2000). Non-indigenous land and freshwater molluscs in the islands of the Pacific: Conservation impacts and threats. pp. 143–172 *in* G. Sherley (Ed.) *Invasive Species in the Pacific: A Technical Review and Draft Regional Strategy.* Apia, Samoa, South Pacific Regional Environment Programme.

Cowie, R. H. (2001a). Invertebrate invasions on Pacific islands and the replacement of unique native faunas: A synthesis of the land and freshwater snails. *Biological Invasions* 3: 119–136.

Cowie, R. H. (2001b). Decline and homogenization of Pacific faunas: The land snails of American Samoa. *Biological Conservation* 99: 207–222.

Cowie, R. H. (2001c). Can snails ever be effective and safe biocontrol agents? *International Journal of Pest Management* 47: 23–40.

Cowie, R. H. & Robinson, D. G. (2003). Pathways of introduction of nonindigenous land and freshwater snails and slugs. pp. 93–122 *in* G. M. Ruiz & Carlton, J. T. (Eds.) *Invasive Species: Vectors and Management Strategies* Washington, DC, Island Press.

Cowie, R. H. & Holland, B. S. (2006). Dispersal is fundamental to biogeography and the evolution of biodiversity on oceanic islands. *Journal of Biogeography* 33: 193–198.

Cowie, R. H., Hayes, K. A., Tran, C. T. & Meyer III, W. M. (2008). The horticultural industry as a vector of alien snails and slugs: Widespread invasions in Hawaii. *International Journal of Pest Management* 54: 267–276.

Cowie, R. H., Dillon, R. T., Robinson, D. G. & Smith, J. W. (2009). Alien non-marine snails and slugs of priority quarantine importance in the United States: A preliminary risk assessment. *American Malacological Bulletin* 27: 113–132.

Cowie, R. H. (2011). *Cornu* Born, 1778 (Mollusca: Gastropoda: Pulmonata: Helicidae): Request for a ruling on the availability of the generic name. *Bulletin of Zoological Nomenclature* 68: 97–104.

Cowie, R. H. & Hayes, K. A. (2012). Apple snails. pp. 207–221 *in* R. A. Francis (Ed.) *A Handbook of Global Freshwater Invasive Species.* Oxford, UK, Earthscan Publications (from Routledge).

Cowie, R. H. (2013a). Biology, systematics, life cycle, and distribution of *Angiostrongylus cantonensis*, the cause of rat lungworm disease. *Hawai'i Journal of Medicine & Public Health* 72: 6.

Cowie, R. H. (2013b). Pathways for transmission of angiostrongyliasis and the risk of disease associated with them. *Hawai'i Journal of Medicine & Public Health* 72: 70.

Cox, L. R. (1960). Gastropoda. General characteristics of Gastropoda. pp. 84–169 *in* R. C. Moore (Ed.) *Treatise on Invertebrate Paleontology Part I. Mollusca. Mollusca.* Vol. 1. Lawrence, KS, Geological Society of America and University of Kansas Press.

Cox, L. R., Newell, N. D., Boyd, D. W., Branson, C. C., Casey, R., Chavan, A., Coogan, A. H. et al. (1969). Bivalvia. pp. i–xxxviii, N1–N489; N491–N868 *in* R. C. Moore & Teichert, C. (Eds.) *Treatise on Invertebrate Paleontology Part N [part 1 and 2] Bivalvia. Treatise on Invertebrate Paleontology.* Vol. Mollusca 6. Lawrence, KS, Geological Society of America and University of Kansas Press.

Cox, P. A. & Sethian, J. A. (1985). Gamete motion, search, and the evolution of anisogamy, oogamy, and chemotaxis. *American Naturalist*: 74–101.

Cox, T. E. & Murray, S. N. (2006). Feeding preferences and the relationships between food choice and assimilation efficiency in the herbivorous marine snail *Lithopoma undosum* (Turbinidae). *Marine Biology* 148: 1295–1306.

Coyne, V. E. (2011). The importance of ATP in the immune system of molluscs. *Invertebrate Survival Journal* 8: 48–55.

Craddock, C., Lutz, R. A. & Vrijenhoek, R. C. (1997). Patterns of dispersal and larval development of archaeogastropod limpets at hydrothermal vents in the eastern Pacific. *Journal of Experimental Marine Biology and Ecology* 210: 37–51.

Cragg, S. M. & Gruffydd, L. D. (1975). The swimming behaviour and the pressure responses of the veliconcha larvae of *Ostrea edulis* L. pp. 43–57 *in* H. Barnes (Ed.) *Proceedings of the 9th European Marine Biological Symposium Oban, Scotland, October 2–8, 1974.* Aberdeen, UK, Aberdeen University Press.

Cragg, S. M. & Crisp, D. J. (1991). The biology of scallop larvae. pp. 75–132 *in* S. E. Shumway (Ed.) *Biology, Ecology and Aquaculture of Scallops. Developments in Aquaculture and Fisheries Science.* Amsterdam, the Netherlands, Elsevier.

Crame, J. A. (2000). Evolution of taxonomic diversity gradients in the marine realm: Evidence from the composition of Recent bivalve faunas. *Paleobiology* 26: 188–214.

Crampton, D. M. (1977). Functional anatomy of the buccal apparatus of *Onchidoris bilamellata* (Mollusca: Opisthobranchia). *Transactions of the Zoological Society of London* 34: 45–86.

Cranfield, H. J. (1975). The ultrastructure and histochemistry of the larval cement of *Ostrea edulis* L. *Journal of the Marine Biological Association of the United Kingdom* 55: 497–503.

Creek, G. A. (1951). The reproductive system and embryology of the snail *Pomatias elegans* (Müller). *Proceedings of the Zoological Society of London* 121: 599–640.

Creek, G. A. (1953). The morphology of *Acme fusca* (Montague) with special reference to the genital system. *Proceedings of the Malacological Society of London* 29: 228–240.

Creese, R. G. & Underwood, A. J. (1976). Observations on the biology of the trochid gastropod *Austrocochlea constricta* (Lamarck) (Prosobranchia). I. Factors affecting shell-banding pattern. *Journal of Experimental Marine Biology and Ecology* 23: 211–228.

Creese, R. G. (1980). Reproductive cycles and fecundities of two species of *Siphonaria* (Mollusca, Pulmonata) in south-eastern Australia. *Australian Journal of Marine and Freshwater Research* 31: 37–47.

Creese, R. G. (1981). Patterns of growth, longevity and recruitment of intertidal limpets in New South Wales. *Journal of Experimental Marine Biology and Ecology* 51: 145–171.

Creese, R. G. (1982). Distribution and abundance of the acmaeid limpet *Patelloida latistrigata*, and its interaction with barnacles. *Oecologia* 52: 85–96.

Creese, R. G. & Underwood, A. J. (1982). Analysis of inter- and intra-specific competition amongst intertidal limpets with different methods of feeding. *Oecologia* 53: 337–346.

Creese, R. G. (1988). Ecology of molluscan grazers and their interactions with marine algae in northeastern New Zealand: A review. *New Zealand Journal of Marine and Freshwater Research* 22: 427–444.

Cretchley, R., Hodgson, A. N., Gray, D. R. & Reddy, K. (1997). Variation in foraging activity of *Acanthochitona garnoti* (Mollusca : Polyplacophora) from different habitats. *South African Journal of Zoology* 32: 59–63.

Cribb, T. H., Bray, R. A. & Littlewood, D. T. J. (2001a). The nature and evolution of the association among digeneans, molluscs and fishes. *International Journal for Parasitology* 31: 997–1011.

Cribb, T. H., Bray, R. N., Littlewood, D. T. J., Pichelin, S. P. & Herniou, E. A. (2001b). The Digenea. pp. 168–185 *in* D. T. J. Littlewood & Bray, R. N. (Eds.) *Interrelationships of the Platyhelminthes.* London, UK, Taylor & Francis Group.

Cribb, T. H., Bray, R. A., Olson, P. D. & Littlewood, D. T. J. (2003). Life cycle evolution in the Digenea: A new perspective from phylogeny. *Advances in Parasitology* 54: 197–254.

Crisp, D. J., Burfitt, A., Rodrigues, K. & Budd, M. D. (1983). *Lasaea rubra*: An apomictic bivalve. *Marine Biology Letters* 4: 127–136.

Crisp, M. (1971). Structure and abundance of receptors of the unspecialized external epithelium of *Nassarius reticulatus* (Gastropoda, Prosobranchia). *Journal of the Marine Biological Association of the United Kingdom* 51: 865–890.

Crisp, M. (1972). Photoreceptive function of an epithelial receptor in *Nassarius reticulatus* (Gastropoda, Prosobranchia). *Journal of the Marine Biological Association of the United Kingdom* 52: 437–442.

Crisp, M. (1973). Fine structure of some prosobranch osphradia. *Marine Biology* 22: 231–240.

Crisp, M. (1981). Epithelial sensory structures of trochids. *Journal of the Marine Biological Association of the United Kingdom* 61: 95–106.

Crofts, D. R. (1929). *Haliotis*. Liverpool Marine Biology Committee: Memoirs on typical British Marine Plants and Animals. Liverpool, UK, Liverpool Marine Biology Committee.

Crofts, D. R. (1937). The development of *Haliotis tuberculata*, with special reference to organogenesis during torsion. *Philosophical Transactions of the Royal Society B* 228: 219–268.

Crofts, D. R. (1955). Muscle morphogenesis in primitive gastropods and its relation to torsion. *Proceedings of the Zoological Society of London* 125: 711–750.

Croghan, P. C. (1983). Osmotic regulation and the evolution of brackish- and fresh-water faunas. *Journal of the Geological Society* 140: 39–46.

Croll, R. P. & Chase, R. (1977). A long-term memory for food odors in the land snail, *Achatina fulica*. *Behavioral Biology* 19: 261–268.

Croll, R. P. (1983). Gastropod chemoreception. *Biological Reviews* 58: 293–319.

Croll, R. P. (1985). Sensory control of respiratory pumping in *Aplysia californica*. *Journal of Experimental Biology* 117: 15–28.

Croll, R. P. & Lo, R. Y. S. (1986). Distribution of serotonin-like immunoreactivity in the central nervous system of the periwinkle, *Littorina littorea* (Gastropoda, Prosobranchia, Mesogastropoda). *Biological Bulletin* 171: 426–440.

Croll, R. P. & Van Minnen, J. (1992). Distribution of the peptide Ala-Pro-Gly-Trp-NH2 (APGWamide) in the nervous system and periphery of the snail *Lymnaea stagnalis* as revealed by immunocytochemistry and in situ hybridization. *Journal of Comparative Neurology* 324: 567–574.

Croll, R. P., Too, C. K. L., Pani, A. K. & Nason, J. (1995). Distribution of serotonin in the sea scallop *Placopecten magellanicus*. *Invertebrate Reproduction & Development* 28: 125–135.

Croll, R. P. & Voronezhskaya, E. E. (1996). Early neurodevelopment in *Aplysia*, *Lymnaea* and *Helisoma*. *Society for Neuroscience Abstracts* 22: 1948.

Croll, R. P., Boudko, D. Y., Pires, A. & Hadfield, M. G. (2003). Transmitter contents of cells and fibers in the cephalic sensory organs of the gastropod mollusc *Phestilla sibogae*. *Cell and Tissue Research* 314: 437–448.

Croll, R. P. & Dickinson, A. J. G. (2004). Form and function of the larval nervous system in molluscs. *Invertebrate Reproduction & Development* 46: 173–187.

Croll, R. P. (2007). Reproductive endocrinology in bivalves: We know less than you might think. *Journal of Shellfish Research* 26: 1302.

Croll, R. P. & Wang, C. (2007). Possible roles of sex steroids in the control of reproduction in bivalve molluscs. *Aquaculture* 272: 76–86.

Crook, R. J. & Basil, J. A. (2008). A role for *Nautilus* in studies of the evolution of brain and behavior. *Communicative & Integrative Biology* 1: 18–19.

Crook, R. J., Hanlon, R. T. & Basil, J. A. (2009). Memory of visual and topographical features suggests spatial learning in nautilus (*Nautilus pompilius* L.). *Journal of Comparative Psychology* 123: 264–274.

Crook, R. J., Lewis, T., Hanlon, R. T. & Walters, E. T. (2011). Peripheral injury induces long-term sensitization of defensive responses to visual and tactile stimuli in the squid *Loligo pealeii*, Lesueur 1821. *Journal of Experimental Biology* 214: 3173–3185.

Crook, R. J. & Walters, E. T. (2011). Nociceptive behavior and physiology of molluscs: Animal welfare implications. *Institute of Laboratory Animal Resources (US) Journal* 52: 185–195.

Crook, R. J., Hanlon, R. T. & Walters, E. T. (2013). Squid have nociceptors that display widespread long-term sensitization and spontaneous activity after bodily injury. *Journal of Neuroscience* 33: 10021–10026.

Crook, R. J., Dickson, K., Hanlon, R. T. & Walters, E. T. (2014). Nociceptive sensitization reduces predation risk. *Current Biology* 24: 1121–1125.

Crosby, N. D. & Reid, R. G. B. (1971). Relationships between food, phylogeny, and cellulose digestion in Bivalvia. *Canadian Journal of Zoology* 49: 617–622.

Crossland, M. R., Alford, R. A. & Collins, J. D. (1991). Population dynamics of an ectoparasitic gastropod, *Hypermastus* sp. (Eulimidae), on the sand dollar, *Arachnoides placenta* (Echinoidea). *Australian Journal of Marine and Freshwater Research* 42: 69–76.

Crossland, M. R. & Alford, R. A. (1998). Evaluation of the toxicity of eggs, hatchlings and tadpoles of the introduced toad *Bufo marinus* (Anura: Bufonidae) to native Australian aquatic predators. *Australian Journal of Ecology* 23: 129–137.

Crosson, L. M., Wight, N., VanBlaricom, G. R., Kiryu, I., Moore, J. D. & Friedman, C. S. (2014). Abalone withering syndrome: Distribution, impacts, current diagnostic methods and new findings. *Diseases of Aquatic Organisms* 108: 261–270.

Crowe, T. P., Lee, C. L., McGuinness, K. A., Amos, M. J., Dangeubun, J., Dwiono, S. A. P., Makatipu, P. C. et al. (2002). Experimental evaluation of the use of hatchery-reared juveniles to enhance stocks of the topshell *Trochus niloticus* in Australia, Indonesia and Vanuatu. *Aquaculture* 206: 175–197.

Croxall, J. P. & Prince, P. A. (1994). Dead or alive, night or day: How do albatrosses catch squid? *Antarctic Science* 6: 155–162.

Crozier, W. J. & Pilz, G. F. (1924). The locomotion of *Limax*: 1: Temperature coefficient of pedal activity. *Journal of General Physiology* 6: 711–721.

Crump, R. G. (1968). The flight response in *Struthiolaria papulosa gigas* Sowerby. *New Zealand Journal of Marine and Freshwater Research* 2: 390–397.

Cruz, A. K. M. da, Pereira, W. O., Santos, E. A. dos, Carvalho, M. G. F., Medeiros, A., da Cunha & de Oliveira, F. W. (2004). Comparative study between the effects of hyaluronic acid and acid galactan purified from eggs of the mollusk *Pomacea* sp. in wound healing. *Acta Cirurgica Brasileira* 19: 13–17.

Cruz, R., Lins, U. & Farina, M. (1998). Minerals of the radular apparatus of *Falcidens* sp. (Caudofoveata) and the evolutionary implications for the phylum Mollusca. *Biological Bulletin* 194: 224–230.

Cruz, R. & Farina, M. (2005). Mineralization of major lateral teeth in the radula of a deep-sea hydrothermal vent limpet (Gastropoda: Neolepetopsidae). *Marine Biology* 147: 163–168.

Cruz, S., Calado, R., Serôdio, J. & Cartaxana, P. (2013). Crawling leaves: Photosynthesis in sacoglossan sea slugs. *Journal of Experimental Botany* 64: 3999–4009.

Csaba, G. & Bierbauer, J. (1979). Effect of oestrogenic, androgenic and gestagenic hormones on the gametogenesis (oogenesis and spermatogenesis) in the snail *Helix pomatia*. *Acta biologica et medica Germanica* 38: 1145.

Cuccu, D., Mereu, M., Porcu, C., Follesa, M. C., Cau, A. & Cau, A. (2013). Development of sexual organs and fecundity in *Octopus vulgaris* Cuvier, 1797 from the Sardinian waters (Mediterranean Sea). *Mediterranean Marine Science* 14: 270–277.

Cudney-Bueno, R., Prescott, R. & Hinojosa-Huerta, O. (2008). The black Murex snail, *Hexaplex nigritus* (Mollusca, Muricidae), in the Gulf of California, Mexico: I. Reproductive ecology and breeding aggregations. *Bulletin of Marine Science* 83: 285–298.

Cuervo-González, R. (2017). *Rhodope placozophagus* (Heterobranchia) a new species of turbellarian-like Gastropoda that preys on placozoans. *Zoologischer Anzeiger* 270: 43–48.

Cui, L. B., Liu, C. L., Liu, X. & Lu, Y. H. (2001a). The cell types and secretion of the digestive gland in *Haliotis discus hannai* Ino. *Acta Zoologica Sinica* 47: 32–37.

Cui, L. B., Liu, C. L., Liu, X. & Lu, Y. H. (2001b). Structure and function of mucous epithelium of the intestine in *Haliotis discus hannai*. *Acta Zoologica Sinica* 47: 324–328.

Cumin, R. (1972). Normentafel zur Organogenese von *Lymnaea stagnalis* (Gastropoda, Pulmonata), mit besonderer Berücksichtigung der Mitteldarmdrüse. *Revue suisse de Zoologie* 79: 709–774.

Cummins, S. F., Degnan, B. M. & Nagle, G. T. (2008). Characterization of *Aplysia* Alb-1, a candidate water-borne protein pheromone released during egg laying. *Peptides* 29: 152–161.

Cummins, S. F., Boal, J. G., Buresch, K. C., Kuanpradit, C., Sobhon, P., Holm, J. B., Degnan, B. M., Nagle, G. T. & Hanlon, R. T. (2011). Extreme aggression in male squid induced by a β-MSP-like pheromone. *Current Biology* 21: 322–327.

Cunningham, A. A. & Daszak, P. (1998). Extinction of a species of land snail due to infection with a microsporidian parasite. *Conservation Biology* 12: 1139–1141.

Current, W. L. (1980). *Cryptobia* sp. in the snail *Triadopsis multilineata* (Say): Fine structure of attached flagellates and their mode of attachment to the spermatheca. *The Journal of Protozoology* 27: 278–287.

Currey, J. D. & Kohn, A. J. (1976). Fracture in the crossed lamellar structure of *Conus* shells. *Journal of Materials Science* 11: 443–463.

Currey, J. D. (1988). Shell form and strength. pp. 183–210 *in* E. R. Trueman & Clarke, M. R. (Eds.) *Form and Function. The Mollusca*. Vol. 11. New York, Academic Press.

Currey, J. D. (1990). Biomechanics of mineralized skeletons. pp. 11–25 *in* J. G. Carter (Ed.) *Skeletal Biomineralization: Patterns, Processes and Evolutionary Trends*. Vol. 1. New York, Van Nostrand Reinhold.

Curtis, L. A. (1980). Daily cycling of the crystalline style in the omnivorous, deposit-feeding estuarine snail *Ilyanassa obsoleta*. *Marine Biology* 59: 137–140.

Curtis, L. A. & Hurd, L. E. (1981). Nutrient procurement strategy of a deposit-feeding estuarine neogastropod, *Ilyanassa obsoleta*. *Estuarine, Coastal and Shelf Science* 13: 277–285.

Cutignano, A., Calado, G., Gaspar, H., Cimino, G. & Fontana, A. (2011). Polypropionates from *Bulla occidentalis*: Chemical markers and trophic relationships in cephalaspidean molluscs. *Tetrahedron Letters* 52: 4595–4597.

Cuvier, G. (1803). Mémoire sur le genre *Laplysia*, vulgairement nommé Lièvre marin, sur son anatomie, et sur quelques-unes de ses espèces. *Annales du Muséum d'histoire naturelle Paris* 2: 287–314.

Cyrus, A. Z., Rupert, S. D., Silva, A. S., Graf, M., Rappaport, J. C., Paladino, F. V. & Peters, W. S. (2012). The behavioural and sensory ecology of *Agaronia propatula* (Caenogastropoda: Olividae), a swash-surfing predator on sandy beaches of the Panamic faunal province. *Journal of Molluscan Studies* 78: 235–245.

D'Arcy Thompson, W. (1917). *On Growth and Form*. Cambridge, UK, Cambridge University Press.

D'Asaro, C. N. (1965). Organogenesis, development, and metamorphosis in the Queen Conch, *Strombus gigas*, with notes on breeding habits. *Bulletin of Marine Science* 15: 359–416.

D'Asaro, C. N. (1966). The egg capsules, embryogenesis, and early organogenesis of a common oyster predator, *Thais haemastoma floridana* (Gastropoda: Prosobranchia). *Bulletin of Marine Science* 16: 884–914.

D'Errico, F., Vanhaeren, M. & Wadley, L. (2008). Possible shell beads from the Middle Stone Age layers of Sibudu Cave, South Africa. *Journal of Archaeological Science* 35: 2675–2685.

D'Asaro, C. N. (1969). The egg capsules of *Jenneria pustulata* (Lightfoot, 1786) with notes on spawning in the laboratory. *The Veliger* 11: 182–184.

da Costa, F., Darriba, S. & Martinez-Patino, D. (2008). Embryonic and larval development of *Ensis arcuatus* (Jeffreys, 1865) (Bivalvia: Pharidae). *Journal of Molluscan Studies* 74: 103–109.

Da Silva, J. J. R. F. & Williams, R. J. P. (1991). *The Biological Chemistry of the Elements*. Oxford, UK, Clarendon Press.

Dahms, H.-U. & Lee, J.-S. (2010). UV radiation in marine ectotherms: Molecular effects and responses. *Aquatic Toxicology* 97: 3–14.

Dakin, W. J. (1912). *Buccinum (the Whelk)*. Liverpool Marine Biology Committee: Memoirs on typical British marine plants and animals. London, UK, Williams and Norgate.

Dalby, J. E. (1989). Predation of ascidians by *Melongena corona* (Neogastropoda: Melongenidae) in the northern Gulf of Mexico. *Bulletin of Marine Science* 45: 708–712.

Dale, B. (1974). Extrusion, retraction and respiratory movements in *Helix pomatia* in relation to distribution and circulation of the blood. *Journal of Zoology* 173: 427–439.

Dall, W. H. (1882). On certain limpets and chitons from the deep waters off the western coast of the United States. *Proceedings of the United States National Museum* 81: 407–414.

Dallinger, R., Berger, B., Triebskorn-Köhler, R. & Köhler, H. R. (2001). Soil biology and ecotoxicology. pp. 489–525 *in* G. M. Barker (Ed.) *The Biology of Terrestrial Molluscs*. Wallingford, UK, CABI Publishing.

Dalziel, A. C. & Stewart, D. T. (2002). Tissue-specific expression of male-transmitted mitochondrial DNA and its implications for rates of molecular evolution in *Mytilus* mussels (Bivalvia: Mytilidae). *Genome* 45: 348–355.

Damborenea, C., Brusa, F. & Paola, A. (2006). Variation in worm assemblages associated with *Pomacea canaliculata* (Caenogastropoda, Ampullariidae) in sites near the Río de la Plata estuary, Argentina. *Biocell* 30: 457–468.

Dame, R. F., Zingmark, R. G. & Haskin, E. (1984). Oyster reefs as processors of estuarine materials. *Journal of Experimental Marine Biology and Ecology* 83: 239–247.

Dame, R. F., Spurrier, J. D. & Wolaver, T. G. (1989). Carbon, nitrogen and phosphorus processing by an oyster reef. *Marine Ecology Progress Series* 54: 249–256.

Dame, R. F. (1996). *Ecology of Marine Bivalves: An Ecosystem Approach*. Boca Raton, FL, CRC Press.

Dame, R. F. & Prins, T. C. (1997). Bivalve carrying capacity in coastal ecosystems. *Aquatic Ecology* 31: 409–421.

Dame, R. F. (2012). *Ecology of Marine Bivalves: An Ecosystem Approach*. 2nd ed. Boca Raton, FL, CRC Press.

Damen, P. & Dictus, W. (2002). Newly-discovered muscle in the larva of *Patella caerulea* (Mollusca, Gastropoda) suggests the presence of a larval extensor. *Contributions to Zoology* 71: 37–45.

Dance, S. P. (1966). *Shell Collecting: An Illustrated History*. London, UK, Faber and Faber.

Dance, S. P. (1986). *A History of Shell Collecting*. Leiden, the Netherlands, Brill Academic Publishers.

Dando, P. R., Southward, A. J. & Southward, E. C. (1986). Chemoautotrophic symbionts in the gills of the bivalve mollusc *Lucinoma borealis* and the sediment chemistry of its habitat. *Proceedings of the Royal Society B* 227: 227–247.

Daniel, M. J. (1973). Seasonal diet of the ship rat (*Rattus r. rattus*) in lowland forest in New Zealand. *Proceedings of the New Zealand Ecological Society* 20: 21–30.

Daraio, C., Nesterenko, V. F., Jin, S., Wang, W. & Rao, A. M. (2006). Impact response by a foamlike forest of coiled carbon nanotubes. *Journal of Applied Physics* 100: 064309 (1–4).

Darmaillacq, A.-S., Dickel, L., Chichery, M.-P., Agin, V. & Chichery, R. (2004). Rapid taste aversion learning in adult cuttlefish, *Sepia officinalis*. *Animal Behaviour* 68: 1291–1298.

Darmaillacq, A. S., Dickel, L. & Mather, J. A., Eds. (2014). *Cephalopod Cognition*. Cambridge, UK, Cambridge University Press.

Darwin, C. R. (1839). *Narrative of the Surveying Voyages of His Majesty's Ships Adventure and Beagle between the Years 1826 and 1836, Describing Their Examination of the Southern Shores of South America, and the Beagle's Circumnavigation of the Globe. Journal and Remarks. 1832–1836.* Vol. 3. London, UK, Henry Colburn.

Daston, M. M. & Copeland, J. (1993). The luminescent organ and sexual maturity in *Dyakia striata*. *Malacologia* 35: 9–19.

Dattagupta, S., Martin, J., Liao, S.-M., Carney, R. S. & Fisher, C. R. (2007). Deep-sea hydrocarbon seep gastropod *Bathynerita naticoidea* responds to cues from the habitat-providing mussel *Bathymodiolus childressi*. *Marine Ecology* 28: 193–198.

Daume, S., Brand-Gardner, S. & Woelkerling, W. J. (1999a). Preferential settlement of abalone larvae: Diatom films vs. non-geniculate coralline red algae. *Aquaculture* 174: 243–254.

Daume, S., Brand-Gardner, S. & Woelkerling, W. J. (1999b). Settlement of abalone larvae (*Haliotis laevigata* Donovan) in response to non-geniculate coralline red algae (Corallinales, Rhodophyta). *Journal of Experimental Marine Biology and Ecology* 234: 125–143.

Daume, S., Krsinich, A., Farrell, S. & Gervis, M. (2000). Growth and survival of *Haliotis rubra* post-larvae feeding on different algal species (Abstract). *Journal of Shellfish Research* 19: 506.

Dauphin, Y. (1996). The organic matrix of coleoid cephalopod shells: Molecular weights and isoelectric properties of the soluble matrix in relation to biomineralization processes. *Marine Biology* 125: 525–529.

Dauphin, Y. & Denis, A. (2000). Structure and composition of the aragonitic crossed lamellar layers in six species of Bivalvia and Gastropoda. *Comparative Biochemistry and Physiology Part A* 126: 367–377.

Davenport, D. (1950). Studies in the physiology of commensalism. I. The polynoid genus *Arctonoë*. *Biological Bulletin* 98: 81–93.

Davenport, J. (1979). Is *Mytilus edulis* a short term osmoregulator? *Comparative Biochemistry and Physiology Part A* 64: 91–95.

Davenport, J. (1981). The opening response of mussels (*Mytilus edulis* L.) exposed to rising sea-water concentrations. *Journal of the Marine Biological Association of the United Kingdom* 61: 667–668.

Davenport, J. (1988). Oxygen consumption and ventilation rate at low temperatures in the Antarctic protobranch bivalve mollusc *Yoldia eightsi* (Courthouy). *Comparative Biochemistry and Physiology Part A* 90: 511–514.

Davidson, D. H. (1976). Assimilation efficiencies of slugs on different food materials. *Oecologia* 26: 267–273.

Davidson, E. H. (1990). How embryos work: A comparative view of diverse modes of cell fate specification. *Development* 108: 365–389.

Davidson, S. K., Koropatnick, T. A., Kossmehl, R., Sycuro, L. & McFall-Ngai, M. J. (2004). NO means 'yes' in the squid-vibrio symbiosis: Nitric oxide (NO) during the initial stages of a beneficial association. *Cellular Microbiology* 6: 1139–1151.

Davies, A., Schertler, G. F., Gowen, B. E. & Saibil, H. R. (1996). Projection structure of an invertebrate rhodopsin. *Journal of Structural Biology* 117: 36–44.

Davies, A., Gowen, B. E., Krebs, A. M., Schertler, G. F. & Saibil, H. R. (2001). Three-dimensional structure of an invertebrate rhodopsin and basis for ordered alignment in the photoreceptor membrane. *Journal of Molecular Biology* 314: 455–463.

Davies, M. S., Hawkins, S. J. & Jones, H. D. (1992). Pedal mucus and its influence on the microbial food supply of two intertidal gastropods, *Patella vulgata* L. and *Littorina littorea* (L.). *Journal of Experimental Marine Biology and Ecology* 161: 57–77.

Davies, M. S. & Williams, G. A. (1995). Pedal mucus of a tropical limpet, *Cellana grata* (Gould): Energetics, production and fate. *Journal of Experimental Marine Biology and Ecology* 186: 77–87.

Davies, M. S. & Hawkins, S. J. (1998). Mucus from marine molluscs. *Advances in Marine Biology* 34: 1–71.

Davis, A. J. R. & Fleure, H. J. (1903). *Patella. Proceedings and Transactions of The Liverpool Biological Society* 17: 193–268.

Davis, E. C. (2004). Odour tracking to a food source by the gastropod *Meridolum gulosum* (Gould, 1864) from New South Wales, Australia (Camaenidae: Eupulmonata: Mollusca). *Molluscan Research* 24: 187–191.

Davis, G. M. & Lindsay, G. K. (1964). Disc electrophoresis in the study of molluscan systematics. *Bulletin of the American Malacological Union* 31: 20–21.

Davis, G. M. (1967). The systematic relationship of *Pomatiopsis lapidaria* and *Oncomelania hupensis formosana* (Prosobranchia, Hydrobiidae). *Malacologia* 6: 1–143.

Davis, G. M. (1981). Different modes of evolution and adaptive radiation in the Pomatiopsidae (Prosobranchia, Mesogastropoda). *Malacologia* 21: 209–262.

Davis, M. & Stoner, A. W. (1994). Trophic cues induce metamorphosis of queen conch larvae (*Strombus gigas* Linnaeus). *Journal of Experimental Marine Biology and Ecology* 180: 83–102.

Davis, W. J., Villet, J., Lee, D., Rigler, M., Gillette, R. & Prince, E. (1980). Selective and differential avoidance learning in the feeding and withdrawal behavior of *Pleurobranchaea californica*. *Journal of Comparative Physiology A* 138: 157–165.

Davison, A. (2002). Land snails as a model to understand the role of history and selection in the origins of biodiversity. *Population Ecology* 44: 129–136.

Davison, A. & Blaxter, M. (2005). Ancient origin of glycosyl hydrolase family 9 cellulase genes. *Molecular Biology and Evolution* 22: 1273–1284.

Davison, A., Wade, C. M., Mordan, P. B. & Chiba, S. (2005). Sex and darts in slugs and snails (Mollusca: Gastropoda: Stylommatophora). *Journal of Zoology* 267: 329–338.

Dawson, J. F. & Holmes, C. F. (1999). Molecular mechanisms underlying inhibition of protein phosphatases by marine toxins. *Frontiers of Bioscience* 4: D646–D658.

Day, J. A. (1969). Feeding of the cymatiid gastropod, *Argobuccinum argus*, in relation to the structure of the proboscis and secretions of the proboscis gland. *American Zoologist* 9: 909–916.

Day, R., Dowel, A., Sant, G., Klemke, J. & Shaw, C. (1995). Patchy predation: Foraging behavior of *Coscinasterias calamaria* and escape response of *Haliotis rubra*. *Marine and Fresh Water Behaviour and Physiology* 26: 11–33.

Day, R. W. & Fleming, A. E. (1992). The determinants and measurement of abalone growth. pp. 141–168 *in* S. A. Shepherd, Tegner, M. J. & Guzman Del Proo, S. A. (Eds.) *Abalone of the World: Biology, Fisheries and Culture*. Oxford, UK, Fishing News Books.

Dayrat, B. A. & Tillier, S. (2002). Evolutionary relationships of euthyneuran gastropods (Mollusca): A cladistic re-evaluation of morphological characters. *Zoological Journal of the Linnean Society* 135: 403–470.

Dayrat, B. A. (2009). Review of the current knowledge of the systematics of Onchidiidae (Mollusca: Gastropoda: Pulmonata) with a checklist of nominal species. *Zootaxa* 2068: 1–26.

Dayrat, B. A., Conrad, M., Balayan, S., White, T. R., Albrecht, C., Golding, R. E., Gomes, S. R., Harasewych, M. G. & de Frias Martins, A. M. (2011). Phylogenetic relationships and evolution of pulmonate gastropods (Mollusca): New insights from increased taxon sampling. *Molecular Phylogenetics and Evolution* 59: 425–437.

Dayton, P. K. (1971). Competition, disturbance, and community organization: The provision and subsequent utilization of space in a rocky intertidal community. *Ecological Monographs* 41: 351–389.

Dazo, B. C. & Moreno, R. G. (1962). Studies on the food and feeding habits of *Oncomelania quadrasi*, the snail intermediate host of *Schistosoma japonicum* in the Philippines. *Transactions of the American Microscopical Society* 81: 341–347.

De-Carli, B. P., Rotundo, M. M., Paschoal, L. R. P., Andrade, D. P. & Cavallari, D. C. (2014). First record of the association between the leech *Helobdella triserialis* (Hirudinea, Glossiphoniidae) and two species of *Pomacea* (Gastropoda, Ampullariidae) in Brazil. *Pan-American Journal of Aquatic Sciences* 9: 136–140.

de Bello, F., Lavorel, S., Díaz, S., Harrington, R., Cornelissen, J. H. C., Bardgett, R. D., Berg, M. P., Cipriotti, P., Feld, C. K. & Hering, D. (2010). Towards an assessment of multiple ecosystem processes and services via functional traits. *Biodiversity and Conservation* 19: 2873–2893.

De Lange, R. P. J., Van Minnen, J. & Boer, H. H. (1994). Expression and translation of the egg-laying neuropeptide hormone genes during post-embryonic development of the pond snail *Lymnaea stagnalis*. *Cell and Tissue Research* 275: 369–375.

De Lisa, E., Carella, F., Vico, G. D. & Cosmo, A. D. (2013). The gonadotropin releasing hormone (GnRH)-like molecule in prosobranch *Patella caerulea*: Potential biomarker of endocrine-disrupting compounds in marine environments. *Zoological Science* 30: 135–140.

De Nie, H. W. (1982). A note on the significance of larger bivalve molluscs (*Anodonta* spp. and *Dreissena* sp.) in the food of the eel (*Anguilla anguilla*) in Tjeukemeer. *Hydrobiologia* 95: 307–310.

de Oliveira, A. L., Wollesen, T., Kristof, A., Scherholz, M., Redl, E., Todt, C., Bleidorn, C. & Wanninger, A. (2016). Comparative transcriptomics enlarges the toolkit of known developmental genes in mollusks. *BMC Genomics* 17: 905.

Dean, D. (1958). New property of the crystalline style of *Crassostrea virginica*. *Science* 128: 837.

Deaton, L. E. (1981). Ion regulation in freshwater and brackish water bivalve molluscs. *Physiology and Zoology* 54: 109–121.

Deaton, L. E. (1990). Potentiation of hypoosmotic cellular volume regulation in the quahog, *Mercenaria mercenaria*, by 5-hydroxytryptamine, FMRFamide, and phorbol esters. *Biological Bulletin* 178: 260–266.

Deaton, L. E., Felgenhauer, B. E. & Duhon, D. W. (2001). Bulbus arteriosus of the bivalve mollusc *Mercenaria mercenaria*: Morphology and pharmacology. *Journal of Morphology* 250: 185–195.

Deaton, L. E. (2008). Osmotic and ionic regulation in molluscs. pp. 107–133 *in* D. H. Evans (Ed.) *Osmotic and Ionic Regulation*. Boca Raton, FL, CRC Press.

Deaton, L. E. (2009). The effect of 5-hydroxytryptamine and molluscan neuropeptides on the hearts of gastropod molluscs: Mechanical activity and second messengers. *Journal of Experimental Marine Biology and Ecology* 377: 43–47.

Debruyne, L., On, S. L. W., De Brandt, E. & Vandamme, P. (2009). Novel *Campylobacter lari*-like bacteria from humans and molluscs: Description of *Campylobacter peloridis* sp. nov., *Campylobacter lari* subsp. *concheus* subsp. nov. and *Campylobacter lari* subsp. *lari* subsp. nov. *International Journal of Systematic and Evolutionary Microbiology* 59: 1126–1132.

DeBusk, B. C., Chimote, S. S., Rimoldi, J. M. & Schenk, D. (2000). Effect of the dietary brominated phenol, lanasol, on chemical biotransformation enzymes in the gumboot chiton *Cryptochiton stelleri* (Middendorf, 1846). *Comparative Biochemistry and Physiology Part C* 127: 133–142.

Decker, C., Zorn, N., Le Bruchec, J., Caprais, J. C., Potier, N., Leize-Wagner, E., Lallier, F. H., Olu, K. & Andersen, A. C. (2017). Can the hemoglobin characteristics of vesicomyid clam species influence their distribution in deep-sea sulfide-rich sediments? A case study in the Angola Basin. *Deep-Sea Research Part II-Topical Studies in Oceanography* 142: 219–232.

Decker, H., Schweikardt, T. & Tuczek, F. (2006). The first crystal structure of tyrosinase: All questions answered? *Angewandte Chemie (International Edition)* 45: 4546–4550.

Decker, H., Jaenicke, E., Hellmann, N., Lieb, B., Meissner, U. & Markl, J. (2007). Minireview: Recent progress in hemocyanin research. *International Comparative Biology* 47: 631–644.

Declerck, C. H. (1995). The evolution of suspension feeding in gastropods. *Biological Reviews* 70: 549–569.

Deeds, J. R., Landsberg, J. H., Etheridge, S. M., Pitcher, G. C. & Longan, S. W. (2008). Non-traditional vectors for paralytic shellfish poisoning. *Marine Drugs* 6: 308–348.

Deffit, S. N. & Hundley, H. A. (2016). To edit or not to edit: Regulation of ADAR editing specificity and efficiency. *Wiley Interdisciplinary Reviews: RNA* 7: 113–127.

Degnan, B. M. & Morse, D. E. (1993). Identification of eight homeobox-containing transcripts expressed during larval development and at metamorphosis in the gastropod mollusc *Haliotis rufescens*. *Molecular Marine Biology and Biotechnology* 2: 1–9.

Degnan, B. M., Degnan, S. M., Fentenany, G. & Morse, D. E. (1997). A Mox homeobox gene in the gastropod mollusc *Haliotis rufescens* is differentially expressed during larval morphogenesis and metamorphosis. *Federation of European Biochemical Societies (FEBS) Letters* 411: 119–122.

Deheyn, D. D. & Wilson, N. G. (2011). Bioluminescent signals spatially amplified by wavelength-specific diffusion through the shell of a marine snail. *Proceedings of the Royal Society B* 278: 2112–2121.

Dele-Dubois, M. L. & Merlin, J. C. (1981). Étude par spectroscopie Raman de la pigmentation du squelette calcaire du corail. *Revue de Gemmologie* 68: 10–13.

Delhaye, W. & Bouillon, J. (1972a). L'évolution et l'adaptation de l'organe excréteur chez les mollusques gastéropodes pulmonés: I: Introduction générale et histophysiologie comparée du rein chez les basommatophores. *Bulletin biologique de France et de Belgique* 106: 45–77.

Delhaye, W. & Bouillon, J. (1972b). L'évolution et l'adaptation de l'organe excréteur chez les mollusques gastéropodes pulmonés: III: Histophysiologie comparée du rein chez les soléolifêres et conclusions générales pour tous les pulmonés. *Bulletin biologique de France et de Belgique* 106: 295–314.

Delhaye, W. & Bouillon, J. (1972c). L'évolution et l'adaptation de l'organe excréteur chez les mollusques gastéropodes pulmonés: II: Histophysiologie comparée du rein chez les stylommatophores. *Bulletin biologique de France et de Belgique* 106: 123–142.

Delhaye, W. (1974a). Histophysiologie comparée des organes excréteurs chez quelques Neritacea (Mollusca: Prosobranchia). *Archives de Biologie* 85: 235–262.

Delhaye, W. (1974b). Histophysiologie comparée du rein chez les Mésogastéropodes Archaeotaenioglossa et Littorinoidea (Mollusca: Prosobranchia). *Archives de Biologie* 85: 461–507.

Delhaye, W. (1974c). Recherches sur les glandes pédieuses des gastéropodes prosobranches, principalement les formes terrestres et leur rôle possible dans l'osmorégulation chez les Pomatiasidae et les Chondropomidae. *Forma et Functio* 7: 181–200.

Delhaye, W. (1974d). Contribution a l'étude des glandes pédieuses de *Pomatias elegans* (Mollusque, Gastéropode, Prosobranche). *Annales des sciences naturelles. Zoologie* 16: 97–110.

Delhaye, W. (1977). Histophysiologie comparée du rein chez les Mésogastéropodes et Cerithoidea (Mollusca: Prosbranchia). *Archives de Biologie* 86: 355–375.

Deliagina, T. G., Orlovsky, G. N., Selverston, A. I. & Arshavsky, Y. I. (2000). Neuronal mechanisms for the control of body orientation in *Clione*. 2. Modifications in the activity of postural control system. *Journal of Neurophysiology* 83: 367–373.

Deline, B., Baumiller, T., Kaplan, P., Kowalewski, M. & Hoffmeister, A. P. (2003). Edge-drilling on the brachiopod *Perditocardinia* aff. *P. dubia* from the Mississippian of Missouri (USA). *Palaeogeography, Palaeoclimatology, Palaeoecology* 201: 211–219.

Dell'Angelo, B. & Laghi, G. F. (1980). *Hippopodinella lata* (Busk, 1856) (Bryozoa, Cheilostomata) Epizoica su *Chiton olivaceus* Spengler, 1797. *Oebalia* 6: 25–30.

deMaintenon, M. J. (1999). Phylogenetic analysis of the Columbellidae (Mollusca: Neogastropoda) and the evolution of herbivory from carnivory. *Invertebrate Biology* 118: 258–288.

deMaintenon, M. J. (2001). Ontogeny of the pseudohermaphroditic reproductive system in *Nassarius vibex* (Gastropoda: Buccinidae: Nassariinae). *Journal of Molluscan Studies* 67: 51–58.

Demarchi, B., Williams, M. G., Milner, N., Russell, N., Bailey, G. & Penkman, K. E. H. (2011). Amino acid racemization dating of marine shells: A mound of possibilities. *Quaternary International* 239: 114–124.

Demian, E. S. & Yousif, F. (1975). Embryonic development and organogenesis in the snail *Marisa cornuarietis* (Mesogastropoda: Ampulariidae). V. Development of the nervous system. *Malacologia* 15: 29–42.

Demopoulos, A. W. J., Gualtieri, D. & Kovacs, K. (2010). Food-web structure of seep sediment macrobenthos from the Gulf of Mexico. *Deep Sea Research Part II: Topical Studies in Oceanography* 57: 1972–1981.

Demopulos, P. A. (1975). Diet, activity and feeding in *Tonicella lineata* (Wood, 1815). *The Veliger* 18: 42–46.

Denny, M. (1995). Predicting physical disturbance: Mechanistic approaches to the study of survivorship on wave-swept shores. *Ecological Monographs* 65: 371–418.

Denny, M. W. (1980a). The role of gastropod pedal mucus in locomotion. *Nature* 285: 160–161.

Denny, M. W. (1980b). Locomotion: The cost of gastropod crawling. *Science* 208: 1288–1290.

Denny, M. W. (1981). A quantitative model for the adhesive locomotion of the terrestrial slug, *Ariolimax columbianus*. *Journal of Experimental Biology* 91: 195–217.

Denny, M. W. & Gosline, J. M. (1981). The physical properties of the pedal mucus of the terrestrial slug, *Ariolimax columbianus*. *Journal of Experimental Biology* 88: 375–395.

Denny, M. W. (1983). Molecular biomechanics of molluscan mucous secretions. pp. 431–465 *in* P. W. Hochachka (Ed.) *Metabolic Biochemistry and Molecular Biomechanics*. Vol. 1. New York, Academic Press.

Denny, M. W. (1985). Wave-forces on intertidal organisms—A case study. *Limnology and Oceanography* 30: 1171–1187.

Denny, M. W. (1988). *Biology and the Mechanics of the Wave-Swept Environment*. Vol. 1. Princeton, NJ, Princeton University Press.

Denny, M. W. (1989). Invertebrate mucous secretions: Functional alternatives to vertebrate paradigms. pp. 337–366 *in* E. Chantler & Ratcliffe, N. A. (Eds.) *Mucus and Related Topics*. Manchester, UK, University of Manchester.

Denny, M. W., Daniel, T. L. & Koehl, M. A. R. (1985). Mechanical limits to size in wave-swept organisms. *Ecology Monographs* 55: 69–102.

Denny, M. W. & Shibata, M. F. (1989). Consequence of surf-zone turbulence for settlement and external fertilization. *American Naturalist* 134: 859–889.

Denny, M. W. & Gaines, S. D. (Eds.) (2007). *Encyclopedia of Tidepools and Rocky Shores*. Encyclopedias of the Natural World, No. 1. Berkeley, CA, University of California Press.

Denny, M. W. & Gaylord, B. (2010). Marine ecomechanics. *Annual Review of Marine Science* 2: 89–114.

Denton, E. J. & Taylor, D. W. (1964). The composition of the gas in the chambers of the cuttlebone of *Sepia officialis*. *Journal of the Marine Biological Association of the United Kingdom* 44: 203–207.

Denton, E. J. & Gilpin-Brown, J. B. (1966). On the buoyancy of the pearly nautilus. *Journal of the Marine Biological Association of the United Kingdom* 46: 723–759.

Denton, E. J. & Warren, F. J. (1968). Eyes of the Histioteuthidae. *Nature* 219: 400–401.

Denton, E. J., Gilpin-Brown, J. B. & Shaw, T. I. (1969). A buoyancy mechanism found in cranchid squid. *Proceedings of the Royal Society B* 174: 271–279.

Denton, E. J. (1974). On buoyancy and the lives of modern and fossil cephalopods. *Proceedings of the Royal Society B* 185: 273–299.

Depledge, M. H., Galloway, T. S. & Billinghurst, Z. (1999). Endocrine disruption in invertebrates. *Environmental Science & Technology* 12: 49–60.

Derby, C. D. (2007). Escape by inking and secreting: Marine molluscs avoid predators through a rich array of chemicals and mechanisms. *Biological Bulletin* 213: 274–289.

Derby, C. D., Kicklighter, C. E., Johnson, P. M. & Zhang, X. (2007). Chemical composition of inks of diverse marine molluscs suggests convergent chemical defenses. *Journal of Chemical Ecology* 33: 1105–1113.

Derby, C. D. & Aggio, J. F. (2011). The neuroecology of chemical defenses. *Integrative and Comparative Biology* 51: 771–780.

Derby, C. D. (2014). Cephalopod ink: Production, chemistry, functions and applications. *Marine Drugs* 12: 2700–2730.

Dermott, R. & Munawar, M. (1993). Invasion of Lake Erie offshore sediments by *Dreissena*, and its ecological implications. *Canadian Journal of Fisheries and Aquatic Sciences* 50: 2298–2304.

Desbruyères, D., Segonzac, M. & Bright, M., Eds. (2006). *Handbook of Deep-sea Hydrothermal Vent Fauna*. Linz, Austria, Land Oberösterreich, Biologiezentrum der Oberösterreichische Landesmuseen.

Deslous-Paoli, J. M. & Heral, M. (1986). *Crepidula fornicata* L. (Gastropoda, Calyptraeidae) in the Bay of Marennes-Oléron: Biochemical-composition and energy value of individuals and spawning. *Oceanologica Acta* 9: 305–311.

Desrosiers, R. R. & Dubé, F. (1993). Flowing seawater as an inducer of spawning in the sea scallop, *Placopecten magellanicus* (Gmelin, 1791). *Journal of Shellfish Research* 12: 263–265.

DeVantier, L. M. & Endean, R. (1988). The scallop *Pedum spondyloideum* mitigates the effects of *Acanthaster planci* predation on the host coral *Porites*: Host defence facilitated by exaptation? *Marine Ecology Progress Series* 47: 293–301.

Dewi, A. S., Cheney, K. L., Urquhart, H. H., Blanchfield, J. T. & Garson, M. J. (2016). The sequestration of oxy-polybrominated diphenyl ethers in the nudibranchs *Miamira magnifica* and *Miamira miamirana*. *Marine Drugs* 14: 198 (191–198).

Dewilde, S., Angelini, E., Kiger, L., Marden, M. C., Beltramini, M. & Salvato, B. (2003). Structure and function of the globin and globin gene from the Antarctic mollusc *Yoldia eightsi*. *Biochemical Journal* 370: 245–253.

Deyoung, R. W., Honeycutt, R. L. & Brennan (2005). The molecular toolbox: Genetic techniques in wildlife ecology and management. *Journal of Wildlife Management* 69: 1362–1384.

Deyrup-Olsen, I., Luchtel, D. L. & Martin, A. W. (1983). Components of mucus of terrestrial slugs (Gastropoda). *American Journal of Physiology* 245: R448–R452.

Deyrup-Olsen, I. & Martin, A. W. (1982). Exudation in terrestrial slugs. *Comparative Biochemistry and Physiology Part C* 72C: 45–51.

Deyrup-Olsen, I., Martin, A. W. & Sawyer, W. H. (1981). Fluid transfer by the body wall of terrestrial slugs. *American Zoologist* 21: 930.

Di Cosmo, A. & Di Cristo, C. (2006). Molluscan peptides and reproduction. pp. 241–246 *in* A. J. Kastin (Ed.) *Handbook of Biologically Active Peptides*. San Diego, CA, Elsevier.

Di Cosmo, A. & Polese, G. (2013). Molluscan bioactive peptides. pp. 276–286 *in* A. J. Kastin (Ed.) *Handbook of Biologically Active Peptides*. Vol. 2. San Diego, CA, Elsevier.

Di Cosmo, A., Maselli, V. & Polese, G. (2018). *Octopus vulgaris*: An Alternative in Evolution. pp. 585–598 *in* M. Kloc & J. Z. Kubiak (Eds.). *Marine Organisms as Model Systems in Biology and Medicine*, Cham, Switzerland, Springer.

Di Cosmo, A. & Polese, G. (2018). Cephalopod Olfaction. in: *Oxford Research Encyclopedia of Neuroscience*. Oxford, UK, Oxford University Press.

Di Cristo, C. (2013). Nervous control of reproduction in *Octopus vulgaris*: A new model. *Invertebrate Neuroscience* 13: 27–34.

Di, G., Ni, J., Zhang, Z., You, W., Wang, B. & Ke, C. (2012). Types and distribution of mucous cells of the abalone *Haliotis diversicolor*. *African Journal of Biotechnology* 11: 9127–9140.

Di Lellis, M. A., Sereda, S., Geissler, A., Picot, A., Arnold, P., Lang, S., Troschinski, S. et al. (2014). Phenotypic diversity, population structure and stress protein-based capacitoring in populations of *Xeropicta derbentina*, a heat-tolerant land snail species. *Cell Stress and Chaperones* 19: 791–800.

Di Marzo, V., Marin, A., Vardaro, R. R., de Petrocellis, L., Villani, G. & Cimino, G. (1993). Histological and biochemical bases of defense mechanisms in four species of Polybranchoidea ascoglossan [sic] molluscs. *Marine Biology* 117: 367–380.

Dias, R. J. P., D'Ávila, S. & D'Agosto, M. (2006). First record of epibionts peritrichids and suctorians (Protozoa, Ciliophora) on *Pomacea lineata* (Spix, 1827). *Brazilian Archives of Biology and Technology* 49: 807–812.

Dias, R. J. P., D'Ávila, S., Wieloch, A. H. & D'Agosto, M. (2008). Protozoan ciliate epibionts on the freshwater apple snail *Pomacea figulina* (Spix, 1827) (Gastropoda, Ampullariidae) in an urban stream of south-east Brazil. *Journal of Natural History* 42: 1409–1420.

Dias, T. L. P. (2009). First field study of the Brazilian endemic marine gastropod *Voluta ebraea* (Mollusca: Volutidae). *Marine Biodiversity Records* 2: e10.

Diaz, R. J. & Rosenberg, R. (2008). Spreading dead zones and consequences for marine ecosystems. *Science* 321: 926–929.

Dickinson, M. H. (1996). Unsteady mechanisms of force generation in aquatic and aerial locomotion. *American Zoologist* 36: 537–554.

Dickinson, P. S. (2006). Neuromodulation of central pattern generators in invertebrates and vertebrates. *Current Opinion in Neurobiology* 16: 604–614.

Dictus, W. J. A. G. & Damen, P. (1997). Cell-lineage and clonal-contribution map of the trochophore larva of *Patella vulgata* (Mollusca). *Mechanisms of Development* 62: 213–226.

Dietl, G. P. (2003). Coevolution of a marine gastropod predator and its dangerous bivalve prey. *Biological Journal of the Linnean Society* 80: 409–436.

Dietl, G. P. & Herbert, G. S. (2005). Influence of alternative shell-drilling behaviours on attack duration of the predatory snail, *Chicoreus dilectus*. *Journal of Zoology* 265: 201–206.

Dietz, T. H. (1974). Body fluid composition and aerial oxygen consumption in the freshwater mussel, *Ligumia subrostrata* (Say): Effects of dehydration and anoxic stress. *Biological Bulletin* 147: 560–572.

Dietz, T. H., Wilcox, S. J., Byrne, R. A., Lynn, J. W. & Silverman, H. G. (1996). Osmotic and ionic regulation of North American zebra mussels (*Dreissena polymorpha*). *Integrative and Comparative Biology* 36: 364–372.

Dijkstra, H. H. & Knudsen, J. (1998). Some Pectinoidea (Mollusca: Bivalvia: Propeamussiidae, Pectinidae) of the Red Sea. *Molluscan Research* 19: 43–103.

Dillaman, R. M., Saleuddin, A. S. M. & Jones, G. M. (1976). Neurosecretion and shell regeneration in *Helisoma duryi* (Mollusca: Pulmonata). *Canadian Journal of Zoology* 54: 1771–1778.

Dillen, L., Jordaens, K., Van Dongen, S. & Backeljau, T. (2010). Effects of body size on courtship role, mating frequency and sperm transfer in the land snail *Succinea putris*. *Animal Behaviour* 79: 1125–1133.

Dillon, R. T. (2000). *The Ecology of Freshwater Molluscs*. Cambridge, UK, Cambridge University Press.

Dilly, P. N., Gray, E. G. & Young, J. Z. (1963). Electron microscopy of optic nerves and optic lobes of *Octopus* and *Eledone*. *Proceedings of the Royal Society B* 158: 446–456 + plts 447–455.

Dilly, P. N. & Messenger, J. B. (1972). The branchial gland: A site of haemocyanin synthesis in *Octopus*. *Zeitschrift für Zellforschung und Mikroskopische Anatomie* 132: 193–201.

Dilly, P. N., Nixon, M. & Young, J. Z. (1977). *Mastigoteuthis*: Whiplash squid. *Journal of Zoology* 181: 527–559.

DiMichele, W. A., Hook, R. W., Beerbower, R., Boy, J. A., Gastaldo, R. A., Hotton, N., Phillips, T. L., Scheckler, S. E., Shear, W. A. & Sues, H. D. (1992). Paleozoic terrestrial ecosystems. pp. 205–325 *in* A. K. Behrensmeyer (Ed.) *Terrestrial Ecosystems through Time*. Chicago, IL, University of Chicago Press.

Dimitriadis, V. K. & Hondros, D. (1992). Effect of starvation and hibernation on the fine structural morphology of digestive gland cells of the snail *Helix lucorum*. *Malacologia* 34: 63–73.

Dimitriadis, V. K. & Liosi, M. (1992). Ultrastructural localization of complex carbohydrates and phosphatases in digestive gland cells of feeding and hibernating snails, *Helix lucorum* (Gastropoda: Helicidae). *Journal of Molluscan Studies* 58: 233–243.

Dimitriadis, V. K. & Andrews, E. B. (1999). Ultrastructural and cytochemical study of the kidney and nephridial gland cells of the marine prosobranch mollusc *Nucella lapillus* (L.) in relation to function. *Malacologia* 41: 187–195.

Dimitriadis, V. K. & Andrews, E. B. (2000). Ultrastructural and cytochemical study of the digestive gland cells of the marine prosobranch mollusc *Nucella lapillus* (L.) in relation to function. *Malacologia* 42: 103–112.

Dimitriadis, V. K. (2001). Structure and function of the digestive system in Stylommatophora. pp. 237–257 *in* G. M. Barker (Ed.) *The Biology of Terrestrial Molluscs*. Wallingford, UK, CABI Publishing.

Dinamani, P. (1963). Feeding in *Dentalium conspicuum*. *Proceedings of the Malacological Society of London* 36: 1–5.

Dinamani, P. (1964). Burrowing behaviour of *Dentalium*. *Biological Bulletin* 126: 28–32.

Dinapoli, A., Zinssmeister, C. & Klussmann-Kolb, A. (2011). New insights into the phylogeny of the Pyramidellidae (Gastropoda). *Journal of Molluscan Studies* 77: 1–7.

Disalvo, L. H. & Carriker, M. R. (1994). Planktonic, metamorphic, and early benthic behavior of the chilean loco *Concholepas concholepas* (Muricidae, Gastropoda. Mollusca). *Journal of Shellfish Research* 13: 57–66.

Distaso, A. (1908). Studi sull'embrione di *Seppia*. *Zoologische Jahrbucher. Abteilung fur Anatomie und Ontogenie der Tiere* 26: 565–650.

Distel, D. L., DeLong, E. F. & Waterbury, J. B. (1991). Phylogenetic characterization and in situ localization of the bacterial symbiont of shipworms (Teredinidae: Bivalvia) by using 16S rRNA sequence analysis and oligodeoxynucleotide probe hybridization. *Applied and Environmental Microbiology* 57: 2376–2382.

Distel, D. L. & Roberts, S. J. (1997). Bacterial endosymbionts in the gills of the deep-sea wood-boring bivalves *Xylophaga atlantica* and *Xylophaga washingtona*. *Biological Bulletin* 192: 253–261.

Distel, D. L., Baco, A. R., Chuang, E., Morrill, W., Cavanaugh, C. & Smith, C. R. (2000). Marine ecology: Do mussels take wooden steps to deep-sea vents? *Nature* 403: 725–726.

Distel, D. L., Beaudoin, D. J. & Morrill, W. (2002a). Coexistence of multiple proteobacterial endosymbionts in the gills of the wood-boring bivalve *Lyrodus pedicellatus* (Bivalvia: Teredinidae). *Applied and Environmental Microbiology* 68: 6292–6299.

Distel, D. L., Morrill, W., MacLaren-Toussaint, N., Franks, D. & Waterbury, J. B. (2002b). *Teredinibacter turnerae* gen. nov., sp. nov., a dinitrogen-fixing, cellulolytic, endosymbiotic gamma-proteobacterium isolated from the gills of wood-boring molluscs (Bivalvia: Teredinidae). *International Journal of Systematic and Evolutionary Microbiology* 52: 2261–2269.

Distel, D. L. (2003). The biology of marine wood boring bivalves and their bacterial endosymbionts. pp. 253–271 *in* B. Goodel, Nicholas, D. D. & Schultz, T. P. (Eds.) *Wood Deterioration and Preservation (Advances in Our Changing World). ACS Symposium Series*. Washington, DC, American Chemical Society.

Distel, D. L., Amin, M., Burgoyne, A., Linton, E., Mamangkey, G., Morrill, W., Nove, J., Wood, N. & Yang, J. (2011). Molecular phylogeny of Pholadoidea Lamarck, 1809 supports a single origin for xylotrophy (wood feeding) and xylotrophic bacterial endosymbiosis in Bivalvia. *Molecular Phylogenetics and Evolution* 61: 245–254.

Distel, D. L., Altamia, M. A., Lin, Z., Shipway, J. R., Han, A., Forteza, I., Antemano, R., Limbaco, M. G. J. P., Tebo, A. G. & Dechavez, R. (2017). Discovery of chemoautotrophic symbiosis in the giant shipworm *Kuphus polythalamia* (Bivalvia: Teredinidae) extends wooden-steps theory. *Proceedings of the National Academy of Sciences of the United States of America* 114: E3652–E3658.

Dixon, D. R., Dixon, L. R. J., Pascoe, P. L. & Wilson, J. T. (2001). Chromosomal and nuclear characteristics of deep-sea hydrothermal-vent organisms: Correlates of increased growth rate. *Marine Biology* 139: 251–255.

Diz, A. P., Dudley, E., Cogswell, A., MacDonald, B. W., Kenchington, E. L. R., Zouros, E. & Skibinski, D. O. F. (2013). Proteomic analysis of eggs from *Mytilus edulis* females differing in mitochondrial DNA transmission mode. *Molecular and Cellular Proteomics* 12: 3068–3080.

Doak, D. F., Estes, J. A., Halpern, B. S., Jacob, U., Lindberg, D. R., Lovvorn, J., Monson, D. H., Tinker, M. T., Williams, T. M. & Wootton, J. T. (2008). Understanding and predicting ecological dynamics: Are major surprises inevitable. *Ecology* 89: 952–961.

Dobberteen, R. A. & Ellmore, C. S. (1986). Embryonic expression of shell dimorphism in *Margarites vorticifera* (Gastropoda: Trochidae). *Malacological Review* 19: 45–52.

Dodds, A. W. & Matsushita, M. (2007). The phylogeny of the complement system and the origins of the classical pathway. *Immunobiology* 212: 233–243.

Dodds, W. K. (2002). *Freshwater Ecology: Concepts and Environmental Applications*. San Diego, CA, Academic Press.

Dodou, D., Breedveld, P., de Winter, J. C. F., Dankelman, J. & Van Leeuwen, J. L. (2011). Mechanisms of temporary adhesion in benthic animals. *Biological Reviews* 86: 15–32.

Doeller, J. E., Kraus, D. W., Colacino, J. M. & Wittenberg, J. B. (1988). Gill hemoglobin may deliver sulfide to bacterial symbionts of *Solemya velum* (Bivalvia: Mollusca). *Biological Bulletin* 175: 388–396.

Doepke, H., Herrmann, K. & Schuett, C. (2012). Endobacteria in the tentacles of selected cnidarian species and in the cerata of their nudibranch predators. *Helgoland Marine Research* 66: 43–50.

Dogelya, V. A. & Zenkevicha, L. A., Eds. (1940). *Manual of Zoology* [in Russian]. Moscow, Russia, Academy of Science of the USSR.

Dogterom, G. E., Bohlken, S. & Joosse, J. (1983). Effect of the photoperiod on the time schedule of egg mass production in *Lymnaea stagnalis*, as induced by ovulation hormone injections. *General and Comparative Endocrinology* 49: 255–260.

Dohmen, M. R. (1983). Gametogenesis. pp. 1–48 *in* N. H. Verdonk, van den Biggelaar, J. A. M., & Tompa, A. S. (Eds.) *Development. The Mollusca*. Vol. 3. New York, Academic Press.

Doi, H., Yurlova, N. I., Kikuchi, E., Shikano, S., Yadrenkina, E. N., Vodyanitskaya, S. N. & Zuykova, E. I. (2010). Stable isotopes indicate individual level trophic diversity in the freshwater gastropod *Lymnaea stagnalis*. *Journal of Molluscan Studies* 76: 384–388.

Dolashka, P. & Voelter, W. (2013). Antiviral activity of hemocyanins. *Invertebrate Survival Journal* 2013: 120–127.

Dominguez, L., Villalba, A. & Fuentes, J. (2010). Effects of photoperiod and the duration of conditioning on gametogenesis and spawning of the mussel *Mytilus galloprovincialis* (Lamarck). *Aquaculture Research* 41: e807–e818.

Donaghy, L., Hong, H.-K., Jauzein, C. & Choi, K.-S. (2015). The known and unknown sources of reactive oxygen and nitrogen species in haemocytes of marine bivalve molluscs. *Fish and Shellfish Immunology* 42: 91–97.

Dong, F. M. (2009). *The Nutritional Value of Shellfish. Washington Sea Grant Publication*. pp. 1–3. Seattle, WA, University of Washington.

Dong, Y., Yao, H., Lin, Z. & Zhu, D. (2012). The effects of sperm-egg ratios on polyspermy in the blood clam, *Tegillarca granosa*. *Aquaculture Research* 43: 44–52.

Dong, Y. W., Miller, L. P., Sanders, J. G. & Somero, G. N. (2008). Heat-shock protein 70 (Hsp70) expression in four limpets of the genus *Lottia*: Interspecific variation in constitutive and inducible synthesis correlates with in situ exposure to heat stress. *Biological Bulletin* 215: 173–181.

Donovan, D. A., Baldwin, J. & Carefoot, T. H. (1999a). The contribution of anaerobic energy to gastropod crawling and a re-estimation of minimum cost of transport in the abalone, *Haliotis kamtschatkana* (Jonas). *Journal of Experimental Marine Biology and Ecology* 235: 273–284.

Donovan, D. A., Danko, J. P. & Carefoot, T. H. (1999b). Functional significance of shell sculpture in gastropod molluscs: Test of a predator-deterrent hypothesis in *Ceratostoma foliatum* (Gmelin). *Journal of Experimental Marine Biology and Ecology* 236: 235–251.

Donovan, D. A., Elias, J. P. & Baldwin, J. (2004). Swimming behavior and morphometry of the file shell *Limaria fragilis*. *Marine and Freshwater Behaviour and Physiology* 37: 7–16.

Donovan, S. K. & Paul, C. R. C. (2010). Land snails from the late Pleistocene lithified sand dunes of Great Pedro Bluff, southwest Jamaica. *Caribbean Journal of Science* 46: 1–11.

Dornellas, A. P. S. & Simon, L. R. L. (2011). Bivalves in the stomach contents of *Calliostoma coppingeri* (Calliostomatidae: Gastropoda). *Strombus* 18: 10–12.

Dorsett, D. A. & Roberts, J. B. (1980). A transverse tubular system and neuromuscular junctions in a molluscan unstriated muscle. *Cell and Tissue Research* 206: 251–260.

Doucet-Beaupré, H., Breton, S., Chapman, E. G., Blier, P. U., Bogan, A. E., Stewart, D. T. & Hoeh, W. R. (2010). Mitochondrial phylogenomics of the Bivalvia (Mollusca): Searching for the origin and mitogenomic correlates of doubly uniparental inheritance of mtDNA. *BMC Evolutionary Biology* 10: 50 (1–19).

Dougherty, L. F., Johnsen, S., Caldwell, R. L. & Marshall, N. J. (2014). A dynamic broadband reflector built from microscopic silica spheres in the 'disco' clam *Ctenoides ales*. *Journal of the Royal Society Interface* 11: 20140407.

Dougherty, L. F., Dubielzig, R. R., Schobert, C. S., Teixeira, L. B. & Li, J. (2017). Do you see what I see? Optical morphology and visual capability of 'disco' clams (*Ctenoides ales*). *Biology Open* 6: 648–653.

Dove, M. C. & Sammut, J. (2007). Histological and feeding response of Sydney rock oysters, *Saccostrea glomerata*, to acid sulfate soil outflows. *Journal of Shellfish Research* 26: 509–518.

Downey, P. & Jahan-Parvar, B. (1972). Chemosensory function of the osphradium in *Aplysia*. *Physiologist* 15: 122.

Downing, W. L. & Downing, J. A. (1993). Molluscan shell growth and loss. *Nature* 362: 506.

Doyle, P. & Bennett, M. R. (1995). Belemnites in biostratigraphy. *Palaeontology* 38: 815–829.

Drew, G. A. (1899). Some observations on the habits, anatomy and embryology of members of the Protobranchia. *Anatomischer Anzeiger* 15: 493–518.

Drew, G. A. (1901). The life history of *Nucula delphinodonta* (Mighels). *Quarterly Journal of Microscopical Science* 44: 313–391.

Drewes, R. C. & Roth, B. (1981). Snail-eating frogs from the Ethiopian highlands: A new anuran specialization. *Zoological Journal of the Linnean Society* 73: 267–287.

Driedzic, W. R., Sidell, B. D., Stewart, J. M. & Johnston, I. A. (1990). Maximal activities of enzymes of energy-metabolism in cephalopod systemic and branchial hearts. *Physiological Zoology* 63: 615–629.

Drinnan, R. E. (1957). The winter feeding of the Oyster catcher (*Haematopus ostralegus*) on the edible cockle (*Cardium edule*). *Journal of Animal Ecology* 26: 441–469.

Driscoll, E. G. (1964). Accessory muscle scars, an aid to protobranch orientation. *Journal of Paleontology* 38: 61–66.

Driscoll, R. J. (2011). Freshwater limpets on a water beetle. *Mollusc World* 2011: 8–9.

Du Buit, M. H. (1978). Remarques sur la denture des rajes et sur leur alimentation. *Vie Milieu* 28–29: 165–174.

Dubilier, N., Bergin, C. & Lott, C. (2008). Symbiotic diversity in marine animals: The art of harnessing chemosynthesis. *Nature Reviews Microbiology* 6: 725–740.

Dubois, R. (1909). Recherches sur la pourpre et sur quelques autres pigments animaux. *Archives de Zoologie Paris* 2: 471–590.

Dubois, S., Blin, J.-L., Bouchaud, B. & Lefebvre, S. (2007). Isotope trophic-step fractionation of suspension-feeding species: Implications for food partitioning in coastal ecosystems. *Journal of Experimental Marine Biology and Ecology* 351: 121–128.

Duda, T. F. & Palumbi, S. R. (1999). Developmental shifts and species selection in gastropods. *Proceedings of the National Academy of Sciences of the United States of America* 96: 10272–10277.

Duda, T. F., Kohn, A. J. & Palumbi, S. R. (2001). Origins of diverse feeding ecologies within *Conus*, a genus of venomous marine gastropods. *Biological Journal of the Linnean Society* 73: 391–409.

Duda, T. F., Bolin, M. B., Meyer, C. P. & Kohn, A. J. (2008). Hidden diversity in a hyperdiverse gastropod genus: Discovery of previously unidentified members of a *Conus* species complex. *Molecular Phylogenetics and Evolution* 49: 867–876.

Dudley, E. C. & Vermeij, G. J. (1978). Predation in time and space: Drilling in the gastropod *Turritella*. *Paleobiology* 4: 436–441.

Dufour, S. C. (2005). Gill anatomy and the evolution of symbiosis in the bivalve family Thyasiridae. *Biological Bulletin* 208: 200–212.

Dullo, W.-C. & Bandel, K. (1988). Diagenesis of molluscan shells: A case study from cephalopods. pp. 719–730 *in* J. Wiedmann & Kullmann, J. (Eds.) *Cephalopods Present and Past. 2nd International Cephalopod Symposium: O. H. Schindewolf Symposium, Tubingen 1985*. Stuttgart, Germany, E. Schweizerbart'sche Verlagsbuchhandlung.

Dumdei, E. J., Kubanek, J., Coleman, J. E., Pika, J., Andersen, R. J., Steiner, J. R. & Clardy, J. (1997). New terpenoid metabolites from the skin extracts, an egg mass, and dietary sponges of the Northeastern Pacific dorid nudibranch *Cadlina luteomarginata*. *Canadian Journal of Chemistry* 75: 773–789.

Dunmore, R. & Schiel, D. (2003). Demography, competitive interactions and grazing effects of intertidal limpets in southern New Zealand. *Journal of Experimental Marine Biology and Ecology* 288: 17–38.

Dunn, C. W., Hejnol, A., Matus, D. Q., Pang, K., Browne, W. E., Smith, S. A., Seaver, E. C. et al. (2008). Broad phylogenomic sampling improves resolution of the animal tree of life. *Nature* 452: 745.

Dunn, C. W., Giribet, G., Edgecombe, G. D. & Hejnol, A. (2014). Animal phylogeny and its evolutionary implications. *Annual Review of Ecology, Evolution, and Systematics* 45: 371–395.

Dunstan, A. J., Alanis, O. & Marshall, J. (2010). *Nautilus pompilius* fishing and population decline in the Philippines: A comparison with an unexploited Australian *Nautilus* population. *Fisheries Research* 106: 239–247.

Dunstan, D. J. & Hodgson, A. N. (2014). Snails home. *Physica Scripta* 89: 068002 (068001–068010).

Dunstan, S. L., Sala-Newby, G. B., Fajardo, A. B., Taylor, K. M. & Campbell, A. K. (2000). Cloning and expression of the bioluminescent photoprotein pholasin from the bivalve mollusc *Pholas dactylus*. *Journal of Biological Chemistry* 275: 9403–9409.

Duperron, S., Bergin, C., Zielinski, F., McKiness, Z. P., Pernthaler, A., DeChaine, E. G., Cavanaugh, C. M. & Dubilier, N. (2006). A dual symbiosis shared by two mussel species, *Bathymodiolus azoricus* and *Bathymodiolus puteoserpentis* (Bivalvia: Mytilidae), from hydrothermal vents along the northern Mid-Atlantic Ridge. *Environmental Microbiology* 8: 1441–1447.

Duperron, S., Sibuet, M., MacGregor, B. J., Kuypers, M. M. M., Fisher, C. R. & Dubilier, N. (2007). Diversity, relative abundance and metabolic potential of bacterial endosymbionts in three *Bathymodiolus* mussel species from cold seeps in the Gulf of Mexico. *Environmental Microbiology* 9: 1423–1438.

Duperron, S., Halary, S., Lorion, J., Sibuet, M. & Gaill, F. (2008). Unexpected co-occurrence of six bacterial symbionts in the gills of the cold seep mussel *Idas* sp. (Bivalvia: Mytilidae). *Environmental Microbiology* 10: 433–445.

Duperron, S. (2010). The diversity of deep-sea mussels and their bacterial symbioses. pp. 137–167 in S. Kiel (Ed.) *The Vent and Seep Biota: Aspects from Microbes to Ecosystems. Topics in Geobiology*. Dordrecht, the Netherlands, Springer.

Duperron, S., Pottier, M.-A., Léger, N., Gaudron, S. M., Puillandre, N., Le Prieur, S., Sigwart, J. D., Ravaux, J. & Zbinden, M. (2012). A tale of two chitons: Is habitat specialisation linked to distinct associated bacterial communities? *FEMS Microbiology Ecology* 83: 552–567.

Duplessis, M. R., Dufour, S. C., Blankenship, L. E., Felbeck, H. & Yayanos, A. A. (2004). Anatomical and experimental evidence for particulate feeding in *Lucinoma aequizonata* and *Parvilucina tenuisculpta* (Bivalvia: Lucinidae) from the Santa Barbara Basin. *Marine Biology* 145: 551–561.

Dutertre, M., Barille, L., Beninger, P. G., Rosa, P. & Gruet, Y. (2009). Variations in the pallial organ sizes of the invasive oyster, *Crassostrea gigas*, along an extreme turbidity gradient. *Estuarine, Coastal and Shelf Science* 85: 431–436.

Dutertre, S., Jin, A. H., Kaas, Q., Jones, A., Alewood, P. F. & Lewis, R. J. (2013). Deep venomics reveals the mechanism for expanded peptide diversity in cone snail venom. *Molecular and Cellular Proteomics* 12: 312–329.

Dutta, S., Hartkopf-Fröder, C., Mann, U., Wilkes, H., Brocke, R. & Bertram, N. (2010). Macromolecular composition of Palaeozoic scolecodonts: Insights into the molecular taphonomy of zoomorphs. *Lethaia* 43: 334–343.

Duval, A. & Runham, N. W. (1981). The arterial system of six species of terrestrial slugs. *Journal of Molluscan Studies* 47: 43–52.

Duval, A. (1983). Heartbeat and blood pressure in terrestrial slugs. *Canadian Journal of Zoology* 61: 987–992.

Dwivedy, R. C. (1973). A study of chemo-receptors on labial palps of the American oyster using microelectrodes. *Proceedings of the National Shellfisheries Association* 63: 20–26.

Dykens, J. A. & Mangum, C. P. (1979). The design of cardiac muscle and the mode of metabolism in molluscs. *Comparative Biochemistry and Physiology Part A* 62: 549–554.

Dykens, J. A., Mangum, C. P. & Arnold, J. (1982). A note on the structural organization of the cardiac myofiber in *Nautilus pompilius*. *Pacific Science* 36: 267–271.

Eakin, R. M. & Brandenburger, J. L. (1975). Understanding a snail's eye at a snail's pace. *American Zoologist* 15: 851–863.

Eales, N. B. (1921). *Aplysia*. Liverpool Marine Biology Committee: Memoirs on typical British marine plants and animals. UK, Liverpool University Press.

Eales, N. B. (1949). The food of the dogfish, *Scyliorhinus caniculus* L. *Journal of the Marine Biological Association of the United Kingdom* 28: 791–793.

Eales, N. B. (1950). Torsion in Gastropoda. *Proceedings of the Malacological Society of London* 28: 53–61.

Ebel, R., Marin, A. & Proksch, P. (1999). Organ-specific distribution of dietary alkaloids in the marine opisthobranch *Tylodina perversa*. *Biochemical Systematics and Ecology* 27: 769–777.

Ebert, T. A. & Russell, M. P. (1988). Latitudinal variation in size structure of the west coast purple sea urchin: A correlation with headlands. *Limnology and Oceanography* 33: 286–294.

Eble, A. F. (2001). Anatomy and histology of *Mercenaria mercenaria*. pp. 117–220 in J. N. Kraeuter & Castagna, M. (Eds.) *Biology of the Hard Clam*. New York, Elsevier.

Echevarria, J. (2016). Ontogeny and evolution within the Myophorellidae (Bivalvia): Paedomorphic trends. *Geobios* 49: 177–189.

Edelstam, C. & Palmer, C. (1950). Homing behavior in gastropods. *Oikos* 2: 259–270.

Edgar, G. J. & Shaw, C. (1995). The production and trophic ecology of shallow-water fish assemblages in southern Australia. II. Diets of fishes and trophic relationships between fishes and benthos at Western Port, Victoria. *Journal of Experimental Marine Biology and Ecology* 194: 83–106.

Edgell, T. C., Brazeau, C., Grahame, J. W. & Rochette, R. (2008). Simultaneous defense against shell entry and shell crushing in a snail faced with the predatory shorecrab *Carcinus maenas*. *Marine Ecology Progress Series* 371: 191–198.

Edgell, T. C. & Miyashita, T. (2009). Shell shape and tissue withdrawal depth in 14 species of temperate intertidal snail. *Journal of Molluscan Studies* 75: 235–240.

Edlinger, K. (1980). Beiträge zur Anatomie, Histologie, Ultrastruktur und Physiologie der chemischen Sinnesorgane einiger Cephalaspidea (Mollusca, Opisthobranchia). *Zoologischer Anzeiger* 205: 90–112.

Edlinger, K. (1982). Colour adaptation in *Haminea navicula* (Da Costa) (Mollusca, Opisthobranchia). *Malacologia* 22: 593–600.

Edlinger, K. (1988a). Torsion in gastropods: A phylogenetic model. pp. 241–250 in W. F. Ponder, Ernisse, D. J. & Waterhouse, J. H. (Eds.) *Prosobranch Phylogeny. Malacological Review Supplement*. Ann Arbor, MI, Malacological Review.

Edlinger, K. (1988b). Contributions to torsion and early evolution of the gastropods. *Zeitschrift für Zoologische Systematik und Evolutionsforschung* 26: 27–50.

Edmondson, C. H. (1962). *Teredinidae, Ocean Travelers*. Vol. 23. Honolulu, HI, Bernice P. Bishop Museum.

Edmunds, M. (1968a). On the swimming and defensive response of *Hexabranchus marginatus* (Mollusca, Nudibranchia). *Journal of the Linnean Society of London, Zoology* 47: 425–429.

Edmunds, M. (1968b). Acid secretion in some species of Doridacea (Mollusca, Nudibranchia). *Proceedings of the Malacological Society of London* 38: 121–133.

Edmunds, M. (1991). Does warning coloration occur in nudibranchs? *Malacologia* 32: 241–255.

Edmunds, M. (2009). Do nematocysts sequestered by aeolid nudibranchs deter predators? – A background to the debate. *Journal of Molluscan Studies* 75: 203–205.

Edsinger-Gonzales, E., van der Zee, M., Dictus, W. J. A. G. & van den Biggelaar, J. A. M. (2007). Brefeldin A or monensin inhibits the 3D organizer in gastropod, polyplacophoran, and scaphopod molluscs. *Development Genes and Evolution* 217: 105–118.

Edwards, C. (1965). The hydroid and the medusa *Neoturris pileata*. *Journal of the Marine Biological Association of the United Kingdom* 45: 443–468.

Edwards, D. D. & Vidrine, M. F. (2013). Patterns of species richness among assemblages of *Unionicola* spp. (Acari: Unionicolidae) inhabiting freshwater mussels (Bivalvia: Unionoida) of North America. *The Journal of Parasitology* 99: 212–217.

Edwards, D. E. (1968). Reproduction in *Olivella biplicata*. *The Veliger* 10: 197–304.

Edwards, S. & Condon, C. (2001). Digestive protease characterization, localization and adaptation in blacklip abalone (*Haliotis rubra* Leach). *Aquaculture Research* 32: 95–102.

Eekhout, S., Raubenheimer, C. M., Branch, G. M., Bosman, A. L. & Bergh, M. O. (1992). A holistic approach to the exploitation of intertidal stocks: Limpets as a case study. *South African Journal of Marine Science* 12: 1017–1029.

Eernisse, D. J. (1986). The genus *Lepidochitona* Gray, 1821 (Mollusca, Polyplacophora) in the northeastern Pacific Ocean (Oregonian and Californian provinces). *Zoologische Verhandelingen (Leiden)* 228: 1–52.

Eernisse, D. J. (1988a). Brooding in a chiton: Why synchronize to self-fertilization? *American Zoologist* 28: 170.

Eernisse, D. J. (1988b). Reproductive patterns in six species of *Lepidochitona* (Mollusca: Polyplacophora) from the Pacific coast of North America. *Biological Bulletin* 174: 287–302.

Eernisse, D. J. & Kerth, K. (1988). The initial stages of radular development in chitons (Mollusca: Polyplacophora). *Malacologia* 28: 95–103.

Eernisse, D. J. & Reynolds, P. D. (1994). Polyplacophora. pp. 55–110 *in* F. W. Harrison & Kohn, A. J. (Eds.) *Microscopic Anatomy of Invertebrates. Mollusca 1*. Vol. 5. New York, Wiley-Liss.

Efford, I. E. & Tsumura, K. (1973). Uptake of dissolved glucose and glycine by *Pisidium*, a freshwater bivalve. *Canadian Journal of Zoology* 51: 825–832.

Ehara, T., Kitajima, S., Kanzawa, N., Tamiya, T. & Tsuchiya, T. (2002). Antimicrobial action of achacin is mediated by L-amino acid oxidase activity. *Federation of European Biochemical Societies (FEBS) Letters* 531: 509–512.

Ehlers, U. & Ehlers, B. (1978). Paddle cilia and discocilia—Genuine structures? *Cell and Tissue Research* 192: 489–501.

Eigenbrodt, H. (1941). Untersuchungen über die Funktion der Radula einiger Schnecken. *Zeitschrift für Morphologie und Ökologie der Tiere* 37: 735–791.

Ekaratne, S. U. K. & Crisp, D. J. (1984). Seasonal growth studies of intertidal gastropods from shell micro-growth band measurements, including a comparison with alternative methods. *Journal of the Marine Biological Association of the United Kingdom* 64: 183–210.

Ekaratne, S. U. K. & Goonewardena, D. P. (1994). Behaviour and consumption of *Thais mutabilis* preying on *Cerithidia cingulata* (Mollusca) and its effect on structuring estuarine soft-bottom communities. *Journal of South Asian Natural History* 1: 49–60.

Ekendahl, A. & Johannesson, K. (1997). Shell colour variation in *Littorina saxatilis* Olivi (Prosobranchia: Littorinidae): A multi-factor approach. *Biological Journal of the Linnean Society* 62: 401–419.

Elder, H. Y. (1979). Studies on the host parasite relationship between the parasitic prosobranch *Thyca crystallina* and the asteroid starfish *Linckia laevigata*. *Journal of Zoology* 187: 369–391.

Eldredge, L. G. (1994). *Introductions of Commercially Significant Aquatic Organisms to the Pacific Islands*. Noumea, Australia, South Pacific Commission.

Ellington, W. R. (1985). Metabolic impact of experimental reductions of intracellular pH in molluscan cardiac muscle. pp. 356–366 *in* R. Gilles (Ed.) *Circulation, Respiration and Metabolism*. Berlin, Germany, Springer-Verlag.

Elliot, A. (1979). Structure of molluscan thick filaments: A common origin for diverse appearances. *Journal of Molecular Biology* 132: 323–340.

Elliott, C. J. H. & Susswein, A. J. (2002). Review: Comparative neuroethology of feeding control in molluscs. *Journal of Experimental Biology* 205: 877–896.

Ellis, D. & Aganpattisina, L. (1990). Widespread neogastropod imposex: A biological indicator of global TBT contamination? *Marine Pollution Bulletin* 21: 248–253.

Ellis, I. & Kempf, S. C. (2011). Characterization of the central nervous system and various peripheral innervations during larval development of the oyster *Crassostrea virginica*. *Invertebrate Biology* 130: 236–250.

Elner, R. W. (1978). The mechanics of predation by the shore crab, *Carcinus maenas* (L.), on the edible mussel, *Mytilus edulis* L. *Oecologia* 36: 333–344.

Elner, R. W. & Campbell, A. (1981). Force, function and mechanical advantage in the chelae of the American lobster *Homarus americanus* (Decapoda: Crustacea). *Journal of Zoology* 193: 269–286.

Elston, R. (1997). Bivalve mollusc viruses. *World Journal of Microbiology and Biotechnology* 13: 393–403.

Elston, R. & Wilkinson, M. T. (1985). Pathology, management and diagnosis of oyster velar virus disease (OVVD). *Aquaculture* 48: 189–210.

Elston, R. A., Moore, J. D. & Brooks, K. (1992). Disseminated neoplasia of bivalve molluscs. *Reviews in Aquatic Sciences* 6: 405–466.

Emerson, W. K. (1956). Upwelling and associated marine life along Pacific Baja California, Mexico. *Journal of Paleontology* 30: 393–397.

Emery, D. G. (1975). The histology and fine structure of the olfactory organ of the squid *Lolliguncula brevis* (Blainville). *Tissue and Cell* 7: 357–368.

Emery, D. G. (1976a). Taste receptors in the lip of *Aplysia*. *American Zoologist* 16: 241.

Emery, D. G. (1976b). Observations on the olfactory organ of adult and juvenile *Octopus joubini*. *Tissue and Cell* 8: 33–46.

Emilia, R., Debora, B., Stefania, A., Nicola, B. & Roberto, B. (2016). *Papillifera papillaris* (O.F. Müller), a small snail living on stones and monuments, as indicator of metal deposition and bioavailability in urban environments. *Ecological Indicators* 69: 360–367.

Emson, R. H., Morritt, D., Andrews, E. B. & Young, C. M. (2002). Life on a hot dry beach: Behavioural, physiological, and ultra-structural adaptations of the littorinid gastropod *Cenchritis* (*Tectarius*) *muricatus*. *Marine Biology* 140: 723–732.

Endean, R. (1972). Aspects of molluscan pharmacology. pp. 421–466 *in* M. Florkin & Scheer, B. T. (Eds.) *Chemical Zoology. Mollusca*. Vol. 7. New York, Academic Press.

Endtz, H. P., Vliegenthart, J. S., Vandamme, P., Weverink, H. W., Van den Braak, N. P., Verbrugh, H. A. & Belkum, A. V. (1997). Genotypic diversity of *Campylobacter lari* isolated from mussels and oysters in The Netherlands. *International Journal of Food Microbiology* 34: 79–88.

Ennis, G. P. (1973). Food, feeding, and condition of lobsters, *Homarus americanus*, throughout the seasonal cycle in Bonavista Bay, Newfoundland. *Journal of the Fisheries Board of Canada* 30: 1905–1909.

Eno, N. C. (1994). The morphometrics of cephalopod gills. *Journal of the Marine Biological Association of the United Kingdom* 74: 687.

Enriquez-Diaz, M., Pouvreau, S., Chavez-Villalba, J. & Le Pennec, M. (2009). Gametogenesis, reproductive investment, and spawning behavior of the Pacific giant oyster *Crassostrea gigas*: Evidence of an environment-dependent strategy. *Aquaculture International* 17: 491–506.

Epstein, S. & Lowenstam, H. A. (1953). Temperature-shell-growth relations of recent and interglacial Pleistocene shoal-water biota from Bermuda. *The Journal of Geology* 61: 424–438.

Erasmus, J. H., Cook, P. A. & Coyne, V. E. (1997). The role of bacteria in the digestion of seaweed by the abalone *Haliotis midae*. *Aquaculture* 155: 377–386.

Erben, H. K., Flajs, G. & Siehl, A. (1969). Die frühontogene-tische Entwicklung der Schalenstruktur ectocochleater Cephalopoden. *Palaeontographica: Abteilung A* 132: 1–54.

Erlandson, J. M., Braje, T. J., Rick, T. C., Jew, N. P., Kennett, D. J., Dwyer, N., Ainis, A. F., Vellanoweth, R. L. & Watts, J. (2011). 10,000 years of human predation and size changes in the owl limpet (*Lottia gigantea*) on San Miguel Island, California. *Journal of Archaeological Science* 38: 1127–1134.

Erlandsson, J. & Johannesson, K. (1994). Sexual selection on female size in a marine snail, *Littorina littorea* (L.). *Journal of Experimental Marine Biology and Ecology* 181: 145–157.

Ermentrout, B., Campbell, J. & Oster, G. (1986). A model for shell patterns based on neural activity. *The Veliger* 28: 369–388.

Esch, G. W., Curtis, L. A. & Barger, M. A. (2001). A perspective on the ecology of trematode communities in snails. *Parasitology* 123: S57–S75.

Esch, G. W. (2002). The transmission of digenetic trematodes: Style, elegance, complexity. *Integrative and Comparative Biology* 42: 304–312.

Eshleman, W. P., Wilkens, J. L. & Cavey, M. J. (1982). Electrophoretic and electron microscopic examination of the adductor and diductor muscles of an articulate brachiopod, *Terebratalia transversa*. *Canadian Journal of Zoology* 60: 550–559.

Eskin, A. & Harcombe, E. (1977). Eye of *Navanax*—Optic activity, circadian-rhythm and morphology. *Comparative Biochemistry and Physiology Part A* 57: 443–449.

Eskin, A. & Takahashi, J. S. (1983). Adenylate cyclase activation shifts the phase of a circadian pacemaker. *Science* 220: 82–84.

Eskin, A., Takahashi, J., Zatz, M. & Block, G. (1984). Cyclic guanosine 3′: 5′-monophosphate mimics the effects of light on a circadian pacemaker in the eye of *Aplysia*. *The Journal of Neuroscience* 4: 2466–2471.

Esmaeelian, B., Benkendorff, K., Johnston, M. & Abbott, C. (2013). Purified brominated indole derivatives from *Dicathais orbita* induce apoptosis and cell cycle arrest in colorectal cancer cell lines. *Marine Drugs* 11: 3802–3822.

Espinosa, E. P., Perrigault, M. & Allam, B. (2010). Identification and molecular characterization of a mucosal lectin (MeML) from the blue mussel *Mytilus edulis* and its potential role in particle capture. *Comparative Biochemistry and Physiology Part A* 156: 495–501.

Espinosa, F., González, A. R., Maestre, M. J., Fa, D. A., Guerra-García, J. M. & García-Gómez, J. C. (2007). Responses of the endangered limpet *Patella ferruginea* to reintroduction under different environmental conditions: Survival, growth rates and life history. *Italian Journal of Zoology* 75: 371–384.

Espoz, C., Guzman, G. & Castilla, J. C. (1995). The lichen *Thelidium litorale* on shells of intertidal limpets: A case of lichen-mediated cryptic mimicry. *Marine Ecology Progress Series* 119: 191–197.

Espoz, C., Lindberg, D. R., Castilla, J. C. & Simison, W. B. (2004). Los patelogastrópodos intermareales de Chile y Perú [Intertidal limpets of Chile and Peru]. *Revista Chilena de Historia Natural* 77: 257–283.

Estabrooks, W. A., Kay, E. A. & McCarthy, S. A. (1999). Structure of the excretory system of Hawaiian nerites (Gastropoda: Neritoidea). *Journal of Molluscan Studies* 65: 61–72.

Estebenet, A. L. & Martín, P. R. (2002). *Pomacea canaliculata* (Gastropoda: Ampullariidae): Life-history traits and their plasticity. *Biocell* 26: 83–89.

Estes, J. A. & Steinberg, P. D. (1988). Predation, herbivory, and kelp evolution. *Paleobiology* 14: 19–36.

Etter, R. J. (1988a). Asymmetrical developmental plasticity in an intertidal snail. *Evolution* 42: 322–334.

Etter, R. J. (1988b). Physiological stress and color polymorphism in the intertidal snail *Nucella lapillus*. *Evolution* 42: 660–680.

Evans, F. (1961). Responses to disturbance of the periwinkle *Littorina punctata* (Gmelin) on a shore in Ghana. *Proceedings of the Zoological Society of London* 137: 393–402.

Evans, J. P. & Marshall, D. J. (2005). Male-by-female interactions influence fertilization success and mediate the benefits of polyandry in the sea urchin *Heliocidaris erythrogramma*. *Evolution* 59: 106–112.

Evans, L. A., Macey, D. J. & Webb, J. (1990). Characterization and structural organization of the organic matrix of the radula teeth of the chiton *Acanthopleura hirtosa*. *Philosophical Transactions of the Royal Society of London B Biological Sciences* 329: 87–96.

Evans, S., Camara, M. D. & Langdon, C. J. (2009). Heritability of shell pigmentation in the Pacific oyster, *Crassostrea gigas*. *Aquaculture* 286: 211–216.

Evers, B. N., Madsen, H. & Stauffer, J. R. (2011). Crush-resistance of soft-sediment gastropods of Lake Malawi: Implications for prey selection by molluscivorous fishes. *Journal of Freshwater Ecology* 26: 85–90.

Evgenèv, M. B., Garbuz, D. G. & Zatsepina, O. G. (2014). Heat shock proteins and adaptation to variable and extreme environments. pp. 59–115 *in* M. B. Evgenèv, Garbuz, D. G. & Zatsepina, O. G. (Eds.) *Heat Shock Proteins and Whole Body Adaptation to Extreme Environments*. Dordrecht, the Netherlands, Springer.

Ewers, W. H. & Rose, C. R. (1966). Polymorphism in *Velacumantus australis* (Gastropoda: Potamididae) and its relationship to parasitism. *Australian Journal of Zoology* 14: 49–64.

Extavour, C. G. & Akam, M. (2003). Mechanisms of germ cell specification across the metazoans: Epigenesis and preformation. *Development* 130: 5869–5884.

Extavour, C. G. M. (2007). Evolution of the bilaterian germ line: Lineage origin and modulation of specification mechanisms. *Integrative and Comparative Biology* 47: 770–785.

Fabricius, K. E. (2005). Effects of terrestrial runoff on the ecology of corals and coral reefs: Review and synthesis. *Marine Pollution Bulletin* 50: 125–146.

Fagbuaro, O., Oso, J. A., Edward, J. B. & Ogunleye, R. F. (2006). Nutritional status of four species of giant land snails in Nigeria. *Journal of Zhejiang University Science B* 7: 686–689.

Fahrner, A. & Haszprunar, G. (2000). Microanatomy and ultrastructure of the excretory system of two pelagic opisthobranch species (Gastropoda: Gymnosomata and Thecosomata). *Journal of Submicroscopic Cytology and Pathology* 32: 185–194.

Fahrner, A. & Haszprunar, G. (2001). Anatomy and ultrastructure of the excretory system of a heart-bearing and a heartless sacoglossan gastropod (Opisthobranchia, Sacoglossa). *Zoomorphology* 121: 85–93.

Fahrner, A. (2002). Comparative microanatomy and ultrastructure of the excretory systems of opisthobranch Gastropoda (Mollusca). PhD, Ludwig Maximilians Universität.

Fahrner, A. & Haszprunar, G. (2002a). Ultrastructure of the renopericardial complex in *Hypselodoris tricolor* (Gastropoda, Nudibranchia). *Zoomorphology* 121: 183–194.

Fahrner, A. & Haszprunar, G. (2002b). Microanatomy, ultrastructure, and systematic significance of the excretory system and mantle cavity of an acochlidian gastropod (Opisthobranchia). *Journal of Molluscan Studies* 68: 87–94.

Faillard, H. & Schauer, R. (1972). Glycoproteins as lubricants, protective agents, carriers, structural proteins and as participants in other functions. pp. 1246–1267 *in* A. Gottschalk (Ed.) *Glycoproteins. Their Composition, Structure and Function. Part B, Volume 5*. Amsterdam, the Netherlands, Elsevier.

Fain, A. (2004). Mites (Acari) parasitic and predaceous on terrestrial gastropods. pp. 505–524 *in* G. M. Barker (Ed.) *Natural Enemies of Terrestrial Molluscs*. Oxford, UK/Cambridge, MA, CABI Publishing.

Fairweather, P. G. (1988). Consequences of supply-side ecology: Manipulating the recruitment of intertidal barnacles affects the intensity of predation upon them. *Biological Bulletin* 175: 349–354.

Falese, L. E., Russell, M. P. & Dollahon, N. R. (2011). Spermcasting of spermatozeugmata by the bivalves *Nutricola confusa* and *N. tantilla*. *Invertebrate Biology* 130: 334–343.

Faller, S., Rothe, B. H., Todt, C., Schmidt-Rhaesa, A. & Loesel, R. (2012). Comparative neuroanatomy of Caudofoveata, Solenogastres, Polyplacophora and Scaphopoda (Mollusca) and its phylogenetic implications. *Zoomorphology* 131: 149–170.

Falniowski, A., Szarowska, M., Glöer, P., Pešïc, V., Georgiev, D., Horsák, M. & Sirbu, I. (2012). Radiation in *Bythinella* Moquin-Tandon, 1856 (Mollusca: Gastropoda: Rissooidea) in the Balkans. *Folia Malacologica* 20: 1–10.

Fan, H., Peng, J., Hamann, M. T. & Hu, J. F. (2008). Lamellarins and related pyrrole-derived alkaloids from marine organisms. *Chemical Reviews* 108: 264–287.

Fan, X., Croll, R. P., Wu, B., Fang, L., Shen, Q., Painter, S. D. & Nagle, G. T. (1997). Molecular cloning of a cDNA encoding the neuropeptides APGWamide and cerebral peptide 1: Localization of APGWamide-like immunoreactivity in the central nervous system and male reproductive organs of *Aplysia*. *Journal of Comparative Neurology* 387: 53–62.

Fänge, R. (1960). The salivary gland of *Neptunea antiqua*. *Annals of the New York Academy of Sciences* 90: 689–694.

Fänge, R. & Lidman, U. (1976). Secretion of sulfuric acid in *Cassidaria echinophora* Lamarck (Mollusca: Mesogastropoda, marine carnivorous snail). *Comparative Biochemistry and Physiology Part A* 53: 101–103.

Fankboner, P. V. (1971). Intracellular digestion of symbiontic zooxanthellae by host amoebocytes in giant clams (Bivalvia: Tridacnidae), with a note on the nutritional role of the hypertrophied siphonal epidermis. *Biological Bulletin* 141: 222–234.

Fankboner, P. V. (1977). The eyes and zooxanthellae of giant clams (Bivalvia, Tridacnidae) (abstract). pp. 16–17. *58th Annual Meeting of the Western Society of Naturalists, Santa Cruz, California*. Santa Cruz, CA, Western Society of Naturalists.

Fankboner, P. V. (1981). Siphonal eyes of giant clams (*Tridacna*; Bivalvia; Tridacnidae) and their relationship to adjacent zooxanthellae. *The Veliger* 23: 245–249.

Fankboner, P. V. & Reid, R. G. B. (1990). Nutrition in giant clams (Tridacnidae). pp. 195–209 *in* B. Morton (Ed.) *The Bivalvia. Proceedings of a Memorial Symposium in Honour of Sir Charles Maurice Yonge (1899–1986), Edinburgh, 1986*. Hong Kong, Hong Kong University Press.

FAO Fisheries. (2006). Food and Agriculture Organisation of the United Nations. Fisheries and Aquaculture Department. Statistics., from http://www.fao.org/fishery/statistics/en.

FAO. (2014). *The State of World Fisheries and Aquaculture. Part 1. World Review of Fisheries and Aquaculture*. Rome, Italy, Food and Agriculture Organization of the United Nations. pp. 1–96.

FAO. (2017). GLOBEFISH—Analysis and information on world fish trade. Retrieved April 16, 2018.

FAO Fisheries. (2016). Food and Agriculture Organisation of the United Nations. Fisheries and Aquaculture Department. Statistics., from http://www.fao.org/fishery/statistics/en.

FAO. (2018). Food and Agriculture Organisation of the United Nations. Fisheries and Aquaculture Department. Statistics—Introduction, from http://www.fao.org/fishery/statistics/en.

Fariña, J. M., Castilla, J. C. & Ojeda, F. P. (2013). The 'idiosyncratic' effect of a 'sentinel' species on contaminated rocky intertidal communities. *Ecological Applications* 13: 1533–1552.

Farley, G. S. (2002). Helical nature of sperm swimming affects the fit of fertilization-kinetics models to empirical data. *Biological Bulletin* 203: 51–57.

Farnie, W. C. (1919). The structure of *Amphibola crenata* Martyn. *Transactions and Proceedings of the New Zealand Institute* 51: 69–85.

Farto, R., Armada, S. P., Montes, M., Guisande, J. A., Perez, M. J. & Nieto, T. P. (2003). *Vibrio lentus* associated with diseased wild octopus (*Octopus vulgaris*). *Journal of Invertebrate Pathology* 83: 149–156.

Fässler, S. M. M. & Kaiser, M. J. (2008). Phylogenetically mediated anti-predator responses in bivalve molluscs. *Marine Ecology Progress Series* 363: 217–225.

Faubel, A., Sluys, R. & Reid, D. G. (2007). A new genus and species of polyclad flatworm found in the mantle cavities of gastropod molluscs in the high-intertidal zone of the Pacific coast of Central America. *Journal of the Marine Biological Association of the United Kingdom* 87: 429–434.

Faucci, A., Toonen, R. J. & Hadfield, M. G. (2007). Host shift and speciation in a coral-feeding nudibranch. *Proceedings of the Royal Society B* 274: 111–119.

Fauchald, K. & Jumars, P. A. (1979). The diet of worms: A study of polychaete feeding guilds. *Oceanography and Marine Biology Annual Review* 17: 193–284.

Faulkner, D. J. & Ghiselin, M. T. (1983). Chemical defense and the evolutionary ecology of dorid nudibranchs and some other opisthobranch gastropods. *Marine Ecology Progress Series* 13: 295–301.

Faulkner, D. J., Molinski, T. F., Andersen, R. J., Dumdei, E. J. & De Silva, E. D. (1990). Geographical variation in defensive chemicals from Pacific coast dorid nudibranchs and some related marine molluscs. *Comparative Biochemistry and Physiology Part C* 97: 233–240.

Faulkner, D. J. (1992). Chemical defenses of marine molluscs. pp. 119–163 in V. J. Paul (Ed.) *Ecological Roles of Marine Natural Products*. Ithaca, NY, Comstock Publishing Associates, Cornell University Press.

Feare, C. J. (1971). The adaptive significance of aggregation behaviour in the dogwhelk *Nucella lapillus* (L.). *Oecologia* 7: 117–126.

Fedosov, A. E., Tiunov, A., Kiyashko, S. & Kantor, Y. I. (2014). Trophic diversification in the evolution of predatory marine gastropods of the family Terebridae as inferred from stable isotope data. *Marine Ecology Progress Series* 497: 143–156.

Fedosov, A. E. (2008). Reduction of the alimentary system structures in predatory gastropods of the superfamily Conoidea (Gastropoda: Neogastropoda). *Doklady Biological Sciences* 419: 136–138.

Fedosov, A. E. & Kantor, Y. I. (2010). Evolution of carnivorous gastropods of the family Costellariidae (Neogastropoda) in the framework of molecular phylogeny. *Ruthenica* 20: 117–139.

Feely, R. A., Sabine, C. L., Hernandez-Ayon, J. M., Ianson, D. & Hales, B. (2008). Evidence for upwelling of corrosive 'acidified' water onto the Continental Shelf. *Science* 320: 1490–1492.

Felbeck, H., Childress, J. J. & Somero, G. N. (1983). Biochemical interactions betweeen molluscs and their algal and bacterial symbionts. pp. 331–358 in P. W. Hochachka (Ed.) *Environmental Biochemistry and Physiology. The Mollusca.* Vol. 2. New York, Academic Press.

Fenberg, P. B., Hellberg, M. E., Mullen, L. & Roy, K. (2010). Genetic diversity and population structure of the size-selectively harvested owl limpet, *Lottia gigantea*. *Marine Ecology* 31: 574–583.

Fenchel, T. (1965). Ciliates from Scandinavian molluscs. *Ophelia* 2: 71–174.

Feng, D. D., Li, Q., Yu, H., Zhao, X. L. & Kong, L. F. (2015). Comparative transcriptome analysis of the Pacific Oyster *Crassostrea gigas* characterized by shell colors: Identification of genetic bases potentially involved in pigmentation. *PLoS ONE* 10: e0145257.

Feranec, R. S. (2007). Ecological generalization during adaptive radiation: Evidence from Neogene mammals. *Evolutionary Ecology Research* 9: 555–577.

Ferguson, G. P. & Messenger, J. B. (1991). A countershading reflex in cephalopods. *Proceedings of the Zoological Society of London* 243: 63–67.

Ferguson, G. P., Messenger, J. B. & Budelmann, B.-U. (1994). Gravity and light influence the countershading reflexes of the cuttlefish *Sepia officinalis*. *Journal of Experimental Biology* 191: 247–256.

Ferguson, R. (1999). *The Effectiveness of Australia's Response to the Black Striped Mussel Incursion in Darwin, Australia. A Report of the Marine Pest Incursion Management Workshop. Number 75.* August 27–28, 1999, Environment Australia, 2000, Australian Government, Department of the Environment and Energy.

Fernandez-Piquer, J., Bowman, J. P., Ross, T. L. & Tamplin, M. L. (2012). Molecular analysis of the bacterial communities in the live Pacific oyster (*Crassostrea gigas*) and the influence of postharvest temperature on its structure. *Journal of Applied Microbiology* 112: 1134–1143.

Fernandez, M. S., Valenzuela, F., Arias, J. I., Neira-Carrillo, A. & Arias, J. L. (2016). Is the snail shell repair process really influenced by eggshell membrane as a template of foreign scaffold? *Journal of Structural Biology* 196: 187–196.

Fernandez, R., Rodriguez, J., Quinoa, E., Riguerra, R., Munoz, L., Fernandez-Suarez, M. & Debitus, C. (1996). Onchidin B: A new cyclodepsipeptide from the mollusc *Onchidium* sp. *Journal of the American Chemical Society* 118: 11635–11643.

Fernandez, R. I., Lyons, L. C., Levenson, J., Khabour, O. & Eskin, A. (2003). Circadian modulation of long-term sensitization in *Aplysia*. *Proceedings of the National Academy of Sciences of the United States of America* 100: 14415–14420.

Ferreira-Cravo, M., Welker, A. F. & Hermes-Lima, M. (2010). The connection between oxidative stress and estivation in gastropods and anurans. pp. 47–61 in C. A. Navas & Carvalho, J. E. (Eds.) *Aestivation: Molecular and Physiological Aspects. Progress in Molecular and Subcellular Biology*. Berlin, Germany, Springer-Verlag.

Ferreira, M. V. R., Alencastro, A. C. R. & Hermes-Lima, M. (2003). Role of antioxidant defenses during estivation and anoxia exposure in the freshwater snail *Biomphalaria tenagophila* (Orbigny, 1835). *Canadian Journal of Zoology* 81: 1239–1248.

Ferriter, A., Doren, B., Thayer, D., Miller, B., Thomas, B., Barrett, M., Pernas, T., Hardin, S., Lane, J. & Kobza, M. (2009). The status of nonindigenous species in the South Florida environment. pp. 9.1–9.101. *2008 South Florida Environmental Report*. Miami, FL, South Florida Water Management District.

Ferro, R. & Cretella, M. (1993). Osservazioni sulla biologia di *Calliostoma granulatem* (Born) (Gastropoda: Trochidae). *Bollettino Malacologico* 29: 49–56.

Fiala-Médioni, A., Michalski, J. C., Jollès, J., Alonso, C. & Montreuil, J. (1994). Lysosomic and lysozyme activities in the gill of bivalves from deep hydrothermal vents. *Comptes rendus de l'Académie des sciences* 317: 239–244.

Fichi, G., Cardeti, G., Perrucci, S., Vanni, A., Cersini, A., Lenzi, C., Wolf, T. De, Fronte, B., Guarducci, M. & Susini, F. (2015). Skin lesion-associated pathogens from *Octopus vulgaris*: First detection of *Photobacterium swingsii*, *Lactococcus garvieae* and betanodavirus. *Diseases of Aquatic Organisms* 115: 147–156.

Fiedler, A. & Schipp, R. (1987). The role of the branchial heart complex in circulation of coleoid cephalopods. *Experientia* 43: 544–553.

Fiedler, A. & Schipp, R. (1990). The effects of biogenic monoamines and related agonists and antagonists on the isolated perfused branchial heart of *Sepia officinalis* L. (Cephalopoda). *Comparative Biochemistry and Physiology Part C* 97: 71–78.

Field, K. G., Olsen, G. J., Lane, D. J., Giovannoni, S. J., Ghiselin, M. T., Raff, E. C., Pace, N. R. & Raff, R. A. (1988). Molecular phylogeny of the animal kingdom. *Science* 239: 748–753.

Field, L. H. (1977). An experimental analysis of the escape response of the gastropod *Strombus maculatus*. *Pacific Science* 31: 1–11.

Fields, W. G. & Thompson, K. A. (1976). Ultrastructure and functional morphology of spermatozoa of *Rossia pacifica* (Cephalopoda, Decapoda). *Canadian Journal of Zoology* 54: 908–932.

Filippov, A. & Riedel, F. (2009). The late Holocene mollusc fauna of the Aral Sea and its biogeographical and ecological interpretation. *Limnologica: Ecology and Management of Inland Waters* 39: 67–85.

Fink, A. L. (1999). Chaperone-mediated protein folding. *Physiological Reviews* 79: 425–449.

Finn, J. K., Hochberg, F. G. & Norman, M. D. (2005). Phylum Dicyemida in Australian waters: First record and distribution across diverse cephalopod hosts. *Phuket Marine Biological Center Research Bulletin* 66: 83–96.

Finn, J. K. & Norman, M. D. (2010). The argonaut shell: Gas-mediated buoyancy control in a pelagic octopus. *Proceedings of the Royal Society B* 277: 2967–2971.

Fiorito, G. & Scotto, P. (1992). Observational-learning in *Octopus vulgaris*. *Science* 256: 545–547.

Fiorito, G. & Chichery, R. (1995). Lesions of the vertical lobe impair visual discrimination learning by observation in *Octopus vulgaris*. *Neuroscience Letters* 192: 117–120.

Fioroni, P. (1966). Zur Morphologie und Embryogenese des Darmtraktes und der transitorischen Organe bei Prosobranchien (Mollusca, Gastropoda). *Revue suisse de Zoologie* 73: 621–876.

Fioroni, P. (1982a). Developmental types and the larval stage in mollusks—Some general findings. *Malacologia* 22: 601–609.

Fioroni, P. (1982b). Larval organs, larvae, metamorphosis and types of development of Mollusca—A comprehensive review. *Zoologische Jahrbücher. Abteilung für Anatomie und Ontogenie der Tiere* 108: 375–420.

Fischer-Piette, E. & Franc, A. (1960a). Classe des Polyplacophores. pp. 1701–1785 *in* P.-P. Grassé (Ed.) *Traité de Zoologie*. Vol. 5, Fascicule 2. Paris, France, Masson.

Fischer-Piette, E. & Franc, A. (1960b). Classe des Aplacophores. pp. 1655–1700 *in* P.-P. Grassé (Ed.) *Traité de Zoologie*. Vol. 5, Fascicule 2. Paris, France, Masson.

Fischer-Piette, E. & Franc, A. (1968). Classe des Scaphopodes. pp. 987–1017 *in* P.-P. Grassé (Ed.) *Traité de Zoologie*. Vol. 5, Fascicule 3. Paris, France, Masson.

Fischer, F. P. (1880–1887). *Manuel de conchyliologie et de paléontologie conchyliologique où histoire naturelle des mollusques vivants et fossiles*. Paris, France, Savy.

Fischer, F. P., Alger, M., Cieslar, D. & Krafczyk, H. U. (1990). The chiton gill: Ultrastructure in *Chiton olivaceus* (Mollusca, Polyplacophora). *Journal of Morphology* 204: 75–87.

Fishelson, L. & Qidron-Lazar, G. (1966). Foot autotomy in the gastropod *Gena varia* (Prosobranchia, Trochidae). *The Veliger* 9: 8.

Fishelson, L. & Kidron, G. (1968). Experiments and observations on the histology and mechanism of autotomy and regeneration in *Gena varia* (Prosobranchia, Trochidae). *Journal of Experimental Zoology Part A* 169: 93–105.

Fisher, C. R., Childress, J. J., Oremland, R. S. & Bidigare, R. R. (1987). The importance of methane and thiosulfate in the metabolism of the bacterial symbionts of two deep-sea mussels. *Marine Biology* 96: 59–71.

Fisher, J. A. D., Rhile, E. C., Liu, H. P. & Petraitis, P. S. (2009). An intertidal snail shows a dramatic size increase over the past century. *Proceedings of the National Academy of Sciences of the United States of America* 106: 5209–5212.

Fisher, W. K. (1904). The anatomy of *Lottia gigantea* Gray. *Zoologische Jahrbücher. Abteilung für Anatomie und Ontogenie der Thiere* 20: 1–66.

Fishlyn, D. A. & Phillips, D. W. (1980). Chemical camouflaging and behavioral defenses against a predatory seastar by three species of gastropods from the surfgrass *Phyllospadix* community. *The Biological Bulletin* 158: 34–48.

Fitt, W. K. & Trench, R. K. (1980). Uptake of zooxanthellae by veliger and juvenile stages of *Tridacna squamosa*. *American Zoologist* 20: 777.

Fitt, W. K. & Trench, R. K. (1981). Spawning, development, and acquisition of zooxanthellae by *Tridacna squamosa* (Mollusca, Bivalvia). *Biological Bulletin* 161: 213–235.

Fitt, W. K., Rees, T. A. V., Braley, R. D., Lucas, J. S. & Yellowlees, D. (1993). Nitrogen flux in giant clams: Size-dependency and relationship to zooxanthellae density and clam biomass in the uptake of dissolved inorganic nitrogen. *Marine Biology* 117: 381–386.

FitzGibbon, S. & Franklin, C. (2010). The importance of the cloacal bursae as the primary site of aquatic respiration in the freshwater turtle, *Elseya albagula*. *Australian Zoologist* 35: 276–282.

Flajnik, M. F. & Du Pasquier, L. (2004). Evolution of innate and adaptive immunity: Can we draw a line? *Trends Immunological* 25: 640–644.

Flari, V. & Charrier, M. (1992). Contribution to the study of carbohydrases in the digestive tract of the edible snail *Helix lucorum* L. (Gastropoda: Pulmonata: Stylommatophora) in relation to its age and its physiological state. *Comparative Biochemistry and Physiology Part A* 102: 363–372.

Flari, V. & Lazaridou-Dimitriadou, M. (1996). Evolution of digestion of carbohydrates in the separate parts of the digestive tract of the edible snail *Helix lucorum* (Gastropoda: Pulmonata: Stylommatophora) during a complete 24-hour cycle and the first days of starvation. *Journal of Comparative Physiology B* 165: 580–591.

Flari, V. A. & Edwards, J. P. (2003). The role of the endocrine system in the regulation of reproduction in terrestrial pulmonate gastropods. *Invertebrate Reproduction & Development* 44: 139–161.

Fleming, P. A., Muller, D. & Bateman, P. W. (2007). Leave it all behind: A taxonomic perspective of autotomy in invertebrates. *Biological Reviews* 82: 481–510.

Fleure, H. J. (1904). On the evolution of topographical relations among the Docoglossa. *Transactions of the Linnean Society of London 2nd Series: Zoology* 9: 269–290.

Fleury, C., Marin, F., Marie, B., Luquet, G., Thomas, J. D., Josse, C., Serpentini, A. & Lebel, J. M. (2008). Shell repair process in the green ormer *Haliotis tuberculata*: A histological and microstructural study. *Tissue and Cell* 40: 207–218.

Flores, E. E. C. (1983). Visual discrimination testing in the squid *Todarodes pacificus*: Experimental evidence for lack of color vision. *Memoirs of the National Museum of Victoria* 44: 213–227.

Flores, E. E. C., Igarashi, S. & Mikami, T. (1978). Studies on squid behavior in relation to fishing: III. On the optomotor response of squid, *Todarodes pacificus* Steenstrup, to various colors. *Bulletin of the Faculty of Fisheries Hokkaido University* 29: 131–140.

Florkin, M. (1966). Nitrogen metabolism. pp. 309–351 *in* K. M. Wilbur & Yonge, C. M. (Eds.) *Physiology of Mollusca*. Vol. 2. New York, Academic Press.

Florkin, M. & Scheer, B. T., Eds. (1972). *Mollusca*. Chemical Zoology. New York, Academic Press.

Focardi, S. & Chelazzi, G. (1990). Ecological determinants of bioeconomics in three intertidal chitons (*Acanthopleura* spp.). *Journal of Animal Ecology* 59: 347–362.

Focarelli, R., Renieri, T. & Rosati, F. (1988). Polarized site of sperm entrance in the egg of a fresh-water bivalve, *Unio elongatulus*. *Developmental Biology* 127: 443–451.

Foden, W., Mace, G., Vié, J.-C., Angulo, A., Butchart, S., DeVantier, L. M., Dublin, H., Gutsche, A., Stuart, S. & Turak, E. (2008). Species susceptibility to climate change impacts. pp. 77–88 *in*

J.-C. Vié, Hilton-Taylor, C. & Stuart, S. N. (Eds.) *Wildlife in a Changing World: An Analysis of the 2008 IUCN Red List of Threatened Species.* Gland, Switzerland, I.U.C.N.

Fokin, S. I., Giamberini, L., Molloy, D. P. & de Vaate, A. B. (2003). Bacterial endocytobionts within endosymbiotic ciliates in *Dreissena polymorpha* (Lamellibranchia: Mollusca). *Acta Protozoologica* 42: 31–39.

Foley, D. A. & Cheng, T. C. (1977). Degranulation and other changes of molluscan granulocytes associated with phagocytosis. *Journal of Invertebrate Pathology* 29: 321–325.

Follesa, M. C., Mulas, A., Cabiddu, S., Porcu, C., Deiana, A. M. & Cau, A. (2010). Diet and feeding habits of two skate species, *Raja brachyura* and *Raja miraletus* (Chondrichthyes, Rajidae) in Sardinian waters (central-western Mediterranean). *Italian Journal of Zoology* 77: 53–60.

Fontana, A., Muniaín, C. & Cimino, G. (1998). First chemical study of patagonian nudibranchs: A new seco-11, 12-spongiane, tyrinnal, from the defensive organs of *Tyrinna nobilis. Journal of Natural Products* 61: 1027–1029.

Fontana, A. (2006). Biogenetical proposals and biosynthetic studies on secondary metabolites of opisthobranch molluscs. pp. 303–328 *in* G. Cimino & Gavagnin, M. (Eds.) *Molluscs. Progress in Molecular and Subcellular Biology.* Berlin, Germany, Springer-Verlag.

Fontoura-da-Silva, V., Dantas, R. & Caetano, C. (2013). Foraging tactics in mollusca: A review of the feeding behavior of their most obscure classes (Aplacophora, Polyplacophora, Monoplacophora, Scaphopoda and Cephalopoda). *Oecologia Australis* 17: 358–373.

Foon, J. K., Clements, G. R. & Liew, T.-S. (2017). Diversity and biogeography of land snails (Mollusca, Gastropoda) in the limestone hills of Perak, Peninsular Malaysia. *ZooKeys* 682: 1–94.

Ford, L. A., Alexander, S. K., Cooper, K. M. & Hanlon, R. T. (1986). Bacterial populations of normal and ulcerated mantle tissue of the squid, *Lolliguncula brevis. Journal of Invertebrate Pathology* 48: 13–26.

Forester, A. J. (1977). The function of the intestine in the pulmonate mollusc *Helix pomatia* (L.). *Experientia* 33: 465–467.

Forester, A. J. (1979). The association between the sponge *Halichondria panicea* (Pallas) and scallop *Chlamys varia* (L.): A commensal-protective mutualism. *Journal of Experimental Marine Biology and Ecology* 36: 1–10.

Fornůsek, L. & Větvička, V. (1992). *Immune System Accessory Cells.* Boca Raton, FL, CRC Press.

Forrest, J. E. (1953). On the feeding habits and the morphology and mode of functioning of the alimentary canal in some littoral dorid nudibranchiate Mollusca. *Proceedings of the Linnean Society of London* 164: 225–235.

Forsythe, J., Kangas, N. & Hanlon, R. T. (2004). Does the California market squid (*Loligo opalescens*) spawn naturally during the day or at night? A note on the successful use of ROVs to obtain basic fisheries biology data. *Fishery Bulletin* 102: 389–392.

Forsythe, J. W., Hanlon, R. T., Bullis, R. A. & Noga, E. J. (1991). *Octopus bimaculoides* (Pickford & McConnaughey, 1949): A marine invertebrate host for ectoparasitic protozoans. *Journal of Fish Diseases* 14: 431–442.

Fortunato, H., Penchaszadeh, P. E. & Miloslavich, P. A. (1998). Observations on the reproduction of *Bifurcium bicanaliferum* (Sowerby, 1832) (Gastropoda: Columbellidae: Strombina-group) from the Pacific coast of Panama. *The Veliger* 41: 208–211.

Fortunato, H. (2002). Reproduction and larval development of the *Strombina*-group (Buccinoidea: Columbellidae) and related gastropods: Testing the use of the larval shell for inference of development in fossil species. *Bolettino Malacologico* 38: 111–126.

Forys, E. A., Allen, C. R. & Wojcik, D. P. (2001). The likely cause of extinction of the tree snail *Orthalicus reses reses* (Say). *Journal of Molluscan Studies* 67: 369–376.

Fotheringham, N. (1971). Life history patterns of the littoral gastropods *Shaskyus festivus* (Hinds) and *Ocenebra poulsoni* Carpenter (Prosobranchia: Muricidae). *Ecology* 52: 743–757.

Fournić, J. & Chetail, M. (1984). Calcium dynamics in land gastropods. *American Zoologist* 24: 857–870.

Fox, D. L. (1953). *Animal Biochromes and Structural Colours.* London, UK, Cambridge University Press.

Fox, D. L. (1966). Pigmentation of molluscs. pp. 249–274 *in* K. M. Wilbur & Yonge, C. M. (Eds.) *Physiology of Mollusca.* Vol. 2. New York, Academic Press.

Fox, D. L. (1976). *Animal Biochromes and Structural Colours: Physical, Chemical, Distributional & Physiological Features of Coloured Bodies in the Animal World.* Berkeley, CA, University of California Press.

Fox, D. L. (1983). Biochromy of the Mollusca. pp. 281–303 *in* P. W. Hochachka (Ed.) *Environmental Biochemistry and Physiology. The Mollusca.* Vol. 2. New York, Academic Press.

Franc, A. (1960). Classe des Bivalves. pp. 1845–2133 *in* P.-P. Grassé (Ed.) *Traité de Zoologie.* Vol. 5, Fascicule 2. Paris, France, Masson.

Franc, A. (1968). Classe des gastéropodes (Gastropoda Cuvier 1798). pp. 1–986 *in* P.-P. Grassé (Ed.) *Traité de Zoologie.* Vol. 5, Fascicule 3. Paris, France, Masson.

Franchini, A. & Ottaviani, E. (1990). Fine structure and acid phosphatase localization of hemocytes in the freshwater snail *Viviparus ater* (Gastropoda, Prosobranchia). *Journal of Invertebrate Pathology* 55: 28–34.

Franchini, A., Kletsas, D. & Ottaviani, E. (1996). Immunocytochemical evidence of PDGF- and TGF-beta-like molecules in invertebrate and vertebrate immunocytes: An evolutionary approach. *Histochemical Journal* 28: 599–605.

Franchini, A. & Ottaviani, E. (2000). Repair of molluscan tissue injury: Role of PDGF and TGF-β. *Tissue and Cell* 32: 312–321.

Francisco, C. J., Almeida, A., Castro, A. M. & Santos, M. J. (2010). Development of a PCR-RFLP marker to genetically distinguish *Prosorhynchus crucibulum* and *Prosorhynchus aculeatus. Parasitology International* 59: 40–43.

Frank, P. W. (1975). Latitudinal variation in the life history features of the black turban snail *Tegula funebralis* (Prosobranchia: Trochidae). *Marine Biology* 31: 181–192.

Frankenberg, D. & Smith, K. L. (1967). Coprophagy in marine animals. *Limnology and Oceanography* 12: 443–450.

Franklin, J. B., Fernando, S. A., Chalke, B. A. & Krishnan, K. S. (2007). Radular morphology of *Conus* (Gastropoda: Caenogastropoda: Conidae) from India. *Molluscan Research* 27: 111–122.

Franz, D. R. (1971). Population age structure, growth and longevity of the marine gastropod *Urosalpinx cinerea* Say. *Biological Bulletin* 140: 63–72.

Franzén, Å. (1955). Comparative morphological investigations into the spermiogenesis among Mollusca. *Zoologiska Bidrag från Uppsala* 30: 399–456.

Franzén, Å. (1983). Ultrastructural studies of spermatozoa in three bivalve species with notes on evolution of elongated sperm types in primitive spermatozoa. *Gamete Research* 7: 199–214.

Franzén, Å. & Rice, S. A. (1988). Spermatogenesis, male gametes and gamete interactions. *Microfauna Marina* 4: 309–333.

Fraser, C. M. L., Coleman, R. A. & Seebacher, F. (2015). Inter-individual variation partially explains patterns of orientation on steeply sloped substrata in a keystone grazer, the limpet *Cellana tramoserica. Aquatic Ecology* 49: 189–197.

Fredensborg, B. L., Mouritsen, K. N. & Poulin, R. (2005). Impact of trematodes on host survival and population density in the intertidal gastropod *Zeacumantus subcarinatus*. *Marine Ecology Progress Series* 290: 109–117.

Fredriksen, S. (2003). Food web studies in a Norwegian kelp forest based on stable isotope ($\delta^{13}C$ and $\delta^{15}N$) analysis. *Marine Ecology Progress Series* 260: 71–81.

Freeman, A. S. & Byers, J. E. (2006). Divergent induced responses to an invasive predator in marine mussel populations. *Science* 313: 831–833.

Freeman, G. (1979). The multiple roles which cell division can play in the localization of developmental potential. pp. 53–76 *in* S. S. Subtelny & Konigsberg, I. R. (Eds.) *Determinants of Spatial Organization*. New York, Academic Press.

Freeman, G. & Lundelius, J. W. (1982). The developmental genetics of dextrality and sinistrality in the gastropod *Lymnaea peregra*. *Development Genes and Evolution* 191: 69–83.

Freeman, G. & Lundelius, J. W. (1992). Evolutionary implications of the mode of D quadrant specification in coelomates with spiral cleavage. *Journal of Evolutionary Biology* 5: 205–247.

Freeman, R. F. H. & Rigler, F. H. (1957). The responses of *Scrobicularia plana* (Da Costa) to osmotic pressure changes. *Journal of the Marine Biological Association of the United Kingdom* 36: 553–567.

Freidl, F. E. (1974). Nitrogen excretion by the freshwater pulmonate snail, *Lymnaea stagnalis jugularis* Say. *Comparative Biochemistry and Physiology Part A* 49: 617–622.

Freidl, F. E. (1979). Some aspects of amino acid catabolism in the freshwater pulmonate snail, *Lymnaea stagnalis*. *Malacologia* 18: 595–604.

French, J. R. P. (1997). Pharyngeal teeth of the freshwater drum (*Aplodinotus grunniens*) a predator of the zebra mussel (*Dreissena polymorpha*). *Journal of Freshwater Ecology* 12: 495–498.

Frenkiel, L. & Mouëza, M. (1980). Ciliated receptors in the cruciform muscle sense organ of *Scrobicularia plana* (Da Costa) (Mollusca Lamellibranchia Tellinacea). *Zeitschrift für mikroskopisch-anatomische Forschung* 94: 881–894.

Frenkiel, L., Gros, O. & Mouëza, M. (1996). Gill structure in *Lucina pectinata* (Bivalvia: Lucinidae) with reference to hemoglobin in bivalves with symbiotic sulphur-oxidizing bacteria. *Marine Biology* 125: 511–524.

Frescura, M. & Hodgson, A. N. (1990). The fine structure of the shell muscle of patellid prosobranch limpets. *Journal of Molluscan Studies* 56: 435–447.

Frescura, M. & Hodgson, A. N. (1992). The fine structure of the columellar muscle of some gastropod mollusks. *The Veliger* 35: 308–315.

Fretter, V. (1937). The structure and function of the alimentary canal of some species of Polyplacophora (Mollusca). *Transactions of the Royal Society of Edinburgh* 59: 119–164.

Fretter, V. (1939). The structure and function of the alimentary canal of some tectibranch molluscs, with a note on excretion. *Transactions of the Royal Society of Edinburgh* 59: 599–646.

Fretter, V. (1943). Studies on the functional morphology and embryology of *Onchidella celtica* (Forbes & Hanley) and their bearing on its relationships. *Journal of the Marine Biological Association of the United Kingdom* 25: 685–720.

Fretter, V. (1948). The structure and life history of some minute prosobranchs of rock pools: *Skeneopsis planorbis* (Fabricius), *Omalogyra atomus* (Philippi), *Rissoella diaphana* (Alder) and *Rissoella opalina* (Jeffreys). *Journal of the Marine Biological Association of the United Kingdom* 27: 597–632.

Fretter, V. (1951). Observations on the life history and functional morphology of *Cerithiopsis tubercularis* (Montagu) and *Triphora perversa* (L.). *Journal of the Marine Biological Association of the United Kingdom* 29: 567–586.

Fretter, V. & Graham, A. L. (1949). The structure and mode of life of the Pyramidellidae, parasitic opisthobranchs. *Journal of the Marine Biological Association of the United Kingdom* 28: 493–532.

Fretter, V. & Graham, A. L. (1954). Observations on the opisthobranch mollusc *Acteon tornatilis* (L.). *Journal of the Marine Biological Association of the United Kingdom* 33: 565–585.

Fretter, V. & Graham, A. L. (1962). *British Prosobranch Molluscs: Their Functional Anatomy and Ecology*. London, UK, Ray Society.

Fretter, V. & Graham, A. L. (1964). Reproduction. pp. 127–164 *in* K. M. Wilbur & Yonge, C. M. (Eds.) *Physiology of Mollusca*. Vol. 1. New York, Academic Press.

Fretter, V. (1965). Functional studies of the anatomy of some neritid prosobranchs. *Journal of Zoology* 147: 46–74.

Fretter, V. (1968). *Studies in the Structure, Physiology, and Ecology of Molluscs*. London, UK, Academic Press.

Fretter, V. (1975a). *Umbonium vestiarium*, a filter-feeding trochid. *Journal of Zoology* 177: 541–552.

Fretter, V. (1975b). Introduction. pp. xi–xxix *in* V. Fretter & Peake, J. (Eds.) *Pulmonates: Functional Anatomy and Physiology*. Vol. 1. London, UK, Academic Press.

Fretter, V. (1982). An external vascular opening in *Littorina* spp. *Journal of Molluscan Studies* 48: 105.

Fretter, V. (1984a). Prosobranchs. pp. 1–45 *in* A. S. Tompa, Verdonk, N. H. & van den Biggelaar, J. A. M. (Eds.) *Reproduction*. *The Mollusca*. Vol. 7. New York, Academic Press.

Fretter, V. (1984b). The functional anatomy of the neritacean limpet *Phenacolepa somanensis* (Biggs) and some comparisons with *Septaria*. *Journal of Molluscan Studies* 50: 8–18.

Fretter, V. (1988). New archaeogastropod limpets from hydrothermal vents; superfamily Lepetodrilacea. II. Anatomy. *Philosophical Transactions of the Royal Society B* 318: 33–82.

Fretter, V. (1989). The anatomy of some new archaeogastropod limpets (Superfamily Peltospiracea) from hydrothermal vents. *Journal of Zoology* 218: 123–169.

Fretter, V. (1990). The anatomy of some new archaeogastropod limpets (Order Patellogastropoda, Suborder Lepetopsina) from hydrothermal vents. *Journal of Zoology* 222: 529–555.

Fretter, V. & Montgomery, M. C. (1968). The treatment of food by prosobranch veligers. *Journal of the Marine Biological Association of the United Kingdom* 48: 499–520.

Fretter, V. & Peake, J. F., Eds. (1975). *Pulmonates: Functional Anatomy and Physiology*. London, UK, Academic Press.

Fretter, V., Graham, A. L. & McLean, J. H. (1981). The anatomy of the Galápagos rift limpet, *Neomphalus fretterae*. *Malacologia* 21: 337–361.

Fretter, V. & Graham, A. L. (1994). *British Prosobranch Molluscs: Their Functional Anatomy and Ecology*. London, UK, Ray Society.

Fried, B., Graczyk, T. K. & Tamang, L. (2004). Food-borne intestinal trematodiases in humans. *Parasitology Research* 93: 159–170.

Friedländer, P. (1909). Über den Farbstoff des antiken Purpurs aus *Murex brandaris*. *Berichte der Deutschen Chemischen Gesellschaft* 42: 765–770.

Friedman, C. S. & Finley, C. A. (2003). Anthropogenic introduction of the etiological agent of withering syndrome into northern California abalone populations via conservation efforts. *Canadian Journal of Fisheries and Aquatic Sciences* 60: 1424–1431.

Friedrich, S., Wanninger, A., Bruckner, M. & Haszprunar, G. (2002). Neurogenesis in the mossy chiton, *Mopalia muscosa* (Gould) (Polyplacophora): Evidence against molluscan metamerism. *Journal of Morphology* 253: 109–117.

Fries, C. R. & Grant, D. M. (1991). Rickettsiae in gill epithelial cells of the hard clam, *Mercenaria mercenaria*. *Journal of Invertebrate Pathology* 57: 166–171.

Fritsch, M., Wollesen, T., de Oliveira, A. L. & Wanninger, A. (2015). Unexpected co-linearity of Hox gene expression in an aculiferan mollusk. *BMC Evolutionary Biology* 15: 1–17.

Fritsch, M., Wollesen, T. & Wanninger, A. (2016). Hox and ParaHox gene expression in early body plan patterning of polyplacophoran mollusks. *Journal of Experimental Zoology Part B* 326: 89–104.

Froesch, D. & Messenger, J. B. (1978). On leucophores and the chromatic unit of *Octopus vulgaris*. *Journal of Zoology* 186: 163–173.

Frogley, M. R. & Preece, R. C. (2004). A faunistic review of the modern and fossil molluscan fauna from Lake Pamvotis, Ioannina, an ancient lake in NW Greece; implications for endemism in the Balkans. pp. 243–260 *in* H. I. Griffiths, Krystufek, B. & Reed, J. M. (Eds.) *Balkan Biodiversity: Pattern and Process in the European Hotspot.* Dordrecht, the Netherlands, Springer.

Frost, W. N. & Kandel, E. R. (1995). Structure of the network mediating siphon-elicited siphon withdrawal in *Aplysia*. *Journal of Neurophysiology* 73: 2413–2427.

Fry, B. (2006). *Stable Isotope Ecology.* New York, Springer.

Fryer, J. L. & Lannan, C. N. (1994). Rickettsial and chlamydial infections of freshwater and marine fishes, bivalves, and crustaceans. *Zoological Studies* 33: 95–107.

Fu, X., Palomar, A. J., Hong, E. P., Schmitz, F. J. & Valeriote, F. A. (2004). Cytotoxic lissoclimide-type diterpenes from the molluscs *Pleurobranchus albiguttatus* and *Pleurobranchus forskalii*. *Journal of Natural Products* 67: 1415–1418.

Fujii, T., Ahn, J.-Y., Kuse, M., Mori, H., Matsuda, T. & Isobe, M. (2002). A novel photoprotein from oceanic squid (*Symplectoteuthis oualaniensis*) with sequence similarity to mammalian carbon–nitrogen hydrolase domains. *Biochemical and Biophysical Research Communications* 293: 874–879.

Fujii, Y., Kawsar, S. M. A., Matsumoto, R., Yasumitsu, H., Kojima, N. & Ozeki, Y. (2009). Purification and characterization of a D-galactoside-binding lectin purified from bladder moon shell (*Glossaulax didyma* Röding). *Journal of Biological Sciences* 9: 319–325.

Fujii, Y., Kawsar, S. M. A., Matsumoto, R., Yasumitsu, H., Ishizaki, N., Dogasaki, C., Hosono, M. et al. (2011). A D-galactose-binding lectin purified from coronate moon turban, *Turbo* (*Lunella*) *coreensis*, with a unique amino acid sequence and the ability to recognize lacto-series glycosphingolipids. *Comparative Biochemistry and Physiology Part B* 158: 30–37.

Fujikura, K., Kojima, S., Tamaki, K., Maki, Y., Hunt, J. & Okutani, T. (1999). The deepest chemosynthesis-based community yet discovered from the hadal zone, 7326 m deep, in the Japan Trench. *Marine Ecology Progress Series* 190: 17–26.

Fujikura, K., Amaki, K., Barry, J. P., Fujiwara, Y., Furushima, Y., Iwase, R., Yamamoto, H. & Maruyama, T. (2007). Long-term in situ monitoring of spawning behavior and fecundity in *Calyptogena* spp. *Marine Ecology Progress Series* 333: 185–193.

Fujimoto, K., Yanase, T., Okuno, Y. & Iwata, K. (1966). Electrical response in the *Onchidium* eyes. *Memoirs of Osaka Gakugei University* 15: 98–108.

Fujiwara-Sakata, M. & Kobayashi, M. (1992). Neuropeptides regulate the cardiac activity of a prosobranch mollusc, *Rapana thomasiana*. *Cell and Tissue Research* 269: 241–247.

Fukuda, H. & Ponder, W. F. (2003). Australian freshwater assimineids, with a synopsis of the Recent genus-group taxa of the Assimineidae (Mollusca: Caenogastropoda: Rissooidea). *Journal of Natural History* 37: 1977–2032.

Fukuda, H. & Ponder, W. F. (2004). A protandric assimineid gastropod: *Rugapedia androgyna* n. gen. and n. sp. (Mollusca: Caenogastropoda: Rissooidea) from Queensland, Australia. *Molluscan Research* 24: 75–88.

Fukuyama, A. & Nybakken, J. (1983). Specialized feeding in mitrid gastropods: Evidence from a temperate species, *Mitra idae* Melvill. *The Veliger* 26: 96–100.

Full, R. J., Caldwell, R. L. & Chow, S. W. (1989). Smashing energetics: Prey selection and feeding efficiency of the stomatopod, *Gonodactylus bredini*. *Ethology* 81: 134–147.

Fulton, F. T. (1975). The diet of the chiton *Mopalia lignosa* (Gould, 1846). *The Veliger* 18: 38–41.

Fulton, W. (1983). Macrobenthic fauna of Great Lake, Arthur's Lake and Lake Sorell, Tasmania. *Australian Journal of Marine and Freshwater Research* 34: 775–785.

Funabara, D., Kanoh, S., Siegman, M. J., Butler, T. M., Hartshorne, D. J. & Watabe, S. (2005). Twitchin as a regulator of catch contraction in molluscan smooth muscle. *Journal of Muscle Research and Cell Motility* 26: 455–460.

Funke, W. (1968). Heimfindevermogen und Ortstreue bei *Patella* L. (Gastropoda, Prosobranchia). *Oecologia* 2: 19–42.

Furgal, S. M. & Brownell, P. H. (1987). Ganglionic circulation and its effects on neurons controlling cardiovascular functions in *Aplysia californica*. *Journal of Experimental Zoology Part A* 244: 347–364.

Furuhashi, T., Beran, A., Blazso, M., Czegeny, Z., Schwarzinger, C. & Steiner, G. (2009a). Pyrolysis GC/MS and IR Spectroscopy in chitin analysis of molluscan shells. *Bioscience, Biotechnology and Biochemistry* 73: 93–103.

Furuhashi, T., Schwarzinger, C., Miksik, I., Smrz, M. & Beran, A. (2009b). Molluscan shell evolution with review of shell calcification hypothesis. *Comparative Biochemistry and Physiology Part B* 154: 351–371.

Furukawa, Y. & Kobayashi, M. (1987). Neural control of heart beat in the African giant snail, *Achatina fulica* Férussac. II: Interconnections among the heart regulatory neurones. *Journal of Experimental Biology* 129: 295–307.

Furuya, H., Hochberg, F. G. & Tsuneki, K. (2003a). Reproductive traits in dicyemids. *Marine Biology* 142: 693–706.

Furuya, H., Hochberg, F. G. & Tsuneki, K. (2003b). Calotte morphology in the phylum Dicyemida: Niche separation and convergence. *Journal of Zoology* 259: 361–373.

Furuya, H., Ota, M., Kimura, R. & Tsuneki, K. (2004). Renal organs of cephalopods: A habitat for dicyemids and chromidinids. *Journal of Morphology* 262: 629–643.

Fusetani, N., Wolstenholme, H. J., Shinoda, K., Asai, N., Matsunaga, S., Onuki, H. & Hirota, H. (1992). Two sesquiterpene isocyanides and a sesquiterpene thiocyanate from the marine sponge *Acanthella* cf. *cavernosa* and the nudibranch *Phyllidia ocellata*. *Tetrahedron Letters* 33: 6823–6826.

Gabe, M. (1966). *Neurosecretion.* Oxford, UK, Pergamon Press.

Gäde, G., Weeda, E. & Gabbott, P. A. (1978). Changes in the level of octopine during the escape responses of the scallop, *Pecten maximus* (L.). *Journal of Comparative Physiology B* 124: 121–127.

Gaest, A. H. Van, Young, C. M., Young, J. J., Helms, A. R. & Arellano, S. M. (2007). Physiological and behavioral responses of *Bathynerita naticoidea* (Gastropoda: Neritidae) and *Methanoaricia dendrobranchiata* (Polychaeta: Orbiniidae) to hypersaline conditions at a brine pool cold seep. *Marine Ecology* 28: 199–207.

Gage, J. D. & Tyler, P. A. (1991). *Deep-sea Biology: A Natural History of Organisms at the Deep-sea Floor.* Cambridge, UK, Cambridge University Press.

Gagnon, Y. L., Sutton, T. T. & Johnsen, S. (2013). Visual acuity in pelagic fishes and mollusks. *Vision Research* 92: 1–9.

Gaines, P. & Carlson, J. R. (1995). The olfactory and visual systems are closely related in *Drosophila. Brazilian Journal of Medical and Biological Research* 28: 161–167.

Gainey, L. F. (1972). The use of the foot and the captacula in the feeding of *Dentalium* (Mollusca: Scaphopoda). *The Veliger* 15: 29–34.

Gainey, L. F. & Greenberg, M. J. (1977). Physiological basis of the species abundance-salinity relationship in molluscs: A speculation. *Marine Biology* 40: 41–49.

Gál, J., Bobkova, M. V., Zhukov, V. V., Shepeleva, I. P. & Meyer-Ruchow, V. B. (2004). Fixed focal-length optics in pulmonate snails (Mollusca, Gastropoda): Squaring phylogenetic background and ecophysiological needs (II). *Invertebrate Biology* 123: 116–127.

Gallardo, C. S. (1977). *Crepidula philippiana* n. sp., nuevo gastropodo Calyptraeidae de Chile con especial referencia al patron de desarrollo *Studies on Neotropical Fauna and Environment* 12: 177–185.

Gallardo, C. S. (1979). Developmental pattern and adaptations for reproduction in *Nucella crassilabrum* and other muricacean gastropods. *Biological Bulletin* 157: 453–463.

Gallardo, C. S. (1981). Egg masses and hatching stage of the muricid gastropod *Chorus giganteus* (Lesson, 1829). *Studies on Neotropical Fauna and Environment* 16: 35–44.

Gallardo, C. S. & Sanchez, K. A. (2001). Induction of metamorphosis and its effect on the growth and survival of postmetamorphic juveniles of *Chorus giganteus* (Gastropoda: Muricidae). *Aquaculture* 201: 241–250.

Gallardo, W. G., Siar, S. V. & Encena, V. (1995). Exploitation of the Window Pane Shell *Placuna placenta* in the Philippines. *Biological Conservation* 73: 33–38.

Galleni, L., Tongiorgi, P., Ferroro, E. & Salghetti, U. (1980). *Stylochus mediterraneus* (Turbellaria: Polycladida), predator on the mussel *Mytilus galloprovincialis. Marine Biology* 55: 317–326.

Galli, D. R. & Giese, A. C. (1959). Carbohydrate digestion in a herbivorous snail, *Tegula funebralis. Journal of Experimental Zoology Part A* 140: 415–440.

Gallien, L. & de Larambergue, M. (1939). Biologie et sexualité de *Lacuna pallidula* da Costa (Littorinidae). *Travaux de la Station Zoologique de Wimereux* 13: 293–306.

Galtsoff, P. S. (1960). The three hearts of the oyster. *Proceedings of the National Shellfisheries Association* 51: 7–11.

Galtsoff, P. S. (1964). The American oyster. *United States Fish and Wildlife Service Fishery Bulletin* 64: 1–480.

Gamarra-Luques, C., Winik, B., Vega, I., Albrecht, E., Catalan, N. & Castro-Vazquez, A. (2006). An integrative view to structure, function, ontogeny and phylogenetical significance of the male genital system in *Pomacea canaliculata* (Caenogastropoda, Ampullariidae). *Biocell* 30: 345–357.

Gamarra-Luques, C., Castro-Vazquez, A., Winik, B. C. & Catala, M. (2009). Ectaquasperm-like parasperm in an internally fertilizing gastropod. *Invertebrate Biology* 128: 223–231.

Gantsevich, M. M., Strelkov, P. P., Basova, L. A. & Malakhov, V. V. (2016). Parasitizing of trematodes provokes warts on the hinge plate of the bivalve mollusk *Macoma balthica* Linnaeus, 1758 (Veneroida, Tellinidae). *Doklady Biological Sciences; General Biology* 466: 8–11.

Gao, J., Liu, J., Yang, Y., Liang, J., Xie, J., Li, S., Zheng, G., Xie, L. & Zhang, R. (2016). Identification and expression characterization of three Wnt signaling genes in pearl oyster *(Pinctada fucata). Comparative Biochemistry and Physiology Part B* 196: 92–101.

Gao, W., Wiederhold, M. L. & Hejl, R. (1997). Development of the statocyst in the freshwater snail *Biomphalaria glabrata* (Pulmonata, Basommatophora). *Hearing Research* 109: 125–134.

Gao, Y. M. & Natsukari, Y. (1990). Karyological studies on seven cephalopods. *Venus* 49: 126–145.

Gapasin, R. S. J. & Polohan, B. B. (2005). Response of the tropical abalone, *Haliotis asinina*, larvae on combinations of attachment cues. *Hydrobiologia* 548: 301–306.

Garcia-Alvarez, O. & von Salvini-Plawen, L. (2007). Species and diagnosis of the families and genera of Solenogastres (Mollusca). *Iberus* 25: 73–143.

Garcia-Esquivel, Z. & Felbeck, H. (2006). Activity of digestive enzymes along the gut of juvenile red abalone, *Haliotis rufescens*, fed natural and balanced diets. *Aquaculture* 261: 615–625.

Garcia, C., Robert, M., Arzul, I., Chollet, B., Joly, J.-P., Miossec, L., Comtet, T. & Berthe, F. (2006). Viral gametocytic hypertrophy of *Crassostrea gigas* in France: From occasional records to disease emergence? *Diseases of Aquatic Organisms* 70: 193–199.

García, F. J. & García-Gómez, J. C. (1988). The anatomy of the nervous system of *Armina maculata* (Rafinesque, 1814) (Gastropoda, Opisthobranchia, Arminacea). *Iberus* 8: 75–88.

García, F. J. & García-Gómez, J. C. (1990). The anatomy of the circulatory system in the arminid nudibranch *Armina maculata* (Rafinesque, 1814) (Gastropoda: Opisthobranchia). *Acta Zoologica* 71: 33–35.

García, G. O. & Biondi, L. M. (2011). Kleptoparasitism by the Caracara Chimango (*Milvago chimango*) on the American Oystercatcher (*Haematopus palliatus*) at Mar Chiquita Lagoon, Argentina. *Ornitología Neotropical* 22: 453–457.

Gardner, R. N. & Campbell, H. I. J. (2002). Middle to Late Jurassic bivalves of the subfamily Astartinae from New Zealand and New Caledonia. *New Zealand Journal of Geology and Geophysics* 45: 1–51.

Garefalaki, M. E., Triantafyllidis, A., Abatzopoulos, T. J. & Staikou, A. (2010). The outcome of sperm competition is affected by behavioural and anatomical reproductive traits in a simultaneously hermaphroditic land snail. *Journal of Evolutionary Biology* 23: 966–976.

Garland, C. D., Cooke, S. L., Grant, J. F. & McMeekin, T. A. (1985). Ingestion of the bacteria on the cuticle of crustose (non-articulated) coralline algae by post-larval and juvenile abalone (*Haliotis ruber* Leach) from Tasmanian waters. *Journal of Experimental Marine Biology and Ecology* 91: 137–149.

Garnerot, F., Pellerin, J., Blaise, C. & Mathieu, M. (2006). Immunohistochemical localization of serotonin (5-hydroxytryptamine) in the gonad and digestive gland of *Mya arenaria* (Mollusca: Bivalvia). *General and Comparative Endocrinology* 149: 278–284.

Garrard, T. A. (1975). A Revision of Australian Cancellariidae (Gastropoda, Mollusca). *Records of the Australian Museum* 30: 1–62.

Garrido-Ramos, M. A., Stewart, D. T., Sutherland, B. W. & Zouros, E. (1998). The distribution of male-transmitted and female-transmitted mitochondrial DNA types in somatic tissues of blue mussels: Implications for the operation of doubly uniparental inheritance of mitochondrial DNA. *Genome* 41: 818–824.

Garrity, S. D. (1984). Some adaptations of gastropods to physical stress on a tropical rocky shore. *Ecology* 65: 559–574.

Garson, M. J., Jones, D. D., Small, C. J., Liang, J. & Clardy, J. (1994). Biosynthetic studies on polypropionates: A stereochemical model for siphonarins A and B from the pulmonate limpet *Siphonaria zelandica*. *Tetrahedron Letters* 35: 6921–6924.

Garstang, W. (1928). Origin and evolution of larval forms. *Nature* 122: 366.

Garstang, W. (1929). The origin and evolution of larval forms. *Report of the British Association for the Advancement of Science* 96: 77–98.

Gascoigne, T. (1977). Sacoglossan teeth. *Malacologia* 16: 101–105.

Gattuso, J.-P. & Hansson, L. (2011). *Ocean Acidification*. Oxford, UK, Oxford University Press.

Gauslaa, Y. (2008). Mollusc grazing may constrain the ecological niche of the old forest lichen *Pseudocyphellaria crocata*. *Plant Biology* 10: 711–717.

Gavagnin, M., Spinella, A., Cimino, G. & Sodano, G. (1990). Stereochemistry of ichthyotoxic diacylglycerols from opisthobranch molluscs. *Tetrahedron Letters* 31: 6093–6094.

Gavagnin, M., Vardaro, R. R., Avila, C., Cimino, G. & Ortea, J. (1992). Ichthyotoxic diterpenoids from the Cantabrian nudibranch *Chromodoris luteorosea*. *Journal of Natural Products* 55: 368–371.

Gavagnin, M. & Fontana, A. (2000). Diterpenes from marine opisthobranch molluscs. *Current Organic Chemistry* 4: 1201–1248.

Gazeau, F., Quiblier, C., Jansen, J. M., Gattuso, J. P., Middelburg, J. J. & Heip, C. H. R. (2007). Impact of elevated CO_2 on shellfish calcification. *Geophysical Research Letters* 34: L07603.

Geary, D. (1988). Heterochrony in gastropods: A paleontological view. pp. 183–196 in M. L. McKinney, Stehli, F. G. & Jones, D. S. (Eds.) *Heterochrony in Evolution: A Multidisciplinary Approach. Topics in Geobiology*. New York, Plenum Publishing Corp.

Geddes, M. C. & Butler, A. J. (1984). Physicochemical and biological studies on the Coorong lagoons, South Australia, and the effect of salinity on the distribution of the macrobenthos. *Transactions of the Royal Society of South Australia* 108: 51–62.

Gegenbauer, C. (1878). *Grundriss der vergleichenden Anatomie. Zweite verbesswerte Auflage*. Leipzig, Germany, W. Engelmann.

Gehring, W. J. & Ikeo, K. (1999). Pax 6: Mastering eye morphogenesis and eye evolution. *Trends in Genetics* 15: 371–377.

Gehring, W. J. (2001). The genetic control of eye development and its implications for the evolution of the various eye-types. *Zoology (Jena)* 104: 171–183.

Gehring, W. J. (2005). New perspectives on eye development and the evolution of eyes and photoreceptors. *Journal of Heredity* 96: 171–184.

Geiger, D. L., Marshall, B. A., Ponder, W. F., Sasaki, T. & Warén, A. (2007). Techniques for collecting, handling, preparing, storing and examining small molluscan specimens. *Molluscan Research* 27: 1–50.

Geiger, D. L., Nützel, A. & Sasaki, T. (2008). Vetigastropoda. pp. 297–330 in W. F. Ponder & Lindberg, D. R. (Eds.) *Phylogeny and Evolution of the Mollusca*. Berkeley, CA, University of California Press.

Gelperin, A. (1974). Olfactory basis of homing behavior in the giant garden slug, *Limax maximus*. *Proceedings of the National Academy of Sciences of the United States of America* 71: 966–970.

Gendron, L. (1992). Determination of the size at sexual maturity of the waved whelk *Buccinum undatum* Linnaeus, 1758, in the Gulf of St. Lawrence, as a basis for the establishment of a minimum catchable size. *Journal of Shellfish Research* 11: 1–7.

Gendron, R. P. (1977). Habitat selection and migratory behaviour of the intertidal gastropod *Littorina littorea* (L.). *Journal of Animal Ecology* 46: 79–92.

Genner, M. J., Todd, J. A., Michel, E., Erpenbeck, D., Jimoh, A., Joyce, D. A., Piechocki, A. & Pointier, J.-P. (2007). Amassing diversity in an ancient lake: Evolution of a morphologically diverse parthenogenetic gastropod assemblage in Lake Malawi. *Molecular Ecology* 16: 517–530.

Geraerts, W. P. M. (1976a). The role of the lateral lobes in the control of growth and reproduction in the hermaphrodite freshwater snail *Lymnaea stagnalis*. *General and Comparative Endocrinology* 29: 97–108.

Geraerts, W. P. M. (1976b). Control of growth by the neurosecretory hormone of the light green cells in the freshwater snail *Lymnaea stagnalis*. *General and Comparative Endocrinology* 29: 61–71.

Geraerts, W. P. M., Bohlken, S. & Joosse, J. (1978). The endocrine control of reproduction in the hermaphroditic snail *Lymnaea stagnalis*. pp. 21–24 in P. J. Gaillard & Boer, H. H. (Eds.) *Comparative Neuroendocrinology*. Amsterdam, the Netherlands, Elsevier/North Holland Biomedical Press.

Geraerts, W. P. M. (1987). Environmental and neuroendocrine control of seasonal reproduction in molluscs. pp. 101–120 in P. Pévet (Ed.) *Comparative Physiology of Environmental Adaptations. 3. Adaptations to Climatic Changes*. Vol. 3. Basel, Switzerland, Karger.

Geraerts, W. P. M., Vreugdenhil, E. & Ebberink, R. H. M. (1988). Bioactive peptides in molluscs. pp. 261–282 in M. C. Thorndyke (Ed.) *Neurohormones in Invertebrates*. Cambridge, UK, Cambridge University Press.

Geraerts, W. P. M. (1992). Neurohormonal control of growth and carbohydrate metabolism by the light green cells in *Lymnaea stagnalis*. *General and Comparative Endocrinology* 86: 433–444.

Gerhart, D. J., Rittschof, D. & Mayo, S. W. (1988). Chemical ecology and the search for marine antifoulants. *Journal of Chemical Ecology* 14: 1905–1917.

Gerhart, J. (1999). 1998 Warkany Lecture: Signaling pathways in development. *Teratology* 60: 226–239.

Gestal, C., Abollo, E. & Pascual, S. (2002). Observations on associated histopathology with *Aggregata octopiana* infection (Protista: Apicomplexa) in *Octopus vulgaris*. *Diseases of Aquatic Organisms* 50: 45–49.

Gestal, C., Pascual, S. & Hochberg, F. G. (2010). *Aggregata bathytherma* sp. nov. (Apicomplexa: Coccidea: Aggregatidae), a new coccidian parasite associated with a deep-sea hydrothermal vent octopus. *Diseases of Aquatic Organisms* 91: 237–242.

Gestal, C. & Castellanos-Martínez, S. (2015). Understanding the cephalopod immune system based on functional and molecular evidence. *Fish and Shellfish Immunology* 46: 120–130.

Geusz, M. E., Foster, R. G., DeGrip, W. J. & Block, G. D. (1997). Opsin-like immunoreactivity in the circadian pacemaker neurons and photoreceptors of the eye of the opisthobranch mollusc *Bulla gouldiana*. *Cell and Tissue Research* 287: 203–210.

Ghiretti, F. (1966). Respiration. pp. 175–208 in K. M. Wilbur & Yonge, C. M. (Eds.) *Physiology of Mollusca*. Vol. 2. New York, Academic Press.

Ghiselin, M. T. (1965). Reproductive function and the phylogeny of opisthobranch gastropods. *Malacologia* 3: 327–378.

Ghiselin, M. T. (1966). The adaptive significance of gastropod torsion. *Evolution* 20: 337–348.

Ghiselin, M. T. (1969). The evolution of hermaphroditism among animals. *Quarterly Review of Biology* 44: 189–208.

Ghiselin, M. T. (1974). *The Economy of Nature and the Evolution of Sex*. Berkeley, CA, University of California Press.

Ghiselin, M. T. (1987). Evolutionary aspects of marine invertebrate reproduction. pp. 609–665 *in* A. C. Giese, Pearse, J. S. & Pearse, V. B. (Eds.) *Reproduction of Marine Invertebrates. General Aspects: Seeking Unity in Diversity*. Vol. 9. Palo Alto, CA, Blackwell Scientific Publications & Boxwood Press.

Ghose, K. C. (1963). Morphogenesis of the shell gland, lung, mantle and mantle cavity of the giant snail *Achatina fulica*. *Proceedings of the Malacological Society of London* 35: 119–126.

Gianguzza, P., Badalamenti, F., Jensen, K. R., Chemello, R., Cannicci, S. & Riggio, S. (2004). Body size and mating strategies in the simultaneous hermaphrodite *Oxynoe olivacea* (Mollusca, Opisthobranchia, Sacoglossa). *Functional Ecology* 18: 899–906.

Gibbs, P. E. (1984). The population cycle of the bivalve *Abra tenuis* and its mode of reproduction. *Journal of the Marine Biological Association of the United Kingdom* 64: 791–800.

Gibbs, P. E. (1993). Phenotypic changes in the progeny of *Nucella lapillus* (Gastropoda) transplanted from an exposed shore to sheltered inlets. *Journal of Molluscan Studies* 59: 187–194.

Gibbs, P. E., Nott, J. A., Nicolaidou, A. & Bebianno, M. J. (1998). The composition of phosphate granules in the digestive glands of marine prosobranch gastropods: Variation in relation to taxonomy. *Journal of Molluscan Studies* 64: 423–433.

Gibson, G. (1995). Why be choosy? Temporal changes in larval sensitivity to several naturally-occurring metamorphic inducers in the opisthobranch *Haminaea callidegenita*. *Journal of Experimental Marine Biology and Ecology* 194: 9–24.

Gibson, G. D. (2003). Larval development and metamorphosis in *Pleurobranchaea maculata*, with a review of development in the Notaspidea (Opisthobranchia). *Biological Bulletin* 205: 121–132.

Gibson, R., Thompson, T. E. & Robilliard, G. A. (1970). Structure of the spawn of an Antarctic dorid nudibranch *Austrodoris macmurdensis* Odhner. *Journal of Molluscan Studies* 39: 221–225.

Gibson, R. N., Atkinson, R. J. A. & Gordon, J. D. M. (2006). Taxonomy, ecology and behaviour of the cirrate octopods. *Oceanography and Marine Biology Annual Review* 44: 277–322.

Gibson, R. N., Atkinson, R. J. A., Gordon, J. D. M., Smith, I. P. & Hughes, D. J. (2011). Impact of ocean warming and ocean acidification on marine invertebrate life history stages: Vulnerabilities and potential for persistence in a changing ocean. *Oceanography and Marine Biology Annual Review* 49: 1–42.

Giere, O. (1985). Structure and function of bacterial endosymbionts in the gill filaments of *Lucinidae* from Bermuda (Mollusca, Bivalvia). *Zoomorphology* 105: 296–301.

Giese, A. C. (1969). A new approach to the biochemical composition of the mollusc body. *Oceanography and Marine Biology: An Annual Review* 7: 175–229.

Giese, A. C. & Pearse, J. S. (1974). Introduction: General principles. pp. 1–49 *in* A. C. Giese & Pearse, J. S. (Eds.) *Reproduction of Marine Invertebrates. Acoelomate and Pseudocoelomate Metazoans*. Vol. 1. New York, Academic Press.

Giese, A. C. & Kanatani, H. (1987). Maturation and spawning. pp. 252–313 *in* A. C. Giese & Pearse, J. S. (Eds.) *Reproduction of Marine Invertebrates. General Aspects: Seeking Unity in Diversity*. Vol. 9. Palo Alto, CA, Blackwell Scientific Publications & Boxwood Press.

Giesel, J. T. (1970). On the maintenance of a shell pattern and behavior polymorphism in *Acmaea digitalis*, a limpet. *Evolution* 24: 98–119.

Gilbert, P. U., Bergmann, K. D., Myers, C. E., Marcus, M. A., DeVol, R. T., Sun, C.-Y., Blonsky, A. Z., Tamre, E., Zhao, J. & Karan, E. A. (2017). Nacre tablet thickness records formation temperature in modern and fossil shells. *Earth and Planetary Science Letters* 460: 281–292.

Gilbert, S. F. (2001). Ecological developmental biology: Developmental biology meets the real world. *Developmental Biology* 233: 1–12.

Gilbert, S. F. & Epel, D. (2008). *Ecological Developmental Biology*. Sunderland, MA, Sinauer Associates.

Gillary, H. L. (1974). Light-evoked electrical potentials from eye and optic-nerve of *Strombus*: Response waveform and spectral sensitivity. *Journal of Experimental Biology* 60: 383–396.

Gillary, H. L. & Gillary, E. W. (1979). Ultrastructural features of the retina and optic nerve of *Strombus luhunaus*, a marine gastropod. *Journal of Morphology* 159: 89–116.

Gillette, M. U. & Gillette, R. (1983). Bursting neurons command consummatory feeding behavior and coordinated visceral receptivity in the predatory mollusk *Pleurobranchaea*. *Journal of Neuroscience* 3: 1791–1806.

Gillette, M. U. & Sejnowski, T. J. (2005). Biological clocks coordinately keep life on time. *Science* 309: 1196–1198.

Gillette, R. (1991). On the significance of neuronal gigantism in gastropods. *Biological Bulletin* 180: 234–240.

Gillette, R., Saeki, M. & Huang, R. C. (1991). Defensive mechanisms in notaspid snails: Acid humor and evasiveness. *Journal of Experimental Biology* 156: 335–347.

Gillikin, D. P. & Dehairs, F. (2013). Uranium in aragonitic marine bivalve shells. *Palaeogeography, Palaeoclimatology, Palaeoecology* 373: 60–65.

Gilmer, R. W. (1974). Some aspects of feeding in thecosomatous pteropod molluscs. *Journal of Experimental Marine Biology and Ecology* 15: 127–144.

Gilmer, R. W. (1986). Preservation artifacts and their effects on the study of euthecosomatous pteropod mollusks. *The Veliger* 29: 48–52.

Gilmer, R. W. & Harbison, G. R. (1986). Morphology and field behavior of pteropod molluscs: Feeding methods in the families Cavoliniidae, Limacinidae and Peraclididae (Gastropoda: Thecosomata). *Marine Biology* 91: 47–57.

Gilmour, T. H. J. (1963). A note on the tentacles of *Lima hians* (Gmelin) (Bivalvia). *Proceedings of the Malacological Society of London* 35: 82–85.

Gilmour, T. H. J. (1967). The defensive adaptations of *Lima hians* (Mollusca, Bivalvia). *Journal of the Marine Biological Association of the United Kingdom* 47: 209–221.

Gilmour, T. H. J. (1990). The adaptive significance of foot reversal in the Limoida. pp. 249–263 *in* B. Morton (Ed.) *The Bivalvia. Proceedings of a Memorial Symposium in Honour of Sir Charles Maurice Yonge (1899–1986), Edinburgh, 1986*. Hong Kong, Hong Kong University Press.

Giménez-Casalduero, F., Thacker, R. W. & Paul, V. J. (1999). Association of color and feeding deterrence by tropical reef fishes. *Chemoecology* 9: 33–39.

Giménez, J., Brey, T., Mackensen, A. & Penchaszadeh, P. E. (2004). Age, growth, and mortality of the prosobranch *Zidona dufresnei* (Donovan, 1823) in the Mar del Plata area, south-western Atlantic Ocean. *Marine Biology* 145: 707–712.

Ginzburg, A. S. (1975). Fertilization of the eggs of bivalve mollusks with different insemination conditions. *The Soviet Journal of Developmental Biology* 5: 300–308.

Giribet, G., Okusu, A., Lindgren, A. R., Huff, S. W., Schrödl, M. & Nishiguchi, M. K. (2006). Evidence for a clade composed of molluscs with serially repeated structures: Monoplacophorans are related to chitons. *Proceedings of the National Academy of Sciences of the United States of America* 103: 7723–7728.

Giribet, G. (2008). Assembling the lophotrochozoan (=spiralian) tree of life. *Philosophical Transactions of the Royal Society of London B* 363: 1513–1522.

Gissi, C., Iannelli, F. & Pesole, G. (2008). Evolution of the mitochondrial genome of Metazoa as exemplified by comparison of congeneric species. *Heredity* 101: 301–320.

Gittenberger, A. & Gittenberger, E. (2005). A hitherto unnoticed adaptive radiation: Epitoniid species (Gastropoda: Epitoniidae) associated with corals (Scleractinia). *Contributions to Zoology* 74: 125–203.

Gittenberger, E. (1988). Sympatric speciation in snails: A largely neglected model. *Evolution* 42: 826–828.

Gittenberger, E. (1991). What about non-adaptive radiation? *Biological Journal of the Linnean Society* 43: 263–272.

Gittenberger, E. (2004). Radiation and adaptation, evolutionary biology and semantics. *Organisms Diversity & Evolution* 4: 135–136.

Giusti, A. F., Hinman, V. F., Degnan, S. M., Degnan, B. M. & Morse, D. E. (2000). Expression of a Scr/Hox5 gene in the larval central nervous system of the gastropod *Haliotis*, a non-segmented spiralian lophotrochozoan. *Evolution & Development* 2: 294–302.

Glass, K., Ito, S., Wilby, P. R., Sota, T., Nakamura, A., Bowers, C. R., Vinther, J., Dutta, S., Summons, R. & Briggs, D. E. (2012). Direct chemical evidence for eumelanin pigment from the Jurassic period. *Proceedings of the National Academy of Sciences of the United States of America* 109: 10218–10223.

Glaubrecht, M., von Rintelen, T. & Korniushin, A. V. (2003). Toward a systematic revision of brooding freshwater Corbiculidae in Southeast Asia (Bivalvia, Veneroida): On shell morphology, anatomy and molecular phylogenetics of endemic taxa from islands in Indonesia. *Malacologia* 45: 1–40.

Glaubrecht, M. (2006). Independent evolution of reproductive modes in viviparous freshwater Cerithioidea (Gastropoda, Sorbeoconcha)—A brief review. *Basteria* 3: 23–28.

Glaubrecht, M. (2008). Adaptive radiation of thalassoid gastropods in Lake Tanganyika, East Africa: Morphology and systematization of a paludomid species flock in an ancient lake. *Zoosystematics and Evolution* 84: 71–122.

Glaubrecht, M. & von Rintelen, T. (2008). The species flocks of lacustrine gastropods: *Tylomelania* on Sulawesi as models in speciation and adaptive radiation. *Hydrobiologia* 615: 181–199.

Glaubrecht, M. (2009). On "Darwinian Mysteries" or molluscs as models in evolutionary biology: From local speciation to global radiation. *American Malacological Bulletin* 27: 3–23.

Glaubrecht, M., Brinkmann, N. & Pöppe, J. (2009). Diversity and disparity 'down under': Systematics, biogeography and reproductive modes of the 'marsupial' freshwater Thiaridae (Caenogastropoda, Cerithioidea) in Australia. *Zoosystematics and Evolution* 85: 199–275.

Glaubrecht, M. (2011). Towards solving Darwin's 'mystery': Speciation and radiation in lacustrine and riverine freshwater gastropods. *American Malacological Bulletin* 29: 187–216.

Glavinic, A., Wilson, N. G., Benkendorff, K. & Rouse, G. W. (2010). Doubly uniparental inheritance of mitochondrial DNA in *Neotrigonia margaritacea* (Bivalvia: Palaeoheterodonta). *Tropical Natural History, Supplement* 3: 256.

Glover, E. A. & Taylor, J. D. (2007). Diversity of chemosymbiotic bivalves on coral reefs: Lucinidae (Mollusca, Bivalvia) of New Caledonia and Lifou. *Natural History* 29: 109–182.

Glusker, J. P., Katz, A. K. & Bock, C. W. (1999). Metal ions in biological systems. *Rigaku Journal* 16: 8–16.

Glynn, P. W. (1968). Ecological studies on the associations of chitons in Puerto Rico, with special reference to sphaeromid isopods. *Bulletin of Marine Science* 18: 572–626.

Gnyubkin, V. F. (2010). The circadian rhythms of valve movements in the mussel *Mytilus galloprovincialis*. *Russian Journal of Marine Biology* 36: 419–428.

Göbbeler, K. & Klussmann-Kolb, A. (2007). A comparative ultrastructural investigation of the cephalic sensory organs in Opisthobranchia (Mollusca, Gastropoda). *Tissue and Cell* 39: 399–414.

Göbbeler, K. & Klussmann-Kolb, A. (2011). Molecular phylogeny of the Euthyneura (Mollusca, Gastropoda) with special focus on Opisthobranchia as a framework for reconstruction of evolution of diet. *Thalassas* 27: 121–154.

Goddard, C. K. & Martin, A. W. (1966). Carbohydrate metabolism. pp. 275–308 *in* K. M. Wilbur & Yonge, C. M. (Eds.) *Physiology of Mollusca*. Vol. 2. New York, Academic Press.

Goddard, J. H. R., Gosliner, T. M. & Pearse, J. S. (2011). Impacts associated with the recent range shift of the aeolid nudibranch *Phidiana hiltoni* (Mollusca, Opisthobranchia) in California. *Marine Biology* 158: 1095–1109.

Goewert, A. E. & Surge, D. (2008). Seasonality and growth patterns using isotope sclerochronology in shells of the Pliocene scallop *Chesapecten madisonius*. *Geo-Marine Letters* 28: 327–338.

Gofas, S. (2000). Systematics of *Planktomya*, a bivalve genus with teleplanic larval dispersal. *Bulletin of Marine Science* 67: 1013–1023.

Goff, R. L. & Daguzan, J. (1991). Growth and life cycles of the cuttlefish *Sepia officinalis* L. (Mollusca: Cephalopoda) in South Brittany (France). *Bulletin of Marine Science* 49: 341–348.

Goffredi, S. K., Barry, J. P. & Buck, K. R. (2004a). Vesicomyid symbioses from Monterey Bay (central California) cold seeps. *Symbiosis* 36: 1–27.

Goffredi, S. K., Warén, A., Orphan, V. J., Van Dover, L. & Vrijenhoek, R. C. (2004b). Novel forms of structural integration between microbes and a hydrothermal vent gastropod from the Indian Ocean. *Applied and Environmental Microbiology* 70: 3082–3090.

Goggin, C. L. & Lester, R. J. G. (1987). Occurrence of *Perkinsus* species (Protozoa, Apicomplexa) in bivalves from the Great Barrier Reef. *Diseases of Aquatic Organisms* 3: 113–117.

Goggin, C. L. & Lester, R. J. G. (1995). *Perkinsus*, a protistan parasite of abalone in Australia: A review. *Marine and Freshwater Research* 46: 639–646.

Gohar, H. A. F. & Soliman, G. N. (1963). On the biology of three coralliophilids boring in living corals. *Publications of the Marine Biological Station Al Ghardaqa (Red Sea)* 12: 99–126.

Goldberg, J. I., Garofalo, R., Price, C. J. & Chang, J. P. (1993). Presence and biological activity of a GnRH-like factor in the nervous system of *Helisoma trivolvis*. *Journal of Comparative Neurology* 336: 571–582.

Goldberg, W. M. (1971). A note on the feeding behavior of the snapping shrimp *Synalpheus fritzmuelleri* Coutière (Decapoda, Alpheidae). *Crustaceana* 21: 318–320.

Golding, R. E., Ponder, W. F. & Byrne, M. (2007). Taxonomy and anatomy of Amphiboloidea (Gastropoda: Heterobranchia: Archaeopulmonata). *Zootaxa* 50: 1–50.

Golding, R. E., Ponder, W. F. & Byrne, M. (2009a). Three-dimensional reconstruction of the odontophoral cartilages of Caenogastropoda (Mollusca: Gastropoda) using micro-CT: Morphology and phylogenetic significance. *Journal of Morphology* 270: 558–587.

Golding, R. E., Ponder, W. F. & Byrne, M. (2009b). The evolutionary and biomechanical implications of snout and proboscis morphology in Caenogastropoda (Mollusca: Gastropoda). *Journal of Natural History* 43: 2723–2763.

Golding, R. E. & Ponder, W. F. (2010). Homology and morphology of the neogastropod valve of Leiblein (Gastropoda: Caenogastropoda). *Zoomorphology* 129: 81–91.

Golding, R. E., Bieler, R., Rawlings, T. A. & Collins, T. M. (2014). Deconstructing *Dendropoma*: A systematic revision of a world-wide worm-snail group, with descriptions of new genera (Caenogastropoda: Vermetidae). *Malacologia* 57: 1–97.

Golikov, A. N. & Kussakin, O. G. (1972). Sur la biologie de la reproduction des Patellides de la familie Tecturidae (Gastropoda, Docoglossa) et sur la position systématique de ses subdivisions. *Malacologia* 11: 287–294.

Gomez-Pinchetti, J. L. & Garcia-Reina, G. (1993). Enzymes from marine phycophages that degrade cell walls of seaweeds. *Marine Biology* 116: 553–558.

Gomot de Vaufleury, A. (2001). Regulation of growth and reproduction. pp. 331–355 in G. M. Barker (Ed.) *The Biology of Terrestrial Molluscs*. Wallingford, UK, CABI Publishing.

Goncalves, P., Anderson, K., Thompson, E. L., Melwani, A., Parker, L. M., Ross, P. M. & Raftos, D. A. (2016). Rapid transcriptional acclimation following transgenerational exposure of oysters to ocean acidification. *Molecular Ecology* 25: 4836–4849.

Gong, Z., Matzke, N. J., Ermentrout, B., Song, D., Vendetti, J. E., Slatkin, M. & Oster, G. (2012). Evolution of patterns on *Conus* shells. *Proceedings of the National Academy of Sciences of the United States of America* 109: E234–E241.

Gonor, S. J. (1965). Predator-prey reactions between two marine prosobranch gastropods. *The Veliger* 7: 228–232.

González-Santander, R. & Garcia-Blanco, E. S. (1972). Ultrastructure of the obliquely striated or pseudostriated muscle fibres of the cephalopods *Sepia*, *Octopus* and *Eledone*. *Journal of Submicroscopic Cytology* 4: 233–245.

Goodacre, S. L. (2002). Population structure, history and gene flow in a group of closely related land snails: Genetic variation in *Partula* from the Society Islands of the Pacific. *Molecular Ecology* 11: 55–68.

Goodfriend, G. A. (1992). The use of land snail shells in paleoenvironmental reconstruction. *Quaternary Science Reviews* 11: 665–685.

Goodhart, C. B. (1987a). Why are some snails visibly polymorphic, and others not? *Biological Journal of the Linnean Society* 31: 35–58.

Goodhart, C. B. (1987b). Garstang's hypothesis and gastropod torsion. *Journal of Molluscan Studies* 53: 33–36.

Gooding, M. P. & LeBlanc, G. A. (2001). Biotransformation and disposition of testosterone in the eastern mud snail *Ilyanassa obsoleta*. *General and Comparative Endocrinology* 122: 172–180.

Goodwill, R. H., Fautin, D. G., Furey, J. & Daly, M. (2009). A sea anemone symbiotic with gastropods of eight species in the Mariana Islands. *Micronesica* 41: 117–130.

Goodwin, D. H., Anderson, L. C. & Roopnarine, P. D. (2008). Evolutionary origins of novel conchologic growth patterns in tropical American corbulid bivalves. *Evolution & Development* 10: 642–656.

Goodwin, T. W. (1972). Pigments of Mollusca. pp. 187–199 in M. Florkin & Scheer, B. T. (Eds.) *Chemical Zoology: Mollusca*. Vol. 7. New York, Academic Press.

Gorbman, A., Whiteley, A. & Kavanaugh, S. (2003). Pheromonal stimulation of spawning release of gametes by gonadotropin releasing hormone in the chiton, *Mopalia* sp. *General and Comparative Endocrinology* 131: 62–65.

Gorbushin, A. M. (1997). Field evidence of trematode-induced gigantism in *Hydrobia* spp. (Gastropoda: Prosobranchia). *Journal of the Marine Biological Association of the United Kingdom* 77: 785–800.

Gorbushin, A. M. & Levakin, I. A. (1999). The effect of trematode parthenitae on the growth of *Onoba aculeus*, *Littorina saxatilis* and *Littorina obtusata* (Gastropoda: Prosobranchia). *Journal of the Marine Biological Association of the United Kingdom* 79: 273–279.

Gordon, M. S. & Olson, E. C. (2013). *Invasions of the Land: The Transitions of Organisms from Aquatic to Terrestrial Life*. New York, Columbia University Press.

Goreau, T. F., Goreau, N. I., Soot-Ryen, T. & Yonge, C. M. (1969). On a new commensal mytilid (Mollusca: Bivalvia) opening into the coelenteron of *Fungia scutaria* (Coelenterata). *Journal of Zoology* 158: 171–195.

Goreau, T. F., Goreau, N. I., Yonge, C. M. & Yeumann, Y. (1970). On feeding and nutrition in *Fungiacava eilatensis* (Bivalvia, Mytilidae), a commensal living in fungiid corals. *Journal of Zoology* 160: 159–172.

Goreau, T. F., Goreau, N. I. & Yonge, C. M. (1972). On the mode of boring in *Fungiacava eilatensis* (Bivalvia: Mytilidae). *Journal of Zoology* 166: 55–60.

Goreau, T. F., Goreau, N. I. & Yonge, C. M. (1973). On the utilisation of photosynthetic products from zooxanthellae and of a dissolved amino-acid in *Tridacna maxima* f. *elongata* (Mollusca: Bivalvia). *Journal of Zoology* 169: 417–454.

Goreau, T. F., Goreau, N. I. & Goreau, T. J. (1979). Corals and coral reefs. *Scientific American* 241: 124–136.

Gorf, A. (1961). Untersuchungen über Neurosekretion bei der Sumpfdeckelschnecke *Vivipara vivipara* L. *Zoologische Jahrbucher. Abteilung für Allgemeine Zoologie und Physiologie der Tiere* 69: 379–404.

Gorf, A. (1963). Der Einfluss des sichtbaren Lichtes auf die Neurosekretion der Sumpfdeckelschnecke *Vivipara vivipara*. *Zoologische Jahrbücher. Abteilung für Allgemeine Zoologie und Physiologie der Tiere* 70: 266–277.

Gorham, W. T. (1990). Uptake of dilute nutrients by marine invertebrates. pp. 84–95 in J. Mellinger (Ed.) *Animal Nutrition and Transport Processes. I. Nutrition in Wild and Domestic Animals*. Vol. 5. Basel, Switzerland, Karger.

Gosline, J. M. & Shadwick, R. E. (1982). The biomechanics of the arteries of *Nautilus*, *Nototodarus*, and *Sepia*. *Pacific Science* 36: 283–296.

Gosline, J. M. & Demont, M. E. (1985). Jet-propelled swimming in squids. *Scientific American* 256: 96–103.

Gosliner, T. M. & Liltved, W. R. (1982). Comparative morphology of three South African Triviidae (Gastropoda: Prosobranchia) with the description of a new species. *Zoological Journal of the Linnean Society* 74: 111–132.

Gosliner, T. M. & Ghiselin, M. T. (1984). Parallel evolution in opisthobranch gastropods and its implications for phylogenetic methodology. *Systematic Zoology* 33: 255–274.

Gosliner, T. M. (1985). Parallelism, parsimony and the testing of phylogenetic hypotheses: The case of opisthobranch gastropods. *Transvaal Museum Monographs* 4: 105–107.

Gosliner, T. M. (1987). *Nudibranchs of Southern Africa. A Guide to Opisthobranch Molluscs of Southern Africa.* Leiden, the Netherlands, E.J. Brill.

Gosliner, T. M. (1991). Morphological parallelism in opisthobranch gastropods. *Malacologia* 32: 313–327.

Gosliner, T. M. (1994). Gastropoda: Opisthobranchia. pp. 253–355 *in* F. W. Harrison & Kohn, A. J. (Eds.) *Microscopic Anatomy of Invertebrates. Mollusca 1.* Vol. 5. New York, Wiley-Liss.

Gosliner, T. M. (1996). The Opisthobranchia. pp. 161–213 *in* P. H. Scott, Blake, J. A. & Lissner, A. L. (Eds.) *Taxonomic Atlas of the Benthic Fauna of the Santa Maria Basin and Western Santa Barbara Channel.* Vol. 9. San Diego, CA, Science Applications International Corporation.

Gosliner, T. M. (2001). Aposematic coloration and mimicry in opisthobranch mollusks: New phylogenetic and experimental data. *Bollettino Malacologico* 37: 163–170.

Gosliner, T. M. & Behrens, D. W. (1990). Special resemblance, aposematic coloration and mimicry in opisthobranch gastropods. pp. 127–138 *in* M. Wicksten (Ed.) *Adaptive Coloration in Invertebrates.* College Station TX, Texas A&M University Sea Grant College Program.

Gosliner, T. M. & Johnson, R. F. (1999). Phylogeny of *Hypselodoris* (Nudibranchia: Chromodorididae) with a review of the monophyletic clade of Indo-Pacific species, including descriptions of twelve new species. *Zoological Journal of the Linnean Society* 125: 1–114.

Gosling, E. M. (2003). *Bivalve Molluscs: Biology, Ecology and Culture.* Oxford, UK, Fishing News Books.

Gotliv, B. A., Kessler, N., Sumerel, J. L., Morse, D. E., Tuross, N., Addadi, L. & Weiner, S. (2005). Asprich: A novel aspartic acid-rich protein family from the prismatic shell matrix of the bivalve *Atrina rigida. ChemBioChem* 6: 304–314.

Goto, R., Hamamura, Y. & Kato, M. (2007). Obligate commensalism of *Curvemysella paula* (Bivalvia: Galeommatidae) with hermit crabs. *Marine Biology* 151: 1615–1622.

Goto, R., Kawakita, A., Ishikawa, H., Hamamura, Y. & Kato, M. (2012). Molecular phylogeny of the bivalve superfamily Galeommatoidea (Heterodonta, Veneroida) reveals dynamic evolution of symbiotic lifestyle and interphylum host switching. *BMC Evolutionary Biology* 12: 172.

Gotow, T. (1986). Decrease of K^+ conductance underlying a depolarizing photoresponse of a molluscan extraocular photoreceptor. *Experientia* 42: 52–54.

Gotow, T. & Nishi, T. (2008). Simple photoreceptors in some invertebrates: Physiological properties of a new photosensory modality. *Brain Research* 1225: 3–16.

Goudet, J., De Meeüs, T., Day, A. J. & Gliddon, C. J. (1994). The different levels of population structuring of the dogwhelk, *Nucella lapillus,* along the south Devon coast. pp. 81–95 *in* A. R. Beaumont (Ed.) *Genetics and Evolution of Aquatic Organisms.* London, UK, Chapman & Hall.

Gould, M. C. & Stephano, J. L. (2003). Polyspermy prevention in marine invertebrates. *Microscopy Research and Technique* 61: 379–388.

Gould, S. J. (1968). Ontogeny and the explanation of form: An allometric analysis. *Journal of Paleontology, Paleontological Society Memoir 2* 42: 81–98.

Gould, S. J. (1969). An evolutionary microcosm: Pleistocene and Recent history of the land snail *P. (Poecilozonites)* in Bermuda. *Bulletin of the Museum of Comparative Zoology* 138: 407–532.

Gould, S. J. (1970). Land snail communities and Pleistocene climates in Bermuda: A multivariate analysis of microgastropod diversity. pp. 486–521 *in* E. L. Yochelson (Ed.) *Proceedings of the North American Paleontology Convention: Field Museum of Natural History, Chicago, September 5–7, 1969 [Part E].* Vol. 1. Lawrence, KS, Allen Press.

Gould, S. J. (1971). Muscular mechanics and the ontogeny of swimming in scallops. *Paleontology* 14: 61–94.

Gould, S. J. (1977). *Ontogeny and Phylogeny.* Cambridge, MA, Harvard University Press.

Gould, S. J. (1984). Morphological channeling by structural constraint: Convergence in styles of dwarfing and gigantism in *Cerion,* with a description of two new fossil species and a report on the discovery of the largest *Cerion. Paleobiology* 10: 172–194.

Gould, S. J. (1985). The sinister and the trivial. *Natural History* 94: 16–26.

Gould, S. J. (1993). Poe's greatest Hit. *Natural History* 102: 10–19.

Gould, S. J. & Robinson, B. A. (1994). The promotion and prevention of recoiling in a maximally snail-like vermetid gastropod: A case study for the centenary of Dollo's Law. *Paleobiology* 20: 368–390.

Gould, S. J. (1997). The exaptive excellence of spandrels as a term and prototype. *Proceedings of the National Academy of Sciences of the United States of America* 94: 10750–10755.

Gould, S. J. (2002). *The Structure of Evolutionary Theory.* Cambridge, MA, Belknap Press of Harvard University Press.

Goulding, M. Q. & Lambert, J. D. (2016). Mollusc models I. The snail *Ilyanassa. Current Opinion in Genetics & Development* 39: 168–174.

Govenar, B., Fisher, C. R. & Shank, T. M. (2015). Variation in the diets of hydrothermal vent gastropods. *Deep Sea Research Part II: Topical Studies in Oceanography* 121: 193–201.

Govindan, K. & Natarajan, R. (1974). Studies on Neritidae (Neritacea: Prosobranchia) from Peninsular India. *Indian National Science Academy. Part B. Biological Sciences. Proceedings* 38: 225–239.

Govindarajan, A. F., Piraino, S., Gravili, C. & Kubota, S. (2005). Species identification of bivalve-inhabiting marine hydrozoans of the genus *Eugymnanthea. Invertebrate Biology* 124: 1–10.

Govorin, I. A. (2007). Allochthonous bacteria in the ecosystem: Marine environment-aquatic organisms-bottom sediment. *Hydrobiological Journal* 43: 48–58.

Gowland, F. C., Boyle, P. R. & Noble, L. R. (2002). Morphological variation provides a method of estimating thermal niche in hatchlings of the squid *Loligo forbesi* (Mollusca: Cephalopoda). *Journal of Zoology* 258: 505–513.

Graf, D. L. & Ó Foighil, D. (2000). The evolution of brooding characters among the freshwater pearly mussels (Bivalvia : Unionoidea) of North America. *Journal of Molluscan Studies* 66: 157–170.

Graham, A. L. (1932). On the structure and function of the alimentary system of the limpet. *Transactions of the Royal Society of Edinburgh* 57: 287–308.

Graham, A. L. (1934). The cruciform muscle of lamellibranchs. *Proceedings of the Royal Society of Edinburgh* 54: 17–30.

Graham, A. L. (1939). On the structure of the alimentary canal of style-bearing prosobranchs. *Proceedings of the Zoological Society of London* 109: 75–112.

Graham, A. L. (1941). The oesophagus of the stenoglossan prosobranchs. *Proceedings of the Royal Society of Edinburgh B* 61: 1–23.

Graham, A. L. (1949). The molluscan stomach. *Transactions of the Royal Society of Edinburgh* 61: 737–778.

Graham, A. L. (1955). Molluscan diets. *Proceedings of the Malacological Society of London* 31: 144–159.

Graham, A. L. (1964). The functional anatomy of the buccal mass of the limpet (*Patella vulgata*). *Proceedings of the Zoological Society of London* 143: 301–329.

Graham, A. L. (1965). The buccal mass of ianthinid prosobranchs. *Proceedings of the Malacological Society of London* 36: 323–338.

Graham, A. L. (1966). The fore-gut of some marginellid and cancellariid prosobranchs. *Studies in Tropical Oceanography (Miami)* 4: 134–151.

Graham, A. L. (1973). The anatomical basis of function in the buccal mass of prosobranch and amphineuran molluscs. *Journal of Zoology* 169: 317–348.

Graham, A. L. (1985). Evolution within the Gastropoda: Prosobranchia. pp. 151–186 *in* E. R. Trueman & Clarke, M. R. (Eds.) *Evolution. The Mollusca*. Vol. 10. New York, Academic Press.

Graham, F. J., Runham, N. W. & Ford, J. B. (1996). Long-term effects of *Riccardoella limacum* living in the lung of *Helix aspersa*. pp. 359–364 *in* I. F. Henderson (Ed.) *BCPC Symposium Proceedings: Slug and Snail Pests in Agriculture*. Vol. 66. Thornton Heath, UK, Food and Agriculture Organisation of the United Nations.

Grahame, J. (1969). The biology of *Berthellina caribbea* Edmunds. *Bulletin of Marine Science* 19: 868–879.

Grahame, J. & Branch, G. M. (1985). Reproductive patterns of marine invertebrates. *Oceanography and Marine Biology* 23: 373–398.

Grande, C., Templado, J. & Zardoya, R. (2008). Evolution of gastropod mitochondrial genome arrangements. *BMC Evolutionary Biology* 8: 1–15.

Grande, C. & Patel, N. H. (2009). Nodal signalling is involved in left-right asymmetry in snails. *Nature* 457: 1007–1011.

Grande, C. (2010). Left-right asymmetries in Spiralia. *Integrative and Comparative Biology* 50: 744–755.

Grande, C., Martín-Durán, J. M., Kenny, N. J., Truchado-García, M. & Hejnol, A. (2015). Evolution, divergence and loss of the Nodal signalling pathway: New data and a synthesis across the Bilateria. *International Journal of Developmental Biology* 58: 521–532.

Grant, G. S. (1996). Prey of the introduced *Bufo marinus* on American Samoa. *Herpetological Review* 27: 67–69.

Grant, W. S., Dempster, Y. L. & Da Silva, F. M. (1988). Use of protein electrophoresis in evolutionary systematics. *Transactions of the Royal Society of South Africa* 46: 295–311.

Grant, W. S. & Utter, F. M. (1988). Genetic heterogeneity on different geographic scales in *Nucella lamellosa* (Prosobranchia, Thaididae). *Malacologia* 28: 275–288.

Grantham, B. A., Eckert, G. L. & Shanks, A. L. (2003). Dispersal potential of marine invertebrates in diverse habitats. *Ecological Applications* 13: 108–116.

Grassle, J. F. (1987). The ecology of deep-sea hydrothermal vent communities. *Advances in Marine Biology* 23: 301–362.

Grasso, F. W. (2008). Octopus sucker-arm coordination in grasping and manipulation. *American Malacological Bulletin* 24: 13–23.

Graus, R. R. (1974). Latitudinal trends in the shell characteristics of marine gastropods. *Lethaia* 7: 303–314.

Graves, S. Y. & Dietz, T. H. (1979). Prostaglandin E2 inhibition of sodium transport in the freshwater mussel. *Journal of Experimental Zoology Part A* 210: 195–202.

Gray, C. A., McQuaid, C. D. & Davies-Coleman, M. T. (2005). A symbiotic shell-encrusting bryozoan provides subtidal whelks with chemical defence against rock lobsters. *African Journal of Marine Science* 27: 549–556.

Gray, D. R. & Hodgson, A. N. (1997). Temporal variation in foraging behaviour of *Patella granularis* (Patellogastropoda) and *Siphonaria concinna* (Basommatophora) on a South African shore. *Journal of Molluscan Studies* 63: 121–130.

Gray, D. R. & Hodgson, A. N. (1998). Foraging and homing behaviour in the high-shore, crevice-dwelling limpet *Helcion pectunculus* (Prosobranchia: Patellidae). *Marine Biology* 132: 283–294.

Gray, E. G. (1970). The fine structure of the vertical lobe of *Octopus* brain. *Philosophical Transactions of the Royal Society B* 258: 379–394.

Gray, J. (1955). The movement of sea-urchin spermatozoa. *Journal of Experimental Biology* 32: 775–801.

Gray, J. (1968). *Animal Locomotion*. New York, Norton.

Gray, M. W., Burger, G. & Lang, B. F. (1999). Mitochondrial evolution. *Science* 283: 1476–1482.

Graziadei, P. (1964). Receptors in the sucker of the cuttlefish. *Nature* 203: 384–386.

Graziadei, P. (1965). Sensory receptor cells and related neurons in cephalopods. *Cold Spring Harbor Symposia on Quantitative Biology* 30: 45–57.

Graziadei, P. P. C. & Gagne, H. T. (1976). An unusual receptor in the octopus. *Tissue and Cell* 8: 229–240.

Green, R. H. (1971). A multivariate statistical approach to the Hutchinsonian niche: Bivalve molluscs of central Canada. *Ecology* 52: 544–556.

Green, T. J. & Barnes, A. C. (2010). Bacterial diversity of the digestive gland of Sydney rock oysters, *Saccostrea glomerata* infected with the paramyxean parasite, *Marteilia sydneyi*. *Journal of Applied Microbiology* 109: 613–622.

Greenberg, S. G., Drake, P. F. & Lasek, R. J. (1983). Differential synthesis of heat-shock proteins by connective cells and by neurons of *Aplysia californica*. *Journal of Cell Biology* 97: A152.

Greene, R. V., Griffin, H. L. & Freer, S. N. (1988). Purification and characterization of an extracellular endoglucanase from the marine shipworm bacterium. *Archives of Biochemistry and Biophysics* 267: 334–341.

Greene, R. V., Cotta, M. A. & Griffin, H. L. (1989). A novel, symbiotic bacterium isolated from marine shipworm secretes proteolytic activity. *Current Microbiology* 19: 353–356.

Greenfield, L. M. (1972). Feeding and gut physiology in *Acanthopleura spinigera* (Mollusca). *Journal of Zoology* 166: 37–47.

Greenwald, L. & Ward, P. D. (1987). Buoyancy in *Nautilus*. pp. 547–562 *in* W. B. Saunders & Landman, N. H. (Eds.) *Nautilus: The Biology and Paleobiology of a Living Fossil. Topics in Geobiology*. New York, Springer.

Greenwood, P. G. (1988). Nudibranch nematocysts. pp. 445–462 *in* D. A. Hessinger & Lenhoff, H. M. (Eds.) *The Biology of Nematocysts*. San Diego, CA, Academic Press.

Greenwood, P. G. (2009). Acquisition and use of nematocysts by cnidarian predators. *Toxicon* 54: 1065–1070.

Grégoire, C., Duchateau, G. & Florkin, M. (1955). La trame protidique des nacres et des perles. *Annales de l'Institut océanographique* 31: 1–36.

Grégoire, C. (1966). On organic remains in shells of Paleozoic and Mesozoic cephalopods (nautiloids and ammonoids). *Bulletin de l'Institut royal des sciences naturelles de Belgique* 42: 1–36.

Gregory, T. R. (2004). Macroevolution, hierarchy theory, and the C-value enigma. *Paleobiology* 30: 179.

Gregory, T. R., Nicol, J. A., Tamm, H., Kullman, B., Kullman, K., Leitch, I. J., Murray, B. G., Kapraun, D. F., Greilhuber, J. & Bennett, M. D. (2007). Eukaryotic genome size databases. *Nucleic Acids Research* 35: D332–D338.

Gregory, T. R., Ed. (2011). *The Evolution of the Genome*. Burlington, MA, Elsevier Academic Press.

Gregory, T. R. (2018). Animal Genome Size Database. from http://www.genomesize.com.

Grenon, J.-F. & Walker, G. (1978). The histology and histochemistry of the pedal glandular system of two limpets, *Patella vulgata* and *Acmaea tessulata* (Gastropoda: Prosobranchia). *Journal of the Marine Biological Association of the United Kingdom* 58: 803–816.

Grenon, J.-F. & Walker, G. (1980). Biochemical and rheological properties of the pedal mucus of the limpet, *Patella vulgata* L. *Comparative Biochemistry and Physiology Part B* 66: 451–458.

Grenon, J.-F. & Walker, G. (1981). The tenacity of the limpet *Patella vulgata* L.: An experimental approach. *Journal of Experimental Marine Biology and Ecology* 54: 277–308.

Grewal, P. S., Grewal, S. K., Tan, L. & Adams, B. J. (2003). Parasitism of molluscs by nematodes: Types of associations and evolutionary trends. *Journal of Nematology* 35: 146–156.

Grewal, P. S., Ehlers, R.-U. & Shapiro-Ilan, D. I. (2005). *Nematodes as Biological Control Agents*. Wallingford, UK, CABI Publishing.

Grey, M., Lelievre, P. G. & Boulding, E. G. (2005). Selection for prey shell thickness by the naticid gastropod *Euspira lewisii* (Naticidae) on the bivalve *Protothaca staminea* (Veneridae). *The Veliger* 48: 1–6.

Grieshaber, M. K. (1978). Breakdown and formation of high-energy phosphates and octopine in the adductor muscle of the scallop, *Chlamys opercularis* (L.), during escape swimming and recovery. *Journal of Comparative Physiology A* 126: 269–276.

Griffin, L. E. (1900). The anatomy of *Nautilus pompilius*. *Memoirs of the National Academy of Sciences* 8: 101–230.

Griffith, G. W. & Castagna, M. (1962). Sexual dimorphism in oyster drills of Chincoteague Bay, Maryland-Virginia. *Chesapeake Science* 3: 215–217.

Griffiths, C. L. & Seiderer, L. J. (1980). Rock lobsters and mussels—Limitations and preferences in a predator-prey interaction. *Journal of Experimental Marine Biology and Ecology* 44: 95–109.

Griffiths, C. L. (1990). Spatial gradients in predation pressure and their influence on the dynamics of two littoral bivalve populations. pp. 321–332 in B. Morton (Ed.) *The Bivalvia. Proceedings of a Memorial Symposium in Honour of Sir Charles Maurice Yonge (1899–1986), Edinburgh, 1986*. Hong Kong, Hong Kong University Press.

Griffiths, C. L. & Klumpp, D. W. (1996). Relationships between size, mantle area and zooxanthellae numbers in five species of giant clam (Tridacnidae). *Marine Ecology Progress Series* 137: 139–147.

Griffiths, O., Cook, A. & Wells, S. M. (1993). The diet of the introduced carnivorous snail *Euglandina rosea* in Mauritius and its implications for threatened island gastropod faunas. *Journal of Zoology* 229: 79–89.

Griffiths, O. L. & Florens, V. F. B. (2006). *A Field Guide to the Non-Marine Molluscs of the Mascarene Islands (Mauritius, Rodrigues and Reunion) and the Northern Dependencies of Mauritius*. Mauritius, Bioculture Press.

Griffiths, R. (1961). Sexual dimorphism in Cypraeidae. *Proceedings of the Malacological Society of London* 34: 79–92.

Griffond, B. & Bride, J. (1985). Contribution à l'étude de l'embryogenèse de l'appareil génital d'*Helix aspersa*; mise en évidence de la participation de deux types cellulaires à l'édification de la gonade. *Reproduction Nutrition Développement* 25: 141–152.

Griffond, B. & Bolzoni-Sungur, D. (1986). Stages of oogenesis in the snail, *Helix aspersa*: Cytological, cytochemical and ultrastructural studies. *Reproduction Nutrition Développement* 26: 461–474.

Grimaldi, A., Tettamanti, G., Acquati, F., Bossi, E., Guidali, M. L., Banfi, S., Monti, L., Valvassori, R. & de Eguileor, M. (2008). A hedgehog homologue is involved in muscle formation and organization of *Sepia officinalis* (Mollusca) mantle. *Developmental Dynamics* 237: 659–671.

Grimm-Jørgensen, Y., Ducor, M. E. & Piscatelli, J. (1986). Surface mucus production in gastropods is dependent on environmental salinity and humidity. *Comparative Biochemistry and Physiology Part A* 83: 415–419.

Grinich, N. P. & Terwilliger, R. C. (1980). The quaternary structure of an unusual high-molecular-weight intracellular haemoglobin from the bivalve mollusc *Barbatia reeveana*. *Biochemical Journal* 189: 1–8.

Grizzle, J. M. & Brunner, C. J. (2009). Infectious diseases of freshwater mussels and other freshwater bivalve mollusks. *Reviews in Fisheries Science* 17: 425–467.

Groombridge, B. & Jenkins, M. D. (2002). *World Atlas of Biodiversity. Prepared by the UNEP World Conservation Monitoring Centre*. Berkeley, CA, University of California Press.

Gros, O., Frenkiel, L. & Mouëza, M. (1997). Embryonic, larval, and post-larval development in the symbiotic clam *Codakia orbicularis* (Bivalvia: Lucinidae). *Invertebrate Biology* 116: 86–101.

Gros, O., Durand, P., Frenkiel, L. & Mouëza, M. (1998a). Putative environmental transmission of sulfuroxidising gill endosymbionts in four tropical lucinid bivalves, inhabiting various environments. *FEMS Microbiology Ecology Letters* 160: 257–262.

Gros, O., Frenkiel, L. & Mouëza, M. (1998b). Gill filament differentiation and experimental colonization by symbiotic bacteria in aposymbiotic juveniles of *Codakia orbicularis* (Bivalvia: Lucinidae). *Invertebrate Reproduction & Development* 34: 219–231.

Gros, O., Liberge, M., Heddi, A., Khatchadourian, C. & Felbeck, H. (2003). Detection of free-living forms of sulfide-oxidising gill endosymbionts in the lucinid habitat *Thalassia testudinum* environment. *Applied and Environmental Microbiology* 69: 6264–6267.

Gros, O., Frenkiel, L. & Aranda, D. A. (2009). Structural analysis of the digestive gland of the queen conch *Strombus gigas* Linnaeus, 1758 and its intracellular parasites. *Journal of Molluscan Studies* 75: 59–68.

Gros, O., Elisabeth, N. H., Gustave, S. D. D., Caro, A. & Dubilier, N. (2012). Plasticity of symbiont acquisition throughout the life cycle of the shallow-water tropical lucinid *Codakia orbiculata* (Mollusca: Bivalvia). *Environmental Microbiology* 14: 1584–1595.

Grosberg, R. K., Vermeij, G. J. & Wainwright, P. C. (2012). Biodiversity in water and on land. *Current Biology* 22: R900–R903.

Gross, C. G. (2002). Genealogy of the 'Grandmother Cell'. *Neuroscientist* 8: 512–518.

Grove, D. I. (1990). *A History of Human Helminthology*. Wallingford, UK, CAB International.

Grubert, M. A., Mundy, C. N. & Ritar, A. J. (2005). The effects of sperm density and gamete contact time on the fertilization success of blacklip (*Haliotis rubra*; Leach, 1814) and greenlip (*H. laevigata*; Donovan, 1808) abalone. *Journal of Shellfish Research* 24: 407–413.

Grubich, J. (2003). Morphological convergence of pharyngeal jaw structure in durophagous perciform fish. *Biological Journal of the Linnean Society* 80: 147–165.

Gründel, J. (2008). Remarks to the classification and phylogeny of the Ataphridae Cossmann, 1915 (Gastropoda, Archaeogastropoda) in the Jurassic. *Neues Jahrbuch für Geologie und Paläontologie Abhandlungen* 250: 177–197.

Grüneberg, H. & Nugaliyadde, L. (1976). Population studies on a polymorphic prosobranch snail (*Clithon* (*Pictoneritia*) *oualaniensis* Lesson). *Philosophical Transactions of the Royal Society B* 275: 385–437.

Grüneberg, H. (1980). On pseudo-polymorphism. *Proceedings of the Royal Society B* 210: 533–548.

Grüneberg, H. (1981). Pseudo-polymorphism in *Nerita polita* (Neritacea, Archaeogastropoda). *Proceedings of the Royal Society B* 212: 53–63.

Grüneberg, H. (1982). Pseudo-polymorphism in *Clithon oualaniensis*. *Proceedings of the Royal Society B* 216: 147–157.

Guderley, H. E. & Tremblay, I. (2013). Escape responses by jet propulsion in scallops. *Canadian Journal of Zoology* 91: 420–430.

Guerreiro, M., Phillips, R. A., Cherel, Y., Ceia, F. R., Alvito, P., Rosa, R. & Xavier, J. C. (2015). Habitat and trophic ecology of Southern Ocean cephalopods from stable isotope analyses. *Marine Ecology Progress Series* 530: 119–134.

Guerriero, A., D'Ambrosio, M. & Pietra, F. (1988). Slowly interconverting conformers of the briarane diterpenoids Verecynarmin B, C, and D, isolated from the nudibranch mollusc *Armina maculata* and the pennatulacean octocoral *Veretillum cynomorium* of East Pyrenean waters. *Helvetica Chimica Acta* 71: 472–485.

Guerriero, A., D'Ambrosio, M. & Pietra, F. (1990). Isolation of the cembranoid preverecynarmin alongside some briaranes, the verecynarmins, from both the nudibranch mollusc *Armina maculata* and the octocoral *Veretillum cynomorium* of the East Pyrenean Mediterranean sea. *Helvetica Chimica Acta* 73: 277–283.

Guiart, J. (1901). *Contribution à l'étude des gastéropodes opisthobranches et en particulier des céphalaspides*. Vol. 14. Lille, Bigot Frères.

Guida, V. G. (1976). Sponge predation in the oyster reef community as demonstrated with *Cliona celata* Grant. *Journal of Experimental Marine Biology and Ecology* 25: 109–122.

Guilford, T. & Cuthill, I. (1991). The evolution of aposematism in marine gastropods. *Evolution* 45: 449–451.

Guilford, T. & Dawkins, M. S. (1993). Are warning colors handicaps? *Evolution* 47: 400–416.

Gunji, K., Ishizaki, S. & Shiomi, K. (2010). Cloning of complementary and genomic DNAs encoding echotoxins, proteinaceous toxins from the salivary gland of marine gastropod *Monoplex echo*. *The Protein Journal* 29: 487–492.

Guo, X. & Allen, S. (1994). Sex determination and polyploid gigantism in the dwarf surfclam (*Mulinia lateralis* Say). *Genetics* 138: 1199–1206.

Guralnick, R. P. & Lindberg, D. R. (1999). Integrating developmental evolutionary patterns and mechanisms: A case study using the gastropod radula. *Evolution* 53: 447–459.

Guralnick, R. P. & Smith, K. (1999). Historical and biomechanical analysis of integration and dissociation in molluscan feeding, with special emphasis on the true limpets (Patellogastropoda: Gastropoda). *Journal of Morphology* 241: 175–195.

Guralnick, R. P. & Lindberg, D. R. (2001). Reconnecting cell and animal lineages: What do cell lineages tell us about the evolution and development of Spiralia? *Evolution* 55: 1501–1519.

Guralnick, R. P. (2004). Life-history patterns in the brooding freshwater bivalve *Pisidium* (Sphaeriidae). *Journal of Molluscan Studies* 70: 341–351.

Gurney, L. J., Mundy, C. & Porteus, M. C. (2005). Determining age and growth of abalone using stable oxygen isotopes: A tool for fisheries management. *Fisheries Research* 72: 353–360.

Gusmão, L. C. & Daly, M. (2010). Evolution of sea anemones (Cnidaria: Actiniaria: Hormathiidae) symbiotic with hermit crabs. *Molecular Phylogenetics and Evolution* 56: 868–877.

Gustafson, K. & Andersen, R. J. (1985). Chemical studies of British Columbia nudibranchs. *Tetrahedron* 41: 1101–1108.

Gustafson, R. G. & Reid, R. G. B. (1986). Development of the pericalymma larva of *Solemya reidi* (Bivalvia: Cryptodonta: Solemyidae) as revealed by light and electron microscopy. *Marine Biology* 93: 411–428.

Gustafson, R. G. & Reid, R. G. B. (1988). Larval and post-larval morphogenesis in the gutless protobranch bivalve *Solemya reidi* (Cryptodonta: Solemyidae). *Marine Biology* 97: 373–387.

Gustafson, R. G. & Lutz, R. A. (1992). Larval and early post-larval development of the protobranch bivalve *Solemya velum* (Mollusca: Bivalvia). *Journal of the Marine Biological Association of the United Kingdom* 72: 383–402.

Guthrie, M. J. & Anderson, J. M. (1957). *General Zoology*. New York, Wiley.

Gutiérrez, J. L., Jones, C. G., Strayer, D. L. & Iribarne, O. O. (2003). Mollusks as ecosystem engineers: The role of shell production in aquatic habitats. *Oikos* 101: 79–90.

Gutnick, T., Shomrat, T., Mather, J. A. & Kuba, M. J. (2017). The Cephalopod brain: Motion control, learning, and cognition. pp. 137–178 *in* S. Saleuddin & Muka, S. (Eds.) *Physiology of Molluscs*. Vol. 2. Oakville, ON, Apple Academic Press.

Haas, F. (1935–1955). Bivalvia. pp. 545–704 *in* H. G. Bronn (Ed.) *H. G. Bronn's Klassen und Ordnungen des Tier-Reichs wissenschaftlich dargestellt in Wort und Bild*. Vol. 3, Mollusca, Abt. 3. Leipzig, Germany, C. F. Winter.

Haas, W. (1972). Untersuchungen über die Mikro- und Ultrastruktur der Polyplacophorenschale. *Biomineralisation* 5: 3–52.

Haas, W., Kriesten, K. & Watabe, N. (1979). Notes on the shell formation in the larvae of the *Placophora* (Mollusca). *Biomineralization Research Reports* 10: 1–8.

Haas, W. (1981). Evolution of calcareous hardparts in primitive molluscs. *Malacologia* 21: 403–418.

Haas, W. (2002). The evolutionary history of the eight-armed Coleoidea. *Abhandlungen der Geologischen Bundesanstalt* 57: 341–351.

Haase, M. & Bouchet, P. (1998). Radiation of crenobiontic gastropods on an ancient continental island: The *Hemistomia*-clade in New Caledonia (Gastropoda: Hydrobiidae). *Hydrobiologia* 367: 43–129.

Haase, M. & Karlsson, A. (2000). Mating and the inferred function of the genital system of the nudibranch, *Aeolidiella glauca* (Gastropoda: Opisthobranchia: Aeolidioidea). *Invertebrate Biology* 119: 287–298.

Haase, M. (2005). Rapid and convergent evolution of parental care in hydrobiid gastropods from New Zealand. *Journal of Evolutionary Biology* 18: 1076–1086.

Haase, M. & Bouchet, P. (2006). The radiation of hydrobioid gastropods (Caenogastropoda, Rissooidea) in ancient Lake Poso, Sulawesi. *Hydrobiologia* 556: 17–46.

Haase, M. (2008). The radiation of hydrobiid gastropods in New Zealand: A revision including the description of new species based on morphology and mtDNA sequence information. *Systematics and Biodiversity* 6: 99–159.

Haber, M., Cerfeda, S., Carbone, M., Calado, G., Gaspar, H., Neves, R., Maharajan, V., Cimino, G., Gavagnin, M. & Ghiselin, M. T. (2010). Coloration and defense in the nudibranch gastropod *Hypselodoris fontandraui*. *Biological Bulletin* 218: 181–188.

Haddock, S. H. D., Moline, M. A. & Case, J. F. (2010). Bioluminescence in the sea. *Annual Review of Marine Science* 2: 443–493.

Hadfield, M. G. (1970). Observations on the anatomy and biology of two California vermetid gastropods. *The Veliger* 12: 301–309.

Hadfield, M. G. (1978). Metamorphosis in marine molluscan larvae: An analysis of stimulus and response. pp. 165–175 *in* F.-S. Chia & Rice, M. E. (Eds.) *Settlement and Metamorphosis of Marine Invertebrate Larvae*. New York, Elsevier.

Hadfield, M. G. (1979). Aplacophora. pp. 1–25 *in* A. C. Giese & Pearse, J. S. (Eds.) *Reproduction of Marine Invertebrates. Molluscs: Pelecypods and Lesser Classes*. Vol. 5. New York, Academic Press.

Hadfield, M. G. & Switzer-Dunlap, M. (1984). Opisthobranchs. pp. 209–350 *in* A. S. Tompa, Verdonk, N. H. & van den Biggelaar, J. A. M. (Eds.) *Reproduction. The Mollusca*. Vol. 7. New York, Academic Press.

Hadfield, M. G. (1994). Extinction in Hawaiian achatinelline snails. pp. 320–334 *in* E. A. Kay (Ed.) *A Natural History of the Hawaiian Islands: Selected Readings II*. Honolulu, University of Hawai'i.

Hadfield, M. G. & Strathmann, M. F. (1996). Variability, flexibility and plasticity in life histories of marine invertebrates. *Oceanologica Acta* 19: 323–334.

Hadfield, M. G., Meleshkevitch, E. A. & Boudko, D. Y. (2000). The apical sensory organ of a gastropod veliger is a receptor for settlement cues. *Biological Bulletin* 198: 67–76.

Hadfield, M. G., Carpizo-Ituarte, E. J., del Carmen, K. & Nedved, B. T. (2001). Metamorphic competence, a major adaptive convergence in marine invertebrate larvae. *American Zoologist* 41: 1123–1131.

Hadfield, M. G. & Koehl, M. A. R. (2004). Rapid behavioral responses of an invertebrate larva to dissolved settlement cue. *Biological Bulletin* 207: 28–43.

Haga, T. & Kase, T. (2013). Progenetic dwarf males in the deep-sea wood-boring genus *Xylophaga* (Bivalvia: Pholadoidea). *Journal of Molluscan Studies* 79: 90–94.

Hagadone, M. R., Burreson, B. J., Scheuer, P. J., Finer, J. S. & Clardy, J. (1979). Defense allomones of the nudibranch *Phyllidia varicosa* Lamarck 1801. *Helvetica Chimica Acta* 62: 2484–2494.

Hagenau, A. & Scheibel, T. (2010). Towards the recombinant production of mussel byssal collagens. *Journal of Adhesion* 86: 10–24.

Hahn, K. O. (1994). The neurosecretory staining in the cerebral ganglia of the Japanese abalone (ezoawabi), *Haliotis discus hannai*, and its relationship to reproduction. *General and Comparative Endocrinology* 93: 295–303.

Hahn, T. & Denny, M. (1989). Tenacity-mediated selective predation by oystercatchers on intertidal limpets and its role in maintaining habitat partitioning by 'Collisella' scabra and Lottia digitalis. *Marine Ecology Progress Series* 53: 1–10.

Hai, T. (2006). The ATF transcription factors in cellular adaptive responses. pp. 329–340 *in* J. Ma (Ed.) *Gene Expression and Regulation*. New York, Springer.

Halanych, K. M. & Kocot, K. M. (2014). Repurposed transcriptomic data facilitate discovery of innate immunity toll-like receptor (TLR) genes across lophotrochozoa. *Biological Bulletin* 227: 201–209.

Haldane, J. B. S. (1932). *The Causes of Evolution*. London, Harpers.

Haldeman, S. S. (1845). *A Monograph of the Freshwater Univalve Mollusca of the United States, Including Notices of Species in Other Parts of North America*. Philadelphia, PA, Conchological Section of the Academy of Natural Sciences.

Hall-Spencer, J. M. & Moore, P. G. (2000). *Limaria hians* (Mollusca: Limacea): A neglected reef-forming keystone species. *Aquatic Conservation: Marine and Freshwater Ecosystems* 10: 267–277.

Hall, J. R. (1970). Description of egg capsules and embryos of squid, *Lolloguncula brevis*, from Tampa Bay, Florida. *Bulletin of Marine Science* 20: 762–768.

Hall, K. C. & Hanlon, R. T. (2002). Principal features of the mating system of a large spawning aggregation of the giant Australian cuttlefish *Sepia apama* (Mollusca : Cephalopoda). *Marine Biology* 140: 533–545.

Hallinan, N. & Lindberg, D. R. (2011). Comparative analysis of chromosome counts infers three paleopolyploidies in the Mollusca. *Genome Biology and Evolution* 3: 1150–1163.

Hallmann, N., Burchell, M., Schöne, B. R., Irvine, G. V. & Maxwell, D. (2009). High-resolution sclerochronological analysis of the bivalve mollusk *Saxidomus gigantea* from Alaska and British Columbia: techniques for revealing environmental archives and archaeological seasonality. *Journal of Archaeological Science* 36: 2353-2364.

Halton, D. W. & Owen, G. (1968). The fine structure and histochemistry of the gastric cuticle of the proto-branchiate bivalve, *Nucula sulcata* Brohn. *Proceedings of the Malacological Society of London* 38: 71–81.

Halwart, M. (1994). The golden apple snail *Pomacea canaliculata* in Asian rice farming systems: Present impact and future threat. *International Journal of Pest Management* 40: 199–206.

Hamada, F. N., Rosenzweig, M., Kang, K., Pulver, S. R., Ghezzi, A., Jegla, T. J. & Garrity, P. A. (2008). An internal thermal sensor controlling temperature preference in *Drosophila*. *Nature* 454: 217–220.

Hameed, P. S. & Paulpandian, A. L. (1987). Quantitative analysis of carbohydrases in the crystalline style of some intertidal bivalve molluscs. *Proceedings of the Indian Academy of Sciences (Animal Sciences)* 96: 41–47.

Hamilton, P. V. (1977). Daily movements and visual location of plant stems by *Littorina irrorata* (Mollusca: Gastropoda). *Marine Behaviour and Physiology* 4: 293–304.

Hamilton, P. V. (1978). Adaptive visually-mediated movements of *Littorina irrorata* (Mollusca: Gastropoda) when displaced from their natural habitat. *Marine Behaviour and Physiology* 5: 255–271.

Hamilton, P. V. & Russell, B. J. (1982a). Experiments involved on the sense organs and directional cues in offshore-oriented swimming by *Aplysia brasiliana* Rang (Mollusca: Gastropoda). *Journal of Experimental Marine Biology and Ecology* 56: 123–143.

Hamilton, P. V. & Russell, B. J. (1982b). Celestial orientation by surface-swimming *Aplysia brasiliana* Rang (Mollusca: Gastropoda). *Journal of Experimental Marine Biology and Ecology* 56: 145–152.

Hamilton, P. V. & Winter, M. A. (1982). Behavioural responses to visual stimuli by the snail *Littorina irrorata*. *Animal Behaviour* 30: 752–760.

Hamilton, P. V., Ardizzoni, S. C. & Penn, J. S. (1983). Eye structure and optics in the intertidal snail, *Littorina irrorata*. *Journal of Comparative Physiology A* 152: 435–445.

Hamilton, P. V. & Winter, M. A. (1984). Behavioural responses to visual stimuli by the snails *Tectarius muricatus*, *Turbo castanea*, and *Helix aspersa*. *Animal Behaviour* 32: 51–57.

Hamilton, W. J. (1948). The food and feeding behavior of the green frog, *Rana clamitans* Latreille, in New York State. *Copeia* 1948: 203–207.

Hammarsten, O. D. & Runnström, J. (1926a). Ein Beitrag zur Diskussion über die Verwandtschaftsbeziehungen der Mollusken. *Acta Zoologica* 7: 1–67.

Hammarsten, O. D. & Runnström, J. (1926b). Zur Embryologie von *Acanthochiton discrepans* Brown. *Zoologische Jahrbücher. Abteilung für Anatomie und Ontogenie der Tiere* 47: 261–318.

Han, G.-Y., Sun, D.-Y., Liang, L.-F., Yao, L.-G., Chen, K.-X. & Guo, Y.-W. (2018). Spongian diterpenes from Chinese marine sponge *Spongia officinalis*. *Fitoterapia* 127: 159–165.

Han, T. S., Teichert, R. W., Olivera, B. M. & Bulaj, G. (2008). *Conus* venoms-a rich source of peptide-based therapeutics. *Current Pharmaceutical Design* 14: 2462–2479.

Hancock, A. & Embleton, D. (1852). On the anatomy of *Doris*. *Philosophical Transactions of the Royal Society* 142: 207–252.

Hancock, D. A. (1956). The structure of the capsule and the hatching process in *Urosalpinx cinerea* (Say). *Proceedings of the Zoological Society of London* 127: 565–571.

Handl, C. H. & Todt, C. (2005). Foregut glands of Solenogastres (Mollusca): Anatomy and revised terminology. *Journal of Morphology* 265: 28–42.

Haneda, Y. (1939). Luminosity in *Rocellaria grandis* (Deshayes) (Lamellibranchia) [in Japanese]. *Kagaku Nanyâo* 2: 36–39.

Haneda, Y. (1955). Luminous organisms of Japan and the Far East. pp. 335–386 in F. H. Johnson (Ed.) *The Luminescence of Biological Systems*. Washington, DC, American Association for the Advancement of Science.

Haneda, Y. (1958). Studies on luminescence in marine snails. *Pacific Science* 12: 152–156.

Haneda, Y. (1981). Luminescence activity of the land snail *Quantula striata*. pp. 252–265 in M. A. DeLuca & McElroy, W. D. (Eds.) *Bioluminescence and Chemiluminescence*. New York, Academic Press.

Hanington, P. C. & Zhang, S. M. (2010). The primary role of fibrinogen-related proteins in invertebrates is defense, not coagulation. *Journal of Innate Immunity* 3: 17–27.

Hanken, J. & Wake, D. B. (1993). Miniaturization of body size: Organismal consequences and evolutionary significance. *Annual Review of Ecology and Systematics* 24: 501–519.

Hanlin, H. G. (1978). Food habits of the greater siren, *Siren lacertina*, in an Alabama coastal plain pond. *Copeia* 1978: 358–360.

Hanlon, R. T. (1982). The functional organization of chromatophores and iridescent cells in the body patterning of *Loligo pealei* (Cephalopoda, Myopsida). *Malacologia* 23: 89–119.

Hanlon, R. T. & Budelmann, B.-U. (1987). Why cephalopods are probably not 'deaf'. *American Naturalist* 129: 312–317.

Hanlon, R. T. & Messenger, J. B. (1988). Adaptive coloration in young cuttlefish (*Sepia officinalis* L.): The morphology and development of body patterns and their relation to behaviour. *Philosophical Transactions of the Royal Society B* 320: 437–487.

Hanlon, R. T. & Forsythe, J. W. (1990). Introduction: Cephalopoda, Annelida, Crustacea, Chaetognatha, Echinodermata, Urochordata. pp. 23–46 in O. Kinne (Ed.) *Diseases of Mollusca. Cephalopoda: Diseases Caused by Microorganisms*. Vol. 3. Hamburg, Biologisches Anstalt Helgoland.

Hanlon, R. T. (1991). Integrative Neurobiology and Behavior of Mollusks Symposium: Introduction, perspectives, and round-table discussion. *Biological Bulletin* 180: 197–199.

Hanlon, R. T. & Messenger, J. B. (1996). *Cephalopod Behaviour*. Cambridge, UK, Cambridge University Press.

Hanlon, R. T. (1998). Mating systems and sexual selection in the squid *Loligo*: How might commercial fishing on spawning squids affect them? *California Cooperative Oceanic Fisheries Investigations Report: (CalCOFI Report)* 39: 92–100.

Hanlon, R. T., Smale, M. J. & Sauer, W. H. H. (2002). The mating system of the squid *Loligo vulgaris reynaudii* (Cephalopoda, Mollusca) off South Africa: Fighting, guarding, sneaking, mating and egg laying behavior. *Bulletin of Marine Science* 71: 331–345.

Hanlon, R. T. & Shashar, N. (2003). Aspects of the sensory ecology of cephalopods. pp. 266–282 in S. P. Collin & Marshall, N. J. (Eds.) *Sensory Processing in Aquatic Environments*. New York, Springer.

Hanlon, R. T., Naud, M. J., Forsythe, J. W., Hall, K., Watson, A. C. & McKechnie, J. (2007). Adaptable night camouflage by cuttlefish. *American Naturalist* 169: 543–551.

Hannides, A. K. & Aller, R. C. (2016). Priming effect of benthic gastropod mucus on sedimentary organic matter remineralization. *Limnology and Oceanography* 61: 1640–1650.

Hansen, B. (1953). Brood protection and sex ratio of *Transennella tantilla* (Gould), a Pacific bivalve. *Videnskabelige meddelelser fra Dansk naturhistorisk forening i Kjøbenhavn* 115: 313–324.

Hansen, T. A. (1978). Larval dispersal and species longevity in Lower Tertiary gastropods. *Science* 199: 885–887.

Hansen, T. A. (1983). Modes of larval development and rates of speciation in Early Tertiary neogastropods. *Science* 220: 501–502.

Hanumante, M. M., Deshpande, U. D. & Nagabhushanam, R. (1978). Susceptibility of urea and uric acid levels to neurohormone influence in the freshwater pulmonate, *Lymnaea acuminata*. *Hydrobiologia* 60: 125–128.

Hanumante, M. M. & Deshpande, U. D. (1980). Regulation of ammonia (NH_3-N) metabolism in the marine pulmonate, *Onchidium verruculatum*. 1. Effects of a neuroendocrine centre on tissue ammonia levels. *Hydrobiologia* 68: 265–268.

Hanumante, M. M., Deshpande, U. D. & Nagabhushanam, R. (1980). Changes in the neurosecretory cells of the freshwater snail, *Lymnaea acuminata* induced by urea administration. *Hydrobiologia* 69: 7–10.

Hara, T. & Hara, R. (1976). Distribution of rhodopsin and retinochrome in the squid retina. *Journal of General Physiology* 67: 791–805.

Hara, T., Hara, R., Kishigami, A., Koshida, Y., Horiuchi, S. & Raj, U. (1995). Rhodopsin and retinochrome in the retina of a tetrabranchiate cephalopod, *Nautilus pompilius*. *Zoological Science* 12: 195–201.

Harasewych, M. G. & Petit, R. E. (1986). Notes on the morphology of *Admete viridula* (Gastropoda: Cancellariidae). *The Nautilus* 100: 85–91.

Harasewych, M. G. (2002). Pleurotomarioidean gastropods. *Advances in Marine Biology* 42: 237–294.

Harasewych, M. G. (2009). Anatomy and biology of *Mitra cornea* Lamarck, 1811 (Mollusca, Caenogastropoda, Mitridae) from the Azores. *Açoreana* Suplemento 6: 121–135.

Harding, J. M. & Mann, R. (2001). Oyster reefs as fish habitat: Opportunistic use of restored reefs by transient fishes. *Journal of Shellfish Research* 20: 951–959.

Harper, E. M. (1991). The role of predation in the evolution of cementation in bivalves. *Palaeontology* 34: 455–460.

Harper, E. M. (1992). Post-larval cementation in the Ostreidae and its implications for other cementing bivalves. *Journal of Molluscan Studies* 58: 37–47.

Harper, E. M. (1994a). Molluscivory by the asteroid *Coscinasterias acutispina* (Stimpson). pp. 339–355 *in* B. Morton (Ed.) *The Malacofauna of Hong Kong and Southern China, III: Proceedings of the Third International Workshop on the Malacofauna of Hong Kong and Southern China, Hong Kong, 13 April–1 May 1992.* Vol. 3. Hong Kong, Hong Kong University Press.

Harper, E. M. (1994b). Are conchiolin sheets in corbulid bivalves primarily defensive? *Palaeontology* 37: 551–578.

Harper, E. M. (1997a). Attachment of mature oysters (*Saccostrea cucullata*) to natural substrata. *Marine Biology* 127: 449–453.

Harper, E. M. (1997b). The molluscan periostracum: An important constraint in bivalve evolution. *Palaeontology* 40: 71–97.

Harper, E. M. (2012). Cementing bivalves. *Treatise Online* Part N, Revised, Volume 1, Chapter 21: 1–12.

Harper, G. L., King, R. A., Dodd, C. S., Harwood, J. D., Glen, D. M., Bruford, M. W. & Symondson, W. O. C. (2005). Rapid screening of invertebrate predators for multiple prey DNA targets. *Molecular Ecology* 14: 819–827.

Harris, J. M. (1993). The presence, nature, and role of gut microflora in aquatic invertebrates: A synthesis. *Microbial Ecology* 25: 195–231.

Harris, J. M., Branch, G. M., Elliott, B. L., Currie, B., Dye, A. H., McQuaid, C. D., Tomalin, B. J. & Velasquez, C. (1998a). Spatial and temporal variability in recruitment of intertidal mussels around the coast of southern Africa. *South African Journal of Zoology.* 33: 1–11.

Harris, J. O., Burke, C. M. & Maguire, G. B. (1998b). Characterization of the digestive tract of Greenlip abalone, *Haliotis laevigata* Donovan. II. Microenvironment and bacterial flora. *Journal of Shellfish Research* 17: 989–994.

Harris, J. R. & Markl, J. (1999). Keyhole limpet hemocyanin (KLH): A biomedical review. *Micron* 30: 597–623.

Harris, S. A., Da Silva, F. M., Bolton, J. J. & Brown, A. C. (1986). Algal gardens and herbivory in a scavenging sandy-beach nassariid whelk. *Malacologia* 27: 299–305.

Harris, W. A. (1997). *Pax-6*: Where to be conserved is not conservative. *Proceedings of the National Academy of Sciences of the United States of America* 94: 2098–2100.

Hart, A. M., Bell, J. D. & Foyle, T. P. (1998). Growth and survival of the giant clams, *Tridacna derasa, T. maxima* and *T. crocea,* at village farms in the Solomon Islands. *Aquaculture* 165: 203–220.

Harte, M. E. (1992). A new approach to the study of bivalve evolution. *American Malacological Union Bulletin* 9: 199–206.

Hartenstein, V. (2006). The neuroendocrine system of invertebrates: A developmental and evolutionary perspective. *Journal of Endocrinology* 190: 555–570.

Hartline, P. H., Hurley, A. C. & Lange, G. D. (1979). Eye stabilization by statocyst mediated oculomotor reflex in *Nautilus. Journal of Comparative Physiology A* 132: 117–126.

den Hartog, J. C., (1987). Observations on the wentletrap *Epitonium clathratulum* (Kanmacher, 1797) (Prosobranchia, Epitoniidae) and the sea anemone *Bunodosoma biscayensis* (Fischer, 1874) (Actiniaria, Actiniidae). *Basteria* 51: 95–108.

Hashimoto, N., Kurita, Y., Murakami, K. & Wada, H. (2015). Cleavage pattern and development of isolated D blastomeres in bivalves. *Journal of Experimental Zoology Part B* 324: 13–21.

Haszprunar, G. (1983). Comparative analysis of the abdominal sense organs of Pteriomorpha (Bivalvia). *Journal of Molluscan Studies* 12A: 47–50.

Haszprunar, G. (1985a). The fine morphology of the osphradial sense organs of the Mollusca. Part 1: Gastropoda—Prosobranchia. *Philosophical Transactions of the Royal Society B* 307: 457–496.

Haszprunar, G. (1985b). The fine structure of the abdominal sense organs of Pteriomorpha (Mollusca, Bivalvia). *Journal of Molluscan Studies* 51: 315–319.

Haszprunar, G. (1985c). On the anatomy and fine structure of a peculiar sense organ in *Nucula* (Bivalvia, Protobranchia). *The Veliger* 28: 52–62.

Haszprunar, G. (1985d). Zur Anatomie und systematischen Stellung der Architectonicidae (Mollusca, Allogastropoda). *Zoologica Scripta* 14: 25–43.

Haszprunar, G. (1985e). On the innervation of gastropod shell muscles. *Journal of Molluscan Studies* 51: 309–314.

Haszprunar, G. (1985f). The Heterobranchia: A new concept of the phylogeny and evolution of the higher Gastropoda. *Zeitschrift für Zoologische Systematik und Evolutionsforschung* 23: 15–37.

Haszprunar, G. (1985g). The fine morphology of the osphradial sense organs of the Mollusca. Part 2: Allogastropoda (Architectonicidae and Pyramidellidae). *Philosophical Transactions of the Royal Society B* 307: 497–505.

Haszprunar, G. (1986a). Fine morphology of gastropod osphradia. pp. 101–104 *in* L. S. Pínter (Ed.) *Proceedings of the 8th International Malacological Congress (Budapest, Hungary) 28th August–4th September, 1983.* Budapest, Hungarian Natural History Museum.

Haszprunar, G. (1986b). Fine morphological investigations on sensory structures of primitive Solenogastres (Mollusca). *Zoologischer Anzeiger* 217: 345–362.

Haszprunar, G. (1987a). Anatomy and affinities of cocculinid limpets (Mollusca: Archaeogastropoda). *Zoologica Scripta* 16: 305–324.

Haszprunar, G. (1987b). The fine morphology of the osphradial sense organs of the Mollusca. Part 3. Placophora and Bivalvia. *Philosophical Transactions of the Royal Society B* 315: 37–61.

Haszprunar, G. (1987c). The fine morphology of the osphradial sense organs of the Mollusca. Part 4. Caudofoveata and Solenogastres. *Philosophical Transactions of the Royal Society B* 315: 63–73.

Haszprunar, G. (1987d). The fine structure of the ctenidial sense organs (bursicles) of Vetigastropoda (Zeugobranchia, Trochoidea) and their functional and phylogenetic significance. *Journal of Molluscan Studies* 53: 46–51.

Haszprunar, G. (1988a). On the origin and evolution of major gastropod groups, with special reference to the Streptoneura (Mollusca). *Journal of Molluscan Studies* 54: 367–441.

Haszprunar, G. (1988b). Comparative anatomy of cocculiniform gastropods and its bearing on archaeogastropod systematics. pp. 64–84 *in* W. F. Ponder, Eernisse, D. J. & Waterhouse, J. H. (Eds.) *Prosobranch Phylogeny. Malacological Review Supplement.* Ann Arbor, MI, Malacological Review.

Haszprunar, G. (1988c). *Sukashitrochus* sp., a scissurellid with heteropod-like locomotion (Mollusca, Archaeogastropoda). *Annalen des Naturhistorischen Museums in Wien Serie B* 90: 367–371.

Haszprunar, G. (1989a). Die Torsion der Gastropoda: ein biomechanischer prozeß? [Torsion in gastropoda—A biomechanical process?]. *Zeitschrift für Zoologische Systematik und Evolutionsforschung* 27: 1–7.

Haszprunar, G. (1989b). New slit-limpets (Scissurellacea and Fissurellacea) from hydrothermal vents. Part 2: Anatomy and relationships. *Natural History Museum of Los Angeles County. Contributions in Science* 408: 1–17.

Haszprunar, G. & Huber, G. (1990). On the central nervous system of Smeagolidae and Rhodopidae, two families questionably allied with the Gymnomorpha (Gastropoda: Euthyneura). *Journal of Zoology* 220: 185–199.

Haszprunar, G. (1992a). Ultrastructure of the osphradium of the tertiary relict snail, *Campanile symbolicum* Iredale (Mollusca, Streptoneura). *Philosophical Transactions of the Royal Society B* 337: 457–469.

Haszprunar, G. (1992b). The first molluscs: Small animals. *Bollettino di Zoologia* 59: 1–16.

Haszprunar, G., Schaefer, K., Warén, A. & Hain, S. (1995). Bacterial symbionts in the epidermis of an Antarctic neopilinid limpet (Mollusca, Monoplacophora). *Philosophical Transactions of the Royal Society B* 347: 181–185.

Haszprunar, G. (1996a). The Mollusca: Coelomate turbellarians or mesenchymate annelids? pp. 1–28 *in* J. D. Taylor (Ed.) *Origin and Evolutionary Radiation of the Mollusca*. Oxford, UK, Oxford University Press.

Haszprunar, G. (1996b). The molluscan rhogocyte (pore-cell, blasenzelle, cellule nucale), and its significance for ideas on nephridial evolution. *Journal of Molluscan Studies* 62: 185–211.

Haszprunar, G. & Kunz, E. (1996). Ultrastructure and systematic significance of the epidermis and haemocoel of *Rhodope* (Gastropoda, Nudibranchia, Doridoidea?). *Journal of Submicroscopic Cytology and Pathology* 28: 485–497.

Haszprunar, G. & McLean, J. H. (1996). Anatomy and systematics of bathyphytophilid limpets (Mollusca, Archaeogastropoda) from the northeastern Pacific. *Zoologica Scripta* 25: 35–49.

Haszprunar, G. & Schaefer, K. (1996). Anatomy and phylogenetic significance of *Micropilina arntzi* (Mollusca, Monoplacophora, Micropilinidae Fam. nov.). *Acta Zoologica* 77: 315–334.

Haszprunar, G. (1997). Ultrastructure of the pseudo-protonephridium of the enigmatic opisthobranch, *Rhodope transtrosa* (Gastropoda, Nudibranchia). *Journal of Submicroscopic Cytology and Pathology* 29: 371–378.

Haszprunar, G. & Schaefer, K. (1997). Monoplacophora. pp. 415–457 *in* F. W. Harrison & Kohn, A. (Eds.) *Microscopic Anatomy of Invertebrates. Mollusca 2.* Vol. 6B. New York, Wiley-Liss.

Haszprunar, G. (1998). Superorder Cocculiniformia. pp. 653–664 *in* P. L. Beesley, Ross, G. J. B. & Wells, A. (Eds.) *Mollusca: The Southern Synthesis. Part B. Fauna of Australia.* Melbourne, Australia, CSIRO Publishing.

Haszprunar, G. & Ruthensteiner, B. (2000). Microanatomy and ultrastructure of the protonephridial system in the larva of the limpet, *Patella vulgata* L. (Mollusca, Patellogastropoda). *Journal of Submicroscopic Cytology and Pathology* 32: 59–67.

Haszprunar, G. & Wanninger, A. (2000). Molluscan muscle systems in development and evolution. *Journal of Zoological Systematics and Evolutionary Research* 38: 157–163.

Haszprunar, G., Friedrich, S., Wanninger, A. & Ruthensteiner, B. (2002). Fine structure and immunocytochemistry of a new chemosensory system in the *Chiton* larva (Mollusca: Polyplacophora). *Journal of Morphology* 251: 210–218.

Haszprunar, G., Schander, C. & Halanych, K. M. (2008). Relationships of higher molluscan taxa. pp. 19–32 *in* W. F. Ponder & Lindberg, D. R. (Eds.) *Phylogeny and Evolution of the Mollusca*. Berkeley, CA, University of California Press.

Haszprunar, G. & Wanninger, A. (2008). On the fine structure of the creeping larva of *Loxosomella murmanica*: Additional evidence for a clade of Kamptozoa (Entoprocta) and Mollusca. *Acta Zoologica* 89: 137–148.

Haszprunar, G., Speimann, E., Hawe, A. & Hess, M. (2011). Interactive 3D anatomy and affinities of the Hyalogyrinidae, basal Heterobranchia (Gastropoda) with a rhipidoglossate radula. *Organisms Diversity & Evolution* 11: 201–236.

Haszprunar, G., Kunze, T., Brückner, M. & Hess, M. (2016). Towards a sound definition of Skeneidae (Mollusca, Vetigastropoda): 3D interactive anatomy of the type species, *Skenea serpuloides* (Montagu, 1808) and comments on related taxa. *Organisms Diversity & Evolution* 16: 577–595.

Haszprunar, G., Kunze, T., Warén, A. & Hess, M. (2017). A reconsideration of epipodial and cephalic appendages in basal gastropods: Homologies, modules and evolutionary scenarios. *Journal of Molluscan Studies* 83: 363–383.

Hatakeyama, D., Fujito, Y., Sakakibara, M. & Ito, E. (2004). Expression and distribution of transcription factor CCAAT/enhancer-binding protein in the central nervous system of *Lymnaea stagnalis*. *Cell and Tissue Research* 318: 631–642.

Hatcher, N. G. & Sweedler, J. V. (2008). *Aplysia* bag cells function as a distributed neurosecretory network. *Journal of Neurophysiology* 99: 333–343.

Hathaway, J. M., Adema, C. M., Stout, B. A., Mobarak, C. D. & Loker, E. S. (2010). Identification of protein components of egg masses indicates parental investment in immunoprotection of offspring by *Biomphalaria glabrata* (Gastropoda, Mollusca). *Developmental and Comparative Immunology* 34: 425–435.

Hatschek, B. (1888–1891). *Lehrbuch der Zoologie, eine morphologische Übersicht des Thierreiches zur Einführung in das Studium dieser Wissenschaft* Vol. 1 [1888], 2 [1889], 3 [1891] Jena, Gustav Fischer.

Hauck, M. & Gallardo-Fernández, G. L. (2013). Crises in the South African abalone and Chilean loco fisheries: Shared challenges and prospects. *Maritime Studies* 12: 3.

Hauffe, T., Albrecht, C., Schreiber, K., Birkhofer, K., Trajanovski, S. & Wilke, T. (2011). Spatially explicit analyses of gastropod biodiversity in ancient Lake Ohrid. *Biogeosciences* 8: 175–188.

Hausdorf, B. (2001). Macroevolution in progress: Competition between semislugs and slugs resulting in ecological displacement and ecological release. *Biological Journal of the Linnean Society* 74: 387–395.

Hausdorf, B. (2002). Introduced land snails and slugs in Colombia. *Journal of Molluscan Studies* 68: 127–131.

Haven, N. (1973). Reproduction, development and feeding of the Australian marine pulmonate, *Trimusculus (Gadinia) conica*. *The Veliger* 16: 61–65.

Haven, N. (1977). Cephalopoda: Nautiloidea. pp. 227–241 *in* A. C. Giese & Pearse, J. S. (Eds.) *Reproduction of Marine Invertebrates. Molluscs: Gastropods and Cephalopods. Reproduction of Marine Invertebrates.* Vol. 4. New York, Academic Press.

Haven, S. B. (1971). Niche differences in the intertidal limpets *Acmaea scabra* and *Acmaea digitalis* (Gastropoda) in central California. *The Veliger* 13: 231–248.

Havenhand, J. N. (1991). On the behaviour of opisthobranch larvae. *Journal of Molluscan Studies* 57: 119–131.

Hawe, A., Gensler, H. & Haszprunar, G. (2014). Bacteriocytes in the mantle cavity of *Lurifax vitreus* Warén & Bouchet, 2001 (Orbitestellidae): The first case among heterobranch gastropoda. *Journal of Molluscan Studies* 80: 337–340.

Hawkins, A. J. S. & Bayne, B. L. (1992). Physiological interrelations, and the regulation of production. pp. 171–222 *in* E. Gosling (Ed.) *The Mussel Mytilus: Ecology, Physiology, Genetics and Culture*. Amsterdam, the Netherlands, Elsevier.

Hawkins, A. J. S. & Klumpp, D. W. (1995). Nutrition of the giant clam *Tridacna gigas* (L.). II. Relative contributions of filter-feeding and the ammonium-nitrogen acquired and recycled by symbiotic alga towards total nitrogen requirements for tissue growth and metabolism. *Journal of Experimental Marine Biology and Ecology* 190: 263–290.

Hawkins, C. M. (1981). Efficiency of organic matter absorption by the tropical echinoid *Diadema antillarum* Philippi fed non-macrophytic algae. *Journal of Experimental Marine Biology and Ecology* 49: 245–253.

Hawkins, R. D., Kandel, E. R. & Bailey, C. H. (2006). Molecular mechanisms of memory storage in *Aplysia*. *Biological Bulletin* 210: 174 -191.

Hay, K. B., Fredensborg, B. L. & Poulin, R. (2005). Trematode-induced alterations in shell shape of the mud snail *Zeacumantus subcarinatus* (Prosobranchia: Batillariidae). *Journal of the Marine Biological Association of the United Kingdom* 85: 989–992.

Hay, M. E., Fenical, W. & Gustafson, K. (1987). Chemical defense against diverse coral-reef herbivores. *Ecology* 68: 1581–1591.

Hay, M. E. & Fenical, W. (1988). Marine plant-herbivore interactions: The ecology of chemical defense. *Annual Review of Ecology and Systematics* 19: 111–145.

Hay, M. E., Pawlik, J. R., Duffy, J. E. & Fenical, W. (1989). Seaweed-herbivore-predator interactions: Host-plant specialization reduces predation on small herbivores. *Oecologia* 81: 418–427.

Hay, M. E., Dufy, J. E., Paul, V. J., Renaud, P. E. & Fenical, W. (1990). Specialist herbivores reduce their susceptibility to predation by feeding on the chemically defended seaweed *Avrainvillea longicaulis*. *Limnology and Oceanography* 35: 1734–1743.

Hay, M. E. & Steinberg, P. D. (1992). The chemical ecology of plant-herbivore interactions in marine versus terrestrial communities. pp. 371–413 *in* M. Berenbaum & Rosenthal, G. A. (Eds.) *Herbivores: Their Interactions with Secondary Plant Metabolites. Ecological and Evolutionary Processes.* Vol. 2. Cambridge, MA, Academic Press.

Hay, M. E. (2009). Marine chemical ecology: Chemical signals and cues structure marine populations, communities, and ecosystems. *Annual Review of Marine Science* 1: 193–212.

Hayami, I. (1991). Living and fossil scallop shells as airfoils: An experimental study. *Paleobiology* 17: 1–18.

Hayashi, M. & Chiba, S. (2004). Enhanced colour polymorphisms in island populations of the land snail *Euhadra peliomphala*. *Biological Journal of the Linnean Society* 81: 417–425.

Hayden, E. C. (2014). The $1,000 genome. *Nature* 507: 294.

Hayes, K. A., Joshi, R. C., Thiengo, S. C. & Cowie, R. H. (2008). Out of South America: Multiple origins of non-native apple snails in Asia. *Diversity and Distributions* 14: 701–712.

Hayes, K. A., Yeung, N. W., Kim, J. R. & Cowie, R. H. (2012). New records of alien Gastropoda in the Hawaiian Islands: 1996–2010. *Bishop Museum Occasional Papers* 112: 21–28.

Hayes, K. A., Burks, R. L., Castro-Vazquez, A., Darby, P. C., Heras, H., Martín, P. R., Qiu, J. W., Thiengo, S. C., Vega, I. A. & Wada, T. (2015). Insights from an integrated view of the biology of Apple Snails (Caenogastropoda: Ampullariidae). *Malacologia* 58: 245–302.

Hayes, T. (1983). The influence of diet on local distributions of *Cypraea*. *Pacific Science* 37: 27–36.

Haynes, A. (1992). The reproductive patterns of the five Fijian species of *Septaria* (Prosobranchia: Neritidae). *Journal of Molluscan Studies* 58: 13–20.

Healy, J. M. (1986a). An ultrastructural study of euspermatozoa, paraspermatozoa and nurse cells of the cowrie *Cypraea erones* (Gastropoda, Prosobranchia, Cypraeidae). *Journal of Molluscan Studies* 52: 125–137.

Healy, J. M. (1986b). Ultrastructure of paraspermatozoa of cerithiacean gastropods (Prosobranchia, Mesogastropoda). *Helgoländer Meeresuntersuchungen* 40: 177–199.

Healy, J. M. (1986c). Euspermatozoa and paraspermatozoa of the relict cerithiacean gastropod *Campanile symbolicum* (Prosobranchia, Mesogastropoda). *Helgoländer Meeresuntersuchungen* 40: 201–218.

Healy, J. M. (1988a). Sperm morphology in *Serpulorbis* and *Dendropoma* and its relevance to the systematic position of the Vermetidae (Gastropoda). *Journal of Molluscan Studies* 54: 295–308.

Healy, J. M. (1988b). Ultrastructural observations on the spermatozoa of *Pleurotomaria africana* Tomlin (Gastropoda). *Journal of Molluscan Studies* 54: 309–316.

Healy, J. M. (1988c). Sperm morphology and its systematic importance in the Gastropoda. pp. 251–266 *in* W. F. Ponder, Eernisse, D. J. & Waterhouse, J. H. (Eds.) *Prosobranch Phylogeny. Malacological Review Supplement.* Ann Arbor, MI, Malacological Review.

Healy, J. M. (1990). Euspermatozoa and paraspermatozoa in the trochid gastropod *Zalipais laseroni* (Trochoidea: Skeneidae). *Marine Biology* 105: 497–507.

Healy, J. M., Schaefer, K. & Haszprunar, G. (1995). Spermatozoa and spermatogenesis in a monoplacophoran mollusc, *Laevipilina antarctica*: Ultrastructure and comparison with other mollusca. *Marine Biology* 122: 53–65.

Healy, J. M. (1996). Molluscan sperm ultrastructure: Correlation with taxonomic units within the Gastropoda, Cephalopoda and Bivalvia. pp. 99–113 *in* J. D. Taylor (Ed.) *Origin and Evolutionary Radiation of the Mollusca.* Oxford, UK, Oxford University Press.

Healy, J. M., Lamprell, K. & Keys, J. L. (2001). The 'water-trap' spiny oyster, *Spondylus varius* G.B. Sowerby I, 1827 (Mollusca: Bivalvia: Spondylidae) from Australia. *Memoirs of the Queensland Museum* 46: 577–588.

Hearty, P. J. (2010). Chronostratigraphy and morphological changes in *Cerion* land snail shells over the past 130 ka on Long Island, Bahamas. *Quaternary Geochronology* 5: 50–64.

Heath, D. (1977). Simultaneous hermaphroditism; cost and benefit. *Journal of Theoretical Biology* 64: 363–373.

Heath, D. J. (1975). Colour, sunlight and internal temperatures in the land-snail *Cepaea nemoralis* (L.) *Oecologia* 19: 29–38.

Heath, H. (1899). The development of *Ischnochiton*. *Zoologische Jahrbücher. Abteilung für Anatomie und Ontogenie der Tiere* 12: 567–656.

Heath, H. (1905). The morphology of a solenogastre. *Zoologische Jahrbücher* 21: 701–734.

Heath, H. (1911a). The Solenogastres. *Memoirs of the Museum of Comparative Zoology* 45: 9–79.

Heath, H. (1911b). Reports on the scientific results of the expedition to the tropical Pacific. XIV. The Solenogastres. *Museum of Comparative Zoology. Memoirs* 45: 17–179.

Heath, H. (1916). The conjugation of *Ariolimax californicus*. *The Nautilus* 30: 22–24.

Heath, H. (1937). The anatomy of some protobranch mollusks. *Mémoires du Musée royal d'histoire naturelle de Belgique* 10: 1–26.

Heatwole, H. (1999). *Sea Snakes.* Sydney, University of New South Wales Press.

Hecht, S. (1920). The photochemical nature of the photosensory process. *Journal of General Physiology* 2: 229–246.

Hecht, S. (1927). The kinetics of dark adaptation. *Journal of General Physiology* 10: 781–809.

Hedegaard, C. (1990). *Shell Structures of the Recent Archaeogastropoda*. Cand. scient Thesis, University of Århus, Denmark.

Hedegaard, C. (1996). *Molluscs: Phylogeny and Biomineralization* PhD, University of Århus, Denmark.

Hedegaard, C. (1997). Shell structures of the Recent Vetigastropoda. *Journal of Molluscan Studies* 63: 369–377.

Hedegaard, C. & Wenk, H. R. (1998). Microstructure and texture patterns of mollusc shells [Part 1]. *Journal of Molluscan Studies* 64: 133–136.

Hedegaard, C., Bardeau, J. F. & Chateigner, D. (2006). Molluscan shell pigments: An in situ resonance Raman study [Part 2]. *Journal of Molluscan Studies* 72: 157–162.

Hedgecock, D. & Okazaki, N. B. (1984). Genetic diversity within and between populations of American oysters (*Crassostrea*). *Malacologia* 25: 535–549.

Hedgecock, D. (1986). Is gene flow from pelagic larval dispersal important in the adaptation and evolution of marine invertebrates? *Bulletin of Marine Science* 39: 550–564.

Hedgecock, D. & Pudovkin, A. I. (2011). Sweepstakes reproductive success in highly fecund marine fish and shellfish: A review and commentary. *Bulletin of Marine Science* 87: 971–1002.

Hedtke, S. M., Stanger-Hall, K., Baker, R. J. & Hillis, D. M. (2008). All-male asexuality: Origin and maintenance of androgenesis in the Asian clam *Corbicula*. *Evolution* 62: 1119–1136.

Heeger, T., Piatkowski, U. & Möller, H. (1992). Predation on jellyfish by the cephalopod *Argonauta argo*. *Marine Ecology Progress Series* 88: 293–296.

Hejnol, A., Martindale, M. Q. & Henry, J. Q. (2007). High-resolution fate map of the snail *Crepidula fornicata*: The origins of ciliary bands, nervous system, and muscular elements. *Developmental Biology* 305: 63–76.

Hejnol, A. & Martindale, M. Q. (2009). The mouth, the anus, and the blastopore: Open questions about questionable openings. pp. 33–40 *in* M. J. Telford & Littlewood, D. T. J. (Eds.) *Animal Evolution: Genomes, Fossils, and Trees*. Oxford, UK, Oxford University Press.

Helama, S. & Valovirta, I. (2008). The oldest recorded animal in Finland: Onto-genetic age and growth in *Margaritifera margaritifera* (L. 1758) based on internal shell increments. *Memoranda Societatis pro Fauna et Flora Fennica* 84: 20–30.

Heldin, C. H. & Westermark, B. (1999). Mechanism of action and in vivo role of platelet-derived growth factor. *Physiology Review* 79: 1283–1316.

Hellberg, M. E. (1996). Dependence of gene flow on geographic distance in two solitary corals with different larval dispersal capabilities. *Evolution*: 1167–1175.

Hellberg, M. E. & Vacquier, V. D. (1999). Rapid evolution of fertilization selectivity and lysin cDNA sequences in teguline gastropods. *Molecular Biology and Evolution* 16: 839–848.

Hellberg, M. E., Burton, R. S., Neigel, J. E. & Palumbi, S. R. (2002). Genetic assessment of connectivity among marine populations. *Bulletin of Marine Science* 70: 273–290.

Hellberg, M. E. (2009). Gene flow and isolation among populations of marine animals. *Annual Review of Ecology, Evolution, and Systematics* 40: 291–310.

Heller, J. (1979). Visual versus non-visual selection of shell colour in an Israeli freshwater snail. *Oecologia* 44: 98–104.

Heller, J. (1990). Longevity in molluscs. *Malacologia* 31: 259–295.

Heller, J. & Farstey, V. (1990). Sexual and parthenogenetic populations of the freshwater snail *Melanoides tuberculata* in Israel. *Israel Journal of Zoology* 37: 75–87.

Heller, J. (1993). Hermaphroditism in molluscs. *Biological Journal of the Linnean Society* 48: 19–42.

Heller, J. (2001). Life history strategies. pp. 413–445 *in* G. M. Barker (Ed.) *The Biology of Terrestrial Molluscs*. Wallingford, UK, CABI Publishing.

Heller, N. E. & Zavaleta, E. S. (2009). Biodiversity management in the face of climate change: A review of 22 years of recommendations. *Biological Conservation* 142: 14–32.

Hellou, J. & Andersen, R. J. (1981). Luteone, a twenty three carbon terpenoid from the dorid nudibranch *Cadlina luteomarginata*. *Tetrahedron Letters* 22: 4173–4176.

Helmuth, B., Broitman, B. R., Yamane, L., Gilman, S. E., Mach, K., Mislan, K. & Denny, M. W. (2010). Organismal climatology: Analyzing environmental variability at scales relevant to physiological stress. *Journal of Experimental Biology* 213: 995–1003.

Helmuth, B. S. T., Veit, R. R. & Holberton, R. (1994). Long-distance dispersal of a subantarctic brooding bivalve (*Gaimardia trapesina*) by kelp-rafting. *Marine Biology* 120: 421–426.

Hemingway, G. T. (1978). Evidence for a paralytic venom in the intertidal snail, *Acanthina spirata* (Neogastropoda: Thaisidae). *Comparative Biochemistry and Physiology Part C* 60: 79–81.

Hemmen, J. (2007). *Annotated and Illustrated Catalogue of Recent Cancellariidae*. Wiesbaden, Germany, Jens Hemmen.

Hemminga, M. A., Maaskant, J. J., Van der Plas, J., & Gabbott, P. A. (1985). The hyperglycemic factor of the CNS of the freshwater snail *Lymnaea stagnalis*: Interaction with glucose stimulation of glycogen synthesis and evidence for its release during anaerobiosis. *General and Comparative Endocrinology* 59: 301–307.

Hennig, W. (1950). *Grundzüge einer Theorie der phylogenetischen Systematik*. Berlin, Deutscher Zentralverlag.

Hennig, W. (1980). *Wirbellose I: Ausgenommen Gliedertiere. Taschenbuch der Zoologie Vol. 2 (5th Ed.)*. Jena, Gustav Fischer.

Henriques, P., Delgado, J. & Sousa, R. (2017). Patellid limpets: An overview of the biology and conservation of keystone species of the rocky shores. pp. 71–95 *in* S. Ray (Ed.) *Organismal and Molecular Malacology*. Rijeka, Croatia, InTech.

Henry, J., Cornet, V., Bernay, B. & Zatylny-Gaudin, C. (2013). Identification and expression of two oxytocin/vasopressin-related peptides in the cuttlefish *Sepia officinalis*. *Peptides* 46: 159–166.

Henry, J. Q., Okusu, A. & Martindale, M. Q. (2004). The cell lineage of the polyplacophoran, *Chaetopleura apiculata*: Variation in the spiralian program and implications for molluscan evolution. *Developmental Biology* 272: 145–160.

Henry, J. Q., Perry, K. J. & Martindale, M. Q. (2006). Cell specification and the role of the polar lobe in the gastropod mollusc *Crepidula fornicata*. *Developmental Biology* 297: 295–307.

Henry, J. Q., Hejnol, A., Perry, K. J. & Martindale, M. Q. (2007). Homology of ciliary bands in spiralian trochophores. *Integrative and Comparative Biology* 47: 865–871.

Henry, J. Q. & Lyons, D. C. (2016). Molluscan models: *Crepidula fornicata*. *Current Opinion in Genetics & Development* 39: 138–148.

Henry, M. (1984a). Ultrastructure des tubules digestifs d'un mollusque bivalve marin, la palourde *Ruditapes decussatus* L., en métabolisme de routine. I: La cellule digestive. *Vie marine* 6: 7–15.

Henry, M. (1984b). Ultrastructure des tubules digestifs d'un mollusque bivalve marin, la palourde *Ruditapes decussatus* en métabolisme de routine. II: La cellule sécrétice. *Vie marine* 6: 17–23.

Henry, M., Benlinmame, N., Belhsen, O. K., Jule, Y. & Mathieu, M. (1995). Immunohistochemical localization of FMRFamide-containing neurons and nerve fibers in the ganglia and the gonad wall of the scallop, *Pecten maximus* (L). *Neuropeptides* 28: 79–84.

Henssen, A. & Jahns, H. M. (1974). *Lichenes: eine Einführung in die Flechtenkunde*. Stuttgart, G. Thieme.

Heras, H., Dreon, M. S., Ituarte, S. & Pollero, R. J. (2007). Egg carotenoproteins in neotropical Ampullariidae (Gastropoda: Arquitaenioglossa). *Comparative Biochemistry and Physiology Part C* 146: 158–167.

Herbert, D. G. (1982). Fine structural observations on the juxtaganglionar organ of *Gibbula umbilicalis* (Da Costa). *Journal of Molluscan Studies* 48: 226–228.

Herbert, D. G. (1991). Foraminiferivory in a *Puncturella* (Gastropoda: Fissurellidae). *Journal of Molluscan Studies* 57: 127–140.

Herbert, D. G., Hamer, M. L., Mander, M., Mkhize, N. & Prins, F. (2003). Invertebrate animals as a component of the traditional medicine trade in KwaZulu-Natal, South Africa. *African Invertebrates* 44: 327–344.

Herbert, G. S. & Portell, R. W. (2004). First paleontological record of larval brooding in the calyptraeid gastropod genus *Crepidula* Lamarck, 1799. *Journal of Paleontology* 78: 424–429.

Herbert, G. S., Merle, D. & Gallardo, C. S. (2007). A developmental perspective on evolutionary innovation in the radula of the predatory neogastropod family Muricidae. *American Malacological Bulletin* 23: 17–32.

Herman, K. G. & Strumwasser, F. (1984). Regional specializations in the eye of *Aplysia*, a neuronal circadian oscillator. *Journal of Comparative Neurology* 230: 593–613.

Hermann, H. T. (1968). Optic guidance of locomotor behavior in the land snail, *Otala lactea*. *Vision Research* 8: 601–612.

Hermann, P. M., Genereux, B. & Wildering, W. C. (2009). Evidence for age-dependent mating strategies in the simultaneous hermaphrodite snail, *Lymnaea stagnalis* (L.). *Journal of Experimental Biology* 212: 3164–3173.

Hermans, C. O. & Satterlie, R. A. (1992). Fast-strike feeding behavior in a pteropod mollusk, *Clione limacina* Phipps. *Biological Bulletin* 182: 1–7.

Hernroth, L. & Gröndahl, F. (1985). On the biology of *Aurelia aurita* (L.) 3. Predation by *Coryphella verrucosa* (Gastropoda, Opisthobranchia), a major factor regulating the development of *Aurelia* populations in the Gullmar Fjord, western Sweden. *Ophelia* 24: 37–45.

Herring, P. J. (1977). Luminescence in cephalopods and fish. pp. 127–159 *in* M. Nixon & Messenger, J. B. (Eds.) *The Biology of Cephalopods. Symposia of The Zoological Society of London*. London, Academic Press.

Herring, P. J. (1987). Systematic distribution of bioluminescence in living organisms. *Luminescence* 1: 147–163.

Herring, P. J., Widder, E. A. & Haddock, S. H. D. (1992). Correlation of bioluminescence emissions with ventral photophores in the mesopelagic squid *Abralia veranyi* (Cephalopoda: Enoploteuthidae). *Marine Biology* 112: 293–298.

Herring, P. J. (2000). Species abundance, sexual encounter and bioluminescent signalling in the deep sea. *Philosophical Transactions of the Royal Society B* 355: 1273–1276.

Herring, P. J. (2002). *The Biology of the Deep Ocean*. Oxford, UK, Oxford University Press.

Hershler, R. (1985). Systematic revision of the Hydrobiidae (Gastropoda: Rissoacea) of the Cuatro Cienegas Basin, Coahuila, Mexico. *Malacologia* 26: 31–123.

Hershler, R. & Sada, D. W. (1987). Springsnails (Gastropoda: Hydrobiidae) of Ash Meadows, Amargosa Basin, California-Nevada. *Proceedings of the Biological Society of Washington* 100: 776–843.

Hershler, R. & Holsinger, J. R. (1990). Zoogeography of North American hydrobiid cavesnails. *Stygologia* 5: 5–16.

Hershler, R. (1998). A systematic review of the hydrobiid snails (Gastropoda: Rissooidea) of the Great Basin, western United States. Part I. Genus *Pyrgulopsis*. *The Veliger* 41: 1–132.

Hershler, R. (1999). A systematic review of the hydrobiid snails (Gastropoda: Rissooidea) of the Great Basin, western United States, Part II: Genera *Colligyrus, Eremopyrgus, Fluminicola, Pristinicola* and *Tryonia*. *The Veliger* 42: 306–337.

Hershler, R., Mulvey, M. & Liu, H.-P. (1999). Biogeography in the Death Valley region: Evidence from springsnails (Hydrobiidae: *Tryonia*). *Zoological Journal of the Linnean Society* 126: 335–354.

Hershler, R. & Liu, H.-P. (2008). Ancient vicariance and recent dispersal of springsnails (Hydrobiidae: *Pyrgulopsis*) in the Death Valley system, California-Nevada. *Geological Society of America Special Papers* 439: 91–101.

Hervio, D., Bower, S. M. & Meyer, G. R. (1993). Detection, isolation, and host specificity of *Mikrocytos mackini*, the cause of Denman Island disease in Pacific oysters *Crassostrea gigas*. *Journal of Shellfish Research* 12: 136.

Hesbacher, S., Baur, B., Baur, A. & Proksch, P. (1995). Sequestration of lichen compounds by three species of terrestrial snails. *Journal of Chemical Ecology* 21: 233–246.

Hescheler, K. (1900). Mollusca. pp. viii, 1–509 *in* A. Lang (Ed.) *Lehrbuch der vergleichenden Anatomie der wirbellosen Thiere*. Lief 1. Jena, Gustav Fischer.

Heslinga, G. A. (1981). Larval development, settlement and metamorphosis of the tropical gastropod *Trochus niloticus*. *Malacologia* 20: 349–357.

Hess, S. D. & Prior, D. J. (1985). Locomotor activity of the terrestrial slug, *Limax maximus*: Response to progressive dehydration. *Journal of Experimental Biology* 116: 323–330.

Hesse, R. (1900). Untersuchungen über die Organe der Lichtempfindung bei niederen Tieren. VI. Die Augen einiger Mollusken. *Zeitschrift für wissenschaftliche Zoologie* 68: 379–477.

Hesse, R. (1908). *Das Sehen der niederen Tiere*. Jena, Fischer.

Heyder, P. (1909). Zur Entwicklung der Lungenhöhle bei *Arion*. Nebst Bemerkungenüber die Entwicklung der Urniere, des Pericards und Herzens. *Zeitschrift für wissenschaftliche Zoologie* 93: 90–156.

Heyer, C. B., Kater, S. B. & Karlsson, U. (1973). Neuromuscular systems in molluscs. *American Zoologist* 13: 247–270.

Heyerdahl, T. (1950). *Kon-Tiki: Across the Pacific By Raft*. Chicago, Rand McNally & Company.

Heyland, A., Vue, Z., Voolstra, C. R., Medina, M. & Moroz, L. L. (2011). Developmental transcriptome of *Aplysia californica*. *Journal of Experimental Zoology Part B* 316B: 113–134.

Hickman, C. S. (1974). Characteristics of bathyal mollusk faunas in the Pacific Coast Tertiary. *Annual Report of the Western Society of Malacologists* 7: 41–50.

Hickman, C. S. (1980). Gastropod radulae and the assessment of form in evolutionary paleontology. *Paleobiology* 6: 276–294.

Hickman, C. S. (1981). Evolution and function of asymmetry in the archaeogastropod radula. *The Veliger* 23: 189–194.

Hickman, C. S. & Lipps, J. H. (1983). Foraminiferivory; selective ingestion of foraminifera and test alterations produced by the neogastropod *Olivella*. *The Journal of Foraminiferal Research* 13: 108–114.

Hickman, C. S. (1984). Form and function of the radulae of pleurotomariid gastropods. *The Veliger* 27: 29–36.

Hickman, C. S. (1985a). Gastropod morphology and function. pp. 138–156 in T.W. Broadhead (Ed.) *Mollusks: Notes for a Short Course Organized by D.J. Bottjer, C.S. Hickman, and P.D. Ward. University of Tennessee Studies in Geology.* Knoxville, TN, University of Tennessee, Department of Geological Sciences.

Hickman, C. S. (1985b). Comparative morphology and ecology of free-living suspension-feeding gastropods from Hong Kong. pp. 217–234 in B. Morton & Dudgeon, D. (Eds.) *Proceedings of the Second International Workshop on the Malacofauna of Hong Kong and of Southern China, Hong Kong, 6–24 April 1983.* Vol. 1 and 2. Hong Kong, Hong Kong University Press.

Hickman, C. S. & McLean, J. H. (1990). Systematic revision and suprageneric classification of trochacean gastropods. *Science Series: Natural History Museum of Los Angeles County* Issue 35: i–vi, 1–169.

Hickman, C. S. (1992). Reproduction and development of trochacean gastropods. *The Veliger* 35: 245–272.

Hickman, C. S. (1996). Tracing gastropod phylogeny through a suspension feeding design space. p. 293. *Geological Society of America, 28th Annual Meeting October 28–31 1996. Abstracts with Programs: Geological Society of America.* Vol. 28. Denver, CO, Geological Society of America.

Hickman, C. S. (1999). Larvae in invertebrate development and evolution. pp. 21–59 in B. K. Hall & Wake, M. H. (Eds.) *The Origin and Evolution of Larval Forms.* San Diego, CA, Academic Press.

Hickman, C. S. (2001). Evolution and development of gastropod larval shell morphology: Experimental evidence for mechanical defense and repair. *Evolution & Development* 3: 18–23.

Hickman, C. S. & Hadfield, M. G. (2001). Larval muscle contraction fails to produce torsion in a trochoidean gastropod. *Biological Bulletin* 200: 257–260.

Hickman, C. S. (2003a). Evaporites, water, and life. Part I: Mollusc–microbe mutualisms extend the potential for life in hypersaline systems. *Astrobiology* 3: 631–644.

Hickman, C. S. (2003b). Functional morphology and mode of life of *Isanda coronata* (Gastropoda: Trochidae) in an Australian macrotidal sandflat. pp. 69–88 in F. E. Wells, Walker, D. I. & Jones, D. S. (Eds.) *The Marine Flora and Fauna of Dampier, Western Australia: Proceedings of the Eleventh International Marine Biological Workshop held in Dampier 24 July–11 August 2000.* Perth, WA, Western Australian Museum.

Hickman, C. S. (2005a). Evolution on flexible hard substrates: Metazoan adaptations for life on seagrasses. p. 405. *Geological Society of America 2005 Annual Meeting: Abstracts.* Vol. 37. Salt Lake City, UT, Geological Society of America.

Hickman, C. S. (2005b). Seagrass fauna of the temperate southern coast of Australia: [Part] 1: The cantharidine trochid gastropods. pp. 199–220 in F. E. Wells, Walker, D. I. & Kendrick, G. A. (Eds.) *The Marine Flora and Fauna of Esperance, Western Australia: Proceedings of the Twelfth International Marine Biological Workshop held in Esperance 3–21 February 2003.* Perth, WA, Western Australian Museum.

Hickman, C. S. (2005c). How have bacteria contributed to the evolution of multicellular animals? pp. 3–33 in M. J. McFall-Ngai, Henderson, B. & Ruby, E. G. (Eds.) *The Influence of Cooperative Bacteria on Animal Host Biology. Advances in Molecular and Cellular Biology.* Cambridge, UK, New York, Cambridge University Press.

Hickman, C. S. & Porter, S. S. (2007). Nocturnal swimming, aggregation at light traps, and mass spawning of scissurellid gastropods (Mollusca: Vetigastropoda). *Invertebrate Biology* 126: 10–17.

Hickok, J. F. & Davenport, D. (1957). Further studies in the behavior of commensal polychaetes. *Biological Bulletin* 113: 397–406.

Higgins, W. J., Price, D. A. & Greenberg, M. J. (1978). FMRFamide increases the adenylate cyclase activity and cyclic AMP level of molluscan heart. *European Journal of Pharmacology* 48: 425–430.

Highsmith, R. C. (1985). Floating and algal rafting as potential dispersal mechanisms in brooding invertebrates. *Marine Ecology Progress Series* 25: 169–179.

Hillis, D., M, Rosenfeld, D. S. & Sanchez, M. (1987). Allozymic variability and heterozygote deficiency within and among morphologically polymorphic populations of *Liguus fasciatus* (Mollusca: Pulmonata: Bulimulidae). *American Malacological Bulletin* 5: 153–157.

Hillman, R. E. (1969). Histochemistry of mucosubstances in the mantle of the clam, *Mercenaria mercenaria.* II. Mucosubstances in the second marginal fold. *Transactions of the American Microscopical Society* 88: 420–425.

Himes, J. E., Riffell, J. A., Zimmer, C. A. & Zimmer, R. K. (2011). Sperm chemotaxis as revealed with live and synthetic eggs. *Biological Bulletin* 220: 1–5.

Himmelman, J. H. (1975). Phytoplankton as a stimulus for spawning in three marine invertebrates. *Journal of Experimental Marine Biology and Ecology* 20: 199–214.

Hinckley, A. D. (1963). Diet of the giant toad, *Bufo marinus* (L.), in Fiji. *Herpetologica* 18: 253–259.

Hine, P. M. (1999). The inter-relationships of bivalve haemocytes. *Fish and Shellfish Immunology* 9: 367–385.

Hine, P. M. & Thorne, T. (2000). A survey of some parasites and diseases of several species of bivalve mollusc in northern Western Australia. *Diseases of Aquatic Organisms* 40: 67–78.

Hinegardner, R. (1974). Cellular DNA content of the Mollusca. *Comparative Biochemistry and Physiology Part A* 47: 447–460.

Hinman, V. F., O'Brien, E. K., Richards, G. S. & Degnan, B. M. (2003). Expression of anterior Hox genes during larval development of the gastropod *Haliotis asinina. Evolution & Development* 5: 508–521.

Hinzmann, M. F., Lopes-Lima, M., Gonçalves, J. & Machado, J. (2013). Antiaggregant and toxic properties of different solutions on hemocytes of three freshwater bivalves. *Toxicological and Environmental Chemistry* 95: 790–805.

Hiong, K. C., Loong, A. M., Chew, S. F. & Ip, Y. K. (2005). Increases in urea synthesis and the ornithine-urea cycle capacity in the giant African snail, *Achatina fulica*, during fasting or aestivation, or after the injection with ammonium chloride. *Journal of Experimental Zoology Part A* 303A: 1040–1053.

Hirohashi, N. & Iwata, Y. (2013). The different types of sperm morphology and behavior within a single species. Why do sperm of squid sneaker males form a cluster? *Communicative & Integrative Biology* 6: e26729.

Hirose, E., Iwai, K. & Maruyama, T. (2005). Where are zooxanthellae distributed in the larvae and juveniles of giant clams? *Zoological Science (Tokyo)* 22: 1515.

Hirose, E., Iwai, K. & Maruyama, T. (2006). Establishment of the photosymbiosis in the early ontogeny of three giant clams. *Marine Biology* 148: 551–558.

Hiroshi, N. (1991). A new species of the genus *Stylactaria* (Cnidaria, Hydrozoa) from Hokkaido, Japan (Systematics and Taxonomy). *Zoological Science* 8: 805–812.

Hirota, H., Okino, T., Yoshimura, E. & Fusetani, N. (1998). Five new antifouling sesquiterpenes from two marine sponges of the genus *Axinyssa* and the nudibranch *Phyllidia pustulosa*. *Tetrahedron* 54: 13971–13980.

Hisano, N., Tateda, H. & Kuwabara, M. (1972). Photosensitive neurones in the marine pulmonate mollusc *Onchidium verruculatum*. *Journal of Experimental Biology* 57: 651–660.

Hoagland, E. (1975). Reproductive strategies and evolution in the genus *Crepidula* (Gastropoda: Prosobranchia). PhD, Harvard University.

Hoagland, K. E. (1978). Protandry and the evolution of environmentally mediated sex change: A study of the Mollusca. *Malacologia* 17: 365–391.

Hoagland, K. E. (1984). Use of the terms protandry, protogyny, and hermaphroditism in malacology. *American Malacological Union Bulletin* 3: 85–88.

Hoagland, K. E. (1986). Patterns of encapsulation and brooding in the Calyptraeidae (Prosobranchia: Mesogastropoda). *American Malacological Union Bulletin* 4: 173–183.

Hobday, A. J., Tegner, M. J. & Haaker, P. L. (2000). Over-exploitation of a broadcast spawning marine invertebrate: Decline of the white abalone. *Reviews in Fish Biology and Fisheries* 10: 493–514.

Hochachka, P. W., Hartline, P. H. & Fields, J. H. (1977). Octopine as an end product of anaerobic glycolysis in the chambered nautilus. *Science* 195: 72–74.

Hochachka, P. W., Fields, J. H. A. & Mommsen, T. P. (1983a). Metabolic and enzyme regulation during rest-to-work transition: A mammal versus mollusc comparison. pp. 55–89 *in* P. W. Hochachka (Ed.) *Metabolic Biochemistry and Molecular Biomechanics. The Mollusca*. Vol. 1. New York, Academic Press.

Hochachka, P. W., Mommsen, T. P., Storey, J., Storey, K. B., Johansen, K. & French, C. J. (1983b). The relationship between arginine and proline metabolism in cephalopods. *Marine Biology Letters* 4: 1–21.

Hochachka, P.W. (1994). Oxygen efficient design of cephalopod muscle metabolism. *Marine and Freshwater Behaviour and Physiology* 25: 61–67.

Hochberg, F. G. (1983). The parasites of cephalopods: A review. *Memoirs of the National Museum of Victoria* 44: 109–145.

Hochberg, F. G. (1990). Introduction: Cephalopoda, Annelida, Crustacea, Chaetognatha, Echinodermata, Urochordata. pp. 47–227 *in* O. Kinne (Ed.) *Diseases of Mollusca. Cephalopoda: Diseases caused by Protistans and Metazoans. Diseases of Marine Animals*. Vol. 3. Hamburg, Biologisches Anstalt Helgoland.

Hochner, B., Shomrat, T. & Fiorito, G. (2006). The octopus: A model for a comparative analysis of the evolution of learning and memory mechanisms. *Biological Bulletin* 210: 308–317.

Hochner, B. (2008). Octopuses. *Current Biology* 18: R897–R898.

Hockey, P. A. R. & Branch, G. M. (1984). Oystercatchers and limpets: Impact and implications: A preliminary assessment. *Ardea* 72: 199–206.

Hockey, P. A. R., Bosman, A. L. & Ryan, P. G. (1987). The maintenance of polymorphism and cryptic mimesis in the limpet *Scurria variabilis* by two species of *Cinclodes* (Aves: Furnariinae) in central Chile. *The Veliger* 30: 5–10.

Hodges, M. H. & Allendorf, F. W. (1998). Population genetics and pattern of larval dispersal of the endemic Hawaiian freshwater amphidromous gastropod *Neritina granosa* (Prosobranchia: Neritidae). *Pacific Science* 52: 237.

Hodgson, A. N. (1984). Use of the intrinsic musculature for siphonal autotomy in the Solenacea (Mollusca: Bivalvia). *Transactions of the Royal Society of South Africa* 45: 129–137.

Hodgson, A. N., Hawkins, S. J., Cross, R. H. M. & Dower, K. (1987). Comparison of the structure of the pallial tentacles of seven species of South African patellid limpets. *Journal of Molluscan Studies* 53: 229–240.

Hodgson, A. N., Baxter, J. M., Sturrock, M. G. & Bernard, R. T. F. (1988). Comparative spermatology of 11 species of Polyplacophora (Mollusca) from the suborders Lepidopleurina, Chitonina and Acanthochitonina. *Proceedings of the Royal Society B* 235: 161–178.

Hodgson, A. N., Ridgway, S., Branch, G. M. & Hawkins, S. J. (1996). Spermatozoan morphology of 19 species of prosobranch limpets (Patellogastropoda) with a discussion of patellid relationships. *Philosophical Transactions of the Royal Society B* 351: 339–347.

Hodgson, A. N., Healy, J. M. & Tunnicliffe, V. (1997). Spermatogenesis and sperm structure of the hydrothermal vent prosobranch gastropod *Lepetodrilus fucensis* (Lepetodrilidae, Mollusca). *Invertebrate Reproduction & Development* 31: 87–97.

Hodgson, A. N. & Eckelbarger, K. J. (2000). Ultrastructure of the ovary and oogenesis in six species of patellid limpets (Gastropoda : Patellogastropoda) from South Africa. *Invertebrate Biology* 119: 265–277.

Hodgson, A. N. (2010). Reproductive seasonality of southern African inshore and estuarine invertebrates: A biogeographic review. *African Zoology* 45: 1–17.

Hoegh-Guldberg, O. & Hinde, R. (1986). Studies on a nudibranch that contains zooxanthellae. I. Photosynthesis, respiration and the translocation of newly fixed carbon by zooxanthellae in *Pteraeolidia ianthina*. *Proceedings of the Royal Society B* 228: 493–509.

Hoegh-Guldberg, O., Hinde, R. & Muscatine, L. (1986). Studies on a nudibranch that contains zooxanthellae. II. Contribution of zooxanthellae to animal respiration (CZAR) in *Pteraeolidia ianthina* with high and low densities of zooxanthellae. *Proceedings of the Royal Society B* 228: 511–521.

Hoeh, W. R., Stewart, D. T., Sutherland, B. W. & Zouros, E. (1996). Multiple origins of gender-associated mitochondrial DNA lineages in bivalves (Mollusca: Bivalvia). *Evolution* 50: 2276–2286.

Hoeh, W. R., Stewart, D. T., Saavedra, C., Sutherland, B. W. & Zouros, E. (1997). Phylogenetic evidence for role-reversals of gender-associated mitochondrial DNA in *Mytilus* (Bivalvia: Mytilidae). *Molecular Biology and Evolution* 14: 959–967.

Hoeh, W. R., Stewart, D. T. & Guttman, S. I. (2002). High fidelity of mitochondrial genome transmission under the doubly uniparental mode of inheritance in freshwater mussels (Bivalvia: Unionoidea). *Evolution* 56: 2252–2261.

Hoffmann, H. (1922). Über die Entwicklung der Geschlechtsorgane bei *Limax maximus*. *Zeitschrift für wissenschaftliche Zoologie* 119: 493–538.

Hoffmann, H. (1929–1930). Amphineura und Scaphopoda. Nachträge. *H. G. Bronn's Klassen und Ordnungen des Tier-Reichs wissenschaftlich dargestellt in Wort und Bild* 3, Abteilung 1–3: 1–511.

Hoffmann, H. (1932–1939). Opisthobranchia. Teil 1. pp. i–xi, 1–1247 *in* H. G. Bronn (Ed.) *H. G. Bronn's Klassen und Ordnungen des Tier-Reichs wissenschaftlich dargestellt in Wort und Bild*. Vol. 3, Mollusca. Abteilung 2, Gastropoda. Buch 3. Leipzig, Akademische Verlagsgesellschaft.

Hoffmann, H. (1940). Opisthobranchia, Teil 2. pp. 1–90 *in* H. G. Bronn (Ed.) *H. G. Bronn's Klassen und Ordnungen des Tier-Reichs wissenschaftlich dargestellt in Wort und Bild*. Vol. 3, Band Mollusca, II. Abteilung Gastropoda, Buch 3.

Hoffmann, S. (1949). Studien über das Integument der Solenogastres, nebst Bemerkungen über die Verwandtschaft zwischen den Solenogastres und Placophoren. *Zoologiska Bidrag från Uppsala* 27: 293–427.

Hofkin, B., Mkoji, G. M., Koech, D. & Loker, E. S. (1991). Control of schistosome-transmitting snails in Kenya by the North American crayfish *Procambarus clarkii*. *American Journal of Tropical Medicine and Hygiene* 45: 339–344.

Hofmann, G. E., O'Donnell, M. J. & Todgham, A. E. (2008). Using functional genomics to explore the effects of ocean acidification on calcifying marine organisms. *Marine Ecology Progress Series* 373: 219–226.

Höglund, N. G. & Rahemtulla, F. (1977). The isolation and characterization of acid mucopolysaccharides of *Chiton tuberculatus*. *Comparative Biochemistry and Physiology Part B* 56: 211–214.

Hohagen, J. & Jackson, D. J. (2013). An ancient process in a modern mollusc: Early development of the shell in *Lymnaea stagnalis*. *BMC Developmental Biology* 13: 27.

Hohenlohe, P. A. (2002). Life history of *Littorina scutulata* and *L. plena*, sibling gastropod species with planktotrophic larvae. *Invertebrate Biology* 121: 25–37.

Van Holde, K. E., Miller, K. I. & Decker, H. (2001). Hemocyanins and invertebrate evolution. *Journal of Biological Chemistry* 276: 15563–15566.

Holland, L. Z. (2005). Non-neural ectoderm is really neural: Evolution of developmental patterning mechanisms in the non-neural ectoderm of chordates and the problem of sensory cell homologies. *Journal of Experimental Zoology Part B* 304: 304–323.

Hollande, A. & Pesson, P. (1945). Biologie et cycle évolutif de *Dimoeriopsis destructor* nov. gen. nov. sp., flagelle parasite des pontes de Limnée. *Bulletin biologique de France et de Belgique* 79: 1–16.

Hollande, A. & Chabelard, R. (1953). Essai de lutte biologique par *Dimoeriopsis destructor* Hollande et Pesson (Protozoaire Flagelle) contra les bilharzioses et les distomatoses. *Minerva Urologica* 5: 145.

Holmes, S. J. (1997). Notes on gamete release and fertilisation in *Calliostoma zizyphinum* (L.) (Gastropoda; Trochidae). *Journal of Molluscan Studies* 63: 471–473.

Holmes, S. P., Cherrill, A. & Davies, M. S. (2002). The surface characteristics of pedal mucus: A potential aid to the settlement of marine organisms? *Journal of the Marine Biological Association of the United Kingdom* 82: 131–139.

Holt, A. L., Vahidinia, S., Gagnon, Y. L., Morse, D. E. & Sweeney, A. M. (2014). Photosymbiotic giant clams are transformers of solar flux. *Journal of The Royal Society Interface* 11: 1–14.

Holten-Andersen, N., Harrington, M. J., Birkedal, H., Lee, B. P., Messersmith, P. B., Lee, K. Y. C. & Waite, J. H. (2011). pH-induced metal-ligand cross-links inspired by mussel yield self-healing polymer networks with near-covalent elastic moduli. *Proceedings of the National Academy of Sciences of the United States of America* 108: 2651–2655.

Holterman, M., Van der Wurff, A., Van den Elsen, S., Van Megen, H., Bongers, T., Holovachov, O., Bakker, J. & Helder, J. (2006). Phylum-wide analysis of SSU rDNA reveals deep phylogenetic relationships among nematodes and accelerated evolution toward crown clades. *Molecular Biology and Evolution* 23: 1792–1800.

Holthuis, B. V. (1995). Evolution between marine and freshwater habitats: A case study of the gastropod suborder Neritopsina. PhD, University of Washington.

Holznagel, W. E., Colgan, D. J. & Lydeard, C. (2010). Pulmonate phylogeny based on 28S rRNA gene sequences: A framework for discussing habitat transitions and character transformation. *Molecular Phylogenetics and Evolution* 57: 1017–1025.

Hommay, G. (2002). Agriolimacidae, Arionidae and Milacidae as pests in west European sunflower and maize. pp. 245–254 *in* G. M. Barker (Ed.) *Molluscs as Crop Pests*. Wallingford, UK, CABI Publishing.

Hooper, A. K., Wegener, B. J. & Wong, B. B. M. (2016). When should male squid prudently invest sperm? *Animal Behaviour* 112: 163–167.

Hooper, S. L. & Thuma, J. B. (2005). Invertebrate muscles: Muscle specific genes and proteins. *Physiological Reviews* 85: 1001–1060.

Hooper, S. L., Hobbs, K. H. & Thuma, J. B. (2008). Invertebrate muscles: Thin and thick filament structure; molecular basis of contraction and its regulation, catch and asynchronous muscle. *Progress in Neurobiology* 86: 72–127.

Horiuchi, S. & Lane, C. E. (1965). Digestive enzymes of the crystalline style of *Strombus gigas* Linné. I. Cellulase and some other carbohydrases. *Biological Bulletin* 129: 273–281.

Horiuchi, S. & Lane, C. E. (1966). Carbohydrases of the crystalline style and hepatopancreas of *Strombus gigas* Linné. *Comparative Biochemistry and Physiology* 17: 1189–1197.

Horiuchi, Y., Kimura, R., Kato, N., Fujii, T., Seki, M., Endo, T., Kato, T. & Kawashima, K. (2003). Evolutional study on acetylcholine expression. *Life Sciences* 72: 1745–1756.

Van Horn, D. J., Garcia, J. R., Loker, E. S., Mitchell, K. R., Mkoji, G. M., Adema, C. M. & Takacs-Vesbach, C. D. (2012). Complex intestinal bacterial communities in three species of planorbid snails. *Journal of Molluscan Studies* 78: 74–80.

Horn, P. L. (1986). Energetics of *Chiton pelliserpentis* (Quoy & Gaimard, 1835) (Mollusca: Polyplacophora) and the importance of mucus in its energy budget. *Journal of Experimental Marine Biology and Ecology* 101: 119–141.

Horne, F. B. & Beck, S. (1979). Purine production during fasting in the slug, *Limax flavus* Linné. *Journal of Experimental Zoology Part A* 209: 309–316.

Horne, F. R. & Boonkoom, V. (1970). The distribution of the ornithine cycle enzymes in twelve gastropods. *Comparative Biochemistry and Physiology* 32: 141–153.

Horne, F. R. (1971). Accumulation of urea by a pulmonate snail during aestivation. *Comparative Biochemistry and Physiology Part A* 38: 565–570.

Horne, F. R. (1973a). Urea metabolism in an estivating terrestrial snail *Bulimulus dealbatus*. *American Journal of Physiology* 224: 781–787.

Horne, F. R. (1973b). The utilization of foodstuffs and urea production by a land snail during estivation. *Biological Bulletin* 144: 321–330.

Horne, F. R. (1977a). Regulation of urea biosynthesis in the slug *Limax flavus* Linné. *Comparative Biochemistry and Physiology Part B* 38: 63–69.

Horne, F. R. (1977b). Ureotelism in the slug, *Limax flavus* Linné. *Journal of Experimental Zoology Part A* 199: 227–232.

Horohov, J., Silverman, H. G., Lynn, J. W. & Dietz, T. H. (1992). Ion transport in the freshwater zebra mussel, *Driessena polymorpha*. *Biological Bulletin* 183: 297–303.

Horwood, J. W. & Goss-Custard, J. D. (1977). Predation by the oystercatcher, *Haematopus ostralegus* (L.), in relation to the cockle, *Cerastoderma edule* (L.), fishery in the Burry Inlet, South Wales. *Journal of Applied Ecology* 14: 139–158.

Hoskin, M. G. (1997). Effects of contrasting modes of larval development on the genetic structures of populations of three species of prosobranch gastropods. *Marine Biology* 127: 647–656.

Hoso, M., Asami, T. & Hori, M. (2007). Right-handed snakes: Convergent evolution of asymmetry for functional specialization. *Biology Letters* 3: 169–172.

Hoso, M. & Hori, M. (2008). Divergent shell shape as an antipredator adaptation in tropical land snails. *American Naturalist* 172: 726–732.

Hoso, M., Kameda, Y., Wu, S. P., Asami, T., Kato, M. & Hori, M. (2010). A speciation gene for left-right reversal in snails results in anti-predator adaptation. *Nature Communications* 1: 133.

Hoso, M. (2012). Cost of autotomy drives ontogenetic switching of anti-predator mechanisms under developmental constraints in a land snail. *Proceedings of the Royal Society of London B* 279: 4811–4816.

Hou, D. F., Zhou, G. S. & Zheng, M. (2004). Conch shell structure and its effect on mechanical behaviors. *Biomaterials* 25: 751–756.

Houbrick, J. R. & Fretter, V. (1969). Some aspects of the functional anatomy and biology of *Cymatium* and *Bursa*. *Proceedings of the Malacological Society of London* 38: 415–429.

Houbrick, R. S. (1984). Revision of higher taxa in genus *Cerithidea* (Mesogastropoda: Potamididae) based on comparative morphology and biological data. *American Malacological Bulletin* 2: 1–20.

Houbrick, R. S. (1989). *Campanile* revisited: Implications for cerithioidean phylogeny. *American Malacological Bulletin* 7: 1–6.

Houlihan, D. F., Innes, A. J. & Dey, D. G. (1981). The influence of mantle cavity fluid on the aerial oxygen consumption of some intertidal gastropods. *Journal of Experimental Marine Biology and Ecology* 49: 57–68.

Houlihan, D. F. & Innes, A. J. (1982). Oxygen consumption, crawling speeds, and cost of transport in four Mediterranean intertidal gastropods. *Journal of Comparative Physiology A* 147: 113–121.

Houlihan, D. F., Innes, A. J., Wells, M. J. & Wells, J. (1982). Oxygen consumption and blood gases of *Octopus vulgaris* in hypoxic conditions. *Journal of Comparative Physiology B* 148: 35–40.

Houlihan, D. F., Duthie, G., Smith, P. J., Wells, M. J. & Wells, J. (1986). Ventilation and circulation during exercise in *Octopus vulgaris*. *Journal of Comparative Physiology B* 156: 683–689.

Houlihan, D. F., Agnisola, C., Hamilton, N. M. & Trara Genoino, I. (1987). Oxygen consumption of the isolated heart of *Octopus vulgaris*: Effects of power output and hypoxia. *Journal of Experimental Biology* 131: 137–157.

Houssay, F. (1884). Recherches sur l'opercule et les glandes du pied des Gastéropodes. *Archives de zoologie expérimentale et générale* 2: 171–288.

Hoving, H. J. T., Lipiński, M. R., Videler, J. J. & Bolstad, K. S. R. (2010). Sperm storage and mating in the deep-sea squid *Taningia danae* Joubin, 1931 (Oegopsida: Octopoteuthidae). *Marine Biology* 157: 393–400.

Howard, A. D. & Anson, B. J. (1922). Phases in the parasitism of the Unionidae. *Journal of Parasitology* 9: 68–82.

Howell, T. R. W. (1989). The response of juvenile *Pecten maximus* (L.) to light and water currents. *International Council for the Exploration of the Sea* 7: 1–10.

Howells, H. H. (1942). The structure and function of the alimentary canal of *Aplysia punctata*. *Quarterly Journal of Microscopical Science* 83: 357–397.

Hsiao, S.-T., Chuang, S.-C., Chen, K.-S., Ho, P.-H., Wu, C.-L. & Chen, C. A. (2016). DNA barcoding reveals that the common cupped oyster in Taiwan is the Portuguese oyster *Crassostrea angulata* (Ostreoida; Ostreidae), not *C. gigas*. *Scientific Reports* 6: 34057.

Hu, M. Y., Yan, H. Y., Chung, W. S., Shiao, J. C. & Hwang, P. P. (2009). Acoustically evoked potentials in two cephalopods inferred using the auditory brainstem response (ABR) approach. *Comparative Biochemistry and Physiology Part A* 153: 278–283.

Hu, M. Y., Guh, Y.-J., Stumpp, M., Lee, J.-R., Chen, R.-D., Sung, P.-H., Chen, Y.-C., Hwang, P.-P. & Tseng, Y.-C. (2014). Branchial NH_4^+-dependent acid–base transport mechanisms and energy metabolism of squid (*Sepioteuthis lessoniana*) affected by seawater acidification. *Frontiers in Zoology* 11: 55.

Hu, M. Y., Lee, J.-R., Lin, L.-Y., Shih, T.-H., Stumpp, M., Lee, M.-F., Hwang, P.-P. & Tseng, Y.-C. (2013). Development in a naturally acidified environment: Na^+/H^+-exchanger 3-based proton secretion leads to CO_2 tolerance in cephalopod embryos. *Frontiers in Zoology* 10: 51.

Huan, P., Wang, H. & Liu, B. (2012). Transcriptomic analysis of the clam *Meretrix meretrix* on different larval stages. *Marine Biotechnology* 14: 69–78.

Huang, C. L. & Mir, G. N. (1972). Pharmacological investigation of salivary gland of *Thais haemastoma* (Clench). *Toxicon* 10: 111–117.

Huang, Z.-X., Chen, Z.-S., Ke, C.-H., Zhao, J., You, W.-W., Zhang, J., Dong, W.-T. & Chen, J. (2012). Pyrosequencing of *Haliotis diversicolor* transcriptomes: Insights into early developmental molluscan gene expression. *PLoS ONE* 7: e51279.

Huang, Z. & Satterlie, R. A. (1989). Smooth muscle fiber types and a novel pattern of thick filaments in the wing of the pteropod mollusc *Clione limacina*. *Cell and Tissue Research* 257: 405–414.

Huang, Z., Li, H., Pan, Z., Wei, Q., Chao, Y. J. & Li, X. (2011). Uncovering high-strain rate protection mechanism in nacre. *Nature. Scientific Reports* 1: 148.

Huang, Z. B., Guo, F., Zhao, J., Li, W. D. & Ke, C. H. (2010). Molecular analysis of the intestinal bacterial flora in cage-cultured adult small abalone, *Haliotis diversicolor*. *Aquaculture Research* 41: e760–e769.

Huaquin, L. G., Guerra, R. & Campo, A. T. del (2004). Ovary cell differentiation and morphology of mature oocytes in *Fissurellacrassa* Lamarck 1822 (Mollusca: Archaeogastropoda). *Invertebrate Reproduction & Development* 46: 103–110.

Hubendick, B. (1958). On the molluscan adhesive epithelium. *Arkiv för Zoologi* 11: 31–36.

Hubendick, B. (1978). Systematics and comparative morphology of the Basommatophora. pp. 1–47 *in* V. Fretter & Peake, J. (Eds.) *Pulmonates: Systematics, Evolution and Ecology*. Vol. 2A. London, Academic Press.

Huber, G. (1993). On the cerebral nervous system of marine Heterobranchia (Gastropoda). *Journal of Molluscan Studies* 59: 381–420.

Hubricht, L. (1943). Notes on the sex ratios in *Campeloma*. *The Nautilus* 56: 138–139.

Huelsken, T. (2011). First evidence of drilling predation by *Conuber sordidus* (Swainson, 1821) (Gastropoda: Naticidae) on soldier crabs (Crustacea : Mictyridae). *Molluscan Research* 31: 125–132.

Huete-Perez, J. A. & Quezada, F. (2013). Genomic approaches in marine biodiversity and aquaculture. *Biological Research* 46: 353–361.

Huffard, C. L., Boneka, F. & Full, R. J. (2005). Underwater bipedal locomotion by octopuses in disguise. *Science* 307: 19–27.

Huffard, C. L. (2006). Locomotion by *Abdopus aculeatus* (Cephalopoda: Octopodidae): Walking the line between primary and secondary defenses. *Journal of Experimental Biology* 209: 3697–3707.

Huffard, C. L., Caldwell, R. L. & Boneka, F. (2008). Mating behavior of *Abdopus aculeatus* (d'Orbigny 1834) (Cephalopoda: Octopodidae) in the wild. *Marine Biology* 154: 353–362.

Huffard, C. L., Saarman, N., Hamilton, H. & Simison, W. B. (2010). The evolution of conspicuous facultative mimicry in octopuses: An example of secondary adaptation? *Biological Journal of the Linnean Society* 101: 68–77.

Hughes, G. M. & Morgan, M. (1973). The structure of fish gills in relation to their respiratory function. *Biological Reviews* 48: 419–475.

Hughes, H. P. I. (1970). A light and electron microscope study of some opisthobranch eyes. *Zeitschrift für Zellforschung und mikroskopische Anatomie* 106: 79–98.

Hughes, H. P. I. (1976). Structure and regeneration of the eyes of strombid gastropods. *Cell and Tissue Research* 171: 259–271.

Hughes, H. P. I. (1985a). Feeding in *Gymnodoris inornata* (Bergh) and *Gymnodoris alba* (Bergh) (Opisthobranchia). pp. 627–633 *in* B. S. Morton & Dudgeon, D. (Eds.) *Proceedings of the Second International Workshop on the Malacofauna of Hong Kong and of Southern China, Hong Kong, 6–24 April 1983*. Vol. 2. Hong Kong, Hong Kong University Press.

Hughes, J. M. & Mather, P. B. (1986). Evidence for predation as a factor in determining shell color frequencies in a mangrove snail *Littorina* sp. (Prosobranchia: Littorinidae). *Evolution* 40: 68–77.

Hughes, M. H. & Parmalee, P. W. (1999). Prehistoric and modern freshwater mussel (Mollusca: Bivalvia: Unionoidea) faunas of the Tennessee River: Alabama, Kentucky, and Tennessee. *River Research and Applications* 15: 25–42.

Hughes, R. N. & Lewis, A. H. (1974). On the spatial distribution, feeding and reproduction of the vermetid gastropod *Dendropoma maximum*. *Journal of Zoology* 172: 531–547.

Hughes, R. N. (1978). The biology of *Dendropoma corallinaceum* and *Serpulorbis natalensis*, two South African vermetid gastropods. *Zoological Journal of the Linnean Society* 64: 111–127.

Hughes, R. N. & Hughes, H. P. I. (1981). Morphological and behavioural aspects of feeding in the Cassidae (Tonnacea, Mesogastropoda). *Malacologia* 20: 385–402.

Hughes, R. N. & Dunkin, S. d. B. (1984). Behavioural components of prey selection by dogwhelks, *Nucella lapillus* (L.), feeding on mussels, *Mytilus edulis* L., in the laboratory. *Journal of Experimental Marine Biology and Ecology* 77: 45–68.

Hughes, R. N. (1985b). Predatory behaviour of *Natica unifasciata* feeding intertidally on gastropods. *Journal of Molluscan Studies* 51: 331–335.

Hughes, R. N. & Emerson, W. K. (1987). Anatomical and taxonomic characteristics of *Harpa* and *Morum* (Neogatropoda: Harpidae). *The Veliger* 29: 349–358.

Hulbert, G. C. E. B. & Yonge, C. M. (1937). A possible function of the osphradium in the Gastropoda. *Nature* 139: 840–841.

Hull, S. L. (1998). Assortative mating between two distinct micro-allopatric populations of *Littorina saxatilis* (Olivi) on the northeast coast of England. *Hydrobiologia* 378: 79–88.

Humes, A. G. & Cressey, R. F. (1960). Seasonal population changes and host relationships of *Myocheres major* (Williams), a cyclopoid copepod from pelecypods. *Crustaceana* 1: 307–325.

Humes, A. G. & Voight, J. R. (1997). *Cholidya polypi* (Copepoda: Harpacticoida: Tisbidae), a parasite of deep-sea octopuses in the North Atlantic and northeastern Pacific. *Ophelia* 46: 65–81.

Humphries, J. E. & Yoshino, T. P. (2003). Cellular receptors and signal transduction in molluscan hemocytes: Connections with the innate immune system of vertebrates. *Integrative and Comparative Biology* 43: 305–312.

Hunsicker, M. E., Essington, T. E., Watson, R. & Sumaila, U. R. (2010). The contribution of cephalopods to global marine fisheries: Can we have our squid and eat them too? *Fish and Fisheries* 11: 421–438.

Hunt, B. P. V., Pakhomov, E. A., Hosie, G. W., Siegel, V., Ward, P. & Bernard, K. H. (2008). Pteropods in southern ocean ecosystems. *Progress in Oceanography* 78: 193–221.

Hunt, S. (1976). The gastropod operculum: A comparative study of the composition of gastropod opercular proteins. *Journal of Molluscan Studies* 42: 251–260.

Hunter, W. R. (1953). The conditions of the mantle cavity of two pulmonate snails living in Loch Lomond. *Proceedings of the Royal Society of Edinburgh* 65: 143–165.

Huntley, J. W., Kaufman, D. S., Kowalewski, M., Romanek, C. S. & Neves, R. J. (2012). Sub-centennial resolution amino acid geochronology for the freshwater mussel *Lampsilis* for the last 2000 years. *Quaternary Geochronology* 9: 75–85.

Huntley, J. W. & Scarponi, D. (2012). Evolutionary and ecological implications of trematode parasitism of modern and fossil northern Adriatic bivalves. *Paleobiology* 38: 40–51.

Huntley, J. W., Fürsich, F. T., Alberti, M., Hethke, M. & Liu, C. (2014). A complete Holocene record of trematode–bivalve infection and implications for the response of parasitism to climate change. *Proceedings of the National Academy of Sciences of the United States of America* 111: 18150–18155.

Huntley, J. W. & De Baets, K. (2015). Trace fossil evidence of trematode-bivalve parasite-host interactions in deep time. *Advances in Parasitology* 90: 201–231.

Hurlbert, S. H. (1984). Pseudoreplication and the design of ecological field experiments. *Ecological Monographs* 54: 187–211.

Hurtado, L. A., Mateos, M., Lutz, R. A. & Vrijenhoek, R. C. (2003). Coupling of bacterial endosymbiont and host mitochondrial genomes in the hydrothermal vent clam *Calyptogena magnifica*. *Applied and Environmental Microbiology* 69: 2058–2064.

Huryn, A. D. & Denny, M. W. (1997). A biomechanical hypothesis explaining upstream movements by the freshwater snail *Elimia*. *Functional Ecology* 11: 472–483.

Husmann, G., Gerdts, G. & Wichels, A. (2010). Spirochetes in crystalline styles of marine bivalves: Group-specific PCR detection and 16S rRNA sequence analysis. *Journal of Shellfish Research* 29: 1069–1075.

Hutson, K. S., Styan, C. A., Beveridge, I., Keough, M. J., Zhu, X., EL-Osta, Y. G. A. & Gasser, R. B. (2004). Elucidating the ecology of bucephalid parasites using a mutation scanning approach. *Molecular and Cellular Probes* 18: 139–146.

Huxham, M., Raffaelli, D. & Pike, A. (1993). The influence of *Cryptocotyle lingua* (Digenea: Platyhelminthes) infections on the survival and fecundity of *Littorina littorea* (Gastropoda: Prosobranchia); an ecological approach. *Journal of Experimental Marine Biology and Ecology* 168: 223–238.

Huxham, M., Raffaelli, D. & Pike, A. W. (1995). The effect of larval trematodes on the growth and burrowing behaviour of *Hydrobia ulvae* (Gastropoda: Prosobranchiata) in the Ythan estuary, north-east Scotland. *Journal of Experimental Marine Biology and Ecology* 185: 1–17.

Huxley, J. (1942). *Evolution: The Modern Synthesis*. London, UK, Allen and Unwin.

Huxley, T. H. (1853). On the morphology of the cephalous Mollusca, as illustrated by the anatomy of certain Heteropoda and Pteropoda collected during the voyage of H. M. S. Rattlesnake in 1846–1850. *Philosophical Transactions of the Royal Society* 143: 29–66.

Huys, R., Lopez-González, P. J., Roldan, E. & Luque, A. A. (2002). Brooding in cocculiniform limpets (Gastropoda) and familial distinctiveness of the Nucellicolidae (Copepoda): Misconceptions reviewed from a chitonophilid perspective. *Biological Journal of the Linnean Society* 75: 187–217.

Hvorecny, L. M., Grudowski, J. L., Blakeslee, C. J., Simmons, T. L., Roy, P. R., Brooks, J. A., Hanner, R. M., Beigel, M. E., Karson, M. A., Nichols, R. H., Holm, J. B. & Boal, J. G. (2007). Octopuses (*Octopus bimaculoides*) and cuttlefishes (*Sepia pharaonis, S. officinalis*) can conditionally discriminate. *Animal Cognition* 10: 449–459.

Hwang, D. F., Arakawa, O., Saito, T., Noguchi, T., Simidu, U., Tsukamoto, K., Shida, Y. & Hashimoto, K. (1989). Tetrodotoxin-producing bacteria from the blue-ringed octopus *Octopus maculosus*. *Marine Biology* 100: 327–332.

Hwang, D. F., Tai, K. P., Chueh, C. H., Lin, L. C. & Jeng, S. S. (1991). Tetrodotoxin and derivatives in several species of the gastropod Naticidae. *Toxicon* 29: 1019–1024.

Hwang, D. F., Lin, L. C. & Deng, S. S. (1992). Occurence of a new toxin and tetrodotoxin in two species of the gastropod mollusk Nassariidae. *Toxicon* 30: 41–46.

Hwang, P. A., Tsai, Y. H., Lu, Y. H. & Hwang, D. F. (2003). Paralytic toxins in three new gastropod (Olividae) species implicated in food poisoning in southern Taiwan. *Toxicon* 41: 529–533.

Hwang, P. A., Noguchi, T. & Hwang, D. F. (2004). Neurotoxin tetrodotoxin as attractant for toxic snails. *Fisheries Science* 70: 1106–1112.

Hyatt, A. (1894). Phylogeny of an acquired characteristic. *American Philosophical Society Proceedings* 32: 350–647.

Hylander, B. L. & Summers, R. G. (1977). An ultrastructural analysis of the gametes and early fertilization in 2 bivalve mollusks *Chama macerophylla* and *Spisula solidissima* with special reference to gamete binding. *Cell and Tissue Research* 184: 469–490.

Hylleberg, J. & Gallucci, V. F. (1975). Selectivity in feeding by the deposit-feeding bivalve *Macoma nasuta*. *Marine Biology* 32: 167–178.

Hyman, L. H. (1967). *Mollusca I*. Vol. 6. New York, McGraw-Hill.

Ibáñez, C. M., Pardo-Gandarillas, M. C. & George-Nascimento, M. (2005). Uso del microhábitat por el protozoo parásito *Aggregata patagonica* Sardella, Ré & Timi, 2000 (Apicomplexa: Aggregatidae) en su hospedador definitivo, el pulpo *Enteroctopus megalocyathus* (Gould, 1852) (Cephalopoda: Octopodidae) en el sur de Chile. *Revista Chilena de Historia Natural* 78: 441–450.

Ibrahim, M. M. (2007). Population dynamics of *Chaetogaster limnaei* (Oligochaeta: Naididae) in the field populations of freshwater snails and its implications as a potential regulator of trematode larvae community. *Parasitology Research* 101: 25–33.

Ieyama, H. & Inaba, A. (1974). Chromosome numbers of 10 species in 4 families of Pteriomorphia Bivalvia. *Venus* 33: 129–137.

Ieyama, H. (1993). Karyotypes of two species in the Dentaliidae (Mollusca: Scaphopoda). *Venus* 52: 245–248.

Iglesias, J. & Castillejo, J. (1999). Field observations on feeding of the land snail *Helix aspersa* Müller. *Journal of Molluscan Studies* 65: 411–423.

von Ihering, H. (1876). Versuch eines natürlichen Systemes der Mollusken. *Deutsche Malakozoologische Gesellschaft Jahrbücher* 3: 97–147.

Iijima, M., Akiba, N., Sarashina, I., Kuratani, S. & Endo, K. (2006). Evolution of Hox genes in molluscs: A comparison among seven morphologically diverse classes. *Journal of Molluscan Studies* 72: 259–266.

Iijima, M., Takeuchi, T., Sarashina, I. & Endo, K. (2008). Expression patterns of engrailed and dpp in the gastropod *Lymnaea stagnalis*. *Development Genes and Evolution* 218: 237–251.

Iijima, R., Kisugi, J. & Yamazaki, M. (2003). A novel antimicrobial peptide from the sea hare *Dolabella auricularia*. *Developmental & Comparative Immunology* 27: 305–311.

Ikeda, Y. (2009). A perspective on the study of cognition and sociality of cephalopod mollusks, a group of intelligent marine invertebrates. *Japanese Psychological Research* 51: 146–153.

Ilano, A. S., Fujinaga, K. & Nakao, S. (2004). Mating, development and effects of female size on offspring number and size in the neogastropod *Buccinum isaotakii* (Kira, 1959). *Journal of Molluscan Studies* 70: 277–282.

Illert, C. (1987). Formulation and solution of the classical seashell problem. Part 1. Seashell geometry. *Il Nuovo Cimento D* 9: 791–814.

Illert, C. (1989). Formulation and solution of the classical seashell problem. Part II. Tubular three-dimensional seashell surfaces. *Il Nuovo Cimento D* 11: 761–780.

Illingworth, J. F. (1902). The anatomy of *Lucapina crenulata* Gray. *Zoologische Jahrbücher. Abteilung für Anatomie und Ontogenie der Tiere* 16: 449–490.

Innes, A. J. & Houlihan, D. F. (1981). A review of the effects of temperature on the oxygen consumption of intertidal gastropods. *Journal of Thermal Biology* 6: 249–256.

Innes, A. J., Marsden, I. D. & Wong, P. P. S. (1984). Bimodal respiration of intertidal pulmonates. *Comparative Biochemistry and Physiology Part A* 77: 441–446.

Innes, A. J. & Houlihan, D. F. (1985). Aerobic capacity and cost of locomotion of a cool temperate gastropod: A comparison with some Mediterranean species. *Comparative Biochemistry and Physiology Part A* 80: 487–493.

Inoue, M. & Oba, T. (1980). Circuli on the shell as an age determination and growth of the Japanese abalone *Haliotis gigantea* [in Japanese]. *Bulletin of the Kanagawa Prefectual Fisheries Experiment Station* 1: 107–113.

Inoue, T., Watanabe, S. & Kirino, Y. (2001). Serotonin and NO complementarily regulate generation of oscillatory activity in the olfactory CNS of a terrestrial mollusk. *Journal of Neurophysiology* 85: 2634–2638.

Ip, H. S. & Desser, S. S. (1984). Apicornavirus-like pathogen of *Cotylogaster occidentalis* (Trematoda: Aspidogastrea), an intestinal parasite of freshwater mollusks. *Journal of Invertebrate Pathology* 43 197–206.

Ireland, C. & Faulkner, D. J. (1978). The defensive secretion of the opisthobranch mollusc *Onchidella binneyi*. *Bioorganic Chemistry* 7: 125–131.

Ireland, C. M., Biskupiak, J. E., Hite, G. J., Rapposch, M., Scheuer, P. J. & Ruble, J. R. (1984). Ilikonapyrone esters, likely defense allomones of the mollusc *Onchidium verruculatum*. *Journal of Organic Chemistry* 49: 559–561.

Isaacs, J. D. & Schwartzlose, R. A. (1975). Active animals of the deep-sea floor. *Scientific American* 233: 84–91.

Iseto, T. (2005). A review of non-commensal loxosomatids: Collection, culture, and taxonomy, with new implications to the benefit of commensalism (Entoprocta: Loxosomatidae). pp. 133–140 *in* G. I. Moyano, Cancino, J. M. & Wyse

Jackson, P. N. (Eds.) *Bryozoan Studies 2004: Proceedings of the Thirteenth International Bryozoology Association Conference, Concepción Chile, 11–16 January 2004.* London, UK, Taylor & Francis Group.

Isgrove, A., Ed. (1909). *Eledone.* Liverpool Marine Biology Committee: Memoirs on typical British marine plants and animals. London, Williams & Norgate.

Ishibashi, M., Yamaguchi, Y. & Hirano, Y. J. (2006). Bioactive natural products from nudibranchs. pp. 513–535 *in* M. Fingerman & Nagabhushanam, R. (Eds.) *Biomaterials from Aquatic and Terrestrial Organisms.* Enfield, NH, Science Publishers.

Ishikura, M., Hagiwara, K., Takishita, K., Haga, M., Iwai, K. & Maruyama, T. (2004). Isolation of new *Symbiodinium* strains from tridacnid giant clam *Tridacna crocea* and sea slug *Pteraeolidia ianthina* using culture medium containing giant clam tissue homogenate. *Marine Biotechnology* 6: 378–385.

Ismail, N. S. & Elkarmi, A. Z. (1999). Age, growth, and shell morphometrics of the limpet *Cellana radiata* (Born, 1778) from the Gulf of Aqaba, Red Sea. *Japanese Journal of Malacology* 58: 61–69.

Isobe, M., Uyakul, D., Sigurdsson, J. B., Goto, T. & Lam, T. J. (1991). Fluorescent substance in the luminous land snail, *Dyakia striata. Agricultural and Biological Chemistry* 55: 1947–1951.

Ito, E., Kojima, S., Lukowiak, K. & Sakakibara, M. (2013). From likes to dislikes: Conditioned taste aversion in the great pond snail (*Lymnaea stagnalis*). *Canadian Journal of Zoology* 91: 405–412.

Ito, N., Kamitaka, R., Takahashi, K. G. & Osada, M. (2010). Identification and characterization of multiple b-glucan binding proteins in the Pacific oyster, *Crassostrea gigas. Developmental and Comparative Immunology* 34: 445–454.

Itoh, N., Momoyama, K. & Ogawa, K. (2005). First report of three protozoan parasites (a haplosporidian, *Marteilia* sp. and *Marteilioides* sp.) from the Manila clam, *Venerupis* (= *Ruditapes*) *philippinarum* in Japan. *Journal of Invertebrate Pathology* 88: 201–206.

Ituarte, C. F. (2009). Unusual modes of oogenesis and brooding in bivalves: The case of *Gaimardia trapesina* (Mollusca: Gaimardiidae). *Invertebrate Biology* 128: 243–251.

IUCN. (2018). The IUCN Red List of Threatened Species. Version 2018–2. Retrieved 15 December, 2018, from https://www.iucnredlist.org/.

Ivanov, D. L. & Tsetlin, A. B. (1981). Origin and evolution of the cuticular pharyngeal armature in trochophore animals having the ventral pharynx [in Russian, English summary]. *Zoologicheskii Zhurnal* 60: 1445–1454.

Ivanov, D. L. (1990a). The origins and early stages of evolutionary transformation of the radula [in Russian]. *Sbornik Trudov Zoologicheskogo Muzeya MGU* 28: 5–37.

Ivanov, D. L. (1990b). The radula in the class Aplacophora [in Russian]. *Sbornik Trudov Zoologicheskogo Muzeya MGU* 28: 159–198.

Ivanov, D. L. (1996). Origin of Aculifera and problems of monophyly of higher taxa in molluscs. pp. 59–66 *in* J. D. Taylor (Ed.) *Origin and Evolutionary Radiation of the Mollusca.* Oxford, UK, Oxford University Press.

Ivanov, D. L. & Sysoev, A. (2009). *Molluscs in the World Cookery.* Moscow, KMK Publications.

Ivanova-Kazas, O. M. (1985a). Origin and phylogentic sigificance of the trochophoran larvae. 2: Evolutionary significance of the larvae of coelomate worms and molluscs [In Russian, English summary]. *Zoologicheskii Zhurnal* 64: 650–660.

Ivanova-Kazas, O. M. (1985b). Origin and phylogenetic significance of the trochophoran larvae. 1: The larvae of coelomate worms and molluscs [in Russian, English summary]. *Zoologicheskii Zhurnal* 64: 485–497.

Ivanovici, A. M. (1960). The adenylate energy charge in the estuarine mollusc, *Pyrazus ebeninus*: Laboratory studies of responses to salinity and temperature. *Comparative Biochemistry and Physiology Part A* 66: 43–55.

Ivany, L. C. (2012). Reconstructing paleoseasonality from accretionary skeletal carbonates - challenges and opportunities. Pp. 133-165 *in* L. C. Ivany & Huber, B. T. *Reconstructing Earth's Deep-Time Climate - the State of the Art in 2012. [Paleontological Society Short Course, November 3, 2012]. The Paleontological Society Papers.* Boulder CO, The Paleontological Society.

Iwakoshi, E., Takuwa-Kuroda, K., Fujisawa, Y., Hisada, M., Ukena, K., Tsutsui, K. & Minakata, H. (2002). Isolation and characterization of a GnRH-like peptide from *Octopus vulgaris. Biochemical and Biophysical Research Communications* 291: 1187–1193.

Iwamoto, M., Ueyama, D. & Kobayashi, R. (2014). The advantage of mucus for adhesive locomotion in gastropods. *Journal of Theoretical Biology* 353: 133–141.

Iwanowicz, D. D., Sanders, L. R., Schill, W. B., Xayavong, M. V., da Silva, A. J., Qvarnstrom, Y. & Smith, T. (2015). Spread of the rat lungworm (*Angiostrongylus cantonensis*) in giant African land snails (*Lissachatina fulica*) in Florida, USA. *Journal of Wildlife Diseases* 51: 749–753.

Iwasa, Y. (1991). Sex change evolution and cost of reproduction. *Behavioral Ecology* 2: 56–68.

Iwata, K. (1975). Ultrastructure of the conchilin matrices in molluscan nacreous layer. *Journal of the Faculty of Science, Hokkaido University. Series IV Geology and Mineralogy* 17: 173–229.

Iwata, Y., Shaw, P., Fujiwara, E., Shiba, K., Kakiuchi, Y. & Hirohashi, N. (2011). Why small males have big sperm: Dimorphic squid sperm linked to alternative mating behaviours. *BMC Evolutionary Biology* 11: 236.

Iyengar, E. V. & Harvell, C. D. (2002). Specificity of cues inducing defensive spines in the bryozoan *Membranipora membranacea. Marine Ecology Progress Series* 225: 205–218.

Iyengar, E. V. (2004). Host-specific performance and host use in the kleptoparasitic marine snail *Trichotropis cancellata. Oecologia* 138: 628–639.

Iyengar, E. V. (2005). Seasonal feeding-mode changes in the marine facultative kleptoparasite *Trichotropis cancellata* (Gastropoda, Capulidae): Trade-offs between trophic strategy and reproduction. *Canadian Journal of Zoology* 83: 1097–1111.

Iyengar, E. V. (2007). Suspension feeding and kleptoparasitism within the genus *Trichotropis* (Gastropoda: Capulidae). *Journal of Molluscan Studies* 74: 55–62.

Iyengar, E. V. (2008). Kleptoparasitic interactions throughout the animal kingdom and a re-evaluation, based on participant mobility, of the conditions promoting the evolution of kleptoparasitism. *Biological Journal of the Linnean Society* 93: 745–762.

Jablonski, D. (1982). Evolutionary rates and modes in Late Cretaceous gastropods: Role of larval ecology. pp. 257–262 *in* B. L. Mamet & Copeland, M. J. (Eds.) *Proceedings of the Third North American Paleontological Convention August 5–7 1982, Montreal, Quebec, Canada.* Vol. 1. Montreal, University of Montreal.

Jablonski, D. & Lutz, R. A. (1983). Larval ecology of marine benthic invertebrates: Paleobiological implications. *Biological Reviews* 58: 21–89.

Jablonski, D. (1986). Larval ecology and macroevolution in marine invertebrates. *Bulletin of Marine Science* 39: 565–587.

Jablonski, D. J. & Lutz, R. A. (1980). Molluscan larval shell morphology: Ecological and paleontological applications. pp. 323–377 *in* D. C. Rhoads & Lutz, R. A. (Eds.) *Skeletal Growth of Aquatic Organisms.* New York, Plenum Publishing Corporation.

Jacklet, J. W. (1972). Circadian locomotor activity in *Aplysia*. *Journal of Comparative Physiology A* 79: 325–341.

Jacklet, J. W. & Colquhoun, W. (1983). Ultrastructure of photoreceptors and circadian pacemaker neurons in the eye of a gastropod, *Bulla*. *Journal of Neurocytology* 12: 673–696.

Jacklet, J. W. & Barnes, S. (1993). Photoresponsive pacemaker neurons from the dissociated retina of *Aplysia*. *NeuroReport* 5: 209–212.

Jackson, A. P., Vincent, J. F. V. & Turner, R. M. (1988). The mechanical design of nacre. *Proceedings of the Royal Society B* 234: 415–440.

Jackson, A. R., MacRae, T. H. & Croll, R. P. (1995). Unusual distribution of tubulin isoforms in the snail *Lymnaea stagnalis*. *Cell and Tissue Research* 281: 507–515.

Jackson, D. J., Wörheide, G. & Degnan, B. M. (2007). Dynamic expression of ancient and novel molluscan shell genes during ecological transitions. *BMC Evolutionary Biology* 7: 160.

Jackson, D. J., McDougall, C., Woodcroft, B., Moase, P., Rose, R. A., Kube, M., Reinhardt, R., Rokhsar, D. S., Montagnani, C., Joubert, C., Piquemal, D. & Degnan, B. M. (2010). Parallel evolution of nacre building gene sets in molluscs. *Molecular Biology and Evolution* 27: 591–608.

Jackson, D. J. & Degnan, B. M. (2016). The importance of evo-devo to an integrated understanding of molluscan biomineralisation. *Journal of Structural Biology* 196: 67–74.

Jackson, G. D. (1994). Application and future potential of statolith increment analysis in squids and sepioids. *Canadian Journal of Fisheries and Aquatic Sciences* 51: 2612–2625.

Jackson, J. B. C. (1974). Biogeographic consequences of eurytopy and stenotopy among marine bivalves and their evolutionary significance. *The American Naturalist* 108: 541–560.

Jackson, R. R. & Barrion, A. (2004). Heteropteran predation on terrestrial gastropods. pp. 483–496 in G. M. Barker (Ed.) *Natural Enemies of Terrestrial Molluscs*. Oxford, UK, Cambridge, MA, CABI Publishing.

Jackson, R. T. (1890). Phylogeny of the Pelecypoda, the Aviculidae and their allies. *Memoirs of the Boston Society of Natural History* 4: 277–400.

Jacob, J. (1957). Cytological studies on Melaniidae (Mollusca) with special reference to parthenogenesis and polyploidy. I. Oogenesis of the partheogenetic species of *Melanoides* (Prosobranchia-Gastropoda). *Transactions of the Royal Society of Edinburgh* 63: 341–352.

Jacob, J. (1958). Cytological studies on Melaniidae (Mollusca) with special reference to parthenogenesis and polyploidy. II. A study of meiosis in the rare males of the polyploid race of *Melanoides tuberculatus* and *Melanoides lineatus*. *Transactions of the Royal Society of Edinburgh* 63: 433–444.

Jacobs, D. K., Wray, C. G., Wedeen, C. J., Kostriken, R., DeSalle, R., Staton, J. L., Gates, R. D. & Lindberg, D. R. (2000). Molluscan engrailed expression, serial organization, and shell evolution. *Evolution & Development* 2: 340–347.

Jacobs, D. K., Hughes, N. C., Fitz-Gibbon, S. T. & Winchell, C. J. (2005). Terminal addition, the Cambrian radiation and the Phanerozoic evolution of bilaterian form. *Evolution & Development* 7: 498–514.

Jacobsen, H. P. & Stabell, O. B. (1999). Predator-induced alarm responses in the common periwinkle, *Littorina littorea*: Dependence on season, light conditions, and chemical labelling of predators. *Marine Biology* 134: 551–557.

Jacobsen, H. P. & Stabell, O. B. (2004). Antipredator behaviour mediated by chemical cues: The role of conspecific alarm signalling and predator labelling in the avoidance response of a marine gastropod. *Oikos* 104: 43–50.

Jagla, K., Bellard, M. & Frasch, M. (2001). A cluster of *Drosophila* homeobox genes involved in mesoderm differentiation programs. *BioEssays* 23: 125–133.

James, J. F. (1889). On variation, with special reference to certain Palaeozoic genera. *American Naturalist* 276: 1071–1087.

James, M. A. (1997). Brachiopoda: Internal anatomy, embryology, and development. pp. 297–407 in F. W. Harrison & Woollacott, R. M. (Eds.) *Microscopic Anatomy of Invertebrates*. Vol. 13. New York, Wiley-Liss.

James, R., Nguyen, T., Arthur, W., Levine, K. & Williams, D. C. (1997). Hydrolase (β-glucanase, α-glucanase, and protease) activity in *Ariolimax columbianus* (banana slug) and *Arion ater* (garden slug). *Comparative Biochemistry and Physiology Part B* 118: 275–283.

Jamieson, B. G. M. (1987). A biological classification of sperm types, with special reference to annelids and molluscs, and an example of spermiocladistics. pp. 311–332 in H. Mohri (Ed.) *New Horizons in Sperm Cell Research*. Tokyo & New York, Japan Scientific Society Press and Gordon & Breach.

Jamieson, G. S. (1993). Marine invertebrate conservation: Evaluation of fisheries' over-exploitation concerns. *American Zoologist* 33: 551–567.

Jander, R., Daumer, K. & Waterman, T. H. (1963). Polarized light orientation by two Hawaiian decapod cephalopods. *Zeitschrift für vergleichende Physiologie* 46: 383–394.

Janer, G., Lyssimachou, A., Bachmann, J., Oehlmann, J., Schulte-Oehlmann, U. & Porte, C. (2006). Sexual dimorphism in esterified steroid levels in the gastropod *Marisa cornuarietis*: The effect of xenoandrogenic compounds. *Steroids* 71: 435–444.

Janson, K. & Ward, R. D. (1984). Microgeographic variation in allozyme and shell characters in *Littorina saxatilis* Olivi (Prosobranchia: Littorinidae). *Biological Journal of the Linnean Society* 22: 289–307.

Janson, K. (1987). Allozyme and shell variation in two marine snails (*Littorina*, Prosobranchia) with different dispersal abilities. *Biological Journal of the Linnean Society* 30: 245–256.

Janssen, A. W. (1985). Evidence for the occurrence of a 'skinny' or 'minute stage' in the ontogenetical development of Miocene *Vaginella* (Gastropoda, Euthecosomata) from the North Sea and Aquitaine Basins. *Mededelingen van de Werkgroep voor Tertiaire en Kwartaire Geologie* 21: 193–204.

Janssen, W. A. (1974). Oysters: Retention and excretion of three types of human waterborne disease bacteria. *Health and Laboratory Science* 11: 20–24.

Janssens, F. A. & Elrington, G. A. (1904). L'element nuclénien pendant les divisions de maturation dans l'oeuf de l'*Aplysia punctata*. *Cellule* 21: 315–326.

Janz, H. (1999). Hilgendorf's planorbid tree: The first introduction of Darwin's theory of transmutation into paleontology. *Palaeontological Research* 3: 287–293.

Japanese Association of Benthology, Ed. (2012). *Threatened Animals of Japanese Tidal Flats: Red Data Book of Seashore Benthos* [in Japanese]. Hadano, Tokai University Press.

Jaremovic, R. & Rollo, C. D. (1979). Tree climbing by the snail *Cepaea nemoralis* (L.): A possible method for regulating temperature and hydration. *Canadian Journal of Zoology* 57: 1010–1014.

Jarne, P. & Charlesworth, D. (1993). The evolution of the selfing rate in functionally hermaphrodite plants and animals. *Annual Review of Ecology and Systematics* 24: 441–466.

Järnegren, J., Tobias, C. R., Macko, S. A. & Young, C. M. (2005). Egg predation fuels unique species association at deep-sea hydrocarbon seeps. *Biological Bulletin* 209: 87–93.

Jarocki, J. (1935). Studies on ciliates from freshwater molluscs: I: General remarks on protozoan parasites of Pulmonata; transfer experiments with species of *Heterocineta* and *Chaetogaster limnaei,* their additional host; some new hypocomid ciliates. *Bulletin international de l'Académie des sciences de Cracovie* 2: 201–230.

Jarrige, P. & Henry, R. (1952). Étude de l'activité glucuronidasique de suc digestif de *Helix pomatia* (L.) et de sa phenolsulfatase. *Bulletin de la Société de chimie biologique* 34: 872–885.

Jass, J. & Glenn, J. (2004). Sexual dimorphism in *Lampsilis siliquoidea* (Barnes, 1823) (Bivalvia: Unionidae). *American Malacological Bulletin* 18: 45–47.

Jeffery, B., Barlow, T., Moizer, K., Paul, S. & Boyle, C. (2004). Amnesic shellfish poison. *Food and Chemical Toxicology* 42: 545–557.

Jeffreys, J. G. (1882). On the Mollusca procured during the 'Lightning' and 'Porcupine' Expeditions, 1868–70. *Proceedings of the Zoological Society of London* 50: 656–688.

Jékely, G. (2011). Origin and early evolution of neural circuits for the control of ciliary locomotion. *Proceedings of the Royal Society B* 278: 914–922.

Jenkins, C., Hick, P., Gabor, M., Spiers, Z., Fell, S. A., Gu, X., Read, A., Go, J., Dove, M. & Connor, W. O. (2013). Identification and characterisation of an ostreid herpesvirus-1 microvariant (OsHV-1 μ-var) in *Crassostrea gigas* (Pacific oysters) in Australia. *Diseases of Aquatic Organisms* 105: 109–126.

Jenkins, M. (2003). Prospects for biodiversity. *Science* 302: 1175–1177.

Jenner, R. A. (2006). Challenging received wisdoms: Some contributions of the new microscopy to the new animal phylogeny. *Integrative and Comparative Biology* 46: 93–103.

Jennings, K. H. (1984). The organization, fine structure and function of the excretory system of the estuarine bivalve, *Scrobicularia plana* (da Costa) and the freshwater bivalve *Anodonta cygnea* (Linn) and other selected species. PhD, University of London UK.

Jensen, H. (1974). Ultrastructural studies of the hearts in *Arenicola marina* L. (Annelida: Polychaeta). *Cell and Tissue Research* 156: 127–144.

Jensen, K. R. (1984). Defensive behavior and toxicity of ascoglossan opisthobranch *Mourgona germineae* Marcus. *Journal of Chemical Ecology* 10: 475–486.

Jensen, K. R. (1997). Evolution of the Sacoglossa (Mollusca, Opisthobranchia) and the ecological associations with their food plants. *Evolutionary Ecology* 11: 301–335.

Jensen, K. R. (2011). Comparative morphology of the mantle cavity organs of shelled Sacoglossa, with a discussion of relationships with other Heterobranchia. *Thalassas* 27: 169–192.

Jensen, S., Duperron, S., Birkeland, N.-K. & Hovland, M. (2010). Intracellular Oceanospirillales bacteria inhabit gills of *Acesta* bivalves. *FEMS Microbiology Ecology* 74: 523–533.

Jernakoff, P., Phillips, B. F. & Fitzpatrick, J. J. (1993). The diet of post-puerulus western rock lobster, *Panulirus cygnus* George, at Seven Mile Beach, Western Australia. *Marine and Freshwater Research* 44: 649–655.

Jespersen, Å., Kosuge, T. & Lützen, J. (2001). Sperm dimorphism and spermatozeugmata in the commensal bivalve *Pseudopythina macrophthalmensis* (Galeommatoidea, Kelliidae). *Zoomorphology* 120: 177–189.

Jespersen, Å., Lützen, J. & Morton, B. (2002). Ultrastructure of dimorphic sperm and seminal receptacle in the hermaphrodites *Barrimysia siphonosomae* and *Pseudopythina ochetostomae* (Bivalvia, Galeommatoidea). *Zoomorphology* 121: 159–171.

Jespersen, Å. & Lützen, J. (2006). Reproduction and sperm structure in Galeommatidae (Bivalvia, Galeommatoidea). *Zoomorphology* 125: 157–173.

Jespersen, Å. & Lützen, J. (2009). Structure of sperm, spermatozeugmata and lateral organs in the bivalve *Arthritica* (Galeommatoidea: Leptonidae). *Acta Zoologica* 90: 51–67.

Jewett, S. C. & Feder, H. M. (1982). Food and feeding habits of the king crab *Paralithodes camtschatica* near Kodiak Island, Alaska. *Marine Biology* 66: 243–250.

Jiang, W., Li, J., Gao, Y., Mao, Y., Jiang, Z., Du, M., Zhang, Y. & Fang, J. (2016). Effects of temperature change on physiological and biochemical responses of Yesso scallop, *Patinopecten yessoensis*. *Aquaculture* 451: 463–472.

Jing, J. & Gillette, R. (1999). Central pattern generator for escape swimming in the notaspid sea slug *Pleurobranchaea californica*. *Journal of Neurophysiology* 81: 654–667.

Jobling, M. (1994). *Fish Bioenergetics*. Vol. 13. London, New York, Chapman & Hall.

Joester, D. & Brooker, L. R. (2016). The chiton radula: A model system for versatile use of iron oxides. pp. 177–206 *in* D. Faivre (Ed.) *Iron Oxides: From Nature to Applications*. Weinhelm, Germany, Wiley-VCH Verlag GmbH & Co.

Johannessen, L. E. & Solhøy, T. (2001). Effects of experimentally increased calcium levels in the litter on terrestrial snail populations. *Pedobiologia* 45: 234–242.

Johannesson, K. (1988). The paradox of Rockall: Why is a brooding gastropod (*Littorina saxatilis*) more widespread than one having a planktonic larval dispersal stage (*L. littorea*)? *Marine Biology* 99: 507–513.

Johannesson, K. (1992). Genetic variability and large scale differentiation in two species of littorinid gastropods with planktotrophic development, *Littorina littorea* (L.) and *Melarchaphe neritoides* (L.) (Prosobranchia: Littorinacea), with notes on a mass occurrence of *Melarchaphe neritoides* in Sweden. *Biological Journal of the Linnean Society* 47: 285–299.

Johannesson, K. & Ekendahl, A. (2002). Selective predation favouring cryptic individuals of marine snails (*Littorina*). *Biological Journal of the Linnean Society* 76: 137–144.

Johansen, K. & Martin, A. W. (1962). Circulation in the cephalopod, *Octopus dofleini*. *Comparative Biochemistry and Physiology* 5: 161–176.

Johnsen, S., Balser, E. J., Fisher, E. C. & Widder, E. A. (1999). Bioluminescence in the deep-sea cirrate octopod *Stauroteuthis syrtensis* Verrill (Mollusca: Cephalopoda). *Biological Bulletin* 197: 26–39.

Johnsen, S., Widder, E. A. & Mobley, C. D. (2004). Propagation and perception of bioluminescence: Factors affecting counterillumination as a cryptic strategy. *Biological Bulletin* 207: 1–16.

Johnsen, S. (2005). The red and the black: Bioluminescence and the color of animals in the deep sea. *Integrative and Comparative Biology* 45: 234–246.

Johnson, J. G., Paul, M. R., Kniffin, C. D., Anderson, P. E., Burnett, L. E. & Burnett, K. G. (2015). High CO_2 alters the hypoxia response of the Pacific whiteleg shrimp *(Litopenaeus vannamei)* transcriptome including known and novel hemocyanin isoforms. *Physiological Genomics* 47: 548–558.

Johnson, K. S., Childress, J. J. & Beehler, C. L. (1988). Short-term temperature variability in the Rose Garden hydrothermal vent field: An unstable deep-sea environment. *Deep Sea Research Part A: Oceanographic Research Papers* 35: 1711–1721.

Johnson, L. J. (1999). Size assortative mating in the marine snail *Littorina neglecta*. *Journal of the Marine Biological Association of the United Kingdom* 79: 1131–1132.

Johnson, M. E. (1988). Why are ancient rocky shores so uncommon? *The Journal of Geology* 96: 469–480.

Johnson, M. P., Burrows, M. T., Hartnoll, R. G. & Hawkins, S. J. (1997). Spatial structure on moderately exposed rocky shores: Patch scales and the interactions between limpets and algae. *Marine Ecology Progress Series* 160: 209–215.

Johnson, M. S. (1976). Allozymes and area effects in *Cepaea nemoralis* on western Berkshire Downs. *Heredity* 36: 105–121.

Johnson, M. S., Clarke, B. & Murray, J. (1977). Genetic-variation and reproductive isolation in *Partula*. *Evolution* 31: 116–126.

Johnson, M. S. & Black, R. (1982). Chaotic genetic patchiness in an intertidal limpet, *Siphonaria* sp. *Marine Biology* 70: 157–164.

Johnson, M. S. & Black, R. (1984a). The Wahlund effect and the geographical scale of variation in the intertidal limpet *Siphonaria* sp. *Marine Biology* 79: 295–302.

Johnson, M. S. & Black, R. (1984b). Pattern beneath the chaos: The effect of recruitment on genetic patchiness in an intertidal limpet. *Evolution* 38: 1371–1383.

Johnson, P. M. & Willows, A. O. D. (1999). Defense in sea hares (Gastropoda, Opisthobranchia, Anaspidea): Multiple layers of protection from egg to adult. *Marine and Freshwater Behaviour and Physiology* 32: 147–180.

Johnson, P. M., Kicklighter, C. E., Schmidt, M., Kamio, M., Yang, H., Elkin, D., Michel, W. C., Tai, P. C. & Derby, C. D. (2006). Packaging of chemicals in the defensive secretory glands of the sea hare *Aplysia californica*. *Journal of Experimental Biology* 209: 78–88.

Johnson, R. F. (2010). Breaking family ties: Taxon sampling and molecular phylogeny of chromodorid nudibranchs (Mollusca, Gastropoda). *Zoologica Scripta* 40: 137–157.

Johnson, R. F. & Gosliner, T. M. (2012). Traditional taxonomic groupings mask evolutionary history: A molecular phylogeny and new classification of the chromodorid nudibranchs. *PLoS ONE* 7: e33479.

Johnson, S. G. & Bragg, E. (1999). Age and polyphyletic origins of hybrid and spontaneous parthenogenetic *Campeloma* (Gastropoda: Viviparidae) from the southeastern United States. *Evolution* 53: 1769–1781.

Johnstone, J. (1899). *Cardium*. Liverpool Marine Biology Committee: Memoirs on typical British marine plants and animals. Liverpool, T. Dobb & Co.

Joll, L. M. (1989). Swimming behavior of the saucer scallop *Amusium balloti* (Mollusca: Pectinidae). *Marine Ecology* 102: 299–305.

Jones, C. G. & Shachak, M. (1990). Fertilization of the desert soil by rock-eating snails. *Nature* 346: 839–841.

Jones, C. G., Lawton, J. H. & Shachak, M. (1994). Organisms as ecosystem engineers. *Oikos* 69: 373–386.

Jones, C. G., Lawton, J. H. & Shachak, M. (1997). Positive and negative effects of organisms as physical ecosystem engineers. *Ecology* 78: 1946–1957.

Jones, D. L. (1961). Muscle attachment impressions in a Cretaceous ammonite. *Journal of Paleontology* 35: 502–504.

Jones, D. S. (1980). Annual cycle of shell growth increment formation in two continental shelf bivalves and its paleoecologic significance. *Paleobiology* 6: 331–340.

Jones, D. S. (1983a). Sclerochronology: Reading the record of the molluscan shell. *American Scientist* 71: 384–391.

Jones, D. S. (1985). Growth increments and geochemical variations in the molluscan shell. pp. 72–87 in T. W. Broadbent (Ed.) *Mollusks: Notes for a Short Course Organized by D. J. Bottjer, C. S. Hickman, and P. D. Ward*. *University of Tennessee Studies in Geology*. Orlando, FL, University of Tennessee, Department of Geological Sciences.

Jones, D. S., Williams, D. F. & Romanek, C. S. (1986). Life history of symbiont-bearing giant clams *Tridacna maxima* from stable isotope profiles. *Science* 231: 46–48.

Jones, D. S. & Gould, S. J. (1999). Direct measurement of age in fossil *Gryphaea*: The solution to a classic problem in heterochrony. *Paleobiology* 25: 158–187.

Jones, E. C. (1963). *Tremoctopus violaceus* uses *Physalia* tentacles as weapons. *Science* 139: 746–766.

Jones, H. D. (1968). Some aspects of the heart function in *Patella vulgata* L. *Nature* 217: 1170–1172.

Jones, H. D. & Trueman, E. R. (1970). Locomotion of the limpet, *Patella vulgata* L. *Journal of Experimental Biology* 52: 201–216.

Jones, H. D. (1971). Circulatory pressures in *Helix pomatia* L. *Comparative Biochemistry and Physiology Part A* 39: 289–295.

Jones, H. D. (1975). Locomotion. pp. 1–32 in V. Fretter & Peake, J. (Eds.) *Pulmonates: Functional Anatomy and Physiology*. Vol. 1. London, UK, Academic Press.

Jones, H. D. (1983b). Circulatory systems of gastropods and bivalves. pp. 189–238 in A. S. M. Saleuddin & Wilbur, K. M. (Eds.) *Physiology, Part 2. The Mollusca*. Vol. 5. New York, Academic Press.

Jones, H. D. (1988). In vivo cardiac pressure, heart rate and heart mass of *Busycon canaliculatum* (L.). *Journal of Experimental Biology* 140: 257–271.

Jones, J. B. & Marshall, B. A. (1986). *Cocculinika myzorama*, new genus, new species, a parasitic copepod from a deep-sea wood-ingesting limpet. *Journal of Crustacean Biology* 6: 166–169.

Jones, J. S. (1973). Ecological genetics and natural selection in mollusks *Science* 182: 439–453.

Jones, J. S., Leith, B. H. & Rawlings, P. (1977). Polymorphism in *Cepaea*: A problem with too many solutions? *Annual Review of Ecology and Systematics* 8: 109–143.

Jones, J. S., Selander, R. K. & Schnell, G. D. (1980). Patterns of morphological and molecular polymorphism in the land snail *Cepaea nemoralis*. *Biological Journal of the Linnean Society* 14: 359–381.

Jones, N. S. (1948). Observations and experiments on the biology of *Patella vulgata* at Port St. Mary, Isle of Man. *Proceedings and Transactions of the Liverpool Biological Society* 56: 60–77.

Jones, P. & Silver, J. (1979). Red and blue-green bile pigments in the shell of *Astraea tuber* (Mollusca: Archaeogastropoda). *Comparative Biochemistry and Physiology Part B* 63: 185–188.

Jones, S. (2001). Cocksure Jones. *Trends in Genetics* 17: 107–108.

Jones, W. J., Won, Y. J., Maas, P. A. Y., Smith, P. J., Lutz, R. A. & Vrijenhoek, R. C. (2006). Evolution of habitat use by deep-sea mussels. *Marine Biology* 148: 841–851.

de Jong-Brink, M., Boer, H. H. & Joosse, J. (1983). Mollusca. pp. 297–355 in K. D. Adiyodi & Adiyodi, R. D. (Eds.) *Oogenesis, Oviposition and Oosorption. Reproductive Biology of Invertebrates*. Vol. 1. Chichester, UK, John Wiley & Sons.

Joordens, J. C., d'Errico, F., Wesselingh, F. P., Munro, S., De Vos, J., Wallinga, J., Ankjærgaard, C., Reimann, T., Wijbrans, J. R., Kuiper, K. F., Mücher, H. J., Coqueugniot, H., Prié, V., van Joosten, I., Os, B., Schulp, A. S., Panuel, M., van der Haas, V., Lustenhouwer, W., Reijmer, J. J. G. & Roebroekset, W. (2015). *Homo erectus* at Trinil on Java used shells for tool production and engraving. *Nature* 518: 228–231.

Joosse, J. (1975). Structural and endocrinological aspects of hermaphroditism in pulmonate snails, with particular reference to *Lymnaea stagnalis* (L.). pp. 158–169 in R. Reinboth (Ed.) *Intersexuality in the Animal Kingdom*. Berlin and New York, Springer-Verlag.

Joosse, J. (1976). Endocrinology of Molluscs. pp. 107–123 *in* M. Durchon & Joly, P. (Eds.) *Actualités sur les hormones d'invertébrés.* Paris, France, Centre national de la Recherche scientifique.

Joosse, J. & Geraerts, W. P. M. (1983). Endocrinology. pp. 317–406 *in* A. S. M. Saleuddin & Wilbur, K. M. (Eds.) *Physiology, Part 1. The Mollusca.* Vol. 4. London, Academic Press.

Joosse, J. & Van Elk, R. (1986). *Trichobilharzia ocellata*: Physiological characterization of giant growth, glycogen depletion, and absence of reproductive activity in the intermediate snail host, *Lymnaea stagnalis. Experimental Parasitology* 62: 1–13.

Joosse, J. (1988). The hormones of molluscs. pp. 89–140 *in* H. Laufer & Downer, R. G. H. (Eds.) *Endocrinology of Selected Invertebrate Types.* New York, Alan R. Liss, Inc.

Jordaens, K., Dillen, L. & Backeljau, T. (2007). Effects of mating, breeding system and parasites on reproduction in hermaphrodites: Pulmonate gastropods (Mollusca). *Animal Biology* 57: 137–195.

Jordaens, K., Dillen, L. & Backeljau, T. (2009). Shell shape and mating behaviour in pulmonate gastropods (Mollusca). *Biological Journal of the Linnean Society* 96: 306–321.

Jordan, P. J. & Deaton, L. E. (1999). Osmotic regulation and salinity tolerance in the freshwater snail *Pomacea bridgesii* and the freshwater clam *Lampsilis teres. Comparative Biochemistry and Physiology Part A* 122: 199–205.

Jørgensen, C. B. (1946). Lamellibranchia: Reproduction and larval development of Danish marine bottom invertebrates, with special reference to the planktonic larvae in the Sound (Øresund). *Meddelelser fra Kommissionen for Danmarks Fiskeriog Havundersøgelser Serie: Plankton* 4: 277–311.

Jørgensen, C. B., Famme, P., Saustrup, K., Larsen, P. S., Möhlenberg, F. & Riisgård, H. U. (1986). The bivalve pump. *Marine Ecology Progress Series* 34: 69–77.

Jorgensen, D. D., Ware, S. K. & Redmond, J. R. (1984). Cardiac output and tissue blood flow in the abalone, *Haliotis cracherodii* (Mollusca, Gastropoda). *Journal of Experimental Zoology Part A* 231: 309–324.

Jörger, K. M., Neusser, T. P., Haszprunar, G. & Schrödl, M. (2008). Undersized and underestimated: 3D visualization of the Mediterranean interstitial acochlidian gastropod *Pontohedyle milaschewitchii* (Kowalevsky, 1901). *Organisms Diversity & Evolution* 8: 194–214.

Jörger, K. M., Hess, M., Neusser, T. P. & Schrödl, M. (2009). Sex in the beach: Spermatophores, dermal insemination and 3D sperm ultrastructure of the aphallic mesopsammic *Pontohedyle milaschewitchii* (Acochlidia, Opisthobranchia, Gastropoda). *Marine Biology* 156: 1159–1170.

Jörger, K. M., Stöger, I., Kano, Y., Fukuda, H., Knebelsberger, T. & Schrödl, M. (2010). On the origin of Acochlidia and other enigmatic euthyneuran gastropods, with implications for the systematics of Heterobranchia. *BMC Evolutionary Biology* 10: 323.

Joseph, S., Poriya, P., Vakani, B., Singh, S. P. & Kundu, R. (2014). Identification of a group of cryptic marine limpet species, *Cellana karachiensis* (Mollusca: Patellogastropoda) off Veraval Coast, India, using mtDNA COI sequencing. *Mitochondrial DNA: The Journal of DNA Mapping, Sequencing, and Analysis* 27: 1328–1331.

Joseph, S. J., Fernández-Robledo, J. A., Gardner, M. J., El-Sayed, N. M., Kuo, C.-H., Schott, E. J., Wang, H., Kissinger, J. C. & Vasta, G. R. (2010). The alveolate *Perkinsus marinus*: Biological insights from EST gene discovery. *BMC Genomics* 11: 228.

Joshi, R. C., Cowie, R. H. & Sebastian, L. S., Eds. (2017). *Biology and Management of Invasive Apple Snails.* Muñoz, Nueva Ecija, Philippine Rice Research Institute.

Joubin, L. (1885). La branchie des céphalopodes. *Archives de zoologie expérimentale et générale* 3: 75–150.

Judd, W. (1979). The secretions and fine structure of bivalve crystalline style sacs. *Ophelia* 18: 205–233.

Judd, W. (1987). Crystalline style proteins from bivalve molluscs. *Comparative Biochemistry and Physiology Part B* 88: 333–339.

Judge, J. & Haszprunar, G. (2014). The anatomy of *Lepetella sierrai* (Vetigastropoda, Lepetelloidea): Implications for reproduction, feeding, and symbiosis in lepetellid limpets. *Invertebrate Biology* 133: 324–339.

Jullien, A. (1948). Recherches sur les fonctions de la glande hypobranchiale chez *Murex trunculus. Comptes rendus des séances de la Société de biologie et de ses filiales* 142: 102–103.

Jumars, P. A. (1993). Gourmands of mud: Diet choice in deposit feeders. pp. 124–143 *in* R. N. Hughes (Ed.) *Diet Selection: An Interdisciplinary Approach to Foraging Behaviour.* Boston, MA, Blackwell, Oxford University Press.

Jung, L. H., Kavanaugh, S. I., Sun, B. & Tsai, P. S. (2014). Localization of a molluscan gonadotropin-releasing hormone in *Aplysia californica* by in situ hybridization and immunocytochemistry. *General and Comparative Endocrinology* 195: 132–137.

Juniper, S. K. (1986). Deposit feeding strategy in *Amphibola crenata:* Feeding behaviour, selective feeding and digestion. *Mauri Ora* 13: 103–115.

Jurberg, P., Cunha, R. A. & Rodrigues, M. L. (1997). Behavior of *Biomphalaria glabrata* Say, 1818 (Gastropoda: Planorbidae). 1. Morphophysiology of the mantle cavity. *Memórias do Instituto Oswaldo Cruz* 92: 287–295.

Kaas, Q., Yu, R., Jin, A.-H., Dutertre, S. & Craik, D. J. (2012). ConoServer: Updated content, knowledge, and discovery tools in the conopeptide database. *Nucleic Acids Research* 40: D325–D330.

Kabat, A. R. (1990). Predatory ecology of naticid gastropods with a review of shell boring predation. *Malacologia* 32: 155–193.

Kaifu, K., Akamatsu, T. & Segawa, S. (2008). Underwater sound detection by cephalopod statocyst. *Fisheries Science* 74: 781–786.

Kaim, A., Kobayashi, Y., Echizenya, H., Jenkins, R. G. & Tanabe, K. (2008). Chemosynthesis-based associations on Cretaceous plesiosaurid carcasses. *Acta Palaeontologica Polonica* 53: 97–104.

Kakoi, S., Kin, K., Miyazaki, K. & Wada, H. (2008). Early development of the Japanese spiny oyster (*Saccostrea kegaki*): Characterization of some genetic markers. *Zoological Science* 25: 455–464.

Kamardin, N. N. (1976). Response of osphradium of the mollusk *Limnea stagnalis* to different oxygen concentrations in water. *Journal of Evolutionary Biochemistry and Physiology* 12: 427–428.

Kamardin, N. N. (1988). Le rôle probable de l'osphradium dans le homing des mollusques marins littoraux *Acanthopleura gemmata* Blainv. (Polyplacophora), *Siphonaria grisea* L. et *Siphonaria* sp. (Gastropoda, Pulmonata). *Bulletin du Musée d'histoire naturelle de Marseille* 48: 125–130.

Kamel, S. J. & Grosberg, R. K. (2012). Exclusive male care despite extreme female promiscuity and low paternity in a marine snail. *Ecology Letters* 15: 1167–1173.

Kamenev, G. M., Nadtochy, V. A. & Kuznetsov, A. P. (2001). *Conchocele bisecta* (Conrad, 1849) (Bivalvia: Thyasiridae) from cold-water methane-rich areas of the Sea of Okhotsk. *The Veliger* 44: 84–94.

Kamio, M., Grimes, T. V., Hutchins, M. H., Van Dam, R., & Derby, C. D. (2010). The purple pigment aplysioviolin in sea hare ink deters predatory blue crabs through their chemical senses. *Animal Behaviour* 80: 89–100.

Kamio, M., Kicklighter, C. E., Nguyen, L., Germann, M. W. & Derby, C. D. (2011). Isolation and structural elucidation of novel mycosporine-like amino acids as alarm cues in the defensive ink secretion of the sea hare *Aplysia californica*. *Helvetica Chimica Acta* 94: 1012–1018.

Kamiya, H., Muramoto, K. & Ogata, K. (1984). Antibacterial activity in the egg mass of a sea hare. *Experientia* 40: 947–949.

Kamiya, H., Muramoto, K., Goto, R., Sakai, M., Endo, Y. & Yamazaki, M. (1989). Purification and characterization of an antibacterial and antineoplastic protein secretion of a sea hare, *Aplysia juliana*. *Toxicon* 27: 1269–1277.

Kanda, A., Takahashi, T., Satake, H. & Minakata, H. (2006). Molecular and functional characterization of a novel gonadotropin-releasing-hormone receptor isolated from the common octopus (*Octopus vulgaris*). *Biochemical Journal* 395: 125–135.

Kandel, E. R. (1976). *Cellular Basis of Behavior. An Introduction to Behavioral Neurobiology*. San Francisco, CA, W. H. Freeman and Co.

Kandel, E. R. (1979). *Behavioral Biology of Aplysia: A Contribution to the Comparative Study of Opisthobranch Molluscs*. San Francisco, CA, W. H. Freeman and Co.

Kandel, E. R. (2001). The molecular biology of memory storage: A dialogue between genes and synapses. *Science* 294: 1030–1038.

Kano, Y., Chiba, S. & Kase, T. (2002). Major adaptive radiation in neritopsine gastropods estimated from 28S rRNA sequences and fossil records. *Proceedings of the Royal Society B* 269: 2457–2465.

Kano, Y. & Kase, T. (2002). Anatomy and systematics of the submarine-cave gastropod *Pisulina* (Neritopsina: Neritiliidae): Part 4. *Journal of Molluscan Studies* 68: 365–383.

Kano, Y. & Haga, T. (2006). Sulphide rich environments. pp. 373–375 *in* P. Bouchet, Le Guyader, H. & Pascal, O. (Eds.) *The Natural History of Santo. Patrimoines Naturels*. Paris, Muséum national d'Histoire naturelle.

Kano, Y. (2008). Vetigastropod phylogeny and a new concept of Seguenzioidea: Independent evolution of copulatory organs in the deep-sea habitats. *Zoologica Scripta* 37: 1–21.

Kano, Y. (2009). Hitchhiking behaviour in the obligatory upstream migration of amphidromous snails. *Biology Letters* 5: 465–468.

Kantha, S. S. (1989). Carotenoids of edible molluscs: A review. *Journal of Food Biochemistry* 13: 429–442.

Kantor, I. & Taylor, J. D. (1991). Evolution of the toxoglossan feeding mechanism: New information on the use of the radula. *Journal of Molluscan Studies* 57: 129–134.

Kantor, I. & Taylor, J. D. (2000). Formation of marginal radular teeth in Conoidea (Neogastropoda) and the evolution of the hypodermic envenomation mechanism. *Journal of Zoology* 252: 251–262.

Kantor, I. (2002). Morphological prerequisites for understanding neogastropod phylogeny. *Bollettino Malacologico* 38: 161–174.

Kantor, I. & Taylor, J. D. (2002). Foregut anatomy and relationships of Raphitomine gastropods (Gastropoda: Conoidea: Raphitominae). *Bollettino Malacologico* 38: 83–110.

Kantor, I. & Sysoev, A. V. (2005). A preliminary analysis of biodiversity of molluscs of Russia and adjacent territories. *Ruthenica* 14: 107–118.

Kantor, Y. I. (1991). On the morphology and relationships of some oliviform gastropods. *Ruthenica* 1: 17–52.

Kantor, Y. I. (1996). Phylogeny and relationships of Neogastropoda. pp. 221–230 *in* J. D. Taylor (Ed.) *Origin and Evolutionary Radiation of the Mollusca*. Oxford, UK, Oxford University Press.

Kantor, Y. I. (2003). Comparative anatomy of the stomach of Buccinoidea (Neogastropoda). *Journal of Molluscan Studies* 69: 203–220.

Kantor, Y. I. (2007). How much can *Conus* swallow? Observations on molluscivorous species. *Journal of Molluscan Studies* 73: 123–127.

Kappes, H. & Haase, P. (2012). Slow but steady: Dispersal of freshwater molluscs. *Aquatic Sciences* 74: 1–14.

Kappner, I., Al-Moghrabi, S. M. & Richter, C. (2000). Mucus-net feeding by the vermetid gastropod *Dendropoma maxima* in coral reefs. *Marine Ecology Progress Series* 204: 309–313.

Karatayev, A. Y., Burlakova, L. E., Molloy, D. P. & Volkova, L. K. (2000). Endosymbionts of *Dreissena polymorpha* (Pallas) in Belarus. *International Revue der gesamten Hydrobiologie und Hydrographie* 85: 543–560.

Karlson, R. H. & Cariolou, M. A. (1982). Hermit crab shell colonization by *Crepidula convexa* Say. *Journal of Experimental Marine Biology and Ecology* 65: 1–10.

Karlson, R. H. & Cornell, H. V. (1999). Integration of local and regional perspectives on the species richness of coral assemblages. *American Zoologist* 39: 104–112.

Karnik, V., Braun, M., Dalesman, S. & Lukowiak, K. (2012a). Sensory input from the osphradium modulates the response to memory-enhancing stressors in *Lymnaea stagnalis*. *Journal of Experimental Biology* 215: 536–542.

Karnik, V., Dalesman, S. & Lukowiak, K. (2012b). Input from a chemosensory organ, the osphradium, does not mediate aerial respiration in *Lymnaea stagnalis*. *Aquatic Biology* 15: 167–173.

Karp, G. C. & Whiteley, A. H. (1973). DNA-RNA hybridization studies of gene activity during the development of the gastropod, *Acmaea scutum*. *Experimental Cell Research* 78: 236–241.

Karp, K. & Thomas, R. (2014). *Summary of Laboratory and Field Experiments to Evaluate Predation of Quagga Mussel by Redear Sunfish and Bluegill. Research and Development Office Science and Technology Program Final Report*. https://www.usbr.gov/research/, Research and Development Office, US Department of the Interior, Bureau of Reclamation: 1–24.

Karpov, K., Haaker, P., Taniguchi, I. & Rogers-Bennett, L. (2000). Serial depletion and the collapse of the California abalone (*Haliotis* spp.) fishery. *Canadian Special Publication of Fisheries and Aquatic Sciences* 130: 11–24.

Karr, T. L. (2007). Application of proteomics to ecology and population biology. *Heredity* 100: 200–206.

Karson, M. A., Boal, J. G. & Hanlon, R. T. (2003). Experimental evidence for spatial learning in cuttlefish (*Sepia officinalis*). *Journal of Comparative Psychology* 117: 149–155.

Karuso, P. (1987). Chemical ecology of the nudibranchs. pp. 31–50 *in* P. J. Scheuer (Ed.) *Bioorganic Marine Chemistry*. Vol. 1. Berlin, Springer-Verlag.

Kasahara, M. (2007). The 2R hypothesis: An update. *Current Opinion on Immunology* 19: 547–552.

Kasai, A., Horie, H. & Sakamoto, W. (2004). Selection of food sources by *Ruditapes philippinarum* and *Mactra veneriformis* (Bivalvia: Mollusca) determined from stable isotope analysis. *Fisheries Science* 70: 11–20.

Kasai, A. & Nakata, A. (2005). Utilization of terrestrial organic matter by the bivalve *Corbicula japonica* estimated from stable isotope analysis. *Fisheries Science* 71: 151–158.

Kase, T. & Kano, Y. (2002). *Trogloconcha*, a new genus of larocheine Scissurellidae (Gastropoda: Vetigastropoda) from tropical Indo-Pacific submarine caves. *The Veliger* 45: 25–32.

Kasinathan, R. (1975). Some studies of five species of cyclophorid snails from Peninsular India. *Proceedings of the Malacological Society of London* 41: 379–394.

Kasschau, M. R. (1975). The relationship of free amino acids to salinity changes and temperature-salinity interactions in the mudflat snail, *Nassarius obsoletus*. *Comparative Biochemistry and Physiology Part A* 51: 301–308.

Kasugai, T. (2001). Feeding behaviour of the Japanese pygmy cuttlefish *Idiosepius paradoxus* (Cephalopoda: Idiosepiidae) in captivity: Evidence for external digestion? *Journal of the Marine Biological Association of the United Kingdom* 81: 979–981.

Katagiri, N., Terakita, A., Shichida, Y. & Katagiri, Y. (2001). Demonstration of a rhodopsin-retinochrome system in the stalk eye of a marine gastropod, *Onchidium*, by immunohistochemistry. *Journal of Comparative Neurology* 433: 380–389.

Katagiri, N., Suzuki, T., Shimatani, Y. & Katagiri, Y. (2002). Localization of retinal proteins in the stalk and dorsal eyes of the marine gastropod, *Onchidium*. *Zoological Science* 19: 1231–1240.

Katagiri, Y., Katagiri, N. & Fujimoto, K. (1985). Morphological and electrophysiological studies of a multiple photoreceptive system in a marine gastropod, *Onchidium*. *Neuroscience Research* 2: S1–S15.

Kataoka, S. (1976). Fine structure of the epidermis of the optic tentacle in a slug, *Limax flavus* L. *Tissue and Cell* 8: 47–60.

Kataoka, S. & Yamamoto, Y. (1981). Diurnal changes in the fine structure of photoreceptors in an abalone, *Nordotis discus*. *Cell and Tissue Research* 218: 181–189.

Kater, S. B. (1974). Feeding in *Helisoma trivolvis*: The morphological and physiological bases of a fixed action pattern. *Integrative and Comparative Biology* 14: 1017–1036.

Kato, K. (1935). *Stylochoplana parasitica* sp. nov., a polyclad parasitic in the pallial groove of the chiton [in Japanese]. *Annotationes Zoologicae Japonenses* 15: 123–127.

Kato, K. (1949). Luminous organ of *Kalophlocamus ramosum*. *Dobutsugaku zasshi* 58: 163–164.

Kato, K. (1951). *Convoluta*, an acoelous turbellarian, destroyed the edible Clam. *Miscellaneous Reports of the Research Institute of Natural Resources* 19–21: 64–67.

Kato, K. & Kubomura, K. (1954). On the origins of the crystalline style of lamellibranchs. *Science Reports of the Saitama University. Series B* 1: 135–152.

Kato, M. & Itani, G. (1995). Commensalism of a bivalve, *Peregrinamor ohshimai*, with a thalassinidean burrowing shrimp, *Upogebia major*. *Journal of the Marine Biological Association of the United Kingdom* 75: 941–947.

Kato, M. (1998). Morphological and ecological adaptations in montacutid bivalves endo- and ecto-symbiotic with holothurians. *Canadian Journal of Zoology* 76: 1403–1410.

Katoh, M. (1989). Life history of the golden ring cowry *Cypraea annulus* (Mollusca: Gastropoda) on Okinawa Island, Japan. *Marine Biology* 101: 227–234.

Katsuno, S. & Sasaki, T. (2008). Comparative histology of radula-supporting structures in Gastropoda. *Malacologia* 50: 13–56.

Katz, P. S. (1998). Neuromodulation intrinsic to the central pattern generator for escape swimming in *Tritonia*. *Annals of the New York Academy of Sciences* 860: 181–188.

Katz, P. S., Fickbohm, D. J. & Lynn-Bullock, C. P. (2001). Evidence that the central pattern generator for swimming in *Tritonia* arose from a non-rhythmic neuromodulatory arousal system: Implications for the evolution of specialized behavior. *American Zoologist* 41: 962–975.

Katz, S., Cavanaugh, C. M. & Bright, M. (2006). Symbiosis of epi-and endocuticular bacteria with *Helicoradomenia* spp. (Mollusca, Aplacophora, Solenogastres) from deep-sea hydrothermal vents. *Marine Ecology Progress Series* 320: 89–99.

Katzoff, A., Ben-Gedalya, T., Hurwitz, I., Miller, N., Susswein, Y. Z. & Susswein, A. J. (2006). Nitric oxide signals that *Aplysia* have attempted to eat, a necessary component of memory formation after learning that food is inedible. *Journal of Neurophysiology* 96: 1247–1257.

Kauffman, E. G. & Buddenhagen, C. H. (1969). Protandric sexual dimorphism in Paleocene *Astarte* (Bivalvia) of Maryland. pp. 76–93 *in* G. E. G. Westermann (Ed.) *Sexual Dimorphism in Fossil Metazoa and Taxonomic Implications*. Vol. 1. Stuttgart, E. Schweizerbart'sche Verlagsbuchhandlung.

Kaufman, M. R., Ikeda, Y., Patton, C., Van Dykhuizen, G. & Epel, D. (1998). Bacterial symbionts colonize the accessory nidamental gland of the squid *Loligo opalescens* via horizontal transmission. *Biological Bulletin* 194: 36–43.

Kaupp, U. B., Hildebrand, E. & Weyand, I. (2006). Sperm chemotaxis in marine invertebrates: Molecules and mechanisms. *Journal of Cellular Physiology* 208: 487–494.

Kavaliers, M. (1980). A circadian rhythm of behavioural thermoregulation in a freshwater gastropod, *Helisoma trivolis*. *Canadian Journal of Zoology* 58: 2152–2155.

Kavaliers, M. (1981). Circadian and ultradian activity rhythms of a freshwater gastropod, *Helisoma trivolis*: The effects of social factors and eye removal. *Behavioral and Neural Biology* 32: 350–363.

Kavaliers, M. & Tepperman, F. S. (1988). Exposure to novel odors induces opioid-mediated analgesia in the land snail, *Cepaea nemoralis*. *Behavioral and Neural Biology* 50: 285–299.

Kavaliers, M. & Ossenkopp, K. P. (1991). Opioid systems and magnetic field effects in the land snail, *Cepaea nemoralis*. *Biological Bulletin* 180: 301–309.

Kawaguti, S. (1959). Formation of the bivalve shell in a gastropod, *Tamanovalva limax*. *Proceedings of the Japan Academy* 35: 607–611.

Kawashima, Y., Nagashima, Y. & Shiomi, K. (2002). Toxicity and tetramine contents of salivary glands from carnivorous gastropods [in Japanese]. *Journal of the Food Hygienic Society of Japan* 43: 385–388.

Kawashima, Y. (2003). Primary structure of echotoxin 2, an actinoporin-like hemolytic toxin from the salivary gland of the marine gastropod *Monoplex echo*. *Toxicon* 42: 491–497.

Kawashima, Y., Nagashima, Y. & Shiomi, K. (2004). Determination of tetramine in marine gastropods by liquid chromatography/electrospray ionization-mass spectrometry. *Toxicon* 44: 185–191.

Kawsar, S. M. A., Matsumoto, R., Fujii, Y., Yasumitsu, H., Dogasaki, C., Hosono, M., Nitta, K., Hamako, J., Matsui, T., Kojima, N. & Ozeki, Y. (2009). Purification and biochemical characterization of a D-galactose binding lectin from Japanese sea hare (*Aplysia kurodai*) eggs. *Biochemistry* 74: 709–716.

Kay, A. E. (1990). Cypraeidae of the Indo-Pacific: Cenozoic fossil history and biogeography. *Bulletin of Marine Science* 47: 23–34.

Kay, E. A. & Young, D. K. (1969). The Doridacea (Opisthobranchia; Mollusca) of the Hawaiian Islands. *Pacific Science* 23: 172–231.

Kay, M. C. & Emlet, R. B. (2002). Laboratory spawning, larval development, and metamorphosis of the limpets *Lottia digitalis* and *Lottia asmi* (Patellogastropoda, Lottiidae). *Invertebrate Biology* 121: 11–24.

Kays, W. T., Silverman, H. G. & Dietz, T. H. (1990). Water channels and water canals in the gill of the freshwater mussel, *Ligumia subrostrata*: Ultrastructure and histochemistry. *Journal of Experimental Zoology Part A* 254: 256–269.

Keas, B. E. & Esch, G. W. (1997). The effects of diet and reproductive maturity on the growth and reproduction of *Helisoma anceps* (Pulmonata) infected by *Halipegus occidualis* (Trematoda). *Journal of Parasitology* 83: 96–104.

Keen, M. (1971). *Sea Shells of Tropical West America: Marine Mollusks from Lower California to Columbia*. 2nd ed. Stanford, Stanford University Press.

Keferstein, W. M. (1862). Malacozoa, Abt. 2: Malacozoa Cephalomalacia [gastropods, scaphopods, cephalopods]. pp. 523–1500 *in* H. G. Bronn (Ed.) *Die Klassen und Ordnungen des Thier-Reichs*. Vol. 3. Leipzig, Germany, C. F. Winter.

Kelaher, B. P., Castilla, J. C., Prado, L., York, P., Schwindt, E. & Bortolus, A. (2007). Spatial variation in molluscan assemblages from coralline turfs of Argentinean Patagonia. *Journal of Molluscan Studies* 73: 139–146.

Keller, H. W. & Snell, K. L. (2002). Feeding activities of slugs on Myxomycetes and macrofungi. *Mycologia* 94: 757–760.

Kellert, S. R. (1993). Values and perceptions of invertebrates. *Conservation Biology* 7: 845–855.

Kelly, S. R. A., Ditchfield, P. W., Doubleday, P. A. & Marshall, J. D. (1995). An Upper Jurassic methane-seep limestone from the fossil Bluff Group forearc basin of Alexander Island, Antarctica. *Journal of Sedimentary Research* 65: 274–282.

Kemp, P. & Bertness, M. D. (1984). Snail shape and growth rates: Evidence for plastic shell allometry in *Littorina littorea*. *Proceedings of the National Academy of Sciences of the United States of America* 81: 811–813.

Kempf, S. C. (1981). Long-lived larvae of the gastropod *Aplysia juliana*: Do they disperse and metamorphose or just slowly fade away. *Marine Ecology Progress Series* 6: 61–65.

Kempf, S. C. & Hadfield, M. G. (1985). Planktotrophy by the lecithotrophic larvae of a nudibranch, *Phestilla sibogae* (Gastropoda) days after hatching in the absence of any external food source. *Cultures* 169: 119–130.

Kempf, S. C., Page, L. R. & Pires, A. (1997). Development of serotonin-like immunoreactivity in the embryos and larvae of nudibranch mollusks with emphasis on the structure and possible function of the apical sensory organ. *Journal of Comparative Neurology* 386: 507–528.

Kennedy, D. (1960). Neural photoreception in a lamellibranch mollusc. *Journal of General Physiology* 44: 277–299.

Keough, M. J., Quinn, G. P. & King, A. (1993). Correlations between human collecting and intertidal mollusc populations on rocky shores. *Conservation Biology* 7: 378–390.

Kerkhoven, R. M., Croll, R. P., Ramkema, M. D., Minnen, J. Van, Bogerd, J. & Boer, H. H. (1992). The VD_1/RPD_2 neuronal system in the central nervous system of the pond snail *Lymnaea stagnalis* studied by in situ hybridization and immunocytochemistry. *Cell and Tissue Research* 267: 551–559.

Kerkhoven, R. M., Ramkema, M. D., Minnen, J. Van, Croll, R. P., Pin, T. & Boer, H. H. (1993). Neurons in a variety of molluscs react to antibodies raised against the VD_1/RPD_2 α-neuropeptide of the pond snail *Lymnaea stagnalis*. *Cell and Tissue Research* 273: 371–379.

Kerth, K. (1979). Phylogenetische Aspekte der Radulamorphogenese von Gastropoden. *Malacologia* 19: 103–108.

Kerth, K. (1983). Radulaapparat und Radulabildung der Mollusken. II. Zahnbildung, Abbau und Radulawachstum. *Zoologische Jahrbücher. Abteilung für Anatomie und Ontogenie der Tiere* 110: 239–269.

Kessel, E. (1942). Über Bau und Bildung des Prosobranchier Deckels. *Zeitschrift fur Morphologie und Okologie der Tiere* 38: 197–250.

Kessel, R. G. (1982). Differentiation of *Acmaea digitalis* oocytes with special reference to lipid-endoplasmic reticulum-annulate lamellae-polyribosome relationships. *Journal of Morphology* 171: 225–243.

Van Kesteren, R. E., Smit, A. B., Dirks, R. W., de With, N. D., Geraerts, W. P. M. & Joosse, J. (1992). Evolution of the vasopressin-oxytocin superfamily: Characterization of a cDNA encoding a vasopressin-related precursor, preproconopressin, from the mollusc *Lymnaea stagnalis*. *Proceedings of the National Academy of Sciences of the United States of America* 89: 4593–4597.

Ketata, I., Denier, X., Hamza-Chaffai, A. & Minier, C. (2008). Endocrine-related reproductive effects in molluscs. *Comparative Biochemistry and Physiology Part C* 147: 261–270.

Kever, L., Colleye, O., Lugli, M., Lecchini, D., Lerouvreur, F., Herrel, A. & Parmentier, E. (2014). Sound production in *Onuxodon fowleri* (Carapidae) and its amplification by the host shell. *Journal of Experimental Biology* 217: 4283–4294.

Kew, H. W. (1893). *The Dispersal of Shells*. London, K. Paul, Trench, Trubner & Company.

Keyes, C. R. (1889). Variations exhibited by a Carbonic gasteropod [sic]. *American Geologist* 3: 330–333.

Keyl, M. J., Michaelson, I. A. & Whittaker, V. P. (1957). Physiologically active choline esters in certain marine gastropods and other invertebrates. *Journal of Physiology* 139: 434–454.

Khan, H. R. & Saleuddin, A. S. M. (1979). Effects of osmotic changes and neurosecretory extracts on kidney ultrastructure in the freshwater pulmonate *Helisoma*. *Canadian Journal of Zoology* 57: 1256–1270.

Khan, H. R. & Saleuddin, A. S. M. (1981a). Involvement of actin and Na^+-K^+ ATPase in urine formation of the freshwater pulmonate *Helisoma*. *Journal of Morphology* 169: 243–251.

Khan, H. R. & Saleuddin, A. S. M. (1981b). Cell contacts in the kidney epithelium of *Helisoma* (Mollusca: Gastropoda): Effects of osmotic pressure and brain extracts—A freeze-fracture study. *Journal of Ultrastructure Research* 75: 23–40.

Khan, H. R., Ashton, M. L. & Saleuddin, A. S. M. (1990). Changes in the fine structure of the endocrine dorsal body cells of *Helisoma duryi* (Mollusca) induced by mating. *Journal of Morphology* 203: 41–53.

Khan, H. R., Price, D. A., Doble, K. E., Greenberg, M. J. & Saleuddin, A. S. M. (1999). Osmoregulation and FMRFamide-related peptides in the salt marsh snail *Melampus bidentatus* (Say) (Mollusca: Pulmonata). *Biological Bulletin* 196: 153–162.

Kicklighter, C. E., Shabani, S., Johnson, P. M. & Derby, C. D. (2005). Sea hares use novel antipredatory chemical defenses. *Current Biology* 15: 549–554.

Kicklighter, C. E. & Derby, C. D. (2006). Multiple components in ink of the sea hare *Aplysia californica* are aversive to the sea anemone *Anthopleura sola*. *Journal of Experimental Marine Biology and Ecology* 334: 256–268.

Kicklighter, C. E., Kamio, M., Gemann, M. & Derby, C. D. (2007). Pyrimidines and mycosporine-like amino acids function as alarm cues in the defensive secretions of the sea hare *Aplysia californica*. *Chemical Senses* 32: A30.

Kido, J. S. & Murray, S. N. (2003). Variation in owl limpet *Lottia gigantea* population structures, growth rates, and gonadal production on southern California rocky shores. *Marine Ecology Progress Series* 257: 111–124.

Kiel, S. & Little, C. T. S. (2006). Cold-seep mollusks are older than the general marine mollusk fauna. *Science* 313: 1429–1431.

Kier, W. M. (1985). Tongues, tentacles and trunks: The biomechanics of movement in muscular-hydrostats. *Zoological Journal of the Linnean Society* 83: 307–324.

Kier, W. M. (1988). The arrangement and function of molluscan muscle. pp. 211–252 *in* E. R. Trueman & Clarke, M. R. (Eds.) *Form and Function. The Mollusca*. Vol. 11. New York, Academic Press.

Kier, W. M. & Van Leeuwen, J. L. (1997). A kinematic analysis of tentacle extension in the squid *Loligo pealei*. *Journal of Experimental Biology* 200: 41–53.

Kier, W. M. & Thompson, J. T. (2003). Muscle arrangement, function and specialization in Recent Coleoids. *Berliner Paläobiologische Abhandlungen* 3: 141–162.

Kier, W. M. & Schachat, F. H. (2008). Muscle specialization in the squid motor system. *Journal of Experimental Biology* 211: 164–169.

Kilburn, R. N. & Herbert, D. G. (1999). Mollusca. pp. 8–12 *in* M. J. Gibbons (Ed.) *The Taxonomic Richness of South Africa's Marine Fauna: A Crisis at Hand*. Vol. 95. Johannesburg, South African Journal of Science.

Kilian, E. F. (1951). Untersuchungen zur Biologie von *Pomatias elegans* (Müller) und ihrer "Konkrementdrüse". *Archiv für Molluskenkunde* 80: 1–16.

Killeen, I. J., Seddon, M. B. & Holmes, A. H., Eds. (1998). *Molluscan Conservation: A Strategy for the 21st Century. Proceedings of the Molluscan Conservation Conference, Cardiff, U.K., November 1996*. Journal of Conchology: Special Publication. London, Conchological Society of Great Britain and Ireland.

Kim, B. H., Moon, T. S., Park, K. Y., Jo, P. G. & Kim, M. C. (2010). Study on spawning induction and larvae breeding of the hard clam, *Meretrix petechiails* (Lamark). *Korean Journal of Malacology* 26: 151–156.

Kim, J. R., Hayes, K. A., Yeung, N. W. & Cowie, R. H. (2014). Diverse gastropod hosts of *Angiostrongylus cantonensis*, the rat lungworm, globally and with a focus on the Hawaiian Islands. *PloS ONE* 9: e94969.

Kim, J. R., Hayes, K. A., Yeung, N. W. & Cowie, R. H. (2016). Identity and distribution of introduced slugs (Veronicellidae) in the Hawaiian and Samoan Islands. *Pacific Science* 70: 477–493.

Kim, Y. H., Nachman, R. J., Pavelka, L., Mosher, H. S., Fuhrman, F. A. & Fuhrman, G. J. (1981). Doridosine, 1-methylisoguanosine, from *Anisodoris nobilis;* structure, pharmacological properties and synthesis. *Journal of Natural Products* 44: 206–214.

Kim, Y. M., Park, K. I., Choi, K. S., Alvarez, R. A., Cummings, R. D. & Cho, M. (2006). Lectin from the Manila clam *Ruditapes philippinarum* is induced upon infection with the protozoan parasite *Perkinsus olseni*. *Journal of Biological Chemistry* 281: 26854–26864.

Kin, K., Kakoi, S. & Wada, H. (2009). A novel role for dpp in the shaping of bivalve shells revealed in a conserved molluscan developmental program. *Developmental Biology* 329: 152–166.

King, A. J. & Adamo, S. A. (2008). Short-term pain for long-term gain: A hypothetical role for the mantle in coleoid cephalopod circulation. *American Malacological Bulletin* 24: 25–29.

King, G. M., Judd, C., Kuske, C. R. & Smith, C. (2012). Analysis of stomach and gut microbiomes of the eastern oyster (*Crassostrea virginica*) from coastal Louisiana, USA. *PLoS ONE* 7: e51475.

Kingsley-Smith, P. R., Richardson, C. A. & Seed, R. (2005). Growth and development of the veliger larvae and juveniles of *Polinices pulchellus* (Gastropoda: Naticidae). *Journal of the Marine Biological Association of the United Kingdom* 85: 171–174.

Kingston, A. C. N., Chappell, D. R. & Speiser, D. I. (2018). Evidence for spatial vision in *Chiton tuberculatus*, a chiton with eyespots. *Journal of Experimental Biology* 221: jeb183632.

Kingston, R. (1968). Anatomical and oxygen electrode studies of respiratory surfaces and respiration in *Acmaea* (Mollusca: Gastropoda: Prosobranchia). *The Veliger* 11: 75–78.

Kingtong, S., Kellner, K., Bernay, B., Goux, D., Sourdaine, P. & Berthelin, C. H. (2013). Proteomic identification of protein associated to mature spermatozoa in the Pacific oyster *Crassostrea gigas*. *Journal of Proteomics* 82: 81–91.

Kinne, O. (1970). Temperature: Animals—Invertebrates. *Marine Ecology* 1: 407–514.

Kirchner, C., Krätzner, R. & Welter-Schultes, F. W. (1997). Flying snails: How far can *Truncatellina* (Pulmonata: Vertiginidae) be blown over sea? *Journal of Molluscan Studies* 63: 479–487.

Kirkendale, L. (2009). Their Day in the Sun : Molecular phylogenetics and origin of photosymbiosis in the 'other' group of photosymbiotic marine bivalves (Cardiidae: Fraginae). *Biological Journal of the Linnean Society* 97: 448–465.

Kirkendale, L. & Paulay, G. (2017). Part N, Revised, Vol. 1, Chapter 9: Photosymbiosis in Bivalvia. *Treatise Online* 89: 1–31.

Kirtland, J. P. (1834). Observations on the sexual characteristics of the animals belonging to Lamarck's family of Naiades. *American Journal of Science and Arts* 26: 117–220.

Kisker, L. G. (1923). Über Anordnung und Bau der interstitiellen Bindesubstanzen von *Helix pomatia* L. *Zeitschrift für wissenschaftliche Zoologie* 121: 65–125.

Kiss, T. & Pirger, Z. (2013). Multifunctional role of PACAP-like peptides in molluscs. *Protein and Peptide Letters* 20: 628–635.

Kisugi, J., Ohye, H., Kamiya, H. & Yamazaki, M. (1989). Biopolymers from marine invertebrates. [Part 10] Mode of action of an antibacterial glycoprotein, aplysianin E, from eggs of a sea hare, *Aplysia kurodai*. *Chemical and Pharmaceutical Bulletin* 37: 3050–3053.

Kitajima, E. W. & Paraense, W. L. (1983). The ultrastructure of the spermatheca of *Biomphalaria glabrata* (Gastropoda, Pulmonata). *Journal of Morphology* 176: 211–220.

Kitching, J. A. (1977). Shell form and niche occupation in *Nucella lapillus* (L.) (Gastropoda). *Journal of Experimental Marine Biology and Ecology* 26: 275–287.

Kito, Y., Narita, K., Seidou, M., Michinomae, M., Yoshihara, K., Partridge, J. C. & Herring, P. J. (1992). A blue-sensitive visual pigment based on 4-hydroxyretinal is found widely in mesopelagic cephalopods. pp. 411–414 *in* J.-L. Rigaud (Ed.) *Structures and Functions of Retinal Proteins Colloque Inserm*. Vol. 221. London, John Libbey.

Klages, N. T. (1996). Cephalopods as prey. II. Seals. *Philosophical Transactions of the Royal Society B* 351: 1045–1052.

Klecka, J. & Boukal, D. S. (2012). Who eats whom in a pool? A comparative study of prey selectivity by predatory aquatic insects. *PLoS ONE* 7: e37741.

Kleemann, K. H. (1990). Coral associations, biocorrosion, and space competition in *Pedum spondyloideum* (Gmelin) (Pectinacea, Bivalvia). *Marine Ecology* 11: 77–94.

Kleemann, K. H. (1996). Biocorrosion by bivalves. *Marine Ecology* 17: 145–158.

Kleeton, G. (1961). New host and distribution record of the copepod *Trochicola entericus*. *Crustaceana* 3: 172.

Klinbunga, S., Amparyup, P., Khamnamtong, B., Hirono, I., Aoki, T. & Jarayabhand, P. (2009). Identification, characterization, and expression of the genes *TektinA1* and *Axonemal Protein 66.0* in the tropical abalone *Haliotis asinina*. *Zoological Science* 26: 429–436.

Kling, G. & Jakobs, P. M. (1987). Cephalopod myocardial receptors: Pharmacological studies on the isolated heart of *Sepia officinalis* (L.). *Experientia* 43: 511–524.

Kling, G. & Schipp, R. (1987). Comparative ultrastructural and cytochemical analysis of the cephalopod systemic heart and its innervation. *Experientia* 43: 502–511.

Klitgaard, A. B. (1995). The fauna associated with outer shelf and upper slope sponges (Porifera, Demospongiae) at the Faroe Islands, northeastern Atlantic. *Sarsia* 80: 1–22.

Klussmann-Kolb, A., Croll, R. P. & Staubach, S. (2013). Use of axonal projection patterns for the homologisation of cerebral nerves in Opisthobranchia (Mollusca, Gastropoda). *Frontiers in Zoology* 10: 1–13.

Knight, J. B. (1952). Primitive fossil gastropods and their bearing on gastropod classification. *Smithsonian Miscellaneous Collections* 117: 1–56.

Kniprath, E. (1979). The functional morphology of the embryonic shell-gland in the conchiferous molluscs. *Malacologia* 18: 549–552.

Kniprath, E. (1980). Ontogenetic plate and plate field development in two chitons, *Middendorffia* and *Ischnochiton*. *Roux's Archives of Developmental Biology* 189: 97–106.

Kniprath, E. (1981). Ontogeny of the molluscan shell field: A review. *Zoologica Scripta* 10: 61–79.

Knorre, H. V. (1925). Schale und Rückensinnesorgane von *Trachydermon* (*Chiton*) *cinereus* und die ceylonesischen Chitonen der Sammlung Plate. *Jenaische Zeitschrift für Naturwissenschaft* 61: 469–632.

Knowlton, N., Brainard, R. E., Fisher, R., Moews, M., Plaisance, L. & Caley, M. J. (2010). Coral reef biodiversity. pp. 65–74 *in* A. D. McIntyre (Ed.) *Life in the World's Oceans: Diversity Distribution and Abundance*. Hoboken, NJ, Wiley-Blackwell.

Knox, G. A. (2000). *The Ecology of Seashores*. Boca Raton, FL, CRC Press.

Knudsen, J. (1991). Observations on *Hipponix australis* (Lamarck, 1819) (Mollusca, Gastropoda, Prosobranchia) from the Albany area, Western Australia. pp. 641–660 *in* F. E. Wells, Walker, D. I., Kirkman, H. & Lethbridge, R. (Eds.) *Proceedings of the Third International Marine Biological Workshop: The Marine Flora and Fauna of Albany, Western Australia*. Vol. 2. Albany, Western Australia, Western Australian Museum.

Knudsen, J. (1993). Observations on *Antisabia foliacea* (Quoy and Gaimard, 1835) (Mollusca, Gastropoda, Prosobranchia, Hipponicidae) from off Rottnest Island, Western Australia. pp. 481–495 *in* F. E. Wells, Walker, D. I., Kirkman, H. & Lethbridge, R. (Eds.) *Proceedings of the Fifth International Marine Biological Workshop: 'The Marine Flora and Fauna of Rottnest Island, Western Australia'*. Perth, WA, Western Australian Museum.

Knutson, L. V. (1966). Biology and immature stages of malacophagous flies: *Antichaeta analis*, *A. atriseta*, *A. brevipennis*, and *A. obliviosa* (Diptera: Sciomyzidae). *Transactions of the American Entomological Society* 92: 67–101.

Knutson, L. V., Berg, C. J., Edwards, L. J., Bratt, A. D. & Foote, B. A. (1967). Calcareous septa formed in snail shells by larvae of snail-killing flies. *Science* 156: 522–523.

Knutson, L. V. & Vala, J.-C. (2011). *Biology of Snail-killing Sciomyzidae Flies*. Cambridge, UK, Cambridge University Press.

Ko, G. W. K., Dineshram, R., Campanati, C., Chan, V. B. S., Havenhand, J. N. & Thiyagarajan, V. (2014). Interactive effects of ocean acidification, elevated temperature, and reduced salinity on early-life stages of the Pacific oyster. *Environmental Science & Technology* 48: 10079–10088.

Ko, R. C., Morton, B. & Wong, P. S. (1975). Prevalence and histopathology of *Echinocephalus sinensis* (Nematoda: Gnathostomatidae) in natural and experimental hosts. *Canadian Journal of Zoology* 53: 550–559.

Ko, R. C. (1976). Experimental infection of mammals with larval *Echinocephalus sinensis* (Nematoda: Gnathostomatidae) from oysters (*Crassostrea gigas*). *Canadian Journal of Zoology* 54: 597–609.

Kobak, J. (2001). Light, gravity and conspecifics as cues to site selection and attachment behaviour of juvenile and adult *Dreissena polymorpha* Pallas, 1771. *Journal of Molluscan Studies* 67: 183–189.

Kobluk, D. R. & Mapes, R. H. (1989). The fossil record, function, and possible origins of shell color patterns in Paleozoic marine invertebrates. *Palaios* 4: 63–85.

Kocot, K. M. (2013). Recent advances and unanswered questions in deep molluscan phylogenetics. *American Malacological Bulletin* 31: 195–208.

Kocot, K. M., Aguilera, F., McDougall, C., Jackson, D. J. & Degnan, B. M. (2016). Sea shell diversity and rapidly evolving secretomes: Insights into the evolution of biomineralization. *Frontiers in Zoology* 13: Article No. 23.

Kocot, K. M., McDougall, C. & Degnan, B. M. (2017). Developing perspectives on molluscan shells, part 1: Introduction and molecular biology. pp. 1–41 *in* S. Saleuddin & Mukai, S. (Eds.) *Physiology of molluscs, a collection of selected reviews*. Vol.1. Oakville, Ontario, Apple Academic Press.

Kodirov, S. A. (2011). The neuronal control of cardiac functions in molluscs. *Comparative Biochemistry and Physiology Part A* 160: 102–116.

Koehl, M. A. R. & Hadfield, M. G. (2010). Hydrodynamics of larval settlement from a larva's point of view. *Integrative and Comparative Biology* 50: 539–551.

Koene, J. M., Jansen, R. F., ter Maat, A., & Chase, R. (2000). A conserved location for the central nervous system control of mating behaviour in gastropod molluscs: Evidence from a terrestrial snail. *Journal of Experimental Biology* 203: 1071–1080.

Koene, J. M. & ter Maat, A. (2005). Sex role alternation in the simultaneously hermaphroditic pond snail *Lymnaea stagnalis* is determined by the availability of seminal fluid. *Animal Behaviour* 69: 845–850.

Koene, J. M. & ter Maat, A. (2007). Coolidge effect in pond snails: Male motivation in a simultaneous hermaphrodite. *BMC Evolutionary Biology* 7: 212 (211–216).

Koene, J. M., Montagne-Wajer, K. & ter Maat, A. (2007). Aspects of body size and mate choice in the simultaneously hermaphroditic pond snail *Lymnaea stagnalis*. *Animal Biology* 57: 247–259.

Koene, J. M. (2010). Neuro-endocrine control of reproduction in hermaphroditic freshwater snails: Mechanisms and evolution. *Frontiers in Behavioral Neuroscience* 4: 167.

Koester, J., Dieringer, N. & Mandelbaum, D. E. (1979). Cellular neuronal control of molluscan heart. *American Zoologist* 19: 103–116.

Koester, J. & Koch, U. T. (1987). Neural control of the circulatory system in *Aplysia*. *Experientia* 43: 972–977.

Koh, H. Y., Lee, J. H., Han, S. J., Park, H., Shin, S. C. & Lee, S. G. (2015). A transcriptomic analysis of the response of the Arctic pteropod *Limacina helicina* to carbon dioxide-driven seawater acidification. *Polar Biology* 38: 1727–1740.

Kohn, A. J. (1959). The ecology of *Conus* in Hawaii. *Ecological Monographs* 29: 47–90.

Kohn, A. J. (1961a). Chemoreception in gastropod molluscs. *American Zoologist* 1: 291–308.

Kohn, A. J. (1961b). Studies on spawning behavior, egg masses and larval development in the gastropod genus *Conus*. II. Observations in the Indian Ocean during the Yale Seychelles Expedition. *Bulletin of the Bingham Oceanographic Collection, Yale University* 17: 1–51.

Kohn, A. J. (1966). Food specialization in *Conus* in Hawaii and California. *Ecology* 47: 1041–1043.

Kohn, A. J. (1970). Food habits of the gastropod *Mitra litterata* in relation to trophic structure of the intertidal marine bench community in Hawaii. *Pacific Science* 24: 483–486.

Kohn, A. J. & Nybakken, J. W. (1975). Ecology of *Conus* on eastern Indian Ocean fringing reefs: Diversity of species and resource utilization. *Marine Biology* 29: 211–234.

Kohn, A. J. (1990). Tempo and mode of evolution in Conidae. *Malacologia* 32: 55–67.

Kohn, A. J. & Peron, F. E. (1994). *Life History and Biogeography: Patterns in Conus*. Oxford, UK, Oxford University Press.

Kohn, A. J., Nishi, M. & Pernet, B. (1999). Snail spears and scimitars: A character analysis of *Conus* radular teeth. *Journal of Molluscan Studies* 65: 461–481.

Kohn, A. J. (2012). Egg size, life history, and tropical marine gastropod biogeography. *American Malacological Bulletin* 30: 163–174.

Køie, M., Karlsbakk, E. & Nylund, A. (2002). A cystophorous cercaria and metacercaria in *Antalis entalis* (L.) (Mollusca, Scaphopoda) in Norwegian waters, the larval stage of *Lecithophyllum botryophorum* (Olsson, 1868) (Digenea, Lecithasteridae). *Sarsia* 87: 302–311.

Koike, H., Takashima, C. & Kano, A. (2012). The Aleut cockle *Clinocardium nuttallii* (Conrad, 1837) from Adk-Oll, Adak Island, Alaska. pp. 167–175 *in* D. West, Hatfield, V., Wilmerding, E., Lefevre, C. & Gualtieri, L. (Eds.) *The People Before: The Geology, Paleoecology and Archaeology of Adak Island, Alaska*. Oxford, UK, Archaeopress.

Kollmann, H. A. (2014). The extinct Nerineoidea and Acteonelloidea (Heterobranchia, Gastropoda): A palaeobiological approach. *Geodiversitas* 36: 349–383.

Kolmann, M. A., Crofts, S. B., Dean, M. N., Summers, A. P. & Lovejoy, N. R. (2015). Morphology does not predict performance: Jaw curvature and prey crushing in durophagous stingrays. *Journal of Experimental Biology* 218: 3941–3949.

Komak, S., Boal, J. G., Dickel, L. & Budelmann, B. U. (2005). Behavioural responses of juvenile cuttlefish (*Sepia officinalis*) to local water movements. *Marine and Freshwater Behaviour and Physiology* 38: 117–125.

Korpelainen, H. (2004). The evolutionary processes of mitochondrial and chloroplast genomes differ from those of nuclear genomes. *Naturwissenschaften* 91: 505–518.

Korschelt, E. (1891). Über die Entwicklung von *Dreissena polymorpha* Pallas. *Sitzungsberichte der Gesellschaft Naturforschender Freunde zu Berlin* 21: 131–146.

Kosnik, M. A., Hua, Q., Kaufman, D. S. & Wüst, R. A. (2009). Taphonomic bias and time-averaging in tropical molluscan death assemblages: Differential shell half-lives in Great Barrier Reef sediment. *Paleobiology* 35: 565–586.

Kosuge, S. (1966). The family Triphoridae and its systematic position. *Malacologia* 4: 297–324.

Kosuge, T. (2007). A specimen of *Neritodryas cornea* (Mollusca, Gastropoda, Neritidae) recaptured after 8 years interval in the wild, in Yonaguni Island. *Nanki Seibutu* 49: 9–10.

Kotrla, M. B. & James, F. C. (1987). Sexual dimorphism of shell shape and growth of *Villosa villosa* (Wright) and *Elliptio icterina* (Conrad) (Bivalvia, Unionidae). *Journal of Molluscan Studies* 53: 13–23.

Kouchinsky, A. (2000). Shell microstructures in Early Cambrian molluscs. *Acta Palaeontologica Polonica* 45: 119–150.

Kouchinsky, A. V., Bengtson, S., Runnegar, B. N., Skovsted, C. B., Steiner, M. & Vendrasco, M. J. (2012). Chronology of early Cambrian biomineralization. *Geological Magazine* 149: 221–251.

Koukouras, A., Russo, A., Voultsiadou-Koukoura, E., Arvanitidis, C. & Stefanidou, D. (1996). Macrofauna associated with sponge species of different morphology. *Marine Ecology* 17: 569–582.

Kozloff, E. N. (1961). A new genus and two new species of ancistrocomid ciliates (Holotricha: Thigmotricha) from sabellid polychaetes and from a chiton. *Journal of Protozoology* 8: 60–63.

Kozloff, E. N. (1992). The genera of the phylum Orthonectida (Mesozoa). *Cahiers de Biologie Marine* 33: 377–406.

Kraeuter, J. N. (2001). Predators and predation. pp. 441–590 *in* J. N. Kraeuter & Castagna, M. (Eds.) *Biology of the Hard Clam*. Amsterdam, the Netherlands, Elsevier Science.

Kraus, D. W. & Wittenberg, J. B. (1990). Hemoglobins of the *Lucina pectinata*/bacteria symbiosis. I. Molecular properties, kinetics and equilibria of reactions with ligands. *Journal of Biological Chemistry* 265: 16043–16053.

Kreeger, D. A. (1993). Seasonal patterns in utilization of dietary protein by the mussel *Mytilus trossulus*. *Marine Ecology Progress Series* 95: 215–215.

Kresge, N., Vacquier, V. D. & Stout, C. D. (2001). Abalone lysin: The dissolving and evolving sperm protein. *BioEssays* 23: 95–103.

Kriegstein, A. R. (1977a). Stages in the post-hatchling development of *Aplysia californica*. *Journal of Experimental Zoology* 199: 275–288.

Kriegstein, A. R. (1977b). Development of the nervous system of *Aplysia californica*. *Proceedings of the National Academy of Sciences of the United States of America* 74: 375–378.

Krijgsman, B. J. (1925). Arbeitsrhythmus der Verdauungsdrüsen bei *Helix pomatia*. I. Teil: Die natürlichen Bedingungen. *Journal of Comparative Physiology A* 2: 264–296.

Krijgsman, B. J. (1928). Arbeitsrhythmus der Verdauungsdrüsen bei *Helix pomatia*: II. Teil: Sekretion, resorption und phagocytose. *Zeitschrift für vergleichende Physiologie* 8: 187–280.

Kristensen, J. H. (1972). Structure and function of crystalline styles of bivalves. *Ophelia* 10: 91–108.

Kristof, A. & Klussmann-Kolb, A. (2010). Neuromuscular development of *Aeolidiella stephanieae* Valdéz, 2005 (Mollusca, Gastropoda, Nudibranchia). *Frontiers in Zoology* 7: 5 (1–24).

Kroll, O., Hershler, R., Albrecht, C., Terrazas, E. M., Apaza, R., Fuentealba, C., Wolff, C. & Wilke, T. (2012). The endemic gastropod fauna of Lake Titicaca: Correlation between molecular evolution and hydrographic history. *Ecology and Evolution* 2: 1517–1530.

Krueger, D. M., Dubilier, N. & Cavanaugh, C. M. (1996a). Chemoautotrophic symbiosis in the tropical clam *Solemya occidentalis* (Bivalvia: Protobranchia): Ultrastructural and phylogenetic analysis. *Marine Biology* 126: 55–64.

Krueger, D. M., Gustafson, R. G. & Cavanaugh, C. M. (1996b). Vertical transmission of chemoautotrophic symbionts in the bivalve *Solemya velum* (Bivalvia: Protobranchia). *Biological Bulletin* 190: 195–202.

Krueger, K. K. (1974). The use of ultraviolet light in the study of fossil shells. *Curator* 17: 36–49.

Krug, P. J. & Manzi, A. E. (1999). Waterborne and surface-associated carbohydrates as settlement cues for larvae of the specialist marine herbivore *Alderia modesta*. *Biological Bulletin* 197: 94–103.

Krug, P. J., Ellingson, R. A., Burton, R. & Valdés, Á. (2007). A new poecilogonous species of sea slug (Opisthobranchia: Sacoglossa) from California: Comparison with the planktotrophic congener *Alderia modesta* (Lovén, 1844). *Journal of Molluscan Studies* 73: 29.

Krug, P. J. (2009). Not my 'type': Larval dispersal dimorphisms and bet-hedging in opisthobranch life histories. *Biological Bulletin* 216: 355–372.

Krug, P. J., Riffell, J. A. & Zimmer, R. K. (2009). Endogenous signaling pathways and chemical communication between sperm and egg. *Journal of Experimental Biology* 212: 1092–1100.

Krug, P. J. (2011). Patterns of speciation in marine gastropods: A review of the phylogenetic evidence for localized radiations in the sea. *American Malacological Bulletin* 29: 169–186.

Krug, P. J., Gordon, D. & Romero, M. R. (2012). Seasonal polyphenism in larval type: Rearing environment influences the development mode expressed by adults in the sea slug *Alderia willowi*. *Integrative and Comparative Biology* 52: 161–172.

Krug, P. J., Vendetti, J. E., Ellingson, R. A., Trowbridge, C. D., Hirano, Y. M., Trathen, D. Y., Rodriguez, A. K., Swennen, C., Wilson, N. G. & Valdes, Á. (2015). Species selection favors dispersive life histories in sea slugs, but higher per-offspring investment drives shifts to short-lived larvae. *Systematic Biology* 64: 983–999.

Krull, H. (1934). Die Aufhebung der Chiastoneurie bei den Pulmonaten. *Zoologischer Anzeiger* 105: 173–183.

Krylova, E. M. & Sahling, H. (2006). Recent bivalve molluscs of the genus *Calyptogena* (Vesicomyidae). *Journal of Molluscan Studies* 72: 359–395.

Kryvi, H. (1977). The fine structure of the cartilage in the annelid *Sabella penicillum*. *Protoplasma* 91: 191–200.

Kuanpradit, C., Cummins, S. F., Degnan, B. M., Sretarugsa, P., Hanna, P. J., Sobhon, P. & Chavadej, J. (2010). Identification of an attractin-like pheromone in the mucus-secreting hypobranchial gland of the abalone *Haliotis asinina* Linnaeus. *Journal of Shellfish Research* 29: 699–704.

Kuanpradit, C., Stewart, M. J., York, P. S., Degnan, B. M., Sobhon, P., Hanna, P. J., Chavadej, J. & Cummins, S. F. (2012). Characterization of mucus-associated proteins from abalone (*Haliotis*): Candidates for chemical signaling. *Federation of European Biochemical Societies (FEBS) Journal* 279: 437–450.

Kubanek, J., Graziani, E. I. & Andersen, R. J. (1997). Investigations of terpenoid biosynthesis by the dorid nudibranch *Cadlina luteomarginata*. *Journal of Organic Chemistry* 62: 7239–7246.

Kubanek, J. & Andersen, R. J. (1999). Evidence for *de novo* biosynthesis of the polyketide fragment of diaulusterol A by the northeastern Pacific dorid nudibranch *Diaulula sandiegensis*. *Journal of Natural Products* 62: 777–779.

Kubanek, J., Faulkner, D. J. & Andersen, R. J. (2000). Geographic variation and tissue distribution of endogenous terpenoids in the northeastern Pacific dorid nudibranch *Cadlina marginata*: Implications for the regulation of *de novo* biosynthesis. *Journal of Chemical Ecology* 26: 377–389.

Kubo, M. (1977). The formation of a temporary-acrosome in the spermatozoon of *Laternula limicola* (Bivalvia, Mollusca). *Journal of Ultrastructure Research* 61: 140–148.

Kubodera, T. & Mori, K. (2005). First-ever observations of a live giant squid in the wild. *Proceedings of the Royal Society of London B* 272: 2583–2586.

Kubota, S. (2000). Parallel, paedomorphic evolutionary processes of the bivalve-inhabiting hydrozoans (Leptomedusae, Eirenidae) deduced from the morphology, life cycle and biogeography, with special reference to taxonomic treatment of *Eugymnanthea*. *Scientia Marina* 64: 241–247.

Kubota, S. (2004). Some new and reconfirmed biological observations in two species of *Eugymnanthea* (Hydrozoa, Leptomedusae, Eirenidae) associated with bivalves. *Biogeography* 6: 1–5.

Kuchel, R. P., Raftos, D. A., Birch, D. & Vella, N. (2010). Haemocyte morphology and function in the Akoya pearl oyster, *Pinctada imbricata*. *Journal of Invertebrate Pathology* 105: 36–48.

Kuchel, R. P., O'Connor, W. A. & Raftos, D. A. (2011). Environmental stress and disease in pearl oysters, focusing on the Akoya pearl oyster (*Pinctada fucata* Gould 1850). *Reviews in Aquaculture* 3: 138–154.

Kudo, R. R. (1954). *Protozoology*. 4th ed. Springfield, IL, Charles C. Thomas.

Kuga, H. & Matsuno, A. (1988). Ultrastructural investigations on the anterior adductor muscle of a Brachiopoda, *Lingula unguis*. *Cell Structure and Function* 13: 271–279.

Kuhajek, J. M. & Schlenk, D. (2003). Effects of the brominated phenol, lanosol, on cytochrome P-450 and glutathione transferase activities in *Haliotis rufescens* and *Katharina tunicata*. *Comparative Biochemistry and Physiology Part C* 134: 473–479.

Kuhara, T., Kano, Y., Yoshikoshi, K. & Hashimoto, J. (2014). Shell morphology, anatomy and gill histology of the deep-sea bivalve *Elliptiolucina ingens* and molecular phylogenetic reconstruction of the chemosynthetic family Lucinidae. *Venus* 72: 13–27.

Kulicki, C. & Doguzhaeva, L. A. (1994). Development and calcification of the ammonitella shell. *Acta Palaeontologica Polonica* 39: 17–44.

Kunigelis, S. C. & Saleuddin, A. S. M. (1986). Reproduction in the freshwater gastropod, *Helisoma*: Involvement of prostaglandin in egg production. *International Journal of Invertebrate Reproduction* 10: 159–167.

Künz, E. & Haszprunar, G. (2001). Comparative ultrastructure of gastropod cephalic tentacles: Patellogastropoda, Neritaemorphi and Vetigastropoda. *Zoologischer Anzeiger* 240: 137–165.

Kupfermann, I. & Kandel, E. (1970). Electrophysiological properties and functional interconnections of two symmetrical neurosecretory clusters (bag cells) in abdominal ganglion of *Aplysia*. *Journal of Neurophysiology* 33: 865–876.

Kupfermann, I., Carew, T. J. & Kandel, E. R. (1974). Local, reflex, and central commands controlling gill and siphon movements in *Aplysia*. *Journal of Neurophysiology* 37: 996–1019.

Kupriyanova, E. & Havenhand, J. N. (2002). Variation in sperm swimming behaviour and its effect on fertilization success in the serpulid polychaete *Galeolaria caespitosa*. *Invertebrate Reproduction & Development* 41: 21–26.

Kurabayashi, A. & Ueshima, R. (2000). Complete sequence of the mitochondrial DNA of the primitive opisthobranch gastropod *Pupa strigosa*: Systematic implication of the genome organization. *Molecular Biology and Evolution* 17: 266–277.

Kurahashi, H. (1970). Tribe Calliphorini from Australian and Oriental Regions. I. Melinda-Group (Diptera: Calliphoridae). *Pacific Insects* 12: 519–542.

Kurihara, H., Kato, S. & Ishimatsu, A. (2007). Effects of increased seawater pCO2 on early development of the oyster *Crassostrea gigas*. *Aquatic Biology* 1: 91–98.

Kurihara, H. (2008). Effects of CO_2-driven ocean acidification on the early developmental stages of invertebrates. *Marine Ecology Progress Series* 373: 275–284.

Kuris, A. M. & Culver, C. S. (1999). An introduced sabellid polychaete pest infesting cultured abalones and its potential spread to other California gastropods. *Invertebrate Biology* 118: 391–403.

Kurita, Y. & Wada, H. (2011). Evidence that gastropod torsion is driven by asymmetric cell proliferation activated by TGF-β signalling. *Biology Letters* 7: 759–762.

Kurita, Y., Hashimoto, N. & Wada, H. (2016). Evolution of the molluscan body plan: The case of the anterior adductor muscle of bivalves. *Biological Journal of the Linnean Society* 119: 420–429.

Kuroda, R., Endo, B., Abe, M. & Shimizu, M. (2009). Chiral blastomere arrangement dictates zygotic left-right asymmetry pathway in snails. *Nature* 462: 790–794.

Kuroda, R. (2015). A twisting story: How a single gene twists a snail? Mechanogenetics. *Quarterly Reviews of Biophysics* 48: 445–452.

Kuroda, S., Kunita, I., Tanaka, Y., Ishiguro, A., Kobayashi, R. & Nakagaki, T. (2014). Common mechanics of mode switching in locomotion of limbless and legged animals. *Journal of the Royal Society Interface* 11: 20140205.

Kuwamura, T., Fukao, R., Nishida, M., Wada, K. & Yanagisawa, Y. (1983). Reproductive biology of the gastropod *Strombus luhuanus* (Strombidae). *Publications of the Seto Marine Biological Laboratory* 28: 433–443.

Kuzirian, A. M., Alkon, D. L. & Harris, L. G. (1981). An infraciliary network in statocyst hair cells. *Journal of Neurocytology* 10: 497–514.

Kuzmina, T. V. & Malakhov, V. V. (2015). The accessory hearts of the articulate brachiopod *Hemithyris psittacea*. *Zoomorphology* 134: 25–32.

Kvitek, R. & Bretz, C. (2004). Harmful algal bloom toxins protect bivalve populations from sea otter predation. *Marine Ecology Progress Series* 271: 233–243.

Kvitek, R. & Bretz, C. (2005). Shorebird foraging behavior, diet, and abundance vary with harmful algal bloom toxin concentrations in invertebrate prey. *Marine Ecology Progress Series* 293: 303–309.

Kvitek, R. G. (1991a). Sequestered paralytic shellfish poisoning toxins mediate glaucous-winged gull predation on bivalve prey. *Auk* 108: 381–392.

Kvitek, R. G. (1991b). Paralytic shellfish toxins sequestered by bivalves as a defense against siphon-nipping fish. *Marine Biology* 111: 369–374.

Kvitek, R. G. & Beitler, M. K. (1991). Relative insensitivity of butter clam neurons to saxitoxin: A pre-adaptation for sequestering paralytic shellfish poisoning toxins as a chemical defense. *Marine Ecology Progress Series* 69: 47–54.

Kvitek, R. G., Degange, A. R. & Beitler, M. K. (1991). Paralytic shellfish poisoning toxins mediate sea otter feeding behavior. *Limnology and Oceanography* 36: 393–404.

Kwong, K. L., Chan, R. K. Y. & Qiu, J.-W. (2009). The potential of the invasive snail *Pomacea canaliculata* as a predator of various life-stages of five species of freshwater snails. *Malacologia* 51: 343–356.

Kyle, C. J. & Boulding, E. G. (1998). Molecular genetic evidence for parallel evolution in a marine gastropod, *Littorina subrotundata*. *Proceedings of the Royal Society B* 265: 303–308.

Kyle, C. J. & Boulding, E. G. (2000). Comparative population genetic structure of marine gastropods (*Littorina* spp.) with and without pelagic larval dispersal. *Marine Biology* 137: 835–845.

LaBella, A. L., Van Dover, C. L., Jollivet, D. & Cunningham, C. W. (2017). Gene flow between Atlantic and Pacific Ocean basins in three lineages of deep-sea clams (Bivalvia: Vesicomyidae: Pliocardiinae) and subsequent limited gene flow within the Atlantic. *Deep-Sea Research Part II -Topical Studies in Oceanography* 137: 307–317.

de Lacaze-Duthiers, F. J. H. (1856–1858). Histoire de l'organization et du développment du Dentale. I–II. *Annales des sciences naturelles. Zoologie* 6: 225–228, 319–385.

de Lacaze-Duthiers, F. J. H. (1872). Du système nerveux des Mollusques gastéropodes pulmonés aquatiques et d'un nouvel organe d'innervation. *Archives de zoologie expérimentale et générale* 1: 437–500.

Lacchini, A. H., Davies, A. J., Mackintosh, D. & Walker, A. J. (2006). β-1, 3-glucan modulates PKC signalling in *Lymnaea stagnalis* defence cells: A role for PKC in H_2O_2 production and downstream ERK activation. *Journal of Experimental Biology* 209: 4829–4840.

Laffy, P. W., Benkendorff, K. & Abbott, C. A. (2013). Suppressive subtractive hybridisation transcriptomics provides a novel insight into the functional role of the hypobranchial gland in a marine mollusc. *Comparative Biochemistry and Physiology Part D* 8: 111–122.

LaFont, R. (2000). The endocrinology of invertebrates. *Ecotoxicology* 9: 41–57.

LaFont, R. & Mathieu, M. (2007). Steroids in aquatic invertebrates. *Ecotoxicology* 16: 109–130.

LaForge, N. L. & Page, L. R. (2007). Development in *Berthella californica* (Gastropoda: Opisthobranchia) with comparative observations on phylogenetically relevant larval characters among nudipleuran opisthobranchs. *Invertebrate Biology* 126: 318–334.

Lagadic, L., Coutellec, M. A. & Caquet, T. (2007). Endocrine disruption in aquatic pulmonate molluscs: Few evidences, many challenges. *Ecotoxicology* 16: 45–59.

Laihonen, P. & Furman, E. R. (1986). The site of settlement indicates commensalism between blue mussel and its epibiont. *Oecologia* 71: 38–40.

Lalli, C. M. & Gilmer, R. W. (1989). *Pelagic Snails: The Biology of Holoplanktonic Gastropod Mollusks.* Stanford, CA, Stanford University Press.

Lamarck, J. B. P. A. (1818). *Histoire Naturelle des Animaux Sans Vertèbres, Présentant les Caractères Généraux et Particuliers de ces Animaux, leur Distribution, leurs Classes, leurs Familles, leurs Genres, et la Citation des Principales Espèces qui s'y rapportent; Précédée d'une Introduction offrant la Détermination des Caractères Essentiels de l'Animal, sa Distinction du Végétal et des Autres Corps Naturelles, enfin, l'Exposition des Principes Fondamentaux de la Zoologie.* Vol. 5. Paris, Verdière.

Lamb, E. J., Boxshall, G. A., Mill, P. J. & Grahame, J. (1996). Nucellicolidae: A new family of endoparasitic copepods (Poecilostomatoida) from the dog whelk *Nucella lapillus* (Gastropoda). *Journal of Crustacean Biology* 16: 142–148.

Lambert, J. D. & Nagy, L. M. (2002). Asymmetric inheritance of centrosomally localized mRNAs during embryonic cleavages. *Nature* 420: 682–686.

Lambert, J. D. (2008). Mesoderm in spiralians: The organizer and the 4d cell. *Journal of Experimental Zoology Part B* 310: 15–23.

Lambert, J. D., Chan, X. Y., Spiecker, B. & Sweet, H. C. (2010). Characterizing the embryonic transcriptome of the snail *Ilyanassa*. *Integrative and Comparative Biology* 50: 768–777.

Lambert, T. C. & Farley, J. (1968). The effect of parasitism by the trematode *Cryptocotyle lingua* (Creplin) on zonation and winter migration of the common periwinkle, *Littorina littorea* (L.). *Canadian Journal of Zoology* 46: 1139–1147.

Lambert, W. J. (1991). Coexistence of hydroid eating nudibranchs: Do feeding biology and habitat use matter? *Biological Bulletin* 181: 248–260.

Lambrix, P., Tan, H., Jakoniene, V. & Strömbäck, L. (2007). Biological ontologies. pp. 85–99 in C. J. O. Baker & Cheung, K.-H. (Eds.) *Semantic Web: Revolutionizing Knowledge Discovery in the Life Sciences*. New York, Springer.

Lamoral, B. H. (1971). Predation on terrestrial molluscs by scorpions in the Kalahari Desert. *Annals of the Natal Museum* 21: 17–20.

Lampert, W. & Sommer, U. (2007). *Limnoecology: The Ecology of Lakes and Streams*. Oxford, UK, Oxford University Press.

Lamprell, K. L. & Healy, J. M. (1998). A revision of the Scaphopoda from Australian waters (Mollusca). *Records of the Australian Museum*, Supplement 24: 1–189.

Lamptey, E. & Armah, A. K. (2008). Factors affecting macrobenthic fauna in a tropical hypersaline coastal lagoon in Ghana, West Africa. *Estuaries and Coasts* 31: 1006–1019.

Lance, J. R. (1962). Two new opisthobranch mollusks from southern California. *The Veliger* 4: 155–159.

Land, M. F. (1968). Functional aspects of the optical and retinal organisation of the mollusc eye. pp. 75–96 in J. D. Carthy & Newell, G. E. (Eds.) *Invertebrate Receptors. Proceedings of a Symposium held at the Zoological Society of London on 30 and 31 May 1967*. New York, Academic Press.

Land, M. F. (1981). Optics and vision in invertebrates. pp. 471–592 in H. Autrum (Ed.) *Comparative Physiology and Evolution of Vision in Invertebrates. Handbook of Sensory Physiology*. Vol.VII/6B. Berlin, Springer.

Land, M. F. (1982). Scanning eye movements in a heteropod mollusc. *Journal of Experimental Biology* 96: 427–430.

Land, M. F. (1984). Molluscs. pp. 699–725 in M. A. Ali (Ed.) *Photoreception and Vision in Invertebrates. NATO Advanced Science Institutes Series. Series A, Life Sciences*. New York, Plenum Press.

Land, M. F. (2003). The spatial resolution of the pinhole eyes of giant clams (*Tridacna maxima*). *Proceedings of the Royal Society B* 270: 185–188.

Land, M. F. (2012). The evolution of lenses. *Ophthalmic and Physiological Optics* 32: 449–460.

Land, M. F. & Nilsson, D. E. (2012). *Animal Eyes*. Oxford, UK, Oxford University Press.

Landau, M., Biggers, W. J. & Laufer, H. (1997). Invertebrate endocrinology. pp. 1291–1388 in W. H. Dantzler (Ed.) *Handbook of Physiology Section 13: Comparative Physiology*. Vol. 2. New York, Oxford University Press.

Landman, N. H. (1988). Heterochrony in ammonites. pp. 159–182 in M. L. McKinney (Ed.) *Heterochrony in Evolution: A Multidisciplinary Approach. Topics in Geobiology*. New York, Springer Science + Business Media.

Landman, N. H., Mikkelsen, P. M., Bieler, R. & Bronson, B. (2001). *Pearls: A Natural History*. New York, Harry N. Abrams.

Landman, N. H., Plint, A. G. & Walaszczyk, I. (2017). Chapter 3: Scaphitid Ammonites from the Upper Cretaceous (Coniacian-Santonian) Western Canada Foreland Basin. *Bulletin of the American Museum of Natural History* 414: 105–172.

Lane, H. S., Webb, S. C. & Duncan, J. (2016). *Bonamia ostreae* in the New Zealand oyster *Ostrea chilensis*: A new host and geographic record for this haplosporidian parasite. *Diseases of Aquatic Organisms* 118: 55–63.

Lane, N. J. (1964). Semper's organ, a cephalic gland in certain gastropods. *Quarterly Journal of Microscopical Science* 105: 331–342.

Lane, N. J. (1966). The fine-structural localization of phosphatases in neurosecretory cells within the ganglia of certain gastropod snails. *American Zoologist* 6: 139–158.

Lang, A. (1900). *Lehrbuch der Vergleichenden Anatomie der Wirbellosen Thiere*. Jena, Fischer.

Lang, K. (1954). On a new orthonectid, *Rhopalura philinae* n. sp., found as a parasite in the opisthobranch *Philine scabra* Müller. *Arkiv för Zoologi* 6: 603–610.

Lang, R. C., Britton, J. C. & Metz, T. (1998). What to do when there is nothing to do: The ecology of Jamaican intertidal Littorinidae (Gastropoda; Prosobranchia) in repose. *Hydrobiologia* 378: 161–185.

Lang, W. (1932). Synbiosen von *Serpula* mit Ammoniten im unteren Lias Norddeutschlands. *Zeitschrift der Deutschen Gesellschaft* 84: 229–234.

Langston, W. J., Bebianno, M. J. & Burt, G. R. (1998). Metal handling strategies in molluscs. pp. 219–283 in W. J. Langston & Bebianno, M. J. (Eds.) *Metal Metabolism in Aquatic Environments*. London, UK, Chapman & Hall.

Langston, W. J., Burt, G. R. & Chesman, B. S. (2007). Feminisation of male clams *Scrobicularia plana* from estuaries in Southwest UK and its induction by endocrine-disrupting chemicals. *Marine Ecology Progress Series* 333: 173–184.

Langton, R. W. & Gabbott, P. A. (1974). The tidal rhythm of extracellular digestion and the response to feeding in *Ostrea edulis*. *Marine Biology* 24: 181–187.

Langton, R. W. (1975). Synchrony in the digestive diverticula of *Mytilus edulis* L. *Journal of the Marine Biological Association of the United Kingdom* 55: 221–229.

Lankester, E. R. (1883). Mollusca. pp. 632–695. *Encyclopaedia Britannica*. London, Encyclopaedia Britannica.

Lannig, G., Eilers, S., Pörtner, H. O., Sokolova, I. M. & Bock, C. (2010). Impact of ocean acidification on energy metabolism of oyster, *Crassostrea gigas*: Changes in metabolic pathways and thermal response. *Marine Drugs* 8: 2318–2339.

Lapan, E. A. (1975). Inositol polyphosphate deposits in the dense bodies of mesozoan dispersal larvae. *Experimental Cell Research* 83: 143–151.

Laporta-Ferreira, I. L. & Da Graca Salomäo, M. (2004). Reptilian predators of terrestrial gastropods. pp. 427–481 in G. M. Barker (Ed.) *Natural Enemies of Terrestrial Molluscs*. Oxford, UK, Cambridge, MA, CABI Publishing.

Laptikhovsky, V., Nikolaeva, S. & Rogov, M. (2018). Cephalopod embryonic shells as a tool to reconstruct reproductive strategies in extinct taxa. *Biological Reviews* 93: 270–283.

Lartillot, N., Le Gouar, M. & Adoutte, A. (2002). Expression patterns of *fork head* and *goosecoid* homologues in the mollusc *Patella vulgata* supports the ancestry of the anterior mesendoderm across Bilateria. *Development Genes and Evolution* 212: 551–561.

Lau, D. C. P. & Leung, K. M. Y. (2004). Feeding physiology of the carnivorous gastropod *Thais clavigera* (Küster): Do they eat 'soup'? *Journal of Experimental Marine Biology and Ecology* 312: 43–66.

Lauckner, G. (1980). Diseases of Mollusca: Gastropoda. pp. 311–424 in O. Kinne (Ed.) *Diseases of Marine Animals. General Aspects, Protozoa to Gastropoda*. Chichester, UK, John Wiley & Sons.

Lauckner, G. (1983a). Diseases of Mollusca: Scaphopoda. pp. 979–983 *in* O. Kinne (Ed.) *Diseases of Marine Animals. Introduction, Bivalvia to Scaphopoda.* Vol. 2. Hamburg, Biologische Anstalt Helgoland.

Lauckner, G. (1983b). Diseases of Mollusca: Bivalvia. pp. 477–961 *in* O. Kinne (Ed.) *Diseases of Marine Animals. Introduction, Bivalvia to Scaphopoda.* Vol. 2. Hamburg, Biologische Anstalt Helgoland.

Lauckner, G. (1983c). Diseases of Mollusca: Amphineura. pp. 963–977 *in* O. Kinne (Ed.) *Diseases of Marine Animals. Introduction, Bivalvia to Scaphopoda.* Vol. 2. Hamburg, Biologische Anstalt Helgoland.

Laursen, D. (1953). The genus *Ianthina*, a monograph. *Dana-Report* 38: 1–40.

Laursen, J. R., di Liu, H., Wu, X.-J. & Yoshino, T. P. (1997). Heat-Shock response in a molluscan cell line: Characterization of the response and cloning of an inducible HSP70 cDNA. *Journal of Invertebrate Pathology* 70: 226–233.

Lavm, B. & Nevo, E. (1981). Genetic diversity in marine molluscs: A test of the niche-width variation hypothesis. *Marine Ecology* 2: 335–342.

Laws, H. M. (1971). Reproductive biology and shell site preference in *Hipponix conicus* (Schumacher). *The Veliger* 13: 115–121.

Laxton, J. H. (1971). Feeding in some Australasian Cymatiidae (Gastropoda: Prosobranchia). *Zoological Journal of the Linnean Society* 50: 1–9.

Lazareth, C. E., Lasne, G. & Ortlieb, L. (2006). Growth anomalies in *Protothaca thaca* (Mollusca, Veneridae) shells as markers of ENSO conditions. *Climate Research* 30: 263–269.

Le Gall, S. & Streiff, W. (1975). Protandric hermaphroditism in prosobranch gastropods. pp. 170–178 *in* R. Reinboth (Ed.) *Symposium on Intersexuality in the Animal Kingdom.* Berlin, New York, Springer-Verlag.

Le Gall, S. (1981). Étude expérimentale du facteur morphogénétique contrôlant la différenciation du tractus génital mâle externe chez *Crepidula fornicata* L. (Mollusque hermaphrodite protandre). *General and Comparative Endocrinology* 43: 51–62.

Le Gall, S., Le Gall, P. & Féral, C. (1983). The neuroendocrine mechanism responsible for penis differentiation in *Crepidula fornicata*. pp. 169–173 *in* J. Lever & Boer, H. H. (Eds.) *Molluscan Neuroendocrinology. Proceedings of the International Minisymposium on Molluscan Endocrinology, held in the Department of Biology, Free University, Amsterdam, August 16–20, 1982.* Amsterdam, the Netherlands, New York, North Holland Publishing Co.

Le Guyader, F. S. & Atmar, R. L. (2007). Viruses in shellfish. pp. 205–226 *in* A. Bosch (Ed.) *Human Viruses in Water. Perspectives in Medical Virology.* Amsterdam, the Netherlands, Elsevier.

Le Guyader, F. S., Le Saux, J.-C., Ambert-Balay, K., Krol, J., Serais, O., Parnaudeau, S., Giraudon, H., Delmas, G., Pommepuy, M. & Pothier, P. (2008). Aichi virus, norovirus, astrovirus, enterovirus, and rotavirus involved in clinical cases from a French oyster-related gastroenteritis outbreak. *Journal of Clinical Microbiology* 46: 4011–4017.

Le Pabic, C., Safi, G., Serpentini, A., Lebel, J. M., Robin, J. P. & Koueta, N. (2014). Prophenoloxidase system, lysozyme and protease inhibitor distribution in the common cuttlefish *Sepia officinalis*. *Comparative Biochemistry and Physiology Part B* 172: 96–104.

Le Pennec, M., Beninger, P. G. & Herry, A. (1995). Feeding and digestive adaptations of bivalve molluscs to sulphide-rich habitats. *Comparative Biochemistry and Physiology Part A* 111: 183–189.

Le Quesne, W. J. F. & Hawkins, S. J. (2006). Direct observations of protandrous sex change in the patellid limpet *Patella vulgata*. *Journal of the Marine Biological Association of the United Kingdom* 86: 161–162.

Leal, J. H. & Bouchet, P. (1991). Distribution patterns and dispersal of prosobranch gastropods along a seamount chain in the Atlantic Ocean. *Journal of the Marine Biological Association of the United Kingdom* 71: 11–25.

Lebour, M. V. (1937). The eggs and larvae of the British prosobranchs with special reference to those living in the plankton. *Journal of the Marine Biological Association of the United Kingdom* 22: 105–166.

Lechanteur, R. G. & Prochazka, K. (2001). Feeding biology of the giant clingfish *Chorisochismus dentex*: Implications for limpet populations. *African Zoology* 36: 79–86.

Lee, B. P., Messersmith, P. B., Israelachvili, J. N. & Waite, J. H. (2011). Mussel-inspired adhesives and coatings. *Annual Review of Materials Research* 41: 99–132.

Lee, H. O., Davidson, J. M. & Duronio, R. J. (2009). Endoreplication: Polyploidy with purpose. *Genes & Development* 23: 2461–2477.

Lee, J. S., Joo, J. A. Y. & Park, J. J. (2007). Histology and ultrastructure of the mantle epidermis of the Equilateral Venus, *Gomphina veneriformis* (Bivalvia: Veneridae). *Journal of Shellfish Research* 26: 413–421.

Lee, J. S., Lee, Y. G., Park, J. J. & Shin, Y. K. (2012). Microanatomy and ultrastructure of the foot of the infaunal bivalve *Tegillarca granosa* (Bivalvia: Arcidae). *Tissue and Cell* 44: 316–324.

Lee, O. H. K. & Williams, G. A. (2002). Locomotor activity patterns of the mangrove littorinids, *Littoraria ardouiniana* and *L. melanostoma*, in Hong Kong. *Journal of Molluscan Studies* 68: 235–241.

Lee, P. G., Lu, L. J. W., Salazar, J. J. & Holoubek, V. (1994). Absence of formation of benzo pyrene/DNA adducts in the cuttlefish *Sepia officinalis*, (Mollusca: Cephalopoda). *Environmental and Molecular Mutagenesis* 23: 70–73.

Lee, P. N., Callaerts, P., de Couet, H. G., & Martindale, M. Q. (2003a). Cephalopod Hox genes and the origin of morphological novelties. *Nature* 424: 1061–1065.

Lee, R. W., Robinson, J. J. & Cavanaugh, C. M. (1999). Pathways of inorganic nitrogen assimilation in chemoautotrophic bacteria-marine invertebrate symbioses: Expression of host and symbiont glutamine synthetase. *Journal of Experimental Biology* 202: 289–300.

Lee, S. E. & Jacobs, D. K. (1998). Expression of distal-less in molluscan gills: An evolutionary relationship between limbs and gills? *Developmental Biology* 198: 202.

Lee, S. E., Gates, R. D. & Jacobs, D. K. (2003b). Gene fishing: The use of a simple protocol to isolate multiple homeodomain classes from diverse invertebrate taxa. *Journal of Molecular Evolution* 56: 509–516.

Lee, T. (1999). Polyploidy and meiosis in the freshwater clam *Sphaerium striatinum* (Lamarck) and chromosome numbers in the Sphaeriidae (Bivalvia, Veneroida). *Cytologia (Tokyo)* 64: 247–252.

Lee, Y. S., Yang, H. O., Shin, K. H., Choi, H. S., Jung, S. H., Kim, Y. M., Oh, D. K., Linhardt, R. J. & Kim, Y. S. (2003c). Suppression of tumor growth by a new glycosaminoglycan isolated from the African giant snail *Achatina fulica*. *European Journal of Pharmacology* 465: 191–198.

Van Leeuwen, J. L. & Kier, W. M. (1997). Functional design of tentacles in squid: Linking sarcomere ultrastructure to gross morphological dynamics. *Philosophical Transactions of the Royal Society B* 352: 551–571.

Leggat, W., Buck, B. H., Grice, A. & Yellowlees, D. (2003). The impact of bleaching on the metabolic contribution of dinoflagellate symbionts to their giant clam host. *Plant Cell and Environment* 26: 1951–1961.

Lehoux, J. & Williams, E. (1971). Metabolism of progesterone by gonadal tissue of *Littorina littorea* (L.) (Prosobranchia, Gastropoda). *Journal of Endocrinology* 51: 411–412.

Leighton, D. L. (1961). Observations on the effect of diet on shell coloration on the red abalone *Haliotis rufescens* Swainson. *The Veliger* 4: 29–32.

Lemche, H. & Wingstrand, K. G. (1959). The anatomy of *Neopilina galatheae* Lemche, 1957 (Mollusca, Tryblidiacea). *Galathea Report* 3: 9–71.

Lemouellic, S., Chauvet, C., Hoang, D. H., Tuyen, H. T., Lu, H. D., Foale, S., Teitelbaum, A., Gereva, S. & Friedman, K. (2008). *Trochus niloticus* (Linnae 1767) growth in Wallis Island. *SPC Trochus Information Bulletin* 14: 2–6.

Leonard, G. H., Bertness, M. D. & Yund, P. O. (1999). Crab predation, waterborne cues, and inducible defenses in the blue mussel, *Mytilus edulis. Ecology* 80: 1–14.

Leonard, J. L. (1990). The hermaphrodite's dilemma. *Journal of Theoretical Biology* 147: 361–371.

Leonard, J. L. (1991). Sexual conflict and the mating systems of simultaneously hermaphroditic gastropods. *American Malacological Bulletin* 9: 45–58.

Leonard, J. L. & Lukowiak, K. (1991). Sex and the simultaneous hermaphrodite: Testing models of male-female conflict in a sea slug, *Navanax inermis* (Opisthobranchia). *Animal Behaviour* 41: 255–266.

Leonard, J. L. (1992). The 'Love-dart' in helicid snails: A gift of calcium or a firm commitment? *Journal of Theoretical Biology* 159: 513–521.

Leonard, J. L., Pearse, J. S. & Harper, A. B. (2002). Comparative reproductive biology of *Ariolimax californicus* and *A. dolichophallus* (Gastropoda; Stylommatophora). *Invertebrate Reproduction & Development* 41: 83–93.

Leonard, J. L. (2006). Sexual selection: Lessons from hermaphrodite mating systems. *Integrative and Comparative Biology* 46: 349–367.

Leonard, J. L., Westfall, J. A. & Pearse, J. S. (2007). Phally polymorphism and reproductive biology in *Ariolimax* (*Ariolimax*) *buttoni* (Pilsbry and Vanatta, 1896) (Stylommatophora: Arionidae). *American Malacological Bulletin* 23: 121–135.

Lesel, M., Charrier, M. & Lesel, R. (1990). Some characteristics of the bacterial flora housed by the brown garden snail, *Helix aspersa* (Gastropoda Pulmonata). Preliminary results. pp. 149–152 in R. Lesel (Ed.) *Microbiology in Poecilotherms: Proceedings of the International Symposium on Microbiology in Poecilotherms. Paris, 10–12 July 1989.* Amsterdam, Elsevier Science.

Lesoway, M. P., Abouheif, E. & Collin, R. (2015). The development of viable and nutritive embryos in the direct developing gastropod *Crepidula navicella. International Journal of Developmental Biology* 58: 601–611.

Lesoway, M. P., Abouheif, E. & Collin, R. (2016). Comparative transcriptomics of alternative developmental phenotypes in a marine gastropod. *Journal of Experimental Zoology Part B* 326: 151–167.

Lespinet, O., Nederbragt, A. J., Cassan, M., Dictus, W., van Loon, A. E. & Adoutte, A. (2002). Characterisation of two snail genes in the gastropod mollusc *Patella vulgata*. Implications for understanding the ancestral function of the snail-related genes in Bilateria. *Development Genes and Evolution* 212: 186–195.

Lesser, W. & Greenberg, M. J. (1993). Cardiac regulation by endogenous SCPs and FMRFamide-related peptides in the snail *Helix aspersa. Journal of Experimental Biology* 178: 205–230.

Leung, K. M. Y. & Furness, R. W. (1999). Induction of metallothionein in dogwhelk *Nucella lapillus* during and after exposure to cadmium. *Ecotoxicology and Environmental Safety* 43: 156–164.

Levasseur, M., Couture, J. Y., Weise, A. M., Michaud, S., Elbrächter, M., Sauvé, G. & Bonneau, E. (2003). Pelagic and epiphytic summer distributions of *Prorocentrum lima* and *P. mexicanum* at two mussel farms in the Gulf of St. Lawrence, Canada. *Aquatic Microbial Ecology* 30: 283–293.

Lever, J., Boer, H. H., Duiven, R. J. T., Lammens, J. J. & Wattel, J. (1959). Some observations on follicle glands in pulmonates. *Proceedings of the Koninklijke Nederlandse Akademie van Wetenschappen. Series C* 62: 139–144.

Lever, J. & Bekius, R. (1965). On the presence of an external hemal pore in *Lymnaea stagnalis* L. *Experientia* 21: 395–396.

Lever, J. (1979). On torsion in gastropods. pp. 5–23 *in* S. van der Spoel, van Bruggen, A. C., & Lever, J. (Eds.) *Pathways in Malacology. 6th International Congress of Unitas Malacologica Europaea, 15–20 August, 1977.* Utrecht and The Hague, Bohn, Scheltema and Holkema.

Lever, J. & Boer, H. H., Eds. (1983). *Molluscan neuroendocrinology.* Proceedings of the International Mini-Symposium on Molluscan Endocrinology, held in the Department of Biology, Free University, Amsterdam August 16–20, 1982. Amsterdam, New York, North Holland Publishing Co.

Levin, L. A. (2003). Oxygen minimum zone benthos: Adaptation and community response to hypoxia. *Oceanography and Marine Biology Annual Review* 41: 1–45.

Levin, M. (2005). Left-right asymmetry in embryonic development: A comprehensive review. *Mechanisms of Development* 122: 3–25.

Levin, S. A. & Lubchenco, J. (2008). Resilience, robustness, and marine ecosystem-based management. *Bioscience* 58: 27–32.

Levine, T. D., Lang, B. K. & Berg, D. J. (2009). Parasitism of mussel gills by dragonfly nymphs. *American Midland Naturalist* 162: 1–6.

Levinton, J. S. (1971). Control of tellinacean (Mollusca: Bivalvia) feeding behavior by predation. *Limnology and Oceanography* 16: 660–662.

Levinton, J. S. & Suchanek, T. H. (1978). Geographic variation, niche breadth and genetic differentiation at different geographic scales in the mussels *Mytilus californianus* and *M. edulis. Marine Biology* 49: 363–375.

Levinton, J. S. (1979). Effect of density upon deposit-feeding populations: Movement, feeding and floating of *Hydrobia ventrosa* Montagu (Gastropoda: Prosobranchia). *Oecologia* 43: 27–39.

Levinton, J. S. (1991). Variable feeding behavior in three species of *Macoma* (Bivalvia, Tellinacea) as a response to water-flow and sediment transport. *Marine Biology* 110: 375–385.

Levitan, D. R. (1993). The importance of sperm limitation to the evolution of egg size in marine invertebrates. *American Naturalist* 141: 517–536.

Levitan, D. R. (1995). The ecology of fertilization in free-spawning invertebrates. pp. 123–156 *in* L. R. McEdward (Ed.) *Ecology of Marine Invertebrate Larvae.* Boca Raton, FL, CRC Press.

Levitan, D. R. & Petersen, C. (1995). Sperm limitation in the sea. *Trends in Ecology & Evolution* 10: 228–231.

Levitan, D. R. (1998). Sperm limitation, gamete competition, and sexual selection in external fertilizers. pp. 175–218 *in* T. R. Birkhead & Møller, A. P. (Eds.) *Sperm Competition and Sexual Selection*. San Diego, CA, Academic Press.

Levitan, D. R. (2000). Sperm velocity and longevity trade off each other and influence fertilization in the sea urchin *Lytechinus variegatus*. *Proceedings of the Royal Society B* 267: 531–534.

Levitan, D. R. (2006). The relationship between egg size and fertilization success in broadcast-spawning marine invertebrates. *Integrative and Comparative Biology* 46: 298–311.

Levri, E. P. & Lively, C. M. (1996). The effects of size, reproductive condition, and parasitism on foraging behaviour in a freshwater snail, *Potamopyrgus antipodarum*. *Animal Behaviour* 51: 891–901.

Levri, E. P. (1999). Parasite-induced change in host behavior of a freshwater snail: Parasitic manipulation or byproduct of infection? *Behavioral Ecology* 10: 234–241.

Levy, M., Blumberg, S. & Susswein, A. J. (1997). The rhinophores sense pheromones regulating multiple behaviors in *Aplysia fasciata*. *Neuroscience Letters* 225: 113–116.

Lewbart, G. (2006). *Invertebrate Medicine*. Chichester, UK, Wiley-Blackwell.

Lewin, R. A. (1970). Toxin secretion and tail autotomy by irritated *Oxynoe panamensis* (Opisthobranchiata; Sacoglossa). *Pacific Science* 24: 356–358.

Lewin, R. A. (1986). Supply-side ecology. *Science* 234: 25–27.

Lewis, R. J., Dutertre, S., Vetter, I. & Christie, M. J. (2012). *Conus* venom peptide pharmacology. *Pharmacological Reviews* 64: 259–298.

Lewy, Z. & Samtleben, C. (1979). Functional morphology and palaeontological significance of the conchiolin layers in corbulid pelecypods. *Lethaia* 12: 341–351.

Li, C.-H., Zhao, J.-M. & Song, L. (2009). A review of advances in research on marine molluscan antimicrobial peptides and their potential application in aquaculture. *Molluscan Research* 29: 17–26.

Li, C. & Guo, X. (2015). Transcriptome analysis of the response to air exposure and cold stress by the eastern oyster. *Journal of Shellfish Research* 34: 652–653.

Li, F., Wu, N., Lu, H., Zhang, J., Wang, W., Ma, M., Zhang, X. & Yang, X. (2013). Mid-Neolithic exploitation of mollusks in the Guanzhong Basin of Northwestern China: Preliminary results. *PLoS ONE* 8: e58999.

Li, H., Parisi, M. G., Parrinello, N., Cammarata, M. & Roch, P. (2011). Molluscan antimicrobial peptides, a review from activity-based evidences to computer assisted sequences. *Invertebrate Survival Journal* 8: 85–97.

Li, J., Zhang, Y., Zhang, Y., Xiang, Z., Tong, Y., Qu, F. & Yu, Z. (2014a). Genomic characterization and expression analysis of five novel IL-17 genes in the Pacific oyster, *Crassostrea gigas*. *Fish and Shellfish Immunology* 40: 455–465.

Li, L., Connors, M. J., Kolle, M., England, G. T., Speiser, D. I., Xiao, X., Aizenberg, J. & Ortiz, C. (2015a). Multifunctionality of chiton biomineralized armor with an integrated visual system. *Science* 350: 952–956.

Li, L., Kolle, S., Weaver, J. C., Ortiz, C., Aizenberg, J. & Kolle, M. (2015b). A highly conspicuous mineralized composite photonic architecture in the translucent shell of the blue-rayed limpet. *Nature Communications* 6: 1–11.

Li, Q., Osada, M., Suzuki, T. & Mori, K. (1998). Changes in vitellin during oogenesis and effect of estradiol-17β on vitellogenesis in the Pacific oyster *Crassostrea gigas*. *Invertebrate Reproduction & Development* 33: 87–93.

Li, S., Liu, Y., Liu, C., Huang, J., Zheng, G., Xie, L. & Zhang, R. (2015c). Morphology and classification of hemocytes in *Pinctada fucata* and their responses to ocean acidification and warming. *Fish and Shellfish Immunology* 45: 194–202.

Li, S., Huang, J., Liu, C., Liu, Y., Zheng, G., Xie, L. & Zhang, R. (2016). Interactive effects of seawater acidification and elevated temperature on the transcriptome and biomineralization in the pearl oyster *Pinctada fucata*. *Environmental Science & Technology* 50: 1157–1165.

Li, X., Bai, Z., Luo, H., Wang, G. & Li, J. (2014b). Comparative analysis of total carotenoid content in tissues of purple and white inner-shell color pearl mussel, *Hyriopsis cumingii*. *Aquaculture International* 22: 1577–1585.

Li, Y., Qin, J. G., Abbott, C. A., Li, X. C. & Benkendorff, K. (2007). Synergistic impacts of heat shock and spawning on the physiology and immune health of *Crassostrea gigas*: An explanation for summer mortality in Pacific oysters. *American Journal of Physiology* 293: R2353–2362.

Liberge, M., Gros, O. & Frenkiel, L. (2001). Lysosomes and sulfide-oxidizing bodies in the bacteriocytes of *Lucina pectinata*, a cytochemical and microanalysis approach. *Marine Biology* 139: 401–409.

Liddiard, K. J., Hockridge, J. B., Macey, D. J., Webb, J. & Bronswijk, W. V. (2004). Mineralisation in the teeth of the limpets *Patelloida alticostata* and *Scutellastra laticostata* (Mollusca: Patellogastropoda). *Molluscan Research* 24: 21–31.

Lieb, B., Dimitrova, K., Kang, H. S., Braun, S., Gebauer, W., Martin, A., Hanelt, B., Saenz, S. A., Adema, C. M. & Markl, J. (2006). Red blood with blue-blood ancestry: Intriguing structure of a snail hemoglobin. *Proceedings of the National Academy of Sciences of the United States of America* 103: 12011–12016.

Lieb, B. & Todt, C. (2008). Hemocyanin in mollusks: A molecular survey and new data on hemocyanin genes in Solenogastres and Caudofoveata. *Molecular Phylogenetics and Evolution* 49: 382–385.

Lillvis, J. L., Gunaratne, C. A. & Katz, P. S. (2012). Neurochemical and neuroanatomical identification of central pattern generator neuron homologues in Nudipleura molluscs. *PLoS ONE* 7: e31737.

Lillywhite, H. B. (2014). *How Snakes Work. Structure, Function, and Behavior of the World's Snakes*. Oxford & New York, Oxford University Press.

Lim, S. S. L. (2008). Body posturing in *Nodilittorina pyramidalis* and *Austrolittorina unifasciata* (Mollusca: Gastropoda: Littorinidae): A behavioural response to reduce heat stress. *Memoirs of the Queensland Museum* 54: 339–347.

Limén, H., Levesque, C. & Juniper, S. K. (2007). POM in macro-/meiofaunal food webs associated with three flow regimes at deep-sea hydrothermal vents on Axial Volcano, Juan de Fuca Ridge. *Marine Biology* 153: 129–139.

Lindberg, D. R., Kellogg, M. G. & Hughes, W. E. (1975). Evidence of light reception through the shell of *Notoacmea persona* (Rathke, 1833) (Archaeogastropoda: Acmaeidae). *The Veliger* 17: 383–386.

Lindberg, D. R. (1978). Note on changes in marine intertidal fungus taxonomy. *The Veliger* 20: 399.

Lindberg, D. R. (1979). *Problacmaea moskalevi* (Golikov and Kussakiin): New record: A new addition to the Eastern Pacific Limpet Fauna (Archaeogastropoda Acmaeidae). *The Veliger* 22: 57–61.

Lindberg, D. R. (1981). Rhodopetalinae, a new subfamily of Acmaeidae from the boreal Pacific: Anatomy and systematics. *Malacologia* 20: 291–305.

Lindberg, D. R. & Dobberteen, R. A. (1981). Umbilical brood protection and sexual dimorphism in the boreal Pacific trochid gastropod *Margarites vorticiferus* Dall. *International Journal of Invertebrate Reproduction* 3: 347–355.

Lindberg, D. R. (1982). Tertiary biogeography and evolution of the genus *Problacmaea* in the North Pacific (Acmaeidae: Patelloidinae). *Journal of Paleontology* 56: 16.

Lindberg, D. R. & Kellogg, M. G. (1982). A note on the structure and pigmentation of the shell of *Notoacmea persona* (Rathke) (Docoglossa: Acmaeidae). *The Veliger* 25: 173–174.

Lindberg, D. R. (1983). Anatomy, systematics and evolution of brooding acmaeid limpets. PhD thesis, University of California, Santa Cruz.

Lindberg, D. R. & Dwyer, K. R. (1983). The topography, formation and role of the home depression of *Collisella scabra* (Gould) (Gastropoda: Acmaeidae). *The Veliger* 25: 229–234.

Lindberg, D. R. (1984a). A recent specimen of *Collisella edmitchelli* from San Pedro, California (Mollusca, Acmaeidae). *Bulletin of the Southern California Academy of Sciences* 83: 148–151.

Lindberg, D. R. (1984b). *Fossil Brooding Bivalve Molluscs from the Neogene of Western North America. Geological Society of America, 97th Annual Meeting.* Reno, NV, Geological Society of America. 16: 576.

Lindberg, D. R. (1985a). Aplacophorans, monoplacophorans, polyplacophorans, scaphopods: The lesser classes. pp. 230–247 *in* T. W. Broadhead (Ed.) *Mollusks: Notes for a Short Course organized by D. J. Bottjer, C. S. Hickman and P. D. Ward. University of Tennessee Studies in Geology.* Vol. 13. Knoxville, TN, University of Tennessee Department of Geological Sciences.

Lindberg, D. R. (1985b). Shell sexual dimorphism of *Margarites vorticifera*: Multivariant analysis and taxonomic implications. *Malacological Review* 18: 1–8.

Lindberg, D. R. & Wright, W. G. (1985). Patterns of sex change of the protandric patellacean limpet *Lottia gigantea* (Mollusca: Gastropoda). *The Veliger* 27: 261–265.

Lindberg, D. R. (1987a). Evolution of the patellogastropod fauna of the Caribbean Sea during the Neogene. *American Malacological Bulletin* 4: 115.

Lindberg, D. R. (1987b). Recent and fossil species of the genus *Erginus* from the North Pacific Ocean (Patellogastropoda: Mollusca). *PaleoBios* 12: 1–8.

Lindberg, D. R., Warheit, K. I. & Estes, J. A. (1987). Prey preference and seasonal predation by oystercatchers on limpets at San Nicolas Island, California, USA. *Marine Ecology Progress Series* 39: 105–113.

Lindberg, D. R. (1988a). Heterochrony in gastropods, a neontological view. pp. 197–216 *in* M. L. McKinney, Stehli, F. G. & Jones, D. S. (Eds.) *Heterochrony in Evolution: A Multidisciplinary Approach. Topics in Geobiology.* New York, Plenum Publishing Corporation.

Lindberg, D. R. (1988b). The Patellogastropoda. pp. 35–63 *in* W. F. Ponder, Eernisse, D. J. & Waterhouse, J. H. (Eds.) *Prosobranch Phylogeny. Malacological Review Supplement.* Ann Arbor, MI, Malacological Review.

Lindberg, D. R. (1990a). Movement patterns of the limpet *Lottia asmi* (Middendorff): Networking in California rocky intertidal communities [USA]. *The Veliger* 33: 375–383.

Lindberg, D. R. (1990b). Systematics of *Potamacmaea fluviatilis* (Blanford), a brackish water patellogastropod (Patelloidinae: Lottiidae). *Journal of Molluscan Studies* 56: 309–316.

Lindberg, D. R. (1990c). Morphometrics and the systematics of marine plant limpets (Mollusca: Patellogastropoda). pp. 301–310 *in* F. J. Rohlf & Bookstein, F. L. (Eds.) *Proceedings of the Michigan Morphometrics Workshop, University of Michigan, May 16–28, 1988.* Ann Arbor, MI, University of Michigan.

Lindberg, D. R. & Pearse, J. S. (1990). Experimental manipulation of shell color and morphology of the limpets *Lottia asmi* (Middendorff) and *Lottia digitalis* (Rathke) (Mollusca, Patellogastropoda). *Journal of Experimental Marine Biology and Ecology* 140: 173–186.

Lindberg, D. R. & Hedegaard, C. (1996). A deep water patellogastropod from Oligocene waterlogged wood of Washington State, USA (Acmaeoidea, Pectinodonta). *Journal of Molluscan Studies* 62: 299–314.

Lindberg, D. R. & Ponder, W. F. (1996). An evolutionary tree for the Mollusca: Branches or roots? pp. 67–75 *in* J. D. Taylor (Ed.) *Origin and Evolutionary Radiation of the Mollusca.* Oxford, UK, Oxford University Press.

Lindberg, D. R. (1998). William Healey Dall: A neo-Lamarckian view of molluscan evolution. *The Veliger* 41: 227–238.

Lindberg, D. R., Estes, J. A. & Warheit, K. I. (1998). Human influences on trophic cascades along rocky shores. *Ecological Applications* 8: 880–890.

Lindberg, D. R. & Ponder, W. F. (2001). The influence of classification on the evolutionary interpretation of structure: A re-evaluation of the evolution of the pallial cavity of gastropod molluscs. *Organisms Diversity & Evolution* 1: 273–299.

Lindberg, D. R. & Ghiselin, M. T. (2003). Fact, theory and tradition in the study of molluscan origins. *Proceedings of the California Academy of Sciences* 54: 663–686.

Lindberg, D. R. & Guralnick, R. P. (2003a). Lineages: Cell and phyletic. pp. 243–249 *in* B. K. Hall & Olson, W. M. (Eds.) *Key Concepts and Approaches in Evolutionary Developmental Biology.* Cambridge, MA, Harvard University Press.

Lindberg, D. R. & Guralnick, R. P. (2003b). Phyletic patterns of early development in gastropod molluscs. *Evolution & Development* 5: 494–507.

Lindberg, D. R., Ponder, W. F. & Haszprunar, G. (2004). The Mollusca: Relationships and patterns from their first half-billion years. pp. 252–278 *in* J. Cracraft & Donoghue, M. J. (Eds.) *Assembling the Tree of Life.* New York, Oxford University Press.

Lindberg, D. R. & Pyenson, N. D. (2006). Evolutionary patterns in Cetacea: Fishing up prey size through deep time. pp. 68–82 *in* J. A. Estes, DeMaster, D. P., Brownell, R. L., Doak, D. F. & Williams, T. M. (Eds.) *Whales, Whaling and Ocean Ecosystems.* Berkeley, CA, University of California Press.

Lindberg, D. R. & Pyenson, N. D. (2007). Things that go bump in the night: Evolutionary interactions between cephalopods and cetaceans in the tertiary. *Lethaia* 40: 335–343.

Lindberg, D. R. (2007a). Patellogastropoda. pp. 753–761 *in* J. T. Carlton (Ed.) *The Light & Smith Manual: Intertidal Invertebrates of the Central California Coast.* Vol. 18. Berkeley, CA, University of California Press.

Lindberg, D. R. (2007b). Reproduction, ecology, and evolution of the Indo-Pacific limpet *Scutellastra flexuosa*. *Bulletin of Marine Science* 81: 219–234.

Lindberg, D. R. (2008). Patellogastropoda, Neritimorpha and Cocculinoidea. The low diversity clades. pp. 271–296 *in* W. F. Ponder & Lindberg, D. R. (Eds.) *Phylogeny and Evolution of the Mollusca.* Berkeley, CA, University of California Press.

Lindberg, D. R. & Sigwart, J. D. (2015). What is the molluscan osphradium? A reconsideration of homology. *Zoologischer Anzeiger* 256: 14–21.

Linsley, R. M. (1977). Some 'laws' of gastropod shell form. *Paleobiology* 3: 196–206.

Linsley, R. M. (1978a). Shell form and the evolution of gastropods. *American Scientist* 66: 432–441.

Linsley, R. M. (1978b). Locomotion rates and shell form in the Gastropoda. *Malacologia* 17: 193–206.

Linsley, R. M. (1978c). The Omphalocirridae, a new family of Paleozoic Gastropoda which exhibits sexual dimorphism. *Memoirs of the National Museum of Victoria* 39: 33–54.

Lipiński, M. R., Butterworth, D. S., Augustyn, C. J., Brodziak, J. K. T., Christy, G., Des Clers, S., Jackson, G. D., O'Dor, R. K., Pauly, D. & Purchase, L. V. (1998). Cephalopod fisheries: A future global upside to past overexploitation of living marine resources? Results of an International Workshop, 31 August–2 September 1997, Cape Town, South Africa. *South African Journal of Marine Science* 20: 463–469.

Lipton, C. S. & Murray, J. (1979). Courtship of land snails of the genus *Partula*. *Malacologia* 19: 129–146.

Liscovitch-Brauer, N., Alon, S., Porath, H. T., Elstein, B., Unger, R., Ziv, T., Admon, A., Levanon, E. Y., Rosenthal, J. J. C. & Eisenberg, E. (2017). Trade-off between transcriptome plasticity and genome evolution in cephalopods. *Cell* 169: 191–202.

Lissmann, H. W. (1945). The mechanism of locomotion in gastropod molluscs II. Kinetics. *Journal of Experimental Biology* 22: 37–50.

List, T. (1902). *Die Mytiliden des Golfes von Neapel und der Angrenzenden Meeres-Abschnitte.* Berlin, R. Friedländer & Sohn.

Little, C. (1965a). Notes on the anatomy of the queen conch, *Strombus gigas*. *Bulletin of Marine Science* 15: 338–358.

Little, C. (1965b). The formation of urine by the prosobranch gastropod mollusc *Viviparus viviparus* Linn. *Journal of Experimental Biology* 43: 39–54.

Little, C. (1965c). Osmotic and ionic regulation in the prosobranch gastropod mollusc *Viviparus viviparus* Linn. *Journal of Experimental Biology* 43: 23–37.

Little, C. (1967). Ionic regulation in the queen conch, *Strombus gigas* (Gastropoda, Prosobranchia). *Journal of Experimental Biology* 46: 459–474.

Little, C. (1968). Aestivation and ionic regulation in two species of *Pomacea* (Gastropoda, Prosobranchia). *Journal of Experimental Biology* 48: 569–585.

Little, C. (1972). The evolution of kidney function in the Neritacea (Gastropoda, Prosobranchia). *Journal of Experimental Biology* 56: 249–261.

Little, C. & Andrews, E. B. (1977). Some aspects of excretion and osmoregulation in assimineid snails. *Journal of Molluscan Studies* 43: 263–285.

Little, C. (1979). Reabsorption of glucose in the renal system of *Viviparus*. *Journal of Molluscan Studies* 45: 207–208.

Little, C. (1981). Osmoregulation and excretion in prosobranch gastropods. Part I: Physiology and biochemistry. *Journal of Molluscan Studies* 47: 221–247.

Little, C. (1983). *The Colonisation of Land: Origins and Adaptations of Terrestrial Animals.* Cambridge, UK, Cambridge University Press.

Little, C., Pilkington, M. C. & Pilkington, J. B. (1984). Development of salinity tolerance in the marine pulmonate *Amphibola crenata* (Gmelin). *Journal of Experimental Marine Biology and Ecology* 74: 169–177.

Little, C. (1985). Renal adaptations of prosobranchs to the freshwater environment. *American Malacological Bulletin* 3: 223–231.

Little, C., Stirling, P., Pilkington, M. C. & Pilkington, J. B. (1985). Larval development and metamorphosis in the marine pulmonate *Amphibola crenata* (Mollusca: Pulmonata). *Journal of Zoology* 205: 489–510.

Little, C. & Kitching, J. A. (1996). *Biology of Rocky Shores.* Oxford, UK, Oxford University Press.

Little, C. (2000). *The Biology of Soft Shores and Estuaries.* New York, Oxford University Press.

Little, C. T. S., Campbell, K. A. & Herrington, R. J. (2002). Why did ancient chemosynthetic seep and vent assemblages occur in shallower water than they do today? *International Journal of Earth Sciences* 91: 149–153.

Little, C. T. S. & Vrijenhoek, R. C. (2003). Are hydrothermal vent animals living fossils? *Trends in Ecology & Evolution* 18: 562–588.

Littler, M. M., Littler, D. S. & Taylor, P. R. (1995). Selective herbivore increases biomass of its prey: A chiton-coralline reef-building association. *Ecology* 76: 1666–1681.

Littlewood, D. T. J. & Marsbe, L. A. (1990). Predation on cultivated oysters, *Crassostrea rhizophorae* (Guilding) by the polyclad turbellarian flatworm *Stylochus* (*Stylochus*) *frontalis* Verrill. *Aquaculture* 88: 145–150.

Liu, D. & Chen, Z. (2013). The expression and induction of heat shock proteins in molluscs. *Protein and Peptide Letters* 20: 601–606.

Liu, F.-M., Fu, Y.-M. & Shih, D. Y.-C. (2004). Occurrence of tetrodotoxin poisoning in *Nassarius papillosus* Alectrion and *Nassarius gruneri* Niotha. *Journal of Food and Drug Analysis* 12: 189–192.

Liu, J., Zhang, Z., Zhang, L., Liu, X., Yang, D. & Ma, X. (2014). Variations of estradiol-17 beta and testosterone levels correlated with gametogenesis in the gonad of Zhikong scallop (*Chlamys farreri*) during annual reproductive cycle. *Canadian Journal of Zoology* 92: 195–204.

Liu, J. W., Lee, I. K., Tang, H. J., Ko, W. C., Lee, H. C., Liu, Y. C., Hsueh, P. R. & Chuang, Y. C. (2006). Prognostic factors and antibiotics in *Vibrio vulnificus* septicemia. *Archives of Internal Medicine* 166: 2117–2123.

Liu, L. L., Foltz, D. W. & Stickle, W. B. (1991). Genetic population structure of the southern oyster drill *Stramonita haemostoma* (equals *Thais haemostoma*). *Marine Biology* 111: 71–80.

Liu, L. L. & Wang, S.-P. (2002). Histology and biochemical composition of the autotomy mantle of *Ficus ficus* (Mesogastropoda: Ficidae). *Acta Zoologica* 83: 111–116.

Liu, X., Wu, F., Zhao, H., Zhang, G. & Guo, X. (2009). A novel shell color variant of the Pacific abalone *Haliotis discus hannai* Ino subject to genetic control and dietary influence. *Journal of Shellfish Research* 28: 419–424.

Liu, Y., Huang, Z., He, S., Fu, J., Gong, X., Ye, Y., Han, X., Zhang, J. & Wan, X. (1993). Preliminary study of the causative agent of the freshwater mussel *Hyriopsis cumingii* plague [in Chinese with English summary]. *Freshwater Fish* 23: 3–7.

Liu, Y., Shenouda, D., Bilfinger, T. V., Stefano, M. L., Magazine, H. I. & Stefano, G. B. (1996). Morphine stimulates nitric oxide release from invertebrate microglia. *Brain Research* 722: 125–131.

Lively, C. M. & Johnson, S. G. (1994). Brooding and the evolution of parthenogenesis: Strategy models and evidence from aquatic invertebrates. *Proceedings of the Royal Society B* 256: 89–95.

Livett, B. G., Gayler, K. R. & Khalil, Z. (2004). Drugs from the sea: Conopeptides as potential therapeutics. *Current Medicinal Chemistry* 11: 1715–1723.

Livingstone, D. A. (1983). Studies on the phylogenetic distribution of pyruvate oxidoreductases. *Biochemical Systematics and Ecology* 11: 415–425.

Livingstone, D. R., de Zwaan, A. & Thompson, R. J. (1981). Aerobic metabolism, octopine production and phosphoarginine as sources of energy in the phasic and catch adductor muscles of the giant scallop *Placopecten magellanicus* during swimming and the subsequent recovery period. *Comparative Biochemistry and Physiology Part B* 70: 35–44.

Livingstone, D. R. & de Zwaan, A. (1983). Carbohydrate metabolism of gastropods. pp. 177–242 *in* P. W. Hochachka (Ed.) *Metabolic Biochemistry and Molecular Biomechanics. The Mollusca.* Vol. 1. New York, Academic Press.

Livingstone, D. R., Stickle, W. B., Kapper, M. A., Wang, S. & Zurburg, W. (1990). Further studies on the phylogenetic distribution of pyruvate oxidoreductase activities. *Comparative Biochemistry and Physiology Part B* 97: 661–666.

Llinás, R. R. (1999). *The Squid Giant Synapse: A Model for Chemical Transmission.* New York, Oxford, UK, Oxford University Press.

Lobo-da-Cunha, A., Batista, C. & Oliveira, E. (1994). The peroxisomes of the hepatopancreas in marine gastropods. *Biology of the Cell* 82: 67–74.

Lobo-da-Cunha, A. (2001). Ultrastructural and histochemical study of the salivary glands of *Aplysia depilans* (Mollusca, Opisthobranchia). *Acta Zoologica* 82: 201–212.

Lobo-da-Cunha, A. (2002). Cytochemical localisation of lysosomal enzymes and acidic mucopolysaccharides in the salivary glands of *Aplysia depilans* (Opisthobranchia). *Journal of Submicroscopic Cytology and Pathology* 34: 217–225.

Lobo-da-Cunha, A., Ferreira, I., Coelho, R. & Calado, G. (2009). Light and electron microscopy study of the salivary glands of the carnivorous opisthobranch *Philinopsis depicta* (Mollusca, Gastropoda). *Tissue and Cell* 41: 367–375.

Lobo-da-Cunha, A., Santos, T., Oliveira, E., Alves, Â., Coelho, R. & Calado, G. (2011). Microscopical study of the crop and oesophagus of the carnivorous opisthobranch *Philinopsis depicta* (Cephalaspidea: Aglajidae). *Journal of Molluscan Studies* 77: 322–331.

Lobo-da-Cunha, A., Pereira-Sousa, J., Oliveira, E., Alves, Â., Guimarães, F. & Calado, G. (2014). Calcium detection and other cellular studies in the esophagus and crop of the marine slug *Aglaja tricolorata* (Euopisthobranchia, Cephalaspidea). *Malacologia* 57: 365–376.

Lobo-da-Cunha, A., Alves, Â., Oliveira, E. & Calado, G. (2015). Comparative study of salivary glands in carnivorous and herbivorous cephalaspideans (Gastropoda: Euopisthobranchia). *Journal of Molluscan Studies* 82: 43–54.

Lobo-da-Cunha, A., Alves, Â., Oliveira, E., Guimarães, F. & Calado, G. (2018). Endocytosis, lysosomes, calcium storage and other features of digestive-gland cells in cephalaspidean gastropods (Euopisthobranchia). *Journal of Molluscan Studies* 84: 451–462.

Lobo-da-Cunha, A. & Calado, G. (2008). Histology and ultrastructure of the salivary glands in *Bulla striata* (Mollusca, Opisthobranchia). *Invertebrate Biology* 127: 33–44.

Lobo-da-Cunha, A., Oliveira, E., Alves, Â., Coelho, R. & Calado, G. (2010). Light and electron microscopic study of the anterior oesophagus of *Bulla striata* (Mollusca, Opisthobranchia). *Acta Zoologica* 91: 125–138.

Loch, I. (1987). Peanuts for breakfast. *Australian Shell News* 58: 6–8.

Lockyer, A. E., Jones, C. S., Noble, L. R. & Rollinson, D. (2004). Trematodes and snails: An intimate association. *Canadian Journal of Zoology* 82: 251–269.

Lodge, S. M. (1948). Algal growth in the absence of *Patella* on an experimental strip of foreshore, Port St. Mary, Isle of Man. *Proceedings of the Liverpool Biological Society* 56: 78–83.

Loest, R. A. (1979a). Ammonia volatilization and absorption by terrestrial gastropods: A comparison between shelled and shell-less species. *Physiological Zoology* 52: 461–469.

Loest, R. A. (1979b). Ammonia-forming enzymes and calcium-carbonate deposition in terrestrial pulmonates. *Physiological Zoology* 52: 470–483.

Logan, J. M., Toppin, R., Smith, S., Galuardi, B., Porter, J. & Lutcavage, M. (2012). Contribution of cephalopod prey to the diet of large pelagic fish predators in the central North Atlantic Ocean. *Deep Sea Research Part II: Topical Studies in Oceanography* 95: 1–9.

Loh, W. K. W., Cowlishaw, M. & Wilson, N. G. (2006). Diversity of *Symbiodinium* dinoflagellate symbionts from the Indo-Pacific sea slug *Pteraeolidia ianthina* (Gastropoda: Mollusca). *Marine Ecology Progress Series* 320: 177–184.

Lohmann, K. J. & Willows, A. O. D. (1987). Lunar-modulated geomagnetic orientation by a marine mollusk. *Science* 235: 331–334.

Loker, E. S. & Kepler, T. B. (2004). Invertebrate immune systems – not homogeneous, not simple, not well understood. *Immunological Reviews* 198: 10–24.

Loker, E. S. (2010). Gastropod immunobiology. pp. 17–43 *in* K. Söderhäll (Ed.) *Invertebrate Immunity.* Vol. 708. New York, Landes Bioscience and Springer Science + Business Media.

Lombardo, P. & Cooke, G. D. (2004). Resource use and partitioning by two co-occurring freshwater gastropod species. *Archiv für Hydrobiologie* 159: 229–251.

Lombardo, R. C., Takeshita, F., Abe, S. & Goshima, S. (2012). Mate choice by males and paternity distribution in offspring of triple-mated females in *Neptunea arthritica* (Gastropoda: Buccinidae). *Journal of Molluscan Studies* 78: 283–289.

Lomnicki, A. (1969). Individual differences among adult members of a snail population. *Nature* 223: 1073–1074.

Long, E. O. & Dawid, I. B. (1980). Repeated genes in eukaryotes. *Annual Review of Biochemistry* 49: 727–764.

Long, J. D. & Hay, M. E. (2006). Fishes learn aversions to a nudibranch's chemical defense. *Marine Ecology Progress Series* 307: 199–208.

Longhurst, A. R. & Pauly, D. (1987). *Ecology of Tropical Oceans.* San Diego, CA, Academic Press.

Longo, F. J. (1973). Fertilization: A comparative ultrastructural review. *Biology of Reproduction* 9: 149–215.

Longo, F. J. & Kunkle, M. (1978). Tranformations of sperm nuclei upon insemination. pp. 149–184 *in* A. A. Moscona & Monroy, A. (Eds.) *Current Topics in Developmental Biology.* Vol. 12. New York, Academic Press.

Longo, F. J. (1983). Meiotic maturation and fertilization. pp. 49–89 *in* N. H. Verdonk, van den Biggelaar, J. A. M. & Tompa, A. S. (Eds.) *Development. The Mollusca.* Vol. 3. New York, Academic Press.

Lonsdale, P. (1977). Clustering of suspension-feeding macrobenthos near abyssal hydrothermal vents at oceanic spreading centers. *Deep Sea Research Part II: Topical Studies in Oceanography* 24: 857–863.

Loot, G., Aldana, M. & Navarrete, S. A. (2005). Effects of human exclusion on parasitism in intertidal food webs of central Chile. *Conservation Biology* 19: 203–212.

López-Vera, E., Aguilar, M. B. & de la Cotera, E. P. H. (2008). FMRFamide and related peptides in the phylum Mollusca. *Peptides* 29: 310–317.

Lopez, G. R. & Levinton, J. S. (1987). Ecology of deposit-feeding animals in marine sediments. *Quarterly Review of Biology* 62: 235–260.

Lopez, J. L., Abalde, S. L. & Fuentes, J. (2005). Proteomic approach to probe for larval proteins of the mussel *Mytilus galloprovincialis*. *Marine Biotechnology* 7: 396–404.

Lord, J. P. (2011). Larval development, metamorphosis and early growth of the gumboot chiton *Cryptochiton stelleri* (Middendorff, 1847) (Polyplacophora: Mopaliidae) on the Oregon coast. *Journal of Molluscan Studies* 77: 182–188.

Lorenzen, S. (2007). The limpet *Patella vulgata* L. at night in air: Effective feeding on *Ascophyllum nodosum* monocultures and stranded seaweeds. *Journal of Molluscan Studies* 73: 267–274.

Lorion, J., Halary, S., Do Nasciment, J., Samadi, D. S., Couloux, A. & Duperron, S. (2012). Evolutionary history of *Idas* sp. Med (Bivalvia: Mytilidae), a cold seep mussel bearing multiple symbionts. *Cahiers de Biologie Marine* 53: 77–87.

Love, A. C. (2009). Marine invertebrates, model organisms, and the modern synthesis: Epistemic values, evo-devo, and exclusion. *Theory in Biosciences* 128: 19–42.

Love, A. C. (2015). Conceptual Change and Evolutionary Developmental Biology. pp. 1–54 in A. C. Love (Ed.) *Conceptual Change in Biology. Scientific and Philosophical Perspectives on Evolution and Development.* Dordrecht, the Netherlands, Springer Science + Business Media.

Lowenstam, H. A. & Weiner, S. (1989). Mollusca. pp. 88–110. *On Biomineralization.* New York, Oxford University Press.

Lozano, V., Martínez-Escauriaza, R., Bernardo-Castiñeira, C., Mesías-Gansbiller, C., Pazos, A. J., Sánchez, J. L. & Pérez-Parallé, M. L. (2014). A novel class of *Pecten maximus POU* gene, *PmaPOU-IV*: Characterization and expression in adult tissues. *Journal of Experimental Marine Biology and Ecology* 453: 154–161.

Lubchenco, J. & Gaines, S. D. (1981). A unified approach to marine plant-herbivore interactions. I. Populations and communities. *Annual Review of Ecology and Systematics* 12: 405–437.

Lucas, A., Chebabchalabi, L. & Aranda, D. (1986). Passage de l'endotrophie à l'exotrophie chez les larves de *Mytilus edulis*. *Oceanologica Acta* 9: 97–103.

Lucas, M. I. & Newell, R. C. (1984). Utilization of saltmarsh grass detritus by two estuarine bivalves: Carbohydrase activity of crystalline style enzymes of the oyster *Crassostrea virginica* (Gmelin) and the mussel *Geukensia demissa* (Dillwyn). *Marine Biology Letters* 5: 275–290.

Luchtel, D. L., Martin, A. W. & Deyrup-Olsen, I. (1984). The channel cell of the terrestrial slug *Agriolimax columbianus* (Stylommatophora, Arionidae). *Cell and Tissue Research* 235: 143–151.

Luchtel, D. L., Martin, A. W., Deyrup-Olsen, I. & Boer, H. H. (1997). Gastropoda: Pulmonata. pp. 459–718 in F. W. Harrison & Kohn, A. J. (Eds.) *Microscopic Anatomy of Invertebrates. Mollusca 2.* New York, Wiley-Liss.

Luchtel, D. L. & Deyrup-Olsen, I. (2001). Body wall: Form and function. pp. 147–178 in G. M. Barker (Ed.) *The Biology of Terrestrial Molluscs.* Wallingford, UK, CABI Publishing.

Ludbrook, N. H. & Gowlett-Holmes, K. L. (1989). Chitons, gastropods, and bivalves. pp. 504–724 in S. A. Shepherd & Thomas, I. M. (Eds.) *Marine Invertebrates of Southern Australia. Part 2.* Adelaide, South Australia, South Australian Government Printing Division.

Lukowiak, K., Ringseis, E., Specer, G., Wildering, W. C. & Syed, N. I. (1996). Operant conditioning of aerial respiratory behaviour in *Lymnaea stagnalis*. *Journal of Experimental Biology* 199: 683–691.

Lukowiak, K., Adatia, N., Krygier, D. & Syed, N. I. (2000). Operant conditioning in *Lymnaea*: Evidence for intermediate- and long-term memory. *Learning and Memory* 7: 140–150.

Lukowiak, K. (2017). Stress, memory, forgetting and what *Lymnaea* can tell us about a stressful world. pp. 67–102 in S. Saleuddin & Mukai, S. (Eds.) *Physiology of Molluscs.* Vol. 2. Oakville, ON, Apple Academic Press Inc.

Lundin, K. & Schander, C. (1999). Ultrastructure of gill cilia and ciliary rootlets of *Chaetoderma nitidulum* Loven 1844 (Mollusca, Chaetodermomorpha). *Acta Zoologica* 80: 185–191.

Lundin, K. & Schander, C. (2001a). Ciliary ultrastructure of protobranchs (Mollusca, Bivalvia). *Invertebrate Biology* 120: 350–357.

Lundin, K. & Schander, C. (2001b). Ciliary ultrastructure of polyplacophorans (Mollusca, Amphineura, Polyplacophora). *Journal of Submicroscopic Cytology and Pathology* 33: 93–98.

Lundin, K. & Schander, C. (2001c). Ciliary ultrastructure of neomeniomorphs (Mollusca: Neomeniomorpha = Solenogastres). *Invertebrate Biology* 120: 342–349.

Lundin, K., Schander, C. & Todt, C. (2009). Ultrastructure of epidermal cilia and ciliary rootlets in Scaphopoda. *Journal of Molluscan Studies* 75: 69–73.

Luo, Y. J., Takeuchi, T., Koyanagi, R., Yamada, L., Kanda, M., Khalturina, M., Fujie, M., Yamasaki, S. I., Endo, K. & Satoh, N. (2015). The *Lingula* genome provides insights into brachiopod evolution and the origin of phosphate biomineralization. *Nature Communications* 6: 9301.

Luoma, S. N. & Rainbow, P. S. (2005). Why is metal bioaccumulation so variable? Biodynamics as a unifying concept. *Environmental Science & Technology* 39: 1921–1931.

Luoma, S. N. & Rainbow, P. S. (2008). *Metal Contamination in Aquatic Environments: Science and Lateral Management.* Cambridge, UK, Cambridge University Press.

Lüter, C. (2016). The Lophophorates. Phylum Brachiopoda. pp. 657–668 in R. C. Brusca, Moore, W. & Shuster, S. (Eds.) *Invertebrates.* Sunderland, MA, Sinauer Associates, Inc.

Lutfy, R. G. & Demian, E. S. (1967). The histology of the alimentary system of *Marisa cornuarietis* (Mesogastropoda: Ampullariidae). *Malacologia* 5: 375–422.

Luttikhuizen, P. C. & Drent, J. (2008). Inheritance of predominantly hidden shell colours in *Macoma balthica* (L.) (Bivalvia: Tellinidae). *Journal of Molluscan Studies* 74: 363–371.

Lutz, R. A. (1976). Annual growth patterns in the inner shell layer of *Mytilus edulis* L. *Journal of the Marine Biological Association of the United Kingdom* 56: 723–731.

Lutz, R. A., Bouchet, P., Jablonski, D., Turner, R. D. & Warén, A. (1986). Larval ecology of mollusks at deep-sea hydrothermal vents. *American Malacological Union Bulletin* 4: 49–54.

Lutz, R. A. & Kennish, M. J. (1993). Ecology of deep-sea hydrothermal vent communities: A review. *Reviews of Geophysics* 31: 211–242.

Lützen, J. (1968). Unisexuality in the parasitic family Entoconchidae (Gastropoda, Prosobranchia). *Malacologia* 7: 7–15.

Lützen, J. & Nielsen, K. (1975). Contributions to the anatomy and biology of *Echineulima* n. g. (Prosobranchia: Eulimidae), parasitic on sea urchins. *Videnskabelige Meddelelser fra Dansk naturhistorisk Forening i Kjøbenhavn* 138: 171–199.

Lützen, J. (1979). Studies on the life-history of *Enteroxenos* Bonnevie, a gastropod endoparasitic in aspidochirote holothurians. *Ophelia* 18: 1–51.

Lützen, J., Sakamoto, H., Taguchi, A. & Takahashi, T. (2001a). Reproduction, dwarf males, sperm dimorphism, and life cycle in the commensal bivalve *Peregrinamor ohshimai* Shoji (Heterodonta: Galeommatoidea: Montacutidae). *Malacologia* 43: 313–325.

Lützen, J., Takahashi, T. & Yamaguchi, T. (2001b). Morphology and reproduction of *Nipponomysella subtruncata* (Yokoyama), a galeommatoidean bivalve commensal with the sipunculan *Siphonosoma cumanense* (Keferstein) in Japan. *Journal of Zoology* 254: 429–440.

Lützen, J., Jespersen, Å., Takahashi, T. & Kai, T. (2004). Morphology, structure of dimorphic sperm, and reproduction in the hermaphroditic commensal bivalve *Pseudopythina tsurumaru* (Galeommatoidea: Kellidae). *Journal of Morphology* 262: 407–420.

Lützen, J., Kato, M., Kosuge, T. & Ó Foighil, D. (2005). Reproduction involving spermatophores in four bivalve genera of the superfamily Galeommatoidea commensal with holothurians. *Molluscan Research* 25: 99–112.

Lützen, J., Berland, B. & Bristow, G. A. (2011). Morphology of an endosymbiotic bivalve, *Entovalva nhatrangensis* (Bristow, Berland, Schander & Vo, 2010) (Galeommatoidea). *Molluscan Research* 31: 114–124.

Luyten, Y. A., Thompson, J. R., Morrill, W., Polz, M. F. & Distel, D. L. (2006). Extensive variation in intracellular symbiont community composition among members of a single population of the wood-boring bivalve *Lyrodus pedicellatus* (Bivalvia: Teredinidae). *Applied and Environmental Microbiology* 72: 412–417.

Lydeard, C. & Lindberg, D. R., Eds. (2003). *Molecular Systematics and Phylogeography of Mollusks.* Smithsonian Series in Comparative Evolutionary Biology. Washington DC, London, Smithsonian Press.

Lydeard, C., Cowie, R. H., Ponder, W. F., Bogan, A. E., Bouchet, P., Clark, S. A., Cummings, K. S., Frest, T. J., Gargominy, O., Herbert, D. G., Hershler, R., Perez, K. E., Roth, B., Seddon, M. B., Strong, E. E. & Thompson, F. G. (2004). The global decline of nonmarine mollusks. *BioScience* 54: 321–330.

Lynn, J. W. (1994). The ultrastructure of the sperm and motile spermatozeugmata released from the freshwater mussel *Anodonta grandis* (Mollusca, Bivalvia, Unionidae). *Canadian Journal of Zoology* 72: 1452–1461.

Lyons, D. C., Perry, K. J., Lesoway, M. P. & Henry, J. Q. (2012). Cleavage pattern and fate map of the mesentoblast, 4d, in the gastropod *Crepidula*: A hallmark of spiralian development. *EvoDevo* 3: 21.

Lyons, D. C. & Henry, J. Q. (2014). Ins and outs of spiralian gastrulation. *International Journal of Developmental Biology* 58: 413–428.

Lyons, D. C., Perry, K. J. & Henry, J. Q. (2015). Spiralian gastrulation: Germ layer formation, morphogenesis, and fate of the blastopore in the slipper snail *Crepidula fornicata*. *EvoDevo* 6: 24.

Lyons, L. C., Rawashdeh, O., Katzoff, A., Susswein, A. J. & Eskin, A. (2005). Circadian modulation of complex learning in diurnal and nocturnal *Aplysia*. *Proceedings of the National Academy of Sciences of the United States of America* 102: 12589–12594.

Lyons, L. C., Rawashdeh, O. & Eskin, A. (2006). Non-ocular circadian oscillators and photoreceptors modulate long term memory formation in *Aplysia*. *Journal of Biological Rhythms* 21: 245–255.

Lyons, L. C. (2011). Critical role of the circadian clock in memory formation: Lessons from *Aplysia*. *Frontiers in Molecular Neuroscience* 4: 1–6.

Ma, H. M., Mai, K. S., Xu, W. & Liufu, Z. G. (2005). Molecular cloning of alpha2-macroglobulin in sea scallop *Chlamys farreri* (Bivalvia, Mollusca). *Fish and Shellfish Immunology* 18: 345–349.

Maas, A. E., Wishner, K. F. & Seibel, B. A. (2012). The metabolic response of pteropods to acidification reflects natural CO_2-exposure in oxygen minimum zones. *Biogeosciences* 9: 747–757.

Maas, A. E., Lawson, G. L. & Tarrant, A. M. (2015). Transcriptome-wide analysis of the response of the thecosome pteropod *Clio pyramidata* to short-term CO_2 exposure. *Comparative Biochemistry and Physiology Part D* 16: 1–9.

Maas, D. (1965). Anatomische und histologische untersuchungen am Mundapparat der Pyramidelliden. *Zeitschrift für Morphologie und Ökologie der Tiere* 54: 566–642.

Maboloc, E. A. & Mingoa-Licuanan, S. S. (2013). Spawning in *Trochus maculatus:* Field observations from Bolinao, Pangasinan (Philippines). *Coral Reefs* 32: 1141.

MacClintock, C. (1967). Shell structure of patelloid and bellerophontoid gastropods (Mollusca). *Peabody Museum of Natural History Yale University Bulletin* 22: 1–140.

MacDonald, I. R., Guinasso, N. L., Reilly, J. F., Brooks, J. M., Callender, W. R. & Gabrielle, S. G. (1990). Gulf of Mexico hydrocarbon seep communities: VI. Patterns in community structure and habitat. *Geo-Marine Letters* 10: 244–252.

MacDonald, J. & Manio, C. B. (1964). Observations on the epipodium, digestive tract, coelomic derivates, and nervous system of the trochid gastropod *Tegula funebralis*. *The Veliger* 6 (Supplement): 50–55.

Macey, D. J. & Brooker, L. R. (1996). The junction zone: Initial site of mineralization in radula teeth of the chiton *Cryptoplax striata* (Mollusca: Polyplacophora). *Journal of Morphology* 230: 33–42.

Machado, J., Reis, M. L., Coimbra, J. & Sá, C. (1991). Studies on chitin and calcification in the inner layers of the shell of *Anodonta cygnea. Journal of Comparative Physiology B* 161: 413–418.

Machin, J. (1968). The permeability of the epiphragm of terrestrial snails to water vapour. *Biological Bulletin* 134: 87–95.

Machin, J. (1972). Water exchange in the mantle of a terrestrial snail during periods of reduced water loss. *Journal of Experimental Biology* 57: 103–111.

Machin, J. (1975). Water relationships. pp. 105–163 *in* V. Fretter & Peake, J. (Eds.) *Pulmonates: Functional Anatomy and Physiology.* Vol. 1. London, UK, Academic Press.

Machkevsky, V. K. (1997). Endosymbionts of mangrove oyster in nature and under cultivation. *Scientia Marina* 61: 99–107.

Mackensen, A. K., Brey, T., Bock, C. & Luna, S. (2012). *Spondylus crassisquama* Lamarck, 1819 as a microecosystem and the effects of associated macrofauna on its shell integrity: Isles of biodiversity or sleeping with the enemy? *Marine Biodiversity* 42: 443–451.

Mackenstedt, U. & Märkel, K. (1987). Experimental and comparative morphology of radula renewal in pulmonates (Mollusca, Gastropoda). *Zoomorphology* 107: 209–239.

Mackenstedt, U. & Märkel, K. (2001). Radular structure and function. pp. 213–236 *in* G. M. Barker (Ed.) *The Biology of Terrestrial Molluscs.* Wallingford, UK, CAB Publishing.

MacKenzie, C. L. (2007). Causes underlying the historical decline in eastern oyster (*Crassostrea virginica* Gmelin, 1791) landings. *Journal of Shellfish Research* 26: 927–938.

Mackie, G. L. (1984). Bivalves. pp. 351–418 *in* A. S. Tompa, Verdonk, N. H. & van den Biggelaar, J. A. M. (Eds.). *Reproduction. The Mollusca.* Vol. 7. New York, Academic Press.

Macy, R. W., Cook, W. A. & DeMott, W. R. (1960). Studies on the life cycle of *Halipegus occidualis* Stafford, 1905 (Trematoda: Hemiuridae). *Northwest Science* 34: 1–17.

Maddock, L. & Young, J. Z. (1987). Quantitative differences among the brains of cephalopods. *Journal of Zoology* 212: 739–767.

Madrid, K. P., Price, D. A., Greenberg, M. J., Khan, H. R. & Saleuddin, A. S. M. (1994). FMRFamide-related peptides from the kidney of the snail *Helisoma trivolvis*. *Peptides* 15: 31–36.

Madsen, H. & Hung, N. M. (2014). An overview of freshwater snails in Asia with main focus on Vietnam. *Acta Tropica* 140: 105–117.

Maes, V. O. & Raeihle, D. (1975). Systematics and biology of *Thala floridana* (Gastropoda: Vexillidae). *Malacologia* 15: 43–67.

Mahmud, S., Mladenov, P. V., Chakraborty, S. C. & Faruk, M. A. R. (2014). Relationship between gonad condition and neurosecretory cell activity in the green-lipped mussel, *Perna canaliculus*. *Progressive Agriculture* 18: 135–148.

Mair, J. & Port, G. R. (2002). The influence of mucus production by the slug, *Deroceras reticulatum*, on predation by *Pterostichus madidus* and *Nebria brevicollis* (Coleoptera: Carabidae). *Biocontrol Science and Technology* 12: 325–335.

Maitland, P. (1965). The feeding relationships of salmon, trout, minnows, stone loach and three spined sticklebacks in the River Endrick, Scotland. *Journal of Animal Ecology* 34: 109–133.

Mäkinen, T., Panova, M. & André, C. (2007). High levels of multiple paternity in *Littorina saxatilis*: Hedging the bets? *Journal of Heredity* 98: 705–711.

Makoni, P., Chimbari, M. J. & Madsen, H. (2005). Interactions between fish and snails in a Zimbabwe pond, with particular reference to *Sargochromis codringtonii* (Pisces: Cichlidae). *African Journal of Aquatic Science* 30: 45–48.

Makowski, H. (1962). Problem of sexual dimorphism in ammonites. *Palaeontologia Polonica* 12: 1–92.

Malaquias, M. A. E. & Reid, D. G. (2008). Systematic revision of the living species of Bullidae (Mollusca: Gastropoda: Cephalaspidea), with a molecular phylogenetic analysis. *Zoological Journal of the Linnean Society* 153: 453–543.

Malek, E. A. & Cheng, T. C. (1974). *Medical and Economic Malacology*. New York and London, Academic Press.

Malik, A. A., Aremu, A., Bayode, G. B. & Ibrahim, B. A. (2011). A nutritional and organoleptic assessment of the meat of the giant African land snail (*Archachatina marginata* Swaison [sic]) compared to the meat of other livestock. *Livestock Research for Rural Development* 23: article 60.

Malkowsky, Y. & Götze, M.-C. (2014). Impact of habitat and life trait on character evolution of pallial eyes in Pectinidae (Mollusca: Bivalvia). *Organisms Diversity & Evolution* 14: 173–185.

Malusa, J. R. (1985). Attack mode in a predatory gastropod: Labial spine length and the method of prey capture in *Acanthina angelica* Oldroyd. *The Veliger* 28: 1–5.

Manahan, D. T. (1990). Adaptations by invertebrate larvae for nutrient acquisition from seawater. *American Zoologist* 30: 147–160.

Mandrioli, M., Mola, L., Cuoghi, B. & Sonetti, D. (2010). Endoreplication: A molecular trick during animal neuron evolution. *Quarterly Review of Biology* 85: 159–169.

Maneveldt, G. W., Wilby, D., Potgieter, M. & Hendricks, M. G. J. (2006). The role of encrusting coralline algae in the diets of selected intertidal herbivores. *Journal of Applied Phycology* 18: 619–627.

Manga-González, M. Y., González-Lanza, C., Cabanas, E. & Campo, R. (2001). Contributions to and review of dicrocoeliosis, with special reference to the intermediate hosts of *Dicrocoelium dendriticum*. *Parasitology* 123: 91–114.

Mangold, K. M. & Froesch, D. (1977). A reconsideration of factors associated with sexual maturation. pp. 541–555 *in* M. Nixon & Messenger, J. B. (Eds.) *The Biology of Cephalopods. Symposia of The Zoological Society of London*. London, UK, Academic Press.

Mangold, K. M. (1989). Céphalopodes. pp. 1–804 *in* P. P. Grassé (Ed.) *Traité de Zoologie*. Vol. 5, Fascicule 4. Paris, Masson.

Mangum, C. P., Dykens, J. A., Henry, R. P. & Polites, G. (1978). The excretion of NH_4^+ and its ouabain sensitivity in aquatic annelids and molluscs. *Journal of Experimental Zoology* 203: 151–157.

Mangum, C. P. (1979). A note on blood and water mixing in large marine gastropods. *Comparative Biochemistry and Physiology Part A* 63: 389–391.

Manker, D. C., Garson, M. J. & Faulkner, D. J. (1988). *De novo* biosynthesis of polypropionate metabolites in the marine pulmonate *Siphonaria denticulata*. *Journal of the Chemical Society, Chemical Communications* 16: 1061–1062.

Manker, D. C. & Faulkner, D. J. (1989). Vallartanones A and B, polypropionate metabolites of *Siphonaria maura* from Mexico. *Journal of Organic Chemistry* 54: 5374–5377.

Manker, D. C., Faulkner, D. J., Stout, T. J. & Clardy, J. (1989). The baconipyrones: Novel polypropionates from the pulmonate *Siphonaria baconi*. *Journal of Organic Chemistry* 54: 5371–5374.

Manker, D. C. & Faulkner, D. J. (1996). Investigation of the role of diterpenes produced by marine pulmonates *Trimusculus reticulatus* and *T. conica*. *Journal of Chemical Ecology* 22: 23–36.

Mann, K. & Jackson, D. J. (2014). Characterization of the pigmented shell-forming proteome of the common grove snail *Cepaea nemoralis*. *BMC Genomics* 15: 249.

Mann, R. & Wolf, C. C. (1983). Swimming behaviour of larvae of the ocean quahog *Arctica islandica* in response to pressure and temperature. *Marine Ecology Progress Series* 13: 211–218.

Mannino, M. A. & Thomas, K. D. (2002). Depletion of a resource? The impact of prehistoric human foraging on intertidal mollusc communities and its significance for human settlement, mobility and dispersal. *World Archaeology* 33: 452–474.

Manríquez, P. H., Navarrete, S. A., Rosson, A. & Castilla, J. C. (2004). Settlement of the gastropod *Concholepas concholepas* on shells of conspecific adults. *Journal of the Marine Biological Association of the United Kingdom* 84: 651–658.

Manríquez, P. H., Lagos, N. A., Jara, M. E. & Castilla, J. C. (2009). Adaptive shell color plasticity during the early ontogeny of an intertidal keystone snail. *Proceedings of the National Academy of Sciences of the United States of America* 106: 16298–16303.

Manseau, F., Sossin, W. S. & Castellucci, V. F. (1998). Long-term changes in excitability induced by protein kinase C activation in *Aplysia* sensory neurons. *Journal of Neurophysiology* 79: 1210–1218.

Manseau, F., Fan, X., Hueftlein, T., Sossin, W. S. & Castellucci, V. F. (2001). Ca^{2+}-independent protein kinase C Apl II mediates the serotonin-induced facilitation at depressed *Aplysia* sensorimotor synapses. *Journal of Neuroscience* 21: 1247–1256.

Manwell, C. (1963). The chemistry and biology of hemoglobin in some marine clams. I. Distribution of the pigment and properties of the oxygen equilibrium. *Comparative Biochemistry and Physiology* 8: 209–218.

Mapstone, B. D., Underwood, A. J. & Creese, R. G. (1984). Experimental analyses of the commensal relation between intertidal gastropods *Patelloida mufria* and the trochid *Austrocochlea constricta*. *Marine Ecology Progress Series* 17: 85–100.

Marchand, C. R., Griffond, B., Mounzih, K. & Colard, C. (1991). Distribution of methionine-enkephalin-like and FMRFamide-like immunoreactivities in the central nervous system (including dorsal bodies) of the snail *Helix aspersa* (Müller). *Zoological Science* 8: 905–913.

Marcus, E. D. B.-R. & Marcus, E. (1960). On *Tricolia affinis cruenta*. *Boletim, Faculdade se Filosofia Ciências e Letras da Universidade de São Paulo* 260: 171–211.

Marder, E., Bucher, D., Schulz, D. J. & Taylor, A. L. (2005). Invertebrate central pattern generation moves along. *Current Biology* 15: R685–R699.

Marie, B., Marin, F., Marie, A., Bédouet, L., Dubost, L., Alcaraz, G., Milet, C. & Luquet, G. (2009). Evolution of nacre: Biochemistry and proteomics of the shell organic matrix of the cephalopod *Nautilus macromphalus*. *ChemBioChem* 10: 1495–1506.

Marie, B., Le Roy, N., Zanella-Cleon, I., Becchi, M. & Marin, F. (2011). Molecular evolution of mollusc shell proteins: Insights from proteomic analysis of the edible mussel *Mytilus*. *Journal of Molecular Evolution* 72: 531–546.

Marie, B., Joubert, C., Tayale, A., Zanella-Cleon, I., Belliard, C., Piquemal, D., Cochennec-Laureau, N., Marin, F., Gueguen, Y. & Montagnani, C. (2012). Different secretory repertoires control the biomineralization processes of prism and nacre deposition of the pearl oyster shell. *Proceedings of the National Academy of Sciences of the United States of America* 109: 20986–20991.

Marie, B., Jackson, D. J., Ramos-Silva, P., Zanella-Cléon, I., Guichard, N. & Marin, F. (2013). The shell-forming proteome of *Lottia gigantea* reveals both deep conservations and lineage specific novelties. *Federation of European Biochemical Societies (FEBS) Journal* 280: 214–232.

Marigómez, I., Soto, M., Cajaraville, M. P., Angulo, E. & Giamberini, L. (2002). Cellular and subcellular distribution of metals in molluscs. *Microscopy Research and Technique* 56: 358–392.

Marin, A., Belluga, M. D. L., Scognamiglio, G. & Cimino, G. (1997). Morphological and chemical camouflage of the Mediterranean nudibranch *Discodoris indecora* on the sponges *Ircinia variabilis* and *Ircinia fasciculata*. *Journal of Molluscan Studies* 63: 431–439.

Marin, A., Alvarez, L. A., Cimino, G. & Spinella, A. (1999). Chemical defence in cephalaspidean gastropods: Origin, anatomical location and ecological roles. *Journal of Molluscan Studies* 65: 121–131.

Marin, A. & Belluga, M. D. L. (2005). Sponge coating decreases predation on the bivalve *Arca noae*. *Journal of Molluscan Studies* 71: 1–6.

Marín, A. & Ros, J. (2004). Chemical defenses in sacoglossan opisthobranchs : Taxonomic trends and evolutive implications. *Scientia Marina* 68: 227–241.

Marin, F., Corstjens, P., de Gaulejac, B., de Vrind-De Jong, E. & Westbroek, P. (2000). Mucins and molluscan calcification: Molecular characterization of mucoperlin, a novel mucin-like protein from the nacreous shell layer of the fan mussel *Pinna nobilis* (Bivalvia, Pteriomorphia). *Journal of Biological Chemistry* 275: 20667–20675.

Marin, F. & Luquet, G. (2004). Molluscan shell proteins. *Comptes Rendus Palevol* 3: 469–492.

Marin, F. & Luquet, G. (2005). Molluscan biomineralization: The proteinaceous shell constituents of *Pinna nobilis* L. *Materials Science and Engineering C* 25: 105–111.

Marin, F., Le Roy, N. & Marie, B. (2012). The formation and mineralization of mollusk shell. *Frontiers of Bioscience* 4: 1099–1125.

Marin, F., Le Roy, N., Marie, B., Ramos-Silva, P., Wolf, S., Benhamada, S., Guichard, N. & Immel, F. (2014). Synthesis of calcium carbonate biological materials: How many proteins are needed? *Key Engineering Materials* 614: 52–61.

Marinesco, S., Wickremasinghe, N. & Carew, T. J. (2006). Regulation of behavioral and synaptic plasticity by serotonin release within local modulatory fields in the CNS of *Aplysia*. *Journal of Neuroscience* 26: 12682–12693.

Mariottini, P., Di Giulio, A., Appolloni, M. & Smriglio, C. (2013). Phenotypic diversity, taxonomic remarks and updated distribution of the Mediterranean *Jujubinus baudoni* (Monterosato, 1891) (Gastropoda Trochidae). *Biodiversity Journal* 4: 343–354.

Märkel, K. (1957). Bau und Funktion der Pulmonaten-Radula. *Zeitschrift für wissenschaftliche Zoologie* 160: 213–289.

Markich, S. J., Jeffree, R. A. & Burke, P. T. (2002). Freshwater bivalve shells as archival indicators of metal pollution from a copper-uranium mine in tropical northern Australia. *Environmental Science & Technology* 36: 821–832.

Markl, H. (1974). The perception of gravity and of angular acceleration in invertebrates. pp. 17–74 *in* H. H. Kornhuber (Ed.) *Handbook of Sensory Physiology. Vestibular System Part 1: Basic Mechanisms*. Vol. 6. Berlin, Germany, Springer-Verlag.

Markl, J. (2013). Evolution of molluscan hemocyanin structures. *Biochimica et Biophysica Acta* 1834: 1840–1852.

Marko, P. B. (1998). Historical allopatry and the biogeography of speciation in the prosobranch snail genus *Nucella*. *Evolution* 52: 757–774.

Marko, P. B. & Vermeij, G. J. (1999). Molecular phylogenetics and the evolution of labral spines among eastern Pacific ocenebrine gastropods. *Molecular Phylogenetics and Evolution* 13: 275–288.

Marlow, H., Tosches, M. A., Tomer, R., Steinmetz, P. R., Lauri, A., Larsson, T. & Arendt, D. (2014). Larval body patterning and apical organs are conserved in animal evolution. *BMC Biology* 12: 7.

Marsh, H. (1971). The foregut glands of some vermivorous cone shells. *Australian Journal of Zoology* 19: 313–326.

Marshall, B. A. (1978). Cerithiopsidae (Mollusca: Gastropoda) of New Zealand, and a provisional classification of the family. *New Zealand Journal of Zoology* 5: 47–120.

Marshall, B. A. (1980). The systematic position of *Triforis* Deshayes (Mollusca: Gastropoda). *New Zealand Journal of Zoology* 7: 85–88.

Marshall, B. A. (1981). The genus *Williamia* in the western Pacific (Mollusca: Siphonariidae). *New Zealand Journal of Zoology* 8: 487–492.

Marshall, B. A. (1983). A revison of the Recent Triphoridae of southern Australia (Mollusca: Gastropoda). *Records of the Australian Museum, Supplement* 2: 1–119.

Marshall, B. A. (1984). Adelacerithiinae: A new subfamily of the Triphoridae (Mollusca: Gastropoda). *Journal of Molluscan Studies* 50: 78–84.

Marshall, B. A. (1985). Recent and Tertiary Cocculinidae and Pseudococculinidae (Mollusca: Gastropoda) from New Zealand and New South Wales [Australia]. *New Zealand Journal of Zoology* 12: 505–546.

Marshall, B. A. (1988a). Thysanodontinae: A new subfamily of the Trochidae (Gastropoda). *Journal of Molluscan Studies* 54: 215–229.

Marshall, B. A. (1988b). Skeneidae, Vitrinellidae and Orbitestellidae (Mollusca: Gastropoda) associated with biogenic substrata from bathyal depths off New Zealand and New South Wales. *Journal of Natural History* 22: 949–1004.

Marshall, B. A. (1993). The systematic position of *Larochea* Finlay, 1927, and introduction of a new genus and two new species (Gastropoda: Scissurellidae). *Journal of Molluscan Studies* 59: 285–294.

Marshall, B. A. (1995). Recent and Tertiary Trochaclididae from the southwest Pacific (Mollusca: Gastropoda: Trochoidea). *The Veliger* 38: 92–115.

Marshall, B. A. (1997). A luminescent eulimid (Mollusca: Gastropoda) from New Zealand. *Molluscan Research* 18: 69–72.

Marshall, D. J. & Hodgson, A. N. (1990). Structure of the cephalic tentacles of some species of prosobranch limpet (Patellidae and Fissurellidae). *Journal of Molluscan Studies* 56: 415–424.

Marshall, D. J. & McQuaid, C. D. (1992). Comparative aerial metabolism and water relations of the intertidal limpets *Patella granularis* L. (Mollusca: Prosobranchia) and *Siphonaria oculus* Kr. (Mollusca: Pulmonata). *Physiological Zoology* 65: 1040–1056.

Marshall, D. J., Dong, Y. W., McQuaid, C. D. & Williams, G. A. (2011). Thermal adaptation in the intertidal snail *Echinolittorina malaccana* contradicts current theory by revealing the crucial roles of resting metabolism. *Journal of Experimental Biology* 214: 3649–3657.

Marshall, D. J. & McQuaid, C. D. (2011). Warming reduces metabolic rate in marine snails: Adaptation to fluctuating high temperatures challenges the metabolic theory of ecology. *Proceedings of the Royal Society B* 278: 281–288.

Marshall, D. J., Krug, P. J., Kupriyanova, E. K., Byrne, M. & Emlet, R. B. (2012). The biogeography of marine invertebrate life histories. *Annual Review of Ecology, Evolution, and Systematics* 43: 97–114.

Marshall, H. C. (2012). Notes on the veliger larvae and settlement in the anaspidean *Akera bullata*. *The Veliger* 51: 97–101.

Martel, A., Larrivée, D. H. & Himmelman, J. H. (1986). Behaviour and timing of copulation and egg-laying in the neogastropod *Buccinum undatum* L. *Journal of Experimental Marine Biology and Ecology* 96: 27–42.

Martel, A. & Chia, F.-S. (1991). Drifting and dispersal of small bivalves and gastropods with direct development. *Journal of Experimental Marine Biology and Ecology* 150: 131–147.

Martel, A. & Diefenbach, T. (1993). Effects of body size, water current and microhabitat on mucous-thread drifting in post-metamorphic gastropods *Lacuna* spp. *Marine Ecology Progress Series* 99: 215.

Marthy, H. J. (1968). Die Organogenese des Cölomsystems von *Octopus vulgaris* Lam. *Revue suisse de Zoologie* 75: 723–763.

Marthy, H. J. (1987). Organogenesis of the nervous system in cephalopods. pp. 443–459 in M. A. Ali (Ed.) *Nervous Systems in Invertebrates. NATO ASI series: Life sciences.* New York, Plenum Press.

Martin, A. W., Stewart, D. M. & Harrison, F. M. (1965). Urine formation in a pulmonate land snail, *Achatina fulica*. *Journal of Experimental Biology* 42: 99–123.

Martin, A. W. & Harrison, F. M. (1966). Excretion. pp. 353–386 in K. M. Wilbur & Yonge, C. M. (Eds.) *Physiology of Mollusca.* Vol. 2. New York, Academic Press.

Martin, A. W., Catala-Stucki, I. & Ward, P. D. (1978). The growth rate and reproductive behaviour of *Nautilus macromphalus*. *Neues Jahrbuch für Geologie und Paläontologie Abhandlungen* 156: 207–225.

Martin, A. W. & Deyrup-Olsen, I. (1982). Blood venting through the pneumostome in terrestrial slugs. *Comparative Biochemistry and Physiology Part C* 72: 53–58.

Martin, A. W. (1983). Excretion. pp. 353–405 in A. S. M. Saleuddin & Wilbur, K. M. (Eds.) *Physiology, Part 2. The Mollusca.* Vol. 5. New York, Academic Press.

Martin, A. W. & Deyrup-Olsen, I. (1986). Function of the epithelial channel cells of the body wall of a terrestrial slug, *Ariolimax columbianus*. *Journal of Experimental Biology* 121: 301–314.

Martin, A. W., Deyrup-Olsen, I. & Stewart, D. M. (1990). Regulation of body volume by the peripheral nervous system of the terrestrial slug *Ariolimax columbianus*. *Journal of Experimental Zoology Part A* 253: 121–131.

Martin, D., Gil, J., Abgarian, C., Evans, E., Turner, E. M. & Nygren, A. (2014). *Coralliophila* from Grand Cayman: Specialized coral predator or parasite? *Coral Reefs* 33: 1017.

Martin, G. G., Oakes, C. T., Tousignant, H. R., Crabtree, H. & Yamakawa, R. (2007). Structure and function of haemocytes in two marine gastropods, *Megathura crenulata* and *Aplysia californica*. *Journal of Molluscan Studies* 73: 355–365.

Martin, G. G., Martin, A. J., Tsai, W. & Hafner, J. C. (2011). Production of digestive enzymes along the gut of the giant keyhole limpet *Megathura crenulata* (Mollusca: Vetigastropoda). *Comparative Biochemistry and Physiology Part A* 160: 365–373.

Martin, R., Hess, M., Schrödl, M. & Tomaschko, K.-H. (2009). Cnidosac morphology in dendronotacean and aeolidacean nudibranch molluscs: From expulsion of nematocysts to use in defense? *Marine Biology*: 261–268.

Martin, R., Tomaschko, K.-H., Hess, M. & Schrödl, M. (2010). Cnidosac-related structures in *Embletonia* (Mollusca, Nudibranchia) compared with dendronotacean and aeolidacean species. *Open Marine Biology Journal* 4: 96–100.

Martin, W. A. (1974). Physiology of the excretory organ of cephalopods. *Fortschritte der Zoologie* 23: 112–123.

Martínez-Pita, I., Guerra-García, J. M., Sánchez-España, A. N. A. I. & García, F. J. (2006). Observations on the ontogenetic and intraspecific changes in the radula of *Polycera aurantiomarginata* García-Gómez and Bobo, 1984 (Gastropoda: Opistobranchia) from Southern Spain. *Scientia Marina* 70: 227–234.

Martínez, G., Mettifogo, L., Lenoir, R. & Olivares, A. (2000). Prostaglandins and reproduction of the scallop *Argopecten purpuratus*: II. Relationship with gamete release. *Journal of Experimental Zoology Part A* 287: 86–91.

de Frias Martins, A. M. (1996). Relationships within the Ellobiidae. pp. 285–294 in J. D. Taylor (Ed.) *Origin and Evolutionary Radiation of the Mollusca.* Oxford, UK, Oxford University Press.

de Frias Martins, A. M. (2001). Ellobiidae: Lost between land and sea. *Journal of Shellfish Research* 20: 441–446.

Martoja, M. (1964). Développement de l'appareil reproducteur chez les Gastéropodes pulmonés. *Annales Biologiques* 3: 14–232.

Martoja, M. (1965a). Données relatives à l'organe juxta-ganglionaire des Prosobranches Diotocardes. *Comptes rendus de l'Académie des sciences* 261: 3195–3196.

Martoja, M. (1965b). Existence d'une organe juxta-ganglionaire chez *Aplysia punctata* Cuv. (Gastéropode, Opisthobranchie). *Comptes rendus de l'Académie des sciences* 260: 4615–4617.

Martoja, M. (1965c). Sur l'incubation et l'existence possible d'une glande endocrine chez *Hydromyles globulosa* Rang (*Halopsyche gaudichaudi* Keferstein), Gastéropode, Gymnosome. *Comptes rendus de l'Académie des sciences* 260: 2907–2909.

Martoja, M. (1975). Le rein de *Pomatias* (= *Cyclostoma*) *elegans* (Gastéropode, Prosobranche): Données structurales et analytiques. *Annales des sciences naturelles. Zoologie* 17: 535–557.

Maruyama, T. & Heslinga, G. A. (1997). Fecal discharge of zooxanthellae in the giant clam *Tridacna derasa*, with reference to their *in situ* growth rate. *Marine Biology* 127: 473–477.

Maruyama, T., Ishikura, M., Yamazaki, S. & Kanai, S. (1998). Molecular phylogeny of zooxanthellate bivalves. *Biological Bulletin* 195: 70–77.

Masefield, J. R. B. (1899). The economic use of some British Mollusca. *Journal of Conchology* 9: 153–157, 161–164.

Mason, A. Z. & Nott, J. A. (1981). Role of intracellular biomineralized granules in the regulation and detoxification of metals in gastropods with special reference to the marine prosobranch *Littorina littorea. Aquatic Toxicology* 1: 239–256.

Mason, M. J., Watkins, A. J., Wakabayashi, J., Buechler, J., Pepino, C., Brown, M. & Wright, W. G. (2014). Connecting model species to nature: Predator-induced long-term sensitization in *Aplysia californica. Learning and Memory* 21: 363–367.

Matera, E. M. & Davis, W. J. (1982). Paddle cilia (discocilia) in chemosensitive structures of the gastropod mollusk *Pleurobranchaea californica. Cell and Tissue Research* 222: 25–40.

Mather, J. A. (1991). Navigation by spatial memory and use of visual landmarks in octopuses. *Journal of Comparative Physiology A* 168: 491–497.

Mather, J. A. & Anderson, R. C. (1993). Personalities of octopuses (*Octopus rubescens*). *Journal of Comparative Psychology* 107: 336–340.

Mather, J. A. (1995). Cognition in cephalopods. *Advances in the Study of Behavior* 24: 317–353.

Mather, J. A. (2008a). Cephalopod consciousness: Behavioural evidence. *Consciousness and Cognition* 17: 37–48.

Mather, J. A. (2008b). To boldly go where no mollusc has gone before: Personality, play, thinking, and consciousness in cephalopods. *American Malacological Bulletin* 24: 51–58.

Mather, J. A. (2010). Vigilance and antipredator responses of Caribbean reef squid. *Marine and Freshwater Behaviour and Physiology* 43: 357–370.

Mather, J. A., Griebel, U. & Byrne, R. A. (2010). Squid dances: An ethogram of postures and actions of *Sepioteuthis sepioidea* squid with a muscular hydrostatic system. *Marine and Freshwater Behaviour and Physiology* 43: 45–61.

Mather, J. A. & Kuba, M. J. (2013). The cephalopod specialties: Complex nervous system, learning, and cognition. *Canadian Journal of Zoology* 91: 431–449.

Mathers, N. F. (1972). The tracing of a natural algal food labelled with a carbon 14 isotope through the digestive tract of *Ostrea edulis* L. *Proceedings of the Malacological Society of London* 40: 115–124.

Mathers, N. F. (1973). A comparative histochemical survey of enzymes associated with the process of digestion in *Ostrea edulis* and *Crassostrea angulata* (Mollusca: Bivalvia). *Journal of Zoology* 169: 169–179.

Mäthger, L. M. & Hanlon, R. T. (2006). Anatomical basis for camouflaged polarized light communication in squid. *Biology Letters* 2: 494–496.

Mäthger, L. M. & Hanlon, R. T. (2007). Malleable skin coloration in cephalopods: Selective reflectance, transmission and absorbance of light by chromatophores and iridophores. *Cell and Tissue Research* 329: 179–186.

Mäthger, L. M., Chiao, C. C., Barbosa, A. & Hanlon, R. T. (2008). Color matching on natural substrates in cuttlefish, *Sepia officinalis. Journal of Comparative Physiology A* 194: 577–585.

Mäthger, L. M., Roberts, S. B. & Hanlon, R. T. (2010). Evidence for distributed light sensing in the skin of cuttlefish, *Sepia officinalis. Biology Letters* 6: 600–603.

Mäthger, L. M., Bell, G. R., Kuzirian, A. M., Allen, J. J. & Hanlon, R. T. (2012). How does the blue-ringed octopus (*Hapalochlaena lunulata*) flash its blue rings? *Journal of Experimental Biology* 215: 3752–3757.

Mathieu, M., Robbins, I. & Lubet, P. (1991). The neuroendocrinology of *Mytilus edulis. Aquaculture* 94: 213–223.

Matos, E., Matos, P. & Azevedo, C. (2005). Observations on the intracytoplasmic microsporidian *Steinhausia mytilovum*, a parasite of mussel (*Mytella guyanensis*) oocytes from the Amazon river estuary. *Brazilian Journal of Morphological Sciences* 22: 183–186.

Matozzo, V., Ballarin, L. & Marin, M. G. (2004). Exposure of the clam *Tapes philippinarum* to 4-nonylphenol: Changes in anti-oxidant enzyme activities and re-burrowing capability. *Marine Pollution Bulletin* 48: 563–571.

Matricon-Gondran, M. (1990). The site of ultrafiltration in the kidney sac of the pulmonate gastropod *Biomphalaria glabrata. Tissue and Cell* 22: 911–923.

Matsui, S., Seidou, M., Horiuchi, S., Uchiyama, I. & Kito, Y. (1988). Adaptation of a deep sea cephalopod to the photic environment: Evidence for 3 visual pigments. *Journal of General Physiology* 92: 55–66.

Matsukuma, A. (1978). Fossil boreholes made by shell-boring predators or commensals. 1. Boreholes of capulid gastropods. *Venus* 37: 29–45.

Matsukuma, A. (1996). Transposed hinges: A polymorphism of bivalve shells. *Journal of Molluscan Studies* 62: 415–431.

Matsunaga, S., Fusetani, N., Hashimoto, K., Koseki, K. & Noma, M. (1986). Bioactive marine metabolites. Part 13. Kabiramide C, a novel antifungal macrolide from nudibranch eggmasses. *Journal of the American Chemical Society* 108: 847–849.

Matsunaga, S. (2006). Trisoxazole macrolides from *Hexabranchus* nudibranchs and other marine invertebrates. pp. 241–260 *in* G. Cimino & Gavagnin, M. (Eds.) *Molluscs: From Chemo-ecological Study to Biotechnological Application. Progress in Molecular and Subcellular Biology.* Berlin, Germany, Springer.

Matsuno, A. (1987). Ultrastructural classification of smooth muscle cells in invertebrates and vertebrates. *Zoological Science* 4: 15–22.

Matsuno, A. & Kuga, H. (1989). Ultrastructure of muscle cells in the adductor of the boring clam *Tridacna crocea* Matsuno. *Journal of Morphology* 200: 247–253.

Matsuno, A., Ishida, H. & Hori, H. (1993). Two kinds of thick filament in smooth muscle cells in the adductor of a clam, *Chlamys nobilis. Tissue and Cell* 25: 325–332.

Matsuno, T. (2001). Aquatic animal carotenoids. *Fisheries Science* 67: 771–783.

Matsuoka, A., Ohie, Y., Imai, K. & Shikama, K. (1996). The dimer-monomer conversion of *Cerithidea* myoglobin coupled with the heme iron oxidation. *Comparative Biochemistry and Physiology Part B* 115: 483–492.

Matthews-Cascon, H., Pereira Alencar, H. A., Guimaraes Rabay, S. & Mota, R. M. S. (2005). Sexual dimorphism in the radula of *Pisania pusio* (Linnaeus, 1758) (Mollusca, Gastropoda, Buccinidae). *Thalassas* 21: 29–33.

Matthiessen, P. & Gibbs, P. E. (1998). Critical appraisal of the evidence for tributyltin-mediated endocrine disruption in mollusks. *Environmental Toxicology and Chemistry* 17: 37–43.

Matthiessen, P. (2008). An assessment of endocrine disruption in mollusks and the potential for developing internationally standardized mollusk life cycle test guidelines. *Integrated Environmental Assessment and Management* 4: 274–284.

Maurand, J. & Loubes, C. (1979). Les microsporidies parasites de mollusques. *Haliotis* 8: 39–48.

Maxfield, B. Y. M. (1953). Axoplasmic proteins of the squid giant nerve fiber with particular reference to the fibrous protein. *Journal of General Physiology* 37: 201–217.

Maxwell, S. L., Fuller, R. A., Brooks, T. M. & Watson, J. E. M. (2016). Biodiversity: The ravages of guns, nets and bulldozers. *Nature* 536: 143–145.

Maxwell, W. L. (1974). Spermiogenesis of *Eledone cirrhosa* Lamarck (Cephalopoda, Octopoda). *Proceedings of the Royal Society B* 186: 181–190.

Maxwell, W. L. (1983). Mollusca. pp. 275–319 *in* K. D. Adiyodi & Adiyodi, R. G. (Eds.) *Reproductive Biology of Invertebrates, Spermatogenesis and Sperm Function*. Vol. 2. Chichester, UK, John Wiley & Sons.

Mayr, E. (1963). *Animal Species and Evolution*. Cambridge, MA, Belknap Press of Harvard University.

McAlister, R. O. & Fisher, F. M. (1968). Responses of the false limpet, *Siphonaria pectinata* Linnaeus (Gastropoda, Pulmonata) to osmotic stress. *Biological Bulletin* 134: 96–117.

McBee, R. H. (1971). Significance of intestinal microflora in herbivory. *Annual Review of Ecology and Systematics* 2 : 165–176.

McBride, C. J., Williams, A. H. & Henry, R. P. (1989). Effects of temperature on climbing behavior of *Littorina irrorata*: On avoiding a hot foot. *Marine Behaviour and Physiology* 14: 93–100.

McCaffrey, K. & Johnson, P. T. J. (2017). Drivers of symbiont diversity in freshwater snails: A comparative analysis of resource availability, community heterogeneity, and colonization opportunities. *Oecologia* 183: 927–938.

McClain, C. R., Stegen, J. C. & Hurlbert, A. H. (2011). Dispersal, environmental niches and oceanic-scale turnover in deep-sea bivalves. *Proceedings of the Royal Society of London B* 279: 2011–2166.

McClain, C. R., Balk, M. A., Benfield, M. C., Branch, T. A., Chen, C., Cosgrove, J., Dove, A. D., Gaskins, L. C., Helm, R. R. & Hochberg, F. G. (2015). Sizing ocean giants: Patterns of intraspecific size variation in marine megafauna. *PeerJ* 3: e715.

McClary, A. (1964). Surface inspiration and ciliary feeding in *Pomacea paludosa* (Prosobranchia: Mesogastropoda: Ampullariidae). *Malacologia* 2: 87–104.

McClintock, J. B. & Janssen, J. (1990). Pteropod abduction as a chemical defense in a pelagic Antarctic amphipod [*Hyperiella dilata*: *Clione limacina*]. *Nature* 346: 462–464.

McClintock, J. B. (1994). Trophic biology of Antarctic shallow-water echinoderms. *Marine Ecology Progress Series* 111: 191–202.

McClintock, J. B., Baker, B. J., Slattery, M., Heine, J. N., Bryan, P. J., Yoshida, W., Davies-Coleman, M. T. & Faulkner, D. J. (1994). Chemical defense of common Antarctic shallow-water nudibranch *Tritoniella belli* Eliot (Mollusca: Tritonidae) and its prey, *Clavularia frankliniana* Rouel (Cnidaria: Octocorallia). *Journal of Chemical Ecology* 20: 3361–3372.

McClymont, H. E., Dunn, A. M., Terry, R. S., Rollinson, D., Littlewood, D. T. J. & Smith, J. E. (2005). Molecular data suggest that microsporidian parasites in freshwater snails are diverse. *International Journal for Parasitology* 35: 1071–1078.

McConnaughey, T. A. & Gillikin, D. P. (2008). Carbon isotopes in mollusk shell carbonates. *Geo-Marine Letters* 28: 287–299.

McDermott, J. J. & Roe, P. (1985). Food, feeding behavior and feeding ecology of nemerteans. *American Zoologist* 25: 113–125.

McDougall, C., Aguilera, F. & Degnan, B. M. (2013). Rapid evolution of pearl oyster shell matrix proteins with repetitive, low-complexity domains. *Journal of the Royal Society Interface* 10: 20130041.

McDowall, R. M. (1992). Diadromy: Origins and definitions of terminology. *Copeia* 1: 248–251.

McDowall, R. M. (2007). On amphidromy, a distinct form of diadromy in aquatic organisms. *Fish and Fisheries* 8: 1–13.

McDowell, I. C., Modak, T. H., Lane, C. E. & Gomez-Chiarri, M. (2016). Multi-species protein similarity clustering reveals novel expanded immune gene families in the eastern oyster *Crassostrea virginica*. *Fish and Shellfish Immunology* 53: 13–23.

McFadien-Carter, M. (1979). Scaphopoda. pp. 95–111 *in* A. C. Giese & Pearse, J. S. (Eds.) *Reproduction of Marine Invertebrates. Molluscs: Pelecypods and Lesser Classes*. Vol. 5. New York, Academic Press.

McFall-Ngai, M. J., Henderson, B. & Ruby, E. G. (2005). *The Influence of Cooperative Bacteria on Animal Host Biology*. Cambridge, UK, Cambridge University Press.

McFarland, F. K. & Muller-Parker, G. (1993). Photosynthesis and retention of zooxanthellae and zoochlorellae within the aeolid nudibranch *Aeolidia papillosa*. *Biological Bulletin* 184: 223–229.

McFarlane, I. D. (1980). Trail-following and trail-searching behaviour in homing of the intertidal gastropod mollusc, *Onchidium verruculatum*. *Marine Behaviour and Physiology* 7: 95–108.

McGee, B. L. & Targett, N. M. (1989). Larval habitat selection in *Crepidula* (L.) and its effect on adult distribution patterns. *Journal of Experimental Marine Biology and Ecology* 131: 195–214.

McGhee, G. R. (1978). Analysis of shell torsion phenomenon in Bivalvia. *Lethaia* 11: 315–329.

McGladdery, S. E. (1999). Shellfish diseases (viral, bacterial and fungal). pp. 723–842 *in* P. T. K. Woo & Bruno, D. W. (Eds.) *Viral, Bacterial and Fungal Infections. Fish Diseases and Disorders*. Wallingford, UK, CABI Publishing.

McGladdery, S. E. (2011). Shellfish diseases (viral, bacterial and fungal). pp. 748–854 *in* P. T. K. Woo & Bruno, D. W. (Eds.) *Viral, Bacterial and Fungal Infections 2nd ed. Fish Diseases and Disorders*. Vol. 3. Wallingford, UK, CABI Publishing.

McIntosh, J. M., Foderaro, T. A., Li, W. Z., Ireland, C. M. & Olivera, B. M. (1993). Presence of serotonin in the venom of *Conus imperialis*. *Toxicon* 31: 1561–1566.

McIntosh, J. M. & Jones, R. M. (2001). Cone venom: From accidental stings to deliberate injection. *Toxicon* 39: 1447–1451.

McKaye, K. R., Stauffer, J. R. & Louda, S. M. (1985). Fish predation as a factor in the distribution of Lake Malawi gastropods. *Experimental Biology* 45: 279–289.

McKee, A. E. & Wiederholt, M. L. (1974). *Aplysia* statocyst receptor cells: Fine structure. *Brain Research* 81: 310–313.

McKillup, S. C., McKillup, R. V. & Pape, T. (2000). Flies that are parasitoids of a marine snail: The larviposition behaviour and life cycles of *Sarcophaga megafilosia* and *Sarcophaga meiofilosia*. *Hydrobiologia* 439: 141–149.

McKillup, S. C. & McKillup, R. V. (2002). Flies that attack polymorphic snails on coloured backgrounds: Selection for crypsis by a sarcophagid parasitoid of *Littoraria filosa*. *Biological Journal of the Linnean Society* 77: 367–377.

McKinney, M. L. (1988). *Heterochrony in Evolution: A Multidisciplinary Approach*. Vol. 7. New York, Springer Science + Business Media.

McKinney, M. L. & McNamara, K. J. (1991). Heterochrony. pp. 1–12. *Heterochrony: The Evolution of Ontogeny*. New York, Springer.

McLachlan, A. & Brown, A. (2006). *The Ecology of Sandy Shores*. 2nd ed. Amsterdam, the Netherlands, Academic Press.

McLean, J. H. (1962). Feeding behavior of the chiton *Placiphorella*. *Proceedings of the Malacological Society of London* 35: 23–26.

McLean, J. H. (1966). West American prosobranch Gastropoda: Superfamilies Patellacea, Pleurotomariacea and Fissurellacea. PhD Dissertation, Stanford University.

McLean, J. H. (1984). Shell reduction and loss in fissurellids: A review of genera and species in the *Fissurellidea* group. *American Malacological Bulletin* 2: 21–34.

McLean, J. H. (1990). Neolepetopsidae, a new docoglossate limpet family from hydrothermal vents and its relevance to patellogastropod evolution. *Journal of Zoology* 222: 485–528.

McLean, J. H. (2008). Three new species of the family Neolepetopsidae (Patellogastropoda) from hydrothermal vents and whale-falls in the northeastern Pacific. *Journal of Shellfish Research* 27: 15–20.

McLean, J. H. (2012). New species and genera of colloniids from Indo-Pacific coral reefs, with the definition of a new subfamily Liotipomatinae n. subfam. (Turbinoidea, Colloniidae). *Zoosystema* 34: 343–376.

McLean, N. (1970). Digestion in *Haliotis rufescens* Swainson (Gastropoda: Prosobranchia). *Journal of Experimental Zoology Part A* 173: 303–318.

McLusky, D. S., Bryant, V. & Campbell, R. (1986). The effects of temperature and salinity on the toxicity of heavy metals to marine and estuarine invertebrates. *Oceanography and Marine Biology* 24: 481–520.

McMahon, B. R., Burggren, W. W., Pinder, A. W. & Wheatly, M. G. (1991). Air exposure and physiological compensation in a tropical intertidal chiton, *Chiton stokesii* (Mollusca: Polyplacophora). *Physiological Zoology* 64: 728–747.

McMahon, B. R., Wilkens, J. L. & Smith, P. J. S. (1997). Invertebrate circulatory systems. pp. 931–1008 in W. H. Dantzler (Ed.) *Handbook of Physiology Section 13, Comparative Physiology.* New York, Oxford University Press.

McMahon, D. G. & Block, G. D. (1982). Organized photoreceptor layer is not required for light responses in three opisthobranch eyes. *Society for Neuroscience, Abstracts* 8: 33.

McMahon, D. G., Wallace, S. F. & Block, G. D. (1984). Cellular analysis of the *Bulla* circadian pacemaker system: Neurophysiological basis of circadian rhythmicity. *Journal of Comparative Physiology* 155: 379–385.

McMahon, R. F. (1985). Interspecific relationships between morphometric parameters and the vertical distribution patterns of seven species of turbinate gastropods on mangrove trees in Hong Kong. pp. 199–215 in B. Morton & Dudgeon, D. (Eds.) *Proceedings of the Second International Workshop on the Malacofauna of Hong Kong and of Southern China, Hong Kong, 6–24 April 1983.* Hong Kong, Hong Kong University Press.

McMahon, R. F. & Britton, J. C. (1985). The relationship between vertical distribution, thermal tolerance, evaporative water loss rate, and behaviour on emergence in six species of mangrove gastropods from Hong Kong. pp. 563–582 in B. Morton & Dudgeon, D. (Eds.) *Proceedings of the Second International Workshop on the Malacofauna of Hong Kong and of Southern China, Hong Kong, 6–24 April 1983. Vol. 2.* Hong Kong, Hong Kong University Press.

McMahon, R. F. (1988). Respiratory response to periodic emergence in intertidal molluscs. *American Zoologist* 28: 97–114.

McMahon, R. F. (1990). Thermal tolerance, evaporative water loss, air-water oxygen consumption and zonation of intertidal prosobranchs: A new synthesis. *Hydrobiologia* 193: 241–260.

McMahon, R. F. & Cleland, J. D. (1990). Thermal tolerance, evaporative water loss and behavior during prolonged emergence in the high zoned mangrove gastropod *Cerithidea ornata*: Evidence for atmospheric water uptake. pp. 1123–1140 in B. Morton (Ed.) *The Marine Flora and Fauna of Hong Kong and Southern China II: Proceedings of the Second International Marine Biological Workshop held in Hong Kong April 2–24, 1986.* Hong Kong, Hong Kong University Press.

McMullin, E. R., Bergquist, D. C. & Fisher, C. R. (2000). Metazoans in extreme environments: Adaptations of hydrothermal vent and hydrocarbon seep fauna. *Gravitational and Space Research* 13: 13–24.

McNamara, K. J. (1978). Symbiosis between gastropods and bryozoans in the late Ordovician of Cumbria, England. *Lethaia* 11: 25–40.

McNeil, B. I. & Matear, R. J. (2008). Southern Ocean acidification: A tipping point at 450-ppm atmospheric CO_2. *Proceedings of the National Academy of Sciences of the United States of America* 105: 18860–18864.

McQuaid, C. D., Cretchley, R. & Rayner, J. L. (1999). Chemical defence of the intertidal pulmonate limpet *Siphonaria capensis* (Quoy & Gaimard) against natural predators. *Journal of Experimental Marine Biology and Ecology* 237: 141–154.

McQuiston, R. W. (1969). Cyclic activity in the digestive diverticula of *Lasaea rubra* (Montagu) (Bivalvia: Eulamellibranchia). *Proceedings of the Malacological Society of London* 38: 483–492.

McQuiston, R. W. (1970). Fine structure of the gastric shield in the lamellibranch bivalve, *Lasaea rubra* (Montagu). *Proceedings of the Malacological Society of London* 39: 69–75.

McShane, P. E. & Smith, M. G. (1992). Shell growth checks are unreliable indicators of age of the abalone *Haliotis rubra* (Mollusca: Gastropoda). *Australian Journal of Marine and Freshwater Research* 43: 1215–1219.

McShane, P. E., Gorfine, H. K. & Knuckey, I. A. (1994). Factors influencing food selection in the abalone *Haliotis rubra* (Mollusca, Gastropoda). *Journal of Experimental Marine Biology and Ecology* 176: 27–37.

McVean, A. R. (1984). *Octopus* extraocular muscle. *Comparative Biochemistry and Physiology Part A* 78: 711–718.

Mead, A. R. (1943). Revision of the giant West Coast land slugs of the genus *Ariolimax* Moerch (Pulmonata: Arionidae). *The American Midland Naturalist* 30: 675–717.

Mead, A. R. (1961). *The Giant African Snail.* Chicago, University of Chicago Press.

Mead, A. R. (1979). *Economic Malacology with Particular Reference to Achatina fulica. Vol. 2B.* London, Academic Press.

Medina, B., Guzman, H. M. & Mair, J. M. (2007). Failed recovery of a collapsed scallop *Argopecten ventricosus* fishery in Las Perlas Archipelago, Panama. *Journal of Shellfish Research* 26: 9–15.

Medina, M., Lal, S., Vallès, Y., Takaoka, T. L., Dayrat, B. A., Boore, J. L. & Gosliner, T. M. (2011). Crawling through time: Transition of snails to slugs dating back to the Paleozoic, based on mitochondrial phylogenomics. *Marine Genomics* 4: 51–59.

Medzhitov, R. & Janeway, C. A. (2002). Decoding the patterns of self and nonself by the innate immune system. *Science* 296: 298–300.

Meenakshi, V. R. (1957). Anaerobiosis in the south Indian applesnail *Pila virens* (Lamarck) during aestivation. *Journal of the Zoological Society of India* 9: 62–71.

Meenakshi, V. R., Harpe, P. E., Watabe, N., Wilbur, K. M. & Menzies, R. J. (1970). Ultrastructure, histochemistry and amino acid composition of the shell of *Neopilina*: Scientific Results of the Southeast Pacific Expedition. II. pp. 1–12 in E. Chin (Ed.) *Scientific Research of the Southeast Pacific Expedition (A. Bruun Report). Vol. 2.* Galveston, TX.

Meeuse, B. J. D. & Fluegel, W. (1958). Carbohydrases in the sugargland juice of *Cryptochiton* (Polyplacophora, Mollusca). *Nature* 181: 699–700.

Meeuse, B. J. D. & Fluegel, W. (1959). Carbohydrate-digesting enzymes in the sugar gland juice of *Cryptochiton stelleri* Middendorff (Polyplacophora, Mollusca). *Archives Néerlandaises de Zoologie* 13: 301–313.

Megina, C. & Cervera, J. L. (2003). Diet, prey selection and cannibalism in the hunter opisthobranch *Roboastra europaea*. *Journal of the Marine Biological Association of the United Kingdom* 83: 489–495.

Van der Meij, S. E. T. & Reijnen, B. T. (2012). First observations of attempted nudibranch predation by sea anemones. *Marine Biodiversity* 42: 281–283.

Meinhardt, H. & Klingler, M. (1987). A model for pattern formation on the shells of molluscs. *Journal of Theoretical Biology* 126: 63–89.

Meisenheimer, J. (1896). Entwicklungsgeschichte von *Limax maximus* L. 1. Theil. Furchung und Keimblätterbildung. 2. Theil. Die Larvenperiode. *Zeitschrift für wissenschaftliche Zoologie* 62: 415–468, 573–664.

Meisenheimer, J. (1912). Die Weinbergschnecke, *Helix pomatia* *Monographien einheimischer Tiere* 4: 1–140.

Melone, G. (1986). Sex changes in *Opalia crenata* (L.) (Gastropoda: Epitoniidae). pp. 54 *in* D. Heppell (Ed.) *9th International Malacology Congress*. Edinburgh, UK, National Museum of Scotland.

Melzner, F., Bock, C. & Pörtner, H. (2007a). Thermal sensitivity of the venous return system in the cephalopod *Sepia officinalis*. *Comparative Biochemistry and Physiology Part A* 146: S165.

Melzner, F., Mark, F. C. & Pörtner, H. O. (2007b). Role of blood-oxygen transport in thermal tolerance of the cuttlefish, *Sepia officinalis*. *Integrative and Comparative Biology* 47: 645–655.

Meng, J., Zhu, Q., Zhang, L., Li, C., Li, L., She, Z., Huang, B. & Zhang, G. (2013). Genome and transcriptome analyses provide insight into the euryhaline adaptation mechanism of *Crassostrea gigas*. *PLoS ONE* 8: 1–14.

Menge, B. A. (1975). Brood or broadcast? The adaptive significance of different reproductive strategies in the two intertidal sea stars *Leptasterias hexactis* and *Pisaster ochraceus*. *Marine Biology* 31: 87–100.

Menge, B. A. (1976). Organization of the New England rocky intertidal community: Role of predation, competition, and environmental heterogeneity. *Ecological Monographs* 46: 355–393.

Menge, B. A. & Sutherland, J. P. (1976). Species diversity gradients: Synthesis of the roles of predation, competition, and temporal heterogeneity. *American Naturalist* 110: 351–369.

Menge, B. A., Berlow, E. L., Blanchette, C. A., Navarrete, S. A. & Yamada, S. B. (1994). The keystone species concept: Variation in interaction strength in a rocky intertidal habitat. *Ecological Monographs* 64: 249–286.

Menshutkin, V. V. & Natochin, V. (2007). Coupled evolution of digestive, respiratory, circulatory, and excretory systems: A model investigation. *Journal of Evolutionary Biochemistry and Physiology* 43: 335–341.

Mercier, A. & Hamel, J.-F. (2008). Nature and role of newly described symbiotic associations between a sea anemone and gastropods at bathyal depths in the NW Atlantic. *Journal of Experimental Marine Biology and Ecology* 358: 57–69.

Mercier, A., Schofield, M. & Hamel, J.-F. (2011). Evidence of dietary feedback in a facultative association between deep-sea gastropods and sea anemones. *Journal of Experimental Marine Biology and Ecology* 396: 207–215.

Merdsoy, B. & Farley, J. (1973). Phasic activity in the digestive gland cells of the marine prosobranch gastropod *Littorina littorea* (L.). *Proceedings of the Malacological Society of London* 40: 473–482.

Merkel, J., Wollesen, T., Lieb, B. & Wanninger, A. (2012). Spiral cleavage and early embryology of a loxosomatid entoproct and the usefulness of spiralian apical cross patterns for phylogenetic inferences. *BMC Developmental Biology* 12: 11.

Mesías-Gansbiller, C., Pazos, A. J., Sánchez, J. L. & Pérez-Parallé, M. L. (2013). First evidence of the presence of *NK2* and *Tlx* genes in bivalve molluscs. *Canadian Journal of Zoology* 91: 275–280.

Messenger, J. B. (1967). The effects on locomotion of lesions to the visuo-motor system in octopus. *Proceedings of the Royal Society B* 167: 252–281.

Messenger, J. B. (1968). The visual attack of the cuttlefish, *Sepia officinalis*. *Animal Behaviour* 16: 342–357.

Messenger, J. B. & Sanders, G. D. (1972). Visual preference and two-cue discrimination learning in *Octopus*. *Animal Behaviour* 20: 580–585.

Messenger, J. B. (1973). Learning in the cuttlefish, *Sepia*. *Animal Behaviour* 21: 801–826.

Messenger, J. B., Wilson, A. P. & Hedge, A. (1973). Some evidence for color blindness in *Octopus*. *Journal of Experimental Biology* 59: 77–94.

Messenger, J. B. (1977a). Evidence that *Octopus* is colour blind. *Journal of Experimental Biology* 70: 49–55.

Messenger, J. B. (1977b). Prey-capture and learning in the cuttlefish, *Sepia*. pp. 347–376 *in* M. Nixon & Messenger, J. B. (Eds.) *The Biology of Cephalopods*. *Symposia of The Zoological Society of London*. London, UK, Academic Press.

Messenger, J. B. (1983). Multimodal convergence and the regulation of motor programs in cephalopods. pp. 77–98 *in* E. Horn (Ed.) *Multimodal Convergences in Sensory Systems*. *Fortschritte der Zoologie*. Vol. 28. Stuttgart, Gustav Fischer.

Messenger, J. B. (1991). Photoreception and vision in molluscs. pp. 364–397 *in* J. R. Cronly-Dillon & Gregory, R. L. (Eds.) *Vision and Visual Dysfunction*. Vol. 2. Houndmills, UK, Macmillan Press.

Messenger, J. B. (1996). Neurotransmitters of cephalopods. *Invertebrate Neuroscience* 2: 95–114.

Messenger, J. B. & Young, J. Z. (1999). The radular apparatus of cephalopods. *Philosophical Transactions of the Royal Society B* 354: 161–182.

Metzger, M. J., Reinisch, C., Sherry, J. & Goff, S. P. (2015). Horizontal transmission of clonal cancer cells causes leukemia in soft-shell clams. *Cell* 161: 255–263.

Meyer-Rochow, V. B. & West, S. (1999). The molluscan radula as a source of information about metal uptake from the environment. *Mémoires de Biospéologie* 26: 85–89.

Meyer, C. P. (2003). Molecular systematics of cowries (Gastropoda: Cypraeidae) and diversification patterns in the tropics. *Biological Journal of the Linnean Society* 79: 401–459.

Meyer, C. P. (2004). Toward comprehensiveness: Increased molecular sampling within Cypraeidae and its phylogenetic implications. *Malacologia* 46: 127–156.

Meyer, C. P., Geller, J. B. & Paulay, G. (2005). Fine scale endemism on coral reefs: Archipelagic differentiation in turbinid gastropods. *Evolution* 59: 113–125.

Meyer, W. M. & Cowie, R. H. (2010). Feeding preferences of two predatory snails introduced to Hawaii and their conservation implications. *Malacologia* 53: 135–144.

Meyer, W. M., Yeung, N. W., Slapcinsky, J. & Hayes, K. A. (2017). Two for one: Inadvertent introduction of *Euglandina* species during failed bio-control efforts in Hawaii. *Biological Invasions* 19: 1399–1405.

Meyhöfer, E., Morse, M. P. & Robinson, W. E. (1985). Podocytes in bivalve molluscs: Morphological evidence for ultrafiltration. *Journal of Comparative Physiology B* 156: 151–161.

Meyhöfer, E. & Morse, M. P. (1996). Characterization of the bivalve ultrafiltration system in *Mytilus edulis*, *Chlamys hastata*, and *Mercenaria mercenaria*. *Invertebrate Biology* 115: 20–29.

Mi, J., Orbea, A., Syme, N., Ahmed, M., Cajaraville, M. P. & Cristobal, S. (2005). Peroxisomal proteomics, a new tool for risk assessment of peroxisome proliferating pollutants in the marine environment. *Proteomics* 5: 3954–3965.

Michaelidis, B. & Beis, I. (1990). Studies on the anaerobic energy metabolism in the foot muscle of marine gastropod *Patella caerulea* (L.). *Comparative Biochemistry and Physiology Part B* 95: 493–500.

Michaelidis, B. (2002). Studies on the extra- and intracellular acid-base status and its role on metabolic depression in the land snail *Helix lucorum* (L.) during estivation. *Journal of Comparative Physiology B* 172: 347–354.

Michel, E. (1994). Why snails radiate: A review of gastropod evolution in long-lived lakes, both recent and fossil. *Ergebnisse der Limnologie* 44: 285–317.

Michel, S., Geusz, M. E., Zaritsky, J. J. & Block, G. D. (1993). Circadian rhythm in membrane conductance expressed in isolated neurons. *Science* 259: 239–241.

Michelson, E. H. (1957). Studies on the biological control of schistosome-bearing snails: Predators and parasites of freshwater Mollusca—A review of the literature. *Parasitology* 47: 413–426.

Michelson, E. H. (1961). An acid-fast pathogen of fresh water snails. *American Journal of Tropical Medicine and Hygiene* 10: 423–433.

Michelson, E. H. (1963). *Plistophora husseyi* sp. n., a microsporidian parasite of aquatic pulmonate snails. *Journal of Insect Pathology* 5: 28–38.

Michelson, E. H. & Richards, C. S. (1975). Neoplasms and tumor-like growths in the aquatic pulmonate snail *Biomphalaria glabrata*. *Annals of the New York Academy of Sciences* 266: 411–425.

Michiels, N. K. (1998). Mating conflicts and sperm competition in simultaneous hermaphrodites. pp. 219–253 *in* T. R. Birkhead & Møller, A. P. (Eds.) *Sperm Competition and Sexual Selection*. San Diego, CA, Academic Press.

Michinomae, M., Masuda, H., Seidou, M. & Kito, Y. (1994). Structural basis for wavelength discrimination in the banked retina of the firefly squid *Watasenia scintillans*. *Journal of Experimental Biology* 193: 1–12.

Middelfart, P. A. & Craig, M. (2004). Description of *Austrodevonia sharnae* n. gen. n. sp. (Galeommatidae: Bivalvia), an ectocommensal of *Taeniogyrus australianus* (Stimpson, 1855) (Synaptidae: Holothuroidea). *Molluscan Research* 24: 211–219.

Mienis, H. K. (1990). Records of slug mites (*Riccardoella* spec.) from terrestrial gastropods in Israel. *Soosiana* 18: 42–46.

Miglavs, I. J., Sneli, J.-A. & Warén, A. (1993). Brood protection of *Oenopota* (Gastropoda) *turridae* eggs by the shrimp *Sclerocrangon boreas* (Phipps). *Journal of Molluscan Studies* 59: 363–365.

Mikkelsen, P. M. & Bieler, R. (1992). Biology and comparative anatomy of three new species of commensal Galeommatidae, with a possible case of mating behavior in bivalves. *Malacologia* 34: 1–24.

Mikkelsen, P. M. (1996). The evolutionary relationships of Cephalaspidea s.l. (Gastropoda: Opisthobranchia): A phylogenetic analysis. *Malacologia* 37: 375–442.

Miletic, I., Miric, M., Lalic, Z. & Sobajic, S. (1991). Composition of lipids and proteins of several species of molluscs, marine and terrestrial, from the Adriatic Sea and Serbia. *Food Chemistry* 41: 303–308.

Millar, R. H. & Scott, J. M. (1967). The larva of the oyster *Ostrea edulis* during starvation. *Journal of the Marine Biological Association of the United Kingdom* 47: 475–484.

Miller, A. C., Ponder, W. F. & Clark, S. A. (1999). Freshwater snails of the genera *Fluvidona* and *Austropyrgus* (Gastropoda, Hydrobiidae) from northern New South Wales and southern Queensland, Australia. *Invertebrate Taxonomy* 13: 461–493.

Miller, A. M., Ramirez, T., Zuniga, F. I., Ochoa, G. H., Gray, S. M., Kelly, S. D., Matsumoto, B. & Robles, L. J. (2005). Rho GTPases regulate rhabdom morphology in octopus photoreceptors. *Visual Neuroscience* 22: 295–304.

Miller, J. A. & Byrne, M. (2000). Ceratal autotomy and regeneration in the aeolid nudibranch *Phidiana crassicornis* and the role of predators. *Invertebrate Biology* 119: 167–176.

Miller, L. P. & Denny, M. W. (2011). Importance of behavior and morphological traits for controlling body temperature in littorinid snails. *Biological Bulletin* 220: 209–223.

Miller, M. F. & Labandeira, C. L. (2002). Slow crawl across the salinity divide: Delayed colonization of freshwater ecosystems by invertebrates. *Geological Society of America Today* 12: 4–10.

Miller, R. L. (1977). Chemotactic behavior of the sperm of chitons (Mollusca: Polyplacophora). *Journal of Experimental Zoology Part A* 202: 203–211.

Miller, R. L., Mojares, J. J. & Ram, J. L. (1994). Species-specific sperm attraction in the Zebra Mussel, *Dreissena polymorpha*, and the Quagga Mussel, *Dreissena bugensis*. *Canadian Journal of Zoology* 72: 1764–1770.

Miller, S. & Hadfield, M. (1990). Developmental arrest during larval life and life-span extension in a marine mollusc. *Science* 248: 356–358.

Miller, S. (1993). Larval period and its influence on post-larval life history: Comparison of lecithotrophy and facultative planktotrophy in the aeolid nudibranch *Phestilla sibogae*. *Marine Biology* 117: 635–645.

Miller, S. E. & Hadfield, M. G. (1986). Ontogeny of phototaxis and metamorphic competence in larvae of the nudibranch *Phestilla sibogae* Bergh (Gastropoda: Opisthobranchia). *Journal of Experimental Marine Biology and Ecology* 97: 95–112.

Miller, S. L. (1974a). Adaptive design of locomotion and foot form in prosobranch gastropods. *Journal of Experimental Marine Biology and Ecology* 14: 99–156.

Miller, S. L. (1974b). The classification, taxonomic distribution, and evolution of locomotor types among prosobranch gastropods. *Journal of Molluscan Studies* 41: 233–261.

Miller, W., Nishioka, R. S. & Bern, H. A. (1973). The 'juxtaganglionic' tissue and the brain of the abalone *Haliotis rufescens* (Swainson). *The Veliger* 16: 125–129.

Milligan, B. J., Curtin, N. A. & Bone, Q. (1997). Contractile properties of obliquely striated muscle from the mantle of squid (*Alloteuthis subulata*) and cuttlefish (*Sepia officinalis*). *Journal of Experimental Biology* 200: 2425–2436.

Millott, N. (1937). On the morphology of the alimentary canal, process of feeding, and physiology of digestion of the nudibranch mollusc *Jorunna tomentosa* (Cuvier). *Philosophical Transactions of the Royal Society B* 228: 173–217.

Mills, L. S., Soulé, M. E. & Doak, D. F. (1993). The keystone-species concept in ecology and conservation. *BioScience* 43: 219–224.

Mills, S. C. & Cote, I. M. (2003). Sex-related differences in growth and morphology of blue mussels. *Journal of the Marine Biological Association of the United Kingdom* 83: 1053–1057.

Milyutina, I. A. & Petrov, N. B. (1989). Divergence of unique DNA sequences in Mytilinae (Bivalvia Mytilidae). *Molekulyarnaya Biologiya (Moscow)* 23: 1373–1381.

Mincher, R. (2008). New Zealand's Challenger Scallop Enhancement Company: From reseeding to self-governance. *FAO fisheries technical paper* 504: 307.

Minchin, D. (1992). Multiple species, mass spawning events in an Irish Sea lough: The effect of temperatures on spawning and recruitment of invertebrates. *Invertebrate Reproduction & Development* 22: 229–238.

Minchinton, T. E. & Ross, P. M. (1999). Oysters as habitat for limpets in a temperate mangrove forest. *Australian Journal of Ecology* 24: 157–170.

Mingo-Licuanan, S. S. & Lucas, J. S. (1995). Bivalves that 'feed' out of water: Phototrophic nutrition during emersion in the giant clam, *Tridacna gigas* Linne. *Journal of Shellfish Research* 14: 283–286.

Minichev, S. & Sirenko, B. J. (1974). Development and evolution of the radula in Polyplacophora [in Russian, English summary]. *Zoologicheskii Zhurnal* 53: 1133–1139.

Van Minnen, J., Reichelt, D. & Lodder, J. C. (1979). An ultrastructural study of the neurosecretory canopy cell of the pond snail *Lymnaea stagnalis* (L.), with the use of the horseradish peroxidase tracer technique. *Cell and Tissue Research* 204: 453–462.

Minniti, F. (1986). Morphological and histochemical study of Pharynx of Leiblein, salivary glands and Gland of Leiblein in the carnivorous Gastropoda *Amyclina tinei* (Maravigna) and *Cyclope neritea* (Lamarck) (Nassariidae: Prosobranchia Stenoglossa). *Zoologischer Anzeiger* 217: 14–22.

Minton, R. L. & Wang, L. L. (2011). Evidence of sexual shape dimorphism in *Viviparus* (Gastropoda: Viviparidae). *Journal of Molluscan Studies* 77: 315–317.

Miranda, R. M., Lombardo, R. C. & Goshima, S. (2008). Copulation behaviour of *Neptunea arthritica*: Baseline considerations on broodstocks as the first step for seed production technology development. *Aquaculture Research* 39: 283–290.

Miron, G., Boudreau, B. & Bourget, E. (1995). Use of larval supply in benthic ecology: Testing correlations between larval supply and larval settlement. *Marine Ecology Progress Series* 124: 301–305.

Mironenko, A. A. & Rogov, M. A. (2015). First direct evidence of ammonoid ovoviviparity. *Lethaia* 49: 245–260.

Mischor, B. & Märkel, K. (1984). Histology and regeneration of the radula of *Pomacea bridgesi* (Gastropoda, Prosobranchia). *Zoomorphology (Berlin)* 104: 42–66.

Mitta, G., Vandenbulcke, F., Hubert, F., Salzet, M. & Roch, P. (2000a). Involvement of mytilins in mussel antimicrobial defense. *Journal of Biological Chemistry* 275: 12954–12962.

Mitta, G., Vandenbulcke, F. & Roch, P. (2000b). Original involvement of antimicrobial peptides in mussel innate immunity. *Federation of European Biochemical Societies (FEBS) Letters* 486: 185–190.

Mitton, J. B. (1977). Shell color and pattern variation in *Mytilus edulis* and its adaptive significance. *Chesapeake Science* 18: 387–390.

Miura, O., Nishi, S. & Chiba, S. (2007). Temperature-related diversity of shell colour in the intertidal gastropod *Batillaria*. *Journal of Molluscan Studies* 73: 235–240.

Miyamoto, H., Endo, H., Hashimoto, N., Limura, K., Isowa, Y., Kinoshita, S., Kotaki, T., Masaoka, T., Miki, T., Nakayama, S., Nogawa, C., Notazawa, A., Ohmori, F., Sarashina, I., Suzuki, M., Takagi, R., Takahashi, J., Takeuchi, T., Yokoo, N., Satoh, N., Toyohara, H., Miyashita, T., Wada, H., Samata, T., Endo, K., Nagasawa, H., Asakawa, S. & Watabe, S. (2013). The diversity of shell matrix proteins: Genome-wide investigation of the pearl oyster, *Pinctada fucata*. *Zoological Science* 30: 801–816.

Miyazawa, K. & Noguchi, T. (2001). Distribution and origin of tetrodotoxin. *Toxin Reviews* 20: 11–33.

Mizzaro-Wimmer, M. & von Salvini-Plawen, L. (2001). *Praktische Malakologie. Beiträge zur vergleichend-anatomischen Bearbeitung der Mollusken: Caudofoveata bis Gastropoda—Streptoneura*. Vienna, Austria, Springer.

Mkoji, G. M., Smith, J. M. & Pritchard, R. K. (1988). Antioxidant systems in *Schistosoma mansoni*: Correlation between susceptibility to oxidant killing and the levels of scavengers of hydrogen peroxide and oxygen free radicals. *International Journal for Parasitology* 18: 661–666.

Modica, M. V. & Holford, M. (2010). The Neogastropoda: Evolutionary innovations of predatory marine snails with remarkable pharmacological potential. pp. 249–270 in P. Pontarotti (Ed.). *Evolutionary Biology: Concepts, Molecular and Morphological Evolution*. Berlin, Germany, Springer-Verlag.

Modica, M. V., Bouchet, P., Cruaud, C., Utge, J. & Oliverio, M. (2011). Molecular phylogeny of the nutmeg shells (Neogastropoda, Cancellariidae). *Molecular Phylogenetics and Evolution* 59: 685–697.

Modica, M. V., Lombardo, F., Franchini, P. & Oliverio, M. (2015). The venomous cocktail of the vampire snail *Colubraria reticulata* (Mollusca, Gastropoda). *BMC Genomics* 16: 441.

Moens, L., Vanfleteren, J. R., Van der Peer, Y., Peeters, K., Kapp, O., Czeluzniak, J., Goodman, M., Blaxter, M. & Vinogradov, S. N. (1996). Globins in nonvertebrate species: Dispersal by horizontal gene transfer and evolution of the structure-function relationships. *Molecular Biology and Evolution* 13: 324–333.

Moffett, S. B. (1991). Neural control of male reproductive function in pulmonate and opisthobranch gastropods. pp. 140–153 in B. Scharrer, Florey, E. & Stefano, G. B. (Eds.) *Comparative Aspects of Neuropeptide Function. Studies in Neuroscience*. New York, Manchester University Press.

Møhlenberg, F. & Riisgård, H. U. (1978). Efficiency of particle retention in 13 species of suspension feeding bivalves. *Ophelia* 17: 239–246.

Van Mol, J.-J. (1967). *Étude morphologique et phylogénétique du ganglion cérébroïde des Gastéropodes pulmonés (mollusques)*. Vol. 37. Bruxelles, Palais des Académies.

Van Mol, J.-J. (1974). Evolution phylogénétique du ganglion cérébroïde chez les gastéropodes pulmonés. *Haliotis* 4: 77–86.

Mølgaard, P. (1986). Food plant preferences by slugs and snails: A simple method to evaluate the relative palatability of the food plants. *Biochemical Systematics and Ecology* 14: 113–121.

Molina, F. R., Paggi, J. C. & Devercelli, M. (2010). Zooplanktophagy in the natural diet and selectivity of the invasive mollusk *Limnoperna fortunei*. *Biological Invasions* 12: 1647–1659.

Molis, M., Scrosati, R. A., El-Belely, E. F., Lesniowski, T. J. & Wahl, M. (2015). Wave-induced changes in seaweed toughness entail plastic modifications in snail traits maintaining consumption efficacy. *Journal of Ecology* 103: 851–859.

Mollo, E., Gavagnin, M., Carbone, M., Guo, Y. W. & Cimino, G. (2005). Chemical studies on Indopacific *Ceratosoma* nudibranchs illuminate the protective role of their dorsal horn. *Chemoecology* 15: 31–36.

Molloy, D. P., Karatayev, A. Y., Burlakova, L. E., Kurandina, D. P. & Laruelle, F. (1997). Natural enemies of zebra mussels: Predators, parasites, and ecological competitors. *Review of Fisheries Science* 5: 27–97.

Moltschaniwskyj, N. A., Pecl, G. T. & Lyle, J. M. (2002). An assessment of the use of short-term closures to protect spawning southern calamary aggregations from fishing pressure in Tasmania, Australia. *Bulletin of Marine Science* 71: 501–514.

Moltschaniwskyj, N. A., Hall, K., Lipiński, M. R., Marian, J. E., Nishiguchi, M., Sakai, M., Shulman, D., Sinclair, B., Sinn, D. L. & Staudinger, M. (2007). Ethical and welfare considerations when using cephalopods as experimental animals. *Reviews in Fish Biology and Fisheries* 17: 455–476.

Moment, G. B. (1962). Reflexive selection: A possible answer to an old puzzle. *Science* 136: 202–203.

Mommsen, T. P., Ballantyne, J. S., Macdonald, D., Gosline, J. & Hochachka, P. W. (1981). Analogs of red and white muscle in squid mantle. *Proceedings of the National Academy of Sciences of the United States of America* 78: 3274–3278.

Monaco, C. J., McQuaid, C. D. & Marshall, D. J. (2017). Decoupling of behavioural and physiological thermal performance curves in ectothermic animals: A critical adaptive trait. *Oecologia* 185: 583–593.

Monod, T. & Dollfus, R. P. (1932). Les copépodes parasites de mollusques. *Annales de parasitologie humaine et comparée* 10: 129–204.

Monteiro, W., Almeida, J. M. G. & Dias, B. S. (1984). Sperm sharing in *Biomphalaria* snails: A new behavioural strategy in simultaneous hermaphroditism. *Nature* 308: 727–729.

Montes, J. F., Durfort, M. & García-Valero, J. (1995). Cellular defence mechanism of the clam *Tapes semidecussatus* against infection by the protozoan *Perkinsus* sp. *Cell and Tissue Research* 279: 529–538.

Montes, J. F., Dufort, M. & García-Valero, J. (1996). When the venerid clam *Tapes decussatus* is parasitized by the protozoan *Perkinsus* sp. it synthesizes a defensive polypeptide that is closely related to p225. *Diseases of Aquatic Organisms* 26: 149–157.

Montes, J. F., Del Río, J. A., Durfort, M. & García-Valero, J. (1997). The protozoan parasite *Perkinsus atlanticus* elicits a unique defensive response in the clam *Tapes semidecussatus*. *Parasitology* 114: 339–349.

Montory, J. A., Pechenik, J. A., Diederich, C. M. & Chaparro, O. R. (2014). Effects of low salinity on adult behavior and larval performance in the intertidal gastropod *Crepipatella peruviana* (Calyptraeidae). *PLoS ONE* 9: e103820.

Moody, M. F. & Parriss, J. R. (1960). The visual system of *Octopus*. 2. Discrimination of polarized light by *Octopus*. *Nature* 186: 839–840.

Moody, M. F. & Parriss, J. R. (1961). The discrimination of polarized light by *Octopus*: A behavioural and morphological study. *Zeitschrift für vergleichende Physiologie* 44: 268–291.

Moor, B. (1977). Zur Embryologie von *Bradybaena* (*Eulota*) *fruticum* Müller (Gastropoda, Pulmonata, Stylommatophora). *Zoologische Jahrbücher, Abteilung Anatomie* 97: 323–399.

Moor, B. (1983). Organogenesis. pp. 123–177 in N. H. Verdonk, van den Biggelaar, J. A. M., & Tompa, A. S. (Eds.) *Development. The Mollusca.* Vol. 3. New York, Academic Press.

Moore, H. B. (1936). The biology of *Purpura lapillus*. I. Shell variation in relation to environment. *Journal of the Marine Biological Association of the United Kingdom* 21: 61–89.

Moore, H. B. (1957). Cephalopoda: Ammonoidea. pp. i–xxii, L1–L490 in R. C. Moore (Ed.) *Treatise on Invertebrate Paleontology Part L.* Vol.L 4. Boulder, CO and Lawrence, KS, Geological Society of America & University of Kansas Press.

Moore, H. B., Ed. (1996). *Cretaceous Ammonoidea*. Treatise on Invertebrate Paleontology Part L. Boulder, CO and Lawrence, KS, Geological Society of America and University of Kansas Press.

Moore, J. (2002). *Parasites and the Behaviour of Animals*. New York, Oxford University Press.

Moore, R. C., Ed. (1964). *Mollusca: General features, Scaphopoda, Amphineura, Monoplacophora, Gastropoda—general features, Archaeogastropoda and some (mainly Paleozoic) Caenogastropoda and Opisthobranchia.* Treatise on Invertebrate Paleontology Part I. Lawrence, KS, Geological Society of America and University of Kansas Press.

Moore, S. & Meyer-Rochow, V. B. (1988). Observations on habitat and reproduction in the pulmonate, basommatophoran gastropod *Latia neritoides* Gray 1850: The only bioluminescent freshwater mollusc in the world. *Verhandlungen des Internationalen Verein Limnologie* 23: 2189–2192.

de Moraes, D. T. & Lopes, S. G. B. C. (2003). The functional morphology of *Neoteredo reynei* (Bartsch, 1920) (Bivalvia, Teredinidae). *Journal of Molluscan Studies* 69: 311–318.

Moran, A. L. & Woods, H. A. (2007). Oxygen in egg masses: Interactive effects of temperature, age, and egg-mass morphology on oxygen supply to embryos. *Journal of Experimental Biology* 210: 722–731.

Morand, S. (1989). Influence du nématode *Nemhelix bakeri* Morand et Petter sur la reproduction de l'escargot *Helix aspersa* Müller. *Comptes rendus de l'Académie des sciences* 309: 367–385.

Morand, S. & Faliex, E. (1994). Study on the life cycle of a sexually transmitted nematode parasite of a terrestrial snail. *Journal of Parasitology* 80: 1049–1052.

Morand, S., Wilson, M. J. & Glen, D. M. (2004). Nematodes (Nematoda) parasitic in terrestrial gastropods. pp. 525–558 in G. M. Barker (Ed.) *Natural Enemies of Terrestrial Molluscs.* Oxford, UK/Cambridge, MA, CABI Publishing.

Mordan, P. B. & Wade, C. M. (2008). Heterobranchia II: The Pulmonata. pp. 409–426 in W. F. Ponder & Lindberg, D. R. (Eds.) *Phylogeny and Evolution of the Mollusca.* Berkeley, CA, University of California Press.

Morishita, F., Furukawa, Y., Matsushima, O. & Minakata, H. (2010). Regulatory actions of neuropeptides and peptide hormones on the reproduction of molluscs. *Canadian Journal of Zoology* 88: 825–845.

Moritz, C. E. (1939). Organogenesis in the gastropod *Crepidula adunca* Sowerby. *University of California Publications in Zoology* 43: 217–248.

Moroz, L. L. & Gillette, R. (1995). From Polyplacophora to Cephalopoda: Comparative analysis of nitric oxide signalling in Mollusca. *Acta Biologica Hungarica* 46: 169–182.

Moroz, L. L., Edwards, J. R., Puthanveettil, S. V., Kohn, A. B., Hla, T., Heyland, A., Knudsen et al. (2006). Neuronal transcriptome of *Aplysia*: Neuronal compartments and circuitry. *Cell* 127: 1453–1467.

Moroz, L. L. (2011). *Aplysia. Current Biology* 21: R60–R61.

Morris, C. E. & Sigurdson, W. J. (1989). Stretch-inactivated ion channels coexist with stretch-activated ion channels. *Science* 243: 807–809.

Morris, M. C. (1950). Dilation of the foot in *Uber* (*Polinices*) *strangei* (Mollusca, Class Gastropoda). *Proceedings of the Linnean Society of New South Wales* 75: 70–81.

Morris, R. H., Putnam, D., Abbott, E. & Haderlie, C. (1980). *Intertidal Invertebrates of California.* Stanford, CA, Stanford University Press.

Morrison, C. M. & Odense, P. H. (1974). Ultrastructure of some pelecypod adductor muscles. *Journal of Ultrastructure Research* 49: 228–251.

Morse, A. N. C., Froyd, C. A. & Morse, D. E. (1984). Molecules from cyanobacteria and red algae that induce larval settlement and metamorphosis in the mollusc *Haliotis rufescens*. *Marine Biology* 81: 293–298.

Morse, D. E., Duncan, H., Hooker, N. & Morse, A. (1978). Hydrogen peroxide induces spawning in molluscs, with activation of prostaglandin endoperoxide synthetase. *Science* 196: 298–300.

Morse, M. P. & Norenburg, J. L. (1983). A molluscan interstitial assemblage at Fort Pierce, Florida (USA). *American Zoologist* 23: 931.

Morse, M. P. (1984). Functional adaptations of the digestive system of the carnivorous mollusc *Pleurobranchaea californica* MacFarland, 1966. *Journal of Morphology* 180: 253–269.

Morse, M. P. (1987a). Distribution and ecological adaptation of interstitial molluscs in Fiji. *American Malacological Union Bulletin* 5: 281–286.

Morse, M. P. (1987b). Comparative functional morphology of the bivalve excretory system. *American Zoologist* 27: 737–746.

Morse, M. P. & Reynolds, P. D. (1996). Ultrastructure of the heart-kidney complex in smaller classes supports symplesiomorphy of molluscan coelomic characters. pp. 89–97 in J. D. Taylor (Ed.) *Origin and Evolutionary Radiation of the Mollusca.* Oxford, UK, Oxford University Press.

Morse, M. P. & Zardus, J. D. (1997). Bivalvia. pp. 7–118 in F. W. Harrison & Kohn, A. J. (Eds.) *Microscopic Anatomy of Invertebrates. Mollusca 2.* Vol. 6A. New York, Wiley-Liss.

Mortensen, S. (2000). Scallop introductions and transfers, from an animal health point of view. *Aquaculture International* 8: 123–138.

Mortensen, T. (1928). *A Monograph of the Echinoidea. I. Cidaroidea.* Copenhagen, C. A. Reitzel.

Morton, B. S. (1969). Studies on the biology of *Dreissena polymorpha* Pall. 1. General anatomy and morphology. *Proceedings of the Malacological Society of London* 38: 301–321.

Morton, B. S. (1971). The diurnal rhythm and tidal rhythm of feeding and digestion in *Ostrea edulis. Biological Journal of the Linnean Society* 3: 329–342.

Morton, B. S. (1972a). Some aspects of the functional morphology and biology of *Pseudopythina subsinata* (Bivalvia, Leptonacea) commensal on stomatopod crustaceans. *Journal of Zoology* 166: 79–96.

Morton, B. S. (1973a). A new theory of feeding and digestion in the filter feeding Lamellibranchia. *Malacologia* 14: 63–79.

Morton, B. S. (1973b). Some factors affecting the location of *Arthritica crassiformis* (Bivalvia: Leptonacea) commensal upon *Anchomasa similis* (Bivalvia: Pholadidae). *Journal of Zoology* 170: 463–473.

Morton, B. S. (1973c). The biology and functional morphology of *Galeomma (Paralepida) takii* (Bivalvia, Leptonacea). *Journal of Zoology* 169: 135–150.

Morton, B. & McQuiston, R. W. (1974). The daily rhythm of activity in *Teredo navalis* Linnaeus correlated with the functioning of the digestive system. *Forma et Functio* 7: 59–80.

Morton, B. S. (1976). Secondary brooding of temporary dwarf males in *Ephippodonta (Ephippodontina) oedipus* sp. nov. (Bivalvia: Leptonacea). *Journal of Conchology* 29: 31–39.

Morton, B. S. (1977a). The hypobranchial gland in Bivalvia. *Canadian Journal of Zoology* 55: 1225–1234.

Morton, B. S. (1977b). The tidal rhythm of feeding and digestion in the Pacific oyster, *Crassostrea gigas* (Thunberg). *Journal of Experimental Marine Biology and Ecology* 26: 135–151.

Morton, B. S. (1978a). Feeding and digestion in shipworms. *Oceanography and Marine Biology Annual Review* 16: 107–144.

Morton, B. S. (1978b). The biology and functional morphology of *Claudiconcha japanica* (Bivalvia-Veneracea). *Journal of Zoology* 184: 35–52.

Morton, B. S. (1978c). The diurnal rhythm and the processes of feeding and digestion in *Tridacna crocea* (Bivalvia: Tridacnidae). *Journal of Zoology* 185: 371–387.

Morton, B. S. (1979). A comparison of lip structure and function correlated with other aspects of the functional morphology of *Lima lima, Limaria (Platilimaria) fragilis,* and *Limaria (Platilimaria) hongkongensis* sp. nov. (Bivalvia: Limacea). *Canadian Journal of Zoology* 57: 728–742.

Morton, B. S. (1980a). Some aspects of the biology and functional morphology (including the presence of a ligamental lithodesma) of *Montacutona compacta* (Bivalvia: Leptonacea) associated with coelenterates in Hong Kong. *Journal of Zoology* 192: 431–455.

Morton, B. S. (1980b). Selective site segregation in *Patelloida (Chiazamea) pygmaea* (Dunker) and *P.(C.) lampanicola* Habe (Gastropoda: Patellacea) on a Hong Kong shore. *Journal of Experimental Marine Biology and Ecology* 47: 149–171.

Morton, B. & Scott, P. J. B. (1980). Morphological and functional specializations of the shell, musculature and pallial glands in the Lithophaginae (Mollusca: Bivalvia). *Journal of Zoology* 192: 179–203.

Morton, B. S. (1981a). The Anomalodesmata. *Malacologia* 21: 35–60.

Morton, B. S. (1981b). The biology and functional morphology of *Chlamydoconcha orcutti* with a discussion on the taxonomic status of the Chlamydoconchacea (Mollusca: Bivalvia). *Journal of Zoology* 195: 81–121.

Morton, B. S. (1981c). Prey capture in the carnivorous septibranch *Poromya granulata* (Bivalvia: Anomalodesmata: Poromyacea). *Sarsia* 66: 241–256.

Morton, B. S. (1982). The biology, functional morphology and taxonomic status of *Fluviolanatus subtorta* (Bivalvia: Trapeziidae), a heteromyarian bivalve possessing 'zooxanthellae'. *Journal of the Malacological Society of Australia* 5: 113–140.

Morton, B. S. (1983a). Feeding and digestion in Bivalvia. pp. 65–147 in A. S. M. Saleuddin & Wilbur, K. M. (Eds.) *Physiology, Part 2. The Mollusca.* Vol. 5. New York, Academic Press.

Morton, B. S. (1983b). The biology and functional morphology of the twisted ark *Trisidos semitorta* (Bivalvia: Arcacea) with a discussion on shell 'torsion' in the genus. *Malacologia* 23: 375–396.

Morton, B. & Morton, J. (1983). *The Sea Shore Ecology of Hong Kong.* Hong Kong, Hong Kong University Press.

Morton, B. S. (1984). Adventitious tube construction in *Brechites vaginiferus* (Bivalvia: Anomalodesmata: Clavagellacea) with an investigation of the juvenile of 'Humphreyia strangei'. *Journal of Zoology* 203: 461–484.

Morton, B. S. (1985). Statocyst structure in the Anomalodesmata (Bivalvia). *Journal of Zoology* 206: 23–34.

Morton, B. S. (1986). The diet and prey capture mechanism of *Melo melo* (Prosobranchia: Volutidae). *Journal of Molluscan Studies* 52: 156–160.

Morton, B. S. (1988a). *Partnerships in the Sea: Hong Kong's Marine Symbioses.* Hong Kong, Hong Kong University Press.

Morton, B. S. & Thurston, M. H. (1989). The functional morphology of *Propeamussium lucidum* (Bivalvia: Pectinacea), a deep-sea predatory scallop. *Journal of Zoology* 218: 471–496.

Morton, B. S. (1990). Corals and their bivalve borers: The evolution of a symbiosis. pp. 11–46 in B. Morton (Ed.) *The Bivalvia. Proceedings of a Memorial Symposium in Honour of Sir Charles Maurice Yonge (1899–1986), Edinburgh, 1986.* Hong Kong, Hong Kong University Press.

Morton, B. S. (1996). The evolutionary history of the Bivalvia. pp. 337–359 in J. D. Taylor (Ed.) *Origin and Evolutionary Radiation of the Mollusca.* Oxford, UK, Oxford University Press.

Morton, B. & Chan, K. (1997). First report of shell boring predation by a member of the Nassariidae (Gastropoda). *Journal of Molluscan Studies* 63: 476–478.

Morton, B. S. (2000a). The pallial eyes of *Ctenoides floridanus* (Bivalvia: Limoidea). *Journal of Molluscan Studies* 66: 449–455.

Morton, B. S. (2000b). The function of pallial eyes within the Pectinidae, with a description of those present in *Patinopecten yessoensis*. pp. 247–255 *in* E. M. Harper, Taylor, J. D. & Crame, J. A. (Eds.) *Evolutionary Biology of the Bivalvia*. Vol. 177. London, UK, The Geological Society of London.

Morton, B. S. (2000c). The biology and functional morphology of *Fragum erugatum* (Bivalvia: Cardiidae) from Shark Bay, Western Australia: The significance of its relationship with entrained zooxanthellae. *Journal of Zoology* 251: 39–52.

Morton, B. S. (2001). The evolution of eyes in the Bivalvia. *Oceanography and Marine Biology Annual Review* 39: 165–205.

Morton, B. & Jones, D. S. (2001). The biology of *Hipponix australis* (Gastropoda: Hipponicidae) on *Nassarius pauperatus* (Nassariidae) in Princess Royal Harbour, Western Australia. *Journal of Molluscan Studies* 67: 247–255.

Morton, B. & Jones, D. S. (2003). The dietary preferences of a suite of carrion-scavenging gastropods (Nassariidae, Buccinidae) in Princess Royal Harbour, Albany, Western Australia. *Journal of Molluscan Studies* 69: 151–156.

Morton, B. S. (2004a). The biology and functional morphology of *Nipponoclava gigantea*: Clues to the evolution of tube dwelling in the Penicillidae (Bivalvia: Anomalodesmata: Clavagelloidea). *Journal of Zoology* 264: 355–369.

Morton, B. S. (2005). Biology and functional morphology of a new species of endolithic *Bryopa* (Bivalvia: Anomalodesmata: Clavagelloidea) from Japan and a comparison with fossil species of *Stirpulina* and other Clavagellidae. *Invertebrate Biology* 124: 202–219.

Morton, B. S. (2006a). Diet and predation behaviour exhibited by *Cominella eburnea* (Gastropoda: Caenogastropoda: Neogastropoda) in Princess Royal Harbour, Albany, Western Australia, with a review of attack strategies in the Buccinidae. *Molluscan Research* 26: 39–50.

Morton, B. S. (2006b). Scavenging behaviour by *Ergalatax contractus* (Gastropoda: Muricidae) and interactions with *Nassarius nodifer* (Gastropoda: Nassariidae) in the Cape d'Aguilar Marine Reserve, Hong Kong. *Journal of the Marine Biological Association of the United Kingdom* 86: 141–152.

Morton, B. S. (2006c). Structure and formation of the adventitious tube of the Japanese watering-pot shell *Stirpulina ramosa* (Bivalvia, Anomalodesmata, Clavagellidae) and a comparison with that of the Penicillidae. *Invertebrate Biology* 125: 233–249.

Morton, B. S. (2007). The evolution of the watering pot shells (Bivalvia: Anomalodesmata: Clavagellidae and Penicillidae). *Records of the Western Australian Museum* 24: 19–64.

Morton, B., Peharda, M. & Harper, E. M. (2007). Drilling and chipping patterns of bivalve prey shell penetration by *Hexaplex trunculus* (Mollusca: Gastropoda: Muricidae). *Journal of the Marine Biological Association of the United Kingdom* 87: 933–940.

Morton, B. S. (2008a). The evolution of eyes in the Bivalvia: New insights. *American Malacological Bulletin* 26: 35–45.

Morton, B. S. (2008b). The biology of sympatric species of *Scintillona* (Bivalvia: Galeommatoidea) commensal with *Pilumnopeus serratifrons* (Crustacea: Decapoda) in Moreton Bay, Queensland, Australia, with description of a new species. *Memoirs of the Queensland Museum* 54: 323–338.

Morton, B. & Harper, E. M. (2009). Drilling predation upon *Ditrupa arietina* (Polychaeta: Serpulidae) from the mid-Atlantic Azores, Portugal. *Açoreana Supplement*: 157–165.

Morton, B. S. (2011). Behaviour of *Nassarius bicallosus* (Caenogastropoda) on a north western Western Australian surf beach with a review of feeding in the Nassariidae. *Molluscan Research* 31: 90–94.

Morton, B., Peharda, M. & Petrić, M. (2011). Functional morphology of *Rocellaria dubia* (Bivalvia: Gastrochaenidae) with new interpretations of crypt formation and adventitious tube construction, and a discussion of evolution within the family. *Biological Journal of the Linnean Society* 104: 786–804.

Morton, J. E. (1950). Feeding mechanisms in the Vermetidae (Order Mesogastropoda). *Nature* 165: 923–924.

Morton, J. E. (1951). The ecology and digestive system of the Struthiolariidae (Gastropoda). *Quarterly Journal of Microscopical Science* 92: 1–25.

Morton, J. E. (1952a). The role of the crystalline style. *Proceedings of the Malacological Society of London* 29: 85–92.

Morton, J. E. (1952b). A preliminary study of the land operculate *Murdochia pallidum* (Cyclophoridae, Mesogastropoda). *Transactions of the Royal Society of New Zealand* 80: 69–79.

Morton, J. E. (1953). The functions of the gastropod stomach. *Proceedings of the Linnean Society of London* 164: 240–246.

Morton, J. E. (1954). The biology of *Limacina retroversa*. *Journal of the Marine Biological Association of the United Kingdom* 33: 297–312.

Morton, J. E. (1955a). The evolution of the Ellobiidae with a discussion on the origin of the Pulmonata. *Proceedings of the Zoological Society of London* 125: 127–168.

Morton, J. E. (1955b). The functional morphology of *Otina otis*, a primitive marine pulmonate. *Journal of the Marine Biological Association of the United Kingdom* 34: 113–150.

Morton, J. E. (1955c). The evolution of vermetid gastropods. *Pacific Science* 9: 3–15.

Morton, J. E. (1955d). The structure and function of the stomach and sorting caecum in *Lunella smaragda* (Martyn) (Turbinidae). *Proceedings of the Malacological Society of London* 31: 123–137.

Morton, J. E. (1956a). The vascular and nervous system of *Struthiolaria* (Prosobranchia, Mesogastropoda). *Transactions of the Royal Society of New Zealand* 83: 721–743.

Morton, J. E. (1956b). The tidal rhythm and action of the digestive system of the lamellibranch *Lasaea rubra*. *Journal of the Marine Biological Association of the United Kingdom* 35: 563–586.

Morton, J. E. (1958a). *Molluscs*. 1st ed. London, UK, Hutchinson.

Morton, J. E. (1958b). Torsion and the adult snail: A re-evaluation. *Proceedings of the Malacological Society of London* 33: 2–10.

Morton, J. E. (1958c). The adaptations and relationships of the Xenophoridae (Mesogastropoda). *Proceedings of the Malacological Society of London* 33: 89–101.

Morton, J. E. (1959). The habits and feeding organs of *Dentalium entalis*. *Journal of the Marine Biological Association of the United Kingdom* 38: 225–238.

Morton, J. E. (1960a). The functions of the gut in ciliary feeders. *Biological Reviews* 35: 92–140.

Morton, J. E. (1960b). The responses and orientation of the bivalve *Lasaea rubra* Montagu. *Journal of the Marine Biological Association of the United Kingdom* 39: 5–26.

Morton, J. E. (1960c). The habits of *Cyclope neritea*, a style-bearing stenoglossan gastropod. *Proceedings of the Malacological Society of London* 34: 96–105.

Morton, J. E. (1963). The molluscan pattern: Evolutionary trends in a modern classification. *Proceedings of the Linnean Society of London* 174: 53–72.

Morton, J. E. (1964). Locomotion. pp. 383–423 *in* K. M. Wilbur & Yonge, C. M. (Eds.) *Physiology of Mollusca*. Vol. 1. New York, Academic Press.

Morton, J. E. & Yonge, C. M. (1964). Classification and structure of the Mollusca. pp. 1–58 *in* K. M. Wilbur & Yonge, C. M. (Eds.) *Physiology of Mollusca*. Vol. 1. New York, Academic Press.

Morton, J. E. (1967). *Molluscs*. 4th ed., revised). London, UK, Hutchinson.

Morton, J. E. & Miller, M. C. (1968). *The New Zealand Sea Shore*. London/Auckland, Collins.

Morton, J. E. (1972). The form and functioning of the pallial organs in the opisthobranch *Akera bullata*, with a discussion on the nature of the gill in Notaspidea and other tectibranchs. *The Veliger* 14: 337–349.

Morton, J. E. (1988b). The pallial cavity. pp. 253–286 *in* E. R. Trueman & Clarke, M. R. (Eds.) *Form and Function. The Mollusca*. Vol. 11. New York, Academic Press.

Morton, J. E. (2004b). *Seashore Ecology of New Zealand and the Pacific*. Auckland, David Bateman Ltd.

Moshel, S. M., Levine, M. & Collier, J. R. (1998). Shell differentiation and engrailed expression in the *Ilyanassa* embryo. *Development Genes and Evolution* 208: 135–141.

Moskalev, L. I. (1964). Distribution of the Acmaeidae (Gastropoda. Prosobranchia) in the northern portion of the Pacific Ocean. *Doklady Akademii Nauk SSSR* 158: 1221–1222.

Moss, B. R. (2010). *Ecology of Freshwaters: A View for the Twenty-first Century*. Chichester, UK, John Wiley & Sons.

Mouëza, M., Gros, O. & Frenkiel, L. (2006). Embryonic development and shell differentiation in *Chione cancellata* (Bivalvia, Veneridae): An ultrastructural analysis. *Invertebrate Biology* 125: 21–33.

Mount, A. S., Wheeler, A. P., Paradkar, R. P. & Snider, D. (2004). Hemocyte-mediated shell mineralization in the eastern oyster. *Science* 304: 297–300.

Moura, K. R. S., Terra, W. R. & Ribeiro, A. F. (2004). The functional organization of the salivary gland of *Biomphalaria straminea* (Gastropoda: Planorbidae): Secretory mechanisms and enzymatic determinations. *Journal of Molluscan Studies* 70: 21–29.

Moy, G. W., Springer, S. A., Adams, S. L., Swanson, W. J. & Vacquier, V. D. (2008). Extraordinary intraspecific diversity in oyster sperm bindin. *Proceedings of the National Academy of Sciences of the United States of America* 105: 1993–1998.

Moy, G. W. & Vacquier, V. D. (2008). Bindin genes of the Pacific oyster *Crassostrea gigas*. *Gene* 423: 215–220.

Moyer, C. L. & Emerson (1982). Massive destruction of scleractinian corals by the muricid gastropod *Drupella* in Japan and the Philippines. *The Nautilus* 96: 69–82.

Moynihan, M. (1975). Conservatism of displays and comparable stereotyped patterns among cephalopods. pp. 276–291 *in* G. Baerends, Beer, C. & Manning, A. (Eds.) *Function and Evolution in Behavior*. Oxford, UK, Oxford University Press.

Moynihan, M. & Rodaniche, A. F. (1982). *The Behaviour and Natural History of the Caribbean Reef Squid Sepioteuthis sepioidea; with a Consideration of Social, Signal, and Defense Patterns for Difficult and Dangerous Environments*. Berlin, Hamburg, Verlag Paul Parey.

Moynihan, M. (1985). *Communication and Noncommunication by Cephalopods*. Bloomington, IN, Indiana University Press.

Mozley, A. (1954). *An Introduction to Molluscan Ecology. Distribution and Population Studies of Fresh-Water Molluscs*. London, UK, H. K. Lewis & Co.

Mugglin, F. (1939). Beiträge zur Kenntnis der Anatomie von *Nautilus macromphalus* G.B. Sow. *Vierteljahrsschrift der Naturforschenden Gesellschaft in Zürich* 84: 25–118.

Mujer, C. V., Andrews, D. L., Manhart, J. R., Pierce, S. K. & Rumpho, M. E. (1996). Chloroplast genes are expressed during intracellular symbiotic association of *Vaucheria litorea* plastids with the sea slug *Elysia chlorotica*. *Proceedings of the National Academy of Sciences of the United States of America* 93: 12333–12338.

Mukai, S. T., Steel, C. G. H. & Saleuddin, A. S. M. (2001). Partial characterization of the secretory material from the dorsal bodies in the snail *Helisoma duryi* (Mollusca: Pulmonata), and its effects on reproduction. *Invertebrate Biology* 120: 149–161.

Mukai, S. T. & Morishita, F. (2017). Physiological functions of gastropod peptides and neurotransmitters. pp. 379–476 *in* S. Saleuddin & Mukai, S. T. (Eds.) *Physiology of Molluscs*. Vol. 2. Oakville, Ontario, Canada, Apple Academic Press.

Muller, P. A. L., Geary, D. H. & Magyar, I. (1999). The endemic molluscs of the Late Miocene Lake Pannon: Their origin, evolution, and family-level taxonomy. *Lethaia* 32: 47–60.

Muneoka, Y. & Twarog, B. M. (1983). Neuromuscular transmission and excitation: Contraction coupling in molluscan muscle. pp. 35–76 *in* A. S. M. Saleuddin & Wilbur, K. M. (Eds.) *Physiology, Part 1. The Mollusca*. Vol. 4. New York, Academic Press.

Muneoka, Y., Fujisawa, Y., Matsuura, M. & Ikeda, T. (1991). Neurotransmitters and neuromodulators controlling the anterior byssus retractor muscle of *Mytilus edulis*. *Comparative Biochemistry and Physiology Part C* 98: 105–114.

Munoz, A. A. & Ojeda, F. P. (1997). Feeding guild structure of a rocky intertidal fish assemblage in central Chile. *Environmental Biology of Fishes* 49: 471–479.

Muñoz, R. C. & Warner, R. R. (2003). Alternative contexts of sex change with social control in the bucktooth parrotfish, *Sparisoma radians*. *Environmental Biology of Fishes* 68: 307–319.

Muntz, W. R. A. & Johnson, M. S. (1978). Rhodopsins of oceanic decapods. *Vision Research* 18: 601–602.

Muntz, W. R. A. & Raj, U. (1984). On the visual system of *Nautilus pompilius*. *Journal of Experimental Biology* 109: 253–263.

Muntz, W. R. A. (1986). The spectral sensitivity of *Nautilus pompilius*. *Journal of Experimental Biology* 126: 513–517.

Muntz, W. R. A. (1987). Visual behaviour and visual sensitivity of *Nautilus pompilius*. pp. 231–244 *in* W. B. Saunders & Landman, N. H. (Eds.) *Nautilus: The Biology and Paleobiology of a Living Fossil. Topics in Geobiology*. New York, Springer.

Muntz, W. R. A. & Gwyther, J. (1988). Visual acuity in *Octopus pallidus* and *Octopus australis*. *Journal of Experimental Biology* 129: 119–129.

Muntz, W. R. A. (1994). Spatial summation in the phototactic behaviour of *Nautilus pompilius*. *Marine Behaviour and Physiology* 24: 183–187.

Muntz, W. R. A. (1999). Visual systems, behaviour, and environment in cephalopods. pp. 467–483 *in* S. N. Archer, Djamgoz, M. B. A., Loew, E. R., Partridge, J. C. & Vallerga, S. (Eds.) *Adaptive Mechanisms in the Ecology of Vision*. Dordrecht, the Netherlands, Kluwer Academic Publishers.

Murakami, M. & Kouyama, T. (2008). Crystal structure of squid rhodopsin. *Nature* 453: 363–367.

Murdock, G. R. & Vogel, S. L. (1978). Hydrodynamic induction of water flow through a keyhole limpet (Gastropoda, Fissurellidae). *Comparative Biochemistry and Physiology Part A* 61: 227–231.

Murphy, D. J. & Pierce, S. K. (1975). The physiological basis for changes in the freezing tolerance of intertidal molluscs. I. Response to subfreezing temperatures and the influence of salinity and temperature acclimation. *Journal of Experimental Zoology Part A* 193: 313–321.

Murphy, M. (2002). Mollusc conservation and the New South Wales Threatened Species Conservation Act 1995: The recovery program for Mitchell's Rainforest Snail *Thersites mitchellae*. *Australian Zoologist* 32: 1–11.

Murray, J. & Clarke, B. (1976). Supergenes in polymorphic land snails. I. *Partula taeniata. Heredity* 37: 253–269.

Murray, J., Murray, E., Johnson, M. S. & Clarke, B. (1988). The extinction of *Partula* on Moorea [French Polynesia]. *Pacific Science* 42: 150–153.

Murray, R. W. (1966). The effect of temperature on the membrane properties of neurons in the visceral ganglion of *Aplysia*. *Comparative Biochemistry and Physiology* 18: 291–303.

Mutlu, E. (2004). Sexual dimorphisms in radula of *Conomurex persicus* (Gastropoda: Strombidae) in the Mediterranean Sea. *Marine Biology* 145: 693–698.

Mutvei, H. (1957). On the relations of the principal muscles to the shell in *Nautilus* and some fossil nautiloids. *Arkiv för Mineralogi och Geologi* 2: 219–254.

Mutvei, H. (1964). Remarks on the anatomy of Recent and fossil Cephalopoda: With description of the minute shell structure of belemnoids. *Stockholm Contributions in Geology* 11: 79–102.

Mutvei, H. (1967). On the microscopic shell structure in some Jurassic ammonoids. *Neues Jahrbuch für Geologie und Paläontologie Abhandlungen* 129: 157–166.

Mutvei, H. (1978). Ultrastructural characteristics of the nacre in some gastropods. *Zoologica Scripta* 7: 287–296.

Mutvei, H. (1980). The nacreous layer in molluscan shells. pp. 49–56 *in* M. Ōmori & Watabe, N. (Eds.) *The Mechanisms of Biomineralization in Animals and Plants. Proceedings of the 3rd International Biomineralization Symposium Ago-chō, Japan 1977*. Tokyo, Tokai University Press.

Mutvei, H. (1983). Flexible nacre in the nautiloid *Isorthoceras*, with remarks on the evolution of cephalopod nacre. *Lethaia* 16: 234–240.

Mydlarz, L. D., Jones, L. E. & Harvell, C. D. (2006). Innate immunity, environmental drivers, and disease ecology of marine and freshwater invertebrates. *Annual Review of Ecology, Evolution, and Systematics* 37: 251–288.

Myers, F. L. & Northcote, D. H. (1958). A survey of the enzymes from the gastro-intestinal tract of *Helix pomatia*. *Journal of Experimental Biology* 35: 639–648.

Myers, G. S. (1949). Usage of anadromous, catadromous and allied terms for migratory fishes. *Copeia* 2: 89–97.

Myers, R. A., Cruz, L. J., Rivier, J. E. & Olivera, B. M. (1993). *Conus* peptides as chemical probes for receptors and ion channels. *Chemical Reviews* 93: 1923–1936.

Naef, A. (1909). Die Organogenese des Cölomsystems und der zentralen Blutgefässe von *Loligo*. Kritische Darstellung nach eigenen Untersuchungen. *Jenaische Zeitschrift für Medizin und Naturwissenschaft* 45: 221–266.

Naef, A. (1911). Studien zur generellen Morphologie der Mollusken. I. Teil: Über Torsion und Asymmetrie der Gastropoden. *Ergebnisse und Fortschritte der Zoologie* 3: 73–164.

Naef, A. (1913). Studien zur generellen Morphologie der Mollusken. II. Teil: Das Cölomsystem in seinen topographischen Beziehungen. *Ergebnisse und Fortschritte der Zoologie* 3: 329–462.

Naef, A. (1926). Studien zur generellen Morphologie der Mollusken. III. Teil: Die typischen Beziehungen der Weichtiere untereinander und das verhältnis ihrer Urformen zu anderen Cälomaten. *Ergebnisse und Fortschritte der Zoologie* 6: 27–124.

Naef, A. (1928). *Die Cephalopoden. Embryologie*. Vol. 35. Berlin, Germany, R. Friedlander & Sohn.

Nagahama, T. & Takata, M. (1988). Food-induced firing patterns in motoneurons producing jaw movements in *Aplysia kurodai*. *Journal of Comparative Physiology A* 162: 729–738.

Naganuma, T., Ogawa, T., Hirabayashi, J., Kasai, K., Kamiya, H. & Muramoto, K. (2006). Isolation, characterization and molecular evolution of a novel pearl shell lectin from a marine bivalve *Pteria penguin. Molecular Diversity* 10: 607–618.

Nagasawa, K., Tomita, K., Fujita, N. & Sasaki, R. (1993). Distribution and bivalve hosts of the parasitic copepod *Pectenophilus ornatus* Nagasawa, Bresciani, and Lützen in Japan. *Journal of Crustacean Biology* 13: 544–550.

Nagata, T., Sameshima, M., Uchikawa, T., Osafune, N. & Kitano, T. (2017). Molecular cloning and expression of the heat shock protein 70 gene in the Kumamoto oyster *Crassostrea sikamea*. *Fisheries Science* 83: 273–281.

Nagel, R. L. (1985). Molluscan hemoglobins. pp. 227–247 *in* W. D. Cohen (Ed.) *Blood Cells of Marine Invertebrates. MBL Lectures in Biology*. New York, Alan R. Liss.

Nagle, G. T., Knock, S. L., Painter, S. D., Blankenship, J. E., Fritz, R. R. & Kurosky, A. (1989). I. *Aplysia californica* neurons R3–R14: Primary structure of the myoactive histidine-rich basic peptide and peptide I. *Peptides* 10: 849–857.

Nair, N. B. & Ansell, A. D. (1968). Characteristics of penetration of the substratum by some marine bivalve molluscs. *Proceedings of the Malacological Society of London* 38: 179–197.

Nair, N. B. & Saraswathy, M. (1971). The biology of wood-boring teredinld molluscs. *Advances in Marine Biology* 9: 335–509.

Nakagawa, M., Iwasa, T., Kikkawa, S., Tsuda, M. & Ebrey, T. G. (1999). How vertebrate and invertebrate visual pigments differ in their mechanism of photoactivation. *Proceedings of the National Academy of Sciences of the United States of America* 96: 6189–6192.

Nakagawa, S. & Takai, K. (2008). Deep-sea vent chemoautotrophs: Diversity, biochemistry and ecological significance. *FEMS Microbiology Ecology* 65: 1–14.

Nakahara, H. & Bevelander, G. (1970). An electron microscope study of the muscle attachment in the mollusc *Pinctada radiata*. *Texas Reports on Biology and Medicine* 28: 279–286.

Nakamura-Kusakabe, I., Nagasaki, T., Kinjo, A., Sassa, M., Koito, T., Okamura, K., Yamagami, S., Yamanaka, T., Tsuchida, S. & Inoue, K. (2016). Effect of sulfide, osmotic, and thermal stresses on taurine transporter mRNA levels in the gills of the hydrothermal vent-specific mussel *Bathymodiolus septemdierum*. *Comparative Biochemistry and Physiology Part A: Molecular & Integrative Physiology* 191: 74–79.

Nakamura, H. K. (1985). A review of molluscan cytogenetic information based on the CISMOCH [computerized index system for molluscan chromosomes]: Bivalvia, Polyplacophora and Cephalopoda. *Venus* 44: 193–225.

Nakamura, H. K. (1986). Chromosomes of Archaeogastropoda (Mollusca: Prosobranchia), with some remarks on their cytotaxonomy and phylogeny. *Publications of the Seto Marine Biological Laboratory* 31: 191–267.

Nakamura, M., Masaki, M., Maki, S., Matsui, R., Hieda, M., Mamino, M., Hirano, T., Ohmiya, Y. & Niwa, H. (2004). Synthesis of *Latia* luciferin benzoate analogues and their bioluminescent activity. *Tetrahedron Letters* 45: 2203–2205.

Nakamura, S., Osada, M. & Kijima, A. (2007). Involvement of GnRH neuron in the spermatogonial proliferation of the scallop, *Patinopecten yessoensiss*. *Molecular Reproduction and Development* 74: 108–115.

Nakamura, Y. (2013). Secretion of a mucous cord for drifting by the clam *Meretrix lusoria* (Veneridae). *Plankton and Benthos Research* 8: 31–45.

Nakano, H. & Stevens, J. D. (2008). The biology and ecology of the blue shark, *Prionace glauca*. pp. 140–151 *in* M. D. Camhi, Pikitch, E. K. & Babcock, E. A. (Eds.) *Sharks of the Open Ocean: Biology, Fisheries and Conservation. Fish and Aquatic Resources.* Oxford, UK, Blackwell Publishing.

Nakano, T. & Spencer, H. G. (2007). Simultaneous polyphenism and cryptic species in an intertidal limpet from New Zealand. *Molecular Phylogenetics and Evolution* 45: 470–479.

Nakano, T., Sasaki, T. & Kase, T. (2010). Color polymorphism and historical biogeography in the Japanese patellogastropod limpet *Cellana nigrolineata* (Reeve) (Patellogastropoda: Nacellidae). *Zoological Science* 27: 811–820.

Nakao, Y., Yoshida, W. Y., Szabo, C. M., Baker, B. J. & Scheuer, P. J. (1998). More peptides and other diverse constituents of the marine mollusk *Philinopsis speciosa*. *Journal of Organic Chemistry* 63: 3272–3280.

Nalepa, T. F. & Schloesser, D. W., Eds. (2014). *Quagga and Zebra Mussels; Biology, Impacts, and Control.* Boca Raton, FL, CRC Press.

Namigai, E. K. O., Kenny, N. J. & Shimeld, S. M. (2014). Right across the tree of life: The evolution of left–right asymmetry in the Bilateria. *Genesis* 52: 458–470.

Napolitano, J. G., Souto, M. L., Fernández, J. J. & Norte, M. (2008). Micromelones A and B, noncontiguous polypropionates from *Micromelo undata*. *Journal of Natural Products* 71: 281–284.

Narain, A. S. (1976). A review of the structure of the heart of molluscs, particularly bivalves, in relation to cardiac function. *Journal of Molluscan Studies* 42: 46–62.

Naranjo-Garcia, E. & Castillo-Rodriguez, Z. G. (2017). First inventory of the introduced and invasive mollusks in Mexico. *The Nautilus* 131: 107–126.

Naraoka, H., Naito, T., Yamanaka, T., Tsunogai, U. & Fujikura, K. (2008). A multi-isotope study of deep-sea mussels at three different hydrothermal vent sites in the northwestern Pacific. *Chemical Geology* 255: 25–32.

Narchi, W. (1969). On *Pseudopythina rugifera* (Carpenter, 1864) (Bivalvia). *The Veliger* 12: 43–52.

Narita, H., Noguchi, T., Maruyama, J., Ueda, Y., Hashimoto, K., Watanabe, Y. & Hida, K. (1981). Occurrence of tetrodotoxin in a trumpet shellfish 'boshubora' *Charonia sauliae*. *Bulletin of the Japanese Society of Scientific Fisheries* 47: 934–941.

Nathan, C. & Xie, Q. W. (1994). Nitric oxide synthases: Roles, tolls, and controls. *Cell* 78: 915–918.

Natochin, V. (1966). Reaction of mussels on separate changes of osmotic concentration and salinity in the environment. *Zhurnal Obshchei Biologii* 27: 473–479.

Navarrete, S. A. & Castilla, J. C. (1993). Predation by Norway rats in the intertidal zone of central Chile. *Marine Ecology Progress Series* 92: 187.

Navarrete, S. A. & Manzur, T. (2008). Individual- and population-level responses of a keystone predator to geographic variation in prey. *Ecology* 89: 2005–2018.

Navarro, J., Coll, M., Somes, C. J. & Olson, R. J. (2013). Trophic niche of squids: Insights from isotopic data in marine systems worldwide. *Deep Sea Research Part II: Topical Studies in Oceanography* 95: 93–102.

Navet, S., Bassaglia, Y., Baratte, S., Martin, M. C. & Bonnaud, L. (2008). Somatic muscle development in *Sepia officinalis* (Cephalopoda-Mollusca): A new role for NK4. *Developmental Dynamics* 237: 1944–1951.

Naylor, E. (2001). Marine animal behaviour in relation to lunar phase. *Earth, Moon and Planets* 85: 291–302.

Nederbragt, A. J., van Loon, A. E. & Dictus, J. A. G. (2002). Evolutionary biology: *hedgehog* crosses the snail's midline. *Nature* 417: 811–812.

Negrete, L. H., Gullo, B. S. & Martín, S. M. (2007). First record of *Helobdella hyalina* (Hirudinea: Glossiphoniidae) in the mantle cavity of Planorbidae from lentic environments in a Buenos Aires province, Argentina. *Brazilian Journal of Biology* 67: 377–378.

Negus, M. R. S. (1968). Oxygen consumption and amino acid levels in *Hydrobia ulvae* (Pennant) in relation to salinity and behaviour. *Comparative Biochemistry and Physiology* 24: 317–325.

Neigel, J. E. (1997). A comparison of alternative strategies for estimating gene flow from genetic markers. *Annual Review of Ecology and Systematics* 28: 105–128.

Nekola, J. C. (2002). Effects of fire management on the richness and abundance of central North American grassland land snail faunas. *Animal Biodiversity and Conservation* 25: 53–66.

Nekola, J. C. (2014). Overview of the North American terrestrial gastropod fauna. *American Malacological Bulletin* 32: 225–235.

Nelson, B. V. (1980). Adaptive value of aggregation in a sessile gastropod. *American Zoologist* 20: 746.

Nelson, L. & Morton, J. E. (1979). Cyclic activity and epithelial renewal in the digestive gland tubules of the marine prosobranch *Maoricrypta monoxyla* (Lesson). *Journal of Molluscan Studies* 45: 262–283.

Neuckel, W. (1985). Anal uptake of water in terrestrial pulmonate snails: Metabolic and transport functions. *Journal of Comparative Physiology B* 156: 291–296.

Neumeister, H. & Budelmann, B.-U. (1997). Structure and function of the *Nautilus* statocyst. *Philosophical Transactions of the Royal Society B* 352: 1565–1588.

Neusser, T. P., Hess, M., Haszprunar, G. & Schrödl, M. (2006). Computer-based three-dimensional reconstruction of the anatomy of *Microhedyle remanei* (Marcus, 1953), an interstitial acochlidian gastropod from Bermuda. *Journal of Morphology* 267: 231–247.

Neusser, T. P., Hess, M. & Schrödl, M. (2009). Tiny but complex: Interactive 3D visualization of the interstitial acochlidian gastropod *Pseudunela cornuta* (Challis, 1970). *Frontiers in Zoology* 6: 20.

Neusser, T. P. & Schrödl, M. (2009). Between Vanuatu tides: 3D anatomical reconstruction of a new brackish water acochlidian gastropod from Espiritu Santo. *Zoosystema* 31: 453–469.

Neusser, T. P., Fukuda, H., Jörger, K. M., Kano, Y. & Schrödl, M. (2011). Sacoglossa or Acochlidia? 3D reconstruction, molecular phylogeny and evolution of Aitengidae (Gastropoda: Heterobranchia). *Journal of Molluscan Studies* 77: 332–350.

Neusser, T. P., Hanke, F., Haszprunar, G. & Jörger, K. M. (2018). 'Dorsal vessels'? 3D-reconstruction and ultrastructure of the renopericardial system of *Elysia viridis* (Montagu, 1804) (Gastropoda: Sacoglossa), with a discussion of function and homology. *Journal of Molluscan Studies* 85: 79–91.

Neuweger, D., White, P. & Ponder, W. F. (2001). Land snails from Norfolk Island sites. *Records of the Australian Museum* 27: 115–122.

Neves, R. (1999). Conservation and commerce: Management of freshwater mussel (Bivalia: Unionoidea) resources in the United States. *Malacologia* 41: 461–474.

Neves, R. J. & Odom, M. C. (1989). Muskrat predation on endangered freshwater mussels in Virginia. *The Journal of Wildlife Management* 53: 934–941.

Neves, R. J., Bogan, A. E., Williams, J. D., Ahlstedt, S. A. & Hartfield, P. W. (1997). Status of aquatic mollusks in the southeastern United States: A downward spiral of diversity. pp. 43–85 *in* G. W. Benz & Collins, D. E. (Eds.) *Aquatic Fauna in Peril: The Southeastern Perspective.* Chattanooga, Tennessee, Lenz Design & Communications for Southeast Aquatic Research Institute.

Newcomb, J. M., Fickbohm, D. J. & Katz, P. S. (2006). Comparative mapping of serotonin-immunoreactive neurons in the central nervous systems of nudibranch molluscs. *Journal of Comparative Neurology* 499: 485–505.

Newcomb, J. M. & Katz, P. S. (2007). Homologues of serotonergic central pattern generator neurons in related nudibranch molluscs with divergent behaviors. *Journal of Comparative Physiology A* 193: 425–443.

Newcomb, J. M., Sakurai, A., Lillvis, J. L., Gunaratne, C. A. & Katz, P. S. (2012). Homology and homoplasy of swimming behaviors and neural circuits in the Nudipleura (Mollusca, Gastropoda, Opisthobranchia). *Proceedings of the National Academy of Sciences of the United States of America* 109: 10669–10676.

Newcomb, J. M., Kirouac, L. E., Naimie, A. A., Bixby, K. A., Lee, C., Malanga, S., Raubach, M. & Watson, W. H. (2014). Circadian rhythms of crawling and swimming in the nudibranch mollusc *Melibe leonina. Biological Bulletin* 227: 263–273.

Newell, G. E. (1958a). The behaviour of *Littorina littorea* (L) under natural conditions and its relation to position on the shore. *Journal of the Marine Biological Association of the United Kingdom* 37: 229–239.

Newell, G. E. (1958b). An experimental analysis of the behaviour of *Littorina littorea* (L) under natural conditions and in the laboratory. *Journal of the Marine Biological Association of the United Kingdom* 37: 241–266.

Newell, G. E. (1965). The eye of *Littorina littorea. Proceedings of the Zoological Society of London* 144: 75–86.

Newell, P. F. & Newell, G. E. (1968). The eye of the slug, *Agriolimax reticulatus* (Müll). pp. 97–111 *in* J. D. Carthy & Newell, G. E. (Eds.) *Invertebrate Receptors. Proceedings of a Symposium held at the Zoological Society of London on 30 and 31 May 1967.* New York, Academic Press.

Newell, P. F. & Skelding, J. M. (1973). Structure and permeability of the septate junction in the kidney sac of *Helix pomatia* L. *Zeitschrift für Zellforschung und Mikroskopische Anatomie* 147: 31–39.

Newell, P. F. (1977). The structure and enzyme histochemistry of slug skin. *Malacologia* 16: 183–195.

Newell, R. C. (1970). *Biology of Intertidal Animals.* London, UK, Logos Press.

Newell, R. C. & Branch, G. M. (1980). The influence of temperature on the maintenance of metabolic energy balance in marine invertebrates. *Advances in Marine Biology* 17: 329–396.

Newell, R. C., Parker, I. & Cook, P. A. (1980). A possible rôle of α-amylase isoenzymes from the style of the mussel *Choromytilus meridionalis* (Krauss) following thermal acclimation. *Journal of Experimental Marine Biology and Ecology* 47: 1–8.

Newman, K. C. & Thomas, R. E. (1975). Ammonia, urea and uric acid levels in active and estivating snails, *Bakerilymnaea cockerelli. Comparative Biochemistry and Physiology Part A* 50: 109–112.

Newman, L. J., Cannon, L. R. G. & Govan, H. (1993). *Stylochus (Imogene) matatasi* n. sp. (Platyhelminthes, Polycladida): Pest of cultured giant clams and pearl oysters from Solomon Islands. *Hydrobiologia* 257: 185–189.

Newman, L. J., Cannon, L. R. G. & Brunckhorst, D. J. (1994). A new flatworm (Platyhelminthes: Polycladida) which mimics a phyllidiid nudibranch (Mollusca, Nudibranchia). *Zoological Journal of the Linnean Society* 110: 19–25.

Newman, R. M., Hanscom, Z. & Kerfoot, W. C. (1992). The watercress glucosinolate-myrosinase system: A feeding deterrent to caddisflies, snails and amphipods. *Oecologia* 92: 1–7.

Newman, R. M., Kerfoot, W. C. & Hanscom, Z. (1996). Watercress allelochemical defends high-nitrogen foliage against consumption: Effects on freshwater invertebrate herbivores. *Ecology* 77: 2312–2323.

Newman, W. A. (1985). The abyssal hydrothermal vent invertebrate fauna: A glimpse at antiquity? *Bulletin of the Biological Society of Washington* 6: 231–242.

Nezlin, L. & Voronezhskaya, E. (2017). Early peripheral sensory neurons in the development of trochozoan animals. *Russian Journal of Developmental Biology* 48: 130–143.

Nezlin, L. P. (1997). The osphradium is involved in the control of egg-laying in the pond snail *Lymnaea stagnalis. Invertebrate Reproduction & Development* 32: 163–166.

Ng, J. S., Wai, T.-C. & Williams, G. A. (2007). The effects of acidification on the stable isotope signatures of marine algae and molluscs. *Marine Chemistry* 103: 97–102.

Ng, P. K. L. & Tan, L. W. H. (1984). The 'shell peeling' structure of the box crab *Calappa philargius* (L.) and other crabs in relation to mollusc shell architecture. *Journal of the Singapore National Academy of Science* 13: 195–199.

Ng, P. K. L. & Tan, L. W. H. (1985). 'Right handedness' in heterochelous calappoid and xanthoid crabs: Suggestion for a functional advantage. *Crustaceana* 49: 98–100.

Ng, T. P. T., Saltin, S. H., Davies, M. S., Johannesson, K., Stafford, R. & Williams, G. A. (2013). Snails and their trails: The multiple functions of trail-following in gastropods. *Biological Reviews* 88: 683–700.

Ng, T. P. T., Davies, M. S., Stafford, R. & Williams, G. A. (2016). Fighting for mates: The importance of individual size in mating contests in rocky shore littorinids. *Marine Biology* 163: 1–9.

Nhan, H. T., Jung, L. H., Ambak, M. A., Watson, G. J. & Siang, H. Y. (2010). Evidence for sexual attraction pheromones released by male tropical Donkey's ear Abalone (*Haliotis asinina*, L.). *Invertebrate Reproduction & Development* 54: 169–176.

Nicaise, G. (1973). The gliointerstitial system of molluscs. *International Review of Cytology* 34: 251–332.

Nicaise, G. & Amsellem, J. (1983). Cytology of muscle and neuromuscular junction. pp. 1–33 *in* A. S. M. Saleuddin & Wilbur, K. M. (Eds.) *Physiology, Part 1. The Mollusca.* Vol. 4. New York, Academic Press.

Nicol, D. (1969). The number of living species of molluscs. *Systematic Zoology* 18: 251–254.

Nie, H., Liu, L., Huo, Z., Chen, P., Ding, J., Yang, F. & Yan, X. (2017). The HSP70 gene expression responses to thermal and salinity stress in wild and cultivated Manila clam *Ruditapes philippinarum. Aquaculture* 470: 149–156.

Nielsen, C. (1991). The development of the brachiopod *Crania (Neocrania) anomala* (O.F. Müller) and its phylogenetic significance. *Acta Zoologica* 72: 7–28.

Nielsen, C. (1995). Phylum Mollusca. pp. i, 110–123. *Animal Evolution: Interrelationships of the Living Phyla.* Oxford, UK, Oxford University Press.

Nielsen, C. (2001). *Animal Evolution: Interrelationships of the Living Phyla*. 2nd ed. Oxford, UK, Oxford University Press.

Nielsen, C. (2004). Trochophora larvae: Cell-lineages, ciliary bands, and body regions. 1. Annelida and Mollusca. *Journal of Experimental Zoology Part B* 302: 35–68.

Nielsen, C., Haszprunar, G., Ruthensteiner, B. & Wanninger, A. (2007). Early development of the aplacophoran mollusc *Chaetoderma*. *Acta Zoologica* 88: 231–247.

Nielsen, C. (2015). Larval nervous systems: True larval and precocious adult. *Journal of Experimental Biology* 218: 629–636.

Nielsen, R. (1972). A study of the shell-boring marine algae around the Danish island Laesø. *Botanisk Tidsskrift* 67: 245–269.

Nierstrasz, H. F. & Hoffmann, H. (1929). Aculifera. pp. 1–64 *in* G. Grimpe & Wagler, E. (Eds.) *Die Tierwelt der Nord- und Ostsee*. Leipzig, Akademische Verlagsgesellschaft.

Nilsson-Ehle, H. (1914). Vilka erfarenheter hava hittillis vunnits rörande möjligheten av växters acklimatisering? *Kungliga Landtbruks Akademiens Handlingar och Tidskrift* 53: 537–572.

Nilsson, D. E. (1994). Eyes as optical alarm systems in fan worms and ark clams. *Philosophical Transactions of the Royal Society B* 346: 195–212.

Nilsson, D. E., Warrant, E. J., Johnsen, S., Hanlon, R. T. & Shashar, N. (2012). A unique advantage for giant eyes in giant squid. *Current Biology* 22: 683–688.

Nilsson, D. E. (2013). Eye evolution and its functional basis. *Visual Neuroscience* 30: 5–20.

Nilsson, D. E., Warrant, E. J., Johnsen, S., Hanlon, R. T. & Shashar, N. (2013). The giant eyes of giant squid are indeed unexpectedly large, but not if used for spotting sperm whales. *BMC Evolutionary Biology* 13: 187.

Nisbet, R. H. (1973). The role of the buccal mass in the trochid. *Proceedings of the Malacological Society of London* 40: 435–468.

Nishi, T. & Gotow, T. (1992). A neural mechanism for processing colour information in molluscan extra-ocular photoreceptors. *Journal of Experimental Biology* 168: 77–91.

Nishiguchi, M. K., Lopez, J. E. & von Boletzky, S. (2004). Enlightenment of old ideas from new investigations: More questions regarding the evolution of bacteriogenic light organs in squids. *Evolution & Development* 6: 41–49.

Nishiguchi, M. K. & Mapes, R. H. (2008). Cephalopoda. pp. 163–199 *in* W. F. Ponder & Lindberg, D. R. (Eds.) *Phylogeny and Evolution of the Mollusca*. Berkeley, CA, University of California Press.

Nishikawa, J.-I., Mamiya, S., Kanayama, T., Nishikawa, T., Shiraishi, F. & Horiguchi, T. (2004). Involvement of the retinoid X receptor in the development of imposex caused by organotins in gastropods. *Environmental Science & Technology* 38: 6271–6276.

Nishikawa, J. (2006). Imposex in marine gastropods may be caused by binding of organotins to retinoid X receptor. *Marine Biology* 149: 117–124.

Nishimura, S. (1976). *Dynoidella conchicola*, gen. et sp. nov. (Isopoda, Sphaeromatidae), from Japan, with a note on its association with intertidal snails. *Publications of the Seto Marine Biological Laboratory* 23: 275–282.

Nishioka, R. S., Bern, H. A. & Golding, D. W. (1970). Innervation of the cephalopod optic gland. pp. 47–54 *in* W. Bargmann & Scharrer, B. (Eds.) *Aspects of Neuroendocrinology*. Berlin, Germany, Springer.

Nishiwaki, S. (1964). Phylogenetic study on the type of the dimorphic spermatozoa in Prosobranchia. *Science Reports. Tokyo Kyoiku Daigaku: Section B* 11: 237–275.

Nixon, M. (1969). The life span of *Octopus vulgaris* (Lamarck). *Proceedings of the Malacological Society of London* 38: 529–540.

Nixon, M. (1979a). Has *Octopus vulgaris* a second radula? *Journal of Zoology* 187: 291–296.

Nixon, M. (1979b). Hole-boring in shells by *Octopus vulgaris* Cuvier in the Mediterranean. *Malacologia* 18: 431–443.

Nixon, M. (1980). The salivary papilla of *Octopus* as an accessory radula for drilling shells. *Journal of Zoology* 190: 53–57.

Nixon, M. (1984). Is there external digestion by *Octopus*? *Journal of Zoology* 202: 441–447.

Nixon, M. (1987). Cephalopod diets. pp. 201–219 *in* P. R. Boyle (Ed.) *Cephalopod Life Cycles. Volume 2. Comparative Reviews*. Orlando, FL, Academic Press.

Nixon, M. (1995). A nomenclature for the radula of the Cephalopoda (Mollusca): Living and fossil. *Journal of Zoology* 236: 73–81.

Nixon, M. & Young, J. Z. (2003). *The Brains and Lives of Cephalopods*. Oxford, UK, Oxford University Press.

Noguchi, F., Kawato, M., Yoshida, T., Fujiwara, Y., Fujikura, K. & Takishita, K. (2013). A novel alveolate in bivalves with chemosynthetic bacteria inhabiting deep-sea methane seeps. *Journal of Eukaryotic Microbiology* 60: 158–165.

Noguchi, T., Maruyama, T., Narita, H. & Hashimoto, K. (1984). Occurrence of tetrodotoxin in the gastropod mollusk *Tutufa lissostoma* (frog shell). *Toxicon* 22: 219–226.

Nolen, T. G., Johnson, P. M., Kicklighter, C. E. & Capo, T. (1995). Ink secretion by the marine snail *Aplysia californica* enhances its ability to escape from a natural predator. *Journal of Comparative Physiology A* 176: 239–254.

Nolen, T. G. & Johnson, P. M. (2001). Defensive inking in *Aplysia* spp.: Multiple episodes of ink secretion and the adaptive use of a limited chemical resource. *Journal of Experimental Biology* 204: 1257–1268.

Nordsieck, H. (1985). The system of the Stylommatophora (Gastropoda) with special regard to the systematic position of the Clausiliidae. I. Importance of the excretory and genital systems. *Archiv für Molluskenkunde* 116: 1–24.

Norman, M. D. (1992). *Ameloctopus litoralis*, gen. et sp. nov. (Cephalopoda: Octopodidae), a new shallow-water octopus from tropical Australian waters. *Invertebrate Systematics* 6: 567–582.

Norman, M. D. (2000). Hitchhiking argonaut. pp. 189 *in* M. D. Norman & Debelius, H. (Eds.) *Cephalopods: A World Guide*. Hackenheim (Germany), ConchBooks.

Norman, M. D. & Reid, A. L. (2000). *A Guide to Squid, Cuttlefish and Octopuses of Australasia*. Melbourne, Australia, CSIRO Publishing, The Gould League of Australia.

Norman, M. D. & Finn, J. (2001). Revision of the *Octopus horridus* species-group, including erection of a new subgenus and description of two member species from the Great Barrier Reef, Australia. *Invertebrate Systematics* 15: 13–35.

Norman, M. D., Paul, D., Finn, J. K. & Tregenza, T. (2002). First encounter with a live male blanket octopus: The world's most sexually size-dimorphic large animal. *New Zealand Journal of Marine and Freshwater Research* 36: 733–736.

Norman, M. D. & Hochberg, F. G. (2005). The 'mimic octopus' (*Thaumoctopus mimicus* n. gen. and sp.), a new octopus from the tropical Indo-West Pacific (Cephalopoda: Octopodidae). *Molluscan Research* 25: 57–70.

Norte, M., Fernández, J. J. & Padilla, A. (1994). Isolation and synthesis of siphonarienal a new polypropionate from *Siphonaria grisea*. *Tetrahedron Letters* 35: 3413–3416.

Norton, J. H., Shepherd, M. A., Long, H. M. & Fitt, W. K. (1992). The zooxanthellal tubular system in the giant clam. *Biological Bulletin* 183: 503–506.

Norton, J. H., Shepherd, M. A. & Prior, H. C. (1993). Papovavirus-like infection of the golden-lipped pearl oyster, *Pinctada maxima*, from the Torres Strait, Australia. *Journal of Invertebrate Pathology* 62: 198–200.

Norton, J. H., Prior, H. C., Baillie, B. & Yellowlees, D. (1995). Atrophy of the zooxanthellal tubular system in bleached giant clams *Tridacna gigas*. *Journal of Invertebrate Pathology* 66: 307–310.

Noshita, K. (2014). Quantification and geometric analysis of coiling patterns in gastropod shells based on 3D and 2D image data. *Journal of Theoretical Biology* 363: 93–104.

Nowikoff, M. (1926). Über die Komplexaugen der Gattung Arca. *Zoologischer Anzeiger* 67: 277–289.

Noy, R., Lavie, B. & Nevo, E. (1987). The niche-width variation hypothesis revisited: Genetic diversity in the marine gastropods *Littorina punctata* (Gmelin) and *L. neritoides* (L.). *Journal of Experimental Marine Biology and Ecology* 109: 109–116.

Nuñez, J. D., Laitano, M. V. & Cledon, M. (2012). An intertidal limpet species as a bioindicator: Pollution effects reflected by shell characteristics. *Ecological Indicators* 14: 178–183.

Nüske, H. (1973). Cytologische Untersuchungen an der Säuredrüse der Meeresschnecke *Cassidaria echinophora*. *Cytobiologie* 7: 164–180.

Nusnbaum, M. & Derby, C. D. (2010). Ink secretion protects sea hares by acting on the olfactory and nonolfactory chemical senses of a predatory fish. *Animal Behaviour* 79: 1067–1076.

Nützel, A. (1998). Über die Stammesgeschichte der Ptenoglossa (Gastropoda). *Berliner Geowissenschaftliche Abhandlungen: Reihe E* 26: 1–229.

Nützel, A. & Frýda, J. (2003). Paleozoic plankton revolution: Evidence from early gastropod ontogeny. *Geology* 31: 829–831.

Nützel, A., Lehnert, O. & Frýda, J. (2006). Origin of planktotrophy: Evidence from early molluscs. *Evolution & Development* 8: 325–330.

Nützel, A. (2014). Larval ecology and morphology in fossil gastropods. *Palaeontology* 57: 479–503.

Nuurai, P., Poljaroen, J., Tinikul, Y., Cummins, S. F., Sretarugsa, P., Hanna, P. J., Wanichanon, C. & Sobhon, P. (2010). The existence of gonadotropin-releasing hormone-like peptides in the neural ganglia and ovary of the abalone, *Haliotis asinina* L. *Acta Histochemica* 112: 557–566.

Nuwayhid, M. A., Davies, P. S. & Elder, H. Y. (1978). Gill structure in the common limpet *Patella vulgata*. *Journal of the Marine Biological Association of the United Kingdom* 58: 817–823.

Nuzzo, G., Cutignano, A., Moles, J., Avila, C. & Fontana, A. (2016). Exiguapyrone and exiguaone, new polypropionates from the Mediterranean cephalaspidean mollusc *Haminoea exigua*. *Tetrahedron Letters* 57: 71–74.

Nybakken, J. & Perron, F. (1988). Ontogenetic change in the radula of *Conus magus* (Gastropoda). *Marine Biology* 98: 239–242.

Nybakken, J. (1990). Ontogenetic change in the *Conus* radula, its form, distribution among the radula types, and significance in systematics and ecology. *Malacologia* 32: 35–54.

Nyffeler, M. & Symondson, W. O. C. (2001). Spiders and harvestmen as gastropod predators. *Ecological Entomology* 26: 617–628.

Nyholm, S. V. & Nishiguchi, M. K. (2008). The evolutionary ecology of a sepiolid squid-vibrio association: From cell to environment. *Vie et Milieu. Life and Environment* 58: 175–184.

O'Brien, E. K. & Degnan, B. M. (2000). Expression of *POU*, *Sox*, and *Pax* genes in the brain ganglia of the tropical abalone *Haliotis asinina*. *Marine Biotechnology* 2: 545–557.

O'Brien, E. K. & Degnan, B. M. (2002). Developmental expression of a class IV *POU* gene in the gastropod *Haliotis asinina* supports a conserved role in sensory cell development in bilaterians. *Development Genes and Evolution* 212: 394–398.

O'Brien, E. K. & Degnan, B. M. (2003). Expression of *Pax258* in the gastropod statocyst: Insights into the antiquity of metazoan geosensory organs. *Evolution & Development* 5: 572–578.

O'Connor, W. A. & Lawler, N. F. (2004). Salinity and temperature tolerance of embryos and juveniles of the pearl oyster *Pinctada imbricata* Röding. *Aquaculture* 229: 493–506.

O'Dor, R., Stewart, J., Gilly, W., Payne, J., Borges, T. C. & Thys, T. (2013). Squid rocket science: How squid launch into air. *Deep Sea Research Part II: Topical Studies in Oceanography* 95: 113–118.

O'Dor, R. K. & Wells, M. J. (1973). Yolk protein synthesis in ovary of *Octopus vulgaris* and its control by optic gland gonadotropin. *Journal of Experimental Biology* 59: 665–674.

O'Dor, R. K. & Wells, M. J. (1975). Control of yolk protein synthesis by *Octopus* gonadotropin *in vivo* and *in vitro* (effects of *Octopus* gonadotropin). *General and Comparative Endocrinology* 27: 129–135.

O'Dor, R. K. & Webber, D. M. (1986). The constraints on cephalopods: Why squid aren't fish. *Canadian Journal of Zoology* 64: 1591–1605.

O'Dor, R. K. & Wells, M. J. (1987). Energy and nutrient flow. pp. 109–133 *in* P. R. Boyle (Ed.) *Cephalopod Life Cycles. Volume 2. Comparative Reviews.* London, Orlando FL, Academic Press.

O'Dor, R. K. & Shadwick, R. E. (1989). Squid, the Olympian cephalopods. *Journal of Cephalopod Biology* 1: 33–55.

O'Dor, R. K., Pörtner, H. & Shadwick, R. E. (1990). Squid as elite athletes: Locomotory, respiratory, and circulatory integration. pp. 481–503 *in* D. L. Gilbert, Adelman, W. J. & Arnold, J. M. (Eds.) *Squid as Experimental Animals.* New York, Plenum Press.

O'Dor, R. K. & Webber, D. M. (1991). Invertebrate athletes: Trade-offs between transport efficiency and power density in cephalopod evolution. *Journal of Experimental Biology* 160: 93–112.

O'Sullivan, J. B., McConnaughey, P. R. & Huber, M. E. (1987). A blood-sucking snail: The Cooper's Nutmeg, *Cancellaria cooperi* Gabb, parasitizes the California Electric Ray, *Torpedo californica* Ayres. *Biological Bulletin* 172: 362–366.

Ó Foighil, D. (1985). Fine structure of *Lasaea subviridis* and *Mysella tumida* sperm (Bivalvia, Galeommatacea). *Zoomorphology* 105: 125–132.

Ó Foighil, D. (1988). Global survey of developmental mode in the brooding bivalve *Lasea*. *American Zoologist* 28: 138.

Ó Foighil, D. (1989). Role of the spermatozeugmata in the spawning ecology of the brooding oyster *Ostrea edulis*. *Gamete Research* 24: 219–228.

Ó Foighil, D. & Thiriot-Quiévreux, C. (1991). Ploidy and pronuclear interaction in Northeastern Pacific *Lasaea* clones (Mollusca, Bivalvia). *Biological Bulletin* 181: 222–231.

Ó Foighil, D. & Taylor, D. J. (2000). Evolution of parental care and ovulation behavior in oysters. *Molecular Phylogenetics and Evolution* 15: 301–313.

Obata, M., Kamiya, C., Kawamura, K. & Komaru, A. (2006). Sperm mitochondrial DNA transmission to both male and female offspring in the blue mussel *Mytilus galloprovincialis*. *Development, Growth & Differentiation* 48: 253–261.

Ochoa, G. H., Clark, Y. M., Matsumoto, B., Torres-Ruiz, J. A. & Robles, L. J. (2002). Heat shock protein 70 and heat shock protein 90 expression in light- and dark-adapted adult *Octopus* retinas. *Journal of Neurocytology* 31: 161–174.

Ockelmann, K. W. (1958). Marine Lamellibranchiata: The zoology of East Greenland. *Meddelelser om Grønland* 122: 1–256.

Ockelmann, K. W. (1964). *Turtonia minuta* (Fabricius), a neotenous veneracean bivalve. *Ophelia* 1: 121–146.

Ockelmann, K. W. & Dinesen, G. E. (2011). Life on wood: The carnivorous deep-sea mussel *Idas argenteus* (Bathymodiolinae, Mytilidae, Bivalvia). *Marine Biology Research* 7: 71–84.

Odhner, N. H. (1914). Zwei Notitzen über die Fauna der Adria. Beiträge zur Kenntnis der Molluskenfauna von Rovigno. *Zoologischer Anzeiger* 44: 156–170.

Odhner, N. H. (1919). Studies on the morphology, the taxonomy and the relations of recent Chamidae. *Kungliga Svenska Vetenskapsakademiens Handlingar* 59: 1–102.

Odierna, G., Aprea, G., Barucca, M., Biscotti, M. A., Canapa, A., Capriglione, T. & Olmo, E. (2008). Karyology of the Antarctic chiton *Nuttallochiton mirandus* (Thiele, 1906) (Mollusca: Polyplacophora) with some considerations on chromosome evolution in chitons. *Chromosome Research* 16: 899–906.

Odiete, W. O. (1978). The cruciform muscle and its associated sense organ in *Scrobicularia plana* (da Costa). *Journal of Molluscan Studies* 44: 180–189.

Oehlmann, J., Stroben, E. & Fioroni, P. (1991). The morphological expression of imposex in *Nucella lapillus* (Linnaeus) (Gastropoda: Muricidae). *Journal of Molluscan Studies* 57: 375–390.

Oehlmann, J. & Schulte-Oehlmann, U. (2003). Endocrine disruption in invertebrates. *Pure and Applied Chemistry* 75: 2207–2218.

Oehlmann, J., Di Benedetto, P., Tillmann, M., Duft, M., Oetken, M. & Schulte-Oehlmann, U. (2007). Endocrine disruption in prosobranch molluscs: Evidence and ecological relevance. *Ecotoxicology and Environmental Safety* 16: 29–43.

Oguma, H., Shimizu, M., Ito, Y. & Mugiya, Y. (1998). Induction and identification of heat shock proteins in scallops, *Patinopecten yessoensis*, exposed to high temperatures. *Bulletin of the Faculty of Fisheries Hokkaido University* 49: 71–83.

Ogura, A., Ikeo, K. & Gojobori, T. (2004). Comparative analysis of gene expression for convergent evolution of camera eye between octopus and human. *Genome Research* 14: 1555–1561.

Ogura, A., Yoshida, M.-A., Moritaki, T., Okuda, Y., Sese, J., Shimizu, K. K., Sousounis, K. & Tsonis, P. A. (2013). Loss of the six3/6 controlling pathways might have resulted in pinhole-eye evolution in *Nautilus*. *Nature. Scientific Reports* 3: 1432.

Ohgaki, S.-I. (1993). Locomotive activity patterns in the four species of Littorinidae on a rocky shore on Ishigaki Island, Okinawa. *Venus* 52: 69–75.

Ohno, S. (1967). *Sex Chromosomes and Sex-linked Genes*. Vol. 1. Berlin, Germany, Springer Science + Business Media.

Okada, K. (1935). Some notes on *Musculium heterodon* (Pilsbry), a freshwater bivalve. I: The genital system and the gametogenesis. II. The gill, the breeding habits and the marsupial sac. *Science Reports of the Tohoku Imperial University* 9: 315–328,373–391.

Okada, K. (1939). The development of the primary mesoderm in *Sphaerium japonicum biwaense* Mori. *Science Reports of the Tohoku Imperial University* 14: 25–87.

Økland, J. & Økland, K. A. (1986). The effects of acid deposition on benthic animals in lakes and streams. *Experientia* 42: 471–486.

Økland, S. (1980). The heart ultrastructure of *Lepidopleurus asellus* (Spengler) and *Tonicella marmorea* (Fabricius) (Mollusca: Polyplacophora). *Zoomorphology* 96: 1–19.

Økland, S. (1981). Ultrastructure of the pericardium in chitons (Mollusca: Polyplacophora), in relation to filtration and contraction mechanisms. *Zoomorphology* 97: 193–203.

Økland, S. (1982). The ultrastructure of the heart complex in *Patella vulgata* L. (Archaeogastropoda: Prosobranchia). *Journal of Molluscan Studies* 48: 331–341.

Okoshi, K. & Ishii, T. (1996). Concentrations of elements in the radular teeth of limpets, chitons and other marine mollusks. *Journal of Marine Biotechnology* 3: 252–257.

Okuda, S. (1947). Notes on the postlarval development of the giant chiton *Cryptochiton stelleri* (Middendorff). *Journal of the Faculty of Science, Hokkaido University* 9: 267–275.

Okumura, S. I., Arai, K., Harigaya, Y., Eguchi, H., Sakai, M., Senbokuya, H., Furukawa, S. & Yamamori, K. (2007). Highly efficient induction of triploid Pacific abalone *Haliotis discus hannai* by caffeine treatment. *Fisheries Science* 73: 237–243.

Okusu, A. (2002). Embryogenesis and development of *Epimenia babai* (Mollusca Neomeniomorpha). *Biological Bulletin* 203: 87–103.

Okutani, T. (1994). A new discovery of streak-like flashing luminescence in *Ctenoides ales* (Finley, 1927) (Bivalvia: Limidae). *Venus* 53: 57–59.

Olden, J. D., Poff, N. L., Douglas, M. R., Douglas, M. E. & Fausch, K. D. (2004). Ecological and evolutionary consequences of biotic homogenization. *Trends in Ecology & Evolution* 19: 18–24.

Oldfield, E. (1964). The reproduction and development of some members of the Erycinidae and Monacutidae (Mollusca, Eulamellibranchia). *Proceedings of the Malacological Society of London* 36: 79–120.

Oliva, M. E. & Sánchez, M. F. (2005). Metazoan parasites and commensals of the northern Chilean scallop *Argopecten purpuratus* (Lamarck, 1819) as tools for stock identification. *Fisheries Research* 71: 71–77.

Olive, P. J. W. (1992). The adaptive significance of seasonal reproduction in marine invertebrates: The importance of distinguishing between models. *Invertebrate Reproduction & Development* 22: 165–174.

Oliver, J. & Babcock, R. (1992). Aspects of the fertilization ecology of broadcast spawning corals: Sperm dilution effects and in situ measurements of fertilization. *Biological Bulletin* 183: 409–417.

Oliver, P. G. (1982). A new species of cancellariid gastropod from Antarctica with a description of the radula. *British Antarctic Survey Bulletin* 57: 15–20.

Oliver, P. G., Southward, E. C. & Dando, P. R. (2013). Bacterial symbiosis in *Syssitomya pourtalesiana* Oliver, 2012 (Galeommatoidea: Montacutidae), a bivalve commensal with the deep-sea echinoid *Pourtalesia*. *Journal of Molluscan Studies* 79: 30–41.

Oliver, P. G. & Drewery, J. (2014). New species of chemosymbiotic clams (Bivalvia: Vesicomyidae and Thyasiridae) from a putative 'seep' in the Hatton-Rockall Basin, north-east Atlantic. *Journal of the Marine Biological Association of the United Kingdom* 94: 389–403.

Oliver, W. R. B. (1923). Marine littoral plant and animal communities in New Zealand. *Transactions of the New Zealand Institute* 54: 495–545.

Olivera, B. M. (2002). *Conus* venom peptides: Reflections from the biology of clades and species. *Annual Review of Ecology and Systematics* 33: 25–47.

Olivera, B. M., Showers Corneli, P., Watkins, M. & Fedosov, A. (2014). Biodiversity of cone snails and other venomous marine gastropods: Evolutionary success through neuropharmacology. *Annual Revue of Animal Biosciences* 2: 487–513.

Olmsted, J. M. (1917). Notes on the locomotion of certain Bermudan mollusks. *Journal of Experimental Zoology Part A* 24: 223–236.

Olson, P. D., Cribb, T. H., Tkach, V. V., Bray, R. A. & Littlewood, D. T. J. (2003). Phylogeny and classification of the Digenea (Platyhelminthes: Trematoda). *International Journal of Parasitology* 33: 733–755.

Olu-Le Roy, K., Sibuet, M., Fiala-Médioni, A., Gofas, S., Salas, C., Mariotti, A., Foucher, J. P. & Woodside, J. (2004). Cold seep communities in the deep eastern Mediterranean Sea: Composition, symbiosis and spatial distribution on mud volcanoes. *Deep Sea Research Part I: Oceanographic Research Papers* 51: 1915–1936.

Onishi, T., Suzuki, M. & Kikuchi, R. (1984). The distribution of polysaccharide hydrolase activity in gastropods and bivalves. *Nippon Suisan Gakkaishi* 51: 301–308.

Onitsuka, C., Yamaguchi, A., Kanamaru, H., Oikawa, S., Takeda, T. & Matsuyama, M. (2009). Molecular cloning and expression analysis of a GnRH-like dodecapeptide in the swordtip squid, *Loligo edulis. Zoological Science* 26: 203–208.

Onitsuka, T., Kawamura, T. & Horii, T. (2010). Reproduction and early life ecology of abalone *Haliotis diversicolor* in Sagami Bay, Japan. *Japan Agricultural Research Quarterly* 44: 375–382.

Orange, C. J. & Freeman, R. F. H. (1988). Intracellular sodium, potassium and chloride in the marine pulmonate, *Amphibola crenata. Marine Behaviour and Physiology* 13: 125–153.

Oren, U., Brickner, I. & Loya, Y. (1998). Prudent sessile feeding by the corallivore snail *Coralliophila violacea* on coral energy sinks. *Proceedings of the Royal Society of London B* 265: 2043–2050.

Orensanz, J. M. L., Hand, C. M., Parma, A. M., Valero, J. & Hilborn, R. (2004). Precaution in the harvest of Methuselah's clams: The difficulty of getting timely feedback from slow-paced dynamics. *Canadian Journal of Fisheries and Aquatic Sciences* 61: 1355–1372.

Orlov, A. M. (1999). Capture of especially large sleeper shark *Somniosus pacificus* (Squalidae) with some notes on its ecology in northwestern Pacific. *Journal of Ichthyology (Voprosy Ikhtiologii)* 39: 548–553.

Orr, J. C., Fabry, V. J., Aumont, O., Bopp, L., Doney, S. C., Feely, R. A., Gnanadesikan, A., Gruber, N., Ishida, A. & Joos, F. (2005). Anthropogenic ocean acidification over the twenty-first century and its impact on calcifying organisms. *Nature* 437: 681–686.

Orr, V. (1962). The drilling habit of *Capulus danieli* (Crosse) (Mollusca: Gastropoda). *The Veliger* 5: 63–67.

Örstan, A. (1999). Drill holes in land snail shells from western Turkey. *Schriften zur Malakozoologie* 13: 31–36.

Örstan, A. (2010). A possible function of the parietal tooth of *Pedipes* (Gastropoda, Pulmonata, Ellobiidae). *Basteria* 74: 111–114.

Orton, J. H. (1912). The mode of feeding of *Crepidula*, with an account of the current-producing mechanism in the mantle cavity, and some remarks on the mode of feeding in gastropods and lamellibranchs. *Journal of the Marine Biological Association of the United Kingdom* 9: 444–478.

Orton, J. H. (1914). On ciliary mechanisms in brachiopods and some polychaetes, with a comparison of the ciliary mechanisms on the gills of Molluscs, Protochordata, Brachiopods, and cryptocephalous polychaetes, and an account of the endostyle of *Crepidula* and its allies. *Journal of the Marine Biological Association of the United Kingdom* 10: 283–311.

Orton, J. H. (1920). Sex-phenomena in the common limpet (*Patella vulgata*). *Nature* 104: 373–374.

Orton, J. H. (1927). A note on the physiology of sex and sex-determination. *Journal of the Marine Biological Association of the United Kingdom* 14: 1047–1055.

Orton, J. H. (1928). Observations on *Patella vulgata*: I. Sex phenomena, breeding and shell-growth. II. Rate of growth of shell. III. Habitats and habits. *Journal of the Marine Biological Association of the United Kingdom* 15: 851–888.

Orton, J. H. (1949). Notes on the feeding habit of *Capulus ungaricus. Report of the Marine Biological Station, Port Erin, Isle of Man* 61: 29–30.

Orton, J. H., Southward, A. J. & Dodd, J. M. (1956). Studies on the biology of limpets. I. The breeding of *Patella vulgata* L. in Britain. *Journal of the Marine Biological Association of the United Kingdom* 35: 149–176.

Osorio, C., Jara, F. & Valdez Ramirez, M. E. (1993). Diet of *Cypraea caputdraconis* (Mollusca: Gastropoda) as it relates to food availability on Easter Island. *Pacific Science* 47: 34–42.

Osorio, C., Brown, D., Donoso, L. & Atan, H. (1999). Aspects of the reproductive activity of *Cypraea caputdraconis* from Easter Island (Mollusca: Gastropoda: Cypraeidae). *Pacific Science* 53: 15–23.

Ostergaard, J. M. (1950). Spawning and development in some Hawaiian marine gastropods. *Pacific Science* 4: 75–115.

Ostrovskaya, R. M., Sitnikova, T. J., Yakovleva, Y. & Finogenko, A. (1996). Polyploidy and mixoploidy in endemic molluscs of Baikal. pp. 54–56 *in* V. E. Gokhman & Kuznetsova, V. G. (Eds.) *Karyosystematics of the Invertebrate Animals. 3. Volume of Scientific Papers.* Moscow, Lomonosov State University.

Ostrovsky, A. N., Lidgard, S., Gordon, D. P., Schwaha, T., Genikhovich, G. & Ereskovsky, A. V. (2016). Matrotrophy and placentation in invertebrates: A new paradigm. *Biological Reviews of the Cambridge Philosophical Society* 91: 673–711.

Otsuka-Fuchino, H., Watanabe, Y., Hirakawa, C., Tamiya, T., Matsumoto, J. J. & Tsuchiya, T. (1992). Molecular cloning of the antibacterial protein of the giant African snail, *Achatina fulica* Férussac. *European Journal of Biochemistry* 209: 1–6.

Ott, J., Bright, M. & Bulgheresi, S. (2004). Marine microbial thiotrophic ectosymbioses. *Oceanography and Marine Biology Annual Review* 42: 95–118.

Ottaviani, E., Franchini, A., Kletsas, D., Bernardi, M. & Genedani, S. (1997). Involvement of PDGF and TGF-β in cell migration and phagocytosis in invertebrate and human immunocytes. *Animal Biology* 6: 91–95.

Ottaviani, E., Franchini, A. & Hanukoglu, I. (1998). In situ localization of ACTH receptor-like mRNA in molluscan and human immunocytes. *Cellular and Molecular Life Sciences* 54: 139–142.

Ottaviani, E., Franchini, A., Malagoli, D. & Genedani, S. (2000). Immunomodulation by recombinant human interleukin-8 and its signal transduction pathways in invertebrate hemocytes. *Cellular and Molecular Life Sciences* 57: 506–513.

Ottaviani, E., Franchini, A. & Kletsas, D. (2001). Platelet-derived growth factor and transforming growth factor-beta in invertebrate immune and neuroendocrine interactions: Another sign of conservation in evolution. *Comparative Biochemistry and Physiology Part C* 129: 295–306.

Ottaviani, E. (2004). The mollusc as a suitable model for mammalian investigations. *Comparative and General Pharmacology* 1: 2–4.

Ottaviani, E. (2006). Molluscan immunorecognition. *Invertebrate Survival Journal* 3: 50–63.

Ottaviani, E. (2011). Is the distinction between innate and adaptive immunity in invertebrates still as clear-cut as thought? *Italian Journal of Zoology* 78: 274–278.

Otter, G. W. (1937). *Rock Destroying Organisms in Relation to Coral Reefs. Scientific Report of the Great Barrier Reef Expedition 1928–29*. Vol. 1. London, UK, British Museum.

Ovchinnikov, Y. A., Abdulaev, N. G., Zolotarev, A. S., Artamonov, I. D., Bespalov, I. A., Dergachev, A. E. & Tsuda, M. (1988). *Octopus* rhodopsin: Amino-acid sequence deduced from C-DNA. *Federation of European Biochemical Societies (FEBS) Letters* 232: 69–72.

Owen, D. F. (1969). Ecological aspects of polymorphism in the African land snail *Limicolaria motensia*. *Journal of Zoology* 159: 79–96.

Owen, G. (1955). Observations on the stomach and the digestive diverticula of the Lamellibranchia: 1. The Anisomyaria and Eulamellibranchia. *Quarterly Journal of Microscopical Science* 96: 517–537.

Owen, G. (1956). Observations on the stomach and the digestive diverticula of the Lamellibranchia. II. The Nuculidae. *Quarterly Journal of Microscopical Science* 97: 541–567.

Owen, G. (1958). Observations on the stomach and digestive glands of *Scutus breviculus* (Blainville). *Proceedings of the Malacological Society of London* 33: 103–114.

Owen, G. (1966). Digestion. pp. 53–96 in K. M. Wilbur & Yonge, C. M. (Eds.) *Physiology of Mollusca*. Vol. 2. New York, Academic Press.

Owen, G. (1970). Fine structure of digestive tubules of marine bivalve *Cardium edule*. *Philosophical Transactions of the Royal Society B* 258: 245–260.

Owen, G. (1974). Feeding and digestion in the Bivalvia. pp. 1–35 in O. Lowenstein (Ed.) *Advances in Comparative Physiology and Biochemistry*. Vol. 5. New York, Academic Press.

Oyeleke, S. B., Egwim, E. C., Oyewole, E. C. & John, E. E. (2012). Production of cellulase and protease from microorganisms isolated from gut of *Archachatina marginata* (Giant African Snail). *Science and Technology* 2: 15–20.

Oz'go, M. (2011). Rapid evolution in unstable habitats: A success story of the polymorphic land snail *Cepaea nemoralis* (Gastropoda: Pulmonata). *Biological Journal of the Linnean Society* 102: 251–262.

Ozaki, K., Terakita, A., Hara, R. & Hara, T. (1986). Rhodopsin and retinochrome in the retina of a marine gastropod, *Conomulex luhuanus*. *Vision Research* 26: 691–705.

Ozeki, Y. (1997). Purification of a 63 kDa β-D-galactoside binding lectin from cuttlefish, *Todarodes pacificus*. *Biochemistry and Molecular Biology International* 41: 633–640.

Packard, A. (1972). Cephalopods and fish: The limits of convergence. *Biological Reviews* 47: 241–307.

Packard, A. & Trueman, E. R. (1974). Muscular activity of the mantle of *Sepia* and *Loligo* (Cephalopoda) during respiration and jetting and its physiological interpretation. *Journal of Experimental Biology* 61: 411–420.

Packard, A. (1988). The skin of cephalopods (coleoids): General and special adaptations. pp. 37–67 in E. R. Trueman & Clarke, M. R. (Eds.) *Form and Function. The Mollusca*. Vol. 11. New York, Academic Press.

Packard, A., Karlsen, H. E. & Sand, O. (1990). Low-frequency hearing in cephalopods. *Journal of Comparative Physiology A* 166: 501–505.

Padilla, D. K. (1998). Inducible phenotypic plasticity of the radula in *Lacuna* (Gastropoda: Littorinidae). *The Veliger* 41: 201–204.

Padilla, D. K. (2001). Food and environmental cues trigger an inducible offence. *Evolutionary Ecology Research* 3: 15–25.

Pafort-van Iersel, T. & van der Spoel, S. (1986). Schizogamy in the planktonic opisthobranch *Clio*:A previously undescribed mode of reproduction in the Mollusca. *International Journal of Invertebrate Reproduction & Development* 10: 43–50.

Page, A. J. & Willan, R. C. (1988). Ontogenetic change in the radula of the gastropod *Epitonium billeeana* (Prosobranchia: Epitoniidae). *The Veliger* 30: 222–229.

Page, L. R. (1994). The ancestral gastropod larval form is best approximated by hatching-stage opisthobranch larvae: Evidence from comparative developmental studies. pp. 206–223 in W. H. Wilson, Stricker, S. A. & Shinn, G. L. (Eds.) *Reproduction and Development of Marine Invertebrates*. Baltimore, MD, Johns Hopkins University Press.

Page, L. R. (1995). Similarities in form and developmental sequence for three larval shell muscles in nudibranch gastropods. *Acta Zoologica* 76: 177–191.

Page, L. R. (1997a). Larval shell muscles in the abalone *Haliotis kamtschatkana*. *Biological Bulletin* 193: 30–46.

Page, L. R. (1997b). Ontogenetic torsion and protoconch form in the archaeogastropod *Haliotis kamtschatkana*: Evolutionary implications. *Acta Zoologica* 78: 227–245.

Page, L. R. (1998). Sequential developmental programmes for retractor muscles of a caenogastropod: Reappraisal of evolutionary homologues. *Proceedings of the Royal Society B* 265: 2243–2250.

Page, L. R. (2000). Development and evolution of adult feeding structures in caenogastropods: Overcoming larval functional constraints. *Evolution & Development* 2: 25–34.

Page, L. R. & Parries, S. C. (2000). Comparative study of the apical ganglion in planktotrophic caenogastropod larvae: Ultrastructure and immunoreactivity to serotonin. *Journal of Comparative Neurology* 418: 383–401.

Page, L. R. (2002a). Comparative structure of the larval apical sensory organ in gastropods and hypotheses about function and developmental evolution. *Invertebrate Reproduction & Development* 41: 193–200.

Page, L. R. (2002b). Ontogenetic torsion in two basal gastropods occurs without shell attachments for larval retractor muscles. *Evolution & Development* 4: 212–222.

Page, L. R. (2002c). Larval and metamorphic development of the foregut and proboscis in the caenogastropod *Marsenina* (*Lamellaria*) *stearnsii*. *Journal of Morphology* 252: 202–217.

Page, L. R. (2002d). Apical sensory organ in larvae of the patellogastropod *Tectura scutum*. *Biological Bulletin* 202: 6–22.

Page, L. R. (2003). Gastropod ontogenetic torsion: Developmental remnants of an ancient evolutionary change in body plan. *Journal of Experimental Zoology* 297B: 11–26.

Page, L. R. (2005). Development of foregut and proboscis in the buccinid neogastropod *Nassarius mendicus*: Evolutionary opportunity exploited by a developmental module. *Journal of Morphology* 264: 327–338.

Page, L. R. (2006a). Modern insights on gastropod development: Re-evaluation of the evolution of a novel body plan. *Integrative and Comparative Biology* 46: 134–143.

Page, L. R. (2006b). Early differentiating neuron in larval abalone (*Haliotis kamtschatkana*) reveals the relationship between ontogenetic torsion and crossing of the pleurovisceral nerve cords. *Evolution & Development* 8: 458–467.

Page, L. R. & Kempf, S. C. (2009). Larval apical sensory organ in a neritimorph gastropod, an ancient gastropod lineage with feeding larvae. *Zoomorphology* 128: 327–338.

Page, L. R. (2011). Developmental modularity and phenotypic novelty within a biphasic life cycle: Morphogenesis of a cone snail venom gland. *Proceedings of the Royal Society B* 279: 77–83.

Page, L. R. & Ferguson, S. J. (2013). The other gastropod larvae: Larval morphogenesis in a marine neritimorph. *Journal of Morphology* 274: 412–428.

Page, L. R. & Hookham, B. (2017). The gastropod foregut – evolution viewed through a developmental lens. *Canadian Journal of Zoology* 95: 227–238.

Paine, R. T. (1966). Food web complexity and species diversity. *American Naturalist* 100: 65–75.

Paine, R. T. (1969a). The *Pisaster-Tegula* interaction: Prey patches, predator food preference, and intertidal community structure. *Ecology* 50: 950–961.

Paine, R. T. (1969b). A note on trophic complexity and community stability. *American Naturalist* 103: 91–93.

Paine, R. T. (1971). The measurement and application of the calorie to ecological problems. *Annual Review of Ecology and Systematics* 2: 145–164.

Paine, R. T. (1974). Intertidal community structure. *Oecologia* 15: 93–120.

Paine, R. T. (1977). Controlled manipulations in the marine intertidal zone and their contributions to ecological theory. pp. 245–270 in C. E. Goulden (Ed.) *Changing Scenes in the Natural Sciences, 1776–1976.* Vol. 12. Philadelphia, PA, Academy of Natural Sciences.

Paine, R. T. & Palmer, A. R. (1978). *Sicyases sanguineus*: A unique trophic generalist from the Chilean intertidal zone. *Copeia* 1978: 75–81.

Paine, R. T. (1980). Food webs: Linkage, interaction strength and community infrastructure. *Journal of Animal Ecology* 49: 667–685.

Paine, R. T. (1995). A conversation on refining the concept of keystone species. *Conservation Biology* 9: 962–964.

Painter, S. D., Clough, B., Garden, R. W., Sweedler, J. V. & Nagle, G. T. (1998). Characterization of *Aplysia* attractin, the first waterborne peptide pheromone in invertebrates. *Biological Bulletin* 194: 120–131.

Painter, S. D., Cummins, S. F., Nichols, A. E., Akalal, D.-B. G., Schein, C. H., Braun, W., Smith, J. S., Susswein, A. J., Levy, M. & De Boer, P. A. C. M. (2004). Structural and functional analysis of *Aplysia* attractins, a family of water-borne protein pheromones with interspecific attractiveness. *Proceedings of the National Academy of Sciences of the United States of America* 101: 6929–6933.

Pairett, A. N. & Serb, J. A. (2013). *De novo* assembly and characterization of two transcriptomes reveal multiple light-mediated functions in the scallop eye (Bivalvia: Pectinidae). *PLoS ONE* 8: e69852 (69851–69812).

Pakarinen, E. (1994). Autotomy in arionid and limacid slugs. *Journal of Molluscan Studies* 60: 19–23.

Pal, P. & Hodgson, A. N. (2002). An ultrastructural study of oogenesis in a planktonic and a direct-developing species of *Siphonaria* (Gastropoda: Pulmonata). *Journal of Molluscan Studies* 68: 337–344.

Pal, P. & Hodgson, A. N. (2005). Reproductive seasonality and simultaneous hermaphroditism in two species of *Siphonaria* (Gastropoda: Pulmonata) from the southeast coast of South Africa. *Journal of Molluscan Studies* 71: 33–40.

Pal, P., Erlandsson, J. & Sköld, M. (2006). Size-assortative mating and non-reciprocal copulation in a hermaphroditic intertidal limpet: Test of the mate availability hypothesis. *Marine Biology* 148: 1273–1282.

Palais, F., Jubeaux, G., Dedourge-Geffard, O., Biagianti-Risbourg, S. & Geffard, A. (2010). Amylolytic and cellulolytic activities in the crystalline style and the digestive diverticulae of the freshwater bivalve *Dreissena polymorpha* (Pallas, 1771). *Molluscan Research* 30: 29–36.

Páll-Gergely, B., Naggs, F. & Asami, T. (2016). Novel shell device for gas exchange in an operculate land snail. *Biology Letters* 12: 20160151.

Palmer, A. R. (1977). Function of shell sculpture in marine gastropods: Hydrodynamic destabilization in *Ceratostoma foliatum*. *Science* 197: 1293–1295.

Palmer, A. R. (1979). Fish predation and the evolution of gastropod shell sculpture: Experimental and geographic evidence. *Evolution* 33 697–719.

Palmer, A. R. (1983). Relative cost of producing skeletal organic matrix versus calcification: Evidence from marine gastropods. *Marine Biology* 75: 287–292.

Palmer, A. R. (1984). Species cohesiveness and genetic control of shell color and form in *Thais emarginata* (Prosobranchia, Muricacea): Preliminary results. *Malacologia* 25: 477–491.

Palmer, A. R. (1985a). Adaptive value of shell variation in *Thais lamellosa*: Effect of thick shells on vulnerability to and preference by crabs. *The Veliger* 27: 349–356.

Palmer, A. R. (1985b). Genetic basis of shell variation in *Thais emarginata* (Prosobranchia, Muricacea). I. Banding in populations from Vancouver Island. *Biological Bulletin* 169: 638–651.

Palmer, A. R. (1988). Feeding biology of *Ocenebra lurida* (Prosobranchia: Muricacea): Diet, predator-prey size relations, and attack behavior. *The Veliger* 31: 192–203.

Palmer, A. R. (1990). Effect of crab effluent and scent of damaged conspecifics on feeding, growth, and shell morphology of the Atlantic dogwhelk *Nucella lapillus* (L.). *Hydrobiologia* 193: 155–182.

Palmer, A. R. (1992). Calcification in marine molluscs: How costly is it? *Proceedings of the National Academy of Sciences of the United States of America* 89: 1379–1382.

Palmer, A. R. (1996). From symmetry to asymmetry: Phylogenetic patterns of asymmetry variation in animals and their evolutionary significance. *Proceedings of the National Academy of Sciences of the United States of America* 93: 14279–14286.

Palmer, B. A., Taylor, G. J., Brumfeld, V., Gur, D., Shemesh, M., Elad, N., Osherov, A., Oron, D., Weiner, S. & Addadi, L. (2017). The image-forming mirror in the eye of the scallop. *Science* 358: 1172–1175.

Palmer, J. D. (1995). *The Biological Rhythms and Clocks of Intertidal Animals.* Oxford, UK, Oxford University Press.

Palumbi, S. R. (1994). Genetic divergence, reproductive isolation, and marine speciation. *Annual Review of Ecology and Systematics* 25: 547–572.

Palumbi, S. R. (1997). Population structure and molecular biogeography of the Pacific. *Coral Reefs* 16: 47–52.

Panchin, V., Popova, L. B., Deliagina, T. G., Orlovsky, G. N. & Arshavsky, I. (1995). Control of locomotion in marine mollusk *Clione limacina*: VIII. Cerebropedal neurons. *Journal of Neurophysiology* 73: 1912–1923.

Panchin, V. (1997). Cellular mechanism for the temperature sensitive spatial orientation in *Clione*. *Neuroreport* 8: 3345–3347.

Pancucci-Papadopoulou, M. A., Zenetos, A., Corsini-Foka, M. & Politou, C. Y. (2005). Update of marine alien species in Hellenic waters. *Mediterranean Marine Science* 6: 147–158.

Pani, A. K., Roghani, A. & Croll, R. P. (2005). Identification of peptide-like substances in the *Placopecten*: Possible role in growth and reproduction. *Biogenic Amines* 19: 47–67.

Paniagua, R., Royuela, M., Garcia-Anchuelo, R. M. & Fraile, B. (1996). Ultrastructure of invertebrate muscle cell types. *Histology and Histopathology* 11: 181–201.

Pankey, S., Sunada, H., Horikoshi, T. & Sakakibara, M. (2010). Cyclic nucleotide-gated channels are involved in phototransduction of dermal photoreceptors in *Lymnaea stagnalis*. *Journal of Comparative Physiology B* 180: 1205–1211.

Panova, M., Boström, J., Hofving, T., Areskoug, T., Eriksson, A., Mehlig, B., Mäkinen, T., André, C. & Johannesson, K. (2010). Extreme female promiscuity in a non-social invertebrate species. *PLoS ONE* 5: e9640.

Pansini, M., Cattaneo-Vietti, R. & Shiaporelli, S. (1999). Relationship between sponges and a taxon of obligatory ingilines: The siliquariid molluscs. *Memoirs of the Queensland Museum* 44: 427–437.

Paps, J., Baguñà, J. & Riutort, M. (2009). Lophotrochozoa internal phylogeny: New insights from an up-to-date analysis of nuclear ribosomal genes. *Proceedings of the Royal Society B* 276: 1245–1254.

Paps, J., Xu, F., Zhang, G. & Holland, P. W. H. (2015). Reinforcing the egg-timer: Recruitment of novel lophotrochozoa homeobox genes to early and late development in the Pacific Oyster. *Genome Biology and Evolution* 7: 677–688.

Pardy, R. L. (1980). Symbiotic algae and ^{14}C incorporation in the freshwater clam, *Anodonta*. *Biological Bulletin* 158: 349–355.

Parent, C. E. & Crespi, B. J. (2006). Sequential colonization and diversification of Galápagos endemic land snail genus *Bulimulus* (Gastropoda, Stylommatophora). *Evolution* 60: 2311–2328.

Parent, C. E. & Crespi, B. J. (2009). Ecological opportunity in adaptive radiation of Galápagos endemic land snails. *American Naturalist* 174: 898–905.

Park, G. (2008). Polyploidy in three sphaeriids (Bivalvia: Veneroida) from Korea. *Molluscan Research* 28: 133.

Park, G. M., Yong, T. S., Im, K. I. & Chung, E.-Y. (2000). Karyotypes of three species of *Corbicula* (Bivalvia: Veneroida) in Korea. *Journal of Shellfish Research* 19: 979–982.

Park, J. J., Lee, J. S., Lee, Y. G. & Kim, J. W. (2012). Micromorphology and ultrastructure of the foot of the equilateral venus *Gomphina veneriformis* (Bivalvia: Veneridae). *Open Journal of Cell Biology* 1: 11–16.

Park, K.-I. & Choi, K.-S. (2001). Spatial distribution of the protozoan parasite *Perkinsus* sp. found in the Manila clams, (sic) *Ruditapes philippinarum*, in Korea. *Aquaculture* 203: 9–22.

Park, K., Lee, J. S., Kang, J.-C., Kim, J. W. & Kwak, I.-S. (2015). Cascading effects from survival to physiological activities, and gene expression of heat shock protein 90 on the abalone *Haliotis discus hannai* responding to continuous thermal stress. *Fish and Shellfish Immunology* 42: 233–240.

Park, M.-S. & Chun, S.-K. (1989). Study on *Marteilioides chungmuensis* Comps *et al.*, 1986 parasite of the Pacific oyster, *Crassostrea gigas* Thunberg. *Journal of Fish Pathology* 2: 53–70.

Parker, G. (1914). The locomotion of chiton. *Contributions from the Bermuda Biological Station for Research* 3: 1–2.

Parker, G. H. (1917). The pedal locomotion of the sea hare *Aplysia californica*. *Journal of Experimental Zoology Part A* 24: 139–145.

Parker, L. M., Ross, P. M., Raftos, D. A., Thompson, E. & O'Connor, W. A. (2011). The proteomic response of larvae of the Sydney rock oyster, *Saccostrea glomerata* to elevated pCO$_2$. *Australian Zoologist* 35: 1011–1023.

Parker, L. M., O'Connor, W. A., Raftos, D. A., Pörtner, H.-O. & Ross, P. M. (2015). Persistence of positive carryover effects in the oyster, *Saccostrea glomerata*, following transgenerational exposure to ocean acidification. *PLoS ONE* 10: e0132276.

Parkin, D. T. (1972). Climatic selection in the land snail *Arianta arbustorum* in Derbyshire, England. *Heredity* 28: 49–56.

Parnell, P. E. (2002). Larval development, precompetent period, and a natural spawning event of the pectinacean bivalve *Spondylus tenebrosus* (Reeve, 1856). *The Veliger* 45: 58–64.

Parra, M., Sellanes, J., Dupré, E. & Krylova, E. (2009). Reproductive characteristics of *Calyptogena gallardoi* (Bivalvia: Vesicomyidae) from a methane seep area off Concepción, Chile. *Journal of the Marine Biological Association of the United Kingdom* 89: 161–169.

Parsons, K. E. (1996). Discordant patterns of morphological and genetic divergence in the 'Austrocochlea constricta' (Gastropoda: Trochidae) species complex. *Marine and Freshwater Research* 47: 981–990.

Pascoal, S., Carvalho, G., Creer, S., Rock, J., Kawaii, K., Mendo, S. & Hughes, R. (2012). Plastic and heritable components of phenotypic variation in *Nucella lapillus*: An assessment using reciprocal transplant and common garden experiments. *PLoS ONE* 7: e30289.

Pascual, S., Gestal, C., Estévez, J. M., Rodriguez, H., Soto, M., Abollo, E. & Arias, C. (1996). Parasites in commercially-exploited cephalopods (Mollusca, Cephalopoda) in Spain: An updated perspective. *Aquaculture* 142: 1–10.

Pascual, S. & Hochberg, F. G. (1996). Marine parasites as biological tags of cephalopod hosts. *Parasitology Today* 12: 324–327.

Pasic, M. & Kartelija, G. (1995). The reaction of *Helix* photosensitive neurons to light and cyclic GMP. *Neuroscience* 69: 557–565.

Passamaneck, Y. J., Furchheim, N., Hejnol, A., Martindale, M. Q. & Lüter, C. (2011). Ciliary photoreceptors in the cerebral eyes of a protostome larva. *EvoDevo* 2: 6.

Passamaneck, Y. J. & Martindale, M. Q. (2013). Evidence for a phototransduction cascade in an early brachiopod embryo. *Integrative and Comparative Biology* 53: 17–26.

Passamonti, M. & Ghiselli, F. (2009). Doubly uniparental inheritance: Two mitochondrial genomes, one precious model for organelle DNA inheritance and evolution. *DNA and Cell Biology* 28: 79–89.

Pastorino, G. (2007). Sexual dimorphism in shells of the southwestern Atlantic gastropod *Olivella plata* (Ihering, 1908) (Gastropoda: Olividae). *Journal of Molluscan Studies* 73: 283–285.

Patek, S. N. & Caldwell, R. L. (2005). Extreme impact and cavitation forces of a biological hammer: Strike forces of the peacock mantis shrimp *Odontodactylus scyllarus*. *Journal of Experimental Biology* 208: 3655–3664.

Paterson, I. G., Partridge, V. & Buckland-Nicks, J. A. (2001). Multiple paternity in *Littorina obtusata* (Gastropoda, Littorinidae) revealed by microsatellite analyses. *Biological Bulletin* 200: 261–267.

Patil, A. D., Feyer, A. J., Eggleston, D. S., Haltiwanger, R. C., Bean, M. F., Taylor, P. B., Caranfa et al. (1993). The inophyllums, novel inhibitors of HIV-1 reverse transcriptase isolated from Malaysian tree *Calophyllum inophyllum*. *Journal of Medicinal Chemistry* 36: 4131–4138.

Patil, A. M. (1958). The occurrence of a male of the prosobranch *Potamopyrgus jenkinsi* (Smith) var. *carinata* Marshall in the Thames at Sonning, Berkshire. *Journal of Natural History* 1: 232–240.

Patterson, C. M. (1969). Chromosomes of molluscs. pp. 635–686. *Proceedings of the Symposium on Mollusca, Cochin, January 1968*. Vol. 2. Cochin, India, Marine Biological Association of India.

Patterson, C. M. & Burch, J. B. (1978). Chromosomes of pulmonate molluscs. pp. 171–217 *in* V. Fretter & Peake, J. (Eds.) *Pulmonates: Systematics, Evolution and Ecology*. Vol. 2A. London, UK, Academic Press.

Paul, C. R. C. (1981). The function of the spines in *Murex* (*Murex*) *pecten* Lightfoot and related species (Prosobranchia: Muricidae). *Journal of Conchology* 30: 285–294.

Paul, C. R. C. (1991). The functional morphology of gastropod apertures. pp. 127–140 *in* N. Schmidt-Kittler & Vogel, K. (Eds.) *Constructional Morphology and Evolution*. Berlin, Germany, Springer-Verlag.

Paul, M. C., Zubía, E., Ortega, M. J. & Salvá, J. (1997). New polypropionates from *Siphonaria pectinata*. *Tetrahedron* 53: 2303–2308.

Paul, V. J. & Van Alstyne, K. L. (1988). Use of ingested algal diterpenoids by *Elysia halimedae* Macnae (Opisthobranchia: Ascoglossa) as antipredator defenses. *Journal of Experimental Marine Biology and Ecology* 119: 15–29.

Paul, V. J., Lindquist, N. & Fenical, W. (1990). Chemical defenses of the tropical ascidian *Atapozoa* sp. and its nudibranch predators *Nembrotha* spp. *Marine Ecology Progress Series* 59: 109–118.

de Paula, S. M. & Silveira, M. (2009). Studies on molluscan shells: Contributions from microscopic and analytical methods. *Micron* 40: 669–690.

Pauletti, P. M., Cintra, L. S., Braguine, C. G., Cunha, W. R. & Januário, A. H. (2010). Halogenated indole alkaloids from marine invertebrates. *Marine Drugs* 8: 1526–1549.

Pauletto, M., Milan, M., de Sousa, J. T., Huvet, A., Joaquim, S., Matias, D., Leitão, A., Patarnello, T. & Bargelloni, L. (2014). Insights into molecular features of *Venerupis decussata* oocytes: A microarray-based study. *PLoS ONE* 9: e113925.

Pauly, D., Trites, A. W., Capuli, E. & Christensen, V. (1998). Diet composition and trophic levels of marine mammals. *ICES Journal of Marine Science* 55: 467–481.

Pavlicek, A., Schwaha, T. & Wanninger, A. (2018). Towards a ground pattern reconstruction of bivalve nervous systems: Neurogenesis in the zebra mussel *Dreissena polymorpha*. *Organisms Diversity & Evolution* 18: 101–114.

Pavlova, G. A., Glantz, R. M. & Willows, A. O. D. (2011). Responses to magnetic stimuli recorded in peripheral nerves in the marine nudibranch mollusk *Tritonia diomedea*. *Journal of Comparative Physiology A* 197: 979–986.

Pawlicki, J. M., Pease, L. B., Pierce, C. M., Startz, T. P., Zhang, Y. & Smith, A. M. (2004). The effect of molluscan glue proteins on gel mechanics. *Journal of Experimental Biology* 207: 1127–1135.

Pawlik, J. R., Albizati, K. F. & Faulkner, D. J. (1986). Evidence of a defensive role for limatulone, a novel triterpene from the intertidal limpet *Collisella limatula*. *Marine Ecology Progress Series* 30: 251–260.

Pawlik, J. R., Kernan, M. R., Molinski, T. F., Harper, M. K. & Faulkner, D. J. (1988). Defensive chemicals of the Spanish dancer nudibranch *Hexabranchus sanguineus* and its egg ribbons: Macrolides derived from a sponge diet. *Journal of Experimental Marine Biology and Ecology* 119: 99–109.

Pawlik, J. R. (1989). Larvae of the sea hare *Aplysia californica* settle and metamorphose on an assortment of macroalgal species. *Marine Ecology Progress Series* 51: 195–199.

Pawlik, J. R. (1992). Chemical ecology of the settlement of benthic marine invertebrates. *Oceanography and Marine Biology Annual Review* 30: 273–335.

Pawlik, J. R. (1993). Marine invertebrate chemical defenses. *Chemical Reviews* 93: 1911–1922.

Pawlik, J. R. (2012). Antipredatory defensive roles of natural products from marine invertebrates. pp. 677–710 *in* E. Fattorusso, Gerwick, W. H. & Taglialatela-Scafati, O. (Eds.) *Handbook of Marine Natural Products*. Dordrecht, the Netherlands, Springer.

Paxton, J. R. (1972). Osteology and relationships of the lanternfishes (Family Myctophidae). *Bulletin of the Natural History Museum of Los Angeles County. Science* 13: 1–81.

Payne, C. M. & Crisp, M. (1989). Ultrastructure and histochemistry of the posterior oesophagus of *Nassarius reticulatus* (Linnaeus). *Journal of Molluscan Studies* 55: 313–321.

Payne, D. W., Thorpe, N. A. & Donaldson, E. M. (1972). Cellulolytic activity and a study of the bacterial population in the digestive tract of *Scrobicularia plana* (Da Costa). *Proceedings of the Malacological Society of London* 40: 147–160.

Payne, J. L., Bush, A. M., Heim, N. A., Knope, M. L. & McCauley, D. J. (2016). Ecological selectivity of the emerging mass extinction in the oceans. *Science*: aaf2416.

Pazos, A. J. & Mathieu, M. (1999). Effects of five natural gonadotropin-releasing hormones on cell suspensions of marine bivalve gonad: Stimulation of gonial DNA synthesis. *General and Comparative Endocrinology* 113: 112–120.

Peake, J. F. (1981). The land snails of islands: A dispersalist view. pp. 247–263 *in* P. L. Forey (Ed.) *The Evolving Biosphere*. Cambridge, UK, British Museum (Natural History), London and Cambridge University Press.

Pearce, S. S. (1889). On the Varieties of our banded Snails, especially those of *Helix caporata*. *Journal of Conchology* 6: 123–135.

Pearce, T. A. (1989). Loping locomotion in terrestrial gastropods. *Walkerana: Transactions of the POETS Society* 3: 229–237.

Pearce, T. A. & Örstan, A. (2006). Terrestrial Gastropoda. pp. 261–285 *in* C. F. Sturm, Pearce, T. A. & Valdes, A. (Eds.) *The Mollusks: A Guide to their Study, Collection, and Preservation*. Boca Raton, FL, Universal Publishers.

Pearse, J. S. (1979). Polyplacophora. pp. 27–86 *in* A. C. Giese & Pearse, J. S. (Eds.) *Reproduction of Marine Invertebrates. Molluscs: Pelecypods and Lesser Classes. Reproduction of Marine Invertebrates*. Vol. 5. New York, Academic Press.

Pearse, J. S., Mooi, R., Lockhart, S. J. & Brandt, A. (2009). Brooding and species diversity in the Southern Ocean: Selection for brooders or speciation within brooding clades. pp. 181–196 *in* I. Krupnik, Lang, M. A. & Miller, S. E. (Eds.) *Smithsonian at the Poles: Contributions to International Polar Year Science*. Washington, DC, Smithsonian Institute.

Pearson, D. A., Greenlees, M., Ward-Fear, G. & Shine, R. (2009). Predicting the ecological impact of cane toads (*Bufo marinus*) on threatened camaenid land snails in north-western Australia. *Wildlife Research* 36: 533–540.

Pechenik, J. A. (1979). Encapsulation in invertebrate life histories. *American Naturalist* 114: 859–870.

Pechenik, J. A. (1982). Ability of some gastropod egg capsules to protect against low-salinity stress. *Journal of Experimental Marine Biology and Ecology* 63: 195–208.

Pechenik, J. A. (1986). The encapsulation of eggs and embryos by molluscs: An overview. *American Malacological Union Bulletin* 4: 165–172.

Pechenik, J. A. (1990). Delayed metamorphosis by larvae of benthic marine invertebrates: Does it occur? Is there a price to pay? *Ophelia* 32: 63–94.

Pechenik, J. A., Hadfield, M. G. & Eyster, L. S. (1995). Assessing whether larvae of the opisthobranch gastropod *Phestilla sibogae* Bergh become responsive to 3 chemical cues at the same age. *Journal of Experimental Marine Biology and Ecology* 191: 1–17.

Pechenik, J. A. (1999). On the advantages and disadvantages of larval stages in benthic marine invertebrate life cycles. *Marine Ecology Progress Series* 177: 269–297.

Pechenik, J. A., Li, W. & Cochrane, D. E. (2002). Timing is everything: The effects of putative dopamine antagonists on metamorphosis vary with larval age and experimental duration in the prosobranch gastropod *Crepidula fornicata*. *Biological Bulletin* 202: 137–147.

Pedrozo, H. A., Schwartz, Z., Luther, M., Dean, D. D., Boyan, B. D. & Wiederhold, M. L. (1996). A mechanism of adaptation to hypergravity in the statocyst of *Aplysia californica*. *Hearing Research* 102: 51–62.

Peel, J. S. (2006). Scaphopodization in Palaeozoic molluscs. *Palaeontology* 49: 1357–1364.

Peile, A. J. (1939). Radula notes–VI. *Journal of Molluscan Studies* 23: 270–273.

Pelseneer, P. (1894). Recherches sur divers opisthobranches. *Mémoires couronnés et mémoires des savants étrangers publiés par l'Académie royale des sciences, des lettres, et des beaux-arts de Belgique. Extrait du tome LIII* 53: 1–157.

Pelseneer, P. (1896). Pulmonés à branchies. *Bulletin de la Société Malacologique de Belgique* 29: 65–66.

Pelseneer, P. (1900). Les yeux céphaliques chez les lamellibranches. *Archives de Biologie* 16: 97–103.

Pelseneer, P. (1906). Mollusca. pp. 1–355 *in* E. R. Lankester (Ed.) *A Treatise on Zoology*. London, UK, V. Adam and Charles Black.

Pelseneer, P. (1907). La concentration du système nerveux chez les Lamellibranches. *Bulletin de la Classe des sciences, Académie royale de Belgique* 9–10: 874–878.

Pelseneer, P. (1911a). *Les Lamellibranches de l'expédition du 'Siboga'. Partie anatomique*. Vol. 53A. Leiden, E. J. Brill.

Pelseneer, P. (1911b). Recherches sur l'embryologie des Gastéropodes. Première partie. Embryologie spéciale. *Académie royale des sciences, des lettres et des beaux-arts de Belgique. Classe des sciences: Mémoires de la Classe des sciences* 3: 1–167.

Pelseneer, P. (1914). Éthologie de quelques *Odostomia* et d'un Monstrillide parasite de l'un d'eux. *Bulletin scientifique de la France et de la Belgique* 48: 1–14.

Pelseneer, P. (1926a). Notes on molluscan embryology. Egg-laying and development. *Bulletin biologique de France et de Belgique* 60: 88–112.

Pelseneer, P. (1926b). La proportion rélative des sexes chez les animaux et particulièrement chez les mollusques. *Mémoires de l'Académie royale de Belgique: Classe des sciences* 8: 1–258.

Pelseneer, P. (1928). La variabilité relative des sexes d'après des variations chez *Patella*, *Trochus* et *Nassa*. *Mémoires de l'Académie royale de Belgique: Classe des sciences* 10: 1–52.

Pelseneer, P. (1931). Quelques particularités d'organisation chez des Pectinacea. *Annales de la Société royale zoologique de Belgique* 61: 12–17.

Pelseneer, P. (1935). *Essai d'éthologie zoologique d'après l'étude des mollusques*. Bruxelles, Palais des Académies.

Pemberton, A. J., Hughes, R. N., Manríquez, P. H. & Bishop, J. D. D. (2003a). Efficient utilization of very dilute aquatic sperm: Sperm competition may be more likely than sperm limitation when eggs are retained. *Proceedings of the Royal Society B* 270 Suppl: S223–S226.

Pemberton, A. J., Noble, L. R. & Bishop, J. D. D. (2003b). Frequency dependence in matings with water-borne sperm. *Journal of Evolutionary Biology* 16: 289–301.

Penkman, K. E. H., Preece, R. C., Bridgland, D. R., Keen, D. H., Meijer, T., Parfitt, S. A., White, T. S. & Collins, M. J. (2013). An aminostratigraphy for the British Quaternary based on *Bithynia opercula*. *Quaternary Science Reviews* 61: 111–134.

Penney, B. K. (2004). Individual selection and the evolution of chemical defence in nudibranchs: Experiments with whole *Cadlina luteomarginata* (Nudibranchia: Doridina). *Journal of Molluscan Studies* 70: 399–402.

Penney, B. K. (2006). Morphology and biological roles of spicule networks in *Cadlina luteomarginata* (Nudibranchia, Doridina). *Invertebrate Biology* 125: 222–232.

Penney, B. K. (2009). A comment on F. Aguado & A. Marin: 'Warning coloration associated with nematocyst-based defences in aeolidioidean nudibranchs'. *Journal of Molluscan Studies* 75: 199–200.

Pennings, S. C. (1991). Reproductive behavior of *Aplysia californica* Cooper: Diel patterns, sexual roles and mating aggregations. *Journal of Experimental Marine Biology and Ecology* 149: 249–266.

Pennings, S. C. & Paul, V. J. (1993). Sequestration of dietary secondary metabolites by three species of sea hares: Location, specificity and dynamics. *Marine Biology* 117: 535–546.

Pennings, S. C., Weiss, A. M. & Paul, V. J. (1996). Secondary metabolites of the cyanobacterium *Microcoleus lyngbyaceus* and the sea hare *Stylocheilus longicauda*: Palatability and toxicity. *Marine Biology* 126: 735–743.

Pennington, J. T. (1985). The ecology of fertilization of echinoid eggs: The consequence of sperm dilution, adult aggregation, and synchronus spawning. *Biological Bulletin* 169: 417–430.

Pennington, J. T. & Chia, F.-S. (1985). Gastropod torsion: A test of Garstang's hypothesis. *Biological Bulletin* 169: 391–396.

Pentreath, V. W. & Cottrell, G. A. (1970). The blood supply to the central nervous system of *Helix pomatia*. *Zeitschrift für Zellforschung und Mikroskopische Anatomie* 111: 160–178.

Pereira, R. B., Andrade, P. B. & Valentão, P. (2016). Chemical diversity and biological properties of secondary metabolites from sea hares of *Aplysia* genus. *Marine Drugs* 14: 39.

Perez-Parallé, M. L., Pazos, A. J., Mesias-Gansbiller, C. & Sanchez, J. L. (2016). Hox, Parahox, EHGbox, and NK genes in bivalve molluscs: Evolutionary implications. *Journal of Shellfish Research* 35: 179–190.

Perkins, F. O., Zwerner, D. E. & Dias, R. K. (1975). The hyperparasite, *Urosporidium spisuli* sp. n. (Haplosporea), and its effects on the surf clam industry. *Journal of Parasitology* 61: 944–949.

Pernice, M., Deutsch, J. S., Andouche, A., Boucher-Rodoni, R. & Bonnaud, L. (2006). Unexpected variation of Hox genes' homeodomains in cephalopods. *Molecular Phylogenetics and Evolution* 40: 872–879.

Pernice, M., Wetzel, S., Gros, O., Boucher-Rodoni, R. & Dubilier, N. (2007). Enigmatic dual symbiosis in the excretory organ of *Nautilus macromphalus* (Cephalopoda: Nautiloidea). *Proceedings of the Royal Society B* 274: 1143–1152.

Pernice, M. & Boucher-Rodoni, R. (2012). Occurrence of a specific dual symbiosis in the excretory organ of geographically distant nautiloid populations. *Environmental Microbiology Reports* 4: 504–511.

Perrins, R. & Weiss, K. R. (1996). A cerebral central pattern generator in *Aplysia* and its connections with buccal feeding circuitry. *The Journal of Neuroscience* 16: 7030–7045.

Perron, F. E. & Turner, R. D. (1978). The feeding behaviour and diet of *Calliostoma occidentale*, a coelenterate-associated prosobranch gastropod. *Journal of Molluscan Studies* 44: 100–103.

Perron, F. E. (1981). The partitioning of reproductive energy between ova and protective capsules in marine gastropods of the genus *Conus*. *American Naturalist* 118: 110–118.

Perron, F. E. & Corpuz, G. C. (1982). Costs of parental care in the gastropod *Conus pennaceus*: Age-specific changes and physical constraints. *Oecologia* 55: 319–324.

Perry, K. J. & Henry, J. Q. (2015). CRISPR/Cas9-mediated genome modification in the Mollusc, *Crepidula fornicata*. *Genesis* 53: 237–244.

Perry, K. J., Lyons, D. C., Truchado-Garcia, M., Fischer, A. H. L., Helfrich, L. W., Johansson, K. B., Diamond, J. C., Grande, C. & Henry, J. Q. (2015). Deployment of regulatory genes during gastrulation and germ layer specification in a model spiralian mollusc *Crepidula*. *Developmental Dynamics* 244: 1215–1248.

Peruzzi, E., Fontana, G. & Sonetti, D. (2004). Presence and role of nitric oxide in the central nervous system of the freshwater snail *Planorbarius corneus*: Possible implication in neuron-microglia communication. *Brain Research* 1005: 9–20.

Peruzzi, E. & Sonetti, D. (2004). Microglia proliferation as a response to activation in the freshwater snail *Planorbarius corneus*: A BrdU incorporation study. *Acta Biologica Hungarica* 55: 287–291.

Petersen, J. A. & Johansen, K. (1973). Gas exchange in the giant sea cradle *Cryptochiton stelleri* (Middendorff). *Journal of Experimental Marine Biology and Ecology* 12: 27–43.

Peterson, C. H. & Black, R. (1995). Drilling by buccinid gastropods of the genus *Cominella* in Australia. *The Veliger* 38: 37.

Petit, R. E. & Harasewych, M. G. (1986). New Philippine Cancellariidae (Gastropoda: Cancellariacea), with notes on the fine structure and function of the nematoglossan radula. *The Veliger* 28: 436–443.

Petkevičiūtė, R., Stanevičiūtė, G., Stunzenas, V., Lee, T. & Ó Foighil, D. (2007). Pronounced karyological divergence of the North American congeners *Sphaerium rhomboideum* and *S. occidentale* (Bivalvia: Veneroida: Sphaeriidae). [Part 4]. *Journal of Molluscan Studies* 73: 315–321.

Petrov, N. B., Aleshin, V. V., Pegova, A. N., Ofitserov, M. V. & Slyusarev, G. S. (2010). New insight into the phylogeny of Mesozoa: Evidence from the 18S and 28S rRNA genes. *Moscow University Biological Sciences Bulletin* 65: 167–169.

Pettit, G. R. (1997). The dolastatins. pp. 1–79 *in* W. Herz, Grisebach, H. & Kirby, G. W. (Eds.) *Fortschritte der Chemie organischer Naturstoffe [Progress in the Chemistry of Organic Natural Products]*. Vienna, Austria, Springer-Verlag.

Peyer, S. M., Pankey, M. S., Oakley, T. H. & McFall-Ngai, M. J. (2014). Eye-specification genes in the bacterial light organ of the bobtail squid *Euprymna scolopes*, and their expression in response to symbiont cues. *Mechanisms of Development* 131: 111–126.

Pfeiffer, C. J. (1992). Intestinal ultrastructure of *Nerita picea* (Mollusca: Gastropoda), an intertidal marine snail of Hawaii. *Acta Zoologica* 73: 39–47.

Pfenninger, M., Hrabáková, M., Steinke, D. & Dèpraz, A. (2005). Why do snails have hairs? A Bayesian inference of character evolution. *BMC Evolutionary Biology* 5: 1–11.

Pfister, C. A., Meyer, F. & Antonopoulos, D. A. (2010). Metagenomic profiling of a microbial assemblage associated with the California mussel: A node in networks of carbon and nitrogen cycling. *PLoS ONE* 5: e10518.

Philipp, E. E. R. & Abele, D. (2010). Masters of longevity: Lessons from long-lived bivalves–a mini-review. *Gerontology* 56: 55–65.

Philippe, H., Brinkmann, H., Copley, R. R., Moroz, L. L., Nakano, H., Poustka, A. J., Wallberg, A., Peterson, K. J. & Telford, M. J. (2011). Acoelomorph flatworms are deuterostomes related to *Xenoturbella*. *Nature* 470: 255–258.

Phillips, D. J. H. (1980). *Quantitative Aquatic Biological Indicators; their use to Monitor Trace Metal and Organochlorine Pollution*. London, UK, Applied Science Publishers.

Phillips, D. J. H. (1990). Use of macroalgae and invertebrates as monitors of metal levels in estuaries and coastal waters. pp. 81–99 *in* R. W. Furness & Rainbow, P. S. (Eds.) *Heavy Metals in the Marine Environment*. Boca Raton, FL, CRC Press.

Phillips, D. W. (1975). Distance chemoreception-triggered avoidance behavior of the limpets *Acmaea* (*Collisella*) *limatula* and *Acmaea* (*Notoacmea*) *scutum* to the predatory starfish *Pisaster ochraceus*. *Journal of Experimental Zoology* 191: 199–209.

Phillips, D. W. (1976). The effect of a species-specific avoidance response to predatory starfish on the intertidal distribution of two gastropods. *Oecologia* 23: 83–94.

Phillips, D. W. (1977). Avoidance and escape responses of the gastropod mollusc *Olivella biplicata* (Sowerby) to predatory asteroids. *Journal of Experimental Marine Biology and Ecology* 28: 77–86.

Phillips, D. W. (1978). Chemical mediation of invertebrate defensive behaviors and the ability to distinguish between foraging and inactive predators. *Marine Biology* 49: 237–243.

Phillips, D. W. & Chiarappa, M. L. (1980). Defensive responses of gastropods to the predatory flatworms *Freemania litoricola* (Heath and McGregor) and *Notoplana acticola* (Boone). *Journal of Experimental Marine Biology and Ecology* 47: 179–189.

Philpott, D. E. & Person, P. (1970). The biology of cartilage. II. Invertebrate cartilages: Squid head cartilage. *Journal of Morphology* 131: 417–430.

Piatigorsky, J., Kozmik, Z., Horwitz, J., Ding, L.-F., Carosa, E., Robison, W. G., Steinbach, P. J. & Tamm, E. R. (2000). Ω-crystallin of the scallop lens: A dimeric aldehyde dehydrogenase class 1/2 enzyme-crystallin. *Journal of Biological Chemistry* 275: 41064–41073.

Piatigorsky, J. (2003). Crystallin genes: Specialization by changes in gene regulation may precede gene duplication. *Journal of Structural and Functional Genomics* 3: 131–137.

Pichon, D., Gaia, V., Norman, M. D. & Boucher-Rodoni, R. (2005). Phylogenetic diversity of epibiotic bacteria in the accessory nidamental glands of squids (Cephalopoda: Loliginidae and Idiosepiidae). *Marine Biology* 147: 1323–1332.

Pichon, Y., Mouëza, M. & Frenkiel, L. (1980). Mechanoreceptor properties of the sense organ of the cruciform muscle in a tellinacean lamellibranch, *Donax trunculus* L., an electrophysiological approach. *Marine Biology Letters* 1: 273–284.

Picken, G. B. & Allan, D. (1983). Unique spawning behavior by the Antarctic limpet *Nacella* (*Patinigera*) *concinna* (Strebel, 1908). *Journal of Experimental Marine Biology and Ecology* 71: 283–288.

Pickford, G. E. (1957). Vampyromorpha. *Galathea Reports* 161: 243–253.

Picos-García, C., Carcia-Carreno, F. L. & Serviere-Zaragoza, E. (2000). Digestive proteases in juvenile Mexican green abalone, *Haliotis fulgens*. *Aquaculture* 181: 157–170.

Piel, J. (2004). Metabolites from symbiotic bacteria. *Natural Product Reports* 21: 519–538.

Piel, W. H. (1991). Pycnogonid predation on nudibranchs and ceratal autotomy. *The Veliger* 34: 366–367.

Pierce, S. K. & Greenberg, M. J. (1972). The nature of cellular volume regulation in marine bivalves. *Journal of Experimental Biology* 57: 681–692.

Pierce, S. K. & Amende, L. M. (1981). Control mechanisms of amino acid-mediated cell volume regulation in salinity-stressed molluscs. *Journal of Experimental Zoology Part A* 215: 247–257.

Pierce, S. K. (1982). Invertebrate cell volume control mechanisms: A coordinated use of intracellular amino acids and inorganic ions as osmotic solute. *Biological Bulletin* 163: 405–419.

Pierce, S. K., Biron, R. & Rumpho, M. E. (1996). Endosymbiotic chloroplasts in molluscan cells contain proteins synthesized after plastid capture. *Journal of Experimental Biology* 199: 2323–2330.

Pierce, S. K., Massey, S. E., Hanten, J. J. & Curtis, N. E. (2003). Horizontal transfer of functional nuclear genes between multicellular organisms. *Biological Bulletin* 204: 237–240.

Pigneur, L.-M., Hedtke, S. M., Etoundi, E. & van Doninck, K. (2012). Androgenesis: A review through the study of the selfish shellfish *Corbicula* spp. *Heredity* 108: 581–591.

Pika, J. & Faulkner, D. J. (1994). Four sesquiterpenes from the South African nudibranch *Leminda millecra*. *Tetrahedron* 50: 3065–3070.

Pila, E. A., Sullivan, J. T., Wu, X. Z., Fang, J., Rudko, S. P., Gordy, M. A. & Hanington, P. C. (2016). Haematopoiesis in molluscs: A review of haemocyte development and function in gastropods, cephalopods and bivalves. *Developmental and Comparative Immunology* 58: 119–128.

Pilkington, J. B., Little, C. & Stirling, P. E. (1984). A respiratory current in the mantle cavity of *Amphibola crenata* (Mollusca, Pulmonata). *Journal of the Royal Society of New Zealand* 14: 327–334.

Pinder, L. C. V., Pottinger, T. G., Billinghurst, Z. & Depledge, M. H. (1999). *Endocrine Function in Aquatic Invertebrates and Evidence for Disruption by Environmental Pollutants*. Bristol, Environment Agency.

Pipe, R. K. (1987). Oogenesis in the marine mussel *Mytilus edulis*: An ultrastructural study. *Marine Biology* 95: 405–414.

Piraino, S., Todaro, C., Geraci, S. & Boero, F. (1994). Ecology of the bivalve-inhabiting hydroid *Eugymnanthea inquiline* in the coastal sounds of Taranto (Ionian Sea, SE Italy). *Marine Biology* 118: 695–703.

Pires, A., Croll, R. P. & Hadfield, M. G. (2000). Catecholamines modulate metamorphosis in the opisthobranch gastropod *Phestilla sibogae*. *Biological Bulletin* 198: 319–331.

Pitcher, C. R. & Butler, A. J. (1987). Predation by asteroids, escape response, and morphometrics of scallops with epizoic sponges. *Journal of Experimental Marine Biology and Ecology* 112: 233–249.

Plagányi, E. E. & Branch, G. M. (2000). Does the limpet *Patella cochlear* fertilize its own algal garden? *Marine Ecology Progress Series* 194: 113–122.

Plante, C. & Jumars, P. (1992). The microbial environment of marine deposit-feeder guts characterized via microelectrodes. *Microbial Ecology* 23: 257–277.

Plate, L. H. (1897). Die anatomie und phylogenie der Chitonen. *Zoologisches Jahrbuch Supplement* Suppl. 4: 1–243.

Plate, L. H. (1901). Die Anatomie und Phylogenie der Chitonen (Fortsetzung). *Zoologisches Jahrbuch Supplement. Band V. Fauna Chilensis* 2: 281–600.

Plaziat, J.-C. (1993). Modern and fossil potamids (Gastropoda) in saline lakes. *Journal of Paleolimnology* 8: 163–169.

Plesch, B., Janse, C. & Boer, H. H. (1975). Gross morphology and histology of the musculature of the freshwater pulmonate *Lymnaea stagnalis* (L.). *Netherlands Journal of Zoology* 25: 332–352.

Plesch, B. (1976). Shell attachment in the pond snail *Lymnaea stagnalis* (L.). *Cell and Tissue Research* 171: 389–396.

Plesch, B. (1977). An ultrastructural study of the musculature of the pond snail *Lymnaea stagnalis* (L.). *Cell and Tissue Research* 180: 317–340.

Pluth, M. D., Tomat, E. & Lippard, S. J. (2011). Biochemistry of mobile zinc and nitric oxide revealed by fluorescent sensors. *Annual Review of Biochemistry* 80: 333–355.

Pohunkova, H. (1967). The ultrastructure of the lung of the snail *Helix pomatia*. *Folia Morphologica* 15: 250–257.

Poizat, C. H. (1985). Interstitial opisthobranch gastropods as indicator organisms in sublittoral sandy habitats. *Stygologia* 1: 26–42.

Poizat, C. H. (1991). New data on an interstitial opisthobranch assemblage and other meiofauna from the Skagerrak, Sweden. *Journal of Molluscan Studies* 57: 167–177.

Pojeta, J. & Runnegar, B. N. (1976). The paleontology of rostroconch mollusks and the early history of the phylum Mollusca. *Geological Survey Professional Paper* 968: 1–88.

Pokora, Z. (1989). Strobilation as a form of asexual reproduction in certain pteropods (Gastropoda, Opisthobranchia): Functional significance of this process in their life cycle. *Przeglad Zoologiczny* 33: 397–410.

Pollard, E. (1975). Aspects of the ecology of *Helix pomatia* L. *Journal of Animal Ecology* 44: 305–329.

Pollard, S. D. & Jackson, R. R. (2004). Gastropod predation in spiders (Araneae). pp. 497–504 *in* G. M. Barker (Ed.) *Natural Enemies of Terrestrial Molluscs*. Oxford, UK, Cambridge, MA, CABI Publishing.

Polner, A., Paul, V. J. & Scheuer, P. J. (1989). Kumepaloxane, a rearranged trisnor sesquiterpene from the bubble shell *Haminoea cymbalum*. *Tetrahedron* 45: 617–622.

Pomiankowski, A. & Reguera, P. (2001). The point of love. *Trends in Ecology & Evolution* 16: 533–534.

Ponder, W. F. (1965). The biology of the genus *Arthritica*. *Transactions of the Royal Society of New Zealand: Zoology* 6: 75–86.

Ponder, W. F. (1966). A new family of the Rissoacea from New Zealand. *Records of the Dominion Museum* 5: 177–184.

Ponder, W. F. (1970a). The morphology of *Alcithoe arabica* (Gastropoda: Volutidae). *Malacological Review* 3: 127–165.

Ponder, W. F. (1970b). Some aspects of the morphology of four species of the neogastropod family Marginellidae with a discussion on the evolution of the toxoglossan poison gland. *Journal of the Malacological Society of Australia* 2: 55–81.

Ponder, W. F. (1971). *Montacutona ceriantha* n. sp., a commensal leptonid bivalve living with *Cerianthus*. *Journal de Conchyliologie* 109: 15–25.

Ponder, W. F. (1972). The morphology of some mitriform gastropods with special reference to their alimentary and reproductive systems (Neogastropoda). *Malacologia* 11: 295–342.

Ponder, W. F. (1974). The origin and evolution of the Neogastropoda. *Malacologia* 12: 295–338.

Ponder, W. F. & Yoo, E. K. (1977). A revision of the Australian species of the Rissoellidae (Mollusca: Gastropoda). *Records of the Australian Museum* 31: 133–186.

Ponder, W. F. (1982). Hydrobiidae of Lord Howe Island (Mollusca: Gastropoda: Prosobranchia). *Australian Journal of Marine and Freshwater Research* 33: 89–159.

Ponder, W. F. (1983). Review of the genera of the Barleeidae (Mollusca: Gastropoda: Rissoacea). *Records of the Australian Museum* 35: 231–281.

Ponder, W. F. (1986). Glacidorbidae (Glacidorbacea: Basommatophora), a new family and superfamily of operculate freshwater gastropods. *Zoological Journal of the Linnean Society* 87: 53–83.

Ponder, W. F. (1987). The anatomy and relationships of the pyramidellacean limpet *Amathina tricarinata* (Mollusca: Gastropoda). *Asian Marine Biology* 4: 1–34.

Ponder, W. F. (1988a). The Truncatelloidean (= Rissoacean) radiation: A preliminary phylogeny. pp. 129–166 in W. F. Ponder, Eernisse, D. J. & Waterhouse, J. H. (Eds.) *Prosobranch Phylogeny. Malacological Review Supplement*. Ann Arbor, MI, Malacological Review.

Ponder, W. F. (1988b). *Potamopyrgus antipodarum*: A molluscan colonizer of Europe and Australia. *Journal of Molluscan Studies* 54: 271–285.

Ponder, W. F. (1988c). Bioluminescence in *Hinea braziliana* Lamarck (Gastropoda Planaxidae). *Journal of Molluscan Studies* 54: 361–362.

Ponder, W. F. & Clark, G. A. (1988). A morphological and electrophoretic examination of *Hydrobia buccinoides*, a variable brackish-water gastropod from temperate Australia (Mollusca: Hydrobiidae). *Australian Journal of Zoology* 36: 661–689.

Ponder, W. F., Hershler, R. & Jenkins, B. (1989). An endemic radiation of Hydrobiidae from artesian springs in northern South Australia: Their taxonomy, physiology, distribution and anatomy. *Malacologia* 31: 1–140.

Ponder, W. F. (1990a). The anatomy and relationships of a marine valvatoidean (Gastropoda: Heterobranchia). *Journal of Molluscan Studies* 56: 533–555.

Ponder, W. F. (1990b). The anatomy and relationships of the Orbitestellidae (Gastropoda: Heterobranchia). *Journal of Molluscan Studies* 56: 515–532.

Ponder, W. F. (1990c). A gravel beach shelled micro-gastropod assemblage from Ceuta, Strait of Gibraltar, with the description of a new truncatelloidean genus. *Bulletin du Muséum national d'histoire naturelle Série A Zoologie biologie et écologie animales* 12: 291–312.

Ponder, W. F. & Clark, G. A. (1990). A radiation of hydrobiid snails in threatened artesian springs in western Queensland. *Records of the Australian Museum* 42: 301–363.

Ponder, W. F. & Taylor, J. D. (1992). Predatory shell drilling by two species of *Austroginella* (Gastropoda: Marginellidae). *Journal of Zoology* 228: 317–328.

Ponder, W. F., Clark, G. A., Miller, A. C. & Toluzzi, A. (1993). On a major radiation of freshwater snails in Tasmania and eastern Victoria: A preliminary overview of the Beddomeia group (Mollusca: Gastropoda: Hydrobiidae). *Invertebrate Taxonomy* 7: 501–750.

Ponder, W. F., Colgan, D. J., Clark, G. A., Miller, A. C. & Terzis, T. (1994). Microgeographic, genetic and morphological differentiation of freshwater snails: The Hydrobiidae of Wilson's Promontory, Victoria, south-eastern Australia. *Australian Journal of Zoology* 42: 557–678.

Ponder, W. F. & Lindberg, D. R. (1997). Towards a phylogeny of gastropod molluscs: An analysis using morphological characters. *Zoological Journal of the Linnean Society* 119: 83–265.

Ponder, W. F. & Grayson, J. E. (1998). *The Australian marine molluscs considered to be potentially vulnerable to the shell trade. Report to Environment Australia*. Sydney, Environment Australia: 1–56.

Ponder, W. F. & Lunney, D., Eds. (1999). *The Other 99%: The Conservation and Biodiversity of Invertebrates [Proceedings of a four-day meeting held at the Australian Museum, Sydney]*. Transactions of the Royal Zoological Society of New South Wales. Sydney, Australia, Royal Zoological Society of New South Wales.

Ponder, W. F. & Avern, G. J. (2000). The Glacidorbidae (Mollusca: Gastropoda: Heterobranchia) of Australia. *Records of the Australian Museum* 52: 307–353.

Ponder, W. F. & Colgan, D. J. (2002). What makes a narrow range taxon? Insights from Australian freshwater snails. *Invertebrate Systematics* 16: 571–582.

Ponder, W. F., Hutchings, P. A. & Chapman, R. (2002). *Overview of the Conservation of Australian Marine Invertebrates [Report to Environment Australia]*. Sydney, Australia, Environment Australia, Australian Museum: 1–588.

Ponder, W. F. (2003). Narrow range endemism in the sea and its implications for conservation. pp. 89–102 in P. Hutchings & Lunney, D. (Eds.) *Conserving Marine Environments: Out of Sight, Out of Mind*. Mosman NSW, Royal Zoological Society of New South Wales.

Ponder, W. F., Clark, S. A., Eberhard, S. & Studdert, J. B. (2005). A radiation of hydrobiid snails in the caves and streams at Precipitous Bluff, southwest Tasmania, Australia (Mollusca: Caenogastropoda: Rissooidea: Hydrobiidae s.l.). *Zootaxa* 1074: 1–66.

Ponder, W. F. & Haase, M. (2005). A new genus of hydrobiid gastropods with Australian affinities from Lake Poso, Sulawesi (Gastropoda: Caenogastropoda: Rissooidea). *Molluscan Research* 25: 27–36.

Ponder, W. F., Colgan, D. J., Healy, J. M., Nützel, A., de Simone, L. R. L. & Strong, E. E. (2008). Caenogastropoda. pp. 331–383 in W. F. Ponder & Lindberg, D. R. (Eds.) *Phylogeny and Evolution of the Mollusca*. Berkeley, CA, University of California Press.

Ponder, W. F. & Lindberg, D. R., Eds. (2008). *Phylogeny and Evolution of the Mollusca*. Berkeley, CA, University of California Press.

Popescu, I. R. & Willows, A. O. D. (1999). Sources of magnetic sensory input to identified neurons active during crawling in the marine mollusc *Tritonia diomedea*. *Journal of Experimental Biology* 202: 3029–3036.

Popham, J. (1974). Comparative morphometrics of the acrosomes of the sperms of "externally" and "internally" fertilizing sperms of the shipworms (Teredinidae, Bivalvia, Mollusca). *Cell and Tissue Research* 150: 291–297.

Popham, J. D. & Dickson, M. R. (1973). Bacterial associations in the teredo *Bankia australis* (Lamellibranchia: Mollusca). *Marine Biology* 19: 338–340.

Popham, J. D. & Marshall, B. A. (1977). The fine structure of the spermatozoan of the protobranch bivalve, *Nucula hartvigiana* Pfeiffer. *The Veliger* 19: 431–433.

Port, G. & Ester, A. (2002). Gastropods as pests in vegetable and ornamental crops in Western Europe. pp. 337–352 in G. M. Barker (Ed.) *Molluscs as Crop Pests*. Wallingford, UK, CABI Publishing.

Porte, C., Janer, G., Lorusso, L. C., Ortiz-Zarragoitia, M., Cajaraville, M. P., Fossi, M. C. & Canesi, L. (2006). Endocrine disruptors in marine organisms: Approaches and perspectives. *Comparative Biochemistry and Physiology Part C* 143: 303–315.

Portmann, A. (1926). Der embryonale Blutkreislauf und die Dotterresorption bei *Loligo vulgaris*. *Zeitschrift für Morphologie und Ökologie der Tiere* 5: 406–423.

Pörtner, H.-O. (2002). Climate variations and the physiological basis of temperature dependent biogeography: Systemic to molecular hierarchy of thermal tolerance in animals. *Comparative Biochemistry and Physiology Part A* 132: 739–761.

Pörtner, H. O., O'Dor, R. K. & Macmillan, D. L. (1994). *Physiology of Cephalopod Molluscs: Lifestyle and Performance Adaptations*. Basel, Switzerland, Gordon and Breach.

Pörtner, H. O., Bennett, A. F., Bozinovic, F., Clarke, A., Lardies, M. A., Lucassen, M., Pelster, B., Schiemer, F. & Stillman, J. H. (2006a). Trade-offs in thermal adaptation: The need for a molecular to ecological integration. *Physiological and Biochemical Zoology* 79: 295–313.

Pörtner, H. O., Peck, L. S. & Hirse, T. (2006b). Hyperoxia alleviates thermal stress in the Antarctic bivalve, *Laternula elliptica*: Evidence for oxygen limited thermal tolerance. *Polar Biology* 29: 688–693.

Postuma, F. A. & Gasalla, M. A. (2015). Ethogram analysis reveals new body patterning behavior of the tropical Arrow Squid *Doryteuthis plei* off the São Paulo coast. *Biological Bulletin* 229: 143–159.

Potts, D. C. (1975). Persistence and extinction of local populations of the garden snail *Helix aspersa* in unfavorable environments. *Oecologia* 21: 313–334.

Potts, G. W. (1981). The anatomy of respiratory structures in the dorid nudibranchs, *Onchidoris bilamellata* and *Archidoris pseudoargus*, with details of the epidermal glands. *Journal of the Marine Biological Association of the United Kingdom* 61: 959–982.

Potts, W. T. W. (1954). The energetics of osmotic regulation in brackish- and fresh-water animals. *Journal of Experimental Biology* 31: 618–630.

Poulin, R. & Mouritsen, K. N. (2006). Climate change, parasitism and the structure of intertidal ecosystems. *Journal of Helminthology* 80: 183–191.

Poulin, R. (2007). *Evolutionary Ecology of Parasites*. Princeton, NJ, Princeton University Press.

Powell, A. W. B. (1941). Seven new species of New Zealand land Mollusca. *Records of the Auckland Institute and Museum* 2: 260–264.

Powell, A. W. B. (1966). The molluscan families Speightiidae and Turridae. *Bulletin of the Auckland Institute and Museum* 5: 1–184.

Powell, E. N., Cummins, H., Stanton, R. J. & Staff, G. (1984). Estimation of the size of molluscan larval settlement using the death assemblage. *Estuarine, Coastal and Shelf Science* 18: 367–384.

Powell, E. N. & Cummins, H. (1985). Are molluscan maximum life spans determined by long-term cycles in benthic communities? *Oecologia* 67: 177–182.

Powell, M. A. & Somero, G. N. (1985). Sulfide oxidation occurs in the animal tissue of the gutless clam, *Solemya reidi*. *Biological Bulletin* 169: 164–181.

Power, A. J., Keegan, B. F. & Nolan, K. (2002). The seasonality and role of the neurotoxin tetramine in the salivary glands of the red whelk *Neptunea antiqua* (L.). *Toxicon* 40: 419–425.

Power, M. E., Tilman, D., Estes, J. A., Menge, B. A., Bond, W. J., Mills, L. S., Daily, G., Castilla, J. C., Lubchenco, J. & Paine, R. T. (1996). Challenges in the quest for keystones: Identifying keystone species is difficult: But essential to understanding how loss of species will affect ecosystems. *BioScience* 46: 609–620.

Prado, S., Montes, J. F., Romalde, J. L. & Barja, J. L. (2009). Inhibitory activity of *Phaeobacter* strains against aquaculture pathogenic bacteria. *International Microbiology* 12: 107–114.

Prashad, B. (1925). Anatomy of the common Indian apple snail, *Pila globosa*. *Memoirs of the Indian Museum* 8: 91–152.

Prashanth, J. R., Lewis, R. J. & Dutertre, S. (2012). Towards an integrated venomics approach for accelerated conopeptide discovery. *Toxicon* 60: 470–477.

Prasopdee, S., Tesana, S., Cantacessi, C., Laha, T., Mulvenna, J., Grams, R., Loukas, A. & Sotillo, J. (2015). Proteomic profile of *Bithynia siamensis* goniomphalos snails upon infection with the carcinogenic liver fluke *Opisthorchis uiverrini*. *Journal of Proteomics* 113: 281–291.

Prat, N., Añón-Suárez, D. & Rieradevall, M. (2004). First record of Podonominae larvae living phoretically on the shells of the water snail *Chilina dombeyana* (Diptera: Chironomidae/Gastropoda: Lymnaeidae). *Aquatic Insects* 26: 147–152.

Prato, F. S., Kavaliers, M., Thomas, A. W. & Ossenkopp, K. P. (1998). Modulatory actions of light on the behavioural responses to magnetic fields by land snails probably occur at the magnetic field detection stage. *Philosophical Transactions of the Royal Society of London B* 265: 367–373.

Prescott, S. A., Gill, N. & Chase, R. (1997). Neural circuit mediating tentacle withdrawal in *Helix aspersa*, with specific reference to the competence of the motor neuron C3. *Journal of Neurophysiology* 78: 2951–2965.

Preston, R. J. & Lee, R. M. (1973). Feeding behavior in *Aplysia californica*: Role of chemical and tactile stimuli. *Journal of Comparitive and Physiological Psychology* 83: 368–381.

Preuss, T. & Budelmann, B.-U. (1995). Proprioceptive hair cells on the neck of the squid *Lolliguncula brevis*: A sense organ in cephalopods for the control of head-to-body position. *Philosophical Transactions of the Royal Society B* 349: 153–178.

Prezant, R. S. (1990). Form, function and phylogeny of bivalve mucins. pp. 83–95 *in* B. Morton (Ed.) *The Bivalvia. Proceedings of a Memorial Symposium in Honour of Sir Charles Maurice Yonge (1899–1986), Edinburgh, 1986*. Hong Kong, Hong Kong University Press.

Price, C. H. (1977). Morphology and histology of the central nervous system and neurosecretory cells in *Melampus bidentatus* Say (Gastropoda: Pulmonata). *Transactions of the American Microscopical Society* 96: 295–312.

Price, C. H. (1979). Physical factors and neurosecretion in the control of reproduction in *Melampus* (Mollusca: Pulmonata). *Journal of Experimental Zoology* 207: 269–282.

Price, C. H. (1980). Water relations and physiological ecology of the salt marsh snail, *Melampus bidentatus* Say. *Journal of Experimental Marine Biology and Ecology* 45: 51–67.

Price, D. A. & Greenberg, M. J. (1977). Structure of a molluscan cardioexcitatory neuropeptide. *Science* 197: 670.

Price, D. A. (1986). The evolution of a molluscan cardioregulatory neuropeptide. *American Zoologist* 26: 1007–1015.

Price, D. A., Lesser, W., Lee, T. D., Doble, K. E. & Greenberg, M. J. (1990). Seven FMRFamide-related and two SCP-related cardioactive peptides from *Helix*. *Journal of Experimental Biology* 154: 421–437.

Price, R. M. (2003). Columellar muscle of neogastropods: Muscle attachment and the function of columellar folds. *Biological Bulletin* 205: 351–366.

Prieur, D. G., Mevel, J. L., Nicolas, A., Plusquellec, A. & Vigneulle, M. (1990). Interactions between bivalve molluscs and bacteria in the marine environment. *Oceanography and Marine Biology Annual Review* 28: 277–352.

Prince, J. D., Nolen, T. G. & Coelho, L. (1998). Defensive ink pigment processing and secretion in *Aplysia californica*: Concentration and storage of phycoerythrobilin in the ink gland. *Journal of Experimental Biology* 201: 1595–1613.

Prince, J. S. & Johnson, P. M. (2006). Ultrastructural comparison of *Aplysia* and *Dolabrifera* ink glands suggests cellular sites of anti-predator protein production and algal pigment processing. *Journal of Molluscan Studies* 72: 349–357.

Prince, J. S. & Young, D. (2010). Ultrastructural study of skin coloration in *Aplysia californica*. *Bulletin of Marine Science* 86: 803–812.

Prince, J. S. & Johnson, P. M. (2013). Role of the digestive gland in ink production in four species of sea hares: An ultrastructural comparison. *Journal of Marine Biology* 2013: 1–5.

Prior, D. J. (1984). Analysis of contact-rehydration in terrestrial gastropods: Osmotic control of drinking behaviour. *Journal of Experimental Biology* 111: 63–73.

Prior, D. J. & Uglem, G. L. (1984). Analysis of contact-rehydration in terrestrial gastropods: Absorption of 14C-inulin through the epithelium of the foot. *Journal of Experimental Biology* 111: 75–80.

Prior, D. J. (1985). Water-regulatory behaviour in terrestrial gastropods. *Biological Reviews* 60: 403–424.

Proćków, M., Strzala, T., Kuźnik-Kowalska, E., Proćków, J. & Mackiewicz, P. (2017). Ongoing speciation and gene flow between taxonomically challenging *Trochulus* species complex (Gastropoda: Hygromiidae). *PLoS ONE* 12: e0170460.

Proestou, D. A., Goldsmith, M. R. & Twombly, S. (2008). Patterns of male reproductive success in *Crepidula fornicata* provide new insight for sex allocation and optimal sex change. *Biological Bulletin* 214: 194–202.

Proksch, P. (1994). Defensive roles for secondary metabolites from marine sponges and sponge-feeding nudibranchs. *Toxicon* 32: 639–656.

Protasov, A., Sylaieva, A., Morozovska, I., Lopes-Lima, M. & Sousa, R. (2015). A massive freshwater mussel bed (Bivalvia: Unionidae) in a small river in Ukraine. *Folia Malacologica* 23: 273–277.

Prowse, T. A. A. & Pile, A. J. (2005). Phenotypic homogeneity of two intertidal snails across a wave exposure gradient in South Australia. *Marine Biology Research* 1: 176–185.

Prozorova, L. A. & Bogatov, V. V. (2006). Large bivalve molluscs (Bivalvia, Unioniformes) of Lake Baikal. *Hydrobiologia* 568: 201–205.

Pruvot-Fol, A. (1937). Étude d'un prosobranche d'eau douce: *Helicostoa sinensis* Lamy. *Bulletin de la Société zoologique de France* 62: 250–257.

Pruvot, G. (1890). Sur le développement d'un Solénogastre. *Comptes rendus hebdomadaire des séances de l'Académie des sciences* 111: 689–692.

Przeslawski, R. (2004a). Chemical sunscreens in intertidal gastropod egg masses. *Invertebrate Reproduction & Development* 46: 119–124.

Przeslawski, R. (2004b). A review of the effects of environmental stress on embryonic development within intertidal gastropod egg masses. *Journal of Molluscan Studies* 24: 43–63.

Przeslawski, R., Davis, A. R. & Benkendorff, K. (2004). Effects of ultraviolet radiation and visible light on the development of encapsulated molluscan embryos. *Marine Ecology Progress Series* 268: 151–160.

Przeslawski, R., Davis, A. R. & Benkendorff, K. (2005). Synergistic effects associated with climate change and the development of rocky shore molluscs. *Global Change Biology* 11: 515–522.

Przeslawski, R., Byrne, M. & Mellin, C. (2015). A review and meta-analysis of the effects of multiple abiotic stressors on marine embryos and larvae. *Global Change Biology* 21: 2122–2140.

Puce, S., Cerrano, C., Di Camillo, C. G. & Bavestrello, G. (2008). Hydroidomedusae (Cnidaria: Hydrozoa) symbiotic radiation. *Journal of the Marine Biological Association of the United Kingdom* 88: 1715–1721.

Pugh, D. (1963). The cytology of the digestive and salivary glands of the limpet, *Patella*. *Quarterly Journal of Microscopical Science* 104: 23–37.

Puillandre, N., Kantor, I., Sysoev, A. V., Couloux, A., Meyer, C., Rawlings, T., Todd, J. A. & Bouchet, P. (2011). The dragon tamed? A molecular phylogeny of the Conoidea (Gastropoda). *Journal of Molluscan Studies* 77: 259–272.

Puinean, A. M., Labadie, P., Hill, E. M., Osada, M., Kishida, M., Nakao, R., Novillo, A., Callard, I. P. & Rotchell, J. M. (2006). Laboratory exposure to 17β-estradiol fails to induce vitellogenin and estrogen receptor gene expression in the marine invertebrate *Mytilus edulis*. *Aquatic Toxicology* 79: 376–383.

Pujalte, M. J., Ortigosa, M., Macián, M. C. & Garay, E. (1999). Aerobic and facultative anaerobic heterotrophic bacteria associated to Mediterranean oysters and seawater. *International Microbiology* 2: 259–266.

Purcell, J. E. (1984). Predation on fish larvae by *Physalia physalis*, the Portuguese man of war. *Marine Ecology Progress Series* 19: 189–191.

Purcell, S. W. (2004). Management options for restocked *Trochus* fisheries. pp. 233–243 *in* K. M. Leber, Kitada, S., Blankenship, H. L. & Svåsand, T. (Eds.) *Stock Enhancement and Sea Ranching: Developments, Pitfalls and Opportunities*. Oxford, UK, Blackwell Publishing.

Purchon, R. D. (1941a). On the biology and relationships of the lamellibranch *Xylophaga dorsalis* (Turton). *Journal of the Marine Biological Association of the United Kingdom* 25: 1–39.

Purchon, R. D. (1941b). A note on the biology of *Martesia striata* L. (Lamellibranchia). *Journal of Zoology* 126: 245–258.

Purchon, R. D. (1955). The structure and function of the British Pholadidae (rock-boring Lamellibranchia). *Proceedings of the Zoological Society of London* 124: 859–911.

Purchon, R. D. (1959). Phylogenetic classification of the Lamellibranchia, with special reference to the Protobranchia. *Proceedings of the Malacological Society of London* 33: 224–230.

Purchon, R. D. (1968). *The Biology of the Mollusca*. Oxford, UK, Pergamon Press.

Purchon, R. D. (1977a). Feeding methods and evolution in the Bivalvia. pp. 101–145 *in* R. D. Purchon (Ed.) *The Biology of the Mollusca*. Amsterdam, the Netherlands, Pergamon Press.

Purchon, R. D. (1977b). Digestion. pp. 207–268 *in* R. D. Purchon (Ed.) *The Biology of the Mollusca*. Amsterdam, the Netherlands, Pergamon Press.

Purchon, R. D. (1987a). The stomach in the Bivalvia. *Philosophical Transactions of the Royal Society B* 316: 183–276.

Purchon, R. D. (1987b). Classification and evolution of the Bivalvia: An analytical study. *Philosophical Transactions of the Royal Society B* 316: 277–302.

Purschke, G., Arendt, D., Hausen, H. & Müller, M. C. M. (2006). Photoreceptor cells and eyes in Annelida. *Arthropod Structure & Development* 35: 211–230.

Puttick, M. N., Morris, J. L., Williams, T. A., Cox, C. J., Edwards, D., Kenrick, P., Pressel, S., Wellman, C. H., Schneider, H. & Pisani, D. (2018). The interrelationships of land plants and the nature of the ancestral embryophyte. *Current Biology* 28: 733–745.

Putz, A., König, G. M. & Wägele, H. (2010). Defensive strategies of Cladobranchia (Gastropoda, Opisthobranchia). *Natural Product Reports* 27: 1386–1402.

Quayle, D. B. & Newkirk, G. F. (1989). *Farming Bivalve Molluscs: Methods for Study and Development*. Vol. 1. Baton Rouge, LA, World Aquaculture Society in association with the International Development Research Centre.

Quesada, H., Wenne, R. & Skibinski, D. O. F. (1999). Interspecies transfer of female mitochondrial DNA is coupled with role-reversals and departure from neutrality in the mussel *Mytilus trossulus*. *Molecular Biology and Evolution* 16: 655–665.

Quick, H. E. (1920). Notes on the anatomy and reproduction of *Paludestina* (*Hydrobia*) *stagnalis*. *Journal of Conchology* 16: 96–97.

Quinn, J. F. (1991). Systematic position of *Basilissopsis* and *Guttula*, and a discussion of the phylogeny of the Seguenzioidea (Gastropoda: Prosobranchia). *Bulletin of Marine Science* 49: 575–598.

Radermacher, P., Schöne, B. R., Gischler, E., Oschmann, W., Thébault, J. & Fiebig, J. (2009). Sclerochronology: A highly versatile tool for mariculture and reconstruction of life history traits of the queen conch, *Strombus gigas* (Gastropoda). *Aquatic Living Resources* 22: 307–318.

Radoman, P. (1983). Hydrobioidea, a superfamily of Prosobranchia (Gastropoda). 1. Systematics. *Serbian Academy of Sciences and Arts Monographs, 547, Department of Science* 57: 1–256.

Raffaelli, D. & Hawkins, S. J. (1996). *Intertidal Ecology*. Dordrecht, the Netherlands, Kluwer Academic Publishers.

Rahat, M. & Monselise, B.-I. (1979). Photobiology of the chloroplast hosting mollusc *Elysia timida* (Opisthobranchia). *Journal of Experimental Biology* 79: 225–233.

Raikow, D. F. & Hamilton, S. K. (2001). Bivalve diets in a midwestern US stream: A stable isotope enrichment study. *Limnology and Oceanography* 46: 514–522.

Rainbow, P. S. (1995). Biomonitoring of heavy metal availability in the marine environment. *Marine Pollution Bulletin* 31: 183–192.

Rainbow, P. S. (2002). Trace metal concentrations in aquatic invertebrates: Why and so what? *Environmental Pollution* 120: 497–507.

Rainbow, P. S. (2007). Trace metal bioaccumulation: Models, metabolic availability and toxicity. *Environment International* 33: 576–582.

Rainbow, P. S. & Luoma, S. N. (2011). Metal toxicity, uptake and bioaccumulation in aquatic invertebrates: Modelling zinc in crustaceans. *Aquatic Toxicology* 105: 455–465.

Rainer, S. F., Ivanovici, A. M. & Wadley, V. A. (1979). Effect of reduced salinity on adenylate energy charge in three estuarine molluscs. *Marine Biology* 54: 91–99.

Rajasekariah, G. R. (1978). *Chaetogaster limnaei* K. von Baer 1872 on *Lymnaea tomentosa*: Ingestion of *Fasciola hepatica* cercariae. *Experientia* 34: 1458–1459.

Ralph, M. R. & Block, G. D. (1990). Circadian and light-induced conductance changes in putative pacemaker cells of *Bulla gouldiana*. *Journal of Comparative Physiology A* 166: 589–595.

Ram, J. L. (1977). Hormonal control of reproduction in *Busycon*: Laying of egg capsules caused by nervous system extracts. *Biological Bulletin* 152: 221–232.

Ramasamy, M. S. & Murugan, A. (2007). Fouling deterrent chemical defence in three muricid gastropod egg masses from the Southeast coast of India. *Biofouling* 23: 259–265.

Ramirez, M. D., Speiser, D. I., Pankey, M. S. & Oakley, T. H. (2011). Understanding the dermal light sense in the context of integrative photoreceptor cell biology. *Visual Neuroscience* 28: 265–279.

Rammelmeyer, H. (1925). Zur morphologie der *Puncturella noachina*. *Zoologischer Anzeiger* 64: 105–114.

Ramnanan, C. J., Groom, A. G. & Storey, K. B. (2007). Akt and its downstream targets play key roles in mediating dormancy in land snails. *Comparative Biochemistry and Physiology Part B* 148: 245–255.

Ramnanan, C. J., Allan, M. E., Groom, A. G. & Storey, K. B. (2009). Regulation of global protein translation and protein degradation in aerobic dormancy. *Molecular and Cellular Biochemistry* 323: 9–20.

Ramnanan, C. J., Bell, R. A. & Hughes, J.-D. M. (2017). Key molecular regulators of metabolic rate depression in the estivating snail *Otala lactea*. pp. 275–302 *in* S. Saleuddin & Mukai, S. (Eds.) *Physiology of Molluscs*. Vol. 1. Oakville, Ontario, Canada, Apple Academic Press.

Ramon, M. (1991). Spawning and development characteristics of *Cymatium cutaceum* and *C. corrugatum* (Gastropoda, Prosobranchia) in the laboratory. *Ophelia* 33: 31–43.

Rampersad, J. N., Agard, J. B. R. & Ammons, D. (1994). Effects of gamete concentration on the in vitro fertilization of manually extracted gametes of the oyster (*Crassostrea rhizophorae*). *Aquaculture* 123: 153–162.

Ramsell, J. & Paul, N. D. (1990). Preferential grazing by molluscs of plants infected by rust fungi. *Oikos* 58: 145–150.

Randall, J. E. (1967). Food habits of reef fishes of the West Indies. *Studies in Tropical Oceanography* 5: 655–847.

Randles, W. B. (1905). Some observations on the anatomy and affinities of the Trochidae. *Quarterly Journal of Microscopical Science* 48: 33–78.

Ranzi, S. (1931a). Sviluppo di parti isolate di embrioni di Cefalopodi. *Pubblicazioni della Stazione Zoologica di Napoli* 11: 104–146.

Ranzi, S. (1931b). Risultati di ricerche di embriologia sperimentate sui Cefalopodi. *Archivio Zoologico Italiano* 16: 403–408.

Rao, B. M. (1976). Studies on the growth of the limpet *Cellana radiata* (Born) (Gastropoda: Prosobranchia). *Journal of Molluscan Studies* 42: 136–144.

Rao, T. & Ramakrishna (1993). Population growth rate of the parasitic rotifer *Proales gigantea*, and susceptibility to parasitization in the snail *Lymnaea acuminata* at different stages of embryonic development. *Hydrobiologia* 254: 1–6.

Rasser, M. W. & Covich, A. P. (2014). Predation on freshwater snails in Miocene Lake Steinheim: A trigger for intralacustrine evolution? *Lethaia* 47: 524–532.

Rath, E. (1988). Organization and systematic position of the Valvatidae. pp. 194–204 *in* W. F. Ponder, Eernisse, D. J. & Waterhouse, J. H. (Eds.) *Prosobranch Phylogeny. Malacological Review Supplement*. Ann Arbor, MI, Malacological Review.

Ratner, S. C. & Jennings, J. W. (1968). Magnetic fields and orienting movements in mollusks. *Journal of Comparative and Physiological Psychology* 65: 365.

Ratner, S. C. (1976). Kinetic movements in magnetic fields of chitons with ferro magnetic structures. *Behavioral Biology* 17: 573–578.

Ratté, S. & Chase, R. (2000). Synapse distribution of olfactory interneurons in the procerebrum of the snail *Helix aspersa*. *Journal of Comparative Neurology* 417: 366–384.

Rauch, C., Vries, J. D., Rommel, S., Rose, L. E., Woehle, C., Christa, G., Laetz, E. M., Wägele, H., Tielens, A. G. M. & Nickelsen, J. (2015). Why it is time to look beyond algal genes in photosynthetic slugs. *Genome Biology and Evolution* 7: 2602–2607.

Raup, D. M. (1962). Computer as aid in describing form in gastropod shells. *Science* 138: 150–152.

Raup, D. M. & Michelson, A. (1965). Theoretical morphology of the coiled shell. *Science* 147: 1294–1295.

Raup, D. M. (1966). Geometric analysis of shell coiling: General problems. *Journal of Paleontology* 40: 1178–1190.

Raup, D. M. (1967). Geometric analysis of shell coiling: Coiling in ammonoids. *Journal of Paleontology* 41: 43–65.

Raut, S. & Barker, G. (2002). *Achatina fulica* Bowdich and other Achatinidae as pests in tropical agriculture. pp. 55–114 *in* G. M. Barker (Ed.) *Molluscs as Crop Pests*. Wallingford, UK, CABI Publishing.

Raven, C. P. (1952). Morphogenesis in *Lymnaea stagnalis* and its disturbance by lithium. *Journal of Experimental Zoology* 121: 1–78.

Raven, C. P. (1958). *Morphogenesis: The Analysis of Molluscan Development*. Oxford, UK, Pergamon Press.

Raven, C. P. (1961). *Oogenesis: The Storage of Developmental Information*. Vol. 10. Oxford, UK, Pergamon Press.

Raven, C. P. (1964). Development. pp. 165–195 *in* K. M. Wilbur & Yonge, C. M. (Eds.) *Physiology of Mollusca*. Vol. 1. New York, Academic Press.

Raven, C. P. (1966). *Morphogenesis: The Analysis of Molluscan Development*. 2nd ed. Oxford, UK, Pergamon Press.

Raven, C. P. (1972). Chemical embryology of Mollusca. pp. 155–185 *in* M. Florkin & Scheer, B. T. (Eds.) *Chemical Zoology. Mollusca*. Vol. 7. New York, Academic Press.

Raw, J. L., Perissinotto, R., Miranda, N. A. F. & Peer, N. (2016). Diet of *Melanoides tuberculata* (Müller, 1774) from subtropical coastal lakes: Evidence from stable isotope ($\delta^{13}C$ and $\delta^{15}N$) analyses. *Limnologica-Ecology and Management of Inland Waters* 59: 116–123.

Rawlings, T. A., Collins, T. M. & Bieler, R. (2001). A major mitochondrial gene rearrangement among closely related species. *Molecular Biology and Evolution* 18: 1604–1609.

Rawlings, T. A., Hayes, K. A., Cowie, R. H. & Collins, T. M. (2007). The identity, distribution, and impacts of non-native apple snails in the continental United States. *BMC Evolutionary Biology* 97: 1–14.

Ray, G. L. (2005). *Invasive Marine and Estuarine Animals of the Gulf of Mexico*. ANSRP Technical Notes Collection (ERDC/TN ANSRP-05-4), U.S. Army Engineer Research and Development Center, Vicksburg, MS. http://el.erdc.usace.army.m il/ansrp/.

Read, K. R. H. (1965). The characterization of the hemoglobins of the bivalve mollusc *Phacoides pectinatus* (Gmelin). *Comparative Biochemistry and Physiology* 15: 137–158.

Read, K. R. H. (1966). Molluscan hemoglobin and myoglobin. pp. 209–232 *in* K. M. Wilbur & Yonge, C. M. (Eds.) *Physiology of Mollusca*. Vol. 2. New York, Academic Press.

Reade, P. & Reade, E. (1972). Phagocytosis in invertebrates. II. The clearance of carbon particles by the clam, *Tridacna maxima*. *Journal of the Reticuloendothelial Society* 12: 349–360.

Redfearn, P. (1974). Biology and distribution of toheroa, *Paphies (Mesodesma) ventricosa* (Gray). *Fisheries Research Bulletin* 11: 1–49.

Rees, B. B. & Hand, S. C. (1993). Biochemical correlates of estivation tolerance in the mountain snail *Oreohelix* (Pulmonata: Oreohelicidae). *Biological Bulletin* 184: 230–242.

Rees, W. J. (1967). A brief survey of the symbiotic associations of Cnidaria with Mollusca. *Proceedings of the Malacological Society of London* 37: 213–232.

Régnier, C., Fontaine, B. & Bouchet, P. (2009). Not knowing, not recording, not listing: Numerous unnoticed mollusk extinctions. *Conservation Biology* 23: 1214–1221.

Reichelt, R. E. (1982). Space: A non-limiting resource in the niches of some abundant coral reef gastropods. *Coral Reefs* 1: 3–11.

Reid, D. G. (1985). Habitat and zonation patterns of *Littoraria* species (Gastropoda: Littorinidae) in Indo-Pacific mangrove forests. *Biological Journal of the Linnean Society* 26: 39–68.

Reid, D. G. (1986a). *The Littorinid Molluscs of Mangrove Forests in the Indo-Pacific Region: The Genus Littoraria (Mollusca, Gastropoda, Littorinidae)*. London, UK, Butler & Tanner Ltd.

Reid, D. G. (1986b). *Mainwaringia* Nevill, 1885, a littorinid genus from Asiatic mangrove forests, and a case of protandrous hermaphroditism. *Journal of Molluscan Studies* 52: 225–242.

Reid, D. G. (1987). Natural selection for apostasy and crypsis acting on the shell colour polymorphism of a mangrove snail, *Littoraria filosa* (Sowerby) (Gastropoda: Littorinidae). *Biological Journal of the Linnean Society* 30: 1–24.

Reid, D. G. & Mak, Y. M. (1999). Indirect evidence for ecophenotypic plasticity in radular dentition of *Littoraria* species (Gastropoda: Littorinidae). *Journal of Molluscan Studies* 65: 355–370.

Reid, D. G., Dyal, P. & Williams, S. T. (2010). Global diversification of mangrove fauna: A molecular phylogeny of *Littoraria* (Gastropoda: Littorinidae). *Molecular Phylogenetics and Evolution* 55: 185–201.

Reid, L. & Clamp, J. R. (1978). The biochemical and histochemical nomenclature of mucus. *British Medical Bulletin* 34: 5–8.

Reid, R. G. B. (1965). The structure and function of the stomach in bivalve molluscs. *Journal of Zoology* 147: 156–184.

Reid, R. G. B. & Reid, A. M. (1974). The carnivorous habit of members of the septibranch genus *Cuspidaria* (Mollusca, Bivalvia). *Sarsia* 56: 47–56.

Reid, R. G. B. & Rauchert, K. (1976). Catheptic endopeptidases and protein digestion in the horse clam *Tresus capax* (Gould). *Comparative Biochemistry and Physiology Part B* 54: 467–472.

Reid, R. G. B. (1977). Gastric protein digestion in the carnivorous septibranch *Cardiomya planetica* Dall; with comparative notes on deposit and suspension-feeding bivalves. *Comparative Biochemistry and Physiology Part A* 56A: 573–575.

Reid, R. G. B. & Crosby, S. P. (1980). The raptorial siphonal apparatus of the carnivorous septibranch *Cardiomya planetica* Dall (Mollusca: Bivalvia), with notes on feeding and digestion. *Canadian Journal of Zoology* 58: 670–679.

Reid, R. G. B. & Friesen, J. A. (1980). The digestive system of the moon snail *Polinices lewisii* (Gould, 1847) with emphasis on the role of the oesophageal gland. *The Veliger* 23: 25–34.

Reid, R. G. B. & Porteous, S. (1980). Aspects of the functional morphology and digestive physiology of *Vulsella vulsella* (Linné) and *Crenatula modiolaris* (Lamarck), bivalves associated with sponges. pp. 291–310 *in* B. Morton (Ed.) *Proceedings of the First International Conference on the Malacofauna of Hong Kong and Southern China 23 March–8 April 1977*. Hong Kong, Hong Kong University Press.

Reid, R. G. B. & Sweeney, B. (1980). The digestibility of the bivalve crystalline style. *Comparative Biochemistry and Physiology Part B* 65: 451–453.

Reid, R. G. B. (1982). Aspects of bivalve feeding and digestion relevant to aquaculture nutrition. pp. 231–251 *in* G. D. Pruder, Langdon, C. J. & Conklin, D. E. (Eds.) *Proceedings of the Second International Conference on Aquaculture Nutrition: Biochemical and Physiological Approaches to Shellfish Nutrition*. Special Publication No. 2. Baton Rouge, LA, Louisiana State University Division of Continuing Education.

Reid, R. G. B., Fankboner, P. V. & Brand, D. G. (1984a). Studies on the physiology of the giant clam *Tridacna gigas* Linné I. Feeding and digestion. *Comparative Biochemistry and Physiology Part A* 78: 95–102.

Reid, R. G. B., Fankboner, P. V. & Brand, D. G. (1984b). Studies on the physiology of the giant clam *Tridacna gigas* Linné II. Kidney function. *Comparative Biochemistry and Physiology Part A* 78: 103–108.

Reid, R. G. B., McMahon, R. F., Ó Foighil, D. & Finnigan, R. (1992). Anterior inhalant currents and pedal feeding in bivalves. *The Veliger* 35: 93–104.

Reid, R. G. B. (1998). Subclass Protobranchia. pp. 235–247 *in* P. L. Beesley, Ross, G. J. B. & Wells, A. (Eds.) *Mollusca: The Southern Synthesis. Part A. Fauna of Australia*. Melbourne, Australia, CSIRO Publishing.

Reidenbach, M. A., Koseff, J. R. & Koehl, M. A. R. (2009). Hydrodynamic forces on larvae affect their settlement on coral reefs in turbulent, wave-driven flow. *Limnology and Oceanography* 54: 318–330.

Reimchen, T. E. (1979). Substratum heterogeneity, crypsis, and colour polymorphism in an intertidal snail (*Littorina mariae*). *Canadian Journal of Zoology* 57: 1070–1085.

Reimchen, T. E. (1989). Shell color ontogeny and tubeworm mimicry in a marine gastropod *Littorina mariae*. *Biological Journal of the Linnean Society* 36: 97–110.

Reindl, S. & Haszprunar, G. (1994). Light and electron microscopical investigations on shell pores (caeca) of fissurellid limpets (Mollusca: Archaeogastropoda). *Journal of Zoology* 233: 385–404.

Reindl, S. & Haszprunar, G. (1996a). Fine structure of caeca and mantle of arcoid and limopsoid bivalves (Mollusca: Pteriomorpha). *The Veliger* 39: 101–116.

Reindl, S. & Haszprunar, G. (1996b). Shell pores (caeca, aesthetes) of Mollusca: A case of polyphyly. pp. 115–118 *in* J. D. Taylor (Ed.) *Origin and Evolutionary Radiation of the Mollusca*. Oxford, UK, Oxford University Press.

Reinecke, A. & Harrington, M. J. (2017). The role of metal ions in the mussel byssus. pp. 113–152 *in* S. Saleuddin & Mukai, S. (Eds.) *Physiology of Molluscs*. Vol. 1. Oakville, Ontario, Canada, Apple Academic Press.

Reise, H. & Hutchinson, J. M. (2002). Penis-biting slugs: Wild claims and confusions. *Trends in Ecology & Evolution* 17: 163.

Reise, H. (2007). A review of mating behavior in slugs of the genus *Deroceras* (Pulmonata: Agriolimacidae). *American Malacological Bulletin* 23: 137–156.

Reise, K. (1985). *Tidal Flat Ecology: An Experimental Approach to Species Interactions*. Berlin, Springer-Verlag.

Remigio, E. A. & Duda, T. F. (2008). Evolution of ecological specialization and venom of a predatory marine gastropod. *Molecular Ecology* 17: 1156–1162.

Renault, T. & Novoa, B. (2004). Viruses infecting bivalve molluscs. *Aquatic Living Resources* 17: 397–409.

Renault, T., Moreau, P., Faury, N., Pépin, J.-F., Segarra, A. & Webb, S. (2012). Analysis of clinical ostreid herpesvirus 1 (Malacoherpesviridae) specimens by sequencing amplified fragments from three virus genome areas. *Journal of Virology* 86: 5942–5947.

Renzoni, A. (1968). Osservazioni istologice, istochimiche ed ultrastrutturali sui tentacoli di *Vaginulus borellianus* (Colosi), Gastropoda, Soleolifera. *Zeitschrift für Zellforschung und Mikroskopische Anatomie* 87: 350–376.

Reuner, A., Brümmer, F. & Schill, R. O. (2008). Heat shock proteins (Hsp70) and water content in the estivating Mediterranean Grunt Snail (*Cantareus apertus*). *Comparative Biochemistry and Physiology Part B* 151: 28–31.

Rex, M. A. (1973). Deep-sea species diversity: Decreased gastropod diversity at abyssal depths. *Science* 181: 1051–1053.

Rex, M. A. & Boss, K. J. (1976). Open coiling in Recent gastropods. *Malacologia* 15: 289–297.

Rex, M. A., Crame, J. A., Stuart, C. T. & Clarke, A. (2005). Large-scale biogeographic patterns in marine mollusks: A confluence of history and productivity? *Ecology* 86: 2288–2297.

Rex, M. A. & Etter, R. J. (2010). *Deep-sea Biodiversity: Pattern and Scale*. Cambridge, MA. London, Harvard University Press.

Reyes, F., Fernández, R., Rodríguez, A., Bueno, S., de Eguilior, C., Francesch, A. & Cuevas, C. (2008). Cytotoxic staurosporines from the marine ascidian *Cystodytes solitus*. *Journal of Natural Products* 71: 1046–1048.

Reynolds, K. C., Watanabe, H., Strong, E. E., Sasaki, T., Uematsu, K., Miyake, H., Kojima, S., Suzuki, Y., Fujikura, K., Kim, S. K. & Young, C. M. (2010). New molluscan larval form: Brooding and development in a hydrothermal vent gastropod, *Ifremeria nautilei* (Provannidae). *Biological Bulletin* 219: 7–11.

Reynolds, P. D. (1990a). Fine structure of the kidney and characterization of secretory products in *Dentalium rectius* (Mollusca, Scaphopoda). *Zoomorphology* 110: 53–62.

Reynolds, P. D. (1990b). Functional morphology of the perianal sinus and pericardium of *Dentalium rectius* (Mollusca: Scaphopoda) with a reinterpretation of the scaphopod heart. *American Malacological Bulletin* 7: 137–146.

Reynolds, P. D. (1992). Distribution and ultrastructure of ciliated sensory receptors in the posterior mantle epithelium of *Dentalium rectius* (Mollusca, Scaphopoda). *Acta Zoologica* 73: 263–270.

Reynolds, P. D., Morse, M. P. & Norenburg, J. L. (1993). Ultrastructure of the heart and pericardium of an aplacophoran mollusc (Neonemiomorpha): Evidence for ultrafiltration of blood. *Proceedings of the Royal Society B* 254: 147–152.

Reynolds, P. D. (2002). The Scaphopoda. *Advances in Marine Biology* 42: 137–236.

Reynolds, P. D. (2006). Scaphopoda: The tusk shells. pp. 229–237 *in* C. F. Sturm, Pearce, T. A. & Valdes, A. (Eds.) *The Mollusks: A Guide to their Study, Collection, and Preservation*. Boca Raton, FL, Universal Publishers.

Rhoads, D. C. & Lutz, R. A. (1980). *Skeletal Growth of Aquatic Organisms: Biological Records of Environmental Change*. New York, Plenum Press.

Rhode, B. (1992). Development and differentiation of the eye in *Platynereis dumerilii* (Annelida, Polychaeta). *Journal of Morphology* 212: 71–85.

Riascos, J. M. & Guzman, P. A. (2010). The ecological significance of growth rate, sexual dimorphism and size at maturity of *Littoraria zebra* and *L. variegata* (Gastropoda: Littorinidae). *Journal of Molluscan Studies* 76: 289–295.

Ricciardi, A. (1994). Occurrence of chironomid larvae (*Paratanytarsus* sp.) as commensals of dreissenid mussels (*Dreissena polymorpha* and *D. bugensis*). *Canadian Journal of Zoology* 72: 1159–1162.

Rice, S. H. (1985). An anti-predator chemical defense of the marine pulmonate gastropod *Trimusculus reticulatus* (Sowerby). *Journal of Experimental Marine Biology and Ecology* 93: 83–89.

Rice, S. H. (1998). The bio-geometry of mollusc shells. *Paleobiology* 24: 133–149.

Richards, C. S. & Sheffield, H. G. (1970). Unique host relations and ultrastructure of a new microsporidian of the genus *Coccospora* infecting *Biomphalaria glabrata*. pp. 439–452 *in* R. A. Sampson, Vlak, J. M. & Peters, D. (Eds.) *Proceedings of the IV International Colloquium of Insect Pathology (1970)*. College Park, MD, Society for Invertebrate Pathology.

Richardson, C. A., Crisp, D. J., Runham, N. W. & Gruffydd, L. D. (1980). The use of tidal growth bands in the shell of *Cerastoderma edule* to measure seasonal growth rates under cool temperate and sub-arctic conditions. *Journal of the Marine Biological Association of the United Kingdom* 60: 977–989.

Richardson, C. A. (1989). An analysis of the microgrowth bands in the shell of the common mussel *Mytilus edulis*. *Journal of the Marine Biological Association of the United Kingdom* 69: 477–491.

Richardson, C. A. (2001). Molluscs as archives of environmental changes. *Oceanography and Marine Biology Annual Review* 39: 103–164.

Richling, I. (2004). Classification of the Helicinidae: Review of morphological characteristics based on a revision of the Costa Rican species and application to the arrangement of the Central American mainland taxa (Mollusca: Gastropoda: Neritopsina). *Malacologia* 45: 195–440.

Richling, I., Malkowsky, Y., Kuhn, J., Niederhofer, H.-J. & Boeters, H. D. (2017). A vanishing hotspot: The impact of molecular insights on the diversity of Central European *Bythiospeum* Bourguignat, 1882 (Mollusca: Gastropoda: Truncatelloidea). *Organisms Diversity & Evolution* 17: 67–85.

Richter, G. & Thorson, G. (1975). Pelagische Prosobranchier-larven des Golfes von Neapel. *Ophelia* 13: 109–185.

Ricketts, E. F., Calvin, J., Hedgpeth, J. W. & Phillips, D. W. (1985). *Between Pacific Tides*. Stanford, CA, Stanford University Press.

Rico, C., Antonio Cuesta, J., Drake, P., Macpherson, E., Bernatchez, L. & Marie, A. D. (2017). Null alleles are ubiquitous at microsatellite loci in the Wedge Clam (*Donax trunculus*). *PeerJ* 5: e3188.

Riddle, W. A. (1983). Physiological ecology of land snails and slugs. pp. 431–461 *in* W. D. Russell-Hunter (Ed.) *Ecology. The Mollusca*. Vol. 6. New York, Academic Press.

Ridgway, I. D. & Richardson, C. A. (2011). *Arctica islandica*: The longest lived non colonial animal known to science. *Reviews in Fish Biology and Fisheries* 21: 297–310.

Riedl, R. (1960). Beiträge zur Kenntnis der *Rhodope veranii*. 2. Entwicklung. *Zeitschrift für wissenschaftliche Zoologie* 163: 237–316.

Riegel, J. A. (1972). *Comparative Physiology of Renal Excretion*. Edinburgh, Oliver and Boyd.

Rieger, R. M. & Rieger, G. E. (1976). Fine structure of the archiannelid cuticle and remarks on the evolution of the cuticle within the Spiralia. *Acta Zoologica* 57: 53–68.

Rieger, R. M. (1984). Evolution of the cuticle in the lower Eumetazoa. pp. 389–399 in Bereiter-Hahn, J., Matoltsy, A. G., Richards, K. S. (Eds.) *Biology of the Integument*, Invertebrates. Springer, Berlin, Germany.

Riera, P. & Richard, P. (1996). Isotopic determination of food sources of *Crassostrea gigas* along a trophic gradient in the Estuarine Bay of Marennes-Oléron. *Estuarine, Coastal and Shelf Science* 42: 347–360.

Riera, P. (2010). Trophic plasticity of the gastropod *Hydrobia ulvae* within an intertidal bay (Roscoff, France): A stable isotope evidence. *Journal of Sea Research* 63: 78–83.

Riese, K. (1930). Phylogenetische Betrachtungen über das Nervensystem von *Cypraea moneta* auf Grund seiner Morphologie und Histologie. *Jenaische Zeitschrift für Naturwissenschaft* 65: 361–486.

Riley, L. A., Dybdahl, M. F. & Hall, R. O. (2008). Invasive species impact: Asymmetric interactions between invasive and endemic freshwater snails. *Journal of the North American Benthological Society* 27: 509–520.

Rio, M., Roux, M., Renard, M. & Schein, E. (1992). Chemical and isotopic features of present day bivalve shells from hydrothermal vents or cold seeps. *Palaios* 7: 351–360.

Riquelme, C., Hayashida, G., Araya, R., Uchida, A., Satomi, M. & Ishida, Y. (1996). Isolation of a native bacterial strain from the scallop *Argopecten purpuratus* with inhibitory effects against pathogenic vibrios. *Journal of Shellfish Research* 15: 369–374.

Riquelme, C. E., Jorquera, M. A., Rosas, A. I., Avendano, R. E. & Reyes, N. (2001). Addition of inhibitor-producing bacteria to mass cultures of *Argopecten purpuratus* larvae (Lamarck, 1819). *Aquaculture* 192: 111–119.

Ritson-Williams, R., Shjegstad, S. & Paul, V. (2003). Host specificity of four corallivorous *Phestilla* nudibranchs (Gastropoda: Opisthobranchia). *Marine Ecology Progress Series* 255: 207–218.

Ritson-Williams, R., Yotsu-Yamashita, M. & Paul, V. J. (2006). Ecological functions of tetrodotoxin in a deadly polyclad flatworm. *Proceedings of the National Academy of Sciences of the United States of America* 103: 3176–3179.

Ritson-Williams, R., Shjegstad, S. M. & Paul, V. J. (2009). Larval metamorphosis of *Phestilla* spp. in response to waterborne cues from corals. *Journal of Experimental Marine Biology and Ecology* 375: 84–88.

Riutort, M., Álvarez-Presas, M., Lázaro, E., Solà, E. & Paps, J. (2012). Evolutionary history of the Tricladida and the Platyhelminthes: An up-to-date phylogenetic and systematic account. *International Journal of Developmental Biology* 56: 5–17.

Riva, C. & Binelli, A. (2014). Analysis of the *Dreissena polymorpha* gill proteome following exposure to dioxin-like PCBs: Mechanism of action and the role of gender. *Comparative Biochemistry and Physiology Part D* 9: 23–30.

Rivera, L., López-Garriga, J. & Cadilla, C. L. (2008). Characterization of the full length mRNA coding for *Lucina pectinata* HbIII revealed an alternative polyadenylation site. *Gene* 410: 122–128.

Rivest, B. R. (1983). Development and the influence of nurse egg allotment on hatching size in *Searlesia dira* (Reeve, 1846) (Prosobranchia: Buccinidae). *Journal of Experimental Marine Biology and Ecology* 69: 217–241.

Robb, M. F. (1975). The diet of the chiton *Cyanoplax hartwegii* in three intertidal habitats. *The Veliger* 18: 34–37.

Robbins, I., Lenoir, F. & Mathieu, M. (1990). A putative neuroendocrine factor that stimulates glycogen mobilization in isolated glycogen cells from the marine mussel *Mytilus edulis*. *General and Comparative Endocrinology* 79: 123–129.

Roberts, A. B. & Sporn, M. B. (1996). Transforming growth factor-β. pp. 275–308 in R. A. F. Clark (Ed.) *The Molecular and Cellular Biology of Wound Repair*. New York, Plenum Press.

Roberts, M. H. & Block, G. D. (1982). Dissection of circadian organization of *Aplysia* through connective lesions and electrophysiological recording. *Journal of Experimental Zoology Part A* 219: 39–50.

Roberts, M. H., Block, G. D. & Lusska, A. E. (1987). Comparative studies of circadian pacemaker coupling in opisthobranch molluscs. *Brain Research* 423: 286–292.

Roberts, R. D. & Lapworth, C. (2001). Effect of delayed metamorphosis on larval competence, and post-larval survival and growth, in the abalone *Haliotis iris* Gmelin. *Journal of Experimental Marine Biology and Ecology* 258: 1–13.

Robertson, R. (1961). The feeding of *Strombus* and related herbivorous marine gastropods. *Notulae Naturae* 343: 1–9.

Robertson, R. (1966). Coelenterate-associated prosobranch gastropods. *American Malacological Union Annual Report* 1965: 6–8.

Robertson, R. (1967). *Heliacus* (Gastropoda, Architectonicidae) symbiotic with *Zoantharia* (Coelenterata). *Science* 156: 246–248.

Robertson, R. (1970a). Review of the predators and parasites of stony corals, with special reference to symbiotic prosobranch gastropods. *Pacific Science* 24: 43–54.

Robertson, R. (1970b). Systematics of Indo-Pacific *Philippia* (*Psilaxis*), architectonicid gastropods with eggs and young in the umbilicus. *Pacific Science* 24: 66–83.

Robertson, R., Scheltema, R. S. & Adams, F. W. (1970). The feeding, larval dispersal, and metamorphosis of *Philippia* (Gastropoda: Architectonicidae). *Pacific Science* 24: 55–65.

Robertson, R. (1980). Gastropods symbiotic with zoantharian sea-anemones (abstract). *Bulletin of the American Malacological Union* 46: 69.

Robertson, R. (1983). Extraordinarily rapid postlarval growth of a tropical wentletrap (*Epitonium albidum*). *The Nautilus* 97: 60–66.

Robertson, R. (1985a). Four characters and the higher category systematics of gastropods. *American Malacological Bulletin* 1: 1–22.

Robertson, R. (1985b). Archaeogastropod biology and the systematics of the genus *Tricolia* (Trochacea: Tricoliidae) in the Indo-West-Pacific. *Monographs of Marine Mollusca* 3: 1–103.

Robertson, R. (1989). Spermatophores of aquatic non-stylommatophoran gastropods: A review with new data on *Heliacus* (Architectonicidae). *Malacologia* 30: 341–364.

Robertson, R. (1993a). Snail handedness. *New York Shell Club Notes* 329: 12–18.

Robertson, R. (1993b). Snail handedness: The coiling directions of gastropods. *National Geographic Research and Exploration* 9: 104–119.

Robertson, R. (2007). Taxonomic occurrences of gastropod spermatozeugmata and non-stylommatophoran spermatophores updated. *American Malacological Bulletin* 23: 11–16.

Robilliard, G. A. (1970). The systematics and some aspects of the ecology of the genus *Dendronotus* (Gastropoda: Nudibranchia) *The Veliger* 12: 433–479.

Robinson, D. G. (1999). Alien invasions: The effects of the global economy on non-marine gastropod introductions into the United States. *Malacologia* 41: 413–438.

Robinson, S. D., Safavi-Hemami, H., McIntosh, L. D., Purcell, A. W., Norton, R. S. & Papenfuss, A. T. (2014). Diversity of conotoxin gene superfamilies in the venomous snail, *Conus victoriae*. *PLoS ONE* 9: e87648.

Robinson, W. E. & Langton, R. W. (1980). Digestion in a subtidal population of *Mercenaria mercenaria* (Bivalvia). *Marine Biology* 58: 173–179.

Robison, B., Seibel, B. A. & Drazen, J. (2014). Deep-sea octopus (*Graneledone boreopacifica*) conducts the longest-known egg-brooding period of any animal. *PLoS ONE* 9: e103437.

Robison, B. H., Reisenbichler, K. R., Hunt, J. C. & Haddock, S. H. D. (2003). Light production by the arm tips of the deep-sea cephalopod *Vampyroteuthis infernalis*. *Biological Bulletin* 205: 102–109.

Robledo, J. A. F., Caceres-Martinez, J., Sluys, R. & Figueras, A. (1994). The parasitic turbellarian *Urastoma cyprinae* (Platyhelminthes: Urastomidae) from blue mussel *Mytilus galloprovincialis* in Spain: Occurrence and pathology. *Diseases of Aquatic Organisms* 18: 203–210.

Robles, C. (1987). Predator foraging characteristics and prey population structure on a sheltered shore. *Ecology* 68: 1502–1514.

Robles, C. D., Desharnais, R. A., Garza, C., Donahue, M. J. & Martinez, C. A. (2009). Complex equilibria in the maintenance of boundaries: Experiments with mussel beds. *Ecology* 90: 985–995.

Roch, G. J., Busby, E. R. & Sherwood, N. M. (2011). Evolution of GnRH: Diving deeper. *General and Comparative Endocrinology* 171: 1–16.

Rochette, R., Morissette, S. & Himmelman, J. H. (1995). A flexible response to a major predator provides the whelk *Buccinum undatum* L. with nutritional gains. *Journal of Experimental Marine Biology and Ecology* 185: 167–180.

Rochette, R., Arsenault, D. J., Justome, B. & Himmelman, J. H. (1998). Chemically-mediated predator-recognition learning in a marine gastropod. *Écoscience* 5: 353–360.

Rochette, R., Doyle, S. P. & Edgell, T. C. (2007). Interaction between an invasive decapod and a native gastropod: Predator foraging tactics and prey architectural defenses. *Marine Ecology Progress Series* 330: 179–188.

Rochman, C. M., Tahir, A., Williams, S. L., Baxa, D. V., Lam, R., Miller, J. T., Teh, F.-C., Werorilangi, S. & Teh, S. J. (2015). Anthropogenic debris in seafood: Plastic debris and fibers from textiles in fish and bivalves sold for human consumption. *Scientific Reports* 5: article 14340.

Rodet, F., Lelong, C., Dubos, M. P. & Favrel, P. (2008). Alternative splicing of a single precursor mRNA generates two subtypes of Gonadotropin-Releasing Hormone receptor orthologues and their variants in the bivalve mollusc *Crassostrea gigas*. *Gene* 414: 1–9.

Rodgers, J. K., Sandland, G. J., Joyce, S. R. & Minchella, D. J. (2005). Multi-species interactions among a commensal (*Chaetogaster limnaei limnaei*), a parasite (*Schistosoma mansoni*), and an aquatic snail host (*Biomphalaria glabrata*). *Journal of Parasitology* 91: 709–712.

Rodrigues, C. F., Webster, G., Cunha, M. R. & Weightman, A. J. (2010). Chemosynthetic bacteria found in bivalve species from mud volcanoes of the Gulf of Cadiz. *FEMS Microbiology Ecology* 73: 486–499.

Rodrigues, C. F., Cunha, M. R., Génio, L. & Duperron, S. (2013). A complex picture of associations between two host mussels and symbiotic bacteria in the Northeast Atlantic. *Die Naturwissenschaften* 100: 21–31.

Rodrigues, L. R. G. & Absalão, R. S. (2005). Shell colour polymorphism in the chiton *Ischnochiton striolatus* (Gray, 1828) (Mollusca: Polyplacophora) and habitat heterogeneity. *Biological Journal of the Linnean Society* 85: 543–548.

Rodrigues, M., Guerra, Á. & Troncoso, J. S. (2012). Reproduction of the Atlantic bobtail squid *Sepiola atlantica* (Cephalopoda: Sepiolidae) in northwest Spain. *Invertebrate Biology* 131: 30–39.

Rodriguez-Ortega, M. J., Grosvik, B. E., Rodriguez-Ariza, A., Goksoyr, A. & Lopez-Barea, J. (2003). Changes in protein expression profiles in bivalve molluscs (*Chamaelea gallina*) exposed to four model environmental pollutants. *Proteomics* 3: 1535–1543.

Rodríguez, J., Riguerra, R. & Debitus, C. (1992). The natural polypropionate-derived esters of the mollusc *Onchidium* sp. *Journal of Organic Chemistry* 57: 4624–4632.

Roffe, T. (1975). Spectral perception in *Octopus*: A behavioural study. *Vision Research* 15: 353–356.

Rogers, D. C. (1971). Surface specializations of the epithelial cells at the tip of the optic tentacle, dorsal surface of the head and ventral surface of the foot in *Helix aspersa*. *Zeitschrift für Zellforschung und Mikroskopische Anatomie* 114: 106–116.

Rogers, S. D. & Paul, V. J. (1991). Chemical defenses of three *Glossodoris* nudibranchs and their dietary *Hyrtios* sponges. *Marine Ecology Progress Series* 77: 221–232.

Rohde, K. (2005). *Marine Parasitology*. Melbourne, Australia, CSIRO Publishing.

Rohde, K. (2006). *Nonequilibrium Ecology*. Cambridge, UK, Cambridge University Press.

Rohfritsch, A., Bierne, N., Boudry, P., Heurtebise, S., Cornette, F. & Lapegue, S. (2013). Population genomics shed light on the demographic and adaptive histories of European invasion in the Pacific oyster, *Crassostrea gigas*. *Evolutionary Applications* 6: 1064–1078.

Rokop, F. J. (1974). Reproductive patterns in deep-sea benthos. *Science* 186: 743–745.

Rokop, F. J. (1977). Seasonal reproduction of the brachiopod *Frieleia halli* and the scaphopod *Cadulus californicus* at bathyal depths in the deep sea. *Marine Biology* 43: 237–246.

Rolán-Alvarez, E., Austin, C. & Boulding, E. G. (2015). The contribution of the genus *Littorina* to the field of evolutionary ecology. *Oceanography and Marine Biology Annual Review* 53: 157–214.

Rolán, E., Guerra-Varela, J., Colson, I., Hughes, R. N. & Rolán-Alvarez, E. (2004). Morphological and genetic analysis of two sympatric morphs of the dogwhelk *Nucella lapillus* (Gastropoda: Muricidae) from Galicia (northwestern Spain). *Journal of Molluscan Studies* 70: 179–185.

Rollo, C. D. & Wellington, W. G. (1979). Intra- and inter-specific agonistic behavior among terrestrial slugs (Pulmonata: Stylommatophora). *Canadian Journal of Zoology* 57: 846–855.

Romanenko, L. A., Uchino, M., Kalinovskaya, N. I. & Mikhailov, V. V. (2008). Isolation, phylogenetic analysis and screening of marine mollusc-associated bacteria for antimicrobial, hemolytic and surface activities. *Microbiological Research* 163: 633–644.

Romero, A., Dios, S., Poisa-Beiro, L., Costa, M. M., Posada, D., Figueras, A. & Novoa, B. (2011). Individual sequence variability and functional activities of fibrinogen-related proteins (FREPs) in the Mediterranean mussel (*Mytilus galloprovincialis*) suggest ancient and complex immune recognition models in invertebrates. *Developmental and Comparative Immunology* 35: 334–344.

Romero, J., Garcia-Varela, M., Laclette, J. P. & Espejo, R. T. (2002). Bacterial 16S rRNA gene analysis revealed that bacteria related to *Arcobacter* spp. constitute an abundant and common component of the oyster microbiota (*Tiostrea chilensis*). *Microbial Ecology* 44: 365–371.

Roots, E. B. I. (2008). The phylogeny of invertebrates and the evolution of myelin. *Neuron Glia Biology* 4: 101–109.

Roper, C. F. E. & Hochberg, F. G. (1988). Behavior and systematics of cephalopods from Lizard Island, Australia, based on color and body patterns *Malacologia* 29: 153–194.

Roper, C. F. E. & Vecchione, M. (1997). *In situ* observations test hypotheses of functional morphology of *Mastigoteuthis* (Cephalopoda, Oegopsida). *Vie et Milieu* 47: 87–93.

Ros, J. (1977). La defensa en los opistobranquios. *Investigación y Ciencia* 12: 48–60.

Rosa, M., Ward, J. E., Holohan, B. A., Shumway, S. E. & Wikfors, G. H. (2017). Physicochemical surface properties of microalgae and their combined effects on particle selection by suspension-feeding bivalve molluscs. *Journal of Experimental Marine Biology and Ecology* 486: 59–68.

Rosa, M., Ward, J. E. & Shumway, S. E. (2018). Selective capture and digestion of particles by suspension-feeding bivalve molluscs: A review. *Journal of Shellfish Research* 37: 1–20.

de Rosa, R., Pereira, J. & Nunes, M. L. (2005). Biochemical composition of cephalopods with different life strategies, with special reference to a giant squid, *Architeuthis* sp. *Marine Biology* 146: 739–751.

Roseghini, M., Severini, C., Erspamer, G. F. & Erspamer, V. (1996). Choline esters and biogenic amines in the hypobranchial gland of 55 molluscan species of the neogastropod Muricoidea superfamily. *Toxicon* 34: 33–55.

Rosen, M. D., Stasek, C. R. & Hermans, C. O. (1978). The ultrastructure and evolutionary significance of the cerebral ocelli of *Mytilus edulis*, the Bay Mussel. *The Veliger* 21: 10–18.

Rosen, M. D., Stasek, C. R. & Hermans, C. O. (1979). The ultrastructure and evolutionary significance of the ocelli in the larva of *Katharina tunicata* (Mollusca, Polyplacophora). *The Veliger* 22: 173–178.

Rosen, S. C., Weiss, K. R., Cohen, J. L. & Kupfermann, I. (1982). Interganglionic cerebral-buccal mechanoafferents of *Aplysia*: Receptive fields and synaptic connections to different classes of neurons involved in feeding behavior. *Journal of Neurophysiology* 48: 271–288.

Rosenberg, G. (1996). Independent evolution of terrestriality in Atlantic truncatellid gastropods. *Evolution* 50: 682–693.

Rosenfield, A. (1976). Recent environmental studies of neoplasms in marine shellfish. *Progress in Experimental Tumor Research* 20: 263–274.

Rosenthal, R. J. (1971). Tropic interaction between the sea star *Pisaster giganteus* and the gastropod *Kelletia kelletii*. *Fishery Bulletin of the Fish & Wildlife Service of the United States* 69: 669–679.

Rosewater, J. (1965). The family Tridacnae in the Indo-Pacific. *Indo-Pacific Mollusca* 1: 3–17.

Rossiter, A. & Kawanabe, H. (2000). *Ancient Lakes: Biodiversity, Ecology and Evolution*. Vol. 31. San Diego, CA, Academic Press.

Roth, B. & Bogan, A. E. (1984). Shell color and banding parameters of the *Liguus fasciatus* phenotype (Mollusca: Pulmonata). *Bulletin of the American Malacological Union Incorporated* 3: 1–10.

Rothman, B. S., Weir, G. & Dudek, F. E. (1983). Egg-laying hormone: Direct action on the ovotestis of *Aplysia*. *General and Comparative Endocrinology* 52: 134–141.

Rothschild, L. & Swann, M. M. (1951). The fertilization reaction in the sea-urchin: The probability of a successful sperm-egg collision. *Journal of Experimental Biology* 28: 403–416.

Roubal, F. R., Masel, J. & Lester, R. J. G. (1989). Studies on *Marteilia sydneyi*, agent of QX disease in the Sydney rock oyster, *Saccostrea commercialis*, with implications for its life cycle. *Marine and Freshwater Research* 40: 155–167.

Rouse, G. W. & Jamieson, B. G. M. (1987). An ultrastructural study of the spermatozoa of the polychaetes *Eurythoe complanata* (Amphinomidae), *Clymenella* sp. and *Micromaldane* sp. (Maldanidae), with definition of sperm types in relation to reproductive biology. *Journal of Submicroscopic Cytology* 19: 573–584.

Rouse, G. W. (2000). The epitome of hand waving? Larval feeding and hypotheses of metazoan phylogeny. *Evolution & Development* 2: 222–233.

Rouse, G. W. (2016). Phylum Annelida. The segmented (and some unsegmented) worms. pp. 531–602 *in* R. C. Brusca, Moore, W. & Shuster, S. (Eds.) *Invertebrates*. Sunderland, MA, Sinauer Associates, Inc.

Roussel, S., Huchette, S., Clavier, J. & Chauvaud, L. (2011). Growth of the European abalone (*Haliotis tuberculata* L.) *in situ*: Seasonality and ageing using stable oxygen isotopes. *Journal of Sea Research* 65: 213–218.

Rouzaud, H. P. J. B. (1885). Recherches sur le développement des organes génitaux de quelques gastéropodes hermaphrodites. PhD, Laboratoire zoologique de la Faculté des sciences de Montpellier et de la Station maritime de Cette.

Roy, A. B. (1955). The steroid sulphatase of *Patella vulgata*. *Biochemical Journal* 62: 41–50.

Roy, K., Collins, A. G., Becker, B. J., Begovic, E. & Engle, J. M. (2003). Anthropogenic impacts and historical decline in body size of rocky intertidal gastropods in southern California. *Ecology Letters* 6: 205–211.

Roy, S., Datta, U., Ghosh, D., Dasgupta, P. S., Mukherjee, P. & Roychowdhury, U. (2010). Potential future applications of spermatheca extract from the marine snail *Telescopium telescopium*. *Turkish Journal of Veterinary and Animal Science* 34: 533–540.

Rozkošný, R. (1984). *The Sciomyzidae (Diptera) of Fennoscandia and Denmark*. Leiden, the Netherlands, Brill.

Rózsa, K. S. (1979). Analysis of the neural network regulating the cardio-renal system in the central nervous system of *Helix pomatia* L. *American Zoologist* 19: 117–128.

Rózsa, K. S. (1983). The role of identified central neurons in the regulation of visceral functions in *Helix pomatia*. pp. 132–138 *in* J. Lever & Boer, H. H. (Eds.) *Molluscan Neuroendocrinology. Proceedings of the International Minisymposium on Molluscan Endocrinology held in the Department of Biology, Free University, Amsterdam, the Netherlands, August 16–20, 1982*. Amsterdam, the Netherlands, New York, North Holland Publishing.

Rózsa, K. S. (1987). Organization of the multifunctional neural network regulating visceral organs in *Helix pomatia* L. (Mollusca, Gastropoda). *Experientia* 43: 965–972.

Rubinoff, D. & Haines, P. (2005). Web-spinning caterpillar stalks snails. *Science* 309: 575.

Ruby, E. G., Henderson, B. & McFall-Ngai, M. J. (2004). We get by with a little help from our (little) friends. *Science* 303: 1305–1307.

Ruddle, F. H., Bartels, J. L., Bentley, K. L., Kappen, C., Murtha, M. T. & Pendleton, J. W. (1994). Evolution of Hox genes. *Annual Review of Genetics* 28: 423–442.

Rudin-Bitterli, T. S., Spicer, J. I. & Rundle, S. D. (2016). Differences in the timing of cardio-respiratory development determine whether marine gastropod embryos survive or die in hypoxia. *Journal of Experimental Biology* 219: 1076–1085.

Rudman, W. B. (1972). Studies on the primitive opisthobranch genera *Bullina* (Férussac) and *Micromelo* (Pilsbry). *Zoological Journal of the Linnean Society* 51: 105–119.

Rudman, W. B. (1978). A new species and genus of the Aglajidae and the evolution of the philinacean opisthobranch molluscs. *Zoological Journal of the Linnean Society* 62: 89–107.

Rudman, W. B. (1981). Further studies on the anatomy and ecology of opisthobranch molluscs feeding on the scleractinian coral *Porites*. *Zoological Journal of the Linnean Society* 71: 373–412.

Rudman, W. B. (1984). The Chromodorididae (Opisthobranchia: Mollusca) of the Indo-West Pacific: A review of the genera. *Zoological Journal of the Linnean Society* 81: 115–273.

Rudman, W. B. (1987). Solar-powered animals. *Australian Natural History* 96: 50–53.

Rudman, W. B. (1991a). Purpose in pattern: The evolution of colour in chromodorid nudibranchs. *Journal of Molluscan Studies* 57: 5–21.

Rudman, W. B. (1991b). Further studies on the taxonomy and biology of the octocoral-feeding genus *Phyllodesmium* (Ehrenberg, 1831) (Nudibranchia, Aeolidoidea). Part 2. *Journal of Molluscan Studies* 57: 167–203.

Rudman, W. B. (1998). Suborder Doridina. pp. 990–1001 *in* P. L. Beesley, Ross, G. J. B. & Wells, A. (Eds.) *Mollusca: The Southern Synthesis. Part B. Fauna of Australia*. Melbourne, Australia, CSIRO Publishing.

Rudman, W. B. & Willan, R. C. (1998). Opisthobranchia: Introduction. pp. 915–942 *in* P. L. Beesley, Ross, G. J. B. & Wells, A. (Eds.) *Mollusca: The Southern Synthesis. Part B. Fauna of Australia*. Vol. 5. Melbourne, Australia, CSIRO Publishing.

Rudman, W. B. (2004). Mimicry. *Sea Slug Forum* Retrieved April, 2016, from http://www.seaslugforum.net/factsheet/mimicry

Rudman, W. B. & Bergquist, P. R. (2007). A review of feeding specificity in the sponge-feeding Chromodorididae (Nudibranchia: Mollusca). *Molluscan Research* 27: 60–88.

Rueda, A., Caballero, R., Kaminsky, R. & Andrews, K. L. (2002). Vaginulidae in Central America, with emphasis on the bean slug *Sarasinula plebeia* (Fischer). pp. 115–144 *in* G. M. Baker (Ed.) *Molluscs as Crop Pests*. Wallingford, UK, CABI Publishing.

Ruggiero, M. A., Gordon, D. P., Orrell, T. M., Bailly, N., Bourgoin, T., Brusca, R. C., Cavalier-Smith, T., Guiry, M. D. & Kirk, P. M. (2015a). A higher level classification of all living organisms. *PLoS ONE* 10: e0119248.

Ruggiero, M. A., Gordon, D. P., Orrell, T. M., Bailly, N., Bourgoin, T., Brusca, R. C., Cavalier-Smith, T., Guiry, M. D. & Kirk, P. M. (2015b). Correction: A higher level classification of all living organisms. *PLoS ONE* 10: e0130114.

Ruiz, G. M. & Lindberg, D. R. (1989). A fossil record for trematodes: Extent and potential uses. *Lethaia* 22: 431–438.

Ruiz, G. M. (1991). Consequences of parasitism to marine invertebrates: Host evolution? *American Zoologist* 31: 831–839.

Rumi, A., Sánchez, J. & Ferrando, N. S. (2010). *Theba pisana* (Müller, 1774) (Gastropoda, Helicidae) and other alien land molluscs species in Argentina. *Biological Invasions* 12: 2985–2990.

Rumpho, M. E., Summer, E. J. & Manhart, J. R. (2000). Solar-powered sea slugs: Mollusc/algal chloroplast symbiosis. *Plant Physiology* 123: 29–38.

Rumpho, M. E., Worful, J. M., Lee, J., Kannan, K., Tyler, M. S., Bhattacharya, D., Moustafa, A. & Manhart, J. R. (2008). Horizontal gene transfer of the algal nuclear gene *psbO* to the photosynthetic sea slug *Elysia chlorotica*. *Proceedings of the National Academy of Sciences of the United States of America* 105: 17867–17871.

Rumsey, T. J. (1972). Osmotic and ionic regulation in a terrestrial snail, *Pomatias elegans* (Gastropoda, Prosobranchia) with a note on some tropical Pomatiasidae. *Journal of Experimental Biology* 57: 205–216.

Runham, N. W., Thornton, P. R., Shaw, D. A. & Wayte, R. C. (1969). The mineralization and hardness of the radular teeth of the limpet *Patella vulgata* L. *Zeitschrift für Zellforschung und Mikroskopische Anatomie* 99: 608–626.

Runham, N. W. & Hunter, P. J. (1970). *Terrestrial Slugs*. London, UK, Hutchinson University Library.

Runham, N. W. (1988). Accessory sex glands. pp. 113–188 *in* K. G. Adiyodi & Adiyodi, R. G. (Eds.) *Reproductive Biology of Invertebrates*. Vol. 3. Chichester UK, John Wiley & Sons.

Runham, N. W. (1993). Mollusca. pp. 311–383 *in* K. G. Adiyodi & Adiyodi, R. G. (Eds.) *Reproductive Biology of Invertebrates. Asexual Propagation and Reproductive Strategies. Reproductive Biology of Invertebrates*. Vol. 6. Chichester, UK, John Wiley & Sons.

Runnegar, B. N. & Pojeta, J. (1985). Origin and diversification of the Mollusca. pp. 1–57 *in* E. R. Trueman & Clarke, M. R. (Eds.) *Evolution. The Mollusca*. Vol. 10. New York, Academic Press.

Rupert, S. D. & Peters, W. S. (2011). Autotomy of the posterior foot in *Agaronia* (Caenogastropoda: Olividae) occurs in animals that are fully withdrawn into their shells. *Journal of Molluscan Studies* 77: 437–440.

Ruppert, E. E. & Smith, P. R. (1988). The functional organization of filtration nephridia. *Biological Reviews* 63: 231–258.

Russell, C. W. & Evans, B. K. (1989). Cardiovascular anatomy and physiology of the black-lip abalone, *Haliotis ruber*. *Journal of Experimental Zoology Part A* 252: 105–117.

Russell, J. M. (2008). Effects of Ultraviolet Radiation (UVR) and Other Environmental Stressors on the Development of Intertital Mollusc Embryos. MSc, Victoria University of Wellington.

Russell, M. P. & Lindberg, D. R. (1988). Real and random patterns associated with molluscan spatial and temporal distributions. *Paleobiology* 14: 322–330.

Russell, M. P., Lindberg, D. R. & Jablonski, D. (1988). Estimates of species duration: Discussion and reply. *Science* 240: 969.

Russell, M. P. & Huelsenbeck, J. P. (1989). Seasonal variation in brood structure of *Transennella confusa* (Bivalvia: Veneridae). *The Veliger* 32: 288–295.

Russell, M. P., Huelsenbeck, J. P. & Lindberg, D. R. (1992). Experimental taphonomy of embryo preservation in a Cenozoic brooding bivalve. *Lethaia* 25: 353–359.

Russell-Hunter, W. D. (1949). The structure and behavior of *Hiatella gallicana* (Lamarck) and *H. arctica* (L.), with special reference to the boring habit. *Proceedings of the Royal Society of Edinburgh B* 63: 271–289.

Russell-Hunter, W. D. (1954). The condition of the mantle cavity in two pulmonate snails living in Loch Lomond. *Proceedings of the Royal Society of Edinburgh B* 65: 143–165.

Russell-Hunter, W. D. & Grant, D. C. (1962). Mechanics of the ligament in the bivalve *Spisula solidissima* in relation to mode of life *Biological Bulletin* 122: 369–379.

Russell-Hunter, W. D. & Brown, S. (1965). Ctenidial number in relation to size in certain chitons, with a discussion of its phyletic significance. *Biological Bulletin* 128: 508–521.

Russell-Hunter, W. D. (1968). *A Biology of Lower Invertebrates*. London, Collier-Macmillan.

Russell-Hunter, W. D. & Apley, M. L. (1968). Pedal expansion in the naticid snails. II. Labelling experiments using insulin. *Biological Bulletin* 135: 563–573.

Russell-Hunter, W. D. & Russell-Hunter, M. (1968). Pedal expansion in the naticid snails. I. Introduction and weighing experiments. *Biological Bulletin* 135: 548–562.

Russell-Hunter, W. D., Apley, M. L. & Hunter, R. D. (1972). Early life-history of *Melampus* and the significance of semilunar synchrony. *Biological Bulletin* 143: 623–656.

Russell-Hunter, W. D. (1978). Ecology of freshwater pulmonates. pp. 335–383 in V. Fretter & Peake, J. (Eds.) *Pulmonates: Systematics, Evolution and Ecology*. Vol. 2A. London, UK, Academic Press.

Russell-Hunter, W. D. (1983). Overview: Planetary distribution of and ecological constraints upon the Mollusca. pp. 1–27 in W. D. Russell-Hunter (Ed.) *Ecology. The Mollusca*. Vol. 6. New York, Academic Press.

Ruthensteiner, B. (1991). Development of *Ovatella* (*Myosotella*) *myosotis* (Draparnaud) (Pulmonata, Ellobiidae). pp. 45–46 in C. Meier-Brook (Ed.) *Proceedings of the 10th International Malacological Congress, Tübingen, 27 August–2 September 1989*. Tübingen, University of Tübingen.

Ruthensteiner, B. (1997). Homology of the pallial and pulmonary cavity of gastropods. *Journal of Molluscan Studies* 63: 353–367.

Ruthensteiner, B. (1999). Nervous system development of a primitive pulmonate (Mollusca: Gastropoda) and its bearing on comparative embryology of the gastropod nervous system. *Bollettino Malacologico* 34: 1–22.

Ruthensteiner, B., Wanninger, A. & Haszprunar, G. (2001). The protonephridial system of the tusk shell, *Antalis entalis* (Mollusca, Scaphopoda). *Zoomorphology* 121: 19–26.

Ruthensteiner, B., Schropel, V. & Haszprunar, G. (2010). Anatomy and affinities of *Micropilina minuta* (Warén, 1989) (Monoplacophora: Micropilinidae). *Journal of Molluscan Studies* 76: 323–332.

Ruxton, G. (2004). What, if anything, is the adaptive function of countershading? *Animal Behaviour* 68: 445–451.

Ryder, T. A. & Bowen, I. D. (1977). Studies on transmembrane and paracellular phenomena in the foot of the slug, *Agriolimax reticulatus* (Mu). *Cell and Tissue Research* 183: 143–152.

Ryland, J. S. & Porter, J. S. (2006). The identification, distribution and biology of encrusting species of *Alcyonidium* (Bryozoa: Ctenostomatida) around the coasts of Ireland. *Biology and Environment: Proceedings of the Royal Irish Academy* 106B: 19–33.

Rymzhanov, T. (1994). Courtship display and copulation mechanism in the slugs of the genus *Deroceras* (Mollusca, Gastropoda Terrestria Nuda) at Transili Alatau mountain range. *Izvestiya Natsional'noi Akademii Nauk Respubliki Kazakhstan Seriya Biologicheskaya* 4: 28–33.

Saavedra, C. & Bachere, E. (2006). Bivalve genomics. *Aquaculture* 256: 1–14.

Saavedra, L., Leonardi, M., Morin, V. & Quiñones, R. A. (2012). Induction of vitellogenin-like lipoproteins in the mussel *Aulacomya ater* under exposure to 17β-estradiol. *Revista de Biología Marina y Oceanografía* 47: 429–438.

Sacarrao, G. F. (1952). Remarks on gastrulation in Cephalopoda. *Arquivos do Museu Bocage* 23: 43–47.

Sadzikowski, M. R. & Wallace, D. C. (1976). A comparison of the food habits of size classes of three sunfishes (*Lepomis macrochirus* Rafinesque, *L. gibbosus* (Linnaeus) and *L. cyanellus* Rafinesque). *American Midland Naturalist* 95: 220–225.

Safavi-Hemami, H., Siero, W. A., Gorasia, D. G., Young, N. D., MacMillan, D., Williamson, N. A. & Purcell, A. W. (2011). Specialisation of the venom gland proteome in predatory cone snails reveals functional diversification of the conotoxin biosynthetic pathway. *Journal of Proteome Research* 10: 3904–3919.

Safavi-Hemami, H., Hu, H., Gorasia, D. G., Bandyopadhyay, P. K., Veith, P. D., Young, N. D., Reynolds, E. C., Yandell, M., Olivera, B. M. & Purcell, A. W. (2014). Combined proteomic and transcriptomic interrogation of the venom gland of *Conus geographus* uncovers novel components and functional compartmentalization. *Molecular & Cellular Proteomics* 13: 938–953.

Safriel, U. N. (1969). Ecological segregation, polymorphism and natural selection in two intertidal gastropods of the genus *Nerita* at Elat (Red Sea, Israel). *Israel Journal of Zoology* 18: 205–231.

Safriel, U. N. (1975). The role of vermetid gastropods in the formation of Mediterranean and Atlantic reefs. *Oecologia* 20: 85–101.

Sagrista, E., Bozzo, M. G., Bigas, M., Poquet, M. & Dufort, M. (1998). Developmental cycle and ultrastructure of *Steinhausia mytilovum*, a microsporidian parasite of oocytes of the mussel, *Mytilus galloprovincialis* (Mollusca, Bivalvia). *European Journal of Protistology* 34: 58–68.

Saibil, H. R. & Hewat, E. (1987). Ordered transmembrane and extracellular structure in squid photoreceptor microvilli. *Journal of Cell Biology* 105: 19–28.

Saidel, W. M. (1979). Relationship between photoreceptor terminations and centrifugal neurons in the optic lobe of *Octopus*. *Cell and Tissue Research* 204: 463–472.

Saidel, W. M., Lettvin, J. Y. & McNichol, E. F. (1983). Processing of polarized light by squid photoreceptors. *Nature* 304: 534–536.

Saidel, W. M., Shashar, N., Schmolesky, M. T. & Hanlon, R. T. (2005). Discriminative responses of squid (*Loligo pealei*) photoreceptors to polarized light. *Biochemistry* 142: 340–346.

Saito, H. & Okutani, T. (1992). Carnivorous habits of two species of the genus *Craspedochiton* (Polyplacophora: Acanthochitonidae). *Journal of the Malacological Society of Australia* 13: 55–63.

Saito, H. & Osako, K. (2007). Confirmation of a new food chain utilizing geothermal energy: Unusual fatty acids of a deep-sea bivalve, *Calyptogena phaseoliformis*. *Limnology and Oceanography* 52: 1910–1918.

Saitongdee, P., Apisawetakan, S., Anunruang, N., Poomthong, T., Hanna, P. & Sobhon, P. (2005). Egg-laying-hormone immunoreactivity in the neural ganglia and ovary of *Haliotis asinina* Linnaeus. *Invertebrate Neuroscience* 5: 165–172.

Sakakibara, M., Aritaka, T., Iizuka, A., Suzuki, H., Horikoshi, T. & Lukowiak, K. (2005). Electrophysiological responses to light of neurons in the eye and statocyst of *Lymnaea stagnalis*. *Journal of Neurophysiology* 93: 493–507.

Sakamoto, K., Touhata, K., Yamashita, M., Kasai, A. & Toyohara, H. (2007). Cellulose digestion by common Japanese freshwater clam *Corbicula japonica*. *Fisheries Science (Tokyo)* 73: 675–683.

Sakamoto, K. & Toyohara, H. (2009). A comparative study of cellulase and hemicellulase activities of brackish water clam *Corbicula japonica* with those of other marine Veneroida bivalves. *Journal of Experimental Biology* 212: 2812–2818.

Sakker, E. R. (1984). Sperm morphology, spermatogenesis and spermiogenesis of three species of chitons (Mollusca, Polyplacophora). *Zoomorphology* 104: 111–121.

Sakurai, A., Gunaratne, C. A. & Katz, P. S. (2014). Two interconnected kernels of reciprocally inhibitory interneurons underlie alternating left-right swim motor pattern generation in the mollusk *Melibe leonina*. *Journal of Neurophysiology* 112: 1317–1328.

Salánki, J. & van Bay, T. (1975). Sensory input characteristics at the chemical stimulation of the lip in the *Helix pomatia* L. *Annales Instituti Biologici (Tihany) Hungaricae Academiae Scientiarium* 42: 115–128.

Salerno, J. L., Macko, S. A., Hallam, S. J., Bright, M., Won, Y. J., McKiness, Z. & Van Dover, C. L. (2005). Characterization of symbiont populations in life-history stages of mussels from chemosynthetic environments. *Biological Bulletin* 208: 145–155.

Saleuddin, A. S. M. (1964). The gonads and reproductive cycle of *Astarte sulcata* (Da Costa) and sexuality in *A. elliptica* (Brown). *Journal of Molluscan Studies* 36: 141.

Saleuddin, A. S. M. & Dillaman, R. M. (1976). Direct innervation of the mantle edge gland by neurosecretory axons in *Helisoma duryi*. *Cell and Tissue Research* 171: 397–401.

Saleuddin, A. S. M. & Kunigelis, S. C. (1984). Neuroendocrine control mechanisms in shell formation. *American Zoologist* 24: 911–916.

Saleuddin, A. S. M., Mukai, S. T. & Khan, H. R. (1990). Hormonal control of reproduction and growth in the freshwater snail *Helisoma* (Mollusca: Pulmonata). pp. 163–182 *in* B. G. Loughton & Saleuddin, A. S. M. (Eds.) *Neurobiology and Endocrinology of Selected Invertebrates*. Toronto, Captus Press.

Saleuddin, A. S. M., Griffond, B. & Ashton, M.-L. (1991). An ultrastructural study of the activation of the endocrine dorsal bodies in the snail *Helix aspersa* by mating. *Canadian Journal of Zoology* 69: 1203–1215.

Saleuddin, A. S. M., Khan, H. R., Ashton, M. L. & Griffond, B. (1992a). Immunocytochemical localization of FMRFamide in the central nervous system and the kidney of *Helisoma duryi* (Mollusca): Its possible antidiuretic role. *Tissue and Cell* 24: 179–189.

Saleuddin, A. S. M., Sevala, V. M., Sevala, V. L., Mukai, S. T. & Khan, H. R. (1992b). Involvement of mammalian insulin and insulin-like peptides in shell growth and shell regeneration in molluscs. pp. 149–169 *in* S. Suga & Watabe, N. (Eds.) *Hard Tissue Mineralization and Demineralization*. Tokyo, Springer-Verlag.

Saleuddin, A. S. M., Mukai, S. T. & Khan, H. R. (1994). Molluscan endocrine structures associated with the central nervous system. pp. 256–263 *in* K. G. Davey, Peter, R. E. & Tobe, S. S. (Eds.) *Perspectives in Comparative Endocrinology: Invited papers from XII International Congress of Comparative Endocrinology, Sponsored by the International Federation of Comparative Endocrinological Societies. Toronto, Ontario, Canada, 16–21 May 1993*. Ottawa, National Research Council of Canada.

Saleuddin, A. S. M., Ashton, M. L. & Khan, H. R. (1997). An electron microscopic study of the endocrine dorsal bodies in reproductively active and inactive *Siphonaria pectinata* (Pulmonata: Mollusca). *Tissue and Cell* 29: 267–275.

Saleuddin, S. & Mukai, S., Eds. (2017). *Physiology of Molluscs: A Collection of Selected Reviews*. Oakville, Canada, Apple Academic Press.

Sallam, A. & El-Wakeil, N. (2012). Biological and ecological studies on land snails and their control. pp. 413–444 *in* S. Soloneski & Larramendy, M. L. (Eds.) *Integrated Pest Management and Pest Control: Current and Future Tactics*. Shanghai, InTech Publishing.

Salvi, D. & Mariottini, P. (2016). Molecular taxonomy in 2D: A novel ITS2 rRNA sequence-structure approach guides the description of the oysters' subfamily Saccostreinae and the genus *Magallana* (Bivalvia: Ostreidae). *Zoological Journal of the Linnean Society* 179: 263–276.

von Salvini-Plawen, L. (1968a). Über Lebendbeobachtungen an Caudofoveata (Mollusca, Aculifera), nebst Bemerkungen zum system der Klasse. *Sarsia* 31: 105–126.

von Salvini-Plawen, L. (1968b). Über einige Beobachtung an Solenogastres (Mollusca, Aculifera). *Sarsia* 31: 131–142.

von Salvini-Plawen, L. (1969). Solenogastres and Caudofoveata (Mollusca, Aculifera): Organisation and phylogenetic significance. *Malacologia* 9: 191–216.

von Salvini-Plawen, L. (1971). *Schild- und Furchenfüßer (Caudofoveata und Solenogastres)*. Vol. 441. Lutherstadt Wittenberg, Ziemsen Verlag.

von Salvini-Plawen, L. (1972). Zur morphologie und phylogenie der Mollusken: Die Beziehung der Caudofoveata und der Solenogastres als Aculifera, als Mollusca und als Spiralia. *Zeitschrift für Wissenschaftliche Zoologie* 184: 205–394.

von Salvini-Plawen, L. & Mayr, E. (1977). On the evolution of photoreceptors and eyes. *Evolutionary Biology* 10: 207–263.

von Salvini-Plawen, L. (1978a). Different blood-cells in species-pairs of Solenogastres (Mollusca). *Zoologischer Anzeiger* 200: 27–30.

von Salvini-Plawen, L. (1978b). Antarktische und subantarktische Solenogastres (eine Monographie 1898–1974). *Zoologica (Stuttgart)* 128: 1–315.

von Salvini-Plawen, L. (1980a). A reconsideration of systematics in the Mollusca (phylogeny and higher classification). *Malacologia* 19: 249–278.

von Salvini-Plawen, L. (1980b). Was ist eine Trochophora? Eine analyse der Larventypen mariner Protostomier [What is a Trochophora?]. *Zoologisches Jahrbücher Anatomie* 103: 398–423.

von Salvini-Plawen, L. (1981a). On the origin and evolution of the Mollusca. *Atti dei Convegni Lincei (Roma)* 49: 235–293.

von Salvini-Plawen, L. (1981b). The molluscan digestive system in evolution. *Malacologia* 21: 371–401.

von Salvini-Plawen, L. (1982). On the polyphyletic origin of photoreceptors. pp. 137–154 *in* J. Westfall (Ed.) *Visual Cells in Evolution*. New York, Raven Press.

von Salvini-Plawen, L. & Haszprunar, G. (1982). On the affinities of Septibranchia (Bivalvia). *The Veliger* 25: 83–85.

von Salvini-Plawen, L. (1984). Die Cladogenese der Mollusca. *Mitteilungen der Deutschen Malakozoologischen Gesellschaft* 37: 89–118.

von Salvini-Plawen, L. (1985a). Early evolution and the primitive groups. pp. 59–150 *in* E. R. Trueman & Clarke, M. R. (Eds.) *Evolution. The Mollusca*. Vol. 10. New York, Academic Press.

von Salvini-Plawen, L. & Haszprunar, G. (1987). The Vetigastropoda and the systematics of streptoneurous Gastropoda (Mollusca). *Journal of Zoology* 211: 747–770.

von Salvini-Plawen, L. (1988). The structure and function of molluscan digestive systems. pp. 301–379 *in* E. R. Trueman & Clarke, M. R. (Eds.) *Form and Function. The Mollusca*. Vol. 11. New York, Academic Press.

von Salvini-Plawen, L. & Steiner, G. (1996). Synapomorphies and plesiomorphies in higher classification of Mollusca. pp. 29–51 *in* J. D. Taylor (Ed.) *Origin and Evolutionary Radiation of the Mollusca*. Oxford, Oxford University Press.

von Salvini-Plawen, L. (2008). Photoreception and the polyphyletic evolution of photoreceptors (with special reference to Mollusca). *American Malacological Bulletin* 26: 83–100.

Samadi, L. & Steiner, G. (2009). Involvement of Hox genes in shell morphogenesis in the encapsulated development of a top shell gastropod (*Gibbula varia* L.). *Development Genes and Evolution* 219: 523–530.

Samadi, L. & Steiner, G. (2010). Expression of Hox genes during the larval development of the snail, *Gibbula varia* (L.): Further evidence of non-colinearity in molluscs. *Development Genes and Evolution* 220: 161–172.

Samadi, S., Quéméré, E., Lorion, J., Tillier, A., von Cosel, R., Lopez, P., Cruaud, C., Couloux, A. & Boisselier-Dubayle, M. C. (2007). Molecular phylogeny in mytilids supports the wooden steps to deep-sea vents hypothesis. *Comptes rendus Biologies* 330: 446–456.

San-Martin, A., Quezada, E., Soto, P., Palacios, Y. & Rovirosa, L. (1996). Labdane diterpenes from the marine pulmonate gastropod *Trimusculus peruvianus*. *Canadian Journal of Chemistry* 74: 2471–2475.

Sanders, B. M., Hope, C., Pascoe, V. M. & Martin, L. S. (1991). Characterization of the stress protein response in two species of *Collisella* limpets with different temperature tolerance. *Physiological Zoology* 64: 1471–1489.

Sanders, B. M. (1993). Stress proteins in aquatic organisms: An environmental perspective. *Critical Reviews in Toxicology* 23: 49–75.

Sanders, G. D. (1970). Long-term memory of a tactile discrimination in *Octopus vulgaris* and the effect of vertical lobe removal. *Brain Research* 20: 59–73.

Sanders, H. L. & Hessler, R. R. (1969). Diversity and composition of abyssal benthos. *Science* 166: 1034–1035.

Sanders, H. L. & Allen, J. A. (1973). Studies on deep-sea Protobranchia (Bivalvia): Prologue and the Pristiglomidae. *Bulletin of the Museum of Comparative Zoology* 145: 237–261.

Sandison, E. E. (1967). Respiratory response to temperature and temperature tolerance of some intertidal gastropods. *Journal of Experimental Marine Biology and Ecology* 1: 271–281.

Sanger, J. W. (1979). Cardiac fine structure in selected arthropods and molluscs. *American Zoologist* 19: 9–27.

Santagata, S. (2015). Phoronida. pp. 231–245 *in* A. Wanninger (Ed.) *Evolutionary Developmental Biology of Invertebrates 2*. Wien, Springer-Verlag.

Santagata, S. (2016). The Lophophorates. Phylum Phoronida. pp. 638–644 *in* R. C. Brusca, Moore, W. & Shuster, S. (Eds.) *Invertebrates*. Sunderland, MA, Sinauer Associates.

Santos, M. B., Clarke, M. R. & Pierce, G. J. (2001). Assessing the importance of cephalopods in the diets of marine mammals and other top predators: Problems and solutions. *Fisheries Research* 52: 121–139.

Santos, S., Ayres-Peres, L., Cardoso, R. C. F. & Sokolowicz, C. C. (2008). Natural diet of the freshwater anomuran *Aegla longirostri* (Crustacea, Anomura, Aeglidae). *Journal of Natural History* 42: 1027–1037.

Santos, X., Bros, V. & Miño, À. (2009). Recolonization of a burned Mediterranean area by terrestrial gastropods. *Biodiversity and Conservation* 18: 3153–3165.

Sarà, G., Romano, C., Widdows, J. & Staff, F. J. (2008). Effect of salinity and temperature on feeding physiology and scope for growth of an invasive species (*Brachidontes pharaonis*: Mollusca—Bivalvia) within the Mediterranean Sea. *Journal of Experimental Marine Biology and Ecology* 363: 130–136.

Sardet, C., Prodon, F., Dumollard, R., Chang, P. & Chênevert, J. (2002). Structure and function of the egg cortex from oogenesis through fertilization. *Developmental Biology* 241: 1–23.

Sartori, A. F., Passos, F. D. & Domaneschi, O. (2006). Arenophilic mantle glands in the Laternulidae (Bivalvia: Anomalodesmata) and their evolutionary significance. *Acta Zoologica* 87: 265–272.

Sasaki, T. (1998). Comparative anatomy and phylogeny of the Recent Archaeogastropoda (Mollusca: Gastropoda). *Bulletin of the University Museum, University of Tokyo* 38: i–vi, 1–223.

Sasaki, T. & Saito, H. (2005). Feeding of *Neomenia yamamotoi* Baba, 1975 (Mollusca: Solenogastres) on a sea anemone. *Venus* 64: 191–194.

Sasaki, T., Shigeno, S. & Tanabe, K. (2010a). Anatomy of living *Nautilus*: Re-evaluation of primitiveness and comparison with Coleoidea. pp. 35–66 *in* K. Tanabe, Shigeta, Y., Sasaki, T. & Hirano, H. (Eds.) *Cephalopods Present and Past*. Tokyo, Tokai University Press.

Sasaki, T., Warén, A., Kano, Y., Okutani, T. & Fujikura, K. (2010b). Gastropods from Recent hot vents and cold seeps: Systematics, diversity and life strategies. pp. 169–254 *in* S. Kiel (Ed.) *The Vent and Seep Biota: Aspects from Microbes to Ecosystems—Topics in Geobiology*. Dordrecht, the Netherlands, Springer.

Sastry, A. N. (1968). Relationships among food temperature and gonad development of bay scallops *Aequipecten irradians* Lamarck. *Physiological Zoology* 41: 44–53.

Sastry, A. N. (1979). Pelecypoda (excluding Ostreidae). pp. 113–292 *in* A. C. Giese & Pearse, J. S. (Eds.) *Reproduction of Marine Invertebrates: Molluscs—Pelecypods and Lesser Classes*. Vol. 5. New York, Academic Press.

Sato, S., Owada, M., Haga, T., Hong, J. S., Lützen, J. & Yamashita, H. (2011). Genus-specific commensalism of the galeommatoid bivalve *Koreamya arcuata* (A. Adams, 1856) associated with lingulid brachiopods. *Molluscan Research* 31: 95–105.

Satterlie, R. A., Labarbera, M. & Spencer, A. N. (1985). Swimming in the pteropod mollusc, *Clione limacina*. 1. Behavior and morphology. *Journal of Experimental Biology* 116: 189–204.

Satterlie, R. A., Norekian, T. P., Jordan, S. & Kazilek, C. J. (1995). Serotonergic modulation of swimming speed in the pteropod mollusc *Clione limacina*. I. Serotonin immunoreactivity in the central nervous system and wings. *Journal of Experimental Biology* 198: 895–904.

Sauer, W. H. H., Smale, M. J. & Lipiński, M. R. (1992). The location of spawning grounds, spawning and schooling behavior of the squid *Loligo vulgaris reynaudii* (Cephalopoda, Myopsida) off the eastern cape coast, South Africa. *Marine Biology* 114: 97–107.

Sauer, W. H. H., Roberts, M. J., Lipiński, M. R., Smale, M. J., Hanlon, R. T., Webber, D. M. & O'Dor, R. K. (1997). Choreography of the squid's 'nuptial dance'. *Biological Bulletin* 192: 203–207.

Saunders, A. M. C. & Poole, M. (1910). The development of *Aplysia punctata*. *Quarterly Journal of Microscopical Science* 55: 497–539.

Saunders, S. E., Kellett, E., Bright, K., Benjamin, P. R. & Burke, J. F. (1992). Cell-specific alternative RNA splicing of an FMRFamide gene transcript in the brain. *The Journal of Neuroscience* 12: 1033–1039.

Saunders, W. B. & Spinosa, C. (1978). Sexual dimorphism in *Nautilus* from Palau. *Paleobiology* 4: 349–358.

Saunders, W. B. (1984). *Nautilus* growth and longevity: Evidence from marked and recaptured animals. *Science* 224: 990–992.

Saur, M. (1990). Mate discrimination in *Littorina littorea* L. and *L. saxatilis* Olivi (Mollusca: Prosobranchia). *Hydrobiologia* 193: 261–270.

Saura, M., Rivas, M. J., Diz, A. P., Caballero, A. & Rolán-Alvarez, E. (2012). Dietary effects on shell growth and shape in an intertidal marine snail, *Littorina saxatilis*. *Journal of Molluscan Studies* 78: 213–216.

Sauriau, P.-G. & Kang, C.-K. (2000). Stable isotope evidence of benthic microalgae-based growth and secondary production in the suspension feeder *Cerastoderma edule* (Mollusca, Bivalvia) in the Marennes-Oléron Bay. *Hydrobiologia* 440: 317–329.

Savazzi, E. (1982). Adaptations to tube dwelling in the Bivalvia. *Lethaia* 15: 275–297.

Savazzi, E. (1984). Adaptive significance of shell torsion in mytilid bivalves. *Palaeontology* 27: 307–314.

Savazzi, E. (1985). SHELLGEN: A BASIC program for the modeling of molluscan shell ontogeny and morphogenesis. *Computers and Geosciences* 11: 521–530.

Savazzi, E. (1989). Burrowing mechanisms and sculptures in Recent gastropods. *Lethaia* 22: 31–48.

Savazzi, E. (1994). Adaptations to burrowing in a few Recent gastropods. *Historical Biology* 7: 291–311.

Savazzi, E. (1996). Adaptations of vermetid and siliquariid gastropods. *Palaeontology* 39: 157–177.

Savazzi, E. (1998). The colour patterns of cypraeid gastropods. *Lethaia* 31: 15–27.

Savazzi, E. (2001a). A review of symbiosis in the Bivalvia, with special attention to macrosymbiosis. *Paleontological Research* 5: 55–73.

Savazzi, E. (2001b). Morphodynamics of an endolithic vermetid gastropod. *Paleontological Research* 5: 3–11.

Savazzi, E. (2005). The function and evolution of lateral asymmetry in boring endolithic bivalves. *Paleontological Research* 9: 169–187.

Savin, K. W., Cocks, B. G., Wong, F., Sawbridge, T., Cogan, N., Savage, D. & Warner, S. (2010). A neurotropic herpes virus infecting the gastropod abalone shares ancestry with oyster herpes virus and a herpes virus associated with the amphioxus genome. *Virology Journal* 7: 301–309.

Sawyer, W. H., Deyrup-Olsen, I. & Martin, A. W. (1984). Immunological and biological characteristics of the vasotocin-like activity in the head ganglia of gastropod molluscs. *General and Comparative Endocrinology* 54: 97–108.

Scannell, D. R., Frank, A. C., Conant, G. C., Byrne, K. P., Woolfit, M. & Wolfe, K. H. (2007). Independent sorting-out of thousands of duplicated gene pairs in two yeast species descended from a whole-genome duplication. *Proceedings of the National Academy of Sciences of the United States of America* 104: 8397–8402.

Scaps, P. (2011). Associations between the scallop *Pedum spondyloideum* (Bivalvia, Pteriomorphia, Pectinidae) and hard corals on the west coast of Thailand. *Zoological Studies* 50: 466–474.

Scardino, A., Nys, R. de, Ison, O., O'Connor, W. A. & Steinberg, P. D. (2003). Microtopography and antifouling properties of the shell surface of the bivalve molluscs *Mytilus galloprovincialis* and *Pinctada imbricata*. *Biofouling* 19: 221–230.

Scemes, E., Salomão, L. C., McNamara, J. C. & Cassola, A. C. (1991). Lack of osmoregulation in *Aplysia brasiliana*: Correlation with response of neuron R15 to osphradial stimulation. *American Journal of Physiology* 260: R777–R784.

Schaefer, K. (1996). Development and homologies of the anal gland in *Haminaea navicula* (Da Costa, 1778) (Opisthobranchia, Bullomorpha). pp. 249–260 *in* J. D. Taylor (Ed.) *Origin and Evolutionary Radiation of the Mollusca*. Oxford, UK, Oxford University Press.

Schaefer, K. & Haszprunar, G. (1997). Organisation and fine structure of the mantle of *Laevipilina antarctica* (Mollusca, Monoplacophora). *Zoologischer Anzeiger* 236: 13–23.

Schaefer, K. (2000). The adoral sense organ in protobranch bivalves (Mollusca): Comparative fine structure with special reference to *Nucula nucleus*. *Invertebrate Biology* 119: 188–214.

Schaefer, K. & Ruthensteiner, B. (2001). The cephalic sensory organ in pelagic and intracapsular larvae of the primitive Opisthobranch genus *Haminoea* (Mollusca: Gastropoda). *Zoologischer Anzeiger* 240: 69–82.

Schäfer, H. (1952). Ein Beitrag zur Ernährungsbiologie von *Bithynia tentaculata* L. (Gastropoda Prosobranchia). *Zoologischer Anzeiger* 148: 299–303.

Van der Schalie, H. (1965). Observations on the sex in *Campeloma* (Gastropoda: Viviparidae). *Occasional Papers. Museum of Zoology, University of Michigan* 641: 1–15.

Schechtman, A. & Knight, P. F. (1955). Transfer of proteins from the yolk to the chick embryo. *Nature* 176: 786.

Scheibel, A. B. & Schopf, J. W., Eds. (1997). *The Origin and Evolution of Intelligence*. Sudbury, MA, Jones & Bartlett Publishers.

Scheibling, R. E. (1994). Molluscan grazing and macroalgal zonation on a rocky intertidal platform at Perth, Western Australia. *Australian Journal of Ecology* 19: 141–149.

Scheide, J. I. & Dietz, T. H. (1986). Serotonin regulation of gill cAMP production, Na, and water uptake in freshwater mussels. *Journal of Experimental Zoology Part A* 240: 309–314.

Scheil, A. E., Hilsmann, S., Triebskorn, R. & Köhler, H. R. (2013). Shell colour polymorphism, injuries and immune defense in three helicid snail species, *Cepaea hortensis*, *Theba pisana* and *Cornu aspersum maximum*. *Results in Immunology* 3: 73–78.

Schejter, L., Escolar, M. & Bremec, C. (2011). Variability in epibiont colonization of shells of *Fusitriton magellanicus* (Gastropoda) on the Argentinean shelf. *Journal of the Marine Biological Association of the United Kingdom* 91: 897–906.

Scheller, R. H. (1985). Gene expression in *Aplysia* peptidergic neurons. pp. 513–529 *in* G. M. Edelman, Gall, W. E. & Cowan, W. M. (Eds.) *Molecular Bases of Neural Development*. New York, Neurosciences Research Foundation.

Scheltema, A. H. (1973). Heart, pericardium, coelomoduct openings, and juvenile gonad in *Chaetoderma nitidulum* and *Falcidens caudatus* (Mollusca, Aplacophora). *Zeitschrift für Morphologie der Tiere* 76: 97–107.

Scheltema, A. H. (1978a). Position of the class Aplacophora in the phylum Mollusca. *Malacologia* 17: 99–109.

Scheltema, A. H. (1981). Comparative morphology of the radulae and alimentary tracts in the Aplacophora. *Malacologia* 20: 361–383.

Scheltema, A. H. (1988a). Ancestors and descendants: Relationships of the Aplacophora and Polyplacophora. *American Malacological Bulletin* 6: 57–68.

Scheltema, A. H. (1992). The Aplacophora: History, taxonomy, phylogeny, biogeography, and ecology. PhD, University of Oslo.

Scheltema, A. H. (1993). Aplacophora as progenetic aculiferans and the coelomate origin of mollusks as the sister taxon of Sipuncula. *Biological Bulletin* 184: 57–78.

Scheltema, A. H. & Jebb, M. (1994). Natural history of a solenogaster mollusk from Papua New Guinea, *Epimenia australis* (Thiele) (Aplacophora, Neomeniomorpha). *Journal of Natural History* 28: 1297–1318.

Scheltema, A. H., Tscherkassy, M. & Kuzirian, A. M. (1994). Aplacophora. pp. 13–54 *in* F. W. Harrison & Kohn, A. J. (Eds.) *Microscopic Anatomy of Invertebrates. Mollusca 1*. Vol. 5. New York, Wiley-Liss.

Scheltema, A. H. (1996). Phylogenetic position of Sipuncula, Mollusca and the progenetic Aplacophora. pp. 53–58 *in* J. D. Taylor (Ed.) *Origin and Evolutionary Radiation of the Mollusca*. Oxford, UK, Oxford University Press.

Scheltema, A. H. (1998). Class Aplacophora. pp. 145–159 *in* P. L. Beesley, Ross, G. J. B. & Wells, A. (Eds.) *Mollusca: The Southern Synthesis. Part A. Fauna of Australia*. Melbourne, Australia, CSIRO Publishing.

Scheltema, A. H. & Ivanov, D. L. (2000). Prochaetodermatidae of the eastern Atlantic Ocean and Mediterranean Sea (Mollusca : Aplacophora). *Journal of Molluscan Studies* 66: 313–362.

Scheltema, A. H. & Ivanov, D. L. (2002). An aplacophoran postlarva with iterated dorsal groups of spicules and skeletal similarities to Paleozoic fossils. *Invertebrate Biology* 121: 1–10.

Scheltema, A. H. & Schander, C. (2006). Exoskeletons: Tracing molluscan evolution. *Venus* 65: 19–25.

Scheltema, R. S. (1968). Dispersal of larvae by equatorial ocean currents and its importance to the zoogeography of shoal-water tropical species. *Nature* 217: 1159–1162.

Scheltema, R. S. (1971). The dispersal of the larvae of shoal-water benthic invertebrate species over long distances by ocean currents. pp. 7–28 *in* D. J. Crisp (Ed.) *Fourth European Marine Biology Symposium held at Bangor, North Wales in September 1969*. Cambridge, UK, Cambridge University Press.

Scheltema, R. S. (1974). Biological interactions determining larval settlement of marine invertebrates. *Thalassia Jugoslavica* 10: 263–296.

Scheltema, R. S. (1977). Dispersal of marine invertebrate organisms: Paleobiogeographic and biostratigraphic implications. pp. 73–108 *in* E. G. Kauffman, Hazel, J. E. & Heffernan, B. D. (Eds.) *Concepts and Methods of Biostratigraphy*. Stroudsburg, PA, Dowden, Hutchinson & Ross.

Scheltema, R. S. (1978b). On the relationship between dispersal of pelagic veliger larvae and the evolution of marine prosobranch gastropods. pp. 303–322 *in* B. Battaglia & Beardmore, J. A. (Eds.) *Marine Organisms: Genetics, Ecology, and Evolution. Proceedings of a NATO Advanced Study Research Institute on the Genetics, Evolution, and Ecology of Marine Organisms. NATO conference series, IV, Marine sciences*. Vol. 2. New York, Plenum Press.

Scheltema, R. S. (1979). Dispersal of pelagic larvae and the zoogeography of Tertiary marine benthic gastropods. pp. 391–398 *in* J. Gray & Boucot, A. J. (Eds.) *Historical Biogeography, Plate Tectonics, and the Changing Environment. Proceedings of the 37th Annual Biology Colloquium (Oregon State University), and selected papers*. Vol. 37. Corvallis, OR, Oregon State University Press.

Scheltema, R. S. & Williams, I. P. (1983). Long-distance dispersal of planktonic larvae and the biogeography and evolution of some Polynesian and Western Pacific mollusks. *Bulletin of Marine Science* 33: 545–565.

Scheltema, R. S. (1986). Long-distance dispersal by planktonic larvae of shoal-water benthic invertebrates among central Pacific islands. *Bulletin of Marine Science* 39: 241–256.

Scheltema, R. S. (1988b). Initial evidence for the transport of teleplanic larvae of benthic invertebrates across the East Pacific Barrier. *Biological Bulletin* 174: 145–152.

Scheltema, R. S. (1994). Adaptations for reproduction among deep-sea benthic molluscs: An appraisal of the existing evidence. pp. 44–75 *in* C. M. Young & Eckelbarger, K. J. (Eds.) *Reproduction, Larval Biology, and Recruitment of the Deep-sea Benthos*. New York, Columbia University Press.

Scherholz, M., Redl, E., Wollesen, T., Todt, C. & Wanninger, A. (2015). From complex to simple: Myogenesis in an aplacophoran mollusk reveals key traits in aculiferan evolution. *BMC Evolutionary Biology* 15: 1–17.

Scherholz, M., Redl, E., Wollesen, T., de Oliveira, A. L., Todt, C. & Wanninger, A. (2017). Ancestral and novel roles of Pax family genes in mollusks. *BMC Evolutionary Biology* 17: 81.

Schiaparelli, S., Cattaneo-Vietti, R. & Chiantore, M. (2000). Adaptive morphology of *Capulus subcompressus* Pelseneer, 1903 (Gastropoda: Capulidae) from Terra Nova Bay, Ross Sea (Antarctica). *Polar Biology* 23: 11–16.

Schiaparelli, S., Barucca, M., Olmo, E., Boyer, M. & Canapa, A. (2005). Phylogenetic relationships within Ovulidae (Gastropoda: Cypraeoidea) based on molecular data from the 16S rRNA gene. *Marine Biology* 147: 411–420.

Schiemann, S. M., Martin-Duran, J. M., Børve, A., Vellutini, B. C., Passamaneck, Y. J. & Hejnol, A. (2017). Clustered brachiopod Hox genes are not expressed collinearly and are associated with lophotrochozoan novelties. *Proceedings of the National Academy of Sciences* 114: E1913–E1922.

Schill, R. O., Gayle, P. M. H. & Köhler, H. R. (2002). Daily stress protein (hsp70) cycle in chitons (*Acanthopleura granulata* Gmelin, 1791) which inhabit the rocky intertidal shoreline in a tropical ecosystem. *Comparative Biochemistry and Physiology Part C* 131: 253–258.

Schilthuizen, M. & Rutjes, H. A. (2001). Land snail diversity in a square kilometre of tropical rainforest in Sabah, Malaysian Borneo. *Journal of Molluscan Studies* 67: 417–423.

Schilthuizen, M. & Davison, A. (2005). The convoluted evolution of snail chirality. *Naturwissenschaften* 92: 504–515.

Schilthuizen, M., Craze, P. G., Cabanban, A. S., Davison, A., Stone, J., Gittenberger, E. & Scott, B. J. (2007). Sexual selection maintains whole-body chiral dimorphism in snails. *Journal of Evolutionary Biology* 20: 1941–1949.

Schilthuizen, M., Looijestijn, S., Chua, S. C. & Cajas, R. F. C. (2014). Mapping of dextral/sinistral proportions in the chirally dimorphic land snail *Amphidromus inversus*. *PeerJ* 3: 1–9.

Schindel, D. E. (1990). Unoccupied morphospace and the coiled geometry of gastropods: Architectural constraint or geometric covariation? pp. 270–304 *in* R. M. Ross & Allmon, W. D. (Eds.) *Causes of Evolution: A Paleontological Perspective*. Chicago IL, University of Chicago Press.

Schipp, R., Höhn, P. & Schäfer, A. (1971). Elektronenmikroskopische und histochemische Untersuchungen zur Funktion des Kiemenherzanhanges (Percardialdrüse) von *Sepia officinalis*. *Zeitschrift für Zellforschung und Mikroskopische Anatomie* 274: 252–274.

Schipp, R. & von Boletzky, S. (1975). Morphology and function of the excretory organs in dibranchiate cephalopods. *Fortschritte der Zoologie* 23: 89–111.

Schipp, R. & von Boletzky, S. (1976). The pancreatic appendages of dibranchiate cephalopods. *Zoomorphologie* 86: 81–98.

Schipp, R., Mollenhauer, S. & von Boletzky, S. (1979). Electron microscopical and histochemical studies of differentiation and function of the cephalopod gill (*Sepia officinalis* L.). *Zoomorphology* 93: 193–207.

Schipp, R. & Hevert, F. (1981). Ultrafiltration in the branchial heart appendage of dibranchiate cephalopods: A comparative ultrastructural and physiological study. *Journal of Experimental Biology* 92: 23–35.

Schipp, R. & Martin, A. W. (1981). Cytology of the renal appendages of *Nautilus* (Cephalopoda, Tetrabranchiata). *Cell and Tissue Research* 219: 585–596.

Schipp, R., Martin, A. W., Liebermann, H. & Magnier, Y. (1985). Cytomorphology and function of the pericardial appendages of *Nautilus* (Cephalopoda, Tetrabranchiata). *Zoomorphology* 105: 16–29.

Schipp, R. (1987a). The blood vessels of cephalopods: A comparative morphological and functional survey. *Experientia* 43: 525–537.

Schipp, R. (1987b). General morphological and functional characteristics of the cephalopod circulatory system: An introduction. *Experientia* 43: 474–477.

Schipp, R., Chung, Y. S. & Arnold, J. M. (1990). Symbiotic bacteria in the coelom of *Nautilus* (Cephalopoda, Tetrabranchiata). *Cell and Tissue Research* 219: 585–604.

Schipp, R., Jakobs, P. M. & Fiedler, A. (1991). Monoaminergic peptidergic interactions in neuroregulatory control of the cephalic aorta in *Sepia officinalis* L. (Cephalopoda). *Comparative Biochemistry and Physiology Part C* 99: 421–430.

Schluter, D. (2000). *The Ecology of Adaptive Radiation*. Oxford, UK, Oxford University Press.

Schmale, H. & Becker, W. (1977). Studies on the urea cycle of *Biomphalaria glabrata* during normal feeding activity, in starvation and with infection of *Schistosoma mansoni*. *Comparative Biochemistry and Physiology Part B* 58: 321–330.

Schmekel, L. & Portmann, A. (1982). *Opisthobranchia des Mittelmeeres, Nudibranchia und Saccoglossa*. Berlin, Germany, Springer-Verlag.

Schmekel, L. (1985). Aspects of evolution within the opisthobranchs. pp. 221–267 *in* E. R. Trueman & Clarke, M. R. (Eds.) *Evolution. The Mollusca*. Vol. 10. New York, Academic Press.

Schmidt-Rhaesa, A. (2007). *The Evolution of Organ Systems*. Oxford, UK, Oxford University Press.

Schmitz, L., Motani, R., Oufiero, C. E., Martin, C. H., McGee, M. D., Gamarra, A. R., Lee, J. J. & Wainwright, P. C. (2013a). Allometry indicates giant eyes of giant squid are not exceptional. *BMC Evolutionary Biology* 13: 45 (41–49).

Schmitz, L., Motani, R., Oufiero, C. E., Martin, C. H., McGee, M. D. & Wainwright, P. C. (2013b). Potential enhanced ability of giant squid to detect sperm whales is an exaptation tied to their large body size. *BMC Evolutionary Biology* 13: 226.

Schnabel, H. (1903). Über die embryonalentwicklung der radula bei den Mollusken. II. Die entwicklung der radula bei den Gastropoden. *Zeitschrift für wissenschaftliche Zoologie* 74: 616–655.

Schneider, D. W. & Frost, T. M. (1986). Massive upstream migrations by a tropical freshwater neritid snail. *Hydrobiologia* 137: 153–157.

Schneider, D. W. & Lyons, J. (1993). Dynamics of upstream migration in two species of tropical freshwater snails. *Journal of the North American Benthological Society* 12: 3–16.

Schneider, J. A. (1995). Phylogeny of the Cardiidae (Mollusca, Bivalvia): Protocardiinae, Laevicardiinae, Lahilliinae, Tulongocardiinae subfam. n. and Pleuriocardiinae subfam. n. *Zoologica Scripta* 24: 321–346.

Schöne, B. R., Flessa, K. W., Dettman, D. L. & Goodwin, D. H. (2003). Upstream dams and downstream clams: growth rates of bivalve mollusks unveil impact of river management on estuarine ecosystems (Colorado River Delta, Mexico). *Estuarine, Coastal and Shelf Science* 58: 715–726.

Schöne, B. R. & Fiebig, J. (2009). Seasonality in the North Sea during the Allerød and Late Medieval Climate Optimum using bivalve sclerochronology. *International Journal of Earth Sciences* 98: 83–98.

Schories, D., Anibal, J., Chapman, A. S., Herre, E., Lillebo, A. I., Reise, K., Sprung, M. & Thiel, M. (2000). Flagging greens: Hydrobiid snails as substrata for the development of green algal mats (*Enteromorpha* spp.) on tidal flats of North Atlantic coasts. *Marine Ecology Progress Series* 199: 127–136.

Schot, L. P. C., Boer, H. H., Swaab, D. F. & Van Noorden, S. (1981). Immunocytochemical demonstration of peptidergic neurons in the central nervous system of the pond snail *Lymnaea stagnalis* with antisera raised to biologically active peptides of vertebrates. *Cell and Tissue Research* 216: 273–291.

Schrag, S. J. & Rollinson, D. (1994). Effects of *Schistosoma haematobium* infection on reproductive success and male outcrossing ability in the simultaneous hermaphrodite, *Bulinus truncatus* (Gastropoda: Planorbidae). *Parasitology* 108: 27–34.

Schreiber, E. S. G., Lake, P. S. & Quinn, G. P. (2002). Facilitation of native stream fauna by an invading species? Experimental investigations of the interaction of the snail, *Potamopyrgus antipodarum* (Hydrobiidae) with native benthic fauna. *Biological Invasions* 4: 317–325.

Schreiber, E. S. G., Quinn, G. P. & Lake, P. S. (2003). Distribution of an alien aquatic snail in relation to flow variability, human activities and water quality. *Freshwater Biology* 48: 951–961.

Schrödl, M. & Neusser, T. P. (2010). Towards a phylogeny and evolution of Acochlidia (Mollusca: Gastropoda: Opisthobranchia). *Zoological Journal of the Linnean Society* 158: 124–154.

Schubart, C. D. & Ng, P. K. L. (2008). A new molluscivore crab from Lake Poso confirms multiple colonization of ancient lakes in Sulawesi by freshwater crabs (Decapoda: Brachyura). *Zoological Journal of the Linnean Society* 154: 211–221.

Schulhof, M. & Lindberg, D. R. (2013). The ontogeny of the lower reproductive tract of the landsnail *Helix aspersa* (Gastropoda: Mollusca). *Organisms Diversity & Evolution* 13: 559–568.

Schulte-Oehlmann, U., Oehlmann, J., Bauer, B., Fioroni, P. & Leffler, U. S. (1998). Toxico-kinetic and -dynamic aspects of TBT-induced imposex in *Hydrobia ulvae* compared with intersex in *Littorina littorea* (Gastropoda, Prosobranchia). *Hydrobiologia* 378: 215–225.

Schultz, J. H. & Adema, C. M. (2017). Comparative immunogenomics of molluscs. *Developmental and Comparative Immunology* 75: 3–15.

Schultz, M. C. (1983). A correlated light and electron microscopic study of the structure and secretory activity of the accessory salivary glands of the marine gastropods, *Conus flavidus* and *C. vexillum* (Neogastropoda, Conacea). *Journal of Morphology* 176: 89–111.

Schüpbach, H. U. & Baur, B. (2008). Experimental evidence for a new transmission route in a parasitic mite and its mucus-dependent orientation towards the host snail. *Parasitology* 135: 1679–1684.

Schwabe, E., Holtheuer, J. & Schories, D. (2014). First record of a mesoparasite (Crustacea, Copepoda) infesting a polyplacophoran (Mollusca, Polyplacophora) in Chilean waters, with an overview of the family Chitonophilidae. *Spixiana* 37: 165–182.

Schwartz, J. A., Curtis, N. E. & Pierce, S. K. (2014). FISH labeling reveals a horizontally transferred algal (*Vaucheria litorea*) nuclear gene on a sea slug (*Elysia chlorotica*) chromosome. *The Biological Bulletin* 227: 300–312.

Scott, A. P. (2012). Do mollusks use vertebrate sex steroids as reproductive hormones? Part I: Critical appraisal of the evidence for the presence, biosynthesis and uptake of steroids. *Steroids* 77: 1450–1468.

Scott, A. P. (2013). Do mollusks use vertebrate sex steroids as reproductive hormones? Part II . Critical review of the evidence that steroids have biological effects. *Steroids* 78: 268–281.

Scott, K. M. (2005). Allometry of gill weights, gill surface areas, and foot biomass $\delta^{13}C$ values of the chemoautotroph-bivalve symbiosis *Solemya velum*. *Marine Biology* 147: 935–941.

Seamone, C. B. & Boulding, E. G. (2011). Aggregation of the northern abalone *Haliotis kamtschatkana* with respect to sex and spawning condition. *Journal of Shellfish Research* 30: 881–888.

Seapy, R. R. (1980). Predation by the epipelagic heteropod mollusk *Carinaria cristata* forma *japonica*. *Marine Biology* 60: 137–146.

Seaward, D. R. (2001). The spaces in between: Interstitial molluscs in Chesil shingle. pp. 377–378 in J. R. Packham, Randall, R. E., Barnes, R. S. K. & Neal, A. (Eds.) *Ecology & Geomorphology of Coastal Shingle*. Otley, West Yorkshire, UK, Westbury Academic and Scientific Publishing.

Seed, R. & Hughes, R. N. (1995). Criteria for prey size-selection in molluscivorous crabs with contrasting claw morphologies. *Journal of Experimental Marine Biology and Ecology* 193: 177–195.

Seed, R. (1996). Patterns of biodiversity in the macro-invertebrate fauna associated with mussel patches on rocky shores. *Journal of the Marine Biological Association of the United Kingdom* 76: 203–210.

Seelemann, U. (1968a). Zur Überwindung der biologischen Grenze Meer-Land durch Mollusken. *Oecologia* 1: 130–154.

Seelemann, U. (1968b). Zur Überwindung der biologischen Grenze Meer-Land durch Mollusken II. Untersuchungen an *Limapontia capitata*, *Limapontia depressa* und *Assiminea grayana*. *Oecologia* 1: 356–368.

Seeley, R. H. (1986). Intense natural selection caused a rapid morphological transition in a living marine snail. *Proceedings of the National Academy of Sciences of the United States of America* 83: 6897–6901.

Segade, P., García, N., Estévez, J. M. G., Arias, C. & Iglesias-Blanco, R. (2016). Encystment/excystment response and serotypic variation in the gastropod parasite *Tetrahymena rostrata* (Ciliophora, Tetrahymenidae). *Parasitology Research* 115: 771–777.

Segal, E. (1961). Acclimation in molluscs. *American Zoologist*: 235–244.

Segarra, A., Pépin, J. F., Arzul, I., Morga, B., Faury, N. & Renault, T. (2010). Detection and description of a particular *Ostreid herpesvirus 1* genotype associated with massive mortality outbreaks of Pacific oysters, *Crassostrea gigas*, in France in 2008. *Virus Research* 153: 92–99.

Segura, C. J., Pechenik, J. A., Montory, J. A., Navarro, J. M., Paschke, K. A., Cubillos, V. M. & Chaparro, O. R. (2016). The cost of brooding in an estuary: Implications of declining salinity for gastropod females and their brooded embryos. *Marine Ecology Progress Series* 543: 187–199.

Seibel, B. A., Dymowska, A. & Rosenthal, J. (2007). Metabolic temperature compensation and coevolution of locomotory performance in pteropod molluscs. *Integrative and Comparative Biology* 47: 880–891.

Seibel, B. A. (2011). Critical oxygen levels and metabolic suppression in oceanic oxygen minimum zones. *Journal of Experimental Biology* 214: 326–336.

Seibel, B. A. (2013). The jumbo squid, *Dosidicus gigas* (Ommastrephidae), living in oxygen minimum zones II: Blood-oxygen binding. *Deep Sea Research Part II: Topical Studies in Oceanography* 95: 139–144.

Seiderer, L. J. & Newell, R. C. (1979). Adjustment of the activity of α-amylase extracted from the style of the black mussel *Choromytilus meridionalis* (Krauss) in response to thermal acclimation. *Journal of Experimental Marine Biology and Ecology* 39: 79–86.

Seiderer, L. J., Davis, C. L., Robb, F. T. & Newell, R. C. (1984). Utilisation of bacteria as nitrogen resource by kelp-bed mussel *Chloromytilus meridionalis*. *Marine Ecology Progress Series* 15: 109–116.

Seiderer, L. J., Newell, R. C., Schultes, K., Robb, F. T. & Turley, C. M. (1987). Novel bacteriolytic activity associated with the style microflora of the mussel *Mytilus edulis* (L.). *Journal of Experimental Marine Biology and Ecology* 110: 213–224.

Seidou, M., Sugahara, M., Uchiyama, H., Hiraki, K., Hamanaka, T., Michinomae, M., Yoshihara, K. & Kito, Y. (1990). On the three visual pigments in the retina of the firefly squid, *Watasenia scintillans*. *Journal of Comparative Physiology A* 166: 769–773.

Seilacher, A. (1959). Schnecken im Brandungssand. *Natur und Volk* 89: 359–366.

Seilacher, A. (1968). Swimming habits of belemnites: Recorded by boring barnacles. *Palaeogeography, Palaeoclimatology, Palaeoecology* 4: 279–285.

Seilacher, A. (1972). Divaricate patterns in pelecypod shells. *Lethaia* 5: 325–343.

Seilacher, A. (1985). Bivalve morphology and function. pp. 88–101 in T. W. Broadhead (Ed.) *Mollusks: Notes for a short course organized by D. J. Bottjer, C. S. Hickman, and P. D. Ward. University of Tennessee Studies in Geology*. Orlando, FL, University of Tennessee Department of Geological Sciences.

Seilacher, A. (1990). Aberrations in bivalve evolution related to photo- and chemosymbiosis. *Historical Biology* 3: 289–311.

Seillière, G. (1906). Sur un cas d'hydrolyse diastasique de la cellulose du coton, après dissolution dans la liqueur de Schweitzer. *Comptes rendus des séances de la Société de biologie et de ses filiales* 61: 205–206.

Sekeroglu, E. & Colkesen, T. (1989). Prey preference and feeding capacity of the larvae of *Ablattaria arenaria* (Coleoptera: Silphidae), a snail predator. *Entomophaga* 34: 227–236.

Selden, R., Johnson, A. S. & Ellers, O. (2009). Waterborne cues from crabs induce thicker skeletons, smaller gonads and size-specific changes in growth rate in sea urchins. *Marine Biology* 156: 1057–1071.

Selkoe, K. A. & Toonen, R. J. (2011). Marine connectivity: A new look at pelagic larval duration and genetic metrics of dispersal. *Marine Ecology Progress Series* 436: 291–305.

Sellius, G. (1733). *Historia naturalis Teredinis seu Xylophagi marini, tubuli-conchoidis, speciatim belgici.* Trajecti ad Rhenum [Utrecht], H. Besseling.

Sellmer, G. P. (1967). Functional morphology and ecological life history of the Gem Clam *Gemma gemma* (Eulamellibranchia, Veneridae). *Malacologia* 5: 137–223.

Selman, B. J. & Jones, A. A. (2004). Microsporidia (*Microspora*) parasitic in terrestrial gastropods. pp. 579–598 *in* G. M. Barker (Ed.) *Natural Enemies of Terrestrial Molluscs.* Oxford, UK, Cambridge, MA, CABI Publishing.

Selman, K. & Arnold, J. M. (1977). An ultrastructural and cytochemical analysis of oogenesis in the squid, *Loligo pealei.* *Journal of Morphology* 152: 381–400.

Selmi, M. G. & Giusti, F. (1980). Structure and function in typical and atypical spermatozoa of Prosobranchia (Mollusca): 1. *Cochlostoma montanum* (Issel) (Mesogastropoda). Atti dell'Accademia dei Fisiocritici, Siena, IV Congresso (Society Malacologica Italiana, Siena), 1978: 115–147.

Selwood, L. (1986). Interrelationships between developing oocytes and ovarian tissues in the chiton *Sypharochiton septentriones* (Ashby) (Mollusca, Polyplacophora). *Journal of Morphology* 125: 71–104.

Seminoff, J. A., Resendiz, A. & Nichols, W. J. (2002). Diet of East Pacific green turtles (*Chelonia mydas*) in the Central Gulf of California, México. *Journal of Herpetology* 36: 447–453.

Semmens, J. M., Jarvis, D., Piasente, M., Schubert, M., Sen, S., Moore, A., Stobutzki, I. & Marton, N. (2012). Commercial scallop *Pecten fumatus.* pp. 88–93. *Status of Key Australian Fish Stocks Reports 2012.* Canberra, Australian Government.

Semmens, J. M., Ewing, G. & Keane, J. P. (2018). *Tasmanian Scallop Fishery Assessment 2017*, University of Tasmania.

Seréne, R. (1957). A megalopa commensal in a squid. pp. 35–36. *Proceedings of the Ninth Pacific Science Congress of the Pacific Science Association: Held at Chulalongkorn University, Bangkok, Thailand, November 18th to December 9th, 1957.* Vol. 10. Bangkok, Secretariat, Ninth Pacific Science Congress.

Serrano, T., Gómez, B. J. & Angulo, E. (1996). Light and electron microscopy study of the salivary gland secretory cells of Helicoidea (Gastropoda, Stylommatophora). *Tissue and Cell* 28: 237–251.

Serviere-Zaragoza, E., Navarrete del Toro, M. A. & García-Carreño, F. L. (1997). Protein-hydrolyzing enzymes in the digestive systems of the adult Mexican blue abalone, *Haliotis fulgens* (Gastropoda). *Aquaculture* 157: 325–336.

Seuront, L. & Spilmont, N. (2015). The smell of sex: Water-borne and air-borne sex pheromones in the intertidal gastropod *Littorina littorea.* *Journal of Molluscan Studies* 81: 96–103.

Seydel, E. (1909). Untersuchungen über den Byssusapparat der Lamellibranchier. *Zoologische Jahrbücher. Abteilung für Anatomie und Ontogenie der Tiere* 27: 465–582.

Seyer, J.-O. (1992). Resolution and sensitivity in the eye of the winkle *Littorina littorea.* *Journal of Experimental Biology* 170: 57–69.

Seyer, J.-O. (1994). Structure and optics of the eye of the hawk-wing conch, *Strombus raninus* (L.). *Journal of Experimental Zoology* 268: 200–207.

Seyer, J.-O. (1998). Comparative optics of prosobranch eyes. PhD, University of Lund, Sweden.

Seyer, J.-O., Nilsson, D. E. & Warrant, E. J. (1998). Spatial vision in the prosobranch gastropod *Ampularia* sp. *Journal of Experimental Biology* 201: 1673–1679.

Seymour, R. S. & Matthews, P. G. D. (2013). Physical gills in diving insects and spiders: Theory and experiment. *Journal of Experimental Biology* 216: 164–170.

Shachak, M., Jones, C. G. & Granot, Y. (1987). Herbivory in rocks and the weathering of a desert. *Science* 236: 1098–1099.

Shachak, M., Jones, C. G. & Brand, S. (1995). The role of animals in an arid ecosystem: Snails and isopods as controllers of soil formation, erosion and desalinization. *Advances in Geoecology* 28: 37–50.

Shadwick, R. E. & Nilsson, E. K. (1990). The importance of vascular elasticity in the circulatory system of the cephalopod *Octopus vulgaris.* *Journal of Experimental Biology* 152: 471–484.

Shadwick, R. E., O'Dor, R. K. & Gosline, J. M. (1990). Respiratory and cardiac function during exercise in squid. *Canadian Journal of Zoology* 68: 792–798.

Shadwick, R. E. (1994). Mechanical organization of the mantle and circulatory system of cephalopods. *Marine and Freshwater Behaviour and Physiology* 25: 69–85.

Shaffer, J. F. & Kier, W. M. (2012). Muscular tissues of the squid *Doryteuthis pealeii* express identical myosin heavy chain isoforms: An alternative mechanism for tuning contractile speed. *Journal of Experimental Biology* 215: 239–246.

Shaffer, J. F. & Kier, W. M. (2016). Tuning of shortening speed in coleoid cephalopod muscle: No evidence for tissue-specific muscle myosin heavy chain isoforms. *Invertebrate Biology* 135: 3–12.

Shalack, J. D., Power, A. J. & Walker, R. L. (2011). Hand harvesting quickly depletes intertidal whelk populations. *American Malacological Bulletin* 29: 37–50.

Shanks, A. L. (2002). Previous agonistic experience determines both foraging behavior and territoriality in the limpet *Lottia gigantea* (Sowerby). *Behavioral Ecology* 13: 467–471.

Shanks, A. L. & Brink, L. (2005). Upwelling, downwelling, and cross-shelf transport of bivalve larvae: Test of a hypothesis. *Marine Ecology Progress Series* 302: 1–12.

Shao, J., Xiang, L., Zhang, M. & Mao, S. (1995). Histopathological studies on the plague disease of *Hyriopsis cumingii* Lea [in Chinese with English summary]. *Journal of Fisheries of China* 191: 1–7.

Sharara, F. I., Seifer, D. B. & Flaws, J. A. (1998). Environmental toxicants and female reproduction [part 4]. *Fertility and Sterility* 70: 613–622.

Sharman, M. (1956). Note on *Capulus ungaricus* (L.). *Journal of the Marine Biological Association of the United Kingdom* 35: 445–450.

Shashar, N., Rutledge, P. S. & Cronin, T. W. (1996). Polarization vision in cuttlefish in a concealed communication channel? *Journal of Experimental Biology* 199: 2077–2084.

Shashar, N., Hanlon, R. T. & Petz, A. de M. (1998). Polarization vision helps detect transparent prey. *Nature* 393: 222–223.

Shashar, N., Hagan, R., Boal, J. G. & Hanlon, R. T. (2000). Cuttlefish use polarization sensitivity in predation on silvery fish. *Vision Research* 40: 71–75.

Shashar, N., Milbury, C. A. & Hanlon, R. T. (2002). Polarization vision in cephalopods: Neuroanatomical and behavioral features that illustrate aspects of form and function. *Marine and Freshwater Behaviour and Physiology* 35: 57–68.

Shasky, D. (1968). Observations on *Rosenia nidorum* (Pilsbry) and *Arene socorroensis* (Strong). *Bulletin of the American Malacological Union* 1967: 74.

Shau-Hwai, A. T., Hui, S. L. M. & Yasin, Z. (2010). *The feeding behaviour of Volutidae snail, Melo melo (Lightfoot, 1786)*. *BioScience for the future. Proceedings of the 7th IMT-GT UNINET and The 3rd International PSU-UNS Conferences on Bioscience, 7–8 October 2010*. Prince of Songkla University, Hat Yai, Songkla, Thailand, Prince of Songkla University.

Shaw, J. A., Macey, D. J. & Brooker, L. R. (2008). Radula synthesis by three species of iron mineralizing molluscs: Production rate and elemental demand. *Journal of the Marine Biological Association of the United Kingdom* 88: 597–601.

Shaw, J. A., Macey, D. J., Brooker, L. R., Stockdale, E. J., Saunders, M. & Clode, P. L. (2009). The chiton stylus canal: An element delivery pathway for tooth cusp biomineralization. *Journal of Morphology* 270: 588–600.

Sheehan, D. & McDonagh, B. (2008). Oxidative stress and bivalves: A proteomic approach. *Invertebrate Survival Journal* 5: 110–123.

Shepherd, S. A. (1973). Studies on southern Australian abalone (genus *Haliotis*). I. Ecology of five sympatric species. *Australian Journal of Marine and Freshwater Research* 24: 217–257.

Shepherd, S. A., Al-Wahaibi, D. & Al-Azri, A. R. (1995). Shell growth checks and growth of the Omani abalone *Haliotis mariae*. *Marine and Freshwater Research* 46: 575–582.

Shepherd, S. A. & Avalos-Borja, M. (1997). The shell microstructure and chronology of the abalone *Haliotis corrugata*. *Molluscan Research* 18: 197–207.

Shepherd, S. A. & Turrubiates, J. R. (1997). A practical chronology for the abalone *Haliotis fulgens*. *Molluscan Research* 18: 219–226.

Shepherd, S. A., Woodby, D., Rumble, J. M. & Avalos-Borja, M. (2000). Micro-structure, chronology and growth of the pinto abalone, *Haliotis kamtschatkana* in Alaska. *Journal of Shellfish Research* 19: 219–228.

Shepherd, S. A. (2008). Abalone of Gulf St Vincent. pp. 448–455 *in* S. A. Shepherd, Bryars, S., Kirkegaarde, I. R., Harbison, P. & Jennings, J. T. (Eds.) *Natural History of Gulf St Vincent*. Adelaide, S. A., Royal Society of South Australia.

Sheybani, A., Nusnbaum, M., Caprio, J. & Derby, C. D. (2009). Responses of the sea catfish *Ariopsis felis* to chemical defenses from the sea hare *Aplysia californica*. *Journal of Experimental Marine Biology and Ecology* 368: 153–160.

Shibata, T., Yamaguchi, K., Nagayama, K., Kawaguchi, S. & Nakamura, T. (2002). Inhibitory activity of brown algal phlorotannins against glycosidases from the viscera of the turban shell *Turbo cornutus*. *European Journal of Phycology* 37: 493–500.

Shick, J. M. & Dunlap, W. C. (2002). Mycosporine-like amino acids and related gadusols: Biosynthesis, accumulation, and UV-protective functions in aquatic organisms. *Annual Review of Physiology* 64: 223–262.

Shigemiya, Y. (2003). Does the handedness of the pebble crab *Eriphia smithii* influence its attack success on two dextral snail species? *Journal of Zoology* 260: 259–265.

Shigeno, S., Kidokoro, H., Tsuchiya, K., Segawa, S. & Yamamoto, M. (2001a). Development of the brain in the oegopsid squid, *Todarodes pacificus*: An atlas from hatchling to juvenile. *Zoological Science* 18: 1081–1096.

Shigeno, S., Tsuchiya, K. & Segawa, S. (2001b). Conserved topological patterns and heterochronies in loliginid cephalopods: Comparative developmental morphology of the oval squid *Sepioteuthis lessoniana*. *Invertebrate Reproduction & Development* 39: 161–174.

Shigeno, S., Tsuchiya, K. & Segawa, S. (2001c). Embryonic and paralarval development of the central nervous system of the loliginid squid *Sepioteuthis lessoniana*. *Journal of Comparative Neurology* 437: 449–475.

Shigeno, S., Sasaki, T. & Haszprunar, G. (2007). Central nervous system of *Chaetoderma japonicum* (Caudofoveata, Aplacophora): Implications for diversified ganglionic plans in early molluscan evolution. *Biological Bulletin* 213: 122–134.

Shigeno, S., Sasaki, T., Moritaki, T., Kasugai, T., Vecchione, M. & Agata, K. (2008). Evolution of the cephalopod head complex by assembly of multiple molluscan body parts: Evidence from *Nautilus* embryonic development. *Journal of Morphology* 269: 1–17.

Shigeno, S., Takenori, S. & von Boletzky, S. (2010). The origins of cephalopod body plans: A geometrical and developmental basis for the evolution of vertebrate-like organ systems. pp. 23–34 *in* K. Tanabe, Shigeta, Y., Sasaki, T. & Hirano, H. (Eds.) *Cephalopods Present and Past*. Tokyo, Tokai University Press.

Shigeno, S., Parnaik, R., Albertin, C. B. & Ragsdale, C. W. (2015). Evidence for a cordal, not ganglionic, pattern of cephalopod brain neurogenesis. *Zoological Letters* 1: 26.

Shikov, E. V. (2011). *Haemopis sanguisuga* (Linnaeus, 1758) (Hirudinea): The first observation of a leech predation on terrestrial gastropods. *Folia Malacologica* 19: 103–106.

Shikuma, N. J. & Hadfield, M. G. (2010). Marine biofilms on submerged surfaces are a reservoir for *Escherichia coli* and *Vibrio cholerae*. *Biofouling* 26: 39–46.

Shimatani, Y., Katagiri, Y. & Katagiri, N. (1986). Light-evoked peripheral reflex of isolated mantle papilla of *Onchidium verruculatum* (Gastropoda, Mollusca). *Zoological Science (Tokyo)* 3: 993.

Shimek, R. L. (1990). Diet and habitat utilization in a Northeastern Pacific Ocean scaphopod assemblage. *American Malacological Bulletin* 7: 147–169.

Shimek, R. L. (1997). A new species of Eastern Pacific *Fissidentalium* (Mollusca: Scaphopoda) with a symbiotic sea anemone. *The Veliger* 40: 178–191.

Shimek, R. L. & Steiner, G. (1997). Scaphopoda. pp. 719–781 *in* F. W. Harrison & Kohn, A. J. (Eds.) *Microscopic Anatomy of Invertebrates. Mollusca 2*. Vol. 6B. New York, Wiley-Liss.

Shimizu, K., Sarashina, I., Kagi, H. & Endo, K. (2011). Possible functions of *dpp* in gastropod shell formation and shell coiling. *Development Genes and Evolution* 221: 59–68.

Shimizu, K., Iijima, M., Setiamarga, D. H. E., Sarashina, I., Kudoh, T., Asami, T., Gittenberger, E. & Endo, K. (2013). Left-right asymmetric expression of *dpp* in the mantle of gastropods correlates with asymmetric shell coiling. *EvoDevo* 4: 1–7.

Shine, R. (1989). Ecological causes for the evolution of sexual dimorphism: A review of the evidence. *Quarterly Review of Biology* 64: 419–461.

Shine, R. (2010). The ecological impact of invasive cane toads (*Bufo marinus*) in Australia. *The Quarterly Review of Biology* 85: 253–291.

Shiomi, K., Mizukami, M., Shimakura, K. & Nagashima, Y. (1994). Toxins in the salivary gland of some marine carnivorous gastropods. *Comparative Biochemistry and Physiology Part B* 107: 427–432.

Shiomi, K., Ishii, M., Shimakura, K., Nagashima, Y. & Chino, M. (1998). Tigloylcholine: A new choline ester toxin from the hypobranchial gland of two species of muricid gastropods (*Thais clavigera* and *Thais bronni*). *Toxicon* 36: 795–798.

Shiomi, K., Kawashima, Y., Mizukami, M. & Nagashima, Y. (2002). Properties of proteinaceous toxins in the salivary gland of the marine gastropod (*Monoplex echo*). *Toxicon* 40: 563–571.

Shipway, J. R., Borges, L. M. S., Müller, J. & Cragg, S. M. (2014). The broadcast spawning Caribbean shipworm, *Teredothyra dominicensis* (Bivalvia, Teredinidae), has invaded and become established in the eastern Mediterranean Sea. *Biological Invasions* 16: 2037–2048.

Shipway, J. R., Altamia, M. A., Haga, T., Velásquez, M., Albano, J., Dechavez, R., Concepcion, G. P., Haygood, M. G. & Distel, D. L. (2018). Observations on the life history and geographic range of the giant chemosymbiotic shipworm *Kuphus polythalamius* (Bivalvia: Teredinidae). *The Biological Bulletin* 235: 167–177.

Shirbhate, R. & Cook, A. (1987). Pedal and opercular secretory glands of *Pomatias, Bithynia* and *Littorina*. *Journal of Molluscan Studies* 53: 79–96.

Shirtcliffe, N. J., McHale, G. & Newton, M. I. (2012). Wet adhesion and adhesive locomotion of snails on anti-adhesive non-wetting surfaces. *PLoS ONE* 7: e36983.

Shivji, M., Parker, D., Hartwick, B., Smith, M. J. & Sloan, N. A. (1983). Feeding and distribution study of the sunflower sea star *Pycnopodia helianthoides* (Brandt, 1835). *Pacific Science* 37: 133–140.

Shomrat, T., Zarrella, I., Fiorito, G. & Hochner, B. (2008). The octopus vertical lobe modulates short-term learning rate and uses LTP to acquire long-term memory. *Current Biology* 18: 337–342.

Shomrat, T., Graindorge, N., Bellanger, C., Fiorito, G., Loewenstein, Y. & Hochner, B. (2011). Alternative sites of synaptic plasticity in two homologous 'fan-out fan-in' learning and memory networks. *Current Biology* 21: 1773–1782.

Shomrat, T., Turchetti-Maia, A. L., Stern-Mentch, N., Basil, J. A. & Hochner, B. (2015). The vertical lobe of cephalopods: An attractive brain structure for understanding the evolution of advanced learning and memory systems. *Journal of Comparative Physiology A* 201: 947–956.

Shtein, G. A. (1954). Ortonektidy roda *Rhopalura* Giard nekotorykh molliuskow *Barentsova moria*. *Uchenye Zapiski Karelo-Finskogo Universiteta, Biologicheskie Nauki* 5: 171–206.

Shukhgalter, O. A. & Nigmatullin, C. M. (2001). Parasitic helminths of jumbo squid *Dosidicus gigas* (Cephalopoda: Ommastrephidae) in open waters of the central east Pacific. *Fisheries Research* 54: 95–110.

Shumway, S. E. & Marsden, I. D. (1982). The combined effects of temperature, salinity, and declining oxygen tension on oxygen consumption in the marine pulmonate *Amphibola crenata* (Gmelin, 1791). *Journal of Marine Biology and Ecology* 61: 133–146.

Shumway, S. E. (1995). Phycotoxin-related shellfish poisoning: Bivalve molluscs are not the only vectors. *Reviews in Fisheries Science* 3: 1–31.

Shumway, S. E., Davis, C., Downey, R., Karney, R., Kraeuter, J., Parsons, J., Rheault, R. & Wikfors, G. (2003). Shellfish aquaculture–in praise of sustainable economies and environments. *World Aquaculture* 34: 8–10.

Shumway, S. E. & Parsons, G. J., Eds. (2006). *Scallops: Biology, Ecology, Aquaculture, and Fisheries*. Developments in Aquaculture and Fisheries Science. Amersterdam, the Netherlands, Elsevier.

Shumway, S. E. (2011). *Shellfish Aquaculture and the Environment*. Chichester, UK, John Wiley & Sons.

Shumway, S. E., Blaschik, N., Kallansrude, L. & Boland, N. (2014a). *Simply Shellfish: A Collection of Recipies by the National Shellfisheries Association*. Kearney, NE, Morris Press Cook Books.

Shumway, S. E., Ward, J. E., Heupel, E., Holohan, B. A., Heupel, J., Heupel, T. & Padilla, D. K. (2014b). Observations of feeding in the common Atlantic slippersnail *Crepidula fornicata* L., with special reference to the "mucus net". *Journal of Shellfish Research* 33: 279–291.

Shumway, S. E. & Parsons, G. J., Eds. (2016). *Scallops: Biology, Ecology, Aquaculture, and Fisheries*. 3rd ed. Developments in Aquaculture and Fisheries Science. Amersterdam, the Netherlands, Elsevier.

Shumway, S. E., Burkholder, J. M. & Morton, S. L. (2018). *Harmful Algal Blooms: A Compendium Desk Reference*. New York, John Wiley & Sons.

Shuto, T. (1974). Larval ecology of prosobranch gastropods and its bearing on biogeography and paleontology. *Lethaia* 7: 239–256.

Sigler, M. F., Hulbert, L. B., Lunsford, C. R., Thompson, N. H., Burek, K., O'Corry-Crowe, G. & Hirons, A. C. (2006). Diet of Pacific sleeper shark, a potential Steller sea lion predator, in the north-east Pacific Ocean. *Journal of Fish Biology* 69: 392–405.

Signor, P. W. (1982a). Constructional morphology of gastropod ratchet sculpture. *Neues Jahrbuch für Geologie und Paläontologie Abhandlungen* 163: 349–368.

Signor, P. W. (1982b). Growth-related surficial resorption of the penultimate whorl in *Terebra dimidiata* (Linnaeus, 1758) and other marine prosobranch gastropods. *The Veliger* 25: 79–82.

Signor, P. W. & Kat, P. W. (1984). Functional significance of columellar folds in turritelliform gastropods. *Journal of Paleontology* 58: 210–216.

Signor, P. W. (1993). Ratchet riposte: More on gastropod burrowing sculpture. *Lethaia* 26: 379–383.

Signor, P. W. & Vermeij, G. J. (1994). The plankton and the benthos: Origins and early history of an evolving relationship. *Paleobiology* 20: 297–319.

Sigurdsson, J. B., Titman, C. W. & Davies, P. A. (1976). The dispersal of young post-larval bivalve molluscs by byssus threads. *Nature* 262: 386–387.

Sigwart, J. D. (2009). The deep-sea chiton *Nierstraszella* (Mollusca: Polyplacophora: Lepidopleurida) in the Indo-West Pacific: Taxonomy, morphology and a bizarre ectosymbiont. *Journal of Natural History* 43: 447–468.

Sigwart, J. D., Schwabe, E., Saito, C. H., Samadi, D. S. & Giribet, E. G. (2010). Evolution in the deep sea: A combined analysis of the earliest diverging living chitons (Mollusca: Polyplacophora: Lepidopleurida). *Invertebrate Systematics* 24: 560–572.

Sigwart, J. D., Sumner-Rooney, L. H., Schwabe, E., Hess, M., Brennan, G. P. & Schrödl, M. (2014). A new sensory organ in 'primitive' molluscs (Polyplacophora: Lepidopleurida), and its context in the nervous system of chitons. *Frontiers in Zoology* 11: 7.

Sigwart, J. D. & Sumner-Rooney, L. H. (2015). Mollusca: Caudofoveata, Monoplacophora, Polyplacophora, Scaphopoda, and Solenogastres. pp. 172–189 *in* A. Schmidt-Rhaesa, Harzsch, S. & Purschke, G. (Eds.) *Structure and Evolution of Invertebrate Nervous Systems*. Oxford, UK, Oxford University Press.

Silberfeld, T., Leigh, J. W., Verbruggen, H., Cruaud, C., De Reviers, B. & Rousseau, F. (2010). A multi-locus time-calibrated phylogeny of the brown algae (Heterokonta, Ochrophyta, Phaeophyceae): Investigating the evolutionary nature of the 'brown algal crown radiation'. *Molecular Phylogenetics and Evolution* 56: 659–674.

Silliman, B. R., Van de Koppel, J., Bertness, M. D., Stanton, L. E. & Mendelssohn, I. A. (2005). Drought, snails, and large-scale die-off of southern U.S. salt marshes. *Science* 310: 1803–1806.

Silva, E. P. & Russo, C. A. M. (2000). Techniques and statistical data analysis in molecular population genetics. *Hydrobiologia* 420: 119–135.

Silva, F. M. Da & Hodgson, A. N. (1987). Fine structure of the pedal muscles of the whelk *Bullia rhodostoma* Reeve: Correlation with function. *Comparative Biochemistry and Physiology Part A* 87A: 143–149.

Silverman, K. (1991). *Edgar A. Poe: Mournful and Never-ending Remembrance*. New York, Harper Collins.

Simakov, O., Marletaz, F., Cho, S.-J., Edsinger-Gonzales, E., Havlak, P., Hellsten, U., Kuo, D.-H. et al. (2013). Insights into bilaterian evolution from three spiralian genomes. *Nature* 493: 526–531.

Simberloff, D. (1998). Flagships, umbrellas, and keystones: Is single-species management passé in the landscape era? *Biological Conservation* 83: 247–257.

Simison, W. B. & Lindberg, D. R. (1999). Morphological and molecular resolution of a putative cryptic species complex: A case study of *Notoacmea fascicularis* (Menke, 1851) (Gastropoda: Patellogastropoda). *Journal of Molluscan Studies* 65: 99–109.

Simison, W. B., Lindberg, D. R. & Boore, J. L. (2006). Rolling circle amplification of metazoan mitochondrial genomes. *Molecular Phylogenetics and Evolution* 39: 562–567.

Simison, W. B. & Boore, J. L. (2008). Molluscan evolutionary genomics. pp. 447–461 *in* W. F. Ponder & Lindberg, D. R. (Eds.) *Phylogeny and Evolution of the Mollusca*. Berkeley, CA, University of California Press.

Simkiss, K. & Wilbur, K. M. (1977). The molluscan epidermis and its secretions. *Proceedings of the Zoological Society of London* 39: 35–76.

Simkiss, K., Jenkins, K. G. A., McLellan, J. & Wheeler, E. (1982). Methods of metal incorporation into intracellular granules. *Experientia* 38: 333–335.

Simkiss, K. & Mason, A. Z. (1983). Metal ions: Metabolic and toxic effects. pp. 101–164 *in* P. W. Hochachka (Ed.) *Environmental Biochemistry and Physiology. The Mollusca*. Vol. 2. New York, Academic Press.

Simkiss, K. (1985). Prokaryote-eukaryote interactions in trace element metabolism. *Desulfovibrio* sp. in *Helix aspersa*. *Experientia* 41: 1195–1197.

Simkiss, K. (1988). Molluscan skin (excluding cephalopods). pp. 11–35 *in* E. R. Trueman & Clarke, M. R. (Eds.) *Form and Function. The Mollusca*. Vol. 11. New York, Academic Press.

Simkiss, K. & Watkins, B. (1990). The influence of gut microorganisms on zinc uptake in *Helix aspersa*. *Environmental Pollution* 66: 263–271.

Simkiss, K. (2017). Developing perspectives on molluscan shells, Part 2: Cellular aspects. pp. 43–76 *in* S. Saleuddin & Mukai, S. *Physiology of Molluscs*. Vol.1, Apple Academic Press & CRC Press.

Simmons, T. L., Andrianasolo, E., McPhail, K., Flatt, P. & Gerwick, W. H. (2005). Marine natural products as anticancer drugs. *Molecular Cancer Therapeutics* 4: 333–342.

Simo, J. A. T., Scott, R. W. & Masse, J. P. (1993). *Cretaceous Carbonate Platforms*. Tulsa, OK, American Association of Petroleum Geologists: 1–479.

de Simone, L. R. L. (2002). Comparative morphological study and phylogeny of representatives of the Superfamily Calyptraeoidea (including Hipponicoidea) (Mollusca, Caenogastropoda). *Neotropica* 2: 2–137.

de Simone, L. R. L. (2004). Comparative morphology and phylogeny of representatives of the Superfamilies of architaenioglossans and the Annulariidae (Mollusca, Caenogastropoda). *Arquivos do Museu Nacional (Rio de Janeiro)* 62: 387–504.

de Simone, L. R. L. (2006). *Land and Freshwater Molluscs of Brazil: An Illustrated Inventory of the Brazilian Malacofauna, including Neighbor regions of the South America, Respect to the Terrestrial and Freshwater Ecosystems*. São Paulo, Brazil, Museu de Zoologia Universidade de São Paulo.

de Simone, L. R. L. (2009). Comparative morphology among representatives of main taxa of Scaphopoda and basal protobranch Bivalvia (Mollusca). *Papéis Avulsos de Zoologia (São Paulo)* 49: 405–458.

de Simone, L. R. L. (2011). Phylogeny of the Caenogastropoda (Mollusca), based on comparative morphology. *Arquivos de Zoologia* 42: 83–323.

Simpson, G. G. (1944). *Tempo and Mode in Evolution*. New York, Columbia University Press.

Simpson, G. G. (1953). *The Major Features of Evolution*. New York, Columbia University Press.

Simroth, H. (1892–1894a). Amphineura und Scaphopoda. pp. i–vii, 1–1056 *in* H. G. Bronn (Ed.) *H. G. Bronn's Klassen und Ordnungen des Tier-Reichs wissenschaftlich dargestellt in Wort und Bild (1892–1894)*. Band 3, Mollusca, Abteilung 1. Leipzig, C. F. Winter.

Simroth, H. (1892–94b). Gastropoda. Prosobranchia. pp. 1–1056 *in* H. G. Bronn (Ed.) *H. G. Bronn's Klassen und Ordnungen des Tier-Reichs wissenschaftlich dargestellt in Wort und Bild (1892–1894)*. Band 3, Mollusca, Abteilung 1. Leipzig und Heidelberg, C. F. Winter.

Simroth, H. & Hoffmann, H. (1896). Pulmonata. pp. 913–1354 *in* H. G. Bronn (Ed.) *H. G. Bronn's Klassen und Ordnungen des Tier-Reichs wissenschaftlich dargestellt in Wort und Bild (1896–1907). Mollusca*. Band 3, Mollusca, Abteilung 2. Leipzig, C.F. Winter.

Simroth, H. (1896–1907). Gastropoda. Buch 1, Prosobranchia. pp. i–vii, 1–1056 *in* H. G. Bronn (Ed.) *H. G. Bronn's Klassen und Ordnungen des Thier-Reichs*. Band 3, Mollusca, Abteilung 2. Leipzig, C. F. Winter.

Sindermann, C. J. & Rosenfield, A. (1967). Principal diseases of commercially important marine bivalve Mollusca and Crustacea. *Fishery Bulletin* 66: 335–385.

Sindermann, C. J. (1990). *Principal Diseases of Marine Fish and Shellfish*. Vol. 2. San Diego, CA, Academic Press.

Singer, S., van Fleet, A. L., Viel, J. J. & Genevese, E. E. (1997). Biological control of the zebra mussel *Dreissena polymorpha* and the snail *Biomphalaria glabrata*, using Gramicidin S and D and molluscicidal strains of *Bacillus*. *Journal of Industrial Microbiology and Biotechnology* 18: 226–231.

Singley, C. T. (1977). An analysis of gastrulation in *Loligo pealei*. PhD, University of Hawaii.

Sinha, S. K. & Nandi, B. C. (2010). Life history of *Liosarcophaga choudhuri* (Sinha and Nandi) (Diptera: Sarcophagidae), a parasitoid of the intertidal snail *Littoraria melanostoma* (Gray) (Gastropoda: Littorinidae). *Tijdschrift voor Entomologie* 153: 85–90.

Sinha, V. R. P. & Jones, J. W. (1967). On the food of the freshwater eels and their feeding relationship with the salmonids. *Journal of Zoology* 153: 119–137.

Sirenko, B. & Minichev, S. (1975). Développement ontogénétique de la radula chez les polyplacophores. *Cahiers de Biologie Marine* 16: 425–433.

Sitnikova, T. J., Kiyashko, S. I., Maximova, N., Pomazkina, G. V., Roepstorf, P., Wada, E. & Michel, E. (2012a). Resource partitioning in endemic species of Baikal gastropods indicated by gut contents, stable isotopes and radular morphology. *Hydrobiologia* 682: 75–90.

Sitnikova, T. J., Michel, E., Tulupova, Y., Khanaev, I., Parfenova, V. & Prozorova, L. A. (2012b). Spirochetes in gastropods from Lake Baikal and North American freshwaters: New multi-family, multi-habitat host records. *Symbiosis* 56: 103–110.

Skelding, J. M. (1973). The fine structure of the kidney of *Achatina achatina* (L.). *Zeitschrift für Zellforschung und Mikroskopische Anatomie* 147: 1–29.

Skelton, M. E., Alevizos, A. & Koester, J. (1992). Control of the cardiovascular system of *Aplysia* by identified neurons. *Experientia* 48: 809–817.

Skelton, M. E. & Koester, J. (1992). The morphology, innervation and neural control of the anterior arterial system of *Aplysia californica*. *Journal of Comparative Physiology* 171: 141–155.

Skinner, T. L. & Peretz, B. (1989). Age sensitivity of osmoregulation and of its neural correlates in *Aplysia*. *American Journal of Physiology* 256: R989–R996.

Skovsted, C. B., Brock, G. A., Lindström, A., Peel, J. S., Paterson, J. R. & Fuller, M. K. (2007). Early Cambrian record of failed durophagy and shell repair in an epibenthic mollusc. *Biology Letters* 3: 314–317.

Sleurs, W. J. M. (1985). Marine microgastropods from the Republic of Maldives. 1. Genus *Ammonicera* Vayssière, 1893, with description of four new species (Prosobranchia: Omalogyridae). *Basteria* 49: 19–27.

Slijepcevic, P. (1998). Telomeres and mechanisms of Robertsonian fusion. *Chromosoma* 107: 136–140.

Sliusarev, G. S. (2008). Phylum Orthonectida: Morphology, biology, and relationships to other multicellular animals [in Russian]. *Zhurnal Obshchei Biologii* 69: 403–427.

Sloan, N. A. (1980). Aspects of the feeding biology of asteroids. *Oceanography and Marine Biology Annual Review* 18: 57–124.

Slootweg, R. (1987). Prey selection by molluscivorous cichlids foraging on a schistosomiasis vector snail, *Biomphalaria glabrata*. *Oecologia* 74: 193–202.

Slootweg, R., Malek, E. A. & McCullough, F. S. (1994). The biological control of snail intermediate hosts of schistosomiasis by fish. *Reviews in Fish Biology and Fisheries* 4: 67–90.

Slugina, Z. V. (2006). Endemic Bivalvia in ancient lakes. *Hydrobiologia* 568: 213–217.

Smaldon, P. R. & Duffus, J. H. (1985). An ultrastructural study of gametes and fertilization in *Patella vulgata*. *Journal of Molluscan Studies* 51: 116–132.

Smale, M. J. (1996). Cephalopods as prey. IV. Fishes. *Philosophical Transactions of the Royal Society B* 351: 1067–1081.

Smallwood, W. M. (1905). Some observations on the chromosome vesicles in the maturation of nudibranchs. *Morphologisches Jahrbuch* 33: 87–105.

Sminia, T., Boer, H. H. & Niemantsverdriet, A. (1972). Haemoglobin producing cells in freshwater snails. *Zeitschrift für Zellforschung und Mikroskopische Anatomie* 135: 563–568.

Sminia, T. & Boer, H. H. (1973). Haemocyanin production in pore cells of the freshwater snail *Lymnaea stagnalis*. *Zeitschrift für Zellforschung und Mikroskopische Anatomie* 145: 443–445.

Sminia, T., Pietersma, K. & Scheerboom, J. E. M. (1973). Histological and ultrastructural observations on wound healing in the freshwater pulmonate *Lymnaea stagnalis*. *Zeitschrift für Zellforschung und Mikroskopische Anatomie* 141: 561–573.

Smit, A. B., Thijsen, S. F. & Geraerts, W. P. M. (1993). cDNA cloning of the sodium-influx-stimulating peptide in the mollusc, *Lymnaea stagnalis*. *European Journal of Biochemistry* 215: 397–400.

Smit, A. B., Spijker, S., Van Minnen, J., Burke, J. F., DeWinter, F., Van Elk, R., & Geraerts, W. P. M. (1996). Expression and characterization of molluscan insulin-related peptide VII from the mollusc *Lymnaea stagnalis*. *Neuroscience* 70: 589–596.

Smith, A. E. (1885). Report on the Lamellibranchiata collected by H.M.S. 'Challenger', during the years 1873–76. *Reports on the Scientific Results of the Voyage of the H. M. S. 'Challenger' (1873–1876) Zoology* 13: 1–341.

Smith, A. M. (1991). The role of suction in the adhesion of limpets. *Journal of Experimental Biology* 161: 151–169.

Smith, A. M. (1992). Alternations between attachment mechanisms by limpets in the field. *Journal of Experimental Marine Biology and Ecology* 160: 205–220.

Smith, A. M. (2002). The structure and function of adhesive gels from invertebrates *Integrative and Comparative Biology* 42: 1164–1171.

Smith, A. M. & Morin, M. C. (2002). Biochemical differences between trail mucus and adhesive mucus from marsh periwinkle snails. *Biological Bulletin* 203: 338–346.

Smith, A. M. (2006). The biochemistry and mechanics of gastropod adhesive gels. pp. 167–182 in A. M. Smith & Callow, J. A. (Eds.) *Biological Adhesives*. Berlin, Germany, Springer-Verlag.

Smith, A. M. (2010). Gastropod secretory glands and adhesive gels. pp. 41–51 in I. Grunwald (Ed.) *Biological Adhesive Systems: From Nature to Technical and Medical Application*. Vienna, Austria, Springer.

Smith, A. M. (2013). Multiple metal-based cross-links: Protein oxidation and metal coordination in a biological glue. pp. 3–15 in R. Santos, Aldred, N., Gorb, S. & Flammang, P. (Ed.) *Biological and Biometric Adhesives: Challenges and Opportunities*. Cambridge, UK, Royal Society of Chemistry.

Smith, B. D. (1987a). Growth rate, distribution and abundance of the introduced topshell *Trochus niloticus* Linnaeus on Guam, Marianna Islands. *Bulletin of Marine Sciences* 41: 466–474.

Smith, C., Papadopoulou, N., Sevastou, K., Franco, A., Teixeira, H., Piroddi, C., Katsanevakis, S., Fürhaupter, K., Beauchard, O. & Cochrane, S. (2015). *Report on Identification of Keystone Species and Processes across Regional Seas*, DEVOTES Project: 1–105.

Smith, C. R. & Baco, A. R. (2003). Ecology of whale falls at the deep-sea floor. *Oceanography and Marine Biology Annual Review* 41: 311–354.

Smith, D. A. S. (1975). Polymorphism and selective predation in *Donax faba* Gmelin (Bivalvia: Tellinacea). *Journal of Experimental Marine Biology and Ecology* 17: 205–219.

Smith, D. G. (1983). On the so-called mantle muscle scars on shells of the Margaritiferidae (Mollusca, Pelecypoda), with observations on mantle-shell attachment in the Unionoida and Trigonioida. *Zoologica Scripta* 12: 67–71.

Smith, E. H. (1969). Functional morphology of *Penitella conradi* relative to shell penetration. *American Zoologist* 9: 869–888.

Smith, F. G. W. (1935). The development of *Patella vulgata*. *Philosophical Transactions of the Royal Society of Edinburgh* 225: 95–125.

Smith, J. M. (1982a). *Evolution and the Theory of Games*. Cambridge, UK, Cambridge University Press.

Smith, L. D. & Palmer, A. R. (1994). Effects of manipulated diet on size and performance of brachyuran crab claws. *Science* 264: 710–712.

Smith, L. D. & Jennings, J. A. (2000). Induced defensive responses by the bivalve *Mytilus edulis* to predators with different attack modes. *Marine Biology* 136: 461–469.

Smith, L. H. (1981a). Quantified aspects of pallial fluid and its effects on the duration of locomotor activity in the terrestrial gastropod *Triodopsis albolabris*. *Physiological Zoology* 54: 407–414.

Smith, L. H. & Pierce, S. K. (1987). Cell volume regulation by molluscan erythrocytes during hypoosmotic stress: Ca^{2+} effects on ionic and inorganic osmolyte effluxes. *Biological Bulletin* 173: 407–418.

Smith, L. S. (1963). Circulatory anatomy of the octopus arm. *Journal of Morphology* 113: 261–266.

Smith, M. & Quayle, D. B. (1963). Deoxyribonucleic acids of marine Mollusca. *Nature* 200: 676.

Smith, M., Quick, T. J. & St. Peter, R. L. (1999). Differences in the composition of adhesive and non-adhesive mucus from the limpet *Lottia limatula*. *Biological Bulletin* 196: 34–44.

Smith, P. J. S. (1981b). The role of venous pressure in regulation of output from the heart of the octopus *Eledone cirrhosa* (Lam.). *Journal of Experimental Biology* 93: 243–255.

Smith, P. J. S. (1982b). The contribution of the branchial heart to the accessory branchial pump in the Octopoda. *Journal of Experimental Biology* 98: 229–237.

Smith, P. J. S. & Boyle, P. R. (1983). The cardiac innervation of *Eledone cirrhosa* (Lamarck) (Mollusca: Cephalopoda). *Philosophical Transactions of the Royal Society B* 300: 493–511.

Smith, P. J. S. (1985). Cardiac performance in response to loading pressures for two molluscan species, *Busycon canaliculatum* (L.) (Gastropoda) and *Mercenaria mercenaria* (L.) (Bivalvia). *Journal of Experimental Biology* 119: 301–320.

Smith, P. J. S. (1987b). Cardiac output in the Mollusca: Scope and regulation. *Experientia* 43: 956–965.

Smith, S. A., Nason, J. & Croll, R. P. (1997). Detection of APGWamide-like immunoreactivity in the sea scallop, *Placopecten magellanicus*. *Neuropeptides* 31: 155–165.

Smith, S. A., Wilson, N. G., Goetz, F., Feehery, C., Andrade, S. C. S., Rouse, G. W., Giribet, G. & Dunn, C. W. (2011). Resolving the evolutionary relationships of molluscs with phylogenomic tools. *Nature* 480: 364–367.

Smith, S. E., Brittain, T. & Wells, R. M. G. (1988). A kinetic and equilibrium study of ligand binding to the monomeric and dimeric haem-containing globins of two chitons. *Biochemical Journal* 252: 673–678.

Smith, S. T. (1967). The development of *Retusa obtusa* (Montagu) (Gastropoda, Opisthobranchia). *Canadian Journal of Zoology* 45: 737–764.

Smits, J. D., Witte, F. & Veen, F. G. (1996). Functional changes in the anatomy of the pharyngeal jaw apparatus of *Astatoreochromis alluaudi* (Pisces, Cichlidae), and their effects on adjacent structures. *Biological Journal of the Linnean Society* 59: 389–409.

Smock, T. & Arch, S. (1986). What commands egg laying in *Aplysia*. *Behavioral and Brain Sciences* 9: 734–735.

Smyth, M. J. (1988). *Penetrantia clionoides*, sp. nov. (Bryozoa), a boring bryozoan in gastropod shells from Guam. *Biological Bulletin* 174: 276–286.

Smyth, M. J. (1989). Bioerosion of gastropod shells: With emphasis on effects of coralline algal cover and shell microstructure. *Coral Reefs* 8: 119–126.

Snegin, E. A., Sychev, A. A. & Shapovalov, A. S. (2016). Estimating the impact of ungulates on Holocene steppe ecosystems by analyzing repaired injuries in land snail shells. *Russian Journal of Ecology* 47: 514–517.

Snyder, M. J., Girvetz, E. & Mulder, E. P. (2001). Induction of marine mollusc stress proteins by chemical or physical stress. *Archives of Environmental Contamination and Toxicology* 41: 22–29.

Snyder, N. F. R. & Snyder, H. A. (1969). A comparative study of mollusk predation by limpkins, everglade kites, and boat-tailed grackles. *Living Bird* 8: 177–223.

Sobrino-Figueroa, A. S., Cáceres-Martínez, C., Botello, A. V. & Nunez-Nogueira, G. (2007). Effect of cadmium, chromium, lead and metal mixtures on survival and growth of juveniles of the scallop *Argopecten ventricosus* (Sowerby II, 1842). *Journal of Environmental Science and Health Part A* 42: 1443–1447.

Sokolov, V. A., Kamardin, N. N., Zaĭtseva, O. V. & Tsirulis, T. P. (1980). Sensory system of osphradium in gastropod mollusks. *Sensory Systems* 1980: 158–174.

Sokolov, V. A. & Zaĭtseva, O. V. (1982). Osphradial chemoreceptors in lamellibranch mollusks, *Unio pectorum* and *Anodonta cygnea*. *Journal of Evolutionary Biochemistry and Physiology* 18: 56–61.

Sokolova, I. M. & Berger, V. (2000). Physiological variation related to shell colour polymorphism in White Sea *Littorina saxatilis*. *Journal of Experimental Marine Biology and Ecology* 245: 1–23.

Sokolova, I. M. & Pörtner, H. O. (2003). Metabolic plasticity and critical temperatures for aerobic scope in a eurythermal marine invertebrate (*Littorina saxatilis*, Gastropoda: Littorinidae) from different latitudes. *Journal of Experimental Biology* 206: 195–207.

Sokolova, I. M. (2009). Apoptosis in molluscan immune defence. *Invertebrate Survival Journal* 6: 49–58.

Sokolove, P. G., Beiswanger, C. M., Prior, D. J. & Gelperin, A. (1977). A circadian rhythm in the locomotive behaviour of the giant garden slug *Limax maximus*. *Journal of Experimental Biology* 66: 47–64.

Soledad Avaca, M., Narvarte, M., Martin, P. & Van der Molen, S. (2013). Shell shape variation in the nassariid *Buccinanops globulosus* in northern Patagonia. *Helgoland Marine Research* 67: 567–577.

Solem, A. C. (1974). *The Shell Makers. Introducing Mollusks*. New York, John Wiley & Sons.

Solem, A. C. & Yochelson, E. L. (1979). North American Paleozoic land snails, with a summary of other Paleozoic nonmarine snails. *United States Geological Survey. Professional Paper* 1072: 1–42.

Solem, A. C., Climo, F. M. & Roscoe, D. J. (1981). Sympatric species diversity of New Zealand land snails. *New Zealand Journal of Zoology* 8: 453–485.

Solem, A. C. (1983). Lost or kept internal whorls: Ordinal differences in land snails. *Journal of Molluscan Studies: Supplement* 12A: 172–178.

Solem, A. C. (1984). A world model of land snail diversity and abundance. pp. 6–22 in A. Solem & van Bruggen, A. C. (Ed.) *World Wide Snails. Biogeographical Studies on Non Marine Mollusca*. Leiden, E.J. Brill/W. Backhuys.

Solem, A. C. (1985). Origin and diversification of pulmonate land snails. pp. 269–293 in E. R. Trueman & Clarke, M. R. (Eds.) *Evolution. The Mollusca*. Vol. 10. New York, Academic Press.

Solem, A. C. (1990). How many Hawaiian land snail species are left? And what we can do for them. *Bishop Museum Occasional Papers* 30: 27–40.

Somero, G. N., Childress, J. J. & Anderson, A. E. (1989). Transport metabolism and detoxification of hydrogen sulfide in animals from sulfide-rich marine environments. *Reviews in Aquatic Sciences* 1: 591–614.

Somero, G. N. (2002). Thermal physiology and vertical zonation of intertidal animals: Optima, limits, and costs of living. *Integrative and Comparative Biology* 42: 780–789.

Sommerville, B. A. (1973). The circulatory physiology of *Helix Pomatia*: I. Observations on the mechanism by which *Helix* emerges from its shell and on the effects of body movement on cardiac function. *Journal of Experimental Biology* 59: 275–282.

Son, M. H. & Hughes, R. N. (2000). Sexual dimorphism of *Nucella lapillus* (Gastropoda: Muricidae) in North Wales, UK. *Journal of Molluscan Studies* 66: 489–498.

Sone, E. D., Weiner, S. & Addadi, L. (2007). Biomineralization of limpet teeth: A cryo-TEM study of the organic matrix and the onset of mineral deposition. *Journal of Structural Biology* 158: 428–444.

Sone, H., Kondo, T., Kiryu, M., Ishiwata, H., Ojika, M. & Yamada, K. (1995). Dolabellin, a cytotoxic bisthiazole metabolite from the sea hare *Dolabella auricularia*: Structural determination and synthesis. *The Journal of Organic Chemistry* 60: 4774–4781.

Sonetti, D., Ottaviani, E. & Stefano, G. B. (1997). Opiate signaling regulates microglia activities in the invertebrate nervous system. *General Pharmacology* 29: 39–47.

Sonetti, D., Mola, L., Casares, F., Bianchi, E., Guarna, M. & Stefano, G. B. (1999). Endogenous morphine levels increase in molluscan neural and immune tissues after physical trauma. *Brain Research* 835: 137–147.

Sonetti, D. & Peruzzi, E. (2004). Neuron-microglia communication in the CNS of the freshwater snail *Planorbarius corneus*. *Acta Biologica Hungarica* 55: 273–285.

Song, Y., Miao, J., Cai, Y. & Pan, L. (2015). Molecular cloning, characterization, and expression analysis of a gonadotropin-releasing hormone-like cDNA in the clam, *Ruditapes philippinarum*. *Comparative Biochemistry and Physiology Part B* 189: 47–54.

Song, Y. F., Lu, Y., Ding, H. B., Lu, H. B., Gao, G. H. & Sun, C. H. (2013). Structural characteristics at the adductor muscle and shell interface in mussel. *Applied Biochemistry and Biotechnology* 171: 1203–1211.

Sorensen, F. E. & Lindberg, D. R. (1991). Preferential predation by American black oystercatchers on transitional ecophenotypes of the limpet *Lottia pelta* (Rathke). *Journal of Experimental Marine Biology and Ecology* 154: 123–136.

Sorensen, R. E. & Minchella, D. J. (2001). Snail-trematode life history interactions: Past trends and future directions. *Parasitology* 123: S3–S18.

Soto, L. A. (2009). Stable carbon and nitrogen isotopic signatures of fauna associated with the deep-sea hydrothermal vent system of Guaymas Basin, Gulf of California. *Deep Sea Research Part II: Topical Studies in Oceanography* 56: 1675–1682.

Soucier, C. P. & Basil, J. A. (2008). Chambered nautilus (*Nautilus pompilius pompilius*) responds to underwater vibrations. *American Malacological Bulletin* 24: 3–11.

Soudant, P., Chu, F. L. E. & Volety, A. K. (2013). Host-parasite interactions: Marine bivalve molluscs and protozoan parasites, *Perkinsus* species. *Journal of Invertebrate Pathology* 114: 196–216.

Sousa, R., Antunes, C. & Guilhermino, L. (2008). Ecology of the invasive Asian clam *Corbicula fluminea* (Müller, 1774) in aquatic ecosystems: An overview. *Annales de Limnologie* 44: 85–94.

Sousa, R., Gutiérrez, J. L. & Aldridge, D. C. (2009). Non-indigenous invasive bivalves as ecosystem engineers. *Biological Invasions* 11: 2367–2385.

Sousa, W. P. (1983). Host life history and the effect of parasitic castration on growth: A field study of *Cerithidea californica* Haldeman (Gastropoda: Prosobranchia) and its trematode parasites. *Journal of Experimental Marine Biology and Ecology* 73: 273–296.

Sousa, W. P. & Gleason, M. (1989). Does parasitic infection compromise host survival under extreme environmental conditions? The case for *Cerithidea californica* (Gastropoda: Prosobranchia). *Oecologia* 80: 456–464.

Sousounis, K., Ogura, A. & Tsonis, P. A. (2013). Transcriptome analysis of *Nautilus* and Pygmy Squid developing eye provides insights in lens and eye evolution. *PLoS ONE* 8: e78054.

South, A. (1992). *Terrestrial Slugs: Biology, Ecology and Control*. London, UK, Chapman & Hall.

Southward, A. J. (1958). The zonation of plants and animals on rocky sea shores. *Biological Reviews* 33: 137–177.

Spagnolia, T. & Wilkens, L. A. (1983). Neurobiology of the Scallop. II. Structure of the parietovisceral ganglion lateral lobes in relation to afferent projections from the mantle eyes. *Marine and Freshwater Behaviour and Physiology* 10: 23–55.

Spalding, M. D., Fox, H. E., Allen, G. R., Davidson, N., Ferdaña, Z. A., Finlayson, M., Halpern, B. S., Jorge, M. A., Lombana, A. L. & Lourie, S. A. (2007). Marine ecoregions of the world: A bioregionalization of coastal and shelf areas. *BioScience* 57: 573–583.

Spann, N., Harper, E. M. & Aldridge, D. C. (2010). The unusual mineral vaterite in shells of the freshwater bivalve *Corbicula fluminea* from the UK. *Naturwissenschaften* 97: 743–751.

Speeg, K. V. & Campbell, J. W. (1968). Purine biosynthesis and excretion in *Otala* (= *Helix*) *lactea*: An evaluation of the nitrogen excretory potential. *Comparative Biochemistry and Physiology* 26: 579–595.

Speight, M. C. D. (1974). Food and feeding habits of the frog *Rana temporaria* in bogland habitats in the west of Ireland. *Journal of Zoology* 172: 67–79.

Speiser, B. (2001). Food and feeding behaviour. pp. 259–288 *in* G. M. Barker (Ed.) *The Biology of Terrestrial Molluscs*. Wallingford, UK, CABI Publishing.

Speiser, B., Zaller, J. G. & Neudecker, A. (2001). Size-specific susceptibility of the pest slugs *Deroceras reticulatum* and *Arion lusitanicus* to the nematode biocontrol agent *Phasmarhabditis hermaphrodita*. *BioControl* 46: 311–320.

Speiser, D. I. & Johnsen, S. (2008). Scallops visually respond to the size and speed of virtual particles. *Journal of Experimental Biology* 211: 2066–2070.

Speiser, D. I. & Johnsen, S. (2011). Seeing through rocks: Chitons use calcium carbonate lenses to form images. *Society for Integrative and Comparative Biology* 51: E130.

Spencer, B. E. (2002). *Molluscan Shellfish Farming*. Oxford, UK, Blackwell Publishing.

Spencer, G. E., Rothwell, C. M. & Benjamin, P. R. (2017). Associative memory mechanisms in the pond snail *Lymnaea stagnalis*. pp. 1–42 *in* S. Saleuddin & Mukai, S. T. (Eds.) *Physiology of Molluscs*. Vol. 2. Oakville, ON, Apple Academic Press.

Spencer, H. G., Marshall, B. A., Maxwell, P. A., Grant-Mackie, J. A., Stilwell, J. D., Willan, R. C., Campbell, H. J., Crampton, J. S., Henderson, R. A. & Bradshaw, M. A. (2009). Phylum Mollusca: Chitons, clams, tusk shells, snails, squids, and kin. pp. 161–209 *in* D. Gordon (Ed.) *New Zealand Inventory of Biodiversity, Volume one. Kingdom Animalia: Radiata, Lophotrochozoa, Deuterostomia*. Christchurch, New Zealand, Canterbury University Press.

Spengel, J. W. (1881). Die Geruchsorgane und das Nervensystem der Mollusken. *Zeitschrift für wissenschaftliche Zoologie* 35: 333–383.

Spiers, Z. B., Gabor, M., Fell, S. A., Carnegie, R. B., Dove, M., Connor, W. O., Frances, J., Go, J., Marsh, I. B. & Jenkins, C. (2014). Longitudinal study of winter mortality disease in Sydney rock oysters *Saccostrea glomerata*. *Diseases of Aquatic Organisms* 110: 151–164.

Spight, T. M., Birkland, C. & Lyons, A. (1974). Life histories of large and small murexes (Prosobranchia: Muricidae). *Marine Biology* 24: 229–242.

Spight, T. M. (1975). Extending gastropod embryonic development and their selective cost. *Oecologia* 21: 1–16.

Spight, T. M. (1976a). Colors and patterns of an intertidal snail, *Thais lamellosa*. *Researches on Population Ecology* 17: 176–190.

Spight, T. M. (1976b). Hatching size and the distribution of nurse eggs among prosobranch embryos. *Biological Bulletin* 150: 491–499.

Spinella, A., Mollo, E., Trivellone, E. & Cimino, G. (1997). Testudinariol A and B, two unusual triterpenoids from the skin and the mucus of the marine mollusc *Pleurobrancus testudinarius*. *Tetrahedron Letters* 53: 16891–16896.

van der Spoel, S. (1967). *Euthecosomata: A Group with Remarkable Developmental Stages (Gastropoda, Pteropoda)*. Gorinchem, the Netherlands, J. Noorduijn en Zoon N.V.

van der Spoel, S. (1973). Strobilation in a mollusk: The development of aberrant stages in *Clio pyramidata* Linnaeus, 1767 (Gastropoda, Pteropoda). *Bijdragen tot de Dierkunde* 43: 202–215.

van der Spoel, S. (1979). Strobilation in a pteropod (Gastropoda, Opisthobranchia). *Malacologia* 18: 27–30.

van der Spoel, S., Van Bruggen, A. C., & Lever, J., Eds. (1979). *Pathways in Malacology*. 6th International Congress of Unitas Malacologica Europaea, Utrecht, August 15–20 1977. Amsterdam, the Netherlands, Scheltema and Holkema B. V.

Sprague, V. (1978). Comments on trends in research on parasitic diseases of shell fish and fish. *Marine Fisheries Review* 40: 26–29.

Spray, D. C. (1986). Cutaneous temperature receptors. *Annual Review of Physiology* 48: 625–638.

Springer, J., Ruth, P., Beuerlein, K., Palus, S., Schipp, R. & Westermann, B. (2005). Distribution and function of biogenic amines in the heart of *Nautilus pompilius* L. (Cephalopoda, Tetrabranchiata). *Journal of Molecular Histology* 36: 345–353.

Srilakshmi, G. (1991). The hypobranchial gland in *Morula granulata* (Gastropoda: Prosobranchia). *Journal of the Marine Biological Association of the United Kingdom* 71: 623–634.

St-Onge, P., Tremblay, R. & Sevigny, J. M. (2015). Tracking larvae with molecular markers reveals high relatedness and early seasonal recruitment success in a partially spawning marine bivalve. *Oecologia* 178: 733–746.

Stahlschmidt, V. & Wolff, H. G. (1972). The fine structure of the statocyst of the prosobranch mollusc *Pomacea paludosa*. *Zeitschrift für Zellforschung und Mikroskopische Anatomie* 133: 529–537.

Stallard, M. O. & Faulkner, D. J. (1974). Chemical constituents of the digestive gland of the sea hare *Aplysia californica*: I. Importance of diet. *Comparative Biochemistry and Physiology Part B* 49B: 25–35.

Stanisic, J. & Ponder, W. F. (2004). Forest snails in eastern Australia: One aspect of the other 99%. pp. 127–149 *in* D. Lunney (Ed.) *Conservation of Australia's Forest Faunas*. Mosman, New South Wales, Australia, Royal Zoological Society of New South Wales.

Stanley, D. W. (2014). *Eicosanoids in Invertebrate Signal Transduction Systems*. Princeton, NJ, Princeton University Press.

Stanley, S. M. (1970). *Relation of Shell Form to Life Habits of the Bivalvia (Mollusca)* Vol. 125. Boulder, CO, Geological Society of America.

Stanley, S. M. (1972). Functional morphology and evolution of byssally attached bivalved molluscs. *Journal of Paleontology* 46: 165–212.

Stanley, S. M. (1981). Infaunal survival: Alternative functions of shell ornamentation in the Bivalvia (Mollusca). *Paleobiology* 7: 384–393.

Stanwell-Smith, D. & Clarke, A. (1998). The timing of reproduction in the Antarctic limpet *Nacella concinna* (Strebel, 1908) (Patellidae) at Signy Island, in relation to environmental variables. *Journal of Molluscan Studies* 64: 123–127.

Staras, K., Kemenes, G. Y. & Benjamin, P. R. (1999). Electrophysiological and behavioral analysis of lip touch as a component of the food stimulus in the snail *Lymnae*. *Journal of Neurophysiology* 81: 1261–1273.

Starliper, C. E. & Morrison, P. (2000). Bacterial pathogen contagion studies among freshwater bivalves and salmonid fishes. *Journal of Shellfish Research* 19: 251–258.

Starr, M., Himmelman, J. H. & Therriault, J. C. (1990). Direct coupling of marine invertebrate spawning with phytoplankton blooms. *Science* 247: 1071–1074.

Stasek, C. R. (1965). Feeding and particle-sorting in *Yoldia ensifera* (Bivalvia: Protobranchia), with notes on other nuculanids. *Malacologia* 2: 349–366.

Stasek, C. R. (1967). Autotomy in the Mollusca. *Occasional Papers. California Academy of Science* 59: 1–44.

Stasek, C. R. (1972). The molluscan framework. pp. 1–44 *in* M. Florkin & Scheer, B. T. (Eds.) *Chemical Zoology. Mollusca*. Vol. 7. New York, Academic Press.

Stasek, C. R. & McWilliams, W. R. (1973). The comparative morphology and evolution of the molluscan mantle edge. *The Veliger* 16: 1–19.

Staub, R. & Ribi, G. (1995). Size-assortative mating in a natural population of *Viviparus ater* (Gastropoda: Prosobranchia) in Lake Zürich, Switzerland. *Journal of Molluscan Studies* 61: 237–247.

Staubach, S. & Klussmann-Kolb, A. (2007). The cephalic sensory organs of *Acteon tornatilis* (Linnaeus, 1758) (Gastropoda Opisthobranchia) cellular innervation patterns as a tool for homologisation. *Bonner Zoologische Beiträge* 55: 311–318.

Staubach, S. (2008). The evolution of the Cephalic Sensory Organs within the Opisthobranchia. PhD, Johann Wolfang Goethe Universität.

Stead, R. A. & Thompson, R. J. (2003). Physiological energetics of the protobranch bivalve *Yoldia hyperborea* in a cold ocean environment. *Polar Biology* 26: 71–78.

Stead, R. A., Thompson, R. J. & Jaramillo, J. R. (2003). Absorption efficiency, ingestion rate, gut passage time and scope for growth in suspension- and deposit-feeding *Yoldia hyperborea*. *Marine Ecology Progress Series* 252: 159–172.

Stearns, S. C. (1976). Life-history tactics: A review of the ideas. *Quarterly Review of Biology* 51: 3–47.

Stearns, S. C. (1977). The evolution of life-history tactics. *Annual Review of Ecology, Evolution, and Systematics* 8: 145–171.

Stearns, S. C. (1980). A new view of life-history evolution. *Oikos* 35: 266–281.

Steele, T. E. & Klein, R. G. (2008). Intertidal shellfish use during the Middle and Later Stone Age of South Africa. *Archaeofauna* 17: 63–76.

Stefaniak, L. M., McAtee, J. & Shulman, M. J. (2005). The costs of being bored: Effects of a clionid sponge on the gastropod *Littorina littorea* (L). *Journal of Experimental Marine Biology and Ecology* 327: 103–114.

Stein, R. A., Kitchell, J. F. & Knežzevic, B. (1975). Selective predation by carp (*Cyprinus carpio* L.) on benthic molluscs in Skadar Lake, Yugoslavia. *Journal of Fish Biology* 7: 391–399.

Steinberg, P. D. (1985). Feeding preferences of *Tegula funebralis* and chemical defenses of marine brown algae. *Ecological Monographs* 53: 333–349.

Steiner, G. (1991). Observations on the anatomy of the scaphopod mantle and the description of a new family, the Fustiariidae. *American Malacological Bulletin* 9: 1–20.

Steiner, G. (1992). Phylogeny and classification of Scaphopoda. *Journal of Molluscan Studies* 58: 385–400.

Steiner, G. & Dreyer, H. (2003). Molecular phylogeny of Scaphopoda (Mollusca) inferred from 18S rDNA sequences: Support for a Scaphopoda-Cephalopoda clade. *Zoologica Scripta* 32: 343–356.

Steiner, H. (2004). Peptidoglycan recognition proteins: On and off switches for innate immunity. *Immunological Reviews* 198: 83–96.

Stella, J. S. (2012). Evidence of corallivory by the keyhole limpet *Diodora galeata*. *Coral Reefs* 31: 579.

Steneck, R. S. & Watling, L. (1982). Feeding capabilities and limitations of herbivorous molluscs: A functional group approach. *Marine Biology* 68: 299–320.

Steneck, R. S. & Carlton, J. T. (2001). Human alteration of marine communities: Students beware! pp. 445–468 *in* M. D. Bertness, Gains, S. D. & Hay, M. E. (Eds.) *Marine Community Ecology*. Sunderland, MA, Sinaur Associates.

Steneck, R. S., Graham, M. H., Bourque, B. J., Corbett, D., Erlandson, J. M., Estes, J. A. & Tegner, M. J. (2002). Kelp forest ecosystems: Biodiversity, stability, resilience and future. *Environmental Conservation* 29: 436–459.

Stenseng, E., Braby, C. E. & Somero, G. N. (2005). Evolutionary and acclimation-induced variation in the thermal limits of heart function in congeneric marine snails (genus *Tegula*): Implications for vertical zonation. *The Biological Bulletin* 208: 138–144.

Stenyakina, A., Walters, L., Hoffman, E. & Calestani, C. (2010). Food availability and sex reversal in *Mytella charruana*, an introduced bivalve in the southeastern United States. *Molecular Reproduction and Development* 77: 222–230.

Stenzel, H. B. (1971). *Oysters*. Vol. 6. Lawrence, KS, Geological Society of America and University of Kansas Press.

Stephano, J. L. & Gould, M. (1988). Avoiding polyspermy in oyster (*Crassostrea gigas*). *Aquaculture* 73: 295–307.

Stephano, J. L. (1992). A study of polyspermy in abalone. pp. 518–526 *in* S. A. Shepherd, Tegner, M. J. & Guzman del Proo, S. A. (Eds.) *Abalone of the World. Biology, Fisheries and Culture*. Oxford; Carlton, Vic, Fishing News Books.

Stephens, P. J. (1978). The sensitivity and control of the scallop mantle edge. *Journal of Experimental Biology* 75: 203–221.

Stephens, P. R. & Young, J. Z. (1978). Semicircular canals in squids. *Nature* 271: 444–445.

Stephenson, T. A. & Stephenson, A. (1949). The universal features of zonation between tide-marks on rocky coasts. *The Journal of Ecology* 37: 289–305.

Stephenson, T. A. & Stephenson, A. (1972). *Life Between Tidemarks on Rocky Shores*. San Francisco, CA, Freeman.

Sterki, V. (1893). Growth changes of the radula in land-mollusks. *Proceedings of the Academy of Natural Sciences of Philadelphia* 45: 388–400.

Sternberg, R. M., Hotchkiss, A. K. & LeBlanc, G. A. (2008). Synchronized expression of retinoid X receptor mRNA with reproductive tract recrudescence in an imposex-susceptible mollusc. *Environmental Science & Technology* 42: 1345–1351.

Sternberg, R. M., Gooding, M. P., Hotchkiss, A. K. & LeBlanc, G. A. (2010). Environmental-endocrine control of reproductive maturation in gastropods: Implications for the mechanism of tributyltin-induced imposex in prosobranchs. *Ecotoxicology* 19: 4–23.

Steuber, T. (2003). Strontium isotope stratigraphy of Cretaceous hippuritid rudist bivalves: Rates of morphological change and heterochronic evolution. *Palaeogeography, Palaeoclimatology, Palaeoecology* 200: 221–243.

Steusloff, U. (1942). Weitere Beiträge zur Kenntnis der Verbreitung und Lebensansprüche der *Vertigo genesii-parcedentata* im Diluvium und Alluvium. (Polyploidie wahrend des Periglazials?). *Archiv für Molluskenkunde* 74: 192–212.

Stevens, A. J., Stevens, N. M., Darby, P. C. & Percival, H. F. (1999). Observations of fire ants (*Solenopsis invicta* Buren) attacking apple snails (*Pomacea paludosa* Say) exposed during dry down conditions. *Journal of Molluscan Studies* 65: 507–510.

Stewart, D. T., Saavedra, C., Stanwood, R. R., Ball, A. O. & Zouros, E. (1995). Male and female mitochondrial DNA lineages in the blue mussel (*Mytilus edulis*) species group. *Molecular Biology and Evolution* 12: 735–747.

Stewart, F. J., Newton, I. L. G. & Cavanaugh, C. M. (2005). Chemosynthetic endosymbioses: Adaptations to oxic-anoxic interfaces. *Trends in Microbiology* 13: 439–448.

Stewart, F. J. & Cavanaugh, C. M. (2006). Bacterial endosymbioses in *Solemya* (Mollusca: Bivalvia): Model systems for studies of symbiont-host adaptation. *Antonie van Leeuwenhoek* 90: 343–360.

Stewart, F. J., Young, C. R. & Cavanaugh, C. M. (2008). Lateral symbiont acquisition in a maternally transmitted chemosynthetic clam endosymbiosis. *Molecular Biology and Evolution* 25: 673–687.

Stewart, H., Westlake, H. E. & Page, L. R. (2014a). Rhogocytes in gastropod larvae: Developmental transformation from protonephridial terminal cells. *Invertebrate Biology* 133: 47–63.

Stewart, M. J., Favrel, P., Rotgans, B., Wang, T., Zhao, M., Sohail, M., Wayne, A. O., Elizur, A., Henry, J. & Cummins, S. F. (2014b). Neuropeptides encoded by the genomes of the Akoya pearl oyster *Pinctata fucata* and Pacific oyster *Crassostrea gigas*: A bioinformatic and peptidomic survey. *BMC Genomics* 15: 840.

Stickle, W. B. (1973). The reproductive physiology of the intertidal prosobranch *Thais lamellosa* (Gmelin). I. Seasonal changes in the rate of oxygen consumption and body component indexes. *Biological Bulletin* 144: 511–524.

Stieglitz, R. R. (1994). The Minoan origin of Tyrian purple. *Biblical Archaeologist* 57: 47.

Stilkerich, J., Smrecak, T. A. & De Baets, K. (2017). 3D-Analysis of a non-planispiral ammonoid from the Hunsruck Slate: Natural or pathological variation? *PeerJ* 5: e3526.

Stillman, J. H. (2003). Acclimation capacity underlies susceptibility to climate change. *Science* 301: 65.

Stimson, J. (1973). The role of the territory in the ecology of the intertidal limpet *Lottia gigantea* (Gray). *Ecology* 54: 1020–1030.

Stine, O. C. (1989). *Cepaea nemoralis* from Lexington, Virginia: The isolation and characterization of their mitochondrial DNA, the implications for their origin and climatic selection. *Malacologia* 30: 305–315.

Stinnacre, J. & Tauc, L. (1969). Central nervous responses to activation of osmoreceptors in the osphradium of *Aplysia*. *Journal of Experimental Biology* 51: 347–361.

Stobbs, R. E. (1980). Feeding habits of the giant clingfish *Chorisochismus dentex* (Pisces: Gobiesocidae). *South African Journal of Zoology* 15: 146–149.

Stoecklin, R., Michalet, S., Menin, L., Bulet, P. & Favreau, P. (2004). Venom proteomics and drug discovery. pp. 478–479 *in* M. Chorev & Sawyer, T. K. (Eds.) *Peptide Revolution: Genomics, Proteomics & Therapeutics. The Proceedings of the 18th American Peptide Symposium*. New York, Springer-Verlag.

Stoll, C. J. (1972). Sensory systems involved in the shadow response of *Lymnaea stagnalis*, as studied with the use of habituation phenomena. *Proceedings of the Koninklijke Nederlandse Akademie van Wetenschappen* 75C: 342–351.

Stoll, C. J. (1973). On the role of eyes and non-ocular light receptors in orientational behaviour of *Lymnaea stagnalis* (L.). *Proceedings of the Koninklijke Nederlandse Akademie van Wetenschappen. Series C. Biological and Medical Sciences* 76: 203–214.

Stone, B. A. & Morton, J. E. (1958). The distribution of cellulases and related enzymes in Mollusca. *Journal of Molluscan Studies* 33: 127–141.

Stone, H. M. I. (1998). On predator deterrence by pronounced shell ornament in epifaunal bivalves. *Palaeontology* 41: 1051–1068.

Stone, J. R. (1996). The evolution of ideas: A phylogeny of shell models. *American Naturalist* 148: 904–929.

Storey, K. B. & Hochachka, P. W. (1975). Alpha-glycerophosphate dehydrogenase: Its role in the control of the cytoplasmic arm of the alpha-glycerophosphate cycle in squid muscle. *Comparative Biochemistry and Physiology Part B* 52: 169–174.

Storey, K. B. & Storey, J. M. (1983). Carbohydrate metabolism in cephalopod molluscs. pp. 91–136 *in* P. W. Hochachka (Ed.) *Metabolic Biochemistry and Molecular Biomechanics. The Mollusca*. Vol. 1. New York, Academic Press.

Storey, K. B. & Storey, J. M. (2010). Metabolic regulation and gene expression during aestivation. pp. 25–45 *in* C. A. Navas & Carvalho, J. E. (Eds.) *Aestivation: Molecular and Physiological Aspects. Progress in Molecular and Subcellular Biology*. Vol. 49. Berlin, Germany, Springer-Verlag.

Stothard, J. R. & Kristensen, T. K. (2000). Medical and veterinary malacology in Africa. *Parasitology Today* 16: 85–86.

Stott, K., Austin, W., Sayer, M., Weidman, C., Cage, A. & Wilson, R. (2010). The potential of *Arctica islandica* growth records to reconstruct coastal climate in north west Scotland, UK. *Quaternary Science Reviews* 29: 1602–1613.

Stramma, L., Johnson, G. C., Sprintall, J. & Mohrholz, V. (2008). Expanding oxygen-minimum zones in the tropical oceans. *Science* 320: 655–658.

Stramma, L., Prince, E. D., Schmidtko, S., Luo, J., Hoolihan, J. P., Visbeck, M., Wallace, D. W. R., Brandt, P. & Körtzinger, A. (2012). Expansion of oxygen minimum zones may reduce available habitat for tropical pelagic fishes. *Nature Climate Change* 2: 33–37.

Strasdine, G. A. & Whitaker, D. R. (1963). On the origin of the cellulase and chitinase of *Helix pomatia*. *Canadian Journal of Biochemistry and Physiology* 41: 1621–1626.

Strathmann, M. F. (1980). Why does a larva swim so long? *Paleobiology* 6: 373–376.

Strathmann, M. F. (1987a). Phylum Mollusca, class Bivalvia. pp. 309–353 *in* M. F. Strathmann (Ed.) *Reproduction and Development of Marine Invertebrates of the Northern Pacific Coast. Data and Methods for the Study of Eggs, Embryos, and Larvae*. Seattle, WA, University of Washington Press.

Strathmann, M. F. (1987b). Phylum Mollusca, class Gastropoda, subclass Prosobranchia. pp. 220–267 *in* M. F. Strathmann (Ed.) *Reproduction and Development of Marine Invertebrates of the Northern Pacific Coast. Data and Methods for the Study of Eggs, Embryos, and Larvae*. Seattle, WA, University of Washington Press.

Strathmann, M. F. & Strathmann, R. R. (2007). An extraordinarily long larval duration of 4.5 years from hatching to metamorphosis for teleplanic veligers of *Fusitriton oregonensis*. *Biological Bulletin* 213: 152–159.

Strathmann, R. R. (1982). Selection for retention or export of larvae in estuaries. pp. 521–536 *in* V. S. Kennedy (Ed.) *Estuarine Comparisons*. New York, Academic Press.

Strathmann, R. R. & Strathmann, M. F. (1982). The relationship between adult size and brooding in marine invertebrates. *American Naturalist* 119: 91–101.

Strathmann, R. R. & Chaffee, C. (1984). Constraints on egg masses. II. Effect of spacing, size, and number of eggs on ventilation of masses of embryos in jelly, adherent groups, or thin-walled capsules. *Journal of Experimental Marine Biology and Ecology* 84: 85–93.

Strathmann, R. R. (1985). Feeding and nonfeeding larval development and life-history evolution in marine invertebrates. *Annual Review of Ecology and Systematics* 16: 339–361.

Strathmann, R. R. & Strathmann, M. F. (1989). Evolutionary opportunities and constraints demonstrated by artificial gelatinous egg masses. pp. 201–209 *in* J. S. Ryland & Tyler, P. A. (Eds.) *Reproduction, Genetics and Distributions of Marine Organisms*. Fredensborg, Denmark, Olsen & Olsen.

Strathmann, R. R. (1993). Hypothesis on the origins of marine larvae. *Annual Review of Ecology and Systematics* 24: 89–117.

Straub, V. A., Staras, K., Kemenes, G. Y. & Benjamin, P. R. (2002). Endogenous and network properties of *Lymnaea* feeding: Central pattern generator interneurons. *Journal of Neurophysiology* 88: 1569–1583.

Straub, W. (1901). Zur Physiologie des Aplysienherzen. *Pflügers Archiv: European Journal of Physiology* 86: 504–531.

Straub, W. (1904). Fortgesetzte Studien am Aplysienherzen (Dynamik Kreislauf und dessen Innervation) nebst Bemerkungen zur vergleichenden Muskelphysiologie. *Pflügers Archiv: European Journal of Physiology* 103: 429–449.

Straw, J. & Rittschof, D. (2004). Responses of mud snails from low and high imposex sites to sex pheromones. *Marine Pollution Bulletin* 48: 1048–1054.

Strayer, D. L. (2008). *Freshwater Mussel Ecology: A Multifactor Approach to Distribution and Abundance*. Berkeley, CA, University of California Press.

Streftaris, N. & Zenetos, A. (2006). Alien marine species in the Mediterranean - the 100 'Worst Invasives' and their impact. *Mediterranean Marine Science* 7: 87–118.

Strong, E. E. (2003). Refining molluscan characters: Morphology, character coding and a phylogeny of the Caenogastropoda. *Zoological Journal of the Linnean Society* 137: 447–554.

Strong, E. E., Harasewych, M. G. & Haszprunar, G. (2003). Phylogeny of the Cocculinoidea (Mollusca, Gastropoda). *Invertebrate Biology* 122: 114–125.

Strong, E. E., Gargominy, O., Ponder, W. F. & Bouchet, P. (2008). Global diversity of gastropods (Gastropoda: Mollusca) in freshwater. *Hydrobiologia* 595: 149–166.

Struck, T. H. (2017). Phylogeny of Annelida. pp. 399–413 *in* P. Gopalakrishnakone & Malhotra, A. (Eds.) *Evolution of Venomous Animals and their Toxins*. Dordrecht, the Netherlands, Springer.

Strumwasser, F. (1984). The structure of the commands for a neuropeptide-mediated behavior: Egg-laying in an opisthobranch mollusk. pp. 36–43 in J. Hoffmann & Porchet, M. (Eds.) *Proceedings in Life Sciences: Biosynthesis Metabolism and Mode of Action of Invertebrate Hormones*. Berlin, Germany, Springer.

Stuart, J. A. & Ballantyne, J. S. (1996). Correlation of environment and phylogeny with the expression of β-hydroxybutyrate dehydrogenase in the Mollusca. *Comparative Biochemistry and Physiology Part B* 114: 153–160.

Stuart, J. A., Ooi, E. L., McLeod, J., Bourns, A. E. & Ballantyne, J. S. (1998). D- and L-beta-hydroxybutyrate dehydrogenases and the evolution of ketone body metabolism in gastropod molluscs. *Biological Bulletin* 195: 12–16.

Sturm, C. F., Pearce, T. A. & Valdés, Á., Eds. (2006). *The Mollusks: A Guide to their Study, Collection, and Preservation*. Boca Raton, FL, Universal Publishers.

Stützel, R. (1984). Anatomische und ultrastrukturelle Untersuchungen an der Napfschnecke *Patella* L. unter besonderer Berücksichtigung der Anpassung an den Lebensraum. *Zoologica (Stuttgart)* 46: 1–54.

Suess, E. (1888). *Das Antlitz der Erde*. Vol. 2. Wien, F. Tempsky.

Sugawara, K., Katagiri, Y. & Tomita, T. (1971). Polarized light responses from *Octopus* single retinular cells. *Journal of the Faculty of Science, Hokkaido University. Series VI Zoology* 17: 581–586.

Sugimoto, C. & Ikeda, Y. (2012). Ontogeny of schooling behavior in the oval squid *Sepioteuthis lessoniana*. *Fisheries Science* 78: 287–294.

Sugiura, S. & Okochi, I. (2006). High predation pressure by an introduced flatworm on land snails on the oceanic Ogasawara Islands. *Biotropica* 38: 700–703.

Sugiura, S. (2010). Prey preference and gregarious attacks by the invasive flatworm *Platydemus manokwari*. *Biological Invasions* 12: 1499–1507.

Sultan, S. E. (2007). Development in context: The timely emergence of eco-devo. *Trends in Ecology & Evolution* 22: 575–582.

Sumbre, G., Gutfreund, Y., Fiorito, G., Flash, T. & Hochner, B. (2001). Control of octopus arm extension by a peripheral motor program *Science* 293: 1845–1848.

Sumbre, G., Fiorito, G., Flash, T. & Hochner, B. (2005). Neurobiology: Motor control of flexible octopus arms. *Nature* 433: 595–596.

Summer, A. T. (1969). The distribution of some hydrolytic enzymes in the cells of the digestive gland of certain lamellibranchs and gastropods. *Journal of Zoology* 158: 277–291.

Summers, A. P. (2000). Stiffening the stingray skeleton: An investigation of durophagy in myliobatid stingrays (Chondrichthyes, Batoidea, Myliobatidae). *Journal of Morphology* 243: 113–126.

Sumner-Rooney, L. H., Murray, J. A., Cain, S. D. & Sigwart, J. D. (2014). Do chitons have a compass? Evidence for magnetic sensitivity in Polyplacophora. *Journal of Natural History* 48: 3033–3045.

Sumner-Rooney, L. H. & Sigwart, J. D. (2015). Is the Schwabe organ a retained larval eye? Anatomical and behavioural studies of a novel sense organ in adult *Leptochiton asellus* (Mollusca, Polyplacophora) indicate links to larval photoreceptors. *PLoS ONE* 10: e0137119.

Sumner-Rooney, L. & Sigwart, J. D. (2018). Do chitons have a brain? New evidence for diversity and complexity in the polyplacophoran central nervous system. *Journal of Morphology* 279: 936–949.

Sumner, A. T. (1965). The cytology and histochemistry of the digestive gland cells of *Helix*. *Quarterly Journal of Microscopical Science* 106: 173–192.

Sumner, A. T. (1966). The fine structure of digestive-gland cells of *Helix, Succinea* and *Testacella*. *Journal of the Royal Microscopical Society* 85: 181–192.

Sun, B., Kavanaugh, S. I. & Tsai, P.-S. (2012). Gonadotropin-releasing hormone in protostomes: Insights from functional studies on *Aplysia californica*. *General and Comparative Endocrinology* 176: 321–326.

Sun, S. E., Li, Q., Kong, L. & Yu, H. (2017). Multiple reversals of strand asymmetry in molluscs mitochondrial genomes, and consequences for phylogenetic inferences. *Molecular Phylogenetics and Evolution* 118: 222–231.

Sunday, J. M., Calosi, P., Dupont, S., Munday, P. L., Stillman, J. H. & Reusch, T. B. H. (2014). Evolution in an acidifying ocean. *Trends in Ecology & Evolution* 29: 117–125.

Sundberg, P. (1988). Microgeographic variation in shell characters of *Littorina saxatilis* Olivi: A question mainly of size? *Biological Journal of the Linnean Society* 35: 169–184.

Sunnucks, P. (2000). Efficient genetic markers for population biology. *TREE* 15: 199–203.

Suntornchashwej, S., Chaichit, N., Isobe, M. & Suwanborirux, K. (2005). Hectochlorin and morpholine derivatives from the Thai sea hare, *Bursatella leachii*. *Journal of Natural Products* 68: 951–955.

Suphamungmee, W., Engsusophon, A., Vanichviriyakit, R., Sretarugsa, P., Chavadej, J., Poomtong, T., Linthong, V. & Sobhon, P. (2010). Proportion of sperm and eggs for maximal *in vitro* fertilization in *Haliotis asinina* and the chronology of early development. *Journal of Shellfish Research* 29: 757–763.

Suresh, P. G., Reju, M. K. & Mohandas, A. (1993). Haemolymph phosphatase activity levels in two fresh-water gastropods exposed to copper. *Science of the Total Environment* 134: 1265–1277.

Susswein, A. J., Schwarz, M. & Feldman, E. (1986). Learned changes of feeding behavior in *Aplysia* in response to edible and inedible foods. *Journal of Neuroscience* 6: 1513–1527.

Susswein, A. J., Katzoff, A., Miller, N. & Hurwitz, I. (2004). Nitric oxide and memory. *The Neuroscientist* 10: 153–162.

Sutcharit, C., Asami, T. & Panha, S. (2007). Evolution of whole-body enantiomorphy in the tree snail genus *Amphidromus*. *Journal of Evolutionary Biology* 20: 661–672.

Sutherland, B., Stewart, D., Kenchington, E. L. R. & Zouros, E. (1998). The fate of paternal mitochondrial DNA in developing female mussels, *Mytilus edulis*: Implications for the mechanism of doubly uniparental inheritance of mitochondrial DNA. *Genetics* 148: 341–347.

Sutherland, N. S. (1957a). Visual discrimination of orientation and shape by *Octopus*. *Nature* 179: 11–13.

Sutherland, N. S. (1957b). Visual discrimination of orientation by *Octopus*. *British Journal of Psychology* 48: 55–71.

Sutherland, N. S. (1959). A test of a theory of shape discrimination in *Octopus vulgaris* Lamarck. *Journal of Comparative and Physiological Psychology* 52: 135–141.

Sutherland, N. S. (1960). Visual discrimination of shape by *Octopus*: Open and closed forms. *Journal of Comparative and Physiological Psychology* 53: 104–112.

Sutton, M. D., Briggs, D. E. G. & Siveter, D. J. (2006). Fossilized soft tissues in a Silurian platyceratid gastropod. *Proceedings of the Royal Society B* 273: 1039–1044.

Suzuki, K.-I., Ojima, T. & Nishita, K. (2003). Purification and cDNA cloning of a cellulase from abalone *Haliotis discus hannai*. *European Journal of Biochemistry* 270: 771–778.

Suzuki, M., Iwashima, A., Kimura, M., Kogure, T. & Nagasawa, H. (2013). The molecular evolution of the Pif family proteins in various species of Mollusks. *Marine Biotechnology* 15: 145–158.

Suzuki, M. & Nagasawa, H. (2013). Mollusk shell structures and their formation mechanism. *Canadian Journal of Zoology* 91: 349–366.

Suzuki, T., Furukohri, T. & Okamoto, S. (1993). Amino acid sequence of myoglobin from the chiton *Liolophura japonica* and a phylogenetic tree for molluscan globins. *Journal of Protein Chemistry* 12: 45–50.

Suzuki, T. & Imai, K. (1998). Evolution of myoglobin. *Cellular and Molecular Life Sciences* 54: 979–1004.

Suzuki, T., Kawamichi, H. & Imai, K. (1998a). A myoglobin evolved from indoleamine 2,3-dioxygenase, a tryptophan-degrading enzyme. *Comparative Biochemistry and Physiology Part B* 121: 117–128.

Suzuki, T., Kawamichi, H. & Imai, K. (1998b). Amino acid sequence, spectral, oxygen-binding, and autoxidation properties of indoleamine dioxygenase-like myoglobin from the gastropod mollusc *Turbo cornutus*. *Journal of Protein Chemistry* 17: 817–826.

Suzuki, T. (2003). Comparison of the sequences of *Turbo* and *Sulculus* indoleamine dioxygenase-like myoglobin genes. *Gene* 308: 89–94.

Suzuki, T. G., Ogino, K., Tsuneki, K. & Furuya, H. (2010). Phylogenetic analysis of dicyemid mesozoans (phylum Dicyemida) from innexin amino acid sequences: Dicyemids are not related to Platyhelminthes. *Journal of Parasitology* 96: 614–625.

Suzuki, Y., Sasaki, T., Suzuki, M., Nogi, Y., Miwa, T., Takai, K., Nealson, K. H. & Horikoshi, K. (2005). Novel chemo-autotrophic endosymbiosis between a member of the Epsilonproteobacteria and the hydrothermal vent gastropod *Alviniconcha* aff. *hessleri* (Gastropoda: Provannidae) from the Indian Ocean. *Applied and Environmental Microbiology* 71: 5440–5450.

Suzuki, Y., Kojima, S., Sasaki, T., Suzuki, M., Utsumi, T., Watanabe, H., Urakawa, H. et al. (2006a). Host-symbiont relationships in hydrothermal vent gastropods of the genus *Alviniconcha* from the southwest Pacific. *Bulletin of the Japanese Society of Scientific Fisheries* 72: 1388–1393.

Suzuki, Y., Kopp, R. E., Kogure, T., Suga, A., Takai, K., Mizota, C., Hirata, T., Chiba, H., Nealson, K. H., Horikoshi, K. & Kirschvink, J. L. (2006b). Sclerite formation in the hydro-thermal-vent 'scaly-foot' gastropod: Possible control of iron sulfide biomineralization by the animal. *Earth and Planetary Science Letters* 242: 39–50.

Sverdrup, H. U., Johnson, M. W. & Fleming, R. H. (1942). *The Oceans: Their Physics, Chemistry, and General Biology.* Vol. 7. New York, Prentice Hall.

Swanson, W. J. & Vacquier, V. D. (2002). The rapid evolution of reproductive proteins. *Nature Reviews Genetics* 3: 137–144.

Swedmark, B. (1967). The biology of interstitial Mollusca. pp. 135–149 in V. Fretter (Ed.) *Symposium of the Zoological Society of London. Symposia of The Zoological Society of London and The Malacological Society of London.* London, UK, Academic Press.

Sweeney, A. M., Haddock, S. H. D. & Johnsen, S. (2007). Comparative visual acuity of coleoid cephalopods. *Integrative and Comparative Biology* 47: 808–814.

Swennen, C. K. & Buatip, S. (2009). *Aiteng ater*, new genus, new species: An amphibious and insectivorous sea slug that is difficult to classify (Mollusca: Gastropoda: Opisthobranchia: Sacoglossa (?): Aitengidae, new family. *Raffles Bulletin of Zoology* 57: 495–500.

Switzer-Dunlap, M., Meyer-Schulte, K. & Gardner, E. A. (1984). The effect of size, age, and recent egg laying on copulatory choice of the hermaphroditic mollusc *Aplysia juliana*. *International Journal of Invertebrate Reproduction & Development* 7: 217–225.

Switzer-Dunlap, M. (1987). Ultrastructure of the juxtaganglionar organ, a putative endocrine gland associated with the cerebral ganglion of *Aplysia juliana*. *International Journal of Invertebrate Reproduction* 11: 295–304.

Switzer, A. D., Sloss, C. R., Jones, B. G. & Bristow, C. S. (2010). Geomorphic evidence for mid-late Holocene higher sea level from southeastern Australia. *Quaternary International* 221: 13–22.

Symondson, W. O. C. (2004). Coleoptera (Carabidae, Staphylinidae, Lampyridae, Drilidae and Silphidae) as predators of terrestrial gastropods. pp. 37–84 in G. M. Barker (Ed.) *Natural Enemies of Terrestrial Molluscs.* Oxford, UK, Cambridge, MA, CABI Publishing.

Sysoev, A. V. & Kantor, I. (1987). Deep-sea gastropods of the genus *Aforia* (Turridae) of the Pacific: Species composition, systematics and functional morphology of the digestive system. *The Veliger* 30: 105–126.

Sysoev, A. V. & Kantor, I. (1989). Anatomy of molluscs of genus *Splendrillia* (Gastropoda: Toxoglossa: Turridae) with descriptions of two new bathyal species of the genus from New Zealand. *New Zealand Journal of Zoology* 16: 205–214.

Szal, R. A. (1971). 'New' sense organ of primitive gastropods. *Nature* 229: 490–492.

Taïb, N. T. & Vicente, N. (1999). Histochemistry and ultrastructure of the crypt cells in the digestive gland of *Aplysia punctata* (Cuvier, 1803). *Journal of Molluscan Studies* 65: 385–398.

Taib, N. T. (2001). Distribution of digestive tubules and fine structure of digestive cells of *Aplysia punctata* (Cuvier, 1803). *Journal of Molluscan Studies* 67: 169–182.

Tajika, A., Morimoto, N., Wani, R., Naglik, C. & Klug, C. A. (2015). Intraspecific variation of phragmocone chamber volumes throughout ontogeny in the modern nautilid *Nautilus* and the Jurassic ammonite *Normannites*. *PeerJ* 3: e1306.

Tajima, M., Ikemori, M. & Arasaki, S. (1980). Abalone pigments originated from algal food: 1. Chromatographic analysis of pigments in the shell of the abalone fed with green algae *Bulletin of the Japanese Society of Scientific Fisheries* 46: 445–450.

Takahashi, J., Takagi, M., Okihana, Y., Takeo, K., Ueda, T., Touhata, K., Maegawa, S. & Toyohara, H. (2012). A novel silk-like shell matrix gene is expressed in the mantle edge of the Pacific oyster prior to shell regeneration. *Gene* 499: 130–134.

Takahashi, J. S., Nelson, D. E. & Eskin, A. (1989). Immunocytochemical localization of serotonergic fibers innervating the ocular circadian system of *Aplysia*. *Neuroscience* 28: 139–148.

Takami, H., Kawamura, T. & Yamashita, Y. (2002). Effects of delayed metamorphosis on larval competence, and postlarval survival and growth of abalone *Haliotis discus hannai*. *Aquaculture* 213: 311–322.

Takeuchi, T., Koyanagi, R., Gyoja, F., Kanda, M., Hisata, K., Fujie, M., Goto, H., Yamasaki, S., Nagai, K. & Morino, Y. (2016). Bivalve-specific gene expansion in the pearl oyster genome: Implications of adaptation to a sessile lifestyle. *Zoological Letters* 2: 3.

Takuwa-Kuroda, K., Iwakoshi-Ukena, E., Kanda, A. & Minakata, H. (2003). *Octopus*, which owns the most advanced brain in invertebrates, has two members of vasopressin/oxytocin superfamily, as in vertebrates. *Regulatory Peptides* 115: 139–149.

Talbot, C. F., Lewis, C. A. & Vacquier, V. D. (1980). Isolation of an egg vitelline layer, lysin, from abalone sperm. *Journal of Cell Biology* 87: A138.

Tamamaki, N. & Kawai, K. (1983). Ultrastructure of the accessory eye of the giant snail, *Achatina fulica* (Gastropoda, Pulmonata). *Zoomorphology* 102: 205–214.

Tamamaki, N. (1989a). Visible light reception of accessory eye in the giant snail, *Achatina fulica*, as revealed by an electrophysiological study. *Zoological Science (Tokyo)* 6: 867–876.

Tamamaki, N. (1989b). The accessory photosensory organ of the terrestrial slug, *Limax flavus* L. (Gastropoda: Pulmonata): Morphological and electrophysiological study. *Zoological Science (Tokyo)* 6: 877–886.

Tan, C. K., Nowark, B. F. & Hodson, S. L. (2002). Biofouling as a reservoir of *Neoparamoeba pemaquidensis* (Page, 1970), the causative agent of amoebic gill disease in Atlantic salmon. *Aquaculture* 210: 49–58.

Tan, K. S. & Lee, S. S. C. (2009). Neritid egg capsules: Are they all that different? *Steenstrupia* 30: 115–125.

Tanabe, K. & Fukuda, Y. (1999). Morphology and function of cephalopod buccal mass. pp. 245–262 in E. Savazzi (Ed.) *Functional Morphology of Invertebrate Skeletons*. London, UK, John Wiley & Sons.

Tanaka, H., Suzukia, H., Ohtsuki, I. & Ojima, T. (2008). Structure–function relationships of molluscan troponin T revealed by limited proteolysis. *Biochimica et Biophysica Acta* 1784: 1037–1042.

Tanaka, R., Sugimura, I., Sawabe, T., Yoshimizu, M. & Ezura, Y. (2003). Gut microflora of abalone *Haliotis discus hannai* in culture changes coincident with a change in diet. *Fisheries Science* 69: 951–958.

Tanaka, R., Ootsubo, M., Sawabe, T., Ezura, Y. & Tajima, K. (2004). Biodiversity and in situ abundance of gut microflora of abalone (*Haliotis discus hannai*) determined by culture-independent techniques. *Aquaculture* 241: 453–463.

Tang, B., Liu, B., Wang, X., Yue, X. & Xiang, J. (2010). Physiological and immune responses of Zhikong scallop *Chlamys farreri* to the acute viral necrobiotic virus infection. *Fish and Shellfish Immunology* 29: 42–48.

Tardy, J. (1970). Organogenèse de l'appareil génital chez les mollusques. *Bulletin de la Société zoologique de France* 95: 407–428.

Tardy, J. (1991). Types of opisthobranch veligers: Their notum formation and torsion. *Journal of Molluscan Studies* 57: 103–112.

Tardy, J. & Dongard, S. (1993). Le complexe apical de la véligère de *Ruditapes philippinarium* (Adams and Reeve, 1850) (Mollusque, Bivalve, Veneride). *Comptes rendus de l'Académie des sciences* 316: 177–184.

Tasaki, K. & Karita, K. (1966). Intraretinal discrimination of horizontal and vertical planes of polarized light by *Octopus*. *Nature* 209: 934–935.

Tascedda, F. & Ottaviani, E. (2014). Tumors in invertebrates. *Invertebrate Survival Journal* 11: 197–203.

Tasumi, S. & Vasta, G. R. (2007). A galectin of unique domain organization from hemocytes of the Eastern oyster (*Crassostrea virginica*) is a receptor for the protistan parasite *Perkinsus marinus*. *Journal of Immunology* 179: 3086–3098.

Taylor, B. E. & Lukowiak, K. (2000). The respiratory central pattern generator of *Lymnaea*: A model, measured and malleable. *Respiration Physiology* 122: 197–207.

Taylor, D. J. & Ó Foighil, D. (2000). Transglobal comparisons of nuclear and mitochondrial genetic structure in a marine polyploid clam (*Lasaea*, Lasaeidae). *Heredity* 84: 321–330.

Taylor, D. W. (1966). A remarkable snail fauna from Coahuila, Mexico. *The Veliger* 9: 152–228.

Taylor, E. L., Taylor, T. N. & Krings, M. (2009). *Paleobotany: The Biology and Evolution of Fossil Plants*. Burlington, MA, Academic Press.

Taylor, G. M. (2000). Maximum force production: Why are crabs so strong? *Proceedings of the Royal Society of London B* 267: 1475–1480.

Taylor, H. H. & Ragg, N. L. C. (2005). The role of body surfaces and ventilation in gas exchange of the abalone, *Haliotis iris*. *Journal of Comparative Physiology B* 175: 463–478.

Taylor, J. D., Kennedy, W. J. & Hall, A. (1969). The shell structure and mineralogy of the Bivalvia. 1. Introduction, Nuculacea: Trigoniacea. *British Museum (Natural History) Bulletin. Zoology* 3: 1–125.

Taylor, J. D. (1973). The structural evolution of the bivalve shell. *Palaeontology* 16: 519–534.

Taylor, J. D., Kennedy, W. J. & Hall, A. (1973). The shell structure and mineralogy of the Bivalvia. II. Lucinacea: Clavagellacea. Conclusions. *British Museum (Natural History) Bulletin. Zoology* 22: 255–294.

Taylor, J. D. (1978). Habitats and diet of predatory gastropods at Addu Atoll, Maldives. *Journal of Experimental Marine Biology and Ecology* 31: 83–103.

Taylor, J. D. & Morris, N. J. (1988). Relationships of neogastropods. pp. 167–179 in W. F. Ponder, Eernisse, D. J. & Waterhouse, J. H. (Eds.) *Prosobranch Phylogeny. Malacological Review Supplement*. Ann Arbor, MI, Malacological Review.

Taylor, J. D. (1989). The diet of coral-reef Mitridae (Gastropoda) from Guam (West Pacific); with a review of other species of the family. *Journal of Natural History* 23: 261–278.

Taylor, J. D. & Miller, J. A. (1989). The morphology of the osphradium in relation to feeding habits in meso- and neo-gastropods. *Journal of Molluscan Studies* 55: 227–237.

Taylor, J. D. & Miller, J. A. (1990). A new type of gastropod proboscis: The foregut of *Hastula bacillus* (Gastropoda: Terebridae). *Journal of Zoology* 220: 603–617.

Taylor, J. D. & Reid, D. G. (1990). Shell microstructure and mineralogy of the Littorinidae: Ecological and evolutionary significance. *Hydrobiologia* 193: 199–215.

Taylor, J. D. (1993). Dietary and anatomical specialization of mitrid gastropods (Mitridae) at Rottnest Island, Western Australia. pp. 583–599 in F. E. Wells, Walker, D. I., Kirkman, H. & Lethbridge, R. (Eds.) *The Marine Flora and Fauna of Rottnest Island, Western Australia. Proceedings of the Fifth International Marine Biological Workshop held at Rottnest Island (January 1991)*. Perth, WA, Western Australian Museum.

Taylor, J. D., Kantor, I. & Sysoev, A. V. (1993). Foregut anatomy, feeding mechanisms, relationships and classification of the Conoidea (=Toxoglossa) (Gastropoda). *Bulletin of the British Museum (Natural History). Zoology* 59: 125–170.

Taylor, J. D. & Wells, F. E. (1994). A revision of the crassispirine gastropods from Hong Kong (Gastropoda: Turridae). pp. 101–116 in B. Morton (Ed.) *The Malacofauna of Hong Kong and Southern China, III: Proceedings of the Third International Workshop on the Malacofauna of Hong Kong and Southern China, Hong Kong*. Vol. 3. Hong Kong, Hong Kong University Press.

Taylor, J. D., Ed. (1996). *Origin and Evolutionary Radiation of the Mollusca*. Oxford, UK, Oxford University Press.

Taylor, J. D. (1998). Understanding biodiversity: Adaptive radiations of predatory marine gastropods. pp. 187–206 in B. Morton (Ed.) *The Marine Biology of the South China Sea 3. Proceedings of the Third International Conference on the Marine Biology of the South China Sea, Hong Kong, 28 October–1 November 1996*. Vol. 3. Hong Kong, Hong Kong University Press.

Taylor, J. D. & Glover, E. A. (2000). Functional anatomy, chemosymbiosis and evolution of the Lucinidae. pp. 207–225 *in* E. M. Harper, Taylor, J. D. & Crame, J. A. (Eds.) *Evolutionary Biology of the Bivalvia*. London, UK, The Geological Society (Special Publication No. 177).

Taylor, J. D. & Glover, E. A. (2003). Food of giants: Field observations on the diet of *Syrinx aruanus* (Linnaeus, 1758) (Turbinellidae) the largest living gastropod. pp. 217–223 *in* F. E. Wells, Walker, D. I. & Jones, D. S. (Eds.) *The Marine Flora and Fauna of Dampier, Western Australia*. Perth, WA, Western Australian Museum.

Taylor, J. D., Glover, E. A. & Williams, S. T. (2005). Another bloody bivalve: Anatomy and relationships of *Eucrassatella donacina* from south western Australia (Mollusca: Bivalvia: Crassatellidae). pp. 261–288 *in* F. E. Wells, Walker, D. I. & Kendrick, G. A. (Eds.) *Proceedings of the Twelfth International Marine Biological Workshop: The Marine Flora and Fauna of Esperance, Western Australia*. Esperance, WA, Western Australian Museum.

Taylor, J. D. & Glover, E. A. (2006). Lucinidae (Bivalvia): The most diverse group of chemosymbiotic molluscs. *Zoological Journal of the Linnean Society* 148: 421–438.

Taylor, J. D. (2008). Ancient chemosynthetic bivalves: Systematics of Solemyidae from eastern and southern Australia (Mollusca: Bivalvia). *Memoirs of the Queensland Museum* 54: 75–104.

Taylor, J. D. & Glover, E. A. (2010). Chemosymbiotic bivalves. pp. 107–135 *in* S. Kiel (Ed.) *The Vent and Seep Biota: Aspects from Microbes to Ecosystems. Topics in Geobiology*. Dordrecht, the Netherlands, Springer.

Taylor, J. S. & Raes, J. (2004). Duplication and divergence: The evolution of new genes and old ideas. *Annual Review of Genetics* 38: 615–643.

Taylor, P. D., Schembri, P. J. & Cook, P. L. (1989). Symbiotic associations between hermit crabs and bryozoans from the Otago region, southeastern New Zealand. *Journal of Natural History* 23: 1059–1085.

Taylor, P. M. & Andrews, E. B. (1987). Tissue adenosine triphosphate activities of the gill and excretory system in mesogastropod molluscs in relation to osmoregulatory capacity. *Comparative Biochemistry and Physiology Part A* 86: 693–696.

Taylor, P. M. & Andrews, E. B. (1988). Osmoregulation in the intertidal gastropod *Littorina littorea*. *Journal of Experimental Marine Biology and Ecology* 122: 35–46.

Tedesco, S., Mullen, W. & Cristobal, S. (2014). High-throughput proteomics: A new tool for quality and safety in fishery products. *Current Protein & Peptide Science* 15: 118–133.

Teichert, C., Kummel, B., Sweet, W. C., Stenzel, H. B., Furnish, W. M., Glenister, B. F., Erben, H. K., Moore, R. C. & Nodine Zeller, D. E. (1964). Cephalopoda: General features: Endoceratoidea, Actinoceratoidea, Nautiloidea, Bactritoidea. pp. K4–K505 *in* R. C. Moore (Ed.) *Treatise on Invertebrate Paleontology Part K*. Vol. 3. Lawrence, KS, Geological Society of America and University of Kansas Press.

Teichert, S. & Nützel, A. (2015). Early Jurassic anoxia triggered the evolution of the oldest holoplanktonic gastropod *Coelodiscus minutus* by means of heterochrony. *Acta Palaeontologica Polonica* 60: 269–276.

Teichmann, E. (1903). Die frühe Entwicklung der Cephalopoden. *Verhandlungen der Deutschen Zoologischen Gesellschaft* 1903: 42–52.

Telford, M. J. (2008). Xenoturbellida: The fourth deuterostome phylum and the diet of worms. *Genesis* 46: 580–586.

Templado, J. & Ortea, J. (2001). The occurrence of the shell-less neritacean gastropod *Titiscania limacina* in the Galápagos Islands. *The Veliger* 44: 404–406.

Tendal, O. S. (1985). Xenophyophores (Protozoa, Sarcodina) in the diet of *Neopilina galathea* (Mollusca, Monoplacophora). *Galathea Report* 16: 95–99.

Tenger, M. J. (1989). The California abalone fishery: Production, ecological interactions, and prospects for the future. pp. 401–420 *in* J. F. Caddy (Ed.) *Marine Invertebrate Fisheries: Their Assessment and Management*. New York, John Wiley & Sons.

Tensen, C. P., Cox, K. J. A., Smit, A. B., Van der Schors, R. C., Meyerhof, W., Richter, D., Planta, R. J. et al. (1998). The *Lymnaea* cardioexcitary peptide (LyCEP) receptor: A G-protein-coupled receptor for a novel member of the RFamide neuropeptide family. *Journal of Neuroscience* 18: 9812–9821.

Tentori, L. (1970). Myoglobin, with particular reference to the myoglobin of *Aplysia*. *Biochemical Journal* 119: 33P–34P.

Terio, V., Martella, V., Moschidou, P., Pinto, P. Di, Tantillo, G. & Buonavoglia, C. (2010). Norovirus in retail shellfish. *Food Microbiology* 27: 29–32.

Terlau, H. & Olivera, B. M. (2004). *Conus* venoms: A rich source of novel ion channel-targeted peptides. *Physiology Reviews* 84: 41–68.

Terlizzi, C. M., Ward, M. E. & Van Dover, C. L. (2004). Observations on parasitism in deep-sea hydrothermal vent and seep limpets. *Diseases of Aquatic Organisms* 62: 17–26.

Termier, H. & Termier, G. (1968). *Évolution et Biocinèse, les Invertébrés dans l'histoire du monde vivant*. Paris, Masson et Cie.

Terwilliger, N. B. & Ryan, M. C. (2006). Functional and phylogenetic analyses of phenoloxidases from brachyuran (*Cancer magister*) and branchiopod (*Artemia franciscana*, *Triops longicaudatus*) crustaceans. *Biological Bulletin* 210: 38–50.

Terwilliger, R. C., Terwilliger, N. B. & Schabtach, E. (1978). Extracellular hemoglobin of the clam, *Cardita borealis* (Conrad): An unusual polymeric haemoglobin. *Comparative Biochemistry and Physiology Part B* 59: 9–14.

Terwilliger, R. C. & Terwilliger, N. B. (1985). Molluscan hemoglobins. *Comparative Biochemistry and Physiology Part B* 81: 255–261.

Test, A. R. G. (1945). Ecology of California *Acmaea*. *Ecology* 26: 395–405.

Tettelbach, S. T. & Rhodes, E. W. (1981). Combined effects of temperature and salinity on embryos and larvae of the northern bay scallop *Argopecten irradians irradians*. *Marine Biology* 63: 249–256.

Teyke, T., Weiss, K. R. & Kupfermann, I. (1992). Orientation of *Aplysia californica* to distant food sources. *Journal of Comparative Physiology A* 170: 281–289.

Thayer, C. W. (1971). Fish-like crypsis in swimming Monomyaria. *Proceedings of the Malacological Society of London* 39: 371–376.

Theologidis, I., Fodelianakis, S., Gaspar, M. B. & Zouros, E. (2008). Doubly uniparental inheritance (DUI) of mitochondrial DNA in *Donax trunculus* (Bivalvia: Donacidae) and the problem of its sporadic detection in Bivalvia. *Evolution* 62: 959–970.

Thiele, J. (1891–1893). *Das Gebiß der Schnecken, zur Begründung einer natürlichen Classification (continuation of work by F. H. Troschel)*. Vol. 2 (7–8). Berlin, Nicolai.

Thiele, J. (1892). Beiträge zur Kenntnis der Mollusken. *Zeitschrift für wissenschaftliche Zoologie* 53: 578–590.

Thiele, J. (1894). Beiträge zur vergleichenden Anatomie der Amphineuren. 1. über einige Neapler Solenogastres. *Zeitschrift für wissenschaftliche Zoologie* 58: 222–302.

Thiele, J. (1895). Zur Phylogenie der Gastropoden. *Biologisches Zentralblatt* 15: 220–236.

Thiele, J. (1929–1935). *Handbuch der systematischen Weichtierkunde. Volume 1, Teil 1, Loricata and Gastropoda I: Prosobranchia (1929), Teil 2, Opisthobranchia & Pulmonata (1931); Volume 2, Teil 3, Scaphopoda, Bivalvia, Cephalopoda, Tiel 4, general and corrections.* Jena, Gustav Fischer Verlag.

Thiele, J. (1992a). *Handbook of Systematic Malacology. Part 2 (Gastropoda: Opisthobranchia and Pulmonata).* Washington, DC, Smithsonian Institutional Libraries & The National Science Foundation.

Thiele, J. (1992b). *Handbook of Systematic Malacology. Part 1 (Loricata: Gastropoda: Prosobranchia).* Washington, DC, Smithsonian Institutional Libraries & The National Science Foundation.

Thiele, J. (1998). *Handbook of Systematic Malacology. Part 3 (Scaphopoda/Bivalvia/Cephalopoda). Part 4 (Comparative Morphology/Phylogeny/Geographical Distribution).* Washington, DC, Smithsonian Institutional Libraries & The National Science Foundation.

Thieltges, D. W. (2006). Parasite induced summer mortality in the cockle *Cerastoderma edule* by the trematode *Gymnophallus choledochus*. *Hydrobiologia* 559: 455–461.

Thiem, H. (1917a). Beiträge zur Anatomie und Phylogenie der Docoglossen. I. Zur Anatomie von *Helcioniscus ardosiaeus* Hombron & Jaquinot unter Bezugnahme auf die Bearbeitung von Erich Schusterinden. *Jenaische Zeitschrift für Naturwissenschaft* 54: 333–404.

Thiem, H. (1917b). Beiträge zur Anatomie und Phylogenie der Docoglossen. II. Die Anatomie und Phylogenie der Monobranchen (Akmäiden und Scurriden nach der Sammlung Plates). *Jenaische Zeitschrift für Naturwissenschaft* 54: 405–630.

Thiengo, S. C., Faraco, F. A., Salgado, N. C., Cowie, R. H. & Fernandez, M. A. (2007). Rapid spread of an invasive snail in South America: The giant African snail, *Achatina fulica*, in Brasil. *Biological Invasions* 9: 693–702.

Thiriot-Quiévreux, C. & Martoja, M. (1976). Appareil génital femelle des Atlantidae (Mollusca, Heteropoda): 2—Étude histologique des structures larvaires, juvéniles et adultes: données sur la fécondation et la pointe. *Vie et Milieu Serie A: Biologie Marine* 26: 201–233.

Thiriot-Quiévreux, C., Insua Pombo, A. M. & Albert, P. (1989). Polyploidy in a brooding bivalve, *Lasaea rubra* (Montagu). *Comptes rendus de l'Académie des sciences* 308: 115–120.

Thiriot-Quiévreux, C. (2002). Review of the literature on bivalve cytogenetics in the last ten years. *Cahiers de Biologie Marine* 43: 17–26.

Thiriot-Quiévreux, C. (2003). Advances in chromosomal studies of gastropod molluscs. *Journal of Molluscan Studies* 69: 187–202.

Thomas, F. I. M. & Poulin, R. (1998). Manipulation of a mollusc by a trophically transmitted parasite: Convergent evolution or phylogenetic inheritance? *Parasitology* 116: 431–436.

Thomas, J. D. (1982). Chemical ecology of the snail hosts of Schistosomiasis: Snail-snail and snail-plant interactions. *Malacologia* 22: 81–91.

Thomas, M. V. & Day, R. W. (1995). Site selection by a small drilling predator: Why does the gastropod *Haustrum baileyanum* drill over muscle tissue of the abalone *Haliotis rubra*? *Marine and Freshwater Research* 46: 647–655.

Thomas, R. D. K. (1975). Functional morphology, ecology, and evolutionary conservatism in the Glycymerididae (Bivalvia). *Palaeontology* 18: 217–254.

Thomas, R. F. (1973). Homing behaviour and movement rhythms in the pulmonate limpet, *Siphonaria pectinata* Linnaeus. *Proceedings of the Malacological Society of London* 40: 303–311.

Thompson, F. G. (1968). *The Aquatic Snails of the Family Hydrobiidae of Peninsular Florida.* Gainsville, FL, University of Florida Press.

Thompson, J. E., Walker, R. P., Wratten, S. J. & Faulkner, J. D. (1982). A chemical defense mechanism for the nudibranch *Cadlina luteomarginata*. *Tetrahedron Letters* 23: 1865–1873.

Thompson, R. C., Crowe, T. P. & Hawkins, S. J. (2002). Rocky intertidal communities: Past environmental changes, present status and predictions for the next 25 years. *Environmental Conservation* 29: 168–191.

Thompson, R. J., Livingstone, D. R. & De Zwamm, A. (1980). Physiological and biochemical aspects of the valve snap and valve closure responses in the giant scallop, *Plactopecten magellanicus*: I: Physiology. *Journal of Comparative Physiology B* 137: 97–104.

Thompson, S. H. & Watson, W. H. (2005). Central pattern generator for swimming in *Melibe*. *Journal of Experimental Biology* 208: 1347–1361.

Thompson, T. E. (1958). The natural history, embryology, larval biology and post-larval development of *Adalaria proxima* (Alder and Hancock) (Gastropoda Opisthobranchia). *Philosophical Transactions of the Royal Society B* 242: 1–58.

Thompson, T. E. & Slinn, S. J. (1959). On the biology of the opisthobranch *Pleurobranchus membranaceus*. *Journal of the Marine Biological Association of the United Kingdom* 38: 507–524.

Thompson, T. E. (1960). The development of *Neomenia carinata* Tullberg (Mollusca, Aplacophora). *Proceedings of the Royal Society B* 153: 263–278.

Thompson, T. E. (1962). Studies on the ontogeny of *Tritonia hombergi* Cuvier (Gastropoda Opisthobranchia). *Philosophical Transactions of the Royal Society B* 245: 171–218.

Thompson, T. E. (1966). Studies on the reproduction of *Archidoris pseudoargus* (Rapp) (Gastropoda. Opisthobranchia). *Philosophical Transactions of the Royal Society B* 250: 343–374.

Thompson, T. E. (1969). Acid secretion in Pacific Ocean gastropods. *Australian Journal of Zoology* 17: 755–764.

Thompson, T. E. & Bebbington, A. (1969). Structure and function of the reproductive organs of three species of *Aplysia* (Gastropoda: Opisthobranchia). *Malacologia* 7: 347–380.

Thompson, T. E. & Bennett, I. (1970). Observations on Australian Glaucidae (Mollusca: Opisthobranchia). *Journal of the Linnean Society of London* 49: 187–197.

Thompson, T. E. (1973). Euthyneuran and other molluscan spermatozoa: General characteristics of molluscan spermatozoa. *Malacologia* 14: 167–206.

Thompson, T. E. (1976). *Biology of Opisthobranch Molluscs.* Vol. 1. London, UK, Ray Society.

Thompson, T. E. & Brown, G. H. (1984). *Biology of Opisthobranch Molluscs.* Vol. 2. London, UK, Ray Society.

Thompson, T. E. (1986). Investigation of the acidic allomone of the gastropod mollusc *Philine aperta* by means of ion chromatography and histochemical localisation of sulphate and chloride ions. *Journal of Molluscan Studies* 52: 38–44.

Thongkukiatkul, A., Sobhon, P., Upatham, E. S., Kruatrachue, M., Wanichanon, C., Chitramvong, Y. P. & Pumthong, T. (2001). Ultrastructure of neurosecretory cells in the cerebral and pleuropedal ganglia of *Haliotis asinina* (Linnaeus). *Journal of Shellfish Research* 20: 733–741.

Thonig, A., Oellermann, M., Lieb, B. & Mark, F. C. (2014). A new haemocyanin in cuttlefish (*Sepia officinalis*) eggs: Sequence analysis and relevance during ontogeny. *EvoDevo* 5: 6 (1–12).

Thorson, G. (1935). Studies on the egg-capsules and development of Arctic marine bottom invertebrates. *Meddelelser om Grønland* 100: 1–71.

Thorson, G. (1946). Reproduction and larval development of Danish marine bottom invertebrates, with special reference to the planktonic larvae in the Sound (Øresund). *Meddelelser fra Kommissionen for Danmarks Fiskeri-og Havunder Søgelser. Serie Plankton* 4: 1–523.

Thorson, G. (1950). Reproductive and larval ecology of marine bottom invertebrates. *Biological Reviews* 25: 1–45.

Tillier, S. (1984). Patterns of digestive tract morphology in the limacisation of helicarionid, succineid and athoracophorid snails and slugs (Mollusca: Pulmonata). *Malacologia* 25: 173–192.

Tillier, S. (1989). Comparative morphology, phylogeny and classification of land snails and slugs (Gastropoda: Pulmonata: Stylommatophora). *Malacologia* 30: 1–304.

Tillier, S. & Ponder, W. F. (1992). New species of *Smeagol* from Australia and New Zealand, with a discussion of the affinities of the genus (Gastropoda: Pulmonata). *Journal of Molluscan Studies* 58: 135–155.

Tills, O., Rundle, S. D. & Spicer, J. I. (2013). Variance in developmental event timing is greatest at low biological levels: Implications for heterochrony. *Biological Journal of the Linnean Society* 110: 581–590.

Tochimoto, T. (1967). Comparative histochemical study on the dimorphic spermatozoa of the Prosobranchia with special reference to polysaccharides. *Science Reports. Tokyo Kyoiku Daigaku: Section B* 13: 75–109.

Todd, C. D., Bentley, M. G. & Havenhand, J. N. (1991). Larval metamorphosis of the opisthobranch mollusc *Adalaria proxima* (Gastropoda: Nudibranchia): The effects of choline and elevated potassium ion concentration. *Journal of the Marine Biological Association of the United Kingdom* 71: 53–72.

Todd, C. D. (1998). The genetic structure of intertidal populations of two species of nudibranch molluscs with planktotrophic and pelagic lecithotrophic larval stages: Are pelagic larvae 'for' dispersal? *Journal of Experimental Marine Biology and Ecology* 228: 1–28.

Todd, M. E. (1964). Osmotic balance in *Hydrobia ulvae* and *Potamopyrgus jenkinsi* (Gastropoda: Hydrobiidae). *Journal of Experimental Biology* 41: 665–677.

Todt, C. & von Salvini-Plawen, L. (2004). Ultrastructure and histochemistry of the foregut in *Wirenia argentea* and *Genitoconia rosea* (Mollusca: Solenogastres). *Zoomorphology* 123: 65–80.

Todt, C. & von Salvini-Plawen, L. (2005). The digestive tract of *Helicoradomenia* (Solenogastres, Mollusca), aplacophoran molluscs from the hydrothermal vents of the East Pacific Rise. *Invertebrate Biology* 124: 230–253.

Todt, C. (2006). Ultrastructure of multicellular foregut glands in selected solenogastres (Mollusca). *Zoomorphology* 125: 119–134.

Todt, C., Büchinger, T. & Wanninger, A. (2008). The nervous system of the basal mollusk *Wirenia argentea* (Solenogastres): A study employing immunocytochemical and 3D reconstruction techniques. *Marine Biology Research* 4: 290–303.

Todt, C., Cárdenas, P. & Rapp, H. T. (2009). The chiton *Hanleya nagelfar* (Polyplacophora, Mollusca) and its association with sponges in the European Northern Atlantic. *Marine Biology Research* 5: 408–411.

Todt, C. & Wanninger, A. (2010). Of tests, trochs, shells, and spicules: Development of the basal mollusk *Wirenia argentea* (Solenogastres) and its bearing on the evolution of trochozoan larval key features. *Frontiers in Zoology* 7: 1–17.

Toevs, L. A. S. (1970). Identification and characterization of the egg-laying hormone from the neurosecretory bag cells of *Aplysia*. PhD, California Institute of Technology.

Togo, T., Osanai, K. & Morisawa, M. (1995). Existence of three mechanisms for blocking polyspermy in oocytes of the mussel *Mytilus edulis*. *Biological Bulletin* 189: 330–339.

Tomanek, L. & Somero, G. N. (1997). The effect of temperature on protein synthesis in snails of the genus *Tegula* from the sub- and intertidal zone. *American Zoologist* 37: 188A.

Tomanek, L. & Somero, G. N. (1998). Features of a lethal heat shock: Impairment of synthesis of heat shock proteins 70 and 90 during recovery in snails of the genus *Tegula* from the sub- and intertidal zone. *American Zoologist* 38: 159A.

Tomanek, L. & Somero, G. N. (1999). Evolutionary and acclimation-induced variation in the heat-shock responses of congeneric marine snails (genus *Tegula*) from different thermal habitats: Implications for limits of thermotolerance and biogeography. *Journal of Experimental Biology* 202: 2925–2936.

Tomanek, L. (2002). The heat-shock response: Its variation, regulation and ecological importance in intertidal gastropods (genus *Tegula*). *Integrative and Comparative Biology* 42: 797–807.

Tomanek, L. & Helmuth, B. S. T. (2002). Physiological ecology of rocky intertidal organisms: A synergy of concepts. *Integrative and Comparative Biology* 42: 771–775.

Tomanek, L. & Sanford, E. (2003). Heat-shock protein 70 (Hsp70) as a biochemical stress indicator: An experimental field test in two congeneric intertidal gastropods (genus: *Tegula*). *Biological Bulletin* 205: 276–284.

Tomanek, L., Zuzow, M. J., Ivanina, A. V., Beniash, E. & Sokolova, I. M. (2011). Proteomic response to elevated PCO_2 level in eastern oysters, *Crassostrea virginica*: Evidence for oxidative stress. *Journal of Experimental Biology* 214: 1836–1844.

Tomanek, L., Zuzow, M. J., Hitt, L., Serafini, L. & Valenzuela, J. J. (2012). Proteomics of hyposaline stress in blue mussel congeners (genus *Mytilus*): Implications for biogeographic range limits in response to climate change. *Journal of Experimental Biology* 215: 3905–3916.

Tomanek, L. (2014). Proteomics to study adaptations in marine organisms to environmental stress. *Journal of Proteomics* 105: 92–106.

Tomarev, S. I. & Piatigorsky, J. (1996). Lens crystallins of invertebrates. *European Journal of Biochemistry* 235: 449–465.

Tomarev, S. I., Callaerts, P., Kos, L., Zinovieva, R. D., Halder, G., Gehring, W. J. & Piatigorsky, J. (1997). Squid Pax-6 and eye development. *Proceedings of the National Academy of Sciences of the United States of America* 94: 2421–2426.

Tomić, N. & Meyer-Rochow, V. B. (2011). Atavisms: Medical, genetic, and evolutionary implications. *Perspectives in Biology and Medicine* 54: 332–353.

Tomlinson, J., Reilly, D. & Ballering, R. (1980). Magnetic radular teeth and geomagnetic responses in chitons. *The Veliger* 23: 167–170.

Tompa, A. S. & Watabe, N. (1976). Ultrastructural investigation of the mechanism of the muscle attachment to the gastropod shell. *Journal of Morphology* 149: 339–352.

Tompa, A. S. (1979). Oviparity, egg retention and ovoviviparity in pulmonates. *Journal of Molluscan Studies* 45: 155–160.

Tompa, A. S. (1980). Studies on the reproductive biology of gastropods: Part III. Calcium provision and the evolution of terrestrial eggs among gastropods. *Journal of Conchology* 30: 145–154.

Tompsett, D. H. (1939). *Sepia*. Liverpool Marine Biology Committee: Memoirs on typical British Marine Plants and Animals. Liverpool; London, UK, University Press of Liverpool.

Tong, H., Ma, W., Wang, L., Wan, P., Hu, J. & Cao, L. (2004). Control over the crystal phase, shape, size and aggregation of calcium carbonate via a L-aspartic acid inducing process. *Biomaterials* 25: 3923–3929.

Too, C. K. L. & Croll, R. P. (1995). Detection of FMRFamide-like immunoreactivities in the sea scallop *Placopecten magellanicus* by immunohistochemistry and Western blot analysis. *Cell and Tissue Research* 281: 295–304.

Torres, S. C., Camacho, J. L., Matsumoto, B., Kuramoto, R. T. & Robles, L. J. (1997). Light/dark-induced changes in rhabdom structure in the retina of *Octopus bimaculoides*. *Cell and Tissue Research* 290: 167–174.

Towns, D. R. (2009). Eradications as reverse invasions: Lessons from Pacific rat (*Rattus exulans*) removals on New Zealand islands. *Biological Invasions* 11: 1719–1733.

Townsend, C. R. (1973). The role of the osphradium in chemoreception by the snail *Biomphalaria glabrata* (Say). *Animal Behaviour* 21: 549–556.

Tramell, P. R. & Campbell, J. W. (1970). Nitrogenous excretory products of the giant South American land snail, *Strophocheilus oblongus*. *Comparative Biochemistry and Physiology* 32: 569–571.

Trappmann, W. (1916). Die Muskulatur von *Helix pomatia*. *Zeitschrift für wissenschaftliche Zoologie* 115: 489–585.

Treen, N., Itoh, N., Miura, H., Kikuchi, I., Ueda, T., Takahashi, K. G., Ubuka, T., Yamamoto, K., Sharp, P. J. & Tsutsui, K. (2012). Mollusc gonadotropin-releasing hormone directly regulates gonadal functions: A primitive endocrine system controlling reproduction. *General and Comparative Endocrinology* 176: 167–172.

Tremblay, I., Guderley, H. E. & Frechette, M. (2006). Swimming performance, metabolic rates, and their correlates in the Iceland scallop *Chlamys islandica*. *Physiological and Biochemical Zoology* 79: 1046–1057.

Trench, R. K., Wethey, D. S. & Porter, J. W. (1981). Observations on the symbiosis with zooxanthellae among the Tridacnidae (Mollusca, Bivalvia). *Biological Bulletin* 161: 180–198.

Treude, T., Smith, C. R., Wenzhöfer, F., Carney, E., Bernardino, A. F., Hannides, A. K., Krüger, M. & Boetius, A. (2009). Biogeochemistry of a deep-sea whale fall: Sulfate reduction, sulfide efflux and methanogenesis. *Marine Ecology Progress Series* 382: 1–21.

Triantis, K. A., Economo, E. P., Guilhaumon, F. & Ricklefs, R. E. (2015). Diversity regulation at macro-scales: Species richness on oceanic archipelagos. *Global Ecology and Biogeography* 24: 594–605.

Triantis, K. A., Rigal, F., Parent, C. E., Cameron, R. A., Lenzner, B., Parmakelis, A., Yeung, N. W., Alonso, M. R., Ibáñez, M. & de Frias Martins, A. M. (2016). Discordance between morphological and taxonomic diversity: Land snails of oceanic archipelagos. *Journal of Biogeography* 43: 2050–2061.

Trigg, C., Harries, D., Lyndon, A. & Moore, C. G. (2011). Community composition and diversity of two *Limaria hians* (Mollusca: Limacea) beds on the west coast of Scotland. *Journal of the Marine Biological Association of the United Kingdom* 91: 1403–1412.

Tripp, M. R. (1975). Humoral factors and molluscan immunity. pp. 201–223 in K. Maramorosch & Shope, R. E. (Eds.) *Invertebrate Immunity: Mechanisms of Invertebrate Vector-parasite Relations.* New York, Academic Press.

Tritt, S. H. & Byrne, J. H. (1980). Motor controls of opaline secretion in *Aplysia californica*. *Journal of Neurophysiology* 43: 581–594.

Troschel, F. H. (1856–1879). *Das Gebiß der Schnecken, zur Begründung einer natürlichen Classification.* Vol. 1 (1–5), 2 (1–6). Berlin, Nicolai.

Trowbridge, C. D., Hirano, Y. M. & Hirano, Y. J. (2009). Interaction webs of marine specialist herbivores on Japanese shores. *Journal of the Marine Biological Association of the United Kingdom* 89: 277–286.

Townsley, P., Richy, R. & Trussell, P. (1966). The laboratory rearing of the shipworm, *Bankia setacea* (Tryon). *National Shellfisheries Association, Proceedings* 56: 49–52.

Trueblood, L. A. & Seibel, B. A. (2014). Slow swimming, fast strikes: Effects of feeding behavior on scaling of anaerobic metabolism in epipelagic squid. *Journal of Experimental Biology* 217: 2710–2716.

Trueman, E. R., Brand, A. R. & Davis, P. (1966). The dynamics of burrowing of some common littoral bivalves. *Journal of Experimental Biology* 44: 469–492.

Trueman, E. R. (1968). The mechanism of burrowing of some naticid gastropods in comparison with that of other molluscs. *Journal of Experimental Biology* 48: 663–678.

Trueman, E. R. & Packard, A. (1968). Motor performances of some cephalopods. *Journal of Experimental Biology* 49: 495–507.

Trueman, E. R. (1975). *The Locomotion of Soft-bodied Animals.* London, UK, Edward Arnold.

Trueman, E. R. & Brown, A. C. (1976). Locomotion, pedal retraction and extension, and the hydraulic systems of *Bullia* (Gastropoda: Nassaridae). *Journal of Zoology* 178: 365–384.

Trueman, E. R. (1983). Locomotion in molluscs. pp. 155–198 in A. S. M. Salauddin & Wilbur, K. M. (Eds.) *Physiology, Part 1. The Mollusca.* Vol. 4. New York, Academic Press.

Trueman, E. R. & Brown, A. C. (1985). Dynamics of burrowing and pedal extension in *Donax serra* (Mollusca: Bivalvia). *Journal of Zoology* 207: 345–355.

Trueman, E. R. & Brown, A. C. (1987). Proboscis extrusion in *Bullia* (Nassariidae): A study of fluid skeletons in Gastropoda. *Journal of Zoology* 211: 505–513.

Trussell, G. C., Johnson, A. S., Rudolph, S. G. & Gilfillan, E. S. (1993). Resistance to dislodgement: Habitat and size-specific differences in morphology and tenacity in an intertidal snail. *Marine Ecology Progress Series* 100: 135–144.

Trussell, G. C. (1996). Phenotypic plasticity in an intertidal snail: The role of a common crab predator. *Evolution* 50: 448–454.

Trussell, G. C. (1997). Phenotypic plasticity in the foot size of an intertidal snail. *Ecology* 78: 1033–1048.

Trussell, G. C., Ewanchuk, P. J. & Bertness, M. D. (2003). Trait-mediated effects in rocky intertidal food chains: Predator risk cues alter prey feeding rates. *Ecology* 84: 629–640.

Tsai, P.-S., Maldonado, T. A. & Lunden, J. B. (2003). Localization of gonadotropin-releasing hormone in the central nervous system and a peripheral chemosensory organ of *Aplysia californica*. *General and Comparative Endocrinology* 130: 20–28.

Tsai, P.-S. (2006). Gonadotropin-releasing hormone in invertebrates: Structure, function, and evolution. *General and Comparative Endocrinology* 148: 48–53.

Tsubaki, R. & Kato, M. (2012). Host specificity and population dynamics of a sponge-endosymbiotic bivalve. *Zoological Science* 29: 585–592.

Tsubata, N., Iizuka, A., Horikoshi, T. & Sakakibara, M. (2003). Photoresponse from the statocyst hair cell in *Lymnaea stagnalis*. *Neuroscience Letters* 337: 46–50.

Tsuji, F. I. (1983). Molluscan bioluminescence. pp. 257–279 *in* P. W. Hochachka (Ed.) *Metabolic Biochemistry and Molecular Biomechanics. The Mollusca*. Vol. 2. New York, Academic Press.

Tsuji, F. I. (2002). Bioluminescence reaction catalyzed by membrane-bound luciferase in the 'firefly squid,' *Watasenia scintillans*. *Biochimica et Biophysica Acta* 1564: 189–197.

Tuersley, M. D. & McCrohan, C. R. (1987). Food arousal in the pond snail, *Lymnaea stagnalis*. *Behavioral and Neural Biology* 48: 222–236.

Tunnicliffe, V., Rose, J. M., Bates, A. E. & Kelly, N. E. (2008). Parasitization of a hydrothermal vent limpet (Lepetodrilidae, Vetigastropoda) by a highly modified copepod (Chitonophilidae, Cyclopoida). *Parasitology* 135: 1281–1293.

Turekian, K. K., Cochran, J. K., Kharkar, D. P., Cerrato, R. M., Vaisnys, J. R., Sanders, H. L., Grassle, J. F. & Allen, J. A. (1975). Slow growth rate of a deep-sea clam determined by ^{228}Ra chronology. *Proceedings of the National Academy of Sciences* 72: 2829–2832.

Turner, A. M. & Chislock, M. F. (2007). Dragonfly predators influence biomass and density of pond snails. *Oecologia* 153: 407–415.

Turner, E. (1978). Brooding of chitons in Tasmania. *Journal of the Malacological Society of Australia* 4: 43–47.

Turner, R. D. (1966). *A Survey and Illustrated Catalogue of the Teredinidae (Mollusca: Bivalvia)*. Cambridge, UK, Museum of Comparative Zoology.

Turner, R. D. & Johnson, A. C. (1971). Biology of marine wood-boring molluscs. pp. 259–301 *in* E. B. G. Jones & Eltringham, S. K. (Eds.) *Marine Borers, Fungi and Fouling Organisms of Wood*. Paris, O.E.C.D. Paris.

Turner, R. D. (1977). Genetic relations of deep-sea wood-borers. *Bulletin of the American Malacological Union Incorporated* 1977: 19–24.

Turner, R. D. & Roberts, T. (1978). Mollusks as prey of arid catfish in the Fly River. *Bulletin of the American Malacological Union* 33: 33–40.

Turner, R. D. & Yakovlev, M. (1983). Dwarf males in the Teredinidae (Bivalvia, Pholadacea). *Science* 219: 1077–1078.

Turner, R. D., Pechenik, J. A. & Calloway, C. B. (1987). The language of benthic marine invertebrate development patterns: Problems and needs. pp. 227–235 *in* M. F. Thompson, Sarojini, R. & Nagabhushanam, R. (Eds.) *Biology of Marine Benthic Organisms: Techniques and Methods as Applied to the Indian Ocean*. New Delhi, Oxford & IBH Publishing Co.

Turnipseed, M., Knick, K. E., Lipcius, R. N., Dreyer, J. & Van Dover, C. L. (2003). Diversity in mussel beds at deep-sea hydrothermal vents and cold seeps. *Ecology Letters* 6: 518–523.

Tutschulte, T. C. & Connell, J. H. (1988). Feeding behavior and algal food of three species of abalones (*Haliotis*) in southern California. *Marine Ecology Progress Series* 49: 57–64.

Twarog, B. M. (1967). The regulation of catch in molluscan muscle. *Journal of General Physiology* 50: 157–169.

Twede, V. D., Miljanich, G., Olivera, B. M. & Bulaj, G. (2009). Neuroprotective and cardioprotective conopeptides: An emerging class of drug leads. *Current Opinion in Drug Discovery & Development* 12: 231.

Twomey, E. & Mulcahy, M. F. (1989). Transmission of a sarcoma in the cockle *Cerastoderma edule* (Bivalvia; Mollusca) using cell transplants. *Developmental and Comparative Immunology* 12: 195–200.

Tyler, A. (1949). Properties of fertilizin and related substances of eggs and sperm of marine animals. *American Naturalist* 83: 195–215.

Tyler, A. V. (1972). Food resource division among northern, marine, demersal fishes. *Journal of the Fisheries Board of Canada* 29: 997–1003.

Tyler, P. A. & Young, C. M. (1999). Reproduction and dispersal at vents and cold seeps. *Journal of the Marine Biological Association of the United Kingdom* 79: 193–208.

Tyndale, E., Avila, C. & Kuzirian, A. M. (1994). Food detection and preferences of the nudibranch mollusc *Hermissenda crassicornis*: Experiments in a Y-Maze. *Biological Bulletin* 187: 274–275.

Tyrakowski, T., Kaczorowski, P., Pawlowicz, W., Ziółkowski, M., Smuszkiewicz, P., Trojanowska, I., Marszalek, A., Żebrowska, M., Lutowska, M., Kopczyńska, E., Lampka, M., Holyńska-Iwan, I. & Piskorska, E. (2012). Discrete movements of foot epithelium during adhesive locomotion of a land snail. *Folia Biologica (Cracow)* 60: 99–106.

Udaka, H. & Numata, H. (2008). Short-day and low-temperature conditions promote reproductive maturation in the terrestrial slug, *Lehmannia valentiana*. *Comparative Biochemistry and Physiology Part A* 150: 80–83.

Ulloa, O., Canfield, D. E., DeLong, E. F., Letelier, R. M. & Stewart, F. J. (2012). Microbial oceanography of anoxic oxygen minimum zones. *Proceedings of the National Academy of Sciences of the United States of America* 109: 15996–16003.

Underwood, A. J. (1972). Spawning, larval development and settlement behaviour of *Gibbula cineraria* (Gastropoda: Prosobranchia) with a reappraisal of torsion in gastropods. *Marine Biology* 17: 341–349.

Underwood, A. J. & Creese, R. G. (1976). Observations on the biology of the trochid gastropod *Austrocochlea constricta* (Lamarck) (Prosobranchia). II. The effects of available food on shell-banding pattern. *Journal of Experimental Marine Biology and Ecology* 23: 229–240.

Underwood, A. J. (1977). Movements of intertidal gastropods. *Journal of Experimental Marine Biology and Ecology* 26: 191–201.

Underwood, A. J. (1978). A refutation of critical tidal levels as determinants of the structure of intertidal communities on British shores. *Journal of Experimental Marine Biology and Ecology* 33: 261–276.

Underwood, A. J. (1979). The ecology of intertidal gastropods. *Advances in Marine Biology* 16: 111–210.

Underwood, A. J. (1980). The effects of grazing by gastropods and physical factors on the upper limits of distribution of intertidal macroalgae. *Oecologia* 46: 201–213.

Underwood, A. J. & Jernakoff, P. (1981). Effects of interactions between algae and grazing gastropods on the structure of a low-shore intertidal algal community. *Oecologia* 48: 221–233.

Underwood, A. J., Denley, E. J. & Moran, M. J. (1983). Experimental analyses of the structure and dynamics of mid-shore rocky intertidal communities in New South Wales. *Oecologia* 56: 202–219.

Underwood, A. J. (1985). Experimental ecology, environmental management and the approaching dark ages for fundamental research. AES working paper 2/85 (School of Australian Environmental Studies). Nathan, Qld., School of Australian Environmental Studies, Griffith University: 1–37.

Underwood, A. J. & Verstegen, P. H. (1988). Experiments on the association between the intertidal amphipod *Hyale media* (Dana) and the limpet *Cellana tramoserica* (Sowerby). *Journal of Experimental Marine Biology and Ecology* 119: 83–98.

Underwood, A. J. & Fairweather, P. G. (1989). Supply-side ecology and benthic marine assemblages. *Trends in Ecology & Evolution* 4: 16–20.

Underwood, A. J. & Chapman, M. G. (1995). *Coastal Marine Ecology of Temperate Australia.* Sydney, Australia, UNSW Press.

Underwood, A. J. (1997). *Experiments in Ecology: Their Logical Design and Interpretation using Analysis of Variance.* Cambridge, UK, Cambridge University Press.

Underwood, A. J. (1998). Molluscs on rocky shores. pp. 29–32 *in* P. L. Beesley, Ross, G. J. B. & Wells, A. (Eds.) *Mollusca: The Southern Synthesis. Part A. Fauna of Australia.* Melbourne, Australia, CSIRO Publishing.

Underwood, A. J. (2000). Experimental ecology of rocky intertidal habitats: What are we learning? *Journal of Experimental Marine Biology and Ecology* 250: 51–76.

Underwood, A. J. & Keough, M. J. (2001). Supply-side ecology: The nature and consequences of variations in recruitment of intertidal organisms. pp. 183–200 *in* M. D. Bertness (Ed.) *Marine Community Ecology and Conservation.* Sunderland, MA, Sinauer Associates.

Urdy, S. (2015). Theoretical modelling of the molluscan shell: What has been learned from the comparison among molluscan taxa? pp. 207–251 *in* C. Klug, Korn, D., De Baets, K., Kruta, I. & Mapes, R. H. (Eds.) *Ammonoid Paleobiology: From Anatomy to Ecology. Topics in Geobiology.* Dordrecht, the Netherlands, Springer.

Uribe, R. & Otaïza, R. D. (1996). *Primer registro de parasitismo por larvas de dípteros en patelogastrópodos de la zona intermareal rocosa.* Book of Abstracts, 16th Seminar of Marine Sciences, Concepción, Chile.

Uyeno, T. A. & Kier, W. M. (2005). Functional morphology of the cephalopod buccal mass: A novel joint type. *Journal of Morphology* 264: 211–222.

Vaccaro, E. & Waite, J. H. (2001). Yield and post-yield behavior of mussel byssal thread: A self-healing biomolecular material. *Biomacromolecules* 2: 906–911.

Vacquier, V. D. (1998). Evolution of gamete recognition proteins. *Science* 281: 1995–1998.

Vader, W. (1972). Associations between amphipods and molluscs: A review of published records. *Sarsia* 48: 13–18.

Vagvolgyi, J. (1975). Body size, aerial dispersal, and origin of the Pacific land snail fauna. *Systematic Zoology* 24: 465–488.

Vail, V. A. (1977). Comparative reproductive anatomy of three viviparid gastropods. *Malacologia* 16: 519–540.

Valdés, Á. & Gosliner, T. M. (2001). Systematics and phylogeny of the caryophyllidia-bearing dorids (Mollusca, Nudibranchia), with descriptions of a new genus and four new species from Indo-Pacific deep waters. *Zoological Journal of the Linnean Society* 133: 103–198.

Valdés, Á., Blanchard, L. & Marti, W. (2013). Caught naked: First report [of] a nudibranch sea slug attacked by a cone snail. *American Malacological Bulletin* 31: 337–338.

Valentine, J. W. (1955). Upwelling and thermally anomalous Pacific coast Pleistocene molluscan faunas. *American Journal of Science* 253: 462–474.

Valentine, J. W. (1997). Cleavage patterns and the topology of the metazoan tree of life. *Proceedings of the National Academy of Sciences of the United States of America* 94: 8001-8005.

Valle-Esquivel, M. (2003). Aspects of the population dynamics, stock assessment, and fishery management strategies of the queen conch, *Strombus gigas*, in the Caribbean. PhD, University of Miami.

Vallès, Y. & Gosliner, T. M. (2006). Shedding light onto the genera (Mollusca: Nudibranchia) *Kaloplocamus* and *Plocamopherus* with description of new species belonging to these unique bioluminescent dorids. *The Veliger* 48: 178–205.

Van Dongen, C. A. M. & Van Geilenkirchen, W. L. M. (1974). The development of *Dentalium* with special reference to the significance of the polar lobe. Parts I–III. Division chronology and development of the cell pattern in *Dentalium dentale* (Scaphopoda). *Proceedings of the Koninklijke Nederlandse Akademie van Wetenschappen. Series C* 77: 57–100.

Van Dover, C. L. (2002). Trophic relationships among invertebrates at the Kairei hydrothermal vent field (Central Indian Ridge). *Marine Biology* 141: 761–772.

Van Dover, C. L., German, C. R., Speer, K. G., Parson, L. M. & Vrijenhoek, R. C. (2002). Evolution and biogeography of deep-sea vent and seep invertebrates. *Science* 295: 1253–1257.

Van Dover, S. L., Aharon, P., Bernhard, J. M., Caylor, E., Doerries, M., Flickinger, W., Gilhooly, W. et al. (2003). Blake Ridge methane seeps: Characterization of a soft-sediment chemosynthetically based ecosystem. *Deep Sea Research Part I: Oceanographic Research Papers* 50: 281–300.

Van Dover, C. L. (2014). Impacts of anthropogenic disturbances at deep-sea hydrothermal vent ecosystems: A review. *Marine Environmental Research* 102: 59–72.

Van Duivenboden, Y. A. (1982). Non-ocular photoreceptors and photo-orientation in the pond snail *Lymaea stagnalis* L. *Journal of Comparative Physiology A* 149: 363–368.

van Gils, J. A., van der Geest, M., Jansen, E. J., Govers, L. L., de Fouw, J. & Piersma, T. (2012). Trophic cascade induced by molluscivore predator alters pore-water biogeochemistry via competitive release of prey. *Ecology* 93: 1143–1152.

Van Golen, F. A., Li, K. W., De Lange, R. P., Van Kesteren, R. E., Van der Schors, R. C. & Geraerts, W. P. (1995). Co-localized neuropeptides conopressin and Ala-Pro-Gly-Trp-NH 2 have antagonistic effects on the vas deferens of *Lymnaea. Neuroscience* 69: 1275–1287.

Van Noord, J. E. (2012). Diet of five species of the family Myctophidae caught off the Mariana Islands. *Ichthyological Research* 60: 89–92.

Vandermeer, J. H. (1972). Niche theory. *Annual Review of Ecology and Systematics* 3: 107–132.

Vanfleteren, J. R. (1982). A monophyletic line of evolution? Ciliary induced photoreceptor membranes. pp. 107–136 *in* J. A. Westfall (Ed.) *Visual Cells in Evolution.* New York, Raven Press.

Vanhaeren, M., D'Errico, F., Stringer, C., James, S. L., Todd, J. A. & Mienis, H. K. (2006). Middle Paleolithic shell beads in Israel and Algeria. *Science* 312: 1785–1788.

Vannini, M., Mrabu, E., Cannicci, S., Rorandelli, R. & Fratini, S. (2008). Rhythmic vertical migration of the gastropod *Cerithidea decollata* in a Kenyan mangrove forest. *Marine Biology* 153: 1047–1053.

Varaksina, G. S. & Varaksin, A. A. (1991). Effect of estradiol, progesterone and testosterone on oogenesis of the Japanese scallop *Mizuhopecten yessoensis. Biologiya Morya* 3: 61–68.

Varjú, D. (2004). Polarization sensitivity in cephalopods and marine snails. pp. 267–275 *in* G. Horváth & Varju, D. (Eds.) *Polarized Light in Animal Vision.* Berlin, Germany, Springer-Verlag.

Varona, P., Rabinovich, M. I., Selverston, A. I. & Arshavsky, Y. I. (2002). Winnerless competition between sensory neurons generates chaos: A possible mechanism for molluscan hunting behavior. *Chaos* 12: 672–677.

Vaughn, C. C. & Taylor, C. M. (2000). Macroecology of a host-parasite relationship. *Ecography* 23: 11–20.

Vaughn, C. C., Gido, K. B. & Spooner, D. E. (2004). Ecosystem processes performed by unionid mussels in stream mesocosms: Species roles and effects of abundance. *Hydrobiologia* 527: 35–47.

Vayssière, A. (1895). Sur le dimorphisme sexuel des Nautiles. *Comptes rendus de l'Académie des sciences* 120: 1431–1434.

Vazquez, N. N., Ituarte, C., Navone, G. T. & Cremonte, F. (2006). Parasites of the stout razor clam *Tagelus plebeius* (Psammobiidae) from the southwestern Atlantic Ocean. *Journal of Shellfish Research* 25: 877–886.

Vecchione, M. & Young, R. E. (1997). Aspects of the functional morphology of cirrate octopods: Locomotion and feeding. *Vie et Milieu* 47: 101–110.

Vecchione, M., Roper, C. F. E., Widder, E. A. & Frank, T. M. (2002). In situ observations on three species of large-finned deep-sea squids. *Bulletin of Marine Science* 71: 893–901.

Veenstra, J. A. (2010). Neurohormones and neuropeptides encoded by the genome of *Lottia gigantea*, with reference to other mollusks and insects. *General and Comparative Endocrinology* 167: 86–103.

Vega, I. A., Amborenea, M. C. D., Gamarra-Luques, C., Koch, E., Cueto, J. A. & Castro-Vazquez, A. (2006). Facultative and obligate symbiotic associations of *Pomacea canaliculata* (Caenogastropoda, Ampullariidae). *Biocell* 30: 367–375.

Velez-Arellano, N., Del Proo, S. G. & Ordonez, E. O. (2009). Gonadal cycle of *Tegula eiseni* (Jordan 1936) (Mollusca: Gastropoda) in Bahia Asuncion, Baja California Sur, Mexico. *Journal of Shellfish Research* 28: 577–580.

Vellutini, B. C. & Hejnol, A. (2016). Expression of segment polarity genes in brachiopods supports a non-segmental ancestral role of engrailed for bilaterians. *Nature. Scientific Reports* 6: 32387.

Vendrasco, M. J., Checa, A. G. & Kouchinsky, A. V. (2011). Shell microstructure of the early bivalve *Pojetaia* and the independent origin of nacre within the mollusca. *Palaeontology* 54: 825–850.

Vendrasco, M. J. & Checa, A. G. (2015). Shell microstructure and its inheritance in the calcitic helcionellid *Mackinnonia*. *Estonian Journal of Earth Sciences* 64: 99.

Venetis, C., Theologidis, I., Zouros, E. & Rodakis, G. C. (2006). No evidence for presence of maternal mitochondrial DNA in the sperm of *Mytilus galloprovincialis* males. *Proceedings of the Royal Society B* 273: 2483–2489.

Venier, P., Domeneghetti, S., Sharma, N., Pallavicini, A. & Gerdol, M. (2016). Immune-related signaling in mussel and bivalves. pp. 93–105 *in* L. Ballarin & Cammarata, M. (Eds.) *Lessons in Immunity: From Single-Cell Organisms to Mammals.* London, UK, Academic Press.

Vercammen-Grandjean, P. H. (1951). Considerations on the habits of *Xenopus laevis victorianus* Ahl. *Annales de la Société belge de médecine tropicale* 31: 409–414.

Verdonk, N. H. & Van den Biggelaar, J. A. M. (1983). Early development and the formation of the germ layers. pp. 91–122 *in* N. H. Verdonk, Van den Biggelaar, J. A. M., & Tompa, A. S. (Eds.) *Development. The Mollusca.* Vol. 3. New York, Academic Press.

Verdonk, N. H. & Cather, J. N. (1983). Morphogenetic determination and differentiation. pp. 215–252 *in* N. H. Verdonk, Van den Biggelaar, J. A. M. & Tompa, A. S. (Eds.) *Development. The Mollusca.* Vol. 3. New York, Academic Press.

Vergés, A., Becerro, M. A., Alcoverro, T. & Romero, J. (2007). Variation in multiple traits of vegetative and reproductive seagrass tissues influences plant-herbivore interactions. *Oecologia* 151: 675–686.

Vergés, A., Alcoverro, T. & Romero, J. (2011). Plant defences and the role of epibiosis in mediating within-plant feeding choices of seagrass consumers. *Oecologia* 166: 381–390.

Vermeij, G. J. (1971). Gastropod evolution and morphological diversity in relation to shell geometry. *Journal of Zoology* 163: 15–23.

Vermeij, G. J. (1972a). Endemism and environment: Some shore molluscs of the tropical Atlantic. *American Naturalist* 106: 89–101.

Vermeij, G. J. (1972b). Intraspecific shore-level size gradients in intertidal molluscs. *Ecology* 53: 693–700.

Vermeij, G. J. (1973a). Morphological patterns in high-intertidal gastropods: Adaptive strategies and their limitations. *Marine Biology* 20: 319–346.

Vermeij, G. J. (1973b). Adaptation, versatility, and evolution. *Systematic Zoology* 22: 466–477.

Vermeij, G. J. (1974). Marine faunal dominance and molluscan shell form. *Evolution* 28: 656–664.

Vermeij, G. J. (1975). Evolution and distribution of left-handed and planispiral coiling in snails. *Nature* 254: 419–420.

Vermeij, G. J. (1976). Interoceanic differences in vulnerability of shelled prey to crab predation. *Nature* 260: 135–136.

Vermeij, G. J. (1977a). Patterns in crab claw size: The geography of crushing. *Systematic Zoology* 26: 138–151.

Vermeij, G. J. (1977b). The Mesozoic marine revolution: Evidence from snails, predators and grazers. *Paleobiology* 3: 245–258.

Vermeij, G. J. (1978). *Biogeography and Adaptation.* Cambridge, MA, Belknap Press.

Vermeij, G. J. & Covich, A. P. (1978). Coevolution of freshwater gastropods and their predators. *American Society of Naturalists* 112: 833–843.

Vermeij, G. J. (1979). Shell architecture and causes of death of Micronesian reef snails. *Evolution* 33: 686–696.

Vermeij, G. J. & Currey, J. D. (1980). Geographical variation in the strength of thaidid snail shells. *Biological Bulletin* 158: 383–389.

Vermeij, G. J., Zipser, E. & Dudley, E. C. (1980). Predation in time and space: Peeling and drilling in terebrid gastropods. *Paleobiology* 6: 352–364.

Vermeij, G. J., Schindel, D. E. & Zipser, E. (1981). Predation through geological time: Evidence from gastropod shell repair. *Science* 214: 1024–1026.

Vermeij, G. J. (1982). Phenotypic evolution in a poorly dispersing snail after arrival of a predator. *Nature* 299: 349–350.

Vermeij, G. J. (1983). Shell-breaking predation through time. pp. 649–669 *in* M. J. S. Tevesz & McCall, P. L. (Eds.) *Biotic Interactions in Recent and Fossil Benthic Communities.* New York, Springer Science + Business Media.

Vermeij, G. J. & Dudley, E. C. (1985). Distribution of adaptations: A comparison between the functional shell morphology of freshwater and marine pelecypods. pp. 461–478 *in* E. R. Trueman & Clarke, M. R. (Eds.) *Evolution. The Mollusca.* Vol. 10. New York, Academic Press.

Vermeij, G. J. & Zipser, E. (1986). A short-term study of growth and death in a population of the gastropod *Strombus gibberulus* in Guam. *The Veliger* 28: 314–317.

Vermeij, G. J. (1987). *Evolution and Escalation.* Princeton, NJ, Princeton University Press.

Vermeij, G. J. (1990). Tropical Pacific pelecypods and productivity: A hypothesis. *Bulletin of Marine Science* 47: 62–67.

Vermeij, G. J., Palmer, A. R. & Lindberg, D. R. (1990). Range limits and dispersal of mollusks in the Aleutian Islands, Alaska. *The Veliger* 33: 346–354.

Vermeij, G. J. (1992). Time of origin and biogeographical history of specialized relationships between northern marine plants and herbivorous molluscs. *Evolution* 46: 657–664.

Vermeij, G. J. & Signor, P. W. (1992). The geographic, taxonomic and temporal distribution of determinate growth in marine gastropods. *Biological Journal of the Linnean Society* 47: 233–247.

Vermeij, G. J. (1993). *A Natural History of Shells*. Princeton, NJ, Princeton University Press.

Vermeij, G. J. (1996a). Adaptations of clades: Resistance and response. pp. 363–380 *in* M. R. Rose & Lauder, G. V. (Eds.) *Adaptation*. San Diego, CA, Academic Press.

Vermeij, G. J. (1996b). An agenda for invasion biology. *Biological Conservation* 78: 3–9.

Vermeij, G. J. & Dudley, R. (2000). Why are there so few evolutionary transitions between aquatic and terrestrial ecosystems? *Biological Journal of the Linnean Society* 70: 541–554.

Vermeij, G. J. & Lindberg, D. R. (2000). Delayed herbivory and the assembly of marine benthic ecosystems. *Paleobiology* 26: 419–430.

Vermeij, G. J. (2001a). Edge-drilling: History and distribution of a novel method of predation. *Paleobios* 21 (Suppl. 2): 130.

Vermeij, G. J. (2001b). Innovation and evolution at the edge: Origins and fates of gastropods with a labral tooth. *Biological Journal of the Linnean Society* 72: 461–508.

Vermeij, G. J. & Raven, H. (2009). Southeast Asia as the birthplace of unusual traits: The Melongenidae (Gastropoda) of northwest Borneo. *Contributions to Zoology* 78: 113–127.

Vermeij, G. J. (2010). Sound reasons for silence: Why do molluscs not communicate acoustically? *Biological Journal of the Linnean Society* 100: 485–493.

Vermeij, G. J. (2016). The limpet form in gastropods: Evolution, distribution, and implications for the comparative study of history. *Biological Journal of the Linnean Society* 120: 22–37.

Vernberg, F. J. & Vernberg, W. B. (1975). Adaptations to extreme environments. pp. 165–180 *in* F. J. Vernberg (Ed.) *Physiological Ecology of Estuarine Organisms*. Columbia SC, University of South Carolina Press.

Vernberg, W. B. & Vernberg, F. J. (1972). The intertidal zone. pp. 58–160 *in* W. B. Vernberg & Vernberg, F. J. (Eds.) *Environmental Physiology of Marine Animals*. Berlin, Heidelberg, Springer.

Verrill, A. E. (1884). Notice of the remarkable marine fauna occupying the outer banks off the southern coast of New England; No. 9. *American Journal of Science* 28: 213–220.

Versen, B., Gokorsch, S., Fiedler, A. & Schipp, R. (1999). Monoamines and the isolated auricle of *Sepia officinalis*: Are there β-like receptors in the heart of a cephalopod? *Journal of Experimental Biology* 202: 1067–1079.

Vershinin, A. (1996). Carotenoids in mollusca: Approaching the functions. *Comparative Biochemistry and Physiology Part B* 113: 63–71.

Vetter, R. D. (1985). Elemental sulfur in the gills of three species of clams containing chemoautotrophic symbiontic bacteria: A possible inorganic energy storage compound. *Marine Biology* 88: 33–42.

Vetter, R. D. (1991). Symbiosis and the evolution of novel trophic strategies: Thiotrophic organisms at hydrothermal vents. pp. 219–245 *in* L. Margulis & Fester, R. (Eds.) *Symbiosis as a Source of Evolutionary Innovation: Speciation and Morphogenesis*. Cambridge, MA, MIT Press.

Vicente, N. & Gasquet, M. (1970). Étude du système nerveux et de la neurosécrétion chez quelques mollusques polyplacophores. *Téthys* 2: 515–546.

Viçose, G. C. D., Viera, M., Bilbao, A. & Izquierdo, M. (2010). Larval settlement of *Haliotis tuberculata coccinea* in response to different inductive cues and the effect of larval density on settlement, early growth, and survival. *Journal of Shellfish Research* 29: 587–591.

Vijver, M. G., Gestel, C. A. M. van, Lanno, R. P., van Straalen, N. M. & Peijnenburg, W. J. G. M. (2004). Internal metal sequestration and its ecotoxicological relevance: A review. *Environmental Science & Technology* 38: 4705–4712.

Villalba, A., Reece, K. S., Ordás, M. C., Casas, S. M. & Figueras, A. (2004). Perkinsosis in molluscs: A review. *Aquatic Living Resources* 17: 411–432.

Villanueva, R. & Guerra, A. (1991). Food and prey detection in two deep-sea cephalopods: *Opisthoteuthis agassizii* and *O. vossi* (Octopoda: Cirrata). *Bulletin of Marine Science* 49: 288–299.

Villanueva, R. (1992). Continuous spawning in the cirrate octopods *Opisthoteuthis agassizii* and *O. vossi*: Features of sexual maturation defining a reproductive strategy in cephalopods. *Marine Biology* 275: 265–275.

Villanueva, R. & Norman, M. D. (2008). Biology of the planktonic stages of benthic octopuses. *Oceanography and Marine Biology* 46: 105–202.

Villanueva, R., Sykes, A. V., Vidal, E. A. G., Rosas, C., Nabhitabhata, J., Fuentes, L. & Iglesias, J. (2014). Current status and future challenges in cephalopod culture. pp. 479–489 *in* J. Iglesias, Fuentes, L. & Villanueva, R. (Eds.) *Cephalopod Culture*. Dordrecht, the Netherlands, Springer.

de Villiers, C. J. & Hodgson, A. N. (1987). The structure of the secondary gills of *Siphonaria capensis* (Gastropoda: Pulmonata). *Journal of Molluscan Studies* 53: 129–138.

Vincent, C., Griffond, B., Wijdenes, J. & Gomot, L. (1984). Contrôle d'une glande endocrine: les corps dorsaux par le système nerveux central chez *Helix aspersa*. *Comptes rendus de l'Académie des sciences* 299: 4231–4236.

Vinn, O. & Zatoń, M. (2012). Phenetic phylogenetics of tentaculitoids: Extinct, problematic calcareous tube-forming organisms. *GFF: Journal of the Geological Society of Sweden* 134: 145–156.

Vinogradov, A. E. (2000). Larger genomes for molluskan land pioneers. *Genome* 43: 211–212.

Visser, M. H. C. (1988). The significance of terminal duct structures and the role of neoteny in the evolution of the reproductive system of Pulmonata. *Zoologica Scripta* 17: 239–252.

Vitturi, R., Colombera, D., Catalano, E. & Arnold, J. M. (1990). Spermatocyte chromosome study of eight species of the class Cephalopoda (Mollusca). *Journal of Cephalopod Biology* 1: 101–112.

Vlés, F. (1907). Sur les ondes pédieuses des Mollusques reptateurs. *Comptes rendus hebdomadaire des séances de l'Académie des sciences* 145: 276–278.

Vogel, H., Czihak, G., Chang, P. & Wolf, W. (1982). Fertilization kinetics of sea urchin eggs. *Mathematical Biosciences* 58: 189–216.

Vogel, S. (1987). Flow-assisted mantle cavity refilling in jetting squid. *Biological Bulletin* 172: 61–68.

Vogel, S. (1994). *Life in Moving Fluids: The Physical Biology of Flow*. Princeton, NJ, Princeton University Press.

Voight, J. R., Pörtner, H. O. & O'Dor, R. K. (1994). A review of ammonia-mediated buoyancy in squids (Cephalopoda: Teuthoidea). *Marine and Freshwater Behaviour and Physiology* 25: 193–203.

Voight, J. R. (1995). Sexual dimorphism and niche divergence in a mid-water octopod (Cephalopoda: Bolitaenidae). *Biological Bulletin* 189: 113–119.

Voight, J. R. (2000). A deep-sea octopus (*Graneledone* cf. *boreopacifica*) as a shell-crushing hydrothermal vent predator. *Journal of Zoology* 252: 335–341.

Voight, J. R., Lee, R. W., Reft, A. J. & Bates, A. E. (2012). Scientific gear as a vector for non-native species at deep-sea hydrothermal vents. *Conservation Biology* 26: 938–942.

Voight, J. R. (2015). Xylotrophic bivalves: Aspects of their biology and the impacts of humans. *Journal of Molluscan Studies* 81: 1–12.

Volland, J. M., Frenkiel, L., Aranda, D. & Gros, O. (2010). Occurrence of Sporozoa-like microorganisms in the digestive gland of various species of Strombidae. *Journal of Molluscan Studies* 76: 196–198.

Voltzow, J. (1985). Morphology of the pedal circulatory system of the marine gastropod *Busycon contrarium* and its role in locomotion (Gastropoda, Buccinacea). *Zoomorphology* 105: 395–400.

Voltzow, J. (1986). Changes in pedal intramuscular pressure corresponding to behavior and locomotion in the marine gastropods *Busycon contrarium* and *Haliotis kamtschatkana*. *Canadian Journal of Zoology* 64: 2288–2293.

Voltzow, J. (1987). Torsion reexamined: The timing of the twist. *American Zoologist* 27: 84.

Voltzow, J. (1988). The organization of limpet pedal musculature and its evolutionary implications for the Gastropoda. pp. 273–283 in W. F. Ponder, Eernisse, D. J. & Waterhouse, J. H. (Eds.) *Prosobranch Phylogeny. Malacological Review Supplement*. Ann Arbor, MI, Malacological Review.

Voltzow, J. (1992). Gastropod pedal architecture: Predicting function from structure and structure from function. pp. 83–84 in C. Meier-Brook (Ed.) *Proceedings of the 10th International Malacological Congress, Tübingen, 27 August–2 September 1989*. Tübingen, University of Tübingen.

Voltzow, J. (1994). Gastropoda: Prosobranchia. pp. 111–252 in F. W. Harrison & Kohn, A. J. (Eds.) *Microscopic Anatomy of Invertebrates. Mollusca 1*. Vol. 5. New York, Wiley-Liss.

Voltzow, J. & Collin, R. (1995). Flow through mantle cavities revisited: Was sanitation the key to fissurellid evolution? *Invertebrate Biology* 114: 145–150.

von Rintelen, T., von Rintelen, K., Glaubrecht, M., Schubart, C. D. & Herder, F. (2012). Aquatic biodiversity hotspots in Wallacea: The species flocks in the ancient lakes of Sulawesi, Indonesia. pp. 290–315 in D. J. Gower, Johnson, K., Richardson, J., Rosen, B., Rüber, L. & Williams, S. (Eds.) *Biotic Evolution and Environmental Change in Southeast Asia*. Cambridge, UK, Cambridge University Press.

Vonk, H. J. & Western, J. R. H. (1984). *Comparative Biochemistry and Physiology of Enzymatic Digestion*. London, UK, Academic Press.

Voogt, P. A. (1983). Lipids: Their distribution and metabolism. pp. 329–370 in P. W. Hochachka (Ed.) *Metabolic Biochemistry and Molecular Biomechanics. The Mollusca*. Vol. 1. New York, Academic Press.

Voronezhskaya, E. E., Tyurin, S. A. & Nezlin, L. P. (2002). Neuronal development in larval chiton *Ischnochiton hakodadensis* (Mollusca: Polyplacophora). *The Journal of Comparative Neurology* 444: 25–38.

Vorwohl, G. (1961). Zur Funktion der Exkretionsorgane von *Helix pomatia* L. und *Archachatina ventricosa* Gould. *Zeitschrift für vergleichende Physiologie* 45: 12–49.

Voss, M., Kottowski, K. & Wünnenberg, W. (2000). Reactions of central neurons to peripheral thermal stimuli in the snail *Helix pomatia* L. *Journal of Thermal Biology* 25: 431–436.

Voss, M. & Schmidt, H. (2001). Electrophysiological responses to thermal stimuli in peripheral nerves of the African giant snail, *Archachatina marginata* S. *Journal of Thermal Biology* 26: 21–27.

Vrijenhoek, R. C. (1997). Gene flow and genetic diversity in naturally fragmented metapopulations of deep-sea hydrothermal vent animals. *Journal of Heredity* 88: 285–293.

Vrijenhoek, R. C. (2010). Genetics and evolution of deep-sea chemosynthetic bacteria and their invertebrate hosts. pp. 15–49 in S. Kiel (Ed.) *The Vent and Seep Biota: Aspects from Microbes to Ecosystems. Topics in Geobiology*. Dordrecht, the Netherlands, Springer Netherlands.

Vrolijk, N. H. & Targett, N. M. (1992). Biotransformation enzymes in *Cyphoma gibbosum* (Gastropoda: Ovulidae): Implications for detoxification of gorgonian allelochemicals. *Marine Ecology Progress Series* 88: 237–246.

Wabnitz, R. W. (1975). Functional states and fine structure of the contractile apparatus of the penis retractor muscle (PRM) of *Helix pomatia* L. *Cell and Tissue Research* 156: 253–265.

Wada, S. K. (1954). Spawning in the tridacnid clams. *Japanese Journal of Zoology* 11: 273–285.

Wada, S. K., Collier, J. R. & Dan, J. C. (1956). Studies on the acrosome: V. An egg-membrane lysin from the acrosomes of *Mytilus edulis* spermatozoa. *Experimental Cell Research* 10: 168–180.

Wada, T., Takegaki, T., Mori, T. & Natsukari, Y. (2005). Alternative male mating behaviors dependent on relative body size in captive oval squid *Sepioteuthis lessoniana* (Cephalopoda, Loliginidae). *Zoological Science* 22: 645–651.

Waddington, C. H. & Cowe, R. J. (1969). Computer simulation of a mulluscan [sic] pigmentation pattern. *Journal of Theoretical Biology* 25: 219–225.

Wade, C. M., Mordan, P. B. & Clarke, B. (2001). A phylogeny of the land snails (Gastropoda: Pulmonata). *Proceedings of the Royal Society B* 268: 413–422.

Wägele, H. & Willan, R. C. (2000). Phylogeny of the Nudibranchia. *Zoological Journal of the Linnean Society* 130: 83–181.

Wägele, H. & Johnsen, G. (2001). Observations on the histology and photosynthetic performance of 'solar-powered' opisthobranchs (Mollusca, Gastropoda, Opisthobranchia) containing symbiotic chloroplasts or zooxanthellae. *Organisms Diversity & Evolution* 1: 193–210.

Wägele, H. (2004). Potential key characters in Opisthobranchia (Gastropoda, Mollusca) enhancing adaptive radiation. *Organisms Diversity & Evolution* 4: 175–188.

Wägele, H., Ballesteros, M. & Avila, C. (2006). Defensive glandular structures in opisthobranch molluscs—From histology to ecology. *Oceanography and Marine Biology* 44: 197–276.

Wägele, H., Klussmann-Kolb, A., Vonnemann, V. & Medina, M. (2008). Heterobranchia I: The Opisthobranchia. pp. 385–408 in W. F. Ponder & Lindberg, D. R. (Eds.) *Phylogeny and Evolution of the Mollusca*. Berkeley, CA, University of California Press.

Wägele, H., Deusch, O., Händeler, K., Martin, R., Schmitt, V., Christa, G., Pinzger, B. et al. (2011). Transcriptomic evidence that longevity of acquired plastids in the photosynthetic slugs *Elysia timida* and *Plakobranchus ocellatus* does not entail lateral transfer of algal nuclear genes. *Molecular Biology and Evolution* 28: 699–706.

Wägele, H., Klussmann-Kolb, A., Verbeek, E. & Schrödl, M. (2014). Flashback and foreshadowing: A review of the taxon Opisthobranchia. *Organisms Diversity & Evolution* 14: 133–149.

Wägele, H. & Martin, W. F. (2014). Endosymbioses in sacoglossan seaslugs: Plastid-bearing animals that keep photosynthetic organelles without borrowing genes. pp. 291–324 in W. Löffelhardt (Ed.) Endosymbiosis. Vienna, Austria, Springer.

Wagner, P. J. (2001). Gastropod phylogenetics: Progress, problems, and implications. Journal of Paleontology 75: 1128–1140.

van der Wal, P., Videler, J. J., Havinga, P. & Pel, R. (1989). Architecture and chemical composition of the magnetite-bearing layer in the radula teeth of Chiton livaceus (Polyplacophora). pp. 153–166 in R. E. Crick (Ed.) Origin, Evolution, and Modern Aspects of Biomineralization in Plants and Animals. New York, Plenum Press.

van der Wal, P., Giesen, H. J. & Videler, J. J. (2000). Radular teeth as models for the improvement of industrial cutting devices. Materials Science and Engineering: C 7: 129–142.

Waldichuk, M. (1989). Amnesic shellfish poison. Marine Pollution Bulletin 20: 359–360.

Walker, A. J., Glen, D. M. & Shewry, P. R. (1996a). Proteolytic enzymes present in the digestive system of the field slug, Deroceras reticulatum, as a target for novel methods of control. pp. 305–312 in I. F. Henderson (Ed.) Slug and Snail Pests in Agriculture. Farnham, UK, British Crop Protection Council, Monograph No. 66.

Walker, A. J., Miller, A. J., Glen, D. M. & Shewry, P. R. (1996b). Determination of pH in the digestive system of the slug Deroceras reticulatum (Müller) using ion-selective microelectrodes. Journal of Molluscan Studies 62: 390–392.

Walker, A. J., Glen, D. M. & Shewry, P. R. (1999). Bacteria associated with the digestive system of the slug Deroceras reticulatum are not required for protein digestion. Soil Biology and Biochemistry 31: 1387–1394.

Walker, C. & Böttger, S. (2008). A naturally occurring cancer with molecular connectivity to human diseases. Cell Cycle 7: 2286–2289.

Walker, C. G. (1968). Studies on the jaw, digestive system, and coelomic derivatives in representatives of the genus Acmaea. The Veliger 11: 88–97.

Walker, C. H. (2008). Organic Pollutants: An Ecotoxicological Perspective. Boca Raton, FL, CRC Press.

Walker, G. (1970a). Light and electron microscope investigations on the salivary glands of the slug, Agriolimax reticulatus (Müller). Protoplasma 71: 111–126.

Walker, G. (1970b). The cytology, histochemistry and ultrastructure of the cell types found in the digestive gland of the slug Agriolimax reticulatus (Müller). Protoplasma 71: 91–109.

Walker, G., Dorrell, R. G., Schlacht, A. & Dacks, J. B. (2011). Eukaryotic systematics: A user's guide for cell biologists and parasitologists. Parasitology 138: 1638–1663.

Walker, K. J. (2003). Recovery Plans for Powelliphanta Land Snails: 2003–2013. Threatened Species Recovery Plan 49. Wellington, NZ, Department of Conservation, pp. 1–208.

Walker, R. P. & Faulkner, D. J. (1981). Chlorinated acetylenes from the nudibranch Diaulula sandiegensis. Journal of Organic Chemistry 46: 1475–1478.

Walker, T. D. & Valentine, J. W. (1984). Equilibrium models of evolutionary species diversity and the number of empty niches. American Naturalist 124: 887–899.

Wallace, C. (1992). Parthenogenesis, sex and chromosomes in Potamopyrgus. Journal of Molluscan Studies 58: 93–107.

Wallace, J. B., Webster, J. R. & Woodall, W. R. (1977). The role of filter feeders in flowing waters. Archiv für Hydrobiologie 79: 506–532.

Waller, D. L. & Lasee, B. A. (1997). External morphology of spermatozoa and spermatozeugmata of the freshwater mussel Truncilla truncata (Mollusca: Bivalvia: Unionidae). American Midland Naturalist 138: 220–223.

Waller, T. R. (1976). The origin of foliated-calcite shell microstructure in the subclass Pteriomorphia (Mollusca: Bivalvia). Bulletin of the American Malacological Union 1975: 57–58.

Waller, T. R. (1980). Scanning electron microscopy of shell and mantle in the order Arcoida (Mollusca: Bivalvia). Smithsonian Contributions to Zoology 313: 1–58.

Waller, T. R. (1990). The evolution of ligament systems in the Bivalvia. pp. 49–71 in B. Morton (Ed.) The Bivalvia. Proceedings of a Memorial Symposium in Honour of Sir Charles Maurice Yonge (1899–1986), Edinburgh, 1986. Edinburgh, UK, Hong Kong University Press.

Waller, T. R. (1998). Origin of the molluscan class Bivalvia and a phylogeny of major groups. pp. 1–45 in P. A. Johnston & Haggart, J. W. (Eds.) Bivalves: An Eon of Evolution. Paleobiological Studies Honoring Norman D. Newell. Calgary, Canada, University of Calgary Press.

Walsby, J. R., Morton, J. E. & Croxall, J. P. (1973). The feeding mechanism and ecology of the New Zealand pulmonate limpet Gadinalea nivea. Journal of Zoology 171: 257–283.

Walsby, J. R. (1975). Feeding and the radula in the marine pulmonate limpet Trimusculus reticulatus. The Veliger 18: 139–145.

Walters, E. T. (1991). A functional, cellular, and evolutionary model of nociceptive plasticity in Aplysia. Biological Bulletin 180: 241.

Wang, C. & Croll, R. P. (2003). Effects of sex steroids on 'in vitro' gamete release in the sea scallop Placopecten magellanicus. Invertebrate Reproduction & Development 44: 89–100.

Wang, D.-Z. (2008). Neurotoxins from marine dinoflagellates: A brief review. Marine Drugs 6: 349–371.

Wang, J. (2003). A monovalent anion affected multi-functional cellulase EGX from the mollusca [sic], Ampullaria crosseana. Protein Expression and Purification 31: 108–114.

Wang, J., Ding, M., Li, Y.-H., Chen, Q.-X., Xu, G.-J. & Zhao, F.-K. (2003a). Isolation of a multi-functional endogenous cellulase gene from mollusc, Ampullaria crosseana. Acta Biochimica et Biophysica Sinica (Chinese Edition) 35: 941–946.

Wang, J. H., Cain, S. D. & Lohmann, K. J. (2003b). Identification of magnetically responsive neurons in the marine mollusc Tritonia diomedea. Journal of Experimental Biology 206: 381–388.

Wang, L., Wang, L., Huang, M., Zhang, H. & Song, L. (2011). The immune role of C-type lectins in molluscs. Invertebrate Survival Journal 8: 241–246.

Wang, L., Yue, F., Song, X. & Song, L. (2015a). Maternal immune transfer in mollusc. Developmental & Comparative Immunology 48: 354–359.

Wang, X., Li, L., Zhu, Y., Song, X., Fang, X., Huang, R., Que, H. & Zhang, G. (2014). Aragonite shells are more ancient than calcite ones in bivalves: New evidence based on omics. Molecular Biology Reports 41: 7067–7071.

Wang, Y., Wang, M., Yin, S., Jang, R., Wang, J., Xue, Z. & Xu, T. (2015b). NeuroPep: A comprehensive resource of neuropeptides. Database: The Journal of Biological Databases and Curation 2015.

Wanninger, A., Ruthensteiner, B., Dictus, W. J. A. G. & Haszprunar, G. (1999a). The development of the musculature in the limpet Patella with implications on its role in the process of ontogenetic torsion. Invertebrate Reproduction & Development 36: 211–215.

Wanninger, A., Ruthensteiner, B., Lobenwein, S., Salvenmoser, W., Dictus, W. J. A. G. & Haszprunar, G. (1999b). Development of the musculature in the limpet *Patella* (Mollusca, Patellogastropoda). *Development Genes and Evolution* 209: 226–238.

Wanninger, A., Ruthensteiner, B. & Haszprunar, G. (2000). Torsion in *Patella caerulea* (Mollusca, Patellogastropoda): Ontogenetic process, timing, and mechanisms. *Invertebrate Biology* 119: 177–187.

Wanninger, A. & Haszprunar, G. (2001). The expression of an engrailed protein during embryonic shell formation of the tuskshell, *Antalis entalis* (Mollusca, Scaphopoda). *Evolution & Development* 3: 312–321.

Wanninger, A. & Haszprunar, G. (2002a). Muscle development in *Antalis entalis* (Mollusca: Scaphopoda) and its significance for scaphopod relationships. *Journal of Morphology* 254: 53–64.

Wanninger, A. & Haszprunar, G. (2002b). Chiton myogenesis: Perspectives for the development and evolution of larval and adult muscle systems in molluscs. *Journal of Morphology* 251: 103–113.

Wanninger, A. & Haszprunar, G. (2003). The development of the serotonergic and FMRF-amidergic nervous system in *Antalis entalis* (Mollusca, Scaphopoda). *Zoomorphology* 122: 77–85.

Wanninger, A., Fuchs, J. & Haszprunar, G. (2007). The anatomy of the serotonergic nervous system of an entoproct creeping-type larva and its phylogenetic implications. *Invertebrate Biology* 126: 268–278.

Wanninger, A. & Wollesen, T. (2015). Mollusca. pp. 103–153 *in* A. Wanninger (Ed.) *Evolutionary Developmental Biology of Invertebrates 2: Lophotrochozoa (Spiralia)*. Vol. 2. Vienna, Austria, Springer.

Wanninger, A. & Wollesen, T. (2019). The evolution of molluscs. *Biological Reviews* 94: 102–115.

Ward, D. V. (1972). Locomotor function of the squid mantle. *Journal of Zoology* 167: 487–499.

Ward, D. V. & Wainwright, S. A. (1972). Locomotory aspects of squid mantle structure. *Journal of Zoology* 167: 437–449.

Ward, J. E. (1965). The digestive tract and its relation to feeding habits in the stenoglossan prosobranch *Coralliophila abbreviata* (Lamarck). *Canadian Journal of Zoology* 43: 447–464.

Ward, J. E. (1966). Feeding, digestion and histology of the digestive tract in the keyhole limpet, *Fissurella barbadensis* Gmelin. *Bulletin of Marine Science* 16: 668–684.

Ward, J. E. & Targett, N. M. (1989). Influence of marine microalgal metabolites on the feeding behavior of the blue mussel *Mytilus edulis*. *Marine Biology* 101: 313–321.

Ward, J. E., Beninger, P. G., Macdonald, B. A. & Thompson, R. J. (1993). Suspension-feeding mechanisms in bivalves: Resolution of current controversies using endoscopy. *Journal of Shellfish Research* 12: 157.

Ward, J. E., Newell, R. I. E., Thompson, R. J. & MacDonald, B. A. (1994). In vivo studies of suspension-feeding processes in the eastern oyster, *Crassostrea virginica* (Gmelin). *Biological Bulletin* 186: 221–240.

Ward, J. E. (1996). Biodynamics of suspension-feeding in adult bivalve molluscs: Particle capture, processing, and fate. *Invertebrate Biology* 115: 218–231.

Ward, J. E. & Shumway, S. E. (2004). Separating the grain from the chaff: Particle selection in suspension- and deposit-feeding bivalves. *Journal of Experimental Marine Biology and Ecology* 300: 83–130.

Ward, M. E., Shields, J. D. & Van Dover, C. L. (2004). Parasitism in species of *Bathymodiolus* (Bivalvia: Mytilidae) mussels from deep-sea seep and hydrothermal vents. *Diseases of Aquatic Organisms* 62: 1–16.

Ward, P. (1981). Shell sculpture as a defensive adaptation in ammonoids. *Paleobiology* 7: 96–100.

Ward, P., Carlson, B. A., Weekly, M. & Brumbaugh, B. (1984). Remote telemetry of daily vertical and horizontal movement of *Nautilus* in Palau. *Nature* 309: 248–250.

Ward, R. D. & Warwick, T. (1980). Genetic differentiation in the molluscan species *Littorina rudis* and *Littorina arcana* (Prosobranchia: Littorinidae). *Biological Journal of the Linnean Society* 14: 417–428.

Ward, R. D. (1990). Biochemical genetic variation in the genus *Littorina* (Prosobranchia: Mollusca). *Hydrobiologia* 193: 53–69.

Warén, A. (1980a). Sexual strategies in Eulimidae (Prosobranchia). *Journal of the Malacological Society of Australia* 4: 231.

Warén, A. (1980b). Revision of the genera *Thyca, Stilifer, Scalenostoma, Mucronalia* and *Echineulima* (Mollusca, Prosobranchia, Eulimidae). *Zoologica Scripta* 9: 187–210.

Warén, A. (1983). A generic revision of the family Eulimidae (Gastropoda, Prosobranchia). *Journal of Molluscan Studies* 49: 1–95.

Warén, A., Burch, B. L. & Burch, T. A. (1984). Description of five new species of Hawaiian Eulimidae. *The Veliger* 26: 170–178.

Warén, A. (1990). Ontogenetic changes in the trochoidean (Archaeogastropoda) radula, with some phylogenetic interpretations. *Zoologica Scripta* 19: 179–187.

Warén, A. & Bouchet, P. (1993). New records, species, genera, and a new family of gastropods from the hydrothermal vents and hydrocarbon seeps. *Zoologica Scripta* 22: 1–90.

Warén, A. & Bouchet, P. (1990). Laubierinidae and Pisanianurinae (Ranellidae), two new deep-sea taxa of the Tonnoidea (Gastropoda: Prosobranchia). *The Veliger* 33: 56–102.

Warén, A., Gofas, S. & Schander, C. (1993). Systematic position of three European heterobranch gastropods. *The Veliger* 36: 1–15.

Warén, A. (1994). Systematic position and validity of *Ebala* Gray, 1847 (Ebalidae fam. n., Pyramidelloidea, Heterobranchia). *Bollettino Malacologico* 30: 203–210.

Warén, A., Norris, D. R. & Templado, J. (1994). Descriptions of four new eulimid gastropods parasitic on irregular sea urchins. *The Veliger* 37: 141–154.

Warén, A. & Gofas, S. (1996a). *Kaiparapelta askewi* McLean and Harasewych, 1995 (Gastropoda: Pseudococculinidae): A spongivorous cocculiniform limpet and a case of remarkable convergence in radular morphology. *Haliotis* 25: 107–116.

Warén, A. & Gofas, S. (1996b). A new species of Monoplacophora, redescription of the genera *Veleropilina* and *Rokopella*, and new information on three species of the class. *Zoologica Scripta* 25: 215–232.

Warén, A. & Bouchet, P. (2001). Gastropoda and Monoplacophora from hydrothermal vents and seeps: New taxa and records. *The Veliger* 44: 116–231.

Warén, A., Bengtson, S., Goffredi, S. K. & Van Dover, C. L. (2003). A hot-vent gastropod with iron sulfide dermal sclerites. *Science* 302: 1007.

Warén, A. (2011). Molluscs on biogenic substrates. pp. 438–448 *in* P. Bouchet, Le Guyader, H. & Pascal, O. (Eds.) *The Natural History of Santo*. Vol. 70. Paris, Muséum national d'Histoire naturelle.

Warham, J. (1990). *The Petrels: Their Ecology and Breeding Systems*. London, UK, Academic Press.

Warner, G. F. & Jones, A. R. (1976). Leverage and muscle type in crab chelae (Crustacea: Brachyura). *Journal of Zoology* 180: 57–68.

Warner, R. R. (1975). The adaptive significance of sequential hermaphroditism in animals. *American Naturalist* 109: 61–82.

Warrant, E. J. & Locket, N. A. (2004). Vision in the deep sea. *Biological Reviews* 79: 671–712.

Watabe, N. (1983). Shell repair. pp. 289–316 in A. S. M. Saleuddin & Wilbur, K. M. (Eds.) *Physiology, Part 1. The Mollusca*. Vol. 4. London, UK, Academic Press.

Watabe, N. (1988). Shell structure. pp. 69–104 in E. R. Trueman & Clarke, M. R. (Eds.) *Form and Function. The Mollusca*. Vol. 11. New York, Academic Press.

Watanabe, H. & Tokuda, G. (2001). Animal cellulases. *Cellular and Molecular Life Sciences* 58: 1167–1178.

Watanabe, J. M. & Cox, L. R. (1975). Spawning behavior and larval development in *Mopalia lignosa* and *Mopalia mucosa* (Mollusca, Polyplacophora) in central California. *The Veliger* 18: 18–27.

Watanabe, T., Suzuki, A., Kawahata, H., Kan, H. & Ogawa, S. (2004). A 60-year isotopic record from a mid-Holocene fossil giant clam (*Tridacna gigas*) in the Ryukyu Islands; physiological and paleoclimatic implications. *Palaeogeography, Palaeoclimatology, Palaeoecology* 212: 343–354.

Waterbury, J. B., Calloway, C. B. & Turner, R. D. (1983). A cellulytic nitrogen-fixing bacterium cultured from the gland of Deshayes in shipworms (Bivalvia: Teredinidae). *Science* 221: 1401–1403.

Watkins, B. & Simkiss, K. (1990). Interactions between soil bacteria and the molluscan alimentary tract. *Journal of Molluscan Studies* 56: 267–274.

Watkins, S. M., Reich, A., Fleming, L. E. & Hammond, R. (2008). Neurotoxic shellfish poisoning. *Marine Drugs* 6: 431–455.

Watling, L., Guinotte, J., Clark, M. R. & Smith, C. R. (2013). A proposed biogeography of the deep ocean floor. *Progress in Oceanography* 111: 91–112.

Watson, M. E. & Signor, P. W. (1986). How a clam builds windows: Shell microstructure in *Corculum* (Bivalvia: Cardiidae). *The Veliger* 28: 348–355.

Watson, W. H. & Trimarchi, J. R. (1992). A quantitative description of *Melibe* feeding behavior and its modification by prey density. *Marine Behaviour and Physiology* 19: 183–194.

Wayne, N. L. & Block, G. D. (1992). Effects of photoperiod and temperature on egg-laying behavior in a marine mollusk, *Aplysia californica*. *Biological Bulletin* 182: 8.

Wayne, N. L. (2001). Regulation of seasonal reproduction in mollusks. *Journal of Biological Rhythms* 16: 391–402.

Wear, R. G. (1966). Physiological and ecological studies on the bivalve mollusk *Arthritica bifurca* (Webster, 1908) living commensally with the tubicolous polychaete *Pectinaria australis* Ehlers, 1905. *Biological Bulletin* 130: 141–149.

Weaver, J. C., Wang, Q., Miserez, A., Tantuccio, A., Stromberg, R., Bozhilov, K. N., Maxwell, P., Nay, R., Heier, S. T. & DiMasi, E. (2010). Analysis of an ultra hard magnetic biomineral in chiton radular teeth. *Materials Today* 13: 42–52.

Weaver, J. C., Milliron, G. W., Miserez, A., Evans-Lutterodt, K., Herrera, S., Gallana, I., Mershon, W. J., Swanson, B., Zavattieri, P. & DiMasi, E. (2012). The stomatopod dactyl club: A formidable damage-tolerant biological hammer. *Science* 336: 1275–1280.

Webber, H. H. (1977). Gastropoda: Prosobranchia. pp. 1–97 in A. C. Giese & Pearse, J. S. (Eds.) *Reproduction of Marine Invertebrates. Molluscs: Gastropods and Cephalopods*. Vol. 4. New York, Academic Press.

Weber, H. (1927). Der Darm von *Dolium galea* L., eine vergleichend anatomische Untersuchung unter besonderer Berücksichtigung der *Tritonium*-arten. *Zeitschrift für Morphologie und Ökologie der Tiere* 8: 663–804.

Webster, J. P., Hoffman, J. I. & Berdoy, M. (2003). Parasite infection, host resistance and mate choice: Battle of the genders in a simultaneous hermaphrodite. *Proceedings of the Royal Society B* 270: 1481–1485.

Webster, N. B. & Vermeij, G. J. (2017). The varix: Evolution, distribution, and phylogenetic clumping of a repeated gastropod innovation. *Zoological Journal of the Linnean Society* 180: 732–754.

Wedemeyr, H. & Schild, D. (1995). Chemosensitivity of the osphradium of the pond snail *Lymnaea stagnalis*. *Journal of Experimental Biology* 198: 1743–1754.

Van Weel, P. B. (1950). Contribution to the physiology of the glandula media intestini of the African Giant Snail, *Achatina fulica* Fér., during the first hours of digestion. *Fisiologia Comparata at Oecologica* 2: 1–19.

Van Weel, P. B. (1957). Observations on the osmoregulation in *Aplysia juliana* Pease (Aplysiidae, Mollusca). *Zeitschrift für Vergleichende Physiologie* 39: 492–506.

Wefer, G. & Killingley, J. S. (1980). Growth histories of strombid snails from Bermuda recorded in their O-18 and C-13 profiles. *Marine Biology* 60: 129–135.

Wehmiller, J. F. & Belknap, D. F. (1978). Alternative kinetic models for interpretation of amino acid enantiomeric ratios in Pleistocene Mollusks: Examples from California, Washington and Florida. *Quaternary Research* 9: 330–348.

Wehmiller, J. F. (1990). Amino acid racemization: Applications in chemical taxonomy and chronostratigraphy of Quaternary fossils. pp. 583–608 in J. G. Carter (Ed.) *Skeletal Biomineralization: Patterns, Processes and Evolutionary Trends*. Vol. 1. New York, Van Nostrand Reinhold.

Wehmiller, J. F. (2013). United States Quaternary coastal sequences and molluscan racemization geochronology: What have they meant for each other over the past 45 years? *Quaternary Geochronology* 16: 3–20.

Wei, L., Wang, Q., Wu, H., Ji, C. & Zhao, J. (2015). Proteomic and metabolomic responses of Pacific oyster *Crassostrea gigas* to elevated pCO_2 exposure. *Journal of Proteomics* 112: 83–94.

Wei, X., Yang, J., Yang, D., Xu, J., Liu, X., Yang, J., Fang, J. & Qiao, H. (2012). Molecular cloning and mRNA expression of two peptidoglycan recognition protein (PGRP) genes from mollusk *Solen grandis*. *Fish and Shellfish Immunology* 32: 178–185.

Weiner, R. M., Walch, M., Labare, M. P., Bonar, D. B. & Colwell, R. R. (1989). Effect of biofilms of the marine bacterium *Alteromonas colwelliana* on set of the oysters *Crassostrea gigas* and *Crassostrea virginica*. *Journal of Shellfish Research* 8: 117–123.

Weiner, S. & Addadi, L. (2011). Crystallization pathways in biomineralization. *Annual Review of Materials Research* 41: 21–40.

Weinstein, J. E. (1995). Fine structure of the digestive tubule of the eastern oyster, *Crassostrea virginica* (Gmelin, 1791). *Journal of Shellfish Research* 14: 97–103.

Weiser, W. & Schuster, M. (1975). The relationship between water content, activity, and free amino acids in *Helix pomatia* L. *Journal of Comparative Physiology B* 98: 169–181.

Weiss, K. R., Bayley, H., Lloyd, P. E., Tenenbaum, R., Gawinowicz Kolks, M. A., Buck, L., Cropper, E. C., Rosen, S. C. & Kupfermann, I. (1989). Purification and sequencing of neuropeptides contained in neuron R15 of *Aplysia californica*. *Proceedings of the National Academy of Sciences of the United States of America* 86: 2913–2917.

Welladsen, H. M., Southgate, P. C. & Heimann, K. (2010). The effects of exposure to near-future levels of ocean acidification on shell characteristics of *Pinctada fucata* (Bivalvia: Pteriidae). *Molluscan Research* 30: 125–130.

Wells, H. W. (1958). Predation of pelecypods and gastropods by *Fasciolaria hunteria* (Perry). *Bulletin of Marine Science* 8: 152–166.

Wells, M. J. & Wells, J. (1957). The function of the brain of *Octopus* in tactile discrimination. *Journal of Experimental Biology* 34: 131–142.

Wells, M. J. (1959). A touch-learning centre in *Octopus*. *Journal of Experimental Biology* 36: 590–612.

Wells, M. J. (1960). Optic glands and the ovary of *Octopus*. *Symposia of the Zoological Society of London* 2: 87–107.

Wells, M. J. (1961). What the octopus makes of it: Our world from another point of view. *American Scientist* 49: 215–227.

Wells, M. J. (1962). *Brain and Behaviour in Cephalopods*. London, UK, Heinemann Educational Books.

Wells, M. J. (1963). Taste by touch: Some experiments with *Octopus*. *Journal of Experimental Biology* 40: 187–193.

Wells, M. J. (1966). Cephalopod sense organs. pp. 523–545 *in* K. M. Wilbur & Yonge, C. M. (Eds.) *Physiology of Mollusca*. Vol. 2. New York, Academic Press.

Wells, M. J. & Young, J. Z. (1975). The subfrontal lobe and touch learning in the octopus. *Brain Research* 92: 103–121.

Wells, M. J. & Wells, J. (1977a). Cephalopoda: Octopoda. pp. 291–336 *in* A. C. Giese & Pearse, J. S. (Eds.) *Reproduction of Marine Invertebrates. Molluscs: Gastropods and Cephalopods. Reproduction of Marine Invertebrates*. Vol. 4. New York, Academic Press.

Wells, M. J. & Wells, J. (1977b). Optic glands and the endocrinology of reproduction. pp. 525–540 *in* M. Nixon & Messenger, J. B. (Eds.) *The Biology of Cephalopods. [Papers from] Symposium held –April 10–11 1975. Symposia of The Zoological Society of London*. London, UK, Academic Press.

Wells, M. J. (1978). *Octopus: Physiology and Behaviour of an Advanced Invertebrate*. London, UK, Chapman & Hall.

Wells, M. J. (1979). Heartbeat of *Octopus vulgaris*. *Journal of Experimental Biology* 78: 87–104.

Wells, M. J. & Wells, J. (1982). Ventilatory currents in the mantle of cephalopods. *Journal of Experimental Biology* 99: 315–330.

Wells, M. J. (1983). Circulation in cephalopods. pp. 239–290 *in* A. S. M. Saleuddin & Wilbur, K. M. (Eds.) *Physiology, Part 2. The Mollusca*. Vol. 5. New York, Academic Press.

Wells, M. J., O'Dor, R. K., Mangold, K. M. & Wells, J. (1983a). Oxygen consumption in movement by *Octopus*. *Marine Behaviour and Physiology* 9: 289–303.

Wells, M. J., Duthie, G. G., Houlihan, D. F. & Smith, P. J. S. (1987). Blood flow and pressure changes in exercising octopuses (*Octopus vulgaris*). *Journal of Experimental Biology* 131: 175–188.

Wells, M. J. & Smith, P. J. S. (1987). The performance of the octopus circulatory system: A triumph of engineering over design. *Experientia* 43: 487–499.

Wells, M. J. & Wells, J. (1989). Water uptake in a cephalopod and the function of the so-called 'pancreas'. *Journal of Experimental Biology* 145: 215–226.

Wells, M. J. (1992a). The cephalopod heart: The evolution of a high performance invertebrate heart. *Experientia* 48: 800–808.

Wells, M. J. (1992b). The evolution of a circulatory system. pp. 5–14 *in* R. B. Hill, Kuwasawa, K., McMahon, B. R. & Kuramoto, R. T. (Eds.) *Phylogenetic Models in Functional Coupling of the CNS and the Cardiovascular System. Comparative Physiology*. Vol. 11. Basel, Switzerland, S. Karger.

Wells, M. J., Wells, J. & O'Dor, R. K. (1992). Life at low oxygen tensions: The behaviour and physiology of *Nautilus pompilius* and the biology of extinct forms. *Journal of the Marine Biological Association of the United Kingdom* 72: 313–328.

Wells, S. M., Pyle, R. M. & Collins, N. M. (1983b). *IUCN Invertebrate Red Data Book*. Gland, Switzerland, IUCN.

Wells, S. M. (1989). Impacts of the precious shell harvest and trade: Conservation of rare or fragile resources. pp. 443–454 *in* J. F. Caddy (Ed.) *Marine Invertebrate Fisheries: Their Assessment and Management*. New York, John Wiley & Sons.

Welsh, K., Elliot, M., Tudhope, A., Ayling, B. & Chappell, J. (2011). Giant bivalves (*Tridacna gigas*) as recorders of ENSO variability. *Earth and Planetary Science Letters* 307: 266–270.

Welter-Schultes, F. W. (2012). *European Non-marine Molluscs, a Guide for Species Identification*. Göttingen, Planet Poster Editions.

Weltje, L. & Schulte-Oehlmann, U. (2007). The seven year itch: Progress in research on endocrine disruption in aquatic invertebrates since 1999. *Ecotoxicology* 16: 1–3.

Wendelaar-Bonga, S. E. (1970). Ultrastructure and histochemistry of neurosecretory cells and heurohaemal areas in the pond snail *Lymnaea stagnalis* (L.). *Zeitschrift für Zellforschung und Mikroskopische Anatomie* 108: 190–224.

Wendelaar Bonga, S. W. (1971). Formation, storage, and release of neurosecretory material studied by quantitative electron microscopy in the fresh water snail *Lymnaea stagnalis* (L.). *Zeitschrift für Zellforschung und Mikroskopische Anatomie* 113: 490–517.

Wenz, W. (1938–1944). Gastropoda: Teil I: Allgemeiner Teil und Prosobranchia. pp. i–viii, 1–1639 *in* O. H. Schindewolf (Ed.) *Handbuch der Paläontologie*. Vol. 6, Teil 1. Berlin, Germany, Gebrüder Borntraeger.

Werneke, S. W., Swann, C., Farquharson, L. A., Hamilton, K. S. & Smith, A. M. (2007). The role of metals in molluscan adhesive gels. *Journal of Experimental Biology* 210: 2137–2145.

Werner, B. (1951). Über die Bedeutung der Wasserstromerzeugung und Wasserstromfiltration für die Nahrungsaufnahme der ortsgebundenen Meeresschnecke *Crepidula fornicata* L. (Gastropoda, Prosobranchia). *Zoologischer Anzeiger* 146: 97–113.

Werner, B. (1953). Über der Nahrungserwerb der Calyptraeidae (Gastropoda, Prosobranchia): Morphologie, Histologie und Funktion der am Nahrungserwerb beteiligten Organe. *Helgoländer Wissenschaftliche Meeresuntersuchungen* 4: 260–315.

Werner, G. D., Gemmell, P., Grosser, S., Hamer, R. & Shimeld, S. M. (2013). Analysis of a deep transcriptome from the mantle tissue of *Patella vulgata* Linnaeus (Mollusca: Gastropoda: Patellidae) reveals candidate biomineralising genes. *Marine Biotechnology* 15: 230–243.

Wertz, A., Roessler, W., Obermayer, M. & Bickmeyer, U. (2007). Functional neuroanatomy of the rhinophore of *Archidoris pseudoargus*. *Helgoland Marine Research* 61: 135–142.

Wesołowska, W. & Wesołowski, T. (2014). Do *Leucochloridium* sporocysts manipulate the behaviour of their snail hosts? *Journal of Zoology* 292: 151–155.

West-Eberhard, M. J. (2003). *Developmental Plasticity and Evolution*. New York, Oxford University Press.

West, D. J., Andrews, E. B., McVean, A. R., Osborne, D. J. & Thorndyke, M. C. (1994). Isolation of serotonin from the accessory salivary glands of the marine snail *Nucella lapillus*. *Toxicon* 32: 1261–1264.

West, D. J., Andrews, E. B., McVean, A. R., Thorndyke, M. C. & Taylor, J. D. (1998). Presence of a toxin in the salivary glands of the marine snail *Cymatium intermedius* that targets nicotinic acetylcholine receptors. *Toxicon* 36: 25–29.

West, T. L. (1990). Feeding behavior and functional morphology of the epiproboscis of *Mitra idae* (Mollusca, Gastropoda, Mitridae). *Bulletin of Marine Science* 46: 761–779.

West, T. L. (1991). Functional morphology of the proboscis of *Mitra catalinae* (Dall 1920) (Mollusca: Gastropoda: Mitridae) and the evolution of the mitrid epiproboscis. *Bulletin of Marine Science* 48: 702–718.

Westermann, B., Ruth, P., Litzlbauer, H. D., Beck, I., Beuerlein, K., Schmidtberg, H., Kaleta, E. F. & Schipp, R. (2002). The digestive tract of *Nautilus pompilius* (Cephalopoda, Tetrabranchiata): An X-ray analytical and computational tomography study on the living animal. *Journal of Experimental Biology* 205: 1617–1624.

Westermann, B., Schmidtberg, H. & Beuerlein, K. (2005). Functional morphology of the mantle of *Nautilus pompilius* (Mollusca, Cephalopoda). *Journal of Morphology* 264: 277–285.

Westermann, G. E. G. (1969). *Sexual Dimorphism in Fossil Metazoa and Taxonomic Implications*. Stuttgart, Germany, E. Schweizerbart'sche Verlagsbuchhandlung.

Westfall, K. M. & Gardner, J. P. A. (2010). Genetic diversity of southern hemisphere blue mussels (Bivalvia: Mytilidae) and the identification of non-indigenous taxa. *Biological Journal of the Linnean Society* 101: 898–909.

Westheide, W. & Rieger, R., Eds. (1996). *Spezielle Zoologie – Erster Teil: Einzeller und Wirbellose*. Stuttgart, Germany, Gustav Fischer.

Westley, C. B., Vine, K. L. & Benkendorff, K. (2006). A proposed functional role for indole derivatives in reproduction and defense of the Muricidae (Neogastropoda: Mollusca). pp. 31–44 *in* L. Meijer, Guyard, N., Skaltsounis, L. & Eisenbrand, G. (Eds.) *Indirubin, the Red Shade of Indigo*. Roscoff, France, Life in Progress Editions.

Westley, C. B. & Benkendorff, K. (2008). Sex-specific Tyrian purple genesis: Precursor and pigment distribution in the reproductive system of the marine mollusc, *Dicathais orbita*. *Journal of Chemical Ecology* 34: 44–56.

Wetzel, R. G. (2001). *Limnology. Lake and River Ecosystems*. 3rd ed. San Diego, CA, Academic Press.

Weymouth, F. W. (1921). The edible clams, mussels and scallops of California. *State of California Fish and Game Commission Fish Bulletin* 4: 1–114.

Whalen, K. E., Starczak, V. R., Nelson, D. R., Goldstone, J. V. & Hahn, M. E. (2010). Cytochrome P450 diversity and induction by gorgonian allelochemicals in the marine gastropod *Cyphoma gibbosum*. *BMC Ecology* 10: 24.

Wheeler, A. P. (1992). Mechanisms of molluscan shell formation. pp. 179–216 *in* E. Bonucci (Ed.) *Calcification in Biological Systems*. Boca Raton, FL, CRC Press.

Whelan, H. A. & McCrohan, C. R. (1996). Food-related conditioning and neuronal correlates in the freshwater snail *Lymnaea stagnalis*. *Journal of Molluscan Studies* 62: 483–494.

Whitaker, J. R. (1994). *Principles of Enzymology for the Food Sciences*. New York, Marcel Dekker.

Whitaker, M. B. (1951). On the homologies of the oesophageal glands of *Theodoxus fluviatilis* (L.). *Journal of Molluscan Studies* 29: 21–34.

White, K. M. (1942). The pericardial cavity and the pericardial gland of the Lamellibranchia. *Proceedings of the Malacological Society of London* 25: 37–88.

White, T. R., Pagels, A. K. W. & Fautin, D. G. (1999). Abyssal sea anemones (Cnidaria: Actiniaria) of the northeast Pacific symbiotic with molluscs: *Anthosactis nomados*, a new species, and *Monactis vestita* (Gravier, 1918). *Proceedings of the Biological Society of Washington* 112: 637–651.

Whiteley, D. A. A., Owen, D. F. & Smith, D. A. S. (1997). Massive polymorphism and natural selection in *Donacilla cornea* (Poli, 1791) (Bivalvia: Mesodesmatidae). *Biological Journal of the Linnean Society* 62: 475–494.

Whitmore, N. (2015). Harnessing local ecological knowledge for conservation decision making via Wisdom of Crowds: The case of the Manus green tree snail *Papustyla pulcherrima*. *Oryx* 50: 684–689.

Whitney-Desautels, N. A. & Beer, R. A. (1999). *Franchthi Cave Riverine and Terrestrial Molluscs*. Bloomington, IN, Indiana University Press.

Widdows, J. & Hawkins, A. J. S. (1989). Partitioning of rate of heat dissipation by *Mytilus edulis* into maintenance, feeding, and growth components. *Physiological Zoology* 62: 764–784.

Wiederhold, M. L., MacNichol, E. F. & Bell, A. L. (1973). Photoreceptor spike responses in the hardshell clam, *Mercenaria mercenaria*. *Journal of General Physiology* 61: 24–55.

Wiederhold, M. L., Sharma, J. S., Driscoll, B. P. & Harrison, J. L. (1990). Development of the statocyst in *Aplysia californica*. I. Observations on statoconial development. *Hearing Research* 49: 63–78.

Wiens, B. L. & Brownell, P. H. (1990). Characterization of cardiac innervation in the nudibranch *Archidoris montereyensis*. *Journal of Comparative Physiology* 167: 51–60.

Wieser, W. (1981). Responses of *Helix pomatia* to anoxia: Changes of solute activity and other properties of the haemolymph. *Journal of Comparative Physiology* 141: 503–509.

Wijdenes, J., Schluter, N. C. M., Gomot, L. & Boer, H. H. (1987). In the snail *Helix aspersa* the gonadotropic hormone-producing dorsal bodies are under inhibitory nervous control of putative growth hormone-producing neuroendocrine cells. *General and Comparative Endocrinology* 68: 224–229.

Wilbur, K. M. & Yonge, C. M., Eds. (1964). *Physiology of Mollusca*. New York, Academic Press.

Wilbur, K. M. & Yonge, C. M., Eds. (1966). *Physiology of Mollusca*. New York, London, UK, Academic Press.

Wilbur, K. M., Ed. (1983–1988). *The Mollusca*, Vol. 1–12. New York, Academic Press.

Wild, S. & Lawson, A. (1937). Enemies of the land and freshwater Mollusca of the British Isles. *Journal of Conchology* 20: 351–361.

Wilke, T., Väinölä, R. & Riedel, F., Eds. (2006). *Patterns and Processes of Speciation in Ancient Lakes. Proceedings of the Fourth Symposium on Speciation in Ancient Lakes, Berlin, Germany, September 4–8, 2006*. Developments in Hydrobiology. Dordrecht, the Netherlands, Springer Science + Business Media.

Wilke, T., Albrecht, C., Anistratenko, V. V., Sahin, S. K. & Yildirim, M. Z. (2007). Testing biogeographical hypotheses in space and time: Faunal relationships of the putative ancient Lake Egirdir in Asia Minor. *Journal of Biogeography* 34: 1807–1821.

Wilke, T., Väinölä, R. & Riedel, F. (2008). Speciation in ancient lakes. *Hydrobiologia* 615: 1–3.

Wilkens, L. A. (1986). The visual system of the giant clam *Tridacna*: Behavioral adaptations. *Biological Bulletin* 170: 393–408.

Wilkens, L. A. (2008). Primary inhibition by light: A unique property of bivalve photoreceptors. *American Malacological Bulletin* 26: 101–109.

Willcox, M. A. (1898). Zur Anatomie von *Acmaea fragilis* Chemnitz. *Jenaische Zeitschrift für Naturwissenschaft* 32: 411–456.

Willem, V. (1892). Contributions à l'étude physiologique des organes des sens chez les mollusques. III. Observations sur la vision et les organes visuels de quelques mollusques prosobranches et opisthobranches. *Archives de Biologie* 12: 123–149.

Willey, A. (1902). Contribution to the natural history of the pearly Nautilus. pp. 691–830. *Zoological Results based on Material from New Britain, New Guinea, Loyalty Islands and Elsewhere, Collected During the Years 1895, 1896 and 1897 by Arthur Willey*. Vol. 6. Cambridge, UK, Cambridge University Press.

Williams, D. E., Ayer, S. W. & Andersen, R. J. (1986). Diaulusterols A and B from the skin extracts of the dorid nudibranch *Diaulula sandiegensis. Canadian Journal of Chemistry* 64: 1527–1529.

Williams, G. A. & Morritt, D. (1995). Habitat partitioning and thermal tolerance in a tropical limpet, *Cellana grata. Marine Ecology Progress Series* 124: 89–103.

Williams, G. A., Pirro, M., de Leung, K. M. Y. & Morritt, D. (2005). Physiological responses to heat stress on a tropical shore: The benefits of mushrooming behaviour in the limpet *Cellana grata. Marine Ecology Progress Series* 292: 213–224.

Williams, J. D., Warren, M. L., Cummings, K. S., Harris, J. L. & Neves, R. J. (1993). Conservation status of freshwater mussels of the United States and Canada. *Fisheries* 18: 6–22.

Williams, J. W. (1889). Variation in the Mollusca and its probable cause. *Science Gossip*: 174–176, 200–203, 245–248.

Williams, S. I. & Walker, D. I. (1999). Mesoherbivore-macroalgal interactions: Feeding ecology of sacoglossan sea slugs (Mollusca, Opisthobranchia) and their effects on their food algae. *Oceanography and Marine Biology Annual Review* 37: 87–128.

Williams, S. L., Crafton, R. E., Fontana, R. E., Grosholz, E. D., Pasari, J. & Zabin, C. (2012). *Aquatic Invasive Species Vector Risk Assessments: A Vector Analysis of the Aquarium and Aquascape (for California Ocean Science Trust)*. Bodega Marine Laboratory, University of California at Davis.

Williams, S. T. & Reid, D. G. (2004). Speciation and diversity on tropical rocky shores: A global phylogeny of snails of the genus *Echinolittorina. Evolution* 58: 2227–2251.

Williams, S. T., Taylor, J. D. & Glover, E. A. (2004). Molecular phylogeny of the Lucinoidea (Bivalvia): Non-monophyly and separate acquisition of bacterial chemosymbiosis. *Journal of Molluscan Studies* 70: 187–202.

Williams, S. T. (2017). Molluscan shell colour. *Biological Reviews* 92: 1039–1058.

Williams, T. (1856). On the mechanisms of aquatic respiration and on the structure of the organs of breathing in invertebrate animals. *Annals and Magazine of Natural History* 17: 28–41.

Williams, W. D. (1986). Limnology, the study of inland waters: A comment on perceptions of salt lake studies, past and present. pp. 471–484 *in* P. De Deckker & Williams, W. D. (Eds.) *Limnology in Australia*. Melbourne and Dordrecht, CSIRO and Dr. W. Junk Publishers.

Williams, W. D. & Mellor, M. W. (1991). Ecology of *Coxiella* (Mollusca, Gastropoda, Prosobranchia), a snail endemic to Australian salt lakes. *Palaeogeography, Palaeoclimatology, Palaeoecology* 84: 339–355.

Williamson, P. G. (1981). Palaeontological documentation of speciation in Cenozoic molluscs from Turkana Basin. *Nature* 293: 437–443.

Williamson, R., Ichikawa, M. & Matsumoto, G. (1994). Neuronal circuits in cephalopod vision. *Netherlands Journal of Zoology* 44: 272–283.

Williamson, R. & Chrachri, A. (2004). Cephalopod neural networks. *NeuroSignals* 13: 87–98.

Willis, K. J. & Bhagwat, S. A. (2009). Biodiversity and climate change. *Science* 326: 806–807.

Willis, T. J., Berglöf, K. T. L., McGill, R. A. R., Musco, L., Piraino, S., Rumsey, C. M., Fernández, T. V. & Badalamenti, F. (2017). Kleptopredation: A mechanism to facilitate planktivory in a benthic mollusc. *Biology Letters* 13: 0447.

Willman, S. (2007). Testing the role of spines as predatory defense. *Journal of Shellfish Research* 26: 261–266.

Willows, A. O. D. (1980). Physiological basis of feeding behavior in *Tritonia diomedea*. II. Neuronal mechanisms. *Journal of Neurophysiology* 44: 849–861.

Willows, A. O. D. (1999). Shoreward orientation involving geomagnetic cues in the nudibranch mollusc *Tritonia diomedea. Marine and Freshwater Behaviour and Physiology* 32: 181–192.

Wilmot, N. V., Barber, D. J., Taylor, J. D. & Graham, A. L. (1992). Electron microscopy of molluscan crossed-lamellar microstructure. *Philosophical Transactions of the Royal Society of London B* 337: 21–35.

Wilson, B. R. (1985). Direct development in southern Australian cowries (Gastropoda: Cypraeidae). *Australian Journal of Marine and Freshwater Research* 36: 267–280.

Wilson, E. B. (1892). The cell-lineage of *Nereis. Journal of Morphology* 6: 361–480.

Wilson, E. B. (1898). Considerations on cell-lineage and ancestral reminiscence based on a re-examination of some points in the early development of annelids and polyclades. *Annals of the New York Academy of Sciences* 11: 1–27.

Wilson, J. H. & La Touche, R. W. (1978). Intracellular digestion in two sublittoral populations of *Ostrea edulis* (Lamellibranchia). *Marine Biology* 47: 71–77.

Wilson, M. A. (1987). Ecological dynamics on pebbles, cobbles, and boulders. *Palaios* 2: 594–599.

Wilson, N. G., Rouse, G. W. & Giribet, G. (2010). Assessing the molluscan hypothesis Serialia (Monoplacophora + Polyplacophora) using novel molecular data. *Molecular Phylogenetics and Evolution* 54: 187–193.

Wilson, S. P. & Hyne, R. V. (1997). Toxicity of acid-sulfate soil leachate and aluminum to embryos of the Sydney rock oyster. *Ecotoxicology and Environmental Safety* 37: 30–36.

Wilt, F. H., Killian, C. E. & Livingston, B. T. (2003). Development of calcareous skeletal elements in invertebrates. *Differentiation* 71: 237–250.

Windoffer, R. & Giere, O. (1997). Symbiosis of the hydrothermal vent gastropod *Ifremeria nautilei* (Provannidae) with endobacteria—Structural analyses and ecological considerations. *Biological Bulletin* 193: 381–392.

Wingstrand, K. G. (1985). On the anatomy and relationships of recent Monoplacophora. *Galathea Report* 16: 1–94.

Winkler, F. M., Estévez, B. F., Jollán, L. B. & Garrido, J. P. (2001). Inheritance of the general shell color in the scallop *Argopecten purpuratus* (Bivalvia: Pectinidae). *Journal of Heredity* 92: 521–525.

Winkler, L. R. & Wagner, E. D. (1959). Filter paper digestion by the crystalline style in *Oncomelania. Transactions of the American Microscopical Society* 78: 262–268.

Winsor, L., Johns, P. M. & Barker, G. M. (2004). Terrestrial planarians (Platyhelminthes: Tricladida: Terricola) predaceous on terrestrial gastropods. pp. 227–278 *in* G. M. Barker (Ed.) *Natural Enemies of Terrestrial Molluscs*. Oxford, UK, Cambridge, MA, CABI Publishing.

de Winter, A. J. & Gittenberger, E. (1998). The land snail fauna of a square kilometer patch of rainforest in southwestern Cameroon, high species richness, low abundance and seasonal fluctuations. *Malacologia* 40: 231–250.

Winter, J. E. (1978). A review on the knowledge of suspension-feeding in lamellibranchiate bivalves, with special reference to artificial aquaculture systems. *Aquaculture* 13: 1–33.

Winterbourn, M. J. (1980). The distribution and biology of the freshwater gastropods *Physa* and *Physastra* in New Zealand. *Journal of the Malacological Society of Australia* 4: 233–234.

Winters, A. D., Marsh, T. L. & Faisal, M. (2011). Heterogeneity of bacterial communities within the zebra mussel (*Dreissena polymorpha*) in the Laurentian Great Lakes Basin. *Journal of Great Lakes Research* 37: 318–324.

Winterstein, H., Ed. (1910–1925). *Handbuch der vergleichenden Physiologie*. Jena, Germany, Gustav Fischer.

de With, N. D. & van den Berg, H. A. (1992). Neuroendocrine control of hydromineral metabolism in molluscs, with special emphasis on the pulmonate freshwater snail, *Lymnaea stagnalis*. *Advances in Comparative Endocrinology* 1: 83–99.

de With, N. D., Boer, H. H., Smit, A. B. & Schors, R. C. Van der (1994). Neurosecretory yellow cells and hydromineral regulation in the pulmonate freshwater snail *Lymnaea stagnalis*. Pp. 81–84 *in* K. G. Davey, Peter, R. E. & Tobe, S. S. (Eds.) *Perspectives in Comparative Endocrinology: invited papers from XII International Congress of Comparative Endocrinology, sponsored by The International Federation of Comparative Endocrinological Societies, Toronto, Ontario, Canada 16–21 May 1993*. Ottawa, National Research Council of Canada.

de With, N. D. (1996). Oral water ingestion in the pulmonate freshwater snail, *Lymnaea stagnalis*. *Journal of Comparative Physiology B* 166: 337–343.

Withers, P., Pedler, S. & Guppy, M. (1997). Physiological adjustments during aestivation by the Australian land snail *Rhagada tescorum* (Mollusca: Pulmonata: Camaenidae). *Australian Journal of Zoology* 45: 599–611.

Withers, P. C. (1992). *Comparative Animal Physiology*. Fort Worth, TX, Saunders College Publishing.

Witman, J. D. & Suchanek, T. H. (1984). Mussels in flow-drag and dislodgement by epizoans. *Marine Ecology Progress Series* 16: 259–268.

Witmer, V. A. (1974). Die Feinstruktur der Kiemenherzen des Cephalopodan *Octopus joubini*. *Zoologische Beiträge* 20: 459–487.

Witte, V., Janssen, R., Eppenstein, A. & Maschwitz, U. (2002). *Allopeas myrmekophilos* (Gastropoda, Pulmonata), the first myrmecophilous mollusc living in colonies of the ponerine army ant *Leptogenys distinguenda* (Formicidae, Ponerinae). *Insectes Sociaux* 49: 301–305.

Wittenberg, J. B. & Stein, J. L. (1995). Hemoglobin in the symbiont-harboring gill of the marine gastropod *Alviniconcha hessleri*. *Biological Bulletin* 188: 5–7.

Wojcik, D. P., Allen, C. R., Brenner, R. J., Forys, E. A., Jouvenaz, D. P. & Lutz, R. S. (2001). Red imported fire ants: Impact on biodiversity. *American Entomologist* 47: 16–23.

Wojtowicz, M. B. (1972). Carbohydrases of the digestive gland and the crystalline style of the Atlantic deep-sea scallop (*Placopecten magellanicus*, Gmelin). *Comparative Biochemistry and Physiology Part A* 43: 131–141.

Wolcott, T. G. (1973). Physiological ecology and intertidal zonation in limpets *Acmaea*: A critical look at 'limiting' factors. *Biological Bulletin* 145: 389–432.

Wolff, W. J. & Montserrat, F. (2005). *Cymbium* spp. (Gastropoda: Mollusca) as bivalve predators at the tidal flats of the Banc d'Arguin, Mauritania. *Journal of the Marine Biological Association of the United Kingdom* 85: 949–953.

Wolfson, F. H. (1968). Spawning notes. I. *Hexaplex erythrostomus*. *The Veliger* 10: 292.

Wollemann, M. & Rózsa, K. S. (1975). Effects of serotonin and catecholamines on the adenylate cyclase of the molluscan heart. *Comparative Biochemistry and Physiology Part C* 51: 63–66.

Wollesen, T., Wanninger, A. & Klussmann-Kolb, A. (2007). Neurogenesis of cephalic sensory organs of *Aplysia californica*. *Cell and Tissue Research* 330: 361–379.

Wollesen, T., Wanninger, A. & Klussmann-Kolb, A. (2008). Myogenesis in *Aplysia californica* (Cooper, 1863) (Mollusca, Gastropoda, Opisthobranchia) with special focus on muscular remodeling during metamorphosis. *Journal of Morphology* 269: 776–789.

Wollesen, T., Rodríguez Monje, S. V., McDougall, C., Degnan, B. M. & Wanninger, A. (2015). The ParaHox gene *Gsx* patterns the apical organ and central nervous system but not the foregut in scaphopod and cephalopod mollusks. *EvoDevo* 6: 41.

Wölper, C. (1950). Das Osphradium der *Paludina vivipara*. *Zeitschrift für vergleichende Physiologie* 32: 272–286.

Woltereck, R. (1909). Weitere experimentelle Untersuchungen über Artveränderung, speziell über das Wesen quantitativer Artunterschiede bei Daphniden. *Verhandlungen der Deutschen Zoologischen Gesellschaft* 19: 110–173.

Wolters, V. & Ekschmitt, K. (1997). Gastropods, isopods, diplopods, and chilopods: Neglected groups of the decomposer food web. pp. 265–306 *in* G. Benckiser (Ed.) *Fauna in Soil Ecosystems: Recycling Processes, Nutrient Fluxes, and Agricultural Production*. New York, Marcel Dekker.

Won, Y.-J., Hallam, S. J., O'Mullan, G. D., Pan, I. L., Buck, K. R. & Vrijenhoek, R. C. (2003). Environmental acquisition of thiotrophic endosymbionts by deep-sea mussels of the genus *Bathymodiolus*. *Applied and Environmental Microbiology* 69: 6785–6792.

Won, Y.-J., Jones, W. J. & Vrijenhoek, R. C. (2008). Absence of cospeciation between deep-sea mytilids and their thiotrophic endosymbionts. *Journal of Shellfish Research* 27: 129–138.

Wondrak, G. (1968). Elektronenoptische Untersuchungen der Körperdecke von *Arion rufus* L. (Pulmonata). *Protoplasma* 66: 151–171.

Wondrak, G. (1969). Die Ultrastruktur der Zellen aus dem Interstitiellen Bindegewebe von *Arion rufus* (L.), Pulmonata, Gastropoda. *Zeitschrift für Zellforschung und Mikroskopische Anatomie* 95: 249–262.

Wondrak, G. (1981). Ultrastructure of the supporting cells in the chemoreceptor areas of the tentacles of *Pomatias elegans* (Mollusca, Prosobranchia) and the ommatophore of *Helix pomatia* (Mollusca, Pulmonata). *Journal of Morphology* 167: 211–230.

Wondrak, G. (2012). Monotypic gland-cell regions on the body surface of two species of *Arion*: Ultrastructure and lectin-binding properties. *Journal of Molluscan Studies* 78: 364–376.

Wong, V. & Saleuddin, A. S. M. (1972). Fine structure of normal and regenerated shell of *Helisoma duryi duryi*. *Canadian Journal of Zoology* 50: 1563–1568.

Wood, J. B., Maynard, A. E., Lawlor, A. G., Sawyer, E. K., Simmons, D. M., Pennoyer, K. E. & Derby, C. D. (2010). Caribbean reef squid, *Sepioteuthis sepioidea*, use ink as a defense against predatory French grunts, *Haemulon flavolineatum*. *Journal of Experimental Marine Biology and Ecology* 388: 20–27.

Wood, T. M. & Bhat, K. M. (1988). Methods for measuring cellulase activities. *Methods in Enzymology* 160: 87–112.

Wood, T. S., Anurakpongsatorn, P., Chaichana, R., Mahujchariyawong, J. & Satapanajaru, T. (2006). Heavy predation on freshwater bryozoans by the Golden Apple Snail, *Pomacea canaliculata* Lamarck, 1822 (Ampullariidae). *The Natural History Journal of Chulalongkong University* 6: 31–36.

Woodward, M. F. (1901). The anatomy of *Pleurotomaria beyrichii*, Hilg. *Quarterly Journal of Microscopical Science* 44: 215–268.

Wootton, J. T., Pfister, C. A. & Forester, J. D. (2008). Dynamic patterns and ecological impacts of declining ocean pH in a high-resolution multi-year dataset. *Proceedings of the National Academy of Sciences of the United States of America* 105: 18848–18853.

Worms, J., Bouchard, N., Cormier, R., Pauley, K. E. & Smith, J. C. (1993). New occurrences of paralytic shellfish poisoning toxins in the southern Gulf of St. Lawrence, Canada. pp. 353–358 *in* T. J. Smayda & Shimizu, Y. (Eds.) *Toxic Phytoplankton Blooms in the Sea. Proceedings of the Fifth International Conference on Toxic Marine Phytoplankton. Newport, Rhode Island, USA 28 October to 1 November 1991*. New York, Elsevier.

Wourms, J. P. (1987). Oogenesis. pp. 49–178 *in* A. C. Giese, Pearse, J. S. & Pearse, V. B. (Eds.) *Reproduction of Marine Invertebrates. General Aspects: Seeking Unity in Diversity.* Vol. 9. Palo Alto, CA, Blackwell Scientific Publications & Boxwood Press.

Wouters-Tyrou, D., Martin-Ponthieu, A., Briand, G., Sautiére, P. & Biserte, G. (1982). The amino-acid-sequence of histone H2A from cuttlefish *Sepia officinalis*. *European Journal of Biochemistry* 124: 489–498.

Wray, C. G., Jacobs, D. K., Kostriken, R., Vogler, A. P., Baker, R. & DeSalle, R. (1995). Homologues of the *engrailed* gene from five molluscan classes. *Federation of European Biochemical Societies (FEBS) Letters* 365: 71–74.

Wray, G. A. (1994). The evolution of cell lineage in echinoderms. *American Zoologist* 34: 353–363.

Wrede, C., Dreier, A., Kokoschka, S. & Hoppert, M. (2012). *Archaea* in symbioses. *Archaea* 2012: 1–11.

Wright, A. C., Fan, Y. & Baker, G. L. (2018a). Nutritional value and food safety of bivalve molluscan shellfish. *Journal of Shellfish Research* 37: 695–708.

Wright, B. R. (1974). Sensory structure of the tentacles of the slug *Arion ater* (Pulmonata, Mollusca). 1. Ultrastructure of the distal epithelium receptor cells and tentacular ganglion. 2. Ultrastructure of the free nerve endings in the distal epithelium. *Cell and Tissue Research* 151: 229–244; 245–257.

Wright, J. M., Parker, L. M., O'Connor, W. A., Scanes, E. & Ross, P. M. (2018b). Ocean acidification affects both the predator and prey to alter interactions between the oyster *Crassostrea gigas* (Thunberg, 1793) and the whelk *Tenguella marginalba* (Blainville, 1832). *Marine Biology* 165: 46.

Wright, W. G. (1982). Ritualized behavior in a territorial limpet. *Journal of Experimental Marine Biology and Ecology* 60: 245–251.

Wright, W. G. (1988). Sex change in the Mollusca. *Trends in Ecology & Evolution* 3: 137–140.

Wright, W. G. (1989). Intraspecific density mediates sex-change in the territorial patellacean limpet *Lottia gigantea*. *Marine Biology* 100: 353–364.

Wright, W. G. & Lindberg, D. R. (1982). Direct observation of sex change in the patellacean limpet *Lottia gigantea*. *Journal of the Marine Biological Association of the United Kingdom* 62: 737–738.

Wright, W. G. & Nybakken, J. W. (2007). Effect of wave action on movement in the owl limpet, *Lottia gigantea*, in Santa Cruz, California. *Bulletin of Marine Science* 81: 235–244.

Wright, W. G. & Shanks, A. L. (1993). Previous experience determines territorial behavior in an archaeogastropod limpet. *Journal of Experimental Marine Biology and Ecology* 166: 217–229.

Wu, H., Ji, C., Wei, L., Zhao, J. & Lu, H. (2013). Proteomic and metabolomic responses in hepatopancreas of *Mytilus galloprovincialis* challenged by *Micrococcus luteus* and *Vibrio anguillarum*. *Journal of Proteomics* 94: 54–67.

Wu, Q., Li, L. & Zhang, G. F. (2011). *Crassostrea angulata* bindin gene and the divergence of fucose-binding lectin repeats among three species of *Crassostrea*. *Marine Biotechnology* 13: 327–335.

Wu, S.-K. (1965). Comparative functional studies of the digestive system of the muricid gastropods *Drupa ricina* and *Morula granulata*. *Malacologia* 3: 211–233.

Wu, S., Chiang, C.-Y. & Zhou, W. (2017). Formation mechanism of $CaCO_3$ spherulites in the myostracum layer of limpet shells. *Crystals* 7: 319.

Wu, X. Z. & Pan, J. P. (2000). An intracellular prokaryotic microorganism associated with lesions in the oyster, *Crassostrea ariakensis* Gould. *Journal of Fish Diseases* 23: 409–414.

Wulff, J. L. (2006). Ecological interactions of marine sponges. *Canadian Journal of Zoology* 84: 146–166.

Wyatt, H. V. (1960). Protandry and self-fertilization in the Calyptraeidae. *Nature* 187: 520.

Wyatt, T. (1838). *A Manual of Conchology, according to the system laid down by Lamarck, with the late improvements by de Blainville. Exemplified and arranged for the use of students.* New York, Harper & Brother.

Wyatt, T. D. (2014). Proteins and peptides as pheromone signals and chemical signatures. *Animal Behaviour* 97: 273–280.

Wyeth, R. C. & Willows, A. O. D. (2006). Odours detected by rhinophores mediate orientation to flow in the nudibranch mollusc *Tritonia diomedea*. *Journal of Experimental Biology* 209: 1441–1453.

Wyeth, R. C., Woodward, O. M. & Willows, A. O. D. (2006). Orientation and navigation relative to water flow, prey, conspecifics, and predators by the nudibranch mollusc *Tritonia diomedea*. *Biological Bulletin* 210: 97–108.

Xu, B., Zhang, Y., Jing, Z. & Fan, T. (2017). Molecular characteristics of hemoglobins in blood clam and their immune responses to bacterial infection. *International Journal of Biological Macromolecules* 99: 375–383.

Xu, F., Domazet-Loso, T., Fan, D., Dunwell, T. L., Li, L., Fang, X. & Zhang, G. (2016). High expression of new genes in trochophore enlightening the ontogeny and evolution of trochozoans. *Scientific Reports* 6: 34664.

Xu, M., Bijoux, H., González, P. & Mounicou, S. (2014). Investigating the response of cuproproteins from oysters (*Crassostrea gigas*) after waterborne copper exposure by metallomic and proteomic approaches. *Metallomics* 6: 338–346.

Xue, D.-X., Zhang, T. & Liu, J.-X. (2016). Influences of population density on polyandry and patterns of sperm usage in the marine gastropod *Rapana venosa*. *Scientific Reports* 6: 23461.

Xue, D., Zhang, T. & Liu, J.-X. (2014). Microsatellite evidence for high frequency of multiple paternity in the marine gastropod *Rapana venosa*. *PLoS ONE* 9: e86508.

Yager, T. D., Terwilliger, N. B., Terwilliger, R. C., Schabtach, E. & Van Holde, K. E. (1982). Organization and physical properties of the giant extracellular homoglobin of the clam, *Astarte castanea*. *Biochimica et Biophysica Acta* 709: 194–203.

Yahel, G., Marie, D., Beninger, P. G., Eckstein, S. & Genin, A. (2009). In situ evidence for pre-capture qualitative selection in the tropical bivalve *Lithophaga simplex*. *Aquatic Biology* 6: 235–246.

Yamada, A., Yoshio, M. & Oiwa, K. (2013). Myosin Mg-ATPase of molluscan muscles is slightly activated by F-actin under catch state in vitro. *Journal of Muscle Research and Cell Motility* 34: 115–123.

Yamada, K., Ojika, M., Kigoshi, H. & Suenaga, K. (2010). Cytotoxic substances from two species of Japanese sea hares: Chemistry and bioactivity. *Proceedings of the Japan Academy, Series B* 86: 176–189.

Yamada, S. B. & Boulding, E. G. (1996). The role of highly mobile crab predators in the intertidal zonation of their gastropod prey. *Journal of Experimental Marine Biology and Ecology* 204: 59–83.

Yamada, S. B. & Boulding, E. G. (1998). Claw morphology, prey size selection and foraging efficiency in generalist and specialist shell-breaking crabs. *Journal of Experimental Marine Biology and Ecology* 220: 191–211.

Yamaguchi, K., Furuta, E. & Nakamura, H. (1999). Chronic skin allograft rejection in terrestrial slugs. *Zoological Science* 16: 485–495.

Yamaguchi, K., Seo, N. & Furuta, E. (2000). Histochemical and ultrastructural analyses of the epithelial cells of the body surface skin from the terrestrial slug, *Incilaria fruhstorferi*. *Zoological Science* 17: 1137–1146.

Yamaguchi, S., Sawada, K., Yusa, Y. & Iwasa, Y. (2013). Dwarf males and hermaphrodites can coexist in marine sedentary species if the opportunity to become a dwarf male is limited. *Journal of Theoretical Biology* 334: 101–108.

Yamamichi, M. & Innan, H. (2012). Estimating the migration rate from genetic variation data. *Heredity* 108: 362–363.

Yamaura, K., Takahashi, K. G. & Suzuki, T. (2008). Identification and tissue expression analysis of C-type lectin and galectin in the Pacific oyster, *Crassostrea gigas*. *Comparative Biochemistry and Physiology Part B* 149: 168–175.

Yanase, T. & Sakamoto, S. (1965). Fine structure of the visual cells of the dorsal eye in molluscan, *Onchidium verruculatum*. *Zoological Magazine (Tokyo)* 74: 238–242.

Yanes, Y., Asta, M. P., Ibáñez, M., Alonso, M. R. & Romanek, C. S. (2013). Paleoenvironmental implications of carbon stable isotope composition of land snail tissues. *Quaternary Research* 80: 596–605.

Yang, C., Wang, L., Zhang, H., Wang, L., Huang, M., Sun, Z., Sun, Y. & Song, L. (2014a). A new fibrinogen-related protein from *Argopecten irradians* (AiFREP-2) with broad recognition spectrum and bacteria agglutination activity. *Fish and Shellfish Immunology* 38: 221–229.

Yang, H., Johnson, P. M., Ko, K. C., Kamio, M., Germann, M. W., Derby, C. D. & Tai, P. C. (2005). Cloning, characterization and expression of escapin, a broadly antimicrobial FAD-containing L-amino acid oxidase from ink of the sea hare *Aplysia californica*. *Journal of Experimental Biology* 208: 3609–3622.

Yang, M., Xu, F., Liu, J., Que, H., Li, L. & Zhang, G. (2014b). Phylogeny of forkhead genes in three spiralians and their expression in Pacific oyster *Crassostrea gigas*. *Chinese Journal of Oceanology and Limnology* 32: 1207–1223.

Yao, H., Dao, M., Imholt, T., Huang, J., Wheeler, K., Bonilla, A., Suresh, S. & Ortiz, C. (2010). Protection mechanisms of the iron-plated armor of a deep-sea hydrothermal vent gastropod. *Proceedings of the National Academy of Sciences of the United States of America* 107: 987–992.

Yarnall, J. L. (1964). The responses of *Tegula funebralis* to starfishes and predatory snails (Mollusca: Gastropoda). *Veliger* 6: 56–58.

Yarra, T., Gharbi, K., Blaxter, M., Peck, L. S. & Clark, M. S. (2016). Characterization of the mantle transcriptome in bivalves: *Pecten maximus*, *Mytilus edulis* and *Crassostrea gigas*. *Marine Genomics* 27: 9–15.

Yasman, Edrada, R. A., Wray, V. & Proksch, P. (2003). New 9-Thiocyanatopupukeanane sesquiterpenes from the nudibranch *Phyllidia varicosa* and its sponge prey *Axinyssa aculeata*. *Journal of Natural Products* 66: 1512–1514.

Yearsley, J. M. & Sigwart, J. D. (2011). Larval transport modeling of deep-sea invertebrates can aid the search for undiscovered populations. *PLoS ONE* 6: e23063.

Yekutieli, Y., Sumbre, G., Flash, T. & Hochner, B. (2002). How to move with no rigid skeleton? The octopus has the answers? *Biologist (London)* 49: 250–254.

Yekutieli, Y., Sagiv-Zohar, R., Aharonov, R., Engel, Y., Hochner, B. & Flash, T. (2005a). Dynamic model of the octopus arm. I. Biomechanics of the octopus reaching movement *Journal of Neurophysiology* 94: 1443–1458.

Yekutieli, Y., Sagiv-Zohar, R., Hochner, B. & Flash, T. (2005b). Dynamic model of the octopus arm. II. Control of reaching movements *Journal of Neurophysiology* 94: 1459–1468.

Yin, C., Zhang, L.-L., Wang, G.-D. & Wang, Y.-L. (2015). Identification of metamorphosis related genes and development of molecular networks of *Lottia gigantea* larval [sic]. *Jimei Daxue Xuebao Ziran Kexue Ban* 20: 339–347.

Yochelson, E. L. & Bridge, J. (1957). The Lower Ordovician gastropod *Ceratopea*. *Geological Survey Professional Paper* 294-H: 280–304.

Yochelson, E. L. & Linsley, R. M. (1972). Opercula of two gastropods from the Lilydale Limestone (Early Devonian) of Victoria, Australia. *Memoirs of the National Museum of Victoria* 33: 1–14.

Yochelson, E. L. & Wise, O. A. (1972). A life association of shell and operculum in the early Ordovician gastropod *Ceratopea unguis*. *Journal of Paleontology* 46: 681–685.

Yokobori, S. I., Fukuda, N., Nakamura, M., Aoyama, T. & Oshima, T. (2004). Long-term conservation of six duplicated structural genes in cephalopod mitochondrial genomes. *Molecular Biology and Evolution* 21: 2034–2046.

Yonge, C. M. (1923). Studies on the comparative physiology of digestion. I. The mechanism of feeding, digestion and assimilation in the lamellibranch *Mya*. *Journal of Experimental Biology* 1: 15–63.

Yonge, C. M. (1925a). The digestion of cellulose by invertebrates. *Science Progress* 20: 242–248.

Yonge, C. M. (1925b). The hydrogen ion concentration in the gut of certain lamellibranchs and gastropods. *Journal of the Marine Biological Association of the United Kingdom* 13: 938–952.

Yonge, C. M. (1926a). Structure and physiology of the organs of feeding and digestion in *Ostrea edulis*. *Journal of the Marine Biological Association of the United Kingdom* 14: 295–386.

Yonge, C. M. (1926b). The digestive diverticula in the Lamellibranchs. *Transactions of the Royal Society of Edinburgh* 54: 703–718.

Yonge, C. M. (1930). The crystalline styles of the Mollusca and a carnivorous habit cannot normally co-exist. *Nature* 125: 444–445.

Yonge, C. M. (1932). *Notes on Feeding and Digestion in Pterocera and Vermetus, with a Discussion on the Occurrence of the Crystalline Style in the Gastropoda*. Scientific Reports of the Great Barrier Reef Expedition. London, British Museum (Natural History). 1: 259–281.

Yonge, C. M. (1936). *Mode of Life, Feeding, Digestion and Symbiosis with Zooxanthellae in the Tridacnidae*. Scientific Reports of the Great Barrier Reef Expedition. London, British Museum (Natural History). 1: 283–321.

Yonge, C. M. (1937). Circulation of water in the mantle cavity of *Dentalium entalis*. *Proceedings of the Malacological Society of London* 22: 333–337.

Yonge, C. M. (1938). Evolution of ciliary feeding in the Prosobranchia, with an account of feeding in *Capulus ungaricus*. *Journal of the Marine Biological Association of the United Kingdom* 22: 453.

Yonge, C. M. (1939a). On the mantle cavity and its contained organs in the *Loricata* (Placophora). *Quarterly Journal of Microscopical Science* 81: 367–390.

Yonge, C. M. (1939b). The protobranchiate Mollusca: A functional interpretation of their structure and evolution. *Philosophical Transactions of the Royal Society B* 230: 79–147.

Yonge, C. M. & Iles, E. J. (1939). On the mantle cavity, pedal gland, and evolution of mucous feeding in the Vermetidae. *Annals and Magazine of Natural History* 11: 536–556.

Yonge, C. M. (1947). The pallial organs in the aspidobranch Gastropoda and their evolution throughout the Mollusca. *Philosophical Transactions of the Royal Society B* 232: 443–518.

Yonge, C. M. (1949). The structure and adaptations of the Tellinacea, deposit feeding Eulamellibranchia. *Philosophical Transactions of the Royal Society B* 234: 29–76.

Yonge, C. M. (1952). The mantle cavity in *Siphonaria alternata* Say. *Proceedings of the Malacological Society of London* 29: 190–199.

Yonge, C. M. (1953a). Observations on *Hipponix antiquatus* (Linnaeus). *Proceedings of the California Academy of Sciences* 28: 1–24.

Yonge, C. M. (1953b). Mantle chambers and water circulation in the Tridacnidae (Mollusca). *Proceedings of the Zoological Society of London* 123: 551–561.

Yonge, C. M. (1958). Observations in life on the pulmonate limpet *Trimusculus (Gadinia) reticulatus* (Sowerby). *Proceedings of the Malacological Society of London* 33: 31–37.

Yonge, C. M. (1960a). Further observations on *Hipponix antiquatus* with notes on North Pacific pulmonate limpets. *Proceedings of the California Academy of Sciences* 31: 111–119.

Yonge, C. M. (1960b). *Oysters.* London, UK, Collins.

Yonge, C. M. (1960c). General characters of Mollusca. pp. I3–I36 *in* R. C. Moore (Ed.) *Treatise on Invertebrate Paleontology Part I. Mollusca.* Lawrence, KS, Geological Society of America and University of Kansas Press.

Yonge, C. M. (1962). On the primitive significance of the byssus in the Bivalvia and its effects in evolution. *Journal of the Marine Biological Association of the United Kingdom* 42: 113–125.

Yonge, C. M. (1967). Observations on *Pedum spondyloideum* (Chemnitz) Gmelin, a scallop associated with reef building corals. *Proceedings of the Malacological Society of London* 37: 311–323.

Yonge, C. M. (1969). Functional morphology and evolution within the Carditacea (Bivalvia). *Journal of Molluscan Studies* 38: 493–527.

Yonge, C. M. & Thompson, T. E. (1976). *Living Marine Molluscs.* London, UK, Collins.

York, B. & Twarog, B. M. (1973). Evidence for release of serotonin by relaxing nerves in molluscan muscle. *Comparative Biochemistry and Physiology Part A* 44: 423–430.

York, P. S., Cummins, S. F., Degnan, S. M., Woodcroft, B. J. & Degnan, B. M. (2012). Marked changes in neuropeptide expression accompany broadcast spawnings in the gastropod *Haliotis asinina. Frontiers in Zoology* 9: 9.

Yoshida, M.-A., Ishikura, Y., Moritaki, T., Shoguchi, E., Shimizu, K. K., Sese, J. & Ogura, A. (2011). Genome structure analysis of molluscs revealed whole genome duplication and lineage specific repeat variation. *Gene* 483: 63–71.

Yoshida, M.-A., Yura, K. & Ogura, A. (2014). Cephalopod eye evolution was modulated by the acquisition of Pax-6 splicing variants. *Scientific Reports* 4: 4256 (4251–4256).

Yoshida, W. Y., Bryan, P. J., Baker, B. J. & McClintock, J. B. (1995). Pteroenone: A defensive metabolite of the abducted Antarctic pteropod *Clione antarctica. Journal of Organic Chemistry* 60: 780–782.

Yoshioka, E. (1989). Experimental analysis of the diurnal and tidal spawning rhythm in the chiton *Acanthopleura japonica* (Lischke) by manipulating conditions of light and tide. *Journal of Experimental Marine Biology and Ecology* 133: 81–91.

Young, C. M. (2002). A brief history and some fundamentals. pp. 1–19 *in* C. M. Young, Sewell, M. A. & Rice, M. E. (Eds.) *Atlas of Marine Invertebrate Larvae.* San Diego, CA, Academic Press.

Young, D. K. (1969). *Okadaia elegans*, a tube-boring nudibranch mollusc from the central and west Pacific. *American Zoologist* 9: 903.

Young, J. Z. (1960). The statocysts of *Octopus vulgaris. Proceedings of the Royal Society B* 152: 3–29.

Young, J. Z. (1962a). The optic lobes of *Octopus vulgaris. Philosophical Transactions of the Royal Society B* 245: 19–58.

Young, J. Z. (1962b). Why do we have two brains? pp. 7–24 *in* V. Mountcastle (Ed.) *Interhemispheric Relations and Cerebral Dominance.* Baltimore, MD, Johns Hopkins University Press.

Young, J. Z. (1962c). The retina of cephalopods and its degeneration after optic nerve section. *Philosophical Transactions of the Royal Society B* 245: 1–19.

Young, J. Z. (1963a). The number and sizes of nerve cells in *Octopus. Proceedings of the Zoological Society of London* 140: 229–254.

Young, J. Z. (1963b). Light- and dark-adaptation in the eyes of some cephalopods. *Proceedings of the Zoological Society of London* 140: 255–272.

Young, J. Z. (1965a). The central nervous system of *Nautilus. Philosophical Transactions of the Royal Society B* 249: 1–25.

Young, J. Z. (1965b). The buccal nervous system of *Octopus. Philosophical Transactions of the Royal Society B* 249: 27–67.

Young, J. Z. (1971). *The Anatomy of the Central Nervous System of Octopus vulgaris.* Oxford, UK, Clarendon Press.

Young, J. Z. (1977). Brain, behaviour and evolution of cephalopods. pp. 377–434 *in* M. Nixon & Messenger, J. B. (Eds.) *The Biology of Cephalopods. Symposia of The Zoological Society of London.* London, UK, Academic Press.

Young, J. Z. (1988a). Evolution of the cephalopod brain. pp. 215–228 *in* M. R. Clarke & Trueman, E. R. (Eds.) *Paleontology and Neontology of Cephalopods. The Mollusca.* Vol. 12. New York, Academic Press.

Young, J. Z. (1988b). Evolution of the cephalopod statocyst. pp. 229–239 *in* M. R. Clarke & Trueman, E. R. (Eds.) *Paleontology and Neontology of Cephalopods. The Mollusca.* Vol. 12. New York, Academic Press.

Young, J. Z. (1989). The angular acceleration receptor system of diverse cephalopods. *Philosophical Transactions of the Royal Society B* 325: 189–238.

Young, J. Z. (1991). Computation in the learning system of cephalopods. *Biological Bulletin* 180: 200–208.

Young, J. Z. (1994). The muscular-hydrostatic radula supports of *Octopus, Loligo, Sepia,* and *Nautilus. Journal of Cephalopod Biology* 2: 65–93.

Young, J. Z. (1995). Multiple matrices in the memory system of *Octopus.* pp. 431–443 *in* J. N. Abbot, Williamson, R. & Maddock, L. (Eds.) *Cephalopod Neurobiology, Neuroscience Studies in Squid, Octopus and Cuttlefish.* Oxford, UK, Oxford University Press.

Young, J. Z. (2010). The central nervous system. pp. 215–222 *in* W. B. Saunders & Landman, N. H. (Eds.) *Nautilus: The Biology and Paleobiology of a Living Fossil. Reprint with additions (Topics in Geobiology).* New York, Springer.

Young, R. E. (1973). Information feedback from photophores and ventral countershading in mid-water squid. *Pacific Science* 27: 1–7.

Young, R. E., Roper, C. F. & Walters, J. F. (1979). Eyes and extraocular photoreceptors in midwater cephalopods and fishes: Their roles in detecting downwelling light for counterillumination. *Marine Biology* 51: 371–380.

Young, R. E. & Mencher, F. M. (1980). Bioluminescence in mesopelagic squid: Diel color change during counterillumination. *Science* 208: 1286–1288.

Young, R. E. & Vecchione, M. (1996). Analysis of morphology to determine primary sister-taxon relationships within coleoid cephalopods. *American Malacological Bulletin* 12: 91–112.

Young, R. E., Vecchione, M. & Mangold, K. M. (1999). Tree of Life Cephalopod Brain Terminology. Retrieved February 2015, from http://tolweb.org/accessory/Cephalopod_Brain_Terminology?acc_id=1944.

Young, R. T. (1942). Spawning season of the Californian Mussel, *Mytilus californicus*. *Ecology* 23: 490–492.

Ysseling, M. A. (1930). Über die Atmung der Weinbergschnecke (*Helix pomatia*). *Zeitschrift für Vergleichende Physiologie* 13: 1–60.

Yu, X., He, W., Gu, J.-D., He, M. & Yan, Y. (2008). The effect of chemical cues on settlement of pearl oyster *Pinctada fucata martensii* (Dunker) larvae. *Aquaculture* 277: 83–91.

Yuasa, H. J., Hasegawa, T., Nakamura, T. & Suzuki, T. (2007). Bacterial expression and characterization of molluscan IDO-like myoglobin. *Comparative Biochemistry and Physiology Part B* 146: 461–469.

Yue, F., Zhou, Z., Wang, L., Ma, Z., Wang, J., Wang, M., Zhang, H. & Song, L. (2013). Maternal transfer of immunity in scallop *Chlamys farreri* and its trans-generational immune protection to offspring against bacterial challenge. *Developmental & Comparative Immunology* 41: 569–577.

Yue, X., Nie, Q., Xiao, G. & Liu, B. (2015). Transcriptome analysis of shell color-related genes in the clam *Meretrix meretrix*. *Marine Biotechnology* 17: 364–374.

Yusa, Y. (2007). Causes of variation in sex ratio and modes of sex determination in the Mollusca—An overview. *American Malacological Bulletin* 23: 89–98.

Yusa, Y. (2008). Size-dependent sex allocation and sexual selection in *Aplysia kurodai*, a hermaphrodite with nonreciprocal mating. *Invertebrate Biology* 127: 291–298.

Yusa, Y. & Kumagai, N. (2018). Evidence of oligogenic sex determination in the apple snail *Pomacea canaliculata*. *Genetica* 146: 265–275.

Zaady, E., Shachak, M. & Groffman, P. (1996). Release and consumption of nitrogen by snail feces in Negev Desert soils. *Biology and Fertility of Soils* 23: 399–404.

Zabludovskaya, S. À. & Badanin, I. V. (2010). The Slug Mite *Riccardoella* (*Proriccardoella*) *oudemansi* (Prostigmata, Ereynetidae) from Ukraine. *Vestnik zoologii* 44: 163–166.

Zack, S. (1975). A description and analysis of agonistic behavior patterns in an opisthobranch mollusc, *Hermissenda crassicornis*. *Behaviour* 53: 238–267.

Zahradnik, T. D., Lemay, M. A. & Boulding, E. G. (2008). Choosy males in a littorinid gastropod: Male *Littorina subrotundata* prefer large and virgin females. *Journal of Molluscan Studies* 74: 245–251.

Zaitseva, O. V., Shumeev, A. N., Korshunova, T. A. & Martynov, A. V. (2015). Heterochronies in the formation of the nervous and digestive systems in early postlarval development of opisthobranch mollusks: Organization of major organ systems of the arctic dorid *Cadlina laevis*. *Biology Bulletin* 42: 186–195.

Zaĭtseva, O. V. & Bocharova, L. S. (1981). Sensory cells in the head skin of pond snails. *Cell and Tissue Research* 220: 797–807.

Zaĭtseva, O. V. (1997). Structural organization of the receptor elements and organs of the terrestrial mollusk *Pomatia elegans* (Prosobranchia). *Neuroscience and Behavioral Physiology* 27: 83–89.

Zal, F., Leize, E., Oros, D. R., Hourdez, S., van Dorsselaer, A., & Childress, J. J. (2000). Haemoglobin structure and biochemical characteristics of the sulphide-binding component from the deep-sea clam *Calyptogena magnifica*. *Cahiers de Biologie Marine* 41: 413–423.

Zaldibar, B., Cancio, I. & Marigómez, I. (2004). Circatidal variation in epithelial cell proliferation in the mussel digestive gland and stomach. *Cell and Tissue Research* 318: 395–402.

Zande, J. M. (1999). An ascomycete commensal on the gills of *Bathynerita naticoidea*, the dominant gastropod at Gulf of Mexico hydrocarbon seeps. *Invertebrate Biology* 118: 57–62.

Zann, L. P. (1973). Interactions of the circadian and circatidal rhythms of the littoral gastropod *Melanerita atramentosa* Reeve. *Journal of Experimental Marine Biology and Ecology* 11: 249–261.

Zapata, F., Wilson, N. G., Howison, M., Andrade, S. C. S., Jörger, K. M., Schrödl, M., Goetz, F. E., Giribet, G. & Dunn, C. W. (2014). Phylogenomic analyses of deep gastropod relationships reject Orthogastropoda. *Proceedings of the Royal Society B* 281: 2014739.

Zardus, J. D. & Morse, M. P. (1998). Embryogenesis, morphology and ultrastructure of the pericalymma larva of *Acila castrensis* (Bivalvia: Protobranchia: Nuculoida). *Invertebrate Biology* 117: 221–244.

Zardus, J. D. (2002). Protobranch bivalves. *Advances in Marine Biology* 42: 1–65.

Zardus, J. D. & Martel, A. L. (2002). Phylum Mollusca: Bivalvia. pp. 289–325 in C. M. Young (Ed.) *Atlas of Marine Invertebrate Larvae*. San Diego, CA, Academic Press.

Zbinden, M., Pailleret, M., Ravaux, J., Gaudron, S. M., Hoyoux, C., Lambourdière, J., Warén, A., Lorion, J., Halary, S. & Duperron, S. (2010). Bacterial communities associated with the wood-feeding gastropod *Pectinodonta* sp. (Patellogastropoda, Mollusca). *FEMS Microbiology Ecology* 74: 450–463.

Zenetos, A., Çinar, M. E., Pancucci-Papadopoulou, M. A., Harmelin, J., Furnari, G., Andaloro, F., Bellou, N., Streftaris, N. & Zibrowius, H. (2005). Annotated list of marine alien species in the Mediterranean with records of the worst invasive species. *Mediterranean Marine Science* 6: 63–118.

Zenetos, A., Gofas, S., Verlaque, M., Çinar, M. E., Garcia Raso, J. E., Bianchi, C. N., Morri, C., Azzurro, E., Bilecenoglu, M. & Froglia, C. (2010). Alien species in the Mediterranean Sea by 2010. A contribution to the application of European Union's Marine Strategy Framework Directive (MSFD). Part I. Spatial distribution. *Mediterranean Marine Science* 11: 381–493.

Zenetos, A., Gofas, S., Morri, C., Rosso, A., Violanti, D., García Raso, J. E., Çinar, M. E., Almogi-Labin, A., Ates, A. S. & Azzurro, E. (2012). Alien species in the Mediterranean Sea by 2012. A contribution to the application of European Union's Marine Strategy Framework Directive (MSFD). Part 2. Introduction trends and pathways. *Mediterranean Marine Science* 13: 328–352.

Zhadan, P. M. & Semen'kov, P. G. (1984). An electrophysiological study of the mechanoreceptory function of the abdominal sense organ of the scallop *Patinopecten yessoensis* (Jay). *Comparative Biochemistry and Physiology Part A* 78: 865–870.

Zhadan, P. M. & Sizov, A. V. (2000). Ultrastructure of abdominal sensory organ of the scallop *Mizuchopecten yessoensis* (Jay). *Sensornye Sistemy* 14: 118–121.

Zhang, G. F., Fang, X., Guo, X., Li, L., Luo, R., Xu, F. S., Yang, P. et al. (2012a). The oyster genome reveals stress adaptation and complexity of shell formation. *Nature* 490: 49–54.

Zhang, H., Wang, L., Song, L., Song, X., Wang, B., Mu, C. & Zhang, Y. (2009). A fibrinogen-related protein from bay scallop *Argopecten irradians* involved in innate immunity as pattern recognition receptor. *Fish and Shellfish Immunology* 26: 56–64.

Zhang, L., Wayne, N. L., Sherwood, N. M., Postigo, H. R. & Tsai, P.-S. (2000). Biological and immunological characterization of multiple GnRH in an opisthobranch mollusk, *Aplysia californica*. *General and Comparative Endocrinology* 118: 77–89.

Zhang, L., Tello, J. A., Zhang, W. & Tsai, P.-S. (2008). Molecular cloning, expression pattern, and immunocytochemical localization of a gonadotropin-releasing hormone-like molecule in the gastropod mollusk, *Aplysia californica*. *General and Comparative Endocrinology* 156: 201–209.

Zhang, L. L., Li, L., Zhu, Y. B., Zhang, G. F. & Guo, X. M. (2014a). Transcriptome analysis reveals a rich gene set related to innate immunity in the Eastern Oyster (*Crassostrea virginica*). *Marine Biotechnology* 16: 17–33.

Zhang, S.-M. & Coultas, K. A. (2011). Identification and characterization of five transcription factors that are associated with evolutionarily conserved immune signaling pathways in the schistosome-transmitting snail *Biomphalaria glabrata*. *Molecular Immunology* 48: 1868–1881.

Zhang, S., Han, G.-D. & Dong, Y.-W. (2014b). Temporal patterns of cardiac performance and genes encoding heat shock proteins and metabolic sensors of an intertidal limpet *Cellana toreuma* during sublethal heat stress. *Journal of Thermal Biology* 41: 31–37.

Zhang, T., Ma, Y., Chen, K., Kunz, M., Tamura, N., Qiang, M., Xu, J. & Qi, L. (2011). Structure and mechanical properties of a pteropod shell consisting of interlocked helical aragonite nanofibers. *Angewandte Chemie (International Edition)* 50: 10361–10365.

Zhang, W., Gavagnin, M., Guo, Y. W. & Mollo, E. (2006). Chemical studies on the South China Sea nudibranch *Dermatobranchus ornatus* and its suggested prey gorgonian *Muricella* sp. *Chinese Journal of Organic Chemistry* 12: 1667–1672.

Zhang, X., Mao, Y., Huang, Z. X., Qu, M., Chen, J., Ding, S. X., Hong, J. N. & Sun, T. T. (2012b). Transcriptome analysis of the *Octopus vulgaris* central nervous system. *PLoS ONE* 7: e40320.

Zhang, Y., Sun, J., Chen, C., Watanabe, H. K., Feng, D., Zhang, Y., Chiu, J. M., Qian, P.-Y. & Qiu, J.-W. (2017). Adaptation and evolution of deep-sea scale worms (Annelida: Polynoidae): Insights from transcriptome comparison with a shallow-water species. *Scientific Reports* 7: 46205.

Zhang, Z.-Q., Ed. (2011). *Animal Biodiversity: An Outline of Higher-level Classification and Survey of Taxonomic Richness*, Auckland, New Zealand, Magnolia Press.

Zhao, J., Shi, B., Jiang, Q. & Ke, C. (2012). Changes in gut-associated flora and bacterial digestive enzymes during the development stages of abalone (*Haliotis diversicolor*). *Aquaculture* 338: 147–153.

Zheng, H., Zhang, T., Sun, Z., Liu, W. & Liu, H. (2013). Inheritance of shell colours in the noble scallop *Chlamys nobilis* (Bivalve: Pectinidae). *Aquaculture Research* 44: 1229–1235.

Zhu, Q., Zhang, L., Li, L., Que, H. & Zhang, G. (2016). Expression characterization of stress genes under high and low temperature stresses in the Pacific Oyster, *Crassostrea gigas*. *Marine Biotechnology* 18: 176–188.

Zhukov, V. V., Borissenko, S. L., Zieger, M. V., Vakoliuk, I. A. & Meyer-Rochow, V. B. (2006). The eye of the freshwater prosobranch gastropod *Viviparus viviparus*: Ultrastructure, electrophysiology and behaviour. *Acta Zoologica* 87: 13–24.

Zhuravlev, V. L. & Safonova, T. A. (1984). Regulation of the heart rate by the visceral ganglion units in the snail *Helix pomatia* [in Russian]. *Physiology Journal of the USSR* 70: 425–429.

Zieger, M. V. & Meyer-Rochow, V. B. (2008). Understanding the cephalic eyes of pulmonate gastropods: A review. *American Malacological Bulletin* 26: 47–66.

Zielinski, F. U., Pernthaler, A., Duperron, S., Raggi, L., Giere, O., Borowski, C. & Dubilier, N. (2009). Widespread occurrence of an intranuclear bacterial parasite in vent and seep bathymodiolin mussels. *Environmental Microbiology* 11: 1150–1167.

Zielske, S., Glaubrecht, M. & Haase, M. (2011). Origin and radiation of rissooidean gastropods (Caenogastropoda) in ancient lakes of Sulawesi. *Zoologica Scripta* 40: 221–237.

Zielske, S. & Haase, M. (2015). Molecular phylogeny and a modified approach of character-based barcoding refining the taxonomy of New Caledonian freshwater gastropods (Caenogastropoda, Truncatelloidea, Tateidae). *Molecular Phylogenetics and Evolution* 89: 171–181.

Zilch, A. (1959–1960). Gastropoda. Euthyneura. pp. 1–824 *in* O. H. Schindewolf (Ed.) *Handbuch der Paläontologie*. Vol. 6, Teil 2. Berlin, Germany, Gebrüder Borntraeger.

Zimmermann, M. R., Luth, K. E. & Esch, G. W. (2011). Complex interactions among a nematode parasite (*Daubaylia potomaca*), a commensalistic annelid (*Chaetogaster limnaei limnaei*), and trematode parasites in a snail host (*Helisoma anceps*). *Journal of Parasitology* 97: 788–791.

Zouros, E. (2000). The exceptional mitochondrial DNA system of the mussel family Mytilidae. *Genes and Genetic Systems* 75: 313–318.

Zugmayer, E. (1904). Über Sinnesorgane an den Tentakeln des Genus *Cardium*. *Zeitschrift für wissenschaftliche Zoologie* 76: 478–508.

Zukowski, S. A. (2001). An exotic snail that 'mimics' a native species—The comparative ecology of *Physa acuta* (Draparnaud 1805) and *Glyptophysa gibbosa* (Walker 1988) (Gastropoda: Pulmonata) in the River Murray, South Australia. BSc (Hons) thesis, University of Adelaide.

Zukowski, S. A. & Walker, K. F. (2009). Freshwater snails in competition: Alien *Physa acuta* (Physidae) and native *Glyptophysa gibbosa* (Planorbidae) in the River Murray, South Australia. *Marine and Freshwater Research* 60: 999–1005.

Zullo, L. & Hochner, B. (2011). A new perspective on the organization of an invertebrate brain. *Communicative & Integrative Biology* 4: 26–29.

de Zwaan, A. & Wijsman, T. C. M. (1976). Anaerobic metabolism in Bivalvia (Mollusca): Characteristics of anaerobic metabolism. *Comparative Biochemistry and Physiology Part B* 54: 313–324.

de Zwaan, A. (1983). Carbohydrate catabolism in bivalves. pp. 137–175 *in* P. W. Hochachka (Ed.) *Metabolic Biochemistry and Molecular Biomechanics. The Mollusca*. Vol. 1. New York, Academic Press.

de Zwaan, A. (1991). Molluscs. pp. 186–217 *in* C. Bryant (Ed.) *Metazoan Life Without Oxygen*. London, UK, Chapman & Hall.

Zylstra, U. (1972). Distribution and ultrastructure of epidermal sensory cells in the freshwater snails *Lymnaea stagnalis* and *Biomphalaria pfeifferi*. *Netherlands Journal of Zoology* 22: 283–298.

Index

Note: Page numbers in italic refer to figures.